High-Performance Backbone Network Technology

OPTICAL ENGINEERING

Founding Editor

Brian J. Thompson

University of Rochester
Rochester, New York

1. Electron and Ion Microscopy and Microanalysis: Principles and Applications, *Lawrence E. Murr*
2. Acousto-Optic Signal Processing: Theory and Implementation, *edited by Nor man J. Berg and John N. Lee*
3. Electro-Optic and Acousto-Optic Scanning and Deflection, *Milton Gottlieb, Clive L. M. Ireland, and John Martin Ley*
4. Single-Mode Fiber Optics: Principles and Applications, *Luc B. Jeunhomme*
5. Pulse Code Formats for Fiber Optical Data Communication: Basic Principles and Applications, *David J. Morris*
6. Optical Materials: An Introduction to Selection and Application, *Solomon Musikant*
7. Infrared Methods for Gaseous Measurements: Theory and Practice, *edited by Joda Wormhoudt*
8. Laser Beam Scanning: Opto-Mechanical Devices, Systems, and Data Storage Optics, *edited by Gerald F. Marshall*
9. Opto-Mechanical Systems Design, *Paul R. Yoder, Jr.*
10. Optical Fiber Splices and Connectors: Theory and Methods, *Calvin M. Miller with Stephen C. Mettler and Ian A. White*
11. Laser Spectroscopy and Its Applications, *edited by Leon J. Radziemski, Richard W. Solarz, and Jeffrey A. Paisner*
12. Infrared Optoelectronics: Devices and Applications, *William Nunley and J. Scott Bechtel*
13. Integrated Optical Circuits and Components: Design and Applications, *edited by Lynn D. Hutcheson*
14. Handbook of Molecular Lasers, *edited by Peter K. Cheo*
15. Handbook of Optical Fibers and Cables, *Hiroshi Murata*
16. Acousto-Optics, *Adrian Korpel*
17. Procedures in Applied Optics, *John Strong*
18. Handbook of Solid-State Lasers, *edited by Peter K. Cheo*
19. Optical Computing: Digital and Symbolic, *edited by Raymond Arrathoon*
20. Laser Applications in Physical Chemistry, *edited by D. K. Evans*
21. Laser-Induced Plasmas and Applications, *edited by Leon J. Radziemski and David A. Cremers*
22. Infrared Technology Fundamentals, *Irving J. Spiro and Monroe Schlessinger*
23. Single-Mode Fiber Optics: Principles and Applications, Second Edition, Re vised and Expanded, *Luc B. Jeunhomme*
24. Image Analysis Applications, *edited by Rangachar Kasturi and Mohan M. Trivedi*
25. Photoconductivity: Art, Science, and Technology, *N. V. Joshi*
26. Principles of Optical Circuit Engineering, *Mark A. Mentzer*
27. Lens Design, *Milton Laikin*
28. Optical Components, Systems, and Measurement Techniques, *Rajpal S. Sirohi and M. P. Kothiyal*

29. Electron and Ion Microscopy and Microanalysis: Principles and Applications, Second Edition, Revised and Expanded, *Lawrence E. Murr*
30. Handbook of Infrared Optical Materials, *edited by Paul Klocek*
31. Optical Scanning, *edited by Gerald F. Marshall*
32. Polymers for Lightwave and Integrated Optics: Technology and Applications, *edited by Lawrence A. Hornak*
33. Electro-Optical Displays, *edited by Mohammad A. Karim*
34. Mathematical Morphology in Image Processing, *edited by Edward R. Dougherty*
35. Opto-Mechanical Systems Design: Second Edition, Revised and Expanded, *Paul R. Yoder, Jr.*
36. Polarized Light: Fundamentals and Applications, *Edward Collett*
37. Rare Earth Doped Fiber Lasers and Amplifiers, *edited by Michel J. F. Digonnet*
38. Speckle Metrology, *edited by Rajpal S. Sirohi*
39. Organic Photoreceptors for Imaging Systems, *Paul M. Borsenberger and David S. Weiss*
40. Photonic Switching and Interconnects, *edited by Abdellatif Marrakchi*
41. Design and Fabrication of Acousto-Optic Devices, *edited by Akis P. Goutzoulis and Dennis R. Pape*
42. Digital Image Processing Methods, *edited by Edward R. Dougherty*
43. Visual Science and Engineering: Models and Applications, *edited by D. H. Kelly*
44. Handbook of Lens Design, *Daniel Malacara and Zacarias Malacara*
45. Photonic Devices and Systems, *edited by Robert G. Hunsberger*
46. Infrared Technology Fundamentals: Second Edition, Revised and Expanded, *edited by Monroe Schlessinger*
47. Spatial Light Modulator Technology: Materials, Devices, and Applications, *edited by Uzi Efron*
48. Lens Design: Second Edition, Revised and Expanded, *Milton Laikin*
49. Thin Films for Optical Systems, *edited by Francoise R. Flory*
50. Tunable Laser Applications, *edited by F. J. Duarte*
51. Acousto-Optic Signal Processing: Theory and Implementation, Second Edition, *edited by Norman J. Berg and John M. Pellegrino*
52. Handbook of Nonlinear Optics, *Richard L. Sutherland*
53. Handbook of Optical Fibers and Cables: Second Edition, *Hiroshi Murata*
54. Optical Storage and Retrieval: Memory, Neural Networks, and Fractals, *edited by Francis T. S. Yu and Suganda Jutamulia*
55. Devices for Optoelectronics, *Wallace B. Leigh*
56. Practical Design and Production of Optical Thin Films, *Ronald R. Willey*
57. Acousto-Optics: Second Edition, *Adrian Korpel*
58. Diffraction Gratings and Applications, *Erwin G. Loewen and Evgeny Popov*
59. Organic Photoreceptors for Xerography, *Paul M. Borsenberger and David S. Weiss*
60. Characterization Techniques and Tabulations for Organic Nonlinear Optical Materials, *edited by Mark G. Kuzyk and Carl W. Dirk*
61. Interferogram Analysis for Optical Testing, *Daniel Malacara, Manuel Servin, and Zacarias Malacara*
62. Computational Modeling of Vision: The Role of Combination, *William R. Uttal, Ramakrishna Kakarala, Spiram Dayanand, Thomas Shepherd, Jagadeesh Kalki, Charles F. Lunskis, Jr., and Ning Liu*
63. Microoptics Technology: Fabrication and Applications of Lens Arrays and Devices, *Nicholas Borrelli*
64. Visual Information Representation, Communication, and Image Processing, *edited by Chang Wen Chen and Ya-Qin Zhang*
65. Optical Methods of Measurement, *Rajpal S. Sirohi and F. S. Chau*
66. Integrated Optical Circuits and Components: Design and Applications, *edited by Edmond J. Murphy*

67. Adaptive Optics Engineering Handbook, *edited by Robert K. Tyson*
68. Entropy and Information Optics, *Francis T. S. Yu*
69. Computational Methods for Electromagnetic and Optical Systems, *John M. Jarem and Partha P. Banerjee*
70. Laser Beam Shaping, *Fred M. Dickey and Scott C. Holswade*
71. Rare-Earth-Doped Fiber Lasers and Amplifiers: Second Edition, Revised and Expanded, *edited by Michel J. F. Digonnet*
72. Lens Design: Third Edition, Revised and Expanded, *Milton Laikin*
73. Handbook of Optical Engineering, *edited by Daniel Malacara and Brian J. Thompson*
74. Handbook of Imaging Materials: Second Edition, Revised and Expanded, *edited by Arthur S. Diamond and David S. Weiss*
75. Handbook of Image Quality: Characterization and Prediction, *Brian W. Keelan*
76. Fiber Optic Sensors, *edited by Francis T. S. Yu and Shizhuo Yin*
77. Optical Switching/Networking and Computing for Multimedia Systems, *edited by Mohsen Guizani and Abdella Battou*
78. Image Recognition and Classification: Algorithms, Systems, and Applications, *edited by Bahram Javidi*
79. Practical Design and Production of Optical Thin Films: Second Edition, Revised and Expanded, *Ronald R. Willey*
80. Ultrafast Lasers: Technology and Applications, *edited by Martin E. Fermann, Almantas Galvanauskas, and Gregg Sucha*
81. Light Propagation in Periodic Media: Differential Theory and Design, *Michel Nevière and Evgeny Popov*
82. Handbook of Nonlinear Optics: Second Edition, Revised and Expanded, *Richard L. Sutherland*
83. Polarized Light: Second Edition, Revised and Expanded, *Dennis Goldstein*
84. Optical Remote Sensing: Science and Technology, *Walter Egan*
85. Handbook of Optical Design: Second Edition, *Daniel Malacara and Zacarias Malacara*
86. Nonlinear Optics: Theory, Numerical Modeling, and Applications, *Partha P. Banerjee*
87. Semiconductor and Metal Nanocrystals: Synthesis and Electronic and Optical Properties, *edited by Victor I. Klimov*
88. High-Performance Backbone Network Technology, *edited by Naoaki Yamanaka*
89. Semiconductor Laser Fundamentals, *Toshiaki Suhara*
90. Handbook of Optical and Laser Scanning, *edited by Gerald F. Marshall*

Additional Volumes in Preparation

High-Performance Backbone Network Technology

edited by
Naoaki Yamanaka
NNT Network Innovation Laboratories
Tokyo, Japan

THE INSTITUTE OF ELECTRONICS, INFORMATION, AND COMMUNICATION ENGINEERS — TOKYO

MARCEL DEKKER, INC. — NEW YORK • BASEL

Library of Congress Cataloging-in-Publication Data
A catalog record for this book is available from the Library of Congress.

ISBN: 0-8247-5321-6

This book is printed on acid-free paper.
The Institute of Electronics, Information, and Communication Engineers (IEICE),
Kikai-Shinko-Kaikan Bldg.,
3-5-8, Shibakoen, Minato-ku, Tokyo, 105-0011 Japan
Phone: +81-3-3433-6692
Fax: +81-3-3433-6659
http://www.ieice.org

Headquarters
Marcel Dekker, Inc., 270 Madison Avenue, New York, NY 10016, U.S.A.
tel: 212-696-9000; fax: 212-685-4540

Distribution and Customer Service
Marcel Dekker, Inc., Cimarron Road, Monticello, New York 12701, U.S.A.
tel: 800-228-1160; fax: 845-796-1772

Eastern Hemisphere Distribution
Marcel Dekker AG, Hulgasse 4, Postfach 812, CH-4001 Busel, Switzerland
tel: 41-61-260-6300; fax: 41-61-260-6333

World Wide Web
http://www.dekker.com

The publisher offers discounts on this book when ordered in bulk quantities. For more information, write to Special Sales/Professional Marketing at the headquarters address above.

Current printing (last digit):

10 9 8 7 6 5 4 3 2 1

PRINTED IN THE UNITED STATES OF AMERICA

To my wife, Asako,
and my three children, Yuki, Takuto, and Noe

IEICE Transactions on Communications, a well-received international publication, was founded in 1975 by the IEICE Communication Society. It aims at providing communications researchers and professionals with a continuous source of high-quality peer-reviewed papers, letters, invited papers, and survey papers. In addition, very important special issues are also published. This journal is published monthly and in 2002 it provided 366 articles contained in over 2895 pages. Its rigorous acceptance policies (the acceptance rate is under 40%) guarantees high-quality papers. The Transactions are becoming more globalized, and about 73.3% of the contributions are now submitted from outside Japan, especially from Korea, Taiwan, and the U.S.A.

The IEICE Communication Society had 13,994 members as of the end of September 2003. *IEICE Transactions on Communications* issues more than 3500 copies per month.

The topics of the transactions include communication networks, network systems, mobile communication, satellites, optical networks, communication quality, and communication devices.

IEICE Transactions on Communication provides both a science citation index and an expanded science citation index as directed by the Information Science Institute (ISI).

Preface

The last 10 years have seen an enormous growth in broadband access line bandwidth and the number of broadband users. These users have severely loaded the communications infrastructure by their enthusiastic adoption of broadband services and multimedia traffic. This has driven the introduction of advanced wavelength division multiplexing (WDM) technologies into the backbone network. Loads have increased by a factor of about 100 since the commencement of asymmetric digital subscriber line (ADSL) service. In an associated trend, Internet protocol (IP) services are morphing from just e-mail into mission-critical services such as voice, data, and streaming image transmission. Quality of service (QoS) assurance has become a critical goal for service providers. To cope with these two factors, high-throughput and high-quality backbone networking technologies are needed. Any careful study of backbone networking technologies must cover both sophisticated traffic control techniques and advanced hardware technologies. Many researchers are now enthusiastically tackling the next-generation IP network that offers QoS, high security, high reliability, and high performance.

This book covers all of the important contributions in these areas. To realize high-quality and high-speed backbone networks, the focus of traffic control was asynchronous transfer mode (ATM) technology starting in 1988. Next, the focus switched from ATM-oriented technology to IP-oriented technology, where network equipment such as switches were assumed to be controlled in a distribution manner. Multi-protocol label switching (MPLS) was a breakthrough technology for the realization of reliable and manageable IP centric networks. Therefore, many important studies have addressed the use of ATM and its application to MPLS/IP and MPLS/photonic networks. We note that many studies have examined the creation of high-speed and high-performance networks on electrical VLSI (very large scale integrated circuit) and optical technologies. They not only address hardware technologies but also describe many important contributions with regard to network and system architectures. Performance studies have examined architectures, experimental prototypes, and field trials, all of which suggest the creation of a new network paradigm.

This book consists of selected papers from *IEICE* (the Institute of Electronics, Information and Communication Engineers) *Transactions on Communications*. The IEICE publishes four well-established international transactions in English and five transactions in Japanese. Each transaction is issued monthly. All of the papers published in the *IEICE Transactions on Communications* are edited by the IEICE Communications Society. *IEICE Transactions on Communication* receives more than 2000 submissions per year from all over the world. This book presents a selection of the best papers published during the 1997–2002 period. It will allow all readers, including students and researchers, to easily discover the most important study results.

Chapter 1 presents an overview of the whole area of transport and switching technologies. It starts with ATM switch architecture reviews with emphasis on the high-speed backbone switch. ATM switch architectures have been studied for the last 10 years. Switch design is moving away from pure theory to more realistic and more application-oriented issues. This chapter goes on to describe ATM switch architectures from the viewpoints of QoS and traffic control performance. QoS service categories are introduced and characteristics are described in detail. This chapter then deals with the traffic issues faced by ATM switches; it pays particular attention to the ATM-LAN case since it raises many issues including IP over ATM, multi-protocol over ATM, voice over ATM, and MPEG-2 over ATM. This is followed by a presentation of the operation and management of ATM systems. The operation system (OpS) is one of the keys to realizing flexible and dynamic telecommunication services. Next, an ATM system that can be used as the backbone is examined. Any future backbone network will require Tbit/s throughput. An innovatively structured ATM-based switch architecture that uses optical WDM technology is then introduced. It is followed by an examination of photonic switching and networking technologies. These technologies are also very important for achieving the backbone network. In addition, the new concept of the photonic router as applied to IP/WDM networks is presented. The photonic router has both IP packet switching capability and wavelength switching capability. These switching capabilities are selectively used according to traffic demand fluctuations. Finally, this chapter describes next-generation Internet and the role of IP over photonic networks.

Chapter 2 presents services and broadband network applications. High-performance backbone networks will stimulate future multimedia services. The most important applications are IP services, voice and telephony service, real-time video transfer, and data transfer. This chapter first describes ATM network services. It also covers standardization issues and some development results. Virtual path (VP) network design methods in ATM networks including VP restoration are then described. A critical network service attribute is reliability. ATM is very effective in realizing high-performance network failure protection. Network reliability and efficiency are determined for a particular VP network topology. Next, QoS control and failure-resistant transport studies are presented. These studies provide important information on the support of guaranteed services such as voice, image, and TV program transmission. Network scalability is also a crucial issue for designing public networks. A scalable, private network node interface (PNNI) interworking architecture is also introduced. It is followed by a description of real-time MPEG transmission in ATM networks. Finally, a layer3 switching technology is introduced. The layer3 switch enables IP datagrams to be rapidly transmitted through the use of the cut-through technique.

Many studies on ATM architectures have emphasized buffer performance and speed. Chapter 3 addresses ATM switch architectures and systems. This chapter starts with a presentation of switch architectures such as input-buffered and input/output-buffered types. To expand switch capacity, multi-stage switch architectures have been advocated. An effective multi-stage switch architecture is introduced. Another important topic in discussing ATM switch architectures is multi-casting. Several different multi-cast architectures are presented. Next, a system-level demonstration of a digital MPEG transport network is presented. A wide-area network can be realized by ATM with local networks based on HIPPI. An experimental system for ATM access networks is also introduced. A subscriber system based on ATM-PDS (passive double star) is described. ATM-PDS is a very cost-effective way of realizing many ATM transfer capabilities. When line speeds become high, the allowable scheduling time decreases and can become a bottleneck. A scheduling scheme that relaxes the constraint of the scheduling time is then introduced. A pipeline-based maximal-sized matching scheduling approach is proposed. Output-buffered-type switches are attractive because they provide ideal delay performance. However, it is difficult to implement high-speed switches. To avoid this problem, the design and analysis of a packet concentrator are presented. The chapter continues on to introduce a high-capacity switching scheme for ATM/WDM networks. The proposed architecture is contention-free, scalable, easy to implement and requires no internal "speedup." Finally, an innovative distributed traffic control scheme for a large multi-stage ATM switching system is proposed. Operation, administration, and maintenance (OAM) cells are defined for the system. They carry route congestion information within the system.

Chapter 4 addresses ATM and IP integration issues. Demand for Internet service is doubling every year, and many kinds of services such as voice, data, leased lines, and video, all of which are associated with multimedia services, are being carried by IP protocols. The problem is how to carry IP protocols that need QoS support functions. IP over WDM or IP over XXX over WDM are potential problem solvers. With regard to the XXX approach, ATM technology was originally established for the subscriber network transport because of its flexible multi-bearer multiplexing capability and QoS support; QoS is also very important for backbone transport capability. This chapter starts by presenting experimental results on an IP + ATM integrated system and networks. Mobile communication based on IP over ATM technology is then introduced. Mobile networks need hand-off and rerouting functions, and location management is also needed for IMT2000. This topic is followed by several performance studies on IP-over-ATM wide area and access networks. These studies include TCP/IP over ATM, switched virtual connection (SVC)-based IP over ATM, and TCP/IP over an asymmetric digital subscriber line (ADSL). Controlling higher-layer traffic flows is very difficult and is a very hot topic. Next, a newly developed traffic measuring tool that has been applied to the real-time simulation of a network is described. This chapter then proceeds to introduce a new extension to classical IP and address resolution protocol over ATM protocol. Support for the real-time and multimedia applications emerging in IP over ATM systems is becoming indispensable in the networking world. Accordingly, a developed ATM switching system with a built-in IP processing mechanism and results of experiments on the accommodation of TCP traffic by an ATM wide-area network are described. This chapter concludes with suggestions on future requirements and directions for technologies.

ATM is a very hot topic because it offers QoS support. For this, it needs traffic control, bandwidth management, and provisioning. Chapter 5 presents studies on the performance of ATM. This chapter starts by describing performance studies on input-buffered ATM switches. This is followed by reports on the delay and performance of the AAL2 switch, which will be used for low bit-rate voice services. Voice services are delay sensitive. This means that the time out incurred by cell assembly must be considered. This chapter continues by presenting a new jitter-controlled traffic scheduling algorithm that provides an upper bound for the delay jitter in the case of rate-controlled connections. It is applied to packet video streams and IP telephony. It is followed by studies on ATM performance for higher-layer protocols such as available bit rate (ABR), guaranteed frame rate (GFR), and fast reservation protocol (FRP). End-to-end delay performances using these protocols are examined in detail. In addition, connection admission control (CAC) and routing problems are also addressed. Next, as a scalable IP-QoS architecture, differentiated services (Diffserv) is considered to be useful within the service provider's backbone and enterprise networks. A Diffserv over ATM architecture that provides Diffserv-based QoS over ATM access networks is studied. Finally, this chapter introduces an integrated scheme for scheduling real-time traffic. It achieves cell loss control in high-speed ATM networks with multiple priorities.

Traffic control studies are presented in Chapter 6. One important function offered by ATM networks is traffic control to guarantee QoS requirements. Several studies have addressed the traffic control issues of connection admission control (CAC), usage parameter control (VPC), bandwidth allocation, traffic measurement, flow control and traffic shaping. This chapter describes the cell traffic patterns that are enforced by the VPC circuit. This is followed by studies on measurement techniques that are used to estimate the residual ATM bandwidth and traffic shaping techniques for VPC circuits. Scheduling algorithms such weighted round robin (WRR) and static priority scheduling are then examined. Next, an ATM flow control method and its design scheme are introduced. The ATM layer protocols are emulated at the edge node and link utilization is monitored periodically between edges. This chapter then presents reliable VP control strategies, an efficient fuzzy traffic policer, and multicast service. In IP networks, the sharing of network resources among the many flows is on a best-effort basis. A first-in-first-out (FIFO)-based mechanism that achieves fairness in terms of throughput is then introduced. Its performance in terms of both fairness and link utilization is examined using extensive simulations. This chapter concludes by addressing end-to-end TCP flow control mechanisms that utilize available bit rate (ABR) information in ATM networks. Goodput comparisons and fairness issues are described.

The next three chapters describe advanced technologies with special emphasis on photonic technologies. These three chapters basically examine networks, architectures, and devices. Very advanced devices are needed to realize photonic systems. Photonic networks are considered in Chapter 7. Photonic technologies are critical to realizing the future backbone network. Recent progress in WDM has led to a breakthrough in network performance. This chapter starts by showing an IP over WDM network architecture. An ultra-fast bus architecture using TDM technology is then described. The remote bridge technique is used in a hierarchical WAN architecture. Some optical TDM transmission experimental results are also presented. This is followed by studies on a backbone crossconnect and one study on a local-area network. The optical technology requirements to ensure application to the backbone and those related to the access network are quite different. This chapter continues by introducing an optical bridge and switch function for protection networks and an all-optical code division switch. Next, network survivability and reliability are addressed. Networks carrying huge transmission loads must provide survivability, since even a very short outage will cause a large amount of information to be lost. Accordingly, optimal configurations for optical cross connect (OXC) systems and optical add-drop-multiplexer (OADM) systems, and a photonic switching system called WAPS (wavelength assignment photonic switching system), are presented. Finally, this chapter addresses the issue of IP over glass (IOG). IOG is a link-layer protocol specially designed for high speed and bandwidth efficient transmission of IPv4 and IPv6 datagrams over optical fibers.

A photonic switch architecture system and developments are described in Chapter 8. Photonic switches are the key to realizing high-performance systems. Photonic switch architectures can be classified into three types: space division, time division packet, and add drop multiplexer (ADM). This chapter first reports an architectural evaluation of photonic switches, and then focuses on a number of optical devices. An architectural performance study on a wavelength division multiplexed network is then described. Next, two prototype switching systems are reported. One is a photonic core node based on a 2.56-Tbit/s opto-electronic switching fabric, and the other is a large-capacity photonic packet switch prototype based on wavelength routing techniques with a throughput of 320Gbit/s. This chapter then presents an all-optical label swapping technique for the photonic label switching router (LSR), an optical code division multiplexing (CDM) self-routing network system, and an experimental 5 Tbit/s packet-by-packet wavelength switching system. Finally, this chapter addresses optical label switching. A novel concept for an optical label switching network and the configuration and function of the core nodes are described.

The final chapter reports on photonic and advanced switching devices. The chapter covers hardware technologies to realize backbone networks. There are several contributions, all of which use state-of-the-art technologies. For realizing a very high-speed backbone system, photonic devices are key components and optical wavelength conversion is the next breakthrough needed. This chapter first reports several important optical wavelength conversion technologies. It continues with a discussion of a broadcast-and-select-type optical switch element, a Tb/s ATM switch architecture that uses both electrical and photonic technologies, and a 10 Gb/s very high-speed packet receiver module. This is followed by a description of the design of micromachined temperature-insensitive filters. Temperature-insensitive add/drop filters and wavelength tunable lasers are indispensable devices for future optical networks. Next, this chapter deals with VCSELs (vertical cavity surface emitting laser diodes). This decade has seen a continuous increase in the level of research on VCSELs. To increase interconnection speeds, optical WDM and parallel interconnections are being implemented using VCSELs. As is well known, WDM technologies are suitable for long-haul transmission. However, for short-reach transmission, a cost limitation occurs because multiple laser sources are expensive. Therefore, optical parallel interconnection is a key technology for cost-effective short-reach transmission. Photonic crystals (PCs) have seen major advances in the last few years. Experiments on a variety of structures and devices, as well as modeling tools, are then described. Finally, this chapter reviews the recent progress made by those working on optical filters and switches for photonic networks based on dense WDM. Three kinds of optical filters—thin-film interference filters, fiber gratings and AWGs—are described in detail.

ACKNOWLEDGMENTS

I thank Dr. Eiji Oki of NTT Network Innovation Laboratories for selecting several key papers from *IEICE Transactions on Communications* without which this book would not be complete. Finally, I would like to acknowledge Mr. Taisuke Soda and other employees of Marcel Dekker, Inc., and IEICE executives for supporting this project.

Naoaki Yamanaka

Contents

Chapter 1

Overview

The level of IP traffic is rising dramatically due to the increasing popularity of WWW access, e-mail, and IP-VPN service. However, the IP network is still in its first generation, i.e., it remains a best effort, point-to-point based computer network. Many researchers are now enthusiastically tackling the next generation IP network that offers quality of service, high-security, high-reliability, and high-performance. The network would well support business use and mission-critical services as well as digital video stream broadcasting.

These requirements can only be satisfied by grounding network transport and switching technologies on IP, ATM and WDM. In other words, combining these three technologies appropriately will realize the multi-Quality-of-Service (QoS), multi-functional backbone network.

This chapter examines all of these transport technologies. Paper 1.1 presents ATM switch architecture reviews with emphasis on the high-speed backbone switch. The target is classified as the buffering type, ATM switch architectures studied in the last 10 years. Switch design is moving away from pure theory to addressing more realistic and more application-oriented issues.

Technical background topics such as CMOS VLSI technology are also explored in detail. A classification based on general buffer location is introduced and performance is discussed. Next, a multi-stage architecture and its performance are described. In addition, sorting network architectures are also explained in detail. Finally, realization of the multicasting function is shown.

Paper 1.2 also describes the ATM switch architecture but from the viewpoints of QoS and traffic control performance. The paper first introduces the QoS service categories and describes characteristics in detail. Next, switching architectures are briefly described. In addition, a strategy for controlling the QoS is shown together with a discussion of switch implementation.

Paper 1.3 deals with traffic issues for ATM switches, for the ATM-LAN case in particular. The LAN area covers many issues such as IP over ATM, Multi-protocol over ATM, Voice over ATM and MPEG-2 over ATM. The ATM transfer technologies are also detailed in the paper. Emphasis is placed on standardization issues. For those items, traffic control methods, such as rate control and packet level control, are described.

Papers 1.4 and 1.5 address the operation and management of ATM systems. The operation system (OpS) is one of the key technologies for realizing flexible and dynamic telecommunication services. Paper 1.4 overviews operation and management systems and some development results. Paper 1.5 considers a different viewpoint. The network operator designs and controls the network through the construction management system. The real network is controlled using this system through the OpS.

Paper 1.6 examines an ATM system that can be used as the backbone. Any future backbone network will require Tbit/s throughput. The paper shows an innovatively structured ATM-based switch architecture that uses optical WDM technology. The architecture is based on a 3-stage switch network. Each stage is connected by optical WDM interconnection. The switch has several Tbit/s switching capabilities.

Photonic switching technologies are also very important in achieving the same goal of the backbone network. Papers 1.7 and 1.8 detail photonic network and switch technologies. Paper 1.7 describes an information network architecture that uses photonic technologies. The network consists of a two-layer switch. The local network is based on an optical WDM broadcasting and selection network. The network can realize point-to-point and multicasting with more than 10 Mb/s transport capacity. The transit network is based on photonic ATM over WDM cross connects. Combining those two networks will yield a future backbone network.

Paper 1.8 overviews photonic switches. Realizing the photonic switch will require advances in device technologies and architectures, the paper describes the switch architecture and some basic signal processing technologies.

Papers 1.9 and 1.10 address reliable transport ATM networks. Paper 1.9 compares the performance attributes of the VP transport network with those of the conventional SDH network and future optical path-based transport network. It is shown that the VP-based network is superior in terms of path-grooming efficiency and connection capability, and will still play an important role when WDM technologies are introduced to carrier networks. Paper 1.10 proposes a self-healing scheme that achieves multiple-reliability levels on ATM networks. The scheme proposed realizes a multiple-reliability level, virtual path network that efficiently integrates and logically manages several VP network services requiring different reliability level.

Paper 1.11 introduces a new concept of the photonic router as applied to IP/WDM networks. The photonic router has both IP packet switching capability and wavelength switching capability. These switching capabilities are adaptively used according to traffic demand fluctuations. The photonic router handles multi-protocol lambda switching, which is an extension of MPLS. The proposed concept is now being standardized as Generalized MPLS in Internet Engineering Task Force (IETF).

Paper 1.12 reviews different IP/WDM networking architectural models and their tradeoffs. The authors outline and discuss a number of management and control issues and possible solutions related to the configuration, fault, and performance management of IP over dynamic WDM networks. Analytical and simulation results demonstrating the potential benefits of dynamic IP over WDM networks are described.

Paper 1.13 describes next-generation Internet and the role of IP over photonic networks. It discusses QoS metrics of the data networks, and elucidates the problems challenging the next-generation Internet. It also addresses how the WDM technology can be used to resolve those problems.

High-Performance Backbone Network Technology edited by Naoaki Yamanaka
ISBN: 0-8247-5321-6 DOI: 10.1081/(120025647)
Published 2004 by Marcel Dekker, Inc.

Architectural Choices in Large Scale ATM Switches

Jonathan Turner and Naoaki Yamanaka

SUMMARY The rapid development of Asynchronous Transfer Mode technology in the last 10–15 years has stimulated renewed interest in the design and analysis of switching systems, leading to new ideas for system designs and new insights into the performance and evaluation of such systems. As ATM moves closer to realizing the vision of ubiquitous broadband ISDN services, the design of switching systems takes on growing importance. This paper seeks to clarify the key architectural issues for ATM switching system design and provides a survey of the current state-of-the-art.

key words: *ATM, switch, B-ISDN, buffer, architecture*

1. Intoduction

Switching systems are central components in communications networks and serve two key purposes. First, they allow a reduction in overall network costs by reducing the number and/or cost of the transmission links required to enable a given population of users to communicate. Second, they enable heterogeneity among terminals and transmission links, by providing a variety of interface types. ATM networks have introduced a range of new issues into switching system design, but the principal ones arise from two key features of ATM. The first is its support of multi-rate virtual circuits with statistical multiplexing for efficient transmission of multimedia and data traffic. The second is its support for multicast virtual circuits which enable efficient transmission of information from a source to many receivers. At the same time, switching systems for ATM, must address the fundamental issues of scalability, reliability and cost-effectiveness. The pressure of the competing requirements on switching systems, together with new opportunities created by advances in underlying technologies, have helped to spark a remarkable period of creativity in switching system design.

As the title of the paper suggests, we attach considerable importance to the issue of scalability in ATM switch design. While one can build communication networks of arbitrary size using switching systems of limited capacity, it turns out to be far more cost-effective to use switching systems that have sufficient capacity to handle traffic from tens or hundreds of thousands of individual users. As ATM networks are more widely deployed and approach the sort of ubiquitous usage that telephone networks enjoy, there will be a compelling need for high capacity systems. Just as telephone switches require capacities of 1–10 Gb/s to provide cost/effective service to large user populations, ATM switches supporting applications using thousands of times the bandwidth of traditional voice services, will need terabit per second capacities. This in turn demands switch architectures with linear or near-linear scaling characteristics.

There have been several prior surveys of ATM switching that provide useful insights into this field. The papers by Ahmadi and Denzel [1] and Tobagi [101] provide good coverage of the state-of- the-art as of the 1990 time frame. The paper by Zegura [117] introduces a systematic method for comparing architectural alternatives in a quantitative fashion and uses it to demonstrate substantial differences in the cost/performance of different switch architectures. In addition, the book by DePrycker [29] includes a survey of a representative set of ATM switching system designs and the collection of papers by Dhas, Konangi and Sreetharan [31] provides a good introduction to the literature.

This paper seeks to place the key architectural issues raised by the design of ATM switching systems into perspective. While we do not provide a comprehensive survey, we have sought to capture the major ideas that have emerged from the switching research community in the last decade, and to fit them into a framework that facilitates the comparison of different system designs. The paper does not make any explicit technology assumptions, but we do tend to focus on systems that can be effectively implemented using CMOS integrated circuits. CMOS is clearly the dominant commodity circuit technology today and is consequently the standard of comparison against which other circuit technologies must compete. Higher speed technologies, such as GaAs, do allow relatively high capacities to be achieved with single stage designs, providing some potential advantage, but this advantage has yet to be demonstrated. We group systems into three major categories: single stage switching systems, buffered multistage systems and unbuffered multistage systems. The succeeding sections of the paper are organized by these categories and within

Manuscript received May 6, 1997.
Manuscript revised July 27, 1997.

[†]The author is with the Department of Computer Science, Washington University, St. Louis, MO 63130–4899, USA.
[††]The author is with NTT Network Service Systems Laboratories, Musashino-shi, 180–8585 Japan.

IEICE TRANS. COMMUN., Vol. E81-B, No. 2, pp. 120–137, February 1998

each section, a variety of sub-categories are detailed.

2. Single Stage Switching Systems

Most ATM switches today use one of several single stage switching techniques. Single stage switches are relatively simple, but are limited in the number of ports and total throughput that they can support effectively. This section surveys the major categories of single stage switches and discusses the key design and performance issues.

2.1 Buses

The most common form of switching system is built around a simple passive bus, as illustrated in Fig. 1. Arriving cells first pass through an *Input Transmission Interface* (ITI), which converts the (typically optical signal) to a parallel electronic form. This involves recovering timing information from the encoded signal and presenting the decoded signal to the subsequent components. The *Input Port Processor* (IPP) implements the input processing functions at the ATM level. This typically involves synchronizing the arriving data stream to the internal timing of the switch, checking for ATM header errors and performing a routing table lookup based on the arriving cells VPI and VCI. Given this information, the lookup yields an outgoing port number and a new VPI and VCI which replace the original values of these fields as the cell is forwarded on the outgoing link. To transfer data to outputs, IPPs contend for access to the bus (using one of a variety of bus-contention techniques) and then transmit cells on the bus in parallel form, using all w signal lines. Along with the received data, the IPP transmits the number of the output port on which the cell is to be forwarded. The *Output Port Processors* (OPP) compare the output port numbers in cells observed on the bus to their own addresses and buffer matching cells in a queue prior to transmission. The OPP may implement a simple fifo, a collection of several fifos with differing priorities or per virtual circuit queues. It may also implement one or more congestion control mechanisms to improve queuing performance in the presence of bursty data traffic.

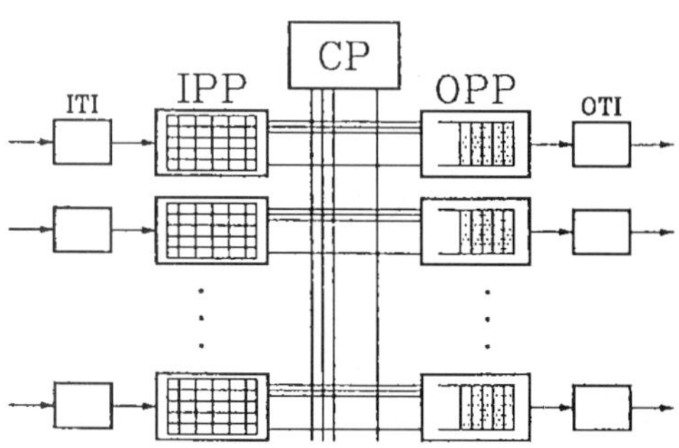

Fig. 1 Simple bus-based switch.

The *Output Transmission Interfaces* (OTI) encode data for transmission and convert it to optical form for transmission to distant switches and terminals.

In addition to the data path components emphasized in Fig. 1 a switch must also have a *Control Processor* (*CP*) which configures the IPP routing tables in response to user requests. The CP is typically implemented as a general-purpose processor with the appropriate software to perform terminal-to-switch and switch-to-switch signaling, in addition to switch control and maintenance functions. In some systems, the CP is connected directly to the bus and may also have a separate control bus connecting it to the memory implementing each of the routing tables. Alternatively, the CP may be connected to the system through one of its external ports, and communicate control information using ATM cells carried on a designated VPI/VCI. This requires that the IPPs and OPPs have the ability to interpret and respond to control cells appropriately. In larger switching systems, multiple control processors may be required. The most straightforward approach in this case is use a conventional shared memory multiprocessor. To provide non-blocking communication, a bus supporting n ports, operating at a data rate of R bits per second each, must provide a bandwidth of at least Rn bits per second. If the clock frequency used for the bus is r Hz, the bus width, w, must be at least Rn/r. For example, a system supporting 16 OC-3 links with an internal clock rate of 40 MHz, requires a bus width of about 64 bits. Notice that as the number of ports in a system increases, both the number of ports connecting to the bus and the width of the bus must increase. This yields a quadratic growth characteristic, making it uneconomical to implement very large systems. Another problem with bus systems is that as the number of ports connecting to a bus increases, the capacitive loading on the signal lines grows, reducing the maximum clock frequency that can be achieved (that is r shrinks with the capacitive loading). This means that the bus width must grow faster than the port count in order to maintain sufficient bus bandwidth. There are techniques to reduce or eliminate the impact of this capacitive loading effect, but they are more complex to implement.

For smaller capacity switches (up to a few Gb/s), the bus-based architecture is very common. A typical commercial system is the ForeRunner ASX-200 made by Fore Systems [37]. TranSwitch produces and sells an integrated circuit that implements the IPP and OPP functions and the required bus- interface logic for a bus-based switch. A detailed description can be found in Ref. [106].

2.2 Rings

Figure 2 shows a single stage system implemented using a ring. This system contains the same components

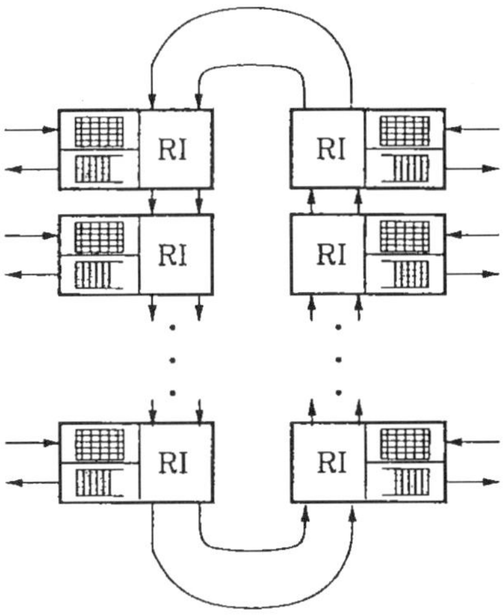

Fig. 2 Ring-based switch.

as the bus system described above, but replaces the bus with a ring-based interconnect. Each IPP and OPP has an associated *Ring Interface* (RI), which is responsible for inserting data onto the ring and removing data from the ring. The simplest ring protocol for ATM switches uses a time slotted approach in which cells are sent on the ring synchronously during specified time slots, and a busy/idle bit is used to indicate the availability or unavailability of a given time slot. An IPP with a cell to send waits for the start of a time slot in which the busy/idle bit indicates that the time slot is available, changes the value of the bit to make the time slot busy and then transmits the cell within the time slot. OPPs compare the output port number of transmitted cells to their own address and copy matching cells off the ring, changing the value of the busy/idle bit to indicate that the slot is now available.

Ring systems, like buses, have quadratic growth characteristics, but unlike buses, they are not further limited by capacitive loading effects. Because all data transmission is point-to-point, one can typically operate rings with higher clock frequencies than bus systems implemented using the same technology. This can become a significant factor in systems with large numbers of ports. Rings do introduce some additional latency, relative to buses, but for switching applications these latencies are typically fairly modest. For example, a ring supporting 16 OC-3 links with a ring clock rate of 80 MHz and a latency of two clock ticks per ring interface would have a latency of under one microsecond, which is about one third the time required to transmit the cell on the external OC-3 link. While rings are popular in local area networks and have advantages over buses in switch design, they are not often used for this purpose. One system that does use a ring for switching is described in [96].

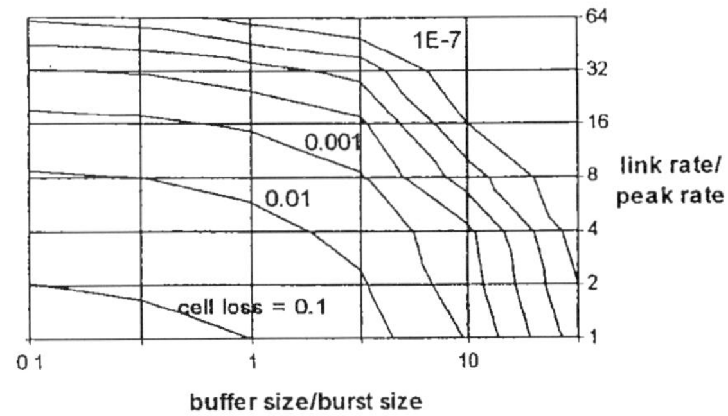

Fig. 3 Contour plot of cell loss rate (peak-to-avg = 10, offered load = 0.6).

2.3 Queuing Issues in Single Stage Switches

ATM networks are designed to handle a variety of different types of applications, including applications with time-varying data rates. For applications in which the peak data rate is much larger than the average data rate, statistical multiplexing is used to share the link bandwidth most efficiently among the competing data streams. In single stage systems based on buses and rings, the queuing occurs primarily at the OPPs. The amount of memory needed for efficient statistical multiplexing is highly dependent on the ratio of the link rate to the peak transmission rate of individual streams. If this ratio is high enough (say 20-to-1 or more), small buffers are sufficient. If the ratio is small, the total buffer requirement at an OPP may need to be as high as ten times the size of the average sustained data burst, in order to achieve acceptable buffer overflow probabilities. Note that when we say burst size, we mean the size of an application-level burst of data, not the size of a transport level packet. For an application such as interactive, high resolution image retrieval, burst sizes of 1 MB are typical, leading to required buffer capacities of up to 10 MB, if the data transfer rates are high, relative to the link rates.

Figure 3 is a contour plot showing how the cell loss rate varies as a function of the ratio of link rate to peak virtual circuit rate, and as a function of the ratio of burst size to buffer size. This plot was produced by simulation of on-off bursty traffic with exponentially distributed holding times in the on and off states and an average traffic intensity equal to 60% of the link's capacity. Similar plots can be produced for different holding time distributions, such as a heavy-tailed distributions like the Pareto distribution. The effect of changing to a Pareto distribution is to shift the place where the contours intercept the horizontal axis to the right. There is no change in where the contours intercept the vertical axis.

For systems in which large buffers are needed to obtain efficient statistical multiplexing, delay can become a significant factor. This leads to a requirement for multiple priority levels, to allow real-time traffic with limited rate variability to obtain access to the link without substantial queuing delays. As a result, switches will often have two or more service class specific queues in the OPP. Some systems go further and provide per virtual circuit queues to ensure fair access to the output link for competing bursty data sources. This adds some complexity to the output port processor but can be important for good performance when the ratio of link rate to peak virtual circuit rate is small. This is most often the case when the link rates are relatively low (150 Mb/s or less). Higher bandwidth links can often get by with smaller amounts of buffering and less sophisticated buffering techniques.

In a bus or ring system in which the bus/ring bandwidth is equal to the sum of the input link bandwidths, all queueing occurs at the outputs. Consequently, the overall queueing performance is determined entirely be the OPP design. Other switch architectures can approximate this characteristic by making the bandwidth of the data path between interconnection network and the output port processor larger than the output link bandwidth. In most cases, a factor of two is adequate and a factor of eight to ten is ample under any realistic traffic scenario.

2.4 Shared Memory

In bus or ring based switches in which large queuing memories are provided at the output ports to ensure good statistical multiplexing performance, one observes that the average memory utilization is usually quite low. This suggests that there is an opportunity for reducing the overall system cost by sharing the memory among the various output ports, instead of dedicating it to each output. This can be done by introducing a large shared memory connected to the bus or ring containing per-link or per-VC queues, implemented as linked lists. Because this implies that cells must pass first from the input ports to the shared memory and from there to the outputs, the bus/ring bandwidth must be doubled for non-blocking switching and the memory bandwidth must be at least as large as the bus/ring bandwidth. While this can be difficult to achieve in larger, high speed systems, the cost-reduction available through the sharing of the memory can be substantial. In typical cases, the amount of memory needed in a shared buffer system is just two to three times larger than the amount needed for a single output port in the case of dedicated buffering. So in a 16 port OC-3 switch, we could reduce the buffer memory requirement by a factor of 5 to 8, at the cost of increasing the parallelism of the bus/ring from 64 to 128 and using a memory subsystem with an aggregate bandwidth of at least 600 MB/s.

MMC Networks makes a set of chips that can be used for shared memory switches. A detailed architectural description of a shared memory switch using these devices can be found in [79].

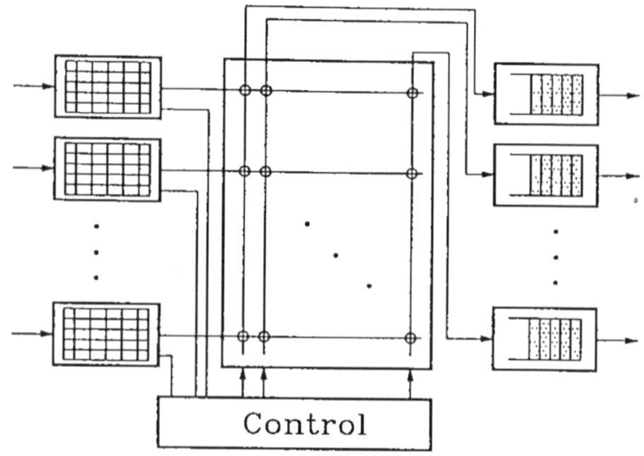

Fig. 4 Crossbar-based switch.

2.5 Crossbars

The last major class of single stage switch is the crossbar, shown in Fig. 4. A crossbar comprises an array of crosspoints, which when closed, connect a horizontal data path to a vertical data path, thereby creating a connection between an input port and an output port. Crossbar-based systems can be significantly less expensive than bus or ring systems with equivalent performance because the crossbar allows multiple data transfers to take place simultaneously between disjoint input/output pairs. This means that each IPP (OPP) need only have the ability to send (receive) data at the rate of its input (output) link times a small constant (called the speed advantage), while in a bus or ring system, it must be able to send (receive) data at n times the link rate. The overall bandwidth of the crossbar must be at least as large as the comparable bus or ring, but each IPP and OPP need not interface directly at the full system bandwidth, allowing the overall system cost to be significantly reduced.

The speed advantage required in a crossbar system depends on the nature of the traffic and the desired queuing behavior. If the queuing is to be done primarily at the outputs, the speed advantage must be large enough so that contention for outputs, caused by the natural random timing of cell arrivals, does not cause too many cells to accumulate at input ports. For bursty traffic with peak rates that are small relative to the link rates, a modest speed advantage (say 2) is sufficient. For larger peak rates, the speed advantage may need to be increased to prevent significant queuing at the input ports. References [32], [41], [114] describe crossbar-based switches that use a speed advantage to reduce or eliminate output contention. The knockout switch described in [115] also addresses output contention by increasing the bandwidth of the data path from the crossbar to the OPPs. In this case however, there is no central controller. Instead, there is a knockout concentrator for each output, which accepts cells from any of the inputs and concentrates them onto a smaller number of outputs (typically 8), which in turn feed into the output buffer in the OPP.

Some switch designers seek to avoid the requirement for an output speedup by providing cell buffers at each crosspoint of the crossbar. If sufficient buffering is provided in the crossbar, head-of-line blocking can be avoided in the IPPs and the buffers storing cells for any particular output compete for that output's bandwidth, independently of all other outputs. Examples of such designs can be found in [82], [83], [103]–[105]. Because this approach dramatically increases the complexity of the crosspoints, the number of which grows quadratically with the number of switch ports, it becomes prohibitively expensive (in comparison to alternative designs) for all but the smallest configurations.

Another way to avoid the need for a speed advantage in the crossbar is to queue at the input ports instead of at the outputs. However, to achieve efficient system operation, one cannot use a single fifo input queue, since this leads to head-of-line blocking. There are several alternatives to a single input queue. The most effective is to maintain a separate linked-list queue for each output port, or each virtual circuit. Another benefit of input buffering is that it does allow a certain amount of buffer sharing, reducing the memory requirements, relative to a system using output queuing or crosspoint queueing. Examples of designs of this sort can be found in [55], [56], [69], [71], [72], [92].

The control of crossbar systems is significantly more complex than for bus or ring systems. Typically, IPPs signal to a central crossbar controller, which output or outputs they have cells for, and the controller matches inputs to outputs so as to maximize the number of cells that can be switched in the upcoming data transfer cycle, configures the crosspoints appropriately and informs each selected IPP of which output it has been connected to. The IPPs may include a priority for each requested output (typically these are dynamic priorities that depend on the number of cells that each IPP has waiting for the requested output). In this case, the controller seeks to maximize the sum of the priorities of the cells selected for transmission. The problem that must be solved by the controller is a maximum matching problem. While exact algorithms for both the weighted and unweighted versions of the problem are known, they are computationally too expensive to use in most switching applications. Consequently, various approximate matching algorithms are generally substituted. While these produce good results in typical situations, they can produce matchings that are significantly sub-optimal in the worst-case [71], [72], [92].

2.6 Multicasting in Single Stage Switches

An important feature of ATM networks is their ability to support multicast virtual circuits for communication among groups of arbitrary size. A simple one-to-many multicast enables data transmission from a source to many receivers. Many-to-many multicasts are also possible, although not currently supported by standards. For switches with a moderate number of ports, multicast can be implemented by including a bit vector in the routing tables in the IPP, with one bit for each output port. When an IPP transfers a cell on the bus or ring, it includes the bit vector with the cell. OPP_i, upon receiving the cell, checks bit i of the bit vector and copies the cell to its output queue if the bit is set. Otherwise it ignores the cell. This approach, while very simple, has the drawback that it results in every copy of the cell having the same VPI/VCI combination. This complicates the switch control software considerably and can lead to blocking due to clashes on the VPI/VCI values. This can be corrected by adding routing tables to each of the output ports, containing new VPI/VCI values for the outgoing cells. Typically, these tables are indexed using a *Multicast Identifier* obtained from the IPP routing table. In this case, each output routing table contains an entry for every multicast identifier, including those for which it does not actually forward cells. This leads to inefficient use of memory (potentially, this approach uses nearly n times the memory required for point-to-point switching), which can be avoided by using a *Content Addressable Memory* (CAM) instead of a simple lookup table.

In shared memory switches, arriving multicast cells must be included in the linked list output queues for all the output links on which the cell is to be sent. When a cell arrives at an IPP, it is transferred on the bus or ring to the shared buffer, copied into memory and then a pointer to the stored cell is included in the linked list for each of the outputs which is to get a copy. The translation from multicast identifier to outgoing VPI/VCI can be done when the cell is read from each output queue on which it is placed. Again, this can be done with either a simple lookup table or a CAM.

For crossbar switches, multicast forwarding raises some more difficult issues. First, the crossbar controller must be able to accept requests to send a single cell to an arbitrary subset of the outputs rather than to just a single output. This requires that an IPP with a multicast cell to send, provide a bit vector indicating which outputs it needs. If two contending multicast cells have some but not all outputs in common, efficient system operation requires immediate forwarding of as many copies as can be sent, without waiting for those that must be held back due to the conflict. This in turn, means that IPPs must keep track of which outputs a given multicast cell has been forwarded to, so that subsequent transmission attempts target only the ones that have yet to receive a copy. In systems using input queuing, queue management for multicast connections can be complex, if one seeks to prevent a busy output port

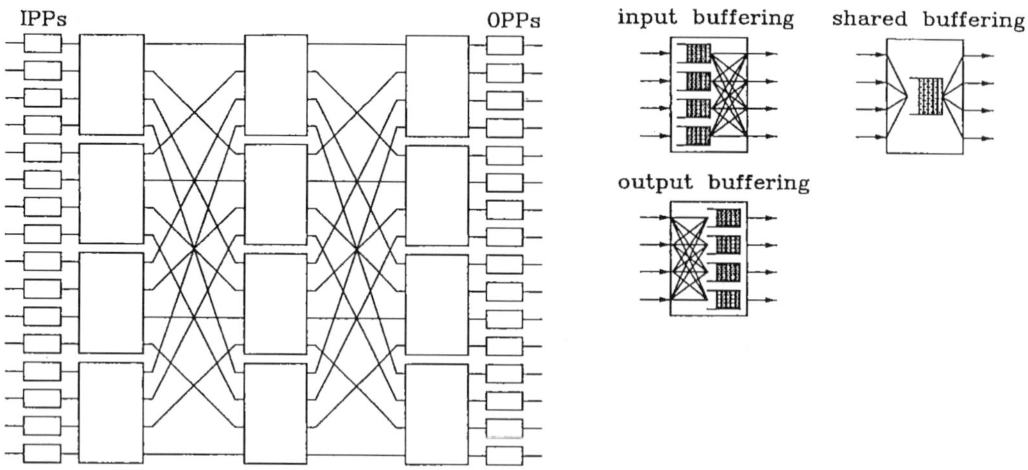

Fig. 5 Buffered multistage networks.

from slowing the transmission of copies of multicast cells to less busy ports.

2.7 Scaling Up Using Port Multiplexing

One way to build switching systems handling larger numbers of external links is to multiplex multiple external links onto a single higher speed link and then use a single stage switch whose ports operate at the higher speed. This approach is used in a number of ATM switches, which are designed for external links of 150 Mb/s but which internally use single stage switches capable of data rates of 600 Mb/s. One example is the Fore Systems ASX 200 [37]. For any given circuit technology, there is a limit to the applicability of this technique, but within its range of applicability it offers the most cost- effective way to scale up to larger sizes.

3. Buffered Multistage Switches

Single stage switching techniques are inherently limited by their quadratic complexity growth. While the use of a higher speed core to handle lower speed ports provides some relief, for any given technology, there comes a point where greater amounts of parallelism are needed to obtain higher throughput. For systems implemented using modern CMOS integrated circuits, buffered multistage networks are among the most attractive. An example of a buffered multistage network is shown in Fig. 5. Each of the switch elements used to construct the network can store some (usually modest) number of cells in its internal buffers. Queues can be located on the input side, output side or can be shared. There are two main classes of systems; those that use dynamic routing and those that use static routing. In systems that use dynamic routing, each cell is routed independently, spreading the traffic as evenly as possible among all the different paths between any given pair of input and output ports. In static routing systems, all cells in a given virtual circuit follow the same path through the multistage network. Static routing networks maintain cell ordering directly, but require explicit path selection and are far more subject to virtual circuit blocking. Dynamic routing networks require the use of some other mechanism to restore cell ordering, after cells emerge from the interconnection network. Typically this involves some form of *resequencing buffer*.

3.1 Dynamic Routing

Interconnection networks based on dynamic routing are among the most cost-effective alternatives for large switches. Perhaps the most popular topology for dynamic routing networks is the Beneš network, $B_{n,d}$ (the subscript n is the number of input and output ports that the interconnection network has, the subscript d is the number of inputs and outputs of the switch elements used to construct the network). The Beneš network is constructed recursively. For $n = d$, it consists of a single d input, d output switch element. For $n = d^2$, it has three stages, with d switch elements in each stage and a single link connecting each pair of switches in consecutive stages (the network in Fig. 5 is a Beneš network with $n = 16$ and $d = 4$). When used to implement an ATM interconnection network with dynamic routing, the first stage distributes arriving cells across all switches in the middle stage to balance the load. The second and third stages route cells to the output, using an output port number inserted in the cell header by the input port processor. This output port number is interpreted as a pair of base d digits, reducing the routing process to a straightforward selection of one of d outputs in each of the last two stages. The Beneš network topology generalizes naturally to five, seven or more stages. To construct a network with $n = d^k$, one first constructs d networks with $n = d^{k-1}$; these subnetworks form the central stages of the network being constructed. The central subnetworks are then preceded by a first stage with d^{k-1} switch elements, each of which is linked to all of the central subnetworks. The central subnetworks are similarly followed by a last stage with d^{k-1} switch elements, each of which is linked to all of the central subnetworks. The resulting network has $2k-1$ stages, altogether. The first $k-1$ of these stages distribute cells to balance the overall system load, while the last k route cells using successive base d digits of the cells' output port numbers.

The Beneš network is particularly attractive for dynamic routing networks because under dynamic routing, the average load on its internal links is guaranteed to be no larger than the maximum load on its inputs or outputs. This means that excellent performance can be obtained with internal link bandwidths that are not substantially larger than those of the external links. As with a single stage crossbar, some speed advantage may be needed to handle bursty traffic with high peak data rates. Because the traffic distribution reduces the effect of high peak rates on the Beneš network's internal links, it is usually sufficient to provide this speed advantage only at the very last stage of the network, on the data paths joining the last stage switch elements with the OPPs.

One drawback of the Beneš network is that it must be expanded in fairly large increments (factors of d). When d is larger than 2, this can be a substantial constraint. Fortunately, the topology can be easily generalized to allow expansion in smaller steps. Moreover, with some care in the design of the switch elements, these networks can be designed to allow expansion while the system is in operation. One issue with networks that use dynamic routing is that cells can get out of order as they pass through the system. This requires the addition of a *resequencing buffer* to each OPP, to restore the correct cell ordering. While a variety of resequencer designs are possible, the simplest approach is to use time-based resequencing in which cells are time-stamped when they enter the interconnection network and the OPPs use the time stamp information to determine the time cells have spent in the interconnection network, allowing them to reorder the arriving cells back into the correct sequence. Cells that pass through the network quickly are delayed in the resequencing buffer in order to give slower cells an opportunity to catch up. This means that the resequencing buffer does introduce some added delay, increasing the minimum latency of the system, but simultaneously reducing the cell delay variation. For systems with up to about 1000 ports, a resequencing buffer of between 50 and 100 cells can be sufficient to reduce the probability of out-of-sequence cells to very low levels.

Systems using buffered multistage networks with dynamic routing have been built by Alcatel [26]–[28], [46], [47], CSELT [6], [15], [17], [65] and Washington University [12], [108], [111].

3.2 Static Routing

Static routing networks operate similarly to dynamic routing networks, except that all cells in a given virtual circuit are constrained to follow the same path. The routing lookup performed in IPPs of static routing networks, inserts a path specification into the cell headers of arriving cells, rather than just output port numbers. The successive digits of a cell's path specification are used to select output ports in the sequence of switch elements that the cell passes through. A variety of interconnection topologies can be used, including the Beneš network. One of the most popular alternatives is a three stage Clos network which is similar to the three stage Beneš network but uses asymmetric switch elements in the first and third stages. In particular, the switch elements in the first stage have d inputs and r outputs, while those in the third stage have r inputs and d outputs. The r middle stage switches have an equal number of inputs and outputs, typically d.

When a new virtual circuit is to be added from some input port x to some output port y of a static routing network, the control processor managing the switching system must find some path through the network with sufficient unused bandwidth on each of its links to accommodate the new virtual circuit. For the three stage Clos network, this can require checking r pairs of links, to determine if they can accommodate the addition of the new virtual circuit. If none of the potential paths has enough unused capacity, the virtual circuit request is blocked.

If one abstracts the bandwidth requirements of a virtual circuit as a single scalar value, one can derive conditions under which a network is nonblocking. For the three stage Clos network, blocking can be avoided if the following inequality is satisfied.

$$r > 2\left\lceil \frac{\beta d - B}{1-B} \right\rceil - 2$$

In this expression, β is the ratio of the bandwidth of the external links to the bandwidth of the internal data paths (so the speed advantage is $1/\beta$) and B is the ratio of the maximum rate of any single virtual circuit to the bandwidth of the internal data paths. (For example, a system with 2.4 Gb/s external interfaces and internal data paths of 4.8 Gb/s will have $\beta = 1/2$. If the users whose virtual circuits are routed through this switching system all have access links that operate at 150 Mb/s, then $B = 1/32$.) If β and B are both 1/2, the number of middle stage switches needed to avoid blocking is $2d-3$. Similar results can be obtained for networks with more than three stages. In particular, a Beneš network is nonblocking for static routing if

$$(1/\beta) \geq (1-2/d)(B/\beta)$$

$$+ (2/3)(1 + (d-1)(k-1))$$

where $k = \log_d n$ and n is the number of ports. In general, the speed advantage needed to make the Beneš network nonblocking is between k and $2k-1$ and approaches $(B/b) + 2k - 2$ for larger values of d. This contrasts sharply with the case of dynamic routing which requires no speed advantage to make the network nonblocking, although, as discussed in the next subsection, some speed advantage (generally much smaller) is needed for good queuing performance in both cases. In systems that are already operating at the maximum clock rate that the given technology can support, increasing speed advantage must be obtained by increasing the amount of parallelism in the system (typically by increasing data path widths). This leads to a proportional increase in cost, making it advantageous to keep the required speed advantage as low as possible.

CNET [30] designed and implemented the first ATM switch using static routing. Similar systems were later produced by NEC [94], Fujitsu [116], Hitachi [59], [60] and IBM [25]. Virtual circuit blocking analyses for this class of systems can be found in [73]–[75].

3.3 Queuing Performance of Buffered Multistage Switches

The queuing performance of buffered multistage interconnection networks is the second key performance issue for this class of systems. The design parameters that have the most impact on performance are the speed advantage, the dimensions of the switch elements, their queue organization and queue size and the presence or absence of inter-stage flow control. In general, large switch elements with shared queues and inter-stage flow control give the best performance for a given total amount of buffering. For uniform random traffic, a system with shared buffer switch elements with eight or more inputs and outputs and four buffer slots per input and output, can handle fully loaded input and output links if the interconnection network has a speed advantage of about 1.25 or more.

There has been only limited study of the performance of this class of systems for time-dependent traffic, and the models used for these analyses are not completely realistic. However, there are some general observations that can be made. Systems using a Beneš network with dynamic routing can generally handle any time-dependent traffic for which the instantaneous load does not exceed the bandwidth of the interfaces between the last stage switch elements and the OPPs. Consequently, if one can ensure that such overloads occur very rarely or not at all, one can be assured of excellent queuing performance with only a small speed advantage. Determining the probability that an interface is overloaded at any random instant is easy to determine for a wide variety of multiplexed data streams with time-dependent behavior, using direct numerical convolution of the individual sources' rate distributions. In general, overload probabilities are small, even for average external link loads close to 100%, when the ratio of the peak bandwidths of individual virtual circuits is small relative to the interface rate. Where this condition is not met, one must increase the speed advantage at the output interface to reduce the overload probability to acceptably small levels. However, it is not necessary to increase the speed advantage throughout the interconnection network. One can eliminate the possibility of overloading the interface between the last stage and the OPP by implementing per virtual circuit flow control mechanisms between the OPPs and the IPPs. This adds significant complexity to the IPP and OPP, although the memory needed for the flow control state information and IPP buffering need not be impractically large in absolute terms.

For networks that use static routing, the queuing performance in the presence of time-dependent traffic is somewhat more problematical. In these systems, congestion can occur on any inter-stage link in the interconnection network. There are essentially three ways to address the congestion issue. First, one can ensure that it occurs rarely by limiting both the average total traffic on any link and/or the ratio of the link rate to the peak rates of individual virtual circuits. The second way to address congestion is to provide sufficient buffering at each switch element to handle typical data bursts. For virtual circuits whose peak rates are large, the buffer sizes must be at least comparable to the size of average length bursts. For large buffers, one must have separate queues for different classes of traffic, in order to prevent real-time traffic from being excessively delayed by bursty data traffic. The third approach to controlling congestion in multistage networks using static routing is to employ more sophisticated flow control mechanisms between stages. The most effective approach is to have a separate queue for each virtual circuit that passes through the given switch element. In this approach, a switch element with a congested output link signals to its upstream neighbors, blocking new cell transmissions for all virtual circuits using the congested link. These signals can be propagated backward through successive stages of the network to the IPPs, where larger buffers are typically available for storing bursts. References [5], [16], [40], [51], [70], [86], [95], [110] provide a sampling of the literature on the queuing performance of buffered multistage networks.

3.4 Multicasting in Buffered Multistage Switches

Multistage networks using dynamic or static routing can be extended in a fairly direct way to support multicast communication. For systems that are not too large, or that involve only small multicast connections, the set of

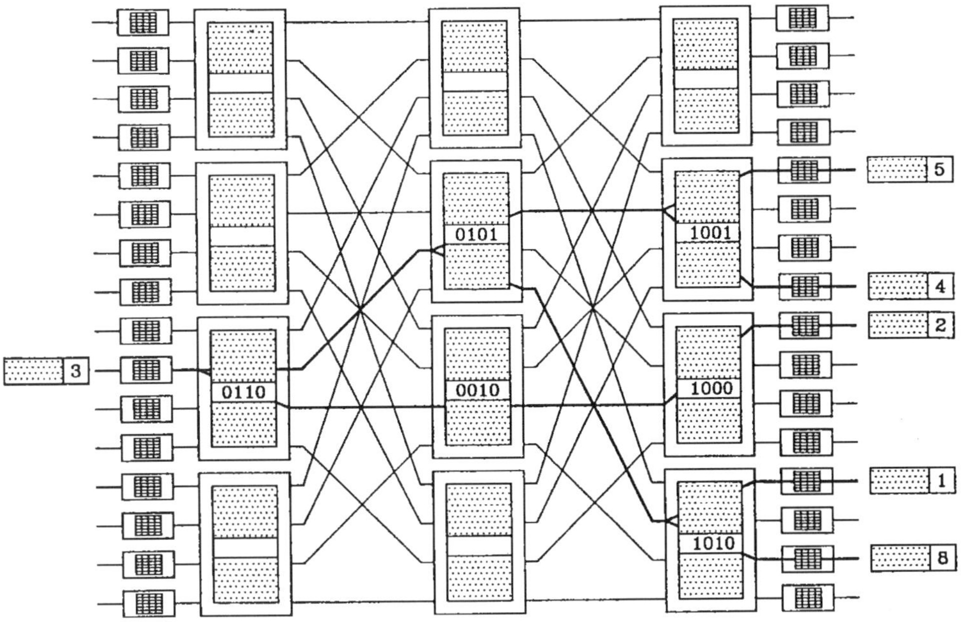

Fig. 6 Multicasting in buffered multistage networks.

all outputs for a cell can be encoded and placed in the cell header at an IPP and used by switch elements to guide the process of routing and copying cells to the required outputs. In general, multicasts for any collection of output sets that can be conveniently encoded using a modest number of bits can be handled in this way. For greatest flexibility, systems need to be capable of labeling different output copies of a given multicast cell with distinct VPI/VCI values. This requires that IPPs insert Multicast Identifiers (MI) into cell headers when cells belonging to multicast virtual circuits arrive at input ports. These are used by OPPs to select the proper outgoing VPI/VCI values from multicast routing tables.

For large systems that must be able to support arbitrary multicasts, a more general technique is needed. For dynamic networks, one method is to augment switch elements with multicast routing tables, indexed by the Multicast Identifier (MI). Switch elements use MIs to select routing table entries that specify the set of switch element outputs an arriving cell should be forwarded to. If a Beneš interconnection network is used, routing tables are needed in the last k stages of a $2k-1$ stage network. The first $k-1$ stages need only do dynamic load balancing, as in the case of point-to-point cells. Because dynamic routing networks can route cells belonging to a given multicast virtual circuit through any part of the interconnection network, the routing information must be replicated multiple times. This redundancy increases the routing table memory requirements significantly and limits the number of distinct multicast virtual circuits that it is practical to implement using this approach. The system developed by Alcatel [26]–[28], [46], [47] uses this method.

Similar techniques can be used for static routing networks, as shown in Fig. 6. In this case routing tables are needed at every stage of a multistage network. As in the case of dynamic routing networks, an MI can be used to determine the set of output ports a given multicast cell should be forwarded to. Alternatively, the routing table entries may be selected using arriving cells' VPI/VCI values, and the entries can specify both the set of outputs to receive copies and the VPI/VCI values to be assigned to the outgoing copies on each of that switch element's outgoing links. This latter approach can require fewer total routing table entries in large configurations but is somewhat more complex to manage. For static routing networks, the introduction of multicast can have a large impact on virtual circuit blocking performance. The best possible blocking performance is obtained in these networks when the branching required for multicast connections occurs as close as possible to the OPPs. This minimizes the use of interconnection bandwidth by multicast connections. Unfortunately, under worst-case conditions, endpoint additions and deletions can lead to a situation in which many multicast virtual circuits branch at early stages, making it more difficult to add new connections or augment existing multicast connections. To prevent blocking, one must increase the interconnection network's speed advantage by a large factor. For the Clos network, the following inequality must be satisfied to make the network nonblocking.

$$r > \left\lceil \min\{r, n/d\} \frac{\beta d - B}{1-B} \right\rceil + \left\lceil \frac{\beta d - B}{1-B} \right\rceil - 2$$

A speed advantage of

$$(1/d)\,[n + (1 + (d-1)(k-1)) - (B/\beta)]$$

is needed to make the Beneš network nonblocking. In some situations, lower costs can be realized by sacrificing the strict nonblocking property. Networks that are *reroutably nonblocking* allow any new virtual circuit to be setup without rearranging any existing ones, but may require rerouting a given multicast virtual circuit when adding a new endpoint to it. Networks satisfying this property require much smaller speed advantages than strictly nonblocking networks and may be acceptable in many practical situations.

For both static and dynamic routing networks, the amount of memory needed for multicast routing can become excessive, particularly in larger system configurations. For this reason, alternative approaches have been developed that can reduce the total amount of routing memory required and simplify the operational aspects of the system. One class of approaches uses a two part interconnection network in which the first part (the *copy network*) produces copies of multicast cells, while the latter part (the *routing network*) routes the copies to the desired outputs. Multicast routing tables are positioned between these two sections, but are not needed at every switch element. See [108] for an example of a system of this type. One way to implement this class of systems is to have IPPs insert an MI and a pair of copy network output numbers in the headers of cells belonging to multicast virtual circuits. The copy network uses the pair of specified output numbers to control the copy process, delivering copies to all output ports in the range bounded by the specified pair of outputs. When cells arrive at the multicast routing tables between the copy and routing networks, the MI is used to select a table entry specifying the OPP that a particular copy is to be forwarded to and the VPI/VCI to be included with the cell when it is forwarded on the outgoing link. In the case of dynamic routing networks, the copies of different multicast cells in the same virtual circuit can be sent to different sets of copy network outputs in order to balance the load across all the copy network outputs. Load balancing makes it possible to build systems with a smaller internal speed advantage than would otherwise be required, but comes at the expense of larger numbers of multicast routing table entries than would be needed without load balancing. One can trade off these two factors to provide the best combination for a given system configuration.

In some situations, one can reduce the cost of the system by combining the functions of the copy and routing networks into a single network that is capable of doing both functions. In such a system, cells belonging to point-to-point virtual circuits are routed through the network to the desired output in a single pass, exactly as in a system supporting only point-to-point virtual circuits. Cells belonging to multicast virtual circuits are handled in two passes through the network. The required number of copies is produced on the first pass, and then the copies are sent through the network again, with each copy forwarded to its required output link. This requires a *recycling data path* between each OPP and its corresponding IPP as shown in Fig. 7. Multicast cells arriving at an OPP are passed along the recycling path to the IPP, which includes a multicast routing table that specifies the output port and VPI/VCI to be used for a particular copy. This multicast routing table can be combined with the virtual path/circuit translation table that the IPP uses for cells arriving on its external link.

The most cost effective system architectures for large-scale multicasting extend this two pass processing to multiple passes, with a bounded number of copies being made on each pass. In the case of *binary copying*, two copies of an original multicast cell are produced in one pass through the network. When cells belonging to a multicast virtual circuit arrive at an IPP from the external link, the IPP table lookup produces two output port numbers and two VPI/VCI fields. These are inserted in the cell header, as the cell is sent to the interconnection network. The interconnection network (which can use either static or dynamic routing), delivers a copy to each of the specified outputs and then each copy is either forwarded on its external link or passed along the recycling path to the IPP, where it passes through another lookup table, acquiring two new output port numbers and VPI/VCI fields. If both copies made in the first pass are recycled to their corresponding IPPs, then they can each produce two copies in the second pass, leading to four copies of the original cell, after two passes through the network. If all four of these copies are recycled, eight copies are produced on the third pass and in general F copies can be produced in $\log_2 F$ passes. The combination of binary copying, cell recycling and a dynamic routing Beneš network yields a system with optimal scaling characteristics. That is, the complexity of the interconnection network grows in proportion to $n \log n$ and the number of routing table entries required for multicast is no larger than is required for point-to-point virtual circuits (although the entries are somewhat larger). This technique for multicasting is used in the system described in Refs. [12], [111].

4. Non-Buffered Multistage Switches

The cell buffers used to resolve contention in buffered multistage networks contribute significantly to their cost, motivating different approaches that attempt to resolve contention in other ways. For ATM, these approaches have not proved to be competitive with the buffered multistage networks, when both are implemented with a high density circuit technology like CMOS. This is because while systems in this class generally require less circuit area for logic devices, they also require either substantially higher bandwidths or

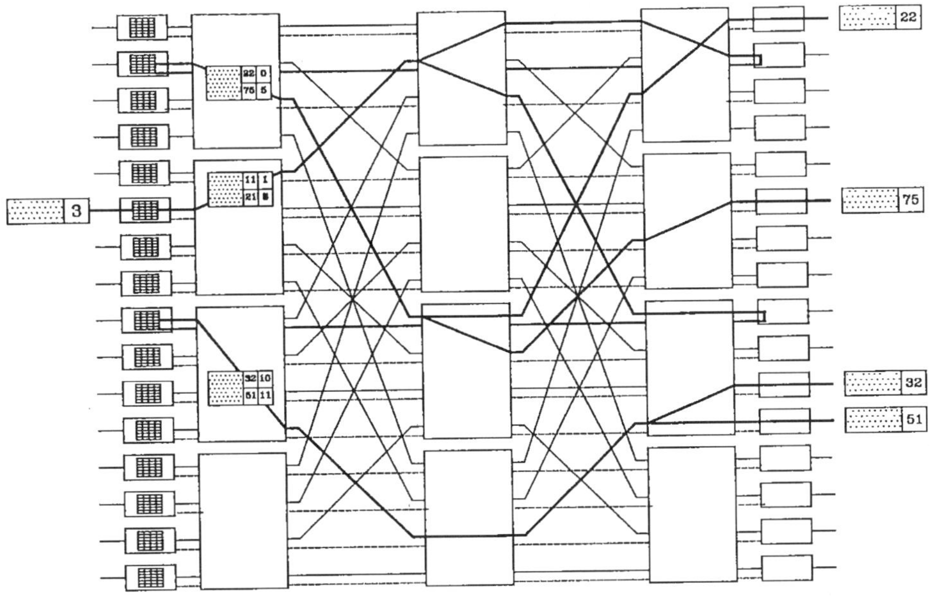

Fig. 7 Multicasting using cell recycling.

increased interconnection wiring, both within and between chips. However they could become competitive in the future, in the context of either higher speed, lower density circuit technologies or larger cell sizes that increase the relative cost of buffered multistage networks.

4.1 Contend and Repeat

The simplest class of non-buffered multistage switches uses a simple contend-and-repeat strategy. As an example, consider a system with a $2k-1$ stage Beneš network topology, in which the individual switch elements include a crossbar and crossbar controller. In the first $k-1$ stages, the crossbar controllers simply configure the crossbar randomly so that arriving cells are distributed evenly across the whole network. In the last k stages, cells are routed to the output specified in their cell headers. If two cells arriving at a particular switch element contend for the same output, one of them is forwarded to the requested output, while the other is simply discarded. Cells are typically delayed by just a few clock ticks at each stage, allowing the first part of a cell to emerge from the interconnection network while the last part is still being sent by the IPP. When a cell arrives at its destination OPP, the OPP returns an acknowledgement signal which flows back through the network along the path taken by the cell, but in the reverse direction, using a circuit path set up as a side effect of the cell routing process. If an IPP fails to receive an acknowledgement signal when expected, it concludes that the cell was discarded and tries again on the next operational cycle. In order to handle fully loaded external links, such systems must be engineered with a substantial speed advantage relative to the external links. For typical configurations, the required speed advantage is in the range of three to five. While this can represent a significant cost factor, it can be acceptable in the context of higher speed circuit technologies. Newman describes a system of this type in [81].

There are many variations on the basic approach described above. All require some mechanism for increasing the overall system bandwidth. In some cases, this is done by linking switch elements with multiple parallel links, rather than a single higher speed link (networks of this sort are called *dilated* networks). In other cases, parallel switch planes are used, with input traffic being distributed over the different parallel switch planes. While these alternatives may offer advantages in particular situations, none is clearly superior to the simpler approach described above.

4.2 Deflection Routing

In systems using the contend-and-repeat paradigm, only one of multiple competing cells is propagated beyond a contention point. Deflection routing networks allow both winning and losing cells to propagate past a contention point, making it possible for the losers to reach the desired output by an alternative path. A simple example of a network of this type is the tandem banyan network, illustrated in Fig. 8. In this system, several banyan networks are cascaded in series, and cells can exit to the desired OPP at the output of any one of the component networks. When a cell loses a contention in any of the banyan networks, a marker bit in its header is set and subsequent switch elements pass it on, using any available output. A cell arriving at the output of a banyan network with its marker bit cleared is passed to the OPP at that point. Cells with the marker bit set are propagated to the next banyan network, with the marker bit is cleared. As cells propagate through the series of banyan networks, the traffic drops, increasing each cell's chance of successfully reaching the required OPP. Of course, even if a large number of banyan networks are used, there remains some non-zero probability that a cell will lose a contention in the last banyan network. If this probability is low enough, it may be acceptable to drop the cell at that point. Alternatively, it can be recirculated back to the input of the last network, and retried at that point.

The tandem banyan network is described in [102]. Another deflection routing architecture, called Shuffle-out is described in [23], [24].

4.3 Sorting Networks

Sorting networks can be used to implement ATM switches that are contention-free, so long as the arriving cells are all addressed to distinct outputs. The most popular sorting network is Batcher's bitonic sorter [3], an example of which is illustrated in Fig. 9. In this figure, cells are indicated by the numbers whose values represent the output address for the given cell. The square boxes in the figure represent *sorting elements,* which propagate arriving cells to the output, based on the relative values of their output addresses; some sorting elements send cells with larger values to the upper output, others to the lower output, as indicated by the arrows in the figure. The sorting network is constructed using a double recursive construction. Some of the components of this construction are highlighted in the figure.

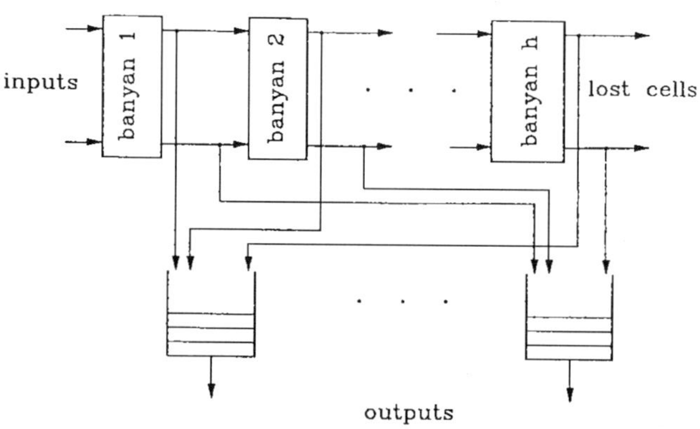

Fig. 8 Tandem Banyan network.

If one happens to have a set of input cells with exactly one cell addressed to each output, the sorting network will deliver each cell to the right output. To handle arbitrary sets of input cells in which some outputs may have zero or more than one cell addressed to them, additional network components are needed. The simplest scaleable approach is to follow the sorting network by a *filter* block and a *routing network.* The filter block simply compares the output port number of a cell arriving on input i to that of a cell arriving on input $i-1$ and propagates the cell on input i to output i, only if the compared port numbers are unequal. When this filter is applied to a sequence of cells that have been sorted by output port number, the result is to propagate at most one cell for each output port. The filters can acknowledge propagated cells by sending a signal back along the reverse path taken through the sorting network. IPPs retry cells that are not acknowledged, during subsequent operational cycles. Cells that pass

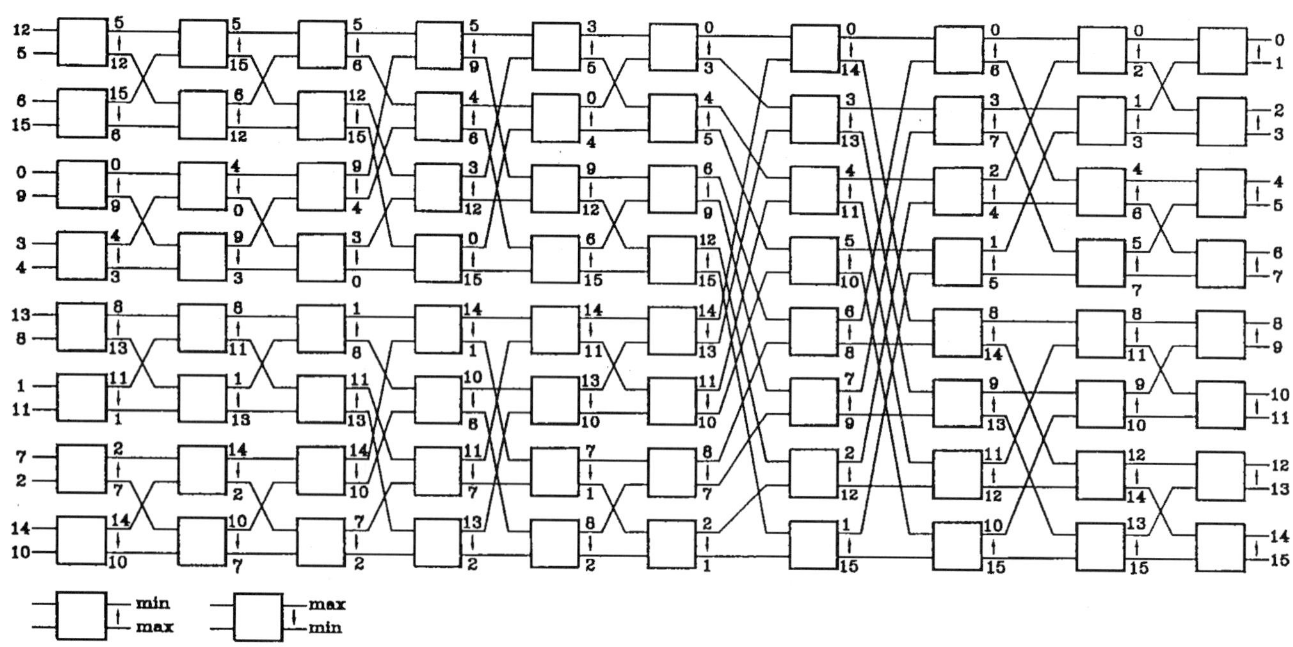

Fig. 9 Bitonic sorting network.

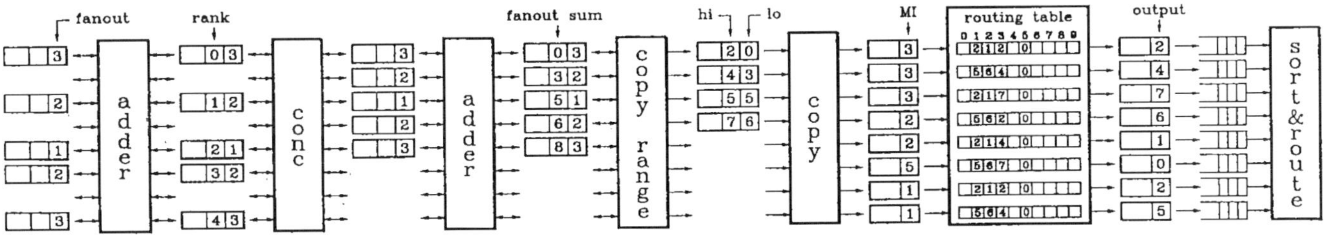

Fig. 10 Copy network.

the filter then go through to the routing network. The most commonly used routing network is built around a banyan network. Banyan networks have the property that a sorted sequence of cells on consecutive inputs can be routed to the outputs without any internal data path contention. A routing network can be constructed by preceding a banyan network with a *concentrator* that places cells on consecutive outputs without changing their relative order.

A system comprising a sorting network, filters and routing network provides a general and scalable switching system. However, the queueing performance can be inadequate to handle fully loaded external links. This limitation can be eliminated by increasing the output bandwidth of the system. One way to do this is to operate the system at a higher internal bandwidth than the external links; for most realistic traffic configurations a two-to-one speed advantage is sufficient. Another approach is to have more than one link from the routing network to each OPP, so that OPPs can receive multiple cells in parallel. For a system that allows r cells to be forwarded to any output in a single cycle, the filter block is modified to propagate up to r cells per output and following the concentrator with r "interleaved" banyan networks.

The major drawback of this class of systems is that they require substantially more interconnection wiring between individual integrated circuits and circuit modules than is needed for buffered multistage networks. This is a consequence of the fact that practical sorting networks require $(1/2)(\log_2 n)(1+\log_2 n)$ stages of sorting elements. Due to pin constraints on integrated circuit packages a larger number of integrated circuits are needed to implement a system with a given number of ports than with alternative architectures. As one example, a 4096 port sorting network implemented with components having 64 inputs and 64 outputs, requires 13 ranks of components, with 64 components per rank. Seven additional ranks of components are required for the filter and routing network. For comparison, a 4096 port Beneš network with 64 port switch elements using internal buffering to resolve contention requires just 3 columns of components.

The first ATM switch design based on the use of sorting networks was the Starlite switch [49]. These ideas were developed further in the Sunshine switch, described in [22], [39], [44], [45], [48]. A system based on a similar approach is described in [76]. Eng et. al. [33]–[36] have developed a hybrid architecture which combines an unbuffered multistage network with buffered switch elements in the last stage. These buffered switch elements have k inputs and $d < k$ outputs (typically $k = 32$, $d = 16$) and a large shared buffer. The unbuffered switching network preceding this rank of buffered switch elements performs an arbitration function, discarding excess cells if more than k arriving cells are destined for any group of d outputs. This approach is feasible when the memory bandwidth of the shared buffer switch elements is high enough to support d simultaneous output ports, but under these conditions, port multiplexing is also an option and can be a more cost-effective approach. The authors suggest a variety of choices for the unbuffered network, including the use of a sorting network and a three stage Clos network (in which the last stage is replaced by the buffered switch elements). Both of these choices have suboptimal scaling characteristics, weakening the authors' assertion of growability.

4.4 Multicasting in Non-Buffered Multistage Switches

Multicasting can be implemented in non-buffered multistage networks, by preceding a point-to-point network with a copy network, capable of producing the requested number of copies of any given input cell. One example of such a system is shown in Fig. 10. A *fanout* field is inserted into cell headers by the IPPs before sending cells to the network. The first two components of the network implement a concentration function; the first *adder* block computes the *rank* of each arriving cell (essentially just numbering cells consecutively) and the *concentrator* block uses the rank to route cells to consecutive outputs. The second adder block sums the fanout fields of cells; the computed fanout sum of a departing cell on output i equals the sum of all fanouts of cells on outputs 0 through $i-1$. Using the fanout sum and fanout fields, the *copy range* component computes the range of outputs that each cell should be copied to in the succeeding *copy* component. The copy component does the actual copying of cells based on the supplied ranges. The routing tables that follow the copy network specify an output port and outgoing VPI/VCI for each copy of an arriving cell (the VPI/VCI fields are not shown in the example). Each table is indexed by the multicast identifier of the arriving cell. The routing tables are followed by a set of cell buffers, which are in turn followed by a point-to-point network. The copy range component of this system filters out cells if the computed copy range extends beyond the number of outputs of the copy component. Acknowledgments are returned to the IPPs along the reverse of the paths followed by the arriving cells and unacknowledged cells are retried on the next operational cycle. The performance of this design can be improved by expanding the number of outputs of the copy network and modifying the copy range block to distribute the load evenly across the outputs of the copy block. The principal disadvantage of this design is that a multicast virtual circuit consumes an entry in every routing table even when it has a small fanout. This leads to excessive memory requirements in the common case of many virtual circuits with small fanout.

This approach to implementing multicast is described in [63] and was later refined in [109]. Similar approaches are developed in [4], [77].

5. Conclusion

Cost is a key factor in evaluating architectural alternatives for ATM switching systems. Throughout the paper, we have touched on issues that contribute to overall system costs. We close with a discussion of a simple cost model for ATM switching systems and a comparison of some of the different design choices, based upon this model. The cost of a single stage switch can be modeled by an expression of the form

$$(C_1 + C_2 m + C_3 B)n + C_4 n^2$$

where the C_i are implementation-dependent constants, n is the number of ports on the switch, m is the number of virtual circuits per port and B is the number of cell buffers per port. The last term in the expression is the part that is determined by the interconnection network. For multistage interconnection networks, this term is replaced by another. For a dynamic routing Beneš network, the last term would be $C_4 n \log n$ while for a static routing, three stage Clos network configured to be nonblocking, it would be $C_4 n^{3/2}$. In each of these cases the constant may differ. The plot in Fig. 11 illustrates the influence of the different components on system cost. The constants were chosen to approximate the cost of systems with 2.4 Gb/s ports. The values chosen are somewhat low at the time of writing, but should be a reasonable reflection of actual costs over the next two to three years. However, they should be viewed as illustrative, not authoritative. A value of $500 was chosen for the per port component (C_1) of the cost. This captures the cost of optical transceivers, transmission interface circuitry and per port ATM circuits, exclusive of

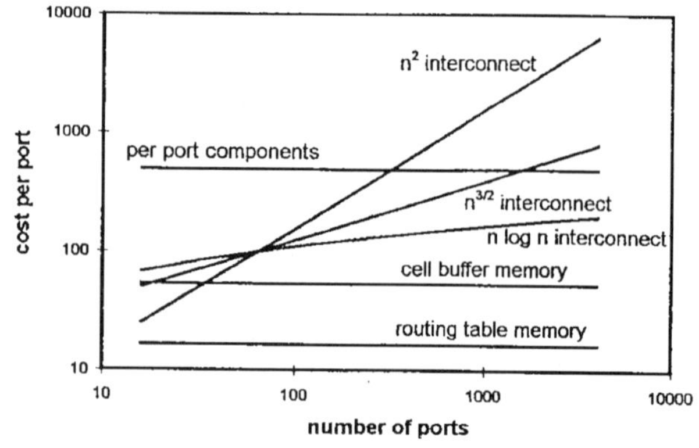

Fig. 11 Switching system cost components.

the memory needed for routing tables and cell buffers. The costs for the cell buffers and routing table memory are based on an assumed 4096 routing table entries per port with 20 bytes per entry, 4096 cell buffers per port with 64 bytes per entry and $200 per megabyte of memory (static RAM with 10 ns cycle time). The constants for the interconnect components of the cost were chosen so that all three resulted in a cost of $100 per port in a 64 port system. In most cases, the per port cost component is dominant, but the interconnect cost of the single stage systems becomes important as the systems grow beyond a few hundred ports (note that at 2.4 Gb/s per port, 512 ports are needed for a 1 Tb/s capacity). For the routing table and buffer sizes chosen here, memory cost is not a major component, but if one were to increase the amount of buffering to 32,768 cells (1.6 Mbytes in cell payloads), the memory cost would increase to over $400 per port, making it comparable in cost to the other per port components.

The design and analysis of modern switching systems has a long tradition, extending back to early work by Shannon and Clos. Changes in circuit technologies have opened up new opportunities, enabling the introduction of new services, expanded functionality and vastly greater bandwidths. The pace of change in technology continues unabated and new innovations in switching system design will be needed to keep pace with these changes, in order to realize the promise that the new technologies have to offer. In this paper, we have tried to put recent developments in a common context and provide an intellectual framework in which past and future developments can be related and compared. We hope that it can help future designers learn from the past, while working to create the switching systems that will take us into the future.

References

[1] H. Ahmadi and W.E. Denzel, "A Survey of Modern High Performance Switching," IEEE J. Sel. Areas Commun., vol.7, no.7, pp.1091–1103, Sept. 1989.

[2] T.R. Banniza, et al., "Design and technology aspects of

VLSIs for ATM switches," IEEE J. Sel. Areas Commun., vol.9, pp.1255–1264, Oct. 1991.

[3] K. Batcher, "Sorting networks and their applications," Proc. Spring Joint Computer Conference, pp.307–314, 1968.

[4] R. Bianchini, Jr. and H. Kim, "Design of a nonblocking shared memory copy network for ATM," Proc. Infocom, pp.876–885, 1992.

[5] G. Bianchi and J.S. Turner, "Improved queueing analysis of shared buffer switching networks," IEEE/ACM Trans. Networking, vol.1, no.4, Aug. 1993.

[6] B. Bostica and L. Licciardi, "Merging electronics and photonics towards tera bit/s ATM switching," Proc. Globecom, pp.943–947, 1996.

[7] J.W. Byun and T.T. Lee, "The design and analysis of an ATM multicast switch with adaptive traffic controller," IEEE/ACM Trans. Networking, June 1994.

[8] H.J. Chao, "A recursive modular terabit/second ATM switch," IEEE J. Sel. Areas Commun., vol.9, no.8, pp.1161–1172, Oct. 1991.

[9] H.J. Chao and B.-S. Choe, "Design and analysis of a large-scale multicast output buffered ATM switch," IEEE/ACM Trans. Networking, vol.3, no.2, pp.126–138, April 1995.

[10] H.J. Chao, "An ATM queue manager handling multiple delay and loss priorities," IEEE/ACM Trans. Networking, vol.3, no.6, Dec. 1995.

[11] A.K. Chatterjee and V.K. Konangi, "Asymmetric ATM switch modules with imbalanced traffic," Proc. Infocom, pp.802–809, 1996.

[12] T. Chaney, J.A. Fingerhut, M. Flucke, and J.S. Turner, "Design of a gigabit ATM switch," Proc. Infocom, 1997.

[13] A.K. Choudhury and E.L. Hahne, "Dynamic queue length thresholds in a shared memory ATM switch," Proc. Infocom, March 1996.

[14] G.L. Choudhury, et al., "Squeezing the most out of ATM," IEEE Trans. Commun., vol.44, no.2, Feb. 1996.

[15] M. Collivignarello, et al., "A complete set of VLSI circuits for ATM switching," Proc. Globecom, pp.134–138, 1994.

[16] B. Collier and H.S. Kim, "Efficient analysis of shared buffer management strategies in ATM networks under non-uniform bursty traffic," Proc. Infocom, 1996.

[17] S. Claretto, et al., "A low power ATM switching element for broadband applications," Proc. International Teletraffic Congress, pp.291–295, 1994.

[18] P. Coppo, et al., "Optimal cost/performance design of ATM switches," IEEE/ACM Trans. Networking, vol.1, no.5, pp.566–575, Oct. 1993.

[19] G. Corazza and C. Raffaelli, "Performance evaluation of input-buffered replicated Banyan networks," IEEE Trans. Commun., vol.41, no.6, pp.841–845, June 1993.

[20] R. Cusani and F. Sestini, "A recursive multistage architecture for multicast ATM switching," Proc. Infocom, pp.1289–1295, April 1991.

[21] G.E. Daddis and H.C. Torng, "A taxonomy of broadband integrated switching architectures," IEEE Commun. Magazine, May 1989.

[22] C. Day, J. Giacopelli, and J. Hickey, "Applications of self-routing switches to data fiber optic networks," Proc. International Switching Symposium, pp.519–423, 1987.

[23] M. Decina, P. Giacomazzi, and A. Pattavina, "Shuffle interconnection networks with deflection routing for ATM switching: The closed-loop shuffleout," Proc. Infocom'91, pp.1254–1263, April 1991.

[24] M. Decina, P. Giacomazzi, and A. Pattavina, "Shuffle interconnection networks with deflection routing for ATM switching: The open-loop shuffleout," June 1991.

[25] W.E. Denzel, A.P.J. Engbersen, and I. Iliadis, "A highly modular packet switch for Gb/s rates," Proc. International Switching Symposium, Oct. 1992.

[26] M. De Prycker and M. De Somer, "Performance of a service independent switching network with distributed control," IEEE J. Sel. Areas Commun., vol.5, no.8, pp.1293–1301, Oct. 1987.

[27] M. De Prycker and J. Bauwens, "A switching exchange for an asynchronous time division based network," Proc. International Communications Conference (ICC), June 1987.

[28] M. De Prycker, et al., "An ATM switching architecture with intrinsic multicast capabilities for the Belgian broadband experiment," Proc. International Switching Symposium, Stockholm, May 1990.

[29] M. DePrycker, "Asynchronous Transfer Mode: Solution for Broadband ISDN," Prentice-Hall, 1995.

[30] M. Devault, J.-Y. Cochennec, and M. Servel, "The prelude ATD experiment: Assessments and future prospects," IEEE J. Sel. Areas Commun., vol.6, no.9, pp.1528–1537, Dec. 1988.

[31] C. Dhas, V. Konangi, and M. Sreetharan, "Broadband Switching, Architectures, Protocols, Design and Analysis," IEEE Computer Society Press, 1991.

[32] Y. Doi, et al., "A very high-speed ATM switch with input and output buffers," Proc. International Switching Symposium, 1992.

[33] K. Eng and M. Kard, "The growable switch architecture: A self-routing implementation for large ATM applications," Proc. International Communications Conference, pp.1014–1020, 1991.

[34] K.Y. Eng, et al., "A growable packet (ATM) switch architecture: Design principles and applications," IEEE Trans. Commun., vol.40, no.2, pp.423–429, Feb. 1992.

[35] K. Eng, et al., "A prototype growable 2.5 Gb/s ATM switch for broadband applications," Journal of High Speed Networks, vol.1, no.3, pp.237–253, 1992.

[36] K. Eng, et al., "Design and prototype of a terabit ATM switch using a concentrator-based growable switch architecture," Proc. International Switching Symposium, March 1995.

[37] Fore Systems, "ForeRunner ATM Switches," http://www.fore.com/products/swtch/index.html, 1997.

[38] S.W. Fuhrmann, "Performance of a packet switch with crossbar architecture," IEEE Trans. Commun., vol.41, no.3, March 1993.

[39] J. Giacopelli, J. Hickey, W. Marcus, W. Sincoskie, and M. Littlewood, "Sunshine: A high- performance self-routing broadband packet switch architecture," IEEE J. Sel. Areas Commun., vol.9, no.8, pp.1289–1298, Oct. 1991.

[40] S. Gianatti and A. Pattavina, "Performance analysis of ATM Banyan networks with shared queueing Part I: Random offered traffic," IEEE/ACM Trans. Networking, vol.2, no.4, pp.398–410, Aug. 1994.

[41] K. Genda, et. al., "A 160 Gb/s ATM switching system using an internal speed-up crossbar switch," Proc. Globecom, Conf., pp.123–133, 1994.

[42] K. Habara, Y. Yamada, A. Misawa, K. Sasayama, M. Tsukada, T. Matsunaga, and K-I Yukimatsu, "Demonstration of frequency-routing type photonic ATM switch (FRONTIERNET) prototype," 22nd ECOC 96, ThC.3.4, 1996.

[43] K. Hajikano, et al., "Asynchronous transfer mode switching architecture for broadband ISDN multistage self-routing switching (MSSR)," Proc. International Communications Conference, 1988.

[44] G. Hayward, et al., "A broadband ISDN local access system using emerging technology components," Proc. International Switching Symposium, 1987.

[45] G. Hayward, et al., "CMOS VLSI applications in broadband circuit switching," IEEE J. Sel. Areas Commun., pp.1231–1241, 1987.

[46] M.A. Henrion, K.J. Schrodi, et al., "Switching network architecture for ATM based broadband communications," Proc. International Switching Symposium, June 1990.

[47] M.A. Henrion, G.J. Eilenberger, et al., "Technology, distributed control and performance of a multipath self-routing switch," Proc. International Switching Symposium, Oct. 1992.

[48] J.Y. Hui and E. Arthers, "A broadband packet switch for integrated transport," IEEE J. Sel. Areas Commun., vol.5, no.8, pp.1264–1273, Oct. 1987.

[49] A. Huang and S. Knauer, "Starlite: A wideband digital switch," Proc. Globecom, pp.121–125, 1984.

[50] A. Itoh, et al., "Practical implementation and packaging technologies for a large-scale ATM switching system," IEEE J. Sel. Areas Commun., vol.9, pp.1280–1288, Oct. 1991.

[51] Y.-C. Jenq, "Performance analysis of a packet switch based on single-buffered Banyan network," IEEE J. Sel. Areas Commun., vol.1, no.6, pp.1014–1021, Dec. 1983.

[52] M.J. Karol and C.-Lin, "Performance analysis of a growable architecture for broadband packet (ATM) switching," IEEE Trans. Commun., vol.40, no.2, pp.431–439, Feb. 1992.

[53] M. Karol and K. Eng, "Performance of hierarchical multiplexing in ATM switch designs," Proc. International Communications Conference, pp.269–275, 1992.

[54] M. Katevenis, et al., "Pipelined memory shared buffer for VLSI switches," Proc. ACM SIGCOMM, pp.39–48, 1995.

[55] M. Katevenis, et al., "ATLAS I: A general-purpose, single-chip ATM switch with credit-based flow control," Proc. IEEE Hot Interconnects, Aug. 1996.

[56] M. Katevenis, et al., "VC-level flow control and shared buffering in the telegraphos switch," Proc. IEEE Hot Interconnects, Aug. 1996.

[57] H.S. Kim, "Design and performance of multinet switch: A multistage ATM switch architecture with partially shared buffers," IEEE/ACM Trans. Networking, vol.2, no.6, pp.571–580, Dec. 1994.

[58] Y.M. Kim and K.Y. Lee, "PR-Banyan: A packet switch with a pseudorandomizer for nonuniform traffic," IEEE Trans. Commun., vol.41, no.7, pp.1039–1042, July 1993.

[59] T. Kozaki, et al., "32 x 32 shared buffer type ATM switch VLSIs for B-ISDNs," IEEE J. Sel. Areas Commun., vol.9, no.8, Oct. 1991.

[60] H. Kuwahara, N. Endo, M. Ogino, and T. Kozaki, "A shared buffer memory switch for an ATM exchange," Proc. Int. Conf. on Commun., June 1989.

[61] K.L.E. Law and A. Leon-Garcia, "ATM multiplexers and output port controllers with distributed control and flexible queueing disciplines," Proc. Infocom, pp.663–670, 1996.

[62] C.-T. Lea, "A multicast broadband packet switch," IEEE Trans. Commun., vol.41, no.4, pp.621–630, April 1993.

[63] T.T. Lee, "Nonblocking copy networks for multicast packet switching," IEEE J. Sel. Areas Commun., vol.6, no.9, Dec. 1988.

[64] T.T. Lee, "A modular architecture for very large packet switches," IEEE Trans. Commun., vol.38, no.7, pp.1097–1106, July 1990.

[65] L. Liccrardi, et al., "A multichip module solution for high performance ATM switching," MCMC96 Conf.

[66] S.C. Liew and K.W. Lu, "A 3-stage interconnection structure for very large packet switches," Proc. International Communications Conference, 1990.

[67] J.-F. Lin and S.-D. Wang, "High-performance low-cost non-blocking switch for ATM," Proc. Infocom, pp.818–821, 1996.

[68] Y.-S. Lin and C.B. Shung, "Queue management for shared buffer and shared multi- buffer ATM switches," Proc. Infocom, pp.688–695, 1996.

[69] J. Lockwood, H. Duan, J. Morikuni, S.-M. Kang, S. Akkineni, and R. Campbell, "Scalable optoelectronic ATM networks: The iPOINT fully functional testbed," Journal of Lightwave Technology, June 1995.

[70] J.V. Luciani and C.Y.R. Chen, "An analytical model for partially blocking finite- buffered switching networks," IEEE Trans. Commun., vol.2, no.5, pp.533–540, Oct. 1994.

[71] N. NcKeown, et al., "The tiny tera: A packet switch core," Proc. Hot Interconnects, 1996.

[72] N. Mckeown, "Achieving 100% throughput in an input-queued switch," Proc. Infocom, 1996.

[73] R. Melen and J.S. Turner, "Nonblocking networks for fast packet switching," Proc. Infocom, March 1989.

[74] R. Melen and J. Turner, "Nonblocking multirate networks," SIAM Journal on Computing, March 1989.

[75] R. Melen and J.S. Turner, "Nonblocking multirate distribution networks," IEEE Trans. Commun., vol.41, no.2, pp.362–369, Feb. 1993.

[76] P.S. Min, H. Saidi, and M.V. Hegde, "A nonblocking architecture for broadband multichannel switching," IEEE/ACM Trans. Networking, vol.3, no.2, pp.181–198, March 1995.

[77] P.S. Min, M.V. Hegde, et al., "Nonblocking copy networks in multi-channel switching," IEEE/ACM Trans. Networking, vol.3, no.6, pp.857–871, Dec. 1995.

[78] A. Misawa, M. Tsukada, Y. Yamada, K. Sasayama, K. Habara, T. Matsunaga, and K-I Yukimatsu, "40-Gbit/s broadcast-and-select photonic ATM switch prototype with FDM output buffers," 22nd ECOC 96, ThD.1.2, 1996.

[79] MMC Networks, "ATMS2000: 5 Gb/s Switch Engine Chip Set," http://www.mmcnet.com/atms2000.html, 1996.

[80] E. Munter, et al., "A high capacity ATM switch based on advanced electronic and optical technologies," Proc. International Switching Symposium, 1995.

[81] P. Newman, "A fast packet switch for the integrated services backbone network," IEEE J. Sel. Areas Commun., vol.6, no.9, pp.1468–1479, Dec. 1988.

[82] S. Nojima, E. Tsutsui, H. Fukuda, and M. Hashimoto, "Integrated services packet network using bus matrix switch," IEEE J. Sel. Areas Commun., vol.5, no.8, pp.1284–1292, Oct. 1987.

[83] E. Oki and N. Yamanaka, "Scalable crosspoint buffering ATM switch architecture using distributed arbitration scheme," Proc. IEEE ATM Workshop, May 1997.

[84] A. Pattavina, "Multichannel bandwidth allocation in a broadband packet switch," IEEE J. Sel. Areas Commun., vol.6, no.9, pp.1489–1499, Dec. 1988.

[85] A. Pattavina, "A broadband packet swtich with input and output queueing," Proc. International Switching Symposium, pp.11–16, 1990.

[86] A. Pattavina and S. Gianatti, "Performance analysis of ATM Banyan networks with shared queueing Part II: Correlated/unbalanced offered traffic," IEEE/ACM Trans. Networking, vol.2, no.4, pp.411–424, Aug. 1994.

[87] B. Prabhakar, N. Mckeown, and R. Ahuja, "Multicast scheduling for input-queued switches," Proc. Princeton Conf., 1996.
[88] K. Padmanabhan, "An efficient architecture for fault-tolerant ATM switches," IEEE/ACM Trans. Networking, vol.3, no.5, pp.527–537, Oct. 1995.
[89] E. Rathgeb, T. Theimer, and M. Huber, "Buffering concepts for ATM switching networks," Proc. Globecom, Nov. 1988.
[90] C. Qiao, "Analysis of space-time tradeoffs in photonic switching networks," Proc. Infocom 96, pp.822–829, 1996.
[91] K. Sasayama, Y. Yamada, K. Habara, and K-I Yukimatsu, "FRONTIERNET: Frequency-routing-type time-division interconnection network," Journal of Lightwave Technology, vol.15, no.3, pp.1–13, 1997.
[92] M.D. Schroeder, et al., "Autonet: A high-speed, self-reconfiguring local area network using point-to-point links," IEEE JSAC, vol.9, no.8, p.1318, Oct. 1991.
[93] F. Sestini, "Recursive copy generation for multicast ATM switching," IEEE/ACM Trans. Networking, pp.329–335, June 1997.
[94] H. Suzuki, et al., "Output-buffer switch architecture for asynchronous transfer mode," Proc. International Communications Conference (ICC), pp.4.1.1–4.1.5., June 1989.
[95] T. Szymanski and S. Shaikh, "Markov chain analysis of packet-switched Banyans with arbitrary switch sizes, queue sizes, link multiplicities and speedups," Proc. Infocom, March 1989.
[96] T. Takeuchi, T. Yamaguchi, H. Niwa, H. Suzuki, and S. Hayano, "Synchronous composite packet switching - A switching architecture for broadband ISDN," IEEE J. Sel. Areas Commun., pp.1365–1376, Oct. 1987.
[97] A. Takase, et. al., "ATM transport node for flexible and robust access networks," Proc. Globecom, pp.1481–1487, 1993.
[98] A. Takase, et. al., "Data-path architecture and technology for large scale ATM switching system," Proc. Globecom, pp.1395–1399, 1996.
[99] C.L. Tarng and J.S. Meditch, "A high performance copy network for B-ISDN," Proc. Infocom, 1991.
[100] K. Thilakam and A. Jhunjhunwala, "Proposed high speed packet switch for broadband integrated networks," Computer Commun., vol.12, no.6, Dec. 1989.
[101] F.A. Tobagi, "Fast packet switch architectures for broadband integrated services digital networks," Proc. of IEEE, vol.78, no.1, pp.133–167, Jan. 1990.
[102] F. Tobagi, T. Kwok, and F. Chiussi, "Architecture, performance, and implementation of the tandem Banyan fast packet switch," IEEE J. Sel. Areas Commun., vol.9, no.8, pp.1173–1193, Oct. 1991.
[103] H. Tomonaga, et. al., "High-speed switching module for large capacity ATM switching system," Proc. Globecom, 1992, pp.123–127, 1992.
[104] H. Tomonaga, N. Matsuoka, Y. Kato, A. Hakata, and K. Murakami, "A structure of ultra high-speed ATM switch," IEICE Trans., vol.J76-B-I, no.11, pp.793–800, 1993.
[105] H. Tomonaga, N. Matsuoka, T. Katoh, M. Kawai, H. Yamashita, and A. Hakata, "Small size and large capacity ATM switching system," Proc. 1995 International Symposium on Communication, pp.374–381, 1995.
[106] TranSwitch, "CUBIT: ATM Cell Bus Switch Device," http://www.txc.com/cubit.html, 1996.
[107] S.C. Tu and W.H. Leung, "Multicast connection-oriented packet switching networks," Proc. Infocom, 1990.
[108] J.S. Turner, "Design of a broadcast packet switching network," IEEE Trans. Commun., vol.36, no.6, pp.734–743, June 1988.
[109] J.S. Turner, "A practical version of Lee's multicast switch architecture," IEEE Trans. Commun., vol.41, no.8, pp.1166–1169, Aug. 1993.
[110] J.S. Turner, "Queuing analysis of buffered switching networks," IEEE Trans. Commun., vol.41, no.2, pp.412–420, Feb. 1993.
[111] J.S. Turner, "An optimal nonblocking multicast virtual circuit switch," Proc. Infocom, pp.298–305, June 1994.
[112] W.D. Zhong, et al., "A copy network with shared buffers for large-scale multicast ATM switching," IEEE/ACM Trans. Networking, vol.1, no.2, pp.157–165, April 1993.
[113] P.C. Wong and M.S. Yeung, "Design and analysis of a novel fast packet switch pipleline Banyan," IEEE/ACM Trans. Networking, vol.3, no.1, pp.63–69, Feb. 1995.
[114] N. Yamanaka, et. al., "320 Gb/s high-speed ATM switching system hardware technologies based on copper-polyimide MCM," 44th ECTE, pp.776–785, 1994.
[115] Y.S. Yeh, M.G. Hluchyj, and A.S. Acampora, "The knockout switch: A simple, modular architecture for high-performance packet switching," IEEE J. Sel. Areas Commun., vol.5, no.8, pp.1274–1283, Oct. 1987.
[116] Y. Kamigaki, T. Nara, S. Machida, A. Hakata, and K. Yamaguchi, "160 Gb/s ATM switching system for public networks," Proc. Globecom, pp.1380–1387, 1996.
[117] E.W. Zegura, "Architectures for ATM switching systems," IEEE Commun. Magazine, pp.28–37, Feb. 1993.
[118] W.D. Zhong, Y. Onozato, and J. Kaniyil, "A copy network with shared buffers for large-scale multicast ATM switching," IEEE/ACM Trans. Networking, March 1993.

High-Performance Backbone Network Technology edited by Naoaki Yamanaka
ISBN: 0-8247-5321-6 DOI: 10.1081/(120025648)
Published 2004 by Marcel Dekker, Inc.

Requirements on ATM Switch Architectures for Quality-of-Service Guarantees

Masayuki Murata

SUMMARY While active researches have been continuously made on the ATM switch architectures and the QoS service guarantees, most of them have been treated independently in the past. In this paper, we first explain the architectural requirement on the ATM switches to implement the mechanism of QoS guarantees in the context of ATM congestion control. Then we discuss how a vital link between two should be built, and remaining problems are pointed out.

key words: *ATM switch architecture, QoS service guarantee, congestion control, buffering strategy, QoS service class*

1. Introduction

The capability of ATM (Asynchronous Transfer Mode) networks to provide large bandwidth and to handle multiple quality-of-service (QoS) guarantees can only be realized by preparing effective traffic management mechanisms [1]–[4]. Traffic management includes congestion control, call admission control, and virtual path (VP)/virtual channel (VC) routing. Especially, essential for stable and efficient operation of ATM networks is congestion control, which is performed between ATM end systems. The ATM end system is a point where an ATM connection is terminated, and the connection goes up to the ATM adaptation layer (AAL). Therefore, each ATM segment – a private local area network (LAN) or a public wide area network (WAN) – may, depending on its characteristics, adopt a different congestion control scheme.

The purpose of congestion control is to maintain the guaranteed network performance to meet QoS required by the user. The set of QoS parameters are defined in the standard, including cell loss ratio (CLR), cell transfer delay (CTD) and cell delay variation (CDV). Several service classes have been standardized to support multiple types of QoS guarantees [2], [5], [6]. Those are UBR (Unspecified Bit Rate), CBR (Constant Bit Rate), VBR (Variable Bit Rate), and ABR (Available Bit Rate) service classes. In each of service classes, the users are requested to specify several *traffic parameters* such as peak cell rate (PCR), sustainable cell rate (SCR), cell delay variation tolerance (CDVT), and burst tolerance (BT). This mechanism allows the contractual QoS guarantees from the network to be met and at the same time to protect the network from unintentional misbehavior of users.

A part of the network resource management function should be included within ATM switches to prepare the above service classes. Since ATM is based on the packet switching technique in principle, the performance of an ATM switch heavily depends on its architecture for the queueing of cells. In other words, there may exist the case where some functions required by the resource management cannot be implemented in the ATM switch, dependent on its architecture. In consequence, the implemented ATM switch may be quite different from a conceptual model of the switch architecture explicitly or implicitly assumed by resource management functions.

In this paper, we first review a service architecture for QoS guarantees to explain what functions are needed within the ATM switch in Sect. 2. We next investigate the effects of buffer location of ATM switches on switch performance in Sect. 3. Recent advancements of the ATM switching architectures are then described in Sect. 4. Purposes of Sects. 3 and 4 are that we examine functional requirements on ATM switch architectures for establishing QoS service architectures. We also investigate how the resource management function should take into account performance limitation posed by the switching architectures. A main subject of this paper is to mention that a vital link between QoS and ATM switch architectures should be paid more attention for ATM to be successfully deployed in the future.

2. QoS Service Class Categories

Two congestion control strategies for ATM networks have been discussed; preventive control and reactive control [7]. With preventive control, the usable bandwidth for each connection is limited based on the notion of a *traffic contract* [2], [7]. To assign bandwidth to the connection, each end system has to declare its traffic parameters (one or more of traffic parameters, i.e., PCR, CDVT, SCR and/or BT, dependent on the service class) to the network before the connection is established. Once the connection request is admitted, its QoS (in terms of CLR, CTD and CDV) is guaranteed throughout the session. A lack of network resources may cause a newly requested connection to be rejected. This sort

Manuscript received May 6, 1997.
Manuscript revised July 31, 1997.

† The author is with the Graduate School of Engineering Science, Osaka University, Toyonaka-shi, 560–8531 Japan.

IEICE TRANS. COMMUN., Vol. E81-B, No. 2, pp. 138–151, February 1998

of congestion control scheme can be applied to real-time communications generating constant-bit-rate (CBR) or variable-bit-rate (VBR) traffic, by which audio and motion video can be accommodated†.

However, the preventive congestion control adopted in CBR and VBR service classes is insufficient for data communications because each connection can never emit cells exceeding its negotiated rate, not even when there is unused bandwidth in the network [1]. Furthermore, bandwidth requirements for data traffic are not likely to be known at connection setup time. Instead, the cell transmission rate can be adjusted if up-to-date information about the congestion status of the network is used appropriately. These are the reasons that reactive congestion control is promising for data communications and is being applied to the ABR (Available Bit Rate) service class in the ATM Forum. It dynamically regulates the cell emission process of each connection by using feedback information from the network. It is therefore especially suitable for data transfer service.

In what follows, we briefly summarize features of CBR, VBR and ABR service classes in Sects. 2.1, 2.2 and 2.3, respectively.

2.1 CBR Service Class

An essential part of preventive congestion control is called CAC (connection admission control), which is performed as follows. In the CBR service class, the user requests an allocation of PCR (peak cell rate) to the network at the connection setup time. If there is enough room in the link capacity, the network allocates a fixed amount of bandwidth, PCR, to the connection. Suppose that the number n of connections have already been accepted on the link. By letting PCR_i be the PCR value of i th connection on the link, the newly arriving $(n+1)$ st connection can be admitted if and only if

$$PCR_{n+1} \leq C - \sum_{i=1}^{n} PCR_i,$$

where C is the bandwidth of the output link of the switch. As far as the sender conforms cell emission rate to allocated PCR, the queueing delay at the switch buffer can be negligible, and the cell loss does not occur in principle in the CBR service class. Or, the Poisson arrival approximation of cell arrivals on the switch buffer may be utilized for safer operation when CDV should also be taken into account.

One might find the above description elsewhere in the introductory papers or books for explaining CAC of the CBR service class. However, it is implicitly (or explicitly) assumed that maximum attainable utilization of the switch is 100%. The problem is that in real implementation, it is difficult to find such an ideal switch as will be shown in the next section.

From a viewpoint of effective resource management, some difficulties arise when the video transfer is applied to the CBR service class. One problem is that in the CBR service class, the user needs to determine the PCR value before the connection setup while it is not likely that the user can estimate it especially when real time encoding of video is performed. Such a problem can be relaxed to some extent by introducing some control mechanism such as a play-out buffer and/or error concealing techniques [8], [9]. Another problem is that if the amount of the allocated bandwidth is set to the peak rate of VBR traffic, an effective bandwidth usage is never expected. For example, almost two thirds of the allocated bandwidth would be unused and wasted in the MPEG-2 coded video [10]. However, the unused bandwidth can be utilized by the data communication which is one of major applications in the real world Thus, an ineffectiveness of the CBR service class can be alleviated if the appropriate means is equipped within the switch. See Sect. 4.7 for discussions on multiple QoS support.

2.2 VBR Service Class

The VBR service class has been introduced for transferring VBR traffic effectively [2]. In the VBR service class, effective bandwidth usage is intended by means of statistical multiplexing of the VBR traffic. When a number of VBR traffic streams is multiplexed onto the link, a total of the required capacities of individual connections becomes lower than the sum of peak rates. Of course, QoS parameters such as CLR and/or CTD are guaranteed only statistically, and QoS are degraded when compared with the CBR service class. A rational of the VBR service class is that by introducing statistical multiplexing, the number of simultaneously accepted connections can be increased.

Since a compressed video exhibits cell generation of the variable-bit-rate, it has been considered that the VBR service class is well suited for video data transfer, and its effectiveness has been investigated through experimental studies (see, e.g., [10], [11]). Those papers mainly focused on the statistical multiplexing gain (SMG) as a measure to show the superiority of the VBR service class over the CBR service class. For example, the authors in [10] show that by applying the video transfer to the VBR service class the number of multiplexed connections can be increased to be over three times larger than that of the CBR service class.

To obtain a statistical multiplexing gain, however, each user needs to declare SCR and BT in addition to PCR as traffic parameters, and to specify QoS parame-

† In ITU-T, CBR and VBR service classes are now called as DBR (Deterministic Bit Rate) and SBR (Statistical Bit Rate) service classes, respectively, to discriminate service classes from traffic characteristics. However, we follow the conventional terminologies in this paper.

ters in terms of CLR, CDV and CTD. Then the network can decide whether a new connection can be accepted or not. The new connection is accepted only if the requested QoS of the new connection is guaranteed and the newly added connection does not affect QoS guarantees of existing connections.

Quite a lot of papers have been published to investigate a theoretical aspect of the statistical multiplexing gain. It is because the CAC mechanism requires to predict the statistical multiplexing gain while guaranteeing the QoS parameters. As an example, cell loss probability for the bursty traffic can be estimated by the equivalent bandwidth method [12]. The basic idea of the equivalent bandwidth method is that the required bandwidth of the bursty source must fall into some value between SCR and PCR. Suppose that the output link bandwidth C and the switch buffer size B are given. Then, the maximum allowable number of connections n_{VBR} is determined by the following equation [12]:

$$n_{VBR} = \max\left\{ \frac{C}{PCR\dfrac{y - B + \sqrt{(y-B)^2 + 4B\,SCR\,y/PCR}}{2y}}, \frac{C}{SCR + SCR(PCR - SCR)\sqrt{2\ln\dfrac{1}{\varepsilon} - \ln 2\pi}} \right\},$$

where ε denotes the requested cell loss probability, and

$$y = \ln(1/\varepsilon)BT(PCR - SCR).$$

Namely, if the current number of connections on the link is less than n_{VBR}, then the newly arriving connection can be admitted. While identical traffic sources are assumed in the above equation, the result has recently been extended to more general cases, but essential is that the output link capacity and the buffer size should be known a priori. As indicated in the above description, a difficulty of the VBR service class is that traffic parameters should be estimated in advance to the connection request. This is the reason that implementation of the VBR service class is still questionable. One exceptional approach that the user does not need to declare the SCR value can be found in [13]–[15], where the measurement-based CAC is used to acquire the traffic load condition of the link.

We last note that a recent advancement of traffic modeling shows that traffic in the real world exhibits a heavy tail distribution and a self-similarity property [16]–[18]. It implies that the above-mentioned CAC does lead to the unexpected performance degradation. However, inapplicability of the Markovian modeling of the traffic source is still controversial [19], and should be left to further researches in this area. The purpose of this subsection is to point out that the output link capacity and the buffer size should be given parameters to implement CAC of the VBR service class.

2.3 ABR Service Class

The rate-based congestion control scheme adopted in the ABR service class controls the cell emission rate of each connection, dependent on the congestion status of the switch within the network. At the switch, congestion occurrence is detected by high and low threshold values associated with the queue length of the buffer. These threshold values are denoted by Q_H and Q_L, respectively. When the queue length exceeds the high threshold value Q_H, the switch detects congestion and notifies it to the source end system via the destination end system by marking the EFCI (Explicit Forward Congestion Indication) bit in the data cells, or by marking the CI (Congestion Indication) bit in the forward RM (Resource Management) cells. After then, when the queue length goes under the low threshold value Q_L, it is regarded that its congestion is terminated. The EFCI bit in data cells or the CI bit in forward RM cells is then cleared to indicate congestion relief to source end systems via the destination end system. This sort of switch operation is often referred to as an EFCI bit marking switch, or simply as a binary switch [20].

A behavior of end systems is standardized in the ATM Forum, and is specified in [21]. Each source end system periodically generates forward RM cells in proportion to its rate; that is, it sends one forward RM cell per $(N_{RM} - 1)$ data cells. As soon as the receipt of the forward RM cell, the corresponding destination end system sends it back as the backward RM cell to the source end system. Allowed cell rate ACR of the source end system is changed in response to backward RM cells. When it receives the CI bit cleared cell, rate increase is performed as

$$ACR \leftarrow \min(ACR + RIF \times PCR, PCR),$$

where PCR is the maximum rate for this connection negotiated at its establishment, and RIF (Rate Increase Factor) is the ratio for proportional rate increase. The CI bit marked cell decreases ACR as

$$ACR \leftarrow \max(ACR - ACR \times RDF, MCR),$$

where MCR (Minimum Cell Rate) is the minimum rate (or guaranteed rate) for this connection but in usual case MCR may be set to zero. RDF (Rate Decrease Factor) is a ratio for exponential rate decrease.

Performance degradation of ABR is mostly caused by an inappropriate setting of control parameters, but selection of appropriate parameters is not an easy task if we use a simulation technique through trial and error. Therefore, an analytical result is derived in [22], [23], where the following two conditions are posed to achieve no cell loss at the buffer and full link utilization by assuming that all connections behave identically.

Condition 1: For the maximum queue length (denoted by Q_{max}) to be less than the buffer size B, we should have

$$\begin{aligned} B &\geq Q_{max} \\ &= Q + \frac{N_{VC}}{RDF}\sqrt{\frac{2N_{RM}\,Q\,RIF \times PCR}{C}} \\ &\quad - \frac{N_{RM}\,N_{VC}}{RDF}\log\left(1 + \sqrt{\frac{2Q\,RIF \times PCR}{C\,N_{RM}}} + \frac{\tau\,RIF \times PCR}{N_{RM}}\right) \\ &\quad + \tau\left(\sqrt{\frac{2C\,Q\,RIF \times PCR}{N_{RM}}} + \frac{N_{VC}\,RIF \times PCR}{RDF}\right) \\ &\quad + \frac{\tau^2\,C \times RIF \times PCR}{N_{RM}} \end{aligned}$$

where Q represents a threshold value at the cell buffer, i.e., $Q = Q_H = Q_L$, and τ does the end-to-end propagation delay between end systems. N_{VC} is the number of active connections on the link. We notice that the above equation is somewhat lengthy without proof. Our intention here is that if we expect no cell loss at the switch, the output link capacity C and the buffer size B should be known to properly set the control values RDF and RIF, and the queue threshold value Q.

Condition 2: The minimum queue length should not reach zero for avoiding "under utilization" of the output link. Since the required condition is more complicated, we omit the equation, but the result shows that the output link capacity C and the buffer size B should also be known a priori.

There may exist many combinations of control parameters that satisfy the above two conditions. In that case, we must choose larger RIF and RDF, which leads to the more rapid rate increase and decrease. It can achieve a better transient behavior.

In the ABR service class, temporal congestion must be absorbed at the switch buffer since it takes at least round-trip time to notify the switch congestion of the end systems. That is, the switch buffer should be moderately large. An order of hundred buffer size is required if the switch is applied to the local area network (LAN) environment (see, e.g., [22], [23]). In the WAN environment, on the other hand, cell buffer even with an order of thousands is not sufficient to assure no cell loss while keeping the link utilization high. Of course, when we consider the ABR service class, some error recovery control is equipped with the upper layer protocol. Then, non-zero cell loss may be accepted in some cases. A typical example is TCP over ABR service class. See, e.g., [24] and references therein.

In the ATM Forum, another type of switch operation, called an explicit-rate (ER) mode, has been considered. While its implementation is rather complex, the explicit-rate switch has a potential to obtain stable operation (i.e., low queue oscillation). A typical operation of the explicit-rate switch is to compute an appropriate bandwidth allocation to the connections based on the available bandwidth and the degree of congestion. The switch then updates the ER value of forward and/or backward RM cells as

$$ER \leftarrow \min(FS, ER),$$

where ER is the ER value in the RM cell written by some other upstream switch, and FS is a computed bandwidth allocation for the connection. A typical example to compute FS is

$$FS = C/N_{VC},$$

where N_{VC} is the number of active connections on the link. When the source end system receives the backward RM cell, it updates its ACR as

$$ACR \leftarrow \min(ACR + PCR \times RIF, PCR, ER).$$

Thus, bandwidth allocation for all connections can be finished within one round-trip time if RIF value is appropriately chosen. The brightness of the above equation is that the source end system does not necessarily know the switch type (i.e., binary-mode or explicit-rate switch). In other words, an effectiveness of explicit-rate switches is fully dependent on the determination method of the FS value.

In the ATM Forum, several switch algorithms with explicit-rate marking have been proposed through standardization process of the rate-based congestion control algorithm [21], [25]. Those include EPRCA (Enhanced Proportional Rate Control Algorithm) [26], CAPC (Congestion Avoidance using Proportional Rate Control) [27], APRC2 (Adaptive Proportional Rate Control) [28] and ERICA (Explicit Rate Indication for Congestion Avoidance) [29] (see also [4]). Each algorithm has its own advantages and disadvantages in terms of effectiveness, robustness, fairness and configuration simplicity. However, the common objective is to share the available bandwidth equally among the active connections. That is, an explicit-rate mode switch requires the knowledge on the bandwidth and the number of active connections for its operation. In the LAN environment, the ER-mode operation requires the cell buffer with an order of tens while an effectiveness is dependent on the implementation method of the algorithm (see, e.g., [30]). In the WAN environment, on the other hand, the stability of queues and the required number of cell buffers have not been fully discussed yet except [31] where the authors study the applicability of the ER-mode operation as well as the binary-mode operation.

3. ATM Switching Architectures

There are so many switching architectures known in the literature, and excellent survey papers have already been published [32]–[36]. In this section, we briefly review various switching architectures by taking account of an applicability to QoS guarantees.

If an ATM switch is nonblocking, there will be a path from every input to every output port. Even if the switch is nonblocking internally, however, contention for output ports would take place since cells arriving at two or more inputs may try to go to the same output simultaneously. If the time that the contention persists is short, contention can only be handled by buffers. On the other hand, long term congestion must be handled by some appropriate congestion control mechanisms such as CAC in CBR/VBR service classes and the rate-based congestion control in the ABR service class in conjunction with switch buffers. Time duration of the congestion depends on the implementation method of congestion control mechanisms as described in the previous section, and henceforth, the amount of cell buffers should be carefully prepared according to the service classes supported by the switch.

With nonblocking switch fabrics, a major focus of design attention is the buffer structure. Since ATM is based on a packet switching technique in principle, the most important traffic issue in designing the switch architecture is the choice of queueing discipline. The position of buffers within a switch greatly influences the switch performance, and at a high level of abstraction three possible disciplines can be used;

- input buffering
- output buffering
- shared buffering

The combination of these three variants can also be considered to optimize the implementation cost and performance.

The physical realization requirement of the above three alternatives are different due to speed requirements and additional control logic for the buffer memory. Three parameters affects the switch complexity;

Memory size depends on the performance requirements of the system.

Memory speed depends on the buffering strategy. It also depends on the size and the speed of the input and output links.

Memory control is required in order to control the queues dependent on the queueing principle. If input and output buffers use the FIFO discipline, only a simple control logic is required, whereas the shared buffering requires an additional control for dynamic buffer management. Various techniques for performance improvement of the input buffering described in the next section also needs a special control logic.

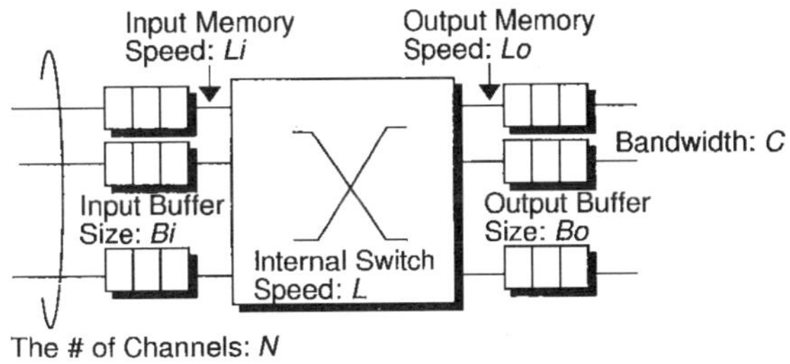

Fig. 1 Generic performance model of ATM switches.

Before proceeding into investigation on performances of different switching architectures, we describe the required speed of buffer memories. The memory can be accessed in parallel (e.g., byte, word, or even cell). We represent the switch size and the link capacity as N and C, respectively. Further let W be the width of memory in bits. Required (dual ported) memory speeds are then given as W/C, $W/(N\,C)$, and $W/(N\,C)$ for input, output, and shared buffering strategies. For example, if we let $W = 16\,[\text{bit}]$, $C = 155\,[\text{Mbps}]$ and $N = 16$, the corresponding memory speeds become 106.6 ns, 6.7 ns and 6.7 ns, respectively.

Keeping those values in mind, we now review the performance aspects of the above three buffering strategies. In what follows, we focus on the class of nonblocking switches, where internal blocking does not occur. We assume the switch fabrics of size $N \times N$. See Fig. 1. Buffers are placed according to the buffering strategy; B_i and B_o denote the size of buffer on an input and an output, respectively. B denotes the buffer size in the case of the shared buffering system. Unless otherwise stated, cells in the buffer are served on a FIFO basis. The internal switching speed is L times faster than the link bandwidth C. That is, the switch can deliver at most L cells to the same output port from the different input ports during each cell time. The speed of output channel L_o is identical to the switching speed L, whereas the input channel speed L_i may be 1 or L (see Sects. 4.2 and 4.3).

Arrivals of cells at each of N input ports are assumed to follow a Bernoulli process; a cell arrives with probability p during cell time, which corresponds to the traffic load at the input port. Uniform traffic refers to the situation where incoming cells are destined to N output ports with a uniform probability of $1/N$. The first concern of the switch performance is the maximum throughput of the switch by assuming the infinite buffer size. Further, the switch size is often assumed to be infinity ($N = \infty$) for simplification of the analysis. For infinite switch size, the cell arrival becomes a Poisson process.

In applying the CBR service class, the Bernoulli

or Poisson arrival assumptions can estimate the performance as a worst case. On the other hand, it is well known that the Bernoulli or Poisson processes cannot model the bursty traffic accurately. However, the implementation of the VBR service class is a still controversial issue as we have already discussed in Sect. 2.2.

Note that other issues including scalability of ATM switches, supports for multicast and broadcast are also important, and should be considered at the architectural design of the switch. However, it is out of scope in this paper, and interested readers should refer to, e.g., [37] and related papers in this issue. See also [38] for commercially available ATM switches.

3.1 Output Buffering

In output buffering, we have the cell buffer only at the output ports (see Fig. 2 (a)). The cell arriving at the input port is lost if it is not successfully delivered to the output port in the arriving slot. If we have $L = N$ switching capacity, every cells at the input ports can be delivered to the same output port.

By assuming a Bernoulli arrival process at each input port, the probability generating function $X(z)$ of the arrival process at tagged output port can be represented as [39]

$$X(z) = \left(1 - \frac{p}{N} + \frac{p}{N}z\right)^N.$$

Then, the probability generating function of the number of cells at the output port can be obtained as

$$Q(z) = \frac{(1-p)(1-z)}{X(z) - z}. \quad (1)$$

The average queue size $\overline{Q}$ can be directly calculated by differentiating Eq. (1) with respect to z and substituting $z = 1$ as

$$\overline{Q} = \frac{N-1}{N}\frac{p^2}{2(1-p)}.$$

Since the mean number of cells of a well-known M/D/1 queueing system with a Poisson arrival process is given by

$$\overline{Q}_{M/D/1} = \frac{p^2}{2(1-p)}.$$

We have a relation

$$\overline{Q} = \frac{N-1}{N}\overline{Q}_{M/D/1},$$

which shows that the average queue size converges to that of an M/D/1 queueing system for $N \to \infty$. It can be verified by the fact that superposition of N independent Bernoulli arrival processes becomes a Poisson process as N reaches ∞. Also, the M/D/1/K queueing system can be utilized for finite buffer case.

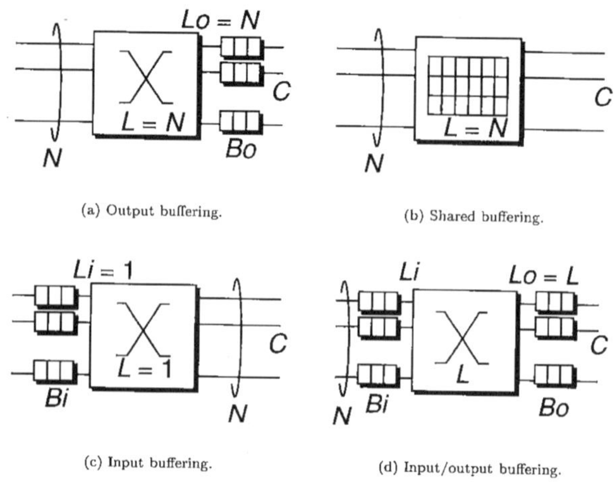

Fig. 2 Buffering strategies of ATM switches.

Since the output link bandwidth and the buffer size can be known a priori, the direct application of service classes described in the previous section can be performed. The output buffering is therefore ideal in that sense when comparing with other buffering strategies introduced in the following subsections.

A problem in the output buffering is that it requires very fast switching speed, and thus limits the buffer size. When we consider the CBR service class, such a limitation may be acceptable. The above results obtained by Bernoulli or Poisson arrival assumption can directly be applied in the case of the CBR service class. When the load p (the sum of PCR's) is restricted to, say, around 0.8, required buffer size is below an order of tens to guarantee 10^{-9} of cell loss probability. It becomes apparent by direct application of the M/D/1/K queueing system, or see Fig. 3 where the result of shared buffering case is also depicted. The applicability of the CBR service class can be confirmed by the following example. When applying circuit emulation to the CBR service class, only a very low cell delay is tolerated within the switch. In the switching fabric running at 155 Mbps, a cell time is 2.7 μs. It is then obvious that the strong timing requirement of circuit emulation can be reached when buffers are smaller than 40 cells.

On the other hand, speed limitation of the memory becomes an obstacle in the case of the VBR service class. The large buffer is necessary in the VBR service class when we want to enjoy the statistical multiplexing gain. The buffer size should be an order of Mbits (tens of thousands in cells) to obtain the statistical multiplexing gain of motion videos while keeping the cell loss ratio to be 10^{-12} [10]. Even if the large buffer is equipped with the switch, possible cell delays experienced at the switch may become an order of hundred milliseconds, and it cannot be applied to interactive applications such as teleconferencing systems.

In the ABR service class, the buffer size should be large enough. As described in the previous section, an

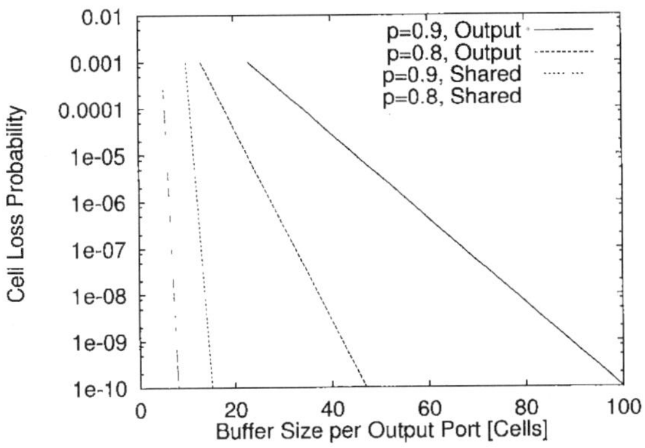

Fig. 3 Cell loss probability of output and shared buffering.

order of hundreds is necessary if the switch is applied to the LAN environment. However, it becomes difficult to be applied to the WAN environment because cell buffer with an order of thousands is required to obtain zero cell loss while keeping the link utilization high.

We last note that in the literature, the output buffering is sometimes referred to as a best solution for performance objectives. It is not true. Actually, cell loss probability can be reduced by introducing the shared buffering, as will be demonstrated in the next subsection. The output buffering would be best if our concern is an easy implementation of traffic control because the bandwidth available on the output link and an associated buffer size can directly be determined. We further notice that even when output buffering is compared with input buffering, one cannot claim that output buffering is better than input buffering under all traffic situations [40]. It will be shown partly in Sect. 4.5.

3.2 Shared Buffering

The shared buffering (Fig. 2 (b)) behaves exactly as the output queueing. However, since multiple queues are combined in a single physical memory, a major advantage of the shared buffering is reflected in the number of cells to be stored in the buffer, but a more complicated control logic is required to ensure that the FIFO discipline for each output link is performed in the single shared memory. The queue size can be calculated as a convolution of N individual output queues.

The number of cells in the shared buffer, Y, equals the sum of cells stored in each output queue, X. That is, the probability distribution of the number of cells in the shared buffer is obtained by the convolution of the queue distribution of each output buffer. By letting $M_X(z)$ be the moment generating function of X, we have $M_X(z)$ by utilizing the M/D/1 queue as

$$M_X(z) = \frac{(1-p)(1-e^z)}{1-e^z\, e^{(1-e^z)p}}.$$

By convolving N independent queues, and applying the Chernoff bound, we have an approximate cell loss probability for the shared buffering system as

$$\mathrm{P}[Y \geqq N \log M_X(z)] \leqq e^{N(\log M_X(z) - z \log M_X(z))}.$$

The relations between the buffer size and cell loss probability in cases of output buffering and shared buffering are shown in Fig. 3, where it is shown that the shared buffering can guarantee the same loss probability by a smaller buffer size. For example, with $p = 0.9$, the queue size of an output queueing system is around 90 cells for a cell loss probability of 10^{-9}, whereas in the shared buffering, only around 15 cells per output port are sufficient. The figure shows a large gain in required memory size for the shared buffering at the expense of its complicated memory control logic. This gain depends on the switch size N; the larger N, the larger the statistical multiplexing effect of the memory is. Note that more accurate approaches can be found in [41]. Further, the approach taken here can also be applied to the bursty traffic case.

The problem is that the buffer should be logically divided for each output port. Let B denote a shared buffer size, and b_i be the assigned buffer size for output port i. If $\sum_i b_i = B$, then the benefit of shared buffer cannot be attained at all. The obtained cell loss ratio becomes identical to the output buffering. On the other hand, if we set b_i's such that $\sum_i b_i > B$, then QoS cannot be guaranteed. Therefore, the benefit of the shared buffering is only its flexibility in that the buffer size assigned to each output port can be statically or dynamically changed. This aspect of the shared buffering will be further discussed in the next section.

3.3 Input Buffering

The input buffering strategy (Fig. 2 (c)) suffers from the head-of-line (HOL) blocking. That is, unless the first cell in the queue is served, other cells behind it cannot be served. It is because with input buffering, only the HOL cells of FIFO input buffers are involved in the arbitration. This reduces the maximum switch throughput. Under the uniform traffic assumption, the throughput of an input-buffered switch with internal switch speed $L = 1$ is limited to 58.6% of output line capacity [39]. We first derive this value.

Suppose that B_n^i cells are destined to output port i, but blocked at the head of the line during cell time n. By serving some number of cells to transmit to their respective output port, a same number of cells proceed into the head of the line. Let A_n^i be the number of new cells appearing at the head of the line destined for output port i during cell time n. Since only one cell can be served for each output port ($L = 1$), the number of blocked cells at the HOL equals the number of blocked cells during the previous cell time B_{n-1}^i minus 1 (served) plus the new cells A_n^i, i.e.,

$$B_n^i = \max(B_{n-1}^i - 1 + A_n^i, 0).$$

In input buffers, new cells may progress to the head of the line if the HOL cells are served. A total number of such cells at cell time n, L_n, is represented as

$$L_n = \sum_{i=1}^{N} A_n^i.$$

Since this value also represents the total number of cells to be served by the switch during cell time $n+1$, L_n can also be written by using the number of blocked cells, B_{n-1}^i, as

$$L_n = N - \sum_{i=1}^{N} B_{n-1}^i. \tag{2}$$

The average of L_n in steady state, $\overline{L}$, equals pN, where p represents the switch throughput per output link, i.e.,

$$\overline{L} = pN. \tag{3}$$

By letting $N \to \infty$, the number of cells moving to the head of the line in steady state destined for output port i is Poisson distributed with rate p. This means that the behavior with $N \to \infty$ is like that of an M/D/1 system. Then, we have

$$\overline{B}^i = \frac{p^2}{2(1-p)}. \tag{4}$$

By averaging Eq. (2) and taking Eq. (3) into account, we have

$$\overline{L} = N - \sum_{i=1}^{N} \overline{B}^i.$$

Then, the following equation holds from Eq. (3);

$$\overline{B}^i = 1 - p. \tag{5}$$

By combining Eq. (4) and Eq. (5) we finally have

$$p_{max} = 2 - \sqrt{2} = 0.5858.$$

That is, switches based on the input buffering strategy are very limited in performance. The maximal achievable throughput is only 58.6%.

We note that by replacing Eq. (4) with the result of an M/M/1 queueing system;

$$\overline{B}^i = \frac{p^2}{(1-p)},$$

we have $p = 0.5$. Its direct interpretation is that the maximum throughput is only a half of the link bandwidth when the cell length follows an exponential distribution [42]. However, it can also be viewed that the upper layer PDU (Protocol Data Unit) follows an exponential distribution. In a more strict sense, the number of cells arriving at some input queue follows the geometric distribution and destined for the same output port [43]. See Sect. 4.5.

In either case, the cell loss probabilities for a given offered load close to the throughput limitation are very high regardless of buffer size. Then, one way to improve the loss probability is to increase the maximum throughput of the switch by modifying the original input buffered-switch, which will be discussed in the next section.

4. Performance Improvement of Input Buffering Strategy

As shown in the previous section, input buffering is difficult to be used as a unique solution due to its low performance, and various techniques to improve the switch performance have been proposed and investigated. Two primary objectives can be found in recent developments of input-buffered switches. One direction is to develop the switch with very high throughput (in an order of hundred Gbps, and towards Tbps). Another is that as has been discussed in the previous section, a large amount of cell buffer is provided for the VBR and ABR service classes. While *pure* input buffering suffers from poor link utilization, it can operate in lower internal switching and memory speeds. Input buffering then becomes a good choice.

4.1 Window Policy

In the pure input buffering, cells are transmitted only from the head of each input queue. Cells behind the head of the queue are blocked regardless of whether or not their destination output port is available. The throughput can be increased by adopting a window policy [44], [45], which reduces HOL blocking by checking the first w cells in each FIFO queue module. When the HOL blocking occurs, a cell destined for the different output port is sought and transmitted. This policy corresponds to $L = L_i = L_o = 1$, but the complex memory control is necessary, and the throughput improvement is limited. For large N, the throughput is improved from 0.586 ($w = 1$) to 0.70 with $w = 2$ or 0.88 even with $w = 8$ [46] in the case of uniform and Bernoulli cell arrivals. Further, when the bursty traffic is applied, cells behind the HOL cell is likely to be destined for the same output port, which inhibits the performance improvement.

Note that if the appropriate arbitration is assumed, the 100% throughput can be attained even in the input buffering by eliminating HOL blocking perfectly [47]. Of course, its applicability would be limited in a real situation due to its algorithm complexity.

4.2 Speedup of Internal Switching Capacity

A direct solution to improve the switch performance is an increase of internal switching capacity $L(1 \leq L \leq N)$ [42], [48], where cases of $L = 1$ and $L = N$

Table 1 Maximum throughput of the switch with speedup ($N = \infty$).

L	Maximum Throughput
1	0.58579
2	0.88454
3	0.97550
4	0.99559
5	0.99931
6	0.99991
7	0.99999
8	≈ 1

correspond to previously-mentioned *pure* input buffering and *pure* output buffering, respectively. In the switch with input bus, the switching speed is equal to that of the input and output channels, but can transfer more than one cell from inputs to a particular output ($1 \leq L = L_o \leq N$). On the other hand, it can deliver only one cell from each input to outputs during cell time ($L_i = 1$). An example of the switch with input bus is the knockout switch [49]. Since at most L cells can be sent to the same output, the output buffer is also necessary in this case (Fig. 2 (d)). As will be shown in Sect. 4.4, a class of the switch with parallel planes is also classified into this category.

For given L, the number of at most L cells can be transferred to the same output port, and other $N - L$ cells are stored at the input buffer if any. By assuming $N = \infty$, the HOL queue behaves like an M/D/L queueing system, where L corresponds to the number of servers. Then the maximum switch throughput can be obtained numerically. Results are shown in Table 1 [48]. As we can see in the table, $L = 3$ or 4 can offer the reasonable high throughput.

4.3 Speedup of Switching Capacities and Input Memories

When the input memory speed (L_i) is also increased, more performance improvement can be obtained. We now treat this case; $1 < L = L_i = L_o \leq N$. Differently from the previous case, the switch transmits the maximum of L cells from *each* input to outputs. Such a switch can be realized by, e.g., the Batcher-Banyan switch or the crossbar switch operating L times faster than the input/output channels. In [50], the crossbar switch is used. This switch also has a desirable feature from a viewpoint of CAC as shown in the below.

Since the switch capacity is L times faster than the input channels, the service rate in Eq. (4) can simply be replaced with

$$\overline{B}^i = \frac{(p/L)^2}{2(1 - (p/L))}.$$

Since the average number of cells moving into the head of queues is represented as p/L, the maximum throughput is derived as

$$p = \left(2 - \sqrt{2}\right) L.$$

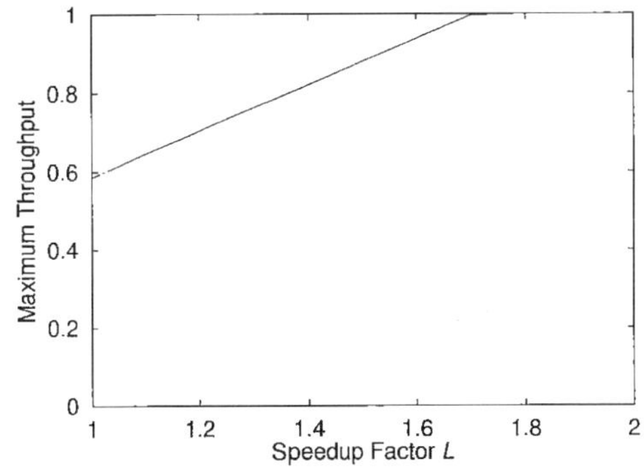

Fig. 4 Speedup effect on maximum switch throughput.

Figure 4 depicts the maximum throughput for given L. Note that in [50], similar results are obtained using more realistic conditions.

Another advantage of this switch can be found in its simple traffic control implementation. If the switch with $L = 2$ is used, the HOL blocking can almost be eliminated [50]. Then, the switch may be viewed as if the output buffering strategy is utilized.

4.4 Parallel Switches

So far, we have seen techniques to improve the performance of a single plane switch. Another type is called a parallel switch, which uses a multiple number of switch planes (denoted as K) to achieve better performance. The maximum throughput of the switch must be identical to the sum of the throughput of each switch plane if we neglect the control overhead. Since the pure input buffering offers 58.6% of throughput, we have 100% of throughput by two parallel switch planes, each of which is implemented even by the pure input buffering. The idea itself is not new. See, e.g., [51].

One problem exists in the case where two input buffers are equipped with each input port, and incoming cells are randomly (or alternatively) placed in one of two buffers (Fig. 5 (a)). In this way, the maximum throughput of 100% can be realized by two parallel planes, since by letting the number of planes K to L, the performance model becomes identical to the one in the previous subsection. However, cells at the output buffer may be in out-of-order. Then, in [45], the arbitration method to choose the plane for the cell in the head of the input buffer is proposed (Fig. 5 (b)). In this case, however, the maximum throughput becomes identical to the case with $L_i = 1$ as having been described in Sect. 4.2.

Another way is to route cells on a VC basis by assuming that each plane is equipped with its own input

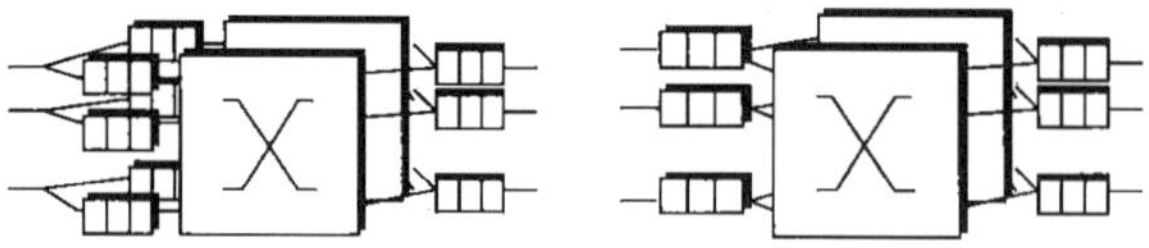

(a) Parallel switch with separate input buffers.

(b) Parallel switch with common input buffers.

Fig. 5 Configuration of parallel switches.

buffer as noted in [34], [35]. Since ATM is connection-oriented, one can easily implement CAC by performing it separately in each plane.

4.5 Backpressure Function

At most L cells at the input ports can be transferred to the output ports during cell time in the previous examples. However, cells are lost when the output port buffer is full. Since the output port buffer requires high speed memory to accept L cells simultaneously, the large amount of the buffer cannot be prepared at the output side. On the other hand, the memory speed of the input buffer can be low and therefore the large buffer at the input side avoids the cell loss. This fact naturally leads to an idea of a backpressure function, where the backpressure signal is sent from the output to the input when the output buffer is full [43], [52]–[55].

This option is useful for the ATM LAN switch since its main application is data communication, where low cell loss probabilities are expected for efficient resource usages. Once the backpressure mechanism is provided, however, the HOL blocking occurs even with $L = N$. In [43], the performance of the switch with backpressure function is analyzed in the case of $L = L_o = N$ and $L_i = 1$. That is, the switch corresponds to the one presented in Sect. 4.2. In Fig. 6, we depict the maximum throughput dependent on the *packet* length for different values of the output buffer size B_o. In the model, it is assumed that cells forming the packet arrive at the input port consecutively, and are destined for the same output port. The maximum throughput can be improved by a small amount of output buffers; 0.9 for $B_o = 20$ and around 0.95 for $B_o = 50$. Note that the case of $B_o = 1$ corresponds to the pure input buffering strategy since it means that at most one cell can be accepted at each output port, and that other cells destined for the same output port are waited in the input port buffer due to the backpressure signal. Another observation can be made from this example. When the packet length becomes large, the maximum throughput reaches 0.5 regardless of the output buffer size as in the case of pure input buffering. This fact makes it difficult to apply the switch to the LAN environment where the main traffic source is bursty computer data.

This is the reason that the authors in [55] proposed RIRO (Random In Random Out) scheduling at the input port side where the logical queue is formed for each

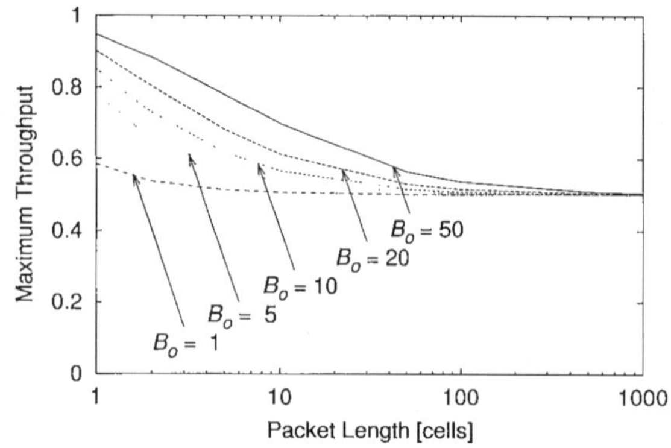

Fig. 6 Maximum throughput of the switch with backpressure function.

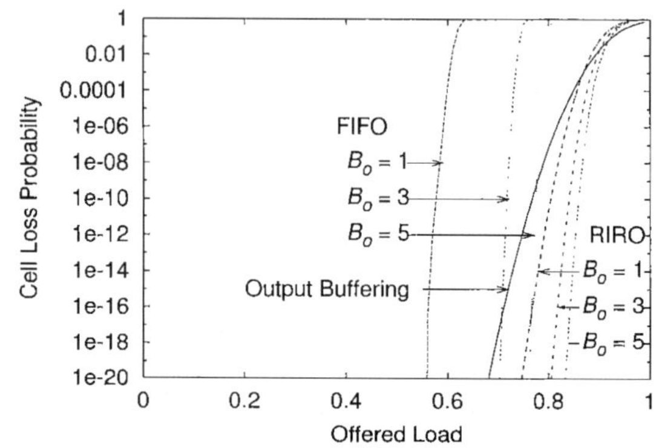

Fig. 7 Improvement of cell loss probability by backpressure function.

of output ports. That is, each input port buffer can be viewed as the shared buffer. Further, by introducing the appropriate arbitration mechanism to select the next cell to be transmitted to the output port, the entire switch can be built as the shared-buffered switch economically by low-speed and large input buffer. In Fig. 7, we compare cell loss probabilities of two scheduling disciplines; FIFO and RIRO. The horizontal axis shows the offered load, and three different values of the output buffer sizes are used. In the figure, the case of output buffering is also shown. For fair comparison, the sum of input and output buffer sizes are always kept to 30.

From the figure, we can see that the RIRO scheduling policy offers better performance even than the output buffering. A more interesting point observed in the figure is that even the FIFO policy is better than the output buffering when the offered load is kept moderately low. It is because at the input port side, cells destined for the same output port are distributed at different input buffers, and henceforth, we have a same effect having been observed in the shared buffering method. See the next section for more general discussion on how to form the shared buffer structure at each input port.

We last note that the backpressure function is useful even in the case of the switch introduced in Sect. 4.3. Actually, the backpressure function is utilized in [50].

By this mechanism, sizes of input and output buffer can be kept fairly small. However, it introduces another problem; a difficulty of CAC implementation, which is a common problem arising when the shared buffering is used somewhere within the switch.

4.6 Shared Buffer Structure at the Input Port

FIFO queueing at the input side of the switch suffers from HOL blocking. Then, non-FIFO queueing becomes important to improve the performance. Namely, multiple logical queues are built in each input buffer. In [56], three different queueing methods are introduced;

1. per-VC queueing
2. per-output queueing
3. per-priority queueing

RIRO scheduling described in the previous subsection corresponds to the second one; per-output queueing. The control complexity must depend on the switch size, the number of VCs the switch can support, and the number of QoS classes. Then it affects the design decision.

Per-priority queueing is mandatory when multiple QoS classes are supported. Priority queueing in the context of QoS guarantees have been extensively studied in the literature. However, its implementation is still questionable. We will be back to this problem in the next subsection. When we are concerned with CAC, per-output queueing becomes necessary because CAC should be implemented for every output link. A remaining problem is how to relate the maximum throughput of the switch to the bandwidth used in CAC. More researches are required in this field. Per-VC queueing is recently considered in [57], where it is used for fair share of bandwidth among the connections in the context of the ABR service class. However, it must be difficult to be applied to support different QoS guarantees for connections due to lack of an effective CAC mechanism.

4.7 Support of Multiple QoS Classes

The problem on how to manage multiple QoS service classes has not been fully resolved yet in the literature. Only an agreement seems to prepare two priority classes; CBR and ABR (or UBR) service classes. The CBR service class cells are given higher priority than the ABR service class cells [30], [55], [58]. As a consequence, two separated queues for CBR and ABR service classes are required physically or logically. When two service classes co-exist, we may be able to utilize a simple *head-of-line* priority scheduling [30], [55], [58]. That is, a combination of per-priority and per-output queueing is necessary in this case.

When we want more service classes, on the other hand, a multiple QoS handler is necessary to manage all queues on a per output link basis. Its task is to decide which one of the waiting cells has to be delivered to ensure QoS of waiting cells. Queue management may be necessary using per-VC queueing. However, its implementation requires a further research. Such a challenging switching architecture can be found in [56]. The proposed switch has non-FIFO input queues and the buffer management can allow multiple QoS classes by preparing per-VC queues. However, the relation to CAC is not clear, especially for the VBR service class, and its implementation would become more difficult when the switching speed is increased.

Several researchers often claim the lack of evaluation using the bursty traffic since VBR traffic such as a compressed video exhibits the bursty nature. However, it is not always necessary. When only the CBR service class is supported, evaluation using the random traffic assumption is sufficient. The bursty traffic would be applied to the ABR service class, but its effectiveness is owing to the congestion control algorithm used in the ABR service class. Implementation of the VBR service class supporting the VBR traffic seems to be difficult as having been described in Sect. 2.2.

5. Summaries and Further Discussions

Before we summarize requirements on ATM switch architectures for QoS guarantees, recall that each of service classes requires the following information:

CBR service class: the bandwidth of the output link on which the connections are multiplexed.

VBR service class: the buffer size of the output port as well as the bandwidth.

ABR service class (binary-mode): the buffer size and the bandwidth.

ABR service class (ER-mode): some implementations require the buffer size as well as the bandwidth.

As described in Sect. 3.1, output buffering is promising for CAC because almost all of past researches on CAC explicitly or implicitly assumed the output buffering mainly due to its simplicity. The bandwidth is assumed to be identical to the bandwidth of the output link, and the buffer is placed at the output port. However, output buffering requires high-speed switching fabric. The high-speed output buffer memory is also necessary, which limits the cell buffer size. When we only consider the CBR service class, it would be sufficient since we need only a few tens of cell buffers to support the CBR service class. On the other hand, the VBR and ABR binary-mode service classes need a large amount of buffers, and output buffering is not appropriate for these service classes.

As described in Sect. 3.2, the shared buffering strategy can offer much smaller cell loss probability for given total buffer size, while it also requires a high-speed operation. The shared buffering introduces the possibility of a small group of output ports hogging the buffer space to the detriment of ports supporting more well-behaved connections. Therefore, the virtual queue for each output port should be maintained. Several shared buffer management strategies have been proposed recently [59]–[61]. However, the relation to CAC is not clear in those papers. Since only the information on an available bandwidth is necessary in the CBR service class, it can be applied to the shared buffering. However, meaning of introducing the shared buffering is not clear in that case. In the VBR service class, on the other hand, the information on the buffer size as well as the output bandwidth is required. When the shared buffer is completely partitioned into the number N of buffers, the advantage of the shared buffering is lost, which is a main problem of this method. Note that there exists the buffering method in which output buffers are shared. The above discussions on the shared buffering strategy can also be applied to this case.

In the output and shared buffering policies, the high-speed memory is required, which is the reason that a large amount of buffers is difficult to be equipped. As has been described before, the VBR service class requires much buffers to gain the statistical multiplexing gain. The binary-mode operation of the ABR service class also requires the large buffer in an order of hundreds/thousands cells because the cell loss ratio is a main performance objective. Only an ER-mode of the ABR service class seems to be applicable to the switch with small buffer when a stability issue is enough resolved [30]. However, the applicability to the WAN environment has not been known yet.

When we consider the input buffering, the story becomes complicated. When the pure input buffering is employed, the available bandwidth is 58% of the output link capacity. The HOL queue formed in the input buffers behaves as the M/D/1 queue. It suggests that the CBR service class can be supported by regarding as 58% of the output port bandwidth as the available bandwidth. A method for realizing more precise bandwidth guarantees can be found in [62]. For the other service classes, however, it is not clear how the available buffer size can be estimated.

The input buffering with speedup of memories of the input buffer as well as the switching fabric has a desirable feature that only two times faster speed is sufficient to obtain 100% throughput (Sect. 4.3). However, it implies that almost all cells are stored in the output buffer. It means that the large amount of memories should be equipped with the output port for supporting the ABR and VBR service classes. Otherwise, the backpressure function is useful to store cells at input buffers. In this case, however, some appropriate means to determine the *logical* buffer size for the designated output ports becomes essential, but it requires further researches.

We last note that the feasibility checking of memory control logic for given memory speed is an important issue to design a high performance switch. The design optimization under the performance constraint is therefore an important problem. One such investigation can be found in [63], but only the model described in Sect. 4.2 is treated there. Researches on other switching architectures in line with recent advancements of the high-speed switching technologies are highly expected.

Acknowledgment

The author would like to thank anonymous reviewers for their helpful comments.

References

[1] P. Newman, "Traffic management for ATM local area networks," IEEE Commun. Mag., pp.44–50, Aug. 1994.

[2] M.W. Garrett, "A service architecture for ATM: From applications to scheduling," IEEE Network, vol.10, no.3, pp.6–14, May/June 1996.

[3] Special issue on "Flow and congestion control," IEEE Commun. Mag., vol.34, no.11, Nov. 1997.

[4] Special issue on "Bandwidth allocation in ATM networks," IEEE Commun. Mag., vol.35, no.5, May 1997.

[5] ITU-T, "ITU-T recommendation I.371, traffic control and congestion control in B-ISDN," International Telecommunication Union, 1992 revised in 1995.

[6] The ATM Forum, "ATM user network interface (UNI) specification version 3.1," Prentice Hall Professional Technical Reference, 1995.

[7] G. Woodruff and R. Ksitpaiboon, "Multimedia traffic management principles for guaranteed ATM network performance," IEEE J. Sel. Areas Commun., vol.8, no.3, pp.437–446, April 1990.

[8] N. Wakamiya, M. Murata, and H. Miyahara, "Quality of MPEG transfer over ATM DBR service class with effective bandwidth usage," Proc. of IEEE GLOBECOM'96, vol.3, pp.1942–1948, Nov. 1996.

[9] S. Aign and K. Fazel, "Temporal & spatial error concealment techniques for hierarchical MPEG-2 video codec," Proc. of IEEE ICC '95, pp.1778–1783, June 1995.

[10] H. Heeke, "Statistical multiplexing gain for variable bit rate video codecs in ATM networks," Inter. J. Digital & Analog Commun. Syst., vol.4, pp.261–268, 1991.

[11] A.R. Reibman and A.W. Berger, "Traffic descriptors for VBR video teleconferencing over ATM networks," IEEE/ACM Trans. Networking, vol.3, no.3, pp.329–339, June 1995.

[12] R. Guerin, H. Ahmadi, and M. Naghshineh, "Equivalent capacity and its application to bandwidth allocation in high-speed networks," IEEE J. Sel. Areas Commun., vol.9, no.7, pp.968–981, Sept. 1991.

[13] H. Saito and K. Shiomoto, "Dynamic call admission control in ATM networks," IEEE J. Sel. Areas Commun., vol.9, no.7, pp.982–989, Sept. 1991.

[14] T.E. Tedijanto and L. Gun, "Effectiveness of dynamic bandwidth management mechanisms in ATM networks," Proc. of IEEE INFOCOM '93, (San Francisco), pp.358–367, March 1993.

[15] H. Saito, "Dynamic resource allocation in ATM networks," IEEE Commun. Mag., vol.35, no.5, pp.146–153,

May 1997.
[16] W. Leland, M. Taqqu, W. Willinger, and D. Wilson, "On the self-similar nature of Ethernet traffic," IEEE/ACM Trans. Networking, vol.2, pp.1–15, 1994.
[17] M. Garett and W. Willinger, "Analysis, modeling and generation of self-similar VBR video traffic," Proc. of SIGCOMM '94, pp.269–280, Oct. 1994.
[18] D. Heyman and T. Lakshman, "What are the implications of long-range dependence for VBR-video traffic engineering?," IEEE/ACM Trans. Networking, vol.4, no.3, pp.301–317, June 1996.
[19] B.K. Ryu and A. Elwalid, "The importance of long-range dependence of VBR video traffic in ATM traffic engineering: Myths and realities," Proc. of SIGCOMM '96, pp.3–14, Aug. 1996.
[20] H. Ohsaki, M. Murata, H. Suzuki, C. Ikeda, and H. Miyahara, "Rate-based congestion control for ATM networks," ACM SIGCOMM Comp. Commun. Review, vol.25, no.2, pp.60–72, April 1995.
[21] The ATM Forum Technical Committee, "Traffic management specification version 4.0 (draft version)," ATM Forum Contribution 95-0013R10, Feb. 1996.
[22] H. Ohsaki, M. Murata, H. Miyahara, C. Ikeda, and H. Suzuki, "Parameter tuning for binary mode switch - analysis," ATM Forum Contribution 95-1483, Dec. 1995.
[23] H. Ohsaki, G. Hasegawa, M. Murata, and H. Miyahara, "Parameter tuning of rate-based congestion control algorithms and its application to TCP over ABR," First Workshop on ATM Traffic Management IFIP WG 6.2, pp.383–390, Dec. 1995.
[24] G. Hasegawa, H. Ohsaki, M. Murata, and H. Miyahara, "Performance of TCP over ABR service class," Proc. of IEEE GLOBECOM '96, pp.1935–1941, Nov. 1996.
[25] R. Jain, "Congestion control and traffic management in ATM networks: recent advances and a survey," ATM Forum Contribution 95-0177, 1995.
[26] L. Roberts, "Enhanced PRCA (proportional rate-control algorithm)," ATM Forum Contribution 94-0735R1, Aug. 1994.
[27] A.W. Barnhart, "Explicit rate performance evaluations," ATM Forum Contribution 94-0983R1, Oct. 1994.
[28] K.-Y. Siu and H.-Y. Tzeng, "Limits of performance in rate-based control schemes," ATM Forum Contribution 94-1077, Nov. 1994.
[29] R. Jain, S. Kalyanaraman, R. Viswanathan, and R. Goyal, "A sample switch algorithm," ATM Forum Contribution 95-0178, Feb. 1995.
[30] Y. Zhao, S. Li, and S. Sigarto, "A linear dynamic model for design of stable explicit-rate ABR control schemes," Proc. of IEEE INFOCOM '97, Kobe, pp.283–292, April 1997.
[31] H. Saito, K. Kawashima, H. Kitazume, A. Koike, M. Ishizuka, and A. Abe, "Performance issues in public ATM service," IEEE Commun. Mag., vol.34, no.11, pp.40–48, Nov. 1996.
[32] H. Ahmadi and W. Denzel, "A survey of modern high–performance switching techniques," IEEE J. Sel. Areas Commun., vol.7, no.7, pp.1091–1103, Sept. 1989.
[33] F. Tobagi, "Fast packet switch architectures for broadband integrated services digital network," Proc. IEEE, vol.78, no.1, pp.133–167, Jan. 1990.
[34] Y. Oie, T. Suda, M. Murata, D. Kolson, and H. Miyahara, "Survey of switching techniques in high-speed networks and their performance," Proc. of IEEE INFOCOM '90, San Francisco, pp.1242–1251, June 1990.
[35] Y. Oie, T. Suda, M. Murata, and H. Miyahara, "Survey of switching techniques in high-speed networks and their performance," Inter. J. of Satellite Commun., vol.9, no.5, pp.285–303, Sept.-Oct. 1991.
[36] E.E. Witte, "A quantitative comparison of architectures for ATM switching systems," Technical Report, WUCS-91-47, Washington University, Sept. 1981. Available at http://wuarchive.wustl.edu/doc/techreports/-wustl.edu/cs/1992/wucs-92-37.ps.Z.
[37] R.O. Onvural, "Asynchronous transfer mode networks: Performance issues," Artech House, Inc., 1994.
[38] E.R. Coover, "ATM switches," Artech House, 1997.
[39] M.J. Karol, M.G. Hluchyj, and S.P. Morgan, "Input versus output queueing on a space-division packet switch," IEEE Trans. Commun., vol.COM-35, no.12, pp.1347–1356, Dec. 1987.
[40] S.C. Liew, "Performance of input-buffered and output-buffered ATM switches under bursty traffic: Simulation study," Proc. of IEEE GLOBECOM '90, pp.1919–1922, Nov. 1990.
[41] J.Y. Hui, "Switching and traffic theory for integrated broadband networks," Kluwer Academic Publishers, 1990.
[42] J.S.-C. Chen and T.E. Stern, "Throughput analysis, optimal buffer allocation, and traffic imbalance study of a generic nonblocking packet switch," IEEE J. Sel. Areas Commun., vol.9, no.3, pp.439–449, March 1991.
[43] H. Ohsaki, N. Wakamiya, M. Murata, and H. Miyahara, "Performance of an input/output buffered type ATM LAN switch with back-pressure function," IEEE/ACM Trans. on Networking, pp.278–290, April 1997.
[44] R. Bubenik and J. Turner, "Performance of a broadcast packet switch," IEEE Trans. Commun., vol.37, no.1, pp.60–69, Jan. 1989.
[45] J.Y. Hui and E. Arthurs, "A broadband packet switch for integrated transport," IEEE J. Sel. Areas Commun., vol.SAC-5, no.8, pp.1264–1273, Oct. 1987.
[46] M. Hluchyj and M. Karol, "Queueing in high-performance packet switching," IEEE J. Sel. Areas Commun., vol.6, no.9, pp.1587–1597, Dec. 1988.
[47] N. McKeown, V. Anantharam, and J. Walrand, "Achieving 100 % throughput in an input-queued switch," Proc. of IEEE INFOCOM '96, San Francisco, pp.296–302, March 1996.
[48] Y. Oie, M. Murata, K. Kubota, and H. Miyahara, "Effect of speed-up in nonblocking packet switch," Proc. of IEEE ICC '89, Boston, pp.410–414, June 1989.
[49] Y.-S. Yeh, M.G. Hluchyj, and A.S. Acampora, "The knockout switch: A simple, modular architecture for high-performance packet switching," IEEE J. Sel. Areas Commun., vol.SAC-5, no.8, pp.1274–1282, Oct. 1987.
[50] K. Genda, Y. Doi, K. Endo, T. Kawamura, and S. Sasaki, "A 160-Gb/s ATM switching system using an internal speed-up crossbar switch," Proc. of IEEE GLOBECOM '94, pp.123–133, Nov. 1994.
[51] J.S. Turner, "Design of a broadcast packet switching network," IEEE Trans. Comput., vol.36, no.6, pp.734–743, June 1988.
[52] I. Iliadis, "Performance of a packet switch with input and output queueing under unbalanced traffic," Proc. of IEEE INFOCOM '92, Florence, pp.743–752, May 1992.
[53] I. Iliadis, "Synchronous versus asynchronous operation of a packet switch with combined input and output queueing," Perfor. Eval., vol.14, pp.241–250, 1992.
[54] A. Pattavina and G. Bruzzi, "Analysis of input and output queueing for nonblocking ATM switches," IEEE/ACM Trans. Networking, vol.1, no.3, pp.314–328, June 1993.
[55] R. Fan, H. Suzuki, K. Yamada, and N. Matsuura, "Expandable ATOM switch architecture (XATOM) for ATM LANs," Proc. of IEEE ICC '94, pp.402–409, May 1994.

[56] H. Duan, J. Lockwood, S. Kang, and J. Will, "A high-performance OC-12/OC-48 queue design prototype for input–buffered ATM switches," Proc. of IEEE INFOCOM '97, Kobe, pp.20–28, April 1997.
[57] H. Li, K.-Y. Siu, H.-Y. Tzeng, C. Ikeda, and H. Suzuki, "TCP performance over ABR and UBR services in ATM," International Phoenix Conference on Computers and Communications '96, 1996.
[58] H. Ohsaki, M. Murata, H. Suzuki, C. Ikeda, and H. Miyahara, "Performance evaluation of rate-based congestion control algorithms in multimedia ATM networks," Proc. of IEEE GLOBECOM '95, pp.1243–1248, Nov. 1995.
[59] B.R. Coller and H.S. Kim, "Efficient analysis of shared buffer management strategies in ATM networks under non-uniform bursty traffic," Proc. of IEEE INFOCOM '96, San Francisco, pp.671–678, March 1996.
[60] G.-L. Wu and J.W. Mark, "A buffer allocation scheme for ATM networks: Complete sharing based on virtual partition," IEEE/ACM Trans. Networking, vol.3, no.6, pp.660–670, Dec. 1995.
[61] A.K. Choudhury and E.L. Hahne, "Dynamic queue length thresholds in a shared memory ATM switch," Proc. of IEEE INFOCOM '96, San Francisco, pp.679–687, March 1996.
[62] D. Stiliadis and A. Varma, "Providing bandwidth guarantees in an input–buffered crossbar switch," Proc. of IEEE INFOCOM '95, San Francisco, pp.960–968, March 1995.
[63] H. Shi, C. Zukowski, and O. Wing, "VLSI design optimization of input/output-buffered broadband ATM switches," Proc. of IEEE INFOCOM '96, San Fransisco, March 1996.

High-Performance Backbone Network Technology edited by Naoaki Yamanaka
ISBN: 0-8247-5321-6 DOI: 10.1081/(120025649)
Published 2004 by Marcel Dekker, Inc.

Traffic Issues in ATM LAN: Principle and Practice

Teruyuki Kubo

SUMMARY Originally intended for application to B-ISDN, which is carrier oriented, ATM technology has been actively studied for application to LAN based environments since the beginning of the 1990s. One of the most notable things in LAN area is development of a rich set of application services. A number of technical specifications for major application services have been developed, which include LAN Emulation, IP over ATM, Multi-protocol over ATM, Voice and Telephony over ATM, as well as Native ATM services such as MPEG2 over ATM. Development of these new services raises new challenges related to traffic management. Keeping pace with the development, a number of traffic control mechanisms have also been developed to maximize the performance of these services. Traffic control and management techniques, however, are still in the early stage of their learning curve. Network engineers are facing challenging problems related to traffic management. This paper reviews major service-related technologies and discusses traffic management issues associated with these services. Especially, it describes the real world traffic management as practiced by average network engineers with state-of-the-art products. Although the thechnology developments have advanced through many research works, there seems to be a considerable gaps between the practice and principles. This paper discusses the traffic issues of ATM LAN from this perspective and points out some challenges for the future. Most of the difficulties in handling traffic issues stems from the differences in implementation details. To alleviate this difficulty, the introduction of a unified node model which describes the traffic handling capability of ATM nodes in sufficient detail is suggested.

key words: *ATM LAN, traffic management, network engineering, traffic model*

1. Introduction

It has been more than a decade since the first conception of ATM technology. For the first several years, the main application area of this technology was considered to be carrier backbone networks to provide integrated transport mechanism for multiple types of services.

In the early 1990s, internetworking vendors and computer vendors realized that ATM is also suitable in LAN environment. Based on this recognition, the ATM Forum was established in 1991. Since then, extensive work has been done to develop technical specifications for new services built on ATM. The scope originally covered by the ATM Forum was centered around LAN-related technologies. However, as the time evolved, the scope has been expanded to include wide technical areas that include enterprise backbone and inter-carrier connections.

As for the traffic issues, a lot of work has been done so far. A number of significant achievements have been made in defining traffic control mechanisms both at ITU-T and the ATM Forum. Most of the results have been implemented in commercial products and services and became available within one year after the approval of agreed specifications. Developed mechanisms are deemed to be effective in principle. However, in practice, achieving the intended benefits sometimes tend to be difficult due to implementation complexity and difficulty in operating the systems.

In Japan, it was only about two years ago when massive introduction of ATM products was started by a number of universities and research institutes building high speed communications test-beds with ATM technology in their campuses. In this sense, we are in the very early stage of learning curve in dealing with ATM traffic issues.

ATM is a flexible technology in the sense that it has a number of programmable parameters. In some cases, it tends to be regarded as the complexity of the technology. There seems to be some gaps between what are built-in by design and achievable in principle and what are actually used in practice. This paper discusses the traffic related gaps observed with ATM LAN and points out some issues to be solved for the future.

2. Background

2.1 ATM LAN and ATM Internetwork

There are several different ways to define ATM LAN. In the narrowest sense, it means a network that provide MAC (Media Access Control) layer service as defined by the IEEE 802. Specifically, it is a service that transfer connectionless data units according to the destination MAC addresses. Services of this type are provided by LAN Emulation [1], [2].

In a broader sense, it means a network that provide legacy LAN-based services such as TCP/IP service as defined by IEFT. Services of this type are provided by IP over ATM [3] and MPOA (Multi-

Manuscript received October 17, 1997.
Manuscript revised December 15, 1997.
† The author is with NTT Business Communications Headquarters, Tokyo, 108-8019 Japan.

IEICE TRANS. COMMUN., Vol. E81-B, No. 5, pp. 840–848, May 1998

protocol over ATM) [4] technologies.

In more broader sense, it means a network on user's premises. This may provide native ATM services in addition to the legacy LAN based services mentioned above. In most cases, ATM LAN refers to the systems which provide legacy data transfer services such as TCP/IP and IEEE802. There are, however, another category of networking which provide native ATM services over ATM. Though this category of services are not yet mature at this point of time, they are expected to play key roles in implementing multimedia services.

In the broadest sense, it also means a private network which covers a wide geographical area based on TCP/IP technology.

2.2 Traffic Management Framework

Typical users' requirements on their enterprise networking environments relate to QoS guarantee, efficiency, and fairness. Traffic management techniques play a key role in meeting these user requirements. Towards this end, extensive work has been done in traffic management field [5]. Especially, a set of service categories have been defined and various traffic management mechanisms have been developed to implement these services. The ATM Forum has defined five service categories (Table 1) to support distinct quality of services.

The objective of supporting multiple service categories is to optimize the resource utilization by trading the balance between two traffic management policies: sharing and isolation. To achieve efficient use of link bandwidth, multiple traffic flows should share the common bandwidth. On the contrary, to assure the desired quality of service to each traffic flow, bandwidth should be divided into pieces in some way and each piece should be allocated to each flow. The characteristics of the divided pieces differ depending upon the types of service categories they support. Harmonized operation of the two conflicting policies is the key in making the most of the ATM's traffic management capability.

Based on the theoretical work, the following mechanisms have been implemented in various types of networking equipment: scheduling, buffer management, flow control, routing, and shaping.

With a rich set of traffic management mechanisms provided, network engineers are experiencing difficulty in designing and operating their networks; what technologies to chose for particular applications; how to optimally configure the system parameters, how to estimate the resulting performance.

The scope of traffic management is rather broad and it can be seen at three hierarchical levels as shown in Fig. 1.

1) Node level controls: The lowest in the hierarchy is the node level controls. Controls at this level operate locally within each node independent of the states of other nodes. This includes buffer scheduling, traffic shaping, and packet discarding.

2) Network level controls: The next level in the hierarchy is the network level controls. Controls at this level operate in coordination of the intelligence within other nodes. This includes routing, call admission control, and flow control.

3) Network engineering: This is the level of traffic management where most field engineers are engaged in. This includes topology design, capacity assignment, routing design, and server location.

The design and operations taken at these three levels are interrelated and each can not be discussed in complete isolation. Dealing with the higher level controls requires knowledge of the lower level mechanisms. One of the reasons why network engineers find difficulty in designing and operating networks lies here. But, to reduce the complexity, they have to be separated as much as possible. This paper discusses

Table 1 ATM layer service categories.

<table>
<tr><td colspan="2">Attributes</td><td>CBR</td><td>rt-VBR</td><td>nrt-VBR</td><td>ABR</td><td>UBR</td></tr>
<tr><td rowspan="2">QoS Parameters</td><td>CLR</td><td colspan="3">specified</td><td>network specific</td><td>unspecified</td></tr>
<tr><td>CTD and CDV</td><td colspan="2">specified</td><td colspan="3">unspecified</td></tr>
<tr><td rowspan="3">Traffic Descriptors</td><td>PCR and CDVT</td><td colspan="5">specified</td></tr>
<tr><td>SCR and MBS</td><td>n/a</td><td colspan="2">specified</td><td colspan="2">n/a</td></tr>
<tr><td>MCR</td><td colspan="3">n/a</td><td>specified</td><td>n/a</td></tr>
<tr><td colspan="2">Feedback</td><td colspan="3">unspecified</td><td>specified</td><td>unspecified</td></tr>
</table>

CLR: Cell Loss Ratio, CTD: Cell Transfer Delay, CDV: Cell Delay Variation
PCR: Peak Cell Rate, CDVT: Cell Delay Variation Tolerance,
SCR: Sustained Cell Rate, MBS: Maximum Burst Size, MCR: Minimum Cell Rate

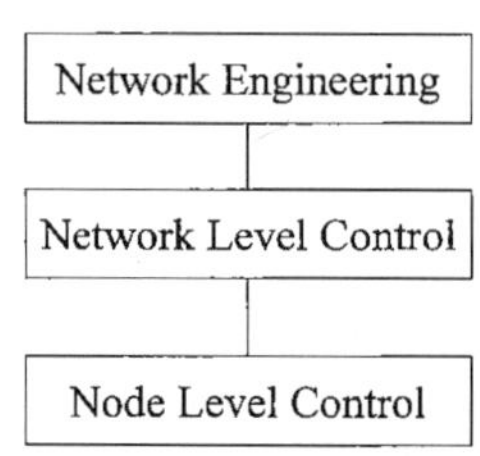

Fig. 1 Traffic management hierarchy.

traffic management issues of ATM LAN from the network engineering aspect; i.e., how to best use the implemented mechanisms through optimal design of the network and optimal configuration of the equipment.

3. Traffic Management Mechanisms

Following several years of experiences in the practical applications, traffic management mechanisms embedded in the network nodes have been improved. This section discusses the achievements done by industries and issues to be solved for the future.

3. 1 Connection Admission Control

Connection admission control (CAC) is a mechanism which determines a newly offered connection should be accepted by the network or not. CAC uses connection traffic descriptors and requested QoS, which are presented by a user, to determine whether a VC connection can be accepted while meeting the requested QoS and without affecting the QoS of connections already established. Traffic descriptors pertain to Peak Cell Rate (PCR), SCR (Sustainable Cell Rate), MBS (Maximum Burst Size), MCR (Minimum Cell Rate), and CDVT (Cell Delay Variation Tolerance) where as QoS parameters pertain to CLR (Cell Loss Ratio), CTD (Cell Transfer Delay), and CDV (Cell Delay Variation).

To make the decisions, offered the traffic descriptors, logic in a switching node performs CAC calculation to evaluate the effect of accepting the new connection to the quality of existing connections. With the most typical technique, traffic descriptors presented by a user is translated into an equivalent capacity. The equivalent capacity for a connection is calculated using vendor-specific calculation formula which typically takes the following parameters as the inputs: peak rate, burst size, sustainable rate, cell loss ratio, buffer size, and link bandwidth. This capacity is compared to the remaining bandwidth of the link on which the new connection is multiplexed.

This calculation usually consumes a significant amount of computation power. Since there is no standard way to perform this calculation, design, operation and evaluation of networks is a complicated work. Network engineers are required to use different metrics for different types of nodes in determining the required network capacity. To alleviate this complexity, a universal metric for bandwidth should be established, with which engineers can determine the adequacy of network resources in a unified manner regardless of vendor-specific implementations. This will facilitate the bandwidth management of inter-network links and will foster user understanding and acceptance.

One of the most significant challenge for this mechanism to work effectively is for users to correctly characterize the source traffic. In many cases, users tend to present larger values in order to ensure satisfactory quality. Actual usage of the link bandwidth is expected to be lower than estimated from user requests. To increase the link efficiency, measurement-based adaptive evaluation of link utilization should be effective.

3. 2 UPC and Shaping

UPC (Usage Parameter Control) and traffic shaping is the basis for guaranteeing the quality of service for the connections after they are admitted to the network by the connection admission control. Guarantees are only valid if the traffic conforms to the negotiated traffic contract. Source devices shape, or adjust, their cell transmission rates on a per-connection (virtual path connection and/or virtual channel connection) basis to meet the contract. For CBR, VBR, and UBR, conformance is defined by the GCRA (Generic Cell Rate Algorithm). Conformance is monitored at the entrance to the network and non-conforming cells may be discarded on the spot or tagged with CLP (Cell Loss Priority) = 1 for lower priority. This process is called UPC. CLP is indicated in a field defined in the ATM cell header. Cells with CLP = 1 may eventually be discarded as they are forwarded into the network and encounter a congested point giving preference to other normal cells.

The basic idea is simple and it should work fine in principle. There are, however, subtle implementation issues for these mechanisms to function effectively.

First, the granularity of control is an issue. GCRA follows a continuous leaky bucket model. But, real products have some limitations on how fine the implementation can be in regulating the cell transmission sequence or measuring the cell arrival sequence. Mismatch of granularity between the UPC and shaping may make it difficult for the user to select optimal bandwidth for connections. It also may result in unexpected cell discards subject to UPC.

Hierarchical traffic control at both VP level and VC level also raises an issue. For the economic reasons, in LAN environment, ATM nodes do not apply UPC at private UNIs. When entering into the public network, gateway nodes to the public network performs traffic shaping at the public UNI. In this case, even if the individual original flows sent out from the terminals conform to the traffic contract, after the multiplexing at the entrance to the public UNI, the resulting flows may be distorted and no longer conform to the contract. If the public network monitors the conformance with respect to both VP and VCs at the same time, and customer premises equipment were to meet the both restrictions at the same time, shaping operation of the traffic becomes extremely complicated. To cope with this problem, some compromising mechanism must be devised which minimizes the traffic fluctuation for both VP and VCs at the same time.

Traffic shaping is taken into consideration only for the flows in user-to-public network direction and between carrier networks/nodes. In order for the simplification of the customer premises equipment, shaping of network-to-user direction flows should also be considered, which would minimize the buffer requirement with customer facilities.

3.3 Flow Control

ABR service has been gaining an increasing attention as the desirable and efficient scheme for supporting LAN data traffic. ABR is a category of ATM services that offers negligibly small cell loss rate while providing guaranteed minimum rate and allowing sending out to the physical limit of the interface. In addition, ABR was developed with the goal of reducing switch buffer requirements. To achieve such service requirements, ABR uses congestion control at the ATM layer which requires network switches to constantly monitor the traffic load and feed congestion information back to the sources. The sources are expected to dynamically adjust their input to the network based on the congestion status of the network. Therefore, the service advantages of ABR over UBR (MCR, fairness, and minimal cell loss) come at the expense of increased switch and end-system complexity.

Functional diagram of ABR control system is illustrated in Fig. 2. There are three types of components in the system: source, switch (es), and destination. In order to provide a mechanism for the network to provide congestion and rate change feedback to the source, the ABR source is required to insert Resource Management (RM) cells periodically within the data flow. The RM cells are turned around by the destination and returned to the source within the data flow in reverse direction. There are three modes of operations for switches to give feedback:

- EFCI binary rate operation: a switch marks EFCI (explicit forward congestion indication) bit in the ATM header of data cells,
- Relative rate binary operation: a switch marks CI (congestion indication) and/or NI (no increase) bits in RM cell, and
- Explicit rate operation: a switch marks ER (Explicit rate) field in RM cell.

These three modes have different characteristics, as shown in Fig. 3 and should be applied in an appropriate manner.

The major part of an RM cell is shown in Fig. 4. Behaviors of the three components are summarized below.

1) ABR source behavior: On receiving a returning RM cell, if CI=1, source decreases its Available Cell Rate (ACR) proportionally:

$$ACR=(1-RDF)*ACR$$

Else if NI=0, source allows linear increase of ACR:

$$ACR=ACR*RIF*PCR$$

It also looks at ER and set the actual ACR the smallest of the two.

2) ABR switch behavior: A switch implements at least one of the following methods to control congestion at queuing points:

a) EFCI marking: The switch may set the EFCI state in the data cell headers;

b) Relative Rate Marking: The switch may set CI=1 or NI=1 in forward and/or backward RM cells;

c) Explicit Rate Marking: The switch may reduce the ER field of forward and/or backward RM cells (Explicit Rate Marking);

3) ABR destination behavior: When a data cell is received, its EFCI indicator is saved as the EFCI state of the connection. On receiving a forward RM cell, the destination turns around the cell to return to the source.

a) If the saved EFCI state is set (=1), then the destination sets CI=1 in the RM cell, and the saved

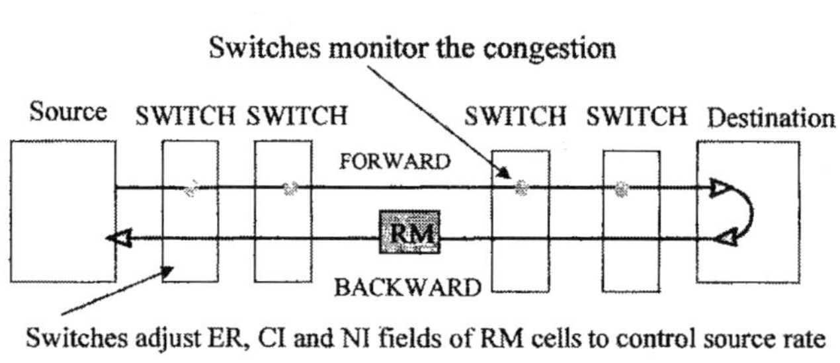

Fig. 2 ABR control system.

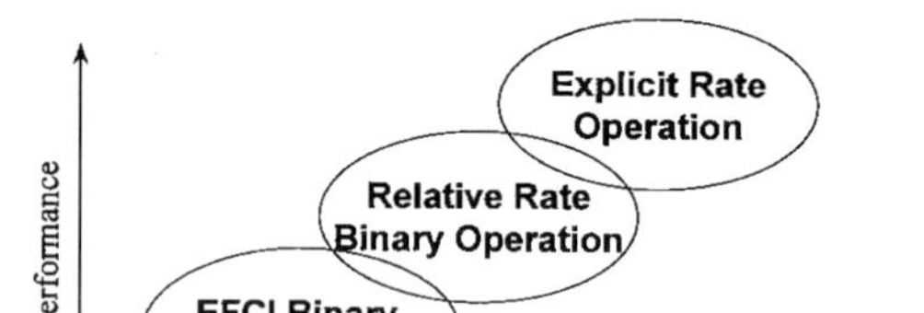

Fig. 3 Relationships among ABR operation modes.

Standard ATM header with PTI = 6	ID	DIR BN CI NI	ER	CCR	MCR		...	CRC10

byte # 1-5 6 7 8-9 10-11 12-13

PTI: Payload Type Identifier, ID: Protocol Identifier, DIR: Direction, BN: BECN, CI: Congestion Indication, NI: No Increase, ER: Explicit Rate, CCR: Current Cell Rate, MCR: Minimum Cell Rate, CRC: Cyclic Redundancy Check

Fig. 4 RM cell format.

EFCI state be reset.
b) The destination (having internal congestion) reduces ER to whatever rate it can support and/or set CI=1 or NI=1.

ABR is expected to lead to superior throughput performance for the TCP-like traffic, attainable performance is not yet clear in practical applications. It is reported that, in certain conditions, a good implementation achieves 1 to 100 improvement in goodput. On the contrary, there are some skepticism that good UBR operation with proper cell discarding capability performs as well as ABR does. Since UBR can be implemented with much simpler hardware, for the ABR to be widely accepted and deployed, clever ways must be devised to make the most of ABR capabilities.

Moreover, the ABR flow control mechanisms may introduce undesirable interoperability issues. There are several important parameters which determines the performance of ABR connections. If these parameters are not properly determined for the nodes, the resulting performance may be inferior than expected. When using binary feedback based on EFCI marking, RIF (Rate Increase Factor) and RDF (Rate Decrease Factor) determines the performance. Values for these parameters are preset at each ATM node taking into account the amount of available buffer capacity or buffer allocation policy. RIF and RDF for an ABR connection are negotiated among the participating nodes and come down to the minimum value encountered along the connection. The resulting RIF and RDF are used by the source device to determine the cell rate available to it, ACR (Available Cell Rate), according to the following logic:

- when source is required to decrease rate:
  ACR=(1−RDF)*ACR,
- when source is allowed to increase rate:
  ACR=ACR*RIF*PCR.

If RIF and RDF take small values, ACR can change only slowly and system can not adapt to the network traffic change quickly. Thus, the available resources in the network may not be utilized efficiently. For nodes to allow RIF and RDF to take large values, they have to be equipped with large buffers. As the industry has not acquired enough experiences to determine on the optimal value for these parameters, it is likely that they are set to conservative values. In such cases, ACR does not change so dynamically that allow the maximum use of unused bandwidth.

With explicit rate control scheme, this problem can be avoided. But there is a different type of interoperability problem. When source receives RM (Resource Management) cell, it checks the ER (Explicit Rate) information in it and adjust the transmission rate accordingly. The ATM node at the source side UNI is expected to police the resulting flow. The control process within the source device may be implemented by hardware or software. Depending on the implementation type, the response time varies significantly; order of microsecond with hardware and order of millisecond with software. This gives the uncertainty what is the best time point for the ATM node to change the UPC parameter values for the connection. If these two control processes are not synchronized, cells may be discarded unnecessarily.

For the engineers to be able to take full advantage of ABR, the following points have to be clarified.
1) Interworking with legacy LANs: while ABR assumes that the end systems comply with the ABR source behavior, most current applications are only connected to ATM via legacy networks such as Ethernet. Therefore, ABR may only push congestion to the network edges and cannot provide flow control on an end-to-end basis.
2) Interworking with applications: ATM level flow control may have some interaction with the flow control implemented in higher level protocol. If the controls are not properly synchronized, the resulting performance may degrade. Since most commonly used applications operate over TCP/IP, control algorithm optimized for TCP would be preferable.
3) ABR performance under multi-vendor environment: There is high expectation that products from various vendors conform to a common specification and exchange control information correctly. However, even in this situation, internal policy taken by each products in choosing the control parameters or other implementation details may have some differences as discussed above. This might result in unexpected performance degradation with specific combination of products. To achieve satisfactory performance levels, network engineers should accumulate sufficient experiences how to choose reasonable values for the parameters.

3.4 Packet Level Processing

It is well known that a single cell loss results in the whole packet loss and it causes severe impact on packet level performance. This is because cells that belong to the same packet as the one the lost cell belongs are transported through the network only to be discarded at the destination. The lost packets are usually retransmitted by higher layer protocols (such as TCP) eventually. This phenomenon add more traffic to the congestion and exacerbate the situation.

To alleviate the problem, packet level processing schemes are used to enhance packet level performance. With these schemes, cells belonging to a packet are selectively discarded at the congested nodes. In general, cell loss ratio degrades when the schemes of this type are applied. There are several specific schemes: EPD (Early Packet Discard), PPD (Partial Packet Discard) and RED (Random Early Detection).

With PPD, ATM node is required to track the cell

Table 2 Packet processing schemes.

	PPD	EPD	RED	ABR
May degrade cell loss ratio	Yes	Yes	Yes	No
Transmission of incomplete packets	Yes	No	No	No
Link left unused	No	Yes	Yes	Yes
Fairness	No	No	Yes	Yes

flow to identify the beginning and end of higher layer packets. When traffic becomes heavy and a node is forced to discard a cell in the middle of a packet, the node also discards every succeeding cell in the packet, except for the last cell. With this scheme, bandwidth is still wasted by sending the preceding part of the packet.

EPD improves this inefficiency by predicting the congestion. In a congested situation, a node begins to drop the entire packet; i.e., all the cells from the first to the last of the packet. The node predicts, at the time of the arrival of the first cell, that a cell in the middle of a packet is very likely to be discarded. The decision may be based on any measure; threshold is used most often. With EPD, however, bandwidth may be left unused when the prediction is too pessimistic. In addition to that, EPD may tend to discard packets belonging to the same connection and thus degrade the fairness among connections.

RED improves the fairness of EPD by randomly selecting the packet to be discarded. When the congested state is encountered, ATM node randomly picks the connections from which a packet is subject to be discarded.

Their characteristics are shown in Table 2. These schemes are typically applied to AAL5 services with UBR category. With the latest products which are equipped with buffers of sufficient capacity, combined with packet discarding schemes, in certain cases, achievable throughput can be as good as the one observed with systems using ABR. Therefore, it is desirable that some criteria be established to determine when to use ABR or packet discarding schemes.

3.5 Queuing Algorithms

In designing queue structures for ATM nodes, there are a lot of factors which affect the performance of the node: buffer capacity, the number of queues, the number of service classes per queue, the number of thresholds per queue, and queue service disciplines (cyclic, strict priority, round-robin, etc.).

After several years of experience in the commercial applications and based on the development of hardware technologies, buffer implementation has been greatly advanced. The buffer sizes are steadily increasing and nodes equipped with 16,000 cells buffer per link of 155 Mbps speed are becoming available. With this level of buffer capacity, very low cell loss rate can be expected with simple UBR operations enhanced by packet level discarding.

In terms of the number of queues, per-VC queuing is also gaining strong attention and being implemented in leading edge products. Coupled with weighted fair queuing discipline, this would be the sound basis for providing solid QoS guarantee and fairness for individual connections, which would be the main differentiator of ATM from other technologies.

When faced with congestion, ATM nodes have to react to the situation either by discarding cells or giving feedback to sources. In order to initiate these actions, nodes identify the following situations happening by observing the amount of traffic exceeding certain thresholds:

- UPC violation
- EFCI marking
- Packet discarding

Optimal values for these thresholds are different with each other and should be determined to optimize the performance.

With all these sophisticated mechanisms, it is very hard for the network engineers to tell how effective those different mechanisms are to the individual applications in practice.

It is also difficult to determine the optimal values for various parameters which determine the resulting queue performances. To overcome these difficulties, it is desirable that some integrated measure be defined to select the proper functions for the nodes and to determine the parameter values for the nodes.

4. Network Services for ATM LAN

4.1 Virtual Networking

Virtual networking offers powerful means to facilitate the operation, administration and management of corporate networks. It allows users to define and maintain logically separate networks over a common network infrastructure. Within each logical network, it is made easy to add, change and move stations. These modifications are taken care of from network management console or traced automatically by the built-in network intelligence by itself without requiring physical rewiring nor configuration change at the stations. Among logical networks, traffic flows are isolated from each other which contribute to lessen the traffic load on the infrastructure and improve the security.

ATM offers a good foundation for virtual networking. Several schemes have been developed which include LAN Emulation (LANE), Classical IP over ATM (CIP), and Multi-Protocol over ATM (MPOA). These schemes are all based on a client-server model. In the case of LANE, there are two types of services that might affect the system performance. One is LANE Services (LESs) which, upon request from a client, translates an offered LAN destination address to a corresponding ATM address and sends the result back, to allow for the client to setup an ATM connection to the destination. This operation is initiated when a new communication is started.

The other is Broadcast and Unknown Services (BUSs) which receives a packet from a client and distribute it to all the other clients belonging to the same emulated LAN (ELAN) as the sending client. This operation is in effect in two cases. One is when the offered LAN frame is destined to broadcast or multicast address. The other is for the case that a client sends out frames prior to the completion of connection setup up to the destination following the address translation operation mentioned above.

One each of these services is defined and used for each ELAN. In many cases, a single physical server support services for multiple ELANs, and these services are located in a single server. As a result, traffic tends to concentrate to these servers and servers themselves or links which connect servers to ATM switches become congested. This situation degrades the throughput and response time of the system. Network engineers have to decide where is the optimal locations to place the servers given the network topology and sets of ELAN members (desktop devices and application servers). Although the location of the servers has a significant impact on the overall performance, there is no established design rules to this design problem.

4. 2 Routing

The ATM Forum completed the specification of PNNI (Private Network-Network Interface) [6] in 1996. This specification defines both signaling and routing aspects of inter node interface. Signaling specifies how to set up connections through a sequence of switching nodes. Routing specifies the mechanisms with which network topology information is collected and shared among switching nodes. What characterize the PNNI routing are dynamic routing and source routing. These two features have significant impact on network design.

1) Dynamic routing

With PNNI, all the nodes in the network exchange topology information with each other and every node creates its topology database. The database describes how the nodes are interconnected, what are the link capacities, and how much the links are loaded. The information exchanges are performed regularly and each node maintains latest status of the network all the time. In this way, changes in the network topology and traffic loading are reflected to the database in a timely manner. Routing decisions are made based on the information in this topology database.

This dynamic nature of PNNI routing has some implications on network design. Network topology and link capacity design usually assume static routing. The results obtained by these designs are not necessarily optimal under the dynamic environment. New approach, integrated routing and topology design, may be required.

2) Source routing:

PNNI adopts source routing in which decisions of the path connections are to follow are made at the originating node. This mechanism allows the network administrators to enforce operations policies due to security and/or cost reasons.

When source routing is implemented in an internetwork environment where element networks operated by different administrations are interconnected, routing decisions may not be consistent among element networks. The routing decisions made at one element networks are not necessarily optimal as seen from other element networks. This may complicate the design of internetwork. It is desirable to define a common framework for solving the trade-off between policy and optimality.

4. 3 Short-Lived Communications Support

In running legacy applications, significant part of the traffic is of transaction type, with which a communication between a source and a destination comprises only a small number of packets and short response time is required. In most cases, there are very large number of source-destination pairs on the network that are to exchange information. For this type of communications, connectionless approaches, such as those using IP (Internet Protocol), are well suited. However, since ATM is a connection oriented technology, it has several challenges to support the communications of this type, which include:

- to minimize the latency until the first packet is delivered to the destination,
- to minimize the amount of wasted resources due to advanced reservations, and
- to minimize the probability of information loss due to resource contentions.

One way to meet these goals is to have permanent connections between all possible source-destination pairs. This eliminates the connection setup delay and helps reduce the response time. But this approach requires a certain amount of bandwidth be allocated to each connection and we may end up with very many small-capacity VCs. To avoid this deficiency, network engineers opt to select shared media networks, such as FDDI, which resolve the resource allocation problem in real time; in the order of one network round trip time, and allow end systems to transmit data at wire speed. ABR is expected to solve this problem, but it may not be sufficient to support this type of traffic, because it reacts to the traffic demand more slowly than required.

One way for the ATM to meet this requirement is to provide sufficient amount of buffers at switching nodes. But it is uncertain if the buffer requirement stays within feasible size. Another way is the introduction of very fast resource management mechanisms. For local area applications, shared media type mechanisms seem to be effective. Related work includes DQDB (Distributed Queue Dual Buffer) [7] and ATMR (ATM Ring) [8]. For wide area applications, several schemes, called FRPs (Fast Reservation Protocols) [9], have been discussed at ITU-T with which connection setup and resource reservation are treated as separate processes. Resource management is done using in-channel signaling where as connection setup is done through normal signaling channel.

This technical area is gaining attention recently in the ATM Forum. This is certainly a challenging topic for further study.

5. Network Design Issues

Network design problem is stated in the following way: "minimize the network cost subject to performance and quality constraints." In designing ATM networks, the main challenges relate to the fact that the target constraints are expressed as higher level specifications, such as frame loss rate or voice delay. These performance figures must be translated into those of ATM. For this purpose systematic way of mapping between those figures are required.

With the latest ATM corporate networks, the ATM service categories that are in practical use are CBR, VBR, and UBR. In this circumstance, the important performance measures are Cell Delay Variation (CDV) and the statistical multiplexing gain. Achievable CDV values are critical with CBR services to assure sufficient quality for voice communications.

5.1 Network Design Processes

To design corporate networks, a typical process takes the phases shown in Fig. 5.

(Phase 1) Requirement Analysis

In this phase, target design objectives are established based on the real user requirements. The first thing is identification of applications to be run on the network and location of sites to be interconnected. The outputs of this phase include traffic demands and target performance.

In general, it is very difficult for users and network engineers to figure out exact values for these items. Support tools that give the first order estimate of the values derived from the service descriptions are being developed.

(Phase 2) Network synthesis

This phase starts with basic topology design in which the level of hierarchies and connections among sites are

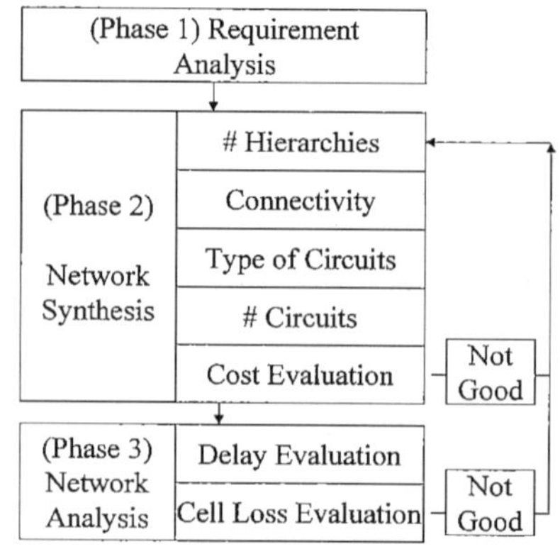

Fig. 5 Network design process.

determined. This is followed by network services selection in which most appropriate network services, i.e., dedicated line, frame relay, as well as ATM, are selected. Finally, the number of circuits and circuit capacities for each link are determined so as to meet the cost criteria.

The operations in this phase are usually supported by network topology design tools. A number of tools have been developed and are commercially available.

(Phase 3) Network analysis

In this stage, the performance and quality achievable by the synthesis are evaluated and compared to the target values. If the conditions met, the whole design process ends at this point. Otherwise, the process goes back to phase 2.

This part of the design cycle is supported by network simulation tools. A number of tools are commercially available for this purpose.

To complete the design, the last two phases must be carried out repeatedly until the target design objectives are met. When going back to phase 2, a new increased cost is assumed for the next round of synthesis.

Since this two phase approach is cumbersome, development of integrated approaches is desirable, in which a single design tool can determine the network design that satisfies both cost and performance/quality goal in a single step.

5.2 Generic Node Model for Traffic Management

During the network analysis phase, achievable values for performance and quality have to be calculated. This process requires detailed descriptions of network elements. Resulting performances differ greatly depending on the details of the traffic management mechanisms of each node. For example, for the same link bandwidth, switches with different buffer sizes show different performances in terms of cell loss ratio. The key mechanisms that affect the performances of ATM systems are discussed in the previous section: buffer size; queue management, flow control, call admission control, UPC/NPC and shaping, packet level processing.

Network designers have to decide on the appropriate products which satisfy the demands. In order to do this, they have to know the differences among products and express the specific characteristics in the simulation model. This process can be facilitated when unified abstract models of the network elements are commonly defined. The models should be able to be customized to meet the specific cases by simply specifying necessary parameter values.

More importantly, performance indexes should be developed for the network engineers to decide how suitable the specific products are for the respective applications. The engineers are experiencing difficulty in figuring out good ways to evaluate their designed systems. Problems of this kind can be solved to some extent when common test conditions are developed.

6. Conclusions

This paper has reviewed the latest status of traffic management issues in ATM LAN environment. More specifically, it has given the status as perceived by network engineers and looked at the gaps between traffic management in practice and in principle. In spite of the extensive work done by researchers and developers, there are some gaps remaining. Network engineers can not always get the clear cut ideas how to best design and configure ATM LANs. These gaps must be filled to bring the full advantages of ATM technologies to practical applications.

Network engineers are required to design networks making most of the rapidly developing technologies and products from multiple vendors. The products offer the same functionality based on different implementations. For the engineers to get a consistent view of networks in designing them, a universal model of ATM nodes should be established. The model is expected to describe the traffic control capability to the degree detailed enough to express difference in performance while hiding the details of individual implementations. GCRA is a good example and the first step toward this.

Topology design and traffic engineering is one of the most challenging issues in designing ATM LANs. Most of the cases, engineers opt to apply rule of thumb, i.e., first estimate the average bandwidth for VBR traffic and determine link capacity so as to keep the link utilization under 70 percent. Tools for automating the design process is desirable. The universal modeling is also effective in developing the network design tools.

Traffic management mechanisms developed for ATM are the very basis for providing high quality services. The promising features, such as ABR, however, have had only limited experiences in practical applications so far. They are in an early stage of their learning curves. As more experiences be accumulated, more benefits will be brought into real world applications.

References

[1] The ATM Forum, "LAN emulation over ATM version 1.0 specification," af-lane-0021.000, Jan. 1995.

[2] The ATM Forum, "LAN emulation over ATM version 2.0—LUNI specification," af-lane-0084.000, July 1997.

[3] IETF, "Classical IP and ARP over ATM," rfc 1577, 1994.

[4] The ATM Forum, "Multi-protocol over ATM v.1.0 specification," af-lane-0087.000, July 1997.

[5] The ATM Forum, "Traffic management specification version 4.0," af-tm-0056.000, 1996.

[6] The ATM Forum, "Private network-network interface version 1.0 specification," af-pnni-0055.000, March 1996.

[7] M. Conti, et al, "A comprehensive analysis of DQDB," ETT, vol. 2, no. 4, July–Aug. 1991.

[8] T. Enomoto, "High-speed shared-media ATM-LAN for multimedia communciations," 5th IEEE COMSOC International Workshop on Multimedia Communications, 1994.

[9] P. E. Boyer and D. P. Tranchier, "A reservation principle with applications to the ATM traffic control," Computer Networks and ISDN Systems, vol. 24, 1992.

High-Performance Backbone Network Technology edited by Naoaki Yamanaka
ISBN: 0-8247-5321-6 DOI: 10.1081/(120025650)
Published 2004 by Marcel Dekker, Inc.

Competitive Telecommunications Management and System Development

Masayoshi Ejiri

SUMMARY Rapid advancing technology and customer demand for sophisticated services are driving the global telecommunications environment into fully competitive and multi service providers environment. To cope with this new environment successfully, open network and open telecommunication management are essential. Telecommunication management is becoming more and more important to realize flexible and dynamic telecommunication services. This paper first gives a view of the new environment in the telecommunication industry and discusses the direction to be taken by service providers. Then, a concept of multi-domain management is proposed to meet a dynamically changing environment. Finally, the paper describes the subjects for the development of telecommunication management system and suggests that a competition based on harmony between the users, service providers and vendors is needed to make customers satisfied with telecommunication services.

key words: *telecommunication management, service provider, open network, management system, interface, multi-domain management, business process, platform, common application*

1. Introduction

The telecommunications industry is now facing great changes driven by two factors. One is the technological innovation represented by digital and distributed processing technologies. This innovation is drastically changing the conventional concepts of services and networks. The fusion of telecommunications, broadcasting, and computing services, as well as that of networks and terminals technologies, removes the border of existing services and networks. The industry is now creating new businesses such as contents business, electronics commerce and so on based on open and borderless services/networks environment.

The other factor is that end-user demand is changing. End users are demanding not only enhanced and manifold services but greater flexibility in service selection. This trend indicates that services in the near future will become tailored based on service negotiations [1].

To keep up with these changes, the liberalization of telecommunications services is accelerating in the world wide telecommunications industries [2]. Service providers (SPs) and vendors are groping for coexsistence and alliances to counter the fierce competition. Existing borders are vanishing and new borders are appearing in telecommunication industries.

As borderless environments in sophisticated and diversified services increase, service and network management becomes more important in providing end users with smooth, seamless services at low cost [1].

The telecommunications management excellence is becoming the differentiator to determining the value of each service, and the service competition becomes the competition of service/network operations.

Under these circumstances, the development and introduction of an efficient management system is becoming an important subject that the service providers and vendors cannot overlook.

To study telecommunications management strategies, this paper first gives a general overview of the telecommunications industry and discusses the direction in which service providers progress under this environment. The paper then proposes a concept of telecommunications management to cope with the trends flexibly. Finally, the paper describes considerations in the development of a management system and indicate that a competition based on the harmony between the users, service providers, and vendors will realize better telecommunications management.

2. New Structure for Providing Telecommunications Services

As mentioned above, the telecommunications industry is about to provide telecommunications services based

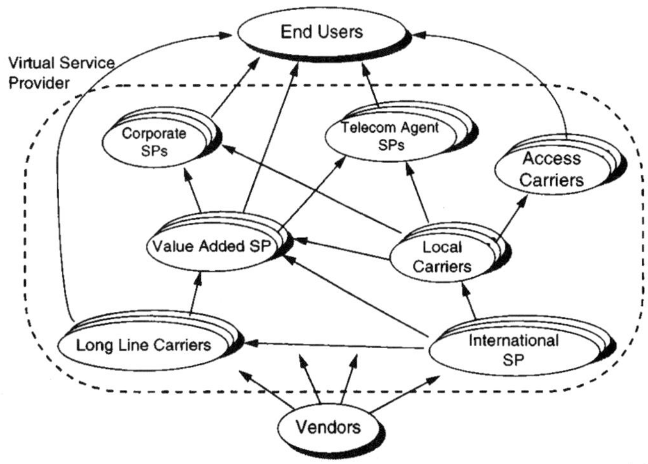

Fig. 1 Service delivery chain.

Manuscript received March 3, 1997.

† The author is with FUJITSU LTD., Kawanaki-shi, 211 Japan.

IEICE TRANS. COMMUN., Vol. E80-B No. 6, pp. 805–810, June 1997 on a new paradigm [1]. In this paradigm, the market of both telecommunications services and networks will be open for end users. Many service providers will provide a variety of value-added services combining the open networks and compete with other service providers.

This phenomenon may be inevitably due to the aforementioned technological innovations and ever changing user demands, especially for free service selection that is not under the control of the service providers or vendors [2].

As Fig. 1 shows, users will be able to select and enjoy services freely, regarding many service providers as a virtual SP. To realize the virtual SP, cooperation among service providers is essential. As we can see from the figure, one service provider may be a customer of another as well as a service provider for others [3]. Since relationships between services providers are complicated, the cooperation between them is inevitable in order to jointly provide a total solution to the customers.

Service providers compete by offering users services of differing quality and price, but also maintain links with their competitors. This is similar to the nature of competition among travel agents who coordinate many package tours with other agents to meet their users' need [1].

As the needs of users grow in quality and variety, the users will themselves get directly involved in SP services. Then the users and service providers may begin to negotiate service conditions (charge, quality, communication time and so on) on an equal footing in the future [1], [2].

3. Direction of Service Providers Business

With the aforementioned changes in the telecommunications service environment, the subjects and objectives of service providers are also changing. In short, these changes are from hardware to software.

In other words, network elements (NEs) may be purchased from the market in which the vendors compete, like other commodities in the market. In order to survive in this competitive market, service provider must establish a strategy in telecommunication management which combines the NEs technologies and customer care to provide a good quality and economical service.

To realize an economical service, the following three most important factors should be considered in addition to a low cost NE procurement [1].

1. Minimizing the network resources by optimum network management
2. Making operations economical and efficient by effectively using operation support systems
3. Providing services well tailored to individual needs (store and forward services, dynamic pricing by negotiation, etc.) to ensure the almost complete use of resources (24 hours a day, 365 days a year).

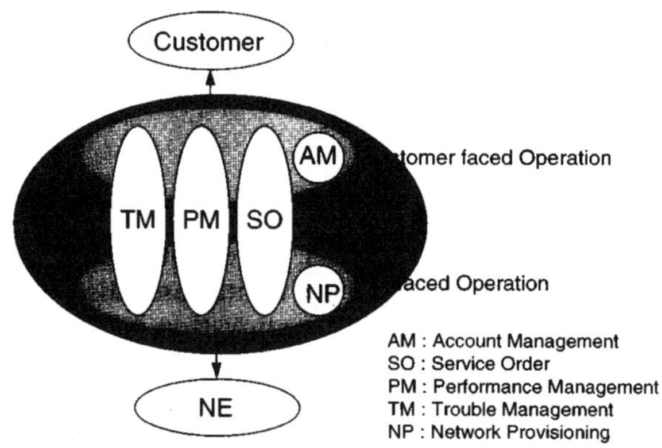

Fig. 2 Basic operations architecture.

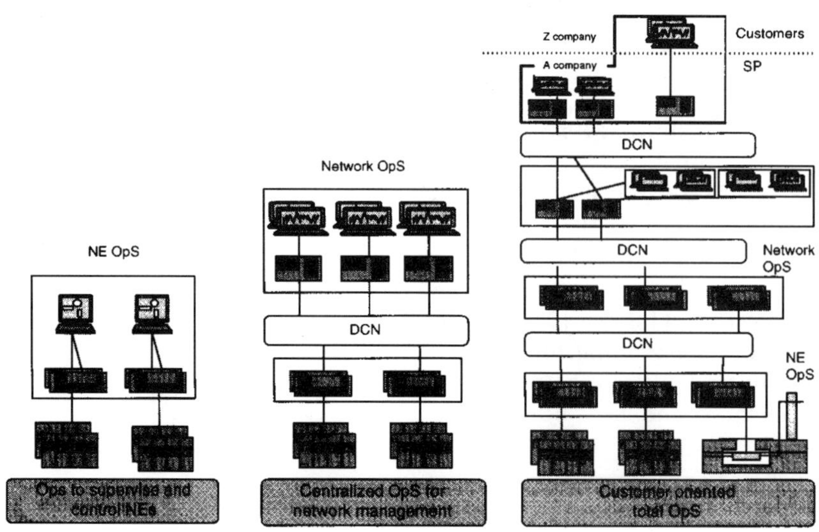

Fig. 3 Trend of operation systems development.

These efforts can be promoted only under appropriate telecommunications management. The depth and the width of an efficient management will directly affect the pricing and the success in competition.

The service providers will also orient their telecommunications management systems toward customers and services by creating new services from the same resources. Enhancing customer care means satisfying customer needs quickly and appropriately through the window of the customer in the basic architecture of telecommunications management shown in Fig. 2 [4].

To enhance customer care, customer faced operations and NE faced operations must be linked in each business process (SO, PM, and others) for information distribution [5]. In addition, inter-process linkage is required for horizontal operations.

The nature of telecommunications management is changing from the conventional network supervision and control to the production of inexpensive, high-quality competitive services that satisfy customers. As Fig. 3 shows, this tendency is reflected in the development of operation systems [4], [6]. Now service and customer management systems based on NE and network management systems are being developed and introduced.

4. Flexible Telecommunications Management and Interface

As mentioned at the beginning of this paper, telecommunications management by service providers should change flexibly with the customer and market changes under the dynamically changing service structure. Conversely, a flexible attitude toward these changes will be the key to satisfactory telecommunications management and efficient management systems.

For this form of telecommunications management, several objects to be managed should be formed into a small group (Domain) and several domains should be integrated into a large one to establish a hierarchical architecture.

This architecture should have a flexible interface (TMN-X interface [7]) that allows one-stop shopping for customers and other service providers. In the world of the virtual SP based on the cooperation and linkage of service providers as shown in Fig. 1, the X interface should be considered as a service management interface.

Figure 4 shows a multilayer management structure of this type as multi-domain management. The definition of a management domain will change dynamically with SP management strategies and market trends. Table 1 shows a representative definition of a management domain.

To allow the timely and flexible reorganization and integration of management domains, the business

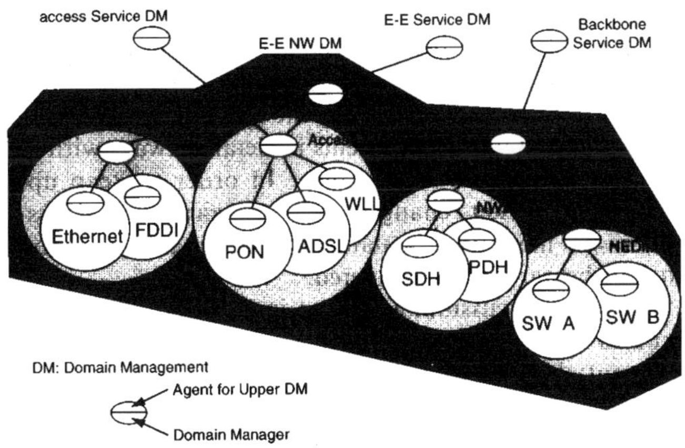

Fig. 4 Multi domain management.

Table 1 Example of managed domain definition.

Managed Domain	Example
Service Domain	POTS, ISDN, Leased Line, FR, ATM
Functional Domain	Access, Backbone, CPN
System Domain	PON, ADSL, SDH, PDH
NE Domain	SW, Transmission, Cable, Power
Vender Domain	A Vendor, B Vendor,,,
Area Domain	Metropolitan, Northern Area,,,

processes and interfaces for telecommunications management must be clarified, shared, and standardized for use in management systems.

Here, "Interface" should be defined by considering three factors: communication protocol, content (information) and service level agreement (SLA). At the beginning of TMN activities, "Communication protocol" caused hot discussions to hide the difference of multi-vendor NE interface. For service-oriented and user-oriented telecommunications management, however, the followings are discussed:

-What kind of SLA should be realized?
-What content should be exchanged in the SLA?
-What is the most suitable communication protocol for content communication?

Through these discussions, it is important to note that SLAs are the potential service differentiators and multiple standardized interface must be shared openly depend on SLAs. Each service providers will select an interface suitable for their business strategy.

5. Subjects for Management System Development

When developing a telecommunications management system, a great emphasis is given on realizing an X interface that ensures interoperability with other systems.

Therefore we had better avoid the problem of system development only meeting occasional and individual requirements, clarifying the system position and relationship with other systems in detail, always with a management overview in mind. In order to keep up with the market changes and reorganize business processes, flexibility, scalability, and interoperability should be carefully considered.

In this discussion, as shown in Fig. 5, the business process reengineer (BPR) of telecommunications management should be analyzed to determine what to be created, as well as, how to make it and a functional architecture should be established according to the results of this analysis. Figure 6 shows an example of functional architecture from the NE management up to service management.

The concept of a virtual SP mentioned in Sect. 1 gave a relative distinction to service provider and customer. This distinction also applies to the management function. As shown in Fig. 7, service management for a service provider may mean network management for another [3], [8]. The classification of management layers is effective when discussing a functional architecture. However, when discussing the system development and interface implementation, this classification may have less meaning because the software is applicable in a borderless way.

In actual system development, an important thing is to promote software distribution and reuse among management systems as far as possible, which will realize inexpensive, stable, and easy to use management systems.

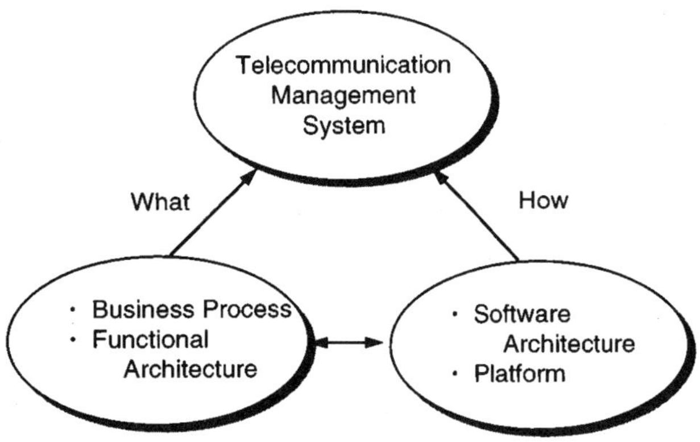

Fig. 5 System development for business solution.

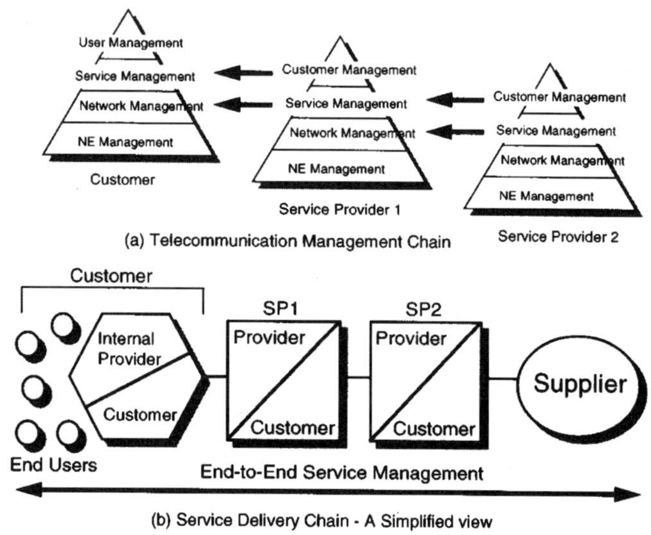

Fig. 7 Telecommunication management for customer/service provider.

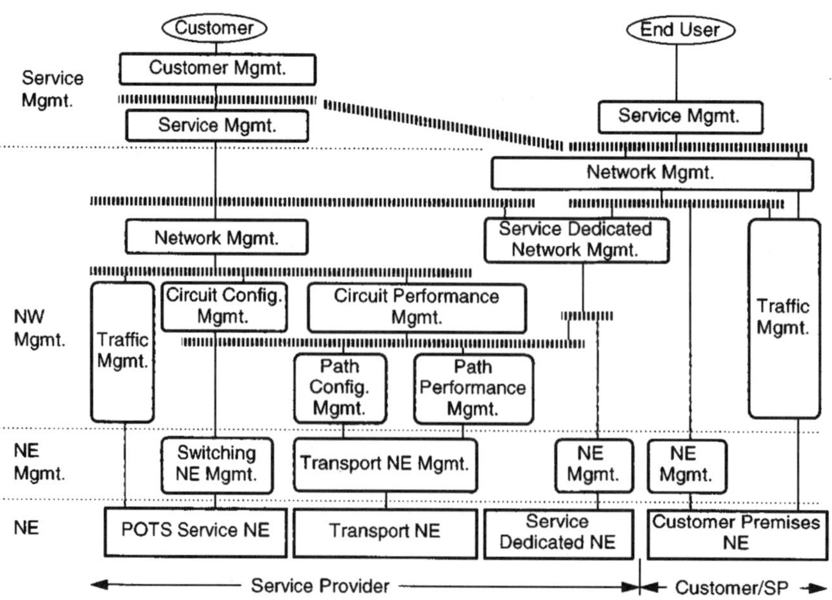

Fig. 6 Example of management functional architecture.

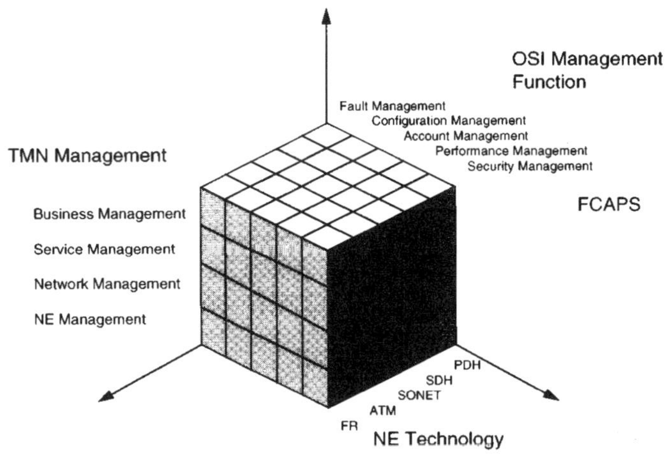

Fig. 8 Management system application architecture.

With regard to software development for telecommunications management, the software platforms for generating Managed Objects, packages for communication protocols and GUI tools are available [9]-[11], but application software is still being developed individually.

Since the processes to be realized differ, it seems that many kinds of application software need to be developed. However, the application architecture can be structured as shown in Fig. 8, so that it can be safely assumed that an application program at a cross-point cube of three axes contains many functions common to application programs expected at other cubes. Therefore, packaging the common functions as a common application component may promote software distribution and reuse and will greatly contribute to economical and stable system development.

The new era of software development age is now coming when new software developed must be kept minimal and the reuse of software maximal. System differentiation will depend much on how to combine and customize software on the market. The main issues for the era will be to create a common understanding of distributing many platforms and common applications, to make product levels standards and to establish a distribution mechanism.

6. Conclusion

The telecommunications world is now in the midst of drastic changes. The changes led by end-users are driving the industry to make services and networks free and open thanks to technological innovations.

When studying telecommunications management, we should understand the changes well and take a top-down approach from the user needs by discussing the objects, purposes, and management methods. For system development and implementation, a bottom-up approach is required to ensure steady system construction, while ascertaining the directions of technological development and standardization [2].

These approaches should not be taken independently by a service provider or vendor, but by the users, service providers, and vendors through cooperation and competition.

We expect that steady approaches for coexistence and coprosperity will realize attractive telecommunications.

References

[1] M. Ejiri, "The paradigm shift in telecommunication services and management," ISINM'95 Keynote Speech, Proc. IFIP, May 1995.

[2] M. Ejiri, "For Whom the Advancing Service/Network Management," NOMS'94 Keynote Speech, Proc. IEEE, Feb. 1994.

[3] E. Adams, "Making it easier to do business across the service delivery chain," Network Management Seminar'95, FPT Japan/NMF, Dec. 1995.

[4] M. Ejiri, "Advancind service operations and operation systems," Special Feature, NTT REVIEW, vol. 6, no. 3, May 1994.

[5] M. Maeda and M. Ejiri, "Enhancement of service front operations," Special Feature, NTT REVIEW, vol. 6, no. 3, May 1994.

[6] M. Ejiri and M. Yoshida, "Progress in telecommunications services and management," IEICE Trans. vol. E78-B, no. 1, Jan. 1995.

[7] ITU-T Rec. M 3010, "Principles for a Telecommunications Management Network."

[8] E. Adams and K. Willetts, "The Lean Communications Provider," McGraw-Hill, May 1996.

[9] L. S. Auld, "Managing the optical access network—The Fujitsu solution—FENS 3000," NOMS'96, Proc. IEEE, April 1996.

[10] A. Dittrich and M. Hoft, "Integration of a TMN-based management platform into a CORBA-based environment," NOMS'96, Proc. IEEE, April 1996.

[11] N. Fujii and T. Yamamura, "ATM transport network operation system in Japan," IEEE Communications, vol. 34, no. 9, Sept. 1996.

High-Performance Backbone Network Technology edited by Naoaki Yamanaka
ISBN: 0-8247-5321-6 DOI: 10.1081/(120025651)
Published 2004 by Marcel Dekker, Inc.

A Construction Manager System over an ATM Transport Network Operation System: Verification of the Basic Technique of Flow through Operation in Configuration Management

Hiroshi Tohjo and Tetsuya Yamamura

SUMMARY We have studied the flowthrough operation to simplify operation processes and to promote the efficiency of operation flows. In this paper, we introduce the Construction Manager System(CMS) that cooperates with the ATM Transport Network Operation System (ATMOS). This system directs the construction processes using the construction tickets newly defined by the GDMO(Guidelines for the Definitions of Managed Objects); these tickets and action scenarios are prepared inside the SMS(Scenario Management System). We confirm that CMS can perform construction management using the construction tickets instead of the traditional off-line communications such as the telephone and the facsimile. Furthermore, because CMS cooperates with ATMOS, it is able to manage actual network elements (NEs) and conduct board construction processes by using construction tickets. Moreover, CMS can confirm that the construction processes of the actual NEs have actually been executed through the Q3 interface. As a result, we verify the basic technique for flowthrough operation for configuration management.

***key words:** ATM, TMN, OpS, construction manager system, flowthrough operation, configuration management*

1. Introduction

The network management category of configuration management is considered herein to include the processes of network planning, resource provisioning, and service provisioning[1]-[3]. Several reports on configuration management have been made at conferences[4]-[9]. Most implemented the configuration management system using a proprietary management protocol or SNMP[4]-[9]. The Network Management Forum (NMF)[13] is considering configuration management. In "A Service Management Business Process Model(SMBPM)"[13], this management was defined as the "Service Configuration Process." Research related to SMBPM[10]-[12] has been reported. However, these definitions or research results only provide outlines of the Service Configuration Process. Recent reports have addressed both configuration management and flowthrough operation[4]-[6], [10]-[12]. However, all these reports contain no information on TMN based configuration management and flowthrough operation, the subject of this paper.

To verify the basic technique for flowthrough operation, this paper examines the automatic operation of resource provisioning and service provisioning through the ATM Transport Network Operation System (ATMOS)[14]-[23]. Resource provisioning and service provisioning include setting-up the NE, the board insertion process, and equipment installation. Such processes are conventionally controlled by using independent media including the telephone and the facsimile. These conventional techniques are, however, not really efficient.

In this paper, the conventional media are replaced by sending construction tickets based on TMN[24] across the ATMOS. Moreover, we describe a Construction Manager System(CMS) developed to verify the basic technique of flowthrough operation. CMS over the ATMOS manages construction process ordering, task progression and completion using the standard interface based on TMN. This CMS handles the construction ticket Managed Object (MO)[25] which is implemented in the OpS and directs the construction processes of NEs and boards as necessary.

First, the concepts of the construction manager system are presented. Next, experimental results from an implemented CMS are presented. Finally, the feasibility of flowthrough operation is confirmed for configuration management.

1.1 Increasing the Efficiency of Configuration Management Processes

It is considered that configuration management processes account for more than fifty percent of all standard network management processes[1]. Since they strongly impact budgets and profits, it is necessary to raise their efficiency[14]-[16]. This paper considers resource and service provisioning as the main configuration management processes. Their efficiency is increased by utilizing ATMOS. These processes include the fundamental process that realize network services such as NE set-up, board insertion, and equipment installation.

The conventional approach is to control such processes by paper documents. The construction workers perform the construction processes specified on the papers. Figure 1 compares the proposed and conventional methods. In this figure, the "Performing Construction Processes on NEs and boards" are the actual activities performed by the construction worker on NEs and boards.

The purpose of this paper is to verify the basic technique of flowthrough operation and to raise the efficiency of configuration management. A CMS based on TMN that uti-

Manuscript received November 25, 1996.
Manuscript revised February 3, 1997.

† The authors are with NTT Optical Network Systems Laboratories, Yokosuka-shi, 239 Japan.

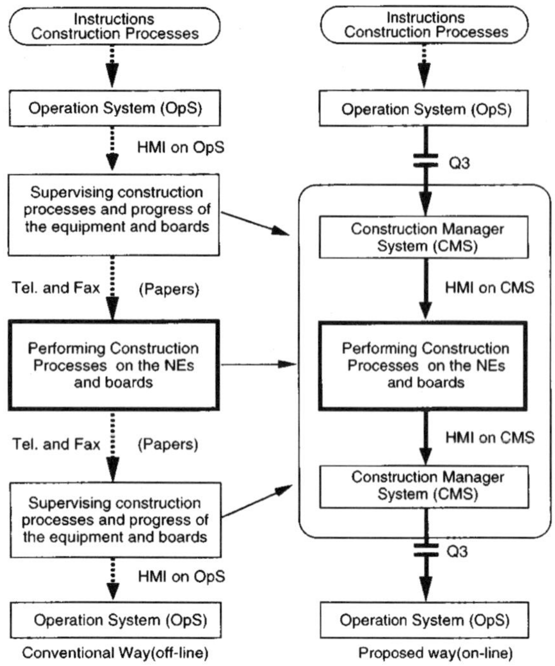

Fig. 1 Promoting efficiency of the configuration management processes.

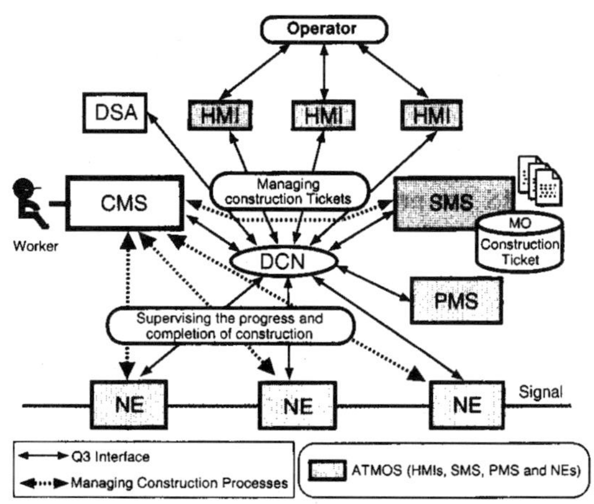

Fig. 2 Construction Manager System(CMS) and ATM transport network Operation System (ATMOS).

lizes ATMOS was built and tested for these purposes.

2. Construction Manager System (CMS)

2.1 Outline and System Component

ATM Operation System (ATMOS) consists of Human Machine Interface (HMI), Scenario Management System (SMS), Path Management System (PMS) and Network Element (NE)[17]-[23]. These systems communicate through the Data Communication Network (DCN) and Directory Service Agent (DSA)[17]-[23]. CMS and ATMOS are shown in Fig. 2. CMS cooperates with ATMOS and provides the construction process indications appropriate for network equipment such as NEs.

The characteristics of this system are described as follows:

(1) Cooperating with SMS [19] by using construction tickets. These MOs and behaviors are implemented in SMS.

(2) Managing the construction processes via the Q3 interface[26].

At the beginning and end of a construction task, CMS directly gets the attributes of the target NE's MO. This allows CMS to manage all phases of the task. Communication between SMS and network element is realized via the Q3 interface. SMS manages the construction tickets as shown in Fig. 2 and ATMOS's HMI issues new construction tickets. CMS receives the Notification(m-event-report) [26]-[28] issued from the new tickets. This Notification is created when the construction status value(ConstructionStatus) is changed. CMS gets the new ticket from SMS and directs the progress of construction in response to the orders in the ticket. The ConstructionStatus is changed to Completion when the task is completed. When SMS receives notice that the ConstructionStatus was changed, it then resumes the work on the network paths that was suspended by reason of the construction processes.

As a result, this system is able to cooperate with TMN-based operation systems.

Table 1 Construction tickets.

Construction Tickets	Information
neConstructionTicket	Directions for setting and removing of the equipment (SLT, XC)
boardConstructionTicket	Directions for inserting and pulling the boards (IF, CONT, VPX and etc.)

2.2 Construction Tickets

Table 1 shows the construction tickets used by SMS. The construction tickets are newly defined in this system. The two types are neConstructionTicket and boardConstructionTicket. The former directs the setting-up and/or removal of NE such as SLT (Subscriber Line Terminal) and XC (Cross Connect)[19] system. The boardConstructionTicket directs the insertion and/or pulling of boards in the network element.

2.3 Construction Ticket Status

The state of a construction task is reflected in the status of the construction ticket (ConstructionStatus). The information is described in Table 2. CMS uses ConstructionStatus to direct the construction worker. Figure 3 shows the transitions in status. For example, the "nothingToDo" condition shows the normal condition for construction processes and the "constructing" and "unconstructing" conditions show that construction processes are in progress.

Table 2 ConstructionStatus.

ConstructionStatus	Information
nothingToDo	Normal Condition
testing	Testing
switching	Switching
softStrapChanging	Doing Soft-Strap
createWait	Order Set-up or Insert
deleteWait	Order Removal or Pulling
constructing	Order Setting or Inserting
unconstructing	Order Removal or Pulling

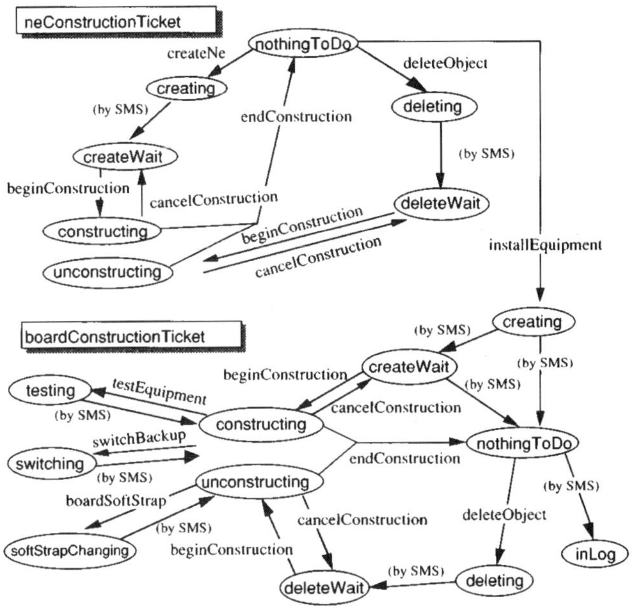

Fig. 3 Construction ticket's status.

2.4 Management Information Tree of Construction Ticket

The neConstructionTicket and the boardConstructionTicket are managed by SMS. The management information tree(MIT) of these construction tickets is shown in Fig. 4. The boardConstructionTicket is inherited from the neConstructionTicket because the boards are located inside the network element.

These construction tickets exist with another MO instances in the SMS.

2.5 Construction Processes Controlled with Construction Tickets

The use of construction tickets is outlined in Fig. 5. CMS manages the following functions for construction.

(1) Give the direction to issue a new construction ticket and complete and manage the processes.

(2) Prepare the original scenario programs to direct the construction processes in CMS.

(3) Q3 interface is used for passing directions.

In addition, CMS handles two attributes: operationalState and availabilityStatus[28]. These attributes indicate the various conditions of the NE or board. CMS manages these attributes in the construction processes.

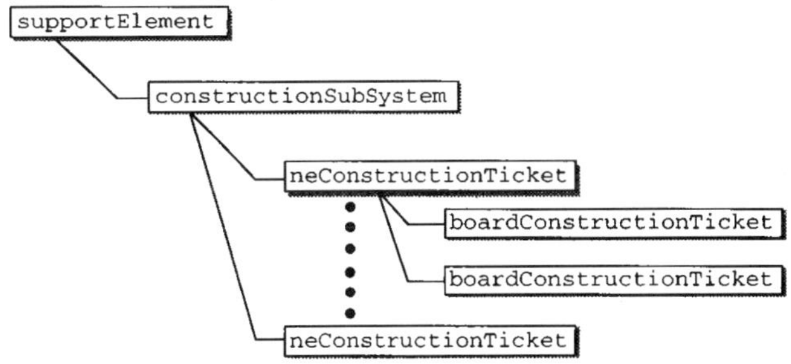

Fig. 4 Management information tree of the construction ticket.

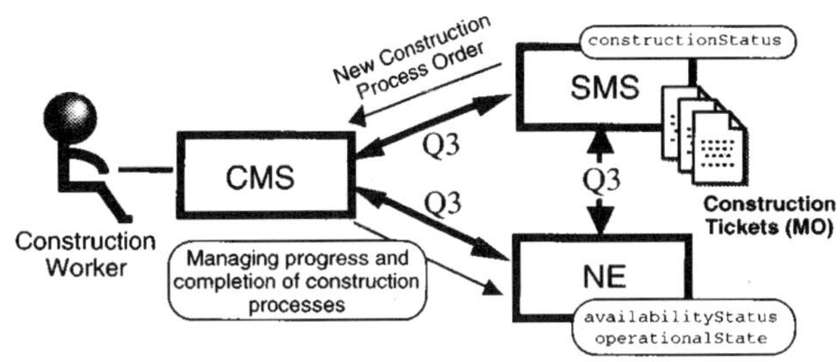

Fig. 5 Managing construction processes with construction tickets.

These ASN.1 definitions[28] are shown below.

```
AvailabilityStatus ::=
    SET OF INTEGER {
        inTest          (0),
        failed          (1),
        powerOff        (2),
        offLine         (3),
        offDuty         (4),
        dependency      (5),
        degraded        (6),
        notInstalled    (7),
        logFull         (8)
    }

OperationalState ::=
    ENUMERATED {
        disabled        (0),
        enabled         (1)
    }
```

2.6 Construction Process Scenarios

Table 3 shows the construction process scenarios which are necessary for construction processes. The construction manager must give directions to the construction worker in these scenarios. For example, the following scenarios between the SMS and NE(Network Element) are described using a sequence diagram [19].

(1) Issuing and accepting construction tickets

This scenario consists of the action messages of the related new construction tickets and acceptance of the new construction processes. Figure 6 shows this scenario.

(2) Beginning and Ending construction processes

This scenario consists of the action messages of the Beginning and Ending the construction processes. Figure 7 shows this scenario.

Table 3 Construction scenarios.

Construction Scenario	Information
Issue and accept construction tickets.	Processes to issue new construction ticket and accept it.
Beginning and ending construction processes.	Indication of the beginning and ending of the construction processes.
Testing the boards inside equipment.	Processes to test boards.
Block double acceptance	Processes to prevent tickets from being accepted twice.
Canceling an issued construction ticket.	Processes to cancel ticket already issued.
Canceling an accepted construction ticket.	Processes to cancel ticket already accepted.
Switching a board.	Switching the boards.

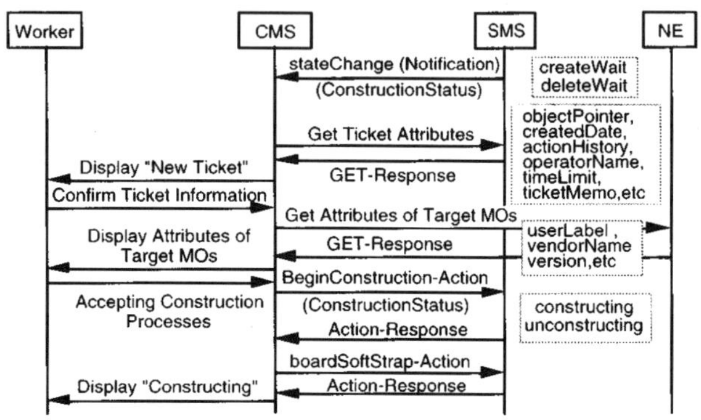

Fig. 6 Issue and accept construction tickets.

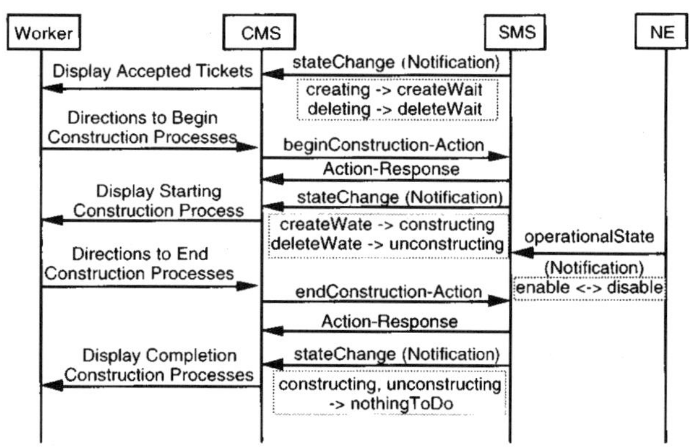

Fig. 7 Beginning and ending construction processes.

(3) Testing boards inside network element

This scenario tests the boards inside the network element using the construction ticket. Figure 8 shows this scenario.

3. Flowthrough Construction Operation with Construction Tickets

By using these construction tickets, CMS directs the construction processes efficiently. Figure 9 shows both the conventional and the proposed process with CMS.

This example is for setting the network paths from the ATMOS's HMI. In the conventional approach, when these

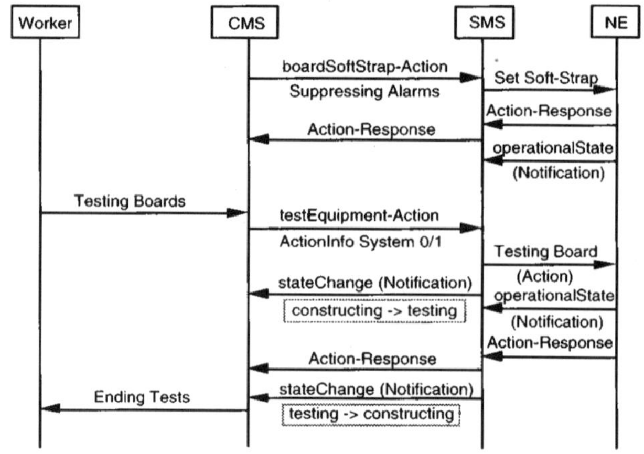

Fig. 8 Testing the boards inside the equipment.

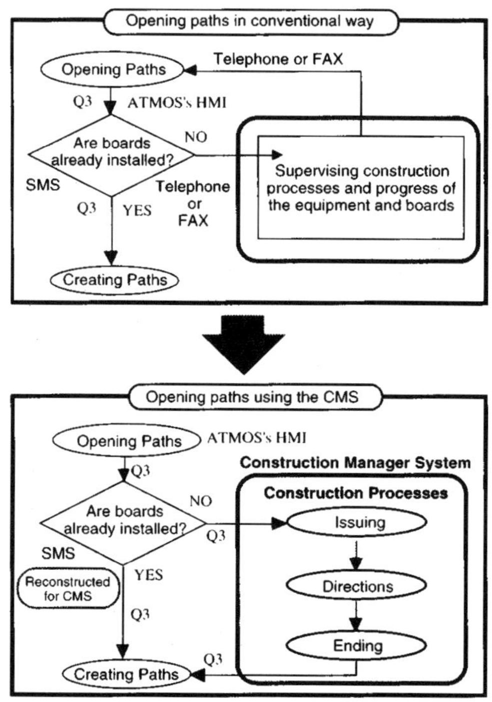

Fig. 9 Opening paths using CMS.

paths are set by HMI and the boards and NEs are not installed or not set-up yet, these paths are never set automatically. Making contact with construction workers through off-line communications such as the telephone and the facsimile is necessary.

By adopting CMS based on TMN, we can direct the construction processes automatically through the Q3 interface even if the boards and network element needed to establish the paths are not set-up. The ATMOS's HMI operator doesn't have to give new directions to set the network paths. This promotes the efficiency of the construction processes.

The construction processes are smoothly and automatically directed.

4. Construction Manager System Implementation

4.1 HMI Specifications of Construction Manager System

Special Human Machine Interfaces (HMIs) were prepared to build a prototype CMS. Table 4 shows these HMIs. These HMIs direct the construction process specified within the construction tickets through Notification(m-event-report) message from the SMS. Moreover, these HMIs also manage the beginning and the ending of the construction processes. As an example, the un-constructing process ticket HMI is explained.

This HMI looks for the un-constructing tickets and displays these ticket lists on the HMI. The construction workers can confirm the new tickets and look at the information of the new tickets. Moreover, the workers are able to give directions on the beginning of the construction process.

The various construction tickets can be viewed using the ticket-finder HMI. Figure 10 shows the un-construction process ticket HMI.

Table 4 CMS's HMI.

Special HMI	Information
Login Window	Starting CMS
Menu Window	Selecting Construction Processes Menu
Un-Constructing Ticket Window	Displaying Un-Constructiong Tickets and Displaying these Ticket lists
Constructing Window	Displaying Constructing Tickets and Displaying these Ticket lists
Completion Ticket Window	Displaying Completion Tickets and Displaying these Ticket lists
Ticket Information Window	Displaying Tickets Information
Direction Construction Processes Window	Confirming Tickets Information of Accepted Tickets
Figure of board installation Window	Displaying Figure of Board Installations
Board Image Display Window	Displaying Board images
Boards Information Window	Displaying Board Information
Display Guidance Window	Displaying Guidance of Construction Processes
Exchange Worker's name Window	Exchanging Construction Worker's Name
Display Log Window	Displaying Event-Log and Construction-Log
Find Ticket Window	Finding Construction Tickets
Set-up Environment Window	Setting up the Environment of CMS
Exit Window	Exiting CMS

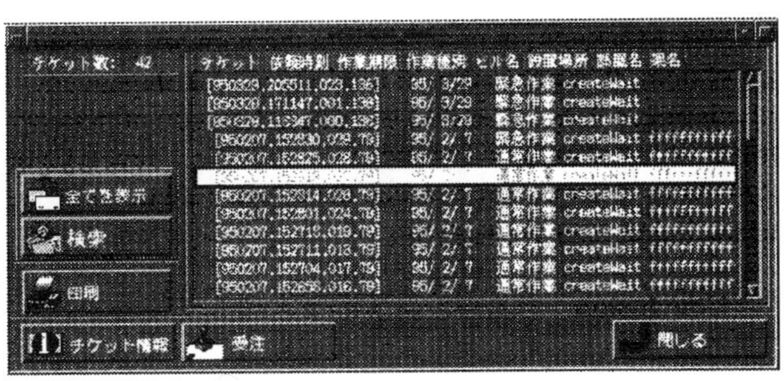

Fig. 10 Un-constructing ticket window HMI.

5. Experimental Results

5.1 Experimental Environments

A CMS was developed by the agents and managers. TNMS Kernel [29]-[31] was used to build CMS as the development platform based on TMN. The construction tickets and the behaviors programs were implemented in addition to the SMS. CMS consists of operation scenario programs for construction processes (11K steps) and GUI handling programs (51K steps: source code was generated automatically). These programs were written in C and C++ program languages. These agents and managers[32] environment information are shown in Table 5.

5.2 Construction Processes Implemented on ATMOS

A series of experiments into construction processes, which were issuing and accepting, testing boards and construction process completion, was performed using CMS and ATMOS. Various construction processes such as inserting and removing boards were performed.

The NEs and the boards were used as follows.

(1) SLT(A Maker) : SIF, VPX, LIF, CONT
(2) XC (A Maker) : SV, LIF, FILE
(3) XC (B Maker) : LIF, FILE

Figure 11 shows the experimental flow. New construction tickets were generated in accordance with this flow as actual construction processes. At result, the status of the network paths changed to normal from suspended, and the SMS was able to manage the operations of these paths. We confirmed that CMS could cooperate with ATMOS.

6. Discussion

6.1 Merit and Demerit Using Q3 Interface

We adopted the Q3 interface for building the Construction Manager System. This has two merits and one demerit.

(1) Merits

(a) CMS uses construction tickets instead of the traditional off-line communications such as the telephone and the

Table 5 Experimental environment.

<table>
<tr><th>System</th><th>Hardware</th><th>OS</th><th>TMN Dev. Tool</th><th>Another Dev. Tool</th></tr>
<tr><td>CMS (M)</td><td rowspan="4">SUN SPARC Station20</td><td rowspan="4">Solaris2.4</td><td rowspan="6">TNMS Kernel Libraries</td><td>Motif1.2.3 UIM/X2.6</td></tr>
<tr><td>SuperManager (M)</td><td>X11R6</td></tr>
<tr><td>SMS (M/A)</td><td rowspan="4"></td></tr>
<tr><td>PMS (M/A)</td></tr>
<tr><td>ATM-SLT (A)</td><td rowspan="2">Real Time OS</td><td rowspan="2">Real Time OS</td></tr>
<tr><td>ATM-XC (A)</td></tr>
</table>

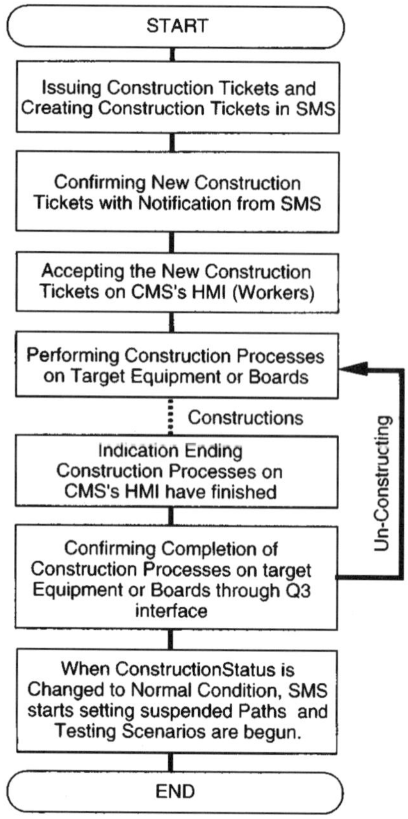

Fig. 11 Experimental flow.

facsimile.

(b) At the beginning and end of a construction task, CMS can directly get the attributes of the target NE's MO. This allows CMS to manage all phases of the task. Communication between SMS and NE is realized via the Q3 interface.

(2) Demerit

NEs have to be equipped with the Q3 interface. If this interface is not available, devices such as the Q3 adapter must be added. This demerit is not critical, and in no way offsets the benefits gained thereby; network management is becoming more complex and closed proprietary system management systems are seen as too rigid and expensive.

In order to realize the NEs easily, a proprietary management protocol or SNMP will be adopted in NEs, because the protocols are simpler than the CMIS/CMIP protocol and implementation is easy in NEs. However, the proprietary management protocol or SNMP don't have the exclusive functions via these protocols. Thus, the OpSs manage NEs more difficulty.

On the other side, in order to manage NEs from the OpSs more easily, the Q3 interface will be adopted in NEs for the operation interface instead of the proprietary management protocol or SNMP. This is because the Q3 interface is more open than existing alternatives and has more exclusive functions than proprietary management protocols or SNMP.

6.2 Automating Resource and Service Provisioning

We examined the automated process of resource provisioning and service provisioning based on the proposed construction tickets. The processes, originally based on off-line communication through papers, were improved and automated. Various construction processes could be directed through the Q3 interface.

Moreover, not only does this system supply the directions needed to control of the construction processes, but it also cooperates with ATMOS. Furthermore, the operator can, through the ATMOS's HMI, set new paths whether the boards are installed or not.

In order to verify the feasibility of automating resource provisioning and service provisioning for configuration management, we confirmed the following requirements.

(1) CMS can cooperate with ATMOS or another TMN-based OpS.

(2) CMS can manage the construction processes using construction tickets based on GDMO.

(3) CMS can confirm and manage the beginning and the ending of the construction processes using the Q3 interface.

(4) SMS can automatically resume the service scenario upon completion of the construction processes.

(5) CMS can manage the construction processes via Q3 interface because processing time is rapid enough.

We also measured the change in the stateChange-Notification processing time with the insertion and/or pulling of boards in the NE. The response time was of the order of a few seconds. Furthermore, the processing time of operations through the Q3 interface was 80-200msec[20]. In effect, we verified the feasibility of automating resource provisioning and service provisioning in configuration management.

6.3 Verification of Basic Technique for Flowthrough Operation on Configuration Management

We examined and verified flowthrough operation for resource provisioning and service provisioning. By managing the construction processes needed to set paths and establish various network services, we verified that these processes were performed in cooperation with the operation systems based on TMN.

Complete configuration management requires the establishment of a network planning system, called network provisioning, that cooperates with TMN-based operation systems. Moreover, in order to build the network provisioning system, it is necessary to implement task dependency, scheduling, roll back, and so on. The efficiency of network provisioning should be enhanced by creating a new type of construction ticket that handles the above functions in the network provisioning system.

7. Conclusion

This paper has introduced the construction manager system(CMS). This system automates and simplifies the construction processes through flowthrough operation. This is achieved by using the construction tickets defined by the GDMO and implemented in SMS. These construction tickets indicate the construction status and the action scenarios possible with the CMIS services. This construction manager system manages the construction processes using the defined status and scenarios. By controlling the real network element such as SLTs and XCs managed by the ATM Transport Network Operation System(ATMOS), we confirmed that CMS could smoothly handle the construction processes.

Moreover, we also verified the basic technique for flowthrough operation for the Configuration Management through the construction tickets based on the GDMO.

Acknowledgments

We thank Dr. Naohisa Ohta of the NTT Optical Network Systems Laboratories for supporting us during this study. We also thank Ko-ichi Sano of NTT Optical Network System Laboratories for valuable discussions.

References

[1] S. Aidarous and T. Plevyak, "Telecommunications Network Management into the 21st Century," IEEE Press, 1995.

[2] ITU-T, "ITU-T Recommendation M.3400, TMN Management Functions," 1992.

[3] H.-G. Hegering and S. Abeck, "Integrated Network and System Management," Addison-Wesley, 1994.

[4] K. Heiler and R. Wies, "Policy driven configuration management of network devices," IEEE Network Operations and Management Symposium, Kyoto, April 1996.

[5] P. Krishnan, S. Nag, and B. Sygla, "An automated technique for designing and improving provisioning processes," IEEE Network Operations and Management Symposium, Kyoto, April 1996.

[6] Q. Kong and G. Chen, "Transactional workflow for telecommunication service management," IEEE Network Operations and Management Symposium, Kyoto, April 1996.

[7] K. Willets, "Service management – The drive for re-engineering," IEEE Network Operations and Management Symposium, Florida, 1994.

[8] R. Morin, "Bell Canada's direction in service provisioning," IEEE Network Operations and Management Symposium, Memphis, TN, April 1992.

[9] G. Rainey and D. Coombs, "An Architecture for Service Management Evolution," IEEE Network Operations and Management Symposium, Memphis, TN, April 1992.

[10] T. Hisazumi and F. Kisaku, "Intelligent dynamic service provisoning architecture in the multimedia era," Proc. SUPERCOMM/ICC '96 (Dallas), pp. 1117-1122, June 1996.

[11] N. Masayoshi, T. Hisazumi, M. Masahiko, and F. Kisaku, "Interlligent dynamic service provisioning," IEEE Network Operations and Management Symposium, Kyoto, pp. 711-722, April 1996.

[12] M. Ejiri and M. Yoshida, "Progress in telecommunications services and management," IEICE Trans. Commun., vol.E78-B, no.1, Jan. 1994.

[13] NMF, "A service management business process model," Issue 1.0, Network Management Forum, 1995.

[14] D. Nakamura and I. Yoda, "Managed information design for the ATM transport network operation system," Proc. The 1994 IEICE Autumn Conference, B-757, 1995.

[15] H. Tohjo and T. Yamamura, "A construction manager system over an ATM transport network operation system," IEICE Technical Report, CQ96-45, Dec. 1996.

[16] H. Tohjo, I. Yoda, and T. Yamamura, "A study on construction manager system for ATM network operation system," Proc. The 1995 IEICE Autumn Conference, B-549, 1995.

[17] I. Yoda and K. Minato, "Development system for network operations systems," NTT R&D, vol.44, no.12, 1995.

[18] N. Fujii and K. Yata, "Technologies for transport network operations system development," NTT R&D, vol.44, no.12, 1995.

[19] T. Kimura, D. Nakamura, H. Tohjo, and M. Tsubokura, "ATM transport network operation system (ATMOS)," NTT R&D, vol.44, no.12, 1995.

[20] K. Minato, I. Yoda, T. Shimizu, and Y. Ohno, "Implementation technologies for operation systems," NTT R&D, vol.44, no.12, 1995.

[21] T. Kimura, Y. Kawakami, and T. Yoshida, "Techniques for management information definitions," NTT R&D, vol.44, no.12, 1995.

[22] K. Yata, I. Yoda, K. Minato, T. Shimizu, H. Tohjo, T. Kimura, and N. Fujii, "Implementation of ATM transport network OpS," IEICE Technical Report, CS93-135, Nov. 1993.

[23] N. Fujii, I. Yoda, and K. Minato, "An ATM transport network operation system based on TMN," IEEE Network Operations and Management Symposium, Kyoto, Japan, April 1996.

[24] ITU-T, "ITU-T Recommendation M.3010, Principles for a Telecommunications Management Network (TMN)," 1992.

[25] ITU-T, "ITU-T Recommendation X.722 1992 I ISO/IEC 10165-4: OSI-Structure of Management Information : Guidelines for the Definition of Managed Objects," 1992.

[26] ITU-T, "ITU-T Recommendation X.711 I ISO/IEC 9596-1: 1991 (E), Information Technology – Open Systems Interconnection – Common Management Information Protocol Specification – Part 1: Specification, Edition 2," 1992.

[27] ITU-T, "ITU-T Recommendation X.710 I ISO/IEC 9595:1991: Common Management Information Service Definition," 1991.

[28] ITU-T, "ITU-T Recommendation M.3100, Generic Network Information Model," Sept. 1992.

[29] I. Yoda, K. Minato, and N. Fujii, "Development of transmission networks operation systems programs by GDMO translator," IEICE Technical Report, CS92-54, 1992.

[30] I. Yoda, K. Yata, and N. Fujii, "TNMS kernel," IEICE Technical Report, CS93-79, Sept. 1993.

[31] I. Yoda, K. Yata, and N. Fujii, "Object oriented TMN based operations systems development platform," Proc. SUPERCOMM/ICC '94 (New Orleans), pp. 1824-1829, May 1994.

[32] K. Yata and I. Yoda, "Implementation of a common management information editor," Proc. The 1993 IEICE Spring Conference, SB-6-1, 1993.

High-Performance Backbone Network Technology edited by Naoaki Yamanaka
ISBN: 0-8247-5321-6 DOI: 10.1081/(120025652)
Published 2004 by Marcel Dekker, Inc.

High-Speed Multi-Stage ATM Switch Based on Hierarchical Cell Resequencing Architecture and WDM Interconnection

Seisho Yasukawa, Naoaki Yamanaka, Eiji Oki, and Ryusuke Kawano

SUMMARY This paper proposesd a non-blocking multi-stage ATM switch based on a hierarchical-cell-resequencing (HCR) mechanism and high-speed WDM interconnection and reports on its feasibility study. In a multi-stage ATM switch, cell-based routing is effective to make the switch non-blocking, because all traffic is randomly distributed over intermediate switching stages. But due to the multi-path conditions, cells may arrive out of sequence at the output of the switching fabric. Therefore, resequencing must be performed either at each output of the final switching stage or at the output of each switching stage. The basic HCR switch performs cell resequencing in a hierarchical manner when switching cells from an input-lines to a output-line. As a result, the cell sequence in each output of the basic HCR switch is recovered. A multi-stage HCR switch is constructed by interconnecting the input-lines and output-lines of these basic HCR switches in a hierarchical manner. Therefore, the cell sequence in each final output of the switching fabric is conserved in a hierarchical manner. In this way, cell-based routing becomes possible and a multi-stage ATM switch with the HCR mechanism can achieve 100% throughput without any internal speed-up techniques. Because a large-capacity multi-stage HCR switch needs a huge number of high-speed signal interconnections, a breakthrough in compact optical interconnection technology is required. Therefore, this paper proposes a WDM interconnection system with an optical router arrayed waveguide filter (AWGF) that interconnects high-speed switch elements effectively and reports its feasibility study. In this architecture, each switch element is addressed by a unique wavelength. As a result, a switch in a previous stage can transmit a cell to any switch in the next stage by only selecting its cell transmission wavelength. To make this system feasible, we developed a wide-channel-spacing optical router AWGF and compact 10-Gbit/s optical transmitter and receiver modules with a compact high-power electroabsorption distributed feedback (EA-DFB) laser and a new bit decision circuit. Using these modules, we confirmed stable operation of the WDM interconnection. This switch architecture and WDM interconnection system should enable the development of high-speed ATM switching systems that can achieve throughput of over 1 Tbit/s.

key words: *ATM, non-blocking, multi-stage switch, WDM, interconnection*

1. Introduction

The rapid growth of Internet traffic and demands for high-speed real-time-data services require a high-speed multimedia infrastructure. Asynchronous transfer mode (ATM) is a key technology for creating future multimedia networks that support these multimedia services [1]. The increasing demands for these services will require a switching system that can offer 1 Tbit/s throughput [2].

A multi-stage switch architecture is a feasible and cost-effective way to make such a system. A three-stage Clos architecture is the best architecture in terms of switch scalability because it can be expanded easily using the same switching block, but it has two basic problems [3]–[5]. One is that if connection-based routing (in which all cells belonging to the same virtual channel (VC) take the same path through the multi-stage switch) is used, this architecture needs resource management in the switches in the intermediate stage to avoid overloading conditon caused by multiple VCs concentrating on the same output and exceeding its capacity. An effective way to eliminate the need for resource management in the intermediate stage is to use cell-based routing (in which cells belonging to the same VC may take different paths through the multi-stage switch). However, cell-based routing may cause some cells to arrive out of order at the output of the switching fabric. Therefore, a cell resequencing function is necessary to correct this cell sequence disorder. The second basic problem is that a huge number of high-speed signal interconnections are necessary for a high-speed multi-stage ATM switch. For example, $N \times N$ (full mesh) interconnection is necessary in the three-stage Clos architecture. From the viewpoint of switch performance, it is better to use an I/O line speed of 10 Gbit/s for a Tbit/s ATM switch; however, the larger the switch size becomes, the more difficult it becomes to interconnect $N \times N$ unit switches in a feasible and cost effective manner.

This paper proposes a non-blocking multi-stage ATM switch based on a hierarchical-cell-resequencing (HCR) mechanism. An HCR switch employs cell-based routing, so it needs to have a cell resequencing function. The basic HCR switch element is constructed from a crosspoint-buffer-type switch and each crosspoint in the basic HCR switch is interconnected in a hierarchical manner to the output. When both the upper and lower

Manuscript received June 18, 1998.
Manuscript revised September 16, 1998.

†The authors are with NTT Network Service Systems Laboratories, Musashino-shi, 180-8585 Japan.

*This paper is also published in IEICE Trans. Electron., Vol.E82-C, No.2, pp.219–228, February 1999.

IEICE TRANS. COMMUN., Vol. E82-B No. 2, pp. 271–280, February 1999

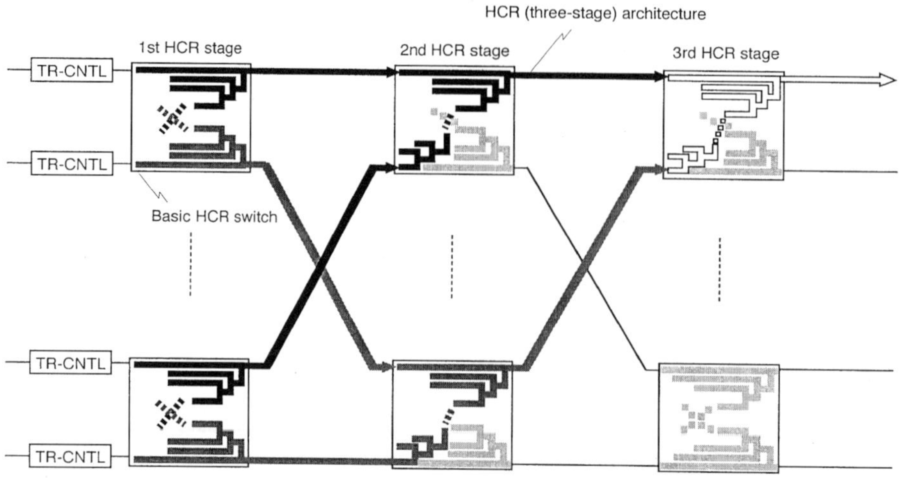

Fig. 1 Logical structure of three-stage HCR switch.

crosspoints have a cell to transmit, the HCR mechanism compares the time stamps in the internal cell headers, which indicates the times at which the cells were is sent to the HCR switch, and switches the cell having the smaller time stamp to the output. This time stamp comparison is repeated for every crosspoint in a hierarchical manner. Thus, a basic HCR switch performs cell resequencing and cell switching in a hierarchical manner at the same time. In a multi-stage HCR switch, each output of an HCR switch is interconnected with an input of the next stage's HCR switch. As a result, a multi-stage hierarchical cell resequencing network can be constructed from these switches. Therefore, perfect cell resequencing is performed at each output of the multi-stage switching fabric and a multi-stage HCR switch can achieve 100% throughput without any internal speed-up techniques.

From the implementation point of view, a high-speed $N \times N$ interconnection system is necessary to interconnect basic HCR switches. To make this multi-stage HCR switch system feasible, a compact interconnection technology is necessary. This paper proposes a high-speed WDM interconnection system and reports on a feasibility study of a 640-Gbit/s system. The proposed WDM interconnection system uses an N-channel optical router arrayed waveguide filter (AWGF) with perfect optical frequency periodicity. The study showed that $N \times N$ optical interconnection was achieved with only one optical router and N signal wavelengths. To make the 640-Gbit/s WDM interconnection system, we developed compact 10-Gbit/s optical transmitter and receiver modules and a wide-channel-spacing optical router AWGF. Using these modules, we demonstrated 640-Gbit/s WDM interconnection.

The remainder of this paper is organized as follows. Section 2 presents the basic-HCR switch architecture and explains how to expand it to a multi-stage one. Switch performance is also reported. Section 3 describes the WDM interconnection system. Section 4 presents a prototype of the 640-Gbit/s WDM interconnection system and the 10-Gbit/s transmitter and receiver modules and optical router AWGF. Finally, Sect. 5 summarizes the key points.

2. High-Speed Multi-Stage ATM Switch with Hierarchical Cell Resequencing Mechanism

2.1 Basic HCR Switch and Multi-Stage HCR Switch Design

The logical structure of a three-stage HCR switch is shown in Fig. 1. This switch uses cell-based routing, and cell resequencinge is performed independently at each basic HCR switch by checking the time-stamps in the internal cell headers. To control the time-stamps and the cell-switching-route, the multi-stage HCR switch has a time-stamp/routing-bit controller (TR-CNTL) at each of its inputs. This has a synchronous counter that is synchronized with the cell time and incremented by one during in each cell time, in order to control the time-stamps. When a cell enters the multi-stage HCR switch, the value of the synchronous counter is written in the header of the cell as a time-stamp. Thus, the time-stamp records the time at which the cell arrives at the multi-stage HCR switch and this time-stamp is used universally throughout the multi-stage HCR switch. When no cells enter the multi-stage HCR switch, TR-CNTL creates a dummy cell and the value of the synchronous counter is written in its header. The dummy cell informs the later basic HCR switches that no cell is arrived at the to multi-stage HCR switch at the time indicated by its time-stamp. This information is used for the sake of correct time comparison in basic HCR switches and to achieve a small cell transfer delay in a multi-stage HCR switch. The TR-CNTL also controls routing bits (RBs) in the internal cell header. The RBs indicate the

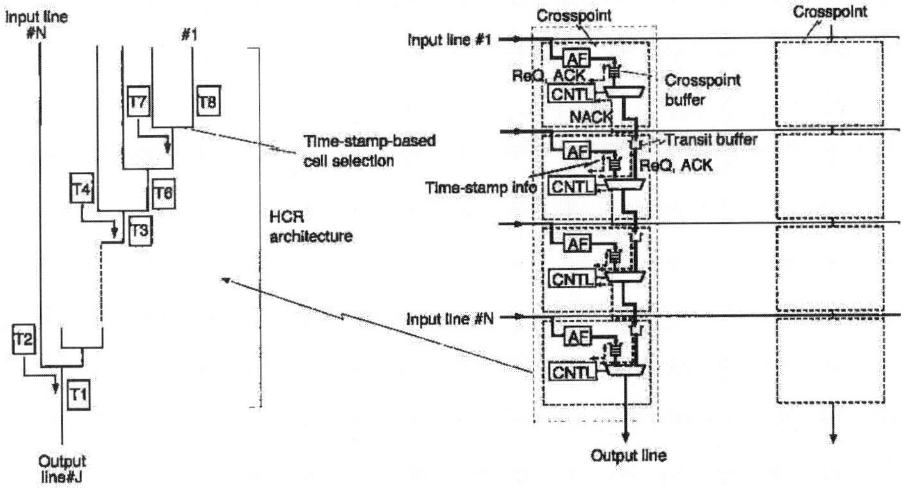

Fig. 2 Basic HCR architecture with time-stamp-based cell sellection.

cell-switching-route that determines the second switch to use when switching a cell. The TR-CNTL cyclically change these RBs in synchronized with the cell time. This balances the load in the second switch, so no internal-speed-up technique is necessary to make the multi-stage HCR switch non-blocking.

The basic HCR switch performs cell switching and resequencing at the same time in switch elements. Its block diagram is shown in Fig. 2. It uses a scalable-distributed-arbitration (SDA) scheme when it switches a cell to an output port [6], [7]. The basic HCR switch is a crosspoint-buffer-type switch. Each crosspoint is constructed from a crosspoint buffer, a transit buffer, an arbitration-control part (CNTL), and a selector. A crosspoint buffer sends a request (REQ) to CNTL if it has at least one cell stored in it. Likewise, the transit buffer also sends REQ to CNTL if it has at least one cell stored in it. A transit buffer stores several cells that are sent from either the upper crosspoint buffer or the upper transit buffer. If the transit buffer is full, it sends a not-acknowledgment (NACK) to the next-upper CNTL. When this upper CNTL receives a NACK signal, it stops sending cells to the lower transit buffer. If a CNTL does not receive NACK from the next lower transit buffer, it selects a cell to transmit within one cell time. The upper-most CNTL always selects a cell in the crosspoint buffer because the upper-most crosspoint does not have a transit buffer. Then the selected cell is sent through a selector to the next lower transit buffer or to the output line in the lowest selector.

The cell selection is performed according to a time-stamp-based cell selection rule, which determines which cell should be sent as follows.

Case 1: If either the transit buffer or the cross-point buffers has no cells to transmit, all of the CNTLs stop selecting a cell to transmit to avoid causing cell sequence disorder in the switch.

Case 2-1: If both the crosspoint and transit buffers have cell that is not a dummy cell to release, the cell with the smaller time stamp is selected. If the time stamp of the cell in the crosspoint buffer equals that in the transit buffer, CNTL determines which cell should be transmitted using a second cell selection rule [7]. Let us consider the k-th crosspoint and transit buffer from the top. The second rule is that the k-th cross-point buffer is selected with probability $1/k$, while the k-th transit buffer is selected with probability $(k-1)/k$ because the k-th transit buffer is interconnected with $k-1$ upper crosspoints, so the probability $(k-1)/k$ represents cell transmission from $k-1$ upper cross-points. Using this second cell selection scheme, this switch can fairly select a cell to transmit when both cells have equal time stamp values.

Case 2-2: If the crosspoint buffer has at least one cell to transmit and the transit buffer has a dummy cell, the cell with the smaller time stamp is selected. If the time stamp of the cell in the crosspoint buffer equals that of the dummy cell in the transit buffer, CNTL selects the cell in the crosspoint buffer for transmission because this cell has a high priority.

Case 2-3: If the transit buffer has at least one cell to transmit and the crosspoint buffer has a dummy cell, the cell with the smaller time stamp is selected. If the time stamp of the cell in the transit buffer equals that of the dummy cell in the crosspoint buffer, CNTL selects the cell in the transit buffer for transmission because this cell has a high priority.

In the above queuing processes, if either crosspoint buffer or transit buffer or both buffers have a dummy cell, then they do not treat it as the tail of the queue, and a transit buffer that has a dummy cell does not send a NACK signal to the upper CNTL. Therefore the dummy cell in the crosspoint buffer or transit buffer is overwritten by a new cell coming from the previous HCR switch's output or upper CNTL to avoid an excessive increase in load.

In this way, the basic HCR switch performs cell resequencing at each crosspoint in a hierarchical manner. The concept of cell resequencing is also shown in Fig.2. After switching, the upper input lines are concentrated one-by-one to lower input lines until the final output line. When two lines are concentrated into one line, cell resequencinge is performed by the time-stamp-based cell selection algorithm. As a result the cell sequence is conserved at each output of the basic HCR switch.

We can expand the switching capacity by interconnecting several basic HCR switches in a multi-stage manner. A three-stage HCR switch architecture is most attractive from the viewpoint of switch scalability. To achieve non-blocking operation without any internal speed-up techniques, cell-based routing is necessary. Thus incoming cells are distributed over switches in the second stage to balance the traffic load of the second stage. As a result, cells belonging to the same VC may take different routes through this switching network, which may cause cell sequence disorder. To avoid cell sequence disorder, this basic HCR switch is interconnected in a hierarchical manner in a three-stage HCR switch. This means that each basic HCR's output is connected to the input of the next stage's basic HCR. Therefore, a larger multi-stage HCR network can be constructed. Because cell resequencing is performed in a hierarchical manner in each stage in this architecture, the cell sequence is perfectly conserved at each output of the three-stage HCR switch. In this way, switching and cell resequencing are performed at the same time in the multi-stage HCR switch.

2.2 Performance of HCR Switch

We evaluated the performance of the HCR switch in terms of average cell transfer delay time by computer simulation. We assumed a three-stage HCR switch (switch size: 64×64) that is composed of basic HCR switch elements (switch size: 8×8). We also assumed that cell arrivals at each input ports follows a Bernoulli process. When the input traffic load is ρ, an incoming cell arrives with probability ρ, during one cell time. And the probability of no cell arriving is $1 - \rho$. The input traffic is assumed to be homogenous and it is distributed uniformly among all input ports. And the destinations of incoming cells are distributed uniformly among all output ports. Figure 3 shows the average cell transfer delay performance of the basic HCR switch and of the three-stage HCR switch. The more the offered load increases, the more the average cell transfer delay increases. This delay performance is identical to the conventional performance of an output-buffer-type switch except for an excess delay time due to the cell resequencing function in the basic HCR switch. This delay time depends on the number of resequencing-steps in the basic HCR switch. In this simulation, we as-

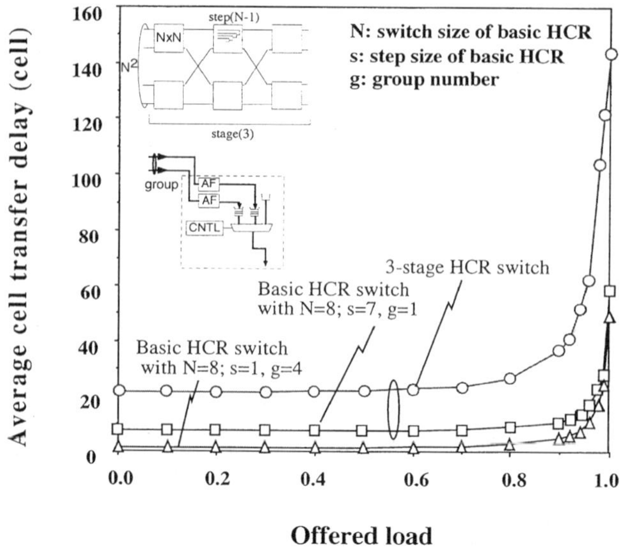

Fig. 3 Average delay performance of HCR switch.

sumed a single-stage HCR network having 7-step time-comparison and a three-stage HCR network having 21-step time-comparison. Therefore, 7 excess cell times and 21 excess cell time are added to the conventional output-buffer-type switch. These results show that a multi-stage HCR switch can achieve 100% throughput and that its delay performance is identical to output-buffer switches connected in series, excluding the cell resequencing delay.

Grouping crosspoint buffers and reducing the number of transit buffers in the basic HCR switch can reduce the resequencing delay. In this architecture, two-phase time-stamp comparison is performed. First, the time-stamps of grouped-crosspoint are compared. When a crosspoint that has a cell having the smallest time-stamp is selected, this crosspoint sends the cell to a lower transit buffer. In the lower crosspoint, the same time-stamp comparison is performed and CNTL compares the time-stamps of cells in the transit buffer and crosspoint buffer. Then resequencing is performed using the time-stamp-based cell selection algorithm mentioned earlier before. Figure 3 also shows the delay performance results for a basic grouped ($g = 4$) HCR switch. In this example, four crosspoint buffers were grouped, and the time-stamp comparison was performed for this group. This grouping reduced the number of transit buffers from 7 to 1. As a result, the average resequencing delay time was reduced from 7 to 1. Using this technique we can construct a multi-stage HCR switch that has a smaller cell transfer delay.

3. WDM Interconnection System with Optical Router

A multi-stage HCR switch is scalable. However, it needs a high-speed $N \times N$ interconnection system to in-

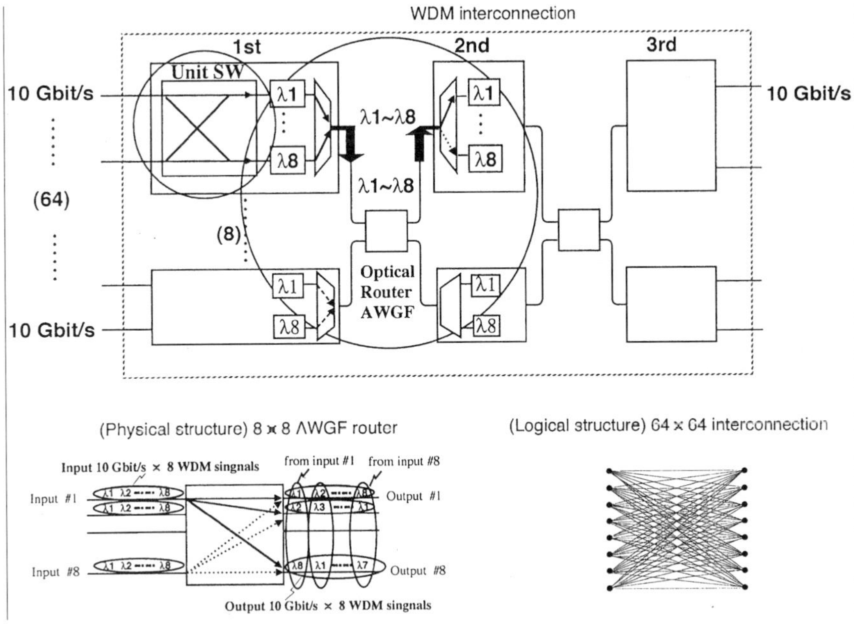

Fig. 4 WDM interconnection.

terconnect N basic HCR switches (switch size: $N \times N$). In this case, the conventional interconnection system, which uses discrete structures such as high-speed coaxial cables, becomes impractical when N approaches ten. The number of intersections in interconnection that must be handled per stage is N^2, which in this case is 100. Therefore, combining wide-bandwidth optical technologies and high-speed electronics technologies will be the key to overcoming the throughput limitation of electrical ATM switch systems [8]–[10], because optical interconnection technology offers a breakthrough in transmission distance, electromagnetic interference, I/O pin area, and so on. To make an optical interconnection system feasible, we need a compact optical interconnection system that can handle a large number of high-speed intersections.

As a step toward creating a Tb/s multi-stage HCR ATM switch system, we propose a WDM interconnection system that interconnects electrical basic HCR switches [11], [12]. Figure 4 shows our proposed WDM interconnection system. This architecture uses N different wavelengths and an $N \times N$ optical router to interconnect N basic switches ($N \times N$) through an optical router [12]. The optical router enables every basic switch in the front stage to address every switch in the next stage by using a different wavelength. For example, a 640-Gbit/s WDM interconnection architecture uses eight different wavelengths to interconnect 80-Gbit/s basic HCR switches through an 8×8 optical router. Every 80-Gbit/s basic HCR switch in the front stage can address every 80-Gbit/s switch in the next stage by using a different wavelength.

The proposed WDM interconnection system is composed of eight transmitter (TX) blocks, one optical router, and eight receiver (RX) blocks. Each TX block is composed of eight 10-Gbit/s optical TX modules, having eight optical sources with different wavelengths, and one optical multiplexer. And each TX block transmits a wavelength-multiplexed 8×10-Gbit/s WDM signal. Each RX block is composed of one optical demultiplexer and eight 10-Gbit/s optical receiver modules. And each RX block receives a wavelength-multiplexed 8×10-Gbit/s WDM signal. The optical router at the center of these modules is composed of an AWGF. This works as an optical router because its free spectral range is eight times the channel spacing. This means that this optical router switches input WDM optical signals to different output ports depending on their wavelength. And the routing relationship between the output channel number and wavelength depends on the input channel number to which the WDM optical signal is introduced. This is done in a cyclic manner as shown in Fig. 4. Thus, a basic HCR switch in the front stage can switch a cell to any basic HCR switch in the next stage using different wavelength optical signals. Thus, this 8×8 interconnection can achieve 80-Gbit/s, 8-WDM throughput. Furthermore, each electronic basic HCR switch is interconnected via a signal of a different wavelength.

From the system point of view and switch scalability, it is better for the operating clock of each unit switch to be independent. Therefore, we prepared an independent clock-data-recovery (CDR) circuit based on a phase-locked loop (PLL) in each 10-Gbit/s highway. They compensate for the phase differences between different 10-Gbit/s highway data. As a result, we obtained stable cell transmission between basic switches in different stages.

This novel system reduces the number of discrete optical fibers needed and makes system simple to operate and can achieve a total system throughput of 640 Gbit/s using only one optical router.

4. Prototype System of WDM Interconnection

4.1 Wide-Channel-Spacing WDM

To introduce WDM technology into switch interconnection, the WDM interconnection system must be simple. The conventional WDM system uses narrow-channel-spacing WDM, because this increases transmission capacity and reduces optical loss in a long fiber transmission line. But this requires strict wavelength control, which makes the WDM system complicated and reduces the system's temperature margin. It is better to widen the passband of the AWGF to attain a wide temperature margin and eliminate the complex temperature control circuits. But a wide AWGF passband causes a large frequency error in the optical router, so wide WDM channel spacing increases the difficulty of selecting optical sources.

Considering these problems, we designed our system to use an AWGF with a wide channel spacing of 525 GHz (center: 193.1 THz (1552.5 nm)). To select the optimal channel condition, we must make the filter loss and frequency error small. Figure 5 shows its loss spectra. The passband of this AWGF (160 GHz) is much wider than that of the conventional AWGF (20–30 GHz). This means that this system can tolerate a temperature range seven times as wide as that of a conventional system, so its temperature circuits can be simpler. As a result, the total system size can be dramatically reduced.

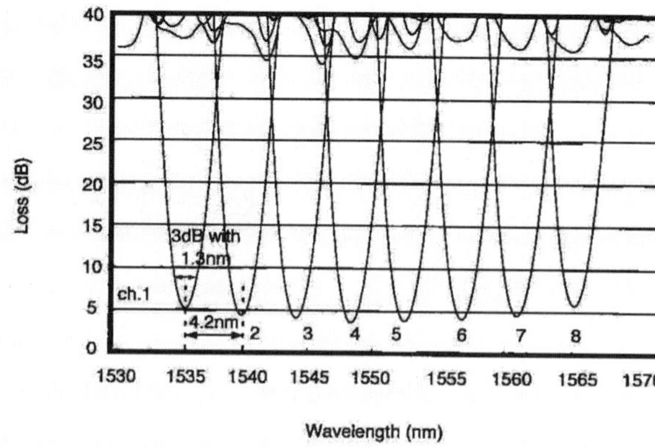

Fig. 5 Transmission loss spectrum of wide-channel-spacing AWGF.

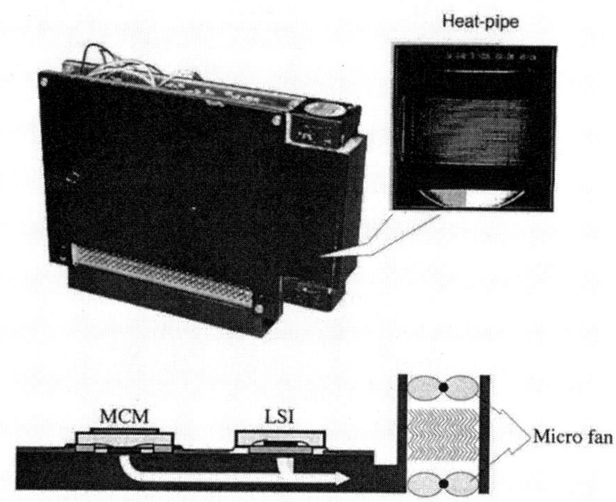

Fig. 6 External view of transmitter/receiver module and new cooling structure.

4.2 Small 10-Gbit/s Transmitter/Receiver Modules

The proposed WDM interconnection system uses a 64-byte ATM cell format. In the WDM interface, 64-byte ATM cells are mapped into a single 10-Gbit/s optical data signal and in the switch interface 64-byte ATM cells are mapped into eighteen 622-Mbit/s datalines (16: data × 32 cycles, 2: frames) and two 622-MHz clock lines. To achieve stable optical transmission, the basic HCR switch has coders and decoders in order to maintain the equilibrium of the signal in the payload of ATM cells in 10-Gbit/s format.

Figure 6 shows an external view of our new transmitter and receiver modules. Both modules are the same size: 80 × 120 × 20 mm. This is about one-twentieth the size of the conventional module [13]. These modules each have a high-speed multi-channel connector and can easily be connected to another board. Figure 7 shows the appearance and block diagrams of the transmitter and receiver modules. Both modules consist of a 10-Gbit/s-class high-frequency printed wiring board (PWB) and a 622-Mbit/s-class low-frequency PWB. In the transmitter and receiver modules, the high- and low-frequency boards are arranged face to face and connected with a fine coaxial cable. These modules have power dissipation of 9.65 and 22.5 W, respectively. Instead of a conventional radiation fin, a heat pipe is used to conduct heat from the heat source to a fin inside the duct on the end of module. Then it is radiated away by push-pull micro-fans. The heat-pipe structure embedded in the substrate of the module is shown in Fig. 6. Using this cooling method, we can fit the 10-Gbit/s optical transmitter and receiver functions into a size of 80 × 120 × 20 mm each with stable operation.

The transmitter uses one silicon bipolar IC, a 16 : 1 multiplexer. This multiplexer IC receives a low speed (622-Mbit/s × 16) ATM cell datastream and performs bit-multiplexing to convert sixteen 622-Mbit/s ATM cell datastreams into one a 10-Gbit/s ATM cell datastream. In this conversion, it compares the phase difference between the input 622-MHz CLK and the divided 622-MHz CLK from the 10-GHz VCO in the

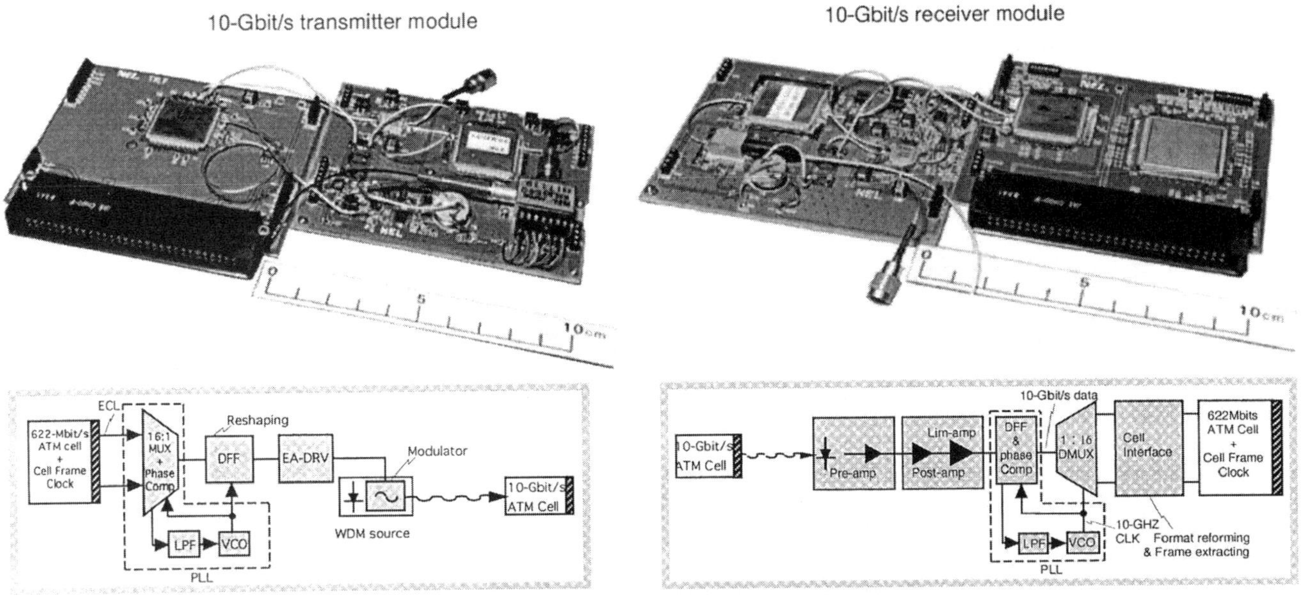

Fig. 7 Appearance and block diagrams of 10-Gbit/s transmitter/receiver module.

phase-locked loop. Thus it can latch 622-Mbit/s input data of arbitrary phase at optimum phase conditions and can multiplex 16 such streams into the 10-Gbit/s data waveform. To obtain a clear eye pattern with an EA-DFB laser, a DFF IC reshapes the 10-Gbit/s data. This DFF IC can attain an output rise time of 40 ps and an output fall time of 30 ps. Following this IC, a modulator driver circuit having GaAs FETs amplifies the signal voltage swing up to $3\,V_{p-p}$, which is applied to the multiple quantum well (MQW) optical modulator of EA-DFB laser. To adjust the extinction characteristics of the modulator wavelength dependency, the driver output is connected to a bias circuit, which can apply optimum bias to the MQW modulator. The output waveform from the EA-DFB transmitter module is shown in Fig. 8. A clear eye pattern was obtained and average optical output power of over 3 dBm was achieved.

In the receiver module, a transimpedance-type preamplifier is packaged together with a PIN-PD. Even though this module is very compact, it can achieve an 8-GHz bandwidth with 52-dBΩ gain. A post-amplifier MCM amplifies the input signal to the SCFL (0: −1 V) level. This MCM is composed of a post-amplifier IC and a limiting amplifier IC. These ICs are fabricated using GaAs MESFETs. The post-amplifier IC amplifies the input signal linearly and achieves a 16-dB gain at 8 GHz. The limiting amplifier IC achieves a 0.8-V_{p-p} output signal at 10 Gbit/s. Clock data recovery is then done by the PLL circuit, which is composed of a decision circuit with a phase comparator and an external low-pass filter and VCO. In this PLL circuit, the phase comparator is made from an EXOR circuit, which compares the phase difference between the raw

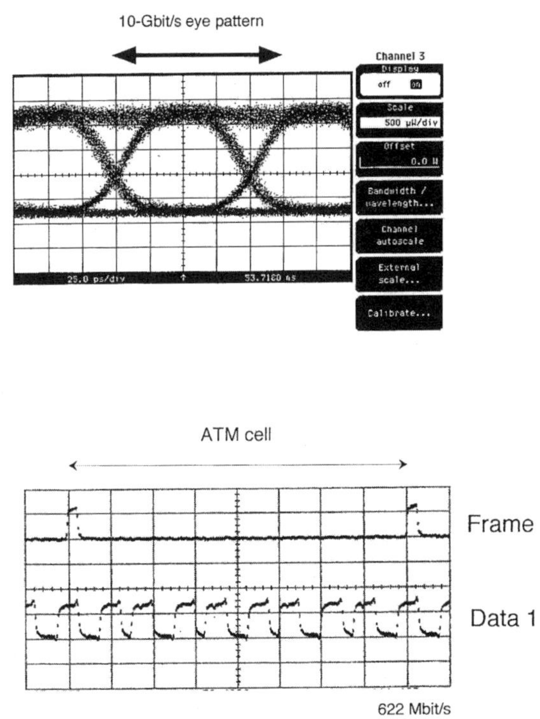

Fig. 8 10-Gbit/s optical output waveform and output waveform of ATM cell interface LSI.

input data and the data output from the decision circuit. The phase error signal is fed back to the VCO via the low-pass filter. Therefore this system can perform perfect clock-data-recovery from an input signal with arbitrary phase. This PLL circuit has a temperature-compensation circuit and can offset the VCO frequency error, which depends on temperature. Thus this mod-

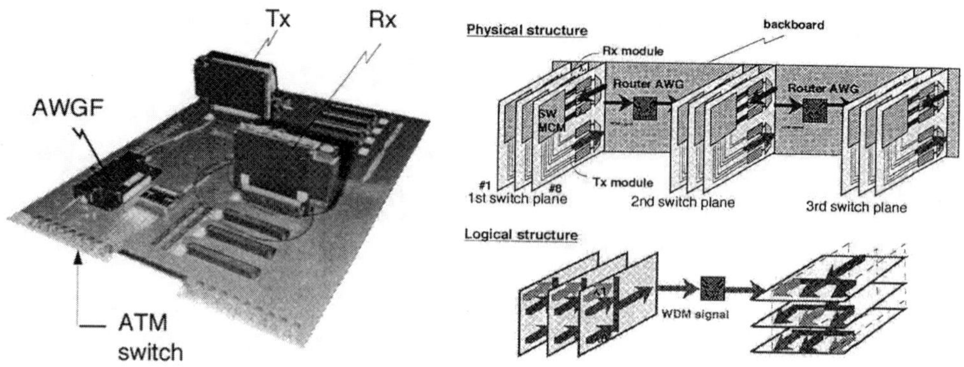

Fig. 9 Overview of optical routing board and system image of three-stage HCR switch.

ule performs stable clock data recovery at junction temperatures up to 70 degrees C. In the 1 : 16 demultiplexer IC, bit-demultiplexing is performed and sixteen 622-Mbit/s datastreams and one 622-Mbit/s clock are demultiplexed from the 10-Gbit/s data. Then 16 signals are introduced into the ATM cell interface synchronization (wordsync) IC, which is fabricated using silicon bipolar technology. This IC protects the ATM cell interface between the basic HCR switch and WDM interconnection. Using this wordsync IC, 10-Gbit/s optical signal interconnection is achieved without considering the ATM cell format because this IC searches for an 8-bit cell synchronization pattern from 16 input-datalines every 32 cycles. Once it finds one in its input datalines, this IC calculates its shift from the ideal ATM cell format and regenerates the ATM cell signal and the frame pulse that indicates the head of the cell. This IC has a triple error protection function, which allows it to ignore a frame error caused by signal jitter until it has detected the error three times. This mechanism enables stable pattern matching and transparent ATM cell transmission between basic HCR switches. Figure 8 also shows the 622-Mb/s ATM cell waveform of the wordsync LSI. Transparent ATM cell transmission is confirmed.

4.3 System Performance

Using these modules, we demonstrated the 640-Gbit/s WDM optical interconnection system. Figure 9 shows an overview of our newly developed optical routing board and illustrated the three-stage HCR switch with WDM interconnection. In this system each switch module is attached to the backboard and interconnected with the optical router via a different wavelength. To achieve stable WDM interconnection, we achieved average optical output power (mark-ratio=1/2) of over 3 dBm and receiver sensitivity of −16.5 dBm at a bit error rate of 10^{-11} (PRBS: 23 N). Using these modules, we can consider a wide optical power level diagram having about 20-dB optical power difference between the front and end of the WDM interconnection system. This optical level margin exceeds the AWGFs loss, which is the sum of each loss at the multiplexer AWGF, router AWGF, and demultiplexer AWGF. As a result, we could construct WDM interconnection without using an optical amplifier. This also simplifies the WDM interconnection system. Moreover, thanks to the sufficient optical power margin and the wide-channel-spacing AWGF, the developed system showed stable operation over a wide temperature range.

5. Conclusion

This paper proposed a high-speed multi-stage ATM switch called a hierarchical cell resequencing (HCR) ATM switch architecture and WDM interconnection architecture. To be non-blocking, the multi-stage HCR switch uses cell-based routing. Even though cell-based routing could cause cell sequence disorder at the output of the switch fabric, each basic HCR switch performs cell resequencing at its crosspoint in a hierarchical manner and conserves the cell sequence order at its output. A hierarchical cell resequencing network is used to construct a multi-stage HCR switch. Therefore, a multi-stage HCR switch performs perfect cell resequencing at each output of the switching fabric. As a result, a multi-stage HCR ATM switch achieves 100% throughput performance at any input traffic pattern without any internal speed-up techniques. Computer simulation results show that this architecture has similar cell transfer delay performance to a conventional output-buffer-type switch; the only difference is the resequencing delay in the basic HCR switch, but this can be reduced by grouping crosspoints in the basic HCR switch.

We also proposed a WDM optical interconnection system to interconnect a huge number of high-speed signal effectively using WDM technology and optical routing technology. Using these technology, we can construct $N \times N$ interconnections with only one optical router AWGF. To confirm the system's feasibility, we developed a 640-Gbit/s WDM interconnection system. This system uses eight wavelengths, each carrying a 10-Gbit/s signal, and 8×8 optical routers to avoid the interconnection limitation. To eliminate complex temperature control circuits, we used wide-channel-spacing WDM technology. Broadening the channel spacing to 525 GHz, compared with the conventional spacing of 70–100 GHz, raised the system's temperature margin about by seven times compared with the conventional system. This dramatically reduced the temperature control circuits and eliminated the need for strict laser selection. Fully integrated 10-Gbit/s optical transmitter and receiver modules were produced using MCM technology and an EA-DFB laser. These small modules are only 80 × 120 × 20 mm each. Using them we demonstrated stable operation of 640-Gbit/s WDM interconnection on the ATM cell level over a wide temperature range and confirmed its feasibility on the system level.

The proposed multi-stage HCR ATM switch and WDM interconnection system is applicable to future broadband ATM networks.

References

[1] H. Ishikawa, "Evolving from narrowband," IEEE Commun. Mag., vol.30, no.8, pp.32–36, 1992.

[2] N. Yamanaka, S. Yasukawa, E. Oki, T. Kawamura, T. Kurimoto, and T. Matsumura, "OPTIMA: Tb/s ATM switching system architecture based on highly statistical optical WDM interconnection," Proc. IEEE ISS'97, Systems Architecture, 1997.

[3] R. Melen and J.S. Turner, "Non-blocking networks for fast packet switching," Proc. IEEE Infocom'89, pp.548–557, 1989.

[4] M. Collivignarelli, A. Daniele, G. Gallassi, F. Rossi, G. Valsecchi, and L. Verri, "System and performance design of the ATM node UT-XC," Proc. ICC'94, 1994.

[5] K.Y. Eng, M.J. Karol, and Y.S. Yeh, "A growable packet (ATM) switch architecture: Design and application," IEEE Trans. Commun., pp.423–230, 1992.

[6] E. Oki and N. Yamanaka, "A high-speed ATM switch based on scalable distributed arbitration," IEICE Trans. Commun., vol.E80-B. no.9, pp.1372–1376, 1997.

[7] E. Oki and N. Yamanaka, "Scalable crosspoint buffering ATM switch architecture using distributed arbitration scheme," Proc. IEEE ATM'97 Workshop, pp. 28–35, 1997.

[8] E. Munter, J. Parker, and P. Kirkby, "A high-capacity ATM switch based on advanced electronic and optical techniques," IEEE Commun. Mag., pp.64–71, 1995.

[9] M. Eiseh, G. Grosskopf, R. Ludwig, W. Pieper, and H.G. Weber, "Photonic ATM switching with semiconductor laser amplifier gates," Electron. Lett., vol.28, no.15, pp.1438–1439, 1992.

[10] M. Tsukada, A. Misawa, J. Nishikido, Y. Shimazu, and H. Nakano, "Experiments on photonic cell switching with an optical input buffer," Electron. Lett., vol.30, no.13, pp.1081–1082, 1994.

[11] C.A. Brackette, "Dense wavelength division multiplexing networks: Principles and applications," J. Sel. Areas Commun., vol.8, pp.948–964, 1990.

[12] H. Takahashi, K. Oda, H. Toba, and Y. Inoue, "Transmission characteristics of arrayed waveguide $N \times N$ wavelength multiplexer," IEEE J. Lightwave Tech., vol.13, no.3, pp.447–455, 1995.

[13] A. Noda, K. Takahashi, T. Isogai, T. Ikeuchi, I. Yoneda, M. Sasagawa, C. Konishi, and S. Fujita, "Fully integrated 10-Gbit/s optical transmitter module and receiver module," Proc. 21st ECOC, pp.669–672, 1995.

High-Performance Backbone Network Technology edited by Naoaki Yamanaka
ISBN: 0-8247-5321-6 DOI: 10.1081/(120025653)
Published 2004 by Marcel Dekker, Inc.

Hyper-Media Photonic Information Networks as Future Network Service Platforms

Masayasu Yamaguchi, Ken-Ichi Yukimatsu, Atsushi Hiramatsu, and Tohru Matsunaga

SUMMARY This paper reviews the hyper-media photonic information network (HM-PIN) concept as a candidate of innovative future networks based on photonic technologies. The HM-PIN having a universal network interface integrates a variety of information services: telecommunications, newspapers, magazines, TV broadcasts and the growing collection of information servers. This network fundamentally offers three items: (1) bi-directional real-time channels with 10-Mbit/s-class or higher bit rate, (2) multipoint connections including multicasting/broadcasting, (3) high accessibility to information. These items are derived from the constraints of the conventional telephone networks and the Internet. By applying photonic technologies, the HM-PIN can be implemented as follows: The local network (the service platform) of the HM-PIN can be achieved by using a wavelength-division-multiplexing (WDM) broadcast-and-select (B&S) architecture that offers broadband multipoint connections (one-to-many, many-to-many) based on an inherent full-mesh topology. The WDM B&S local network will be able to support 10,000 to 100,000 channels (each with 10-Mbit/s or more bandwidth) by using optical and electrical multiplexing techniques. The backbone network can be constructed by combining photonic asynchronous transfer mode (ATM) switching systems and WDM transmission systems (including cross-connects). Two deployment scenarios of the HM-PIN (cost-oriented and service-oriented deployment scenarios) are also described for smoothly introducing the HM-PIN even before the cost issue is solved. The HM-PIN based on photonic technologies will be a future network service platform that greatly enhances communication services.
key words: *hyper-media, photonic network, WDM, multicast, broadcast*

1. Introduction

The Internet has continued its explosive growth supported by the rapid progress of related technologies. State-of-the-art IP routers and ATM switches (key devices for the Internet) have tens of gigabits per second throughput and will have hundreds of gigabits or one terabits per second throughput early in the next century [1]. However, it seems very difficult to keep increasing the throughput if we use only electronic technologies because of their inherent physical limitations (e.g., interconnection bottleneck, heating problem, device speed limit). Therefore, manufacturers have begun to seriously think about the introduction of photonic technologies into routers and ATM switches to overcome these limitations. The first target of photonic technology introduction will be optical interconnections between functional blocks in the equipment. The second target will be photonic switching functions assisted by electric controls. In addition to this replacement of the physical layer structures in the equipment, reconsideration of network architecture including node structures is very important to increase the total network throughput and to enhance network services.

To study new network architectures, we must first understand the potential demand for network services. Although major services (e.g., electronic mail, World Wide Web) on the Internet are still served on the basis of one-to-one (point-to-point) connections, advanced services (e.g., push type information delivery services like 'PointCast,' distant collaboration tools like multipoint video conferencing) need multipoint connections such as one-to-many or many-to-many connections. However, existing networks (the Internet and telephone networks) cannot efficiently support these multipoint connections because they consume too much of the network resources.

As an extended concept of the so-called multimedia network, we proposed the concept of hyper-media photonic information network (HM-PIN) and have been studying its implementation for several years [2]–[17]. The hyper-media network integrates, on a universal network interface, various information services such as telecommunications, newspapers, magazines and TV broadcasts, plus the growing collection of information servers. We believe the HM-PIN concept is a solution to the above mentioned issues.

This paper reviews this concept to clarify an image of future networks based on photonic technologies and shows possible deployment scenarios of the HM-PIN. Section 2 summarizes the characteristics and problems of conventional information media, then derives the requirements for the proposed hyper-media network. In Sect. 3 we discuss the standpoint and basic model for the hyper-media network architecture. Sections 4 and 5 present the local and backbone network technologies. Section 6 describes possible deployment scenarios of the

Manuscript received September 11, 1998.
[†]The authors are with NTT Optical Network Systems Laboratories, Yokosuka-shi, 239-0847 Japan.
[††]The author is with NTT Telecommunication Network Laboratory Group, Tokyo, 160-0022 Japan.
[†††]The author is with NTT Technology Department, Tokyo, 163-8019 Japan.
*This paper is also published in IEICE Trans. Electron., Vol.E82-C, No.2, pp.170–178, February 1999.

IEICE TRANS. COMMUN., Vol. E82-B, No. 2, pp. 222–230, February 1999

HM-PINs.

2. Conventional Communication Media vs. Hyper-Media

In this section we will first summarize the characteristics and problems of conventional information media, then derive the requirements for hyper-media.

2.1 Telecommunication Media

Telecommunication media currently provides a very important function; it enables real-time bi-directional communication between users. However, since it was designed for the speech communication, it has the following constraints.

(a) The bandwidth (or bit-rate) is not sufficient for transmitting high definition still images or motion pictures with good quality (e.g., 80-Mbit/s for MPEG-2 HD pictures {High level} [18]) without noticeable delay (e.g., less than 150 msec).
(b) Because they were designed for communication between humans, networks have an average call-holding period of three minutes. Thus, many computers generating a huge number of very short or very long calls (e.g., few seconds to several hours) will cause traffic congestion, degrading service quality.
(c) Also because of their design for person-to-person communication, networks cannot handle multicast information delivery to a specified group or broadcast information delivery to an anonymous group.

While broadcast media (e.g., groundwave, cable, satellite TVs) are good at efficiently delivering good-quality motion pictures to anonymous people, they do not allow ordinary people to transmit or broadcast their own information because the systems are basically for down-stream communications from a broadcasting center to users. Bi-directional communication is now being tested in cable TV systems, but it is not easy to support an up-stream channel with capacity sufficient for many users. Therefore, these cable TV systems are adopting asymmetric bi-directional communications with a high-bit-rate down-stream channel (e.g., 30 Mbit/s) and a low-bit-rate up-stream channel (e.g., 1 Mbit/s or more) [19], [20].

2.2 The Internet

There is now a very revolutionary communication medium: the Internet. The WWW service on the Internet presents many noteworthy features:

(a) Users can extract various types of information on demand from around the world. Since the directory information is distributed on the same network as the target information, navigation services for finding target information can be easily built (e.g., Yahoo!).
(b) Anyone with a simple personal computer can start a service for delivering information to anonymous people (e.g., personal homepages).
(c) Although character-based information with still images (e.g., homepages with GIF or JPEG pictures) is dominant, multimedia presentations using voice, video and animation (presented by software tools such as QuickTime, RealPlayer and Shockwave) are growing.

Although there are still many problems, such as transmission delays due to insufficient network capacity and insufficient CPU performance, weak security, and an imbalance in network cost-sharing, the WWW is likely to continue to grow rapidly and become a standard information medium.

2.3 Hyper-Media Network

The hyper-media network is a medium (having a universal network interface) that integrates various information services such as telecommunications, newspapers, magazines and TV broadcasts, plus the growing collection of information servers. The aim for the hyper-media network is to overcome the constraints of the conventional telecommunication media and the Internet, and to support all the functions of current media through a unified interface. It thus has to satisfy the following requirements.

(a) It must support bi-directional real-time communication channels with sufficient capacity (e.g., 10-Mbit/s to 1-Gbit/s per user) for good-quality video and provide the same convenience and reliability as the telephone.
(b) Any users must be able to use multicast and broadcast functions on demand.
(c) Users must be able to instantly obtain any kind of information in any amount in any representation form from anywhere.

3. Network Architecture for Hyper-Media

The constraints on conventional media arise from the network/system architecture on which the media are based. Thus, to realize the hyper-media concept, the present network architectures and technologies should be reconsidered from the ground up. In this section we will discuss the standpoint and basic model for the hyper-media network architecture.

3.1 Network Downsizing

To facilitate the penetration of easy-to-use hyper-media, the network architecture should be designed considering the following objectives.

(1) Each connection should be able to maintain a certain quality of service without interference from other connections.
(2) Both point-to-point and multicast/broadcast connections should be available.
(3) Users should be able to flexibly customize their communication services on demand.
(4) The network should be very reliable and secure and should be able to quickly adapt to various changes in the environment.

These objectives have been achieved in the computer world by downsizing. Downsizing through the use of VLSI technology, which makes electrical circuits faster and smaller, has brought about a world in which ordinary people can buy a computer with the performance of early mainframe computers. This gives users the freedom to use computers to meet their own needs. The heavy jobs used to be handled by a mainframe computer can now be handled by a collection of many small personalized computers interconnected via a network.

We propose applying a similar downsizing strategy to global network design. This means that rather than designing a global network as a uniform, inseparable system based on a top-down master plan, it should be designed as a collection of many independent local networks, each with its own policies, functions, and services. By building smaller networks, the number of users on each network is reduced, making it easier for the users to 'personalize' their own local network. The difficult task of achieving a consensus among the users is made easier, and the users can develop new services for themselves more freely and more quickly.

This downsizing strategy is appropriate for the hyper-media network for several reasons:

(1) A smaller network can be built more easily with sufficient resources, so the users can have more flexibility in network resource usage.
(2) Smaller networks have simpler topologies, making network design and management easier. Users can easily customize the network to meet their needs.
(3) Good security and reliability are easier to achieve in a smaller network.
(4) Since updating the network hardware and software is easier, the initial network can be made more simply and less expensively. The strategy is to start small and upgrade later by applying new technologies.

While it is important for a communication network to provide connections to users around the world, the network infrastructure and network services can be developed more quickly if each local network is initially independent. They can be connected later by using broadband links.

3.2 Three-Layered Network Architecture

Present telecommunication networks consist of subscriber loops, subscriber-line switches, toll switches, and network-service databases. This architecture is the outcome of distributing the network functions over entire networks. Subsets of these network elements do not function as a network.

The proposed network architecture, based on network downsizing, is very different. The global network would consist of many self-contained unit networks, called local networks, and a backbone network connecting the local networks. User terminals would be connected to the local networks via access links. All of the basic network services for users would be provided by the local networks; the backbone network would conceal the geographical distribution of the local networks and the distances between them from the users through the use of broadband inter-local network connections. For the access links serving as entry points to the network, only the connectivity and capacity would be important; the physical technologies used, such as whether an optical fiber or a wireless connection is used, would be of no concern to the users.

3.3 Basic Technologies for Hyper-Media Network

Significant breakthroughs in implementation technologies and network architecture concepts are needed to achieve the objectives described in Sect. 3.1. In particular, the total traffic volume would be several thousand times the total bit rate currently carried in a public telephone network. It is impossible to handle this amount of traffic by simply expanding the current network elements; new technologies are needed to dramatically increase the transmission capacity and node processing capability.

The best way to provide the required transmission capacity is to fully utilize the several-teraherz optical spectrum bandwidth of optical fiber. Wavelength-division-multiplexing (WDM) technology makes this possible. A one-Tbit/s signal transmission has already been achieved experimentally by using WDM transmission [21]–[24]. Technologies for controlling and routing optical WDM signals, such as WDM cross-connection, WDM add/drop, and wavelength conversion, are now being actively studied in many research laboratories. It may be possible to increase network capacity 1000-fold by simply combining these WDM technologies.

The required node processing capacity needs optical switching and ultra-high-speed ultra-high-density LSI technologies. While the optical-switching technology relies on opto-electronics, it is not simply the fabrication of switch elements using optical devices. It encompasses all the technologies for utilizing, handling, and managing the broad optical bandwidth de-

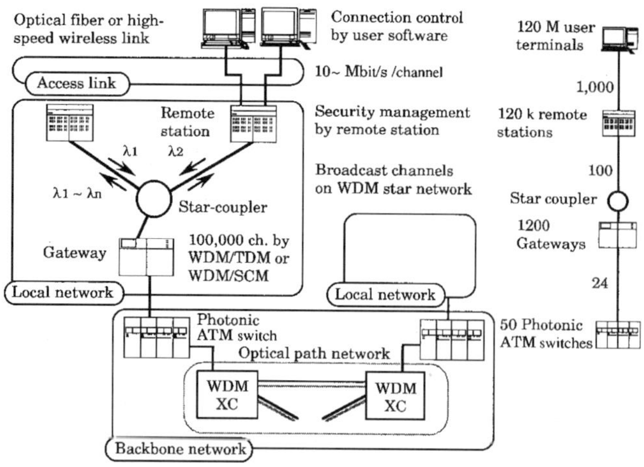

Fig. 1 Hyper-media photonic information network architecture.

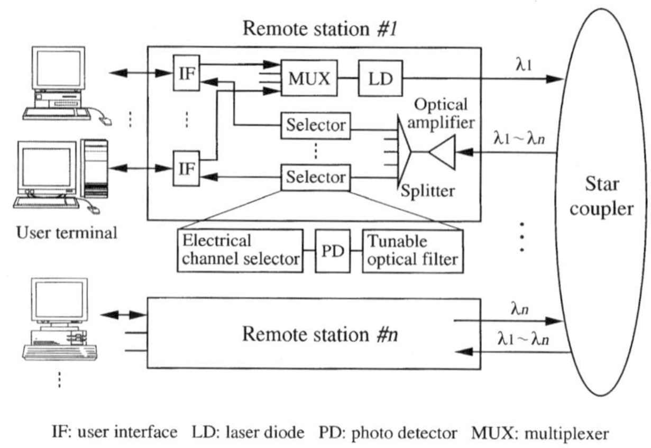

Fig. 2 Structure of a WDM broadcast-and-select network.

scribed above [25]. Ultra-high-speed ultra-high-density LSI technology is key to hyper-media because combined with software and firmware, it provides the flexible high-performance processing capability necessary for handling the various kinds of information that will be sent over the hyper-media network. There is a prospect that CPU performance will reach 100 billion instructions per second (about 250-times faster than today's CPUs) by 2010.

Wireless communication technology is also important for information portability. Although radio frequencies have been mainly used for information broadcasting, radio channels can also be used as mobile-user access links, serving as a significant alternative to optical access fibers. Functions like information broadcasting can then be handled in the wired network.

The other important technologies for achieving a hyper-media network, such as information transfer control, network management, hyper-media application, and distributed database management, will be discussed elsewhere.

3.4 Hyper-Media Photonic Information Network

The proposed hyper-media photonic information network is a concrete network model that integrates the basic concept, network architecture, and key technologies discussed above. It consists of three layers: access links, local networks, and a backbone network as shown in Fig. 1.

(1) Access links: connect users to a local network by using either a wireless or an optical-fiber link depending on user requirements. Their basic function is to establish point-to-point connections with adequate capacity and security.

(2) Local networks: are self-contained and support all network services except inter-local network communication. The architecture is a full-mesh network with an optical WDM broadcast bus supporting many independent broadband broadcast channels. This is a broadcast-and-select WDM local network; it is described in detail in Sect. 4. The independence of local networks promotes innovation in network services and applications in the respective local network. Even the information from outside the local network is provided to users through virtual information servers (like proxy or mirror servers) in the local network, so that users do not have to know the structure of anything outside their local network.

(3) Backbone network: is a global broadband information transfer network connecting many local networks. Since cost and reliability are the most important aspects, its architecture is simplified and dedicated to simple point-to-point information transfer. A combination of photonic ATM switching systems and an optical path network consisting of WDM transmission links and WDM cross-connects seems suitable, because it can provide the capacity sufficient for all those connections as well as the flexibility needed for efficient capacity handling. Details are discussed in Sect. 5.

4. Broadcast-and-Select WDM Local Network

To integrate telecommunication and broadcast services in the local networks of the HM-PIN, we adopted a broadcast-and-select WDM network architecture offering point-to-point and multipoint connections, which is the same as TV studio video signal distribution systems [26]. A star-coupler connects many remote stations, each of which contains an electrical multiplexer and selectors for many users as shown in Fig. 2. A remote station multiplexes all information from the users connected to it and transmits a light signal of a preassigned wavelength. The star-coupler combines all light signals from remote stations and then redistributes the

WDM signal to all remote stations. In a remote station, the WDM signal is divided and fed to each user selector. Each user can receive the information from any other users by tuning to the wavelength and electrical channel of the selector assigned to him. The reason that electrical multiplexing is used in combination with WDM is that the number of available wavelengths is much smaller than that of users. The bandwidth of each wavelength has to be divided into smaller channels to increase the number of channels.

The star-coupler-based WDM B&S network can be replaced by a ring-type WDM B&S local network where remote stations are located on a ring (like add/drop multiplexers). Each remote station adds its own wavelength, and it splits and receives all of the wavelengths multiplexed on the ring fiber. Thus the ring-type WDM B&S network is logically equivalent to the star-type network.

The technologies required for the local networks can be considered as extensions and enhancements of those studied for video distribution systems. A local network needs more channels and the selector cost should be very low. It also requires well-designed functions for connection control and security issues.

We are currently investigating these topics and trying to quickly demonstrate the feasibility of this local network concept. We have already built three small prototype systems with 128 channels (16 electrical channels, 8 wavelengths, 20–150 Mbit/s/channel) and successfully confirmed the operation of the broadcast channel capability and user selector function [15],[16]. Here, aiming at the selector cost reduction employing digital CATV technologies, the combination of WDM and subcarrier multiplexing (SCM) has been tried [2] as well as the well-established WDM/TDM technology.

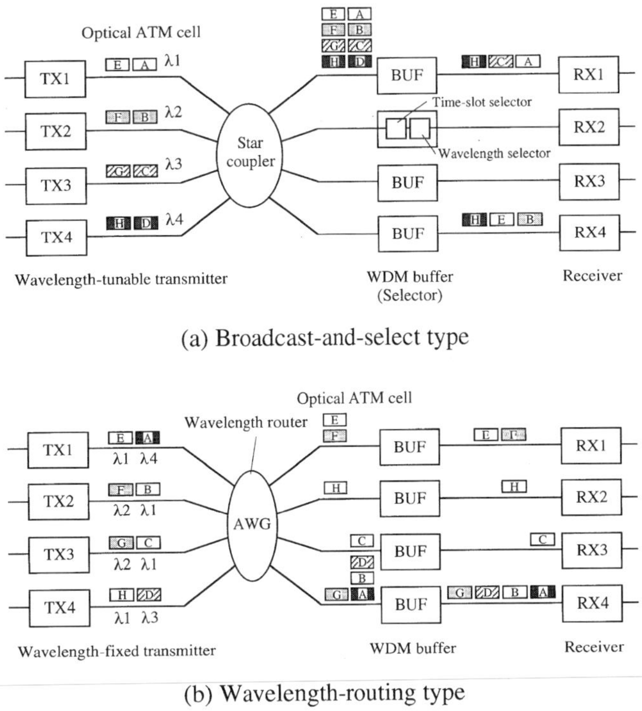

Fig. 3 Basic structure of prototype photonic ATM switching system.

5. Backbone Network Technologies

The key technologies for the backbone network are optical path handling based on WDM routing [27] and virtual channel handling based on photonic ATM switching [28]. These technologies allow large capacity links between any requested locations to be flexibly allocated.

As for WDM routing technology, Tbit/s transmission [21]–[24] and sub-Tbit/s optical cross-connect systems [29] have been reported, and many WDM testbeds have been demonstrated [30]–[32]. It will be feasible to build the core of a backbone network that can handle thousands of optical paths, each of which carries more than 1-Gbit/s traffic. However, thousands of optical paths can only connect several tens of points in a full-mesh network, millions of links are necessary to connect thousands of local networks in full-mesh.

The virtual channel technology offered by ATM is essential to create a lot more links on each wavelength and to dynamically handle link capacity. In particular, photonic ATM switching [33] is important if the backbone network is to handle thousands of Gbit/s optical wavelengths very simply. Accordingly, we are investigating the technologies to build a photonic ATM switching system with more than 1-Tbit/s throughput. Two different approaches employing WDM technologies have been tried to build a high-throughput ATM switch because WDM technologies offer switch structures without internal cell collisions. One approach is based on broadcast-and-select [34],[35] while the other uses wavelength routing [36],[37]. In the broadcast-and-select type, a different wavelength is assigned to each input port, and all cells generated from different input ports are combined by a star-coupler and distributed to all output ports as shown in Fig. 3(a). Each output port then selects only those cells destined to that output port from all distributed cells. In the wavelength routing type, an arrayed waveguide grating is used to connect the input and output ports, and a different wavelength is assigned to each input and output port pairs as shown in Fig. 3(b). When an input port generates a cell of a certain wavelength, the cell is routed to the destined output port according to the wavelength. In both approaches, the WDM cell buffer technology based on fiber delay lines is used to temporarily store optical cells in WDM signal form.

At this moment, the broadcast-and-select type is more practical because wavelength-tunable light sources are not required; the wavelength routing type has good potential for increasing the number of ports because of its better power budget.

6. Deployment Scenarios of the HM-PIN

6.1 Deployment Issue

Since the HM-PIN aims at being a future universal network, it should support a variety of services ranging from conventional point-to-point connection services to new multipoint connection services. Therefore, the HM-PIN adopts WDM B&S local networks that always broadcast any information to all user selectors. However, this broadcast function is redundant and increases the cost for supporting point-to-point connections. Even before solving this cost issue, the proposed HM-PIN concept will be applied to advanced network service platforms by adopting two scenarios: the cost-oriented and service-oriented deployment scenarios (Fig. 4).

6.2 Cost-Oriented Deployment Scenario

To reduce the first-stage deployment cost of the HM-PIN, resource sharing and hybridization are both effective approaches. Figure 4(a) illustrates a modified network architecture employing these methods. In this architecture, traffic concentration functions at access or local nodes enable many users to share expensive network resources and reduce the cost per user. In addition, the hybrid architecture of a WDM B&S function (for multipoint connections) and a legacy function (for point-to-point connections) can effectively reduce the costs for both multipoint and point-to-point services.

Traffic concentration functions increases the number of customers per unit local network. This increase in customer number will accelerate the creation of new applications, and vice versa. Accordingly, this cost-oriented architecture can effectively reduce the first-stage deployment cost and promote smooth deployment. In the future, the hybrid networks will be replaced by large-capacity pure WDM B&S networks as shown in Fig. 4(c).

6.3 Service-Oriented Deployment Scenarios

Another deployment scenario is to offer full service menus supported by the pure WDM B&S systems to high-end regional communities such as campuses, enterprises, and hospitals. This is because high-end communities tend to seek advanced and high-quality services, which demand large-capacity and flexible networks (e.g., broadband multicast/broadcast channels, full-mesh topology, user-oriented controls), even if they are costly. The cost-performance however must be much higher than LAN alternatives such as Gigabit LANs. In this scenario, regional community networks

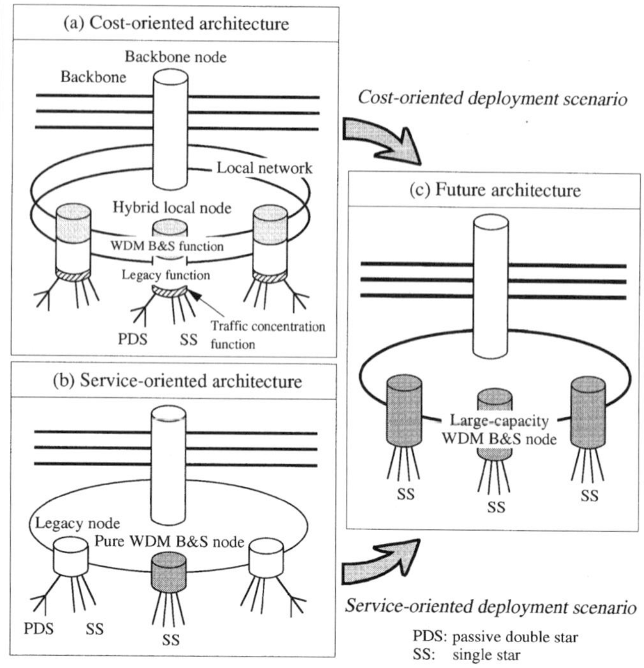

Fig. 4 Deployment scenarios of the HM-PIN. (a) Cost-oriented architecture, (b) Service-oriented architecture, (c) Future architecture.

based on the WDM B&S architecture are first introduced and they are connected via public backbone networks according to the demands for inter-community communications, as shown in Fig. 4(b). The conventional services, such as telephone services, are also supported via the community networks. In the future, all community networks would be replaced by inexpensive large-capacity WDM B&S networks as shown in Fig. 4(c).

6.4 Comparison of Two Scenarios

Table 1 summarizes and compares the features of the two deployment scenarios. The first-stage cost per user in the cost-oriented scenario is much lower than that in the service-oriented scenario because of its resource sharing and hybrid architecture. However, network service menus and performance are limited in the cost-oriented scenario while the other scenario can support a full service menu with high performance. At the first introduction stage, the cost-oriented network needs the development of just a small amount of software for supporting multicast-related network services while the other needs large amount of software for full services. At the last stage, however, these advantages are reversed. Although each deployment scenario has a few drawbacks compared to the ideal case (right most in Table 1), each will promote HM-PIN deployment by offering flexible network environments for creating new applications.

Table 1 Comparison of deployment scenarios.

	Cost-oriented	Service-oriented	Ideal (for comparison)
First target user	mass user	high-end community user	mass user & high-end user
Number of users per unit system	1-10 million/system (local, e.g., prefecture)	1,000 - 10,000/system (community)	1,000 - 100,000/system (local)
Initial cost-per-user	low (resource sharing & hybrid architecture)	high (each user has a selector)	low (optimization & mass production)
Network service menu	limited multipoint service menu conventional services supported by legacy networks	full service menu including conventional services	full service menu including conventional services
Performance	limited (due to traffic concentration)	high	high
Software development volume (first-stage)	small - middle	large	large
Software development volume (last-stage)	middle - large	small	-

7. Conclusion

In this paper, we described the hyper-media photonic network (HM-PIN) concept that integrates various information services on a universal network interface as a candidate of future network service platforms. Its possible implementation was presented by describing the broadcast-and-select WDM local networks and photonic ATM switching systems for a backbone network. Its deployment scenarios were also discussed for smooth HM-PIN introduction even though the cost issue of WDM B&S networks has yet to be solved.

It is quite difficult to properly understand the true needs and to draw specific goals for future information networks. In addition, the targets may dramatically change with later advances in technologies. Therefore, we need much more discussion about practical implementations using photonic technologies. Moreover, discussions about network controls including protocols are also important for making practical network service platforms.

Acknowledgement

The authors would like to thank Dr. Shigehiko Suzuki, Senior Vice President, Executive Manager of NTT Research and Development Headquarters, for his guidance and encouragement.

References

[1] N. McKeown, M. Izzard, A. Mekkittikul, B. Ellersick, and M. Horowitz, "The Tiny Tera: A Packet Switch Core," Hot Interconnects V, Stanford University, Aug. 1996. (http://tiny-tera.Stanford.EDU/tiny-tera/)

[2] A. Hiramatsu, M. Yanagiya, M. Ogawara, K. Hirabayashi, and K. Yukimatsu, "Hyper-media photonic information networks based on WDM/SCM broadcast-and-select switching," Proc. 9th IEEE Lasers and Electro-Optics Society 1996 Annual Meeting, vol.2, no.ThG1, pp.312–313, Nov. 1996.

[3] K. Yukimatsu and A. Hiramatsu, "Hyper-media photonic information networks," in Photonic Networks, ed. G. Prati, pp.67–78, Springer, London, 1997.

[4] K. Yukimatsu, "Photonic technologies for nest generation information networks," Journal of The Information Processing Society of Japan (IPSJ), vol.38, no.8, Aug. 1997.

[5] A. Hiramatsu, M. Yanagiya, M. Yamaguchi, T. Kominato, M. Ogawara, K. Hirabayashi, and K. Yukimatsu, "A signaling method for a broadcast-and-select optical WDM system," IEICE Technical Report, SSE94-197, Feb. 1995.

[6] M. Yanagiya, A. Hiramatsu, and M. Ogawara, "Datagram delivery over a broadcast-and-select WDM system," IEICE Technical Report, SSE95-75, Sept. 1995.

[7] M. Tsukada and A. Hiramatsu, "Switch capacity and the number of channels in broadcast-and-select SCM-WDM switch," IEICE Technical Report, OCS96-78, Nov. 1996.

[8] M. Ogawara, A. Hiramatsu, M. Yanagiya, J. Nishikido, M. Tsukada, and K. Yukimatsu, "Development and experiment of broadcast-and-select WDM systems," IEICE Technical Report, SSE96-145, Feb. 1997.

[9] M. Yanagiya, A. Hiramatsu, and M. Ogawara, "Inter-domain communication and IPv4/v6 implementation over broadcast-and-select WDM networks," IEICE Technical Report, SSE96-146, Feb. 1997.

[10] J. Nishikido, M. Tsukada, M. Ogawara, A. Hiramatsu, and M. Yanagiya, "A study of dense WDM broadcast & select network system," IEICE Technical Report, SSE96-147, Feb. 1997.

[11] K. Yukimatsu, T. Matsunaga, and A. Hiramatsu, "Hypermedia photonic information network," 1997 OSA Tech. Digest Series, Photonic in Switching, vol.10, no.PWC4, pp.62–65, April 1997.

[12] A. J. Keating, M. Tsukada, and A. Hiramatsu, "All-optical selection of 160-Mb/s channels from a 51.2-Gb/s datastream," 1997 OSA Tech. Digest Series, Photonic in Switching, vol.10, no.PThA4, pp.80–83, April. 1997.

[13] K. Yukimatsu, A. Hiramatsu, and T. Matsunaga, "Network downsizing and hyper-media photonic information network," IEICE Technical Report, SSE96-33, July 1997.

[14] J. Nishikido, A. Hiramatsu, M. Ogawara, M. Tsukada, and K. Yukimatsu, "Multiwavelength securely-authenticated broadcast network," Proc. 23rd European Conf. on Opt. Commun., Edinburgh, U.K., vol.4, no.17, pp.17–20, Sept. 1997.

[15] A. Hiramatsu, M. Tsukada, T. Matsunaga, and K. Yukimatsu, "Hyper-media photonic information network: WDM-based solutions and prototype systems," Proc. XVI

World Telecom Congress, International Switching Symposium '97, Toronto, Canada, no.IS2-20, pp.351–358, Sept. 1997.

[16] M. Ogawara, A. Hiramatsu, J. Nishikido, M. Yanagiya, M. Tsukada, and K. Yukimatsu, "Feasibility demonstrations of hyper-media photonic information networks using prototype WDM broadcast-and-select local network systems," IEICE Trans. Commun., vol.E82-B, no.2, pp.317–325, Feb. 1999, IEICE Trans. Electron., vol.E82-C, no.2, pp.265–273, Feb. 1999.

[17] A. J. Keating, M. Tsukada, and K. Wakita, "Switching of 100-Mbit/s channels from 60-Gbit/s data-stream in new broadcast-and-select WDM/TDM local network," Proc. 24th European Conf. on Opt. Commun., Madrid, Spain, vol.1, no.WdD24, pp.613–614, Sept. 1998.

[18] For example: http://www.crs4.it/~luigi/MPEG/mpeg2.html.

[19] For example: http://cabledatacomnews.com/cmic1.htm.

[20] C.A. Eldering, N. Himayat, and F.M. Gardner, "CATV return path characterization for reliable communications," IEEE Commun. Mag., vol.33, no.8, pp.62–69, Aug. 1995.

[21] H. Onaka, H. Miyata, G. Ishikawa, K. Otsuka, H. Ooi, Y. Kai, S. Kinoshita, M. Seino, H. Nishimoto, and T. Chikama, "1.1 Tb/s WDM transmission over a 150 km 1.3 μm zero-dispersion single-mode fiber," 1996 OSA Tech. Digest Series Conf. Edition, Opt. Fiber Commun. '96, San Jose, U.S.A., vol.2, no.PD19, Feb. 1996.

[22] A.H. Gnauck, A.R. Chraplyvy, R.W. Tkach, J.L. Zyskind, J.W. Sulhoff, A.J. Lucero, Y. Sun, R.M. Jopson, F.Forghieri, R.M. Derosier, C. Wolf, and A.R. McCormick, "One terabit/s transmission experiment," 1996 OSA Tech. Digest Series Conf. Edition, Opt. Fiber Commun. '96, San Jose, U.S.A., vol.2, no.PD20, Feb. 1996.

[23] T. Morioka, H. Takara, S. Kawanishi, O. Kamatani, K. Takiguchi, K. Uchiyama, M. Saruwatari, H. Takahashi, M. Yamada, T. Kanamori, and H. Ono, "100 Gbit/s × 10 channel OTDM/WDM transmission using a single supercontinuum WDM source," 1996 OSA Tech. Digest Series Conf. Edition, Opt. Fiber Commun. '96, San Jose, U.S.A., vol.2, no.PD21, Feb. 1996.

[24] Y. Yano, T. Ono, K. Fukuchi, T. Ito, H. Yamazaki, M. Yamaguchi, and K. Emura, "2.6 Terabit/s WDM transmission experiment using optical duobinary coding," Proc. 22nd European Conf. on Opt. Commun., Oslo, Norway, vol.5, no.ThB.3.1, pp.3–6, Sept. 1996.

[25] K. Yukimatsu, T. Matsunaga, M. Yamaguchi, and A. Hiramatsu, "Photonic switching technologies for next-generation flexible networks," Tech. Digest, 10th International Conf. on Integrated Optics and Optical Fibre Communication, vol.2, no.WC2-5, pp.86–87, Hong Kong, June 1995.

[26] M. Misono, N. Shimosaka, T. Shiozawa, M. Murakami, and M. Fujiwara, "Practical demonstration of WD/TD optical network for broadcasting station," Proc. 21st European Conf. on Opt. Commun., Brussels, Belgium, vol.1, no.Tu.B.1.6, pp.247–250, Sept. 1995.

[27] K. Sato, "Photonic transport network OAM technologies," IEEE Commun. Mag. vol.34, no.12, pp.86–94, Dec. 1996.

[28] T. Matsunaga, K. Habara, A. Misawa, Y. Yamada, T. Okugawa, M. Tsukada, S. Hino, K. Sasayama, and K. Yukimatsu, "Design of photonic ATM switch and a rack-mounted prototype," Conf. Record 1997 IEEE International Conf. on Commun., Montreal, Canada, vol.3, no.36.1, pp.1292–1297, June 1997.

[29] M. Koga, A. Watanabe, S. Okamoto, K. Sato, and M. Okuno, "8 × 16 delivery-and-coupling-type optical switches for a 320-Gbit/s throughput optical path cross-connect system," 1996 OSA Tech. Digest Series Conf. Edition, Opt. Fiber Commun. '96, San Jose, U.S.A., vol.2, pp.259–261, Feb. 1996.

[30] R.S. Vodhanel, J.-C. Chiao, J. Gamelin, C.A. Gibbons, H. Shirokmann, M. Rauch, J. Young, G.-K. Chang, R.E. Wagner, A. Luss, M. Maeda, J. Pastor, M. Post, C.-C. Shen, J. Wei, B. Wilson, Y. Tsai, L.D. Garrett, R.M. Derosier, A.H. Gnauck, A.R. McCormick, R.W. Tkach, S.H. Patel, C.L. Allyn, A.R. Chraplyvy, J. Judkins, A.K. Srivastava, J.W. Sulhoff, Y. Sun, A.M. Vengsarkar, C. Wolf, J.L. Zyskind, A. Chester, B. Cornissiong, G.W. Davis, G. Duverney, N.A. Jackman, A. Jozan, V. Nichols, B.H. Lee, R. Vora, A.F. Yorinks, G. Newsome, P. Bhattacharjya, D. Doherty, J. Ellson, C. Hunt, A. Rodriguez-Moral, W. Kraeft, and J. Ippolito, "National-scale WDM networking demonstration by the MONET Consortium," 1997 OSA Tech. Digest Series Conf. Edition, Opt. Fiber Commun. '97, Dallas, U.S.A., no.PD27 1, Feb. 1997.

[31] A.M. Hill and A.J.N. Houghton, "Optical networking in the European ACTS programme," 1996 OSA Tech. Digest Series Conf. Edition, Opt. Fiber Commun. '96, San Jose, U.S.A., vol.2, Thl1, pp.238–239, Feb. 1996.

[32] K. Oda, H. Toba, and T. Matsumoto, "Application of WDM technology to regional optical networks," 1st Opto-electronics and Communications Conf. (OECC '96) Tech. Digest, no.17B1-2, pp.102–103, Chiba, Japan, July 1996.

[33] K. Habara, T. Matsunaga, and K. Yukimatsu, "Large-scale WDM star-based photonic ATM switches," IEEE J. Lightwave Tech., Special Issue on Photonic Packet Switching System Technologies and Techniques, vol.16, no.12, pp.2191–2201, Dec. 1998.

[34] A. Misawa, M. Tsukada, Y. Yamada, K. Sasayama, K. Habara, T. Matsunaga, and K. Yukimatsu, "40-Gbit/s broadcast-and-select photonic ATM switch prototype with FDM output buffers," Proc. 22nd European Conf. on Opt. Commun., Oslo, Norway, vol.4, no.ThD.1.2, pp.107–110, Sept. 1996.

[35] A. Misawa, Y. Yamada, M. Tsukada, K. Sasayama, K. Habara, T. Matsunaga, and K. Yukimatsu, "A prototype broadcast-and-select photonic ATM switches with a WDM output buffer," IEEE J. Lightwave Tech., Special Issue on Photonic Packet Switching System Technologies and Techniques, vol.16, no.12, pp.2202–2211, Dec. 1998.

[36] K. Habara, Y. Yamada, A. Misawa, K. Sasayama, M. Tsukada, T. Matsunaga, and K. Yukimatsu, "Demonstration of frequency-routing type photonic ATM switch (FRONTIERNET) prototype," Proc. 22nd European Conf. on Opt. Commun., Oslo, Norway, vol.4, no.ThC.3.4, pp.41–44, Sept. 1996.

[37] Y. Yamada, K. Sasayama, K. Habara, A. Misawa, M. Tsukada, T. Matsunaga, and K. Yukimatsu, "Optical output buffered ATM switch prototype based on FRONTIERNET architecture," IEEE J. Sel. Areas Commun., Special Issue on High-Capacity Optical Transport Networks, vol.16, no.7, pp.1298–1308, Sept. 1998.

High-Performance Backbone Network Technology edited by Naoaki Yamanaka
ISBN: 0-8247-5321-6 DOI: 10.1081/(120025654)
Published 2004 by Marcel Dekker, Inc.

Photonic Packet Switching: An Overview

Rodney S. Tucker and Wen De Zhong

SUMMARY The application of photonic technologies to packet switching offers the potential of very large switch capacity in the terabit per second range. The merging of packet switching with photonic technologies opens up the possibility of packet switching in transparent photonic media, in which packets remain in optical form without undergoing optoelectronic conversion. This paper reviews recent work on photonic packet switching. Different approaches to photonic packet switching and key design issues are discussed.

key words: *photonic packet switching, photonic packet buffer, wavelength routing, wavelength conversion*

1. Introduction

There is growing motivation for the development of ultra broadband photonic networking technologies. This move to photonic networks arises from the increasing demand for broadband communications and the capability of optical components and devices to provide tremendous bandwidth to satisfy this demand [1], [2]. Asynchronous transfer mode (ATM) switching and other forms of packet switching such as IP routing are emerging as key to the implementation of future high-capacity and multi-function telecommunications networks. Electronic ATM switching techniques, combined with transmission systems using the SDH/SONET standard, provide a platform for flexible transmission and management of broadband traffic from a variety of sources, including video, data, and voice. Meanwhile, IP routing and switching is emerging as a key to future expansion of data and other internet-related services. Electronic packet switching offers great potential, but is ultimately limited in data rate and throughput by the processing capacity of electronic devices [3]–[8]. The ultimate solution is to replace electronic packet switches with optical packet switches in which signals remain in optical form.

The merging of packet switching with photonic technologies opens up the possibility of packet switching in transparent photonic media. Thus, the wide bandwidth of photonic components, combined with the flexibility of wavelength-division multiplexing and wavelength routing techniques [1], [2], and the high-speed capabilities of optical devices such as optical gates and switches [9], [10], optical wavelength converters [11]–[13], and fast multiwavelength lasers [12], [14], provides the potential of packet-switched networks with throughput in the terabit per second range [3]–[8], [15]–[20].

The application of photonic technologies to packet switching have been motivated by a number of driving forces. One of these forces is the anticipated need for high-capacity nodes [3]–[8]. Some studies have shown that the ultimate capacity of photonic-based switching nodes will exceed the capacity of electronic systems [3], [4], [15]–[20]. Thus, high-capacity photonic nodes will be attractive as alternatives to very large electronic switch nodes, even when the individual channels operate at data rates that are compatible with electronic capabilities. Another motivation is the prospect offered by optically transparent self-routing networks [3]–[8], in which optical packets pass through photonic packet switches in a transparent fashion, without undergoing any optical to electrical or electrical to optical conversions. The obvious attraction of such a network is that there is no electronic processing in the data path and the data can ideally pass through the switch without any limitation on bit rate. Also, wavelength-division multiplexing (WDM) can be readily exploited. Thus very large throughput becomes feasible. A third application is in the development of photonic local area and metropolitan networks with a vastly improved throughput [21]–[25].

The purpose of this paper is to give an overview of recent work on photonic packet switching. We will discuss key design issues and different approaches to photonic packet switching. The discussion will focus on photonic approaches to packet routing and buffering, switch fabric architectures, packet header replacement and packet synchronization.

2. Photonic Packet Switches

There have been a number of recent proposals and experimental demonstrations of photonic packet switching in university and industrial laboratories [3]–[8], [15]–[20], [26]–[33]. The switching and/or routing fabric in these demonstrations has generally been opti-

Manuscript received August 31, 1998.

†The authors are with the Australian Photonics Cooperative Research Centre, Department of Electrical and Electronic Engineering, The University of Melbourne, Parkville, Victoria 3052, Australia.

*This paper is also published in IEICE Trans. Electron., Vol.E82-C, No.2, pp.202–212, February 1999.

IEICE TRANS. COMMUN., Vol. E82-B, No. 2, pp. 254–264, February 1999

cal, in order to capitalize on the optical transparency and large bandwidth. However, electronic and opto-electronic components remain attractive alternatives to photonic devices in functions such as address processing, switching and buffing controls [3]–[8]. Large photonic packet switches will probably rely on electronics for control functions, with the packet routing and buffering being carried out by photonic means. This approach provides very large capacity through the transparency of the photonic devices, combined with the functionality and processing power of electronic control circuits.

Figure 1 shows the schematic diagram of a general photonic packet switch. As shown in this diagram, key functions affecting the operation and implementation of photonic packet switches include (a) packet routing, (b) packet buffering, (c) packet header replacement, and (d) packet synchronization and timing recovery. Truly photonic switches require photonic implementations of all of these functions. Packets generally have a fixed duration and consists of a header and a payload. The header contains routing information and other control information. As the header has to be processed at each switch node, it is desirable that the header has a relatively low fixed bit rate suitable for electronic processing (say, less than 10 Gb/s); while the payload could have a variety of bit rates ranging from 10 Gb/s to 100's of Gb/s [3], [4], [7]. The use of fixed length packets can significantly simplify the implementation of packet contention resolution and buffering, packet routing, as well as packet synchronization [3], [4], [7].

Packet header replacement is an important function in ATM-like connection-oriented networks, where it is necessary to change the contents of the packet header as it passes through each switch node. In optically transparent packet switching, it is necessary that the header replacement is carried out photonically [34]. Advances in photonic devices and components such as high-speed optical space switches, semiconductor optical amplifier (SOA) gate switches [9], [10], wavelength routing devices [35], [36], fast tunable lasers and filters [4], [12], [14], and wavelength converters [11], [12], [13], [16] have made it possible to construct photonic packet switch fabrics with a throughput in excess of their electronic counterparts [3]–[8]. Possible routing schemes for photonic packet switch fabrics include wavelength routing, broadcast-and-select routing, space-switch-based routing, and a combination of them. We will discuss the options in detail in Sect. 4.

As shown in Fig. 1, buffering of photonic packets somewhere within the switch fabric is essential to prevent packet contention. The difficulty in implementing fully-functional photonic equivalents to the electronic random access memory (RAM) makes clocked optical buffering and synchronization difficult. One possible approach to the photonic buffering problem is to use fibre delay lines either in recirculating or travelling con-

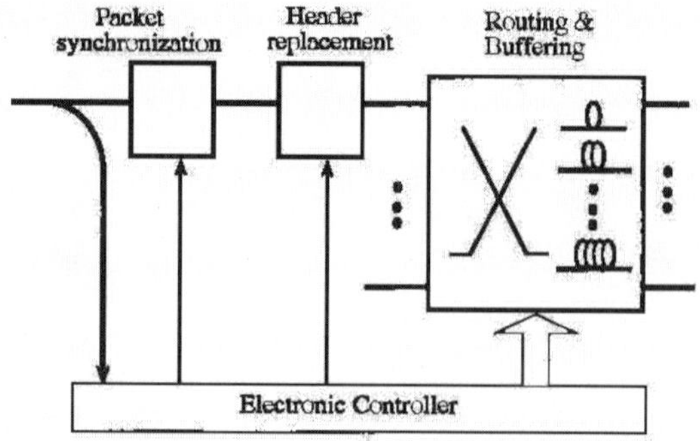

Fig. 1 Schematic of a general photonic packet swith.

figurations [4], [5], [37]. Deflection routing may offer an attractive approach to circumvent packet contention in distributed systems, but this approach provides an effective "connectionless" system, in which the order of packet arrivals cannot be guaranteed [6], [25], [27], [63].

We now consider more details of different approaches to photonic buffering and routing, architectures for switch fabrics, and techniques for photonic packet header replacement and packet synchronization.

3. Buffering

In principle, advanced electronic RAMs can be used for packet buffering. Earlier proposed photonic packet switching systems have used electronic RAMs for packet buffering [23], [38]–[40]. However, electronic RAMs have a limited access speed, which will eventually constrain the speed and capacity of photonic packet switching systems. In addition, this electronic approach requires optical-to-electronic (O/E) and electronic-to-optical (E/O) conversions when packets are written into and read out of electronic RAMs and hence adds to the complexity. An all-optical RAM would avoid the speed bottleneck and O/E and E/O conversions. There has been much effort to investigate and develop optical RAMs [41], [42]. Unfortunately, useful optical RAM suitable for photonic packet switching has not yet been found. The alternative is to use optical fibre-delay-lines incorporating other components such as optical gate switches, optical couplers, optical amplifiers, and wavelength converters to realize photonic buffering.

A number of photonic packet buffers based on optical fibre delay-lines have been proposed and demonstrated [4]–[6], [37]. In general, these optical fibre delay-line based buffers can be classified into two basic categories: travelling-type and recirculating type. A travelling-type buffer generally consists of multiple optical fibre delay-lines whose lengths are equivalent to multiples of a packet duration T, and optical space switches to select delay lines [37], as shown in Fig. 2. Travelling-type buffers can be arranged in serial or parallel structures [20], [37], [43], [44]. In such travelling-type buffers, the storage time of a packet is simply

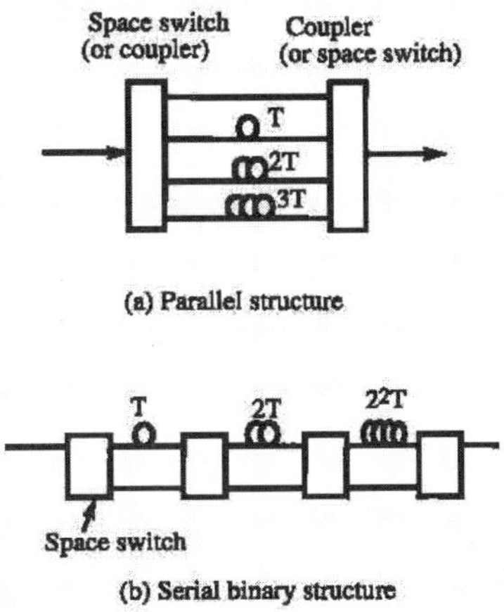

Fig. 2 Travelling-typy packet buffer.

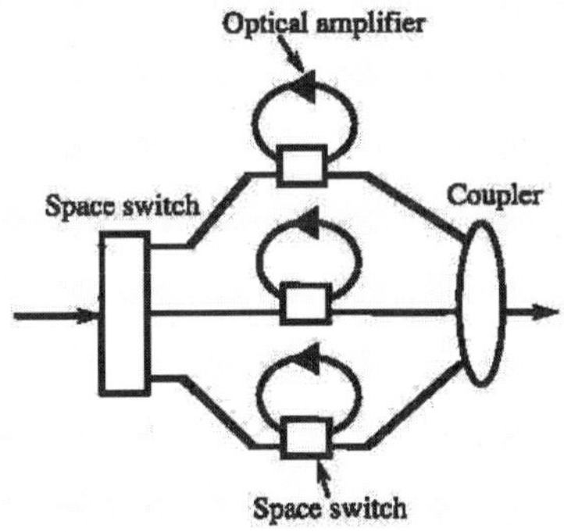

Fig. 3 Recirulating-type packet buffer with a parallel structure.

determined by the duration of a packet propagating through a length of optical fibre. For the parallel configuration, the number of delay-lines and the size of optical space switch is increased as the number of programmable delay amounts is increased. Figure 3 schematically shows a recirculating-type buffer with a parallel configuration [37]. It consists of multiple optical fibre delay-lines each forming a loop with one circulation time equal to one packet duration. The storage time of a packet in such a recirculating-type buffer is determined by the circulating number times the packet duration T. Both of the packet buffers shown in Figs. 2 and 3 are able to store multiple packets with the constraint that only one packet enters and leaves the buffer at a time. This constraint can be removed if WDM is invoked [15], [29], [33], [45].

The recirculating-type buffer is more flexible than the travelling-type buffer in that the packet storage time is adjustable by changing the circulation number [27], [46]. In principle, a recirculating-type buffer offers the capability of random access with storage time depending on the number of recirculations. On the other hand, the storage time of a travelling-type buffer is predetermined by the length of the optical fibre delay-lines. A problem with the recirculating-type buffer is that the signal has to be amplified during each circulation to compensate for the power loss [27], [46]–[48]. This results in accumulated spontaneous emission (ASE) noise from the optical amplifier and signal-level fluctuation, which eventually limits the maximum buffering time [46], [47], [49]. In addition, the gain from the optical amplifier has to be carefully controlled so that the product of gain and the loss in one circulation is slightly less than unity to prevent optical signal lasing.

In packet-switched networks, it is extremely important to ensure a low packet loss probability, while achieving a relatively high throughput [50]. This requires a large buffering capacity, depending on switch architecture. For example, an output-buffered packet switch requires a buffering capacity of more than 50 packets at each output to achieve a packet loss probability less than 10^{-6} at a traffic load of 90% [50]. In the context of photonic packet switching, WDM has been considered to increase the buffering capacity. Based on the principle of travelling-type or recirculating-type buffers, a number of WDM photonic buffers have been reported and demonstrated. These include random wavelength accessible recirculating loop buffers [29], [45], cascaded recirculating loop buffers [18], [49], and wavelength-routed buffers [33], [51]. More details are given in the next section.

4. Switch Architectures

Many different photonic packet switch architectures have been proposed and demonstrated (see [3], [4], [8], [15], [17]–[20], [27]–[33], for example). All of these use photonic means to perform packet buffering and routing; while electronics plays the important role in such functions as address processing, routing and buffering controls. Possible approaches to packet routing within these photonic packet switches include wavelength routing, broadcast-and-select routing, and space-switch based routing.

4.1 Wavelength-Routed Photonic Packet Switch

The first wavelength-routed packet switch discussed here is the photonic matrix switch proposed by Gabriagues and Jacob [30], [52]. This packet switch uses wavelength coding for packet routing and buffering. As shown in Fig. 4, it consists of three functional blocks: a packet encoding block, a buffering block and a packet demultiplexing block. The packet encoding block is composed of N optical tunable wavelength converters (TWCs) which each convert a packet to a new wavelength corresponding to its desired output, i.e. a packet addressed to the i-th output is assigned a wavelength λ_i. The buffering block consists of a N×K SOA-gate switch matrix followed by a set of K optical delay-lines whose lengths range from 0 to (K−1)-packet time. The SOA-gate switch matrix provides wavelength-encoded packets access to appropriate

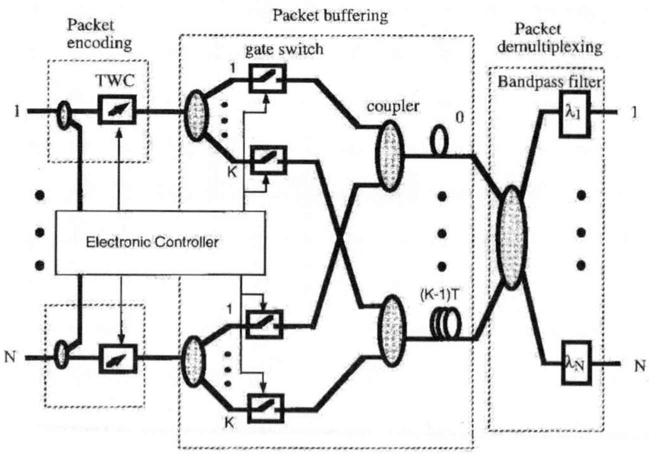

Fig. 4 Wavelength-routed photonic matrix switch [30].

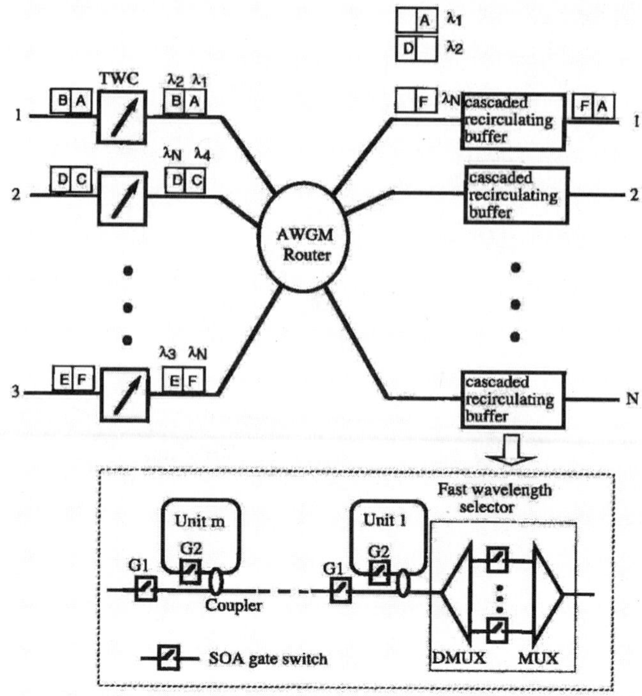

Fig. 5 Frontiernet photonic packet switch[18].

delay-lines in a way that packets that are destined for a given output leave the switch in a first-in first-out (FIFO) manner. The demultiplexing block consists of a K×N star coupler followed by a set of N bandpass filters, one at each output to select packets whose wavelengths are matched with its passband. In addition, there is an electronic controller to control the TWCs and space switches. It is interesting to note that, because packets are wavelength encoded from λ_1 to λ_N, the K delay-lines function as N FIFO buffers, each corresponding to one output and capable of storing up to K packets. One difficulty with this architecture is that the number of SOA gates required in the buffering block increases in proportion to the product of N and K, i.e., the switch size and the buffer size. Also, the optical power loss of a packet is proportional to NK^2 for N<K or N^2K for N>K. The value of K depends on the required packet loss probability and the traffic load.

K. Sasayama, et al., proposed another wavelength-routed packet switch, known as the Frontiernet packet switch [18], [28]. The architecture, as shown in Fig. 5, uses an arrayed-waveguide grating multiplexer (AWGM) [35], [36] in conjunction with wavelength conversion to perform packet routing. For a N×N packet switch, it only requires N wavelengths to fully interconnect N inputs with N outputs. The relationship between the inputs, the outputs and the wavelengths within an 4×4 AWGM is shown in Table 1. With reference to Table 1, this AWGM is capable of simultaneously routing up to N packets of different wavelengths to a given output. This is illustrated in Fig. 5, in which packets A and F from inputs 1 and N are simultaneously routed to output 1 at wavelengths λ_1 and λ_N. The function of the buffer at each output is to regulate packets such that only one packet leave the output at any time slot.

The buffer in the Frontiernet is shown at the bottom of Fig. 5. It consists of multiple recirculating loop units connected in a serial configuration, and followed by a fast wavelength selector (or tunable filter). Each recirculating unit consists of a length of optical fibre forming a loop, and two SOA gate switches and a 3-dB coupler. By appropriately turning on or off the two gate switches G1 and G2 at each loop unit, packets can be either coupled into or cleared from any fibre loop. For example, assuming all loop units are empty and three packets are entering the buffer at the start of time slot 1, these packets are fed into the loop unit 1 by turning on all gate switches G1, and switching off all gate switches G2 except for the gate switch G2 at the loop unit 1. At the same time, one packet is selected out by the fast wavelength selector. In time slot 2, new packets are fed into loop unit 2 only, and so on. The packets in the i-th loop unit are forwarded to the (i−1)-th loop unit if and only if no packet exists in the (i−1)-th loop unit.

Table 1 Relationship between inputs, outputs and wavelengths within a 4 × 4 AWGM.

		Output			
		1	2	3	4
Input	1	λ_1	λ_2	λ_3	λ_4
	2	λ_2	λ_3	λ_4	λ_1
	3	λ_3	λ_4	λ_1	λ_2
	4	λ_4	λ_1	λ_2	λ_3

Simulation has shown that each buffer requires at

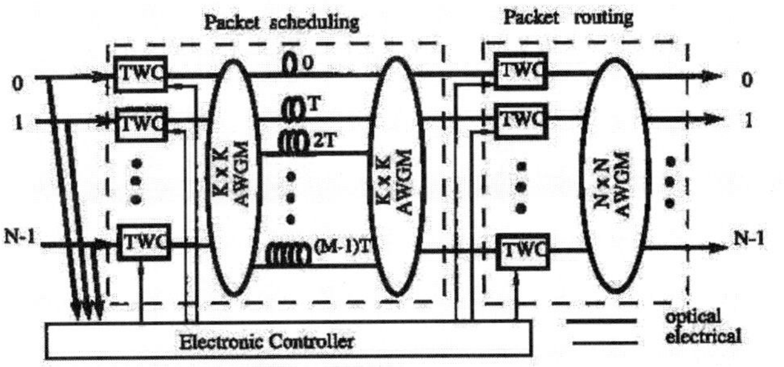

Fig. 6 Input-buffered wavelength-routed packet switch [33].

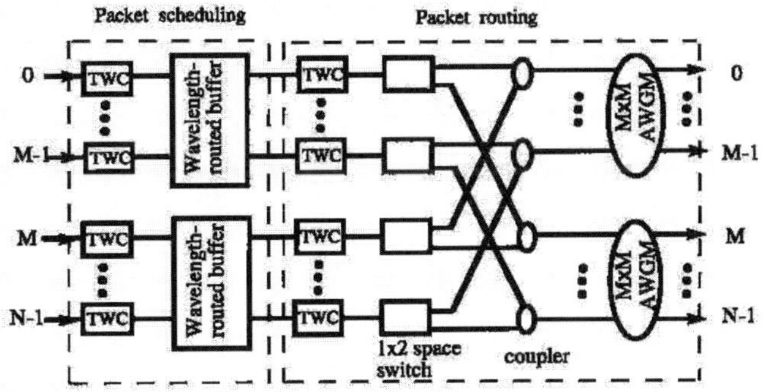

Fig. 7 Modified input-buffered wavelength-routed packet switch [33].

least 18 loop units to achieve a packet loss probability less than 10^{-6} for an 80% traffic load [18]. The complexity of this buffer structure is that an electronic controller (not shown in the figure) has to trace the number of packets in each loop unit, and to carry out forwarding of packets from one loop unit to the next loop unit, which could make the implementation complicated. Forwarding of packets could be avoided if the loop units are arranged in a parallel configuration.

In the above two switch architectures, switching of packets to its desired outputs is carried out before packet buffering. They effectively operate as an output buffered switch and hence achieve the best delay-throughput performance [50]. We have recently proposed a wavelength-routed input-buffered packet switch in which packet buffering is performed before packets are routed to their desired outputs [33], [51]. As a consequence, this packet switch has less complexity, which is in proportion to the switch size N, not N^2 or $N\log_2 N$, in comparison to other photonic packet switches [33]. This is achieved by using a new wavelength-routed packet buffer. The structure is shown in Fig. 6. The switch consists of two blocks: packet scheduling and packet routing. Similar to the Frontiernet packet switch, packet routing is performed by an AWGM in conjunction with a set of N tunable wavelength converters (TWCs), but no more than one packet will be directed to a given output of the AWGM at the routing block. Packet contention is prevented by packet scheduling in a way that each packet is allocated a minimum time delay subject to two conditions in any time slot: (i) no more than one packet is addressed to a given output of the routing block, (ii) only one packet can appear at any input of the routing block.

The packet scheduling block in Fig. 6 contains N TWCs and a wavelength-routed packet buffer. This wavelength-routed packet buffer consists of a pair of AWGMs connected by a set of fibre delay-lines of lengths ranging from 0 to (K−1)T. According to the routing principle of the AWGM (Table 1), a packet entering the buffer at the i-th input port will leave the buffer from the i-th output port after receiving a certain time delay determined by the packet wavelength. In an extreme case, a maximum of N packets from all the inputs can simultaneously access any single delay line without collision. Therefore, this wavelength-routed buffer is equivalent to N single-input single-output packet buffers, which share a common pool of fibre delay-lines through wavelength division multiplexing. Because no couplers or space switches are used, this packet switch has a potentially low power loss, independent of switch size. Simulation has shown that this wavelength-routed input-buffered packet switch can achieve a delay-throughput performance almost as good as the output-buffered packet switch and a very low packet loss probability [33], [51]. It has been shown that a set of 35 delay lines will be sufficient to obtain a packet loss probability less than 10^{-7} at a traffic load of 80%. Figure 7 shows how this architecture can be modified to double its switch capacity by adding additional blocks and 1×2 space switches, but without an increase in the number of wavelengths [33].

4.2 Broadcast-and-Select Packet Switches

The broadcast-and-select approach has been widely employed in photonic networks. With this approach, all the information from signal sources is multiplexed by a star coupler and distributed to all the receivers. Each receiver selects the desired packets. Both time-division multiplexing (TDM) and wavelength-division multiplexing (WDM) can be applied in this broadcast-and-select approach.

Shimatsu and Tsukada [32] have reported an ultrafast photonic ATM switch, known as the ULPHA switch, where an ultra high-speed optical TDM channel was used to broadcast ATM data cells. The architecture of this switch is shown in Fig. 8. The bit pattern of the electronic ATM cells is used to modulate the sequence of ultrashort optical pulses. Shrinking the interval between two consecutive pulses allows data cells from the different inputs to be time-multiplexed on a cell-by-cell basis and broadcast to all the outputs. The corresponding address data of cells are also time-multiplexed and broadcast to all the outputs, but at a different wavelength. Since an address data is much shorter than an ATM cell, the address data need not

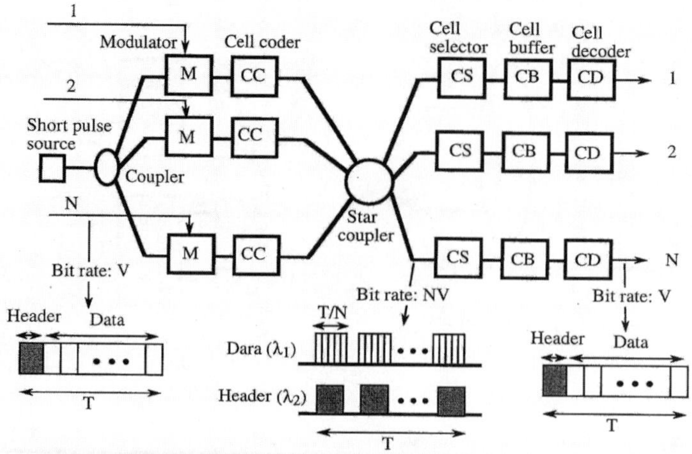

Fig. 8 ULPHA switch [32].

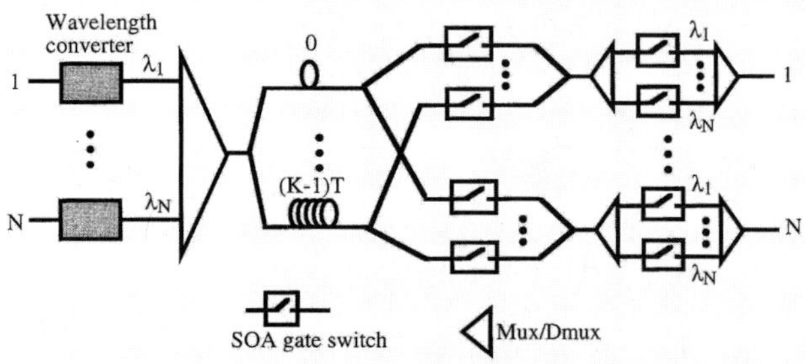

Fig. 9 Broadcast-and-select photonic packet switch [31].

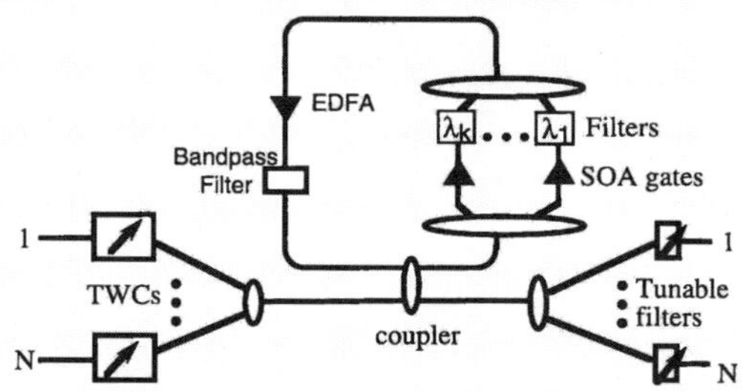

Fig. 10 Fibre loop memory switch [45].

to be time-compressed. This facilitates the detection of address data at each output of the switch. The difficulty with this packet switch is that a high-speed cell coder is required to compress cells at each input, and a cell decoder to decompress cells back into their original form at each output. To overcome this problem, an optical bit-interleaved multiplexing ATM switch was proposed in [53], in which bit-interleaved multiplexing was invoked, rather than cell-interleaved multiplexing. As a result, optical cell compression and decompression are not necessary. It is reported in [17] that by using both ultra high-speed TDM and WDM, the broadcast-and-select architecture can offer a very large switching throughput in the terabit per second range.

Chiaroni, et al. [31], have proposed and demonstrated another broadcast-and-select photonic packet switch, using multiple wavelengths to distinguish packets arriving at different inputs. The switch is shown in Fig. 9. Packets from different inputs are encoded on a different wavelength before they are combined at a star coupler, and then the combined WDM packets are broadcast and fed to a set of K optical delay-lines. Similar to the wavelength-routed packet switches (Figs. 4 and 6), this set of delay-lines is shared by all the packets. After propagating through the set of K delay-lines, each packet can receive any amount of delay ranging from 0 to (K−1) packet-times. Packets emerging at each delay-line are further broadcast to all of the output ports, where two sets of optical gate switches are used for packet selection. That is, at each output, the first set of gate switches selects a particular delay-line through which a desired packet arrives; the second set of gate switches in conjunction with a WDM demultiplexer-multiplexer pair chooses a wavelength at which the desired packet is encoded. In comparison to the above wavelength-routed packet switches, the advantage of this architecture is that it does not require any tunable components. But it has an optical power loss which is proportional to NK^2 and requires N(N+K) gate switches to perform packet selection.

In [45], Bendelli, et al., proposed a simple broadcast-and-select photonic packet switch using a random wavelength accessible recirculating loop buffer. This switch is shown in Fig. 10. It consists of N tunable wavelength converters, one at each input, a random wavelength accessible recirculating loop buffer, and N tunable filters, one at each output. The random wavelength accessible recirculating loop buffer is shared by all the packets from all the inputs. Before entering the buffer, each packet is assigned a wavelength not being used by the packets within the buffer. All the packets in the buffer are broadcast to all the outputs. The tunable filter at each output selects appropriate packets by tuning its wavelength. The packets which are selected by the tunable filters are cleared out of the buffer by turn-

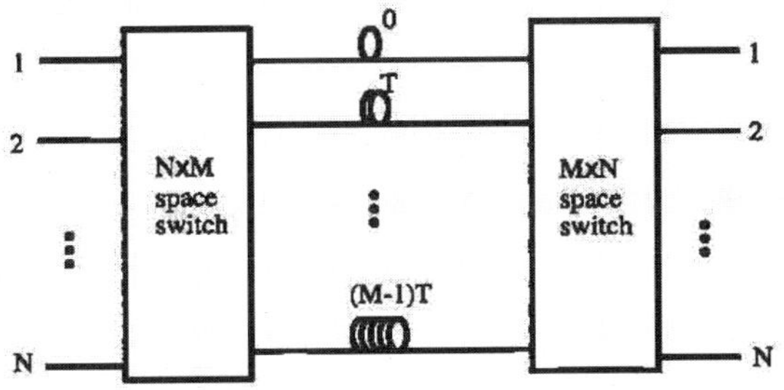

Fig. 11 Staggering switch [20].

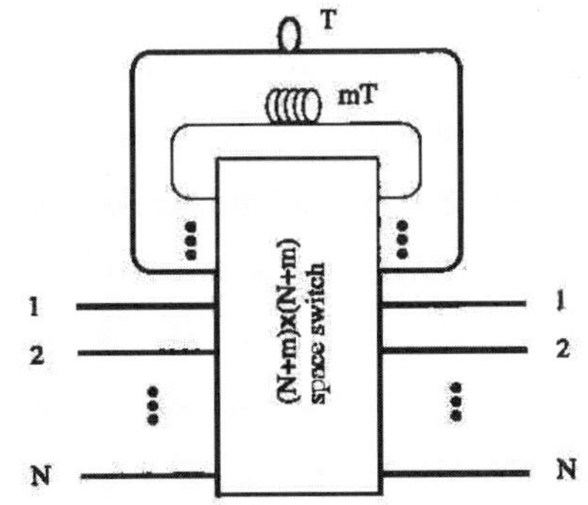

Fig. 12 Shared memory packet switch with recirculating loops [54].

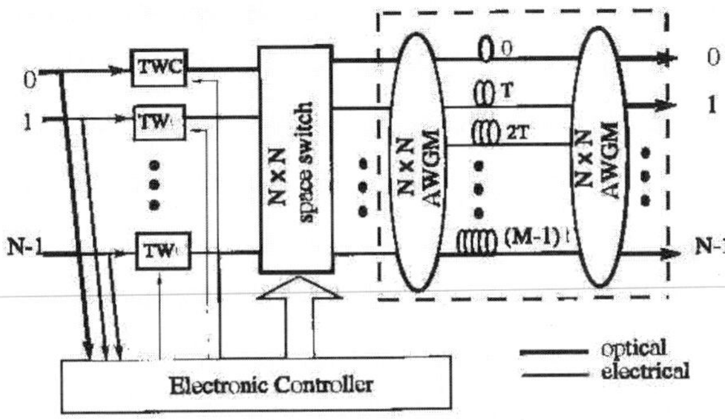

Fig. 13 Output-buffered photonic packet switch using a wavelength-routed packet buffer [33].

ing off the corresponding gate switches contained in the buffer. An electronic controller (not shown in the figure) maintains N packet queues each corresponding to one output of the switch. The capacity of this switch is limited by the tuning range of tunable lasers and filters, and the EDFA bandwidth. The solution is to use multiple loop buffers with a combination of optical space switches [4].

4.3 Space-Switch Based Photonic Packet Switches

Optical space-switches are key components in photonic packet switching systems. With either the wavelength-routed architecture or the broadcast-and-select architecture, optical space-switches are inevitable. A variety of optical space switches have been developed, including $LiNbO_3$ optical switches, InGaAsP/InP-based SOA gate switches, and silicon-silica based waveguide switches. In the context of photonic packet switching, they are required to have a switching speed in a range of several nanoseconds, low crosstalk, and low power loss. Generally, the switching time has to be less than 10% of a packet time to achieve a high utilization in transmission. Another important requirement is that they should be able to be monolithically integrated to reduce the cost and to increase the capacity. It is for these reasons that optical space switches made of SOA gates are most promising [9], [10]. We here describe several photonic packet switch architectures where optical space-switches play the major role in packet routing and buffering.

The Staggering switch [20] uses space switches to perform packet routing and buffering. The architecture of this switch is shown in Fig. 11. It consists of two nonblocking space switches that are interconnected by a set of optical delay-lines of different delay times. The first N×M space switch provides packets with access to the delay lines in a way that, at any time slot, no two packets arrive at the second space switch destined for the same output port, and hence packet contention is resolved. The second space switch routes the packets to their desired outputs. The delay-lines can also be arranged in a recirculating configuration [54], as shown in Fig. 12. A drawback with both of the above architectures is that packets arriving on the same input may leave the switch in the reverse order on the same output [20], [54].

Another space-switch based photonic packet switch, as described in [33], is shown in Fig. 13. Packet routing is performed by an N×N nonblocking space switch capable of broadcasting, while packet buffering is carried out by a wavelength-routed buffer in conjunction with a set of N tunable wavelength converters preceding the space switch. This is an output-buffered packet switch. There may be more than one packet destined for the same output in a time slot, resulting packet contention. However, this packet contention is resolved by wavelength conversion and wavelength-routing based buffering. More specifically, in each time slot, packets that are destined for the same output will be shifted to different wavelengths before they can be routed to the desired output by the optical space switch. Since these packets have been given different wavelengths by the wavelength converters at the input of the switch, they will receive different packet delays at the succeeding wavelength-routed buffer, so that only one packet will emerge at a given output of the switch in any time slot and hence packet contention is resolved. One difficulty with the space-switch based architecture is that the number of crosspoints (or SOA gates) required increases dramatically as the switch size grows.

In order to reduce the complexity of large buffers, a number of photonic packet networks and switches em-

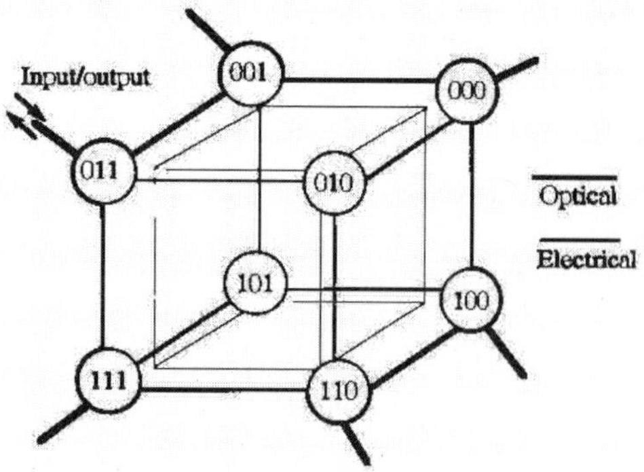

Fig. 14 HiPower photonic ATM switch [55].

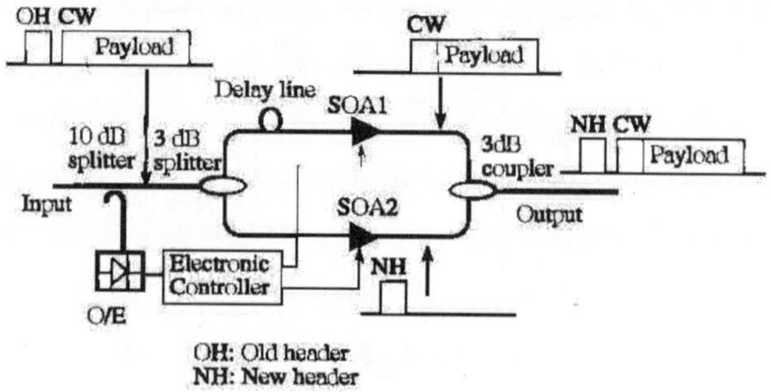

Fig. 15 Packet header replacement module [34].

ploy multihop and deflection routing approaches [23], [39], [63]. Here deflection routing means if more than one packet competes for a given output, only one packet is selected for the desired output, and the remaining packets are routed to other idle outputs, but they will eventually be routed to their desired outputs. Generally these packet networks and switches consist of small packet switch nodes, in which packets may have to traverse multiple nodes before they reach their destination. They can be configured with different topologies such as perfect shuffle or hypercube topologies [23], [39]. The advantage of such packet switches is that they can be implemented by using small space switches and only a small packet buffer is required if deflection routing is used. Figure 14 shows a multihop packet switch architecture reported in [55], using a hypercube topology. For an n-dimensional hypercube packet switch, it consists of $N=2^n$ nodes, each being a small $(n+1)\times(n+1)$ packet switch. Each node is allocated an n-bit binary address. The addresses of adjacent nodes differ by only one bit, which simplifies packet routing control [55]. A link between two adjacent nodes has a length of exactly one or multiple packet times, acting as a packet buffer. There is an electronic layer which has exactly the same topology as the optical layer to control packet routing. Because deflection routing is invoked, only a small packet buffer is required at each local input and output port [55], [56].

5. Packet Header Replacement and Synchronization

With connection-oriented packet-switched networks such as ATM-like networks, the routing information contained in the packet header is known as the virtual circuit number (VCI) and virtual path number (VPI). When packets transverse a switch node, the VCI and VPI are used to identify the output port of the switch by looking up a local routing table. To make the routing table as small as possible and to make the packet header as short as possible, the VCI and VPI contained in the packet header only have a local meaning. This means that the packet header has to be replaced by a new one before the packet is delivered to the next node. In the context of photonic packet switching, optical replacement of packet header is necessary [34]. So far two different methods have been proposed to optically implement the packet header replacement.

In 1993, Spring, et al. proposed and demonstrated a novel technique for optical packet header replacement. The technique introduces a continuous wave (CW) portion between the header and payload when a packet is generated at the source node. As a packet traverses through a switch node, the old packet header is removed from the packet. A copy of the CW portion is modulated by the new header data and is then affixed in the place of the old packet header. Figure 15 shows an implementation of this packet header replacement. It mainly consists of two SOA gates: SOA1 and SOA2. During the packet header period, SOA1 is turned off so as to remove the old packet header, while SOA2 is used to modulate a copy of the CW portion with the new header data and to block the other portions of the packet. An optical delay line of a length equal to the packet header time is inserted before SOA1 to adjust the timing so that the new packet header exiting SOA2 is exactly in the place of the old packet header.

An alternative technique is to use wavelength conversion [60]. With this technique, the payload of the original optical packet is used to optically modulate a CW light at a different wavelength, and while the new header data electronically modulates the same CW light within the place of the old header. Figure 16 shows an example of this technique, where cross-gain compression in SOAs is assumed to perform wavelength conversion. The architecture contains two SOA gates: SOA1 and SOA2. During the packet header period, SOA1 is turned off to block the old header. At the same time, the new header data electrically modulates the CW light at SOA2. The payload of the packet passes through SOA1 and then optically modulates the CW light at SOA2. This wavelength conversion approach does not require an additional CW portion inserted between the packet header and the payload, and hence can more effectively use the transmission bandwidth.

As discussed in Sect. 4, wavelength conversion can play an important role in routing and buffering. By in-

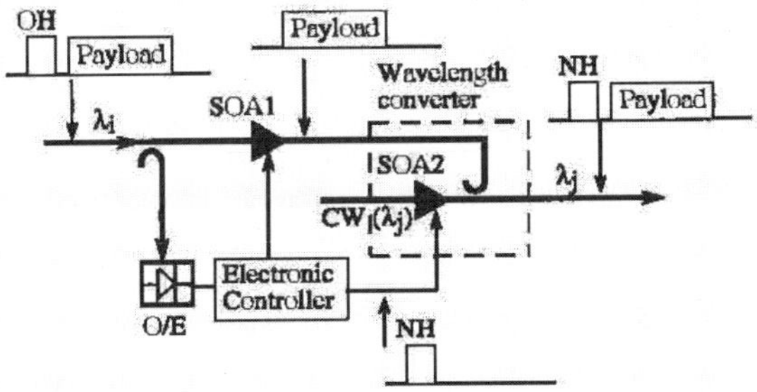

Fig. 16 Packet header replacement using wavelength conversion.

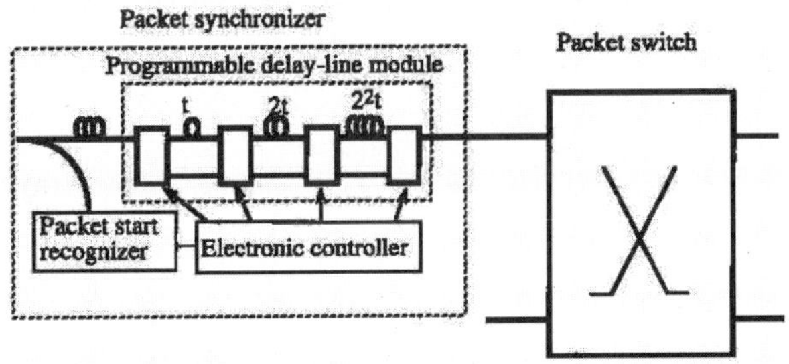

Fig. 17 Schematic of packet synchronization.

tegration of the packet header replacement with wavelength routing or buffering, an exclusive wavelength converter for use in the packet header replacement is not necessary. More importantly, some studies have also shown that optical wavelength conversion capable of regenerating optical pulse signals can significantly remove accumulated noise and crosstalk [58]–[60], and hence improve the cascadability of optically transparent packet switching nodes [57], [60].

Most of the proposed photonic packet switches assume that packets are synchronized at the switch inputs. This is because synchronous operation can simplify the implementation of packet contention resolution, and subsequent buffering and routing. However, all fibre spans interconnecting nodes cannot be designed as integer numbers of the packet duration T. In addition, a slow temperature change would induce variation in path length. Therefore, packet synchronization is necessary for large photonic packet switches. A general implementation of packet synchronization is shown in Fig. 17, which consists of a packet start recognizer and a programmable delay line module [61], [62]. The function of the packet start recognizer is to identify the packet start [61]. Once the packet start is detected, the arrival time of packets can be adjusted by setting the appropriate path within the programmable delay line module [61], [62]. The programmable delay line module can be implemented by a sequence of delay lines and optical space switches, as shown in Fig. 17. The resolution of tunable time is restricted by the number of stages used in the delay line module. If the maximum delay is D and the number of stages is n, the resolution of tunable time is given by $D/2^n$. A fine tuning of delay time can be realized by a tunable wavelength converter and a length of high dispersion fibre [62].

Packet synchronization may substantially increase the hardware complexity. For photonic packet networks that use small switch nodes and multihop routing approaches, asynchronous operation would be useful. This has been demonstrated in a 2×2 input-buffered photonic packet switch [27].

6. Conclusion

Recent advances in photonic technologies and the anticipation of multi-functional telecommunication networks with a very large switch capacity in the terabit per second range have been the driving force in research and development of photonic packet-switched networks. We have reviewed various approaches and techniques to photonic packet switching. It has been shown that the lack of optical equivalents to electronic random access memories and electronic processing capabilities has resulted in the hybrid optoelectronic implementation of photonic packet switching systems. The switching and routing fabric has generally been implemented optically, in order to capitalize on the optical transparency and large bandwidth; while electronic and optoelectronic components remain the important role in such functions as address processing, switching and buffering controls. To achieve satisfactory performance, significant efforts are required in developing key devices and components such as semiconductor optical amplifiers, optical gates and space switches, wavelength converters, fast tunable lasers and filters, and wavelength routers. Along with the developments, new approaches to buffering are needed. Photonic implementation of packet header replacement and synchronization is another key in the implementation of truly optically transparent packet switching systems.

Acknowledgement

This work is supported by the Australian Research Council under the Grant A49702803.

References

[1] Special Issue on Broad-Band Optical Networks, IEEE J. Lightwave Technol., vol.11, no.5/6, May/June 1993.

[2] Special Issue on Multiwavelength Optical Technology and Networks, IEEE J. Lightwave Technol., vol.14., no.6, June 1996.

[3] F. Masetti, et al., "ATMOS (ATM optical switching): Results and conclusions of the RACE R2039 project," Proc. ECOC'95, pp.645–652, 1995.

[4] F. Masetti, et al., "High speed, high capacity ATM optical switches for future telecommunications transport net-

works," IEEE J. Select. Areas Commun., vol.14, pp.979–998, 1996.

[5] R.S. Tucker, "Photonic packet switching," OFC'95, Tutorial, pp.4–28, Feb. 1995.

[6] D.J. Blumenthal, et al., "Photonic packet switches: Architectures and experimental implementations," Proc. IEEE, vol.82, pp.1650–1667, Nov. 1994.

[7] M. Renaud, et al., "Network and system concepts for optical packet switching," IEEE Commun. Mag., April 1997.

[8] P.R. Prucnal, "Optically processed self-routing, synchronization, and contention resolution for 1-d and 2-d photonic switching architectures,"IEEE J. Quantum Electron., vol.29, pp.600–612, Feb. 1993.

[9] R.F. Kalman, L.G. Kazovsky, and J.W. Goodman, "Space division switches based on semiconductor optical amplifiers," IEEE Photon. Technol. Lett., vol.4, pp.1048–1051, Sept. 1992.

[10] E. Almstrom, et al., "Experimental and analytical evaluation of packaged 4×4 InGaAsP/InP semiconductor optical amplifier gate switch matrices for optical networks," IEEE J. Lightwave Technol., vol.14, pp.996–1004, June 1996.

[11] T. Durhuus, et al., "All-optical wavelength conversion by semiconductor optical amplifiers,"IEEE J. Lightwave Technol., vol.14, pp.942–954, June 1996.

[12] B.S. Glance, et al., "New advances on optical components needed for FDM optical networks," IEEE J. Lightwave Technol., vol.11, pp.882–889, May/June 1993.

[13] J. Wiesenfeld, et al., "Wavelength conversion at 10 Gb/s using semiconductor optical amplifier," IEEE Photon. Technol. Lett. vol.5, pp.1300–1303, Nov. 1993.

[14] A. Talneau, et al., "Accurate wavelength spacing and fast switching in a four channel integrated source emitting at 1.55 mm," ECIO'97, Paper EWE3, Stockholm, April 1997.

[15] S.L. Danielsen, et al., "WDM packet switch architectures and analysis of the influence of tunable wavelength converters on performance," IEEE J. Lightwave Technol. vol.15, pp.219–227, Feb. 1997.

[16] M. Renaud, "Key components for optical packet switching," Proc. of Photonics in Switching, EFD1, Stockholm, Sweden, April 1997.

[17] W.D. Zhong, M. Tsukada, K. Yukimatsu, and Y. Shimatsu, "Terahipas: A modular and expandable terabit/second hierarchically multiplexing photonic ATM switch architecture," IEEE J. Lightwave Technol., vol.12, pp.1307–1315, 1994.

[18] K. Sasayama, Y. Yamada, K. Habara, and K. Yukimatsu, "FRONTIERNET: Frequency-routing-type time-division interconnection network," IEEE J. Lightwave Technol., vol.15, pp.417–429, March 1997.

[19] D. Chiaroni, P. Gavignet-Mortin, and A. Jourdan, "Theoretical feasibility analysis of a 256×256 ATM optical switch for broadband applications," Proc. ECOC'93, pp.485–488, 1993.

[20] Z. Hass, "The staggering switch: An electronically controlled optical packet switch," IEEE J. Lightwave Technol., vol.11, pp.925–936, May/June 1993.

[21] R.M. Fortenberry, Y. Cai, and R.S. Tucker, "Optically Transparent Nodes for Photonic Packet-Switched Ring Networks," Electron. Lett., vol.29, pp.417–418, Feb. 1993.

[22] Y. Cai, R.M. Fortenberry, and R.S. Tucker, "Demonstration of photonic packet-switched ring network with optically transparent nodes," IEEE Photon. Technol. Lett., vol.6, pp.1139–1141, Sept. 1994.

[23] A.S. Acampora and M.J. Karol, "An overview of lightwave packet networks," IEEE Network, vol.3, pp.29–41, Jan. 1989.

[24] R.A. Barry, et al., "All-optical network consortium—Ultrafast TDM networks," IEEE J. Select. Areas Commun., vol.14, pp. 999–1013, 1996.

[25] S.W. Seo, K. Bergman, and P.R. Prucnal, "Transparent optical networks with time-division multiplexing," IEEE J. Select. Areas Commun., vol.14, pp.1039–1051, 1996.

[26] A. Jajszczyk and H.T. Mouftah, "Photonic fast packet switching," IEEE Commun. Mag., pp.58–65, Feb. 1993.

[27] J. Spring and R.S. Tucker, "Photonic 2×2 packet switch with input buffers," Electron. Lett. vol.29, pp.284–285, Feb. 1993.

[28] K. Sasayama, K. Habara, W.D. Zhong, and K. Yukimatsu, "Photonic ATM switch using frequency-routing-type time-division interconnection network," Electon. Lett., vol.29, pp.1778–1779, 1993.

[29] A.de Bosio, et al., "ATM photonic switching node architecture based on frequency switching techniques," Proc. Photonic Switching, pp.226–228, 1990.

[30] J.M. Gabriagues and J.B. Jacob, "Photonic ATM switching matrix based on wavelength routing," Proc. Photonic Switching, p.2H2, 1992.

[31] D. Chiaroni, et al., "A novel photonic architecture for high capacity ATM switching applications," Proc. Photonics in Switching, p.PThC3, 1995.

[32] Y. Shimazu and M. Tsukada, "Ultrafast photonic ATM switch with optical output buffers," IEEE J. Lightwave Technol., vol.10, pp.265–272, 1992.

[33] W.D. Zhong and R.S. Tucker, "A wavelength-routing based photonic packet buffer and its application in photonic packet switching systems," J. Lightwave Technol, vol.16, pp.1737–1745, Oct. 1998.

[34] J. Spring, R.M. Fortenberry, and R.S. Tucker, "Photonic Header Replacement for Packet Switching," Electron. Lett., vol.29, pp.1523–1524, Aug. 1993.

[35] H. Takahashi, et al., "Arrayed-waveguide grating for wavelength division multi/demultiplexer with nanometre resolution," Electron. Lett., vol.26, pp.87–88, 1990.

[36] C. Dragone, "An N×N optical multiplexer using a planar arrangement of two star couplers," IEEE Photon. Technol. Lett., vol.3, no.9, pp.896–899, Sept. 1991.

[37] P. Gambini, G. Bendelli, B. Bostica, and M. Puleo, "Photonic packet switching based on fibre memories," Proc. Photonics in Switching, pp.2–3, 1996.

[38] K.Y. Eng, "A photonic knockout switch for high-speed packet networks," IEEE J. Select. Areas Commun., vol.6, pp.1107–1116, Aug. 1988.

[39] M.S. Goodman, "Multiwavelength networks and new approaches to packet switching," IEEE Commun. Mag., vol.27, pp.27–35, Oct. 1989.

[40] A. Cisneros and C.A. Brackett, "A large ATM Switch based on memory switches and optical star couplers," IEEE J. Select. Areas Commun., vol.9, pp.1348–1357, 1991.

[41] K. Kasahara, et al., "Double heterostructure optoelectronic switch as a dynamic memory with low power consumption," Appl. Phys. Lett., vol.52, pp.679–681, Feb. 1988.

[42] S. Suzuki and K. Kasahara, "Electro-photonic devices for gigabit networks," IEEE LTS, pp.36–40, Aug. 1992.

[43] I. Chlamtac and A. Fumagilli, "Quadro-star: a high performance optical WDM star network," IEEE Trans. Commun., vol.42, no.8, pp.2582–2591, Aug. 1994.

[44] T. Maeda, T. Nishi, and S. Kuroyanagi, "An optical fiber traveling-type buffer memory for photonic atm switching networks," Proc. OEC'94, pp.264–265, July 1994.

[45] G. Bendelli, et al., "Photonic ATM switch based on multiwavelength fibre-loop buffer," OFC'95, vol.8, pp.141–142, Feb. 1995.

[46] Y. Yamada, K. Sasayama, and K. Habara, "BER measurements in frequency-division-multiplexed optical loop buffer

at data rate of 2.5 Gbit/s," Electron. Lett., vol.30, pp.1869–1870, Oct. 1994.

[47] A. Dickson and R.S. Tucker, "Noise accumulation in fibre loop memories," Proc. 17th Australian Conference on Optical Fibre Technology, (postdeadline paper), p.4, Nov. 1992.

[48] M. Calzavara, et al., "Resolution of ATM cell contention by multiwavelength fibre loop memory," Proc. ECOC'94, pp.567–570, Sept. 1994.

[49] Y. Yamada, K. Sasayama, K. Habara, and K. Yukimatsu, "FDM optical-loop buffer and its applications to a photonic ATM switch: Frontiernet," OFC'95, vol.8, pp.140, Feb. 1995.

[50] M.G. Hluchyj and M.J. Karol, "Queueing in high-performance packet switching," IEEE Select. Areas Commun., vol.6, pp.1587–1597, Dec. 1988.

[51] W.D. Zhong and R.S. Tucker, "A wavelength-routing based photonic packet buffer for photonic packet switching," Proc. OECC'98, paper 16A3-3, Japan, 1998.

[52] J.M. Gabriagues, et al., "Design and implementation of a gigabit ATM photonic switching matrix," J. High Speed Networks, vol.4, no.4, Oct. 1995.

[53] W.D. Zhong, et al., "Optical bit-interleave multiplexing ATM switch," Electron. Lett., vol.31, pp.995-996, Japan, June 1995.

[54] M. Karol, "Shared memory optical packet (ATM) switch," SPIE, Multigigabit Fibre Communication Systems, vol.2024, pp.212–222, 1993.

[55] T. Matsunaga, "Sorting-based routing algorithms of a photonic ATM switch: HiPower," IEEE Trans. Commun., vol.41, pp.1356–1363, Sept. 1993.

[56] A. Misawa and T. Matsunaga, "Optical input buffers for the HiPower photonic ATM switch: Analysis and experiments," IEEE J. Lightwave Technol., vol.14, pp.2173–2183, 1996.

[57] D. Chiaroni, et al., "New regeneration functionalities of all-optical wavelength converters for packet-switch applications," Proc. Photonics in Switching, pp.10–11, 1996.

[58] K. Inoue and K. Oda, "Noise suppression in wavelength conversion using a light-injected laser diode," IEEE Photon. Technol. Lett., vol.7, pp.500–501, 1995.

[59] T. Durhuus, et al., "Penalty free all-optical wavelength conversion by SOA's in Mach-Zehnder configuration," Proc. ECOC'93, pp.129–132, 1993.

[60] D. Chiaroni, et al., "An all-optical regenerative interface for the cascadeability of photonic packet-switching nodes," Proc. Photonics in Switching, pp.54–57, 1997.

[61] L. Zucchelli, M. Burzio, and P. Gambini, "New solutions for optical packet delineation and synchronization in optical packet switched networks," Proc. ECOC'96, pp.3.301–3.304, 1996.

[62] M. Burzio, et al., "An optical cell synchronizer for packet switched nodes," Proc. Photonics in Switching, pp.12–17, USA, March 1995.

[63] A. Bononi and P.R. Prucnal, "New structures of the optical node in transparent optical multihop networks using deflection routing," Proc. IEEE INFOCOM'94, paper 3d4, 1994.

High-Performance Backbone Network Technology edited by Naoaki Yamanaka
ISBN: 0-8247-5321-6 DOI: 10.1081/(120025655)
Published 2004 by Marcel Dekker, Inc.

ATM VP-Based Economical Transport Network Architecture for Multi-Reliability and Broadband Integrated Service Infrastructure

Yoshihiko Uematsu, Hiroshi Ohta, Ryoichi Iwase, Koji Watanabe, Masayuki Matsuda, Seiichi Takagi, and Toshinori Tsuboi

SUMMARY The ATM Virtual Path(VP)-based transport network is a candidate for the future multi-reliability and broadband integrated service infrastructure. This paper compares the performance attributes of the VP transport network with those of the conventional SDH network and future optical path-based transport network. This paper shows that the VP-based network is superior in terms of path-grooming efficiency and connection capability, and will still play an important role when WDM technologies are introduced to carrier networks. This paper also describes the recently fabricated VP Cross Connect(VP-XC) system configuration and the VP Automatic Protection Switching(APS) performance of the XC.

key words: *ATM, transport network, virtual path, cross-connect, protection switching*

1. Introduction

Network traffic carrying data or video services has explosively increased in recent years and it will continue to grow. Moreover, network users will demand various connection bandwidths, various grades of service reliability and QoS, and various types of connections. Thus the carriers require a unified network infrastructure that accommodates all kinds of services efficiently and economically. ATM is a transport technology that handles all types of traffic using a common cell format[1]. Many studies have looked at applying the ATM transport capability to multimedia traffic. ATM can also support conventional voice services and leased-line services in an economical manner so is a key technology in constructing a unified network infrastructure.

This paper describes the ATM Virtual Path based transport network(VP-NW) architecture that can support various transport services economically. Performance attributes of VP-NW are compared with those of the SDH High-Order-Path based network(HOP-NW) and Optical(Wavelength) Path based network(WP-NW). It is concluded that VP-NW has superior performance in terms of grooming efficiency and connection capability, and so will still play an important role in the future WDM-introduced transport networks. We also describe a scalable structure for the VP Cross Connect system(VP-XC) that we have already fabricated and show VP automatic protection switching(VP-APS) performance of a testbed XC.

Manuscript received May 28, 1999.
Manuscript revised August 26, 1999.

† The authors are with NTT Network Service Systems Labs., Yokosuka-shi, 239-0847 Japan.

2. Future Transport Network Architectures

Future carrier network will support various services with;
- multi-protocol (Native ATM, Native IP, Voice,...)
- multi-reliability (route-redundant services or single-routed services)
- multi-QoS (QoS guaranteed or Best Effort).

The requirements for the transport network are as follows:
- It should support full mesh connections among all service nodes in nation-wide regions.
- Reliability and QoS of each connection are well-controlled.
- It should be a simple one-layered network regardless of native(access) protocols.

ATM-based transport network can satisfy the requirements and so it's a leading candidate for the future unified transport network.

The ATM transport concept has two layered connection types: virtual path connection (VPC) and virtual channel connection(VCC). In the ATM based public network, ATM service nodes(switching systems or subscriber line terminals) will mainly accommodate permanent VPC services, permanent VCC services and switched VCC services. It is conceivable that the permanent services are mainly leased-line services and that enterprise users will use the services to interconnect their intra-enterprise networks. Individuals generally use economical switched services. So each service node must offer VP/VC handling functions.

The role of transport networks is to interconnect the service nodes economically and reliably. To this end, the ATM transport network should establish full-mesh permanent VPCs; there are two main reasons:
- If the transport network supports VCC-level routing, the operation system would have to administer a connection database holding both VPCs and VCCs. The database would be changed frequently since most VCCs are switched connections. Transport network administration would be simpler if it only permanent VPCs were handled.
- If the network supported VCC-level routing, some network elements would need to offer VP/VC handling functions and so would have to be equipped with VP/VC switching functions and VP/VC layered OAM functions. This would lead to higher transport network cost. Meanwhile NE configuration would be quite simple if each

NE handled only permanent VPCs.

This paragraph shows that the full-mesh VPC network can support the sufficient number of connections. In the logical full-mesh VPC network, each service node has VP/VC handling functions and grooms many VCCs having the same destination node into a permanent VPC. A VPC can accommodate 65536 VCCs in terms of 16-bit VCI field. So if a 2.4Gbit/s VPC conveys 65536 VCCs, the average bit rate of each VCC is 37 kbit/s, which is lower than the typical bit-rate required by individuals. So lack of VPC capacity will come sooner than lack of VCC connection capability. As for VPC connectivity, each transmission line can support 4096 VPCs in terms of 12-bit VPI field. Each service node also accommodates permanent VPCs the end points of which are the users' terminals. If there are 101 service nodes in the network, each service node must use 100 VPCs to accommodate VCCs. Thus the service node can offer 3996 VPCs for leased-line services even if it is connected to the transport network via only one transmission line. Moreover, if the network uses a 10 Gbit/s transmission line with fully concatenated payload (STM-64/VC-4-64c) to convey 4096 VPCs, the average bit rate of each VPC is 2.4 Mbit/s. The value is lower than the typical bit-rate of leased-line services for enterprise users. Thus the number of VPCs that the transmission line can support is sufficient, and lack of link capacity will come sooner than lack of connections in any accommodation case. Consequently, the full-mesh VPC network can support sufficiently large number of permanent and switched connections.

There are two basic types of physical transport network topologies: those that directly connect each service node using full-mesh optical fibers, and Cross Connect system (XC) based path grooming networks. The network construction cost of the latter is generally lower than that of the former since the latter offers lower transmission line (optical fibers and regenerators) cost [2]. There are three XC based transport network configurations that accommodate full-mesh permanent VPCs. They are the VP Cross Connect (VP-XC) based network (VP-NW), the SDH High-Order-Path Cross Connect (HOP-XC) based network (HOP-NW), and the Wavelength Path Cross Connect (WP-XC) based network (WP-NW).

The VP-NW proposed herein, shown in Fig. 1, grooms the STM/ATM traffic effectively. VP-XC has STM-based interoffice interfaces and the VP cross connect function. Permanent VPCs are set-up between ATM service nodes. In the development stage of ATM based services, STM based service nodes will already exist, so ATM based service nodes and STM based service nodes will co-exist. In this case, it is not cost effective to construct separated networks. We have already proposed that the path CLAD converts STM paths (VC-3/VC-11) into permanent VPs at the edge of the VP-NW, and VP-XC simply handles STM traffic as VPs [3].

The HOP-NW configuration and the HOP-XC are summarized in Fig. 2. This fully utilizes the conventional SDH High Order Path Cross Connect Systems (HOP-XCs). As shown in Fig. 2(b), HOP-XCs generally connect the conventional STM service nodes to each other and HOPs are set-up between STM service nodes. The basic HOP-XC has the VC-3/VC-4 cross connect function and has STM-based interoffice interfaces. As for ATM traffic, HOPs are set-up between HOP-XCs, and HOP-XCs should also have VP switching functions in order to achieve more efficient VPC grooming [4]. This type of XC, sometimes called the hybrid XC, has the STM based transport capability and additional ATM based grooming capability. It may be a cost-effective configuration in the initial stage of ATM service introduction.

Figure 3 shows the WP-NW and WP-XC configurations; both can be realized by the progress anticipated in WDM technologies, and both will explosively increase the network throughput [5]. Each WP is established between WP-XCs, which have the functions of cross connecting WPs and grooming SDH/ATM traffic into WPs. Since the granularity of WPs is large and WP-NW has poor connectivity (discussed in the next Section), the WP grooming function will be necessary to fully utilize the WP

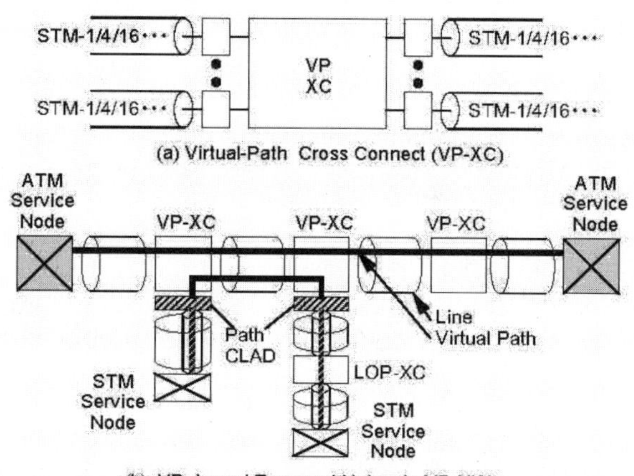

Fig. 1 VP based transport network.

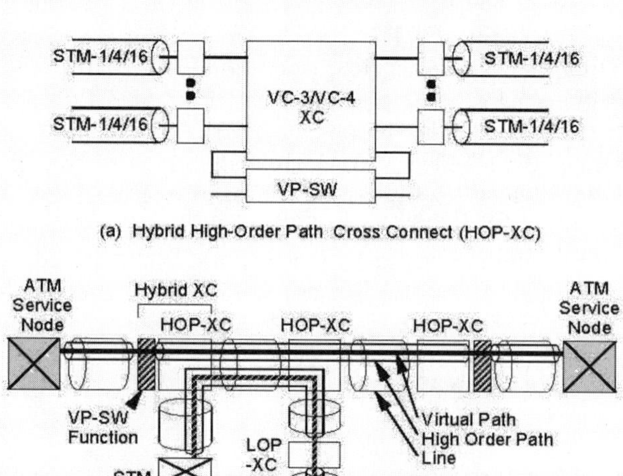

Fig. 2 HOP based transport network.

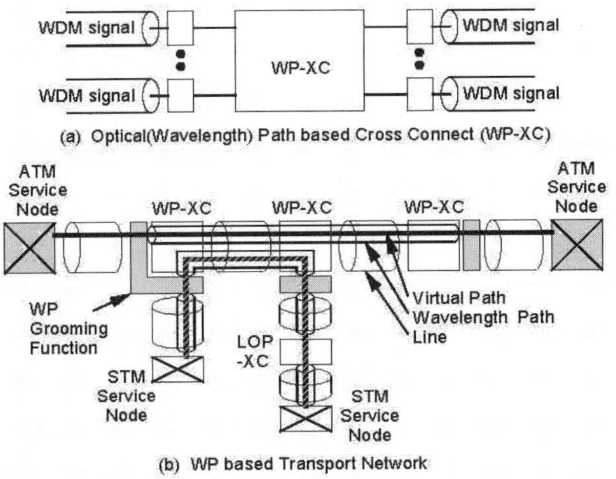

Fig. 3 Optical (Wavelength) path based transport network.

capacities.

Recently the number of internet users has exploded and the amount of native IP traffic that carriers should convey will continue to grow. So IP-packet based network and IP over WDM network may be the important candidates for the unified transport network. If VP-XCs are replaced by IP routers and all the traffic between service nodes is IP-packet based in Fig. 1, that gives the figure of IP based transport network. Also in Fig. 3, if all the traffic between service nodes is IP-packet based, it presents IP over WDM network. Although the IP-based transport networks save the transmission capacity overhead of ATM-layer and ATM adaptation layer, traffic engineering (route control, accommodation design, performance measurement, and so on) is difficult in the present IP-based routing mechanism. Thus they cannot offer QoS guaranteed services if the traffic density is high and so cannot keep the quality of present native ATM or voice services when accommodating the services. This paper presents VP-NW, HOP-NW, and WP-NW as broadband transport network candidates in terms of supporting multi-protocol, multi-reliability and multi-QoS services, and the necessity of timely deployment. Recently multiprotocol label switching (MPLS) with some explicit routing protocols have been recognized as a QoS guaranteed IP traffic transfer scheme, and of course they can run on VP-NW, HOP-NW, and WP-NW described herein. In case of supporting MPLS in those network candidates, it will be necessary for the service nodes to have label switching control components, but it will not be necessary for the XCs if the transport layer supports full-mesh connections among the service nodes.

3. Advantages and Role of ATM-Transport Network

This section shows that VP-NW is the best way to construct economical transport networks by virtue of its higher path grooming efficiency, given that TDM-based transmission technologies are dominant and will continue develop. It also shows that if WP-NW is introduced, VP-NW will still be necessary to compensate its rough WP granularity and poor connection capability.

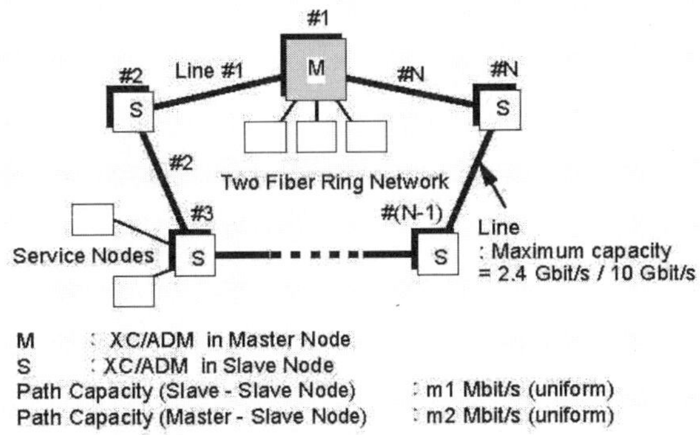

Fig. 4 Model network topology.

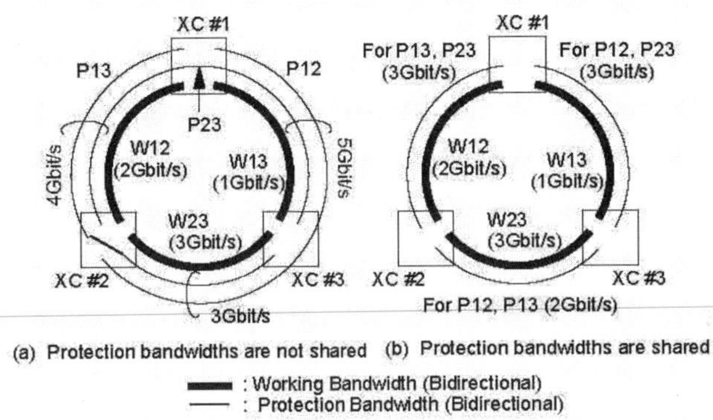

Fig. 5 Required bandwidth reduction effect using shared protection scheme.

3.1 Grooming Efficiency Comparison between SDH and ATM Transport Networks

In the HOP-NW, the finest granularity in path capacity between HOP-XCs is 50 Mbit/s; the VP-NW offers finer granularity and so offers higher link utilization than the HOP-NW. This section confirms this merit with a quantitative estimation. Figure 4 shows the network model that is based on the N node 2-fiber ring topology. Let us consider the situation that paths of m1 Mbit/s capacity are established uniformly between all slave nodes and m2 Mbit/s paths are established between the master node and each slave node. We consider only CBR(Constant Bit Rate) traffic. It's more advantageous to the VP-NW to take other transfer modes such as VBR (Variable Bit Rate), ABR (Available Bit Rate), and UBR (Unspecified Bit Rate) into account. This is because the working connections of those transfer modes share line capacities by statistical multiplexing in the VP-NW, meanwhile in the HOP-NW the connections share each 50Mbit/s path capacity. The VP-NW is assumed to employ VP level automatic protection switching (VP-APS) to recover traffic from transmission line failures [6], [7]. For each VP, protection switching is performed at the end point XCs that add/drop the VP to/from the ring network(path-based switching). All protection connections are routed such that they do not overlap their corresponding working routes. 1:1 protection switching is

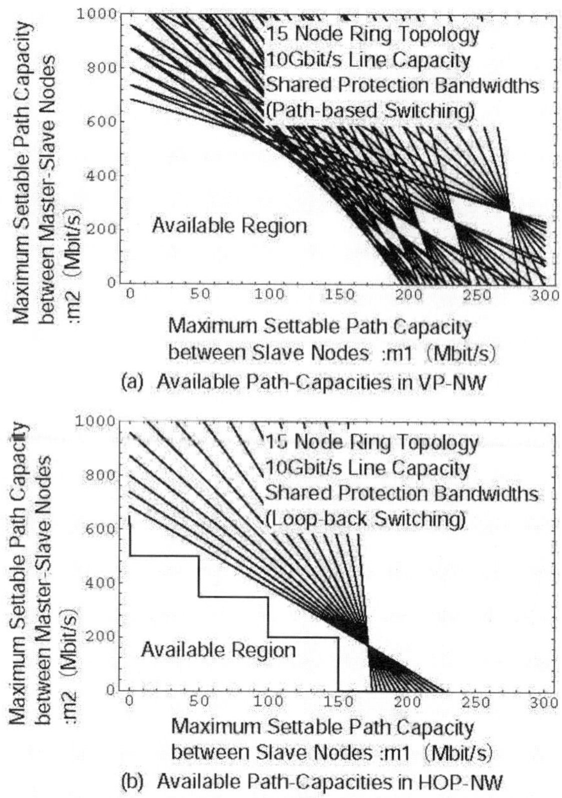

Fig. 6 Maximum settable path capacity comparison.

assumed and a protection bandwidth sharing scheme is adopted to gain grooming efficiency of highly-reliable paths. Figure 5 shows the protection bandwidth sharing scheme for the case of a 3-node ring. In this estimation, the shared protection bandwidth is designed on condition that all the traffic should be saved from any single transmission line failure. HOP-NW adopts the Multiplex Section Shared Protection Ring (MS-SPRING) for network protection [8]. In MS-SPRING, the XC/ADMs nearest to the line failure switch all failed HOPs to the opposite direction (loop-back switching). In MS-SPRING, protection bandwidth sharing scheme can also used to gain grooming efficiency of highly-reliable paths.

The maximum m1, m2 values of VP-NW and HOP-NW are shown in Fig. 6(a) and (b) for the case of 10 Gbit/s transmission line capacity. In Fig. 6(a), each line shows the maximum m1 and m2 relationship on the condition that

$$Wi + Pi(j) \leq \text{Line Capacity} \quad (1 \leq i,j \leq N,\ i \neq j) \quad (1)$$

where Wi is the total bandwidth of all working VPCs in transmission line #i and Pi(j) is the protection bandwidth for transmission line #i required to restore all VPCs when a failure occurs on transmission line #j. Accordingly, in Fig. 6(a), the area under all lines indicates the m1 and m2 values that are available. Maximum m1 and m2 values for HOP-NW, shown in Fig. 6(b), were calculated in the same manner as that for VP-NW and the HOP granularity of 50Mbit/s (VC-3) was taken into account. It is obvious VP-NW offers much larger values than HOP-NW. The reasons for this are as follows:

-Due to the rough granularity, the minimum increment/decrement unit of path-capacity in HOP-NW is 50Mbit/s, meanwhile the practical increment/decrement demands are not such skipping values. So there are some regions in the m1-m2 plane that VP-NW can cover but HOP-NW cannot cover.

-Protection connection in HOP-NW is always longer than that in VP-NW, if working connections are the same length in both cases. According to this fact, Pi(j) in VP-NW is always equals to or smaller than Pi(j) in HOP-NW. Detail proof is omitted here, but main reasons are as follows: In HOP-NW, setting-up a connection means occupying some bandwidth. So in MS-SPRING, protection bandwidth sharing means protection connection sharing. In the connection sharing, it is necessary for all the related XCs to know the other XCs switching status for the purpose of avoiding misconnection. MS-SPRING achieves the connection sharing by the loop-back switching at the simply two XCs connected to the failed line and the switching status notification to the other nodes via K1/K2 bytes of the Section OverHead [8]. In VP-NW, we can design bandwidth allocation independently of connection establishment. If dedicated protection connection is established for each working connection, we do not have to consider the misconnection. So it is easy to implement the path-based switching with sharing protection bandwidths in the VP-NW.

Consequently the path grooming efficiency of VP-NW is much higher than that of HOP-NW. This leads to cost reduction of each accommodated paths in the long run.

3.2 Role of ATM-XCs in Future WDM-Introduced Transport Networks

This subsection discusses WP-NW capability and the roles of VP-NWs and WP-NWs. Let us consider a nation-wide transport network. Traffic is very dense in metropolitan areas and rather sparse in regional areas. The discussion in 3-1 confirmed that the VP-NW is the best solution in the long run, provided that the VP-XC can accommodate the traffic. However in dense traffic areas, existing VP-XC and/or transmission line may have insufficient throughput. In such dense traffic regions, the WP-XC based transport network may be the more economical configuration. In introducing WPs, it is a very expensive configuration for WPs to support full-mesh connections among all the service nodes because all the service nodes should have dense-WDM-based interfaces and the granularity of WP is very rough (2.4Gbit/s or more). Thus Fig. 3 configuration will be a reasonable configuration, where WPs are established between WP-XCs and WP-XCs groom VPs into WPs. It's a two-layered transport network (VP/WP-NW) and VP-layer supports the full mesh connections between the service nodes.

Table 1 shows the minimum number of connections per link in the N node ring network topology required to achieve

Table 1 Full-mesh connection capabilities of ATM/SDH/WP transport systems.

	No. of Working Connections	Total(W+P) No. of Connections(Dedicated)*	Total(W+P) No. of Connections(Shared)**
Required No. of connections per Line(***)	$\frac{N^2-1}{8}$ (28)	$\frac{N(N-1)}{2}$ (105)	$\frac{N^2-1}{4}$ (56)
ATM-XC/ADM	○ Supports Easily	○ Supports Easily	○ Supports Easily
SDH-XC/ADM	○ Supports Easily	○ STM-64 Ring(UPSR) can support the No. of connections.	○ STM-64 Ring(BLSR) can support the No. of connections.
WP-XC/ADM	△ → ? (present) (future) (****)	× → ? (present) (future)	× → ? (present) (future)

(*) Each protection connection is dedicated to its working connection.
(**) Protection connections are shared among some working connections.
(***) present the equations when N(the number of nodes) is odd and show values if N = 15.
(****) Full-mesh connection is possible if 4-fiber ring topology is used[10].

Table 2 Role of VP-NW in WDM-introduced network.

Full Mesh Connection Layer		Protection Layer: Virtual Path	Protection Layer: Wavelength Path
	VP/WP (VP/WP-NW)	-Support Full Mesh Connectivity -Groom VPCs into WPs -Protection Switching	-Support Full Mesh Connectivity -Groom VPCs into WPs
	VP (VP-NW with WDM Line)	-Support Full Mesh Connectivity -Groom VPCs into WDM Lines -Protection Switching	(*) -Support Full Mesh Connectivity -Groom VPCs into WDM Lines

(*) WDM Line -based protecion switching is employed.

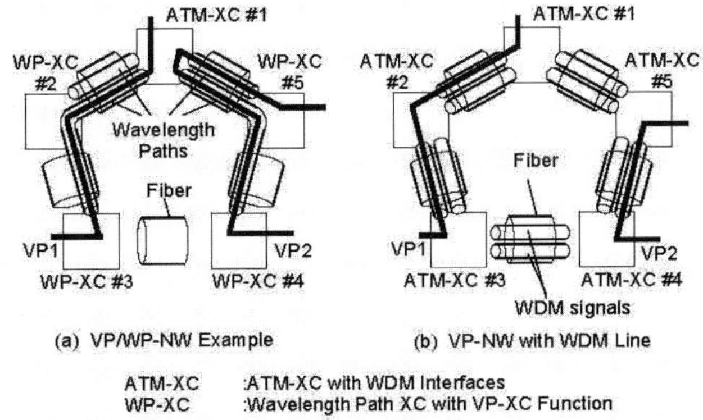

Fig. 7 Role of ATM-XC in WDM-based transport network.

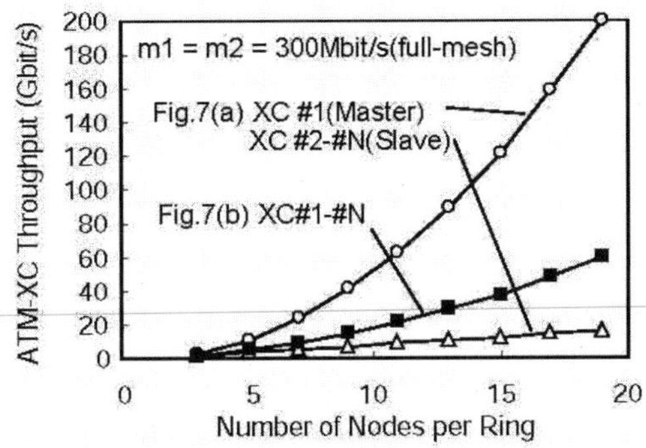

Fig. 8 Required ATM-XC throughput.

full-mesh connections among N nodes and also shows the support offered by ATM/SDH/WDM-based transport systems. Although the maximum number of wavelengths multiplexed in a fiber is about 100 for point-to-point transmission experiment today, current WP-XC system testbeds offer no more than 16 [9]. So existing WP-XC cannot accommodate full-mesh connections for more than 8 node 2-fiber ring if we take the protection WPs into account. There are two ways to compensate this poor connectivity; expanding the network into multiple fiber topology [10], or designing WPs well to reduce the throughput required for VP-XC functions in WP-XCs. In the latter case, VP layer actually supports the full mesh connectivity among WP-XCs. and the role of WP-NW is reduced.

Table 2 summarizes the WDM-introduced network configurations and the roles of VP transport layer. VP/WP-NW is a network that partly introduces WPs, and its example is shown in Fig. 7(a). VP-NW with WDM line introduces WDM technology only to gain line capacities. In the former, WP-XC structure is complicated because it should have two switching fabrics (VP-SW and WP-SW). It also requires future striking progress in WP-XCs connectivity or topological expansion to support full mesh connections among WP-XCs. Moreover, the former requires two-layered network operation and path-accommodation design systems for VPs and WPs. On the other hand, network elements, operation, and path-accommodation of the latter remain simple. Thus the latter would be the preferred configuration in terms of OAM simplicity or path-accommodation design, although the former reduces the required VP-XC throughput to a certain extent.

As for the path-accommodation design in VP/WP-NW, the interworking between the VP-designing and the WP-designing is very important. Figures 7 and 8 show that lack of the interworking results in extra-increase in the required VP-XC throughput. In Fig. 7(a) each WP is established between master node and each slave node and each WP-XC in a slave node terminates the WP destined for the node. ATM-XC in the master node must offer tremendous throughput as shown in Fig.8 because all traffic goes through the master node. In Fig. 8 calculation, it is assumed that in Fig. 7(a) each WP has its redundant route (only working WP appears in the figure) and VP protection switching is employed to protect traffic. Thus the interworking is necessary to place minimum demand on each VP-XC.

As for the protection switching function, the number of paths switched simultaneously is large and this leads to large processing loads on switching if VP protection switching is employed. However, redundancy-required VPs and single-routed VPs are efficiently groomed because network operators can set redundancy on per VP basis. And VP switching time is short enough if many VPs are switched simultaneously (see Sect. 4).

Although quantitative comparison among the configurations in Table 2 remains for further study, it can be concluded that VP-NW will be necessary for full-mesh connections, efficient traffic grooming, and protection switching.

4. VP-XC Structure

4.1 Up to 160 Gbit/s Scalable VP-XC Structure

Figure 9 shows the scalable VP-XC structure we already proposed. It is more cost-effective to install smaller scale XCs in the initial stage of introduction, and extend the installed XCs step by step to match the growth in traffic. The basic XC comprises a 2.4 Gbit/s 8×8 cell switching fabric(M-SW) and STM-1/4/16 inter-office or intra-office interfaces. Cells extracted from several STM-1/4 interfaces are multiplexed into a 2.4 Gbit/s cell stream and introduced into the M-SW. The M-SW can support multi-speed interfaces effectively and can minimize CDV values when accommodating STM-16 interfaces. Since the XC handles only permanent VPCs, no VC handling function is required. The XC has a permanent VP routing table in the M-SW and VP OAM functions in the interfaces.

If the required throughput exceeds 20 Gbit/s, additional basic XCs are installed and a 20 Gbit/s 8×8 SW connects the basic XCs as shown in Fig. 9(b). We have already proposed a 20 Gbit/s 8×8 cross-connect architecture using input/output-buffer switches [2]. The system is designed to cause no errors on the previously established VPs in case of the system expansion.

We have already fabricated a testbed of the basic VP-XC; its specifications are summarized in Table 3. As for system redundancy, the switch and the clock distribution have 1+1 redundant and non-redundant operating mode. So the network operators can select the mode according to reliability and cost required for the system. As for route redundancy, the VP-XC has the Section-APS function and the VP-APS function, either of which the operators can select. The merit of having the two protection switching functions is shown in Fig. 10. Since the segment between XC 2 and XC 3 accommodates redundant VP (VP #1) and non-redundant VPs (#2, #3), the operators had better adopt VP-APS for the segment to achieve high-grooming efficiency. Meanwhile they can adopt Section-APS for the segment between XC 1 and XC 2 to shorten the switching time and reduce the processing loads on switching because the segment does not accommodate non-redundant VPs.

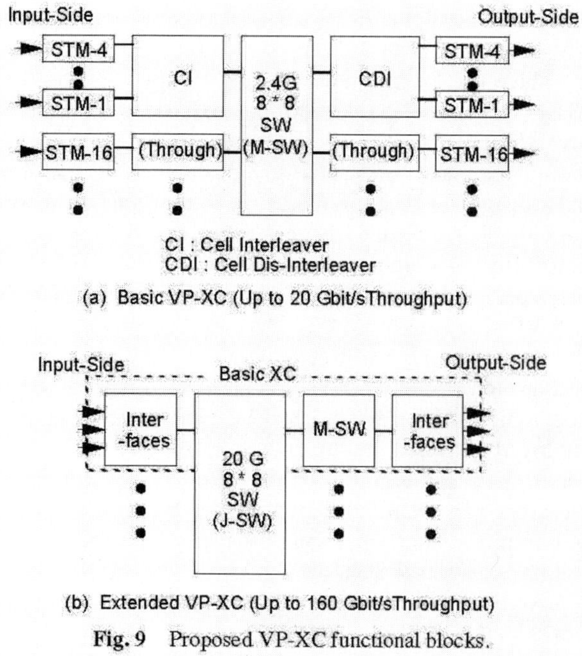

(a) Basic VP-XC (Up to 20 Gbit/sThroughput)

(b) Extended VP-XC (Up to 160 Gbit/sThroughput)

Fig. 9 Proposed VP-XC functional blocks.

4.2 VP Automatic Protection Switching Function

ITU-T SG 13 is now discussing the detail switching mechanisms for VP-APS functions [7]. Recommendation I.630 states that the switching protocol is a 1-phase bi-directional switching protocol and each NE realizes switching with a simple global priority logic. It describes 1+1 and 1:N switching configuration with revertive and non-revertive modes. The testbed VP-XC now employs the NTT proprietary 2-phase switching protocol of 1:1 non-revertive mode, but can be easily updated to the ITU-T-based switching protocol.

Figure 11 shows VP-APS function implementation for the 1:1 switching configuration. Switching is completed by VP Routing Table (VRT) update that allows the establishment of working/protection VP pairs with any VP rate, and between any transmission line interfaces. Detailed

Table 3 Specifications of fabricated VP-XC.

Items	Specifications
XC Capacity	20 Gbit/s per Bay (non-blocking)
Interfaces	STM-1 / 4 / 16 (Intra-office and Inter-office) All IFs can accommodate 4096 VPs.
System Redundancy	XC Switch : 1+1 or non-redundant Clock Distribution : 1+1 or non-redundant Interface : 1+1 Section Protecion or non-redundant mode
Interface Accommodation	STM-1 : 128 STM-4 : 32 x (1+1) Redundancy STM-16 : 8 x (1+1) Redundancy
VP Redundancy	1:1 VP-APS or non-redundant mode 2048 redundant VPs per Bay
OAM Functions	VP-Alarm Transfer, VP-Performance Monitoring, VP Loopback, VP-Test

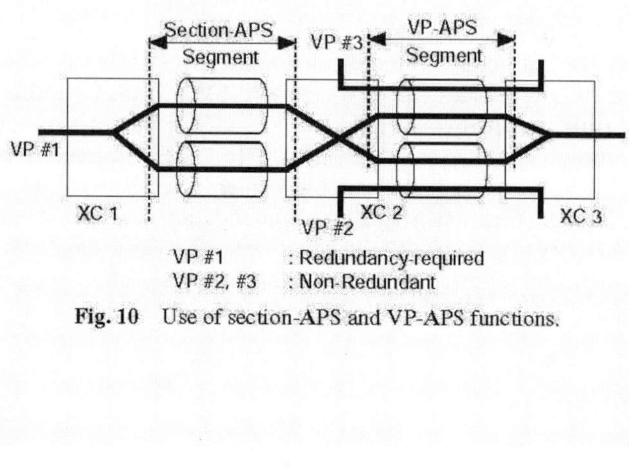

Fig. 10 Use of section-APS and VP-APS functions.

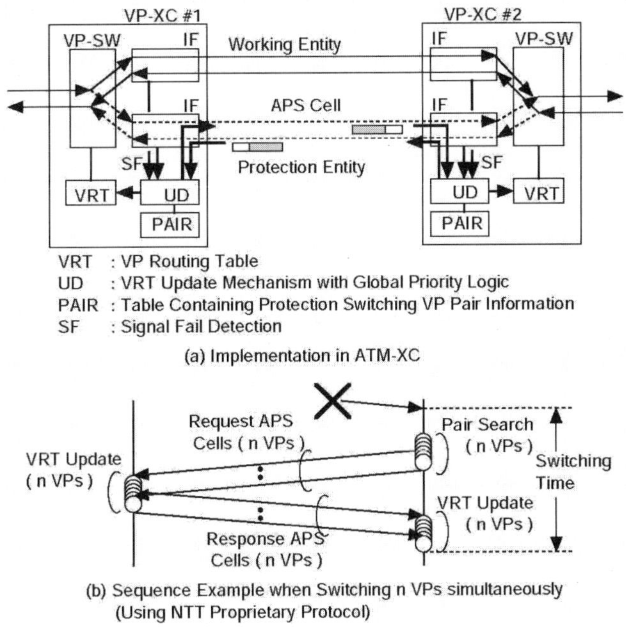

Fig. 11 1:1 VP-APS function implementation.

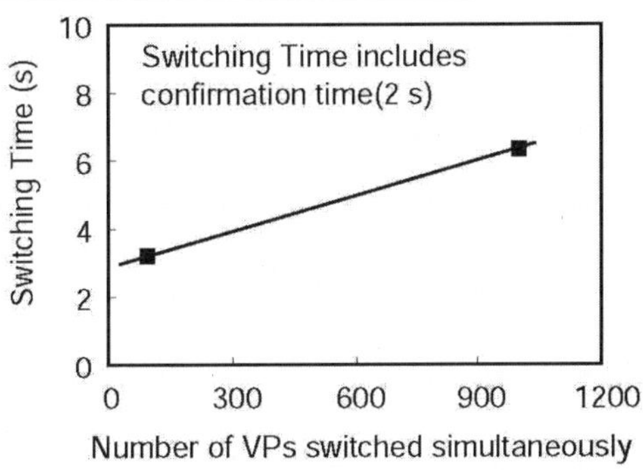

Fig. 12 VP-APS switching time in ATM-XC.

protection switching sequence from working entity to protection entity is summarized in Fig. 11(b), where n VPs are damaged at the same time and n VPs are switched to the protection entities simultaneously. Figure 12 shows the measured VP switching time of the test-bed XC when switching many VPs simultaneously. The switching time is defined in Fig. 11(b).

5. Conclusion

With the goal of offering various future broadband services economically in nation-wide regions, this paper described the VP-XC based, HOP-XC based, and WP-XC based transport network architecture. It was shown that the VP-NW has superior performance, especially in terms of grooming efficiency, to the HOP-XC based transport networks. VP-XC based transport networks will be always more cost-effective provided that increases in the transport network capacity are based on advances in TDM technologies. It was also shown that when WP based network is introduced in the future, the VP-based transport layer will still be necessary to compensate the rough granularity and the poor connectivity of WPs.

A scalable structure was also proposed for a VP-XC that supports the VP-APS function. The ATM-XC described herein will be introduced into NTT's commercial network within 1999 to construct a cost-effective backbone network.

Acknowledgment

Authors wish to thank Mr. Kichiro AKINO, Executive Manager of NTT Network Service Systems Laboratories for encouragement.

References

[1] T.Aoyama, I.Tokizawa, and K.Sato, "Introduction strategy and technologies for ATM VP-based broadband networks," IEEE J. SAC, vol.10, no.9, pp.1434-1447, Dec. 1992.

[2] R.Iwase, K.Watanabe, H.Ohta, M.Hiraki, Y.Kanayama, and T.Tsuboi, "Economical unified platform using ATM cross-connect system for both STM and ATM transport systems," APCC'97 Proceedings, TuK, pp.588-592, 1997.

[3] H.Uematsu and H.Ueda, "STM signal transfer techniques in ATM networks," ICC'92 Conference Proceedings, 311.6, pp.0281-0285, 1992.

[4] Bellcore, "ATM Functionality in SONET Digital Cross-Connect Systems — Generic Criteria," GR-2891-CORE, ISSUE 1, Aug. 1995.

[5] K.Sato, Advances in Transport Network Technologies, Artech House Publishers, 1996.

[6] R.Kawamura and I.Tokizawa, "Self-healing virtual path architecture in ATM networks," IEEE Commun. Mag., pp.72-79, Sept. 1995.

[7] ITU-T Draft Recommendation I.630.

[8] ITU-T Recommendation G.841, 842.

[9] M.Koga, A.Watanabe, T.Kawai, K.Sato, and Y.Ohmori, "Large-capacity optical path cross-connect system for WDM photonic transport network," IEEE J. SAC, vol.16, no.7, pp.1260-1269, Sept. 1998.

[10] N.Nagatsu, A.Watanabe, S.Okamoto, and K.Sato, "Architectural analysis of multiple fiber ring networks employing optical paths," IEEE JLT, vol.15, no.10, pp.1794-1804, Oct. 1997.

High-Performance Backbone Network Technology edited by Naoaki Yamanaka
ISBN: 0-8247-5321-6 DOI: 10.1081/(120025656)
Published 2004 by Marcel Dekker, Inc.

New Self-Healing Scheme that Realizes Multiple Reliability on ATM Networks

Taishi Yahara and Ryutaro Kawamura

SUMMARY This paper proposes a new restoration concept for ATM networks. It realizes the rapid and multiple reliability/cost level restoration required to support many different network services. First, the necessity in realizing rapid and multiple-reliability-level restoration in the future network is shown. The self-healing schemes that is based on distributed restoration mechanism satisfies the rapidity in restoration, but does not satisfy multiple reliability levels. Thus a new self-healing scheme that satisfies them is presented and a Multiple Reliability Level Virtual Path network concept is proposed based on the new self-healing scheme. Next, how to realize the new self-healing scheme is explained as an extension of the existing self-healing scheme with two new simple functions. Finally, evaluations confirm the effectiveness of the proposed scheme. These results show that the proposed new scheme realizes a network that fulfills the rapidity and multiple reliability requirements that are strongly required.

key words: *restoration, self-healing, failure, reliability, ATM*

1. Introduction

Various network services have deeply spread to our society, and dependency to the network have been increasing year by year. In such situation, network reliability will become a crucial issue because even one failure will cause an enormous loss of data. The data volume transmitted through each link increases year by year, so even one failure will significantly damage the information society. Thus rapid restoration from failure is required to minimize data loss and stop network services. Conventional schemes, self-healing schemes that achieve rapid restoration by distributing the control mechanism among Network Elements have received considerable attention for effective restoration scheme, and many self-healing schemes have been proposed and evaluated [1]–[20].

Furthermore, various services will be integrated into the same network in ATM networks, so different levels of reliability, such as restoration rapidity or restoration probability, must be established in the same network. As examples of network services, high reliability services such as tele-medicine require minimal data loss and interruption time because degeneration in service could have serious repercussions on human life. Therefore, these services require sufficient dedicated spare resources to recover simply and rapidly from a failure and incur a high cost. On the other hand, low reliability services such as services that do not require a backup can tolerate data loss and short interruptions which in turn decreases the cost. These examples exemplify the trade-off relationship between reliability and cost, i.e., sufficient spare resources enable a high level of reliability. Thus multiple reliability/cost level should be achieved to satisfy user requirements.

Thus, rapidity and multiple reliability for restoration must be achieved in ATM networks. The conventional self-healing scheme has achieved rapidity, but unfortunately, no existing self-healing scheme provides a complete and sufficient framework for multiple reliability/cost levels needed from the viewpoint of user cost and application. This paper proposes a new self-healing scheme that fulfills the restoration requirements of rapidity and multiple reliability by adding two new functions to the conventional self-healing scheme proposed in [3]–[6].

The remainder of the paper is organized as follows. In the next section, the conventional self-healing scheme and its problems in realizing multiple reliability are described. Section 3 shows the Multiple Reliability Level Virtual Path (VP) network concept based on the new self-healing scheme and its applicable region. Section 4 describes the realization of a new self-healing scheme with the two new functions that extend the conventional self-healing scheme and the realization of a Multiple Reliability Level VP network. Section 5 quantitatively evaluates the proposed scheme through simulations. Concluding remarks are presented in Sect. 6.

2. Existing Self-Healing Schemes

Existing self-healing schemes for ATM networks are classified into the flooding type [10]–[14] and the backup type [3]–[9]. The flooding type restoration looks for a backup route by broadcasting messages after a failure is detected. This scheme can flexibly detect a backup route in any failure pattern, however, this takes an excessive amount of time to achieve restoration because of the time spent in the search process. On the other hand, the backup type restoration preassigns backup routes. The backup type restoration realizes rapid restoration by preassigning backup routes [3]. In ATM networks, they tend to accommodate the increased number of paths in a link, and a lot of paths have to be restored simultaneously when a failure occurs [4]. Thus, simple and rapid restoration scheme is strongly required. Therefore, the backup type restoration is considered to be suitable for restoring ATM networks, and we have developed a scheme of this type called the Backup VP scheme for ATM networks [3]–[6]. It is realized in the VP layer in ATM networks. The controllability

Manuscript received December 1, 1999.
Manuscript revised June 12, 2000.

† The authors are with the NTT Network Innovation Laboratories, Yokosuka-shi, 239-0847 Japan.

IEICE TRANS. COMMUN., Vol. E83-B, No. 12, pp. 2615–2625, December 2000 and flexibility that are possible in ATM networks realizes effective network restoration [21].

Backup VPs are preassigned with zero bandwidth when established. This is feasible because ATM networks allow route establishment to be independent of the bandwidth setting [21]. The spare resources utilized for restoration in a link are shared by the backup VPs accommodated in that link and are managed by the node connected to the link. Spare resources are captured only when a failure occurs. If a network failure occurs, the appropriate bandwidth is captured by distributed message communication between nodes on the backup VP after the failure is detected by VP failure detection (e.g. VP-AIS [22]). After the appropriate bandwidth is captured in all links on the backup VP, the failed working VP is switched to the backup VP (Fig. 1). The Backup VP scheme achieves rapid restoration by preassigning backup routes [3]. The cost is decreased by sharing spare resources and switching at termination nodes compared with restoration schemes utilizing dedicated spare resources and switching at switching nodes [4], [23]–[26]. Furthermore, assuming the utilization of the Backup VP scheme, we can guarantee restoration probability depending on the service requirements in reasonable failure patterns (for example, any 1 link/node failure we define as either "foreseeable failure"). The reason for this is that backup routes are preassigned, so we can estimate the spare resources in each link required to restore all failed VPs against all foreseeable failure patterns by satisfying their bandwidth requirement, i.e., the maximum bandwidth required to counter all foreseeable failure patterns [3].

The Backup VP scheme has many advantages as described above, however, this scheme and the flooding type self-healing scheme did not correspond to multiple reliability levels. Therefore, all VPs have the same reliability level. This leads to two problems. One is the cost increase. It is very expensive to establish dedicated spare resources to provide all VPs with a complete restoration level and satisfy the service requirements of the highest reliability VP. The other problem is no prioritized order control. If there are insufficient spare resources to restore all failed VPs because of an unforeseeable failure (e.g., multiple failures such as any 2 link failures) or for some other reason, the VPs that can be restored are arbitrarily selected. This is because spare resources in a link are captured in the order that the message is received. It is possible to establish the spare resources required to counter an unforeseeable failure, but establishing such increases the cost dramatically [27].

To solve these problems, some schemes that realizes priority control have recently been proposed in [8], [13] and [30], not only ATM networks but also other network layers. In [8], a high priority path precedently captures spare resources canceled by a low priority path which fails in the middle of restoration . In [13], a low priority path waits in specific time, thus a high priority path captures spare resources. In [30], a high priority path preempt a low priority path when setting up PNNI (Private Network-Network Interface) path. But, these schemes do not obtain controllability of reliability that is executed with path reliability defined.

This paper proposes a new Backup VP scheme that realizes a multiple reliability/cost level ATM networks. This new scheme solves above problems and achieves the rapidity and multiple reliability level on restoration. These multiple level is achieved by simple integer value set on each VP. Each reliability level is controlled based on these values.

3. Multiple Reliability Level VP Networks

3.1 Multiple Reliability Level VP Networks Concept

We have proposed the Multiple Reliability Level VP networks concept to realize the multiple reliability level networks (Fig. 2) [5], [28]. This network logically comprises multiple VP networks requiring different levels of reliability on a single physical network. Each service is accommodated in one of these VP networks that satisfies the service requirement. The VP layer on ATM networks is suitable for realizing this concept, since the VP layer achieves several networks logically and realizes controllable and flexible network management of these several networks.

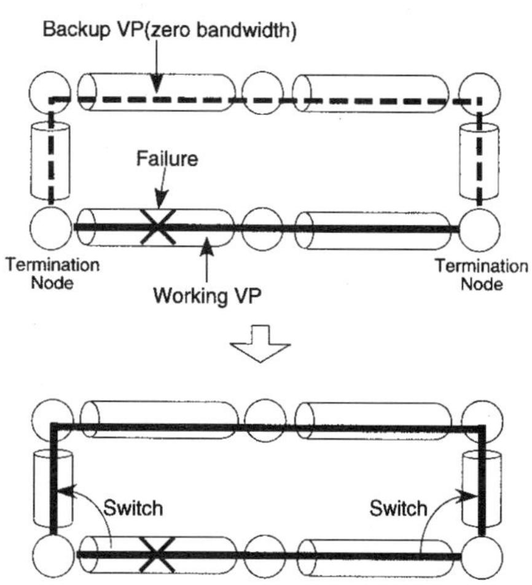

Fig. 1 Backup virtual path scheme.

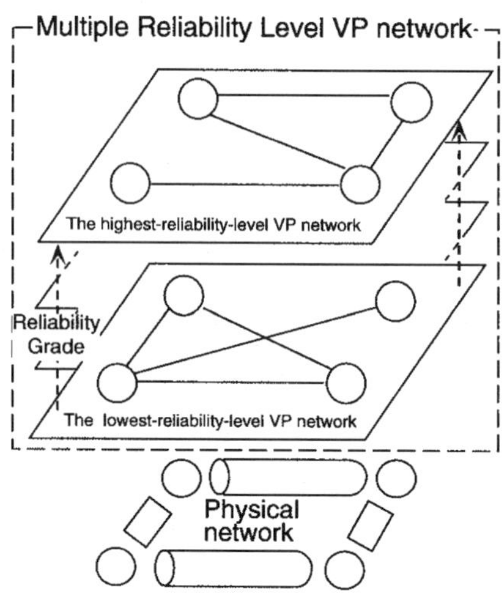

Fig. 2 Multiple reliability level VP network.

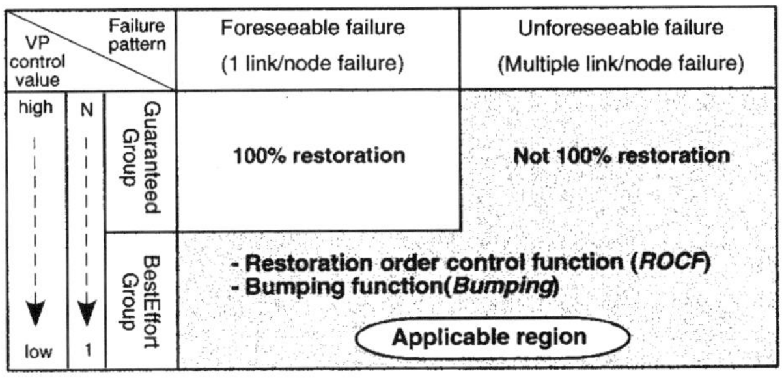

Fig. 3 Applicable region.

As one method to realize this concept, a method that utilizes different restoration schemes in each VP network has been proposed [5]. In this method, three VP networks are established: a VP networks that utilize a 1:1 VP restoration scheme (it utilizes dedicated spare resources and realizes very rapid restoration); a Backup VP scheme; and no restoration scheme. Network resources, however, are not shared between each VP network and each restoration scheme must be developed independently in this method. Thus this method increases the cost in both network establishment and development.

On the other hand, we propose a method that is based on the extended Backup VP scheme. All VP networks utilize the same extended Backup VP scheme and share network resources. All network resources are utilized depending on the reliability level of each VP. Thus, compared width the method in [5], our proposed method realizes Multiple Reliability Level VP networks in cost-effective.

3.2 Applicable Region and Effectiveness

The applicable region of Multiple Reliability Level VP networks is described in Fig. 3. To meet customer demands for low cost, we consider splitting VPs into two groups: Guaranteed Group and BestEffort Group. In Guaranteed Group, VP Automatic Protection Switch (VP-APS) [5], which realizes rapid self-healing scheme offering complete reliability, is also utilized as a restoration scheme realizing high reliability.

Against foreseeable failures, all Guaranteed Group VPs are guaranteed 100% restoration based on the contract. But, BestEffort Group VPs are not guaranteed 100% restoration against any type of failure to minimize cost. Guaranteed Group VPs are not also guaranteed 100% restoration in case of unforeseeable failures, since spare resources assignment to counter the unforeseeable failure increases the cost dramatically [16]. In such cases, reliability performance is achieved by the priority control function (Restoration Order Control Function (*ROCF*) that restores high-reliability-level VPs first and Bumping function (*Bumping*) that restores high-reliability-level VPs by utilizing released bandwidth from low-reliability-level VPs) proposed in this paper.

4. Achievement of Multiple Reliability Level VP Network

This section describes details of two functions that extend the conventional Backup VP scheme and realize the Multiple Reliability Level VP network. To provide details of two new functions, we first define VP reliability levels and VP control values that control VP reliability, and a database in a node. Then the details of two functions that extend the Backup VP scheme are explained.

4.1 Definition of VP Reliability Level

There are plenty of measurement parameter about network reliability, such as availability, Cell Loss Ratio, jitter and more. In this paper, we define "restoration probability" as the reliability measurement. We aim to establish a framework in which high-reliability-level VPs obtain high restoration probability, while low-reliability-level VPs obtain relatively low restoration probability. To realize this framework, we assign a "VP control value (integer)" to all VPs in the network. Higher-reliability-level VPs, such as large bandwidth VPs aggregating lots of customers, pertain higher values, while the lowest reliability VPs pertain the value of 1. This value represents its reliability level, and we define a set of VPs that pertain the same VP control value n which belongs to "VP reliability level n." These VPs in the same VP reliability level comprise one logical VP network. Classification of restoration probability is achieved through the two proposed functions that depend on the VP control value. VPs with a higher VP control value will obtain a higher restoration probability than that of VPs with a lower relative VP control value.

4.2 Link and VP Information

To realize the proposed scheme in a distributed manner, a node that connects to a link must manage information pertaining to the bandwidth utilization in the links and VP data accommodated in the links. In this paper, we define such nodes as "Management nodes." A Management node obtains the Link information described in Table 1, and the VP information containing all data on working VPs and backup VPs in the link, described in Table 2. When a failure happens, restoration processes referring these tables distributively.

4.3 Restoration Order Control Function (*ROCF*)

This function starts restoration in order of the VP control value among all failed VPs detected in a switching node. This function is executed in a distributed manner, so this function does not guarantee a complete restoration order in a global network. If multiple VPs fail simultaneously, the switching node detecting the failure refers to the VP information in Table 2 and decides which VP to restore first among those detected. VPs that obtain a higher VP control value are relatively restored first. *ROCF* prioritizes restoration requests while ini-

Table 1 Link information.

Resources utilized for working VPs
Spare resources

Table 2 VP information.

Data	Contents	
VP classification	Working VP / Backup VP	
VP Identifier(VPI, ingress and egress)	Specific value for VP in a link[21]	
Bandwidth (BW)	Required bandwidth in the Working VP	
VP control value (CV)	The value is established depends on the service requirement. Higher-reliability-level VPs gain a high value. The lowest-reliability-level VPs gain value 1.	
Reserve Condition (RC)	Working VP	Null(Initial state) / Reserved(Reserved to release the bandwidth) / Executing(Now executing to release) / Finish(Finish the releasing bandwidth)
	Backup VP	Null(Initial state) / Release(Requiring bandwidth release for restoration and reserve)
Reserve Information (RI)	Working VP	Null(Initial state) / Reserved bandwidth (Reserve condition = "Reserved")
	Backup VP	Null(Initial state) / VPI of reserved Low-reliability-level VPs (Reserve condition = "Release")

tializing the self-healing process. As an alternative, it is always able to prioritize restoration requests. However, this proposal increases the complexity of network elements, especially if many VPs fail. This complexity will deteriorate the capability of network elements. Thus, we adopt a simple algorithm to decrease the complexity of network elements.

4.4 Bumping Function (*Bumping*)

This function cuts and releases all bandwidth of lower-reliability-level VPs (VP control values of working VPs are lower than that of the VP being restored) to restore higher-reliability-level VPs if the spare resources are insufficient. As candidates for lower-reliability-level VPs, working VPs that do not fail and VPs that are restored in advance are selected. Difference between the algorithm in [29] is that this algorithm completely cut the selected VPs, while the algorithm in [29] decreases the bandwidth of the selected low-reliability-level VPs. *Bumping* improves the restoration probability of VPs that obtain a higher VP control value even if spare resources are insufficient.

Figure 4 shows an example of a VP network and the simple *Bumping* sequence. $Node_i$ indicates the specific node named "*i*." Figure 4(a) shows an example of two VPs, a Backup VP with VP control value *m* (BVP_m) and a VP with VP control value *n* (VP_n). In this figure, *n* is set to be less than *m*. The link between $Node_b$ and $Node_c$ lacks spare resources to restore BVP_m. We assume that Management nodes in Fig. 4 are $Node_a$, $Node_b$, and $Node_c$. The switching node that detects the failure of VP_m (its backup VP is BVP_m) is defined as the Start node. In Fig. 4(a), $Node_a$ is the Start node. The other switching node, $Node_d$, is termed the Destination node. The *Bumping* sequence has 3 phases.

4.4.1 Phase I: Bandwidth Capture Phase

The Start node sends a Capture_message to the next node on the backup VP until that message reaches the Destination node. When a Management node receives a Capture_message, it executes the sequence described in Algorithm 1(Fig. 5) to reserve the appropriate bandwidth on the link. If the spare resources are insufficient, one or more lower-reliability-level VPs (total VP bandwidth released permits restoration) are selected and reserved to release their bandwidth by referring to the VP information. Release of the bandwidth is executed in Phase II to eliminate unnecessary bandwidth release, that is a high-reliability-level VP is not restored after releasing a low-reliability-level VP.

To realize control of the restoration probability dependent on the VP control value and effective network resource utilization, two additional information items W_{sum} and VPI_{bump} are embedded in the Capture_message. W_{sum} describes the summation of the VP control value of VPs reserved to restore the same backup VP being restored. Because realizing control of the restoration probability depends on the VP control value, we should set a limit (*Bumping_limit*) to prevent unlimited bandwidth release of lower-reliability-level VPs, that is unlimited low-reliability-level VPs are released to restore a high-reliability-level VP. Furthermore, *Bumping* is executed in a distributed manner, so the *Bumping_limit* should be checked at each Management node. Thus we set W_{sum} in a Capture_message. We define the *Bumping_limit* such that W_{sum} must not exceed the VP control value of the VP being restored. VPI_{bump} describes the VPI of reserved VPs in previous nodes on the backup VP. *Bumping* is executed in a distributed manner, so more VPs may be released even if they can be restored by releasing less VPs. To minimize released lower-reliability-level VPs, Management nodes receiving a Capture_message refer to VPI_{bump} and preferably select VPs reserved in previous nodes. VPI_{bump} in the Capture_message will be updated to indicate properly the reserved lower-reliability-level VPs in previous Management nodes as long as a backup VP and reserved VPs pass the same links (Fig. 6). If a Management node reserves VPs indicated in VPI_{bump}, W_{sum} is not increased.

When a Management node receives a Capture_message, it first initializes *RB* (required bandwidth for restoration) to the bandwidth of BVP being restored (BVP_c) and checks the VPI_{bump} in the Capture_message (in line 1). If this node manages VPs that are included in VPI_{bump}, it reserves these VPs (in line 2-5). In this case, W_{sum} does not increase. In line 3, if there is sufficient bandwidth that is not reserved in VP_i (the element in VPI_{bump} and LVP_m), bandwidth *RB* is reserved in VP_i. If not, the total unreserved bandwidth in VP_i is reserved. Next, this node checks spare resources. If there are adequate spare resources, this node captures the appropriate bandwidth (*RB*) (in line 6). If not, this node captures all spare resources (in line 7). Finally, this node selects lower-reliability-level VPs from LVP_m if RB exceeds 0. First, the function *Select(Value)* selects the VP that RC (Reserve Condition in Table 2) in VP information is "Reserved" and the total bandwidth is not reserved (in line 9). This VP has been reserved, so utilizing the unreserved bandwidth decreases the total number of released lower-reliability-level VPs. Next, the function *Select(Value)* selects the VP that has the lowest VP control value in LVP_m. To realize more efficient utilization of network resources, a combination of all lower-reliability-level VPs should be considered, but this sequence is complex [30]. Thus we choose the simple sequence to decrease the calculation complexity to select lower-reliability-level VPs, since this complexity will deteriorate the capability of network element. When a VP are reserved, the RC and RI (Reserve Information in Table 2) of VP_i and BVP_c, and W_{sum} are updated. This selection is repeated until conditions in line 8 are fulfilled (in line 8-14).

As a result, this Management node can check whether or not the appropriate bandwidth is captured from RB. If it is captured, this node sends a Capture_message containing updated W_{sum} and VPI_{bump} (in line 16). If not (i.e. $Link_m$ does not have sufficient lower-reliability-level VPs to support backup, or W_{sum} exceeds the VP control value of BVP_c), a Cancel_message is sent back to the Start node (in line 18). The management node receiving a Cancel_message releases the captured spare resources and initializes reserved low-reliability-level VPs by changing Link information and VP information.

When a node that is not a Management node receives a Capture_message, it only transmits it to the next node with an updated VPI_{bump} if necessary.

4.4.2 Phase II: Bandwidth Release Phase

If it is not necessary to release any lower-reliability-level VPs in all links on the backup VP (i.e, $W_{sum} = 0$), the Destination node switches the failed VP to the backup VP and sends a Switch_message to the Start node as in the conventional Backup VP scheme [3]. The Start node receiving the Switch_message switches the failed working VP to the backup VP and completes the restoration process.

If it is necessary to release bandwidth (i.e., $W_{sum} > 0$), it

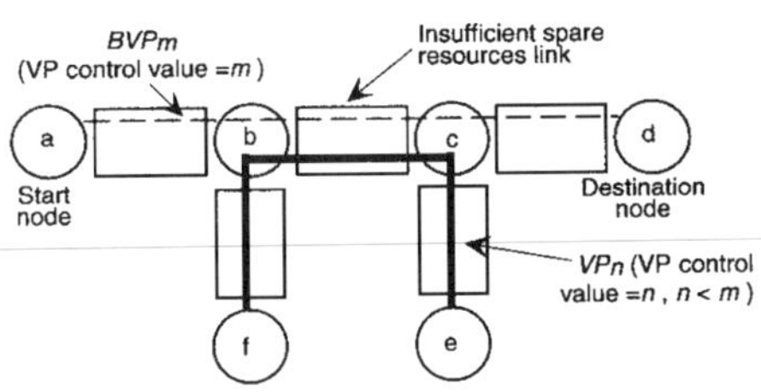

(a)

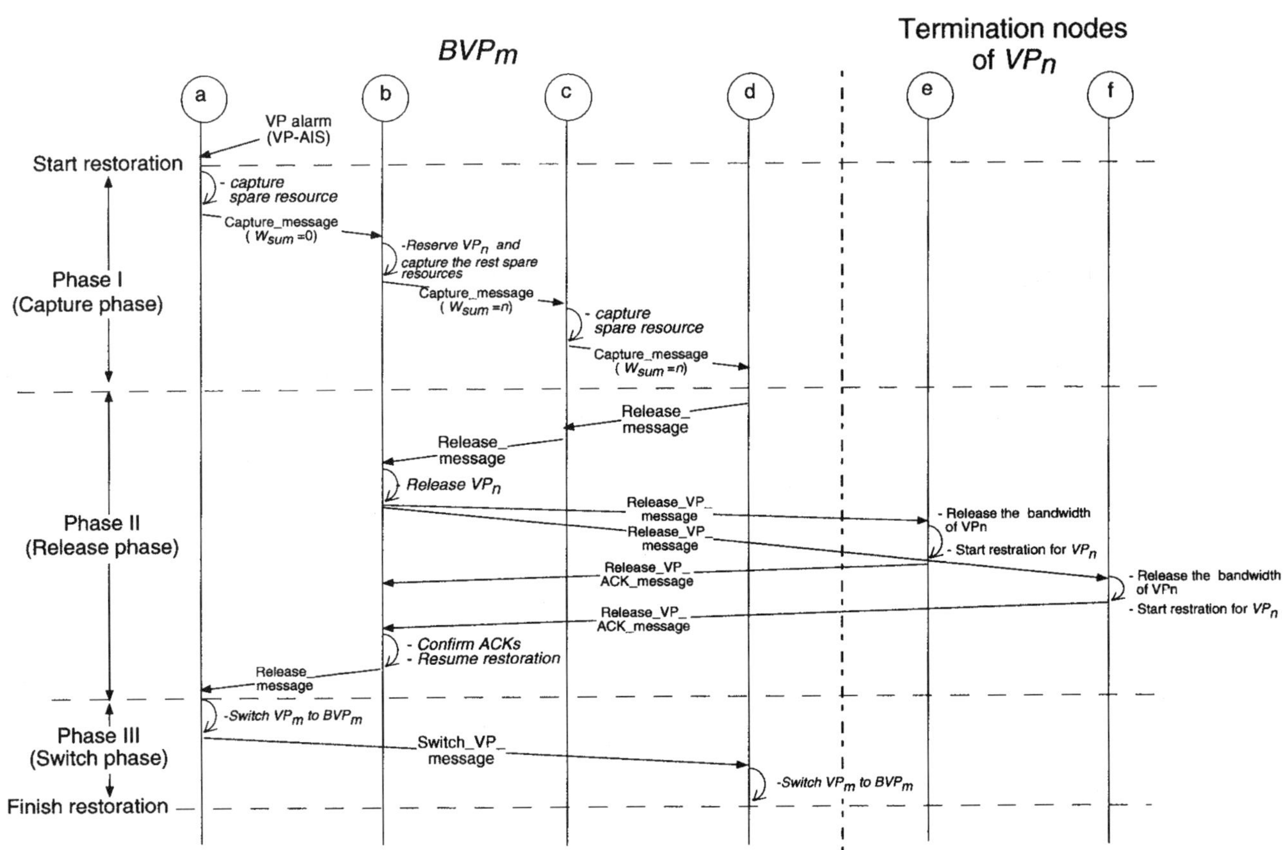

(b)

Fig. 4 Bumping sequence.

sends a Release_message to the Start node via the backup VP. When the Management nodes receive the Release_message, they execute Algorithm 2 (Fig. 7) to release the reserved lower-reliability-level VPs.

This algorithm utilizes the additional information, VPI_{rel}. VPI_{rel} is utilized to indicate VPIs of lower-reliability-level VPs that were released in previous nodes. Management nodes do not deal with VPs included in VPI_{rel}, since previous nodes release the bandwidth. VPs released in a previous node completely stop the cell stream, so VPs included in VPI_{rel} do not require execution of bandwidth release twice. The proposed algorithm guarantees the stoppage of the cell stream. This information enables a decrease in the total number of messages for releasing bandwidth and improves restoration time. VPI_{rel} is also updated to indicate properly as the VPI_{bump} did in Fig. 6. In this algorithm, a Management node first checks the *RC* of BVP_c. If *RC* of BVP_c is "Null," the Management node simply sends a Release_message to the next node on the backup VP (in line 12). If *RC* of BVP_c is "Release" (i.e., it is necessary to release the bandwidth of some lower-reliability-level VPs), this node deals with reserved lower-reliability-level VPs for BVP_c and certainly confirms that all reserved VPs have their bandwidth released. This node can check whether a VP have been reserved or not by referring to *RI* of BVP_c(see line 13 in Algorithm 1)

The Management node checks the VPI_{rel} first. If VP_k (an element in $Bump_m$) is included in VPI_{rel}, it is not necessary to execute disposal for release, since VP_k are released in a previous node. This node only updates *RC* of VP_k to "Finish" (in line 3-4). Next, this node deals with all reserved lower-reli-

Algorithm 1 (Bandwidth Capture Phase)

Variants:

```
SR_m: Spare resources in Link_m;
BVP_c: Backup VP being restored;
RB: Required bandwidth for restoration;
{LVP_m}: Set of VPs with lower VP control value than that of BVP_c in Link_m;

1  RB = {the bandwidth of BVP_c described in VP information};
2  while (there are VPs(VP_i) in LVP_m that are indicated in VPI_bump)
3      Reserve(VP_i, min(RB, unreserved bandwidth of VP_i));
4      RB = RB - min(RB, unreserved bandwidth of VP_i);
5      {LVP_m} = {LVP_m}- {VP_i};
   endwhile
6  if (RB < SR_m) {RB = 0; SR_m = SR_m - RB;} // There are adequate spare resources
7  else {RB = RB - SR_m; SR_m = 0;}
8  while ((RB > 0) or ({LVP_m}  null) or ( W_sum < the VP control value of BVP_c))
9      VP_j = Select(the VP control value of BVP_c)*;
10     Reserve(VP_j, min(RB, unreserved bandwidth of VP_j))**;
11     RB= RB - min(RB, unreserved bandwidth of VP_j);
12     - Update VP information of BVP_c:
           Reserve Condition = "Release"; Reserve Information = the VPI of the VP_j;
13     W_sum = W_sum + (the VP control value of VP_j);
14     {LVP_m} = {LVP_m}- {VP_j};
   endwhile;
15 if (RB = 0)
16     - Send Capture message to the next node with updated W_sum and VPI_bump;
17 else
18     - Send back a Cancel_message;
```

**Select*(*Value*): Select one VP that has a lower VP control value than *Value*. Selection condition is described in order of importance.
1. *RI* in VP information is "Reserved", but total bandwidth is not reserved.
2. It has the lowest VP control value

***Reserve*(VP_i, *BW*): Reserve the VP_i by exchanging its Reserve Condition(*RC*) in VP information "Reserved" and its Bandwidth by adding *BW* to Reserve Information (*RI*) in the VP Information.

Fig. 5 Bandwidth capture algorithm.

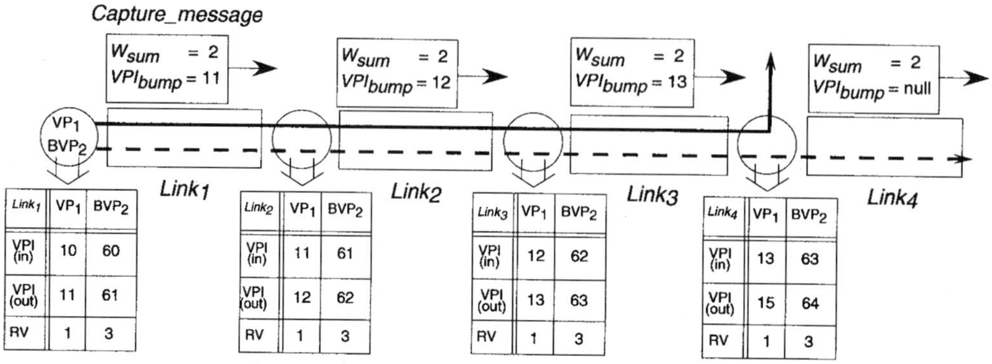

Fig. 6 Transmission of the Capture_message (BVP_2 reserved VP_1 for bumping).

```
Algorithm 2 (Bandwidth Release Phase)

Variants:
  BVP_c : Backup VP being restored;
  Bump_m : Set of VPs reserved for BVP_c in Link_m;

1  if (BVP_c's Reserve Condition = "Release") then
2     while (Bump_m ≠ null)
3        if (VPI_rel ∋ VP_k) then
4           - Update VP information of VP_k:
                Reserve Condition="Finish";
         endif
5        if (Reserve Condition = "Reserved") then
6           - Send Release_VP_messages to both termination nodes of selected lower-reliability-level VPs
7           - Update VP Information of reserved lower-reliability-level VPs:
                Reserve Condition="Executing";
8           - After confirming a Release_VP_ACK_messages, update RC of VP information of a lower-reliability-level VP:
                Reserve Condition = "Finish";
         endif
9        {Bump_m} = {Bump_m} - {VP_k};
      endwhile;
10    - Send Release_message to the next node with updated VPI_rel after RC of all reserved lower-reliability-level VPs becomes "Finish";
11 else // BVP_c's Reserve condition is "Null"
12    - Send Release_message to the next node with updated VPI_rel ;
```

Fig. 7 Bandwidth release algorithm.

ability-level VPs in which *RC* is "Reserved." This node sends Release_VP_messages to both termination nodes of reserved lower-reliability-level VPs and waits for Release_VP_ACK_messages while exchanging its *RC* in "Executing". A termination nodes of a lower-reliability-level VP receiving this message releases its bandwidth and send back a Release_VP_ACK_message. After confirming a Release_VP_ACK_message, Management nodes update *RC* of a lower-reliability-level VP to "Finish" and remove the VP from $Bump_m$ (in line 5-9). Some VPs in $Bump_m$ have been *RC* "Executing" when a Release_message was received. This is because one VP (VP_a) is reserved by multiple backup VPs. When VP_a is reserved by multiple backup VPs, the Management node receiving Release_message first switches the *RC* of VP_a to "Executing." When the Management node receives another Release_message requesting bandwidth release of VP_a, it can not execute the sequence for bandwidth release. This is because the first disposal releases all the bandwidth, so all reserved bandwidths are released at one time. This method decreases the total number of messages to release bandwidth.

After confirming that the *RC* of all reserved lower-reliability-level VPs becomes "Finish," this node sends a Release_message to the next node. This node waits to send the Release_message until the *RC* from all the reserved VPs become "Finish" (in line 10). This sequence is needed to prevent cell overflow from the released VPs. If we simply release a VP and reassign its bandwidth in one link before the *RC* of all reserved VPs become "Finish," cells will continue to arrive at the upstream node which may degrade the cell loss quality of other working VPs.

Released lower-reliability-level VPs also trigger VP restoration which more fully utilize the network resources.

4.4.3 Phase III: Switch Phase

The Start node receiving a Release_message switches a failed VP to its backup VP and sends a Switch_VP_message to the Destination node. The Destination node receiving this message also switches them, and finishes the restoration process.

5. Evaluation of the Proposed Scheme

This section quantitatively examines the effectiveness of our proposed self-healing scheme through simulations. The simulation conditions are described below.

-Network topology:

1. Metropolitan LATA network model [9] (15 nodes joined by 28 links (Fig. 8(a)))
2. Japan long haul network model (56 nodes joined by 85 links (Fig. 8(b)))

-Foreseeable failure pattern: 1 link failure

-VP setting: VPs with three reliability levels are set between all nodes with 10, 20, and 30 Mbps bandwidths (the percentage of each is 33.3%, Total number of working VPs: 315 (Fig. 8(a)) and 4120 (Fig. 8(b)))

-VP routing: Working VP occupies the shortest path. Backup VP occupies the shortest path that is not common (except termination nodes) with working VP.

-Message processing time: 10-12 ms (random)

-Failure detection delay: transmission delay (number of hops from the point where failure occurs in the termination node)

-Alarm detection delay: 10 ms (failure generated time is set to 0)

-Spare resources: The fewest spare resources enabling complete restoration of all VPs against any 1 link failure is set as the spare resources ratio (*SRR*) which is 1 [4]. *SRR* is decreased to test the effectiveness of the proposed scheme. *SRR* = 0 means that there are no spare resources.

Figures 9 and 10 show the average restoration probability at each VP reliability level as *SRR* is decreased; the plots indicate the average values gained by simulating all foreseeable failure patterns. The result using Backup VP scheme shows the total restoration probability in the network. The proposed self-healing scheme (Backup VP scheme + *ROCF* + *Bumping*) is utilized. Restoration probability RP_i at the VP reliability level i is calculated using Eq. (1).

$$RP_i = (RN_i - BN_i + BRN_i)/FN_i \quad (1)$$

RN_j : Restored VP number in VP reliability level j
BN_k : Bumped VP number in VP reliability level k
BRN_l : Bumped and restored VP number in VP reliability level l
FN_m : Failed VP number in VP reliability level m

These figures show the feasibility of multiple reliability in terms of the restoration probability. VPs in reliability level 3 gain higher restoration probability than the conventional Backup VP scheme (maximum difference is about 60% in $SRR = 0.2$, Fig. 9), while VPs in reliability level 1 obtain a lower restoration probability. The effect of the proposed scheme is emphasized especially when *SRR* is low. VPs in reliability level 3 can be restored by utilizing *Bumping* even if *SRR* equals 0. If fewer reliability level 1 VPs are restored than are bumped, the restoration probability of VP reliability level 1 falls below zero ($SRR = 0$–0.2 in Fig. 9). However, topological differences cause differences in the restoration probability even if the proposed scheme is used. The average hop length of the backup VPs in Fig. 8(a) is 3.4 and that in Fig. 8(b) is 8.3, so it is necessary to reserve the bandwidth in more links in Fig. 8(b). Therefore, failed VPs tend not to restore more because of the *Bumping_limit* in Fig. 8(b). It shows the restoration probability strongly depends on the network topology and the VP network in which the proposed scheme is adapted.

Furthermore, simulations were performed at different VP reliability levels (VP reliability levels were 1, 2, and 6). The results are shown in Figs. 11 and 12. The dashed lines in Fig. 11 indicate the results in Fig. 9 and those in Fig. 12 indicate the results in Fig. 10. These graphs show that VP reliability level 6 has a high restoration probability than VP reli-

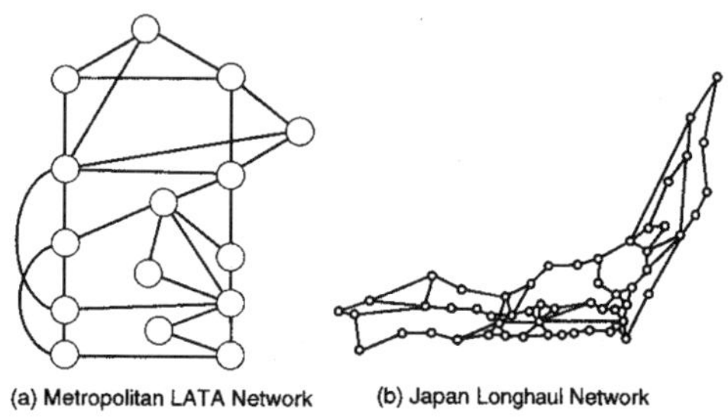

Fig. 8 Network topology.

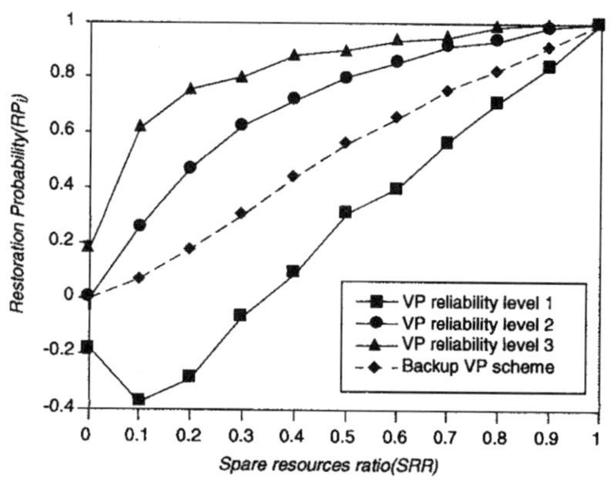

Fig. 9 Restoration probability at each reliability level (Metropolitan LATA Network, VP reliability level: 1, 2, 3).

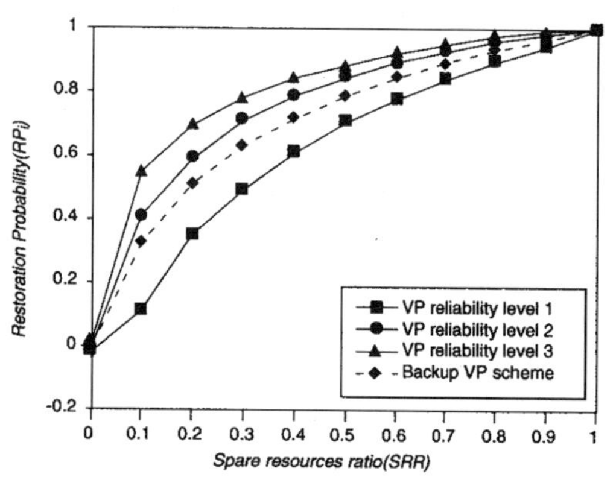

Fig. 10 Restoration probability at each reliability level (Japan long haul Network, VP reliability level: 1, 2, 3).

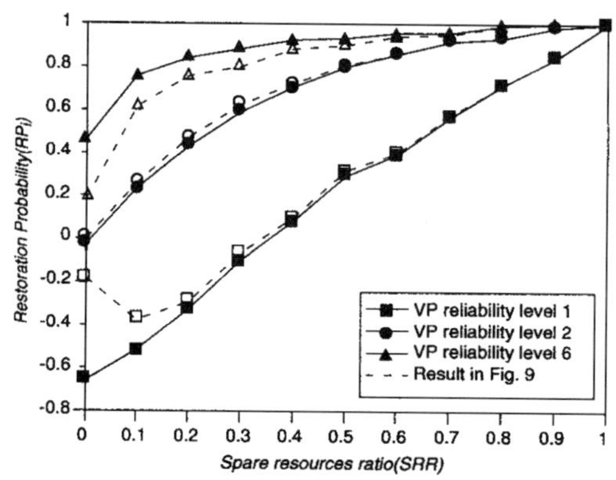

Fig. 11 Restoration probability at each reliability level (Metropolitan LATA network, VP reliability level: 1, 2, 6).

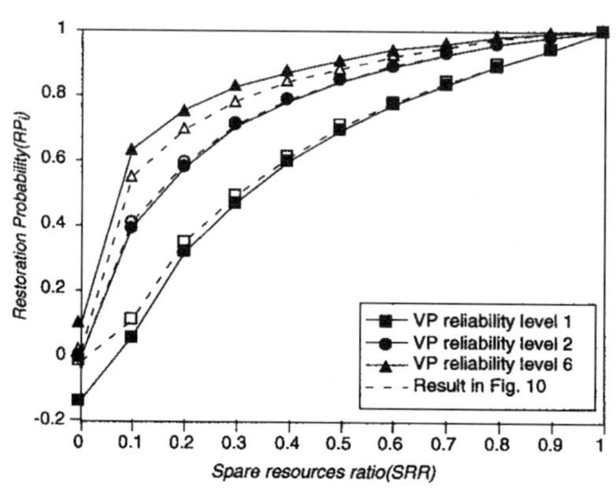

Fig. 12 Restoration probability at each reliability level (Japan long haul network, VP reliability level: 1, 2, 6).

ability level 3, in contrast to VP reliability level 1 which has a lower restoration probability, especially when *SRR* is low. This is because the *Bumping_limit* at level 6 is higher than that at level 3, so the tendency is to restore more often using *Bumping*. This result shows that the restoration probability in each VP reliability level can be controlled using the control value, and also using the *Bumping_limit*.

To evaluate the effect of each function, Figs. 13 and 14 show the average *WRP* in Eq. (2) versus *SRR* when VP reliability levels were 1,2 and 3. Equation (2) shows how many high-reliability VPs are restored in a same failure pattern.

$$WRP=\sum_i CV_i FV_i RP_i / \sum_i CV_i FV_i \qquad (2)$$

$$\left(\begin{array}{l} CV_j : \text{VP control value of VP reliability level } j \\ FV_k : \text{Percentage of failed VP in VP reliability level } k \\ RP_l : \text{Restoration probability in VP reliability level } l \end{array}\right)$$

Three different restoration schemes, method 1 (Backup VP scheme + *ROCF*), method 2 (Backup VP scheme + *ROCF* + *Bumping*), and just the Backup VP scheme, were compared. RP_i in *WRP* was calculated using Eq. (1). These graphs shows that the proposed techniques of *ROCF* and *Bumping* are effective in improving the value of *WRP*, that is VPs with a higher reliability level gain a higher restoration probability. Backup VP scheme and method 1 obtain same restoration probability, but *ROCF* preferably restores higher-reliability-level VP and improves the *WRP*. In this simulation, the average number of failed VPs per one failure is 24.4 in Fig. 8(a) and 272.7 in Fig. 8(b), and Figs. 15 and 16 show the average number of released VPs per failure versus *SRR*. Focusing on *SRR* = 0.4 in Fig. 8(a), method 2 obtains 9.7% higher *WRP* than that of method 1. This result almost matches that 2 more VPs in VP reliability level 3 is restored, in contrast 2 VPs in VP reliability level 1 is not restored influentially in this case. The average number of the released VP is 1.53 in *SRR* = 0.4, so these results show that VPs released are effectively selected by our proposed *Bumping* algorithm. If a VP being restored releases plural VPs, the average number of the released VP will increases over 2. In comparison to Fig. 13, Fig. 14 shows a smaller difference between the Backup VP scheme and method 1. This is because the hop number of the backup VPs in Fig. 14 is larger than that in Fig. 13, so *ROCF* does not have a significant impact. It is because spare resources tend to be captured by lower-reliability-level VPs during a message communication to capture bandwidth for high-reliability-level VPs.

However, the proposed self-healing scheme requires more restoration time than the Backup VP scheme if *Bumping* is executed because of the additional message communication to release bandwidth. Restoration time is also an important factor, so severe degradation should be avoided. Thus we should evaluate the relationship between restoration time and network size. Figures 17 and 18 show the restoration time of each method versus *SRR*. The proposed scheme is set

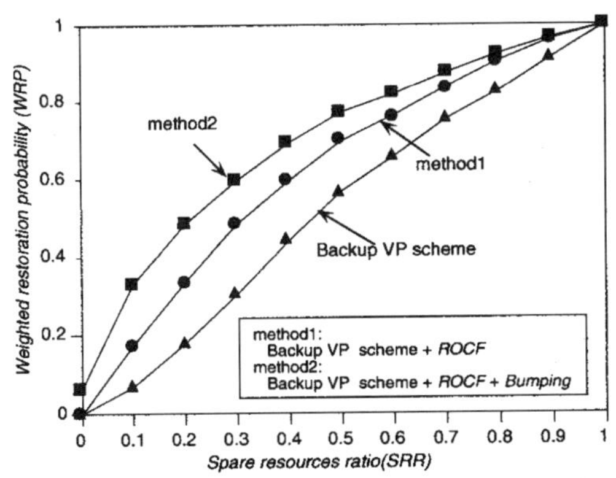

Fig. 13 Weighted restoration probability (Metropolitan LATA network).

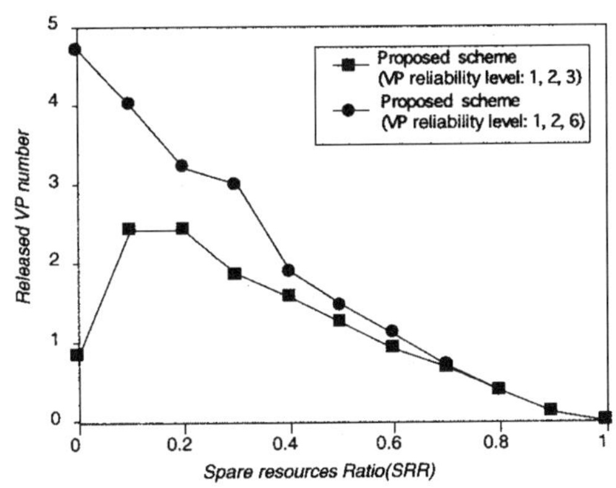

Fig. 15 Released VP number (Metropolitan LATA network).

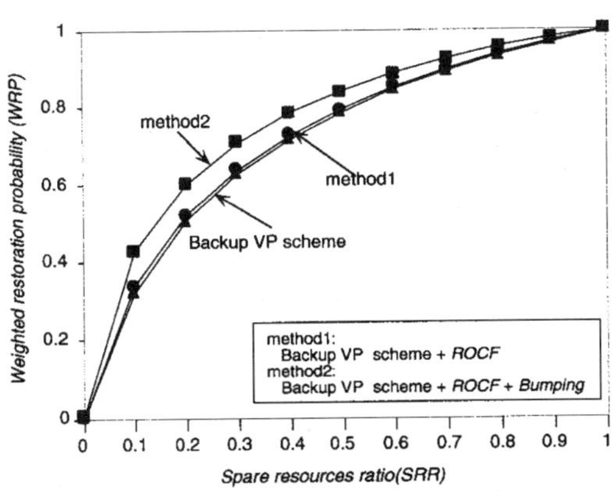

Fig. 14 Weighted restoration probability (Japan long haul network).

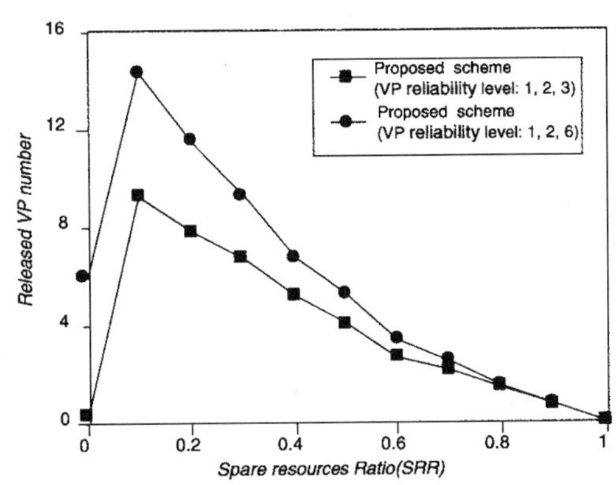

Fig. 16 Released VP number (Japan long haul network).

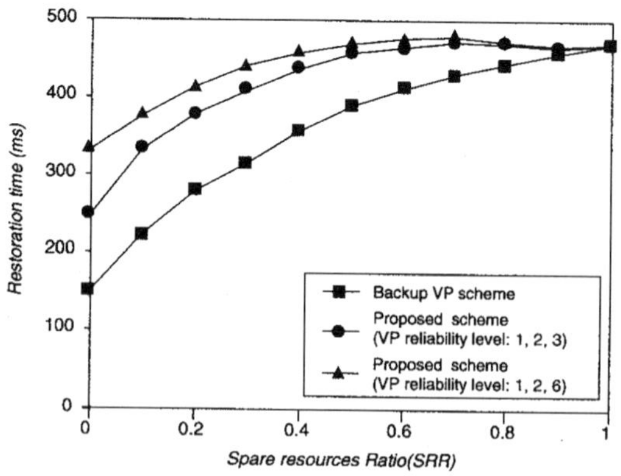

Fig. 17 Comparison of restoration time (Metropolitan LATA network).

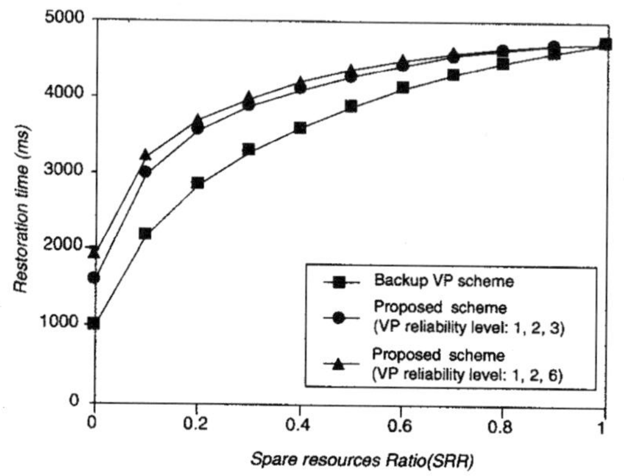

Fig. 18 Comparison of restoration time (Japan long haul network).

in two patterns. One is set to the VP reliability level: 1, 2 and 3. The other is set to the VP reliability level: 1, 2 and 6. Figures 17 and 18 show that restoration time increases with *Bumping*, especially when the highest level is 6, but the impact of *Bumping* on restoration time is relatively small. This is because *Bumping* is executed in a distributed manner, so the time spent is dispersed among the nodes involved. Moreover, Fig. 17 shows a smaller increase in restoration time. There are two reasons. One is that the hop number of the backup VPs in Fig. 18 is larger than that in Fig. 17, so failed VPs tend to interrupt the restoration sequence more. The impact of the interruption reflects in the restoration probability in Fig. 10. Compared with Fig. 9, Fig. 10 shows the small difference between each VP reliability level. The other is that there are larger nodes in Fig. 18, then the disposal for restoration is widely dispersed. The latter mainly impacts to the restoration time. These results on restoration time show that the proposed scheme will satisfy the requirement of rapid restoration.

6. Conclusion

This paper proposed a new self-healing scheme that realizes multiple reliability level on ATM networks. This new self-healing scheme realizes the Multiple Reliability Level Virtual Path network that effectively integrates and logically manages several VP network services requiring different reliability level. We also proposed how to realize the new self-healing scheme as an extension of the conventional self-healing scheme. Two functions were proposed to extend the capabilities. One is the Restoration Order Control Function (*ROCF*) which preferentially restores high-reliability-level VPs. The other is Bumping function (*Bumping*) which captures the bandwidth needed to restore high-reliability-level VPs by releasing low-reliability-level VPs.

Simulation results show that our proposed scheme realizes classification of restoration probability depending on the reliability level, effective utilization of network resources (e.g. improvement of *WRP*) and control of the restoration probability by the setting of the VP control value, and that the increase in the restoration time is relatively small. These results confirm that the proposed scheme realizes networks that fulfill the rapidity and multiple reliability levels for restoration.

References

[1] T.-H. Wu, Fiber Network Service Survivability, Artech House, 1992.

[2] W.D. Grover, "The selfhealing network," Proc. IEEE Globecom '87, pp.1090-1095, 1987.

[3] R. Kawamura and I. Tokizawa, "Self-healing virtual path architecture in ATM networks," IEEE Commun. Mag., vol.33, no.9, pp.72-79, 1995.

[4] R. Kawamura, K. Sato, and I. Tokizawa, "Self-healing ATM networks based on virtual path concept," IEEE J. SAC, vol.12, no.1, pp.120-127, Jan. 1994.

[5] R. Kawamura and S. Ohta, "Self-healing ATM network architecture; Implementation results from NTT's nationwide testbed ATM network," Proc. Networks'96, pp.717-722, 1996.

[6] R. Kawamura, H. Hadama, and I. Tokizawa, "Implementation of self-healing function in ATM networks based on virtual path concept," Proc. IEEE Infocom'95, pp.303-311, 1995.

[7] P. Veitch and I. Hawker, "Administration of restorable virtual path mesh networks," IEEE Commun. Mag., vol.34, no.12, pp.96-101, Dec. 1996.

[8] J. Carlier and D. Nace, "A new distributed restoration algorithm for telecommunication networks," Proc. Networks'94, pp.377-380.

[9] N.D. Lin, A. Zolfaghari, and B. Lusignan, "ATM virtual path self-healing based on a new path restoration protocol," Proc. IEEE Globecom'94, pp.794-798, 1994.

[10] H. Komine, T. Chujo, T. Ogura, K. Miyazaki, and T. Soejima, "A distributed restoration algorithm for multiple-link and node failures of transport networks," Proc. IEEE Globecom'90, pp.459-463.

[11] K. Wipusitwarakun, H. Tode, and H. Ikeda, "Fast failure restoration algorithm with reduced messages based on flooding mechanism," IEICE Trans. Commun., vol.E80-B, no.4 , pp.564-572, April 1997.

[12] H.G. Song, H. Lee, B. Moon, and S.J. Chung, "Dynamic rerouting for ATM virtual path restoration," IEEE Globecom'97, pp.1808-1812, 1997.

[13] H. Saito, M.M. Slominski, and M. Yoshida, "An improved guided restoration algorithm for ATM crossconnect networks," Proc. NOMS'96, pp.225-234, 1996.

[14] E. Ayanoglu, "A fast topology update algorithm for restoration under multiple failures in broadband networks," Proc. IEEE ICC'93, pp.1295-1299.

[15] J. Anderson, B.T. Doshi, S. Dravida, and P. Harshavardhana, "Fast restoration of ATM networks," IEEE J. SAC, vol.12, no.1,

pp.128-138, Jan. 1994.

[16] M.T. Busche, C.M. Lockhart, and C. Olszewski, "Dynamic k-shortest path (DKSP) facility restoration algorithm," Proc. IEEE Globecom'94, pp.536-542, 1994.

[17] K. Struyve and P. Demeester, "Escalation between recovery schemes: Beyond ATM backup recovery," Proc. IEEE Globecom'96, pp.755-761, 1996.

[18] K.W. Ko and K.W. Cheung, "A local map based (LMB) self-healing scheme for arbitrary topology networks," IEEE Globecom'97, pp.1393-1397, 1997.

[19] D.A. Dunn, W.D. Grover, and M.H. MacGregor, "Comparison of k-shortest paths and maximum flow routing for network facility restoration," IEEE J. SAC, vol.12, no.1, pp.88-99, Jan. 1994.

[20] P. Veitch and D. Johnson, "ATM network resilience," IEEE Network, pp.26-33, 1997.9/10.

[21] K. Sato and I. Tokizawa, "Flexible asynchronous transfer mode networks utilizing virtual paths," Proc. IEEE ICC'90, pp.831-838, 1990.

[22] ITU-T Recommendation I.610, "B-ISDN operation and maintenance principles and functions," 1992.

[23] S. Ohta and K. Nunokawa, "Comparison of alternative service protection mechanism for wide area ATM networks," Proc. Networks'96, pp.281-286, 1996.

[24] D. Johnson, "Survivability strategies for broadband networks," IEEE Globecom'96, pp.452-456, 1996.

[25] M. Herzberg and A. Utano, "Optimal assignment of spare capacity to backup VPs in survivable ATM networks," Proc. Networks'96, pp.665-670, 1996.

[26] K. Murakami and H.S. Kim, "Comparative study on restoration schemes of survivable ATM networks," Proc. IEEE Infocom'97, 1997.

[27] K. Miyazaki, T. Ogura, H. Komine, K. Iseda, and T. Chujo, "A study on spare capacity assignment for distributed restoration system," IEICE Technical Report, CS 91-91, 1994.

[28] T. Yahara and R. Kawamura, "Virtual path self-healing scheme based on multi-reliability ATM network concept," IEEE Globecom'97, pp.1803-1807, 1997.

[29] T. Yahara, R. Kawamura, and S. Ohta, "New self-healing scheme that realizes differentiated bandwidth requirement on ATM networks," IEICE Trans. Commun., vol.E83-B, no.3, pp.672-679, March 1999.

[30] M. Peyravian and A.D. Kshemkalyani, "Connection preemption: Issues, algorithms, and a simulation study," Proc. IEEE Infocom'97, pp.143-151, 1997.

High-Performance Backbone Network Technology edited by Naoaki Yamanaka
ISBN: 0-8247-5321-6 DOI: 10.1081/(120025657)
Published 2004 by Marcel Dekker, Inc.

Autonomously Controlled Multiprotocol Wavelength Switching Network for Internet Backbones

Yoshiaki Yamabayashi, Masafumi Koga, and Satoru Okamoto

SUMMARY In order that they fully support human activities, new network services and applications are overwhelming conventional ones, such as telephony, facsimile, and telegraph. Demands for digital networks are exploding, not only in terms of quantity but also quality. Nobody can predict where these demands will lead. Traffic engineering, which is impossible in pure Internet protocol (IP) -based networks, is recognized as being indispensable for quality of service (QoS) control. It includes guaranteed services in terms of bandwidth, delay, delay variation (jitter), and service protection. The "engineered tunnel" through IP network supports virtual private networks (VPNs) and allows us to develop voice-over-IP (VoIP), teleconferencing and other secure private network services. This paper proposes the "photonic router" which makes use of wavelength-based networks for signal routing. IP packets having the same destination are bundled into a wavelength path. Interchange nodes along the path route control path routing on the basis of wavelength information, not on IP headers, which can not be read or processed with current optical techniques. In short, wavelength path routing offers "cut-through" in the photonic layer. This paper shows its feasibility by describing the combination of an optical cross-connect, payload assembler/disassembler, label controller, and IP router. Optical cross-connect systems, which are now being intensively studied worldwide, are deemed to be key equipment for a wavelength-path network with centralized control system. This paper proposes to apply the cross-connect to an IP network with distributed autonomous control.

key words: *optical networks, WDM network architecture, mutiprotocol label switching, optical cross-connects*

1. Introduction

Advanced network services and applications are demanding new infrastructures. E-commerce applications, for example, require the infrastructure network to be secure and cost-effective. Demands for increasing the capacity and quality of digital networks are exploding. State-of-the-art backbone IP routers, which are fully armed with electrical devices, have a maximum switching capacity of about 100 Gbit/s. Traffic engineering, which is impossible in networks based solely on the Internet protocol (IP), is now recognized as being indispensable for quality of service (QoS) control. It yields guaranteed services in terms of bandwidth, delay, delay variation (jitter), and service protection. The "engineered tunnel" through IP network allows us to realize virtual private networks (VPNs) and to develop services for voice-over-IP (VoIP), teleconferencing, and other secure private network services. To achieve this, MPLS (mutiprotocol label switching) technology is becoming the focus of interest [1] . This paper proposes a photonic extension that we call photonic MPLS or mutiprotocol lambda switching [2] . It realizes optical signal routing in a wavelength-based network. A distributed autonomous network control mechanism is introduced for wavelength path control. First, IP packets destined for the same geographical location are bundled into a wavelength path or an optical label switched path (OLSP) in a broad sense. Second, interchange nodes along the path control path routing on the basis of wavelength information, not on IP headers, which can not be read or processed with current optical techniques. In short, wavelength path routing offers "cut-through" in the photonic layer, or photonic routing.

Information processing in computers, Ethernet switches, IP routers is performed by electronic devices. Gordon Moore predicted that the integration level of silicon chips would double every 18 months. Surprisingly the real world has followed this trend for the last 30 years. His prediction is now recognized as "Moore's law." More importantly, it is also applicable to the capacity and size of memory chips, hard-disks, and communication bandwidth [3]. Bit-rate expansion in the backbone TDM (time-division-multiplex) WAN (wide-area network) has followed Moore's law, as shown as the doted line in Fig. 1. This is reasonable because the multi/demultiplexers in TDM systems are made with silicon chips. There is some saturation after 1995. One reason is the chip's

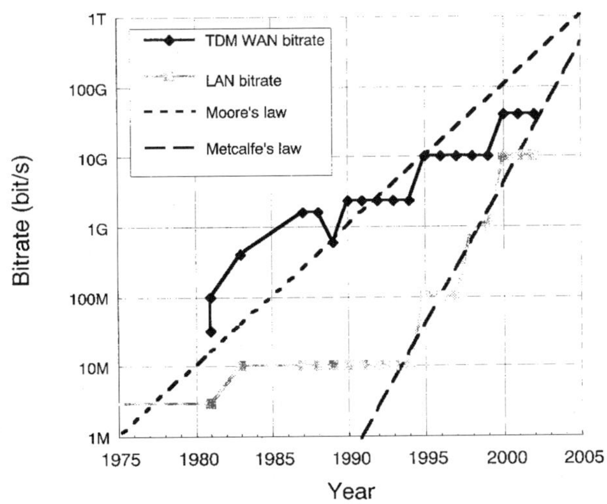

Fig. 1 TDM-WAN and LAN bit-rate growth compared to "Moore's law" and "Metcalfe's law."

Manuscript received May 18, 2000.

† The authors are with NTT Network Innovation Laboratories, Yokosuka-shi, 239-0847 Japan.
* Presently, with NTT Electronics Corp., Yokohama-shi, 221-0052 Japan.

IEICE TRANS. COMMUN., Vol. E83-B, No. 10, pp. 2210–2215, October 2000

material. Moore's law may not be applicable, if the chip material has to be transferred from silicon to another. Only recently, have silicon chips (SiGe) been able to cover the 10 Gbit/s region; GaAs-FETs dominated the high bit-rate region for the 10 year period before 2000. At 40 Gbit/s, even GaAs transistors are unworkable, and InP devices or some new materials have to be developed. Generally speaking, new materials take time to develop because of their unfamiliar physical characteristics. Therefore, the integration level of new material chips can not be increased rapidly.

Robert Metcalfe, one of the Ethernet developers, created another "law," stating that the utilization of any service or network would increase as the square of the number of users. The network traffic growth rate will follow Metcalfe's law: the traffic growth rate will always exceed the growth in communication speed. This seems reasonable since every electronic and office machines and home appliance will contain microchips, and communicate with each other. This trend has started in LAN link-speed after 1994, as the broken line (Metralfe's law) in Fig. 1 indicates. This may result in that LAN speed exceeds that of backbone TDM WAN after 2002. Ultra-fast LAN technologies will lead the development of capacity expansion thereafter, and the transmission format will be controlled by LAN standards.

When the single-channel connection capacity between network nodes is not enough, parallel transmission becomes necessary. Optical TDM (OTDM), which offers faster serial TDM than electrical TDM (ETDM), is not constrained to follow Moore's law and offers better band-efficiency. However, many problems have to be solved in the areas of transmission, fiber-dispersion control including PMD (polarization mode dispersion), as well as circuit board accommodation because of the relatively large optical devices.

One parallel solution is to use fiber ribbons. Since we can not escape skew or delay inequality between fibers, and the high costs incurred by multiple connectors and splices for installation and maintenance, it is not suitable for long-haul applications. On the other hand, WDM (wavelength-division multiplex) shares the same fiber, and makes channel extension easy with minimal cost. Even though there is one difficult issue, wavelength control of semiconductor lasers, especially for the back-up transmitters, many large capacity systems have started to employ WDM technology to achieve ultra-high bit-rates at low cost. A terabit capacity system (1.28 Tbit/s = 10 Gbit/s × 128 λ) has been announced. Due to its analog nature, WDM does not depend on the payload's synchronization or its digital format, and offers different characteristics from multiplexing in the synchronous digital hierarchy.

"IP over WDM" allows transmission systems to transfer IP packets over WDM lines regardless of small variations in protocol stack structure (e.g. IP over SDH over WDM, IP over ATM over WDM, or IP over Ethernet over SDH over WDM, etc.). Since research has solved how to transmit a bunch of IP packets, the next issue is how to route them at each IP router. Commercially available IP routers are reported to have peak processing capacities of several dozens of millions of packets per second. However, Metcalfe's law predicts that one hundred times faster routers, i.e. several billions of packets per second, will be needed in five years.

The first routing concept is to construct a point of presence (POP), as shown in Fig. 2(a), to share the load among clustered IP routers; they are connected to each other by very short reach (VSR) interfaces. However, since processing traffic among the routers tends to be uneven, processing performance will not be proportional to the number of routers, even though the IP routers are densely connected. With the drawback being the high number of links needed for interconnecting the routers, all IP router functions can be achieved with this scheme.

The second concept is to introduce the photonic network, as shown in Fig. 2(b) [4]. IP packets with the same destination are bundled into a wavelength path or OLSP. Second, interchange nodes along the way control the path routing on the basis of wavelength, not IP headers. In short, wavelength path routing offers "cut-through" in the photonic layer. The IP router capacity of the POP is relatively small, only those packets that are being added to or dropped from the

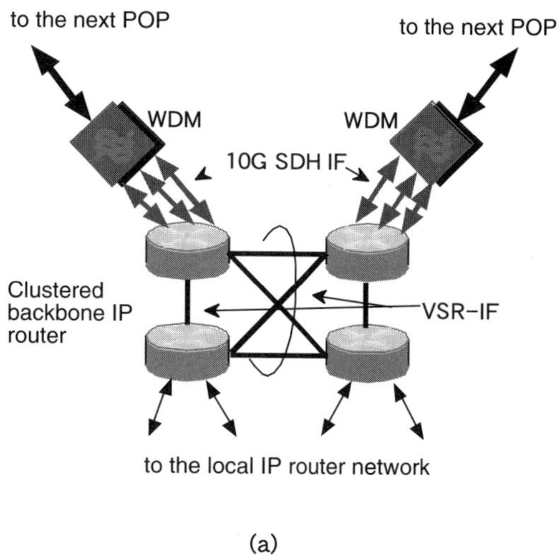

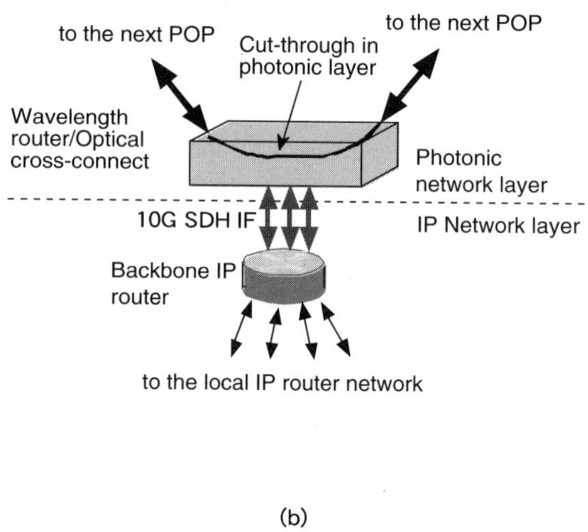

Fig. 2 POP configuration examples: (a) IP over WDM, (b) Photonic router.

transport network need be processed. Therefore, the photonic router makes it possible to minimize the packet relay delay, and simplifies the control of traffic quality with the minimum IP router capacity. The Optical Internetworking Forum (OIF) was established in April 1998 to address IP over WDM technology standards at the international level.

2. Concept of the Photonic Router

This paper proposes a new kind of layer-3 equipment or IP router. As shown in Fig. 3, the photonic router consists of four functional blocks: an electronic IP router and an optical cross-connect (OXC), a photonic transport payload assembler/disassembler (PAD), and a hardware manager. The manager monitors all the circuit elements in the POP. It supervises the dependent repeaters, and controls labeling of LSPs. Labeling includes such functions as generating and canceling, and routing the (optical) label switched paths. The OXC switches and add/drops the wavelength paths. The PAD loads the signal stream onto the wavelength path, converts the signal format, and performs protection/restoration of the wavelength paths. By communicating with other nodes, each node is able to control the wavelength-label switched paths or OLSPs. This router will realize the IP network over photonic network with distributed autonomous control. An MPLS control mechanism has already been developed and can be applied to this autonomous control mechanism with minor modifications; i.e., the logical path attributes (i.e. LSP) must be converted into those the physical wavelength path or OLSP.

In principle, it is not necessary to construct the OXC with 100% optical switches. Electrical switching, however, requires more switches with broadband interconnections. With optical switching, on the other hand, the unique characteristics of lightwave make advanced devices that are also smaller possible. For example, in the case of an n-port cross-connect switch with k-wavelength multiplexed interface, the electrical switch configuration requires, as shown in Fig. 4(a), $2n$ units of $k \times k$ and k of $n \times n$ switches with $2(k \times n)$ lines to connect them. The optical switch version can reduce these numbers to n units of $k \times n$ specially designed switches with $k \times n$ interconnections, and n optical couplers, as shown in Fig. 4(b). An $M \times N$ optical switch consists of M lines with N serially connected optical 1×2 switches, and N optical couplers to combine the M-switch outputs, as shown in Fig. 5. Since, the "delivery and coupling" switch allows all wavelength channels to be multiplexed and output through a single port, the optical cross-connect needs fewer switches than the electrical version. Moreover, k of $n \times n$ electrical switches must be used regardless of which ports will actually be used. In the optical switch, modular blocks, indicated with the dashed rectangles, can be added one by one to expand the port. In other words, the optical switch offers minimal startup costs, and so is suitable for dynamic networks like the Internet.

The proposed photonic router utilizes two switching techniques. One is the electrical processing based conventional label switching, the other is optical processing based wavelength switching (or routing). Being a fermion, electrons are not permitted to exist in the same state. This means that electrons strongly interact with each other. The result is that electrons are suitable to for realizing interacting devices like logical elements for the sophisticated label switching. However, since electrons are easily affected by electric and/or magnetic fields, they are not suited for long-distance transmission applications. This fact is proven by the fact that all long-distance coaxial systems have been replaced by fiber systems. Since photons are bosons, they can share the same

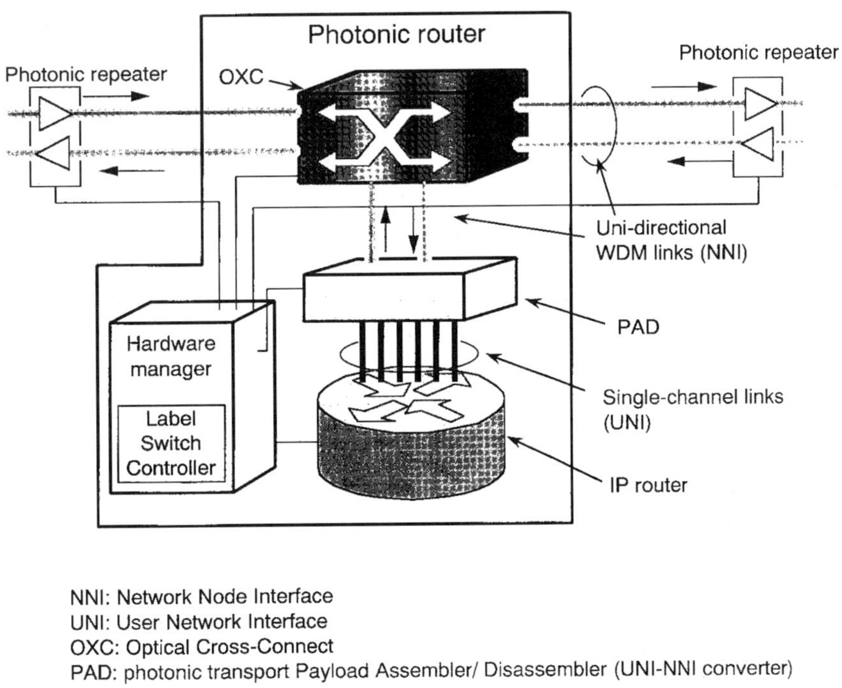

Fig. 3 Photonic router logical configuration.

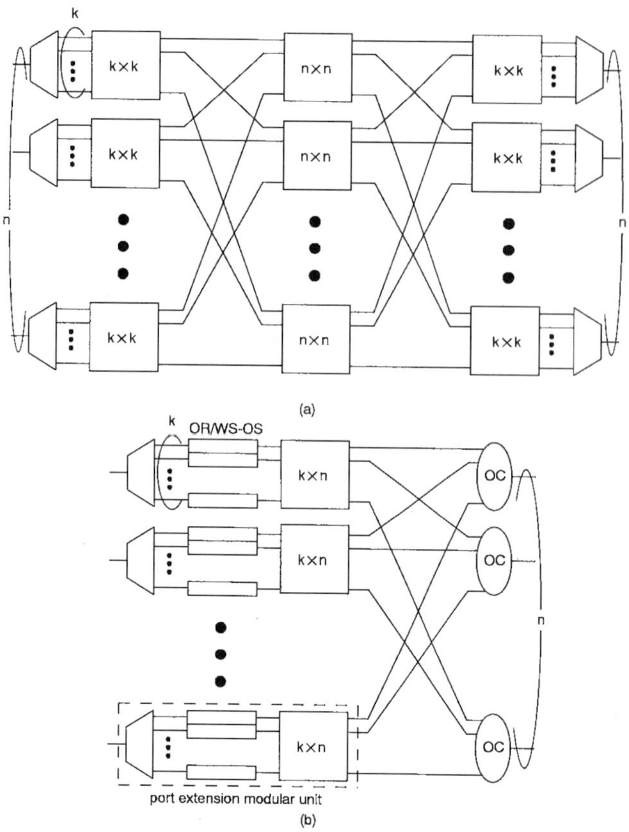

Fig. 4 Optical cross-connect configuration with: (a) space switches, (b) optical switches.

state, and because they do not directly respond to electric or magnetic fields, wavelength-division-multiplexing and long-distance transmission are possible with photons. The basic characteristics of bosons prevent us from using photons to realize complicated signal processing operations. The limited functions possible with photons can be well realized by using currently available miniature optical circuits, such as interferometric switches and filters being formed on a glass or silica plate, have been developed and are commercially available. Optical label discrimination and stream routing should be and can be performed within the optical layer in order to construct large throughput label switched routers (LSRs). From the IP router's viewpoint, the photonic router can be regarded as an MPLS router that uses wavelengths as labels. Therefore, the photonic router can be also called an optical label switched router (OLSR).

As mentioned earlier, the optical cross-connect (OXC) is an important component in the photonic router. Actually, optical cross-connects are deemed the key device in wavelength-path networks with centralized control [5] , and are now being intensively studied worldwide [6]–[9]. By applying them to a distributed autonomous control network, the "photonic IP router" can be constructed. Two types of OXCs have been studied: with wavelength conversion [7] and without [8], [9] . The former is preferred for label switched networks, where the wavelength labels may be changed at the intermediate nodes. Fewer wavelength resources are needed

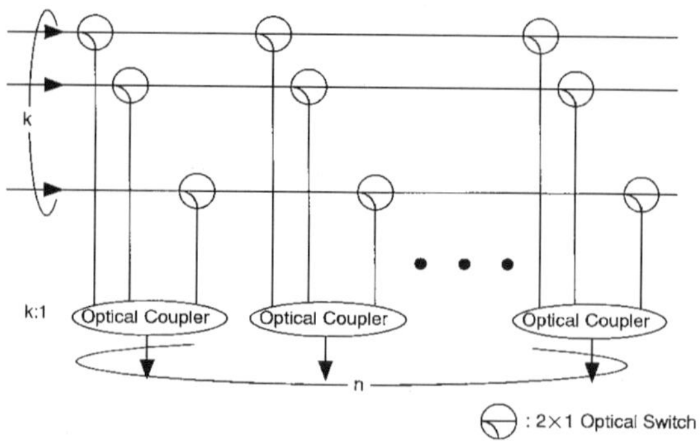

Fig. 5 Delivery-and-coupling optical switch fabric.

than the latter case. With fixed-wavelength paths, all the path labels have to be determined so that there are no conflicts, i.e. paths having the same wavelength.

3. Functions of the Photonic Router

Among the four functional blocks shown in Fig. 3, the IP router block offers IP routing. The label-switch controller performs conventional LSP switching and generation/cancellation of the optical LSPs, and packet multiplexing to make a fat LSP whose bandwidth is equivalent to single optical path bandwidth. The optical block, namely OXC and PAD, performs functions such as multiplexing variable and/or asymmetric streams with changing wavelength numbers or bitrates, and quick network protection and/or restoration through lower-layer supervision. The time-consuming restoration based on IP rerouting can be minimized by giving priority to lower layer restoration or protection.

Since labels are represented by optical wavelength, not digital numbers, the MPLS function is naturally limited. In the near term, only one hundred wavelengths are available for use as optical labels. However, for long-haul transmission, the number is further limited to 16 – 32, because of optical amplifier's gain and noise characteristics and nonlinear effects between wavelength paths. To make full use of the wavelength resource, it is necessary for the intermediate nodes to offer the wavelength conversion function. Wavelength tunable lasers that follow the ITU-T gird should offer the best results, but there some remaining issues. It is a fact that the wavelength band of practical light-sources is narrower than that of the fiber amplifier band, and that the controllability and long-term stability of selected oscillation wavelengths have been not established yet. A virtual tunable laser can be constructed by combining lasers set at different wavelengths and an optical switch to individually select them. The selected lightwave is injected into a modulator that is driven by the path signal. It is obtained from the input path with using optical-electrical conversion, since direct optical-optical modulation is not yet feasible.

The current limitations of the "wavelength" path are:

1. Label merge, that is merging several labels to the same POP to a single label, is not possible. It requires some techniques other than WDM, such as code-division multiplexing or sub-carrier multiplexing.

2. Since label stacking needs layered path identification, it is difficult without introducing other optical multiplication method as mentioned above. Under the current technology, flexible bandwidth control should be left to the conventional MPLS or other layer-2 protocols performed by the IP router.
3. Multicasting is possible for the down-stream path. The up-stream equivalent is currently difficult to achieve since it require label merge. However, this will be possible if uni-directional paths can be redefined in place of the current bi-directional ones.

Despite the above limitations, "photonic router" is the feasible given current optical technologies. A fully-functional optical router would be much simpler to construct if optical memory could be implemented in the network. Revolutionary optical memory devices, or usage of the fiber line as the memory [10], and wavelength resources to avoid signal collision at the receivers are possible solutions. Another approach with no internal memory scheme in the network is also possible, if we can restrict the network topology to a ring [11]. Research into these topics must be accelerated in order to fully support the information society.

4. Conclusion

Demands for digital network are exploding, not only in terms of quantity but also quality. Traffic engineering, which is impossible in Internet protocol (IP) -based networks, is now recognized as being indispensable for quality of service (QoS) control. It includes guaranteed services in terms of bandwidth, delay, delay variation (jitter), and service protection. The "engineered tunnel" through IP network is needed to support virtual private networks (VPNs) to allow us to develop voice-over-IP (VoIP), teleconferencing, and secure other private network services. The "photonic router" with photonic MPLS or multiprotocol lambda switching technology using wavelength-based network for discriminating and routing optically labeled path is feasible and well supports large-capacity IP routers even though their functions are limited compared to the conventional MPLS.

Acknowledgment

The authors are grateful to Ken-ichi Sato for his helpful discussions.

References

[1] G. Swallow, "MPLS advantages for traffic engineering," IEEE Commun. Mag., vol.37, no.12, pp.54-57, 1999.

[2] D.O. Awduche, Y. Rekhter, J. Drake, and R. Coltun, "Multi-protocol lambda switching: Combining MPLS traffic engineering control with optical crossconnects," IETF Internet-draft draft-awduche-mpls-te-optical-01.txt Nov. 1999.

[3] L. Downes and C. Mui, UNLEATHING THE KILLER APP, Harvard Business School Press, Boston, 1998.

[4] A. Watanabe, S. Okamoto, and K. Sato, "Robust IP backbone network utilizing WDM optical paths," IEICE Trans. Commun., vol.E82-B, no.8, pp.1115-1120, 1999.

[5] S. Okamoto, M. Koga, H. Suzuki, and K. Kawai, "Robust photonic transport network implementation with optical cross-connect systems," IEEE Commun. Mag., vol.38, no.3, pp.94-103, 2000.

[6] E.L. Goldstein, L.Y. Lin, and R.W. Tkach, "Multiwavelength opaque optical-crossconnect networks," IEICE Trans. Commun., vol.E82-B, no.8, pp.1095-1104, 1999.

[7] M. Koga, A. Watanabe,T. Kawai, K-I. Sato, and Y. Ohmori, "Large-capacity optical path cross-connect system for WDM photonic transport network," IEEE J. Sel. Areas Commun., vol.16, no.17, pp.1260-1269, Sept. 1998.

[8] M. Teshima, T. Kawai, N. Sakaida, H. Yasaka, H. Ishii, and M. Koga, "Virtual wavelength path cross-connect based on opto-electronic wavelength conversion," IEEE J. Lightwave Technol., vol.16, no.9, pp.1537-1545, 1998.

[9] T. Kawai, M. Teshima, H. Yasaka, M. Kobayashi, and M. Koga, "Wavelength selectable photonic transport system applicable to unequal channnel spacing," ECOC'98, pp.617-618, Sept. 1998.

[10] X. Wang, H. Morikawa, and T. Aoyama, "An optical-burst deflecting routing method for future optical internet," IEICE Technical Reports PNI99-16, Nov. 1999.

[11] J.H. Park, A. Takada, W. Imajuku, and Y. Yamabayashi, "Ultrafast Ring Network with Add-Drop Type Optical Label Switching," to be presented at OECC2000.

High-Performance Backbone Network Technology edited by Naoaki Yamanaka
ISBN: 0-8247-5321-6 DOI: 10.1081/(120025658)
Published 2004 by Marcel Dekker, Inc.

Network Control and Management for the Next Generation Internet

John Y. Wei, Chang-Dong Liu, Sung-Young Park, Kevin H. Liu, Ramu S. Ramamurthy, Hyogon Kim, and Mari W. Maeda

SUMMARY The Next Generation Internet Initiative was launched in the U.S. to advance key networking technologies that will enable a new wave of applications on the Internet. Now, in its third year, the program has launched and fostered over one hundred new research projects in partnership with academic, industrial and government laboratories. One key research area that has been emphasized within the program is the next-generation optical networking. Given the ever increasing demand for network bandwidth, and the recent phenomenal advances in WDM technologies, the Next Generation Internet is expected to be an IP-based optical WDM network. As IP over WDM networking technologies mature, a number of important architectural, management and control issues have surfaced. These issues need to be addressed before a true Next Generation Optical Internet can emerge. This paper provides a brief introduction to the overall goals and activities of DARPA's NGI program and describes the key architectural, management, and control issues for the Optical Internet. We review the different IP/WDM networking architectural models and their tradeoffs. We outline and discuss several management and control issues and possible solutions related to the configuration, fault, and performance management of IP over dynamic WDM networks. We present an analysis and supporting simulation results demonstrating the potential benefits of dynamic IP over WDM networks. We then discuss the issues related to IP/WDM traffic engineering in more detail, and present the approach taken in the NGI SuperNet Network Control and Management Project funded by DARPA. In particular, we motivate and present an innovative integrated traffic-engineering framework for re-configurable IP/WDM networks. It builds on the strength of Multi-Protocol Label Switching (MPLS) for fine-grain IP load balancing, and on the strength of Re-configurable WDM networking for reducing the IP network's weighted-hop-distance, and for expanding the bottleneck bandwidth.

key words: *IP, WDM, MPLS, MPλS, NGI, optical network, IP/WDM, traffic engineering, IP load balancing, WDM reconfiguration, SuperNet, NC&M*

1. Introduction

The Next Generation Internet (NGI) Initiative in the U.S. is a Federal research program that is developing technology and applications for the future growth and scaling of the Internet. This multi-agency effort was initiated in 1998 for the period of five years at a funding level of approximately $100 million per year, and has launched approximately one hundred new projects involving academic, industrial and federal research communities (www.ngi.gov). The Initiative has three high-level goals:

- Research in networking technologies, especially to enable orders of magnitude enhancement in end-to-end flows over the Internet.
- Deployment of national-scale testbeds that provide 100 to 1000 times current end-to-end Internet performance observed today.
- Research and development of discipline applications requiring high performance networks.

While the first goal covers a very broad research area, DARPA's researchers are particularly interested in networking technologies for enabling multi-Gigabit streams to end-applications (www.darpa.mil/ito/research/ngi/index.html). Development of > 10 Gbps network access modules, generation-after-next optical access techniques, and the so-called third generation switched optical networking technologies are being actively pursued. Within the second goal, two classes of networks have been deployed to satisfiy the 100x and 1000x requirements. Approximately 200 sites are now connected to the NGI's 100x testbed with each site being able to access up to approximately 155 Mbps on an IP over ATM or IP over SONET service. In contrast, approximately 30 sites are now connected to the NGI 1000x or SuperNet network, where access rates are Gigabit per second or based on multi-wavelength technology. SuperNet is a dual-purpose testbed, in the sense that it encourages both advanced applications research as well as communications or networking research. The testbed supports ongoing experiments involving prototype optical network elements, network management and control systems, new network and transport layer protocols. Additional

Manuscript received June 14, 2000.

†The authors are with Telcordia Technologies, Red Bank, NJ 07701-5699, U.S.A.

††The author is with Tellium, Inc., Eatontown, NJ 07724, U.S.A.

†††The author is with Ajou University, Seoul 100-095, Korea.

††††The author is with DARPA/ITO, Arlington, VA 22203-1714, U.S.A.

*The author can be contacted at 331 Newman Springs Road, Red Bank, NJ 07701-5699, U.S.A.

**The work reported here was conducted at Telcordia Technologies and funded partly under DARPA NGI SuperNet NC&M Project Contact Number F30602-98-C-0202.

IEICE TRANS. COMMUN., Vol. E83-B, No. 10, pp. 2191–2209, October 2000

information concerning the NGI SuperNet testbed can be found at www.ngi-supernet.org. In summary, the overall goal of the NGI Initiative is to develop the technologies needed, and to lay the foundation for the Next Generation Internet infrastructure.

Given the ever increasing demand in network bandwidth, and the recent phenomenal advances in WDM technologies, there is now a growing consensus that a major component of the Next Generation Internet will be an IP-based optical WDM network. An IP over WDM network combines gigabit and terabit IP routers with WDM switching and transmission systems to create an optimized transport network. As IP over WDM networking technologies mature, several important architectural, management and control issues have surfaced. These issues need to be addressed before a true Next Generation Optical Internet can emerge. In this paper we enumerate some of the most important architectural, management and control issues and discuss some of the approaches taken in the NGI SuperNet NC&M project.

This paper is organized as follows. In Sect. 2, we discuss and illustrate the three different generations of IP/WDM networking architectures, their possible variations, and comment on their tradeoffs and applicability. In Sect. 3, we enumerate several important issues in the management and control of IP/WDM networks, and briefly outline the research pursued in the NGI SuperNet NC&M Project funded under the US Government's Defense Advanced Research Project Agency's (DARPA) NGI Program. Section 4 then presents an analysis of the potential benefits of the re-configurable IP/WDM networking architecture. In this section, we also present a simulation study that highlights the benefits achievable in IP over re-configurable WDM networks. In Sect. 5, we motivate and present an innovative traffic-engineering framework based on the integrated MPLS-based IP routing and MPλS-based WDM reconfiguration architecture, and comment on different implementation possibilities. Finally, we conclude the paper in Sect. 6.

2. IP Over WDM Networking Architecture

The development of IP/WDM technology and networking architecture can be broadly classified into three generations:

- *First Generation*: In the first generation, Dense WDM (DWDM) systems are used for point-to-point high-bandwidth pipes between adjacent IP routers. Typically, SONET is used for framing and transport of overhead information on the WDM channels. IP packets are encapsulated in SONET frames using Packet-over-SONET schemes. Many IP router and WDM equipment vendors have products commercially available today that can support IP over point-to-point DWDM. Point-to-point DWDM systems have seen widespread deployment in long distance carriers.
- *Second Generation*: In second generation IP/WDM systems, WDM channels are routed in the WDM network using WDM cross-connects enabling more efficient WDM bandwidth utilization and IP router interface utilization. Because of the reconfigurability afforded in this generation of products, there is a drive to move protection switching and restoration directly into the optical layer, thereby eliminating the need for the SONET intermediate layer. Many WDM vendors have announced WDM cross-connect products that will enable the second generation of IP over WDM networking.
- *Third Generation*: In the third generation of IP/WDM systems, IP packets are directly transported and switched by WDM packet switches leading to much finer granularity in traffic multiplexing on the respective wavelength channels. WDM packet-switches have been successfully demonstrated in laboratory trials, including in the DARPA funded Optical Label Switching Project [1]. However, it remains to be seen whether such optical packet switching technologies can mature and be made commercially available in the near future.

2.1 IP over Point-to-Point DWDM

In an IP over point-to-point DWDM architecture, IP routers are directly connected to each other via multi-wavelength fiber links. Figure 1 illustrates such an architecture in which the neighboring router for a given router interface is fixed. Figure 2 depicts the more detailed router-to-router interconnections in this architecture. With IP over point-to-point DWDM, the network topology is fixed, and the network configurations are all static. Management systems for such networks are typically centralized, with minimal interactions between the IP and WDM layers.

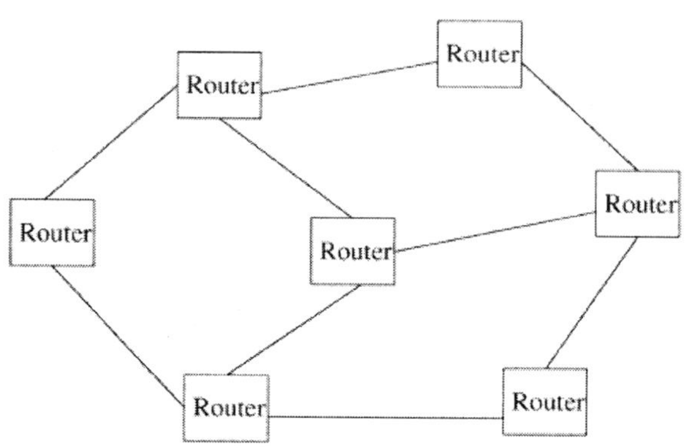

Fig. 1 Architecture of an IP over point-to-point WDM network.

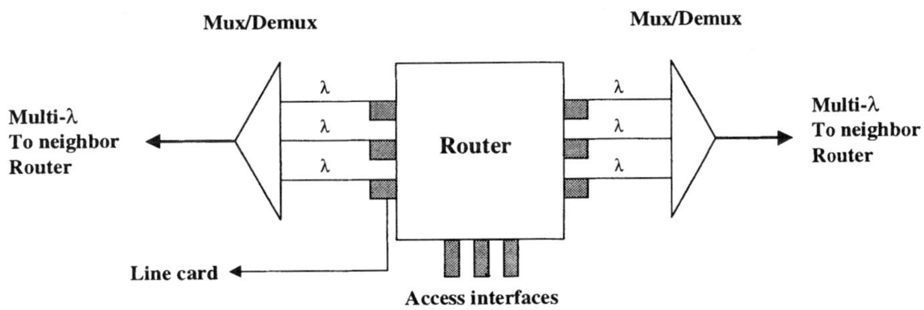

Fig. 2 Router interconnection in a point-to-point WDM network.

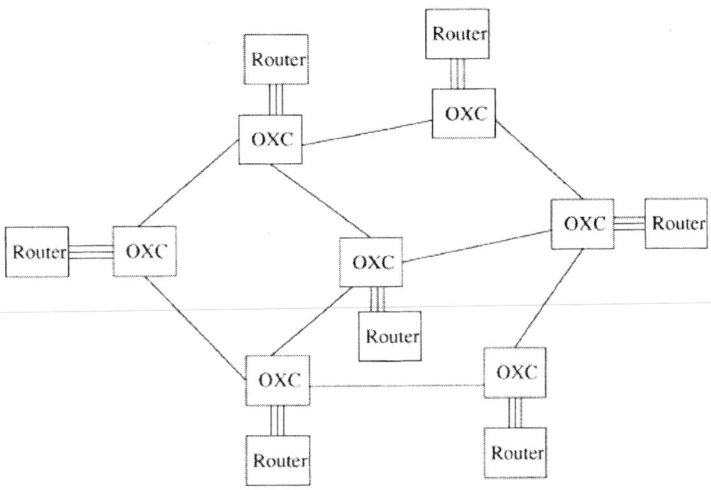

Fig. 3 Architecture of an IP over re-configurable WDM network.

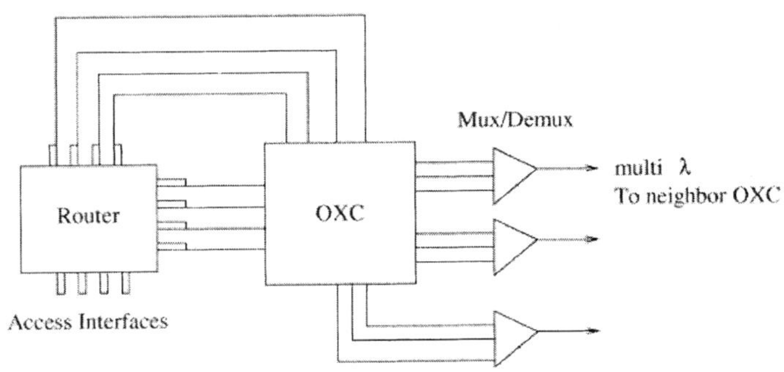

Fig. 4 Router Interconnection in a re-configurable WDM network.

2.2 IP over Re-Configurable WDM

In an IP over Re-configurable WDM architecture, router interfaces from the IP routers are connected to the ports of WDM cross-connects. Figure 3 illustrates such an IP over Re-configurable WDM network. A WDM cross-connect can connect any of its input port to any of its output port. Figure 4 illustrates in detail how an IP router is connected to such a WDM cross-connect. In this architecture, the WDM cross-connects are themselves interconnected in a mesh configuration with multi-wavelength fiber links. By appropriately configuring the WDM cross-connects, a given router interface can be connected to any other router interface at any other router. As a result, the neighboring router for a given router interface is configurable under this architecture. Within this architecture, the IP and the WDM layers can be combined in several different models [2] as illustrated in Fig. 5:

- *Overlay model* where the two layers relate to each other in a *client-server* relationship, with IP being the client layer, and WDM being the server layer. The IP network layer links are realized by the corresponding WDM layer connections. Management and control of the two layers remain separate, with

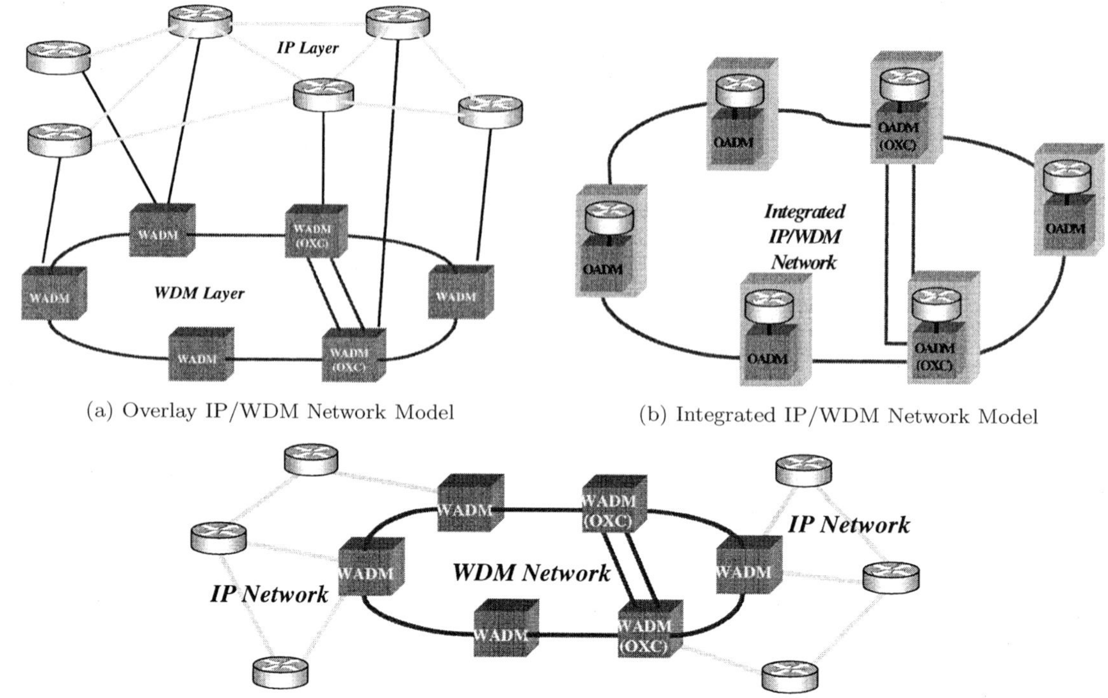

(a) Overlay IP/WDM Network Model

(b) Integrated IP/WDM Network Model

(c) Peering IP/WDM Network Model

Fig. 5 Different IP/WDM architecture possibilities.

well-defined interaction interfaces between them.

- *Integrated model* where the two layers are collapsed into a single layer. In this model, the IP router and the optical cross-connect device are typically integrated together and considered as a single integrated network element. Multi-Protocol Label Switching (MPLS) [3] and its lambda variant (MPλS) [4] are used to provide a single integrated control plane for the collapsed layer over a single network routing and signaling domain employing a common addressing plan.
- *Peering model* where devices from the two layers relate to each other in a peer-to-peer relationship. In this model, MPLS and its lambda variant MPλS again can be used to provide a uniform control plane. However, the IP routers and the optical cross-connects belong to different routing and signaling domains running separate instances of the routing and signaling protocols, and interact with each other on a peer-to-peer fashion using exterior gateway protocols.

The overlay model is likely to be adopted in near-term rapid deployment of relatively static IP/WDM networks. Because of their overall simplified management and control structures, the peering and integrated models are likely to be adopted in the long run for highly dynamic IP/WDM networks, when supporting vendor hardware and software emerge.

2.3 IP over Switched WDM

In an IP over Switched WDM architecture, the WDM infrastructure directly supports a per-packet-switching capability, as opposed to simply providing ingress-to-egress lightpaths. As such, it enables a much finer grain sharing than re-configurable WDM. Various switched WDM approaches have been proposed, including *Optical Burst Switching* (OBS) [5]–[7], and *Optical Label Switching* (OLS) [1]. Since optical logic processing and optical data buffering technologies are still immature at present, switched WDM systems are typically bufferless[†]. Likewise, they rely on electronic processing of the packet header to control the switching actions. We recognize two variants of bufferless switched WDM technology:

- *Optical Burst Switching*: In OBS, the packet control header is sent out along the route path ahead of the actual optical data packet. The idea is that the control header will arrive at the intermediate switching nodes, allowing each switch to perform switching decision computation and install the switch cross-connect setting just in time before the actual arrival of the data packet. In this way, the optical data

[†]Although there are optical packet switching projects [8] that attempt design of optical buffers, those designs remain complex. Therefore most switched WDM system prototype efforts opted for a bufferless design.

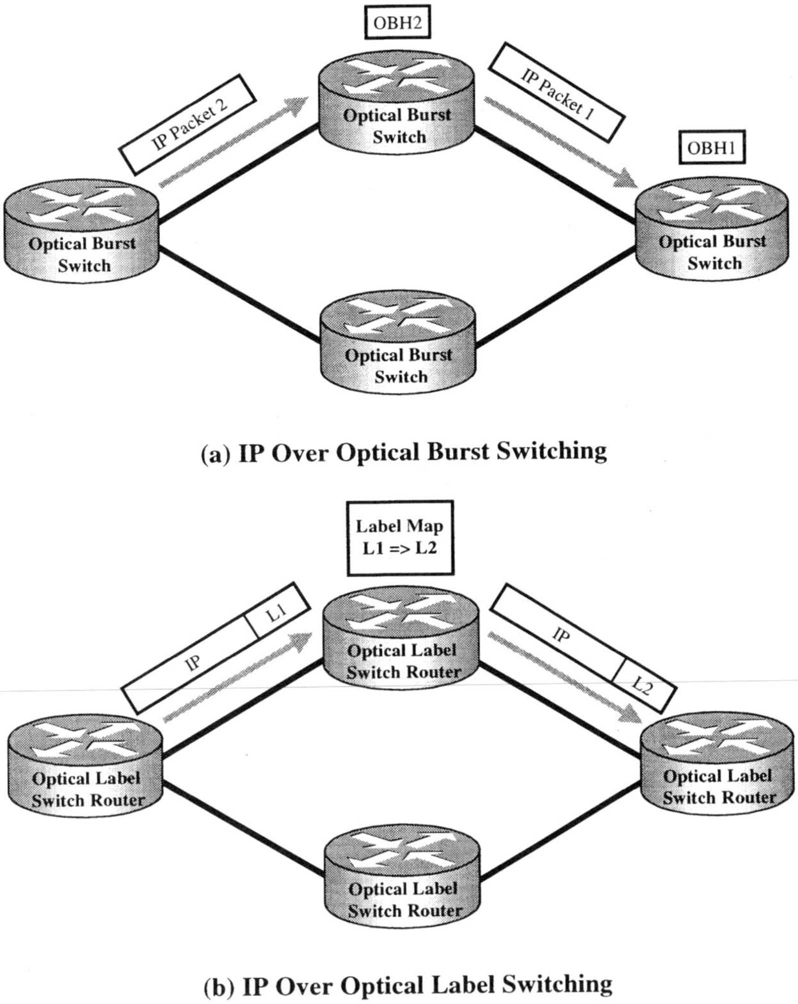

Fig. 6 IP over switched WDM networking technologies.

packet can thread its way across the optical network from ingress to egress. The delay between the control header and the data packet increases with the number of hops and the expected processing delay at the intermediate switches. Figure 6(a) depicts the operation of such an optical burst switched WDM system.

- *Optical Label Switching*: In OLS, the packet control header (called the label) is typically sent together simultaneously with data packet along the route path. To allow for the time needed for switching decision computation and cross-connect setting installation, the data packet is always routed through a local optical delay line upon arrival at an intermediate switch. The delay value is selected so that when the data packet emerges from the delay line, the desired optical cross-connect setting will have been installed. This delay value is local and constant at each intermediate switching node, independent of the particular route paths taken by the packets. Figure 6(b) depicts the operation of such an optical label switched WDM system.

Switched WDM networking technologies are still evolving and it remains to see if they will mature in the next few years to become commercially viable. Similar to the re-configurable WDM approach, IP over switched WDM networks can also follow the overlay model, the integrated model, or the peering model. In the latter cases, an MPLS-based scheme again can be applied to control the IP over Switched WDM network. Such a scheme is proposed in [9] for OBS and a similar scheme is pursued in the NGI SuperNet Project for OLS [10].

3. Network Management and Control Issues

As IP/WDM networking architecture shifted from a static point-to-point architecture towards more dynamic re-configurable and switched architectures, the increased flexibility and agility in networking equipment espoused two important trends:

- A shift away from *static planned* resource alloca-

tion and service provisioning, towards *dynamic on-demand* resource allocation and service provisioning.
- A shift away from *centralized management* and *off-line optimization* strategies towards *distributed control* and *on-line incremental* heuristics in network and traffic management.

There is a desire to endow more intelligence into the IP/WDM network elements to produce a more autonomous network, thereby simplifying and reducing the cost of network operations. In this section, we describe and discuss several management and control issues under this general trend.

Conventional network management discipline divides the management and control issues into the following five functional categories: Fault, Configuration, Accounting, Performance, and Security [11] (FCAPS.) In this section we will enumerate and discuss the issues in the areas of IP/WDM network configuration, fault, and performance management and control. Accounting and security issues are not discussed in this paper.

3.1 IP/WDM Configuration and Routing

3.1.1 Network Configuration and Signaling

Depending on the architectural model, the configuration of IP over dynamic WDM networks raises some new issues. In an overlay model, the WDM layer is a separate layer managed by its own management system and control protocols. In order for the WDM layer to dynamically establish wavelength connection on demand, the WDM layer will require its own addressing scheme and corresponding routing and signaling protocols. An example of such a signaling protocol and its corresponding addressing plan is the *Just-In-Time Signaling Protocol* [12] designed under the DARPA sponsored Multiwavelength Optical Networking [13] (MONET) Project[†]. In order to couple the IP and the WDM layers, either *a WDM Address Resolution Protocol* (WDM-ARP), or a scheme like the ATM Forum's *Multi-Protocol Over ATM* (MPOA) [14] or the IETF's *Next Hop Resolution Protocol* (NHRP) [15] must be developed and adopted. Such schemes would allow the discovery and resolution of IP device addresses and routes across a WDM network.

For the peering and the integrated model, an integrated MPLS-based control plane is adopted that employs the IP addressing scheme uniformly across both IP and WDM devices. Furthermore, MPLS signaling protocols like variants of the *Label Distribution Protocols* (LDP/CR-LDP) [16], [17], and extensions of the *Resource Reservation Protocols* (RSVP/RSVP-TE) [18]–[20], are used in these models. They are used to setup and teardown the coarse-grained *lambda switched paths* (λSP) or lightpath tunnels. Fine-grain IP *label switched paths* (LSP) are then switched over the established lightpath tunnels.

Notice that regardless of the model being adopted, logically a re-configurable IP/WDM network always sees a *virtual topology* that is dynamically re-configurable. In the overlay model, the dynamic virtual topology is the IP link topology. In the peering and the integrated models, the dynamic virtual topology is the lightpath tunnel topology. In all cases, fine-grained IP traffic is routed over the respective virtual topology.

For a small IP/WDM network, given a sufficient number of router interfaces[††] and OXC channels, it is possible to setup a full mesh virtual IP topology allowing *one-hop* IP connectivity (either via an IP link or via a lambda switched path) from any ingress router to any egress router. For larger IP/WDM networks, multi-hop IP routing will be needed to deliver traffic under the overlay model, and two-layer IP MPLS over WDM MPλS stacking will be needed under the peering and the integrated model.

3.1.2 Wavelength Routing and Conversion

Dynamic IP/WDM networks raise new issues concerning the routing of IP wavelength circuits or lambda switched paths. For re-configurable IP/WDM networks, wavelength circuits are established end-to-end to serve either as IP links or as lambda switched paths. Wavelength circuit routing may be performed by link state routing protocols similar to OSPF [21] or IS-IS [22] used for IP routing, with suitable extensions [23]. However, unlike the IP layer where routing can be traffic independent, WDM layer routing must be load dependent. In particular, a WDM fiber that has exhausted all its wavelength channels cannot admit new wavelength circuits. To enable successful routing, wavelength loading must be considered in routing computations. Therefore, routing protocols like OSPF or IS-IS must be extended to propagate wavelength loading in their link state advertisement (LSA) updates.

Another routing issue in IP over re-configurable WDM networks is that wavelength circuit routing is affected by an OXC's ability to perform *wavelength conversion*. Wavelength conversion helps to reduce the WDM connection blocking probabilities. Optical wavelength-conversion technology is still immature and expensive at present. Hence all-optical cross-connects that support wavelength conversion are likely to remain small in size, and be deployed sparingly. Such

[†]Although the JIT Signaling Protocol is designed primarily to support optical burst switching, it is also designed explicitly to support conventional ingress-to-egress lightpath connection establishment.

[††]In practice, the number of interfaces a router can support is limited by the aggregate throughput of the router. E.g., a 40 Gb/s router can support at most 16 (8 in and 8 out) interfaces with 2.5 Gb/s channel speed. This allows at most 8 neighboring routers. Therefore full mesh connectivity will not be possible for networks with more than 9 such IP routers.

capability is perhaps most useful at the edge of a network administrative domain to reduce the information correlation and dependencies across domains. For electro-optic cross-connects, i.e. those that perform *opto-electronic-opto* (OEO) conversions, wavelength conversions can be provided readily. Therefore, depending on the capability of the deployed cross-connects, the wavelength circuit or λSP routing algorithm may have to explicitly cope with the wavelength continuity constraint. In such a case, more detailed wavelength assignment information would have to be included in the link state advertisement updates.

Wavelength conversion capability takes on a different meaning in bufferless IP over switched WDM systems. In such systems, *dynamic wavelength hopping* is relied upon to reduce the switching contention statistics to acceptable levels. Therefore, practical switched WDM systems are likely to incorporate wavelength conversion support in their switches. In such systems, "lambda" switching will not be dictated *a priori* by the LSP setup, rather, the lambda switching decisions will be dynamically made on a per-packet basis to resolve contention.

3.2 IP/WDM Fault Location and Recovery

In order for IP/WDM networks to gain wide deployment in the Next Generation Internet, techniques to guarantee network reliability must be addressed. In this subsection, we will discuss issues concerning the detection and location of faults in IP over dynamic WDM networks, and the corresponding recovery actions when faults have been localized.

3.2.1 Fault Detection and Location

Fault detection and location in IP/WDM networks poses several difficult issues. The problem becomes even more complex in IP over re-configurable WDM networks where the wavelength circuits can be dynamically changed.

Monitored parameters provided by the optical WDM layer are mostly low level analog signal parameters such as optical signal power, optical SNR, and wavelength registration measurements [24] bearing no fixed definitive relations to the high level IP layer observable properties. Likewise, the IP layer's performance monitoring produces soft metrics which bear no simple and direct correlation to the WDM layer measurements. E.g., a low signal power may or may not cause difficulty with the optical receivers, which may or may not result in bringing down an IP link. If the effect is marginal, the IP layer may simply report high bad packet counts. However a high bad IP packet count may also be due to low optical SNR.

Because WDM monitors analog signal parameters, alarms in the WDM layer are typically triggered by the crossing of threshold values set by network management. Because of the uncertainty in setting these thresholds, fault alarms could be generated when the threshold values are set too low. Likewise, alarms that should be generated will be missing when thresholds are set too high. Setting a different threshold could completely change the observable IP layer outcome of the particular fault event. These uncertainties complicate the fault management logic.

An important advantage of optical WDM networks is that the WDM layer can be made transparent, i.e., client signal modulation, format, and bit-rate independent. However this also means that the WDM network has little or no information about its client signal. Threshold values that are tuned for an OC-192 rate signal may be too restrictive for an OC-48 rate signal, generating too many false alarms. Furthermore, because the WDM route path supporting an IP link or lambda switched path can be changed dynamically owing to restoration or traffic engineering activities, threshold values may have to be updated whenever such activities occur. These uncertainties render threshold administration very complex.

Yet another complexity arises from different signal processing and fault propagation characteristics in different WDM equipment. For example, a transparent all-optical cross-connect employing only 1R (re-amplification) regeneration will likely propagate a loss-of-signal fault condition to its downstream nodes. In contrast, most electro-optic cross-connects employing 2R/3R (re-shaping/re-timing) regeneration will emit an idle signal to downstream nodes. If an IP/WDM network includes both types of equipment, anomalous behaviors can result. An example of such was reported in [25] where a tail-end protection mechanism functioned correctly for signals flowing from a transparent all-optical cross-connect towards an electro-optic cross-connect. But the same mechanism failed for signals flowing in the opposite direction because the upstream electro-optic cross-connect masked the lost of signal condition needed to trigger protection switching at the all-optical cross-connect.

Because of these complexities, effective IP/WDM fault detection and location requires tight integration across both IP and WDM. An integrated fault model that encompasses hard and soft metrics from both IP and WDM layers is necessary. This is an area that requires much additional research. A promising approach to integrated IP/WDM fault localization was reported in [26].

3.2.2 Fault Recovery and Network Survivability

The primary objective of network fault recovery is to maintain the end-to-end IP service even under degraded QoS, in spite of network resource failures such as link or node failures. In most IP over point-to-point DWDM systems deployed today, carrier-grade network survivability is provided by the protection and restoration facility in the SONET transport layer. However given the trend and desire to eliminate SONET in the newer re-configurable and switched IP/WDM networks, it has to be replaced by other means. This can be achieved in several different ways:

- *Server or WDM Layer Approach*: In an overlay model, the WDM layer provides wavelength circuits for the routing of IP packets. Therefore it is possible to protect large amount of traffic by protecting and restoring WDM circuits. More specifically, the survivability schemes can be divided into two categories:
  - *Protection schemes*, which pre-provision backup resources that the network can quickly utilize for recovery from a resource failure along the primary path. E.g., link and node disjoint[†] backup circuits can be setup simultaneously with the primary connection to guarantee fast switching to the protection path [24]. Depending on the level of recovery guarantee, the resources along the backup path may be exclusively deployed, or they may be shared among multiple backup paths. E.g., if it is required to tolerate only single-failure, then resources can be shared among all backup paths for which their primary paths have no intersection.
  - *Restoration schemes*, where backup resources are deployed dynamically only when failures occur. Recovery is effected via signaling triggered by the detection of failure. Many schemes exist including both end-to-end and hop-by-hop recovery [27], and possible preemptive and non-preemptive variants.

  Because of their simplicity, protection schemes are typically faster and are most popular in traditional SONET rings. Restoration schemes are much more complex and therefore typically slower. However they are more flexible and adaptive, and may be amenable to a distributed on-line implementation. E.g., ATM signaling protocols have been extended to support this type of dynamic restoration [28], [29]. In practice, a combination of both schemes may be needed.
- *Client or IP Layer Approach*: The IP layer itself employs distributed routing algorithms which are designed to react and adapt to topology changes that may result from resource failures. Likewise, edge routers that employ LSPs can reroute these LSPs when a break is detected. Therefore it is arguable that one may consider relying on the IP layer alone to restore service [30]. For example, a simple IP-based restoration scheme that looks promising has recently been demonstrated in [31]. However such an approach has several concerns:
  - A primary concern for this approach is the slow convergence and response time of IP link failure detection and routing algorithms that renders them unsuitable for use with critical or premium services. E.g., IP link failure detection typically requires multiple rounds of missing *hello* exchanges [32]. Likewise, LSP failure detection can take a long time, since notification messages are propagated and processed hop-by-hop along the LSP back to the ingress router. However, these concerns may be addressed by suitable modification to the IP layer protocol behaviors.
  - Another concern has to do with the scalability and efficiency of the approach. As WDM channels approach OC-192 rates (10 Gb/s), there is a huge amount of traffic to be protected and rerouted in the event of a single wavelength channel (not to mention fiber) failure. Therefore instead of rerouting individual fine grain IP flows, it is much more scalable if a server layer protection mechanism can be engineered. The flip side of the argument of course is that IP layer protection can be made much finer grain and therefore affords much more focused protection, without wasting unnecessary bandwidth for those flows not needing protection.
  - Yet a third concern is that in a purely IP-based approach, there is no way to assure physical WDM layer path diversity between the backup paths setup by an IP/MPLS protection/restoration algorithm. Obtaining such an assurance requires knowledge of the WDM layer physical layout.

  For carrier grade operations, it is certainly desirable (in fact necessary) to have reliable protection and restoration schemes in place. However, for many IP data applications that assume a best effort IP datagram service, many have argued that a pure IP based recovery approach is in fact sufficient. In any case more detailed studies are needed to discern if a pure IP-based approach can compete with WDM layer protection and restoration.
- *Integrated or MPLS Based Approach*: Recently, MPLS-based protection and restoration schemes are being actively pursued in the IETF MPLS working group, and some initial approaches have been already proposed [33]–[35]. For the peering and the integrated models that adopt a unified control plane based on MPLS/MPλS, it would be possible to setup alternate lambda switched paths. This will allow fast backup switching when faults occur to provide uninterrupted end-to-end communications. In essence, the protection and restoration schemes in this category are analogous to the WDM server layer based approach and share the same properties. The main

[†]Simply being logically link and node disjoint may not be sufficient. It may also be necessary to ensure that the WDM links are carried by fibers installed in geographically separated conduits to protect against physical damages.

advantage is that the backup lambda switched paths are now visible to the IP layer, promising more effective coordination across the layers. Furthermore, physical path diversity can be directly enforced if MPλS control has access to the WDM layer's conduit layout information, in addition to the fiber topology.

3.3 IP/WDM Traffic Performance Engineering

In addition to enabling dynamic network configuration and on-demand service provisioning, flexibility and agility in dynamic IP/WDM networks also affords the opportunity to perform *traffic engineering* to optimize network resource utilization and network traffic performance.

3.3.1 IP Routing Considerations

Traditional IP networks employ destination-based shortest path first (SPF) routing such as RIP [36], OSPF [21], and IS-IS [22], to forward traffic from source to destination. This algorithm is simple and stable, and works well when the network is lightly loaded. However, the drawback is that it is traffic independent and does not support diverse routes. The shortest paths from different source-destination pairs in general will overlap at intermediate links causing congestion and packet loss. Although, in theory, it may be possible to adjust the link weights in a routing algorithm to adapt to traffic loading, in practice it is not followed owing to stability and correlation concerns. A recent enhancement to the SPF paradigm is to allow multiple equal-cost paths to a destination, and split the traffic between the multiple paths[†]. E.g., the latest OSPF specification allows for *Equal Cost Multi-Path Routing* (ECMP). However, even ECMP is traffic independent in the sense that there is no load balancing between the multiple paths, because there is also no feedback between the traffic load and the routing algorithm.

A further enhancement to ECMP for load balancing, called *Optimized Multi-Path Routing* (OMP), has been proposed recently. OMP utilizes and extends a link-state routing protocol like OSPF to periodically broadcast link loading information. The routing algorithm utilizes the link-loading information to split the traffic load among multiple near-equal cost paths. Packet forwarding computes a hash on the (source, destination)-tuple, and the hash-space is split among the available paths. Traffic load balancing is achieved by adjusting the hash boundaries among different paths.

Recently MPLS has emerged to support IP traffic engineering needs. In MPLS, the routing and forwarding functions in IP are separated. MPLS supports constraint based routing, including in particular explicit routing, thereby allowing direct control on the exact paths of IP traffic flows. Therefore MPLS can be used to perform multi-path load balancing of IP traffic. Using MPLS, IP traffic load balancing can be effected by rerouting selected LSPs away from congested links. A more elaborate implementation can even manipulate the Forwarding Equivalence Classes (FEC) using a hashing scheme like that in the OMP approach[††] to fine-split the LSPs and achieve more effective load balancing.

In an IP over re-configurable WDM network, traffic engineering can also be effected through wavelength circuit reconfiguration that adapts the IP network (virtual) topology to the evolving traffic pattern. By exploiting the WDM layer's reconfigurability, we can change the IP network's virtual topology to better match the ensuing traffic demand pattern. Below we examine the two mechanisms in more detail.

3.3.2 IP Load Balancing versus WDM Reconfiguration

With MPLS it is possible to attempt load balancing among different possible routes between a router pair. Initially, at low loads, only the shortest-cost path exists. As traffic load between a router pair increases (or if loads on links of the path increases), an MPLS-based traffic engineering algorithm (MPLS-TE) can select less loaded alternate routes, and reroute the different LSPs between the router pair on these alternate routes. The alternate-route selection process works by computing the shortest-cost-path on an auxiliary graph formed by deleting heavily loaded links of the topology.

As network traffic load increases, MPLS-TE keeps rerouting traffic to the alternate paths. However, the maximum traffic loading that can be supported between a node pair is limited by the min-cut-capacity of the IP topology[†††] separating the two nodes. As traffic between a router pair increases, the min-cut-capacity will eventually be saturated, and MPLS-TE can no longer find alternate load balancing paths between the router pair. The cut-edges between the router pair are now all saturated with traffic. The only solution at that point is to expand the capacity of the min-cut (expand the bottleneck) via topology reconfiguration.

As an example, consider the topology and traffic illustrated in Fig. 7(a) where the capacity of each link is assumed to be 1 unit. The *loading* on the links are

---

[†]The splitting algorithm must ensure that packets that belong to the same flow are routed along the same path, to prevent out-of-order arrivals. Simple techniques based on the hashing of source-destination IP addresses can be used for this purpose.

[††]In fact a second proposal called MPLS/OMP was also posted on the IETF Internet Draft site in 1999, but both the OSPF/OMP and the MPLS/OMP drafts have since expired without renewal.

[†††]This is the max-flow min-cut theorem from graph theory.

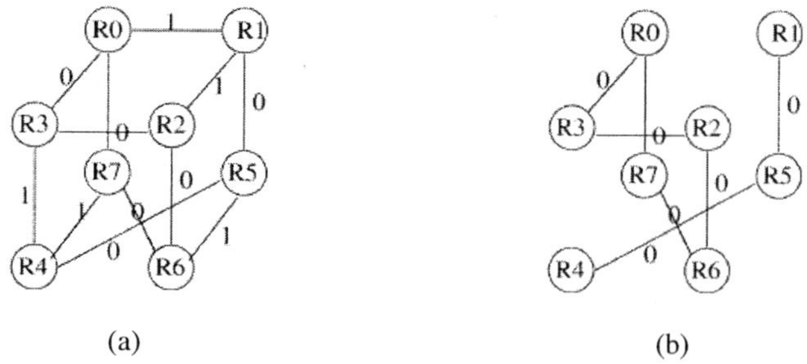

Fig. 7 Illustrative example showing network congestion and capacity exhaustion.

illustrated for some traffic demand. Under this loading, no additional traffic between routers R0 and R1 can be accommodated, because, the remaining capacity across any cut between routers R0 and R1 in this topology is zero. Figure 7(b) illustrates the network topology obtained where links with loading 1 are removed. The resultant topology is partitioned into two disjoint components, with R0 and R1 in different network partitions. Further increase in traffic between the routers R0 and R1 cannot be supported under this traffic loading and network topology.

The re-configurable optical layer can be used to expand the min-cut capacity between a router-pair. Continuing with our example, Fig. 8 illustrates three different topologies† obtained via reconfiguration for which the min-cut capacity between R0-R1 has been increased. In all three cases, an additional unit of spare capacity between R0 and R1 has been created via reconfiguration. These reconfigured topologies can therefore accommodate an additional unit of traffic loading between R0-R1.

For an IP over switched WDM network, the WDM layer also controls the number of wavelengths allocated to each virtual IP link. Such allocations can be dynamic and change over the duration of the connection. This flexibility can be exploited to effect traffic engineering by adapting the link bandwidths available to the IP network. IP/WDM traffic engineering will be discussed in greater detail in the next two sections.

4. Benefits of IP over Re-Configurable WDM Networking

In this section, we present an analysis and some simulation results to illustrate the benefits of running IP over a re-configurable optical layer. Our analysis is by no means exhaustive, but is meant to highlight some of these benefits. In particular, we examine the potential interface resource savings and network throughput improvements. To illustrate our analytical and simulation results, we will utilize the example network depicted in Fig. 9. This example network is derived from a major carrier's IP backbone network. It is representative of a nation-wide long distance IP network.

4.1 Benefit Analysis

The traffic incident at an IP router can be divided into access traffic and transit traffic. Access traffic is the traffic that enters or exits the network through the router. Transit traffic is the traffic that transits across the router to another router. Access interfaces are those router interfaces that carry access traffic, and transit interfaces are those that carry transit traffic.

Let c_{ij} denote the traffic demand from router i to router j, $1 \leq i, j \leq N$. Let $C = \sum_{ij} c_{ij}$, denote the aggregate network traffic demand, and let W be the bandwidth of a wavelength channel††. Assuming that the network can support the entire traffic demand without any loss, the average amount of access traffic at router i is given by $\sum_{j=1}^{N} c_{ij}$. Given a certain routing algorithm such as SPF, ECMP, or OMP, let d_{ij} denote the *average hop distance* traversed by a packet from router i to router j. We define the *average weighted hop distance*, H, for the network as:

$$H = \frac{\sum_{ij} c_{ij} d_{ij}}{\sum_{ij} c_{ij}}$$

The average weighted hop distance H, is determined by three factors: (a) the network connection topology, (b) the network routing function, and (c) the network traffic demand pattern.

4.2 On Router Interface Savings

Because of its expensive high-speed electronics, IP router interfaces constitute a major cost component in the overall cost of an IP router. It is instructive to analyze the potential IP router interface savings possible under a re-configurable IP/WDM architecture. Actual cost savings would be somewhat less due to the cost of the required OXC equipment in such a network. However, the average cost of a high-speed IP router interface is about four to five times that of a corresponding OXC interface. Therefore cost savings in the IP router interfaces will more than compensate for the extra cost incurred by the OXC equipment. Below, we will analyze the number of router interfaces required to carry different IP traffic demands. The purpose of the analysis is not so much to derive exact requirements, but rather to give a qualitative understanding of how reconfiguration can induce resource savings.

†The reconfigured topologies must satisfy the nodal degree constraints dictated by the number of router interfaces.

††The bandwidth of a WDM channel is typically at OC-48 rate of 2.5 Gb/s at present.

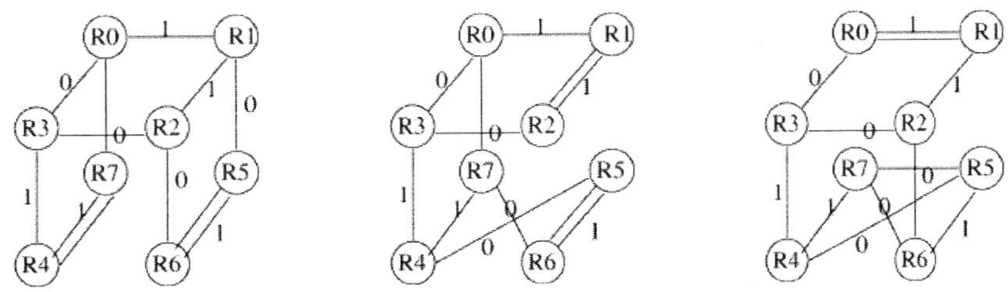

Fig. 8 Reconfigured network topologies.

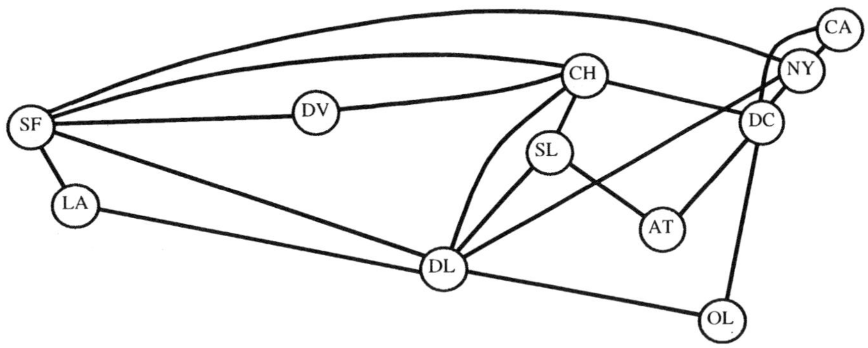

Fig. 9 An example IP backbone network topology.

4.2.1 Given Traffic Demand

The average traffic, T, flowing through a router is computed as follows:

$$T = \frac{\sum_{ij} c_{ij}(d_{ij} + 1)}{N}$$

Therefore, the average number of interfaces at a router is given as $\frac{T}{W}$ which is equal to $\frac{C(H+1)}{NW}$, where W is the bandwidth of a wavelength channel. This quantity is actually the sum of two quantities: (a) the average number of transit interfaces which is equal to $\frac{CH}{NW}$, and (b) the average number of access interfaces at a router which is equal to $\frac{C}{NW}$.

We observe that the number of IP router transit interfaces is directly proportional to the weighted hop distance induced by the routing algorithm on the IP network topologies. Therefore by reducing the average weighted hop distance, H, it is possible to reduce the number of transit interfaces required to support a given traffic.

For our example topology illustrated in Fig. 9, the average hop distance (using min hop routing) is 1.8. Therefore, under uniform traffic, for an IP over point-to-point WDM network architecture with the example topology, the total number of IP router interfaces required is equal to $2.8C/W$. Now consider an IP over re-configurable WDM network architecture. Assume in the ideal case that a full mesh IP topology embedding can be constructed over the re-configurable optical layer, then the total number of IP router interfaces required becomes $2C/W$, because the average hop distance is 1 for a full mesh topology. Therefore, the potential saving in the total number of router interfaces is $(H-1)/(H+1)$. For our example network under uniform traffic, this is equal to $(1.8-1)/(1.8+1)$, which is about 29%.

4.2.2 Locally Constrained Dynamic Traffic

Assume now that we try to size the number of router interfaces for a dynamic traffic demand. Assume that the traffic demand c_{ij} is locally constrained such that $\sum_i c_{ij} \leq nW$ and $\sum_j c_{ij} \leq nW$, where $c_{ij} = c_{ji}$, and $c_{ij} = 0$ or $c_{ij} = W$. This corresponds to the situation where router i is constrained to source and sink at most n bi-directional data flows to and from other routers, and where a flow is a full channel capacity of demand between a router pair. In this case, each router has n access interfaces, and the total traffic that is sourced or sinked at each router cannot exceed nW. Assume again that a full mesh IP topology can be configured over the re-configurable optical layer. A re-configurable IP network that routes traffic in one-hop, would require exactly n transit interfaces at router i, since, depending on the traffic demand, one-hop paths to the appropriate neighbors can be set up. Therefore, the total number of transit interfaces required to fully support this dynamic traffic demand is Nn. Consider now the case for fixed point-to-point WDM. Each fiber would require n interfaces. The total network wide number of transit interfaces required to satisfy the demand is at least

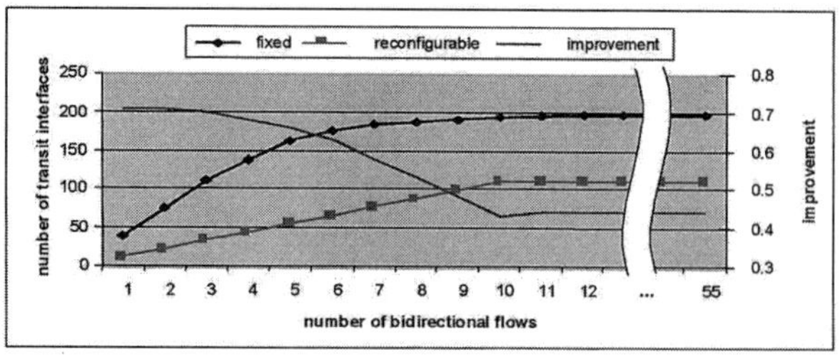

Fig. 10 Potential savings in router transit interfaces for globally constrained dynamic traffic.

$2nL$, where L is the number of links in the IP topology. Therefore to fully accommodate such a dynamic traffic demand, the potential saving in the number of router interfaces is

$$\geq \frac{n(N+2L)-2nN}{n(N+2L)} = \frac{2L-N}{2L+N}.$$

For our example network in Fig. 9 where $N = 11$, $L = 19$, the potential saving in router interfaces is at least 55%.

4.2.3 Globally Constrained Dynamic Traffic

Consider another example traffic demand, where $c_{ij} = 0$ or $c_{ij} = W$, $c_{ij} = c_{ji}$, and $\sum_{ij} c_{ij} = 2nW$. This symmetric traffic corresponds to a network with exactly n bi-directional flows, $1 \leq n \leq N(N-1)/2$, where a flow is, again, a full channel capacity of demand between a router pair. For a re-configurable IP/WDM network, we assume one-hop routing for each flow. Therefore, for n flows, when $1 \leq n \leq N-1$, the number of transit interfaces required per router is equal to n. When $n \geq N$, the number of transit interfaces required per router is equal to $N-1$. For a fixed point-to-point WDM network, the number of transit interfaces required can be computed based on the *convergence degree* of each link in the fixed network induced by its corresponding network routing function. The convergence degree α_k of a network link k, induced by a routing function is the number of source-to-destination routes that are routed over that link. The number of transit interfaces that should be connected to a particular link k is equal to α_k, or to n, whichever is smaller. The total number of transit interfaces needed by a point-to-point IP/WDM network is obtained by summing this quantity over all links. Figure 10 illustrates the router transit interface requirements for our example network topology, when the number of bi-directional traffic flows ranges from 1 through 55.

4.3 On Network Throughput Improvement

In this subsection, we will analyze and estimate the potential improvement in network throughput that may be achieved under a re-configurable IP/WDM network. Let T_f denote the loss-free throughput of the IP over fixed point-to-point network, and T_r denote the *loss-free* throughput of the IP over re-configurable network. In order to make a meaningful comparison, we will adopt the physical constraint of maintaining and utilizing the same number of transit interfaces at each router in both the fixed point-to-point IP/WDM network and in the re-configurable IP/WDM network. Furthermore, we assume that the IP routers all have abundant access bandwidth, so that the *loss-free* traffic demand can be scaled up if there is extra available transit bandwidth in the re-configurable network. Therefore we have $T_f = C_f$ and $T_r = C_r$. Given our previous results on the total number of transit interfaces needed to support a given traffic demand pattern, we have:

$$\frac{T_f H_f}{W} = \frac{T_r H_r}{W} \Rightarrow \frac{T_r - T_f}{T_f} = \frac{H_f - H_r}{H_r}$$

In other words, by decreasing the average weighted hop distance through reconfiguration, the average throughput of a network can be increased correspondingly.

As an example, consider a uniform traffic demand, that is, $c_{ij} = c$, $1 \leq i, j \leq N$, where c is the uniformly constant demand. If we assume that for the uniform traffic, a full mesh between IP routers can be embedded in a re-configurable WDM network, then the improvement in throughput for the resultant one-hop reconfigured network over the original point-to-point network is given by:

$$\frac{T_r - T_f}{T_f} \approx H_f - 1$$

where H_f is the hop-distance of the original fixed point-to-point topology. For our example network in Fig. 9, using min hop routing, where $H_f = 1.8$, the potential improvement in throughput for uniform traffic can be as high as 80%.

4.4 Some Illustrative Simulation Results

To gauge the benefits possible in IP traffic engineering via optical layer reconfiguration, we performed extensive simulation studies. The simulated offered traffic loads captured typical Internet traffic characteristics, such as sudden surge between some node pairs, spatially unbalanced loading and regional congestion. The different loads emulate short-term hot-spot problems, and long term unbalanced resource usage due to poor planning. In this subsection, we present some of the typical simulation results over our representative example network depicted in Fig. 9.

The IP traffic matrix depicted in Table 1 was used as the network load in our simulation study. In this simulation, the links are assumed to be T3 channels with 45 Mb/s bandwidth. Each matrix entry represents a traffic rate in kilobits per second. The representative traffic matrix captures an unbalanced Internet traffic

Table 1 Traffic matrix for the simulation study.

Src\Dst	AT	CA	CH	DC	DL	DV	LA	NY	OL	SF	SL
AT		8250	8250	8250	33	16	88	8250	28490	264	2750
CA	8250		8250	8250	22	44	990	8250	8250	11	2750
CH	8250	8250		8250	22	11	1	24750	8250	176	2750
DC	8250	8250	8250		28	28	506	8250	8250	28	2750
DL	11	22	55	143		22	28	1023	33	22	4
DV	187	550	22	33	33		11	22	55	44	83
LA	44	880	198	11	77	55		358	198	132	16
NY	8250	8250	23100	8250	0	16	16		8250	16	2750
OL	27830	8250	8250	8250	231	33	55	8250		44	2750
SF	1001	22	231	33	11	11	16	22	242		55
SL	2750	2750	2750	2750	22	33	825	2750	2750	231	

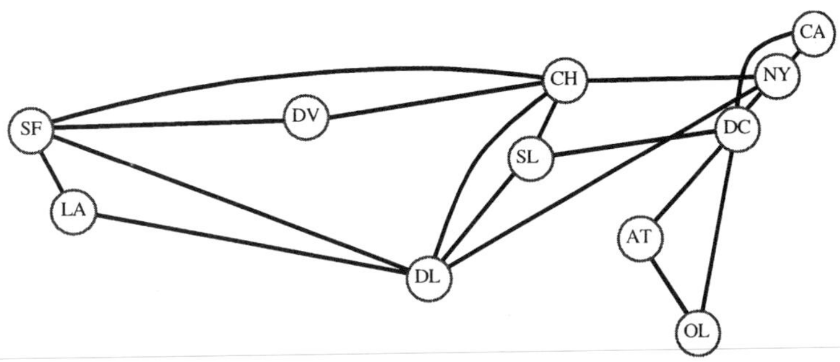

Fig. 11 Reconfigured topology of the simulated network.

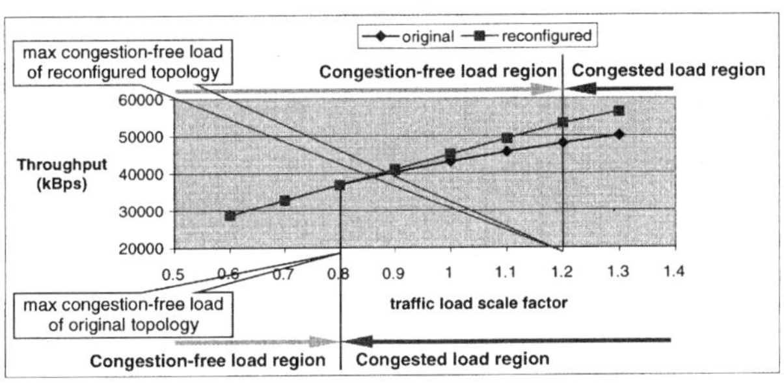

Fig. 12 Network throughput for the orginal and the reconfigured topologies.

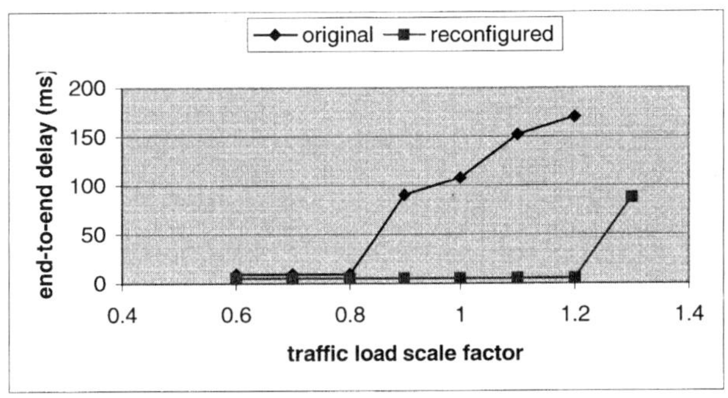

Fig. 13 Packet latency for the original and the reconfigured topologies.

loading. We assume shortest hop path IP routing. A new network topology illustrated in Fig. 11 was constructed by reconfiguring the original topology, subject to the original topology's same nodal degree constraints. The details of the IP topology design and reconfiguration heuristic algorithms used will be reported elsewhere. For the given traffic matrix, the weighted hop distance reduces from 1.72 for the original topology, to 1.56 for the reconfigured topology.

To further stress test these two different network topologies, we scaled the traffic demand matrix proportionally. Figure 12 and Fig. 13 depict the throughput and latency behaviors of the original and the reconfigured networks under the scaled traffic loading. The simulation shows that network performance is improved in many ways. Overall IP traffic throughput is increased by up to 40% over the throughput obtained without reconfiguration under the given skewed traffic demand. The congestion-free load region is also expanded by the same scale. The simulation shows that network wide average packet end-to-end latency is improved in all load regions, although not all flows experience reduced end-to-end latency.

End-to-end latency in an IP network is composed of propagation delay, queuing delay, processing delay, and transmission delay. Packet processing delay is negligible in high performance routers. The dominant delay in light to moderately loaded high bandwidth wide area networks is the propagation delay, which is uniquely determined by the length of the optical path connecting the routers over the physical WDM network. Queuing delay along packet paths is not significant (or at least should not be) until congestion is encountered. Indeed, in our simulation, while the improvement in light load is marginal, significant improvement is observed under heavy load. Figure 14 illustrates the uti-

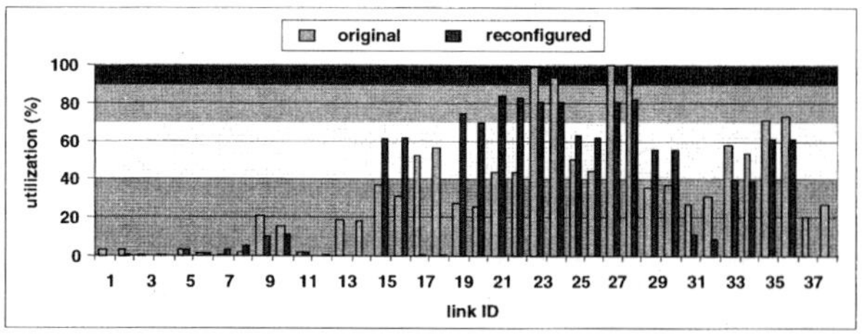

Fig. 14 Link utilization for the original and the reconfigured topologies.

lization for all the links in the original topology and in the reconfigured topology. It clearly illustrates that the effect of reconfiguration is to disperse load from congested links to less congested links. The improvements in average delay are the results of a more efficient network resource allocation, which migrates the loads on congested links over to less utilized ones.

5. IP/WDM Traffic Engineering

5.1 IP/WDM Traffic Engineering Paradigms

Traffic engineering actions in IP over WDM networks can be effected in either the IP layer or in the WDM layer, as well as coordinated across both layers. These correspond to the different paradigms (see Fig. 15) of *overlay traffic engineering* and *integrated traffic engineering* [37], [38].

With overlay traffic engineering, operations in one layer are pursued independent from those in the other layer. Traffic engineering solutions developed for either IP networks or WDM networks are applied directly to their respective layer, with little or no inter-layer coordination. In IP/WDM networks, traffic engineering in the IP layer can theoretically be effected via IP routing algorithms that can adapt the IP packet routes, and perform load balancing. Likewise, traffic engineering in the WDM layer can be effected through light path reconfiguration that adapts the IP network's virtual topology to the evolving traffic pattern. In the case of switched WDM, the WDM layer can also adjust the number of wavelengths allocated to each virtual link, thereby affecting the bandwidth and contention statistics of the virtual link. In practice, because traditional IP routing algorithms are all oblivious to traffic loading, traffic engineering in today's IP networks often relies entirely on link layer adjustments. For example, an IP over ATM network typically maintain a full IP mesh connectivity, and rely entirely on the ATM layer to adjust the bandwidths provisioned for the ATM virtual circuits that support the IP links. These ATM layer actions are completely transparent to the IP layer routing algorithm [38]. Overlay traffic engineering is most likely adopted in near-term deployment of overlay IP/WDM networks.

With the emergence of MPLS-based IP routers, and MPλS-based WDM OXCs, it is now possible to define an integrated control plane for both layers†, which makes it easier to pursue a more integrated approach. With integrated traffic engineering, optimization with respect to selected objectives is pursued coordinately across both the IP and the WDM layers. All of the IP and WDM traffic engineering mechanisms described above still exist, and can now be considered together. Time scale and traffic granularity considerations will determine which specific mechanisms from the different layers are most suitable for different traffic conditions. With more sophisticated hardware that integrates functionality of both IP and WDM at each network element (NE), integrated traffic engineering can be performed even more naturally and efficiently.

In this subsection, we propose and describe an innovative traffic-engineering framework that builds on the strengths of MPLS for fine-grain traffic load balancing, and on optical layer re-configuration for reducing the average weighted-hop-distance, and for expanding the network bottleneck capacity.

5.2 Integrated IP/WDM Traffic Engineering Framework

It is clear that the re-configurable network achieves better flexibility and performance than the fixed network counterpart. We have illustrated the benefits of a re-configurable optical layer in previous sections. However reconfiguration at the optical layer has a special difficulty compared with other technologies such as ATM, which stems from the large and fixed bandwidth granularity of wavelengths. So a challenge is how to exploit WDM reconfiguration in an IP/WDM networking environment despite this coarse granularity.

We have also shown that IP/MPLS has a limitation when it is used without WDM reconfiguration support. As illustrated in Sect. 3, the maximal traffic that MPLS-TE can place between a given node pair is bounded by the maximum flow between the pair which is equal to the minimum cut capacity separating the two nodes. Without WDM reconfiguration, MPLS-TE cannot further engineer traffic in order to accommodate additional load, once the traffic between a node pair reaches the minimum cut between them. WDM reconfiguration however can rearrange the IP virtual topology so that the minimum cut between an overloaded pair is expanded. In effect, reconfiguration at the optical layer can be utilized by MPLS-TE, or any other traffic engineering mechanisms that operates on a higher layer above WDM (i.e. layer 3 (IP) or 2.5 (MPLS).)

Since the WDM reconfiguration and the MPLS load balancing algorithms each have their own limita-

†This possibility applies primarily to the integrated model.

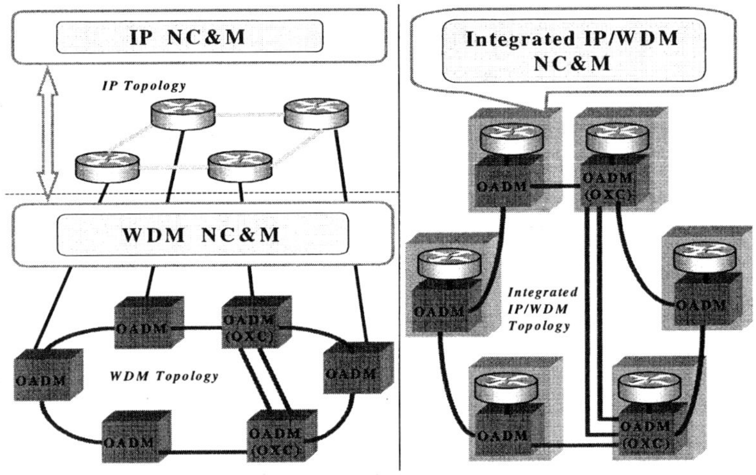

Fig. 15 IP over WDM traffic engineering paradigms.

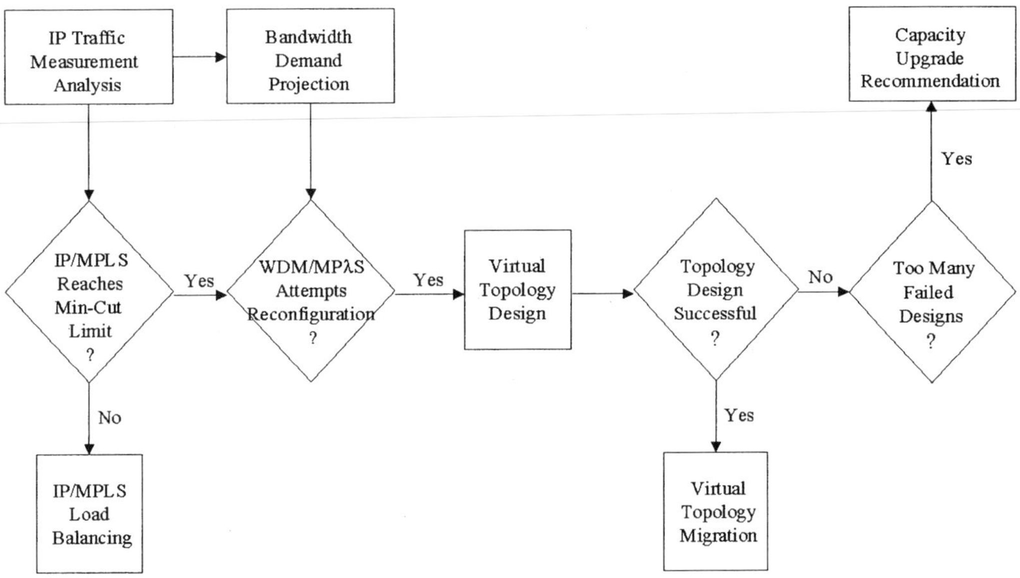

Fig. 16 Integrated traffic engineering workflow with MPLS/MPλS for IP/WDM.

tions when used separately, or used in an uncoordinated way, here we propose an integrated framework. This framework builds on the strengths of MPLS-TE for load balancing and WDM reconfiguration for reducing weighted hop distance, and for expanding bottleneck capacity. Figure 16 illustrates the workflow of our integrated traffic engineering framework.

In our integrated framework, the MPLS-TE algorithm balances load between label switched paths for a given node pair, until it hits the max-flow (i.e. min-cut capacity) for the node pair. Graph-theoretically, this condition is arrived when the source and destination nodes become disconnected in the spare network capacity graph. Upon detecting this condition MPLS-TE stops balancing the load. Instead it requests the WDM reconfiguration algorithm to change the virtual topology and expand the min-cut capacity. The WDM reconfiguration algorithm then derives a new IP virtual topology based on the traffic demand matrix collected from the IP/MPLS router measurements. Various optimization heuristics and additional operational constraints† can be applied in this derivation and the details of these heuristics and their performances will be reported in a separate paper elsewhere. For our present purpose, it suffices to say that the objective is to pick an alternative topology with the smallest average hop distance and a larger min-cut size while simultaneously satisfying the traffic demand of all node pairs, in addition to the currently overloaded node pair.

If no such topology can be found, the network is

†For example, to prevent excessive reconfiguration which could lead to unstable network operations, the reconfiguration system will typically delay reconfiguration until multiple min-cut exhaustion events have been observed.

considered genuinely congested. The WDM reconfiguration algorithm will then declare failure. Typically this would trigger the generation of a *reconfiguration failure alarm*. Furthermore, depending on the threshold set by the network management system, a trouble ticket may be issued if too many such reconfiguration failure alarms have been recorded. This in turn may be routed to the network planning staff, requesting deployment of new capacity. Otherwise, the reconfiguration system goes ahead and changes the IP virtual topology as prescribed by the topology derivation. The IP layer routing algorithm and hence MPLS-TE, will learn about the new topology from its subsequent link state message exchanges.

In essence, we observe that: (1) WDM reconfiguration is not an adequate tool for IP traffic load balancing mainly due to the coarse and fixed wavelength granularity. MPLS-TE does a much better job in load balancing. (2) The load balancing capacity of MPLS-TE alone, however, is bounded by the underlying virtual IP topology. These observations led us to conclude that the traffic engineering on layer 2.5/3 and WDM reconfiguration are mutually complementary. By linking MPLS load balancing and WDM reconfiguration together in a well-coordinated way, we can build a much more powerful traffic-engineering tool than when we use only one of them, or use them independently. In particular, we can make MPLS-TE stop when traffic reaches the graph-theoretic limit and invoke the WDM reconfiguration facility. We expect that such an integrated traffic-engineering approach will greatly enhance the utility and performance of future IP/WDM networks.

5.3 Integrated IP/WDM Traffic Engineering Components

An integrated IP/WDM traffic-engineering system will require extensive support from the overall network management system including IP/WDM configuration and fault management functions. For example, an MPLS signaling functional component such as RSVP or CR-LDP will be needed to establish IP level label switched paths. In this subsection, we enumerate and describe those components that are directly related to traffic engineering activities:

- *Traffic measurement and analysis* is responsible for collecting traffic statistics from the network elements, e.g. IP routers and OXCs. These statistics are analyzed and/or aggregated to prepare for the traffic engineering and network reconfiguration related decision-making.
- *Bandwidth demand projection* attempts to project the bandwidth requirements in the near future based on past and present measurements and the characteristics of the traffic arrival processes. The bandwidth projections are used for subsequent bandwidth allocation.
- *IP/MPLS load balancing and re-routing* consists of the set of policies and algorithms that decide when an IP level LSP re-routing and/or load splitting/balancing action should be performed. This is based on traffic measurements, and on operational issues, e.g., to balance between the desire to react quickly, versus the need to maintain stability and conserve resources.
- *WDM/MPλS reconfiguration trigger* consists of a set of policies and algorithms that decide when a WDM level reconfiguration should be performed. Again, this is based on traffic measurements and bandwidth demand projections, and on operational issues, e.g., to suppress the influence of transitional factors and reserve adequate time for network to converge.
- *Topology design* provides a network topology based on the traffic measurements and projections. Conceptually this can be considered as optimizing a graph (i.e. IP routers connected by light paths in the WDM layer) for specific objectives (e.g. maximizing throughput), subject to certain constraints (e.g. nodal degree, interface capacity), for a given load matrix (i.e. traffic load applied to the network.) This is in general a NP-hard problem. Since reconfiguration is regularly triggered by continually changing traffic patterns, an optimized solution may not be stable. It may be more practical to develop heuristics that emphasize more on factors like fast convergence and less impacts on ongoing traffic, rather than on optimality.
- *Topology migration* consists of algorithms to coordinate the network migration from an old topology to a new topology. As WDM reconfiguration deals with large-capacity channels, changing allocation of channel resources in this coarse granularity has significant effects on a large number of IP MPLS end user flows. These flows have to adapt to the light path changes at and after each migration step. If not handled correctly, these effects can potentially spread over the routing pattern of the network, which in turn may affect more user flows.

An implementation of our traffic-engineering framework can be either *centralized* or *distributed*. A centralized traffic engineering system could interact with conventional IP and WDM network management systems via standard protocols and interfaces[†]. It can collect the global information needed to make proactive bandwidth projections, compute an optimal traffic engineering solution, and can systematically schedule the topology migration steps. Therefore a centralized implementation is more naturally used to provide an *off-line* traffic engineering solution. In contrast, a distributed implementation will likely perform traffic engineering

[†]Such protocols include SNMP, OSPF for IP resource and topology information, CMIP and TL1 for WDM layer information, and in the case of the MONET NC&M [24], object-oriented CORBA APIs.

optimizations based on limited information, and schedule only local incremental topology changes. Therefore, a distributed implementation tends to be more reactive, and is more natural in supporting an *on-line* traffic engineering solution. A distributed implementation requires tight integration with the IP and WDM network control functions. It is also more complex because of the additional synchronization concerns. Therefore centralized implementations are likely to be adopted in near-term traffic engineering solutions.

6. Conclusions

In this paper, we reviewed the different IP/WDM networking architectural models and their tradeoffs. We outlined and discussed a number of management and control issues and possible solutions related to the configuration, fault and performance management of IP over dynamic WDM networks. We presented an analysis and simulation results demonstrating the potential benefits of dynamic IP over WDM networks. Optical layer reconfigurability allows for the reduction of the weighted-hop-distance of the IP topology given a certain traffic demand. A reduction in the weighted hop distance reduces the router interface requirements and improves overall throughput by consuming less network resources.

We discussed the issues related to IP/WDM traffic engineering in detail. WDM reconfiguration is confined to the coarse and fixed bandwidth granularity of wavelength channels. On the other hand, while MPLS-TE achieves fine-grained load balancing, it is limited by the min-cut capacity of the underlying network topology. Based on these observations, we presented an innovative integrated traffic-engineering framework for re-configurable IP/WDM networks that is developed in the NGI SuperNet NC&M Project funded by DARPA. It builds on the strength of Multi-Protocol Label Switching (MPLS) for fine-grain IP load balancing and on the strength of configurable WDM networking for reducing the IP network's weighted-hop-distance as well as expanding the bottleneck bandwidth.

For further investigation and experimentation with this integrated traffic-engineering framework, the NGI SuperNet NC&M project includes the design and implementation of an integrated IP/WDM traffic engineering system prototype. We are currently experimenting with different on-line and off-line algorithms. The details of this design, implementation, and our experimental results will be reported in a separate paper elsewhere.

Finally, note that while we have enumerated many important and fundamental architectural, management, and control issues regarding IP over WDM networks, the list is by no means exhaustive. For example, issues involving the support of IP multicast in IP/WDM networks using optical signal splitting [39], [40], and multi-domain IP/WDM network interworking have not been discussed because of paper length limitations. These are important issues that also need to be addressed before a true Next Generation Optical Internet can emerge.

Acknowledgements

The authors wish to thank Yukun Tsai, Narayanan Natarajan, Tsong-Ho Wu, Deh-Phone Hsing, Michael Post, Arunendu Roy, Young Yong Kim, and Xijun Zhang, for their contributions and participation during various stages of the NGI SuperNet NC&M Project of which part of this paper is based on. The first author also wishes to thank Gee-Kung Chang, Georgios Ellinas, Brian Meagher, Jeffery Young, Arshad Chowdhury, and Brian Wilson for the many stimulating discussions during the course of this work.

References

[1] G.K. Chang, et al., "A proof-of-concept, ultra-low latency optical label switching testbed demonstration for next generation internet networks," OFC'2000, vol.2, Section WD5, pp.56–58, Baltimore, Maryland, March 2000.

[2] J. Luciani, et al., "IP over optical networks, a framework," IETF Internet Draft, http://www.ietf.org/internet-drafts/draft-ip-optical-framework-00.txt.

[3] E.C. Rosen, et al., "Multi-protocol label switching architecture," IETF Internet Draft, http://www.ietf.org/internet-drafts/draft-ietf-mpls-arch-06.txt.

[4] D. Awduche, et al., "Multi-protocol lambda switching: Combining MPLS traffic engineering control with optical cross-connects," IETF Internet Draft, http://www.ietf.org/internet-drafts/draft-awduche-mpls-te-optical-01.txt.

[5] J.S. Turner, "Terabit burst switching," International J. High Speed Networks, 1999.

[6] J.Y. Wei, et al., "Just-in-time optical burst switching for multiwavelength networks," IEEE IFIP'99, Hong Kong, Nov. 1999.

[7] M. Yoo and C. Qiao, "A new optical burst switching protocol for supporting quality of service," Proc. SPIE'98, All Optical Networking: Architecture, Control, and Management Issues, pp.396–405, Boston, Massachusetts, Nov. 1998.

[8] D. Chiaroni and B. Lavigne, "All-optical fast packet-switched networks: Physical and logical limits of operations," Proc. SPIE'98, All Optical Networking: Architecture, Control, and Management Issues, pp.378–389, Boston, Massachusetts, Nov. 1998.

[9] H.M. Chaskar, et al., "A framework to support IP over WDM using optical burst switching," Optical Network Conference 2000, Dallas, Jan. 2000.

[10] J.Y. Wei, "NGI SuperNet network control and management," DARPA/ITO NGI 1999 PI Meeting, http://www.dyncorp-is.com/darpa/meetings/ngi99dec/NGI%20Slides/Wei,%20John.pdf, Dec. 1999.

[11] M.W. Maeda, "Management and control of transparent optical network," J. Sel. Areas Commun., vol.16, no.7, pp.1008–1023, July 1998.

[12] J.Y. Wei and R.I. McFarland, "Just-in-time signaling for WDM optical burst switching networks," To appear in the Special Issue on Optical Networks, Journal of Lightwave

Technology, Dec. 2000.

[13] R.E. Wagner, et al., "MONET: Multiwavelength optical networking," J. Lightwave Technol., vol.14, no.6, pp.1349–1355, June 1996.

[14] "Multi-Protocol over ATM Version 1.1," ATM Forum Technical Specification, AF-MPOA-0114.000, May 1999.

[15] J. Luciani, et al., "NBMA next hop resolution protocol (NHRP)," IETF RFC 2332, April 1998.

[16] L. Andersson, et al., "LDP specification," IETF Internet Draft, http://www.ietf.org/internet-drafts/draft-ietf-mpls-ldp-06.txt, Oct. 1999.

[17] Z.B. Tang, et al., "Extensions to CR-LDP for path establishment in optical networks," IETF Internet Drafts, http://www.ietf.org/internet-drafts/draft-tang-crldp-optical-00.txt, March 2000.

[18] J.P. Lang, et al., "Extensions to RSVP for optical networking," IETF Internet Draft, http://www.ietf.org/internet-drafts/draft-lang-mpls-rsvp-oxc-00.txt, March 2000.

[19] D. Saha, et al., "RSVP extensions for signaling optical paths," IETF Internet Draft, http://www.ietf.org/internet-drafts/draft-saha-rsvp-optical-signaling-00.txt, March 2000.

[20] Y. Fan, et al., "Extensions to CR-LDP and RSVP-TE for optical path set-up," IETF Internet Draft, http://www.ietf.org/internet-drafts/draft-fan-mpls-lambda-signaling-00.txt, March 2000.

[21] J. Moy, "OSPF Version 2," IETF RFC 2328, April 1998.

[22] D. Oran, "OSI IS-IS intra-domain routing protocol," IETF RFC 1142, Feb. 1990.

[23] K. Kompella, et al., "Extensions to IS-IS/OSPF and RSVP in support of MPL(ambda)S," IETF Internet Draft, http://www.ietf.org/internet-drafts/draft-kompella-mpls-optical-00.txt, Feb. 2000.

[24] J.Y. Wei, et al., "Connection management for multiwavelength optical networking," J. Sel. Areas Commun., vol.16, no.7, p.1097, July 1998.

[25] W.T. Anderson, et al., "The MONET project—A final report," OFC'2000, vol.2, Section WJ1, pp.148–149, Baltimore, Maryland, March 2000.

[26] C. Mas and P. Thiran, "An efficient fault localization algorithm for IP/WDM networks," Optical Network Conference 2000, Dallas, Jan. 2000.

[27] T. Wu, Fiber Network Service Survivability, Artech House, 1992.

[28] D. Hsing, et al., "A restoration methodology based on preplanned source routing in ATM networks," ICC'97, June 1997.

[29] G. Goncu, et al., "A restoration strategy based on PNNI routing in ATM networks," GLOBECOM'98, Nov. 1998.

[30] S. Chaudhuri, "Facility management and restoration in the optical vs. data layers," OFC'2000, vol.3, Section ThE, p.67, Baltimore, Maryland, March 2000.

[31] A. Greenberg, et al., "Smart routers—Simple optics: A network architecture for IP over WDM," OFC'2000, vol.3, Section ThU, pp.292–294, Baltimore, Maryland, March 2000.

[32] C. Huitema, Routing in the Internet, Prentice Hall, 1995.

[33] S. Makam, et al., "Framework for MPLS based recovery," IETF Internet Draft. http://www.ietf.org/internet-drafts/draft-makam-mpls-recovery-frmwrk-00.txt, March 2000.

[34] D. Haskin, et al., "A method for setting an alternative label switched paths to handle fast reroute," IETF Internet Draft, http://www.ietf.org/internet-drafts/draft-haskin-mpls-fast-reroute-03.txt, March 2000.

[35] L. Andersson, et al., "Requirement framework for fast reroute with MPLS," IETF Internet Draft, http://www.ietf.org/internet-drafts/draft-andersson-reroute-frmwrk-00.txt.

[36] G. Malkin, "RIP version 2," IETF RFC 2453, Nov. 1998.

[37] J.Y. Wei, et al., "IP over WDM traffic engineering," IEEE/LEOS 2000 Summer Topical Meetings on Broadband Optical Networks, Aventura, Florida, July 2000.

[38] W. Barnes, "Traffic engineering with the overlay model," International Conference on MPLS'98, McLean, Virginia, Nov. 1998.

[39] X. Zhang, et al., "On fundamental issues in IP over WDM multicast," ICCCN'99, Boston, Massachusetts, Nov. 1999.

[40] X. Zhang, et al., "Constrained multicast routing in WDM networks with sparse light splitting," INFOCOM'2000, Tel Aviv, Israel, March 2000.

High-Performance Backbone Network Technology edited by Naoaki Yamanaka
ISBN: 0-8247-5321-6 DOI: 10.1081/(120025659)
Published 2004 by Marcel Dekker, Inc.

Challenges for the Next-Generation Internet and the Role of IP over Photonic Networks

Masayuki Murata

SUMMARY In this article, we first discuss QoS metrics of the data networks, followed by raising the challenging problems for the next-generation Internet with high-performance and high-quality. We then discuss how the WDM technology can be incorporated for resolving those problems. Several research issues for the IP over WDM networks are also identified.
key words: *internet, IP (Internet Protocol), TCP (Transmission Control Protocol), congestion control, WDM (Wavelength Division Multiplexing), network dimensioning*

1. Introduction

Computer networks (i.e., the Internet) and telephone networks (and succeeding Narrowband and Broadband ISDN networks) have evolved in different ways. A fundamental difference of two networks is to offer the connection-oriented service or the connectionless service. The reason of taking such different approaches can be found by observing how the network providers view the network. According to [1], the standpoint of the Internet supporters is as follows;

> *The subnet's job is moving bits around and nothing else. The subnet is inherently unreliable, no matter how it is designed. Therefore, the hosts should accept the fact that it is unreliable and do error control and flow control (and congestion control) by themselves.*

The above argument directly leads to the packet-switched connectionless network. Accordingly, the control complexity is put into the transport layer residing at the end hosts. A rational behind this is;

> *user's computing power has become cheap, so that there is no reason not to put the complexity in the hosts.*

On the other hand, the telephone network puts the control complexity in the network layer because

> *most users are not interested in running complex transport layer protocols in their machines. What they want is reliable, trouble-free service, and this service can be best provided with network layer connections. Furthermore, some services, such as real time audio and video are much easier to provide on top of a connection-oriented network layer than on top of a connectionless network layer.*

Those are reasons why the telephone network adopts the connection-oriented service.

Perhaps, both standpoints above are correct in order to offer their intended services; voice in the telephone networks and data in the computer networks. However, we are now facing with a rapid penetration of the Internet into the civil society as well as the industry, and therefore we have no reasons not to build the network suitable to the data service. It is still not clear whether two networks will be converged, or those will keep on working separately because of reasons of existence. However, it must be true that the next-generation Internet would not be realized without the photonic technology, which was originally developed for telephone networks.

In this article, we will first review the challenging problems for the next-generation Internet to provide the high-performance and high-quality service. We will then discuss how the WDM technology should be contributed to resolve those problems. Our intention is not to mention that all those problems can be resolved by the WDM technology, but several of problems should be coped by the upper-layer protocols than the WDM layer. In other words, we have several problems that the WDM layer should not treat. The above two alternatives of connection-oriented and connectionless services become very important to discuss the network structure of IP over WDM networks. Experiences on ATM networks is also helpful for choosing the appropriate network architecture, which is summarized in the next section.

2. Why ATM is Not Accepted as a Data Transport Technology?

As IP (Internet Protocol) becomes a dominant networking technology, the way for building wide are networks (WAN) is evolving to meet the needs of data traffic. The ATM technology itself has a capability of supporting both of QoS-guaranteed and best-effort services. See, e.g., [2], [3]. For the QoS-guaranteed service, a connection is established before actual communication by allocating physical network resources such as the bandwidth, in addition to the logical network resources

Manuscript received May 11, 2000.
†The author is with Cybermedia Center, Osaka University, Toyonaka-shi, 560-0043 Japan.

IEICE TRANS. COMMUN., Vol. E83-B, No. 10, pp. 2153–2165, October 2000

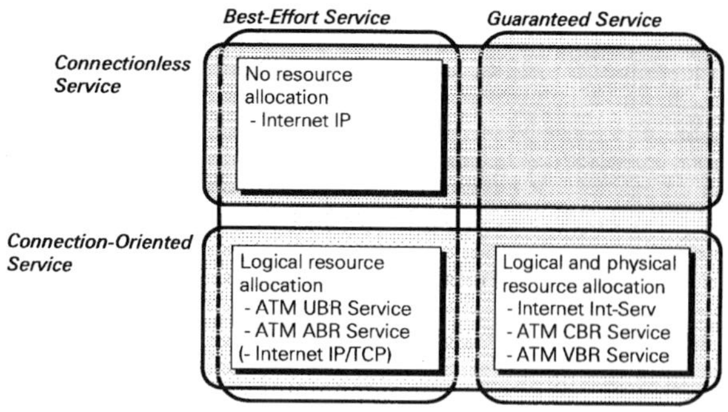

Fig. 1 Classification of network services.

(which correspond to the connection identifier, VPI and VCI, in the ATM networks). Such a mechanism is defined in CBR (Constant Bit Rate) and VBR (Variable Bit Rate) service classes in ATM [2]. Those are intended to be primarily applied for real-time traffic requiring QoS guarantees. See Fig. 1.

For supporting data traffic, ATM defines UBR (Unspecified Bit Rate) and ABR (Available Bit Rate) service classes where no physical network resources are explicitly allocated to the connection. The problem is that the ATM network is essentially connection-oriented; the connection requires the identifier before starting the communication. It inherently inhibits a spread use of the ATM native data communication services. It is because dominant in the current Internet are short-lived connections of Web-based applications. Here, the short-lived connection means that its life-time is very short; after the client sends the request to the server, the connection immediately ceases by receiving small responses (about 10 KByte response is typical in Web document retrievals [4]), only by which we can enjoy *net-surfing*. It is quite different from the conventional telephone network where the connection is assumed to continue during a few minutes or even hours after the connection is established. In the Internet, the connection is established within the transport layer of TCP, but it does not require the connection establishment physically or even logically in the network layer, since its fundamental principle is that TCP/IP does not make any assumption on the lower-layer services (connection-oriented or connectionless).

As an alternative, we can utilize the ATM technology as the link layer protocol, where the ATM link is provided to IP only as a physical link between IP routers. It is called permanent VP or VC (PVP and PVC). In this case, however, we have competitive transport technologies such as SONET. When comparing with SONET, ATM is just a technology that introduces the overhead, which is so-called a *cell tax* [5].

During the standardization effort of the ATM technology, another approach was invented. It utilizes the high-speed switching technology developed for the ATM switch, and the connection path (VP/VC of the ATM) is set up on demand [6]. When the path becomes unnecessary, it is teared down. It seems to be one ideal solution for integrating IP (connectionless service) and ATM (connection-oriented service). However, the signaling protocol to set up the connection path is a burden to support short-lived connections. In the new technology of MPLS (Multiprotocol Label Switching), a cloud constituted by some technology (in the current context, ATM) is offered to the upper layer protocol by permanent connections. See, e.g., [7].

The service suitable to the ATM principle is probably real-time multimedia applications requiring large bandwidth, but it may or may not dominate in the future. At least, we are now going towards the data networking era, and the applications we have already had are best tuned for connectionless IP.

From discussions above, we have found several choices to build the underlying network for IP;

1. to provide the physical link to adjacent IP routers,
2. to provide the cloud of the underlying network; e.g., the current standardization of MPLS,
3. to provide the cloud as the underlying network, and the connection path is set up on demand basis; e.g., the conventional model of IP over ATM switching, or
4. to replace IP by a new networking technology. ATM was believed to have such an ability, but its connection-oriented service became an obstacle.

A successful choice would be dependent on the ability to handle short-lived connections as described before. The only architecture that can handle those connections would survive. From this point of view, the third and fourth choices above are difficult to be applied to the current implementation of Web-based applications.

3. What is QoS for Data Applications?

In this section, we define *QoS (Quality-of-Service)* for data applications. In real-time applications, specifying QoS of communication services is straight-forward. It can be well-defined in terms of, e.g., throughput, packet delays and loss probability, although it is not easy to guarantee the latter two metrics in general [3]. Further, we have a clear relationship between the allocated bandwidth and quality perceived by the user in terms of, e.g., MOS (Mean Opinion Score) values [8].

How about the case of data applications? Do we have some performance metrics to define the quality of data communication services? The answer is probably no. It is because especially in the case of data communication, many elements affect its quality; those include

1. the physical distance between sender and receiver,
2. processing for DNS,
3. the transmission capacity of links,

4. packet switching and forwarding at routers,
5. protocol processing at sender and receiver, and
6. application processing.

One may think that the end-to-end delay (e.g., after clicking the button in the Web page until the related Web document displayed) is the performance metric of the data applications. It would be adequate for the data application of Web browsing. Less than several hundred milliseconds may be an appropriate parameter value by the author's personal feeling. One realization seems to be that the delay is divided and allocated to each component, and that each component is designed to assure the allocated delay. However, such an idea cannot be applied to the heterogeneous network environments consisting of various backbone networks, access networks of ISPs (Internet Service Providers), access lines of users to ISP, and server/client computers.

To further discuss QoS of data applications, consider the bandwidth-guaranteed service for data connections. Some ISP announces that the user can be guaranteed 64 kbps access (in the case of the narrowband ISDN channel). It is apparently impossible;

1. The number of modem ports prepared by ISP limits the number of users simaltaneously accessed. It means that *call blocking* may occur when the user accesses ISP.
2. The capacity of the access line between ISP and the backbone may limit the individual user's throughput.
3. To guarantee the user's throughput even on the access line, the number of users simultaneously admitted should be limited. It also leads to call blocking.
4. Item 3 above implies that the throughput larger than 64 kbps is never allowed.
5. The slow backbone may deteriorate the throughput.

The bandwidth guarantee is only concerned with the access line between the user and ISP, and it is quite far from the guaranteed service of data communications.

It is not too much to say that the current Internet is supported by user's patience. QoS in data communication services is probably not to guarantee some performance metrics such as the end-to-end delay or throughput, but to lower the delay at each component as much as possible with *best-effort*. In other words, we have neither a ultimate nor shorter way to build the next-generation Internet with high-performance and high-quality; all that we can do is to make efforts for improving each network component separately. We then expect that the entire delay becomes improved as a result.

4. Challenging Problems for the Next-Generation Internet

In this section, we summarize challenging problems for the next-generation Internet. An intention of this section is to identify what the WDM layer should support, and what it need not, or even what it should not.

4.1 Supporting QoS for Real-Time Multimedia

By the Internet community, an effort to establish the connection-oriented service analogous to the conventional telephone network has recently been made. It is called ISPN (Integrated Services Packet Network) [9], [10] or intserv (Integrated Services Architecture) in the IETF [11]. An analogy to the telephone network means that intserv requires

- an end-to-end signaling protocol, and
- a bandwidth allocation mechanism at the router.

Those are implemented in a form of RSVP [12] and WFQ [13] (and its variants), respectively. By integrating those two mechanisms, we have a connection-oriented service on the Internet [14], and QoS of real-time multimedia is guaranteed. One important research area in this field is that many researches have been neglecting QoS guarantees at the end hosts. One approach is to utilize the real-time OS to integrate QoS guarantee mechanisms at end systems and networks [15].

As intserv is deployed, however, its problem has also been pointed out. It is scalability. The first scalability problem exists in the signaling protocol, which should be passed by all the routers along the path between end systems. The second is related to the packet scheduling algorithm of WFQ. As the number of active connections on the router increases, processing time at the router is increased.

It follows that differentiated service (or diffserv in short) is now drawn considerable attention [16]. The diffserv architecture is not a mechanism to guarantee QoS, but to discriminate the service grade among classes. Thus, the diffserv architecture is not a final goal for QoS guarantees of real-time multimedia. By diffserv, it can be expected that the user who pays more money can receive better performance. Another possible application of diffserv is to build VPN (Virtual Private Network) by its ability to allocate some portion of the bandwidth to the class.

The question is whether we need an ISPN-like service in the current and future Internet or not? It probably depends on how much real-time multimedia applications will be deployed. If it becomes major in the future Internet, it is valuable to invest in the ISPN architecture. Another scenario we can consider is that at the edge router, we employ the WFQ-like packet

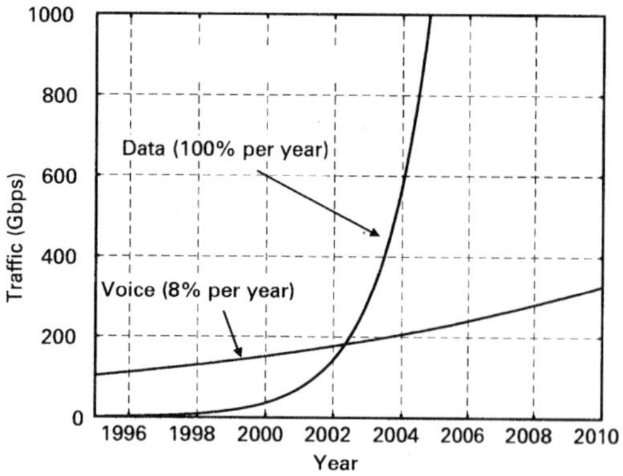

Fig. 2 Traffic growth of the Internet.

scheduling algorithm to ensure better QoS of real-time applications. The core router within the network does not discriminate flows under the assumption that the backbone network becomes sufficiently fast by, e.g., the photonic technology.

4.2 Building High-Speed Packet-Switching Backbone Networks

We are now facing a data explosion in the Internet. One projection states that in the very near future, the data traffic dominates the network [17]. See Fig. 2. Since its projection is based on the statistics of mid 90's when the Internet was rapidly accepted, the projection is likely to overestimate the future growth of data traffic. However, the tendency would continue at least for the time being.

It used to be pointed out that the future killer application is multimedia applications. However, it is now proven to be incorrect. The true story is that the killer is a mass growth of the Internet users. It means that the support of the connection-oriented service (necessary for the real-time multimedia requiring large bandwidth) is not mandatory, but high-speed packet switching for the short-lived connections is important. Especially on the backbone network, the control mechanism such as the active queue management requiring per-flow queueing would not be necessary, and instead a capability of fast packet forwarding is expected. In that sense, WDM (but ATM) is promising for future data transport network. We will be back on this subject in the next section.

4.3 Establishing High Performance Transport Protocols

Problems of TCP have been repeatedly pointed out and in the early 90's, new light-weight transport protocols have been actively discussed [18]. Those include NETBLT [19] and XTP [20]. However, time changes and TCP is now widely deployed in the operating network. Thus, it now becomes difficult for some new protocol

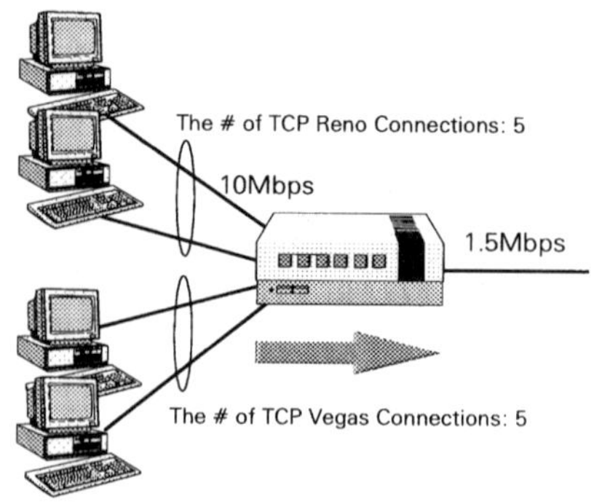

(a) Network model.

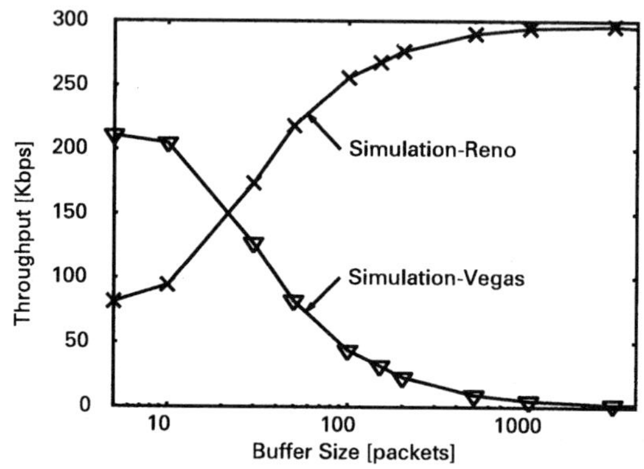

(b) Comparisons of throughput.

Fig. 3 (a) Network model. (b) Comparisons of throughput. The upper figure shows the network model where TCP Vegas and Reno shares the link. The lower figure shows the result. Especially when the router buffer size becomes large, TCP Reno can attain much more throughput than TCP Vegas.

(including the modified version of TCP) to take the place of TCP unless the protocol designer considers a migration path from the original TCP. Otherwise, the new protocol would not be received.

One interesting example can be found in the TCP Vegas version, which was proposed in [21]. In [21], it is shown that TCP Vegas can achieve between 40 and 70% better throughput, with one-fifth to one-half the packet losses, as compared to the existing TCP Reno version [22]. However, the simulation model in [22] assumes that the link is shared only by TCP Vegas connections. When those share the link with TCP Reno connections, however, TCP Reno connections easily dominate the link and receive much more throughput than TCP Vegas connections [23]. See Fig. 3. The reason is that TCP Vegas tries to share the link modestly with other connections, while TCP Reno increases the window size until the packet is lost. That is, an aggressive increase of the window size of TCP Reno results in that TCP Reno connections first grab the link, and TCP Vegas connections miss the chance to increase its window size.

Fortunately, the role of the receiver in the TCP

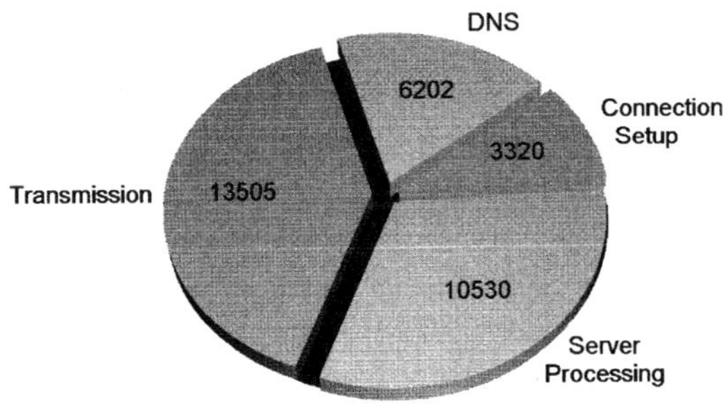

Fig. 4 Delay components of the Web document download. The values in the figure show the 95 percentile delay in milliseconds from randomly chosen 100 sites.

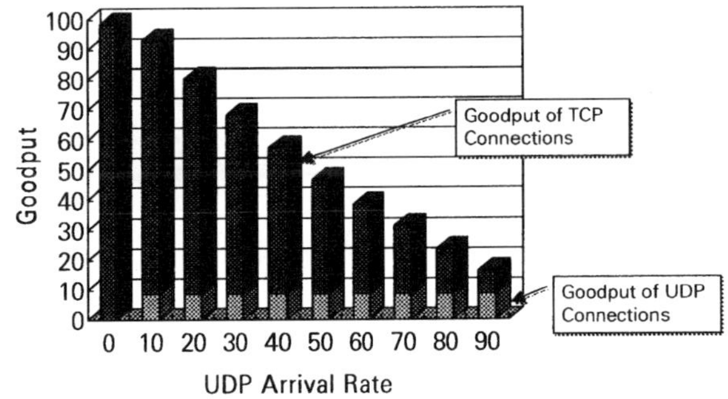

Fig. 5 UDP connections wastes the link and TCP connections receive unfair treatment.

mechanism is just to return ACKs (acknowledgements) as the packet is successfully received. It indicates that we have a chance to incorporate some modification at the sender side of TCP to improve the performance. Such an example can be found in [24], [25], where the socket buffer allocated to the TCP connections is adjusted based on the bandwidth-delay product of connections.

As the number of Internet users grows, the congestion control aspect of TCP has been paid much attention. Namely, the problem is, say, how 1.5 Mbps link can be shared by a thousand of users with 64 kbps access channel. Researches addressing this aspect of TCP can be found in [26].

4.4 Establishing High Performance End Systems

In the previous subsection, we argued that it is important to effectively share the link bandwidth by the number of connections. Another important aspect is how to accomplish the high-speed data transfer between end systems. As we discussed in Sect. 3, the delay is incurred at the end systems in addition to the network. It is partly demonstrated in Fig. 4† while the delay component at the client side is not included in the figure. As the figure clearly shows, no component dominates the end-to-end delay.

Several commercial products such as Gigabit Ethernet [27] and MAPOS [28] have already provided the interface exceeding 1 Gbps to the end system. Especially in such an environment, the role of end systems becomes important in the Internet architecture, because protocol processing of upper layers (transport and application layers) is performed at the end system. In the projection paper of [29], the gigabit processing would be possible even by the conventional Internet protocol hierarchy. Actually, the gigabit order of throughput is reaching by the widely used personal computer [30], if an overhead due to memory copy can be effectively avoided based on a zero copy technique [31]. However, the method in [30] adopts a proprietary transport protocol, and more researches should be performed in this area.

4.5 Keeping Fairness among Connections

In the Internet, the congestion control mechanism is performed at end systems. It gives a flexibility in the Internet, but it also introduces several problems. One such a problem is fairness. A typical example is that the connections cannot receive the same throughput as described below. Such a problem would never occur if the network explicitly allocates the bandwidth to each connection, but the network is not allowed to actively perform congestion control in the Internet. The fairness problem is one of most important problems in the Internet, and we believe that it is often more important than performance issues.

4.5.1 Fairness between TCP and UDP Connections

It is now well known that when TCP and UDP connections share the link, UDP connections tend to occupy the link [32]. It is because the UDP connection does not adequately react against the congestion while the TCP connection does. Figure 5 shows goodput values of UDP and TCP connections [32], where the simulation setting is as follows. Three TCP and one UDP connections compete over a congested 1.5 Mbps link. The access links for all senders are 10 Mbps, except that the access link to the receiver of the UDP flow is 128 kbps. When the UDP source rate exceeds 128 kbps, most of the UDP packets will be dropped at the output port of the link. The vertical axis shows the goodput defined as the rate successfully delivered to the receiver. All values given in the figure are defined as a fraction of the bandwidth of the 10 Mbps link.

We should note here that it is true that most of the currently available real-time multimedia applications are equipped with its proprietary delay- and/or rate-adaptive control mechanisms. See, e.g., [33]. However, it is only for its own purpose of performing better-quality presentation. Accordingly, a notion of

†The figure is produced by using the data available at ftp://ftp.telcordia.com/pub/huitema/stats/

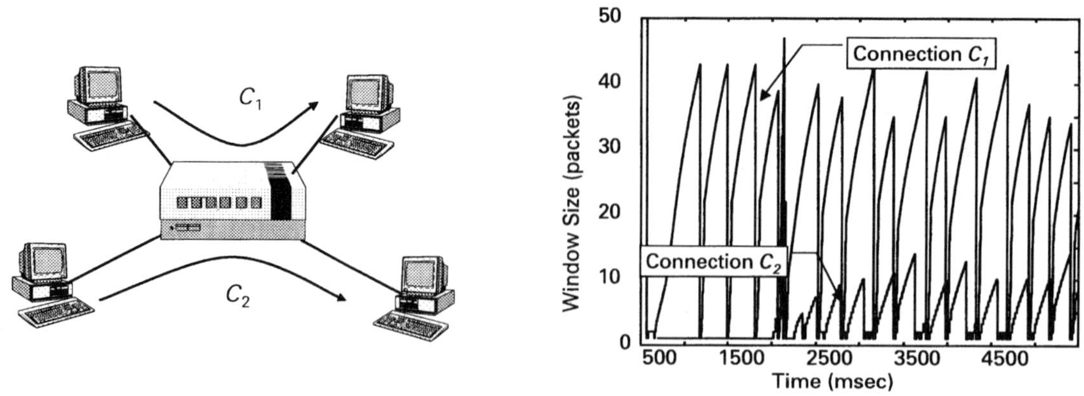

(a) Network model. (b) Comparisons of time-transient throughput.

Fig. 6 Comparisons of instantaneous throughput values of TCP connections with different propagation delays. The propagation delay of connection C_2 has two time larger than that of connection C_1.

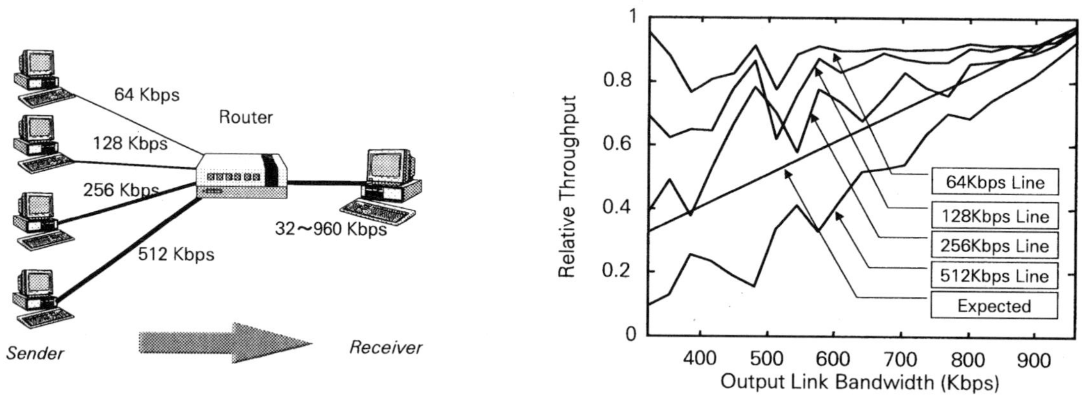

(a) Network model. (b) Comparisons of long-term throughput values.

Fig. 7 Comparisons of long-term throughput values of TCP connections with different values of access link capacities.

TCP friendly congestion control is now proposed [34], where TCP friendly is defined as *a non-TCP connection should receive the same share of bandwidth as a TCP connection if they traverse the same path.* However, its definition still contains ambiguity; an appropriate time scale of the fair share is still not clear, which is important especially when *TCP-friendly* is applied to real-time multimedia.

4.5.2 Fairness among TCP Flows

Due to an intrinsic nature of the window-based congestion control mechanism of TCP [22], TCP connections do not receive the same instantaneous throughput even if connections share the same end-to-end path. If TCP connections have different propagation delays, even the long-term throughput values become different. An experimental example by simulation is shown in Fig. 6 [35], where Connection C_1 with smaller propagation delay clearly obtains the larger throughput than C_2 with longer one.

More importantly, if the bandwidth values of TCP connections are different, the relative throughput of TCP connections against its access link bandwidth are much different. It means that more bandwidth does not help improve the throughput. See Fig. 7 for an example result [36]. The vertical axis of the figure shows the relative throughput of the connection, which is defined as the ratio of the throughput against the capacity of the input link of the connection. The *expected* line shows the ideal values by our definition. Resolving this problem is important since the heterogeneous environment is becoming common as various access methods of the Internet, such as the ADSL and CATV, are realized.

A main cause of the unfairness in TCP is owing to its AIMD (additive increase/multiplicative decrease) window updating algorithm. It was shown in [37] that only AIMD establishes the fairness among connections. However, the window updating algorithm of TCP is self-clocked. That is, the window size is only updated as the sender receives ACK from the receiver, and therefore the window curves of TCP connections with different propagation delays and/or different access link capacities become different. Then, the window updating algorithm is now actively studied [38]. Solution techniques against the unfairness problem mentioned above and possible limitations are also discussed in the literature, which will be introduced in the next subsection.

4.6 Reallocating Network Functions

So far, we have emphasized that the fundamental philosophy of the Internet is that the network is loosely-coupled and the control is performed in a distributed fashion. Then network functions such as congestion control is left to the end systems (i.e., TCP), and the only role of the network is to carry the *bits*. Of course, it is a rather philosophical statement and several control functions are tried to be moved into the network layer.

Actually, fully distributed control makes it difficult to realize the fair service. It is reported that several end systems seem to modify the source code of TCP such that even if the congestion takes place within the network, the window size is not throttled [39]. Since other TCP connections decrease the window size according to the congestion control principle, congestion is terminated and the modified TCP can enjoy better performance than others. The network should take an active queue management for achieving the fair service among TCP connections and eliminating ill-behaved connections.

The first example is the RED router [40] which avoids the burst dropping of packets from the same connection. It results in a fair treatment of connections to some extent. DRR is another active queue management method to process packets with per-flow queueing [41], by which active connections at the router is fairly treated.

However, the scalability problem exists in the per-flow queueing. It seems difficult for the core router to maintain the per-flow queueing and scheduling against thousands of active flows. On the other hand, the edge router could handle per-flow queueing since the number of active flows is not very large. Another extreme is an FIFO router where the packets are simply forwarded in an FIFO manner. As noted earlier, the primary objective of core routers must be fast packet forwarding and switching, and therefore the FIFO scheduling seems to be sufficient. A compromise between per-flow and FIFO queueing may also be possible while discrimination of flows becomes imperfect. Such examples can be found in [42].

An ECN mechanism [43] is another example that the network plays an active role against congestion. In ECN, the router actively returns the congestion information to the sender so that the congestion relief can be performed quickly.

We should last recall that too complicated and restrictive control by the network looses the merit of the Internet and we need more investigation on this aspect.

4.7 Establishing Network Dimensioning Methods

It is important to provide stable performance to the user. For this purpose, adequate network dimensioning is necessary. The Erlang loss formula has been providing a powerful tool for the telephone network, where the following positive feedback loop is established.

1. Set the target call blocking probability (e.g., 1% during busiest hours).
2. Estimate the traffic characteristics (e.g., the offered load and call holding time).
3. Apply the Erlang loss formula in order to determine the required link capacities (the number of telephone lines) such that the target call blocking probability can be fulfilled.
4. Build the network according to the estimated values.
5. Keep the traffic measurement to assess that the target call blocking probability is satisfied. If not, go back to Step 3 or 2 to adjust link capacities.

The advantage of the above procedure is owing to the fact that the Erlang loss formula is robust in the sense that the distribution of call holding times does not matter. Only a finite average is necessary.

On the other hand, dimensioning of the data network is not easy. We have several obstacles.

- We do not have an adequate performance metric to determine the required amount of network resources as we have addressed in Sect. 3.
- Even if the performance metric is known, the network provider has no means to measure it. Only the users can perceive the performance (e.g., response times of the Web document retrieval). It is a quite different point from the traditional telephone network where the network operator can measure the user's QoS by observing the number of lost calls. In the data network, the delay consists of several factors, and the network operator can only know a part of entire delays.
- A rapid growth of data traffic makes it difficult to predict the future traffic demand. The various applications (including the Web-based services and multimedia applications) have different traffic characteristics. Only an exception is characteristics of Web documents. See, e.g., [4] and references therein.

Perhaps, we will not be able to have a theory-based network dimensioning method like the one in the telephone network. The traffic measurement then becomes important to establish the positive feedback loop. In doing so, the following problems should be resolved.

1. The route is instable [44], which implies that the traffic load at the link is changed even if the traffic generation at the source is fixed.
2. Due to the window-based congestion control of TCP, the packet arriving rate at the link is changed even if the link is not congested.
3. The packet arriving rate at the link contains the packet retransmission of TCP. We should eliminate those packets in estimating an adequate traffic load.
4. In the current Internet, real-time application implements its proprietary rate control. Even if we can estimate the traffic rate of each flow, it is not clear how much the user is satisfied with the quality.
5. The low utilization of some link might be caused by
   a. the congestion control of TCP which keeps the packet transmission rate low,
   b. the low bandwidth of the access line of the user, or
   c. the low computing power of end systems.

Many of current research efforts on traffic measurement are devoted just to acquire the traffic characteristics on the link [45]–[48]. More researches are necessary to resolve the above problems. One point that we should note here is that for effective and meaningful positive feedback loop, the flexible bandwidth management is expected. ATM has such an capability, which is well known as dynamic VP bandwidth management [49].

4.8 Establishing a Fundamental Theory for the Network Researches

The fundamental theory in the data network has been a queueing theory during a long time. Its origin can be found in [50], and its usefulness is needless to say in the field. However, the queueing theory only reveals the basic property of a single entity, corresponding to the packet buffer of the router in the case of the Internet. We can find the packet queueing delay and loss probability (for the finite buffer) by applying the queueing theory. However, as described before, the QoS metric of the data network is neither packet delay time nor loss probability at the router. The performance at the router is an only component of the entire delay. It is a quite different point from the teletraffic theory (i.e., the Erlang loss formula). The derived call blocking probability is directly related to the user's perceived performance.

We have another theory, called a queueing network theory, which treats the network of queues (see, e.g., [51]). However, it does not reflect the dynamic behavior of TCP, which is essentially the window-based dynamic congestion control. We thus need another fundamental theory to model and evaluate the data network. One promising approach is a control theory that has an ability to explicitly model the feedback loop of the congestion control. See, e.g., [52], [53].

Another important research area is on what kind of performance model should be used in evaluating network performance. When a new version of TCP is proposed, how can its effectiveness be convinced? We do not have an appropriate modeling approach for the world-wide and scalable Internet. The modeling approach needs to be further researched [54].

We last note that in recent years, a self-similar nature of the network traffic has been paid much attention [55], and a fractal queueing theory becomes one of active research areas [56]. However, the author's feeling is that it is still not clear whether we actually need a fractal queueing theory or not. It is true that the Internet traffic appear to be self-similar or at least to be heavy-tailed according to the past researches. However, most of traffic characterizations are just results on traffic measurements. While the self-similarity was observed on the Ethernet [57], it includes effects of the data link layer protocol (i.e., CSMA/CD of the Ethernet) and the transport layer protocol (i.e., TCP). That is, it does not directly mean that the packet arrival at, e.g., the router shows self-similarity. Other examples can be found in the case of the file transfer time on the Internet [58] and Web documents [4], [59]. Just an indication drawn from those results is that the distribution of file sizes follows the heavy-tailed distribution [60], and is not directly related to the modeling of the self-similarity of the packet arrival. Another attention is necessary; most of past results has shown that the heavy-tailed distribution is only exhibited if we look at the tail distribution. An entire distribution often has a finite mean of log-normal distribution [4].

5. Architectural Considerations on IP over WDM Networks

Keeping challenging problems raised in the previous section in mind, we now discuss IP over WDM networks. Since we have several excellent papers for IP over WDM networks (see, e.g., [5], [61]), our discussion below is mainly devoted to architectural considerations.

5.1 Four Alternatives of IP over WDM Networks

For building the IP over WDM network, we have several alternatives. Typical structures are shown in Fig. 8. The first choice shown in Fig. 8(a) provides the WDM link between adjacent electronic routers. This approach does not utilize the WDM networking technology, but by introducing the wavelength multiplexing technology, the link capacity is certainly increased by the number

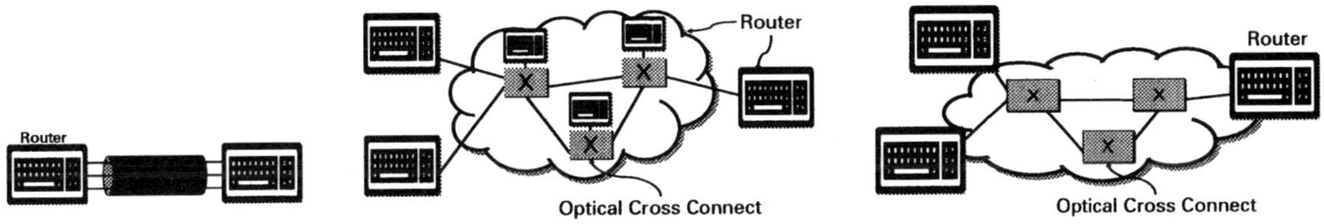

(a) WDM link. (b) WDM path network. (c) MPLS-based path network.

Fig. 8 Structures for IP over WDM networks.

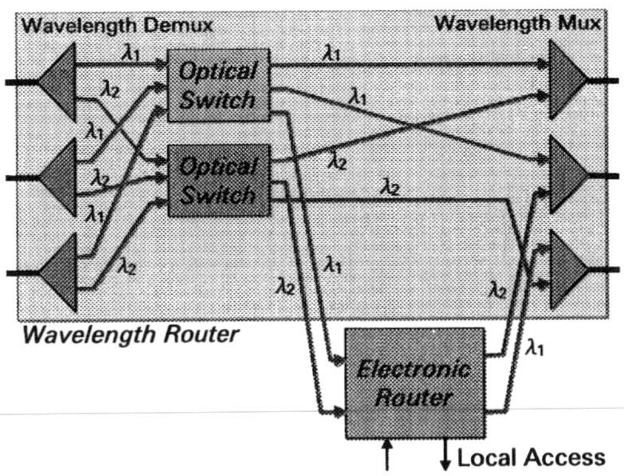

Fig. 9 Structure of wavelength router [62].

of wavelengths on the fiber. However, it is insufficient to resolve an explosion of traffic demands since it is likely to result in that the bottleneck is only shifted to an electronic router.

To relax the bottleneck at the router, an introduction of optical switches has actively been discussed. By utilizing the optical switch, we can have the wavelength routed network, where wavelength paths are established on the physical network by only using optical switches and bypassing electronic routers (Fig. 8(b)). Here, the physical network means an actual network consisting of the optical nodes and the optical-fiber links connecting nodes. Each optical switches directly connects an input wavelength to an output wavelength, by which no electronic packet processing is necessary at that node. Then, the wavelength path can be set up directly between two nodes via one or more optical switches. An example structure of the optical node is shown in Fig. 9.

A set of the wavelength paths and optical nodes constitutes the WDM-based logical network. An example is shown in Fig. 10. The physical network consists of five optical nodes, and each fiber has two wavelengths of λ_1 and λ_2 (Fig. 10(a)). By establishing the wavelength-routed network as in Fig. 10(b), we have the logical topology consisting of wavelength paths as illustrated in Fig. 10(c), showing a logical view of the underlying network to the IP routers. In Fig. 10(c), the direct wavelength path is set up by using wavelength λ_2 from node N_1 to node N_3, by which the processing for packet forwarding at node N_2 is not necessary. There is no direct wavelength path for packets from node N_1 to N_4. It is because the wavelength is not available to establish the wavelength path, and therefore, those packets should be passed to the electronic router at node N_3 for further forwarding the packets destined for node N_4 (see the lower part of Fig. 9). By comparing Figs. 10(a) and 10(b), however, it is apparent that required packet processing at the router can be reduced by introducing the optical switches. Furthermore, the logical topology provided to IP tends to have a larger number of degrees at each node, leading to less hop counts of IP routes. It is desirable since the processing overhead at IP routers can be alleviated.

We note here that in the above example, we do not consider the wavelength changes at the optical switch. If it becomes available, more flexible wavelength routing can be achieved [63]. We further note that the other structure of optical nodes can also be considered, but the above-mentioned node architecture is preferable since there is no need to modify the IP mechanism of the electronic routers.

While a lot of research works have been devoted to the logical topology design method for wavelength-routed WDM networks (see the survey papers [62], [64] and references therein), it is still not clear how much the increased number of wavelengths can reduce the requirement on the processing overhead of the electronic router. See, e.g., [65].

The IP over WDM network based on the MPLS technology is recently discussed in the IETF [66] (Fig. 8(c)). The difference from the previous case (Fig. 8(b)) is that the network is internally constituted only by the wavelength paths. By optical switching, higher performance can be expected by this structure. For this purpose, however, the wavelength path for every node pairs should be established, and too much wavelengths are necessary to establish such a network [63]. In MPLS, the LSR (Label Distribution Protocol) is defined for binding the label and the actual path. In MPLS-based WDM networks, the label and path correspond to the wavelength and the optical path. By comparing with the MPLS-based WDM network, the previous approach takes a compromise; the logical topology consisting of the wavelength paths is established by using the available wavelengths as much as possible. If

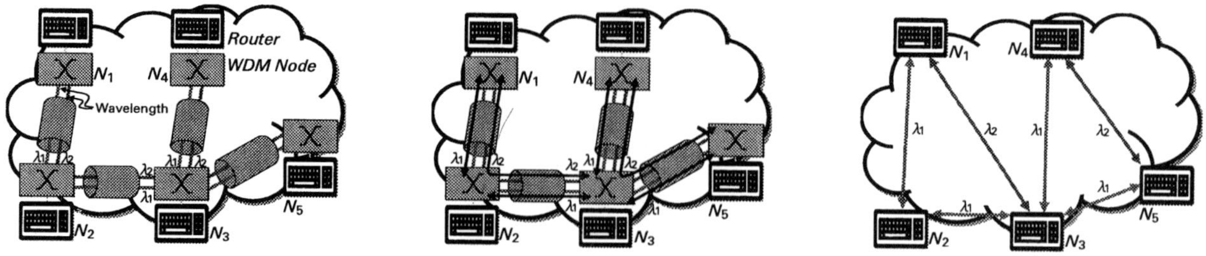

(a) Physical topology

(b) Wavelenth routing on the physical topology

(c) Logical topology

Fig. 10 Physical and logical topologies.

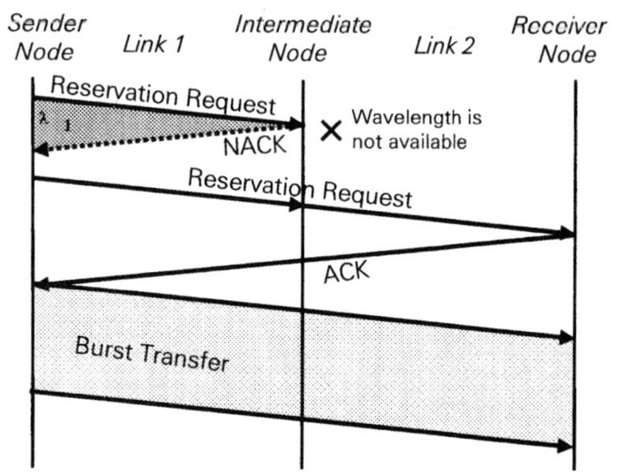

Fig. 11 Optical burst switching: the case of forward reservation.

the direct wavelength path cannot be set up between two nodes, two or more wavelength paths are used for packets to reach the destination.

The last one is to provide the packet switching network within the WDM layer. One promising approach is optical burst switching [67]. When the burst arrives at the sender, it requests the optical path between source and destination nodes. After the path is established, the sender transmits the burst (Fig. 11). In this way, there is no need to buffer the burst within the network, and is suitable to the photonic network where the optical memory is difficult to be implemented. The figure shows the case where the wavelength is reserved in the forward direction. A variant exists; in the forward direction, only the availability of wavelengths are checked and the wavelength is actually reserved in the backward direction. The disadvantage of the optical burst switching is that it is apparently not suitable to the short-lived connections since the two-way propagation delay to reserve the wavelength becomes an overhead. However, it must be one of promising methods for the future data transport.

5.2 Functional Partitioning for IP over WDM Networks

In this subsection, we discuss the appropriate choice of the IP over WDM network architecture by taking account of the challenging problems described in the precious section. Functional partitioning is an essential issue to be taken into account.

The WDM network can provide a solution for the first problem; providing the very large bandwidth and high switching capacity. How about other problems such as the congestion control and fairness control with per-flow queueing? Probably, the WDM network should not provide solutions for those problems. It should be devoted to carry the packets as fast as possible, and the other service-specific processing should be left to the edge routers (or access networks) [68].

To further discuss the above problem, we again recall the history of ATM. ATM has a capability of providing various network functions such as

1. connection establishment by signaling protocols,
2. congestion control,
3. routing control, and
4. reliability control.

Since IP is connectionless, the connection establishment control is not necessary. An exception is the application that needs the connection establishment such as the real-time multimedia. However, it is not likely that each real-time connection needs one wavelength since its bandwidth is too large for the single connection. Further, IP has the routing control mechanism and TCP does the congestion control mechanism.

In the standardization effort in the ATM Forum, the congestion control for the ATM ABR service class became a hot topic, and the rate-based congestion control mechanism was defined as a standard [69]. However, we faced with a difficult problem that if ATM is applied to IP, distinct two layers (ATM ABR service class and TCP) provide the similar mechanism. If both of two mechanisms work well, a very effective use of the network resources can be expected. The true story was different. If we tune the control parameters of rate-based congestion control in ATM properly, the TCP over ABR service works quite well [70]. If it fails, on the other hand, the result is worse than that with no control mechanism. More importantly, in the standardization process of the rate-based congestion control, the majority of short-lived connections in the Internet was not fully taken into account. Since ATM is essentially connection-oriented, the long-lived connection is expected. The frequently changes of the active connections in the actual Internet must give an ill-effect on the network performance.

We therefore conclude that network functionalities such as the routing and congestion control mechanisms should be left to the upper layers (i.e., TCP/IP) and the WDM layer should not provide those functions. Only an exception must be a reliability control. Of course, IP's routing function takes account of the reliability, which is just a reason that ARPANET (the origin of the Internet) was introduced [1]. However, its reaction against the network failure is slow (in the order of 10 seconds) while WDM has an ability to offer the protection mechanism in the order of ten milliseconds. Accordingly, if the number of wavelengths is sufficiently large, some portion can be used as protection paths [71]. Even if all the paths are not protected, IP is able to recover from network failures.

Thus, the next step towards IP over WDM networks must be how the logical topology of the WDM network is established with consideration on network reliability. For this purpose, we have several problems.

1. We need to establish a logical topology design method with reliability. Constraints on the number of wavelengths on the fiber and the technological availability of optical switches should be taken into account.
2. Does the protection mechanism of WDM have to cover all reliability issues? And, should the restoration mechanism [72] be also considered?

In addition to the above problem, network dimensioning should also be carefully treated. In the logical topology design algorithm of WDM networks, the traffic load is assumed to be given a priori in bps or pps (packet per second). We need an effective way of the network dimensioning method suitable for IP over WDM networks. In dimensioning the Internet, the positive feedback loop is mandatory. In IP over WDM networks, it should be implemented in the wavelength and routing assignment mechanism. A flexible bandwidth management mechanism is also necessary to increase/decrease the bandwidth of paths. Those are important future research topics.

6. Concluding Remarks

For concluding remarks, we want to raise several myths for the next-generation Internet, which we have already discussed in this paper.

- Performance metrics of QoS parameters in data networks are packet delay and loss probability at the router.
- QoS of data networks is to guarantee the application level delays.
- We have a queueing theory for the data network, which corresponds to a teletraffic theory for the telephone network.
- If TCP connections follow the standard congestion control mechanism, the fairness can be achieved among those TCP connections.
- Since TCP has a congestion control capability, the network support is not necessary.
- Since TCP is an old protocol, we need a new lightweight high-performance protocol.
- By introducing a WDM technology, the network bandwidth can be increased by the number of wavelengths.
- Multi-layer control of the network can always lead to the high performance network.
- A Poissonian traffic is old and it is always necessary to consider a self-similar nature of the Internet traffic.

Note that above statements are the author's opinions and several statements may not be widely accepted.

References

[1] A.S. Tanenbaum, Computer Networks, Prentice Hall, 1996.

[2] P. Newman, "Traffic management for ATM local area networks," pp.44–50, Aug. 1994.

[3] M. Murata, "Quality of service guarantees in multimedia networks: Approaches and open issues," IEICE Trans., vol.J80-B-1, no.6, pp.296–304, June 1997.

[4] M. Nabe, M. Murata, and H. Miyahara, "Analysis and modeling of World Wide Web traffic for capacity dimensioning of Internet access lines," vol.34, no.4, pp.249–271, Dec. 1998.

[5] J. Anderson, J.S. Machester, Antonio Rodriguez-Moral, and M. Veeraraghavan, "Protocols and architecture of IP optical networking," Bell Labs Tech. J., pp.105–124, Jan.–March 1999.

[6] P. Newman, T. Lyon, and G. Minshall, "Flow labelled IP: A connectionless approach to ATM," Proc. IEEE SIGCOMM '96, Sept. 1996.

[7] "Multiprotocol label switching (mpls) charter," http://www.ietf.org/html.charters/mpls-charter.html.

[8] K. Fukuda, N. Wakamiya, M. Murata, and H. Miyahara. "QoS mapping between user's preference and bandwidth control for video transport," Proc. 15th IFIP International Workshop on Quality of Service, pp.291–302, New York, May 1997.

[9] D. Clark, S. Shenker, and L. Zhang, "Supporting real-time applications in an integrated services packet network: Architecture and mechanisms," Proc. ACM SIGCOMM '92, pp.14–26, Aug. 1992.

[10] R. Braden, D. Clark, and S. Shenker, "Integrated services in the Internet architecture: An overview," RFC 1633, July 1994.

[11] "Integrated services (intserv) charter," http://www.ietf.org/html.charters/intserv-charter.html.

[12] B. Braden, L. Zhang, S. Bersion, S. Herzog, and S. Jamin, "ReSource ReserVation protocol (RSVP)—version 1 functional specification," RFC 2205, Sept. 1997.

[13] A.K. Parekh and R.G. Gallager, "A generalized processor

sharing approach to flow control in integrated services networks: The single-node case," vol.1, no.3, pp.344–357, June 1993.

[14] P.P. White, "RSVP and integrated services in the Internet: A tutorial," vol.34, no.5, pp.100–106, May 1997.

[15] K. Fukuda, N. Wakamiya, M. Murata, and H. Miyahara, "QoS guarantees based on end-to-end resource reservation for real-time video communications," Proc. 16th International Teletraffic Congress, pp.857–866, June 1999.

[16] "Differentiated services (diffserv) charter," http://www.ietf.org/html.charters/diffserv-charter.html.

[17] K.G. Coffman and A.M. Odlyzko, "The size and growth rate of the Internet," available at http://www.research.att.com/ amo.

[18] W. Doeringer, D. Dykeman, M. Kaiserswerth, B. Meister, H. Rudin, and R. Williamson, "A survey of light-weight transport protocols for high-speed networks," vol.38, no.11, Nov. 1990.

[19] D.D. Clark, M.L. Lambert, and L. Zhang, "NETBLT: A high throughput transport protocol," Proc. ACM SIGCOMM '87, pp.353–359, Aug. 1987.

[20] R.M. Sanders and A.C. Weaver, "The Xpress transfer protocol (XTP)—a tutorial," vol.20, no.5, pp.67–80, Oct. 1990.

[21] L.S. Brakmo and S.W. O'Malley, "TCP Vegas: New techniques for congestion detection and avoidance," Proc. ACM SIGCOMM '94, pp.34–35, Aug. 1994.

[22] W.R. Stevens, TCP/IP Illustrated, Volume 1: The Protocols, Addison-Wesley Publishing Company, 1994.

[23] K. Kurata, G. Hasegawa, and M. Murata, "Fairness comparisons between TCP Reno and TCP Vegas for future deployment of TCP Vegas," Proc. INET 2000, June 2000.

[24] J. Semke, J. Mahdavi, and M. Mathis, "Automatic TCP buffer tuning," Proc. ACM SIGCOMM '98, pp.315–323, Aug. 1998.

[25] T. Matsuo, G. Hasegawa, M. Murata, and H. Miyahara, "Scalable automatic buffer tuning to provide high performance and fair service for TCP connections," Proc. INET 2000, Aug. 2000.

[26] R. Morris, "TCP behavior with many flows," "Proc. 4th IEEE International Conference on Network Protocols," Oct. 1997.

[27] "Gigabit Ethernet alliance," available at http://www.gigabit-ethernet.org/.

[28] M. Maruyama and K. Murakami, "MAPOS version 1 assigned numbers," RFC 2172, June 1997.

[29] C. Partridge, "How slow is one gigabit per second?" vol.20, no.1, pp.44–53, Jan. 1990.

[30] X. Xiao, L. Ni, and W. Tang, "Benchmarking and analysis of the user-perceived performance of tcp/udp over myrinet," Technical Report of Michigan State Univ., Jan. 1997.

[31] P. Druschel and L.L. Peterson, "Fbufs: a high-bandwidth cross-domain transfer facility," Proc. 14th ACM Symposium on Operating Systems Principles, pp.189–202, Dec. 1993.

[32] S. Floyd and K. Fall, "Promoting the use of end-to-end congestion control in the Internet," vol.6, no.8, Aug. 1999.

[33] "Real.com," available at http://www.real.com/.

[34] "The TCP-friendly Website," available at http://www.psc.edu/networking/tcp_friendly.html.

[35] G. Hasegawa, M. Murata, and Hideo Miyahara, "Fairness and stability of congestion control mechanisms of TCP," Proc. IEEE INFOCOM '99, March 1999.

[36] G. Hasegawa, T. Matsuo, M. Murata, and H. Miyahara, "Comparisons of packet scheduling algorithms for fair service among connections on the Internet," Proc. IEEE INFOCOM 2000, March 2000.

[37] D. Chiu and R. Jain, "Analysis of the increase/decrease algorithms for congestion avoidance in computer networks," Comput. Networks and ISDN Syst., vol.17, no.1, pp.1–14, June 1989.

[38] P. Hurley, J.-Y.L. Boudec, and P. Thiran, "A note on the fairness of additive increase and multiplicative decrease," Proc. 16th International Teletraffic Congress, pp.336–350, June 1999.

[39] V. Paxson, "Automated packet trace analysis of TCP implementations," Proc. ACM SIGCOMM '97, Sept. 1997.

[40] S. Floyd and V. Jacobson, "Random early detection gateways for congestion avoidance," vol.1, no.4, pp.397–413, Aug. 1993.

[41] M. Shreedhar and G. Varghese, "Efficient fair queueing using deficit round robin," vol.25, no.4, pp.231–242, Oct. 1995.

[42] I. Stoica, S. Shenker, and H. Zhang, "Core-stateless fair queueing: Achieving approximately fair bandwidth allocations in high speed networks," vol.28, no.4, pp.118–130, Sept. 1998.

[43] S. Floyd, "TCP and explicit congestion notification," vol.24, no.5, pp.8–23, Oct. 1994.

[44] C. Labovitz, G.R. Malan, and F. Jahanian, "Internet routing instability," Proc. ACM SIGCOMM '97, Sept. 1997.

[45] V. Paxson, J. Mahdavi, A. Adams, and M. Mathis, "An architecture for large-scale Internet measurement," vol.36, no.8, pp.48–54, Aug. 1998.

[46] "CAIDA: Cooperative association for Internet data analysis," available at http://www.caida.org/.

[47] "IPMA: Internet performance measurement and analysis project," available at http://www.merit.edu/ipma/.

[48] "Surveyer home page," available at http://www.advanced.org/csg-ippm/.

[49] S. Shioda and H. Uose, "Virtual path bandwidth control method for ATM networks: Successive modification method," vol.74, no.12, pp.4061–4068, Dec. 1991.

[50] L. Kleinrock, Queueing Systems, Volume II: Computer Applications, New York, Wiley-Interscience, 1975.

[51] P.G. Harrison and N.M. Patel, Performance Modelling of Communication Networks and Computer Architectures, Addison-Wesley, 1993.

[52] H. Ohsaki, M. Murata, T. Ushio, and H. Miyahara, "Stability analysis of window-based flow control mechanism in TCP/IP networks," Proc. 1999 IEEE International Conference on Control Applications, pp.1603–1606, Aug. 1999.

[53] V. Firoiu and M. Borden, "A study of active queue management for congestion control," Proc. IEEE INFOCOM 2000, March 2000.

[54] V. Paxson and S. Floyd, "Why we don't know how to simulate the Internet," Proc. 1997 Winter Simulation Conference, Dec. 1997.

[55] A. Feldmann, A.C. Gilbert, W. Willinger, and T.G. Kurtz, "The changing nature of network traffic: Scaling phenomena," vol.28, no.2, pp.5–29, April 1998.

[56] W. Willinger and V. Paxson, "Where mathematics meets the Internet," Notices of the American Mathematical Society, vol.45, no.8, Aug. 1998.

[57] W.E. Leland, M.S. Taqqu, W. Willinger, and D.V. Wilson, "On the self-similar nature of Ethernet traffic," vol.2, no.1, pp.1–15, 1994.

[58] V. Paxson, "End-to-end Internet packet dynamics," vol.7, no.3, pp.277–292, March 1999.

[59] M.E. Crovella and A. Bestavros, "Self-similarity in World Wide Web traffic: Evidence and possible causes," Proc. ACM SIGCOMM '96, pp.160–169, Aug. 1996.

[60] G. Irlam, "Unix file size survey," available at http://www.base.com/gordoni/ufs93.html, 1994.

[61] P. Bonenfnt, A. Rodriguez-Moral, and J. Manchester, "IP over WDM: The missing link," available at http://www.lucent-optical.com/resources/whitepapers/wp010.pdf, 2000.

[62] R. Dutta and G.N. Rouskas, "A survey of virtual topology design algorithms for wavelength routed optical networks," Optical Networks, Jan. 2000.

[63] R. Ramaswami and K.N. Sivarajan, "Routing and wavelength assignment in all–optical networks," vol.3, pp.489–500, Oct. 1995.

[64] E. Leonardi, M. Mellia, and M.A. Marsan, "Algorithms for the logical topology design in WDM all-optical networks," Optical Networks, Jan. 2000.

[65] M. Murata, K. Kitayama, and H. Miyahara, "IP over a–thousand–wavelength division multiplexing: Is it useful and possible for resolving the network bottlenecks?" submitted for publication, 2000.

[66] N. Ghani, "Lambda-labeling: A framework for IP-over-WDM using MPLS," to appear in Optical Networks, April 2000.

[67] C. Qiao and M. Yoo, "Choices, features and issues in optical burst switching," to appear in Optical Networks, April 2000.

[68] D.S. Isenberg, "The dawn of the stupid network," ACM Networker, vol.2, no.1, pp.24–31, Feb./March 1998.

[69] H. Ohsaki, M. Murata, H. Suzuki, C. Ikeda, and H. Miyahara, "Rate-based congestion control for ATM networks," vol.25, no.2, pp.60–72, April 1995.

[70] G. Hasegawa, H. Ohsaki, M. Murata, and H. Miyahara, "Performance evaluation and parameter tuning of TCP over ABR service in ATM networks," IEICE Trans. Commun., vol.E79-B, no.5, pp.668–683, May 1996.

[71] O. Gerstel and R. Ramaswami, "Optical layer survivability: A services perspective," vol.38, pp.104–113, March 2000.

[72] B.T. Doshi, S. Dravida, P. Harshavardhana, O. Hauser, and Y. Wang, "Optical network design and restoration," Bell Labs Tech. J., pp.58–84, Jan.–March 1999.

Chapter 2

Broadband Networks

High-performance backbone networks will stimulate future new multimedia services. First of all, this chapter describes services and broadband network applications. The most important applications are IP services, voice and telephony service, real time video transfer, and data transfer.

H. Ohnishi et al. present ATM network technology and service in Paper 2.1. Network services are created step-by-step by using ATM technology. The paper also covers standardization issues and some development results.

J. Roberts presents a network service model based on applications in Paper 2.2. The present Internet service and telephone network service require different transport networks, for example. The paper suggests two service classes, one is stream and the other is elastic traffic. These service classes are provided by a sophisticated ATM traffic control method.

A critical network service attribute is reliability. Paper 2.3 introduces a new VP restoration method. ATM is very effective in realizing high-performance network failure protection. Some experimental network results also described.

Paper 2.4 addresses a design scheme for VP networks. The author has developed a new software tool for automatic network system design. Network reliability and efficiency are determined for a particular VP network topology. The theoretical approaches are described in detail.

B. Mark et al. describe a Quality-of-Service (QoS) control method for performance and network resource control in Paper 2.5. QoS is realized by sophisticated traffic control techniques such as VPC or CAC. The paper models traffic using these traffic control functions and describes network efficiency.

Paper 2.6, written by T. Masaki et al., deals with voice transmission over ATM networks, which is a key issue since voice traffic is very sensitive to packet delay. The paper introduces short cell and a shared Virtual Cell Identifier (VCI) cell architecture. A shared VCI cell is realized by combining several short cells. Using this architecture, a standard 53-byte cell switch can be used to support several shared voice connections. The paper also reports VLSI implementation results.

S-J. Chung et al. propose a virtual topology management scheme for ATM and MPLS networks in Paper 2.7. It purses optimal resource utilization through the unification of VP-connection overlay network configuration, VC routing, and capacity allocation. It supports point-to-multipoint connections as well as point-to-point connections. The authors develop an algorithm to solve the large-scale nonlinear mathematical programming formulation of the proposed model. Experimental results show that the proposed scheme finds an optimal VP management that establishes a trade off between the conflicting attributes of ATM technologies.

Y. Kanayama et al. address an efficient, failure-resistant ATM transport method for high-quality real-time signals in Paper 2.8. The proposed method can prevent not only cell loss but cell delay variation (CDV) even when network failure occurs. Simulation results suggest that the CDV is very small and remains unchanged at network failure. The proposed scheme can be applied to real-time services such as a nationwide TV program transmission service between TV stations.

Network scalability is a crucial issue for designing public networks. M. Kakemizu et al. address a scalable private network node interface (PNNI) interworking architecture in Paper 2.9. It includes a peer group configuration to avoid explosions in the routing load and a method of PNNI emulation in ATM public networks. The proposed PNNI interworking architecture enables the economical, seamless, and scalable interconnection of each ATM LAN with any other ATM-LAN via PNNI internetworking, including public networks and ATM-LANs.

Paper 2.10 by S. Tojo et al. proposes a phase assignment algorithm based on traffic measurement for real-time MPEG sources in ATM networks. MPEG traffic will become the dominant load in future broadband networks. The proposed algorithm decides when a new MPEG source should begin to transmit based on its notification parameters and traffic measurement of ongoing connections. It is shown that the proposed algorithm can provide good QoS and improve network utilization.

K. Kawahara et al. address a layer3 switch in Paper 2.11. The layer3 switch enables us to rapidly transmit IP datagrams using the cut-through technique. The authors analyze the cut-through rate, the datagram waiting time and the mis-ordered rate of a layer3 switch for the case of flow-driven connection setup. They model the arrival process of IP flow and IP datagram from each source.

High-Performance Backbone Network Technology edited by Naoaki Yamanaka
ISBN: 0-8247-5321-6 DOI: 10.1081/(120025660)
Published 2004 by Marcel Dekker, Inc.

Issues in ATM Network Service Development, Standardization, and Deployment

Hirokazu Ohnishi and Kou Miyake

SUMMARY To construct the future multimedia network, ATM network technology and services should support cost-effective, high-speed interconnectivity and a variety of service-providing functions. Furthermore, as the infrastructure of future multimedia service, the ATM architecture should be adaptable to changes without needing replacement of its core functions and platform capabilities. This paper presents an overview of the current state of development, standerdization and deployment of the ATM network service technologies and architecture concept. It also discusses the trend toward the integration of ATM technology and Internet technology. Also reported is the state of development and standardization for the individual ATM technologies and related issues, including access networks, bearer services, signalling, network middleware, and future ATM switching system technology.

key words: *ATM, network, standardization, multimedia, internet*

1. Introduction

The multimedia communications network of the future should cost effectively support a variety of communications services operating at high speed over broadband connections. The key to this support is the uniform and bit-rate-independent asynchronous transfer mode (ATM) protocol with its flexibility in providing a variety of connections, bit rates, and qualities of service.

In the deployment phase for the multimedia network, practical ATM network systems should have flexible and on-demand control and management functionality to support the variety of multimedia services. However, in the future multimedia communications environment, the number of different communication bit rates, service categories, and communication modes are likely to increase to support various kinds of multimedia communications services. Furthermore, it is still unclear what types of services and applications will come to dominate. Therefore, the architecture for the ATM network and services serving as the infrastructure for future multimedia services should be adaptable to changes in requirements without requiring replacement of its core functions and platform capabilities.

To achieve these concepts and network service capabilities, much effort has gone into developing and deploying ATM network services. However, there are still several issues to be solved. In this paper, we overview the development, standardization and deployment of ATM network service technologies and ATM network architecture concept. We also discuss the more towards integrating ATM and Internet technologies. Finally, we report on the development and standardization of the individual ATM technologies, focusing on such related issues as access network, bearer services, signalling, middleware, and switching systems.

2. Overview of ATM Network Service Technologies and Architecture Concept

The ITU-T began its study of the asynchronous transfer mode (ATM) protocol for the next stage of ISDN development in 1985; it outlined the specifications for such items as the cell formats, the ATM adaptation layer, and the signalling protocols in the late 1980s.

2.1 ATM Technology and Service Development and Standardization

The ATM protocol was studied considering in the main characteristics of optical fiber, i.e., its high speed and low bit-error rate transmission capability. The great importance of this standardization is that it is unique and international. This will allow a common interworking, interoperable, and multi-vendor environment in the future multimedia network, with no overhead for adaptation.

In the 1990s, the ATM Forum and IETF took the initiative in applying basic ATM protocols to third-generation LAN technologies. As a result, ATM characteristics such as high speed, assurance of QoS/bandwidth, and seamless interworking with WANs (wide area networks) can be used in a practical LAN environment. These organizations have actively specified new protocols and architectures for incorporating ATM technologies into a LAN environment, thereby establishing a basis for future multimedia networks. Their main achievements are IP over ATM, LAN emulation, and MPOA.

The recently introduced ATM technologies have

Manuscript received June 23, 1997.
Manuscript revised September 10, 1997.

† The author is with NTT Global Business Headquarters, Tokyo, 163-1419 Japan.
†† The author is with NTT Network Service System Laboratories, Musashino-shi, 180-0012 Japan.

IEICE TRANS. COMMUN., Vol. E81-B, No. 2, pp. 152–163, February 1998

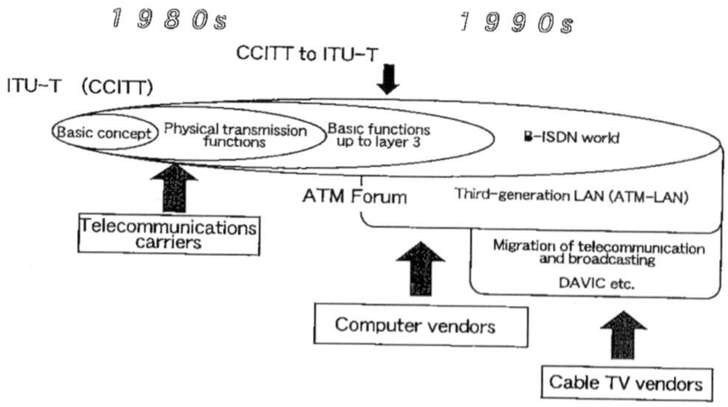

Fig. 1 Expansion of ATM standardization activities.

been targeted at high-speed computer communication or integrated enterprise voice and data backbone networks. They have also been applied as the backbone of large frame-relay networks and the Internet. To effectively provide these services, interworking technologies, such as frame-relay ATM interworking protocol, NHRP (Next Hop Routing Protocol) and IP switching, have also been developed to facilitate the establishment of a higher-speed frame-relay and Internet backbone.

Also, in the area of the enterprise networks, ATM technologies have been developed for implementing high-speed flexible networks to enhance current TDM-based networks for data and voice by unifying their backbones. For this purpose, a new protocol called VTOA (voice and telephony over ATM) has been specified that provides efficient and high-quality voice communications.

The full service network (FSN) is being studied by several organizations, such as DAVIC, to comprehensively provide various services such as VoD, high speed Internet access, telephony, and N-ISDN, through a single subscriber line. The expansion of ATM-related standardization activities is shown in Fig. 1.

Along with the rapid expansion in the range of applications due to new ATM technologies, it is applied to local enterprise networks and their links and to the backbones of public computer networks. For local enterprise networks, ATM technologies have been introduced that simplify constructing an ATM base network, a LAN backbone network (ATM-LAN), and a unified voice and data network. By connecting such networks within branch offices by high-speed leased lines, these networks can be widened into WANs. However, instead of such a network configuration, enterprise networks are gradually shifting to ones using cell-relay service with a native ATM interface for connecting the subnetworks. This ATM-based internetwork configuration meets user requirements for high-speed access.

ATM networks will be used as backbones not only for frame-relay and internet services, but also for existing services, e.g., telephony and N-ISDN, in the future. In addition, future multimedia ATM networks will have to provide native ATM interfaces for end users that can handle the access-speed variations for the various types of multimedia applications.

2.2 ATM Network Architecture

Existing networks have been optimized to support particular types of communication services. Users must have several network interfaces to access the various services that they need. Service providers must build and maintain several networks. These drawbacks should be resolved in future multimedia ATM networks. Furthermore, a network should be able to evolve to provide enhanced services without changing its basic interface and network structure. Our proposed ATM network architecture concept is shown in Fig. 2.

i) user-network interface

The proposed network provides users with end-to-end connectivity and a set of protocols. ATM-based interfaces are defined so that users can access many types of services from a common user-network interface (UNI). Because any new service must be able to be supported through this interface, a high-speed physical interface, such as 156 or 600 Mbps, is needed.

ii) access network

The network consists of an access network and an ATM-based core network. The former would be constructed using optical-fiber technology to support the high-speed UNI. This network is the most expensive relative to the total network cost, and should be able to run for years without interfering with new service deployment. Therefore, the access network functions should be limited to basic transport functions and serve as a common network platform. The technology needed for the access network is described in Sect. 4. Between the access network and the core network, a common interface may be defined: the service node interface (SNI). Through this interface, several types of network service nodes (ATM bearer service, CATV headend, etc.) can access the access network and be connected to the UNI.

iii) core network

The lower-layer core network provides users with the ATM bearer capabilities, such as the ATM connection bit rate, the traffic type, such as constant-bit-rate or variable-bit-rate, and quality of service. To control these capabilities, the signalling channel and local management interface are terminated in the physical layer. These capabilities should be limited to those for basic transport, which can be provided independently of the higher-layer service capabilities. On top of the core network, several types of service-dependent functions can be attached through a well-defined open interface. Therefore, the core network can serve as a common platform for multimedia services.

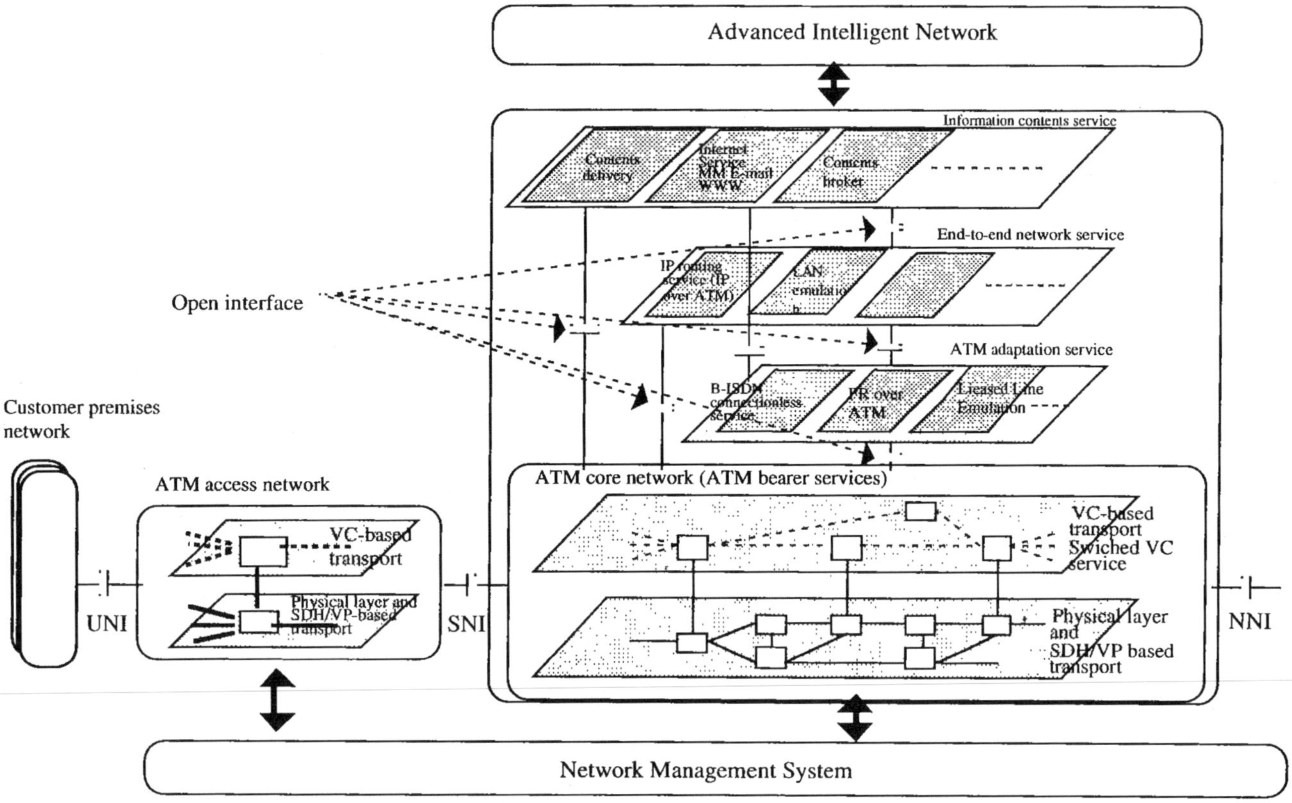

Fig. 2 Proposed ATM network architecture.

iv) higher-layer service capabilities

The higher layer is classified into three sub-layers. The ATM adaptation layer (AAL) service sub-layer provides such AAL related services as connectionless services and interworking services between the ATM network and existing networks. Connectionless service over ATM networks, which is very similar to SMDS, was standardized by the ITU-T as Recommendation I.364 [1]. An interworking service example is frame relay over ATM, which was standardized by the ITU-T [2] and the ATM Forum [3].

These service capabilities are provided either alone or combined with other upper-layer services. The end-to-end network service sub-layer provides end-to-end routing services, such as the IP-over-ATM service studied by the IETF or the LAN emulation service studied by the ATM Forum. In this layer, address-resolution protocols are terminated and end-to-end routing capabilities are provided in conjunction with an ATM bearer capability. Also provided are a contents delivery service, an Internet World-Wide-Web service, a contents broker service, or any other contents-related services.

In addition to these capabilities, advanced intelligent network capabilities and network management capabilities would be supported in accordance with required network functions.

2. 3 Integration of ATM and Internet Technologies

The rapid integration of ATM and Internet technologies will have a big impact on future multimedia networks.

2. 3. 1 Protocol Architecture

The basic protocol for the Internet is TCP/IP. It specifies the fundamental rules over the peer-to-peer network layer so that any protocol can be used below the data-link layer. The IP over ATM protocol developed by the IETF and other organizations enables IP packets to be sent over ATM connections, [4], [5].

The main issue in sending IP packets over an ATM network is how to set the ATM connection for sending IP packets. The protocol used in the network layer (TCP/IP) is a connectionless type, while the corresponding ATM protocol is a connection-oriented type. Therefore, the critical problem is how to establish, control, and select an ATM connection corresponding to the forward-to destinations of IP packets.

There are three basic methods for sending IP packets over ATM networks.

(1) PVC connection: This method uses ATM connections established permanently. It is aimed at connections among routers having ATM-interface cards. Connection selection is done by mapping between the

forward-to destinations of the IP packets and the VPCI/VCIs in the ATM network. Transporting of IP packets requires using routers, so this method is best suited for networks that do not have many forward-to destinations. The ATM connection is used as a pipe for statistical multiplexing.

(2) ATM-connectionless service: This method uses connectionless services specified for ATM. It is represented by the protocol defined in ITU-T Recommendation I.364. The basic idea of this method is to identify a group of ATM cells as an IP packet with a frame identifier in the AAL layer when IP packets are divided into ATM cells to be sent. This makes it possible to collectively transmit ATM cells consisting of IP packets without reconstructing the IP packets at the transit nodes. This is connectionless-type communication because routing is determined by the address at the header of an ATM cell. This transit transportation without reconstructing IP packets results in high-speed connectionless communication over the ATM network; however, achieving high-speed real-time communication is difficult because the required transit bandwidth is not always available.

(3) SVC connection: In this method, the switched virtual connection (SVC) protocol controls the establishment of an ATM connection corresponding to the forward-to destination of an IP packet at the time of the transmission request. The SVC protocol provides ATM-layer address resolved by the IP address. Once an ATM connection is established, the IP packets can be transmitted in the same way as over a PVC connection. This method achieves high-speed real-time communication, because the bearer services and the negotiation of bandwidth can be managed by establishing the ATM connection and because the transit transportation can be controlled without reassembling the IP packets. However, an increasing number of forward-to IP-packet destinations may make SVC management difficult. As a result, this method may not be suitable for transaction-type communication, in which short packets are forwarded to many destinations frequently.

As part of the development of an IP transportation method, the issues of transaction management, high-speed real-time communication, and so on need to be addressed.

2.3.2 System and Network Architecture

The Internet network is strongly characterized by the use of an autonomous distributed architecture. In such an architecture, when nodes or networks are added to an already operating network, the nodes or networks being added can automatically exchange transit-route information, address information, and so on with the existing nodes or networks, so as to identify and configure the new network configuration. With this architecture, new nodes can be connected to an existing network without network operators involvement. However, if the network is expanded, there could be difficulty in transmitting operation information among nodes. Therefore, the scope of application of this architecture is limited to some extent. Achieving routing that corresponds to policy routing and traffic priorities is difficult, because the information transmitted is limited to those of basic ones.

On the other hand, ATM networks, especially carrier networks, usually have a large centralized-control architecture based on POTS and N-ISDN. The routing and network-configuration management are concentrated in the operation systems, enabling ATM networks that can achieve policy routing that is difficult to use in the Internet architecture. ATM networks still need to provide systematic function for their operation.

A comparison of these architecture is shown in Fig. 3. The ATM Forum is now developing new protocols, such as PNNI, that have the merits of both the ATM and Internet protocols. The PNNI protocol provides essential functions for ATM network operation, e.g., policy routing, and autonomous distribution, but is still not well suited for large networks and is thus the subject of further study.

3. Connection Type, Bearer Services and QoS

The proposed ATM network technology is designed to support various kinds of bearer services on the same network. Example ATM bearer services defined by ITU-T Recommendation I.371 and the ATM Forum UNI 3.1 and 4.0 specifications include legacy-type IP interconnection service, non real-time variable-bit-

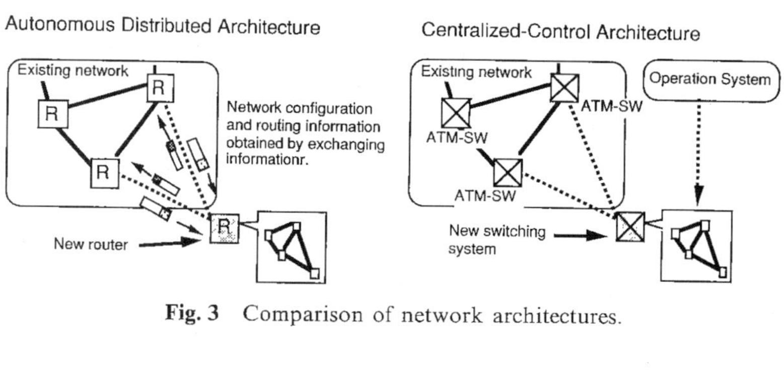

Fig. 3 Comparison of network architectures.

Table 1 Bearer service and QoS class mapping.

(ITU-T I.356, I.371; ATM Forum UNI4.0)

Bearer Services: ITU-T	Bearer Services: ATM Forum		Class 1 (CLR 0+1< 3 *10, CTD < 400ms)	Class 2 (CLR 0+1< 10^{6}, CTD:undefined)	Class 3 (CLR 0< 10^{5})	U class (Unspecified)
DBR PCR,CDVT CLP=0+1)	CBR PCR, CDVT (CLP=0+1)		X	-	-	X
-	rt-VBR PCR,CDVT, SCR,MBS		-	-	-	-
SBR PCR, CDVT, SCR, MBS	nrt-VBR PCR, CDVT, SCR, MBS	CLP=0+1	-	X	-	X same definition
		CLP=1 not tagging	-	-	X	-
		CLP=1 tagging	-	-	X	-
-	UBR PCR,CDVT, CLP=0+1 or CLP=1 and tagging		-	-	-	CLP=0+1 / CLP=1 and tagging
ABR PCR, MCR, etc.	ABR PCR, MCR, etc.	Binary	-	-	X	X
		ER	-	-	X	X

rate/statistical-bit-rate (VBR/SBR) service and unspecified-bit-rate (UBR) service. These services can be used to achieve a statistical multiplexing effect in the network. For voice or real-time video communication, stringent QoS and bandwidth reservation are required. In this case, deterministic-bit-rate/constant-bit-rate (DBR/CBR) services can be used.

The QoS classes of ATM bearer services defined in ITU-T Recommendation I.356 can be used for selecting bearer services and the QoS. Table 1 summarizes the standardized ATM bearer services, QoS classes, and their relationships. The upper limit of the QoS objectives, such as the cell-loss rate (CLR) and the cell-transfer delay (CTD) are specified in I.356. The ATM network should be able to provide these bearer services through a standardized interface according to a user's request. It must be able to handle the traffics using the network resources efficiently.

Some connection types and call/control capabilities are defined for flexibly managing end-to-end communication. Basically, an ATM connection is either a virtual path connection (VPC) or a virtual channel connection (VCC).

From the connection-control point of view, a permanent virtual connection (PVC) and a switched virtual channel (SVC) are defined. The connection setup and release for a VPC are controlled through the network operation system; the connection is held regardless of whether the end-to-end communication is active. In contrast, the connection setup and release for an SVC are directly initiated by the end user, using the signalling channel and protocol. Therefore, the connection is set up when the end user desires communication.

Each type of connection has its own merits. For example, if a PVC connection has already been set up, there is no need to use a signalling method, so a PVC connection is very useful for a connection-less service over an ATM network. On the other hand, for connection-oriented, stringent-QoS, bandwidth-guaranteed services, such as video or voice communication, SVC-based on-demand connection control manages the network resources more efficiently. In conjunction with bearer service capabilities, these connection types and control mechanisms enable the network resources to be used more efficiently.

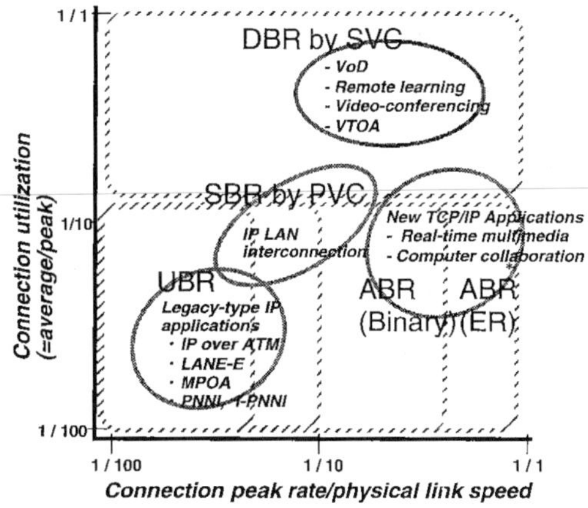

Fig. 4 Applcation areas of bearer service and connection types.

The mapping between applications, the ATM bearer service, and the connection-control method is shown in Fig. 4. An ATM network must therefore be able to provide these capabilities depending on user demand.

4. Access Network Architecture and Technology

Economical high-speed access networks are essential to enlarging ATM network services, not only as the backbone of enterprise networks but also for Broadband public service. The three major transmission methods being developed for access networks are compared in Table 2. HFC uses the coaxial cables that are already being used for CATV. xDSL uses the metallic cables that are used for telephony and ISDN. ATP-PDS uses optical fibers. Both HFC and xDSL are aimed at using existing access networks to create

Table 2 Comparison of alternative access technologies.

	xDSL	HFC	ATM-PDS
Media	Twisted Pair	Coaxial	Optical Fiber
Bandwidth	Up: 16 to 640 kb/s (occupied) Down: 1.5 to 9 Mb/s (occupied) (VDSL) Up: 1.6 to 2.3 Mb/s Down: 13 to 55 Mb/s	Up: 2 to 10 Mb/s (Shared) Down: 10 to 36 Mb/s (Shared)	Up: 156 Mb/s (partly occupied) Down: 156 Mb/s (partly occupied)
Standardization	ADSL: almost finished (ATM soon) VDSL: Planned	Almost finished	Planned
Advantages	• Uses existing copper • Occupied bandwidth • Small system improvement • Higher security	• Cheap cable modem • Integration of broadcast and communications	• Flexible transfer speed • Flexible service • Occupied bandwidth
Disadvantages	• Expensive ADSL modem • Limited Distance	• Shared bandwidth (with tens of other users) • Large system improvement (two-way AMP, headend)	• Expensive compared with ADSL and cable modem

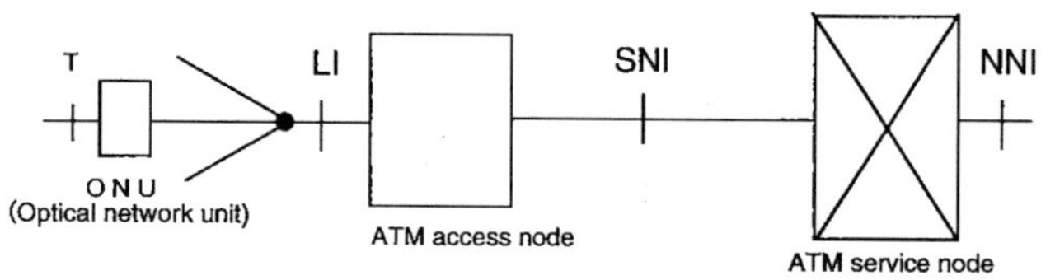

Fig. 5 Basic access network architecture model.

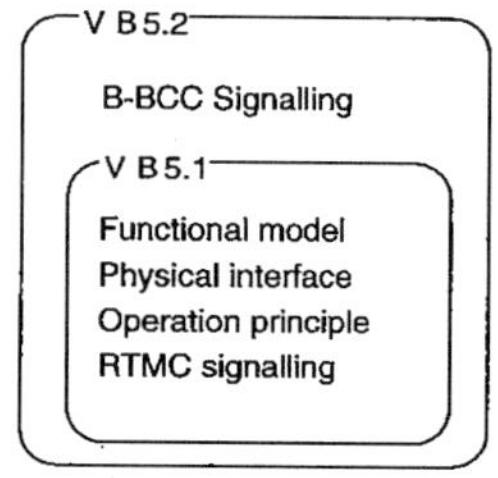

Fig. 6 Relationship between VB5. 1 and VB5. 2 specifications.

economical network configurations. However, their available bandwidths are limited. While the optical fibers for ATM-PDS, provide higher access speeds, they require additional capital investment.

In addition, access networks configuration needs following requirements:

(1) Access networks and ATM services nodes in these networks should be developed respectively.

(2) Any manufacturer's device can construct access networks.

(3) Sharing the same access network with more than one carrier.

The ITU-T has thus been studying the standardization of access network architectures and their interface specifications. The basic model being used is shown in Fig. 5. It uses ATM-PDS to specify the LI reference point. The reference point between the access network and a service node is an SNI. Two specifications are now under discussion.

VB5.1: An access network only has the VP cross-connect function. The operation system establishes paths on a provisional basis.

VB5.2: An access network includes a VCC assignment function on demand. Broadband-bearer connection control (B-BCC) signalling specified among service nodes establishes connections for access networks on demand (See Fig. 6).

For both specifications, Real-time management coordination (RTMC) signalling provides the interface that service nodes should exchange management information in real time.

5. Signalling

In addition to the ATM technology, the key requirements for future ATM networks are a higher layer protocol set that supports various types of media, signaling systems that support various types of connection configuration, and enhanced network management functions that support simple and flexible OAM capabilities. One possible set of protocols that provides all ATM network capabilities is shown in Fig. 7. Each protocol needs to be standardized at the same pace as the whole set.

5. 1 Relationship between ITU-T and ATM Forum

Study Group 11 of the ITU-T and the SOG Working group of the ATM Forum are both studying signalling functions and signalling protocols for ATM multimedia services. Both the ITU-T and ATM Forum are proceeding with developing specifications from the basic signalling functions in a step-by-step manner. However, the difference in their priorities has created a small difference in the protocol specification step (Figure 8 shows the UNI protocol-specification step).

The reason is that the ITU-T, which mainly aims at public networks, consists of network providers, communications equipment manufacturers, govern-

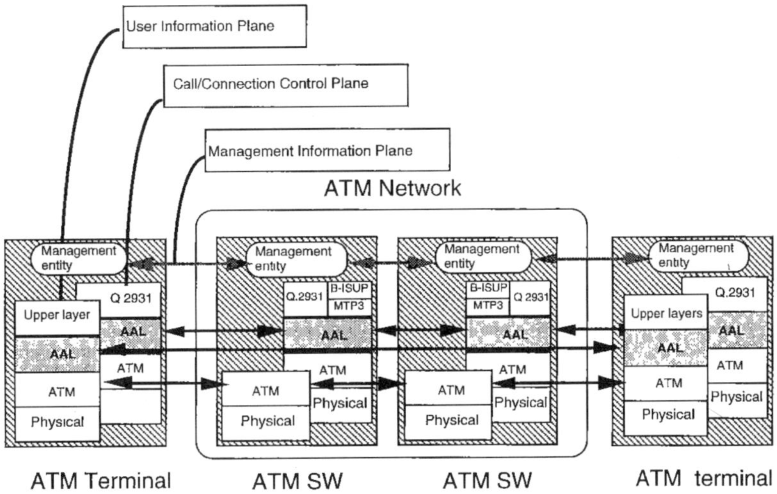

Fig. 7 Example protocol set for an ATM network.

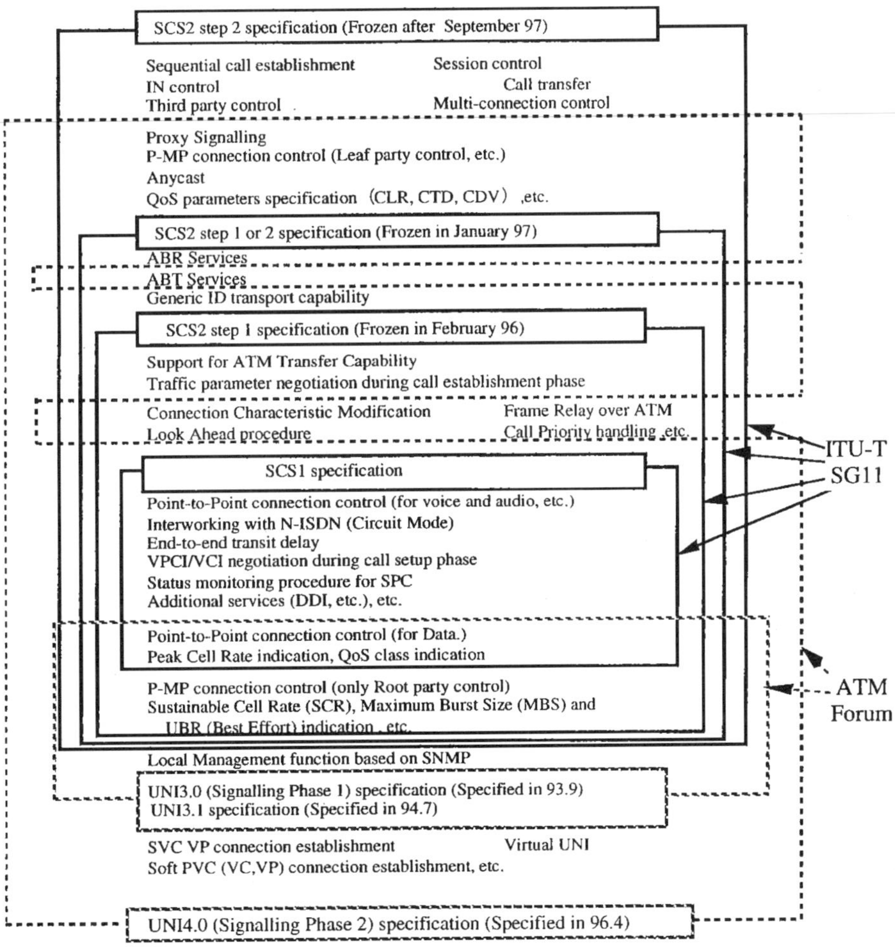

Fig. 8 ATM signalling protocol specification step (UNI).

ment organizations, a small group of computer manufacturers, and so on, while the ATM Forum, which aims at LANs and private networks, consists of network providers, communications equipment manufacturers, computer manufacturers, semiconductor manufacturers, LAN vendors, software developers, and so on. Although there are some differences in the protocol specification steps of the ITU-T and ATM Forum, the key members attend meetings of both groups, and they try to include the capabilities needed for each group. The specifications of the ITU-T and ATM Forum provide the same capabilities in basically the same way thus ensuring compatibility.

5.2 Basic Signalling Functions for Multimedia Services

The ITU-T specified Signalling Capability Set 1 (SCS1) for the first step, followed by SCS2 Step 1 (SCS2 has a wider scope and is thus divided into smaller parts), which was finished in February 1996. The ATM Forum completed the UNI 3.1 specification in July 1994, followed by UNI 4.0 in April 1996. These specifications cover not only point-to-point connections, single connections, call/connection setup and release, and constant bit rate, but also point-to-multipoint connections, traffic parameter negotiation during call setup, connection characteristic modification, additional traffic parameters for SBR, and so on. Therefore, we now have almost all the basic signalling functions needed for multimedia services, such as VOD, LAN connections, and LAN Emulation.

5.3 Current Issues

Interworking between enterprise and carrier networks is necessary for providing wide-area network service. However, because of the difference in the protocol specification step, there are still some issues that should be solved.

5.3.1 LAN Emulation Interconnection

For LAN emulation, the edge switches in private networks will require UBR call/connection with a PCR equal to the line rate, even when they do not need such a wide bandwidth. This is because the edge switches in private networks do not know which PCR is appropriate for its prior to the connection setup. This is not a problem within a private network because the switches in a private network do not concern PCR when UBR is used. However, switches in a public network do need PCR, and too many rescues are reserved even when UBR is required because the resources in a public network are shared by many users (Fig. 9).

In addition, a call/connection request is rejected if the line rate is narrower than the requested PCR on the connection pathway. An originating edge switch can not know prior to connection setup of a line narrower than the line attached to the public network.

This problem can be solved by using ABR and traffic parameter negotiation during call setup. The appropriate resources in the public network can be shared by using ABR and the traffic parameters. For example, PCR may be specified during call setup by using traffic parameter negotiation. However, an algorithm for allocating the bandwidth during traffic parameter negotiation in the call-setup phase requires for further study.

5.3.2 PNNI over B-ISUP

The PNNI protocol over a wide-area network is essential for interworking between two different enterprise networks through a carrier network. Figure 10 shows an example of a connection between PNNI enterprise networks via a B-ISUP carrier network with a PNNI interface (PNNI over B-ISUP).

The PNNI protocol consists of a signalling protocol and a routing protocol, whereas B-ISUP has only the signalling protocol, which can currently interwork only with UNI signalling. In addition, the PNNI routing and that of B-ISUP are different: the former is

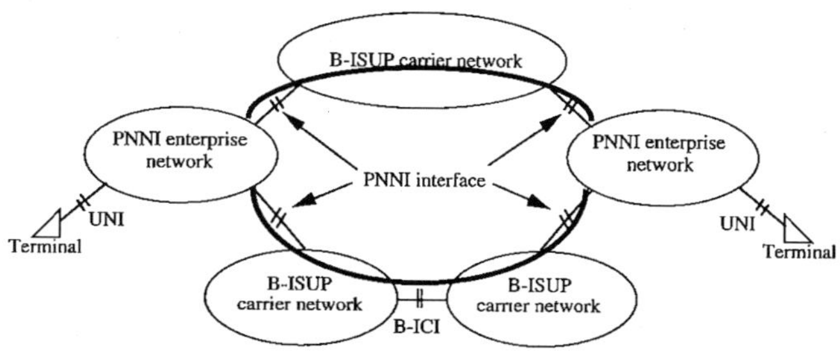

Fig. 10 Example of connection between private networks via carrier network.

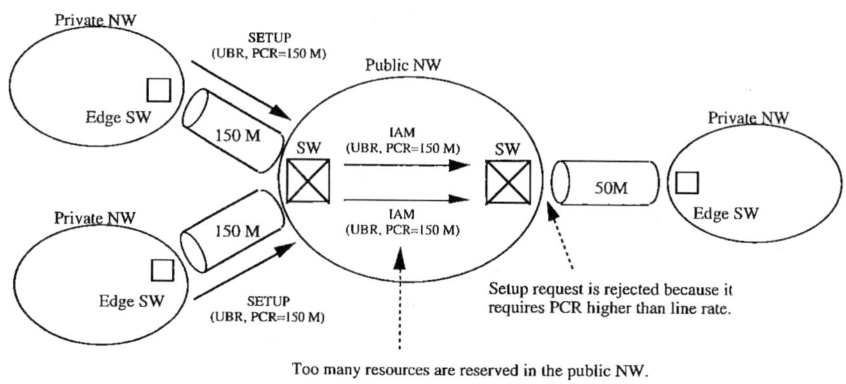

Fig. 9 Example call/connection setup request with PCR equal to the originating line rate.

source routing and the latter is hop-by-hop routing. Therefore, the main requirements for achieving PNNI over B-ISUP are

(1) interworking between B-ISUP signalling and PNNI signalling,

(2) handling the PNNI routing protocol in a B-ISUP network, and

(3) coordinating PNNI (source) routing and B-ISUP (hop-by-hop) routing.

To achieve interworking, it is possible for B-ISUP signalling to interwork with PNNI signalling, with only a small expansion for PNNI-specific signalling (Fig. 11), such as a designated transit list (DTL). This is because PNNI signalling is based on UNI signalling, and B-ISUP signalling originally interworked with UNI signalling.

To achieve PNNI routing in a B-ISUP network, the routing information can be transparently transferred through a B-ISUP network via a PVC as a first step towards PNNI over B-ISUP. To handle routing information more efficiently, a routing server to edit and distribute the routing information can be added to the B-ISUP network.

To achieve routing coordination, the address used for routing in a B-ISUP network is not the called-party number, but the carrier network address (e.g., the E. 164 address) assigned to the interface attached to the egress node of the B-ISUP network because the PNNI and B-ISUP networks have different numbering plans. This should be the address that enables the B-ISUP network to follow the route indicated in the DTL (Fig. 12).

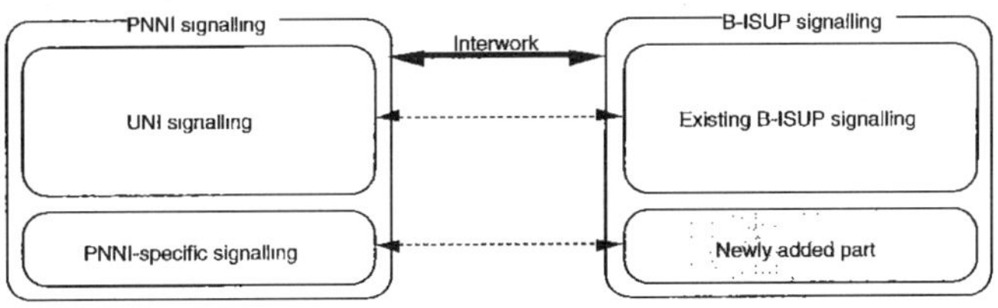

Fig. 11 Interworking between PNNI and B-ISUP signalling.

5.4 Items to be Addressed

The items to be addressed in ATM signalling include
-Session control, call/connection transfer, third-party control and so on for the introduction of VOD;
-interworking of public and private networks, leaf initiated join*, group multicast, anycast*, and so on for ATM LANs and private networks;
-introduction of AIN to provide integrated services.
(* The ATM Forum has specified these UNI signalling protocols in UNI 4.0.)

A session is a global association of clients, servers. and network resources to provide telecommunication service. It can associate multiple connections via multiple networks, support call control without connection

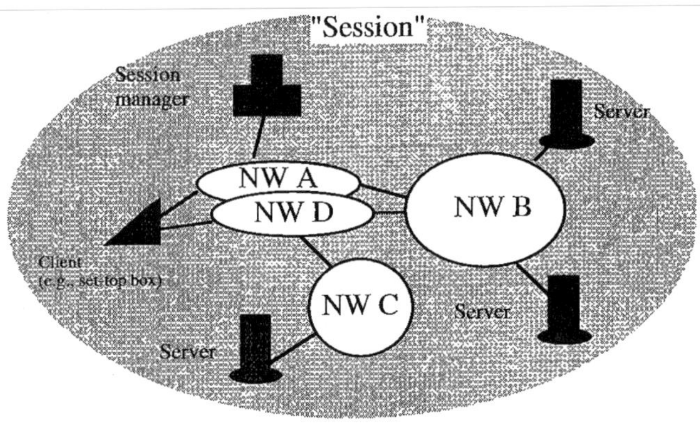

Fig. 13 Example of a session over multiple networks.

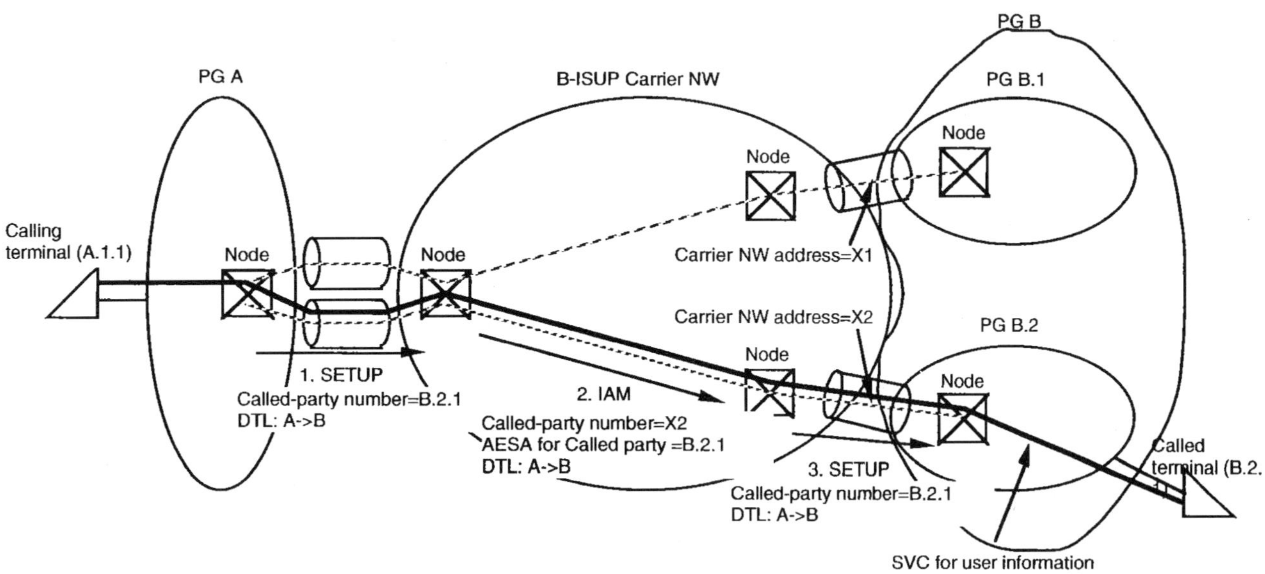

Fig. 12 Example of setup sequence.

(e.g., calling without connection, pre-negotiation, or look ahead) and coordinate multiple parties and multiple connections. Figure 13 shows an example of a session over multiple networks.

6. Network Middleware

To provide multimedia services in ATM networks, the entire network must be controlled in perspective in order to connect end-to-end smoothly. An example of this is the address resolution function. An ATM-LAN generally uses a private addressing system, such as NSAP addressing, in order to ensure independent operation and the quality of extension. On the other hand, public networks generally use E. 164 addressing to ensure interworking between different users. Therefore, mapping between private addresses and public network addresses is indispensable. Figure 14 shows an image of this mapping. This is an example of address resolution in the ATM layer. Address resolution between different layers, like IP over ATM is also needed.

Another example of controlling the entire network is the virtual private network (VPN). Providing the backbone of enterprise networks in public networks is becoming increasingly important. It is therefore becoming more and more necessary to provide not only virtual private networks but also security assurance. Users constructing private networks especially, require a service that grants or denies access according to the policies of the originating and connecting private networks. The closed user group (CUG) service meets this requirement. Basically, servers controlling access management hold a filtering database and in the connection-establishment phase, both the originating and the connecting node inquire of these data to these servers, which realizes CUG.

A third example is point-to-multipoint connection control. Tele-conferencing, distance learning, and so on often uses the P-MP connection in the ATM layer. The basic signalling protocol for this P-MP connection has been studied by the ATM Forum and other organizations, and from the point of view of NW management, a function that selects and controls the best copy point based on the called-party numbers must be provided in order to make the best use of the network resources. That's why a P-MP control module needs a function for deciding the copy point based on the network topology or the copy position information in the network.

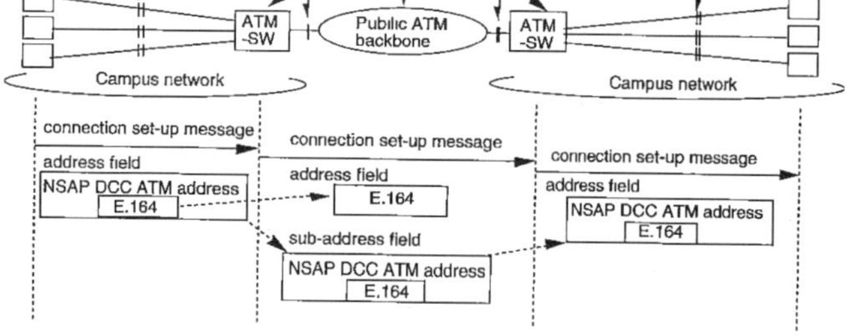

Fig. 14 Mapping between private and public network addresses.

A final example is the need for a bridging function adaptable to various telecommunication media, e.g., an image-merge function for tele-conferencing. As shown in the above discussion, establishing network middleware technology based on high speed, wide bandwidth, and reliable distributed control is needed to flexibly provide the various communication configurations required by future multimedia services.

7. Recent Technologies for Large-Scale ATM Switching Systems

Much larger ATM switches will be required by carriers to provide future multimedia services. Simply to handle the current telephone subscribers, the required capacity will easily exceed 1 Tbit/s (assuming each subscriber has a broadband access line with a capacity of tens or hundreds of Mbit/s). In this section we describe recent trends in electronic and photonic switching technologies toward Tbit/s switching.

7.1 Electronic ATM Switching

When the input/output lines each have a speed exceeding several Gbit/s, an electronic ATM switch usually has a space-division architecture because it is difficult to implement a common memory switch or a common bus switch due to the access-speed limitation. The key technologies for high-speed space-division switching are LSI, packaging, and interconnection.

One way to achieve large-capacity switching is to use a simple switch architecture, such as a matrix switch, and speed up the input/output lines by using ultrahigh-speed LSI technologies [6], [7]. A-160 Gbit/s (16 input and 16 output) switch composed of 20-Gbit/s (twice as high as the input/output line speed) 1×1 non-buffer crosspoint LSI chips fabricated using Si-bipolar super self-aligned process technology (SST) and input/output buffer LSI chips fabricated using Bi-CMOS technology has been reported [1]. Another way to attain both high speed and high integration density is to use new LSI technologies [8]. CMOS/SIMOX (separation by implanted oxygen) technology is a promising new technology. A 40-Gbit/s LSI chips that includes a 5-Gbit/s 8×8 rerouting subnetwork, 20-kgate logic, and 110-kbit 2-port RAM has been reported [8]. These above LSI technologies should make it possible to attain Tbit/s switching when the power consumption problem is overcome.

A promising high-speed LSI packaging technol-

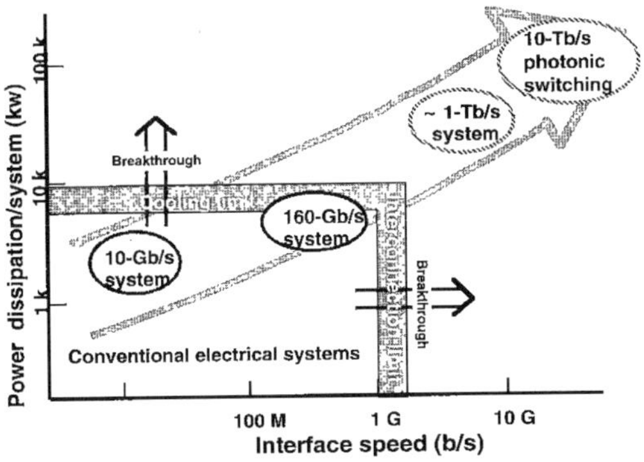

Fig. 15 Key technologies for future ATM switching systems.

ogy is multi-chip module (MCM) technology [6], [9]. An MCM enables shorter-length interconnections between the chips, resulting in compact packaging and higher-speed operation.

A promising interconnection technology is optical interconnections, which is gradually being applied to ATM switches [9]. Optical interconnections will become more common when E/O and O/E devices and optical lasers become more economical and useless power.

The biggest problem to be overcome in the electronics area is power consumption. Cooling limitations within the cabinet require that the power consumption of LSI chips be lowered so they can be packaged more density.

7.2 Photonic ATM Switching

Large-scale photonic ATM switches using frequency multiplexing have been reported [10]–[12]. These switches use a common bus architecture with output buffers because frequency multiplexing yields an effect similar to that of internal link speed-up. Photonic technologies hold the promise of attaining simple and large-scale ATM switching by taking advantage of their ultrahigh speed and large number of frequencies. Two kinds of switch configurations, the star-coupler type [10], [12] and the frequency-router type [11], have been proposed. An output buffer is implemented by an optical delay line to absorb the conflicts for the output line and to select the frequency from among several frequencies. Switch throughputs of 40 Gbit/s (10 Gbit/s 4×4 [10] or 2.4 Gbit/s 16×16 [11]) and 160 Gbit/s (10 Gbit/s 16×16 [12]) have been reported. When compact packaging of optical buffers and effective absorption of optical-level variances are attained with stable and reliable operation, a compact, lower-power, large-scale ATM switching system will no longer be a dream (Fig. 15). Photonic switching is considered to be the most promising approach to a switching capacity greater than several Tbit/s.

8. Conclusion

We have overviewed the development and deployment of ATM network service technologies and the ATM network architecture concept. We discussed the trend towards integrating ATM and Internet technologies. Finally we reported on the development and standardization of individual ATM technologies and related issues, including access networks, bearer services, signalling, middleware, and switching systems.

References

[1] ITU-T Recommendation I.364, "Support of broadband connectionless data service on B-ISDN," March 1993.

[2] ITU-T Recommendation I.555, "Frame relaying bearer service interworking," July 1993.

[3] The ATM Forum, "ATM user-network interface, Version 3.1 (UNI 3.1) specification," Sept. 1994.

[4] J. Heinanen, "Multiprotocol encapsulation over ATM adaptation layer 5," IETF RFC 1483, July 1993.

[5] M. Laubach, "Classical IP and ARP over ATM," IETF RFC 1577, Jan. 1994.

[6] K. Genda, Y. Doi, K. Endo, T. Kawamura, and S. Sasaki, "A 160 Gbit/s ATM switching system using an international speed-up crossbar switch," Globecom '94, pp. 123–133, 1994.

[7] K. Y. Eng, M. J. Karol, G. J. Cyr, and M. A. Pashan, "Design and prototype of a terabit ATM switch using a concentrator-based growable switching architecture," ISS'95, Session 3.7, 1995.

[8] Y. Ohtomo, S. Yasuda, M. Nogawa, J. Inoue, K. Yamakoshi, H. Sawada, M. Ino, S. Hino, Y. Sato, Y. Takei, T. Watanabe, and K. Takeya, "A 40-Gb/s 8×8 ATM switch LSI using 0.25-mm CMOS/SIMOX," ISSCC '97, pp. 154–155, 1997.

[9] S. Hino, S. Yamaguchi, Y. Ohtomo, S. Yasuda, S. Urushidani, and K. Yamasaki, "An implementation of a high performance ATM switching fabric," ISS '95, vol. 1, pp. 399–403, 1995.

[10] A. Misawa, M. Tsukada, Y. Yamada, K. Sasayama, K. Habara, T. Matsunaga, and K. Yukimatsu, "40 Gbit/s broadcast-and-select photonic ATM switch prototype with FDM output buffers," ECOC '96, pp. 4.107–110, 1996.

[11] K. Habara, Y. Yamada, A. Misawa, K. Sasayama, M. Tsukada, T. Matsunaga, and K. Yukimatsu, "Demonstration of frequency-routing type photonic ATM switching (FRONTIERNET) prototype," ECOC '96, pp. 5.41–44, 1996.

[12] D. Chiaroni, D. de Bouard, C. Chauzat, J. C. Jacquinot, D. Bayart, P. Bousselet, M. Bachmann, and M. Sotom, "A 160 Gbit/s throughput photonic switch for fast packet switching systems," PS '97, PWB 3/37–40, 1997.

High-Performance Backbone Network Technology edited by Naoaki Yamanaka
ISBN: 0-8247-5321-6 DOI: 10.1081/(120025661)
Published 2004 by Marcel Dekker, Inc.

Quality of Service Guarantees and Charging in Multiservice Networks

James W. Roberts

SUMMARY Quality of service requirements are satisfied conjointly by the service model, which determines how resources are shared and by network engineering, which determines how much capacity is provided. In this paper we consider the impact of the adopted charging scheme on the feasibility of fulfilling QoS requirements. We identify three categories of charging scheme based respectively on flat rate pricing, congestion pricing and transaction pricing.

key words: *quality of service, multiservice networks, traffic control, pricing, Internet, ATM*

1. Introduction

There are two main candidates for a future universal multiservice network: the Internet and the ATM based B-ISDN. Some would say that the Internet has already won the contest since it is already providing multimedia services to some tens of millions of users throughout the world. However, the network suffers chronically from congestion and this is severely limiting the development of new applications. The need to introduce quality of service guarantees is thus now widely recognized, although there is still no consensus on how this should be accomplished.

The B-ISDN has been designed with quality of service as a prime concern with multiple service classes standardized by the ITU and the ATM Forum with particular usage categories in mind. However, although ATM is increasingly used in the backbone network, the prospects of deploying an end to end ATM network still appear remote.

Quality of service depends on two factors: the network service model, which determines how resources are shared, and the network provisioning strategy, which determines how much capacity is available. Both factors depend significantly on the way users are charged for network services. Fundamental differences between Internet and B-ISDN service models can often be traced to the fact that the former is based on flat rate pricing while the latter is conceived as a generalization of the telephone network with per call charging. Adequate provisioning clearly depends on the existence of an economic incentive: a network provider will only install enough capacity if it is economically viable to do so given the charging scheme.

The economy of the Internet is the subject of active research. The recent book "Internet Economics" [15] gives an excellent introduction to the subject. Web pages maintained by MacKie-Mason [12] and Varian [21] and the bibliography established by Klopfenstein [10] point to further useful references. The survey of "frequently asked questions," "firmly expressed opinions" and "partially baked ideas" compiled by MacKie-Mason and Varian is a particularly useful overview [14].

Telephone network economics has also received a large amount of attention recently but from a largely different community. It is here mainly a question of regulating state or private monopolies and providing the necessary economic conditions for the introduction of fair competition. We have by no means made a comprehensive survey of relevant literature in this area, the author's understanding of the telephone pricing issue being derived mainly from the books by Curien and Gensolen [5] and Baumol and Sidak [1].

The present paper is addressed mainly to readers who, like the author, have a background in network engineering and the design of traffic controls. It is intended to stress the importance of including the economic point of view in these activities. We hope that economists will excuse an imperfect knowledge and understanding of their science and accept the following discussion as a contribution to a necessary multidisciplinary approach to multiservice network design.

We discuss the nature of quality of service requirements for a broad categorization of traffic types. We then consider possible charging schemes, distinguishing flat rate pricing, congestion pricing and transaction pricing, before considering their impact on the choice of service model.

2. Quality of Service Requirements

In discussing quality of service requirements it is useful to distinguish two broad categories of traffic:

- *stream* traffic entities are flows having an intrinsic duration and rate (which may be variable) whose time integrity must be (more or less) preserved by the network; such traffic is generated by applica-

Manuscript received September 29, 1997.
Manuscript revised January 9, 1998.

†The author is with France Télécom, BD/CNET/DAC, 38–40, rue du Général Leclerc, 92794 Issy-Moulineaux, France.

IEICE TRANS. COMMUN., Vol. E81-B, No. 5, pp. 824–831, May 1998

tions like the telephone and interactive video services such as videoconferencing;

- *elastic* traffic entities are digital objects which must be transferred from one place to another; these objects might be files of alphanumeric data, texts or pictures, for example.

Stored video and audio sequences accessed remotely across the network can be considered as stream traffic if they are emitted at their natural playback rate or as elastic traffic if the entire sequence is transferred for storage at the destination prior to playback beginning.

Quality of service depends on the statistical nature of traffic through three main phenomena:

- *transparency*, referring to the time and semantic integrity of transferred data;
- *throughput*, a quality of service measure for elastic traffic defined as the document size divided by the transfer time;
- *accessibility*, the probability of admission refusal and the delay for set up in case of blocking.

Stream traffic requires time integrity while a certain degree of data loss is tolerable. Elastic traffic on the other hand can by definition tolerate delays occasioned by network queueing or flow control but transport should preserve semantic integrity.

Throughput requirements for elastic traffic entities are not well known. It is useful to distinguish traffic emitted for immediate attention, like web pages, and traffic which is essentially deferable such as e-mail. For the latter throughput in the sense defined above is not really relevant since transfers can be delayed for minutes or hours without significant inconvenience. For the great majority of web documents, throughput of some tens or hundreds of kilobits/sec would be sufficient for quasi-immediate local display. The transfer of larger documents (e.g., entire data bases) may require much higher rates.

Accessibility is only relevant to quality of service in a network employing admission control. The probability of blocking depends in general on the rate required by a given transaction and a network provider will only aim to meet a target value for demands up to a certain maximum rate. A blocking probability of around 1% in the busy hour is a typical target in the telephone network.

Quality of service requirements are satisfied jointly by the network service model, which determines how resources are shared, and by network engineering procedures which determine how much capacity is provided. The service model alone allows transparency and throughput guarantees for some users; accessibility and throughput for a given population of users or an entire class of traffic additionally requires adequate network provisioning.

Depending on the definition of the service model, the network may or may not see stream and elastic traffic flows as individual transactions. It is usual presently, for example, to offer network services for LAN interconnection where the traffic entity is in fact an aggregation of flows. The administrative advantages of this approach (billing, absence of flow identification) must be weighed against the difficulty of providing quality of service guarantees for such traffic. These difficulties stem in large part from the particularly complex characteristics of traffic aggregations [11].

3. Charging Options

The feasibility of satisfying quality of service requirements depends significantly on the charging scheme employed. We distinguish three broad categories.

3.1 Flat Rate Pricing

With flat rate pricing, users pay a fixed charge, every month say, independently of the volume of traffic they produce. The price generally depends on the capacity of their network access line. It should be calculated to cover all the network provider's costs including any settlement charges incurred to provide interconnection with another provider's network. This is how the large majority of Internet users are charged today.

The major advantage of flat rate charging is its simplicity leading to lower network operating costs. A weakness is its inherent unfairness, a light user having to pay as much as a heavy user. The level of the flat rate charge excludes potential users having a lower, though positive, evaluation of the value of using the network. Flat rate schemes are also open to abuse by resale of access capacity. A more immediate problem is the absence of restraint inherent in this charging scheme which may be said to contribute to the present state of congestion of the Internet.

3.2 Congestion Pricing

Charging can be used to modulate network traffic dynamically by exploiting the elasticity of demand with respect to price. Congestion pricing allows users to pay for the level of quality of service they judge necessary. The payment is not intended to cover the cost of the network but only the cost of congestion represented by the inconvenience caused to other users who are denied quality of service. Network costs must still be recovered by a flat rate, as in the previous scheme. Congestion pricing relies explicitly on a service model where users can express the worth they attach to their traffic.

A well known congestion pricing scheme is the so-called "smart market" [13]. In the smart market, users include a bid in each packet. In case of congestion, the users offering the lowest bids are discarded first and accepted packets are tariffed at a rate determined by the highest bid among the rejected packets. The cost of carrying each packet is thus related to the marginal value (represented by the bid) of the traffic which is squeezed out.

If the revenue gained by the network provider exceeds the cost of adding extra capacity, he will have an incentive to expand bottleneck links in order to admit more traffic and thus gain even more. The network will cease to expand when the marginal cost of capacity is equal to the marginal cost of congestion, leading to a micro-economic optimum.

The optimal charging paradigm has a number of disadvantages which have been pointed out by Shenker and co-authors [19]. A major problem is that of identifying the value of a unit of traffic such as a packet when the true worth of a communication is associated with the higher level transaction (e.g., a document transfer or a telephone conversation). The complexity of the scheme and the implied billing system may severely limit the practicality of its implementation.

It is argued in [19] that it is necessary to sacrifice economic optimality in order to take account of structural aspects of the underlying service model. The authors suggest that it is sufficient to offer differentially priced service classes with charges increasing with the guaranteed level of quality of service. Users regulate their charge by choosing or not to use a higher quality of service class in times of congestion. The simplest service model fulfilling this objective has just two differently charged service classes.

A simple two-tier charging scheme would result from an evolution of the Internet in which users are required to pay usage charges for reservations (depending on reservation parameters and duration) but only the flat rate for best effort traffic. In case of congestion, users requiring a certain minimum throughput would be obliged to reserve capacity and thus pay more for their communication.

A second possibility is the two-tier best effort Internet service advocated by Clark where users identify their packets as being "in" or "out" with respect to an "expected capacity profile" [3], [4]. Only "in" packets are charged above the flat rate. The profile may be defined as a long term constraint on expected use as part of the data defining a subscription. Alternatively, a user could define an expected profile for a shorter term session or choose to mark packets as in or out according to his own priority criteria. The definition of the expected capacity profile is left open although the token bucket filter is quoted as an example.

The proposition of Songhurst and Kelly [20] for pricing ABR connections in an ATM network is also based on two-tier charging. In this case, the cost of an ABR connection depends on the minimum cell rate parameter MCR. Users pay more per bit transferred as the value of MCR increases allowing the user to pay to maintain a larger share of throughput in case of congestion.

Congestion pricing has the considerable advantage of allowing users the possibility of expressing the value of their traffic and gaining corresponding priority in access to network resources. If the network provider responds appropriately by investing congestion revenue in extra capacity, the scheme should also lead to an optimally dimensioned network with low congestion.

Likely user perception of congestion pricing is unclear, however, since the cost of a given transaction depends on invisible factors: how can users tell if the network provider isn't deliberately causing congestion? why should they pay more to an inefficient provider? Note that congestion pricing is not generally employed in other service industries subject to demand overloads such as electricity supply or public transportation or, indeed, the telephone network.

3.3 Transaction Pricing

In the telephone network, switches and links are generally sized to ensure that demand congestion occurs only exceptionally. This operating model relies on being able to reliably predict demand, given announced charges, and to size the network to avoid congestion when that demand prevails. Since demand fluctuates over time, the representative traffic volume used for provisioning is generally that of an appropriately defined "busy hour." Enough capacity is provided to handle the busy hour traffic with good quality of service (less than 1% of calls blocked, say).

The price must be set at a value allowing the network operator to recover the cost of investment and is determined on a long term basis, ideally at an optimal level such that the revenue from an additional unit of demand, stimulated by further lowering the price, would be equal to the extra cost of carrying that unit. Differential pricing on a time of day basis is used to smooth out the demand profile to some extent but this is not generally viewed as a congestion control mechanism.

The question of charging for the telephone service has been the subject of much study over the last few years as the market has been opened to competition. It is generally recognized that prices for different services (e.g., local, long distance) and market segments (e.g., residential, business) must align more closely with costs. It remains difficult, however, to evaluate these costs taking into account resource sharing and the realization of economies of scale and scope. Certainly, the marginal cost of handling an additional transaction is not sufficient to cover the major part of the costs of a network provider. A more satisfactory basis is to relate charges to an "average incremental cost" per unit of demand, as discussed by Baumol and Sidak [1], taking account of all scale and scope economies attributable to resource sharing. The average incremental cost includes the fixed capital and operational costs of these shared resources.

The average incremental cost of a transaction in a multiservice network depends on the amount of resources necessary to handle the flow concerned. This amount generally depends on many parameters including the traffic characteristics and performance requirements of the transaction and the bandwidth and buffer capacity of the network links it uses. Songhurst and Kelly have discussed a rational transaction pricing scheme for multiservice networks based on the formula:

$$\text{price} = a(x) \times \text{duration} + b(x) \times \text{volume} + c(x)$$

where a, b and c depend in a rather complex way on the traffic and quality of service characteristics of the transaction x [20]. The complexity of this type of scheme may be reduced in the case of a large network where individual transactions only use a small fraction of available resources.

By the scale economies effect, the efficiency of resource sharing increases as network capacity grows. Indeed, networking is generally only economically advantageous when a large number of small users share a large resource. In this case, it may be shown that required bandwidth for a stream flow tends to its mean rate for whatever loss rate while the throughput of an elastic flow remains satisfactory even as links tend to saturation [17].

In this large network limit, it may be natural to replace the above formula by:

$$\text{price} = b \times \text{volume} + c(x)$$

i.e., a flat rate per byte charge b for all transactions plus a set up charge c which may depend of the nature of the connection.

Note that the latter simplified transaction pricing scheme gives no incentive to users to identify elastic traffic flows as deferable rather than intended for immediate delivery. By definition, this type of traffic can use any spare network capacity over and above that required to handle traffic with quality of service guarantees. However, this does not imply that a deferable traffic service class would have zero average incremental cost. This cost must be evaluated in common with that of other elastic traffic with which it would be indistinguishable in the absence of congestion.

Transaction pricing works well for the telephone network because there is only one service whose use can be forecast very accurately. On the other hand, the traffic in any computer network is extremely varied and indeed changes all the time. It is simply not possible to predict future trends based on current usage and, indeed, current usage is itself extremely difficult to characterize and measure. Furthermore, Internet traffic overall is growing at an exceptional rate of more than 100% per annum such that any capacity expansion is rapidly saturated and congestion can hardly be avoided. The following two questions must be addressed:

- is it possible to forecast multiservice traffic sufficiently reliably to be able to provision the network to make demand congestion an exceptional event?
- will transaction pricing perform satisfactorily when demand congestion does occur?

The unpredictability of Internet traffic is undeniable and largely explains and justifies the absence of any established network traffic engineering practice in the Internet. We would argue, however, that this situation is indicative of an initial transitory phase which will disappear as soon as the network has grown sufficiently to absorb latent demand from the population at large. In a mature network serving a large population of "ordinary" users, it is not unreasonable to suppose that demand for stream and elastic traffic can be characterized on a statistical basis with the idiosyncrasies of individual flows being insignificant with respect to a stable ensemble statistical behaviour.

The second question is particularly relevant in the transitory network growth phase when it may be difficult for a network to keep up with demand. The impact of overload depends on whether or not admission control is employed. Indeed, the absence of admission control appears incompatible with transaction pricing: overload would lead to unacceptable transparency and throughput performance degradation for a paid transaction. Use of admission control is intended to ensure that performance is acceptable for admitted transactions with overload manifested by increased blocking probabilities.

3.4 Other Charging Issues

As pointed out in [19], effective cost recovery is not the only issue to be resolved in designing the charging scheme. The following questions are particularly relevant:

- who pays for any data exchange: sender, receiver or some third party?
- who receives the payment, the network provider to which the user is directly connected or all network providers used by the flow?
- should users be charged for goodput or for all emitted data, particularly when congestion results in the need for multiple retransmissions?
- what basis should be used to charge for multipoint communications?

4. Choosing a Service Model

In this section we consider how the adopted charging scheme influences the definition of the service model. We examine successively three major components of the service model: the definition of service classes, the use of admission control and requirements for network mechanisms to allow quality of service guarantees.

4.1 Service Classes

If the network provider offers flat rate pricing there is no case for introducing service classes with different degrees of the same quality of service measure (e.g., expected packet delay) since users have no incentive to choose other than the best. However, it remains useful to distinguish between service classes specialized for stream and elastic flows, respectively, since their quality requirements are different: stream transactions require time integrity for flows with a bounded peak rate; elastic flows require a minimum throughput but both users and network gain by providing a much higher rate whenever possible. Users have a quality of service, rather than financial, incentive to choose the service class corresponding to their application.

Congestion pricing where the individual packets have explicit priority determined by their value may be seen as an enhancement to best effort service which *avoids* the need to introduce distinct service classes. The simplicity of this solution is clearly attractive for the current Internet. To enable quality of service through resource reservation, on the other hand, leads to an indirect definition of service classes through the identity of a flow and the conformity of its packets with respect to the negotiated traffic specification.

Transaction pricing explicitly relies on the definition of service classes and on the characterization of a flow within a service class by a set of traffic and performance parameters. The problem of defining statistical traffic parameters which are relevant for resource allocation and yet can be policed by the network has still not received a satisfactory solution.

If price were essentially determined by the traffic volume realized during the transaction and not by preannounced traffic parameters, a simple distinction between two service classes could be sufficient: one class for stream flows and another for elastic flows. Stream flows would be characterized by their peak rate (for admission control purposes) and elastic streams would be assigned a common minimum throughput guarantee, no further differentiation being necessary or useful under this charging scheme.

4.2 Admission Control

The need for admission control can be avoided with flat rate and congestion pricing at the expense, however, of any real guarantees with respect to transparency or throughput. Transaction pricing on the other hand, as noted previously, relies on admission control. As a general statement, admission control allows transparency and throughput guarantees at the cost of variable accessibility quality of service. The use or not of admission control is one of the most significant differences between Internet and B-ISDN service models.

Reliance on admission control adds considerable complexity to the network, necessitating connection oriented operation with resource reservation. Admission control algorithms can also be very complex, depending on the nature and number of quality of service criteria which must be satisfied for different flows.

For stream flows, admission control is easiest and best understood in the case of so-called "bufferless" multiplexing: delay and loss performance is guaranteed (statistically) by ensuring that the combined input rate of flows multiplexed on a given link is less than the link bandwidth with very high probability. The most promising procedures, like that proposed by Gibbens et al [9], are based on measurements of real traffic combined with sure knowledge of flow peak rates. A stream transaction would not be accepted if, given current traffic levels, a constant rate flow at the stream peak rate would lead to a non-negligible probability of rate overload.

Bufferless multiplexing is a mathematical model based on the interpretation of stream traffic as a superposition of variable rate fluid flows. In practice it is necessary to provide a non-zero buffer to account for the asynchronism of packet arrivals from different flows and for the imprecision of rate definition in the presence of jitter. The ATM multiplexing scheme known as Rate Envelope Multiplexing allows controlled performance if the peak rate of flows can be guaranteed by spacing cells at network ingress [2], [18].

Required precision of admission control can be relaxed when link capacity is shared with a non-negligible amount of elastic traffic. The latter can easily adjust to the rapid rate variations of the combined stream traffic. Admission control for elastic flows could be performed simply by comparing the sum of minimum required throughputs to available link bandwidth, after subtraction of the current estimated requirement for stream flows. Admission of elastic flows would of course also need to take account of memory requirements, as determined by the implemented queue scheduling mechanism.

Note finally that admission control allows a form of congestion pricing to be applied at transaction level. A given transaction might be admitted or rejected depending on whether or not its value, as declared by the user, were greater than an estimated "shadow price" equal to the expected value of hypothetical subsequent transactions which would be rejected if the considered transaction were accepted.

4.3 Mechanisms for Service Differentiation

Different service models impose widely varying requirements on network queueing and flow control mechanisms. In fact, requirements on queue scheduling and protocols for flow control are complementary. Simple FIFO queueing may be adequate to ensure quality of service if end to end protocols closely adjust flow rates to avoid excessive overflow. On the other hand, individual flows may have to be protected by complex scheduling mechanisms, like weighted fair queueing, if the network ensures no rate enforcement.

FIFO works well in the Internet when the rate of elastic flows is adjusted according to the congestion control algorithms of TCP and the volume of stream traffic, generally transported using the non-responsive protocol UDP, remains well within the capacity of network links. However, performance does rely on the cooperative behaviour of users which is less and less sure as the Internet expands. The importance of using queue mechanisms to reduce the impact of flows which are "not TCP friendly" is stressed by Floyd and Fall [6].

The introduction of congestion pricing would require mechanisms of varying complexity, ranging from a simple selective rejection device for the two-class service differentiation scheme of Clark [4] to some unspecified means to implement the "auction" in the case of the smart market.

Head of line priority queueing can provide quality of service differentiation. It is possible to closely control the quality of service of stream traffic in the highest priority service class if admission control is also employed. It may not, however, be feasible to realize guarantees for lower priorities without implementing supplementary flow controls like those of TCP or ABR. The same high priority queue can be used for different service classes if the realized quality of service is sufficient for the most exigent class. To realize different loss rates for different classes, for example, would require a more sophisticated class-based queueing discipline.

In a network where users pay for quality of service, in the case of transaction pricing for example, it appears unreasonable to rely on users executing an end to end protocol like TCP in preference to implementing network level flow control. The ABR service class is a particularly complex flow control protocol devised for ATM [7]. A simpler protocol may be sufficient if rate control is applied to individual elastic transactions rather than to bursty aggregates. The protocol is simpler still if there is no requirement to differentiate between flows because they all have the same per byte charge.

We have in mind a window-based protocol like TCP where, however, the need for congestion avoidance is greatly alleviated by the use of admission control. A fixed window size can provide a reasonably fair share of spare bandwidth over and above a minimum throughput guaranteed by limiting the number of admitted flows. The required queueing mechanism would be a simple, but large, FIFO queue in second priority behind the queue reserved for stream flows.

In Sect. 2, we identified the category of deferable elastic traffic (e-mail, etc.) which requires no throughput guarantees. This could be handled by a third and last priority queue served only in the absence of traffic from the stream and immediate delivery elastic flows. To clearly distinguish two elastic service classes, access to this third priority could be restricted to specialized (mail) servers: users pay to deposit their documents in a local mail server; the mail service provider pays the network provider for transporting the document to the mail server corresponding to one or more destination users.

Priority queues have limited applicability if it is necessary to identify more than two or three service classes or if it is necessary to satisfy individually specified flow quality of service requirements. Much more complex mechanisms, including "weighted fair queueing" [16] and "earliest deadline first" scheduling [8], have been extensively studied in the last few years. There are two main motivations:

- to satisfy deterministic delay guarantees;
- to protect individual flows from the traffic of other users.

In this paper we have not considered the need for absolute delay guarantees, as provided for in the guaranteed service category defined by the IETF. If such delays are necessary, it is clear that simple priority queueing, as considered above, is insufficient. However, we pretend that the stream traffic category only requires that the probability of delay exceeding a given limit be negligibly small. We also believe that this objective can be achieved simply by employing bufferless multiplexing with measurement based admission control, as previously mentioned.

Per flow scheduling as a means to ensure that a flow receives a "fair share" of bandwidth, independently of the traffic on other flows, may be seen as a requirement in a network having no control on input rates. The requirement is less obvious if admission control is employed and the network itself ensures that stream flows respect their traffic contract and that elastic flows are prevented from exceeding their current rate allocation.

5. Conclusion

Quality of service requirements in a multiservice network handling a mixture of stream and elastic traffic concern transparency, throughput and accessibility. The feasibility of fulfilling these requirements depends both on the service model, which defines how resources are shared, and on how much capacity is made available. Charging has a central role with respect to both factors.

We have identified three broad charging schemes: flat rate pricing, congestion pricing and transaction pricing. From the point of view of the business model, however, the options for a network provider reduce to two since the revenue raised by congestion pricing is not intended to cover network infrastructure and operation costs.

The business model of current Internet service providers is generally based on flat rate pricing: users pay only an access capacity dependent connection fee. This fee must be sufficient to cover the cost of both user-dedicated and shared resources. The network provider has limited incentive to invest in additional shared resources to avoid congestion since this would cost more but produce no additional revenue (except by attracting customers from a competitor).

The present Internet best effort service model, with the absence of admission control and reliance on end to end flow control, can provide virtually no quality of service guarantees. Proposed enhancements include the introduction of resource reservation and service differentiation. We interpret the use of these new facilities as a kind of congestion pricing since users only have an incentive to use the premium services in case of congestion. The appearance of non-TCP friendly transport protocols is leading to an additional requirement for more sophisticated queue management mechanisms.

The business model of the telephone network (with some notable exceptions) is based on transaction pricing. The price of transactions is fixed, demand at that price is forecast and the network is provisioned to handle that demand with high quality of service. This model may be generalized to a multiservice network on condition that admission control is systematically employed. It is then feasible to charge for each transaction since required transparency and throughput are guaranteed.

We have suggested a simple service model based on transaction pricing with just two service classes, one for stream traffic and one for elastic traffic. Stream traffic flows, characterized by their peak rate, would be handled using "bufferless" multiplexing with measurement based admission control ensuring negligible probability of rate overload. Admission control would also be employed to ensure minimum guaranteed throughput for elastic flows. A simple network layer flow control, possibly based on a fixed sized window, could be sufficient to ensure efficient and fair sharing of available link bandwidth. Resource sharing would be assured by a simple queue with head of line priority for stream flows and FIFO service in each class. In a large network, the cost related transaction price could be determined, for both stream and elastic flows, by just the volume of data transmitted. It remains to more fully explore the feasibility of this simple model which should for the time being be classified among the "partially baked ideas" of [14].

In this paper, we have developed the idea that the adopted charging scheme has a significant impact on the nature of the service model. An additional "firmly held opinion" derives from the converse of this statement: a service model without systematic use of admission control precludes the use of transaction pricing and may consequently jeopardize the solvability of the network provider.

References

[1] W.J. Baumol and J.G. Sidak, "Toward Competition in Local Telephony," The MIT Press, Cambridge, 1994.

[2] F. Brichet, L. Massoulié, and J. Roberts, "Stochastic ordering and the notion of negligible CDV," Proc. ITC15 (eds. V. Ramaswami and P.E. Wirth, "Teletraffic Contributions for the Information Age"), Elsevier, 1997.

[3] D.D. Clark, "A model for cost allocation and pricing in the Internet," in Internet Economics, eds. L.W. McKnight and J.P. Bailey [15], 1997.

[4] D.D. Clark and W. Fang, "Explicit allocation of best effort packet delivery service," Laboratory for Computer Science, MIT, Preprint, available via URL http://diffserv.lcs.mit.edu/Papers/exp-alloc-ddc-wf.pdf, 1997.

[5] N. Curien and M. Gensollen, "Economie des Télécommunications—Ouverture et Réglementation," ENSPTT/Economica, Paris, 1992.

[6] S. Floyd and K. Fall, "Router mechanisms to support end-to-end congestion control," Lawrence Berkeley National Laboratory, Preprint, 1997.

[7] E.J. Hernandez-Valencia, L. Benmohamed, S. Chong, and R. Nagarajan, "Rate control algorithms for the ATM ABR service," Europ. Trans. Telecom., vol.8, no.1, Jan.–Feb. 1997.

[8] L. Georgiadis, R. Guérin, V. Peris, and R. Rajan, "Efficient support of delay and rate guarantees in an Internet," Proc. ACM SIGCOMM '96, 1996.

[9] R. Gibbens, F. Kelly, and P. Key, "A decision theoretic approach to call admission control in ATM networks," IEEE JSAC, vol.13, no.6, Aug. 1995.

[10] B.C. Klopfenstein, "Internet economics: An annotated bibliography," to be published in Journal of Media Economics, available via URL http://www.bgsu.edu/departments/tcom/annota.html, 1997.

[11] W. Leland, M. Taqqu, W. Willinger, and D. Wilson, "On the self-similar nature of Ethernet traffic," IEEE/ACM Trans. Networking, vol.2, no.1, Feb. 1994.

[12] J.K. MacKie-Mason, "Intenet economics," Web pages via URL http://www.spp.umich.edu/telecom/net-economics.html, 1997.

[13] J. MacKie-Mason and H. Varian, "Pricing the Internet," Public Access to the Internet, eds. B. Kahin and J. Keller, Prentice Hall, 1995.

[14] J.K. MacKie-Mason and H.R. Varian, "Economic FAQs about the Internet," Internet Economics, eds. L.W. McKnight and J.P. Bailey [15], 1997.

[15] L.W. McKnight and J.P. Bailey, eds., "Internet Economics," The MIT Press, Cambridge, (earlier versions of published papers available in the Jounal of Electronic Publishing via URL http://www.press.umich.edu/jep/econTOC.html), 1997.

[16] A.K. Parekh and R.G. Gallager, "A generalized processor sharing approach to flow control in integrated services networks: The single node case," IEEE/ACM Trans. Net-

working, vol.2, no.2, April 1993.

[17] J. Roberts, "Realizing quality of service guarantees in multiservice networks," Proceedings of IFIP Conference PMCCN '97, Chapman and Hall, 1998.

[18] J. Roberts, U. Mocci, and J. Virtamo, eds., "Broadband Network Teletraffic (Final Report of COST 242)," LNCS 1155, Springer Verlag, 1996.

[19] S. Shenker, D. Clark, D. Estrin, and S. Herzog, "Pricing in computer networks: Reshaping the research agenda," Telecommunications Policy, vol.20, no.3, pp.183–201, 1996.

[20] D.J. Songhurst and F.P. Kelly, "Charging schemes for multiservice networks," Proc. ITC15 (eds. V. Ramaswami, P.E. Wirth, "Teletraffic Contributions for the Information Age"), Elsevier, 1997.

[21] H.R. Varian, "Network economics," Web pages via URL http://www.sims.berkeley.edu/resources/infoecon/Networks.html, 1997.

High-Performance Backbone Network Technology edited by Naoaki Yamanaka
ISBN: 0-8247-5321-6 DOI: 10.1081/(120025662)
Published 2004 by Marcel Dekker, Inc.

A Failure-Resistant Self-Healing Scheme in ATM Networks

Ryutaro Kawamura

SUMMARY In B-ISDN, network reliability is a very significant theme. This paper proposes the Failure-Resistant Virtual Path (FRVP) scheme that prevents any information loss even with network failure, to realize a high-end reliability service in B-ISDN. The FRVP scheme is based on simple parallel transmission established using the superior characteristics of ATM. In the FRVP scheme, the transmitter duplicates user cells and transmits them across several VPs simultaneously. The receiver chooses the perfect cells and sends them to the user. As a result, the cell stream output by the receiver is not affected by VP failure if at least one VP remains active. I develop a prototype FRVP system and conduct field trials using NTT's nationwide ATM testbed network. The FRVP scheme is shown to achieve extremely-reliable ATM networks and services.

key words: *self-healing, restoration, failure, ATM, Virtual Path*

1. Introduction

B-ISDN has been actively researched and developed using ATM (Asynchronous Transfer Mode) techniques. In B-ISDN, one of the most important themes is the increased level of network reliability needed. This is because the steady increase in optical transmission bandwidth (e.g. 10 Gbps) suggests that even single failure will seriously impact huge numbers of service users. Moreover, the advance of the information society increases the "mission critical" nature of data communications.

Several restoration scheme based on transmission link switching are being utilized in current telecom networks against network failures. The reliability (e.g. restoration rapidity, restoration probability) secured by these restoration techniques involves a trade-off against network redundancy, and so impacts the cost of services provided.

B-ISDN will integrate a wide variety of services, thus reliability requirements will also vary widely. Several reliability grades have to be realized and integrated into the framework of B-ISDN in order to achieve appropriate service cost [1]. In particular, high-end applications that require very extremely high network reliability continue to be found, for example tele-medicine, electric commerce, TV program transmission and intra-net backbone systems. In these applications, network failure causes significant economic and social damage, even if the interruption period is short.

The goal of this paper is the realization of a failure-resistant transmission scheme that can perfectly prevent information loss at network failure for high-end reliability service in B-ISDN. This paper proposes the Failure-Resistant Virtual Path (FRVP) scheme utilizing benefits of ATM, and reports prototype development results.

Manuscript received August 25, 1997.
Manuscript revised November 19, 1997.

† The author is with the NTT Optical Network Systems Laboratories, Yokosuka-shi, 239-0847 Japan.

2. FRVP Requirements and the Multi-Reliability VP Network Concept

Figure 1 shows the relationship between restoration rapidity and the spare resources required by typical network reliability techniques. The required spare resource is the ratio of required spare capacity to the capacity required by the basic network. Schemes typical of current networks or being researched include centralized control switching [2]-[4], Self-Healing Network (SHN) [1], [5]-[7] and Automatic Protection Switch (APS) [5], [8]. All are based on switching the failed facility (fiber, path etc.) to an alternate route. Switching is initiated by a failure indication signal, so these schemes can not prevent service interruption completely.

Compared to these conventional schemes, the target of this paper is failure restoration with no service interruption. Diversity Coding (DC) techniques have been studied [9]-[11] for this purpose. In DC, the transmitter encodes the signal components as a matrix including error detection and correction codes, and transmits them via multiple independent paths. The receiver reconstructs the matrix from the signals arriving on each path and decodes the original signal. When one transmission path fails, the receiver detects the failure, and

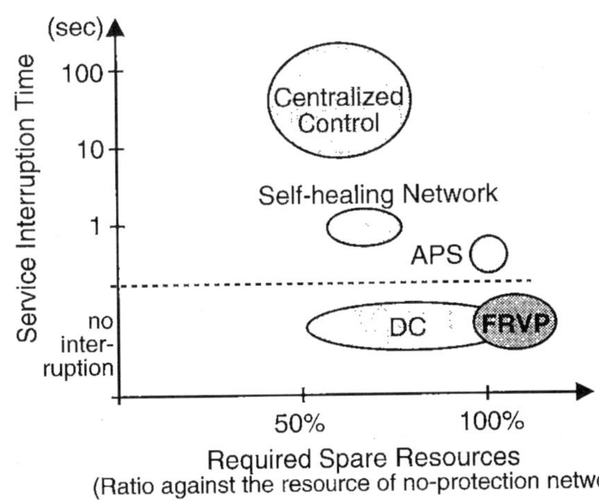

Fig. 1 The FRVP scheme compared conventional network reliability schemes.

restores the lost information from the error correction codes.

If there are many independent routes between source to destination nodes, DC offers better resource efficiency than the proposed FRVP scheme. However, recent physical network topologies tend to be quite simple reflecting the increase of fiber capacity and the trend toward reducing management costs by simplifying the network.

Moreover, DC encounters two problems. One is the increase in transmission delay. The encoding and decoding operations of DC increase the transmission delay. Moreover, the receiver can not respond until all information (matrix) is received on all diversity routes. Therefore, delay is determined by the longest route's delay. Another source of delay is the time needed by the receiver to detect failure. In ATM, cells suffer from jitter, so the receiver must allow for jitter when detecting failure. The increase of transmission delay, the "long-fat pipe problem," leads to a lowering of network performance. Another problem is the size of the buffer in the receiver. A large buffer is needed for each receiver to handle the delay difference between the shortest and the largest route, jitter, failure detection time, and matrix size.

Another technique is hitless switching. The hitless path/line switching schemes were developed for SDH and ATM networks [12], [13]. These schemes realize hitless switching if the network operator initiates switching. However, they can not realize hitless switching against network failure. Therefore, they are not used for failure restoration, only preplanned route alteration.

This paper proposes the Failure-Resistant Virtual Path (FRVP) scheme [14], [15]. It is based on simple parallel transmission instead of matrix coding, and so suits simple physical network topologies while minimizing transmission delay. FRVP realizes failure free transmission using the near resources as the 1:1 or 1+1 APS scheme [5], if the level of FRVP redundancy is 2. The redundancy is defined as the number of parallel transmission VPs and details are described below.

Reliability generally increases network cost through the overhead of redundancy. This tendency is also shown in Fig. 1. Therefore, it impacts the cost of services provided. On the other hand, B-ISDN will integrate a wide variety of services, so the reliability requirements will also vary widely. Therefore several reliability grades have to be realized and integrated into the framework of B-ISDN in order to minimize service costs [16], [17]. One way of realizing multiple reliability grades is the integration of several restoration schemes.

Figure 2 shows the proposed multi-reliability VP network concept for B-ISDN [16], [17]. The logical networks, which have different levels of reliability, and realized using several restoration schemes on the VP layer. Each service is accommodated on the appropriate network according its reliability and cost requirements. As a result, the network can provide multi-reliability VP/VC services. In this architecture, FRVP is a key technique for realizing the most reliable VP network.

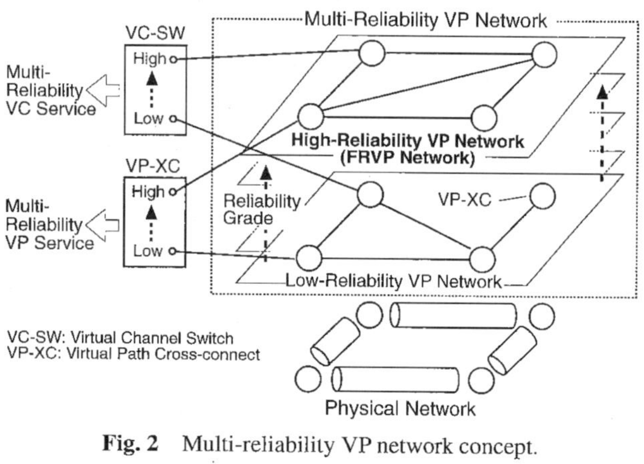

Fig. 2 Multi-reliability VP network concept.

3. FRVP Techniques

3.1 Basic Concept of FRVP

Figure 3 shows the basic concept of the FRVP scheme. FRVP consists of a transmitter, receiver, and several VPs that establish multiple transmission between the transmitter and receiver. The transmitter duplicates the incoming cells from the users and sends the same cells over all assigned VPs. The receiver always chooses the cells without error and sends them to the user side. Therefore, if the level of FRVP redundancy is more than two, no interruption occurs upon VP failure if at least one VP remains. The assigned VPs have completely independent routes to prevent multiple VPs being cut by a single failure. In order to identify the cell order at the receiver, index cells are inserted into the data cells by the transmitter.

FRVP uses very simple parallel transmission techniques to minimize the delay. Moreover it is suitable for simple network topologies. In this paper, I focus on applying FRVP techniques to the VP layer, however, they could also be applied other layers, for example the SDH layer or the VC layer.

In the FRVP scheme, the level of redundancy, that is number of parallel VPs, can be freely set but must be more than two. This value is the most important factor in determining survivability and redundancy of the FRVP scheme. The following discussion considers the redundancy level of two for simplicity, a reasonable level of network redundancy, and simple physical topology.

3.2 Cell Frame

In FRVP, the cell frame is used for synchronization between VPs and error detection in user cells in the receiver [18]. The receiver must match the cells arriving on the VPs even though they have different levels of transmission delay and jitter. If

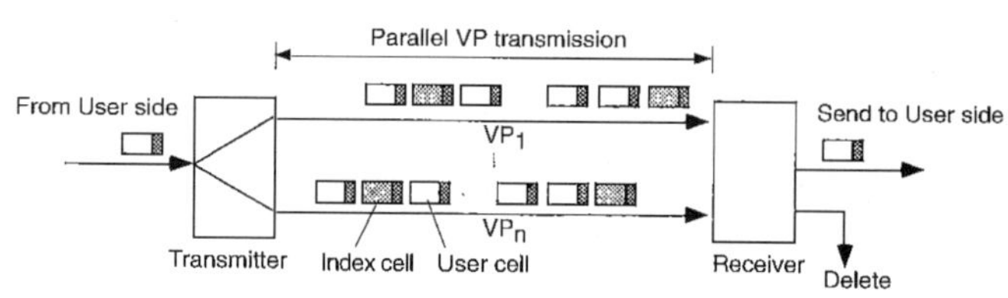

Fig. 3 FRVP concept.

the differences are relatively small, line protection using the same route for example, synchronization can be realized by comparing user cells [19]. However, the differences are very large if we consider a long-haul ATM VP network (order of the maximum delay differences may be 10 ms which corresponds to 10^3 cells in a 150Mbps VP). Moreover, jitter (CDV) must also be considered in ATM. Therefore, FRVP inserts index cells into the user cell stream for synchronization.

The index cells are also used by the receiver to detect errors in user cells. Figure 4 shows the cell frame format. The cell frame consists of a fixed block of user cells followed by an index cell for that block of cells. Figure 5 shows the index cell format used in the current prototype. The sequence number (*SN*) identifies each cell frame and is used for synchronization. The number of user cells is used to detect cell loss and mis-insertion. A error detection code (CRC-32) is used to detect bit error in the user cells in the frame. The receiver stores one cell frame in the buffer and checks for errors in the user cells or the index cell. The VCI value must be reserved; moreover, the position of the index cells in the cell stream must not be changed during transmission. For example, some OAM cells are not suited as the index cell, because they are dropped and re-inserted at transit nodes.

3.3 Transmit Function

Figure 6 shows a functional block diagram of the FRVP node. The transmitter consists of a frame generator and a cell duplicator. The frame generator forms cell frames by inserting index cells. User cells are transmitted as soon as received; buffering is not used. The cell duplicator copies all cells including the index cells, and transmits them to each VP.

Figure 7 shows the index cell insertion procedure. The index cell is inserted, which then completes a frame, when either the limit of the frame cell number (N_f) or the time limit (T_i) is reached. The former limit controls the maximum number of user cells in one cell frame, and this guarantees the error detection capability. On the other hand, the time limit determines the maximum duration of one frame. This limit guarantees the maximum waiting delay in the receiver, because user cells are buffered until the arrival of the index cell. This limit is particularly effective if the traffic is not continuous such as on/off type traffic. The moment the traffic is turned off, the index cell triggered by this limit can flush the waiting cells from the receiver buffer.

The index cell bandwidth ratio (b_i), that is the ratio of index cell traffic to user traffic in a CBR VP, is given by

$$b_i = \frac{1}{\min\left\{N_f, \lfloor B_w T_i / 53 \rfloor\right\}} .$$

(B_w: user's VP bandwidth, T_i: time limit)

3.4 Receive Function

The receiver identifies same cell frames arriving through each VP, chooses those with no error, and sends the user cells in the frames to the user side. There are several alternative algorithms to this cell frame selection [14], however, I use the First-Arrived-Select-Soon (FASS) scheme in order to minimize delay and buffer in receiver. The receiver selects fastest arriving cell frame that does not include error, and then sends user cells in the frame to the user. Duplicates of the same frame that arrive later are deleted in the receiver.

The lower part of Fig. 6 illustrates the function block of

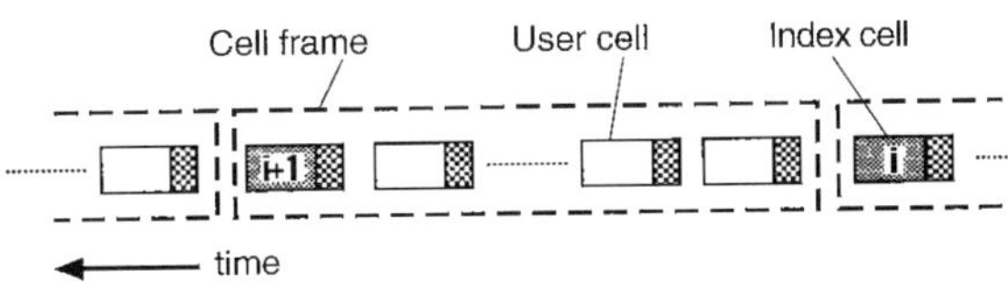

Fig. 4 Definition of the cell frame.

byte \ bit	8	7	6	5	4	3	2	1
1	VPI(12bit:NNI/8bit:UNI): same value as user cell							
2								
3	VCI(16bit): Index cell Identifier (reserved value for index cell)							
4					PT			CLP:0
5	HEC							
6	Index cell payload type (normal ,alarm indication etc.)							
7,8	Sequence number (SN) (0-65535)							
9,10	Number of user cells in this cell frame							
11-14	Error detection code for user cells in the frame (CRC32)							
⋮								
52								
53	Error detection code for this index cell (CRC-10)							

Fig. 5 The index cell format used in the prototype.

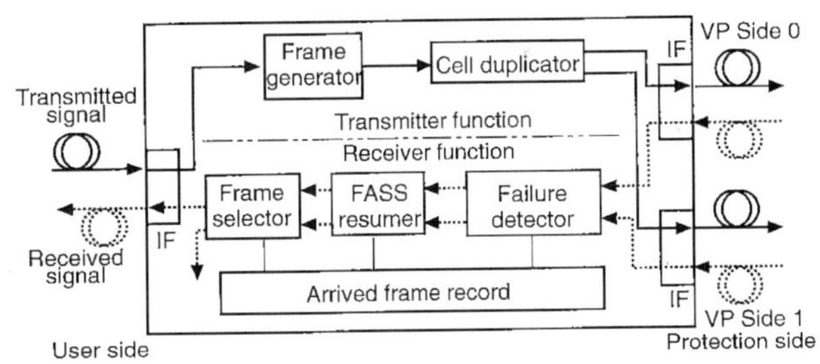

Fig. 6 Function block of the FRVP node.

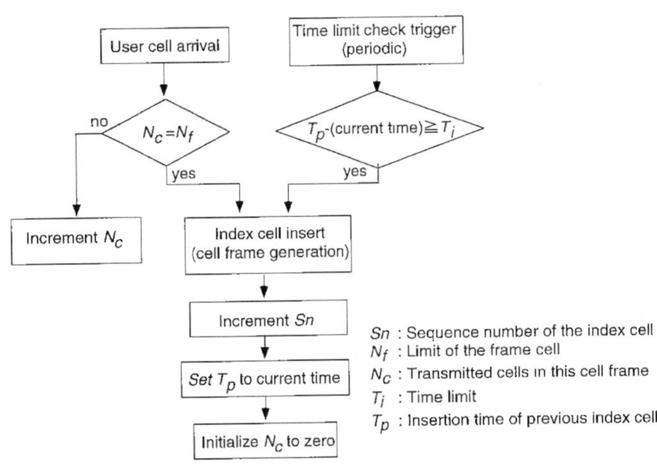

Fig. 7 Index cell insertion (cell frame generation) procedure.

the receiver. The failure detector observes the connectivity and quality of the cell stream using the *SN*, number of cells, error detection code in the index cell and arrival of index cell that arrives more frequent than (T_i +maximum jitter) period even if there is no user cell. The frame selector performs the FASS procedure; it refers to the arrived frame record and the *SN* field in the frame, and transmits user cells to the user if the cells are error-free. If the jitter is larger than the delay difference, the side selected may change frame-by-frame. Details of this sequence are shown in Fig. 8. By adopting FASS, maximum waiting delay in the receiver is held to just one frame and the cell frame buffer for each VP can be reduced to just one frame.

The arrived frame record describes the *SN* of the last cell frame arrived via each VP. The FASS resumer continues with the FASS rule when the failed VP is restored; details of this procedure are described in the next section.

Figure 9 illustrates the typical behavior of the receiver in failure. When both VPs are normal, cell frames arriving on the shorter VP are always selected by the FASS rule. If the shortest delay VP fails, cell frames on the other VP would be selected (Fig. 9(a)). On the other hand, when the longer delay VP fails, frame selection continues from the shorter VP (Fig. 9(b)).

3.5 Handling of Restored VP Traffic

If one of the VPs fails and is then restored, the FASS rule must be resumed to guard against subsequent failure. In particular, if the restored VP has minimum delay, the receiver has to select the cell frames arriving on this VP without any frame jump (cell loss).

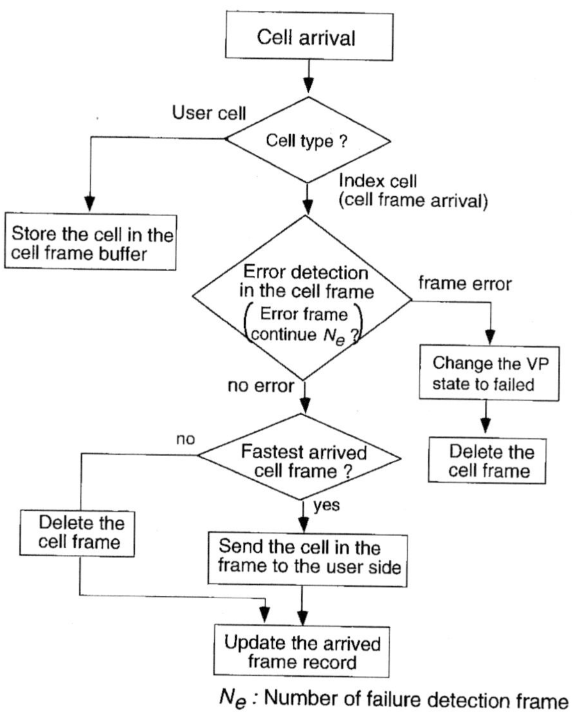

Fig. 8 FASS algorithm in receiver.

Figure 10 shows the flow of the FASS resumer. If the restored VP has larger delay, buffering is not necessary. The resumer permits cell frames of this VP to be selected which reestablishes the FASS process. On the other hand, if the restored VP has minimum delay, delay control (buffering) is necessary to prevent any frame jump. First, the resumer starts by identifying the index cells and buffering the restored VP cells. Buffer length is equivalent to the maximum expected delay difference. When an index cell identical to one held in the buffer arrives on the other VP, the resumer starts to read cells from the cell buffer. When all cells are read out from the buffer, the process is completed and the FASS rule is resumed. However, cell readout must be only slightly faster than the normal cell stream rate. To achieve this, I use the following procedure which is similar to one proposed for a hitless switching scheme [12].

- When a cell arrives on the restored VP, readout one cell

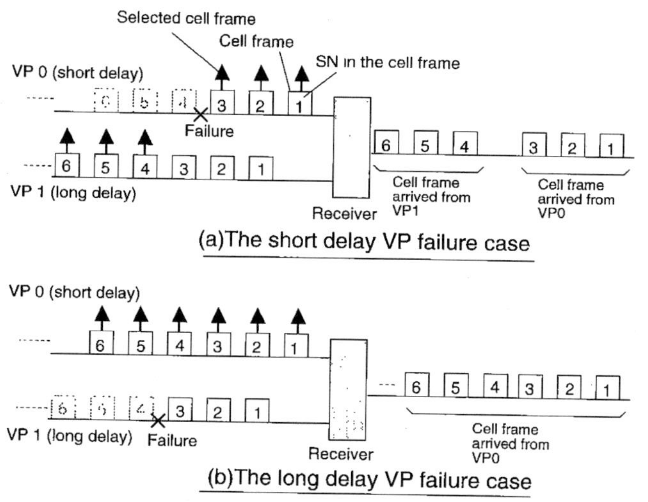

Fig. 9 Receiver behaviors upon failure.

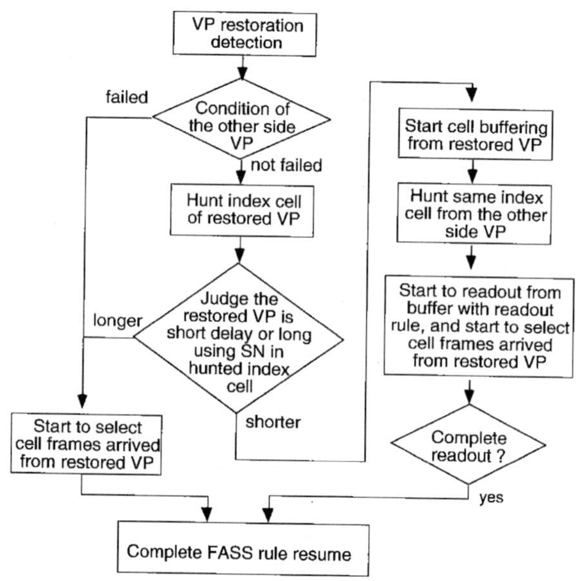

Fig. 10 The FASS rule resume procedure after VP restoration.

from the buffer.
- Every N_a cells, readout one extra cell from buffer.

This procedure increases the readout speed to $1/N_a$ faster than the original cell stream, and buffer readout is gradual. The readout completion period (T_r) is given by

$$T_r = N_a T_d .$$

(N_a : frame readout control, T_d: delay difference)

This procedure assumes the existence of enough idle cells or the increasing readout rate (B_w/N_a) is lower than the bandwidth of index cell. With reasonable parameters including worst case in the Table 1 and 2 (T_d=5msec, N_a=128cell), the resume procedure completes in several seconds. Therefore, FASS resume function and the buffer in it can be shared by all VPs.

4. Prototype Development and Examination Using NTT's Testbed ATM Network

4.1 Prototype Development

Figure 11 shows two typical applications of the FRVP function in ATM networks. One is end-to-end VP service protection. The FRVP acts in a CPN node. The other is an application that runs inside the network. The FRVP is implemented in the NEs (Network Elements; e.g. XC) of the network provider to improve network reliability.

Two prototype FRVP node systems (NE type and CPN type) were developed and evaluated in NTT's testbed ATM network. Figure 6 illustrates the functional block diagram of the prototype systems. It consists of the transmitter function and the receiver function for bidirectional FRVP transmission. Figure 12 shows the prototype node (CPN type).

Table 1 shows the principal specifications of the prototypes. The maximum delay difference is 5ms (equivalent to 1000km of fiber) which is representative of long-haul networks. Moreover, the maximum number of VP pairs supported in this system is eight for the NE type and four for the CPN type.

In the FRVP scheme, transmission characteristics (e.g. transmission efficiency, delay) are influenced by the parameters described in Sect. 3. In the prototype, these parameters were controlled by the OpS (NE type) or a dip switch (CPN type). Table 2 shows the variable parameters. These parameters can be independently defined for each VP.

The failure detection frame (N_e) is the number of frames used to determine VP failure. If N_e corrupted frames arrive continuously, the receiver identifies this VP as having failed. Therefore, if this value is one, the receiver determines VP failure from the first failed frame and does not select this frame for output to user. As a result, the cell stream passed to the user does not include any corrupted cells. On the other hand, if this value is set to more to than one, slight errors are ignored in the receiver, and are output to the user.

Moreover, the prototype uses unique OAM cells for alarm indication between FRVP nodes. This covers the failure of parallel VPs; the protected segment can be processed as if it were just one reliable VP.

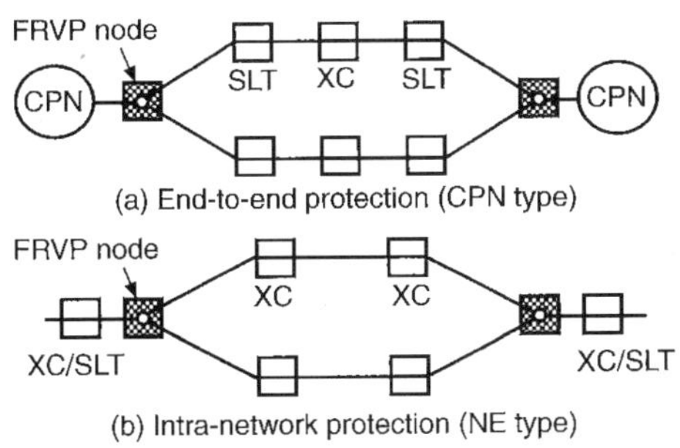

Fig. 11 Typical applications of FRVP function in ATM networks.

4.2 Examination Using NTT's ATM Testbed Network

In order to verify the FRVP scheme and prototype, a trial was performed using NTT's nationwide ATM testbed network [20]. Figure 13 illustrates the test network. The difference in VP length was controlled from 1 m (5 ns) to 1000 km (5 ms). Moreover, a non-protected VP was also established

Fig. 12 Prototype node (CPN type).

Table 1 Principal specifications of prototype system.

specification \ prototype type	CPN type	NE type
Optical Interface	155Mbps(STM-1)	
Level of redundancy	2	
Maximum delay difference	5msec (1000km)	
Maximum jitter	1msec	
Number of VP pairs	4	8
OpS Interface	No	exist
Parameter control	dip switch	OpS-IF

Table 2 Variable parameters.

Limit of the frame cell(N_f)	1 - 240(cell)
Time limit of index cell(T_i)	0.2 - 4(msec)
Number of failure detection frame(N_e)	1 - 12(frame)
Frame readout control(N_a)	1 - 128 (cell)
VCI value of index cell	0 - 65535

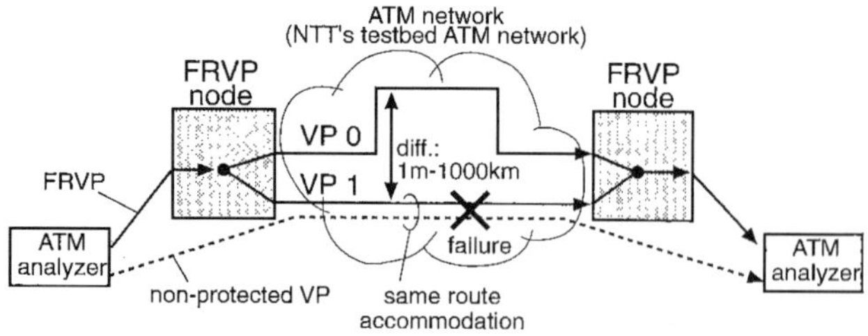

Fig. 13 Test network.

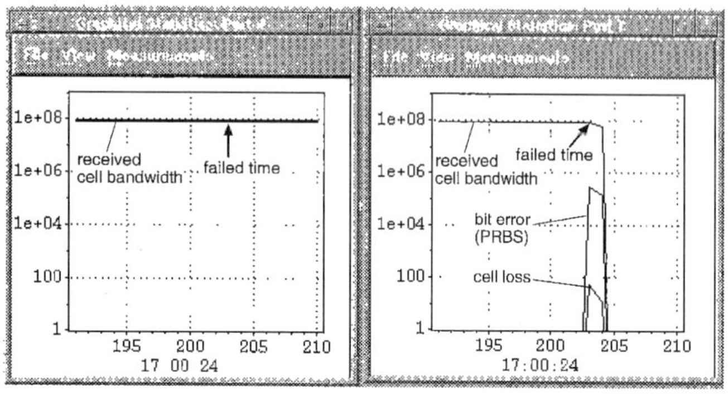

(a) FRVP (b) non-protected VP

Fig. 14 A sample of the examinarion results.

along the shorter route.

Figure 14 shows one of the measured results. This is a dump of an ATM analyzer display. Figure 14(a) shows the output of the FRVP system, and Fig. 14(b) shows the output of the additional VP not protected by FRVP. Each graph shows received cell stream bandwidth, amount of bit error, and cell loss. When the shorter fiber was cut (203 sec in the graphs), the FRVP VP did not degrade, while the non-FRVP VP failed with bit error and cell loss.

5. Conclusion

In order to realize highly reliable services in B-ISDN, I have proposed the FRVP scheme. FRVP is based on simple parallel transmission techniques, but it is a very effective solution in ATM networks. This scheme completely prevents information loss due to network failure. I have developed two types of prototype FRVP nodes, and examined their operation in NTT's testbed ATM network. As a result, I verified that FRVP prevents link failure from impacting the user.

In the future, FRVP can be applied to (1) end-to-end VPs between customer premises networks, and (2) intra-network VPs in the network provider's facilities.

Acknowledgments

I would like to thank Dr. Ichiro Yamashita and Dr. Koichi Sano of NTT Optical Network Systems Laboratories and Dr. Naohisa Ohta and Dr. Toshinori Tsuboi of NTT Network Service Systems Laboratories for their support of this work. I must acknowledge Prof. Tokuhiro Kitami of Meiji University , Dr. Ikuo Tokizawa of NTT Advanced Technologies and Mr. Hiroshi Ohta of NTT Network Service Systems Laboratories for their valuable suggestions.

References

[1] R.Kawamura and I.Tokizawa, "Self-healing virtual path architecture in ATM networks," IEEE Commun. Mag., p.72-79, 1995.

[2] H.S.Burns, C.-W.Chao, P.M.Dollard, R.E.Mallon, H.Eslambolchi, and P.A.Wolfmeyer, "FASTER(SM) operations in the real AT&T transport network," Proc. Globecom'93, pp.229-233, 1993.

[3] C.-W.Chao, G.Fuoco, and D.Kropfl, "FASTER platform gives the network a competitive edge," AT&T TECHNICAL JOURNAL, no.JULY/AUGUST, pp.69-81, 1994.

[4] K.Nishihata, "SDH-based 52Mb/s digital cross-connect systems with hit-less function, TMN-based operation system," Proc. IEEE DCS Workshop VI, Banff, Canada, June 1995.

[5] T.-H.Wu, "Fiber Network Service Survivability," Artech House, 1992.

[6] R.Kawamura, K.Sato, and I.Tokizawa, "Self-healing ATM networks based on virtual path concept," IEEE J. Select. Areas Commun., vol.12, no.1, pp.120-127, 1994.

[7] R.Kawamura, H.Hadama, and I.Tokizawa, "Implementation of self-healing function in ATM networks based on virtual path concept," Proc. IEEE INFOCOM'95, Boston, April 1995.

[8] R.D.Hall and S.Whitt, "Protection of SONET based networks," Proc. Globecom '89, pp.0821-0825, 1989.

[9] E.Ayanoglu, I.Chih-Lin, R.D.Gitlin, and J.E.Mazo, "Diversity coding for transparent self-healing and fault-tolerant communication networks," IEEE Trans. Commun., vol.41, no.11, pp.1677-1686, 1993.

[10] E.Ayanoglu and R.D.Gitlin, "Broadband network restoration," IEEE Commun. Mag., p.110-119, 1996.

[11] T.Onuma and T.Kitami. "Diversity coding," IEICE Technical Report, CS93-150, 1993.

[12] H.Tatsuno and N.Tokura, "Hitless path protection switching techniques for ATM networks," IEICE Trans. Commun., vol.J76-B-I, no.6, pp.421-430, 1993.

[13] H.Ohta and H.Ueda, "ATM hitless line protection switching method on forced switch commands," IEICE Trans. Commun., vol.J78-B-I, no.11, pp.652-663, 1995.

[14] R.Kawamura, "A failure-resistant virtual path scheme for ATM networks," IEICE Technical Report, CS94-196, 1995.

[15] R.Kawamura, "A development of failure-resistant virtual path in ATM networks," IEICE Technical Report, CS96-94, 1996.

[16] R.Kawamura and S.Ohta, "Self-healing ATM network architecture; Implementation results from NTT's nationwide testbed ATM network," Proc. Networks'96, Sydney, pp.717-722, 1996.

[17] T.Yahara and R.Kawamura, "Virtual path self-healing scheme based on multi-reliability ATM network concept," Proc. IEEE Globecom'97, 1997.

[18] R.Kawamura, "A failure-resistant virtual path scheme for ATM networks," IEICE Technical Report, CS94-196, 1995.

[19] H.Ohta and H.Ueda, "Hitless line protection switching method for ATM networks," Proc. IEEE ICC'93, pp.272-276, 1993.

[20] N.Fujii, I.Yoda, and K.Minato, "An ATM transport network operation system based on TMN," Proc. IEEE/IFIP NOMS'96, pp.124-134, Kyoto, Japan, 1996.

High-Performance Backbone Network Technology edited by Naoaki Yamanaka
ISBN: 0-8247-5321-6 DOI: 10.1081/(120025663)
Published 2004 by Marcel Dekker, Inc.

A Study on VP Network Designer

Ayano Yamashita and Satoru Ohta

SUMMARY VP Network Designer, a software tool that automates the tasks involved in the transaction of service orders in VP leased line services, is introduced here in this paper. The tool is composed of two functions: VP Design Explorer which, given a request for VP establishment, determines a disjoint backup and target VP routes to support fault tolerance under VP-APS scheme; Network Resource Administrator which provides data administering functions, useful in maintaining a clear insight into the state of the network. The paper focuses on the implementation of VP Design Explorer. A scheme, proposed as disjoint VP routes search, is used to evaluate a pair of VP routes that guarantee duct level independence. The scheme is based on an integer programming modeling approach, and is proved to be effective for evaluating disjoint paths in a hierarchical network. VP Design Explorer is equipped with an additional feature where, under conditions of resource shortage, it presents a set of additional resources that are necessary to satisfy VP demand, together with the VP routes that become possible by the additions. Formulation of the problem is attained through an extension of the disjoint VP routes search scheme. A prototype of VP Network Designer is presented and its performance tested using computer simulations. The tool is found to achieve a computational time performance in the order of seconds and minutes, for evaluating disjoint VP routes search problems under real life ATM network conditions.

key words: *network design, disjoint paths, integer programming, VP-APS, VP leased line, ATM*

1. Introduction

A major service that will be provided by an ATM network is believed to be a VP (virtual path) leased line service [1]. The service involves provisioning for a VP connection of given bandwidth between a pair of customer sites. Service is made available by the network provider through the following set of actions. Service order (SO), a customer request for a VP connection of certain bandwidth, is passed to the network designer whose job is to determine the network route over which the VP will be accommodated. The task involves the selection of a valid set of TP (transmission path) trails with a capacity enough to accommodate VP demand. VP identifiers (VPI) must then be assigned for the VP over every TP trail. IF (interface) ports of ATM devices through which VP is passed must also be defined. Once these data have been collected, it then becomes possible for the operation system (OpS) [2] to establish and provision the VP leased line service.

This paper introduces VP Network Designer [3]-[5], a software tool designed for the purpose of automating the tasks involved in the transaction of SOs. A prototype of the tool has been developed, and its implementation discussed, together with its level of performance.

The tool supports two prominent features that are vital in providing leased line services effectively. One feature is the support for a VP-APS (Automatic Protection Switch) scheme [6]. With this tool it is possible to search for a pair of disjoint paths for target and backup VPs. Previous studies on disjoint paths algorithms [7] are based on a non-layered network model, which cannot be applied to an ATM network of hierarchical structure. In order to search for a pair of disjoint paths in a hierarchical network, therefore, the tool proposes a scheme called, disjoint VP routes search, based on an integer programming (IP) modeling approach.

The second feature is the capability of the tool to elicit VP routes that require additional resource. When an existing network cannot accommodate VP demand, the amount of additional resource that must be supplemented onto the network is evaluated, together with the VP routes that become feasible as a result. Under such conditions of resource shortage, it is favorable to derive a solution whereby the VPs are made to run over routes that require the least cost of additional resource. In an ATM network which is expected to accommodate a variety of multimedia services, traffic forecast is expected to become highly unreliable, leading to an overconsumption of network resource. To make use of the underutilized resource to the best extent, the tool proceeds to search for VP routes that require the minimum amount of additional resource. This way, major construction works involving the layout of new fibres and building sites requiring prohibitively large capital investments are avoided whenever possible.

In addition to the above features, a function that allows network designers to inspect on the state of network resource is incorporated into the prototype package.

One problem envisaged for the tool is its computational time performance. Because the tool uses integer programming for finding disjoint VP routes pair, the corresponding program that evaluates the problem is expected to run in exponential time order. It is therefore imperative that the tool be verified of its runtime performance against practical situations found in a real sized ATM network. Simulation results shown in this paper do in fact prove that VP Network Designer achieves solutions with acceptable speed.

Manuscript received July 1, 1997.
Manuscript revised October 20, 1997.
† The authors are with the NTT Optical Network Systems Laboratories, Yokosuka-shi, 239-0847 Japan.

IEICE TRANS. COMMUN., Vol. E81-B, No. 3, pp. 503–510, March 1998

The remainder of the paper is organized as follows. Sect. 2 explains the tool architecture. Section 3 outlines its functional framework. Section 4 proposes a unique set of disjoint VP routes search, the performance of which is analyzed in Sect. 5. The final section outlines the conclusions reached.

2. Architecture of VP Network Designer

Figure 1 illustrates the architecture of the prototype tool that has been implemented. Tcl/Tk [8] is used to provide a flexible, easy to use HMI. The program for evaluating VP routes is written in C. IP formulation that is produced by the program is solved using a commercial optimization software package[9]. The RDBMS (Relational Database Management System) maintains information on network topology and its available resource. This includes data on each of the ATM network layers, as well as the tier relationship between these layers.

3. Functional Framework of VP Network Designer

VP Network Designer is composed of two major functions, VP Design Explorer and Network Resource Administrator. A schematic view of the functions is illustrated in Fig. 2.

3.1 VP Design Explorer

VP Design Explorer performs the task of VP route design when SO arises. Service order may be concerned with a customer request to setup a new VP, to reassign VP bandwidth, or to alter VP reliability condition.

To provide a cost effective service, VP routes having the shortest length are selected. For VP services that warrants high reliability, both backup and target routes are evaluated so as to make VP restoration possible under network failures. Under conditions of resource shortage, Explorer will suggest facility additions necessary for VP establishment, and will derive VP routes based on these additions.

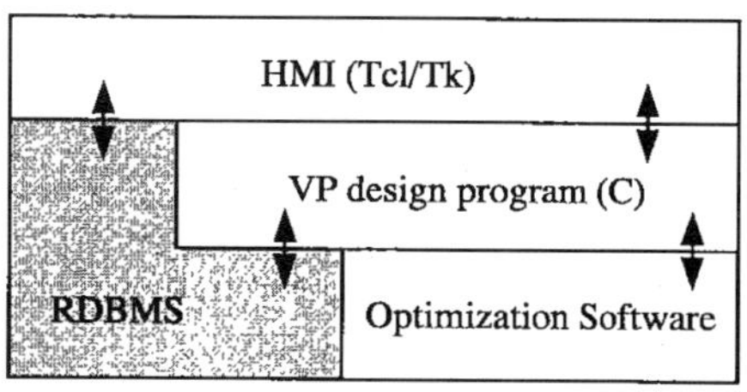

Fig. 1 Architecture of VP Network Designer.

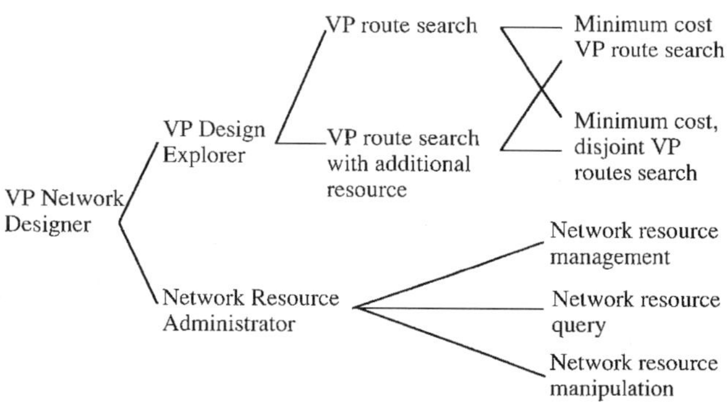

Fig. 2 Functional framework of VP Network Designer.

3.2 Network Resource Administrator

For a network designer to design the VP network effectively, he must be able to maintain a clear insight into the state of the existing ATM network. Network Resource Administrator, through its efficient functions in managing, searching, and editing the immense network data, makes this task realizable.

Apart from the most obvious, basic query functions such as displaying line and equipment data which match certain conditions, Administrator may also provide higher level functions such as triggering alarms when resource runs out after VP establishment. These functions will assist the designer in taking early actions against shortage of resource.

Administrator plays a supporting role in the Designer tool, aiding the human designer to make rational decisions over the design of VP network. Explorer is the function that conducts the evaluation of individual VP routes. The remainder of this paper will focus on the implementation of Explorer.

4. Implementation of VP Design Explorer

Before going into detail about the implementation of VP Design Explorer, it is worth spending a moment to look at the overall structure of an ATM network [10]. Figure 3 illustrates a simple ATM network. A collection of TPs are established across ATM devices of type DSU (Digital Service Unit), SLT (Subscriber Line Terminal), and XC (Cross-Connect). TPs are accommodated onto a physical transmission medium such as optical fibre, which in turn is laid out over a set of cable ducts. Backbone network is comprised of a collection of XC switches which are intercon-

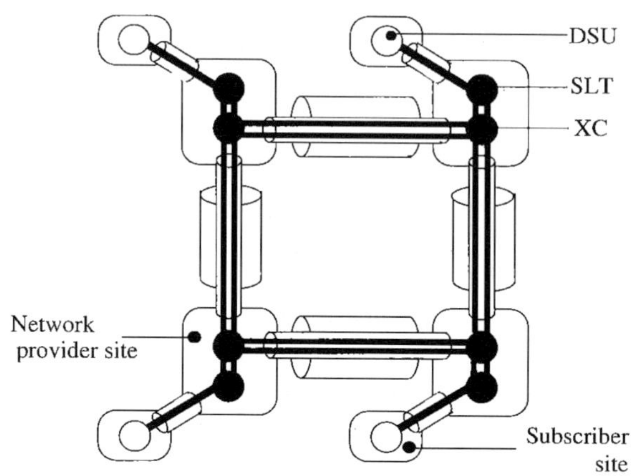

Fig. 3 ATM network topology.

nected by TPs. Subscriber sites are connected to the closest network provider site containing an SLT. The subscriber network and the backbone network are linked together by TPs that run across SLT and XC devices.

When a customer requests for a VP leased line service, network provider must cater for a resource that will enable the establishment of VP across a pair of customer DSUs. The procedure implemented by the prototype of VP Design Explorer in designing the VP route, is illustrated in Fig. 4. Its objective is to search for all possible routes that are capable of accommodating the requested demand at minimum expense.

VP route evaluation is realized in two steps. First, a TP which links the DSU and SLT devices of the given subscriber network, is selected, at both ends of the route. If a TP does not already exist, the program determines the additional facilities necessary to set up a new TP. By defining a valid pair of links across the subscriber networks, one effectively determines the homing nodes for the backbone network.

A VP restoration scheme called VP-APS [6] has been developed for the purpose of ensuring reliability in an ATM network. The scheme proposes to preassign a backup VP at service establishment, with a bandwidth of equal magnitude as the target VP. When a network failure causes the working VP to fail, the service is restored by switching over to the backup VP. Explorer supports the VP-APS scheme by making possible the evaluation of a pair of disjoint VP routes. ATM network which has a multilayered architecture, is prone to failure in any of the layers of VP, TP, transmission medium or the duct. The most serious of all failures would involve duct abeyance, which may easily be caused by underground construction works. In order for the VP service to benefit from VP-APS under such heavy failures, the target and backup VPs must be set over a pair of duct sequences that are disjoint across the backbone network. The disjoint VP routes search scheme which has been adopted to realize the above objective is explained in Sect. 4.1.

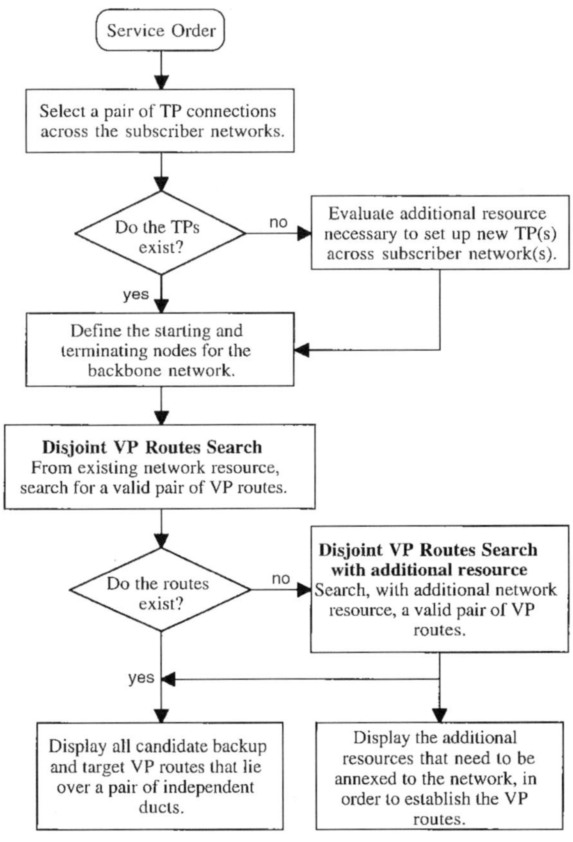

Fig. 4 Flow diagram of VP Design Explorer program.

At the end of the procedure, the tool must display all candidates of backup and target VP routes which satisfy duct level independence. This is achieved as follows. By the method given in Sect. 4.1, a pair of disjoint VP routes is evaluated, for which the corresponding pairs of underlying TP sets and duct sets are defined. Any sequence of TPs, which constitute a path across the duct set that runs over the shorter distance, becomes a candidate for accommodating the target VP. Similarly, TP sequences across the duct set that is running over the longer distance become the candidates of the backup VP.

There may, however, be occasions when the network designer finds himself short of network resource in order to satisfy the SO. The information of no feasible solution under existing resource is of not much help to him. Explorer serves to be effective under such situations by deriving VP routes that will require additional facilities of the smallest cost. Details of the algorithm are discussed in Sect. 4.2.

4.1 Disjoint VP Routes Search

4.1.1 Representation of ATM Network

The special case of finding a pair of VP routes in an ATM network renders a problem that cannot be solved by a conventional, disjoint paths algorithm [7]. The conventional algorithm, intended for a single layer network, has limitations when applied to an ATM network of hierarchical structure. A new approach is therefore imperative, if the dual VP routes problem is to be portrayed correctly under multilayered network conditions. An example illustrated in Fig. 5 clarifies the importance of this approach.

Consider, for a simple ATM backbone network, shown in the figure, a pair of VPs are to be routed across XCs marked *S* and *T*. Applying the conventional algorithm to the TP layers of the network, as in Fig. 5(a), clearly cannot conform to the duct level route independence. Alternatively, applying it to the duct layer, as shown in Fig. 5(b), will provide no guarantee that a valid sequence of TPs exist over the selected routes. A way to search for a disjoint pair of VP routes is therefore necessary, so that they will run over a sequence of TPs laid out over an independent set of ducts. This is shown in Fig. 5(c). The following IP formulation achieves this objective.

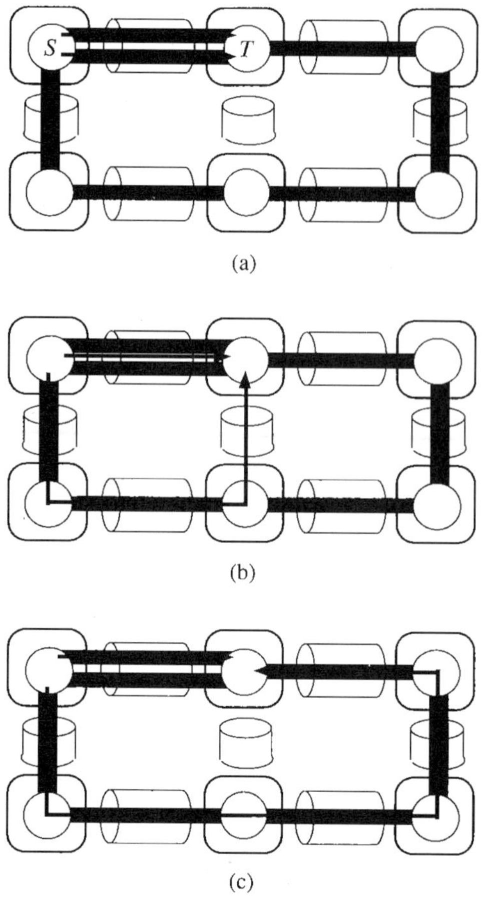

Fig. 5 (a) ATM network with routes over a pair of disjoint TPs. (b) ATM network with routes over a pair of disjoint duct sequences. (c) ATM network with routes over a pair of TP sequences running over independent ducts.

4.1.2 IP Formulation

An ATM network may be represented as a weighted digraph. In the following, node n_i represents a XC or an SLT device. Node N_x represents a network provider site in which ATM devices are placed. Link (n_i, n_j) represents a TP with enough idle capacity to accommodate VP demand. These TPs are contained within ducts (N_x, N_y) across the respective sites. Given the links (n_i, n_j) and their related lengths $a_{i,j}$, the disjoint VP routes search may be expressed as a problem in which a flow of magnitude 2 is run through a network having links of unit capacity. This is formulated into the following IP problem.

$$\min \sum_{i=1}^{n_{max}} \sum_{j=1}^{n_{max}} a_{i,j} f_{i,j} \tag{1}$$

subject to,

$$\sum_{j=1}^{n_{max}} f_{ij} - \sum_{j=1}^{n_{max}} f_{ji} = \begin{cases} 2 \text{ if node of } i \text{ is the source} \\ -2 \text{ if node of } i \text{ is the sink} \\ 0 \text{ otherwise} \end{cases} \tag{2}$$

$$f_{i,j} \leq 1 \tag{3}$$

$$f_{i,j} \geq 0, \text{ integer} \qquad \text{for every } (n_i, n_j) \tag{4}$$

$$\sum_{(n_i,n_j) \text{ included in } (N_X,N_Y)} f_{i,j} + \sum_{(n_i,n_j) \text{ included in } (N_X,N_Y)} f_{j,i} \leq 1 \qquad \text{for every } (N_X, N_Y) \tag{5}$$

The model extends from an ordinary minimum cost flow problem [11] by the following factors. Integer constraint on link flow, $f_{i,j}$, given by Eq. (4), assures the solution to yield a pair of route flows each of unit capacity. Furthermore, defining a capacity constraint of unity over a bundle of links associated with each duct, given by Eq. (5), guarantees duct level independence for the pair of flow routes. Reference [11] provides an alternative approach to formulate flow problems with capacity constraints over a bundle of links. The reference suggests a "route-flow" approach, in which the individual components of flow associated with the possible routes are identified in every link. To take this approach in the disjoint VP routes problem, route-flow variables must be specified with a 0-1 integer constraint. Consequently, a greater number of integer constrained variables will be necessary in formulating a problem under the route-flow approach. This approach is therefore unfavourable for large scale ATM networks that need to be handled by the tool.

Although the integer constraint causes the disjoint VP routes problem to have an exponential time complexity, we will see in the section to come that, with the use of a commercial optimization software, it is possible to solve the model within an acceptable time.

4.2 Disjoint VP Routes Search with Additional Resource

4.2.1 Representation of ATM Network

In Sect. 4.1, the paper uses an IP model to formulate the disjoint VP routes search. The model used for the disjoint VP routes problem involving additional resource is a protraction of it. We can extend on the graph of existing ATM network by defining additional virtual links and nodes that become possible as a result of installing additional facilities. Link length may then indicate the cost of installing the necessary facilities for setting up the link. The problem may then be solved as an IP problem as before. The virtual links having a non-zero flow indicate the necessary additions.

The task of configuring a network graph becomes more complex when virtual nodes and links need to be incorporated. As in the previous section of 4.1, an ATM device is represented by a node and a TP connection by a link. Subsequently, a virtual node and a link are associated with potential ATM device and TP that may be set up over an existing network to establish a VP route. The manner in which virtual links and nodes are laid out over the graph is best understood through a careful review of the properties of network components that constitute an ATM network.

To begin with, a concept of neighbour devices is introduced here in the paper. Every ATM device is associated with a set of neighbour devices, or neighbours, to which a TP connection may be established. Depending on the state of resource consumption, every neighbour of a given device may be categorized as either an associate or a dissociate neighbour. A neighbour with which a TP connection exists, with a capacity to accommodate VP demand, is defined as an associate neighbour. A neighbour with which no such TP connections exist is then a dissociate neighbour.

Thus, an ATM device is a VP switch that is capable of channeling VP routes across its neighbour devices. With preestablished TPs, a device is capable of channeling a VP route between its associate neighbours. A device is also capable of securing a channel to a dissociate neighbour by enacting an IF port onto which a new TP is terminated. Connections to dissociate neighbours where a new TP may be established becomes depicted as virtual links in the network graph.

If a device is to channel a VP route across an associate and a dissociate neighbour, a single IF port needs to be enacted on the device to support the set up of a new TP. If it is to channel a VP route across a pair of dissociate neighbours, then two IF ports becomes necessary. If the device lacks the port capacity to channel a VP route across any pair of its neighbours, then a new ATM device must be replenished. A virtual node representing a new ATM device is defined at a building site if, the site contains a device which has inadequate port capacity to channel a VP route across its neighbours.

The ability of an ATM device to channel a VP route across its neighbours is therefore determined by it's IF port capacity to support existing and new TP connections. When representing a device, therefore, a node may be described by a set of the following type of subnodes, each representing a collection of IF ports that the device may possess.

type 1: represents a collection of IF ports that are occupied by existing TPs with sufficient capacity to accommodate VP demand.

type 2: represents an idle IF port that can be enacted to support a new TP establishment.

type 3: represents a second idle IF port that can be enacted to support a second TP establishment.

A node is constructed by up to three subnodes of type 1, 2, and 3, depending on the state of connectivity of the corresponding device to its neighbours. Type 1 subnode is defined if more than one associate neighbours exist. Type 2 is defined if atleast one dissociate neighbours exist, given that the device has an idle IF port. Type 3 if two or more dissociate neighbours exist, given that the device has a second idle IF port. The collection of subnodes of every device is organized as a strongly connected subgraph.

A network graph may therefore be configured from a set of subnodes, which are connected by a collection of links. The set of subnodes is defined as a set S, and the set of subnodes for each of the types 1, 2 and 3 are defined as S_1, S_2 and S_3 respectively. Preestablished TPs are represented by links across type 1 subnodes. Potential TP connections are represented by virtual links across type 2, type 3 subnodes. Links connecting subnodes n_i and n_j are defined as (n_i,n_j) where $i,j \in S$, and ducts connecting the building sites N_X and N_Y are described as (N_X,N_Y). Link length, $a_{i,j}$, is described by the link addition cost, $p_{i,j}$, and the costs of IF ports, p_i, p_j, at either end, as shown below.

$$a_{i,j}=\frac{p_i}{2}+\frac{p_j}{2}+p_{ij} \tag{6}$$

Decomposing the nodes into a collection of subnodes allows the graph to convey the correct information on the additional resources which are allowed over the network. Consequently the additions necessary to establish the selected VP routes can be derived. Consider an example of Fig. 6(b), which illustrates a network graph of a part of the ATM network given in Fig. 6(a). Suppose a VP flow is to be channeled from node E to A, via D. To pass from node E to D, a unit flow needs to be accommodated in either of the links, (n_5,n_3) and (n_5,n_4). If the cost of enacting the first IF port is cheaper than that of the second IF port, virtual link (n_5,n_3) is selected. Then, to pass from node D to A, flow is accommodated in virtual link (n_4,n_2). Selecting link (n_3,n_2) is forbidden because subnode n_3 is reserved as an IF port for establishing a TP across nodes E and D. Following the above rules, the passage of flow over the two subnodes, n_3, n_4, of node D correctly conveys that two new IF ports need to be enacted on device D, if a VP is to be routed across devices E and A.

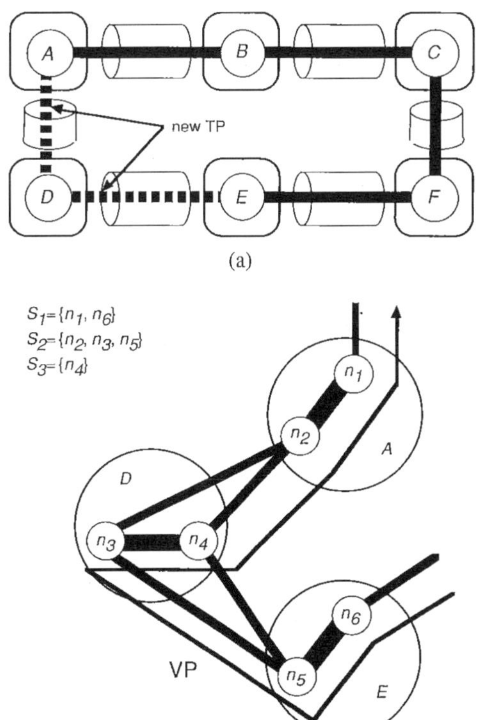

(b)

Fig. 6 (a) Example of an ATM network. (b) Representation of part of ATM network by network graph.

Although subnodes of type S_2 and S_3 both represent an addition of a new IF port, the two may not be abbreviated into a single subnode. The distinction comes from the fact that the cost of the links attached to each of the two subnodes do not necessarily correspond to each other due to the differences in cost of enacting the first IF port and the second.

4.2.2 IP Formulation

IP formula described below defines an objective function that aims to minimize the investment costs involved in establishing the VP routes.

$$\min \sum_{i \in S} \sum_{\substack{j \in S \\ i \neq j}} a_{i,j} f_{i,j} \tag{7}$$

subject to,

$$\sum_{\substack{j \in S \\ i \neq j}} f_{i,j} - \sum_{\substack{j \in S \\ i \neq j}} f_{j,i} = \begin{cases} 2 \text{ if node of } i \text{ is the source and } i \in S_1 \\ -2 \text{ if node of } i \text{ is the sink and } i \in S_1 \\ 0 \text{ otherwise} \end{cases} \tag{8}$$

$$f_{i,j} \leq 1 \tag{9}$$

$$f_{i,j} \geq 0, \text{ integer} \qquad \text{for every } (n_i, n_j) \tag{10}$$

$$\sum_{\substack{j \in S_2 \cup S_3 \\ \text{node of } i \neq \text{node of } j}} f_{i,j} + \sum_{\substack{j \in S_2 \cup S_3 \\ \text{node of } i \neq \text{node of } j}} f_{j,i} \leq 1 \qquad \text{for every } i \in S_2 \cup S_3 \tag{11}$$

$$\sum_{(n_i, n_j) \text{ included in } (N_X, N_Y)} f_{i,j} + \sum_{(n_i, n_j) \text{ included in } (N_X, N_Y)} f_{j,i} \leq 1 \qquad \text{for every } (N_X, N_Y) \tag{12}$$

The formulation, when compared to the case of Sect. 4.1, includes an additional constraint, of Eq. (11). Subnodes S_2 and S_3 each represents an additional IF port which can support at most a single TP. Therefore, the maximum quantity of flow passing into and out of a node from an S_2 and an S_3 subnode must each be limited to unity.

5. Performance of VP Design Explorer

The bete noire commonly felt towards the use of integer programming is that, they are considerably harder to solve when compared to the ordinary, linear programming (LP) problems. IP is generally solved by a method known as branch and bound [9], in which the optimal solution is searched through solving a series of related LP models. Due to this enumeration process, computation time has the tendency to increase dramatically with increase in the number of integer constraints. VP Network Designer is based on an algorithm which cannot, theoretically, be solved in polynomial time. It is often the case, however, that IP problems that are encountered in practical situations perform surprisingly well compared to the worst case scenario. Thus, confirming the performance of the program in solving the disjoint VP routes search problem in an ATM network is vital, if Designer is to prove itself as an effectual tool.

The VP Design Explorer program, which was developed for the prototype of VP Network Designer, was examined for its computational time efficiency under networks of varying shapes and sizes. Computer simulations verified that, under real life ATM network conditions, Explorer is capable of obtaining solutions in the range of seconds and minutes. A Sun Sparc 20 workstation was used to run the simulations.

5.1 Network Models

Two network models have been selected to illustrate the behaviour of the VP Design Explorer program. The simplest configuration that may be found in an ATM network is portrayed by model 1, as shown in Fig. 7(a). Only the backbone topology is illustrated for convenience. Model 1 is characterized by the perfectly lattice shape of both fibre and duct layers. A pair of TPs is assumed to be set up across each of the fibres contained in the backbone network. The lengths of ducts are set uniformly to 100km. Positions of network provider sites are described by the corresponding row and column numbers. Model 2, as shown in Fig. 7(b), describes a more complex configuration. A bypass at the fibre layer, and consequently at the TP layer, is defined across sites ($3m$+2,$2n$+1) and ($3m$+3,$2n$+2), where m and n are integer values. Horizontal ducts along the rows, ($3m$+1), are defined with a longer length of 250km.

Simulations were used to initiate VP demands over the two network models, for which a pair of disjoint VP routes was evaluated using the Explorer program. Demands were generated across a pair of subscriber sites that are linked to the provider sites (1,1) and (1,$2n$+1), for a network of size (4,$2n$+1). Network size was increased by varying the column size using the variable, n. Row number was fixed to a value of 4. It was assumed for the TPs of the network to all have an idle capacity greater than the demanded VP bandwidth.

The Explorer program was expected to perform poorly for the model 2 network, compared to the case of model 1. An ordinary ATM network topology, typified by model 1, is known to generate a naturally integer solution for its IP problems. For example, the shortest pair of routes across sites (1,1) and (1,3) for a Fig. 7(a) network, are given by *P1* and *P2*. Both routes can carry a flow of unit capacity in the corresponding network graph. The exceptional case, illustrated by model 2, is when bypass TPs exist over which a fraction of unit flow may be routed across. LP optimum will derive a flow across passages *P2* and *P3*, to carry a quarter of the net flow each. Only through enumeration will IP optimality be reached, to achieve the required VP routes *P1* and *P3*.

5.2 Simulation Results

Figure 8 illustrates the CPU time performance of VP Design

Explorer for searching disjoint VP routes, against network size. Two CPU times were measured for each model: one expressing the time taken to complete the procedures of Explorer as illustrated in Fig. 4, the other expressing the time taken by the optimization software to solve the associated IP formulae. Each plot represents the mean value obtained from repeating the simulation 50 times.

Contradictory to the behaviour that was expected, no severe degradation in performance was found for model 2. In fact, the performance curve of Explorer program for model 1 shows a poorer result compared to model 2. Irrespective of the network model, increase in network size induces a noticeable degradation in the performance of Explorer program. However, because there are a greater number of link variables involved in depicting model 1 compared to model 2 of the same size, Explorer exhibits a poorer performance with the former model.

The time involved in the evaluation of IP problems within the program, in contrast, increases only linearly with network size. Figure 9 illustrates the number of branch and bounds executed to reach IP optimality. The graph verifies that no enumeration is necessary to reach an IP solution for model 1,and that the solution be achieved in polynomial time scale. For the case of model 2, the optimization software that is employed in the tool is found to be extremely effective in suppressing the number of branch and bounds. The number of branch and bounds increases only linearly with network size. This verifies the unexpectedly high level of performance achieved under the model.

The level of performance obtained by Explorer indicates that the proposed IP modeling approach is indeed

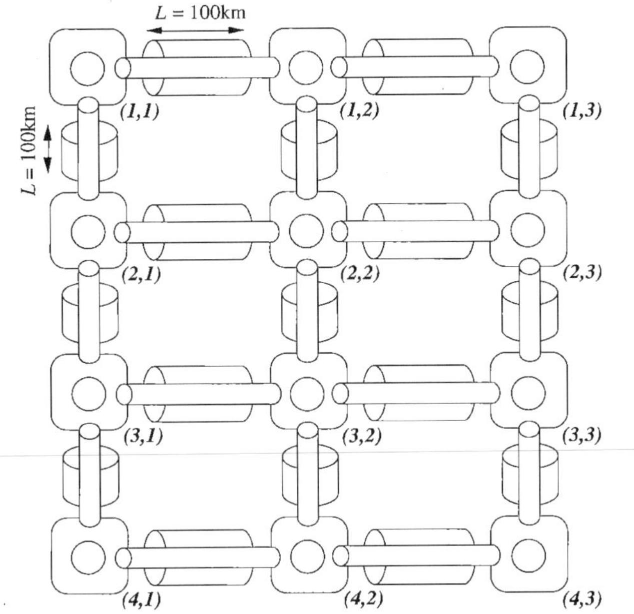

Passage P1: (1,1) → (1,2) → (1,3)
Passage P2: (1,1) → (2,1) → (2,2) → (2,3) →(1,3)

(a)

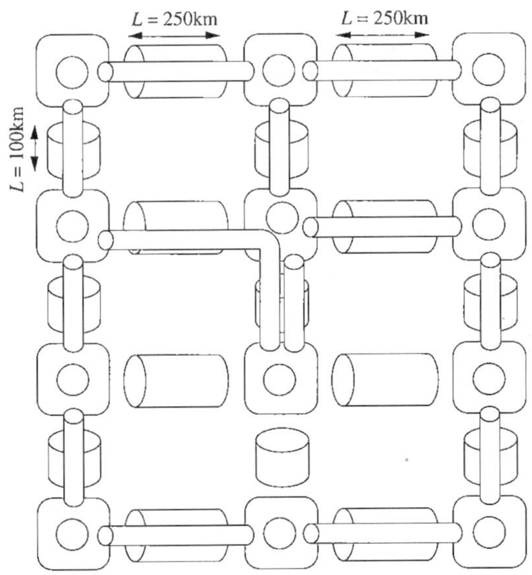

Passage P1: (1,1) → (1,2) → (1,3)
Passage P2: (1,1) → (2,1) → (3,2) → (2,2) →(2,3) →(1,3)
Passage P3: (1,1) → (2,1) → (3,1) → (4,1) →(4,2) →(4,3) → (3,3) → (2,3) → (1,3)

(b)

Fig. 7 (a) Network model 1. (b) Network model 2.

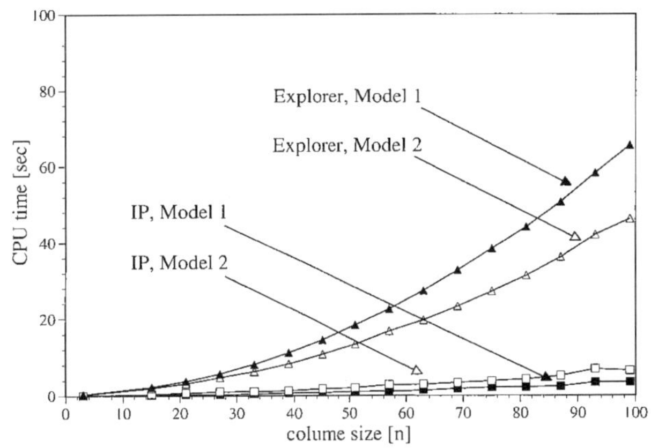

Fig. 8 Run-time performance with network size.

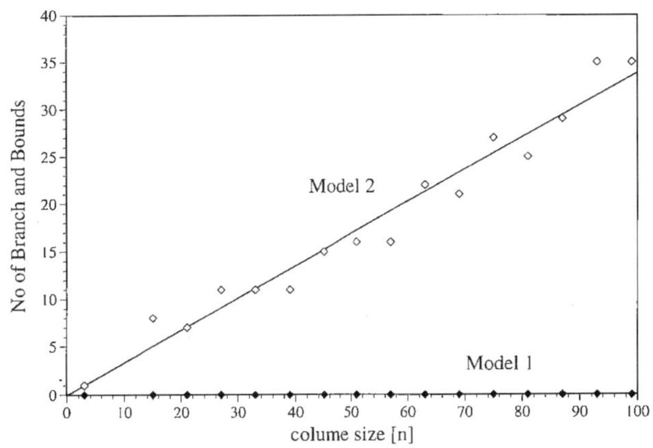

Fig. 9 Number of enumerations with network size.

practical for solving disjoint VP routes in a real size ATM network.

6. Conclusion

In this paper we introduced VP Network Designer, a software tool that automates tasks involved in the transaction of service orders. The tool supports fault tolerant design under VP-APS by a scheme proposed as disjoint VP routes search. The proposed scheme evaluates backup and target VP routes that runs over independent ducts. Formulation of the scheme, based on an IP modeling approach, was presented and was elucidated on its ability to evaluate disjoint paths in a hierarchical network structure. Under conditions of resource shortage, the tool provides information on the additional resource necessary to establish a VP service, together with the VP routes that become possible by the additions. Formulation of the problem was presented as an extension to the disjoint VP routes search scheme proposed earlier.

The prototype of VP Network Designer was presented, and its runtime performance on the evaluation of disjoint VP routes assessed. During the assessment, it was demonstrated that the topology of a typical ATM network was such that the corresponding network graph provided a naturally integer solution to the IP problem. Consequently, solutions to the IP formulae were achieved in the polynomial time scale. The only exception to this model was shown to be when bypass TP links existed in the network over which shortest routes containing noninteger flow was possible. Even in such cases, the optimization software package employed by the tool was found to be very effective in suppressing the number of branch and bounds necessary to achieve IP optimality. Simulation results showed that, under conditions of real life ATM networks, VP Network Designer was capable of obtaining disjoint VP routes solution within the order of seconds and minutes.

Acknowledgments

The author would like to thank Dr. Toshinori Tsuboi and Dr. Koichi Sano of NTT Optical Network Systems Laboratories for their kind assistance.

References

[1] T.Aoyama, I.Tokizawa, and K.Sato, "ATM VP-based broadband networks for multimedia services," IEEE Commun. Mag., vol.31, no.4, pp.30-39, 1992.
[2] N.Fujii and T.Yamamura, "ATM transport network operation system in Japan," IEEE Commun. Mag., vol.34, no.9, pp.70-75, 1996.
[3] A.Yamashita and S.Ohta, "A study on ATM network design tool (Part 1)," Proc. IEICE Fall Conference, vol.2, B-797, 1996.
[4] A.Yamashita and S.Ohta, "A study on ATM network design tool (Part 2)," Proc. IEICE Spring Conference, vol.2, B-7-163, 1997.
[5] A.Yamashita and S.Ohta, "A study on ATM network designer," IEICE Technical Report, IN97-37, April 1997.
[6] R.Kawamura and H.Ueda, "A (1+1) virtual path protection switching scheme in ATM networks," Proc. IEICE Fall Conference, vol.3, B-488, 1992.
[7] J.W.Suurballe, "Disjoint paths in a network," Networks, vol.4, pp.125-145, 1974.
[8] J.K.Ousterhout, "Tcl and the Tk toolkit," Addision-Wesley, 1994.
[9] L.Schrage, "LINDO, an optimization modeling system," Boyd & Fraser, 1991.
[10] "Special feature (1): ATM telecommunications services and networks," NTT Review, vol.6, no.1, pp.20-60, Jan. 1994.
[11] M.Iri and T. Kobayashi, "Network Theory," JULE, 1994.

High-Performance Backbone Network Technology edited by Naoaki Yamanaka
ISBN: 0-8247-5321-6 DOI: 10.1081/(120025664)
Published 2004 by Marcel Dekker, Inc.

Real-Time Traffic Characterization for Quality-of-Service Control in ATM Networks

Brian L. Mark and Gopalakrishnan Ramamurthy

SUMMARY One of the important challenges in the design of ATM networks is how to provide quality-of-service (QoS) while maintaining high network resource utilization. In this paper, we discuss the role of real-time traffic characterization in QoS control for ATM networks and review several approaches to the problem of resource allocation. We then describe a particular framework for QoS control in which real-time measurements of a connection stream are used to determine appropriate parameters for usage parameter control (UPC). Connection admission control (CAC) is based on the characterization of the aggregate stream in terms of the individual stream UPC descriptors, together with real-time measurements.

key words: *traffic characterization, quality-of-service, resource allocation, congestion control*

1. Introduction

The ATM network is anticipated to become a ubiquitous communications medium, capable of supporting a wide range of traffic services. As a successor to traditional packet-switching and circuit-switching technologies, ATM has the potential to support diverse quality-of-service (QoS) requirements in an efficient manner. To fully realize this potential, practical mechanisms for characterizing and controlling ATM traffic are essential.

This paper focuses on the topic of real-time traffic characterization and its role in QoS control for ATM networks. Our main interest in this paper is not in traffic modeling per se, but in schemes that employ traffic characterization for making real-time resource allocation decisions supporting QoS. In Sect. 2 we provide a brief overview of various approaches to traffic characterization and resource allocation that have been investigated in the literature. In Sects. 3, 4, we discuss a particular framework for QoS control that we have investigated which comprises: real-time characterization of an individual stream and admission control via traffic descriptors together with real-time measurements of the aggregate stream at the multiplexing point. Section 5 concludes the paper with a discussion of future research directions.

Manuscript received September 29, 1997.
Manuscript revised December 15, 1997.

[†]The authors are with C&C Research Laboratories, NEC USA, Inc., 4 Independence Way, Princeton, NJ 08540, USA.

2. Traffic Characterization and Control

2.1 Stochastic Modeling of ATM Traffic

The advent of high-speed networking technologies, ATM in particular, has spurred much research on statistical traffic models and their performance analyses. Due to statistical multiplexing, one expects more efficient resource usage in ATM compared with the circuit-switching paradigm. In TDM, for example, each stream must be allocated a sufficient number of time-slots to accommodate the peak rate; this may be extremely wasteful for bursty, variable bit rate (VBR) streams. To achieve statistical multiplexing gain in ATM, however, one requires more sophisticated characterizations of the offered traffic than merely the peak rates of the individual streams.

The purpose of traffic modeling is to capture the statistical characteristics of a traffic stream. Statistical models can be used to generate traffic for simulation studies of network performance or for dimensioning network resources. A substantial literature exists on the subject of traffic modeling (cf. [1], [2] and references cited therein). In the remainder of this subsection, we provide a brief overview of traffic models for high-speed networks. Simplified versions of these models can sometimes be used effectively in traffic control schemes.

Early on, it was recognized that the Poisson process, widely used in teletraffic studies, does not adequately model the bursty traffic streams expected in ATM networks. Anick et al. [3] popularized the Markov on-off fluid source as a model for bursty high-speed network traffic. The fluid abstraction is a reasonable approximation for ATM cell streams, since the cell size is small, relative to ATM link capacities. More general Markov modulated fluid systems have also been studied (cf. [4]). Other authors have investigated point process and discrete-time counterparts of the fluid models, respectively, the Markov Modulated Poisson Process (MMPP) (cf. [5]) and Markov Modulated Bernouilli Process (MMBP) (cf. [6]).

Another class of popular models for high-speed networks, in particular video traffic, is the class of autoregressive models (cf. [7]). The TES mode [8] is an autoregressive model that fits the marginal distribution of the data and allows the user to match the autocorre-

IEICE TRANS. COMMUN., Vol. E81-B, No. 5, pp. 832–839, May 1998

lation function to that of the observed data. Although the autoregressive models are useful for generating random traffic for simulation purposes, they do not lead to simple solutions for the multiplexer model.

Self-similar phenomena were observed in traces of LAN traffic [9] and VBR video services [10]. This led to the proposal of a class of self-similar traffic models with long-range dependence such as fractional Brownian motion and fractional ARIMA processes [11]. Unfortunately, these processes are difficult to analyze in queueing models. Recently, some papers have questioned the importance of long-range dependence on multiplexer performance (cf. [12]).

Nevertheless, the solution of statistical multiplexer models, even with Markovian traffic sources can be a computationally prohibitive task for real-time applications. The concept of effective bandwidth [13], [14] is one approach to simplifying the task of resource allocation. Roughly speaking, the effective bandwidth of a traffic source is the bandwidth needed to achieve a QoS requirement (i.e., cell loss probability) when the source is offered to a constant service rate queue. In the asymptotic regime of large buffer sizes and small loss probabilities, the effective bandwidth of a source typically takes on a relatively simple form. In this regime, the effective bandwidth of a superposition of streams turns out to be the sum of the effective bandwidths of the individual streams.

Effective bandwidth does not capture the statistical multiplexing gain when many sources are multiplexed together. Further, effective bandwidth formulas can, in some cases, lead to optimistic allocations (see [15]), i.e., cell loss constraints may be violated. Nevertheless, the notion of effective bandwidth can be useful in designing resource allocation policies.

2.2 Access and Admission Control

Resource allocation and traffic control schemes generally require some description of the offered traffic. Traffic characterization must be sufficiently simple to be obtained in real-time and used to make control decisions. As such, there is a significant gap in complexity between the models for bursty traffic discussed above and the real-time traffic characterizations which form the basis of schemes for access and admission control.

The ATM Forum [16] uses the term *usage parameter control* (UPC) to refer to an open-loop access control applied to a user stream at the network boundary. The UPC function monitors an incoming cell stream and either discards or tags cells which are judged to be nonconforming. The UPC function provides deterministic limits on the traffic emitted by a given stream, thereby protecting the QoS experienced by other streams sharing common resources. The most popular algorithms for performing the UPC are based on the leaky bucket (cf. [17]). The ATM Forum has proposed a UPC for real-time traffic which consists of two leaky buckets [16]: one to limit the peak rate and the other to limit the sustainable rate (i.e., long-term average rate) and the maximum burst size.

Connection admission control (CAC), is another essential function for networks supporting QoS. The CAC is responsible for deciding whether or not a new connection can be accepted while maintaining QoS in the network. In the process, buffer and bandwidth resources are allocated among the admitted connections. All CACs must be based on some form of traffic characterization in order to make appropriate resource allocations. In the remainder of this subsection, we discuss some representative approaches to traffic control, with an emphasis on the way in which traffic is characterized in each case.

2.2.1 Statistical Model-Based Schemes

A pure model-based approach to QoS control assumes that the traffic streams at a multiplexer are represented by traffic models. To determine whether or not a new traffic stream can be admitted at a multiplexer, the multiplexer model including the new stream is analyzed. If the model satisfies QoS constraints on cell loss, delay, etc., the stream is accepted. Otherwise, the stream is rejected. In practice, it is difficult for users to specify their traffic stream in terms of a statistical model. One approach is to perform measurements on each stream and then to match the empirical measurements to a traffic model. However, it may be difficult to ensure the accuracy of such a matching procedure and the multiplexer model may not be solvable in real-time. Further, the model parameters may change with time.

Guérin and Gün [18] propose a unified approach to bandwidth allocation and access control based on the Markov on-off fluid model. In their scheme, each traffic source is characterized by a set of three parameters that determine an on-off source model. Given the on-off source model, bandwidth allocation is performed using the method described in [19]. The required bandwidth for a set of streams at a multiplexing point is taken as the minimum of:

- the sum of the effective bandwidths and
- the aggregate required bandwidth for a bufferless model in which the source bit-rate distributions are assumed to be Gaussian.

Further, based on the on-off model, the two parameters for leaky bucket access control of the source are determined.

The approach in [18] relies on the assumption that the traffic source obeys the on-off fluid model. The actual traffic stream is matched to an on-off source model through measurements of the mean rate. To account for the deficiency in the on-off model assumption, the burst size parameter is chosen not as the measured mean burst size, but as a modified burst size chosen to capture the impact of more general sources (see [18]).

2.2.2 Deterministic Model-Based Schemes

Rather than modeling a traffic source by a stochastic process, an alternative approach is to specify a deterministic rule-based algorithm to which the traffic stream conforms. Thus, the parameters of the access control method (i.e., the UPC parameters) serve as a traffic descriptor. Cruz [20] studies worst-case delays in networks where the input traffic streams are assumed to conform to the parameters of a leaky bucket access control. By considering worst-case scenarios, Cruz obtains end-to-end delay bounds in terms of the leaky bucket parameters.

Various other authors have used the deterministic approach. Parekh and Gallager [21] obtain end-to-end delay bounds assuming each node schedules cells according to generalized processor sharing. More complex rule-based algorithms for access control which lead to tighter delay bounds have also been considered. For example, by cascading multiple leaky buckets the traffic stream emitted by a source can be characterized by a tighter *envelope* [22].

Given the UPC parameters, the admission of a new connection can be decided in terms of worst-case delay bound computations. To guarantee worst-case delays for all streams, sufficient buffer space must be available at network nodes. A distinct feature of the deterministic approach is that it can yield relatively simple end-to-end delay bounds results. By contrast, the end-to-end performance analysis of statistical traffic streams is a difficult task. The statistical approaches may have the potential to achieve higher resource utilization with probabilistic QoS measures.

In the deterministic approach, the characteristics of the offered traffic are abstracted in the form of UPC parameters. These parameters are often assumed to be given, or else the parameters are assumed to be determined by an independent process. In principle, the UPC parameters should be chosen to accommodate the worst-case behavior of a given traffic stream. However, this requires that the traffic stream pattern must be known a priori. In Sect. 3, we describe a methodology for obtaining leaky bucket parameters for a given traffic stream based on real-time traffic measurements.

2.2.3 Measurement-Based Schemes

Measurement-based approach involves measurements of the individual streams and/or the multiplexed stream. The use of traffic measurements can lead to more aggressive CAC schemes. However, the risk of QoS violation must be minimized in such schemes.

The bandwidth allocation scheme of Guérin and Gün [18] employs real-time measurements of each source to obtain a set of parameters for an *equivalent* on-off source model. In this sense, their scheme may be regarded as measurement-based, although it relies heavily on the on-off source model. Other authors have proposed measurement-based schemes that make less model-specific assumptions on the traffic streams. We mention a framework for resource management proposed by de Veciana et al. [23] which is based on on-line estimations of per-stream effective bandwidth. In their approach, the effective bandwidth can be estimated either by fitting the stream to a parametric model or via real-time buffer measurements.

Saito and Shiomoto [24] propose a measurement-based admission control scheme based on a theoretical bound on cell loss ratio for general arrival processes. For a multiplexer, assume a buffer of size B and consider the aggregate arrival process formed by the set of admitted streams. Assume that time is slotted into ATM cell times and an *observation interval* of length r cell times is chosen. Let $p(k)$ denote the probability of k arrivals to the multiplexer during an observation interval. Suppose a new connection arrives with declared peak rate λ_p and mean rate λ_m in units of cells per observation interval. Define a *worst-case* arrival distribution for the new connection as follows:

$$\theta(k) = \begin{cases} \lambda_m/\lambda_p, & \text{if } k = \lambda_p \\ 1 - \lambda_m/\lambda_p, & \text{if } k = 0 \\ 0, & \text{otherwise} \end{cases} \tag{1}$$

The cell loss ratio that would result by admitting the new connection can be upper bounded as follows [24]:

$$P^*_{CLR} \leq \frac{\sum_{k=0}^{\infty}[k-r]^+ p(k) \star \theta(k)}{\sum_{k=0}^{\infty} kp(k) \star \theta(k)}, \tag{2}$$

where $\star$ denotes convolution, $r \leq B+1$ and the arrival process is assumed to be ergodic and stationary.

By taking measurements over N observation periods, an empirical estimate, $\{\hat{p}(k)\}$, for the arrival distribution can be obtained. A CAC can then be formulated based on (2) with $\{\hat{p}(k)\}$ substituted for $\{p(k)\}$. To smooth out measurement errors in the estimate, $\{\hat{p}(k)\}$, exponential averaging between adjacent measurement intervals is suggested (see [24] for further details).

Jamin et al. [25] propose an ad-hoc measurement-based scheme based on a deterministic delay model. They use empirical estimates of delay to obtain "approximate delay bounds." Although their approach cannot guarantee delay bounds, it may be useful for services that do not require strict real-time guarantees. Papers by Gibbens et al. [26] and Grossglauser and Tse [27] also discuss measurement-based CACs. Both papers assume homogeneous, bufferless multiplexer models and include call-level dynamics in their models. The assumption of homogeneity makes the model conservative relative to a heterogeneous system, leaving room for the possibility of improving resource efficiency in the more general case. In [26], measurements of instantaneous aggregate traffic load at the multiplexer are combined with a prior distribution for the mean cell arrival rate and mean call rate to obtain a threshold-based CAC rule that maximizes a reward function. The CAC scheme in [27] employs estimators of the mean and the variance of the source arrival rates. Given estimates of the mean and variance, the cell loss ratio can be approximated by a Gaussian distribution via an appeal to a central limit-type argument. An interesting aspect of their study is the effect of using memory in their estimators. As one would expect, exponential smoothing over past measurements can greatly improve the performance of an estimator.

3. Traffic Characterization for UPC

In this section, we outline the real-time UPC estimation approach developed in [28], [29]. The end result of the UPC estimation is a UPC descriptor that is used to shape and to characterize the stream entering the network boundary. We assume that sources are shaped to their UPC descriptors to avoid cell dropping at the network edge. Following the ATM Forum [16], we assume a dual leaky bucket descriptor, which we denote by $(\lambda_p, \lambda_s, B_s)$, where λ_p is the peak rate, λ_s is the sustainable rate and B_s is the bucket depth. In Fig. 1, the UPC shaper delays cells (where necessary), such that the shaped stream passes the UPC policing mechanism on the network side without violations. In this case, the delay due to the UPC shaper must be taken into account in the end-to-end QoS for the stream. By allowing shaping on the source side, a "tighter" UPC descriptor can be chosen to characterize the stream.

3.1 UPC-Based Characterization

From observations of the cell stream over a short interval of time T, estimates of the peak rate and mean rate can be obtained. The peak rate estimate can be mapped directly to the UPC parameter λ_p. The peak rate λ_p serves as an upper bound for the choice of sustainable rate, λ_s. Similarly, the mean rate estimate, λ_m, serves as a lower bound for λ_s.

Let μ be a candidate value for the sustainable rate,

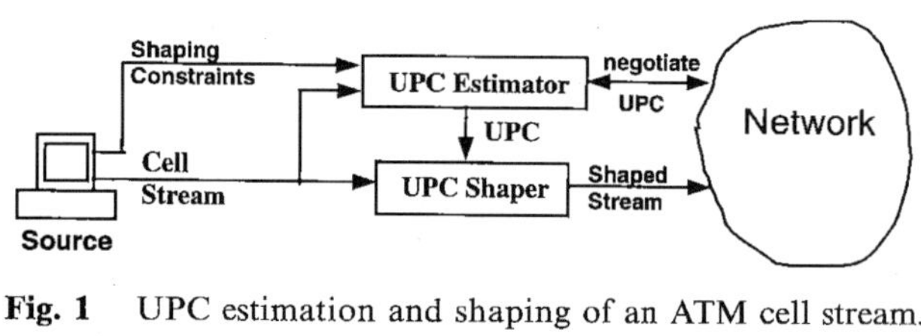

Fig. 1 UPC estimation and shaping of an ATM cell stream.

where $\lambda_m < \mu < \lambda_p$. Suppose that we offer the cell stream to a leaky bucket shaper with leak rate μ. Define $B(\mu, \epsilon)$ as the minimum bucket size necessary to ensure that the shaping probability (i.e., the probability that an arbitrary cell has to wait in the data buffer) does not exceed ϵ, $0 < \epsilon < 1$. Then, the assignment

$$\lambda_s = \mu, \quad B_s = B(\mu, \epsilon). \tag{3}$$

provides a UPC descriptor, $(\lambda_p, \lambda_s, B_s)$, for a user who can tolerate the shaping probability ϵ. In principle, the stream can be characterized as follows:

$$\bar{C}_S(\epsilon) = [\lambda_p; \lambda_m; (\mu, B(\mu, \epsilon)), \lambda_m < \mu < \lambda_p], \tag{4}$$

where ϵ ranges over the interval $(0, 1)$. In the limit, $\epsilon \to 0$, $B(\mu) = \lim_{\epsilon \to 0} B(\mu, \epsilon)$ represents the minimum bucket size needed to ensure that no shaping occurs.

Let us suppose that the cell stream is offered to a *fictitious queue* with constant service rate μ, $\lambda_m < \mu < \lambda_p$. Let Q denote the steady-state queue occupancy. Then,

$$B(\mu, \epsilon) = \inf\{B \geq 1 : P(Q > B) \leq \epsilon\}. \tag{5}$$

Hence, $B(\mu, \epsilon)$ can be determined from the steady-state queue length (or equivalently, waiting time) distribution. We therefore seek to characterize the cell stream in terms of the behavior of a fictitious queue to which it is offered.

To simplify the characterization, we shall assume that the complementary waiting time distribution has the exponential form:

$$P(W > t) \approx a(\mu) e^{-b(\mu)t}, \quad t \geq 0, \tag{6}$$

where a and b are functions of the service rate μ. For a wide class of arrival processes, the exponential form (6) holds asymptotically in t. This assumption gives a reasonable approximation for Markovian traffic models and is exact in the case of on-off Markov sources. For small-to-moderate buffer sizes, Markovian models have been found to give accurate solutions for a wide class of traffic types, including the self-similar traffic models (cf. [12]).

For a given cell stream, the method in [28] chooses values of $a(\mu)$ and $b(\mu)$ such that the queueing behavior is approximated by (6). Let S_a and Q_a denote, respectively, the number of customers in service and the number of customers in queue seen by an arbitrary customer arrival. Further, define $q(\mu) = E[Q_d]$ and $\tau_r(\mu)$ to be the conditional average remaining service time of the customer in service as seen by an arriving customer (given that a customer is in service). Under the assumption (6), one can determine [28]:

$$a(\mu) = E[S_a] = P(W > 0) \tag{7}$$

$$b(\mu) = \frac{\mu a(\mu)}{\mu \tau_r(\mu) a(\mu) + q(\mu)}. \tag{8}$$

3.2 Choice of UPC Descriptor

Given a characterization, C_S, of a stream in terms of the queue parameters $a(\mu)$ and $b(\mu)$, the minimum bucket size needed to achieve a shaping probability no more than ϵ can be determined as:

$$B(\mu, \epsilon) = \frac{\mu}{b(\mu)} \log\left(\frac{a(\mu)}{\epsilon}\right). \tag{9}$$

With respect to the shaping constraint, the space of feasible UPC descriptors is given by:

$$\mathcal{F} = \{(\lambda_p, \mu, B(\mu, \epsilon) : \lambda_m < \mu < \lambda_p\}$$

A unique UPC descriptor in $\mathcal{F}$ is determined by fixing a sustainable rate λ_s from among the candidate rates $\mu \in (\lambda_m, \lambda_p)$. We remark that the shaping constraint can also be formulated in terms of other shaping delay measures, e.g., mean delay, cell delay variation, etc. (see [28])

From the network point of view, a resource cost is associated with a given UPC descriptor. This cost is closely related to the algorithm used for resource allocation. Given the parameters $(\lambda_p, \lambda_s, B_s)$, consider an on-off arrival process with peak rate λ_p and deterministic on and off periods given by

$$T_{on} = \frac{B_s}{\lambda_p - \lambda_s} \text{ and } T_{off} = \frac{B_s}{\lambda_s}. \tag{10}$$

Doshi [30] shows that such a source model maximizes the cell loss probability for the case of a homogeneous superposition at a bufferless multiplexer. Even for the more general heterogeneous buffered multiplexer, empirical evidence indicates that this source model is often the worst-case or nearly the worst-case.

Based on the worst-case model, the following cost function was developed in [28]:

$$c(\lambda_p, \lambda_s, B_s) = \min\left(\lambda_p, \lambda_s + K\frac{\lambda_s(\lambda_p - \lambda_s)B_s}{\lambda_p}\right),$$

where K is a fixed parameter that can be determined as a function of the multiplexer buffer size and the target cell loss probability. In effect, the cost function is an approximate measure of the worst-case bandwidth required for a source characterized by the UPC descriptor $(\lambda_p, \lambda_s, B_s)$. The operating UPC parameter set U^* is chosen to minimize the cost function, i.e., $U^* = \min_{U \in \mathcal{F}} c(U)$. As an example, Fig. 2 (a) shows $B(\mu, \epsilon)$ as a function of μ and with target shaping probability $\epsilon = 0.01$ for the 'mobi' sequence. Figure 2 (b) shows $c(\lambda_p, \mu, B(\mu, \epsilon))$ as a function of μ. From Fig. 2 (b), the rate which minimizes the cost function is approximately 27 cells/ms. Then the corresponding bucket size can be read from Fig. 2 (b) as 900 cells.

3.3 Dynamic Traffic Characterization

The statistical characteristics of a cell stream may

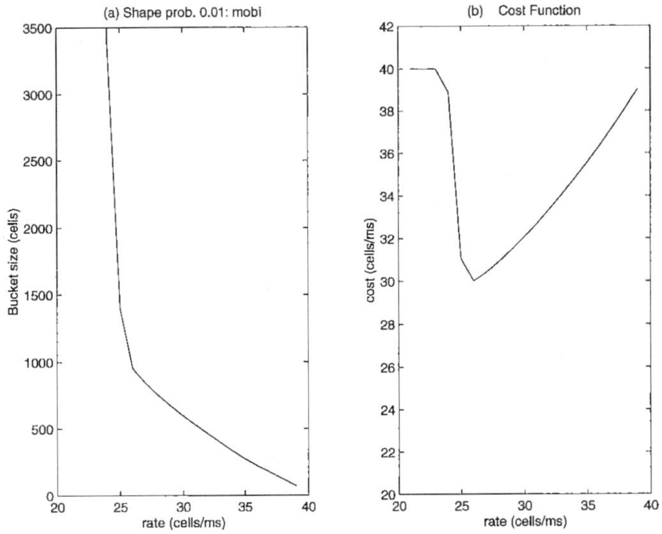

Fig. 2 Shaping constraint and cost curves.

change over time. For example, a video stream may consist of several scene types with very different bit rate characteristics. In this case, a single UPC descriptor may be an inadequate description of the offered traffic. To avoid excessive shaping of the cell stream, a user must choose a conservative UPC descriptor. This may be wasteful of the network bandwidth and result in low utilization. Moreover, since the cost charged to the user will depend on the declared UPC descriptor, the user has an incentive to negotiate different UPC descriptors for different segments of the cell stream. During a low bit rate scene, both the network and the user can benefit by negotiating smaller UPC parameters than during higher bit rate scenes.

The real-time UPC characterization method described above can be made dynamic [29]. Divide the time axis into measurement periods, $P_0, P_1, \cdots$, where $P_n = (nT, (n+1)T]$. Over interval P_0, mean and peak rates, $\lambda_m(0)$ and $\lambda_p(0)$, respectively, are estimated. Over interval P_1, a UPC descriptor $U(1) = (\lambda_p(1), \lambda_s(1), B_s(1))$ is determined. The user then negotiates the UPC descriptor, $U_{neg}(1) = U(1)$, with the network. For the next measurement period, the candidate rates may be spaced nonuniformly, with some rates chosen close to the candidate sustainable rate selected in the previous measurement period. This reduces the need for a large number of candidate rates.

Although a new UPC descriptor $U(n)$ is generated at the end of each measurement period, an attempt to renegotiate the UPC is made only when a significant change occurs. On the source side, the criteria for a significant change could be in terms the change in measured peak rate and/or a violation of the shaping constraint observed through empirical measurements. At the end of measurement period P_n, a renegotiation request for the new UPC parameters $U(n)$ is issued to the network. If the renegotiation is successful, the new set of negotiated UPC parameters becomes $U(n)$. Otherwise, the declared parameters remain the same as in the previous period, P_{n-1}.

4. Traffic Characterization for CAC

In this section, we consider resource allocation at a multiplexing point in the network, where a number of connections share a common buffer pool of size B_{max} cells and link capacity C_{max}. Each connection is specified by a UPC descriptor that may be determined according to the methodology of Sect. 3.

4.1 UPC-Based Model

Let $r(t)$ denote the aggregate arrival rate process for a given class. Appealing to a "central limit theorem" type argument, when the number of connections is large, the stationary distribution of $r(t)$ is approximately normally distributed (cf. [19]). Let m denote the mean and let σ^2 denote variance of the stationary distribution of $r(t)$. For a bufferless fluid multiplexer model with constant service rate C, the cell loss ratio is given by:

$$CLR = \lim_{t\to\infty} \frac{E[r(t) - C]^+}{E[r(t)]}. \tag{11}$$

With the UPC-based aggregate traffic model, the cell loss ratio at the multiplexer can then be approximated by

$$CLR = \int_C^\infty \frac{x - C}{m\sqrt{2\pi\sigma^2}} \exp\left[-\frac{(x-C)^2}{2\sigma^2}\right] dx. \tag{12}$$

Using the UPC descriptors specified for the connections, a model for the aggregate traffic stream can be derived. The aggregate model can then be used to obtain the required bandwidth. Let C denote the set of streams and let $(\lambda_p^{(i)}, \lambda_s^{(i)}, B_s^{(i)})$ denote the UPC descriptor for stream i. As discussed in Sect. 3, a worst-case on-off model can be derived based on the UPC descriptor The mean and variance for the aggregate rate process are as follows:

$$m = \sum_{i\in C} \lambda_s^{(i)}, \; \sigma^2 = \sum_{i\in C} \lambda_s^{(i)}(\lambda_p^{(i)} - \lambda_s^{(i)}). \tag{13}$$

Using Eq. (12), an approximation for the cell loss ratio at the multiplexer can be obtained. The result for CLR as computed above will likely be overly conservative, since a bufferless multiplexer was assumed.

To take into account the effect of buffering, a *modified source model* for a UPC descriptor, $(\lambda_p, \lambda_s, B_s)$, was proposed in [31]. Define the time constant

$$T_* = \frac{B_{max}}{2C_{max}}, \tag{14}$$

where C_{max} denotes the maximum bandwidth that can be allocated to the traffic class. The value of T_* represents the time needed to empty a half-full buffer at the maximum service rate C_{max}. The modified source model alternates between two states, a high state of duration T_{on} and a low state of duration T_{off}, where T_{on} and T_{off} are the durations of the on- and off-states in the worst-case model defined in Eqs. (10). The rates in the high and low state are denoted by λ_H and λ_L, respectively, and defined as follows:

if $T_{on} \geq T_*$ **then**:

$$\lambda_H = \lambda_p, \quad \lambda_L = 0$$

if $T_{on} < T_*$ **then**:

$$\lambda_H = \frac{T_{on}}{T_*}\lambda_p + \left(1 - \frac{T_{on}}{T_*}\right)\lambda_s,$$

$$\lambda_L = \left(1 - \frac{T_{on}}{T_*}\right)\lambda_s.$$

If $T_{on} \geq T_*$, the modified model is identical to the worst-case on-off model. Otherwise, the modified model is a *smoothed* version of the on-off model, which to some extent takes into account the effect of buffering at the multiplexer.

With the modified source model, the mean and variance of the aggregate arrival rate are given by

$$m = \sum_{i\in C} \lambda_s^{(i)}, \tag{15}$$

$$\sigma^2 = \sum_{i\in C} (\lambda_H^{(i)} - \lambda_L^{(i)})^2 p_{on}^{(i)}(1 - p_{on}^{(i)}), \tag{16}$$

where $\lambda_L^{(i)}$ and $\lambda_H^{(i)}$ are the rates for the modified model for source i and

$$p_{on}^{(i)} = \lambda_s^{(i)}/\lambda_p^{(i)}$$

is the probability that the source is transmitting at $\lambda_H^{(i)}$. We now seek the bandwidth required to ensure a cell loss ratio less than ϵ. Setting $CLR = \epsilon$, Eq. (12) can be solved approximately for C [31]:

$$C = m + \eta\sigma, \tag{17}$$

where:

$$\eta = 1.8 - 0.46\log_{10}\left(\frac{m\sqrt{2\pi}}{\sigma}\epsilon\right). \tag{18}$$

4.2 Real-Time Measurements

The UPC-based CAC may lead to an underutilization. In [32], the UPC-based approach is combined with real-time measurements to produce a "hybrid" model with improved accuracy. Empirical estimates of the mean and variance of the aggregate process are obtained.

Let m_{meas} and σ^2_{meas}, denote the empirical estimates for the mean and variance, respectively. Let m_{upc} and σ^2_{upc} denote, respectively, the mean and variance of the aggregate rate process obtained from the UPC descriptors, as discussed in Sect. 4.1. Then the mean and variance of the hybrid model are taken to be:

$$m = \alpha m_{upc} + (1 - \alpha) m_{meas}, \quad (19)$$

$$\sigma^2 = \alpha \sigma^2_{upc} + (1 - \alpha) \sigma^2_{meas}, \quad (20)$$

where $\alpha \in (0,1)$ is an averaging weight between the UPC-based statistics and the measurement-based statistics. When $\alpha = 0$, the hybrid model is purely derived from the measured statistics, whereas for $\alpha = 1$, the hybrid model reduces to the UPC-based aggregate model. Traffic classes with more stringent QoS requirements should be assigned higher values of α.

4.3 CAC Procedure

Given the hybrid model, the required bandwidth for the hybrid model can then be computed from (17) and (18). When a new call arrives with UPC descriptor $(\lambda'_p, \lambda'_s, B'_s)$, the CAC assumes the modified source model for the call. To determine the total required bandwidth to accommodate the new connection, a new aggregate rate process, $r^+(t)$ is constructed, with mean and variance given by

$$m^+ = m + \lambda'_s \quad (21)$$

$$\sigma^{+2} = \sigma^2 + (\lambda'_H - \lambda'_L) p'_{on} (1 - p'_{on}), \quad (22)$$

where $p'_{on} = \lambda'_s / \lambda'_p$ and λ'_H and λ'_L are the high and low rates for the modified source model of the new call. The bandwidth, C^+, needed to accommodate the new connection can then computed. The new call is accepted if and only if

$$\Delta C = C^+ - C \leq FP,$$

where C is the current bandwidth allocation and FP is the *free pool* of bandwidth.

If the call is accepted, the incremental bandwidth, ΔC^-, bandwidth is subtracted from the free pool. The bandwidth allocation is updated to be C^+ and the UPC-based aggregate traffic model parameters are updated to include the new call. When a call departs from the system, the bandwidth requirement, C^-, for the aggregate process *without the call* can be computed. The incremental bandwidth

$$\Delta C^- = C - C^-,$$

can then be released to the free pool.

5. Conclusions

In this paper, we have presented an overview of methodologies for traffic characterization and QoS control in ATM networks. Researchers have studied relatively sophisticated models that capture the statistical characteristics of VBR traffic. Simplified versions of these models, e.g., the Markov on-off fluid model, can often be used to perform resource allocation.

We briefly discussed several different approaches to traffic control. The model-based approach assumes a simple traffic model for each source. The parameters of the model can be estimated through real-time measurements and then used to perform resource allocation. The deterministic approach characterizes each source according to parameters of a UPC mechanism, e.g., the leaky bucket. The CAC function is based on delay bounds which provide deterministic QoS guarantees. Measurement-based approaches can provide higher utilization, but incur the risk of QoS violations,

We then discussed a UPC-based approach to resource allocation. In this approach, an arbitrary source is characterized, using real-time measurements, in terms of a UPC descriptor. The CAC is based on forming a statistical model for each stream based on the stream's UPC descriptor. Since the model is based on the UPC parameters, the CAC can guarantee strict QoS, although it may be conservative in its allocations. To improve utilization further, without sacrificing QoS, real-time measurements of the actual aggregate traffic can be incorporated into the UPC-based CAC.

As we have seen, network resource allocation involves traffic characterization, access control and admission control. The deterministic approach to resource allocation can provide strict QoS guarantees in a relatively straightforward manner. Schemes which employ statistical models and/or real-time measurements have the potential to achieve higher utilization with minimal sacrifices in QoS. The design of such schemes entails the consideration of modeling and measurement errors, but the added complexity may be well-justified by the improvement in network efficiency. Although stochastic traffic models have been studied extensively, there has been less work on methods for characterizing traffic in real-time and their application to real-time traffic control. We anticipate that future resource allocation schemes will integrate real-time traffic characterization with the traffic controls in order to provide QoS in an efficient manner.

References

[1] D.L. Jagerman, B. Melamed, and W. Willinger, "Stochastic modeling of traffic processes," in Frontiers in Queueing: Models, Methods and Problems, ed. J. Dshalalow, pp.271–320, CRC Press, 1997.

[2] J.W. Roberts, U. Mocci, and J. Virtamo, eds., "Broadband network teletraffic: Performance evaluation and design of broadband multiservice networks," No.1155 in Lecture notes in computer science, Springer-Verlag, 1996.

[3] D. Anick, D. Mitra, and M. Sondhi, "Stochastic theory of a data-handling system with multiple sources," The Bell System Tech. J., vol.61, pp.1871–1894, Oct. 1982.

[4] T. Stern and A.I. Elwalid, "Analysis of separable Markov-modulated rate models for information-handling systems," Adv. Appl. Prob., no.23, pp.105–139, 1991.

[5] H. Heffes and D.M. Lucantoni, "A Markov modulated characterization of packetized voice and data and related

statistical multiplexer performance," IEEE J. Sel. Areas in Commun., vol.4, pp.856–868, Sept. 1986.

[6] J. Le Boudec, "An efficient solution method for Markov models of ATM links with loss priorities," IEEE J. Sel. Areas in Commun., vol.9, April 1991.

[7] P. Sen, B. Maglaris, N.E. Rikli, and D. Anastassiou, "Models for packet switching of variable-bit-rate video sources," IEEE J. Sel. Areas in Commun., vol.7, pp.865–869, June 1989.

[8] B. Melamed, "TES: A class of methods for generating autocorrelated uniform variates," ORSA Journal on Computing, vol.3, no.4, pp.317–329, 1991.

[9] W. Leland, M. Taqqu, W. Willinger, and D. Wilson, "On the self-similar nature of ethernet traffic (extended version)," IEEE/ACM Trans. Networking, vol.2, no.1, pp.1–15, 1994.

[10] J. Beran, R. Sherman, M. Taqqu, and W. Willinger, "Variable bit-rate video traffic and long-range dependence," IEEE Trans. Commun., vol.43, pp.1566–1579, 1995.

[11] M. Garrett and W. Willinger, "Analysis, modeling and generation of self-similar VBR traffic," Proc. ACM Sigcomm'94, (London, UK), pp.269–280, 1994.

[12] D.P. Heyman and T.V. Lakshman, "What are the implications of long-range dependence for VBR-video traffic engineering?," IEEE/ACM Trans. Networking, vol.4, pp.301–317, June 1996.

[13] A.I. Elwalid and D. Mitra, "Effective bandwidth of general Markovian traffic sources and admission control of high speed networks," IEEE/ACM Trans. Networking, vol.1, pp.323–329, June 1993.

[14] F. Kelly, "Effective bandwidths at multi-class queues," Queueing Systems, vol.9, pp.5–15, 1991.

[15] G.L. Choudhury, D.M. Lucantoni, and W. Whitt, "Squeezing the most out of ATM," IEEE Trans. Commun., vol.44, pp.203–217, Feb. 1996.

[16] ATM Forum Technical Committee, "Traffic Management Specification, Version 4.0," ATM Forum, 1996.

[17] J.S. Turner, "New directions in communications (or which way to the information age?)," IEEE Communications Magazine, pp.8–15, Oct. 1986.

[18] R. Guérin and L. Gün, "A unified approach to bandwidth allocation and access control in fast packet-switched networks," IEEE INFOCOM '92, (Florence, Italy), May 1992.

[19] R. Guérin, H. Ahmadi, and M. Naghshineh, "Equivalent capacity and its application to bandwidth allocation in high-speed networks," IEEE J. Sel. Areas in Commun., vol.9, pp.968–981, Sept. 1991.

[20] R.L. Cruz, "A calculus for network delay, part II: Network analysis," IEEE Trans. Inf. Theory, vol.37, no.1, pp.132–141, 1991.

[21] A.K. Parekh and R.G. Gallager, "A generalized processor sharing approach to flow control in integrated services networks: The multiple node case," IEEE/ACM Trans. Networking, vol.2, pp.137–150, April 1994.

[22] C.S. Chang, "Stability, queue length and delay of deterministic and stochastic queueing networks," IEEE Trans. Automatic Control, vol.39, pp.913–931, May 1994.

[23] G. de Veciana, G. Kesidis, and J. Walrand, "Resource management in wide-area ATM networks using effective bandwidths," IEEE J. Sel. Areas in Commun., vol.13, pp.1081–1090, Aug. 1995.

[24] H. Saito and K. Shiomoto, "Dynamic call admission control for ATM networks," IEEE J. Sel. Areas in Commun., vol.9, pp.982–989, Sept. 1991.

[25] S. Jamin, P. Danzig, S. Shenker, and L. Zhang, "A measurement-based admission control algorithm for integrated service packet networks," IEEE/ACM Trans. Networking, vol.9, pp.56–70, Feb. 1997.

[26] R. Gibbens, F. Kelly, and P. Key, "A decision-theoretic approach to call admission control in ATM networks," IEEE J. Sel. Areas in Commun., vol.13, pp.1101–1114, Aug. 1995.

[27] M. Grossglauser and D. Tse, "A framework for robust measurement-based admission control," Proc. ACM Sigcomm., pp.237–248, Sept. 1997.

[28] B.L. Mark and G. Ramamurthy, "Real-time estimation of UPC parameters for arbitrary traffic sources in ATM networks," Proc. INFOCOM '96, (San Francisco, California), pp.384–392, March 1996.

[29] B.L. Mark and G. Ramamurthy, "Dynamic renegotiation of UPC parameters for arbitrary traffic sources in ATM networks," Proc. ICC '96, (Dallas, Texas), pp.1707–1711, June 1996.

[30] B.T. Doshi, "Deterministic rule based traffic descriptors for broadband ISDN: Worst case behavior and connection acceptance control," Proc. IEEE GlobeCom'93, pp.1759–1764, Dec. 1993.

[31] G. Ramamurthy and Q. Ren, "Multi-class connection admission control policy for high speed ATM switches," Proc. IEEE INFOCOM '97, Kobe, Japan, April 1997.

[32] Q. Ren and G. Ramamurthy, "A hybrid model and measurement based connection admission control and bandwidth allocation for multi-class ATM networks," NEC Technical Report no.97-C039, July 1997.

High-Performance Backbone Network Technology edited by Naoaki Yamanaka
ISBN: 0-8247-5321-6 DOI: 10.1081/(120025665)
Published 2004 by Marcel Dekker, Inc.

Voice Communication on Multimedia ATM Network Using Shared VCI Cell

Toshihiro Masaki, Yasuhiro Nakatani, Takao Onoye, Nariyoshi Yamai, and Koso Murakami

SUMMARY This paper presents novel multimedia ATM networks which are capable of transmitting voice data efficiently and unify the switching methods among heterogeneous traffic. Fully ATMized multimedia networks are using *fellow cell switches.* The proposed assembly method can pack plural calls which have different virtual channel connection (VCC) into one cell. Every call in cells is able to be dynamically rearranged by the fellow cell switch to achieve an efficient use of network resources. The switching functions are supported by *shared virtual channel identifier (VCI) cells* and *fellow cells* in it. The fellow cell switch for 622 Mbps links is integrated into a single chip. The multimedia ATM networks including voice transmission can be constructed by the fellow cell switches being attached to the standard ATM switches.

key words: *ATM, VTOA, multimedia network, voice communication, VCI*

1. Introduction

Asynchronous Transfer Mode (ATM) is designed to transport all types of traffic such as data, video, and voice. Recently, Voice and Telephony over ATM (VTOA) has become points of interest in the ATM Forum [1], and the International Telecommunication Union-Telecommunication standardization sector (ITU-T) focusing on interworking function between traditional Synchronous Transfer Mode (STM) and ATM networks [2].

The technical challenge in applying the ATM to voice traffic is to transmit voice data efficiently and to switch compatibly with standard ATM cells, even if many narrowband voice connections (e.g. 64 kbps pulse code modulation) are in the broadband ATM highway (e.g. 156 Mbps, 622 Mbps). Moreover, ITU-T recommendations specify the average network delay for speech traffic, and the delay of about 1 ms is acceptable for a node of digital networks. The multimedia ATM networks must be designed to accomplish comparable performance with the digital ones. Therefore, to cut the delay of equipments in ATM networks, such as ATM switches and CLAD's (cell assembly and disassembly), has an good effect on reducting deterioration of voice quality such as echo problems. The additional controls of voice traffic must be reduced in order to standardize voice transmission. The dominant factor of the delay is the cell assembly delay, which is the time to fill up the payload of an ATM cell with voice data. In order to reduce the cell assembly delay, which takes up to 6 ms for 64 kbps data, several B-ISDN trunking methods for narrowband services were already introduced.

[3] proposes the remap interface in the ATM network for efficient data transmission. The remap function receives *composite cells* from one switch and transmits remapped composite cells to the other switch. However, remapping nodes must preserve rearrangement tables of all connections since the voice cells have no information on each call. After calls established, [3] needs many complicated control performed in software. Moreover, ITU-T standardized the AAL-2 (ATM adaptation layer type two) which is an AAL for low bit-rate voice. The AAL-2 is able to reduce the cell assembly delay and obtain high transmission efficiency if the networks have interworking units (IWU). In other words, the AAL-2 is very useful for the ATM networks which are used as the backbone of low bit-rate voice networks.

This paper presents the multimedia ATM networks which serve not only to transmit voice data efficiently but also to unify the switching methods among heterogeneous traffic. There is no need to perform software control. The particular processes for setting up or tearing down the voice connections are almost unnecessary. The end points of proposed networks will be a subscriber terminal as oppose to the end point being IWU accommodating multiple voice channel. The *fellow cell switch* is integrated into a single chip by using CMOS technology.

In the following section, a description of how the multimedia ATM networks serve to integrate heterogeneous traffic and to switch voice cells is proposed. In Sect. 3, the structures of the *shared virtual channel identifier (VCI) cell* and the *fellow cell* are described. The hardware architecture of the fellow cell switch, and its simulation and implementation results are explained in Sects. 4 and 5, respectively.

2. Network Structure and Communication Method

Figure 1 outlines a fully ATMized network for multime-

Manuscript received May 6, 1997.
Manuscript revised July 31, 1997.

†The authors are with the Graduate School of Engineering, Osaka University, Suita-shi, 565–0871 Japan.
††The authors are with the Computation Center, Osaka University, Ibaraki-shi, 567–0047 Japan.

IEICE TRANS. COMMUN., Vol. E81-B, No. 2, pp. 340–346, February 1998

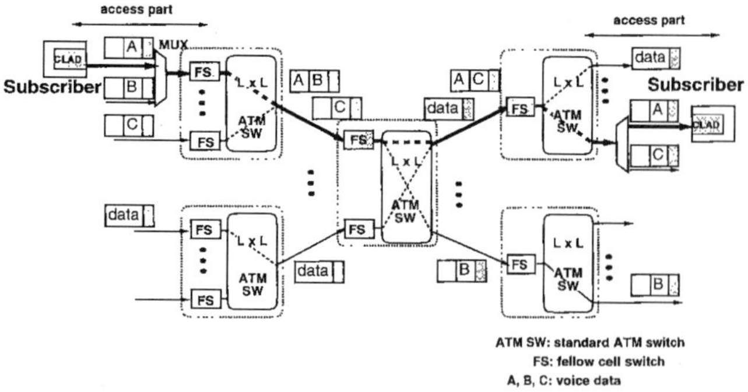

Fig. 1 Constructions of multimedia ATM network.

dia which this paper is going to discuss the development of. The ATM network is wired to every subscriber to transmit moving pictures, high speed data, and voice information with standard ATM cells constituting multimedia ATM networks, the goal of the ATM. In this model, $L \times L$ standard ATM switches are connected with fellow cell switches. Each ATM switch has up to one hundred (L) 156 Mbps or 622 Mbps highways which operate at 19.5 MHz or 77.8 MHz in 8-bit parallel, respectively. The ATM switches of this kind are available with the latest VLSI technology. Once the 3-stage networks are constructed, the multimedia ATM networks can then possibly be extended to multiple-stage networks owing to the repeatability of the 3-stage networks. Expanding switching capacity of transit switch or constructing more than 3-stage networks lead to increase the accommodating capability of a multimedia ATM network.

ATM networks can now provide high speed data communication especially in a local area environment for bringing us a capability of distributed processing among personal computers and workstations. All traffic in the proposed multimedia ATM network must be consistent with standard ATM protocols.

To provide video data transmission capability over the multimedia ATM networks is extremely required. The bandwidth for a subscriber is estimated as follows. Considering that 12,000 subscribers are accommodated in one local switch and 2,400 subscribers (20% of whole subscribers) require to connect simultaneously. Assuming that 80% of highways are connected with subscribers, approximately 5 Mbps bandwidth is assigned to each subscriber on average to transmit video, voice, and high speed data if each ATM switch has one hundred highways per switch with transfer capacity of 156 Mbps. The bandwidth of 5 Mbps per subscriber is adequate to transfer video data since the bandwidth of MPEG2 (MP@ML) is from 3 to 5 Mbps.

The most simple and straightforward method to assign the 64 kbps voice connections to ATM cells is through the homogeneous cell. One ATM virtual channel connection (VCC) is used for each 64 kbps connection with each cell being completely filled with voice data. The main disadvantage of this method is the long cell assembly delay to fill up the payload of an ATM cell. For 64 kbps speech, to fill a ATM payload requires 6 ms. To avoid deterioration of voice quality such as echo problems caused by cell assembly delay, additional delay introduced by CLAD had better be shortend [1], [4].

The concepts of transferring voice data are described as follows. First, CLAD equipments generate the *solitary cells* which contain only a few bytes of voice data (A, B, and C in Fig. 1). Considering cell assembly delay at CLAD, adopting solitary cells in these access parts of the network is feasible since the traffic density is sufficiently low. Then, the fellow cell switch rearranges the calls into one shared VCI cell if all calls go in the direction of the same next fellow cell switch regardless of whether the destinations of calls are the same or not ((A and B) or (A and C) in Fig. 1). To put it briefly, the multimedia ATM networks which this paper describes can assemble plural calls which have different VCC's into an ATM cell and reassemble them in order to obtain high transmission efficiency. The reassignment of calls in the fellow cell switches can be performed in hardware in order not to decrease high speed capability of ATM switching. The equipments existing outside of the fellow cell switch, such as ATM switches, will not be required to discriminate between the standard ATM cells and the shared VCI cells since the latter cells comply with standard ATM protocols. Therefore, the multimedia ATM networks are constructed with the use of conventional equipments such as ATM switches and CLAD's except for fellow cell switches. The multimedia ATM networks are readily realized by attaching the fellow cell switches to the standard ATM switches.

The shared VCI cells are newly introduced to pack calls into an ATM cell in order to realize call switching functions. Following section describes the detailed structure of the shared VCI cell format and its assembly method in the fellow cell switches.

3. Shared VCI Cell Assembly Method

In order to overcome the cell assembly delay, several mechanisms are proposed. The structures of four conventional cell packetization schemes, the partial-filled cell, the short cell, the composite cell, and the layered cell [3], [5], are shown in Figs. 2 (a), (b), (c), (d), respectively. The assembly methods of the composite cell, the layered cell, and AAL-2 (AAL-CU: AAL composite user information), which is a version of the layered cell, are so-called many-to-one mapping approach.

Considering the case where the above cell assembly methods are employed for the multimedia ATM networks. The partial-filled cell for voice information wastes bandwidth and hinders communications of other calls or information sources, especially in the parts of networks where traffic density is high. The short cell can hardly coexist with standard ATM cells because of the

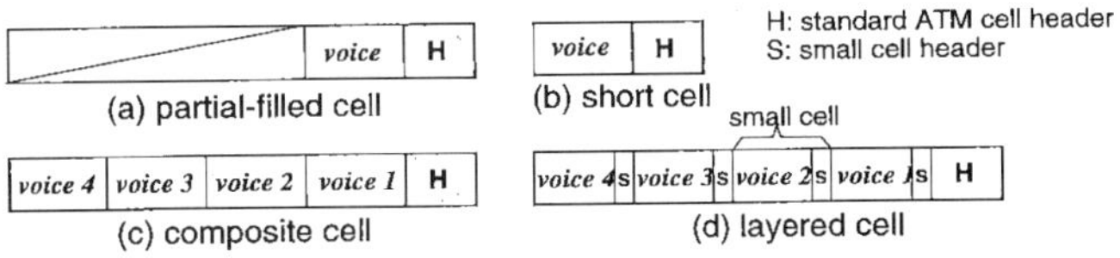

Fig. 2 Structures of conventional cell assembly methods.

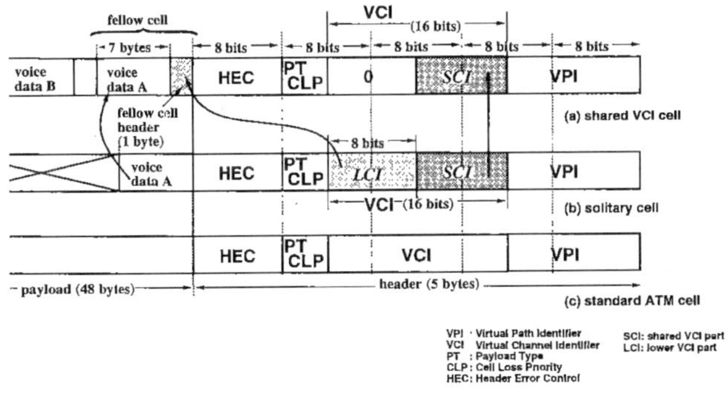

Fig. 3 Structures of ATM cells for multimedia ATM networks.

difference in the cell length. The many-to-one mapping approaches are not available for the networks illustrated in Fig. 1, since there are no Interworking Units (IWU's) which are not just end points of ATM networks but are also points where different calls converge. Both cells are merely suitable for crossconnect of voice data as opposed to the switching for every voice data.

[3] proposes the remapping node in the ATM network, which is used as the backbone of narrowband networks, in order to transmit composite cells efficiently. However, rearrangement information of all connection must be maintained by remapping nodes since the composite cells have no information on each call. At the time an individual connection is set up, signaling is required to specify the slot of the composite cell. After call established, the remap function disassemble arriving composite cells, and remap them with the use of rearrangement table. Therefore, [3] needs much complicated software control.

The concept of a new cell assembly and re-assembly method to transmit voice data efficiently is illustrated. The shared VCI cells can be dynamically rearranged by the fellow cell switches in the ATM networks, in contrast with other cell assembly methods. The shared VCI cell is not only dedicated to transmission of calls but is also compatible with the standard ATM cells. The proposed shared VCI cell fully utilizes the function of standard header converter (HCV). In standard ATM networks, HCV's are established at the front of ATM switches.

The structures of the shared VCI cell, the solitary cell, and the standard ATM cell, which are employed by the proposed voice transmission method, are outlined in Figs. 3 (a), (b), and (c), respectively. The standard ATM cell is utilized for transmission of all multimedia data except for voice. The shared VCI cell is composed of the header and fellow cells to transmit plural calls. The solitary cell has its header and a few bytes data of a call. The most significant bit of VCI is the identifier for voice cells including shared VCI cells and solitary cells. In other words, the multimedia ATM networks reserve a portion of VC for voice transmission. The number of bits for the identifier is not limited to one.

In standard ATM network node interface (NNI) protocol, 28 bits are available for routing. After this, the number of possible virtual path identifier (VPI) in one physical transmission path is limited to one for simplicity without losing generality.

Each solitary cell has only one call data to transmit during the access parts of the network. In these parts, many-to-one mapping approaches cannot be adopted because cells have not reached the assembly point, fellow cell switches. The solitary cell has the identifier for voice cell, although the partial-filled cells do not support it. On the other hand, in order to reduce the number of cells between standard ATM switches, the new cell assembly method can pack plural calls which have different VCC into one shared VCI cell. In these parts, each fellow cell contains its own VCI information, which is up to 16 bits long. As a 2-byte header for a fellow cell is too large to obtain high transmission efficiency, a fellow cell header is one byte or less to hold a part of VCI. In order to assist this method, the VCI field in use is divided into the *shared VCI* part, SCI, and the *lower VCI* one, LCI which is kept in fellow cell header, in the cases of the solitary cell and the shared VCI cell. The remaining bits of VCI omitted from fellow cell header, SCI, are held in shared VCI cell header. Therefore, all SCI missed out from fellow cell headers must be the same bit strings, if the fellow cells are packed into a shared VCI cell. The following is the way to assign the VCI for each connection.

The entire VCI of each fellow cells must be reconstructed from shared VCI header and corresponding fellow cell header. The shared VCI header has SCI and the fellow cell header has LCI. On the occasion of connecting new VCC for a call a, the SCI has the same SCI of previously connected calls x, if and only if a and x go to the same fellow cell switch in a VC link of the VCC. Noting that in shared VCI cell header, LCI field is equal to 0. This means that the shared VCI cell seems to have the VCI, (SCI, 0), which will not correspond to any VCI of existing VCC's. The bit string (SCI, 0) is an imaginary VCI existing only during the VC link. The shared VCI cells do not require any additional bits to identify each call. The fellow cell switches only use the shared VCI bit strings to rearrange the shared VCI cell as well as the HCV's in the standard ATM networks do.

Owing to elaborate cell assembly method, the number of cells transmitted through the network is greatly reduced. However its reduction rate is variable according to the length of fellow cell. Candidates of fellow cell length including 1-byte header are 4, 6, 8, and 12 bytes, since cells of these length fit perfectly in 48-byte payload. Analysis of simulation results shows that the shorter the length of the fellow cell, the shorter the cell assembly delay is. On the other hand, the shorter fellow cell length causes the transmission efficiency to worsen and the number of cells to increase. The 8-byte long fellow cell (including 1-byte header) is selected as the most ideal fellow cell with the constraints that delay at CLAD is not to exceed 1 ms, and the acceptable rate of fellow cell header per an ATM payload is to be approximately 10%.

The AAL enhances the service provided by the ATM layer to support functions required by the next higher layer of protocol stack. AAL type 0, type 1/unstructured data transfer (UDT), and type 5 are acceptable for voice channel at 64 kbps [4]. Since the proposed cell assembly method can have each routing information of the connection which is in the cell, the three types of ATM cell are readily converted to the shared VCI cell. In case of transferring with AAL1/UDT, segmentation and the reassembly-protocol data unit (SAR-PDU) header is maintained in the payload of the fellow cell the same as the SAR-PDU payload.

4. Fellow Cell Switch Architecture

In this section, the hardware architecture to convert standard ATM cell headers and to rearrange shared VCI cells is described. Since shared VCI cells make good use of HCV's, conversion of ATM cell headers and rearrangement of shared VCI cells are integrated in a single fellow cell switch chip inscribed with FS in Fig. 1. The fellow cell switch is composed of a *new header table* with an *address converter*, a *standard cell buffer*, a *fellow cell controller*, and a *fellow cell buffer*, as shown in Fig. 4. The following describes functions of the fellow cell switch.

The address converter is to generate the address of the new header table from incoming header information. If the routing bits of the ATM cell header are input to the address converter, this unit eliminates unused

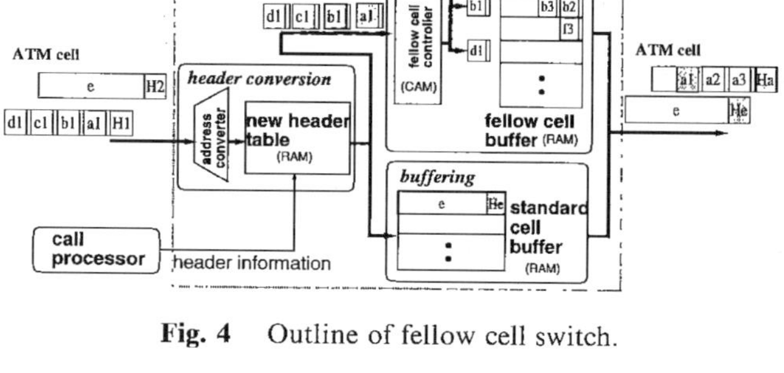

Fig. 4 Outline of fellow cell switch.

bit from routing bit through the use of appropriate control registers. The output of this unit is called *packed route data*. When a new cell arrives at the fellow cell switch, the address converter judges whether the cell is a shared VCI cell or not, by checking the most significant bit(s) of VCI which is(are) identifier(s) for the shared VCI cell. In the case of the shared VCI cell including calls a, the address converter is fed with SCI, which is in the shared VCI cell header, and LCI, which is in the fellow cell header for a. In other words, reconstructed VCI of the call a is fed. One shared VCI cell may cause the unit to look up the table six times, since the cell may have six fellow cells. In the case of standard ATM cell, the address converter run once.

The new header table converts the incoming packed route data, which is the address of this table provided by the address converter, into following packed route data specifying VCC during next VC link. The new header table consists of four 16-bit × 2,048-word RAM's, which supports 2^{13} VCC entries, leading to the total bits within the VPI and VCI fields used for routing, being provided to 13 bits or less.

The standard cell buffer is necessary for the fellow cell switch to avoid output conflicts between standard ATM cells and shared VCI cells. The standard cell buffer is FIFO based to store M standard ATM cells with new header. The packed route data for a standard cell from the new header table is expanded into normal 40-bit header, and then attached to its payload to construct an ATM cell. The ATM cell with new header will be stored in the standard cell buffer, and output unless shared VCI cells require to be output.

For the purpose of reconstructing shared VCI cell, the fellow cell buffer stores fellow cells composed of new fellow cell header from the new header table, and corresponding voice data. The fellow cell buffer is partitioned into N *standby rooms* $(SR_0, SR_1, \ldots, SR_{N-1})$ which are the space to store payloads of shared VCI cells. The standby room occupies 48 bytes to store up to 6 fellow cells with the new fellow cell header.

The fellow cell controller, based on content addressable memory (CAM), switches the fellow cells from input shared VCI cells to new ones. The write address to store the fellow cell now being processed and the read address to output the rearranged shared VCI cell are generated in the controller. The fellow cell controller consists of N 13-bit *packed route registers* $(pr_0, pr_1, \ldots, pr_{N-1})$, N 1-bit *valid registers* $(v_0, v_1, \ldots, v_{N-1})$, N 3-bit *#calls registers* $(c_0, c_1, \ldots, c_{N-1})$, N *tc*-bit *time stamp registers* $(tm_0, tm_1, \ldots, tm_{N-1})$, one 13-bit *mask register*, *read/write address generators*, and a *time counter*, as shown in Fig. 5. One fellow cell reassigned by this fellow cell switch is divided into its payload and header portion in order to keep the payload in a SR_i and the packed route data which is to become new header in corresponding pr_i. The valid register, v_i, is set equal to

1 if at least one fellow cell is stored in SR_i. The number of fellow cells in the standby room, SR_i, is specified by c_i.

The write address generator creates the address that the fellow cell of a call a, now being processed for example, should be stored. If the fellow cell of the call a is found to have the same common packed route data except for LCI as that of any one of the fellow cell already stored in a particular SR_i, then the fellow cell of a will be stored in that particular SR_i. In the case where the SR_i already holds six fellow cells or no matching fellow cells are found, a new SR_i is prepared for the fellow cells of the call a.

The reconstructed shared VCI cell is taken out when the standby room is fully filled with fellow cells. When c_i is equal to six, the read address generator outputs read address of the fellow cell buffer. In order to prevent significant delay produced by fellow cell switch, reconstructed shared VCI cell is taken out after certain constant time, TM, whether the standby room is fully filled with fellow cells or not. A unit of time measurement in terms of TM is the cell time which is normally 53 clock cycle. When the first fellow cell for SR_i arrives, the time counter sets the (current time $+ TM$) to tm_i. When tm_i is equal to the value of time counter, the read address generator also outputs read address of the fellow cell buffer.

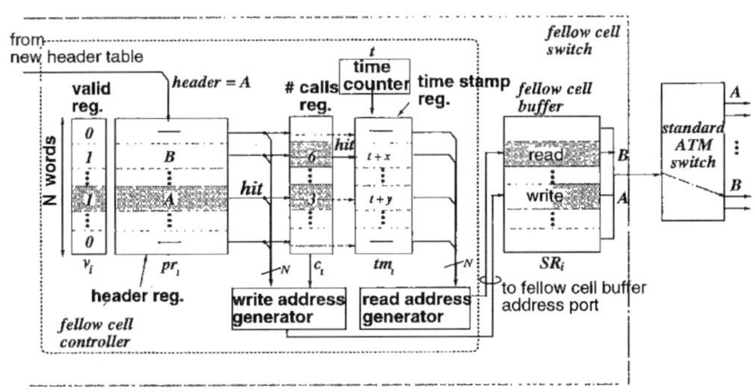

Fig. 5 Outline of fellow cell controller.

With this fellow cell switch, once a VCC is linked over network, shared VCI cells on the connection can be converted to new header solely by hardware processing. Therefore, the shared VCI cell method does not decrease high speed capability of ATM switching. Furthermore, this fellow cell switch can also handle header conversion of standard ATM cells, so that the unit can be replaced with HCV requiring no other hardware.

5. Simulation and Implementation Results

The output priority of the shared VCI cells is superior to that of the standard ATM cells under normal conditions. Therefore, when the fellow cell switch will go on outputting shared VCI cells, the standard ATM cells are not able to be output and finally the standard cell buffer will be flooded. For the purpose of avoiding significant delay or cell loss generated by the buffer, the standard ATM cell kept in the fellow cell switch is also forced to output after certain constant time, time out TM. It is stated that the mean one-way propagation delay for a switch is in the order of 0.5 – 1.0 ms [5]. The time out TM can be set equal to 128 since the additional delay introduced by the buffers is insignificant. It only takes 0.09 ms when 622 Mbps links are fed into fellow cell switch.

The size of standard cell buffer M and that of fellow cell buffer N are determined in order to prevent cell loss in the fellow cell switch. The simulator, which is written in the C language, calculates the maximum buffer sizes, M and N, varying with the both the cell arrival rate and the shared VCI cell occupation rate. The cell arrival rate is the percentage of the valid slots of cells which is randomly coming into the fellow cell switches. The shared VCI cell occupation rate denotes the ratio of shared VCI cells to standard cells. Each buffer size is examined on the assumption that all arriving shared VCI cells have six fellow cells, which is the worst case.

Figures 6 (a), (b) show the maximum standard and fellow cell buffer size to avoid the cell loss, respectively.

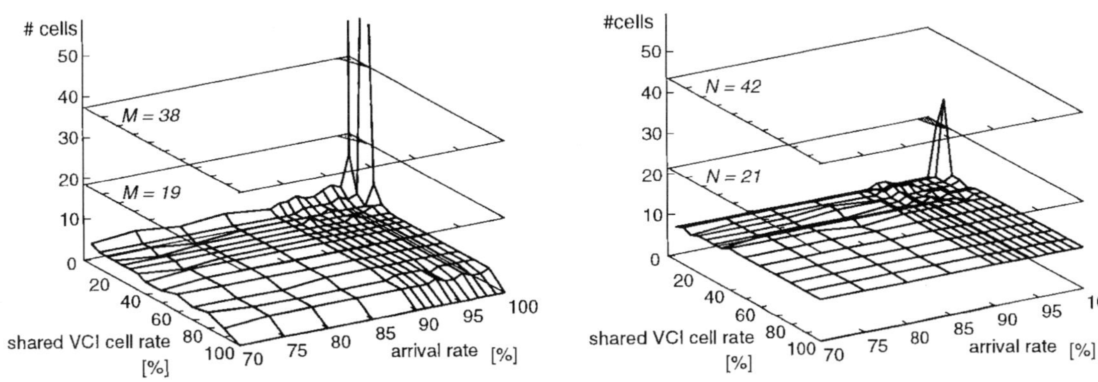

(a) Standard cell buffer (b) Fellow cell buffer

Fig. 6 Simulated results of non-blocking buffer size.

Table 1 Main features of the fellow cell switch.

Technology	0.6 μm CMOS triple-level Al.
chip size	10.4 mm $\times$ 6.82 mm
# gates	307,994
clock frequency	94 MHz

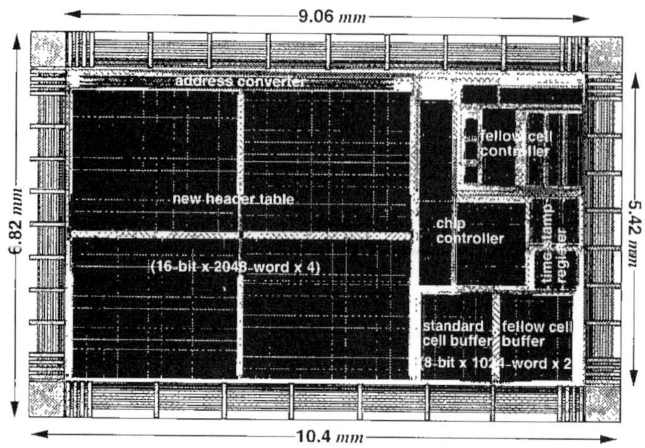

Fig. 7 Layout of fellow cell switch.

With the cell arrival rate set high and the shared VCI cell occupation rate set low, the buffer size to keep standard cell is massively increased. This inexcusable state is brought about by the following cell increment mechanism. An incoming shared VCI cell with six fellow cells may be decomposed into up to six shared VCI cells and output with partially filled, since the time out TM is too short for shared VCI cell arrival rate. Each plane in the graphs is inscribed with buffer sizes constructed by 8-bit $\times$ 1,024-word RAM ($M = 19$, $N = 21$), and 8-bit $\times$ 2,048-word RAM ($M = 38$, $N = 42$) from bottom to top. The hatched areas show the cell loss range that the buffer is flooded because of the cell increment mechanism. Evaluating occupied area for the standard cell buffer and the fellow cell buffer, 8-bit $\times$ 1,024-word RAM ($M = 19$, $N = 21$) proves to be the best solution.

The fellow cell switch has been integrated in a single chip through the use of *COMPASS Design Navigator*. Table 1 lists the main features of the fellow cell switch. Figure 7 depicts the layout of the fellow cell switch obtained by employing 0.6 μm CMOS standard-cell and datapath libraries. The additional function units for shared VCI cell processing are mainly the fellow cell buffer, the fellow cell controller, and the chip controller with some standard-cell. The area increase resulted by having replaced the HCV for standard ATM cells with the fellow cell switch is 54.2%. The maximum clock rate of this unit overcomes 77.8 MHz demanded for 8-bit parallel 622 MHz link.

6. Conclusion

This paper has described new multimedia ATM networks which are capable of efficient transmission of voice data. A new cell assembly and re-assembly method dedicated to voice data transmission are also described to help realize fully ATMized multimedia networks. The proposed shared VCI cell multiplexes fellow cells, which have their own VCC, into one ATM cell as long as every fellow cell go through the same route to the next switch. Moreover, a combination of the shared VCI cell header and the fellow cell header keeps a VCI of the call even if the fellow cell switches reassemble the call in the networks. With this fellow cell switch, once a VCC is linked over network, shared VCI cells on the connection can be converted to new header solely by hardware processing. Therefore, shared VCI cell method does not decrease high speed capability of ATM switching. The hardware architecture of fellow cell switches has been implemented mainly with several 2-port RAM's and CAM based controller in a single chip. The multimedia ATM networks including voice transmission can be constructed by the fellow cell switches being attached to the standard ATM switches. The future works for practical use are suporting low bit-rate voice, variable length fellow cell size, and fellow cell header error protection.

Acknowledgments

The authors would like to express sincere gratefulness to Prof. I. Shirakawa who gave us valuable comments.

References

[1] "Voice and telephony over ATM -ATM Trunking for narrowband services," ATM Forum, 95-0446R5, Feb. 1996.

[2] H. Iwata, H. Yamada, T. Sampei, and S. Kikuchi, "An ATM subscriber line interface circuit for voice and telephony over ATM (VTOA)," Proc. GLOBECOM, pp.270–274, Nov. 1995.

[3] R.A. Spanke and J.M. Adrian, "ATM composite cell switching for DS0 digital switches," Proc. ISS, vol.1, pp.268–272, April 1995.

[4] D.J. Wright, "Voice over ATM: An evaluation of implementation alternatives," IEEE Communications Magazine, pp.72–80, May 1996.

[5] I. Gard and L.-G. Petersen, "Supporting STM traffic with ATM—a switch implementation," Proc. ISS, vol.1, pp.62–66, April 1995.

High-Performance Backbone Network Technology edited by Naoaki Yamanaka
ISBN: 0-8247-5321-6 DOI: 10.1081/(120025666)
Published 2004 by Marcel Dekker, Inc.

An Optimal Virtual Topology Configuration for Multicast in ATM and MPLS Networks

Sung-Jin Chung, Sung-Pil Hong, Sang-Baeg Kim, and Hoo-Sang Chung

SUMMARY This paper has a dual purpose. First, it proposes a virtual path management model for ATM networks. The model unifies VPC overlay network configuration, VCC routing, and capacity allocation in a single framework. It accommodates multiple end-to-end offered traffics of various QoS requirements and traffic types. Especially, it also supports point-to-multipoint as well as point-to-point connections in a resource-efficient manner. The objective is to minimize the overall network resource cost. To do so, it pursues an optimal trade-off among the gains offered by ATM technology. The application of the proposed model is naturally extended to the multiprotocol label switching framework. Second, it proposes an efficient algorithm to solve the model. The mathematical formulation of such a unifying model typically involves a very large-scale intractable optimization which, treated by a straightforward method, requires excessive computational efforts. In this paper, we show how the computational structure of formulation can be exploited to tailor a solution method providing good solutions in dramatically reduced computational efforts.

key words: *ATM, virtual path management model, VCC routing, multicast, optimization*

1. Introduction

We propose a virtual path (VP) management model which unifies, in a single framework, virtual path connection (VPC) overlay network configuration, virtual channel connection (VCC) routing, and capacity allocation including bandwidth and buffer assignment. The model explicitly considers source characteristics, QoS requirements, and connection types of offered traffics. Especially, it supports point-to-multipoint as well as point-to-point connections in a resource-efficient manner. The objective is to minimize the total network resource cost incurred in the management domain.

Such a model typically involves very large-scale intractable optimization: highly nonlinear, non-convex program with tens of million variables. A straightforward nonlinear algorithm would require too excessive computation time and space. In this paper, we apply carefully designed optimization techniques to the special computational structure of the mathematical model to minimize computational efforts while maintaining the quality of the solution. As a result, tested on the networks of various sizes, the algorithm yields good solutions in dramatically reduced computation time.

This paper is organized as follows. In Sect. 2, we propose the model. Then our model is discussed in comparison with previous works. Also, we argue that the proposed model is applicable in the MPLS framework which inherits and extends the characteristics of ATM. We describe the algorithm in Sect. 3. In Sect. 4, we summarize the computational experiments and their results. Finally, in Sect. 5, some concluding remarks are presented. Also the complete set of notation is summarized in Appendix in the order of appearance.

2. The Model

2.1 Input

Our model assumes the following set of input data is given to the management plane.

Backbone

Denote by $G = (N, A)$ the physical network, where $N = \{1, 2, \cdots, n\}$ represents the set of ATM nodes and each physical link $(i, j) \in A$ from node i to node j has capacity c_{ij} bps. From this, we can define the logical network $\bar{G} = (\bar{N}, \bar{A})$ with the candidate VPC set $\bar{A}$ and the VC switching node set $\bar{N}$. B_i is the maximum buffer capacity available at node $i \in \bar{N}$.

QoS requirements

In our model, when end stations want a VCC, they need to specify QoS requirement, the characteristic parameters, and connection type of the offered traffic.

Our model assumes that there are Q possible QoS classes specified by the maximum cell loss ratios, $\epsilon(t)$ $(t = 1, \cdots, Q)$. We conveniently assume that

$$\epsilon(1) < \epsilon(2) < \cdots < \epsilon(Q). \tag{1}$$

On the other hand, our model does not give an explicit consideration to the call-level QoS, or the call blocking probability. However, the call blocking probability can be guaranteed in an implicit manner. As the model assigns a fixed number of VCCs to each offered traffic until the next reconfiguration, we can calculate the call

Manuscript received September 4, 2000.
Manuscript revised February 14, 2001.

[†]The authors are with Seoul National University, Korea.
[††]Corresponding author. The author is with Chung-Ang University, Ansong-Shi, Gyong-gi-Do, 456-756, Korea.
[†††]The author is with Korea Telecom, Korea.
[*]Presently, with LG-EDS Systems, Korea.

[**]This paper contains in part the results in the paper, "An optimal VP-based multicast routing in ATM networks," in Proc. 1996 IEEE INFOCOM [30].

IEICE TRANS. COMMUN., Vol. E84-B, No. 9, pp. 2656–2669, September 2001

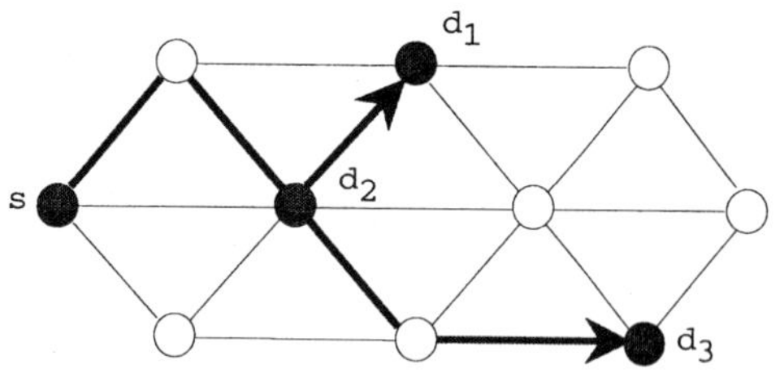

Fig. 1 Steiner tree spanning origin s and destinations d_1, d_2 and d_3.

blocking probability from the call arrival parameters. Thus an inverse calculation will compute the number of VCCs that should be reserved to guarantee a required call blocking probability. So we may assume that the offered traffic intensity is given as the number of VCCs to be reserved for satisfying the call-level QoS.

Traffic parameters

We assume that there are L possible types of traffic and the jth ($j = 1, \cdots, L$) traffic type is characterized by the following standard set of parameters: R_j, the peak bit rate, m_j, the mean bit rate, b_j, the average duration of burst state, and the utilization, $\rho_j = m_j/R_j$.

Connection types

For efficiency of bandwidth utilization, our model also distinguishes the offered traffics based on their connection types: point-to-point (unicast) or point-to-multipoint (multicast). In particular, the proposed model supports the point-to-multipoint connections by a directed *Steiner tree* rooted at the origin and spanning the set of destinations (see Fig. 1). In this paper, we call this *Steiner tree virtual channel connection* (TVCC). Each arc of TVCC is a VPC just the same as each VCC is a concatenation of VPCs.

For each possible origin-destination pair $(s, d) \in N \times N$, we denote by $r(s, d; j, t)$, the number of demanded point-to-point VCCs of jth traffic type and tth QoS class from s to d. This is equivalent to a $Q \times L$ traffic matrix.

Similarly for each multicast group (s, D), where s and D, respectively, are the origin and the set of destinations, $a(s, D; j, t)$ is the intensity of the offered traffic of the multicast expressed in terms of the equivalent number of VCCs of jth traffic type and tth QoS class. Denote by $\mathcal{D}$ the set of offered point-to-multipoint connections.

2.2 Output

Given the input data, we decide the following;

- the VPC overlay network on the physical network,
- the VCC routing on the VPC network, and
- capacity allocation including VP bandwidth assignment to the VPCs and buffer assignment to the VP/VC switching nodes.

This approach is illustrated with an example in Fig. 2. A physical network with four nodes $\{1, 2, 3, 4\}$ and six links $\{(1,4), (4,1), (1,3), (3,1), (2,3), (3,2)\}$ is given in the bottom layer. (For simplicity, the two links connecting a pair of nodes are represented by a single undirected arc.) Suppose that there are three offered traffics (whose QoS classes and traffic types are not specified for brevity): two point-to-point demands, $r(1, 2; ..)$ VCCs from node 1 to node 2, and $r(3, 4; ..)$ VCCs from node 3 to node 4, and one point-to-multipoint demand, $a(1, \{3, 4\}$: ..) TVCCs from node 1 to nodes 3 and 4.

The model may configure the following in the two upper layers: 1) a VPC network in the mid-layer consisting of 6 VPCs: 3 single-hop VPCs, P_1 corresponding to the link $(1, 4)$, P_2 to $(2, 3)$, and P_3 to $(3, 2)$; 2 three-hop VPCs, P_4 corresponding to the physical path $2-3-1-4$, and P_5 to $4-1-3-2$; finally the two-hop VPC, P_6 corresponding to the path $1-3-2$, 2) VCC routing in the top-layer: VC1, the VCC consisting of P_1 and P_5 to accommodate $r(1, 2; ..)$, VC2, the VCC consisting of P_3 and P_4 for $r(3, 4; ..)$, and finally TVC1, the TVCC for $a(1, \{3, 4\}$: ..) consisting of P_6, P_4, and P_2, and 3) the bandwidth reservation for each VPC and the buffer capacity assignment to each VP/VC switching node (which are not specified in the figure).

It should be emphasized that the configurations and capacity allocations in two upper layers are done simultaneously. Most previous works configure a logical network first, and then route the traffics on the logical network [2], [12], [19]. The new approach will enhance the optimality of the solution as it does not fix a logical network *a priori*.

2.3 Constraints

QoS guarantee

Given an offered traffic, our model guarantees the cell-level QoS by enforcing the corresponding VCC to be multiplexed in a VPC supporting an equal or more stringent QoS class. That is, a VCC carrying traffic of tth QoS class can use the kth VPC, P_k if and only if q_k, the QoS class supported by P_k, is al least as stringent as t: $\epsilon(q_k) \leq \epsilon(t)$, or equivalently, $q_k \leq t$ from Eq. (1). This VCC grouping strategy [27] not only guarantees the user's QoS requirements but also improves the resource utilization by increasing the number of choices.

VCC and TVCC routing constraints

In Fig. 2, the offered traffics and the corresponding VCCs are accommodated by the logical network consisting of 9 unidirectional VPCs in the mid-layer. In principle, the candidate VPCs can be any subset of possible VPCs. The larger the set of candidate VPCs is, the more chance we have for better resource utilization. In our model, the candidate VPC set consists of the first three shortest (in hop count) paths of physical links for each QoS type between every (s, d)-pair. Hence the the total number of candidate VPCs is nor-

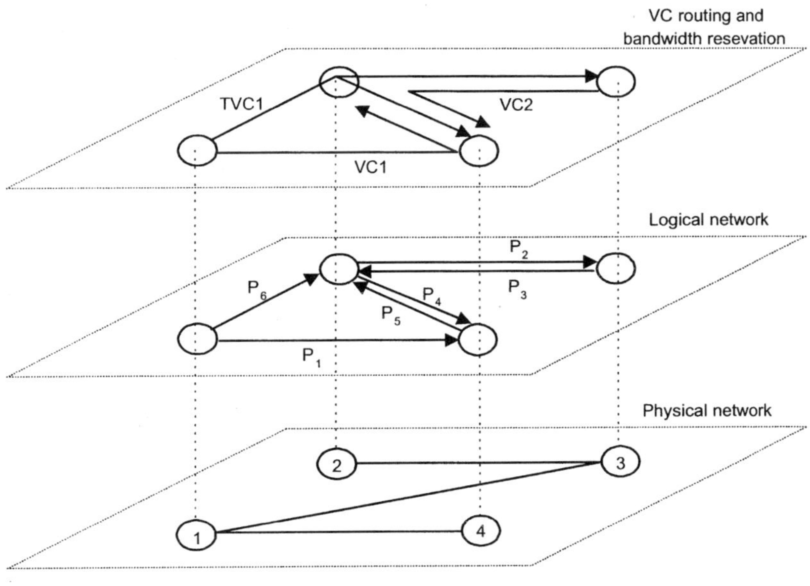

Fig. 2 Output of model: two upper layers are configured simultaneously.

mally $O(Qn^2)$ (where n is the number of nodes in the physical networks).

Once the candidate VPCs have been chosen, for each traffic requirement $r(s,d;j,t)$, we need to establish the same number of VCCs from s to d. This is equivalent to find a flow of size $r(s,d;j,t)$ from the origin node s to the destination node d on the logical network whose nodes and links are the VC switching nodes and the candidate VPCs, respectively. So routing single point-to-point offered traffic is equivalent to a (single-commodity) network flow problem [1] on the logical network of candidate VPCs. As there are multiple point-to-point offered traffics distinguished in terms of the origin (s), destination (d), traffic types (j), and QoS (t), routing the complete set of point-to-point offered traffics is equivalent to the multi-commodity network flow problem (see, for example pp.649–694 of [1])on the logical nework in which each commodity flow corresponds to the VCCs for a traffic requirement $r(s,d;j,t)$ for some fixed s, d, j and t.

At each VC switching node, however, a unit commodity flow (a VCC for $r(s,d;j,t)$) cannot be sent to an arc (a VPC) P_k if P_k supports a lower QoS class: namely, if $q_k > t$. Hence the usual network flow problem should be modified for the QoS requirements. So we introduce $\Gamma^+(i;t)$ ($\Gamma^-(i;t)$, resp.), the set of VPCs, P_k with $q_k \le t$, originating from (terminating at, resp.) VC switching node i.

Denote by $f_k(s,d;j,t)$ the decision variables representing the number of VCCs for a traffic requirement $r(s,d;j,t)$, which are to be multiplexed in P_k. Then the multi-commodity flow formulation guaranteeing QoS requirements is given as follows:

$$\sum_{k\in\Gamma^+(i;t)} f_k(s,d;j,t) - \sum_{k\in\Gamma^-(i;t)} f_k(s,d;j,t) = \begin{cases} r(s,d;j,t) & \text{if } i=s \\ -r(s,d;j,t) & \text{if } i=d \\ 0 & \text{otherwise} \end{cases}$$
$$\forall s,d \in N, \forall j=1,\cdots,L, \forall t=1,\cdots,Q. \quad (2)$$

Similarly, for each $(s,D)\in\mathcal{D}$, we need to find a directed Steiner tree rooted at s and spanning the destinations D in the network of candidate VPCs. The following is a standard and exact integer programming formulation of the Steiner tree problem in the networks, modified similarly for QoS guarantee. Here $z_k(s,D;j,t)$ is the binary variable telling whether the kth VPC, P_k is used in the multicast or not. The other variable $h_k(s,D;d,j,t)$ is an auxiliary one which enforces that the subgraph induced by P_k with $z_k(s,D;j,t)=1$ should be a Steiner tree spanning the source s and all the destinations in D. For more details, the readers are referred to [26] or Appendix.

$$z_k(s,D;j,t) \ge h_k^d(s,D;d,j,t)$$
$$\forall\ k\in\bar{A}\ \forall d\in D,$$
$$\forall (s,D)\in\mathcal{D}, \forall j=1,\cdots,L, \forall t=1,\cdots,Q,$$

$$z_k(s,D;j,t)=0, \text{ or } 1$$
$$\forall\ k\in\bar{A},\ \forall (s,D)\in\mathcal{D}, \forall j=1,$$
$$\cdots,L, \forall t=1,\cdots,Q,$$

$$\sum_{k\in\Gamma^+(i;t)} h_k^d(s,D;d,j,t)$$

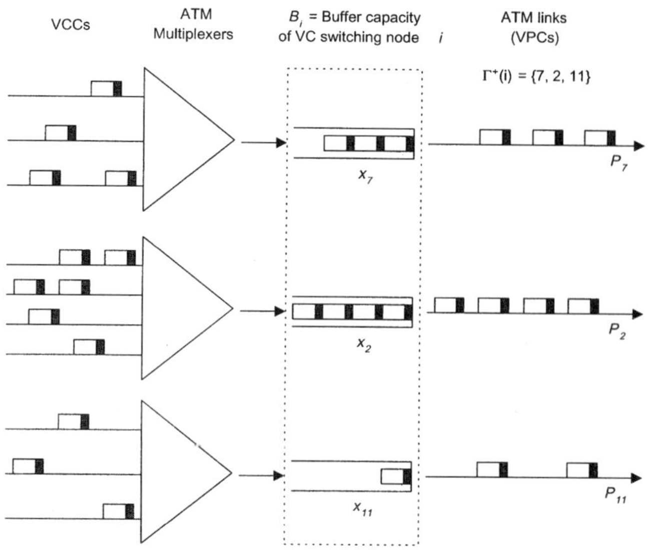

Fig. 3 Multiplexer queues at output ports of an ATM node; each ATM link or VPC consists of a transmission path and a buffer.

$$- \sum_{k\in\Gamma^-(i;t)} h_k^d(s,D;d,j,t) = \begin{cases} 1 & \text{if } i=s \\ -1 & \text{if } i=d \\ 0 & \text{otherwise} \end{cases}$$
$$\forall d\in D, \forall (s,D)\in\mathcal{D}, \forall j=1,\cdots,L, \forall t=1,\cdots,Q. \quad (3)$$

Buffer and link capacity constraints

Let $i\in\bar{N}$ be a VC switching node (or ATM node), and $\Gamma^+(i)$ be the index set of VPCs originating from i. Each VPC, $P_k\in\Gamma^+(i)$ consists of its own buffer and a transmission path. We assume output queueing and each VPC, P_k is reserved the buffer capacity x_k. (See Fig. 3.) Then if B_i is the buffer capacity of i, we have:

$$\sum_{k\in\Gamma^+(i)} x_k \le B_i \quad \forall i\in\bar{N}. \quad (4)$$

For the VCC and TVCC routing along with the bandwidth allocation at the logical level to be valid, we need to add the capacity consideration at the physical level: the sum of bandwidth reserved for the VPCs sharing a common physical link (i,j) should not exceed the capacity of the link c_{ij}. In Fig. 2, the unidirectional physical link $(1,4)$ is shared by unidirectional VPCs, P_3 and P_5 (from node 2 to 4) which, in turn, accommodate VC1 and TVC1, respectively. Hence the sum of bandwidth reserved for P_3 and P_5 should not exceed c_{14}.

A more general situation may look like Fig. 4. When multiple VCCs are multiplexed into a VPC, the equivalent capacity is determined by the types of traffic carried by the VCCs and the QoS supported by the VPC. If we denote by y_k^j the number of jth traffic type VCCs using kth VPC, P_k, then the complete set of VCCs multiplexed in P_k can be written as the vector $\boldsymbol{y}_k=(y_k^1,y_k^2,\cdots,y_k^L)$. From Eqs. (2) and (3), we have

$$y_k^j = \sum_{s,d\in N}\sum_{t=1}^{Q} f_k(s,d;j,t) + \sum_{(s,D)\in\mathcal{D}}\sum_{t=1}^{Q} z_k(s,D;j,t)a(s,D;j,t)$$
$$\forall k\in\bar{A}, \forall j=1,\cdots,L. \quad (5)$$

Suppose P_k supports the q_kth QoS class and is assigned buffer capacity x_k. Let $c(\boldsymbol{y}_k,x_k)$ be the *equivalent capacity* to be reserved at P_k to accommodate the traffic $\boldsymbol{y}_k$. Guérin et al. [21] suggest the following formula.

$$c(\boldsymbol{y}_k,x_k) = \text{Min}\ (c_g(\boldsymbol{y}_k),\ c_f(\boldsymbol{y}_k,x_k)), \quad (6)$$

where, $c_g(\boldsymbol{y}_k)$ and $c_f(\boldsymbol{y}_k,x_k)$ are, respectively, the *Gaussian approximation function* and *the fluid flow approximation function* (see Appendix).

This formula is incorporated into the link capacity constraint so that the statistical multiplexing effect of ATM is explicitly reflected in the model. Let $P(i,j)$ be the set of VPCs that use a physical link (i,j). Then the capacity constraint to be satisfied at each link is

$$\sum_{k\in P(i,j)} c(\boldsymbol{y}_k,x_k) \le c_{ij} \quad \forall (i,j)\in A. \quad (7)$$

The functional form of the *fluid flow approximation*, $c_f(\boldsymbol{y}_k,x_k)$ is quite complicated and makes the mathematical programming formulation highly nonlinear. It is observed in [15] that this function can be modified to a relatively tractable bilinear function (see Eq. (A· 1) in Appendix). The modification is a slight overestimation of the original function. But it facilitates the solution method and hence reduces the computational efforts significantly. This paper also adopts this modification, denoted by $\hat{c}$, in place of c in Eqs. (6) and (7).

2.4 Cost Function

The cost function is a linear combination of the reserved bandwidth, the reserved buffer capacity, and the required switching/control level for the VPC deployment. The coefficients in the linear combination are the cost factors of the relevant resources. The cost function has the following terms: $o_{ij}u$, the cost incurred at physical link (i,j) when the reserved bandwidth is u, and $\zeta_k v$, the cost incurred at VPC, P_k when the buffer reservation level is v.

We assume the switching cost incurred at each VPC, P_k is a linear function of the total accommodated traffic u, $(\Omega_k+(|P_k|-1)\omega_k)u$ where Ω_k and ω_k represent, respectively, the VC and VP switching cost coefficients, and $|P_k|$ is the physical hop length of P_k.

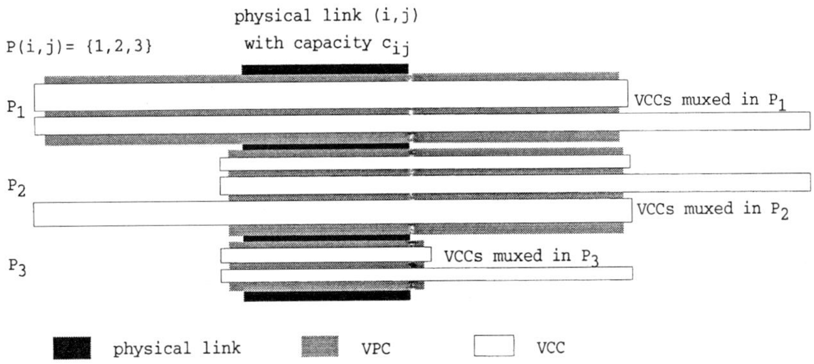

Fig. 4 Bandwidth allocation hierarchy.

In sum, the total switching cost incurred at P_k with the traffic level u is given by $w_k u$, where

$$w_k = \Omega_k + (|P_k| - 1)\omega_k. \tag{8}$$

Let $A(P_k)$ denote the set of physical links of VPC, P_k and define

$$\Phi_k \equiv \sum_{(i,j)\in A(P_k)} o_{ij} + \Omega_k + (|P_k| - 1)\omega_k. \tag{9}$$

Then if the reserved bandwidth and buffer capacity for P_k are u and v respectively, the total cost incurred at P_k is

$$\Phi_k u + \zeta_k v. \tag{10}$$

Suppose P_k supports the q_kth QoS class and is assigned buffer capacity x_k. Then, since we need to reserve bandwidth $\hat{c}(\boldsymbol{y}_k, x_k)$ for P_k to accommodate the traffic $\boldsymbol{y}_k$, Eq. (10) implies that the total cost is given by

$$\theta(\boldsymbol{y}, x) \equiv \sum_{k\in\bar{A}} \Phi_k \hat{c}(\boldsymbol{y}_k, x_k) + \sum_{k\in\bar{A}} \zeta_k x_k. \tag{11}$$

2.5 Mathematical Formulation

Summing up the discussion so far, the mathematical programming formulation of our model is written as follows. The objective function Eq. (12) follows from the modified version of Eqs. (6) and (11). The constraints, Eqs. (13), (14), (15), (16), and (17), respectively, are Eqs. (2), (3), (4), (5), and (7) with $c(\boldsymbol{y}_k, x_k)$ replaced by $\hat{c}(\boldsymbol{y}_k, x_k)$.

$$\text{Min } \theta(\boldsymbol{y}, x) \equiv \sum_{k\in\bar{A}} \Phi_k \hat{c}(\boldsymbol{y}_k, x_k) + \sum_{k\in\bar{A}} \zeta_k x_k \tag{12}$$

$$\text{sub to } \sum_{k\in\Gamma^+(i;t)} f_k(s,d;j,t) - \sum_{k\in\Gamma^-(i;t)} f_k(s,d;j,t) = \begin{cases} r(s,d;j,t) & \text{if } i = s \\ -r(s,d;j,t) & \text{if } i = d \\ 0 & \text{otherwise} \end{cases}$$
$$\forall s,d \in N, \forall j = 1,\cdots,L, \forall t = 1,\cdots,Q \tag{13}$$

$$z_k(s,D;j,t) \ge h_k^d(s,D;d,j,t) \quad \forall\, k \in \bar{A}\ \forall\, d \in D,$$
$$\forall (s,D) \in \mathcal{D}, \forall j = 1,\cdots,L, \forall t = 1,\cdots,Q,$$
$$\sum_{k\in\Gamma^+(i;t)} h_k^d(s,D;d,j,t) - \sum_{k\in\Gamma^-(i;t)} h_k^d(s,D;d,j,t) = \begin{cases} 1 & \text{if } i = s \\ -1 & \text{if } i = d \\ 0 & \text{otherwise} \end{cases}$$
$$\forall d \in D, \forall (s,D) \in \mathcal{D}, \forall j = 1,\cdots,L, \forall t = 1,\cdots,Q, \tag{14}$$

$$\sum_{k\in\Gamma^+(i)} x_k \le B_i \quad \forall i \in \bar{N} \tag{15}$$

$$y_k^j = \sum_{s,d\in N} \sum_{t=1}^{Q} f_k(s,d;j,t) + \sum_{(s,D)\in\mathcal{D}} \sum_{t=1}^{Q} z_k(s,D;j,t) a(s,D;j,t)$$
$$\forall k \in \bar{A}, \forall j = 1,\cdots,L \tag{16}$$

$$\sum_{k\in P(i,j)} \hat{c}(\boldsymbol{y}_k, x_k) \le c_{ij} \quad \forall (i,j) \in A \tag{17}$$

$$f_k(s,d;j,t) \ge 0$$
$$\forall s,d \in N, \forall j = 1,\cdots,L, \forall t = 1,\cdots,Q,$$
$$x_k \ge 0 \quad \forall k \in \bar{A},$$
$$h_k^d(s,D;d,j,t) \ge 0, \quad z_k(s,D;j,t) = 0,\ 1$$
$$\forall d \in D, \forall (s,D) \in \mathcal{D}, \forall j = 1,\cdots,L, \forall t = 1,\cdots,Q.$$

2.6 Discussion on the Model

Trade-off between capacity and control costs: It is well-known that ATM technology offers two fundamental gains, setup and control cost reduction and efficient network resource usage through the multiplexing effects due to its asynchronous nature [9]. However, they are conflicting. The "express bit pipes" [19], [35] (i.e. VPCs) reduce setup and control costs. On the other hand, if VCCs are accommodated by a maximal number of VPCs, then the multiplexing gains will be minimal and hence deteriorate the network performance. Our model pursues an optimal trade-off between these gains to minimize the total cost. Note that the trade-off is realized only if the three problems are solved simultaneously in a single framework. Also for an optimal trade-off, it is essential to base the formulation on a logical network with a maximal number of candidate VPCs. (In the experiment, the number of candidate VPCs used is $O(Qn^2)$.) We note that the model in [37], like ours, adopts as the objective to minimize the sum of capacity cost and control cost with an explicit consideration of multiplexing effects.

Depending on the scope and time scale, there can be other types of trade-off appropriate in ATM networks. For instance, in [4] the model is focused on the tradeoff between call throughput and processing load on the signaling systems. The work of [12] pursues the throughput maximization subject to call setup time limits.

The objective of existing models are mainly focused on the call-level QoS such as throughput or call blocking probability [2], [20], [31], [34], or cell-level QoS such as cell delay [13], [19], [35] or cell loss ratio [25], [32].

Unifying approach: To the best of our knowledge, the proposed model is the first one unifying VPC overlay network configuration, VCC routing and capacity allocation *in a single framework* and therefore enhances the optimality of the solution. This is enabled by the explicit consideration of the logical network consisting of a variety of candidate VPCs (which was called "second order topology" in [19]), as well as by the careful mathematical construction which gets the routing and capacity allocation at the logical level to be also valid in the physical level. The model in [25] performs logical network configuration, routing and capacity allocation in an integrated manner. However, capacity allocation is done only in the physical level.

[14] and [35] propose routing algorithms to meet given VPC capacity demands, whereas [20], [34] and [36] consider capacity allocation to predetermined VPCs. In [4], capacity allocation and routing is performed on a predetermined logical network combined with physical links (single hop VPCs).

In [2], [12], and [19], capacity allocation and routing are done in an iterative manner after a logical network is determined first. There are other models in which routing is performed after VPC capacity allocation [3], [32], [37], or VPC capacity allocated after VCC routing is performed [13].

Multiple QoS classes, traffic types, connection types: In principle, at the expense of problem-size increase, the model allows an arbitrary number of cell-level QoS classes and traffic types. (In the experiment, both are set to 3.) Most of the previous models allow a single traffic type, except [4], [34], a single QoS class per each traffic type.

In our model, as the point-to-point traffic accommodation is based on the multi-commodity flow formulation on the second order topology, the VCC routing can be done by flexible VPC deployment: for the same (s, d)-pair, there can be multiple VCCs of multiple VPC hop count. Also it supports point-to-multipoint connections in a resource-efficient manner by adopting Steiner tree virtual channel connection proposed in [3].

However, this unifying feature seems to be restricted to the following limits.

Centralized and static model: Our model is a centralized and static model. As the model pursues simultaneously the whole set of solutions based on the forecasted state of the management domain, it does not adapt to the dynamic conditions in traffic or network itself. The accuracy of the forecasted data is also critical.

Extensive computational efforts: Since the proposed model integrates ATM network resource management in a single framework while accommodating various requirements for offered traffics, the mathematical programming formulation, which is large-scale and highly nonlinear, requires extensive computational efforts for finding a good solution. We may develop an efficient algorithm by carefully exploiting the structural properties of the model (which is the main topic of the latter half of this paper). However, in any case, the computational efforts are not likely to be comparable to those for the VP management schemes which are usually for dynamic routing and whose management decisions are based on short-term and local state of the network. In this sense, the proposed model may not be suitable for such a practical VP management and the expensive global configuration provided by the model is most useful at startup or after major change in the traffic pattern or network state on a long-term time scale.

However, it is worth noting that our model is compatible and can be combined with dynamic VP management schemes, as in [19]. The model maintains a complete set of information for a dynamic call admission or routing such as the *least loaded routing* [16], [40]: operational VPCs, their reserved bandwidth in terms of equivalent capacity, and the portion of each physical link capacity currently used by the operational VPCs.

Table 1 Comparison of proposed model with others.

capability	[2]	[4]	[12]	[14]	[19]	[25]	[32]	[37]	ours
Static/ Dynamic	static/ dynamic	static/ dynamic	static	static	static	static	static	static	static
Objective function	call blocking	revenue	call blocking	min-max link load	average delay	capacity	cell loss	capacity + control	capacity + control
VP topology	output	input	output	input	output	output	output	output	output
VPC capacity	output	output	output	input	output	none	output	output	output
VPC routing	yes	no	no	yes	no	no	yes	yes	yes
VCC routing	yes	yes	yes	no	yes	yes	no	no	yes
No. traffic types	single	multiple	single	single	single	single	single	single	multiple
No. VPC hops per VCC	multiple	multiple	multiple	single	sinlge	sinlge	sinlge	sinlge	multiple
No. VCCs per (s, d)-pair	multiple	multiple	multiple	single	multiple	multiple	multiple	multiple	multiple
Call blocking	objective	constraint	objective	-	constraint (implicit)	-	-	constraint (implicit)	constraint (implicit)
Statistical multiplexing	no	yes	yes (implicit)	yes (implicit)	no	no	no	yes	yes

This enables a dynamic routing scheme on the available resources.

In Table 1, our model is compared to some previous models.

2.7 Applicability in MPLS

As the concept of MPLS (multiprotocol label switching) [10], [38] inherits and extends the characteristics of ATM, it is rather natural to expect an ATM VP management model to be also applicable in the MPLS framework. It is worth to emphasize that the intended MPLS needs not to be an ATM-integrating framework. For instance, the "MPLS tunnel," or *label switched path* (LSP) [5], [39], [41] of the generic MPLS framework seems to be a pure analogy of the VPC in the ATM network. Both can be managed by, e.g. RSVP [5] signaling procedure depending on the relevant time-scales.

By stretching this line of abstraction, we can argue the applicability of our model in MPLS networks by establishing the one-to-one correspondences between the elements of our model and the MPLS framework. First of all, with VPCs identified with LSPs, which are bundles of IP flows, VCCs are naturally matched to the end-to-end IP flows supported by the connection-oriented MPLS.

An IP flow is a collection of one or more microflows at a standardized interface. A microflow is a sequence of packets passing across a standardized interface in which all packets have the same classification. Packet classification is the process of classifying packets based on the content of their IP header and other relevant information. Traffic parameters of an IP flow may include the peak rate, the mean rate, and the maximum burst size of a flow [29]. For possible QoS classes and their requirements, readers are referred to [28].

Also the assumption on QoS requirements and characteristics of offered traffics of our model is general enough to accommodate the differentiated service in IP-integrating MPLS [8]. Recall that the proposed model carefully considers the bandwidth reservation hierarchy (see Fig. 4) and reserves bandwidth based on an explicit statistical multiplexing calculation. It is not hard to assume that this is also applicable, at least in principle, for an efficient aggregation of traffics in LSPs.

Finally, we note that the proposed model seems particularly advantageous in configuring VPNs (virtual private networks), one of the main applications of MPLS [33], [38]. Existing VPN service models are restricted to point-to-point connections. Our model, due to its unique ability to support point-to-multipoint connections, can configure VPNs in a more resource-efficient form of Steiner trees of LSPs with the explicit bandwidth reservation guaranteeing the QoS requirements of customers.

3. The Algorithm

The mathematical program, (P) in Sect. 2.5 is a concave minimization. More specifically, due to Eq. (7) which restricts the solution to the complement of a convex set, (P) is a global optimization problem with *reversed convex* constraints [22]. It implies that (P) is NP-hard and it is unlikely to obtain a global optimum with reasonable computational efforts. Even for a local minimum, the computational efforts can be enormous. As will be shown in Sect. 4, when the physical network is the "USA Network" with 28 nodes and 90 bidirectional edges, (P) has about 62 million variables and 14 million constraints. Thus, to deal with a instance of realistic size, it is crucial to devise an algorithm which optimally trades off between the computation time/space and the quality of solution.

Figure 5 depicts the computational structure of the constraints in (P). Besides the physical level capacity constraints from Eq. (17), the rest shows a block diago-

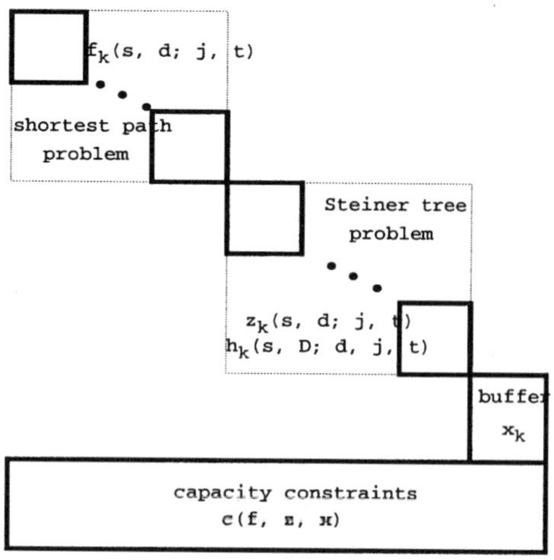

Fig. 5 The computational structure of model.

nal structure: the constraints are separable in terms of the variables in the same block. There are three groups of blocks. The first group is from Eq. (13), the multi-commodity flow formulation for point-to-point traffic accommodation. Without the capacity constraints, it reduces to a set of shortest path problems which are easy to solve. The second group of blocks corresponds to Eq. (14), in which each block is a Steiner tree problem. The Steiner tree problem is NP-hard. There are, however, several fast 2-approximation algorithms which guarantees a solution whose cost is within two times the exact minimum. In practice, they are known to produce near-optimal solutions [6], [26]. The last group of constraints is from the buffer constraints Eq. (15) which are tractable linear constraints. From this, the *Lagrangian relaxation* method which relaxes the capacity constraints, Eq. (17) would be the most plausible ingredient of our solution method.

Our main algorithm is the combination of a sequential linear program approximation of (P) and the Lagrangian relaxation applied to each linear program approximation.

Algorithm 3.1: Main algorithm

Input: Initial feasible solution $(\boldsymbol{y}, \boldsymbol{x}) = (\bar{\boldsymbol{y}}, \bar{\boldsymbol{x}})$.
begin
  $t_{new} \leftarrow \theta(\bar{\boldsymbol{y}}, \bar{\boldsymbol{x}})$;
  compute linear approximation of c_g around $\bar{\boldsymbol{y}}$
  to construct local bilinear program $B(\boldsymbol{y}, \boldsymbol{x})$;
  repeat
    begin
      $t_{old} \leftarrow t_{new}$;
      find an optimal solution $\tilde{\boldsymbol{y}}$ of $B(\boldsymbol{y}, \bar{\boldsymbol{x}})$
      by Lagragian relaxation method;
      find an optimal solution $\tilde{\boldsymbol{x}}$
      of linear program $B(\tilde{y}, x)$;
      $t_{new} \leftarrow \theta(\tilde{\boldsymbol{y}}, \tilde{\boldsymbol{x}})$;
      $(\bar{\boldsymbol{y}}, \bar{\boldsymbol{x}}) \leftarrow (\tilde{\boldsymbol{y}}, \tilde{\boldsymbol{x}})$;
    end
  until $\frac{t_{old} - t_{new}}{t_{old}} < \epsilon$
end

Table 2 Traffic parameters.

	Data	Image	Voice
R_j (Mbps)	10	2	0.064
m_j (Mbps)	1	0.087	0.021
b_j (ms)	14.37	55.20	38.43

Initially, we find a feasible solution $(\bar{\boldsymbol{y}}, \bar{\boldsymbol{x}})$ of (P). Then we compute the linear approximation of Gaussian component c_g of $\hat{c}$ (Eq. (A·1) in Appendix) around $\bar{\boldsymbol{y}}$. Then since the fluid flow component $\hat{c}_f$ is bilinear with respect to $\boldsymbol{x}$ and $\boldsymbol{y}$, we get a local bilinear program approximation $B(\boldsymbol{y}, \boldsymbol{x})$ of (P). Then we fix $\boldsymbol{x}$ to $\bar{\boldsymbol{x}}$ to get the linear program $B(\boldsymbol{y}, \bar{\boldsymbol{x}})$ which is then solved to get an optimal solution $\boldsymbol{y} = \tilde{\boldsymbol{y}}$. Then, we fix $\boldsymbol{y}$ to $\tilde{\boldsymbol{y}}$ to get the linear program $B(\tilde{\boldsymbol{y}}, \boldsymbol{x})$ with respect to $\boldsymbol{x}$. Solve it to get an optimal solution $\boldsymbol{x} = \tilde{\boldsymbol{x}}$. Repeat the procedure with the new feasible solution, $(\tilde{\boldsymbol{y}}, \tilde{\boldsymbol{x}})$, until the objective function value improvement becomes insignificant. (See Algorithm 3.1.) It is not hard to show that this algorithm is descendent, namely, the feasible solutions after the repeat loops have monotone decreasing objective values. Thus the stopping criterion is well-defined.

The linear program $B(\boldsymbol{y}, \bar{\boldsymbol{x}})$ is a large-scale problem. However, as mentioned earlier, by relying on the Lagrangian method which relaxes the capacity constraints, we can reduce the problem to a set of separable and relatively small shortest path problems and Steiner tree problems, which can be solved effectively. Indeed, it is shown in Sect. 4 that the computation time reduction due to this approach is dramatic.

On the other hand, as the size of $\boldsymbol{x}$ variables is relatively small to that of the total variables, the linear program $B(\tilde{\boldsymbol{y}}, \boldsymbol{x})$ is easily solvable by an LP-solver.

4. Experimental Results

Traffic scenario
Each offered traffic may belong to one of three possible QoS classes with the cell loss probability 10^{-8}, 10^{-6}, or 10^{-4}. We use three traffic types in Table 2 whose parameters are adopted from [11].

In case of unicast, for each of 9 possible QoS class-traffic type combinations, the number of VCCs required between each (s, d)-pair are randomly generated from the set $\{1, 2, , \cdots 20\}$ when the traffic type is data or image, and from the set $\{1, 2, \cdots 500\}$ when the traffic type is voice. The same method is used for multicast traffic.

Topology
Two specific physical topologies are adopted. The first one is a "small" network with 6 nodes and 10 bidirectional physical links. The second one is called "USA"

network [18] which has 28 nodes and 45 bidirectional links. Other parameters are summarized in Table 3. Note that in case of USA network which is assigned 6804 unicasts and 252 multicasts each of which has 8 destinations, (P) has more than sixty million variables and thirteen million constraints.

Cost parameters

We set VC and VP switching cost, respectively, $\Omega = 1$ and $\omega = 0.1\Omega$ to reflect the belief that the former is about 10% of the latter [9]. Also we use rather hypothetical buffer cost $\zeta = 0.001$ and capacity cost $o = 2$.

Computation time

First, we compare the computational efforts of our algorithm to the algorithm, called *KT algorithm*, which was developed in [30] and used in *Korea Telecom*. This algorithm is the combination of *Zoutendijk*'s feasible direction method and the exterior penalty technique to maintain the feasibility of search directions [7]. For the USA network instances, we could not execute KT algorithm on our computing device (Sunspark Ultra 1) as it requires an excessive memory. The comparison of computation times on 5 small network instances are summarized in Table 4. It shows that the computation time reduction is dramatic. Also, the quality of solution is maintained with fewer VPC deployment.

Quality of solution

Ideally, the quality of solution should be compared to the exact global optimum. This is not feasible as (P) is highly nonlinear nonconvex minimization. Instead, we test the quality of solution in two respects: "local" duality gap and "robustness."

It is well-known [7], [17] that the Lagragian relaxation offers a lower bound on the global minimum. It implies that we can perform a conservative quality test by comparing the solution to the lower bound. However, this alternative test should be taken cautiously as the lower bound principle is applied only when the Lagrangian relaxation can be solved globally, which is not the case as we solve it locally.

Thus the *duality gap*(= (cost of solution − lower bound)/lower bound) summarized in Table 5 indicates how good is each solution compared to the best solution in "some" neighborhood where the solution is found.

The "robustness" test is complementary to the duality gap test in the sense that how the algorithm is consistent to the expansion of the solution set. If we uniformly increase the link capacities, a global optimum algorithm should provide a lower total cost. We generated 15 base instances and the link capacities of each was uniformly and monotonically increased up to three levels so 45 extra instances were solved. To test the robustness for a more various set of network topologies, we adopt an additional medium size network called "New Jersey LATA" which has 15 nodes and 28 bidirec-

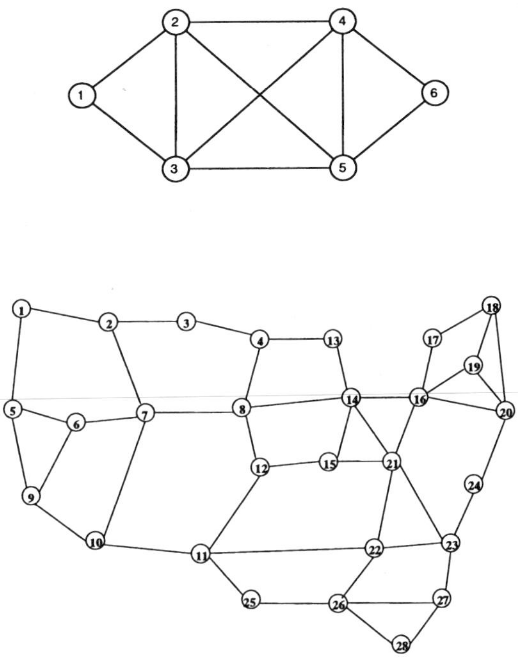

Fig. 6 "Small" and USA network.

Table 3 Network parameters and problem sizes.

	"small"	USA
$\lvert N\rvert$, $\lvert A\rvert$	6, 20	28, 90
$\lvert\bar{N}\rvert$, $\lvert\bar{A}\rvert$	6, 270	28, 6804
c_{ij} (Gbps)	0.40 - 0.60	5-6.5
B_i (Mbits)	10	10
No. unicasts	270	6804
No. multicasts	54	252
$\lvert D\rvert$	3	8
No. f var.	72,900	46,294,416
No. z var.	14,580	1,714,608
No. h var.	43,740	13,716,864
No. x var.	270	6804
Total var.	131,490	61,732,692
No. const.	46,358	13,963,942

Table 4 Computation times.

Instance	KT algorithm			Our algorithm		
	CPU time (sec)	Total cost	No. of used VPs	CPU time (sec)	Total cost	No. of used VPs
1	9,055.31	15,738.45	26	16.87	15,610.90	20
2	35,416.32	20,198.21	63	12.50	19,867.02	26
3	30,586.42	16,088.56	60	13.89	15,315.70	24
4	9,042.65	21,367.57	63	27.87	20,479.10	30
5	34,402.08	23,437.93	30	12.41	23,370.20	22

Table 5 Local duality gap.

instance		CPU time	LB	obj. value	gap (%)
Small	1	23.14	18210.7	18384.8	0.96
	2	17.23	15512.6	15667.7	1.00
	3	14.87	17804.5	17965.1	0.90
	4	20.97	17976.5	18138.1	0.90
	5	17.02	18141.2	18241.6	0.55
USA	1	2166.83	552141	567015	2.69
	2	1980.89	546915	557038	1.85
	3	2176.14	546701	560018	2.43
	4	1964.95	552601	560932	1.51
	5	2563.71	552281	561504	1.59

Table 6 Robustness test.

	c_{ij}	0.45	0.50	0,55	0.60
Small	1	18414.8	18401.4	18384.8	18384.8
	2	15677.5	15667.7	15667.7	15667.7
	3	17955.3	17965.1*	17947.3	17906.5
	4	18333.3	18111.4	18138.1*	18135.0
	5	18307.7	18241.6	18234.5	18232.0
	c_{ij}	0.95	1.00	1.10	1.20
LATA	1	66925.7	66851.3	66365.7	65091.6
	2	66947.8	66650.6	66497.8	66410.6
	3	67575.3	67488.9	67415.2	67486.1*
	4	67061.7	67061.3	66807.7	66745.6
	5	65346.5	65322.3	65029.4	64673.0
	c_{ij}	5.0	5.5	6.0	6.5
USA	1	573757	567339	567015	567158*
	2	565634	561036	557038	558746*
	3	564672	562387	560018	561297*
	4	567956	567189	564031	560932
	5	572228	565582	563825	561504

tional physical links. Among them, 6 instances failed the test: they have larger costs than the instances with smaller solution sets. So 87% of tested instances passed the robustness test. (See Table 6.)

Validity test

Finally, we need to see if the model correctly pursues the trade-off between the gains discussed in Sect. 2.6. We consider a base instance of which the unit VC switching costs Ω_k and unit capacity costs o_{ij} are all set to 3. Then we generate two groups of instances from the base instance. In the first group, with all the unit capacity costs fixed to 3, we consider 5 instances whose unit switching costs are to the same values 1, 2, 3, 4 and 5. In the 5 instances of the second group, all the unit switching costs are fixed and the corresponding ratios are 3:1, 3:2, 3:3, 3:4 and 3:5. (Note that the base instance is common in both groups.)

In the first group, as the ratio of unit switching cost to the fixed unit capacity cost gets larger, the number of deployed VPCs is expected to increase to offset the expensive switching cost. The number of deployed VPCs plotted in Fig. 7 shows that this is indeed the case. Figure 8 plots the resulting trade-off effects in total cost: the percentage of the switching cost to the total cost are successfully restrained despite the increase in the unit switching cost.

For the second instance group, the symmetric observation can be made in terms of the switching and capacity costs. When the capacity cost gets expensive,

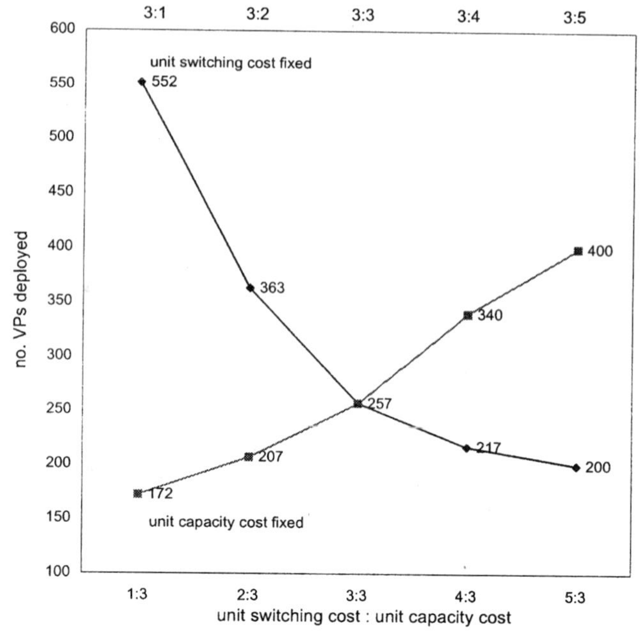

Fig. 7 Trade-off effects in VPC deployment.

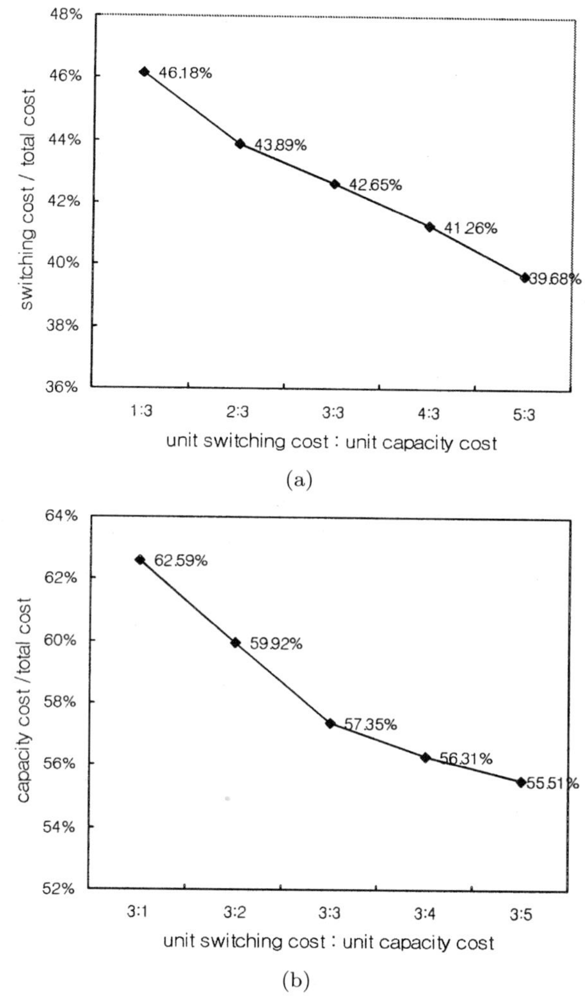

Fig. 8 Trade-off effects in total cost.

the model maintains a minimal total cost restraining the percentage of capacity cost by multiplexing the offered traffics in a smaller number of VPCs to fully utilize statistical gains.

5. Conclusion

We have proposed a VP management model for ATM networks. The model pursues an optimal resource utilization unifying the VPC overlay network configuration, VCC routing, and capacity allocation in a single framework. It accommodates multiple traffic types and QoS classes. Especially, it supports point-to-multipoint as well as point-to-point connections, with Steiner tree virtual channel connection.

Also, we have developed an algorithm to solve the large-scale nonlinear mathematical programming formulation of the model. Tested on the small to large size networks, the algorithm, tailored to the special structure of the problem, finds good solutions in a robust manner and dramatically reduced computation time.

Finally experimental results show that the model finds an optimal VP management which trades off between the conflicting gains of ATM technology.

Acknowledgement

The authors would like to thank two anonymous referees for their helpful comments.

References

[1] R.K. Ahuja, T.L. Magnanti, and J.B. Orlin, Network Flows: Theory, Algorithms, and Applications, Prentice-Hall, New Jersey, 1993.

[2] S.H. Ahn, R.P. Tsang, S.R. Tong, and H.C. Du, "Virtual path layout design on ATM networks," Proc. IEEE INFOCOM'94, pp.192–200, 1994.

[3] M.H. Ammar, S.Y. Cheung, and C.M. Scoglio, "Routing multipoint conections using virtual paths in an ATM network," Proc. IEEE INFOCOM'93, pp.98–105, 1993.

[4] N. Anerousis and A.A. Lazar, "Virtual path control for ATM networks with call level quality of service guarantees," IEEE/ACM Trans. Networking, vol.6, no.2, pp.222–236, 1998.

[5] D.O. Awduche, L. Berger, D.-H. Gan, T. Li, V. Srinivasan, and G. Swallow, "RSVP-TE: Extensions to RSVP for LSP Tunnels," draft-ietf-mpls-framework-07.txt, Aug. 2000.

[6] F. Bauer, Multicast Routing in Point-to-Point Networks under Constraints, Ph.D. Dissertation, Department of Computer Engineering, University of California, Santa Cruz, 1996.

[7] M.S. Bazaraa, H.D. Sherali, and C.M. Shetty, Nonlinear Programming: Theory and Algorithms, John Wiley & Sons, New York, 1993.

[8] D. Black, "Differentiated services and tunnels," draft-ietf-diffserv-tunnels-02.txt, July 2000.

[9] J. Burgin and D. Dorman, "Broadband ISDN resource management: The role of virtual paths," IEEE Commun. Mag., pp.44–48, Sept. 1991.

[10] R. Callon, G. Swallow, N. Feldman, A. Viswanathan, P. Doolan, and A. Fredette, "A framework for multiprotocol label switching," draft-ietf-mpls-framework-05.txt, Sept. 1999.

[11] J.H.S. Chan and D.H.K. Tsang, "Bandwidth allocation of multiple QOS classes in ATM environment," Proc. IEEE INFOCOM'94, pp.360–367, 1994.

[12] K.T. Cheng and F.Y.S. Lin, "On the joint virtual path assignment and virtual circuit routing problems in ATM networks," Proc. IEEE GLOBECOM'94, pp.777–782, 1994.

[13] K.T. Cheng and F.Y.S. Lin, "Minimax end-to-end delay routing and capacity assignment for virtual circuit networks," Proc. IEEE GLOBECOM'95, pp.2134–2138, 1995.

[14] I. Chlamtac, A. Fargo, and T. Zhang, "Optimizing the sytems of virtual paths," IEEE/ACM Trans. Networking, vol.2, pp.581–587, 1994.

[15] S.J. Chung, S.P. Hong, and H.G. Song, "Towards real-time optimal VP management for ATM nets," Telecommunications Systems, vol.14, pp.219–242, 2000.

[16] Z. Dziong, ATM Network Resource Mangement, McGraw-Hill, New York, 1997.

[17] M.L. Fisher, "The Lagrangian relaxation method for solving integer programming problems," Management Science, vol.27, pp.1–18, 1981.

[18] B. Gavish and I. Neuman, "A system for routing capacity assignment in computer communication networks flows," IEEE Trans. Commun., vol.37, pp.360–366, 1989.

[19] M. Gerla, J.A.S. Monteiro, and R. Pazos, "Topological design and bandwidth allocation in ATM nets," IEEE J. Sel. Areas Commun., vol.7, pp.1253–1262, 1989.

[20] G. Gopal, C.K. Kim, and A. Weinrib, "Algorithms for reconfigurable networks," Proc. ITC-13, pp.341–347, 1991.

[21] R. Guérin, H. Ahmadi, and M. Naghshshineh, "Equivalent capacity and its application to bandwidth allocation in high-speed networks," IEEE J. Sel. Areas Commun., vol.9, pp.968–981, 1991.

[22] R. Horst and H. Tuy, Global Optimization: Deterministic Approaches, Springer-Verlag, Berlin Heidelberg 1990.

[23] I. Hsu and J. Walrand, "Admission control for ATM networks," Proc. IMA Workshop on Stochastic Networks, 1994.

[24] I. Hsu and J. Walrand, "Admission control for multi-class ATM traffic with overflow constraints," Technical Report, Dept. of EECS, U.C. Berkeley, 1996, Comp. Networks and ISDN Sys. J. for the special issue on high speed networks and applications, vol.28, pp.1739–1751, 1996.

[25] J.Y. Hui, M.B. Gursoy, N. Moayeri, and R.D. Yates, "A layered broadband switching archtecture with physical or virtual path configurations," IEEE J. Sel. Areas Commun., vol.9, pp.1416–1426, 1991.

[26] F.K. Hwang, D.S. Richards, and P. Winter, The Steiner Tree Problem, Annals of Discrete Mathematics 53, Elsevier Science Publishers B.V., Amsterdam, 1992.

[27] ITU-T, B-ISDN general network aspects I.311, Geneva, 1996.

[28] ITU-T, Internet Protocol Communication Service -IP Performance and Availability Objectives and Allocations, Y.1541(Draft), Geneva, 2000.

[29] ITU-T, Traffic Control and Congestion Control in IP based networks, Y.iptc(Draft), Geneva, 2000.

[30] S.B. Kim, "An optimal VP-based multicast routing in ATM networks," Proc. IEEE INFOCOM'96, pp.1302–1309, 1996.

[31] S.O. Larsson and A. Arvidsson, "Performance evaluation of a local approach for VPC capacity management," IEICE Trans. Commun., vol.E81-B, no.5, pp.870–876, May 1998.

[32] M.J. Lee and J.R. Yee, "A design algorithm for reconfigurable ATM networks," Proc. IEEE INFOCOM'93, pp.144–151, 1993.

[33] T. Li, "MPLS and the evolving Internet architecture,"

IEEE Commun. Mag., pp.38–41, Dec. 1999.

[34] M. Logothetis, S. Shioda, and G. Kokkinakis, "Optimal virtual path management assuring network reliablity," Proc. IEEE ICC'93, pp.30–36, 1993.

[35] S. Mitra and D. Ghosh, "Configuring express pipes in emerging telecommuication networks," Telecommunications System, vol.1, pp.339–363, 1993.

[36] S. Ohta and K. Sato, "Dynamic bandwidth control of the virtual path in an asynchronous transfer mode network," IEEE Trans. Commun., vol.40, pp.1239–1247, 1992.

[37] B.H. Ryu, H. Ohsaki, M. Murata, and H. Miyahara, "Design algorithm for virtual path based ATM networks," IEICE Trans. Commun., vol.E79-B, no.2, pp.97–107, Feb. 1996.

[38] C. Semeria, Multiprotocol label switching: Enhancing routing in the new public network, Juniper networks white paper, 1999. (www.juniper.net)

[39] C. Semeria, Traffic engineering for the new public network, Juniper networks white paper, 1999. (www.juniper.net)

[40] M.E. Streenstrup, Routing in communications networks, Prentice Hall, Englewoodcliffs, New Jersey, 1995.

[41] G. Swallow, "MPLS advantages for traffic engineering," IEEE Commun. Mag., pp.54–57, Dec. 1999.

Appendix A: Notation

The notation is summarized in the order of appearance.

$G = (N, A)$: the physical backbone network;

c_{ij} : the capacity of the physical link $(i, j) \in A$;

B_i : the buffer capacity of the node $i \in N$;

Q : the number of possible QoS classes;

$\epsilon(t)$: the maximum cell loss ratio of the tth QoS class, here we assume that $\epsilon(1) < \epsilon(2) < \cdots < \epsilon(Q)$;

L : the number of possible traffic types;

R_j : the peak bit rate of the jth traffic type;

m_j : the mean bit rate of the jth traffic type;

b_j : the average duration of burst state of the jth traffic type;

ρ_j : the utilization, m_j/R_j of the jth traffic type;

$\bar{G} = (\bar{N}, \bar{A})$: the logical network i.e., $\bar{N}$ is the set of VC switching nodes and $\bar{A}$ is the index set of VPC candidates;

P_k : the kth VPC candidate, where $k \in \bar{A}$;

q_k : the QoS class of the kth VPC, i.e. $q_k \in \{1, \cdots, Q\}$;

$\Gamma^-(i; t)$: the set of VPCs, P_k in $\bar{A}$ terminating at the node $i \in \bar{N}$ with $q_k \leq t$ i.e., the QoS value $\epsilon(q_k)$ supported by P_k is more stringent than or equal to $\epsilon(t)$;

$\Gamma^+(i; t)$: the set of VPCs, P_k in $\bar{A}$ originating from the node $i \in \bar{N}$ with $q_k \leq t$;

$\Gamma^+(i)$: the set of VPCs in $\bar{A}$ originating from the node $i \in \bar{N}$, i.e. $\Gamma^+(i) = \bigcup_{t=1}^{Q} \Gamma^+(i; t)$;

$r(s, d; j, t)$: the number of VCCs of the jth traffic type and tth QoS class required from node s to node d ($j = 1, \cdots, L$, $s, d \in \bar{N}$, $t = 1, \cdots, Q$);

$f_k(s, d; j, t)$: the number of VCCs of the j traffic type and tth QoS class assigned to P_k for the requirement from node s to node d ($k \in \bar{A}$, $j=1, \cdots, L$, $s, d \in \bar{N}$, $t=1, \cdots, Q$);

$\mathcal{D}$: the set of multicast groups;

$a(s, D; j, t)$: the number of VCCs of the jth traffic type and tth QoS class required by a multicast group (s, D) ($j = 1, \cdots, L$, $(s, D) \in \mathcal{D}$, $t = 1, \cdots, Q$);

$z_k(s, D; j, t)$: the 0-1 variable indicating whether the kth VPC is used for routing the VCCs of the jth traffic type and tth QoS class required by a multicast group (s, D) ($k \in \bar{A}$, $j = 1, \cdots, L$, $(s, D) \in \mathcal{D}$, $t = 1, \cdots, Q$);

$h_k(s, D; d, j, t)$: the auxiliary variable telling whether the kth VPC is used to route the VCCs from node s to node $d (\in D)$ for a multicast group (s, D) of the jth traffic type and tth QoS class ($k \in \bar{A}$, $j = 1, \cdots, L$, $(s, D) \in \mathcal{D}$, $t = 1, \cdots, Q$);

y_k^j : the total number of the jth type VCCs assigned to P_k, i.e. $y_k^j = \sum_{s,d \in \bar{N}} \sum_{t=1}^{Q} f_k(s, d; j, t) + \sum_{(s,D) \in \mathcal{D}} \sum_{t=1}^{Q} z_k(s, D; j, t) a(s, D; j, t)$ ($k \in \bar{A}$, $j = 1, \cdots, L$);

$\boldsymbol{y}_k$: the vector of $y_k^j (j = 1, \cdots, L)$, i.e. $\boldsymbol{y}_k = (y_k^1, \cdots, y_k^L)$ ($k \in \bar{A}$), and we write $\boldsymbol{y} = (\boldsymbol{y}_1, \cdots, \boldsymbol{y}_{|\bar{A}|})$;

x_k: the buffer capacity assigned to P_k, $k \in \bar{A}$, and we write $\boldsymbol{x} = (x_1, \cdots, x_{|\bar{A}|})$;

$P(i, j)$: the set of VPCs that use the physical link (i, j), $P(i, j) \subset \bar{A}$;

$A(P_k)$: the set of physical links that the kth VPC, P_k uses, $A(P_k) \subset A$;

$\hat{c}(\boldsymbol{y}_k, x_k)$: the equivalent capacity required at P_k which uses buffer capacity of x_k, supports the q_kth QoS class and accommodate y_k^j VCCs of the jth service type ($j = 1, \cdots, L$);

$o_{ij}[u]$: the opportunity cost incurred at physical link (i, j) when the reserved capacity level is u;

$\zeta_k[v]$: the opportunity cost of buffer capacity v reserved for P_k;

$w_k[u]$: the total switching/control costs incurred at P_k when the (multiplexed) traffic level is u;

$\Omega_k[u]$: the VC switching/control cost incurred at the terminating nodes of P_k when the switched traffic level is u;

$\omega_k[u]$: the VP switching/control cost incurred at each intermediate node of P_k when the switched traffic level is u;

$\theta(\boldsymbol{y}, x)$: the total cost incurred from accommodation of VCCs $\boldsymbol{y}$ and buffering x. It is given by $\sum_{k \in \bar{A}} \Phi_k \hat{c}(\boldsymbol{y}_k, x_k) + \sum_{k \in \bar{A}} \zeta_k x_k$;

Appendix B: Equivalent Capacity Formula

The Gaussian approximation function $c_g(y_k)$ is given by

$c_g(y_k)$

$$= \sum_{j=1}^{L} y_k^j m_j + \delta_k \sqrt{\sum_{j=1}^{L} y_k^j m_j (R_j - m_j)}, \quad (A\cdot 1)$$

where, $\delta_k = \sqrt{-2\log(\epsilon(q_k)) - \log(2\pi)}$.

Also, the fluid flow approximation function, $c_f(y_k, x_k)$ is given by

$$c_f(y_k, x_k) = \sum_{j=1}^{L} \gamma(y_k^j, x_k) \quad (A\cdot 2)$$

where,

$$\gamma(y_k^j, x_k) = y_k^j R_j \frac{\beta_k^j - x_k + \sqrt{(\beta_k^j - x_k)^2 + 4x_k\rho_j\beta_k^j}}{2\beta_k^j}. \quad (A\cdot 3)$$

and $\rho_j = \frac{m_j}{R_j}$, $\beta_k^j = \alpha_k b_j (1-\rho_j) R_j$, $\alpha_k = \log(\frac{1}{\epsilon(q_k)})$.

Note that $\gamma(y_k^j, x_k)$ is a decreasing convex function in terms of x_k. Hence the marginal effect of x_k to $\gamma(y_k^j, x_k)$, tends to be negligible as x_k exceeds a certain threshold value. This implies that we may restrict the range of buffer capacity *a priori* in using fluid flow approximation function.

There is a more fundamental reason to restrict the buffer capacity than this technical difficulty caused by this specific choice of equivalent capacity formula. As the intended services of ATM networks include the real-time traffic which cannot tolerate an extensive buffering. Thus it is indispensable to limit the level of buffering to some upper bound.

Remark : Some authors [24] even propose a call admission policy in which the real-time traffic is assigned no buffer (but, allocated higher service priority to the bandwidth instead). The rationale of this policy is that the small buffers to the real-time traffic can only negligible improvement in the loss rate [23]. Thus real-time traffic relies on the statistical multiplexing gain rather than extensive buffering.

There seems to be no universal solution to this problem as it depends on specific VP management policies. We suggest a bound which seems to be fairly independent of the specific policies and furthermore can be easily calculated from the traffic characteristic parameters offered by the users: namely, $x_k \leq \beta_k^j = \alpha_k b_j (1-\rho_j) R_j$.

The rationale is that β_k^j is much larger than (α_k times) the total amount of traffic arrived at the peak rate minus the mean bit rate during a burst period. That is, β_k^j is larger than the amount of traffic which possibly needs smoothing. In this sense, we may consider that β_k^j is a sufficient upper bound on x_k.

Once we choose the range of buffer capacity as above, for fixed y_k^j, $\gamma(y_k^j, x_k)$ should be very close to linear function of x_k since in Eq. (A·3) the inside of square root is dominated by the square term when x_k

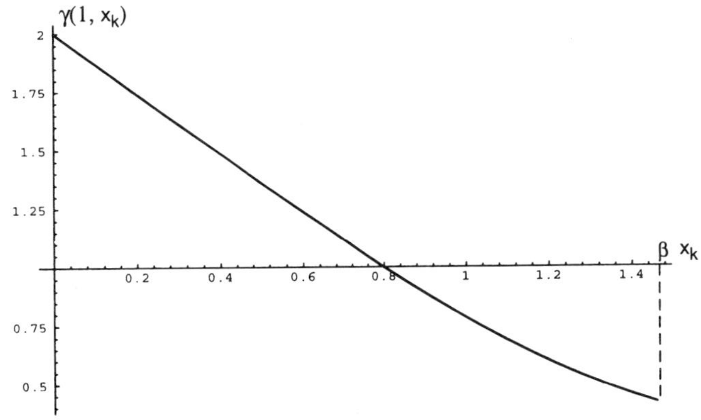

Fig. A·1 $\gamma(y_k^j, x_k)$ of Still image with $y_k^j = 1$.

is small. For example, in Fig. A·1, $\gamma(1, x_k)$ is plotted using the still image characteristic parameters given in Table 2.

We will simplify the function $\gamma(1, x_k)$ to the linear function, $\hat{\gamma}(1, x_k)$ which intersects $\gamma(y_k^j, x_k)$ at $x_k = 0$ and $x_k = \beta_k^j$:

$$\hat{\gamma}(y_k^j, x_k) = y_k^j R_j (1 - \frac{1-\sqrt{\rho_j}}{\beta_k^j} x_k). \quad (A\cdot 4)$$

Then the simplified fluid flow approximation becomes,

$$\hat{c}_f(y_k, x_k) = \sum_{j=1}^{L} \hat{\gamma}(y_k^j, x_k). \quad (A\cdot 5)$$

Of course, this simplification may sacrifice the accuracy of the model. Nevertheless, it should be noticed that the resulting solution is still valid for the original problem, since $\hat{c}_f(y_k, x_k)$ is greater than the convex function $c_f(y_k, x_k)$.

A more important thing is, as shown in the experimental results, that the "quasi-linearity" of $\gamma(1, x_k)$ and the facility for more efficient solution technique in sum produce a solution of higher quality in significantly reduced computational effort than the nonlinear methods applied to the original nonlinear problem. In this sense, the simplification can be justified.

We will denote the fluid flow equivalent capacity function with linear approximation by $\hat{c}_f$ instead of c_f. Then the modified equivalent capacity function can be denoted as follows:

$$\hat{c}(y_k, x_k) = Min(c_g(y_k), \hat{c}_f(y_k, x_k)) \quad (A\cdot 6)$$

Remark : Notice that a similar linear approximation can be constructed for other choice of the bound on x_k.

Appendix C: Integer Program Formulation of Steiner Tree Problem

Suppose we are given a network with 5 nodes, 1, 2, 3, 4

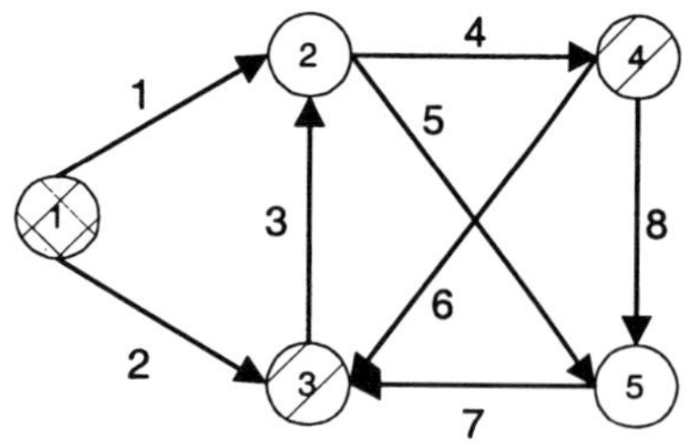

Fig. A· 2 Multicast group with origin $s = 1$ and destinations $D = \{3, 4\}$.

and 5, and the arcs indexed $k = 1, 2, 3, 4, 5, 6,$ and 7 as in Fig. A· 2. We want to configure point-to-multipoint connection as a Steiner tree for the multicast group, with origin, $s = 1$ and destinations, $D = \{3, 4\}$.

Now we give an integer formulation of this problem.

$$z_k = 0, \text{ or } 1 \quad \text{for } k = 1, \cdots 7 \tag{A· 7}$$

$$z_k \geq h_k^3 \; z_k \geq h_k^4 \quad \text{for } k = 1, \cdots 7 \tag{A· 8}$$

$$\sum_{k \in \Gamma^+(i)} h_k^3 - \sum_{k \in \Gamma^-(i)} h_k^3 = \begin{cases} 1 & \text{if } i = 1 \\ -1 & \text{if } i = 3 \\ 0 & \text{otherwise} \end{cases} \tag{A· 9}$$

$$\sum_{k \in \Gamma^+(i)} h_k^4 - \sum_{k \in \Gamma^-(i)} h_k^4 = \begin{cases} 1 & \text{if } i = 1 \\ -1 & \text{if } i = 4 \\ 0 & \text{otherwise} \end{cases} \tag{A· 10}$$

$$h_k^3 \geq 0, \; h_k^4 \geq 0 \quad \text{for } k = 1, \cdots 7$$

where, $\Gamma^+(i)$ ($\Gamma^-(i)$, resp.) are the arcs originating from (arriving at) the node i.

z_k are the binary variable telling whether the kth arc (a VPC, P_k in our model) is used in the multicast or not: $z_k = 1$ if kth arc is in the tree, $z_k = 0$, otherwise (Eq. (A· 7)). h_k^3 and h_k^4 are auxiliary variables which enforce that the subgraph induced by arcs with $z_k = 1$ is indeed a Steiner tree spanning the source s and all the destinations in D. Equation (A· 9) send a unit flow from the origin 1 to the destination 3. Equation (A· 10) do the same thing from the origin 1 to the destination 4. Equation (A· 8) enforce that kth arc is included in the connection if and only if the arc is used in sending flow. Thus the configured tree will span all the destinations as desired.

High-Performance Backbone Network Technology edited by Naoaki Yamanaka
ISBN: 0-8247-5321-6 DOI: 10.1081/(120025667)
Published 2004 by Marcel Dekker, Inc.

A Failure-Resistant ATM Transport Method for the Transmission of High-Quality Real-Time Signals

Yukiharu Kanayama, Kazuhiro Fujihara, and Toshinori Tsuboi

SUMMARY This paper proposes a new failure-resistant ATM transport method for real-time applications (FRAT-RA). This method can prevent not only cell loss but also cell delay variation (CDV) even when a network failure occurs. First, a general overview of the existing fault-resistant method is briefly described and its problems for real-time applications are discussed. Next, the proposed FRAT-RA method is described in detail. Finally, evaluation results on the CDV of FRAT-RA and its QoS improvement are presented. The results show that failure-resistant and high-quality transmission of real-time signals can be achieved with FRAT-RA.

key words: *hitless switching, ATM, CDV, video transmission*

1. Introduction

The ATM network is a promising platform for providing a wide variety of services for multimedia applications. Increasing network reliability and providing several reliability grades of services are becoming key issues. For increasing network and service reliability, many kinds of restoration and self-healing techniques have been intensively studied [1], [2]. There are high-end applications that require extremely highly-reliable services, for example TV program transmission from a key TV station to local stations and tele-medicine. In these applications, service interruptions due to network failures or maintenance activities cause significant economic and social turmoil, even if the interruption period is very short. The failure-resistant virtual path (FRVP) method prevents information loss when network failures occur [3]. However, this method is not applicable to real-time applications such as TV program transmission between TV stations. This is because the method produces large cell delay variations (CDVs) when a failure or cell loss occurs in a network, and the large CDVs may cause cell buffer overflow or underflow in the user's ATM systems. This paper proposes a new failure-resistant ATM transport method for real-time applications (FRAT-RA) that prevents the loss of any information without changing the CDV when a network failure occurs.

High-end video transmission applications may also require the highest quality of service (e.g., cell loss ratio (CLR) and bit error ratio (BER)) as well as excellent reliability. For example, the HDTV signal distribution requires a quality level better than 1×10^{-9} (CLR) and better than 3×10^{-12} (BER) [4]. Some networks may not achieve such levels from their 'basic' ATM layer cell transfer functions. The FRAT-RA method can also greatly improve network CLR and BER. This eliminates the requirement that the user system must have a forward error correction (FEC) function in the ATM Adaptation layer (AAL) or upper layers.

The FRAT-RA method is somewhat similar to the existing FRVP method, however they are quite different in how they handle CDV. In the following sections, first an overview of the existing FRVP method is briefly described and its problem when applied to real-time signal transmission is discussed. Next, the proposed FRAT-RA method is described in detail. The CDV of FRAT-RA is then discussed and evaluation results are presented that prove that it can hold the CDV under a specified value. Finally, the CLR and BER improvements offered by FRAT-RA are evaluated.

2. Existing FRVP Method and Its Problems

2.1 Overview of the FRVP Method

The FRVP method comprises a transmitter, a receiver, and two separate VPs linking them [3]. Figure 1 shows the basic configuration of FRVP. At the transmitter, incoming user cells from the user system are duplicated and then transmitted to the separate VPs. Each VP is accommodated on a physically different route to avoid encountering simultaneous network failures. At the receiver, only one correct cell is selected from the two corresponding cells received from each VP and that cell is transmitted to the user system. Using this method, user cells can be transmitted hitlessly across the network unless both of the corresponding cells are lost or have an error.

In the FRVP method, an index cell is defined in order to select the correct cell from the corresponding cells of each VP at the receiver. This index cell has information for detecting errors and identifying corresponding cells. The index cell is generated and inserted into the user cell stream at the trans-

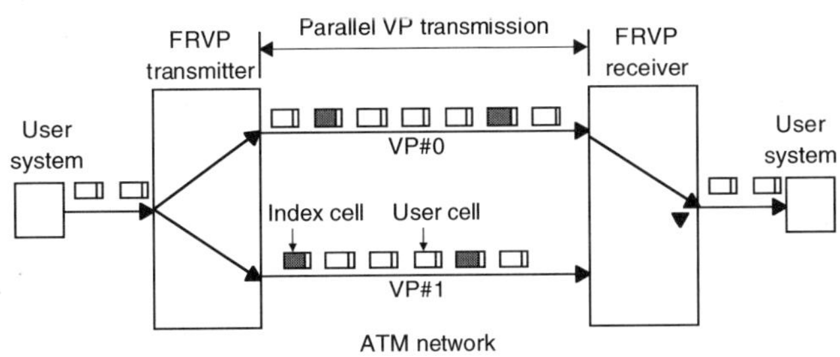

Fig. 1 Overview of FRVP method.

Manuscript received May 21, 1999.
Manuscript revised September 7, 1999.
† The authors are with the NTT Network Service Systems Laboratories, Yokosuka-shi, 239-0847 Japan.

IEICE TRANS. COMMUN., Vol. E83-B, No. 2, pp. 293–298, February 2000

mitter either when the number of user cells after the previous index cell reaches the maximum frame size (N_f cells), or when the time interval after the previous index cell reaches the time limit (T_i seconds). The unit, which consists of an index cell and a number of user cells transmitted before the index cell, is called a cell frame. At the receiver, error detection and identification are performed for each cell frame. The structure of the cell frame and the index cell format are shown in Fig. 2. The 'Frame sequence number' in the payload of the index cell is the sequence number for each cell frame. The receiver identifies the corresponding cell frames with this number. Index cell loss can be also detected from a sequence jump of this number. The 'Number of user cells' indicates the number of user cells in the cell frame. The receiver can detect cell loss or mis-insertion by comparing this number with the number of cells actually received. The 'Frame error detection code' is the CRC code for all payload bits of user cells in the cell frame and the 'Index cell error detection code' is also the CRC code for the payload bits of the index cell. The receiver detects the cell frame error or the index cell error using these CRC codes.

2.2 Problem of the FRVP Method

The FRVP receiver selects a correct frame and transmits it to the user system as soon as the first correct frame arrives. The receiver does not wait until the arrival of the other corresponding frames unless the first frame has an error. This can cause significant changes in the transmission delay time when the difference between the two VP lengths is large. This situation is illustrated in Fig. 3. The figure shows the situation where the fourth and the following frames in VP#0 (shorter delay VP) were lost due to network failure. The result is a long time gap that corresponds to the delay difference of the two VPs. In addition, the output rate from the receiver would increase when the failure of the shorter delay VP is restored [3]. The time gap change and the rate increase imply a large network CDV. This large network CDV may cause cell buffer overflow or underflow in the user's ATM system because the most user's ATM systems can accept network CDV up to around 1.5 msec [5]. The user system cannot absorb time gaps longer than 1.5 msec which corresponds to the difference in VP length of 300 km. Therefore, the existing FRVP method is not applicable to real-time applications.

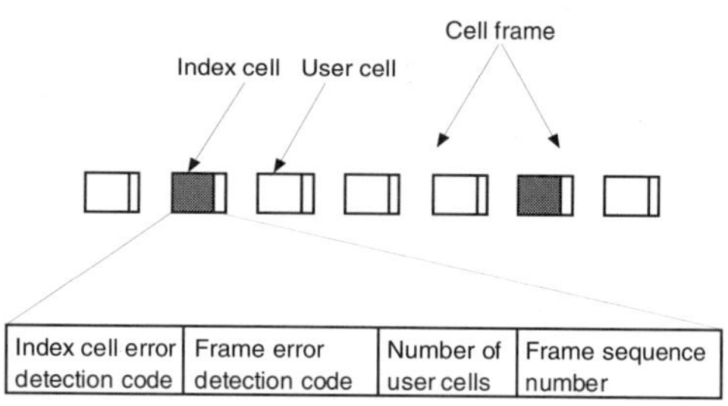

Fig. 2 Cell frame structure and index cell format.

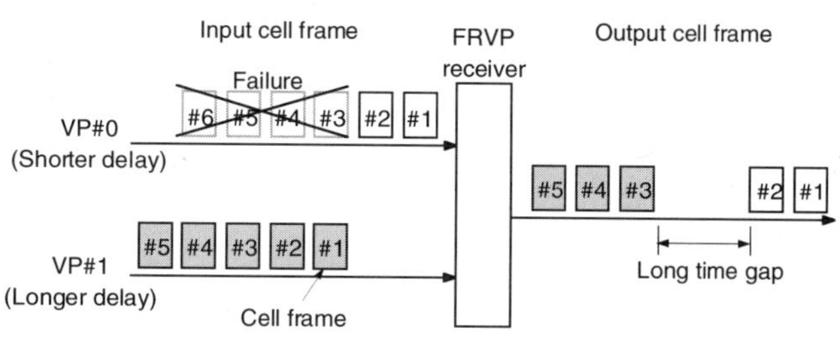

Fig. 3 Receiver output at failure.

3. Proposed FRAT-RA Method

3.1 Requirement for Transmitting Real-Time Signals

For transmitting real-time signals, a new fault-resistant method is required that can hold network CDV under a target value, however there is a large delay difference between two VPs. Those methods that simply select the first arriving frame and transmit it soon after can not prevent the time gap change and the rate increase from occurring. Therefore, the new method must select and transmit a frame at roughly the same time as the corresponding frame on the other VP arrives. The FRAT-RA method realizes this function by delaying frames from the shorter delay VP for a period corresponding to the delay difference between the two VPs.

3.2 FRAT-RA Algorithm

The basic configuration and transmitter function of the FRAT-RA method are similar to those of the FRVP method presented in Sect. 2.1. In the following, only the receiver function of FRAT-RA is described in detail. The block diagram of the FRAT-RA receiver is shown in Fig. 4. Functions and processes in each block are as follows:

(1) Failure detector

This block detects frame errors (bit error or cell loss in a frame) and VP failures of each VP. In the event of a index cell loss or a index cell error, it is detected as two consecutive frame errors when the next correct index cell is received. The VP failure is detected when no index cells are received for the time limit (T_i + *maximum network CDV* seconds) and its recovery is detected when a correct frame is received. When a frame error or a VP failure is detected, failure information is notified to other blocks. Each block then performs a proce-

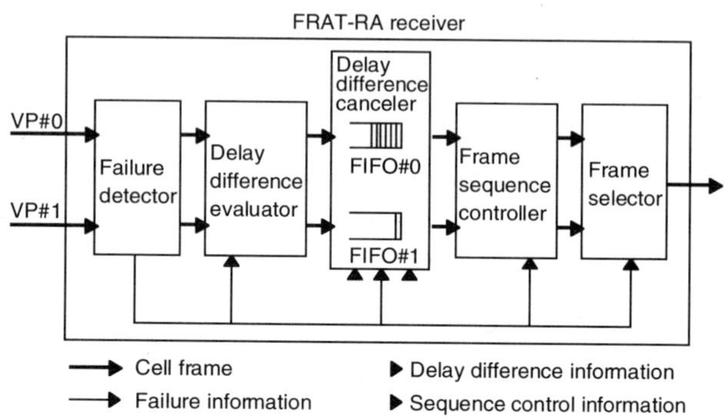

Fig. 4 Block diagram of FRAT-RA receiver.

dure determined by the content of the information.

(2) Delay difference evaluator

This block automatically evaluates the difference in transmission delay between the two VPs. In general, the transmission delay consists of a fixed part and a variable part. The fixed part mainly reflects the transmission delay of long transmission lines and the variable part is due to cell multiplexing in ATM nodes. The block evaluates only the fixed part because the delay variation of each VP can change quickly and independently of each other. For this purpose, the block measures the difference in the arrival times of two corresponding index cells and calculates the delay difference between the two VPs periodically by averaging N_a samples of the measured differences. The value for N_a is selected so that the delay difference can be evaluated precisely. Details will be discussed in Sect. 5.2. Samples from error frames are discarded to avoid mis-calculation. With this block, the delay difference can be automatically adjusted even if the difference is changed due to VP protection switching or a load increase in ATM nodes.

(3) Delay difference canceler

This block delays cells from the shorter delay VP for the period equivalent to the delay difference time calculated at the Delay difference evaluator. The block has FIFO buffers for each VP. Adjusting the delay time is achieved by controlling the readout timing of the buffers and discarding incoming idle cells. An example of delay adjustment is shown in Fig. 5. In this example, in-coming non-target VP cells (e.g., idle cells) are also written into the buffers with the target VP cells. If the delay difference increases compared to the previous value (Case a), the readout operation from the buffer of the shorter VP is stopped for the period corresponding to the delay increase. On the other hand, if the delay difference decreases (Case b), a number of in-coming idle cells corresponding to the delay decrease are discarded before being written into the buffer. When a VP failure is detected, delay adjustment is stopped to keep the delay difference unchanged.

(4) Frame sequence controller

As described above, the delay difference evaluator and canceler cancel only the fixed part of the delay difference; the variable part remains. Due to this delay variation, it is possible for the n-th and the n+1-th (and the following) frames of one VP to arrive from the Delay difference canceler before the n-th frame of the other VP. This situation is illustrated in Fig. 6. In this situation, if the first n-th frame of one VP has an error, the correct n+1-th frame of that VP arrives earlier than the correct n-th frame of the other VP. This causes inversion in the cell frame sequence or untransmitted frames. To avoid these problems, this block sends sequence control messages to the Delay difference canceler to control the readout timing of frames from the buffers when the first n-th frame has an error. In this control operation, readout of the n+1-th

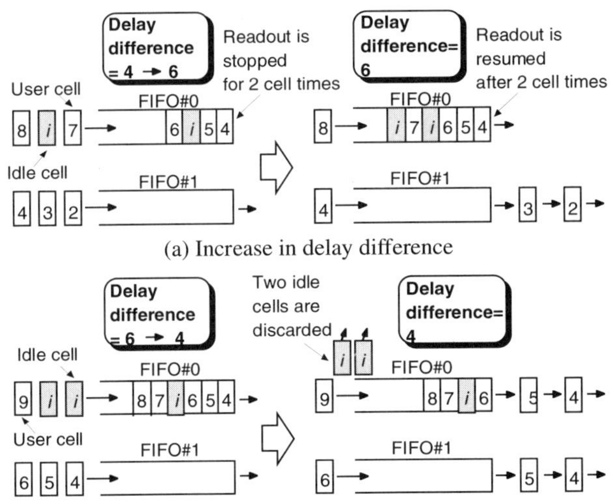

Fig. 5 Examples of delay difference adjustment.

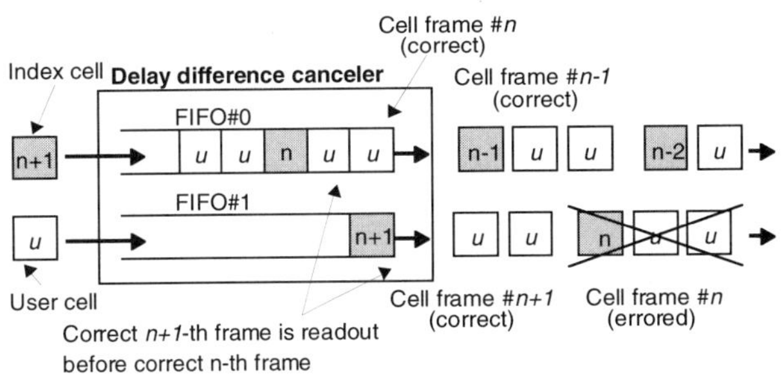

Fig. 6 Frame sequence disorder due to cell delay variation.

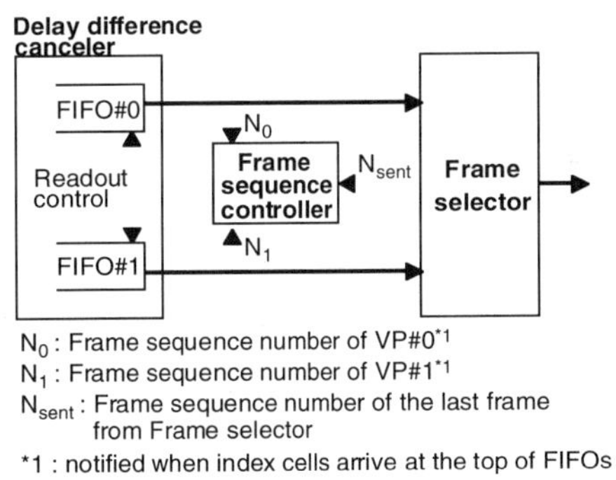

Fig. 7 Relation between Frame sequence controller and other blocks.

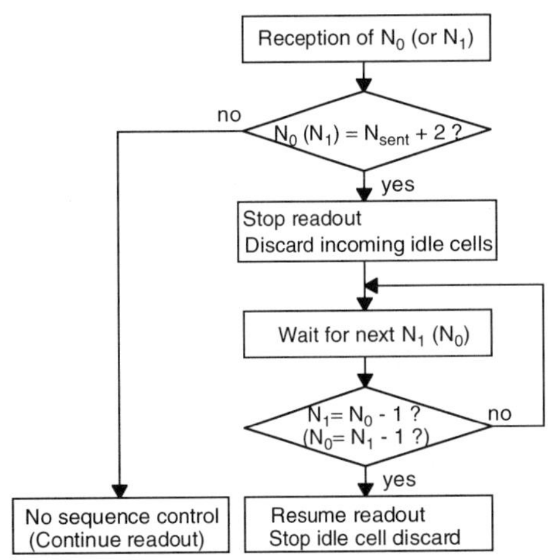

Fig. 8 Frame sequence control procedure.

frame of the same VP from the buffer is stopped until the n-th frame of the other VP is read out, and also a number of idle cells corresponding to the stopped periods are discarded. Details of this procedure is shown in Figs. 7 and 8. When a VP failure is detected, this sequence control is stopped and all cells from the other VP are transferred to the Frame selector without changing the sequence.

(5) Frame selector

This block selects the first-arriving correct frame and soon transmits all user cells in the frame to the user system. The subsequently arriving frame is discarded. In the event that the first-arriving frame has errors, the frame is discarded and the subsequently arriving frame is transmitted, whether or not it is correct, in order to minimize the number of cell losses. When a VP failure is detected, all frames of the other VP are selected and transmitted until the failure is recovered.

3.3 Usage with Other Restoration Methods

Generally FRAT-RA is used without other restoration methods and it alone achieves reliable transmission because the possibility that both of two separate VPs fails simultaneously is very small. It should be avoided to use FRAT-RA together with a restoration method that changes the VP length. It may change the length of longer VP in FRAT-RA. This change causes large CDV of FRAT-RA.

4. Algorithm Verification

To verify whether the proposed FRAT-RA method behaves as intended, a software simulator was constructed, and changes in the CDV of FRAT-RA after VP failure and upon a fall in CLR is present were examined.

4.1 Network Model

The network model used in this verification is shown in Fig. 9. It is assumed that the length of VP#0 is 60 km and that of VP#1 is 560 km (corresponding to a 100-cell time delay and 1000-cell time delay at 150 Mbps), and it is also assumed that these VPs exhibit uniform random CDVs with a maximum value of 30 cell times. VP rate is 30 Mbps and the cell frame size is 10 cells.

4.2 Behavior at VP Failure

Figure 10 shows the delay time changes of the FRAT-RA method, along with those of the FRVP method, where VP#0 (shorter delay VP) failed at 3000 cell times and recovered at 7000 cell times. The results clearly show that the delay time of FRAT-RA was unchanged before and after the failure and recovery. On the other hand, the delay time of the existing method changed greatly.

4.3 Behavior under Degraded CLR

Figure 11 shows the delay time distributions of FRAT-RA and FRVP under the condition that the CLR of both VPs is 1 × 10^{-5}. The delay time of FRAT-RA remained in a confined range at around 1000 cell times. The distribution is equivalent to that seen under no cell loss conditions (not shown in the figure). On the other hand, the delay time of FRVP varied in the range corresponding to the delay difference of the two VPs.

These results confirm that the CDV of the proposed FRAT-RA algorithm remains unchanged, as intended, under failure or degraded CLR conditions. In addition, no cell losses and no inversion in cell sequence were observed through these evaluations.

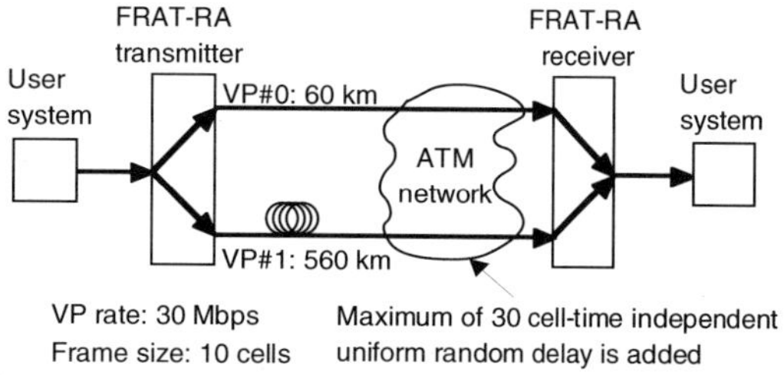

Fig. 9 Network model.

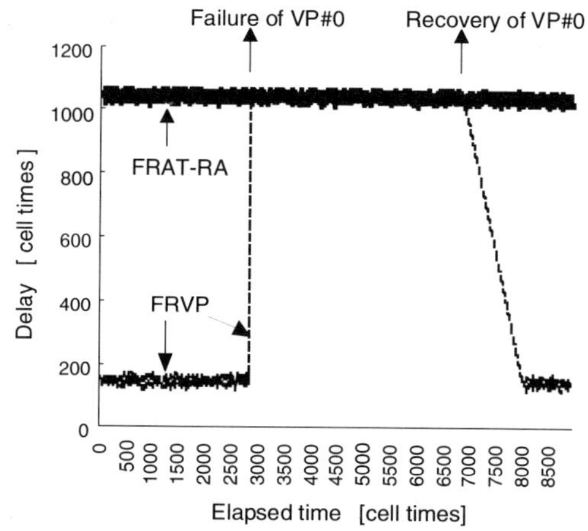

Fig. 10 Delay time changes at VP failure and recovery.

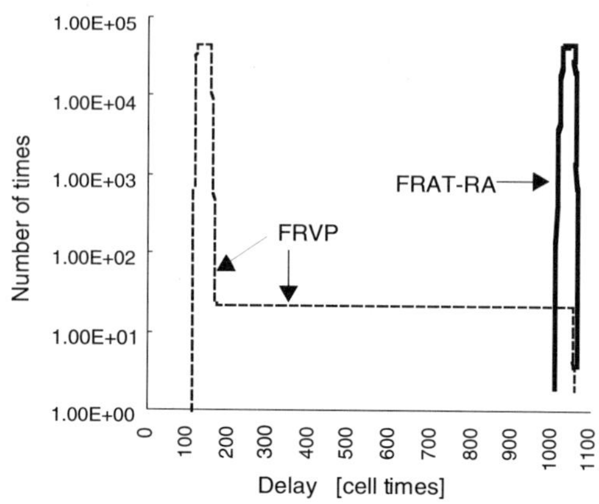

Fig. 11 Delay distribution under degraded cell loss ratio.

5. Consideration on the CDV of FRAT-RA

FRAT-RA must achieve the CDV target to transmit high-quality real-time signals. The CDV of FRAT-RA is influenced by parameters, N_a, the number of samples averaged to evaluate delay difference of VPs; and N_f, the frame size. In addition, the CDV is also influenced by the CDV of the network. The influence of these parameters on the CDV of FRAT-RA is shown below.

5.1 Influence of Parameter N_a

As described in Sect. 3.2, the delay time comprises a fixed part and a variable part. When N_a is not large enough, the calculated delay difference suffers from the impact of the variable part and this may increase the CDV of FRAT-RA. To estimate this influence, the delay time distribution on the same network model described in Sect. 4.1 was evaluated under the condition that N_a was changed from 1 to 100 samples. The results are shown in Fig. 12. The results indicate that the delay distribution is mostly unchanged if N_a is larger than 10. Consequently, N_a should be larger than 10 when FRAT-RA is used in a network whose delay variation matches that of the network model described.

5.2 Influence of Frame Size N_f

The output from the FRAT-RA receiver basically forms a continuous cell burst whose length equals the frame size because the receiver selects and transmits on a frame-by-frame basis. Consequently, the maximum CDV caused by the cell framing $CDV_{max[framing]}$ can be expressed as follows, where N_c denotes the cell interval time of the VP. Therefore, $CDV_{max[framing]}$ can be determined by selecting the frame size.

$$CDV_{max[framing]} = N_c * (N_f - 1) \quad (1)$$

5.3 Influence of Network CDV

To estimate the influence of network CDV, the delay time distribution was evaluated using the maximum random CDV values in the network of 10, 30, and 50 cell times and N_a is 10 samples. The results are presented in Fig. 13. It shows that the delay distribution range widens as the network CDV increases.

The maximum CDV values of FRAT-RA ($CDV_{max[FRAT\text{-}RA]}$), which are found from these results, are listed in Table 1 along with the maximum CDV values caused by framing ($CDV_{max[framing]}$) and the maximum CDV values of the network ($CDV_{max[network]}$). The values in the table show that the following relations hold among these values.

$$CDV_{max[FRAT\text{-}RA]} <= CDV_{max[framing]} + CDV_{max[network]} \quad (2)$$

The CDV of FRAT-RA can be regarded as consisting of the part due to the cell framing and the part due to network CDV, when N_a is sufficiently large. The part due to network

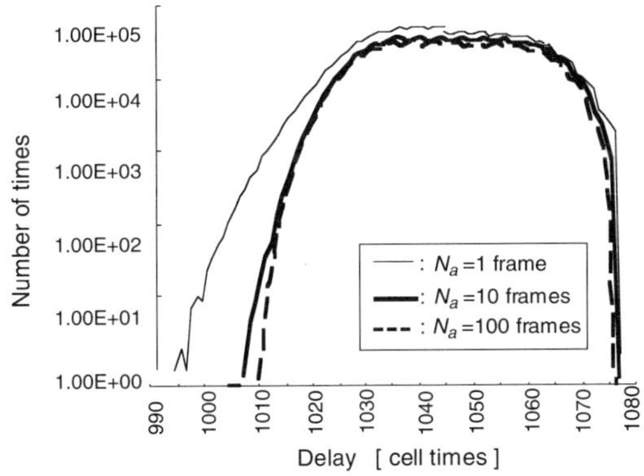

Fig. 12 Influence of N_a on CDV of FRAT-RA.

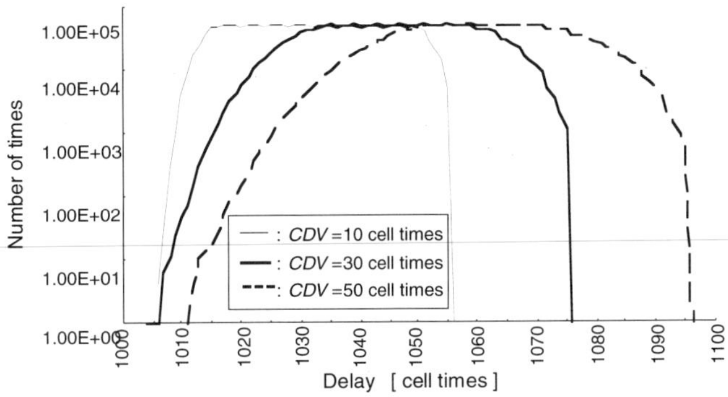

Fig. 13 Influence of network CDV on CDV of FRAT-RA.

Table 1 Evaluated CDV values.

$CDV_{max[network]}$	$CDV_{max[framing]}$	$CDV_{max[FRAT\text{-}RA]}$
10	45	52
30		71
50		87

CDV can be considered to be smaller than the network CDV itself, because selecting the first-arriving frames from two VPs must reduce the CDV. Therefore, relation (2) would hold.

Relation (2) proves that the upper bound value of the FRAT-RA's CDV is determined by $CDV_{max[framing]}$ and $CDV_{max[network]}$. In general, the maximum CDV value of the network is determined from the VP accommodation ratio on the transmission line, node buffer size, and the number of nodes in the VP [6]. Therefore, FRAT-RA can meet the target CDV value by selecting the proper N_f value using Eq. (1). For example, if the objective CDV value is 1.5 msec and the maximum network CDV is 1.0 msec, the maximum CDV due to framing must be less than 0.5 msec and this is realized by setting the frame size to under 36 cells.

6. Evaluation of QoS Improvement

FRAT-RA improves the CLR and BER of the network as well as achieving hitless ATM transport when a failure occurs. Using FRAT-RA, cell losses and bit errors will arise only when they arise simultaneously in both the first-arriving cell frame and the corresponding subsequently arriving cell frame.

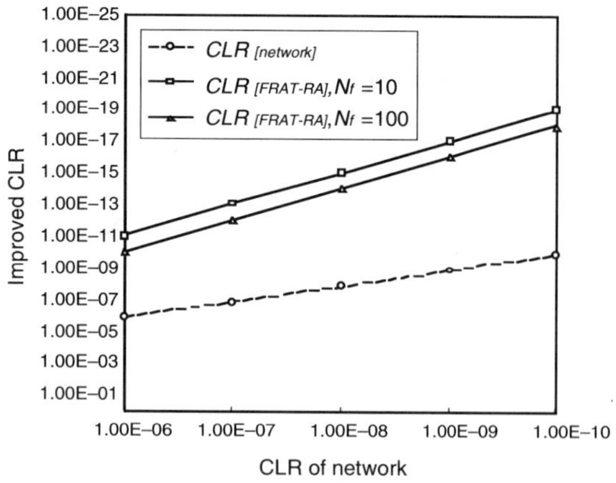

(a) Improvement in CLR

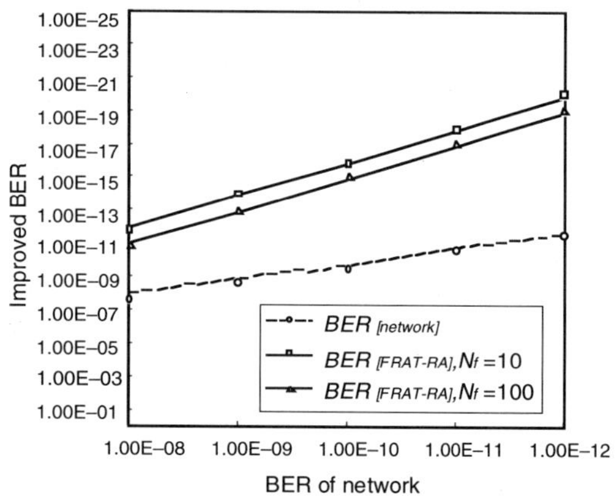

(b) Improvement in BER

Fig. 14 QoS improvement by FRAT-RA.

Therefore, the CLR and BER of FRAT-RA can be expressed as follows:

$$CLR_{[FRAT\text{-}RA]} = \{1 - (1 - CLR_{[network]})^{(Nf+1)}\} * CLR_{[network]}$$

$$BER_{[FRAT\text{-}RA]} = \{1 - (1 - BER_{[network]})^{53*8*(Nf+1)}\} * BER_{[network]}$$

Calculated results for the case of $N_f = 10$ and 100 are shown in Fig. 14. The results show that both CLR and BER are greatly improved by FRAT-RA. If a HDTV signal distribution application requires a CLR of better than 1×10^{-10}, a network CLR of around 1×10^{-6} is acceptable if we apply FRAT-RA.

7. Conclusions and Future Work

This paper proposed a new fault-resistant ATM transport method for high-quality real-time signals (FRAT-RA) that prevents the loss of any information without changing CDV when a network failure occurs. For transmitting real-time signals, the CDV of FRAT-RA was discussed in detail, and simulation results were introduced that proved that the CDV is very small and remains unchanged at network failure. The improvement in QoS offered by FRAT-RA was also evaluated.

At present, prototype FRAT-RA systems are being developed and high-quality video transmission experiments on these systems are planned. If the experimental results are satisfactory, we will conduct further research into applying FRAT-RA to real services such as a nation-wide TV program transmission service between TV stations.

References

[1] T-H.Wu, Fiber Network Service Survivability, Artech House, 1992.

[2] E.Ayanoglu and D.Gitlin, "Broadband network restoration," IEEE Commun. Mag., pp.110-119, July 1996.

[3] R.Kawamura, "A failure-resistant self-healing scheme in ATM networks," IEICE Trans. Commun., vol.E81-B, no.4, pp.699-705, April 1998.

[4] ATM Forum Contribution 94-0640, Quality of Service Requirements for Audio-Visual Multimedia Services.

[5] ITU Recommendation I.356

[6] Y.Sato and K.Sato, "Virtual path and link capacity design for ATM network," IEEE J. SAC, vol.9, no.1, pp.104-111, Jan. 1991.

High-Performance Backbone Network Technology edited by Naoaki Yamanaka
ISBN: 0-8247-5321-6 DOI: 10.1081/(120025668)
Published 2004 by Marcel Dekker, Inc.

PNNI Internetworking Architecture over ATM Public Networks

Mitsuaki Kakemizu, Kazunori Murata, and Masaaki Wakamoto

SUMMARY An increase in high quality of service (QoS) applications such as video conferencing and distribution, and the evolution of the Internet have popularized ATM-LAN use based on the private network node interface (PNNI). Also in public networks, which serve as backbone networks for LANs, ATM technology is being introduced for high-speed and broadband communication. These situations lead to a great demand for economically and seamlessly interconnecting remote ATM-LANs via ATM public networks, which are based on broadband ISDN user part (BISUP). This paper discusses a method of peer group configuration method for such internetworking architecture that can avoid an overload of the PNNI routing processing in each peer group. The paper also proposes a method for seamless interconnection of remote ATM-LANs. In this method, complete PNNI signaling and routing is executed between a local switch (LS) in a public network and each ATM-LAN. It also can reduce the PNNI routing processing load on each public network by emulating PNNI routing and signaling between LSs.

key words: *PNNI, BISUP, routing, signaling, ATM*

1. Introduction

The number of ATM-LANs based on the private network node interface (PNNI) [1] has been increasing since the PNNI specification was standardized at the ATM Forum. Also in public networks, which serve as backbone networks for LANs, ATM technology is being introduced for high-speed and broadband communication. An increase in high quality of service (QoS) applications and the evolution of the Internet will therefore lead to a great demand for economically and seamlessly interconnecting remote ATM-LANs via several ATM public networks based on broadband ISDN user part (BISUP) [2].

PNNI consists of signaling and routing functions for establishing switched virtual connections and paths (SVCs and SVPs). In contrast, BISUP consists of only the signaling function. This implies that ATM public networks must support PNNI emulation to seamlessly interconnect remote ATM-LANs because of avoidance of vast implementation cost in ATM public networks. PNNI routing is characterized by the peer group and the link state protocol. The peer group is a logical sub-network used for routing; the link state protocol is a routing protocol by which the state information of each link in the sub-network is flooded among the switches in the peer group. That is, the peer group configuration directly affects routing efficiency and load. This paper proposes a scalable PNNI internetworking architecture that includes a peer group configuration to avoid explosions in the routing load and a method of PNNI emulation in ATM public networks.

Manuscript received June 4, 1999.
Manuscript revised August 26, 1999.

† The Authors are with Fujitsu Laboratories, Kawasaki-shi, 211-8588 Japan.
†† The Author is with Fujitsu Kyushu Communication Systems, Fukuoka-shi, 814-8588 Japan.

2. Requirements for PNNI Internetworking

The ATM Forum studies PNNI internetworking methods for ATM public networks based on BISUP, and proposes three methods [3], [4], depending on whether ATM public networks emulate the PNNI signaling and routing functions (see Fig. 1).

(a) Interconnection between a specific pair of ATM-LANs via PVP

An ATM public network provides one permanent virtual path (PVP) for a pair of ATM-LANs and emulates neither the PNNI signaling nor the routing function. In this method, ATM public networks are regarded as one link from the perspective of the pair of ATM-LANs.

(b) Interconnection between a specific pair of ATM-LANs via a PVC-based RCC and signaling VC

An ATM public network provides one permanent virtual channel (PVC) for a pair of ATM-LANs and emulates only the PNNI signaling function. This PVC is used as a routing control channel (RCC) that transparently transports topology information generated by executing the PNNI routing function between these ATM-LANs. Also, in this method, ATM public networks are regarded as a single link from the perspective of the pair of ATM-LANs although user data connec-

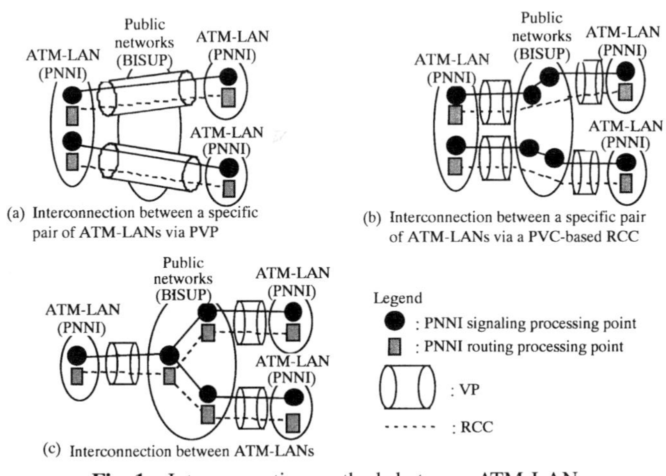

Fig. 1 Interconnecting methods between ATM-LANs.

tions are established dynamically by signaling in the public network.

(c) Interconnection between ATM-LANs

An ATM public network provides one PVP for each ATM-LAN and emulates both the PNNI signaling and routing functions. That is, an ATM public network not only converts between PNNI signaling messages and BISUP signaling messages, but exchanges topology information with each ATM-LAN via one PVC in a PVP.

In this method, ATM public networks are regarded as a PNNI network from the perspective of each ATM-LAN.

In methods (a) and (b), each ATM-LAN gets topology information only for the pair ATM-LANs connected to the RCCs. This means that each ATM-LAN is not able to interconnect with any other ATM-LANs not connected to the RCCs because the designated transit list (DTL) parameter, which is one of the mandatory information elements of PNNI signaling, is based on the topology information for the remote ATM-LANs. In method (c), an ATM public network advertises each ATM-LAN of topology information for any of the other ATM-LANs connected to the ATM public network via the RCCs. This allows each ATM-LAN to economically interconnect with any remote ATM-LAN connected to the ATM public network. In this paper, define such PNNI-based interconnections between remote ATM-LANs via ATM public networks as "PNNI internetwork." For scalable and economical PNNI internetworking, method (c) is, therefore, indispensable. This implies that ATM public networks must emulate both the PNNI signaling and routing functions.

3. Proposed Internetworking Architecture

3.1 Interface between ATM Public Networks and Each ATM-LAN

A user network interface (UNI) in each LS that accommodates one or more ATM-LANs supports the native PNNI routing and signaling protocol in addition to Q.2931 [5]. This LS can recognize the difference between Q.2931 signaling messages and PNNI signaling messages sent via the same virtual channel identifier (VCI =5) by reference to a protocol discriminator. Because this method supports the native PNNI routing and signaling protocol between each ATM-LAN and LS, it enables seamless interconnection of each ATM-LAN with any other ATM-LAN using the PNNI emulation methods between LSs, which are shown in this subsection.

3.2 Peer Group Configuration

Figure 2 shows the ATM end system address (AESA) formats [6] used in ATM-LANs. Each network prefix in each AESA format completely corresponds with peer group (PG) hierarchy. As shown in [7], if the PNNI internetwork were organized in a hierarchy and each of the topology information items were flooded in the PNNI internetwork, the more the ATM-LANs would increase, the heavier the routing processing load on the PNNI internetwork would be. The PNNI internetwork must be, therefore, organized into multiple hierarchical layers. In contrast, most external registration authorities in the world (e.g. ANSI in the USA) allocate each ATM-LAN to an unregistered value into a higher HO-DSP field in order of registration. As a result, several ATM-LANs administered by different, unrelated administrators are accommodated by immediate identical parent PGs. These ATM-LANs are interconnected via links for seamless interconnection, but it is generally difficult for these ATM-LANs to share the costs of the links. As shown in Fig. 3, an approach would be therefore effective, in which the PG (A.1) accommodating one ATM-LAN (#1) and the LS (LS1) interconnects with the other PG (A.2) via these LSs (LS1, LS2). Because multiple switches and physical links between these LSs are owned by public networks, the problem can be solved by charging these ATM-LANs.

For efficient emulation of PNNI signaling and routing in ATM public networks, logical peer group representation of ATM public networks should be considered. Detailed net-

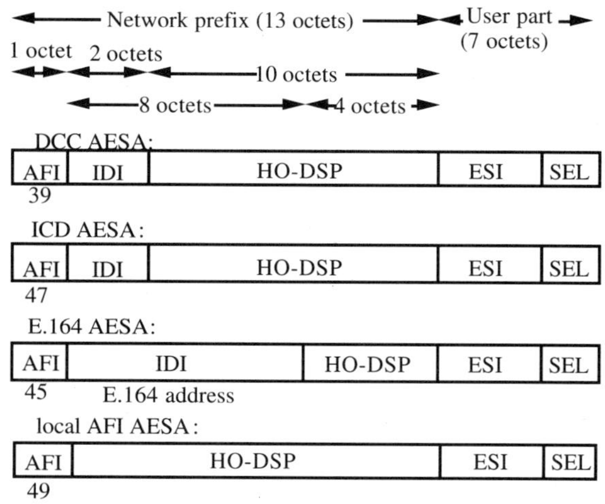

Legend
AFI: Authority and format identifier
IDI: Initial domain identifier
DSP: Domain specific part
HO-DSP: High order DSP
ESI: End system identifier SEL: Selector

Fig. 2 AESA formats.

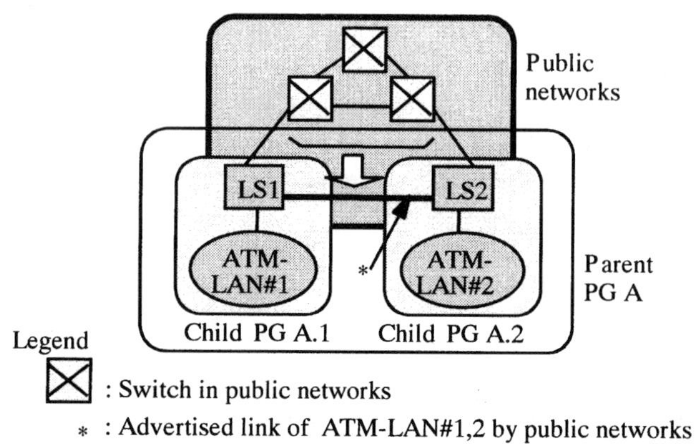

Fig. 3 A low layer peer group configuration in the PNNI internetwork.

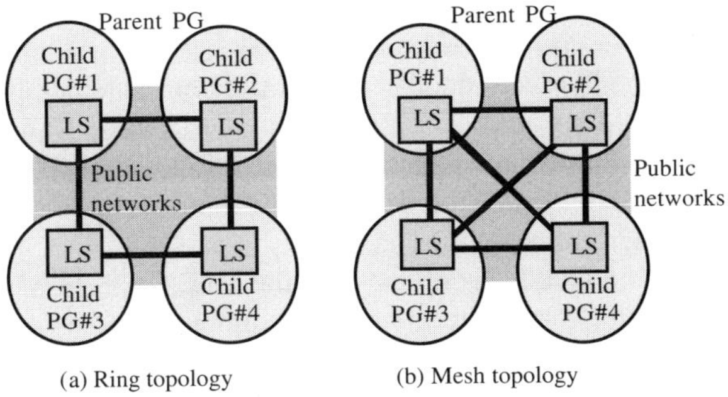

Fig. 4 Peer group configuration in identical parent peer group.

work topology between LSs for public networks is generally secret, and to show each ATM-LAN the detailed network topology in public networks requires that each switch in the public networks fully support PNNI routing protocol. Consequently, as shown in Fig. 3, we propose an approach wherein public networks show each ATM-LAN only one link between the LS accommodating the ATM-LAN and the other LS accommodating the other ATM-LAN. This approach avoids the requirement that all switches in public networks support the PNNI routing protocol.

The remaining issue is how the lowest or higher child PGs in public networks are configured in immediate identical parent PGs. One solution is ring topology, and another is mesh topology, as shown in Fig. 4. Of these alternatives, best choice is mentioned in Sect. 3.4 because this choice depends on the method of PNNI emulation

3.3 PNNI Emulation

3.3.1 PNNI Routing Emulation

Each LS that accommodates ATM-LANs interworks with these ATM-LANs through the PNNI protocol. Each LS must therefore advertise each of these ATM-LANs of topology information based on the PNNI routing protocol. This topology information includes the link state parameters of the links in public networks. There are two methods for calculating the link state parameters of the links in public networks. These methods are given below.

(a) Calculate whenever each link state parameter between the LSs or the network topology is changed

The link between the LSs generally consists of multiple switches and multiple physical links in public networks. In this method, the link state parameter of this link must be periodically calculated from the combined states of these switches and the physical links between the LSs because these are combined dynamically. Each LS advertises each ATM-LAN of the calculated link state parameter. Each LS also advertises each ATM-LAN of the network prefix of PG attached to or removed from the public networks when the network topology is changed. As a result, the public networks incur PNNI emulation overload, although each ATM-LAN can issue the optimal DTL that designates the SVC route.

(b) Calculate only whenever the network topology is changed

The link state parameter is calculated only whenever the network topology is changed. That is, subscription or release of each ATM-LAN to the public networks, or LS faults in the public networks. In this case, each LS advertises each ATM-LAN of the network prefix of PG attached to or removed from the public networks. Each LS is downloaded for the fixed link state parameter which is independent of the combined dynamical states of the switches and the physical links between the LS and each of the other LSs. Each LS also advertises each ATM-LAN of this fixed link state parameter as the link state parameter of the link attached to the public networks. As a result, the PNNI emulation load in this method is lower than that in method (a) although each ATM-LAN cannot issue an optimal DTL.

3.3.2 PNNI Signaling Emulation

The destination native E.164 address must be resolved to establish SVCs between any remote ATM-LANs via the public networks because public networks use BISUP signaling. If an ATM-LAN uses ATM end system address (AESA) based on E.164 AESA format, public networks can use the IDI value as the destination native E.164 address. If an ATM-LAN uses AESA based on DCC or ICD AESA format, public networks can resolve the destination native E.164 address from the network prefix in the AESA by using the ATM name system (ANS) [8] corresponding to the domain name system (DNS). PNNI-specific information elements such as DTL and crankback must be transparently transported over the BISUP signaling messages. The ATM-Forum[3] proposes packing these information elements into the user-to-user information element for the cases covered by method (b) in Sect. 2. New information elements should be introduced to transparently transport these information elements for the cases covered by method (c). This is because the information transported by using the user-to-user information element is subject to charge, and because the DTL must be referenced to select each SVC route on public networks. There are two ways of using PNNI signaling emulation methods in public networks. These methods are divided by the difference between the SVC routes in public networks. The methods are given below.

(A) SVC route based on DTL

The SVC route is selected from the DTL at each PG, even in public networks. As shown in Fig. 5, if the next PG of the PG designated by the current transit pointer in the DTL is located in a public network, each LS (P.2, R.2) resolves the network prefix of the neighboring switch into a native E.164 address. In addition, each LS sends a BISUP IAM message (ii, iii) to the neighboring switch. If the next PG of the PG designated by current transit pointer in the DTL is not located in a public network, each LS (Q.2) sends a PNNI SETUP message (iv) to the neighboring switch without resolving the network prefix of the neighboring switch.

(B) Shortcut SVC route

The public networks take a shortcut from the ingress LS to the egress LS to establish each SVC. Each LS is downloaded for all network prefixes of the PGs that satisfy all the following conditions:

1) The PGs are located in public networks. That is, each of the PGs accommodates one or more LSs.
2) The PGs are accommodated by the identical parent PG that accommodates the downloaded LS.

Using these network prefixes, as shown in Fig. 6, each ingress LS (P.2) receiving a PNNI SETUP message from the PNNI switch specifies an egress PG (Q) that satisfies all the following conditions:

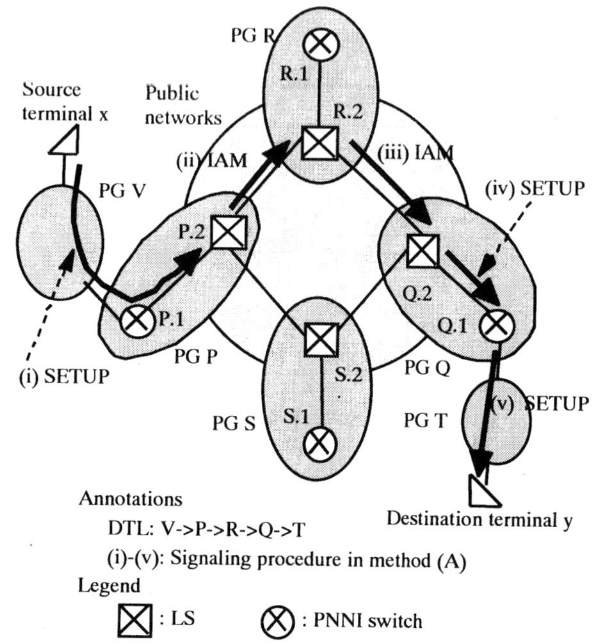

Fig. 5 PNNI signaling emulation procedures in method (A).

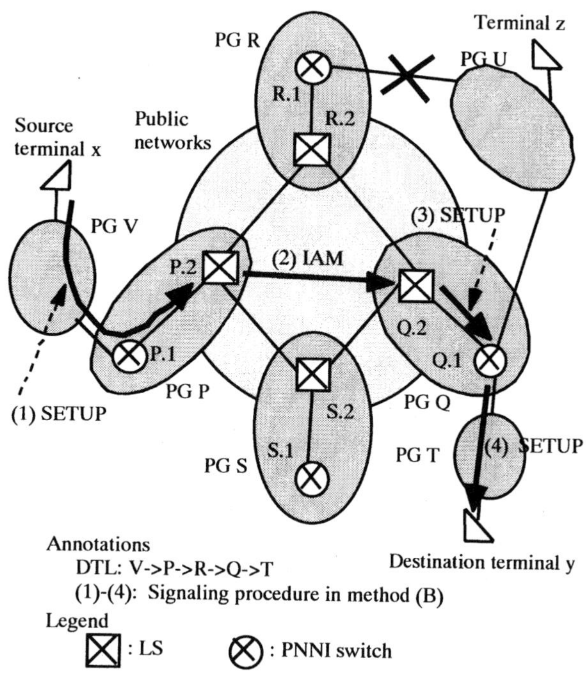

Fig. 6 PNNI signaling emulation procedures in method (B).

1) The PG is located in public networks. That is, The PG accommodates one or more LSs.
2) The PG accommodates the destination terminal. If this PG does not accommodate the destination terminal, the following condition must be satified. If this PG does accommodate the destination terminal, condition 3) does not have to be satisfied.
3) Of all the PGs contained in the DTL, the egress PG is the PG closest to the PG accommodating the destination terminal.

The ingress LS then resolves the network prefix of the egress LS (Q.2) in this egress PG into a native E.164 address, and sends a BISUP IAM message (2) to the egress LS. In this case, the ingress LS does not change the DTL received from the PNNI switch. The egress LS sends a PNNI SETUP message (3) to the neighboring PNNI switch (Q.1). This method allows each ATM-LAN not to have to recognize that the SVC route in the public networks is different from the DTL.

Other method is to simply resolve the destination E.164 address from the network prefix of the destination terminal without reference to the DTL. However, this method forces the public networks to ignore the DTL issued by the PNNI switch. That is, neither of the selected egress LS and the remote ATM-LAN accommodated by the egress LS may be on the DTL. This method thus leads to an increase of the probability of blocking SVC establishment because the remote ATM-LAN not on the DTL can neither issue the DTL parameter in the SETUP message nor select the neighboring switch. Even if public networks change the DTL so that the remote ATM-LAN can issue the DTL and select the neighboring switch, this method leads to an increase of the probability of blocking SVC establishment because the route designated by the originally issued DTL is one of the optimal routes. As shown in Fig. 6, an example based on this simple resolution method is shown below.

Consider that source terminal x sends a PNNI SETUP message to terminal z via several switches and links, and that PG (U), accommodating terminal z, is connected to two PGs (Q, R) but the link between PG (U) and PG (R) is faulty. By executing the PNNI routing protocol, the switch accommodating terminal x issues the DTL as "V->P->S->Q->U" to bypass the faulty link. If the native E.164 address for the network prefix of terminal z is unfortunately registered as one of the terminal addresses of the LS (R.2) in ANS, the ingress LS (P.2) determines the terminal address registered in ANS based on this simple resolution method. When sending an IAM message to the LS (R.2), the ingress LS must also change the DTL into "V->P->R->U" so that each switch in PG (R) can issue the DTL and select the neighboring switch. The LS (R.2) then sends the SETUP message to the PNNI switch (R.1), based on the changed DTL. However, the PNNI switch (R.1) cannot send a SETUP message to a neighboring PNNI switch in PG (U) because of a link failure between PG (R) and PG (U). This simple method of resolution thus increases the probability of blocking SVC establishment.

Table 1 Evaluation of peer group configuration and PNNI emulation.

Scheme	Characteristics			Signaling load	Efficient use of bandwidth	Routing load	Total evaluation
1	(a)	(A)	M	Light	Efficient	Heavy	Bad: Uses bandwidth efficiently, but heavy routing load
2			R	Moderate	Inefficient	Heavy	Bad: Many LSs along SVC, and heavy routing load
3		(B)	M	Light	Fair	Heavy	Bad: Cannot select the route according to optimal DTL, and heavy routing load
4			R	Light	Fair	Heavy	
5	(b)	(A)	R	Moderate	Inefficient	Light	Bad: Light routing load, but many LSs along SVC
6			M	Light	Fair	Moderate	Fair: Though few LSs along SVC, route not optimal, and heavy Hello protocol load
7		(B)	M	Light	Fair	Moderate	
8			R	Light	Fair	Light	Good: Though not optimal route, few LSs along SVC, and light routing load

Legend
(a): Analyzes link information in response to changes in link topology information between LSs.
(b): Determines constant link information in response to changes in link configuration between LSs.
(A): SVC route based on DTL.
(B): Shortcut SVC route.
M: Peer group configuration based on mesh topology.
R: Peer group configuration based on ring topology.

3.4 Considerations

The combination of peer group configurations and emulation methods shown in Sect. 3.2, 3.3 produces eight kinds of schemes for PNNI internetworking architecture. Table 1 lists qualitative evaluations of these schemes. The routing load is the most important evaluation item because this item could create bottlenecks in implementing PNNI internetworking architecture. The signaling load includes resolving the load from the network prefix of AESA to a native E.164 address.

In scheme 1, each topology information item about links between LSs is periodically calculated from the combined states of multiple switches and physical links. These multiple switches and physical links are configured between the LSs in public networks. Scheme 1 therefore enable efficient establishment of SVCs based on an optimal DTL by advertising each ATM-LAN of this accurate topology information about the links between the LSs. Schemes 1-4 would not be appropriate for PNNI internetworking architecture, however, because periodic calculation load in public networks is heavy, and advertising the ATM-LANs of this topology information increases the cost of the PNNI routing protocol. Scheme 5, which adopts ring topology in public networks, causes the establishment of inefficient SVCs and a signaling overload due to the increase of LSs on the SVCs. This is because, in this scheme, selection of each SVC route is based on the DTL, even in public networks. Schemes 6 and 7 enable more efficient establishment of SVCs than scheme 5 due to the reduction of LSs on the SVCs because mesh topology is adopted for the public networks. These schemes can also reduce the signaling load, but each LS incurs the overload of the Hello protocol included in the PNNI routing protocol because the number of links possessed by each LS in mesh topology is more than that in ring topology. The purpose of the Hello protocol is for each switch to specify each of its neighboring switches and to confirm whether each of links possessed by the switch is active. Each switch executes the Hello protocol by periodically sending a Hello packet to each of its neighboring switches and periodically receiving a Hello packet from each of its neighboring switches. Scheme 8 can reduce the PNNI routing emulation load including the load of the Hello protocol because of ring topology. Scheme 8 also enables more efficient establishment of SVCs than scheme 5 and reduces the signaling load. This is because of the reduction of the number of LSs on the SVCs. That is, in this scheme, the public networks take a shortcut from the ingress LS to the egress LS. Consequently, of these schemes, scheme 8 is the most appropriate for PNNI internetworking architecture.

4. Conclusions

This paper proposed a PNNI internetworking architecture that includes peer group configuration in the PNNI internetwork and PNNI emulation methods in public networks. The proposed peer group configuration has three features: interconnection between ATM-LANs via two or more LSs, advertisement each ATM-LAN of only one link between the LSs, and ring topology in the public networks. The proposed PNNI emulation methods are characterized by the following points:

- Native PNNI routing and signaling protocols are used between each ATM-LAN and each LS.
- The link state parameters are calculated only when the network topology is changed.
- Public networks take a shortcut from the ingress LS to the egress LS.

This PNNI internetworking architecture enables the economical, seamless and scalable interconnection of each ATM-LAN with any other ATM-LAN via the PNNI internetwork, including public networks and ATM-LANs.

References

[1] ATM Forum, "PNNI Specification Version 1.0," ATM Forum/

af-pnni-0055.000, 1996.
[2] ITU-T Recommendation Q.2764, B-ISDN, B-ISDN User Part — Basic Call Procedures, 1995.
[3] ATM Forum, "Specification of Interworking among ATM Networks (Draft)," ATM Forum/BTD-RA-IAN-01.07, 1998.
[4] M.Aihara, T.Ohba, and J.Aramomi, "Interworking method between public ATM network and private ATM network(1)," Proc. 1997 Commun. Society Conf. Of IEICE, Sept. 1997.
[5] ITU-T Recommendation Q.2931(Draft), B-ISDN DSS2 Signaling for Basic Call/Connection Control, 1993.
[6] ATM Forum, "ATM forum addressing: Reference guide (Draft)," ATM Forum/BTD-RA-ADDR-01.07, 1998.
[7] J.Aramomi, T.Ohba, and M.Aihara, "Study of interworking between public ATM network and private ATM network(3)," Proc. 1997 Commun. Society Conf. Of IEICE, Sept. 1997.
[8] ATM Forum, "ATM name system specification version 1.0," ATM Forum/96-1284, 1996.

High-Performance Backbone Network Technology edited by Naoaki Yamanaka
ISBN: 0-8247-5321-6 DOI: 10.1081/(120025669)
Published 2004 by Marcel Dekker, Inc.

Phase Assignment Algorithm Based on Traffic Measurement for Real-Time MPEG Sources in ATM Networks

Shinya Tojo, Fumio Ishizaki, and Chikara Ohta

SUMMARY This paper proposes a phase assignment algorithm "Silent Wave Algorithm (SWA)" for real-time MPEG traffic in ATM networks. Our algorithm decides when a new MPEG source should begin to transmit based on its notification parameters and traffic measurement of ongoing connections. Simulation results show that it is hard to accommodate MPEG traffic effectively without any control of phase assignment. On the other hand, the SWA can provide better QOS and improve the network utilization.

key words: *ATM, MPEG video, phase assignment, measurement*

1. Introduction

Cell networks using ATM are expected to support a diverse mix of traffic types such as voice, data and video, which are typically bursty. Thus, the arrival process at switch or multiplexer in cell networks is then essentially a superposition of bursty sources. A considerable amount of bursty traffic also has a periodic nature due to its origin. Such a typical one is MPEG traffic, which will become major traffic in future high-speed networks. MPEG video encoder produces a sequence of three kinds of encoded pictures according to specific pattern at a certain frame rate. As a result, we can regard traffic from a real-time MPEG video source as bursty and periodic (see Figs. 1 and 2).

It is worth while noting that the performance of switch or multiplexer depends on the combination of phase of periodic sources [2], [3]. For instance, the phase synchronization of periodic sources with the same periodicity and burstiness will lead serious degradation of performance measures. Such a situation is observed in Sect. 4. The degradation will last for a certain period because the combination of the periodic sources will be kept during the connection holding times. On the other hand, in the case where the phases of periodic sources are assigned appropriately, we can expect that the statistical multiplexing gain is obtained in usage of network resources, and QOS (Quality Of Service) is improved [3]. In the case of real-time MPEG video transfer, the concentration of transfer of I-picture on a switch leads to congestion of the switch since I-picture has the most amount of data statistically. If users can adjust the phase of MPEG video stream flexibly and avoid the serious situation, QOS of their connection will be improved [11], [12].

In ITU-T and ATM Forum, traffic control for VBR (Variable Bit Rate) and CBR (Constant Bit Rate) traffic is standardized as an open-loop fashion [4]–[6], which is realized by the cooperation of both connection admission control (CAC) [6], [10] and usage parameter control (UPC) [10]. When a source end system (SES) wants a new virtual circuit (VC), it informs its traffic contract and QOS. The CAC mechanism checks whether it is possible or not for the network to handle the new connection without adversely affecting ongoing connections. Every ongoing connection should be observed whether it keeps its traffic contract or not by UPC mechanism since the CAC is performed based on the traffic contract. The CAC algorithms can be divided into two categories: analysis based approach [6] and measurement based approach [7]. The former is based on the queueing analysis where traffic is modeled by traffic contract. In the latter approach, network monitors the condition of network resources and call admission is judged based on the measurement. In this case, the statistical multiplexing gain must also be improved if the bursty periodic traffic is accommodated appropriately.

In this paper, contrary to the analysis based approaches shown in [3], [4], [11], [12], we propose a measurement based phase assignment algorithm which is suitable for measurement based CAC environment. Throughout our paper, we focus on real-time MPEG video traffic as typical periodic VBR traffic. We, however, believe that our algorithm is available for other periodic VBR traffic.

The rest of the paper is organized as follows. In Sect. 2, we briefly observe the characteristics of MPEG video traffic, which is a typical pseudo periodic VBR traffic. In Sect. 3, we propose a phase assignment algorithm based on traffic measurement, which is named "SWA (Silent Wave Algorithm)." Section 4 shows some

Manuscript received February 1, 1999.
Manuscript revised June 21, 1999.

[†]The authors are with the Department of Information Science and Intelligent Systems, Faculty of Engineering, University of Tokushima, Tokushima-shi, 770-8506 Japan.

[*]This paper was in part presented at the Seventh International Conference on Computer Communications and Networks (IC3N '98), October 1998.

simulation results to investigate the effect of the phase assignment algorithm on CLR (cell loss ratio). Finally, conclusions are drawn in Sect. 6.

2. MPEG Video Coding

A video encoder of MPEG 1 or 2 produces the sequence of three types of encoded pictures: I-picture, P-picture and B-picture. I-picture is intra-coded and does not need any data of other pictures. P-picture is predictive-coded and make reference to its previous I-picture or P-picture. B-picture is bi-directively-coded and needs the data of both its previous and following I-pictures or P-pictures. Each picture type has larger amount of data in order of I-picture, P-picture and B-picture statistically due to the different coding methods [9].

In order to facilitate random access, GOP (Group Of Pictures) structure is introduced to MPEG coding. GOP structure is usually predetermined by the two parameters N and M: A GOP consists of N pictures, and I-picture or P-picture appears every M frames† in the GOP. The rest pictures are B-pictures if any.

Note that there is a difference between display sequence and transmission sequence. This is because since each B-picture needs its following reference pictures in encoding and decoding, the latter is encoded and decoded ahead of the former. Then, the reference picture following a group of B-pictures in display sequence is usually transmitted ahead of the group. For instance, if a display sequence is

IBBPBBPBBIBBPBBPBBIBBPBBPBB . . . ,

then the transmission sequence becomes

IPBBPBBIBBPBBPBBIBBPBBPBBIB . . . ,

where $N = 9$ and $M = 3$. In the above case, the transmission sequence repeats the GOP pattern "IBBPBBPBB" indefinitely except the first part according to the GOP structure.

The output of MPEG encoder then becomes pseudo periodic VBR traffic since it produces the different types of pictures at a constant picture rate (e.g. 30 frames/s) according to the GOP structure [1]. Figures 1 and 2 exemplify the amount of data generated in each frame of actual MPEG video streams. MPEG video streams shown in Figs. 1 and 2 consist of 135 and 131 frames, respectively. The MPEG video sequences shown in Fig. 1 is generated from 720×480 pixels ISO test video sequence "mobile" where $N = 15$ and $M = 3$. On the other hand, the MPEG video sequences shown in Fig. 2 is generated from 720×480 pixels ISO test video sequence "multibars" where $N = 12$ and $M = 3$. In both generating sequences, we used the software MPEG-2 TM5 encoder "mpeg2encode" [8] and set the frame rate $1/F$ to 30 frames/sec and the target rate to 5 Mbps. As expected, these MPEG video sequences have not only bursty nature but also periodic one.

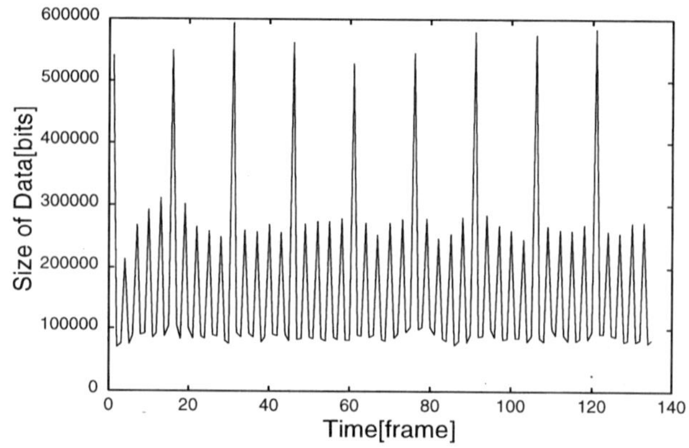

Fig. 1 Fluctuation of MPEG traffic data. [N=15,M=3]

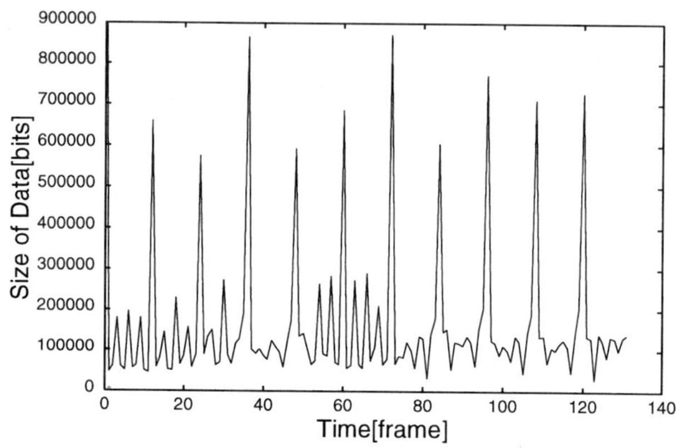

Fig. 2 Fluctuation of MPEG traffic data.[N=12,M=3]

When many MPEG video connections are multiplexed on a link, the QOS of each connection fairly depends on how these connections are superposed on the link as shown in Sect. 4. Once many I-pictures arrive intensively on a output link of an ATM switch, the switch often falls into the congested state and such serious situations continue due to the periodic nature of MPEG video stream. To solve the above problem, we propose a measurement based phase assignment algorithm for real-time MPEG video traffic.

3. Silent Wave Algorithm

The performance of switch or multiplexer fairly depends on the phase combination of periodic VBR sources. To support as many connections satisfying QOS as possible, the phase of periodic sources should be assigned appropriately at CAC operation. However, it is difficult to assign an appropriate phase for every periodic connection through its whole path. We thus assume that the phase assignment control is executed only at the subscriber switch (local switch).

†We use the terms "picture" and "frame" interchangeably.

We here provide a rough description of the pahse assignment control. The detailed explanation of individual procedures will be provided in Sect. 3.1. Roughly speaking, the phase assignment control operates as follows. All the measurement and calculation for the phase assignment control is executed based on "base-slot" which composes of successive T slots[†], where T is a positive integer parameter. The purpose of introducing "base-slot" is to reduce the computational effort of the control. First, when an MPEG source requires a new connection, it declares its frame rate, the mean amount of data of each picture type and GOP parameters M and N. Second, based on the declared parameters, the switch computes the "virtual traffic" $\{v_i\}$ of the new connection. The switch then calculates the "wave value (WV)" $\{w_i\}$ of the new connection from its virtual traffic $\{v_i\}$ and mean traffic λ_{new}, where w_i is the difference between v_i and λ_{new}. On the other hand, the switch maintains the WV $\{W_i\}$ of ongoing connections, which are calculated from the virtual traffic $\{V_i\}$ of ongoing connections and the measurements $\{X_i\}$ of the number of arriving cells. The switch defines an objective function of $\{w_i\}$, $\{W_i\}$ and the phase of the new connection. The switch selects the phase of the new connection such that the objective function minimizes. Based on the selected phase, the switch determines the base-slot in which the source should start transmission, and notifies the source when to start transmission. Finally, the source starts transmission according to the notification.

3.1 Algorithm

In this subsection, we provide a detailed explanation of the algorithm which determines the base-slot in which the new connection should start transmission. We hereafter call the algorithm "Silent Wave Algorithm (SWA)."

In order to reduce computational effort, the SWA is executed based on "base-slot" instead of slot. A base-slot consists of successive T slots. T should be determined by

$$T = \frac{L}{Fr \times Ce \times D}, \tag{1}$$

where L, Fr, Ce denote the link capacity (Mbps), the frame rate (s^{-1}) and the length of a cell (53×8 bits), respectively, and D is a positive integer parameter. D will determine the precision of the control. If D equals to 1, a unit time of the control becomes one frame. Our experience shows that appropriate value of D is 10–20. We hereafter call the parameter D the precision parameter.

Figure 3 shows the fluctuation of the number of cells arriving during a base-slot, when 20 MPEG connections are superposed. Each MPEG source used in this figure is the same one shown in Fig. 1. We assume

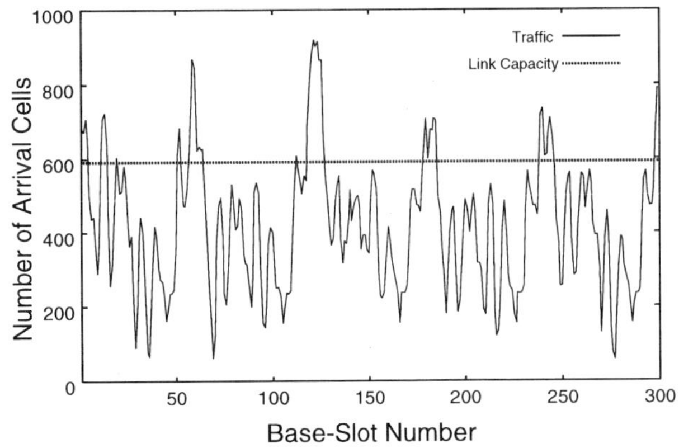

Fig. 3 Fluctuation of traffic volume of 20 connections with their phase chosen randomly.

that each MPEG source randomly selects its transmission starting time. In this figure, we set the precision parameter $D = 20$, $L = 150$ Mbps and $Fr = 30$ frames/s, and thus the base-slot $T = 589$ slots. The amount of the traffic wildly fluctuates, and such traffic behavior leads to the congestion of the switch. Thus, if we can assign a phase to the new connection in such a way that the fluctuation of total traffic minimizes, good performance will be achieved. To do so, we consider the phase assignment of the new connection as an optimization problem and propose SWA to solve it approximately.

In SWA, the management module of subscriber switch counts the number of arriving cells from ongoing connections during each base-slot. Let X_i denote the number of arriving cells from ongoing connections in the ith base-slot (see Fig. 5). We assume that such measured data are stored over the past K base-slots, where K is a positive integer parameter. K should be determined by $K = V_c \times D$, where V_c is a positive integer parameter. We hereafter call the parameter V_c the virtual cycle. Although we can set the virtual cycle V_c to any positive integer, setting V_c to a common multiple of N's or M's is preferable. This will be observed in Sect. 4.

We here assume that the current time is in the tth base-slot and a new MPEG source arrives at the moment. Let C denote a variable representing the cycle of the new connection in base-slot. C is given by

$$C = \begin{cases} M \times D & \text{(if } V_c \text{ is a multiple of } M \\ & \text{of the new connection} \\ & \text{but not a multiple of } N\text{),} \\ N \times D & \text{(otherwise).} \end{cases} \tag{2}$$

Let $\{v_i\}_{i=-\infty}^{\infty}$ denote the "virtual traffic". v_i represents the number of cells arriving from the new connection during the ith base-slot, when the new connection generates cells according to the notification. $\{v_i\}$ is deter-

[†]Throughout our paper, we assume that time is slotted on cell-basis on the transmission link.

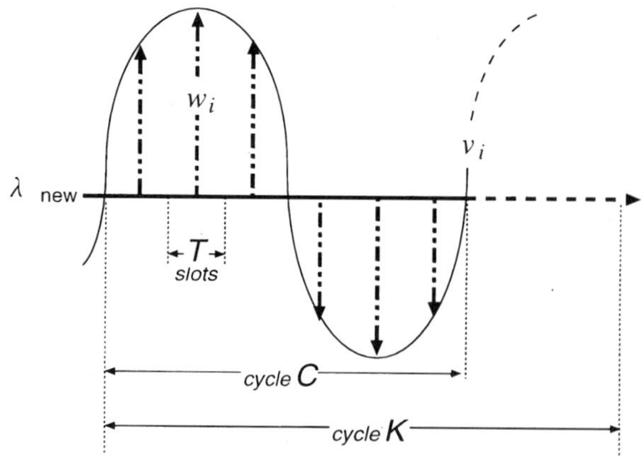

Fig. 4 Parameters of SWA for new connection.

mined as follows. For $i = 0, \ldots, C-1$, v_i is determined by the declaration of the source (i.e., the frame rate, the mean amount of data of each type of picture and GOP parameters M and N). For the other i's, v_i is given by

$$v_i = v_{i \bmod C}. \tag{3}$$

Note that the virtual traffic of the new connection is not real traffic, but it is enough for switch to understand the features of the new connection.

In order to construct the objective function of SWA, we introduce the notion of "wave value (WV)" which denotes the difference between the mean value of a sequence of variables and the value of the corresponding variable. Let w_i denote the wave value of v_i. The w_i is then given by

$$w_i = v_i - \lambda_{\text{new}}, \tag{4}$$

where λ_{new} denotes the average v_i of the new connection during C base-slots, which is declared in the connection-setup phase. Note that the larger variance of wave values of the new connection means the wilder fluctuation of its traffic volume (see Fig. 4).

In order to reflect the influence of newly arriving or released connection in the measured total traffic of ongoing connections, we introduce the total virtual traffic to rectify the value of $\{X_i\}$. Let $\{V_i\}$ denote the "total virtual traffic," where V_i reflects the amount of traffic of newly arriving or released connections in the ith base-slot. The initial value of V_i is set to zero and it is modified by (7), which will be explained later. For the sequence $\{X_i + V_i\}_{i=t-K}^{t-1}$ of the rectified measured traffic, we also define the wave value W_i as

$$W_i = X_i + V_i - \frac{1}{K}\sum_{j=1}^{K}(X_{t-j} + V_{t-j}). \tag{5}$$

Note that the latter term in the right-hand side of Eq. (5) is the mean value of the rectified measured traffic over the past K base-slots. Recall that a wave value represents the difference between the mean value of a sequence of variables and the value of the corresponding variable. Then, in order to reduce the fluctuation of total traffic, we superpose the new connection in such a way that the sum of the abosolute values of the wave values minimizes. Keeping this in mind, we construct an objective function of our optimization problem. Consider the case where the new connection delays the transmission by j base-slots. Let $S_t(j)$ denote the objective function, which is the sum of estimated wave values of total traffic over the past K base-slots given that the newly accommodated connection delays the transmission by j base-slots. We then have

$$S_t(j) = \sum_{k=1}^{K} |W_{t-k} + w_{t-j-k}|. \tag{6}$$

Since the new connection has periodicity with the cycle C, it is enough to find the value of j that minimizes the value of $S_t(j)$ for $0 \le j < C$. Once the value of j is obtained, the management module of subscriber switch makes the new connection delay the transmission by j base-slots from the current time. In the case of $Fr = 30$ (s^{-1}) and $N = 15$, the play-out delay becomes 0.5 s at most.

The above is the basic procedure of phase assignment in SWA. We now return to the consideration of the virtual traffic V_i to rectify the measured traffic X_i. The management module of subscriber switch must suppose the future behavior of traffic based on the past measurement data. However, it is clear that traffic of a newly accepted connection is not taken into account in the measured traffic. The management moodule then needs to reflect the influence of the new connection in the virtual traffic using its notification parameters until the beginning of the transmission. Provided that a newly accommodated connection arrives at the current tth base-slot and it delays the starting time by j base-slots, it is necessary to take account of the virtual traffic of the new connection into the total virtual traffic from the past K base-slot to the $(t+j)$th base-slot when the new connection starts transmission. In this case, the values of V_i are modified as follows:

$$V_k \leftarrow V_k + v_{k-j} \quad \text{for } t-K \le k \le t+j. \tag{7}$$

This procedure is illustrated in Fig. 5.

On the other hand, the effect of released connection should be taken into account. To do so, provided that an ongoing connection is released at the current tth base-slot, we subtract the virtual traffic of the released connection from the total virtual traffic for the past K base-slot in order to remove the effect of the released connection from the measured data. In this case, the values of V_i are modified as follows:

$$V_k \leftarrow V_k - v_{k-j} \quad \text{for } t-K \le k \le t. \tag{8}$$

This procedure is also illustrated in Fig. 5.

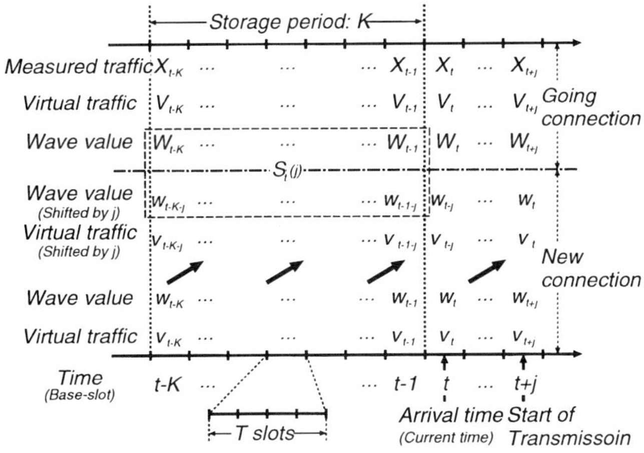

Fig. 5 Relationship between wave values of ongoing connections and a new connection.

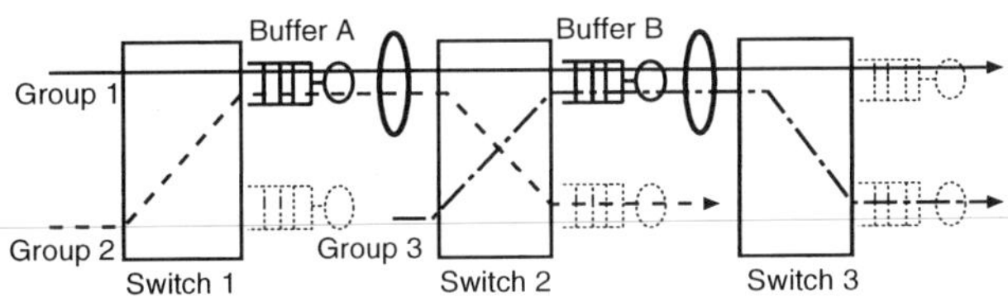

Fig. 6 Simulation model.

We should note that SWA requires a little extra signaling procedure since the local switch equipping SWA and the source end system exchange the information for phase assignment.

4. Simulation Results

In this section, we provide some simulation results to evaluate the effect of the phase assignment control on CLR (Cell Loss Ratio).

Throughout this section, we use a three-stage network model shown in Fig. 6. There are three groups of connections, and the connections of the same group traverse the same path. More precisely, the connections of groups 1 and 2 share the same link between the switches 1 and 2, and the group 1 also shares the links between the switches 2 and 3 with the group 3. The interference then occurs among the groups. Each group consists of 8 connections.

Further, we assume the followings. Time is slotted based on cell transmission duration on a link. The link capacity is 150 Mbps. The length of cell is 53 bytes and its payload is 48 bytes. The buffer capacity $Q_{\max}$ of the switch is set to 300 cells.

As for MPEG sources used in the simulation, we assume the followings if not refered to. The MPEG video trace shown in Fig. 1 is utilized to generate traffic in the simulation. The target coding rate is set to 5 Mbps. The number N of frames in a GOP is 15 frames and the distance M between I-picture (P-picture) and P-picture is 3 frames. Frame rate is set to 30 frame/s.

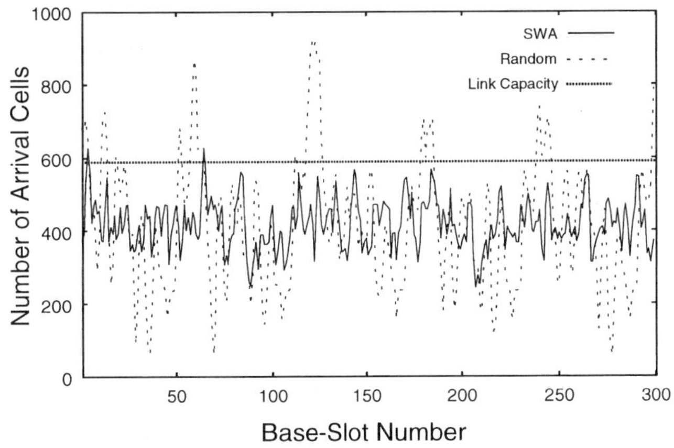

Fig. 7 Fluctuation of traffic volume.

This sequence is used repeatedly since it consists of the limited 135 frames. The number of generated cells in a frame is determined by simply dividing the amount of data in the frame by 48 bytes. The peak rate is set to 20 Mbps. Each connection requests a connection setup to its subscriber switch randomly during the first 135 frames.

Throughout the simulation provided in this section, we set the precision parameter $D = 10$ and thus the base-slot $T = 1179$ slots.

4.1 Homogeneous Case

In this subsection, we consider homogeneous cases.

Figure 7 shows the fluctuation of the total traffic volume arriving at the switch 1 in the cases with and without SWA. Note that the traffic volume is represented in the number of arrival cells during 589 slots. In the case without SWA, the traffic volume grows up frequently. This leads to the congestion of the switch. On the other hand, SWA tries to keep the traffic volume as almost constant as possible. As expected, in the case with SWA, the magnitude of the fluctuation is fairly reduced due to the phase assignment. This will make the QOS and the network utilization improved more.

Figure 8 shows the cumulative frequencies of CLR at the switches 1 and 2. In SWA, we set the virtual cycle $V_c = 15$ and thus $K = 150$. In each case, 500 simulation trials are executed. In the case without SWA, no cell is lost in 226 trials among 500 trials, whereas the value of CLR fairly differs among the rests of trials. In particular, the values of CLR at the switches 1 and 2 exceed 10^{-2} in 25 and 24 trials, respectively. This is because that some MPEG video streams happen to synchronize, which causes the congestion of the switch. As a result, it is hard to accommodate MPEG sources effectively without any control of phase assignment. On the other hand, no cell is lost at the switch 1 in all trials with SWA. At the switch 2, a few cells are

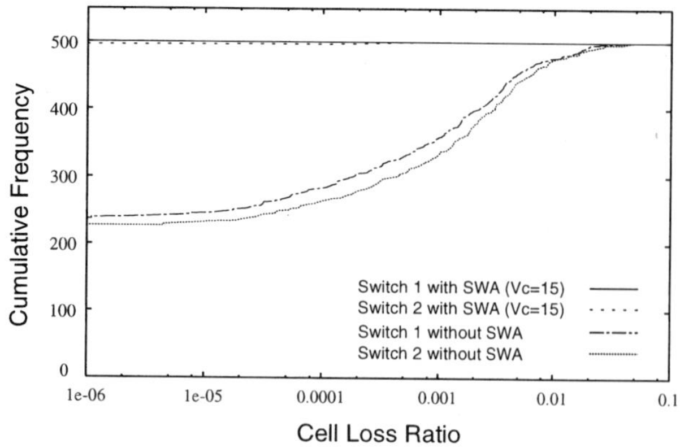

Fig. 8 CLR at the switches 1 and 2. (homogeneous case, $V_c = 15$)

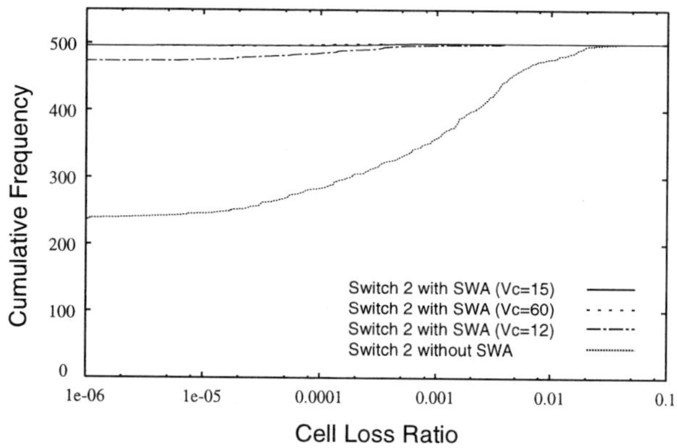

Fig. 9 CLR at the switch 2. (homogeneous case, $V_c = 15, 12, 60$)

lost in 3 trials while the CLR is less than 10^{-3}. This is because that the management module of the switch 2 can assign the phases of only the group 3. In such cases, however, SWA can also provide better QOS.

Next, we observe the effect of values of the virtual cycle V_c on the performance of SWA. Figure 9 shows the cumulative frequencies of CLR at the switch 2 for various values of V_c. V_c is set to 15, 12 and 60. In the first case, V_c is set to N (i.e., the least multiple of N and also a multiple of M). In the second case, V_c is a multiple of M but not a multiple of N. In the third case, V_c is a multiple of N but not the least multiple. In all the cases, no cell is lost at the switch 1 in all trials with SWA.

Also, Fig. 10 shows the cumulative frequencies of CLR at the switches 1 and 2 for various values of V_c. V_c is set to 16, 61 and 70. In these cases, V_c is not a common multiple of N nor M. 500 simulation trials are executed in each case.

In those figures, we observe the following: (a) If V_c is a multiple of N (and thus also a multiple of M), SWA can greatly reduce CLR. (b) If V_c is a multiple of M but not a multiple of N, SWA can also reduce CLR. (c) When V_c is not a multiple of N nor M, SWA may

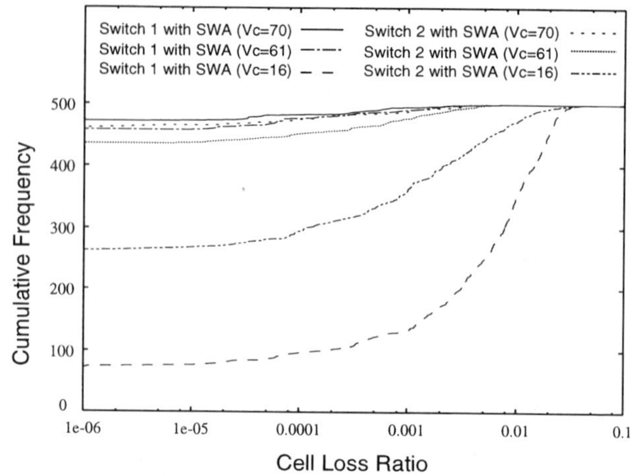

Fig. 10 CLR at the switches 1 and 2. (homogeneous case, $V_c = 16, 61, 70$)

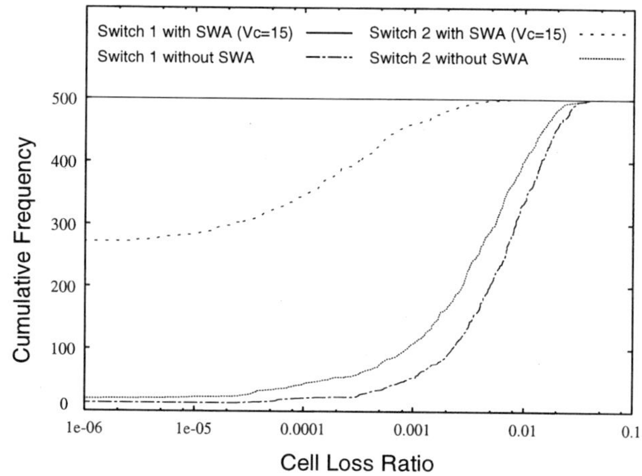

Fig. 11 CLR at switches 1 and 2. (low average rate connections)

or may not reduce CLR. However, even in such cases, if V_c is sufficiently large compared to N, SWA can reduce CLR.

Next, we consider the scenario that many connections at small average rate are multiplexed in ATM networks. The network model is slightly changed from the original network model shown in Fig. 6. The only difference is that each group consists of 45 connections, the MPEG video coding rate is 1.5 Mbps and the peak rate is 4 Mbps.

Figure 11 shows the cumulative frequencies of CLR at the switches 1 and 2. Also, 500 simulation trials are executed. In this figure, we also observe that SWA can also greatly reduce CLR in the case that many connections at small average rate are multiplexed.

We also consider the scenario that 15 connections without any B-pictures are multiplexed in the original network model shown in Fig. 6. In such cases, SWA may become less effective, because traffic from the MPEG sources will be less bursty. In order to generate such

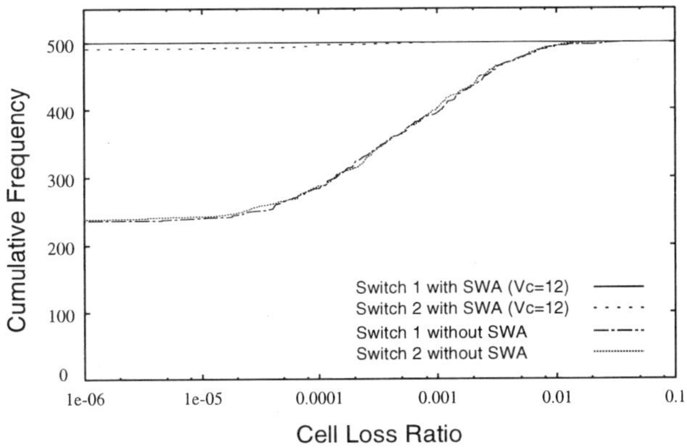

Fig. 12 CLR at switches 1 and 2. (the case without B-pictures)

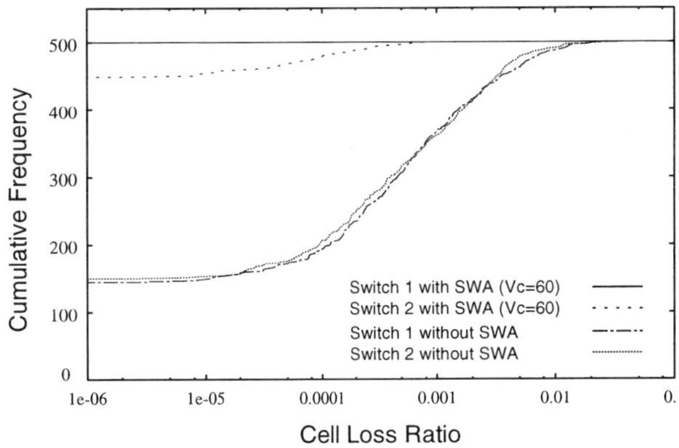

Fig. 13 CLR at swtiches 1 and 2. (heterogeneous case, $V_c = 60$)

an MPEG traffic model, we set to $N = 12$ and $M = 1$. Further, the target coding rate is set to 5 Mbps.

Figure 12 shows the cumulative frequencies of CLR at the switches 1 and 2. Also, 500 simulation trials are executed. In this figure, we also observe that SWA can also greatly reduce CLR even in the case that connections without any B-pictures are multiplexed.

4.2 Heterogeneous Case

In this subsection, we consider heterogeneous cases where MPEG sources with different values of N and M are multiplexed.

To construct heterogeneous environment, in addition to the MPEG source shown in Fig. 1, the MPEG source shown in Fig. 2 is used in the simulation. The characteristic of the MPEG source shown in Fig. 2 is the following. The target coding rate is set to 5 Mbps. The number N of frames in a GOP is 12 frames and the distance M between I-picture (P-picture) and P-picture is 3 frames. Frame rate is set to 30 frame/s. This sequence is used repeatedly since it consists of the limited 131 frames.

Among 8 connections of each group in the network model, 4 connections come from the MPEG source shown in Fig. 1 and the other 4 connections come from the MPEG source shown in Fig. 2. Each source requests a connection setup to its subscriber switch randomly during the first 131 frames.

Figure 13 shows the cumulative frequencies of CLR at the switches 1 and 2 . We set the virtual cycle $V_c = 60$ and thus $K = 600$. In this setting, V_c is the least common multiple of N's (i.e., 12 and 15). No cell is lost at the switch 1 in all trials with SWA. At the switch 2, a few cells are lost in 51 trials and the CLR is less than 10^{-3}. This is because that the management module of the switch 2 can assign the phases of only the group 3. However, even in such cases, SWA can also reduce CLR.

Next, we observe the effect of values of the virtual cycle V_c on the performance of SWA. Figure 14 shows

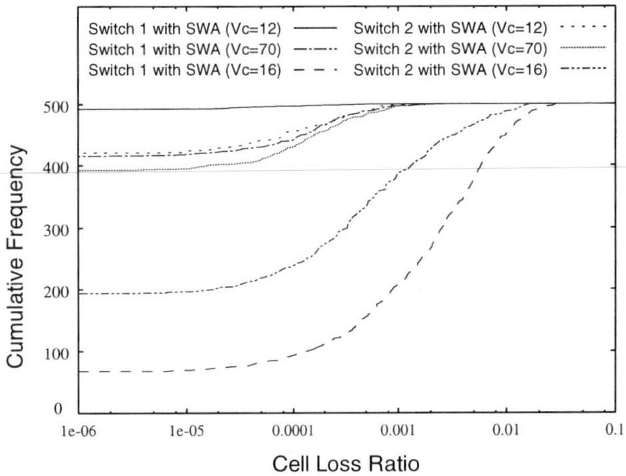

Fig. 14 CLR at the switches 1 and 2. (heterogeneous case, $V_c = 12, 70, 16$)

the cumulative frequencies of CLR at the switches 1 and 2 for various values of V_c. V_c is set to 12, 16 and 70. In the first case, V_c is a common multiple of M's. In the second and third cases, V_c is not a common multiple of M's nor N's. In each case, 500 simulation trials are executed.

Those figures provide us similar observations shown in the homogeneous case: (a) If V_c is a multiple of M's, SWA can greatly reduce CLR. (b) When V_c is not a multiple of N nor M, SWA may or may not reduce CLR. However, even in such cases, if V_c is sufficiently large compared to N's, SWA can reduce CLR.

5. Conclusions

In this paper, we propose a measurement-based phase assignment algorithm "Silent Wave Algorithm (SWA)" for real-time MPEG video transfer in ATM networks. Although we consider only real-time MPEG video sources in this paper, our basic idea can be applicable to other periodic VBR sources. We provide some simulation results to investigate the effect of SWA on the performance of switches. In the simulation results, we observe that it is hard to accommodate real-time MPEG video sources effectively without any control, and SWA can provide better QOS and improve the network utilization. In addition, we examine the effect of values of the parameters, especially, the virtual cycle V_c on the performance of SWA. We observe the following in both homogeneous and heterogeneous cases: (a) If V_c is a multiple of the GOP paramters N's and M's, SWA can reduce CLR. In particular, if V_c is a multiple of N's, SWA can greatly reduce CLR. (b) If V_c is not a multiple of N's nor M's, SWA may or may not reduce CLR. However, even in such cases, if V_c is sufficiently large compared to N's, we can expect that SWA reduces CLR. We thus recommend to set the virtual cycle $V_c = 60$. The reason is as follows: If we set $V_c = 60$, the computational costs for the control are not expensive, because V_c is not so large. (Note that large V_c leads to large amounts of computational efforts for the control.) In addition, V_c then has many diviosrs including 12 and 15, which N's are likely to be set to.

References

[1] K. Ishimaru, T. Fujii, T. Sawabe, J. Suzuki, and S. Ono, "Transmission characteristics of MPEG2 encoded super high definition images," IEICE Trans. Commun., vol.J80B-I, no.12, pp.961–969, Dec. 1997.

[2] F. Ishizaki and T. Takine "Bounds for the tail distribution in a queue with the superposition of general periodic Markov sources," Proc. IEEE INFOCOM '97, pp.1090–1097, 1997.

[3] F. Ishizaki and C. Ohta, "Finding good combinations of initial phases for periodic sources in discrete-time queues," Proc. IEEE ICC '98, pp.1510–1514, 1998.

[4] F. Ishizaki and C. Ohta, "Phase assignment control for MPEG video sources," Proc. IEEE ICIPS '98, pp.138–142, 1998.

[5] ITU-T Recommendation I.371, Traffic Control and Congestion Control in B-ISDN, July 1995.

[6] R. Guerin, H. Ahmadi, and M. Naghshinch, "Equivalent Capacity and Its Application to Bandwidth Allocation in High-Speed Networks," IEEE J. Select. Areas in Commun., vol.9, pp.968–981, 1991.

[7] S. Jamin, P.B. Danzig, S. Shenker, and L. Zhang, "A measurement-based admission control algorithm for integrated services packet networks," ACM Computer Communication Review, vol.25, no.4, pp.2–13, Oct. 1995.

[8] "MPEG Software Simulation Group," available at "http://www.mpeg.org/MSSG/."

[9] M. Krunz, R. Sass, and H. Hughes, "Statistical characteristics and multiplexing of MPEG streams," Proc. IEEE INFOCOM'95, pp.455–462, 1995.

[10] W. Stallings, High-Speed Networks: TCP/IP and ATM Design Principles, Prentice Hall, 1997.

[11] B.-H. Roh, H.-J. Ahn, and J.-K. Kim, "Starting time selection method for minimum cell loss ratio of superposed VBR MPEG video traffic in ATM networks," Proc. ICC '98, pp.573–577, 1998.

[12] M. Krunz and S.K. Tripathi, "Exploiting the temporal structure of MPEG video for the reduction of bandwidth requirements," Proc. INFOCOM '97, pp.67–74, 1997.

High-Performance Backbone Network Technology edited by Naoaki Yamanaka
ISBN: 0-8247-5321-6 DOI: 10.1081/(120025670)
Published 2004 by Marcel Dekker, Inc.

Performance Analysis of Layer3 Switch: Case of Flow-Driven Connection Setup

Kenji Kawahara, Shougo Nakazawa, Tetsuya Takine, and Yuji Oie

SUMMARY The layer3 switch enables us to fast transmit IP datagrams using the cut-through technique. The current layer3 router would become bottleneck in terms of delay performance as the amount of traffic injected into high speed networks gets relatively large. Thus, the layer3 switch should be an important element constructing the next generation Internet backbone. In this paper, we analyze the cut-through rate, the datagram waiting time and the mis-ordered rate of a layer3 switch in case of flow-driven connection setup. In the analysis, by using 3-state Markov modulated Bernoulli process (MMBP), we model the arrival process of IP flow and IP datagram from each source. Furthermore, we investigate impacts of the arrival rate and the average datagram length on the performance.

key words: *layer3 switching, ATM, flow-driven type, 3-state MMBP*

1. Introduction

On the Internet, the number of hosts continues to double approximately every year. Within this overall growth rate, there is also increasing demand for multimedia applications, which require quite stringent quality of service (QoS) delivery in terms of packet delay, loss rate and minimum bandwidth. Furthermore, as the World-Wide Web is increasingly used for business, there are a growing number of users for whom delay-bounded access to information is very important. Thus, the Internet backbone must increase its bandwidth available to end users as well as offer QoS provisioning. However, traditional layer3 IP routers may be bottle-neck in terms of high-speed data transfer, since they carry out forwarding decisions in software based on address resolution with hop-by-hop transmission of IP datagrams of variable length [1].

On the other hand, a layer2 switch based on ATM achieves high bit rates through hardware switching of fixed-length cells and commonly supports QoS by connection oriented cell transfer. Therefore, ATM can be exploited in future IP networks with QoS provisioning. This combination has been prompted by some proposals including classical IP over ATM [2] by Internet Engineering Task Force (IETF) and multi-protocol over ATM (MPOA) [3] by ATM Forum, in which a number of servers mapping next hop's IP address to its ATM one are required. However, this address resolution mechanism is not scalable. In recent years, so-called **layer3 switching** mechanism is employed in order to efficiently combine layer3 routing with layer2 high-speed switching and enables us to fast transmit IP datagrams. It has been implemented in IP switch [4] of Ipsilon, CSR(Cell Switch Router) [5] of Toshiba and Tag Switching [6] of Cisco so far.

The layer3 switch consists of the layer3 datagram routing kernel and the layer2 switching fabric. Hereafter, we will identify the switching fabric with ATM switch. A specific source-destination IP address pair is regarded as a *flow*. When a datagram arrives at the layer3 switch, the routing kernel distinguishes its source and destination IP addresses from the IP header, creates the specific *flow*, and maps it to some virtual connection (VC) on the ATM switch. The subsequent datagrams of the *flow* are directly transmitted on the ATM switch without address resolution on the routing kernel. This will be called cut-through. Therefore, the number of VC on the ATM switch is most important parameter in terms of the performance. If there are not enough VCs for transmitting datagrams via the layer3 switch, the rate of cut-through datagrams becomes small thereby causing the out-of-order transmission of datagrams.

Lin et al. [7] show the impact of the number of VCs on both rates of cut-through and the mis-ordered, which means out-of-order transmission, by means of trace-driven simulation, while datagram transfer delay, which must be important, is not shown. Furthermore, the impact of traffic parameter on the performance can not be investigated in that manner. Niu et al. [8] analyze the datagram loss rate, transfer delay and the VC utilization of the layer3 switch focusing only on single flow/VC. Hence, in this paper, we model the arrival process of both IP flow and datagrams by 3-state Markov modulated Bernoulli process (MMBP), and an-

Manuscript received April 23, 1999.
Manuscript revised September 6, 1999.

†The authors are with the Department of Computer Science and Electronics, Faculty of Computer Science and Systems Engineering, Kyushu Institute of Technology, Iizuka-shi, 820-8502 Japan.

††The author is with Nara Advanced Institute of Science and Technology, Ikoma-shi, 630-0101 Japan.

†††The author is with the Department of Applied Mathematics and Physics, Graduate School of Informatics, Kyoto University, Kyoto-shi, 606-8501 Japan.

*The earlier version of this paper was presented at the Internet Workshop'99 (IWS'99).

IEICE TRANS. COMMUN., Vol. E83-B, No. 2, pp. 130–139, February 2000

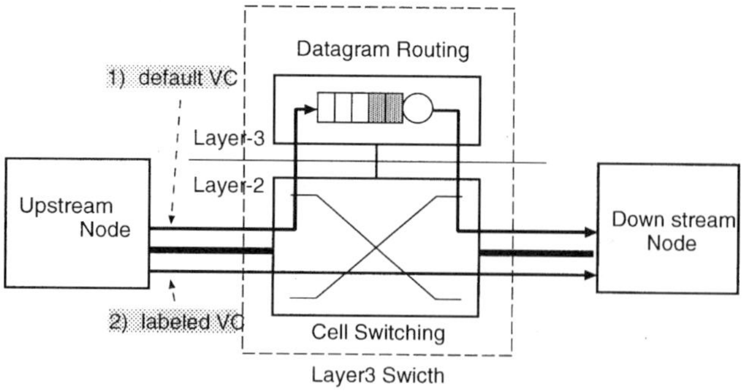

Fig. 1 Concept of layer3 switching.

alyze the datagram waiting time as well as the cut-through rate and the mis-ordered one in case of multiplexing these flows. Furthermore, we can investigate the impact of traffic parameter on the performance.

The remainder of this paper is organized as follows. In Sect. 2, we explain the mechanism of the layer3 switch and 3-state MMBP as a source model. In Sect. 3, we analyze the steady state probability of the layer3 switch and derive performance measures. Numerical results and discussions are given in Sect. 4. Section 5 presents conclusion and future works.

2. Analytical Model and Traffic Model

2.1 Layer3 Switch: Cut-Through Mechanism

Figure 1 illustrates the concept of the layer3 switching mechanism. The layer3 switch consists of the layer3 datagram routing kernel and the layer2 switching fabric or ATM switch. When a datagram from the upstream node via the *default VC* arrives at the layer3 switch, firstly, the datagram is raised to the routing kernel which checks its source and destination IP addresses, and then it is forwarded to the down-stream node which is determined according to its destination address. At the same time, the routing kernel regards this source-destination IP address pair as a *flow* and maps it to some VC, which is called *labeled VC*. Subsequent datagrams of the *flow* are directly transmitted on ATM switch via the labeled VC without address resolution on the routing kernel. This mechanism is referred to as *cut-through* and it allows datagrams to be quickly transmitted. In this case, VC association is flow-driven for cut-through VCs. On the other hand, it is called as topology-driven when cut-through VCs are predetermined according to provided routing information. The former type is IP switch [4] and CSR [5] and the latter is Tag Switching [6]. In this paper, we shall deal with the flow-driven type layer3 switch.

2.2 Source Model Description

Firstly, we consider a discrete-time queuing system with time slot being equal to an ATM cell transmission time. The IP flow model is shown in Fig. 2. The period of on state corresponds to the IP flow established by the source. Furthermore, in on$^+$ state, one cell always arrives in each slot and the period of on state consists of exactly one IP datagram, while no cell arrives in on$^-$ state. In off state, any IP flow is not established or no cell arrives. Hence, this process can be expressed by the following three states: on$^+$, on$^-$ and off, and its state transition in a slot is given as the following matrix†.

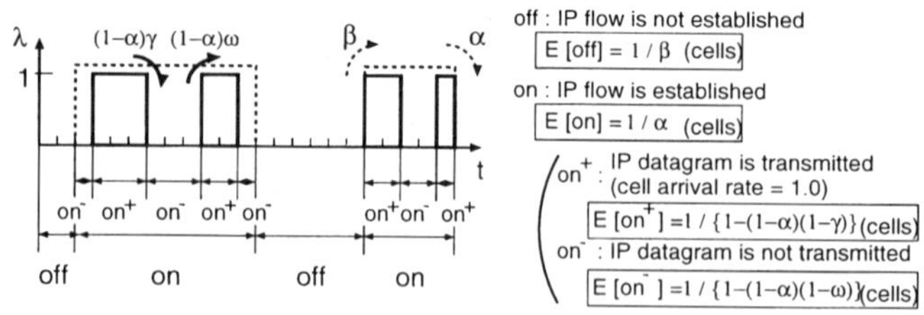

Fig. 2 Source model: 3-state MMBP.

$$\begin{array}{c} \\ \text{on}^+ \\ \text{on}^- \\ \text{off} \end{array} \begin{array}{c} \begin{array}{ccc} \quad\text{on}^+ \quad & \quad\text{on}^- \quad & \quad\text{off} \end{array} \\ \begin{pmatrix} (1-\alpha)(1-\gamma) & (1-\alpha)\gamma & \alpha \\ (1-\alpha)\omega & (1-\alpha)(1-\omega) & \alpha \\ \beta\frac{\omega}{\gamma+\omega} & \beta\frac{\gamma}{\gamma+\omega} & 1-\beta \end{pmatrix} \end{array} \tag{1}$$

From this equation, we find that the duration of the IP flow follows a geometric distribution with parameter $(1-\alpha)$ and the average IP flow length $E[\text{on}]$ is equal to $1/\alpha$, and that the duration of off state follows a geometric distribution with parameter $(1-\beta)$ and the average length $E[\text{off}]$ becomes $1/\beta$. Furthermore, in each slot, the process changes its state from on$^+$ state to on$^-$ state with probability $(1-\alpha)\gamma$, or to off state with probability α. This means that the datagram length, which indicates the number of cells generating in an on$^+$ period, follows a geometric distribution with parameter $(1-\alpha)(1-\gamma)$. Similarly, the duration of an on$^-$ state follows a geometric distribution with parameter $(1-\alpha)(1-\omega)$. Thus, the average duration of on$^+$ state $E[\text{on}^+]$ and on$^-$ state $E[\text{on}^-]$ are equal to $\{1-(1-\alpha)(1-\gamma)\}^{-1}$ and $\{1-(1-\alpha)(1-\omega)\}^{-1}$, respectively. We will call this process the 3-state (namely, on$^+$, on$^-$ and off) Markov modulated Bernoulli process (MMBP) characterized by a parameter set $(\alpha, \beta, \gamma, \omega)$. From Eq. (1), the steady state probability of each state is given by

$$\pi_{\text{off}} = \frac{\alpha}{\alpha+\beta} \tag{2}$$

†Here, we set the probability to α that a state of each source changes from on$^-$ in the current slot to off in the following one. Since both states look same in that no cell arrives there, the transition from on$^-$ to off may be considered to be unneccesary. However, if this transition is not permitted, on$^-$ state will be independent from on state. This does not allow on$^+$ and on$^-$ to alternately change in a on period; i.e., IP flow can not be modeled. For this reason, the transition is required.

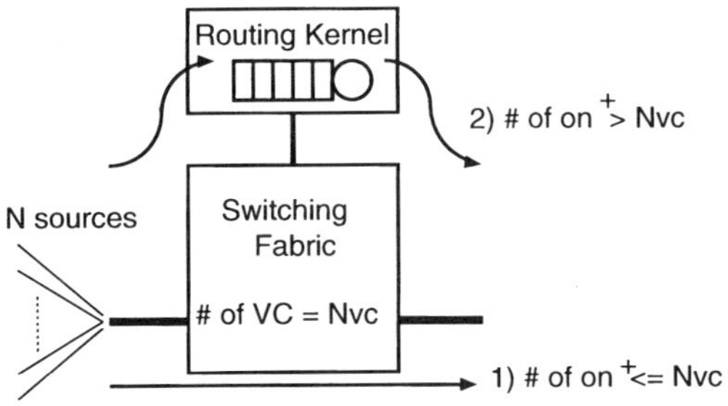

Fig. 3 Analytical model.

$$\pi_{\text{on}^+} = \frac{\beta\omega}{(\alpha+\beta)(\gamma+\omega)} \tag{3}$$

$$\pi_{\text{on}^-} = \frac{\beta\gamma}{(\alpha+\beta)(\gamma+\omega)} \tag{4}$$

Since one cell arrives in each slot of on$^+$ state whereas no cell is generated in both on$^-$ and off states, the average arrival rate λ of a source is given by Eq.(3). In this model, both the IP flow and datagram length follow geometric distributions and there is a correlation in terms of cell-level arrival. If the impact of the upper layer protocol, i.e. TCP, is taken into account, we may introduce the traffic model with long range dependence in terms of packet-level. However, in this paper, we just focus on the layer3 performance, so that the 3-state MMBP model would be suitable for analyzing pure layer3 switches.

2.3 Analytical Model Assumption

Figure 3 shows our queuing model for analysis. We suppose that the switching fabric is an ATM switch with its capacity of N, and that the routing kernel has its capacity of one and its buffer size is infinite. N sources are multiplexed at the layer3 switch and they are homogeneous in terms of traffic characteristics, namely, each source has a parameter set of 3-state MMBP, $(\alpha, \beta, \gamma, \omega)$. When the IP flow is established by some source, the cut-through VC has been assigned during transmission of just one IP datagram; it means that sources are in on$^+$ state. IP datagrams from them thus can be cut through and the waiting time of them becomes 0, since we do not suppose the buffering on the ATM switch. If the number of sources being in on$^+$ state exceeds N_{VC}, then datagrams are lifted up to the routing kernel. In this case, all cells composing one IP datagram arrive at the ATM switch and are stored in the buffer of the routing kernel at once, that is, when the state of each source changes from on$^+$ to on$^-$ or off, cells only arrive at the routing kernel.

3. Analysis

In this section, we will derive the steady state probability and give the cut-through rate, the mis-ordered rate and the average packet waiting time of the layer3 switching system. We first introduce following random variables in the tth slot:

- $D(t)$: the queue length in cells of the buffer at the routing kernel;
- $S_{\text{on}^+}(t)$: # of sources being in the on$^+$ state;
- $S_{\text{on}^-}(t)$: # of sources being in the on$^-$ state.

Since sources are homogeneous in terms of traffic characteristics, we can completely describe the state of our system by the above variables. Furthermore, each source generates cells according to the 3-state MMBP described in the previous section, so that the system state $\{D(t), S_{\text{on}^+}(t), S_{\text{on}^-}(t); t = 0, 1, 2\cdots\}$ can form a discrete time Markov chain. Note that when $S_{\text{on}^+}(t) = j$ and $S_{\text{on}^-}(t) = k$, the number of sources in the off state is given by $N - j - k$, where N denotes the number of sources in our system. Here, we define S as the state space for $(S_{\text{on}^+}(t), S_{\text{on}^-}(t))$, i.e., $S = \{(j,k)|0 \le j, k, j+k \le N\}$. A state (j,k) is labeled using the following one-to-one mapping function $M(j,k)$:

$$M(j,k) = \frac{1}{2}j(2R - j + 3) + k. \tag{5}$$

Therefore, S can be equivalently represented by the state space $S^* = \{0, \cdots, M_s - 1\}$, where $M_s = (N+1)(N+2)/2$.

3.1 Transition Probability of the Number of Sources Being in on State

We assume that cell generations from each source happen just after phase state transitions of 3-state MMBP occur. Thus, we first define the transition probability of the number of sources in each state.

Since we suppose that cells arrive in batch at the routing kernel only when the state of each source changes from on$^+$ to on$^-$ or off, we will define the following random variables:

- $S_{+-}(t)$: the number of sources changing their states from on$^+$ at the $t-1$st slot to on$^-$ at the tth slot;
- $S_{+*}(t)$: the number of sources changing their states from on$^+$ at the $t-1$st slot to off at the tth slot.

By using these variables, we can define the transition probability $P_{M(j,k),M(m,n),p,q}$ as follows.

$$\begin{aligned} &P_{M(j,k),M(m,n),p,q} \\ &\triangleq \Pr\{S_{\text{on}^+}(t) = m, S_{\text{on}^-}(t) = n, S_{+-}(t) = p, \\ &S_{+*}(t) = q | S_{\text{on}^+}(t-1) = j, S_{\text{on}^-}(t-1) = k\} \end{aligned} \tag{6}$$

By supposing that $s(r)$ sources change their states from on$^-$ to on$^+$(off), we can completely give the transition of the number of sources as shown in Table 1.

Table 1 Transition of the number of sources in each state.

	$S_{\mathrm{on}^+}(t)$	$S_{\mathrm{on}^-}(t)$	$S_{\mathrm{off}}(t)$	
$S_{\mathrm{on}^+}(t-1)$	j-p-q	p	q	j
$S_{\mathrm{on}^-}(t-1)$	s	k-r-s	r	k
$S_{\mathrm{off}}(t-1)$	m-j+p+q-s	n-k-p+r+s	N^+-q-r	N^-
	m	n	N^+	N

The rightmost column (the bottom row) represents the number of sources in each state before (after) the transition. In this table, $N^- = N - j - k$ and $N^+ = N - m - n$. From this table and Eq. (1), we thus have $P_{M(j,k),M(m,n),p,q}$ as

$$\begin{aligned}
&P_{M(j,k),M(m,n),p,q}\\
&\quad = \binom{j}{q}\alpha^q(1-\alpha)^{j-q}\binom{j-q}{p}\\
&\quad\times \gamma^p(1-\gamma)^{j-p-q}\sum_{r=\max(0,N^+-N^--q)}^{\min(k,N^+-q)}\\
&\quad\cdot\sum_{s=\max(0,k-n+p-r)}^{\min(k-r,m-j+p+q)}\binom{k}{r}\alpha^r(1-\alpha)^{k-r}\\
&\quad\times\binom{k-r}{s}\omega^s(1-\omega)^{k-r-s}\\
&\quad\times\binom{N^-}{N^+-q-r}(1-\beta)^{N^+-q-r}\\
&\quad\times\left(\frac{\beta}{\gamma+\omega}\right)^{N^--N^++q+r}\\
&\quad\times\binom{N^--N^++q+r}{m-j+p+q-s}\\
&\quad\times\gamma^{n-k-p+r+s}\omega^{m-j+p+q-s}\\
&\qquad\text{if } p+q\le j
\end{aligned}\tag{7}$$

3.2 Definition of Cell Arrival Probability at the Routing Kernel

By letting $a(t)$ denote a random variable which represents the number of cells arriving at the routing kernel, we define the cell arrival probability $a_{M(j,k),M(m,n)}(x)$ and obtain it as follows.

$$\begin{aligned}
&a_{M(j,k),M(m,n)}(x)\\
&\quad=\Pr\{a(t)=x, S_{\mathrm{on}^+}(t)=m,\\
&\qquad S_{\mathrm{on}^-}(t)=n|S_{\mathrm{on}^+}(t-1)=j,\\
&\qquad S_{\mathrm{on}^-}(t-1)=k\}
\end{aligned}$$

$$=\begin{cases}
\displaystyle\sum_{i=0}^{j}\sum_{q=0}^{j-i}P_{M(j,k),M(m,n),i,q} & \text{if } x=0, j\le N_{VC}\\
0 & \text{if } x>0, j\le N_{VC}\\
\displaystyle\sum_{i=0}^{N_{VC}}\sum_{q=0}^{N_{VC}-i}P_{M(j,k),M(m,n),i,q} & \text{if } x=0, j>N_{VC}\\
\displaystyle\sum_{i=N_{VC}+1}^{j}P(x)^{(i-N_{VC})*}\times\sum_{q=0}^{i}P_{M(j,k),M(m,n),i-q,q} & \text{if } x>0, j>N_{VC}
\end{cases}\tag{8}$$

where $P(x)^{i*}$ denotes the probability that i sources change their states from on$^+$ to on$^-$ or off and x cells from i sources are stored in the queue of the routing kernel. Since the duration of on$^+$ states follows the geometric distribution with parameter $(1-\alpha)(1-\gamma)$ in each source, $P(x)^{i*}$ is given by the negative binomial distribution with parameter $(i, 1-(1-\alpha)(1-\gamma))$, i.e.,

$$\begin{aligned}
P(x)^{i*} &= \binom{x-1}{x-i}\{(1-\alpha)(1-\gamma)\}^{x-i}\\
&\quad\times\{1-(1-\alpha)(1-\gamma)\}^i.
\end{aligned}\tag{9}$$

3.3 Steady State Probability

We define the system state probability at the tth slot as

$$\begin{aligned}
&P_t(b, M(j,k))\\
&\quad\triangleq \Pr\{D(t)=b, S_{\mathrm{on}^+}(t)=j, S_{\mathrm{on}^-}(t)=k\},
\end{aligned}$$

and the steady state probability $x(b, M(j,k))$ and its vector representation as

$$\begin{aligned}
&x(b,M(j,k))\triangleq\lim_{t\to\infty}P_t(b,M(j,k))\\
&\boldsymbol{x}_b=[\cdots,x(b,M(j,k)),\cdots]\\
&\qquad 0\le M(j,k)\le M_s-1
\end{aligned}\tag{10}$$

We next give the transition probability matrix $\boldsymbol{A}(x)$ representing that x cells arrive at the routing kernel with changing state of sources, as follows:

$$\boldsymbol{A}(x)=\begin{pmatrix} a_{0,0}(x) & \cdots & a_{0,M_s-1}(x)\\ \vdots & \cdots & \vdots\\ a_{M_s-1,0}(x) & \cdots & a_{M_s-1,M_s-1}(x)\end{pmatrix}.\tag{11}$$

Since the system state forms Markov chain of M/G/1 type queuing system, the steady state probability vector $\boldsymbol{x}_b$ can be related to the transition probability matrix $\boldsymbol{A}(x)$ as follows [9]:

$$\boldsymbol{x}_b = \boldsymbol{x}_0 \boldsymbol{A}(b) + \sum_{i=1}^{b+1} \boldsymbol{x}_i \boldsymbol{A}(b+1-i), \quad \text{for } b \geq 0$$

$$\sum_{b=0}^{\infty} \boldsymbol{x}_b \cdot \boldsymbol{e}[M_s] = 1, \tag{12}$$

where $\boldsymbol{e}[M_s]$ represents a column vector of dimension M_s, whose elements are all equal to one and Eq. (12) can be stably solved by using the algorithm proposed in [10].

3.4 Derivation of Performance Measures

By using the steady state probability obtained in the previous subsection, we can get some performance measures.

3.4.1 Cut-Through Rate, R_c

We shall first derive the cut-through rate R_c. R_c is the ratio of the average number of sources changing their states from on$^+$ to on$^-$ or off when the number j of sources being in on$^+$ states is less than or equal to N_{VC}, to the average number of sources changing their states from on$^+$ in a slot. This is given by

$$R_c = \frac{\displaystyle\sum_{b=0}^{\infty} \sum_{j=1}^{N_{VC}} \sum_{k=0}^{N-j} x(b, M(j,k)) \sum_{i=1}^{j} i D_{M(j,k)}(i)}{\displaystyle\sum_{b=0}^{\infty} \sum_{j=1}^{N} \sum_{k=0}^{N-j} x(b, M(j,k)) \sum_{i=1}^{j} i D_{M(j,k)}(i)} \tag{13}$$

where $D_{M(j,k)}(i)$ is the probability that i datagrams get to the routing kernel on condition that the number of on$^+$ (on$^-$) sources was j (k) in the previous slot. Here, let $D(t)$ denote a random variable which indicates the number of datagrams arriving in the tth slot. $D_{M(j,k)}(i)$ is thus given as follows.

$$\begin{aligned} D_{M(j,k)}(i) &= \Pr\{D(t) = i \\ &\qquad |S_{\text{on}^+}(t-1) = j, S_{\text{on}^-}(t-1) = k\} \\ &= \sum_{m=0}^{N} \sum_{n=0}^{N-m} \sum_{q=0}^{i} P_{M(j,k),M(m,n),i-q,q}. \end{aligned} \tag{14}$$

3.4.2 Average Datagram Waiting Time, W_d

We let the z transform of the steady state probability vector $\boldsymbol{x}_b$ be $\boldsymbol{X}(z)$, The first equations of (12) lead to

$$\boldsymbol{X}(z)[z\boldsymbol{I}(M_s) - \boldsymbol{A}(z)] = \boldsymbol{x}_0(z-1)\boldsymbol{A}(z), \tag{15}$$

where $\boldsymbol{A}(z) = \sum_{i=0}^{\infty} \boldsymbol{A}(i) z^i$ and $\boldsymbol{I}(M_s)$ is the unit matrix of the dimension M_s. Neuts [9] shows that the n-th moment of the probability density $\boldsymbol{x}_b$ is obtained by the n-times differentiation of $\boldsymbol{X}(z)$ with respect to z. Thus, the average queue length L of the buffer at the routing kernel is given by $L = \left.\frac{d\boldsymbol{X}(z)}{dz}\right|_{z=1}$. Since the total arrival rate at the layer3 switch becomes $N\lambda$, by using Little's formula, the average datagram waiting time W_d can be obtained as follows.

$$W_d = \frac{L}{N\lambda}. \tag{16}$$

3.4.3 Mis-Ordered Rate, R_m

Finally, we will obtain the mis-ordered rate R_m. The term "mis-order" means the out-of-order transmission of datagrams in [7] and we also use the term in the paper. The mis-order occurs according to following steps;

1. A datagram of a flow, which will be called *first datagram*, is not cut through on the ATM switch.
2. The subsequent datagrams of the flow is directly transfered on the ATM switch by cut-through before the first datagram is served at the routing kernel.

Namely, focusing on the state transition of a source, we find that it is possible to occur that the state of each source changes from on$^+$ to on$^-$, and again to on$^+$. Here, we will define the probability $D^*_{M(j,k)}(i)$, which indicates that i datagrams from sources arrive at the routing kernel and sources changes their states to on$^-$, as $D^*_{M(j,k)}(i) = \sum_{m=0}^{N} \sum_{n=0}^{N-m} P_{M(j,k),M(m,n),i,0}$. R_m is thus obtained by

$$\begin{aligned} R_m &= \sum_{b=1}^{\infty} \sum_{j=N_{VC}+1}^{N} \sum_{k=0}^{N-j} x(b, M(j,k)) \sum_{i=1}^{j} i D^*_{M(j,k)}(i) \\ &\quad \times \sum_{l=1}^{b} \{(1-\alpha)(1-\omega)\}^{l-1} \{(1-\alpha)\omega\} R_c \\ &\quad / \sum_{b=0}^{\infty} \sum_{j=1}^{N} \sum_{k=0}^{N-j} x(b, M(j,k)) \sum_{i=1}^{j} i D_{M(j,k)}(i) \\ &= R_c(1-\alpha)\omega \sum_{b=1}^{\infty} \sum_{j=N_{VC}+1}^{N} \sum_{k=0}^{N-j} x(b, M(j,k)) \\ &\quad \times \sum_{i=1}^{j} i D^*_{M(j,k)}(i) \left(1 - \{(1-\alpha)(1-\omega)\}^b\right) \\ &\quad / (1 - (1-\alpha)(1-\omega)) \sum_{b=0}^{\infty} \sum_{j=1}^{N} \sum_{k=0}^{N-j} x(b, M(j,k)) \\ &\quad \times \sum_{i=1}^{j} i D_{M(j,k)}(i). \end{aligned} \tag{17}$$

4. Numerical Results and Discussions

In this section, we evaluate the performance using the analysis presented in the preceding section.

4.1 Impact of N_{VC}: Comparison with Simulation Results

First, we will show the impact of the number N_{VC} of the maximum VCs for cut through transmission on performance in Fig. 4. We set the number N of sources to 8 and the average arrival rate λ to 0.125, that is, the total arrival rate $N\lambda$ equals 1. Furthermore, the average IP flow length (i.e. $E[\text{on}]$), the average datagram length (i.e. $E[\text{on}^+]$) and $E[\text{on}^-]$ are set to 100, 10 and 10 in cells, respectively.

It would be decided whether datagrams are cut-through via the ATM switch or stored in the routing kernel when the first cell of datagrams just arrives at the layer3 switch. However, in our assumption, a datagram is forwarded to the routing kernel if the number of sources being in on^+ state is more than N_{VC} in *the* slot just when the last cell of it arrives at the ATM switch. Hence, we should first ascertain the assumption to be valid. For this purpose, we also compare analytical results with simulation ones in Fig. 4.

Figure 4(a) illustrates the cut-through rate R_c. We can see from this figure that analytical results just agree with simulation ones, and that R_c is almost equal to 1 when $N_{VC} \geq 4$. This implies that if the number of VCs provided is more than a half of the number of sources, almost all datagrams are transfered by cut-through.

The average datagram waiting time W_d is shown in Fig. 4(b) from which we have found that analytical results are larger than simulation ones, however these differences are quite small. When $N_{VC} = 0$, namely, the layer3 switch works as pure layer3 router, W_d becomes approximately 1000 (cells) while it is less than 40 (cells) if N_{VC} is set to only 1. Therefore, we can say that the cut-through technique is very effective in reducing transfer delay.

Figure 4(c) shows the mis-ordered rate R_m, and that analytical results are slightly larger than simulation ones. Note that when $N_{VC} = 0$, $R_m = 0$. It is found that R_m becomes the highest value if $N_{VC} = 1$, however, as N_{VC} increases, R_m exponentially decreases.

Through the above discussions, we can neglect the difference between analytical results and simulation ones, hereafter, we will discuss the performance based only upon the analytical results.

4.2 Impact of Total Arrival Rate, $N\lambda$

Figure 5 shows the impact of the total arrival rate $N\lambda$ at the layer3 switch on performance. In this figure,

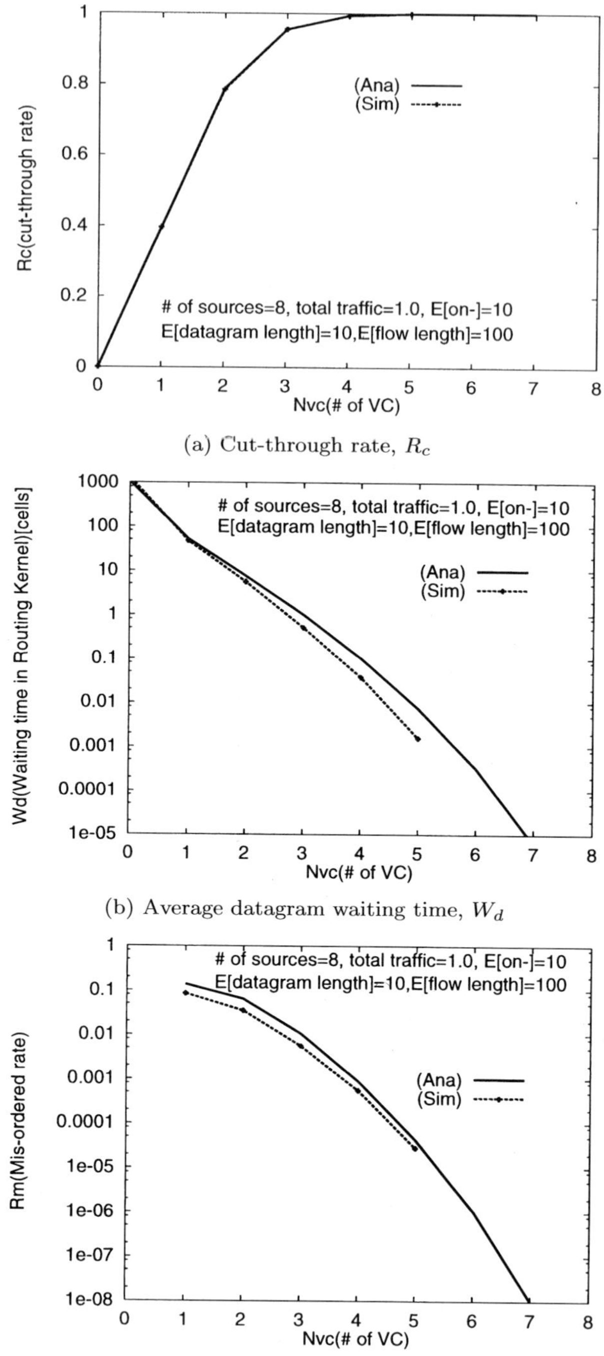

(a) Cut-through rate, R_c

(b) Average datagram waiting time, W_d

(c) Mis-ordered rate, R_m

Fig. 4 Performance Comparison between analysis and simulation.

we set $N\lambda$ to 0.5, 1.0, 2.0 and 3.0. The cut-through rate R_c is illustrated in Fig. 5(a). We can see that the smallest N_{VC} achieving that $R_c = 1$ becomes larger as $N\lambda$ is larger. For example, if $N\lambda = 3.0$, the value of N_{VC} becomes 6 whereas only 3 VCs are required when $N\lambda = 0.5$.

Figure 5(b) shows the average datagram waiting time W_d. Note that the queue length distribution obtained from the steady state probability $\boldsymbol{x}_b$ is unstable when the actual arrival rate at the routing kernel $N\lambda(1-R_c)$ exceeds 1.0 of the capacity of the routing kernel. We thus omit W_d in this range of N_{VC}. When $N\lambda$ doubles, W_d degrades about one order of magnitude.

Figure 5(c) illustrates the mis-ordered rate R_m. From this figure, R_m is maximized at N_{VC} of 1, 1, 2 and 3 when $N\lambda = 0.5, 1.0, 2.0$ and 3.0 respectively. This is due to the followings. The mis-ordered transmission of datagrams occurs when the first arrival datagram of some flow is lifted up to the routing kernel and the subsequent datagram of the flow is directly transmitted via the ATM switch by cut-through. Thus, R_m would be in proportion to $R_c(1-R_c)$, which is maximized when R_c becomes 0.5. From Fig. 4(a), when $N\lambda = 0.5, 1.0, 2.0$ and 3.0, the values of N_{VC} making R_c of about 0.5 become 1, 1, 2 and 3 respectively.

4.3 Impact of the Number N of Sources

Figure 6 shows the impact of the number N of sources on performance. In this figure, N is set to 4, 8 and 12 and the x axis indicates the ratio of N_{VC} to N. Note that the average arrival rate λ of each source gets smaller as N becomes larger since $N\lambda$ is fixed, i.e., set to 1.0 for each N. Therefore, as N_{VC} increases, R_c becomes smaller and both W_d and R_m get larger when N is smaller, however, from a viewpoint in terms of N_{VC}/N, we have an opposite tendency. This implies that in addition to the number N_{VC}, the assigned bandwidth of each VC is very important factor for improving performance of the layer3 switch.

4.4 Impact of Traffic Parameter

First, we show the impact of the average IP datagram length $E[\text{on}^+]$ on performance in Fig. 7. We set $E[\text{on}^+]$ to 2, 5, 10 and 20. Note here that, if the average arrival rate λ is fixed, the cut-through rate R_c is kept some fixed value although some of the MMBP parameters $(\alpha, \beta, \gamma, \omega)$ are quite different. In this subsection, since we suppose λ to be fixed, we omit the figure of R_c which is the same as Fig. 4(a). From Fig. 7(a), the average datagram waiting time W_d gets larger as $E[\text{on}^+]$ becomes larger. From Fig. 7(b), mis-ordered rate R_m is also larger when $E[\text{on}^+]$ is larger. Furthermore, as N_{VC} increases, the differences of R_m become larger.

Next, on condition that the number of datagrams in one flow is set to the fixed value, the impact of $E[\text{on}^+]$ on performance is shown in Fig. 8. In comparison with the case in Fig. 7, the differences of W_d are relatively small, and R_m for any $E[\text{on}^+]$ gets the fixed value except for the case when N_{VC} is large; R_m becomes smaller when $E[\text{on}^+]$ is smaller.

Finally, we also show the impact of the average IP flow length $E[\text{on}]$ in Fig. 9. In this figure, we set $E[\text{on}]$ to 50, 100, 500 and 1000. From Fig. 9(a), W_d becomes larger when $E[\text{on}]$ is larger if N_{VC} is set to a relatively small value. Otherwise W_d takes the fixed value for each $E[\text{on}]$. From Fig. 9(b), R_m is smaller as $E[\text{on}]$ is set to a smaller value, however, the differences of R_m are insensitive to N_{VC}.

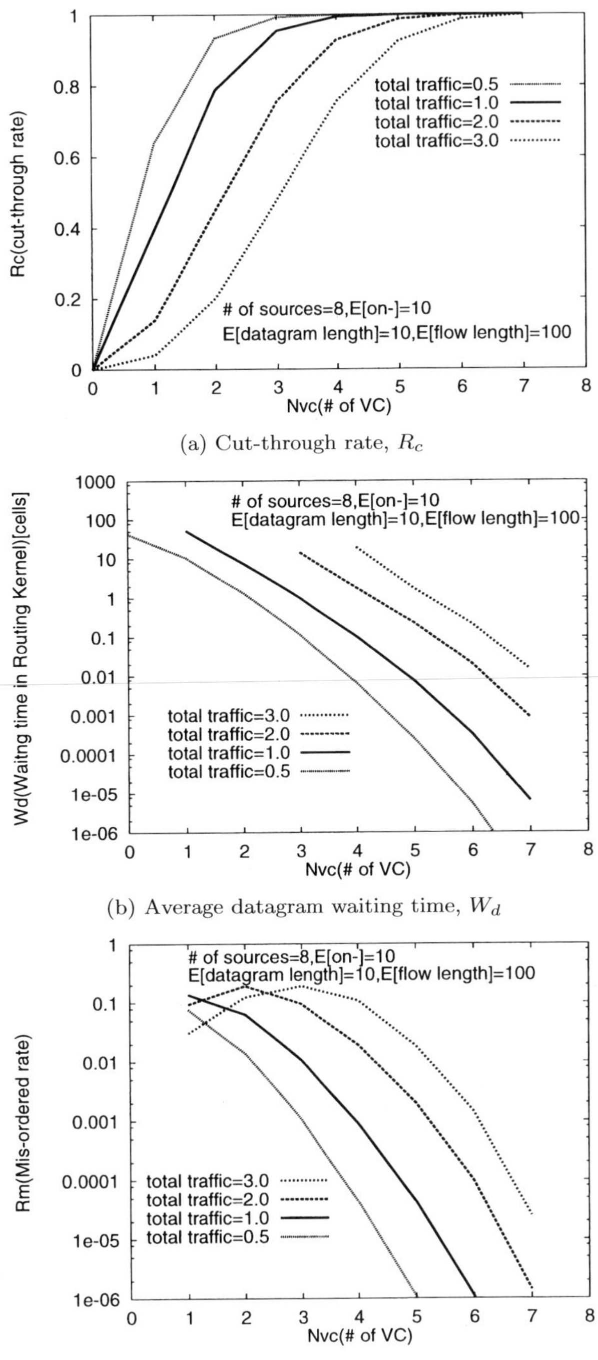

(a) Cut-through rate, R_c

(b) Average datagram waiting time, W_d

(c) Mis-ordered rate, R_m

Fig. 5 Impact on total traffic rate, $N\lambda$.

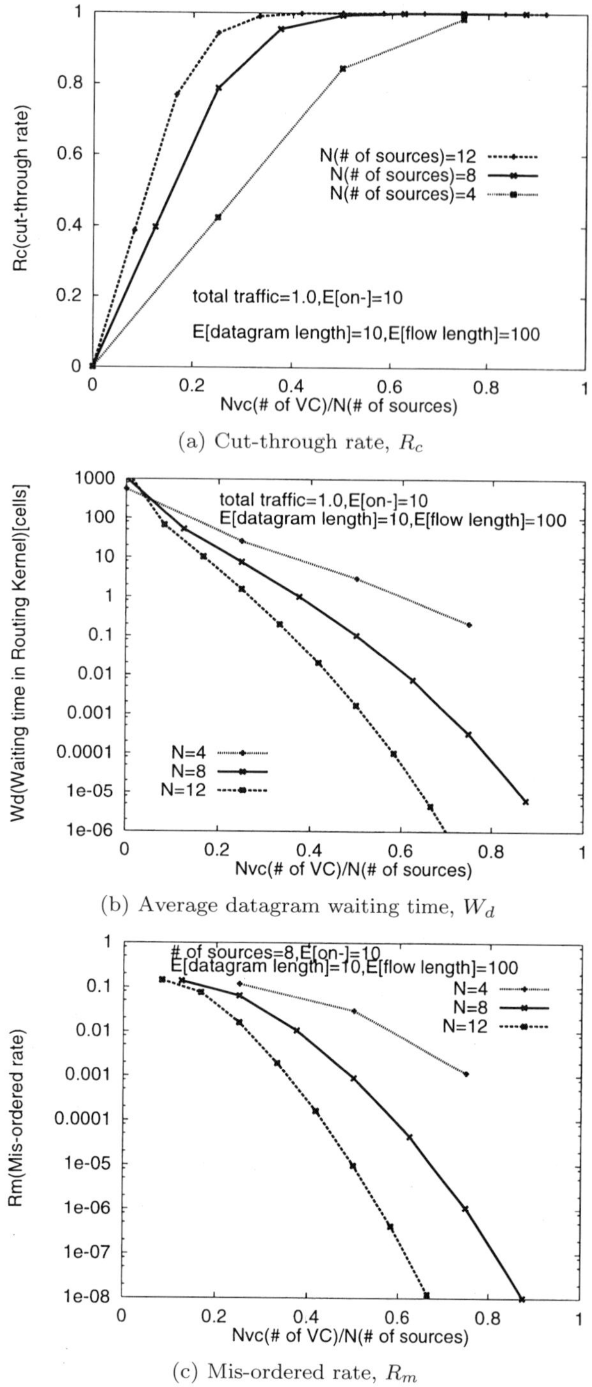

(a) Cut-through rate, R_c

(b) Average datagram waiting time, W_d

(c) Mis-ordered rate, R_m

Fig. 6 Impact of the number of sources, N.

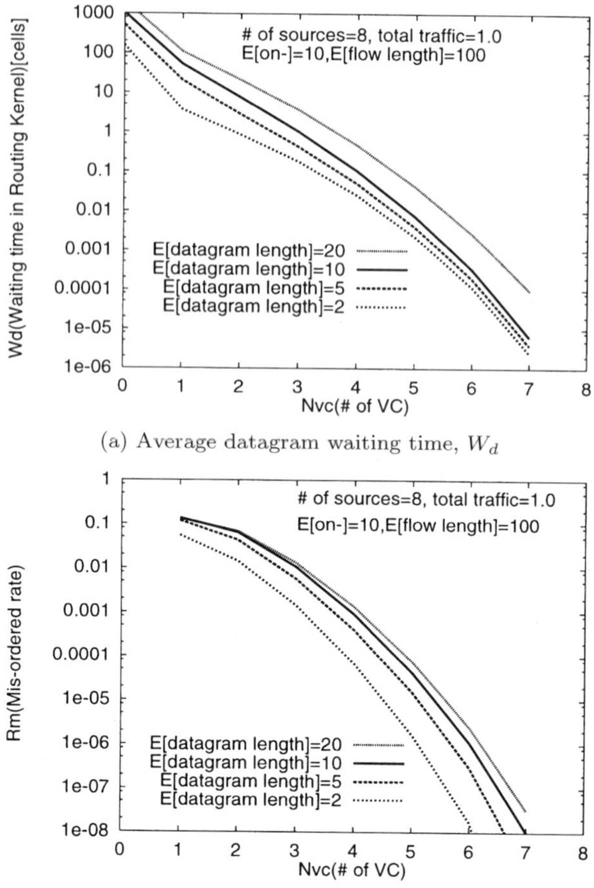

(a) Average datagram waiting time, W_d

(b) Mis-ordered rate, R_m

Fig. 7 Impact of the datagram length, E[on+].

5. Concluding Remarks

The cut-through technique on the layer3 switch has been considered as a way to fast transmit IP datagrams in the Internet backbone. Therefore, in this paper, we have proposed the analysis of the cut-through rate, the average IP datagram waiting time at the routing kernel and the mis-ordered rate of the layer3 switch in case of the flow-driven connection setup. We modeled the arrival process of IP flow and datagram from each source as the 3-state Markov modulated Bernoulli process (MMBP) with a parameter set $(\alpha, \beta, \gamma, \omega)$, which allows us to investigate the relation of the number of multiplexing sources and that of VCs provided cut-through on performance of the layer3 switch.

Through some numerical results, we have found that if the number of VCs provided for cut-through is more than a half of the number of sources almost all datagrams are directly transferred via the ATM switch when the total arrival traffic is equal to 1.0 of the capacity of the routing kernel. Both the average datagram waiting time W_d and the mis-ordered rate R_m exponentially decreases as the number of provided VCs increases. Furthermore, R_m is in proportion to $R_c(1-R_c)$ where R_c indicates the cut-through rate.

The impacts of traffic parameters on performance obtained here are summarized below.

- R_c is kept the fixed value when the total arrival rate at the layer3 switch is fixed.
- Both W_d and R_m increases as the average IP datagram length.
- On condition that we set the number of datagrams

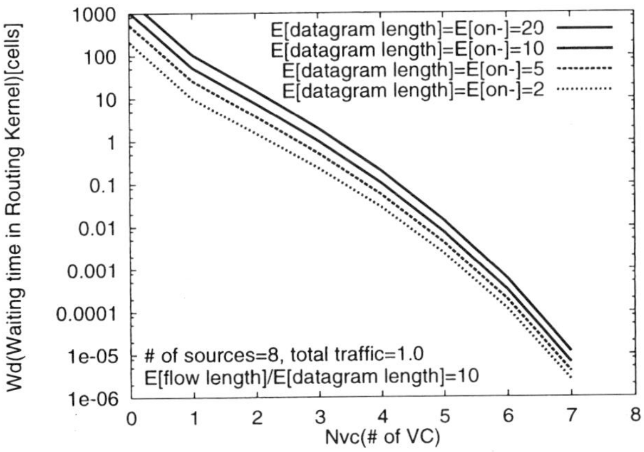

(a) Average datagram waiting time, W_d

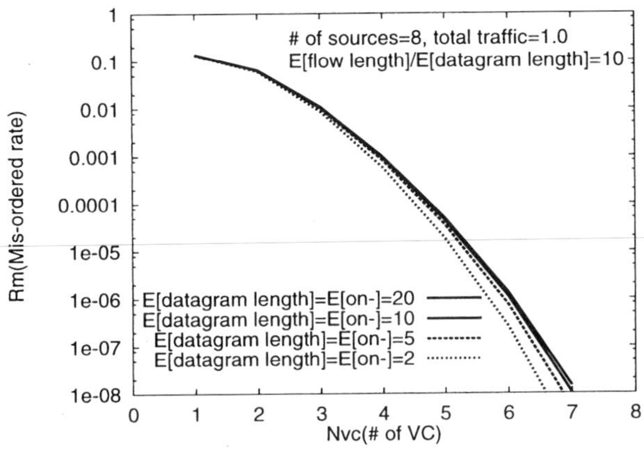

(b) Mis-ordered rate, R_m

Fig. 8 Impact of the datagram length, E[on+], in case that the number of datagrams in a flow is fixed.

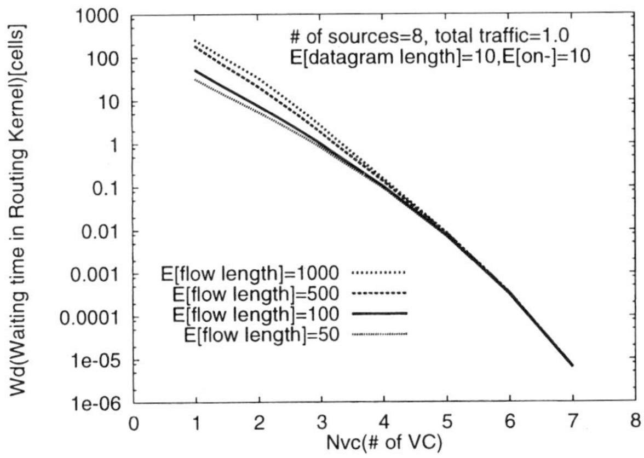

(a) Average datagram waiting time, W_d

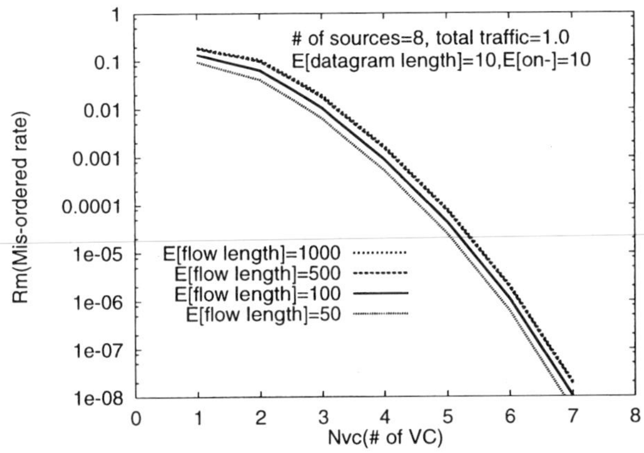

(b) Mis-ordered rate, R_m

Fig. 9 Impact of the flow length, E[on].

in single IP flow to be fixed, R_m is fixed for any value of the average datagram length.

- As the average IP flow length becomes larger, R_m increases, while the differences are quite little.

In the analysis, since the occupied memory space for calculation is in proportion to $O(N^4)$, where N denotes the number of sources, we can calculate the steady state probability of the system with N of up to 12. Although the behavior of the system was qualitatively clarified, we should also deal with the case for larger N (e.g. the order of hundreds or thousands). Thus, we will extend the model for this purpose.

As mentioned in [7], the number of required VCs for backbone network is about 30,000 and this would be explosively increasing as the number of hosts connected via the Internet keeps on growing. One of solutions to "VC explosion" [1] is the VC merging in the topology-driven type layer3 switch such as Aggregated Route-Based IP Switching (ARIS) [11]. Another solution is to selectively cut-through the traffic with stringent delay bound and this would lead to the contribution to some QoS provisioning. Thus, IETF have recently proposed multi-protocol label switching (MPLS) [12] as generalized and expanded layer3 switching mechanism. However, investigation of these topics will be left for further study.

Acknowledgement

This work was supported in part by Research for the Future Program of Japan Society for the Promotion of Science under the Project "Integrated Network Architecture for Advanced Multimedia Application Systems" (JSPS-RFTF97R16301) and Telecommunications Advancement Organization of Japan.

References

[1] P.P. White, "ATM switching and IP routing integration: The next stage in internet evolution?" IEEE Commun. Mag., vol.36, no.4, pp. 79–83, April 1998.
[2] M. Laubach, "Classical IP and ARP over ATM," RFC1577, Jan. 1994.
[3] "Multi-protocol over ATM Version-1," af-mpoa-0087.000, July 1997.
[4] P. Newman, T. Lyon, and G. Minshall, "Flow labeled IP: Connectionless ATM under IP," Networld+Interop, April 1996.
[5] TOSHIBA Corporation, "Cell switch router," whitepaper,

Nov. 1996.

[6] Y. Rekhter, B. Davie, D. Katz, E. Rosen, and G. Swallow, "Cisco systems' tag switching architecture overview," RFC2105, Feb. 1997.

[7] S. Lin and N. McKeown, "A simulation study of IP switching," ACM SIGCOMM'97, Sept. 1997.

[8] Z. Niu, Y. Takahashi, and N. Endo, "Performance evaluation of SVC-based IP-over-ATM networks," IEICE Trans. Commun., vol.E81-B, no.5, pp. 948–957, May 1998.

[9] M.F. Neuts, Structured Stochastic Matrices of M/G/1 Type and Their Applications, Marcel Dekker, New York, 1989.

[10] V. Ramaswami, "Stable recursion for the steady state vector in Markov chains of M/G/1 type," Stochastic Models, vol.4, pp.183–188, 1988.

[11] A. Viswanathan, N. Feldman, R. Boivie, and R. Woundy, "ARIS: Aggregate route-based IP switching," Internet draft, March 1997.

[12] A. Viswanathan, N. Feldman, Z. Wang, and R. Callon, "Evolution of multiprotocol label switching" IEEE Commun. Mag., vol.36, no.5, pp. 165–173, May 1998.

Chapter 3

ATM Switch Architecture and Systems

Many studies on ATM architectures have emphasized buffer performance and speed. Papers 3.1 through 3.4 describe new switch architectures such as input/output buffer types, input buffer and multi-stage switches. Paper 3.1 details the Input/Output buffer type switch with feedback loop. Head-of-line blocking is reduced by employing a feedback loop from the blocked output buffer.

Another approach to reducing head-of-line blocking is described in Paper 3.2, the odd/even switch. This switch dramatically improves the switching performance even if the input buffers are the First-In-First-Out (FIFO) type. Each input port has Odd FIFO for odd destination ports and Even-FIFO for even destination ports. HOL contention is reduced by appropriately selecting the Odd or Even FIFO.

Paper 3.3 proposes an input-buffered switch architecture based on a virtual-output-queued structure supporting QoS. It offers both deterministic, i.e., firm, and stochastic delay bounds per connection for prioritized traffic. This is achieved by a hybrid static and dynamic arbitration scheme.

To expand switch capacity, a multi-stage switch architecture is often used. J. Agrawal et al. show a modified multi-stage ATM switch architecture in Paper 3.4. The architecture combines the multistage and recurrence architectures.

Another important topic in discussing ATM switch architectures is multicasting. Papers 3.5 through 3.8 tackle this field. Multicasting can be realized by the tandem copy network or the recurrence architecture. Papers 3.5 and 3.6 describe the copy network architecture for tandem copy networks. They determine the optimal copy point in a multi-stage switch network. Paper 3.7 provides several theoretical approaches to the multi-stage copy network. The switch proposed in Paper 3.8 consists of non-blocking unicast and extended multicast networks. For multicasting, a simple bus copy network is used. This approach differs from those in Papers 3.5 and 3.6.

A system level demonstration of a digital MPEG transport network is shown in Paper 3.9. The network is built around HIPPI and an ATM connection transmission system. HIPPI and ATM protocol are converted at the network edge point. The study shows that a wide-area network can be realized by ATM with local networks based on HIPPI. A performance study and experimental results are also detailed.

An experimental system for ATM access networks is introduced by H. Kasahara et al. in Paper 3.10. The authors describe a subscriber system based on ATM-PDS (Passive Double Star). ATM-PDS is a very cost-effective way to realize many ATM transfer capabilities. The paper shows the application of ATM switches and the system in detail.

When line speeds become high, the allowable scheduling time decreases and can become a bottleneck. E. Oki et al. address a scheduling scheme that relaxes the constraint of the scheduling time in Paper 3.11. The authors propose a pipelined-based maximal-sized matching scheduling approach. Each sub-scheduler is allowed to take more than one time to complete maximal-size matching. It is shown that the proposed pipelined-based scheme provides 100% throughput as the non-pipelined algorithm does.

An output-buffered-type switch is attractive because it provides ideal delay performance. However, it is difficult to implement high-speed switches. To avoid this problem, Y-W. Leung addresses the design and analysis of a packet concentrator in Paper 3.12. A packet concentrator is useful since it simplifies output-buffered switches. The proposed concentrator has a significantly smaller probability of packet loss than previous alternatives.

Paper 3.13 by I. Elhanany introduces a high-capacity switching scheme for ATM/WDM networks. The proposed architecture is contention-free, scalable, easy to implement and requires no internal "speedup." The authors show that the throughput of one terabit/sec is achieved with low latency. The proposed architecture yields a powerful solution for high-performance packet-switch networks.

The final paper of this chapter studies a distributed traffic control scheme for large multi-stage ATM switching system. K. Nakai et al. in Paper 3.14 define operation, administration, and maintenance (OAM) cells inside the system. They carry route congestion information inside the system. The performance of the proposed scheme depends on the interval between OAM cells and the authors show how an optimal interval can be determined.

High-Performance Backbone Network Technology edited by Naoaki Yamanaka
ISBN: 0-8247-5321-6 DOI: 10.1081/(120025671)
Published 2004 by Marcel Dekker, Inc.

Input and Output Queueing Nonblocking Switch with Feed-Back Loop

Jaewan Choi, Shigeki Shiokawa, and Iwao Sasase

SUMMARY Input and output queueing nonblocking switch with feed-back loop is proposed in order to reduce the system delay. It has b head of the line (HOL)'s and b output ports feed-backed into the HOL's of feed-back loop. The cell which receives backpressure over threshold is switched and enters one of these HOL's through a feed-back loop. It provides the cell, which should wait behind the HOL of input queue until the cell in that HOL is served, with opportunity to be served immediately by vacating the HOL of the input queue. We obtain the mean system delay and consider the effect of the size of b by computer simulation. As the results, it is shown that proposed switch with appropriate size of b can reduce the system delay compared to the conventional switch.

key words: *backpressure mode, HOL, head of the line blocking, feed-back loop, cell flowing*

1. Introduction

Input queueing, output queueing, and input and output queueing nonblocking switches have been studied [1]–[4]. Input queueing switch has the advantages of high speed and low cost. However, only one cell can be transmitted over the output trunk and the throughput is restricted. Output queueing switch provides optimum throughput but all cells destined for a specific output queue simultaneously must be accepted immediately. This multiport property can be realized by running switch fabrics at appropriately higher speed or by a highly paralleled switch implementation. However both methods are complex and expensive [3]. Consequently, only limited amount of buffers can be provided per output port. In order to cope with disadvantages of the input queueing switch and output queueing switch, the switch with input and output queueing has been studied [3], [4]. In input and output queueing switch, sufficiently large amount of input buffers is provided for each input port and small amount of output buffer is provided for each output port. Iliadis et al. proved that in the input and output queueing switch output buffers of relatively small capacity, say, B_0 cells, are sufficient for closely approaching the ideal performance of infinite output queues [3]. Thus, the required speedup or parallelism for the output queues can be limited to a factor B_0. However, when the output buffer is saturated, the classical problem of output port contention occurs. If the cell dispatched to a specific output port loses contention, it must be discarded. Backprssure Mode (BP mode) resolves this problem and avoids cell loss at the output queue. In BP mode, each cell which can not be switched to output port receives backpressure, which is the number of time slots each cell has to wait at the HOL position until the cell will be switched to the corresponding output port. However in BP mode, other cell may be queued behind the cell waiting at a HOL. And this queued cell is blocked from reaching momentarily idle output port. This is referred to as head of the line blocking and it increases waiting time in input queue. We focus our attention on decreasing the effect of head of the line blocking on input and output queueing $N \times N$ nonblocking switch with backpressure mode. Some methods to decrease the effect of head of the line blocking are studied [5], [6]. Input buffer control method in [5] considers speedup factor of the switch with infinite output buffer and needs special control policy in input queueing. Two stages ATM switch in [6] needs an extra switch and intermediate buffer between two switches. It is not wanted that hardware to be grown so much and controlling to be complicated. Several switch architectures in which packets are feedbacked are studied [7]–[11]. M.G. Hluchyj et al. [7] show that *Completely Shared Buffering* which represents *Starlite Digital Switch* [8] requires smaller total buffer memory for large N than that of output queueing switch. However, the switch fabric is enlarged from $N \times N$ to $N(b+1) \times N(b+1)$ and the packets continue to recirculate through the switch fabric and shared buffer until they are routed to output because the switch has neither input buffer nor output buffer. In *Sunshine* [9] that utilizes both internal and output queueing techniques, excess cells due to momentary output overloads, overflow into a shared recirculating queue. Packet loss due to an excessive number of re-entering packets will increase, as well as the possibility of arriving packets out-of-sequence [10]. In *Feedback Input Queueing* system of G. Kbar [11], feedback packets reenter the head of the input queue to maintain the sequence integrity. Therefore, the problem to reduce the head of the line blocking remains. Furthermore, extra routing network is needed in the feedback loops to route feedback packet to original head of the input queue.

In this paper, we propose input and output queue-

Manuscript received April 24, 1997.
Manuscript revised July 28, 1997.

[†]The authors are with the Faculty of Science and Technology, Keio University, Yokohama-shi, 223–0061 Japan.

IEICE TRANS. COMMUN., Vol. E81-B, No. 2, pp. 175–181, February 1998

ing $(N+b) \times (N+b)$ nonblocking switch model with feed-back loop. Feed-back loop consists of b HOLF's and b output trunks which are feed-backed into the HOLF's. Our switch with backpressure mode moves the cell which gets backpressure over threshold among cells in excess of output capacity momentarily from HOL to HOLF, so that the head of the line blocking is solved. This promotes cell flowing through the switch and this promotion produces the same effect as an increase of output capacity. Thus, our switch with enlarging switch fabric from $N \times N$ to $(N+b) \times (N+b)$ can reduce output buffer. The cells which once receive backpressure waits to be dispatched to the destined output in HOL or HOLF until the backpressure becomes 0. Therefore, there is no recirculation and no extra routing network is needed because the cell is switched from HOL to HOLF using main switch fabric. Each cell holds given backpressure and each output port issues backpressure according to the memory that remembers largest backpressure which the output port has issued. And both of them are decreased 1 by 1 as a time slot goes by. Thus, the backpressure mechanism maintains cell sequence over all cells which request to go to the same output port wherever the cell waits in HOL or HOLF. By removing cells in HOL's with large backpressure, the cells waiting behind the HOL's can obtain the opportunities to be served immediately. It decreases waiting time in input queue. Computer simulated performance study of the input and output queueing switch with feed-back loop is presented. In Sect. 2, switch operation and behavior of feed-back loop in the proposed switch model is presented. The problem in conventional switch is also presented. In Sect. 3, non-uniform traffic model used in the simulation is presented. Simulation results are presented in Sect. 4. Finally, Sect. 5 presents conclusion.

2. Switch Model

$(N+b) \times (N+b)$ nonblocking switch model is shown in Fig. 1. We assume that the number of buffer at each input port is infinite, whereas the number of buffer at each output port, B_o is finite. Each of HOL's of feed-back loops (HOLF's) has no buffer and behaves in the same way the HOL's of input queues behave. The switch operates synchronously, in that ATM cells are received at different switch inlets and transmitted to different switch outlets in the same time interval, called slot. One cell (or no cell) per one time slot can be received by each switch inlet and one cell (or no cell when output buffer is empty) can be departed from a switch outlet, whereas up to C cells per one time slot can be transferred from different input queues to each output queue by the interconnection network. C is referred to speedup factor and it is assumed that the speedup factor is equal to B_o+1. The speedup can be realized with switch fabrics which run C times as fast as the input and output trunk does

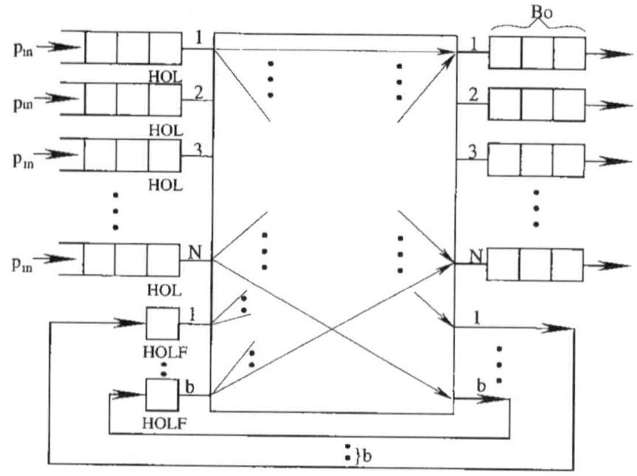

Fig. 1 $(N+b) \times (N+b)$ nonblocking switch.

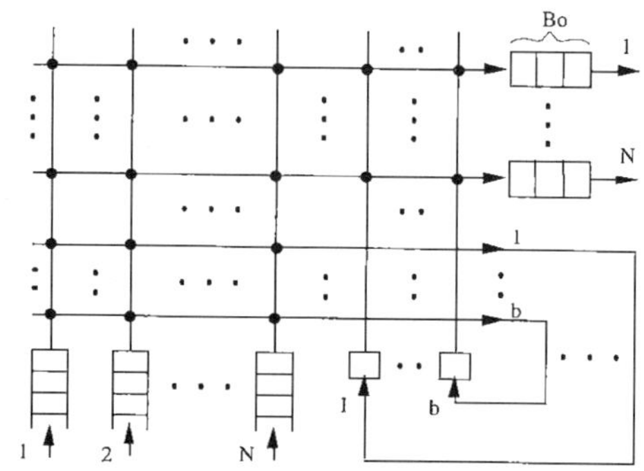

Fig. 2 Crossbar implementation of proposed model.

or a highly paralleled implemented switch architecture. A simple example of a nonblocking switch fabric is the crossbar interconnected with N^2 switch points. In Fig. 2 crossbar implementation of proposed model is shown. Speedup may be satisfied by dispatching C cells from C input ports to a specific output port in a time slot. The arrival process on each switch inlet is Bernoulli with probability p_{in}. The destination of an arbitrary cell can be any of the N output ports with equal probability $1/N$ under the uniform traffic model. We also consider the output ports which an arbitrary cell aims to with higher probability than $1/N$ as non-uniform traffic model in Sect. 3. We explain behaviors of conventional and proposed switches in detail at the next sub chapters. First, output port contention, backprssure mechanism for resolving output port contention and head of the line blocking in the conventional $N \times N$ nonblocking switch with backpressure mode are explained. Secondly, we explain switch operation and behavior of feed-back loop of the proposed $(N+b) \times (N+b)$ nonblocking switch and the process that the feed-back loop resolves the phenomenon of head of the line blocking.

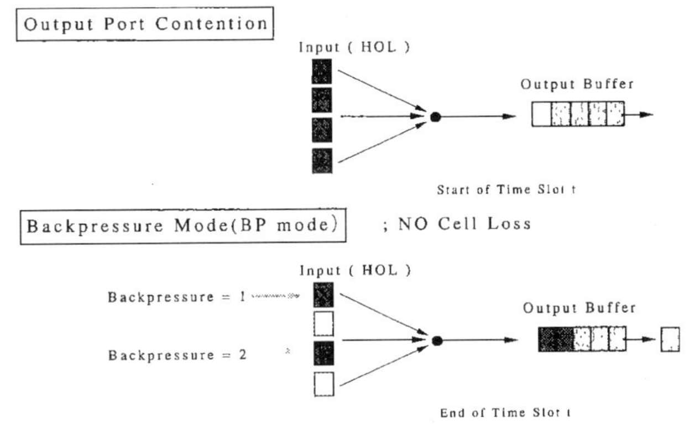

Fig. 3 Output port contention and backpressure mode (BP mode).

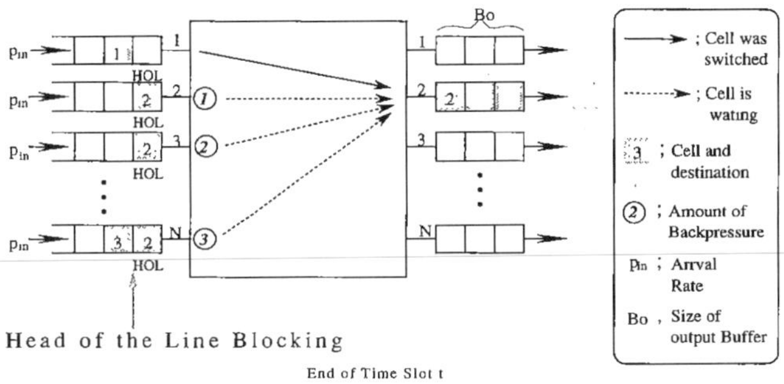

Fig. 4 Cell flowing in the conventional $N \times N$ nonblocking switch with BP mode.

2.1 Head of the Line Blocking in Conventional $N \times N$ Switch with Backpressure Mode

It is assumed that a cell enters HOL position at the instant of the start of time slot. In Fig. 3, output port contention and backpressure mode are shown. Suppose there is one vacancy in the output buffer and 4 cells are destined to this output at each HOL of 4 inputs at the start of time slot t. Since only one cell will depart from the output port, there will be two vacancies in the output buffer. Therefore, two of the 4 cells are switched and stored in the output buffer, and the other cells must be discarded due to output contention in Queue Loss (QL) mode at the end of time slot t. On the other hand, in BP mode, two cells which can not be switched to output receive backpressure 1, 2, respectively, by random manner, and wait to be dispatched to destination output port as long time slots as the amount of backpressure at HOL position. However in BP mode, other cell may be queued behind the cell which is waiting at HOL of port N as shown in Fig. 4. Because the cell which was at HOL position of the input port 1 at the start of time slot t was switched to output port 2 as a solid line represents, the cell which is behind HOL position of the input port 1 enters HOL position to be served and can be switched to destination output 1 at time slot $t + 1$. The cells which were at HOL's of input ports 2, 3 and N lost contention to be dispatched to output buffer 2 as dotted lines represent and received backpressure 1, 2 and 3 by random manner. The cell behind HOL position of the input port N can not enter HOL at the start of time slot $t + 1$, and therefore, it can not be switched to the destination output port 3 even though it may be switched to if it enters HOL at time slot $t+1$. This head of the line blocking causes the increase of waiting time in input queue.

2.2 Proposed $(N + b) \times (N + b)$ Switch Model with Feed-Back Loop

The cells that remain at HOL's will be transferred to the output buffer in the subsequent time slots according to the amount of backpressure. In our proposed $(N + b) \times (N + b)$ switch model with feed-back loop, the threshold of backpressure is set in order to use a feed-back line. If the backpressure applied to a cell is over threshold, that cell is switched into a output trunk connected to a feed-back loop in the $(N + b) \times (N + b)$ switch. It goes through a feed-back line, stored in HOL of a feed-back loop in current time slot and waits to be dispatched according to the amount of backpressure. In order to decrease the effect of head of the line blocking, b HOL's of feed-back loop work as shown in Fig. 5. When a cell reaches HOL position, it is switched to the requesting output port. If requested output port 1 is idle, the cell which came from input 2 flows through the switch. If the output port is busy but it's buffer is not full, the cell is switched and stored in the buffer like the cell which came from input port 1. Suppose the threshold of backpressure is set to be 1. The cell in HOL 3 which requested to go to the output port 3 lost the contention and received backpressure 1 and then waits according to it's amount. The cell in HOL N which requested to go to output port 3 also lost the contention and received backpressure 2. However, in the proposed switch because backpressure 2 is over threshold, this cell was switched to the trunk of feed-back loop 1 in the current time slot and waits at HOLF 1. Consequently, head of the line blocking is solved which would occur in input buffer N. At the end of time slot t in Fig. 5, the cell in HOL 3 receives backpressure 1, and the cell in HOL N receives backpressure 2 and immediately is switched to HOLF 1 due to the threshold value of 1. At the start of time slot $t + 1$ in the Fig. 6, the amount of back-pressure of the cell in HOL 3 becomes 0 and this cell will be dispatched to output buffer 3 soon. The amount of backpressure of 2 in HOLF 1 becomes 1. If the cells entering HOL 1 and HOL N request to go to output 3, they receive backpressure 2, 3 respectively by random manner due to backpressure mechanism since the output 3 is already reserved by the cells in HOL 3 and HOLF 1 with backpressure 0, 1, respectively. These two cells from HOL 1 and HOL N will be switched to HOLF's hence both backpressures

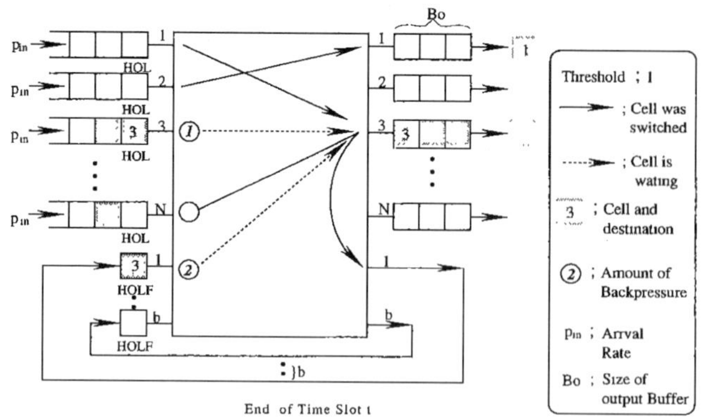

Fig. 5 Cell flowing in the proposed $(N + b) \times (N + b)$ nonblocking switch.

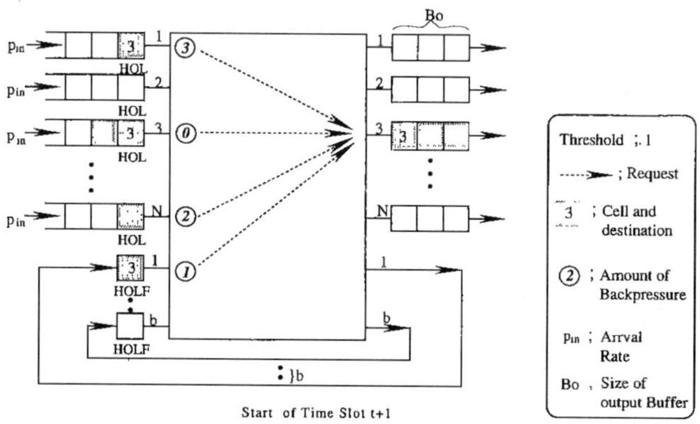

Fig. 6 Cell sequence in the proposed $(N + b) \times (N + b)$ nonblocking switch.

are over the threshold value of 1. At this time, these cells keep the amount of backpressure as like the cell which moved from HOL N to HOLF 1 in time slot t in Fig. 5. If only one of HOLF is empty and the others are fully occupied, one of them is switched to available HOLF by random selection and the other one remains at original HOL position and tries to be switched to HOLF as long as whose back-pressure is still over threshold at the next time slot. Every cell competes for requesting output and receives backpressure only once when it enters HOL position for the first time, and competes for HOLF at every time slot as long as its backpressure is over threshold. The amount of backpressure of each cell is kept wherever the cell is in HOL or HOLF. It decreases one by one as a time slot goes by. The cell which is moved from HOL to HOLF keeps the amount of backpressure and is switch to destined output as soon as its backpressure becomes 0. Each output port issues backpressure according to the memory that remembers largest backpressure which the output port has issued and revises it by decreasing 1 by 1 as a time slot goes by. Consequently, the sequence of HOLFs' cells and other input ports' cells are maintained, and the switch control mechanism is not complicated.

3. Non-Uniform Traffic Model

Under uniform traffic, each cell has equal probability $1/N$ of being addressed to any given output, and successive cells are assumed to independent. However, under non-uniform traffic, there exist single or plural output ports of higher access rate, so called hot-spot. Let p_{ij} be the probability that the cells of input i being addressed to output j and n be the number of hot-spot ports, then we have for all i an j,

$$p_{ij} = \begin{cases} \dfrac{h}{N+n(h-1)}, & \text{for hot-spot} \\ \dfrac{1}{N+n(h-1)}, & \text{for non hot-spot,} \end{cases} \tag{1}$$

here h is the coefficient for non-uniform traffic, which means that the access rate to hot-spot port is h times as the access rate to non hot-spot port. And when p_{in} is the arrival rate at each switch inlet, the traffic load given to tagged output which is defined as the average number of cells destined tagged output per time slot, P_{out}, is obtained as [6].

$$p_{out} = \begin{cases} \dfrac{hN}{N+n(h-1)} \times p_{in}, & \text{for hot-spot} \\ \dfrac{N}{N+n(h-1)} \times p_{in}, & \text{for non hot-spot.} \end{cases} \tag{2}$$

Since the number of cell which can depart from output queue per slot is at most one, if p_{out} for hot-spot port is greater than one, the output buffer is saturated. So, we can compare the delay performances of conventional switch and proposed switch in the range $1 \leq h < \frac{N-n}{NP_{in}-n}$.

4. Simulation Results

We show computer simulated results of system delay under uniform traffic and non-uniform traffic. The results in [3] are used for comparison. A switch size of 120×120 is used in the simulation so that the mismatch between the finite size of the simulated switch and the infinite switch of the conventional model can be neglected [4]. In the simulation, each value is obtained by computing over 200,000 time slots. Over 95% of the values are included in the errorbar. When the size of HOLF $b = 0$, the simulated plots of delay performance coincide with the analytical results of the conventional switch.

Figure 7 shows the delay in input queue versus load(p_{in}) for various thresholds with HOLF size $b = 18$ and $B_o = 3$ comparing with that of conventional $N \times N$ switch with backpressure mode when $B_0 = 3$. When the value of threshold is small, because a lot of cells

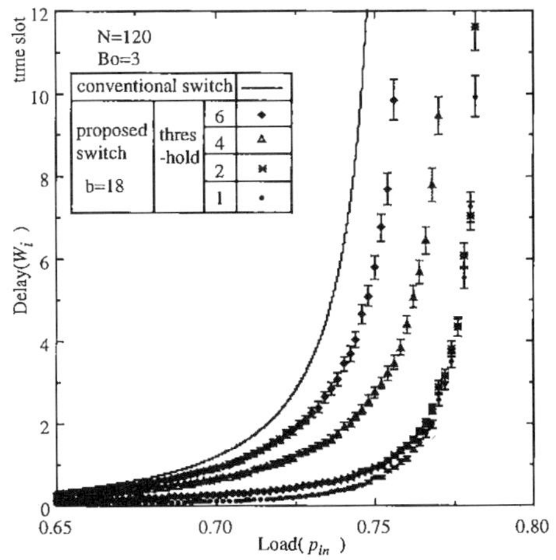

Fig. 7 Delay in input queue versus load(p_{in}) for various thresholds.

are transfered from HOL's of input queues to HOL's of feed-back loops the waiting time in input queue is small. The feed-back loops suppress the increase of delay. When the value of threshold is small, the cells do not stay long in HOL position. Moreover, because the backpressure associated with the cells which enter HOLF's is small, they leave these HOLF's soon. Therefore, flowing of cells is promoted through the feed-back loops. That provides more opportunities of being served to the cells behind HOL's of the input queues. Consequently, delay in the input queue is suppressed. When the value of threshold is large, flowing of cells is not promoted that much. Until getting backpressure over threshold, the delay progresses to a certain extent. Moreover, even if the cells are switched to HOLF's they stay long at those HOLF's. Therefore, the flowing of cells through the feed-back loops is slow. It can not provide many opportunities of being served to the cells behind HOL's of the input queues. Consequently, the delay in the input queue is not suppressed that much. The value of threshold 2 is sufficiently small for improving delay performance when $b = 18$.

Figure 8 shows the system delay versus load(p_{in}) for various values of parameter b of the proposed switch with $B_o = 3$. The values of b are 6, 12, 18, 24, 30 and 36 in the case of threshold 2 and 6, 12, 18 and 24 in the case of threshold 4. The system delay of conventional switch with $B_o = 3$ is also shown. When threshold is 2, because a lot of cells which get backpressure over the threshold are switched into HOLF's, as b increases until pretty large size, e.g. $b = 36$, the delay performance is more improved. That is, when b is large, more cells can be circulated through large feed-back loops. On the other hand, threshold 4 does not enhance the effect of b. Because until getting backpressure over threshold 4, the delay progresses to a certain extent, and even if cells are switched to HOLF's, they leave at least equal to or more than 4 time slots after. With threshold 4, $b = 12$ is effective.

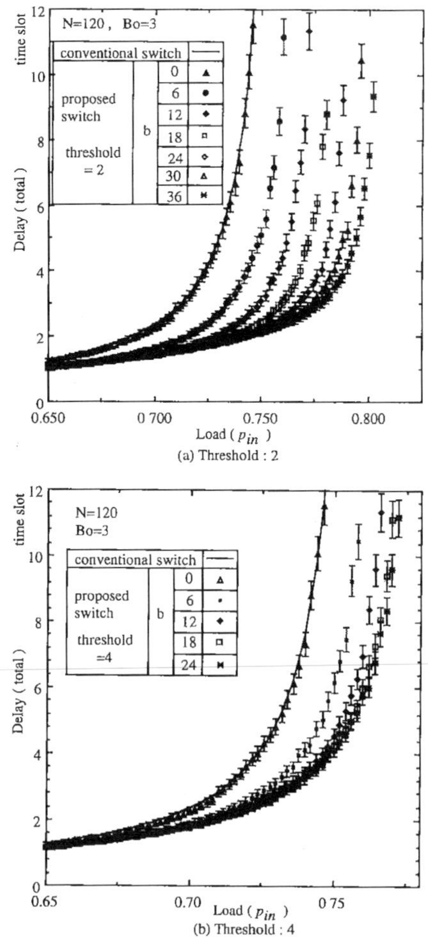

Fig. 8 System delay versus load(p_{in}) for various values of b.

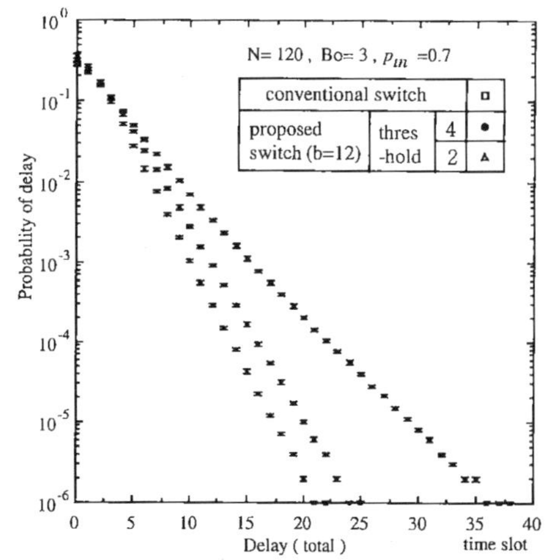

Fig. 9 Probability of delay.

Figure 9 shows the probability of delay, i.e., how many percentage of total arrived cells spend how long time in the switch when arrival rate is 0.7 and $B_o = 3$. It is shown that in the proposed switch with threshold 4

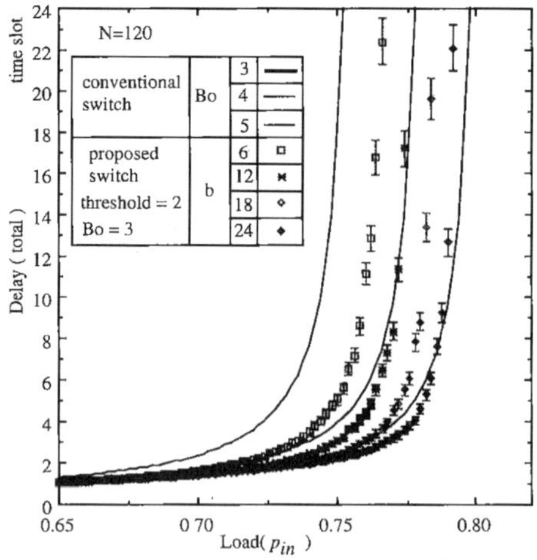

Fig. 10 System delay versus load(p_{in}) under various parameters B_o and b.

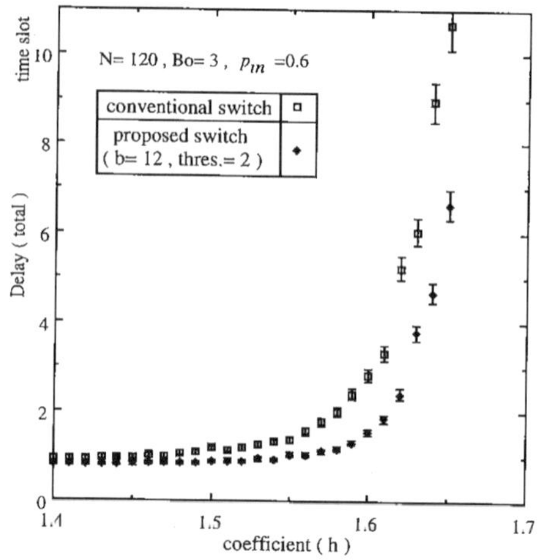

※h = $\frac{\text{access rate to hot-spot port}}{\text{access rate to non hot-spot port}}$

Fig. 11 System delay versus coefficient(h) under non-uniform Traffic.

and 2 when b is 12, the cells go through the switch faster than in the $N \times N$ switch. There are cells spending even 38 time slots in the $N \times N$ switch. In $(N+b) \times (N+b)$ switch, cells spend at most 25 time slots when threshold is 4 and 22 time slots when threshold is 2. It is shown that the feed-back loops accelerate stream of cells.

Figure 10 shows the system delay versus load(p_{in}) under various parameters B_o and b. The solid lines represent the delay of conventional $N \times N$ switch when output buffer size B_o is 3, 4 and 5, respectively. The plotted curves represent the delay of the proposed $(N+b) \times (N+b)$ switch with threshold 2, $B_o = 3$ when HOLF size b is 6, 12, 18 and 24, respectively. The proposed switch has b more switch fabrics, b more HOL's and b feed-back lines compared to conventional $N \times N$ switch. However, in order to have b feed-back loops, it is not needed to increase speedup or parallelism or employ complicated control policy. In the case of 120×120 switch with output buffer size 3, to improve the delay performance to the level of the switch with output buffer size 4, we need not only 120 more buffers but also speedup or parallelism to be increased by 1 unit. But, if we use feed-back loop, almost same delay performance can be obtained with only 12 feed-back loops and better delay performance can be obtained with 18 feed-back loops.

In Fig. 11, the system delay versus coefficient(h) under non-uniform traffic is shown when load$p_{in} = 0.6$ and $B_o = 3$. Under non-uniform traffic our proposed switch when $b = 12$ and threshold is 2 shows better delay performance. Coefficient h represents the ratio of the access rate to hot-spot port to the access rate to non hot-spot port. And hot-spot port means a port which each cell in the input queue has higher probability of being addressed to than the other ports. Each cell in the input queues is easy to receive backpressure over threshold owing to hot-spot. However the feed-back loop promotes cell flowing by vacating HOL's that may be occupied by such cells, and thus, delay is suppressed.

5. Conclusion

We have proposed input and output queueing $(N+b) \times (N+b)$ nonblocking switch with feed-back loop model in order to decrease system delay. Our system removes HOLs' cells with backpressure over the threshold and provides the cells waiting behind the HOL's with the opportunities to be served immediately. Thus, the head of the line blocking is solved. This promotes cell flowing through the switch and this promotion produces the same effect as an increase of output capacity. It is shown that in the case of 120×120 switch with output buffer size 3 under uniform traffic, instead of adding one more output buffer at each 120 output ports, we can obtain better delay performance by using 18 feed-back loops. That is attractive at the aspect of hardware complexity. Furthermore, backpressure mechanism maintains cell sequence over all cells which are destined to the same output port, and therefore, the switch control mechanism is not complicated. It is also shown that under non-uniform traffic HOLF's mitigate the effects that hot-spot severely blunts all traffic streams, not just the traffic streams to hot-spot port. As the results, it is found that our switch with feed-back loop improves the delay performance compared to the conventional switch with backpressure mode.

Acknowledgment

This work is partly supported by Ministry of Education, Yazaki, Memorial Foundation for Science and Technology, Mitsubishi Electric Corporation and Hitachi Ltd.

References

[1] M.J. Karol and M.G. Hluchyj, "Input versus output queueing on a space-division packet switch," IEEE Trans. Commun., vol.COM-35, no.9, pp.1347–1356, Dec. 1987.
[2] M.G. Hluchyj and M.J. Karol, "Queueing in high-performance packet switching," IEEE J. Selet. Areas Commun., vol.6, no.9, pp.1587–1597, Dec. 1988.
[3] I. Iliadis and W.E. Denzel, "Analysis of packet switches with input and output queueing," IEEE Trans. Commun., vol.41, no.5, pp.731–740, May 1993.
[4] A. Pattavina and G. Bruzzi, "Analysis of input and output queueing for nonblocking ATM switch," IEEE Trans. Networking, vol.1, no.3, pp.314–327, June 1993.
[5] N. Yoshihashi and I. Sasase, "Effect of number of packets sent simultaneously to the same output and input buffer control policy on the waiting time in a packet switch," IEICE Trans., vol.J78B-I, no.3, pp.147–156, March 1995.
[6] S. Shiokawa and I. Sasase, "Input and output queueing two stages ATM switch with hot-spot route," Proc. ICT '97, Melbourne, Australia, vol.1, pp.7–12, April 2-5, 1997.
[7] M.G. Hluchyj and M.J. Karol, "Queueing in high-performance packet switching," IEEE J. Select Areas Commun., vol.6, no.9, pp.1587–1597, Dec. 1988.
[8] A. Huang and S. Knauer, "Starlite: A wideband digital switch," Proc. IEEE GLOBECOM '84 Conf. Rec., pp.121–125, Nov. 1984.
[9] J.N. Giacopelli, J. Hickey, W. Marcus, W.D. Sincoskie, and M. Littelwood, "Sunshine: A high-performance self-routing broadband packet switch architecture," IEEE J. Select. Areas Commun., vol.9, no.8, pp.1289–1991, Oct. 1991.
[10] G. Kbar, "Survey on ATM switches," Proc. ICT '97, vol.2, pp.693–699, April 1997.
[11] G. Kbar, "Analysis of ATM switch with feedback input queueing and output queueing under burst traffic," Proc. Int. Conf. on Comput. Communi. 12th, pp.259–264, Aug. 1995.

High-Performance Backbone Network Technology edited by Naoaki Yamanaka
ISBN: 0-8247-5321-6 DOI: 10.1081/(120025672)
Published 2004 by Marcel Dekker, Inc.

The Odd-Even ATM Switch

Christos Kolias and Leonard Kleinrock

SUMMARY This paper introduces and studies the performance of an $N \times N$ space-division, single-stage ATM switch with dual input-queueing. Each input port has two separate FIFO queues, an "odd" and an "even" queue. An incoming cell is stored at the input at either of two FIFOs according its output port destination (output ports are also labeled as "odd" or "even"). Hence we call this scheme the *Odd-Even* switch. We compare the Odd-Even switch to the ordinary input-buffered switch and we find that it can achieve a remarkably higher performance, in terms of throughput, mean delay and cell loss. This is due to the fact that the Head-of-Line effect becomes less problematic under the Odd-Even switch. Our results are based on various traffic models. Finally, we compare the Odd-Even scheme to the Look-ahead (input "window") policy.

key words: *odd-even, ATM switching, input-queueing, head-of-line blocking*

1. Introduction

Asynchronous Transfer Mode (ATM) wide area networks are currently being deployed by telecommunications companies in order to support the forthcoming B-ISDN services. Various blocking and non-blocking, time-division and space-division, single stage and multistage ATM switches have been proposed and studied extensively, in order to provide high performance packet switching for integrated ATM transport. Some of these have been developed and tested under real-time traffic requirements and have eventually become available as commercial products.

Switch architectures are classified [7] according to their buffering methods (e.g., input buffers, output buffers), according to their switch fabric (i.e. crossbar, Batcher-Banyan, Starlite) or according to their internal switching (non-blocking, blocking).

In this paper we study an input-buffered, non-blocking, crossbar, ATM switch with N input ports and N output ports. Incoming cells are stored in (one of the two) dedicated buffers at the input ports, according to their output port destination.

The model we present for this switch ignores any hardware design and specific implementation details; we are mainly concerned with the performance evaluation of the switching system. Achieved throughput and delay are the two main performance metrics of interest. That allows us to compare it in terms of performance with other switches that have different characteristics (i.e. arbitration policy) but with the same philosophy (i.e. input buffering).

The paper is organized as follows. Section 2 describes the Odd-Even switch along with our modeling assumptions. In Sect. 3, we present some performance evaluation results for the Odd-Even switch while we compare it to the ordinary (simple) input-queuing switch - to which we refer as "regular" switch - under various traffic conditions. Section 4 extends the comparison of the Odd-Even model by considering also the Look-ahead scheme. Concluding remarks are provided in Sect. 5.

2. The Odd-Even Switch

2.1 Switch Architecture

The switch under consideration uses input-queueing to buffer a cell before it accesses the switch fabric in the case where there is no available connection to the output port and there are buffers only at the inputs. Cell arbitration among the input buffers is needed in order to control the switch and initiates the contention resolution cycle. The switch has N inputs and N outputs, therefore the switch is of size N. What is novel about this switch (Fig. 1) is the fact that each input is split (expanded) into two FIFO queues which we call Odd and Even; hence we call this switching system the Odd-Even

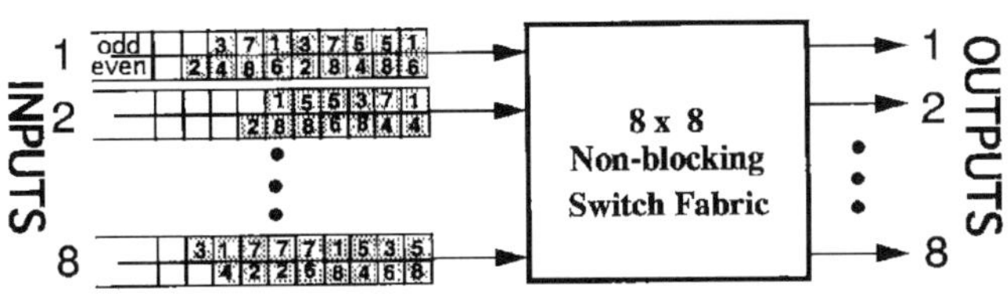

Fig. 1 An 8×8 Odd-Even ATM switch (two FIFO queues per input port).

Manuscript received April 25, 1997.
Manuscript revised August 21, 1997.

[†]The authors are with the Computer Science Department, UCLA, Los Angeles, CA 90095, U.S.A.

[*]This research was supported by ARPA/CSTO under grant MDA-972-91-J-1011, Advanced Networking and Distributed Systems. The first author was also supported by the Lilian Voudouri Fellowship (Greece). Part of this paper was presented at the International Conference on Communications (ICC) '96, Dallas, Texas, June 23–27, 1996. Web URL: http://millennium.cs.ucla.edu/~ck

IEICE TRANS. COMMUN., Vol. E81-B, No. 2, pp. 244–250, February 1998

ATM switch[†]. By convention, we refer to outputs numbered as $1, 3, 5, \ldots$ as odd, and to $2, 4, 6 \ldots$ as even ports. An incoming cell destined to an odd (even) numbered output port joins the odd (even) queue of the input port to which it was fed and awaits its turn to be switched to its output. When more than one cell are present at the heads of the queues contending for the same output address only one can access that output while the others remain blocked until they are finally released in future time slots. This blocking phenomenon is known as *head-of-line* (*HOL*) blocking, a problem from which all input-buffered packet switches suffer. A queued cell that moves up to the head of its queue, after the previous HOL is switched, is called a *fresh* cell. Time is slotted and since cells are packets of fixed size, service time is deterministic and equal to a time slot (which is determined by the actual speed of the switch fabric) and which we assume here to be one time unit.

As mentioned above, cell arbitration is employed in order to resolve any potential contention among the input buffers. The full contention resolution process occurs at the beginning of each time slot, immediately following the cell arrivals, and consists of two (very short) rounds. Arbitration during the first round involves the HOL cells at the even input queues. In the second round cells at the HOL of the odd input queues contend for the odd output addresses. However, those input ports whose even queues could not access an output port in the first round — because they either lost the contention (and therefore were blocked) or simply did not have any HOL cells present — are allowed to participate in a contention among their odd queues in the subsequent second round (see example in Fig. 2). Consequently, at each time slot, an input port always gets the opportunity to send a cell either from an even or an odd queue, but not more than one cell is allowed to be cleared, from any given input port, in both arbitration phases during the course of a time slot.

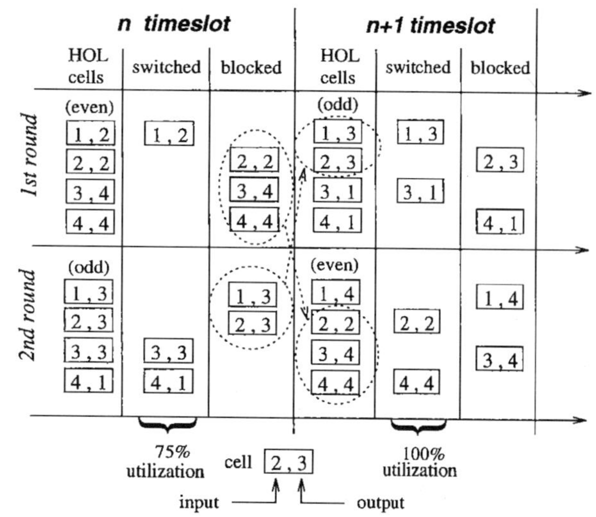

Fig. 2 Arbitration and contention resolution (random selection) example in a 4 × 4 Odd-Even switch.

In order to maintain fairness we allow arbitration cycles to interchange order in every time slot (if it is even-odd in one time slot it is odd-even in the next one), otherwise it would be as if we introduced some kind of priorities. We use a random HOL selection policy[††]. Under a balanced traffic assumption, this arbitration strategy can guarantee stability and fairness. Other possible policies are the "longest queue selection"[†††] and the "longest waiting HOL cell selection"[††††]—if there is a tie, the lowest numbered input queue is the winner. Scheduling among the HOL cells can further improve the throughput. We do not consider any such mechanisms or algorithms in our paper.

Clearly, the proposed architecture has no speed-up features. However, as we demonstrate in the next sections, comparing this switch architecture to the "regular" input-buffered switch it can increase rather significantly the output trunk utilization and decrease delays and switch cell loss. This is achieved as the Odd-Even scheme introduces a certain degree of parallelism. We now proceed with the assumptions supporting our model.

2.2 Modeling Approach

The proposed switch is modeled as a multi-queue (input links), multi-server (output trunks), discrete-time queueing system. Cells arriving at the input links are simultaneously and independently routed to the appropriate output ports. Cells arrivals occur at the beginning of the time slot and incoming cells are queued accordingly in the Odd-Even FIFO queues. Cell departures from the switch occur in a concurrent, independent fashion.

The model ignores any flow or congestion control mechanism or implications by assuming that such control is accounted for in other parts of the network. We assume that the traffic management is efficient enough so that there is no cell loss inside the switch (though it may occur elsewhere as part of a policing mechanism) and that input buffers are infinitely large (we will later relinguish this assumption). Buffers are considered to be FIFO queues, thus preventing any resequencing of cells. It is clear from the arbitration policy, that the throughput of the Odd-Even model is at least as high as that of the regular switch. A packet switch where

[†]We would like to thank Larry Roberts, CTO of *Connectware Inc.*, for introducing the Odd-Even model to us.

[††]When k cells contend for the same output at the beginning of a time slot one of the k is randomly chosen to be switched.

[†††]The buffer with the longest queue (most stored cells) is selected.

[††††]The HOL cell that has been waiting the longest at the HOL position is selected. This obviously requires timestamping.

the input traffic intensity is the same for every input link and the output address of each incoming cell is assigned with equal probability to any output port is said to be characterized as a homogeneous system.

The *throughput* of the switch, denoted by γ and defined as the average number of cells successfully (i.e. not blocked) switched per unit of time (timeslot), will be used as the main performance evaluation measure. Switch throughput is obtained as the average utilization of the N servers, where each server's utilization is obtained as the percentage of time the server is busy. What actually presents interest is finding the maximum achievable switch throughput, γ_{max}. The maximum throughput (also known as saturation throughput) is the maximum applied input load than can be handled by the switch without causing any instability (the switch then operates at its maximum attainable limit). Input traffic exceeding that critical load threshold drives the system to unstable and overloaded situations (e.g., infinitely large queues and delays), while for input load less than γ_{max} throughput is equal to the input load. A throughput of, e.g., $\gamma_{max} = 0.5$ means that on average only half of the input ports route cells in a typical time slot and that throughput gives the probability that a cell at an HOL position gets served during a time slot, provided of course that all HOL positions are occupied, i.e. a heavy traffic situation. Another interpretation is that an HOL cell has to wait, again on average, for two ($1/\gamma_{max}$) timeslots, to be switched, assuming again a heavy traffic case. *Delay* is obtained as the total time experienced by a cell when it is fed into an input port until it is finally switched out from its desired output port. Thus, delay comprises of any waiting time (queueing delay) and the (fixed) switching time (equal to a time slot). The time spent at the HOL position is considered part of the queueing delay.

Our performance study is carried under a variety of traffic patterns: non-bursty, bursty, and non-uniform. Since exact analysis is rather intractable, we present simulation results in order to evaluate the Odd-Even model[†]. In our simulation runs we considered various switch sizes, namely, $N = 1, 2, 4, 8, 16, 32, 64, 128, 256$ and 512. Throughput results, presented as a function of N (on a logarithmic scale), refer to saturation situations. As we have indicated, we are concerned with those the cases where input queues become saturated (saturation may be exhibited at different input load levels for switches of different sizes). For the balanced traffic case we expect all input queues, including odd and even queues in our model, to become saturated at the same input load.

3. Performance Results

3.1 Non-Bursty Traffic

For a quantitative evaluation and comparison of the

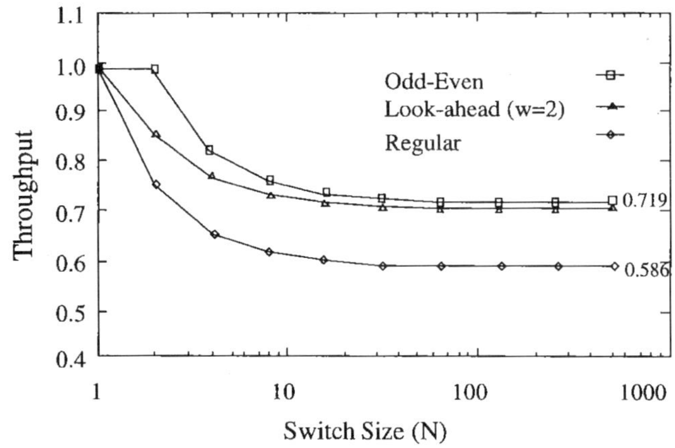

Fig. 3 Throughput of the Odd-Even switch vs. regular switch (Bernoulli arrivals).

Odd-Even switch we consider a homogeneous system. Cell arrivals to each of the N input ports are independent (thus we consider non-bursty traffic) and identically distributed according to a Bernoulli process with parameter λ, that is s cell arrives at an input port with probability λ.

Figure 3 shows the throughput of the Odd-Even switch (upper curve), whose value is (for large N) $\gamma_{max} = 0.719$, a sizable increase compared to that of the ordinary input buffered switch, which is known to be $2 - \sqrt{2} \approx 0.586$ (cf. [2], [3], [6]), as also seen in the same figure. In [4] we show that an approximate mathematical analysis results in a switch throughput of 0.705 for the Odd-Even model, which is in fact a good approximation to 0.719. We see that the Odd-Even scheme considerably outperforms the regular switch (by more than 20%).

The mean waiting time for the Odd-Even model, as compared to the regular input-queueing switch, is shown in Fig. 4, as a function of λ and for the values of $N = 2$, 32 and 512, under a random HOL selection policy. We see again a significant improvement using the Odd-Even switch over the regular one.

Table 1 shows how the Odd-Even model behaves under different policies. While in the regular switch the selection policy has no effect on throughput, this is not true for the Odd-Even. In fact, we notice that the random selection strategy yields a higher throughput than the other two. Similar results (which we do not include in this paper) were obtained with respect to the mean waiting time which showed the random selection policy yields the lowest.

Last but not least we extend our performance evaluation, by considering finite-size buffers as to study the cell drop ratio for both switches, namely the Odd-Even and the regular switch. The *cell drop ratio* is measured as the fraction of incoming cells that are discarded due

[†] See [4] for an analytical treatment of throughput for the non-bursty case.

to insufficient buffer space (buffer overflow) inside the switch. For fairness, we assume the same total buffer capacity, per input port, for both switching systems, i.e., if we assume that a FIFO queue in the regular switch is 32 cells long then the two FIFO queues (per input) in the Odd-Even switch are 16 cells each. Table 2 shows various cell drop ratios for different switch and buffer sizes. The tabulated ratios refer to those saturation cases where the applied input load slightly exceeds the switch's throughput. These input load levels are different for switches of different sizes. We observe from that table that the Odd-Even switch has lower (with a few exceptions) cell drop ratios when the same input load is applied to both switching systems.

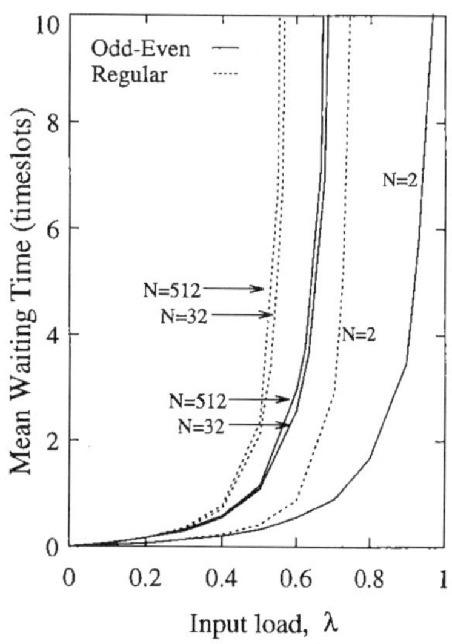

Fig. 4 Mean waiting times (Bernoulli arrivals).

Table 1 Maximum throughput, γ_{max}, for different selection policies.

Scheme	γ_{max}
Regular - Random	0.586
Odd-Even, Longest Waiting HOL	0.713
Odd-Even, Longest Queue	0.718
Odd-Even, Random	0.719

3.2 Interrupted Bernoulli Process (IBP) Traffic

As Variable Bit Rate (VBR) applications characterize a significant portion of the traffic (e.g., packetized voice and video and large data files) carried by ATM networks, we evaluate our model also under a bursty traffic pattern. In this model, input traffic alternates between active (busy) and silent (idle) periods, while output port addresses are still uniformly distributed. In particular, each of the inputs is described by the same *On-Off* model where both busy and idle periods have a geometrically distributed .duration Cells within the same burst are destined to the same output port (i.e. cells belong to the same fragmented packet). Basically, each time slot is governed, again, by a Bernoulli process, where given an input is in an On (busy) state, it will remain in that state (i.e. during the next time slot) with probability $1-p$, while it will switch to the Off (idle) state with probability p (see Fig. 5). An On state corresponds to a burst being transmitted on an input link while the Off state corresponds to a silence period (Fig. 5). The probability of a burst consisting of k time slots is given by:

$$b(k) = p(1-p)^{k-1}, \qquad k \geq 1 \tag{1}$$

If the input is in the Off (idle) state, where no cells arrive, it will stay in the same state with probability $1-q$ and then the probability that an idle period lasts k time

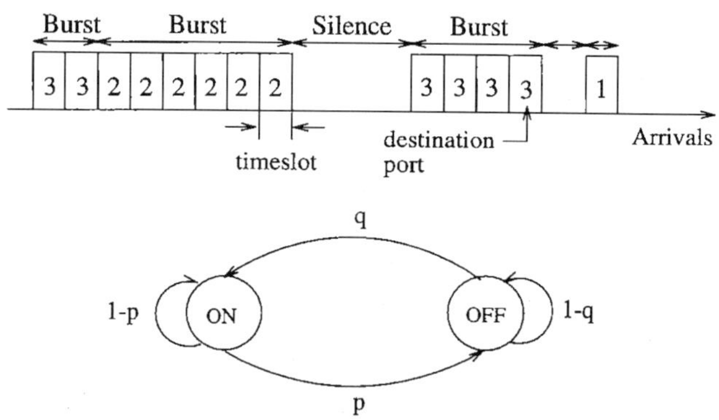

Fig. 5 On (Burst)-Off (Silence) traffic model.

Table 2 Cell drop ratios for the Odd-Even (OE) and the regular switches under various buffer capacities and for different switch sizes.

Switch Size	Cell drop ratio									
	$b=2$		$b=16$		$b=32$		$b=128$		$b=512$	
$N \times N$	OE	regular	OE	regular	OE	regular	OE	regular	OE	regular
2×2	1.94E-1	2.48E-1	2.92E-2	2.49E-1	1.40E-2	2.49E-1	2.62E-3	2.49E-1	1.90E-5	2.49E-1
4×4	2.17E-1	2.43E-1	5.45E-2	2.29E-1	3.83E-2	2.29E-1	2.79E-2	2.29E-1	2.73E-2	2.29E-1
8×8	2.28E-1	2.48E-1	6.71E-2	2.27E-1	5.04E-2	2.27E-1	4.08E-2	2.27E-1	4.04E-2	2.27E-1
16×16	2.20E-1	2.31E-1	4.98E-2	1.98E-1	2.99E-2	1.98E-1	1.39E-2	1.98E-1	1.07E-2	1.97E-1
32×32	2.23E-1	2.33E-1	5.28E-2	1.98E-1	3.22E-2	1.98E-1	1.64E-2	1.98E-1	1.32E-2	1.97E-1
64×64	2.27E-1	2.37E-1	5.84E-2	2.03E-1	3.81E-2	2.03E-1	2.32E-2	2.03E-1	2.09E-2	2.03E-1
128×128	2.23E-1	2.32E-1	5.28E-2	1.95E-1	3.19E-2	1.94E-1	1.50E-2	1.94E-1	1.15E-2	1.94E-1

slots is simply:

$$i(k) = q(1-q)^k, \qquad k \geq 0 \tag{2}$$

which also takes into account the possibility that a burst can be followed immediately by another burst (e.g., two different streams are multiplexed on the same input link) in which case there is no idle period ($k = 0$). This traffic model is referred as the Interrupted Bernoulli Process (IBP). The offered input load is calculated as:

$$\rho = \frac{E[\text{busy period}]}{E[\text{busy period}] + E[\text{idle period}]} = \frac{q}{q+p-pq} \tag{3}$$

A $p = 0.5$ means that the average burst length is $1/p$ = 2 cells. Figure 6 illustrates the saturation throughput for both the Odd-Even and the regular switch, for IBP traffic with $p = 0.05$ (i.e., very bursty), while Fig. 7 gives the maximum throughput for different values of p (the probability of being at the On state) for a 512×512 switch. For $\rho = 1$ then from Eq. (3) we get $q = 1$ which means that there are no silence periods, regardless of the value of p ($p \neq 0$). This leads the system to overloaded situations. If now $p = 1$ (bursts consist of only one cell) then $q = \rho$ which then has the same effect, in terms of throughput, as the Bernoulli ($\lambda = \rho$) case in the previous paragraph. From Fig. 7, which gives the asymptotic behavior, we notice that the absolute difference in throughput between the two switches remains almost constant for any value of p. This means that the percentage gain in throughput using the Odd-Even switch becomes larger as p decreases (hence traffic is more bursty).

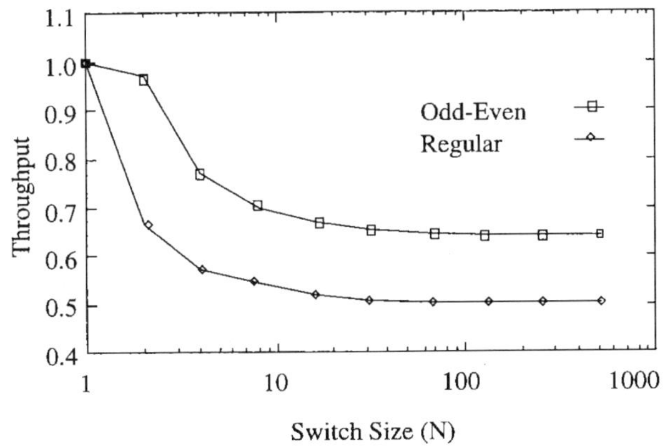

Fig. 6 Throughput of the Odd-Even switch vs. regular switch (IBP arrivals, $p = 0.05$).

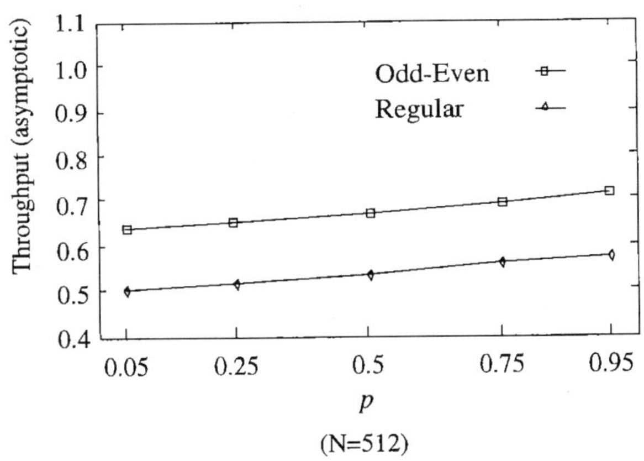

Fig. 7 Asymptotic throughput of the Odd-Even switch vs. regular switch (IBP arrivals).

3.3 Non-uniformly Distributed Destinations

So far we have assumed uniformity for the output port addresses. In this section we evaluate the switch's performance under the assumption of a non-uniform, non-bursty traffic pattern. The output port mapping for each incoming cell is determined by a binomial distribution [8], where for $i = 2, \ldots, N-1$:

$$a_{i,j} = \Pr[\text{a cell arriving at the } i\text{-th input is destined to the } j\text{-th output}]$$
$$= \binom{N}{j} r_i^j (1-r_i)^{N-j}, \qquad j = 1, \ldots, N \tag{4}$$

and r_i is a probability associated with input i, $2 < i < N$. Note that for this binomial distribution the maximum probability occurs for $j = \lfloor N r_i \rfloor$. Now, if we choose $N - i$ to be the output to which the highest percentage of traffic arriving on i goes (therefore it has the largest probability associated with input i), then we have

$$r_i = 1 - \frac{i}{N}, \qquad i = 2, \ldots, N-1. \tag{5}$$

For inputs 1 and N the output address for $j = 1, \ldots, N$ is given by a normalized Poisson-like distribution with rate r,

$$a_{1,j} = \frac{\frac{r^{N-j}}{(N-j)!}}{\sum_{\forall j} \frac{r^{N-j}}{(N-j)!}}, \qquad a_{N,j} = \frac{\frac{r^j}{j!}}{\sum_{\forall j} \frac{r^j}{j!}} \tag{6}$$

One of the advantages of using a binomial distribution for inputs $i = 2, \ldots, N-1$, is that we still retain the balanced traffic effect for the Odd and Even input queues in each input port. However, this observation does not apply for inputs 1 and N, where, depending on the value of r, there is some bias introduced against the even outputs, but as N gets very large this bias becomes negligible. Another reason for using this kind of binomial distribution is it gives us a substantial number of "hot spots"† (instead of one or two), which is more realistic as the number of switch ports increases (and thus the number of potential connections among hosts in an ATM network).

† Destinations that collect most of the incoming traffic.

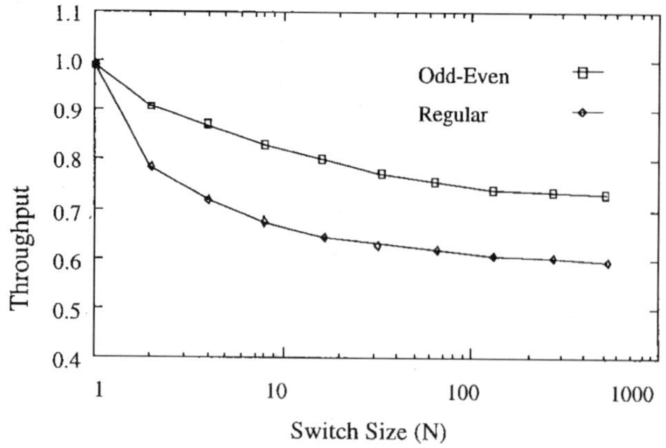

Fig. 8 Throughput of the Odd-Even switch vs. regular switch (non-uniform traffic).

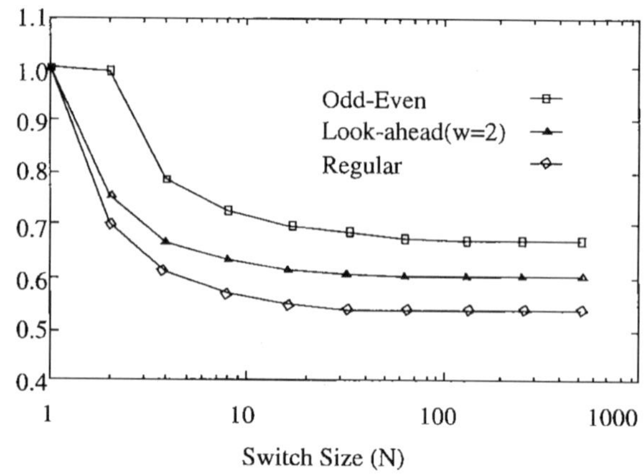

Fig. 9 Throughput of the Odd-Even switch vs. Look-ahead ($w = 2$) and regular switches (IBP arrivals, $p = 0.5$).

In our simulation runs, we chose a value of $r = 0.4$; however r has no significant relevance on estimating the throughput of the system. Using the same input-to-output address assignment functions (4)–(6) for both the Odd-Even and the Regular switch, Fig. 8 shows the throughput of each switch respectively.

4. The Odd-Even and the Look-Ahead Model

Lastly, we compare the Look-ahead scheme [1] (an input "window" policy for dealing with the HOL problem) to the Odd-Even switch. In the Look-ahead strategy (also known as the bypass discipline) the arbitration and cell selection process employs w separate contention resolution rounds. During the first one all the HOL cells at the input queues contend for an output port. Those queues that lost the contention during that first round can participate in a new arbitration where the cells right behind the HOL cells (that were blocked) compete for outputs that are still idle (i.e. they have not received any cells so far in the current time slot). In the same fashion a contention cycle that is comprised of a total of w rounds, allows, during the i-th round, the i-th (deep in the queue) cells (from those input queues that have not sent a cell) to contend for an unclaimed output. It is clear that only one cell is allowed to be switched from a single input and only one cell can be served by an output port during a time slot. For $w = 1$, we have the regular pure input-buffered switch.

In our comparison we consider a Look-ahead scheme with a window of $w = 2$. We feel that a window of two makes it comparable to the Odd-Even model where contention resolution also consists of two rounds (one for polling the odd and one for polling the even queues). Also, note that in both strategies only one cell can be cleared from a single input port (there is not any speed-up feature).

Figure 3 illustrates how all three models, namely the Odd-Even, the Look-ahead and the regular one compare, assuming Bernoulli arrivals. We see that the throughputs of the Odd-Even and the Look-ahead are very close. However as traffic becomes more bursty (which means that cell arrivals are correlated) this is not any longer true. The Odd-Even performs much better, since a window of $w = 2$ for Look-ahead does not provide much of a benefit (see Fig. 9).

5. Conclusion

In this paper we have described an innovative crossbar, input-buffered ATM switch, the *Odd-Even* switch, which allows incoming cells to be stored in either of the two FIFOs placed in every input port, according to their destination, namely odd or even. Its performance has been quite extensively examined and compared it to the simple input-queueing switch under various traffic conditions. Furthermore, we compared the Odd-Even model to the Look-ahead scheme for which we assumed a window of $w = 2$, in order to make the latter comparable to the Odd-Even. In all cases, it has been demonstrated that the Odd-Even can achieve a noteworthy gain of more than 20% in throughput over the regular input-buffered switch. Other performance results (with respect to mean waiting times, cell drop ratios) clearly suggest a superior performance of the Odd-Even switch over the ordinary input-buffered switch. We recognize that the Odd-Even switch may require additional hardware and complexity (e.g. input controller, more complex arbitrator), however it is still simple in its implementation and effective in terms of performance.

We have expanded the idea of the Odd-Even switching scheme to include multiple (instead of just two) input FIFO queues connected to the same input port. This led us to introducing the Multiple Input-Queueing switch which we study in [4]. As an additional application of the Odd-Even switch we mention the use of this dual input-queueing strategy in the buffered-banyan multistage interconnection networks. This has resulted in us proposing a high-performance switching network architecture which we call the Dual-Banyan (DB) switch (cf. [5]).

References

[1] M.G. Hluchyj and M.J. Karol, "Queueing in high-performance packet switching," IEEE J. Sel. Areas Commun., vol.6, no.9, pp.1587–1597, Dec. 1988.

[2] J.Y. Hui and E. Arthurs, "A broadband packet switch for integrated transport," IEEE J. Sel. Areas Commun., vol.SAC-5, no.8, pp.1264–1273, Oct. 1987.

[3] M.J. Karol, M.G. Hluchyj, and S.P. Morgan, "Input versus output queueing on a space-division packet switch," IEEE Trans. Commun., vol.COM-35, no.12, pp.1347–1356, Dec. 1987.

[4] C. Kolias and L. Kleinrock, "Throughput analysis of multiple input-queueing in ATM switches," in Broadband Communications, eds., L. Mason and A. Casaca, pp.382–393, Chapman & Hall, United Kingdom, 1996.

[5] C. Kolias and L. Kleinrock, "The dual-banyan (DB) switch: A high-performance buffered-Banyan ATM switch," Proc. International Conference on Communications (ICC)'97, pp.770–776, Montreal, Canada, June 1997.

[6] C. Kolias and L. Kleinrock, "An analytically simplified performance study of ATM switches with input-queueing," submitted for review to the IEEE Trans. Commun., March 1997.

[7] R. Onvural, "Asynchronous Transfer Mode Networks: Performance Issues," Artech House, Norwood, MA, 1994.

[8] S.Q. Li, "Nonuniform traffic analysis on a nonblocking space-division packet switch," IEEE Trans. Commun., vol.38, no.7, pp.1085–1096, July 1990.

High-Performance Backbone Network Technology edited by Naoaki Yamanaka
ISBN: 0-8247-5321-6 DOI: 10.1081/(120025673)
Published 2004 by Marcel Dekker, Inc.

An Architecture Supporting Quality-of-Service in Virtual-Output-Queued Switches

Rainer Schoenen

SUMMARY Input buffered switches most efficiently use memory and switch bandwidth. With Virtual Output Queueing (VOQ), head-of-line blocking can be avoided, thus breaking the throughput barrier of 58.6%. In this paper a switch architecture based on VOQ is proposed, which offers deterministic and stochastic delay bounds for prioritized traffic. This is achieved by a hybrid static and dynamic arbitration scheme, which matches ports both by a precalculated schedule and realtime calculations. By using weighted dynamic arbitration algorithms 100% throughput with lowest delays under all admissible traffic can be achieved. An integrated global priority scheme allows the multiplexing of realtime and data traffic. Following the arbitration decision, a cell scheduler decides locally in the input ports upon the next connection from which a cell is forwarded. Cell scheduling based on earliest-deadline-first (EDF) is shown to perform similar to its behaviour in an output-queued switch.

key words: *ATM, VOQ, arbitration, allocation, scheduling*

1. Introduction

ATM networks will have to consist of switches supporting the individual per-connection Quality-of-Service (QoS) ATM promised to deliver. Even IPv6 switches offering Diffserv/Intserv need architectures with built-in QoS features. Among the switching architectures, input queued switches are most powerful because the access rate of crossbar and buffer memory is not higher than the line rate of the connected links. With Virtual Output Queueing (VOQ) [1], where each input manages a separate queue for each output, it has been shown that a throughput of 100% can be achieved [2]–[4]. *Arbitration algorithms* resolve the contention for output ports in each time slot. The achievable throughput and delay performance heavily depends the used algorithm.

Arbitration can principially be performed by two ways: Statically by assigning time slots to specific connections or connection groups in advance or dynamically by resolving the contention for the same output port in each time slot anew.

The *static arbitration* (also called *allocation*) reserves time slots for specific connections in advance. This fixed schedule offers bandwidth guarantees, deterministic delay bounds for worst-case traffic and actively shapes (smoothes) the traffic. As shown in this paper, static arbitration can guarantee delay bounds for traffic accepted by the connection admission control (CAC) of ATM by assigning a constant service bandwidth to each connection. This immediately supports CBR and VBR services, whose traffic parameters are known at CAC time. Similarily, IPv6 guaranteed service [5] with RSVP [6] is supported. Due to the boundedness of the departure process, bounds can as well be given end-to-end. Analysis and results are in this paper.

On the other hand, *dynamic arbitration* distributes the left-over bandwidth to contending ports. Weighted algorithms [3] offer the best delay performance compared to unweighted algorithms and they operate much better in more difficult than symmetric load configurations [7]. Additionally, a static priority distinction between service classes must be integrated in a global arbitration method. A local priority decision within each input port alone is not sufficient. In this paper a method for prioritized arbitration and its performance are treated.

After the the centralized arbitration decision, the input ports are notified, which output port they have to send a cell to. A local *cell scheduler* within each input port must select the first queued cell of a suitable connection for transmission to the given output port. Unlike other papers assuming FCFS service it is necessary to provide cell scheduling in the input ports within each priority class. Thus problems of individual QoS, flow separation and fairness can be addressed. The difficulty is the complexity of the system involving the hierarchical decision of both arbiter and scheduler. In this paper EDF scheduling is shown to perform similar in the VOQ configuration as it does in output-queued switches. Together with an approximative analytic model of the arbitration, the delay performance for FCFS and EDF can be obtained in a closed form.

The paper is organized as follows. Section 2 discusses VOQ and arbitration algorithms and related work. The static arbitration is explained and analysed in Sect. 3. Dynamic methods are treated in Sect. 4. Section 4.3 contains performance results for the dynamic arbitration method. Based on an approximative analytic model of the dynamic arbitration in Sect. 5, cell scheduling is treated in Sect. 6.

2. VOQ System Description

The VOQ configuration [1] shown in Fig. 1 consists of

Manuscript received May 28, 1999.
Manuscript revised August 26, 1999.

[†]The author is with the Institute of Integrated Signal Processing Systems (ISS) at Aachen University of Technology (RWTH).

IEICE TRANS. COMMUN., Vol. E83-B, No. 2, pp. 171–181, February 2000

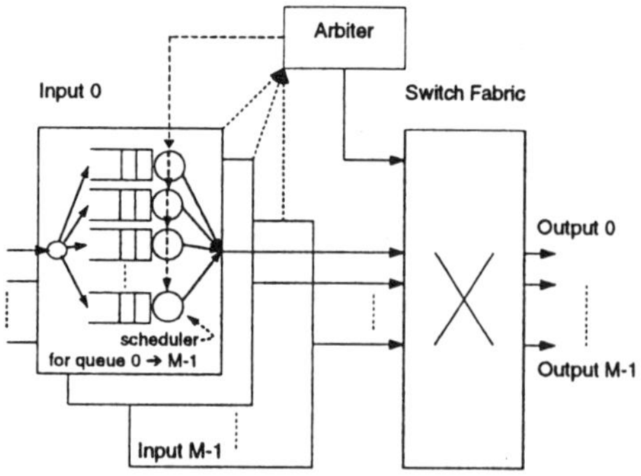

Fig. 1 VOQ architecture.

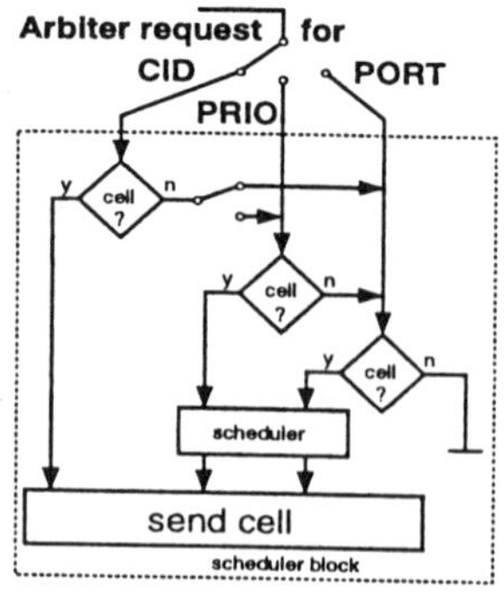

Fig. 3 Input scheduling.

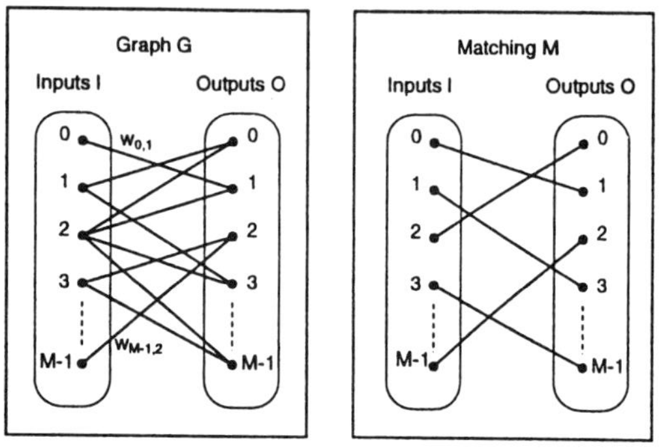

Fig. 2 Bipartite matching.

a nonblocking switch fabric, an arbitration unit and M input and output ports. Each port has a link rate of r_{link}. Arriving cells on input port i are placed into the corresponding queue for their destination port o. This queue does not necessarily maintain cells in FCFS order, as often assumed, it can as well serve cells in weighted round-robin or earliest-deadline-first (EDF) order [8], as shown in Sect. 6. Let its current queue size be $q_{i,o}$. The process of arrivals to this queue is characterized by a mean rate $\lambda_{i,o}$. Let the input and output loads be

$$\rho_o^{out} = \sum_{i=0}^{M-1} \lambda_{i,o} T_{slot}; \qquad \rho_i^{in} = \sum_{o=0}^{M-1} \lambda_{i,o} T_{slot} \tag{1}$$

This is admissible [1] if $\forall o : \rho_o^{out} < 1$ and $\forall i : \rho_i^{in} < 1$.

In each time slot the arbiter selects unique pairs of input and output ports (a "match" (i, o)) either by a lookup in the allocation table (if there is an entry) or by realtime calculations based on weights $w_{i,o}$ sent to it from the input ports. This task is equivalent to a bipartite graph matching problem [1] as shown in Fig. 2.

Static arbitration (allocation) has been used in traditional circuit-switched TDM systems, where the arrival and departure instances of frames are known in advance. TDM switches only have to precompute a periodic schedule to control the crosspoints in a switching network. For packet-switched networks (ATM) there is no frame reference time. Due to the asynchronous nature of packet traffic a periodic structure cannot immediately be exploited. Ideas to integrate precompute schedules for packet switches appeared in [9]–[11]. As shown in Sect. 3, the constant bandwidth which is guaranteed per connection combined with the bounded traffic model in Sect. 3.1 offers delay bounds per switch and end-to-end. Figure 3 shows that when the connection indentifier (CID) is allocated and a cell of that connection is available, no more decisions have to be taken. However, if there is no cell at all, the input-output port match cannot be changed and this slot is unused. Since the distance between service slots is deterministic and usually† larger than with dynamic arbitration (see Fig. 16), the average cell delay can be much lower with dynamic arbitration. This is why in a second stage a dynamic arbitration is performed on all unmatched ports.

A number of algorithms exist to solve the dynamic matching optimally, either as a maximum size matching (MSM) or a maximum weight matching (MWM) [12]. MSM finds the maximum number of input-output connections. MWM maximizes the weight sum of the match. It has been shown that 100% throughput can be achieved for all admissible i.i.d. arrivals [3] with MWM and $w_{i,o} = q_{i,o}$ (LQF).

Algorithms for solving the MWM problem are computationally complex with $O(M^3 \log M)$. A number of alternatives have been proposed, such as iMCFF [4], iLQF [2] and SIMP [7]. Even the unweighted MSM problem is rated $O(M^{2.5})$. MSM approximations exist with PIM [10], iSLIP [2], WFA [13] or FARR [14].

Without weight information instability and unfairness are likely, because queue backlogs can accumulate and cell delays become very large for bursty or asymmetric load [7]. Explicit treatment of priorities [15] is important, because it is required to separate traffic with extremely different QoS requirements (realtime and data traffic). For ABR an additional credit-based flow control mechanism [16] is required to guarantee zero cell loss by avoiding buffer overflows.

Cell scheduling within the input ports to offer dif-

† In an underloaded switch.

ferentiated QoS is a quite new topic. As shown in Fig. 3 the cell scheduler decides on which exact cell of which connection to forward, given the destination port o. The priority is given implicitely, because the arbiter has already chosen globally that this port has the highest priority for output o. With a static priority scheme the requested priority is served. But within the priority level a cell scheduler can decide using any known scheduling algorithm, not only FCFS. An approach for WFQ can be found in [17]. In Sect. 6 we show that given the performance of FCFS which is derived from an arbiter model, the performance of EDF can be calculated.

3. Static Bandwidth Allocation

The delay guarantee concept of allocation is based on the individual (per-VC) guarantee of a service rate. This guarantee is firm, i.e. the probability $Pr\{d > d_{max}\}$[†] to exceed a delay d_{max} is zero. This is in contrast to statistical delay bounds, where $Pr\{d > d_{max}\}$ is never zero but only tends to zero for higher d_{max} (typically with an exponential decay [18]) and the performance heavily depends on the total load of all other connections and their statistical properties (an undesired property). The requirement for firm delay bounds with allocation is that the traffic of the connection-of-interest is bounded, as shown now.

3.1 Bounded Traffic

Traffic bounds are common in integrated services networks. For ATM, the known traffic descriptor (PCR, $CDVT$) bounds the peak cell rate over an interval, (SCR, MBS) bounds the average rate over a longer interval. The contract conformance can easily be policed by the GCRA [19] algorithm. IPv6 offers a TSpec in its RSVP message [20] during connection admission, which is very similar.

The parameters are fitted into the traffic bound model of Cruz [21], [22] which bounds the amount of traffic A (given by its rate function R) in an interval (t_1, t_2) by

$$\int_{t_1}^{t_2} R(t)dt = A(t_1, t_2) \leq \min_{1 \leq i \leq n} \{\sigma_i + \rho_i(t_2 - t_1)\} \quad (2)$$

This can be written $R \sim (\vec{\sigma}, \vec{\rho})$ for n pairs (σ_i, ρ_i). Each pair bounds the traffic rate to ρ_i with a burst tolerance σ_i. ATM offers $n = 1$ for CBR and $n = 2$ for VBR in the traffic contract at the UNI [19]. Using $T_{PCR} = 1/PCR$ and $T_{link} = 1/r_{link}$ we give the following mapping of parameters[††]

$$\begin{aligned} \sigma_1 &= BS = 1 + \lfloor CDVT/(T_{PCR} - T_{link}) \rfloor, \\ \rho_1 &= PCR, \quad \sigma_2 = MBS, \quad \rho_2 = SCR \end{aligned} \quad (3)$$

To calculate firm delay bounds, a worst-case traffic model [23] using these parameters must be applied. It consists of periodically repeated bursts of length BS (MBS f. VBR) cells followed by silence such that the average rate during the period length is PCR (SCR). The benefit of this model is that delay bounds derived for the worst-case cannot be exceeded by any other traffic which also obeyes the traffic contract.

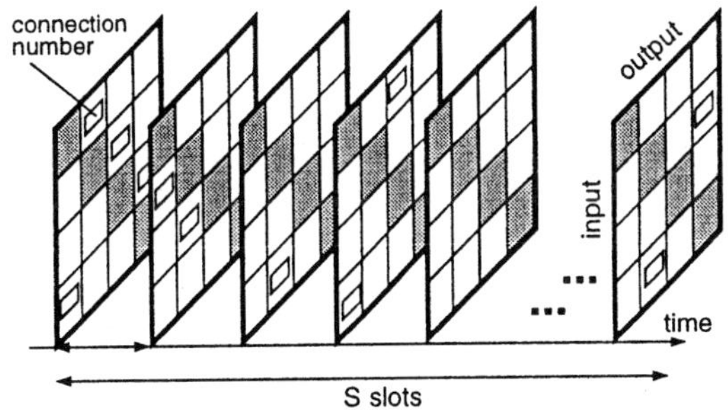

Fig. 4 Allocation table.

3.2 Construction of the Allocation Table

A switch that supports static arbitration maintains an allocation table $T = [\vec{t_k}]$ $(0 \leq k < S)$ that contains the precomputed and periodically repeated schedule for S time slots in a frame [11] ($T_{period} = S \cdot T_{link}$), see Fig. 4. In a slot k, $\vec{t_k}$ contains the input port numbers i_o for each precomputed match $(i_o \rightarrow o)$:

$$(i_o \rightarrow o) \quad \Leftrightarrow \quad t_{k,o} = i_o \quad (4)$$

For each entry a connection identifier ($CID = f(VPI, VCI)$) is specified. Empty entries are readily available for dynamic arbitration. For each CID a number of slots $n(CID)$ are reserved (for CBR: $r = PCR$, VBR: $r = SCR$):

$$n(CID) = \left\lceil \frac{\omega \cdot r}{r_{slot}} \right\rceil \quad (5)$$

Each allocated slot contributes to a bandwidth[†††] of

$$r_{slot} = r_{link}/S = T_{period}^{-1} \quad (6)$$

The overallocation factor ω $(\omega > 1)$ is used to adjust the service rate by a desired amount. This connection is then served with an allocated bandwidth of

$$\mu_s(CID) = T_\mu^{-1} = n(CID) \cdot r_{slot} \quad (7)$$

which determines the individual (per-VC) load ρ_{CID}

$$\rho(CID) = \lambda_a/\mu_s(CID) = T_\mu/\bar{a} \quad (8)$$

where the arrival traffic rate $\lambda_a = \bar{a}^{-1}$ is the mean rate

[†] CDF, a complementary distribution function.

[††] In units of cells and cells/second.

[†††] Corresponds to the smallest supported rate, e.g. 64 kb/s.

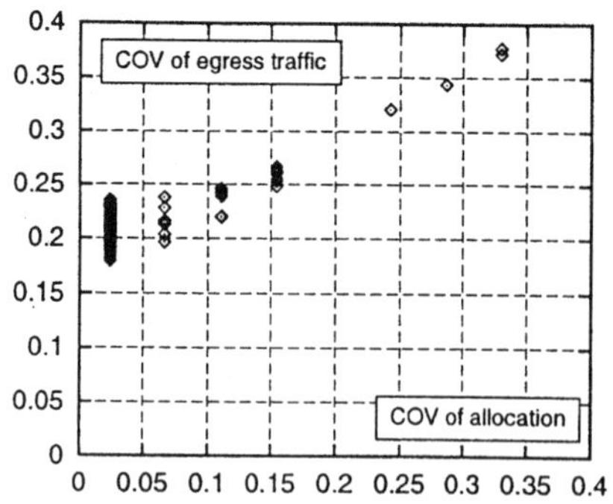

Fig. 5 COV_o vs. COV_s.

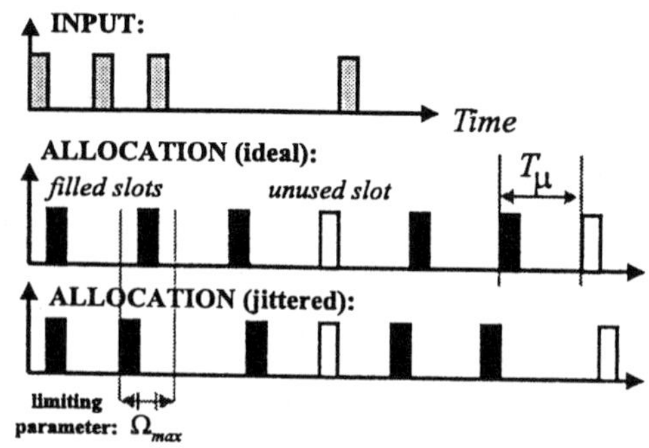

Fig. 6 Allocation jitter.

of the stationary traffic process. Due to the discrete nature of μ_s, the load can only be adjusted in discrete steps (Eq. (5)). For that reason the effective ω changes slightly to

$$\omega(CID)=\mu_s(CID)/PCR(CID)=1/\rho(CID). \quad (9)$$

3.3 Metrics for Allocation

Algorithms for the distribution of $n(CID)$ slots into the allocation table can be obtained from [9]–[11]. The connections are served in a per-VC manner with the service time given by the slot distances. A number of allocated slots for a connection implies an average slot distance of $T_\mu = T_{period}/n(CID)$. However, due to quantization and blocking effects this is not equidistant. The maximum tolerable slot offset is controlled with a parameter Ω_{max}, which bounds the $CDVT_{max}$ of the allocated slots of each connection. The deviation is quantified with the coefficient-of-variation COV_s of the interservice time $\tau_{s,i} = t_{i+1} - t_i$.

$$COV_s = \sqrt{Var(\tau_s)}/E(\tau_s) \quad (10)$$

It is a measure of the quality of the allocation in the sense of a smooth service time distribution. It is desirable to have $COV_s = 0$ $(G/D/1)$. The actual COV_s depends on S, $n(CID)$, the total number of other allocated (blocked) slots, the allocation algorithm and Ω_{max} (see Fig. 6).

The output traffic stream is shaped such that it assimilates the service process. Thus COV_s and the output traffic variation COV_o are strongly correlated. COV_o also depends on $\omega(CID)$: Even with ideal allocation ($COV_s = 0$), unused slots lead to gaps in the output traffic stream (Fig. 6) which adds one component to the COV_o: Let for simplicity the random process *slot used* be i.i.d. and the probability of a used slot be $p = 1 - \rho(CID)$. Then the assumed geometrically distributed interdeparture time has

$$COV_o = \sqrt{p} = \sqrt{1 - \rho(CID)} \quad (11)$$

For example, with $\omega(CID) = 1.05$, $\rho(CID) \approx 0.95$ the contribution is $COV_o = 0.218$, which can be seen in Fig. 5. Here the traffic arrived with $COV_a = 0.4$. The graph has been obtained by allocation of 240 connections. Each dot represents the values for one connection. The COV_s due to quantization is

$$COV_{s,quant} = \sqrt{T_{link}^2/12T_\mu^2} \quad (12)$$

which is 0.015 here. Allocation jitter due to Ω_{max} is at most $COV_{s,\Omega} = 0.33$ here, which can be seen in Fig. 5. It is bounded by

$$COV_{s,\Omega} \le \Omega_{max} \cdot T_{link}/(\sqrt{3} \cdot T_\mu) \quad (13)$$

Table 1 Properties of allocation.

Properties of allocation
isolated, per-VC treatment (no influence of other traffic)
pre-determined constant delay bound → QoS
output traffic is shaped and bounded
self-policing (output conforms to GCRA)
QoS is adjustable with a single parameter ω
simple model for stochastic traffic: $G/E_k/1$
can be used together with dynamic arbitration → 100% utilization
non work-conserving globally → mean delay is higher
hybrid arbitration offers lower mean delay but higher miss ratio

3.4 Allocation Performance

Allocation has a number of positive properties (Table 1) which can be quantified for given traffic bounds, i.e. during CAC the model parameters are known for connections demanding real-time QoS.

When bounded (deterministic) traffic (Sect. 3.1) is applied, an explicit delay bound is guaranteed (Fig. 8). Assuming a fully equidistant service ($COV_s = 0$) with a service rate $1/T_\mu$ higher than PCR for CBR or SCR for VBR, the time between instances where the queue is empty (busy time) is less than the period of the periodic worst case traffic bounded by the given traffic parameters. Thus delay bounds can be given.

A bound for CBR is derived here in continuous time. The delay a cell of bounded stream experiences is composed of the sum of two statistically independent terms d_e =*delay until service if queue is empty* and d_b =*delay due to cell's position in burst*. With a random phase, d_e is equally distributed (rect-shaped) over

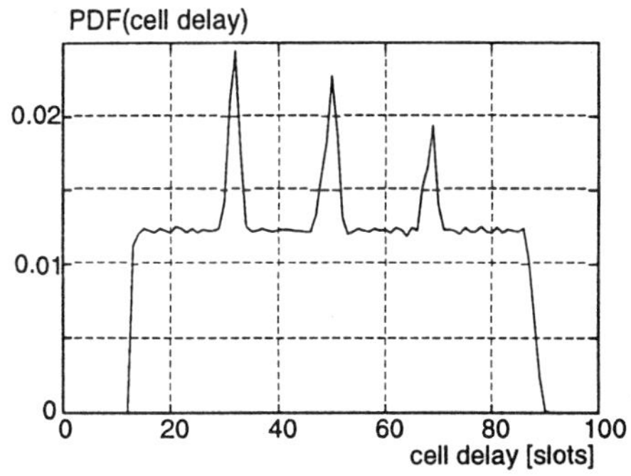

Fig. 7 Example $PDF(d)$.

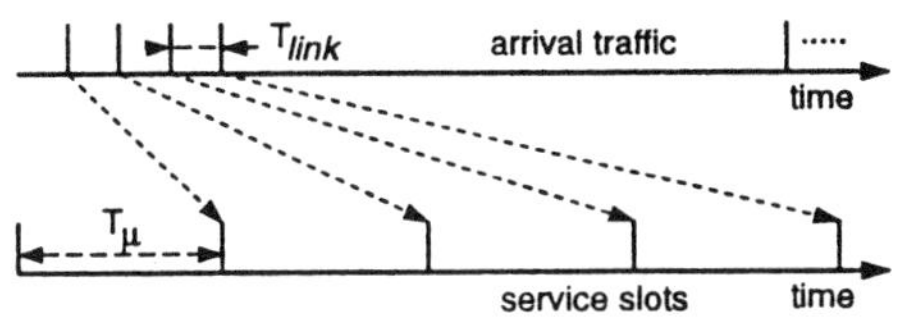

Fig. 8 Worst case traffic.

the interval $[0; T_\mu]$: $PDF(d_e) = p_e(d_e) = \prod(d_e/T_\mu - \frac{1}{2})$. The distribution $PDF(d_b)$ is a function shaped like a sequence of dirac (δ) impulses ($||||$):

$$p_b(d_b) = \frac{1}{BS} \sum_{i=0}^{BS-1} \delta(d_b - i \cdot (T_\mu - T_{link})) \quad (14)$$

The resulting delay distribution is the convolution ($\otimes$) of these components: $p(d) = p_e(d) \otimes p_b(d) = \prod \otimes ||||$. For the example, with $BS = 4$ a sequence of four *diracs* convoluted with a *rect* should have the sandcastle shape, as the simulated result in Fig. 7 shows. The maximum value leads to this delay bound:

$$\max(d_{CBR}) \leq (BS - 1) \cdot (T_\mu - T_{link}) + T_\mu \quad (15)$$

The situation for VBR is similar. One level of burstiness more makes it more complex, however. For a smooth worst case model [23] the same holds:

$$\max(d_{VBR}) \leq (MBS - 1) \cdot (T_\mu - T_{PCR}) + T_\mu \quad (16)$$

For the heavy VBR model† Eq. (17) holds.

$$T_O := BS \cdot (T_\mu - T_{PCR})$$

$$\max(d_{VBR}) \leq \lfloor (MBS - 1)/BS \rfloor \cdot T_O + [(MBS - 1) \bmod BS] \cdot (T_\mu - T_{link}) + T_\mu \quad (17)$$

When the assumption $COV_s = 0$ is relaxed, an additional delay of $CDVT_{max}$ must be added.

3.5 End-to-End Delay Bounds

With allocation an end-to-end delay guarantee d_N over N hops can be given (Fig. 9).

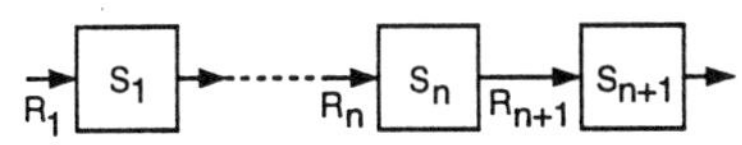

Fig. 9 Traffic bounds.

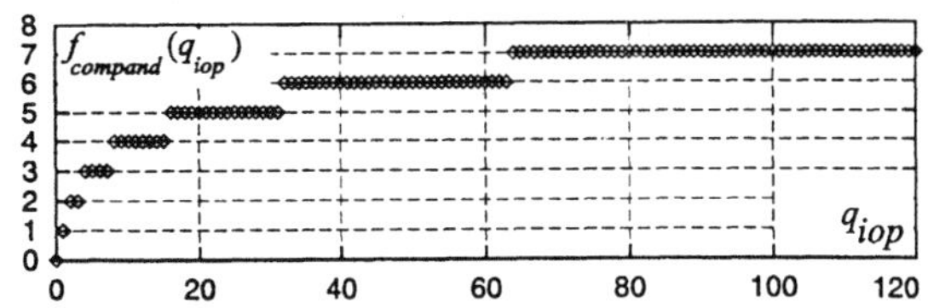

Fig. 10 $f_{compand}$.

Lemma 1: For any switch n on the route and given CBR†† input traffic parameters $R_n \sim (\sigma_n, \rho_n)$ the output is shaped such that its traffic is bounded by pairs $R_{n+1} \sim (\sigma_{n+1}, \rho_{n+1})$. Of the infinitely many possible pairs, two are of interest: Eq. (18) at the service time $\rho_n \cdot \omega_n$ and Eq. (19) at the declared ρ_n.

$$\rho'_{n+1} = PCR_{n+1} = \rho_n \cdot \omega_n \quad (18)$$

$$\sigma'_{n+1} = 1 + \lfloor 2T_{link}\Omega_{max}/(T_{PCR_{n+1}} - T_{link}) \rfloor$$

$$\rho''_{n+1} = PCR_{n+1} = \rho_n \quad (19)$$

$$\sigma''_{n+1} = 1 + \lfloor CDVT''_{max}/(T_{PCR_{n+1}} - T_{link}) \rfloor$$

$$CDVT''_{max} = (\sigma_n - 1)(T_{PCR} - T_\mu) + T_\mu + 2\Omega_{max}T_{link}$$

In fact, after the first allocating switch the output traffic is much smoother due to active shaping. What results are lower delay bounds for any following switch. Overallocation ω is not necessary for switches $n > 1$, because overload is impossible by construction.

Theorem 1: For any limited number of hops n the total cell delay $d_{sum,n}$ and its output traffic R_{i+1} is bounded. Let R_i be the traffic bounds before switch i and $d_i(R_i)$ the resulting delay bound for it. Then holds

$$d_{sum,n} = \sum_{i=1}^{n} d_i(R_i) \quad (20)$$

Proof (by induction): With given bounded traffic parameters R_1, the first switch at the network ingress has a bounded delay $d_{sum,1} = d_1$ given by Eqs. (15) and (17). This is obviously the delay for $N = 1$.

Let the theorem be true for switch n. Switch $n+1$ then has also a bounded delay output.

Proof: The worst-case delay of switch $n + 1$, d_{n+1}, adds to $d_{sum,n}$, the maximum delay guaranteed so far.

†Maximal length bursts on link and PCR rate [23].

††For VBR substitute $SCR \rightarrow PCR$.

Table 2 Delay bounds for example realtime applications (with $maxCTD = 200\,\text{ms}$, $\omega = 1$) for switch 1 and the following (n).

Traffic	Descriptors	$max(d_1)$	$max(d_n)$
CBR	$PCR = 64\,\text{kb/s}$, $CDVT = 5\,\text{ms}$	5.5 ms	5.5 ms
CBR	$PCR = 1\,\text{Mb/s}$, $CDVT = 1.4\,\text{ms}$	1.76 ms	352μs
VBR	$PCR = 10\,\text{Mb/s}$, $SCR = 3.3\,\text{Mb/s}$ $MBS = 1000$ (MPEG-2)	107 ms	400μs

d_{n+1} is bounded because its input traffic R_i is bounded (Lemma 1) and allocation then guarantees Eqs. (15) and 17.

Thus Theorem 1 holds for all n.

In Table 2 example values for typical voice/video traffic show the practical suitability of allocation. CAC rules can be obtained easily [11] by using the bounds above.

4. Dynamic Arbitration

Dynamic arbitration serves as a complementary mechanism to use unallocated slots for connections which do not require strict delay bounds or require a lower mean cell delay, which cannot be provided by static mechanisms.

4.1 Priority Support for Weighted Matching

Weighted matching algorithms can achieve full throughput under admissible load [3]. However, with different priority classes a stable operation cannot be guaranteed. In fact an admissible load of higher priority cell streams can nevertheless lead to unbounded delay if lower priority (best effort) connections experience congestion (high queue weight $w_{i,o}$). When a smaller weight on another port represents real-time cells only, they have to suffer. Only a globally prioritizing arbitration can achieve the correct separation.

The goal is to have no influence of lower priority queue weights over higher priority weights. A three-dimensional extension of the known weighted matching algorithms is discarded because of the complexity. We aim for a method operating in two dimensions only.

Consider a queue organization as shown to Fig. 1 and assume the queues are additionally separated per priority (with P priority levels, 1=highest). Let the queue states on input i, output o and priority level p be $q_{i,o,p}$. The arrivals to these queues are characterized by a mean rate $\lambda_{i,o,p}$. All weighted matching algorithms operate on a two-dimensional weight $w_{i,o}$. Therefore we map $[q_{i,o,p}]$ to $[w_{i,o}]$ in a special way. The simplest case is $w_{i,o} = \sum_{p=1}^{P} q_{i,o,p}$ (no priority support).

The key idea of the simplification proposed in this paper is to use the following mapping function (modified weighted summation) to reduce q_{iop} to a two-dimensional format that can be handled by any weight dependent arbitration algorithm [15].

$$w_{io} = \sum_{p=1}^{P} c_p \cdot f(q_{iop}) \tag{21}$$

$$c_p = \prod_{q=p+1}^{P-1} k_q = \prod_{q=p+1}^{P-1} 2^{b_q} \tag{22}$$

The coefficients c_p are computed in Eq. (22) such that the weight of a queue with priority p is k_{p+1} times higher (b_{p+1} bits) than that of the next lower priority $p+1$. For the function $f(q_{iop})$ the alternatives are

$$f_{eq}(q_{iop}) = q_{iop} \tag{23}$$

$$f_{sat}(q_{iop}, k_p) = \left\{ \begin{array}{ll} q_{iop} & \text{if} \quad q_{iop} < k_p \\ k_p - 1 & \text{if} \quad q_{iop} \geq k_p \end{array} \right\} \tag{24}$$

$$f_{compand}(q_{iop}) = \left\{ \begin{array}{rl} 0 : & q_{iop} = 0 \\ \lfloor \log_2(q_{iop}) + 1 \rfloor : & \text{else} \\ 2^{b_p} - 1 : & q_{iop} \geq 2^{b_p} \end{array} \right. \tag{25}$$

f_{eq} is a simple addition, which still has a problem when due to large values an overflow into the bits reserved for higher priorities occurs. A sufficient priority separation is achieved if the overflow probability is very low. The saturation f_{sat} cuts off any overflow. This has the trade-off that the wordlength (the dynamic range) is small. An example for saturation is $P = 3; b_p = 3 \Rightarrow k_p = 8; q_{io1} = 2 = 010_b, q_{io2} = 5 = 101_b, q_{io3} = 15 \Rightarrow f_{sat}(q_{io3}) = 111_b \Rightarrow w_{io} = 010101111_b$. If a wider dynamic range is necessary (e.g. for bursty traffic) but bits for the wordlength must be saved, the companding characteristic $f_{compand}$[†] is recommended. It requires $2^{b_p} - 1 = \lfloor \log_2(2^{b_v} - 1) + 1 \rfloor$ or $2^{b_p} - 1 = b_v$. The most effective value is $b_p = 3$ bit offering a dynamic range of $b_v = 7$ bit (Fig. 10), which has shown to be sufficient for good results. The operation can very well be implemented in hardware, since only the position of the highest 1 bit must be determined for the $\log_2$ operation.

With this mechanism priorities are supported because queued cells of a higher priority are given a higher weight $w_{i,o}$. The number of bits per priority b_p representing the dynamic range within the class allows scaling of the arbitration performance from MSM ($b_p = 1$) to MWM ($b_p \to \infty$). This has consequences especially for asymmetric load [7].

4.2 An Approximation for Weighted Arbitration

As one low complexity method to achieve weighted matching by approximating MWM, the SIMP [7] algorithm is outlined shortly in Table 3. SIMP approximates MWM by successively choosing the edge with the highest weight for each output node in order (step 4). This resolves the output contention problem for each port in a cyclic manner. It cannot generally find the global maximum MWM would find. Visually we

[†]Similar to the μlaw used for speech coding.

Table 3 The SIMP algorithm.

1	let I be an ordered list of all input ports and O the list of all output ports
2	let $I' \leftarrow I$ and $O' \leftarrow O$
3	choose the first output port o_c out of the ordered list O'
4	choose the input port i_c to match as one with $w_{i_c} = \max_{i \in I'}(w'_i)$, resolve ambiguities (same weights) in round-robin fashion
5	if $w_{i_c} > 0$, match i_c with o_c and let $I' \leftarrow I' \setminus i_c$ (set minus)
6	reduce the match space by letting $O' \leftarrow O' \setminus o_c$
7	repeat steps 3-6 until $O' = \emptyset$ (M repititions)
8	shift the list O cyclically before the next slot to achieve round-robin fairness between outputs
9	start the next time slot at step 2

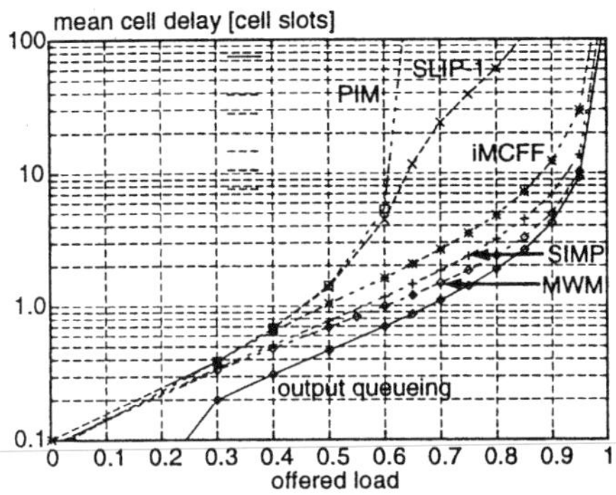

Fig. 11 Std. scenario.

can imagine that SIMP only considers one triangular half of the weight matrix $[w_{io}]$, instead of the whole as with MWM. Due to the cyclic shift of this viewport the number of weights considered for a specific output port changes cyclically from 1 to M. As we see in Sect. 4.3 the approximation yields reasonable results.

4.3 Performance Results for Arbitration

In this section we compare the performance results for reference algorithms output queueing, MWM, iSLIP, PIM, iMCFF with SIMP using simulations for an 16×16 switch with OPNET [24].

With Bernoulli traffic and a symmetrically chosen destination (1 of $M-1$), each input port carries a total offered load of ρ, i.e. $\lambda_{i,o} = \rho/(M-1)$.

Bursty traffic is important for many reasons. According to [25] we use a packet train model producing a burst of $B(t)$ cells at full rate followed by empty slots such that the mean rate $\lambda_{i,o} = \rho/(M-1)$ is the desired fraction of the input rate. B is exponentially distributed here; the *mean burst length* $\bar{B}$ is 32. This corresponds to the mean length of an Ethernet PDU segmented into ATM cells.

In Fig. 11 the results for PIM and iSLIP are as in [25]. The best case is output queueing which we want to approximate with VOQ. The ideal arbitration for VOQ is MWM which most closely approximates output queueing. As we see, iMCFF performs worse than SIMP. The performance of SIMP is in between iMCFF and MWM.

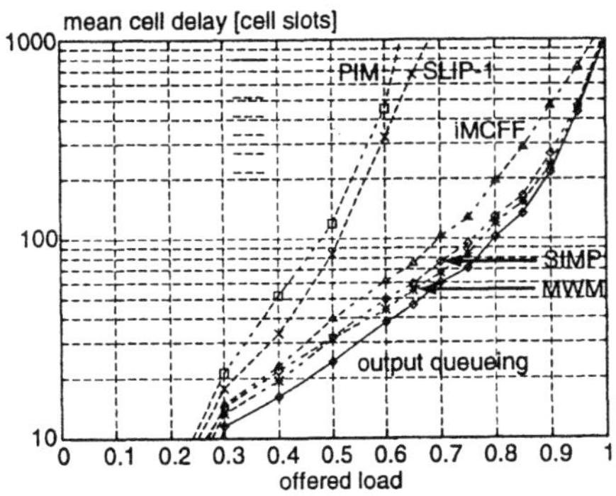

Fig. 12 Burst scenario.

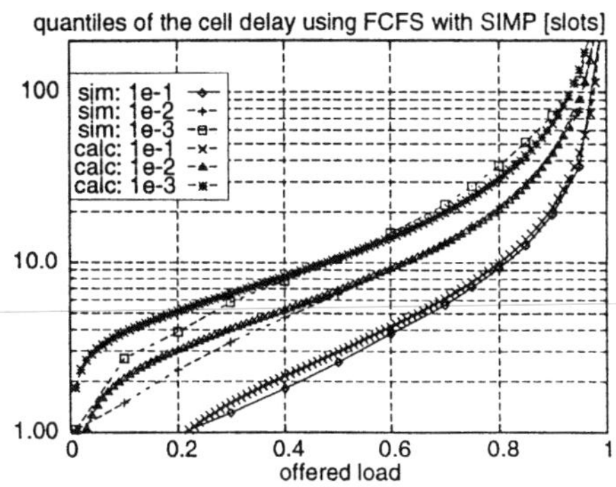

Fig. 13 $d_{quantile(x)}(\rho)$.

In Fig. 12 we see the delay performance for the bursty scenario. The most important characteristics are: (i) the absolute delay is two decades higher than for the previous scenario due to the burst scale queueing effect and the short-term asymmetries. (ii) PIM and iSLIP perform similar with more than 300 cell transmission times above $\rho = 0.6$, i.e. noticably worse than the other algorithms. (iii) As before, iMCFF, SIMP, MWM and output queueing are quite close to each other with improved performance (less delay) in ascending order.

We observe that the algorithms that decide based on weights (iMCFF, SIMP, MWM) perform very well for typical traffic.

When using the first priority mechanism f_{eq} the main parameter for achieving priority separation is the priority weight factor $k_q = 2^{b_q}$ (b_q shift bits). The other two degrees of freedom in the choice of traffic parameters manifest in $\rho_{total} = \rho_1 + \rho_2$ and s_{p1} (share of first priority cells) with $\rho_1 = s_{p1} \cdot \rho_{total}$ and $\rho_2 = (1 - s_{p1}) \cdot \rho_{total}$. The behavior for two priority classes[†] can be seen in Fig. 15, where $\rho_1 = 0.2$ is held constant and ρ_2 is varied as bursty traffic. We see that there is almost no influence of the lower priority load on the higher priority performance when a wordlength of seven or more bits is used.

To analyse quantitatively how many bits b_p are

[†]Higher priority traffic is Bernoulli, lower is bursty.

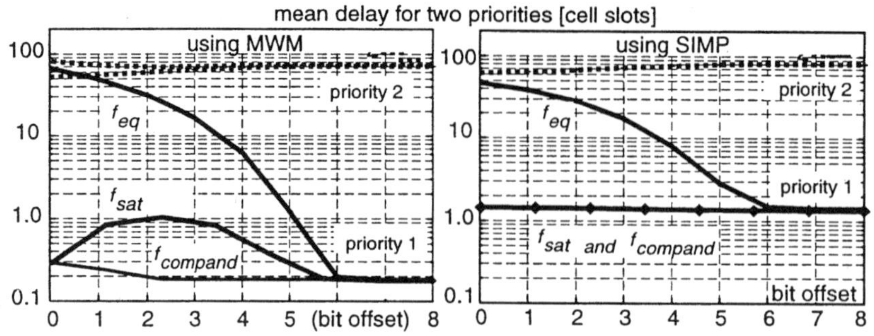

Fig. 14 Priority distinction.

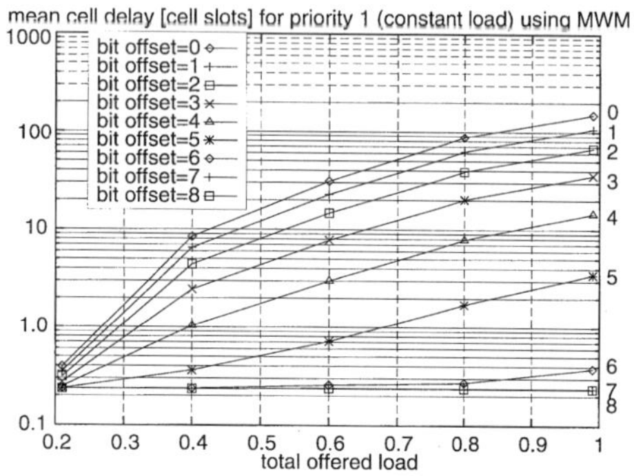

Fig. 15 Priority scenario.

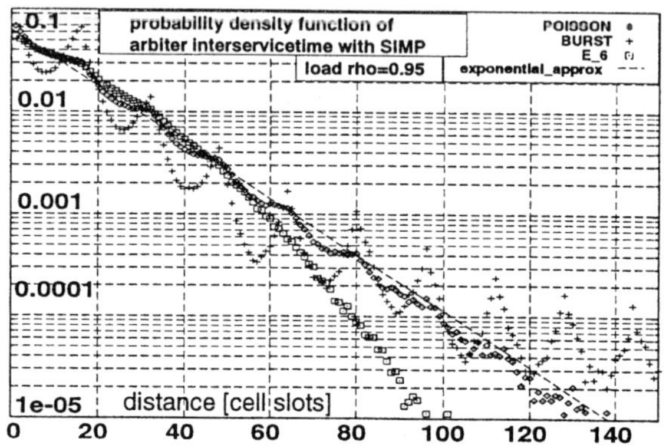

Fig. 16 VOQ switch: PDF for interservice time of SIMP arbiter.

needed we study the mean delay for both priority groups at a fixed load $\rho = 0.7$ and a fixed ratio $s_{p1} = 0.2$ as a function of different b_p. The fixed values used have revealed to be the most expressive.

For bursty traffic Fig. 14 shows that about $b_p = 6$ bits are needed to fully separate the priorities. We can verify that with enough bits $\bar{d}_{prio1} = 0.2$ slots at load $\rho_1 = 0.2$. This is exactly the delay for MWM with $\rho = 0.2$ in Fig. 11. Thus we have an exact priority separation. The higher priority traffic performance does only depend on its total load.

The necessary value of b_p depends on the traffic in priority p. This is not desirable, so we apply the deterministic separation by saturation (Eq. (24)), where no overlapping is possible, i.e. $w_{io,p+1}$ cannot become higher than 2^{b_p}. We lose some of the weight dynamics in this priority because the values possible are in the interval $[0, 2^{b_p} - 1]$.

With saturation we observe in Fig. 14 that now the separation is much improved for $1 \ldots 5$ bits. There is still some influence typical for MWM[†]. We can even choose one bit per priority and achieve a considerable separation. This, however, ignores all weights within a priority level and yields a bad performance under asymmetric workload as MSM does.

With companding (Fig. 14) we finally observe the very best separation. Here three bits are sufficient for a complete separation, whereas the dynamics is as good as pure MWM.

5. Calculating VOQ Performance

In a VOQ configuration there are M virtual schedulers for each priority in each input port, one for each output. Each of these schedulers is only activated to serve a cell if induced by the arbiter. Thus the arbiter appoints the service interval, opposed to an OQ switch, where a cell is served in each time slot. In Fig. 16 the time between successive service events is shown for a number of configurations, e.g. different load and different traffic for SIMP. Observe the periodicity in multiples of $M \cdot T_{slot}$. For lower load or smooth traffic with SIMP an exponential distribution looks very similar, but assuming an $M/M/1$ queueing system is wrong because in the VOQ system there is no statistical independence between arrivals and service. The dependence is exactly what we want weighted matching algorithms to achieve.

For determining the delay performance of the scheduler a performance model for the arbiter is needed. The delay quantiles can be obtained by simulation and analytical modelling. With a detailled stochastic PN system (Petri Net [26]) the arbitration can be modelled

[†]The maximum matching is the one that maximizes the sum of the matched weights. This sum might be still higher for some lower priority matches (e.g. $7 + 7$) than for the alternative match (e.g. $8 + 0$).

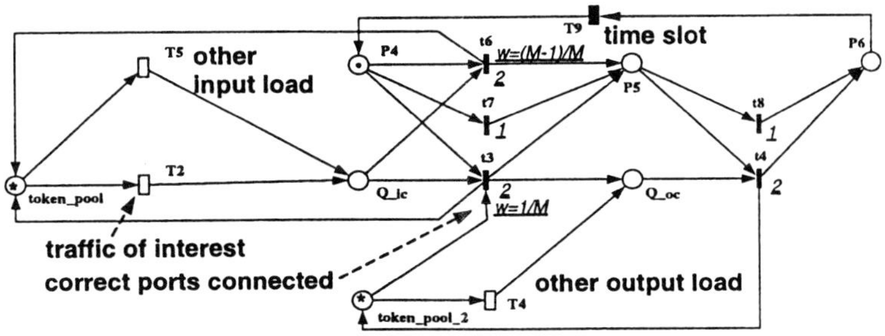

Fig. 17 PN system for VOQ.

most accurately [8], but the stationary solution is computationally very intensive as the number of Markov states grow large. Alternatively a more abstract modelling of the reasons for a higher delay, the input and output conflicts, provides quite good results. For the performance from port i to o this is modelled in Fig. 17 by having the input and output port loaded by ρ_i^{in} and ρ_o^{out} respectively. With an independent probability $1/M$ a token in Qic is transferred to Qoc. The performance is similar to two virtual $M/D/1$-type queues in series. Thus the calculated approximation can be obtained based on Eq. (26).

$$Pr\{w \leq t\} = 1 - ae^{-bt} \tag{26}$$

$$a = \frac{2E[w]^2}{E[w^2]} \quad \text{and} \quad b = \frac{2E[w]}{E[w^2]} \tag{27}$$

where the first and second moment are obtained for the $M/D/1/\infty/FCFS$ system [27]. For realtime applications the QoS demand per switch can be expressed as $Pr\{d > d_{max}\} < \epsilon$ where ϵ and d_{max} are derived from ATM traffic and QoS parameters [19]. The delay quantiles are given in Eq. (28), and especially for Poisson traffic in Eq. (29).

$$d_\epsilon = \frac{E[w^2]}{E[w]} \ln(2\frac{E[w]^2}{\epsilon E[w^2]}) \tag{28}$$

$$d_{\epsilon,Poisson} = \frac{T_{slot}(2+\rho)}{3(1-\rho)} \ln(\frac{3\rho}{\epsilon(2+\rho)}) \tag{29}$$

In Fig. 13 the simulated and calculated results for $\epsilon \in \{10^{-1}, 10^{-2}, 10^{-3}\}$ are compared and show an acceptable accuracy. This has been used to obtain approximations for the FCFS scheduling performance in Fig. 19 at $\rho = 0.95$.

6. Cell Scheduling in VOQ Switches

The reason for sophisticated cell scheduling algorithms is the global provision of an individual per-stream QoS and the decision which concrete cell to send in a slot-by-slot timescale. In traditional output-queued (OQ) switches all cells for a specific egress link are handled by one scheduler in the output port and there is almost no difference between the multiplexer being fed by M input links with a rate of λ/M each or one input with a rate λ. For $M \to \infty$ an FCFS† scheduler with Poisson traffic is appropiately modelled by an $M/D/1$ queueing system. In many cases the performance can be numerically calculated as the cell delay distribution $PDF(d)$ in stationarity.

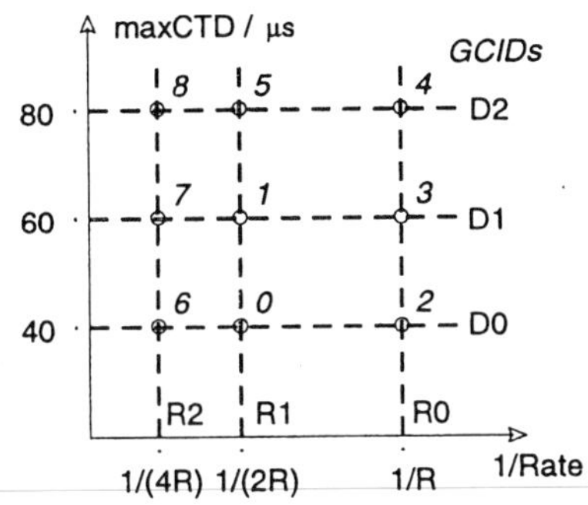

Fig. 18 $Rate[VC]$ and $d_{max}[VC]$.

More complex schedulers have evolved for the need to treat distinct traffic streams differently. Here we concentrate only on the most promising algorithm EDF† [29], [30], because the principal properties of VOQ scheduling are to be shown. EDF offers a parameter $D[VC]$ which has the meaning of deadline. With a properly chosen parameter $D[VC] = d_{max}[VC] + c$ this very successfully guarantees individual statistical delay bounds.

For the graphs shown in this paper the slot time is assumed to be $T = 1\,\mu s$ in a 16×16 switch. The analyzed simulation scenario consists of 2160 streams†† with nine different characteristics in rate and deadline, as shown in Fig. 18. All ports are homogeneously and symmetrically loaded with $\rho = 0.95$ (Poisson traffic).

The results in Figs. 19 and 20 show that in principle the connection separation works the same way as in an OQ architecture. For all loads ρ these results can be obtained using the delay quantiles in Fig. 13.

For EDF, a good approximation for the CDF is given by

†FCFS=first come first served [28], EDF=earliest deadline first.

††16 ports · 15 destinations each · 9 classes.

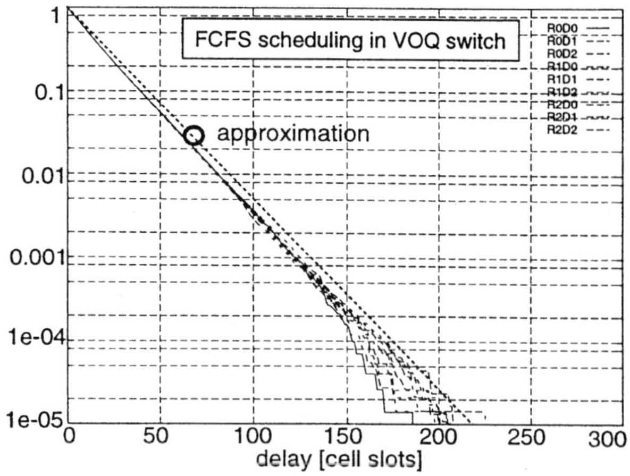

Fig. 19 VOQ switch: $Pr\{d > t\}$ for FCFS.

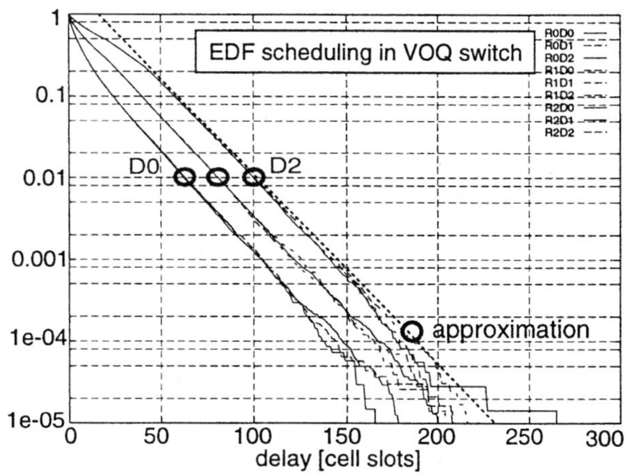

Fig. 20 VOQ switch: $Pr\{d > t\}$ for EDF.

$$Pr\{w \leq t\} = 1 - ae^{-b(t+\bar{D}_0 - D_{0i})} \tag{30}$$

using the mean deadline value [27] and the shape parameters a, b obtained from the VOQ model in Eq. (27).

$$\bar{D}_0 = \frac{\sum_i \lambda_i D_{0i}}{\sum_i \lambda_i} \tag{31}$$

Using Eq. (30) and a, b from Eq. (27), the performance can be calculated. The simulated graphs in Fig. 20 show the typical behaviour: A separate performance for each delay class, the distance is exactly the difference of the deadlines. The slope of the graphs is exactly the same as for FCFS, which validates that our approximation is applicable. As it can be seen, EDF supports deadlines as it does in an output-queued switch.

7. Conclusion

It is shown that in a switch using virtual-output-queueing (VOQ) a number of arbitration algorithms can be applied. Static allocation offers deterministic, i.e. firm, delay bounds for bounded traffic per-VC. However, the mean delays can become quite high. Dynamic algorithms perform quite close to the ideal output queueing. With weighted algorithms priorities can be supported and statistical delay bounds can be given, the performance of which depends on other connections, however. Scheduling algorithms can further separate connections and operate similar to their output-queued variant. With the performance approximations for the SIMP algorithm given in this paper the performance of more complex schedulers can be derived, as shown for EDF here.

Acknowledgement

The autor would like to thank the anonymous reviewers for their valuable comments.

References

[1] A. Mekkittikul and N. McKeown, "A starvation-free algorithm for achieving 100% throughput in an input-queued switch," Proc. IEEE International Conference on Communications Networks, 1996.

[2] N. McKeown, "Scheduling algorithms for input-queued cell switches," Ph.D. thesis, UC Berkeley, 1995.

[3] N. McKeown, A. Mekkittikul, V. Anantharam, and J. Walrand, "Achieving 100% throughput in an input-queued switch," IEEE Trans. Commun., 1998.

[4] A. Mekkittikul and N. McKeown, "A practical scheduling algorithm to achieve 100% throughput in input-queued switches," Proc. IEEE INFOCOM, 1998.

[5] S. Shenker, C. Partridge, and R. Guerin, "Specification of guaranteed quality of service," IETF RFC 2212, Sept. 1997.

[6] J. Wroclawski, "The use of RSVP with IETF integrated services," IETF RFC 2210, Sept. 1997.

[7] R. Schoenen, G. Post, and G. Sander, "Weighted arbitration algorithms with priorities for input-queued switches with 100% throughput," Proc. IEEE Broadband Switching Systems, 1999.

[8] R. Schoenen, R. Hying, and J. Kozinc, "Distributed cell scheduling algorithms for virtual-output-queued switches," Proc. IEEE GLOBECOM, 1999.

[9] J.Y. Hui, Switching and Traffic Theory for Integrated Broadband Networks, Kluwer, 1990. ISBN 0-7923-9061-X.

[10] T. Anderson, S. Owicki, J. Saxe, and C. Thacker, "High speed switch scheduling for local area networks," ACM Trans. Computer Systems, vol.11, Nov. 1993.

[11] R. Schoenen and G. Post, "Static bandwidth allocation for input-queued switches with strict QoS bounds," Proc. IEEE Broadband Switching Systems, 1999.

[12] C.H. Papadimitriou and K. Steiglitz, Combinatorial Optimization, Prentice-Hall, Inc., 1982.

[13] Y. Tamir and H.-C. Chi, "Symmetric cross bar arbiters for VLSI communication switches," IEEE Trans. Parallel & Distributed Systems, vol.4, no.1, pp.13-27, 1993.

[14] C. Lund, S. Phillips, and N. Reingold, "Fair prioritized scheduling in an input-buffered switch," Proc. IEEE International Conference on Broadband Communications, 1996.

[15] R. Schoenen, G. Post, and G. Sander, "Prioritized arbitration for input-queued switches with 100% throughput," Proc. ATM Workshop '99, 1999.

[16] R. Schoenen, G. Post, and A. Müller, "Analysis and dimensioning of credit-based flow control for the ABR service in ATM networks," Proc. IEEE GLOBECOM, vol.4, p.2399,

1998.
[17] D. Stephens and H. Zhang, "Implementing distributed packet fair queueing in a scalable switch architecture," Proc. IEEE INFOCOM, 1998.
[18] L. Kleinrock, Queueing Systems, Vol.II: Applications, John Wiley & Sons, New York, 1976.
[19] The ATM Forum, ATM user-network interface (UNI) specification version 3.1, Prentice Hall, Englewood Cliffs, N.J., 1994.
[20] S. Shenker and J. Wroclawski, "General characterization parameters for integrated service network elements," IETF RFC 2215, Sept. 1997.
[21] R. Cruz, "A calculus for network delay, Part I: Network elements in isolation," IEEE Trans. Inf. Theory, vol.37, p.114, Jan. 1991.
[22] H. Michiel and K. Laevens, "Teletraffic engineering in a broad-band era," Proc. IEEE, vol.85, p.2007, Dec. 1997.
[23] J. Roberts, U. Mocci, and J. Virtamo, eds., Broadband Network Teletraffic: Performance Evaluation and Design of Broadband Multiservice Networks; final report of action COST 242. LNCS1155, Springer, 1996.
[24] MIL 3, Inc., 3400 International Drive, NW Washington, DC 20008, USA, OPNET Release 6.0, 1999. http://www.mil3.com
[25] N. McKeown and T. Anderson, "A Quantitative Comparison of Scheduling Algorithms for Input-Queued Switches," 1997, unpublished available at http://http.cs.berkeley.edu./%7Etea/atm.html.
[26] M. Marsan, Modelling with Generalized Stochastic Petri Nets, Wiley, 1996. ISBN 0-471-93059-8.
[27] B. Walke and W. Rosenbohm, "Waiting-time distributions for deadline-oriented serving," Performance of Computer Systems, p.214, North-Holland Publishing Company, 1979.
[28] L. Kleinrock, Queueing Systems, Vol.I: Theory, John Wiley & Sons, New York, 1975.
[29] F. Chiussi and V. Sivaraman, "Achieving high utilization in guaranteed services networks using earliest-deadline-first scheduling," Proc. International Workshop on Quality-of-Service, 1998.
[30] S.W. Lee, D.H. Cho, and Y.K. Park "Improved dynamic weighted cell scheduling algorithm based on earliest deadline first scheme for various traffics of ATM switch," Proc. IEEE Globecom '96, pp.1959–1963, London, 1996.

High-Performance Backbone Network Technology edited by Naoaki Yamanaka
ISBN: 0-8247-5321-6 DOI: 10.1081/(120025674)
Published 2004 by Marcel Dekker, Inc.

Bodhi: A Highly Modular Terabit ATM Switch Fabric Architecture

Jagan P. Agrawal and Fa Toh Yap

SUMMARY In this paper, we propose a high performance highly modular ATM switch architecture known as Bodhi which is suitable for small, large, and very large size ATM switch implementations. Its basic configuration consists of two stages: an input stage and an output stage. The input stage consists of input group modules (IGMs) and output stage output group modules (OGMs). Each IGM-OGM pair is connected by multiple paths which carry cells from IGMs to OGMs. Excess cells at the IGMs are recycled to minimize the cell loss probability. Another module called recirculation module is used to couple several IGMs together to create additional routes for recirculating cells which gives this architecture robustness against nonuniform and directed traffic. Multicasting has been implemented by integrating copying and broadcasting techniques, and using some novel techniques to minimize the switch complexity. A shared buffer architecture is employed for OGMs such that it implements multiple priorities dynamically in a weighted manner, requires no speedup, and, can function in standalone mode as small switches. The performance of Bodhi has been evaluated by computer simulation to select design parameters for a $1k \times 1k$ ($k = 1024$) switch fabric. It allows growability from 2.4 gigabit to 150 gigabit switch, and, expandability to 100^{+} Terrabit switch with the complexity increasing approximately linearly with size. Based on the study presented in this paper, it is seen that the Bodhi architecture offers a high degree of modularity, weighted dynamic priority control, robustness against nonuniform traffic conditions, low complexity, low latency, and, supports multicasting efficiently and without any limitation. Bodhi, therefore, has high potential for application in high speed broadband networks.
key words: *ATM switch, multicast, broadcast, modularity, growability*

1. Introduction

The need for high performance ATM switches has been well established during the last decade after ATM was accepted to be the information transfer mode for the future local and long haul high speed broadband networks such as BISDN. In order to fulfill the needs of such networks, an ATM switch fabric should be designed to: i. sustain high input loading (i.e., 90 to 95%) at a cell loss rate below 10^{-9}, ii. have selfrouting and low latency for high speed switching, iii. provide dynamic prioritization in a weighted manner for service quality control, iv. have ability to handle nonuniform and directed traffic effectively, v. have fast and efficient multicasting capability with no limit on the number of destinations per cell, and, with least increase in the switch complexity, vi. have no degradation in the quality of service for unicast cells due to multicast traffic, vii. have minimum number of buffers per port with no head of line (HOL) blocking, viii. require minimal or no switch speedup to accept multiple cells per port in a slot, ix. be highly modular with linear, or nearly linear, growability and expandability, and, x. have low complexity architecture suitable for efficient VLSI implementation. A number of ATM switch architectures have been proposed in the literature [1]–[29] which meet the above requirements in varying degrees. The purpose of this paper is to develop an ATM switch architecture which will meet all the above criteria.

Most ATM switch architectures reported in the literature are based on either a single stage crossbar, multistage crossbar, or common memory configuration. In single stage crossbar architectures [2], [6], each output port accesses all input lines simultaneously in every slot. While simpler to implement, the complexity of single stage $N \times N$ configurations increases exponentially with N. The multistage architectures switches [1], [3], [5], [9], [10], [16], [18], [21], [24], [25] reduce hardware complexity of the switch significantly compared to that of the single stage configurations at the cost of increased implementation complexity. Typical examples of multistage switches are starlite [1], sunshine [5], SCOQ [10], moonshine [4], St. Louis switch [3], TBSF [8] and pipelined banyan switch [24]. The multistage configurations suffer from limitations in introducing one or more of the following capabilities: modularity, dynamic priority mechanism, multicasting and shared buffers. For example, multicasting is generally introduced using a copy network as a preprocessor at the input of the switch which introduces copy-overflow, delay, fairness, and/or head-of-line blocking problems [3], [9], [18], [22], [26], [27].

Some interesting broadband switch architectures are based on common memory switch (CMS) approach [12]–[14], also known as shared buffer architectures. Besides the reduced buffer requirement due to sharing, another key advantage of this configuration is in the implementaion of the multicast function. A multicast cell is stored in only one memory location which is released when all the corresponding output ports have read the cell from it. The most critical function in CMS is the managment of the buffers. As N increases, so does

Manuscript received April 24, 1997.
[†]The authors are with Department of Computer Networking University of Missouri-K.C. 5100 Rockhill Road Kansas City, MO 64110-2499, U.S.A.

IEICE TRANS. COMMUN., Vol. E81-B, No. 2, pp. 182–193, February 1998

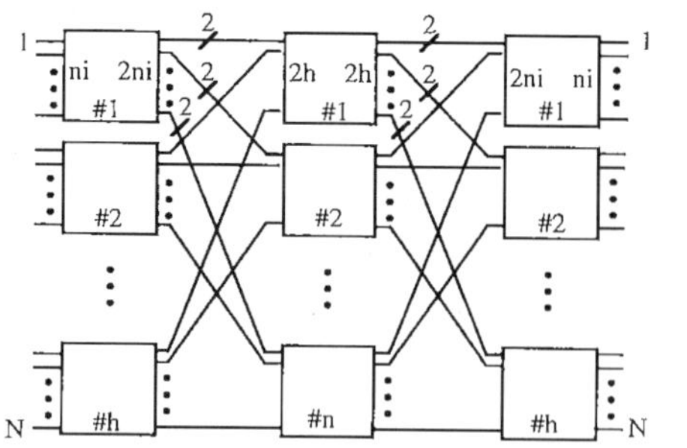

Number of distinct paths between any input-output pair =2ni
h=N/ni

Fig. 1 An implementation of the three-stage clos network.

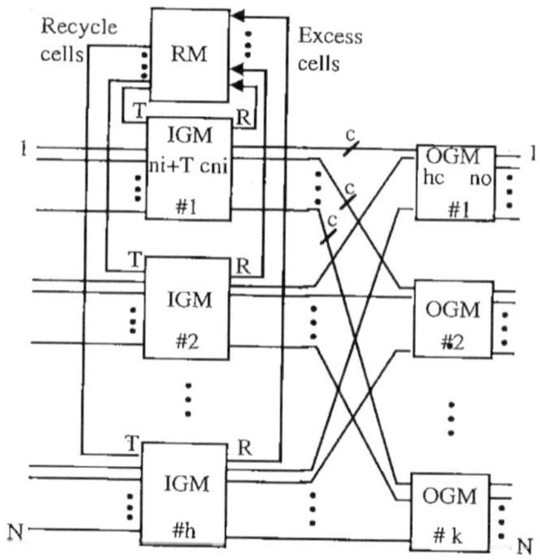

IGM: Input group module
OGM: Output group module
RM: Recirculation module

Fig. 2 Proposed two-stage virtual Clos network.

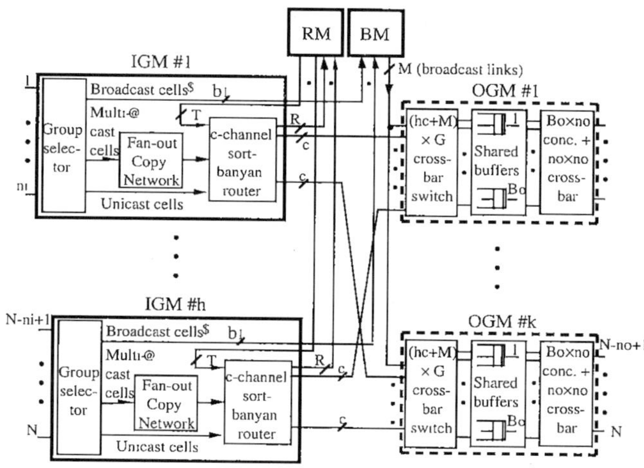

IGM/OGM: Input/output Group Module
Bi ith one-cell buffer \$: > cp copies/cell @ ≤ cp copies/cell G:Number of virtual buffer groups.

Fig. 3 General configuration of the $N \times N$ Bodhi architecture.

the buffer management complexity and the number of write/read cycles required within one slot time. Therefore, the shared buffer architectures are unsuitable when the switch size is large. One of the reasons for limitations in various switch fabric architectures is their development in phases, i.e., first development of the basic architecture, then introducing additional capabilities one at a time. Therefore, the motivation for this study was to develop a comprehensive ATM switch architecture which would exploit the techniques and characteristics of the existing architectures and overcome their limitations to achieve the ten desired features listed earlier. We propose such an architecture in this paper which has been named Bodhi.

After studying various existing architectures, the highly modular three-stage Clos network, shown in Fig. 1, was found to be an ideal base level configuration for our ATM switch design as in [25]. As seen in the above figure (i) the input stage has h modules (called input group modules or IGMs), each with n_i inputs and $2n_i$ outputs, and, (ii) there are $2n_i$ paths between each input port-output port pair. We modified the three-stage Clos network into a two-stage virtual Clos network to facilitate cell switching a cell recirculation mechanism called recirculation module (RM), as shown in Fig. 2, such that the new cells arriving at an IGM have c paths to each of the k output modules (called output group modules or OGMs), and, the recirculating cells have hc paths between every input port-output port pair through h IGMs, respectively. Development of the virtual Clos network is discussed in more detail in Sect. 2.

The c-channel switching with cell recirculation, then, led us to use the basic approach of sunshine switch architecture [5] to implement IGMs. Multicasting was introduced in the IGMs integrating both the conventional multicasting techniques: broadcast [13], [14], [19] and cell replication [3], [18], [26], [27] approaches to exploit the advantages of both as follows: a multicast cell is replicated if its number destinations is less than or equal to some number cp, and, broadcast if greater than cp through a central module called broadcast module (BM) to all the OGMs over an M-link bus. Our above new approach to multicasting allowed us to do cell replication without using a copy network.

Implementation of OGMs was done using shared buffer approach to minimize the number of buffers per port. A buffer configuration and queueing procedure was developed to allow writing all incoming cells in the buffers simultaneously at the internal switching speed and implementation of multiple level weighted dynamic priority control. The resulting Bodhi configuration is shown in Fig. 3.

The use of RM module provides a completely shared buffering at the input stage which reduces the recirculation buffer requirement, and, distribution of recirculating cells to IGMs in a round robin fashion balances the recirculation load (or excess load) between h IGMs, a feature useful in handling nonuniform traffic. The integrated approach to multicasting [33] minimizes the increase in complexity due to mulicasting by copying without using a copy network and reducing the size of the broadcast bus, provides protection against overflow due to multicasting, and, minimizes the memory requirement for VPI/VCI translation and routing tag insertion in the multicasting cells.

Our proposed architecture is suitable for designing ATM switches growable from the size of 16×16 to $16^{\nu} \times 16^{\nu}$ where $\nu \geq 0$ is an integer, for operation at OC-3 bit rate. The OGM is designed to work as a standalone switch for sizes from 16×16 to 64×64. Larger switch sizes up to $1k \times 1k$ require IGMs, OGMs, an RM module and a BM module in two stages configured as shown in Fig. 3. Beyond $1k \times 1k$ size, expansion is done by a factor of 16 at a time, each time adding an additional input stage at approximately linearly increasing cost.

The performance of the $1k \times 1k$ Bodhi architecture has been studied by computer simulation for uniform as well as non-uniform traffic with the objective to dimension various design parameters. Center for Development of Telematics (CDOT), New Delhi, India, has decided to commercialize the Bodhi architecture at all sizes. Currently, it is prototyping the OGM. In this paper, first we present our design approach and basic Bodhi architecture in Sect. 2, then the architecture with multicasting capability in Sect. 3. This is followed by the determination of group sizes for IGMs and OGMs in Sect. 4. The simulation details are, then, presented in Sect. 5 and the simulation results in Sect. 6. This is, then, followed by the estimation of the complexity in Sect. 7, illustration of growability and expandability in Bodhi in Sect. 8, and, estimation of latency in Sect. 9. In the end, some concluding remarks are made in Sect. 10.

2. The Basic Bodhi Architecture

In this section, we first present the rationale behind the development of the two-stage virtual Clos network, then design of its four modules: IGM, OGM, RM and BM.

2.1 The Two-Stage Virtual Clos Network

In a strict sense nonblocking Clos network, shown in Fig. 1, an available path must be found and scheduled for each new call arriving at the IGMs. This works well for the point-to-point circuit switching applications where the number of new call arrivals is only a small fraction of the ongoing calls. In ATM switching environment, each cell arrival will require path finding, and, multipoint-to-multipoint switching will introduce internal blocking in the switching fabric.

Eng, Carol and Yeh [25] introduced the multipoint-to-multipoint switching capability in the Clos network using group routing to develop a three-stage growable ATM switch architecture. However, their architecture still requires path finding for each cell. By using the cell recirculation concept along with the group routing concept, we can eliminate the need for path finding at the cost of a mean queueing delay of about 1% of a slot. Further, the above also reduces the complexity by reducing the number of stages from three to two. This process is illustrated in Fig. 2. There are c paths provided from an ith IGM to each of the k OGMs for group routing. The cells from all the IGMs which can not be routed during a slot are brought into the recirculation module (RM) where they are compacted, sorted, held for one slot time and, then, distributed to all the IGMs in a round robin manner; each IGM receiving upto T cells. Therefore, there are hc possible paths for a recirculating cell to reach its destination. Since paths between each IGM-OGM pair are unique, pathfinding for recirculating cells is not required. However, a small recirculation delay, on the average only a small fraction of a slot, is introduced. Since the resulting two-stage configuration in Fig. 2 continues to provide the alternate routing path capabilities of the Clos network in a delayed manner, we call this configuration a 'two-stage virtual Clos Network.' The dimensioning of various parameters is done in Sect. 5 based on computer simulation results. Next we present development of the basic architecture of an IGM.

2.2 Basic Architecture of the Input Group Modules (IGMs)

First let us consider the routing tag fields to be attached to the cells for routing, sequencing, and implementing multiple priority levels. Let $n = \log_2 N$. Then the n-bit destination address of the cells can be written as $(d_1 \ldots d_g d_{g+1} \ldots d_n)$ where $(d_1 \ldots d_g)$ is its OGM address. Assuming p priority levels, we will have an $\alpha = \log_2 p$-bit priority field $(P_1 \ldots P_\alpha)$. Since there are hT recirculation lines, we will have $s = \log_2 hT$-bit sequence number field $(S_1 \ldots S_s)$. An activity bit 'a' is attached in the beginning to indicate if a cell is present on the link. In all the cases, we assume a lower value of a field indicates higher priority. As such, the sequence field in all the new cells is set to be $(S_1 \ldots S_s) = (1 \ldots 1)$. Then, the routing tag to be attached to each cell is $(a, d_1 \ldots d_g, P_1 \ldots P_a, S_1 \ldots S_s, d_{g+1} \ldots d_n)$. Next we discuss design of an IGM using the sunshine switch architecture [5] with some modifications.

As in a Sunshine switch [5], first the n_i+T inputs at an IGM are sorted in the order of their OGM destinations using $(d_1 \ldots d_g, P_1 \ldots P_\alpha, S_1 \ldots S_s)$ field by the $(n_i+T) \times (n_i+T)$ sorter and up to c cells per OGM destination are identified by the trap network for routing. Then the $(n_i + T) \times (n_i + T)$ concentrator separates the cells to be routed from the cells to be recirculated. As illustrated in [1], the cells to be routed emerge compact and monotone at one end of the concentrator output, and, the cells to be recirculated emerge compact but not monotone at the other end. The first R cells from the recirculation group are sent to the recirculation module dropping any recirculation cells in excess of R. We observe that (i) since sorting of the cells is by their OGM destinations, it does not favour any particular output ports, and, (ii) since the recirculation cells at the concentrator output are not monotone, any cell loss is not confined to specific output ports.

We do cell routing from an IGM to the OGMs using a c-channel router instead of c individual routers used in the sunshine switch [5]. In a c-channel router, the cells on the first c output ports of the router belong to the first OGM, on the next c output ports to the second OGM, and so on. Although using a c-channel router increases the number of switching elements slightly, it has the following advantages over using c individual routers:

- simplified interconnection network between the concentrator and router by avoiding distribution of the cells to the routers in a round robin fashion, and,
- cells at the router output ports appear in the order of priority, i.e., for the first OGM the highest priority cell arrives on the first output port, next highest priority cell on the second output port, and so on. This in turn reduces the cell resequencing problem in the OGMs, as will be seen in the subsection on the OGM architecture.

The c-channel routing is implemented as follows. Let $e = \log_2 c$-bit field $(E_1 \ldots E_e)$ indicate the number i for the ith cell selected by the trap network for a particular OGM destination so that $1 \leq i \leq c$. Then we modify the trap network so that in addition to identifying the cells to be routed, it also adds the corresponding cell number field $(E_1 \ldots E_e)$ into each cell to extend the IGM-to-OGM routing field from $(d_1 \ldots d_g)$ to $(d_1 \ldots d_g E_1 \ldots E_e)$. As a result, format of the routing tags in the cells becomes $(a, d_1 \ldots d_g, E_1 \ldots E_e, P_1 \ldots P_\alpha, S_1 \ldots S_s, d_{g+1} \ldots d_n)$.

Sequencing of the recirculating cells to ensure excess-load balancing between IGMs, and, first in-first out routing is done in the RM as discussed in the following subsection.

2.3 Basic Architecture of the RM

In order to ensure FIFO routing, the recirculation cells should be given priority for routing from an IGM to OGMs over the newly arrived cells. Therefore, we introduce a simple and efficient cell sequencing mechanism as shown in Fig. 4. All newly arrived cells at the IGMs are given the sequence field of $(1 \ldots 1)$. Therefore, after the first slot, the hR cells coming to the RM will have the sequence field values in the range $(0 \ldots 0)$ and $(1 \ldots 1)$. These hR cells are

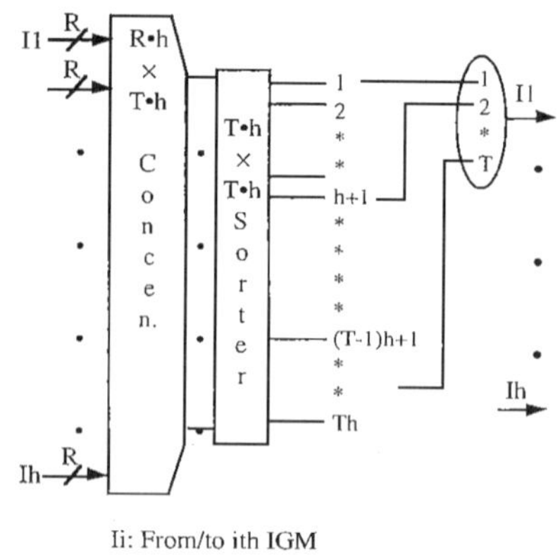

Fig. 4 Architecture of the recirculation module.

first compacted hT lines using an $hR \times hT$ concentrator and, then, sorted by an $hR \times hT$ sorter using the fields $(a, d_1 \ldots d_g, E_1 \ldots E_e, P_1 \ldots P_\alpha, S_1 \ldots S_s)$, as in the IGMs. After sorting, the field $(S_1 \ldots S_s)$ in the cells is substituted by the corresponding sorter output link numbers. For example, if value of the sequence field of a cell is 12 and it appears on the sorter output link #5, the field is changed to 5. The sequenced cells are then distributed to the IGMs in a round robin manner. The sequence number $(1 \ldots 1)$ is not used at the output of the sorter in the RM since it is reserved for the new cells. The above process ensures that (i) for any destination, older the cell higher its priority, and, (ii) the recirculation cells destined to the same OGM are distributed to different IGMs. The latter ensures that all hc paths through h IGMs seen collectively by the cells of the same destination are utilized. Therefore, a burst of cells at an IGM destined for a particular OGM will get distributed evenly to all the IGMs during the next slot, thus handling the burst effectively.

2.4 Architecture of the OGMs

In order to keep the total buffer requirement at the output ports to a minimum, we chose a shared buffer architecture for OGMs as shown in Fig. 5. Each OGM consists of $hc + M$ input ports to accept hc inputs from the h IGMs and M inputs from the broadcast module if multicasting is implemented (to be discussed in the next section). However, the hc input links carry cells for only n_o output ports. Therefore, the OGMs are designed using shared buffer architecture which operate (i) in standalone mode as a $n_o \times n_o$ switch growable to the size $(hc + M) \times (hc + M)$ using OGMs only, and, (ii) as output module for the larger size switches. While OGMs receive cells from the IGMs, no feedback is necessary from the OGMs to IGMs since both operate independently.

We recall that the c-channel routing in the IGMs places older and/or higher priority cells on the lower numbered output ports. Therefore, the inputs from the

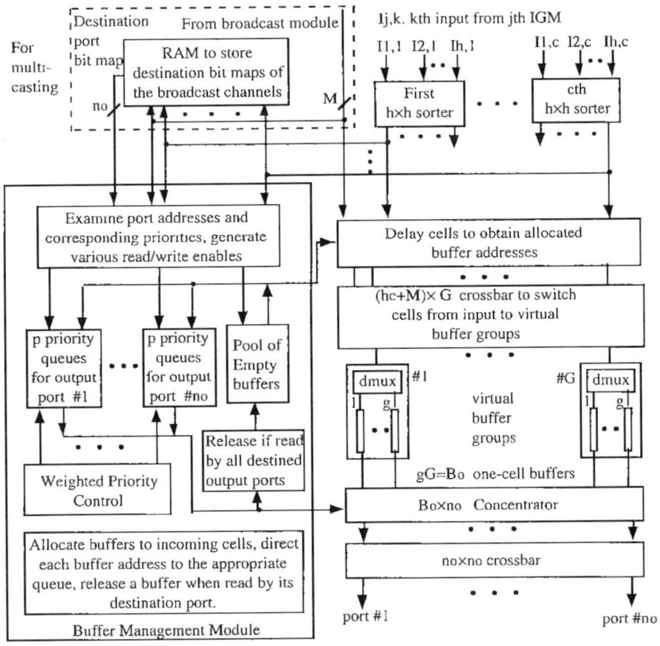

Fig. 5 Block diagram of the OGM.

h IGMs are interleaved so that the first inputs from the IGMs get higher priority over the second inputs from the IGMs, and so on. This creates c interleaved groups at the input of the OGMs as shown in Fig. 5. However, due to the inter-IGM distribution of the recirculation cells, it is possible for a group to have cells arranged in an out-of-sequence order. Therefore, to avoid out-of-sequence situation, the cells in each group are sorted using c $h \times h$ sorters in each OGM, as shown in Fig. 5. Our other considerations in the OGM design were:

- Each input link should have access to each buffer to maximize the buffer sharing.
- All input links with cells should be able to write into buffers simultaneously so that no speedup of the internal switching is required. This implies that there must be at least $hc + M$ distinct buffers, one per input link.
- The buffer organization should not introduce HOL blocking. This consideration implies using B_o single cell buffers where B_o is the number of one-cell buffers required to meet the cell loss rate criteria.

Therefore, the optimal size for the input-to-buffer switch is $(hc + M) \times (hc + M)$, and, buffers-to-output ports switch is $B_o \times n_o$. In order to accomplish this, we organized the B_o buffers in G groups with g buffers per group such that the number of buffers seen by the input links is G while the number of buffers seen by the output ports is $B_o = gG$. Therefore, the input switching is done in two steps: first input links to groups, then to individual cell buffers within a group. From the computer simulation study it was found that $G > (hc + M)$ to avoid cell losses due to the above group switching. Therefore, for switching at the input side we use $(hc{+}M) \times G$ crossbar switch followed by G 1-to-g demultiplexers. Since $(hc{+}M)$ is not expected to be large, and, $G > (hc{+}M)$, a crossbar switch was found to be more efficient than a banyan router. After buffer allocation to each input link with a cell, the switching in the OGM is only poin-to-point. Therefore, there is no collision of the cells at the crosspoints. For switching at the output side we use a $B_o \times n_o$ concentrator, since $n_o \ll B_o$, followed by a $n_o \times n_o$ crossbar switch to reduce the switching complexity. The $n_o \times n_o$ crossbar switch has the ability to route a cell to multiple output ports to support multicasting. However, only one cell can arrive at any output ports.

The buffer management module shown in Fig. 5 examines routing tags in the cells, allocates buffers to each active cell, updates the output port queues, provides addresses of the buffers to be read by the output ports during a slot, releases buffers after they are read, and, maintains a pool of available buffers. Each output port has p priority queues which are controlled externally through a weighted priority control as follows: first x_1 cells are read from the priority 1 queue, then x_2 cells are read from the priority 2 queue, ..., then x_p cells are read from the priority p queue. The values of $x_1, x_2, \ldots, x_p$ are set dynamically by the switch control.

In the next section we introduce multicasting in the IGM and OGM modules and develop the complete Bodhi architecture.

3. The Bodhi Architecture with Multicasting

In order to provide multicasting capability we take

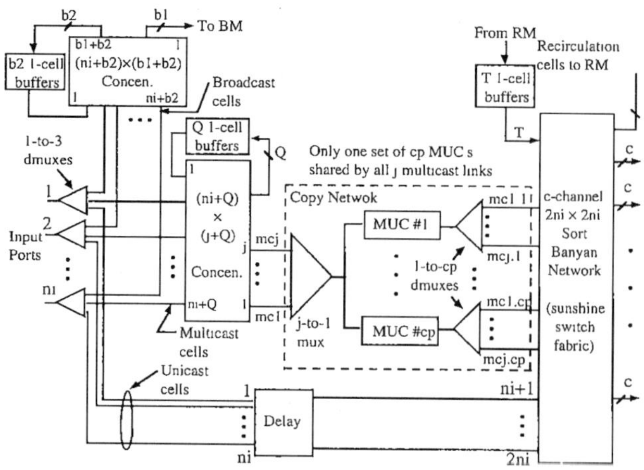

mci: multicast cell on ith output link
mci k: kth copy of the cell #mci
MUC: multicast cell to unicast cell converter
cp: maximum number of copies per multicast cell

Fig. 6 IGM block diagram with integrated multicasting.

an integrated approach to multicasting[33] and exploit both: the cell replication approach and broadcast approach. In this approach, we split the multicasting function in two parts as follows: (i) a cell whose number of copies requirement (i.e., number of destinations) is more than one but less than or equal to *cp* is termed a multicasting cell, and, (ii) a cell whose number of destinations is more than *cp* is termed a broadcasting cell. To implement this, we modify the routing tag fields in the cells as follows:

unicast: $(a, a_M, a_B, d_1 \ldots d_g, P_1 \ldots P_\alpha, S_1 \ldots S_s, d_{g+1} \ldots dn)$; $a_M, a_B = 0, 0$

multicast: $(a, a_M, a_B, m_1 m_2 \ldots m_\gamma, P_1 \ldots P_\alpha, 0 \ldots 0)$; $a_M, a_B = 1, 0$

broadcast: $(a, a_M, a_B, B_1 B_2 \ldots B_\beta, P_1 \ldots P_\alpha, 0 \ldots 0)$; $a_M, a_B = 0, 1$

where $(m_1 m_2 \ldots m_\gamma)$ is the multicast channel number (MCN) and $(B_1 B_2 \ldots B_\beta)$ is the broadcast channel number (BCN). Zeros are appended at the end of the routing tags to make the routing field length in all cases the same. As seen in Fig. 6, the incoming cells in the IGMs are separated into three groups: unicast, multicast and broadcast using n_i one-to-three demultiplexers (demuxes) based on the bits a_M and a_B which are deleted at the output of the demuxes. The unicast group is then delayed until the multicast cells are converted into unicast cells. The multicast and broadcast groups are processed as discussed below.

The multicast cells are concentrated in an $(n_i+Q) \times (j+Q)$ concentrator where n_i are the new inputs from 1-to-3 demuxes, j is the number of multicast cells which are converted into unicast cells per slot, and, Q is the number of multicast cells recirculated to avoid cell loss due to overloads. The above recirculation is done as follows to ensure priority to the older cells. The oldest cell to be recirculated is on $(j+1)$th output port of the concentrator and it is recirculated through its first input port. Similarly, the next oldest cell is on the $(j+2)$th output port which is recirculated through the second input port, and so on. This allows the older cells to appear at the output ports of the concentrator in the same order starting at the first output port. The new multicast cells, then, appear at the subsequent output ports.

Since the maximum number of destinations of a multicast cell is *cp*, the copies are made simply by fanning out each of the above j outputs into *cp* branches as seen in Fig. 6. Each branch has a multicast cell to unicast cell converter (MUC) associated with it in which a RAM is used to store an activity bit a_{cp}, one destination address and the corresponding H-bit VPI/VCI field for header translation in the location indicated by the γ-bit MCN field. a_{cp}, is initially zero in all RAM locations, and is changed to '1' in a RAM in a particular location only if the corresponding branch is needed by an MCN to make a copy. When the a_{cp}, bit is '0' in a RAM, the cell on that branch is marked inactive so that only desired number of copies are made. For example if $cp = 7$ and only four copies are needed by a multicast cell whose $MCN = 138$, then a_{cp}, bit in the RAM location 138 in the first four branches will be set to '1' and will be left to be '0' in the remaining three. Thus, the above copy replication process adds $j.cp$ cells to n_i original unicast links but the number of active cells added is less than or equal to $j.cp$. The MUC on the first branch contains the routing and header translation information for destination #1, the second for destination #2, and so on. As a result of the above fixed destination locations, all the j multicast cells can share one set of *cp* MUCs in time division multiplexed manner to minimize the memory requirement. The VPI/VCI fields in the multicast cells must become available before the header translation can be completed. It will take a delay of at least $(n+\alpha+s+H-\gamma)$ bits afer receiving the γ-bit MCN for the above to happen. Even for the modest size switch, this delay is likely to be at least 32 bits which is sufficient time to perform at least 8 RAM-read operations; as we will see later, the value of j is less than or equal to eight. As such, the above sharing of MUCs by all j multicasting cells does not introduce any additional delay. Further, when operating in standalone mode, either type of cell can arrive on any OGM input link. The bit a_B is, therefore, used in the OGMs to distinguish between unicast and broadcast cells. Therefore, before a cell is sent from an IGM to any OGM, the bit $a_B = 0$ is inserted in its routing tag field following the activity bit a to indicate that it is a unicast cell.

Similar to multicast group, the broadcast group in an IGM is concentrated by an $(n_i+b_2) \times (b_1+b_2)$ concentrator where b_1 is the number of broadcast cells selected per slot for broadcast from an IGM and b_2 is the number of recirculation links provided for overload protection. Since many of the above b_1 links are likely to be inactive, these are concentrated using an $(hb_1 + M_1) \times (M + M_1)$ concentrator called broadcast module (BM) where M is the number of cells broadcast to the OGMs and M_1 is the number of recirculation links for overload protection. FIFO characteristic is preserved the same way as in the multicast group concentrator.

A bit map of the recipient output ports of the broadcast cells is stored in a 2^β-words $\times$ n_o-bit RAM for each BCN as shown in Fig. 5. Distribution of a broadcast cell in an OGM is, therefore, done simply by queueing the buffer address of the cell in the queues of the recipient output ports as indicated by the bit map stored in the cell's BCN location. The bit map is also used to determine the count of a cell's destinations in that OGM for use by the buffer management module to ensure that the corresponding buffer is released only after it has been read that many times.

The MCNs are IGM specific, and, assigned and managed locally for each IGM while the BCNs have switch-wide significance. Therefore, the total multicast capability of the switch is $h \cdot 2^\gamma + 2^\beta$ simultaneous multicast calls, significantly larger than the conventional capacity of $2^\gamma + 2^\beta$. Further, the definition of the multicast and broadcast calls can be changed dynamically by the switch control by changing the value of *cp* for the new calls to regulate the amount of traffic through the unicast portion of the switch if desired.

The values of the parameters b_1, b_2, Q, cp, j, M and M_1 depend on N, n_i, n_o, multicast loading, output loading and the desired cell loss probability. Therefore, in the next section, we present how to determine the values of n_i and n_o, and, select their values for the design of a $1k \times 1k$ switch presented in this paper.

4. The Input and Output Group Sizes

A smaller value of the output group size n_o implies lower OGM complexity and switch growth in smaller increments. On the other hand, a larger n_o implies a more efficient sharing of buffers in an OGM, and, a more efficient sharing of hc, the number of input links from the h IGMs to an OGM, in group routing. The latter also implies a smaller number of recirculating cells in the IGMs since more cells per slot will arrive at the OGMs, and, a greater ability to carry bursty traffic directed towards specific OGMs. Based on the previous studies [12]–[14], [23], [28], the reduction in the output buffer requirement due to sharing is near optimum when $n_o = 16$, which is neither too small or too large. Therefore we select $n_o = 16$ for our design. Then, for a 1024×1024 switch design, the number of OGMs $k = N/n_o = 1024/16 = 64$.

Next, we choose the size of hc. Let $P(x_{OGM} > hc)$ be the probability that the total number of arrivals for an OGM is greater than hc. Then, assuming (i) binomial arrival process at the input ports with all 1024 destinations equally likely, and, (ii) an input loading of 95%, $P(x_{OGM} > 40) = 2.4 \cdot 10^{-8}$, and, $P(x_{OGM} > 48) = 2.8 \cdot 10^{-12}$. If we assume that the traffic on all 1024 input ports is directed towards only 512 output ports with all 512 ports equiprobable, then we have $P(x_{OGM} > 40) = 0.036$, and, $P(x_{OGM} > 48) = 9.7 \cdot 10^{-4}$. Therefore, we select $hc = 48$ so that the overall probability that a new cell will not be routed to its destination OGM as it arrives at an IGM is below 0.1% under uniform as well as directed traffic with twice the group loading.

If $c = 2$, as in the Clos networks, it will lead to a large number of recirculation lines from an IGM to RM [5] in multipoint to multipoint switching applications. This problem will be more critical when there are sudden bursts of cells in an IGM for specific OGMs. Therefore, based on the sunshine switch study in [5], and, our simulation results presented in the following section we choose $c = 3$ for our design. Therefore, $h(= 48/3) = 16$ and $n_i(= 1024/h) = 64$.

In the next section, we present computer simulation of the Bodhi architecture to determine the size of recirculation lines in IGM, RM and BM and to determine the buffer size in OGMs.

5. Simulation of the Bodhi Architecture

Simulation of the Bodhi architecture was done for the switch size of $1k \times 1k$, input group size $n_i = 64$, output group size $n_o = 16$ with three channel router between IGMs and OGMs to determine the desired size of parameters $Q, j, b_1, b_2, R, T, M, M_1$ and buffers in OGMs to achieve a cell loss rate below 10^{-9}. The software was developed in C in unix environment, and, simulation time was about 1 million slots for each simulation. The simulations were conducted for no cell loss, i.e., none of the parameters was limited. The results were used to compute p.m.f. of each parameter which were, then, used to determine cell loss rate vs. parameter size performances. Simulation of the full switch was conducted for unicast traffic only. The multicast traffic (or, say, B/M traffic) was considered only locally to determine parameters of the multicast and broadcast concentrators. Uniform traffic as well as nonuniform traffic was used in the study. The nonuniform traffic was obtained by generating uniform traffic with destinations between 1 and $N/2$ on the links in the even numbered IGMs, and, between $1 + N/2$ and N on the links in the odd numbered IGMs to test the ability of the recirculation scheme of the design to handle skewed traffic conditions which are also created by a bursty traffic.

The B/M traffic was generated as follows. Let ρ_i be the total (unicast+B/M) input loading, ρ_b (a fraction of ρ_i) the B/M input loading and C_a the average number of destinations requested by a B/M cell. Then, first the traffic at an input port was generated setting $\rho_i = \rho_o/(1 + \rho_b(C_a - 1))$. Then, a cell at an input

Input traffic	Output loading (ρ_o)	Input load. per dest. (ρ_i)	Max. value of R#	Max. value of T$	Mean/max recir. delay in slots	Corresponding Max. size of RM Module
Uniform (all N	0.95		12	4	.020/1	192×50
destinations in	0.90		10	3	.018/1	160×39
each IGM)	0.85	ρ_o/N	9	3	.015/1	144×33
	0.75		8	3	.011/1	128×26
Non-uniform (N/2	0.95		23	10	.111/1	368×156
dest. in each IGM	0.90		21	9	.098/1	336×136
with N destinations	0.85	2ρ_o/N	20	8	.085/1	320×119
every two IGMs)	0.75		16	6	.064/1	256×86

#R. no. of recirculation lines from an IGM to RM, $T. no. of recirculation lines from RM to an IGM.

Fig. 7 Sizes of the recirculation lines and RM module for zero cell loss under various loadings for uniform and nonuniform traffic obtained by simulation (switch configuration: $N = 1024, c = 3, h = 16$ and $k = 64$).

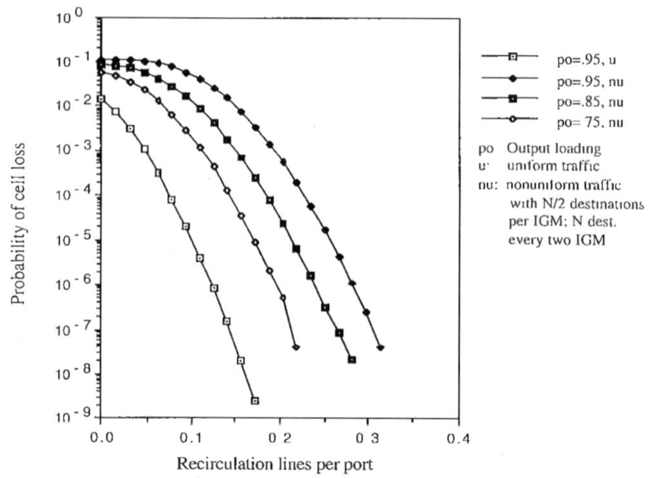

Fig. 8 Cell Loss Rate vs Number of recirculation lines/port performances of the sort-banyan network in an IGM under uniform and nonuniform traffic conditions.

port was designated to be a B/M cell with probability $\rho_i \rho_b$. The number of destinations of a B/M cell, *cp*, was generated assuming a Poisson p.m.f. between 2 and N with a mean of C_a. Then, as per our definition, if the number of destinations was between two and *cp* it was designated to be a multicast cell, otherwise a broadcast cell.

In this study, we assumed $C_a = 7$. Then, at $\rho_b = 0.1$ and $\rho_o = 0.95$, the B/M load accounted for 41.5% and unicast load for 53.5%, and, at $\rho_b = 0.15$ and $\rho_o = 0.95$, the B/M and unicast loads changed to 52.5% and 42.5%, respectively. Also, for the above parameters, it is easily seen that *cp* should be less than or equal to eight so that both multicast and broadcast routes are utilized efficiently.

6. Simulation Results and Selection of Parameters

6.1 Unicast Traffic:

Results of the simulation of unicast $1k \times 1k$ switch are given in Fig. 7 to Fig. 10. Figure 7 presents the minimum sizes of R, T and RM module at various switch loadings for the above two types of traffic required for no cell loss while Fig. 8 to Fig. 10 present cell loss rate performances for various switch parameters. Table 1 also gives the mean and maximum recirculation delay in slots. The maximum recirculation delay of only one slot implies that none of the cells is recirculated more than once. The mean recirculation delays are negligible: only 2% of a slot for uniform traffic and about 10% of a slot for nonuniform traffic at 95% input loading. Further, it is seen that the size of RM increases h times as fast as R, and, the size of T is always between 33% to 45% of R depending upon the type of traffic and input loading. Therefore, R should be chosen such that the size of RM is manageable, and, the desired cell loss rate criteria is met.

The cell loss rate performance at various values of R under a few selected traffic conditions is given in Fig. 8. If the target overall cell loss is less than or equal to 10^{-9}, the losses at various points in the switch must be lower by about an order of magnitude. In Bodhi architecture, there are four sources of cell loss: multicast concentrator, broadcast concentrator, broadcast module, and shared buffers in the OGMs. Therefore, assuming that the cell loss due to buffer overflow in the OGMs is to be lower than 10^{-9}, the cell loss at each of the remaining three points should be lower than 10^{-10}. Further, the total number of inlet+outlets (including power, clock and other requirements) in the RM should be kept to about 500 to keep its VLSI implementation practical. From Fig. 8 we see that choosing 0.3 recirculation lines per port (i) R will be $0.3n_i$ or 20, and, (ii) it will be possible to support the selected nonuniform traffic at about 80% loading with a cell loss rate below 10^{-10} (based on interpolation and extrapolation). Similarly, from Fig. 9 giving the cell loss rate vs. number of the RM outlets performance, we see that 128 outlets in RM with $T = 8$ will provide a cell loss rate below 10^{-10} under the above traffic conditions. Therefore, we select $R = 20$, RM of size 320×128 and $T = 8$.

Figure 10 gives the cell loss rate performance for various buffer sizes at different input loading for uniform traffic. Identical results were obtained for the nonuniform traffic also since the overall loading at each output port remained to be uniform. As seen from this figure, a buffer size of about 14 and 28 per port will be needed to achieve a cell loss rate of 10^{-8} at 90% and 95% output loadings, respectively. To achieve a cell loss rate of 10^{-10} at 95% loading, the number of buffers per port should be at least 32 based on the linear extrapolation of the results in Fig. 10. The mean queueing delay was seen to increase exponentially with loading as expected. For example it was about 4.25 slots at

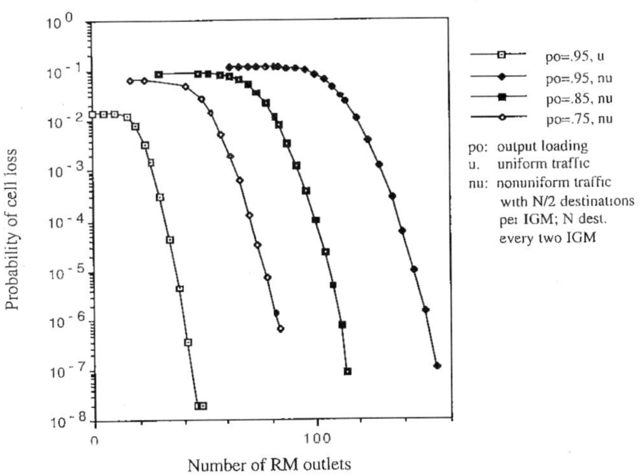

Fig. 9 Cell loss rate vs Output size of the recirculation module performances under uniform and nonuniform traffic conditions and various input loadings.

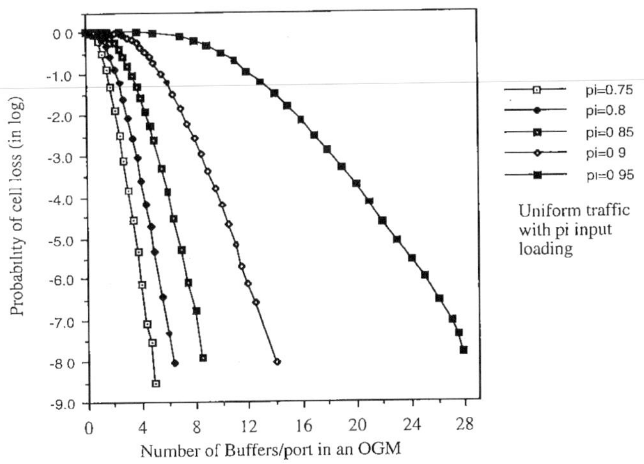

Fig. 10 Cell loss rate vs buffer size/port in an OGM under uniform traffic conditions with input loading as a parameter.

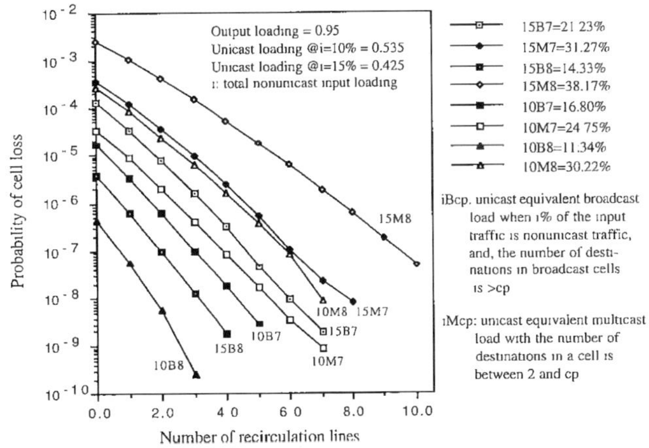

Fig. 11 Cell loss vs number of recirculation lines performance of the multicast and broadcast concentrators in an IGM.

90% and 9.75 slots at 95% loading. The above results agree with the results presented in the previous studies [2], [8], [10], [12]–[14], [28]. The grouped shared buffer configuration was simulated separately based on the above results. It was found that 576 buffers (36 buffers per port) organized in 72 buffer groups with 8 buffers in each group provided a cell loss rate below 10^{-10}. Therefore, the above buffer configuration was adopted for the OGM implementation.

6.2 Multicast Traffic

The results of the simulation study using unicast traffic were used to set the values of parameter j in the multicast concentrator and b_1 in the broadcast concentrator as follows. With $T+j\cdot cp \leqq n_i(=64)$, $T=8$ and $cp \leqq 8$ as the constraints, j should be less than or equal to 7. To obtain a conservative estimate of the parameter Q, we set $j=6$. Further, since the number of destinations in the broadcast cells will always be larger than cp, the broadcast concentrator will receive a smaller number of cells compared to the multicast concentrator. Hence we set $b_1=5$, one less than j.

The results of the simulation using multicast traffic are presented in Fig. 11. As one would expect, the minimum number of recirculation lines (b_2 or Q) required to achieve a desired cell loss rate increases as the load in the corresponding concentrator increases. For example when load in the broadcast concentrator increases from 10B8 (11.34%) to 10B7 (16.8%), b_2 increases from 2 to 5 to achieve a cell loss rate below 10^{-8}. In all cases, the mean recirculation delay was found to be less than 1.5% of a slot which is negligible compared to other delays in the switch. Also, in nearly all cases, the cell loss rate drops by about two orders of magnitude if the number of recirculation lines is increased by three. Therefore, we chose $Q=b_2=16$ for our design since it will provide us with a cell loss rate below 10^{-10} in each concentrator and also give us a greater flexibility in splitting B/M load between the two and/or handling a larger B/M load.

We used p.m.f. of b_1 as the basis to determine the values of parameters M and M_1 in the broadcast module. Convolving b_1 with itself 15 $(=h-1)$ times, we obtained p.m.f. of the input loading of the BM. This p.m.f. was then used to determine M and M_1 as in the other cases. It was found that $M=12$ and $M_1=24$ were sufficient to obtain a cell loss rate below 10^{-10}. Again, for greater flexibility and capability, we chose $M=16$ and $M_1=48$ making the BM a 128×64 concentrator.

In the next section, we compare the complexities of Bodhi and sunshine architectures.

7. Complexities of the Bodhi and Sunshine Architectures

In this section we determine complexity of the input

stage of the $1k \times 1k$ Bodhi architecture and compare it with that of the sunshine architecture. Since a shared buffer is highly desirable to minimize the number of buffers required at the output ports, it is assumed that a module similar to OGM will be used at the output ports of the sunshine architecture also and is not considered here. We consider complexities of the architectures without multicasting as well as with multicasting. In the case of sunshine architecture, it is assumed that multicasting is done using nonblocking copy network described in [18] with one buffer per input port to hold the cells in case of copy overflow. Further, the complexities are determined in terms of 2×2 switching/sorting elements (SEs) assuming they have the same level of complexity, 1-cell buffers for recirculation and memory requirements to implement multicasting. Also, we assume that (i) the total number of multicasting channels is 4k in the sunshine switch and 5k in Bodhi (0.25 k multicast channels per IGM+1k broadcast channels), (ii) 4 bytes of memory per channel is needed to store header and routing tag information in the copy network, and, (iii) eight outputs of the copy network can share the above memory. For sorter complexity, it is assumed that an arbitrary size sorter $(N+N_1) \times (N+N_1)$ where $N = 2^n > N_1 = 2^{n_1}$ can be implemented using $0.25[n(n+1)N+n_1(n_1+1)N_1]+0.5[nN+n_1N_1]+N_1$ sorting elements [32]. The complexity of an $X \times Y$ or $Y \times X$ banyan network where X and/or Y is not a power of two has been estimated by pruning the unused switching elements.

Then, the complexities of the two architectures determined as above are as follows:
(i) No Multicasting
Bodhi: 37600 SEs + 128 Buffers
Sunshine: 63104 SEs + 256 Buffers
(ii) Multicasting
Bodhi: 60896 SEs + 688 Buffers + 2 MB memory
Sunshine: 78464 SEs + 1280 Buffers + 16 MB memory

Therefore, as seen above, the complexity of the Bodhi architecture is significantly lower than that of the sunshine architecture. This can be attributed mainly to the following three features in Bodhi: (i) modularity, (ii) recirculation module, and, (iii) integrated approach to multicasting.

8. Growability and Expandability in Bodhi

The Bodhi architecture is highly modular and expandable. We discuss how these capabilities are implemented using the design parameters chosen above.

8.1 Growability

The first level of growability is from 16×16 to $16\eta \times 16\eta$ where $1 \leq \eta \leq 4$, simply by using η OGMs in standalonemode. In order to facilitate this, an externally controlled two-bit mode control $(\mu_1\mu_2)$ is introduced in the OGMs which is used to program an OGM for output port numbers $16(\eta - 1) + 1$ to 16η by setting them to (00), (01), (10) and (11) in the first, second, third and fourth OGM, respectively. The mode control bits in an OGM filter out the cells destined for the other OGMs.

In order to increase the size beyond 64×64, the two stage configuration presented in this paper must be used. Growability is in increments of 16×16 by adding one OGM at a time as needed, and using IGMs in the steps of 8, 12 and 16 for the switch sizes of up to 128×128, 512×512 and $1k \times 1k$, respectively, and distributing the input links among the IGMs equally. These numbers are based on additional simulation and computational results. The IGMs can not be added one at a time since it will not provide sufficient number of links between the IGMs and OGMs that are in use. When fully configured, the number of IGMs per OGM in a $1k \times 1k$ switch is only 0.25. Further, the cost of an IGM is expected to be only a fraction of the cost of an OGM since the OGMs contain buffering and buffer management. Therefore, the cost of Bodhi increases nearly linearly as its size increases from 16×16 to $1k \times 1k$.

8.2 Expandability

The above growth principle is extended to expand the switch size further 16-fold at a time. Each 16-fold expansion requires addition of one more stage at the input side. We consider here expansion from $1k \times 1k$ configuration to $16k \times 16k$. The 16k input links are considered as 16 groups, each consisting of 1k links. Let the input stage of the $1k \times 1k$ switch be called a plane. Then the input stage of the $16k \times 16k$ switch consists of 16 planes. Then, its second stage should be designed such that each input link sees 64 destinations, as in the case of two stage switch. This will then imply that each group in the second stage will receive the following number of inputs: 3 inputs from each IGM in the input stage (a total of (3×16 IGMs per plane) $\times$(16 planes)=768) + 16 inputs from each BM (a total of 16 per $BM \times 16$ planes=256) = 1024. Hence, the second stage of the $16k \times 16k$ switch consists of 64 planes. For multicasting, the definition of the multicast in the first stage is modified so that multicast cells are those cells whose destinations are in up to cp planes.

In the cases of larger expansion, it is easily seen that there is no increase in the number of planes beyond the second stage since the growth in the number of planes in a stage is neutralized by the equivalent reduction in the number of destinations of a plane in that stage. Therefore, when the switch size is extended to $16^{\nu}k \times 16^{\nu}k$ size where $\nu \geq 0$, it will have $(\nu+2)$ stages such that its input or the first stage consists of 16^{ν} planes, the last stage consists of $16^{\nu-1}k$ OGMs and each intermediate stage consists of $4 \cdot 16^{\nu}$ planes. Therefore, the number of IGMs per OGM is given by the ratio: total number of

IGMs/total number of OGMs which is easily computed as $(0.25+\nu)$. And since the cost of an OGM is expected to be several times the cost of an IGM, the expansion cost factor is nearly constant for expansion to very large size switches. Operating at OC-3 bit rates with $\nu=3$, the switch throughput will be 650 Terabit per second. Using the above expansion technique, therefore, the Bodhi architecture can be expanded to any size at a linearly increasing cost, subject to only physical and practical considerations.

9. The Switch Latency

In determining the latency, we assume that the switching/sorting time per stage in the routing/sorting networks = 1 bit, and, one RAM read or write time = 2 bits. Then the maximum amount of time taken by a multicast cell, assuming zero queueing delay, is determined to be 122 bits through an IGM and 140 bits through an OGM. Therefore, the latency for a switch size less than equal to 64×64 is 140 bits, for a switch size greater than 64×64 but less than or equal to $1k\times1k$ is 262 bits, and, for a switch size of $16^{\nu}k\times16^{\nu}k$ where $\nu\geq0$ is $(\nu+1)\cdot122+140$ bits. Thus we see that the switch latency increases only by about 29% of a slot for every expansion by a factor of 16. Therefore, Bodhi is a low latency switch architecture.

10. Conclusion

We have presented a high performance highly modular architecture which implements multicasting efficiently using some novel approaches and exploiting various approaches presented in the literature. The basic architecture presented is for a $1k\times1k$ configuration showing how it can be used for the switch sizes from 16×16 to $16^{\nu}k\times16^{\nu}k$ where $\nu\geq0$ with switch complexity increasing linearly with size. Latency of the switch is less than one slot duration for sizes up to $16k\times16k$ and increases only by about 25% of a slot for every expansion in size by a factor of 16. Using recirculation module, robustness against nonuniform and skewed traffic has been introduced. It is seen that its complexity is significantly lower than that of the sunshine switch [5]. Shared buffering between 16 ports has been employed at the outputs to minimize the output buffer requirement at a cell loss rate below 10^{-9}. Also, multiple weighted priority levels have been implemented in a dynamic manner to allow high degree of service quality control.

Therefore, we conclude that Bodhi is a novel highly modular switch fabric architecture well suited for small to very large ATM switches for applications ranging from 2.4 gigabit/s to 100^{+} Terabit/s in broadband networks providing the desired capabilities and robustness.

References

[1] A. Huang and S. Knauer, "Starlite: A wideband digital switch," Proc. GLOBECOM'84, Nov. 1984.

[2] Y.S. Yeh, M.G. Hluchyj, and A.S. Acompora, "The knockout switch: A simple, modular architecture for high performance packet switching," IEEE J. Sel. Areas Commun., vol.SAC-5, no.8, Oct. 1987.

[3] J.S. Turner, "Design of a broadcast packet Switch network," IEEE Trans. Commun., vol.36, no.6, 1988.

[4] J.Y. Hui and E. Arthurs, "A broadband packet switch for integrated transport," IEEE J. Sel. Areas Commun., vol.5, no.8, Dec. 1988.

[5] J.N. Giacopelli, et al., "Sunshine: A high performance self-routing broadband Packet Switch Architecture," IEEE J. Sel. Areas Commun., vol.9, no.8, pp.1289–1298, 1991.

[6] S. Nojima, et al, "Integrated services packet network using bus matrix switch," IEEE J. Sel. Areas Commun., vol.5, no.8, 1987.

[7] S. Urushidani, "Rerouting network: A high performance self-routing switch for B-ISDN," IEEE J. Sel. Areas Commun., vol.9, no.8, 1194–1204, 1991.

[8] F.A. Tobagi, T. Kwok, and F.M. Chiussi, "Architecture, performance, and implementation of the Tandem Banyan fast packet switch," IEEE J. Sel. Areas Commun., vol.9, no.8, pp.1173–1193, 1991.

[9] T.T. Lee, "A modular architecture for very large packet switches," IEEE Trans. Commun., vol.38, no.7, 1990.

[10] D.X. Chen and J.W. Mark, "SCOQ: A fast packet switch with shared concentration and output queueing," IEEE/ACM Trans. on Networking, vol.1, no.1, pp.142–151, 1993.

[11] H. Suzuki, H. Nagano, T. Suzuki, T. Takeuchi, and S. Iwasaki, "Output-buffer switch architecture for asynchronous transfer mode," Proc. IEEE ICC '89, pp.99–103, June 11–14, 1989.

[12] H. Kuwahara, et al., "A shared buffer memory switch for an ATM exchange," Proc. IEEE ICC '89, vol.1, pp.118–122, June 1989.

[13] T. Kozaki, et al., "32432 shared buffer type ATM switch VLSI's for B-ISDN's," IEEE J. Sel. Areas Commun., vol.9, no.8, pp.1239–1247, 1991.

[14] Y. Shobatake, et al., "A one chip scalable 8∗8 ATM switch LSI employing shared buffer architecture," IEEE J. Sel. Areas Commun., vol.9, no.8, pp.1248–1254, 1991.

[15] T. Koinuma, et al., "An ATM switching system based upon a distributed control architecture," Proc. Int'l Switching Symp. (ISS'90), vol.5, 1990.

[16] Y. Sakurai, et al., "Large scale ATM multistage switching network with shared buffer memory switches," IEEE Commun. Mag., Jan. 1991.

[17] C.L. Tang and J.S. Meditch, "A high performance copy network for B-ISDN," Proc. INFOCOM'91.

[18] T.T. Lee, "Nonblocking copy networks for multicast packet switching," IEEE J. Sel. Areas Commun., vol.6, no.9, 1988.

[19] K.Y. Eng, M.G. Hluchyj, and Y.S. Yeh, "Multicast and broadcast services in a knockout packet switch," Proc. INFOCOM'88, New Orleans, pp.29–34, March 1988.

[20] H.J. Chao, "Design and analysis of a large-scale multicast output buffered ATM switch," IEEE/ACM Trans. Networking, vol.3, no.2, pp.126–138, April 1995.

[21] H.J. Chao, "A recursive modular Terabit/second ATM switch," IEEE J. Sel. Areas Commun., vol.9, no.8, pp.1161–1172, 1991.

[22] T.T. Lee, "A modular architecture for very large packet switches," IEEE Trans. Commun., vol.38, no.7, pp.1097–

1106, July 1990.
[23] D.X. Chen and J.W. Mark, "A buffer management scheme for the SCOQ switch under nonuniform traffic loading," IEEE Trans. Commun., vol.42, no.10, pp.2899–2907, Oct. 1994.
[24] P.C. Wong and M.S. Yeung, "Design and analysis of a novel fast packet switch - pipeline banyan," IEEE/ACM Trans. networking, vol.3, no.1, pp.63–69, 1995.
[25] K.Y. Eng, M.J. Karol, and Y.S. Yeh, "A growable packet (ATM) switch architecture: design principles and applications," IEEE Trans. Commun., vol.40, no.2, pp.423–430, Feb. 1992.
[26] C.T Lea, "A multicast broadband packet switch," IEEE Trans. Commun., vol.41, no.4, pp.621–630, April 1993.
[27] J.S. Turner, "An optimal nonblocking multicasting virtual circuit switch," Proc. Infocom '94, pp.298–305, June 1994.
[28] M.J. Karol, M.G. Hluchyj, and S.P. Morgan, "Input verses output queueing on a space-division packet switch," IEEE Trans. Commun., vol.35, Dec. 1987.
[29] K.E. Batcher, "Sorting networks and their applications," presented at the Proc. 1968 Sprint Joint Comput. Conf., 1968.
[30] F.T. Yap, "A study and performance evaluation of Bodhi: A high performance ATM switch," Ph.D. dissertation, University of Missouri-Kansas City, Aug. 1995.
[31] J.P. Agrawal, "Asynchronous Transfer Mode Packet Switch," U.S. Patent no.5, 636, 210, June 3, 1997, US Patent Office, Washington, D.C.
[32] J.P. Agrawal, "Arbitrary size bitonic sorters and their applications in broadband ATM switching," Proc. IPCCC '96, pp.454–458, March 27–29, 1996.
[33] J.P. Agrawal, S.V. Sridhar, R. Halker, and F.T. Yap, "An integrated approach to multicasting in ATM switches," Proc. IEEE IPCCC '97, Tempe, AZ, pp.238–244, Feb. 5–7, 1997.

High-Performance Backbone Network Technology edited by Naoaki Yamanaka
ISBN: 0-8247-5321-6 DOI: 10.1081/(120025675)
Published 2004 by Marcel Dekker, Inc.

Multicast Packet Switch Based on Dilated Network

Pierre U. Tagle and Neeraj K. Sharma

SUMMARY Multicasting is an important feature for any switching network being intended to support broadband integrated services digital networks (B-ISDN). This paper proposes an improved multicast packet switch based on Lee's nonblocking copy network [1]. The improved design retains the desirable features of Lee's network including its nonblocking property while adopting techniques to overcome the various limitations mentioned in various literature. The proposed network architecture utilizes d-dilated banyan networks to increase the amount of cells that can be replicated within the copy network. Cell splitting is used to optimize the utilization of the network's available bandwidth. Furthermore, the proposed architecture allows for the modular expansion in capacity to accomodate changing traffic patterns. The modular design of the proposed switch likewise offers easy handling and replacement of faulty modules.

key words: *ATM Switching, cell splitting, copy networks, dilated banyan networks, fault-tolerance, multicasting*

1. Introduction

Multicasting or the ability to provide point-to-multipoint connections is an important feature for switching networks intended to support broadband integrated services digital networks (B-ISDN). Among the various communication services which require information from a source to be delivered to multiple destinations are teleconferencing, entertainment video, LAN bridging and distributed data processing.

The most common approach used to implement a multicast switch is to employ a copy network in tandem with a point-to-point routing network. The copy network replicates the inputs to the requested number of outputs, and the routing network then routes each replicated cell to the required destination. Many of the previously proposed multicast networks follow this basic approach [1], [2], [7]–[11].

Among the mentioned networks is the *nonblocking copy network for multicast packet switching* proposed by T. Lee [1]. Lee's copy network is based on a banyan network architecture with broadcast switch nodes capable of packet replications based on two-bit header information. The processing of the cells prior to entering the broadcast banyan network guarantees that the copy network is nonblocking. One problem in Lee's copy network and in many multicast packet switches based on a similar design is the problem of overflow. This happens when the total number of copy requests exceeds the total number of output ports. The problem becomes more serious as the input load increases as more and more cells would have to be dropped. A multicast packet switch supporting one-to-one; one-to-many; and one-to-all connections should ideally be able to handle N request from each of the N inputs or a total of $N \times N$ copies. Another problem occurs when the next cell's fanout is larger than the remaining free network outputs. In such a situation, this and the other cells in the higher numbered inputs are dropped. This evidently results in the underutilization of the network bandwidth. Lastly, some multicast packet switches are designed to support a particular expected traffic pattern. A particular network for example may be designed so that cell loss rates will be within acceptable limits provided that maximum fanout per cell is limited to some number, say $N/4$. Problem arises when the traffic pattern changes to such a degree that the performance of the network suffers. There is a need to allow for modular expansion in input load handling capability.

There have been a number of efforts to address these problems. To handle the overflow problem, Turner proposed to use an expanding copy network which allows for more than N output ports [2]. However the size of the expanding copy network would be fixed to a certain value and therefore would not be flexible to handle any changes in the traffic patterns. If it is to handle the maximum possible requests of $N \times N$ then a network with N^2 output ports is needed. While this is ideal, not all applications may need such a capability. This also heighten another problem in multicast packet switches based on Lee's original design. As each output requires the use of an address translation table, the increased number of outputs likewise increases the requirement for more memory. Turner addressed this problem by proposing to use shared tables which allows for a certain number of outputs to share one translation table [2]. Another proposal is the use of cell-splitting [3]. This addresses the problem of network utilization as well as overflow. The solution requires the use of an input cell-splitting section and shared memory buffers. The cell-splitting

Manuscript received March 26, 1997.
Manuscript revised July 31, 1997.

†The authors are with the School of Computer Science and Computer Engineering, La Trobe University, Bundoora, Victoria 3083, Australia.

*This paper was presented at the 1996 IEEE Global Telecommunications Conference held in London, UK on 18–22 November 1996.

IEICE TRANS. COMMUN., Vol. E81-B, No. 2, pp. 258–265, February 1998

section has its own running adder separate from another running adder used to form queueing addresses. This results in high hardware costs.

This paper proposes an improved multicast packet switch based on Lee's design [1]. The proposed architecture utilizes d-dilated copy network in place of the non-dilated copy network used by Lee. In routing networks it has already been shown that dilated banyan networks reduce output conflict since up to d cells can be transmitted simultaneously [4]–[6], [12], [13]. In our multicast packet switch it is used to increase the number of available output ports. The new design thus offers $N \times d$ outputs where N is the network size and d is the dilation. The increased number of available paths offered by dilated networks also provides fault-tolerance at link level in a natural way. The new design retains the desirable features of Lee's designs and adapts desirable techniques proposed in [2], [3] to overcome the previously mentioned problems. The proposed network also allows for easy modular expansion to handle changing traffic patterns.

The rest of the paper is organized as follows. A review of dilated banyan networks and Lee's multicast switch is done in Sect. 2. The proposed copy network is discussed in Sect. 3. Extensions to this basic design is then presented in Sect. 4. The paper is concluded in Sect. 5.

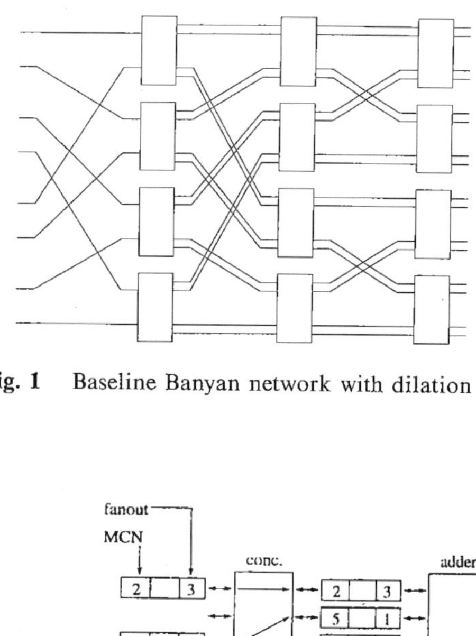

Fig. 1 Baseline Banyan network with dilation 2.

2. Review of Dilated Network and Lee's Multicast Switch

2.1 Dilated Banyan Network

Cells in non-dilated banyan networks are transferred from a switch in one stage to the switch in the next stage using only one link. In a d-dilated banyan network, there are d such links connecting the input ports of one switch to the output ports of another. This allows for up to d cells to be simultaneously transferred to the next stage [4]. A 2-dilated banyan network is shown in Fig. 1. The non-dilated banyan network is hence a special case of the dilated banyan with the dilation equal to one.

Nonblocking Property of a Dilated Banyan Network. It is well-known that the banyan network is nonblocking if the incoming cells are ordered according to their destination addresses and that there are no output conflicts [10]. A d-dilated banyan network is also nonblocking given the following conditions [4].

1) outputs are monotone, i.e. $y_1 \leqq y_2 \leqq \ldots \leqq y_k$ or $y_1 \geqq y_2 \geqq \ldots \geqq y_k$

2) any input x between two active inputs x_i and x_j is also active, i.e. $x_i \leqq x \leqq x_j$

3) no more than d cells have the same destination address

where $x_1 \ldots x_k$ and $y_1 \ldots y_k$ specify the active inputs and outputs respectively.

2.2 Lee's Copy Network

Lee's copy network is composed of a concentrator network, a running adder network, dummy address encoders (DAE), a broadcast banyan network, and trunk number translators (TNT). Each input cell is assigned a multicast channel number and a fanout prior to entering the copy network. Lee's copy network is shown in Fig. 2 [1].

The concentrator network places the incoming cells in consecutive positions to ensure nonblocking operation in the subsequent stages of the network. The cells then pass through a running adder network which as-

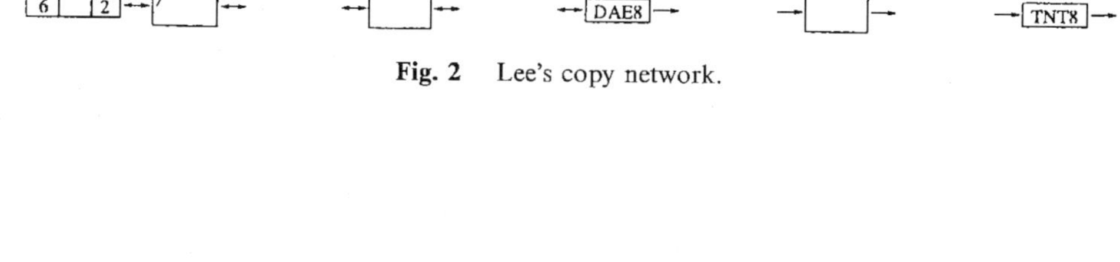

Fig. 2 Lee's copy network.

signs output addresses for each cell according to the requested fanout. This is done by computing for a cell in port position i the fanouts of all cells on ports 0 to $i-1$. The dummy address encoders (DAE) then use the running sum to specify a dummy address interval $[LO, HI]$ for each cell. For a cell in port i, LO is equal to the running sum of the fanouts of all cells in ports 0 to $i-1$ while HI is equal to the sum of LO and fanout -1 of the ith cell. This ensures that the output addresses are monotone, another condition for nonblocking operation. This information is then used by the succeeding broadcast banyan network to replicate the cells using the boolean splitting algorithm [1]. The trunk number translators (TNT) then assigns a final destination address using each cell's multicast channel number and its copy index, i.e. the $LO = HI$ value when it exits the broadcast banyan network. This address is to be used by the succeeding point-to-point routing network to guide it to its final destination.

3. Proposed Copy Network

In Lee's copy network, only cells whose running sum is equal or less than N can be replicated, the rest are blocked. To increase the number of cells that can successfully pass through the copy network, a dilated banyan network can be used instead. A non-dilated banyan would have N possible outputs while a dilated banyan with d dilation would have Nd outputs. Hence, modification is needed to Lee's original design.

Cells entering the proposed copy network would also have a multicast channel number (MCN) and a fanout. These cells pass through the concentrator network and running adder as they would in Lee's copy network. The dummy address encoders would now utilize this information to form the headers to replicate these cells when they pass through the dilated banyan network as shown in Fig. 3.

As in Lee's design the cell header would include a maximum (HI') and minimum (LO') addresses. However, these fields are divided into two subfields one of length $\log_2 N$ bits and the other of length $\lceil \log_2 d \rceil$ bits. Hence HI' is actually $HI + Hextra$ and LO' is $LO + Lextra$. For a cell in port i, let LO' be equal to the running sum of the fanouts of all cells in ports 0 to $i-1$ while HI' is equal to the sum of LO' and the fanout -1 of the ith cell. It is divided into the two subfields wherein the upper $\log_2 N$ bits of the HI' field becomes the value for HI and the lower $\lceil \log_2 d \rceil$ bits is the $Hextra$. The same process is true for LO'. The extra bits are needed to specify which among the d lines of the output will the address interval begin and end. For a non-dilated network, the outputs can be labeled from 0 to $N-1$. For a d-dilated network, the outputs would have to be labeled from 0 to $Nd-1$. However, by excluding the lower $\lceil \log_2 d \rceil$ bits it is possible to use the same routing algorithm as that used in a non-dilated network except in the last stage. In the last stage, the lower $\lceil \log_2 d \rceil$ bits are examined to ascertain which among the d lines will the cell be sent to. For example, in the 2-dilated broadcast banyan network of size $N = 8$ shown in Fig. 4 the first cell has a minimum address $LO' = 0000$ and a maximum address $HI' = 0010$. Thus, $LO = 000$, $Lextra = 0$, $HI = 001$, and $Hextra = 0$. Looking at LO' and HI', the cell is to be sent to output lines 0000, 0001 and 0010. This is done by routing the cell up to last stage by simply using LO and HI as in any non-dilated banyan network. In the last stage $Hextra$ and $Lextra$ are considered in addition to LO and HI which would then give the proper information on which links are included in the interval, i.e. from line 0 of output 000 to line 0 of output 001 for a total of three outputs.

Routing through the dilated broadcast banyan will follow the same procedure as in the boolean splitting algorithm. A switching element in stage j, where $2 \leq j < n$ (where $n = \log_2 N$) need only examine the jth bit of both the HI and LO fields to ascertain its destination and whether it is to be replicated. If the jth bit of the two fields are the same then the cell is sent to the output corresponding to that bit. If the bits are different then the cell is sent through both outputs and the minimum-maximum interval is modified according to the boolean splitting algorithm presented earlier [1].

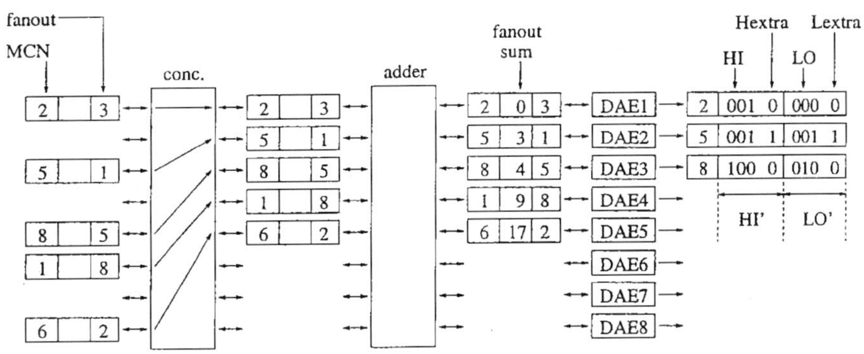

Fig. 3 Formation of dummy cell headers in proposed copy network.

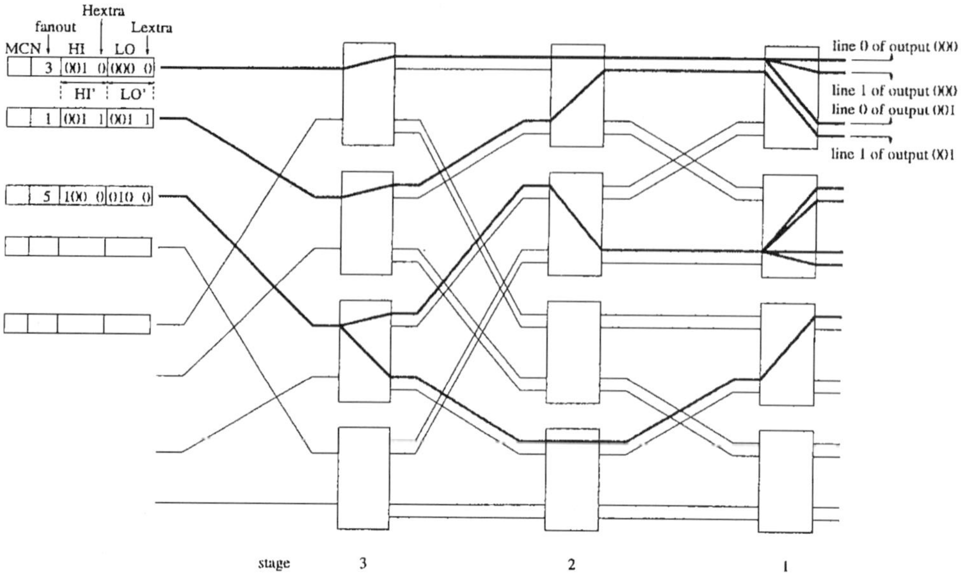

Fig. 4 Routing in a 2-dilated broadcast Banyan.

However, unlike in the non-dilated banyan network the new intervals are formed by applying the splitting algorithm on the interval $[LO',HI']$ and not $[LO,HI]$. This is despite the fact that routing through the stages with the exception of the last one depend solely on the latter. At the last stage, the least significant $\lceil \log_2 d \rceil$ bits of HI and LO are examined together with the extra bits, i.e. $Lextra$ and $Hextra$. This will then allow the cell to be sent to a subset of the output lines or to all of them.

As in Lee's copy network, the proposed network would have a large memory requirement because of the needed multicast address translation to be done by the trunk number translators (TNTs). We adopt the approach proposed by Turner for the sharing of TNTs among several output lines [2]. Turner has pointed out that for a clock rate of 150 MHz, serial data paths and an ATM cell size of 424 bits, 32 lines can share a single TNT while still allowing for 80 ns for each memory access.

4. Utilizing Multiple Dilated Broadcast Banyan Networks

For a multicast packet switch, the worst case would be when all N inputs are sending to all N destinations. For such a situation in Lee's network, all except the first would have to be dropped and resent at a later time slot. In a d-dilated network, up to d input cells can be accomodated while the remaining $N-d$ are dropped. Therefore a N-dilated network would be the ideal to handle the worst case traffic. However, as N increases such a network would be too complex to implement. Furthermore, not all installations would require the capability that all N inputs be able to broadcast to all N destinations simultaneously. A better solution is to adopt a modular design which could be expanded as the need for bandwidth increases.

To achieve such a multicast network, we present a modification of the architecture described in the previous section. The modification needed comes in three parts, i.e. the running adder, the dummy header, and the inputs to the dilated broadcast banyan network.

In the discussion in the previous section, the number of copies that can be replicated by a d-dilated broadcast banyan network is $d \times N$. Hence any cell whose running sum goes above the said value would have to be dropped. The same is true with any other cell after it. Instead of doing this, the running adder is allowed to sum up the fanouts of all the cells. The maximum running sum would therefore be less than or equal to $N \times N$ or N^2 or a value requiring $2\log_2 N$ bits. The d-dilated broadcast banyan requires $\log_2 N + \lceil \log_2 d \rceil$ bits for proper replication. As in [3], the excluded most significant bits gives the degree of overflow above dN. To put it simply, if the cells were to be grouped in terms of those whose running sum would be equal to dN, these bits would provide the information on what group a cell would belong to.

There are two possible ways of doing the above. The first is to adopt a running adder which upon encountering a cell k whose running sum is larger than dN, would reset the lower $\log_2 N + \lceil \log_2 d \rceil$ bits and increment the most significant bits. The dummy address encoder would then simply place the value of these bits into an additional field in the dummy headers, i.e. a group field. An example is shown in Fig. 5.

The cells are then sent through a bus to which one or more dilated banyan networks are connected as shown in Fig. 6 (with 2-dilated banyan networks). Each dilated banyan network has been assigned a group

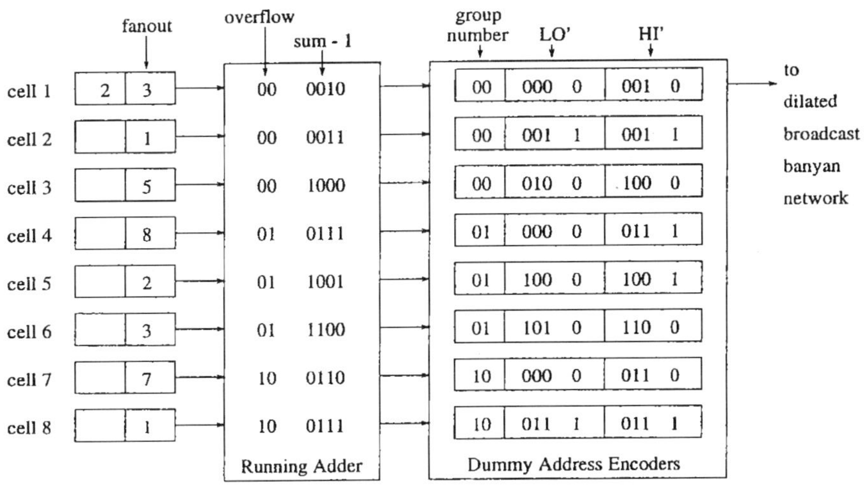

Fig. 5 Group assignment without splitting.

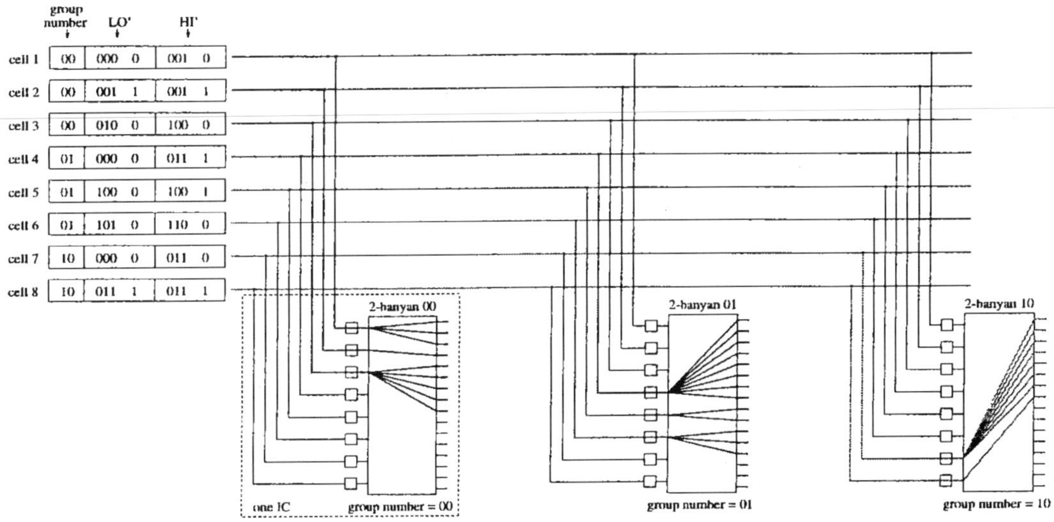

Fig. 6 Multiple dilated networks without cell splitting.

number during installation. Each input of the dilated banyan network has a comparator which simply compares the group number of the cell passing on the bus. If the group number matches with its group number then it is allowed to enter the network otherwise it is ignored.

The disadvantage of using this approach is that use of the dilated broadcast banyan networks would not be optimized. The reason for this being that it is not always the case that enough cells whose total fanout would equal to Nd will be assigned to a single group. It is highly possible that a number of the available Nd output lines will not be used at all. For example, take the situation in Fig. 6. Three dilated banyan networks were needed though numerous outputs were still free in 2-dilated banyan networks 00 and 01.

A more efficient approach is to perform cell splitting. An algorithm for cell splitting was presented in [3] for nondilated broadcast banyan network. In this approach, the running adder would simply calculate the running sum for all incoming input cells. A dummy address encoder unit with cell-splitting capability would then take this information to calculate the required field values including the group field. The process is shown in Fig. 7 for dilated broadcast banyan network.

In the inputs to the dilated broadcast banyan, an input controller would examine the group field of each incoming cell by simply comparing it with its own group number, which like in the previous configuration is assigned to it during installation. If the contents of the group field and the group number are equal then the cell is routed to the dilated broadcast banyan. Note that in the event of cell splitting and when there are two cells present in the inputs of the input controller, only one will have a group number equal to that input controller's setup number. An example is shown in Fig. 8. It can be seen that the concentration of the cells

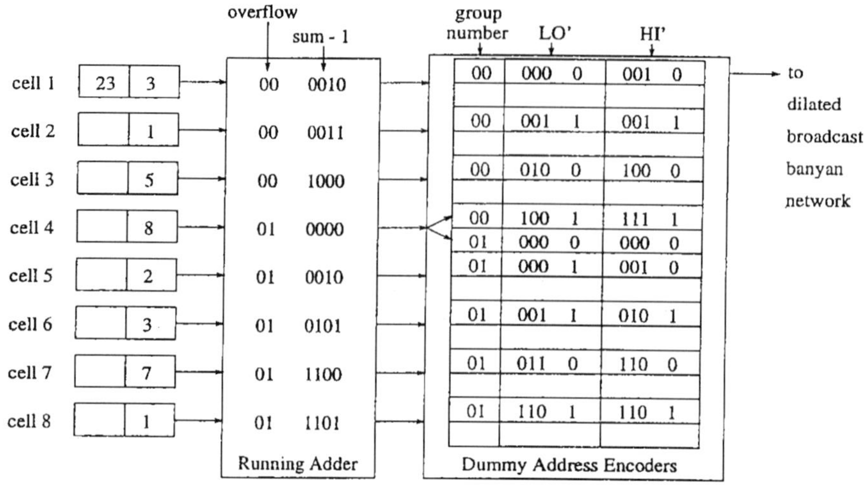

Fig. 7 Group assignment with splitting for dilated broadcast banyan network.

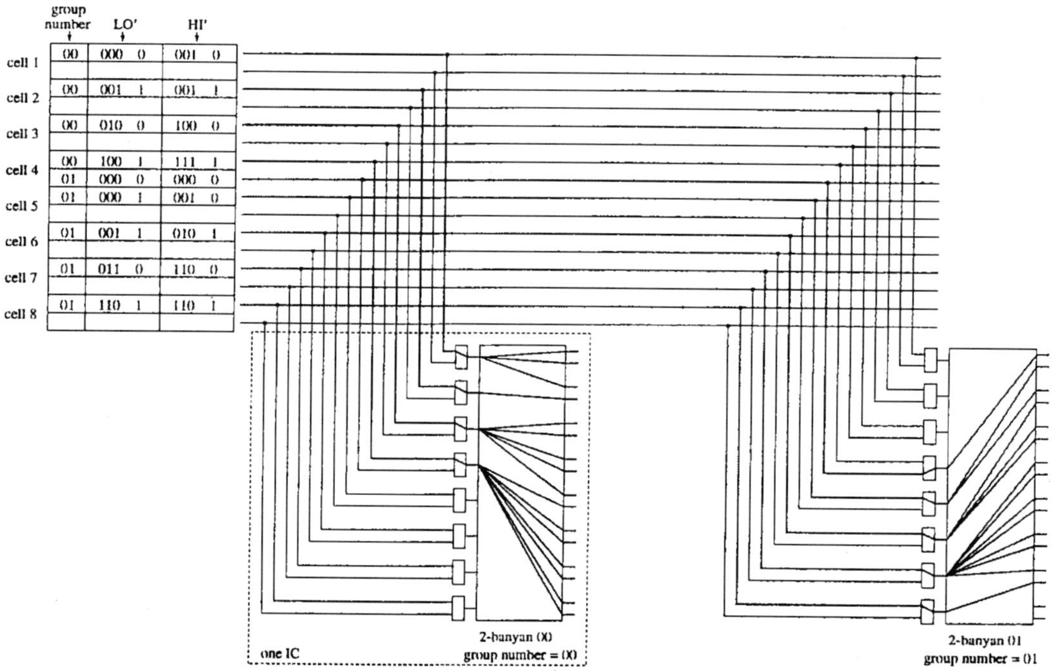

Fig. 8 Multiple dilated networks with cell splitting.

are preserved for the succeeding networks. This guarantees that every network in the system is non-blocking. In [3], cell splitting is likewise used but in contrast to the technique proposed above the cells are kept in shared-memory buffers for a time delay equal to the cells DT value. The DT value corresponds to the group number in the proposed technique. This extra delay is avoided in the proposed network. Furthermore, in the proposed copy network output conflicts are avoided as cells assigned the same output intervals are sent to different dilated networks. However, the number of buses needed in the proposed network with cell splitting increases by a factor of two compared to the implementation without cell splitting.

The outputs of all networks are linked to the trunk number translators. We adopt Turner's approach of sharing TNTs to allow the memory requirement to be kept manageable [2].

Each set of input controllers and a d-dilated broadcast banyan network can be incorporated in one chip. Each chip would have provisions for it to be assigned a setup or network number during installation. By using such an approach, it is possible to simply add more chips if there is a need for more bandwidth. This modular design allows for easy expansion of the copy network. The dilation d and network size N are chosen

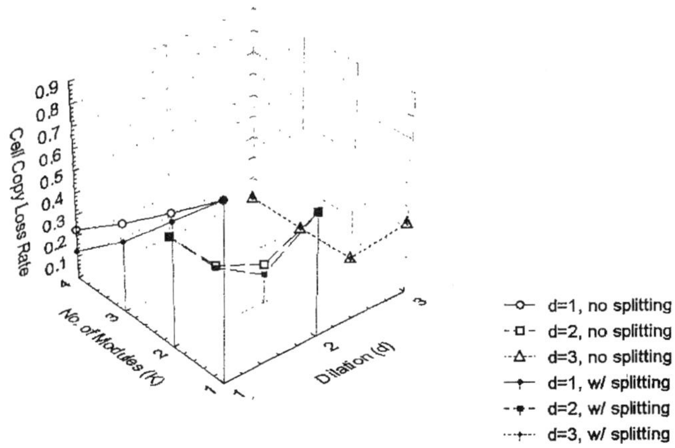

Fig. 9 Cell copy loss rate for $N = 8$ at 400% output load.

prior to chip production. Once these have been set then the number of dilated network chips needed (denoted as k) can be determined using the following equation: $\sum_{i=1}^{N} averagefanout_i \leq kdN$.

As an example, consider a copy network of size $N = 8$. Using simulations, a full input load composed of 25% unicast and 75% multicast with average fanout of 5 is applied to the network. Using this randomly generated mixed unicast-multicast traffic that produces a combined output load of 400% (25%+75%×5), the cell copy loss rate for various number of modules ($k = 1$ to 4) and network dilation ($d = 1$ to 3) is obtained. The results are shown in Fig. 9. Take note that the case $k = 1$ and $d = 1$ with no splitting is simply Lee's copy network. It can easily be seen from the said graph that increasing the dilation reduces the need for more modules. Furthermore, the benefit of using cell splitting is only evident if there are more than one module utilised in the copy network. Performance is therefore dependent on both the amount of traffic expected and the network installed. If enough d-dilated banyan networks are installed then no cells will be lost in the copy network as shown in Fig. 9.

The proposed copy network is also suited to easy handling of faulty networks. As each network is packaged as a separate IC, a faulty network can be detected by a fault-detection mechanism and be easily repaired by replacing it with a new IC. The rest of the network will remain intact and in operation.

To complete the multicast packet switch a routing network is needed to work in tandem with the copy network described in the previous two sections. With k d-dilated banyan networks of size N, the maximum traffic expected from the copy network is kdN. For this, a routing network with a $k \times d$ speedup is required.

5. Conclusion

This paper described an improved multicast packet switch based on Lee's original nonblocking copy network design. It retained some of the desirable features in Lee's design and incorporated new and improved techniques to overcome the various limitations of the said network.

The proposed architecture utilized d-dilated banyan network which increased the number of output ports by a factor of d. Routing is kept simple and only requires the addition of $2 \times \lceil \log_2 d \rceil$ bits to the dummy header used in the broadcast banyan network. The nonblocking property of the copy network is maintained in the improved copy network architecture. The proposed copy network also allowed for easy modular expansion to handle changing traffic patterns by allowing for the connection of up to k d-dilated banyan networks which further increases the copy network output from dN to kdN, where N is the network size. Using the degree of overflow obtained from the running adder it was shown that multicast cells can be sent to the appropriate dilated banyan network while satisfying the conditions for nonblocking within the broadcast banyan network.

The number of trunk number translators (TNTs) needed is kept low by adapting the technique proposed in [2]. Furthermore, the bandwidth is maximized by using cell-splitting. The said technique also help minimize the number k of d-dilated banyan networks needed by maximizing the use of each of the said network. Cell loss in the proposed multicast packet switch will be kept low as long as there are k d-dilated networks that will satisfy the equation $\sum_{i=1}^{N} averagefanout_i \leq kdN$.

Moreover, the modular nature of the architecture allows for the easy handling and replacement of faulty networks.

References

[1] T.T. Lee, "Nonblocking copy networks for multicast packet switching," IEEE J. Sel. Areas Commun., vol.6, no.9, pp.1455–1467, Dec. 1988.

[2] J. Turner, "A Practical Version of Lee's Multicast Architecture," IEEE Trans. Commun., vol.41, no.8, pp.1166–1169, Aug. 1993.

[3] X. Liu and H.T. Mouftah, "A dynamic cell splitting copy network design for ATM multicast switching," IEEE GLOBECOM 1994, pp.458–462.

[4] T.T. Lee and S.C. Liew, "Broadband packet switches based on dilated interconnection networks," IEEE ICC 1992, pp.255–261.

[5] T.T. Lee and S.C. Liew, "Broadband packet switches based on dilated interconnection networks," IEEE Trans. Commun., vol.42, no.2/3/4, pp.732–744, Feb./March/April 1994.

[6] I. Widjaja, "Tandem banyan switching fabric with dilation," Electronic Letters, vol.27, no.19, pp.1770–1772, Sept. 12, 1991.

[7] T.T. Lee, R. Boorstyn, and E. Arthurs, "The architecture of a multicast broadband packet switch," IEEE INFOCOM

1988, pp.1–8.

[8] T.-H. Lee and S.-J. Liu, "A fair high-speed copy network for multicast packet switch," IEEE INFOCOM 1992, pp.886–894.

[9] J. Turner, "Design of a broadcast packet network," IEEE INFOCOM 1986, pp.667–675.

[10] A. Huang and S. Knauer, "Starlite: A wideband digital switch," IEEE GLOBECOM 1984, pp.121–125.

[11] P.U. Tagle and N.K. Sharma, "A multicast switching network for B-ISDN," IEEE ICAAAPP 1995, pp.916–919.

[12] M. Kumar and J.R. Jump, "Performance of unbuffered shuffle-exchange networks," IEEE Trans. Comput., vol.C-35, no.6, pp.573–578, June 1986.

[13] C.P. Kruskal, "The performance of multistage interconnection networks for multiprocessors," IEEE Trans. Comput., vol.C-32, no.12, pp.1091–1098, Dec. 1983.

High-Performance Backbone Network Technology edited by Naoaki Yamanaka
ISBN: 0-8247-5321-6 DOI: 10.1081/(120025676)
Published 2004 by Marcel Dekker, Inc.

PPCN: A High-Performance Copy Network for Large Scale ATM Switching Systems

Wen-Tsuen Chen and Yao-Wen Deng

SUMMARY In this paper a high-performance copy network named PPCN is proposed for large scale ATM switching systems. The proposed copy network consists of multiple planes of the *P2I Copy Networks* (PCN) arranged in parallel. The PCN planes are designed based on the P2I multistage interconnection networks (MINs). A single PCN plane is itself a preliminary self-routing copy network which, however, is not a non-blocking one. A novel dispatcher is designed to dispatch input cells to the PCN planes such that no internal blocking nor output contention arises during the cell replication procedure and the offered load can be shared in an efficient way. The architecture of the PPCN provides -exibilit y for the maximum fanout for an input cells. In a PPCN system, the maximum fanout for an input cells is determined only by the number of interconnection stages within the PCN planes, independent of the input size of the system. The performance of the PPCN is studied under uniform traffic. It is shown that a small constant number of PCN planes are sufficient for a PPCN system to achieve an acceptable low over-o w probability regardless of the system size. The hardware complexity of an $N \times N$ PPCN is $O(N \log_2 K)$ and the length of the routing tag is $O(\log_2 K)$ bits, where K is the maximum fanout for an input cell. The storage complexity of the translation tables adopted in an N-inlet PPCN is $O(N)$, which is much lower than that of the previously proposed ones.

key words: *copy networks, ATM switching systems, multicast*

1. Introduction

In order to support advanced multimedia communication services such as video on demand, distant learning and video conferencing, the ATM Forum has added the support for multicast connections into the ATM standards [1] as one of the many attractive new features of the ATM networks. A multicast connection is a uni-directional one-to-many connection with which the source node can deliver a packet to multiple destinations simultaneously by sending only a single copy of the packet onto the network. The use of multicast connections greatly improves the efficiency of multipoint communications.

Hardware support for multicast connections has been proposed recently. Cell replication function is built into the switching systems. According to the cell replication mechanism adopted, the proposed multicast switching systems (MSS) are classified into two major categories [2]: (1) *multicast tree type* (2) *broadcast type*. Multicast tree type MSS are designed based on multistage interconnection networks (MINs) such as the Banyan networks [3] and the shuffle-exchange networks [4]. Broadcast type MSS are built based on broadcast media such as the crossbar networks [2], shared bus systems, and rings [5].

Broadcast type MSS are suitable for small-scale systems. They can be built with efficiency. Optimal throughput performance can be achieved if output buffering scheme is adopted. However, most of the broadcast media have scalability problem which makes it difficult to implement a large-scale system. Moreover, since bit map addressing scheme is commonly adopted by the broadcast type MSS, the length of the routing tags used in an $N \times N$ broadcast type MSS is $O(N)$ bits. If $N \geq 512$ then the length of the routing tag could be longer than that of an ATM cell (424 bits.) Chao [2] proposed a multiple-stage structure that can be applied to the broadcast type MSS and reduce the length of the routing tags from $O(N)$-bit to $O(\sqrt[t]{N})$-bit, where t is the number of stages. However, similar to the multicast tree type MSS, which will be explained later, extra storage space for translation tables are required to translate routing tags between cascading stages.

Multicast tree type MSS are scalable and suitable for VLSI implementation because the MINs have regular structures and most of them can be constructed recursively. The advantages of using multicast tree type MSS include less complex switching fabric, shorter switching delay, and better scalability. As shown in Fig. 1, a multicast tree type MSS performs cell replication function and switching function by two separate hardware modules. Cell copies are generated by the copy network and then forwarded to the point-to-point routing network which switches them toward the correct output ports. Both of the copy network and the point-to-point routing network are designed based on some kind of MINs such as the Banyan networks and the shuffle-exchange networks. These MINs have $O(\log_2 N)$ stages and $O(N \log_2 N)$ switching elements (SEs). They operate in a self-routing manner and the routing tags used is only $O(\log_2 N)$ bits long. Cell replication paths or the paths along which cell copies traverse through the copy network are constructed like binary trees rooted at the input ports. This is the reason behind the name multicast tree type MSS.

Manuscript received December 6, 1997.
Manuscript revised May 27, 1998.

†The authors are with the Department of Computer Science, National Tsing Hua University, Hsin-Chu, Taiwan 30043, R.O.C.

*A preliminary version of this paper is presented at the TJCOM'98, Hsin Chu, Taiwan, R.O.C., January 1998.

IEICE TRANS. COMMUN., Vol. E82-B, No. 1, pp. 1–13, January 1999

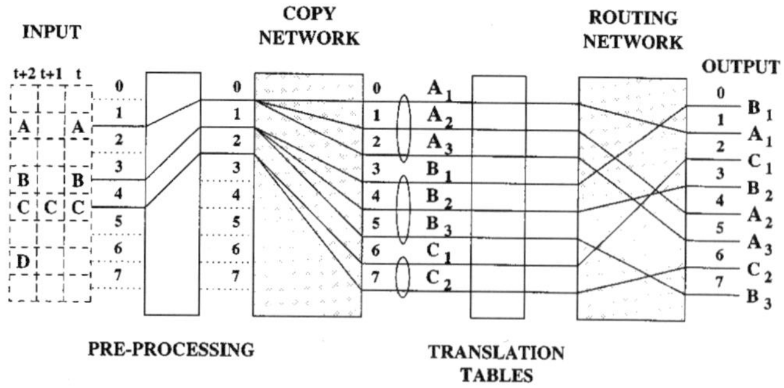

Fig. 1 The general architecture of a multicast tree type MSS.

One of the most serious problems for multicast tree type MSS is that they require very large storage space for translation tables which translate routing information between the copy network and the routing network. To illustrate this problem, let us consider the non-blocking copy network proposed by Lee [3]. In Lee's design, the copy network is constructed by a Broadcast Banyan Network and the point-to-point routing network can be constructed by a Batcher-Banyan network. In order to satisfy the non-blocking conditions of the Broadcast Banyan Network, the input cells must be pre-processed such that their dummy addresses are cyclically compacted and monotonically increasing or decreasing. Because the total number of input cells changes from time slot to time slot, cell copies traverse through the copy network via different paths even their VPI/VCI values are the same. In other words, the input cells generated from the same connection may have different replication paths as the system load varies. Therefore, the translation table at each outlet of the copy network must be large enough to maintain global translation information for all the cells from every inlet.

If an input cell can request a maximum of K cell copies, the total storage space required by the translation table is $KNBW$ bits where N is the number of outlets, B is the total number of VCIs available in the system, and W is the width (in bits) of an entry in the translation table. If $K = N$ then the storage complexity of the translation table is $O(N^2)$. For an output expanded 256×512 copy network [3] where $K = 256$, $N = 512$, $B = 2^{16}$, and $W = 16$, a total of 128 Gb (128×2^{30} bits) are required to implement the translation table at the outlets. It is clear that the implementation of a copy network with such a high storage complexity is not practical.

High storage complexity is a critical problem not only for the copy networks designed based on the Broadcast Banyan Networks but also for copy networks designed based on other MINs. In fact, any multiple-stage switch architecture which does not have a fixed or predictable output pattern at the preceding stage requires large storage space for the translation tables between cascading stages.

Turner [6] proposed some methods to reduce the storage space required by the translation tables used in Lee's copy network. By sharing a table with s lines and properly partitioning the tables, the storage space requirement can be reduced to $2\sqrt{KNBW/s}$ bit. That is, if $K = N$ the storage complexity is reduced from $O(N^2)$ to $O(N\sqrt{N})$. Liew [4] proposed a Companion Copy Network which sends translation information from the inlets to the outlets simultaneously as the ordinary multicast cells are passing through the original copy network. In this way the translation tables are moved from the outlets to the inlets and thus the storage complexity can be reduced by a factor of N. However, complicated synchronization mechanism is required to make the Companion Copy Network and the original copy network work together.

Another critical problem on the design of copy networks is the over-o w control problem. Over-o w occurs when the total number of cell copies generated during a time slot exceeds the output capacity of the copy network. Although some over-o w control schemes have been proposed to avoid system over-o w and improve the throughput of the copy network [7], [8], throughput degradation is inevitable as the integrated performance of the whole MSS is evaluated. Many of the previously proposed over-o w control schemes adopt input queuing approach. An over-o w controller schedules the input cells before they are forwarded to the copy network. By using one of the the over-o w control schemes, a copy network can achieve 100% throughput performance if fanout[†] splitting is permitted. However, output port contentions may exist among the cell copies such that the cell copies generated by the copy network in a time slot may then be delayed for two or more time slots before they can be switched to their corresponding output ports. In other words, the throughput of the MSS degrades because the over-o w control scheme considers only the throughput of the copy network rather than the throughput of the whole MSS system.

An intuitive remedy for the throughput degradation problem mentioned above is integrating output port contention resolution algorithms into the over-o w control scheme. However, it is shown that scheduling a set of multicast cells at the input queue of an MSS to optimize the throughput is an NP-complete problem no matter fanout splitting is permitted or not [9]. Some heuristic solutions and alternative scheduling schemes can be found in [9]–[11]. On the other hand, if we can design an output-queued copy network which can accept all the copy requests from the input ports and store the cell copies into its output queues, which is also the input queues of the point-to-point routing network, then optimal MSS system throughput could be achieved if a simple output contention resolution scheme with

†The *fanout* of an input cell is the total number of cell copies generated by the copy network for the cell.

lookahead window [12] is adopted by the point-to-point routing network.

In this paper a high-performance copy network architecture for ATM switching systems is proposed. The proposed copy network architecture adopts multiple copy network planes and the output queuing scheme to achieve high throughput performance. As the number of copy network planes increases, the over‑o w probability decreases. It can be shown that the relationship between the number of copy network planes and the over‑o w probability converges and that under uniform traffic fifteen copy network planes are sufficient for the proposed copy network system to achieve a reasonably low over‑o w probability ($\leq 10^{-9}$) regardless of the system size.

To reduce the storage complexity of the translation tables, a special MIN named P2I network is used to construct the copy network planes. A P2I network is a variation of the *Plus-Minus* 2^i (PM2I) network proposed in [13]. The copy network planes are hence named PCN which stands for *P2I Copy Network*. Cell replication paths on the PCN planes are determined at connection setup time. At run time, cells generated from the same connection traverse through the PCN planes along the same path. Therefore, the size of the translation tables at the outlets can be reduced significantly.

The proposed copy network architecture is named PPCN which stands for *Parallel P2I Copy Network*. PPCN† is adaptive in that it can be designed to fulfill different fanout requirement. In a PPCN, the maximum fanout for an input cell is determined only by the number of interconnection stages of the PCN. A PCN with k stages can generate a maximum of $K = 2^{k-1}$ cell copies for an input cell in a time slot with $k \geq 2$. Therefore, we can design an $N \times N$ PPCN which provides a maximum fanout of K cell copies for each input cell, no matter $K \geq N$ or $K < N$. The hardware complexity of an $N \times N$ PPCN with maximum fanout K is $O(N \log_2 K)$ and the length of the routing tag is $O(\log_2 K)$ bits. The storage complexity of the translation tables is $O(N)$.

The rest of this paper is organized as follows. The architecture of the PPCN is described in Sect. 2. In Sect. 3 the performance of the PPCN is studied. Section 4 presents the conclusions.

2. The Proposed Copy Network Architecture

The proposed copy network architecture is shown in Fig. 2. An $N \times N$ PPCN consists of a set of input processors, a dispatcher, N distributors, several PCN planes, N concentrators, output buffers, and a set of trunk number translators (TNT).

The input processors add to each of the input cells a new cell header containing necessary information for the cell replication procedure. Cell replication function

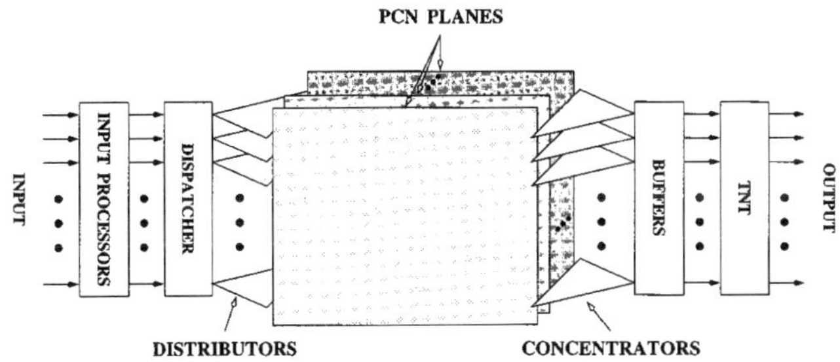

Fig. 2 The architecture of the PPCN.

is performed by the PCN planes. The dispatcher dispatches each input cells to an appropriate PCN plane such that no replication path contention exists among the input cells assigned to the same PCN plane. The distributors distribute the input cells to their corresponding PCN planes as they are assigned by the dispatcher. The concentrators collect cell copies from the PCN planes and store them into the output buffers. The TNT translates the cell header into another format which will be used by the following routing network.

With a simple encoding scheme, the PCN planes can perform cell replication function in a self-routing manner. Unlike many of the previously proposed copy networks in which cell replication paths change according to system load, cell replication paths on a PCN plane are predetermined. Let I_i be the inlet with label i and O_j be the outlet with label j, on a PCN plane. Let F_i be the fanout request of an input cell arrived at I_i. If $F_i = k$, then k cell copies are generated over the outlets $O_i, O_{i+1}, \ldots, O_{i+k-1}$, consecutively. The closed interval $[i, i+k-1]$ is defined to be the *output range* of the input cell from I_i with fanout k.

The deterministic output range property greatly reduces the storage complexity of the translation tables. However, it also makes the PCN a blocking system. Contention occurs when two or more input cells with con‑icting output ranges are sent to a PCN plane simultaneously. On a PCN plane, two or more input cells are said to have con‑icting output ranges if their output ranges overlap with each other.

To avoid this kind of contentions, multiple PCN planes are arranged in parallel and the dispatcher is used to dispatch input cells with con‑icting output ranges to different PCN planes. A dispatcher is an $N \times M$ mesh of processing elements (PEs) for an N-inlet PPCN with M PCN planes. In Sect. 2.2, a simple dispatch algorithm is developed with which the dispatcher can dispatch the input cells to appropriate PCN planes such that no con‑icting output range exists among the input cells forwarded to a PCN plane.

†In this paper, we use PPCN to name the whole copy network system and PCN to name a single copy network plane.

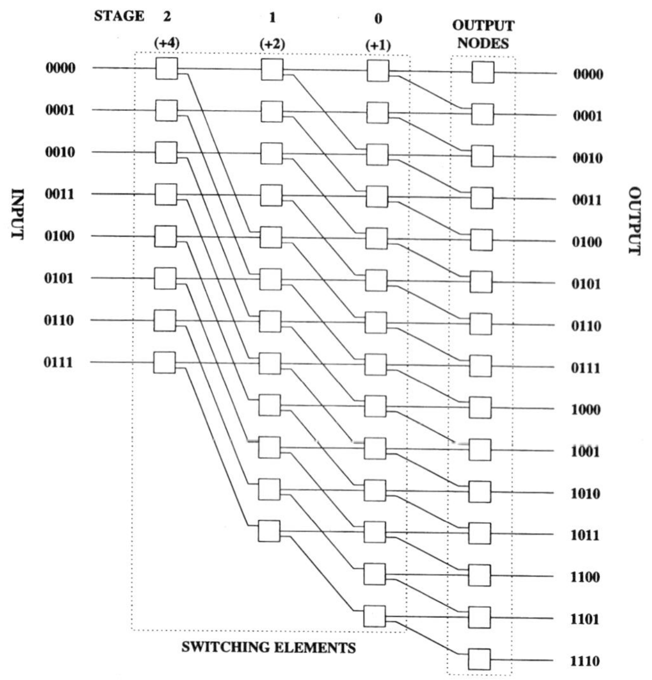

Fig. 3 Architecture of the PCN: an example 8-inlet PCN.

2.1 The Architecture of the PCN Planes

An N-inlet PCN plane can be viewed as the combination of N full binary trees. An example 8-inlet PCN plane is shown in Fig. 3 to illustrate the architecture of the PCN.

On a PCN plane, stages of SEs are labeled from left to right in descending order. The right most stage is the output stage which consists of output nodes. The primary function of the output nodes is to generate *index reference* (IR). which is used to identify a cell copy during the cell header translation procedure. Detailed description on the use of index reference will be provided in Sect. 2.5.

SEs at neighboring stages are connected by a horizontal link and an oblique link. The interconnection function of the oblique links at stage q is given by

$$P2_q(p) = p + 2^q. \tag{1}$$

Let $\mathrm{SE}_{p,q}$ denotes the SE at line p, stage q. The oblique output link connects $\mathrm{SE}_{p,q}$ to $\mathrm{SE}_{p+2^q,q-1}$ and the horizontal output link connects $\mathrm{SE}_{p,q}$ to $\mathrm{SE}_{p,q-1}$.

Each SE has two input links and two output links. If one of the input links is not used, it is connected to the logical ground such that a sequence of 0s are generated to indicate that the link is not active. If both of the input links are used, under normal operation state, only one of them should be active. If both of the input links are active, a contention arises and the output result is unpredictable.

In the following, we describe the self-routing cell

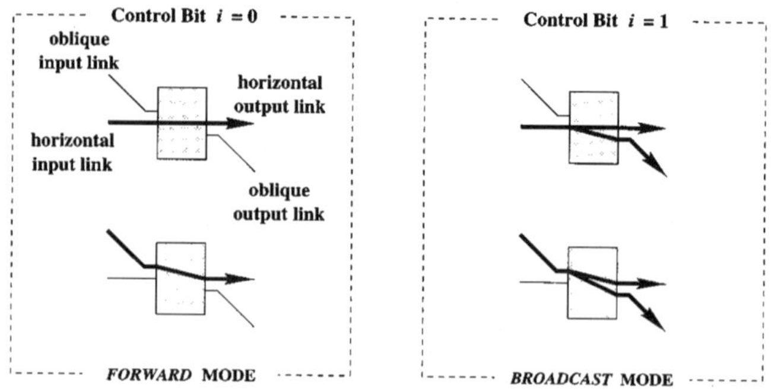

Fig. 4 Operation modes of the SE.

replication (SRCR) scheme for the PCN planes under the assumption that all SEs operate at normal state, i.e., no contention exists. In Sect. 2.2 a contention avoidance scheme is presented which ensures that no contention arises on the PCN planes and all SEs operate at normal state.

2.1.1 The Self-Routing Cell Replication Scheme

As shown in Fig. 4, an SE can be designed to change its operating mode according to a control bit. If the value of the control bit equals 0, the SE enters the "forward" mode and the input cell is forwarded to the horizontal output link, leaving the oblique output link idle. Otherwise, the SE enters the "broadcast" mode and the input cell is duplicated and forwarded to both of the output links.

Instead of using the term "routing tag" which should be dedicated to the usage on "routing," we define the term "replication tag" which is used by the copy networks during the cell replication procedure. A replication tag is a binary bit pattern with which a PCN plane can perform cell replication function in a self-routing manner. If the PCN plane has T stages (excluding the output stage) then the replication tag is T bits long. The replication tag for an input cell with fanout request $F_i = k$ is given by $\mathrm{RT}^i(k) = k - 1$.

Let $\mathrm{RT}^i_{p,q} = b_{T-1} \cdots b_q b_{q-1} \cdots b_0$ be the replication tag seen by $\mathrm{SE}_{p,q}$. $\mathrm{SE}_{p,q}$ examines bit b_q to determine its operation mode. If $b_q = 0$, $\mathrm{SE}_{p,q}$ operates in the "forward" mode and the input cell is forwarded to $\mathrm{SE}_{p,q-1}$ via the horizontal output link. If $b_q = 1$, $\mathrm{SE}_{p,q}$ operates in the "broadcast" mode, the input cell is duplicated and forwarded to $\mathrm{SE}_{p,q-1}$ and $\mathrm{SE}_{p+2^q,q-1}$ via the horizontal output link and the oblique output link, respectively. When $\mathrm{SE}_{p,q}$ operates in the "broadcast" mode, the last q bits of the replication tag sent out on the horizontal output link is replaced by a sequence of 1s such that

$$\mathrm{RT}^i_{p,q-1} = b_{T-1} \cdots b_q \underbrace{1 \cdots 1}_{q \text{ bits}}.$$

The decision rule described above is summarized in

Fig. 5.

Two cell replication examples are shown in Fig. 6 which illustrates the SRCR scheme of the PCN planes. In Fig. 6, cell A arrived at I_0 with fanout request $F_0 = 3$. The initial replication tag for cell A is $\mathrm{RT}^A(3) = 011 - 001 = 010$. At $\mathrm{SE}_{0,2}$ we have $\mathrm{RT}^A_{0,2} = \breve{0}10$. Since $b_2 = 0$, $\mathrm{SE}_{0,2}$ operates in the "forward" mode and a copy of cell A is forwarded to $\mathrm{SE}_{0,1}$. At $\mathrm{SE}_{0,1}$, $\mathrm{RT}^A_{0,1} = 0\breve{1}0$ and $\mathrm{SE}_{0,1}$ switches to the "broadcast" mode such that cell A is duplicated and forwarded to $\mathrm{SE}_{0,0}$ and $\mathrm{SE}_{2,0}$. The replication tag of the cell replica forwarded to $\mathrm{SE}_{0,0}$ is modified to be $01\underline{1}$ such that $\mathrm{SE}_{0,0}$ will be operating in the "broadcast" mode.

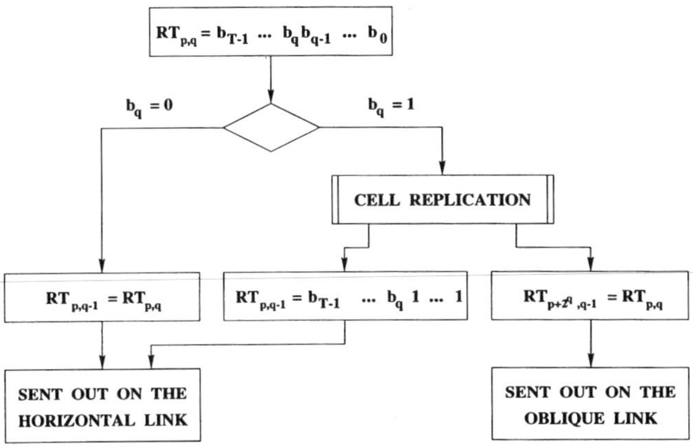

Fig. 5 The decision rule for $\mathrm{SE}_{p,q}$.

The initial replication tag for cell B is $\mathrm{RT}^B(7) = 110$. $\mathrm{SE}_{6,2}$ operates in the "broadcast" mode and the replication path for cell B branches over there. The replication tag of the cell replica forwarded to $\mathrm{SE}_{6,1}$ via the horizontal output link becomes $1\underline{11}$ such that $\mathrm{SE}_{6,1}$ and all the visiting SEs that follows will all be operating in the "broadcast" mode; while the replication tag of the cell replica forwarded to $\mathrm{SE}_{10,1}$ via the oblique output link of $\mathrm{SE}_{6,2}$ is kept intact.

2.1.2 Implementation Parameters

The logic design of the SEs that implements the SRCR scheme is illustrated in Fig. 7. At the entrance of $\mathrm{SE}_{p,q}$, signals from both of the input linksare sent into an OR gate. Recall that under normal operation modes, at most one of the input links should be active and the idle one generates a series of 0s to indicate that it is not active. The cell controller extracts control bits from the copy cell header and stores them into registers. The *active bit* A is the first bit in the copy cell header. If $A = 0$, it means that the input cell is an idle cell, i.e., both of the input links are not active. In this case, both of the output links should carry a series of 0s to inform the SEs at the next stage ($\mathrm{SE}_{p,q-1}$ and $\mathrm{SE}_{p+2^q,q-1}$) that the links are not active.

The route controller controls the selectors to generate appropriate output according to the control bits

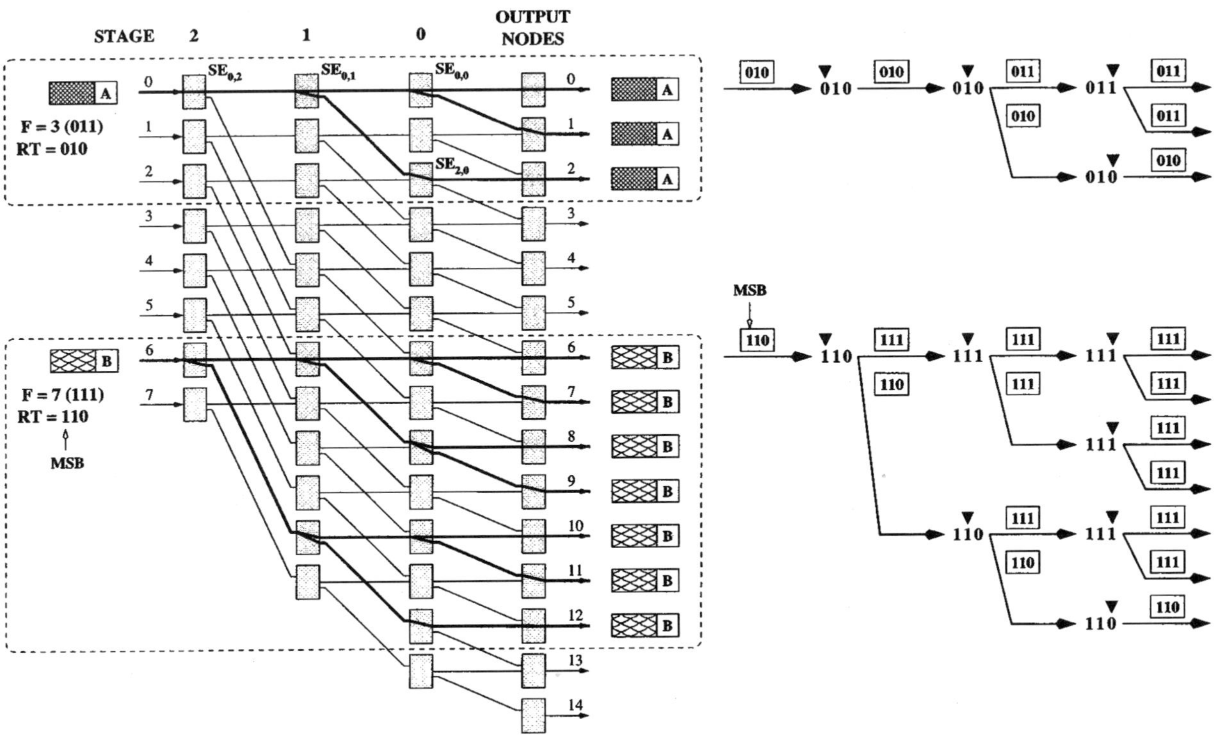

Fig. 6 The self-routing cell replication scheme of the PCN.

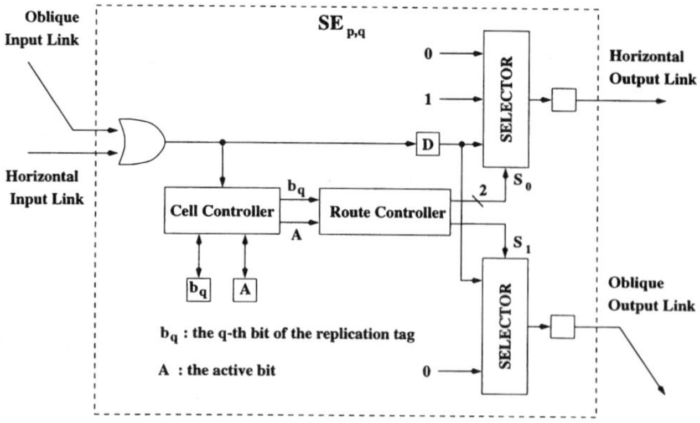

Fig. 7 Logic design of $SE_{p,q}$.

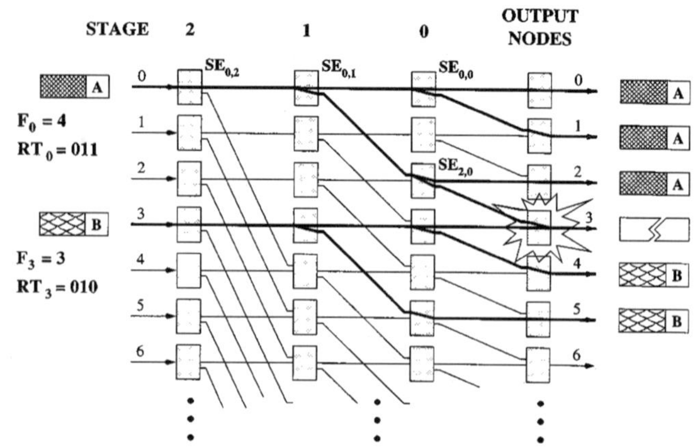

Fig. 8 A contention example in a PCN.

A and b_q. If $b_q = 0$, $SE_{p,q}$ operates in the "forward" mode; the input data is copies to the horizontal output and the oblique output should generate a series of 0s. If $b_q = 1$, $SE_{p,q}$ operates in the "broadcast" mode; the input data should be copies to both of the outputs. In this case, the replication tag $RT_{p,q-1}$ sent out on the horizontal output link should be modified to be $b_{T-1} \cdots b_q 1 \cdots 1$. The route controller must maintain a finite state machine to copy the bit sequence $b_{T-1} \cdots b_q$ from the input data and then generate a sequence of 1s that is q bits long.

For an N-inlet PCN with T stages, the number of output nodes equals

$$N' = N + \sum_{i=0}^{T-1} 2^i = N + 2^T - 1. \tag{2}$$

The total number of SEs in the PCN equals

$$\begin{aligned} S &= N \cdot T + T \cdot (2^T - 1) - \sum_{i=0}^{T-1} \sum_{j=0}^{i} 2^j \\ &= N \cdot T + (T-2) \cdot 2^T + 2. \end{aligned} \tag{3}$$

The maximum fanout for an input cell is

$$K = 1 + \sum_{i=0}^{T-1} 2^i = 2^T \tag{4}$$

Note that the maximum fanout for an input cell on an N-inlet PCN plane depends only on the number of stages. Therefore, it is -exible for the PCN to adjust its stage number to meet different fanout request. That is, we can customize a PCN plane to best fit a small switch with large per cell fanout or a large switch with small per cell fanout. For example, given $N = 1024$ and $K = 32$ the PCN that best fit this specification has $\log_2 32 = 5$ stages, $1024 + 32 - 1 = 1055$ output nodes and a total of $1024 \times 5 + (5-2) \times 2^5 + 2 = 5218$ SEs. For the nonblocking copy network proposed by Lee [3], it requires $\log_2 1024 = 10$ stages and a total of $0.5 \times 1024 \times \log_2 1024 = 524288$ SEs to meet the specification[†].

The hardware complexity of an N-inlet PCN is summarized as follows. If the maximum fanout for an input cell is K then the PCN has $N + K - 1$ output nodes, $\log_2 K$ stages, and the total number of SEs is of $O(N \log_2 K)$.

2.2 The Dispatcher

A single PCN plane is not a non-blocking system; contention occurs in a PCN when two active inlets are too close to each other as shown in Fig. 8. More precisely, contentions take place when an active inlet is within the output range of another active inlet. In the example shown in Fig. 8, cell A arrives at I_0 with fanout $F_0 = 4$. The output range of cell A is thus $[0, 3]$. When inlet I_3 is also active due to the arrival of cell B, a contention within the PCN is destine to occur because cell copies for cell A and cell B will be sent to outlet O_3 at the same time. It is easy to see that the output range of cell B, which is $[3, 5]$, is overlapped with that of cell A.

To avoid contentions from occurring, it must be ensured that no two active inlets have overlapped output range. Some room has to be saved between two neighboring active inlets. In PPCN, multiple PCN planes are arranged in parallel. Whenever it is detected that two cells with overlapping output ranges will be sent to the same PCN plane, such as the case shown in Fig. 8, one of them will be deferred and sent to a different PCN plane.

The *dispatcher* is designed to prevent contentions from occurring in a PPCN system by dispatching incoming cells with overlapping output ranges to different PCN planes. The basic idea behind the design of the dispatcher is simple. A brief description of the dispatch procedure is given as follows.

The dispatcher scans linearly from I_0 to I_{N-1} for

[†]For an $N \times N$ Broadcast Banyan Network, $K = N$, $T = \log_2 N$ and $S = \frac{N}{2} \log_2 N$.

Algorithm 1 (Dispatch Procedure):

```
1: for P = 1 to M do
   /* PCN planes are made available in tern */
2:     L = 0; /* initialize L */
3:     for i = 0 to (N − 1) do
       /* scan the inlets in sequence */
4:         if P_i = 0 and L ≤ i then
5:             P_i = P;       /* mark the cell */
6:             L = i + F_i;   /* update L */
7:         endif
8:     enddo
9: enddo
```

Fig. 9 The dispatch procedure.

active cells. Once an active cell is found, it is marked by writing the label of the currently available PCN plane into the cell header. Then the scan continues by skipping the inlets which are in the output range of the marked cell. As the scan ends, the dispatcher picks up the next free PCN plane and makes it the currently available one and then repeats the scan. Marked cells will be ignored during the following scan procedure. The scan procedure is repeated until no more free PCN plane exists.

When the dispatch procedure completes, all active cells left unmarked must be discarded because the PPCN does not employ an input queuing strategy. These cells are lost due to short of available PCN planes. In other words, they are discarded because the total fanout of the incoming cells exceeds the capacity of the copy network under the given dispatch policy. This is called the *overflow* problem as defined in Lee's paper [3].

A formal description of the dispatch procedure is shown in Fig. 9. Assume that the PPCN system has M PCN planes. Let F_i be the fanout of the input cell from I_i. Initially, if I_i is not active, $F_i = 0$; otherwise, $1 \leq F_i \leq K$ where K is the maximum fanout an active cell can do. Let P_i be the label of the PCN plane to which the input cell from I_i is dispatched. The initial value of P_i depends on that of F_i. If $F_i > 0$, $P_i = 0$; otherwise, P_i is assigned an illegal large value, say $M + 1$. Let P be the label of the currently available PCN plane, $1 \leq P \leq M$. A cell is marked by assigning $P_i = P$ for some P. Let L be the last available outlet at the current PCN plane, $0 \leq L \leq N - 1$, that is, O_L is the first outlet that is outside the output range of the last marked cell.

A dispatcher is implemented as an $N \times M$ mesh of PEs for an N-inlet PPCN system with M PCN planes. Each PE performs part of the dispatch procedure. The PE at row i column P of the mesh performs the dispatch procedure steps, as shown in Algorithm 1, from line 4 to line 7.

A PE can be implemented by simple combinational circuit as shown in Fig. 10(a). The truth table of the circuit is shown in Fig. 10(b).

An example 8 × 3 dispatcher is shown in Fig. 11 to illustrate the architecture and operation of the dispatcher. As shown in Fig. 11, the input cell from I_0 has fanout $F_0 = 3$. The *init* nodes initialize the value

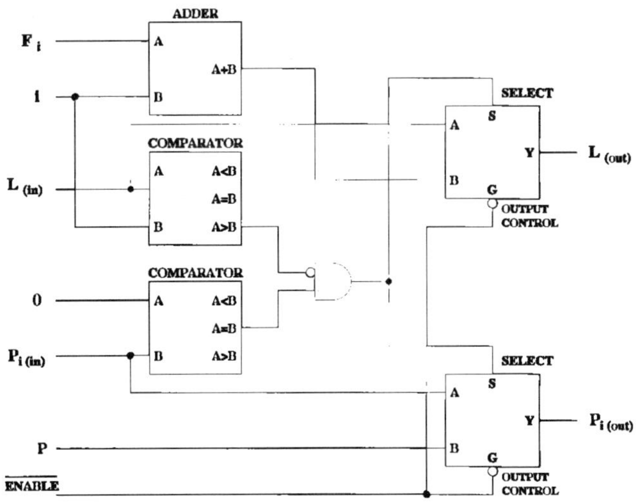

(a) Logic design of the circuit.

Inputs/State			Outputs	
ENABLE	$L_{(\text{in})} > 0$	$P_{i(\text{in})} = 0$	$L_{(\text{out})}$	$P_{i(\text{out})}$
1	X	X	Z	Z
0	0	0	$L_{(\text{in})}$	$P_{i(\text{in})}$
0	0	1	$F_i + i$	P
0	1	0	$L_{(\text{in})}$	$P_{i(\text{in})}$
0	1	1	$L_{(\text{in})}$	$P_{i(\text{in})}$

(b) The truth table of the circuit. X = irrelevant, Z = high impedance (off)

Fig. 10 The logic design of the PE.

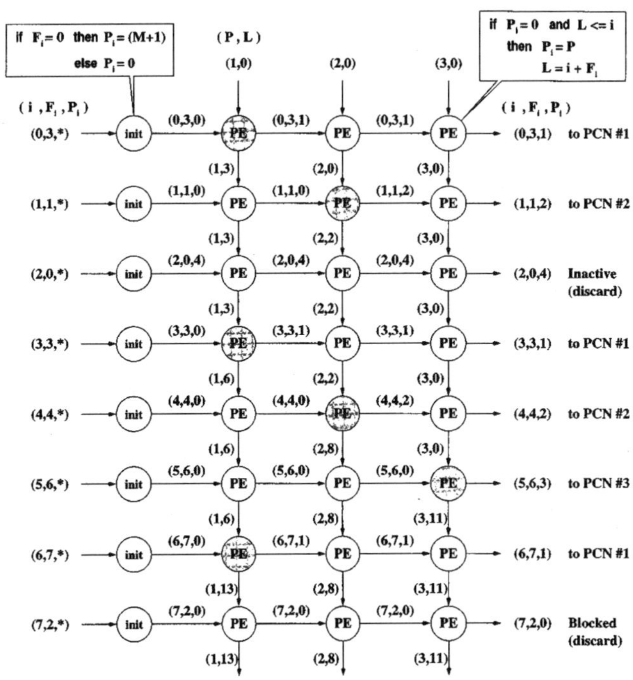

Fig. 11 The architecture of the dispatcher: an $N = 8$, $M = 3$ example.

of P_i. Let $\mathrm{PE}_{i,P}$ denotes the PE at row i column P of the PE mesh. In Fig. 11 the inputs to $\mathrm{PE}_{1,1}$ are $(i = 0, F_i = 3, P_i = 0)$ and $(P = 1, L = 0)$. Because $P_i = 0$ and $L \leq i$, the outputs generated by $\mathrm{PE}_{1,1}$ are $(i = 0, F_i = 3, P_i = 1)$ and $(P = 1, L = 3)$. That is, the cell from I_0 is dispatched to the 1st PCN plane and that at the 1st PCN plane the last available outlet to the rest of the input cells is O_3.

If the value of some P_i remains 0 at the exit of the dispatcher mesh, it means that no appropriate PCN plane is available for the input cell from I_i. In this case the input cell from I_i will be discarded. As shown in Fig. 11, the input cell from I_7 with fanout $F_7 = 2$ will be discarded because $P_7 = 0$ at the exit of the dispatcher.

It is easy to see that the dispatcher prefers active cells from upper inlets than those from lower inlets. That is, the dispatch procedure is unfair. For example, cells from inlet I_0 will never suffer from discarding because it is always the first one to scan. The probability that the cell from I_{N-1} be discarded is the largest one amongst all the cells arrived during the same time slot. Moreover, cells are discarded though some PCN planes may still have idle outlets but none of the discarded cells can utilize them without causing a contention.

It is intuitive that adding more PCN planes into the PPCN system can reduce the probability that incoming cells be discarded. However, hardware cost increases as more PCN planes are used. In Sect. 3 we will show that at most 15 PCN planes is sufficient for a PPCN system to reduce the over-o w probability to be lower than 10^{-9}, under uniform traffic, regardless of the size of the PPCN system. Therefore, the cell lost probability of I_{N-1} will be small enough to an acceptable level although it is treated unfairly.

The design of the dispatcher is simple and effective. Although the time complexity of the dispatch procedure (Fig. 9) turns out to be $O(NM)$, the dispatcher can be operated in a pipeline fashion. Moreover, since $M = 15$ is a large enough value for most systems, it can be treated as a constant. For 155 Mbps ATM links, the length of a time slot is about 2.7μs during which the dispatcher must complete a scan from I_0 to I_{N-1}. However, as the PEs can be implemented by very simple combinational circuits (Fig. 10), it is not difficult to implement a thousand-port PPCN system.

2.3 The Distributors

There are N planes of distributors in an N-inlet PPCN system. Each distributor provides service for one inlet; it routes the input cell to the assigned PCN plane according to the P_i value computed by the dispatcher. The distributors are constructed as full binary trees. The architecture of the distributor and a distribution example is shown in Fig. 12.

The distributors also operate in a self-routing manner. Assume that the PPCN system has M PCN planes

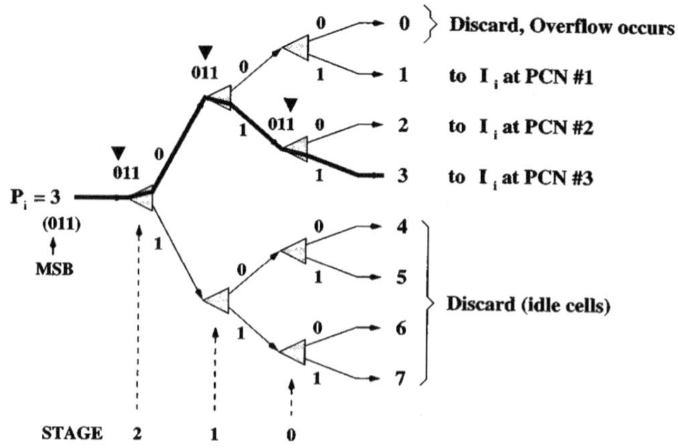

Fig. 12 A distribution example. $M = 3$.

and let $m = \lfloor \log_2 M \rfloor$. Consider the distributor serves inlet I_i. The binary representation of P_i can be written as $b_m b_{m-1} \cdots b_0$. The SEs of the distributors have one input link and two output links. An SE at stage j of the dispatcher examines bit b_j of P_i to determine the route. If the $b_j = 0$, the cell is sent out on the upper output link; otherwise it is sent out on the lower output link. It is easy to see that with the self-routing scheme the cell with $P_i = p$ will be sent to outlet p of the distributor, which is connected to inlet I_i of the p-th PCN plane.

In an N-inlet PPCN system with M PCN planes, there are N planes of $1 \times 2^{m+1}$ distributors. Since a $1 \times 2^{m+1}$ distributor has $2^{m+1} - 1$ SEs, the total number of distributor SEs in the PPCN system is $N \times (2^{m+1} - 1)$ which is of $O(N \cdot M)$. However, since M can be treated as a constant, the hardware complexity of the distributors is $O(N)$.

2.4 Output Concentration

Over the output end of the PCN planes, concentrators are used to reduce the number of outlets back to N if the point-to-point routing network that follows has only N inlets. It is known that a concentrator network can be constructed by a running adder followed by a reverse Banyan network [3] or alternatively constructed by the *Compact Routing Concentrator* [14],[15]. Since there are M PCN planes, if every PCN plane connects to the concentrators in the same way then the input size of the concentrators should be

$$C = \left\lceil \frac{N'}{N} \right\rceil \times M. \tag{5}$$

where N' is the output size of the PCN given by Eq. (2). The interconnection between the outlets of the PCN and the inlets of the concentrators is shown in Fig. 13.

Let $r = \left\lceil \frac{N'}{N} \right\rceil$, O_i of the j-th PCN plane is connected to the $(r \times (j-1) + \lceil \frac{i}{N} \rceil)$-th inlet of the concentrator at line $(i \mod N)$. For example, if $N = 8$,

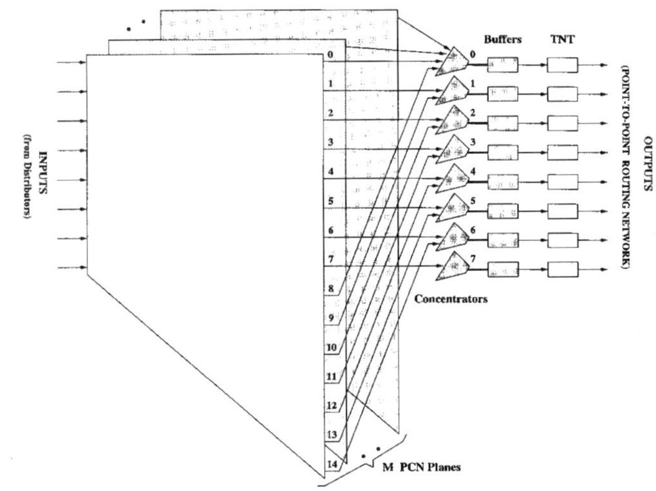

Fig. 13 The interconnection between the PCN and the concentrators.

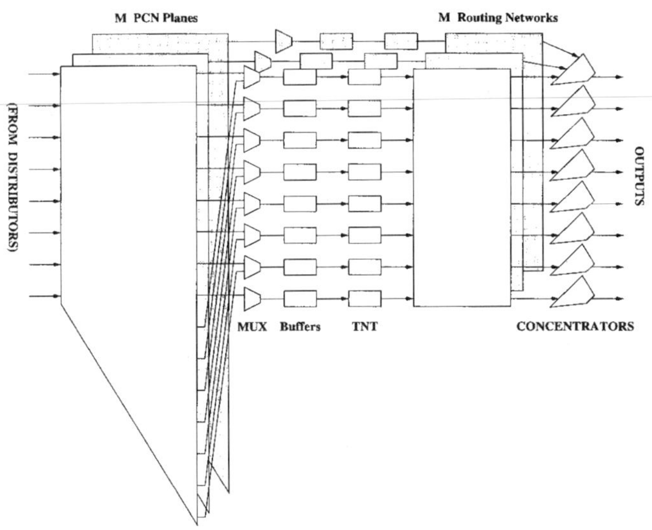

Fig. 14 Connecting each PCN plane to a point-to-point routing network.

$r = 2$, $i = 14$ and $j = 1$ then $r \times (j-1) + \lceil \frac{i}{N} \rceil = 2$, i.e., O_{14} of the 1st PCN plane is connected to the 2nd inlet of the concentrator at line $(14 \mod 8) = 6$. If $N = 8$, $r = 2$, $i = 7$ and $j = 2$ then $r \times (j-1) + \lceil \frac{i}{N} \rceil = 3$ and hence O_7 of the 2nd PCN plane is connected to the 3rd inlet of the concentrator at line 7.

The use of concentrators as the interconnection between the PCN planes and the point-to-point routing network may degrade the throughput of the copy network. In Fig. 13, as there are M $N \times (N + K - 1)$ PCN planes and only one $N \times N$ point-to-point routing network, the concentrator at line i may has $2M$ inputs $(0 \le i \le K-2)$ or M inputs $(K-1 \le i \le N-1)$ but only one output. In this case, the concentrators should be running at a speed that is thirty times faster than the line speed if $M = 15$.

Alternatively, we may connect each PCN plane to one dedicate point-to-point routing network and perform Knockout concentration [16] at the output port of the multicast switching system as shown in Fig. 14. In this case, only 2-to-1 multiplexors (MUX) are used between the PCN planes and the point-to-point routing networks.

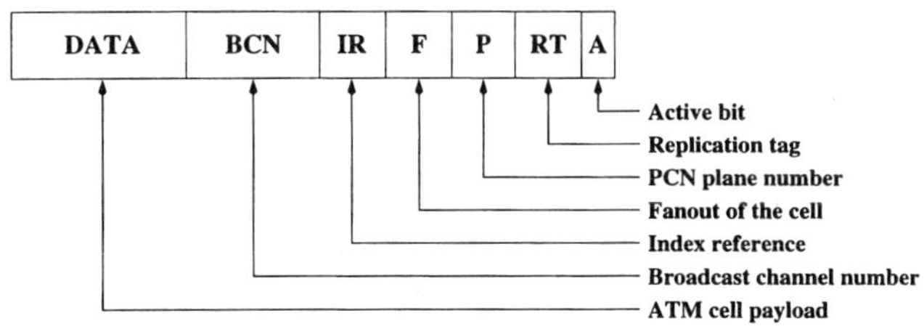

Fig. 15 Cell header format used within the PPCN.

2.5 Cell Header Translation

ATM cells can be identified by the unique VPI/VCI number in the cell header. As the cells enter a switching system, a table lookup procedure is performed by the input processors a new cell header named *copy cell header* is added to the input cells. The copy cell header contains necessary information for the cell replication procedure. The contents of the copy cell header is shown in Fig. 15.

The *broadcast channel number* (BCN) is the identifier used within the copy network to identify the cell of a connection. The F field records the fanout request of the input cell. In Sect. 2.2, we described that the dispatcher dispatches the input cell to an appropriate PCN plane for a contention-free cell replication procedure. As shown in Algorithm 1, the dispatch procedure computes the PCN plane number for an input cell based on the label of the inlet and the fanout of the cell. The result of the dispatch procedure is stored in the P field of the copy cell header and then be used by the distributors to forward the input cell to the dispatched PCN plane.

After cell copies are generated, the BCN value is no longer a unique identifier for the cells because all the cell copies generated based on the same input cell have the same BCN value. To uniquely identify a copied cell, the assistance of the IR field is required. The value of the IR field is filled by the output nodes of the PCN plane. It contains the label of the PCN outlet from which the cell copy left the PCN plane. Because the output range of an input cell is deterministic and each PCN outlet generates at most one cell copy during a time slot, the value of the BCN field together with that of the IR field must be unique among all the cell copies during the time slot.

Over the output ports of the PPCN, the TNT translates the copy cell header into the routing cell header which will be used by the following point-to-point routing network. The point-to-point routing net-

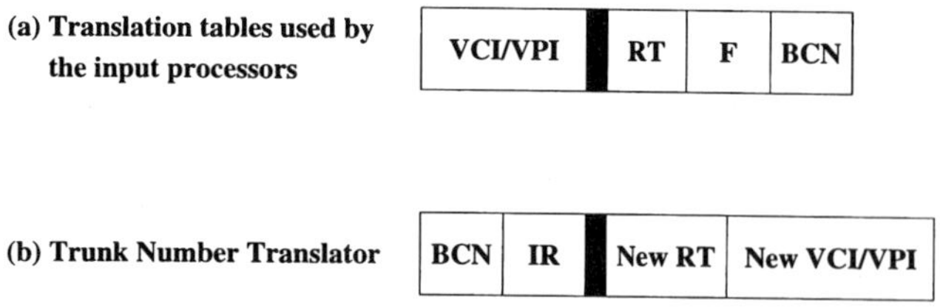

Fig. 16 The structure of the translation tables.

work can be constructed, for example, by a Batcher-Banyan switching network. Detailed design issues on the point-to-point routing network for ATM switching systems can be found in [17]–[21].

The content of the translation tables used by the input processors and that of the TNT are computed in advance at connection setup time. At run time, only table lookup operations are needed. The translation tables used by the input processors and the TNT can both be implemented by content addressable memory (CAM). The structure of the TNT and that of the translation tables used by the input processors are shown in Fig. 16.

Let B be the total number of BCNs available in a PPCN system and W be the width of an entry in the TNT. If all the available BCN numbers can be shared evenly by the inlets, the storage spaced required by the TNT is BW bits per outlet and the total storage space required is NBW bits which is of $O(N)$ space complexity.

3. Performance Study

In this section the performance of the PPCN is studied. The primary concern is on the over-o w probabilities and the system throughput. The *throughput* of a PPCN system is defined as the expected number of cell copies that an outlet can generate when the system is under full load. The *overflow probability* of a PPCN system is defined as the probability that there is at least one incoming cell being discarded during one run of the dispatch procedure.

3.1 Over-o w Probability

It is assumed that cell arrivals over the inlets are independent and identically distributed (i.i.d.) from time slot to time slot. If the offered (average) traffic load is σ then the probability that an inlet is active during a given time slot is σ and the probability that it is idle is $(1-\sigma)$.

Let F_i be a random variable representing the fanout of the input cell from I_i. Assume that the fanout between input cells are i.i.d. and have a truncated geometric distribution. The probability density function of F_i is

$$f(k) = \Pr\{F_i = k\} = \frac{(1-r)r^{k-1}}{1-r^K}, 1 \le k \le K, \tag{6}$$

where K is the maximum fanout per input cell. The moment generating function of F_i is

$$\mathrm{G}(s) = \sum_{k=1}^{K} f(k)e^{sk} = \frac{e^s(1-r)(1-r^K e^{sK})}{(1-r^K)(1-re^s)}. \tag{7}$$

The expected value of the fanout is

$$\overline{F} = E[F_i] = \dot{\mathrm{G}}(0) = \frac{1}{1-r} - \frac{Kr^K}{1-r^K} \tag{8}$$

and the effective system load is

$$\rho = \sigma\overline{F}. \tag{9}$$

For a stationary system the effective system load should not exceed 1.0.

Let $\alpha_{i,j}$ be the probability that a copied cell is generated and sent from I_i to O_j. We have

$$\alpha_{i,j} = \begin{cases} 0 & \text{if } i > j, \\ \sigma & \text{if } i = j, \\ \sigma \cdot \Pr\{F_i > (j-i)\} & \text{if } i < j. \end{cases} \tag{10}$$

Let Z_j be a random variable representing the total number of cell copies arrive at O_j during a time slot. Assume that cell arrivals over the outlets are i.i.d. The probability that z cells simultaneously arrive at O_j is

$$\Pr\{Z_j = z\} = \sum_{\mathcal{A} \subset \{1,2,\ldots,j\}} \left(\prod_{i \in \mathcal{A}} \alpha_{i,j} \times \prod_{k \in \{1,2,\ldots,j\} - \mathcal{A}} (1-\alpha_{k,j}) \right) \tag{11}$$

where $\mathcal{A}$ is a subset of $\{1,2,\ldots,j\}$ with its size $|\mathcal{A}| = z$, $z \le j$.

To simplify the analysis, let β_j be the probability that the total number of cell copies destined for O_j exceeds the number total number of PCN planes, i.e., over-o w occurs at O_j. Then we have

$$\begin{aligned} \beta_j &= \Pr\{Z_j > M\} \\ &= \begin{cases} 0 & \text{if } j \le M \\ \sum_{z=M+1}^{j} \Pr\{Z_j = z\} & \text{if } j > M \end{cases} \end{aligned}$$

and the overall over-o w probability of the system is thus

$$\beta = \sum_{j=0}^{N+K-1} \beta_j = \sum_{j=M+1}^{N+K-1} \sum_{z=M+1}^{j} \Pr\{Z_j = z\}. \tag{12}$$

Figure 17 shows a comparison between the results obtain by computer simulation of the PPCN system and the evaluation of Eq. (12) when $\rho = 1.0$ and

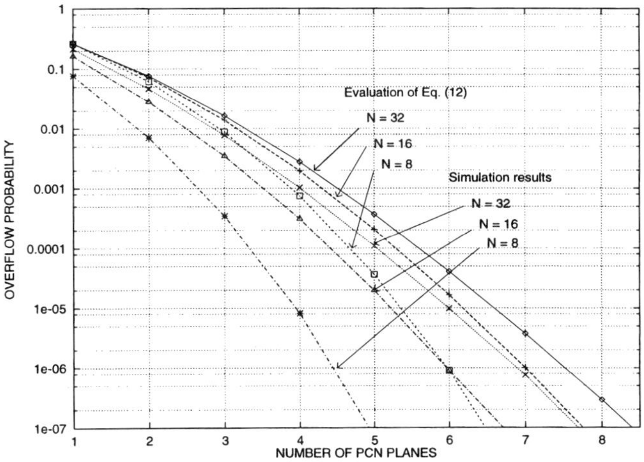

Fig. 17 The evaluation results of Eq. (12) and the computer simulation results. $\rho = 1.0$, $\overline{F} = N/2$.

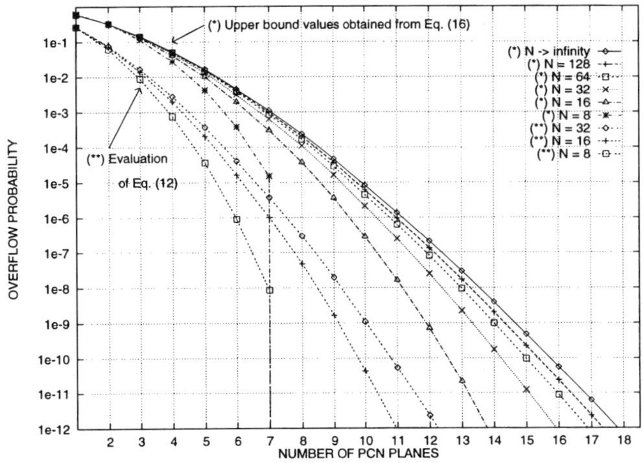

Fig. 18 Eq. (16) is a looser upper bound than Eq. (12).

$\overline{F} = N/2$. It is observed that the values of the simulation results are smaller than that of the evaluation results under the same condition. One possible explanation for this outcome is that Eq. (12) is obtained under the i.i.d. assumption but correlation between the cell arrivals over the outlets do exist. After all, we learn from Fig. 17 that Eq. (12) is an acceptable approximation for the evaluation of over-o w probabilities.

The evaluation of Eq. (12) is time consuming. A looser but easier obtained upper bound is desirable. When $i \leq j$, it is easy to see from Eq. (10) that $\alpha_{i,j} \leq \sigma$ and hence Eq. (11) can be bounded by

$$\Pr\{Z_j = z\} \leq \binom{j}{z} \sigma^z (1-\sigma)^{j-z}. \tag{13}$$

Therefore,

$$\beta_j = \sum_{z=M+1}^{j} \Pr\{Z_j = z\} \tag{14}$$

$$\leq \sum_{z=M+1}^{j} \binom{j}{z} \sigma^z (1-\sigma)^{j-z}. \tag{15}$$

Since β_j is an increasing function of j, $\beta_j \leq \beta_{j+1} \leq \cdots \leq \beta_N$ and β_N may be a dominant term in the over-o w probability. When N is large we have

$$\beta_N \leq 1 - \sum_{z=0}^{M} \frac{e^{-2\rho}(2\rho)^z}{z!} \tag{16}$$

as $\overline{F} = N/2$ and $\sigma = \rho/\overline{F}$.

Figure 18 shows a comparison between the results obtained by evaluating Eq. (12) and Eq. (16). Curves marked with (*) are results obtained by the evaluation of Eq. (16) and curves marked with (**) are results obtain from Eq. (12), $\rho = 1.0$ and $\overline{F} = N/2$.

The result shown in Fig. 18 reveals that the over-o w probability given by Eq. (12) is bounded by Eq. (16) and that fifteen PCN planes are sufficient for a PPCN system to achieve an acceptable low over-o w probability ($\leq 10^{-9}$) regardless of the size of the system.

Besides the over-o w probability, we may also want to know the *cell loss ratio* (CLR) of the system. Let N_e be the total number of cell copies that are expected to be generated, i.e., $N_e = \sum_{i=0}^{N-1} F_i$ and N_a be the total number of cell copies that the copy network actually generates. The CLR for a PPCN can be defined as

$$\text{CLR} = \frac{N_e - N_a}{N_e} \tag{17}$$

Since the dispatcher discards incoming cells in case of over-o w before cell copies are generated, the CLR of a PPCN can be estimated by the over-o w probability times the expected fanout. That is,

$$\text{CLR} \approx \beta \times \overline{F}. \tag{18}$$

For $M \geq 16$ and $\overline{F} < 1024$ the CLR can be made smaller than 10^{-9}.

3.2 Throughput

The throughput of a PPCN is determined by the over-o w probability and the way it is connected to the following point-to-point routing network. In Section 2.4 two approaches for the interconnection between the PPCN and the point-to-point routing network have been described. The first approach is using N concentrators to reduce the number of outlets from $N+K-1$ to N and then connects the PPCN to an $N \times N$ point-to-point routing network as shown in Fig. 13. The second one is connecting each PCN plane to a point-to-point routing network as shown in Fig. 14.

In this section the throughput performance of the two interconnection approaches are evaluated through simulation. Uniform traffic is assumed and cell arrivals

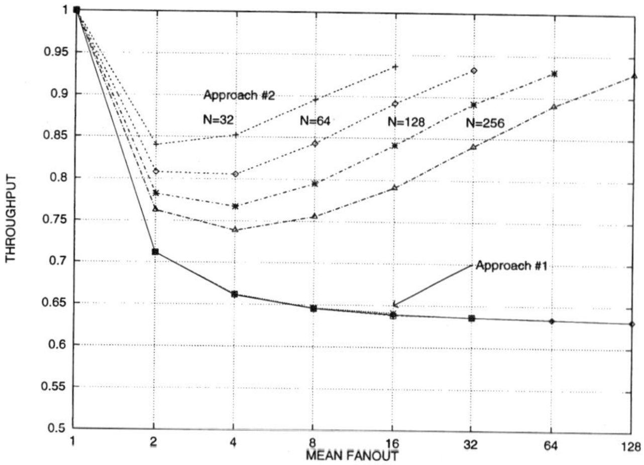

Fig. 19 Throughput performance of the PPCN. $M = 15$.

are assumed to be an i.i.d. process. Cell fanout is assumed to has a truncated geometric distribution as defined by Eq. (6). The simulation results are shown in Fig. 19.

In Fig. 19, maximal throughput is achieved when $\overline{F} = 1$. In this case, the system is reduced to a unicast one. With the first interconnection approach, throughput degrades as $\overline{F}$ increases. The minimal value is about 0.63. One explanation for this phenomenon is that the cell copies generated by the PCN planes are not evenly distributed among the outlets. Some cell copies build up a queue in the buffer between the PPCN and the routing network. Because each outlet of the PPCN forwards only one cell copy to an inlet of the routing network during a time slot, a cell copy at the rear of the queue must wait for its turn until all the cells in front of it be forwarded — a phenomenon similar to the head-of-line blocking in some input queuing switching systems. The longest queue length observed during the simulation equals 8.

The throughput performance for the second interconnection approach is much better than that of the first approach. An interesting result is observed that larger mean fanout makes better throughput. When $\overline{F} = N/2$ the throughput is about 0.93. However, when $N = 256$ and $\overline{F} = 4$ the throughput is only 0.74. This result re-ects the in-uence of the granularity of output ranges on the throughput. Because cell copies for an incoming cell are generated consecutively within its output range, when $\overline{F}$ is small the expected size of output ranges is small and the distribution of cell copies are more random than that when $\overline{F}$ is large.

Although the second interconnection approach requires M point-to-point routing networks. Since $M = 15$ is a sufficient large value for most systems, we may consider M as a constant. The hardware complexity of a multicast switch built of a PPCN and M point-to-point routing networks can be kept as low as $O(N \log N)$ if $O(N \log N)$-complexity routing networks, e.g. the Batcher Banyan Network, are employed. The overall hardware complexity of the PPCN is equivalent to that of a single PCN plane. If the maximum fanout per input cell in an N-inlet PPCN is K then the hardware complexity of the PPCN is of $O(N \log K)$. If $K = N$ then the hardware complexity of the PPCN is $O(N \log N)$ which is the optimal value for switching systems.

4. Conclusions

In this paper a high-performance copy network named PPCN is proposed for large scale ATM switching systems. PPCN stands for *Parallel P2I Copy Network*. A PPCN consists of multiple PCN planes arranged in parallel. A PCN is a preliminary copy network designed based on the P2I multistage interconnection network. A single PCN plane is itself a self-routing copy network which, however, is not a non-blocking system. In PPCN, a novel dispatcher is designed to properly dispatch input cells to the PCN planes such that no internal blocking nor output contention arises and that the offered traffic load can be shared by the PCN planes in an efficient way.

PCN has a regular structure and is constructed by very simple SEs so that it is suitable for large scale switching systems. The output capacity of a PCN is -exible. On a PCN plane, the maximum fanout for an input cell is related only to the number of interconnection stages, independent of the system size. Therefore, it is possible for an N-inlet PCN to generate more than N cell copies for an input cell during a single time slot. The operation of the PCN planes involves only simple operations such as table lookup and self-routing so that it can be designed to operate at high clock rate. Furthermore, according to the self-routing cell replication scheme proposed, the cell copies generated according to a given input cell departs from the PCN over predefined outlets. This deterministic output property greatly reduces the storage complexity of the translation tables. In an N-inlet PPCN system, the total storage space required by the translation tables is of $O(N)$, which is much lower than that of the previously proposed ones.

The performance of the PPCN under uniform traffic is studied. It is shown that 15 PCN planes are sufficient for a PPCN system to achieve an over-o w probability lower than 10^{-9} regardless of the system size. Therefore, as the over-o w probability is specified, the number of PCN planes used in a PPCN system can be treated as a constant. The hardware complexity of an $N \times N$ PPCN is $O(N \log_2 K)$ and the length of the routing tag is $O(\log_2 K)$ bits, where K is the maximum fanout for an input cell.

References

[1] The ATM Forum, "ATM User-network interface specification version 3.1," The ATM Forum Technical Committee,

af-uni-0010.002, Sept 1994.

[2] H.J. Chao and B.S. Choe, "Design and analysis of a large-scale multicast output buffered ATM switch," IEEE/ACM Trans. Networking, vol.3, pp.126–138, April 1995.

[3] T.T. Lee, "Nonblocking copy networks for multicast packet switching," IEEE J. Sel. Areas Commun., vol.6, pp.1455–1467, Dec. 1988.

[4] S.C. Liew, "A general packet replication scheme for multicasting with application to shuffle-exchange networks," IEEE Trans. Commun., vol.44, pp.1021–1033, Aug. 1996.

[5] M. Rahnema, "The fast packet ring switch: A high-performance efficient architecture with multicast capability," IEEE Trans. Commun., vol.38, pp.539–545, April 1990.

[6] J.S. Turner, "A practical version of Lee's multicast switch architecture," IEEE Trans. Commun., vol.41, pp.1166–1169, Aug. 1993.

[7] C.J. Chang and C.J. Ling, "Over-o w control in copy network of broadband packet switch," Elect. Lett, vol.27, pp.927–939, May 1991.

[8] X.Liu and H.T. Moutfah, "Design of a high performance nonblocking copy network for multicast ATM switching," IEE Proc. Commun., vol.141, pp.317–324, Oct. 1994.

[9] P.R. Sheu, "A study on some scheduling problems in multicast switching systems," Ph.D. thesis, National Tsing Hua University, Hsin-Chu, Taiwan, Republic of China, June 1992.

[10] J.W. Byun and T.T. Lee, "The design and analysis of an ATM multicast switch with adaptive traffic controller," IEEE/ACM Trans. Networking, vol.2, pp.288–298, June 1994.

[11] B. Prabhakar, N. McKeown, and R. Ahuja, "Multicast scheduling for input-queued switches," IEEE J. Sel. Areas Commun., vol.15, pp.855–866, June 1997.

[12] W.-T. Chen, H.-J. Liu, and Y.-T. Tsay, "High-throughput cell scheduling for broadband switching systems," IEEE J. Sel. Areas Commun., vol.9, pp.1510–1523, Dec. 1991.

[13] R.R. Seban and H.J. Siegel, "Shuffling with the Illiac and PM2I SIMD networks," IEEE Trans. Comput., vol.C-33, pp.619–625, July 1984.

[14] W.J. Cheng and W.T. Chen, "A new self-routing permutation network," IEEE Trans. Computers, vol.45, pp.630–636, May 1996.

[15] W.J. Cheng, "Design of High-Performance ATM Switching Architectures," Ph.D. thesis, Department of Computer Science, National Tsing Hua University, Hsin Chu, Taiwan, R.O.C., 1996.

[16] Y.S. Yeh, M.G. Hluchyj, and A.S. Acampora, "The knockout switch: A simple, modular architecture for high-performance packet switching," IEEE Trans. Commun., vol.SAC-5, pp.1274–1283, Oct. 1987.

[17] A. Huang and S. Knauer, "Starlite: A wideband digital switch," Proc. IEEE GLOBECOM'84, pp.121–125, 1984.

[18] T.T. Lee, "A modular architecture for very large packet switches," IEEE Trans. Commun., vol.38, pp.1097–1106, July 1990.

[19] J.N. Giacoplli, J.J. Hickey, W.S. Marcus, W.D. Sincoskie, and M.Littlewood, "Sunshine: A high-performance self-routing broadband packet switch architecture," IEEE J. Sel. Areas Commun., vol.9, pp.1289–1298, Oct. 1991.

[20] W.J. Cheng and W.T. Chen, "A simple architecture for ATM switching systems," Proc. IEEE INFOCOM'95, pp.1206–1213, 1995.

[21] R. Kannan, R. Bartoš, K. Lee, and H.F. Jordan, "SXmin: A self-routing high-performance ATM packet switch based on group-knockout principle," IEEE Trans. Commun., vol.45, pp.710–722, June 1997.

High-Performance Backbone Network Technology edited by Naoaki Yamanaka
ISBN: 0-8247-5321-6 DOI: 10.1081/(120025677)
Published 2004 by Marcel Dekker, Inc.

A General Expansion Architecture for Large-Scale Multicast ATM Switches

Sung Hyuk Byun and Dan Keun Sung

SUMMARY This paper proposes a general expansion architecture for constructing large-scale multicast ATM switches with any type of small multicast switch, called the multicast Universal Multistage Interconnection Network (multicast UniMIN). The proposed architecture consists of a buffered distribution network that can perform cell routing and replication simultaneously, and a column of output switch modules (OSMs). The adoption of channel grouping and virtual first-in-first-out (FIFO) buffers results in high delay/throughput performance, and the distributed lookup table scheme for multicast addressing greatly reduces the size of a single lookup table. Analytical and simulation results show that high delay/throughput performance is obtained for both unicast and multicast traffic, and the proposed architecture yields an even better performance for multicast traffic than for unicast traffic. In addition, the multicast UniMIN switch has such good features as modular expandability, simple hardware, and no internal speed-up operation.

key words: *general expansion architecture, large-scale, multicast, ATM switch*

1. Introduction

Various multicast (point-to-multipoint) services, such as teleconferencing, video on demand (VOD), and distributed data processing, are expected to be among major applications of the future Broadband Integrated Services Digital Network (B-ISDN). Multicast function of ATM switches can be implemented either with a separate *copy network* preceding the routing network, or with a single switching fabric performing both replication and routing functions. The latter will be called as the *unified* approach in this paper.

The copy network approach is widely employed in many space-division switches [1]–[5]. The copy network replicates each incoming multicast cell (called a master cell) according to its fanout value (the number of destinations). The header of each copied cell is translated to its destined output port address through a translation table lookup at the trunk number translator (TNT) [2]; the copied cells are then routed to the destination output ports through the following point-to-point switch. One problem in copy network architectures is overflow, which occurs when the number of copy requests exceeds the number of output ports of the copy network. To overcome this problem, Lee [2] and Turner [4] proposed asymmetric copy networks, which has a larger number of output ports than that of input ports. Another related problem is head-of-line (HOL) fanout blocking [6]. When the fanout of an incoming cell is greater than the number of remaining idle output ports, the cell is blocked; further, the rest of the cells entering at the same time slot are also blocked. To resolve this problem, a fanout limiting scheme [4] and the cell-splitting method [6] has been proposed. In addition, the memory size of TNT tables increases significantly as the fanout and the switch size increase, because each table of output port needs to have the routing data for all the copied cells of each multicast connection [2].

The unified approach is preferred in the shared medium output buffering [7] or shared buffering [8], [9] ATM switches, because cell replication function can be implemented by simply modifying their switching fabrics. As a rule, the unified approach is simpler than the copy network approach, and it does not suffer from HOL fanout blocking or overflow problems.

Both approaches may encounter some difficulties as the switch size grows. In the copy network approach, they include the implementation of large copy network and large translation tables. To alleviate such problems, Zhong [5] proposed a modular copy network to implement large-scale multicast switches with moderate-size copy modules and small translation tables, by restricting the multicast calls processed in a copy module to a subset of all multicast calls. However, this network still requires a large-scale point-to-point switch to construct a large-scale multicast switch. The switching capacities of the unified type switches are limited by the bandwidth of shared medium or shared buffers. Furthermore, all switch architectures suffer fundamental physical limitations in the number of input/output ports and the circuit density that a VLSI chip or a PBA can accommodate. Therefore, large-scale multicast switches should have multistage structures with moderately sized modules.

We define the *general expansion architecture* as a multistage architecture that can be utilized in building large-scale switches with any type of small ATM switch. The Clos network [10] is one of the most popular general expansion architectures [7], [8], [11], [12], because it can be built by simply interconnecting switch mod-

Manuscript received May 30, 1997.

† The authors are with the Department of Electrical Engineering, Korea Advanced Institute of Science & Technology, Taejon, Korea.

* This paper was supported in part by the Ministry of Information and Communications, Korea.

IEICE TRANS. COMMUN., Vol. E80-B, No. 11, pp. 1671–1679, November 1997 ules in a regular pattern. However, its multipath property requires efficient path hunting methods[13], and performance degradation due to the multistage configuration is inevitable[8]. Besides, internal routing for multicast connections is even more complicated [14].

In this paper, we apply a unified multicast approach to our general expansion architecture for large-scale, point-to-point ATM switches, called the *UniMIN* (Universal Multistage Interconnection Network) [15]–[17]. The UniMIN switch can be classified as a *distribution network-based architecture*, which is another general expansion architecture class[17]. In the distribution network-based ATM switches, output ports are divided into multiple groups accommodated by small ATM switch modules. Designing a high-performance distribution network with manageable hardware complexity is one of the main issues in the distribution network-based ATM switch architectures. Growable switches of AT&T[18],[19] and Chao's MOBAS[20] can be classified into this category. Their distribution networks have bufferless architectures that yield optimal performance at the expense of high hardware complexity. On the other hand, the proposed distribution network of the UniMIN switch has a buffered multistage structure. However, queueing delay in the distribution network is minimized to several cell times, and its throughput can be increased up to 100%. Therefore, performance degradation due to the multistage configuration can be negligible. The *multicast UniMIN* switch architecture inherits such good features of unicast UniMIN[17] as self-routing capability, modular expandability, no internal speed-up, low hardware complexity, and high delay/throughput performances. In addition, a multicast capability is added by simply modifying the distribution modules without an additional large copy network.

The remainder of this paper is organized as follows. In Sect. 2, the architecture of the multicast UniMIN switch and its distribution module are introduced, and the multicasting scheme is described. In Sect. 3, the performance of the proposed multicast switch is evaluated, and is then verified by computer simulations. Finally, conclusions are given in Sect. 4.

2. Structure of the Multicast UniMIN Switch

2.1 Distribution Network-Based General Expansion Architecture

The $N \times N (N = 2^s n, s = 0, 1, 2, \ldots)$ UniMIN switch is comprised of a distribution network with s stages of 2^s $n \times n$ output-buffered modules and 2^s output switch modules (OSMs) at the last stage. Figure 1 shows an example of the UniMIN switch for $N = 4n$ and $s = 2$. The internal links of the UniMIN switch are grouped with $n/2$ links, and the links are shared within each link group. This is the concept of *channel grouping*[22], and

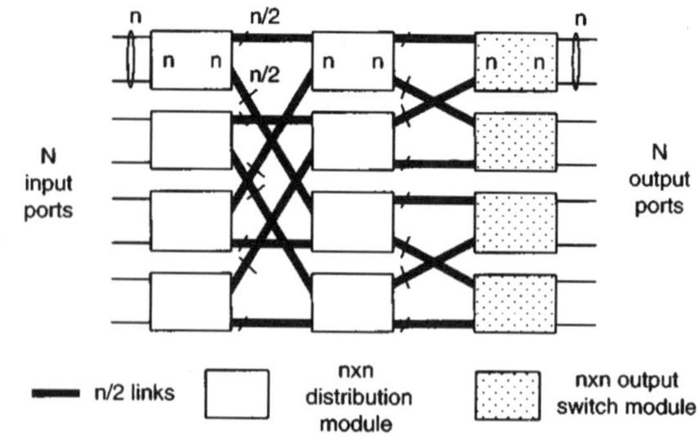

Fig. 1 Multicast UniMIN switch architecture for $N = 4n$.

results in a small delay and reduced buffer size in the distribution modules. Though the successive cells of a VC can take different links within a link group, cell sequences in the distribution network can be preserved during cell switchings. This cell sequence integrity can be guaranteed by assigning a higher priority on the upper input link when buffering the cells and by transmitting the first cell of the virtual FIFO on the top output link, the second cell on the next link, and so on. It is also maintained at OSMs by processing the uppermost incoming cell first, if OSMs have no input buffers. The input queueing OSM may require a resequencing operation.

Since the link groups follow the interconnection pattern of a delta network with 2×2 modules, this network can inherently support the self-routing feature based on the destination output port address. Unicast cells utilize the self-routing feature according to their destination output port addresses. The first s bits of the destination address are used as a routing tag at the distribution network; the k-th bit in the routing tag determines the outgoing link group of the k-th stage, i.e., "0" for upper link group and "1" for lower link group. The remaining $\log_2 n$ bit address is utilized in the routing at the OSM.

Multicast cells have multiple destinations belonging to one or more OSMs. In the proposed architecture, cell replications take place as far away as possible from the input ports. Based on this scheme, only one copy of a multicast cell is sent to each destined OSM, and each OSM performs further cell replication and routing within the module. This scheme can significantly reduce actual cell traffic in the distribution network, and thus can improve the delay/cell loss performances of multicast traffic. Multicast cells are replicated and routed, based on the lookup tables of each distribution module and OSM. More detailed explanations on this multicasting scheme are given in the following subsections.

2.2 Distribution Module

An $n \times n$ distribution module consists of n input port controllers (IPCs), two multicast lookup table control

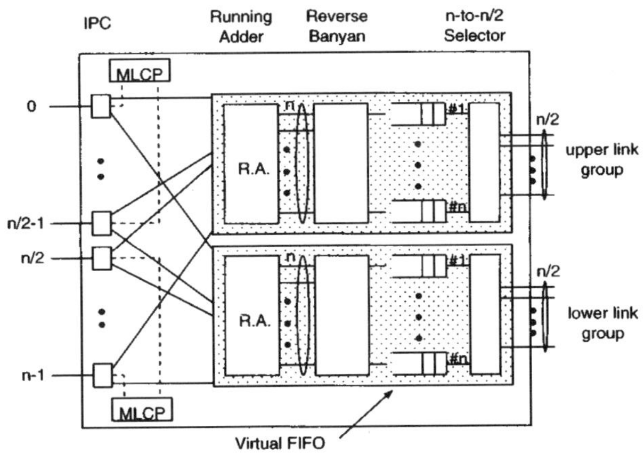

Fig. 2 $n \times n$ distribution module.

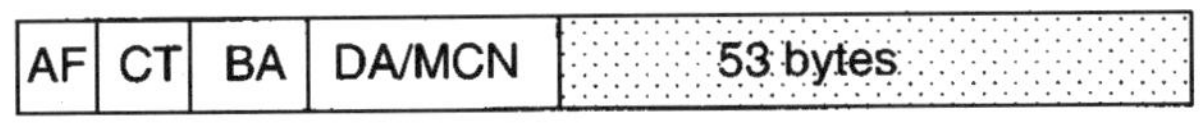

AF : activity flag (1bit) ('1' : active cell, '0' : idle cell)
CT : cell type
'00' : unicast user cell
'01' : OSM multicast control cell
'10' : multicast user cell
'11' : DN multicast control cell
BA : buffer address at the virtual FIFO
DA : destination address for CT = '00' and '01'
MCN : multicast channel number for CT='10' and '11

Fig. 3 Internal cell format.

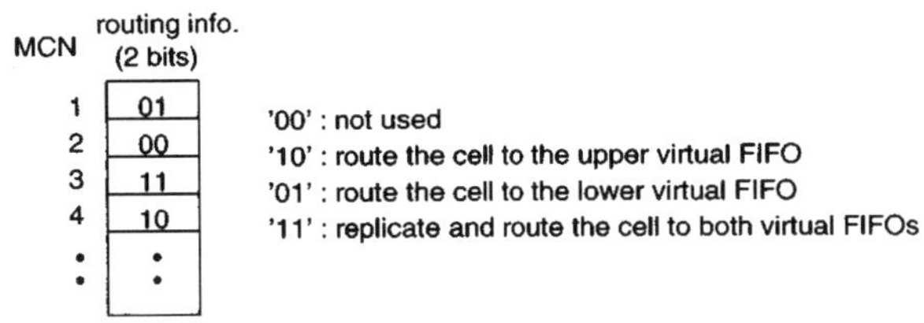

Fig. 4 Lookup table at the distribution network.

processors (MLCPs), and two virtual FIFO blocks, as shown in Fig. 2. Output links are divided into two groups of $n/2$ links each, and each output link group has a virtual FIFO block shared by all the links of a link group. The distribution module routes each incoming cell to one or both of the output link groups, according to its destination address for a unicast cell or lookup table entry for a multicast cell. IPC performs such a routing decision. Each IPC has a multicast lookup table, and the table contents of $n/2$ IPCs are updated by an MLCP. Though $n/2$ IPCs can share a single lookup table, we utilize $n/2$ separate lookup tables with identical contents to relax the timing constraints in table lookup operations. Each lookup table requires only $2B$ bits for B multicast channels.

The internal cell format of the multicast UniMIN switch is shown in Fig. 3. It has four additional fields in front of the 53-byte cell body: activity flag (AF), cell type (CT), buffer address (BA), and destination address (DA) or multicast channel number (MCN), determined by the cell type. AF is set to "1" and "0" for active cells or idle cells, respectively. CT has two-bit information CT_0CT_1, which is "00" for unicast user cells, "01" for OSM multicast control cells that carry an OSM lookup table entry of a multicast channel, "10" for multicast user cells, and "11" for distribution network (DN) multicast control cells that contain the DN lookup table contents. BA is used for temporary storage of the buffer address in which the cell should be stored inside the virtual FIFO of each stage distribution network, and it is calculated by the running adder of each stage. The fourth field has different meanings according to the cell type. If the cell is a unicast cell, it is the destination output port address used as a routing tag. For an OSM multicast control cell, it is the destined OSM address. In the case of multicast user cells and DN multicast control cells, it corresponds to the multicast channel number to which the cells belong.

IPC comprises a multicast lookup table and a 1×2 switch and some control logic circuits. The lookup table contains information on whether a multicast cell should be replicated to both output link groups, or just routed to one of two output link groups. The lookup table entry for each MCN has two bits, as shown in Fig. 4. If the routing information for an MCN is "10" or "01," the cell is routed to the upper virtual FIFO or the lower virtual FIFO, respectively. Cells are replicated out when the lookup table entry has "11." If a multicast connection is released, the entry is set to "00."

IPC performs different operations based on the cell type of the incoming cell. If CT_0 = "0," IPC routes the cell to one of the virtual FIFOs according to the routing bit of DA. If CT_0 = "1," it regards the fourth field as MCN. When CT = "10," it obtains the multicast information through a table lookup, and then performs a cell replication and routing based on the lookup table entry. When CT = "11," it sends the DN multicast control cell to the corresponding MLCP and both virtual FIFOs. The DN multicast control cell is delivered to all the MLCPs lying in the possible routing paths of the multicast channel within the distribution network. MLCP should process up to $n/2$ DN multicast control cells in a cell time. It extracts the lookup table entry of the distribution module from the control cell, and updates corresponding $n/2$ lookup tables.

The virtual FIFO is a kind of output buffer for each output link group and can accept up to n cells and transmit up to $n/2$ cells in a cell time. Though the time-multiplexed shared buffer architecture with internal speed-up n times [8] can be an alternative, the virtual FIFO architecture is employed to avoid the internal speed-up operation. Each virtual FIFO consists of a running adder (RA), a reverse banyan network, n separate FIFO buffers, and an n-to-$n/2$ selector. The n separate buffers are operated externally as a virtually single FIFO buffer by storing and transmitting cells to and from n buffers in a cyclic order. The running adder and reverse banyan perform cyclic-order cell bufferings, and the n-to-$n/2$ selector ensures cyclic cell transmissions. Such a "virtual FIFO" concept was first applied in the output buffer design of the Knockout switch to implement output buffer without internal speed-up [21], and several implementation examples can be found in [20] and [23]. More detailed descriptions on the virtual FIFO operation can be found in [17].

2.3 Multicasting Scheme

Determining a multicast addressing scheme is one of the design issues of multicast switches. One simple approach is to carry all the destination addresses in the cell header. However, this scheme requires the address information of maximum $N \log_2 n$ bits for an $N \times N$ switch. The routing header can be reduced to N bits by using bitmap addressing, in which each bit corresponds to one output port, and indicates whether the cell should be replicated or not. However, for a large N, the header size becomes impractically large. The copy network approaches have tried to avoid such a problem by placing the destination addresses in TNT tables.

The multicast UniMIN switch utilizes a *distributed lookup table scheme* as a multicast addressing method. The multicast cells carry only their multicast channel numbers, and each distribution module has two bit information for an MCN. OSMs also have multicast lookup tables for their own output ports, and perform their own internal multicast functions. Since multicast routing information is distributed to all the modules, each lookup table of distribution modules can be built with a much smaller memory size ($2B$ bits where B is the number of multicast channels of the whole switch), compared with NBW (N: switch size, W: address size) of Lee's copy network [2].

Lookup table entries are updated when a new multicast call is established, or when any changes such as addition or deletion of a leaf and call release occur in the existing multicast call. The multicast call control processor (MCCP) of the switching system constructs a multicast tree of the multicast channel, based on the principle that cell replication should take place as far away as possible from the input ports. Figure 5 a) shows a multicast tree example with a source module $DM_{1,3}$ and the destination set {2, 6, $n+3$, $2n$, $2n+2$, $2n+3$}, where $DM_{i,j}$ denotes the j-th distribution module of stage i and OSM_j the j-th OSM. From the multicast tree, information fields of DN and OSM multicast control cells can be deduced. A DN multicast control cell has the lookup table entries of all the DMs for a multicast channel, $LT_{1,1} \cdots LT_{1,k} LT_{2,1} \cdots LT_{s,k}$, where $k (= 2^s)$ is the number of modules of a stage, s the number of DN stages, and $LT_{i,j}$ the two bit lookup table

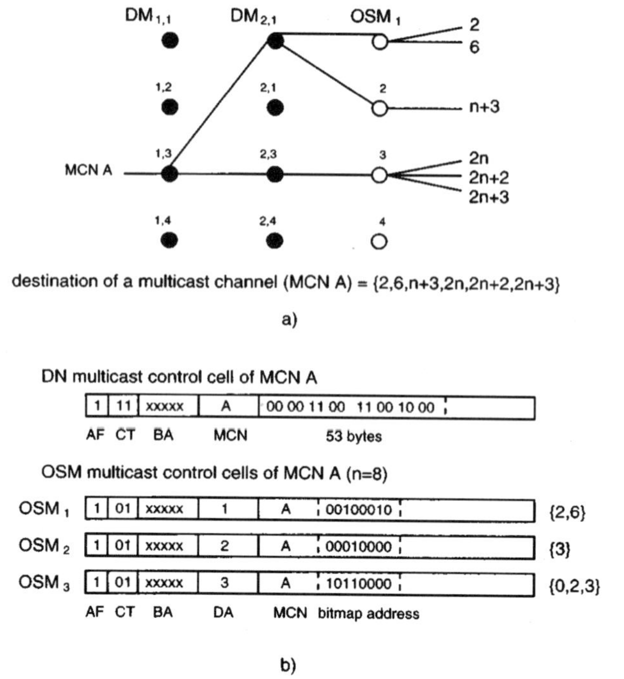

Fig. 5 a) A multicast tree of MCN A when the destination set is $\{2, 6, n+3, 2n, 2n+3, 2n+2, 2n+3\}$ in the $4n \times 4n$ multicast UniMIN swith, b) OSM and DN multicast control cells for MCN A.

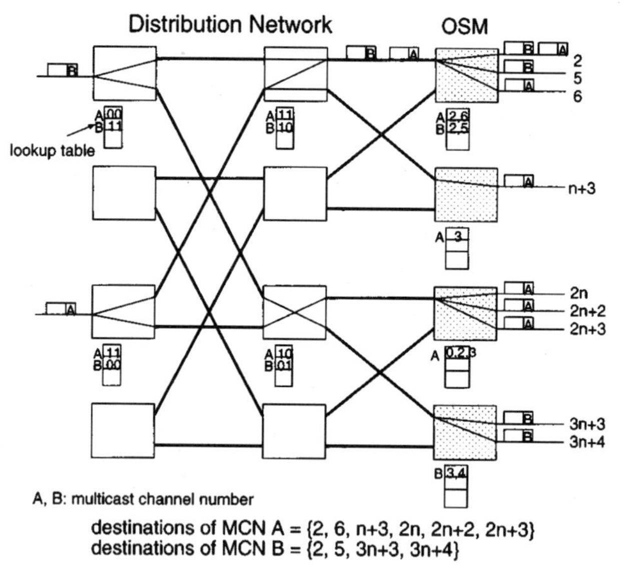

Fig. 6 Routing example of multicast cells in the $4n \times 4n$ multicast UniMIN.

entry for $DM_{i,j}$ $(1 \leqq i \leqq s, 1 \leqq j \leqq k)$. An OSM multicast control cell is generated for each OSM that contains more than one destination port, and it has two fields of MCN and an n-bit bitmap address of destination ports. For a multicast channel (MCN=A) with destination set {2, 6, $n+3$, $2n$, $2n+2$, $2n+3$}, a DN multicast control cell and three OSM multicast control cells are illustrated in Fig. 5 b). A multicast routing example in a $4n \times 4n$ multicast UniMIN switch is also given in Fig. 6.

3. Performance Evaluation of the Proposed Multicast Switch

We consider an $N \times N$ ($N = 2^s n$) UniMIN switch that consists of s stages of $n \times n$ distribution modules, and $n \times n$ OSMs of shared medium type ATM switch modules such as the ATOM switch [7]. The switching system is assumed to operate synchronously. In the first half of a time slot, incoming cells of each stage are stored at the corresponding buffers; the buffered cells are then transferred to the next stage in the next half time slot. If the virtual FIFO has no room to accept a cell, a cell loss occurs. When a multicast cell with k copy cells is lost, it is counted as k cell losses. Under the assumption that the destinations of a multicast cell are uniformly distributed to all the output ports in the symmetric architecture, the performance of a virtual FIFO represents that of the stage to which the virtual FIFO belongs, without loss of generality. Therefore, we consider the upper virtual FIFO of the uppermost distribution module of each stage.

We first analyze the distribution of the output process for the first stage distribution network. Since this is also the input process of the second stage, we can evaluate the performance of the second stage. Continuing this procedure up to the final OSM stage, we can obtain the cell loss probability and mean delay of the distribution network and the UniMIN switch. In this analysis, we assume that the incoming cells are multicast cells. However, the performance of unicast traffic can also be deduced, because the unicast can be seen as a special case of multicast (constant fanout of one), and we have confirmed that it is consistent with the analytical results for unicast traffic in [17].

3.1 Distribution Network

We assume that the arrivals of multicast cells follow a Bernoulli process with load λ, and consider three fanout models for multicast cells as follows:

i) *Constant distribution*

$$F_{const}(k) = \begin{cases} 1, & k = C \\ 0, & \text{otherwise} \end{cases} \qquad (1)$$

$$E[F_{const}] = C.$$

ii) *Uniform distribution*

$$F_{uni}(k) = \begin{cases} 1/U, & 1 \leqq k \leqq U \\ 0, & \text{otherwise} \end{cases} \qquad (2)$$

$$E[F_{uni}] = \frac{U+1}{2}.$$

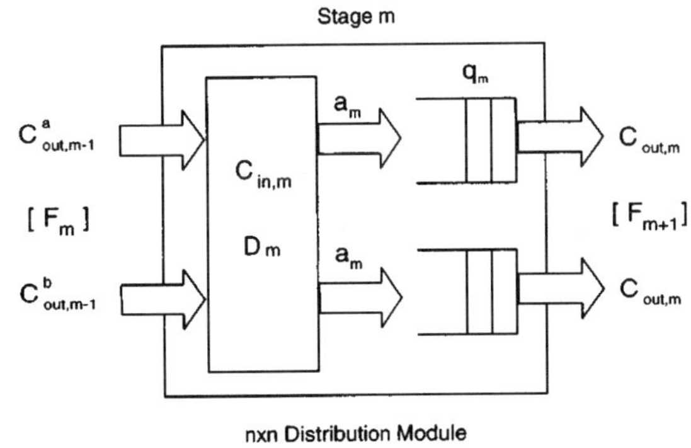

Fig. 7 Analytical model of the m-th stage module.

iii) *Truncated geometric distribution*

$$F_{tgeom}(k) = \begin{cases} \dfrac{(1-q)q^{k-1}}{1-q^N}, & 1 \leqq 1 \leqq N \\ 0, & \text{otherwise} \end{cases} \qquad (3)$$

$$E[F_{tgeom}] = \frac{Nq^{N+1} - (N+1)q^N + 1}{(1-q)(1-q^N)}.$$

For the traffic load λ of incoming master cells, the effective offered load of multicast traffic with fanout F is given by $\lambda E[F]$.

Figure 7 illustrates the analytical model of the m-th stage module. In order to derive queue length distribution of the virtual FIFO of each stage, we need to know the actual cell arrival process to the virtual FIFOs. A multicast cell carries multiple copy cells which are virtually self-routed to their destinations, and is divided into two cells at an IPC when any one of the copy cells should take different path to the other's ones. Such branching nodes are specified in the multicast tree, and therefore in the lookup tables of DMs.

We first introduce some notations related to copy cells as follows:

$F_m(k) =$ Pr{An incoming cell of stage m carries k copy cells.}

$P_m(l) =$ Pr{An incoming cell of stage m carries l copy cells destined to the upper virtual FIFO.}

$D_m =$ Pr{An incoming cell of stage m has at least one copy cell destined to the upper virtual FIFO.}

F_m is the fanout distribution of incoming cells at stage m, and P_m denotes how many copy cells in an incoming multicast cell are going to the upper virtual FIFO. $F_1(k)$ takes one of the three fanout models Eqs. (1)–(3). In the m-th stage, $F_m(k)$ is written as

$$F_m(k) = \frac{P_{m-1}(k)}{D_{m-1}}, \qquad 1 \leq k \leq \frac{N}{2^{m-1}}, \qquad m \geqq 2. \qquad (4)$$

As the destinations of copy cells in a multicast cell are uniformly distributed, $P_m(l)$ can be obtained as

$$P_m(l) = \sum_{j=l}^{N/2^{m-1}} \frac{\binom{N/2^m}{l}\binom{N/2^m}{j-l}}{\binom{N/2^{m-1}}{j}} F_m(j),$$

$$0 \le l \le \frac{N}{2^m}. \quad (5)$$

The copy cell whose destination address is one of $[0, N/2^m - 1]$ is routed to the upper virtual FIFO in the m-th stage. D_m is the probability that an incoming cell is routed or duplicated to the upper virtual FIFO, and is given by

$$D_m = \sum_{j=1}^{N/2^m} P_m(j). \quad (6)$$

Next, we derive the arrival and departure processes of the virtual FIFO. Let $C_{in,m}(k)$ denote the probability of k cell arrivals during a time slot in the m-th module, and $C_{out,m}(k)$ denote the probability that k cells depart from the m-th stage virtual FIFO during a time slot. $C_{in,m}(k)$ $(m \neq 1)$ is the sum of two output processes of the previous stage virtual FIFOs, $C^a_{out,m-1}$ and $C^b_{out,m-1}$, which are assumed to have independent, identical distributions. It is written as

$$C_{in,m}(k) = \sum_{\substack{i+j=k,\\ 0\le i,j \le \frac{n}{2}}} C_{out,m-1}(i) C_{out,m-1}(j),$$

$$0 \le k \le n,\ m \ge 2. \quad (7)$$

In the first stage, $C_{in,1}(k)$ has a binomial distribution

$$C_{in,1}(k) = \binom{n}{k} \lambda^k (1-\lambda)^{n-k}. \quad (8)$$

As the virtual FIFO has n inputs and $n/2$ outputs with a finite buffer size M, we can model it as a finite buffer multi-server queueing system. Let $A_{m,t}$ and $Q_{m,t}$ denote the number of cell arrivals during the t-th time slot and the total number of buffered cells just after the t-th time slot in the m-th stage, respectively. Then $Q_{m,t}$ is given by

$$Q_{m,t} = \min\{\max\{Q_{m,t-1} + A_{m,t-1} - n/2, 0\}, M - n/2\}, \quad (9)$$

and $Q_{1,t}$ can be modeled as a discrete time Markov chain with finite state space $\{0, 1, \ldots, M - n/2\}$. The steady state probability of k cell arrivals at the m-th stage virtual FIFO during a time slot, $a_m(k)$ is expressed as

$$a_m(k) = \sum_{l=k}^{n} \binom{l}{k} (D_m)^k (1-D_m)^{l-k} C_{in,m}(l),$$

$$0 \le k \le n. \quad (10)$$

A set of balance equations for the steady state queue length distribution at the departure epoch, $q_m(k)$, can be derived from Eq. (9). Finally, we obtain $\{q_m(k),\ 0 \le k \le M - n/2\}$ by solving the balance equations with a normalization equation $\sum_{i=0}^{M-n/2} q_m(i) = 1$.

The output process of a virtual FIFO, $C_{out,m}(k)$ can be obtained as

$$C_{out,m}(k) = \begin{cases} \displaystyle\sum_{\substack{i+j=k,\\ 0\le i<\frac{n}{2},\\ j\ge 0}} a_m(i) q_m(j) & \text{if } 0 \le k < \frac{n}{2}, \\ \displaystyle\sum_{\substack{i+j\ge\frac{n}{2},\\ 0\le i\le n,\\ 0\le j\le M-\frac{n}{2}}} a_m(i) q_m(j) & \text{if } k = \frac{n}{2}. \end{cases} \quad (11)$$

Repeating this procedure up to the final stage distribution network, we have the PDFs of queue length of all the virtual FIFOs in the distribution network. Therefore, we can obtain the mean delay and the cell loss probability in the distribution network. The delay of a copy cell at a virtual FIFO corresponds to that of a multicast cell carrying the copy cell. Hence, the mean delay at the m-th stage distribution module, W_m is given by

$$W_m = \frac{\bar{Q}_m}{\lambda'_m} + 1 = \frac{\sum_{k=1}^{M-n/2} k q_m(k)}{\sum_{k=1}^{n/2} k C_{out,m}(k)} + 1, \quad (12)$$

where $\bar{Q}_m$ is the mean queue length of the virtual FIFO, and λ'_m the actual input traffic load except for the blocked cells due to buffer overflow, which is identical to the output link load in the steady state. W_{DN} is the total delay at the s-stage distribution network, and it is written as

$$W_{DN} = \sum_{m=1}^{s} W_m. \quad (13)$$

The cell loss probability at the distribution network $P_{loss,DN}$ can be obtained as

$$P_{loss,DN} = 1 - \frac{E[C_{out,s}]E[F_s]}{E[a_1]E[F_1]}, \quad (14)$$

where $E[a_1]E[F_1]$ is the effective input load to the first stage virtual FIFO, and $E[C_{out,s}]E[F_s]$ is the effective output link group load at the final stage of the distribution network. In Eq. (14), a loss of a multicast cell with k copy cells is counted as the loss of k cells. W_{DN} and $P_{loss,DN}$ are performance indices of the distribution network of UniMIN switches, regardless of the OSM architecture.

3.2 Output Switch Module

We assume that OSM is an $n \times n$ output buffering ATM switch with output buffer size B. The fanout distribution F_{om} has the same form as F_m of the distribution network.

$$F_{om}(k) = \frac{P_s(k)}{D_s}, \qquad 1 \leq k \leq n. \qquad (15)$$

In the OSM, the copy cells of a multicast cell are routed to their destined output ports. If D^*_{om} is defined as Pr{the tagged output port is one of the destinations of the incoming cell}, it can be represented as

$$D^*_{om} = \sum_{j=1}^{n} \left[1 - \frac{\binom{n-1}{j}}{\binom{n}{j}}\right] F_{om}(j), \qquad (16)$$

and the probability of k cell arrivals to the tagged output port in a time slot, $a_{om}(k)$ is given by

$$a_{om}(k) = \sum_{l=k}^{n} \binom{l}{k} (D^*_{om})^k (1 - D^*_{om})^{l-k} C_{in,om}(l),$$
$$0 \leq k \leq n, \qquad (17)$$

where $C_{in,om}(l)$ is the probability of l cells arriving at the given OSM in a time slot, and it can be obtained from $C_{out,s}(k)$ using Eq. (11). Letting $Q_{om,t}$ and $A_{om,t}$ denote the number of cells in the tagged output buffer just after time slot t and the number of cell arrivals during time slot t, respectively, we have

$$Q_{om,t} = \min\{\max\{Q_{om,t-1} + A_{om,t-1} - 1, 0\}, B - 1\}. \qquad (18)$$

After modeling Q_{om} as a discrete time Markov chain with a finite state space $\{0, 1, \ldots, B-1\}$, we can obtain $\{q_{om}(k),\ 0 \leq k \leq B-1\}$, which is the steady state queue length distribution of the output buffer.

The mean OSM delay W_{om} is derived using Little's Law.

$$W_{om} = \frac{\bar{Q}_{om}}{\lambda'_{om}} + 1 = \frac{\sum_{k=1}^{B-1} k q_{om}(k)}{1 - q_{om}(0) a_{om}(0)} + 1, \qquad (19)$$

where $1 - q_{om}(0)a_{om}(0)$ is the output link utilization. The total cell switching delay in the UniMIN switch is

$$W_{total} = W_{DN} + W_{om} = \sum_{m=1}^{s} W_m + W_{om}. \qquad (20)$$

Finally, the cell loss probability of the UniMIN switch, $P_{loss,total}$ is written as

$$P_{loss,total} = 1 - \frac{1 - q_{om}(0) a_{om}(0)}{E[F_1]\lambda}. \qquad (21)$$

3.3 Results

We now present various numerical results of delay/cell loss analysis of the multicast UniMIN switch. Figure 8 compares the mean delay of multicast traffic with that of unicast traffic. It shows that the delay of multicast traffic is smaller than that of unicast traffic, which is a result contrary to that of the buffered copy network-based switches [3], [5]. This is because actual cell traffic in the distribution network is reduced as cell replications take place as close as possible to their destinations. However, both cases show very small delay (11 cell times for load = 0.9) in the 1024 × 1024 switch, and the delay of multicast traffic in the distribution network remains at the minimum value (5 cell times for a five-stage distribution network), even for a high traffic load of 0.95.

We investigate the effect of fanout distribution on delay characteristics in Fig. 9, and the results show that different fanout models do not incur any noticeable difference in the mean delay of multicast UniMIN switches. Therefore, the delay characteristic obtained from any one of the three fanout models can be a good estimate of multicast UniMIN switch delay performance for real multicast traffic with general fanout distributions.

The effects of mean fanout on the delay and cell

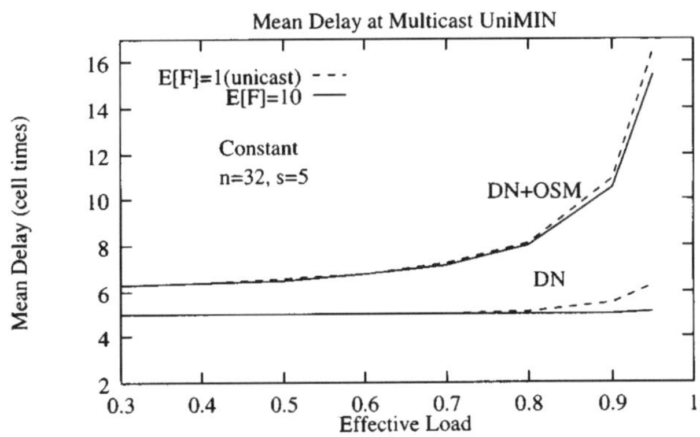

Fig. 8 Mean delay at the distribution network (DN) and whole switch (DN+OSM) of 1024×1024 UniMIN switch.

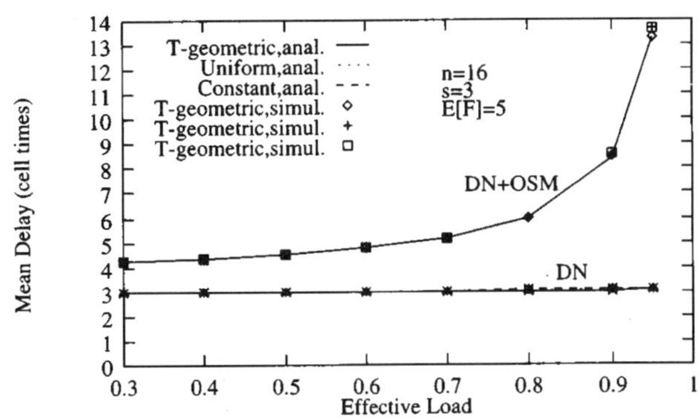

Fig. 9 Mean delay for various fanout distribution models ($n = 16$, $s = 3$, E[F] = 5).

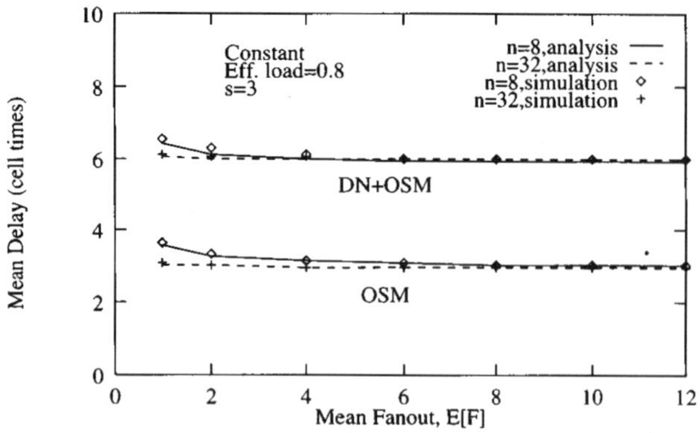

Fig. 10 Mean delay for various mean fanout (constant fanout, $s = 3$, eff. load = 0.8).

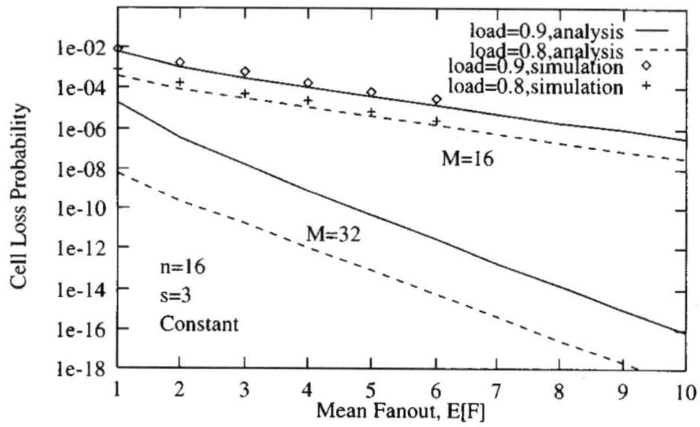

Fig. 11 Cell loss probability of distribution network in the multicast UniMIN switch (constant fanout, $s = 3$, $n = 16$).

loss performances are illustrated in Figs. 10 and 11, respectively. Many conventional multicast switches [1]–[3], [5] yield worse delay/cell loss performances for the multicast traffic of larger fanouts. However, the proposed multicast UniMIN switch yields better delay/cell loss performances for large fanouts, as shown in Figs. 10 and 11. This results from the reduction of cell arrival rate in the distribution network as the mean fanout increases.

Figure 12 shows the cell loss probability of the 1024×1024 UniMIN switch with 32×32 output buffering OSMs. Cell loss probability decreases exponentially as the output buffer size B increases. When the output buffer size B exceeds a certain value, cell loss probability remains at the constant value that is determined by the buffer size of the virtual FIFO at the distribution module. This figure also illustrates that cell loss performance for multicast traffic is better than that for unicast traffic, due to reduced actual cell traffic in the distribution network for multicast traffic. From the above results, we can expect that 1024×1024 or even larger multicast ATM switches meeting the given performance objectives in terms of cell loss and delay can be constructed by adopting the proposed UniMIN switch architecture.

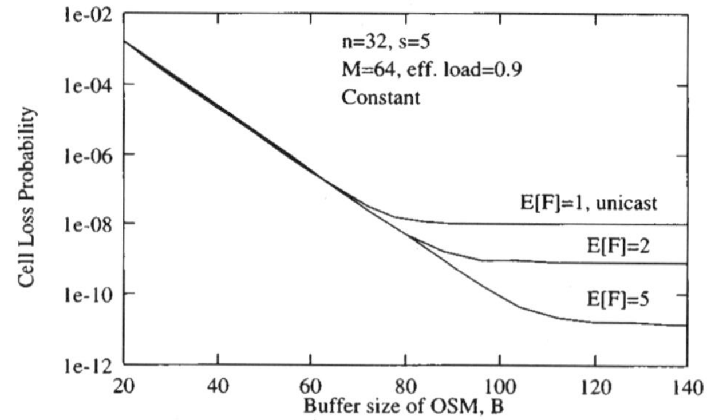

Fig. 12 Cell loss probability of the 1024×1024 multicast UniMIN switch (constant fanout, $n = 32$, $s = 5$, $M = 64$, eff. load = 0.9).

4. Conclusions

In this paper, we proposed a large-scale multicast UniMIN switch architecture. It consists of a buffered distribution network performing both cell replications and routings simultaneously and a column of OSMs. The proposed architecture does not require an additional large copy network to support multicast traffic, and does not suffer from overflow or HOL fanout blocking problems of conventional copy network architectures. The distributed lookup table scheme of the multicast UniMIN also greatly reduces the size of the single lookup table to $2B$ bits where B is the number of multicast channels. The evaluated performance for multicast traffic is even better than that of the unicast traffic, which is the opposite result compared with that of conventional copy network type multicast switches. The multicast UniMIN switch can support a multicast traffic ratio up to 100% with more improved delay/loss performances. It has an excellent multicast capability in addition to such good features as a general expansion architecture for large scale ATM switches [17], and thus is expected to be a promising solution for large-scale ATM switches for future public B-ISDN.

References

[1] J.S. Turner, "Design of a broadcast packet network," IEEE INFOCOM'86, pp.667–675, 1986.

[2] T.T. Lee, "Nonblocking copy networks for multicast packet switching," IEEE J. Select. Areas Commun., vol.6, pp.1455–1467, Dec. 1988.

[3] P. Newman and M. Doar, "A slotted ring copy network for a multicast fast packet switch," ISS'92, Yokohama, Japan, vol.2, pp.161–165, Oct. 1992.

[4] J.S. Turner, "A practical version of Lee's multicast switch architecture," IEEE Trans. Commun., vol.41, pp.1166–1169, Aug. 1993.

[5] W.D. Zhong, Y. Onozato, and J. Kaniyil, "A copy network with shared buffers for large-scale multicast ATM switching," IEEE/ACM Trans. Networking, vol.1, pp.157–165, April 1993.

[6] X. Liu and H.T. Mouftah, "A dynamic cell-splitting

copy network design for ATM multicast switching," IEEE GLOBECOM'94, San Francisco, California, USA, pp.458–462, Nov. 1994.

[7] H. Suzuki, H. Nagano, T. Suzuki, T. Takeuchi, and S. Iwasaki, "Output-buffer switch architecture for asynchronous transfer mode," IEEE ICC'89, vol.1, pp.99–103, 1989.

[8] Y. Shobatake, M. Motoyama, E. Shobatoke, T. Kamitake, S. Shimizu, M. Noda, and K. Sakaue, "A one-chip scalable ATM switch LSI employing shared buffer architecture," IEEE J. Select. Areas in Commun., vol.9, pp.1248–1254, Oct. 1991.

[9] T. Kozaki, N. Endo, Y. Sakurai, O. Matsubara, M. Mizukami, and K. Asano, "32 × 32 shared buffer type ATM switch VLSIs for B-ISDN's," IEEE J. Select. Areas in Commun., vol.9, pp.1239–1247, Oct. 1991.

[10] C. Clos, "A study of non-blocking switching networks," Bell Syst. Tech. J., vol.32, pp.406–424, March 1953.

[11] Y. Sakurai, N. Ido, S. Gohar, and N. Endo, "Large-scale ATM multistage switching network with shared buffer memory switches," IEEE Commun. Magazine, vol.29, pp.90–104, Jan. 1991.

[12] A. Itoh, W. Takahashi, H. Nagano, M. Kurisaka, and S. Iwasaki, "Practical implementation and packaging technologies for large-scale ATM switching system," IEEE J. Select. Areas Commun., vol.9, pp.1248–1254, Oct. 1991.

[13] M. Collier and T. Curran, "Path allocation in a three-stage broadband switch with intermediate channel grouping," IEEE INFOCOM'93, San Francisco, California, USA, pp.927–934, March 1993.

[14] S.C. Liew, "Multicast routing in 3-stage Clos ATM switching networks," IEEE Trans. Commun., vol.42, pp.1380–1390, Feb./March/April 1994.

[15] S.H. Byun and D.K. Sung, "A universal multistage interconnection network for large scale ATM switches," IEEE GLOBECOM'93, Houston, Texas, USA, pp.19–23, Nov. 1993.

[16] S.H. Byun and D.K. Sung, "A UniMIN switch architecture with shared output link type interconnection modules," JC-CNSS'94, Taejon, Korea, pp.263–268, July 1994.

[17] S.H. Byun and D.K. Sung, "The UniMIN switch architecture for large-scale ATM switches," Submitted to IEEE/ACM Trans. Networking.

[18] K.Y. Eng, M.J. Karol, and Y.S. Yeh, "A growable packet (ATM) switch architecture: design principles and applications," IEEE GLOBECOM'89, Dallas, Texas, USA, pp.1159–1165, Nov. 1989.

[19] K.Y. Eng, M.J. Karol, G.J. Cyr, and M.A. Pashan, "A 160-Gb/s ATM switch prototype using the concentrator-based growable switch architecture," IEEE ICC'95, Seattle, Washington, USA, pp.550–554, June 1995.

[20] H.J. Chao and B.-S. Choe, "Design and analysis of a large-scale multicast output buffered ATM switch," IEEE/ACM Trans. Networking, vol.3, no.2, pp.126–138, April 1995.

[21] K.Y. Eng, M.G. Hluchyj, and A.S. Acampora, "The Knockout switch: A simple, modular, architecture for high-performance packet switching," IEEE J. Select. Areas Commun., vol.5, pp.1274–1283, Oct. 1987.

[22] A. Pattavina, "Multichannel bandwidth allocation in a broadband packet switch," IEEE J. Select. Areas Commun., vol.SAC-9, pp.1489–1499, Dec. 1988.

[23] H.S. Kim, "Multichannel ATM switch with preserved packet sequence," IEEE ICC'92, Chicago, Michigan, USA, pp.1634–1638, June 1992.

High-Performance Backbone Network Technology edited by Naoaki Yamanaka
ISBN: 0-8247-5321-6 DOI: 10.1081/(120025678)
Published 2004 by Marcel Dekker, Inc.

A Multicast ATM Switch Based on Broadcast Buses

Ming-Huang Guo and Ruay-Shiung Chang

SUMMARY Multicasting functionality is an important criterion in judging the powerfulness and extendibility of ATM switches. Recently many multicast ATM switches have been proposed. They are mostly modified from unicasting switches. For such switches to support multicasting, they often need some additional components, e.g., copy network, and numerous modifications on the original unicasting network. Although the multicasting is supported, the method is often inefficient and the architecture is complicated and costly. In this paper, we propose a new multicast ATM switch. In the proposed architecture, the nonblocking unicast network will not be modified. All that need is a simple external nonblocking multicast module. The external nonblocking multicast module will not delay the original nonblocking unicast network module, and the cost in hardware complexity for the external nonblocking multicasting module will be $O(NK \log K)$, where N is the switch size and K is the number of broadcast buses used in the multicast module.

key words: *multicast, ATM switch, broadcast bus, nonblocking unicast network*

1. Introduction

ATM technology [16], [30], as it evolves, will offer a global linkage of voice, video, and data all running together on flexible, high speed networks. There are many applications currently developed or envisioned over ATM. For example, video-on-demand, video conferencing, distant learning, and so on. These applications all require a piece of information (a cell) to be multicast to several users. Therefore, it is very important to design an efficient multicast ATM switch.

For a multicast ATM switch, the most important issue is the replication facility [3], [4], [6], [7], [9]–[11], [13]–[15], [17], [19], [20], [22]–[29], [31], [32], [34]. In general, there are two approaches to replicate cells in a multicast ATM switch: One is using a special component for replicating [9], [13], [15], [17], [19], [27], [28], [34], and the other is combining the replication function within the routing elements [3], [4], [6], [7], [10], [11], [13], [14], [19], [20], [22]–[26], [29], [31], [32]. The former can simply the design in switch architecture, but the transmitting cycles, the time needed to transmit cells through the switch, will be longer. The latter does not have the former's disadvantage, but it will make the routing elements more complex. Whatever the method is, unicast cells have to go through the replicating and routing elements as the multicasting cells do. This will complicate the routing and cell forwarding in internal switching elements.

In this paper, we propose a simple design to separate the unicast and multicast traffic. The multicast traffic is transported using broadcast buses. The broadcast bus is used in multicast Knockout switch [13] and some of its variations [4], [6], [7], [9], [20], [31]. All output ports attached to the broadcast bus can read a cell at the same time. Therefore, routing and replicating of multicasting cells will be done at the same time, and no modification is needed in the part of unicasting networks.

But previous switches [4]–[9], [13], [20], [31], [33] that contain broadcast buses are often based on crossbar network. The hardware complexities or crosspoints in the switch are very high. The major reason for using the crossbar network is that it can deal with unicasting cells and multicasting cells in the same way at the same time. However, as said before, mixing unicast and multicast traffic is not a good approach.

The multicast Knockout switch [13] uses one $N \times N$ crossbar network to transmit the unicasting cells and one $N \times M$ crossbar network to transfer multicasting cells, where M is the number of broadcast bus used for transmitting multicast cells. Although the hardware complexity is very high, it shows a good idea for the transmission of the unicasting cells and multicasting cells. Afterwards, in [4]–[7], [9], [20], the researches build new switch architectures based on combining the Knockout switch and the multistage interconnection network (MIN). But in these switches, due to the serial combination of shared buffers, a major problem is the queueing delay in those shared buffers, and the internal blocking of MIN also occurs.

Following the ideas in [4]–[9], [13], [20], [31], [33], in this paper, we will use an MIN, instead of the $N \times N$ crossbar network, to carry the unicasting cells, and broadcast buses, instead of the $N \times M$ crossbar network, to deal with multicasting cells. By carefully designing the number of broadcast buses used, the per-

Manuscript received May 11, 1998.
Manuscript revised September 28, 1998.

†The author is with the Department of Information Management, National Taiwan University of Science and Technology, Taipei, Taiwan, ROC.

††The author is with the Department of Computer Science and Informantion Engineering, National Dong Hwa University, Hualien, Taiwan, ROC.

*This research is supported in part by NSC under contract numbers NSC86-2622-E-007-001 and NSC86-2213-E-011-056.

IEICE TRANS. COMMUN., Vol. E82-B, No. 6, pp. 915–922, June 1999

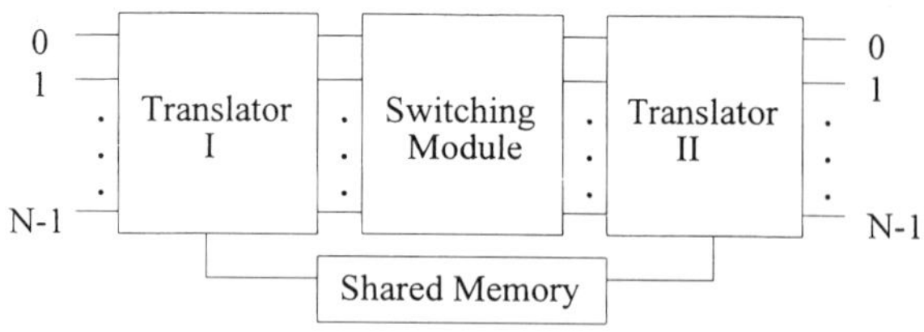

Fig. 1 Basic architecture for a multicast ATM switch.

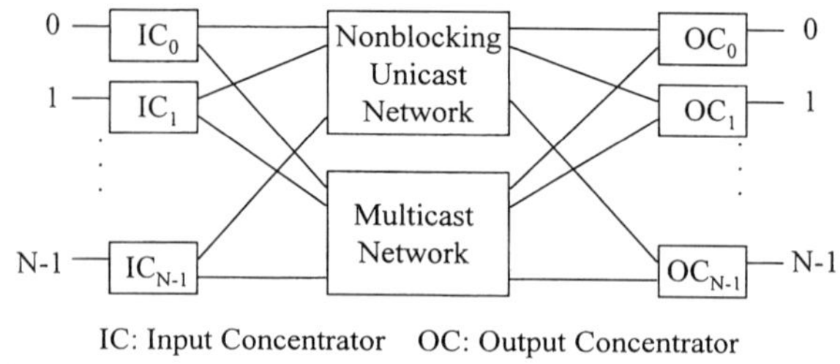

Fig. 2 Architecture for the switching module in Fig. 1.

formance of the proposed multicast switch will be the same as the nonblocking unicast networks. Because the proposed switch adopts output buffering, the transmission delay problem in [4]–[7], [20] will not occur in our switch. Furthermore, the hardware complexity will not be as costly as that in [13].

When a multicasting or unicasting cell is transmitted through the switch, the whole cells are usually processed while routing. That is, for a standard ATM cell, 53 bytes and some additional bytes used by the switch are transferred through the switch. In our design, it is not necessary to transmit the whole cell through the switch. Only bytes containing routing information are needed to travel through the switch. This will reduce the bytes in transmitting and hence will improve the performance of the switch.

The rest of this paper is organized as follows. Section 2 will introduce the switch architecture and its detailed operations. Hardware complexities and transmission time for the switch are analyzed in Sect. 3. Section 4 is the statistical analysis and simulation results. Conclusions and future research directions are discussed in Sect. 5.

2. Architecture and Operations

2.1 Basic Architecture

The basic architecture of the proposed multicast switch is depicted in Fig. 1. When a cell comes into the switch, it is stored temporarily in the shared memory first. An index to identify the cell in the shared memory will be given. Because we transfer the multicasting cells and unicasting cells in different components, the requested output ports of the multicasting cells and unicasting cells will be represented by an N-bit and a $\log N$-bit destination pattern, respectively. For a multicasting cell, if the output port i is in one of the destinations, the i-th bit is set to 1 in the pattern; otherwise, it is set to 0. But for an unicasting cell, the destination output port will be expressed by the $\log N$-bit directly. The cell storing, index assigning and destination pattern translating are done by translator I.

For example, when $N = 8$, assume a cell in input port 1 wants to multicast to output ports 0, 2, 4, and 6. If translator I uses the input port number as the index to the cell in the shared memory, then the destination pattern and index of the multicasting cell are 10101010 and 001, respectively. These $N+\log N$ ($2\log N$ for unicast cells) bits contain enough information for the switch to transmit the cell.

After the cell reaches the output port, its original contents will be retrieved from the shared memory according to the index. The translator II is used to retrieve the cell contents from the shared memory and makes the cell ready for its next hop.

2.2 Switching Module

To prevent the multicasting cells from interfering the transmission of the unicasting cells, they will be processed in a different routing module. Figure 2 shows the architecture.

In Fig. 2, the incoming cells will reach the corresponding input concentrator (IC) first. The IC examines the cell and decides where the cell should go as its next step. If the cell does not request multicasting, it will be sent to the nonblocking unicast network to transmit to its destination output port. Otherwise, it will be sent to the multicast network. After being transmitted, the cell will arrive its destination output port concentrator (OC). OC will then pass the received cell(s) to the output port.

From previous researches, there are many alternatives for the nonblocking unicast network [1], [2], [5], [8], [12], [20], [23], [32] and the multicast network [3], [4], [6], [7], [9]–[11], [13]–[15], [17], [19], [20], [22], [24]–[28], [30], [31], [33]. In this paper, we will only focus on the multicast network. In the previous multicast network fabrics proposed [3], [4], [6], [7], [9]–[11], [14], [15], [17], [19], [20], [22], [24]–[28], [30], [31], [33], they often process unicasting cells and multicasting cells in the same way. Only the multicast switch in [13] uses the different paths to transmit the unicasting cells and multicasting cells to the destination ports. This is a very good idea, and we will apply the idea in our switch.

To implement the above idea in our switch, we need a special component which serves only the multicasting cells and performs the cell replication while routing. There are some candidates [3], [4], [6], [7], [9]–[11], [13]–[15], [17], [19], [20], [22], [24]–[28], [30], [31], [33] for the required multicasting module, but we choose the simplest fabric: the broadcast bus. Under the characteristic of broadcast bus, any output port that attaches to

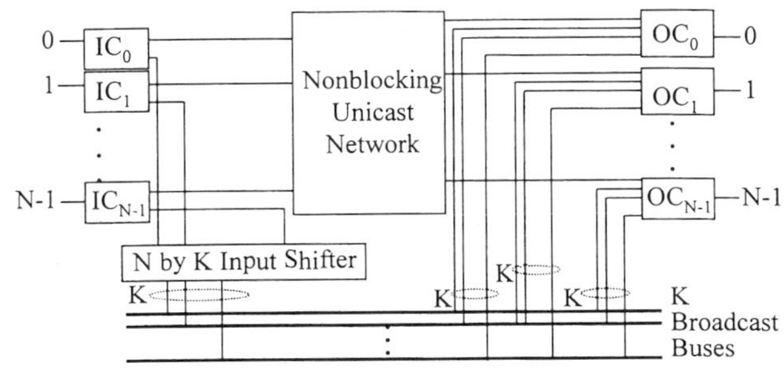

Fig. 3 The proposed switching module.

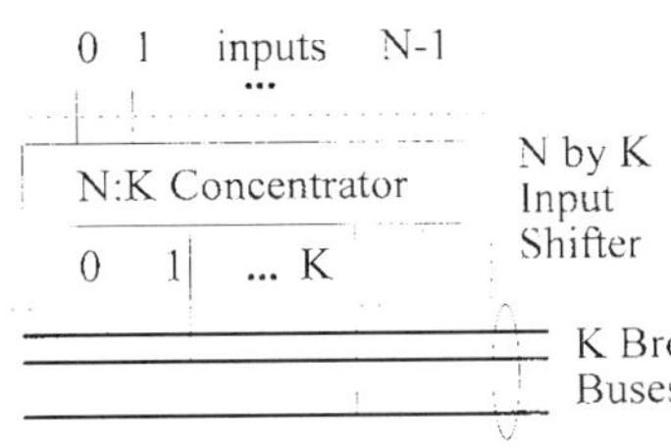

Fig. 4 The architecture of the $N \times K$ input shifter.

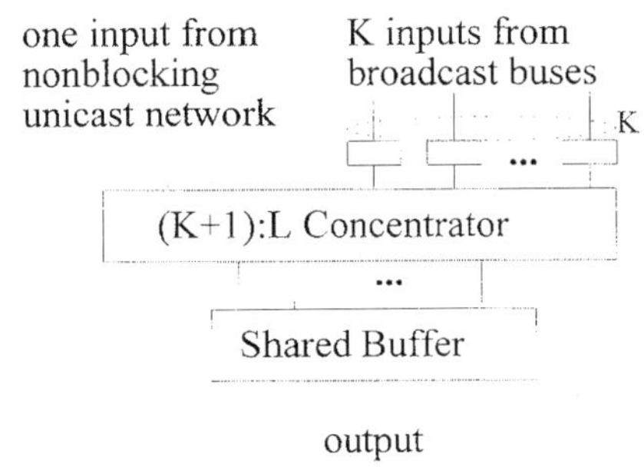

Fig. 5 The architecture of each output concentrator.

the broadcast bus can simultaneously receive a copy of the cell transmitting in the bus. Hence, the cell replication and routing can be done at the same time by the broadcast bus. Figure 3 shows the detailed architecture of the switching module.

In Fig. 3, the unicasting cells will go through the nonblocking unicast network to the requested destination output port, and the multicasting cells will be injected into the $N \times K$ input shifter, which is shown in Fig. 4. In this shifter, to prevent sending conflicts which might occur in the broadcast buses, K multicasting cells are allowed to reach the broadcast buses each time, and every bus will only deal with one multicasting cell. Hence, $\lceil \frac{N}{K} \rceil$ cycles will be needed for K broadcast buses to transmit N multicasting cells to the output ports in the worst case.

After the above operations, cells will reach requested output ports. In every output port, there may be unicasting cell and multicasting cells coming from different links. For the links from broadcast buses, every output concentrator (OC) can read the cells at the same time. The architecture of each OC is illustrated in Fig. 5. In each OC, K output filters are used to check the destination pattern of the cells coming from the broadcast buses. If the output port that the output filter attaches to is one of the destinations required, the filter will inject the cell into the concentrator. Otherwise, the filter will discard the cell. After that, the $(K+1) \times L$ concentrator selects up to L cells by the Knockout contention scheme [32] and feeds them into the shared buffer which acts as an FIFO queue with L inputs and 1 output.

One problem remained to be solved is the cell ordering problem. Since unicasting and multicasting take different paths in the switch, if the same source of data sometimes unicast and sometimes multicast, it is hard to maintain cell sequence. Therefore, we assume that when a connection begins unicasting or multicasting, it will keep doing this until the connection is closed and a new connection is built. In this way, the cell sequence can be maintained by the mechanisms originally used by the unicast module and the multicast module.

3. Complexity Analysis

3.1 Transmission Time Analysis

The total transmission time needed in the switch is the summation of processing time in every component. In the switch, the critical part of the transmission time will be either in the nonblocking unicasting network, or in the input shifter. The reason for this claim is that in these two components, they might transmit cells in several cycles if some cells request the same output port. Therefore, only the time cost in the nonblocking unicasting network and the input shifter are considered in this analysis. Time took in other components is assumed negligible.

Assume T_U and T_M represent the transmission cycles needed for one cycle of unicasting cells and one cycle of multicasting cells, respectively. Because the nonblocking unicasting network and the input shifter work in parallel, the transmission time T of the proposed switch should be the maximum of T_U and T_M, denoted by Max(T_U, T_M). First, assume that there are only unicasting cells injected into the switch. For an $N \times N$ nonblocking unicasting network, assume its transmission cycle time is t_U. In the worst case, if N transmission cycles are needed to deal with a cycle of input cells, the time needed will be Nt_U.

Next, assume that there are only multicasting input cells. In this case, the input shifter will take $\lceil \frac{N}{K} \rceil$ cycles to transmit N multicasting cells to the output ports by K broadcast buses. Assume that each cycle of the input shifter needs time t_M. The above two cases are the worst case inputs for the proposed switch, so the total transmission time of the proposed switch is

$\text{Max}(Nt_U,\, t_M\lceil \frac{N}{K} \rceil)$.

In the proposed switch, the input shifter is an $N \times K$ concentrator, and it can be constructed by an $N \times N$ multistage interconnection network. For the nonblocking unicast network based on MIN [1], [2], [12], their internal stages are always not smaller than an $N \times N$ MIN. Hence, we can assume that $t_m \leq t_u$, and

$$T = \text{Max}\left(Nt_U, t_M\left\lceil \frac{N}{K} \right\rceil\right) = Nt_U. \tag{1}$$

The complexity of the total transmission time of the proposed network, therefore, will be no worse than a nonblocking unicasting network.

3.2 Hardware Complexity Analysis

In the proposed switch fabric, there are N 1×2 input port controllers, an $N \times K$ input shifter, K broadcast buses, $N((K+1) \times L + L \times 1)$ output port controllers, and a nonblocking unicasting network. Hence, the total hardware complexity of the proposed network is the summation of the above. If Batcher-Banyan network [2] is used as the nonblocking unicast network, there are $O(\frac{1}{4}N\log N^2 + N\log N)$ 2×2 switching elements, which is $O(N\log N^2)$. Assume that the $N \times L$ concentrator in the input shifter is built by an $N \times N$ MIN, and the $(K+1) \times L$ output port concentrator is built by a $(K+1) \times (K+1)$ MIN. By counting the number of the 2×2 switching elements used in the proposed switch, the total hardware complexity of the proposed switch is

$$O\big(N + N\log N + N(K+1)\log K + 1 + N\log N^2\big). \tag{2}$$

In Eq. (2), $O(N\log N^2 + N(K+1)\log K + 1)$ is the hardware complexity of the whole multicast switch, and K is the critical number in the equation. In Sect. 4.2, we will discuss what value of K will be for the multicast module to be economical and efficient.

The Knockout switch supporting multicasting [13] has hardware complexity $O(N^2 + NM)$, where M is the number of broadcast buses used for transmitting multicasting cells. When $(K+1)\log K + 1$ is less than N, $O(N^2 + NM)$ will be larger than Eq. (2). Even for the non-multicasting Knockout switch [33], its total hardware complexity is $O(N^2)$, and it is still larger than that of our network in the same case. Besides that, counting the number of broadcast buses used in the switch, there are $N + M$ and N broadcast buses in [13] and [33], respectively. But in our switch, there are only K broadcast buses, and as explained later, K will be smaller than N.

According to the above reasoning, the transmission time of the proposed network is in the same degree as that of the nonblocking unicasting network adopted. When K is not greater than $\log N^2$, the hardware complexity of the proposed multicast module will become $O(N\log N^2 + 2N\log N^2 \log\log N)$. In this case, the total hardware complexity of the proposed switch will be $O(N\log N^2 \log\log N)$, and it is the same in hardware complexity as the Batcher-Banyan network.

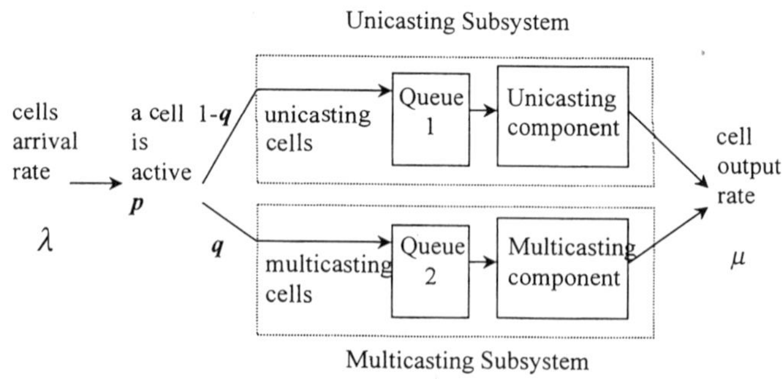

Fig. 6 System processing diagram with respective parameters for the proposed multicast switch.

4. Performance Analysis and Simulation Results

4.1 Queueing Analysis

Assume that the cells arriving process for the proposed system is a Poisson process and the system service time is exponentially distributed. Because there are two different components in the proposed multicast ATM switch, we can use two M/M/1 queues to simulate the performance of the proposed switch. The system architecture with the statistical parameters is depicted in Fig. 6.

In Fig. 6, λ is the cells arrival rate for the whole system, p is the probability that an incoming cell is active, q is the probability that an active cell requires multicasting, and μ_U and μ_M are the service rates of the unicasting component and multicasting component, respectively. Let the actual cells arrival rate for the unicasting component be λ_U, and for the multicasting component λ_M. In the following analysis, assume p is 1, and λ_U is $\lambda(1-q)$, and λ_M is λq. Next, we define the following terms:

$\lambda(\lambda_U, \lambda_M)$ = the steady-state arrival rate of cells in the system (nonblocking unicast network, multicasting module),

$\mu(\mu_U, \mu_M)$ = steady-stage departure rate of cells in the system (nonblocking unicast network, multicasting module),

$N(N_U, N_M)$ = the average number of cells in the system (unicasting component, multicasting component),

$W(W_U, W_M)$ = the average cells waiting time in the system (unicasting component, multicasting component),

$NQ(NQ_U, NQ_M)$ = the average number of cells waiting in the queue (unicasting component, multicast-

$WQ(WQ_U, WQ_M)$ = the average cells waiting time in the queue (unicasting component, multicasting component).

By [21], the values for the above variables can be derived as:

$$N = N_U + N_M = \frac{\lambda_U}{\mu_U - \lambda_U} + \frac{\lambda_M}{\mu_M - \lambda_M},$$

$$\begin{aligned} W &= \max(W_U, W_M) = \max\left(\frac{N_U}{\lambda_U}, \frac{N_M}{\lambda_M}\right) \\ &= \max\left(\frac{1}{\mu_U - \lambda_U}, \frac{1}{\mu_M - \lambda_M}\right), \end{aligned}$$

$$\begin{aligned} WQ &= \max(WQ_U, WQ_M) \\ &= \max\left(W_U - \frac{1}{\mu_U}, W_M - \frac{1}{\mu_M}\right) \\ &= \max\left(\frac{1}{\mu_U - \lambda_U}, \frac{1}{\mu_M - \lambda_M}\right), \end{aligned}$$

$$\begin{aligned} NQ &= NQ_U + NQ_M = \lambda_U WQ_U + \lambda_M WQ_M \\ &= \frac{\lambda_U}{\mu_U - \lambda_U} + \frac{\lambda_M}{\mu_M - \lambda_M} - \frac{\lambda_M}{\mu_M}, \end{aligned}$$

and

$$\lambda = [\lambda p(1-q) + \lambda pq]_{p=1} = \lambda(1-q) + \lambda q,$$

$$\mu = \mu_U + \mu_M.$$

4.2 Number of Broadcast Buses Needed

It is well-known that for an $N \times N$ crossbar network, it has the best transmission time O(1), but the worst hardware complexity $O(N^2)$. To decrease the cross-points in the crossbar network, one solution is partitioning the output ports into multiple output port groups [4]–[7], [20]. As shown in Fig. 7, m output ports are grouped into an output port group, and there are $\frac{N}{m}$ output port groups in the $N \times N$ switch. In this way, the crosspoints in the switch can be reduced to $O(\frac{N^2}{m})$ [7]. However, each output port group needs some additional time to realize which output port in the group is the destination of the incoming cells, and this increases transmission time.

In our switch, because the fabric between the multicasting module and the output ports is very similar to crossbar network, there is also a trade-off problem in our switch. However, the trade-off here is the number of broadcast buses used. To find out the best combination, we run simulations to test the performance. To simply the simulation, we only find the combination when the switch size is 128.

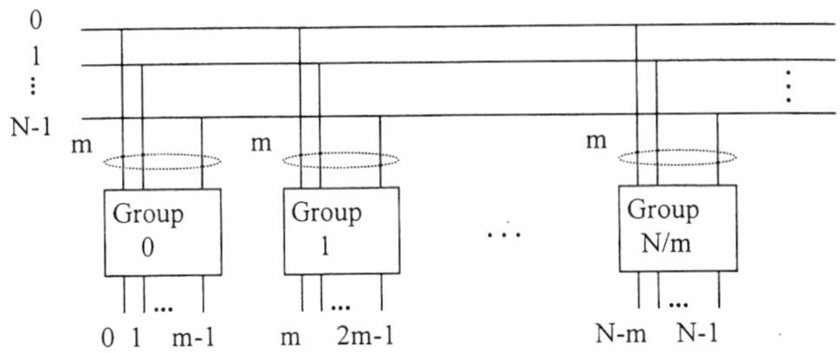

Fig. 7 Partitioning output ports into multiple output port groups.

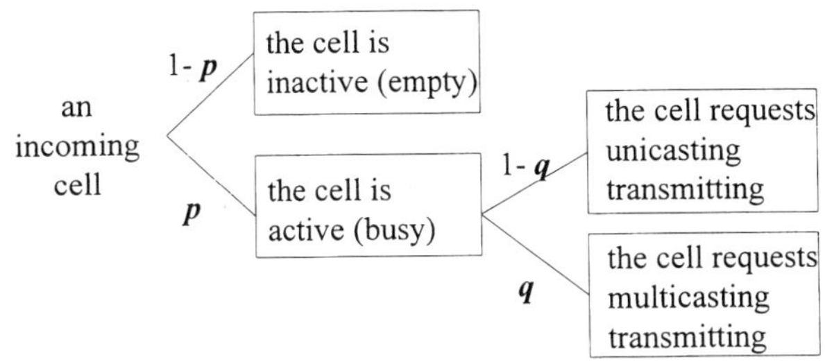

Fig. 8 The decision tree used in the simulation.

In the simulation, there are three important parameters: the cast ratio, the timeout ratio, and the utilization factor ρ. The cast ratio is the probability that an active cell requests multicasting, the timeout ratio is defined as the number of cycles allowed to transmit one cycle of input cells, and the utilization factor ρ is $\frac{\lambda}{\mu}$, where λ and μ are the same as described in Sect. 4.1. According to [21], ρ can be viewed as a measure of the average use of the service facility. When a working system is in the steady state, $0 < \rho < 1$. In the simulations, the cast ratio is assigned 0.1, 0.5, and 0.9, the timeout ratio is assigned 1, 2, and 4, and the utilization factor is varied from 0.1 to 0.9 with step 0.1. From the simulations with varying parameters, we can find out the variations in the number of broadcast buses needed to provide a no cell loss transmitting. In the simulation, we use Batcher-Banyan network as the nonblocking unicast network.

The input traffic pattern used in the simulations is described as follows. Let A be the event that an incoming cell is active or inactive, and let B be the event that an active cell requests unicasting or multicasting transmitting. In the simulation, assume Pro(A = active) = p, Pro(B = multicasting transmitting) = q and these two events are all Bernoulli processes. Figure 8 depicts the relationship of these two events and the decision tree used in the simulation. Moreover, assume every output port in the switch has the same probability of being chosen.

In Table 1, we show the number of broadcast buses needed for switch size $N = 128$ and $p = 1$ under the given cast ratio and timeout ratio. When the timeout ratio is 1, all cells belong to the same transmission cycle

Table 1 Broadcast bus needed for switch size $N = 128$ and $p = 1$ under the given q and timeout ratio.

Cast ratio	Timeout ratio	utilization factor								
		0.1	0.2	0.3	0.4	0.5	0.6	0.7	0.8	0.9
0.1	1	1	2	4	4	4	8	8	8	8
0.1	2	1	1	1	1	1	2	4	4	4
0.1	4	1	1	1	1	1	1	1	1	1
0.5	1	4	8	16	16	16	32	32	32	32
0.5	2	2	4	4	8	8	8	16	16	16
0.5	4	1	1	2	2	2	4	4	4	4
0.9	1	16	32	32	32	32	64	64	64	64
0.9	2	8	8	16	16	16	32	32	32	32
0.9	4	2	4	8	8	8	8	8	16	16

are not allowed to be transmitted in the next cycle. But when the timeout ratio is k, $k > 1$, all cells belong to the same transmission cycle are allowed to be transmitted in k cycles.

According to Table 1, when timeout ratio is 1, the number of broadcast buses, K, will exceed $\log N^2$, where N is the switch size. But when timeout ratio is greater than 1, K is smaller than $\log N^2$, and from Sect. 3.2 the hardware complexity of the whole multicast switch will be in the same level as a Batcher-Banyan network.

Figures 9 to 11 show the average queue size in each broadcast bus adopted with given cast ratio and timeout ratio under ρ factor from 0.1 to 0.9. Intuitively, the average queue size in each bus should increase when ρ increases. According to above three figures, when the buses used in each point on the curves in Figs. 9 to 11 are the same, the previous "intuition" is true. But actually in the above three figures, the number of buses used in each point on the curves is from to Table 1, and are not the same. Furthermore, when ρ is increased, because the number of buses used is increased to prevent cells loss at the same time, the corresponding average queue size in each bus might not increase. In Table 2, we show the total average queue size for a given cast ratio and timeout ratio under different ρ from 0.1 to 0.9 when switch size is 128. In the table, (0.1, 1) in the first column means the cast ratio, q, is 0.1 and the timeout ratio is 1. The value in the first column is ρ.

From Table 2, the maximum average queue size among the entire table is close to 161. The corresponding simulation parameters for that value are $q = 0.9$, timeout ratio $= 4$, ρ is 0.9, no cell loss, and 16 broadcast buses are used, according to Table 1. This means that when there are 90% multicasting cells in the arrival traffic, timeout ratio is 4, and ρ is 0.9, 16 broadcast buses with total average queue size 161 are needed in the proposed multicast module to provide no cell loss transmitting. By reducing the timeout ratio to 2, according to Table 1 and Table 2, 32 buses with total average queue size 55 are needed for no cell loss transmitting. In this case, the proposed multicast switch has the same level in hardware complexity as Batcher-Banyan network ($N = 128$, $\log N = 7$, $\log N^2 = 49 > 32$), but the total buffer needed in the proposed multicast module is less than 64, half of switch size.

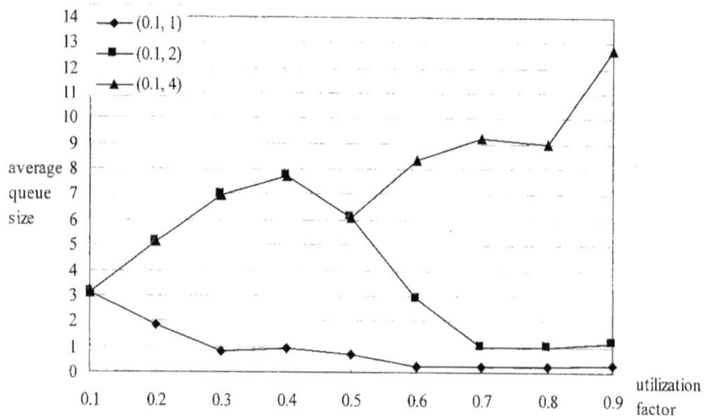

Fig. 9 Average queue size with ($q = 0.1$, timeout ratio) for different utilization factor.

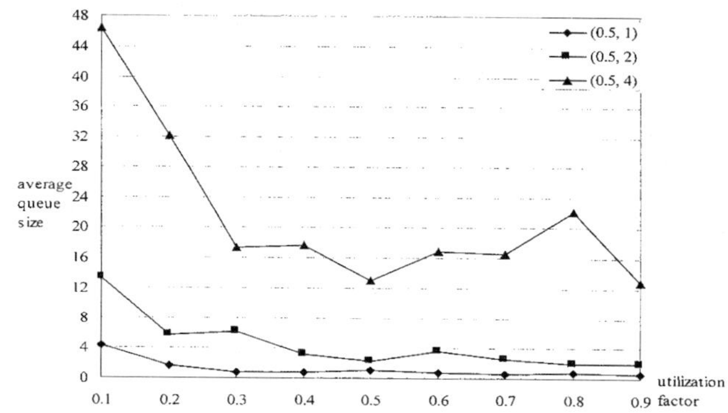

Fig. 10 Average queue size with ($q = 0.5$, timeout ratio) for different utilization factor.

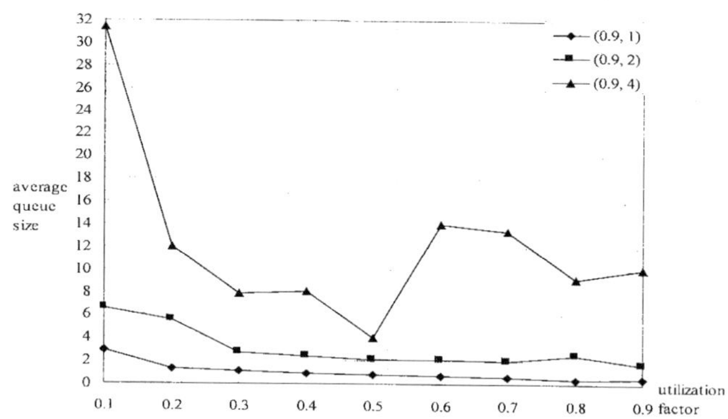

Fig. 11 Average queue size with ($q = 0.9$, timeout ratio) for different utilization factor.

5. Conclusions

In this paper, a new prototype for multicast ATM switches is proposed. It divides the whole switch into two sub-networks: one is the nonblocking unicast network, and the other is the multicast network. To simply

Table 2 Total average queue size with (q, timeout ratio) for different utilization factor from 0.1 to 0.9.

utilization factor	(q, timeout ratio)								
	(0.1, 1)	(0.1, 2)	(0.1, 4)	(0.5, 1)	(0.5, 2)	(0.5, 4)	(0.9, 1)	(0.9, 2)	(0.9, 4)
0.1	3.10027	3.10027	3.10027	17.34785	26.51508	46.39252	45.99624	52.30676	62.73081
0.2	3.74358	5.10712	5.10712	13.61330	22.95080	32.20919	40.77089	44.30432	47.95288
0.3	3.17916	6.97022	6.97022	13.11073	24.63390	34.72897	35.54543	42.92361	62.86678
0.4	3.78548	7.69765	7.69765	12.82516	25.61305	35.22335	30.32859	39.32481	64.61079
0.5	2.69380	6.10320	6.10420	16.31880	17.75800	26.10388	25.60640	33.41775	32.36285
0.6	1.63845	5.74641	8.35278	22.63351	28.68089	67.43371	42.76058	68.71416	112.20190
0.7	1.62317	4.00833	9.22218	19.24270	39.82128	66.33402	39.19458	64.13894	107.02652
0.8	1.64540	3.93568	9.01547	21.78983	30.55658	88.59522	26.13050	81.93359	145.92312
0.9	2.25509	4.71547	12.73679	20.33227	30.59926	51.53501	29.10857	54.66839	160.66941

the architecture, we use a broadcast bus as the multicast network. It can replicate and route the multicasting cells in the switch at the same time. For a cell to route in the switch, only index for identifying itself in the shared memory and its destination pattern are used and transmitted, rather than the whole cell contents. Although switching in this way will need larger memories for cells content storing, switch routing will be easier and transmission time can be decreased. The transmission time of the proposed switch is comparable to the Batcher-Banyan network. The hardware complexity of whole multicast switch can be made to be the same as Batcher-Banyan network, too. Moreover, when there are 90% multicasting cells in the arrival traffic, utilization factor is 0.9, timeout ratio is 2, and switch size is 128, only 32 buses with total buffer size 55 are needed to provide no cell loss transmitting. Therefore, our switch not only supports multicasting transmission, but also has the same complexity in time and hardware as the Batcher-Banyan network, and does not needed very larger buffer size.

To design an efficient multicast ATM switch, the trade-off between transmission time and hardware complexity of the switch is the basic design consideration. Another interesting research direction is the fault tolerant capabilities of multicast switches. That is, if some of the components failed, would its multicasting function be impaired? Would it still support unicasts?

References

[1] S. Bassi, M. Decina, P. Giacomazzi, and A. Pattavina, "Multistage shuffle networks with shortest path and deflection routing for high performance ATM switching: The open-loop shuffleout," IEEE Trans. Commun., vol.42, pp.2881–2888, Oct. 1994.

[2] K.E. Batcher, "Sorting networks and their applications," AFIPS Proc. Spring Joint Comput. Conf., pp.307–314, 1968.

[3] J.W. Byun and T.T. Lee, "The design and analysis of an ATM multicast switch with adaptive traffic controller," IEEE/ACM Trans. Networking, vol.2, no.3, pp.288–298, June 1994.

[4] K.S. Chan, S.Chan, K.L. Yeung, K.T. Ko, and E.W.M. Wong, "Clos-Knockout: A large-scale modular multicast ATM switch," IEEE GLOBECOM, pp.1358–1362, 1996.

[5] H.J. Chao, "A recursive modular terabit/second ATM switch," IEEE J. Sel. Areas Commun., vol.9, no.8, pp.1161–1172, Oct. 1991.

[6] H.J. Chao and B.-S. Choe, "Design and analysis of a large-scale multicast output buffered ATM switch," IEEE/ACM Trans. Networking, vol.3, no.2, pp.126–138, April 1995.

[7] H.J. Chao, B.S. Choe, J.S. Park, and N.Uzun, "Design and implementation of Abacus switch: A scalable multicast ATM switch," IEEE J. Sel. Areas Commun., vol.15, no.5, pp.830–843, June 1997.

[8] D.X. Chen and J. W. Mark, "SCOQ: A fast packet switch with shared concentration and output queueing," IEEE/ACM Trans. Networking, vol.1, pp.142–151, Feb. 1993.

[9] D.X. Chen and J.W. Mark, "Multicasting in the SCOQ switch," IEEE INFOCOM, pp.290–297, 1994.

[10] X. Chen and V. Kumer, "Multicast routing in self-routing multistage networks," IEEE INFOCOM, pp.306–314, 1994.

[11] R. Cusani and F. Sestini, "A recursive multistage structure for multicast ATM switching," IEEE INFOCOM, pp.1289–1295, 1991.

[12] M. Decina, P. Giacomazzi, and A. Pattavina, "Multistage shuffle networks with shortest path and deflection routing for high performance ATM switching: The closed-loop shuffleout," IEEE Trans. Commun., vol.42, pp.3034–3044, Nov. 1994.

[13] K.Y. Eng, M.G. Hluchyj, and Y.S. Yeh, "Multicast and broadcast services in a knockout packet switch," IEEE INFOCOM, pp.29–34, 1988.

[14] J.N. Giacopelli, J.J. Hickey, W.S. Marcus, W.D. Sincoskie, and M. Littlewood, "Sunshine: A high-performance self-routing broadband packet switch architecture," IEEE, J. Sel. Areas Commun., vol.9, no.8, pp.1289–1298, Oct. 1991.

[15] M.H. Guo and R.S. Chang, "Multicast scheduling for ATM switches," Proc. National Computer Symposium, vol.1, pp.251–258, Dec. 1995.

[16] R. Handel, M. N. Huber, and S. Schroder, "ATM Networks: Concepts, Protocols, Applications," Addison-Wesly, 1994.

[17] A. Huang and S. Knauer, "Starlite: A wideband digital switch," IEEE GLOBECOM, pp.121–125, 1984.

[18] J.Y. Hui, "Switching and traffic theory for integrated broadband networks," Norwell, Kluwer, MA, 1990.

[19] H.S. Kim, "Design and performance of multinet switch: A multistage ATM switch architecture with partially shared buffers," IEEE/ACM Trans. Networking, vol.2, pp.571–580, Dec. 1994.

[20] Y.M. Kim and K.Y. Lee, "KSMIN's: Knockout switch-based multistage interconnection networks for high-speed packet switching," IEEE Trans. Commun., vol.43, no.8, pp.2391–2398, Aug. 1995.

[21] L. Kleinrock, "Queueing systems volume 1: Theory," Wiley-Interscience, 1975.

[22] T.T. Lee, "Nonblocking copy networks for multicast packet switching," IEEE J. Sel. Areas Commun., vol.6, no.9, pp.1455–1467, Dec. 1988.

[23] T.T. Lee, R. Boorstyn, and E. Arthurs, "The architecture of a multicast broadband packet switch," IEEE INFOCOM, pp.1–8, 1988.

[24] T.T. Lee and S.C. Liew, "Broadband packet switches based on dilated inter-connection networks," IEEE Trans. Commun., vol.42, pp.732–744, Feb. 1994.

[25] D.J. Marchok, C.E. Rohrs, and R.M. Schafer, "Multicasting in a growable packet (ATM) switch," IEEE INFOCOM, pp.850–858, 1991.

[26] S. Shimamoto, W.D. Zhong, Y. Onozato, and J. Kaniyil, "Recursive copy networks for large multicast ATM switches," IEICE Trans. Commun., vol.E75-B, no.11, pp.1208–1219, Nov. 1992.

[27] P.U. Tagle and N.K. Sharma, "Multicast packet switch based on dilated network," IEEE GLOBECOM, pp.849–

853, 1996.

[28] J.S. Turner, "Design of a broadcast packet switching network," IEEE Trans. Commun., vol.36, pp.734–743, June 1988.

[29] J.S. Turner, "A practical version of Lee's multicast switch architecture," IEEE Trans. Commun., vol.4, pp.1166–1169, Aug. 1993.

[30] R. Vetter, "ATM concepts, architectures, and protocols," CACM, vol.38, no.2, p.109, 1995.

[31] R.J.F. Vries, "ATM multicast connections using the GAUSS switch," IEEE GLOBECOM, pp.211–217, 1990.

[32] Y. Yang and G.M. Masson, "Broadcast ring sandwich networks," IEEE Trans. Computers, vol.44, pp.1169–1180, Oct. 1995.

[33] Y.S. Yeh, M.G. Hluchyj, and A.S. Acampora, "The Knockout switch: A simple, modular architecture for high-performance packet switching," IEEE, J. Sel. Areas Commun., vol.SAC-5, pp.1274–1283, Oct. 1987.

[34] W.D. Zhong and K. Yukimatsu, "Design requirements and architectures for multicast ATM switching," IEICE Trans. Commun., vol.E77-B, no.11, pp.1420–1428, Nov. 1994.

High-Performance Backbone Network Technology edited by Naoaki Yamanaka
ISBN: 0-8247-5321-6 DOI: 10.1081/(120025679)
Published 2004 by Marcel Dekker, Inc.

Hybrid Transmission Scheme on HIPPI-ATM Connection and Its Performance

Hideki Tode, Shuhei Takimoto, and Hiromasa Ikeda

SUMMARY For the realization of the very high-speed data transmission over wide area network, the HIPPI-ATM conversion for connecting two remote HIPPI end systems is specified by ANSI. Since the credit scheme is adopted for the flow control on the ATM part in the specification, the propagation delay influences the remarkable transfer throughput degradation with the longer distance between end systems. In this paper, two priority types of HIPPI bursts are proposed, which include the flag indicating the priority, high priority or low priority. We propose the hybrid scheme that the credit scheme is adopted only for sending the high priority HIPPI bursts and the source can send low priority HIPPI bursts without reserving buffers in the receiver. In order to confirm the effectiveness of the proposed scheme, we show the characteristics of transfer performance and low priority bursts loss ratio v.s. transmission distance by simulation. Furthermore, we introduce the implemental example of the proposed scheme in the MPEG video transmission application, and show it's good performance.

key words: *HIPPI-ATM conversion, hybrid flow control, credit scheme*

1. Introduction

HIPPI (High-Performance Parallel Interface) is the very-high speed (800 Mbps/1.6 Gbps) parallel interface standardized in ANSI (American National Standard for Information Systems) [1], [2]. HIPPI, inherently, was assumed to be for the interface between a supercomputer and peripherals as well as the Fiber Channel [3]. Accordingly, the connection distance was limited to 25 m at the maximum (currently, 50 m). The serial-HIPPI can extend the connection distance up to 10 km by using a pair of fiber optical cables [4] instead of the parallel cables. However, the serial-HIPPI will not be applied to longer distance communication than 10 km. Furthermore, it is an implementation agreement, not an ANSI standard.

While, recently, the standardization for applying this existing high-speed HIPPI interface to the long-distance communication over 10 km is recently discussed in many organizations, as one of desiring technologies [5] for the realization of wide area applications requiring very large volume of high-quality and high-resolution video and image data transmission, such as broadcasts, movies and medical applications, and furthermore, requiring the transmission of excessively large data files in science field studies, such as meteorological studies, seismological observations, etc.

In particular, as one of most effective solutions to overcome the distance limitation, the technology of tunneling through ATM (Asynchronous Transfer Mode) network between remotely located high-speed terminals (computers) employing HIPPI interfaces is considered. Currently, ANSI has defined specifications for a HIPPI-ATM conversion method [6]. In these specifications, credit-based flow control scheme is adopted as a flow control on the ATM network. It is reported that this credit scheme also leads to the throughput fairness of the multiplexed HIPPI connections in the ATM network in addition to the essential flow control function [7].

As another alternative, it is considered to use commercial available IP routers, e.g. the GigaRouter from NetStar, Inc. that has both HIPPI and ATM ports instead of HIPPI-ATM converter to extend the HIPPI's connectivity [8], [9]. In Ref. [9], it is reported that both HIPPI-ATM conversion and IP routing can attain a high degree of bandwidth utilization in the OC-3 (135.632 Mbps) link transmission experiment. However, in the near future, more higher speed transmission such as Gbps order will be required in several applications. In general, IP routing must forward data packets between HIPPI networks and other networks at the network layer, while HIPPI-ATM converter can forward them at the physical layer. Therefore, for such more higher transmission requirement, HIPPI-ATM conversion will have higher performance potential than IP router. Accordingly, in this paper, we study the HIPPI-ATM conversion defined by ANSI as an distance extention method of the HIPPI connection.

On the other hand, as for the performance evaluation of the HIPPI-ATM conversion, it is confirmed in [10], [11] that when long-distance data transmission beyond 100 km (0.5 ms one-way propagation delay when 5 μs/km) via a wide area network is executed under the conventional credit scheme, the throughput is rapidly degraded. This is because the receive waiting time in the receiving end system increases with the larger propagation delay of the backward control signal to notify

Manuscript received Feburary 12, 1998.
Manuscript revised May 22, 1998.

†The author is with the Department of Information Systems Engineering, Osaka University, Suita-shi, 565–0871 Japan.
††The authors are with the Department of Communications Engineering, Osaka University, Suita-shi, 565–0871 Japan.

IEICE TRANS. COMMUN., Vol. E81-B, No. 10, pp. 1859–1867, October 1998

the receiving side's buffer status to the sending side, and in addition, the credit size corresponding to the size of receiving buffer with the very high speed access ability will be limited because of the realization of efficient congestion avoidance and the hardware cost problem [12]. In practice, it is reported that in a certain HIPPI terminal adapter, the initial credit size was set to 63 [13].

In this paper, so as to solve this throughput degradation problem in the high speed data transmission using the HIPPI-ATM conversion defined by ANSI, the hybrid transmission scheme using the intra-traffic priority class is proposed and the system configuration for the realization of the proposed scheme is presented. Proposed scheme adds the priority flag (High/Low) on every "HIPPI burst" of a certain connection which is the HIPPI minimum transmission unit, according to the indication from the upper layer. As for the ATM-part flow control for high-reliability traffic (high-priority traffic), the conventional credit scheme is adopted. While, for the loss tolerant traffic (low-priority traffic), the data are transmitted without taking account of the receiving side's buffer. However, if, on a certain connection, the low priority traffic can be interleaved among the high-priority traffic, the improvement effect of the loss performance can be expected in this proposed scheme.

In the paper, this scheme is evaluated by the computer simulation comparing to the credit scheme defined by ANSI, from the viewpoint of the relation between the transmission performance and the distance, and low-priority HIPPI burst loss probability on condition that the processing speed of the receiving end system reduces.

2. HIPPI-ATM Conversion

Figure 1 shows the protocol stack of the HIPPI-ATM conversion. In the HIPPI-ATM conversion specified by ANSI, HIPPI Protocol Data Unit (H-PDU) is configured by appending the header information called HB_Header into the HIPPI burst. Then, the H-PDU is mapped into the CPCS-PDU in AAL 5.

2.1 Ready Pulse Scheme in HIPPI Part

As for the flow control, HIPPI part and ATM part are independently controlled. As HIPPI-part flow control, the READY pulse scheme is adopted, that is the conventional flow control in the HIPPI layer (HIPPI-PH). In the READY pulse scheme, immediately after the connection establishment in the HIPPI-PH, receiving end of the HIPPI part sends the READY pulses to the sending end of the HIPPI part, whose number corresponds to that of receive-ready HIPPI burst, i.e, the receiving end's buffer size. The sending end can transmit the HIPPI burst according to the number of received pulses. Also, whenever the processing of a received HIPPI burst finishes and as a result, one buffer space becomes empty, the receiving end sends one READY pulse to the sending end. By this procedure, the READY pulse scheme can avoid the receiving end's buffer overflow.

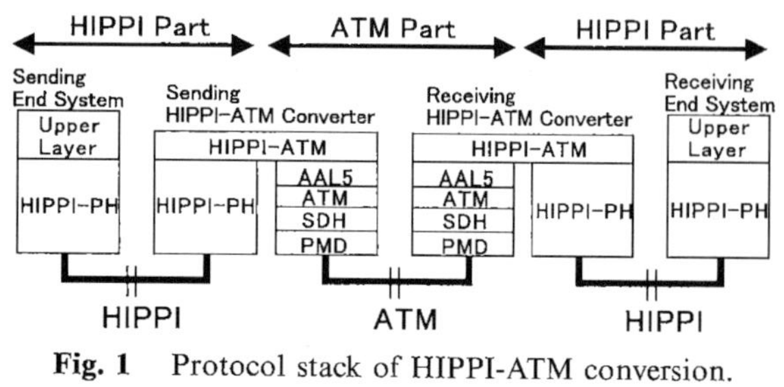

Fig. 1 Protocol stack of HIPPI-ATM conversion.

2.2 Conventional Flow Control Scheme in ATM Part

In the standardization by ANSI, the credit scheme, a kind of window flow control [14] is adopted as the ATM-part flow control. In the credit scheme the credits are returned from the receiving converter to the sending one once in a certain time period called credit period. The sending converter can be permitted to transmit the H-PDUs corresponding to the credit size at most within a credit period.

Flow control sequence diagram between HIPPI-ATM converters is illustrated in Fig. 2.

The sending-side HIPPI-ATM converter (sending converter) has a counter (Credit Count) and a register (Old Credit) and can send as many H-PDUs as the value of the Credit Count. While, the receiving-side HIPPI-ATM converter (receiving converter) has a counter (New Credit) and a register (Initial Credit).

Basic procedure of the credit scheme is as follows [10], [11].

(1) The sending converter initially sets the Credit Count to 0 and sends the H-PDU with the RESET bit in the HB_Header set to 1, to the receiving converter.

(2) When the receiving converter receives the H-PDU,

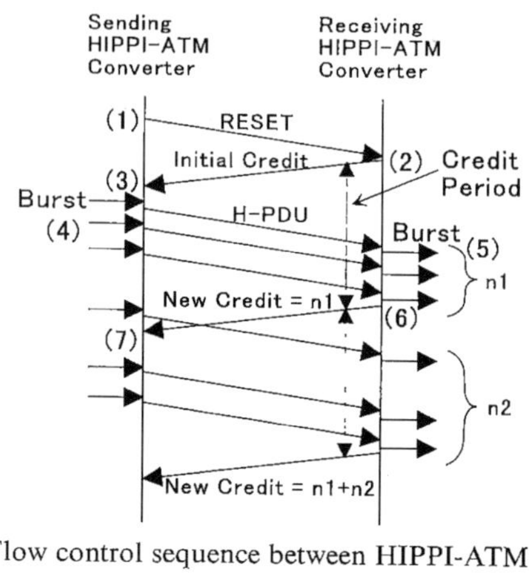

Fig. 2 Flow control sequence between HIPPI-ATM converters.

it initializes the receiving buffer and its state and sets the New Credit to 0. Furthermore, the receiving converter sets the Initial Credit to the number of H-PDUs which can be stored in the receiving buffer, and sends a specific H-PDU which contains the Initial Credit value on the Credit Information field in the HB_Header, to the sending converter.

(3) When the sending converter receives the H-PDU including the Initial Credit value, it sets the Credit Count to the Initial Credit value and sets the Old Credit to 0. Then, the sending converter can transmit H-PDUs as many as the Initial Credit value.

(4) The sending converter decreases the Credit Count by 1 whenever it sends one H-PDU. If the Credit Count becomes 0, the sending process is stopped.

(5) The receiving converter increases the New Credit by 1 whenever it finishes processing one H-PDU space of the receiving buffer.

(6) The receiving converter periodically send a specific H-PDU to the sending converter, which has the New Credit value on the Credit Information field in the HB_Header.

(7) When the sending converter receives the H-PDU including the New Credit value, it rewrites the Credit Count value according to Eq. (1).

$$\textit{Credit Count} := \textit{New Credit} - \textit{Old Credit} + \textit{Credit Count} \qquad (1)$$

Where, the descriptor ":=" means new substitution of the counter value. (*New Credit* – *Old Credit*) represents the number of H-PDUs which can be newly transmitted.

Then, in the sending converter the Old Credit is updated to the New Credit value.

The ATM-part flow control is performed by repeating the steps (1) to (7).

The credit scheme has the benefit that the receiver-side buffer overflow can be avoided completely since the data are transmitted after holding the buffer space in the receiving converter. Because the credit scheme, however, belongs to the window flow control whose window size corresponds to the receiving buffer size, i.e. the constant value, the influence of the propagation delay is not negligible as the transmission distance and transmission speed are larger [15]. Accordingly, the credit scheme is not suitable for long distance and high speed communication using the HIPPI-ATM conversion because of the inefficient utilization of the connection bandwidth.

As for the performance evaluation of the HIPPI-ATM conversion, it is confirmed in [10], [11] that when long-distance data transmission beyond 0.5 ms one-way propagation delay (100 km if 5 μs/km and intra-node delay is ignored) via a wide area network is executed, the throughput is rapidly degraded, under the condition of a certain buffer size (64 kByte) in the receiving converter. Where, the receiving buffer size that corresponds to the credit size will be limited due to the hardware difficulty of the very high-speed data access and the productive cost [12] in addition to obtaining the efficient effect for network congestion avoidance. In this evaluation, 64 kByte receiving buffer size was determined by the credit size, 64 [13] multiplied by the H-PDU size, 1 kByte. As shown in Ref. [11], as the credit size increases, the distance beyond which throughput performance start to degrade is also linearly longer. However, the shape of performance degradation curve does not change.

3. A Proposal of Hybrid Flow Control Scheme

If the flow control is not implemented in the ATM part, the connection bandwidth can be efficiently used. However, the buffer overflow in the receiving converter can occur just when the process in the receiving end system is congested. On the other hand, the credit scheme avoids the buffer overflow, but can not attain the high-throughput transmission in the wide-area high-speed network. Therefore, in this paper, we propose the following hybrid flow control scheme based on intra-traffic priority classes [16], [17].

In the HIPPI layer (HIPPI-PH), the originated HIPPI bursts are assumed to be classified into high-priority burst and low-priority one according to the indication from the upper layer in the proposed scheme. We classify the HIPPI bursts requiring high reliability into the high-priority HIPPI bursts (HP-bursts), and those relatively tolerating loss or bit error into the low-priority HIPPI bursts (LP-bursts). The HP-bursts are transmitted according to the credit scheme defined by ANSI. While, the LP-bursts are transmitted in no-control manner without holding the buffer space in the receiving converter. That means the best-effort transmission for LP-bursts. This hybrid scheme enables HP-bursts to guarantee the reliability by applying the credit scheme. On the other hand, it can be regarded that LP-bursts are interleaved into the HP-burst's flow. Accordingly, the proposed scheme can utilize the connection bandwidth more efficiently.

Let me explain this effect by using Fig. 3. For simplicity, the credit size is set to 5. In the credit scheme (Fig. 3 (a)), periodically, the empty buffer space is reported to the sending converter as the New credit which is accumulative value. If the transmission becomes higher speed and longer distance, the credit waiting time in the sending converter becomes longer. As a result, it causes the throughput degradation. On the other hand, in the proposed scheme, we assume the mixed priority generation pattern as shown in Fig. 3 (b). In this case, in the first period, totally 5 PDUs have arrived, 1 High

H-PDU and 4 Low H-PDUs, Thus, only 1 credit is returned. In the second period, in addition to receiving 5 high H-PDUs, 8 low H-PDUs are also received by interleaving the high H-PDU stream. Due to such interleaving, the throughput improvement is expected. In this period, since 5 high H-PDUs were already sent, the sixth high H-PDU has to be waited. And it is transmitted in the head of the next period just when the new credits are returned to the sending side.

From the point of throughput performance, though the proposed scheme has the worst performance when all HIPPI bursts belong to the HP-bursts, at least, it can guarantee the same throughput performance as the conventional credit scheme defined by ANSI. Furthermore, if the low priority traffic can be enough interleaved by the high-priority traffic in this proposed scheme, the improvement effect of the loss performance compared with totally no-control transmission case can also be expected.

The system configuration to realize the hybrid flow control scheme with intra-traffic priority classes is depicted in Fig. 4. In particular, the detail structure of the receiving side HIPPI-ATM converter is illustrated in Fig. 5. The difference from the conventional system employing the specification of ANSI is that two buffers for high-priority H-PDU (High H-PDU) and low-priority H-PDU (Low H-PDU), respectively, are implemented in the receiving converter. On the contrary, we considered single buffer should be implemented in the sending converter. In the upper layer, the data are classified into two priority categories according to importance of the information contents, and as a result, each H-PDU in the HIPPI-ATM layer has its own priority. Thus, the H-PDU stream with mixed priorities has the pre-determined transmission sequence. Even if the two buffer is implemented in the sending converter, the determined sequence must be maintained. In addition, the buffer over-flow control in the sending converter, named READY pulse scheme, is implemented. Accordingly, single buffer in the sending converter is implemented in Fig. 4.

In the credit scheme for the HP-bursts, the buffer size of High H-PDH becomes the Initial Credit value.

The strict procedure of the hybrid scheme is as follows.

(1) In the sending end sytem, originated data and its priority are passed from the upper layer to the HIPPI-PH layer.

(2) HIPPI bursts and their priority information are sent to the sending converter. The priority information can be passed through one of non-reserved links in the HIPPI parallel interface.

(3) In the sending converter, A H-PDU is configured by a passed HIPPI burst and its priority informa-

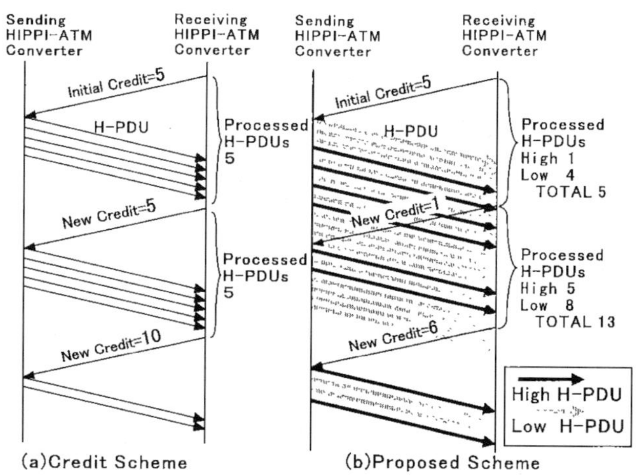

Fig. 3 Comparison of the credit scheme and the hybrid scheme.

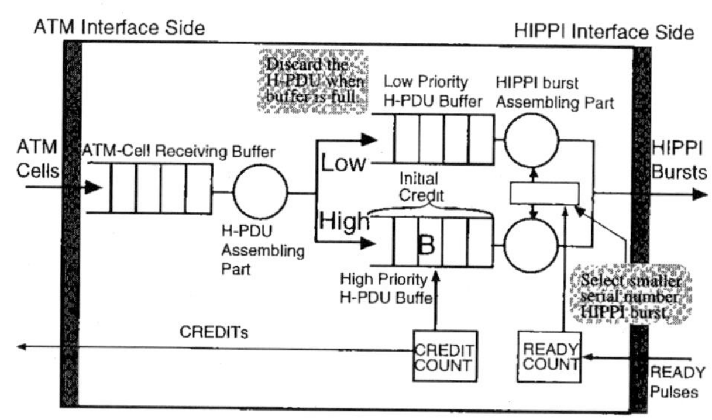

Fig. 5 Configuration of the receiving HIPPI-ATM converter.

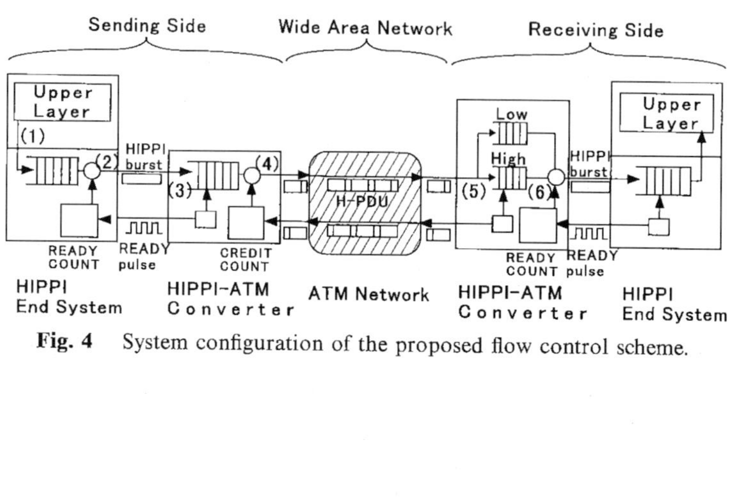

Fig. 4 System configuration of the proposed flow control scheme.

tion. Where, the priority information is written on Rsv field in the HB_Header.

(4) The priority of the H-PDU on the head of the buffer is detected in the sending converter. When it is the High H-PDU, if the Credit Count value equals to 0, the transmission process is stopped. Otherwise, the transmission is performed and then the Credit Count is decreased by 1. While, when it is the Low H-PDU, the transmission is performed, independent of the Credit Count value. In order to transmit the H-PDU to ATM network, exactly speaking, first, the H-PDU is segmented into multiple ATM cells after the conversion into the CPCS-PDU in the AAL 5. Next, these cells are transmitted sequentially.

(5) After the receiving converter reassembles the arrived ATM cells into the H-PDU, the H-PDU is stored in either buffer, high-priority buffer or low-priority buffer according to the priority information in the HB_Header. The high-priority buffer can prevent the H-PDU overflow by the credit scheme. While, the low-priority buffer can occur the H-PDU overflow.

(6) The receiving converter selects a H-PDU with the smaller sequence number among H-PDUs in the heads of two buffers, just when it receives the READY pulse from the receiving end system. Then, it converts the H-PDU into the HIPPI burst and transmit the burst to the receiving end system.

Where, the discarded LP-bursts are not retransmitted in the HIPPI layer but done in the upper layer. This is the same process as the case of normal HIPPI communication [2].

4. Performance Evaluation

We evaluated the performance of the proposed scheme through the computer simulation under the following common conditions.

- Transmission rate in ATM part is 622.08 Mbps, where, effective rate without overhead becomes 542.54 Mbps. The transmission rate in HIPPI part is 800 Mbps.
- Both the HIPPI bursts and H-PDUs have the constant length, 1 kByte.
- The cell loss in the ATM network does not occur.
- The credit transmission period is 1 ms.
  This value is selected as the one that throughput degradation is very slight in the long-distance transmission environment where one-way propagation delay becomes more than 1 ms in Ref. [11].

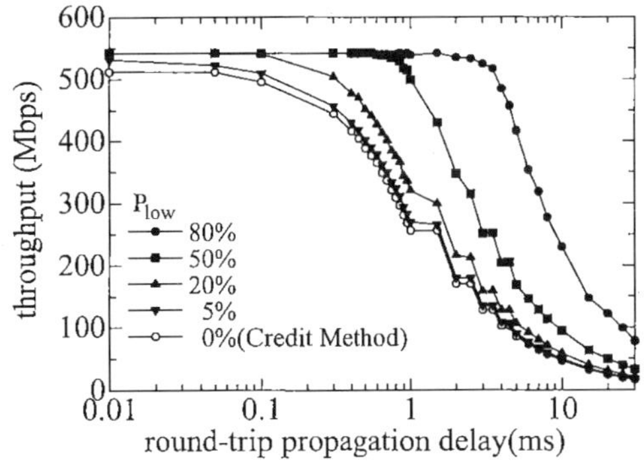

Fig. 6 Throughput and transmission distance.

- According to two priorities, two 64 kByte buffers are set up in the receiving HIPPI-ATM converter. While, one 64 kByte buffer is set up in the converter for the conventional system defined by ANSI.
- The sending end system always generates the data (Greedy generation). Thus, its buffer is always filled with the data.

4.1 Total Throughput

The relation between the throughput in the receiving end system and the transmission distance is presented in this section. Figure 6 shows the throughput and the distance as a function of the LP-burst generation probability, P_{low}. Where, the priority of the HIPPI bursts is randomly determined based on P_{low}. Also, for the purpose of maximum throughput evaluation, the processing rate in the receiving end system is assumed to be enough faster than the transmission rate in the ATM network. Accordingly, the LP-bursts can not be discarded in the receiving HIPPI-ATM converter in this simulation.

From Fig. 6, the larger P_{low} can attain the high throughput even in the longer transmission distance. Also, in the small P_{low} cases, especially, the throughput discontinuously decreases with the longer distance. This phenomena are caused by the relation between the credit transmission period and propagation delay. This phenomena are alleviated in the larger P_{low} cases because of the less influence of the credit scheme.

In this scheme, $\frac{1}{1-P_{low}}$ H-PDUs can be transmitted per one Credit Count on average. Accordingly, it is confirmed in Fig. 6 that the proposed scheme can accomplish the throughput $\frac{1}{1-P_{low}}$ times larger than that of the credit scheme. However, since the effective data transmission rate is 542.53 Mbps, the throughput is less than its value.

Next, the impact of the successive generation of the same priority HIPPI bursts is illustrated in Fig. 7. Where, P_{low} is set up to the constant value. Average succession number of the HP-bursts is defined as H_{av}, and that of the LP-bursts is defined as L_{av}. Both periods are assumed to follow the geometrical distribution that

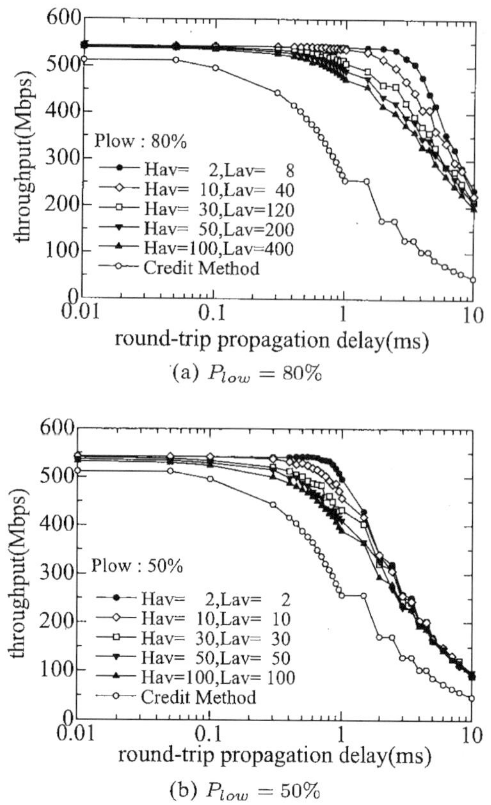

Fig. 7 Impact of succession number of same kind of bursts.

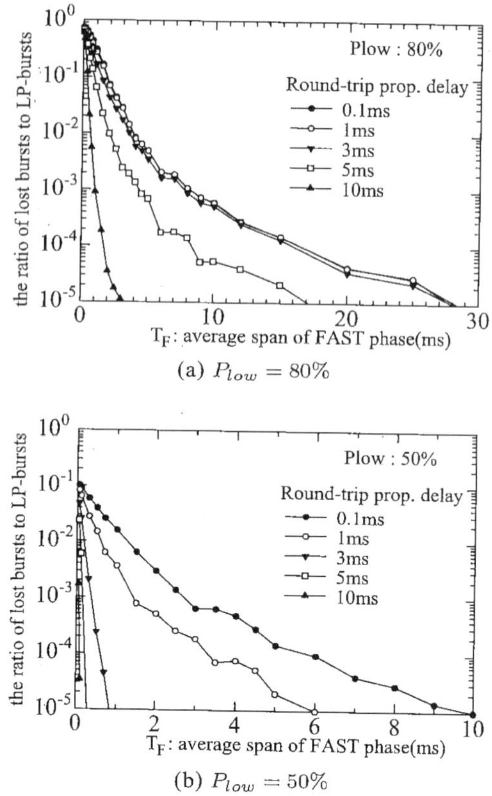

Fig. 8 Ratio of lost bursts to LP-bursts and FAST phase span.

has the memory-less property. These values H_{av}, L_{av} depends on information contents of the target application and the way of the priority assignment. Namely, they are not tuning parameters if the information contents and the priority assignment policy are determined beforehand. We have evaluated the system performance in several different H_{av}, L_{av} cases. From Fig. 7, the following results are confirmed. The proposed scheme has better performance than the credit scheme when the H_{av} is less than 100. Also, this scheme effectively behaves in case that $H_{av} \ll$ *Initial Credit*.

4.2 Low Priority HIPPI-Burst Loss Rate

In the section, we evaluate the low-priority HIPPI burst loss rate on condition that the processing speed in the receiving end system reduces. Its processing speed is classified into two phases, "FAST phase" where the process is performed at higher speed compared to the ATM network and "STOP phase" where the process is stopped. The succession time of the STOP phase has the constant value 1 ms, which is equal to the credit transmission period. While, the succession time of the FAST phase exponentially distributes with mean T_F. The relation between the LP-bursts loss rate and parameter T_F is depicted in Fig. 8 as a function of the transmission distance.

The LP-burst loss rate is defined as the ratio of the lost LP-bursts to all LP-bursts. The priority of the HIPPI bursts is randomly given in this simulation.

Figure 8 shows that larger P_{low} leads to the higher loss rate. In Fig. 8 (a), the same loss performance can be shown under the transmission distance less than 3 ms round-trip propagation delay (300 km end-to-end distance if intra-node delay is ignored). This is because the maximum throughput is the same in Fig. 6 when the round-trip propagation delay is less than 3 ms. The H-PDU loss happens when the process of the receiving end system enters the STOP phase. Once the successive H-PDU loss initiates, these Low H-PDU continues to be discarded until the process of the receiving end system returns to the FAST phase. Accordingly, since the discarded sequence has complete burstness, the retransmission control can be performed in the upper layer effectively.

It is considered that the LP-burst loss rate rapidly degrades with the longer STOP phase succession time. In this case, a certain control must be implemented to prevent the buffer overflow. One example is the feedback congestion notification of the Low H-PDU buffer to the sending converter when its buffer length exceeds a certain threshold. The detail consideration about this improvement is further study issue.

4.3 Application for the MPEG Compressed Video File Transmission

When the application data can be classified into two classes, the proposed scheme can attain the higher throughput communication. Let me consider the MPEG compressed video file communication for example. We assume the stored MPEG file transmission whose data can be classified into two priority classes, and for simplicity, whose rate is set to the interface speed as CBR in the ATM part. The multiplexed realtime video transmission like CATV is out of concern.

Furthermore, we assume the stored MPEG video data specification for the throughput evaluation as follows [18], [19].

- Frame size is 512 × 480 (pixel).
- GOP (Group Of Picture) is "IBBPBB."
- Compression rate corresponds to the NTSC video quality.

In this application, the following two priority assignment policies, "Inter-frame priority assignment policy" and "Intra-Frame priority assignment policy" can be considered.

4.3.1 Inter-Frame Priority Assignment Policy

In the MPEG compression, three frames (called pictures), I (Intra) picture, P (Predictive) picture and B (Bidirectionally predictive) picture are defined. If I pictures are discarded, the video quality is greatly degraded since I picture has a great influence to reconstruct both P and B pictures. Thus, the Inter-frame priority assignment policy is defined as the policy that the HIPPI bursts composed of I picture information are classified into high-priority bursts (HP-bursts), and those composed of P or B picture information are classified into low-priority bursts (LP-bursts) in this paper. In other words, only I picture information is controlled by the credit scheme in this policy.

Figure 9 (a) shows an example of the priority assignment of this policy. In this figure, the GOP with "IBBPBB" pattern is illustrated. The large square means each frame (picture) which is composed of small squares called "slices." Based on the information generation pattern of the "flowg.mpg" file in Fig. 1 of Ref. [18], we assumed that $P_{low} = 63\%$, $H_{av} = 120$, $L_{av} = 200$ in this policy. Namely, the successive generation of each priority HIPPI bursts becomes larger. However, at least, throughput improvement will be expected by the hybrid scheme to a degree even in such a inter-frame priority assignment policy.

4.3.2 Intra-Frame Priority Assignment Policy

As shown in Fig. 7, in order for the proposed scheme

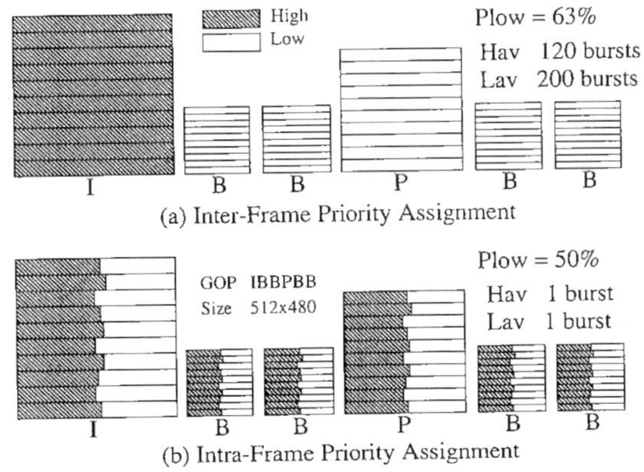

Fig. 9 Priority assignment policies for MPEG compressed video data.

to behave effectively, the small succession of the HP-bursts is suitable. The intra-frame priority assignment policy is defined as the policy that assigns the priority to each data fraction within a frame according to the importance of contents. In this paper, especially, the policy called "Slice-based decision [19]" is adopted as the representative of the Intra-frame assignment policies, which assigns the different priority to each data fraction within a "slice," smaller information unit to construct a frame. In this policy, very small H_{av} can be realized. Strictly speaking, according to the Ref. [19], under the condition of the HP ratio more than 16/64 where the degradation of image quality was found to be very small, each priority class information per slice will be almost accommodated into a HIPPI burst (Fig. 9 (b)). In other words, the HP-burst and LP-burst can be generated almost alternatively. Accordingly, we assumed that $P_{low} = 50\%$, $H_{av} = 1$, $L_{av} = 1$ in the intra-frame assignment policy. Due to the small succession of specific priority bursts, it is considered that the intra-frame priority assignment policy has better performance than the inter-frame one in our proposed scheme.

4.3.3 Numerical Results

We depict the throughput performance based on the above two priority assignment policies in Fig. 10. The QOS performance can also be important characteristics. However, QOS depends on the performance of priority assignment policies themselves. In this paper, to evaluate the performance of the transmission scheme itself, we compared the throughput performance. From this graph, in less than 2 ms round-trip propagation delay region, intra-frame policy can realize the higher throughput performance. However, on the contrary, in more than 2 ms round-trip propagation delay region, inter-frame policy can realize the higher one.

In general, as shown in Fig. 7, in case of the same P_{low}, the smaller H_{av} is, the more effectively the priority control performs. However, in the longer distance re-

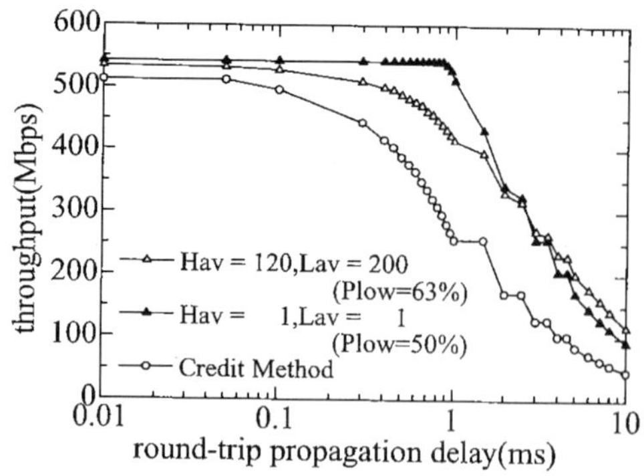

Fig. 10 Throughput in the MPEG compressed video transmission.

gion, the throughput performance converges to the same curve without regard to the each priority HIPPI burst generation pattern. Accordingly, in the short distance region, intra-frame policy having smaller H_{av} obtains the better performance. While, in the longer distance region, the parameter P_{low} becomes more dominant factor than H_{av}, as a result, inter-frame policy having a little larger P_{low} shows better performance. In both cases, the hybrid scheme has twice as high throughput as the conventional credit scheme in more than 1 ms round-trip propagation delay region.

5. Conclusions

In this paper, the hybrid transmission scheme using the intra-traffic priority class is proposed, in order to solve the throughput degradation problem when the credit flow control is implemented in the high-speed and long-distance HIPPI-ATM communication. The priority flag (High/Low) is introduced in every "HIPPI burst," based on the indication from the upper layer. As for the high priority HIPPI bursts, it adopts the conventional credit scheme as the ATM-part flow control. While, as for the low priority HIPPI bursts, it transmits the data without taking account of the buffer capacity in the receiving HIPPI-ATM converters. The simulation results show that the proposed scheme can improve the throughput performance, especially under the condition of the larger ratio of the low-priority HIPPI bursts, and smaller succession of the high-priority HIPPI bursts. Moreover, the low priority burst loss rate was clarified when the processing at the receiving end system is occasionally congested. As a result, unless the receiving end system is severely congested, quite good loss performance can be attained. Finally, implemental example in case of the MPEG compressed video transmission has been shown, and their performance improvement effect was confirmed.

The further research issues include the modification of the proposed scheme to improve the low-priority loss characteristics and the evaluation taking account of the cell loss process in the ATM network.

Acknowledgement

The authors wish to express their deep appreciation to Mr. Kazuhiro Sato, NTT Multimedia Network Laboratory, for his useful information, and to Ultra High Speed Network and Computer Technology Laboratories for financial support.

References

[1] D. Tolmie and J. Renwick, "HIPPI: Simplicity yields success," IEEE Network, pp.28–32, Jan. 1993.
[2] ANSI X3.183-1991, "High-performance parallel interface—mechanical, electrical, and signalling protocol specification (HIPPI-PH)," Nov. 1994.
[3] ANSI, "Fiber channel phisical and signalling interface (FC-PH)," -REV4.3.
[4] Serial HIPPI Implementators Group, "Serial HIPPI specification, revision 1.0," May 1991.
[5] C. Partridge, "Gigabit Networking," Addison-Wesley, 1994.
[6] ANSI X3.299-199x, "High-performance parallel interface—mapping to asynchronous transfer mode (HIPPI-ATM)," April 1996.
[7] H. Gotoh, Y. Hori, K. Kawahara, H. Sunahara, and Y. Oie, "Performance of HIPPI over ATM high speed networks," IEICE Technical Report, SSE96-118, Dec. 1996.
[8] NetStar Inc., "Gigarouter system description, revision 2," Dec. 1994.
[9] J. Hsieh, D. Du, and J. MacDonald, "Experimental study of extended HIPPI connections over ATM networks," IEEE INFOCOM'96, 10b.3, pp.1261–1268, March 1996.
[10] K. Sato and S. Takimoto, "Evaluating performance of high-speed and long distance communications on HIPPI-ATM networks," 21st Conference on Local Computer Networks, pp.424–432, Oct. 1996.
[11] K. Sato and S. Takimoto, "Performance evaluation of flow control method on HIPPI-ATM conversion," IEICE Technical Report, IN95-69, Oct. 1995.
[12] K. Arai, "A study on interworking technology between HIPPI and ATM," IEICE Technical Report, CS95-127, Oct. 1995.
[13] J. Kawamata, H. Denpou, and C. Ikeda, "A consideration of credit and throughput on the HIPPI terminal adapter," Proc. 1996 IEICE General Conference, B-733, March 1996.
[14] D. Bertsekas and R. Gallager, "Data Networks," Prentice-Hall, 1987.
[15] L. Kleinrock, "The latency/bandwidth tradeoff in Gigabit networks," IEEE Commun. Mag., April 1992.
[16] S. Takimoto, H. Tode, and H. Ikeda, "A study on flow control method with the order of priority on HIPPI-ATM connection," IEICE Technical Report, SSE96-181, OCS96-119, Feb. 1997.
[17] H. Tode, S. Takimoto, and H. Ikeda, "Hybrid flow control scheme using the priority class on HIPPI-ATM connection," Proc. the third Asia-Pacific Conference on Communications (APCC'97), pp.377–381, Dec. 1997.
[18] N. Wakamiya, Y. Kobayashi, M. Murata, and H. Miyahara, "Study on MPEG transfer over ATM CBR service," IEICE Technical Report, IN95-142, 1996.
[19] P. Pancha and M. El Zarki, "Prioritized transmission of variable bit rate MPEG video," IEEE GLOBECOM'92, pp.1135–1139, 1992.

High-Performance Backbone Network Technology edited by Naoaki Yamanaka
ISBN: 0-8247-5321-6 DOI: 10.1081/(120025680)
Published 2004 by Marcel Dekker, Inc.

ATM Based Broadband Access System Using Bearer Connection Control

Hideki Kasahara, Shinichiro Chaki, Hiroaki Sato, and Hiromi Ueda

SUMMARY This paper discusses an ATM based access system for a broadband access network that provides flexible and cost effective multimedia services for mass consumption including households. The access system proposed herein uses ATM-PDS subscriber transmission technologies, thus enabling multiple users to share a single fiber resource while transmitting or receiving multimedia information within the broadband network. The system also has virtual channel concentration function by using the bearer connection control protocol being studied for $V_B5.2$ interface in the ITU-T. We show that the virtual channel concentration function reduces the access cost per user. We also discuss multiple QoS control methods in the system to provide multiple services efficiently. This paper evaluates two methods to handle ABR class and UBR class traffic: shared bandwidth with preferred ABR method, and guaranteed bandwidth with overriding method. The result indicates that when using the shared bandwidth with preferred ABR method for the access system, ABR throughput and UBR throughput per VC depend on the proportion of the number of ABR-VC connections to the total active VCs and on the each buffer size in the access system. And it is difficult to control ABR-VC and UBR-VC throughputs in the access network by using the shared bandwidth with preferred ABR method, which simple mechanism. With the guaranteed bandwidth with overriding method, while ABR-VC throughput and UBR-VC throughput also depend on the proportion of ABR-VCs to the total VCs and on the buffer size, it can offer the minimum guaranteed throughput to the UBR traffic. The result shows the method is effective for easy service provisioning.
key words: *ATM, access NW, multi-QoS, PDS, SNI*

1. Introduction

In order to provide multimedia communication services that include home access services, it is necessary to realize cost-effective broadband access networks. FTTH (Fiber To The Home) in which optical fiber cables are extended to individual homes is one of the most powerful infrastructure for realizing such a network. The ATM-PDS (Passive Double Star) technology is a cost-effective approach to realize this FTTH configuration because it enables multiple users to share a single fiber resource while transmitting multimedia information within a broadband network [1]-[3].

This paper describes an ATM based access system for the broadband access network. To provide various types of multimedia service to residential users cost-effectively, the access system will be required to have not only VP (Virtual Path) multiplexing function but also VC (Virtual Channel) concentration and VC based statistical multiplexing functions. First, we discuss the basic model of a broadband access network and the required functions of the access system. We then propose a desirable system schematic for the ATM access system and evaluate its features.

Manuscript received April 25, 1997.
Manuscript revised July 28, 1997.
† The authors are with NTT Access Network Systems Laboratories, Chiba-shi, 261-0023 Japan.

2. Basic Model for a Broadband Access Network

Standards organizations such as the ITU-T have been discussing the establishment of a basic model for broadband access networks [4]. Figure 1 shows a connection model between a broadband access network and a service node. As shown in Fig. 1, an SNI (Service Node Interface) is specified between the broadband access network and the service node which provides various types of services. The broadband access network is connected to the broadband service nodes through a V_B5 interface as an ATM-based broadband SNI [5],[6]. Furthermore, it is connected to the existing telephone/ISDN service nodes through a V5 interface which has already been standardized [7].

To share the resource of access network by multiple service nodes, the V_B5 interface discussed in the ITU-T allows multiple service nodes to manage an access network simultaneously in real time [8],[9]. The V_B5 interface has been

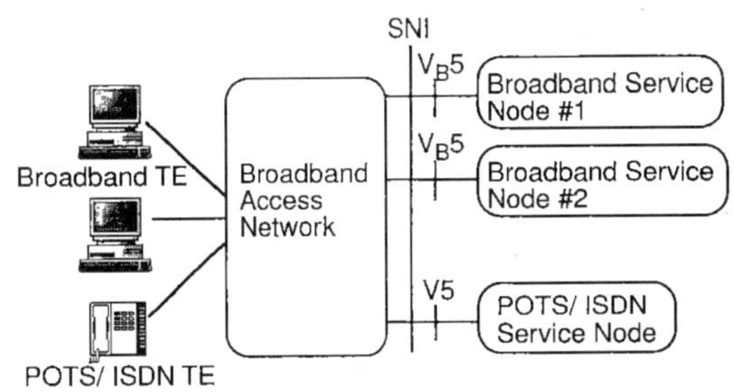

Fig. 1 Basic model for a broadband access network.

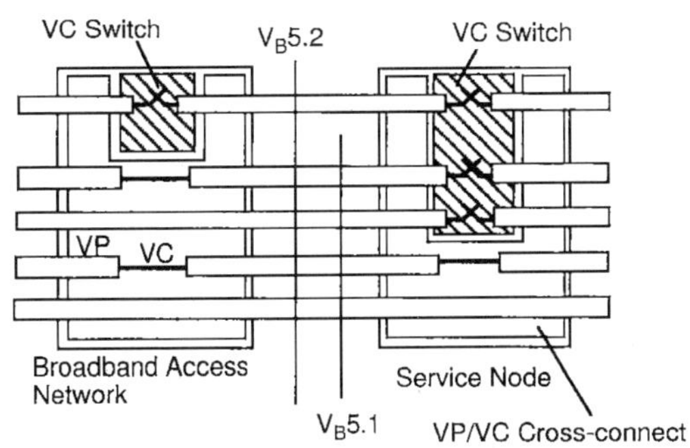

Fig. 2 Function model for V_B5 interface.

classified into two types, $V_B5.1$ and $V_B5.2$, depending on the functions of the access network. Hence, there are two kinds of network capabilities offered by the broadband access network. As shown in Fig. 2, in case of $V_B5.1$, the access network is required to have VP and VC cross-connecting function. On the other hand, in case of $V_B5.2$ the access network is required to have VC concentration function in which each VC is dynamically switched as well as VP and VC based cross-connecting function.

3. Required Functions and System Configuration of a Broadband Access System

3.1 Required Functions of a Broadband Access System

To provide more cost-effective multimedia services to users, an access system for a broadband access network is required to have remote traffic concentration function. The service node connected to the access network is often located in a place distant from the user. In such cases, remote traffic concentration function in the access network will be effective in reducing the access cost. To provide remote traffic concentration function to the service node, following functions; VC concentration function and multiple service classes handling function are required to the access system.

(1) VC Concentration Function: VC concentration allows the cost of the access network per user to be reduced by sharing of the access network resource among multiple users on a call by call basis. VC concentration function should be implemented using a connection control signaling for SNI used to set up or release a connection from the service node to the access system in real time [10].

As shown in Fig. 3, in the case of handling an SVC (Switched Virtual Channel) connection, the access system transmits a 'SETUP' message transparently from the user to the service node. When the service node receives the 'SETUP' message, it performs necessary tasks such as analyzing subscriber data, and transmits control signals using a B-BCC (Broadband Bearer Connection Control) channel back to the access system. Upon receiving these control signals from the service node, the access system sets up a VC connection.

Though the VC concentration function of the access network can be realized by small service nodes in close to users, it is not possible to limit the changes necessitated by additional services within the service node. When the access system is based on $V_B5.2$, new services can be introduced easily and quickly.

(2) Multiple Service Classes Handling Function: To reduce the access cost of services provided by each service node, the access system should have service multiplexing function. The access system is required to transport various kinds of information such as IP and FR packets, image data such as VOD, and POTS/ISDN signals as well as native ATM service signals.

Table 1 shows relationship between ATCs (ATM Transfer Capability) and the services. The access system accommodates non-ATM services by converting Non-ATM signals into ATM cells, and the ATM cells are transported by using corresponding ATCs in the ATM network. In order to perform service multiplexing, it is necessary for the access system to handle different ATCs simultaneously and guarantee QoS (Quality of Service) requirement of each ATC. Furthermore, to reduce the bandwidth of the transport line connecting the access system to the service node, cell based statistical multiplexing and cell based priority control within different ATCs are required to the access system.

Table 1 Broadband services and ATM transfer capabilities.

	DBR/ CBR	SBR/ VBR	UBR	ABR
ATM bearer Services	✓	✓	✓	✓
FR,IP Services		✓	✓	✓
VOD/SVB	✓			
POTS/ISDN	✓			

SVB: Switched Video Broadcast

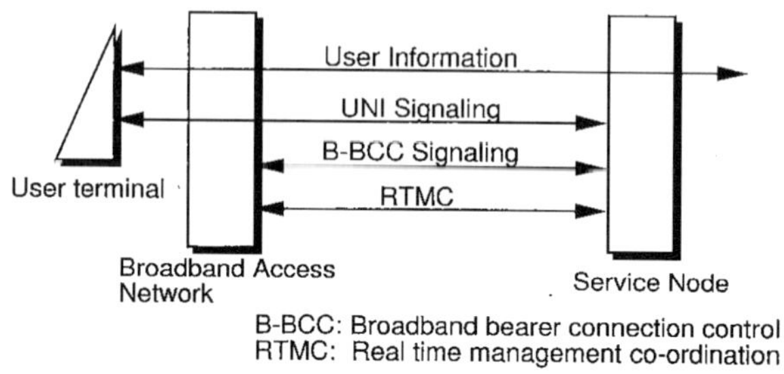

Fig. 3 Signal types via $V_B5.2$ interface.

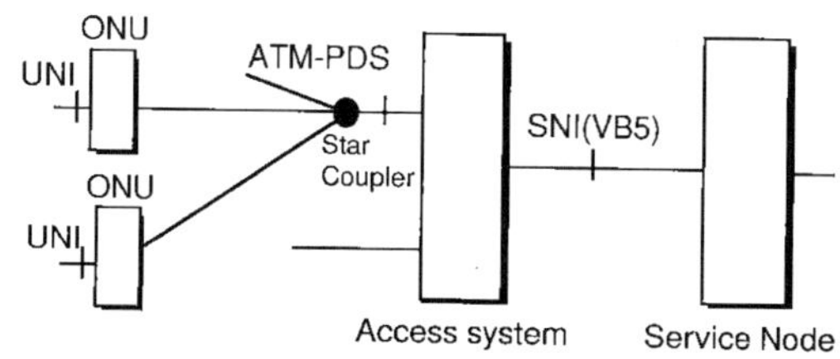

Fig. 4 System model for a broadband access network.

3.2 System Configuration

Figure 4 shows a system model of the broadband access network. The access system terminates the PDS access lines, concentrates the VC traffic, and transmits the user traffic to the service node through the SNI. The ONU (Optical Network Unit) terminates the PDS access line and provides the UNI to users.

4. VC Concentration in Access Network

4.1 Evaluation Model

In this section, we evaluate the effect of the VC concentration function in the access system on the basis of the number of users accommodated to the access network, using the cost of the $V_B5.1$ access system with VP multiplexing function (Model 1) and the cost of the $V_B5.2$ access system with VC concentration function (Model 2).

In Model 1, the VPs between the user and service node are statically set up and relieved according to commands via operation system. Therefore, it is possible to reduce the cost of the common functions such as system control part and switch fabric in the system. In Model 2, the VCs are set up and release in real time through the B-BCC signaling from the service node. In this case, the real time processing of B-BCC signaling increases the cost of common part of the access system. As the number of accommodated users increase, cost reductions in transport line interface board and transport lines becomes significant since VC concentration enables the reduction in the number of transport lines.

Figure 5 shows the system model used for this evaluation. It is assumed that the access line interface boards (PDSIF) which terminate the ATM-PDS access line are not duplicated, and that the common part and transport line interface boards (TLIF) are duplicated. To evaluate the access cost we divide the network into four parts. The relative costs of each part are assumed as shown in Fig. 5. The costs of the access lines and the transport lines do not include the cost of conduits and construction. Considering the size of the control part, the maximum number of accommodated interface boards is assumed to be 40 for model 1 and 30 for Model 2, because B-BCC signaling termination function has to be installed to Model 2. If the number of both required PDSIF and TLIF to handle user traffic exceeds the maximum number of accommodated interface boards of the system, it is assumed to install an additional system in the access network.

For simplicity, the number of accommodated users per PDSIF is assumed to be 5, and the PCR (Peak Cell Rate) of each VC is assumed to be the same as the VP bandwidth of each user. The traffic characteristic is assumed to be DBR (Deterministic Bit Rate). The call blocking ratio at the access system is assumed to be 0.001 or less.

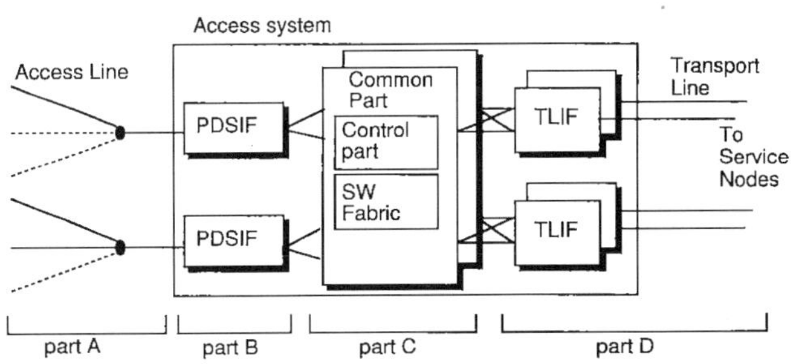

(a) System model for evaluation

	part A	part B	part C	part D
Model 1	2	5	100	90
Model 2	2	5	200	90

(b) Relative cost parameters for evaluation

Fig. 5 Access system evaluation model.

4.2 Evaluation Results

Figure 6 shows the simulated results when taking 10 Mb/s as PCR, and 0.1 and 0.01 Erlang as traffic intensity per user. Owing to the effect of VC concentration, in the cases of both 0.1 and 0.01 for traffic intensities, the network cost per user using Model 2 becomes lower than the cost using Model 1 when the number of users accommodated to the access network exceeds 30 as shown in Fig. 6.

Simulation using the same assumption shows that a system based on Model 1 can accommodate a maximum of 120 users, while one based on Model 2 can accommodate a maximum of 130 (when 0.1 Erlang) and 140 (when 0.01 Erlang) users. Although fewer interface boards can be accommodated in the model 2, this does not affect the difference in cost per user between Model 2 and Model 1.

Figure 7 shows the results when 0.1 Erlang as traffic intensity per user, and when 10 Mb/s or 15 Mb/s is assumed for the PCR. When the PCR is 15 Mb/s. The cost per user using Model 2 becomes lower than the cost using Model 1 when the number of users exceeds 20. The difference in cost between the two peak rates is not very significant for Model 2, while the peak rate significantly affects the cost per user for Model 1. Therefore, it can be said that under the conditions assumed in this paper, it is effective to have the access network feature the VC concentration function in the case of

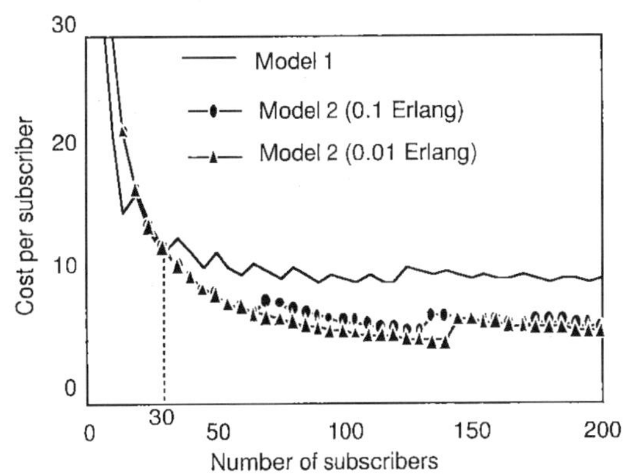

Fig. 6 Cost vs. number of users.

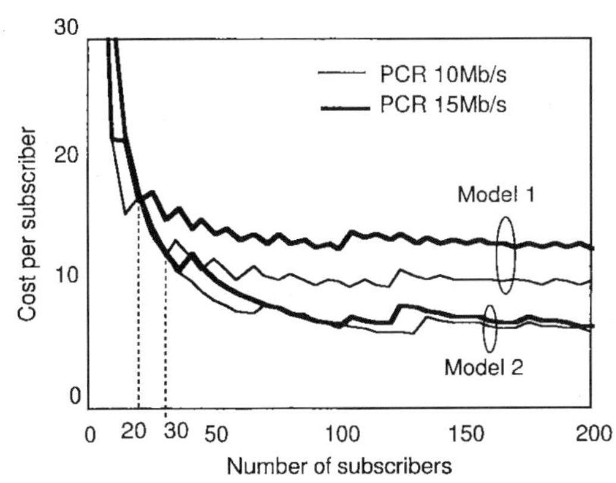

Fig. 7 Cost vs. number of users.

30 or more users in the access network. Moreover, this effect becomes more apparent at higher PCRs.

5. Handling of Multiple Service Classes

5.1 Multiple Service Classes Handling Method

As described in Sect. 3, the access system must support multiple service classes efficiently and also must obtain cell based statistical gain. The access system has to handle four ATCs: DBR/CBR (Constant Bit Rate), SBR (Statistical Bit Rate)/VBR (Variable Bit Rate), ABR (Available Bit Rate) and UBR (Unspecified Bit Rate). These ATCs can be categorized into two groups: a high-quality group (DBR/CBR and SBR/VBR) and a best-effort group (ABR and UBR).

The DBR/CBR and SBR/VBR ATCs are required to support quality of each service based on usage parameter declared by each user. On the other hand, The ABR and UBR ATCs are not so sensitive for quality, because these services are used for data communication, such as IP transmission service [11],[12].

To assure the quality of the high-quality group services easily, we propose high-quality group traffic and best-effort group traffic are allocated to different deterministic bandwidth VPs in the access network. In this method, the quality of the high-quality group traffic can be estimated easily, because deterministic bandwidth is allocated to the high-quality group traffic.

As a means of allocating bandwidth to ABR and UBR within the VP, we propose two methods; a Shared Bandwidth with Preferred ABR method (SB/PA), and a Guaranteed Bandwidth with OverRiding method (GB/OR) [13].

The SB/PA method outputs ABR cells at the highest priority for any cell emission timing and outputs UBR cells only when there are no ABR cells in the output buffer (see Fig. 8). This method provides ABR traffic with the maximum throughput, but very few UBR cells are transmitted when the ABR load is heavy. The GB/OR method assigns a minimum bandwidth to UBR traffic. It means that the method gives priority cyclically in a certain period to UBR. When UBR has priority, UBR cells can be transmitted even if there are ABR cells in the ABR buffer. If UBR has priority but it has no UBR cells to transmit, ABR cell can be transmitted. If there are no ABR cells to transmit, UBR cells can be also transmitted (See Fig. 9).

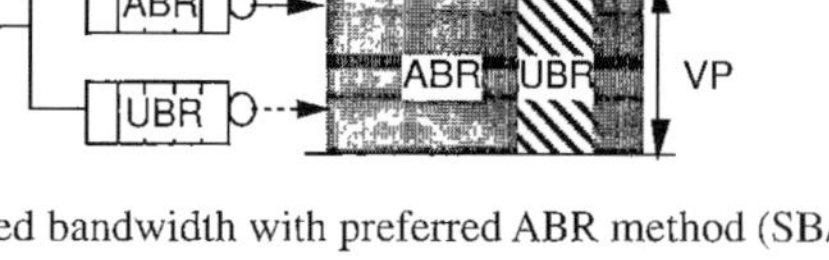

Fig. 8 Shared bandwidth with preferred ABR method (SB/PA).

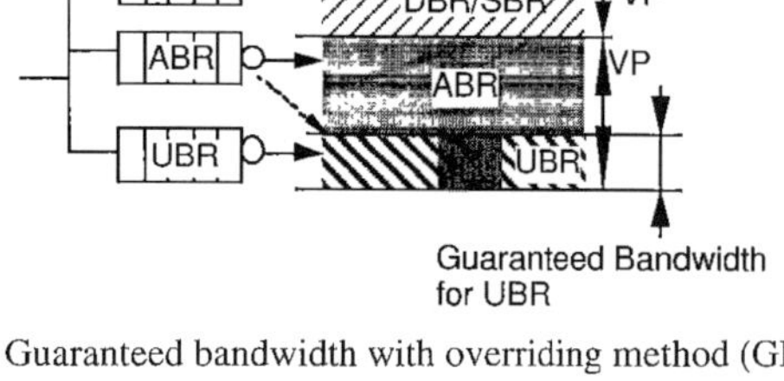

Fig. 9 Guaranteed bandwidth with overriding method (GB/OR).

5.2 Evaluation Model

The SB/PA and GB/OR methods are used to performed transmit cells to the PDS access line (downstream) or to the transport line connected to the service nodes (upstream). This paper evaluates both cases. Evaluation model is shown in Fig. 10.

(1) Upstream direction: Many VCs on low speed access lines are multiplexed into a high speed transport line in the access system, as shown on the left hand side of Fig. 10. The access system has buffers to multiplex the cells. We assume that the VP bandwidth for best-effort class is 10 Mb/s and the VP bandwidth for the transport line is 150 Mb/s. Simultaneous active VCs at the transport line are 30 VCs. The PCR of each VC is equal to the VP bandwidth.

(2) Downstream direction: Several VCs in the transport line VP are connected to the same user via a 10 Mb/s access line VP, as shown on the right hand side of Fig. 10. The access system has buffers to transmit the cells into 10 Mb/s access line VP. We assume that 8 active VCs are connected to each access line simultaneously.

5.3 Evaluation Parameters Related to Access System

The access system has 4k cell buffers for each output port in our simulation. The buffers are divided into two queues, one for ABR and one for UBR. The output algorithm for each queue is based on the SB/PA or the GB/OR methods.

To perform ABR service, the access system counts the number of active ABR-VCs. Then, to divide the bandwidth among the active ABR-VCs fairly, the system changes the value of ER (Explicit Rate) in the arriving RM (Resource Management) cells to control the ABR user traffic. The ABR queue has a congestion threshold, and if the ABR queue length exceeds the congestion threshold, the system sets the congestion indication in the user cells [14]-[16].

5.4 Evaluation Parameters Related to a Terminal

ABR sources as user terminals shape the cell streams based on whether the received ER or calculated ACR (Available

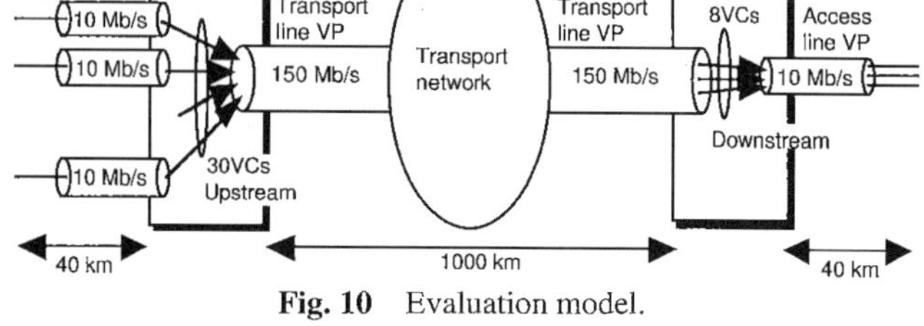

Fig. 10 Evaluation model.

Cell Rate) is smaller. UBR sources always transfer cells at the PCR (Peak Cell Rate). Both ABR and UBR sources transmit cells in these manners and have enough data to transmit. All terminals transfer data segments by using TCP as the transport layer protocol because most of data traffic is carried using TCP. TCP includes the congestion window control and the retransmission mechanism based on round trip time measurements [17].

For our evaluation, we adopt the congestion window control and the retransmission mechanism based on the 4.3 BSD Tahoe UNIX operating system. The congestion window size (cwnd) at the source is initially set to one segment. The slow start threshold (ssthres) is set to half the maximum window size. The cwnd is increased in two phases; the slow start phase and congestion avoidance phase. The source increases the cwnd by one segment for each received acknowledge segment (ACK) in the slow start phase. If the cwnd becomes larger than the ssthres, the source changes to the congestion avoidance phase. While in the congestion avoidance phase, the cwnd is increased by 1/cwnd for each received ACK. If congestion is indicated by time-out, the cwnd is set to one segment and the source becomes the slow start phase. The round trip time granularity is assumed to be 40 μsec to match the 10 Mb/s VP. The receiving terminals have no congestion and send an ACK as soon as a TCP segment is reconstruct. The parameters about ABR and TCP are shown in Table 2.

5.5 Evaluation Results

(1) Upstream direction: Figure 11 shows TCP throughput over ABR-VCs and UBR-VCs using the SB/PA method. In this evaluation, TCP throughput is defined as the ratio of the amount of the TCP segments to the measured time, where cells comprising the segment are not lost in the network and can be reassembled completely at the destination terminal. The TCP throughput can be regarded as the maximum throughput per VC using TCP, because we assume there is enough data to transmit at the source terminal.

The total number of VCs is 30. When there are fewer than 15 ABR-VCs, ABR-VC throughput is at the almost PCR (10 Mb/s) because the sum of the PCRs of the ABR-VCs is lower than the VP bandwidth of the transport line. The UBR-VCs can use the remaining bandwidth. When there are 15 to 20 ABR-VCs, the UBR throughput is almost nil because the ABR traffic occupies almost the entire bandwidth. When there are 25 ABR-VCs, the UBR-VC throughput increases to 2 Mb/s, because the ABR congestion mechanisms excessively reduce the cell rate of the ABR-VCs and the bandwidth for the UBR is remaining.

If the ABR buffer is too small, TCP throughput over ABR falls. In Fig. 12, the ABR buffer size is 100 cells and the UBR buffer size is 3900 cells. With over 20 ABR-VCs, the ABR-VC throughput is lower than UBR-VC throughput. Moreover, ABR-VC throughput drops to nil when there are more than 25 ABR-VCs. The loss of even one cell causes complete TCP segment loss. When the ABR buffer is too small, few complete TCP segments over ABR can be reconstructed at the destination. Therefore the TCP at the ABR sources can't receive an ACK and then the ABR sources are stalled. On the other hand, TCP over UBR can transfer TCP segments by using the available bandwidth. If the ABR buffer is 500 cells, as shown in Fig. 13, ABR-VC throughput is always higher than UBR throughput. To keep ABR-VC throughput high, the ABR buffer size must be more than 500 cells.

Table 2 Traffic parameters for simulation.

TCP	Average segment size	1 kB
	Max window size	64 kB
ABR	Peak Cell Rate	10 Mb/s
	Minimum Cell Rate	64 kb/s
	Rate Increase Factor	1/16
	Rate Decrease Factor	1/16
UBR	Peak Cell Rate	10 Mb/s
LINK	Access link speed	10 Mb/s
	Transit link speed	150 Mb/s

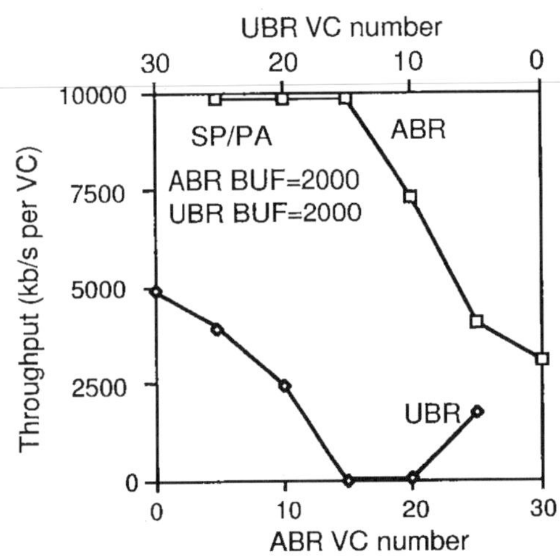

Fig. 11 TCP throughput with SB/PA (ABR BUF=2000, UBR BUF=2000).

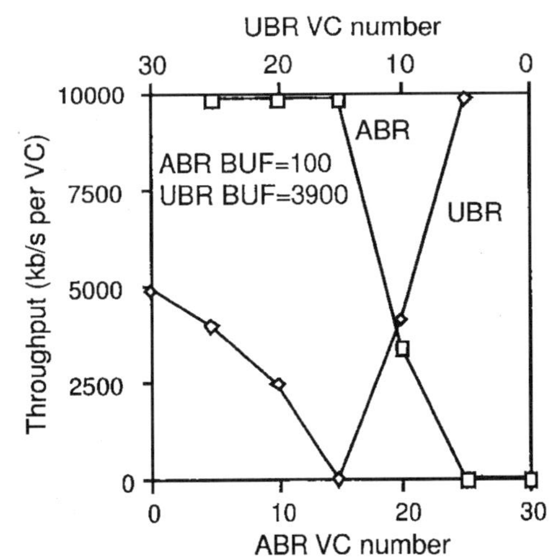

Fig. 12 TCP throughput with SB/PA (ABR BUF=100, UBR BUF=3900).

Figure 14 shows TCP throughput with GB/OR. The minimum bandwidth of 1 Mb/s is allocated to each UBR-VC. Compared to SB/PA case (Fig. 11), UBR traffic can obtain 800 kb/s throughput per VC when the number of active UBR-VCs is 15, because the preferred bandwidth for UBR traffic is allocated. In this result, UBR traffic obtains only 800 kb/s throughput per VC, even though a bandwidth of 1 Mb/s is allocated for UBR-VC. The difference is caused by the loss of cells in the UBR buffer in the access system.

(2) Downstream direction: Figure 15 shows TCP throughput over ABR-VC and UBR-VC using the SB/PA method in downstream direction. The total number of VCs is 8. Even if there is only one ABR-VC, the UBR-VCs can not transfer. The ABR-VC traffic occupies all of the access line VP bandwidth because ABR congestion control does not work. If the ABR buffer is too small, the UBR-VC throughput rapidly increases.

Figure 16 shows 100 cells are allocated for ABR and 3900 cells for UBR. When there are 7 ABR-VCs, ABR-VC throughput is nil. No complete TCP segments arrive at the destination because many cells are lost. If 500 cell buffers are allocated to ABR, as shown in Fig. 17, ABR-VC throughput is always higher than UBR-VC throughput. The ABR buffer has to be more than 500 cells to achieve reasonable

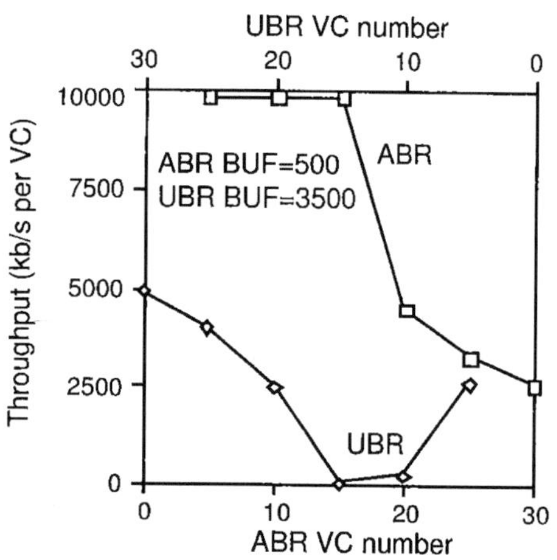

Fig. 13 TCP throughput with SB/PA (ABR BUF=500, UBR BUF=3500).

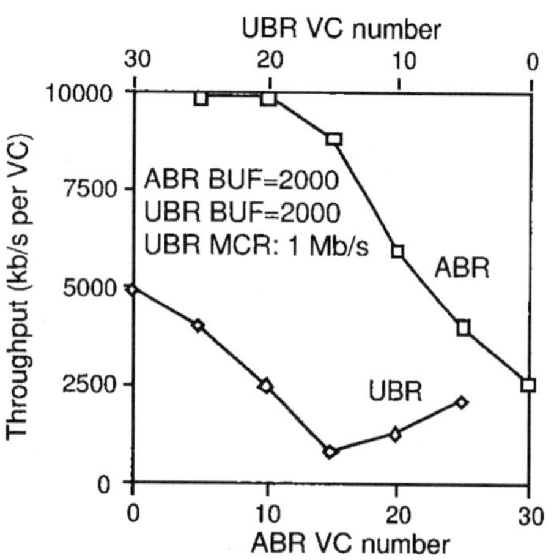

Fig. 14 TCP throughput with GB/OR (ABR BUF=2000, UBR BUF=2000).

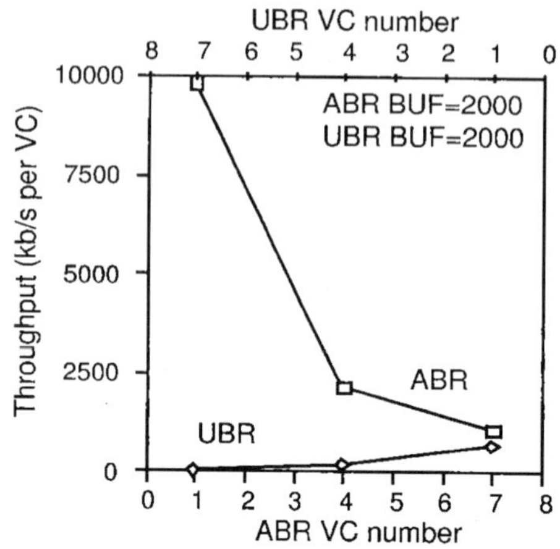

Fig. 15 TCP throughput with SB/PA (ABR BUF=2000, UBR BUF=2000).

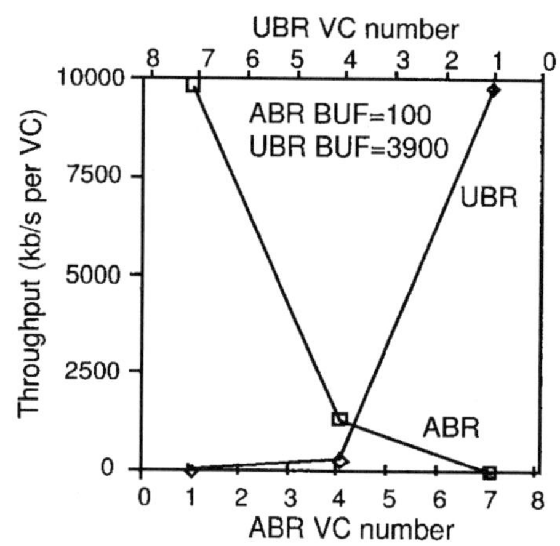

Fig. 16 TCP throughput with SB/PA (ABR BUF=100, UBR BUF=3900).

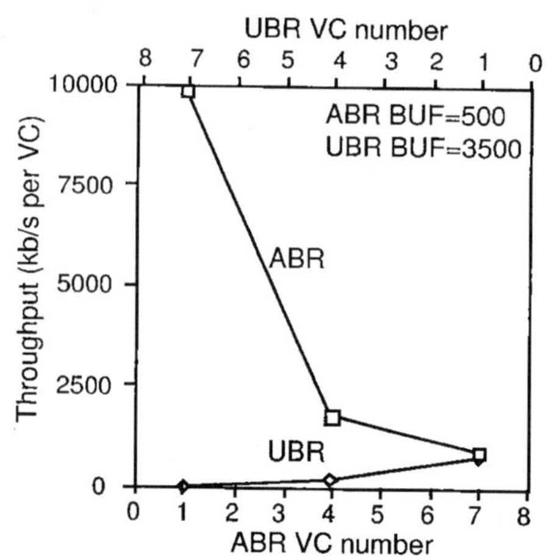

Fig. 17 TCP throughput with SB/PA (ABR BUF=500, UBR BUF=3500).

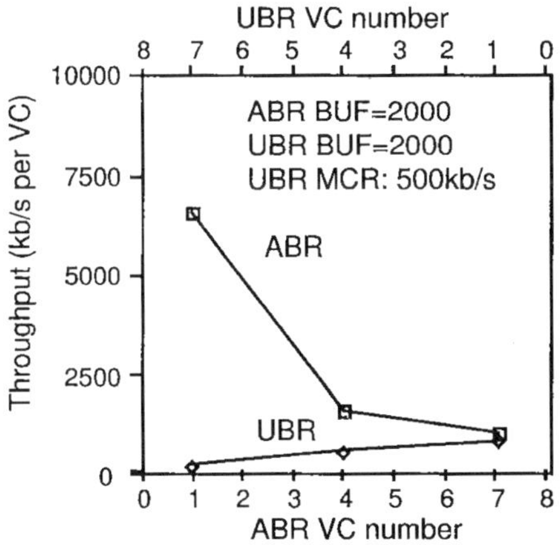

Fig. 18 TCP throughput with GB/OR (ABR BUF=2000, UBR BUF=2000).

ABR throughput with the SB/PA method.

Figure 18 shows throughput using the GB/OR method. The minimum bandwidth of 500 kb/s is allocated to each UBR-VC. Compared to the SB/PA case, UBR can achieve at least 400 kb/s per VC when the number of active UBR-VCs is 7.

With the SB/PA method, the ABR-VC and UBR-VC throughputs depend on the proportion of the number of ABR-VC connections to the total active VCs and the buffer size. Therefore, it is difficult to control ABR-VC and UBR-VC throughput on transport line and access line, but the method is simple mechanism. The minimum bandwidth allocation for UBR, the GB/OR method is more effective for providing a minimum guaranteed TCP throughput for UBR traffic. The result shows the GB/OR method is effective for easy service provisioning when the access network is required to maintain the throughput of UBR traffic.

6. Conclusion

A broadband access system for realizing a cost-effective broadband access network was discussed in this paper. It was shown that the access system with VC concentration and statistical multiplexing functions is effective in terms of access cost. We proposed and evaluated two methods for supporting multiple service classes in the access system: shared bandwidth with preferred ABR method (SB/PA), and guaranteed bandwidth with overriding method (GB/OR). The result indicated that when using the SB/PA method for the access system, ABR throughput and UBR throughput per VC depend on the proportion of the number of ABR-VC connections to the total active VCs and on the each buffer size. And the result showed it is difficult to control ABR-VC and UBR-VC throughputs in the access network by using the SB/PA method, which simple mechanism. With GB/OR method, while ABR-VC throughput and UBR-VC throughput also depend on the proportion of ABR-VCs to the total VCs and on the buffer size, it can offer the minimum guaranteed throughput to the UBR traffic. The result showed the method is effective for easy service provisioning.

Acknowledgments

The authors wish to thank Mr. Ichiro Yamashita of the NTT Optical Network Systems Laboratories, Dr. Shigehiko Suzuki and Dr. Tatsuro Takahashi of the NTT Network Service Systems Laboratories for their valuable comments.

References

[1] T.Kanada, S.Ishikawa, Y.Okumura, and N.Terada, "ATM-PON based FTTH access network and its application to multimedia service experiments," ISSLS96, pp.251-256, 1996.

[2] Y.Takigawa, S.Aoyagi, and E.Maekawa, "ATM based passive double star system offering B-ISDN, N-ISDN and POTS," IEEE GLOBECOM '93, p.14, Nov. 1993.

[3] O.Casals, J.Garcia, and C.Blondia, "A MAC protocol for a Passive Optical Network with Traffic Spacing," Proc. 4th IFIP Conf. High Perf. Networking, 1993.

[4] ITU-T Recommendation G.902, "Framework Recommendation on Functional Access Networks (AN)."

[5] A.Profumo, L.D.Ascoli, M.Costa, and P.Casaschi, "Interoperability in Broadband Access Networks using ATM PONs," Global Information Infrastructure (GII) Evolution, pp.322-331, IOS press, 1996.

[6] A.Gillepie, B.Orth, A.Profumo, and S.Webster, "Evolving access networks: A european perspective," IEEE Communication Magazine, vol.35, pp.47-54, March 1997.

[7] ITU-T Recommendation G.965, "V interfaces at the digital Local Exchange (LE) V5.2 interface (based on 2048 kbit/s) for the support of Access Network (AN)."

[8] ITU-T Draft Recommendation G.VB51, "V-interfaces at the service node (SN)- VB5.1 Reference Point Specification," June 1997.

[9] ITU-T Draft Recommendation G.VB52, "V-interfaces at the service node (SN)- VB5.2 Reference Point Specification," June 1997.

[10] T.Oh-ishi, H.Matsumura, S.Shiraishi, and Y.Hiramatsu, "A study on the structure of the broadband access network," IEICE Technical Report, SSE96-43, Aug. 1996.

[11] P.Newman, "Traffic management for ATM local area networks," IEEE Communication Magazine, pp.44-50, Aug. 1994.

[12] M.W.Garret, "A service architecture for ATM: From applications to scheduling," IEEE Network, vol.10, pp.6-14, May/June 1996.

[13] H.Sato, S.Chaki, H.Kasahara, and H.Ueda, "A Traffic Control Method for ABR/UBR in ATM Node Systems," APSITT'97, pp.12.2.1-12.2.4, 1997.

[14] The ATM Forum Technical Committee, "Traffic management specification version 4.0," ATM Forum Contribution 95-0013R10, 1995.

[15] H.Saito, K.Kawashima, H.Kitazume, A.Koike, M.Ishizuka, and A.Abe, "Performance issue in public ABR service," IEEE Communication Magazine, pp.40-48, Nov. 1996.

[16] T.M.Chen, S.T.Liu, and V.K.Samalam, "The available bit rate service for data in ATM networks," IEEE Communication Magazine, pp.56-71, March 1996.

[17] W.R.Stevens, "TCP/IP Illustrated," Volume 1, Addison-Wesley, Massachusetts, 1996.

High-Performance Backbone Network Technology edited by Naoaki Yamanaka
ISBN: 0-8247-5321-6 DOI: 10.1081/(120025681)
Published 2004 by Marcel Dekker, Inc.

A Pipelined Maximal-Sized Matching Scheme for High-Speed Input-Buffered Switches

Eiji Oki, Roberto Rojas-Cessa, and H. Jonathan Chao

SUMMARY This paper proposes an innovative Pipeline-based Maximal-sized Matching scheduling approach, called PMM, for input-buffered switches. It dramatically relaxes the limitation of a single time slot for completing a maximal matching into any number of time slots. In the PMM approach, arbitration is operated in a pipelined manner, where K subschedulers are used. Each subscheduler is allowed to take more than one time slot for its matching. Every time slot, one of the subschedulers provides the matching result. We adopt an extended version of Dual Round-Robin Matching (DRRM), called iterative DRRM (iDRRM), as a maximal matching algorithm in a subscheduler. PMM maximizes the efficiency of the adopted arbitration scheme by allowing sufficient time for the number of iterations. We show that PMM preserves 100% throughput under uniform traffic and fairness for best-effort traffic of the non-pipelined adopted algorithm, while ensuring that cells from the same virtual output queue (VOQ) are transmitted in sequence. In addition, we confirm that the delay performance of PMM is not significantly degraded by increasing the pipeline degree, or the number of subschedulers, when the number of outstanding requests for each subscheduler from a VOQ is limited to 1.

key words: *scheduling, pipeline, input-buffered switch, maximal-sized matching*

1. Introduction

The explosion of Internet traffic has led to a greater need for high-speed switches and routers that have over 1-Tbit/s throughput [1]. Most high-speed packet switching systems adopt a fixed-size cell in the switch fabric. Variable-length packets are segmented into several fixed-sized cells when they arrive, are switched through the switch fabric, and are reassembled into packets before they depart.

There are various types of buffering strategies in switch architectures: input buffering, output buffering, or crosspoint buffering [2]–[6]. Input buffering is a cost-effective approach for high-speed switches. This is because input-buffered switches do not require internal speedup or allocation of buffers at each crosspoint. It relaxes memory-bandwidth and memory-size constraints.

In input-buffered switches, it is well known that head-of-line (HOL) blocking limits the maximum throughput to 58.6% in a input-buffered switch with the First-In-First-Out (FIFO) structure [7]. A Virtual-Output-Queue (VOQ) structure is used to overcome the HOL-blocking problem [10]. Consider an $N \times N$ input-buffered switch with VOQs at the inputs and a crossbar switch fabric, as shown in Fig. 1. A fixed-size cell is sent from any input to any output, provided that no more than one cell is sent from the same input and no more than one cell is received by the same output. Each input i has N VOQs, each of which is denoted as $VOQ(i,j)$, where cells that are destined for output j are stored. The HOL cell in each VOQ can be selected for transmission across the switch in each time slot. Therefore, every time slot, a scheduler has to determine one matching set.

For an input-buffered switch with VOQs, maximum-sized matching algorithms have been proposed to achieve 100% throughput [8], [9]. Although the maximum-sized matching algorithms provide a maximum match, they suffer from high-computing time complexity. Therefore, it is difficult to implement such algorithms for high-speed switching systems.

Maximal-sized matching algorithms have been proposed as an alternative to maximum matching ones, such as iSLIP [10] and Dual Round-Robin Matching (DRRM) [11], [12]. Both algorithms reduce their computing complexity compared with maximum matching ones, and provide 100% throughput under uniform traffic and complete fairness for best-effort traffic. How-

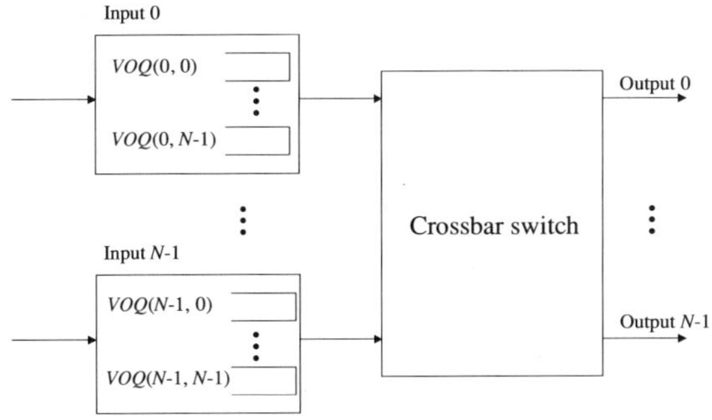

Fig. 1 Input-buffered switch with VOQs.

Manuscript received October 30, 2001.
Manuscript revised February 5, 2002.

†The author is with NTT Network Innovation Laboratories, NTT Corporation, Musashino-shi, 180-8585 Japan. This work was done while he was a Visiting Scholar at Polytechnic University, Brooklyn, NY.
††The authors are with the Department of Computer and Electrical Engineering, Polytechnic University, 6 Metrotech Center, Brooklyn, NY 11201 USA.

*This research is supported by NSF Grant 9906673.

IEICE TRANS. COMMUN., Vol. E85-B, No. 7, pp. 1302–1311, July 2002

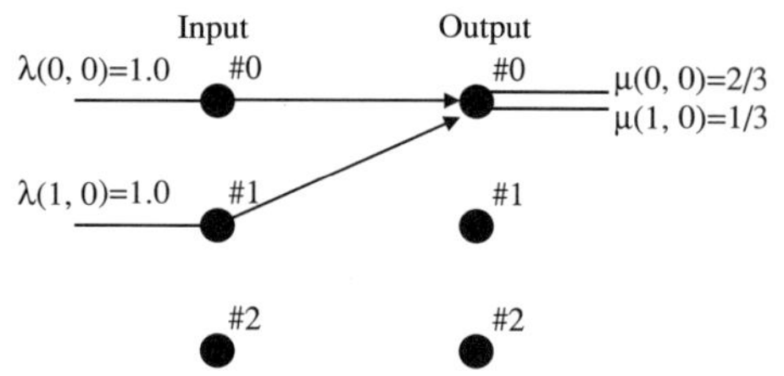

Fig. 2 Example of unfairness for RRGS under unbalanced traffic.

ever, they still have the strict constraint that the maximal matching has to be completed within one time slot. The constraint is a bottleneck when the switch size increases or a port speed becomes high, because the arbitration time becomes longer than one time slot or the time slot shrinks, respectively. Consider a 64-byte fixed-length cell at a port speed of 40 Gbit/s (OC-768). The computation time given for maximal-sized matching is only 12.8 ns.

To relax the strict scheduling timing constraint, a pipeline-based scheduling algorithm called Round-Robin Greedy Scheduling (RRGS) was proposed by Smiljanić et al. [13]. The RRGS switch has N round-robin arbiters, each of which is associated with an input. Each input performs one round-robin arbitration to select one VOQ in one time slot. N round-robin operations that select their own candidates to be transmitted at time slot T are performed during the previous N time slots $\{T-N, T-N+1, \ldots, T-1\}$. Each round-robin operation is assigned into different time slots in a simple pre-determined cyclic manner so that RRGS can avoid output contention. (See Appendix A.)

However, RRGS is not able to provide max-min fair share for a best-effort service. Let $\lambda(i,j)$ and $\mu(i,j)$ be input offered load to $VOQ(i,j)$ and acceptable transmission rate from $VOQ(i,j)$, respectively. Consider a 3×3 switch, and $\lambda(0,0) = \lambda(1,0) = 1.0$ and other input offered loads $\lambda(i,j) = 0$, as shown in Fig. 2. According to the RRGS algorithm described in [13], the acceptable transmission rate is obtained as $\mu(0,0) = 2/3$ and $\mu(1,0) = 1/3$. This is due to the simple cyclic allocation mechanism. The RRGS operation in this example is also described in Appendix A. Thus, when traffic is not balanced, some inputs can unfairly send more cells than others. Smiljanić also proposed weighted-RRGS (WRRGS), which guarantees pre-reserved bandwidth [14]. In WRRGS, however, the fairness problem is not yet solved for best-effort traffic. In addition, every N time-slot cycle, an idle time slot is produced when N is an even number. This means that RRGS does not efficiently utilize the switching capacity.

In addition, RRGS causes a fixed latency of N time slots at worst case. The average fixed latency is $N/2$ time slots. When the switch size is large, the latency will be the dominant delay. Thus, RRGS is not scalable in terms of the switch size.

It is a challenge to find a maximal matching scheduling scheme to meet the following requirements.

- The scheduling time should be relaxed into more than one time slot.
- High throughput should be provided.
- Fairness should be maintained for best-effort traffic.
- Cell sequence should be kept to avoid a cell re-sequence function at an output port.

To our knowledge, no study has been presented to satisfy these requirements.

This paper presents a solution to these requirements. We introduce an innovative Pipeline-based approach for Maximal-sized Matching scheduling in input-buffered switches, called PMM. Within the scheduler, more than one subscheduler operate in a pipelined manner. Each subscheduler is allowed to take more than one time slot. Every time slot, a subscheduler provides a matching result. The subschedulers can adopt a maximal matching algorithm such as *i*SLIP or DRRM, while preserving the properties of the non-pipelined versions. In this paper, we adopt an extended version of DRRM, called iterative DRRM (*i*DRRM), as a maximal matching algorithm in each subscheduler by considering the hardware implementation. *i*DRRM is easier to be implemented than *i*SLIP when round-robin arbiters are implemented in a distributed manner. As a result, PMM provides 100% throughput under uniform traffic, and maintains fairness for best-effort traffic. To use the advantage of the features of the adopted maximal matching algorithm, PMM allows one subscheduler to grant a younger request earlier than another subscheduler to grant an older request. However, PMM ensures that cells from the same VOQ are transmitted in sequence as only the HOL cell is transmitted when a grant is received. Therefore, PMM does not need a re-sequence function at an output port. We investigated the impact of the allowable number of outstanding requests distributed to subschedulers on the PMM delay performance. Furthermore, we show that the delay performance is not significantly degraded by increasing the number of subschedulers when the number of outstanding requests for each subscheduler is limited to 1.

The remainder of this paper is organized as follows. Section 2 describes the PMM scheme. Section 3 describes the performance of the PMM scheme. Section 4 summarizes the key points.

2. Pipeline-Based Maximal-Sized Matching (PMM)

The PMM scheme is able to relax the computation time for maximal-sized matching into more than one time slot. The PMM scheduler consists of N^2 request counters and K subschedulers, as shown in Fig. 3. Each

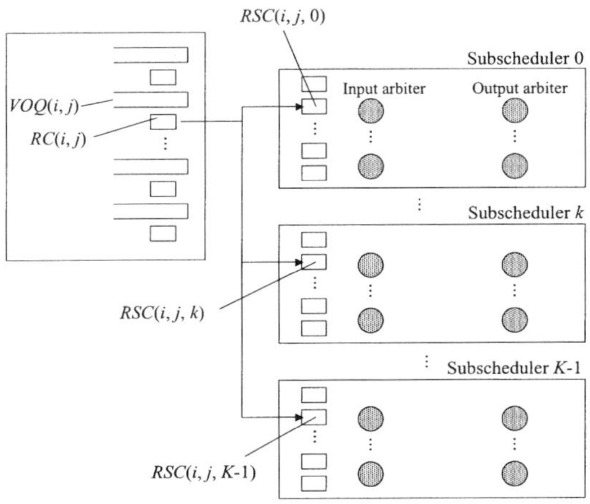

Fig. 3 Structure of PMM scheduler.

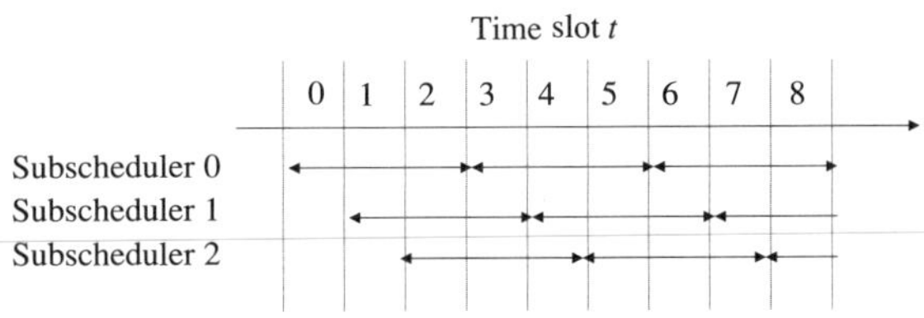

Fig. 4 Timing diagram of PMM with $K = 3$.

subscheduler has N^2 request subcounters. Detailed notations in Fig. 3 are described below. Each subscheduler operates the maximal-sized matching in a pipelined manner, and takes K time slots to complete the matching, as shown in Fig. 4. We assume that PMM adopt an extended version of DRRM in each subscheduler.

DRRM was originally intended to use input and output round-robin arbitration with one iteration. The brief description of the DRRM algorithm is described in Appendix B. In the extended version of DRRM, the number of iterations are equal to or more than one. We call it iterative DRRM (iDRRM). In iDRRM, input and output round-robin arbitration is iterated to find matching input-output pairs among unmatched pairs. The input and output round-robin pointers are updated only when a grant is issued at the first iteration†. iDRRM mainly improves the delay performance of DRRM by considering more iterations.

From the hardware implementation point of view, iDRRM is easier to implement than iSLIP when input and output round-robin arbiters are distributed at the input/output ports [12], [15]. We describe the decentralized implementation comparison of DRRM and iSLIP and clarify why iDRRM is easier than iSLIP in Appendix C††. When a scheduling algorithm is implemented in a decentralized manner, the amount of communication signals between printed circuit boards (PCBs) is sensitive due the limitation of the number of PCB pins and the transmission delay at a signal speed. In iSLIP, a request from $VOQ(i,j)$ at input port i is first sent to an output arbiter at output port j. On the other hand, iDRRM, a request from $VOQ(i,j)$ at input port i is first sent to an input arbiter at input port i. Thus, iSLIP first performs the round-robin arbitration for output ports and then does for input ports, while iDRRM first performs the round-robin arbitration for input ports and then does for output ports. As a result, iSLIP needs multiple requests and multiple grants between an input PCB and the crossbar PCB, while iDRRM needs only one request and one grant between them. In addition, iSLIP has an additional signal compared with iDRRM. Therefore, iDRRM has less communication signals between PCBs than iSLIP.

We note that iDRRM is logically equivalent to iSLIP, although the implementation of iDRRM is simpler than that of iSLIP [12]. The logical equivalence between iDRRM and iSLIP can be easily derived in the same ways as a logical equivalence between Parallel Iterative Matching (PIM) and Logical-Equivalent PIM (LE-PIM) was derived by Nong et al. in [16]. Let Θ be an input-traffic matrix. Each element $\theta(i,j)$ expresses input traffic to $VOQ(i,j)$. The performance of iDRRM with Θ is equivalent to that of iSLIP with Θ^T, where Θ^T is a transposed matrix of Θ.

Several notations used here are defined in the following. A request counter $RC(i,j)$ is associated with $VOQ(i,j)$, as shown in Fig. 3. The value of $RC(i,j)$ is denoted as $C(i,j)$, where $0 \leq C(i,j) \leq L_{max}$. L_{max} is the maximum VOQ occupancy. $C(i,j)$ expresses the number of accumulated requests associated with $VOQ(i,j)$ that have not been sent to any subscheduler. Each request subcounter $RSC(i,j,k)$ is associated with $VOQ(i,j)$ and subscheduler k, where $0 \leq k \leq K-1$. The value of $RSC(i,j,k)$ is denoted as $SC(i,j,k)$, where $0 \leq SC(i,j,k) \leq SC_{max}$. SC_{max} has to be set carefully to provide the best performance. We found that when $SC_{max} = 1$, the delay performance is the best. The impact of SC_{max} will be discussed in detail in Sect. 3. $SC(i,j,k) > 0$ means that input i has at least one request for output j in subscheduler k. $SC(i,j,k) = 0$ means that input i has no request for output j in subscheduler k. At initial time, $C(i,j)$ and $SC(i,j,k)$ are set to zero.

PMM operates as follows.

- Phase 1: When a new cell enters $VOQ(i,j)$, the counter value of $RC(i,j)$ is increased: $C(i,j) = C(i,j) + 1$.
- Phase 2: At the beginning of every time slot t, if $C(i,j) > 0$ and $SC(i,j,k) < SC_{max}$, where $k = t \mod K$, $C(i,j) = C(i,j) - 1$ and $SC(i,j,k) =$

†This operation of iDRRM ensures max-min fair share as that of iSLIP.

††Although Appendix C focuses on the non-pipelined schemes with one iteration for simplicity, the discussion is easily applied to the pipelined schemes with multiple iterations.

$SC(i,j,k)+1$. Otherwise, $C(i,j)$ and $SC(i,j,k)$ are not changed.

- Phase 3: At $Kl+k \leq t < K(l+1)+k$, where l is an integer, subscheduler k operates the maximal-sized matching according to the adopted algorithm.
- Phase 4: By the end of every time slot t, subscheduler k, where $k = (t-(K-1)) \mod K$, completes the matching. When input-output pair (i,j) is matched, $SC(i,j,k) = SC(i,j,k)-1$. The HOL cell in $VOQ(i,j)$ is sent to output j at the next time slot. This ensures that cells from the same VOQ are transmitted in sequence, even if $L(i,j)-C(i,j) > 1$, where $L(i,j)$ is the occupancy of $VOQ(i,j)$. Note that $L(i,j)-C(i,j) = \sum_{k=0}^{K-1} SC(i,j,k) \leq K \times SC_{max}$. In PMM, a request from $RC(i,j)$ belongs to only $VOQ(i,j)$, but not the HOL cell of $VOQ(i,j)$. Therefore, any request that is granted, independent of its age, is used by the HOL cell, thus cells are always transmitted in sequence. A VOQ can accumulate up to $K \times SC_{max}$ issued requests to the PMM scheduler. When input-output pair (i,j) is not matched, $SC(i,j,k)$ is not changed.

Whenever a condition associated with any phase is satisfied, the phase is executed by the main scheduler.

To apply the iDRRM algorithm as a matching algorithm in subscheduler k in PMM, we use $SC(i,j,k)$ instead of VOQ requests as described in the DRRM scheme [11], [12]. Each subscheduler has its own round-robin pointers. The position of pointers in subscheduler k is modified by the results only from subscheduler k. The operation of iDRRM in subscheduler k is the same as that of the non-pipelined iDRRM scheme. Therefore, it may happen that one subscheduler grants to a younger request earlier than another subscheduler does to an older request, but PMM avoids the out-of-sequence problem. This is because a request is not related to a cell but to a VOQ in PMM.

3. Performance

This section describes delay performance and throughput of PMM under uniform traffic, and the fairness for best-effort traffic. In addition, the effect of the PMM scheduling relaxation is described.

3.1 Delay

First, we discuss the dependency of SC_{max}. Let us consider two extreme cases, which are $SC_{max} = L_{max}$ and $SC_{max} = 1$. Here, we assume that L_{max} is large enough to avoid cell loss at a VOQ to simplify the discussion.

Figures 5 and 6 show the average delay with $SC_{max} = L_{max}$ and $SC_{max} = 1$, respectively. Simulation results are obtained with a 95% confidence interval, not greater than 5% for the average cell delay.

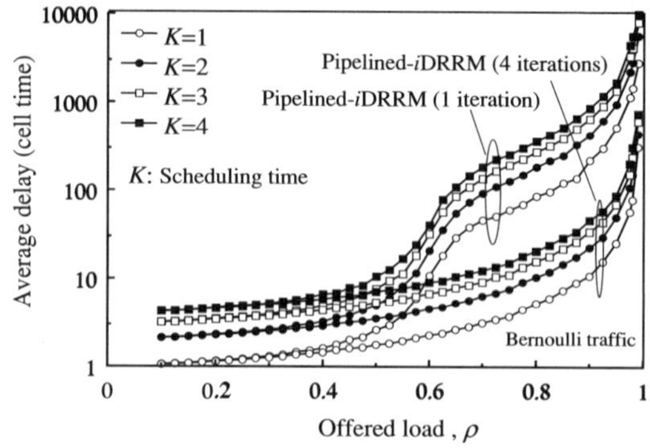

Fig. 5 Delay dependency on scheduling time K for Bernoulli traffic with $SC_{max} = L_{max}$. (switch size $N = 32$)

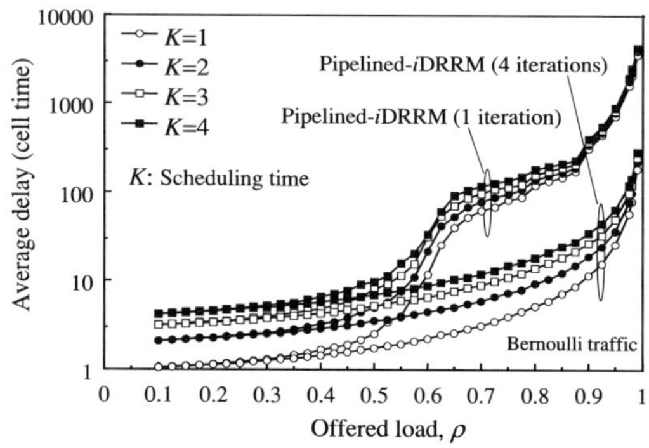

Fig. 6 Delay dependency on scheduling time K for Bernoulli traffic with $SC_{max} = 1$. (switch size $N = 32$)

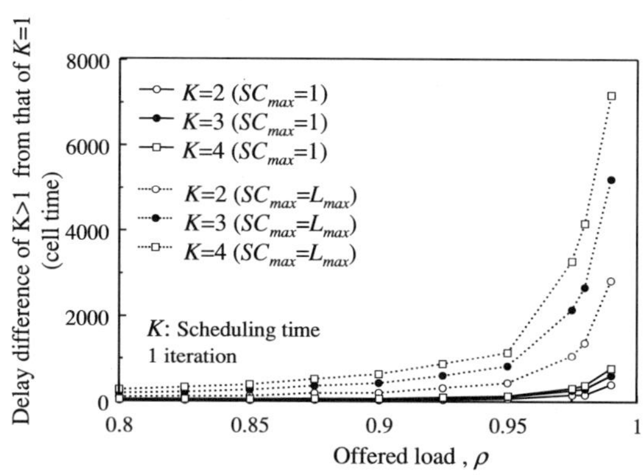

Fig. 7 Delay difference of $K > 1$ from that of $K = 1$. (switch size $N = 32$)

In this evaluation, we include absolute delay caused by the scheduling time in the delay performance.

These two figures show that delay performance degrades with $SC_{max} = L_{max}$, while it does not significantly degrade with $SC_{max} = 1$. Figure 7 shows the delay difference of $K > 1$ from that of $K = 1$. For example, with ρ=0.95, where ρ is the offered load, and $K = 4$, the delay difference of SC_{max}=1 is only 126 cell times, while that of $SC_{max} = L_{max}$ is 1155 cell times. When the input line speed is 10 Gbit/s, 126 cell times are equal to 6.45 μs when assuming a 64-byte

	Initail	t=0 (r)	t=0 (g)	t=1 (r)	t=1 (g)	t=2 (r)	t=2 (g)	t=3 (r)	t=3 (g)	t=4 (r)	t=4 (g)	t=5 (r)	t=5 (g)	t=6 (r)	t=6 (g)	t=7 (r)	t=7 (g)
C (0, 0)	1	0	0	0	0	0	0	0	0	0	0	0	0	0	0	0	0
C (0, 1)	1	0	0	0	0	0	0	0	0	0	0	0	0	0	0	0	0
C (0, 2)	3	2	2	1	1	0	0	0	0	0	0	0	0	0	0	0	0
SC (0, 0, 0)	0	1	1	1	0	0	0	0	0	0	0	0	0	0	0	0	0
SC (0, 1, 0)	0	1	1	1	1	1	1	1	0	0	0	0	0	0	0	0	0
SC (0, 2, 0)	0	1	1	1	1	2	2	2	2	2	2	2	1	1	1	1	0
SC (0, 0, 1)	0	0	0	0	0	0	0	0	0	0	0	0	0	0	0	0	0
SC (0, 1, 1)	0	0	0	0	0	0	0	0	0	0	0	0	0	0	0	0	0
SC (0, 2, 1)	0	0	0	1	1	1	0	0	0	0	0	0	0	0	0	0	0

A request is granted.

Fig. 8 Example of counter status with $SC_{max} = L_{max}$.

	Initail	t=0 (r)	t=0 (g)	t=1 (r)	t=1 (g)	t=2 (r)	t=2 (g)	t=3 (r)	t=3 (g)	t=4 (r)	t=4 (g)	t=5 (r)	t=5 (g)
C (0, 0)	1	0	0	0	0	0	0	0	0	0	0	0	0
C (0, 1)	1	0	0	0	0	0	0	0	0	0	0	0	0
C (0, 2)	3	2	2	1	1	1	1	0	0	0	0	0	0
SC (0, 0, 0)	0	1	1	1	0	0	0	0	0	0	0	0	0
SC (0, 1, 0)	0	1	1	1	1	1	1	1	0	0	0	0	0
SC (0, 2, 0)	0	1	1	1	1	1	1	1	1	1	1	1	0
SC (0, 0, 1)	0	0	0	0	0	0	0	0	0	0	0	0	0
SC (0, 1, 1)	0	0	0	0	0	0	0	0	0	0	0	0	0
SC (0, 2, 1)	0	0	0	1	1	1	0	1	1	1	0	0	0

A request is granted.

Fig. 9 Example of counter status with $SC_{max} = 1$.

cell. Thus, K does not affect the delay performance with $SC_{max} = 1$ for a practical use.

We explain why the delay performance of $SC_{max} = L_{max}$ is significantly degraded although that of $SC_{max} = 1$ is not. When the value of SC_{max} is large, some subschedulers have more requests, while others have less requests. The large value of SC_{max} causes unbalance of the number of requests among subschedulers. The unbalanced request distribution makes delay performance worse.

Figures 8 and 9 show examples of counter status for $C(i,j)$ and $SC(i,j,k)$ with $SC_{max} = L_{max}$ and $SC_{max} = 1$, respectively. We consider $N = 3$ and $K = 2$. In both cases, $C(0,0) = 1$, $C(0,1) = 1$, and $C(0,2) = 3$ are set and other counter values of $C(i,j)$ and $SC(i,j,k)$ are zero at the initial condition. We assume that no additional new cell enters a VOQ and all round-robin pointers point at zero. In these figures, $t = T$ (r) means a request phase, which is phase 2 as described in Sect. 2. $t = T$ (g) means a grant phase, which is phase 4. The counter status follows the PMM algorithm as described in Sect. 2.

In Fig. 8, at $t = 0$ (r), $C(0,0)$, $C(0,1)$, and $C(0,2)$ are decremented and $SC(0,0,0)$, $SC(0,1,0)$, and $SC(0,2,0)$ are incremented. At $t = 1$ (r), $C(0,2)$ is decremented to 1 and $SC(0,2,1)$ is incremented. At $t = 1$ (g), $SC(0,0,0)$ is granted by subscheduler 0, because the input round-robin pointer points at zero. The pointer is updated to 1. At $t = 2$ (r), $C(0,2)$ is decremented and $SC(0,2,0)$ is incremented. At this time, all the values of $C(i,j)$ are zero. At $t = 2$ (g), $SC(0,2,1)$ is granted by subscheduler 1. $SC(0,1,0)$, $SC(0,2,0)$, and $SC(0,2,0)$ are granted by subscheduler 0 at $t = 3$ (g), $t = 5$ (g), and $t = 7$ (g), respectively. The main problem of $SC_{max} = L_{max}$ is that there is no request in subscheduler 1 from $t = 3$ (r), although subscheduler 0 has some requests until $t = 7$ (g).

On the other hand, in Fig. 9, at $t = 2$ (r), $SC(0,2,0)$ is not fed by $C(0,2)$. This is because $SC(0,2,0)$ is equal to $SC_{max}(= 1)$. $C(0,2)$ is decremented and $SC(0,2,1)$ is incremented at $t = 3$ (r). At this time, the values of all $C(i,j)$ are zero. $SC(0,0,0)$, $SC(0,2,1)$, $SC(0,1,0)$, $SC(0,2,1)$, and $SC(0,2,0)$ are granted at $t = 1$ (g), $t = 2$ (g), $t = 3$ (g), $t = 4$ (g), and $t = 5$ (g), respectively. Thus, all requests of $SC_{max} = 1$ are granted two cell time slots earlier than that of $SC_{max} = L_{max}$, although the values of all $C(i,j)$ of $SC_{max} = 1$ are zero one time slot later than those of $SC_{max} = L_{max}$.

From these observations, when we set SC_{max} to 1, the requests from $RC(i,j)$ are distributed more effectively and the delay degradation is suppressed. When K increases, the requests from $RC(i,j)$ are dispersed more to each $RSC(i,j,k)$ associated with each subscheduler k. As a result, the desynchronization effect becomes less efficient as K increases. The delay degradation with $SC_{max} = 1$ is suppressed better than when $SC_{max} = L_{max}$ and, in this case, the delay is slightly degraded with K.

Since we know that when SC_{max} is set to 1, the delay performance is the best, we discuss the PMM performance with $SC_{max} = 1$ in the following.

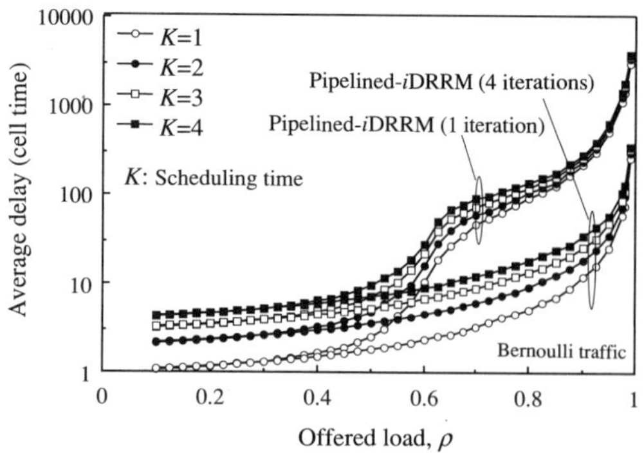

Fig. 10 Delay dependency on scheduling time K for Bernoulli traffic. (switch size $N = 64$)

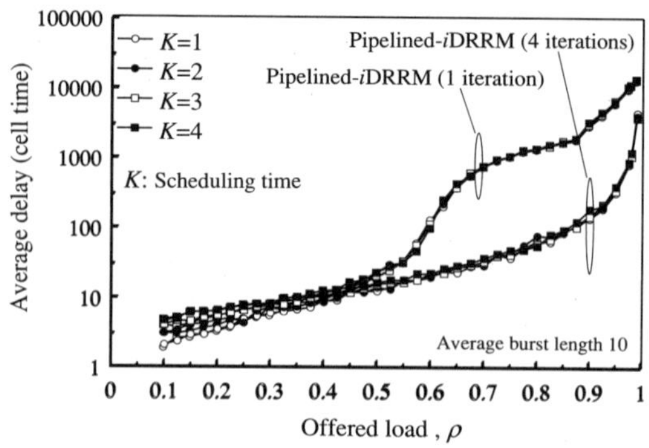

Fig. 11 Delay dependency on scheduling time K for bursty traffic. (switch size $N = 32$)

In Fig. 6, the delay with one iteration increases slightly at $0.65 \leq \rho < 0.9$. At the light offered load, $\rho < 0.5$, the contention among input requests is not significant. At $0.5 \leq \rho < 0.65$, the contention becomes significant. When ρ reaches 0.65, the pointer desynchronization is more easily obtained compared with the light offered load [10], [12]. However, when ρ exceeds 0.9, the contention among input requests significantly affects the delay performance.

The delay performance is improved with more iterations. Since PMM relaxes the scheduling timing constraint, a large number of iterations is not a bottleneck even when the switch size increases or a port speed becomes fast, compared with the non-pipelined algorithm, as will be described in Sect. 3.4. Note that we showed the delay performance up to four iterations because there is no measurable improvement with more iterations.

The above observations for $N = 32$ are applied to different switch sizes. Figure 10 shows the same delay tendency as that in Fig. 6 for a switch with $N = 64$.

We also consider an ON-OFF source model in which an arrival process to an input port alternates between ON (active) and OFF (idle) periods. A traffic source, during the ON period, continues sending cells in every time slot, but stops sending cells in the OFF period. Both the durations on the ON and OFF periods are assumed to be geometrically distributed. The average burst length used in this paper corresponds to the average ON period. In this evaluation, the average burst length is set to 10.

Figure 11 shows that K does not affect the delay performance for a practical use even when a bursty arrival process is considered. The dependency on K under bursty traffic is lighter than under Bernoulli traffic. This is because the arrival process of bursty traffic is less independent than that of Bernoulli traffic.

3.2 Throughput

As we saw the delay performance of PMM in Sect. 3.1, as PMM adopts the *i*DRRM algorithm, PMM provides 100% throughput under uniform traffic.

The reason is as follows. Consider the offered load as 1.0. If some inputs cannot send cells, outstanding requests are maintained in each subscheduler. In other words, $SC(i,j,k) = 1$. As a result, $C(i,j)$ is not decremented in phase 2 and increased in phase 1. Since $C(i,j)$ reaches a large enough value to be always satisfied with $C(i,j) > 0$, $SF(i,j,k) = 1$ is kept in phase 2[†]. In this situation with $SC(i,j,k) = 1$ at any t in phase 3, subscheduler k provides a complete matching result every K time slots due to the desynchronization effect of input and output round-robin arbiters, as described in [12], [17][††].

Thus, PMM preserves the throughput advantage of *i*DRRM.

3.3 Fairness

Since we adopt a round-robin-based algorithm in subscheduler k, subscheduler k can maintain a fair scheduling. Therefore, PMM provides max-min fair share for best-effort traffic.

3.4 Scheduling Timing Relaxation

Figure 12 shows the effect of the PMM scheduling timing relaxation. We assume that a cell size, L_{cell}, is 64×8 bits. Let the allowable arbitration time per iteration, a port speed, and the number of iterations be T_{arb}, C, and I, respectively. T_{arb} is given by,

$$T_{arb} = \frac{KL_{cell}}{CI}. \quad (1)$$

T_{arb} decreases with I and C, but increases with K. In the non-pipelined *i*DRRM scheme, K is 1 as a special case of PMM in Eq. (1). In the non-pipelined

[†]Although $SC(i,j,k)$ becomes 0 in phase 4 when a request is granted, $SC(i,j,k)$ is always changed to 1 in phase 2.

[††]We note that the pointer desynchronization is achieved within the same subscheduler. There is no relationship between pointers in different subschedulers.

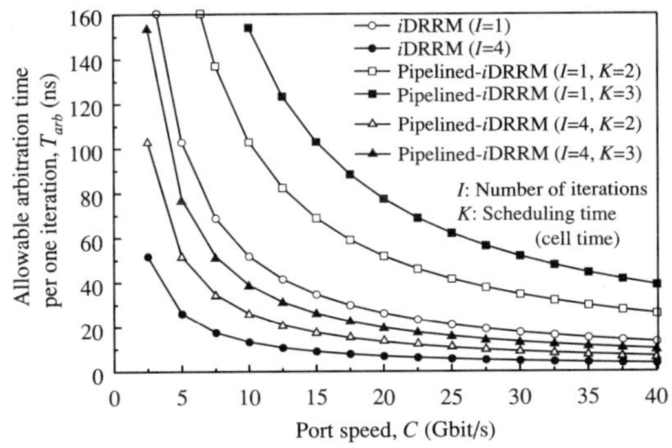

Fig. 12 Effect of PMM scheduling timing relaxation.

iDRRM, when C=40 Gbit/s and $I = 4$, T_{arb}=3.2 ns. Under this timing constraint, it is difficult to implement a round-robin arbiter that supports a large N in hardware by using available CMOS technologies, in which, for example, a typical gate-delay time is about 100 ps [18]. In this case, PMM can be used to expand T_{arb} by increasing K. When C=40 Gbit/s, $I = 4$ and $K = 3$, T_{arb}=9.6 ns. Therefore, PMM achieves the desired number of iterations even when N increases or C becomes fast.

4. Conclusions

This paper has proposed a Pipeline-based Maximal-sized Matching scheduling approach, called PMM, for input-buffered switches. It dramatically relaxes the scheduling timing constraint. Each subscheduler is allowed to take more than one time slot to complete a maximal-size matching. Every time slot, one subscheduler provides the matching results. The subschedulers adopt an extended version of DRRM, called iterative DRRM (iDRRM). iDRRM offers flexibility for implementation in a centralized or distributed manner. PMM maximizes the efficiency of the adopted arbitration scheme by allowing sufficient time for a number of iterations. We showed that PMM provides 100% throughput under uniform traffic and fairness for best-effort service as the non-pipelined algorithm does, while ensuring that cells from the same VOQ are transmitted in sequence. In addition, we investigated the impact of SC_{max} on the PMM delay performance. The delay performance is not significantly degraded by increasing the number of subschedulers when SC_{max} is set to 1.

The hardware in PMM is proportional to K, but PMM provides the flexibility to implement the schedulers in a distributed manner so that the increasing of hardware has a slight impact in the overall hardware amount.

Thus, PMM is a solution to a maximal-sized matching scheduler for input-buffered switches when the switch size increases or a port speed becomes high, such as OC-768.

References

[1] N. Yamanaka, E. Oki, S. Yasukawa, R. Kawano, and K. Okazaki, "OPTIMA: Scalable, multi-stage, 640-Gbit/s ATM switching system based on advanced electronic and optical WDM technologies," IEICE Trans. Commun., vol.E83-B, no.7, pp.1488-1496, July 2000.

[2] G. Nong and M. Hamdi, "On the provision of quality-of-service guarantees for input queued switches," IEEE Commun. Mag., vol.38, no.12, pp.62–69, Dec. 2000.

[3] E. Oki, N. Yamanaka, Y. Ohtomo, K. Okazaki, and R. Kawano, "A 10-Gb/s (1.25 Gb/s × 8) 4 × 2 0.25-μm CMOS/SIMOX ATM switch based on scalable distributed arbitration," IEEE J. Solid-State Circuits, vol.34, no.12, pp.1921–1934, Dec. 1999.

[4] R. Rojas-Cessa, E. Oki, Z. Jing, and H.J. Chao, "CIXB-1: Combined input-one-cell-crosspoint buffered switch," Proc. 2001 IEEE Workshop on High Performance Switching and Routing (HPSR), pp.324–329, May 2001.

[5] J. Turner and N. Yamanaka, "Architectural choices in large scale ATM switches," IEICE Trans. Commun., vol.E81-B, no.2, pp.120–137, Feb. 1998.

[6] H.J. Chao, C.H. Lam, and E. Oki, Broadband Packet Switching Technologies: A Practical Guide to ATM Switches and IP Routers, John Wiley & Sons, New York, 2001.

[7] M.J. Karol, M.G. Hluchyj, and S.P. Morgan, "Input versus output queueing on a space-division packet switch," IEEE Trans. Commun., vol.COM-35, no.12, pp.1347–1356, 1987.

[8] J.E. Hopcroft and R.M. Karp, "An algorithm for maximum matching in bipartite gragphs," Soc. Ind. Appl. Math. J. Comptation, vol.2, pp.225–231, 1973.

[9] N. Mckeown, A. Mekkittikul, V. Anantharam, and J. Walrand, "Achieving 100% throughput in an input-queued switches," IEEE Trans. Commum., vol.47, no.8, pp.1260–1267, Aug. 1999.

[10] N. Mckeown, "The iSLIP scheduling algorithm for input-queued switches," IEEE/ACM Trans. Networking, vol.7, no.2, pp.188–200, April 1999.

[11] H.J. Chao and J.-S. Park, "Centralized contention resolution schemes for a large-capacity optical ATM switch," Proc. IEEE ATM Workshop'97, Fairfax, VA, May 1998.

[12] H.J. Chao, "Saturn: A terabit packet switch using dual round-robin," IEEE Commun. Mag., vol.38, no.12 pp.78–84, Dec. 2000.

[13] A. Smiljanić, R. Fan, and G. Ramamurthy, "RRGS-round-robin greedy scheduling for electronic/optical terabit switches," Proc. IEEE Globecom'99, pp.1244–1250, 1999.

[14] A. Smiljanić, "Flexible bandwidth allocation in terabit packet switches," Proc. IEEE Workshop on High Performance Switching and Routing 2000, pp.233–239, 2000.

[15] K. Genda, Y. Doi, K. Endo, T. Kawamura, and S. Sasaki, "A 160-Gb/s ATM switching system using an internal speed-up crossbar switch," Proc. IEEE GLOBECOM'94, pp.123–133, 1994.

[16] G. Nong, J.K. Muppala, and M. Hamdi, "Analysis of nonblocking ATM switches with multiple input queues," IEEE/ACM Trans. Networking, vol.7, no.1, pp.60–74, Feb. 1999.

[17] Y. Li, S. Panwar, and H.J. Chao, "On the performance of a dual round-robin switch," Proc. IEEE INFOCOM 2001, pp.1688–1697, 2001.

[18] Texas Instruments, "GS40 0.15-μm CMOS, Standard Cell/Gate Array," http://www.ti.com/, version 0.2, May 2000.

Appendix A: Unfairness of RRGS

Before we explain why RRGS suffers from the unfairness problem, we describe briefly the RRGS algorithm. Figure A·1 shows an example of the RRGS algorithm. Let us consider a 3×3 switch. We denote the round-robin arbiter associated with input i as I_i. At time slot T_0, I_0 chooses a candidate to be transmitted at time slot T_3. At time slot T_1, I_1 and I_2 choose their own candidates to be transmitted at time slot T_3 and T_4, respectively. At time slot T_2, I_2, I_0, and I_1 choose their own candidates to be transmitted at time slot T_3, T_4, and T_5, respectively. In this way, at each time slot, each input arbiter chooses one candidate to be transmitted at different time slot from other arbiters. If we focus on a certain transmission time slot, for example, T_3, the candidates are selected at T_0, T_1, and T_2.

Next, we explain the unfairness problem of RRGS. We consider input traffic as shown in Fig. 2. The results obtained by RRGS are shown in Fig. A·2. At time slot T_0, I_0 chooses $VOQ(0,0)$ as a candidate to be transmitted at time slot T_3. Output 0 for T_3 is reserved. At time slot T_1, I_1 and I_2 try to choose their own candidates to be transmitted at time slot T_3 and T_4, respectively. However, I_1 cannot select any candidate for T_3 because $VOQ(1,0)$ is the only non-empty VOQ at input 1 and output 0 for T_3 is already reserved. I_2 cannot select any candidate for T_4 because all the VOQs at input 2 are empty. As a result, output 0 for T_4 is not reserved at T_1. At time slot T_2, I_0 and I_1 choose $VOQ(0,0)$ and $VOQ(1,0)$ as candidates to be transmitted at time slot T_4 and T_5, respectively. As can be seen in Fig. A·2, every three-time-slot cycle, $VOQ(0,0)$ is selected twice while $VOQ(1,0)$ is once. Thus, when traffic is not balanced, as in the example shown in Fig. 2, input 0 can unfairly send twice more cells than input 1.

T_0	T_1	T_2	T_3	T_4	T_5	T_6
		I_1 for T_5	I_2 for T_5	I_0 for T_5	I_1 for T_8	I_2 for T_8
	I_2 for T_4	I_0 for T_4	I_1 for T_4	I_2 for T_7	I_0 for T_7	I_1 for T_7
I_0 for T_3	I_1 for T_3	I_2 for T_3	I_0 for T_6	I_1 for T_6	I_2 for T_6	I_0 for T_9

Fig. A·1 Timing diagram of RRGS switch.

T_0	T_1	T_2	T_3	T_4	T_5	T_6
		VOQ(1,0) for T_5	*Not selected*	*Not selected*	*VOQ(1,0)* for T_8	*Not selected*
	Not selected	*VOQ(0,0)* for T_4	*Not selected*	*Not selected*	*VOQ(0,0)* for T_7	*Not selected*
VOQ(0,0) for T_3	*Not selected*	*Not selected*	*VOQ(0,0)* for T_6	*Not selected*	*Not selected*	*VOQ(0,0)* for T_9

Fig. A·2 Candidates selected by RRGS.

Appendix B: DRRM Algorithm

We briefly describe the DRRM algorithm. The more details are described in [11], [12]. Although the scheduling algorithm with one iteration is presented here, it is easy to apply it for *i*DRRM in the same way as *i*SLIP [9].

DRRM arbitration has four steps in cycle:

- Each input arbiter performs request selection.
- The input arbiters send their requests to the output arbiters.
- Each output arbiter performs grant arbitration.
- The output arbiters send grant signal to input arbiters.

Figure A·3 shows an example of the DRRM arbitration algorithm. In a request phase, each input chooses a VOQ and sends a request to an output arbiter. Assume input 0 has cells destined for outputs 0 and 1. Since its round-robin pointer is pointing 0, input arbiter 0 sends a request to output 0. Let us consider output 2 in the grant phase. Since its round-robin pointer is pointing to input 2, the output arbiter grants input 2. The input and output pointers are updated only when a grant is issued.

Similar to *i*SLIP, the DRRM scheme has the desynchronization effect. Consider the fully loaded situation in which every VOQ always has cells. Figure A·4 shows the head-of-line cells chosen from each input port in different time slots. In time slot 1, each input port chooses a cell destined for output A. Among those three cells, only one (the first in this example) is granted and the other two have to wait at HOL. The round-robin pointer of the first input advances to point to output B

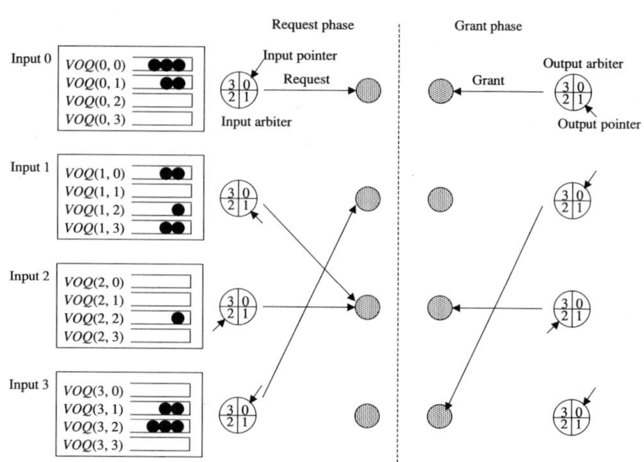

Fig. A·3 An example of the dual round-robin scheduling algorithm.

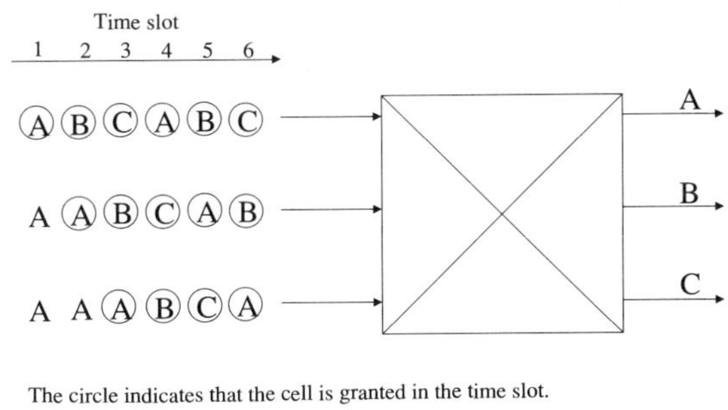

Fig. A· 4 The desynchronization effect of DRRM under the fully loaded situation.

in time slot 2, and a cell destined for B is chosen and then granted because of no contenders. The other two inputs have their HOL cells unchanged, both destined for output A. Only one of them (the one from the second input) is granted; the other has to wait until the third time slot. At that time, the round-robin pointers among the three inputs have been desynchronized and point of C, B, and A, respectively. As a result, all three cells chosen are granted.

Appendix C: Decentralized Implementation of DRRM and *i*SLIP

The *i*SLIP scheduler is implemented in a centralized manner [10]. However, when the switch size increases, the scheduler cannot be implemented in a single chip due the limitations of the gate and the pin number. Even if a multiple-chip implementation in a centralized manner is employed, the bottleneck of the number of pins in a chip still exits. The DRRM scheduler has also the same problem as *i*SLIP when it is implemented in a centralized manner. Therefore, a decentralized implementation should be adopted for a large-size switch.

We describe why DRRM is easier to implement than *i*SLIP when input and output round-robin arbiters are distributed at the input/output ports. Figures A· 5 and A· 6 show schematic examples of DRRM and *i*SLIP when they are implemented in a decentralized manner, respectively. We consider one iteration in these figures for simplicity.

In Fig. A· 5, input arbiter $A_I(i)$ for input port i is located in input port controller (IPC) i and output arbiter $A_O(j)$ for output port j is located in a crossbar part [12]. An example of the distributed round-robin arbiters was presented in [6], [15]. IPC i is usually implemented in a single PCB. In DRRM, $A_I(i)$ first selects one request out of several ones. IPC i sends the selected request to the crossbar. $A_O(j)$ selects one grant and the crossbar sends the grant to IPC i. These communications has to be done between different PCBs.

In Fig. A· 6, $A_I(i)$ and $A_O(j)$ of *i*SLIP are implemented in the same way as DRRM. In *i*SLIP, IPC i first sends multiple requests at most N to the crossbar. $A_O(j)$ selects one grant and the crossbar sends multiple

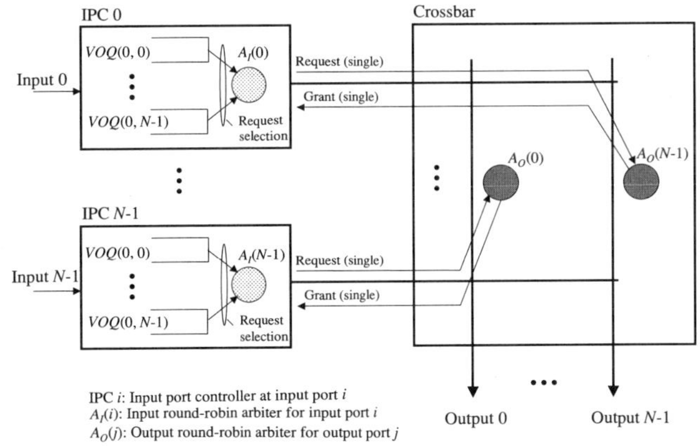

Fig. A· 5 Decentralized implementation of DRRM.

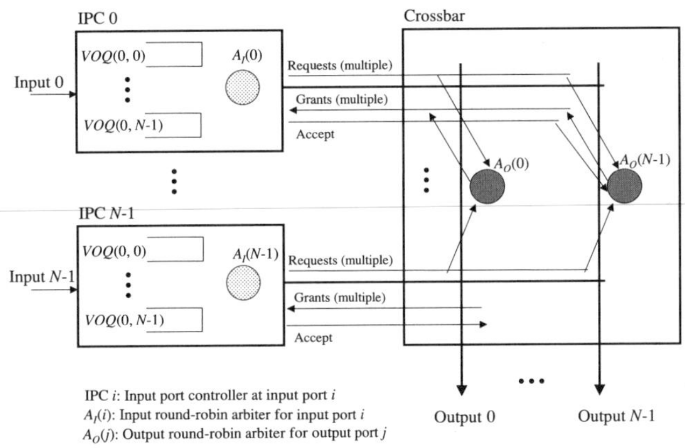

Fig. A· 6 Decentralized implementation of *i*SLIP.

grants up to N to IPC i. $A_I(i)$ accepts one grant and sends the accept to the crossbar.

Thus, *i*SLIP needs multiple requests and multiple grants between IPC i and the crossbar, while DRRM needs only one request and one grant between them. In addition, *i*SLIP has an additional accept signal compared with DRRM. Since DRRM has less communication signals between PCBs than *i*SLIP, DRRM is easier to implement in a decentralized manner than *i*SLIP.

High-Performance Backbone Network Technology edited by Naoaki Yamanaka
ISBN: 0-8247-5321-6 DOI: 10.1081/(120025682)
Published 2004 by Marcel Dekker, Inc.

Design and Analysis of a Packet Concentrator

Yiu-Wing Leung

SUMMARY Packet concentrators are used in many high-speed computer communication systems such as fast packet switches. In these systems, the time available for concentration is very short. It is therefore desirable to realize the packet concentrators as hardware chips for fast concentration. The knockout concentrator was proposed for hardware realization. In this paper, we improve this concentrator to reduce the probability of packet loss, and the improved concentrator is called *wraparound knockout concentrator*. This concentrator has several wraparound paths within it, and it does not require any additional pin per chip. After contention among the packets in a slot, each winner goes to a distinct output, some losers circulate along the wraparound paths for contention in the subsequent slot, and the remaining losers are discarded. In this manner, some losers are not discarded immediately and they still have the chance to go to the outputs in the subsequent slot, thereby reducing the probability of packet loss. We analyze the number of logic gates required and the probability of packet loss. The numerical results show that if the proposed concentrator has a few wraparound paths, the probability of packet loss can already be reduced by orders of magnitude.

key words: *computer communication systems, switches, concentrators, design and analysis*

1. Introduction

Consider a packet concentrator with N inputs and L outputs where $N > L$. The time of each input is divided into time slots, where each slot can accommodate one data unit called packet. When L inputs or less than L inputs have incoming packets in a slot, the concentrator conveys each packet to a distinct output. When more than L inputs have incoming packets in a slot, the concentrator conveys L packets to L respective outputs and discards the other packets. In this case, there is packet loss.

Packet concentrators are used in many high-speed computer communication systems such as fast packet switches [1]–[8]. In these systems, the time available for concentration is very short and hence it is desirable to realize the packet concentrators as hardware chips for fast concentration. The knockout concentrator was proposed for hardware realization [2]. Figure 1(a) shows a knockout concentrator with eight inputs and four outputs. It is composed of 2×2 contention elements. When a contention element has one incoming packet, it conveys this packet to the left output; when it has two incoming packets, it selects a packet as the winner randomly, conveys the winner to the left output and conveys the loser to the right output (see Fig. 1(b)). A knockout concentrator is constructed by properly interconnecting the contention elements, so that it can select up to L packets for output through contention.

A knockout concentrator requires about NL contention elements, where each contention element can be realized using 16 logic gates [2]. When this concentrator is realized as a chip, the limitation is the number

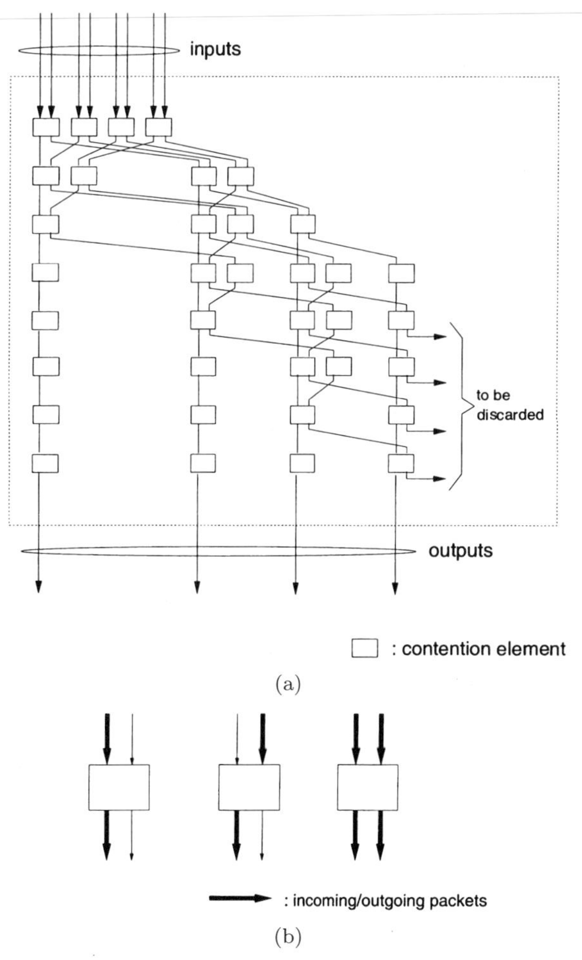

Fig. 1 A knockout concentrator. (a) A knockout concentrator is composed of 2×2 contention elements. (b) The operation of the 2×2 contention element.

Manuscript received July 1, 1999.
Manuscript revised December 13, 1999.

[†]The author is with the Department of Computer Science, Hong Kong Baptist University, Kowloon Tong, Hong Kong.

IEICE TRANS. COMMUN., Vol. E83-B, No. 5, pp. 1115–1121, May 2000

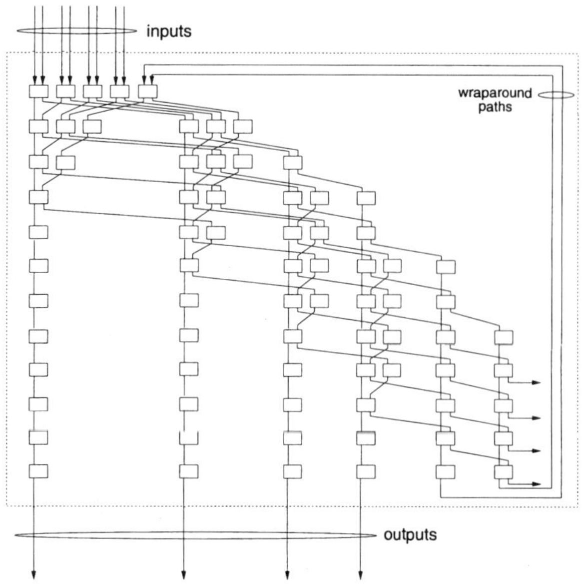

Fig. 2 A wraparound knockout concentrator with two wraparound paths. It requires the same number of pins as the knockout concentrator.

of pins per chip [2], [9]–[11]. For example, if a chip can provide 200 pins for the inputs and outputs, a knockout concentrator with 180 inputs and 20 outputs only require 3994 contention elements or 63904 logic gates on a chip and it can be realized using the current technology [11]–[13].

In this paper, we improve the knockout concentrator to reduce the probability of packet loss, and the improved concentrator is called *wraparound knockout concentrator*. The main idea is to add several wraparound paths within the concentrator, such that some packets that lose in a contention in a slot are not discarded immediately but they circulate along the wraparound paths for contention in the subsequent slot. We demonstrate that the proposed concentrator has a significantly smaller probability of packet loss than the knockout concentrator and it can be realized using the same number of pins per chip.

2. Wraparound Knockout Concentrator

2.1 Design

The wraparound knockout concentrator is an improved version of the knockout concentrator. We add M wraparound paths within the concentrator where M is a design parameter. For example, Fig. 2 shows a concentrator with two wraparound paths. Suppose n packets are to be concentrated in a slot. They contend for going to the outputs, and we distinguish the consequence of contention into the following three cases:

- Case 1: $n \leq L$. All the n packets are winners. The concentrator conveys each winner to a distinct output (e.g., see Fig. 3(a)).
- Case 2: $L < n \leq L+M$. L packets are winners and $n-L$ packets are losers. The concentrator conveys each winner to a distinct output, and circulates each of the $n-L$ losers along a distinct wraparound path for contention in the subsequent slot (e.g., see Fig. 3(b)).
- Case 3: $n > L + M$. L packets are winners and $n - L$ packets are losers. The concentrator conveys each winner to a distinct output, circulates M losers along M respective wraparound paths for contention in the subsequent slot, and discards the remaining $n - L - M$ losers (e.g., see Fig. 3(c) and Fig. 3(d)).

In cases 2 and 3, some losers are not discarded immediately and they still have the chance to go to the outputs in the subsequent slot. Therefore, the probability of packet loss is smaller.

After circulating along a wraparound path, the packet reaches a contention element. It stays in this contention element for a short duration in order to synchronize with the newly incoming packets. In the subsequent slot, the new packets and the circulated packets contend for going to the outputs.

The wraparound knockout concentrator does not require any additional pin per chip, because the wraparound paths can be located within the concentrator. However, it requires additional logic gates to realize the additional contention elements required. The original concentrator requires a moderate number of logic gates per chip because of the constraint on the number of pins per chip (see Sect. 1), and the number of additional logic gates required is also moderate (see Sect. 3). Consequently, the wraparound knockout concentrator requires a moderate number of logic gates and it can be realized using the current VLSI technology. For example, a wraparound knockout concentrator with 256 inputs, 12 outputs, and 4 wraparound paths requires 71200 logic gates (see Sect. 3).

Remark 1: When it is necessary to preserve the packet sequence, we can increase the priority of a packet when it is circulated for contention in the subsequent slot. When two packets of different priority enter a contention element, the high priority packet always wins. When $M \leq L$, two priority levels are sufficient because the number of high priority packets is less than or equal to the number of low priority packets in any slot.

Remark 2: The maximum size of a wraparound knockout concentrator on a chip is limited by the number of pins on this chip. Once a wraparound knockout concentrator has been fabricated on a chip, we can cascade multiple chips to construct a larger concentrator. For example, if we want to construct a 512×32 concentrator, we can connect the 512 inputs to four 128×32 concentrators, and then connect the outputs of these concentrators to a 128×32 concentrator. In

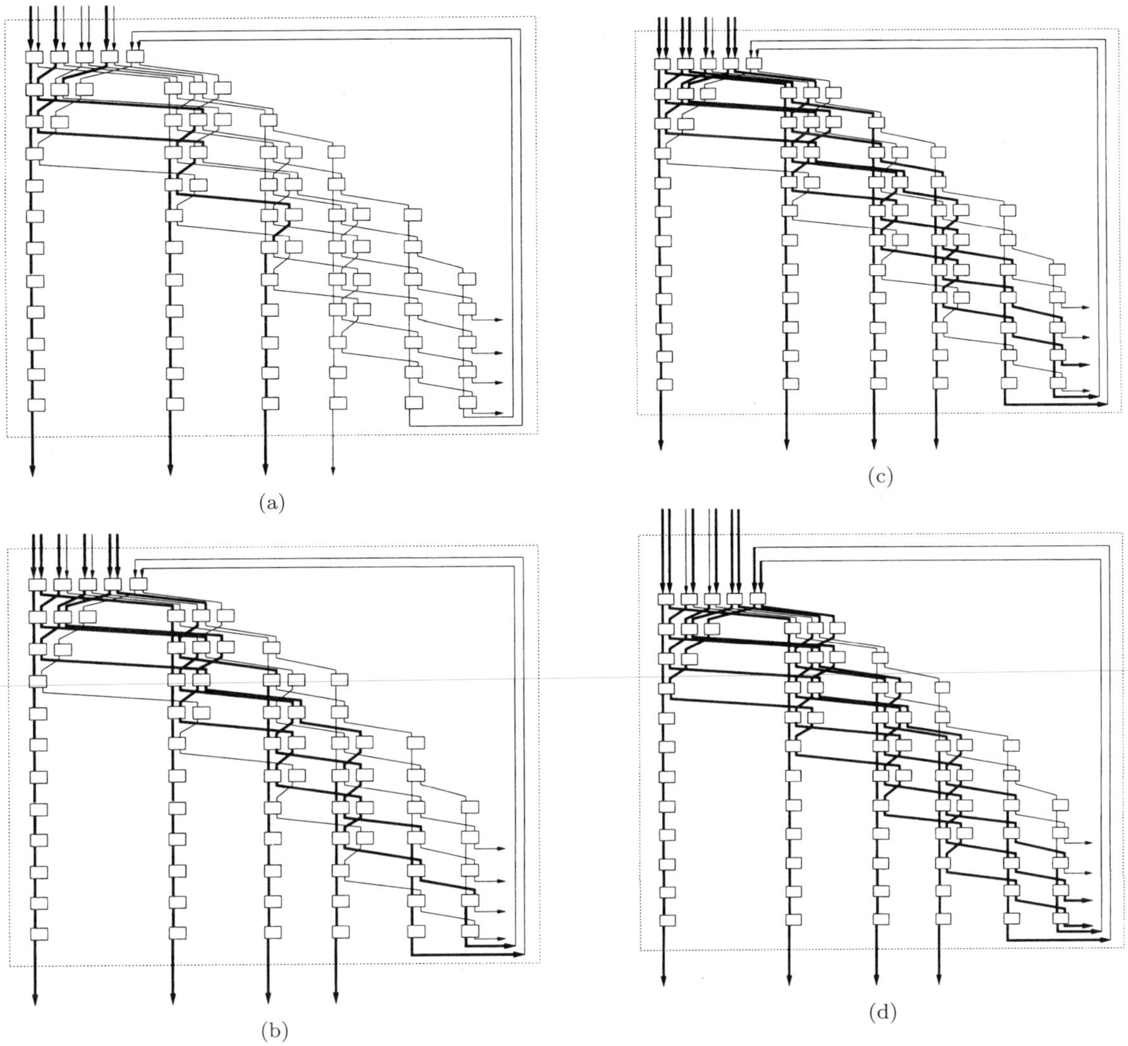

Fig. 3 Operation of the wraparound knockout concentrator. (a) There are three new packets. In this slot, no packet is circulated or discarded. (b) There are six new packets. In this slot, two packets are circulated and no packet is discarded. (c) There are seven new packets. In this slot, two packets are circulated and one packet is discarded. (d) There are six new packets and two circulated packets from the previous slot. In this slot, two packets are circulated and two packets are discarded.

other words, we can use five 128×32 concentrators to construct a 512×32 concentrator.

2.2 Number of Contention Elements

A knockout concentrator with N inputs and L outputs has about NL contention elements, where this number is only an approximation [2]. In this subsection, we compute the exact number of contention elements in the wraparound knockout concentrator and in the knockout concentrator. Since a contention element can be realized using 16 logic gates, the total number of logic gates required is then equal to 16 times the number of contention elements.

We divide a wraparound knockout concentrator into multiple *stages* and multiple *outlets* as shown in Fig. 4. We let $\mathbf{I} = [I_{i,j}]$ where $I_{i,j}$ is equal to the number of inputs to stage i of outlet j. $\mathbf{I}$ has $L+M$ columns because there are $L+M$ outlets. For the concentrator shown in Fig. 4, $\mathbf{I}$ is given by:

$$\mathbf{I} = \begin{bmatrix} 8 & 0 & 0 & 0 \\ 4 & 4 & 0 & 0 \\ 2 & 4 & 2 & 0 \\ 1 & 3 & 3 & 1 \\ 1 & 2 & 3 & 2 \\ 1 & 1 & 3 & 2 \\ 1 & 1 & 2 & 2 \\ 1 & 1 & 1 & 2 \end{bmatrix}$$

We determine the rows of $\mathbf{I}$ one after the other as follows. To determine the 1st row of $\mathbf{I}$, we note that the N inputs and the M wraparound paths only enter

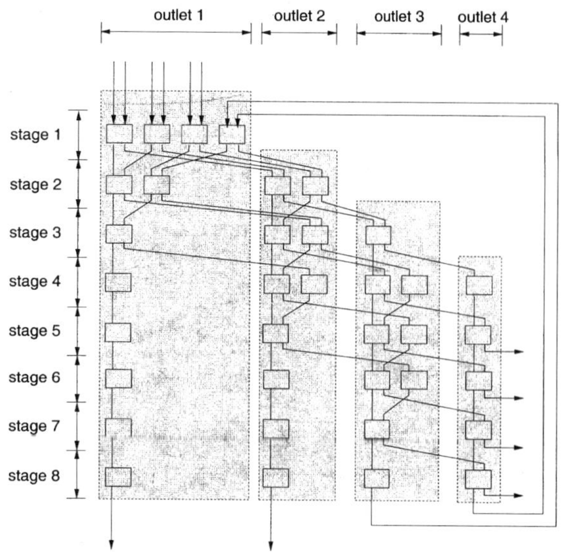

Fig. 4 A wraparound knockout concentrator is divided into multiple stages and multiple outlets.

stage 1 of outlet 1. Therefore, we have

$$I_{1,j} = \begin{cases} N+M & j=1 \\ 0 & 1<j\le L+M \end{cases} \tag{1}$$

We now determine the ith row for $i>1$. For outlet 1, the inputs of stage i are connected to half the number of outputs from stage $i-1$ of the same outlet. For outlet j $(1<j\le L+M)$, the inputs of stage i are connected to: (1) half the number of outputs from stage $i-1$ of the same outlet and (2) half the number of outputs from stage $i-1$ of the previous outlet. Therefore, $I_{i,j}$ is given by $(i>1)$:

$$I_{1,j} = \begin{cases} \left\lceil \frac{I_{i-1,1}}{2} \right\rceil & j=1 \\ \left\lfloor \frac{I_{i-1,j-1}}{2} \right\rfloor + \left\lceil \frac{I_{i-1,j}}{2} \right\rceil & 1<j\le L+M \end{cases} \tag{2}$$

We can apply the above equation to compute the ith row based on the $(i-1)$th row of **I**. When the total number of inputs to stage i of all the outlets is equal to $L+M+1$, the total number of outputs will be $L+M$ and hence concentration is completed at this stage. In other words, when $\sum_{j=1}^{L+M} I_{i,j} = L+M+1$, the ith row is the last row of **I**.

Stage i of outlet j has $I_{i,j}$ inputs and hence it requires $\lceil I_{i,j}/2 \rceil$ contention elements. The total number of contention elements in the wraparound knockout concentrator is therefore equal to $\sum_i \sum_j \lceil I_{i,j}/2 \rceil$. The details for computing the total number of contention elements in a wraparound knockout concentrator are given as follows:

Step 1. Apply (1) to compute the 1st row of **I**. Initialize i to one.

Step 2. Increment i by one. Apply (2) to compute the ith row based on the $(i-1)$th row of **I**. Repeat this step until $\sum_{j=1}^{L+M} I_{i,j} = L+M+1$.

Step 3. The total number of contention elements is equal to $\sum_i \sum_j \lceil I_{i,j}/2 \rceil$.

For the knockout concentrator, there is no wraparound path. To compute the total number of contention elements in this concentrator, we can set $M=0$ and execute the above steps.

2.3 Probability of Packet Loss

The concentrator has N input lines, and therefore at most N new packets can arrive simultaneously in a slot. Within the concentrator, the new and the circulated packets compete with each other for going to the output lines. When the total number of these packets is larger than $L+M$, there is packet loss. This concentrator model is similar to but different from a discrete-time G/D/L/M-FIFO queueing model, in which the packets arrive one after the other and the packets are served in a FIFO manner. In this subsection, we analyze the probability of packet loss.

Let $\widetilde{A}$ be a random variable denoting the number of newly arrived packets in a slot, and $\widetilde{Z}$ be a random variable denoting the number of circulated packets in a slot. We denote $\lambda_i \equiv P(\widetilde{A}=i)$ and $\pi_j \equiv P(\widetilde{Z}=j)$ for $0\le i\le N$ and $0\le j\le M$. In the followings, we derive λ_i and π_j and then derive the probability of packet loss.

We assume that a packet arrives at an input in a slot with probability ρ, where the packet arrivals at different inputs or in different slots are independent. Then λ_i is given by:

$$\lambda_i = \binom{N}{i} \rho^i (1-\rho)^{N-i} \tag{3}$$

To derive π_j, we distinguish two cases: (1) $M\le L$ and (2) $M>L$. We first consider the case $M\le L$. When $\widetilde{A}+\widetilde{Z}\le L$, the concentrator can output all the packets and it does not re-circulate any packet. Therefore, π_0 is equal to $P(\widetilde{A}+\widetilde{Z}\le L)$ and it is given by:

$$\begin{aligned} \pi_0 &= \sum_{j=0}^{M} P(\widetilde{A}\le L-j)\cdot P(\widetilde{Z}=j) \\ &= \sum_{j=0}^{M} \Big(\sum_{i=0}^{L-j} \lambda_i\Big)\pi_j \end{aligned} \tag{4}$$

When $\widetilde{A}+\widetilde{Z} = L+i$ where $1\le i\le M-1$, the concentrator conveys L packets to the outputs and re-circulates the other i packets. Therefore, π_i is equal to $P(\widetilde{A}+\widetilde{Z}=L+i)$ and it is given by:

$$\begin{aligned} \pi_i &= \sum_{j=0}^{M} P(\widetilde{A}=L+i-j)\cdot P(\widetilde{Z}=j) \\ &= \sum_{j=0}^{M} \lambda_{L+i-j}\cdot \pi_j \qquad 1\le i\le M-1 \end{aligned} \tag{5}$$

Since $\sum_{i=0}^{M} \pi_i = 1$, π_M is equal to $1-\sum_{i=0}^{M-1} \pi_i$.

Based on this equation and (4),(5), we have $M+1$ simultaneous linear equations with $M+1$ unknowns $\pi_0, \pi_1, \pi_2, \cdots, \pi_M$. We apply numerical methods to solve these linear equations to get $\pi_0, \pi_1, \pi_2, \cdots, \pi_M$. (e.g., see [14]).

We now consider the case $M > L$. π_0 is equal to $P(\widetilde{A}+\widetilde{Z} \le L)$ and it is given by

$$\pi_0 = \sum_{j=0}^{L}\left(\sum_{i=0}^{L-j}\lambda_i\right)\pi_j \tag{6}$$

For $1 \le i \le M-1$, π_i is equal to $P(\widetilde{A}+\widetilde{Z} = L+i)$ and it is given by

$$\pi_i = \begin{cases} \sum_{j=0}^{L+i} \lambda_{L+i-j}\cdot\pi_j & 1 \le i \le M-L \\ \sum_{j=0}^{M} \lambda_{L+i-j}\cdot\pi_j & M-L+1 \le i \le M-1 \end{cases} \tag{7}$$

π_M is equal to $1-\sum_{i=0}^{M-1}\pi_i$. Based on this equation and (6),(7), we have $M+1$ simultaneous linear equations with $M+1$ unknowns $\pi_0, \pi_1, \cdots, \pi_M$. We apply numerical methods to solve these linear equations to get $\pi_0, \pi_1, \cdots, \pi_M$.

The mean number of packet arrivals in a slot is $\sum_{i=0}^{N} i\cdot P(\widetilde{A}=i)$. When the total number of new packets and circulated packets in a slot is larger than $L+M$, the concentrator conveys L packets to the outputs, circulates M packets for contention in the subsequent slot, and discards the other packets. The mean number of discarded packets in a slot is $\sum_{i=L+M+1}^{N+M}(i-L-M)\cdot P(\widetilde{A}+\widetilde{Z}=i)$ where

$$P(\widetilde{A}+\widetilde{Z}=i) = \begin{cases} \sum_{j=0}^{M} P(\widetilde{A}=i-j)\cdot P(\widetilde{Z}=j) \\ \qquad L+M+1 \le i \le N \\ \sum_{j=i-N}^{M} P(\widetilde{A}=i-j)\cdot P(\widetilde{Z}=j) \\ \qquad N+1 \le i \le N+M \end{cases} \tag{8}$$

The probability of packet loss is then given by:

$$\begin{aligned} &P(packet\ loss) \\ &= \frac{1}{N\rho}\left\{\sum_{i=L+M+1}^{N}(i-L-M)\sum_{j=0}^{M}\lambda_{i-j}\pi_j \right. \\ &\left. + \sum_{i=N+1}^{N+M}(i-L-M)\sum_{j=i-N}^{M}\lambda_{i-j}\pi_j\right\} \end{aligned} \tag{9}$$

3. Numerical Results

We consider a wraparound knockout concentrator with N inputs and L outputs. Suppose this concentrator is used in every output adaptor of an $N\times N$ packet switch (for details, please refer to Fig. 12 in [1]). We consider a heavy and uniform traffic, such that the probability that a packet arrives at an input of the switch in a slot is 1.0 and the probability that an incoming packet is destined for any one of the N outputs of the switch is $\frac{1}{N}$. Therefore, ρ is equal to $1.0\times\frac{1}{N}=\frac{1}{N}$.

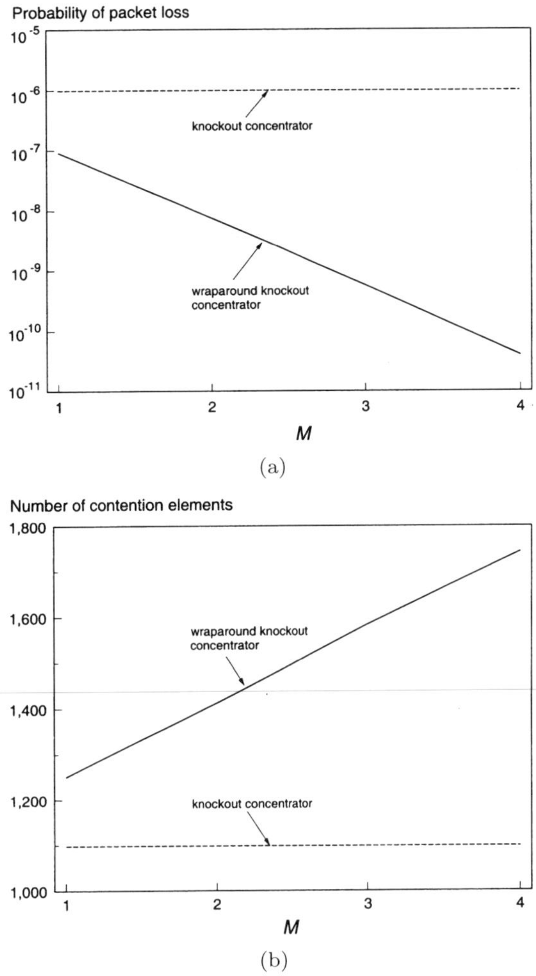

Fig. 5 Performance of the knockout concentrator and the wraparound knockout concentrator; $N = 128$ and $L = 8$. (a) Probability of packet loss versus the design parameter M. (b) Number of contention elements versus the design parameter M.

Figure 5(a) shows the probability of packet loss versus the design parameter M for $N = 128$ and $L = 8$. The wraparound knockout concentrator has a significantly smaller probability of packet loss than the knockout concentrator. When M is increased by one, the probability of packet loss is further reduced by about one order of magnitude. For example, the probability of packet loss in the knockout concentrator is 9.865×10^{-7}, while that in the wraparound knockout concentrator with $M = 1,2,3,4$ are 9.065×10^{-8}, 7.534×10^{-9}, 5.706×10^{-10}, and 3.962×10^{-11} respectively. Figure 5(b) shows the number of contention elements required versus M. When M is increased by one, the wraparound knockout concentrator requires

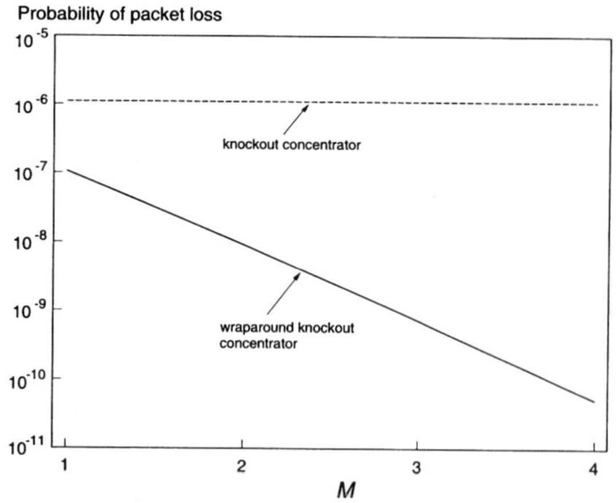

(a)

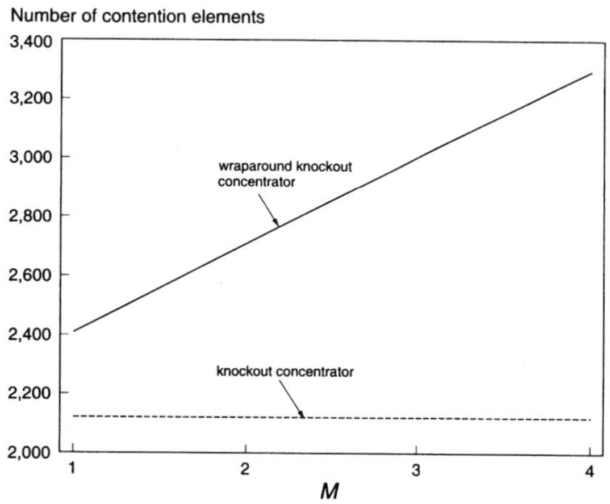

(b)

Fig. 6 Performance of the knockout concentrator and the wraparound knockout concentrator; $N = 256$ and $L = 8$. (a) Probability of packet loss versus the design parameter M. (b) Number of contention elements versus the design parameter M.

about 10% additional contention elements. For example, the knockout concentrator requires 1098 contention elements, while the wraparound knockout concentrator with $M = 1, 2, 3, 4$ requires 1251, 1412, 1584, 1744 contention elements respectively. We see that the probability of packet loss can be reduced significantly while the number of additional logic gates required is only moderate.

When N is increased to 256, Fig. 6(a) shows the probability of packet loss versus M, and Fig. 6(b) shows the number of contention elements required versus M. We see again that when M is increased by one, the probability of packet loss in the wraparound knockout concentrator is reduced by about one order of magnitude and it requires about 10% additional contention elements.

4. Conclusions

We designed the wraparound knockout concentrator, and analyzed the number of logic gates required and the probability of packet loss. This concentrator has a significantly smaller probability of packet loss than the knockout concentrator while it can be realized using the same number of pins per chip. The numerical results show that if the proposed concentrator adopts a few wraparound paths, the probability of packet loss can already be reduced by orders of magnitude.

References

[1] R.Y. Awdeh and H.T. Mouftah, "Survey of ATM switch architectures," Comput. Net. ISDN Syst., vol.27, pp.1567–1613, 1995.

[2] Y.S. Yeh, M.G. Hluchyj, and A.S. Acampora, "The knockout switch: A simple, modular architecture for high-performance packet switching," IEEE J. Sel. Areas Commun., vol.5, no.8, pp.1274–1283, Oct. 1987.

[3] D.X. Chen and J.W. Mark, "SCOQ: A fast packet switch with shared concentration and output queueing," IEEE/ACM Trans. Network., vol.1, no.1, pp.142–151, Feb. 1993.

[4] W. Wang and F.A. Tobagi, "The Christmas-tree switch: An output queueing space-division fast packet switch based on interleaving distribution and concentration functions," Comput. Net. ISDN Syst., vol.25, pp.631–644, 1993.

[5] Y.M. Kim and K.Y. Lee, "KSMIN's: Knockout switch-based multistage interconnection networks for high speed packet switching," IEEE Trans. Commun., vol.43, no.8, pp.2391–2398, Aug. 1995.

[6] K.L.E. Law and A. Leon-Garcia, "A large scalable ATM multicast switch," IEEE J. Sel. Areas Commun., vol.15, no.5, pp.844–854, June 1997.

[7] R. Kannan, R. Bartos, K.Y. Lee, and H.F. Jordan, "SXmin: A self routing high-performance ATM packet switch based on group-knockout principle," IEEE Trans. Commun., vol.45, no.6, pp.710–722, June 1997.

[8] K.S. Chan, S. Chan, K.L. Yeung, K.T. Ko, and E.W.M. Wong, "Clos-knockout: A large-scale modular multicast ATM switch," IEICE Trans. Commun., vol.E81-B, no.2, pp.266–275, Feb. 1998.

[9] K.J. Schultz and P.G. Gulak, "Physical performance limits for shared buffer ATM switches," IEEE Trans. Commun., vol.45, no.8, pp.997–1007, Aug. 1997.

[10] T.C. Banwell, R.C. Estes, S.H. Habiby, G.A. Hayward, T.K. Helstern, G.R. Lalk, D.D. Mahoney, D.K. Wilson, and K.C. Young, "Physical design issues for very large ATM switching systems," IEEE J. Sel. Areas Commun., vol.9, no.8, pp.1227–1238, Oct. 1991.

[11] T.R. Banniza, G.J. Eilenberger, B. Pauwels, and Y. Therasse, "Design and technology aspects of VLSI's for ATM switches," IEEE J. Sel. Areas Commun., vol.9, no.8, pp.1255–1264, Oct. 1991.

[12] L. Geppert, "Technology 1997 analysis and forecast: Solid state," IEEE Spectrum, vol.34, no.1, Jan. 1997.

[13] L. Geppert, "Technology 1998 analysis and forecast: Solid state," IEEE Spectrum, vol.35, no.1, pp.23–28, Jan. 1998.

[14] S. Yakowitz and F. Szidarovszky, An Introduction to Numerical Computations, Chapter 2, Macmillan, 1989.

High-Performance Backbone Network Technology edited by Naoaki Yamanaka
ISBN: 0-8247-5321-6 DOI: 10.1081/(120025683)
Published 2004 by Marcel Dekker, Inc.

A Contention-Free Tbit/sec Packet-Switching Architecture for ATM over WDM Networks

Itamar Elhanany and Dan Sadot

SUMMARY Future high-speed switches and routers will be expected to support a large number of ports at high line rates carrying traffic with diverse statistical properties. Accordingly, scheduling mechanisms will be required to handle Tbit/sec aggregated capacity while providing quality of service (QoS) guarantees. In this paper a novel high-capacity switching scheme for ATM/WDM networks is presented. The proposed architecture is contention-free, scalable, easy to implement and requires no internal "speedup." Non-uniform destination distribution and bursty cell arrivals are examined when studying the switching performance. Simulation results show that at an aggregated throughput of 1 Tbit/sec, low latency is achieved, yielding a powerful solution for high-performance packet-switch networks.

key words: *Tbit routers, scheduling, QoS (quality-of-service), global considerations*

1. Introduction

The exponential growth of Internet and datacom traffic encourages research and development of high-performance networking infrastructure consisting of both increased media transmission speeds and performance enhancement of switches and routers [1]–[5], [8]. Optical technologies, such as wavelength division multiplexing (WDM), are widely accepted as the technology for such high capacity, scalable and cost effective networks. Based on WDM technology, Tbit/sec point-to-point transmission speeds have been demonstrated. Hence, the main bottleneck of the network has shifted towards designing high-speed switches and routers, which are expected to efficiently regulate large amounts of diversely characterized traffic. Accordingly, considerable attention has been paid to WDM contention resolution and wavelength assignment issues, commonly targeted at packet-switched networks such as asynchronous transfer mode (ATM). It is generally acknowledged that the two main goals of network switches are (1) to optimally utilize the available bandwidth in order to achieve maximum throughput, and (2) to support quality of service (QoS) requirements. Constraints derived from these goals typically contradict in the sense that the maximal bandwidth provision may not necessarily correlate with the most urgent traffic flow.

Manuscript received May 28, 1999.
Manuscript revised August 27, 1999.

†The authors are with the Electrical and Computers Engineering Department, Ben-Gurion University, Beer-Sheva 84105, Israel.

Recently, several switching schemes were designed in order to support high capacity, large number of ports and low latency requirements [1], [3], [5]. Many of these scheme employ output-queuing mechanisms, which means that cells arriving at the input node are transmitted through the cross-connect fabric to a designated output queue. In order to avoid transmission contention in an $N \times N$ switch, either N^2 independent channels or N times faster circuitry are required by the switch core. Considering today's high line rates, N times faster circuitry is becoming infeasible. Typical designs implement either centralized or output queuing mechanisms in order to maximize switch bandwidth. However, as the line rates and number of ports increase, output queuing is found impractical for high-performance switches.

An alternative to output queuing is input queuing, whereby cell buffering is located at the input nodes. It has been shown that an input-queued switch employing a single FIFO at each input, may achieve a maximum of 58.6% throughput due to the head-of-line (HOL) blocking phenomenon [6]. An alternative technique, which entirely eliminates the HOL blocking, is known as the *virtual output queuing* (VOQ). In VOQ each input node contains a separate queue for each output. Several scheduling algorithms have been proposed for VOQ switches [6], [7], [9]. As indicated by Chuang et al. [5], most maximum matching algorithms known to-date are too complex to be implemented in hardware on real-time, therefore, are found unsuitable for switches with a large number of nodes at high line rates. Moreover, the algorithms proposed are frequently examined under uniform traffic assumptions, which clearly does not represent real life traffic.

One method of enhancing VOQ based switching is to increase the "speedup" of the switch. A switch with a speedup of L can transmit L cells in a single cell-time. However, the switch-core speed is a paramount resource making speedup a drawback of any scheduling approach. In this paper, we introduce a new packet-switching architecture that offers Terabit/sec capacity along with support of over 100 ports and low switching latency. Scheduling is based on a sequential-reservation scheme with prioritized-matching that was initially introduced in [13], [14], and is broadened here to comply with diverse traffic characteristics. We show that the new architecture is contention-free, requires no speedup

IEICE TRANS. COMMUN., Vol. E83-B, No. 2, pp. 225–230, February 2000

and has the advantages of low implementation complexity and high scalability. In addition, the architecture can easily be adapted to comply with diverse quality of service (QoS) requirements.

Although this work focuses on packets having fixed length, many network protocols, such as IP, have variable length packets. Most switches and routers today segment these packets into fixed-length packet (or "cells") prior to entering the switch fabric. The original packets are reconstructed at the output stage. This methodology is commonly practiced in order to achieve high switching performance. Hence, the techniques described in this paper can be applied to both fixed and variable length packet sizes.

In Sect. 2 the proposed switch architecture is described. Section 3 focuses on the WDM perspective, channel (wavelength) allocation discipline and its VLSI implementation considerations. Traffic modeling is described in Sect. 4, while simulation results and switch metrics are presented and discussed in Sect. 5. Section 6 draws the main conclusions.

2. Switch Architecture

2.1 The WDM Perspective

In the proposed switch we focus on a tunable-transmitter fixed-receiver (TT-FR) configuration whereby a node tunes its transmitter to a predefined wavelength according to the desired destination. By the same token, a fixed-transmitter tunable-receiver (FT-TR) configuration can be applied with little changes in the proposed switching architecture. Employed scheduling schemes for Tbit/sec WDM-based switches are required to possess the following properties:

- *Maximal bandwidth utilization*: the algorithm should maximize the use of the switch internal bandwidth in order to provide maximal switching throughput.
- *Contention-resolution*: Since the wavelengths constitute an expensive resource, no more than N internal channels should be assumed. Two or more simultaneous transmissions over the same wavelength result in collision. Hence, packet losses due to optical collision should be avoided.
- *No speedup*: In the optical context, speeding up internal transmission rates is a wasteful act. Schemes that require no speedup (speedup factor of 1) are more pragmatic for scalable networks.
- *Non-blocking*, *Single-stage*: WDM cross-connect technology is commonly deployed in a single-stage, non-blocking constellation whereby optical signals are not re-routed or re-transmitted within the switching core. The ultra-fast decision processes require that minimal latency is contributed by the cross-connecting element.
- *Large number of ports*: The diversity of communication protocols and line speeds imply that future switches will have port densities of hundreds and even thousands of ports. Therefore, a good switch architecture must support a large number of asymmetric ports.
- *Simple to implement*: The switching architecture and respective algorithm should be plainly implemented in designated custom hardware. Performance in terms of speed is directly affected by the simplicity of the hardware design.

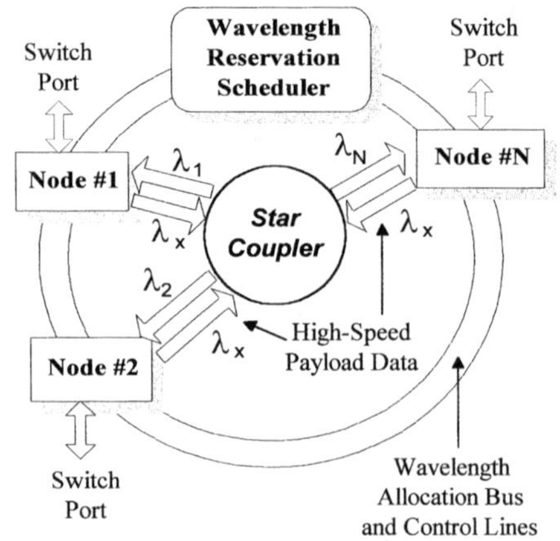

Fig. 1 Terabit/sec ATM over WDM switch architecture.

2.2 The Scheduling Algorithm

Figure 1 depicts the proposed switch architecture. The nodes, corresponding to the switch ports, have bi-directional optical data links interconnected via an optical passive star coupler. The passive star topology acts as a single-stage non-blocking cross-connect fabric. Each transmitter can be tuned to any of the N wavelengths, while each receiver is assigned a fixed and unique wavelength. ATM traffic received at each port is distributed to various buffers within the node on a cell-by-cell basis, where each buffer relates to a single wavelength according to the desired destination node.

All nodes are connected to a central wavelength reservation scheduler via a common electronic wavelength reservation bus and individual control lines. The N bus lines are accessible to all nodes and indicate the reservation status of each of the N wavelengths. The individual control lines are used by the scheduler to signal each node, in turn, to commence the wavelength reservation procedure.

3. Wavelength Reservation

Upon receiving a control signal from the wavelength reservation scheduler, each node performs wavelength reservation according to two major guidelines: (a) global switch resources status, i.e., available wavelengths at the reservation time, and (b) local consid-

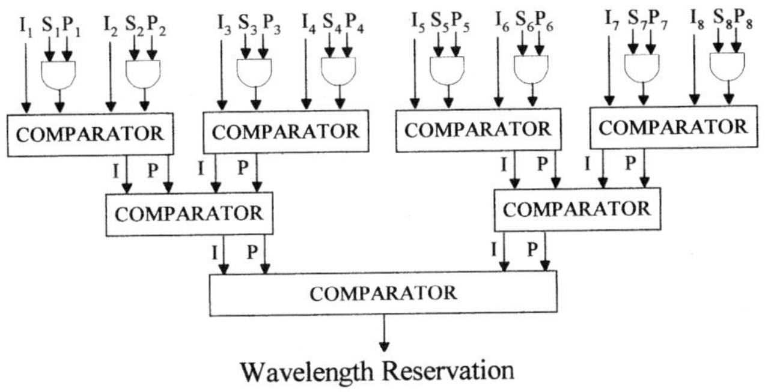

Fig. 2 Block diagram of node wavelength reservation logic for an 8×8 switch.

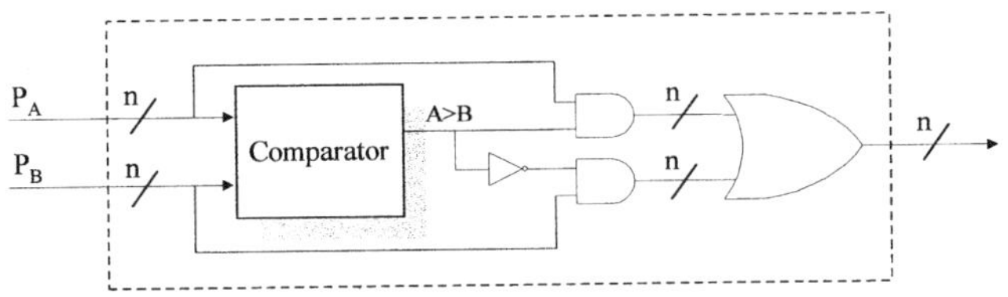

Fig. 3 Single-level comparator logic with n-bits per priority value.

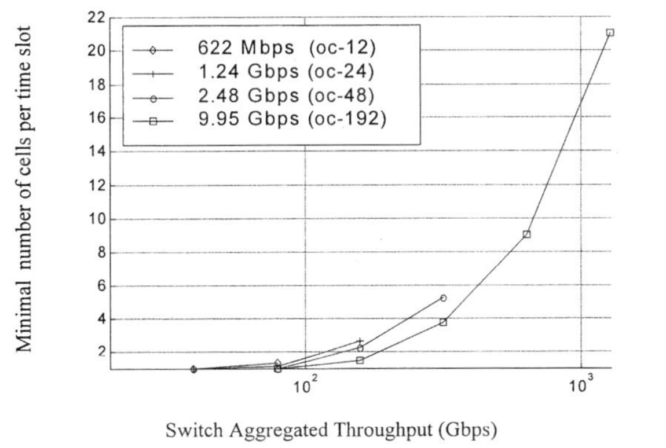

Fig. 4 Minimal number of cells per time slot vs. switch aggregated throughput.

erations, i.e., the status and priorities of the node's internal buffers. Figure 2 illustrates a block diagram of the wavelength reservation hardware of a single node in an 8×8 switch.

At the highest level, the wavelength reservation status lines, denoted by S_i, either grant or discard buffer priorities, denoted by P_i, via designated AND logic. Consequently, only buffer indices, denoted by I_i, relating to available wavelengths advance to the lower levels. Each level consists of a set of comparators, which concurrently receive as input a pair of indices, along with their respective priorities, and output the higher priority and its corresponding index. Figure 3 depicts the logic comprising each comparator unit. The output of the last comparator determines the "prevailing" buffer, which held the highest priority out of the subset of buffers relating to unreserved wavelengths. Any number of parameters, such as buffer load, accumulated delay and required QoS can affect the buffer priority metrics. Node wavelength selection is instantly followed by assertion of the relevant line within the wavelength reservation bus. At that point, utilizing a weighted Round Robin procedure, the central scheduler signals the next node to begin wavelength reservation. Contention is inherently avoided, since at any given time only one node attempts to reserve a wavelength. After all N nodes complete wavelength reservation, high-speed data is optically transmitted via the star coupler.

The wavelength reservation and data transmission are conducted in a time slot discipline. Nodes transmit data concurrently at wavelengths reserved during the previous time slot. Assuming N nodes, the time slot period can be calculated as

$$t_{ts} = N \cdot \log_2(N) \cdot t_c \tag{1}$$

where t_c is the propagation delay of a single-level comparator logic. Accordingly, t_{ts} dictates the minimal number of cells required to be transmitted during each time slot and hence the queuing time delays.

Utilizing current CMOS technology $t_c = 1\,\text{ns}$ is found feasible. Consequently, extremely short processing time for resource allocation is attained, yielding high switching performance. Figure 4 shows the relationship between the minimal number of cells to be transmitted during each time slot and the switch aggregated throughput. Various port bitrates are presented as a parameter. The marks on each curve represent standard number of ports, e.g., 8, 16, 32, 64 and 128.

4. Traffic Modeling

Next generation networks will carry diverse traffic embodying a wide range of statistical properties. Two principal criteria that strongly affect switching performance are the packet arrival statistics and destinations distribution. Typically, it is assumed that arriving packets obey a binomial process and are uniformly distributed to all destinations. Since current high-speed routers and switches are limited to an aggregated capacity of several hundred Gbit/sec, it is difficult to predict how valid will these assumptions be in future multi-Tbit/sec interconnections. It is also well known that real-life traffic tends to burstiness due to modern applications such as compressed video and multimedia services. Accordingly, packet destinations vary dynamically and hence are non-uniformly distributed.

4.1 Traffic Arriving Process

4.1.1 Binomial Arrivals

Networks that contain IP packets or ATM cells are regarded as discrete systems. At each cell time there is a probability p that a cell will arrive, consequently a probability $q = 1 - p$ that no cell arrives. According to the Binomial i.i.d. distribution model, the probability of k packets arrivals during n cell slots is given by:

$$P\{k\} = \frac{n!}{k!(n-k)!} p^k q^{n-k}, \tag{2}$$

It should be noted that for $n \to \infty$ and $p \ll 1$, the Binomial distribution converges to the continuous Poisson distribution.

4.1.2 Bursty Arrivals

It is well known that real-life traffic tends to burstiness due to the prevalence of modern multi-media applications such as compressed video and sound. Correspondingly, packet destinations vary dynamically and are non-uniformly distributed. Many models for bursty traffic have been proposed [15]–[17]. In [17] it has been claimed that network traffic is bursty at any level making bursty models an essential testbed for switching performance. We employ an on-off arrival process modulated by a two-state Markov chain. The result is a train of bursty cell arrivals, each containing cells with identical destination, followed by periods of empty cell slots. The expression for the mean offered load under such bursty traffic is:

$$\textit{Offerd Load} = \frac{q}{p+q} \tag{3}$$

where p and q are the transition probabilities with regard to the active and idle periods, respectively. Due to the geometric distribution of the duration of active and idle periods, the average burst length is $1/p$. From the average burst length and offered load, an idle parameter q can easily be determined.

4.2 Destination Distribution

Real-life traffic destinations are not uniformly distributed; traffic tends to be focused on preferred or popular destinations. Unfortunately, the performance of many scheduling algorithms degrades under non-uniform traffic conditions, where not all queues are evenly and heavily loaded. Maximum matching algorithms are known to perform poorly and cause queue starvation [2] under these conditions. We introduce here a destination distribution model named Zipf's law. The Zipf law was proposed by G. K. Zipf [10]–[12]. This model may be used as a reference to investigate and compare the performance of different scheduling algorithms. The Zipf law states that frequency of occurrence of some events (P), as a function of the rank (i), where the rank is determined by the above frequency of occurrence, is a power-law function: $P_I \sim 1/i^k$, with the exponent k close to unity. The most famous example of Zipf's law is the frequency of English words in a given text. Most common is the word "the," then "of,"

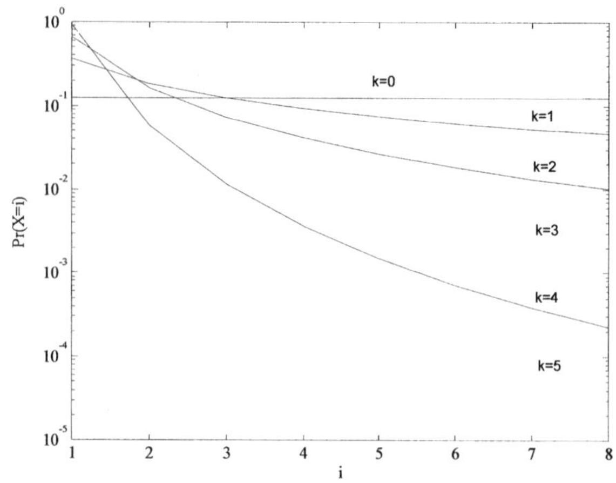

Fig. 5 The Zipf distribution function for various orders.

"to" etc. When the number of occurrence is plotted as the function of the rank ($i = 1$ most common, $i = 2$ second most common, etc.), the functional form is a power-law function with exponent close to 1. Figure 5 illustrates the Zipf distribution for different values of the k parameter. It was shown that many natural and human phenomena such as Web access statistics, company size and biomolecular sequences all obey the Zipf law with k close to 1 [11]. We use the Zipf distribution to model packet destination distribution. The probability that an arriving packet is heading destination i is given by:

$$\textit{Zipf}(i) = \frac{\frac{1}{i^k}}{\sum_{j=0}^{N} \frac{1}{j^k}} \tag{4}$$

where i is the packet destination, k is the Zipf order and N is the system order, i.e., the number of switch ports. While $k = 0$ corresponds to uniform distribution, as k increases the distribution becomes more biased towards preferred destinations. In order to generate a stable and realistic traffic model, the average steady state load at each input port must not exceed 100%. Similarly, the average steady state aggregated traffic rate arriving from all input ports to any destination port must not exceed 100%.

5. Simulation Results and Discussion

Simulations were carried out for various switch sizes (N), each assumed to receive data at a line rate of 10 Gbps (oc-192). The WDM-based switch fabric is single-staged, non-blocking with N available internal channels (wavelengths), as described earlier. It is important to note that in this case the fabric speedup is 1. Destination distributions both uniform and Zipf were examined under Bernoulli and bursty cell arrival

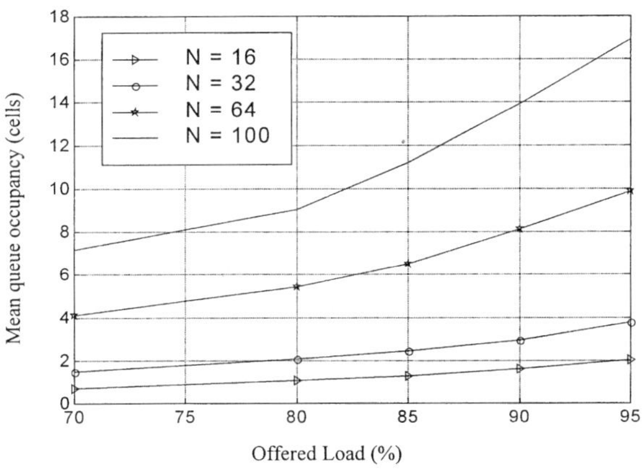

Fig. 6 Mean queue occupancy for uniformly distributed Bernoulli traffic loads.

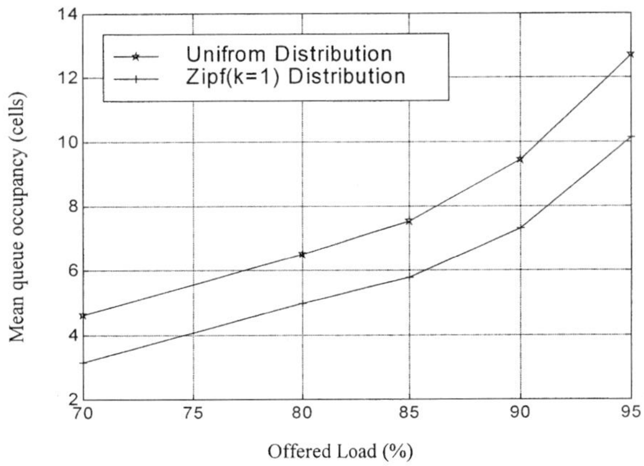

Fig. 8 Mean queue occupancy for Zipf and uniformly distributed bursty traffic loads (average burst length is 10 cells) in a 100 × 100 switch.

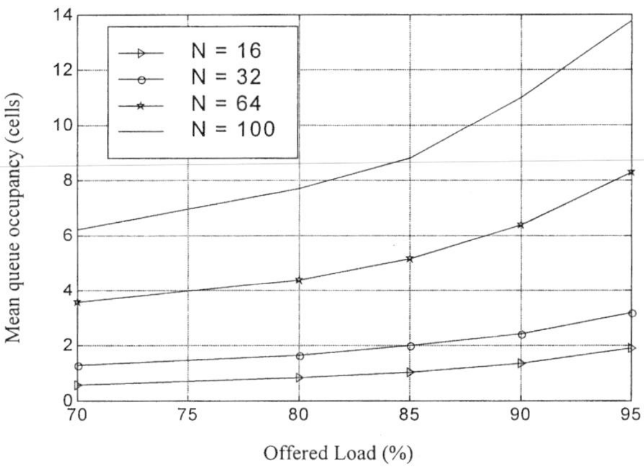

Fig. 7 Mean queue occupancy for Zipf distributed Bernoulli traffic loads.

conditions. For the Zipf distribution, $k = 1$ was selected in accordance with the claims of Sect. 4.2. Any other order of k has lead to improved switching performance results, therefore $k = 1$ can be regarded as the worst case analysis. Each simulation experiment included transmission of two million cells obeying the specific arrival process and destination distribution that was initially selected. It was observed that simulation results have converged to a steady-state value significantly prior to accomplishing the two million events, leading to the conclusion the simulation error was essentially negligible.

Figures 6 and 7 depict, respectively, the mean queue occupancy (MQO) for uniformly and Zipf distributed cells, with Bernoulli arrival characteristics, as a function of offer traffic load. As can be seen, regardless of the destination distribution behavior, the resulting queue occupancies amount to several cell times. A switch of 100 × 100 ($N = 100$) under 90% traffic load, has an MQO of less than 20 cells. The metrics are mildly affected by the traffic destination characteristics, as can be derived from Fig. 7.

Figure 8 depicts the MQO for bursty traffic (mean burst length is 10 cells) arriving at a 100 × 100 switch, under both distribution conditions as a function of the offered load. The results show that bursty traffic, which typically degrades switching performance, has no degrading impact. On the contrary, due to the time slot discipline applied by the architecture, bursty traffic yields better results. The MQO directly relates, and hence may be translated, to the mean cell delay (MCD). Under non-uniform destination distribution conditions, translation to the MCD should refer to a specific queue/traffic channel.

Our simulations can be compared to other high-speed buffer management approaches such as in [1] and [2]. In such architectures, various types of real-time Round Robin algorithms are employed and are typically limited in performance under non-uniform and bursty traffic. In [2], for example, a 16 × 16 switch under 95% offered load of uniform Bernoulli arrivals yields an average latency of approximately 100 cells, as compared to a latency of 2 cells under the same conditions using our proposed scheme. The fact that no Round Robin pointers are employed in our scheme results in low variance of the service-time distribution and hence the lower mean cell delays.

Although in this paper we discuss WDM as the physical layer switching fabric and fixed-size packets, the proposed method can be extended to other multiplexing technology, e.g., optical SDM (space division multiplexing), and to variable-size packets, e.g., IP traffic.

6. Conclusions

A novel Tbit/sec switch architecture for ATM over WDM packet-switched networks has been proposed. By applying an ultra-high speed scheduling discipline, extremely high capacity and low latency switching is achieved for fabrics of up to 100 × 100. Simulation results demonstrate that bounded queue lengths and low latency can be achieved utilizing the proposed scheme, even when applying complex traffic models with non-uniform destination distributions and bursty cell arrivals. Using current CMOS technology, Tbit/sec switching throughput can be attained with delay in the range of tens of microseconds. The design is contention-free, simple to implement, scalable and can easily be adapted to various multiplexing technologies.

Acknowledgement

This work was supported by the Israeli Ministry of Science under contract no. 9404-1-97, the Wolfson foundation and the Rector's office at the Ben-Gurion University of the Negev, Israel.

References

[1] F.M. Chiussi, J.G. Kneuer, and V.P. Kumar, "Low-cost scalable switching solutions for broadband networking: The ATLANTA architecture and chipset," IEEE Commun. Mag., vol.25, no.12, pp.44–53, Dec. 1997.

[2] N. McKeown, "The iSLIP scheduling algorithm for input-queued switched," IEEE/ACM Trans. Networking, vol.7, no.2, pp.188–201, April 1999.

[3] V. Sivaraman and G.N. Rouskas, "HiPeR-1: A high performance reservation protocol with look-ahead for broadcast WDM networks," Proc. IEEE INFOCOM'97, vol.3, pp.1270–1277, April 1997.

[4] C. Salisbury and R. Melhem, "A high speed scheduler/controller for unbuffered Banyan networks," Proc. IEEE ICC'98, Session S18-6, June 1998.

[5] S. Chuang, A. Goel, N. McKeown, and B. Prabhakar, "Matching output queuing with a combined input output queued switch," Proc. IEEE INFOCOM'99, March 1999.

[6] M. Karol, M. Hluchyj, and S. Morgan, "Input versus output queuing on a space division switch," IEEE Trans. Commun., vol.12, no.35, pp.1347–1356, 1987.

[7] A. Mekkittikul and N. McKeown, "A practical scheduling algorithm to achieve 100% throughput in input-queued switches," Proc. IEEE INFOCOM'98.

[8] H. Obara, "Optimum architecture for input queueing ATM switches," IEE Electronic Letters, pp.555–557, March 1991.

[9] N. McKeown, V. Anantharam, and J. Walrand, "Achieving 100% throughput in an input-queued switch," Proc. IEEE INFOCOM'96, San Francisco, March 1996.

[10] G.K. Zipf, Psycho-Biology of languages, MIT Press, Houghton-Miffin, 1965.

[11] R.E. Wyllis, "Measuring scientific prose with rank-frequency ('Zipf') curves: A new use for an old phenomenon," Proc. American Society for Information Science, vol.12, pp.30–31, Washington, DC, 1975.

[12] B.M. Hill, "A simple general approach to inference about the tail of a distribution," Anals. of Statistics, vol.3, pp.1163–1174, 1975.

[13] J. Nir, I. Elhanany, and D. Sadot, "A new Tbit/sec switching scheme for ATM/WDM networks," Electron. Lett., vol.35, no.1, pp.30–31, Jan. 1999.

[14] I. Elhanany and D. Sadot, "A novel Tbit/sec switch architecture for ATM/WDM high-speed networks," Proc. IEEE ATM Workshop'99, Kochi City, Japan, 1999.

[15] H. Heffes and D.M. Lucantoni, "A Markov modulated characterization of packetized voice and data traffic and related statistical multiplexer performance," IEEE J. Sel. Areas Commun., vol.4, pp.856–868, 1988.

[16] R. Jain and S.A. Routhier, "Packet trains: Measurements and a new model for computer network traffic," IEEE J. Sel. Areas Commun., vol.4, pp.986–995, April 1986.

[17] W.E. Leland, W. Willinger, M. Taqqu, and D. Wilson, "On the self-similar nature of ethernet traffic," Proc. SIGCOMM, pp.183–193, San Francisco, CA, Sept. 1993.

High-Performance Backbone Network Technology edited by Naoaki Yamanaka
ISBN: 0-8247-5321-6 DOI: 10.1081/(120025684)
Published 2004 by Marcel Dekker, Inc.

A Distributed Traffic Control Scheme for Large-Scale Multi-Stage ATM Switching Systems

Kohei Nakai, Eiji Oki, and Naoaki Yamanaka

SUMMARY This paper describes a distributed traffic control scheme for large multi-stage ATM switching systems. When a new virtual circuit is to be added from some source line-interface unit (LU) to a destination LU, the system must find an optimal path through the system to accommodate the new circuit. Conventional systems have a central control processor and control lines to manage the bandwidth of all the links in the systems. The central control processor handles all the virtual circuits, but have trouble doing this when the switching system becomes large because of the limited ability of the central processor to handle the number of virtual circuits. A large switching system with Tbit/s-class throughput requires a distributed traffic control scheme. In our proposed switching system, each port of the basic switches has its own traffic monitor. Operation, administration, and maintenance (OAM) cells that are defined inside the system carry the path-congestion information to the LUs, enabling each LU to route new virtual circuits independently. A central control processor and control lines are not required. The performance of the proposed system depends on the interval between OAM cells. This paper shows how an optimal interval can be determined in order to maximize the bandwidth for user cells. This traffic control scheme will suit future Tbit/s ATM switching systems.

key words: *ATM, switch, multi-stage, large-scale, distributed control*

1. Introduction

Asynchronous transfer mode (ATM) is expected to yield the best high-speed multimedia infrastructure [1]. Recently, data traffic in the public network has been expanding explosively, thus necessitating an expansion of the switching system. In this situation, ATM networks will require switching systems that can offer Tbit/s throughput in a cost-effective manner [2]–[4].

Most ATM switches today use several single-stage switching techniques. Single-stage switches are relatively simple, but are limited in the number of ports and total throughput that they can support effectively. For large systems, multi-stage switching systems are needed. Three-stage switch architectures using unique basic switch elements are attractive. This is because they can be expanded easily using the same functional blocks [5], [6].

We have previously proposed a scalable 3-stage ATM switching architecture that uses optical WDM (wavelength division multiplexing) grouped links and dynamic bandwidth sharing [7], [8]. The former reduces the number of cables necessary. The latter prevents the statistical multiplexing gain from falling as the size of the switching system increases.

Figure 1 shows the scalable 3-stage ATM switching architecture. Each basic switch has N input ports and N output ports. Each port multiplexes M links corresponding to the M wavelengths that are multiplexed into one optical fiber. The total throughput of this system is MN times that of the basic switch. For example, a system capacity of 5.2 Tbit/s can be achieved using 8×8 80-Gbit/s basic switches and 8 wavelengths ($N = 8, M = 8$).

However, multi-stage switching systems need a scheme for controlling traffic. When a new virtual circuit is to be added from some source line-interface unit (LU) to a destination LU, the system must find an optimal path through the system to accommodate the new circuit. Conventional systems have a central control processor and control lines to manage the bandwidth of all the links in the system. The central control processor handles all the virtual circuits, but it has trouble doing this when the switching system becomes large. A large switching system with Tbit/s-class throughput requires a distributed traffic control scheme.

This paper proposes a new distributed traffic control scheme for large multi-stage switching systems in which each port of the basic switches has its own traffic monitor. Operation, administration, and maintenance (OAM) cells that are defined inside the system carry the path-congestion information to the LUs, enabling each LU to route new virtual circuits independently. The performance of the proposed system depends on the interval between OAM cells. This paper shows how an optimal interval can be determined.

The remainder of this paper is organized as follows. Section 2 briefly reviews conventional traffic control schemes for multi-stage switching systems. Section 3 presents a distributed traffic control scheme. Section 4 describes the optimal design, and Sect. 5 concludes the paper.

2. Conventional Traffic Control Schemes for Multi-Stage Switching Systems

Multi-stage switching systems need a scheme for controlling traffic. A 3-stage switching system needs to bal-

Manuscript received June 2, 1999.
Manuscript revised August 26, 1999.

†The authors are with the NTT Network Service Systems Laboratories, Musashino-shi, 180-8585 Japan.

IEICE TRANS. COMMUN., Vol. E83-B, No. 2, pp. 231–237, February 2000

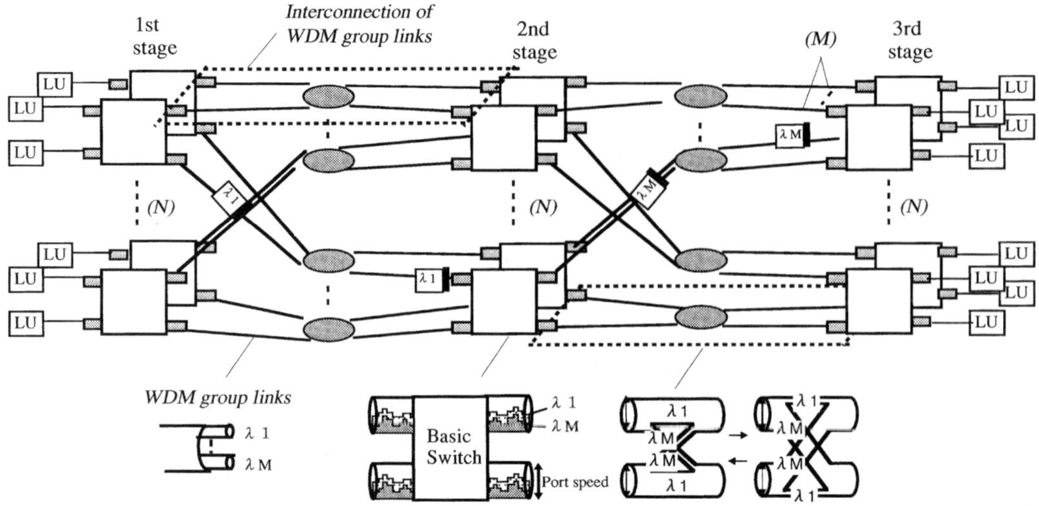

Fig. 1 Scalable 3-stage ATM switch architecture using WDM grouped links.

ance the 2nd-stage switching load. There are two types of traffic control scheme for multi-stage switching systems: those that use cell-based routing and those that use connection-based routing. Here, we assume that the internal link speed between basic switches is not increased. This is because all the same basic switches may be used in the system assuming the modularity of the basic switches.

If the system uses cell-based routing, each cell is routed independently in each input LU. Thus traffic is spread as evenly as possible among all the different paths [3], [9]. Some cells may arrive out of sequence in the wrong order at an output LU because cells belonging to the same virtual circuit may take different paths through the multi-stage switching system. Therefore, cell-resequencing is needed at the output LU. In a conventional cell-resequencing function, cells are time-stamped at the input LUs and the function uses the time-stamp information to reorder the cells arriving at an LU into the correct sequence.

However, the higher the speed of the output link, the harder this cell-resequencing function is to implement. This is because the cell time is shorter. For example, if the link speed of a switching system is 10 Gbit/s, the cell time is only 42.4 ns. This cell-resequencing function will be problematic since it will raise the cost of implementation.

If the system uses connection-based routing, on the other hand, all cells in a given virtual circuit follow the same path through the multi-stage switching system. The system maintains cell ordering directly, but requires explicit path selection. When a new virtual circuit is to be added from some input LU to some output LU, the system must find an optimal path through the system to accommodate the new circuit [10], [11]. Conventional systems have a central control processor and control lines to manage the bandwidth of all the links in the system.

However, conventional systems have trouble doing this when the switching system becomes large. The central processor can only handle a limited number of connections. The centralized control scheme has to find a new path in a very short time, and the system need to be able to handle a lot of information, such as the traffic characteristics of accommodated connections and their routes. This information is needed to manage the bandwidth of all the internal links. To find a new path, the cell loss ratio must be estimated after a new connection is accommodated. For example, assuming that each LU get 500 requests a second, the average period to find a new path is only 4 μs in a system with 512 LUs. Within only 4 μs, therefore, the system has to estimate the cell loss ratio and search for appropriate available routes. To do these heavy tasks, a very high-speed processor would be needed. We think that this centralized control approach would be too expensive. Thus, a centralized control scheme would not allow the switching system to be expanded in a cost-effective manner.

Therefore, we need a new traffic control scheme that does not require the cell-resequencing function used in the cell-based routing and that does not need a central control processor to manage all the links in connection-based routing. A traffic control scheme that works in a distributed manner is required.

3. Proposed Traffic Control Scheme

Our distributed traffic control scheme uses connection-based routing. We think that this scheme can handle CBR (constant bit rate) and VBR (variable bit rate) traffic. Our scheme does not require a central control processor or control lines. Instead, each port of the basic switches has its own traffic monitor. OAM cells carry the path-congestion information to the LUs, enabling each LU to route new virtual circuits independently.

One way to estimate the residual bandwidth is to count the cells arriving at a port. We can use an easy-to-implement simple traffic monitor described in [12]. If the traffic descriptors of connections such as average bit rate and peak bit rate are given, we can also employ a bandwidth management scheme [15] instead of the measurement-based estimation scheme. This scheme estimates the residual bandwidth that satisfies the specified cell-loss probability by generating a series of virtual requests for connection. Both of these estimation schemes support VBR traffic take into account statistical multiplexing effects as well as CBR traffic.

The threshold bandwidth, which is defined as B_r, is set for the residual bandwidth to judge whether the port is congested, as shown in Fig. 2. B_r is designed by taking into account the traffic characteristics of connections to satisfy QoS of connections and overbooking probability. The overbooking probability will be described in the next section.

Note that we use the term "port" to refer to the port of each basic switch. B_r must be set to an optimal bandwidth. If it is set too wide, the bandwidth assigned by LUs decreases; if it is set too narrow, many cells are lost in the port. Section 4 describes this in more detail.

If the residual bandwidth is lower than the threshold, the congestion-information (CI) bit of arriving cells is set to 1. LUs route virtual circuits using this information. The congestion probability at port i, which is defined as $P_c(i)$ is given by the traffic condition and

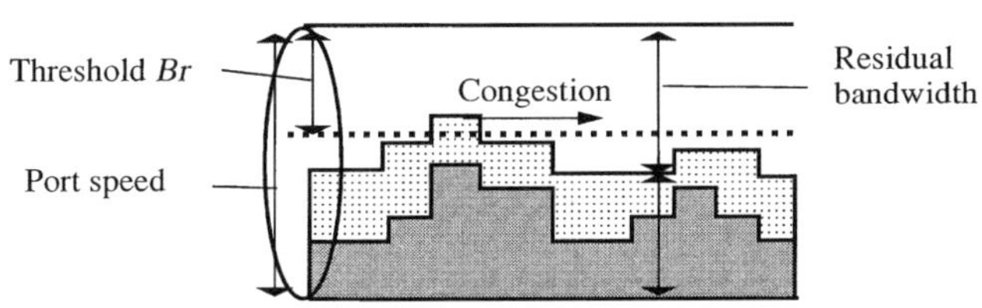

Fig. 2 Traffic monitor in port of basic switches.

switching system structure.

OAM cells are used in the system to check route performance. An LU has an input module and an output module. A virtual circuit is routed from the input module of a source LU to the output module of a destination LU. There are many routes from the source LU to the destination LU, so forward OAM cells are sent from the source LU along each route in turn. The destination LU returns the path-congestion information using backward OAM cell to the source LU through the reverse route, as shown in Fig. 3. The system does not require control lines because it uses OAM cells.

The CI bit of the OAM cell sent from a source LU is set to 0. If forward OAM cells arrive at a congested port, the CI bit is set to 1. When a load is low, we can assume that each $P_c(i)$ is independent. If a load increases to some degree, we can also assume that it is independent when there are many alternative routes in the system. However, when a load becomes very heavy, the traffic does not follow the Poisson distribution and $P_c(i)$ is not independent. As a first step in this study, we assume that the traffic follows the Poisson distribution and $P_c(i)$ is independent. In the next study, we should take into account the dependence of $P_c(i)$ on the heavy traffic condition. If there are P ports on a route, the worst congestion probability of the route P_r is given by Eq. (1). In Fig. 3, there are 4 ports on a route that may be congested.

$$P_r = 1 - \prod_{i}^{P} (1 - P_c(i)). \tag{1}$$

Each LU sends forward OAM cells along each route every T_{oam} seconds, as shown in Fig. 4. If a route is available, the LU rewrites the routing-lookup table which indicates currently available routes. If a route is congested, it immediately sends an OAM cell along another route. The probability that a route is congested is less than P_r. If an LU can not get the backward OAM

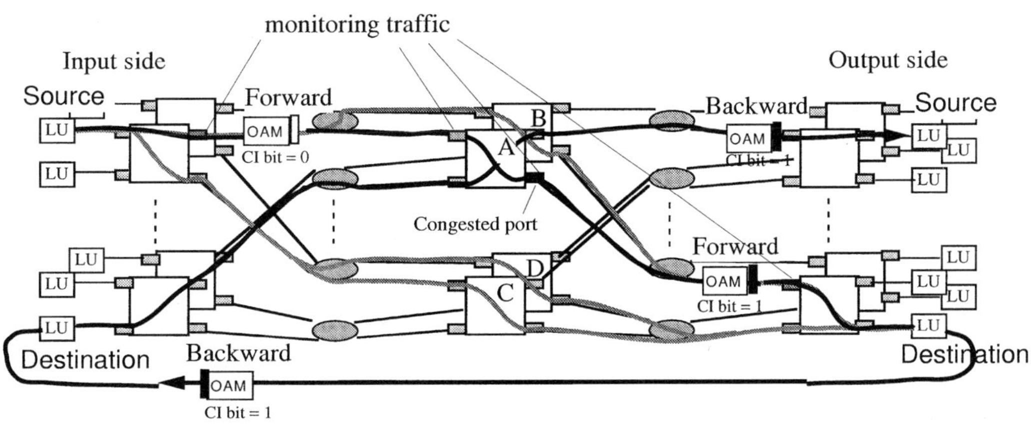

Fig. 3 Route-performance check using OAM cells.

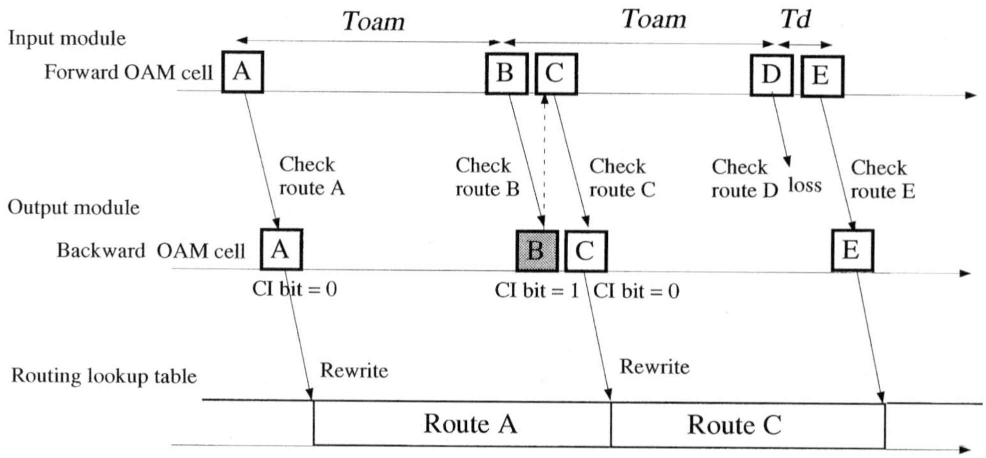

Fig. 4 Path selection by an LU.

cell within a maximum delay time (T_d), the OAM cell is considered to have been dropped or to still be waiting in a long queue. This situation means that the route is not available. In this case, the LU sends an OAM cell along another route.

Each LU is allowed to search available routes R times. If R routes are congested, the LU refuses a new requested virtual circuit. We call this a internal blocking. The internal blocking ratio in the system is less than P_r^R. The maximum time when an LU sends new forward OAM cells until it change the routing-lookup table is $R\ T_d$.

We explain how connection admission control (CAC) is executed using the distributed traffic control scheme described above. When a new connection is requested to be connected, the switching system has to check whether cell-loss probabilities at both the external link and the internal links of the switching system after the new connection is accepted are smaller than each specified probability. If both estimated cell-loss probabilities are satisfied with the conditions, the connection is accepted. The check for the cell loss probability at the external link is executed using well-known cell-loss estimation schemes, for example, those presented in [16]–[18], and [19]. For the internal links, on the other hand, each LU manages its own routing-lookup table which expresses whether available routes that satisfied the specified cell-loss probability exist between the source LU and the destination LU. We note that the cell loss probability is taken into account by using the residual bandwidth estimation scheme presented in [12] and [15].

4. Optimal Design Scheme for the Proposed System

Switching systems need to be designed in a cost-effective manner [13]. We should design bandwidth for user cells in a port (B_{usr} bit/s) to maximize the system

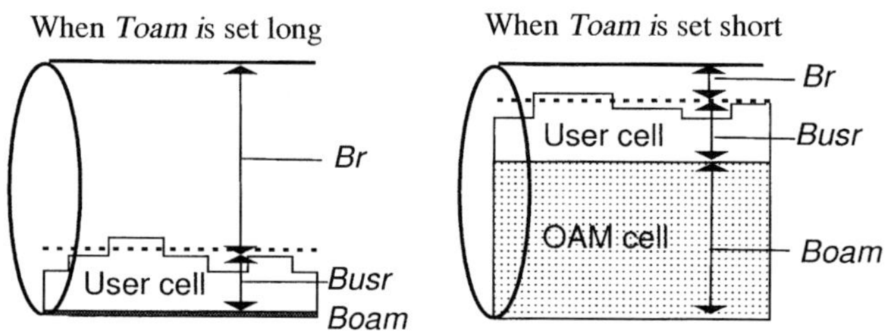

Fig. 5 Bandwidth for user cells in a port.

efficiency. B_{usr} is given by

$$B_{usr} = C - B_r - B_{oam}. \tag{2}$$

B_{usr} depends on the interval T_{oam} between OAM cells. Figure 5 shows the bandwidth for user cells in a port. If we set it too long, we must set B_r to a wide bandwidth. But if we set it too short, the bandwidth for OAM cells in a port (B_{oam} bit/s) increases while B_r is reduced. We show how an optimal interval between OAM cells can be determined.

If we set T_{oam} too long, LUs assign many virtual circuits into a currently "available" route that was given when LUs received the last OAM cell. However, before the next available route is given, the currently "available" route may be congested, and it will thus not be an available route. Therefore, the total bandwidth allocated in a port exceeds the speed of the switch port. We say that the port is overbooked [14]. Figure 6 shows the relationship between the residual bandwidth and the overbooking probability. Here, we assume that T_d is much smaller than T_{oam} in order to simplify the discussion on the relationship between T_{oam} and the overbooking probability. We must set the threshold B_r to an optimal bandwidth to ensure the specified overbooking probability. This is because if a port is overbooked, many cells are lost in the port. Figure 7 shows B_r for assuring that the overbooking probability is less than 10^{-9}. As T_{oam} becomes longer, the optimal B_r be-

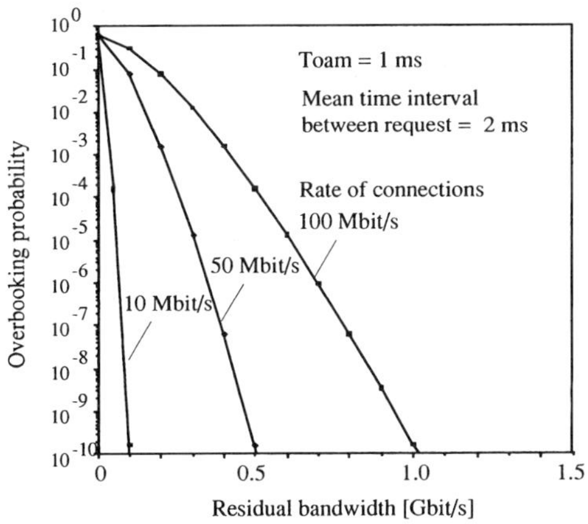

Fig. 6 The overbooking probability of CBR traffic.

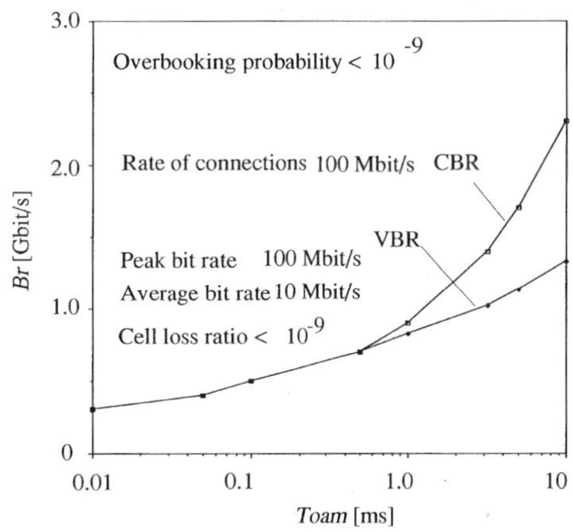

Fig. 8 B_r for CBR and VBR traffic.

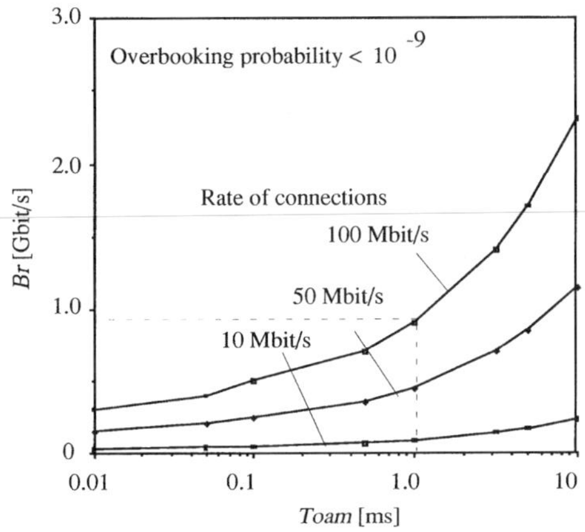

Fig. 7 B_r to ensure overbooking probability $< 10^{-9}$.

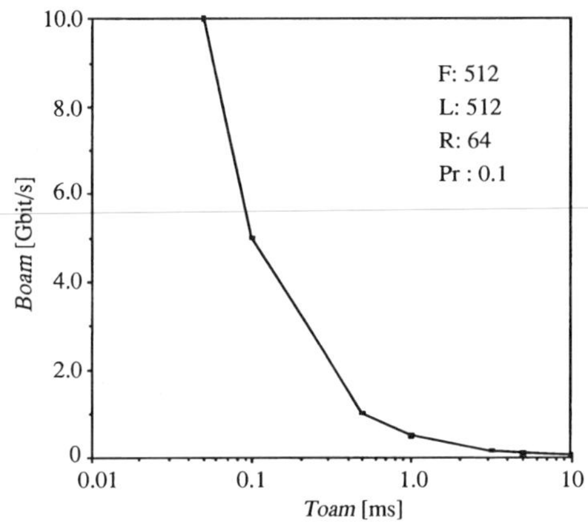

Fig. 9 Relationship between T_{oam} and B_{oam}.

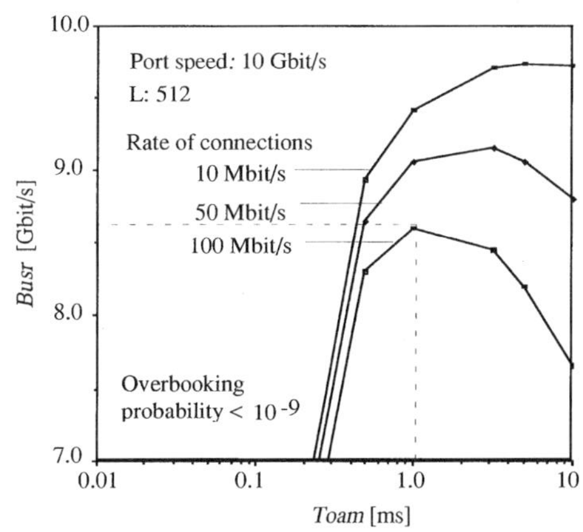

Fig. 10 Relationship between T_{oam} and B_{usr}.

comes wider. If we set T_{oam} to 1 ms, then B_r must be set to 0.9 Gbit/s, assuming that the rate of connections is 100 Mbit/s. The overbooking probability is estimated by using maximum peak bit rate in terms of a conservative design. However, if we can predict the traffic characteristic of new connections that will requested to be connected, we can design the B_r more effectively due to multiplexing effect of VBR connections. The B_r for VBR traffic is estimated less than that of CBR traffic, as shown in Fig. 8, assuming that a certain cell loss ratio is allowed. If VBR connections and CBR connections coexist, B_r can be designed between the B_r for CBR and the B_r for VBR.

On the other hand, if we set T_{oam} too short, the number of OAM cells in the system increases. B_{oam} is inversely proportional to T_{oam}, as shown in Fig. 9. B_{oam} is given by

$$B_{oam} = \frac{2FL\sum_{i=0}^{R} Pr^i}{T_{oam}} \tag{3}$$

where F is the cell length in bits and L is the number of output LUs. On the right side of Eq. (3), the coefficient of 2 means both forward and backward OAM cells, and $\sum_{i=0}^{R} Pr^i$ is the average number of transmissions until an available route is found up to a maximum of R times.

B_{usr} has a maximum at a certain value of T_{oam}. Figure 10 shows the relationship between T_{oam} and B_{usr}. When the rate of connections is 100 Mbit/s, we can maximize B_{usr} (8.6 Gbit/s) if we set T_{oam} to 1 ms.

5. Conclusions

We have proposed a distributed traffic control scheme for large multi-stage ATM switching systems. When a new virtual circuit is to be added from some source LU to a destination LU, the system must find an optimal path through the system to accommodate the new circuit. Conventional systems have a central control processor and control lines to manage the bandwidth of all the links in the system. The central control processor handle all the virtual circuits, but it has trouble doing this when the switching system becomes large. A large switching system with Tbit/s-class throughput requires a distributed traffic control scheme. In our switching system, each port of the basic switches has its own traffic monitor. OAM cells carry the path-congestion information to the LUs, enabling each LU to route new virtual circuits independently. Our system does not require a central control processor and control lines. The performance of the proposed system depends on the interval between OAM cells. We showed how an optimal interval can be determined in order to maximize the bandwidth for user cells. This control scheme will suit future Tbit/s ATM switching systems.

References

[1] H. Ohnishi and K. Miyake, "Issues in ATM network service development, standardization and deployment," IEICE Trans. Commun., vol.E81-B, no.2, pp.152–163, Feb. 1998.

[2] J. Turner and N. Yamanaka, "Architectural choices in large scale ATM switches," IEICE Trans. Commun., vol.E81-B, no.2, pp.120-137, Feb. 1998.

[3] T. Chaney, J.A. Fingerhut, M. Flucke, and J.S. Turner, "Design of a gigabit ATM switch," Proc. IEEE INFOCOM'97, pp.2–11.

[4] K. Genda and N. Yamanaka, "TORUS: Terabit-per-second ATM Switching system architecture based on distributed internal speed-up ATM switch," IEEE J. Sel. Areas Commun., vol.15, no.5, 1997.

[5] Y. Kamigaki, T. Nara, S. Machida, A. Hakata, and K. Yamaguchi, "160 Gbit/s ATM switching system for public network," Proc. IEEE GLOBCOM'96, pp.1380–1387.

[6] N. Yamanaka, S. Yasukawa, E. Oki, and T. Kawamura, "OPTIMA: Tb/s ATM switching system architecture: Based on highly statistical optical WDM interconnection," Proc. IEEE ISS'97, System Architecture, 1997.

[7] N. Yamanaka, "Breakthrough technologies for the high-performance electrical ATM switching system" IEEE J. Lightwave Technol., Dec. 1998.

[8] K. Nakai, E. Oki, and N. Yamanaka, "Scalable 3-stage ATM switch architecture using optical WDM grouped links based on dynamic bandwidth sharing," IEICE Trans. Commun., vol.E82-B, no.2, pp.265–270, Feb. 1999.

[9] M. Prycker and M. Somer, "Performance of a service independent switching network with distributed control," IEEE J. Sel. Areas Commun., vol.5, no.8, 1987.

[10] R. Melen and J. Turner, "Nonblocking multirate distribution networks," IEEE Trans. Commun., vol.41, no.2, Feb. 1993.

[11] S.C. Liew, M. Ng, and W.C. Chan, "Blocking and nonblocking multirate clos switching network," IEEE Trans. Networking, vol.6, no.3, June 1998.

[12] K. Shiomoto and N. Yamanaka, "An admission control scheme based on measurement of instantaneous utilization," IEICE Trans., vol.J80-B-I, no.12, pp.950–960, Dec. 1997.

[13] P. Coppo, M. D'Ambrosio, and R. Melen, "Optimal cost/performance design of ATM switches," IEEE Trans. Commun., vol.1, no.5, Oct. 1993.

[14] N. Yamanaka, K. Shiomoto, and H. Hasegawa, "ALPEN: A simple and flexible ATM network based on multiprotocol emulation at edge nodes," IEICE Trans. Commun., vol.E79-B, no.4, pp.611–615, 1996.

[15] E. Oki and N. Yamanaka, "Performance of high-speed admission control in ATM networks based on virtual request generation," IEICE Trans. Commun., vol.E82-B, no.3, pp.473–480, 1999.

[16] T. Murase, H. Suzuki, S. Sato, and T. Takeuchi, "A call admission control scheme for ATM networks using a simple quality estimate," IEEE J. Sel. Areas Commun., vol.9, no.9, pp.1461–1470, April 1991.

[17] H. Saito, "Call admission control in an ATM network using upper bound of cell loss probability," IEEE Trans. Commun., vol.40, no.9, pp.1512–1521, 1992.

[18] H.G. Perros and K.M. Elsayed, "Call admission control schemes: A review," IEEE Commun. Mag., pp.82–91, Nov. 1996.

[19] T. Lee, K. Lai, and S. Duann, "Design of a real-time call admission controller for ATM networks," IEEE/ACM Trans. Networking, vol.4, no.5, pp.758–765, 1996.

Chapter 4

ATM and IP Integration

Demand for Internet service is doubling every year, and many kinds of services such as voice, data, leased lines, and video—all of which are associated with multimedia services—are being carried by IP protocols. The problem is how to carry IP protocols that need QoS support functions. IP over WDM or IP over XXX over WDM are potential problem solvers. With regard to the XXX approach, ATM technology was originally established for the subscriber network transport because of its flexible multi-bearer multiplexing capability and QoS support; QoS is also a very important for backbone transport capability.

Paper 4.1 presents a typical approach to integrating IP and ATM. IP long-life sessions are cut-through at the ATM switches. A prototype system with 2.4 Gb/s throughput has been fabricated and is described in this paper. The results of many performance experiments are also reported. The cell-switch-router (CSR) approach described is very important for high-performance QoS guaranteed IP transport networks.

Paper 4.2 proposes the connectionless switch architecture. IP traffic can be transferred through an ATM VC connection using the network level cut-through approach. The paper provides not only an architectural study but also a network level demonstration and system implementation results. The system configurations and network level experimental results are described in detail. These two papers are very important in understanding architectures, networks, and systems.

Paper 4.3 introduces an experiment on the traffic control issue. The next generation IP network must offer high-performance and QoS capability. This paper studies the performance of congestion control methods. ATM is connection oriented and has a lot of traffic engineering functionality. The combination of IP and ATM traffic control realizes IP level QoS.

Mobile communication can also be based on IP over ATM technology. LAN emulation (LANE) has been studied for a long time as noted in Paper 4.4. Mobile networks need hand off and rerouting functions, and location management is also needed for IMT2000. This last issue is now a very hot topic for next generation mobile networks.

Paper 4.5 details a performance study on an IP over ATM wide area network. TCP-IP offers a traffic control capability that does not depend on lower level control. ATM also has traffic control functions. The paper introduces the results of network simulation and experiments. Controlling this service in a WAN is very difficult and is a very hot topic.

IP traffic arrives in a very bursty manner, and SVC-based IP over ATM needs SVC setup when bursty IP traffic is present. SVC setup in this situation is quite different from that seen in conventional telephony SVC. Paper 4.6 shows a numerical model and performance results of SVC-based IP over ATM.

C. Dou et al. evaluate the performance of packet-level multiple QoS control in IP over ATM networks in Paper 4.7. They provide closed-form solutions for the mean cell waiting time and the mean packet waiting time of individual traffic type. M. Jo et al. address packet discard schemes for TCP over IP over ATM in Paper 4.8. They improve an early packet discard scheme that was developed by using IP packets to well handle TPC connections.

Paper 4.9 by R. Kawahara et al. studies the performance of TCP/IP over an asymmetric digital subscriber line (ADSL). The bandwidth of an ADSL link varies over time due to changes in the link's physical conditions, which degrades TCP performance. They evaluate the performance when using various ATM traffic control schemes, such as available bit rate (ABR) and generic flow control, to handle variations in the ADSL bandwidth.

H. Saito et al. describe a newly developed traffic measuring tool that is applied to the real-time simulation of a network in Paper 4.10. It monitors IP traffic on an ATM link and continuously transfers the length and timestamp of each IP packet to a post-processing system. The developed system enables the implementation of a new approach to network engineering.

Emerging real-time and multimedia applications in IP over ATM are becoming an indispensable part of the networking world. I. Erturk et al. present a new extension to classical IP and address resolution protocol over ATM protocol in Paper 4.11. They show that the proposed scheme provides better QoS support to the IP end users in an IP over ATM environment.

A. Arutaki et al. describe a developed ATM switching system with a built-in IP processing mechanism in Paper 4.12. They verify that the IP function can be directly provided to a high-speed ATM network without using external IP equipment. It is also confirmed that the high reliability, secure maintainability, and testability of a public network switch can be achieved with this IP processing mechanism.

Paper 4.13 by S. Ano et al. describes results of experiments on the accommodation of TCP traffic by an ATM wide-area network. They propose a variable bit rate (VBR) service with selective cell discard (SCD) option. They show that the VBR service with SCD option is appropriate for accommodating best effort TCP/IP traffic.

The final paper of this chapter is provided by K. Okimi. Paper 4.14 encapsulates today's data network configurations, systems, and services. In addition, the paper suggests future requirements and directions for technologies.

High-Performance Backbone Network Technology edited by Naoaki Yamanaka
ISBN: 0-8247-5321-6 DOI: 10.1081/(120025685)
Published 2004 by Marcel Dekker, Inc.

Architecture of Cell Switch Router and Prototype System Implementation

Shigeo Matsuzawa, Ken'ichi Nagami, Akiyoshi Mogi, Tatsuya Jinmei, Hiroshi Esaki, and Yasuhiro Katsube

SUMMARY Overview of Cell Switch Router (CSR) and the CSR prototype system are described. CSR can simultaneously support both connection oriented IP flows and connectionless IP flows. CSR contains cell switch fabric and IP packet switch fabric to achieve high throughput IP forwarding. IP packets are forwarded either through a cut-thru packet transmission, in which packet are forwarded without reassembling IP packet nor IP header processing, or through a conventional hop-by-hop IP packet forwarding. This paper describes and proposes the mechanism to forward the connectionless IP packet flows at the CSR. A CSR prototype system has been developed. The CSR prototype system uses PVC connections to transfer the IP packets. With the CSR prototype system, we can make sure that CSR system can achieve a high throughput, i.e., 2.4 Gbps aggregated throughput. For end-to-end TCP/IP packet transmission, more than 90 Mbps can be achieved and realtime video transmission (30 Mbps video) can be achieved.

key words: *ATM, CSR, cut-thru, hop-by-hop, IP, router, prototype*

1. Introduction

ATM (Asynchronous Transfer Mode) is realized as one of platforms to provide high speed datalink layer service with certain QOS (Quality of Service), with a unique user network interface [1],[6]. Since the user information is handled by the small and fixed sized packet (i.e., called as cell) in the ATM networks, it is said that ATM networks will provide a highly flexible and cost-effective network for handling a wide variety of communications and the applications will be obtain a flexible and cost-effective communication pipe. The cells are transferred through the connection oriented channel, which is called a virtual channel connection (VCC), in order to reduce the protocol overhead. As a result, it is said that an ATM network could provide a high speed data communication channel for end applications.

Sometimes, it is said that ATM network will provide a router-less large cloud datalink platform, which does not have any routers within the large cloud network. However, we will need routers, even large ATM [13] cloud datalink platform will be provided [2]. As discussed in [2], the IP subnet must not be so large size so as to provide a large scale soft-state multicast service [14]. Therefore, physical or logical datalink network segments (i.e., IP subnets) will be interconnected through routers (i.e., through the network layer entities), even when the ATM becomes a major datalink platform.

This paper focuses on how to transport conventional connectionless IP packets [11] over the ATM networks. The hop-by-hop packet forwarding at the router by the *software processing* would be one solution. However, there would be the limitation of processing capability of processors in the routers for the future high speed and large amount of packet flows. In order to scale associated with the required bandwidth by the applications and with the total amount of packet flows handled by the routers, some efficient IP packet forwarding technique (e.g., cut-thru routing) is necessary. [2],[7] and [4] propose a router architecture containing both conventional IP processing module and cell-switching module. The architecture of the router proposed in this paper is further detailed architecture based on the architecture proposed in [2],[7]. [2],[7] discuss the architecture to forward both connection oriented and connectionless IP packet flows. On the contrary, the architecture discussed in this paper focuses on the forwarding of conventional connectionless IP packets.

The following is the structure of the rest of this paper. Section 2 briefly discusses issues of the conventional IP over ATM architecture models. Sections 3 and 4 propose and evaluate the IP over ATM architecture using CSR (Cell Switch Router). In Sect. 5, the overview and specification of CSR prototype system using PVC connections. Section 6 gives a brief conclusion.

2. Related Works of IP over ATM Architecture

In this section, the issues of IP service architectures over ATM proposed in IETF and ATM Forum are discussed.

2.1 Classical IP Model [12], LAN Emulation Model [8]

In general, logical IP subnets (LISes) will overlay on a single NBMA ATM network segment. Figure 1 shows the case where packets are transferred from Host A.1 belonging to subnet A to Host B.1 belonging to subnet B. Logical IP Subnet A and subnet B exist on the NBMA ATM network.

Manuscript received October 1, 1996.
Manuscript revised December 19, 1996.

†The authors are with R&D Center, TOSHIBA CORPORATION, Kawasaki-shi, 210 Japan.

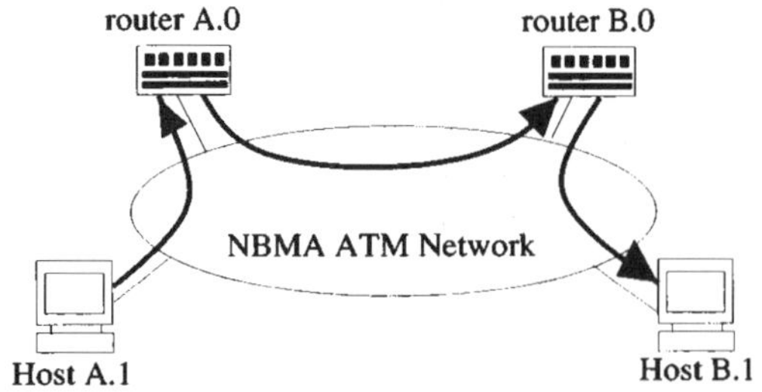

Fig. 1 Classical IP/LAN emulation model.

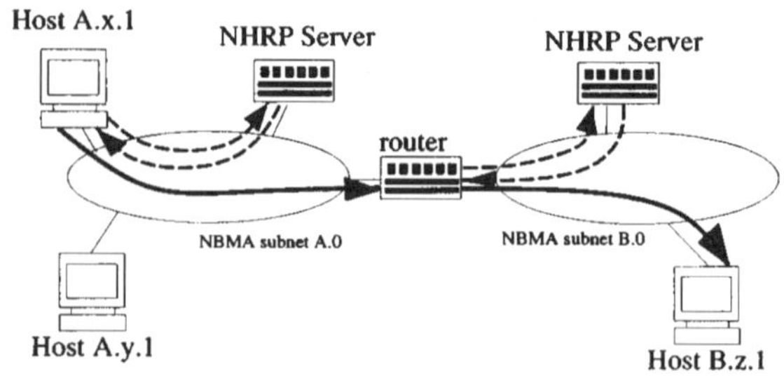

Fig. 2 NBMA with NHRP model.

LISes are interconnected by the conventional routers. For the intra-IP subnet communication, a direct ATM-VCC between the nodes can be provide. On the other hand, for the inter-IP subnet communication, the IP packet must pass through the router(s). This means that ATM-VCC shall be always terminated at the router. In the case of Fig. 1, the packet from Host A.1 to Host B.1 is transferred to router A.0 using the direct ATM-VCC. Router A.0 forwards the received packet to the router B.0, which is the next hop router indicated by routing protocol. There may be the direct ATM-VCC between the router A.0 and router B.0, as shown in the Fig. 1. However, in general, there would be intermediate router(s) between router A.0 and router B.0. The packet received packet by the router B.0 is forwarded to the Host B.0 through the direct ATM-VCC.

Therefore, we can not achieve high throughput IP packet forwarding for inter-subnet communication. This is because, even for the packet transmission within the same NBMA ATM network, the IP packet for inter-subnet communication has to experience more than one IP processing at the router(s).

2.2 NHRP with NBMA Model [9],[10]

LIS will overlay on a single ATM datalink segment. NBMA network is defined as the network domain where end-to-end ATM-VCC can be established. Within the NBMA network, each host could establish a direct ATM-VCC toward any host that is located both at other LIS and at the same LIS without passing through router(s), i.e., short-cut routing. However, when the IP packet should be transferred to outside an NBMA network, IP packet will be transferred to the border router between two NBMA networks. This means that we can not obtain high performed IP packet forwarding beyond the NBMA network.

Host or router will resolve the ATM address for a given destination IP address. If the given destination IP address is not directly reachable, the most appropriate exit point's (i.e., router's) ATM address will be resolved. Here, even if we use NHRP, a packet could experience hop(s) of IP processing (passing through router(s)) within a ATM cloud (i.e., NBMA domain). As well known [10], since a direct ATM-VCC can be established regardless of routing topology/hierarchy defined in the routing protocols, there is a possibility to be created a long period routing loop.

Since the conventional routing protocol for connectionless IP packet routing is still running among the routers, NHRP may require some major modifications of IP packet forwarding for the routers and hosts. The routers attached to the NBMA domain exchange routing information (e.g., topological information and link attributes) among them. When a router receives an IP packet, the appropriate *next* hop node is resolved by the routing table calculated by routing protocol. ARP will be used, if ARP cache for MAC address of the *next* hop router is flushed. As discussed above, in the usual router's procedure, there is no chance to try to resolve the final destination router to reach the destination host. In order to apply NHRP, the IP forwarding procedure, that is usually implemented in kernel space, has to be substantially modified.

The other issue for NHRP model is the scalability for soft-state multicast service using RSVP. Since the transit routers will be physically bypassed using short-cut routes, the number of receivers in the ATM level multicast connection can be large. In RSVP, each receivers (end-hosts or routers) send *Resv* message to the up-stream node, which corresponds to the root node of ATM multicast connection, whenever they receive *Path* message [14]. Therefore, when the number of receivers in the ATM multicast connection is large, the number of *Resv* messages that the up-stream node must handle becomes large. Large number of *Resv* message reception may cause a performance degradation of the up-stream node, due to so many process interruptions.

Figure 2 shows the case where the packet is transferred from Host A.y.1 to Host B.z.1. The two NBMA ATM networks are internetworked through a router. Each NBMA ATM network, subnet A and subnet B, would have multiple LISes. When Host A.x.1 sends a packet to Host B.z.1, Host A.x.1 resolves the ATM address of the egress router leading to Host B.z.1, using the NHRP Server. NHRP Server gives the ATM address of the egress router to forward the packet to Host B.z.1, and Host A.x.1 establishes an ATM-VCC to the egress router to send packets to Host B.z.1. When the egress router receives the packet from Host A.x.1, the egress router analyses the destination IP address of the received packet. The egress router analyses the ATM address of Host B.z.1 using the NHRP Server in subnet B NBMA ATM network. The NHRP Server gives the ATM address of the Host B.z.1 to forward the packet to Host B.z.1, and the egress router establishes an ATM-VCC to Host B.z.1 to send packet to Host B.z.1. As shown in the figure, the packet has to experience the IP packet processing, when the packet is transferred to the host located at the different NBMA ATM network.

2.3 IP Switch Model

In the IP switch model [4], routers containing ATM switch fabric in it are interconnected by transparent links.

The architecture model of IP switch is shown in Fig 3. Data-link network segment is identical to a transparent link between IP switches. Therefore, IP packet beyond IP subnet must always pass router (i.e. IP switch). In order to transfer the IP packet (both connection oriented IP packet and connectionless IP packet) from host A to host B, host A transfer IP packet to the router1 (i.e. IP switch). Routers (1, 2 and 3) relay the received IP packet toward host B. When the IP packet flow from host A to host B is either active flow or connection oriented IP transport flow, the IP packet will be forwarded without IP forwarding process in the router.

This model can provide high throughput and small latency packet forwarding even for inter-subnet packet exchanging. However, since the link among the IP switches must be transparent links (e.g., physical link or SONET link), we can not use ATM switches in the network, as shown in Fig. 4.

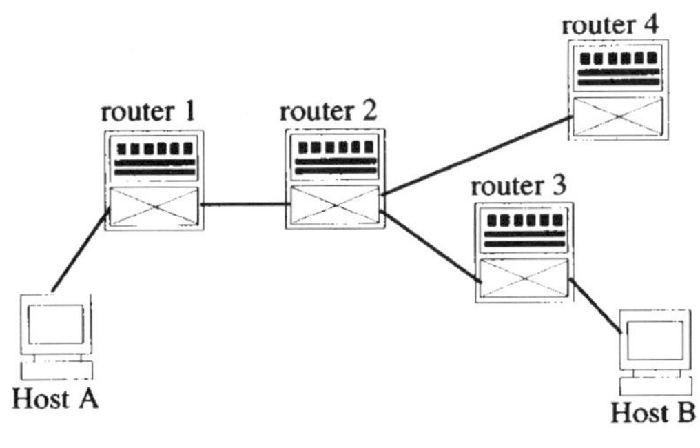

Fig. 3 IP switch model.

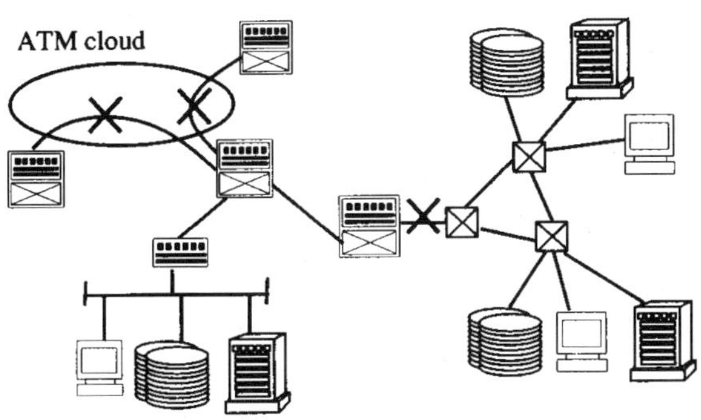

Fig. 4 Issue of IP switch model.

In general, IP switches can not be interconnected through the ATM least lines. IP switched can interconnected through the ATM least line only with the specific configuration, where ingress VPI/VCI value and egress VPI/VCI value of the ATM link (VP) are the same. This particular configuration is not a general configuration in the ATM network. Further, IP switches can not interconnect with standard ATM switches, since IP switch does not support ATM UNI. Since IP switch does not have an interconnectivity with ATM switches, the network with IP switch can not provide a subnet constructed by ATM switch(es).

3. ATM Network Architecture Using Cell Switch Router

3.1 Architecture of Cell Switch Router (CSR)

In this section, the proposed ATM network architecture using CSR is discussed. As described in [2],[7], CSR supports both connection oriented and connectionless IP flows. This paper focuses on further discussion on how CSR supports connectionless IP packet flows and gives some evaluations on the proposed architecture.

Figure 5 shows one example of internet model. ATM-based subnet is composed of switch nodes, and the interface with the adjacent network segment is UNI (User Network Interface), even when the adjacent network is an ATM network. Data-link network segments (i.e., IP subnets) are interconnected by routers, and each data-link network segment corresponds to an IP subnet. Regarding ATM networks, since the interconnecting interface is UNI, each ATM-based subnet can operate independently.

CSR provides a cell switching capability in addition to the conventional IP packet switching. By the use of cell switching capability, high throughput IP packet forwarding can be achieved. CSR can forward some IP packet flows while bypassing the packet assembly/reassembly and IP header processing, i.e., the cut-thru packet forwarding. In the cut-thru packet forward-

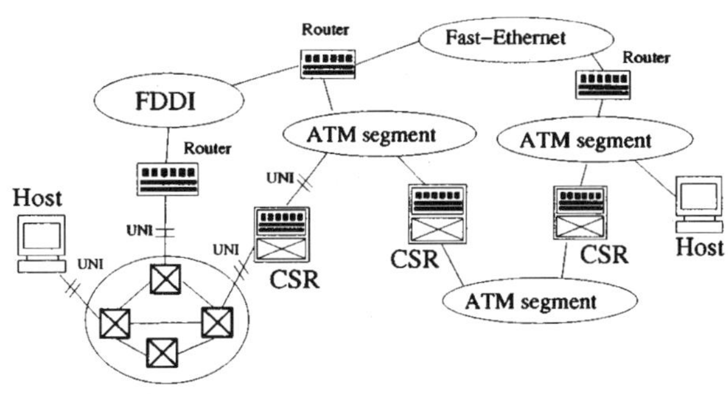

Fig. 5 Large scale internet with cell switch router.

ing, IP packet is forwarded based on VPI/VCI and IP header is not examined at the CSR. On the contrary, the conventional hop-by-hop IP packet forwarding based on IP header can be also applied to the other IP packet flows.

The key functionalities of a router interconnecting ATM networks are as follows.

- Hop-by-hop packet forwarding
  The packets forwarded with hop-by-hop packet forwarding are transferred through the default VC, that is established when CSR turns on. In the hop-by-hop packet forwarding, the IP packet is reassembled into the IP packet to analyze the IP header. The CSR decides the next hop node (i.e., router or host) based on destination IP address in the received IP packet, according to the routing information indicated by routing protocol (e.g., OSPF).

- Cut-thru packet forwarding path establishment and cell relaying
  Cells belonging to the packets, that should and could be forwarded without usual IP processing at the CSR, are relayed as the ATM layer's function, i.e., cut-thru packet forwarding. When the CSR receives IP packet, that should and could be forwarded by cut-thru mode, the CSR tries to establish a dedicated VC between the neighbor nodes (e.g., CSR). When the ingress dedicated VC and the egress dedicated VC, regarding a IP packet flow to be cut-thru, are established, these two dedicated VCs are coupled at the CSR. After the coupling of these two dedicated VCs, the cells belonging to this IP packet flow are forwarded in cut-thru mode. Since egress and ingress VCCs are independently established, CSR has to have the VPI/VCI translation functionality.
  Here, whether an IP packet flow is handled by the cut-thru packet forwarding or is handled by the conventional hop-by-hop packet forwarding will be basically determined by every router's local decision.

Through bypassing the conventional IP forwarding process using cell-relaying, we could dramatically increase the IP processing capacity of the router.

In order to provide the cut-thru packet forwarding by the CSR, the additional protocol (i.e., FANP) messages briefly discussed below must be defined. Neighbor routers must exchange the information how the IP packet flows are aggregated into datalink pipe (i.e., dedicated VC).

3.2 ATM-VCC Management Architecture

Figure 6 shows an example of the configuration of ATM-VCCs and IP packet flows with the cut-thru packet forwarding at CSRs. All ATM-VCCs, i.e., VCCs between CSRs and between CSR and host, are PVC (Permanent Virtual Connection). During the boot procedure of each node (CSR and host), the PVCs are established to the adjacent nodes.

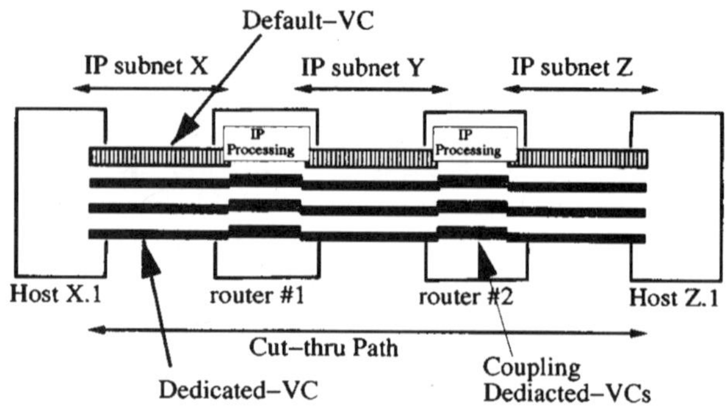

Fig. 6 ATM-VCC configuration example for CSR.

The default-VC transfers both control packets (e.g., the control packets of FANP described in the following subsection) and hop-by-hop based user packets. On the contrary, the dedicated-VC transfers only the user packets that are forwarded through the cut-thru packet forwarding mode. Here, the cell flow arrived at a CSR through a dedicated-VC may be transferred by the default-VC to forward the IP packet to the adjacent CSR, and vice versa.

The control messages to manage the user packet flows using the dedicated-VCs are transferred through the default-VC. The IP packets transferred through the default-VC are always performed the usual IP processing. Therefore, the management of cut-thru path can be hop-by-hop, and the IP header in the cut-thru path does not have to be examined. By the control packets (e.g., ST2, RSVP messages) exchanged among the CSRs, hard-state, soft-state and stateless IP packet forwarding routes are provided. The detail of the protocols and messages for control and management of the user packet flows are not discussed in this paper.

- Hard-state route (e.g., ST2 flow):
  The route of cut-thru path is not changed during a session. Therefore, the control packet may be exchanged only at route setup phase and at route termination phase. Any control packet may not be exchanged for the purpose of state refreshment.

- Soft-state route (e.g., RSVP flow, cut-thru flow):
  The control packet are continuously exchanged among the CSRs for the purpose of state refreshment using the hop-by-hop default-VC. The dedicated VCs are managed hop-by-hop, and the cut-thru path can dynamically reflect to the routing path indicated by the routing protocol.

- Stateless route:
  In stateless route, the route of IP packet can be dynamically changed packet-by-packet. Usually, the

cut-thru path will not be established for the stateless route.

Management of the dedicated-VCs, i.e., establishment, tearing down of the dedicated-VCs, and management of the mapping between IP packet flow and ATM-VCC are matters of every CSR's local decision. Q.2931 will be used to establish or tear-down the dedicated-VCs among the CSRs. When we use digital links (e.g., OC-3 or DS3 SONET pipe) to interconnect CSRs, Q.2931 will not be used to establish/tear-down the dedicated-VCs.

3.3 Flow Attribute Notification Protocol: FANP [3]

The purpose of FANP is establishment and maintenance (i.e., refreshment and flush the state) of dedicated VCs. Using the FANP, the up-stream node indicates to the neighbor down-stream node how the IP packet flows are aggregated into the datalink pipe (i.e., ATM-VCC for ATM networks). By means of this procedure, the corresponding dedicated VC will be established. Here, the information exchanged between the router by the FANP is not only useful for CSR, but also is generally useful for the routers attached to the connection oriented datalink platform (e.g., Frame Relay and TDM circuit network).

By the flow aggregating information exchanged by the FANP, the router could optimize the packet handling. For the CSR, when the flow aggregation is performed based on the destination IP address, the packet forwarding will be able to be optimized using a cut-thru packet forwarding capability. In general, the FANP indicates the following information associated with how the IP packet flows are aggregated to the neighbor down-stream node.

- Destination Address
  The IP packet flows that have a common destination address (i.e., individual destination address or destination address prefix) could be aggregated into a single datalink pipe (e.g., ATM-VCC). The source IP addresses of the IP packet flows may not be identical.
- QOS Class
  The IP packet flows that have a common QOS class could be aggregated into a single datalink pipe (e.g., ATM-VCC). Scheduling of IP packet processing at the router may be optimized based on QOS class.
- Flow Identifier
  The IP packet flow that have a common flow identifier (e.g., IPv6 flow-id [5]) could be aggregated into a single datalink pipe (e.g., ATM-VCC). And, the IP packet flow that is identified by the flow identifier (e.g., RSVP's packet flow) may ought to mapped to a single ATM-VCC.

Above three criteria are the orthogonal parameters. Therefore, in the actual case, the above parameters will be combined to aggregate the IP packet flows. The packet flow aggregation policy based on QOS criteria will not appropriate for CSR. Both the packet flow aggregation based on destination address and flow identifier will be appropriate for CSR architecture.

Here, in the prototype CSR system described in below, the IP packet flow is identified by the pair of source and destination IP addresses.

3.4 Packet Forwarding of Connectionless IP Flow

IP packets beyond the data-link segment are sent to an appropriate router from the source end-host. Routers exchange routing information and forward IP packets to the appropriate router.

The operation of connectionless IP packet forwarding is shown in Fig. 7. When the first IP packet (of

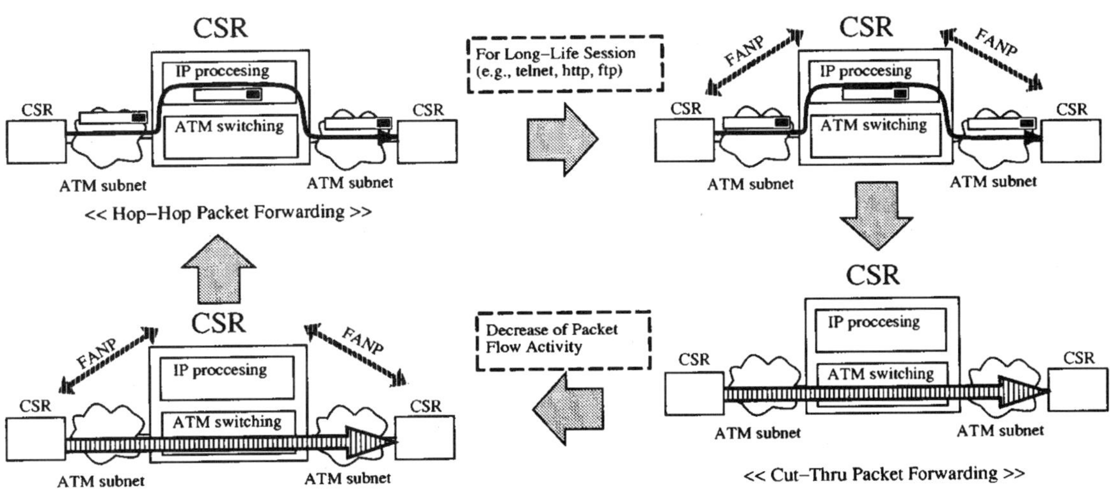

Fig. 7 Operation of hop-by-hop ATM-VCC cacheing for active IP flow.

the session or after the idle period) arrives at a CSR through the dedicated VC, the next hop router is decided using the IP header, i.e., the routing decision will be done based on the IP address (and the flow-ID in IPv6). After the next router is decided, the received IP packet is forwarded through the default VC toward the next hop router. CSR analyzes TCP/UDP header, as well as IP header. When the received IP packet belongs to the IP flow, that is likely to be long life session, a dedicated VC is picked up or is established for the IP packet flow and CSR establishes the mapping table between the incoming VPI/VCI (or some flow-ID in the data-link layer) and the outgoing VPI/VCI. When the packet comes from the conventional platform (e.g., Ethernet/FDDI), the router always examines the IP header in the case of IPv4 (in IPv6, the flow-ID could be mapped to the outgoing VPI/VCI). When the IP packet comes from an ATM network, the successive IP packet is forwarded based on the incoming VPI/VCI value, without any usual IP processing.

If the activity of the IP packet flow becomes low, the mapping information (i.e., the cached information) is pushed out. This means that the pushed out IP packet flow from the cache must be examined by the usual IP processing (looking at the IP header) at the CSR, i.e., transferred through the default VC. As a result, IP packet flow with a high activity can cut-thru a usual IP processing at the CSR.

The above procedure is performed independently at every CSR. The job of CSR is maintaining the VPI/VCI mapping table in the cell switch, or is maintaining the mapping table between the flow-ID and the VPI/VCI.

4. Evaluation and Discussion of Cell Switch Router

In this section, the quantitative and qualitative evaluation of cell switch router and the discussions on cell switch router is performed. Through the evaluations, the effectiveness of cell switch router and the implementation feasibility is shown.

Table 1 shows the comparison of IP over ATM architecture models, regarding some major points. As shown, the functionalities and features achieved by the cell switch router is better than the other architecture models. The meaning of "ATM link (VC/VP) availability" is whether the routers can be interconnected through the *standard* ATM links (VCC/VPC) provided by the ATM network. The meaning of "ATM switch connectivity" is whether the router can be directly connected with the *standard* ATM switch.

4.1 Aggregated System Throughput

The aggregated system throughput of cell switch router depends basically on the product of number and operation clock speed of interface ports. For the larger throughput of cell switching fabric, the required packet processing power of packet processing fabric is larger.

Table 2 shows the maximum aggregated system throughput of cell switch router with some configurations.

For example, with eight ports of OC3 interface, the maximum aggregated system throughput of CSR is about 2.4 million packets per second for 64 bytes average packet length. The throughput of 2.4 million packets per second approximately corresponds to the ten times larger system throughput of current high-end routers.

As shown, the cell switch router can easily achieve larger aggregated system throughput than the current high-end routers can. The expected maximum system throughput of CSR can be easily decades times larger than the maximum system throughput of the current high-end routers.

4.2 Delay Variance with CSR

In CSR, IP packets using different dedicated VCs are transferred in parallel, since IP packets are transferred with a cell-interleaving. On the contrary, in the conventional router, IP packets are transferred packet-by-packet with an FIFO rule. This means that the IP packet has to be in the queue until the all IP packets before it are transferred, i.e., HOL (Head Of Line) blocking. When we compare the packet delay variation (PDV), the PDV with CSR will be smaller than the PDV with the conventional router.

Table 2 Aggregated system throughput of cell switch router.

	CSR System Throughput		
	Bit Level	Pakcet Level	
		64 B packts	100 B packets
OC3–I/F 8 ports	1.2 Gbps	2.4 MPPS	1.5 MPPS
OC3–I/F 16 ports	2.4 Gbps	4.8 MPPS	3.1 MPPS
OC12–I/F 8 ports	4.8 Gbps	9.6 MPPS	6.1 MPPS

(*) MPPS; Million Packets Per Second

Table 1 Comparison of IP over ATM architecture models.

	Software router	NHRP /MPOA	IP Switch	CSR
Throughput	poor	fair	Good	Good
Routing loop possibility	No	Yes	No	No
ATM link (VP/VC) availability	Yes	Yes	No	Yes
ATM switch connectivity	Yes	Yes	No	Yes
System Complexity	Simple	Complex	Simple	Simple

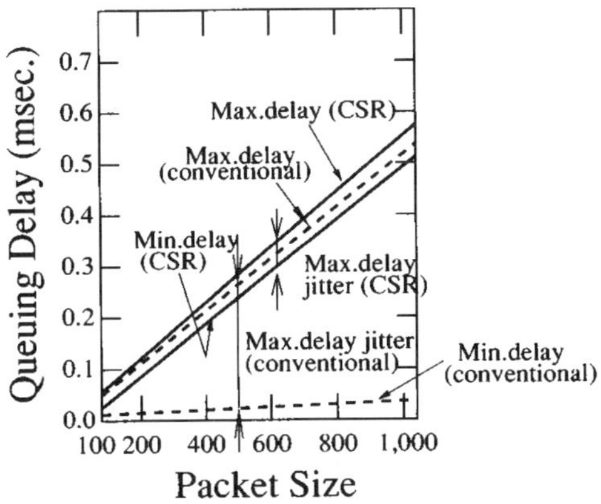

Fig. 8 Delay variation of packet delivery.

In the following evaluations, it is assumed that the switching speed for ingress IP packets is sufficiently fast than the egress link speed both for CSR and for conventional router. That means IP packets (i.e., cells with CSR) are queued at the egress link to the neighbor node.

The minimum packet transmission delay at the conventional router is smaller than that of CSR. No packet is queued before the IP packet experiencing the minimum packet transmission delay in the conventional router. The packet experiencing the maximum packet transmission delay waits in the queue, until all IP packets before the IP packet in the queue are transferred. With CSR, the PDV will be very smaller than that of the conventional router, though the minimum packet transmission delay of CSR would be larger than that of the conventional router and it would be almost the same as the maximum packet transmission delay of conventional router. This is because CSR does not transfer only one IP packet, but the multiple IP packets are transferred simultaneously with cell-interleaving.

Figure 8 shows one example of the maximum and minimum packet transmission delay versus the IP packet size, when 10 packets are simultaneously transferred to the queue at the egress link. Here, the data transmission speed of egress link is 155 Mbps. As shown, the PDV of CSR is quite small, however, the PDV of conventional router is large. With CSR, the maximum delay jitter is always 27 μsec. However, the maximum delay jitter with conventional router increases according to the increase of IP packet size; i.e., the maximum delay jitter with 1 KB packet is 464 μsec.

Some applications (e.g., continuous bit-stream transmission) are sensitive to the PDV rather than to average packet transmission delay. Typical applications, that are sensitive to the PDV, are voice and video signal transmission. With voice and video applications, the required receiving buffer space goes to larger for the larger PDV in order to maintain the playback point. This means that the required receiving buffer space for voice and video signal transmission using CSR will be much smaller than that using very high speed conventional routers.

Since the CSR provides less chance of conventional packet processing at the router than the other architectures (e.g., NHRP, LANE) do, the PDV characteristics using CSR will be better than that using other architectures. This is because LANE always has to perform the conventional packet processing at the router and NHRP has to perform the convetional packet processing at the egress point of NBMA ATM network.

4.3 Effectiveness of Cut-Thru Packet Forwarding

The effectiveness of cut-thru packet forwarding in the CSR mainly depends on the characteristics of IP packet flows. When we have many long-life sessions generating many IP packets, the cut-thru IP packet forwarding is meaningful and the cost to establish cut-thru paths (i.e., dedicated VCs) is acceptable.

Tables 3 and 4 show (a) number of cut-thru sessions that will be simultaneously established at the CSR, (b) total amount of IP packets, belonging to the typical applications that would likely be long life sessions, that will be forwarded with cut-thru mode, and (c) the average cut-thru session life-time for typical applications that would likely be long life sessions. In these evaluations, four applications, i.e., web, ftp, telnet and nntp, are assumed as the sessions to be forwarded with cut-thru mode. When CSR receives the first IP packet belonging to one of these four application, dedicated VC will be established to forward the rest of succeeding IP packets with cut-thru mode. The cut-thru mode will be terminated to go to the hop-by-hop mode, when the application does not generate any IP packet for two minutes. Figure 3 is based on the packet transmission traces, which is available with http://town.hall.org/pub/ITA/html/contrib/DEC-PKT.html, for each one hour's worth of all wide-area packet transmission between Digital Equipment Corporation and the rest of the world. Figure 4 is based on the packet transmission traces for eight hour's worth of all packet transmission between the R&D Center (about thousand peoples) of Toshiba Corporation and the rest of the Toshiba's corporate network.

As shown, the total amount of IP packets forwarded by cut-thru mode will be more than 86% with the DEC's backbone traffic and will be more than 70% in the Toshiba R&D center backbone traffic. Also, the average life time of cut-thru sessions will be more than eight minutes both for DEC's backbone and for Toshiba R&D center backbone. These evaluation results indicate that, even with the *current* IP flows, a large amount of IP packets will be able to be forwarded through cut-thru mode and the life time of cut-thru sessions will be long enough to establish dedicated-VC for them.

In the future, long life sessions will increase in the Internet and in the intranets. Voice and video signal transmission generates large amount of IP packets con-

Table 3 Cut-thru packet forwarding for current packet flows (DEC backbone).

Session	# of Sessions		Traffic Volume	Average Life-Time
	Ave.	Max.		
web	130	161	13.41%	324.7 sec.
telnet	16	22	1.43%	1131.9 sec.
ftp	100	144	32.91%	407.3 sec.
nntp	130	139	39.94%	1058.3 sec.
Total	276	430	86.69%	481.1 sec.

Table 4 Cut-thru IP packet forwarding for current packet flows (Toshiba R&D backbone).

Session	# of Sessions		Traffic Volume	Average Life-Time
	Ave.	Max.		
web	63	98	8.88%	504.2 sec.
telnet	15	27	12.50%	897.0 sec.
ftp	4	19	14.15%	238.5 sec.
nntp	4	12	34.81%	260.5 sec.
Total	86	115	70.34%	515.2 sec.

tinuously for a long time and the traffic volume of these applications will be large and will increase. By means of using proxy servers, web application will generate larger amount of traffic and will have longer session life time to the proxy server. This is because the client host always exchanges the IP packets with its proxy server, even when the client host access the different web site in the Internet. The above tendency will further improve the effectiveness of cut-thru IP packet forwarding provided by CSR.

4.4 Required Number of Dedicated VCs for CSR

The required number of dedicated VCs provided by the CSR depends on the number of cut-thru sessions established simultaneously in the CSR. When the number of cut-thru sessions established simultaneously is too large, CSR could not provide dedicated VCs for all IP flows that potentially require cut-thru packet forwarding.

Tables 3 and 4 shown in the previous subsection indicate the number of dedicated VCs that CSR has to be able to provide. As for DEC's backbone router, CSR has to be able to provide more than 500 dedicated VCs. As for Toshiba R&D center backbone router, CSR has to be able to provide more than 120 dedicated VCs. In the future, the number of dedicated VCs, that can be provided by the CSR, will increase. The maximum number of dedicated VCs, that can be provided by the CSR, seems to be few thousand, and it seems to be possible to provide even by the current cell switching fabric.

4.5 High Performance Firewall Function

CSR may provide three levels of security functions, which are no packet filtering, packet filtering per session, and per packet filtering. With the per packet filtering, i.e., every packet is filtered at the router, we could not achieve a high throughput. With per session filtering, since only the first packet in the session has to be filtered, the rest of the packets can be forwarded in cut-thru mode without any IP packet filtering. This is means that we can apply the cut-thru packet forwarding mechanism to this type of sessions. As a result, when the traffic volume of the sessions, that require the per packet filtering, is not large portion in the aggregated traffic handled by CSR, CSR with a firewall function could achieve a sufficient high throughput.

4.6 Discussion

In the following subsections, the issues, that require further consideration and evaluation, are discussed.

4.6.1 Cut-Thru Parameters

Cut-thru IP packet forwarding is initialized by TCP/UDP information in the received IP packet and it is released when no packet is generated during a given period (T_{term}). The performance and effectiveness of cut-thru packet forwarding depends on these parameters, i.e., (1) which TCP/UDP ports are the trigger to initiate the cut-thru forwarding and (2) how long T_{term} should be.

When we pick up many applications (i.e., many TCP/UDP ports) to be as cut-thru trigger, the total amount of cut-thru packets will be going to be large. However, the dedicated VC establishment overhead may be too large, since many short-life sessions may picked up as a cut-thru trigger. Therefore, we have to choose the appropriate applications to achieve an effective cut-thru packet forwarding. Though we may have some default rule which applications should be cut-thru trigger, each CSR can independently determine which applications are cut-thru trigger. When we introduce a new application, each CSR decides whether it should be treated as a cut-thru trigger or not. Regarding the current applications over the Internet and intranet, ftp, telnet, web, nntp, voice and video seem to be appropriate applications to be cut-thru trigger.

Parameter T_{term} is also important parameter to optimize the effectiveness of cut-thru packet forwarding. When T_{term} is so long, the utilization efficiency of cut-thru pipe (i.e., dedicated VC) will be poor, while the percentage of packets forwarded with cut-thru mode is large enough. On the contrary, when T_{term} is so short, the utilization efficiency of cut-thru pipe may be good. However, the overhead of dedicated VC establishment may be too large. And, the percentage of packets forwarded with cut-thru mode may be small, since the first packet after the termination of dedicated VC is always transferred through the hop-by-hop mode. The optimal value of T_{term} will be unique for each applications and unique for each user. However, we can not define T_{term}

for each applications and for each user in the actual operation, i.e., we have to pick up one T_{term} value for each CSR. This value T_{term} should be determined either by the default value provided by the vendor or by the actual operation (i.e., try and error).

The detailed evaluation on these parameters should be further study item.

4.6.2 Scaling up of IP Processing Capability

Theoretically, the packet processing capacity of CSR is scaleable according to the scalability of cell switch fabric. However, in the actual CSR system, the processing capability of IP processing module will be the bottleneck of system performance. When the cell switching fabric scales, the processing capability of IP processing module has also to scale.

Figure 9 shows an example of implementation image of CSR. In order to scale the IP packet processing capability, multiple IP processing units are implemented in the IP packet switch fabric. The IP packet switch fabric will be able to implement using the multiprocessor technology applied to the high-end workstations. However, since IP packet switch fabric has multiple ingress ATM interfaces, some mechanism (e.g., new routing mechanism within the cell switching fabric) has to be developed to distribute the IP packets to each interface.

An alternative scaleable router architecture, shown in Fig. 10, is that each interface port has the IP router module, and these IP router modules are interconnected by cell switch. In this case, the cell switch is used for the packet transmission module among the IP router modules. When IP packets are handled by software processing, the IP packet processing speed at IP router modules at the interface port may become insufficient. In order to solve this concerning, each IP router module at interface port should have both a cut-thru (i.e., cell-relaying with hardware processing) and a hop-by-hop (i.e., packet forwarding with software processing) packet forwarding functionalities. In this case, we can see the IP router module at the interface port as the CSR having two interface ports.

Yet, another approach to scale the total IP packet processing capability of the CSR system is to interconnecting multiple CSRs, say as the aggregated CSR. By configuring the aggregated CSRs, we can easily achieve larger IP processing capability. Minor problem of this approach would be the physical space that the CSR system requires. The routing among the CSRs could be performed some IP routing protocol (e.g., OSPF), which is effective in the aggregated CSR. There would be a concerning to waste the IP address due to having multiple nodes (i.e., CSRs) in the aggregated CSR. However, we need not waste the IP address space by means of unnumbering technique commonly used for point-to-point links.

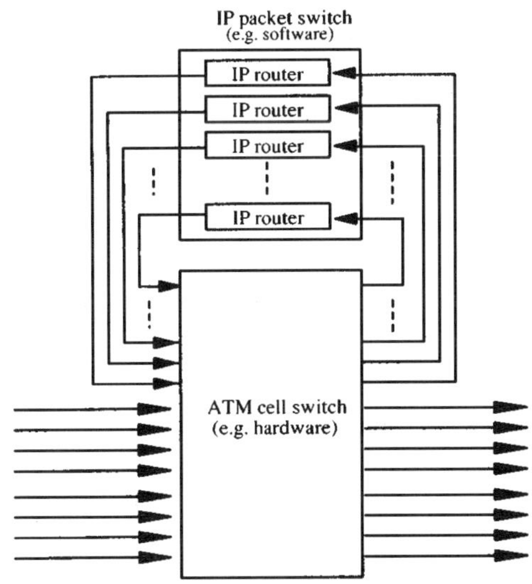

Fig. 9 Scaleable CSR implementation example (I).

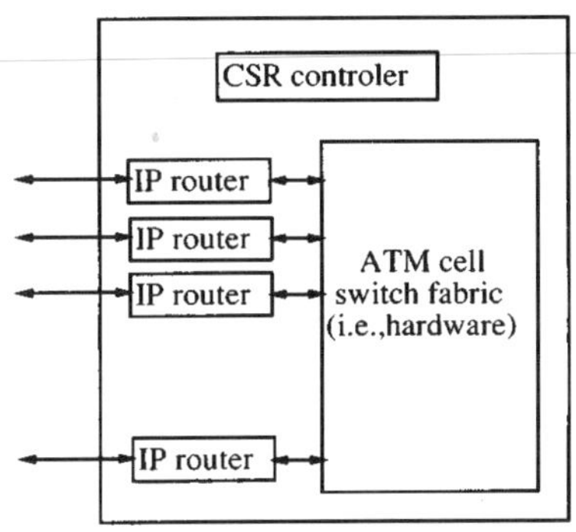

Fig. 10 Scaleable CSR implementation example (II).

4.6.3 Cut-thru IP Packet Flow Aggregation

When the CSRs are interconnected through an ATM link across the wide area networks (WAN), it may be better to multiplex multiple IP flows using the dedicated-VC into a single ATM-VCC, in order to transfer IP packets to the CSR across the WAN. This would happen when the cost of an additional ATM-VCC establishment across the WAN is more expensive, than using an ATM-VCC which has a sufficiently large bandwidth. In this case, multiple cell flows using different dedicated-VCs must be merged into a single dedicated-VC. In general, we must re-construct IP packets in the multiple dedicated-VCs in order to merge them into a single dedicated-VC. However, if we have a cell switching fabric that can schedule a cell transmission from the output port so as to avoid a cell interleaving associated with the packets that belong to the different ingress ATM-VCC but using the same egress ATM-VCC (i.e., using the same VPI/VCI), we do not have to relay on the packet switching fabric with the reconstruction of IP packets. This will be possible when the internal data-unit transferred in the cell switching fabric has the ingress VPI/VCI value and the input port identifier. If we have such a cell switching fabric, the packet switching fabric can transfer the IP packet in pipe-line (i.e., IP packet transmission will be able to start before re-assembly of IP packet is completed) and could aggregate the multiple cut-thru IP packet flows into a single cut-thru path.

4.6.4 Route Change

The IP packet routing path is changed according to the change of network status. Both hop-by-hop path with default VC and cut-thru path with dedicated VC will dynamically reflect to the packet forwarding path indicated by the routing protocol (e.g., OSPF). When the routing path is changed in the CSR system, one or two packets would be dropped at the edge node of cut-thru path. This is because all cells in the packet will not take the same path to be reordered. The one or two packet drops causes the degradation of end-to-end throughput for the old version of TCP due to a TCP windows shrink. However, since the current version of TCP generally implements a fast retransmission mechanism, one or two packet drops will not cause the throughput degradation.

When the routing path is changed, the packet forwarding goes to a hop-by-hop mode and each CSR tries to establish a cut-thru path based on the TCP/UDP port information of the first packet to be transferred to a new path. When the routing path frequently change, we can not achieve the effective operation of cut-thru packet forwarding. However, actually, since many routing protocol (e.g., OSPF, BGP) has a pinning option to avoid frequent path flapping, we can expect that we will be able to achieve effective use of cut-thru forwarding path.

4.6.5 TTL Issue

At the CSR, TTL does not decrease with cut-thru packet forwarding mode. And, the TTL is decreased by one at the egress node of cut-thru path. Therefore, the decreased TTL experienced cut-thru path is smaller than the decreased TTL with conventional hop-by-hop packet forwarding. The actual purposes of TTL would be (1) avoiding permanently existing packet on the routing loop, and (2) define the scope where the packet can travel.

The former purpose would be satisfied with CSR, since TTL is decreased at the egress point of cut-thru path anyway. Only problem associated with item (1) is that the period before discarding the packet on the routing loop will be large due to slow decrease of TTL.

The later purpose will be satisfied with CSR, since routing protocol is performed with hop-by-hop path (default VC) and the first packet in the session is never forwarded with cut-thru mode. Either by the routing protocol or by the first packet's hop-by-hop packet forwarding, the purpose of TTL field to define IP packet's traveling scope will be satisfied.

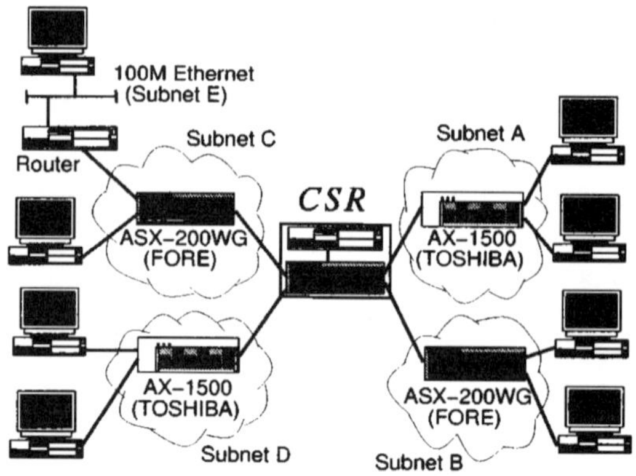

Fig. 11 CSR prototype system overview.

5. CSR Prototype System

Figure 11 shows the overview of the CSR prototype system using PVC connections to transfer the IP packets. Four ATM switched are interconnected through a CSR. Each ATM switch has it's own IP subnet. Therefore, the CSR in the figure internetworks four IP subnets constructed by ATM switch. One of four ATM based subnet is internetworked with CSR and with 100M Ethernet through a PC-based router. As a result, the prototype system has four IP subnets constructed by ATM switch and one 100M Ethernet IP subnet.

The following is the specification of developed CSR prototype system shown in Fig. 11.

- Maximum throughput of CSR; 4.8 MPPS for 64 B packets (2.4 Gbps)
- End-to-end TCP throughput;
  - ATM host to ATM host; 90 – 100 Mbps
  - 100 M Ethernet host to ATM host; 40 Mbps
- Realtime video (NTSC signal) transmission; 30 Mbps
- Remote CSR monitoring and management with SNMP
- Internetworking with 100M Ethernet
- IP packet forwarding using PVCs
- CSR IP packet switch module;
  - CPU; Pentium 100 MHz
  - OS; BSD/OS v2.1
- CSR cell-switch module; FORE ASX-200WG
- Cut-thru Parameters (configurable)

- Applications to be cut-thru: web, ftp, telnet, nntp
- Cut-thru time out (T_{term}): two minutes

6. Conclusion

Over view of CSR architecture and CSR prototype system focusing on the connectionless IP packet transmission is discussed. CSR has both cell switching fabric and packet switching fabric. Since CSR can forward some IP packets cell-by-cell based on VPI/VCI without examining IP header, rather than the conventional packet-by-packet forwarding, CSR can achieve high throughout IP forwarding.

In order to perform the cut-thru IP packet forwarding using the cell relaying capability in the CSR, CSRs exchange the information how the IP packet flows are aggregated into dedicated-VC through FANP. This information exchanging with FANP is hop-by-hop base, and the cut-thru decision is a matter of every router's local decision. When all routers along the path, that an IP packet flow takes, perform the cut-thru packet forwarding, the soft-state and seamless cell-relaying channel is established to get a high throughput IP packet delivery.

Also, we can make sure that CSR can operate correctly and can achieve high throughput IP forwarding by the prototype system. Even with prototype system, The maximum aggregated throughput of prototype system is 2.4 Gbps, which corresponds to 4.8 million packets per second with 64 B packets.

As the future works, CSR will use SVC (Switched VC), as well as PVC, and will support RSVP to provide the end-to-end QoS (Quality of Service) over the Internet. Also, CSR will support the multicast service using the hardware-based multicast within the CSR, in order to provide high performance multicast services.

References

[1] ATM Forum, "ATM user network interface specification version 3.1," Sept. 1994.

[2] H. Esaki, K. Nagami, and M. Ohta, "High speed datagram delivery over internet using ATM technology," IEICE Trans. Commun., vol.E78-B no.8, Aug. 1995.

[3] Y. Katsube, K. Nagami, and H. Esaki, "Mapping of IP flow to datalink layer connection," IETF Internet-Draft, draft-katsube-flow-mapping-dl-conn-00.txt, May 1995.

[4] P. Newman, T. Lyan, and G. Minshall, "Flow labelled: Connectionless ATM under IP," Engineer Conference, Networld+Interop'96 Las Vegas, April 1996.

[5] S. Deering and R. Hinden, "Internet protocol, version 6 (IPv6), specification," IETF Internet-Draft, draft-ietf-ipngwg-ipv6-spec-01.txt, March 1995.

[6] ITU-T Recommendation I.150, "B-ISDN asynchronous transfer mode," 1990.

[7] Y. Katusbe, K. Nagami, and H. Esaki, "Router architecture extensions for ATM: Overview," IETF Internet-Draft, draft-katsube-router-atm-overview-00.txt, March 1995.

[8] ATM Forum, "LAN emulation over ATM specification version 1.0," April 1995, Sept. 1990.

[9] J. Heinanen, "NBMA address resolution protocol," IETF RFC1735, Dec. 1994.

[10] J.V. Luciani, D. Katz, D. Piscitello, and B. Cole, "NBMA next hop resolution protocol (NHRP)," IETF Internet-Draft, draft-ietf-rolc-nhrp-10.txt, Oct. 1996.

[11] ISI-USC, "Internet protocol: DARPA internet program protocol specification," IETF RFC791, Sept. 1981.

[12] M. Laubach, "Classical IP and ARP over ATM," IETF RFC1577, Oct. 1993.

[13] Y. Rekhter and D. Kandlur, "Local/remote forwarding decision in switched datalink subnetworks," IETF RFC1937, May 1996.

[14] L. Zhang, R. Braden, S. Berson, S. Herzog, and S. Jamin, "Resource reservation protocol (RSVP)–Version 1 functional specification," IETF Internet-Draft, draft-ietf-rsvp-spec-13.txt, Aug. 1996.

High-Performance Backbone Network Technology edited by Naoaki Yamanaka
ISBN: 0-8247-5321-6 DOI: 10.1081/(120025686)
Published 2004 by Marcel Dekker, Inc.

Scalable Internet Backbone Using Multi-Gigabit ATM-Based Connectionless Switching

Shigehiko Ushijima, Hiroyuki Ichikawa, Katsunori Noritake, and Naoya Watanabe

SUMMARY We propose a hardware-based packet forwarder for multi-gigabit IP backbone networks. The conventional Internet deploys routers as a key block, but its software-controlled architecture makes it hard to scale up the packet forwarders, especially for table-lookup processes. We propose introducing a pure connectionless (CL) switching approach with a hardware-based forwarder to construct the core part of a scalable IP multi-gigabit backbone. Compared to a software-based forwarder, the table-lookup time is reduced to 100 ns by using content-addressable memory. This hardware-based pipeline implementation easily achieves a maximum forwarding performance, of up to 9.6-Gbps, or 23 million packets per second, for applications ranging from traditional best-effort IP applications to newly emerging time-critical ones. We also consider additional processing when transferring IP packets to enhance best-effort quality. This is done using selective packet-level discarding, including early packet discard and its enhancement, to achieve minimum bandwidth guaranteed service at the packet level. We discusse the IP backbone scalability issue from the viewpoint of new IP-forwarder technologies, paying special attention to connection-oriented (CO) vs. CL switching and hardware vs. software implementation. A pure CL switching solution consisting of a CL server (CLS) and a CL client (CLC) is proposed to balance the hardware- and software-based CL transport functions. As a first step to this solution, a compact CLS has been developed. It supports 600-Mbps throughput and up to 9.6-Gbps forwarding power using a modular architecture. It was evaluated in an ATM field trial using an experimental network. The results show the effectiveness of our approach to providing enhanced best effort services.
key words: *connectionless, Internet, router, ATM, gigabit, IP*

1. Introduction

The exponential growth in Internet traffic has made it necessary to enhance the Internet protocol (IP) packet throughput of existing routers. It has also made it necessary for backbone network capacity to be made easily scalable, irrespective of changes in traffic volumes and characteristics.

A software-based forwarder limits the packet throughput, mainly due to lookups in its routing table. Several approaches have been proposed for applying a hardware-based ATM switching fabric to improve the internal packet forwarding power [4], [5] or to interconnect routers via high-speed ATM backbone networks [6]. Unfortunately, these approaches must handle the connectionless (CL) transfer mode of IP over a connection-oriented (CO) ATM network.

One of the most significant problems in such a network is handling the connection setups, maintenance, and tear-downs needed to handle the quickly changing IP traffic flows. This is due to the bottlenecks in connection control suffered by heavy-weight signalling protocols. An alternative is to keep the switched virtual connections (SVCs) active, thus reducing the signalling load and avoiding the setup delays in IP packet transmission. However, this increases the number of active connections, which raises network management issues and restricts network scalability. The IP- and tag-switch approaches reduce the signalling overhead by using light-weight signalling. However, if the software-based processing power is insufficient, not all of the IP traffic can be switched to on-the-fly transmission. Currently, there is no promising solution for the multi-gigabit IP backbone environment.

Internet service providers (ISPs) need to focus on raising the quality of their best-effort service rather than on developing new services. Various IP-based user applications are now emerging, and multimedia/multicast applications, including Internet phone/desktop video conferencing, are becoming popular. However, most access to the Internet is still through dial-up low-speed dedicated lines without any traffic control mechanism like that in ATM. The ISPs therefore need to provide a better solution for lowering packet latency and packet losses in highly multiplexed IP traffic. To handle such traffic in an IP backbone, a packet-level traffic control mechanism is needed to achieve sustainable throughput even in congested ATM networks.

The IP packets are carried in the form of cells, and once a cell has been lost due to buffer overflow, the parent packet can no longer be reassembled. To reduce the chance of this occurring, several alternatives to early packet discard (EPD) [10] have been studied. The random packet discarding in simple EPD causes an unfair distribution of the available bandwidth among users. To obtain a sustainable throughput for all users, an enhanced EPD scheme is needed that provides a minimum bandwidth to each user.

Our approach is to use the pure hardware-based CL switching fabric for multi-gigabit packet forwarding. We have developed a CL server (CLS) that can forward IP packets on a cell-by-cell basis. Unlike with a software-controlled architecture, IP packet throughput can be increased up to the multi-gigabit level irrespective of packet length, and its no-packet-reassembly feature results in a hop

Manuscript received April 25, 1997.
Manuscript revised July 25, 1997.

† The author is with the NTT PC Communications, Tokyo, 105-0004 Japan.
†† The authors are with the NTT Network Service Systems Laboratories, Musashino-shi, 180-0012 Japan.

IEICE TRANS. COMMUN., Vol. E81-B, No. 2, pp. 324–332, February 1998

delay as low as 20 μs. This fabric is thus applicable to both emerging time-critical multimedia applications and existing TCP/IP applications. In addition, we have developed an enhanced EPD scheme that achieves a minimum-bandwidth (MBW) guaranteed service by selectively discarding those packets whose transfer volume exceeds the MBW.

In Sect. 2 of this paper, we discuss the alternative high-speed IP-forwarding technologies and describe our proposed multi-gigabit IP backbone network architecture. In Sect. 3 we show the practicality of our first-step approach by describing the functional architecture of our compact CLS. We pay special attention to the functional alignment, including the packet routing/copying mechanism, traffic control mechanism, and supported service features. In Sect. 4, we discuss the issues related to the implementation of our compact CLS, which reflect the core part of the backbone architecture discussed in Sect. 3. In Sect. 5, we discuss our evaluation of the performance of our CLS equipment in both laboratory and field trials, and show the feasibility of our proposed solution.

2. High-Speed Packet Forwarding

2.1 Software vs. Hardware Forwarder

A typical IP-packet-processing diagram for a router is shown in Fig. 1. The major tasks in conventional routers often involve memory-intensive operations, such as table lookups. Because, these processes are handled by software, search performance depends directly on the RAM access speed. The effectiveness of processor power upgrades in boosting router-forwarding performance is thus limited. For example, for a RAM access speed of 20 ns, a standard device speed, searching for a routing-table entry is expected to take 1.44 log (number of entries) $\times$ 20 ns, based on the PATRICIA tree algorithm [15]. For 1000 entries, about 15 RAM accesses are needed. Including the related processor overhead, these multiple accesses can consume more than 300 ns, even when cache memory is used. This is why using only software-based processes for improving the IP-packet forwarding power is not enough.

The hardware-logic approach easily overcomes this software processing bottleneck by dividing each single-thread process into several pipelined sub-processes. The IP-forwarding power is determined by the longest processing time of these sub-processes. The most time-consuming tasks, including routing-table lookup, can be done within two memory cycles for any size lookup by using content-addressable memory (CAM). The forwarding power can thus be easily improved. For example, with a 20-MHz system clock, only 100 ns are needed for any routing-table lookup. Additionally, with better organization of the pipeline processes, performance can be increased up to 10 million packet/s.

2.2 ATM-Switching Approaches

Conventional hardware forwarding approaches use an ATM switching fabric. The IP-switch [4] and tag-switch [5] use ATM switching fabrics as an internal hardware packet forwarder to make up for the poor performance of software-based forwarding. The next hop resolution protocol (NHRP) [6] approach uses an ATM wide-area network to connect conventional ATM routers via direct cut-through ATM connections.

Maximizing the use of the ATM switching fabrics as IP traffic forwarders will enable multi-gigabit networking. However, because of the CO nature of the ATM switch, a connection has to be established that can stay current with the dynamic IP packet transmissions. The IP-switch approach supports a light-weight signalling protocol, which improves the SVC handling capability. However, the power of the software-based processor is shared by the signalling and by the short-lived traffic forwarding, so further enhancement is needed to meet the Internet backbone requirements.

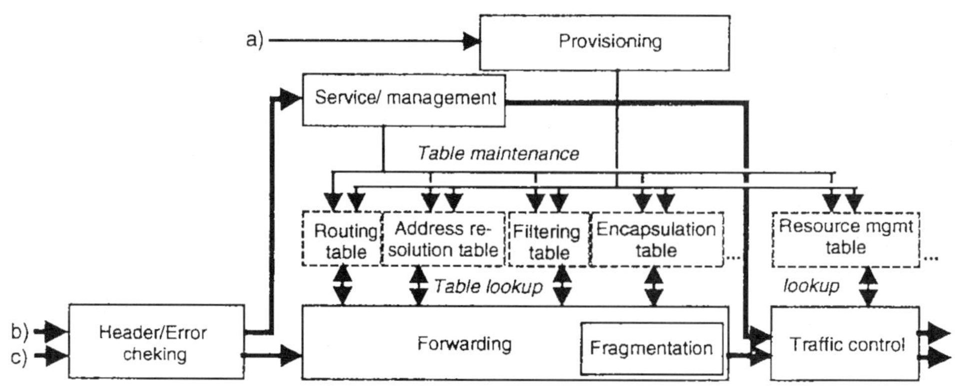

Remarks:
- a): Operation flows, b): Control message flows, c): User message flows
- Header/Error check process: version, TTL, length, framing, check sum
- Forwarder process: forwarding decision based on table lookups, fragmentation
- Service management process: routing, multicast group, resource management driven by IP service layer
- Provisioning process: setting up managed/control data manually
- Traffic control process: congestion/flow control, load balancing

Fig. 1 Typical IP-packet-processing diagrams.

To avoid this limitation in handling dynamic signalling, another alternative is a permanent setup of VCs. However, this raises network scalability problems. Assuming that n routers are connected via VC connections in a full-mesh topology, and that i ATM switching fabrics partially accommodate the same number (n/i) of routers, the number of VC connections required at each ATM switching fabric is

$$\binom{n}{2}-\binom{k}{2},\ k=\frac{n\,(i-1)}{i}\ . \tag{1}$$

The relationship between the number of ATM switching fabrics and the number of connected routers is shown in Fig. 2. If the maximum number of VCs is assumed to be A, then the maximum number of connecting routers is

$$n \approx i\sqrt{\frac{2A}{(2i-1)}}\ . \tag{2}$$

This equation shows that once the maximum number of VC resources has been consumed, adding switches barely increases the number of connecting routers. This implies that the restriction on the maximum number of VCs in an ATM switch causes scalability problems. Additionally, a huge number of VCs would require the switching fabric to support many VC-level functions, including usage parameter control (UPC), VC shaping, buffer management, and network operation. This would make the ATM switching fabric very expensive. In contrast, the CL-based approach can easily scale the network capacity without consuming VC resources and requiring VC-level functions.

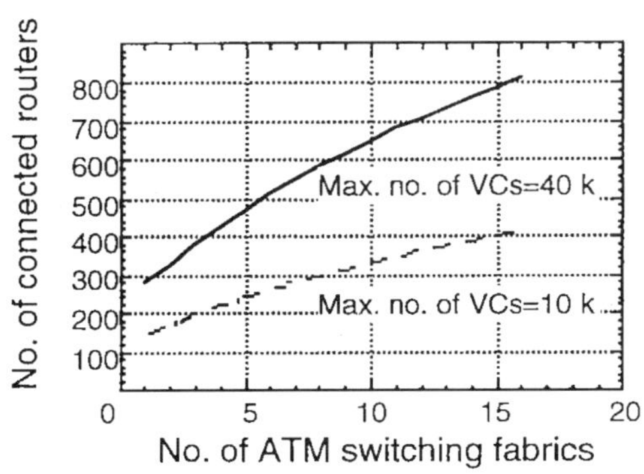

Fig. 2 Number of accommodated routers.

2.3 Our Approach

We take a synergetic approach that uses collaborative elements: a CL server (CLS) at the core and CL clients (CLCs) at the edge. The CLS provides the hardware-based functions, such as high-performance packet forwarding, while the CLCs provides the software-based functions, such as address resolution. Optimizing their functional assignment enhances the network's overall forwarding power and service features.

There is a trade-off between performance and service provision. Although the cell-based packet-forwarding mechanism of the CLS provides high throughput, it has several drawbacks. It requires that the first cell of each packet include complete routing information in order to complete the routing decision. However, the variable-length header, like the optional header in the IP, often exceeds the cell payload capacity. Moreover, services driven by the IP service layer, like IP multicasting, need to reassemble packets used in order to exchange topological information. The IP backbone thus needs a software engine to handle packet reassembly; i.e., a service/management function is needed. However, this software-controlled architecture seriously degrades throughput performance. Because multiple CLCs based on the current software technology are accommodated by a single CLS, the CLS can't supply enough throughput with a software-based architecture.

Therefore, the basic concept of our proposed IP backbone architecture is to support time-critical hardware-based CL switching at the core of the backbone, and to enhance the multiple-service support by using a software-based CLC at the edge of the backbone. The proposed functional architecture is shown in Fig. 3. The CLSs in the core part forward and/or filter the packets in hardware while the CLCs at the edge handle other functions in software. This architecture independently enhances the forwarding capacity and the service-specific functions.

As a first step to demonstrating the feasibility of this backbone architecture, we have developed a compact CLS that uses ITU-T I.364 as the now-available baseline standard. The CLCs use ATM-router products supporting the Connectionless Network Access Protocol (CLNAP) [1] standardized in the ITU-T. This standard requires only AAL3/4 encapsulation of IP packets, so the CLCs only need

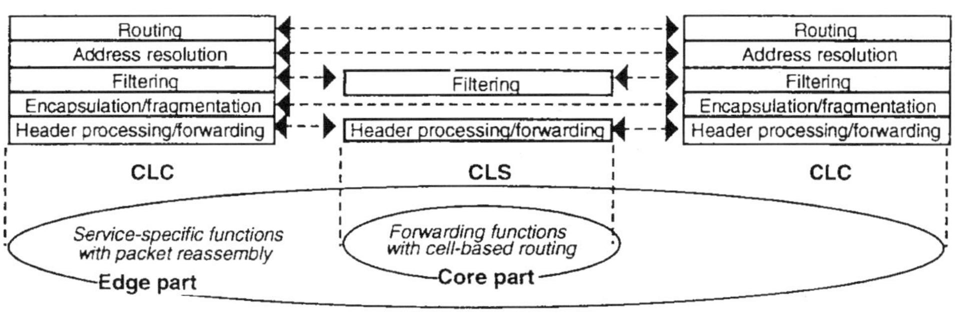

Fig. 3 Proposed IP backbone architecture.

store-and-forward operation. However, using AAL5 would enable the CLCs to perform on-the-fly transmission. The CLSs could then directly handle the IP-based protocol for seamless cell-based interworking with ATM end-user terminals. This is the target of our proposed architecture.

3. Functional Architecture of Compact CLS

3.1 Supported Functions

The primary objective in developing our compact CLS was to obtain a scalable packet forwarder that can handle transmissions up to the multi-gigabit level. It should also be cost-effective and make full use of the CL features. A modular architecture is used to meet the growing IP traffic requirements. To achieve economical usage of the backbone network, multi-stage multicasting with cell-copying and selective-packet discarding is also supported.

3.2 Hardware Configuration

The hardware configuration of our compact CLS is shown in Fig. 4. The components inside the CLS are explained later. This CLS can act as a standalone system or can incorporate an internal ATM switching fabric for interconnection. In the latter case, it is stackable on the Multimedia Handling Node for ATM (MHN-A) system [14], one of NTT's central-office ATM switching fabrics, in a trunk position. Another alternative is to locate the CLS function at the back-end of the switching fabric. In comparison with the server-based approach, the trunk-based one avoids switching bottlenecks and enhances forwarding scalability and reliability.

3.3 Packet Information Management

The cell-based packet forwarding is done based only on the destination address of the CLNAP (denoted here as DA) in the first cell. The following cells in the same packet have no routing information. Therefore, the result of the routing decision has to be retained until the last cell has been sent. Each packet is identified by a set of locally significant identifiers (VPI: virtual path identifier, VCI: virtual channel identifier, and MID: multiplexing identification [2]) that are converted at each CLS. Packet information management thus handles temporary mapping between received and transmitted identifiers based on the routing decision and on the assignment of a new MID value. The depth of its management table depends on the maximum number of packets transferred concurrently. The table size is reduced if the CLC transmits packets without cell interleaving.

This packet-forwarding approach is very effective, even though the actual transfers are done in the form of cells. For example, dropping a packet in the case of error detection or overload is done by simply deleting its temporary mapping entry. Additional attributes describing packet behavior would enhance the multi-purpose transfer capability.

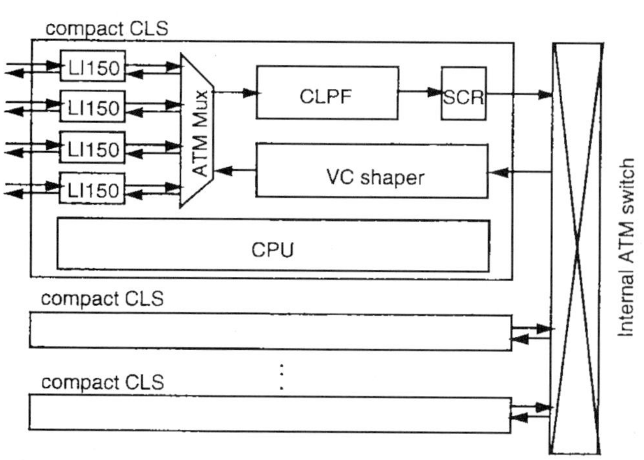

Fig. 4 Hardware configuration of proposed compact CLS.

3.4 Routing and Packet Filtering

The ITU-T standard uses the E.164 [3] address for routing. This layered numbering plan allows each subscriber address (E.164) to be abbreviated to its area code. It also facilitates designing the network accordingly. Routing is done by looking up the entries for the next hop destination in the routing table, which consists of the E.164 address and the VPI/VCI. The area code is a serial short-length number; however, the subscriber address is discretely assigned. To reduce the table size, LSI CAM is used to search for subscriber addresses. For faster searching, both the subscriber address and area code are looked up in parallel.

Connectionless-based routing can share a forwarding route with the next hop although the final destinations are different. A routing tag is defined to summarize the entries. This tag is derived from the DA in the first cell, and the actual VPI/VCI values are selected by referring to it. In comparison with direct management of an E.164 address and VPI/VCI entry pair, the number of entries can be reduced to the number of VPI/VCIs directly connected to the neighboring CLSs. This tag is managed in the packet-information management table described in Sect. 3.3.

Packet filtering is another important requirement for guaranteeing security. Multi-provider IP backbones and virtual private networks for business applications require logically separated network configurations over the backbone network. This requires checking the combination of the source address of the CLNAP (denoted here as SA) and DA in each cell being transferred. However, this combination is too long for a filtering decision to be made with only one lookup. For this reason, a subscriber tag is introduced that normalizes the local subscriber number to a shorter form according to the number of the subscriber (10 bits per 1000 subscribers). Replacing either the DA or the SA with this short tag reduces the combined length to 58 bits (for the Japanese ISDN numbering plan). Thus, the search overhead for packet filtering is reduced.

3.5 Multicasting

Reducing bandwidth usage is important when transferring multicast traffic over a wide-area network. However, the ITU-T based multicasting mechanism requires an extra first cell to contain the CLNIP [1] header that carries the group-address resolution. This extra encapsulation increases the transmission delay and degrades the efficiency of the backbone. We propose a multicasting mechanism that does not need an extra cell.

Our approach is to pre-define a spanning tree network at the VPI/VCI level and to prepare a group-address resolution table containing the group-addresses and the corresponding VPI/VCI values that refer to the partial spanning tree links on the CLS (Fig. 5). When the transit link is resolved as a next-hop destination, only one packet is transferred on the link. The multicasting is done by copying packets to this link and the subscribers assigned to the group. This allows multi-stage copying, which reduces the bandwidth required. This feature is very important for scalable multicasting services.

The multiple-copy mechanism also creates multiple copies inside the CLS. Priority control is used to smooth the traffic peaks. The copied cells are sent during the idle times when no traffic is arriving. This control alleviates the traffic overflow normally generated by multicasting. However, if multicasting became dominant, it would not work very well. Improvements would be needed to control the acceptable bandwidth for singlecast and multicast traffic.

3.6 Buffering and Dynamic Shaper

The CLS generally provides best-effort service when sharing DBR-based VC bandwidth. Therefore, a VC shaper is needed at the final stage of the CLS to absorb the burstiness by using its large-capacity buffers, thus producing a statistical multiplexing gain.

Due to the unpredictable changes in and burstiness of CL traffic, it is often concentrated at a particular port, resulting in cell loss. Our solution is to construct a large-capacity common buffer whose maximum available capacity per VC can be defined so as to localize the overflow to a specific VC. We have developed a dynamic shaper mechanism that dynamically assigns the read-out timing only to VCs carrying actual traffic. This is in contrast to static assignment, which has idle cycles when the corresponding VC has no traffic, and consequently wastes bandwidth. Our solution results in the maximum usage of the available VC resources. Earlier evaluation [8] showed that 6.5 times as much aggregate peak-based bandwidth can be expected assuming the sustainable bit rate is 10% of the peak traffic. The shaper buffer capacity (Table 1) was designed to achieve a five-fold statistical multiplexing gain under an 80% link load and a packet-loss of 10^{-4}.

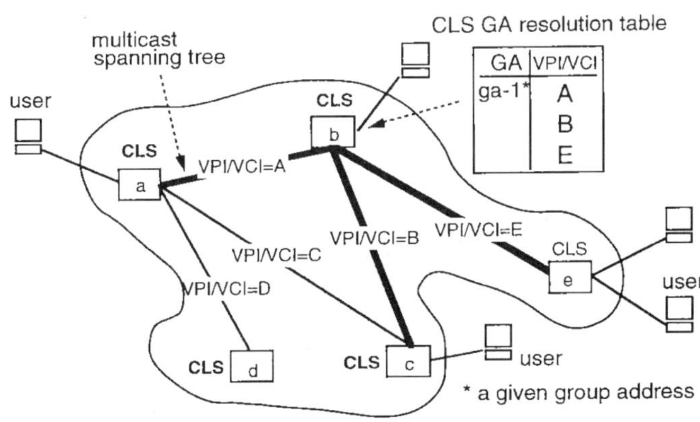

Fig. 5 Multicast transport mechanism.

Table 1 Specification of the compact CLS.

Address entries	2048*
Throughput	600 Mb/s - 9.6 Gb/s
IP-forwarding power	720 kpps - 11.5 Mpps
Packet processing delay	20 μs
Address screening entries	4096
VC shaper scheme	Dynamic R/W
VC shaper buffer capacity	24,000 cells
VC bandwidth	1 - 150 Mb/s
Congestion control	EPD
Operation interface	telnet-based
SNMP-MIB	to be provided
Cabinet size	30(W)x30(H)x40(D) cm

* Both individual and group addresses are included.

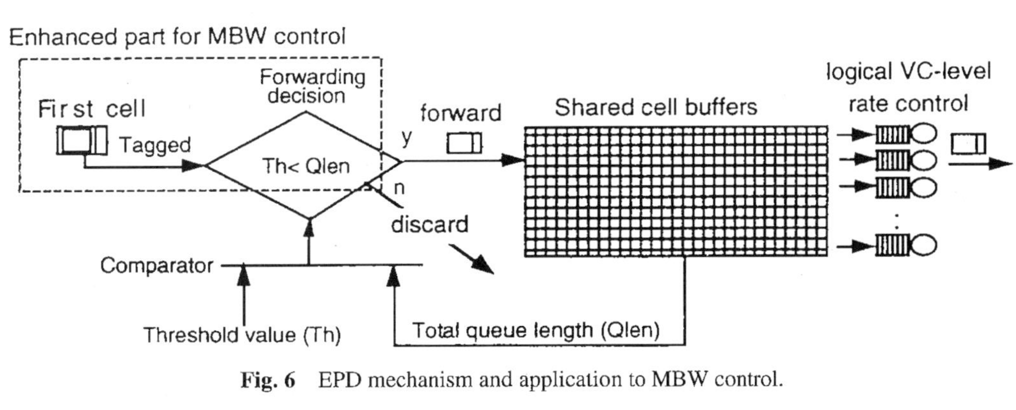

Fig. 6 EPD mechanism and application to MBW control.

3.7 Selective Packet Discarding

Enhancement of a best-effort service like UBR is a controversial issue in carrying IP traffic over ATM. A simple mechanism and a minimum bandwidth guarantee are needed because the performance of IP on top of ATM is directly affected by the ATM-layer transport quality. In spite of its cell-based forwarding, the CLS can perform packet-based handling due to its packet-information management function. Partial packet discard and early packet discard (EPD) [10] are very effective for maintaining the packet-level throughput, even though a congested ATM network discards cells. Current TCP/IP implementations support selective retransmission [12], unlike the previous go-back-N retransmission.

Our approach to EPD is to discard the whole packet when the shaper buffer overflows. This discard mechanism is activated when the total queue length of the shaper buffer exceeds a threshold. If this happens, the temporary packet-information management table, which is described in Sect. 3.3 for packet-level forwarding, is not created (Fig. 6). The remaining cells in the corresponding packet are then discarded. Although this mechanism is very simple, the threshold value has to be carefully fixed. Our experimental evaluation of EPD is discussed in Sect. 5.2.

3.8 Minimum Bandwidth Guarantee

The minimum bandwidth (MBW) guarantee is an appropriate solution to providing 'better' best-effort services. Too many TCP/IP packet losses decreases the throughput to zero, and no other rate-control protocol, such as UDP/IP, interferes the TCP/IP flows. MBW control overcomes this problems so that each user can fairly utilize a minimal bandwidth, irrespective of the load on the backbone network.

In UBR+ [11] service, CLP tagging is used to indicate cell traffic exceeding the MBW, and tagged cells are dropped with a higher probability during congestion. This basic concept can be used in the CLS. However, each cell-level tagging has to be mapped to the packet level because random cell dropping would degrade good-put. A more difficult problem is the difference in dimensions between packet/s and cell/s. Short-length packets are contained more lower-layer overhead than longer ones, so that shorter ones are regarded to degrade good-put.

Figure 6 shows our approach to supporting the MBW guarantee mechanism as an enhancement of EPD. Another packet-discard condition is added: the tagged packets are discarded with higher priority during congestion. Packet tagging is done in the CLC by measuring the octets of the passing packet traffic; packet discarding is done in the CLS by checking the tagging of the first cell during congestion. This approach takes effect only when the occupancy of the tagged packets is dominant. However, if the guaranteed minimum bandwidth is low enough compared to the peak, efficient control can be achieved with this simple function.

4. Implementation

4.1 Pipeline Process

A diagram of the implemented pipelines for the CL packet forwarding (CLPF) and the address screening (SCR) parts

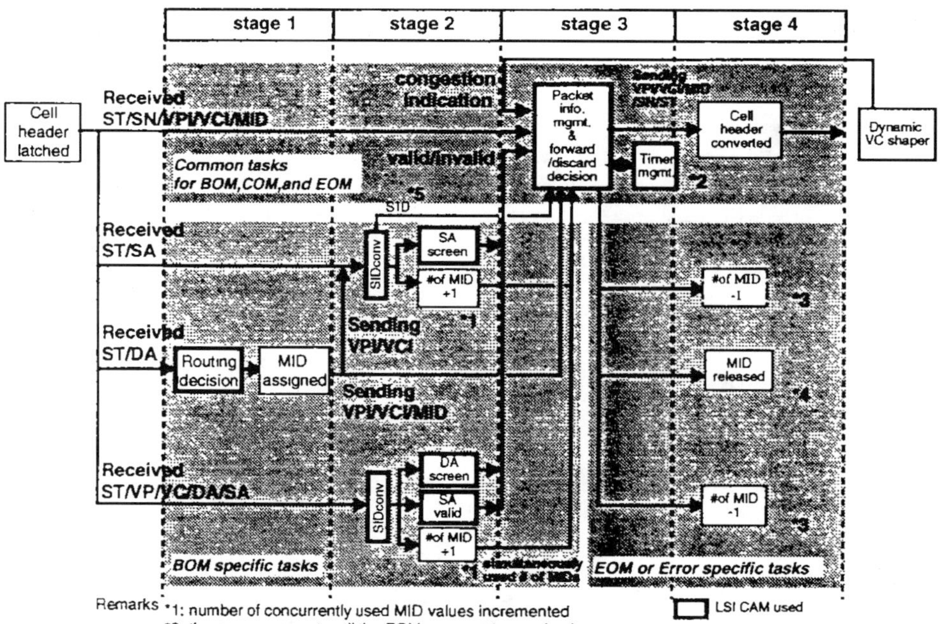

Remarks *1: number of concurrently used MID values incremented
*2: timer management until the EOM segment has arrived
*3: number of concurrently used MID values decremented
*4: the used MID value is released
*5: subscriber identifier

Fig. 7 Pipelined implementation of CLPF and SCR parts.

are shown in Fig. 7. To achieve 600-Mb/s unblocked pipelined operation, each stage is assigned a unit-cell transmission time of 700 ns.

The DA and SA addresses in the first cell (BOM) in the first two stages are handled in parallel. The routing decision, SA/DA screening, and assignment of the new MID value are performed and the results are sent to the packet-information management table. In the third stage, a decision is made between cell-based forwarding and discarding for all cell types in accordance with the results of the previous two stages. A cell is discarded by EPD control when the VC shaper is informed of a congestion state. In the fourth stage, the VPI/VCI/MID of the transmitted cell is actually converted. When the last cell (EOM) has arrived, the table entry used for the packet and its MID value are removed. Timer management releases the table entry if the EOM cell "disappears." Because four stages are needed to perform cell-based forwarding, the time needed is as short as 2.7 μs.

To achieve such time-critical operation, we developed an LSI CAM that enable an entry to be found within 100 ns, with a 20-MHz clock. This CAM is used as shown in Fig. 7 to reduce the table-lookup overhead. We plan to upgrade it to a 40-MHz clock and to 16 k entries to expand the service capacity. A sequencer is implemented in each stage by "full programmable gate array (FPGA)" devices containing firmware. By exchanging this firmware and making a small modification in hardware, other CL protocols like IP over ATM can be supported.

Fig. 8 Front view of compact CLS.

4.2 Specifications

Table 1 shows the specifications of our developed compact CLS. Up to 2048 CLCs can be connected to one CLS. Thus, a packet-forwarding power of up to 23 million packet/s can be achieved. For IP, however, the forwarding power is 11.5 million packet/s because even the shortest IP packet needs two cells.

The front view of our CLS is shown in Fig. 8. It is very compact for a device providing 600-Mb/s throughput. In the current implementation, only three cards are needed to perform the CLPF. To further integrate it into the ATM switching fabric, a higher-density implementation is needed.

5. Experimental Evaluation

5.1 Field Trial

These CLSs have been deployed in NTT's ATM field-trial network since Feb. 1996, working as IP backbone forwarders for 12 ATM routers across Japan (Fig. 9). Hundreds of users are supported by this ATM router backbone via 1.5- or 6-Mbps dedicated lines. The ATM routers operate on a peer-to-peer basis over a logical IP subnet (LIS) configured by two CLSs.

Using this network, several performance experiments [13] were done to determine the effectiveness of this hardware-based CL forwarder. Here we focus on comparing the end-to-end TCP/IP throughput and delay characteristics between the CLS transit and router transit cases. The former corresponds to our proposed solution.

As shown in Table 2, using the CLS reduced the end-to-

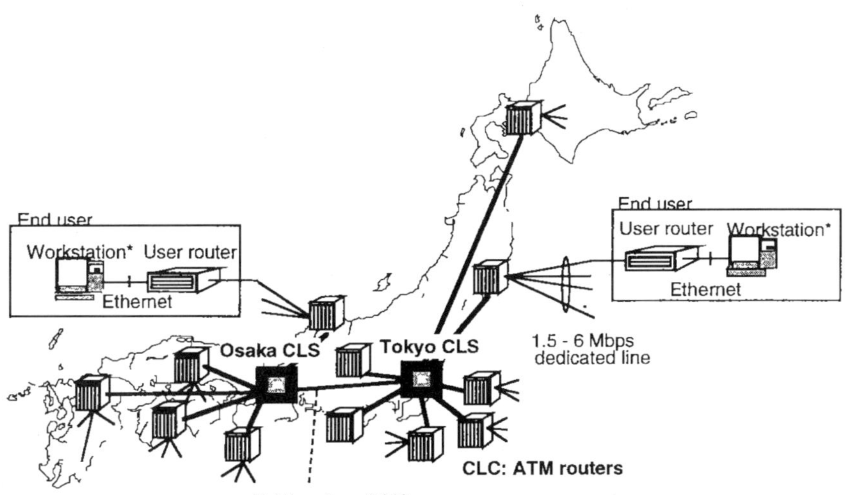

Fig. 9 Field-trial network configuration.

Table 2 End-to-end round-trip time.

Packet length (bytes)	64	512	1500
CLS transit (ms)	24	28	36
Router transit (ms)	26	30	41

note: the delay in the ICMP packet was measured between two routers about 1,000 km apart.

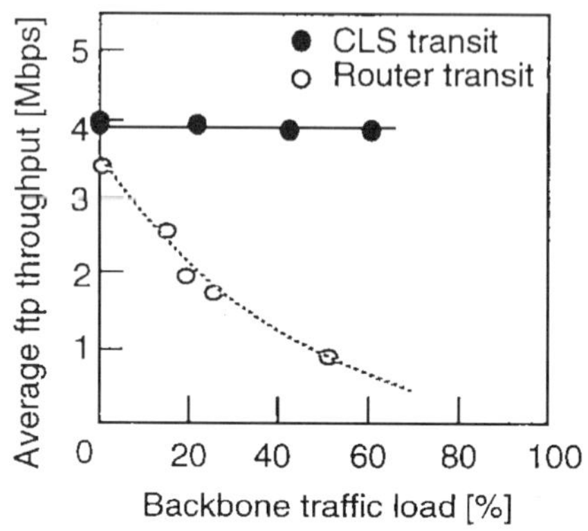

Fig. 10 TCP/IP throughput performance over field- trial network.

end round-trip delay. This improved TCP/IP throughput over an uncongested backbone. An even more remarkable result was for TCP/IP throughput over a congested backbone. Figure 10 shows the file transfer protocol (FTP) throughput between two UNIX workstations under congested conditions. With CLS transit, the TCP/IP throughput remained at about 4 Mbps, similar to that in the uncongested condition. However, with the router, throughput dropped to less than 1 Mbps at a 60% backbone traffic load.

With CLS transit, an overload to a specific destination did not affect the quality of the other IP flows. This is an advantage of an unblocked pipelined process. In contrast, the software-based forwarder degraded the throughput of the other IP flows under a congested backbone because each flow is processed by the same common processor resource. This is a significant difference between software- and hardware-based forwarders.

These experimental results show that applying cell-based packet forwarding increases end-to-end TCP/IP throughput and decreases the round-trip time. Although this result directly depends on the specific implementation of the routers, a general router architecture, such as a single-thread process implementation, and packet reassembly-based forwarding degrade router-forwarding performance.

Consequently, our proposed architecture with its synergetic combination of software- and hardware-based forwarders provides better packet-transmission quality, i.e., 'better' best-effort service.

5.2 EPD Evaluation

To determine the effectiveness of EPD control for actual TCP/IP applications, we compared the throughput for FTP.

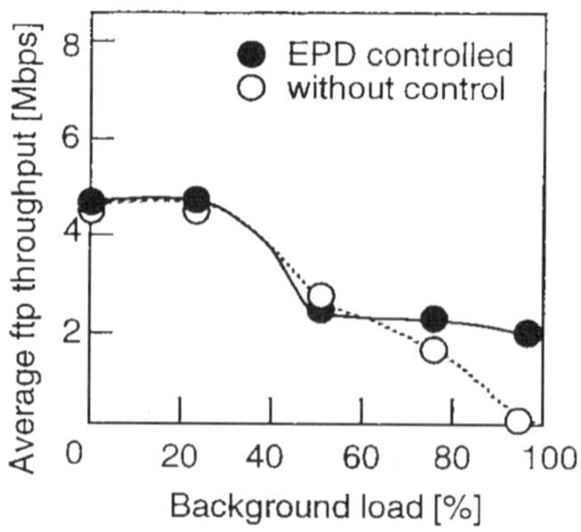

Fig. 11 FTP throughput improvement by EPD control.

In this experiment, we assumed the per-VC buffer capacity and threshold queue length were 1024 and 900 cells, respectively, and that the sender and receiver workstations supported fast-retransmission. Figure 11 shows FTP throughput with and without EPD control under a backbone loaded condition. We found that EPD control improved the FTP throughput in the case of heavily loaded background traffic. Without control, the throughput fell to almost zero. This is because the receiver side could not reassemble the cells into a parent packet due to random cell losses. This resulted in a smaller window size for the TCP protocol.

This example shows that EPD control maintains TCP/IP throughput even when traffic is congested. Because our proposed MBW control is based on the same EPD mechanism, it can guarantee the MBW for each user's IP traffic.

6. Conclusion

We have proposed a pure-hardware connectionless (CL) forwarder and an application architecture for constructing a multi-gigabit IP backbone. This architecture, consisting of a cell-based hardware forwarder and a software-based packet forwarder, provides efficient operation by locating a simple CL transport function at the core of the backbone and the service-specific functions at the edges.

To demonstrate the feasibility of our proposed architecture, we developed a compact CL server (CLS) for the core function of our proposed architecture. With a cell-based pipeline hardware implementation using the developed content addressable memory (LSI CAM), it achieved 600-Mb/s throughput and 720-k packet/s forwarding power as a standalone system. Its modular-node architecture scaled the maximum system capacity up to 9.6-Gb/s throughput. Experimental evaluation showed that in comparison to router transit approaches, our CLS maintains the TCP/IP user throughput during congestion. The throughput is improved even more by using our early packet discard (EPD) mechanism.

Our evaluation showed that a cell-based CL forwarder at the core of an IP backbone increases TCP/IP transmission throughput and decreases the delay, even with conventional

ATM routers deployed at the edge of the backbone. Our proposed IP backbone is thus a better approach to improving the level of service. Along with enhancing the forwarding capacity of the CLSs, the IP backbone can be scaled up to the multi-gigabit level, while maintaining the same topology, without a big increase in the number of connections.

References

[1] ITU-T Rec. I.364, "Support of the broadband connectionless data bearer service by the B-ISDN," July 1995.
[2] ITU-T Rec. I.363, "B-ISDN ATM adaptation layer (AAL) specification," March 1993.
[3] ITU-T Rec. E.164, "Numbering plan for the ISDN era."
[4] P.Newman, T.Lyon, and G.Minshall, "Flow labelled IP: A connectionless approach to ATM," INFOCOM'96, March 1996.
[5] Y.Rekhter, B.Davie, and D.Katz, "Tag switching architecture overview," Internet draft, Sept. 1996.
[6] D.Katz, J.Luciani, and D.Piscitello, "NBMA next hop resolution protocol(NHRP)," Internet draft, 1996.
[7] S.Ushijima, K.Imai, M.Hirano, and K.Noritake, "High-performance connectionless server for public ATM networks and its applications towards the multiple services support in heterogeneous networking environments," International Symposium on Interworking'94, Sophia Antipolis, France.
[8] S.Chaki, "ATM network for high-speed data communication," IEEE Singapore International Conference on Networks/ International Conference on Information Engineering'93, IEEE SICON/ICIE'93", Singapore, 6-11, Sept. 1993.
[9] Report of SWP 2-5/13 on Q.27/13, "Support of broadband connectionless data service on B-ISDN," Temporary document of ITU-T, SG13, April 1996.
[10] A.Romanow and S.Floyd, "Dynamics of TCP traffic over ATM Networks," IEEE JSAC, vol.13, no.4, 1995.
[11] V.K.Samalam, "Traffic Management Working Group Living List," ATM Forum, LTD-TM-01.02, 1997.
[12] W.R.Stevens, "TCP/IP Illustrated," Volume 1, pp.312-313, Addison Wesley, 1994, ISBN 0-201-63346-9.
[13] K.Noritake, S.Ushijima, S.Shimojo, and S.Yamaguchi, "Field trial service using ATM-based connectionless service," International Symposium on Interworking'96, Nara, Japan, Oct. 1996.
[14] H.Ohnishi and M.Nagasawa, "Using the NS8000 series to build multimedia networks—New ATM switching system: Multimedia handling node-ATM(MHN-A)," NTT Review, vol.9, no.1, Jan. 1997.
[15] K.Sklower, "A tree-based packet routing table for Berkeley UNIX," USENIX-Winter'91, Dallas, TX.

High-Performance Backbone Network Technology edited by Naoaki Yamanaka
ISBN: 0-8247-5321-6 DOI: 10.1081/(120025687)
Published 2004 by Marcel Dekker, Inc.

Experimental Demonstrations of IP over ATM with Congestion Avoidance Flow Control: CEFLAR

Yoshio Kajiyama, Hideo Tatsuno, and Nobuyuki Tokura

SUMMARY Experimental demonstrations of IP over ATM with congestion avoidance flow control are described. The results confirm that this network can control the transmission rate without upper layer anomalies, IP packet loss, or TCP retransmission. Efficient and fair sharing of the available bandwidth is realized in local area networks.

key words: *traffic management, flow control, IP over ATM QoS*

1. Introduction

One of the hottest topics in the network community is IP over ATM. Unless the problem of "easy network congestion" is resolved, the Quality of Service (QoS) will be decimated with the rapid spread of computer networks, Internet.

A congestion avoidance technique has been already implemented in TCP layer, that is the upper layer of IP in Internet. However, effective throughput can not be gained in the higher layer without flow control in the ATM layer [1]. Our flow control technique CEFLAR (Congestion Avoidance Estimation Feedback by Limited Acceleration-Rate/Ratio) is based on congestion avoidance and guarantees ATM QoS, cell loss rate, and cell delay time [2]. We've already confirmed its basic feasibility in wide area networks [3].

We implemented the CEFLAR function in both a terminal adapter and a node and demonstrated the characteristics of IP over ATM networks (CEFLAR-ATM).

2. CEFLAR Principle

The congestion control technique for available bit rate services is specified in TM 4.0 of the ATM-Forum [4]. This is also rate based flow control like CEFLAR. This control allows a network to enter the congestion state. Accordingly, the target of the ATM-Forum's control is fast recovery from the congestion state and a high throughput under normal and congestion states.

On the other hand, CEFLAR does not allow a network to enter the congestion state. Logically, the target of CEFLAR is stable management and low transmission delay under the normal state. As a result, the ATM-Forum's control can be applied to small networks such as LAN and CPN while CEFLAR can be applied to large networks such as MAN and WAN.

Figure 1 shows the functional requirements of CEFLAR terminals and networks. A congestion avoiding network is realized if the network state can be estimated accurately. We introduced two new concepts, the bit rate increase per unit time (acceleration-rate $=\alpha$ bit/sec^2) and the bit rate increase ratio (acceleration-ratio $=\exp(\beta)$), and shown that a congestion avoiding network can be attained with these parameters [3]. Rate shaper and rate choker functions are required on the terminal side with rate monitoring and rate choking notification functions on the node side.

When the ATM-Forum's control requires a low transmission delay and a small buffer in MAN or WAN, it can be realized by the CEFLAR concept. In this case, ICR defined in TM 4.0 of the ATM-Forum, is set to MCR and the cell rate envelope is as follows.

$$CCR(t) = ICR^* \exp(PCR^* RIF^* t/(424^* Nrm)) + PCR^* RIF,$$

$$RIF = 424^* Nrm^* \beta / PCR$$

3. Experimental Setup

The experimental setup is shown in Fig. 2. It consists of four CEFLAR-terminals, two ATM-Hubs, and a CEFLAR-node. Four Sun Microsystems' SPARC station 20s running SunOS 4.1.3 were used as CEFLAR terminals and two Fore System's ASX-200 ATM switches as ATM-Hubs. Permanent VPs were preestablished between terminals and ATM-Hubs to

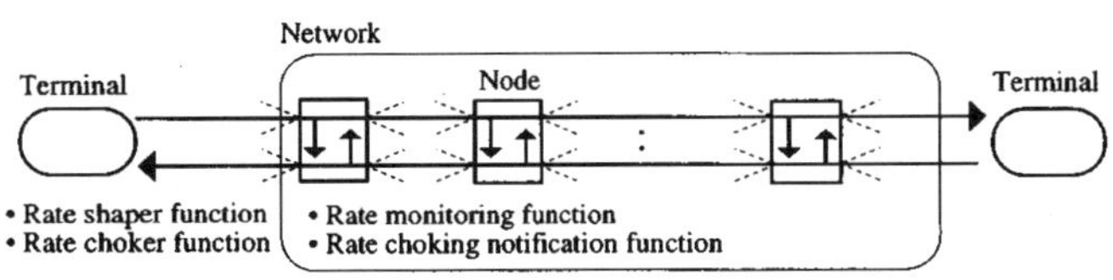

Fig. 1 Functional requirements for CEFLAR terminal and CEFLAR node.

Manuscript received September 2, 1996.
Manuscript revised January 27, 1997.

† The authors are with Optical Subscriber Networks Laboratory, NTT Optical Network Systems Laboratories, Yokosuka-shi, 238-03 Japan.

IEICE TRANS. COMMUN., Vol. E80-B, No. 7, pp. 1098–1100, July 1997

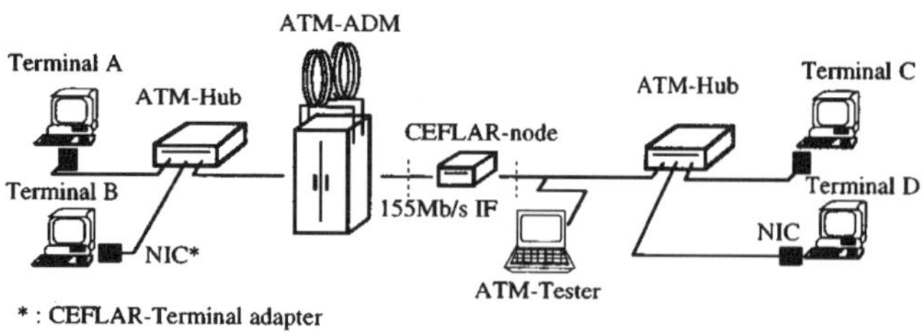

Fig. 2 Experimental set up.

Table 1 Throughput of IP over UBR-ATM.

Network	Protocol	Level	Throughput (Mbit/sec)
UBR-ATM	TCP	Socket	71.3
		ATM	80.9
	UDP	Socket	91.3*
		ATM	148.9

* : Packet loss occurred in the receiving terminal.

eliminate the connection setup time. In demonstrating a local area network, equipment was connected by 5 m fibers and the total length between send and receive terminals was 25 m.

4. Results and Discussion

The characteristics of IP over CEFLAR-ATM were evaluated by measuring the cell delay time, the cell rate, the total cell number, the total IP packet number and IP packet length with an ATM tester running ttcp†. The combination of the send and receive buffer sizes influences TCP throughput and causes throughput anomalies in some conditions [5]. 50 kbyte buffers in both send and receive terminals were used in all experiments to achieve the highest throughput.

We also investigated the throughput of IP over ATM (UBR-ATM) to evaluate terminal adapter performance. Table 1 shows the results. One terminal transmits 16777216 bytes of data using TCP/IP or UDP/IP protocol. (This amount of data is used in all tests.) The size of each write/read system call is set at 8192 bytes and ttcp activates 2048 byte write calls from the user data. The rate was averaged over 10 ttcp runs.

Figure 3 shows the throughput of the ATM layer with TCP protocol in UBR-ATM. The total packet number was 1839–1858 requiring 352134–352185 ATM cells. The average rates obtained were 80.9 Mbit/s (ATM level) and 71.3 Mbit/s (socket level).

Figure 4 also shows the throughput of the ATM layer in CEFLAR-ATM with adequate bandwidth. The initial rate was set at 600 kb/s and the maximum rate, 60 Mb/s was attained in 1 sec (β=4.6 sec^{-1}). The average rates were 44.3 Mbit/s (ATM level) and 39.0 Mbit/s (socket level). Transmission took 0.88 sec longer in the ATM layer than UBR-ATM. However, as the total packet and cell number were the same as in

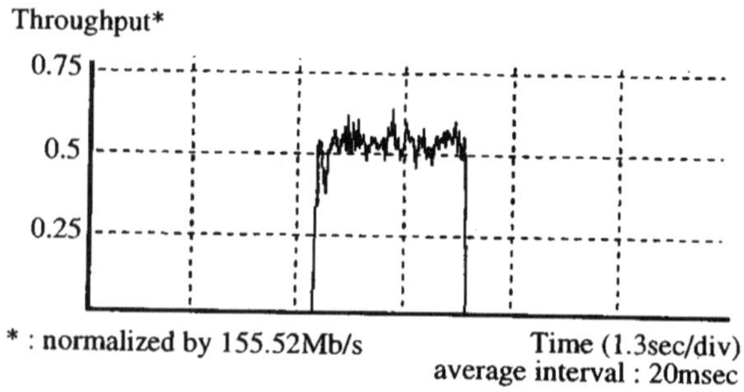

Fig. 3 Hard copy of ATM tester (one terminal throughput of TCP/IP over UBR-ATM).

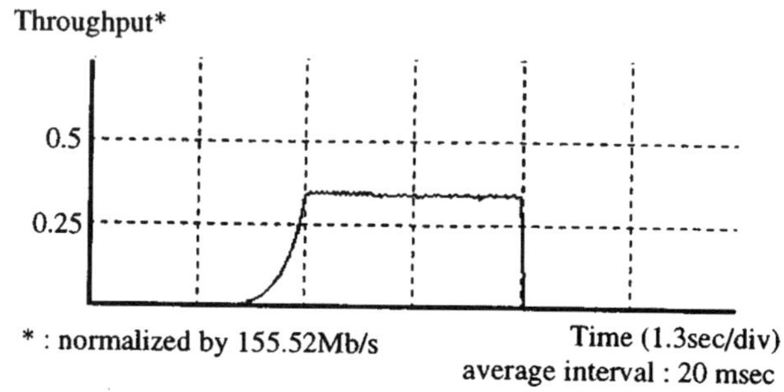

Fig. 4 Hard copy of ATM tester (one terminal throughput of TCP/IP over CEFLAR-ATM).

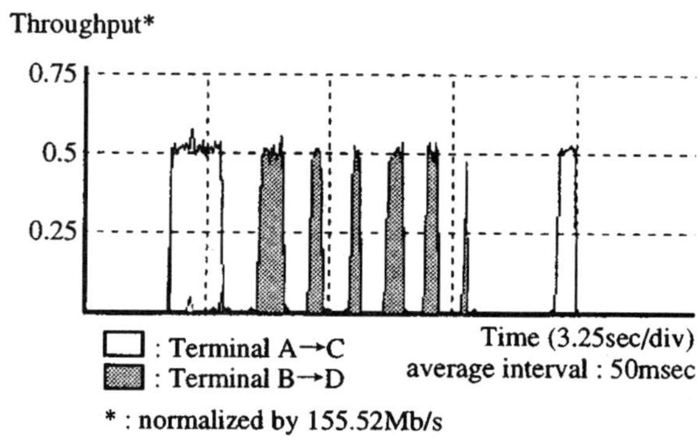

Fig. 5 Hard copy of ATM tester (two terminal throughput of TCP/IP over UBR-ATM).

the case of UBR-ATM, this means that there was no cell loss and the delay time was so small that TCP retransmission was not triggered.

Figure 5 shows the throughput of the ATM layer for two terminal transmission under UBR-ATM. When both terminals begin to send the data at the same time, the total rate exceeds the transmission line rate and cell loss occurs. Re-transmission is required by the receiving terminal whenever this occurs, so cells are transmitted intermittently. Overall, the average throughput of the two terminals were 12.4 Mb/s (A→C) (ATM level) and 18.2 Mb/s (B→D). The times needed to send all data were 10.6 sec and 7.3 sec respectively.

† ttcp is public domain software and is used to monitor the socket level throughput between two IP hosts.

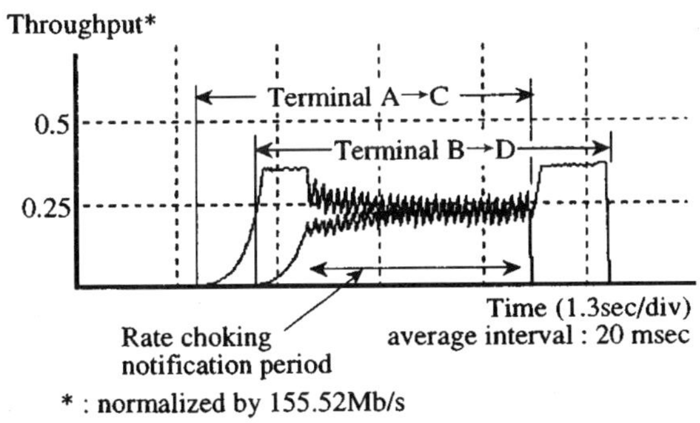

Fig. 6 Hard copy of ATM tester (two terminal throughput of TCP/IP over CEFLAR-ATM with the same line length).

Figure 6 shows the throughput of the ATM layer with two terminal transmission under CEFLAR-ATM. The width of the sliding window was 1000 cell times and the threshold for rate choking notification was set at 0.6 in order to exceed the total multiplexed rate of the two terminals in the CEFLAR node. Each terminal reduces its rate and begins to transfer cells at the low rate pointed by the new address of the look-up table. This is gained by multiplying the present time (address) by 0.9.

One terminal began to transmit data followed by the other terminal after 1 sec delay. When both were transmitting data, the total throughput exceeded the threshold and the network began to require rate choking. In this period, they shared the available bandwidth fairly because the rate of the first terminal was reduced. After the first terminal finished transmission, the second terminal occupied the bandwidth of the network. In this experiment, the data was transmitted at the maximum rate of 60 Mbit/s with the average rate of 34.4 Mb/s sec (A→C) (ATM level) and 32.3 Mb/s (B→D). The times needed to send all data, 4.3 and 4.6 sec, were shorter than for the case of UBR-ATM even if the maximum rate was restricted. The number of cells was also the same as in the other experiments. This means that the data was transmitted without upper layer anomalies.

Figure 7 also shows the experimental results with the same condition shown in Fig. 6 except transmission length. The transmission line between terminal A and C is set to 120 m and the transmission line between terminal B and D is set to 1040 km. To set such a long path in this experiment, the ATM-ADM is used to loop cells through a 65 km long fiber 16 times. In the period of network notification, each terminal also shared the available bandwidth fairly even if they have different transmission line lengths. This shows that the fair bandwidth allocation method is also effective even in wide area networks.

5. Conclusion

We demonstrated IP over CEFLAR-ATM under several conditions. CEFLAR-ATM was confirmed to share the available bandwidth more efficiently and fairly than UBR-ATM.

We confirmed that CEFLAR-ATM transfers IP packets most efficiently.

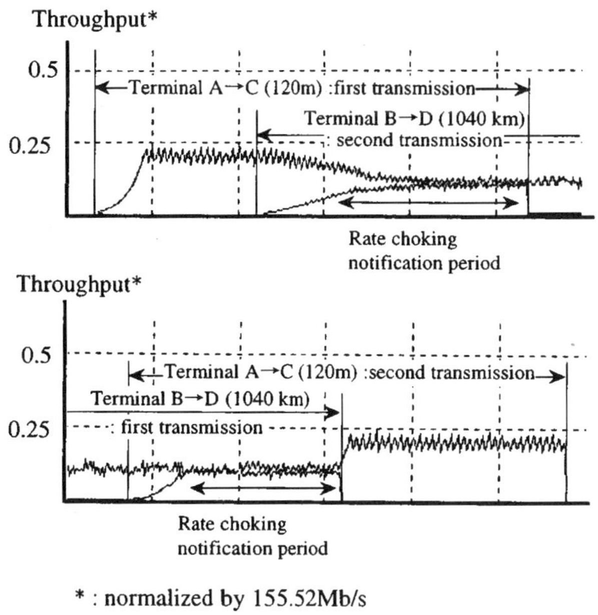

Fig. 7 Hard copy of ATM tester (two terminal throughput of TCP/IP over CEFLAR-ATM with different line lengths).

Acknowledgment

The authors would like to thank Dr. Y. Aoyama, Mr. I. Yamashita, and Mr. K. Yoshida for their helpful encouragement.

References

[1] A. Romanow and S. Floyed, "Dynamics of TCP traffic over ATM networks," IEEE J. Select. Areas Commun., vol. 13, no. 4, pp. 633–641, 1995.

[2] N. Tokura, H. Tatsuno, and Y. Kajiyama, "Congestion avoidance networks: CEFLAR," IEEE ICC'94, pp. 75–79, 1994.

[3] N. Tokura, Y. Kajiyama, and H. Tatsuno, "An experiment on congestion avoidance network: CEFLAR," IEICE Technical Report, CS-94-94, pp. 27–32, 1994.

[4] The ATM-Forum, "ATM forum traffic management specification version 4.0," (95-0013R13).

[5] D. E. Comer and J. C. Lin, "TCP Buffering and Performance over an ATM network," Technical Report CSD-TR 94-026, Purdue University, West Lafayette, Indiana, March 1994.

High-Performance Backbone Network Technology edited by Naoaki Yamanaka
ISBN: 0-8247-5321-6 DOI: 10.1081/(120025688)
Published 2004 by Marcel Dekker, Inc.

ATM LAN Emulation for Mobile Cellular Networks

Nen-Fu Huang and Yao-Tzung Wang

SUMMARY In this paper, the design issues of ATM LAN emulation (LANE) in a mobile environment are addressed and investigated. One of the most important issues is to support the transparent services for mobile stations. We show that the wireline LANE model, defined by the ATM Forum, for the legacy LANs (Ethernet/FDDI/token ring) is not efficient enough to handle a handoff. On occurring a handoff, the ATM network has to maintain network connections and reroute data to the new location of mobile stations. Combined with the conventional cellular handoff schemes, the wireline LANE-based data rerouting is performed at the base station. As a result, a communication path between two mobile stations may become inefficient when the station moves, e.g., the path elongates and incurs extra processing overheads. To overcome this problem, we suggest maintaining a separate connection for each pair of mobile stations instead of each pair of LANs (as defined in the ATM Forum). Following on this suggestion, an extensive cellular handoff scheme is proposed for an ATM-based wireless network. In the proposal, the rerouting decision is furnished at the ATM switch. It shows that this scheme not only meets the basic handoff requirements (data continuity and transparency), but also offers smaller handoff latency and a shorter path. A path migration scheme is also suggested to migrate an inefficient path to a better one, if any. The effectiveness of the proposed handoff scheme for the wireless LAN emulation service is evaluated by analysis. Some implementation issues and the cost/performance tradeoffs between additional connections (one for each pair of mobile stations) required and bandwidth waste caused by path elongation are studied.

key words: *ATM, handoff, LAN emulation, wireless LAN*

1. Introduction

The asynchronous transfer mode (ATM) has emerged as the multiplexing and switching standard for B-ISDN [4]. Differing from conventional shared-medium LAN access methods, ATM adopts a connection-oriented service, i.e., data may not be sent until a virtual channel connection (VCC) is established. Given the vast installation base of the legacy LANs (Ethernet or token ring), ATM has to coexist and interwork with them. LAN emulation (LANE), developed by the ATM Forum [1], is a technique to use ATM's connection-oriented fabric to emulate the connectionless service of a LAN, and enable existing LAN applications to run over ATM networks in the same manner as they are over legacy LANs.

There are several possible models for the provision of a connectionless service over a connection-oriented ATM network. The simplest entails the use of a connectionless server. The server maintains a point-to-multipoint connection, in which the server is the sender of the connection and all stations of the emulated LAN (ELAN) are recipients of the connection. When a station wishes to send a frame, it sends it to the server who in turn forwards it to every station of the ELAN. While this model works, the server acts like a frame relay, forwarding data frames to destination end nodes. Consequently, the server may become a performance bottleneck, resulting in throughput degradation. Such an implementation of LAN emulation fails to directly employ the high-performance switching capability of ATM.

To use the ATM switching capability in a more direct and efficient way, the ATM Forum LANE exploits a combination of two connection models, mesh and star, to carry different types of MAC frames, unicast and multicast. Most unicast frames are sent via a mesh of point-to-point connections. Each sender is required to discover the recipient and set up a VCC between them. To enable the sender to make the discovery of the recipient, a mechanism is needed to determine the corresponding ATM address of the recipient's MAC address. As for sending multicast (or broadcast) frames, the server-based approach described above with a star (point-to-multipoint) connection is exploited by its simplicity. In summary, the ATM LANE defines a so-called star-plus-mesh connection topology to emulate broadcast LAN, while providing most of the advantages of connection-oriented networks [9]. An in-depth overview of the main architectural issues related to the provision of LAN emulation service over ATM and potential solutions can be found in [7], [9], [15].

The ATM LANE provides broadcast and multicast services, address resolution between legacy LANs and ATM, connection setup and data transfer in order to emulate the same services as a MAC layer. In other words, the ATM LANE is to define an ATM MAC layer that is totally independent to upper layer protocols, and

Manuscript received August 25, 1997.
Manuscript revised December 18, 1997.

† Department of Computer Science, National Tsing Hua University, Hsinchu, Taiwan, Republic of China.
* Also, the author is with Computer & Communication Research Laboratories (CCL), ITRI, Hsinchu, Taiwan, Republic of China.

** This work is currently supported by the R.O.C. National Science Council under Contract Number NSC86-2622-E007-009, and by the R.O.C. Ministry of Economic Affairs under Contract Number MOEA86-EC-2-A-15-0146.

IEICE TRANS. COMMUN., Vol. E81-B, No. 6, pp. 1171–1187, June 1998

compatible with legacy LAN MACs. Therefore, well-known bridging techniques can be applied to bridging between Ethernets (token rings) and ATM ELANs.

LAN emulation over ATM has attracted much attention recently across the networking industry, as it allows the installed LAN base to migrate to ATM without major changes and supports interoperability with internetworking devices such as bridges or routers. The ATM LANE is potentially more rewarding in a mobile environment where the constant moving of mobile stations is high anticipated. This is a challenging issue given the ATM network inherently does not handle the mobility very well owing to its connection-oriented nature.

The objective of this paper is to design a mechanism to emulate wireless/mobile LANs (e.g., IEEE 802.11 [6], HIPERLAN [5]) over ATM networks, which can efficiently support transparent LANE service in a mobile environment. Specifically, this mechanism exploits an extensive cellular handoff scheme to reroute user connections from old location to new one which 1) maintains the data continuity without frame loss during a handoff; 2) preserves in-sequence ATM transmission for bi-directional traffic as stations move during a communication; 3) minimizes the handoff time; and 4) keeps the rerouting path as short as possible.

This paper is organized as follows. Design issues of wireless LAN emulation are described in Sect. 2. In Sect. 3, we describe the location management, handoff and connection rerouting algorithms for wireless LAN emulation over ATM networks. The associated data structures and signaling procedures are described in Sect. 4. The performance evaluation and implementation issues of the proposal are investigated in Sect. 5. A brief review of related work is given in Sect. 6. Finally, concluding remarks are given in Sect. 7.

2. Design Issues of Wireless LAN Emulation

2.1 LAN Emulation Architecture

The components of an ELAN consist of LAN emulation clients (LECs), a LAN emulation server (LES), a broadcast and unknown server (BUS) and an optional LAN emulation configuration server (LECS) [1]. The LEC is generally located in ATM end stations or LAN-to-ATM bridges, which performs data forwarding, address resolution, and other control functions. The LES provides a facility for registration and supports the LAN emulation address resolution protocol (LE-ARP), needed by a source LEC to determine the ATM address of the target LEC† responsible for a destination MAC address. The BUS is used to forward all broadcast/multicast traffic to all attached LECs. It may also handle initial unicast data which are sent by a source LEC before a direct VCC to a target LEC is set up. The LECS provides configuration information to LECs to locate LES for joining an ELAN.

The following describes the procedure of transmitting a unicast frame (with an individual MAC address) over an ELAN.

1. A LEC has an LE-ARP cache which maintains MAC address, VCC association. Upon receiving a frame from a high layer, a LEC first attempts to send the frame directly to the target LEC over the associative VCC.

2. If the destination MAC address is not in the LE-ARP cache, the source LEC sends an LE-ARP-Request to the LES for the ATM address of the target LEC.

3. After receiving the LE-ARP-Reply from the LES, the source LEC updates its LE-ARP cache. It then establishes a direct VCC to that ATM address for transmitting the frame.

In the case of a broadcast/multicast frame, the source LEC simply sends it to the BUS for broadcasting to all stations in an ELAN.

From the above descriptions, the following observations hold. Basically, only one direct VCC is required between each pair of LECs. The reason is that it is unrealistic to predict the bandwidth requirements of bursty data traffic in advance; and therefore a service that dynamically shares the available bandwidth between all active end stations is required. This service is referred to

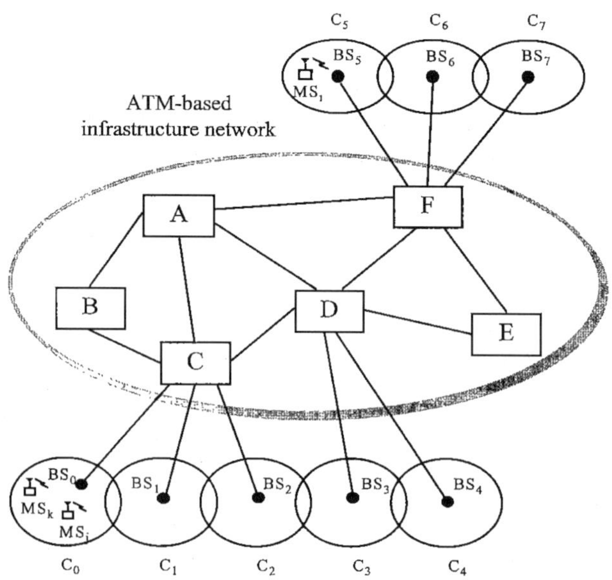

Fig. 1 A typical wireless LAN interconnection via ATM-based infrastructure network.

†If the target LEC is a LAN-to-ATM bridge, it has to serve as a proxy and respond for all LAN stations behind it.

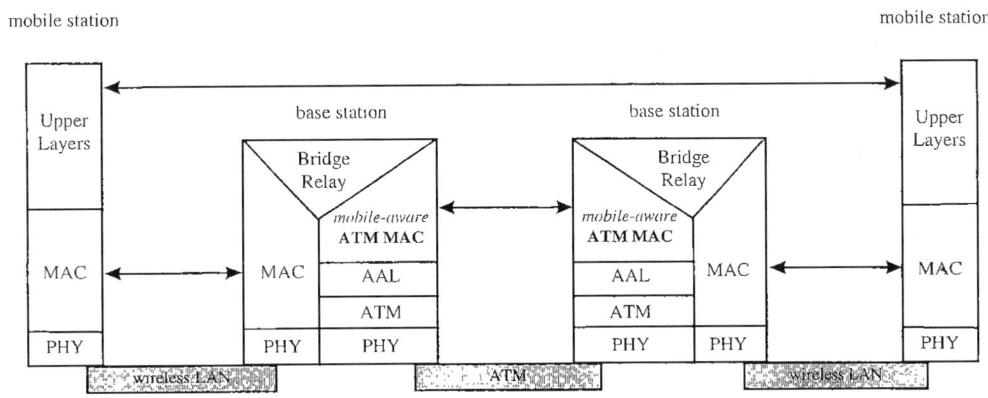

Fig. 2 The system reference model for wireless LAN interconnection over ATM networks.

as available bit rate (ABR) service which has no provisions for guaranteed bandwidth, quality of service, and so forth [2].

In summary, (1) the LANE service is one kind of ABR service class. (2) For data transfer, only one direct VCC is needed between each pair of LECs.

We now turn to the issue of wireless LAN emulation (WLANE) based on the above data transfer model. A typical network configuration of wireless LAN interconnection over an ATM infrastructure is illustrated in Fig. 1†. Mobile stations (MSs), such as MS_i, gain access to the ATM infrastructure via base stations (BSs), such as BS_5 connected to ATM switch F. The base station is to coordinate the access of various mobile stations within its wireless cell. Further, each of the base stations here acts like a wireless LAN-to-ATM bridge. Figure 2 illustrates the system reference model to interconnect two mobile stations located apart from each other over ATM via MAC layer bridging method. It is worth to note that the mobile-aware ATM MAC layer in the base station not only supports all the LANE-related functions, but also is responsible to handle a user connection during a handoff. Two possible approaches can tackle this problem, one is to elongate the connection from an old base station to a new one, the other is to reroute the connection on an end-to-end basis. We next show neither of them is efficient, and better solution has to be sought after.

2.2 BS-to-BS Elongation Rerouting

Consider the case shown in Fig. 1 where one mobile station MS_i is located in cell C_5, and MS_j and MS_k are located in cell C_0. Assume that MS_i is communicating to MS_j and MS_k, simultaneously. Then a direct VCC is established between BS_5 and BS_0, say $VCC_{5,0}$. Assume now MS_j moves into C_1. Clearly, the ATM cells destined to MS_j "should be" forwarded to BS_1 by the ATM switch C. However, since all the ATM cells transferring either from MS_i to MS_j or from MS_i to MS_k are assigned with the same VPI/VCI along the shared connection $VCC_{5,0}$, the ATM switches on the $VCC_{5,0}$ do not have the capability to distinguish which cells should be forwarded. The only way to reroute the cells destined to MS_j is to establish another VCC (say $VCC_{0,1}$) from BS_0 to BS_1††. Then, BS_0 will first reassemble the cells for MS_j, check the destination MAC address (now moved out), and then segment the frame into cells and transmit the cells down the new established $VCC_{0,1}$ again. We call it as the frame-forwarding-based rerouting. Base stations do most data rerouting works. Generally, these works are performed in software.

This approach is quite simple but as we can see, data path (or the $VCC_{5,0}$) is elongated as ($VCC_{5,0}$, $VCC_{0,1}$). The phenomenon of path elongation will be more serious as MS_j continuously moves from left to right. For example, if MS_j further moves into C_2 and C_3, then the data path between MS_i and MS_j might be elongated as ($VCC_{5,0}$, $VCC_{0,1}$, $VCC_{1,2}$, $VCC_{2,3}$), as shown in Fig. 3. After traveling through several wireless cells, an elongated path may consume too much network resources (e.g., bandwidth or buffer) and become very inefficient. Extremely, it might have redundant loops.

2.3 End-to-End Rerouting

To avoid such an elongated path due to the movements of mobile stations, end-to-end VCCs can be dynamically established to reroute cells to new base station. For example, in Fig. 3 we can establish a new direct VCC between BS_5 and BS_3, and the possible path for the new VCC is (BS_5, F, D, BS_3). Obviously, it is the shortest path for the given network topology. Now there are two possible paths between BS_5 and BS_3, one via the elongated path (described above) and another via the new established VCC. Both paths are required, one for data continuity, one for efficiency. They may

† For simplicity, the LANE-related servers and ATM end stations do not appear in the figure.

†† The BS-to-BS connection could be prior-established before a handoff.

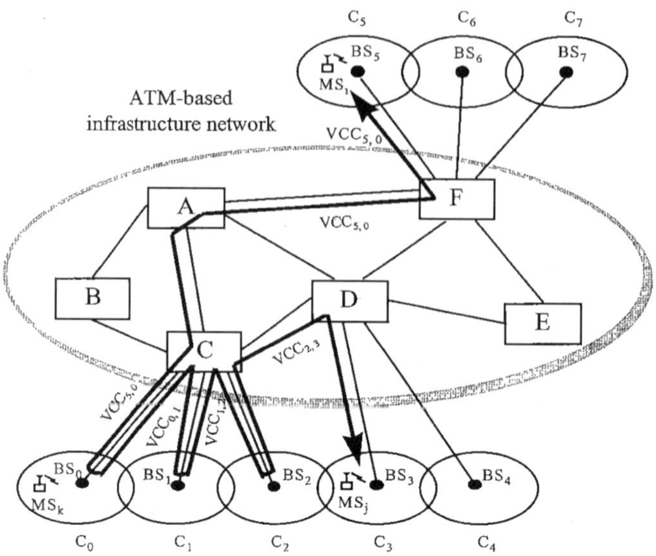

Fig. 3 Data path being elongated due to station (MS_j) moves.

result in frames being delivered out of order. The first frame sent down the new established VCC may arrive sooner than the last frame along the elongated path.

One solution for this problem is to use the flush message procedure [1]. The new base station (BS_3) informs the far-end base station (BS_5) a handoff occur and BS_5 transmits a flush message down the elongated path ($VCC_{5,0}$, $VCC_{0,1}$, $VCC_{1,2}$, $VCC_{2,3}$). Then BS_5 holds further data frames until the flush message is acknowledged. Once the flush message is acknowledged, it ensures that the elongated path is empty of frames before the new established VCC is put into service. As we can see, a three-way handshake is required. That is, it needs three trips over an ATM infrastructure for a handoff. The propagation delays, typically, can be up to multiple seconds. It can lead to a long pause in communication, resulting in large part of frame and throughput losses [8].

This shows that neither BS-to-BS elongation rerouting nor end-to-end rerouting is efficient to support the LAN emulation service in a mobile environment. In the following, we propose an extensive cellular handoff algorithm with cell-forwarding-based rerouting (in contrast with frame-forwarding-based rerouting) for an ATM-based wireless network. The rerouting decision is furnished at the ATM switch and the rerouting overhead is minimized. Also a distributed location management scheme is mentioned.

3. Proposed Algorithms

In this section, we provide a high-level description of the proposed algorithms for location management, handoff and connection rerouting in a WLANE environment. The key contributions of this paper regarding these algorithms are as follows: a fast handoff scheme based on distributed switching/buffering which maintains frame sequence and a connection rerouting scheme based on maintaining a separate connection for each pair of mobile stations to keep rerouting path as short as possible (Sect. 3.1), and a distributed mobile tracking and locating scheme (Sect. 3.2).

3.1 Handoff and Rerouting

Handoff is regarded as the most important operation in a wireless network since it allows communication links to continue without severe interruption. Typically, two types of network links, wireless and wireline, need to be carefully maintained during a handoff. First, a new wireless (radio) link is established as follows. In a cellular topology, a mobile station is communicating with only one base station. Base stations make their presence known by transmitting beacons periodically via wireless channel. An MS decides to change to another cell when it receives a beacon from a new BS (BS_{new}) with a stronger wireless signal than that from the old BS (BS_{old}). To do this, the MS sends a greeting message to BS_{new} and is acknowledged. Then BS_{new} adds the MS to its list of served MSs. The BS_{new} also notifies BS_{old} that the MS has moved out and can be reached via BS_{new}. Then BS_{old} removes the MS from its served list and acknowledges the handoff to BS_{new}. Finally, BS_{new} becomes the default gateway for the MS, routing frames that originate in the MS from the wireless network to the ATM network. Similarly, it forwards frames destined to the MS from the ATM network to the wireless network.

Next, we describe how to retain a continuous wireline link during a handoff by the use of connection rerouting techniques [12]. In order to offer efficient connection rerouting, we recognize that any pair of communicating MSs, instead of any pair of LANs in (wireline) LANE described previously, should establish an individual VCC. Based on this property, the ATM switches along the VCC can distinguish the ATM cells destined to a moving station and reroute them to the new cell's base station. That is, reroute (forward) the ATM cells directly, avoiding unnecessary reassemble/segmentation operation at base stations, as noted earlier.

A so-called "almost-all reuse with shortest elongation (ARSE)" algorithm is proposed for rerouting connections. ARSE attempts to reuse as much of an existing connection as possible, establishing only a new partial connection between the ATM switch and the new cell's base station. The corresponding portion of the connection is then torn down. Basically we can exploit the current signaling protocols [2], [3] to achieve the connection setup and release. An example of such schemes is similar to that of [12].

For illustration, the mobility is further classified into two categories: intra-switch move and inter-switch move. An ATM switch might connect several BSs, each of which covers one wireless cell. Assume that an MS is moving from the cell of BS_i to the cell of BS_j. Then the mobility is called an intra-switch move if BS_i and BS_j are connected to the same ATM switch. Otherwise, it is called an inter-switch move. For an intra-switch move, we can reuse most portion of the VCC and only change the data path from one ATM switch port to another internally. Apparently, the handoff can be performed very quickly and the path is not elongated. As to an inter-switch move, based on the ARSE, the new established partial connection(s) is from BS_{new} to the ATM switch associated with BS_{old}, SW_{BSold}. One important benefit of the ARSE-based rerouting is that the crossover ATM switch (i.e., SW_{BSold}) is definitely located without any sophisticated discovery mechanisms so that establishing the partial connection(s) can be initiated once a handoff occurs, thereby minimizing handoff time.

At the time of handoff, all the cells are followed the BS-to-BS elongation rerouting approach to redirect ATM cells to BS_{new} for data continuity. Cells obviously preserve their in-order sequence. However, there is some inefficiency because cells are always routed to/from base stations. To improve network efficiency, we update the routing database in the crossover switch (SW_{BSold}) to route the ATM cells destined for a moving station, to BS_{new}. This may cause two paths between SW_{BSold} and BS_{new}. For migrating the path to an efficient one, the crossover switch SW_{BSold} sends a migrate message to BS_{old} and BS_{new}, respectively, indicating that the path ($path_{old}$) to BS_{old} is clear and the path ($path_{new}$) to BS_{new} is active. Clearly, BS_{new} has to handle the cell re-sequence work since cells sent from SW_{BSold} down the $path_{new}$ may arrive sooner than those along the $path_{old}$.

3.2 Location Management

In an ATM-based wireless network, there must exist a means to support the location of a mobile station in order to deliver an incoming connection setup request. Consider a simple network configuration where all base stations belong to the same emulated LAN (ELAN). In this model, all base stations connect to the same LAN emulation server (LES). One solution to provide location tracking and query resolution (LE-ARP-Request) functions is that each base station registers with the LES all the mobile stations located in its wireless cell. Movements of mobile stations are detected by base stations which then inform the LES to update the database. This, however, would lead to more registration and deregistration traffic between the LES and the base stations due to station mobility. In our proposal, the LES does not need to track mobile locations. Instead, mobile tracking is done by having each BS maintain a list of the mobile stations located in its wireless cell. For locating a mobile station, the LES broadcasts a query (LE-ARP-Request) to all base stations in the ELAN and waits for the response. This solution avoids centralized location database and eliminates unnecessary (de)registration process.

In practice, each base station potentially needs to connect to multiple ELANs given that it has mobile stations belonging to different ELANs, and a single ELAN can span over multiple base stations' cells. In this model, there may exist multiple LESs and each of them corresponds to one ELAN. To provide efficient mobile tracking and locating for given multiple ELANs over ATM, we recommend that LESs for different ELANs are interconnected into a hierarchical structure. The location information is maintained and determined based on the interactions among mobile stations and LESs. In the design, it provides three advantages: 1) providing a distributed mobile tracking and locating; 2) offering a moderate trade-off between the tracking and locating costs; and 3) supporting ELAN creation and deletion in an easy and fast manner. Space limitations prohibit us from describing these details, but we refer the interested readers to [11] for more information.

4. Signaling Support and the Operation

In this section, we describe the detailed signaling procedures used to support the high-level algorithms for handling handoff and connection rerouting. For an intra-switch move, no path elongation is required. That is, the rerouting overhead due to mobility has been minimized. As for an inter-switch move, the elongated path is kept as short as possible. In addition, a shortest-path based migration is performed, if necessary, to eliminate the elongated path due to an inter-switch move.

First of all, some routing databases, prior-established virtual connections (VC) and handoff messages to support mobility in a WLANE environment are introduced.

Registration database: Each BS has a registration database to record all of the MSs within the served cell. Whenever an MS moves to a new cell, it makes a registration with the 48-bit MAC address [6] to the new BS and the old BS removes the corresponding entry from its registration database. Based upon this database, a BS can determine whether a received MAC frame is local (source and destination MAC addresses are within the same served cell) or remote. The registration database can also be used to reply an LE-ARP-Request (if the requested MAC address is in the registration database).

MS pair-VCC mapping database (MVD): Recall that each pair of communicating MSs has an individual virtual connection, which is identified by a virtual path/channel identifier (VPI/VCI). Thus a BS has to maintain an MS pair-VCC mapping database (MVD) for frame forwarding. If the destination/source MAC address pair of a remote frame arriving from a local MS has a match in the MVD, the BS forwards the frame to the ATM network with the corresponding VPI/VCI value. Otherwise, the frame will be sent to the BUS and the LE-ARP operation will be started to establish a VCC with the serving BS of the destination MAC address.

BS identity database (BID): Each BS also maintains the identity information for all neighboring BSs. For instance, in a regular hexagonal cellular layout, there will be six cells in each neighborhood. Thus each BS keeps six identities of the surrounding BSs in the BID. In an ATM-based wireless network, the ATM address could be treated as the identity of a BS.

ATM switch routing database (SRD): At each switching node, the incoming VCC (IN-VCC) and incoming port (IN-PORT) of an arriving ATM cell are used as indices in the switch routing database (SRD) to uniquely identify the outgoing port (OUT-PORT) and the outgoing VCC (OUT-VCC) associated with the next link. Thus, at each switching unit, the mapping tuple (IN-VCC, IN-PORT, OUT-VCC, OUT-PORT) uniquely defines a VCC. To set up a VCC, a path passing through the ATM network is selected by the signaling protocol, and a new mapping tuple is added to the SRD of each of the switching nodes along the path.

Signaling VC: It is a predefined VPI/VCI value (VPI = 0, VCI = 5) for supporting of demand (switched) channel connections. These connections are established in real time using signaling procedures [2], [3].

Handoff VC: Here we define a handoff VPI/VCI value to carry the messages of handling a move. The handoff VC is a point-to-point connection between a BS and an ATM switch or between two ATM switches.

BS-to-BS VC: We assume that every pair of BSs which serving adjacent wireless cells has a virtual connection between them, used for handoff control.

Hint Message: In the design of handoff scheme, we assume that the topology of the wireless network is overlapping between adjacent cells. Once the moving MS enters the overlap area, it transmits a Hint Message to BS_{old}. It is an advanced warning that a handoff from BS_{old} to BS_{new} is about to occur.

Connection Message: This message is transmitted by BS_{old} to BS_{new} via the BS-to-BS connection. It contains the profile for each MS-associated active connection to be rerouted. The profile may include the MS pair information, the identity of the far-end BS, ATM traffic descriptor and quality of service (QoS) parameters, and so on. For our discussion, only the first two items are used.

Reroute Message: The BS_{old} sends a Reroute Message with (MS pair-VCC correspondence; ATM BS_{new}) to the attached ATM switch via the Handoff VC, in which *ATM* BS_{new} stands for the ATM address of BS_{new}. According to the MS-associated VCCs, the ATM switch alters the corresponding tuples in the SRD for path modification, if necessary.

Deregistration Message: The change-of-BS (handoff) procedure requires a deregistration of the MS with BS_{old}. This is done by having BS_{new} send a deregistration message to BS_{old} on behalf of the MS.

4.1 Intra-Switch Handoff Control Procedures

For the sake of simplicity, we use the term "mobile connection" to refer the virtual connection between any pair of MSs. These mobile connections are dynamically established by the LANE/Signaling protocols. Consider a veracious representation of a mobile connection topology shown in Fig. 4. In this example, we have three ATM switches X, Y, Z and six cells $C_0, C_1, C_2, \ldots, C_5$. Base stations (BS_0, BS_1), (BS_2, BS_3), and (BS_4, BS_5) are connected to ATM switches X, Y, and Z, respectively. Initially, there are three mobile connections (a, c), (a, d), and (b, c).

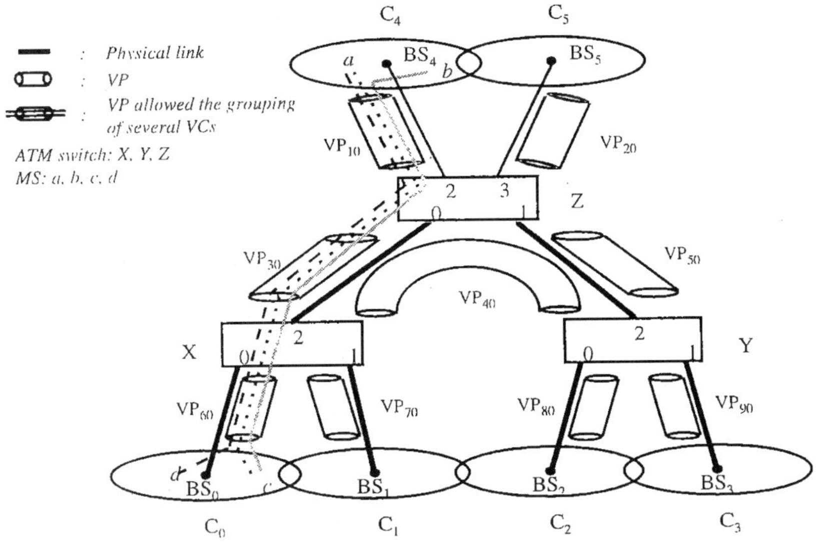

BS0's MS Pair-VCC Mapping Database (MVD)

MS (a, c)-60/100
MS (b, c)-60/200
MS (a, d)-60/300

Switch X's Routing Database (SRD)

IN-VCC	IN-PORT	OUT-VCC	OUT-PORT
60/100	0	30/400	2
30/400	2	60/100	0
60/200	0	30/500	2
30/500	2	60/200	0
60/300	0	30/600	2
30/600	2	60/300	0

BS4's MS Pair-VCC Mapping Database (MVD)

MS (a, c)-10/700
MS (b, c)-10/800
MS (a, d)-10/900

Switch Z's Routing Database (SRD)

IN-VCC	IN-PORT	OUT-VCC	OUT-PORT
30/400	0	10/700	2
10/700	2	30/400	0
30/500	0	10/800	2
10/800	2	30/500	0
30/600	0	10/900	2
10/900	2	30/600	0

Fig. 4 An example of mobile connections (before move).

Each mobile connection is identified by a sequence of VPI/VCI values; each of which is assigned in the physical link between two adjacent switching nodes along the path. That is, the sequences of VPI/VCI values for mobile connections (a, c), (b, c) and (a, d) are $(10/700, 30/400, 60/100)$, $(10/800, 30/500, 60/200)$, and $(10/900, 30/600, 60/300)$, respectively. The VPI value is based on the assignment of virtual paths between two ATM switches or between an ATM switch and a BS. The VCI value here is randomly selected and used to identify an end-to-end mobile connection only. The corresponding contents of the MVDs in BS_0 and BS_4 and the SRDs in ATM switches X and Z are also shown in Fig. 4.

To show the procedure of handling an intra-switch handoff, let us consider the example shown in Fig. 4 again. Assume MS_c is moving from C_0 to C_1. Note that MS_c has two mobile connections with MS_a and MS_b, respectively. Also, the mobile connection is always bi-directional. The rerouting of a bi-directional connection consists of two independent procedures, each of which handles one direction. Here we call the data crossing over the ATM backbone to the wireless network as the downlink data. As for the uplink data it is in the opposite way.

A time diagram showing the exchange of messages for handling an intra-switch handoff is given in Fig. 5, and the control steps are stated as follows.

1. The Hint Message is first delivered to BS_0 identifying the potential new BS, BS_1.

2. BS_0 issues a Connection Message with $MS(a, c)$, $MS(b, c)$ and $ATM\ BS_4$ (far-end BS's identity) to BS_1. Based on the identity information in the BID, BS_0 recognizes the identity of new BS ($ATM\ BS_1$), then sends a Reroute Message ($MS(a, c)$-60/100, $MS(b, c)$-60/200; $ATM\ BS_1$) to ATM switch X.

3. Once the Reroute Message is received, ATM switch X establishes the new partial connections to BS_1 for each of the MS-associated VCCs (60/100, 60/200) using the ATM connection setup procedure. The ATM address of switch X ($ATM\ X$) and the $ATM\ BS_1$ are used as the "calling party" and

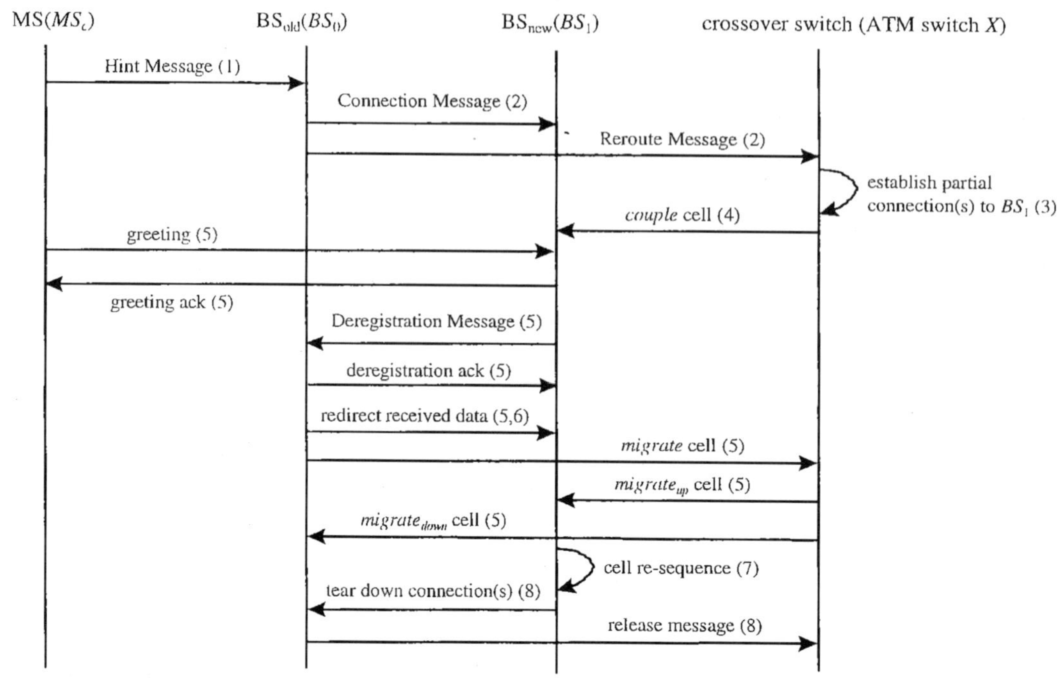

Fig. 5 Intra-switch handoff control procedures.

"called party" identities in the connection setup message. Comparing the *ATM* BS_1 to the routing information, ATM switch X knows that the destination end (BS_1) is connected to it. Thus the new partial connections can be quickly established. During the connection setup, data cells are sent over the old connection.

4. To construct the new MS pair-VCC mapping, ATM switch X will send a special control cell, called couple, down to each new established partial connection†. The couple cell contains the MS pair information, which will be sent along the corresponding new partial connection to new BS. Once receiving the couple control cells, BS_1 knows the ends of the VPI/VCI for the new established partial connections are 70/100 and 70/200, respectively. Therefore, the new MS pair-VCC mapping is constructed, i.e., $MS(a, c)$-70/100 and $MS(b, c)$-70/200. Also BS_1 adds the mapping information to the profile of the MS-associated connections and does some arrangement to make the partial connection as the part of end-to-end connection.

5. When MS_c moves completely into the new cell C_1, it sends a greeting message to BS_1 and is acknowledged. BS_1 also sends a Deregistration Message along the BS-to-BS connection to BS_0. Then BS_0 removes the MS_c from its served list, acknowledges the handoff and redirects the received data to BS_1. At the same time, BS_0 sends a special control cell, called migrate, to the ATM network over the old connections. On receiving it, ATM switch X will replicate the migrate cell to $migrate_{up}$ and $migrate_{down}$ cells; and send them to BS_0 ($migrate_{down}$) and BS_1 ($migrate_{up}$) via the old and new connections, respectively. For downlink traffic, the $migrate_{down}$ is used to identify the traffic "brand point". Before the $migrate_{down}$, the downlink data is along the old connection. After that, it is along the new connection. For a successful reassembly at BS_0, the $migrate_{down}$ control cell should be sent while the end-of-frame receives††. Following the same idea, before the $migrate_{up}$ the uplink data is along the old connection. After that, it is along the new connection. When BS_1 receives the $migrate_{up}$ control cell, it knows that the old connection is clear of uplink data, and may use the new connection to transmit later uplink frames.

6. During the handoff, BS_0 constantly monitors the $migrate_{down}$ control cell. If BS_0 is still providing communication with MS_c, it just forwards the data via wireless channel. Otherwise (MS_c moves to the coverage area C_1), BS_0 redirects the received data (including the $migrate_{down}$ cell) over the prior-established BS-to-BS connection to BS_1 until the $migrate_{down}$ control cell is detected.

†This may be referred to as the in-channel signaling, i.e., control data sent over a user connection.

††This can be achieved by continuously monitoring cells from the connection until the switch sees the ATM-layer-user-to-user (AUU) parameter set in the ATM header, indicating the end of the AAL frame. Because AAL5 does not support the simultaneous multiplexing of frames on a single VCC, the AUU parameter can be used to identify the end of a sequence of cells for the same frame.

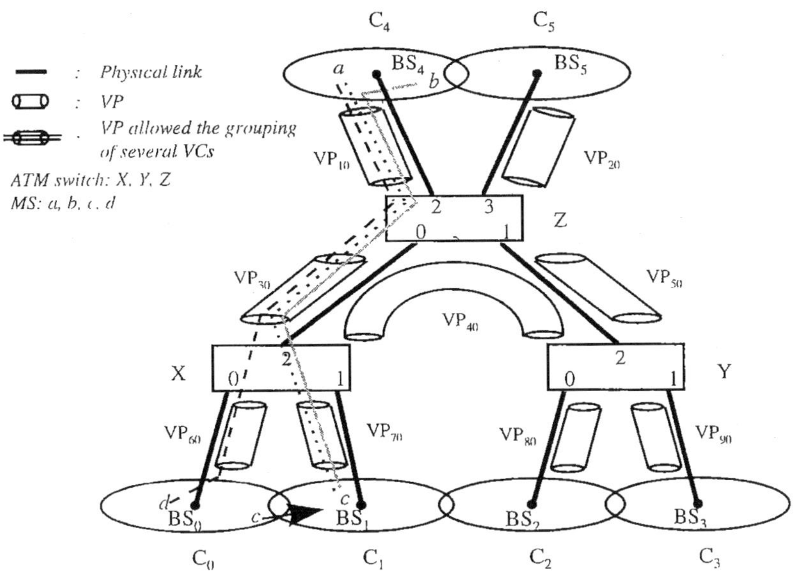

BS$_1$'s MS Pair-VCC Mapping Database (MVD)

MS (a, c)-70/100
MS (b, c)-70/200

BS$_4$'s MS Pair-VCC Mapping Database (MVD)

MS (a, c)-10/700
MS (b, c)-10/800
MS (a, d)-10/900

Switch X's Routing Database (SRD)

IN-VCC	IN-PORT	OUT-VCC	OUT-PORT
60/100->70/100	0->1	30/400	2
30/400	2	60/100->70/100	0->1
60/200->70/200	0->1	30/500	2
30/500	2	60/200->70/200	0->1
60/300	0	30/600	2
30/600	2	60/300	0

Switch Z's Routing Database (SRD)

IN-VCC	IN-PORT	OUT-VCC	OUT-PORT
30/400	0	10/700	2
10/700	2	30/400	0
30/500	0	10/800	2
10/800	2	30/500	0
30/600	0	10/900	2
10/900	2	30/600	0

Fig. 6 An example of mobile connections (after an intra-switch move).

7. To ensure in-order delivery to MS_c during rerouting, all the cells received by BS_1 over the new connection are buffered in the extra buffer space, and not forwarded to MS_c. However, when BS_1 receives the *migrate*$_{down}$ control cell, it processes the buffered cells as if they have been received. Therefore, the in-order delivery can be preserved.

8. On completing the handoff, the new BS (BS_1) informs the old BS (BS_0) via the BS-to-BS connection to tear down the old connection. BS_0 then sends a connection release message to the connected ATM switch X and is acknowledged. According the handoff information, ATM switch X knows that it just needs to tear down the portion from it to BS_0 and the release process then ceases.

From the above descriptions, the following observations can be made. The old BS (BS_0 in the step 5) requires buffering some of the frames until the Deregistration Message (sent from BS_1) is received. As for the new BS (BS_1 in the step 5), it can not start the uplink communications until the *migrate*$_{up}$ control cell is received. The pause time is dependent on the propagation time between the old and new BSs. The mobile connections and the corresponding contents in the MVDs and SRDs after an intra-switch move are depicted in Fig. 6. Here we simply change the data path from one ATM switch port to another internally. Apparently, the rerouting load to ATM networks due to this intra-switch move is minimized and the path (VCC) is not elongated. Moreover, we do not have to change the MVD of BS_4 for such a move. Therefore, it is also transparent to the far-end mobile stations MS_a and MS_b.

Note that the proposed handoff control is based on the cell overlapping topology where new partial connections can be established through the crossover switch to the new BS before an MS enters the new cell. Still, in some cases handoff may occur when cells do not overlap. The proposed handoff procedures can easily adapt to non-overlapping conditions as follows. First, store

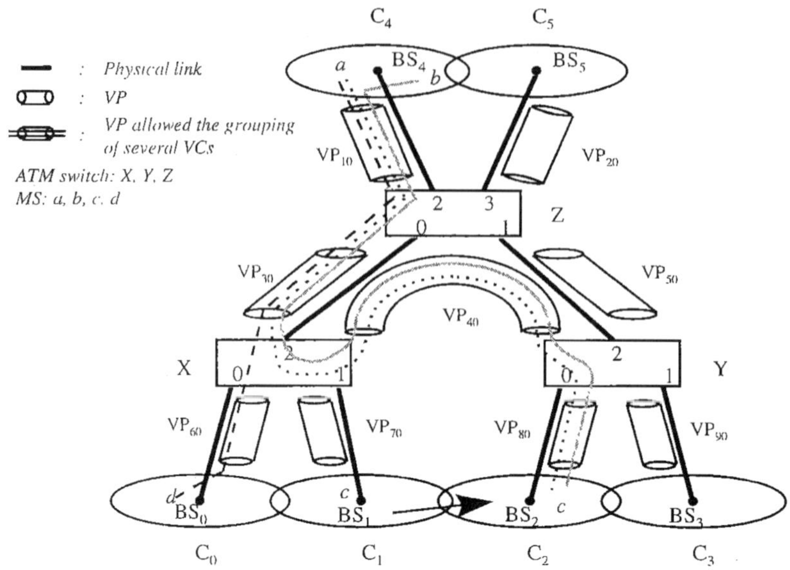

Fig. 7 An example of mobile connections (after an inter-switch move).

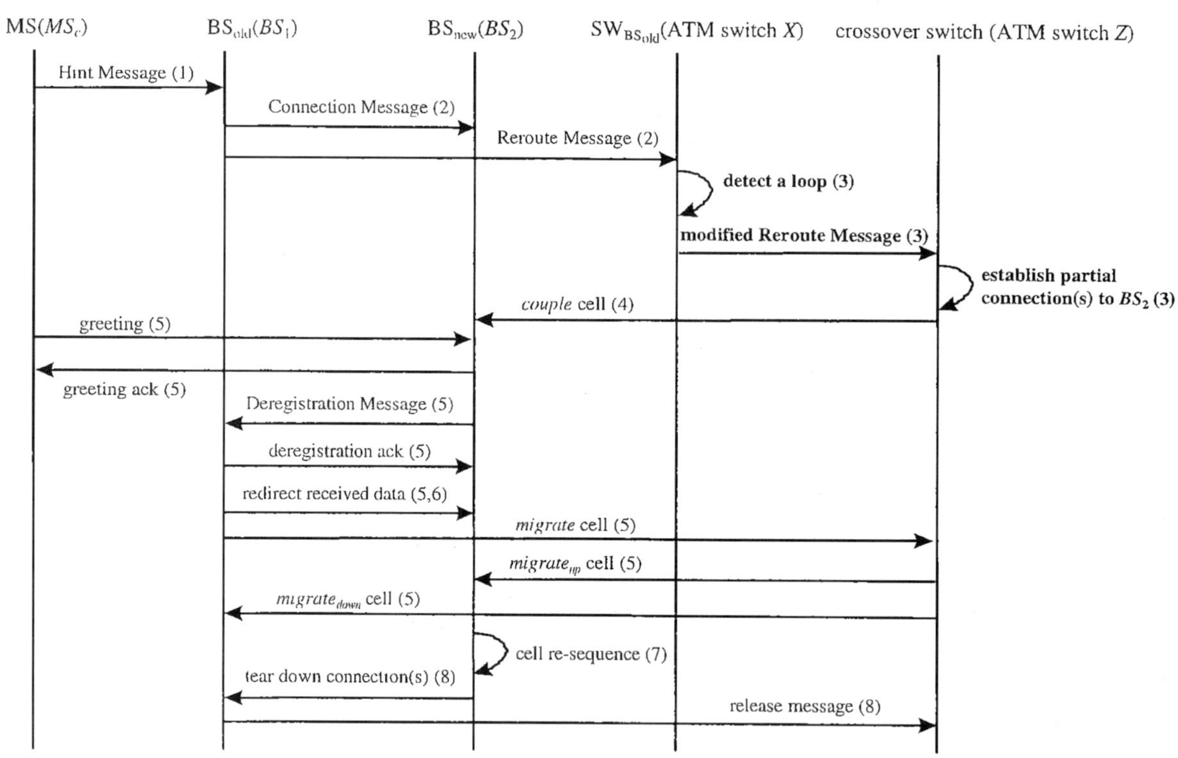

Fig. 8 Inter-switch handoff control procedures while a loop exists.

frames at the old BS until the MS comes within range of new BS. Then, associate the MS with the new BS by sending a greeting message. Finally, follow the steps 2-8 listed above except skipping the MS association to the new BS in step 5.

4.2 Inter-Switch Handoff Control Procedures

So far we have introduced the handoff scheme for intra-switch mobility. Following the same idea, we try to make the elongated path due to an inter-switch move as short as possible.

To demonstrate the procedure of handling an inter-switch handoff, let us consider the example shown in Fig. 6 again. Assume that MS_c in C_1 continues roaming to C_2. Then an inter-switch handoff has to be taken. Following the similar control steps in the intra-switch handoff, the only difference is that the new established partial connections to BS_2 will pass through more than one ATM switch. For example, the partial connections

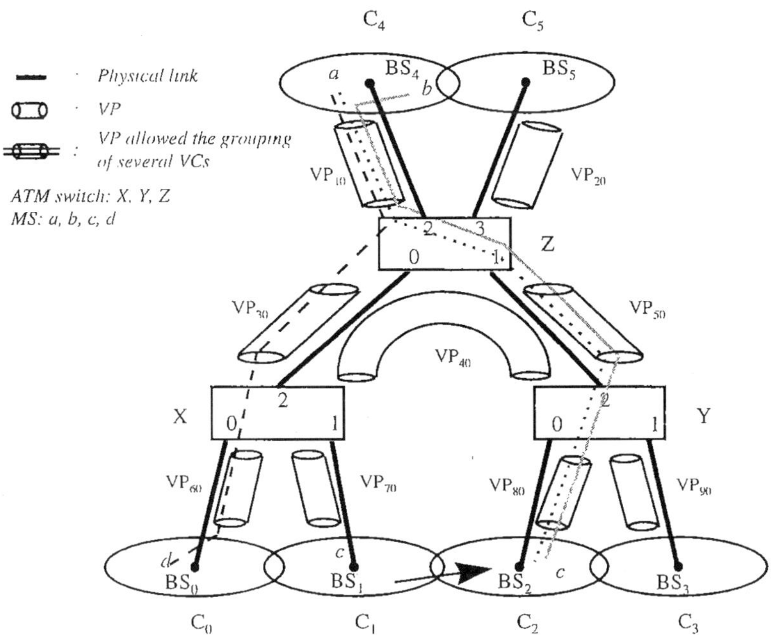

BS$_2$'s MS Pair-VCC Mapping Database (MVD)

MS (a, c)-80/350
MS (b, c)-80/450

Switch Y's Routing Database (SRD)

IN-VCC	IN-PORT	OUT-VCC	OUT-PORT
80/350	0	50/150	2
50/150	2	80/350	0
80/450	0	50/250	2
50/250	2	80/450	0

BS$_4$'s MS Pair-VCC Mapping Database (MVD)

MS (a, c)-10/700
MS (b, c)-10/800
MS (a, d)-10/900

Switch Z's Routing Database (SRD)

IN-VCC	IN-PORT	OUT-VCC	OUT-PORT
30/400->50/150	0->1	10/700	2
10/700	2	30/400->50/150	0->1
30/500->50/250	0->1	10/800	2
10/800	2	30/500->50/250	0->1
30/600	0	10/900	2
10/900	2	30/600	0

Fig. 9 An example of loop elimination (the loop in switch X in Fig. 7 is removed).

are from switch X then switches Z, Y and finally to BS_2, as shown in Fig. 7. We note the elongated path is shorter than the case of BS-to-BS elongation. Nevertheless, we can see that in this example there exists a loop (OUT-PORT = IN-PORT) within the ATM switch X. This is unnecessary and can be eliminated later.

A time diagram showing the exchange of messages for handling an inter-switch handoff is given in Fig. 8. For an inter-switch move, most control steps used for the intra-switch handoff can be applied to it except the step 3 which is replaced by the following one.

3. Whenever an ATM switch (X) detects the output port and input port of a mobile connection are identical, it sends a modified Reroute Message ($MS(a,c)$-30/400, $MS(b,c)$-30/500; $ATM\ BS_2$) to the adjacent ATM switch (Z) on the end-to-end route of the mobile connection. Note that the MS-associated VCCs are replaced with that of the upper link (between switches Z and X). The adjacent ATM switch will establish new partial connections from it to the new BS (BS_2) for the elongated path and detect the loop condition as well. If there is no loop, the handoff procedure is done†; otherwise, the procedure described before is repeated until the loop within an ATM switch is eliminated.

For the loop example shown in Fig. 7, we have a better result after the loop is eliminated as shown in Fig. 9. This not only offers a shorter path but also provides a better bandwidth utilization.

4.3 Path Migration Schemes

Having finished the handoff operation, the elongated

†The ATM switch Z now becomes the crossover switch for the handoff.

path due to an inter-switch move is kept as short as possible. However, it might not be the shortest one between the source MS and the destination MS. A path migration could further be applied to migrate an inefficient path to a better one, if any. The path migration schemes proposed in [10] can be regarded as intentional migrations, which are triggered by recognizing an inefficient path. Basically, these migration schemes can be applied to an ATM-based wireless network. However, we still have to take care the problems of when to find a feasible better path and make a migration. In our design, we suggest that the path migration, if necessary, be done once an inter-switch handoff occurs. This is because based on the proposed handoff scheme, we don't have to take care of the rerouting problem for intra-switch handoffs. When an inter-switch handoff occurs, the BS of the new cell first establishes a VCC to the remote destined BS by executing some routing scheme, e.g., the shortest path scheme. Then makes a migration to the new path. The flush message protocol [1] can be used to switch the path from the old one to the new one for keeping the frame order.

To demonstrate the procedure of a path migration, let us consider the example shown in Fig. 9. On finishing an inter-switch handoff, the path migration mechanism could be initialized, if necessary, and the steps are stated as follows.

1. BS_2 establishes a virtual connection to BS_4 for each of the MS-associated VCCs (80/350, 80/450) using the ATM connection setup procedure. BS_2 and BS_4 (far-end BS's identity recorded in the profile of MS-associated connections) are used as the "calling party" and the "called party" in the connection setup message, respectively. Now there exist two paths between BS_2 and BS_4. The first one is that produced during the inter-switch handoff described before. The other is the new established path.

2. BS_2 initiates the flush message protocol for switching the path from the old one to the new one, if necessary. In this example, since the two paths have the same hop counts, the migration can either be done or just ignored.

5. Analysis and Implementation

5.1 Handoff Latency, Throughput and Bandwidth Waste Analyzes

In this subsection, we provide an analysis of our scheme, by comparing it with others. This analysis assumes perfect delivery of control messages. The processing time for control messages is the same in an MS, BS or ATM switch. Our analysis using hints (advance warning of a handoff) assumes the suboptimal case, where all partial connections have been established by the time the MS enters the new cell, but the wireless channel is just being set up. Below is a given set of network parameters and their definitions partially adopt from [12] that characterize the network environment shown in Figs. 4, 6, 7 and 9.

BW_{wl}	bandwidth of the wireless link	1 Mb/s
BW_w	bandwidth of the wireline (ATM) network	600 Mb/s
t_{wl}	propagation time of the wireless link	7 ms
t_w	propagation time of the wireline link	0.5 ms
t_{ctrl}	protocol processing time for control messages	3 ms
t_{acq}	time for an MS to acquire a wireless channel	20 ms
t_{infra}	cell delivery time between two MSs through the ATM infrastructure network	300 ms
S_{ctrl}	maximum size of a control message	53 bytes (an ATM cell)
S_{data}	maximum size of a data frame	2312 bytes (maximum frame size of IEEE 802.11 wireless LAN)
$L_{old,intra}$	number of links from the crossover switch to the old BS for an intra-switch move	1
$L_{old,inter}$	number of links from the crossover switch to the old BS for an inter-switch move	2 (It means the loop condition always occurs once for an inter-switch move.)
$L_{old,intra}$	number of links from the crossover switch to the new BS for an intra-switch move	1
$L_{old,inter}$	number of links from the crossover switch to the new BS for an inter-switch move	4
$L_{BS,intra}$	number of links of the partial connection connecting the old and new BSs for an intra-switch move	2
$L_{BS,inter}$	number of links of the partial connection connecting the old and new BSs for an inter-switch move	4

We now measure the handoff latency and the bandwidth waste caused by path elongation to show the effectiveness of our scheme. Let us compare three rerouting schemes, namely end-to-end (E2E), BS-to-BS elongation (BSE) and ours. The first two schemes are based on the wireline LANE model which specifies only a VCC between LANs, to handle a mobility. The E2E scheme performs an end-to-end handoff procedure. The new BS tells its far-end BS that it is replacing the old BS. The far-end BS uses the standard flush message protocol to flush its data, and then starts sending to the new BS. As to the BSE scheme, it operates on the basis of BS-to-BS elongation to perform the handoff and data rerouting. In the following, we evaluate the proposed handoff scheme first.

When a station moves to a new cell, it identifies a new BS and sends a greeting message to the BS via the wireless channel. This procedure consists of the time to acquire a wireless channel, the transmission time, the propagation time on the wireless link and the protocol processing time on the new BS. Thus, we have

$$T_1 = t_{acq} + (S_{ctrl}/BW_{wl}) + t_{wl} + t_{ctrl}.$$

The greeting acknowledgment message is then sent back to the MS. Immediately after sending the greeting acknowledgment message, the new BS sends a deregistration message to the old BS for deleting the MS from its served list via the BS-to-BS connection between the two adjacent base stations. Its total latency is the end-to-end delay through the BS-to-BS connection plus the processing time in the old BS. It can be obtained

$$T_2 = \begin{cases} [(S_{ctrl}/BW_w) + t_w] * L_{BS,intra} + t_{ctrl} \\ \qquad \text{for an intra-switch move, or} \\ [(S_{ctrl}/BW_w) + t_w] * L_{BS,inter} + t_{ctrl} \\ \qquad \text{for an inter-switch move.} \end{cases}$$

The deregistration acknowledgment message is then sent from the old BS to the new BS. Its total latency is computed similarly to that of the deregistration message. That is,

$$T_3 = T_2 = \begin{cases} [(S_{ctrl}/BW_w) + t_w] * L_{BS,intra} \\ \qquad + t_{ctrl} \\ \qquad \text{for an intra-switch move, or} \\ [(S_{ctrl}/BW_w) + t_w] * L_{BS,inter} \\ \qquad + t_{ctrl} \\ \qquad \text{for an inter-switch move.} \end{cases}$$

In the proposed handoff scheme, the old BS needs to send a migrate control cell for the clear of uplink data. This migrate cell will be returned to the new BS by the crossover switch. The trip time is dependent on the propagation time between the old and new BSs. The total latency needed for the migrate cell is the end-to-end delay through the crossover switch plus the processing time in the new BS. We obtain

$$T_4 = \begin{cases} [(S_{ctrl}/BW_w) + t_w] * (L_{old,intra} \\ \qquad + L_{new,intra}) + t_{ctrl} \\ \qquad \text{for an intra-switch move, or} \\ [(S_{ctrl}/BW_w) + t_w] * (L_{old,inter} \\ \qquad + L_{new,inter}) + t_{ctrl} \\ \qquad \text{for an inter-switch move.} \end{cases}$$

Also, the old BS has to redirect the received data to the new BS. The delay incurred by the data redirecting is the end-to-end delay through the BS-to-BS connection between two BSs. It is given by

$$T_5 = \begin{cases} [(S_{data}/BW_w) + t_w] * L_{BS,intra} \\ \qquad \text{for an intra-switch move, or} \\ [(S_{data}/BW_w) + t_w] * L_{BS,inter} \\ \qquad \text{for an inter-switch move.} \end{cases}$$

Let $T_{handoff}$ denote handoff latency. If the MS is transmitting frames when it crosses cell boundary, then $T_{handoff}$ is measured from the moment the MS greets the base station in the new cell till the transmission is resumed. In the step 5 of the intra-switch handoff, we know the deregistration acknowledgment message and the migrate control cell are initiated at the same time (when an MS moves completely into a new cell). When the MS is transmitting frames, the handoff latency for the proposed scheme (ours) can be estimated as

$$T_{handoff\_ours(tx)} = T_1 + T_2 + max\{T_3, T_4\}.$$

If the MS is receiving frames when it crosses cell boundary, then $T_{handoff}$ is measured from the moment the MS greets the base station in the new cell till it can begin to receive. We assume that the first redirected data frame from the old BS to the new BS immediately follows the deregistration acknowledgment message. It is given by

$$T_{handoff\_ours(rx)} = T_1 + T_2 + (S_{ctrl}/BW_w) + T_5.$$

Let T_{elong} be the elongated path delay due to handoff. For an intra-switch move, the proposed handoff scheme does not elongate the path, i.e., $T_{elong\_ours} = 0$. However, the path will be elongated from the crossover switch to the new BS for an inter-switch move. Thus, we have

$$T_{elong\_ours} \\ = max\{(L_{new,inter} - L_{old,inter}), 0\} * t_w.$$

Following the similar computations, the handoff latency and the elongated path delay for the schemes E2E and BSE are obtained below.

$$T_{handoff\_E2E} = max\{T_1 + T_2 + T_3, 3 * t_{infra}\},$$
$$T_{handoff\_BSE} = T_1 + T_2 + T_3,$$

and

$$T_{elong\_E2E} = 0,$$

$$T_{elong\_BSE} = \begin{cases} 2 * (S_{data}/BW_w) + L_{BS,intra} * t_w \\ \qquad \text{for an intra-switch move, or} \\ 2 * (S_{data}/BW_w) + L_{BS,inter} * t_w \\ \qquad \text{for an inter-switch move.} \end{cases}$$

As noted in Sect. 2, the E2E scheme needs three trips over the ATM infrastructure network for a handoff. In the BSE scheme, the old BS first reassembles the cells and then segments the frame into cells and forwards the cells into the BS-to-BS connection. Thus the term $2 * (S_{data}/BW_w)$ means the sum of the cells' reception and transmission times in the old BS.

We have derived formula for the performance metrics for E2E, BSE and ours schemes with the availability of hints regarding advance warning of a handoff. It is worth to extend our analysis to the case without hints. For the schemes without using cell overlap hints, the handoff latency ($T_{handoff}$) derived above should add two additional items: one is the roaming time between two adjacent non-overlapping cells, the other is the setup time for partial connection(s). The former parameter mainly depends on the speed of a moving user and the size of non-overlapping area. As for the latter one, it is subject to the connection admission control complexity and the network size. For example, when user's moving speed is in the tens of meters per minute, it will then dominate the handoff latency. To the best of our knowledge, there are no well-justified methods to measure them. We thus do not give any numerical examples for the case without hints in the following analysis.

Identified in Table 1 are the performance results of the handoff latency and elongated path delay of the E2E, BSE and proposed handoff schemes. The values displayed here assume that the moves of a station will not return back to the cell it has been visited. We can find that the wireless channel acquiring time dominates the handoff latency for both the BSE and proposed schemes. However, the trip time over the ATM network contributes significantly for the E2E scheme.

We now investigate the effect of handoff latency and elongated path delay on throughput, respectively. In [8], Caceres and Iftode showed that long handoff latency will reduce throughput compared to that of no handoffs. This is because some reliable transport protocols like TCP interpret the handoff latency (service disruption) as the sign of network congestion and consequently throttle their transmission rate. Their experiment measured the average application-level throughput by transferring four Mbytes of data between a mobile and a fixed station. The mobile station switches to another cell every eight seconds. As a result, multiple handoffs occurs during this data transfer. The throughput performance under different motion scenarios are reported. They indicated that for the cases where a mobile station switches to a new cell but cannot communicate with the new BS for 50 milliseconds (handoff latency), then the throughput drops by 12%. For the cases where the handoff latency reaches one second, the throughput drops by 31%. These cases correspond to 50-millisecond and 1-second handoff latencies respectively. From these we can see that the impact of handoff latency on the throughput is significant.

Table 1 Comparison ($T_{handoff}$ and T_{elong}) of different handoff schemes.

handoff schemes / metrics		 E2E	 BSE	 ours
$T_{handoff}$	*intra-switch move*	900 ms ($3^* t_{infra}$)	38.42 ms ($T_1+T_2+T_3$)	38.42 ms ($T_{handoff\_ours(rx)}$)
				35.48 ms ($T_{handoff\_ours(tx)}$)
	inter-switch move	900 ms ($3^* t_{infra}$)	40.42 ms ($T_1+T_2+T_3$)	41.42 ms ($T_{handoff\_ours(tx)}$)
				37.54 ms ($T_{handoff\_ours(rx)}$)
T_{elong} (*after 100 intra-switch moves*)		0 ms	106 ms	0 ms
T_{elong} (*100 intra-switch moves then another 100 inter-switch moves*)		0 ms	312 ms	100 ms

Without loss of preciseness, the above experiment results can be directly applied to measure the throughput degradation listed in Table 1. In the worst case, the handoff latencies of the proposed, BSE and E2E schemes are 41.42 milliseconds (about 50 milliseconds), 40.42 milliseconds (about 50 milliseconds) and 900 milliseconds (about one second), respectively. We then can conclude that both the proposed and BSE handoff schemes have a better throughput with about 19% (31 minus 12) improvement than the E2E scheme. In [13], the authors demonstrated that the larger elongated path delay, the higher bandwidth-delay product produces. They also showed that the higher bandwidth-delay product, the less throughput obtains. Based on this fact, we can see that the proposed scheme furnishes a better throughput than the BSE scheme. The elongation impact on throughput of the BSE scheme will be worse than the proposed one due to elongated path delays of the proposed one and the BSE scheme are 100 and 312 milliseconds, respectively.

Next we study the bandwidth waste caused by path elongation. As before, the propagation time of a wireline link is 500 microseconds (t_w). During the time interval, a total amount of 37,500-byte ($t_w{}^*BW_w$) data could be sent into the ATM network. The data frame size is measured by the maximum frame of wireless LAN, i.e., 2,312 bytes (S_{data}). Thus sending this data frame will occupy 6% (divide 2,312 by 37,500) of the total bandwidth for a physical link, i.e., 36 Mb/s (multiply 600 Mb/s by 0.06). After 100 intra-switch moves and another 100 inter-switch moves, the elongated paths of the proposed and the BSE handoff schemes take 200 ($100^*(L_{new,inter}$-$L_{old,inter}$)) and 600 ($100^*L_{BS,intra} + 100^*L_{BS,inter}$) links, respectively. The BSE scheme wastes three times network bandwidth of the proposed one. The rescued network bandwidth is up to 14.4 Gb/s (400*36 Mb/s). For example, if a mobile connection is 50-link long and the throughput on each link equals to the wireless link capacity, i.e., 1 Mb/s. Thus, each mobile connection consumes a total bandwidth of 50 Mb/s. Then, surprisingly the rescued network bandwidth (14.4 Gb/s) can furnish up to 288 mobile connections (divide 14.4 Gb/s by 50 Mb/s) as described above.

In summary, from the performance viewpoint, maintaining a VCC for each pair of mobile stations rather than each pair of LANs is justifiable in a mobile environment. However, this takes more network resources and protocol overheads such as a larger routing table in a BS or an ATM switch and the increased efforts to establish and manage all of these additional VCCs.

5.2 Implementation Issues

From now, we will discuss some implementation issues of the proposed handoff scheme. In particular, we will look at how to merge the new established partial connection into an existing connection and buffering requirements while handoffs occur.

5.2.1 Merging the Partial Connection into an Existing Connection

The proposed scheme needs to establish partial connection (elongated path) between a switch and the new BS for a handoff. As a switch receives the Reroute Message, it sends a connection setup message, in which the calling party and called party are the switch and the new BS, respectively. The partial connection is established accordingly. Now the mobile connection after handoff includes two separate parts: one is the reuse portion and the other is the new established (partial) portion. The switch is responsible to make them as one. Thus the switch not only modifies the routing database for connecting them, but also merges the profiles of the two connections into one. Before merging, the two ends of the old mobile connection are the far-end BS and the old BS; and as to that of the partial connection are the switch and the new BS. After merging, the two ends of the new formed mobile connection are the far-end BS and the new BS. Such a connection merging operation is transparent to end users but may cause extra management efforts to an existing switch. Following the same idea, the new BS also needs to do some arrangement to make the partial connection as the part of the end-to-end connection.

5.2.2 Buffering Requirements due to Handoffs

When an MS moves into a new cell, the old BS will redirect the received data frames to the new BS through the BS-to-BS connection. For in-order delivery, all the cells received by the new BS over the new connection are buffered, and not forwarded to the destined MS, until the redirected data frames from the old BS are clear. As described, we can find that during a handoff buffering is only required at BSs, the switch does not need to buffer any data cell.

Let DB_{BSold} denote the amount of buffering required per connection on the old BS for downlink buffering which is determined by the amount of data destined to a moving MS through the BS-to-BS connection. This data is that which has been received while the new BS is arranging for data forwarding[†] from the old BS and waiting for the corresponding acknowledgment. We assume that the first forwarded data frame immediately follows the acknowledgment of the deregistration message. We obtain,

$$DB_{BSold} = BW_{wl} * [T_1 + T_2 + (S_{ctrl}/BW_w)].$$

Let DB_{BSnew} denote the amount of buffering required per connection on the new BS for downlink buffering which is determined by the difference of the delay between two paths, one path is from the crossover switch to the old BS and then to the new BS, the other is from the crossover switch to the new BS. Thus,

$$DB_{BSnew} = \begin{cases} BW_{wl} * [(S_{data}/BW_w) + t_w] * \\ max\{(L_{BS,intra} + L_{old,intra} \\ \quad - L_{new,intra}), 0\} \\ \qquad \text{for an intra-switch move, or} \\ BW_{wl} * [(S_{data}/BW_w) + t_w] * \\ max\{(L_{BS,inter} + L_{old,inter} \\ \quad - L_{new,inter}), 0\} \\ \qquad \text{for an inter-switch move.} \end{cases}$$

Also, the new BS has to buffer the uplink data frames until the $migrate_{up}$ control cell received for in-order delivery. Let UB_{BSnew} denote the amount of buffering required on the new BS for uplink buffering which is given by

$$UB_{BSnew} = BW_{wl} * (T_1 + T_2 + T_4).$$

Using the network parameters given at the beginning of this section, we can obtain the buffering required for an inter-switch move (the worst case we only need to consider) at base stations. The buffering required at the old BS for downlink data (DB_{BSold}) is about 4427.5 bytes per connection. On the other hand, the total buffering required at the new BS for downlink and uplink data ($DB_{BSnew} + UB_{BSnew}$) is about 5302.5 bytes (125 + 5177.5) per connection. For a high-speed internetworking device, the amount of BS buffering required to prevent data loss and to support in-order frame delivery during a handoff is feasible.

6. Related Works

There have been various proposals to support connection rerouting for handoffs in an ATM-based wireless

[†]The arrangement work consists of two parts: receiving a greeting message from the moving MS and sending a deregistration message to the old BS.

network [14], [16], [17]. In [17], the rerouting scheme is based on forwarding of cells after a successful handoff. The ATM switch associated with the old base station acts like a crossover (forwarding switch) for the handoff connection and forwards data cells to the new base station. After the handoff is completed, the crossover ATM switch migrates the user connection to the partial connection (as illustrated before) established due to a handoff. This means that either the crossover ATM switch or the new ATM switch might *need* to buffer some of the cells before the handoff is successfully completed. Following the similar idea, in [16] the initial (home) base station is used as a crossover for the rerouting. This paper proposes a two-stage rerouting scheme. At the time of handoff (stage 1), all the ATM cells are redirected to the new base station along the prior-established (BS-to-BS) connection. Once the handoff is completed, the ATM cells can immediately forwarded to the mobile station. The handoff latency, obviously, appears *much* smaller than that of cell forwarding after a complete handoff. Following the time of handoff later (stage 2), cell in-order delivery and loop removal procedures are performed. Note that in the proposed rerouting mechanism buffering is only required at base stations, requiring no intervention from ATM switch.

The case for locating a suitable crossover switch (rather than the intuitive selections described above) has been advocated by some researchers, for example, Toh [14]. The crossover switch is held responsible for performing connection rerouting during handoffs. Therefore, buffering some of the cells is also required at the crossover ATM switch. It is interesting to note that the general case with route changes over multiple switches is more complex and will require careful protocol software and switch hardware design to avoid handoff-induced cell losses. Our ARSE-based rerouting mechanism attempts to reuse as much of an existing connection as possible, thereby reducing the need for a complex protocol to handle the change of route in a wider network area.

7. Conclusions

In this paper, the main issues of wireless LAN emulation over ATM networks have been addressed and investigated. We showed that the (wireline) LANE service, in which only a VCC is established between each pair of LANs, is not efficient enough to handle the handoff of the wireless LANs due to the fact that the VCC might be elongated when mobile stations move in and out. An elongated path (VCC) not only introduces additional transmission delay but also consumes bandwidth. To overcome this problem, we suggest that a VCC should be established between each pair of communicating mobile stations instead of each pair of LANs as defined in the ATM Forum. Based on this suggestion, an extensive cellular handoff scheme is proposed which not only meets the requirements of data continuity and transparency, but also provides shorter handoff latency and keeps the communication path as short as possible. A path migration scheme is also suggested to further migrate an inefficient elongated path to a better one (due to the roaming of mobile stations), if any. The results from the analytical studies indicate that the proposed solution is suitable to be used in a mobile environment in terms of throughput, transmission delay and wasted network bandwidth. The proposal brings one step closer towards facilitating the mobility management in an ATM network while conforming to the emerging LAN emulation and signaling standards. However, it adds functionality, such as connection merging and data rerouting, to ATM switch to provide station mobility.

References

[1] ATM Forum, "LAN Emulation over ATM Version 2-LUNI Specification Draft 4," Sept. 1996.

[2] ATM Forum, "ATM User-Network Interface Specification Ver. 3.1," Sept. 1994.

[3] ATM Forum, "P-NNI Draft Specification," Contribution 94-0471 (latest version), March 1995.

[4] CCITT, I-Series Recommendation on B-ISDN, Geneva, 1991.

[5] ETSI TC-RES, "Radio Equipment and Systems (RES); High Performance Radio Local Area Network (HIPERLAN); Functional Specification," ETSI, 06921 Sophia Antipolis Cedex, France, draft prETS 300 652, July 1995.

[6] IEEE Standards Dept., Draft Std. IEEE 802.11, "Wireless LANs (P802.11D3)," Jan. 1996.

[7] I.F. Akyildiz and K.L. Bernhardt, "ATM local area network: A survey of requirements, architecture, and standards," IEEE Commun. Mag., vol.35, no.7, pp.72–80, July 1997.

[8] R. Caceres and L. Iftode, "Improving the performance of reliable transport protocols in mobile computing environments," IEEE J. Select. Areas Commun., vol.13, no.5, pp.850–857, June 1995.

[9] N. Finn and T. Mason, "ATM LAN emulation," IEEE Commun. Mag., vol.34, no.6, pp.96–100, June 1996.

[10] N.F. Huang and K.S. Chen, "A distributed paths migration scheme for IEEE 802.6 based personal communication networks," IEEE J. Select. Areas Commun., vol.12, no.8, pp.1415–1425, Oct. 1994.

[11] N.F. Huang, Y.T. Wang, B. Li, and T.L. Liu, "Mobility management of interconnected virtual LANs over ATM networks," Proc. IEEE GLOBECOM, pp.1156–1161, London, Nov. 1996.

[12] K. Keeton, B.A. Mah, S. Seshan, R.H. Katz, and D. Ferrari, "Providing connection-oriented network services to mobile hosts," Proc. of USENIX Symp. on Mobile and Location-Independent Computing, Cambridge, Massachusetts, Aug. 1993.

[13] P. Manzoni, D. Ghosal, and G. Serazzi, "Impact of mobility on TCP/IP: An integrated performance study," IEEE J. Select. Areas Commun., vol.13, no.5, pp.858–867, June 1995.

[14] C.K. Toh, "A Unifying methodology for handovers of heterogeneous connections in wireless ATM networks," ACM Comp. Commun. Rev., vol.27, no.1, pp.12–30, Jan. 1997.

[15] H.L. Truong, W.W. Ellington, J-Y Le Boudec, A.X. Meier, and J.W. Pace, "LAN emulation on an ATM network,"

IEEE Commun. Mag., vol.33, no.5, pp.70–85, May 1995.

[16] M. Veeraraghavan, M.J. Karol, and K.Y. Eng, "Mobility and connection management in a wireless ATM LAN," IEEE J. Select. Areas Commun., vol.15, no.1, pp.50–68, Jan. 1997.

[17] R. Yuan, S.K. Biswas, and D. Raychaudhuri, "Mobility support in a wireless ATM network," Proc. 5th WINLAB Workshop on Third Generation Wireless Info. Networks, pp.335–345, Kluwer, April 1995.

High-Performance Backbone Network Technology edited by Naoaki Yamanaka
ISBN: 0-8247-5321-6 DOI: 10.1081/(120025689)
Published 2004 by Marcel Dekker, Inc.

Simulation and Measurement of TCP/IP over ATM Wide Area Networks

Georgios Y. Lazarou, Victor S. Frost, Joseph B. Evans, and Douglas Niehaus

SUMMARY Predicting the performance of high speed wide area ATM networks (WANs) is a difficult task. Evaluating the performance of these systems by means of mathematical models is not yet feasible. As a result, the creation of simulation models is usually the only means of predicting and evaluating the performance of such systems. In this paper, we use measurements to validate simulation models of TCP/IP over high speed ATM wide area networks. Validation of simulations with measurements is not common; however, it is needed so that simulation models can be used with confidence to accurately characterize the performance of ATM WANs. In addition, the appropriate level of complexity of the simulation models needs to be determined. The results show that under appropriate conditions simulation models can accurately predict the performance of complex high speed ATM wide area networks. This work also shows that the user perceived performance is dependent on host processing demands.

key words: *performance prediction, validation of simulation models, model complexity, TCP/IP over ATM*

1. Introduction

Predicting the performance (that is, throughput) of high speed wide area ATM networks (WANs) is a difficult task. Simulation models enable the systematic performance evaluation of these systems when mathematical models are not available and experiments on the actual systems are impossible or impractical [8]. However, in most cases a very high level of model complexity is required for accurate performance prediction.

In this paper, the level of model complexity required to accurately predict measured performance is determined by using measurements from high speed wide area ATM networks to validate simulation results. We present validation results of simulations of TCP/IP over the MAGIC gigabit testbed and the ACTS ATM Internetwork (AAI) wide area research network. The experimental results presented here also show that today's workstations can fully utilize OC-3 links. These simulation results indicate that network congestion effects can be accurately modeled, under these conditions simulation models can accurately predict the performance of complex high speed ATM wide area networks. However, we observed large discrepancies between measurements and simulations when hosts were required to process multiple TCP/IP streams.

The rest of this paper is organized as follows: Sect. 2 provides an overview of TCP/IP over ATM; Sect. 3 presents background information on the simulation software environment, the TCP model, and the MAGIC and AAI networks; Sect. 4 describes the validation of simulation with experimental results; and Sect. 5 discusses lessons learned.

2. Overview of TCP/IP over ATM

2.1 Importance of TCP/IP over ATM

Probably the most successful idea in data networking over the past twenty years has been the concept of internetworking [15]. It is a method for interconnecting networks, regardless of the particular networking technology (Ethernet, ATM, HIPPI, Frame Relay, FDDI), used by the individual systems. What makes internetworking possible is the development of protocols like TCP/IP. The TCP/IP protocol suite is the internetworking protocol used on the Internet, a global collection of networks connecting millions of computers and users, and incorporating a large variety of different network technologies [15]. It allows computers of all sizes, from many different computer vendors, running totally different operating systems, to communicate with each other [19].

ATM technology is the emerging standard adopted by telecommunications and computer vendors for high speed networks. The fast-cell switching technology employed by ATM helps to provide scalable (in size and speed) networks [5]. Further, ATM technology provides bandwidth on demand and enables the integration of real-time and data traffic over the same physical medium for wide area networks.

Although TCP/IP and ATM often have been viewed as competitors, their complementary strengths and limitations form a natural alliance that combines

Manuscript received March 28, 1997.
Manuscript revised August 22, 1997.

[†]The authors is with the Information & Telecommunication Technology Center at the University of Kansas, 2291 Irving Hill Road, Lawrence, KS 66045, U.S.A.

[††]The authors are with the Faculty of Electrical Engineering and Computer Science Department, the University of Kansas, Lawrence, KS 66045, U.S.A.

the best aspects of both technologies [16]. In the near future, a large portion of the traffic carried by the ATM networks will be generated by applications written to run over a TCP/IP protocol stack [13]. In fact, many of the existing ATM networks employ TCP/IP over ATM technology.

2.2 Previous Studies of TCP/IP over ATM

Numerous simulation and experimental studies have been performed in order to predict the performance of TCP/IP over ATM under congestion and buffer overflow conditions that arise from bandwidth mismatches or multiple sources contending for the same link [2], [5], [9], [13], [17], [18], [21]. The simulation study in [18] examines the performance of TCP over ATM under conditions of network congestion. The results show that the TCP/IP over ATM performance is poor when there is congestion caused by small switch buffers and large TCP segment and window sizes. This performance degradation is caused by a loss-rate multiplier effect caused by the switch dropping cells from multiple packets. In [17], the performance of TCP connections over ATM networks without ATM-level congestion control is investigated. Simulation results of congested networks show that the effective throughput of TCP over ATM can be quite low when cells are dropped at the congested ATM switch. To improve the performance, a mechanism called early packet discard (EPD) which brings throughput performance to its optimal level is proposed.

The work in [13] considers some undesirable interactions between the congestion control scheme used in TCP and the policing mechanisms used in ATM networks that can significantly degrade the throughput of TCP traffic. It is shown that in the presence of policing, once a TCP connection has increased its window size beyond the sustainable cell rate (SCR) times the round–trip time and if the bottleneck capacity exceeds the SCR, the buffer at the site providing the policing mechanism fills up quickly and most of the packets in the TCP window are dropped. This causes the average throughput to be significantly lower than SCR value. In order to improve the performance, the use of smarter policing or cell-level traffic shaping schemes is suggested.

The work in [2], [14] describes performance measurements taken from the MAGIC gigabit testbed relating to the performance of TCP in wide area ATM networks. Results show that the TCP rate control mechanism alone is inadequate for congestion avoidance and control in wide area gigabit networks. It is also illustrated that TCP augmented by cell-level pacing allows the full link capacity to be utilized. Cell level pacing is necessary because the TCP rate control mechanism does not control traffic burstiness sufficiently to avoid congestion-induced cell losses in wide area networks.

These studies separately present simulation and measurement results. However, comparison of performance predictions from measurement techniques and simulation models of TCP/IP-ATM networks is needed to refine our understanding of the network operation. Validation of models against measurements is also required so simulation models can be used with confidence to capture the effects of network control, and to accurately characterize the performance of ATM WANs. However, in order to achieve that goal, the complexity of the models (which is directly proportional to the model execution time) needs to be determined because of the computational demands of simulation.

3. Background

This section provides some background information on the simulation software environment, the TCP BONeS model, and the MAGIC and AAI networks.

3.1 The Simulation Software Environment

The simulation models in this study are implemented using BONeS DESIGNER, a commercial software package for modeling and simulating event-driven systems [20]. Other network simulation software packages are available. One of these is the Lawrence Berkeley National Laboratory (LBNL) Network Simulator, an event-driven simulator embedded into Tcl with an extensible simulation engine implemented in C++ [10]. However, BONeS provides the user with a rich array of library functions, making the task of building our models less complex. A system model can be constructed hierarchically and graphically using building blocks from the BONeS model library, or using models written in C or C++. The relative merits of commercial general simulation systems with respect to special purpose simulations can be found in [6].

3.2 TCP BONeS Primitive Module

A TCP BONeS [20] primitive module was created for this study. The source code for this primitive was based on the MIT Network Simulator (NetSim) TCP module [11], [12]. However, using NetSim module we were unable to match measurement with simulation results in the presence of network congestion because specific TCP timer mechanisms were not modeled. The NetSim TCP module is partially based on the Berkeley Standard Distribution (BSD) 4.3 Tahoe version. All TCP implementations that are 4.3 BSD based include two timer functions: one is called every 200 ms (the fast timer) and the other every 500 ms (the slow timer) [19]. The fast timer is used with the delayed ACK timer and the slow timer is mainly used with the retransmission timer. The NetSim TCP model did not include these timers and thus can not accurately predict the performance of TCP over ATM under congestion.

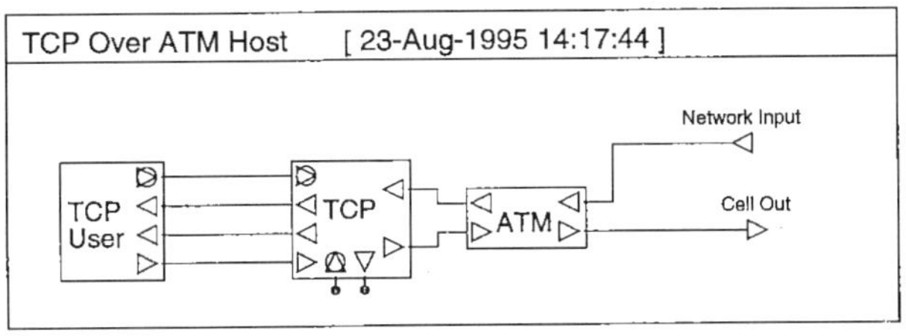

Fig. 1 TCP over ATM BONeS host model.

The modified TCP model developed here is based on the 4.3 BSD Reno version. It supports the following major mechanisms: fast and slow timers, slow start and congestion avoidance, fast retransmit/recovery, and window advertisement. The BONeS model of a TCP over ATM host is shown in Fig. 1. The TCP User block generates the data buffers for transmission and the ATM block accepts TCP packets and generates ATM cells. It also accepts cells from the network and reconstructs the TCP packets.

From Fig. 1 we see that an IP module is not used. IP is the main protocol at the network layer of the TCP/IP protocol suite. It provides an unreliable, connectionless datagram delivery service between hosts attached to an TCP/IP internetwork. In this study, the connectionless IP datagrams are carried only on connection-oriented ATM networks. Hence, the IP routing functionality is redundant in our simulation models.

3.3 MAGIC and AAI Networks

This study focuses on measurements and models of the MAGIC and AAI networks. The Multidimensional Applications and Gigabit Internetwork Consortium (MAGIC) is a group of industrial, academic, and government organizations participating in gigabit networking research. The MAGIC backbone network (see Fig. 2) operates at 2.4 Gb/s and each site on the network includes LANs or hosts communicating at gigabit per second rates [7]. The ACTS ATM Internetwork (AAI) provides wide area Asynchronous Transfer Mode (ATM) connectivity. It connects several DoD High Performance Computing centers and the MAGIC and ATDnet gigabit testbeds. The ATM service is provided by Sprint. The AAI network is depicted in Fig. 3. At the time of these experiments, site access was at DS-3 rates. These links have subsequently been updated to OC-3.

4. Using Measurements to Validate Simulation Models

This section presents the model validation results of TCP over ATM performance for three different cases: over the MAGIC network and varying the TCP window size, over the MAGIC network and under rate mismatch conditions, and over the AAI network. In addition, this section provides information about the simulation system parameters and network traffic.

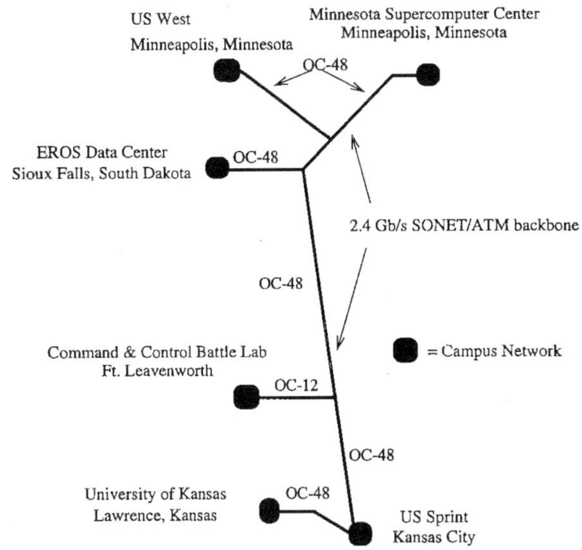

Fig. 2 MAGIC Network.

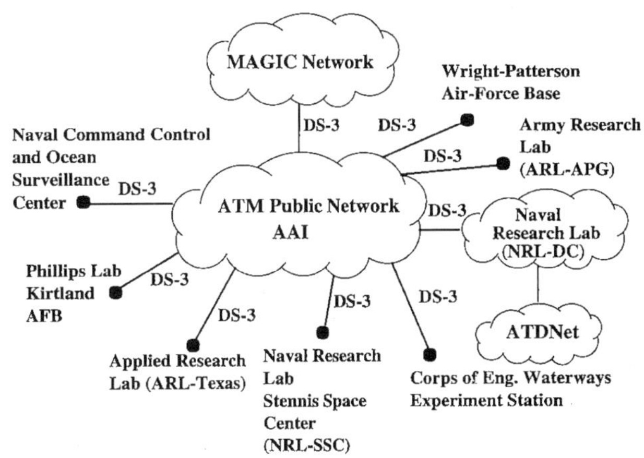

Fig. 3 AAI Network.

4.1 System Parameters

Beside the level of model complexity, the simulation system parameters need to be correctly specified. The system parameters that are used in our simulation systems are listed in Table 1. The maximum size of the TCP segment is specified by the TCP MTU (maximum transmission unit) size parameter. TCP Processing and OS Overhead Time is the overall time needed by the TCP software to create a segment for transmission or process an incoming segment and for the operating system to handle all system calls and I/O operations during transmission or reception of a TCP segment. In Fig. 1, the TCP User module sends data buffers to the TCP module for transmission. The size of these data buffers is designated by the TCP User Send Buffer Size pa-

Table 1 Run-Time Simulation Parameters.

System Parameter	Value
TCP MTU Size	9180 Bytes
TCP Processing & OS Overhead Time	
DEC 3000 AXP	200–300 μs
SGI & SPARC 10	550 μs
SPARC 5	700 μs
TCP User Send Buffer Size	64 KBytes
Slow-Timer Period	0.5 s
Fast-Timer Period	0.2 s
Minimum RTO	1.0 s
AAL5 SAR Processing Time	0.2 μs
AAL5 Cell Payload Size	48 Bytes
Switch Processing Time	4 μs
Switch Output Buffer Size per VC	256 Cells
OC–3c Link Speed	155 Mb/s
TAXI Link Speed	100 Mb/s
DS–3 Link Speed	45 Mb/s

Table 2 Maximum Throughput for Different Link Speeds.

Link	Max SONET Layer Throughput	Max TCP Layer Throughput
OC-3	149 Mb/s	135 Mb/s
100 Mb/s TAXI	98 Mb/s	88 Mb/s
DS-3	40.7 Mb/s	36 Mb/s

rameter. The retransmission timer is decremented every 0.5 seconds (Slow-Timer Period), and only when the timer reaches 0 a retransmission is performed. A delayed ACK is sent every time the 0.2 second delayed ACK timer (Fast-Timer) expires. The retransmission timer is bounded by TCP to be between 1 (Minimum Retransmission Time-Out) and 64 seconds [19].

For this study, the host interfaces use the ATM adaptation layer AAL5 for mapping IP datagrams to ATM cell streams. The AAL5 Segmentation and Reassembly Processing Time is the time required for the AAL5 SAR sublayer to map the IP datagrams to ATM cell streams or to reconstruct cell streams to IP datagrams. The AAL5 cell payload is 48 bytes long. The values of the TCP Processing and OS Overhead Time, AAL5 Segmentation and Reassembly Processing Time, and Switch Processing Time parameters are based on experimental measurements [1], [3]. For reference, Table 2 lists the maximum achievable physical and TCP level throughput for SONET links after excluding the SONET, ATM, and IP overhead.

4.2 Network Traffic Characterization

To measure the maximum end-to-end throughput at the TCP layer, a public domain software tool, *ttcp*, was used in all experiments. This tool transfers TCP packets from local memory to memory on a remote host as fast as the operating system, interfaces, and network allow. In each of the experiments performed for this study, *ttcp* used 64 KBytes user data buffers. In order to match the *ttcp* traffic, the data send buffer at the transmitting host always was kept full in all the simulations presented here.

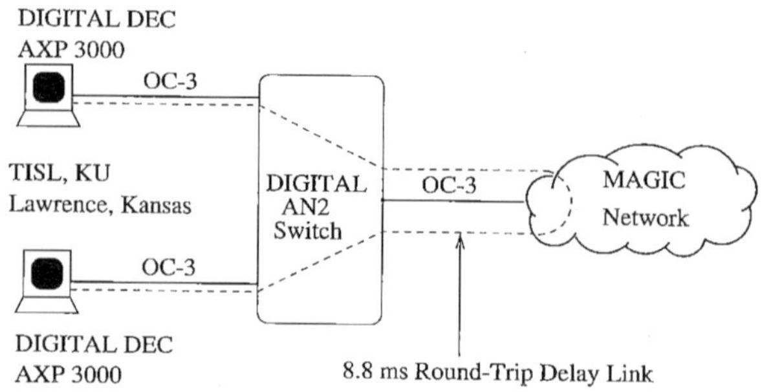

Fig. 4 Experimental Set Up for ATM WAN Performance vs TCP Window Size.

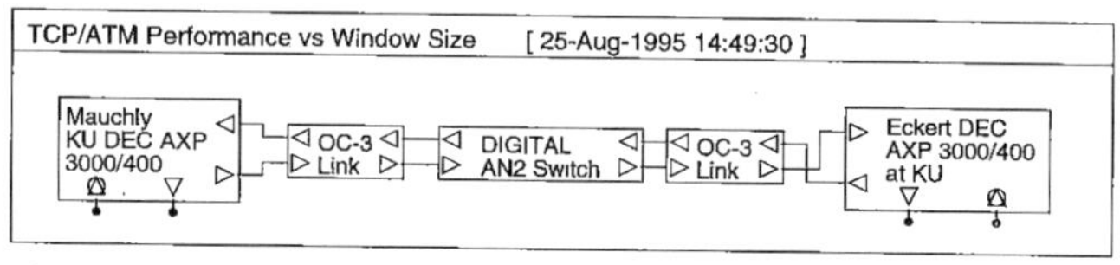

Fig. 5 Simulation Model for ATM WAN Performance vs TCP Window Size.

4.3 Validating ATM WAN vs TCP Window Size Performance

An experiment was carried out by Ewy and Evans [2] using the MAGIC testbed in order to study the ATM WAN performance versus TCP window size. The experimental set up is shown in Fig. 4. A Digital DEC 3000 AXP with an OTTO OC-3 interface transmits to another DEC 3000 AXP over a SONET OC-3 link with 8.8 round-trip delay. A simulation for this experiment is created using our BONeS TCP/ATM host model. Figure 5 shows the top level of the simulation model. The BONeS model for the two DEC AXP 3000 blocks uses the TCP over ATM host model shown in Fig. 1. The two OC-3 links are modeled by delay blocks, while the DIGITAL AN2 Switch is modeled by a simple FIFO queue with a server. The simulation and experimental results are shown in Fig. 6. These results match our expectations given the bandwidth-delay product of this link. The results of this experiment clearly show that today's workstations can fully utilize an OC-3c link.

4.4 Validating TCP/ATM Performance under Rate Mismatch Conditions

Another experiment was conducted by Ewy and Evans [2] in order to study the performance of TCP over ATM under the case of rate mismatch. The configuration for this experiment is shown in Fig. 7. A single host transmits to another host with a 155 Mb/s to 100 Mb/s bandwidth constriction in the path. The simulation system used for this experiment is shown in Fig. 8. The TCP window size is set to 128 KB and the

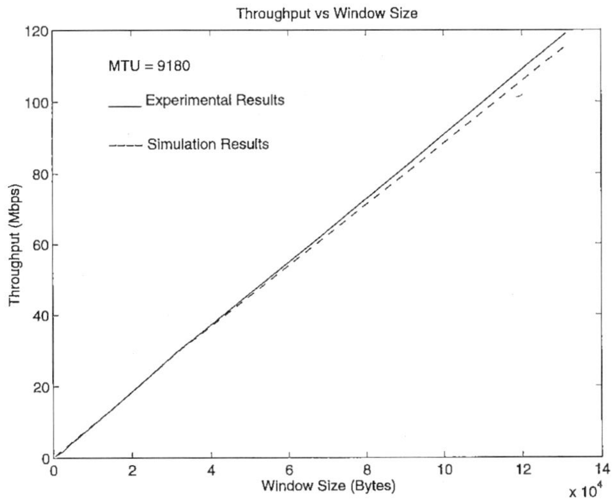

Fig. 6 ATM WAN Performance vs TCP Window Size.

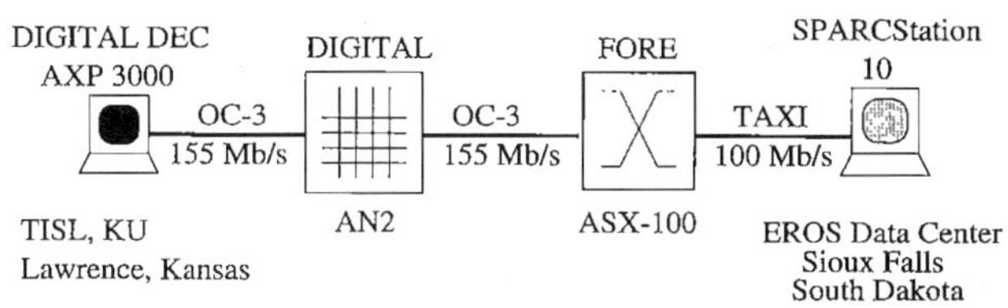

Fig. 7 Experimental Set Up for TCP/ATM Performance under Rate Mismatch Conditions.

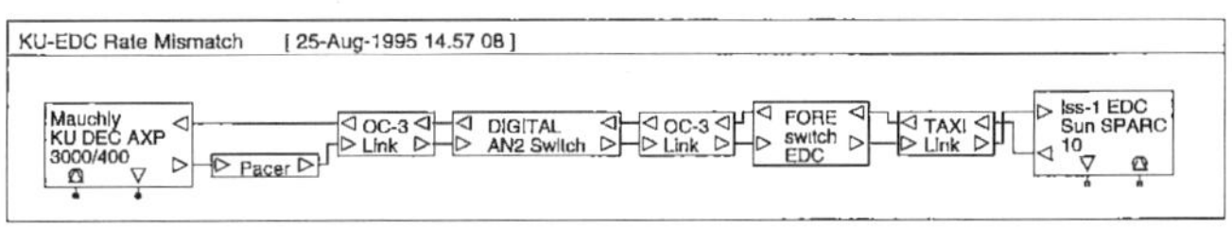

Fig. 8 Simulation Model for TCP/ATM Performance under Rate Mismatch Conditions.

Table 3 Throughput of TCP over ATM with Rate Mismatch.

	No Pacing	Pacing
Experimental Results	1.26 Mb/s	68.20 Mb/s
Simulation Results	1.22 Mb/s	68.59 Mb/s

cell-level pacing is fixed at a rate of 70 Mb/s. Cell-level pacing is the mechanism which reduces the source cell transmission rate. The Pacer functionality is modeled by an infinite size FIFO queue and a server with service rate equal to the pacing rate. The results are shown in Table 3. This confirms that network congestion effects can be accurately modeled.

4.5 Validating TCP/ATM Performance over AAI Network

A simulation model was created for predicting the performance of the AAI Network. The simulation results over a single connection were then validated by measurement. The experimental configuration is shown in Fig. 9. A Digital DEC 3000 AXP with an OTTO OC-3 interface transmits to an SGI Indigo 2 host with a

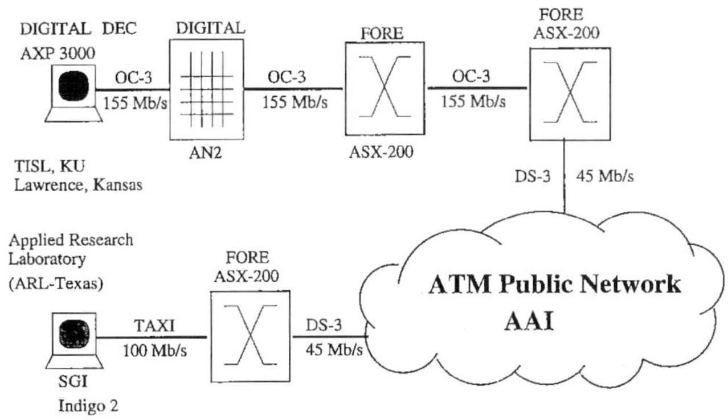

Fig. 9 Experimental Set Up of a Single Connection in the AAI Network.

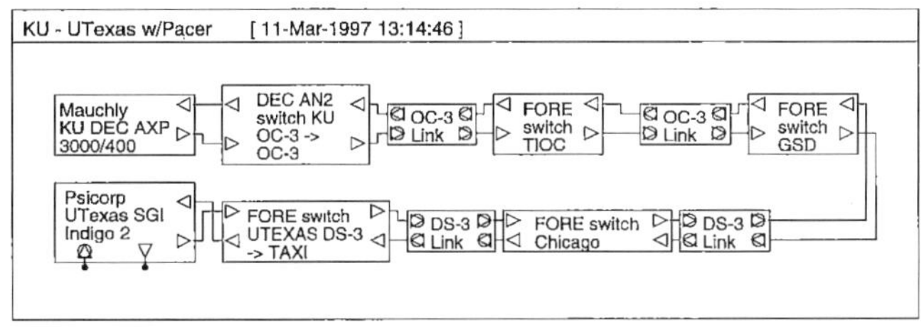

Fig. 10 Simulation Model of a Single Connection in the AAI Network.

Table 4 Throughput of TCP over ATM using AAI Network.

Experimental Results	35.86 Mb/s
Simulation Results	35.97 Mb/s

100 Mb/s TAXI interface. The round-trip time for this connection is about 36 ms. The simulation model for this experiment is shown in Fig. 10. Note in Fig. 9 that the lowest link capacity is DS-3. Hence, a cell-level pacer is used with a rate of 40 Mb/s since the maximum rate of a DS-3 link after excluding the physical layer overhead is 40.7 Mb/s. We used 40 Mb/s at the pacer to avoid any possibility of cell loss due to network congestion. A TCP window size of 256 KB is used. The results are shown in Table 4. Once more, the results show that simulation can accurately predict the network performance of complex high speed ATM wide area networks.

A set of three experiments were carried out by Dasilva [4] over the AAI network in order to study the ATM WAN performance versus multiple simultaneous traffic streams between multiple hosts:

- Experiment 1: an SGI Onyx host with a TAXI interface transmits to a SPARC 10 and SPARC 5 hosts with TAXI interface simultaneously.
- Experiment 2: a SPARC 10 and a SPARC 5 hosts transmit simultaneously to an SGI Onyx host.
- Experiment 3: experiment 1 and 2 carried out simultaneously.

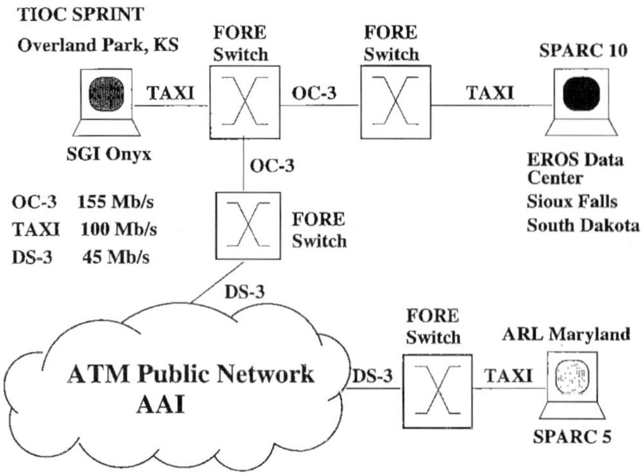

Fig. 11 Experimental Set Up of Simultaneous connections over the AAI Network.

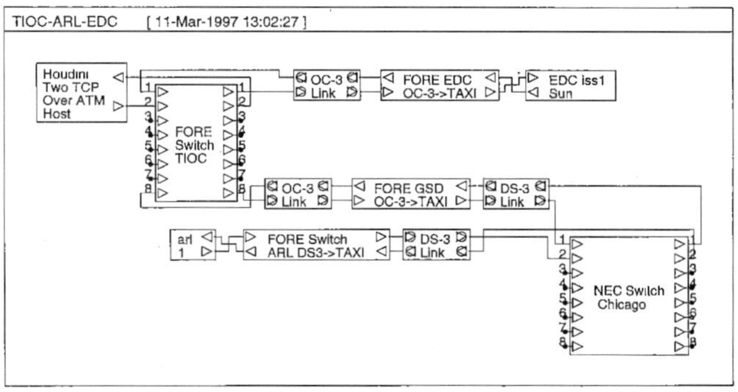

Fig. 12 BONeS model of Simultaneous connections over the AAI Network.

The experimental set up is shown if Fig. 11. In these experiments, the lowest link capacity, DS-3, is between the TIOC and ARL connection. However, a cell-level pacing was not used because the maximum throughput of SPARC 5 with SunOS 4 is less than the DS-3 rate. The round-trip times for TIOC-EDC and TIOC-ARL connections are about 9 ms and 25 ms respectively. For the TIOC-EDC connection, the TCP window size was set to 128 KB, and for the TIOC-ARL it was set to 32 KB. The BONeS simulation model developed for these experiments is shown in Fig. 12. The simulation versus experimental performance results are shown in Table 5.

Table 5 Simulation & Experimental Results of Simultaneous connections over the AAI Network.

Connection	Experimental Results	Simulation Results
Baseline Results: Point-to-Point		
TIOC-to-ARL	4.2 Mb/s	7.18 Mb/s
TIOC-to-EDC	64.2 Mb/s	65.98 Mb/s
Simultaneous traffic streams: single source, two destinations		
TIOC-to-ARL	4.45 Mb/s	4.60 Mb/s
TIOC-to-EDC	64.36 Mb/s	61.37 Mb/s
Simultaneous traffic streams: two sources, single destination		
ARL-to-TIOC	2.15 Mb/s	4.87 Mb/s
EDC-to-TIOC	52.42 Mb/s	65.01 Mb/s
Simultaneous full duplex traffic streams:		
TIOC-to-ARL	4.34 Mb/s	5.16 Mb/s
ARL-to-TIOC	4.3 Mb/s	5.16 Mb/s
TIOC-to-EDC	22.18 Mb/s	41.80 Mb/s
EDC-to-TIOC	31.18 Mb/s	41.30 Mb/s

Table 6 Level of model Complexity.

Model	Level of Complexity
ATM	High
ATM Switch	Medium
IP	Low
Link	Medium
Pacer	Medium
SONET	Low
TCP	High
TCP User (Application Layer)	Low

5. Lessons Learned

TCP imposes heavy processing overhead when individual packets are retransmitted immediately after they time out and subsequently retransmitted. To reduce this overhead, many TCP implementations use the slow and fast timer mechanisms to handle the retransmission and acknowledgment operations. However, the cost of doing that is a performance degradation. The long delay introduced by these timer mechanisms causes a significant reduction in the performance TCP over ATM when there are cell losses due to network congestion. All workstations used for this study employ TCP implementations which use these timers. Therefore, it was necessary to include them in our TCP model in order to match our simulation results with measurements. For example, in our simulation studies, we first used the NetSim TCP model which does not use these timers. The simulation result (without modeling these timers) for the experiment with no pacing presented in Sect. 4.4 was 24 Mb/s, which is 20 times greater than the experimental result.

Our results indicate that simulations can be used to accurately predict the performance of high speed wide area networks. However, to enable feasible simulations of such systems, the minimum level of model complexity for each system element must be used. The simulation's run time is inversely proportional to the level of model complexity. Hence, simulation's run time can be reduced significantly if redundant model precision is avoided. Table 6 shows the minimum required level of model complexity used in our simulation systems. Detailed IP and SONET models are not required in our simulation systems; their impact is captured by simply accounting for their information overhead. We avoided the need for a SONET model at the physical layer by reducing the OC-3 link speed from 155 Mb/s to 149 Mb/s. An IP model is not used because the IP routing functionality is not needed in ATM networks (see Sect. 3.2). Here, the ATM switch is modeled only by a single FIFO queue and a server. High precision for this switch model is unnecessary since a single connection with no cross-traffic was considered. Also, the details of the pacing algorithm were not modeled, since a simple FIFO was sufficient to capture its impact on performance. To obtain the results presented in this study, a high level of complexity was needed for the TCP and ATM modules shown in Fig. 1. In these models, we left out only references to Unix and IP specific functions and the open and close connection operations. By using the different levels of model complexity shown in Table 6, we were able to reduce the simulation time significantly. The run time of the simulation system shown in Fig. 8 on a SPARCstation-10 with 120 MByte of RAM is about 20 to 30 minutes for each second of real-time. If accurate IP, SONET, and switch models were used, the simulation of these high speed wide area ATM networks would be difficult.

The deviation between the simulated and experimental results shown in Table 5 can be explained by the fact that the operating system complexity, such as CPU scheduling and interrupt calls, was not within the scope of this simulation study. This explain why these models did not accurately predict cases where the host was required to process multiple streams. Thus, more complex and computationally intensive models are required to accurately capture host effects. Further, these models are not directly applicable to large and more complex networks such as Internet. Predicting the performance perceived by users over large networks, e.g. the Internet, is a difficult task. In this paper we demonstrate that the impact of network components can be accurately modeled. However, such models fail to capture host effects and can significantly overestimate the expected performance. In some of the cases considered here a difference of 10-20 Mbps was observed between measured and simulated performance.

The ATM Forum is currently in the process of standardizing traffic and congestion control techniques at the cell level for ATM technology. The results presented here show that simulations can be used to capture and evaluate the interactions of these techniques with TCP packet flow and congestion control mechanisms. Questions like "Can TCP packet flow and congestion control techniques cooperate with ATM cell flow and congestion control mechanisms" can be answered using the simulation models developed here.

References

[1] T.E. Anderson, S.S. Owicki, and C.P. Thacker, "High speed switch scheduling for local area network," DEC Internal Publication, 1993.

[2] B.J. Ewy, J.B. Evans, V.S. Frost, and G.J. Minden, "TCP/ATM experiences in the MAGIC tested," Fourth IEEE International Symposium on High Performance Distributed Computing, pp.87–93, Aug. 1995.

[3] H.Y. Chen, J.A. Hutchins, and N. Testi, "TCP performance over wide area networks," SAND93-8243, Sept. 1993.

[4] L. Dasilva, Personal Communication, Dec. 1995.

[5] C. Fang, H. Chen, and J. Hutchins, "Simulation analysis of TCP performance in congested ATM LAN using DEC's flow control scheme and two selective cell-drop schemes," ATM Forum Contribution 94-0119, Jan. 1994.

[6] V.S. Frost, "Modeling and simulation in network management," in Telecommunication Network Management into the 21th Century, eds. S. Aidarous and T. Plevyak, chapter 6, IEEE Press, 1994.

[7] B. Fuller and I. Richer, "An overview of the MAGIC project," MITRE Technical Report M93B0000/73, Dec. 1993.

[8] H. Kobayashi, "Modeling and Analysis: An Introduction to System Performance Evaluation Methodology," Addison-Wesley, Reading, Massachusetts, 1978.

[9] T.V. Lakshman and U. Madhow, "Performance analysis of window-based flow control using TCP/IP: Effect of high bandwidth-delay products and random loss," IFIP Trans. High Performance Networking, V (C-26), 1994.

[10] Lawrence Berkeley National Laboratory (LBNL) Network Simulator (NS), *http://www-nrg.ee.lbl.gov/ns/*

[11] G.Y. Lazarou and V.S. Frost, "BONeS TCP primitive module validation," TISL Technical Report TISL-10980-06, The University of Kansas, 1995.

[12] D. Martin, "Network Simulator User's Manual," MIT, 1988.

[13] P.P. Mishra, "Throughput degradation for TCP over ATM in the presence of traffic policing," IEEE ComSoc TCGN Gigabit Networking Workshop, 1995.

[14] I. Mohammed and V.S. Frost, "Validation measurements for LANSE (DRAFT)," TISL Technical Report TISL-11560-01, The University of Kansas, 1995.

[15] C. Partridge, "Gigabit Networking," Addison-Wesley, Reading, Massachusetts, 1994.

[16] G. Parulkar, D.C. Schmidt, and J.S. Turner, "GIPR: A gigabit IP router," IEEE ComSoc TCGN Gigabit Networking Workshop, 1995.

[17] A. Romanow and S. Floyd, "Dynamics of TCP traffic over ATM networks," IEEE J. Sel. Areas Commun., vol.13, no.4, May 1995.

[18] A. Romanow, "TCP over ATM: Some performance results," ATM Forum Contribution 93-784, July 1993.

[19] W.R. Stevens, "TCP/IP Illustrated, Volume 1,2," Addison-Wesley, Readings, Massachusetts, 1994.

[20] Systems & Networks, "KANSAS BONeS DESIGNER 3.0 Modeling Guide," Lawrence, KS, 1995.

[21] H. Zhu, L.A. Dasilva, J.B. Evans, and V.S. Frost, "Performance evaluation of congestion control mechanisms in ATM networks," Proc. of the 21st International Conf. on the Resource Management & Performance Evaluation of Enterprise Computing Systems (CMG'95), Nashville (TN), Dec. 1995.

High-Performance Backbone Network Technology edited by Naoaki Yamanaka
ISBN: 0-8247-5321-6 DOI: 10.1081/(120025690)
Published 2004 by Marcel Dekker, Inc.

Performance Evaluation of SVC-Based IP-Over-ATM Networks

Zhisheng Niu, Yoshitaka Takahashi, and Noboru Endo

SUMMARY We propose a finite-capacity single-vacation model, with close-down/setup times and a Markovian arrival process (MAP), for SVC-based IP-over-ATM networks. This model considers the SVC processing overhead and the bursty nature of IP packet arrivals. Specifically, the setup time corresponds to the SVC setup time and the vacation time corresponds to the SVC release time, while the close-down time corresponds to the SVC timeout. The MAP is a versatile point process by which typical bursty arrival processes like the IPP (interrupted Poisson process) or the MMPP (Markov modulated Poisson process) is treated as a special case. The approach we take here is the *supplementary variable technique.* Compared with the embedded Markov chain approach, it is more straightforward to obtain the steady-state probabilities at an arbitrary instant and the practical performance measures such as packet loss probability, packet delay time, and SVC setup rate. For the purpose of optimal design of the SVC-based IP-over-ATM networks, we also propose and derive a new performance measure called the *SVC utilization ratio.* Numerical results show the sensitivity of these performance measures to the SVC timeout period as well as to the burstiness of the input process.

key words: *switched virtual connection (SVC), IP-over-ATM, IP switching, traffic control, bursty arrivals, vacation*

1. Introduction

Recently there has been much interest in *IP-over-ATM* technology [1]. This combines the speed of layer-2 ATM switching with the flexibility of layer-3 IP routing. This has been prompted by numerous proposals or de facto standards, including classical IP-over-ATM (IETF) [2], multi-protocol-over-ATM (MPOA) (ATM Forum) [3], and most recently a number of draft proposals [4]–[8] in IETF's MPLS (multi-protocol-label-switching) working group.

Since ATM is a connection-oriented network technology, an end-to-end connection – called virtual connection (VC) – through ATM networks has to be set up first before information can be exchanged. So far, ITU-T has defined two different VC setup mechanisms: (1) PVC (permanent virtual connection) which is set up a priori by network management protocols and left open indefinitely, and (2) SVC (switched virtual connection) which is set up and released dynamically on demand by using signaling protocols. Although the PVC mechanism is superior to the SVC in terms of the processing overhead and the time delay for setting up a VC, some network resources (e.g., bandwidth) are wasted when the VC is not in use. Furthermore, since a PVC is required for each pair of end users connected to the ATM networks, communication using PVC is not adapted to large-scale networks. Therefore, SVC-based IP-over-ATM systems will be paid more attention regarding use in future large-scale ATM wide-area-networks.

As mentioned above, compared with PVC-based networks, setting up and releasing an SVC using signaling protocols involve additional processing overhead and time delay. Therefore, releasing an SVC immediately after the transmission of the last packet in a packet buffer is not a good scheme, especially when the arrival process of IP packets is bursty. One way to reduce the overhead of the SVC-based IP-over-ATM systems is to assign an inactive timer to each SVC. In this way, the SVC is kept for a while even if there are no packets to transmit. The SVC is released only when no packets arrive during this inactive period. This reduces the processing overhead of SVC setup and release at the cost of wasted network resources (e.g., SVC bandwidth) since the SVC is not effectively used during the inactive period. Therefore, the length of the SVC timeout plays a crucial role in the optimal design of the SVC-based IP-over-ATM networks.

So far, there have been several quantitative studies on this topic [9], [10]. In [9] the authors proposed an adaptive prediction algorithm for the timeout period. Here, the new timeout period is calculated based on the mean and the standard derivation of packet interarrival times, while in [10] it is based on the distribution of the previous packet interarrival times. However, both of them used a specific *empirical dataset* of packet arrivals and assumed a very high capacity SVC so that queueing of packets does not have to be considered. Moreover, they did not analyze the network performance as a function of timeout period. In [11], the authors proposed an M/G/1 delayed vacation model with setup time for SVC-based ATM networks by as-

Manuscript received September 26, 1997.
Manuscript revised December 10, 1997.

[†]The author is on his sabbatical leave at Central Research Laboratory, Hitachi, Ltd., Kokubunji-shi, 185–8601 Japan, from Tsinghua University, Beijing, 100084 China.

[††]The author is with the NTT Multimedia Network Laboratories, Musashino-shi, 180–8585 Japan.

[†††]The author is with Central Research Laboratory, Hitachi, Ltd., Kokubunji-shi, 185–8601 Japan.

[*]Presently, with the Department of Electronic Engineering, Tsinghua University, Beijing, 100084 China.

IEICE TRANS. COMMUN., Vol. E81-B, No. 5, pp. 948–957, May 1998

suming a fixed-length timeout period. Later, Kitazume et al.[12] extended the model in [11] to the case of a generally distributed timeout period by also assuming the buffer capacity to be infinite. In [13], the authors considered a finite-capacity single-vacation model, with both setup and close-down times, for the SVC-based IP-over-ATM networks, but the input process was assumed to be Poissonian too. Recent studies on Internet traffic properties have revealed that the arrival process of IP packets are bursty in nature[14]. Although a queueing analysis of a finite-capacity vacation queue with non-renewal input can be found in [15], this work considered the multiple-vacation model only and did not take the server setup and close-down times into account too.

In this paper, we propose a finite-capacity single-vacation queue (with close-down/setup times and a Markovian arrival process (MAP)) for the SVC-based IP-over-ATM networks. This is quite a practical model that can both characterize the effect of SVC processing overhead (setup, release, and timeout) and capture the impact of the burstiness of the input process on the network performance. Specifically, via ATM signaling protocols, the setup time corresponds to the SVC setup time and the vacation time corresponds to the SVC release time, while the close-down time corresponds to the SVC timeout. The MAP is a versatile point process by which typical bursty arrival processes like the IPP (interrupted Poisson process) or the MMPP (Markov-modulated Poisson process) is treated as a special case.

The rest of this paper is organized as follows. In Sect. 2, through two practical examples, we describe the queueing model proposed for SVC-based IP-over-ATM networks. In Sect. 3, we perform an exact queueing analysis by using the supplementary variable method. In Sect. 4, four practically interested performance measures, i.e., packet loss probability, average time delay, SVC setup rate, and SVC utilization ratio, are derived. In Sect. 5, numerical examples demonstrate the applicability of the proposed queueing model and its analysis. Section 6 concludes the paper with some closing remarks.

2. Queueing Model of the SVC-Based IP-Over-ATM Networks

In this section, we propose a queueing model for the SVC-based IP-over-ATM networks by considering two practical examples. The first one is an SVC-based ATM virtual LAN (local area network) system[12] (Fig. 1) which is a typical classical IP-over-ATM network. In this system, IP packets from the ATM virtual terminal A or B are queued in the SVC manager which is responsible for the setup and release of SVC's via ATM signaling protocols. Once an appropriate SVC is successfully established, the IP packets will be segmented into ATM cells by the AAL (ATM adaptation layer) function and forwarded to the destination successively

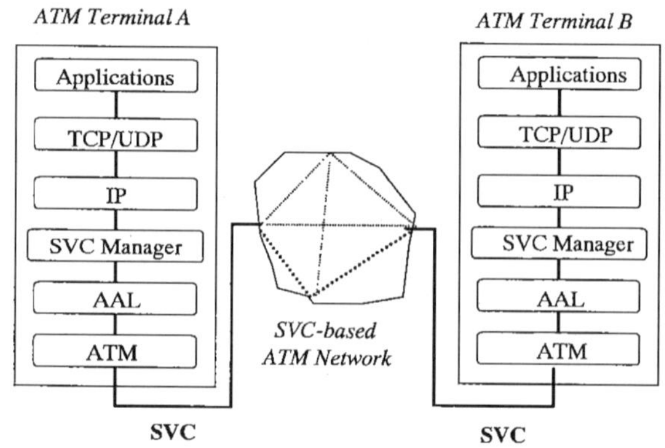

Fig. 1 A functional model of SVC-based ATM virtual LAN systems.

without further setup action. In order to reduce the SVC setup overhead (especially in the case where the arrival process of IP packets is bursty), the SVC manager usually assigns an inactive timer to each SVC. In this way, the SVC is kept for a while even when there are no packets to transmit. The SVC is released only if no packets arrive during the inactive timer.

The other example is the so-called IP switching[4] as shown in Fig. 2. IP packets from an IP-based LAN are queued in the per-flow queues in the gateway according to their source/destination addresses and maybe other flow parameters (see[16] for the concept of IP flow). The first packet from a specific flow has to be forwarded to the IP controller for the setup of an appropriate SVC. Once the SVC is established, all the packets from the same flow will be segmented into ATM cells by CLAD (cell assembly-and-disassembly) and forwarded to the destination through the SVC without further help from the IP controller. As in the case of the ATM virtual LAN systems, the IP controller assigns an inactive timer to each SVC if it sees an empty queue of a particular IP flow, and then the IP controller releases the SVC only when it sees an empty queue again after the timer expires.

As shown in Fig. 3, we propose a finite-capacity single-vacation model, with close-down/setup times and a Markovian arrival process (MAP), for SVC-based IP-over-ATM networks. This is a practical model that can both characterize the effect of the SVC processing overhead (e.g., setup, release, and timeout) and capture the impact of the burstiness of the input process on network performance. The capacity of the queueing system (including the one in the server) is limited to K. Accepted customers (e.g., packets) are served by a single server (e.g., SVC manager or IP controller) on an FIFO (first-in-first-out) basis until the system becomes empty. If the server finds no customers in the system upon a customer departure, it enters the *close-down* phase (e.g., to start an inactive timer of the SVC). If a customer arrives at the system during the close-down period, the server immediately goes back to the *service* phase in order to

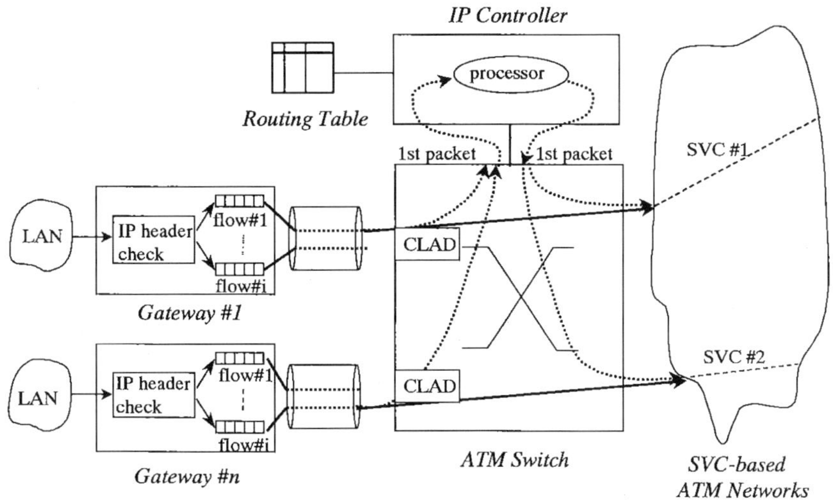

Fig. 2 A functional model of SVC-based IP switching systems.

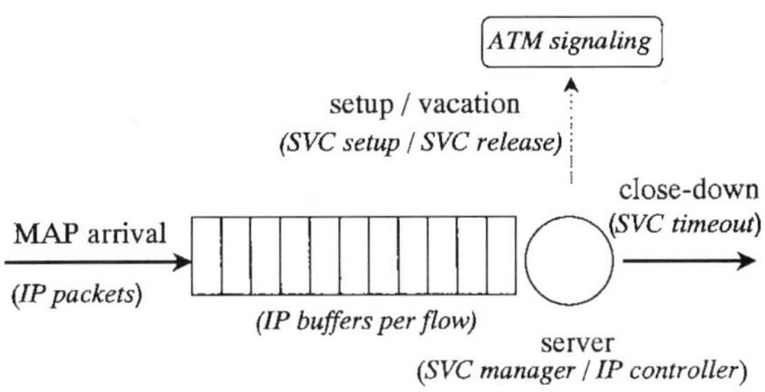

Fig. 3 Queueing model of SVC-based IP-over-ATM networks.

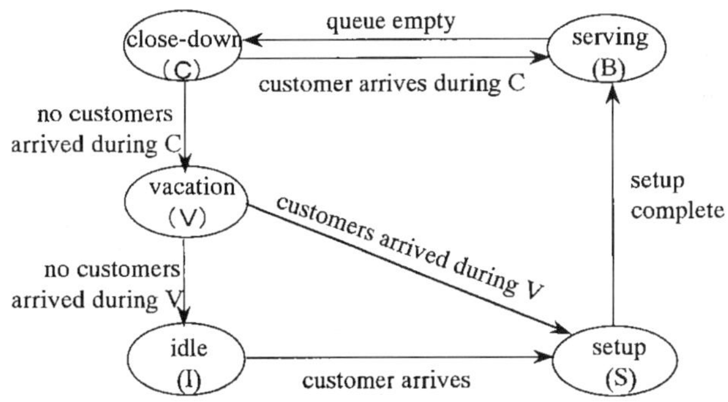

Fig. 4 State transition diagram of MAP/G/1/K vacation model with close-down and setup times.

serve the customer. If no customers arrive during the close-down time period, the server goes to the *vacation* phase (e.g., to release the SVC by signaling protocols). If the server finds one or more customers waiting in the queue at the end of its vacation, it begins to serve them after a *setup* (e.g., to set up a new SVC) period for only the first customer. The subsequent customers who arrive during the setup period of the first customer are served one-by-one without any more setup times. If the server sees an empty system again at the end of its vacation, it enters the *idle* phase. If a customer arrives when the server is in the idle phase, the server goes to the setup phase. The close-down, vacation, setup, and service times are assumed to be generally distributed with probability distribution functions of $C(x)$, $V(x)$, $S(x)$, and $B(x)$, respectively. Figure 4 shows the transitions among the close-down, vacation, setup, service, and idle phases.

Customers are assumed to arrive at the queueing system according to an m-state *Markovian arrival process* (*MAP*) with representation $(D_0, D_1)^{\dagger}$. Here, D_0 is an $m \times m$ matrix, with negative diagonal elements and nonnegative off-diagonal elements, which governs the state transitions that correspond to no customer arrivals, whereas D_1 is an $m \times m$ matrix, with nonnegative elements, which governs the state transitions that correspond to a customer arrival. Specifically, MAP defines a class of non-renewal inputs and by using this process typical bursty input processes like the IPP (interrupted Poisson process) or the MMPP (Markov-modulated Poisson process) is treated as a special case. (For the details of MAP, see [17].)

First, denote by D the sum of D_0 and D_1 and let $\boldsymbol{\pi}$ be the stationary probability vector of D such that

$$\boldsymbol{\pi} D = 0, \qquad \boldsymbol{\pi e} = 1, \tag{1}$$

where $\boldsymbol{e}$ is a column vector of 1's. Then, the average arrival rate of the MAP is given by

$$\bar{\lambda} = \boldsymbol{\pi} D_1 \boldsymbol{e}. \tag{2}$$

Lastly, we denote the LST's of $B(x)$, $V(x)$, $S(x)$, and $C(x)$ by $\tilde{B}(s)$, $\tilde{V}(s)$, $\tilde{S}(s)$, and $\tilde{C}(s)$, respectively, and the corresponding means by m_B, m_V, m_S, and m_C.

† In this paper, we use notation (D_0, D_1) instead of the usual notation (C, D) for the representation of MAP as in [17] for the purpose of consistency with the notation of a more general point process BMAP (batch Markovian arrival process) [18].

3. Queueing Analysis

In this section, we perform an exact analysis to the queueing system shown in Fig. 3 by applying the supplementary variable technique. Compared with the embedded Markov chain approach [15], it is more straightforward to obtain the steady-state probabilities at an *arbitrary* instant and the practical performance measures such as loss probability, server setup rate, and waiting time distribution via the supplementary variable method. To do this, we first define the following random variables at an arbitrary time epoch t,

N_t: number of customers in the system,
J_t: state of the Markovian arrival process,
$\bar{B}_t$: remaining service time of $B(x)$,
$\bar{C}_t$: remaining close-down time of $C(x)$,
$\bar{V}_t$: remaining vacation time of $V(x)$,
$\bar{S}_t$: remaining setup time of $S(x)$,

$$\xi_t = \begin{cases} 1, & \text{if the server is in service phase,} \\ 2, & \text{if the server is in close-down phase,} \\ 3, & \text{if the server is in vacation phase,} \\ 4, & \text{if the server is in setup phase,} \\ 5, & \text{if the server is in idle phase.} \end{cases}$$

It is then clear that the joint distribution of queue length N_t, state J_t, and supplementary variables $\{\bar{B}_t|_{\xi_t=1} \cup \bar{C}_t|_{\xi_t=2} \cup \bar{V}_t|_{\xi_t=3} \cup \bar{S}_t|_{\xi_t=4}\}$ has a Markovian property with $0 \leq N_t \leq K$ and $1 \leq J_t \leq m$.

Next, we denote the joint probability *density* vector of the queue length, the state of MAP, and the remaining close-down, vacation, setup, or service times by

$$\phi_0^{(C)}(x) = (\phi_{01}^{(C)}(x), \cdots\cdots, \phi_{0m}^{(C)}(x)),$$
$$\phi_i^{(V)}(x) = (\phi_{i1}^{(V)}(x), \cdots\cdots, \phi_{im}^{(V)}(x)),$$
$$\phi_i^{(S)}(x) = (\phi_{i1}^{(S)}(x), \cdots\cdots, \phi_{im}^{(S)}(x)),$$

or

$$\phi_i^{(B)}(x) = (\phi_{i1}^{(B)}(x), \cdots\cdots, \phi_{im}^{(B)}(x)),$$

where each component is defined by

$$\phi_{0j}^{(C)}(x)dx \equiv \lim_{t\to\infty} \Pr[N_t = 0, J_t = j, x < \bar{C}_t \leq x + dx, \xi_t = 2],$$
$$\phi_{ij}^{(V)}(x)dx \equiv \lim_{t\to\infty} \Pr[N_t = i, J_t = j, x < \bar{V}_t \leq x + dx, \xi_t = 3],$$
$$\phi_{ij}^{(S)}(x)dx \equiv \lim_{t\to\infty} \Pr[N_t = i, J_t = j, x < \bar{S}_t \leq x + dx, \xi_t = 4],$$

or

$$\phi_{ij}^{(B)}(x)dx \equiv \lim_{t\to\infty} \Pr[N_t = i, J_t = j, x < \bar{B}_t \leq x + dx, \xi_t = 1].$$

If the server is in the idle phase, however, there is no corresponding supplementary variable to be defined since the system transits to the setup phase immediately upon a customer arrival. Hence, we define the joint *probability* vector of the queue length and the state of MAP by

$$\phi_0^{(I)} = (\phi_{01}^{(I)}, \cdots\cdots, \phi_{0m}^{(I)}),$$

where

$$\phi_{0j}^{(I)} \equiv \lim_{t\to\infty} \Pr[N_t = 0, J_t = j, \xi_t = 5].$$

For example, $\phi_1^{(B)}(0)$ is the transition rate vector of the system from the service phase to the close-down phase. Therefore, $\phi_1^{(B)}(0)C_0$ represents the transition rates of the system from the close-down phase to the vacation phase when there is no customer arrivals during the close-down time. This is evidently equal to $\phi_0^{(C)}(0)$. Consequently, by observing the possible phase transitions between two successive (close-down, vacation, setup, or service) completion epochs in a similar manner (Fig. 4), we obtain the following recursive formulas for the probability density vectors as well as $\phi_0^{(I)}$ in terms of $\phi_1^{(B)}(0)$,

$$\phi_0^{(C)}(0) = \phi_1^{(B)}(0)C_0, \tag{3}$$
$$\phi_0^{(I)} = \phi_1^{(B)}(0)C_0V_0(-D_0)^{-1}, \tag{4}$$
$$\phi_i^{(V)}(0) = \phi_0^{(C)}(0)V_i, \quad (0 \leq i \leq K) \tag{5}$$
$$\phi_i^{(S)}(0) = \phi_0^{(I)}D_1S_{i-1} + \sum_{n=1}^{i} \phi_n^{(V)}(0)S_{i-n}, \quad (1 \leq i \leq K-1) \tag{6}$$
$$\phi_K^{(S)}(0) = \phi_0^{(I)}D_1\hat{S}_{K-1} + \sum_{n=1}^{K} \phi_n^{(V)}(0)\hat{S}_{K-n}, \tag{7}$$
$$\phi_i^{(B)}(0) = \phi_1^{(B)}(0)(I - C_0)(-D_0)^{-1}D_1B_{i-1} + \sum_{n=1}^{i}\left(\phi_{n+1}^{(B)}(0) + \phi_n^{(S)}(0)\right)B_{i-n}. \quad (1 \leq i \leq K-1) \tag{8}$$

Here, the (ν, j)th element of C_n (V_n, S_n, or B_n) is the conditional probability that, given that the MAP was in state ν at the beginning of the close-down (vacation, setup, or service) time, n ($n = 0, 1, \cdots, K$) customers arrive during a close-down (vacation, setup, or service) time period and that the MAP is in state j at the end of the close-down (vacation, setup, or service) time. For the derivation of C_n, V_n, S_n, and B_n, see Appendix A. In addition, we denote $\hat{S}_i = \sum_{n=i}^{K} S_n$.

Now, we move on to the calculation of the steady-state probabilities of the queueing system. To do this, we further denote the LST vectors by

$$\tilde{\Phi}^{(C)}{}_0(s) = (\tilde{\Phi}^{(C)}{}_{01}(s), \cdots\cdots, \tilde{\Phi}^{(C)}{}_{0m}(s)),$$

$$\tilde{\boldsymbol{\Phi}}^{(V)}{}_i(s) = (\tilde{\Phi}^{(V)}{}_{i1}(s), \cdots\cdots, \tilde{\Phi}^{(V)}{}_{im}(s)),$$
$$\tilde{\boldsymbol{\Phi}}^{(S)}{}_i(s) = (\tilde{\Phi}^{(S)}{}_{i1}(s), \cdots\cdots, \tilde{\Phi}^{(S)}{}_{im}(s)),$$
$$\tilde{\boldsymbol{\Phi}}^{(B)}{}_i(s) = (\tilde{\Phi}^{(B)}{}_{i1}(s), \cdots\cdots, \tilde{\Phi}^{(B)}{}_{im}(s)),$$

where $\tilde{\Phi}_{0j}^{(C)}(s)$ ($\tilde{\Phi}_{ij}^{(V)}(s)$, $\tilde{\Phi}_{ij}^{(S)}(s)$, or $\tilde{\Phi}_{ij}^{(B)}(s)$) is the Laplace transform of $\phi_{0j}^{(C)}(x)$ ($\phi_{ij}^{(V)}(x)$, $\phi_{ij}^{(S)}(x)$, or $\phi_{ij}^{(B)}(x)$). Then, as shown in Appendix B, the steady-state probabilities at an arbitrary instant are given by

$$\tilde{\boldsymbol{\Phi}}_0^{(C)}(0) = (\boldsymbol{\phi}_1^{(B)}(0) - \boldsymbol{\phi}_0^{(C)}(0))(-D_0)^{-1}, \tag{9}$$

$$\tilde{\boldsymbol{\Phi}}_i^{(V)}(0) = \boldsymbol{\phi}_0^{(C)}(0)Q^i - \sum_{n=0}^{i} \boldsymbol{\phi}_n^{(V)}(0)Q^{i-n}, \quad (0 \leq i \leq K-1) \tag{10}$$

$$\tilde{\boldsymbol{\Phi}}_i^{(S)}(0) = \sum_{n=0}^{i} \boldsymbol{\phi}_n^{(V)}(0)Q^{i-n} - \sum_{n=1}^{i} \boldsymbol{\phi}_n^{(S)}(0)Q^{i-n}, \quad (1 \leq i \leq K-1) \tag{11}$$

$$\tilde{\boldsymbol{\Phi}}_i^{(B)}(0) = (\boldsymbol{\phi}_1^{(B)}(0) - \boldsymbol{\phi}_0^{(C)}(0))Q^i + \sum_{n=1}^{i}(\boldsymbol{\phi}_n^{(S)}(0) + \boldsymbol{\phi}_{n+1}^{(B)}(0) - \boldsymbol{\phi}_n^{(B)}(0))Q^{i-n}, \quad (1 \leq i \leq K-1) \tag{12}$$

$$\tilde{\boldsymbol{\Phi}}_K^{(V)}(0)\boldsymbol{e} = m_V \boldsymbol{\phi}_0^{(C)}(0)\boldsymbol{e} - \sum_{i=0}^{K-1} \tilde{\boldsymbol{\Phi}}_i^{(V)}(0)\boldsymbol{e}, \tag{13}$$

$$\tilde{\boldsymbol{\Phi}}_K^{(S)}(0)\boldsymbol{e} = m_S \sum_{i=0}^{K} \boldsymbol{\phi}_i^{(V)}(0)\boldsymbol{e} - \sum_{i=1}^{K-1} \tilde{\boldsymbol{\Phi}}_i^{(S)}(0)\boldsymbol{e}, \tag{14}$$

$$\tilde{\boldsymbol{\Phi}}_K^{(B)}(0)\boldsymbol{e} = m_B \Big[\sum_{i=1}^{K} \boldsymbol{\phi}_i^{(B)}(0)\boldsymbol{e} + \sum_{i=1}^{K} \boldsymbol{\phi}_i^{(S)}(0)\boldsymbol{e} - \boldsymbol{\phi}_0^{(C)}(0)\boldsymbol{e}\Big] - \sum_{i=1}^{K-1} \tilde{\boldsymbol{\Phi}}_i^{(B)}(0)\boldsymbol{e}, \tag{15}$$

where $Q^i = [(-D_0)^{-1}D_1]^i(-D_0)^{-1}$.

Finally, $\boldsymbol{\phi}_1^{(B)}(0)$ can be determined by using the boundary equation at $i = K$ of $\boldsymbol{\phi}_i^{(B)}(0)$ and the normalizing condition, i.e.,

$$\boldsymbol{\phi}_K^{(B)}(0) = \sum_{n=1}^{K-1} \boldsymbol{\phi}_{n+1}^{(B)}(0)\hat{B}_{K-n} + \sum_{n=1}^{K} \boldsymbol{\phi}_n^{(S)}(0)\hat{B}_{K-n} + \boldsymbol{\phi}_1^{(B)}(0)(I - C_0)(-D_0)^{-1}D_1\hat{B}_{K-1}, \tag{16}$$

$$\boldsymbol{\phi}_0^{(I)}\boldsymbol{e} + \tilde{\boldsymbol{\Phi}}_0^{(C)}(0)\boldsymbol{e} + \sum_{i=0}^{K} \tilde{\boldsymbol{\Phi}}_i^{(V)}(0)\boldsymbol{e} + \sum_{i=1}^{K} \tilde{\boldsymbol{\Phi}}_i^{(S)}(0)\boldsymbol{e} + \sum_{i=1}^{K} \tilde{\boldsymbol{\Phi}}_i^{(B)}(0)\boldsymbol{e} = 1, \tag{17}$$

where we denote $\hat{B}_i = \sum_{n=i}^{K} B_n$.

4. Performance Measures

4.1 Loss Probability

Denote the probability that an arbitrary customer is lost upon arrival by P_L. By applying Little's law to the service process of the server (excluding the waiting room), we have

$$\bar{\lambda}(1 - P_L)m_B = \sum_{i=1}^{K} \tilde{\boldsymbol{\Phi}}_i^{(B)}(0)\boldsymbol{e}, \tag{18}$$

where the right-hand side of the equation signifies the mean number of customers in the server (the probability that the server is busy) and the left-hand side signifies the throughput multiplied by the mean sojourn time in the server. The loss probability P_L is then obtained as

$$P_L = 1 - \frac{\sum_{i=1}^{K} \tilde{\boldsymbol{\Phi}}_i^{(B)}(0)\boldsymbol{e}}{\rho}, \tag{19}$$

where $\rho = \bar{\lambda} m_B$ is the input traffic intensity.

4.2 Average Waiting Time

Denote the *effective* arrival rate of the accepted customers (excluding the customers that were blocked) by λ' and their average waiting time by $\bar{W}$. Since $\sum_{i=1}^{K} \tilde{\boldsymbol{\Phi}}_i^{(B)}(0)\boldsymbol{e}$ represents the probability that the server is busy (in the service phase), we have

$$\lambda' = \frac{\sum_{i=1}^{K} \tilde{\boldsymbol{\Phi}}_i^{(B)}(0)\boldsymbol{e}}{m_B}. \tag{20}$$

Hence, from Little's formula, we have

$$\bar{W} = \left[\sum_{i=1}^{K} i\tilde{\boldsymbol{\Phi}}_i^{(V)}(0)\boldsymbol{e} + \sum_{i=1}^{K} i\tilde{\boldsymbol{\Phi}}_i^{(S)}(0)\boldsymbol{e} + \sum_{i=1}^{K}(i-1)\tilde{\boldsymbol{\Phi}}_i^{(B)}(0)\boldsymbol{e}\right]\Big/\lambda'. \tag{21}$$

4.3 Setup Rate

The server setup rate γ_S is defined as the average number of times the server enters the setup phase per unit time. This is a key measure of the SVC processing overhead in an SVC environment. Similarly, by applying Little's law to the setup process of the server (excluding the waiting room), we have

$$\gamma_S(-\tilde{S}'(0)) = \sum_{i=1}^{K} \tilde{\boldsymbol{\Phi}}_i^{(S)}(0)\boldsymbol{e}, \tag{22}$$

from which we obtain

$$\gamma_S = \frac{\sum_{i=1}^{K} \tilde{\boldsymbol{\Phi}}_i^{(S)}(0)\boldsymbol{e}}{m_S}. \tag{23}$$

4.4 Utilization Ratio

The utilization ratio R_U is defined as the ratio of the time during which the server is serving a customer (busy) to the time during which the server is occupied (either in the close-down phase, on a vacation, or serving a customer). This is a key measure of the network performance and is useful for the optimal design of SVC-based networks. Clearly, it is given by

$$R_U = \frac{\sum_{i=1}^{K} \tilde{\boldsymbol{\Phi}}_i^{(B)}(0)\boldsymbol{e}}{\tilde{\boldsymbol{\Phi}}_0^{(C)}(0)\boldsymbol{e} + \sum_{i=0}^{K} \tilde{\boldsymbol{\Phi}}_i^{(V)}(0)\boldsymbol{e} + \sum_{i=1}^{K} \tilde{\boldsymbol{\Phi}}_i^{(B)}(0)\boldsymbol{e}}. \tag{24}$$

5. Numerical Examples and Discussions

In this section, to demonstrate the applicability of our queueing model and the analytical results to practical SVC-based IP-over-ATM systems, we show numerical examples with the following parameters.

- IP packet size: exponential distribution with a mean of 256 bytes,
- Bandwidth of an SVC: 384 kbit/s,
- IP packet buffer size: 50 packets,
- SVC setup time: exponential distribution with a mean of 100 ms,
- SVC release time: exponential distribution with a mean of 20 ms,
- SVC timeout period: deterministic distribution whose mean is a parameter ($m_C = 10{\sim}500$ ms).

From these parameters, we know $K = 50$, $m_B = 5.3333$ ms, $m_V = 20$ ms, and $m_S = 100$ ms, where m_C varies from 10 to 500 ms.

Two kinds of input processes for IP packet arrivals are compared in the example: a Poisson process (random arrivals) and an interrupted Poisson process (IPP; bursty arrivals). Both are in fact special cases of the MAP, where for the Poisson process $D_0 = -\lambda$, $D_1 = \lambda$, and for the IPP

$$D_0 = \begin{pmatrix} -\lambda - r_1 & r_1 \\ r_2 & -r_2 \end{pmatrix}, \quad D_1 = \begin{pmatrix} \lambda & 0 \\ 0 & 0 \end{pmatrix}. \tag{25}$$

Given the input traffic intensity ρ and squared coefficient of variation C_a^2 of the interarrival time distribution, a possible parameter-matching to the IPP is $\lambda = 2\rho m_B^{-1}$ and $r_1 = r_2 = \lambda/2(C_a^2 - 1)$. For the Poisson process, λ should be set to ρm_B^{-1}.

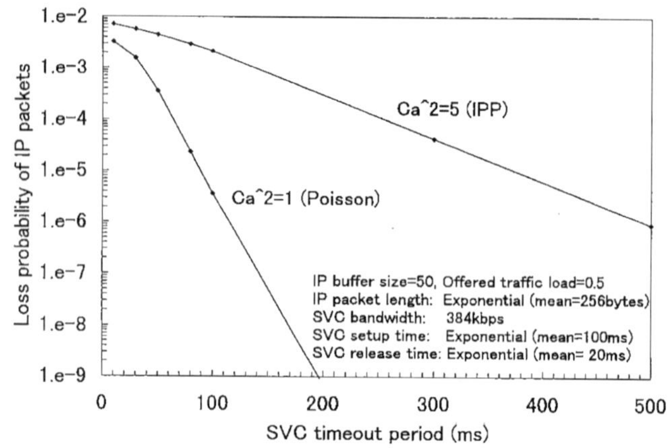

Fig. 5 Loss probability as a function of SVC timeout period.

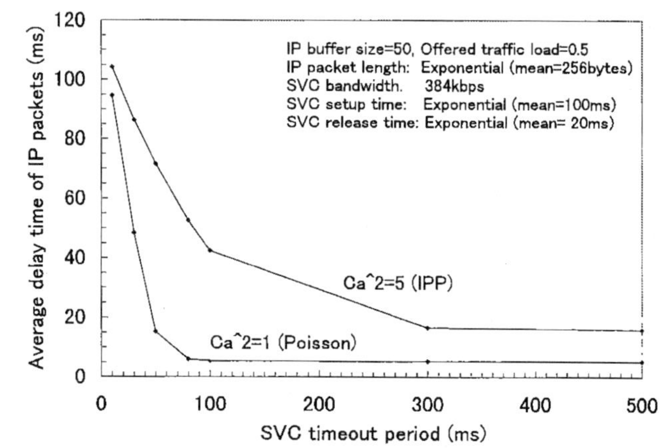

Fig. 6 Average delay time as a function of SVC timeout period.

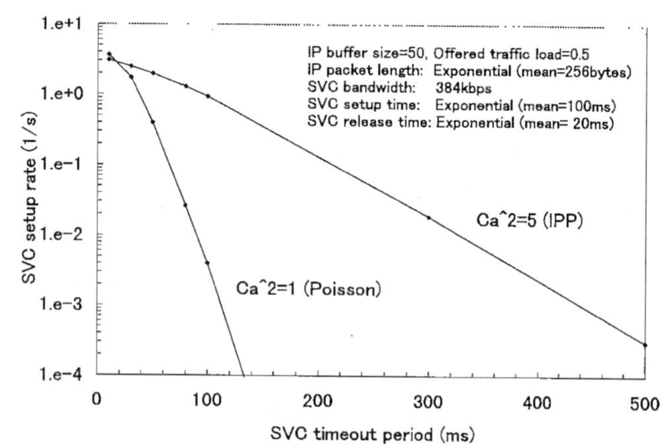

Fig. 7 SVC setup rate as a function of SVC timeout period.

In Figs. 5–8, we show the calculated results of the loss probability of IP packets, average delay time of IP packets, setup rate of SVC connections, and SVC utilization ratio as a function of SVC timeout period. Here, it is worth noting that the loss probability and average time delay of IP packets are measures of a user's

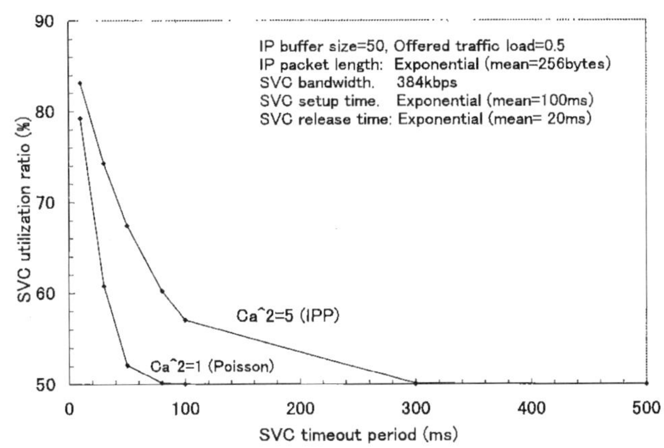

Fig. 8 SVC utilization ratio as a function of SVC timeout period.

quality-of-service (QoS) and the SVC utilization ratio is a key measure of the network performance, while the SVC setup rate measures the SVC processing overhead. A calculation algorithm is summarized in [20]. Also note that the unit time of the SVC setup rate in Fig. 7 has been normalized to one *second* for easy reading.

It is readily seen from Figs. 5–7 that the longer the SVC timeout period, the smaller the loss probability, average delay time, and SVC setup rate. This means that the QoS of IP packets and the SVC processing overhead can be improved by increasing the SVC timeout period (i.e., by reserving the network resources for a longer time). At the same time, however, the network performance deteriorates as shown in Fig. 8. Alternatively speaking, the SVC bandwidth (one kind of network resources) is wasted as the SVC timeout increases. This is the trade-off between a user's QoS and network performance. Therefore, it is crucial to work out an appropriate SVC timeout period for given performance measures (e.g., a combination of the loss probability, average delay time, and setup rate) to maximize the SVC utilization ratio. This is left for future study.

Comparing the results for $C_a^2 = 1$ (Poisson arrivals) with the corresponding ones for $C_a^2 = 5$ (bursty arrivals) in Figs. 5–8, it is seen that the burstiness of the arrival process has a great impact on the key performance measures. Specifically, a Poissonian assumption for the input process underestimates the performance measures and, therefore, results in a dangerous design for the networks. For instance, under the same system parameters and performance measures, our results show that a longer timeout period is needed when the arrival process is getting burstier.

To be more exact, Figs. 5–6 indicate that more IP packets are discarded by the system and the accepted IP packets wait a longer time for transmission if they arrive at the system in a bursty manner. This is a straightforward result and can be easily explained by queueing theory. Similarly, since the tail probability of the interarrival times (the probability that the interarrival time is larger than a relatively large value) of the IPP is larger than that of Poisson process, it is more frequent for the system to release the SVC connection. This results in a higher SVC setup rate as shown in Fig. 7. As for the SVC utilization ratio, however, Fig. 8 shows a reverse phenomenon where the IPP model has a higher utilization ratio than that of the Poisson model. This is because the larger SVC setup rate of the IPP model causes a larger SVC release rate and, therefore, a shorter holding (occupying) time of the SVC is required in order to carry almost the same traffic load.

6. Conclusions

We proposed a finite-capacity single-vacation MAP/G/1/K queue, with server setup and close-down times, for SVC-based IP-over-ATM networks. This is a practical model that can both characterize the effect of the SVC processing overhead (e.g., setup, release, and timeout) and capture the impact of the burstiness of the input process on the network performance. Applying the supplementary variable technique enabled us to measure the performance in terms of loss probability of IP packets, average delay time of IP packets, and SVC setup rate. For the purpose of optimal design, we also introduced a new performance measure called SVC utilization ratio. The numerical results demonstrated that the QoS of IP packets (e.g., loss probability and delay time) and the SVC processing overhead (e.g., SVC setup rate) can be improved by increasing the SVC timeout (i.e., reserving the network resource for a longer time). At the same time, however, the network performance (e.g., SVC utilization ratio) deteriorates. Our results also showed that the burstiness of the input process has a great impact on the performance measures.

It is left for future work to investigate in more details the impact of the SVC timeout period, SVC setup time, SVC release time, and the burstiness of arrival processes on the QoS for users and the network performance. This work is very important for the optimal design of the SVC-based IP-over-ATM networks and, in fact, this paper provides a mathematical basis for this future work.

Acknowledgements

The first author would like to express his great thanks to Dr. Michiharu Nakamura (General Manager of Hitachi Central Research Laboratory (HCRL)) and Shiro Tanabe (Unit Leader of HCRL) for giving him this visiting opportunity and for their continuous encouragement. The authors also wish to thank the reviewers for their helpful comments.

This work is partially supported by the National Natural Science Foundation (NNSF) of China (Grant No.: 69572018).

References

[1] G. Parulkar, D. Schmidt, and J. Turner, "IP/ATM: A strategy for integrating IP with ATM," Proc. ACM SIGCOMM, Cambridge, MA, Sept. 1995.

[2] M. Laubach, "Classical IP and ARP over ATM," IETF Internet Draft RFC 1577, Jan. 1994.

[3] C. Brown, "Baseline text for MPOA," ATM Forum/95-0824r5, Jan. 1996.

[4] P. Newman, T. Lyon, and G. Minshall, "Flow labelled IP: A connectionless approach to ATM," Proc. IEEE Infocom'96, vol.3, pp.1251–1260, 1996.

[5] Y. Rekhter, B. Davie, D. Katz, E. Rosen, and G. Swallow, "Tag switching architecture overview," IETF Internet Draft RFC2105, draft-rekhter-tagswitch-arch-00.txt, Feb. 1997.

[6] A. Viswanathan, N. Feldman, R. Boivie, and R. Woundy, "ARIS: Aggregate route-based IP switching," IETF Internet Draft RFC2107, draft-viswanathan-aris-overview-00.txt, March 1997.

[7] Y. Katsube, K. Nagami, and H. Esaki, "Toshiba's router architecture extensions for ATM: Overview," IETF Internet Draft RFC 2098, 1997.

[8] "The framework of multi-protocol label switching," IETF Internet Draft, draft-ietf-mpls-framework-00.txt, June 1997.

[9] H. Saran and S. Keshav, "An empirical evaluation of virtual circuit holding times in IP-over-ATM networks," IEEE Infocom'94, pp.1132–1140, 1994.

[10] C. Loud, S. Philips, and N. Reingold, "Adaptive holding policies for IP-over-ATM networks," IEEE Infocom'95, pp.80–87, 1995.

[11] M. Hassan and M. Atiquzzaman, "A delayed vacation model of an M/G/1 queue with applications to SVCC-based ATM networks," IEICE Trans. Commun., vol.E80-B, no.2, pp.317–323, Feb. 1997.

[12] H. Kitazume and Y. Takahashi, "A queueing model of SVC mechanism for IP over ATM networks," Symposium on Performance Models for Information Communication Networks, Kamakura, Japan, pp.28–33, 1997.

[13] Y. Sakai, Y. Takahashi, Y. Takahashi, and T. Hasegawa, "A composite queue with vacation/setup/close-down times for IP over ATM networks," Proc. of ORSJ Autumn Conference, 1997.

[14] V. Paxson and S. Floyd, "Wide area traffic: The failure of Poisson modeling," SIGCOMM'94, 1994.

[15] C. Blondia, "Finite capacity vacation models with non-renewal input," J. Appl. Prob., vol.28, pp.174–197, 1991.

[16] K.C. Claffy, H.-W. Braun, and G.C. Polyzos, "A parameterizable methodology for Internet traffic flow profiling," IEEE J. Sel. Areas Commun., vol.13, no.8, pp.1481–1494, 1995.

[17] D.M. Lucantoni, K. Meier-Hellstern, and M.F. Neuts, "A single server queue with server vacations and a class of non-renewal arrival processes," Adv. Appl. Prob., vol.22, pp.676–705, 1990.

[18] D. Lucantoni, "New results on the single server queue with a batch Markovian arrival process," Stoch. Models, vol.7, pp.1–46, 1991.

[19] T. Takine, Y. Matsumoto, T. Suda, and T. Hasegawa, "Mean waiting times in nonpreemptive priority queues with Markovian arrival and i.i.d. service processes," Perform. Eval., vol.20, pp.131–149, 1994.

[20] Z. Niu, Y. Takahashi, and N. Endo, "Queueing model of SVC-based IP over ATM networks," IEICE Technical Reports, SSE97-153, 1997.

[21] Z. Niu and Y. Takahashi, "A finite-capacity queue with exhaustive vacation/close-down/setup times and Markovian arrival processes," submitted to Queueing Systems, 1997.

Appendix A: Derivation of C_n, V_n, S_n, and B_n

Consider the counting process for *accepted* customers in the queueing system during a time interval $(0, x]$, i.e., $P(n,x) = \{p_{ij}(n,x)\}$ $(1 \leq i,j \leq m)$ for $0 \leq n \leq K$, where

$$p_{ij}(n,x) \equiv Pr\{N_x = n, J_x = j | N_0 = 0, J_0 = i\}.$$

Clearly, $P(n,x)$ satisfies the following Chapman-Kolmogorov equations:

$$\left.\begin{array}{l} P(0,0) = I_m, \\ P(n,0) = 0, \qquad (n \geq 1) \\ P'(n,x) = P(n,x)D_0 + P(n-1,x)D_1, \\ \qquad\qquad (1 \leq n \leq K-1) \\ P'(K,x) = P(K,x)D + P(K-1,x)D_1. \end{array}\right\} \quad \text{(A·1)}$$

Here I_m stands for an $m \times m$ unit matrix where the diagonal elements are 1's and the non-diagonal elements are 0's. By repeating the *uniformization* arguments for $P(n,x)$ in a similar manner as shown in [18], it leads to

$$P(n,x) = \sum_{l=n}^{\infty} e^{-\theta x} \frac{(\theta x)^l}{l!} U_n^{(l)}, \quad (0 \leq n \leq K) \text{ (A·2)}$$

where $\theta = \max_i \{(-D_0)_{ii}\}$ and $\{U_n^{(l)}\}$ is given recursively by

$$\left.\begin{array}{l} U_0^{(0)} = I_m, \\ U_n^{(l)} = 0, \quad (1 \leq n \leq K, 0 \leq l \leq n-1) \\ U_0^{(l+1)} = U_0^{(l)}(I_m + \theta^{-1}D_0), \\ U_n^{(l+1)} = U_n^{(l)}(I_m + \theta^{-1}D_0) + \theta^{-1}U_{n-1}^{(l)}D_1, \\ \qquad\qquad (1 \leq n \leq K-1, l \geq n-1) \\ U_K^{(l+1)} = U_K^{(l)}(I_m + \theta^{-1}D) + \theta^{-1}U_{K-1}^{(l)}D_1. \end{array}\right\} \text{(A·3)}$$

In fact, by focusing on the counting process for *accepted* customers in the queueing system, this is an extension of the corresponding results for the infinite-capacity MAP/G/1 queue in [19] to the finite-capacity MAP/G/1/K model.

From the defination of C_n, V_n, S_n, and B_n given in Sect. 3, i.e.,

$$C_n = \int_0^{\infty} P(n,x)dC(x), \quad (0 \leq n \leq K) \qquad \text{(A·4)}$$

$$V_n = \int_0^{\infty} P(n,x)dV(x), \quad (0 \leq n \leq K) \qquad \text{(A·5)}$$

$$S_n = \int_0^{\infty} P(n,x)dS(x), \quad (0 \leq n \leq K) \qquad \text{(A·6)}$$

$$B_n = \int_0^{\infty} P(n,x)dB(x), \quad (0 \leq n \leq K) \qquad \text{(A·7)}$$

we then have

$$C_n = \sum_{l=n}^{\infty} r_l^{(C)} U_n^{(l)}, \quad (0 \leq n \leq K) \qquad \text{(A·8)}$$

$$V_n = \sum_{l=n}^{\infty} r_l^{(V)} U_n^{(l)}, \quad (0 \leq n \leq K) \tag{A·9}$$

$$S_n = \sum_{l=n}^{\infty} r_l^{(S)} U_n^{(l)}, \quad (0 \leq n \leq K) \tag{A·10}$$

$$B_n = \sum_{l=n}^{\infty} r_l^{(B)} U_n^{(l)}, \quad (0 \leq n \leq K) \tag{A·11}$$

where

$$r_l^{(C)} = \int_0^{\infty} e^{-\theta x} \frac{(\theta x)^l}{l!} dC(x), \tag{A·12}$$

$$r_l^{(V)} = \int_0^{\infty} e^{-\theta x} \frac{(\theta x)^l}{l!} dV(x), \tag{A·13}$$

$$r_l^{(S)} = \int_0^{\infty} e^{-\theta x} \frac{(\theta x)^l}{l!} dS(x), \tag{A·14}$$

$$r_l^{(B)} = \int_0^{\infty} e^{-\theta x} \frac{(\theta x)^l}{l!} dB(x). \tag{A·15}$$

For the stopping criteria in Eqs. (A·8)–(A·11), see [19].

As an example, values of B_n for $n = 0, 1, \cdots, K$ are given by the following equations if the service time distribution $B(x)$ is exponential; that is

$$B_n = m_B^{-1}[(m_B^{-1} I_m - D_0)^{-1} D_1]^n (m_B^{-1} I_m - D_0)^{-1}, \quad (0 \leq n \leq K-1) \tag{A·16}$$

$$B_K = m_B^{-1}[(m_B^{-1} I_m - D_0)^{-1} D_1]^K (m_B^{-1} I_m - D)^{-1}. \tag{A·17}$$

Appendix B: Derivation of the Steady-State Probabilities

In this appendix, we outline the derivation of the steady-state probabilities of the MAP/G/1/K single-vacation model with close-down and setup times. More detailed queueing analysis can be found in [21].

Without loss of generality, we consider the possible transitions into the state $\{N_t = 0, J_t = j, x < \bar{C}_t \leq x + dx, \xi_t = 2\}$ during $(t, t+\Delta t]$ as an example. From a basic probability argument, we have

$$\begin{aligned} \phi_{0j}^{(C)}&(t+\Delta t, x-\Delta t) \\ &= [1 - (-D_0(j,j)\Delta t)]\phi_{0j}^{(C)}(t,x) \\ &\quad + \sum_{\substack{\nu=1 \\ \nu \neq j}}^{m} D_0(\nu, j)\Delta t \phi_{0\nu}^{(C)}(t,x) \\ &\quad + \phi_{1j}^{(B)}(t,0)\big(C(x) - C(x-\Delta t)\big) \\ &\quad + o(\Delta t), \quad (1 \leq j \leq m) \end{aligned} \tag{A·18}$$

where $\phi_{0j}^{(C)}(t,x)dx = \Pr[N_t = 0,\ J_t = j,\ x < \bar{C}_t \leq x + dx,\ \xi_t = 2]$. Representing Eq. (A·18) in matrix form, it follows that

$$\begin{aligned} \phi_0^{(C)}&(t+\Delta t, x-\Delta t) \\ &= \phi_0^{(C)}(t,x)(I_m + D_0\Delta t) + \phi_1^{(B)}(t,0)dC(x), \end{aligned} \tag{A·19}$$

or equivalently

$$\frac{\phi_0^{(C)}(t+\Delta t, x-\Delta t) - \phi_0^{(C)}(t, x-\Delta t)}{\Delta t} - \frac{\phi_0^{(C)}(t,x) - \phi_0^{(C)}(t,x-\Delta t)}{\Delta t} = \phi_0^{(C)}(t,x)D_0 + \phi_1^{(B)}(t,0)\frac{dC(x)}{dx}. \tag{A·20}$$

Taking the limits of $\Delta t \to 0$ as well as $t \to \infty$ yields

$$\frac{d\phi_0^{(C)}(x)}{dx} = -\phi_0^{(C)}(x)D_0 - \phi_1^{(B)}(0)\frac{dC(x)}{dx}. \tag{A·21}$$

Using the similar arguments as above to the cases of $\phi_i^{(V)}(x)$, $\phi_i^{(S)}(x)$, and $\phi_i^{(B)}(x)$, we have

$$\frac{d\phi_0^{(V)}(x)}{dx} = -\phi_0^{(V)}(x)D_0 - \phi_0^{(C)}(0)\frac{dV(x)}{dx}, \tag{A·22}$$

$$\frac{d\phi_i^{(V)}(x)}{dx} = -\phi_i^{(V)}(x)D_0 - \phi_{i-1}^{(V)}(x)D_1, \quad (1 \leq i \leq K-1) \tag{A·23}$$

$$\frac{d\phi_K^{(V)}(x)}{dx} = -\phi_K^{(V)}(x)D - \phi_{K-1}^{(V)}(x)D_1, \tag{A·24}$$

$$\frac{d\phi_1^{(S)}(x)}{dx} = -\phi_1^{(S)}(x)D_0 - (\phi_0^{(I)}D_1 + \phi_1^{(V)}(0))\frac{dS(x)}{dx}, \tag{A·25}$$

$$\frac{d\phi_i^{(S)}(x)}{dx} = -\phi_i^{(S)}(x)D_0 - \phi_{i-1}^{(S)}(x)D_1 - \phi_i^{(V)}(0)\frac{dS(x)}{dx}, \quad (2 \leq i \leq K-1) \tag{A·26}$$

$$\frac{d\phi_K^{(S)}(x)}{dx} = -\phi_K^{(S)}(x)D - \phi_{K-1}^{(S)}(x)D_1 - \phi_K^{(V)}(0)\frac{dS(x)}{dx}, \tag{A·27}$$

$$\frac{d\phi_1^{(B)}(x)}{dx} = \phi_1^{(B)}(x)D_0 - \Big(\phi_2^{(B)}(0) + \phi_1^{(S)}(0) + \tilde{\boldsymbol{\Phi}}_0^{(C)}(0)D_1\Big)\frac{dB(x)}{dx}, \tag{A·28}$$

$$\frac{d\phi_i^{(B)}(x)}{dx} = -\phi_i^{(B)}(x)D_0 - \phi_{i-1}^{(B)}(x)D_1 - \Big(\phi_{i+1}^{(B)}(0) + \phi_i^{(S)}(0)\Big)\frac{dB(x)}{dx}, \quad (2 \leq i \leq K-1) \tag{A·29}$$

$$\frac{d\phi_K^{(B)}(x)}{dx} = -\phi_K^{(B)}(x)D - \phi_{K-1}^{(B)}(x)D_1 - \phi_K^{(S)}(0)\frac{dB(x)}{dx}, \tag{A·30}$$

where $\tilde{\boldsymbol{\Phi}}_0^{(C)}(0)D_1 = \lim_{\substack{\Delta t \to 0 \\ t \to \infty}} \int_0^\infty \phi_0^{(C)}(t,x)D_1 dx$ represents the transition (from the close-down phase to the service phase immediately after a customer arrival) rate vector. Therefore, taking the LST for Eqs. (A· 21)–(A· 30) leads to

$$\tilde{\boldsymbol{\Phi}}_0^{(C)}(s)(sI + D_0) = \phi_0^{(C)}(0) - \phi_1^{(B)}(0)\tilde{C}(s), \tag{A· 31}$$

$$\tilde{\boldsymbol{\Phi}}_0^{(V)}(s)(sI + D_0) = \phi_0^{(V)}(0) - \phi_0^{(C)}(0)\tilde{V}(s), \tag{A· 32}$$

$$\tilde{\boldsymbol{\Phi}}_i^{(V)}(s)(sI + D_0) = \phi_i^{(V)}(0) - \tilde{\boldsymbol{\Phi}}_{i-1}^{(V)}(s)D_1, \quad (1 \leqq i \leqq K-1) \tag{A· 33}$$

$$\tilde{\boldsymbol{\Phi}}_K^{(V)}(s)(sI + D) = \phi_K^{(V)}(0) - \tilde{\boldsymbol{\Phi}}_{K-1}^{(V)}(s)D_1, \tag{A· 34}$$

$$\tilde{\boldsymbol{\Phi}}_1^{(S)}(s)(sI + D_0) = \phi_1^{(S)}(0) - (\phi_0^{(I)}D_1 + \phi_1^{(V)}(0))\tilde{S}(s), \tag{A· 35}$$

$$\tilde{\boldsymbol{\Phi}}_i^{(S)}(s)(sI + D_0) = \phi_i^{(S)}(0) - \tilde{\boldsymbol{\Phi}}_{i-1}^{(S)}(s)D_1 - \phi_i^{(V)}(0)\tilde{S}(s), \quad (2 \leqq i \leqq K-1) \tag{A· 36}$$

$$\tilde{\boldsymbol{\Phi}}_K^{(S)}(s)(sI + D) = \phi_K^{(S)}(0) - \tilde{\boldsymbol{\Phi}}_{K-1}^{(S)}(s)D_1 - \phi_K^{(V)}(0)\tilde{S}(s), \tag{A· 37}$$

$$\tilde{\boldsymbol{\Phi}}_1^{(B)}(s)(sI + D_0) = \phi_1^{(B)}(0) - \left(\phi_2^{(B)}(0) + \phi_1^{(S)}(0) + \tilde{\boldsymbol{\Phi}}_0^{(C)}(0)D_1\right)\tilde{B}(s), \tag{A· 38}$$

$$\tilde{\boldsymbol{\Phi}}_i^{(B)}(s)(sI + D_0) = \phi_i^{(B)}(0) - \tilde{\boldsymbol{\Phi}}_{i-1}^{(B)}(s)D_1 - \left(\phi_{i+1}^{(B)}(0) + \phi_i^{(S)}(0)\right)\tilde{B}(s), \quad (2 \leqq i \leqq K-1) \tag{A· 39}$$

$$\tilde{\boldsymbol{\Phi}}_K^{(B)}(s)(sI + D) = \phi_K^{(B)}(0) - \tilde{\boldsymbol{\Phi}}_{K-1}^{(B)}(s)D_1 - \phi_K^{(S)}(0)\tilde{B}(s). \tag{A· 40}$$

By substituting $s = 0$ directly into Eqs. (A· 31)–(A· 33), Eqs. (A· 35)–(A· 36), and Eqs. (A· 38)–(A· 39), and substituting $s = 0$ after differentiation into Eq. (A· 34), Eq. (A· 37), and Eq. (A· 40), we obtain Eqs. (9)–(15).

High-Performance Backbone Network Technology edited by Naoaki Yamanaka
ISBN: 0-8247-5321-6 DOI: 10.1081/(120025691)
Published 2004 by Marcel Dekker, Inc.

Performance Analysis of Packet-Level Scheduling in an IP-over-ATM Network with QoS Control

Chie Dou, Cheng-Tien Lin, Shu-Wei Wang, and Kuo-Cheng Leu

SUMMARY In this paper, we study the performance of packet-level scheduling in IP-over-ATM networks with QoS control. That is, we assume the traffic is composed of multiple classes. We analyze the performance of different queue mappings between traffic types and the number of available traffic classes (priority queues). Since cells of a given packet are not bound to be transmitted back-to-back if multiple traffic classes are used, it is quite interesting to know the packet delay characteristic as well as the cell delay characteristic of respective traffic types in different queue mappings. Closed-form solutions for the mean cell waiting time and the mean packet waiting time of individual traffic types in different queue mappings are presented. The numerical results obtained in this paper can be helpful in understanding the behavior of IP-over-ATM networks which adopting packet-level scheduling and QoS control.
key words: *IP-over-ATM, packet-level scheduling, QoS*

1. Introduction

Recently there has been much interest in IP-over-ATM technology. This combines the speed of layer-2 ATM switching with the flexibility of layer-3 IP routing. In this case, each IP packet may be segmented to multiple 53-byte cells before being switched. Most ATM switch systems today operate in cut-through mode, scheduling service for each cell upon its arrival. While appropriate in an ideal homogeneous native-ATM environment, cell-level scheduling is limiting in the more common multi-protocol environment with legacy applications. The ability to operate in store-and-forward mode, to schedule service for each packet upon arrival of its last cell, can yield a number of advantages [1]. Packet-level scheduling is important to tag and IP switching for multipoint-to-point services; it allows different connections to be merged into a single Virtual Connection (VC) to reduce routing complexity [1], [2]. With packet-level scheduling, consecutive cell arrivals with the same VCI will belong to the same source until the complete packet is constructed. As an additional benefit, packet-level scheduling can enhance intelligent discarding schemes by reducing the number of partial packets that are transmitted; this further reduces wasted bandwidth during congestion. As a result of these advantages, Cisco Systems has incorporated packet-level scheduling in their LightStream 1010 ATM switch [3].

Widjaja and Elwalid had investigated the performance issues in VC-merge capable switches for IP-over-ATM networks in [2]. With VC merging, cells belonging to different packets do not interleave if they are merged to the same VC. Thus, each incoming cell belonging to a given packet needs to be stored in a dedicated buffer (called the reassembly buffer) until the last cell of the packet arrives. When the last cell arrives, all cells in the packet are transferred to a certain output buffer for service scheduling. As stated in [2], if all traffic is best-effort, *full-merging* where all incoming VCs destined for the same destination network are mapped to the same outgoing VC, can be implemented. However, if the traffic is composed of multiple classes, it is desirable to implement *partial merging*, where incoming VCs destined for the same {destination network, QoS} are mapped to the same outgoing VC. With full merging, a single output buffer is used and the switch essentially emulates frame switching. With partial merging, multiple output buffers are used and each corresponds to a {destination network, QoS}. If multiple output buffers are used, VC merging is different from frame switching since cells of a given packet are not bound to be transmitted back-to-back. This implies that cells of a given packet may be interleaved with the other cells of different packets. Priority/Fair queueing can be implemented so that cells from their respective output buffers are served according to some QoS requirements. In [2], the authors analyzed the performance of a VC-merge capable ATM switch under the assumption that full merging with a single FIFO output buffer is used. To compute the steady-state distribution in the output buffer, a discrete-time queueing model with batch Markovian arrival process (D-BMAP/D/1) was used. The result of the D-BMAP/D/1 queue is complex and need to be solved through a rather sophisticated iterative algorithm.

In this paper, we study the performance of packet-level scheduling in an IP-over-ATM network with QoS control. That is, we assume the traffic is composed of multiple classes. Incoming VCs destined for the same

Manuscript received July 9, 1999.
Manuscript revised November 8, 1999.
[†]The authors are with the Department of Electrical Engineering, National Yunlin University of Science and Technology, Touliu, Yunlin, TAIWAN 640, Republic of China.
[††]The authors are with Computer & Communications Research Laboratories, Industrial Technology Research Institute, Taiwan, R.O.C.
[*]The Computer & Communications Research Laboratories, ITRI, Taiwan, R.O.C. supported this research, under the contract G4-88008-m.

IEICE TRANS. COMMUN., Vol. E83-B, No. 7, pp. 1534–1543, July 2000

destination network are mapped to their corresponding output buffers (partial merging) according to the QoS requirements of their associated classes. Since cells of a given packet are not bound to be transmitted back-to-back if multiple output buffers are used, it is quite interesting to know the packet delay characteristic as well as the cell delay characteristic of respective output buffers. In this paper, we present closed-form solutions for the mean cell waiting times and the mean packet waiting times of respective output buffers. In this paper we also consider the mappings between traffic types and the number of available output buffers (traffic classes), if the number of traffic types is greater than the number of available output buffers. As suggested in the Annex of [4], the following list of traffic types, each of which can benefit from simple segregation from the others, seems to command widespread support in IP internetworks: 'Network Control' (NC), 'Voice' (VO), 'Video' (VI), 'Controlled Load' (CL), 'Excellent Effort' (EE), 'Best Effort' (BE) and 'Background' (BK).

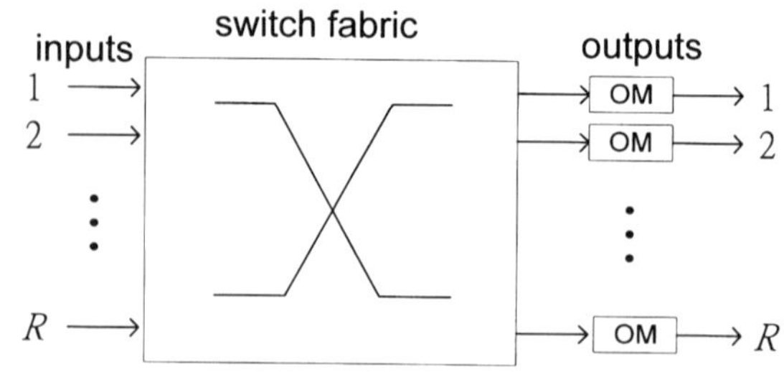

Fig. 1 A generic VC-merge capable ATM switch.

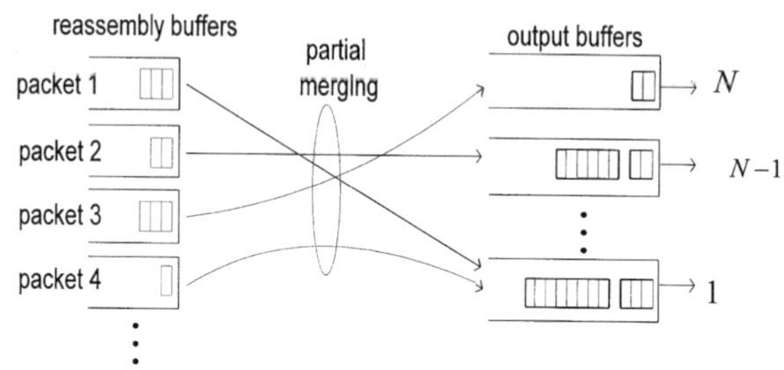

Fig. 2 An output module with QoS control.

2. Switch Architecture

The switch architecture considered in this paper is mainly adopted from [2]. The only difference is that in [2] only single FIFO output buffer is considered for each output module. In this paper we consider each output module can support N output buffers (traffic classes) to achieve better QoS control in VC merging. Figure 1 shows a generic VC-merge capable ATM switch architecture adopted in [2]. The reassembly buffers are located at the output so that the switch fabric can forward each cell independently as in normal ATM switching. Since VC merging with QoS control is considered, the structure of the output module (OM) can be implemented as shown in Fig. 2. The output module contains a number of reassembly buffers (RB), each of which correspond to a particular input port and an incoming VC value. If the VC values are unique to the entire switch rather than to the port, then an RB would correspond to an incoming VC value. The output module also contains a number of output buffers, each of which correspond to a particular QoS requirement (traffic class). Incoming VCs with the same QoS requirement are mapped to the same output buffer. Priority/Fair queueing can be implemented so that the cells from their respective output buffers are served according to some service scheduling disciplines.

With VC merging, all cells belonging to the same packet are delivered to the switch from the same input port and the same incoming VC value. All these cells will be destined for the same reassembly buffer in a certain output module until the last cell of the packet arrived. Before the last cell arrives, all the early arrived cells of a packet will be stored in the reassembly buffer and experiencing the reassembly delay of the packet they belong. This reassembly delay is closely related to the cell arrival rate of the packet. The cell arrival rate of a packet on a given input port depends on the cell transmission rate of the corresponding VC assigned to the packet on that port. For a CBR VC, we will generally give it the highest transmission priority on the link. Thus a CBR VC can always get the required bandwidth it needs. But for an ABR VC or an UBR VC, we will generally assign it lower (or the lowest) transmission priority on the link. Once an ABR VC (or an UBR VC) can not get the enough bandwidth, the cell transmission rate of it will become low. This will cause the reassembly delay of the packet delivered through this VC increases. Clearly, the cell transmission rates of different VCs on different input links have large influence on the reassembly delays of packets reassembled in the reassembly buffers of output modules. In real implementation, cells are usually stored in the data buffer of an output module. And the reassembly buffers are implemented as link lists that are used to link the buffer locations allocated to the cells belonging to the same packet. The output buffers are also implemented as link lists that are used to link all the link lists of packets resided in the same output buffer (priority queue). When the last cell of a packet arrives, the link list of the packet (i.e. the content of the corresponding reassembly buffer) is added to the tail of the link list maintained by the output buffer to which the packet is destined for. Hence, in our study we did not consider the cell transmission rate between reassembly buffers and the output buffers. We assume cells are arrived at the output buffers in batches. In addition, we did not address the reassembly delay mentioned above in our analysis. In this paper we only addressed on the waiting time analysis of the output buffers shown in the latter stage of Fig. 2.

3. Modeling and Analysis

To analyze the queueing behavior of output buffers shown in Fig. 2, we simply consider a priority multi-class VC merged output port as a single server priority queueing model, as shown in Fig. 3, where the arrival process is batched-arrivals (in cell). An output port can support N output buffers, each of which correspond to a particular traffic class. A separate logical queue is reserved for each N output buffers. We assume the server serves cells from N different classes of packets in a fixed priority order (strict priority queue). That is, cells from traffic class i are given preferential treatment over cells from class i-1. Cell-level nonpreemptive service discipline is assumed. In addition, we assume each output port has an infinite-buffer, and a work-conserving server. Since the switch fabric shown in Fig. 1 forwards each cell independently as in normal ATM switching, cells to a certain output module have to be sequential. If an incoming cell is not the last cell of any packet, no packets (in the reassembly buffers) can accomplish their reassembly. At this moment no packets can be transferred from the reassembly buffers to the output buffers. But if the incoming cell is the last cell of a specific packet, this packet and only this packet can accomplish its reassembly. And immediately after that all the cells in the packet are transferred to a certain output buffer as a batch arrival. At the same time, there will be no other packets can be transferred to the output buffers simultaneously. That is, arrival packets to a certain output buffer are sequential. Since packet arrivals (after reassembly) to each output buffer may come from different input ports with different VC values, their arrival times can be rather random on the time axis. For analytical simplicity, we assume N incoming streams of packets shown in Fig. 3 are independent of each other, and their arrival instants can lie at any point on the time axis and constitute an ASTA process. As stated in [5], an important property of the ASTA process is that an arriving customer finds the queue in its steady state. That is, at an arrival instant of an ASTA process, an arriving customer sees the queue behavior as if it were observed at an arbitrary time instant. In [5], it is also stated that ASTA processes include the Poisson process, the Markov modulated Poisson processes, the arrival processes at various queues of Jackson queueing networks, etc. Thus we may also assume the arrival process is Poisson in our analysis.

3.1 Calculating Mean Waiting Times

Let λ_i denote the arrival rate of class-i packets. Moreover, let $\bar{G}_i$ and $\bar{G}_i^{(2)}$ denote the first of two moments of the number of cells in a class-i packet, and x denote the fixed service time of an ATM cell (53-byte). We

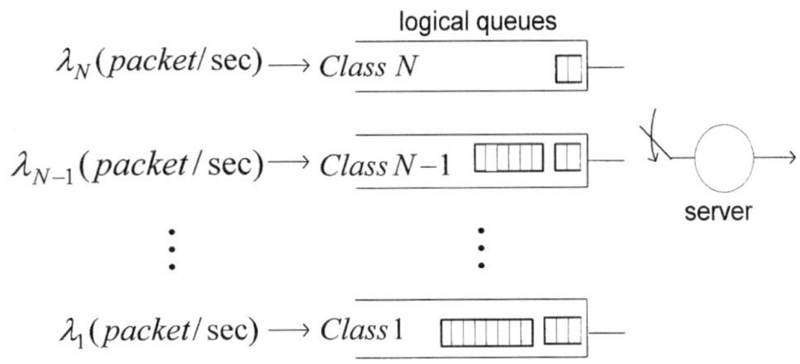

Fig. 3 The queueing model of an output port.

will write $\rho_i = \lambda_i \bar{G}_i x$ and $\rho = \sum_{i=1}^{N} \rho_i$ for the mean offered load to the server due to class-i packets and the total offered load, respectively. In addition, we will write $W_0 = \rho\frac{x}{2}$ for the mean remaining service time of a cell in service at an arbitrary time instant, and write $\bar{M}_i = \frac{\bar{G}_i^{(2)}}{2\bar{G}_i} - \frac{1}{2}$ for the mean number of cells arriving in the same class-i packet but will receive service before the tagged cell does. Due to the fixed (strict) priority service discipline [6], the mean cell waiting time of class-i packets, denoted by $W_{i,cell}$, can be written as

$$W_{N,cell} = W_0 + \lambda_N W_{N,cell} \bar{G}_N x (= \rho_N W_{N,cell}) + \bar{M}_N x, \tag{1}$$

where $\lambda_N \bar{G}_N$ is the mean cell arrival rate of class-N packets. Using Little's formula ($Q = \lambda W$), $\lambda_N \bar{G}_N W_{N,cell}$ denotes the mean number of cells of class-N packets waiting for service at an arbitrary instant. And $\bar{M}_N$ is the mean number of cells arriving in the same class-N packet but will receive service before the tagged cell does.
For $i = 1, \cdots, N$-1, we have

$$W_{i,cell} = W_0 + \left(\sum_{k=i}^{N} \rho_k W_{k,cell}\right) + \bar{M}_i x + \left(\sum_{k=i+1}^{N} \rho_k\right) W_{i,cell}, \quad i = 1, \cdots, N-1, \tag{2}$$

where the second term in the right-hand side represents the delay due to cells found in the system by our tagged cell upon its arrival who receive service before it does. The last term denotes the delay to our tagged cell by later arrivals than it.
Solving recursively, we obtain

$$W_{N,cell} = \frac{W_0 + \bar{M}_N x}{1 - \rho_N}, \tag{3}$$

$$W_{i,cell} = \frac{W_0 + (\sum_{j=i+1}^{N} \rho_j \bar{M}_j) x}{(1 - \sum_{j=i}^{N} \rho_j)(1 - \sum_{j=i+1}^{N} \rho_j)} + \frac{\bar{M}_i x}{1 - \sum_{j=i}^{N} \rho_j}, \quad i = 1, \cdots, N-1. \tag{4}$$

and

$$\sum_{k=i}^{N} \rho_k W_{k,cell} = \frac{\sum_{j=i}^{N} \rho_j W_0 + \sum_{j=i}^{N} \rho_j \bar{M}_j x}{1 - \sum_{j=i}^{N} \rho_j}, \quad i = 1, \cdots, N. \tag{5}$$

Equation (5) is known also by the name of *conservation law formula*.

Now we derive the mean packet waiting time for class-i packets, denoted by $W_{i,pkt}$. That is, a class-i packet spends on the average $W_{i,pkt}$ sec in its corresponding logical queue until its last cell receives service. Similar to (1) and (2), $W_{i,pkt}$ can be written as

$$W_{N,pkt} = W_0 + \rho_N W_{N,cell} + (\bar{G}_N - 1)x, \tag{6}$$

and

$$W_{i,pkt} = W_0 + \left(\sum_{k=i}^{N} \rho_k W_{k,cell} \right) + (\bar{G}_i - 1)x + \left(\sum_{k=i+1}^{N} \rho_k \right) W_{i,pkt}, \; i = 1, \cdots, N-1. \tag{7}$$

Combining (3)–(5) and (6), we obtain

$$W_{N,pkt} = \frac{W_0 + \rho_N \bar{M}_N x}{1 - \rho_N} + (\bar{G}_N - 1)x, \tag{8}$$

and

$$W_{i,pkt} = \frac{W_0 + (\sum_{j=i}^{N} \rho_j \bar{M}_j)x}{(1 - \sum_{j=i}^{N} \rho_j)(1 - \sum_{j=i+1}^{N} \rho_j)} + \frac{(\bar{G}_i - 1)x}{1 - \sum_{j=i+1}^{N} \rho_j}, \; i = 1, \cdots, N-1. \tag{9}$$

3.2 Analysis of Queue Mappings between Traffic Types and Traffic Classes

In this subsection we analyze the queue mappings between traffic types and the number of available output buffers (traffic classes), if the number of traffic types is greater than the number of available output buffers. It is a common practice of today's ATM switches to support multiple priority queues (classes) on each of their output ports. For instance, Cisco LightStream 1010 ATM switch has four priority classes on each of its thirty-two output ports. To have a thorough presentation of our analytical results, we consider N=2, N=3 and N=4 three cases separately. Without loss any generality, we assume the number of different traffic types is seven for each case. In fact, the results obtained in this section can be applied to any value of N and any number of traffic types ($\geq N$). In IP internetworks each traffic type may have its particular QoS requirement. There exists a priority order among different traffic types in IP internetworks as described in [4]. Adjacent traffic types may be mapped to a same priority queue (output buffer) if queue mapping is needed.

Let α_i denote the arrival rate of type-i packets. Moreover, let $\bar{g}_i$ and $\bar{g}_i^{(2)}$ denote the first two moments of the number of cells in a type-i packet. We will write $\eta_i = \alpha_i \bar{g}_i x$ and $\rho = \sum_{i=1}^{7} \eta_i$ for the mean offered load to the server due to type-i packets and the total offered load, respectively. In addition, we will write $W_0 = \rho \frac{x}{2}$ for the mean remaining service time of a cell in service at an arbitrary time instant, and write $\bar{m}_i = \frac{\bar{g}_i^{(2)}}{2\bar{g}_i} - \frac{1}{2}$ for the mean number of cells arriving in the same type-i packet but will receive service before the tagged cell does.

case 1: $N = 2$

In this case, input packet streams belonging to seven traffic types have to be merged to two priority queues. For convenient, we call these seven traffic types NC, VO, VI, CL, EE, BE and BK. If we assume BK, BE and EE three lower priority types (type-1~type-3) are mapped to the lower priority queue. The remaining four higher priority types CL, VI, VO and NC (type-4~type-7) are mapped to the higher priority queue. Due to the fixed (strict) priority service discipline [6], the mean cell waiting time of type-i i=4,$\cdots$, 7 packets in the higher priority queue, denoted by $W_{i,cell}^{(2)}$, can be written as

$$W_{i,cell}^{(2)} = W_0 + \left(\sum_{k=4}^{7} \alpha_k W_{k,cell}^{(2)} \bar{g}_k x \right) + \bar{m}_i x, \quad i = 4, \cdots, 7. \tag{10}$$

From Little's result we know that on the average there will be $\alpha_k W_{k,cell}^{(2)} \bar{g}_k$ cells from the type-k packet stream present in the queue when our tagged packet arrives. Given that $\alpha_k W_{k,cell}^{(2)} \bar{g}_k x = \eta_k W_{k,cell}^{(2)}$. Thus, (10) can be rewritten as

$$W_{i,cell}^{(2)} = W_0 + \left(\sum_{k=4}^{7} \eta_k W_{k,cell}^{(2)} \right) + \bar{m}_i x, \quad i = 4, \cdots, 7. \tag{11}$$

From (11), we can easily obtain

$$\sum_{k=4}^{7} \eta_k W_{k,cell}^{(2)} = \frac{\sum_{j=4}^{7} \eta_j W_0 + \sum_{j=4}^{7} \eta_j \bar{m}_j x}{1 - \sum_{j=4}^{7} \eta_j}. \tag{12}$$

Since we assume our system is work-conserving, the mathematical expression of (12) has the same form as that shown in (5). Substitute (12) into (11), we have

$$W_{i,cell}^{(2)} = \frac{W_0 + \bar{m}_i x + \sum_{j=4, j \neq i}^{7} \eta_j (\bar{m}_j - \bar{m}_i) x}{1 - \sum_{j=4}^{7} \eta_j}, \quad i = 4, \cdots, 7. \tag{13}$$

Similarly, the mean cell waiting time of type-i (i=1, 2

and 3) packets in the lower priority queue, denoted by $W^{(1)}_{i,cell}$, can be written as

$$W^{(1)}_{i,cell} = W_0 + \left(\sum_{k=1}^{3} \eta_k W^{(1)}_{k,cell} + \sum_{k=4}^{7} \eta_k W^{(2)}_{k,cell}\right) + \bar{m}_i x + \sum_{k=4}^{7} \eta_k W^{(1)}_{i,cell},\ i = 1, 2 \text{ and } 3. \tag{14}$$

Combine (11)–(12) and (14), it can be shown that

$$\sum_{k=1}^{3} \eta_k W^{(1)}_{k,cell} + \sum_{k=4}^{7} \eta_k W^{(2)}_{k,cell} = \frac{\rho W_0 + \sum_{k=1}^{7} \eta_k \bar{m}_k x}{1-\rho}, \tag{15}$$

where $\rho = \sum_{k=1}^{7} \eta_k$. Clearly, (15) also satisfies the *conservation law.*
Substitute (15) into (14), we have

$$W^{(1)}_{i,cell} = \frac{W_0 + \bar{m}_i x + \sum_{j=1,j\neq i}^{7} \eta_j (\bar{m}_j - \bar{m}_i)x}{(1-\rho)(1-\sum_{j=4}^{7} \eta_j)}, \quad i = 1,\ 2 \text{ and } 3. \tag{16}$$

Now, we derive the mean packet waiting time for type-i packets ($i = 4, \cdots, 7$) in the higher priority queue, denoted by $W^{(2)}_{i,pkt}$. That is, a type-i packet spends on the average $W^{(2)}_{i,pkt}$ sec in the higher priority queue until its last cell receives service. Similar to (10), $W^{(2)}_{i,pkt}$ can be written as

$$W^{(2)}_{i,pkt} = W_0 + \left(\sum_{k=4}^{7} \eta_k W^{(2)}_{k,cell}\right) + (\bar{g}_i - 1)x, \quad i = 4, \cdots, 7. \tag{17}$$

Substitute (12) into (17), we have

$$W^{(2)}_{i,pkt} = \frac{W_0 + \sum_{j=4}^{7} \eta_j \bar{m}_j x}{1-\sum_{j=4}^{7} \eta_j} + (\bar{g}_i - 1)x, \quad i = 4, \cdots, 7. \tag{18}$$

Similarly, the mean packet waiting time of type-i (i=1, 2 and 3) packets in the lower priority queue, denoted by $W^{(1)}_{i,pkt}$, can be written as

$$W^{(1)}_{i,pkt} = W_0 + \left(\sum_{k=1}^{3} \eta_k W^{(1)}_{k,cell} + \sum_{k=4}^{7} \eta_k W^{(2)}_{k,cell}\right) + (\bar{g}_i - 1)x + \sum_{k=4}^{7} \eta_k W^{(1)}_{i,pkt}, \quad i = 1,\ 2 \text{ and } 3. \tag{19}$$

Substitute (15) into (19), we have

$$W^{(1)}_{i,pkt} = \frac{W_0 + \sum_{j=1}^{7} \eta_j \bar{m}_j x}{(1-\rho)(1-\sum_{j=4}^{7} \eta_j)} + \frac{(\bar{g}_i - 1)x}{1-\sum_{j=4}^{7} \eta_j}, \quad i = 1,\ 2 \text{ and } 3. \tag{20}$$

Case 2: N = 3

In this case, input packet streams belonging to seven traffic types have to be merged to three priority queues. If we assume BK, BE and EE three lower priority types (type-1~type-3) are still mapped to the lowest priority queue. And CL and VI two types are segregated from VO and NC, and are mapped to a medium priority queue. The remaining VO and NC are mapped to the highest priority queue. Similar to the case of N=2, the mean cell waiting time of type-i (i=6 and 7) packets in the highest priority queue, denoted by $W^{(3)}_{i,cell}$, can be written as

$$W^{(3)}_{i,cell} = W_0 + \left(\sum_{k=6}^{7} \eta_k W^{(3)}_{k,cell}\right) + \bar{m}_i x, \quad i = 6 \text{ and } 7. \tag{21}$$

Solve (21), we obtain

$$W^{(3)}_{i,cell} = \frac{W_0 + \bar{m}_i x + \sum_{j=6,j\neq i}^{7} \eta_j (\bar{m}_j - \bar{m}_i)x}{1-\sum_{i=6}^{7} \eta_j}, \quad i = 6 \text{ and } 7. \tag{22}$$

and

$$\sum_{k=6}^{7} \eta_k W^{(3)}_{k,cell} = \frac{\sum_{j=6}^{7} \eta_j W_0 + \sum_{j=6}^{7} \eta_j \bar{m}_j x}{1-\sum_{j=6}^{7} \eta_j}. \tag{23}$$

Clearly, (22) and (23) have the same expressions of (13) and (12), respectively. This is because our system is work conserving. Consequently, the mean cell waiting time of type-i (i=4 and 5) packets in the medium priority queue, denoted by $W^{(2)}_{i,cell}$, can be obtained as

$$W^{(2)}_{i,cell} = \frac{W_0 + \bar{m}_i x + \sum_{j=4,j\neq i}^{7} \eta_j (\bar{m}_j - \bar{m}_i)x}{(1-\eta_6-\eta_7)(1-\sum_{j=4}^{7} \eta_j)}, \quad i = 4 \text{ and } 5. \tag{24}$$

For the mean cell waiting time of type-i (i=1, 2 and 3) packets in the lowest priority queue, denoted by $W^{(1)}_{i,cell}$, it is not difficult to understand that the solution of $W^{(1)}_{i,cell}$ (i=1, 2 and 3) shown in (16) for N=2 is also the solution for N=3.

Now, we derive the mean packet waiting time for type-i packets (i=6 and 7) in the highest priority queue, denoted by $W^{(3)}_{i,pkt}$. Similar to (17), $W^{(3)}_{i,pkt}$ can be written as

$$W^{(3)}_{i,pkt} = W_0 + \left(\sum_{k=6}^{7} \eta_k W^{(3)}_{k,cell}\right) + (\bar{g}_i - 1)x, \quad i = 6 \text{ and } 7. \tag{25}$$

Substitute (23) into (25), we have

$$W^{(3)}_{i,pkt} = \frac{W_0 + \sum_{j=6}^{7} \eta_j \bar{m}_j x}{1 - \sum_{j=6}^{7} \eta_j} + (\bar{g}_i - 1)x, \quad i = 6 \text{ and } 7. \tag{26}$$

Similarly, the mean packet waiting time of type-i (i=4 and 5) packets in the medium priority queue, denoted by $W^{(2)}_{i,pkt}$, can be obtained as

$$W^{(2)}_{i,pkt} = \frac{W_0 + \sum_{j=4}^{7} \eta_j \bar{m}_j x}{(1 - \eta_6 - \eta_7)(1 - \sum_{j=4}^{7} \eta_j)} + \frac{(\bar{g}_i - 1)x}{1 - \eta_6 - \eta_7}, \quad i = 4 \text{ and } 5. \tag{27}$$

For the mean packet waiting time of type-i (i=1, 2 and 3) packets in the lowest priority queue, denoted by $W^{(1)}_{i,pkt}$, again we notice that the solution shown in (20) for N=2 is also the solution for N=3.

Case 3: $N = 4$

In this case, input packet streams belonging to seven traffic types have to be merged to four priority queues. If we assume BE and EE two types are further segregated from BK, and are mapped to an additional medium-low priority queue. And the mapping rules of the other types are the same as those specified in the previous case. Then, we have

$$W^{(4)}_{i,cell} = \frac{W_0 + \bar{m}_i x + \sum_{j=6,j\neq i}^{7} \eta_j (\bar{m}_j - \bar{m}_i)x}{1 - \sum_{j=6}^{7} \eta_j}, \quad i = 6 \text{ and } 7. \tag{28}$$

$$W^{(3)}_{i,cell} = \frac{W_0 + \bar{m}_i x + \sum_{j=4,j\neq i}^{7} \eta_j (\bar{m}_j - \bar{m}_i)x}{(1 - \eta_6 - \eta_7)(1 - \sum_{j=4}^{7} \eta_j)}, \quad i = 4 \text{ and } 5. \tag{29}$$

$$W^{(2)}_{i,cell} = \frac{W_0 + \bar{m}_i x + \sum_{j=2,j\neq i}^{7} \eta_j (\bar{m}_j - \bar{m}_i)x}{(1 - \sum_{j=4}^{7} \eta_j)(1 - \sum_{j=2}^{7} \eta_j)}, \quad i = 2 \text{ and } 3. \tag{30}$$

and

$$W^{(1)}_{1,cell} = \frac{W_0 + \bar{m}_1 x + \sum_{j=2}^{7} \eta_j (\bar{m}_j - \bar{m}_1)x}{(1 - \rho)(1 - \sum_{j=2}^{7} \eta_j)}. \tag{31}$$

Obviously, the results shown in (28) and (29) are exactly the same as the results shown in (22) and (24), respectively.

Also, $W^{(4)}_{i,pkt}$ (i=6 and 7) and $W^{(3)}_{i,pkt}$ (i=4 and 5) have the same results as those shown in (26) and (27), respectively. Furthermore, we can easily obtain

$$W^{(2)}_{i,pkt} = \frac{W_0 + \sum_{j=2}^{7} \eta_j \bar{m}_j x}{(1 - \sum_{j=4}^{7} \eta_j)(1 - \sum_{j=2}^{7} \eta_j)} + \frac{(\bar{g}_i - 1)x}{1 - \sum_{j=4}^{7} \eta_j}, \quad i = 2 \text{ and } 3. \tag{32}$$

and

$$W^{(1)}_{1,pkt} = \frac{W_0 + \sum_{j=1}^{7} \eta_j \bar{m}_j x}{(1 - \rho)(1 - \sum_{j=2}^{7} \eta_j)} + \frac{(\bar{g}_1 - 1)x}{1 - \sum_{j=2}^{7} \eta_j}. \tag{33}$$

4. Numerical Results

In this section, we use the analytical results derived in the previous section to evaluate the performance of individual traffic types in different queue mappings under various traffic conditions. Both mean cell waiting and mean packet waiting are observed for each traffic type. Table 1 describes the characteristics of data packets belonging to each individual traffic type in IP-over-ATM networks. Here we assume seven traffic types as suggested in [4] are considered. We also assume the packet length of each type is uniformly distributed between a minimum value (MIN) and a maximum value (MAX) of that type. The link capacity we used is 155 Mbps. Since the server utilization of the traffic type-i, η_i, can be used to represent the mean offered load of type-i traffic to the server, a specific traffic condition can be denoted by a 7-tuple vector $(\eta_1, \eta_2, \eta_3, \eta_4, \eta_5, \eta_6, \eta_7)$. Figure 4 shows the mean cell waiting times and the mean packet waiting times versus the total server utilization under the condition η_i=$\rho/7$, for all i. That is the server utilization of individual traffic types shares the same fraction of the total server utilization. Here we further assume the switch supports seven priority classes on each output port. This will simplify the traffic type to priority class mapping to have a one-to-one correspondence. In Fig. 4, seven solid curves and seven dotted curves represent $W_{i,cell}$ and $W_{i,pkt}$ i=1,⋯, 7, respectively. We observe that $W_{i,pkt}$ is always greater than $W_{i,cell}$ for all i, and the gap between $W_{i,cell}$ and $W_{i,pkt}$

Table 1 Characteristics of data packets in IP-over-ATM networks.

Priority	Traffic Type	Packet length MIN MAX	$\bar{g}_i$	$\bar{g}_i^{(2)}$	$\bar{m}_i$
7	NC	40 ~64	2	4	0.5
6	VO	64	2	4	0.5
5	VI	180	5	25	2
4	CL	150 ~ 300	5.52	31.44	2.34
3	EE	200 ~ 500	8.21	69.32	3.62
2	BE	300 ~ 1000	14.37	224.37	7.31
1	BK	500 ~ 2000	26.87	803.17	14.45

Note 1: packet length in byte.

Note 2: $g_i = \left\lceil \frac{packet\ length + 16}{48} \right\rceil$

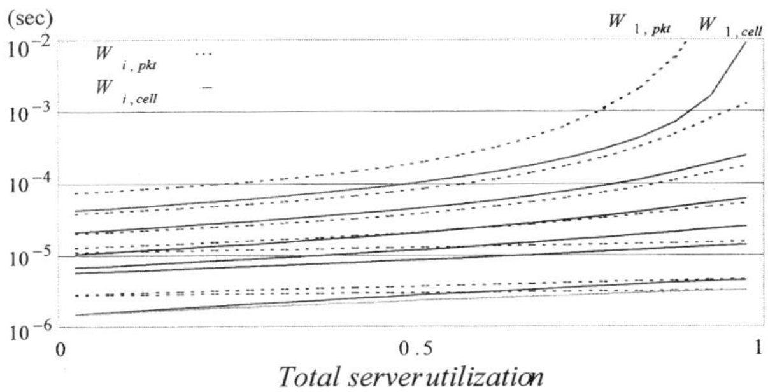

Fig. 4 $W_{i,cell}$ and $W_{i,pkt}$ vs. ρ under $\eta_i = \rho/7$.

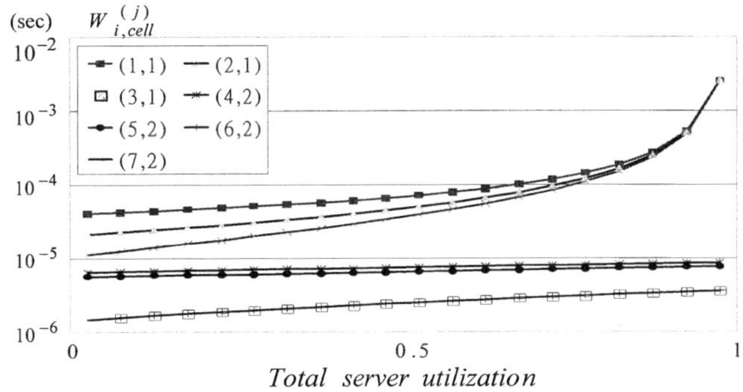

Fig. 5 $W^{(j)}_{i,cell}$ vs. ρ ($\rho^{(I)} = 6\rho/7$, $\rho^{(II)} = \rho/7$).

decreases as the value of i increases. From (1) and (6), we have that $W_{7,pkt} - W_{7,cell} = (\bar{g}_7 - \bar{m}_7 - 1)x = x/2$, which is a constant. Due to the scaling effect, this constant gap is getting smaller as ρ approaches to 1. The gap between $W_{i,cell}$ and $W_{i,pkt} (i \neq 7)$ is getting wider as ρ becomes larger (this phenomenon is not so evident in Fig. 4 due to the scaling effect). The reason of why $W_{7,pkt} = W_{6,pkt}$ and $W_{7,cell} = W_{6,cell}$ as ρ approaches to 0 is because the queueing effect is negligible in this case, and both type-6 and type-7 traffic have the same values of $\bar{g}_i$ and $\bar{m}_i$. As ρ is getting larger, the queueing effect becomes the key dominator of the mean waiting times of every traffic types, especially for those lower priority traffic types.

Figures 5–10 display the performance of mean cell waiting times for N=2 and N=4 under three different traffic conditions. If we let $\rho^{(I)} = \eta_1 + \eta_2 + \eta_3$ denote the aggregate server utilization of the lower three traffic types including BK, BE and EE, and let$\rho^{(II)} = \eta_4 + \eta_5 + \eta_6 + \eta_7$ denote the aggregate server utilization of the higher four traffic types including CL, VI, VO and NC. The first traffic condition considered in Figs. 5 and 8 assumes $\rho^{(I)} = 6\rho/7$, $\eta_1 = \eta_2 = \eta_3 = \rho^{(I)}/3$; and $\rho^{(II)} = \rho/7$, $\eta_4 = \eta_5 = \eta_6 = \eta_7 = \rho^{(II)}/4$. In this case, the total server utilization is heavily dominated by the data of lower priority types. The second traffic condition considered in Figs. 6 and 9 assumes all $\eta_i = \rho/7$, for i=1, 2,$\cdots$, 7. The third traffic condition considered in Figs. 7 and 10 assumes $\rho^{(I)} = \rho/7$, $\eta_1 = \eta_2 = \eta_3 = \rho^{(I)}/3$; and $\rho^{(II)} = 6\rho/7$, $\eta_4 = \eta_5 = \eta_6 = \eta_7 = \rho^{(II)}/4$. In this case, the total server utilization is heavily dominated by the data of higher priority types. Figures 5, 6 and 7 show the results for N=2. Figures 8, 9 and 10 display the results for N=4. In each figures we see only six curves but not seven curves. This is because VO and NC traffic types are mapped to the same queue and the values of $\bar{g}_i$ and $\bar{m}_i$ are the same. Hence the two curves for VO and NC respectively are overlapped. Note that in the legend of each figure we use a (i, j) pair to represent the curve of $W^{(j)}_{i,cell}$.

From Figs. 5, 6 and 7, we observe that for individual traffic types the six curves shown in Fig. 7 performs

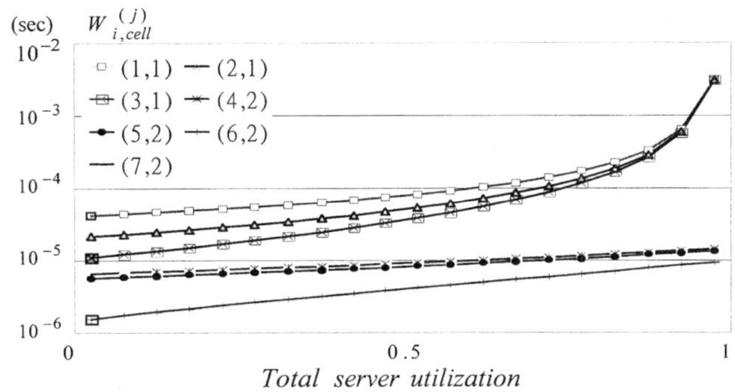

Fig. 6 $W^{(j)}_{i,cell}$ vs. ρ ($\rho^{(I)} = 3\rho/7$, $\rho^{(II)} = 4\rho/7$).

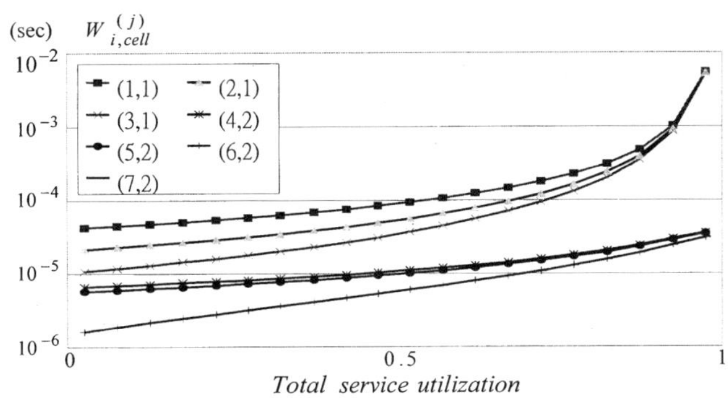

Fig. 7 $W^{(j)}_{i,cell}$ vs. ρ ($\rho^{(I)} = \rho/7$, $\rho^{(II)} = 6\rho/7$).

the worst. The result shown in Fig. 5 performs the best. This implies that the network performance will become worse if the mean offered load to the server is dominated with the higher priority types. Same observation can be obtained from Figs. 8, 9 and 10. Now we consider the effect of queue mappings to the performance of all traffic types. The difference between N=2 and N=4 is that we segregate BK from BE and EE, and segregate CL, VI from VO and NC. From Figs. 5 and 8, we notice that the mean waiting times of type-2 and type-3 traffic are improved dramatically after segregation. But the improvement of type-6 and type-7 traffic is only slightly. This is because in Fig. 5 the total traffic load is dominated with lower priority types. On the

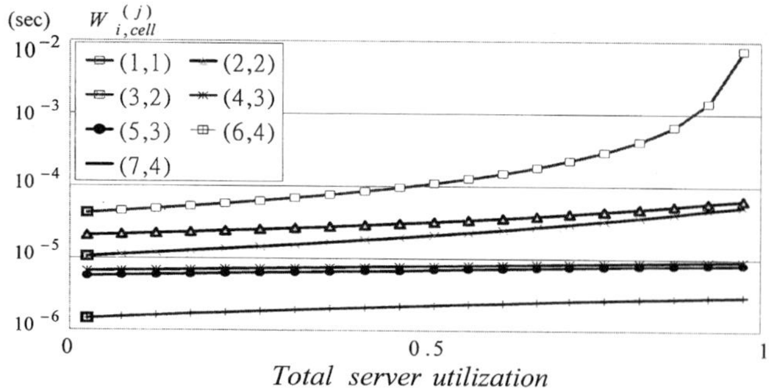

Fig. 8 $W^{(j)}_{i,cell}$ vs. ρ ($\rho^{(I)} = 6\rho/7$, $\rho^{(II)} = \rho/7$).

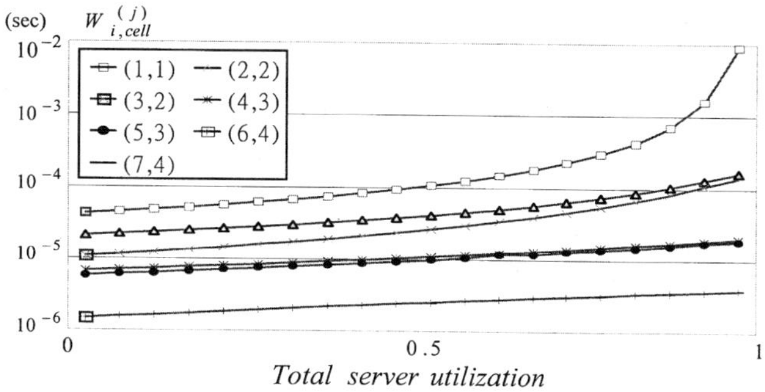

Fig. 9 $W^{(j)}_{i,cell}$ vs. ρ ($\rho^{(I)} = 3\rho/7$, $\rho^{(II)} = 4\rho/7$).

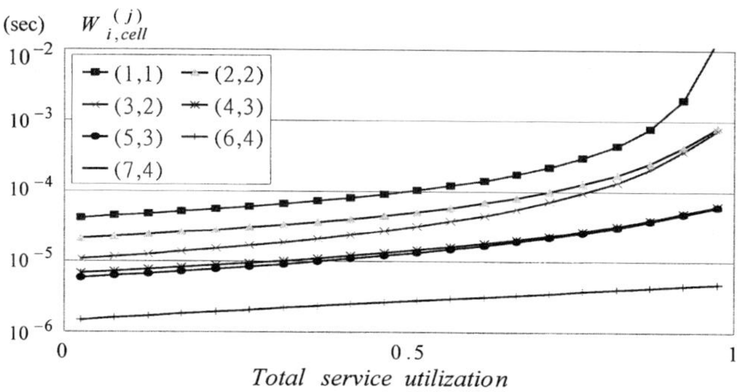

Fig. 10 $W^{(j)}_{i,cell}$ vs. ρ ($\rho^{(I)} = \rho/7$, $\rho^{(II)} = 6\rho/7$).

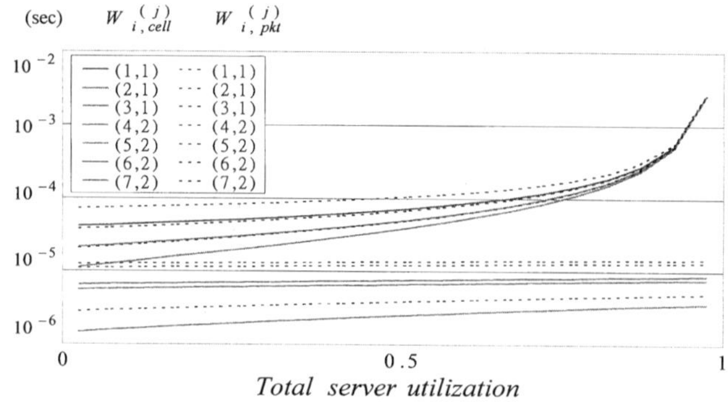

Fig. 11 W vs. ρ for N=2, $\rho^{(I)} = 6\rho/7$, $\rho^{(II)} = \rho/7$.

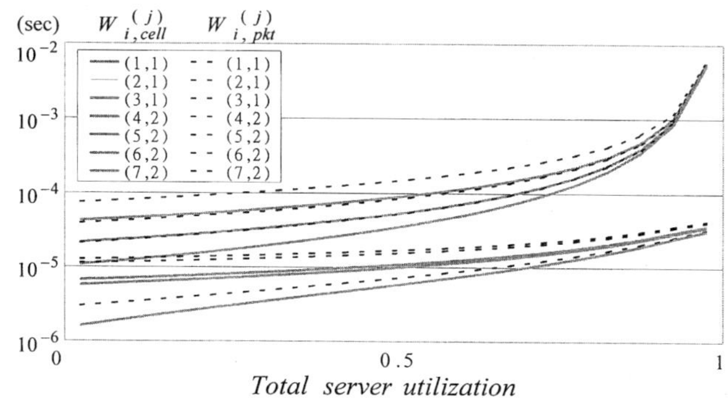

Fig. 12 W vs. ρ for N=2, $\rho^{(I)} = \rho/7$, $\rho^{(II)} = 6\rho/7$.

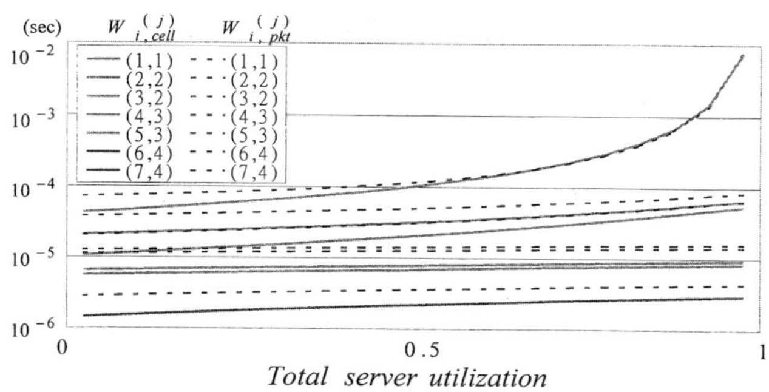

Fig. 13 W vs. ρ for N=4, $\rho^{(I)} = 6\rho/7$, $\rho^{(II)} = \rho/7$.

contrary, in Figs. 7 and 10 the total traffic load is assumed to be dominated with higher priority traffic. We observe that the performance improvement of type-2 and type-3 traffic after segregation is much less than that we observed in Figs. 5 and 8. However, the performance improvement of type-6 and type-7 traffic is more evident than that we found in Figs. 5 and 8.

For N=3, we assume CL and VI are segregated from VO and NC, and BK, BE and EE are still mapped to the lowest priority queue. The numerical result for N=3 can be obtained by putting the three curves of lower priority types in N=2 and the four curves of higher priority types in N=4 together. In fact, we may easily have another mapping rule for N=3. By putting the four curves of higher priority types in N=2 and the three curves of lower priority types in N=4 together, we obtain the numerical result for N=3 with the following new mapping rule. Traffic type BK is segregated from BE and EE, and CL, VI, VO and NC are mapped to the same highest priority queue.

Figures 11–14 display the performance of both mean cell waiting times and mean packet waiting times for N=2 and N=4 under two traffic conditions considered in Fig. 5, respectively. Similar to Fig. 4, here we observe that $W^{(j)}_{i,pkt}$ is always greater than $W^{(j)}_{i,cell}$ for all (i, j) pair in each figure, and both values increase with the same trend as ρ is getting larger. The gap between $W^{(j)}_{i,pkt}$ and $W^{(j)}_{i,cell}$ decreases as the value of i increases. Moreover, we notice that both mean cell waiting time and mean packet waiting time for individual traffic types shown in Fig. 11 perform better than

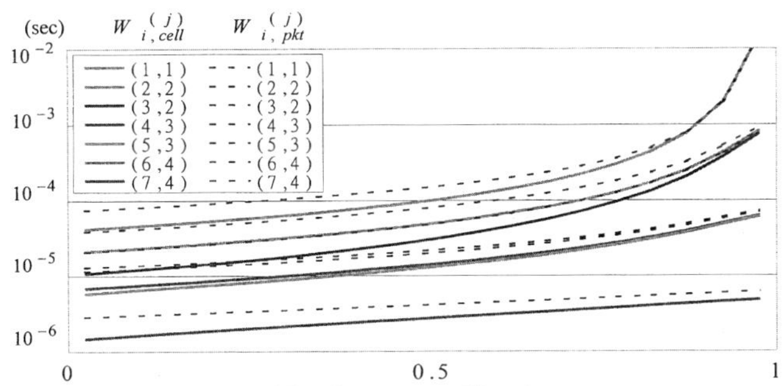

Fig. 14 W vs. ρ for N=4, $\rho^{(I)} = \rho/7$, $\rho^{(II)} = 6\rho/7$.

the corresponding results shown in Fig. 12. This observation also can be seen from Figs. 13 and 14. This implies that the network performance, including mean cell waiting and mean packet waiting for each traffic type, will become worse if the total traffic load is dominated with higher priority types.

5. Conclusions and Discussions

In this paper, we have investigated the performance of packet-level scheduling in IP-over-ATM networks with QoS control. That is, we assume multiple traffic classes are supported on each output port in the switch. We analyzed the performance of different queue mappings between traffic types and the number of available priority queues (traffic classes). For different queue mappings, closed-form solutions for the mean cell waiting time and the mean packet waiting time of individual traffic types are presented. To evaluate the performance of individual traffic types (including mean cell waiting time and mean packet waiting time) in different queue mappings under various traffic conditions. We considered three different traffic conditions and three different queue mappings (N=2, N=3 and N=4) in our numerical study. Performance comparisons and discussions were made for different queue mappings between different traffic conditions. Some remarkable observations are listed as follows:

1. $W^{(j)}_{i,pkt}$ is always greater than $W^{(j)}_{i,cell}$ for all (i, j) pair in each case we considered.
2. Both values of $W^{(j)}_{i,pkt}$ and $W^{(j)}_{i,cell}$ increase with the same trend as ρ is getting larger.
3. The gap between $W^{(j)}_{i,pkt}$ and $W^{(j)}_{i,cell}$ decreases as the value of i increases.
4. For N=4, if the total traffic load is dominated with lower priority types, we notice that the mean waiting times of type-2 and type-3 traffic are improved dramatically after segregation. But the improvement of type-6 and type-7 traffic due to traffic segregation is only slightly.
5. For N=4, if the total traffic load is dominated with higher priority types, we notice that the mean waiting times of type-6 and type-7 traffic are improved dramatically after segregation. But the improvement of type-2 and type-3 traffic is much limited (as compared to the previous observation).
6. The network performance, including mean cell waiting and mean packet waiting for each traffic type, will become worse if the total traffic load is dominated with higher priority types.

References

[1] C. Fulton, S.Q. Li, and A. Lin, "Impact analysis of packet-level scheduling on an ATM shared-memory switch," Proc. INFOCOM'98, pp.947–954, 1998.
[2] I. Widjaja and A.I. Elwalid, "Performance issues in VC-merge capable switches for IP over ATM networks," Proc. INFOCOM'98, pp.372–380, 1998.
[3] LightStream 1010 Multiservice ATM Switch, Cisco Systems, Inc. http://www.cisco.com.
[4] ISO/IEC Final DIS 15802-3, Information Technology—local and metropolitan area networks—common specifications—Part 3: MAC bridges, 1998.
[5] C. Bisdikian, "A note on the conservation law for queues with batch arrivals," IEEE Trans., vol.COM-41, no.6, pp.832–835, 1993.
[6] L. Kleinrock, Queueing Systems, Volume II: Computer Applications, John-Wiley & Sons, Inc., 1976.

High-Performance Backbone Network Technology edited by Naoaki Yamanaka
ISBN: 0-8247-5321-6 DOI: 10.1081/(120025692)
Published 2004 by Marcel Dekker, Inc.

A Study on Generalization of Packet Discard Schemes for TCP over ATM

Manhee Jo and Yoshihiko Ebihara

SUMMARY In order to ease the impact of the packet fragmentation problem and to avoid network congestion in TCP over UBR, packet discard schemes in ATM layer (such as PPD and EPD) have been proposed. These schemes drop packets before they reach their intended destinations if the network is congested and the packets are to be partially discarded. On the other hand, TCP also regulates data flow with its own flow control method. Due to restriction of data flow at the TCP layer, buffer space is not fully used in an ATM switch. In order to make use of more buffer resources, this paper generalizes the PPD and EDP schemes. From this generalization, an optimistic packet discard scheme named the "Probability-Based Delayed Packet Discard" (PDPD) scheme is proposed. Depending on a particular probability, this scheme sets a discard flag to delay actual discard operation. This paper presents the results of several simulated models to find out the potential of improvement of goodput by PDPD. The results of these simulations indicate that PDPD obtains higher goodput than ordinary schemes when the packet size is large and the input load is not light. This author concludes that a PDPD scheme should achieve effective goodput and link utilization while using more buffer resources effectively.

key words: *TCP over ATM, EPD, PDPD, congestion control, UBR*

1. Introduction

ATM has been adopted for B-ISDN as a means to carry a variety of services having different characteristics. It is a connection-oriented, 53-octet-cell-based transport mechanism which enables high speed switching. Because of its features, ATM is one of the most prominently used techniques to support current backbone networks.

TCP is one of the most popular transmission protocols being used by a number of users world wide. To support TCP applications over ATM, ABR service category, as well as UBR, is applied on a best-effort basis. Flow control algorithms in ABR rely on closed-loop feedback information by means of periodical Resource Management (RM) cells. A number of intelligent and systematic congestion control schemes associated with ABR to improve quality of services and/or throughput have been proposed and studied [1]–[4]. While ABR services have been receiving increasing attention in recent years as a research topic concerning quality of services, the function of ABR is much more complex than that of UBR. Therefore, taking into account that the implementation complexity and cost of ABR are significantly higher than UBR, we choose UBR as a basic service category to support TCP applications over ATM in this paper.

A TCP packet should be segmented into fixed sized cells in order to be transmitted over ATM networks. One of the major shortcomings of TCP over ATM is due to this packet fragmentation. If a cell gets lost within the network, the receiver TCP cannot reassemble the original packet. Thus, the source TCP has to retransmit the entire packet because ATM does not support reliable services, while TCP does. Therefore, provided one cell is lost, it is useless to relay its subsequent cells as well as the already-buffered cells of the corresponding packet. Even worse, the transfer of these void cells may lead to network congestion itself.

In order to prevent not only this unnecessary transmission but also network congestion, a number of studies about packet discard policies have been conducted [5]–[8]. Generally, cells are discarded aggressively to avoid and to solve the problem using these policies. However, because of this aggressive discard, network resources sometimes are not fully utilized. In this paper, we studied these type of packet discard schemes to seek the feasibility of better utilization of network resources. We first succeeded in generalizing those schemes in a simple form. As a result of this generalization, we propose a packet discard scheme using a certain probability. This scheme sets a discard flag depending on a particular probability to delay actual discard operation. We believe this scheme will make use of more network resources to obtain an effective transmission rate.

First, in Sect. 2, we describe an overview of TCP congestion control and several packet discard policies. Next, in Sect. 3, we show our preliminary simulation. Thereafter, we generalize packet discard schemes and propose the "Probability-Based Delayed Packet Discard" (PDPD) scheme to prove the feasibility of higher network utilization. We show the simulation results in Sect. 4 and then conclude in Sect. 5.

Manuscript received April 4, 2000.
Manuscript revised July 31, 2000.

[†]The author is with the Doctoral Program in Engineering, University of Tsukuba, Tsukuba-shi, 305-8573 Japan.
[††]The author is with the Institute of Information Sciences and Electronics, University of Tsukuba, Tsukuba-shi, 305-8573 Japan.

IEICE TRANS. COMMUN., Vol. E84-B, No. 2, pp. 275–283, February 2001

2. Overview of Congestion Control Strategies

2.1 TCP Congestion Controls

TCP is a widely used transmission protocol to service diverse user applications. Modern TCP implements end-to-end algorithms based on a congestion window (*cwnd*) not only to support reliable services but also to control network congestion. TCP implementations include *slow start*, *congestion avoidance*, *fast retransmission*, and *fast recovery* [9], [10]. When a connection is newly set up or a transmission restarts after loss of a packet, the TCP source sends one packet and waits for an acknowledgement (ACK) from its peer receiver. With the slow start mechanism, whenever the TCP sender receives an ACK, the *cwnd* is increased by one packet. This makes the congestion window size twice every round-trip time (RTT). When a TCP sender has data to transfer, the amount of data to be transmitted is based on the minimum between the receiver's advertised window size and the *cwnd* size at most. Using the slow start policy, *cwnd* is increased exponentially until it reaches a predefined threshold value (*slow start threshold*). If the *cwnd* size crosses over this threshold value, the congestion avoidance phase begins, which reduces the increasing ratio of *cwnd* to alleviate network congestion. The congestion window is increased one packet each round-trip time at most. In other words, the *cwnd* is increased linearly in this phase. When a fast retransmit algorithm is used, three duplicated ACKs (i.e., acknowledgements with the same sequence number) trigger retransmission of the packet without waiting for the expiration of the retransmission timer. Fast recovery is an intelligent congestion window control scheme to avoid drastic throughput degradation. With this fast recovery, once a packet is detected to be lost, the congestion window is set to half its previous value instead of beginning slow start. Thus, the congestion window size is recovered quickly and effectively. By these intelligent algorithms, TCP can achieve quite good performance for normal TCP applications.

2.2 The Fragmentation Problem of TCP over ATM

In order to relieve the packet fragmentation problem of TCP over ATM, some simple strategies have been studied which save most of packets by sacrificing a small number of certain packets. The Partial Packet Discard scheme is one of them [8]. In this strategy, when a cell is dropped, an ATM switch marks the packet and discards subsequent cells of the same packet except the last cell (i.e., the tail part of the packet). This is why PPD is also called Packet Tail Discard [5]. PPD can partially reduce forwarding of those needless cells. However, since part of the packet is already queued and waiting to be transmitted, the switches still try to relay the cells belonging to the useless packets.

In contrast, EPD (Early Packet Discard) scheme [6] prevents forwarding of needless cells more completely. In the EPD scheme, a certain buffer occupancy level is set as a threshold value which is used by an ATM switch to estimate congestion state of a network. If a new packet arrives when a queue length exceeds the threshold, the packet is not allowed to be buffered. Thus, by this EPD scheme, the entire packet except the last cell is discarded while only part of the packet is discarded by the PPD scheme. In both PPD and EPD schemes a new packet is identified by an end of packet (EOP) flag in a cell.

Numerous studies of the EPD scheme have been conducted. Lapid, Rom, and Sidi have examined the performance of the EPD scheme using an analytical model with different load and different arrival processes [11]. These researchers showed that the EPD policy performs better than the PPD policy and no control policy when the load is not light. Meanwhile, one of the most important parameters of EPD is the threshold value. If the threshold is too large, the EPD scheme cannot take advantage of early discard of packets because the queue length reaches its end too quickly to hold both new packets and the rest of the already received ones. When the threshold is so small that a number of new packets are prohibited from entering, buffer utilization as well as the propagation rate is reduced. Therefore, determination of the threshold value has been an important factor of EPD. In [12], Jaussi, Lorang, and Nelissen evaluated EPD using a simple model. They showed that the spare space of one packet up to three packets is enough to achieve moderate performance. Basic TCP, however, is not said to fully utilize network resources in high-speed networks because of the limitation of its congestion window size. In addition, since TCP regulates its own streams by means of acknowledgements, it cannot send any more packets while it is waiting for a corresponding acknowledgement. Thus, if cells from TCP packets are controlled by an EPD algorithm in the ATM layer, the buffer in the ATM switch is not fully utilized, as Okuda has pointed out [13]. Therefore, if this unused buffer space can be exploited more efficiently, it should result in higher TCP performance.

Here, we show the possibility of utilizing more buffer space through generalization of ordinary packet discard schemes.

3. Probability-Based Delayed Packet Discard Schemes

Reaction to congestion detection should be started as quickly as possible. Thus, complicated schemes are not recommended to control flow and to avoid congestion. EPD meets this requirement for simplicity and is an effective way to avoid congestion. Nevertheless,

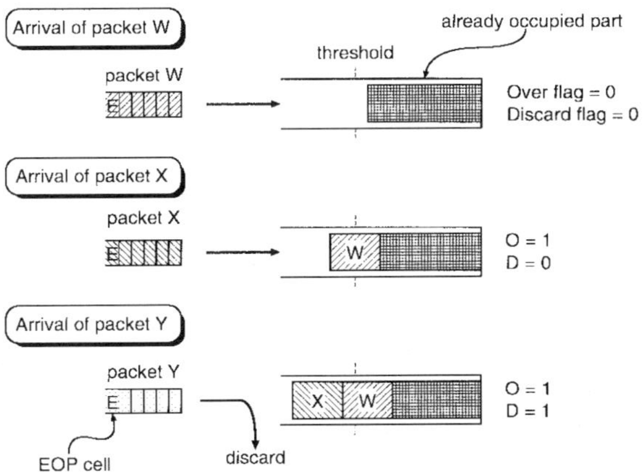

Fig. 1 Introduction of preliminary PDPD scheme.

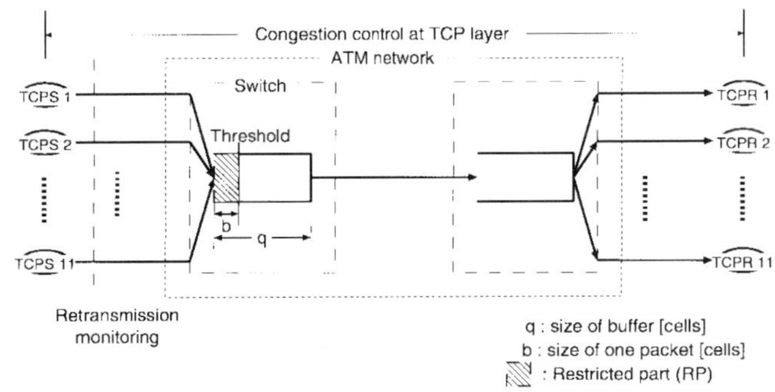

Fig. 2 Simulation model of preliminary PDPD.

TCP monitors traffic and regulates it at an upper layer than the ATM layer. Thus, EPD does not have many chances to work. Although EPD begins to work after the queue length is larger than the threshold, the data stream must have already been controlled by TCP at the higher layer. Therefore, when TCP packets are transmitted over ATM, EPD may regulate more network traffic at the cell level which reduces the utilization of the buffer in a switch. We first generalize these packet discard methods to propose a new scheme which is expected to help exploit more available network resources.

3.1 Preliminary Experiments of the Probability-Based Delayed Packet Discard Scheme

Before we describe our new scheme, we will show the results of a preliminary simulation experiment [14] to show the potential for the availability of buffer space in a switch. In this paper, we call this "pre-PDPD." Figure 1 depicts a brief introduction of the pre-PDPD scheme. In EPD, if a queue length in a switch grows larger than the predefined threshold value, a *Discard* (D) flag is set and newly arrived packets are discarded for the duration this D flag is set. This hasty recourse, however, hinders the network from using more available buffer resources in a switch.

Pre-PDPD provides an additional *Over* (O) flag to indicate whether the queue length is larger than the threshold. If it is larger than the threshold, an O flag is set but a D flag is not. Provided a new packet arrives when an O flag is set and a D flag is not set, the packet is allowed to be buffered setting the D flag on. Actual discard begins when O and D flags are set. Once these flags are set, they will be reset only when the queue length shrinks to less than the threshold. Pre-PDPD is an optimistic algorithm. It appears to be identical with the EPD scheme with a higher threshold value by one packet size. Obviously, both the EPD scheme and the pre-PDPD scheme are identical in the D-flag-setting phase. In the D-flag-resetting phase, however, pre-PDPD has to suspend buffering longer than EPD operating with a higher threshold. This is the tradeoff incurred for using this optimistic scheme.

This pre-PDPD scheme simply permits one more new packet to enter the buffer even when the queue length is larger than the threshold. Considering the regulation of data flow by TCP congestion control schemes at a higher layer, we hypothesize that there remains available space in the buffer for a few more packets. We examined this pre-PDPD scheme in a simulation based on the model shown in Fig. 2. There are 11 TCP sources of a priority level connected to 11 TCP receivers via two ATM switches transmitting segmented cells back-to-back. Various flow controls of the Reno version are executed in the TCP level. User data arrives at TCP sources deterministically and ceaselessly. The maximum transmission rate of each source (from TCPS1 to TCPS11) is 150 Mbps. However, since TCP sources control their streams, the bandwidth does not seem to be fully used, which leaves available space for other streams. The propagation delay between a source and a receiver is about 50 microseconds. We call the buffer space over the threshold a *restricted part* (RP) in this paper because it restricts the entrance of new packets. A buffer can hold q cells, and RP size is b cells. Hence the threshold of buffer is $(q - b)$.

We show throughput of TCP sources when MTU m is 9180 octets for variable buffer size q (Fig. 3). The size of RP or b is about 400 cells, which corresponds to two-packet sizes. This simulation was verified by the repetition method [15]. Our results satisfy a confidence level of 90% and a confidence interval of 2%. Each point represents the total throughput of TCP sources using no specific policy (plain TCP or *ptcp*), EPD and pre-PDPD, respectively. In Fig. 3, as the buffer size increases, throughput is also increased for all policies. In addition, throughput of pre-PDPD is at most 10 Mbps higher than that of EPD when the buffer size is 2000 cells.

Figure 4 shows the snap shots of the queue length when EPD (upper graph) and pre-PDPD are used. The X-axis shows the time slot. One time slot is one cell time in this simulation. Considering the snapshots in

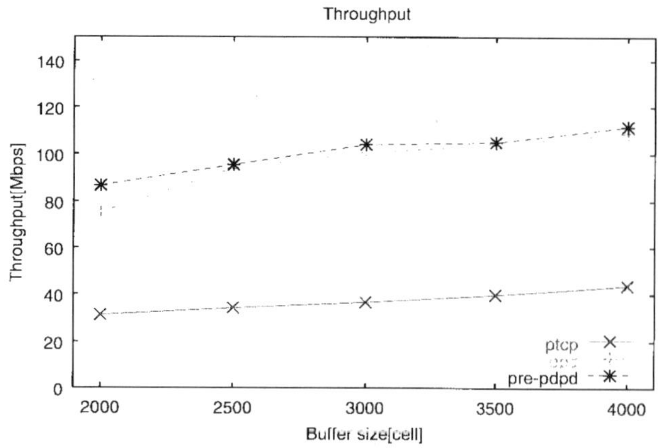

Fig. 3 Total throughput of TCP sources when m is 9180 octets.

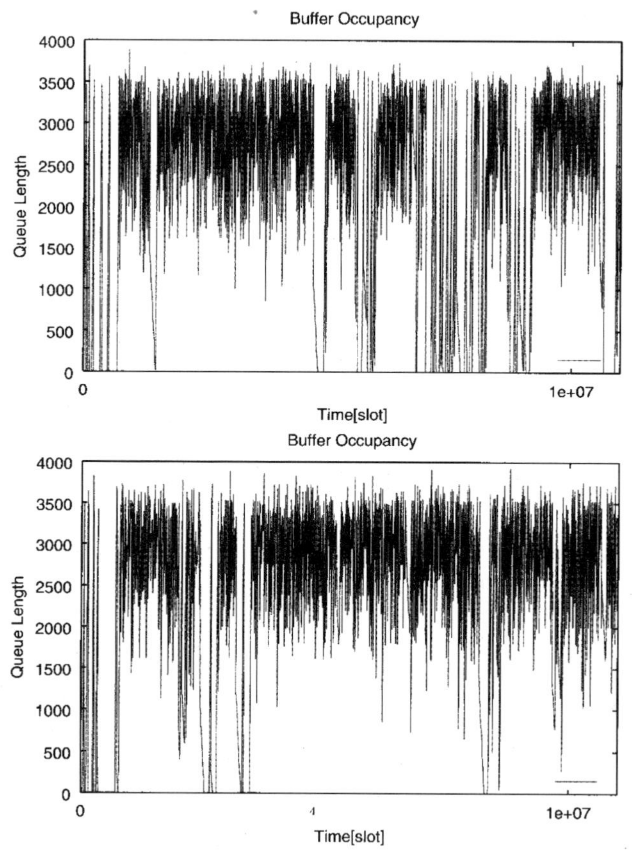

Fig. 4 Snapshots of the queue length in a switch using EPD (upper) and pre-PDPD. The MTU is 9180 octets and the buffer size is 4000 cells with an RP size of approximately 400 cells.

Fig. 4, RP (buffer space from 3600 to 4000) would appear to be more utilized in pre-PDPD than in EPD. Specifically, the average queue length of the pre-PDPD switch in this figure is approximately 16% longer than that of the EPD switch. It is true that there is a greater possibility that buffer overflow will occur in pre-PDPD rather than in EPD. However, since pre-PDPD is basically an optimistic policy to exploit unused network resources, when the higher layer regulates its streams, it is feasible to apply pre-PDPD speculatively. From these results, we learned that by permitting one more packet to enter the buffer, not only buffer utilization but also throughput is improved. Consequently, as TCP regulates the streams, aggressive dropping of cells by EPD makes more available buffer space and leaves it unused. Therefore, if we alleviate this aggressiveness of EPD, it is possible to use more buffer space efficiently, thereby increasing network throughput at the same time. We describe generalizing of ordinary packet discard schemes in the next subsection.

3.2 Generalization of Packet Discard Schemes

In EPD, a D or discard flag is set in two cases. In the first case, if a queue length gets larger than the predefined threshold value, a D flag is set, thereby preventing any cells of a new packet from entering the buffer. In the second case, the flag is set when any cells are dropped due to buffer overflow, for example. In contrast, PPD does not use this threshold value. It basically sets a D flag when a cell is dropped due to buffer overflow. In both PPD and EPD schemes, if the D flag is set, all the subsequent cells of the packet are discarded until the flag is reset. Comparing the characteristics of these two methods, we can understand that PPD is a particular form of EPD whose threshold value is the size of the buffer.

In addition, the PPD scheme can be understood as another specific form of EPD from another point of view. Let us assume that PPD also has the same threshold value as that of EPD. In PPD, even if the queue length is larger than the threshold, a D flag is never set. As the buffer becomes full and overflows, cells are dropped and the D flag is set at last, thereby dropping all cells of the tail part of a packet. Consequently, if a D flag is always set when a queue length is larger than the threshold, it is EPD. On the contrary, if the D flag is never set when a queue length is larger than the threshold, it is PPD. Therefore, if it is possible to set the discard flag with a probability, it is also possible to generalize these ordinary packet discard methods. In other words, PPD and EPD are two extreme cases of the generalized packet discard algorithm. Figure 5 simply shows the generalization of these packet discard schemes. As long as the queue length is less than the threshold, the D flag is a zero. Once a queue length gets larger than the threshold, PPD will set the D flag with a probability 0 while EPD will set the flag with a probability 1. In terms of implementation, it is clear that PDPD is slightly more complex than PPD and EPD. Despite this, however, the advantage PDPD offers is that it is still very simple. As in the EPD, the switch has to monitor the queue length. Thus, the decision of the flag setting in cases when the queue length is larger than the threshold can be calculated beforehand using this monitoring duration with one more instruction.

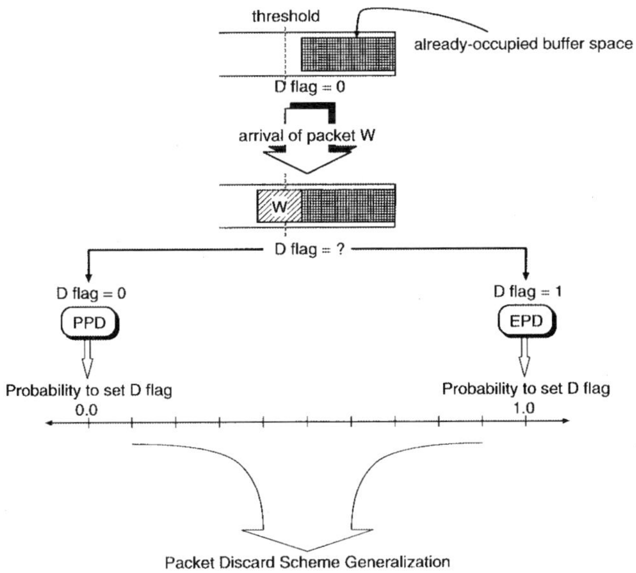

Fig. 5 Generalization of packet discard schemes.

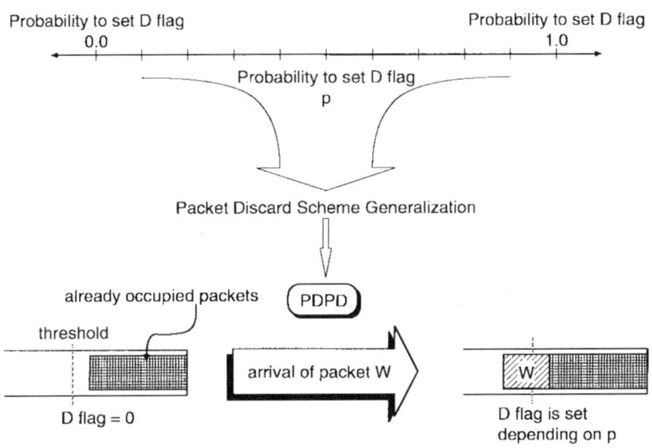

Fig. 6 Probability-Based Delayed Packet Discard scheme. D flag is set with a probability p.

3.3 Probability-Based Delayed Packet Discard Scheme

Based on the above formalization of packet discard schemes, we propose a Probability-Based Delayed Packet Discard or PDPD scheme (Fig. 6). In the PDPD scheme, when a queue length in a buffer is larger than the threshold, the D flag is set with a probability p. In other words, at probability $(1 - p)$, the D flag remains unset. This D flag is reset when a queue length becomes less than the threshold value. Since the actual dropping of the packet begins after the flag is set, as long as the flag is a 0, a new packet is allowed to be buffered. Provided the D flag is a 0 due to the probability p, when a new packet arrives and the queue length is larger than the threshold, the discarding operation is delayed until the D flag is set. This is a major difference between the original EPD and our PDPD scheme. It is also why we call our algorithm the "Probability-Based Delayed

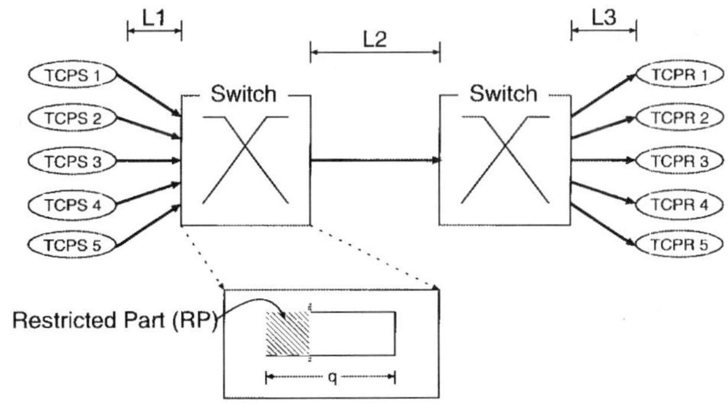

Fig. 7 Simulation model of Probability-Based Delayed Packet Discard scheme.

Packet Discard" scheme.

PDPD is a rather optimistic algorithm which permits a new packet to enter the buffer even when a queue length is larger than the threshold. We examined the PDPD scheme in a simulation with a modified simulator of CLASS, whose original program was developed by CELT (*Centro Studi e Laboratori Telecomunicazioni*) [16]. The simulation environment and results are described in the next section.

4. Simulation

Figure 7 is a simple model of our simulation. Five unidirectional TCP sources (from TCPS1 to TCPS5) send packets to five TCP receivers (from TCPR1 to TCPR5) via two ATM switches ceaselessly. The capacity of each link is 48 Mbps. A TCP packet size or maximum segment size (MSS) is 9140 octets, which corresponds to 192 ATM cells. The user buffer size of each TCP source is 1536 cells. This comes from the maximum size of the congestion window (64 kbytes), which can hold seven TCP packets. The TCP sources send data constantly to the receivers through a bottleneck link between two back-to-back switches. The TCP timer granularity is 500 milliseconds and congestion control policies of the TCP Reno version are also performed in the TCP level. All packets from these TCP sources are fragmented into AAL5 format ATM cells. These segmented cells are transmitted over UBR. We simulated the model for various buffer sizes (q) and various probabilities (p), which vary from 0.0 to 1.0 in increments of 0.25. As we previously described, when p is 0.0 or 1.0, it is PPD or EPD, respectively. Buffer size q varies from 2000 cells to 4000 cells in intervals of 400 cells. The RP size is about two packets (400 cells). The length of each link is 10 km.

In the simulation, *goodput* is defined as the sum of all the packets that reached the receiver without being damaged (excluding retransmitted packets), divided by the connection-set-up duration [16]. This parameter is a net propagation rate that a TCP user would experience during the transmission, which is expressed in bits

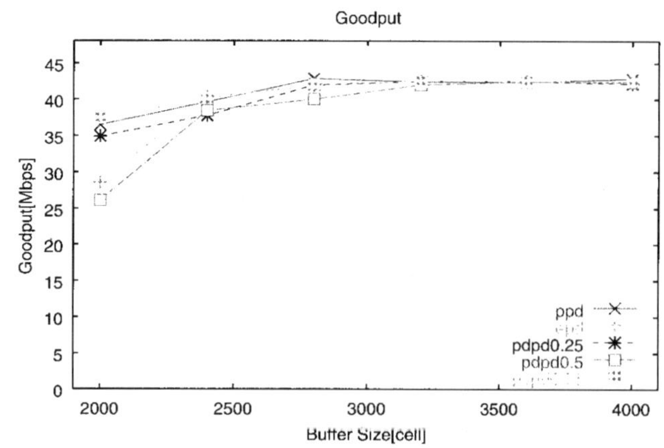

Fig. 8 Total goodput of TCP sources. RP size is 400 cells.

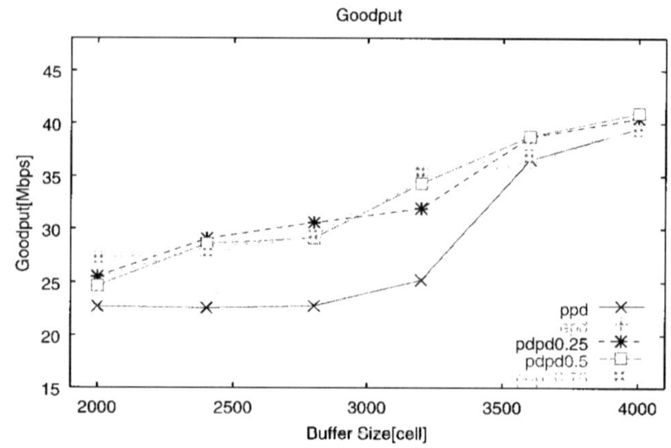

Fig. 10 Goodput (upper) of TCP sources with doubled window scaling.

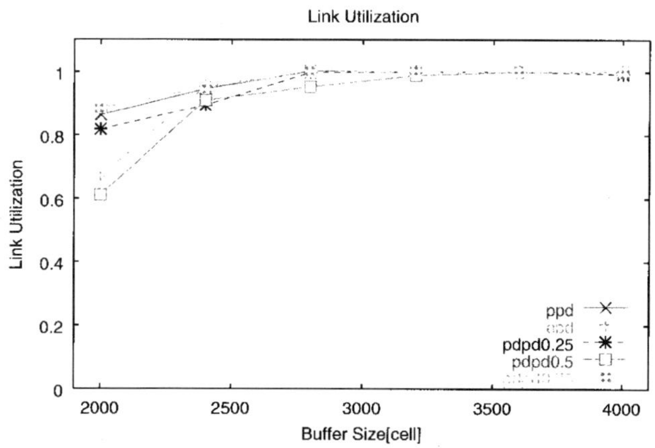

Fig. 9 Link utilization of L2.

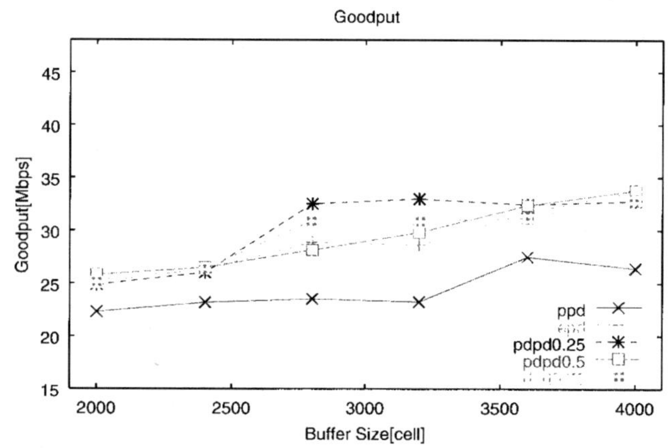

Fig. 11 Goodput (upper) of TCP sources with quadrupled window scaling.

per second. Here, we present the sum of goodputs of five TCP connections. Each experiment was run for approximately 90-seconds for five different seeds. Of the five experimental results, the maximum and the minimum results are excluded. All data shown here are the averages of the rest.

Figure 8 shows the total goodput of TCP sources when the RP size is 400 cells. PDPD with $p = 0.75$ (*pdpd* 0.75) achieves quite high goodput for all buffer size ranges. Note that this is not lower than that of EPD in this figure. In other words, EPD may gain a higher goodput with a more lenient packet discard control.

We show the link utilization of L2 in Fig. 9. It is also the average of three results excluding the maximum and the minimum from the five simulation results. Link utilization is the number of cells passed over the link divided by the link capacity during connection-set-up duration. As cells from the damaged packets are not allowed to be buffered, ineffective cell ratio amounts to a very small portion. In other words, the number of cells passed over the link is almost same as the effective cells. This is why the graphs in Fig. 9 and Fig. 8 are very alike in shape. When the link utilization is a 1, it does not mean that the goodput is equivalent to the link capacity because the cell contains the header. Thus, the relation between goodput and link utilization is as follows:

$$goodput = 48/53 \times link\ utilization \times link\ capacity$$

All of our simulation results satisfy this equation. Therefore, we make no mention of link utilization for some results.

In the Fig. 8, the goodput of all the schemes are saturated from the point when the buffer size becomes 3200 cells. In addition, the goodput of PPD or a PDPD with $p = 0.0$ seems to be high for all the buffer size ranges in the figure. This is because the input load is relatively small compared to the capacity of the links since the congestion window size is basically 64 kbytes. This congestion window regulates data rate of input streams.

In contrast, RFC 1323, however, allows more data to be transmitted using a window scale option. We show the goodput and link utilization of each scheme with window scaling factors of 2 and 4 in Fig. 10 and Fig. 11, respectively. When the window scaling factor

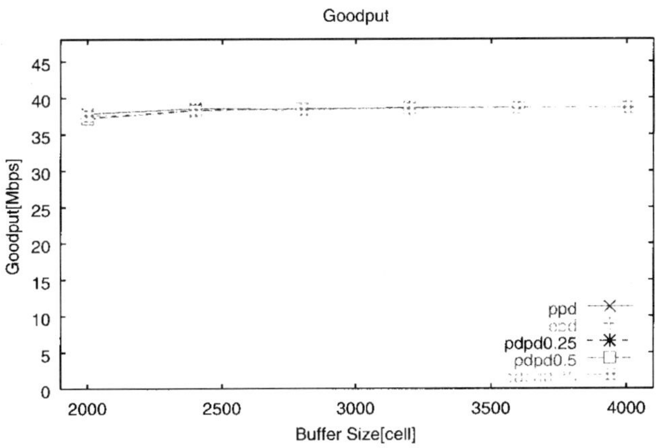

Fig. 12 Goodput of TCP sources when a packet size is 512 octets.

is 2, *pdpd* 0.25 achieves high goodput and high link utilization for small buffer sizes. For large buffer sizes, *pdpd* 0.5 achieves high performance in goodput and in link utilization (Fig. 10). This is because they accept more packets to be buffered and prevent excessive buffering at the same time.

Nevertheless, all schemes including PPD achieve high goodput, as buffer size is large enough. With a window scaling factor of 4, goodput and link utilization obtained by *pdpd* 0.25 are almost the highest for all the buffer size ranges in Fig. 11. This is because, with *pdpd* 0.25, it is possible to buffer more packets than EPD and to avoid more buffer overflows than PPD. In Fig. 10 and Fig. 11, all schemes except PPD achieve high goodput, as buffer size is large enough. This is because as the input load is enlarged, the simulated buffer size is not enough for PPD to obtain high goodput.

In order to test our hypothesis, we tried a variety of packet sizes as a parameter of our simulation. The RP size is two packets as in the other simulation environments. However, for small packet sizes, there is little advantage to using RP. Thus, when a packet size is small, all schemes work very similar to PPD. The goodputs of all schemes are presented in Fig. 12. The packet size is 512 bytes and the window scaling factor is 2. As described above, there is no difference between all schemes for the entire buffer size range. However, as a packet size is increased, influences of not only RP but also our scheme on goodput begin to appear. Figure 13 shows the goodput when the packet size is 1500 bytes (upper) and 4352 bytes, respectively. Window scaling factor is a 2, again. Though there is little difference between goodputs of all schemes, fairly high goodput is obtained by PDPDs when $p = 0.5$ and $p = 0.75$ for the entire range of buffer sizes.

We increased the number of switches to test PDPD in different models (Fig. 14). The length of links ($L1$, $L2$, $L3$ and $L4$) is 10 km. The packet size is 9180 octets and the RP size is about two packets. Figure 15 is the goodput when a window scaling factor is 1. In this

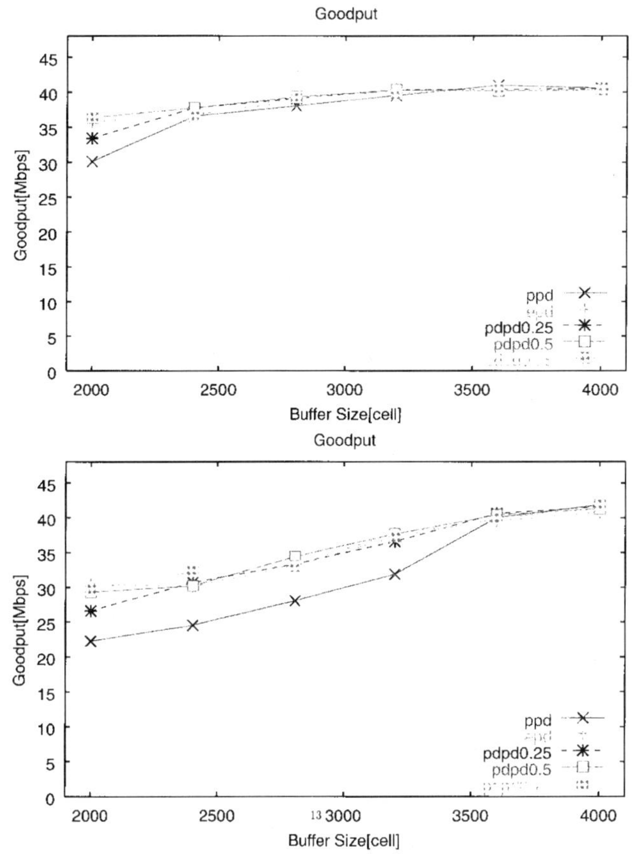

Fig. 13 Goodput of TCP sources when a packet size is 1500 octets (upper) and 4352 octets.

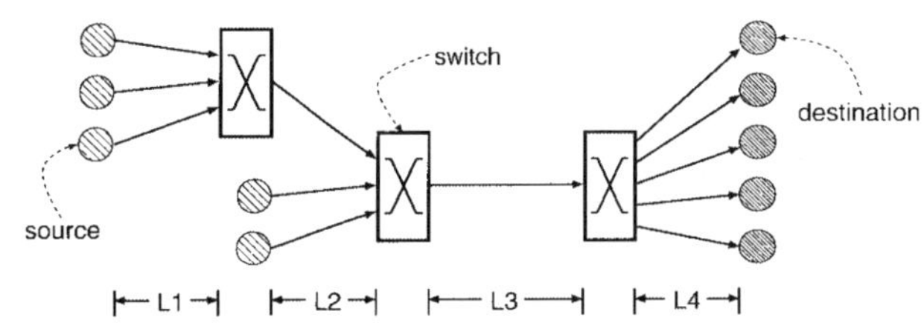

Fig. 14 Simulation model with three switches.

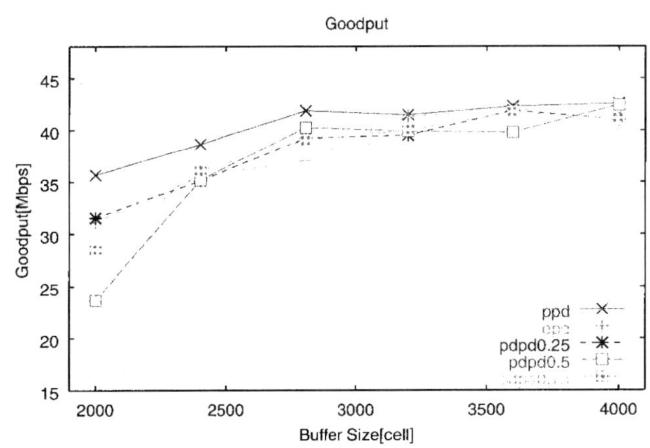

Fig. 15 Goodput of TCP sources for no window scaling.

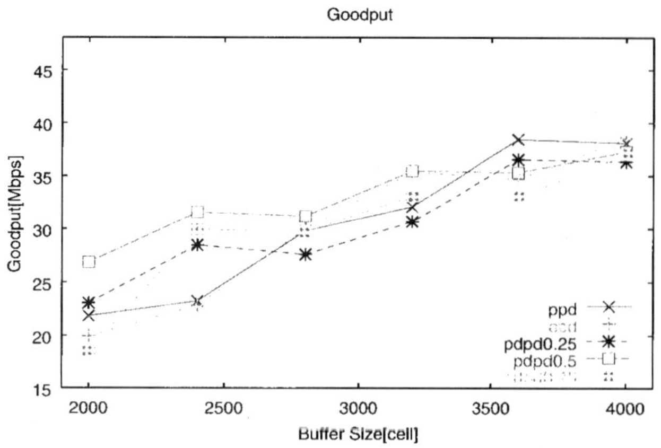

Fig. 16 Goodput of TCP sources for doubled window scaling.

figure, PPD achieves the highest goodput for all buffer sizes. This is because as the load is already regulated by the TCP congestion window to 64 kbytes at most, the buffer resources are maximally utilized. In addition, by using PPD, the load is distributed over many switches and each data stream has more chances to obtain network resources. For small p, the curve of *pdpd* 0.25 also shows quite high goodput for the entire buffer size range. However, there is little difference between goodputs of all schemes when the buffer size becomes large.

Next, we show a simulation result with doubled window scaling in Fig. 16. In this figure, the more the input load is increased, the more the goodput of all the schemes is decreased. This is because as the input data rate is increased, the buffer size is not enough to hold many packets. Nevertheless, random flag setting appears to be fairly useful. This is why the goodput of *pdpd* 0.5 is conspicuous in Fig. 16. Because as the buffer size is 4000 cells, the goodputs of all the schemes are quite high.

5. Conclusion

This paper generalizes EPD-based packet discard schemes. In the generalized form, if the queue length is larger than the threshold value, each scheme sets D (discard) flag based on a probability of p. The actual discard operation begins after the D flag is set. In this generalization, PPD is a special case when $p = 0.0$ and EPD is another special case when $p = 1.0$. By means of this formalization, it has become much easier to understand the packet discard schemes. Moreover, it is possible to change p according to network congestion. Based on this idea, we propose an optimistic packet discard scheme named the "Probability-Based Delayed Packet Discard" (PDPD) scheme in this paper. PDPD alleviates the aggressiveness of packet discard by EPD and tries to use more buffer resources than EPD.

We examined PDPD in a simulation using a modified simulator of CLASS. As the discarding operation in ATM layer is executed optimistically by the PDPD scheme, the possibility that more packets from TCP sources compete the buffer resources is also increased compared with EPD. This may give rise to more discarding of packets as well as decrement of goodput. Nevertheless, PDPD can achieve higher goodput and higher link utilization than EPD and PPD when the packet size is large and the load is high (with the window scaling option of TCP). We posit that the unused buffer space is utilized more effectively with optimistic PDPD methods than with EPD. At the same time, PDPD should succeed in preventing PPD's excessive buffering of packets to avoid network congestion. Moreover, for large buffer sizes, goodputs of all the schemes are quite high and they appear to be saturated. In other words, it can be said that if there is enough buffer space, the goodput is independent of the control schemes used. Thus we can conclude that improvement of buffer utilization is partly related to the improvement of goodput.

This paper attempts to show the potential of PDPD to improve goodput. For future work, we will determine under what conditions the PDPD works most effectively and achieves higher goodput with respect to the probability p.

References

[1] F. Bonomi and K.W. Fendick, "The rate-based flow control framework for the available bit rate ATM service," IEEE Network Magazine, pp.25–39, March/April 1995.

[2] R. Goyal, S. Fahmy, R. Jain, and B. Vandalore, "Effect of RM cell interval on ABR feedback: A simulation study using OPNET," ATM Forum, ATM_Forum/98-0407, July 1998.

[3] R. Jain, S. Kalyanaraman, S. Fahmy, R. Goyal, and S. Kim, "Source behavior for ATM ABR traffic management: An explanation," IEEE Commun. Mag., vol.34, no.11, pp.50–57, Nov. 1996.

[4] S. Kalyanaraman, R. Jain, S. Fahmy, R. Goyal, F. Lu, and S. Srinidhi, "Performance of TCP/IP over ABR service on ATM networks," IEEE GLOBECOM'96, pp.468–475, Nov. 1996.

[5] J.S. Turner, "Maintaining high throughput during overload in ATM switches," INFOCOM'96, pp.287–295, 1996.

[6] A. Romanow and S. Floyd, "Dynamics of TCP traffic over ATM networks," IEEE J. Sel. Areas Commun., vol.13, no.4, pp.633–641, 1995.

[7] M. Shinohara, H. Suzuki, and R. Fan, "Buffer management for best effort traffic in ATM networks," IEICE Technical Report, SSE94-94, IN94-62, CS94-93, Sept. 1994.

[8] G.J. Armitage and K.M. Adams, "Packet reassembly during cell loss," IEEE Network Magazine, vol.7, no.5, pp.26–34, Sept. 1993.

[9] V. Jacobson, "Congestion avoidance and control," SIGCOMM'88, pp.314–329, 1988.

[10] W. Stevens, "TCP slow start, congestion avoidance, fast retransmit, and fast recovery algorithms," IETF, Network Working Group, RFC 2001, Jan. 1997.

[11] Y. Lapid, R. Rom, and M. Sidi, "Analysis of packet discarding policies in high-speed networks," IEEE INFOCOM'97, pp.1191–1198, Kobe, Japan, April 1997.

[12] L. Jaussi, M. Lorang, and J. Nelissen, "A detailed experi-

mental performance evaluation on TCP over UBR," Proc. IEEE ICATM'98, pp.214–223, Colmar, France, June 1998.
[13] M. Okuda and T. Ishihara, "Evaluation of TCP/IP traffic over ATM network," IEICE Technical Report, SSE96-89, IN96-73, CS96-97, Sept. 1996.
[14] M. Jo and Y. Ebihara, "A study on delayed packet discard schemes," Proc. ICDCS 2000 Workshops, pp.B.63–B.70, Taipei, Taiwan, April 2000.
[15] M.H. MacDougall, Translated by Makoto Kobayashi, Simulating Computer Systems: Techniques and Tools, pp.140–151, Kogakusha, Tokyo, 1990.
[16] M. Ajmone Marsan, A. Bianco, T.V. Do, L. Jereb, R. Lo Cigno, and M. Munafò, "ATM simulation with CLASS," Performance Evaluation: Performance Modeling Tools, no.24, pp.137–159, 1995.
[17] A. Li, L.-Y. Siu, H.-Y. Tzeng, C. Ikeda, and H. Suzuki, "A simulation study of TCP performance in ATM networks with ABR and UBR services," IEICE Trans. Commun., vol.E79-B, no.5, pp.658–667, May 1996.

High-Performance Backbone Network Technology edited by Naoaki Yamanaka
ISBN: 0-8247-5321-6 DOI: 10.1081/(120025693)
Published 2004 by Marcel Dekker, Inc.

Performance of TCP/IP over ATM over an ADSL

Ryoichi Kawahara and Hiroshi Saito

SUMMARY The performance of TCP/IP over ATM over an asymmetric digital subscriber line (ADSL) was investigated. Because the bandwidth of an ADSL link can vary over time due to changes in the link's physical conditions, which degrades TCP performance, we performed simulations for various ATM traffic controls, including available bit rate (ABR) and generic flow control, used to handle variations in the ADSL bandwidth. This analysis showed that using an ABR control is effective under various traffic conditions. An ABR switch algorithm that can achieve good performance under any condition was investigated.

key words: *ADSL, ATM, TCP, bandwidth variations, ABR, ABR switch algorithm, GFC*

1. Introduction

The rapid growth in demand for new telecommunication services, such as WWW access through the Internet, has led to a need for subscriber lines with broadband access. The asymmetric digital subscriber line (ADSL) [1] was developed to convert an ordinary telephone subscriber line into a broadband access subscriber line. In North America, for example, ADSL provides 8 Mbps downstream and 640 kbps upstream. Universal ADSL, which is a simplified ADSL that the ITU-T [2] and UAWG (Universal ADSL working group) [3] are aiming to achieve through their standardization efforts, provides 1.5 Mbps downstream and 512 kbps upstream. ADSL is an attractive approach to implementing broadband access for existing telecommunication network operators because it uses their existing ordinary telephone subscriber lines.

From the traffic performance point of view, a particular concern about ADSL is its varying bandwidth. That is, the bandwidth of an ADSL link may vary over time due to changes in the link's physical conditions, caused by temperature variations, electromagnetic interference, and so on. Thus, the high-layer protocols should be robust against these bandwidth variations.

High-speed access to the Internet through ATM over an ADSL is one of the hottest reference models being discussed in the standardization bodies [1]–[3]. We have thus investigated the performance of TCP/IP over ATM over an ADSL by using the reference model shown in Fig. 1 (a detailed explanation is given in Sect. 2). Despite the keen interest in Internet broadband access through ADSL, the performance of this approach has not been well studied.

The performance of TCP has been estimated for networks with different bandwidths between their upstream and downstream links [5]–[8]. However, the work reported so far has neither investigated the use of different ATM traffic controls nor discussed the effects of ADSL bandwidth variations. Giroux and Ramanan [4] investigated the performance of TCP/IP over ATM over an ADSL when available-bit-rate (ABR) control was used for the downstream traffic.

We investigated the performance of TCP/IP over ATM over an ADSL when an ABR local loop [9] was used to control the upstream traffic. We also investigated the applicability of ABR control using various switch algorithms under various conditions. We investigated how well these various types of ABR control could cope with short-term fluctuations in the ADSL bandwidth, which occur frequently in a network having ISDN subscriber lines. We also looked at the applicability of generic flow control (GFC).

In this paper, after briefly describing the reference model we used, we describe various ATM traffic controls that can be used to handle variations in the ADSL bandwidth, including ABR and GFC control. Next, we evaluate the performance of TCP over an ADSL when these traffic controls are used. Finally, we investigate an ABR switch algorithm that achieves good performance under various traffic conditions.

2. Reference Model

The reference model we used for our investigation is shown in Fig. 1. The ATU-C (ADSL transmission unit at the network end) passes traffic to and from the ATM network; it can include buffers and a mechanism to shape the traffic sent to the ADSL, which has a vari-

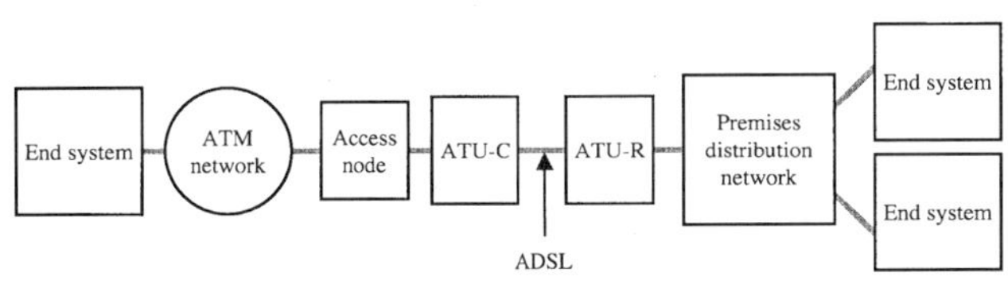

Fig. 1 Network reference model.

Manuscript received May 28, 1999.
Manuscript revised August 26, 1999.

†The author is with NTT Information Sharing Platform Laboratories, Musashino-shi, 180-8585 Japan.
††The author is with NTT Service Integration Laboratories, Musashino-shi, 180-8585 Japan.

IEICE TRANS. COMMUN., Vol. E83-B, No. 2, pp. 140–154, February 2000

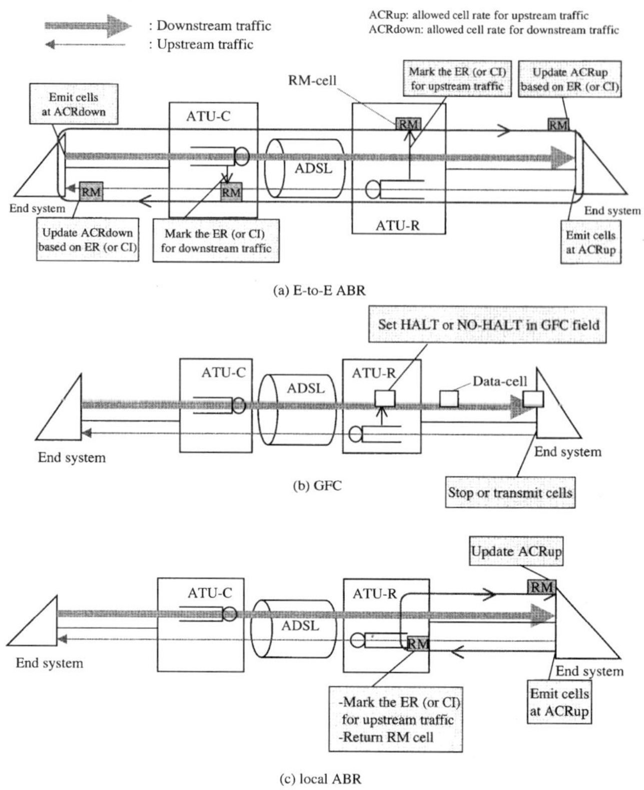

Fig. 2 ATM traffic controls for variations in ADSL bandwidth.

able bandwidth, between the ATU-C and the ATU-R. The ATU-R (ADSL transmission unit at the customer-premises end) terminates the ADSL and provides an interface to the premises network; it can include buffers and a mechanism to shape the traffic sent to the ADSL. The bandwidth of the ADSL varies with the link's physical condition. By measuring the signal-to-noise ratio for each carrier, the ATU-C and ATU-R determine and occasionally change the bit mapping (the number of bits assigned) for each carrier. They notify each other of any changes in the bit mapping. As a result, the bandwidth (the number of bits that will be carried) can vary from moment to moment.

3. ATM Traffic Controls

We considered the ATM traffic controls shown in Fig. 2 for handling the ADSL bandwidth variations (We assumed that the bottleneck was in the ADSL link between the ATU-R and the ATU-C in Fig. 1; hence, the ATM network, the access node, and the premises-distribution network are omitted in Fig. 2. For completeness, we also considered congestion point in the ATM network in simulation evaluation section (condition 4 in Sect. 4.2)).

(a) To handle the variations in both the upstream and downstream directions, we applied ABR service to each virtual channel (VC) between the end system sending the downstream traffic and the one sending the upstream traffic (that is, from end to end). We call this "E-to-E ABR control."

(b) To handle the variations in the upstream direction, we applied GFC (In this case, unspecified-bit-rate (UBR) or variable-bit-rate (VBR) service is assumed to be applied to each VC).

In addition to these controls, we also investigated the application of one more control.

(c) To handle the variations in the upstream direction, we applied an ABR local loop between the ATU-R and the end system sending the upstream traffic. We call this "local ABR control." (The ATM Forum is currently discussing [9] applying ABR locally. That is, using ABR between the ATU-R and the end system connected to it through the premises distribution network. The remaining portion of the VC could use UBR or VBR. The practicality of this approach remains for further study.)

3.1 ABR Control and Its Application to ADSL Reference Model

ABR is an ATM service category that controls the cell emission rate of each VC based on congestion information fed back by the network [10]. Resource management (RM) cells are sent through the feedback control loop. The allowed cell rate (ACR) of the source is updated based on the information carried by the cells. A source-end system (SES) sends a forward RM cell every Nrm (e.g., 32) data cells. When these forward RM cells reach the destination-end system (DES), the DES returns them to the SES as backward RM cells. During this round-trip, the ATM switches mark them. A switch sets the congestion-indication (CI) bits of the cells to "1" if the switch is congested or writes into the explicit-rate (ER) field of the cells the specific rate at which the SESs can transmit cells. The former type of switch is called a binary-mode switch, and the latter type is called an ER-mode switch.

When the SES receives a backward RM cell, it updates the ACR by using the following formulas.

$$\text{if CI} = 1,\ \text{ACR} := \text{ACR} - \text{ACR*RDF} \tag{1}$$

$$\text{if CI} = 0,\ \text{ACR} := \text{ACR} + \text{PCR*RIF} \tag{2}$$

$$\text{ACR} := \min\{\min\{\text{ACR, PCR}\},\ \text{ER}\} \tag{3}$$

$$\text{ACR} := \max\{\text{ACR, MCR}\} \tag{4}$$

The peak cell rate (PCR), minimum cell rate (MCR), rate-increase factor (RIF), and rate-decrease factor (RDF) are negotiation parameters specified at call set-up.

This ABR control mechanism can handle the ADSL bandwidth variations. As shown in Fig. 2(a), when the bandwidth of an ADSL changes in the upstream (or downstream) direction, the ATU-R (or ATU-C) writes the ER into the RM cells corresponding to the currently available bandwidth. Alternatively, the ATU-R (or ATU-C) may set the CI bit in the RM cells to "1" if its queue length exceeds a predetermined threshold. This makes it possible to adjust the ACR of each end system sending upstream (or downstream) traffic according to the ADSL bandwidth variations.

To establish E-to-E ABR control, both end systems must comply with ABR source behavior (e.g., transmission of RM cells and updating of the ACR), but in practice not all the end systems do. In contrast, local ABR control does not require that the end systems sending downstream traffic comply with ABR behavior. It only requires that the ones sending upstream traffic (i.e., ADSL users) do, and that the ATU-R is able to mark and return a forward RM cell from the ATU-R as a backward RM cell (Fig. 2(c)). This return mechanism enables local ABR control without requiring that the end systems sending downstream traffic comply with ABR behavior.

3.2 Use of GFC

Generic flow control (GFC) is executed using the GFC field in the data cells as follows [11][†]. The controlling equipment places a "HALT" or "NO_HALT" into the GFC field of the data cells. When the controlled equipment receives a HALT-marked cell, it stops sending cells to the controlling equipment. When the controlled equipment receives a NO_HALT-marked cell, it resumes sending cells.

In the ADSL reference model shown in Fig. 1, GFC can be used to control the traffic. However, it is not useful in the downstream direction because it can only be used at user network interfaces. Therefore, we use GFC to handle variations in the upstream direction, as shown in Fig. 2(b).

4. Simulation Model and Simulation Scenarios

This section describes the simulation model and various scenarios we used to evaluate whether and how the ATM traffic controls described in Sect. 3 could handle ADSL bandwidth variations. For comparison, we also simulated UBR service without any control mechanism; we refer to this as "no-control." For ABR control, we simulated both binary and ER modes. We call E-to-E ABR control using the binary and ER modes "E-to-E ABR binary control" and "E-to-E ABR ER control," respectively. And we call local ABR control using the binary and ER modes "local ABR binary control" and "local ABR ER control," respectively.

Several kinds of ER control algorithms have been proposed (For example, CAPC2 (congestion avoidance using proportional control version 2 [13]) and EPRCA (enhanced proportional rate-control algorithm [14]). Arulambalam, et al. have surveyed the algorithms that have been proposed [15]). However, it is unclear which kinds of ER algorithm should be applied to ADSL environment. Therefore, for our simulations, we began with the following simple control algorithm for the ER-mode controls and investigated which ER algorithm is suitable for the ADSL environment based on the evaluation of the simple ER control. The SES was sent an ER calculated as

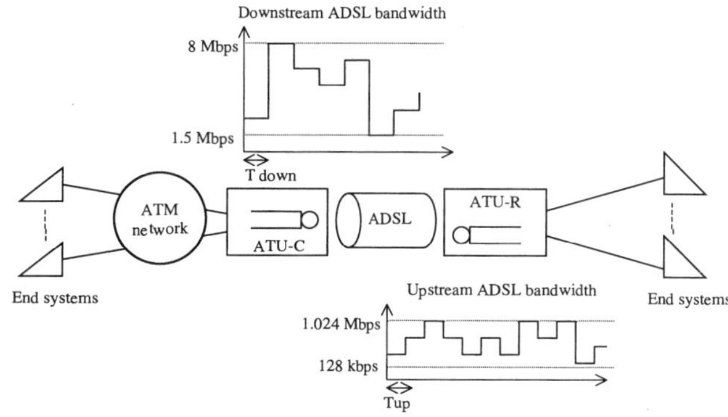

Fig. 3 Simulation model.

$$\text{ER} = \{\text{current ADSL bandwidth}\} / \{\text{number of active VCs}\}. \quad (5)$$

The number of active VCs was estimated every T_est seconds. If a VC was observed to have at least one cell during the preceding T_est-s period, it was judged to be active.

4.1 Models

(1) Network

The network model is shown in Fig. 3. Several TCP connections are multiplexed onto an ADSL link via ATM connections. We assume that one TCP connection is established over one ATM connection (i.e., one VC). To simulate a large-scale ATM network, the distance between the end systems sending the downstream traffic and the ATU-C was set to 10,000 km. However, because the ADSL is a subscriber line, the distance between ATU-C and ATU-R was set to 1 km, while that between ATU-R and the end systems sending the upstream traffic was set to 10 m.

To simulate variations in ADSL bandwidth in the upstream (or downstream) direction, the bandwidth was changed every Tup (or Tdown) seconds. By setting Tup (or Tdown) from 0.01 to 10 s, we simulated bandwidth fluctuations ranging from short- to long-term

[†]The GFC defined in Ref. [11] uses commands other than "HALT" and "NO_HALT." However, for simplicity, here we use only "HALT" and "NO_HALT," which are the most fundamental of the GFC commands.

ones. Every Tup seconds, the upstream ADSL bandwidth was set such that it was uniformly distributed between 1.024 Mbps and 128 kbps. Every Tdown seconds, the downstream ADSL bandwidth was set such that it was uniformly distributed between 8 and 1.5 Mbps.

A FIFO buffer was allocated in front of the ADSL link at the ATU-C (or ATU-R), and it was shared among the downstream (or upstream) traffic flows. It could hold up to 1024 cells.

(2) Source

To emulate TCP behavior, each end system used Reno with fast retransmit recovery implemented in 4.3 BSD (Berkeley Software Distribution) [12]. We set the TCP parameters as follows. The timer granularity was 500 ms, the maximum segment size was 9140 bytes, and the maximum window size was 64 kbytes.

4.2 Scenarios

To evaluate performance under various traffic conditions, we simulated four conditions.

[Condition 1: Upstream data transfer] Two TCP connections were established, and the end systems sent data only in the upstream direction, for which the data size was infinite.

[Condition 2: Downstream data transfer] Two TCP connections were established, and the end systems sent data only in the downstream direction, for which the data size was infinite.

[Condition 3: Coexistence of upstream and downstream data transfers] Four TCP connections were established: two connections sent data in the downstream direction, and the other two sent data in the upstream direction. Each data size was infinite.

[Condition 4: Upstream data transfer with congestion point in backbone network] Unlike condition 1 (and also conditions 2 and 3), there was a congestion point in the ATM network in Fig. 3. In the network, two TCP connections passing through ADSL and the other two connections not passing through ADSL shared a 2.048-Mbps link (that is, the latter connections generated background traffic). The end systems sent an infinite amount of data only in the upstream direction.

We used condition 1 to evaluate the performance when the ADSL bandwidth varied in the upstream direction. We used condition 2 to estimate the downstream TCP throughput. We used condition 3 to evaluate the impact on TCP performance of both ADSL bandwidth variations and network asymmetry. Several reports [5]–[8] have described how network asymmetry causes TCP throughput to degrade in the downstream direction: acknowledgments of downstream data transfers can be delayed or lost due to the smaller bandwidth of the upstream link. We thus looked at how both ADSL bandwidth variations and asymmetry affect the downstream TCP throughput and whether ATM traffic controls can reduce this degradation. We used condition 4 to evaluate the influence of backbone-network-congestion on each ATM control.

4.3 Parameter Values

We set the parameters used for the traffic control methods described in Sect. 3 as follows. As previous studies reported (e.g., [17]), the setting of parameters (RIF, RDF, and threshold) seriously affects the performance of ABR binary control. Therefore, in this paper, we determined these parameters based on an investigation of appropriate parameter values in Appendix A.

(a) No-control
- PCR for upstream traffic (PCRup) = 1.024 Mbps (Equal to the maximum rate of the ADSL link in the upstream direction)
- PCR for downstream traffic (PCRdown) = 8 Mbps (Equal to the maximum rate of the ADSL link in the downstream direction)

(b) GFC
- PCRup = 1.024 Mbps, PCRdown = 8 Mbps
- Threshold for congestion detection (Thigh) = 256 cells
- Threshold for non-congestion detection (Tlow) = 128 cells

(If the queue length exceeds Thigh, HALT continues to be placed in the GFC field until the queue length is less than Tlow. Otherwise, NO_HALT is placed in the GFC field.)

(c) E-to-E ABR ER
- PCRup = 1.024 Mbps, PCRdown = 8 Mbps
- RIF for upstream traffic (RIFup) = 1/16
- RIF for downstream traffic (RIFdown) = 1/16

(The value of RIF was based on the recommendation in the ATM Forum [18]. The RDF was not used because we did not treat the ER algorithm that uses CI bit.)
- Interval for estimating the number of active VCs: T_est = 10 ms

(d) E-to-E ABR binary
- PCRup = 1.024 Mbps, RIFup = 1/16, RDFup = 1/4
- PCRdown = 4 Mbps, RIFdown = 1/512, RDFdown = 1/256

(Based on the observation that a small PCR is effective when binary control is used in a WAN environment [19], we set PCRdown to a small value.)
- Threshold for congestion detection = 256 cells

(e) local ABR ER
- PCRup = 1.024 Mbps, PCRdown = 8 Mbps
- RIFup = 1/16
- Interval for estimating the number of active VCs: T_est = 10 ms

(f) local ABR binary
- PCRup = 1.024 Mbps, PCRdown = 8 Mbps RIFup = 1/16, RDFup = 1/4

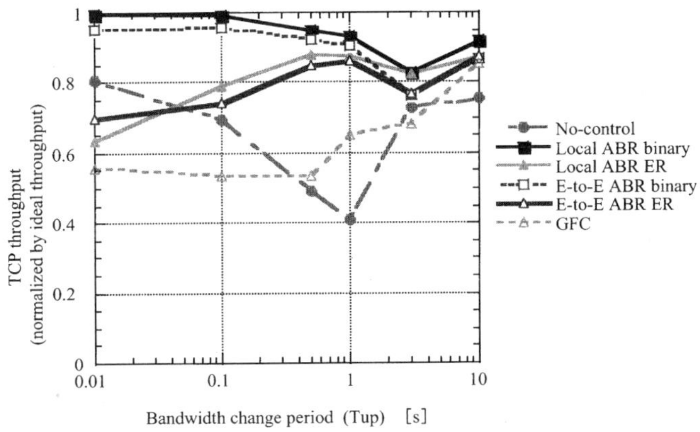

Fig. 4 TCP throughput vs. ADSL bandwidth change period (Tup) in upstream direction.

– Threshold for congestion detection = 256 cells

5. Performance Evaluation

To evaluate the performance of ATM traffic controls under conditions 1–4 in Sect. 4.2, we ran 70-s simulations for each and measured the TCP-level throughput for the interval 10–70 s (the initial bias was removed by starting the measurement at the 10-s mark). The throughput was calculated by dividing the measured amount of data transmitted by the measurement time (60 s).

5.1 TCP Performance for Upstream Data Transfers

Under condition 1, we measured the TCP-level throughput in the upstream direction while varying the frequency of the bandwidth change (Tup) from 0.01 to 10 s (see Fig. 4). The measured throughputs were normalized by the ideal ones[†].

[TCP data loss under no-control and GFC] The ABR controls maintained high throughputs compared with no-control and GFC. This is because the ABR controls avoided cell loss in the ATU-R buffer (Fig. 5) by keeping the input cell rate close to the ADSL bandwidth as it varied (Fig. 6). In contrast, buffer overflow occurred under GFC and no-control (Fig. 5). Generic flow control could not avoid cell loss under condition 1 because the only downstream traffic was acknowledgments of upstream data transfers, though GFC requires frequent downstream cell flows to control the upstream traffic. Compared with no-control, GFC had worse throughput when Tup = 0.01 and 0.1 s, as shown in Fig. 4. This was also due to the scattered downstream cell flows. If there were no cells to be sent in the downstream direction when the queue length was below the threshold for non-congestion detection, the end system could not start emitting cells until the GFC control timer had expired (here, we set this to 5 s, which is the default value [11]). This resulted in lower utilization.

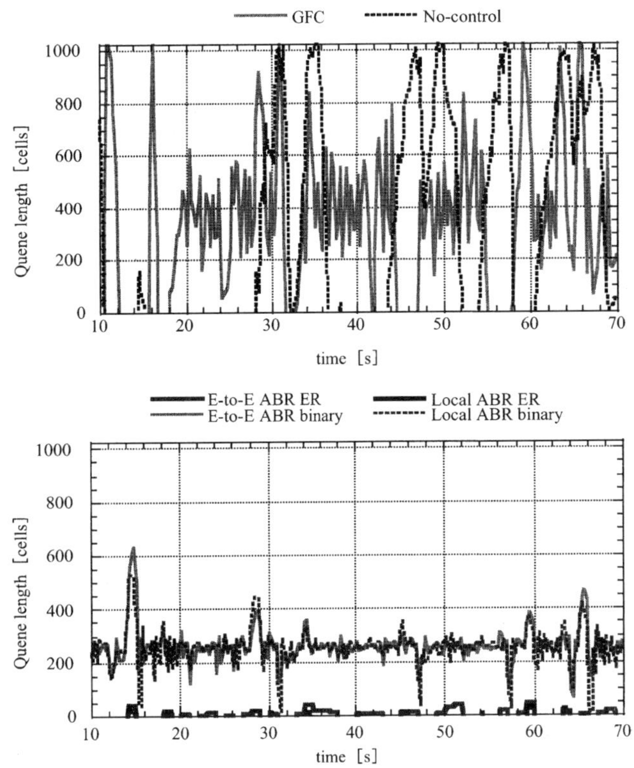

Fig. 5 Queue-length behavior (Tup = 1 s).

For no-control, the worst performance occurred when Tup was 1 s; no-control could cope with a small Tup by absorbing the short-term fluctuations in the ADSL bandwidth at the ATU-R buffer and with a large Tup because the TCP flow control tracked the long-term fluctuations.

[Absorption of short-term fluctuations under ABR binary] Among the ABR controls, local ABR binary control attained high throughputs at all values of Tup. In contrast, both E-to-E ABR ER and local ABR ER control had low throughput when there were short-term fluctuations (i.e., Tup = 0.01 or 0.1 s). This is because ER controls overreact to short-term fluctuations in the bandwidth. The ACR behavior of local ABR ER control when Tup was 0.01 s is shown in Fig. 7. We also show the transient behavior of the bandwidth of ADSL (strictly speaking, the bandwidth of ADSL in Fig. 7 shows the actual bandwidth divided by 2 (the number of VCs), because we want to compare with the ACR of a VC). Once the control reduced the ACR of each end system to correspond to the reduction in the ADSL bandwidth, the frequency of RM cells sent by

[†]The ideal throughput at the TCP level is calculated as follows:
Ideal throughput = average ADSL bandwidth * MSS/(MSS + TCP/IP header size + AAL5trailer + padding)/48*53, where average ADSL bandwidth = average ADSL bandwidth measured in interval [10 s, 70 s], MSS = maximum segment size (here, 9140 bytes), TCP/IP header size = 40 bytes, AAL5 trailer = 8 bytes, and padding = 28 bytes.

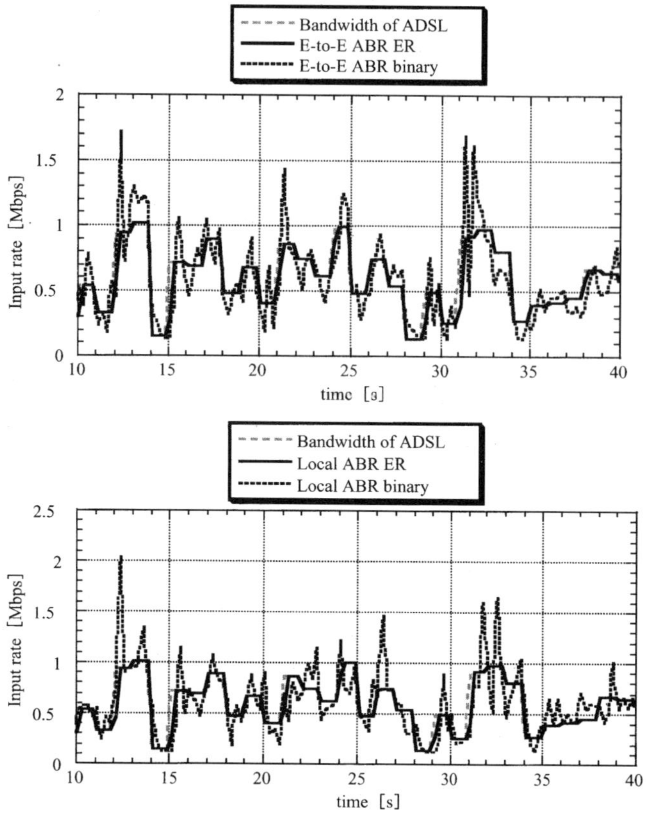

Fig. 6 Input cell-rate behavior (Tup = 1 s).

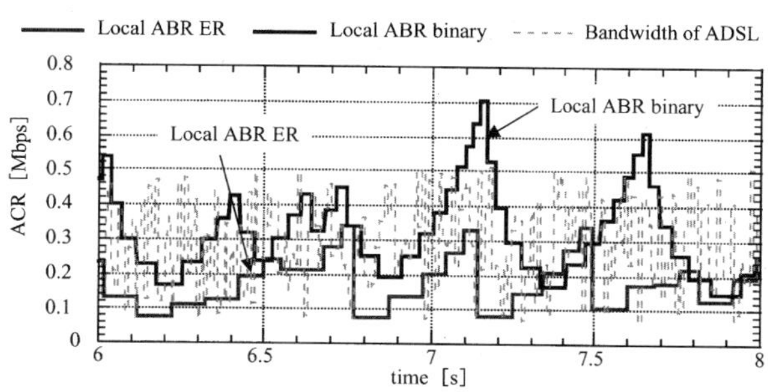

Fig. 7 ACR behaviors when Tup is 0.01 s.

the end system became smaller. Accordingly, it took a long time for the end system to detect an increase in the ADSL bandwidth. Conversely, when the ACR was large, the frequency of RM cells was also large, and the end system responded quickly to the decrease in the ADSL bandwidth. Therefore, ER controls cannot utilize the bandwidth effectively.

In contrast, the binary controls are robust against short-term fluctuations in the bandwidth because they do not detect a decrease in the ADSL bandwidth unless it continues until the queue length exceeds a threshold. Therefore, binary controls can maintain high throughput by absorbing short-term fluctuation at the ATU-R buffer.

[TCP time-out without data loss] Next, we consider long-term changes in the ADSL bandwidth. As shown in Fig. 4, the throughputs under all of the ABR controls were slightly lower than ideal due to the over-reaction of TCP. Figure 8 shows the behaviors of the TCP window size, the round-trip time (RTT), the queue length, and the ADSL bandwidth under local ABR binary control when Tup was 3 s. When the ADSL bandwidth was suddenly reduced (at 42 s in Fig. 8), the acknowledgments were delayed. As a result, the TCP timer expired and the window size was reduced to one segment size based on the assumption that data was lost, though actually none was. This caused the data to be re-transmitted, which degraded the throughput. For comparison, we measured the throughput when we changed the timer granularity from 500 to 100 ms. The normalized throughput under local ABR binary control was 0.772 when Tup was 1 s, while it was 0.930 when the timer granularity was 500 ms. This means that coarser timer granularity helps prevent the TCP from over-reacting. Giroux and Ramanan made this same observation [4].

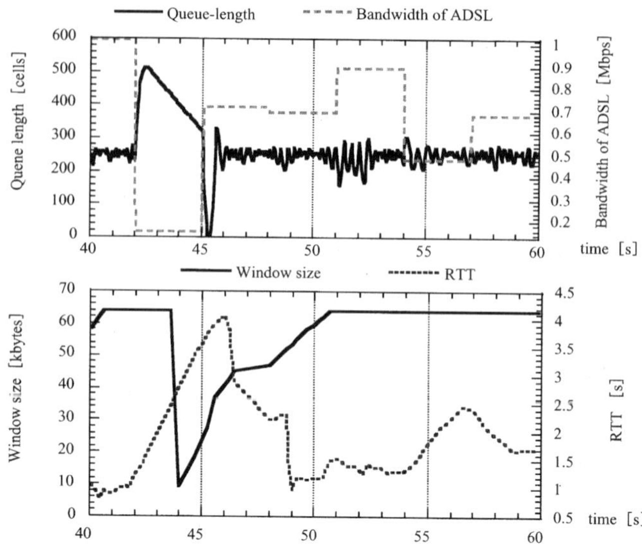

Fig. 8 Behavior of local ABR binary when ADSL bandwidth is suddenly reduced.

5.2 TCP Performance for Downstream Data Transfers

Under condition 2, we measured TCP throughput under no-control and E-to-E ABR binary and ER controls while varying the frequency of bandwidth variation (Tdown) from 0.01 to 10 s (Fig. 9). Local ABR controls and GFC had no effect under this condition because end systems sent data only in the downstream direction, so we omitted the results for these controls. The E-to-E ABR ER control had high throughputs, while E-to-E ABR binary control had low throughputs.

We measured the transient behaviors of the ACR and queue length under E-to-E ABR binary and ER controls when Tdown was 3 s. As shown in Fig. 10,

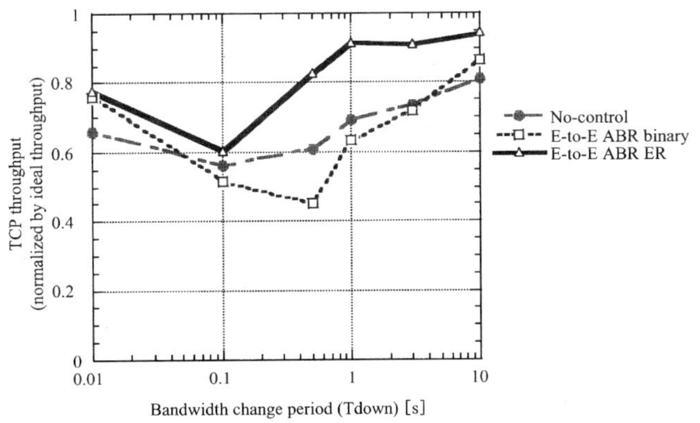

Fig. 9 TCP throughput vs. ADSL bandwidth change period (Tdown) in the downstream direction.

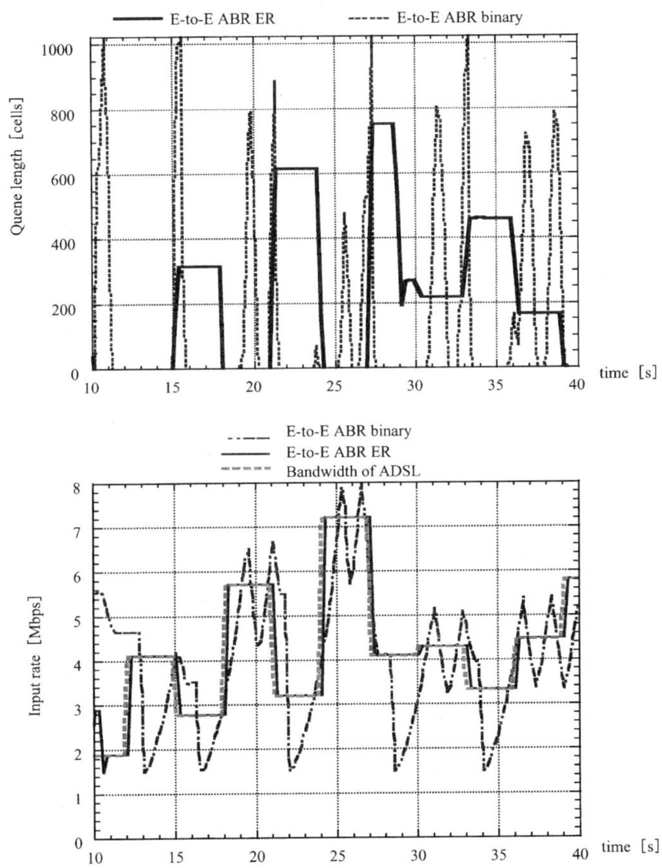

Fig. 10 Transient behaviors of ACR and queue length under E-to-E ABR binary (Tdown = 3 s).

under E-to-E ABR binary control, cell losses occurred or bandwidth was not effectively utilized, because this control could not cope with the ADSL bandwidth variations when the control-delay was long. In contrast, E-to-E ABR ER control could utilize the bandwidth effectively by keeping the input cell rate close to the ADSL bandwidth as it varied, as shown in Fig. 10.

Compared with E-to-E ABR binary control, no-control had better throughputs. Under no-control, the

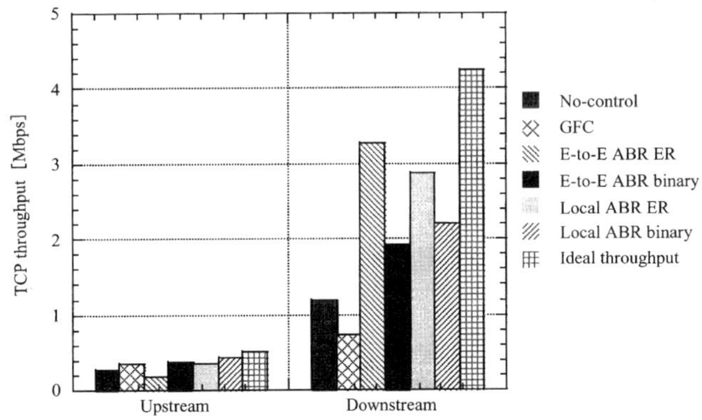

Fig. 11 TCP throughput in the upstream and downstream directions.

performance-degradation point under condition 2 was at Tdown = 0.1 s (Fig. 9), while that under condition 1 was at Tup = 1 s (Fig. 4). This is because the bandwidth in the downstream direction is generally larger than that in the upstream direction, which shifted the degradation point.

5.3 TCP Performance under Coexistence of Upstream and Downstream Data Traffic

Under condition 3, we measured the throughputs for both upstream and downstream data transfers when Tup was 1 s and Tdown was 3 s (Fig. 11). Unlike conditions 1 and 2, the coexistence with upstream (or downstream) traffic affected the downstream (or upstream) TCP throughput.

[ACK loss under no-control and ACK delay under GFC] No-control and GFC had low throughputs in both directions. The reason for the throughput degradation under no-control in the downstream direction was both data loss at the ATU-C and acknowledgment (ACK) loss at the ATU-R. The ACK loss problem in asymmetric networks has been previously reported [7] and [8]. As for GFC, throughput in the downstream direction was degraded, though GFC could avoid loss of ACKs sent upstream. This is because ACKs of downstream data transfers were greatly delayed at the end system sending upstream. This delay was caused by the end systems in the connections sending downstream data traffic being over-regulated. This over-regulation occurred because GFC could not adequately control the end systems in the connections sending upstream data traffic because of scattered downstream cell flow on these connections.

[Over-regulation of upstream traffic under ABR ER] The E-to-E ABR ER control had the highest throughput in the downstream direction because it avoided both data and ACK loss. However, it had the worst throughput in the upstream direction because it over-regulated the upstream data traffic by allocating the same bandwidth to VCs transmitting only back-

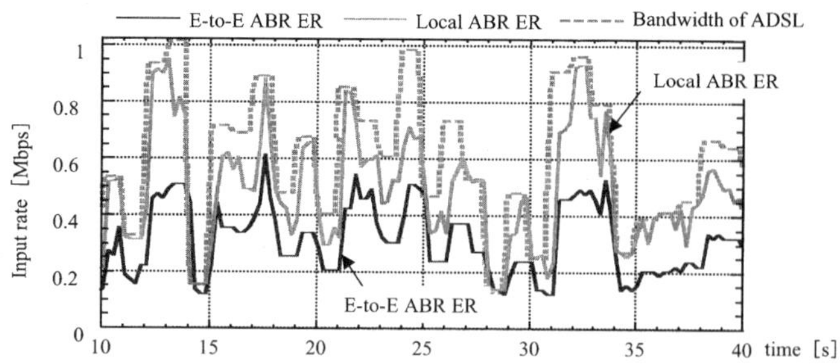

Fig. 12 Upstream input rate under E-to-E ABR ER, and local ABR ER.

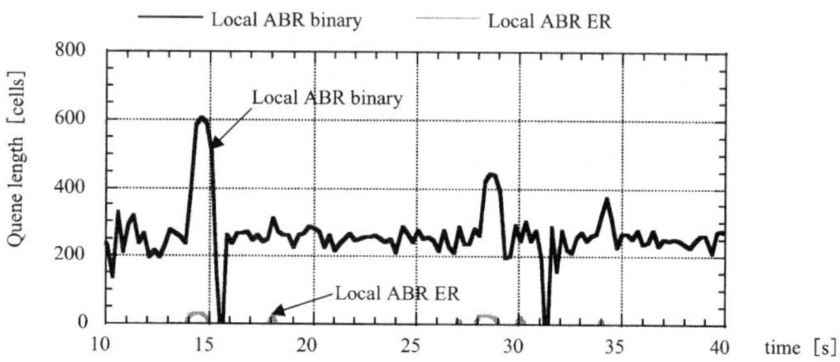

Fig. 13 Queue length in ATU-R under local ABR ER, and local ABR binary.

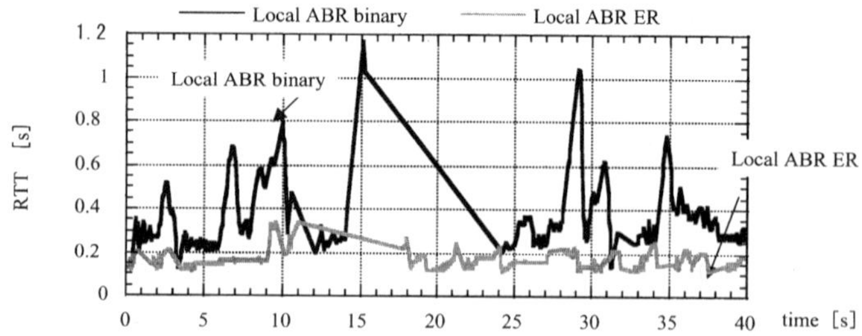

Fig. 14 RTT behavior under local ABR ER, and local ABR binary.

Table 1 Utilization of downstream ADSL bandwidth under condition 3.

No-control	GFC	E-to-E ABR ER control	E-to-E ABR binary control	Local ABR ER control	Local ABR binary control
0.350	0.187	0.830	0.492	0.747	0.538

ward RM cells and ACKs upstream as to those transmitting upstream data traffic. This happened because the control judged the former VCs to be active despite their low cell rate. The result was ineffective utilization of the upstream ADSL bandwidth (see Fig. 12).

In contrast, local ABR ER control had better throughput than E-to-E ABR ER control but worse throughput than local ABR binary control. This was because local ABR ER control also over-regulated the upstream data traffic by allocating the same bandwidth to VCs transmitting only ACKs as to those transmitting upstream data traffic. However, the over-regulation was less than in the case of E-to-E ABR ER control. This was because the frequency at which local ABR ER control judged VCs transmitting only ACKs to be active was low because local ABR ER control does not have the backward RM cells that E-to-E ABR ER control has. Therefore, local ABR ER control utilized the bandwidth more effectively than E-to-E ABR ER control, as shown in Fig. 12.

[Avoidance of ACK loss and delay under ABR ER] For the downstream data transfers under local ABR ER control, high throughput was maintained even though the control did not directly control the downstream traffic. This good performance occurred because local ABR ER control avoided the loss of ACKs of downstream data transfer by controlling the upstream traffic flow. In contrast, though local ABR binary control also avoided ACK loss, as shown in Fig. 13, it did not have high throughput in the downstream direction. This is because the round-trip-time (RTT) under local ABR binary control was much longer than that under local ABR ER control, as shown in Fig. 14. This was due to the queueing delay at the ATU-R buffer caused by the local ABR binary control keeping the queue length at around the congestion threshold (256 cells). In contrast, local ABR ER control returned ACKs quickly to the end systems sending downstream data by avoiding ACK loss and ACK delays in the upstream direction, which enabled it to achieve better TCP performance in the downstream direction. For reference, the link utilization at the cell level in the downstream direction is shown in Table 1.

5.4 TCP Performance under Existence of Backbone-Network-Congestion

Under condition 4, we measured the throughput of two TCP connections passing through an ADSL link when Tup was 1 s. The results are shown in Table 2. For comparison, we also show the throughput under condition 1 as that without backbone-congestion, because condition 4 is equal to condition 1 if there do not exist the two connections not passing through ADSL in the ATM network under condition 4. E-to-E ABR controls could attain almost the same high throughput as without congestion. This is because E-to-E ABR can cope with not only ADSL congestion but also backbone-network-congestion. In contrast, since local ABR controls only cope with ADSL congestion, throughput degradation occurred under backbone-network-congestion. However, local ABR controls had better performance than no-control or GFC.

5.5 Relationship between Buffer Size and TCP Performance

We investigated the relationship between buffer size and the performance of each control. To evaluate

Table 2 TCP throughput under existence of backbone-network-congestion. (normalized by ideal throughput)

(normalized by ideal throughput)

	no-control	GFC	E-to-E ABR ER	E-to-E ABR binary	Local ABR ER	Local ABR binary
With congestion	0.198	0.359	0.832	0.890	0.665	0.618
Without congestion	0.410	0.650	0.862	0.904	0.876	0.930

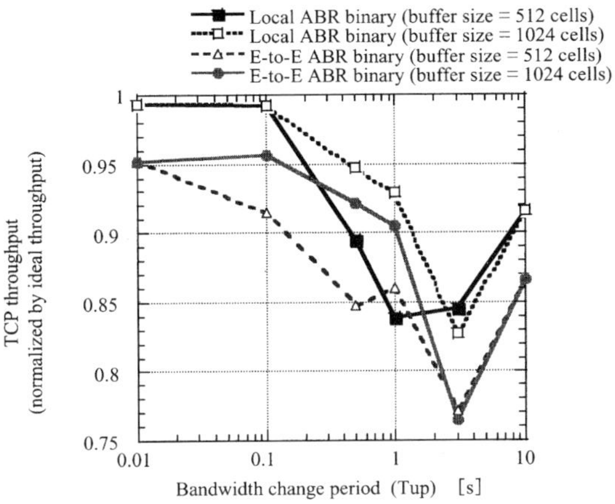

Fig. 15 Upstream throughput under ABR binary controls when we change buffer size.

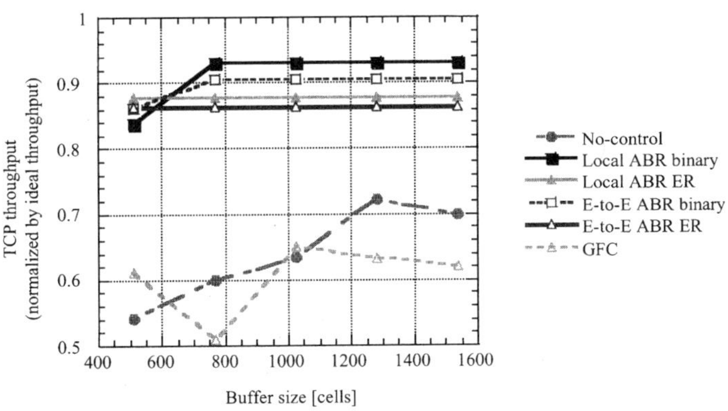

Fig. 16 Upstream throughput vs. buffer size (Tup = 1 s).

whether ABR binary controls with a small buffer could keep the same good performance as in Sect. 5.1, we measured the upstream throughput under condition 1 when the buffer size was reduced to 512 cells (Fig. 15). (We omit the results for ER controls because the maximum queue length under condition 1 was less than 512 cells.) We found that ABR binary controls required a larger buffer than ER controls in order to absorb ADSL bandwidth variations (especially Tup = 0.5 or 1 s). However, as shown in Fig. 16, even with an insufficiently large buffer, the upstream performance of binary controls remained much better than that of GFC or no-control. As for the downstream direction, even with a sufficiently large buffer, E-to-E ABR binary control performed worse than no-control, as shown in Fig. 17

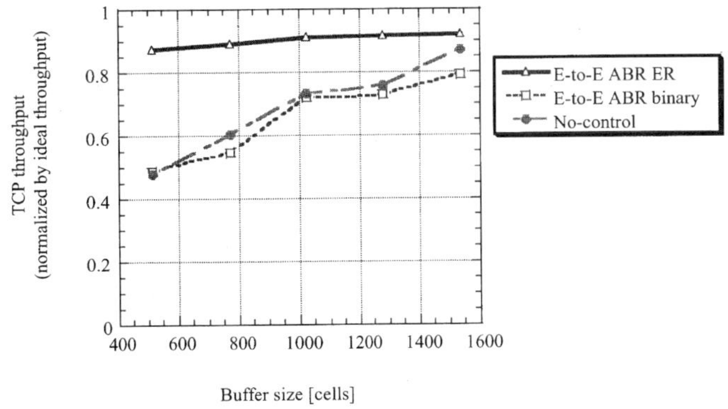

Fig. 17 Downstream throughput vs. buffer size (Tdown = 3 s).

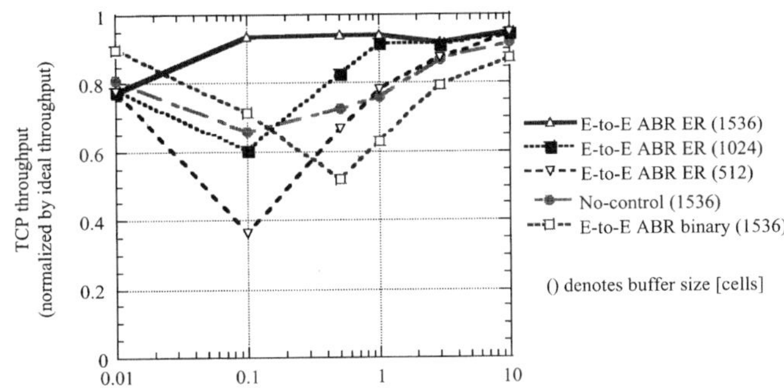

Fig. 18 Downstream throughput under E-to-E ABR ER when we change buffer size.

(we changed the buffer size under condition 2 when Tdown was 3 s). This is because binary control cannot cope with a long-control-delay. To evaluate the downstream performance of E-to-E ABR ER with an insufficiently (or sufficiently) large buffer in more detail, we measured the downstream throughput under condition 2 when the buffer size was changed to 512 cells (or 1536 cells). Figure 18 shows that E-to-E ABR ER with the insufficiently large buffer (512 cells) could not cope with ADSL bandwidth variations (especially Tup = 0.1 or 0.5 s) due to the long-control-delay. Conversely, the sufficiently large buffer (1536 cells) made it possible to achieve high throughput under any Tdown, unlike no-control or binary control with 1536-cell buffer size.

5.6 Summary of Simulation Results

In summary, our simulations clarified the following traffic characteristics.

5.6.1 Upstream Traffic Characteristics

(1) With only upstream data traffic (condition 1), ABR binary controls (both local and E-to-E ABR binary controls) can maintain high TCP throughput even under ADSL short-term bandwidth fluctuations

by absorbing them the ATU-R buffer. In contrast, ABR ER controls (both local and E-to-E ABR ER controls) suffer throughput degradation because they overreact to short-term fluctuations. GFC has low throughput because it requires frequent downstream traffic flows.
(2) When upstream and downstream data transfers coexist (condition 3), ABR binary controls have high upstream throughput. However, ABR ER controls have poor throughput because they over-regulate the upstream traffic. This is due to the large bandwidth allocated to VCs that are transmitting only ACKs or backward RM cells being sent upstream.
(3) Even when there is backbone-network-congestion, E-to-E ABR controls have good performance because they can cope with not only ADSL congestion but also backbone-network-congestion.
(4) With an insufficiently large buffer, ABR binary controls have slightly degraded performance. In contrast, ABR ER controls can keep the same performance as when the buffer is sufficiently large.

5.6.2 Downstream Traffic Characteristics

(1) With only downstream data traffic (condition 2), E-to-E ABR ER control has the best performance. No-control is better than E-to-E ABR binary control.
(2) When upstream and downstream data transfers coexist (condition 3), local ABR ER control has high downstream TCP throughput without directly controlling the downstream traffic. This is because it can quickly return ACKs upstream by avoiding loss and delay of ACKs of downstream data transfer. Though ABR binary controls avoid ACK loss, their downstream throughputs are low because of ACK delays caused by queueing delays at the ATU-R buffer.
(3) E-to-E ABR ER control can attain good performance under any bandwidth-change-period with a large buffer. In contrast, E-to-E ABR binary control has worse throughput than no-control even with a large buffer.

6. Performance Improvement of ABR Controls

In Sect. 5, we showed that ABR controls can be effectively applied to the ADSL environment. However, we also found that ABR binary and ABR ER controls have both advantages and disadvantages. The disadvantages are as follows:
(a) ABR ER control overreacts to short-term fluctuations in the ADSL bandwidth.
(b) ABR binary control causes queueing delays in the upstream direction, which reduces the downstream TCP throughput.
(c) ABR ER control over-regulates the upstream traffic when downstream and upstream data transfers coexist.

To overcome these disadvantages, here we consider how to modify the ABR controls treated in Sect. 5. For binary control, we can greatly reduce the congestion detection threshold (we call this "ABR low-T binary control" or simply "low-T binary"). Low-T binary control should reduce the queueing delay, which overcomes the second disadvantage. We set the threshold to 64 cells. However, lowering the threshold may impair the good performance attained by the original ABR binary control. We will examine this later.

For the ER control, we need an ER algorithm that is suitable for the ADSL environment. To overcome all three disadvantages, the algorithm should be:
(1) robust against short-term fluctuations in the ADSL bandwidth,
(2) able to control the queueing delay, and
(3) able to utilize the bandwidth effectively.
To satisfy the second and third requirements, we use ERICA+ (explicit rate indication for congestion avoidance: complete version [16]) control. This uses the queue length and the measured total cell rate to determine the ER such that the queueing delay remains close to the predetermined target queueing delay (T0). For every measurement of the rate at which cells are arriving at the link (Rate), the ER is calculated as follows:

$$\mathrm{Q0} = \mathrm{ABR\_Capacity} * \mathrm{T0} \tag{6}$$

$$\mathrm{R0} = \mathrm{G(Q,\ Q0)} * \mathrm{ABR\_Capacity} \tag{7}$$

$$\mathrm{G(Q,\ Q0)} = \max[\alpha * \mathrm{Q0}/\{(1-\alpha) * \mathrm{Q} + \mathrm{Q0}\},\ \mathrm{QDLF}],\ \text{for } \mathrm{Q} > \mathrm{Q0} \tag{8a}$$

$$\mathrm{G(Q,\ Q0)} = \beta * \mathrm{Q0}/\{(1-\beta) * \mathrm{Q} + \mathrm{Q0}\},\ \text{for } \mathrm{Q} \leqq \mathrm{Q0} \tag{8b}$$

$$\mathrm{ER} := \max\{\mathrm{R0/m},\ \mathrm{R0/Rate} * \mathrm{ACR}\}, \tag{9}$$

where Q0 is the target queue length, R0 is the target cell rate, Q is the current queue length, G(Q,Q0) is the queue-control function, m is the estimated number of active VCs, and ACR is the current ACR of the end system. ERICA+ control makes it possible to keep the queueing delay at T0 while utilizing the bandwidth effectively (satisfying the last two requirements).

To satisfy the first requirement, instead of using the current ADSL bandwidth, as in Eq. (5), we use the following smoothed ADSL bandwidth as the ABR_Capacity:

$$\mathrm{ABR\_Capacity} := (1-\mathrm{x}) * \mathrm{ABR\_Capacity} + \mathrm{x} * \{\text{current ADSL bandwidth}\}, \tag{10}$$

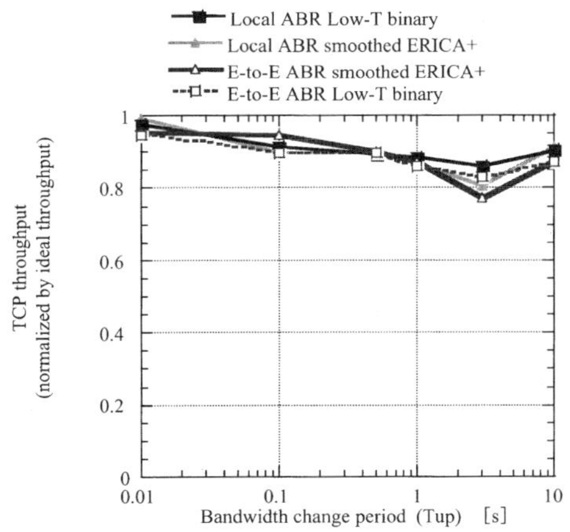

Fig. 19 Upstream throughput under condition 1 for low-T binary and smoothed ERICA+.

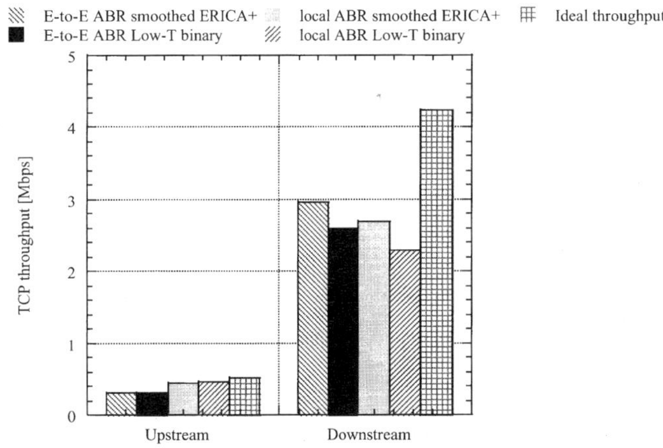

Fig. 20 TCP throughput under condition 3 for low-T binary and smoothed ERICA+.

Table 3 Link utilization and TCP throughput with short-term fluctuations.

	Non-smoothed ERICA+ control	Smoothed ERICA+ control	Simple ER control
Link utilization	0.875	0.997	0.653
TCP throughput	0.872	0.994	0.633

Table 4 Upstream-cell-rate in the VCs transmitting ACKs or b-RM cells [kbps].

	b-RM cell rate	ACK cell rate	Total cell rate
Local ABR	0	19.9	19.9
E-to-E ABR	117.8	19.2	137.0

where x is an averaging factor. ABR_Capacity is updated every Tsm seconds. We call ERICA+ using this smoothed ABR capacity "ABR smoothed ERICA+ control" or simply "smoothed ERICA+." Parameters are set as follows:

Downstream direction:

- target queueing delay: T0 = 200 cell times when the cell rate is 8 Mbps
- parameter used in Eq. (8): $\alpha = 1.15$, $\beta = 1.05$, QDLF = 0.5
- averaging factor in Eq. (10): x = 1/8
- smoothing interval: Tsm = 1 ms

Upstream direction:

- target queueing delay: T0 = 200 cell times when the cell rate is 1.024 Mbps
- parameters used in Eq. (8): $\alpha = 1.5$, $\beta = 1.5$, QDLF = 0
- averaging factor in Eq. (10): x = 1/8
- smoothing interval: Tsm = 10 ms

6.1 Effect of Smoothed ERICA+ under Condition 1

Figure 19 shows that smoothed ERICA+ control had high throughput when there were short-term fluctuations in the bandwidth (i.e., Tup = 0.01 s). To evaluate the effect of smoothing, we measured the link utilization and normalized TCP throughput when ERICA+ was used without smoothing (i.e., we set x to 1) (Table 3). For comparison, we also evaluated local ABR ER control using Eq. (5) (we call this "simple ER" in Table 3). We found that smoothing is necessary to handle the short-term fluctuations.

For low-T binary control, we reduced the threshold to improve the performance of local ABR binary used under condition 3. As we mentioned, this might impair bandwidth utilization. However, compared with Fig. 4, both E-to-E and local ABR low-T binary controls had almost the same high throughput in Fig. 19.

6.2 Effect of Low-T Binary and Smoothed ERICA+ under Condition 3

Figure 20 shows the throughputs under condition 3. Both smoothed ERICA+ and low-T binary controls had high throughput in both the upstream and downstream directions. In both E-to-E and local ABR controls, the downstream throughput under smoothed ERICA+ control was slightly lower than that under simple ER control (cf. Fig. 11) because simple ER control over-regulates the upstream data traffic. When the upstream data traffic is over-regulated, the ACKs sent upstream can return more quickly without waiting in the ATU-R buffer. However, this over-regulation degrades the upstream throughput, as shown in Fig. 11. In contrast, smoothed ERICA+ controls had high throughput in both directions. The reason the upstream throughput under E-to-E ABR controls was slightly less than that under local ABR controls was that E-to-E controls require both ACK and the backward RM cell flow in the upstream direction. Table 4 shows the upstream-cell-rate in the VCs transmitting only ACKs or backward RM (b-RM) cells when smoothed ERICA+ was used.

The backward RM cell rate was almost equal to the "downstream data cell rate / Nrm (32 cells)" (the measured downstream data cell rate was 3.59 Mbps).

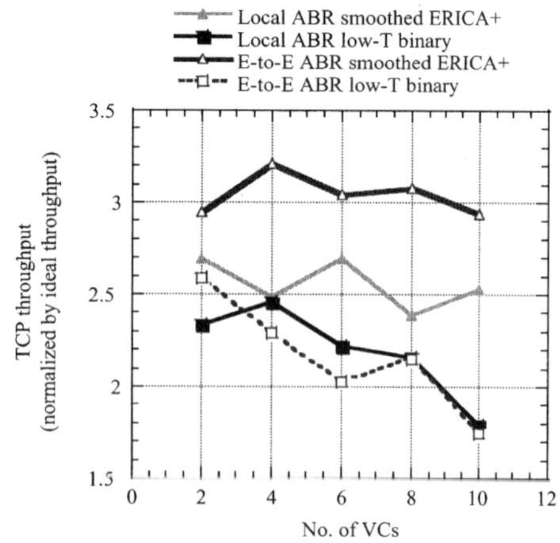

Fig. 21 Downstream throughput vs. no. of VCs under condition 3.

Table 5 Maximum queue length when the number of VCs increases ([cells]).

		2 VCs	8 VCs
Smoothed ERICA+	Tup = 0.01 s	231	258
	Tup = 1 s	436	550
Low-T binary	Tup = 0.01 s	150	391
	Tup = 1 s	288	946

Table 6 TCP throughput under condition 4. (normalized by ideal throughput)

	E-to-E ABR smoothed ERICA+	E-to-E ABR low-T binary	Local ABR smoothed ERICA+	Local ABR low-T binary
With congestion	0.832	0.860	0.720	0.661
Without congestion	0.862	0.890	0.876	0.883

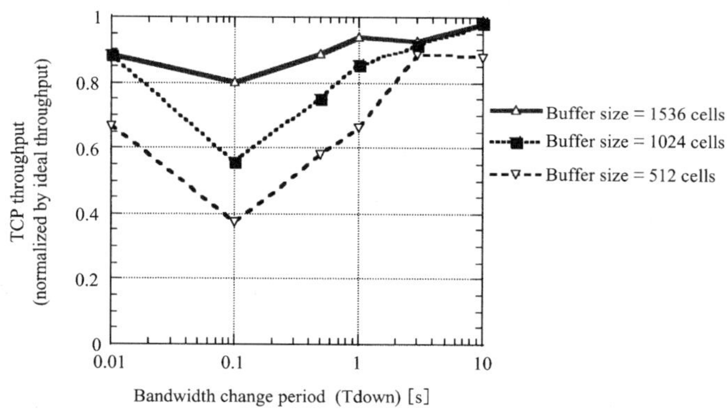

Fig. 22 Downstream throughput under condition 2 for E-to-E ABR smoothed ERICA+.

Due to the network asymmetry, the backward RM cell rate upstream was high compared with the upstream bandwidth. However, ERICA+ enabled E-to-E ABR control to attain higher upstream throughput than simple ER (upstream TCP throughput under ERICA+ was 316 kbps, compared to 184 kbps under simple ER). Here, to increase upstream throughput under E-to-E ABR smoothed ERICA+, we measured throughput when Nrm in the downstream direction was set to 64 cells. When Nrm was increased, the upstream throughput was increased from 316 to 369 kbps, while downstream throughput was maintained around 2.95 Mbps. Hereafter, we use this large Nrm in E-to-E ABR smoothed ERICA+.

Low-T binary controls could improve the downstream throughput compared with the original ABR binary controls. In particular, E-to-E ABR low-T binary control could attain much higher downstream throughput than the original (cf. Fig. 11). Compared with smoothed ERICA+ controls, low-T binary controls had slightly worse performance. As the number of VCs transmitting data in the upstream direction increased, the difference in the downstream throughput between smoothed ERICA+ and low-T binary controls also increased (Fig. 21). This is because smoothed ERICA+ control kept the queue length close to that needed to achieve a predetermined queueing delay even when the number of VCs increased. In contrast, under low-T binary control, the queueing delay and cell loss at the ATU-R buffer became significant when the number of VCs reached 10, which caused delay and loss of ACKs of downstream data transfer.

6.3 Buffer Requirement for Smoothed ERICA+ and Low-T Binary

Under condition 1, the maximum queue length of both smoothed ERICA+ and low-T binary controls was less than 512 cells. Reducing the threshold also made the buffer requirement for ABR binary controls small (in Sect. 5.5, binary controls required a buffer size of more than 512 cells). In this subsection, we investigate the buffer requirement in more detail. As previous studies reported (e.g., [20]), ABR binary requires a larger buffer when the number of VCs increases. Thus, we measured the maximum queue length of local ABR controls under condition 1 when the number of VCs was 8 (Table 5).

Under smoothed ERICA+, the maximum queue length hardly grew at all even when the number of VCs increased. However, low-T binary resulted in a long queue when the number of VCs was 8 and Tup was 1 s. This means that ADSL bandwidth variations require a larger buffer to ABR binary control.

6.4 Other Performance under Smoothed ERICA+

As shown in Table 6 and Fig. 22, for other conditions, smoothed ERICA+ and low-T binary controls had almost the same performance as the original ER and binary controls (cf. Table 2 and Fig. 18). (We omit the results under E-to-E ABR low-T binary in condition 2 because the low-T approach was introduced into the upstream direction in order to reduce queueing delay at the ATU-R buffer).

6.5 Summary

In summary, we found that

- E-to-E ABR smoothed ERICA+ control has the best performance in both directions under any condition. Furthermore, it can cope with an increase in the number of VCs,
- Local ABR smoothed ERICA+ control has high throughput not only in the upstream but also downstream direction by avoiding loss and delay of the ACKs of downstream data transfer, if there is no congestion point in the backbone networks, and
- ABR low-T binary controls have passable performance in spite of their simple control mechanism (they only set the congestion-indication bit in RM cells when the queue length exceeds a threshold), though they need a large buffer when the number of VCs increases.

7. Conclusion

We have evaluated the performance of TCP/IP over ATM over an ADSL when various ATM traffic controls, such as ABR and GFC, were used to handle the variations in the ADSL bandwidth. Our analysis showed that applying an ABR control to the ADSL bandwidth variations is effective under various traffic conditions. In particular, end-to-end ABR control performs well under any condition. If there is no congestion in the backbone networks, then applying an ABR local loop to the upstream traffic also achieves high throughput not only in the upstream direction but also in the downstream direction. This is because local ABR control quickly returns the acknowledgments sent upstream by avoiding loss and delay of the acknowledgments of downstream data transfer. As for buffer requirement for ER and binary controls, binary controls need a large buffer in the upstream direction especially when the number of VCs increases. E-to-E ABR binary control has worse downstream throughput than no-control even with a large buffer. ER controls can keep the same upstream performance with an insufficiently large buffer as with a sufficiently large buffer. As for the downstream performance, E-to-E ABR ER control can attain good performance under any bandwidth-change-period with a large buffer.

We used simulation to determine which of several ABR switch algorithms achieve better performance under any condition, and investigated ABR control methods suitable for ADSL environments. One approach is to apply smoothed ERICA+, which maintains high throughput even when the number of VCs is increased. Another approach is to use ABR binary with a low threshold. This is slightly inferior to the smoothed ERICA+ control, but it is still effective despite its simple control mechanism.

Acknowledgement

We greatly appreciate the cooperation for the simulation evaluation by Miss Mika Ishizuka of NTT Information Sharing Platform Laboratories.

References

[1] ADSL Forum, http://www.adsl.com/
[2] ITU-T, SG15, G.992.2 (G.lite) Splitterless Asymmetrical Digital Subscriber Line (ADSL) Transceivers, 1999.
[3] Universal ADSL working group (UAWG), http://www.uawg.org/
[4] N. Giroux and S. Ramanan, "Handling physical rate changes in ADSL and other technologies," ATM Forum Contribution 98-0825, 1998.
[5] Subir Varma, "Performance and buffering requirements of TCP applications in asymmetric networks," INFOCOM'99, 1999.
[6] T.V. Lakshman, U. Madlhow, and B. Suter, "Window-based error recovery and flow control with a slow acknowledgement channel: A study of TCP/IP performance," INFOCOM'97, pp.1199–1209, 1997.
[7] O. Elloumi, H. Afifi, and M. Hamdi, "Improving congestion avoidance algorithms for asymmetric networks," ICC97, pp.1417–1421, 1997.
[8] G. Hasegawa, M. Murata, and H. Miyahara, "Performance evaluation of HTTP/TCP on asymmetric networks," IEICE Technical Report, IN97-172, Feb. 1998.
[9] The ATM Forum TC Liaison, "The ATMF, ADSL, T1E1 workshop and subsequent actions," ATMF/98L-044, 1998.
[10] The ATM Forum, "Traffic management specification 4.0," af-tm-0056.000, April 1996.
[11] ITU-T, "B-ISDN ATM layer specification—ISDN overall network aspects and functions," ITU-T Recommendation I.361, 1995.
[12] G. Wright and W.R. Stevens, TCP/IP Illustrated Volume 2: The Implementation, Addison Wesley, 1995.
[13] A.W. Barnhart, "Examples switch algorithm for section 5.4 of TM Spec.," ATM Forum Contribution 95-0195, 1995.
[14] L. Roberts, "Enhanced PRCA (proportional rate-control algorithm)," ATM Forum Contribution 94-0735R1, 1994.
[15] A. Arulambalam, X. Chen, and N. Ansari, "Allocating fair rates for available bit rate service in ATM networks," IEEE Commun. Mag., pp.92–100, Nov. 1996.
[16] R.J. Kalyanaraman, R. Goyal, S. Fahmy, and R. Viswanathan, "ERICA switch algorithm: A complete description," ATM Forum Contribution 96-1172, 1996.
[17] H. Osaki, M. Murata, and H. Miyahara, "Parameter tuning for binary mode switch-analysis," ATM-Forum/95-1483.
[18] The ATM Forum, Traffic Management Specification 4.0, atmf95-0013R8, Oct. 1995.
[19] H. Saito, K. Kawashima, H. Kitazume, A. Koike, M. Ishizuka, and A. Abe, "Performance issues in public ABR services," IEEE Commun. Mag., vol.34, no.5, 1996.
[20] H. Osaki, M. Murata, H. Suzuki, C. Ikeda, and H. Miyahara, "Rate-based congestion control for ATM networks," ACM SIGCOMM (Computer Communication Review 95), pp.60–72, 1995.

Appendix A: Parameter Values for ABR Binary Controls

We investigated appropriate values for parameters (RIF, RDF, and threshold) under ABR binary controls in the ADSL environment.

A.1 Parameter Values in the Upstream Direction

First, we set the threshold to the same value (256 cells) as in our group's previous study [19]. For (RIFup, RDFup), we began with four sets (1/64, 1/16), (1/128, 1/32), (1/256, 1/64) and (1/512, 1/128), which are recommended parameter settings in the ATM forum [18]. Figures A·1 and A·2 show the upstream throughput of local and E-to-E ABR binary controls under condition 1. We also measured the link utilization under local ABR binary in Table A·1.

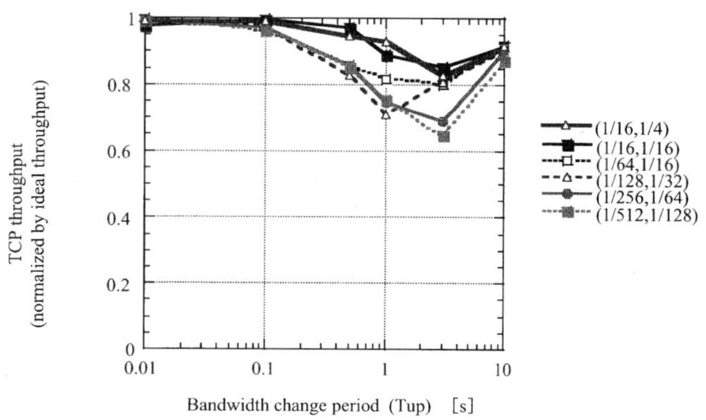

Fig. A·1 Upstream throughput under local ABR binary when we change values of (RIFup, RDFup).

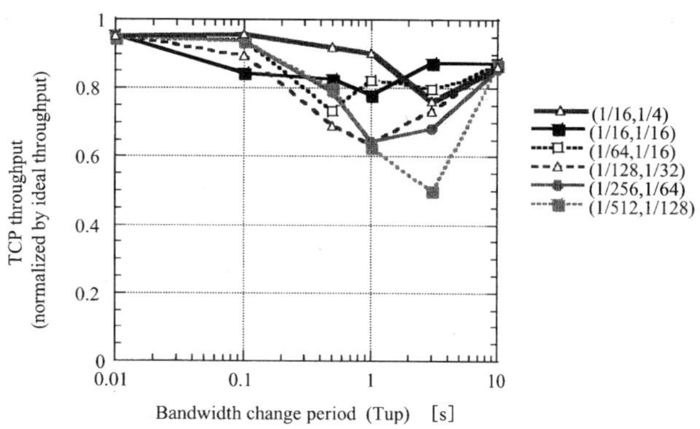

Fig. A·2 Upstream throughput under E-to-E ABR binary when we change values of (RIFup, RDFup).

Table A·1 Link utilization under local ABR binary.

Tup	0.1	0.5	1	3
(1/16,1/16)	1.00	0.993	0.995	0.988
(1/64,1/16)	0.979	0.906	0.922	0.968

From these results, we found that the four sets recommended in the ATM Forum could not utilize bandwidth effectively because they could not keep up with the ADSL bandwidth variations. Therefore, we introduced larger RIFup settings ((1/16, 1/16) and (1/16, 1/4)), and found that (1/16, 1/4) could lead to good performance in both E-to-E and local ABR binary controls. As shown in Figs. A·3 and A·4, even though the congestion thresholds were changed, the original four sets could not achieve high throughput.

A.2 Parameter Values in the Downstream Direction

We measured the downstream throughput under condition 2 when we applied various sets of (RIFdown, RDFdown) to E-to-E ABR binary (Fig. A·5). We also measured the throughput when we changed the threshold (Fig. A·6). These results show that E-to-E ABR binary control performs as badly as or worse than no-control under any set because of the long-control-delay. Thus, at least, in order to cope with very long-term or very short-term bandwidth variations, we evaluate the performance when the ADSL bandwidth was constant (here, we set to be 4.75 Mbps). As shown in Fig. A·7, the four sets recommended in the ATM Forum could not attain high throughput because large RIF caused

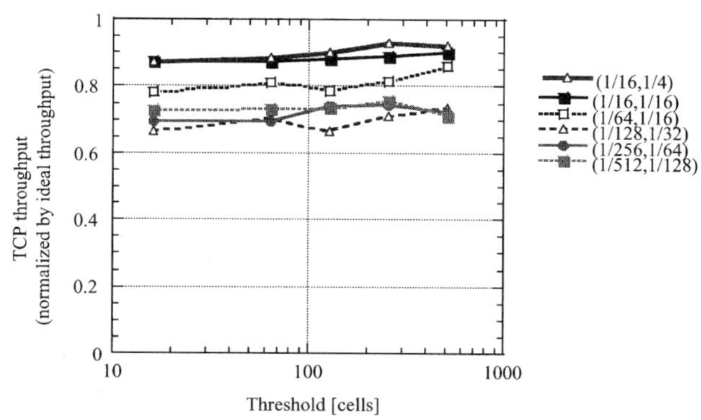

Fig. A·3 Upstream throughput under local ABR binary when we change values of threshold (Tup = 1 s).

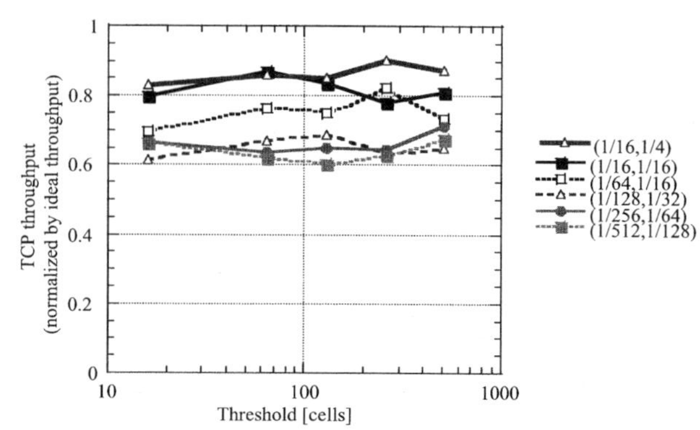

Fig. A·4 Upstream throughput under E-to-E ABR binary when we change values of threshold (Tup = 1 s).

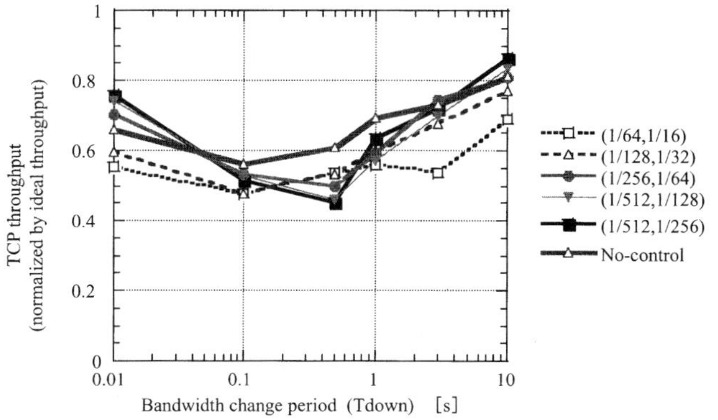

Fig. A· 5 Downstream throughput under E-to-E ABR binary when we change values of (RIFdown, RDFdown).

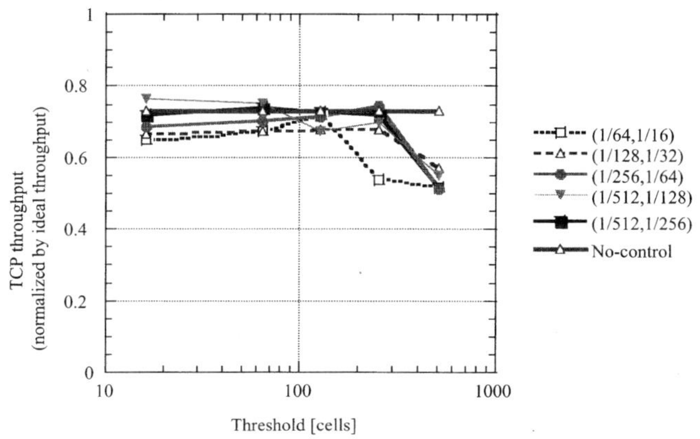

Fig. A· 6 Downstream throughput under E-to-E ABR binary when we change values of threshold (Tdown = 3 s).

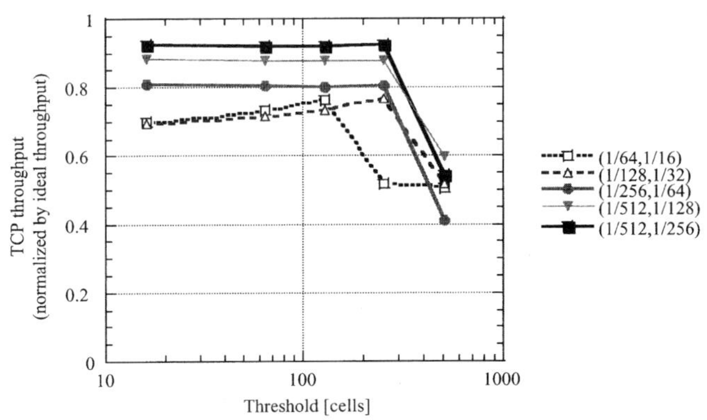

Fig. A· 7 Downstream throughput under E-to-E ABR binary when we bandwidth is constant.

cell loss and large RDF caused ineffective utilization of bandwidth. Therefore, we set (RIFdown, RDFdown) to be (1/512, 1/256), which made it possible to utilize the bandwidth effectively without cell loss.

High-Performance Backbone Network Technology edited by Naoaki Yamanaka
ISBN: 0-8247-5321-6 DOI: 10.1081/(120025694)
Published 2004 by Marcel Dekker, Inc.

Real-Time Cell Arrival Sequence Estimation and Simulation for IP-over-ATM Networks

Hiroshi Saito, Toshiaki Tsuchiya, Daisuke Satoh, Gyula Marosi, Gyorgy Horvath, Peter Tatai, and Shoichiro Asano

SUMMARY We have developed a new traffic measuring tool and applied it to the real-time simulation of a network. It monitors IP traffic on an ATM link and continuously transfers the length and timestamp of each IP packet to a post-processing system. The post-processing system receives the data, estimates the cell's arrival epoch at the transmission queue of the ATM link, and simulates the queueing behavior on-line if conditions differ from those of the actual system. The measuring tool and real-time simulation represent a new approach to traffic engineering. A new estimation problem, the arrival sequence estimation, is shown and some algorithms are proposed and evaluated. Also, a new dimensioning algorithm called the queue decay parameter method, which is expected to be robust and applicable to real-time control, is proposed and evaluated.

key words: *real-time simulation, traffic engineering, ATM, IP*

1. Introduction

Internet traffic is growing rapidly world-wide, and the proliferation of new applications is causing its characteristics to change. Meanwhile, Asynchronous Transfer Mode (ATM) has begun to be deployed in internet backbone networks, several types of WAN services, and fast broadband private networks for companies within a group. In such new telecommunication environments, traffic engineering faces two difficulties. One is the limited capability to monitor and measure traffic in network elements such as ATM switches and routers. While capturing traffic characteristics is one of the most important issues for economical development and evaluation of new technologies, extra functions in network elements to monitor and measure traffic would raise the cost of the network element and reduce the processing power for call control and/or packet forwarding. Thus, existing network elements do not give traffic engineers all the information about the traffic characteristics that they would like, while the traffic characteristics are becoming more and more complicated and the link and processor capacities are increasing. Even when we can monitor traffic, traffic engineering (including traffic control, management, and dimensioning) is still a challenge in an ATM backbone for Internet traffic. This is the second difficulty. In particular, empirical tests (with theoretical background) are more preferable than pure theoretical approaches in traffic engineering these days. Thus, modeling and characterizing traffic may not be enough to show the effectiveness of a proposed traffic engineering scheme. This paper shows how these two challenging issues can be overcome by developing traffic measuring tools outside network elements and passing its data to a real-time on-line simulator.

This kind of approach has received attention recently. DARPA (Defense Advanced Research Projects Agency), through its NGI (Next Generation Internet) project, aims at real-time network simulations as part of network engineering [11]. Hewlett Packard has also announced that NetMetrix, one of its products, can be combined with a simulation tool [12]. In practice, it is different from our tool because we continuously monitor traffic and give every packet a timestamp, and the simulator uses the timestamp sequence continuously.

In addition to the measurement tool and simulator, this paper also proposes a promising link capacity dimensioning algorithm. The algorithm's accuracy is verified by the simulation tool using actual traffic measured by the developed measurement tool.

Overall, we propose traffic engineering steps using real-time on-line simulation: In the first step, the number of network resources such as link capacity or buffer size is dimensioned by a certain method. (The dimensioning algorithm may use traffic data measured at an ATM switching node and may be executed in a network management system. Actually, it is possible to apply a simple dimensioning algorithm that uses and keeps the target utilization. This is because, the third step checks whether the dimensioning was appropriate.) In the second step, the dimensioned number of network resources is set in the simulator using the actual traffic data continuously. Instead of the dimensioned number, the number planned in the following year's budget, for example, can be used. In the third step, the simulator checks whether the dimensioned or planned number of network resources is appropriate. The fourth step yields the number of network resources that was dimensioned/planned, after it has been checked and modified

Manuscript received April 4, 2000.
Manuscript revised July 31, 2000.

†The authors are with NTT Service Integration Laboratories, Musashino-shi, 180-8585 Japan.
††The authors are with Technical University of Budapest, Hungary.
†††The author is with NII (National Institute of Informatics), Tokyo, 101-8430 Japan.

*Parts of this paper were presented at ICCCN'99 [13] and Networking 2000 [14].

IEICE TRANS. COMMUN., Vol. E84-B, No. 3, pp. 634–643, March 2001

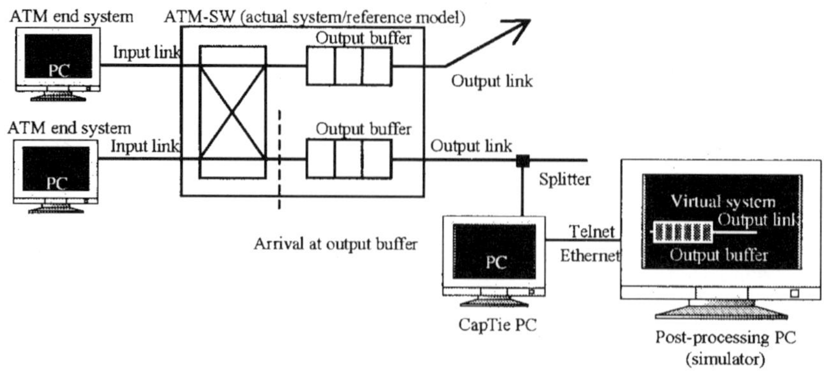

Fig. 1 ATM network with our system.

if necessary. (Normally, the dimensioned or planned number may need a certain margin to accommodate the traffic growth by the time the updated network resources are provided. If we can do these four steps in a short period of time and repeat them, we can implement the self-sizing network concept, which can adapt its network size to the ever-changing traffic [5]–[9].)

2. Measuring Tool

We developed a new traffic measuring tool called CapTie (capturing traffic while being tied to a post-processing computer) on the real-time OS called VxWorks [10]. The first version of CapTie [13] ran under MS-DOS because it was based on the header trace mode of OC3MON [1], [2]. OC3MON is an attractive tool that normally works in a statistical mode, which counts packets at a regular interval, e.g., five minutes. However, for our purpose, finer granularity is needed and multiple processes are convenient for remote operation. In addition, the original header trace mode in OC3MON writes data to a disk and stops working when the disk is full. This is inconvenient for us because we intended to simulate a network continuously. Therefore, we completely rewrote CapTie.

CapTie is a program running on a PC tapped at an ATM link. It monitors traffic on the link by using an ATM network interface card (NIC). CapTie can monitor the IPv4 traffic on ATM adaptation layer type 5 (AAL5). When the ATM NIC receives a cell and processes the physical layer, it judges whether it is the first cell of a IP datagram (packet) by checking that the VPI/VCI of the cell is in the last-cell table in CapTie. (A VPI/VCI of a cell is added in this table if the cell is judged as the last cell of a packet by checking the payload type field in the cell header. If the VPI/VCI of the cell is registered in this table and the cell is not the last cell, the VPI/VCI is removed from the table.)

If a cell is judged as the last cell of a packet, CapTie removes the LLC/SNAP encapsulation header and reads the packet length field in the packet header in the cell and generates a record consisting of the following four items: the CapTie counter value, the ATM NIC slot number, and the packet's timestamp and length. The record generated by CapTie is transferred to another PC that simulates (part of) a network using the record, and CapTie continues working (Fig. 1). As shown later, the first target of the simulated network is the output buffers at the switching node in which the monitored link is accommodated.

The record items are used as follows: (1) CapTie and the post-processing PC are connected via telnet over Ethernet. In our experience, the data rate between them is a few percent of the traffic on the monitored ATM link. Thus, when 100-Mbps traffic on the link is monitored, 1 Mbps or more of data traffic is transferred between the two PCs. Here, there is a danger of some data being lost. The value of CapTie's counter increases one by one when a packet is monitored. Thus, the post-processing PC can check whether any traffic records are lost. (2) CapTie can monitor more than one link simultaneously. To distinguish the record for the ATM link being monitored, an ATM NIC slot number is used. This allows network simulation to be performed if the PC running CapTie has many slots accommodating ATM NICs. (3) The timestamp is a key item of data for simulation. The timestamp uses 64 bits: 32 bits show the number of seconds and another 32 bits express the fraction of a second; this achieves nano-second granularity. (4) Packet length is another key item of data in the simulation. It determines the amount of traffic.

3. Estimation of an Arrival Sequence

The first goal of our traffic measuring tool is to reproduce the cell/packet arrival sequence at the transmission queue (output buffer) of the output link. (Note that CapTie monitors the cell/packet carried sequence of the link.) If this reproduction succeeds, we can analyze the traffic offered to the transmission queue accurately, simulate the transmission queue behavior, and can evaluate some interesting metrics. Otherwise, the result of the traffic analysis is misleading even when the timestamp has fine granularity. (For example, the peak traffic never exceeds the link capacity when we observe the traffic transmitted at the link.) However, the accurate reproduction of the cell/packet arrival sequence is not a straightforward task. Since our measurement device is tapped at a link, what we observe is not the cell arrival sequence or the packet arrival sequence but the cell transmission sequence or the packet transmission sequence. Therefore, our measuring device incorporates an algorithm for estimating the arrival sequence from the transmission sequence.

In this section, we investigate algorithms for estimating an arrival sequence from a transmission sequence. Here, the arrival sequence is defined by the series of cell/packet arrival epochs observed just before the output transmission queue in the ATM switching node that has an ingress end point of the monitored link in Fig. 1. The output buffer in the ATM switching node is sometimes called the reference model when we would like to emphasize the comparison between the reference model and the virtual model, which is a model used in the simulation (Fig. 1). In the remainder of this section, UBR (unspecified bit rate) on the monitored ATM link is implicitly assumed because UBR is used for data transmission on ATM in most cases. As a result, cells of a packet are assumed to be offered to the network at the line speed, while packet transmission from an end system may be controlled by an upper layer.

3.1 Simple Cell Arrival Sequence Estimation Algorithm

Assume that CapTie observes the sequence of cells passing a tapping point in the middle of a link and judges whether a cell is the first cell of a packet based on the payload type indicator of the cell header. If it is the first cell of a packet, CapTie reads the packet length field of the IP packet header in the cell and generates a record consisting of the packet length and the timestamp showing when the cell was detected by CapTie. Actually we can observe the time instance when the first cell of each packet passes a certain point of a link and then observe how many cells belong to the same packet and will pass the point. Based on this information, the simple cell arrival sequence estimation (S-CASE) algorithm estimates the arrival instance of each cell at the arrival observation point.

Let $t(i)$ be the observed timestamp of the first cell of the i-th packet, and let $n(i)$ be the number of cells belonging to the i-th packet. Let H be the capacity of the input link to the ATM switching node that has the ingress end point of the monitored link (Fig. 1) where we assume for simplicity that each input link has the same link capacity. Let L be the length of a cell. The following estimated arrival epoch of the j-th cell belonging to the i-th packet, $t(i,j)$, is provided by the S-CASE algorithm.

$$t(i,j) = t(i) + (j-1)^{*}L/H \tag{1}$$

The S-CASE algorithm assumes implicitly that the first cell does not wait for any time in the transmission queue and that the remaining cells arrive at the input link speed without interruption and do not wait in the transmission queue.

3.2 Busy Period Sensing Cell Arrival Sequence Estimation Algorithm

Assume that we can observe whether or not the output link is busy as well as observing the timestamp of each cell of a packet and the number of cells belonging to the packet. Let b be a busy flag, which is 1 when the output link is busy and 0 when it is idle. This flag can obtained by directly observing the status of the output link or by checking the timestamp of each cell.

The busy period sensing cell arrival sequence estimation (BPS-CASE) algorithm estimates the arrival epoch of the j-th cell belonging to the i-th packet as follows. It maintains x, the number of active flows on the output link, and T, the reference time.

(i) The number of active flows, x, increases by one when the first cell of a packet is detected and decreases by one when the last cell of a packet is detected. That is, the definition of the number of active flows in this paper is the number of simultaneously existing packets on the link, and it is not defined by using the source (destination) address or source (destination) port number.

(ii) If the link is busy at start time s, the reference time T is updated to be s. When the first cell of a packet is detected and if the timestamp is t, the reference time T is updated to be t.

The BPS-CASE algorithm stores the values of b, x, and T for estimating the arrival epoch of each cell of the i-th packet when the first cell of the packet is detected. Let $b(i)$, $x(i)$, and $T(i)$ be their values just before the detection of the first cell of the i-th packet. By using them, the BPS-CASE algorithm estimates the arrival epoch of each cell of the i-th packet when the last cell of the packet is detected.

Consider the case in which the first cell of the i-th packet is detected when the link is idle (i.e., $b(i) = 0$). In this case, this cell does not wait for transmission. Therefore,

$$t(i,1) = t(i),$$

where $t(i,j)$ is the estimated arrival epoch of the j-th cell of the i-th packet and $t(i)$ is the timestamp of the first cell of the i-th packet.

Next, consider the case in which the first cell of the i-th packet is transmitted during a busy period of the link and assume that the link is kept busy while cells from the first one of the $(i-1)$-th packet through the first one of the i-th packet are detected (Fig. 2). In

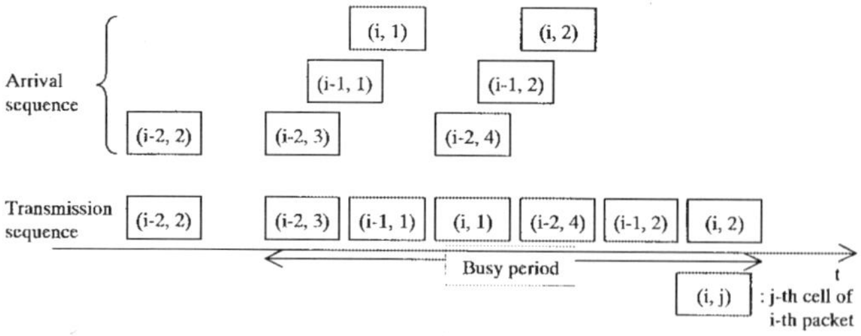

Fig. 2 Example of the event sequence (1).

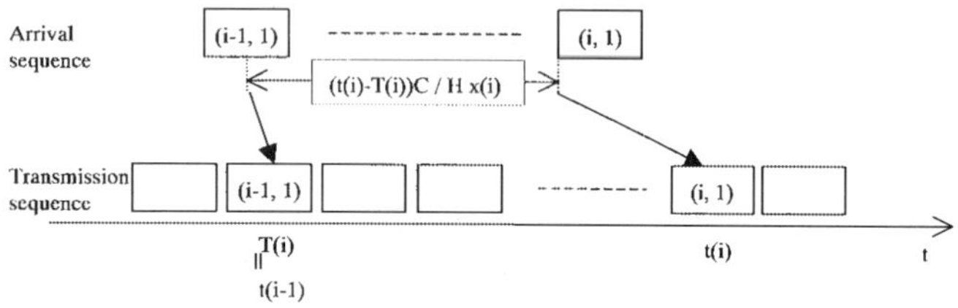

Fig. 3 Example of the event sequence (2).

this case, $T(i)$ is the transmission time (equivalently, the time stamp) of the first cell of the $(i-1)$-th packet. Because the link is busy between $T(i)$ and $t(i)$, the number of cells between the first cell of the $(i-1)$-th packet and the first cell of the i-th packet is given by $(t(i)-T(i))^{*}C/L$, where C is the output link capacity (Fig. 3). Since the number of active flows is $x(i)$, the mean time taken for this number of cells to arrive at the transmission queue of the reference model is $(t(i)-T(i))C/(Hx(i))$ if each active flow offers cells at the input link capacity H and the number of active flows does not change between $T(i)$ and $t(i)$. Therefore, if the estimate $t(i-1,1)$ of the arrival epoch of the first cell of the $(i-1)$-th packet is accurate, the arrival epoch of the first cell of the i-th packet can be estimated as

$$t(i,1) = t(i-1,1) + (t(i)-T(i))C/(Hx(i)). \quad (2)$$

Here we should note that the estimated arrival epoch $t(i,1)$ of the first cell of the i-th packet should be less than or equal to the transmission time $t(i)$ of this cell. To keep this constraint $t(i,1) \le t(i)$ for any $T(i) = t(i-1) \ge t(i-1,1)$, we use $\max(C, Hx(i))$ instead of $Hx(i)$. Consequently, we obtain the following estimation rule that is applicable to both $C > Hx(i)$ and $C \le Hx(i)$.

$$t(i,1) = t(i-1,1) + (t(i)-T(i))C \,/\max(C, Hx(i)). \quad (3)$$

Finally, consider the case in which the first cell of the i-th packet is transmitted during a busy period and the first cell of the $(i-1)$-th packet is not included in this busy period (Fig. 4). In this case, $T(i)$ is the transmission epoch of the first cell forming this busy period. Note that the arrival epoch of this cell is $T(i)$ because this cell does not wait for transmission. Therefore, similarly to the discussion above,

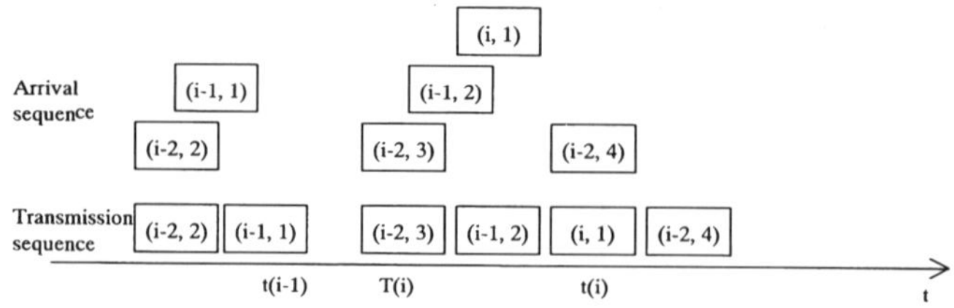

Fig. 4 Example of the event sequence (3).

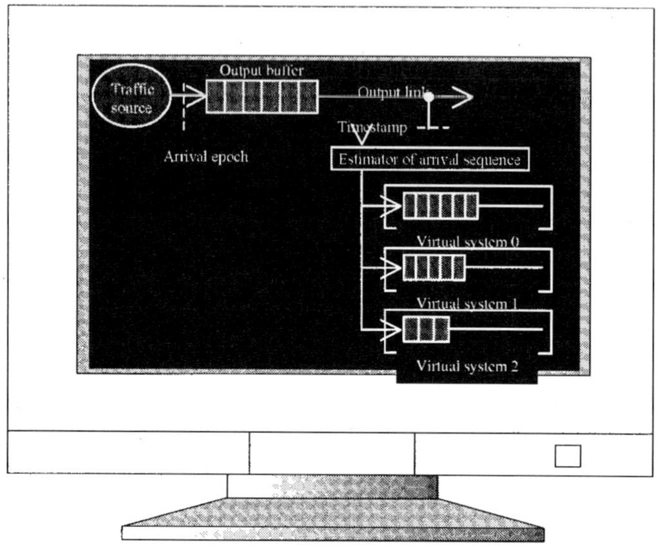

Fig. 5 Computer simulation.

$$t(i,1) = T(i) + (t(i)-T(i))C/\max(C, Hx(i)). \quad (4)$$

The arrival epoch of the j-th cell of the i-th packet is estimated as follows for any of the cases mentioned above.

$$t(i,j) = t(i,1) + (j-1)L/H \quad (5)$$

3.3 Numerical Examples

We investigated the accuracy of the S-CASE and BPS-CASE algorithms through a computer simulation, which simulated the whole system including the reference model and the point where the timestamp is applied (Fig. 5). In this computer simulation, packets were generated according to a Poisson process and their length distribution was assumed to be geometric. They were offered to the transmission queue of the reference model (a FIFO queue). Through observation of the cell transmission sequence, the cell arrival sequence to the transmission queue of the reference model was estimated by the S-CASE and BPS-CASE algorithms. Since we knew the true cell arrival sequence in this computer simulation, we could compare the true and estimated cell arrival sequences. The accuracy was compared in terms of the difference in cell loss ratios (CLRs)

Table 1 Simulation conditions and virtual systems.

Simulation condition	Average packet length (Cell)	Offered load/ source (Mbps)	Number of sources	Input speed (Mbps)	Virtual system (buffer size, output link capacity (Mbps))		
					0	1	2
1	10	1	30	50	(128, 50)	(128, 37)	(128, 30)
2	10	1	30	150	(128, 30)	(128, 37)	(128, 50)
3	50	1	30	50	(128, 50)	(128, 37)	(128, 30)
4	50	1	30	15	(128, 50)	(128, 37)	(128, 30)
5	10	1	30	150	(128, 50)	(128, 37)	(128, 30)
6	50	1	30	50	(256, 50)	(256, 37)	(256, 30)
7	50	1	30	50	(256, 50)	(128, 50)	(384, 50)
8	50	1	30	50	(128, 50)	(128, 75)	(128, 37)

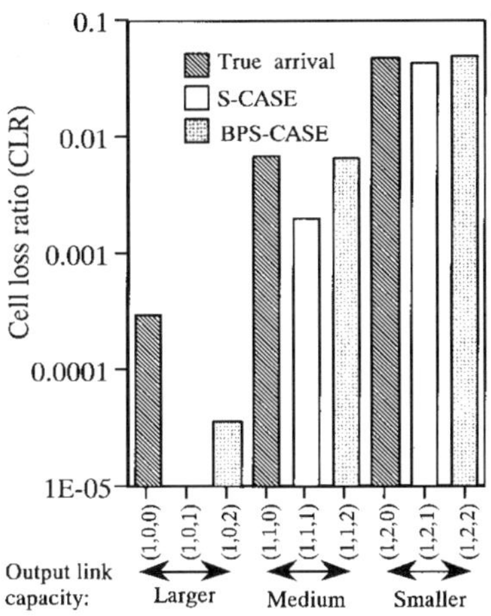

Fig. 6 Cell loss ratio for condition 1.

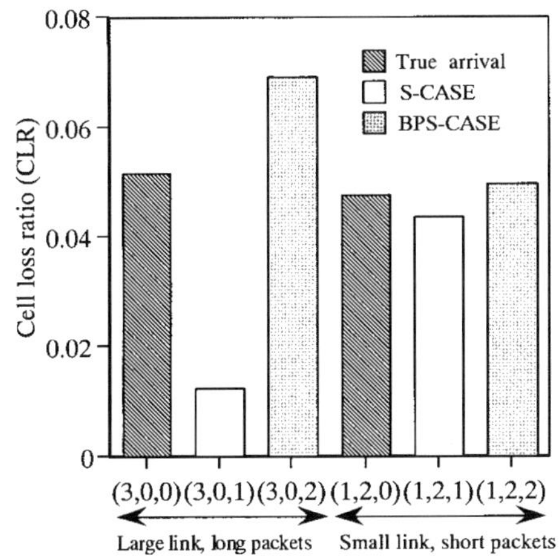

Fig. 7 Errors in estimation for similar CLR.

that occurred when the true and estimated cell arrival sequences were offered to the virtual system and the reference model. This is explained below. One reason for measuring the accuracy in terms of the CLR is that the final target of the cell arrival sequence estimation is to use it for performance analysis.

By using the actual and estimated cell arrival sequences, the CLRs of the reference model and virtual systems in the computer simulation were evaluated, where the virtual systems were defined as a FIFO queue with a different buffer size or a different output link capacity from the reference model. In the following figures, (k, m, n) denotes the k-th simulation condition, the m-th virtual system, and the n-th estimation method. Simulation conditions and virtual systems (their buffer size and output link capacity) are summarized in Table 1. (The 0-th virtual system means the reference model.) The estimation method is 0 when the estimated cell arrival sequence is the true cell arrival sequence; 1 when the estimation method is the S-CASE algorithm; and 2 when it is the BPS-CASE algorithm.

The results for Condition 1 are plotted in Fig. 6. For high CLR or small output link capacity, both algorithms were accurate. As the CLR became lower or the output link capacity became larger, the S-CASE underestimated the CLR.

Figure 7 shows our investigation of Conditions 1 and 3. While the CLRs for the true cell arrival sequence were almost the same for (3, 0, 0) and (1, 2, 0), the estimation accuracies were different. Using either estimation algorithm, the CLRs in (1, 2, 1) and (1, 2, 2) were similar to the true CLR of (1, 2, 0). This means that the estimation in (1, 2, *) was easy. On the other hand, the CLRs in (3, 0, 1) and (3, 0, 2) were different from the true CLR of (3, 0, 0). This means that the estimation in (3, 0, *) was difficult. This seems to be mainly because it was easier to estimate the CLR of a small output link capacity (with short packets) than that of a large one (with long packets). (Item (v) below shows that long packets made estimation easier. Thus, a small output link capacity seems to be one of the main reasons for this result.)

Figure 8 compares the CLRs derived by the estimated and true cell arrival sequences. In addition, the CLR evaluated by the upper bound formula [4], [5] using the mean and peak cell rates of each flow is also shown for comparison. Here, the mean cell rate was

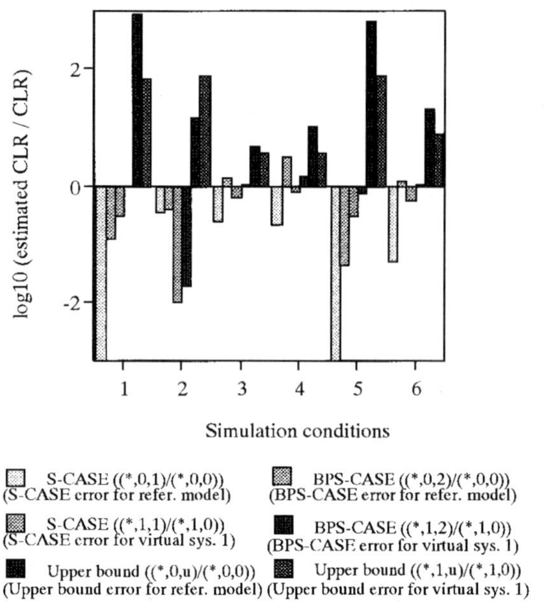

Fig. 8 Errors in estimation.

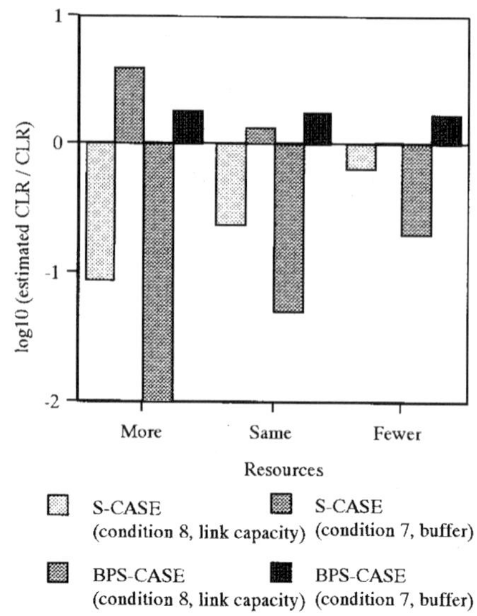

Fig. 9 Errors in estimation for different network resources.

the true mean cell rate and the peak cell rate was the cell rate transmitted by the input link capacity. In this figure, the y-axis denotes the log of the ratio of an estimated CLR to the true CLR. For example, for each simulation condition, S-CASE $((^*,0,1)/(^*,0,0))$ means the log of the ratio of the CLR estimated by S-CASE for the virtual system 0 to the true CLR for the virtual system 0. That is, the y-axis denotes a metric of CLR estimation error.

(i) We should note that the estimation was accurate when the capacity of the output link of the virtual system was smaller than that of the reference model, but may be inaccurate when it is larger. For example, for simulation condition 1, the estimation error for virtual system 0 when S-CASE was used (that is, S-CASE $((^*,0,1)/(^*,0,0))$) was a larger absolute value than the estimation error for virtual system 1 when S-CASE was used (S-CASE $((^*,1,1)/(^*,1,0))$). Here, virtual system 1 had less capacity than virtual system 0 (the reference model).

(ii) BPS-CASE is normally more accurate than S-CASE. For example, for simulation condition 1, S-CASE's estimation error for virtual system 1 (S-CASE $((^*,1,1)/(^*,1,0))$) had a larger absolute value than BPS-CASE's (BPS-CASE $((^*,1,2)/(^*,1,0))$).

(iii) Simulation conditions 1 and 5 for virtual system 0 and simulation condition 2 for virtual system 1 produced low CLRs and the estimation was inaccurate with S-CASE for simulation conditions 1 and 5 and with BPS-CASE for simulation condition 2. See S-CASE $((^*,0,1)/(^*,0,0))$, S-CASE $((^*,1,1)/(^*,1,0))$ and BPS-CASE $((^*,1,2)/(^*,1,0))$. It was difficult to estimate a low CLR accurately.

(iv) The accuracy of BPS-CASE deteriorated when the input link capacity was much larger than the output link capacity of the reference model. See BPS-CASE $((^*,1,2)/(^*,1,0))$ for simulation condition 2 and compare it with that for simulation condition 1.

(v) BPS-CASE was accurate for simulation conditions 3, 4, and 6. There, the input link capacity was less than or equal to the output link capacity of the reference model and the packet size was long. It was more accurate for simulation condition 6 than for simulation condition 3, while the CLR for simulation condition 6 was lower than for simulation condition 3. (Normally, the estimation error becomes large when the CLR is lower.) This seems to be because the virtual system for simulation condition 6 had a longer buffer than that for simulation condition 3.

(vi) The absolute value of the estimation error obtained by BPS-CASE was always smaller (i.e., more accurate) than that by the upper bound formula.

(vii) We also compared the CLR of each virtual system for each simulation condition derived using the Poisson arrival assumption instead of the arrival sequence estimated by S-CASE or BPS-CASE. Here, the Poisson arrival assumption means that the cells were generated according to a Poisson process with mean equal to the mean observed in the simulation. Since the CLR obtained under the Poisson arrival assumption was smaller than -10 in this figure for all simulation conditions, it was not plotted.

Figure 9 shows our investigation of conditions with different network resources (buffer and link capacity). When there were more network resources in the virtual system than in the actual system, the estimation errors were larger than when there were fewer network resources in the virtual system. BPS-CASE was more accurate and better than S-CASE. In particular, BPS-CASE was accurate even when the buffer size of the virtual system was different from that of the actual system.

4. Real-Time Simulation

The post-processing PC, which receives the data from CapTie via Ethernet, uses one of the arrival sequence algorithms mentioned in the previous section. As a result, we can (approximately) reproduce the cell arrival sequence at the multiplexing point (the transmission queue) of a switching node. Therefore, if we provide a simulator in the post-processing PC, provide a virtual system or a model of the transmission queue in the simulator, and give it the reproduced arrival sequence, we can simulate the performance behavior of new controls and new network resource conditions as if they were applied at the transmission queue (Fig. 1).

In this study, we used a simulator to evaluate the accuracy of an output link capacity dimensioning method called the queue decay parameter method, which is described in the following section. This dimensioning method is expected to be applicable to various arrival processes without explicit traffic models and is appropriate even for real-time computation. Therefore, if this dimensioning method is accurate, it will be a good candidate for an algorithm that can be used for the self-sizing network referred in Sect. 1. The simulator in the post-processing PC simulates an output link and its transmission queue in the switching node that has an ingress end-point of the monitored link. We assumed that the transmission queue could be modeled as a single-server FIFO queue with a finite-size waiting buffer where the service was cell transmission.

As a first step, the output link capacity was dimensioned by the proposed dimensioning algorithm, which used the traffic measurement data obtained by the switching node. The dimensioned output link capacity was used for the output link capacity (that is, the service rate of the server) in the simulator and the actual buffer size of the simulated transmission queue was used as the buffer size of the simulator. The estimated and reproduced cell arrival sequences were offered to the simulator in the post-processing PC, and the CLR of the transmission queue (output buffer) of the output link was evaluated in the simulator. In the numerical example below, S-CASE was used in the simulation for simplicity.

5. Queue Decay Parameter Method

Here, we describe a simple and heuristic link capacity dimensioning method based on the buffer occupancy measurement. We call this method the queue decay parameter method. Consider a single server queueing model as a model of the transmission queue in an ATM switching node. From the result of the large deviation principle and related works (e.g., see [6]), the asymptotic behavior of the queue length Q is expressed as

$$p(k) \equiv P\{Q \geq k\} = \alpha \exp(-\beta\nu(k)), \tag{6}$$

where α and β are constants (independent of k) and $\nu(k)$ is a function of k. These parameters depend entirely on the nature of the input traffic. In order to study the input traffic, we made cell-level measurements in a real network in advance and captured the real-time sequence of cell arrivals. At each measurement, we obtained a sample sequence of 1.3 million cell arrivals. The total number of available samples during five days of measurement was 150. With those data, we made a simulation of a FIFO single server queueing model, where the service time was the cell transmission time. (Note that this was not an on-line simulation but one using the recorded cell arrival data repeatedly.) Figure 10 shows the distribution $p(k)$ obtained by the simulation with 44 samples (on the same day). (The x-axis denotes k, the queue length, normalized by the buffer size.) In this case, the arrival rate of each sample was between 6 and 12 Mbps, and the link capacity was set to 15 Mbps. From the figure, we can observe that the decay parameter α is nearly equal to 1. Moreover, the decay of the queue length seems exponential. Therefore, $\nu(k) = k$ and the traffic is considered to have short-term dependence. At least, in this range of queue length, we can dimension the traffic under the assumption that it has short-term dependence. This argument remained valid for other sample sequences, though we omit similar figures. As a result, we have a simple approximation formula

$$p(k) \cong \exp(-\beta k). \tag{7}$$

Equation (7) is similar to the M/M/1 queueing model with different values of β. In the M/M/1 case, $\beta = -\log(\rho)$ where ρ denotes the utilization. In Fig. 10, each sample gives the ratio of $-\log(\rho)$ to β as being from 6 to 11, which is greater than in the M/M/1 case (because of the burstiness of the input traffic). With the same samples, we tested the values of $-\log(\rho)/\beta$ for different link capacities, varying from 15 to 40 Mbps, in Fig. 11. For each sequence of data, the values of

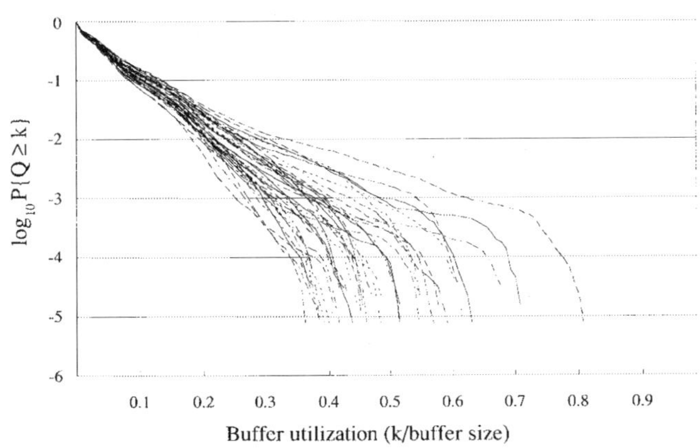

Fig. 10 Queue length distribution.

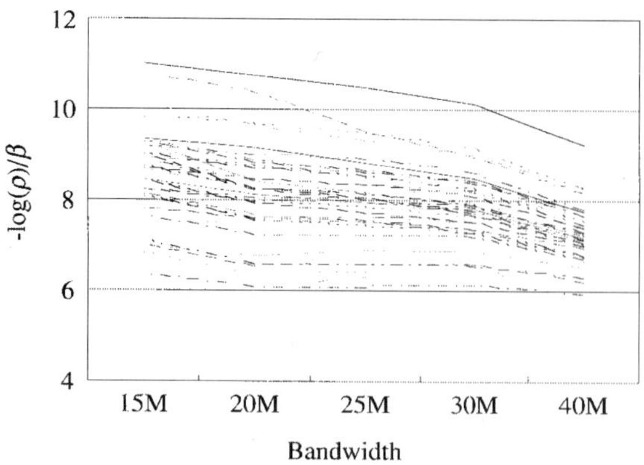

Fig. 11 $-\log(\rho)/\beta$ vs. bandwidth.

$-\log(\rho)\beta$ were almost constant, where the link capacity was in the range of 15–40 Mbps. From those observations, we finally obtained a simple link capacity dimensioning method using a probability $p(k)$ that the queue length exceeds a certain k as follows.

(1) Measure the utilization ρ_m and probability $p(k)$ that the queue length exceeds a certain k.
(2) Estimate the decay rate β_m by $\beta_m = -\log(p(k))/k$ (see Eq. (7)).
(3) With objective cell loss ratio CLRo and buffer size K, estimate the desirable decay rate β^* that satisfies CLRo $= p(K) = \exp(-\beta k)$ by $\beta^* = -\log(CLR_o)/K$.
(4) Assume that $-\log(\rho)/\beta$ is constant and calculate the utilization ρ^* when $p(K) = CLR_o$ by solving the formula

$$-\log(\rho^*)/\beta^* = -\log(\rho_m)/\beta_m. \tag{8}$$

(5) Dimension the link capacity so that the utilization becomes ρ^*.

When the ratio $-\log(\rho_m)/\beta_m$ in Eq. (8) changes in some range, conservative management is possible with the greatest value, because a larger value will lead to lower utilization ρ^*.

6. Numerical Example of Real-Time Simulation

6.1 Example 1

The developed system was applied to a bi-directional link in SINET (the Science Information Network) [3], which is a nationwide large IP network for research organizations in Japan, whose core network is implemented on an ATM network. The monitored link was between the University of Tokyo and the network office of NACSIS (National Center for Science and Information Systems). As mentioned in the previous section, the capacity of each link was dimensioned using the queue decay parameter method based on data measured in the week before this trial by an ATM switching node

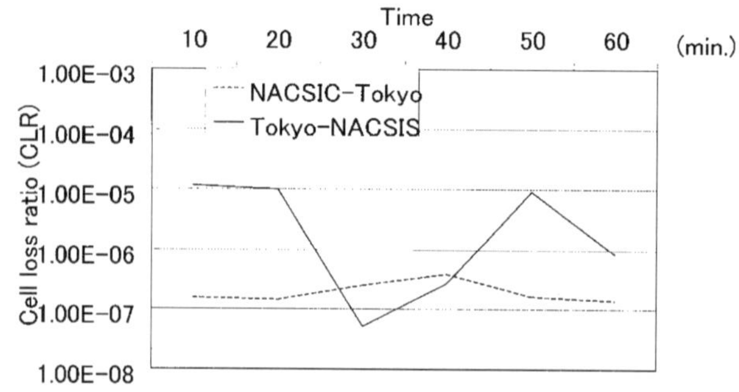

Fig. 12 Cell loss ratio in simulated output link.

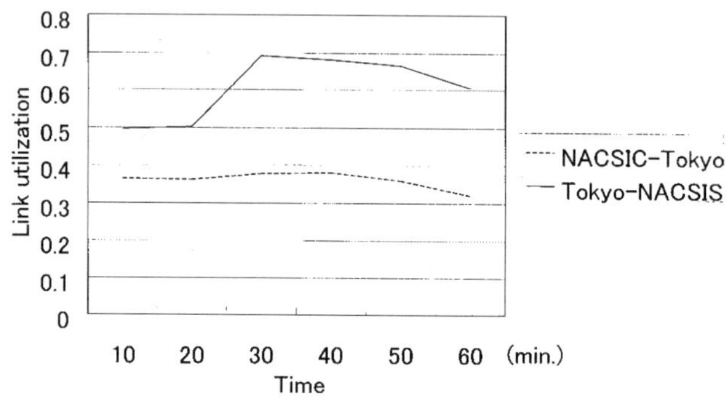

Fig. 13 Link utilization in simulated output link.

accommodating the link. The CLR objective was 10^{-6}.

Figure 12 [13] plots the CLR of each link simulated by the real-time simulator for one hour during a busy hour. The CLR obtained by the simulator agreed closely with the objective. (In our experience, the CLR is a very difficult parameter to manage. Thus, the agreement was actually far better than we expected.)

The direction from the NACSIS network office to the University of Tokyo has a large amount of traffic with long packets because it includes a lot of traffic downloaded from the U.S. The opposite direction has less traffic with shorter packets. The link utilization obtained in the simulation was higher in the direction from the University of Tokyo to the NACSIS than in the opposite direction (Fig. 13). That is, the smaller link had higher utilization than the larger one whereas a larger link can usually achieve higher utilization due to the statistical multiplexing gain. The reason for this unexpected result is that short packets are dominant in the direction from the University of Tokyo to the NACSIS network office (Fig. 14). However, the CLR of this direction had a larger fluctuation than that of the opposite direction because the offered traffic was small (Fig. 12).

Comparing Figs. 12 and 13, we see that high average utilization during ten minutes does not mean a high CLR. This is because traffic fluctuations over a period much shorter than ten minutes strongly affect the CLR. Thus, we cannot estimate the CLR based only on a ten-minute average utilization.

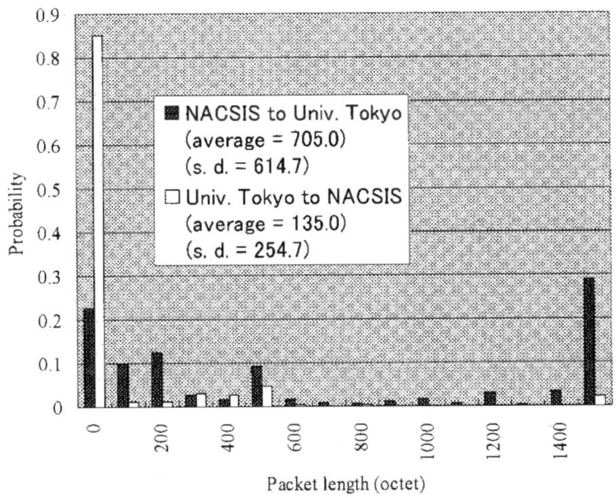

Fig. 14 Packet length distribution.

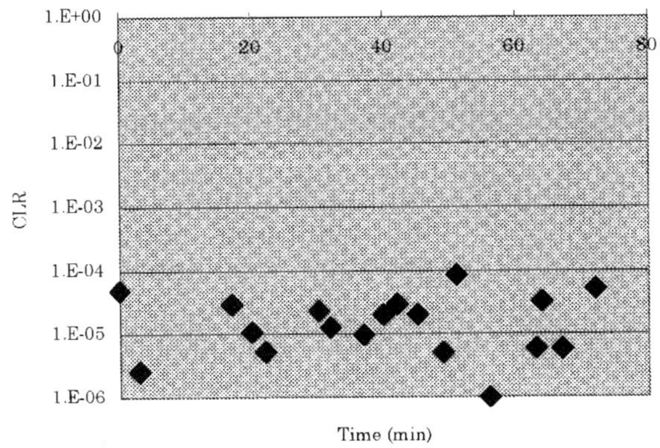

Fig. 15 Cell loss ration in the simulation.

6.2 Example 2

We monitored the carried traffic between the open computer network (OCN) and the network office of NACSIS, which provides a nation-wide internet among research organizations. The OCN is an IP network service provided by NTT Communications. By using the queue simulation implemented in the developed system, we evaluated the VP bandwidth estimated by using the $H2/D/1/K+1$ queue model for the output queue of the VP, where K is the buffer size and the queue has the first-in first-out discipline.

Our system monitored the traffic and calculated the average and the variance of the number of bytes transferred during each minute. Based on this average and this variance, we calculated the loss probability of the $H2/D/1/K+1$ for a fixed service rate and found a service rate for which the loss probability is equal to the cell loss ratio (CLR) objective. (To simplify the analysis, we used the large deviation approximation for $H2/D/1/K+1$.) In the simulator, we set the VP bandwidth to the newly calculated value for every fifteen minutes and evaluated the CLR of the simulated model. Figure 15 shows the CLR of the simulated model in the simulator when the objective CLR is 1.0E-3. (Thus, if the $H2/D/1/K+1$ model is accurate, this CLR should be 1.0E-3.) The approximated $H2/D/1/K+1$ queue analysis gave more bandwidth than the bandwidth which met the objective CLR because the simulated CLR was lower than the objective CLR.

7. Conclusions

We have developed a traffic monitoring tool called CapTie and a real-time simulator that uses data from CapTie. The cell arrival sequence estimation methods used in it were evaluated. Using these two systems, the proposed dimensioning method was evaluated for real IP traffic data on an ATM network and shown to be accurate. The developed system enables us to implement a new approach to network engineering. In the near future, we will apply the system to a self-sizing network.

Acknowledgments

We would like to express our thanks to Joel Apisdorf, who developed OC3MON. Without his help, we would not have started this study. We also thank engineers in NTT Advanced Technology and the staff of NACSIS, who helped us to apply our system to the network. To monitor the traffic between OCN and NACSIS, Kawahara Ryoichi helped us. We thank him.

References

[1] K. Thompson, G.J. Miller, and R. Wilder, "Wide-area internet traffic patterns and characteristics," IEEE Network, pp.10–23, Nov./Dec. 1997.
[2] http://www.nlanr.net/NA/Oc3mon
[3] S. Asano, "Science information network in Japan," IEICE Journal, vol.81, no.4, pp.402–406, 1998.
[4] H. Saito, Teletraffic Technologies in ATM Networks, Artech House, Boston, 1994.
[5] H. Saito, "Call admission control in an ATM network using upper bound of cell loss probability," IEEE Trans. Commun., vol.40, no.9, pp.1512–1521, 1992.
[6] N.G. Duffield and N. O'Connell, "Large deviations and overflow probabilities for the general single-server queue with applications," Proc. Cam. Phil. Soc., vol.118, pp.363–374, 1994.
[7] H. Saito, S. Nakagawa, Y. Kato, and S. Ohta, "Innovations of circuit/path operations in ATM networks—Self-sizing network," NTT Review, vol.8, no.1, pp.56–65, 1996.
[8] S. Nakagawa, Y. Kato, S. Nakai, K. Ogura, and H. Saito, "Self-sizing network operation systems in ATM networks," NOMS'96, Kyoto, 1996.
[9] H. Saito, "Dynamic resource allocation in ATM networks," IEEE Commun. Mag., vol.35, no.5, pp.146–153, 1997.
[10] http://www.wrs.com/products/html/vxworks.html
[11] M. Maeda, "NGI: The next generation internet program," IEEE ATM Workshop'99 Proceedings, Kochi, 1999.
[12] http://www.tmo.hp.com/tmo/ntd/products/products.html

[13] H. Saito, T. Tsuchiya, G. Marosi, G. Horvath, P. Tatai, and S. Asano, "Real-time simulation with traffic monitoring tool," ICCCN'99, Boston, 1999.

[14] H. Saito, T. Tsuchiya, G. Marosi, G. Horvath, P. Tatai, and S. Asano, "Real-time cell arrival sequence estimation and simulator for IP/ATM networks," IFIP Networking 2000, Paris, 2000.

High-Performance Backbone Network Technology edited by Naoaki Yamanaka
ISBN: 0-8247-5321-6 DOI: 10.1081/(120025695)
Published 2004 by Marcel Dekker, Inc.

A New Approach for Real-Time QoS Support in IP over ATM Networks

Ismail Erturk and Elias Stipidis

SUMMARY Emerging real-time and multimedia applications in IP over ATM networks are becoming an indispensable part of the networking world. Considering the different and essential natures of IP and ATM, it seems that they complement each other. Although several approaches have been proposed for IP and ATM integration, so far a key problem (i.e., Quality of Service support) in foreseeable future networking applications remains unsolved. This paper describes a new extension to Classical IP and ARP over ATM (CLIP) protocol, Real-time and Multimedia services using CLIP (RMCLIP), providing ATM service classes to the IP end users. It defines that different service classes are mapped to the associated IP addresses. It enables different IP end user applications (e.g., real-time video, voice and data) to obtain their own dedicated VCs.
key words: *IP over ATM, CLIP, QoS, real-time multimedia*

1. Introduction

In recent years IP (Internet Protocol) has been dominating other protocols deployed in global networks. Considering its primary usage (i.e., digital data communications), IP was initially a prominent success in networking technologies. However, its inadequacy for Quality of Service (QoS) support has resulted in failing to provide emerging multimedia and real-time applications to the desktop satisfactorily. Since 1990, several approaches have been proposed for IP and Asynchronous Transfer Mode (ATM) interoperability so that QoS can also be achieved. Apparently, Classical IP and ARP (Address Resolution Protocol) over ATM is one of the most favourable methods, exploiting the connection-oriented nature of ATM to transport IP traffic. This technique has been brought to reality since IP is already prevalent from an access point of view, and after the indications of ATM's increasing future implementation in Local/Wide Area Networks (LANs/WANs) as well. ATM has universal support for both present and future applications; moreover, it has been designed to cope with even unknown requirements.

Various concepts for QoS support at the IP level have been proposed and/or are being developed, especially by the Internet Engineering Task Force (IETF). A considerable work by IETF in this area has produced two very complicated architectures called the Integrated Services (IntServ - RFC1633) and the Differentiated Services (DiffServ - RFC2475). The QoS requirements of the services are signalled on a per-flow basis through the network and the required network resources are reserved using Resource Reservation Protocol (RSVP - RFC2205) in the former. The latter provides QoS to aggregated traffic on a per-hop basis. It is based on standard behaviours such as expedited forwarding for delay sensitive traffic, assured forwarding for traffic requiring a guaranteed bandwidth and best effort (BE) for the rest.

The IETF MPLS Working Group has recently introduced Multiprotocol Label Switching (MPLS - RFC2702) where extended traffic engineering methods are applied for IP and therefore it also may support QoS. In addition, the IETF Audio/Video Transport Working Group is currently working on a Real-time Transport Protocol (RTP). Although RTP provides end-to-end network transport functions suitable for applications transmitting real-time data, neither does it address resource reservation nor guarantees QoS for real-time services.

Having briefly explained some important characteristics of ATM and IP in Sect. 2, well-known ATM-IP integration methods in Sect. 3, and QoS aspects of multimedia applications in Sect. 4, this paper introduces a new extension to Classical IP and ARP over ATM (CLIP) protocol, called RMCLIP, which provides better QoS support to the IP end users in an IP over ATM environment. This scheme improves the only classical support of IP BE service over one VC, and enables each IP host pair to have multiple VCs with different QoS, where multiple IP addresses are assigned to each host.

2. IP and ATM

Having considered the multi-service nature of the information technologies in present applications as well as deployment of different computer networks worldwide, neither a pure IP network (due to its connectionless nature it can not quarantee QoS for multiple services) nor a pure ATM network (due to its very complicated higher layer signalling and data protocols) is adequate to fully cover all user and network provider requirements. Therefore an IP network adapted to an

Manuscript received July 25, 2001.
Manuscript revised January 15, 2002.
[†]The author is with Technical Education Faculty, Kocaeli University, 41300 Kocaeli, Turkey.
[††]The author is with School of Engineering and Information Technology, University of Sussex, Brighton, UK.

IEICE TRANS. COMMUN., Vol. E85-B, No. 10, pp. 2311–2318, October 2002

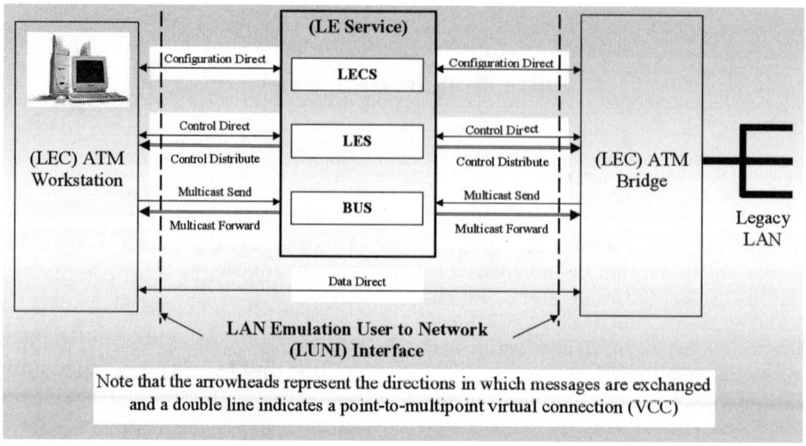

Fig. 1 LAN emulation service.

ATM core is essential to take advantage of the power of ATM for its performance and QoS support. Using both technologies it means that Local/Metro systems can take advantage of ATM's QoS performance in the lower/network layers and the QoS transition to a full optical core IP network (Photonic Transport Systems). The standard bodies, such as ATM Forum and IETF, promote this concept. [1], [2] suggested IP as the common glue for interconnecting heterogeneous networks into a single, large internetwork, where IP runs on top of the various network layers.

One of the strengths of IP is that a vast variety of user applications are available with IP. With its connectionless datagram delivery service, IP can handle bursty traffic and network congestion conditions adequately. Its scalability and application in very large networks with less central management are exceptionally valuable for network designers. On the other hand its datagram-based packets are not suited to real-time traffic and since they are routed independently, out of order delivery is also likely to occur. Furthermore, the number of fields of an IP header and its variable length, which are sources of slow packet processing, are deemed as its setbacks [3]–[5].

In contrast to IP, ATM is intrinsically connection-oriented (hence more complex) and allow QoS guarantees. It has superior features including flexibility, scalability, fast switching and use of statistical multiplexing, which utilise network resources efficiently. Different traffic characteristics are maintained in a scalable manner. However like IP, there are also several problems with ATM. Principally, it must allow interoperability with existing network infrastructure using specially designed signaling, addressing and routing protocols. An adaptation layer has to be used to provide the necessary services (e.g., segmentation/reassembly, adding extra framing and processing overheads) in order to transport variable size data frames over fixed size ATM cells [3], [6], [7].

3. LANE, MPOA, Connectionless ATM, and Classical IP over ATM

The ATM Forum and IETF have been carrying out most of the work on ATM and IP interoperability. The former has emerged with two solutions which have been evolving since 1995. They are namely LAN Emulation (LANE) and Multi Protocol Over ATM (MPOA) [8], [9]. The latter has proposed a different approach in its work, referred to as Classical IP and ARP over ATM, where an ATM network environment is configured as a Logical IP Subnetwork (LIS) [10]. In addition to these specifications (based on software schemes), [11] in an individual attempt, called Connectionless ATM (CATM), suggested a method based on a hardware scheme to provide special support for connectionless traffic in ATM networks. The potential size of the hardware routing tables required for each link or group of links is its main limit to scalability [11].

LANE is a service provided over an ATM network, which operates at the Layer 2 (Data Link Layer) of the Open Systems Interconnection (OSI) reference model and emulates the services of existing Ethernet and Token Ring applications. It hides the features of ATM from applications and therefore applications with QoS are unable to exploit the unique capabilities of ATM, even though they work over ATM through LANE without modification [1], [12]. However, the newly completed LANE Version 2.0 provides some locally administered QoS for communication between ATM-attached end systems [13].

As shown in Fig. 1, an LE Service is comprised of three different server functions, a LAN Emulation Server (LES), a Broadcast and Unknown Server (BUS) and a LAN Emulation Configuration Server (LECS). The LES implements the control co-ordination function for the emulated LAN. It both registers/resolves MAC addresses and/or routes descriptors to ATM addresses. The BUS provides not only broadcast and multicast

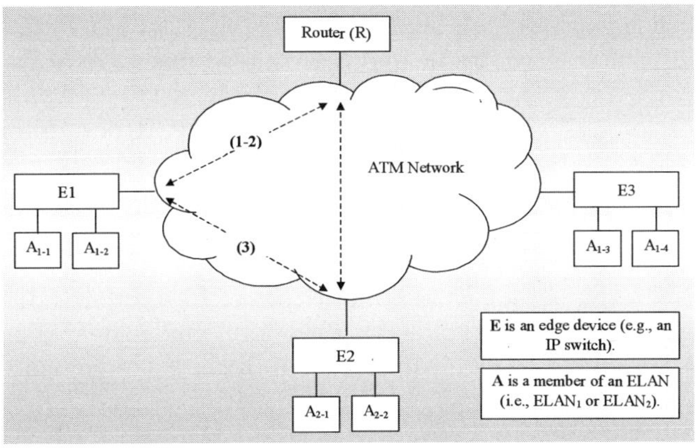

Fig. 2 MPOA connection establishment.

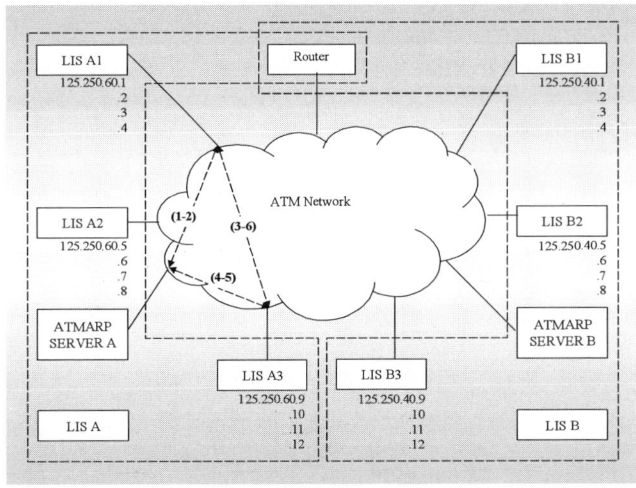

Fig. 3 Connection establishment in CLIP: (1) A1 sends an ATMARP request to the server-A on the LIS A and (2) the server returns ATM address of the A3. (3) A1 establishes an ATM link using this address. (4) When A3 receives the first packet, it also sends an ATMARP request to the server-A to find out the ATM address of the A1 (sender). (5) Once the server returns the appropriate address, (6) A1 and A3 can communicate directly without any further server involvement.

services, but also broadcast of packets whose destination addresses have not been determined by the LES yet. Finally, the assignment of individual LECs to different emulated LANs is implemented by the LECS. The communication paths between the clients, and between the LE Service and clients, are known as ATM Virtual Channel Connections (VCCs). Most traffic flows over the Data Direct VCCs that are established as required and may be released at any time.

MPOA has been created to improve LANE by allowing inter Emulated LAN (ELAN) traffic to go through shortcut connections rather than routers. It is built on top of LANE. The traffic between two end systems on the same ELAN still follows the LANE specification. MPOA allows an end system to obtain the ATM address of another end system that lies in a different ELAN by using a mechanism called flow detection, and establishes end-to-end connection with it. As a result, it significantly reduces slowing packet throughput at intermediate routers, providing a service that is a combination of Layer 2 bridging and Layer 3 (Network Layer) routing. However, MPOA does not support multicast traffic over ATM shortcut connections (but multicast or broadcast services can be provided to the end systems on the same ELAN) and its complexity is a major practical obstacle in the industry (Fig. 2).

As can be seen from Fig. 2, considering two ELANs (i.e., A_1 and A_2) are defined, connections among $A_{1\text{-}1}$, $A_{1\text{-}2}$, $A_{1\text{-}3}$ and $A_{1\text{-}4}$ are accomplished through LANE procedures as explained above. Router (R), on which an entity called MPOA Server is running, is a member of both ELANs. If $A_{1\text{-}1}$ wants to communicate with $A_{2\text{-}2}$ (each one is on a different ELAN), it first sends the traffic to R, as in LANE. However, E1 detects this flow of traffic after a while and (1) contacts R to find which edge device is "nearest" to the destination $A_{2\text{-}2}$. (2) As soon as the source receives the ATM address of E2 replied by the R, (3) E1 builds a connection to E2 and all traffic from $A_{1\text{-}1}$ to $A_{2\text{-}2}$ is sent along this connection, therefore avoiding the router.

The CLIP approach preserves the IP characteristic at the Layer 3 for the existing applications, while switching Layer 2 from Ethernet/Token Ring to ATM. IP packets are carried in ATM Adaptation Layer-5 Payload Data Units (AAL-5 PDUs). The behavioural characteristics of ATM LISs are similar to the well-known IP subnets (Fig. 3).

CLIP can run over both Permanent and Switched Virtual Circuits (PVCs and SVCs), and is based on the notion of logically independent LISs. Similar to an IP subnet, a LIS consists of a group of IP hosts directly connected to a single ATM network and belonging to the same IP subnet. Like in LANE, the traffic between the hosts on different LISs has to go through a router that is also a member of both subnets. Every IP address is resolved to its corresponding ATM address, and vice versa, using ATMARP mechanisms. ATM connections are consequently set up. As a result of this basic mapping definition originally described in RFC1577 that has been replaced by RFC2225 [10], all the traffic with different characteristics is transmitted through only one VC providing only a BE service and set up between two communicating hosts at a given time. In this paper, the IP over ATM protocol is modified so that different traffic types including real-time multimedia, can utilise the QoS guarantees provided by the underlying ATM network (Fig. 4).

4. Real-Time Multimedia Applications and ATM Service Classes

Before looking at the new proposed approach for CLIP, it is necessary to introduce some facts about the multimedia service classes supported by ATM.

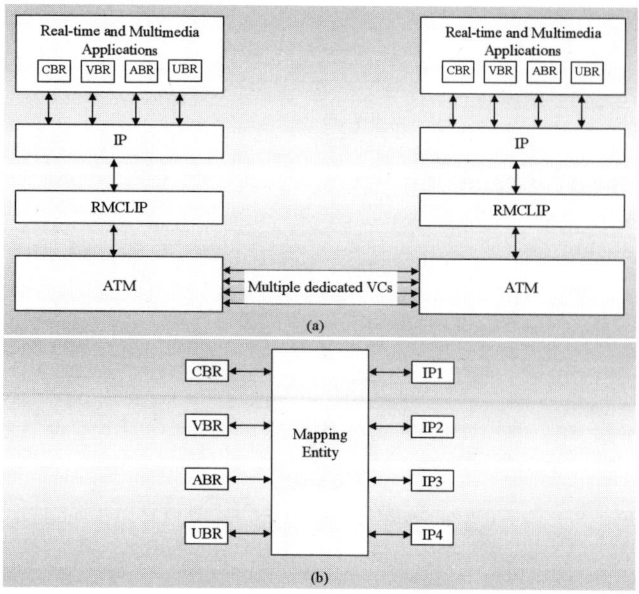

Fig. 4 RMCLIP with multiple VC connections.

Table 1 Advanced TV formats for digital video.

Format	*Lines*	*Horizontal Pixels*	*Frame Rate*
ITU-T 601	480	704	60, 30, 24
MPEG-1	240	320	30, 24
VGA 480	480	640	60, 30, 24

Real-time multimedia is a class of technologies and applications, which covers up two (e.g., data and audio) or more media (e.g., data, audio and video). It can operate with delivery from as little as 56 Kbps to 155 Mbps of information to a user. The 1.5–6 Mbps data rate per user is a basic range designed to support multimedia technology developers. Some video-based application data rates are given in [14], as follows:

- Desktop video-conferencing applications using H.320/H.261 ITU-T standards produce from 128 to 768 kbps per user.
- Multimedia applications using, for example, Intel's Indeo methods produce about 1.5 Mbps per user.
- Entertainment video and distance learning applications using MPEG-1 or MPEG-2 standards produce 1.5 or 6 Mbps, respectively per user (examples of advanced TV formats are shown in Table 1).

Real-time digital audio and video applications require Constant Bit Rate (CBR) and Variable Bit Rate Real-Time (VBR-RT) service categories respectively. For CBR services, the Peak Cell Rate (PCR) descriptor must be provided to the ATM network (called Peak Rate Allocation), whilst for VBR services, Minimum and Sustained Cell Rates (MCR and SCR) are also mandatory (called Dynamic Rate Allocation). Other QoS attributes (e.g., Cell Loss Ratio (CLR), Cell Delay Variation (CDVT) and Cell Transfer Delay (CTD)) can be at a default setting within the ATM network.

Interactive traffic and LAN interconnect traffic are carried over Available Bit Rate (ABR) service. Applications using this service specify the MCR descriptor and can also add PCR, CDVT, and CLR.

Applications where guaranteed packet delivery is not required (e.g., Internet traffic, e-mail, etc.) are provided over Unspecified Bit Rate (UBR) service. Essentially this is the ATM counterpart of Best-Effort (BE) service category of the IP. Therefore, any cells sent on UBR connections have to wait until there is free bandwidth available and not being used by either CBR or VBR traffic.

5. New Approach to Support ATM Service Classes in IP over ATM Networks Using CLIP (RMCLIP)

The goal is to extend the CLIP protocol in order that all ATM Service Classes can be offered to the end users in IP over ATM networks seamlessly (Fig. 4(a)). This approach, called Real-time and Multimedia services using CLIP (RMCLIP), will be used especially to provide IP users with ATM CBR, VBR-RT and ABR service classes, in addition to the UBR (i.e., BE). This scheme improves the only classical support of IP BE service over one VC, enabling each IP host pair to have 4 different dedicated VCs with different QoS.

As shown in Fig. 4(b), the idea behind this approach is that each service category is carried using a different IP address. Thus, this technique requires 3 more IP addresses over which all ATM service classes together with BE can be provided for every RMCLIP end user that demands real-time multimedia applications, considering the fact that for such applications ATM CBR, VBR-RT and ABR service support is necessary. The ATM equivalent service classes are mapped to the associated IP addresses at the source IP user. This process is realised by using a locally (source and destination) configured mapping entity to switch VBR-RT, CBR, ABR and UBR traffic to the IP1, IP2, IP3 and IP4 addresses respectively, and vice versa. As this process is local to the source and destination, the CLIP connection establishment remains the same (Fig. 3). The QoS requested by the client is on the basis of the networking equipment (Routers) to be able to "mask" groups of IP addresses according to their values and mapping these IP addresses to the associated service classes. The recently standardized IPv6 addresses are only used for the moment in backbone structures and IPv4 addresses are only used for end user equipment. If the end user equipment were to be transferred to IPv6 some of the additional bits could be used as a field for the Point-of-Presence (PoP) Routers to identify the service class while translating these fields to IPv6 for the backbone operations. This way the backbone, no matter ATM or IP, can function as before and

Table 2 Priority scheme modifying Options field of IP packets.

ATM Service Class	*Priority (IP PDU Options field)*
CBR	11
VBR	10
ABR	01
UBR	00

the Routers can use the same configuration for the end users as they do for the backbone. The IPv6 addresses are actually assigned to individual interfaces and not to the nodes themselves [5], [10], hence, software interfaces within the end user equipment can hold the different IP addresses and dynamically assign them through the operating system to the hardware network interface. Only one IP address is assigned to a single hardware interface at one point in time. Dynamically assigned IP addresses are already in operation (specially used for laptops) for different local domains irrespective to the operating system of the end user equipment. Therefore, the current configuration for all networking equipment provides the basis of easily migrating to the RMCLIP when IPv6 becomes available with just some minor software additions for

(a) end user dynamically allocated software IP addresses according to the service class requested and
(b) the PoP Routers to include a masking function to discriminate between the IP associated service classes.

The possible setback of RMCLIP having 3 more IP addresses is thought to be insignificant since, with a total number of 2^{128} addresses available (IPv6), address exhaustion ought not to occur for a long time. Moreover, this approach is only provided for the IP users that require real-time multimedia services over ATM. This approach is best suit to applications supporting IPv6 just due to the need for more IP addresses although it can be used for IPv4 (where, respectively, 2^{128} and 2^{32} IP addresses are available) as well.

In addition to this approach, the same goal can also be achieved using the Options (Priority in IPv6) field of the IP PDUs so that each application with different service requirements is assigned a different priority (e.g., Table 2). This scheme, where there is not an end to end ATM VC connection established and every IP packet undergoes normal IP routing operations in the IP/ATM gateway, will introduce extra packet processing for end user equipment, resulting in more end-to-end message delays despite avoiding the use of 3 more IP addresses as in RMCLIP. The extra delays though are relative to the backbone QoS protocol efficiency. If the RMCLIP was to be deployed, the backbone algorithms providing the QoS would have a more simplified task of providing QoS guarantees as the QoS class allocation would be driven by the end user equipment. It is the classical solution of de-centralizing and distributing tasks across the network and simplifies central networking equipment intense processing functions.

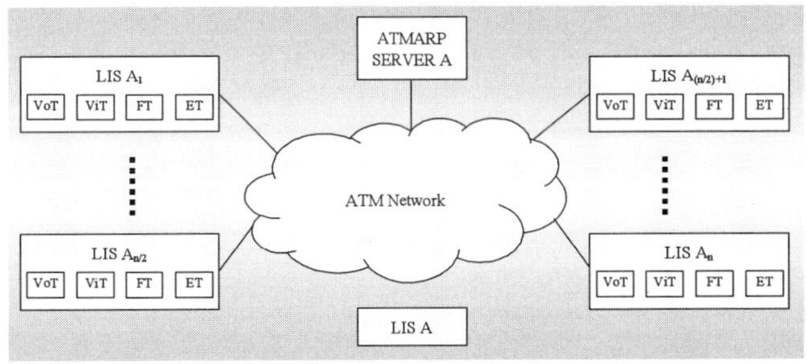

Fig. 5 RMCLIP with multiple VC connections.

5.1 Simulation Models

As ATM is widely known for the QoS performance, the purpose of using IP and ATM within the simulation models with RMCLIP is to illustrate the advantages of providing end-to-end QoS for the end user, by the end user. ATM can provide the best QoS performance in comparison to any other protocol but only if end to end connection is through ATM networking equipment. Figure 5 shows the simulation model created to evaluate the new RMCLIP approach with comparisons to the CLIP, using a commercially available simulation package called COMNET-III. The designed model provides IP end users with CBR, VBR-RT and ABR service classes in addition to CLIP's only UBR (i.e., BE) support. Each user (i.e., A_1–A_n) is connected to an ATM Network. Every user with 4 applications running and requiring different service classes is presented on the network. These are namely a UBR service application (e.g., E-mail Transfer (ET)), an ABR service application (e.g., File Transfer (FT)), a VBR-RT service application (e.g., Video Transfer (ViT)) and finally a CBR service application (e.g., Voice Transfer (VoT)). Although one-way traffic would produce similar results, two-way traffic has been chosen for more realistic and accurate results, from source to destination, both of which are on the same LIS and vice versa. Throughout the simulation time, 10 LIS users (i.e., $n = 10$) with 4 different applications transmit all simultaneously over one ATM physical link (with a line rate of 622,080 Mbps). In order to reduce jitter (for especially CBR services) leaky bucket algorithm as the traffic shaper is used. The service parameter contract of the ATM Network is given in Table 3.

ABR traffic management is accomplished using Resource Management (RM) cells to determine network congestion. To obtain network feedback, the ABR traffic sources send RM cells after every Nrm-1 (the default value for Nrm is 32). Destinations simply return these RM cells with the Current Cell Rate (CCR), the Explicit Rate (ER), the Congestion Indication (CI) flag and the No Increase (NI) flag back to the sources. If upon return of the RM cell no congestion is indicated,

Table 3 ATM service parameters.

Service Class (AAL)	*Data + OH** (bytes)*	*PCR Mbps*	*MCR Mbps*	*SBR Mbps*	*ICR* Mbps*	*AIR Mbps*	*RDF*
CBR (AAL1)	47 + 6	10	-	-	-	-	-
VBR-RT (AAL2)	47 + 6	10	5	8	-	-	-
ABR (AAL3/4)	44 + 9	10	0.5	-	0.5	0.3	2
UBR (AAL5)	48 + 5	10	-	-	-	-	-

* Initial Cell Rate
** Overheads

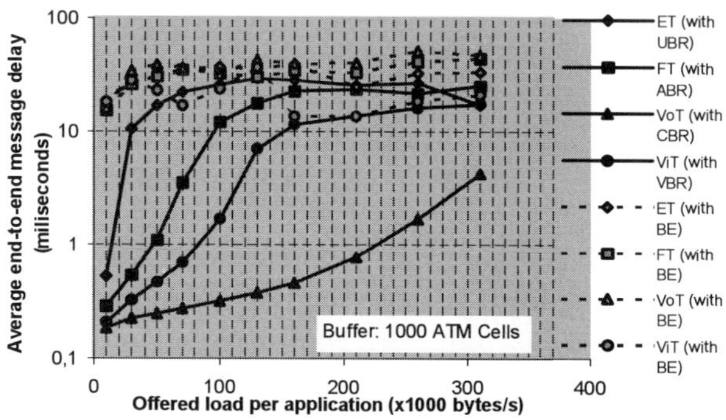

Fig. 6 Average message delays.

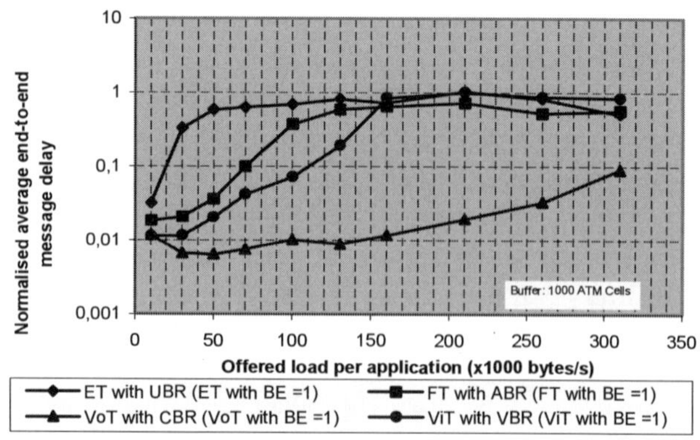

Fig. 7 Normalised average message delays for RMCLIP.

the transmission rate is increased by the Additive Increase Rate (AIR). Otherwise, transmission rate is decreased by the Rate Decrease Factor (RDF) (i.e., the current transmission rate is divided by the RDF).

Throughout the simulation time (i.e., 100 seconds), ET, FT, ViT and VoT data are created and introduced to the network every Inter Arrival Time (IAT) seconds (i.e., 1 second), with an exponential distribution[†] by every IP end user on the LIS. Considering the high volume of user traffic in the ATM network, the delays introduced by the ATMARP mechanisms are considerably low (Fig. 3); hence ignored while measuring end-to-end message delays.

5.2 Simulation Results

Varying the message sizes of the ET, FT, ViT and VoT, end-to-end message delays for both RMCLIP and CLIP end users are provided and explained in the next paragraphs. All of the applications traffic are chosen to be equal so that the justification of the ATM services, each of which is used by a different application of the RMCLIP end users, and CLIP's BE service can easily be done and compared.

Considering the maximum agreeable buffer size is 1000 ATM cells for real-time applications [15], results are presented for the buffer size of 1000 ATM cells which indicates the congestion threshold.

Figures 6 and 7 show the average end-to-end message delays and normalised average end-to-end message delays respectively as a function of offered load per application, for both RMCLIP and classical models. Since the ATM network has offered each application a different and negotiated service as required in the RMCLIP model, VoT (with CBR) has experienced the least average message delay, in spite of the fact that all multimedia applications have been given the same message size. ET (with UBR) and FT (with ABR) have had almost equal average end-to-end delays as the offered load reaches 150,000 bytes/s and above, since the network load increases and consequently network resources start decreasing (effecting the ABR service). As expected, CLIP's BE service has produced not only erratic, but also enormous message delays (plotted as dotted lines), even when the offered load is less than 30,000 bytes/s as a result of its nondeterministic nature.

Having normalised the message delays for the RMCLIP model with their classical model counterparts (Fig. 7), it can be justified that even in the worst case (i.e., when the offered load per application is 210,000 bytes/s), all of the RMCLIP traffic has better end-to-end message delays than those of the classical model. Especially, the VoT (with CBR) traffic experiences 10 to 100 times lower message delays in the RMCLIP model compared to the same traffic carried with the CLIP's only BE service. Considering the fact that UBR and BE service classes are identical, it is not unusual that the message delay for ET (with UBR) traffic has almost the same characteristics as that of the classical model.

The maximum end-to-end message delays of each application traffic for both models are shown in Fig. 8 as a function of offered load. VoT (with CBR) has again outperformed the others. The maximum message delay drop in both ET (with UBR) traffic for offered

[†]Exponential Distribution Function (Exp (A)) takes one positive number which is the mean (A) and returns a positive real number. Samples are likely to be less than the mean and the distribution has a long tail. It occasionally provides a sample value significantly higher than the mean. This behaviour is very useful in modelling random arrival patterns but obviously it might not match to all traffic types.

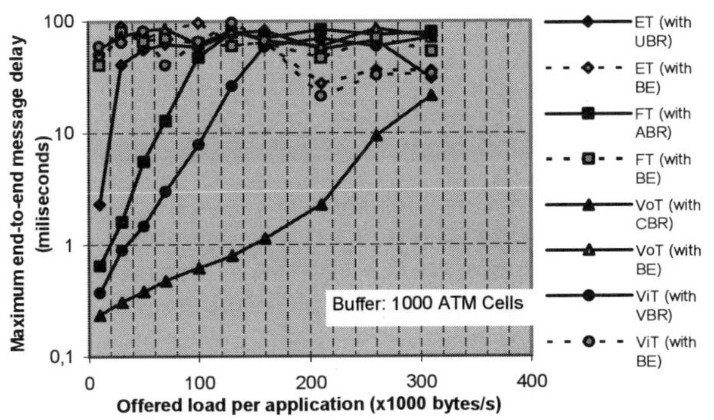

Fig. 8 Maximum message delays.

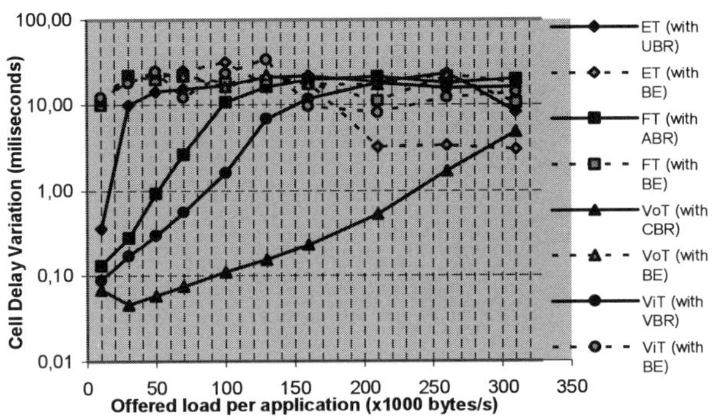

Fig. 9 Cell delay variation.

loads over 250,000 bytes/s in the RMCLIP model, and all traffic for offered loads over 150,000 bytes/s in the classical model, results from the fact that the number of delivered messages decreases dramatically.

Another essential performance metric, i.e. Cell Delay Variation (CDV), comparison of both methods is presented in Fig. 9, which is especially important for both voice traffic and video traffic. It can be seen that there is a considerable difference between CDV results of the classical and RMCLIP approaches for the same type of traffics. The new approach provides VoT CDV values almost negligible when the offered load is less than 100,000 bytes/s and compared to the classical one, the results are much better. Although this VoT traffic is carried using a CBR service, the increase in CDV (seen especially for the offered loads over 210,000 bytes/s) results from the fact that it is multiplexed together with VBR traffic (i.e., ViT) over a single link. Also the ViT CDV values when the offered load is less than 100,000 bytes/s are in acceptable limits considering the fact that the allowable end-to-end jitter for compressed video is 1ms [16]. On the other hand the classical approach is far away from meeting this performance need of the VoT and ViT traffics with its BE service support. Another point to make in the figure is that the CDV drop in all classical model graphics when the offered load for all traffic types is over 130,000 bytes/s is again related to the decrease in the number of delivered messages.

6. Conclusion

A detailed overview of the efforts that have been known to support IP over ATM networks is presented. The LANE method is inherently deficient in terms of QoS support, but it does enable legacy LAN users to run over ATM with no modification. It supports any Layer 3 applications, whilst CLIP only works with IP. Both the inter ELAN traffic and inter LIS traffic have to travel through a router. ATM Forum's MPOA has been created to overcome this crucial negative aspect, providing cut-through connections between ELANs.

QoS support in IP over ATM networks is a hot topic. First implementations of LANE, MPOA and CLIP are deficient in real-time multimedia support. The basic concept developed in this paper shows the establishment of how an improved performance can be obtained in IP over ATM environments, bringing different service classes to desktops. RMCLIP offers a simple approach allowing IP users with real-time multimedia applications to run over ATM networks with QoS. It can be realised between ATM and IP as a mapping layer. Compared with the classical approach, the RMCLIP not only offers all ATM service classes to the IP end users, but also provides over 10 times better end-to-end message delay results for applications traffic requiring ABR, VBR-RT and CBR service class support.

Simulation studies and performance evaluation of RMCLIP together with the classical approach, have shown that CLIP's only BE service (i.e., UBR) support for real-time multimedia applications is no longer a major obstacle in IP over ATM networks.

References

[1] R. Gupta, "The glue of networks: Looking at IP over ATM," 53 Bytes, vol.7, no.1, pp.4–5, March 1999.

[2] M.L. Hankins, "Protocol spawns debate over future of voice, data, video communications," Signal Magazine, p.5, May 1999. (http://www.us.net/signal/Archive/May99/)

[3] Z. Fan, "New trends in ATM networks: A research view," Computer Communications, vol.22, pp.499–515, 1999.

[4] M. Decina and V. Trecorci, "Convergence of telecommunications and computing to networking models for integrated services and applications," Proc. IEEE, pp.1887–1914, Dec. 1997.

[5] W. Stallings, High-speed networks: TCP/IP and ATM design principles, Prentice-Hall, 1998.

[6] I. Erturk, M. Tanrah, E. Stipidis, and A. Kutlu, "Interconnection of ATM LANs with legacy LANs across an ATM backbone using LAN Emulation Service," Proc. 14th International Symposium on Computer and Information Sciences, ISCIS'99, pp.703–710, Izmir, Turkey, 18–20 Oct. 1999.

[7] J. Sterbenz, "Protocols for high speed networks: Life after ATM?" IFIP Conference on Protocols for High Speed Networks, pp.3–18, 1995.

[8] The ATM Forum, "LAN emulation over ATM, version 2," Feb. 1999.

[9] The ATM Forum, "MPOA baseline, version 1.1," May 1999.

[10] M. Laubach and J. Halpern, "Classical IP and ARP over ATM," The Internet Society and IETF, RFC 2225, April 1998.

[11] R. Barnett, "Connectionless ATM," Electronics Communication Engineering Journal, vol.9, no.5, pp.221–230, Oct. 1997.

[12] D.J. Newson, D. Ginsburg, and M.T. Wilkins, "Next generation local area networks," BT Technical Journal, vol.16, no.1, pp.76–87, Jan. 1998.

[13] H. Sun, K. Huang, and L. Li, "Supporting IP on the ATM networks: An overview," Computer Communications, vol.21, no.11, pp.1020–1029, 1998.

[14] D. Minoli and A. Alles, LAN, ATM, and LAN Emulation Technologies, Artech House, 1996.

[15] P.B. Key, "Connection admission control in ATM networks," BT Technology Journal, vol.13, pp.52–65, 1995.

[16] R. Steinmetz and T. Meyer, "Modelling distributed multimedia applications," Proc. IWACA'92, pp.337–349, Munich, 1992.

High-Performance Backbone Network Technology edited by Naoaki Yamanaka
ISBN: 0-8247-5321-6 DOI: 10.1081/(120025696)
Published 2004 by Marcel Dekker, Inc.

ATM and IP Integration by Built-In IP Handling Capability in an ATM Switching System

Akira Arutaki, Hiroshi Ikeda, Masahiko Honda, Kazuhiko Isoyama, Tatsuhiko Amagai, Kenji Yamada, and Tetsurou Nishida

SUMMARY The rapid growth of the Internet impacts ATM networks to be furnished IP handling capability. This paper discusses networking issues for IP and ATM integration. First, it considers function allocation at the boundary of an ATM backbone network and the Internet. As the result, the paper explains the necessity of built-in IP handling capability into an ATM switching system, and summarizes functional requirements for the system architecture. According to the discussion above, the authors propose the system architecture of the IP/ATM integration in the ATM switching system. The implementation of the proposed architecture is evaluated, and the wire-speed IP handling capability in the ATM switch is confirmed.

key word: *high-speed SAR, IPOA, ATM switch, IP handler, H/W routing*

1. Introduction

The surge in the Internet traffic in recent years has placed a great demand for IP processing capability on ATM networks, which are expected to convey IP traffic as a part of multimedia information stream. As is well-known, IP and ATM traffic is handled by IP routers and ATM switches, respectively. This paradigm requires that a network operator, who provides IP routing services via ATM networks, manages IP routers and ATM switches as separate network elements.

Integration of such ATM and IP handling capabilities in a single system is assumed to reduce a great amount of operation costs. To meet this demand, it is necessary to realize an efficient and high-performance IP processing function for an ATM switch which is the primary component of an ATM network.

This paper starts at the point of function allocation in terms of IP handling capabilities in an ATM network. And it discusses the requirements for IP related functions in an ATM switch, and introduces an architecture which satisfies the requirements.

2. Accommodation of the Internet in ATM Network

An ATM network is expected to accommodate a variety of services, such as Cell Relay (ATM), Frame Relay, the Internet, Ethernet LANs, etc. Figure 1 shows the architecture of the

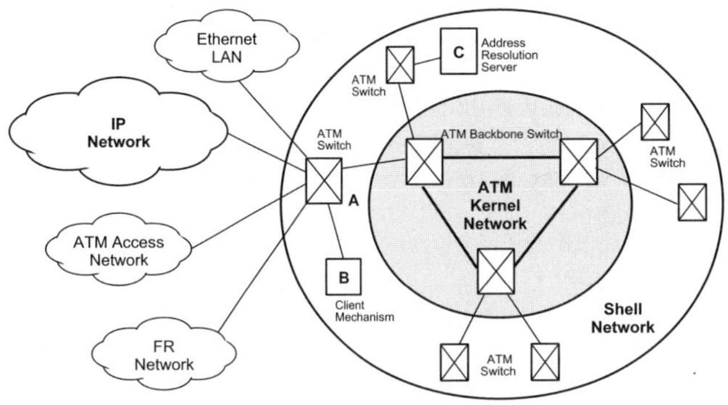

Fig. 1 Architecture of ATM network.

ATM network. The network is composed of three major parts; Access networks, a Shell network, and a Kernel network [9]. The ATM Kernel network is furnished with the characteristics of simplicity/high speed as the UNIX Kernel is. The Shell network has serviceability/flexibility on the analogy of the UNIX shell.

The ATM Kernel network is designed to focus on high-speed ATM transport. It does not recognize any protocol, but ATM. It should be simple, robust, and reliable. As a result, such traffic types as IP, FR (Frame Relay), and CR (Cell Relay) that are integrated on the same transport vehicle, i.e., ATM, are transmitted over the wide area public network at high speed.

The Shell network glues access networks to the Kernel network. It accommodates a variety of traffic including such non-ATM traffic as IP protocol. Sophisticated functions such as security control, QoS guarantee, authentication, multicast, toll/charging, etc. are to be provided in this network. The Shell network is also expected to efficiently provide format and address conversion functions for connecting protocols of other networks (e.g., IP and FR) to the ATM Kernel network. To accommodate IP traffic with this network, the following components are required

(A) ATM switching system to convey IP messages to the Kernel network in an ATM form
(B) Client mechanism that handles protocol processing
(C) Address resolution mechanism that associates each interfaced protocol address with an ATM address/VC (virtual channel)

More specifically, in the MPOA (Multi-Protocol Over ATM) framework [5], (B) is LANE (LAN Emulation) [4]/ an NHRP (Next Hop Resolution Protocol) client and (C) an

Manuscript received June 11, 1999.
Manuscript revised September 2, 1999.

† The authors are with C&C Network Products Development Laboratories, NEC Corporation, Abiko-shi, 270-1198 Japan.
†† The author is with Switching Division, NEC Corporation, Abiko-shi, 270-1198 Japan.
††† The author is with Switching Software Engineering Division, NEC Corporation, Abiko-shi, 270-1198 Japan.

MPOA server. When the ATM network directly interfaces the Internet, (B) is an IP router mechanism and (C) an IPoA server that associates an IP address with an ATM address.

3. Architecture of ATM Switching System in Shell Network

The conventional method is to realize each (A), (B), and (C) with an independent network element. For example, to accommodate the Internet, an IP handling mechanism is provided separately with the ATM switch (A).

The problems with this approach are:

1. It is necessary to provide ATM interface cards in the IP routers, including ATM cell SAR (Segmentation and Reassembly) functions, in parallel with an ATM switch.
2. The IP/ATM conversion speed depends upon, and is therefore restricted by, the software processing speed of the IP router in case of typical implementation.
3. An additional IP router (capital investment) is required to add a subnet.
4. IP router management and operation become more complex as facility grows, because the IP routers and ATM switches are separately managed.

As a solution to these problems, the authors propose an ATM switching system with a built-in IP packet processing function [1]. Figure 2 shows a diagram of the architecture. The switching system is furnished with non-ATM line interface cards, as well as ATM interface cards, which are equipped with a non-ATM protocol termination/RE (Routing Engine) and SAR functionality as clients.

A packet processing server equipped with SAR [3], Routing Engine (RE) [2], an address resolution function and routing protocol handlers are located in the switch fabric's trunk side. Some merits of furnishing protocol servers in the trunk side are as follows:

- The multiple number of upper-layer protocols over ATM can be concentrated on a dedicated server for a specific upper-layer protocol. This protocol aggregation allows a line card to eliminate the multiple number of protocol handlers.
- Even in case that a single upper-layer protocol, typically IP, is to be handled in the system, the system complexity may be reduced by adopting the server cards in the trunk side.

Assuming that;

C_{LA} is complexity of a line card without a packet handler.

C_{LP} is complexity of a line card with a packet handler.

C_T is complexity of a server.

α is % of upper-layer traffic in a line card, which should be examined in the system.

n is the number of line cards.

The system complexity factor, C, is defined as follows:

$$C = n\,(C_{LP} - C_{LA}) - C_T \lceil n\alpha \rceil \quad \text{--- (eq.1)}$$

If $C > 0$, a switching system with $\lceil n\alpha \rceil$ server cards is more efficient than that without servers.

Card complexities shown above are not simple, because they are impacted by size, power consumption, cost, the number of VC's, etc. Also α is not necessarily identical in each input line. But still "$C > 0$" holds, when α is small, which can typically implies the small number of VC's for conveying upper-layer traffic to be terminated in the system.

Integrating these clients and servers into the ATM switching system solves the problems 1, 3, and 4 mentioned above.

4. Architecture of IP Packet Processing Mechanism

Problem 2 described above requires the protocol processing assisted by hardware. A high-speed processing mechanism is required especially in the server side in order to handle concentrated requests from many subscribers as a service trunk. Below, the IP packet processing server is used as an example to discuss its requirements and architecture. Figure 3 shows the architecture.

4.1 Hardware Assisted Protocol Processing (Routing Engine)

Faster packet processing requires an efficient collaboration between hardware and software to replace what had been

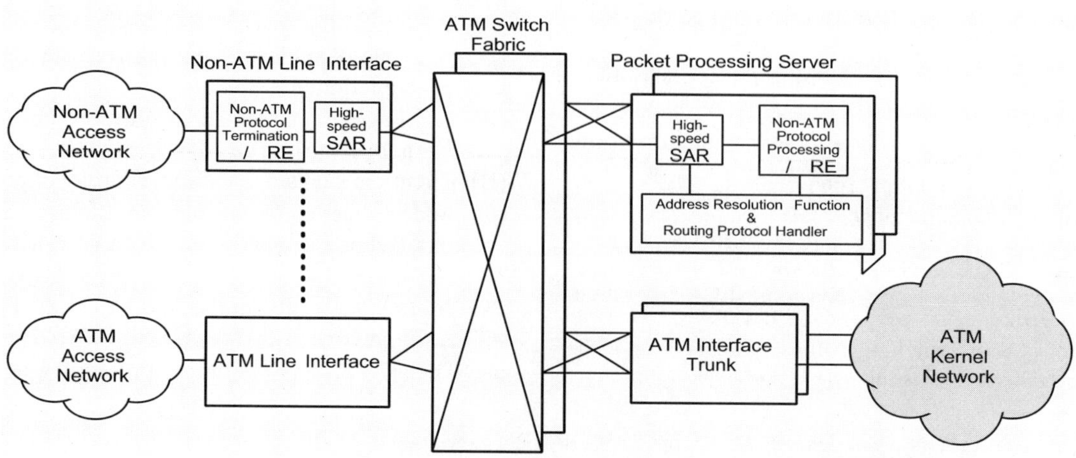

Fig. 2 Architecture of the ATM switching system with built-in packet handling capability.

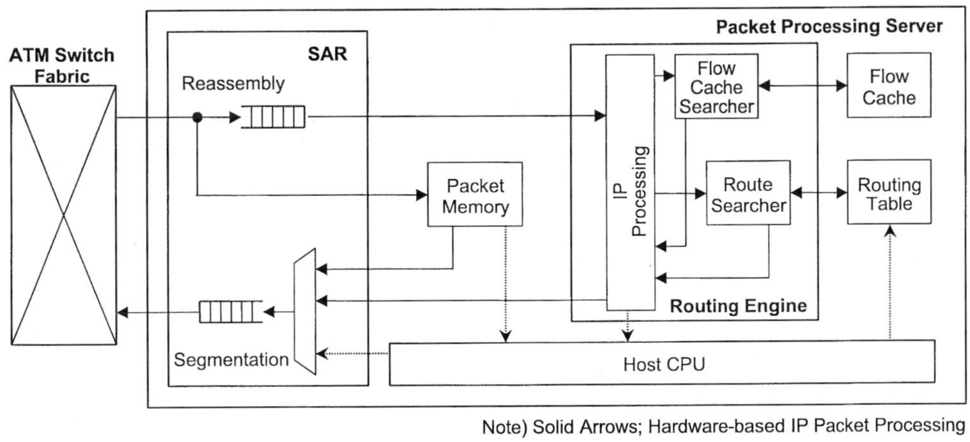

Fig. 3 Architecture of packet processing server.

performed only by software. Routing Engine requires basic functions such as the ones listed below providing high-speed processing power by hardware:

- IP header check and regeneration

Consistency check of LLC/SNAP, IP address, version, and header length, calculation of header check sum, and recalculation after header modification.

- High-speed flow cache

This cache function can be realized by using CAM (Content Addressable Memory), for example, to search destinations of continuously-arriving IP packet flows. As is often observed, it is the typical case that a TCP session generates this type of series of IP packets.

- High-speed large-capacity routing table search

Route search of wire speed (e.g., 735 Kpps for 622 Mbps) is required even when the maximum number of entries (e.g., 128 K) is registered.

- Processing packet judgement

The settings of automatic transfer, discard and handover to software of packets on SAR are possible on the hardware, for each header check error type or flow cache/routing table entry.

4.2 High-Speed SAR

Because the Kernel network is based on a native ATM transport mechanism, a high-speed SAR function should be allocated within the Shell network. For the segmentation and reassembly of ATM cells, the switching system needs an AAL5-SAR processing function supporting the cell rate of the switch fabric's port speed, i.e., wire-speed SAR functionality. To perform high-speed processing, it is also necessary, for example, to eliminate actual packet data movement between the memories.

Additional functions required in high-speed SAR are described below, as well as AAL5 functionality:

- Store and forward using only ATM VCC (Virtual Channel Connection)

It is necessary to realize the packet forwarding of SAR alone using the VCC set in the ATM switch fabric. One of the additional applications of this mechanism is a LANE-BUS (Broadcast and Unknown Server) function [4], where packet data in an ATM cell format is copied from unicast VCC to multicast VCC to perform a broadcast function.

- IP packet marge; VC aggregation

VC aggregation function is also required in the SAR, where IP packets in different VCCs are multiplexed into a single VCC in the form of a packet.

- IP routing and forwarding

The Routing Engine and the SAR must be associated with an Index (in the pointer format) for executing an address search. This makes it possible for the Routing Engine hardware to map a non-ATM protocol address, i.e., IP address, to VPI/VCI of ATM at wire speed.

- Layer 4 and UP routing

It is necessary for the Routing Engine to identify the TCP port number, etc., if the SAR Index is used, for example, to support IP switching [6], [7]. By returning such a VPI/VCI value based on the L4 and UP information to the SAR, it is possible to realize wire-speed QoS-based routing that takes into account the application attributes.

These functions are realized within a single VLSI of CMOS 0.35 μm process. Table 1 shows the specification of the SAR.

Table 1 Specification of SAR.

Speech Path Interface	622Mbps (Redundant)
Host Interface	PCI-Bus
AAL Type	Type 5 (and Type 0)
Packet Forwarding Capability	735 Kpps (64byte packet)
Process/Package/Supply Voltage	0.35μm CMOS / 580pin PBGA / 3.3V ±10%

4.3 Realization of Packet Forwarding with IPv4/IPv6 Conversion

Figure 4 shows a mechanism of packet forwarding with ATM-based IPv4/IPv6 conversion using high-speed SAR.

In this SAR, to support multi-protocol forwarding for a variety of higher layer protocols than ATM (and its conver-

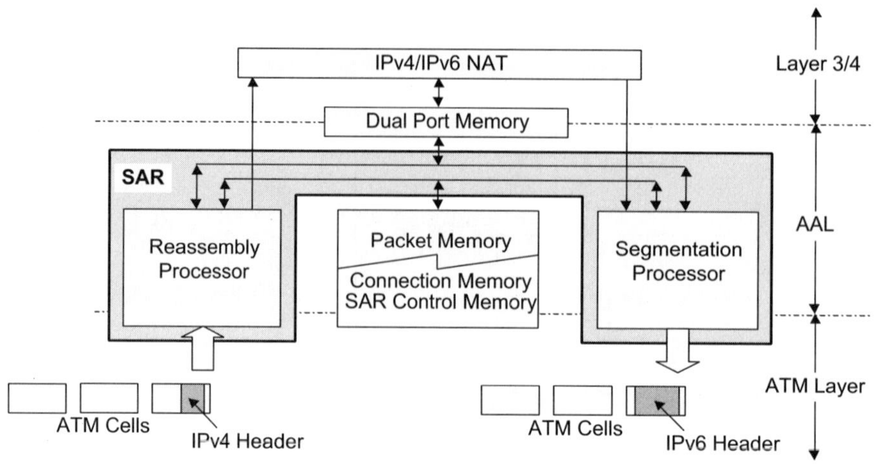

Fig. 4 Mechanism of packet forwarding with IPv4/IPv6 conversion.

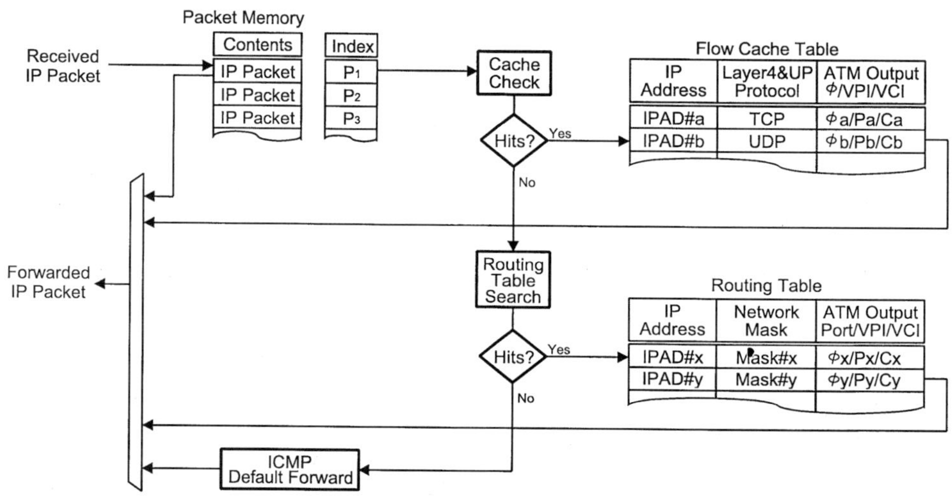

Fig. 5 Address search algorithm and operations.

sion) such as IPv4/IPv6 conversion, data length parameters written in the memory for packet header storage are programmable. The forwarding process is described below.

1. The Reassembly Processor gathers ATM cells into an IP packet and writes the IPv4 packet header data in the Dual Port Memory, which is necessary for upper-layer processing.
2. After reassembling the packet, the processor writes received packet information in the descriptor in the memory and passes pointer nformation of the packet to the Network Address Translator (NAT).
3. The NAT converts the IPv4 address to the IPv6 address, updates the contents of the packet header and descriptor, and returns the pointer to the SAR.
4. The Segmentation Processor refers to the descriptor released by the NAT, and combines the IPv6 header and the packet payload in the Dual Port Memory and in the Packet Memory, respectively. The combined data is segmented into ATM cells.

4.4 High-Speed Search Operation

This section explains one of the important mechanisms to realize high-speed IP packet processing; searching IP addresses by hardware. As is mentioned in the above section, the Flow Cache Table and Routing Table are equipped in the Server. Figure 5 depicts the address search algorithm and operations.

A received IP packet is stored in the Packet Memory. The IP address of the packet associated with the Index is checked by the Flow Cache Searcher in the RE referring to the Flow Cache Table. If the address is found in the table, i.e., "HIT," then ATM output port number, VPI/VCI value, and recalculated IP header are returned to the SAR with the Index number. The SAR picks IP packet contents up from the Packet Memory, and composes IP packet with ATM VCI/VPI to send to the ATM switch fabric.

In case that the Flow Cache Table does not have the IP address of the received IP packet, then the Routing Table is searched by the Route Searcher. When the address is found in the table, then ATM output port number, VPI/VCI value, and recalculated IP header are returned to the SAR. If the address is not found in the Routing Table, then ICMP messages generated by the host CPU, which is a 64-bit RISC CPU running at 200MHz, are sent through the ATM switch fabric. The RADISH search algorithm is known as an efficient table searching algorithm [8].

The entry number of the Flow Cache Table is pre-settable to increase flexibility. This is because the source/destination address distribution is not identical; some end users are apt to send IP packets to frequently-accessed receivers, but some aren't. In case that fewer entry size of the cache table satisfies a given address distribution, then a data width of a single entry can be set longer. In this situation this table can be taken advantage of to specify an ATM cell depending on the received IP packet quality, by specifying IP QoS (e.g., DiffServ, etc.) into the table. Then the IP QoS is related to the ATM class which is specified VCI/VPI in the table.

To realize the above flexibility, the Cache Table and the Routing Table searching functions are implemented in a separate FPGA's. These FPGA's, running at 50 MHz, control the Index tables as well as the Cache/Routing Tables.

4.5 Architecture of Packet Server

The server built into a switch and equipped with the hardware components as described above must provide the following features.

Utmost reliability, or the ability to satisfy its specifications as public network equipment, is the most important feature. To ensure this feature, the server must be capable of redundant card operation. As a switch package, the server must be in the form of a card that can be inserted into a frame slot of the switching system. This implementation realizes an integrated operation and management of the system as a single entity of ATM and IP in the network. Next, the server must allow hot-swap (during redundant operations) to facilitate maintenance, replacement, and testing while the switching system is running. It must be possible to increase or decrease the number of cards according to the demanded service load in the server (e.g., the increase or decrease in the number of accommodated subscribers). In other words, flexibility and scalability are needed.

5. Conclusion

The authors developed an ATM switching system (Fig. 6) with the built-in IP processing mechanism as described above, and evaluated its performance. The evaluation proved that a packet processing server (Fig. 7) installed in a card-slot of the switching system handles 735 K IP packets per second (622 Mbps wire speed) per card. It was also confirmed that, by installing multiple servers on the same switch, the packet processing capacity increases in proportion to the number of servers installed. The authors have therefore verified that the IP function can be directly provided to a high-speed ATM network without using external IP equipment. The authors have also confirmed that the high reliability, secure maintainability and testability of a public network switch can be applied to this IP processing mechanism.

Fig. 6 ATM switching system with the built-in processing mechanism.

Fig. 7 Packet processing server card.

Acknowledgment

The authors would like to show their thanks to those who have been involved in the project.

References

[1] K.Fukumoto, M.Honda, H.Ikeda, A.Arutaki, K.Izawa, and T.Maeda, "Development of packet processor server in ATM switch," IEICE Fall Conference, 1998.

[2] K.Isoyama, T.Amagai, M.Honda, H.Ikeda, and A.Arutaki, "Development of packet forwarding engine," IEICE Fall Conference, 1998.

[3] T.Amagai, M.Yamashita, T.Aramizu, H.Ikeda, and A.Arutaki, "Development of 622Mbps ATM-SAR LSI with high speed packet forwarder," IEICE Fall Conference, 1998.

[4] ATM Forum LAN Emulation Over ATM Specification Version 1.0.

[5] "Multiprotocol Over ATM Version 1.0," ATM Forum/str-mpoa-mpoa-01.00, 28 Feb. 1998.

[6] P.Newman, "Ipsilon flow management protocol specification of IPv4," RFC1954, IETF, May 1996.

[7] R.Callon, "A framework for multiprotocol label switching," draft-ietf-mpls-tramework-02, IETF, Nov. 1997.

[8] K.Yamamoto, "Radish — A simple routing table structure for CIDER," A technical memo of WIDE project, July 1995.

[9] A.Arutaki, "The Internet and ATM," The 1st Workshop of the Internet Study Group, IEICE, Oct. 1997.

High-Performance Backbone Network Technology edited by Naoaki Yamanaka
ISBN: 0-8247-5321-6 DOI: 10.1081/(120025697)
Published 2004 by Marcel Dekker, Inc.

An Experimental Study on Performance During Congestion for TCP/IP Traffic over Wide Area ATM Network Using VBR with Selective Cell Discard

Shigehiro Ano, Toru Hasegawa, and Toshihiko Kato

SUMMARY It is important to establish the technology to accommodate best effort TCP/IP traffic over wide area ATM networks. The UBR (Unspecified Bit Rate) service category is the most typical service category for the best effort traffic, especially in the LAN environment. On the other hand, the VBR (Variable Bit Rate) service category with SCD (Selective Cell Discard) option is considered as the service category which is appropriate for wide area networks due to its fairness and minimum guarantee of the cell transmission using not only PCR (Peak Cell Rate) but SCR (Sustainable Cell Rate) and MBS (Maximum Burst Size). However, there is no actual evaluation for such service. We have, therefore, performed the experimental studies on TCP/IP over VBR with SCD along with UBR and VBR without SCD by VC (Virtual Channel) level policing when each TCP connection is mapped to a different VC. Through these experiments, we measured the link utilization of the effective data and the fairness between each obtained TCP throughput during the congestion of the ATM switch. From the results of the link utilization, the value is over 95% under the various conditions. Therefore, even in the case of the cell losses due to SCD or buffer overflow in ATM switch congestion, average throughput is almost the same as the value which equals the trunk line speed divided by the number of the accommodated TCP connections. From the results of the fairness, VBR with SCD per VC is better than UBR and also obtains better TCP throughput than VBR without SCD. Furthermore, to confirm those characteristics more generally, we adopt the accommodated TCP connections not only with the same TCP send/receive socket buffer size but with different sizes. Finally, we discuss the effectiveness between VBR with SCD and the other service categories, such as UBR and ABR (Available Bit Rate) and GFR (Guaranteed Frame Rate), and conclude that VBR with SCD is one of the most suitable ATM service categories for accommodating best effort traffic.

key words: *ATM network, TCP/IP, variable bit rate, usage parameter control, selective del discard*

1. Introduction

Recently, it becomes important to establish a technology to accommodate TCP/IP traffic over wide area ATM networks. Since the TCP/IP traffic is a best effort traffic, UBR (Unspecified Bit Rate), ABR (Available Bit Rate) and GFR (Guaranteed Frame Rate) are considered to be typical service categories [1] for accommodating it. However, in most of wide area ATM networks which provide commercial ATM services, only CBR (Constant Bit Rate) and VBR (Variable Bit Rate) service categories are supported. In such an actual environment, we consider that VBR with SCD (Selective Cell Discard) option is an appropriate service for accommodating best effort TCP/IP traffic. The best effort communication over VBR with SCD is realized as follows. TCP terminals send ATM cells with conforming to the traffic contract for PCR (Peak Cell Rate), but without conforming to that for SCR (Sustainable Cell Rate) and MBS (Maximum Burst Size). ATM switches for wide area networks (ATM WAN switches for short) perform UPC (Usage Parameter Control) to detect and control non-conforming cells to the traffic contract using dual leaky buckets, one for PCR and another for SCR and MBS. Especially in SCR and MBS conformance checking, the ATM WAN switches with SCD only set CLP (Cell Loss Priority) bits of non-conforming cells (called as *cell tagging*), and do not discard them. The tagged cells (non-conforming cells) are discarded only during the congestion which is the period when the switch output buffer is over certain *SCD threshold*. This accommodation allows ATM networks to pass all incoming cells when there is no congestion, and to guarantee the bandwidth determined by SCR and MBS during the congestion.

However, only a few simulation studies [2]–[4] and none of experimental studies are reported for the performance of TCP/IP traffic over VBR with SCD. So, we have performed an experiment [5] for comparing performance of accommodating TCP/IP traffic using VBR with SCD, UBR and VBR without SCD (where non-conforming cells to contracted SCR and MBS are discarded in ATM WAN switches).

As for VBR with SCD, it is important to evaluate the performance when there is congestion, because all incoming TCP/IP traffic is passed when there is no congestion. So, the experiment focuses on the condition where there is congestion in an ATM trunk line between ATM WAN switches which incoming traffic is aggregated into. In this experiment, the following performance parameters are measured.

- Link Utilization: As for the throughput aspects, we measure the link utilization of an ATM trunk line, which is defined by the ratio of the sum of throughput of TCP/IP traffic to the ATM trunk

Manuscript received June 3, 1999.
Manuscript revised August 30, 1999.

[†]The authors are with KDD R&D Laboratories Inc., Kamifukuoka-shi, 356-8502 Japan.

IEICE TRANS. COMMUN., Vol. E83-B, No. 2, pp. 155–164, February 2000

line speed.

- Fairness among TCP throughput of individual ATM connections: Compared with UBR, VBR is expected to provide better fairness because VBR controls the traffic of each ATM connection using an independent virtual queue of UPC, which is a leaky bucket of GCRA (Generic Cell Rate Algorithm).

This paper describes the result of this experiment. Section 2 describes the overview of experiments. Section 3 shows the results of link utilization and fairness. Based on these results, we discuss the effectiveness of the VBR with SCD in Sect. 4 and finally conclude the results in Sect. 5.

2. Overview of Experiments

2.1 Testing Methods

We performed the experiments according to the following principles.

1. To make a comparison among VBR with SCD, UBR and VBR without SCD, we measure the TCP/IP performance under the following three types of traffic control methods of ATM WAN switch.

    a. VBR with SCD: At the ATM WAN switch, UPC tags the non-conforming cells based on the assigned SCR and MBS. The SCD threshold of the ATM WAN switch cell buffer is varied from 50% to 95%. If the SCD threshold becomes certain percentage, the tagged cells are discarded before the non-tagged cells are discarded by the buffer overflow.
    b. UBR: ATM WAN switch is operated as UBR which require no policing control in the ATM WAN switch. In this case, if congestion occurs, the cell buffer overflow discards the cells randomly.
    c. VBR without SCD: This control method is not intended to accommodate best effort traffic. However, it is useful for the comparison. At the ATM WAN switch, UPC function discards the non-conforming cells based on the assigned SCR. According to the discard, it is possible to suppress the buffer overflow of the switch.

2. In our experiment, one TCP connection is mapped to one ATM VC (Virtual Channel) connection. As for the policing, VBR can operate UPC by the following two levels.

    a. VC level policing: The targets of UPC are individual TCP connections. It is expected that the fairness of each TCP throughput can be achieved more precisely.
    b. VP level policing: The targets of UPC are the aggregated TCP traffic over the same ATM WAN switch input lines. Therefore, the VPI (Virtual Path Identifier) values of the aggregated TCP traffic over the same input lines should be the same number.

    In the experiment discussed in this paper, we adopt the VC level policing, which is different from the principle of our previous experiment reported in [5].
3. As for the number of TCP connections, we performed preliminary testing where the number of TCP connections is gradually increased under the above service categories from the number when no cell loss has occured to the number when many cell losses have occured. As a result, we have selected a certain value for which cell losses have occured due to the ATM WAN switch congestion.
4. For the evaluation of fairness, we evaluated not only the uniform TCP/IP traffic with the same send/receive socket buffer size (S/R buffer size for short) but also the non-uniform TCP/IP traffic with different S/R buffer sizes.
5. In order to determine SCR and MBS values of each VC, we adopt the following approach:

    - Each SCR value is proportional to S/R buffer size of TCP connection included in a VC.
    - The sum of the SCR values of all VCs is equal to the bandwidth of ATM trunk line. This means that there is no overbooking of SCR values in the ATM WAN switch.
    - The MBS value is determined from each S/R buffer size of the TCP connections, according to the previous studies [6], [7]. The actual values must take the various overheads (i.e. cell header, AAL5 (ATM Adaptation Layer Type 5) padding field and trailer, TCP/IP header) into consideration. LLC/SNAP (Logical Link Control/Sub-Network Access Protocol) is not used as an en capsulating method in our experiment.

6. In this experiment, we use eight PCs (Personal Computers) so that the performance of PCs may not become a bottleneck of the throughput. When multiple TCP connections are used, we established them between different send/receive PC pairs, by considering the load balances.
7. To evaluate the best effort service both in the cases of short distance and long distance, we used different values of RTT (Round Trip Time) between PCs.
8. We select the two MSSs (Maximum Segment Sizes) for this experiment. One MSS value is determined by considering use of Ethernet. Another MSS value is determined from the size of user data read

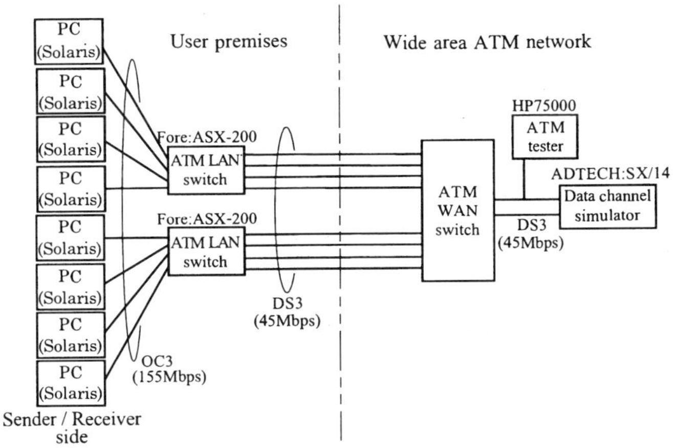

Fig. 1 Experimental configuration.

from or written to the S/R buffer (user data size for short). The latter MSS value is possible to transmit an unit of user data by one TCP segment when we use MTU (Maximum Transmission Unit) size of the ATM NIC (Network Interface Card) device driver.

9. The PCR shaping at send PCs is needed. The following are the reasons.
    a. To evaluate the difference among three types of traffic control methods of ATM WAN switch described above, it is needed that the ATM WAN switch does not discard cells in UPC for PCR, but discard cells in UPC for SCR.
    b. To focus on the behavior of the ATM WAN switch, no cells should be discarded by the user premises, such as output buffer overflow of the ATM LAN switches.

2.2 Experimental Configuration

Figure 1 shows the configuration of experiments. Eight PCs with an ATM NIC (Fore PCA-200EUX/OC3SC) are used. Six PCs are EPSON DIRECT AT-700C and two PCs are MICRON SE440BX2-AT. All PCs are installed with Pentium II processor with 450 MHz and running Solaris 2.6 as an operating system. Each PC is connected to an ATM LAN switch (Fore ASX-200BX) via OC3-c line and the ATM LAN switches are connected to the ATM WAN switch via DS3 line. At the ATM LAN switch, cells over each OC-3c line are switched to the corresponding DS3 line and the PCR shaping at send PC is performed to avoid the output buffer overflow as described above. Therefore, cell loss due to the buffer overflow does not occur at ATM LAN switches. At the ATM WAN switch, eight DS3 lines are multiplexed into one DS3 line as the ATM trunk line.

The output buffer size of the ATM WAN switch is set to default value of 8320 cells. The buffer size of typical ATM LAN switches is ranging from about 15000 cells to 30000 cells. On the other hand, in the wide area ATM networks, delay requirement for VBR might be strict. Therefore, the above value seems to be reasonable values for the practical applications.

Over the trunk line, the data channel simulator (ADTECH SX/14) is connected to insert propagation delay, and the ATM tester (HP75000) is also connected to monitor the physical line usage rate. A free software, *ttcp*, for TCP throughput measurement is used in the TCP/IP communication between PCs. We modified the software so that it can specify the determined measurement period. It can calculate TCP throughput in the case that a greedy transmitter is used like ftp, by varying the values of various TCP parameters, such as S/R buffer size and the user data size.

The experimental parameters are as follows:

- We adopt TCP/IP software supported by Solaris 2.6 for all our measurements. Solaris 2.6 is based on UNIX System V Release 4 (SVR4) whose TCP/IP software is developed from 4.3 BSD (Berkeley Software Distribution) UNIX Reno release [8][†] rather than Tahoe release [9]. Because this TCP/IP software supports not only slow start, congestion avoidance and fast retransmit algorithms but also fast recovery algorithm for TCP congestion control [10].
- The total number of TCP connections is determined to 24.
- As for the S/R buffer size, 48 kbyte is adopted for uniform TCP/IP traffic. For non-uniform TCP/IP traffic, 24 kbyte for 8 connections and 48 kbyte for 16 connections are adopted. As a result, SCR and MBS are set to the following values:
    1. In the case of the uniform traffic
       SCR = 1.69 Mbit/s, MBS = 1078 cells
    2. In the case of the non-uniform traffic
       SCR for 24 kbyte = 1.02 Mbit/s, MBS for 24 kbyte = 539 cells
       SCR for 48 kbyte = 2.03 Mbit/s, MBS for 48 kbyte = 1078 cells
- The value of each PCR shaping is fixed at 35 Mbit/s including the cell header.
- The duration of each TCP throughput measurement is fixed to 300 seconds.
- The RTT value is set to 10 ms as a short distance communication and to 160 ms as a long distance communication.
- The user data size is fixed at 8192 byte. As for MSS, 1460 byte considering the Ethernet MTU size

[†]To be accurate, the TCP/IP software of Solaris 2.6 is developed from 4.4 BSD due to supporting multicasting [11] and long fat pipe modifications [12]. However, it is out of the scope of our evaluation. In that sense, we can regard the software as 4.3 BSD Reno release.

Table 1 Link utilization of trunk line (Number of TCP connections = 24, S/R = 48 kbyte).

RTT	MSS	VBR with SCD	UBR	VBR without SCD
10 ms	1460 byte	95.36%	96.03%	49.90%
	8192 byte	98.37%	98.12%	41.33%
160 ms	1460 byte	96.12%	97.19%	65.17%
	9148 byte	96.89%	97.26%	67.36%

and 8192 byte which is the same as the user data size mentioned above are adopted.

3. Results of Performance Measurement

3.1 Link Utilization

Under the configuration of Fig. 1, we measured the throughput of each TCP connection under VBR with SCD, UBR and VBR without SCD, and calculated the link utilization of the ATM trunk line, which is given by the sum of measured throughput divided by the ATM trunk line speed, as described above. Table 1 shows the results under the three control methods for the uniform TCP/IP traffic with 48 kbyte S/R buffer size. This table shows that VBR with SCD and UBR provide high link utilization over 95%. This means that, for these two control methods, each TCP throughput does not decrease suddenly even if there are cell losses due to congestion. It seems that the reason for it is that the cell losses cause packet losses and consequently the TCP level retransmission and congestion control, only for a few connections at a time.

On the other hand, the link utilization for VBR without SCD is about 65% in the case of RTT = 160 ms, and about 40–50% in the case of RTT = 10 ms. In our experiment, the SCR value is determined by the ATM trunk speed divided by the number of VCs, independent of the characteristics of TCP/IP traffic, and send PCs does not conform to SCR. In the case that, in VBR without SCD, send TCP terminals do not perform the SCR/MBS shaping, the SCR value set in an ATM switch needs to correspond to the original throughput of TCP traffic determined by S/R buffer size and RTT, and if it is smaller than the original TCP throughput, the TCP throughput for VBR without SCD will be degraded due to the policing in the ATM switch [6], [7]. Since the SCR value used in this experiment is smaller than the original TCP throughput, the throughput of each TCP connection in the case of VBR without SCD becomes lower and the link utilization also becomes lower. This is more severe for the case of RTT = 10 ms, which the original TCP throughput is much higher than the SCR value.

Figure 2 shows the relationship between the link utilization and SCD threshold. As shown in the figure,

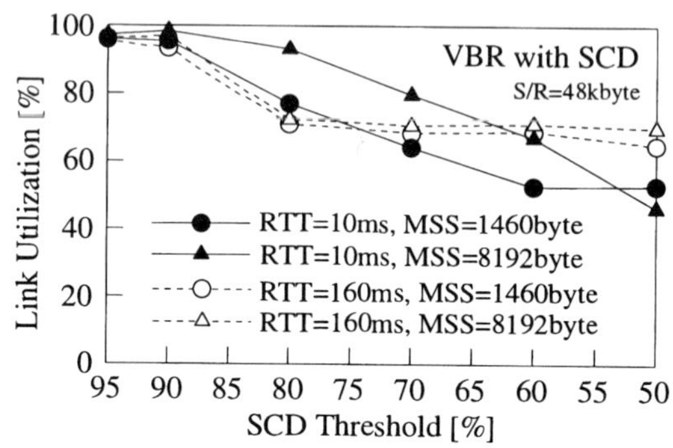

Fig. 2 Relationship between SCD threshold and link utilization.

the link utilization is more than 95% when the SCD threshold is over 90%. The link utilization decreases gradually when the SCD threshold becomes smaller. Among the four combinations of RTT and MSS shown in the figure, in the case of RTT = 10 ms and MSS = 8192 byte, the link utilization remains almost 95% when the SCD threshold is 80%. In this case, the original TCP throughput is high because RTT is small, and the large MSS allows the congestion window to grow quickly in the slow start algorithm after retransmission caused by cell losses. Therefore, it is considered that the throughput of each TCP connection remains high compared with the other cases of RTT and MSS.

3.2 Fairness on Throughput with the Same S/R Buffer Size

Once the congestion occurs, the difference between the maximum and the minimum TCP throughput becomes larger. This section shows the measurement of fairness under the uniform TCP/IP traffic with the same S/R buffer size during congestion. Through the experiment, we obtained the following results.

1. Table 2 shows the maximum, minimum and average value of TCP throughput of one VC and its standard deviation, for VBR with SCD, UBR and VBR without SCD. For the VBR with SCD, the SCD threshold is 90%, which provides high link utilization as described in the previous section. As shown in this table, the difference between maximum and minimum values of TCP throughput (and the standard deviation as a result) is small

Table 2 Fairness on TCP throughput (maximum/minimum/average/standard deviation) of uniform TCP/IP traffic with 48 kbyte S/R buffer size (Number of TCP connections = 24).

RTT	MSS	VBR with SCD(Mbit/s)	UBR(Mbit/s)	VBR without SCD(Mbit/s)
10 ms	1460 byte	1.399/1.385/1.392/3.45e-3	2.500/1.166/1.402/2.89e-1	0.769/0.706/0.729/1.69e-2
	8192 byte	1.886/1.301/1.500/1.39e-1	2.096/1.224/1.496/3.80e-1	0.659/0.609/0.630/9.30e-3
160 ms	1460 byte	1.415/1.389/1.403/6.47e-3	2.011/1.139/1.419/2.51e-1	0.963/0.941/0.951/5.93e-3
	8192 byte	1.483/1.448/1.477/4.36e-3	2.050/1.121/1.383/2.97e-1	1.037/0.996/1.027/8.65e-3

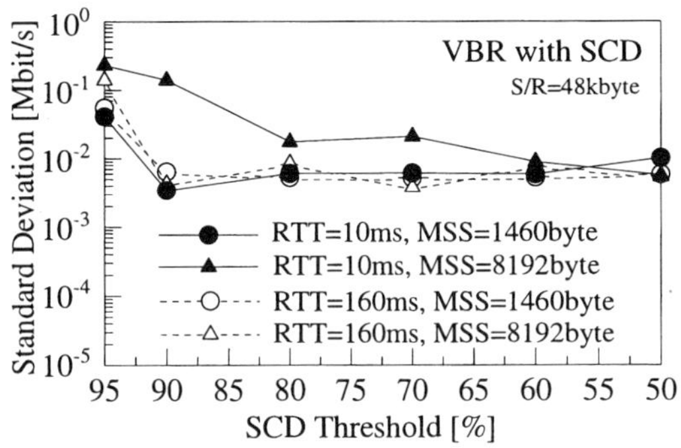

Fig. 3 Relationship between SCD threshold and fairness of uniform traffic.

for VBR with SCD and VBR without SCD, and on the other hand, it is larger for UBR compared with the other two control methods. Therefore, it is considered that VBR based methods can provide more fair cell transmission than UBR. As for the throughput itself, the average value is high for VBR with SCD and UBR, and low for VBR without SCD. The reason for it has been already described in the previous section. From those results, it can be said that VBR with SCD can provide both high link utilization and fairness compared with UBR and VBR without SCD.

2. To evaluate the relationship between the fairness and SCD threshold for VBR with SCD, we measured the standard deviation of TCP throughput for various SCD threshold values. Figure 3 shows the results. As shown in this figure, for the SCD threshold values equal or smaller than 90%, the standard deviation is in the order of 10 kbit/s while the TCP throughput itself is in the order of 1 Mbit/s. Therefore, it is considered that the fairness of VBR with SCD is stable from the viewpoint of SCD threshold. The only exception is the case of RTT = 10 ms and MSS = 8192 byte. In this case, the standard deviation is in the order of 100 kbit/s, which is in the same order as UBR. This is considered to be caused by the following reason. In this case, the throughput of each TCP connection is quickly recovered even if there are cell losses, as described in Sect. 3.1. Therefore, it is considered that the congestion in the ATM WAN switch is more severe than the other cases, and that the output buffer in the ATM WAN switch causes overflow even if SCD is working. So, the behavior of VBR with SCD in this case becomes similar to that of UBR.
3. To confirm the fairness more precisely, we examined the distribution of individual TCP throughput by making the histogram of throughput for VBR with SCD, UBR and VBR without SCD. Figure 4 shows the histogram in the case of RTT = 160 ms and MSS = 8192 byte. In the case of UBR, the obtained throughput is distributed widely and the two peaks are observed at around 1.5 Mbit/s and 2.0 Mbit/s. It is considered that this is caused by the difference between the types of PCs as described in Sect. 2.2. All PCs have the same CPU and the same operating system. However, we used two types of PCs for the experiments. If the sender side uses one type of PC (MICRON), the obtained TCP throughput is about 2.0 Mbit/s, and if sender side uses the other type of PC (EPSON DIRECT), the throughput is about 1.5 Mbit/s. On the other hand, in the case of VBR with SCD and VBR without SCD, the distribution of the TCP throughput is very sharp.

3.3 Fairness on Throughput with Different S/R Buffer Sizes

To evaluate the fairness on TCP throughput with different S/R buffer sizes, eight TCP/IP connections with 24 kbyte of S/R buffer size and sixteen TCP/IP connections with 48 kbyte of S/R buffer size are aggregated into the ATM trunk line. Tables 3 and 4 show the fairness on TCP throughput to the non-uniform TCP/IP traffic with 24 kbyte and 48 kbyte S/R buffer sizes when MSS is 1460 byte and 8192 byte, respectively. The SCD threshold value is 90% with the exception of 80% for the case of RTT = 10 ms and MSS = 8192 byte. Figure 5 shows the distribution of each TCP throughput for the case of RTT = 160 ms and MSS = 8192 byte. From these results, the obtained throughput in UBR is distributed more widely than the other two service category based on VBR, especially in the case of larger S/R buffer size of 48 kbyte. The other characteristics are the same as the case with the uniform S/R buffer

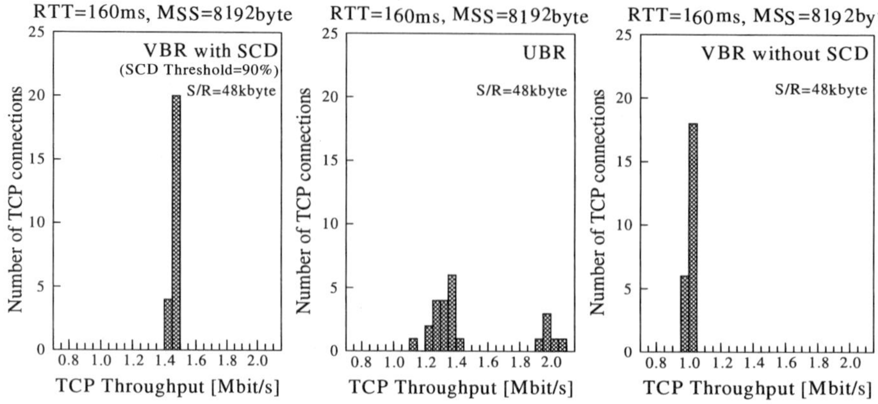

(a) In the case of VBR with SCD (b) In the case of UBR (c) In the case of VBR without SCD

Fig. 4 Distribution of TCP throughput of uniform traffic.

Table 3 Fairness on TCP throughput (maximum/minimum/average/standard deviation) of non-uniform TCP/IP traffic with 24 kbyte and 48 kbyte S/R buffer size (Number of TCP connections = 8 for 24 kbyte S/R buffer size and 16 for 48 kbyte S/R buffer size, MSS = 1460 byte).

RTT	S/R buffer size	VBR with SCD(Mbit/s)	UBR(Mbit/s)	VBR without SCD(Mbit/s)
10 ms	24 kbyte	0.822/0.803/0.810/5.80e-3	1.926/0.801/1.108/3.46e-1	0.648/0.628/0.636/6.44e-3
	48 kbyte	1.722/1.699/1.712/7.00e-3	1.753/1.403/1.560/1.38e-1	1.278/1.259/1.268/6.01e-3
160 ms	24 kbyte	0.839/0.827/0.834/4.44e-3	1.013/0.817/0.881/6.79e-2	0.542/0.519/0.531/7.52e-3
	48 kbyte	1.735/1.724/1.730/2.70e-3	2.075/1.486/1.668/2.06e-1	1.042/1.024/1.036/4.92e-3

Table 4 Fairness on TCP throughput (maximum/minimum/average/standard deviation) to non-uniform TCP/IP traffic with 24 kbyte and 48 kbyte S/R buffer size (Number of TCP connections = 8 for 24 kbyte S/R buffer size and 16 for 48 kbyte S/R buffer size, MSS = 8192 byte).

RTT	S/R buffer size	VBR with SCD(Mbit/s)	UBR(Mbit/s)	VBR without SCD(Mbit/s)
10 ms	24 kbyte	0.821/0.799/0.812/7.74e-3	1.950/0.850/1.196/3.45e-1	0.431/0.416/0.426/4.58e-3
	48 kbyte	1.758/1.729/1.739/4.62e-3	2.014/1.381/1.606/2.24e-1	0.749/0.699/0.730/6.69e-3
160 ms	24 kbyte	0.888/0.874/0.879/3.95e-3	1.013/0.817/0.881/6.79e-2	0.752/0.718/0.723/4.29e-3
	48 kbyte	1.779/1.748/1.768/6.86e-3	2.075/1.486/1.668/2.06e-1	1.503/1.488/1.497/4.82e-3

size.

4. Discussions

(1) Based on the experiment results and the following considerations, it is considered that VBR with SCD is the most realistic solution for the best effort service in wide area ATM networks in the current stage.

1. Compared with UBR, VBR with SCD can provide better fairness on TCP throughput, although the link utilization of ATM trunk line is similar. UBR does not guarantee the fairness at the ATM level, and the fact that the fairness of UBR was at a certain level in our experiment depends on the invocation of TCP congestion control such as the slow start algorithm. If the accommodated traffic includes non-TCP traffic such as a typical audio/video stream traffic, the fairness on throughput will become worse.
2. Compared with VBR without SCD, VBR with SCD can provide better link utilization, although the fairness on TCP throughput is similar. This result comes from the fact that SCR set in ATM WAN switch is smaller than the original TCP throughput. This means that, in VBR without SCD, users need to perform SCR/MBS shaping. However, this shaping restricts the effective accommodation of best effort traffic when network is not congested.
3. ABR [1] is considered as a one candidate to accommodate best effort traffic in ATM network. However, the mechanism for realizing ABR is complicated and hard to implement, and, in the current stage, there are few ATM switches and ATM NICs

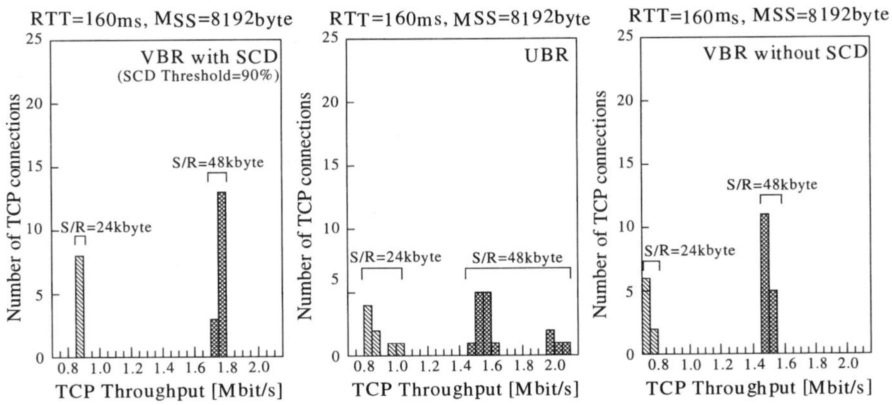

(a) In the case of VBR with SCD (b) In the case of UBR (c) In the case of VBR without SCD

Fig. 5 Distribution of TCP throughput of non-uniform traffic.

supporting it.

4. Recently, GFR, a new service category which guarantees an MCR (Minimum Cell Rate) at a frame level and allows users to send in excess of MCR, is discussed in the ATM Forum and ITU-T [1], [14]. Several simulation studies [3], [4], [13], [15] have pointed out the merits of GFR, such that GFR provides better fairness than VBR with SCD especially when the sum of MCR is smaller than the bandwidth of ATM trunk link. However, GFR is not implemented in the current stage. On the other hand, VBR with SCD is based on simpler mechanism which is implemented in the current ATM switches, and it is considered that the link utilization and the fairness obtained in our experiment will be enough to satisfy actual user demands.

(2) As for VBR with SCD itself, some papers have already mentioned that it is one of the candidates to accommodate best effort TCP/IP traffic. Papers [3], [4] reports the simulation results on the effectiveness of TCP throughput fairness in the case of congestion. They conclude that the obtained TCP throughput is close to the theoretical value and the difference between maximum and minimum throughput is small. Our results are basically similar with the results described in those papers. The contributions of our paper is that we performed an actual experiment using commercial ATM switches, not software simulation, and showed the evaluation results of link utilization and fairness of TCP throughput.

(3) In our experiment, the SCR value is determined from the network's viewpoint. However, we believe that VBR with SCD is also effective in the case that the SCR value is determined based on the user's requirement. It can be also said that VBR with SCD can satisfy the requirements of users of VBR without SCD. In this case, if the users do want to send out their traffic conforming to the subscribed values, they only send out the cells as CLP bit $= 0$, and all of the traffic will be guaranteed to be transmitted without any cell discards.

(4) In the experiment, we adopted the configuration that one TCP connection is mapped to one ATM VC connection. This is because we aimed at evaluating the TCP behavior over VBR with SCD in detail. However, our configuration is not realistic in the real TCP/IP communication, because TCP connections are activated and terminated frequently in contrast of VCs, and because multiple TCP connections are aggregated over one VC. Therefore, from the actual services' viewpoint, we performed additional experiment where multiple TCP connections are aggregated into one ATM connection.

In this configuration, the VBR policing is applied to the aggregated TCP connections, and therefore, the throughput of individual TCP connection takes diverged values even for the case of VBR with SCD [5]. However, since the subscription to ATM network is the ATM connection basis, the fairness among the total throughput of TCP connections aggregated over one ATM connection needs to be evaluated. To confirm the fairness between the aggregated TCP connection over one VP/VC connection, we again measured the TCP throughput under the configuration of Fig. 1. Eight VP connections are set between each ATM LAN switch and ATM WAN switch and the UPC for each VP is performed at the incoming ATM WAN switch port from each ATM LAN switch. Each VP aggregates three TCP connections. Therefore, the total TCP connections for the measurements is 24. Every TCP connection adopts the 48 kbyte S/R buffer size. As for SCR and MBS, 5.08 Mbit/s and 3234 cells are determined, respectively, based on the principle in Sect. 2.2. Other conditions are the same as those descried in Sect. 2.2. Table 5 shows the fairness on aggregated TCP throughput. From the table, the standard deviation of VBR with SCD is much smaller than that of UBR, and on the other hand, it is only a little larger than VBR without SCD. Although the numbers of VP and TCP connections is limited due to the performance of each PC and number of ports of

Table 5 Fairness on aggregated TCP throughput over each VP connection (maximum/minimum/average/standard deviation) of uniform TCP/IP traffic with 48 kbyte S/R buffer size (Number of VP connections = 8, Number of TCP connections over each VP connection = 3).

RTT	MSS	VBR with SCD(Mbit/s)	UBR(Mbit/s)	VBR without SCD(Mbit/s)
10 ms	1460 byte	4.236/4.061/4.162/7.38e-2	5.125/3.632/4.207/4.20e-1	2.255/2.138/2.186/4.17e-2
	8192 byte	4.513/4.261/4.425/1.01e-1	4.674/4.102/4.415/2.21e-1	2.087/1.822/1.958/9.47e-2
160 ms	1460 byte	4.410/4.045/4.202/1.39e-1	4.823/3.733/4.256/3.55e-1	2.968/2.729/2.835/9.49e-2
	8192 byte	4.481/4.214/4.369/9.23e-2	5.027/3.971/4.400/2.87e-1	3.160/3.013/3.057/5.04e-2

Table 6 Fairness on TCP throughput (maximum/minimum/average/standard deviation) of uniform TCP/IP traffic with 48 kbyte S/R buffer size when the sum of the SCR values is set to 80% of the bandwidth of ATM trunk line (Number of TCP connections = 24).

RTT	MSS	VBR with SCD(Mbit/s)	VBR without SCD(Mbit/s)
10 ms	1460 byte	1.403/1.367/1.389/9.81e-3	0.623/0.577/0.599/1.44e-2
	8192 byte	1.796/1.298/1.476/1.13e-1	0.621/0.578/0.597/9.01e-3
160 ms	1460 byte	1.403/1.372/1.392/8.38e-3	0.888/0.861/0.874/7.19e-3
	8192 byte	1.501/1.451/1.467/9.91e-3	0.934/0.896/0.916/1.26e-2

Table 7 Fairness on TCP throughput (maximum/minimum/average/standard deviation) to non-uniform TCP/IP traffic with 24 kbyte and 48 kbyte S/R buffer size when the sum of the SCR values is set to 80% of the bandwidth of ATM trunk line (Number of TCP connections = 8 for 24 kbyte S/R buffer size and 16 for 48 kbyte S/R buffer size, MSS = 8192 byte).

RTT	S/R buffer size	VBR with SCD(Mbit/s)	VBR without SCD(Mbit/s)
10 ms	24 kbyte	0.986/0.883/0.922/3.92e-2	0.364/0.346/0.354/6.59e-3
	48 kbyte	1.621/1.567/1.598/1.96e-2	0.601/0.578/0.591/9.04e-3
160 ms	24 kbyte	0.880/0.869/0.873/4.32e-3	0.615/0.588/0.601/9.82e-3
	48 kbyte	1.701/1.677/1.691/9.12e-3	1.197/1.159/1.175/1.43e-2

ATM switches, the fairness of VBR with SCD shows relatively good results. As the results, it might be concluded that the fairness of aggregated TCP connections over one VP/VC connection, which is more realistic configuration, is better for VBR with SCD, similarly as we indicated in the configuration where one TCP connection is mapped to one ATM connection.

(5) Our evaluation is performed when the sum of the SCR values is the same (i.e. 100%) as the bandwidth of ATM trunk line as described in Sect. 2.1. However, in the practical situation, there might be many cases that the sum of the SCR values is less than the bandwidth of ATM trunk line [4]. Therefore, we performed additional experiment for such a case. In this experiment, the sum of the SCR values is set to 80% of the bandwidth of ATM trunk line. The MBS values remain the same. Tables 6 and 7 show the fairness on TCP throughput with the same S/R buffer size and the different S/R buffer size, respectively. Compared with the 100% case, most of the standard deviation is the same or a little larger for VBR with SCD and VBR without SCD. To evaluate the fairness with the different S/R buffer size in detail, the ratio of the average throughput with 48 kbyte S/R buffer size to that with 24 kbyte S/R buffer size is defined. Table 8 shows the results. Under the condition of RTT = 10 ms, the ratio for the case of VBR with/without SCD becomes smaller and close to that for the case of UBR when the allocation of the residual bandwidth becomes larger. It is because VBR tends to allocate the residual bandwidth to each VC equally independent of the contracted SCR and MBS as described in [4]. On the other hand, under the condition of RTT = 160 ms, the ratio is almost the same.

5. Conclusion

In this paper, we have adopted the VBR service category to accommodate TCP/IP traffic and presented the experimental results on the accommodation of TCP traffic to an ATM WAN of VBR with SCD option. By comparing to the UBR and VBR without SCD, we

Table 8 The ratio of the average TCP throughput with 48 kbyte S/R buffer size to that of 24 kbyte S/R buffer size (Number of TCP connections = 8 for 24 kbyte S/R buffer size and 16 for 48 kbyte S/R buffer size, MSS = 8192 byte).

RTT	VBR with SCD(Mbit/s)		UBR(Mbit/s)	VBR without SCD(Mbit/s)	
	100%	80%		100%	80%
10 ms	2.114	1.733	1.408	1.994	1.669
160 ms	2.074	1.937	1.893	1.951	1.955

showed the TCP throughput, its network efficiency and the fairness of VBR with SCD. From the following results, it is considered that VBR with SCD is a realistic solution for the best effort service in the wide area ATM networks.

1. First, the link utilization of the ATM trunk line is almost near to 100%. This means that each TCP throughput can attain the maximum value under the provided network resources. The value is almost the same as the throughput in the case of UBR.
2. Second, the fairness of TCP throughput is almost the same as the case of VBR without SCD, even in the case that the different types of TCP/IP connections with different S/R buffer sizes are accommodated. This results shows that if the users set the corresponding values to the UPC parameters according to each TCP performance, throughput fairness can be achieved even in the case of non-uniform types of traffic such as different RTTs and S/R buffer sizes.

Acknowledgement

The authors would like to express their gratitude to Dr. T. Muratani, President & CEO of KDD R&D Laboratories Inc., and Dr. K. Suzuki, Executive Vice President, for continuous support of this work.

References

[1] The ATM Forum Technical Committee, Traffic Management Specification Version 4.1, AF-TM-0121.00, March 1999.
[2] M. Perloff and K. Reiss, "Improvements to TCP performance in high speed ATM networks," Commun. ACM, vol.38, no.2, pp.90–109, 1995.
[3] T. Asaka, R. Kawahara, and T. Ozawa, "Performance analysis of ATM traffic control mechanisms with guaranteed minimum cell rate," Proc. 1998 IEICE General Conference, B-7-73, p.194, March 1998.
[4] T. Asaka, R. Kawahara, and T. Ozawa, "Performance analysis of ATM traffic control mechanisms for guaranteed throughput service," IEICE Trans., vol.J81-B-I, no.11, pp.731–740, Nov. 1998.
[5] S. Ano, T. Hasegawa, and T. Kato, "A study on accommodation of TCP/IP best effort traffic to wide area ATM network with VBR service category using selective cell discard," IEEE ATM'99 Workshop, pp.535–540, May 1999.
[6] S. Ano, T. Hasegawa, T. Kato, K. Narita, and K. Hokamura, "Performance evaluation of TCP traffic over wide area ATM network," IEEE ATM'97 Workshop, pp.73–82, May 1997.
[7] S. Ano, T. Hasegawa, T. Kato, K. Narita, and K. Hokamura, "A study on accommodation of TCP/IP traffic using window scale option to international ATM network with VBR service category," First IEEE International Conference on ATM, pp.484–491, June 1998.
[8] G.A. Wright and W.R. Stevens, TCP/IP Illustrated, Volume 2: The Implementation, Addison-Wesley, ISBN 0-201-63354-X, Jan. 1995.
[9] S.J. Leffler, M.K. McKusick, M.J. Karels, and J.S. Quarterman, The Design and Implementation of the 4.3 BSD UNIX Operating System, Addison-Wesley, ISBN 0-201-06196-1, 1989.
[10] M. Allman, V. Paxson, and W.R. Stevens, "TCP congestion control," IETF RFC2581, April 1999.
[11] S.E. Deering, "Host extensions for IP multicasting," IETF RFC1112, Aug. 1989.
[12] V. Jacobson, R. Braden, and D. Borman, "TCP extensions for high performance," IETF RFC1323, May 1992.
[13] R. Goyal, R. Jain, S. Fahmy, B. Vandalore, and S. Kalyanaraman, "Simulation experiments with guaranteed frame rate for TCP/IP traffic," The ATM Forum Contribution, 97-0607, July 1997.
[14] ITU-T, "Annex C to the report of the Q7/13 meeting," SG13 Geneva meeting report, Temporary Document 55 (WP 2/13), pp.1–25, Feb. 1999.
[15] S. Tsuboi, S. Chaki, and H. Kasahara, "Line utilization evaluation of GFR implementation methods," Proc. 1999 IEICE General Conference, B-7-11, p.120, March 1999.

High-Performance Backbone Network Technology edited by Naoaki Yamanaka
ISBN: 0-8247-5321-6 DOI: 10.1081/(120025698)
Published 2004 by Marcel Dekker, Inc.

Developing Customer-Focused IP Network Services

Katsuya Okimi

SUMMARY In pursuing its vision that digital signals will come to dominate not only voice communications but also various services on an integrated basis in future telecommunications, NTT has successfully completed a major part of telephone network preparations by digitizing all of its networks by the end of 1997 [1]. Accompanying digitization, it has also started providing various new services including the initiation of best-effort services through OCN. At the same time, the steady progress in Internet diffusion, the globalization of enterprise networks and inter-company electronic transactions are leading to the full scale usage of IP network communications. Because IP networks are still faced with pressing demand and unresolved issues, NTT believes it is also necessary to structure and provide computer-compatible networks suitable for IP communications by shifting from telephone-focused architecture in order to respond better to these growing needs. This paper overviews the digital network assets including ISDN that NTT has accumulated as a carrier as well as the IP networks such as the Internet that have spread throughout our societies. It also outlines new access services and network offerings based on optical and other technologies and the ATM networks that serve as the infrastructure for highly reliable and economical IP networks, and presents our thinking behind how such networks will be organically combined.

key words: *digitization, ISDN, ATM, IP network, IPv6, optical fiber*

1. Introduction

NTT has carried out the digitization of its networks under the concept that future telecommunications will come to handle not only voice communications but also various services on an integrated basis in the form of digital signals. In pursuit of this view, NTT undertook technological developments aimed at network digitization from the 1970s to the 1980s, started the full-scale introduction of new digital equipment in 1985, and completed network digitization at the end of 1997 [1]. Accompanying digitization, ISDN service was started in 1988, and number of B-channels reached five million at Feb. 1998. A major purpose behind the use of ISDN service today is to provide for Internet access. As the growth of the Internet has been extremely rapid, it was necessary to develop a vision for the service that would come after the telephone. Accordingly, NTT announced its "NTT's Basic Concept and Current Activities for the Coming Multimedia Age" in January 1994, and has accumulated experience in multimedia services utilizing a high-speed ATM environment through joint utilization trials with interested companies, universities and so on. Moreover, NTT started OCN as a best-effort service in the end of 1996, which represented a milestone realizing high-quality line service suitable for both voice and data by using digital technology. From that point forward, NTT has been heading towards the development of access networks based on optical technology and node systems supported by ATM technology.

When we review the progress in communications network developments throughout the world, it becomes obvious that even major carriers such as NTT cannot be content with the status quo, as new enterprises are appearing one after another to provide more economical and more flexible IP-level services. These new offerings are skillfully capturing customer needs under the themes of high-speed Internet and globalization.

At the same time, technologies such as ATM and optical access have finally reached the level of reliability for full-scale use and next-generation IP technologies are also beginning to emerge.

This paper presents an overview of the present communications environment and describes some of our thinking on how NTT will respond to increasingly diversified needs by incorporating these latest technologies in the digital network infrastructure it has structured so far.

2. NTT Networks and Current IP Network Status

2.1 NTT Digital Networks

The objectives of network digitization are to enable the handling of various services in the form of digital signals on an integrated basis, and to improve efficiency in network operations in order to succeed in the era of competition. As NTT has addressed digitization as its most important task after communications liberalization, it has been able to return the results of this progress to customers in the form of reduced telecommunications charges (Fig. 1).

Moreover, digitization has also made it possible to establish interconnections with other carrier networks smoothly by overcoming the many connection limitations of the analog network era. At the same time, dig-

Manuscript received April 9, 1998.
Manuscript revised June 24, 1998.

[†]The author is with NTT Network Strategy Planning Department, Global Business Headquarters, Tokyo, 163-8019 Japan.

IEICE TRANS. COMMUN., Vol. E81-B, No. 11, pp. 1959–1965, November 1998

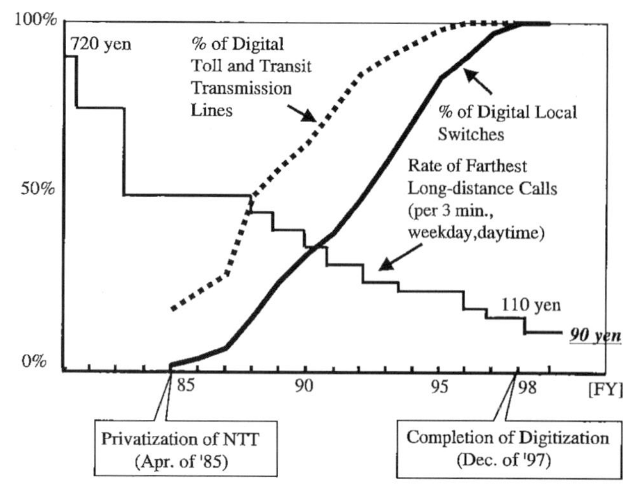

Fig. 1 Progress in digitization.

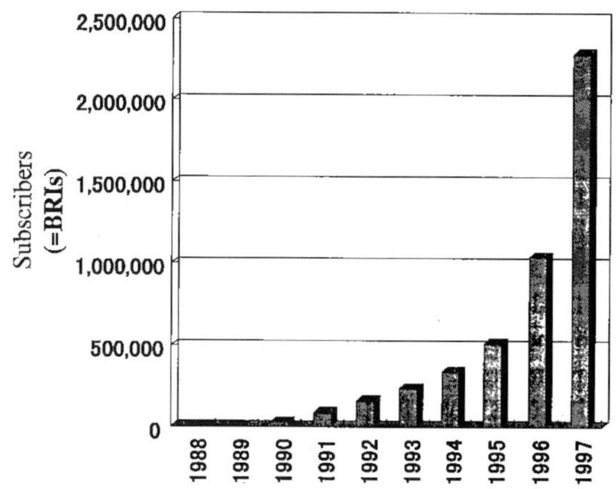

Fig. 2 ISDN subscribers growth.

itization has enabled NTT to provide a host of new services, including detailed charge statements, various discount programs and the recently introduced Number Display. (CLIP: Calling Line Identification Presentation)

Reflecting these changes, NTT's multimedia-related revenues show an increasing trend. In particular, ISDN subscribers have recorded rapid growth largely (Fig. 2), we think, due to the increase in household Internet usage. Similarly, while network digitization has also made it possible to provide IN (Intelligent Network) services, there is still room for making greater use of IN attributes in advanced services when compared to the situation in the US. We will later describe services that represent a convergence of ISDN and Internet functions, which NTT hopes to provide in the future.

2.2 Current IP Network Status

In touching on the differences between Japan and the US, we will describe the Internet usage situation in Japan from the aspects of ISPs, enterprises and resi-

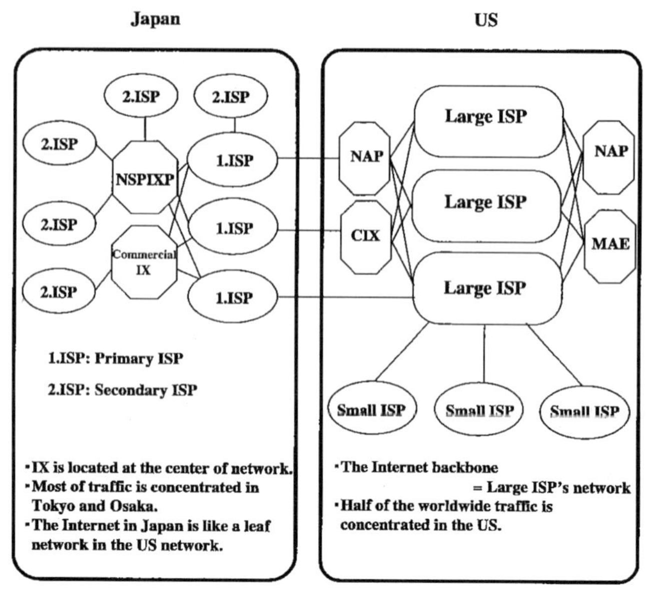

Fig. 3 Difference in Internet structure between Japan and the US.

dential usage.

(1) ISPs

In the midst of rapid growth in global Internet usage in recent years, the number of commercial ISPs has been rapidly growing in Japan as well. There were 900 ISPs and 3 million Internet users in June 1996, then in February 1998, there were about 2,500 ISPs and 10 million Internet users in Japan [2]. Traffic in Japan is concentrated around the Tokyo and Osaka areas, and the network structure is centered on IXs. Japan's major ISPs are secondary providers that use ISPs in the US as backbones (Fig. 3). Moreover, as contents are mainly concentrated in the US, half of the overseas connection traffic flows to the US. While the US has some major ISPs, mergers and consolidations involving carriers are being vigorously pursued in this fast-moving environment that has recently seen the emergence of WorldCom as a major new entrant. Moreover, all of the major ISPs known as Tier1 have come under the umbrella of carriers.

Due to the rapid mushrooming of membership and the expansion of network scale, major ISPs are encountering various difficulties. These problems include bloated routing information, delays caused by increases in the number of hops accompanying multiple router steps, and lowered network performance due to modifications in network setup information. Other problems are also emerging, such as the lack of operating engineers who can manage large-scale networks as well as the paucity of management tools available. Moreover, the assignment of non-consecutive addresses due to the insufficiency of global IP address space also makes routing more complex and otherwise complicates operations.

While Internet users want low-priced dedicated access service by ADSL and other technologies, such moves would not be wholeheartedly welcomed as they would require ISPs to reinforce transmission lines. Line congestion has come to occur in dial-up access as well due to the problem of carrier facility costs and facility restrictions. Accordingly, the present situation is one in which additional facility and line installations cannot catch up with the continuing increase in usage due to the rapid growth of access users (subscribers). In particular, shouldering the line costs between transmission sites and access points has become a heavy burden to nationwide-scale ISPs – creating a predicament in which ISPs cannot increase the number of access points without careful consideration. At the same time, moreover, the competition created by the growing number of ISPs is forcing charge reductions and further complicating management parameters. In addition, in order to secure advantages in areas other than charges, offerings such as multicast service that provides for multiaddress distribution of newspapers articles as well as key-delivery service to improve security are gradually starting to appear. And we are even seeing the emergence of ISPs that have shifted to providing consulting services.

(2) Enterprises

The number of intra-company LANs and leased-line contracts for intra-company networks is showing positive growth, and plant and equipment investment in PCs and workstations is also high. While infocommunications systems using dedicated hosts and terminals as well as dedicated communications protocols were structured in the past, today's environment is shifting towards what are called open systems, or client/server systems by PCs, workstations and TCP/IP networks.

We have seen an increase in enterprise moves to reduce communications costs by connecting what were originally independently structured leased-line networks to the Internet via ISPs and utilizing them for communications between intra-company sites or inter-company communications. As concerns over IP network security still remains high, however, demand for VPN service has been rising. In response to this need, IP network service designed for enterprise users is now being offered. NTT also provides VPN service for enterprises called EBN (Enhanced Business Network), which utilizes frame-relay VCs. Moreover, as communications between intra-company sites and inter-company communications increase, the problem of private IP address overlaps between islands has become more prevalent, leading to a mounting demand for abundant global address space.

As enterprise networks become larger and more complex as described, and as the importance of information flowing through networks increases, the costs for structuring, managing and operating such networks also continues to rise and has started exerting pressures on company bottom lines. Indeed, the line cost itself can account for a high ratio of enterprise infocommunications system expenses. Moreover, factors such as differing access units for varying network services and separate contracts for the lines used are also issues that require improvement.

(3) Residential usage/home businesses

The diffusion of PCs can also be seen in households, half of which are carrying out Internet connections or computer communications. In order to respond to the need for improved performance of peripheral equipment and the diffusion of digital equipment as now seen in networking digital AV equipment at home, users will naturally seek means of communications at speeds higher than the current ISDN (64 kbit/s $\times$ 2). This INS Net 64 service provides two B channels per contracted line, which permits the simultaneous usage of only one unit each for PCs and telephones. For homes where each family member owns a PC or a telephone, we need to provide additional options rather than asking customers to subscribe to multiple lines. These include further increasing the number of available channels, providing higher communications speeds, or using personal mobile terminals such as PHS that permit digital data communications.

The widespread usage of the Internet and PCs indicates that computer communications is no longer a special application that is limited to users with special technical knowledge, and instead requires easy operability similar to that of TVs and VCRs. At the same time, interfaces must also be made easy to understand even by general users because service diversification is increasingly complicating the interfaces that are seen by customers. Moreover, quick response to service demand and modifications is also needed.

As commercial transactions such as mail orders are expected to increase in the future among individuals, home businesses and corporations, it will become important to establish a network environment linking individuals, companies and stores. While large corporations can prepare the network environment for transactions on their own, these markets can be expected to further expand if the environment for EC (electronic commerce) is provided for individuals and self-employed users as well.

2.3 Characteristics and Tasks of Internet in Japan

(1) Autonomous dispersion (bottom-up) development format

- Internet usage was expanded by autonomous dispersion with the initiative taken by research institutes, and was managed and operated on a voluntary basis.
- Network expansion led to complicated routing control and management.
- It is difficult to predict what impact partial modifications of the now greatly expanded network would have on the overall system.

(2) Communications quality and reliability

- The disadvantages of connectionless communications have become prominent, revealing serious weaknesses and congestion.
- Information flow on the Internet now includes multimedia such as voice and video in addition to text, which imposes a greater load on the network.

(3) Security

- In the past there was no expectation of security for the Internet.
- Security equivalent to that for telephones is required for the Internet because of realizing EC and similar services on it.

(4) Human factors in network operation and management

- The design of large-scale networks is an unknown field in designing commercial ISP networks.
- The systematic development of engineers who can operate commercial ISP networks is necessary.

(5) Needs for reduced costs and globalization

- As ISPs in Japan are now entering an era of competitive shake-outs, each ISP is striving to find ways to reduce costs for lines, facilities and operations.
- The Internet in Japan is now like a leaf network in the US network, which currently provides switching functions for some 50% of the worldwide traffic. Moreover, as contents are also concentrated in the US, major ISPs are mainly looking towards the US rather than Japan.

3. Future IP Network (NTT's Approaches)

While the Internet (IP network) has many characteristics that present problems as mentioned in Sect. 2, it is expected to further grow and expand for the time being because of the following advantages.

- It is the largest network in the world and anyone can access it.
- Specifications of IP are open and they are diffused as defactstandard.
- A wide variety of applications are available.
- Information origination and disclosure were started on the Internet, with WWW commonly used as the end-user interface.
- Because network structuring and management are left up to end users, there is some flexibility.

While NTT focused on telephone service in the past, it hopes to provide networks services suitable for computer communications in the future by making full use of its fully digitized network. The next section discusses how NTT would like to contribute to the Internet and IP network.

3.1 Review of IP Networks

IP networks structured by ISPs and enterprises consist of leased lines and various other networks. This requires considerable cost burdens by customers and results in management overlap even from the NTT perspective. If a large-scale IP network is to be structured, substantial delays are expected due to the multi-step router configuration. As a possible solution to these problems, we are considering an ATM-based IP network. Sharing this network among ISPs and enterprises can reduce the costs, and is effective in resolving the problems in which dedicated-access calls such as Internet calls from access lines occupy circuit-switching network facilities. Moreover, a flexible network configuration can be adopted in accordance with customer convenience by monitoring network usage status. In addition, usage of IP over ATM can minimize delays and relieve congestion, and will permit network expansion and QoS service. Figure 4 presents the ATM-based common network for enterprise and ISPs.

With regard to the current problems of complicated IP network routing as well as disordered network configuration, we consider these issues can be resolved by establishing a systematic topology through incorporating the concept adopted for the telephone network and carrying forward with network stratification. Consideration is required for timing any modification of the

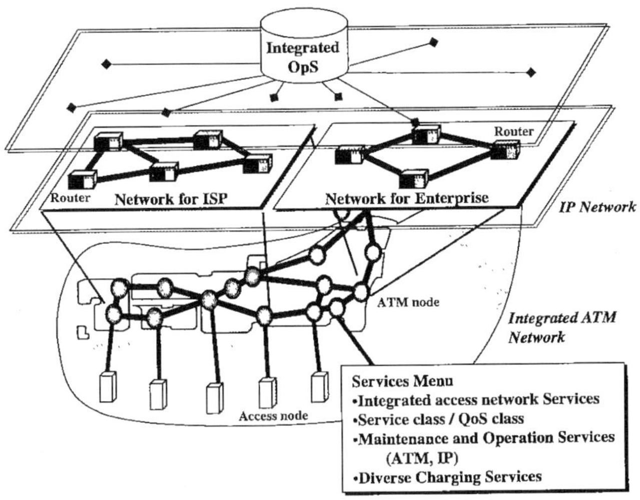

Fig. 4 Utilization of ATM network and integrated network.

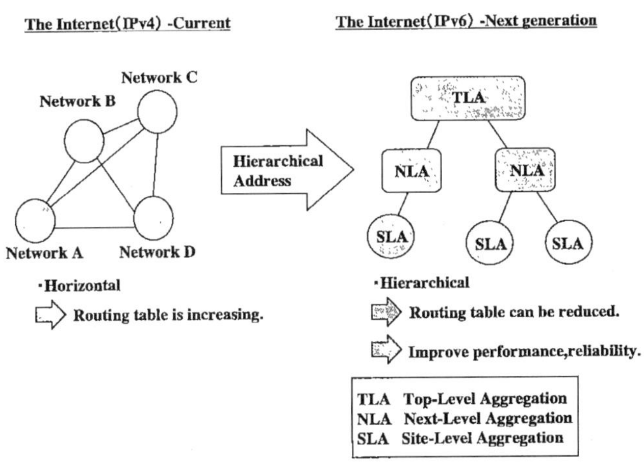

Fig. 5 Simplification and stratification of routes via IPv6.

router network topology now in operation, such as concurrent with the shift to IPv6 [3], which is the next-generation protocol. This new protocol expands address space from the current 32 bits to 128 bits to eliminate the shortage of global address space, and adopt an address architecture based on the concept of hierarchy and aggregation. Accordingly, it will be able to reduce the entries of routing table and minimize router load (Fig. 5). As it will also facilitate the management of the network itself and can limit the range of affected portions when new networks are added, this will provide a good opportunity to re-organize IP networks.

The NTT Laboratories are proposing GMN/CL (Global Mega-media Networks/ConnectionLess) as the architecture that can resolve the problems of current IP networks while still coexisting with these existing networks [4]. GMN/CL supports cut-through transfers in the transmission network, and adopts core protocol which is a new Layer 3 connectionless network architecture. It encapsulates IP protocols and reduces number of transit steps. This architecture is now being imple-

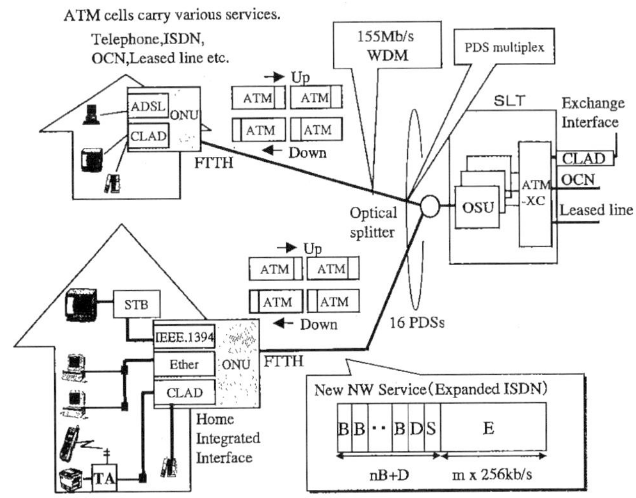

Fig. 6 Access interfaces via ATM-PDS.

mented on the ATM network. QoS control using ATM functions is also possible.

3.2 Access Network and User Interface

Greater sophistication of access lines is needed to provide for such attributes as congestion relief, higher speeds and advanced user interfaces. Access lines must be viewed as the entrance to multimedia communications services rather than simply providing access to the Internet. In particular, when the networking inside the customer premises is taken into account, a variety of user interfaces ranging from the LAN interface such as Ethernet and the digital AV equipment interface such as IEEE. 1394 needs to be supported in accordance with usage purposes. In response to these needs, the ATM-PDS (passive double star) (Fig. 6) system which is now undergoing standardization work at ITU can meet the demands for higher speeds and a greater number of channels at low costs. The ATM-PDS offers a logical speed of 10 Mbps per PDS, and can accommodate 156 BRI (ISDN basic interface) lines [5]. In addition to the expanded ISDN basic interface of nB + D, the system can handle data transmissions such as Internet access and video transmissions by establishing a new "E" channel (enhanced channel). As the user interface for this E-channel, LAN interfaces such as Ethernet and IEEE. 1394 are being considered. Moreover, as the ongoing shift to optical cable permits reductions in the construction period, customer requests for service modifications can be quickly handled. Accordingly, diversified needs from ISDN to IP communications such as the Internet, and video communications such as CATV, can be promptly met.

3.3 Security and Reliability

In order to use IP networks as the infrastructure for the

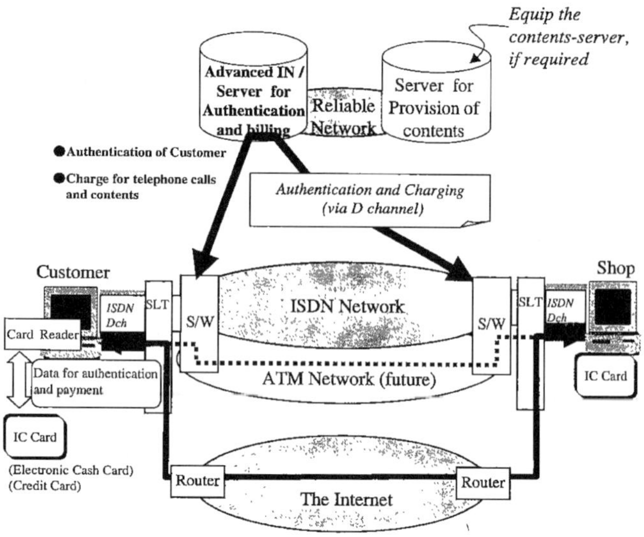

Fig. 7 Overview of information distribution support service.

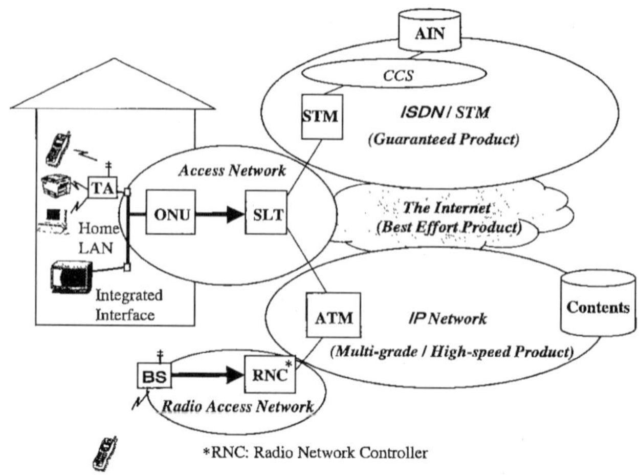

Fig. 8 Future networks described by NTT.

strong growth expected in EC, high security and reliability are essential. Today, we can realize high safety of data by the adoption of encryption and authentication architectures in the IP network. The IPv6 is equipped with expansion functions for QoS control and security functions to allow for needed modifications at the time when security functions on the Internet are generally required [6]. At the same time, the usage of ATM technology will permit the structuring of highly reliable IP networks that can be used for VPN services for enterprises and QoS services for ISPs by setting various service and quality classes and controlling access by minimizing delays and congestion.

If highly reliable networks such as the common-channel signalling system long used by NTT are linked with the Internet, the authentication and charging services offered by an advanced IN are readily available to customers even through the Internet. As indicated in Fig. 7, for example, if the ISDN D-channel is used to exchange authentication data and send charging information to the NTT customer database, any ISDN subscribers can easily use settlement and authentication service anywhere in the country.

3.4 Networks Described by NTT

An outline of networks described by NTT is presented in Fig. 8. Without being aware of which network is connected, customers can select a suitable network simply by choosing the desired service in accordance with quality, price and interface conditions. In other words, the user interface is integrated at the service level, and automatically activated in response to whatever service the customer selects.

We are considering using the Web-based interface widely used on the Internet and IN as the interface be-

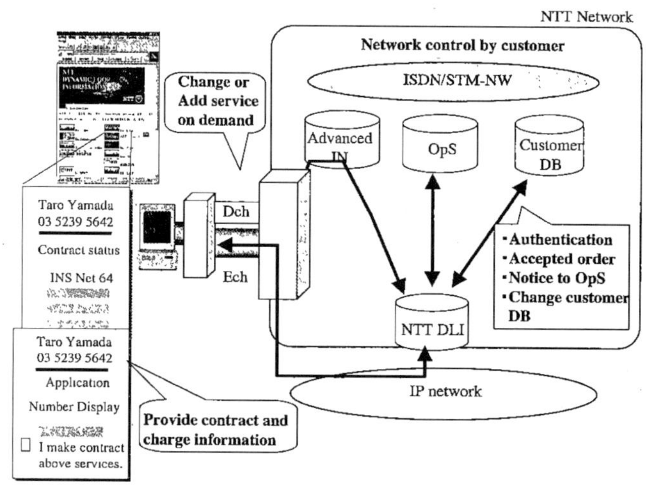

Fig. 9 Service options from DLI (NTT's official homepage).

tween customers and NTT at such times (Fig. 9). In offering a variety of network services, NTT hopes to provide the flexibility to enable customers to directly request network control, charge information presentation, and service options via the Internet.

4. Conclusion

Under the concept so far described, NTT intends to pursue new access services and network offerings based on optical and other technologies and the ATM networks that serve as the infrastructure for highly reliable and economical IP networks, in addition to fully utilizing digital network assets including ISDN that NTT has accumulated as a carrier as well as the IP networks such as the Internet that have come to spread throughout out societies. NTT plans to meet future needs for computer communications by establishing an infrastructure that links these networks organically.

References

[1] "Tokusyu dejitaruka kanryo," NTT Gizyutu Journal, vol.10, no.2, pp.8–33, Feb. 1998.
[2] Internet Association of Japan, "Internet White Book '98," Impress, Tokyo, 1998.
[3] R. Hinden and S. Deering, "IP Version 6 Addressing Architecture," draft-ietf-ipngwg-addr-arch-v2-06.txt, Jan. 1998.
[4] H. Sugioka, K. Koyanagi, T. Adachi, and T. Chiba, "A technological initiative towards next-generation computer networks," NTT R&D, vol.47, no.4, pp.411–414, 1998.
[5] I. Yamashita, T. Kanda, and K. Harikae, "PDS technologies realizing economical full access network opticalization," NTT REVIEW, vol.9, no.5, pp.38–43, Sept. 1997.
[6] S. Kent and R. Atkinson, "Security Architecture for the Internet Protocol," draft-ietf-ipsec-arch-sec-05.txt, May 1998.

Chapter 5

Performance Study on Switch and Network

ATM is a very hot topic because it offers QoS support. For this, it needs traffic control, bandwidth management, and provisioning. Many papers have examined the performance of ATM.

The first paper, Paper 5.1, presents a new switch architecture and a performance study. ATM switch buffer design is one important study issue and reducing head-of-line blocking is another. The paper reports a new crosspoint switch that can reduce the probability of HOL blocking.

Paper 5.2 provides a performance study on the input buffer type switch with window policy. Performance is examined in very fine detail under uniform and non-uniform traffic.

Paper 5.3 reports the delay and performance of AAL2 which will be used for low bit-rate voice service. As described in Chapter 2, voice service is delay sensitive. This means that the time out incurred by cell assembly must be considered. The paper describes AAL2 switch performance in terms of cell delay, cell loss, link bandwidth, and link efficiency (load).

In Paper 5.4, Y. S. Sun et al. propose a new jitter-controlled traffic scheduling algorithm that provides an upper bound for the delay jitter in the case of rate-controlled connection, such as packet video streams and IP telephony. It guarantees bounded delay and worst-case fair weighted fairness.

Available Bit Rate (ABR) service is standardized to handle applications of data traffic in ATM network. As a flow control method, the rate-based flow control is adopted and applied to manage the ABR service. H. Prihanto et al. introduce an ABR flow control scheme based on adaptive buffer thresholds, which solve the beat-down problem in ATM networks in Paper 5.5.

Guaranteed Frame Rate (GFR) service is also standardized as a new ATM service category to provide a minimum cell rate (MCR) guarantee to each connection at the frame level. Paper 5.6 by R. Kawahara et al. proposes a GFR implementation method that uses the ABR control mechanism.

A connection-wise end-to-end delay is crucial to network resource decision or control problems such as call admission control (CAC) and routing. H-W. Ferng et al. present a systematic method for connection-wise end-to-end delay analysis in ATM network in Paper 5.7. Y-C. Wang et al. solve a per-connection end-to-end CAC problem to allocate network resources to an input session to guarantee its QoS requirements in Paper 5.8.

Fast Reservation Protocol (FRP) is a protocol that can achieve efficient resource management for bursty traffic. Y-Z. Cho et al. investigate the performance of FRP for burst-level bandwidth allocation in ATM networks in Paper 5.9. They analyze two schemes, FRP/delayed transmission (DT) and FRP/immediate transmission (IT) methods.

As a scalable IP-QoS architecture, Differentiated Services (Diffserv) is considered to be useful within the service provider's backbone and enterprise networks. T. Ishihara et al. propose a Diffserv over ATM architecture to provide Diffserv-based QoS over ATM access networks in Paper 5.10. They present a developed prototype ATM card to support Diffserv-based QoS.

The final paper of this chapter is provided by C. Lin et al. Paper 5.11 introduces an integrated scheme for scheduling real-time traffic and realizing cell loss control in high-speed ATM networks with multiple priorities. The proposed scheme is based on variable queue length thresholds for scheduling and the partial buffer sharing policy for cell loss control.

High-Performance Backbone Network Technology edited by Naoaki Yamanaka
ISBN: 0-8247-5321-6 DOI: 10.1081/(120025699)
Published 2004 by Marcel Dekker, Inc.

A High-Speed Tandem-Crosspoint ATM Switch Architecture with Input and Output Buffers

Eiji Oki and Naoaki Yamanaka

SUMMARY This paper proposes a high-speed input and output buffering ATM switch, named Tandem-Crosspoint (TDXP) switch. The TDXP switch consists of multiple crossbar switch planes. These switch planes are connected in tandem at every crosspoint. Even if a cell can not be transmitted to an output port on the first plane, it has a chance to be transmitted on the next plane. Cell transmission is executed on each switch plane in a pipeline manner. Therefore, more than one cell can be transmitted to the same output port within one cell time slot, although the internal line speed of each switch is equal to the input /output line speed. The TDXP switch architecture has several advantages in implementation. First, the TDXP switch does not increase the internal line speed in eliminating Head-Of-Line (HOL) blocking. Second, since the TDXP switch employs a simple cell reading algorithm at the input buffer in order to retain the cell sequence, the TDXP switch does not require to rebuild the cell sequences at output buffers using time stamps, as is required by a parallel switch. These merits make implementing the high-speed ATM switch easy. Numerical results show that the TDXP switch can eliminate the HOL blocking effectively and achieve high throughput both for unicasting and multicasting traffic. This switch architecture is expected to enable the development of high-speed ATM switching systems that can realize over 1 Tb/s throughput in a cost-effective way.

key words: *ATM, switch, high-speed, crosspoint, buffer*

1. Introduction

The asynchronous transfer mode (ATM) concept is believed to well support high-speed multimedia infrastructure. ATM networks are able to handle various services, such as high-speed data communications, real-time video conferences, and HDTV(High-Definition Television) broadcasting [1], [2]. The demand for these services is now expanding. In this situation, ATM networks will require switching systems that offer 1 Tb/s throughput and above in a cost-effective manner [3]–[6].

To realize a high-speed ATM switch, several architectures have been addressed so far [7]–[11]. Examples are the shared-memory switch and the input and output buffering switch. In the former, memory resources are effectively utilized because of statistical multiplexing gain. However, high-speed memory access is required. This requirement increases memory cost significantly [8].

The input and output buffering switch is a crossbar-type switch, and has separate buffers in each input line and output line. Head-Of-Line (HOL) blocking can be avoided by using either the internal speed-up technique or the parallel-switch technique [12], [13]. It is very difficult to implement a high-speed switch whose internal lines are sufficiently fast; we have to use ultra-high-speed devices such as GaAs or Si bipolar VLSI and special high-density MCM technologies for the interconnection of the ultra-high-speed VLSIs forming the switch [14]. In addition, even the parallel switch needs a complex scheme to keep cells from being placed onto the output lines out sequence. For example, time stamps can be written in the cell headers and compared at output buffers; the cell sequence is regenerated according to the time stamps. This procedure also makes the cost-effective implementation of high-speed switches difficult to achieve [15]. Furthermore, the hardware resources needed to implement the switch are twice that of a single plane switch.

To implement high-speed ATM switches cost-effectively, we adopt high-density devices such CMOS VLSI rather than ultra-high-speed GaAs or Si bipolar VLSI. CMOS technologies offer more gates and larger memories at the cost of speed, compared to bipolar technologies.

This paper proposes a new input and output buffering switch architecture, named the Tandem-Crosspoint (TDXP) switch, under the above consideration on device technologies. The TDXP switch has, logically, multiple crossbar switch planes. These switch planes are connected in tandem at every crosspoint. The internal speed in each switch plane equals input/output line speed. The TDXP switch has at least one cell buffer at each tandem-crosspoint. A cell at the head of an input buffer tries to be transmitted to the destination tandem-crosspoint by way of the first plane switch. When several cells from different input lines are destined for the same output buffer, only one cell can be transmitted to the output buffer on the first switch plane. Other cells are moved to the next switch plane and try to be accepted in the same way. These operations are executed in a pipeline manner at each cell time. Therefore, more than one cell can be transmitted to the same output buffer within one cell time, even though the internal line speed equals input/output line speed. In order to keep the cell order, we employ a simple algorithm; cells

Manuscript received April 25, 1997.
Manuscript revised July 14, 1997.
†The authors are with NTT Network Service Systems Laboratories, Musashino-shi, 180–8585 Japan.

IEICE TRANS. COMMUN., Vol. E81-B, No. 2, pp. 215–223, February 1998

are not transmitted to the tandem-crosspoint when a cell in the buffer at the tandem-crosspoint exits.

Therefore, the TDXP switch does not need to increase the internal line speed, nor does it use time stamps or cell rearrangement at the output buffers. The TDXP switch avoids HOL blocking by effectively using more gates and memories than the conventional switch architecture.

The remainder of this paper is as follows. Section 2 briefly reviews conventional input and output buffering switch architectures. Section 3 presents the TDXP switch architecture. Section 4 shows its performance. Finally, Sect. 5 summarizes the key points. Throughout this paper, we assume that the switch size is $N \times N$ (N input ports and N output ports). Input and output lines have the same speed.

2. Conventional Switch Architectures with Input and Output Buffers

Input and output buffering switches are one type of crossbar switch. An advantage of the crossbar switch is scalability. The modularity of the switch design allows us to expand the switch easily. One can build a larger switch simply by adding more crosspoint switches. In addition, the cell transmission delay in the switch is smaller than in Banyan-type switches. This is because it has the smallest number of connecting points between arbitrary input/output pairs.

Variants of crossbar-type switches include the input buffering switch and the output buffering switch. The performance of the former was analyzed in [16]. They showed that, when switch size N is infinite, the maximum throughput of the switch is 0.586, assuming that the internal speed of the switch is equal to that of input/output lines. The limitation of the maximum throughput is due to HOL blocking in the input buffers.

Several problems have to be addressed in order to improve the limited throughput of the input buffering switch [15]. One possible solution to HOL blocking is to increase the internal line speed of the switch as shown in Fig. 1 (a). Oie et al. analyzed the performance of the internal speed-up switch with input and output buffers when the speed-up factor is L ($1 \leq L \leq N$) [17]. Yamanaka et al. developed a high-speed switching system that has 160 Gb/s throughput; the internal line speed was twice that of the input/output lines [11], [12], [14]. In the switch reported, input/output speed is 10 Gb/s so the internal line speed is 20 Gb/s. To realize these speeds, the switch adopted ultra-high-speed Si bipolar devices and special high-density MCM techniques [11], [14]. However, for much larger throughputs this internal speed-up crossbar switch architecture is not so cost-effective given the limitation of current hardware technologies.

Another possible approach to improve the performance of crossbar-type switches is to employ a parallel

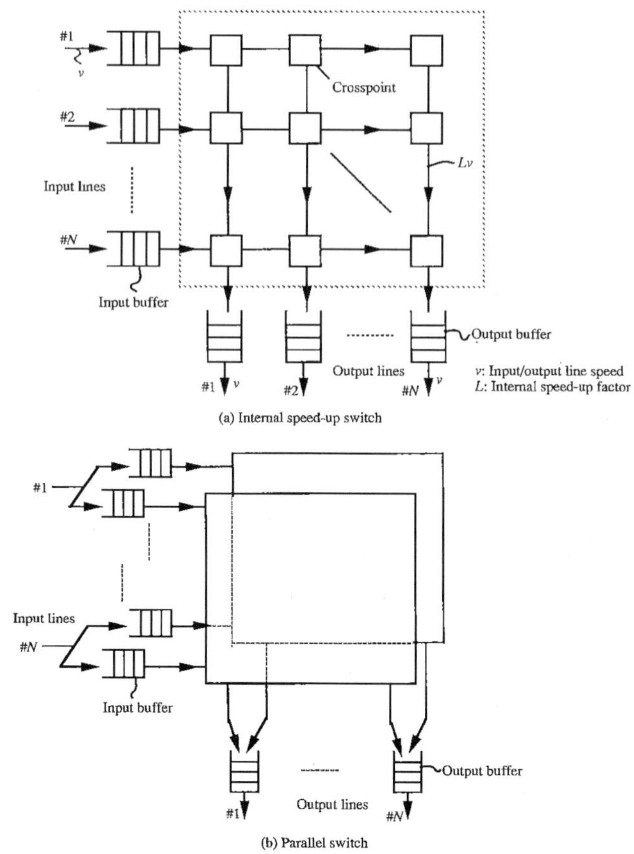

Fig. 1 Conventional input and output buffering switch.

switch architecture as shown in Fig. 1 (b) [20], [21]. The parallel switch consists of K identical switch planes. Each switch plane has its own input buffer and shares output buffers with other planes. The parallel switch with $K = 2$ achieves the maximum throughput of 1.0. This is because the maximum throughput of each switch plane is more than 0.586 for arbitrary switch size N. Using this concept, Balboni et al. developed an industrial 160 Gb/s cross-connect system [13], [18]. At the input buffers, however, time stamp values must be placed in each cell header. At the output ports, cells are buffered by implementing a maximum delay equalization mechanism, in order to rebuild the cell sequences which, due to the internal routing algorithm, can not be guaranteed [19]. Thus, this type of parallel switch requires time stamps, and cell sequence regeneration at the output buffers. In addition, the hardware resources needed to implement the switch are double those of a single plane switch. Considering the implementation for much larger switches, rebuilding of the cell sequences at high-speed also makes cost-effective implementation unlikely.

3. Tandem-Crosspoint (TDXP) Switch Architecture

3.1 Overview of TDXP Switch

To implement ATM switches that have much larger

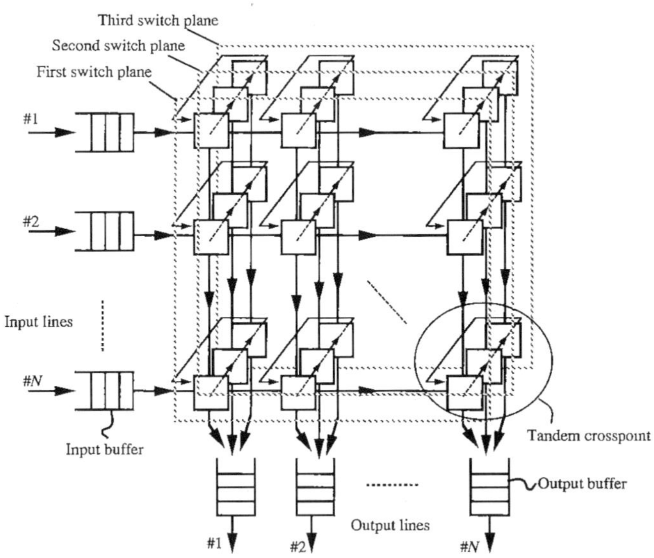

Fig. 2 Tandem-crosspoint switch structure ($K = 3$).

throughput than conventional switches cost-effectively, we need a new switch architecture that does not increase the internal line speed nor does it have to rebuild cell sequences at the output buffers. In addition, we would prefer to use high-density devices such CMOS technologies rather than to use ultra-high-speed GaAs or Si bipolar technologies. CMOS technologies provide more gates and larger memories at the cost of clock speed. For example, an ATM switch LSI with 0.25 μm CMOS that has 220 Kgate logic and 110 Kbit RAM, was presented in [22]. Using CMOS devices relaxes several of the limits imposed by hardware technologies such as VLSI interconnections.

Under the above consideration of the relationship between switch architecture and device technologies, this paper presents a new input and output buffering switch architecture, named the Tandem-Crosspoint (TDXP) switch. Figure 2 shows the structure of the TDXP switch. It has, logically, multiple crossbar switch planes. The number of crossbar switch planes is K in general. The case with $K = 3$ is shown in Fig. 2. These switch planes are connected in tandem at every crosspoint. That is why we call this switch TDXP switch. The internal speed in each plane is the same as the input /output line speed. In other words, each switch plane can transmit only one cell to each output port within one cell time slot. If more than one cell goes to the same output port on the same switch plane, unsuccessful cells that are not transmitted to the output port are stored in the tandem-crosspoint. However, the TDXP switch that has multiple switch planes can transmit up to K cells to each output port within one time slot.

We note that, although Fig. 2 depicts a single output buffer at each output port for simplicity, the TDXP switch can provide multiple quality of service levels if multiple buffers replace the single output buffers and

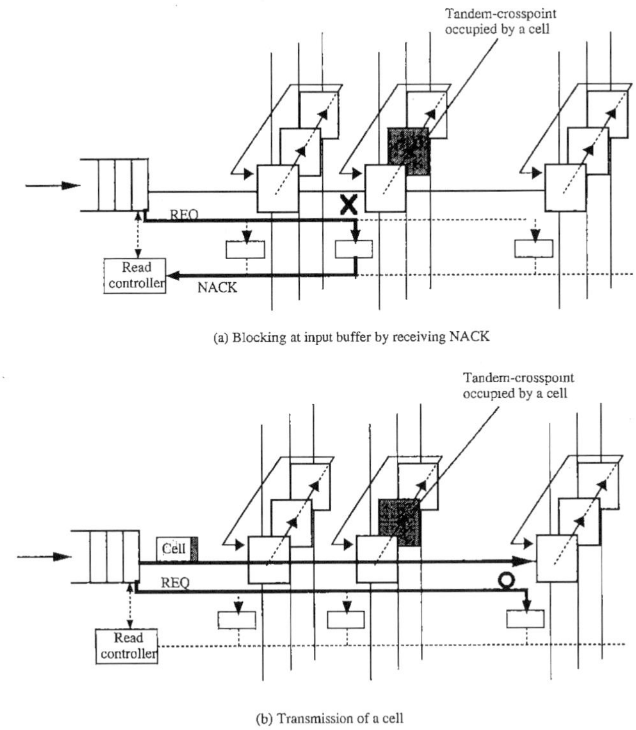

Fig. 3 Cell transmisson mechanism of TDXP switch.

priority controls are employed.

3.2 Unicasting Mechanism

The cell transmission algorithm in the TDXP switch for unicasting is first explained.

Step 1: A cell at the head of the input buffer sends a request signal (REQ) to the destination tandem-crosspoint according to the routing bits written in the cell header. Then go to Step 2.

Step 2: The tandem-crosspoint that receives an REQ sends a not-acknowledge signal (NACK) back to the input buffer if the tandem-crosspoint is already handling or buffering a cell that can not be transmitted to the output line because of contention on the output line, as shown in Fig. 3 (a). Then go to Step 3.

Step 3: The cell at the head of the input buffer that sent the REQ, is sent to the destination crosspoint on the first switch plane, if NACK is not received within a certain time, as shown in Fig. 3 (b). Then go to Step 4, setting $k = 1$. Otherwise, the cell is not sent to the crosspoint and, at the next cell time, go to Step 1.

Step 4: The kth crosspoint sends a request signal to an arbitration controller asking for transmission to the destination output buffer. (We refer to the crosspoint on the kth plane as the kth crosspoint in this paper.) The arbitration control on the kth plane is executed independent of that of the other planes. For example, ring arbitration is one possible approach. If the request is accepted by the arbitration controller, the cell is transmitted to the output buffer. Then, go to Step 5. Otherwise, the cell is moved to the $k+1$th crosspoint if k is not equal to K, and k is set $k = k+1$. If k is equal to K, the cell is moved to the first crosspoint and k is set 1. Then, go back to the beginning of Step 4 after one cell time slot.

Step 5: The cell transmitted from the tandem-crosspoint is stored in the output buffer. The output buffer can receive more than one cell within one cell time.

We note that each tandem-crosspoint need only one cell buffer for arbitrary K. This is because, when one cell is stored in a tandem-crosspoint, the following cell does not go to the same tandem-crosspoint due to the back-pressure mechanism.

To clarify the cell transmission mechanism, we consider that $K+1$ cells request to be transmitted to the same output port on the first switch plane. First, on the first switch plane, only one cell is transmitted to the output port, and other K cells go to the second switch plane. At the next cell time slot, only one cell is transmitted to the output port at the second switch plane, and $K-1$ cells go to the third plane switch. In the same way, on the K th switch plane, one cell out of the remaining two cells is transmitted to the output port. The unsuccessful cell that can not be transmitted to the output port goes back to the first plane switch and tries again to be transmitted to the output port, competing with other cells that request to be transmitted on the first plane switch.

Figure 4 shows the behavior of the cell transmission mechanism with $K = 3$, when 4 cells request to be transmitted to the same output port on the first plane switch at $t = 0$. In Fig. 4, the states of only one output port are depicted. At $t = 0$, the cell at the second input port is transmitted on the first plane switch. Then, the cell at the third input port is transmitted on the second switch plane at $t = 1$, the cell at the first input port is transmitted on the third switch plane at $t = 2$, and the cell at the fifth input port is transmitted on the first plane at $t = 3$.

These procedures are executed in a pipeline manner at every cell time. Therefore, more than one cell can be transmitted to the same output buffer within one cell time slot, even though the internal line speed of each switch plane equals the input/output line speed. When $K = 3$, three cells that come from different input lines can, as the maximum case, be transmitted to the output buffer at the same time slot at $t = 2$, as shown in Fig. 4. Thus, the TDXP switch achieves a similar result to the internal speed-up switch in eliminating HOL blocking. However, the effect on eliminating HOL blocking in the TDXP switch is not exactly same as that of the internal speed-up switch. This is because the TDXP switch has one cell buffer at each tandem-crosspoint and employs a back-pressure mechanism in the input buffers, while the internal speed-up switch does not have any crosspoint buffers. A detailed discussion is given in Sect. 4,

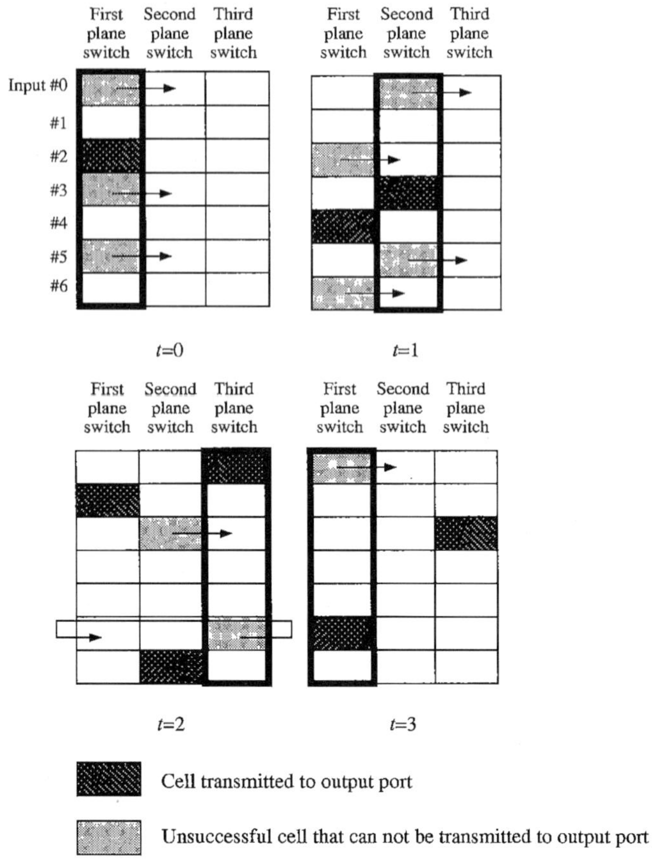

Fig. 4 Behavior of cell transmission mechanism ($K = 3$).

considering the effect such a back-pressure mechanism.

In addition, while a tandem-crosspoint is handling a cell, the input buffer does not send the head-of-line cell to the same tandem-crosspoint. The identical tandem-crosspoint never transmits more than one cell within the same cell time slot. Therefore, cell sequences are completely guaranteed and there is no need to rebuild cell sequences at the output buffers.

Furthermore, another advantage of the TDXP switch is that, although it has, logically, multiple crossbar switch planes, the hardware resources are much less than those of the conventional parallel switch. This is because the parallel switch can not share the hardware resources, while the TDXP switch can shares resources, such as input buffers, internal input lines and so on. This is a significant implementation benefit.

To clarify the advantage of the TDXP switch, several crossbar-type switches are classified according to buffering strategies as shown in Table 1 [15], [23]. Compared to conventional crossbar-type switches, the TDXP switch does not need internal speed-up lines nor rebuilding of cell sequences, and needs only one cell buffer at each tandem-crosspoint. Therefore, the TDXP switch can be easily implemented even if input and output line speed increases.

Table 1 Classification of crossbar-type switches.

Switch type	Buffering strategies	Internal speed-up factor L	Features
Input buffering switch	Input	$L = 1$	• Low throughput, 0.58 with $N=\infty$
Internal speed-up switch	Input/output	$1 < L \leq N$	• High throughput • Difficult Implementation due to internal lines
Parallel switch	Input/output	$L = 1$	• High throughput • Rebuilding cell sequences required
Crosspoint buffering switch	Crosspoint	$L = 1$	• High throughput • Large crosspoint buffers required
TDXP switch	Input/output /tandem-crosspoint	$L = 1$	• High throughput • No need internal speed-up lines • No need to be build cell sequences • Small tandem-crosspoint buffer (only 1 cell)

3.3 Multicasting Mechanism

Next, we explain the multicasting mechanism in the TDXP switch. Unicast is a subset of multicast. For the multicasting, Step 1 and 3 are modified a little. The remaining procedures, Step 2, 4, and 5, are the same as above. The modified procedures, Step 1′ and 3′, for multicast are as follows.

Step 1′: A cell at the head of the input buffer sends request signals (REQ) to all the destination tandem-crosspoints according to the routing bits written in the cell header. The routing bits for multicasting are, for example, written using a bit-map scheme. Then go to Step 2.

Step 3′: The cell at the head of the input buffer that sent REQ before, is sent to all the destination crosspoints on the first plane switch, if *no* NACK is received from *any* of the crosspoints polled to the input buffer within a certain time. Then go to Step 4. If even one NACK is received by the input buffer within a certain time, the cell is not sent to any of the destination crosspoints and, after the next cell time, go to Step 1′.

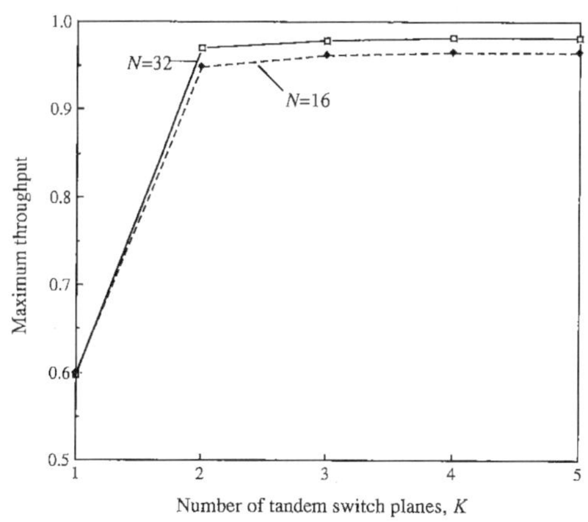

Fig. 5 Impact of number of tandem switch planes on maximum throughput.

4. Performance of TDXP Switch

The performance of the TDXP switch was evaluated by event-driven computer simulation. The simulation programs were written using C language. The performance parameters of interest were the maximum throughput, the delay time, and the required buffer size guaranteeing a specified cell loss ratio objective. The maximum throughput is defined as the ratio of total transmitted cells to output ports to total offered input cells. In the estimation of the maximum throughput, all offered input loads are set at 1.0.

We assume that cell arrival at N input ports follows a Bernoulli process. When the input traffic load is ρ, an incoming cell arrives with probability ρ, in a cell time, and there is no arrival with probability $1-\rho$. The incoming cells are distributed uniformly to all output ports. The input traffic is assumed to be homogeneous, and it is distributed uniformly to all input ports. We focus only on Bernoulli traffic. Performance estimation using Poisson traffic and bursty traffic is for further study. In addition, a simple arbitration mechanism for cell output in the switch, ring arbitration in this paper, is established between the crosspoints belonging to the same output port.

First we present the performance of the TDXP switch for unicasting traffic. Figure 5 shows how many tandem switch planes K are needed to obtain the maximum throughput in the TDXP switch architecture. To evaluate the maximum throughput, we assume that the sizes of the input and output buffers are infinite. We can see that the maximum throughput is almost saturated with $K = 2$.

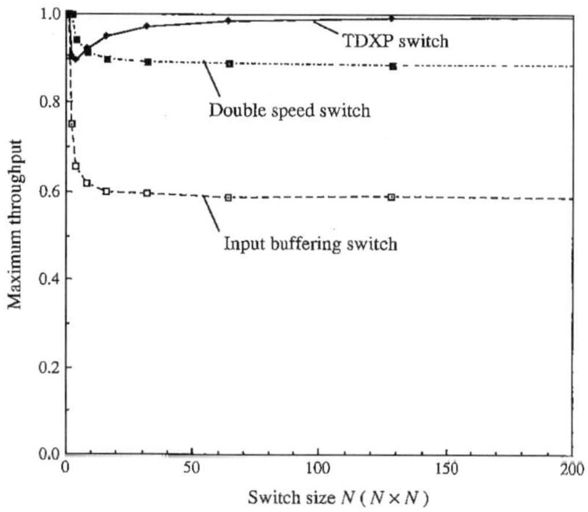

Fig. 6 Comparison of maximum throughput.

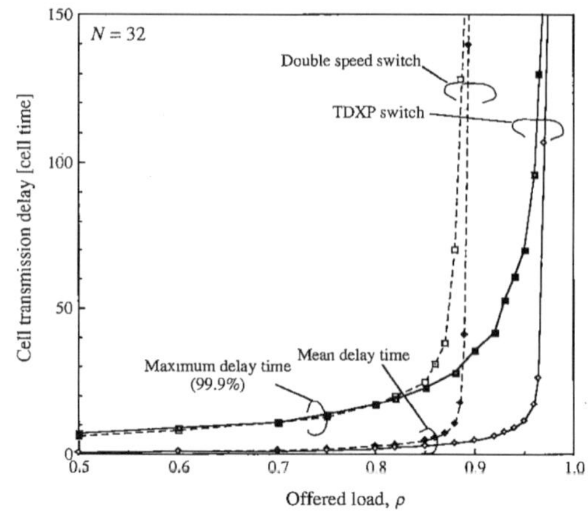

Fig. 7 Delay performance.

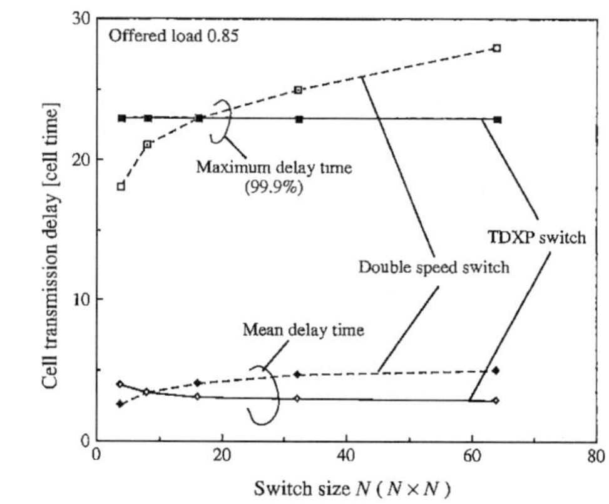

Fig. 8 Delay performance versus switch size.

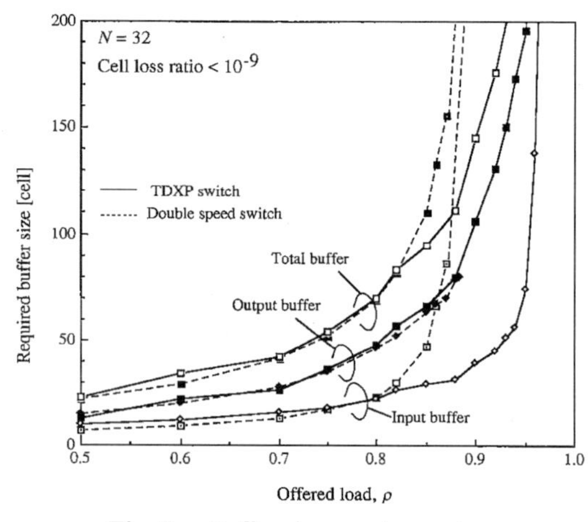

Fig. 9 Buffer size requirements.

We consider that $K = 2$ is large enough to obtain the maximum throughput. Therefore, in the following the performance evaluation of the TDXP switch, we set K to 2. In addition, we also show the performance of the conventional internal speed-up switch with input and output buffers for reference. The internal speed-up factor L is set to 2 for rough equivalency [17].

The maximum throughput of the TDXP switch increases with switch size N, and is higher than that of the double speed switch. Figure 6 compares the maximum throughput with those of the internal double speed-up switch ($L = 2$) and the input buffering switch. The maximum throughput of the TDXP switch decreases with small N, say $N \leqq 8$, while it increases with larger N. The reason is as follows. The probability P_{suc} that two successive cells at the same input buffer are destined to the same output port is $1/N$. When N is small, P_{suc} is large. As a result, the later cell in the input buffer is likely to be blocked (NACK is received) because the first cell is still being handled in the tandem-crosspoint. On the other hand, if N is large, P_{suc} is small, i.e. the blocking probability is small. The maximum throughput approaches 1.0 with $N \to \infty$, because of $P_{suc} \to 0$. In the double speed switch, the maximum throughput decreases to a certain value with N as is true for the input buffering switch. Thus, the maximum throughput of the TDXP switch is higher than that of the double speed switch when N is large.

The cell transmission delay of the TDXP switch is compared with that of the double speed switch with $N = 32$. The maximum delay, defined as the 99.9% value, and the average delay are shown in Fig. 7. The maximum and average delay of the TDXP switch are almost the same as those of the double speed switch with small $\rho(\leqq 0.85)$. However, when the offered load ρ is larger than 0.85, the maximum and average delay values of the double speed switch increase strongly because the limitation of the maximum throughput is smaller than that of the TDXP switch. In the TDXP switch, although the internal speed is the same as that of the input and output lines and a cell is buffered in a tandem-crosspoint before it is transmitted to the output ports, the delay performance is better than that of the double speed switch. This is simply because the TDXP switch has the higher maximum throughput.

Switch size dependency of the cell transmission delay is shown in Fig. 8. We can see that the maximum

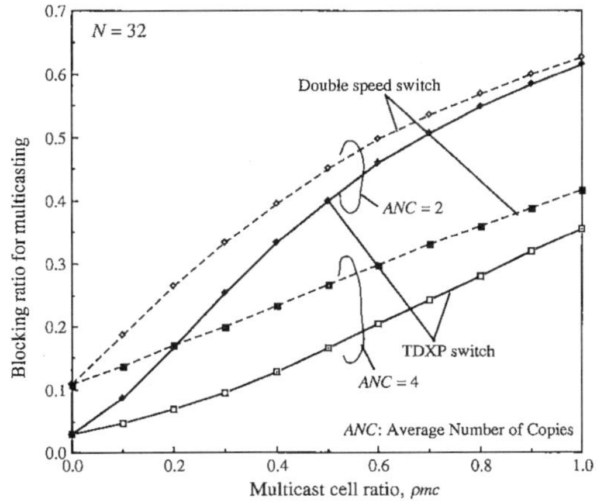

Fig. 10 Blocking ratio for multicasting.

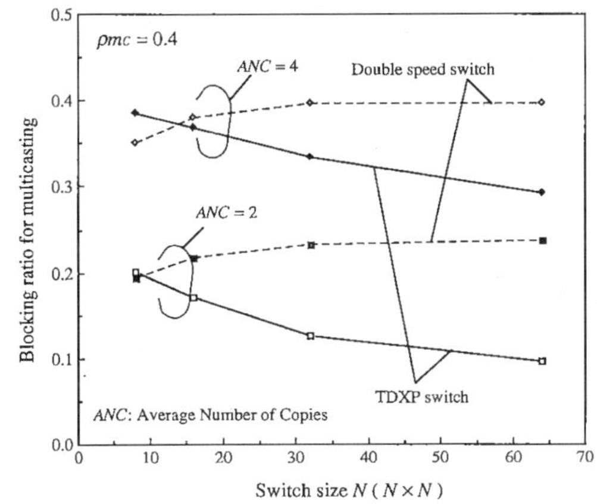

Fig. 11 Blocking ratio for multicasting dependent on switch size and average number of copies.

and average delay of the TDXP switch does not increase with N, while the delay of the double speed switch does. This results from the lower HOL blocking probability of the TDXP switch with large N, as explained in Fig. 6. Thus, the TDXP switch has scalability in terms of N.

Figure 9 shows the required buffer sizes per port, for an input buffer, an output buffer and the sum of the input and output buffers, assuming a cell loss ratio of under 10^{-9}. As is true for the delay performance results in Fig. 7, the total size of the TDXP switch is almost the same as that of the double speed switch, but with offered loads higher than $\rho = 0.85$, that of the double speed switch increases very strongly. Therefore, buffer resources in the TDXP switch are used effectively.

Next, the results of the TDXP switch for multicasting traffic are presented. In the performance evaluation, we define the multicast cell ratio ρ_{mc} as the ratio of offered total input cells to offered multicast cells. The distribution of the number of copies is assumed to follow a binomial distribution.

Figures 10 and 11 show the blocking ratio B_{block} for multicasting traffic. B_{block} is defined as the ratio of the offered input cells to the not-transmitted cells from the input buffer, due to HOL blocking. In this case, the offered input traffic load is set to 1.0. The multicast mechanism employed is that presented in Sect. 3.3. B_{block} of the TDXP switch is smaller than that of the double speed switch with $N = 32$. The reason is as follows. In the double speed switch, if at least one of the destination output ports is occupied by another cell belonging to a different input port, the multicast cell is blocked at the head of line in the input buffer. On the other hand, in the TDXP switch, even when the destined output ports is so occupied, the cell is buffered in the tandem-crosspoint, and an attempt is made to send it through the next switch plane in the next time slot. Therefore, the following multicast cell in the input buffer can be transmitted to all destination tandem-crosspoints as long as none of them are buffering a cell. This benefit of the TDXP switch is large, when N is large and the average number of copies (ANC) is small, in other words, N/ANC is large, as shown in Fig. 11. However, if N/ANC is small, B_{block} of the TDXP switch is higher than that of the double speed switch. For example, we can see such a case with $N \leq 15$ and $ANC = 4$.

5. Conclusions

This paper has proposed the high-speed input and output buffering ATM switch named the Tandem-Crosspoint (TDXP) switch. The TDXP switch consists of multiple crossbar switch planes. These switch planes are connected in tandem at each crosspoint. Even if a cell can not be transmitted to an output port through the first plane, it has a chance to be transmitted through the other planes. Cell transmission is executed on each switch plane in a pipeline manner. Therefore, more than one cell can be transmitted to the same output port within one cell time slot, even though the internal line speed of each switch is equal to the input/output line speed.

There are several implementation advantages to the TDXP switch. First, the TDXP switch does not increase the internal line speed while still eliminating the HOL blocking. That relaxes the limitation of hardware technologies. Second, since the TDXP switch employs a simple linear cell algorithm at the input buffer, the TDXP does not require cell sequence reconstruction at output buffers using time stamps, as is required in the parallel switch. In addition, another advantage of the TDXP switch is that the hardware resources can be used more effectively compared with the parallel switch.

Numerical results have shown the performance of the TDXP switch. The TDXP switch architecture eliminates HOL blocking effectively, and achieves higher maximum throughput than the double speed switch with $K = 2$, where K is the number of multiple planes. Furthermore, the delay time and the required buffer sizes of the TDXP switch are lower than those of the double speed switch.

According to our estimation of the required gates and memories for the TDXP switch, we consider that advanced CMOS device technologies [22] may allow us to implement a 40 Gb/s throughput 4×4 switch LSI with about 200 Kgate logic and 150 Kbit RAM. Therefore, a 1 Tb/s throughput switch can be realized with 625 switch LSIs, using MCM technologies, high-speed optical interconnection technologies and the switch-LSI expansion scheme presented in [24].

Therefore, the proposed TDXP switch will be a suitable architecture for future high-speed cost-effective ATM networks.

References

[1] H. Ishikawa, "Evolving from narrowband," IEEE Commun. Mag., vol.30, no.8, pp.32–36, 1992.

[2] E. Oki, N. Yamanaka, and F. Pitcho, "Multiple-availability-level ATM network architecture," IEEE Commun. Mag., vol.33, no.9, pp.80–88, 1995.

[3] N. Yamanaka, S. Yasukawa, E. Oki, and T. Kurimoto, "OPTIMA: Tb/s ATM switching system architecture," Proc. IEEE ATM'97 Workshop, pp.691–696, 1997.

[4] T. Chaney, J.A. Fingerhut, M. Flucke, and J.S. Turner, "Design of a gigabit ATM switch," Proc. IEEE INFOCOM'97, pp.2–11.

[5] K. Genda and N. Yamanaka, "TORUS-switch: Scalable Tb/s ATM switch architecture based on the internal speed-up ATM switch," Proc. IEEE GLOBECOM'95, pp.1738–1745, 1995.

[6] E. Munter, J. Parker, and P. Kirkby, "A high-capacity ATM switch based on advanced electronic and optical techniques," IEEE Commun. Mag., pp.64–71, 1995.

[7] A. Takase, T. Kozaki, M. Takatori, and H. Abe, "Datapath architecture and technology for large scale ATM switching systems," Proc. IEEE GLOBECOM'96, pp.1395–1399.

[8] H. Kuwahara, N. Endo, M. Ogino, and T. Kozaki, "A Shared buffer memory switch for an ATM exchange," Proc. IEEE ICC'89, pp.118–122, 1989.

[9] H. Yamada, S. Yamada, H. Kai, and T. Takahashi, "A multi-purpose memory switch LSI for ATM-based systems," Proc. IEEE GLOBECOM'90, pp.1602–1607, 1990.

[10] H. Tomonaga, N. Matsuoka, Y. Kato, and Y. Watanabe, "High-speed switching module for a large capacity ATM switching system," Proc. IEEE GLOBECOM'92, pp.123–127, 1992.

[11] N. Yamanaka, K. Endo, K. Genda, H. Fukuda, T. Kishimoto, and S. Sasaki, "320 Gb/s high-speed ATM switching system hardware technologies based on copper-polyimide MCM," IEEE Trans. CPMT, vol.18, pp.83–91, 1995.

[12] K. Genda, Y. Doi, K. Endo, T. Kawamura, and S. Sasaki, "A 160-Gb/s ATM switching system using an internal speed-up crossbar switch," Proc. IEEE GLOBECOM'94, pp.123–133, 1994.

[13] G. Balboni, M. Collivignarelli, L. Licciardi, A. Paglialunga, G. Rigolio, and F. Zizza, "From transport backbone to service platform: facing the broadband switch evolution," Proc. ISS'95, p.6, 1995.

[14] T. Kawamura, H. Ichino, M. Suzuki, K. Genda, and Y. Doi, "Over 20 Gbit/s throughput ATM crosspoint switch large scale integrated circuit using Si bipolar technology," Electron. Lett., vol.30, pp.854–855, 1994.

[15] Y. Oie, T. Suda, M. Murata, D. Kolson, and H. Miyahara, "Survey of switching techniques in high-speed networks and their performance," Proc. IEEE INFOCOM'90, pp.1242–1251.

[16] M.J. Karol, M.G. Hluchyj, and S.P. Morgan, "Input versus output queueing on a space-division packet switch," IEEE Trans. Commun., vol.COM-35, pp.1347–1356, Dec. 1987.

[17] Y. Oie, M. Murata, K. Kubota, and H. Miyahara, "Effect of speedup in nonblocking packet switch," Proc. ICC89, p.410, 1989.

[18] L. Licciardi and F. Serio, "Technical solutions in a flexible and modular architecture for ATM switching," Proc. Int. Symp. on Commun. (ISCOM'95), pp.359–366, 1995.

[19] M. Collivignarelli, A. Daniele, P. De Nicola, L. Licciardi, M. Turolla, and A. Zappalorto, "A complete set of VLSI circuits for ATM switching," Proc. IEEE GLOBECOM'94, pp.134–138, 1994.

[20] J.S. Turner, "Design of a broadcast packet switching networks," IEEE Trans. Commun., vol.36, pp.734–743, 1988.

[21] P. Newman, "A fast packet switch for the integrated services backbone network," IEEE J. Sel. Areas Commun., vol.6, pp.1468–1479, 1988.

[22] Y. Ohtomo, S. Yasuda, M. Nogawa, J. Inoue, K. Yamakoshi, H. Sawada, M. Ino, S. Hino, Y. Sato, Y. Takei, T. Watanabe, and K. Tekeya, "A 40 Gb/s 8×8 ATM Switch LSI using 0.25 μm CMOS/SIMOX," Proc. ISSCC97, FA 9.3, pp.154–155, 1997.

[23] J.W. Byun, "Survey of commercially available ATM switches," Proc. Asia-Pacific Conf. Commun. (APCC95), 5A.1.1, pp.491–495, 1995.

[24] E. Oki and N. Yamanaka, "Scalable crosspoint buffering ATM switch architecture using distributed arbitration scheme," Proc. IEEE ATM'97 Workshop, pp.28–35, 1997.

High-Performance Backbone Network Technology edited by Naoaki Yamanaka
ISBN: 0-8247-5321-6 DOI: 10.1081/(120025700)
Published 2004 by Marcel Dekker, Inc.

The Maximum Throughput of a Nonblocking Packet Switch with Window Policy

Dye-Jyun Ma

SUMMARY It is known that in a nonblocking packet switch with input queueing, head-of-line (HOL) blocking affects significantly the maximum throughput of the packet switch. To alleviate the HOL blocking effect, a window policy has been proposed in that each input queue can scan up to a fixed number of packets (called the window size) to select for transmission on the outputs. However, the performance of the window policy has never been precisely characterized. In this paper, we use a closed queueing network model to characterize the performance of the packet switch with window policy. We obtain explicit closed-form formulae for the maximum throughput of the packet switch as a function of the window size. Both balanced and imbalanced traffic patterns are discussed. The formulae can easily determine the effectiveness of the window policy.

key words: *nonblocking packet switch, throughput, head-of-line blocking, window policy*

1. Introduction

Advances in high speed packet switching for integrated service networks have attracted considerable attention on packet switch architectures that can provide simultaneous packet transmissions from several input links to output links through switch fabric for aggregated throughput (e.g. [2], [4]–[12]). In a nonblocking packet switch with N inputs and N outputs, there are always N simultaneous paths that can be established through the switch fabric. Despite that the switch fabric is nonblocking, the packet switch utilization in general does not achieve 100%, or the throughput does not achieve the capacity of the switch which is basically N times the capacity of one switch fabric. The reason is due to the so-called output conflict, i.e. if more than one inputs have packets to send to the same output at the same time, only one of these contending packets may be allowed to pass through to the output and the others must be queued for later transmission on the output. This type of congestion heavily depends on the input traffic pattern, and determines the throughput that the packet switch can actually have.

In the literature, attention has been focused on the performance study of various packet switch architectures (e.g. [2], [6]–[9], [12], [13]). Among the many well-known schemes are input queueing and output queueing [6]–[8]. Input queueing scheme provides input buffer for each input link to queue packets which can not be immediately transmitted due to output conflicts. Output queueing scheme provides no buffers at inputs, but requires that each switch fabric have a transmission speed which is N times that of an input link, and the switch thus can transmit simultaneously up to N input packets to the same output even with output conflicts. Since the output link is often at the same speed as the input link, the packet arrival rate to an output link may be N times faster than the packet departure rate from an output link, and thus output buffer is provided at each output to queue packets from the inputs. Although the output queueing scheme suffers no congestion due to output conflict, it requires that the switch fabric be N times faster, which may be quite costly and not practically feasible. As such, considerable studies have been devoted to the study of the input queueing scheme. The main question is then how much throughput degradation is (or what the maximum throughput is) with input queueing under various traffic patterns. Note that with input queueing where each input buffer is according to the first-in first-out (FIFO) discipline, if the head-of-line packet of an input queue can not be transmitted becasue of output conflict, other packets which are queued behind it in the FIFO are blocked from reaching possibly idle outputs on the switch. This is the so-called head-of-line (HOL) blocking, and is the source that degrades the throughput of the packet switch. It has been shown that with balanced and uniform input traffic, the maximum normalized throughput (i.e. the utilization), as the number of input/output links goes to infinity, is 0.586 for fixed-size packets [7], and is 0.5 for packets with lengths being exponentially distributed [6].

One method to alleviate the HOL blocking effect, thus to increase the maximum throughput, is to use a window policy to relax the strict FIFO queueing discipline of input buffers [7]. With the window policy, each input queue can now scan, starting with the HOL packet, up to a fixed number of packets, called the window size, to select for transmission on the outputs. In other words, if the HOL packet can not be transmitted because of output conflict, the input queue can check to see if the next packet in line has no output conflicit,

Manuscript received July 30, 1996.
Manuscript revised December 2, 1996.

[†]The author is with the Department of Electrical Engineering, National Chung-Hsing University, Taichung, Taiwan, 402, R.O.C.

[*]The work of the author was supported in part through ROC NSC Grant NSC85-2213-E005-001.

i.e. if the destination output of the next packet in line is free, and if so, the packet can then be transmitted successfully to its output link. If the answer is negative, then the input queue will again go to the next one in line and repeat the same procedure until either a packet can be transmitted or the window size is reached.

It is intuitively clear that the window policy improves the performance of the packet switch. But by how much performace improvement can the window policy achieve? This is an important question to answer, since a clear trade-off between architecture complexity and performance is a very useful information for switch design. Existing studies on the performance of the window policy have been mostly based on simulation [2], [7], [12], [13], and there exist little analytical results, at least to the author's knowledge. An exception is [9], where some analytic/numerical results were obtained for a banyan switch.

In this paper, we develop a closed queueing network model to approximate the behavior of the packet switch with window policy. From the model, we analytically derive the maximum throughput of the packet switch with window policy for packet lengths to be exponentially distributed. Simple and explicit closed-form formula based on the product-form solution [1] is obtained as a function of the window size for balanced and uniform input traffic pattern. For nonuniform or imbalanced traffic, simple recursive formulae based on Buzen's algorithm [3] are derived. The formulae are validated with simulations and the errors are within 5%. The formulae allows one to easily determine the effectiveness of the window policy under various traffic patterns. Note that with the window size equal to 1, our closed-form formula reduces to existing results with no window policy implemented [4], [6].

One distinct feature of the underlying closed queueing network model approach that is developed here is that it works for any number of input/output links. This is unlike existing works on the performance modeling of packet switches (e.g. [6]–[8], [10], [11]) where their analytic results are valid only when the number of input/output links, N, is large, since they use the fundamental fact that as $N \to \infty$, the arrival process of the packets that content for the same output tends to be an independent Poisson process.

Although the model developed here assumes exponentially distributed packet lengths. Interestingly enough, the formula obtained here is also very close to (within 1-2% of errors of) the simulation results of [7] for the window policy with fixed-length packets, especially when the window size W gets large. This kind of coincidence may be traced back to the following fact: In general, the chance of output conflict is always higher in an asynchronous (continuous-time) system than in a synchronous (discrete-time) system; with $W = 1$ output conflict directly results in HOL blocking, but with $W > 1$, an input blocking due to output conflicts has to come from the fact that all the W packets in line have output conflicts with packets in some other inputs, and as W gets large, this blocking impact is further reduced, and thus small performance difference between the continous-time and discrete-time models can be expected. This may also be explained by the work of [4], [5], where a discrete-time packet switch model with no window policy is considered. In [4], [5], it was shown that the steady-state distributions of the discrete-time packet switch have an approximate product-form solution which has the same form as that of a closed queueing network model with exponential service times.

The paper is organized as follows. Section 2 gives the model for the packet switch with window policy. Based on the model, Sect. 3 derives closed-form formulae for the maximum throughput of the packet switch with window policy. Section 4 compares the analytic results and simulations. Finally, the paper is concluded in Sect. 5.

2. The Packet Switch and the Model

We consider a generic $N \times N$ nonblocking packet switch, as shown in Fig. 1, consisting of N input links and N output links. The packets arriving at an input link are first put in the input buffer associated with the link, waiting for transmission to its destination output. Consider first the case with no window policy implemented. The packets in an input queue are moved one by one to the HOL position according to the FIFO service discipline. At the HOL position, a packet contends with possible other packets at the HOL positions of other input queues that have the same destination output. The HOL packets destined to the same output form a (logical) request queue of that output. The packets in a request queue are transmitted according to FIFO queue-

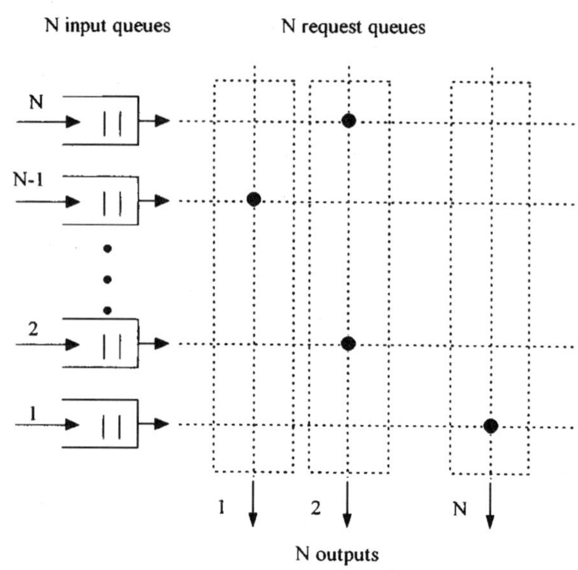

Fig. 1 System model.

ing discipline. For those packets behind the current HOL packet in a request queue, their corresponding input links are blocked from transmissions, i.e. they suffer from the so-called HOL blocking. After the HOL packet from an input queue reaches the output, the next HOL packet in the same input queue is then put in the request queue of the packet's destination output. In Fig. 1, the HOL packets are illustrated by the black dots in the dashed boxes, which represent the request queues.

The behavior of N request queues which correspond to N outputs completely determines the throughput of the packet switch. For example, if packets from all the inputs are destined to a same output, then the request queue of that output always contains N packets from N different inputs (assuming saturation), whereas other remaining $N-1$ request queues are always empty. Thus, the total throughput is only the maximum throughput generated by the only non-empty request queue which is simply the capacity of one switch fabric, and the utilization is thus only $\frac{1}{N}$, since the total capacity of switch fabric is N times the capacity of one switch fabric. In another extreme case, if packets from each input are always destined to a fixed output, with destination outputs for different inputs being different, then at saturation each request queue contains exactly one packet from the corresponding input at all the time, and the total throughput is then the sum of the maximum throughput generated by each of the N request queues. This gives a total throughput which is N times the capacity of one switch fabric, yielding a utilization of 1.

From the above discussion, it should be clear that different input traffic patterns give different behaviors of the request queues, which in turn yield different throughput of the request queues, thus the throughput of the switch. To obtain the maximum throughput of the packet switch, we assume that each input is always saturated in that each input queue has an infinity number of packets waiting to be transmitted. Once a packet from an input link is transmitted by a request queue to the output, a next packet from the same input link is immediately sent to a request queue based on the packet's destination output. This is equivalent to a packet transition from one request queue to another request queue upon service completion. Thus the set of N request queues can be viewed as a closed queueing network containing a fixed number of packets. Each request queue is modeled by a single-server station, with the server corresponding to the transmission fabric to the output link. Let μ be the transmission speed of each switch fabric. We assume that packet lengths from all the inputs are exponentially distributed with mean equal to 1. This is equivalent to the assumption that the service times of packets by the server at every request queue are exponentially distributed with mean equal to $\frac{1}{\mu}$.

To complete the closed queueing network model

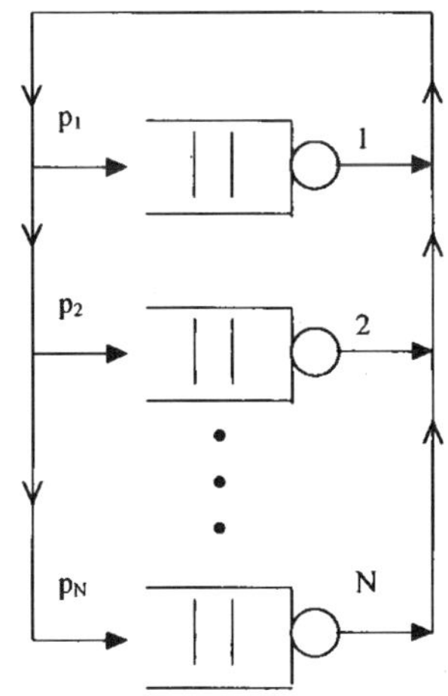

N request queues

Fig. 2 The closed queueing network model.

for the request queues, we need to specify transition probabilities between request queues. The transition takes place when a packet from an input link of the switch has been successfully transmitted to its output. The question as to which request queue the next packet from the same input link should join depends on how the input traffic pattern is specified. Here, we shall assume that the selection of destination output for each packet in each input link is according to the probabilities $\{p_i, i = 1, 2, ..., N\}$ with $\sum_{i=1}^{N} p_i = 1$, where p_i represents the probability that a packet from an input is destined to output i. With the assumption, we have a homogeneous system in which all the inputs are statistically indistinguishable, and the closed queueing network model of the request queues becomes a Jackon network with a single class of customers, and the transition probability $q_{i,j}$ from request i to request queue j is simply equal to p_j. This is illustrated in Fig. 2.

With the above-introduced general output selection probabilities $\{p_i\}$, we consider the case with nonuniform input traffic patterns. If we further assume that $p_i = \frac{1}{N}$ for all $1 \leq i \leq N$, then we consider balanced and uniform traffic patterns. Furthermore, if we allow the output destination selection probabilities to be different for different inputs, then the probability p_i becomes $p_{k;i}$, where k represents input link k, and the closed queueing network model becomes a multiclass closed queueing network with the class representing the input link, and the transition probability for a class k packet from request queue i to request queue j becomes $q_{k;i,j} = p_{k;j}$. In this case, we consider the nonhomogeous and imbalanced system.

The above model description and discussion directly applies to the case with no window policy implemented. As discussed earlier, the behavior of the request queueing network exactly models the behavior of the HOL blocking of the packet switch. In this case, the closed queueing network of the request queues contains exactly a total of N packets, with one packet (the HOL packet) from each input. The situation with window policy is more involved. Let W denote the window size associated the window policy. Since the first W packets in an input queue can now all have the chance to be transmitted, depending on whether or not their destination outputs are being idle, we can view all the first W packets as service position (or HOL) packets and thus to allow all of them to join their corresponding request queues. In this manner, the closed queueing network of the request queues becomes one that contains N classes of packets, with W packets for each class, giving a total of NW customers. The transition probabilities between request queues are the same as before. While for $W = 1$ (the case with no window policy) the model is exact as mentioned earlier, for $W \geq 2$, the model is only an approximation to the actual behavior of the packet switch with window policy. The approximation comes from the following fact: In the closed queuing network model, as long as a request queue is nonempty, it will generate throughput, while in the packet switch, if for each one of the packets waiting in a request queue, there is one packet from the same input that is currently being serviced (i.e. in transmission to output) by some other request queue, this request queue are blocked, not serving any packets and thus producing no throughput (since only one transmission can take place from one input). Thus, the model overestimates the actual throughput of the switch, and essentially provides an upper bound for it. We believe that the approximation should be good, since the above-mentioned event that overestimates the throughput is rare and the probability should be very small. This point is further illustrated and discussed in Sect. 4.

Under the exponential packet length assumption, the closed queueing network model described above will have a product-form solution for the steady-state probabilities [1]. This allows us to analytically derive the maximum throughput of the packet switch. This is done in the next section.

3. The Maximum Throughput

As described in Sect. 2, in general, the state process of the packet switch request queues is that of a closed queueing network with N classes of customers where each class has W customers. To simplify the discussuion, we focus on the case where the destination output selection probabilities $\{p_i, i = 1, 2, ..., N\}$ are the same for each input link. In the terminology of the queuing network, this means that all classes have the same routing probabilities $q_{i,j} = p_j$ between request queues. In this case, there is only one class of packets in the system. The state of the system can be described as $\mathbf{n} = (n_1, n_2, ..., n_N)$ with n_i being the number of packets of request queue i and $\sum_{i=1}^{N} n_i = NW$. Let $p(\mathbf{n})$ be the steady-state probability of the state process being at state $\mathbf{n}$. It is well known that $p(\mathbf{n})$ has a product-form solution [1].

To derive $p(\mathbf{n})$, let v_i be the visit ratio to request queue i, $i = 1, 2, ..., N$, i.e. a nonnegative solution to the equation

$$v_i = \sum_{j=1}^{N} q_{j,i} v_j, \quad j = 1, 2, ..., N. \tag{1}$$

Since $q_{j,i} = p_i$, we have

$$v_i = \sum_{j=1}^{N} p_i v_j = p_i \sum_{j=1}^{N} v_j, \quad j = 1, 2, ..., N. \tag{2}$$

If we set $\sum_{j=1}^{N} v_j = 1$, then we easily have $v_i = p_i$ for $i = 1, 2, ..., N$.

Let $y_i = \frac{v_i}{\mu} = \frac{p_i}{\mu}$ for $i = 1, 2, ..., N$. For every $k = 1, 2, ..., N$ and $n = 1, 2, ..., NW$, set

$$\begin{aligned} G_k(n) &:= \sum_{n_1+\cdots+n_k=n} \prod_{i=1}^{k} y_i^{n_i} \\ &= \frac{1}{\mu^n} \sum_{n_1+\cdots+n_k=n} \prod_{i=1}^{k} p_i^{n_i} \end{aligned} \tag{3}$$

with $G_1(n) = y_1^n$ and $G_k(0) = 1$. The quantity $G_k(n)$ is the normalizing constant for the closed queueing network with k server stations and a total of n packets. We have the product-form solution [3]

$$p(\mathbf{n}) = \frac{1}{G_N(NW)} \prod_{i=1}^{N} y_i^{n_i}. \tag{4}$$

From (3), (4), we can obtain the marginal distribution of the number of packets in the request queue N as

$$\begin{aligned} p(n_N = n) &= \sum_{n_1+\cdots+n_{N-1}=NW-n} p(\mathbf{n}) \\ &= \frac{y_N^n}{G_N(NW)} \sum_{n_1+\cdots+n_{N-1}=NW-n} \prod_{i=1}^{N-1} y_i^{n_i} \\ &= y_N^n \frac{G_{N-1}(NW-n)}{G_N(NW)}. \end{aligned} \tag{5}$$

The throughput of queue N is then

$$\begin{aligned} \eta_N &= \sum_{n=1}^{NW} \mu p(n_N = n) \\ &= \sum_{n=1}^{NW} v_N y_N^{n-1} \frac{G_{N-1}(NW-n)}{G_N(NW)} \\ &= v_N \frac{G_N(NW-1)}{G_N(NW)}. \end{aligned} \tag{6}$$

Here in (6), we use the equality that

$$G_N(NW-1) = \sum_{j=0}^{NW-1} y_N^j G_{N-1}(NW-1-j) \quad (7)$$

which follows easily from (3). Similarly, the throughput of any request queue i is

$$\eta_i = v_i \frac{G_N(NW-1)}{G_N(NW)}. \quad (8)$$

Thus, the total throughput of all the request queues is

$$\eta = \sum_{i=1}^{N} \eta_i = \frac{G_N(NW-1)}{G_N(NW)}. \quad (9)$$

Since all the inputs are identical, the maximum throughput of each input link is

$$\tau = \frac{1}{N}\eta = \frac{1}{N}\frac{G_N(NW-1)}{G_N(NW)} \quad (10)$$

and the maximum utilization of the switch (also the maximum utilization of each input link) is

$$\rho = \frac{1}{N\mu}\eta = \frac{1}{N\mu}\frac{G_N(NW-1)}{G_N(NW)} \quad (11)$$

We now first apply (9)–(11) to the case with balanced and uniform traffic pattern, i.e. $p_i = \frac{1}{N}$ for all $i = 1, 2, ..., N$. In this case, we have $v_i = p_i = \frac{1}{N}$, $y_i = \frac{1}{N\mu}$, and from (3),

$$G_N(NW) = \frac{1}{(N\mu)^{NW}} \sum_{n_1+\cdots+n_N=NW} 1 = \frac{1}{(N\mu)^{NW}} \binom{NW+N-1}{NW} \quad (12)$$

and similar form for $G_N(NW-1)$ with NW being replaced by $NW-1$ in (12).

From (9)–(12), we get the total throughput of the switch

$$\eta = N\mu \frac{NW}{NW+N-1}, \quad (13)$$

the maximum throughput per input link is

$$\tau = \frac{NW}{NW+N-1}\mu, \quad (14)$$

and the maximum utilization of the packet switch is simply

$$\rho = \frac{NW}{N(W+1)-1}. \quad (15)$$

When $W = 1$, the formula (15) coincides with the one with no window policy [4], [6].

It is easy to see that the formula for the maximum utilization of the switch is monotonically decreasing in N and increasing in W. The following limits are easily obtained: First, as the window size W goes to infinity, we have

$$\lim_{W\to\infty} \rho = 1. \quad (16)$$

This is expected, since as one searches through an infinite input queue, one can always find a packet having a destination output that is not busy receiving any packets. Next, as the number of input queues N goes to infinity, we have

$$\lim_{N\to\infty} \rho = \frac{W}{W+1}. \quad (17)$$

The formulae clearly quantify the performance of the window policy under balanced and uniform traffic.

In the case of nonuniform traffic, i.e. with general $\{p_i, i = 1, 2, ..., N\}$, the maximum utilization ρ given in (11) of the switch can be easily computed by computing the normalizing constants $G_N(NW)$ and $G_N(NW-1)$: Note that from (3), the output $G_k(n)$ satisfies the recursion

$$G_k(n) = \sum_{j=0}^{n} y_k^j G_{k-1}(n-j) \quad (18)$$

for every $k = 1, 2, ..., N$ and $n = 0, 1, ..., NW$. The output of the recursion with $k = N$ and $n = NW-1$ gives $G_N(NW-1)$, and with $k = N$ and $n = NW$ gives $G_N(NW)$. This is known as Buzen's algorithm [3].

Other than obtaining numerical results based on Buzen's algorithm (18), it is also possible to derive somewhat more explicit form for the maximum utilization under some specific nonuniform traffic patterns so that some qualitative information can be extracted. Here we present a special case for illustrative purpose. Consider the situation where the output selection probabilities are such that $p_i = p$ for some i, say $i = N$, and $p_j = \frac{q}{N-1}$ for all $j \neq i$ where $p + q = 1$. That is, the output N (which may correspond to, say a file or vedio server) distinguishes itself from other remaining $N-1$ ouputs. In this case, we easily have

$$G_N(NW) = \left(\frac{q}{(N-1)\mu}\right)^{NW} \sum_{j=0}^{NW} \left(\frac{(N-1)p}{q}\right)^j \binom{NW-j+N-2}{NW-j} \quad (19)$$

and similar form for $G_N(NW-1)$ with NW replaced by $NW-1$ in (19). Observe that with $p = \frac{1}{N}$ we have $\frac{(N-1)p}{q} = 1$, and that (19) reduces to (12) for balanced and uniform traffic. Based on (19), it is also possible to show that for each N and W, the maximum utilization ρ is maximized at $p = \frac{1}{N}$, i.e. the balanced and uniform traffic pattern, although the calculations may be quite tedius. In other words, the uniform traffic pattern gives the largest maximum throughput in this case.

Other special cases are also possible. In any case, one can always use Buzen's algorithm (18), which is simple and effective, to obtain the maximum utilization of the switch.

4. Results and Discussion

The maximum utilization obtained from the analytic closed-form formula (15) for the balanced and uniform traffic is first compared with that obtained by simulation. To that end, we consider and simulate packet switches with N equal to 2, 4, 8, 16, 32 and 64, where for each N the window policy with window size W ranging from $W = 1$ to $W = 8$ is included. The results are shown in Fig. 3 through Fig. 8, where each figure gives the results for a particular switch size N. In each figure, the two solid lines represent the results obtained from the formula and simulation, respectively. It can be easily seen that in each case of N, the result from the formula always upper-bounds that from simulation. This is expected, since as discussed in Sect. 2, the analytic queueing model tends to overestimate the actual maximum throughput of the packet switch. It is more interesting to observe that the overestimations (the errors) of the formula are less than 5%. In general, the errors get small and reduce to around 3%, as the window size gets large. Note also that in each case of N, the analytic and simulation results pretty much coincide for $W = 1$. This is also expected, since as discussed in Sect. 2, in the case of $W = 1$ the analytic queueing model gives the exact behavior of the packet switch.

For each case N, we also compare our results with the simulation results for fixed-size packets obtained in [7]. The comparison is also shown in each one of the figures from Fig. 3 to Fig. 8. In each figure, the dashed line represents the corresponding simulation results for switch with fixed-size packets. For switch with fixed-size packets, the queueing networks developed here become discrete-time queueing networks. Generally speaking, a discrete-time queueing network provides better performance results than a continuous-time queueing network with the same means, because less randomness in con-

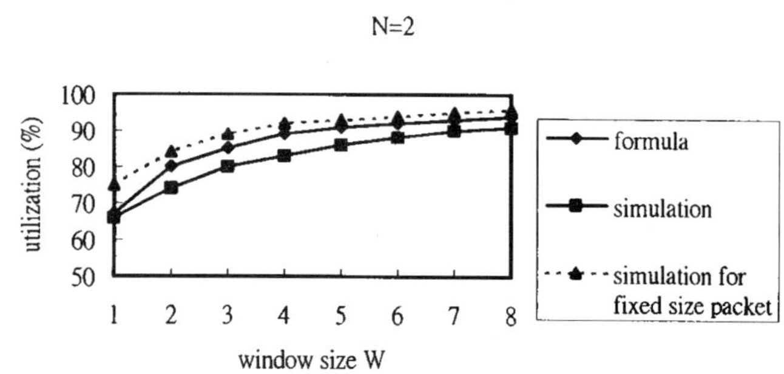

Fig. 3 The maximum utilization for $N = 2$ with uniform traffic.

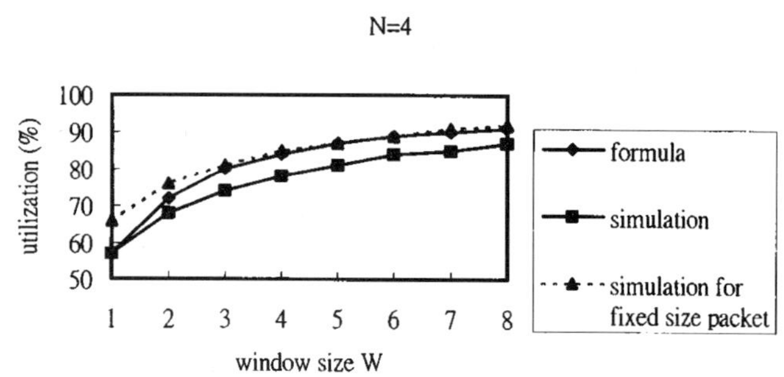

Fig. 4 The maximum utilization for $N = 4$ with uniform traffic.

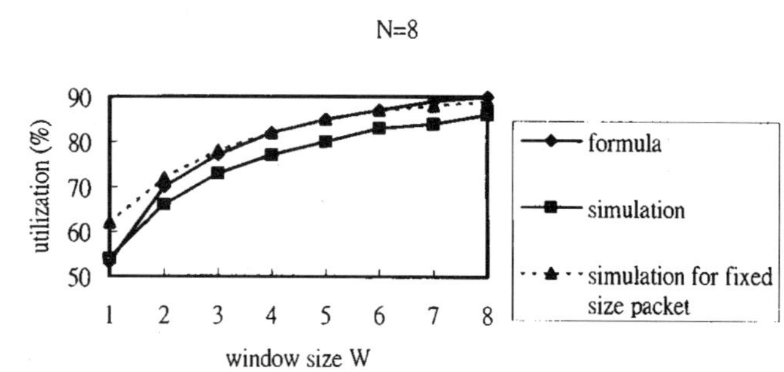

Fig. 5 The maximum utilization for $N = 8$ with uniform traffic.

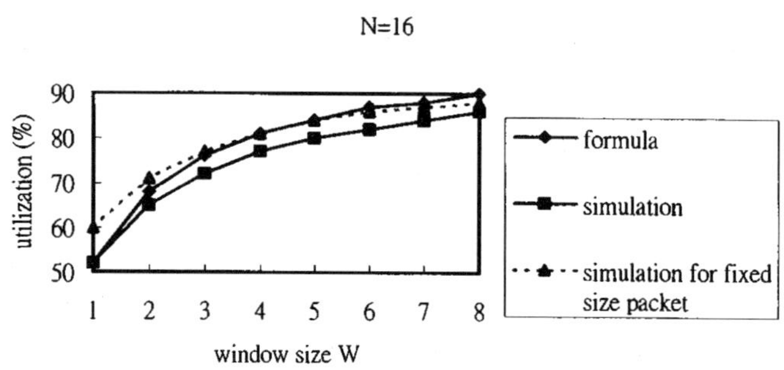

Fig. 6 The maximum utilization for $N = 16$ with uniform traffic.

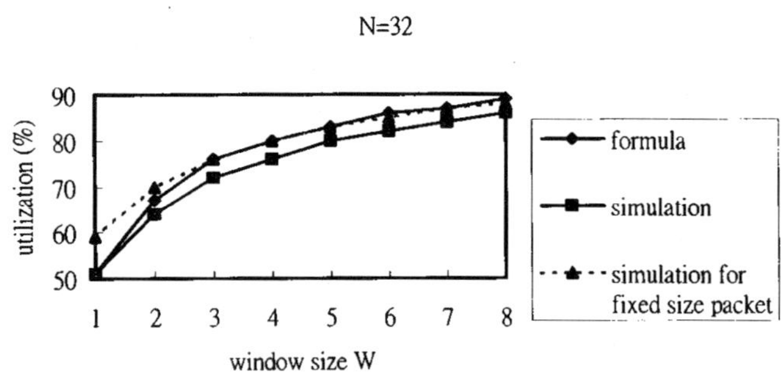

Fig. 7 The maximum utilization for $N = 32$ with uniform traffic.

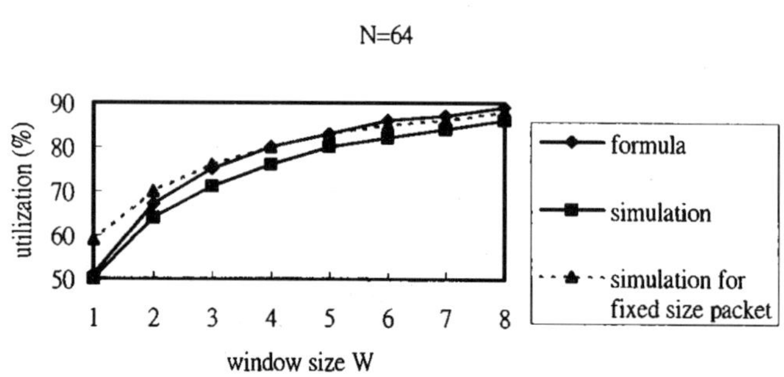

Fig. 8 The maximum utilization for $N = 64$ with uniform traffic.

tention (which causes output conflict) is involved. This is easily seen for small W, especially for $W = 1$. Observe that the formula for $W = 1$ is $\frac{N}{2N-1}$ which as N goes to infinity, converges to 0.5, while the simulation for fixed-size packets converges to 0.586, as they should. However, as W increases, this randomness impact is reduced (because the original output conflict at the HOL packet may now be resolved by one of the next $W - 1$ packets in line), and the performance discrepancy is thus also reduced. As shown in each figure, the errors are within 1% especially for large W. In fact, it has been shown in [4],[5] that the steady-state distributions of a discrete-time packet switch have an approximate product-form solution which has the same form as that of a closed queueing network model with exponential service times. This may also help explain the comparison results that are obtained here.

We also obtain some numerical results based on Buzen's algorithm (18) for nonuniform traffic patterns, and compare them with simulation. The packet switches with N equal to 4 and 8 are considered, where again for each N the window policy with window size W ranging from $W = 1$ to $W = 8$ is included. The results for $N = 4$ are shown in Fig. 9, where two types of input traffic patterns are considered. In both types, we divide the 4 output links into 2 groups with each group containing 2 output links. Each input packet selects the first (resp. second) output group as the destination output group with probability p (resp. $1 - p$), then within the selected group, the input group selects one of the output uniformly as its destination output. In the first type, we assume $p = 0.2$ (this corresponds to $p_1 = 0.1$, $p_2 = 0.1$, $p_3 = 0.4$ and $p_4 = 0.4$), and in the second type we assume $p = 0.4$ (this corresponds to $p_1 = 0.2$, $p_2 = 0.2$, $p_3 = 0.3$ and $p_4 = 0.3$). In other words, the first type of traffic pattern is more nonuniform and imbalanced. The comparison errors are all within 3-4%, and better fit can be observed with the more nonuniform traffic pattern (i.e. the first type). This can be expected, since in such case the queueing network model tends to give less overestimation. Also observe that the maximum utilization is smaller with the more nonuniform traffic pattern (the first type), as one would naturally expect.

The results for $N = 8$ for nonuniform traffic patterns are shown in Fig. 10, where again two types of input traffic patterns are considered. Here, in both types, we divide the 8 output links into 4 groups with each group containing 2 output links. Similar procedure as in the previous case (Fig. 9) is used for an input packet to select its output destination link. In the first (resp. second) type, the probabilities for selecting the four output groups are 0.1, 0.2, 0.3 and 0.4 (resp. 0.2, 0.2, 0.3 and 0.3). Thus, the first type of traffic pattern is more nonuniform and imbalanced. Again, the comparison errors are all within 3%, and similar observation and conclusion as in Fig. 9 are obtained.

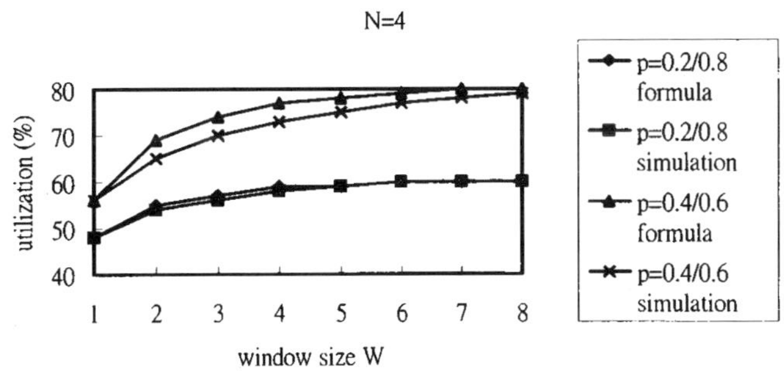

Fig. 9 The maximum utilization for $N = 4$ with nonuniform traffic.

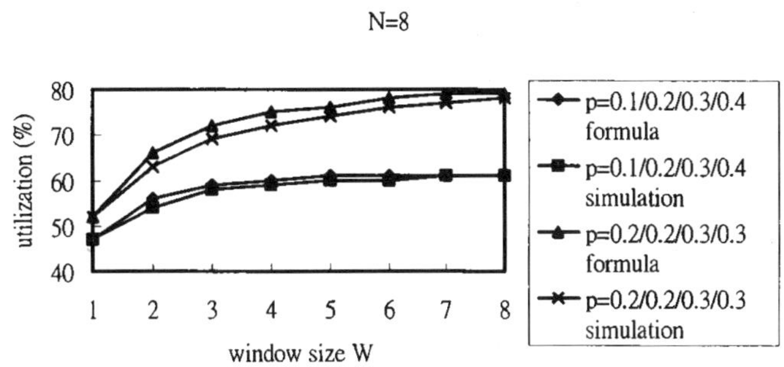

Fig. 10 The maximum utilization for $N = 8$ with nonuniform traffic.

5. Conclusion

We have studied the maximum throughput of a nonblocking packet switch with input queueing with window policy. We have shown that the hehavior of the packet switch with window policy can be approximately modeled as a closed queueing network. Based on the model, we have derived explicit closed-form formulae for the maximum throughput of the packet switch as a function of the window size. Both balanced and imbalanced traffic patterns have been discussed. The analytic results have been compared with simulation, showing good agreement. Our formulae can easily determine the effectiveness of the window policy.

References

[1] F. Baskett, K.M. Chandy, R.R. Muntz, and F.G. Palacios, "Open, closed, and mixed networks of queues with different classes of customers," Journal of ACM, vol.22, pp.284–260, 1975.

[2] R.G. Bubenik and J.S. Turner, "Performance of a broadcast packet switch," IEEE Trans. Commun., vol.37, pp.60–69, 1989.

[3] J.P. Buzen, "Computational algorithm for closed queueing networks with exponential servers," Communications of the ACM, vol.16, pp.527–531, 1973.

[4] X.-R. Cao, "The maximum throughput of a nonblocking space-division packet switch with correlated destinations," IEEE Trans. Commun., vol.43, pp.1898–1901, 1995.

[5] X.-R. Cao and D. Towsley, "A performance model for ATM switches with general packet length distributions," IEEE/ACM Trans. Networking, vol.3, pp.299–309, 1995.

[6] J.S.-C. Chen and T.E. Stern, "Throughput analysis, optimal buffer allocation, and traffic imbalance study of a generic nonblocking packet switch," IEEE J. Sel. Areas Commun., vol.9, pp.439–449, 1991.

[7] M.G. Hluchyj and M.J. Karol, "Queueing in high performance switching," IEEE J. Sel. Areas Commun., vol.6, pp.1587–1597, 1988.

[8] M.J. Karol, M.G. Hluchyj and S.P. Morgan, "Input versus output queueing on a space-division packet switch," IEEE Trans. Commun., vol.35, pp.1347–1356, 1987.

[9] G.Y. Lee and C.K.Un, "Design and performance analysis of an asynchronous banyan network switch with window policy," IEE Proceedings-Communications, vol.142, pp.54–60, 1995.

[10] S.-Q. Li, "Nonuniform traffic analysis on a nonblocking space-division packet switch," IEEE Trans. Commun., vol.38, pp.1085–1096, 1990.

[11] S.-Q. Li, "Performance of a nonblocking space-division packet switch with correlated input traffic," IEEE Trans. Commun., vol.40, pp.97–108, 1992.

[12] P. Newman, "A fast packet switch for the integrated services backbone network," IEEE J. Sel. Areas Commun., vol.6, pp.1468–1479, 1988.

[13] K.W. Sarkies, "The bypass queue in fast packet switching," IEEE Trans. Commun., vol.39, pp.766–774, 1991.

High-Performance Backbone Network Technology edited by Naoaki Yamanaka
ISBN: 0-8247-5321-6 DOI: 10.1081/(120025701)
Published 2004 by Marcel Dekker, Inc.

Performance Evaluation of AAL2 Switch Networks

Hiroshi Saito

SUMMARY A new asynchronous transfer mode adaptation layer (AAL), called AAL2, is being designed mainly for low-bit-rate voice traffic, and nodes that can assemble and disassemble AAL2 cells are being developed to make AAL2 usage efficient. This paper investigates the delay and performance of AAL2 nodes by an analytical method. Then, using the results, it analyzes a network using AAL2 nodes and shows the bandwidth reduction achieved by using AAL2 switching nodes as transit nodes.
key words: *AAL2, IMT-2000, network evaluation*

1. Introduction

The asynchronous transfer mode (ATM) protocol has become widely accepted for networks providing integrated multimedia services, including the voice and telephony over ATM (VTOA) service. With the rapid world-wide growth of mobile telecommunications using low-bit-rate voice codecs increasing the demand for efficient transfer of low-bit-rate voice traffic over ATM networks, third-generation mobile technologies (e.g., International Mobile Telecommunications 2000) are being designed to use ATM as their infrastructure.

ATM adaptation layer 1 (AAL1) is suitable for transferring 64-kbps voice traffic, but it causes an unacceptable delay in cell assembly when it is applied to low-bit-rate voice traffic. For example, using AAL1, a 48-byte cell payload is filled in 6 ms at 64 kbps, but it takes about 50 ms when an 8-kbps voice codec is used. This is too long for intelligible voice communication without the use of echo cancellation. Even with echo cancellation, the transmission delay and the reassembly delay at the receiver make telephone conversations uncomfortable for users.

To resolve this problem, the ITU-T and the ATM Forum started discussing and eventually reached agreement on the use of AAL2 for low-bit-rate voice traffic [1]. The issues are comprehensively summarized in Refs. [2] and [3].

Using AAL2 reduces the cell-assembly delay because it allows multiple voice channels to be carried in a single cell. This enables the payload of a cell to be filled quickly even when each channel has a low bit rate. Under AAL2, the payload of a single cell consists of several short packets, called common-part sublayer (CPS) packets, from different voice sources (Fig. 1). The first byte of the cell payload is the start field. This indicates the position of the header (CPS packet header) of the first short packet; this information is needed because the cell payload may contain the remainder of a CPS packet started in the preceding cell. The start field also shows the sequence number for this cell. A 3-byte CPS packet header is used for each CPS packet in the cell. It indicates the channel to which the CPS packet belongs and the packet's length.

The voice signals are sampled in a standardized manner (for example, [4]) and assembled into a compressed signal burst ranging in length from several bits to several hundred bits (e.g., 2 to 36 bytes for code-division multiple access [2]) in the encoder. Each burst becomes a CPS packet, and an appropriate CPS packet header is attached to it at the AAL2 source node. These CPS packets are accommodated in the cell payload on a first-come first-served (FCFS) basis. When the length of a CPS packet exceeds the remaining cell payload, the packet is divided into two parts. The first part fills the remaining cell payload, and the second part is accommodated in the next cell (Fig. 1). A timer is used to limit the assembly delay, as explained later.

At the AAL2 destination node, the channel that the CPS packet belongs to is identified from the CPS packet header indicated by the start field. The decoder of that channel decodes the compressed voice signals conveyed by the CPS packet.

Typically, the AAL2 source/destination nodes are

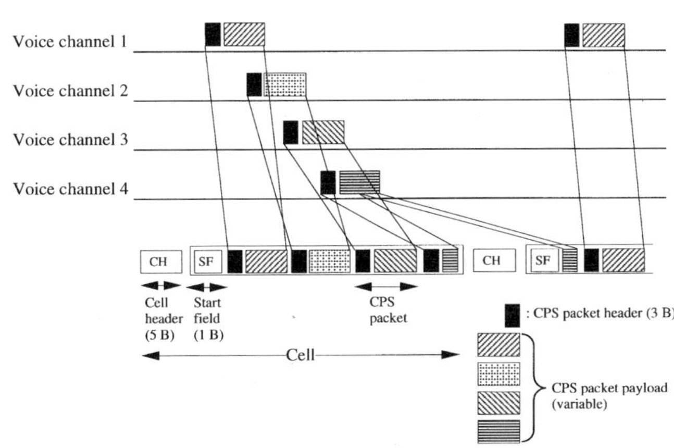

Fig. 1 Composition of AAL2 cell.

Manuscript received December 11, 1998.
†The author is with NTT Service Integration Laboratories, Musashino-shi, 180-8585 Japan.
*A part of this paper is presented at ICC'99.

IEICE TRANS. COMMUN., Vol. E82-B, No. 9, pp. 1411–1423, September 1999

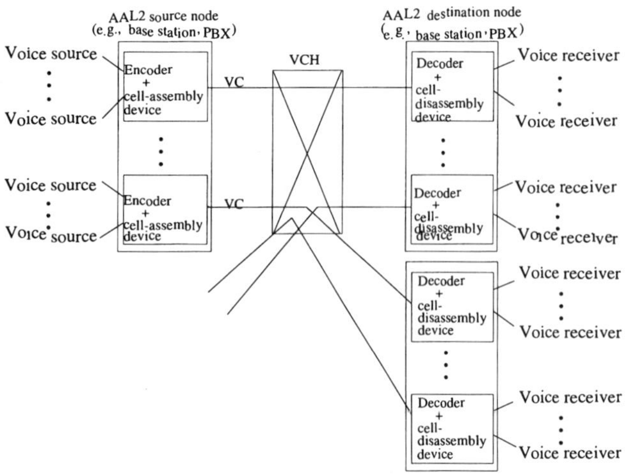

Fig. 2 Typical AAL2 network.

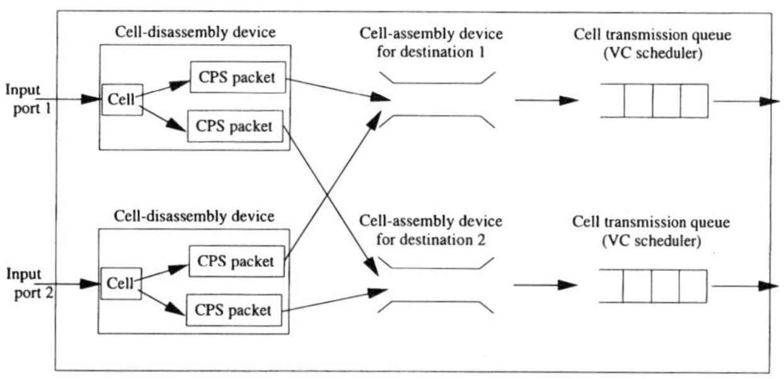

Fig. 3 AAL2 switching node.

base stations of a cellular network or PBXes on an ATM WAN, having ATM interfaces between them. Therefore, the cell-assembly device at the AAL2 source node normally accommodates several voice sources having the same destination node cell-disassembly device into a single VC, and virtual channel handlers (VCHs) in the network transit the VC (Fig. 2).

One problem with the AAL2 network in that a single VC can carry only traffic that has the same pair of the cell-assembly and cell-disassembly devices at the source and destination nodes, and cannot mix traffic that has different cell-assembly devices or different cell-disassembly devices into a single VC.

For example, however, each base station handles a fairly small amount of traffic and the traffic between the same cell assembly device at the source node and the same cell disassembly device at the destination node in each base station is too small to make a non-partially filled cell. To resolve this problem, an AAL2 switching node is being developed that will disassemble cells into CPS packets and reassemble them into ones having packets with the same receiver/destination (Fig. 3). That is, the AAL2 switching node will work as a switching node with cell-disassembly devices, cell-assembly devices, and a CPS packet-based switching function.

While this approach eliminates the problem of each source node having too little traffic to effectively use AAL2, it lengthens the cell-assembly delay in each switching node. In other words, there is a tradeoff between the cell-assembly delay and traffic aggregation.

The performance of AAL2 systems has been analyzed by using both computer simulation [5] and simple experiments [6], [7]. Several papers [2], [3], [32] used rough analyses and/or computer simulation to evaluate AAL2 performance, but their main objective was to investigate AAL2 design. The papers [8], [31] analyzed the effect of bit dropping by using a Poisson-arrival approximation, but their focus was not on the AAL2 switching node and its network. In [33], an end-to-end QoS is evaluated by an analytical model. There, the wireless portion as well as the land portion using AAL2 is considered and a very simple model was used. In [9] and [10], performance of AAL2 CLAD is analyzed. However, both of them focus on a single CLAD and do not investigate the network performance. The paper [30] investigated the bandwidth management methods, but the network evaluation was not the main objective.

The use of voice packets to improve communication began about 30 years ago [11], although original voice-packet communication did not have a layered cell structure. In the early stages, the voice-signal bursts were assumed to be transmitted using a technology called digital speech interpolation [12], mainly over time-division multiple access links. Performance was evaluated using such statistics as talkspurt length and silence-period length [13]–[17]. After that, there were studies on packet voice communication, and numerous papers on its performance evaluation were published [18]–[29]. Although these studies differed from AAL2 performance evaluation, some of their methods, for example their modeling of the voice-packet arrival process, are still significant.

2. Model

In the model used here (Fig. 3), the CPS packets arrive at a cell-assembly device at an AAL2 switching node or at an AAL2 source node in a Poisson process with a mean arrival rate of ρ. (An extension to a general distribution of the number of CPS packets arriving in a slot is straightforward.) A cell is assumed to be assembled as soon as enough packets to form a cell have arrived at the cell assembly device. For simplicity, the length L of the packets and the number of packets in a cell are assumed to be fixed. A timer is used to keep the cell-assembly delay below a certain level. If the timer expires, a partially filled cell is generated and sent. That is, the packets waiting in the cell assembly device are accommodated in the cell payload, and when the timer expires the remaining payload filled with padding.

After cells have been assembled in the cell-assembly device they immediately arrive at a transmission queue, where they await transmission in a buffer having a finite size. Let K be the maximum number

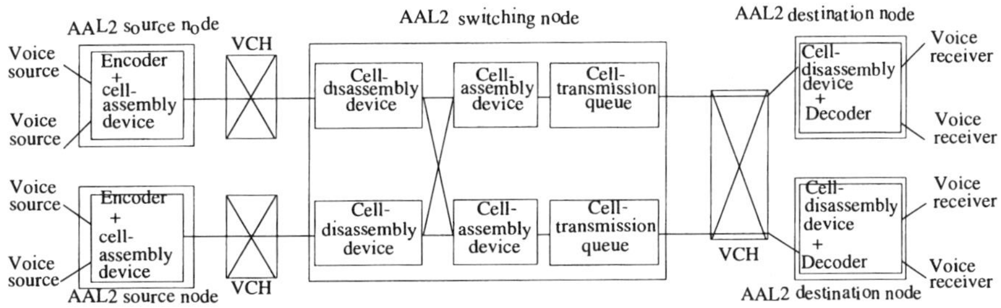

Fig. 4 Network with AAL2 CLADs, switching node, and VCHs.

of cells in the transmission queue. That is, K is the buffer size of the transmission queue plus the cell that is currently being transmitted. There are several ways of transmitting cells in the transmission queue. One approach is to transmit them at the fixed VC bandwidth on a FCFS basis; that is, a VC accommodating voice channels operates at a constant bit rate (CBR). (A CBR VC makes the VCHs simple because only the AAL2 nodes need to judge whether or not additional voice channels can be admitted.) Another approach is to accommodate the AAL2 channels in unspecified-bit-rate (UBR) VCs in a CBR virtual path [2], [10]. While this approach needs additional bandwidth management functions in the VCHs, it can achieve a statistical multiplexing gain without using AAL2 switching nodes. A third option is a variable-bit-rate VC, but this approach makes QoS (quality of service) management more complex. This paper assumes the first approach, that is, a CBR VC.

In the model used here (Fig. 4), a VC is established between an AAL2 source node and an AAL2 switching node, between two AAL2 switching nodes (not shown in the figure), or between an AAL2 switching node and an AAL2 destination node through VCHs. When there is no AAL2 switching node (AAL2 transit node) between an AAL2 source node and an AAL2 destination node, a direct VC is established between them (Fig. 2). At each AAL2 source or switching node, a cell assembly device is provided for each VC and CPS packets are assembled into a cell if they have the same VC destination.

In the model used here, the time axis is discrete, with each unit equal to the time it takes to transmit one cell over a VC (Fig. 5). At the beginning of a time slot, CPS packets arrive at the cell assembly device, and a cell is generated on an FCFS basis as soon as enough packets have arrived. Cells can also be generated due to a timeout (timer expiry) at the beginning of a time slot. The generated cells are placed in the transmission queue. A single cell is transmitted on an FCFS basis at the end of each slot. The timer for cell assembly starts at zero, then increments by one at the middle of the time slot. A partially filled cell is assumed to be generated at the beginning of a time slot if the timer is $T-1$ at the cell assembly (Fig. 5), where T is the cell-assembly timeout.

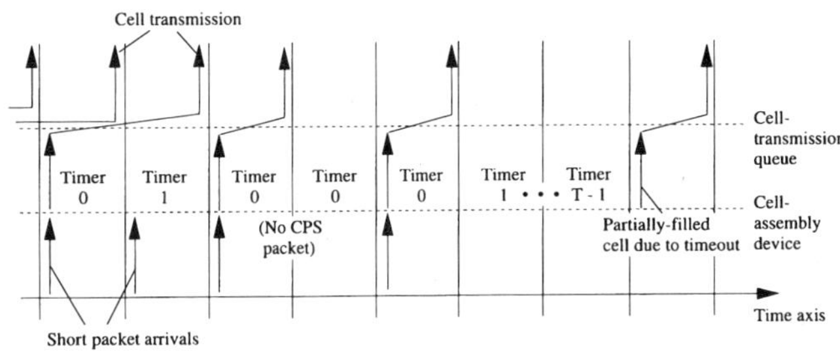

Fig. 5 Discrete time model.

3. Analysis

Reference [10] analyzed the AAL2 CLAD by focusing on the transition between two consecutive cell generation epochs. In the present paper, however, we consider the state of the system at each time slot. This approach enables us to derive the delay distribution in an AAL2 switching node and the cell delay variation, but it needs a large state space. Depending on the objectives, we choose one or other of the following two approaches.

Let D_i $(i = 0, 1, 2, \cdots)$ be the probability that i CPS packets arrive at a cell-assembly device in a slot. Since a Poisson arrival with mean rate ρ is assumed, we have $D_i = e^{-\rho}\rho^i/i!$ but the following analysis is valid for a general D_i.

Consider a state (t, x, y) at the boundary of the time slot where t $(t = 0, 1, \cdots, T-1)$ is a value of the timer, x $(x = 0, 1, ..., K-1)$ is the number of cells in a transmission queue, and y $(y = 0, 1, ..., L-1)$ is the number of CPS packets in the cell-assembly device.

Case 1: Assume that $y > 0$. If m, the number of CPS packets arriving in the slot, is less than $L-y$ (Case 1-i), then state (t, x, y) makes a transition to state $(t+1, [x-1]^+, y+m)$ when $t < T-1$ (Case 1-i-a) or to state $(0, x, 0)$ when $t = T-1$ (Case 1-i-b). This is because, in Case 1-i-a, CPS packets may arrive without a cell being generated so the timer proceeds, while in Case 1-i-b, the number of CPS packets is not enough to make a full cell, so a partially-filled cell is generated due to timeout and the timer resets. If $m \geq L-y$ (Case 1-ii), a cell is generated in the slot. Thus, state (t, x, y) makes a transition to state $(0, min(K, x+\lfloor(y+m)/L\rfloor)-1, y+m-L\lfloor(y+m)/L\rfloor)$. Here, $\lfloor z\rfloor$ is the largest integer smaller than or equal to z. Thus, $\lfloor(y+m)/L\rfloor$ denotes the number of cells assembled by $(y+m)$ CPS packets.

Case 2: Assume that $y=0$. Note $t=0$ if $y=0$. If $0 \le m < L-y$ (Case2-i), then no cell is generated in the slot. Thus, state $(t=0, x, y=0)$ makes a transition to state $(t=0, [x-1]^+, y+m)$. (Note that the timer does not proceed at this slot.) If $m \ge L-y$ (Case 2-ii), a cell is generated in the slot. Therefore, state $(t=0, x, y=0)$ makes a transition to state $(t=0, min(K, x+\lfloor (y+m)/L \rfloor)-1, y+m-L\lfloor (y+m)/L \rfloor)$.

3.1 Stationary Probability

Based on the above discussion, this subsection derives $\{p(t,x,y)\}$, the stationary probability of (t,x,y) at the boundary of a time slot. Set $\mathbf{p}^{(t,x)} = (p(t,x,0), p(t,x,1), \cdots, p(t,x,L-1))$, $\mathbf{p}^{(t)} = (\mathbf{p}^{(t,0)}, \mathbf{p}^{(t,1)}, \cdots, \mathbf{p}^{(t,K-1)})$.

Let us consider the transition of the timer. The timer counts up one by one and resets when a cell (either normal or partially-filled) is assembled. Thus, the state equations of the stationary probability vector $\{p(t,x,y)\}$ have the following structure due to the Markov property.

$$\mathbf{p}^{(0)} = \sum_{i=0}^{T-1} \mathbf{p}^{(i)} T_{i,0} \tag{1}$$

$$\mathbf{p}^{(i)} = \mathbf{p}^{(i-1)} T_{i-1,i} \qquad (i = 1, 2, \cdots, T-1). \tag{2}$$

Here, $T_{i,j}$ is a $(K*L, K*L)$ matrix denoting the transition from (i,x,y) to (j,x',y').

Consider the case in which the timer counts up. (This happens only in Case 1-i-a.) This transition is described by $T_{i-1,i}$ $(i = 1, 2, \cdots, T-1)$. In this case, x (the number of cells in the transmission queue) becomes $[x-1]^+$, and y $(0 < y < L)$ (the number of CPS packets in the cell-assembly device) becomes $y+m$ $(0 < y+m < L)$, where m is the number of CPS packets arriving in the slot. Thus, $T_{i-1,i}$ has the following structure.

$$T_{i-1,i} = \begin{pmatrix} X_- & 0 & 0 & \cdots & 0 & 0 \\ X_- & 0 & 0 & \cdots & 0 & 0 \\ 0 & X_- & 0 & \cdots & 0 & 0 \\ 0 & 0 & X_- & & 0 & 0 \\ \vdots & \vdots & \vdots & \ddots & \vdots & \vdots \\ 0 & 0 & 0 & & X_- & 0 \end{pmatrix}. \tag{3}$$

Here, X_- is an (L,L) matrix given by

$$X_- = \begin{pmatrix} 0 & 0 & 0 & \cdots & 0 \\ 0 & D_0 & D_1 & \cdots & D_{L-2} \\ 0 & 0 & D_0 & \cdots & D_{L-3} \\ 0 & 0 & 0 & \cdots & D_{L-4} \\ \vdots & \vdots & \vdots & & \vdots \\ 0 & 0 & 0 & \cdots & D_0 \end{pmatrix}. \tag{4}$$

Next, let us consider the other cases. Note that, in a slot, x cannot decrease by more than one. Thus, $T_{i,0}$ has the following structure.

$$T_{i,0} = \begin{pmatrix} X^{(i)}_{0,0} & X^{(i)}_{0,1} & X^{(i)}_{0,2} & \cdots & X^{(i)}_{0,K-2} & X^{(i)}_{0,K-1} \\ X^{(i)}_{1,0} & X^{(i)}_{1,1} & X^{(i)}_{1,2} & \cdots & X^{(i)}_{1,K-2} & X^{(i)}_{1,K-1} \\ 0 & X^{(i)}_{2,1} & X^{(i)}_{2,2} & \cdots & X^{(i)}_{2,K-2} & X^{(i)}_{2,K-1} \\ 0 & 0 & X^{(i)}_{3,2} & \cdots & X^{(i)}_{3,K-2} & X^{(i)}_{3,K-1} \\ \vdots & \vdots & & \ddots & \vdots & \vdots \\ 0 & 0 & 0 & 0 & X^{(i)}_{K-1,K-2} & X^{(i)}_{K-1,K-1} \end{pmatrix}. \tag{5}$$

$X^{(i)}_{j,k}$ is an (L,L) matrix denoting a transition from (i,j,y) to $(0,k,y')$.

Consider the cases of $t=0$ to determine $T_{t,0}$. That is, consider $T_{0,0}$. In these cases, the timer stays at 0 either i) because there were no CPS packets in the cell-assembly device in the previous time slot (Case 2), or ii) because a cell was assembled in the previous slot, so there is a CPS packet left and a cell is assembled in this slot (Case 1-ii, $t=0$). As discussed before, in Case 2, state $(t=0, x, y=0)$ makes a transition to state $(t=0, [x-1]^+, y+m=m)$ or state $(t=0, min(K, x+\lfloor (y+m)/L \rfloor)-1, y+m-L\lfloor (y+m)/L \rfloor)$ with probability D_m. The former transition is described by $X^{(0)}_{j,j-1}$ when $x>0$, where $x=j$ and by $X^{(0)}_{0,0}$ when $x=0$ and the latter transition is described by the $X^{(0)}_{j,j+k-1}$, where $x=j$ and $1 \le k = \lfloor (y+m)/L \rfloor < K-j$, or $X^{(0)}_{j,K-1}$ where $x=j$ and $K \le x+\lfloor (y+m)/L \rfloor$. In Case 1-ii with $t=0$, state $(t=0, x, y>0)$ becomes $(0, min(K, x+\lfloor (y+m)/L \rfloor)-1, y+m-L\lfloor (y+m)/L \rfloor)$ with probability D_m. This transition is also described by $X^{(0)}_{j,j+k-1}$ and $X^{(0)}_{j,K-1}$.

Therefore, for $1 \le j < K$,

$$X^{(0)}_{j,j-1} = \begin{pmatrix} D_0 & D_1 & \cdots & D_{L-1} \\ 0 & 0 & \cdots & 0 \\ \vdots & \vdots & & \vdots \\ 0 & 0 & \cdots & 0 \end{pmatrix} \tag{6}$$

and

$$X^{(0)}_{0,0} = \begin{pmatrix} D_0 + D_L & D_1 + D_{L+1} & \cdots & D_{L-1} + D_{2L-1} \\ D_{L-1} & D_L & \cdots & D_{2L-2} \\ D_{L-2} & D_{L-1} & \cdots & D_{2L-3} \\ \vdots & \vdots & & \vdots \\ D_1 & D_2 & \vdots & D_L \end{pmatrix}. \tag{7}$$

In addition, for $0 \le j < K-1$; $k \ge 1$; $j+k-1 < K-1$; except for $\{j=0, k=1\}$,

$$X^{(0)}_{j,j+k-1} = \begin{pmatrix} D_{kL} & D_{kL+1} & \cdots & D_{kL+L-1} \\ D_{kL-1} & D_{kL} & \cdots & D_{kL+L-2} \\ \vdots & \vdots & & \vdots \\ D_{kL-L+1} & D_{kL-L+2} & \cdots & D_{kL} \end{pmatrix} \quad (8)$$

and for $0 \le j < K$; $k \ge 1$ and $j + k - 1 = K - 1$,

$$X^{(0)}_{j,j+k-1} = \begin{pmatrix} \sum_{m=k}^{\infty} D_{mL} & \sum_{m=k}^{\infty} D_{mL+1} & \cdots & \sum_{m=k}^{\infty} D_{mL+L-1} \\ \sum_{m=k}^{\infty} D_{mL-1} & \sum_{m=k}^{\infty} D_{mL} & \cdots & \sum_{m=k}^{\infty} D_{mL+L-2} \\ \vdots & \vdots & & \vdots \\ \sum_{m=k}^{\infty} D_{mL-L+1} & \sum_{m=k}^{\infty} D_{mL-L+2} & \cdots & \sum_{m=k}^{\infty} D_{mL} \end{pmatrix}. \quad (9)$$

Next, let us consider the cases that are described by $T_{i,0}$ $(0 < i < T-1)$. Case 1-ii with $0 < t(= i) < T - 1$ corresponds to these cases. As discussed before, in Case 1-ii, the transition from $(t = i, x, y > 0)$ to $(0, min(K, x + \lfloor (y+m)/L \rfloor) - 1, y + m - L\lfloor (y+m)/L \rfloor)$ happens with probability D_m. Set $x = j$ and $k = \lfloor (y+m)/L \rfloor$. Therefore, for $1 \le k < K - j$; $0 \le j < K$; and $0 < i < T - 1$, we have

$$X^{(i)}_{j,j+k-1} = \begin{pmatrix} 0 & 0 & 0 & \cdots & 0 \\ D_{kL-1} & D_{kL} & D_{kL+1} & \cdots & D_{kL+L-2} \\ D_{kL-2} & D_{kL-1} & D_{kL} & \cdots & D_{kL+L-3} \\ \vdots & \vdots & \vdots & \ddots & \vdots \\ D_{kL-L+1} & D_{kL-L+2} & D_{kL-L+3} & \cdots & D_{kL} \end{pmatrix}, \quad (10)$$

and, for $j = 0, 1, 2, \cdots, K - 1$; $0 < i < T - 1$; and $j + k - 1 = K - 1$, we have

$$X^{(i)}_{j,j+k-1} = \begin{pmatrix} 0 & 0 & 0 & \cdots & 0 \\ \sum_{m=k}^{\infty} D_{mL-1} & \sum_{m=k}^{\infty} D_{mL} & \sum_{m=k}^{\infty} D_{mL+1} & \cdots & \sum_{m=k}^{\infty} D_{mL+L-2} \\ \sum_{m=k}^{\infty} D_{mL-2} & \sum_{m=k}^{\infty} D_{mL-1} & \sum_{m=k}^{\infty} D_{mL} & \cdots & \sum_{m=k}^{\infty} D_{mL+L-3} \\ \vdots & \vdots & \vdots & \ddots & \vdots \\ \sum_{m=k}^{\infty} D_{mL-L+1} & \sum_{m=k}^{\infty} D_{mL-L+2} & \sum_{m=k}^{\infty} D_{mL-L+3} & \cdots & \sum_{m=k}^{\infty} D_{mL} \end{pmatrix}. \quad (11)$$

The number of cells in the transmission queue does not decrease in these cases. Thus, for $0 < i < T - 1$; $1 \le j < K$,

$$X^{(i)}_{j,j-1} = 0. \quad (12)$$

Finally, consider the cases that are described by $T_{T-1,0}$. They are Cases 1-i-b and 1-ii with $t\,(= i) = T - 1$. Note that, in Case 1-i-b, the number of cells assembled in this slot is 1. Thus, for $k = \lfloor (y+m)/L \rfloor > 1$, Case 1-ii is concerned but Case 1-i-b has nothing do to with that. Case 1-ii, however, experiences no effect of timeout in this analysis because only normal cells are generated at timeout. Therefore, for $k = \lfloor (y+m)/L \rfloor > 1$; $0 < i < T-1$; $0 \le j < K-1$; $0 < j+k-1 \le K-1$, we have

$$X^{(T-1)}_{j,j+k-1} = X^{(i)}_{j,j+k-1}. \quad (13)$$

Furthermore, consider the case in which one cell (either a normal or partially-filled cell) is assembled. If it is a normal cell, the timeout has no effect. Thus, its transition probability from $(t = T - 1, x = j, y)$ to $(0, x = j, y + m - L)$ is given by $X^{(i)}_{j,j}$, where $0 < i\,(= t) < T-1$. If the cell is partially-filled, timeout does have an effect and state $(t = T - 1, x = j, y > 0)$ makes a transition to $(0, x = j, 0)$ with probability $\sum_{m=0}^{L-y-1} D_m$. Therefore, for $0 < i < T - 1$, $k = 1$ and $0 \le j < K$, we have

$$X^{(T-1)}_{j,j+k-1} = X^{(i)}_{j,j} + \begin{pmatrix} 0 & 0 & 0 & \cdots & 0 \\ \sum_{m=0}^{L-2} D_m & 0 & 0 & \cdots & 0 \\ \sum_{m=0}^{L-3} D_m & 0 & 0 & \cdots & 0 \\ \vdots & \vdots & \vdots & \ddots & \vdots \\ D_0 & 0 & 0 & \cdots & 0 \end{pmatrix}. \quad (14)$$

The number of cells in the transmission queue does not decrease in these cases. Thus, for $1 \le j < K$,

$$X^{(T-1)}_{j,j-1} = 0. \quad (15)$$

3.2 Delay Distribution

Using on the stationary probability $\{p(t,x,y)\}$, we can derive the delay distribution of CPS packets as follows. Consider a CPS packet. We call this CPS packet a tagged packet and call the cell including it a tagged cell. Note that d, the delay (sojourn time) of the tagged packet in the system is $x' + s$ when s is the time the tagged packet waits for cell assembly and x' is the number of cells in the transmission queue when the tagged cell arrives. x' includes the tagged cell.

Now we can derive the maximum delay. Note that $x' = x'' + \tilde{x}$. Here x'' is the number of cells that are in the transmission queue and arrived at least one unit of time before the tagged cell arrives at the transmission queue, and $\tilde{x}$ is the number of cells that arrive in the same time slot as the tagged cell but that arrive before it. $\tilde{x}$ includes the tagged cell. If the tagged packet waits for cell assembly at the cell-assembly device for at least one unit of time, then the tagged cell is the first cell to arrive at the transmission queue in a time slot. Thus, $\tilde{x} = 1$. While the tagged packet is waiting for cell assembly, no other cells arrive at the transmission queue. Thus, if the number of cells in the transmission queue is x just before the arrivals of CPS packets including the tagged packet at the cell-assembly device, a is the number of cells generated in the slot in which the tagged packet arrives, and the tagged packet waits for cell assembly for $s(>0)$ units of time, then $x' = x'' + 1 = [min(K, x+a) - s]^+ + 1$. Therefore, $d = 1 + s + [min(K, x+a) - s]^+ \le 1 + max(s, K) \le 1 + max(T, K)$ if $s > 0$. If $s = 0$, then $d = x' \le K$. Consequently, $1 + max(T, K)$ is the maximum delay.

Let us start studying the delay distribution. Assume that m CPS packets arrive in the same time slot in which the tagged packet arrives, that the tagged packet is the l-th packet among them, and that the state of the system is (t, x, y) at the boundary of the time slot just before their arrival. (Here, $m \ge 1, l \le m$.) In the following, let us consider $p_l(s, x'|m, t, x, y)$, the l-th CPS packet's conditional probability of (s, x'), given m and (t, x, y). If a tagged cell is lost due to buffer overflow, the delay of the tagged packet is not counted. For simple notation, we set $p_l(s, x'|m, t, x, y) = 0$ when the tagged cell is lost.

If $t < T - 1$ and $0 < y + m < L$, the tagged packet waits for cell assembly for at least one unit of time and keeps waiting for s units of time before enough CPS packets to make a cell have arrived. When s has passed, either enough packets to make a cell have arrived or the timer expires. The first cell that is assembled after the tagged packet arrives is the tagged cell and this cell is never lost due to buffer overflow. Note that the one-step transition of the number of CPS packets in the cell-assembly device during this period is given by the matrix

$$R(y+m) = \begin{pmatrix} D_0 & D_1 & \cdots & D_{L-1-y-m} \\ 0 & D_0 & \cdots & D_{L-2-y-m} \\ & & & \\ 0 & 0 & \ddots & \vdots \\ 0 & 0 & \cdots & D_0 \end{pmatrix}.$$

Thus, for $l \le m$; $t < T - 1$ and $0 < y + m < L$,

$$p_l(s, x'|m, t, x, y)$$
$$= \begin{cases} \sum_{i=L-y-m}^{\infty} D_i, & \text{for } s = 1, x' = [x-1]^+ + 1, \\ Q(y+m)R(y+m)^{s-2} \begin{pmatrix} \sum_{i=L-y-m}^{\infty} D_i \\ \sum_{i=L-y-m-1}^{\infty} D_i \\ \vdots \end{pmatrix} & \\ \text{for } 1 < s < T - t - 1, x' = [x-s]^+ + 1 & \\ Q(y+m)R(y+m)^{s-2} \begin{pmatrix} 1 \\ 1 \\ \vdots \end{pmatrix} & \\ \text{for } s = T - t - 1, x' = [x-s]^+ + 1, & \\ 0 & \text{otherwise} \end{cases} \tag{16}$$

where $Q(y+m) = (D_0 \quad D_1 \quad \cdots \quad D_{L-1-y-m})$.

If $t < T - 1$ and $y + m \ge L$, then $\lfloor (y+m)/L \rfloor$ cells are assembled in the slot in which the tagged packet arrives, and the tagged packet may be included in any of them. However, the tagged packet may be left in the cell-assembly device if it is not included in the first $(L\lfloor (y+m)/L \rfloor - y)$ CPS packets arriving in the slot. If $(K - x)L - y < l \le L\lfloor (y+m)/L \rfloor - y$, the tagged cell is assembled but lost due to overflow. Thus, if $y + m \ge L$, $l \le min(L\lfloor (y+m)/L \rfloor - y, (K-x)L)$, the tagged cell is assembled in the time slot that the tagged packet arrives, and it is not lost. Therefore, we have

$$p_l(s, x'|m, t, x, y)$$
$$= \begin{cases} 1 \text{ for } s = 0, x' = x + 1, 1 \le l \le L - y \\ 1 \text{ for } s = 0, x' = x + k, (k-1)L - y < l \le kL - y, \end{cases} \tag{17}$$

where k is the smallest integer larger than or equal to $(y + l)/L$.

When the tagged packet is left in the cell-assembly device, the number of CPS packets in the cell-assembly device becomes $y' = y + m - L\lfloor (y+m)/L \rfloor$ and the number of cells in the system becomes $x''' = x + \lfloor (y + m)/L \rfloor$. In this case, the tagged packet experiences the same delay as the tagged packet arriving at $(t, x''', y' - 1)$ and $m' = 1$, where m' is the number of packets arriving in the same slot as the tagged packet. This delay was already derived above because $y' - 1 + m' = y' < L$. So, for $l > L\lfloor (y+m)/L \rfloor - y$ and $t < T - 1$,

$$p_l(s, x'|m, t, x, y) = \begin{cases} \sum_{i=L-y'}^{\infty} D_i \text{ for } s = 1, x' = [x''' - 1]^+ + 1 \\ Q(y')R(y')^{s-2} \begin{pmatrix} \sum_{i=L-y'}^{\infty} D_i \\ \sum_{i=L-y'-1}^{\infty} D_i \\ \vdots \end{pmatrix} \\ \text{for } 1 < s < T - t, x' = [x''' - s]^+ + 1, \\ Q(y')R(y')^{s-2} \begin{pmatrix} 1 \\ 1 \\ \vdots \end{pmatrix} \\ \text{for } s = T - t, x' = [x''' - s]^+ + 1, \\ 0 \qquad \text{otherwise.} \end{cases} \tag{18}$$

In addition, if $t = T - 1$ and $y + m \geq L$, then timeout actually has no effect and normal cells are generated. Thus, $p_l(s, x'|m, t, x, y)$ with $t = T - 1$ and $y+m \geq L$ is identical to $p_l(s, x'|m, t, x, y)$ with $t < T-1$ and $y + m \geq L$.

On the other hand, if $t = T - 1$ and $y = m < L$, then timeout actually does have an effect. The tagged cell is a partially-filled cell, so, for $l \leq m$,

$$p_l(s = 0, x' = x + 1|m, t, x, y) = 1. \tag{19}$$

We are now in a position to derive the delay distribution $\Pr(d)$.

$$\Pr(d = s + x') = \sum_{s+x'=d} \sum_{t,x,y} \sum_{m=1}^{\infty} \sum_{l=1}^{m} p(t, x, y) D_m p_l(s, x'|m, t, x, y) \Big/ \left(\sum_m m D_m (1 - PLR) \right), \tag{20}$$

where PLR is the CPS packet loss ratio derived in the following subsection.

3.3 Cell Loss Ratio and CPS Packet Loss Ratio

The cell loss ratio (CLR) and CPS packet loss ratio (PLR) can be derived from the stationary probability $p(t, x, y)$. Note that if the system state is (t, x, y) just before m CPS packets arrive, then $\lfloor (y + m)/L \rfloor$ cells are generated if $t < T - 1$, and $max(1, \lfloor (y + m)/L \rfloor)$ cells are generated if $t = T - 1$.

$$CLR = \left\{ \sum_{t<T-1,x,y} \sum_m [x + \lfloor (y + m)/L \rfloor - K]^+ D_m p(t, x, y) + \sum_{t=T-1,x,y} \sum_m [x + max(1, \lfloor (y + m)/L \rfloor) - K]^+ D_m p(t, x, y) \right\} \Big/ \left\{ \sum_{t<T-1,x,y} \sum_m \lfloor (y + m)/L \rfloor D_m p(t, x, y) + \sum_{t=T-1,x,y} \sum_m max(1, \lfloor (y + m)/L \rfloor) D_m p(t, x, y) \right\}. \tag{21}$$

Note that partially-filled cells are not lost. Thus, one cell loss always means L CPS packet losses. Thus,

$$PLR = \left(\sum_{t<T-1,x,y} \sum_m L[x + \lfloor (y + m)/L \rfloor - K]^+ D_m p(t, x, y) \right. \tag{22}$$

$$\left. + \sum_{t=T-1,x,y} \sum_m L[x + max(1, \lfloor (y+m)/L \rfloor) - K]^+ D_m p(t, x, y) \right) / \rho. \tag{23}$$

4. Numerical Examples: Single Node

This section investigates the performance of AAL2 in a single node. It considers the sensitivity of cell delay distribution and cell loss ratio for various parameters such as load, buffer size, and timeout.

4.1 Load vs. Performance

Figure 6 shows the changes in performance when the offered load changes. As the load increases, the performance (delay and cell loss ratio) deteriorates monotonically. The delay distribution has an nearly-exponential tail but its slope changes near the maximum delay, which is equal to the buffer size when the buffer size is larger than the timeout. A difference in offered load implies a difference in the slope of the delay distribution.

4.2 VC Bandwidth vs. Performance

Figure 7 shows the changes in performance when the VC bandwidth changes (384, 512 and 768 kbps) and the offered load is proportional to the bandwidth. Both delay and loss performance improve as the VC bandwidth increases. This is known as statistical multiplexing gain. Figure 8 shows the delay performance when the VC bandwidth changes and the offered load is determined such that the cell loss ratio is 10^{-4}. When the VC bandwidth is small, the slope of the delay distribution is steep when the buffer size is larger than the timeout (Fig. 8(a)). Thus, for a fixed cell loss ratio and a larger buffer size than the timeout, the delay performance is worse for a large VC than for a small one. When the buffer size is smaller than the timeout

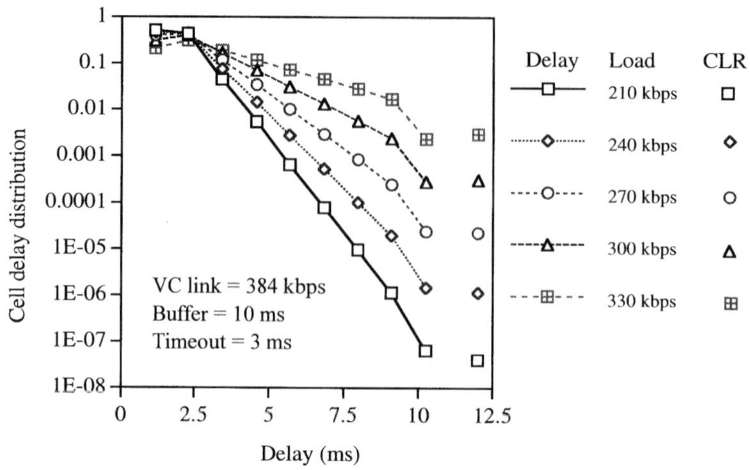

Fig. 6 Load vs. delay and CLR.

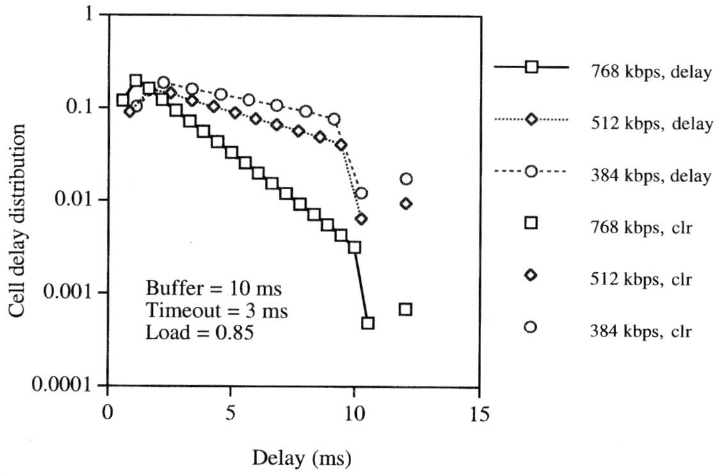

Fig. 7 VC bandwidth vs. delay and CLR.

(Fig. 8(b)), the slope of the delay distribution becomes steeper as the VC bandwidth increases. Thus, the delay performance is worse for a small VC than for a large one. This is because the transmission queue is dominant for both delay and loss when the buffer size is larger than the timeout, but the cell-assembly device is the dominant delay factor when the buffer size is smaller than the timeout.

4.3 Buffer Size and Timeout vs. Performance

Figure 9 shows the changes in performance when the buffer size changes. The result when the buffer size (3, 10 ms) is equal to or less than the timeout (10 ms) is shown in Fig. 9 (a) and when it (5, 10 ms) is larger than or equal to the timeout (5 ms) is shown in Fig. 9(b). Figure 10 shows them when the timeout changes (3, 10 ms) in the case of both a heavy and light load. From these figures, we see that cell loss ratio is very sensitive to the buffer size, whereas the delay is insensitive to it but sensitive to the timeout when traffic is light. The cell loss ratio is insensitive to the timeout regardless of the load. When the load is heavy, cells are assembled quickly, so timeout has no influence on the delay.

5. Numerical Examples: Network

5.1 Network Model

In this subsection, we analyze the network model shown in Fig. 11. Assume that each AAL2 source/destination node is accommodated in an AAL2 switching node (transit node) and that AAL2 switching nodes are fully connected with each other. The network is completely symmetric. The distance between AAL2 switching nodes or between an AAL2 switching node and an AAL2 source/destination node is assumed to be 1000 km. Assume that traffic is also symmetric. Voice traffic generated in an AAL2 source node is transferred uniformly to all other AAL2 destination nodes under

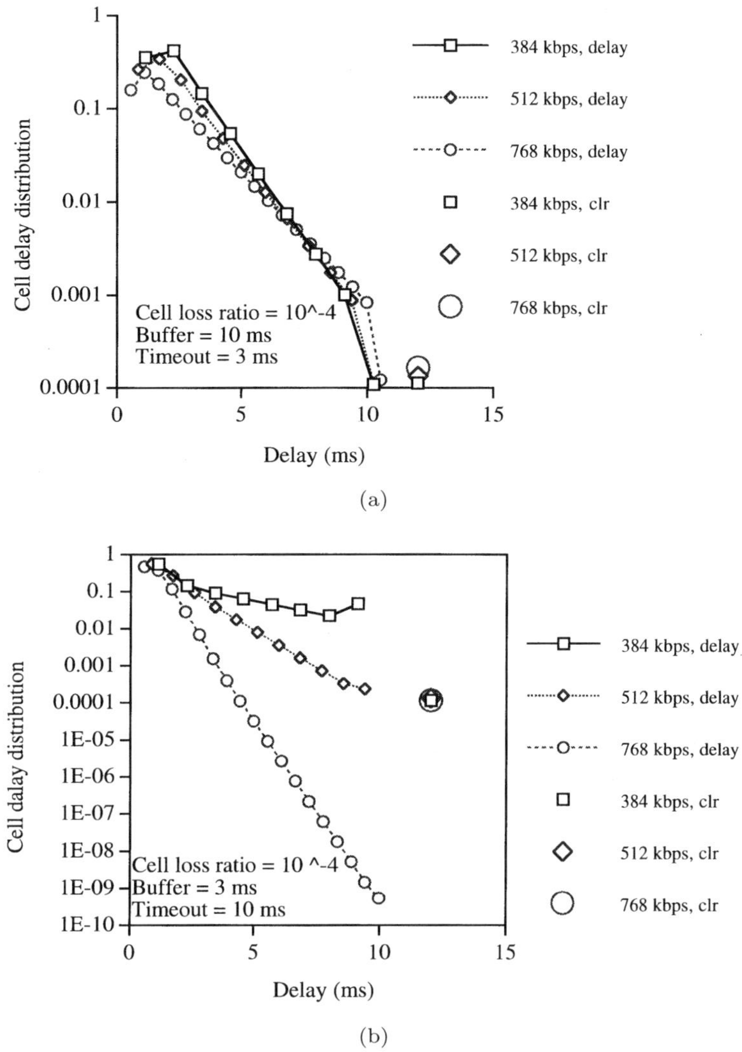

Fig. 8 Delay distribution with fixed CLR.

all AAL2 switching nodes, except for the AAL2 nodes under its own AAL2 switching node. Thus, every cell carried between an AAL2 source node and an AAL2 destination node passes through two AAL2 switching nodes and travels 3000 km.

This subsection assumes that timeout is fixed at 3 ms and buffer space for one cell is provided for every 32 kbps of offered voice traffic in each AAL2 node. The VC link bandwidth is dimensioned to satisfy the cell loss ratio target of 10^{-4} at each AAL2 node. When the offered voice traffic exceeds 780 kbps, multiple VCs are assumed to be used because channel identifier (CID) in the CPS packet header is only 8 bits long. In this case, the offered traffic is equally divided among the VCs.

Let N_1 be the number of AAL2 source/destination nodes under an AAL2 switching node. Let N_2 be the number of AAL2 switching nodes connected to an AAL2 switching node. Thus, N_2+1 is the total number of AAL2 switching nodes. Let ρ' be the total offered voice traffic at each AAL2 source node. The offered load ρ to a single VC between an AAL2 source node and an AAL2 switching node is given by ρ'/(the number of VCs between these two nodes). $N_1\rho'/N_2$ is the offered voice traffic between each pair of AAL2 switching nodes, and ρ' is the offered load between an AAL2 switching node and an AAL2 destination node. Between each pair of AAL2 source and destination nodes, the offered traffic is $\rho'/(N_1N_2)$.

Under the assumption that the delays at nodes are independent of each other, we evaluate the end-to-end delay, where end-to-end denotes the total delay including cell assembly delays, propagation delays, and delays for waiting for transmission, but excluding the delay for cell disassembly at the AAL2 destination nodes.

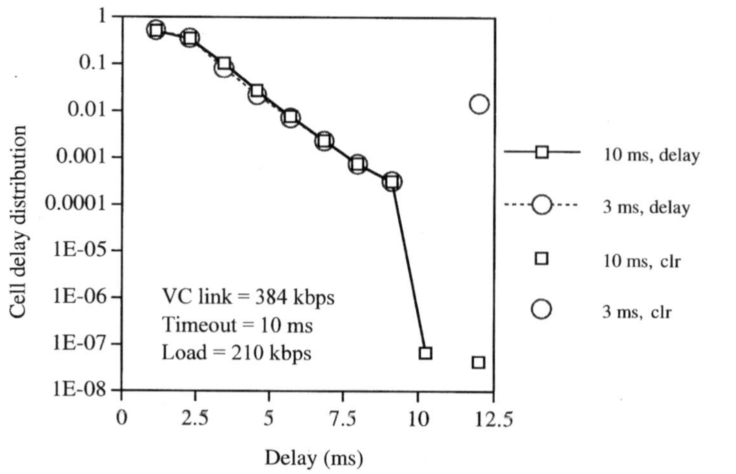

(a)

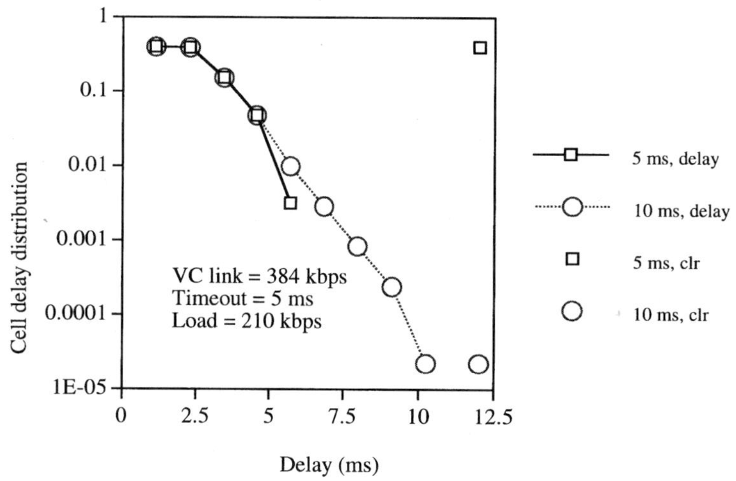

(b)

Fig. 9 Buffer size vs. delay and CLR.

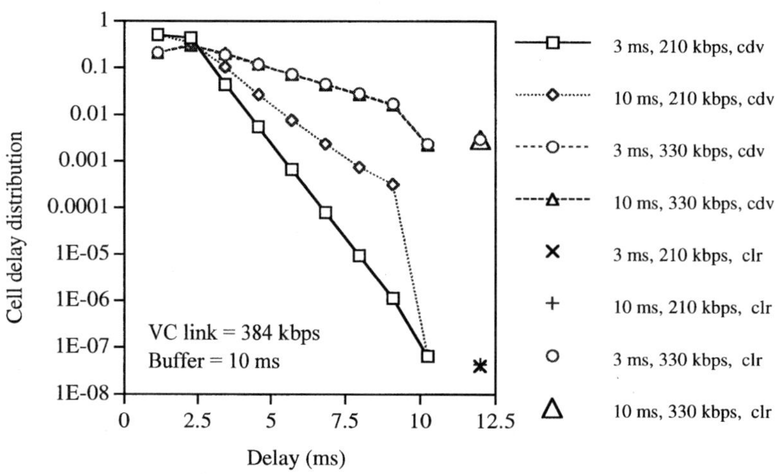

Fig. 10 Timeout vs. delay and CLR.

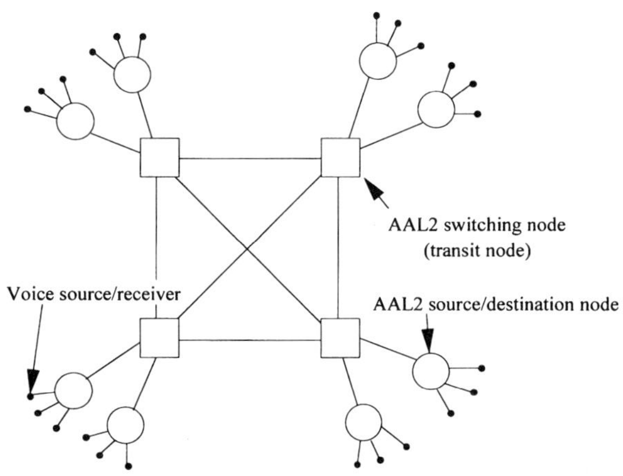

Fig. 11 Network model for numerical examples.

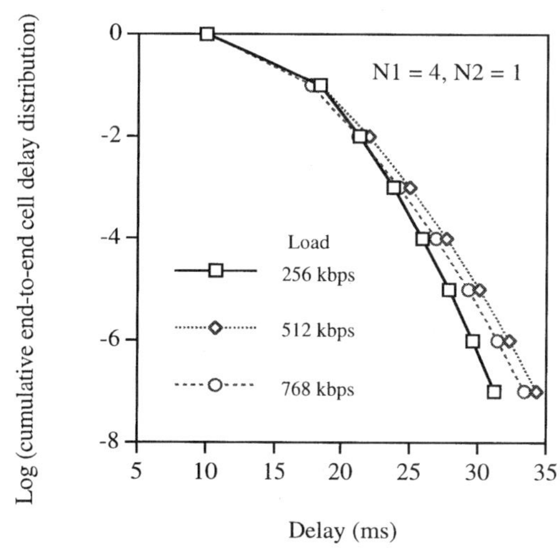

Fig. 12 Load vs. end-to-end delay.

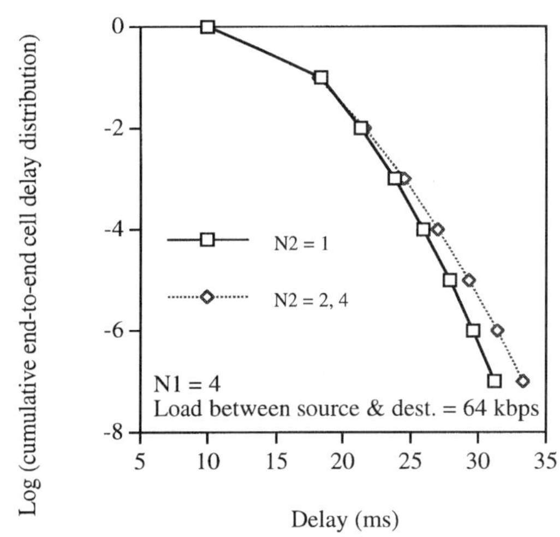

Fig. 13 Number of AAL2 switching nodes vs. end-to-end delay.

5.2 End-to-End Delay Distribution

Figures 12 and 13 show the end-to-end delay distribution for the network shown in Fig. 11. Figure 12 shows the results when the load between an AAL2 source/destination node and an AAL2 switching node is 256 kbps, 512 kbps, or 768 kbps. The difference is not so large. When the number of AAL2 switching nodes changes, the difference is also small (Fig. 13). This is because the load between consecutive AAL2 nodes is fairly large even when the load between each pair of source and destination nodes is small. Moreover, due to the limit of CID, the statistical gain achieved by each VC saturates at a fairly small load. Therefore, the performance of the network is fairly insensitive to the network load and the number of nodes under the buffer and bandwidth dimensioning method used here.

In this example, the minimum delay was fixed to about 10 ms and the delay variation of 15 ms (equivalent to an end-to-end delay of about 25 ms) is about the 10^{-4}-quantile. Thus, if the receiver provides a 15-ms buffer, the maximum cell loss ratio at the receiver is 10^{-4}, and the total end-to-end delay including the delay in the receiver is about 40 ms.

5.2.1 Bandwidth

Assuming that the dimensioning method mentioned in this subsection is used, the bandwidth requested to carry the traffic given in Table 1 is evaluated and plotted in Figs. 14 and 15. For comparison, a network that replaces each AAL2 switching node with a VCH in Fig. 11 is considered. In this network, each VC is directly connected between an AAL2 source node and an AAL2 destination node through VCHs, and it carries traffic that has the AAL2 source node as the origin and the AAL2 destination node as the destination. Thus the traffic that each VC can carry is limited and small.

As expected, using AAL2 switching nodes reduces the bandwidth. This is because each VC carries traffic having the same source node and the same transit node, the same transit nodes, or the same transit node and the same destination node, so more traffic can be carried than when VCHs are used.

Table 1 Conditions of numerical examples.

Case	N1	N2	Offered load (kbps) source/destination-transit	transit-transit	end to end
1	4	1	256	1024	64
2	4	1	512	2048	128
3	4	1	768	3072	192
4	4	2	512	1024	64
5	4	4	1024	1024	64
6	8	1	512	4096	64

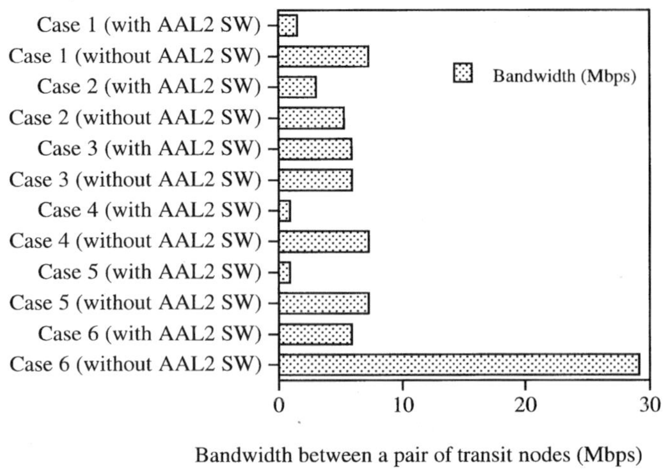

Fig. 14 Bandwidth between transit nodes.

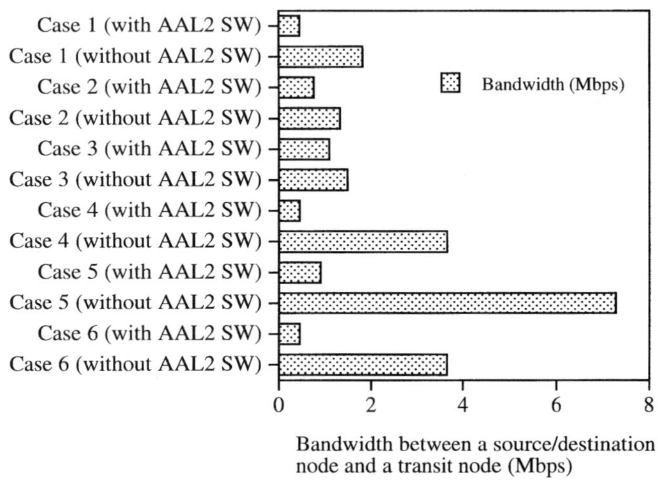

Fig. 15 Bandwidth between a source/destination node and a transit node.

As the end-to-end offered load increases, the bandwidth reduction achieved by AAL2 switching nodes decreases (Cases 1–3). For a fixed end-to-end offered load, as the number ($N_2 + 1$) of AAL2 switching nodes (transit nodes) increases, the reduction in bandwidth between a source (or destination) and a transit node increases (Cases 1, 4, and 5). This is the statistical gain between a source (or a destination) and a transit node due to traffic aggregation. For a fixed end-to-end offered load, as the number N_1 of AAL2 source/destination nodes increases, the reduction in bandwidth between a source (or a destination) and a transit node increases (Cases 1 and 6). Once again, this is the statistical gain due to traffic aggregation. On the other hand, the bandwidth reduction for the link between transit nodes is fairly insensitive to the number of source/destination nodes. This is because, even for a small number of source/destination nodes, the traffic between the transit nodes is large and the statistical gain due to the traffic aggregation saturates for the link between transit nodes for a fairly small number of source/destination nodes.

6. Conclusions

This paper investigated the delay and loss performance of AAL2 nodes by an analytical method. The stationary probability of the system state (the timer, number of cells in the transmission queue, and number of CPS packets in the cell-disassembly device) was derived. This was then used to derive the delay distribution and cell/packet loss probabilities. Numerical results showed that buffer size is critical to loss performance and timeout is critical to delay performance. The performance of AAL2 networks was also investigated and a large reduction in bandwidth in the network was shown to be achieved.

References

[1] ITU-T Recommendation I.363.2, "B-ISDN ATM Adaptation Layer Type 2 Specification," Toronto, 1997.

[2] J.H. Baldwin, B.H. Bharucha, B.T. Doshi, S. Dravida, and S. Nanda, "AAL-2–A new ATM adaptation layer for small packet encapsulation and multiplexing," Bell Labs Technical Journal, vol.2, no.2, pp.111–131, 1997.

[3] J.H. Baldwin, B.H. Bharucha, B.T. Doshi, S. Dravida, and S. Nanda, "AAL-CU: An ATM adaptation layer for small packet encapsulation," The 15th International Teletraffic Congress, pp.1445–1456, 1997.

[4] ITU-T Recommendation G. 729, "Conjugate Structure-Algebraic Code-Excited Linear Prediction," 1996.

[5] K. Nagata and H. Fujiya, "Technologies of AAL-2 for low-bit-rate voice communications," NTT R&D, vol.46, no.11, pp.1181–1188, 1997.

[6] H. Nakamura, H. Tsuboya, M. Nakano, and A. Nakajima, "Applying ATM to mobile infrastructure networks," IEEE Commun. Mag., vol.36, no.1, pp.66–73, 1998.

[7] B. Subbiah and S. Dixit, "ATM adaptation layer 2 (AAL2) for low bit rate speech and data: Issues and challenges," 1998 IEEE ATM Workshop, pp.225–233, 1998.

[8] K. Sriram and Y.T. Wang, "Voice over ATM using AAL2 and bit dropping: Performance and call admission," IEEE ATM Workshop, pp.215–224, 1998.

[9] N. Gerlich and M. Menth, "The performance of AAL-2 carrying CDMA voice traffic," 11th ITC Specialist Seminar, Yokohama, pp.17–24, 1998.

[10] H. Saito, "Performance evaluaton and dimensioning for AAL2 CLAD," INFOCOM'99, New York, 1999.

[11] H. Saito, Teletraffic Technologies in ATM Networks, Artech House, Boston, 1994.

[12] S.J. Campanella, "Digital speech interpolation," COMSAT Tech. Rev., vol.6, no.1, pp.128–157, 1976.

[13] P.T. Brady, "A statistical analysis of on-off patterns in 16 conversations," Bell Syst. Tech. J., vol.47, pp.73–91, 1968.

[14] E. Lyghounis, I. Poretti, and G. Monti, "Speech interpolation in digital transmission systems," IEEE Trans. Commun., vol.22, no.9, pp.1179–1189, 1974.

[15] Y. Yatsuzuka, "Highly sensitive speech detector and high-speed voice band data discriminator in DSI-ADPCM systems," IEEE Trans. Commun., vol.COM-30, no.4, pp.739–750, 1982.

[16] J.G. Gruber, "A comparison of measured and calculated speech temporal parameters relevant to speech activity detection," IEEE Trans. Commun., vol.COM-30, no.4, pp.728–738, 1982.

[17] H. Lee and C. Un, "A study of an on-off characterisation of conversational speech," IEEE Trans. Commun., vol.34, no.6, pp.630–637, 1986.

[18] T.E. Stern, "A queueing analysis of packet voice," GLOBECOM83, pp.2.5.1–6, 1983.

[19] T.E. Stern, "Analytical and computational techniques for packet voice system performance evaluation," Proc. Conf. Inf. Sci. Syst. 1984, pp.415–419, 1984.

[20] Y.C. Jenq, "Approximations for packetized voice traffic," INFOCOM84, pp.256–259, 1984.

[21] J. Daigle and J. Langford, "Queueing analysis of a packet voice communications system," INFOCOM85, pp.18–26, 1985.

[22] J. Daigle and J. Langford, "Models for analysis packet voice communications system," IEEE J. Sel. Areas Commun., vol.4, no.6, pp.847–855, 1986.

[23] K. Sriram and W. Whitt, "Characterizing superposition arrival processes in packet multiplexers for voice and data," IEEE J. Sel. Areas Commun., vol.1, no.6, pp.1124–1132, 1983.

[24] H. Heffes and D.M. Lucantoni, "A Markov modulated characterization of packetized voice and data traffic and related statistical multiplexer performance," IEEE J. Sel. Areas Commun., vol.4, no.6, pp.856–868, 1986.

[25] I. Ide, "Superposition of interrupted Poison process and its application to packetized voice multiplexer," ITC12, 3.1B2, 1988.

[26] S. Li, "Study of packet loss in a packet switched voice system," ICC88, pp.1519–1526, 1988.

[27] R. Nagarajan, J.F. Kurose, and D. Towsley, "Approximation techniques for computing packet loss in finite-buffered voice multiplexers," IEEE J. Sel. Areas Commun., vol.9, no.3, pp.368–377, 1991.

[28] J. Seguel, Y. Tanaka, and M. Akiyama, "Simulation analysis of the waiting time distribution of a packetized voice concentrator," IECE Trans., vol.J65, no.2, pp.115–122, Feb. 1982.

[29] K. Sriram, R.S. Mckinney, and M.H. Sherif, "Voice packetization and compression in broadband ATM networks," IEEE J. Sel. Areas Commun. Networks, vol.9, no.3, pp.294–304, 1991.

[30] H. Saito, "Bandwidth Management in AAL2 Networks," the 16th International Teletraffic Congress, Edinburgh, 1999.

[31] K. Sriram and Y.T. Wang, "Voice over ATM Using AAL2 and bit dropping: Performance and call admission control," IEEE JSAC, vol.17, no.1, pp.18–28, 1999.

[32] K. Sriram, T.G. Lyons, and Y.T. Wang, "Anomalies due to delay and loss in AAL2 packet voice systems: Performance models and methods of mitigation," IEEE JSAC, vol.17, no.1, pp.4–17, 1999.

[33] A.G. Valko, A. Racz, and G. Fodor, "Voice QoS in third-generation mobile systems," IEEE JSAC, vol.17, no.1, pp.109–123, 1999.

High-Performance Backbone Network Technology edited by Naoaki Yamanaka
ISBN: 0-8247-5321-6 DOI: 10.1081/(120025702)
Published 2004 by Marcel Dekker, Inc.

Frame-Based Worst-Case Weighted Fair Queueing with Jitter Control

Yeali S. Sun, Yung-Cheng Tu, and Wei-Kuan Shih

SUMMARY In the past, a number of scheduling algorithms that approximate GPS, such as WFQ, have been proposed and have received much attention. This class of algorithms provides per-flow QoS guarantees in terms of the bounded delay and minimum bandwidth guarantee. However, with $O(\log N)$ computational cost for each new arrival scheduling, where N is the number of backlogged flows, these algorithms are expensive to implement (e.g., in terms of scalability). Moreover, none of them addresses the issues of delay distribution and jitter. In this paper, we propose a new traffic scheduling discipline called *Jitter Control Frame-based Queueing* (JCFQ) that provides an upper bound for delay jitter in the case of rate-controlled connections, such as packet video streams and IP telephony, while guaranteeing bounded delay and worst-case fair weighted fairness, such as in the WF^2Q algorithm, but with $O(1)$ complexity in selecting the next packet to serve, assuming that the number of flows is fixed. Three different algorithms for slot or service order assignment between flows are proposed: Earliest Jitter Deadline First (EJDF), Rate Monotonic (RM) and Maximum Jitter First (MJF). In these algorithms, delay jitter is formulated into the virtual finish time calculation. We compare the fairness, delay and jitter performance of the JCFQ with that of the MJF algorithm with WF^2Q via simulation. The results show that with proper choice of the slot size, JCFQ can achieve better flow isolation in delay distribution than can WF^2Q.

key words: *jitter control, weighted fair queueing, quality of service, frame-based*

1. Introduction

As the Internet has grown in terms of the amount of traffic and the number of users, the amount of interest in providing multimedia communication services, including IP telephony, video teleconferencing, WebTV and other streaming applications, over the Internet has increased. Many of these new applications rely on the ability of networks to provide QoS guarantees, typically in the form of bounded end-to-end delay, bandwidth and delay jitter.

One of the most important issues in the design of QoS networks is the use of proper traffic scheduling algorithms for switches and routers. In the past, a number of scheduling algorithms have been proposed for output-buffered switches (for example, PGPS (the Packet-by-packet version of Generalized Processor Sharing, also known as WFQ (Weighted Fair Queueing)) [1], SCFQ (Self-Clocked Fair Queueing) [2], WF^2Q (Worst-case Fair Weighted Fair Queueing) [3],Weighted Round Robin [4], DRR (Deficit Round Robin) [5], QLWFQ (Queue Length-based Weighted Fair Queueing) [6], FFQ (Frame-based Fair Queueing) [7], Delay-EDD (Delay Earliest Due Date) [8], etc.). Based on their structures, these algorithms can be grouped into two types: *timestamp-based* and *frame-based.* With timestamp-based schedulers, such as PGPS and Delay-EDD, each packet arrival is assigned a timestamp; packets are served according to their timestamps. The computational cost is $O(1)$, where N is the number of backlogged flows. For high-speed backbone networks with large N, scalability becomes a major issue. There are algorithms such as SCFQ that approximate WFQ and reduce the complexity to $O(1)$ [15] of selecting the next packet to serve, assuming that the number of connections or flows is fixed. However, this results in an increase of end-to-end delay bound.

In [1], Parekh showed two important relationships between a fluid GPS system and its corresponding packet WFQ system. First, in terms of delay, a packet will finish service in a WFQ system later than in the corresponding GPS system by no more than the transmission time of one maximum size packet. As long as the input traffic is leaky bucket constrained, the algorithm can guarantee a delay bound to a session. Second, they also showed that in terms of the total number of bits served for each flow, a WFQ system does not fall behind a corresponding GPS system by more than one maximum size packet. In [3], it was further shown that WFQ can also surpass GPS in terms of the number of bits served for a flow. A modification of WFQ called WF^2Q was proposed to ensure that packets do not leave earlier in that system than in a GPS system.

For frame-based scheduling disciplines such as WRR (Weighted Round Robin) [4] and DRR (Deficit Round Robin) [5], the time axis is divided into frames. Packets are served according to a service order list known as the frame table. In WRR, frame slots are assigned according to the weight of a flow. In conventional round-robin service disciplines, the number of packets served in each round is fixed. A major problem raised is the unfairness that may be caused by different packet sizes between flows. To address this issue, in DRR, each flow is assigned a deficit counter that records its service quantum. When the round-robin

Manuscript received June 5, 2000.
Manuscript revised November 13, 2000.

[†]The author is with the Department of Information Management, National Taiwan University, Taipei, Taiwan.
[††]The author is with the Department of Computer Science, Tsing-Hua University, Hsinchu, Taiwan.

IEICE TRANS. COMMUN., Vol. E84-B, No. 8, pp. 2266–2278, August 2001

pointer points to the flow, the value of the quantum size will be added to the deficit counter. The head-of-line packet will be served only when its size is less than the current counter value. After transmission, the value of the packet size is subtracted from the deficit counter. Otherwise, the pointer rounds to the next flow.

A major advantage of this class of algorithms is the $O(1)$ computational complexity of selecting the next packet to serve, compared to $O(\log N)$ in WFQ, assuming that the total number of flows N is fixed. However, algorithms like FFQ [7] that perform dynamic scheduling of the frame table have $O(\log N)$ time complexity. Moreover, the fairness and delay bound performance of these algorithms is generally worse than that of timestamp-based algorithms. Besides, these algorithms do not address the issue of how to minimize the normalized worst-case fairness [3].

In this paper, we propose a new traffic scheduling algorithm called *Jitter Control Frame-based Queueing* (JCFQ) that provides an upper bound for the delay jitter in the case of rate-controlled connections, such as packet video streams and IP telephony, while guaranteeing bounded delay and worst-case fair weighted fairness, such as in the WF^2Q algorithm, but with $O(1)$ complexity. Three different algorithms for slot or service order assignment between flows are proposed. In these algorithms, instead of performing the virtual finish time calculation for every packet arrival based on its arrival time and weighted service time as in WFQ and WF^2Q, the delay jitter is taken into account and formulated into a virtual finish time calculation in time slot assignment within a frame. The number of slots assigned to individual flows (or queues) is based on the principle of weighted fair queueing. Moreover, slots assigned to a flow are spaced out within a frame in order to minimize the Worst-case Fair Index [3] as well as delay jitters. Since the frame table is calculated at the call level, i.e., upon arrival and departure of flows, it has $O(1)$ computational complexity during the packet transfer phase. Therefore, searching for the packet with the smallest timestamp is not necessary, while it is in the case of WFQ or WF^2Q. In this paper, we also show that JCFQ with the proposed slot assignment algorithms can achieve better flow isolation in the delay distribution than can WF^2Q.

The rest of the paper is organized as follows. In Sect. 2, we describe the basic scheme of Jitter Control Frame-based Queueing (JCFQ). In Sect. 3, the three different slot assignment algorithms are presented. In Sect. 4, a simple analysis of the fairness property of the proposed algorithms and a comparison with WF^2Q are presented. In Sect. 5, the fairness, delay and delay jitter performance of the MJF algorithm is evaluated and compared with that of WF^2Q via simulation. The results are analyzed. Finally, Sect. 6 is conclusion.

2. Jitter Control Frame-Based Queueing (JCFQ)

In this section, we present a new traffic scheduling discipline called Jitter Control Frame-based Queueing (JCFQ) that provides not only per-flow bandwidth and delay bound guarantees as in GPS, but also an upper bound on the delay jitter. The algorithm can be viewed as a combination of frame-based and timestamp-based scheduling disciplines. It guarantees worst-case fairness as in WF^2Q but with $O(1)$ complexity in selecting the next packet to serve. Moreover, in Sect. 5, we show that JCFQ with the any of the slot assignment algorithms described in the next section can achieve better performance in terms of flow isolation and delay distribution than can WF^2Q, especially for connections requiring a smaller share of bandwidth.

Similar to other frame-based scheduling disciplines, in JCFQ, the time axis is divided into frames. Queues of flows are served according to the order recorded in the frame table. Rather than performing frame slot scheduling upon arrival of each packet, as in QLWFQ and FFQ, or before each frame starts, as in DRR and WRR, in JCFQ, the service order between flows is scheduled at the call level, i.e., upon the arrival and departure of flows.

The structure of the scheduler is shown in Fig. 1. When a new flow establishment request is received, its service information, including its traffic descriptor and QoS requirements, is forwarded to the slot assignment unit subject to admission control. If admitted, the unit performs resource allocation, including service weight calculation and slot assignment, in the frame table. The frame table records the service order of admitted flows. Packets of different flows are assumed to be stored in separate queues. Within each queue, packets are served on a first-in-first-out (FIFO) basis. Figure 2 shows an

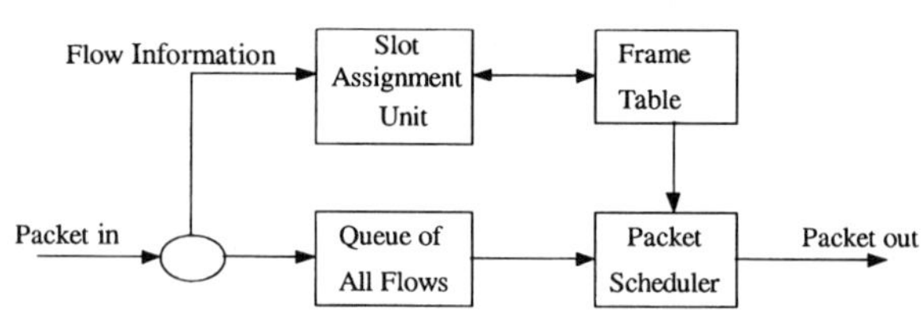

Fig. 1 The structure of the frame-based per-flow QoS guaranteed system.

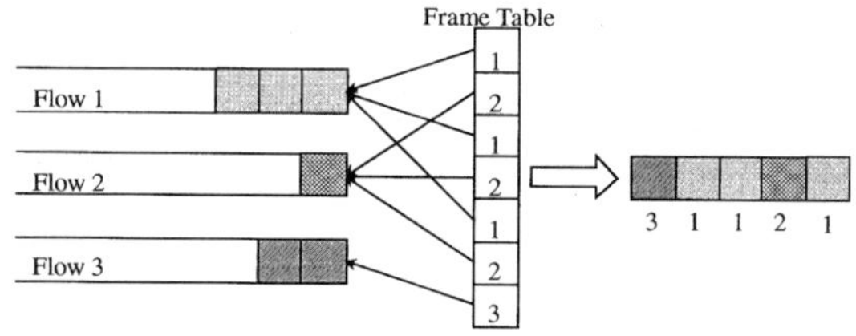

Fig. 2 An example of frame-based packet scheduler.

example of three flows sharing a common link. The frame table has seven slots with the service order set by the slot assignment unit. The packet scheduler reads the number of the next flow to serve from the table and checks whether the corresponding queue has any packet waiting for service. If it does, the scheduler will remove the packet from the head of the queue and transmit it. Otherwise, the packet scheduler will go on to serve the flow indicated in the next slot. After finishing a round of service, the packet scheduler repeats the same service order until the frame table is updated. The table will be updated when a new flow is admitted, when an existing flow departs or when any other resource re-negotiation occurs that might change the assignment.

In the JCFQ algorithm, queues of individual flows are served according to the order specified in the frame table. No search is necessary once the frame table is computed. When time slot k is to be served and the corresponding queue is empty, the server will skip the queue and move on to serve the flow indicated in the next time slot $(k+1)$. The duration of a frame cycle varies with the state of the admitted queues (backlogged or empty). In this case, only constant time computation is necessary for selecting the next queue to serve.

Consider a server serving N flows on a link. The server is characterized by N positive real numbers, $\phi_1, \phi_2, ...\phi_N$. These are the relative amounts of service given to each flow in the sense that if $W_i(T_k)$ is defined as the amount of flow i traffic serves during frame cycle T_k, then when all the flows are continuously backlogged during the cycle, we have

$$\frac{W_i(T_k)}{W_j(T_k)} \geq \frac{\phi_i}{\phi_j}$$

for any flow i and j, $1 \leq i, j \leq N$. Let the increment of flow i, denoted by I_i be defined as

$$I_i = \frac{1}{\phi_i}. \tag{1}$$

I_i is used in the slot assignment of the frame table to control the jitter between service visits to flow i

In JCFQ, at the beginning of a frame cycle, each admitted flow is credited with a service amount, denoted by Q_i, proportional to its service weight. Let N and M be the total number of flows in the system and the maximum amount of traffic (e.g., Bytes) served in a frame cycle, respectively. We have

$$Q_i = \frac{\phi_i}{\sum_{j=1}^{N} \phi_j} M. \tag{2}$$

To control the amount of service a flow receives during a frame cycle, each flow is associated with a pair of counters: a *flow_quantum* counter and a *slot_quantum* counter, denoted by $F_i(T_k)$ and $S_i(T_k^{l_{i,m}})$. $F_i(T_k)$ is the maximum amount of service guaranteed to flow i at cycle T_k. $S_i(T_k^{l_{i,m}})$ is the maximum amount of service allowed to flow i at time slot $l_{i,m}$ of cycle T_k, where $m = 1, 2, \ldots, l_i$ and l_i is the number of slots assigned to flow i. The slot_quantum counter is designed to address the fairness issue for networks with variable size packets. For example, suppose the next queue to serve is flow i; it is now allowed to transmit a maximum amount of traffic as indicated in its slot_quantum counter. Every time a packet is sent, both its flow_quantum and slot_quantum counters are deducted by the value of the packet size. In the case that the value of the slot_quantum counter is less than the size of the head-of-line packet in the queue, the service of the current time slot terminates. The remaining credit will be carried over to the next time slot.

When working with fixed size packets, such as in ATM networks, we assume that the slot_quantum is the packet size. For networks with variable size packets like the Internet, the slot_quantum can take any value. we set the slot_quantumthe to the maximum packet size, denoted by L_{max}. If a flow is backlogged with a number of large packets, the maximum number of packets sent in a frame cycle will be close to the number of slots assigned to the flow in the frame table. On the other hand, if a flow has a large backlog of small packets, then bursts of back-to-back packets are sent in the time slots assigned to the flow. The maximum burst size is bounded by the value of the slot_quantum.

In general, at the beginning of a frame cycle, if a flow is newly admitted or a flow becomes backlogged from an idle state when the server visits the flow, its *slot_quantum* counter will always be initialized with the default value S_{max}. Otherwise, the counter will be updated at the beginning of each time slot assigned to flow i as follows:

$$S_i(T_k^{l_{i,m}}) = \begin{cases} S_i(T_k^{l_{i,m-1}}) + S_{max} & \text{if } 1 < m \leq l_i \\ S_i(T_{k-1}^{l_{i,l_i}}) + S_{max} & \text{if } m = 1. \end{cases} \tag{3}$$

Note that the remainder is carried over from the previous slot. The counter is also credited with a new amount of service of S_{max} at the beginning of a service slot.

JCFQ is a frame-based scheme. The maximum number of slots (i.e., Bytes) in a frame cycle is fixed. Therefore, in each cycle, it may not be able to serve all the flows active in the system, that is at least one packet. The variable slot_quantum is used to record the accumulated credits of a flow. The choice of S_{max} is basically independent of the maximum packet size. However, the value of S_{max} may affect the delay and jitter performance of flows, especially in the case of networks with variable size packets like the Internet. In Sect. 5, we study the effect of S_{max} on the delay and jitter distributions of different types of applications under JCFQ/MJF. In the real world, it is an operational decision to choose a proper value of S_{max} based on the given applications.

Although the service order is pre-calculated for all the flows that are admitted, if a flow is idle with an empty queue and zero slot_quantum, the server will skip the queue and move on to serve the next flow in the service order list. Thus, the JCFQ algorithm is work-conserving because service will not be suspended due to idle flows. If a time slot is assigned to queue i, then the queue was not able to send the packet because the size of its head-of-line packet was too large, so the remainder from this slot or cycle will be carried over to the next slot or cycle. Thus, deficits are kept track off. On the other hand, if a queue becomes empty at the end of a slot, the slot_quantum counter will be cleared. Similarly, if a flow becomes idle at the end of a cycle, the flow_quantum counter will also be cleared. The goal is to avoid any penalty for backlogged queues as in the Virtual Clock algorithm [11]. For the flow_quantum, we have

$$F_i(T_{k-1}) = \begin{cases} 0 & \text{not backlogged at the end of } T_{k-1} \\ F_i(T_{k-1}) & \text{backlogged at the end of } T_{k-1}. \end{cases} \tag{4}$$

At the beginning of a cycle, a new value of the service quantum is added to the flow's flow_quantum counter as follows, regardless of whether the flow is active or idle:

$$F_i(T_k) = F_i(T_{k-1}) + Q_i. \tag{5}$$

If the flow is active, its flow_quantum counter will be reduced by the amount of data served during each server's visit to the queue. When the server or scheduler skips a slot assigned to an idle flow, the value of the slot quantum is also deducted from the counter to maintain a weighted fair share of bandwidth between flows. During a frame cycle, when an idle flow becomes active, it joins the service automatically. As long as its flow_quantum counter has sufficient credit, the queue will be served according to the frame table. By the end of a frame cycle, the remainder of each queue's flow_quantum counter must be either a deficit or zero.

Consider the example depicted in Fig. 3, where three flows share the same link with service weights of 1/2, 1/3 and 1/6, respectively. To illustrate the above concepts, assume packets have only two sizes: large and small packets. Large packets are twice the size of small packets. Without loss of generality, assume that a small packet takes one unit of credit, and that a large packet takes two units for a transmission. For the sake of simplicity, let the default value of the slot quantum be 2. The flow_quantum counters for the three flows are 6, 4 and 2, respectively. The initial backlogs of the three flows are shown in Fig. 3(a). The frame table is shown

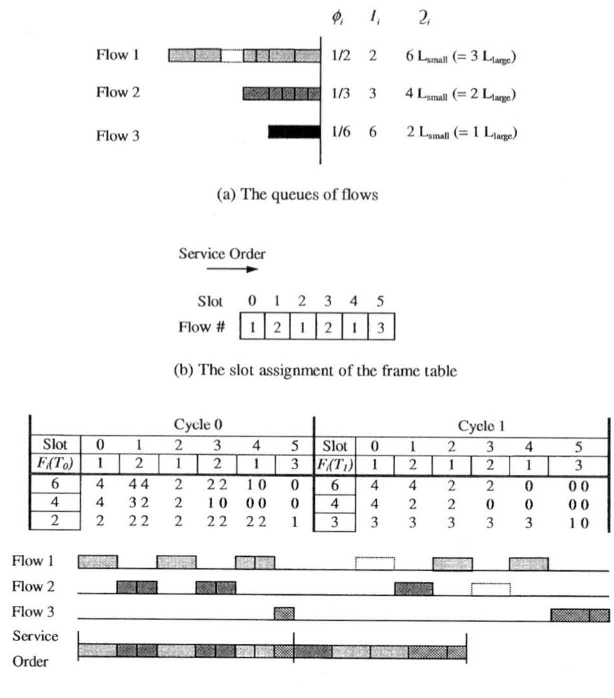

Slot	0	1	2	3	4	5	Slot	0	1	2	3	4	5
$F_i(T_0)$	1	2	1	2	1	3	$F_i(T_1)$	1	2	1	2	1	3
6	4	4 4	2	2 2	1 0	0	6	4	4	2	2	0	0 0
4	4	3 2	2	1 0	0 0	0	4	4	2	2	0	0	0 0
2	2	2 2	2	2 2	2 2	1	3	3	3	3	3	3	1 0

Fig. 3 An example of how packets of flows are served in JCFQ.

in Fig. 3(b). The service sequence is 1, 2, 1, 2, 1 and 3. (This order is generated by using the EJDF algorithm described in the next section.)

Figure 3(c) shows the sequence of packets served and the contents of the flow_quantum counters during the first two cycles. In cycle 0, the first slot is assigned to Flow 1. The head of the queue is a large packet. The packet is served and two units of credit are deducted from the flow's flow_quantum counter. The next slot is assigned to Flow 2. Since the slot quantum allows either one large packet or two small packets to be transmitted, the first two small packets of Flow 2 are transmitted back-to-back. Flow 2's flow_quantum counter is updated after the transmission. Note that at time slot 5 assigned to Flow 3, after the transmission of a small packet, the flow_quantum counter (with only one unit of credit left) does not have sufficient credit for a large packet transmission. The service terminates, and the remaining credit is carried over to the next cycle. The current service cycle ends. Note that at the beginning of cycle 1, Flow 3's flow_quantum counter has a value of three—the sum of the previously carried-over credit of one and a new credit of two. At the first slot, the queue of Flow 1 is empty, the server deducts the two units of the service quantum from the flow's flow_quantum counter and moves on to the next slot. Suppose two large packets arrive at queue of Flow 1 during the second slot. Flow 1 becomes backlogged at slot 2, and the first packet is served. At time slot 3, the queue of Flow 2 is empty, and two units of credit are deducted from its flow_quantum counter. At time slot 5, since both the flow_quantum and slot_quantum counters have sufficient credit, a large packet and a small packet from Flow 3 are served.

3. Service Order Assignment

In this section, we describe how the service order between flows is assigned in JCFQ. The goal of the design is to provide an upper bound on the delay jitter for rate-controlled connections, such as packet video streams and IP telephony, while guaranteeing both worst-case fair weighted fairness and bounded delay and jitter as in WF^2Q. The delay jitter bound guarantee affects the smoothness of the playback quality of these applications. Three algorithms are proposed here. All of them formulate delay jitter into virtual finish time calculations but with different definitions. They all achieve a small value of the Normalized Worse case Fair Index. Let

- N : the total number of flows;
- r_i : the bandwidth requirement of flow i;
- r : the link capacity;
- M : the total number of slots of a frame cycle;
- $m(i)$: the ith slot in a frame cycle;
- $s(i)$: the virtual start time of the head-of-line packet of flow i;
- $t(i)$: the actual start time of the head-of-line packet of flow i;
- $v(i)$: the virtual finish time of the head-of-line packet of flow i;
- $J(i)$: the jitter of the head-of-line packet of flow i.

The first algorithm is called Earliest Jitter Deadline First (EJDF), and is shown in Fig. 4. It is based on the idea of the EDF (Earliest Deadline First) algorithm which has been proven to be optimal in real-time systems. Here, a deadline is defined for each flow in order to control the maximum delay jitter or the time interval that the flow can tolerate before the next visit of the scheduler. The main idea is that when the next flow is chosen for service, we only consider the set of flows that should have started, i.e., the virtual start time is smaller than the current slot time, and select the packet with the smallest virtual finish time as the one to send. In WF^2Q, a similar idea is used. However, the WF^2Q algorithm has higher complexity due to the virtual finish time calculation, which involves sorting upon arrival of each packet, whereas in our algorithms, slot assignment is performed only at call arrivals and departures.

The second algorithm, called the Rate Monotonic (RM) algorithm, is shown in Fig. 5. It is based on the concept behind the RM (Rate Monotonic) algorithm. In our algorithm, flows with higher bandwidth requirements have smaller slot increments according to EQ.(1) and, thus, smaller maximum jitters. Similar to the EJDF algorithm, when the next flow is chosen for service, the scheduler only considers the set of flows that should have started and selects the one that has the smallest slot increment.

The third algorithm, called the Maximum Jitter First (MJF) algorithm, is shown in Fig. 6. In this algorithm, when the next flow is chosen for service, the scheduler selects the flow with the maximal weighted delay jitter. The variable $t(i)$ records the first time slot assigned to flow i in order to ensure that all the flows will be scheduled in the time frame. The variable $s(i)$ records the virtual start time of the head-of-line packet of flow i. The weighted delay jitter is defined as the ratio of the distance of the normalized current time from $s(i)$ to the increment of the flow. If the weighted delay jitter is positive, this means that the head-of-line packet should have started. The larger the value, the further the service to the queue is behind. Otherwise,

```
/* Initialization */
I_i = r / r_i ;
s(i) = 0;
v(i) = I_i ;

/* Start */
j = 0
repeat
{
  /* Select the packet with the minimal virtual finish time
     from the set of flows that should have started */
  v(f) = min{ v(i) | (s(i) <= j) }

  /* flow f is selected; update the information */
  m(j) = f;
  j = j + 1;
  s(f) = v(f);
  v(f) = v(f) + I_f ;

} until (all v(i) >= M);
```

Fig. 4 Scheduler 1—the Earliest Jitter Deadline First (EJDF) algorithm.

```
/* Initiatization */
I_i = r / r_i ;
s(i) = 0;

/* Start */
j = 0
repeat
{
  /* Select the packet with the minimal slot increment
     from the set of flows that should have started */
     I_f = min{ I_i | (s(i) <= j) }

  /* flow f is selected; update the information */
  m(j) = f;
  j = j + 1;
  v(f ) = s(f) + I_f
  s(f) = v(f);

} until (all v(i) >= M);
```

Fig. 5 Scheduler 2—the Rate Monotonic (RM) algorithm.

```
/* Initiatization */
I_i = r/r_i;
s(i) = 0;
J(i) = 0;
t(i) = -1;

/* Start */
j = 0
repeat
{
   for all F_i
   {
      if ( t(i) >= 0)
         J(i) = (j- t(i)-s(i)) / I(i) ;
   }

   /* Select the packet with the the largest weighted delay jitter
      from the set of flows that should have started */
      J(f) = max{ J(i) | (s(i) <= j) }

   /* Update the information */
   if ( t(f) < 0 )
      t(f) = j;
   m(j) = f;
   j = j + 1;
   v(f) = s(f) + I(f);
   s(f) = v(f);

} until (all v(i) >= M);
```

Fig. 6 Scheduler 3—the Maximum Jitter First (MJF) algorithm.

the service to the queue is on track. Since the objective is to minimize the delay jitter of flows, the flow with the largest weighted delay jitter is selected for service. If there is more than one flow, then the flow with the maximum bandwidth requirement is chosen.

4. Fairness Analysis

4.1 Fairness Index

A scheduler is considered to be fair if the difference between the normalized service amounts offered to any two continuously backlogged flows i, j during any time interval (t_1, t_2) is bounded, i.e.,

$$\left|\frac{W_i(t_1,t_2)}{r_i} - \frac{W_j(t_1,t_2)}{r_j}\right| \le FR,$$

where FR is a measure of the fairness of the algorithm. As shown in the previous section, each flow i will be assigned $\lfloor M \times r_i/r \rfloor$ slots in each frame of size M. Assume that the quantum unit of a slot is S. Each flow i will then obtain $S \times \lfloor M \times r_i/r \rfloor$ units of credit. The maximum amount of service that a flow can receive in a frame cycle is $S \times (M \times (r_i/r)) + L_{max}$. L_{max} is the maximum packet size. The additional L_{max} represents the maximum credit that a flow is allowed to carry over from the previous cycle. Similarly, if a flow is always backlogged, then the minimum amount of service received is $S \times (M \times (r_i/r)) - L_{max}$. Thus, we have the Fairness Index computed as follows:

$$\left|\frac{W_i(t_1,t_2)}{r_i} - \frac{W_j(t_1,t_2)}{r_j}\right| \le \left|\frac{S \times (M \times (r_i/r)) + L_{max}}{r_i} - \frac{S \times (M \times (r_i/r)) - L_{max}}{r_j}\right|$$

That is,

$$\left|\frac{W_i(t_1,t_2)}{r_i} - \frac{W_j(t_1,t_2)}{r_j}\right| \le \frac{L_{max}}{r_i} + \frac{L_{max}}{r_j} \tag{6}$$

4.2 Normalized Worst-Case Fair Index

The Normalized Worst-case Fair Index is defined in [3]. For the scheduling method s, let a_i^k be the arrival time of packet k of flow i; the delay of the packet is bounded above by $C_{i,s}$, i.e.,

$$d_{i,s}^k < a_i^k + \frac{Q_{i,s}(a_i^k)}{r_i} + C_{i,s},$$

where $d_{i,s}^k$ is the departure time of the packet, r_i is the bandwidth guarantee to flow i, $Q_{i,s}(a_i^k)$ is the queue size of flow i when the packet arrives at time a_i^k, and $C_{i,s}$ is a constant independent of the queues of other flows sharing the same link. $C_{i,s}$ is called the *Worst-case Fairness Index* for flow i at server s. The Normalized Worst-case Fair Index of flow i at server s is

$$c_{i,s} = \frac{r_i C_{i,s}}{r}.$$

Then, the Normalized Worst-case Fair Index of a server s is defined as

$$c_s = \max_i\{c_{i,s}\}.$$

For networks with fixed size packets, say L, such as ATM networks, let the queue size $Q_{i,s}(a_i^k)$ be measured in bytes. The number of packets in the queue is $Q_{i,s}(a_i^k)/L$. In the proposed algorithms, the ideal interval for service to flow i is r/r_i slots. That is, the ideal departure time interval between two consecutive packets of flow i is $(r/r_i) \times L/r = L/r_i$. Thus, we have

$$d_{i,s}^k - a_i^k = \frac{Q_{i,s}(a_i^k)}{L} \times \frac{L}{r_i} = \frac{Q_{i,s}(a_i^k)}{r_i}.$$

That is,

$$d_{i,s}^k = a_i^k + \frac{Q_{i,s}(a_i^k)}{r_i} + 0. \tag{7}$$

This equation gives the relation between the arrival and departure time of packet k of flow i given scheduler s in an ideal case. Note that the ideal Worst-case Fair Index for flow i is 0. So is the Normalized Worst-case Fair Index of the system. This means that if one can schedule service visits to a queue so that they are equally spaced out in a frame table, then the zero Normalized Worst-case Fairness Index can be achieved. However, the ideal case cannot always be achieved. In order to space out slots assigned to a flow equally within a frame, it is possible that more than one flow will need to be assigned to the same time slot. Hence, the worst-case fair index cannot be zero. This was also shown in [3]. Any scheduler in the real world can serve only one packet at a time, which causes the fairness falling behind the GPS scheduler.

4.3 Normalized Average Fair Index

Note that the Normalized Worst-case Fair Index only shows the worst-case bound on fairness. In the following, we will define a new measure called the *Normalized Average Fairness Index* that takes the average value among all the slots in a frame. Let $C_{i,s}^k$ be the difference between the actual departure time and the departure time in the ideal GPS system. Since in the real world, a packet may leave the system earlier or later than in the GPS system, the value of $C_{i,s}^k$ could be positive (later) or negative (earlier). The Worst-case Fair Index for flow i is obtained by taking the maximum value among all the packets, i.e., $C_{i,s} = \max_k\{C_{i,s}^k\}$. The Normalized Worst-case Fair Index was initially defined in [3]. The focus is on how late a packet may depart from a packet-version system compared to the situation in the ideal GPS system. Therefore, it only shows the worst-case bound on fairness. In this paper, we are interested in the jitter performance. In the definition of the Normalized Average Fair Index for flow i, i.e., EQ(8), we consider all the delay differences among packets by taking the absolute value of $\{C_{i,s}^k\}$. In other words, we take into account the delay differences for both early and late departures.
Let

$$d_{i,s}^k = a_i^k + \frac{Q_{i,s}(a_i^k)}{r_i} + C_{i,s}^k.$$

The Normalized Average Fair Index for flow i is defined as follows:

$$A_{i,s} = \frac{\sum_{k=1}^{n_i} \left|C_{i,s}^k \times \frac{r_i}{r}\right|}{n_i}, \quad (8)$$

where n_i is the number of slots assigned to flow i. We have the Normalized Average Fairness Index for server s as

$$A_s = \max_i\{A_{i,s}\}.$$

By taking the average, the Normalized Average Fair Index provides a better estimate of the fairness for frame-based scheduling algorithms.

In the following, we will use an example to compare the Normalized Worst-case Fair Index with the Normalized Average Fair Index of the proposed three

Table 1 The slot assignment under different scheduling algorithms.

Slot #	0	1	2	3	4	5	6	7	8	9	0	...
WRR	1	1	2	1	2	3	1	1	2	1	1	...
WF^2Q	1	2	1	3	1	1	2	1	2	1	1	...
EJDF	1	2	1	3	1	1	2	1	2	1	1	...
RM	1	2	1	3	1	1	2	1	2	1	1	...
MJF	1	2	1	3	1	2	1	1	2	1	1	...

Table 2 The slot assignment under different scheduling algorithms.

	Normalized Worst-CaseFair Index				Normalized AverageFair Index			
	Flow1	Flow2	Flow3	Max	Flow1	Flow2	Flow3	Max
WRR	0.8	0.2	0	0.8	0.40	0.27	0	0.40
$WF2Q$	0.2	0.5	0	0.5	0.27	0.33	0	0.33
EJDF	0.2	0.5	0	0.5	0.27	0.33	0	0.33
RM	0.2	0.5	0	0.5	0.27	0.33	0	0.33
MJF	0.2	0.2	0	0.2	0.27	0.13	0	0.27

algorithms with WRR and $WF2Q$. Consider three flows that are always backlogged and sharing the same link. The bandwidth allocation weights are as follows: $r_1/r = 6/10$, $r_2/r = 3/10$ and $r_3/r = 1/10$. The frame table size is ten. In Table 1, we show the service order resulting from WRR [4], WF^2Q, EJDF, RM and MJF. The service order for WF^2Q is computed based on the assumption that flows are always backlogged.

Table 2 compares the Normalized Worst-case Fair Index with the Normalized Average Fair Index for the five algorithms. One can see that the proposed three algorithms and WF^2Q outperform WRR. Since EJDF, RM and WF^2Q have the same service sequence, they all have the same fairness indexes. Among the three slot assignment algorithms, MJF has the best performance. This is because the slot assignment of MJF best approximates the ideal case. In this example, the ideal increments for Flows 1, 2 and 3 are 10/6, 10/3 and 10, respectively. In time slot five, MJF assigns the time slot to Flow 2 to keep its jitter from continuing to grow while EJDF picks up Flow 1.

Note that in EQ. 8, the absolute value of $\{C_{i,s}^k\}$ is taken. For those early departures, because the corresponding $\{C_{i,s}^k\}$ has a negative value, it won't affect the value of the Normalized Worst-case Fair Index. However, in the calculation of the Normalized Average Fair Index, every packet jitter $\{|C_{i,s}^k|\}$ is taken into account. Therefore, its value might be greater than the value of the Normalized Worst-case Fair Index.

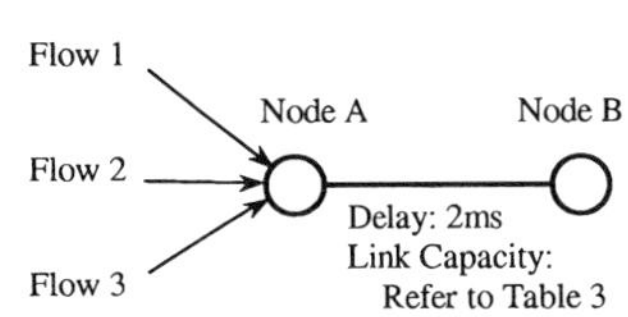

Fig. 7 Network configuration.

5. Performance Evaluation

In this section, we will study and compare the fairness, delay and jitter performance of the JCFQ/MJF algorithm with that of WF^2Q via simulation. All the simulations were performed using *ns* [12]. The network configuration is shown in Fig. 7. Three different types of media transmissions were considered: packet video, packet voice and data traffic. For video flows, an actual video trace (the "Star Wars" movie [13]) was used as the video traffic source in the simulation. For packet-switched voice communications, the G.729 compression algorithm was assumed. A single phone call was digitized at the rate of 8 kbps. Adding up the UDP and IP protocol overhead, the actual data generation rate could range from 6 kbps to 19.2 kbps, depending on how voice codewords were packetized and whether silence suppression is executed. Currently, different products have different ways of packetization. Here, we assumed that voice packets were generated every 20 ms with a size of 48 Bytes [14]. Table 3 shows the information about the traffic sources used in the experiments.

Two sets of experiments were conducted. The goal of the first experiment was to show the weighted fairness property of different scheduling algorithms in terms of the amount of service received by individual flows. We considered three flows sharing a bottleneck 100 kbps link. The flows were assumed to be backlogged at all times and to generate fixed size packets. Each flow had a different bandwidth allocation. The frame size was assumed to be 256 slots; the packet size was 256 Bytes, and so was the slot quantum.

In Fig. 8, we compare the average service rate, i.e., $W_i(t)/t$ received by individual flows under JCFQ/MJF and WF^2Q. Because the link was in the state of congestion, the total input rate of all flows was greater than the link capacity. The service rate shown in Fig. 8 is the departure rate of individual flows. The service amount was sampled every 0.01 second. A moving average with a measurement window t of one second was used. From the figure, one can see that for both algorithms, the service rates converged to the individual's bandwidth allocations and remained stable throughout the course. With JCFQ/MJF, the service rate converged slightly faster than it did with of WF^2Q.

In the second set of experiments, we studied the delay and jitter performance of three different types of media transmissions under JCFQ/MJF and WF^2Q. The characteristics of the traffic sources are shown in Table 3. For JCFQ, the frame size was assumed to be 256 slots, and the size of the slot quantum was 48 Bytes. In Figs. 9, 10 and 11, we show the packet delay profiles of the three flows. Figures 12, 13 and 14 show the corresponding delay jitter profiles. The delay

Table 3 The traffic sources used in the simulation experiments.

	Flow 1	Flow 2	Flow 3	Link Capacity
1	Always Backlogged; Bandwidth allocation is 60 kbps.	Always Backlogged; Bandwidth allocation is 30 kbps.	Always Backlogged; Bandwidth allocation is 10 kbps	100 kbps
2	MPEG file; Max Packet Size: 1024 Bytes; Rate: 1500 kbps.	Voice over IP; Packet size: 48 Bytes; Rate:19.2 kbps (48 Bytes/20ms).	Traffic with Pareto distr.; Burst/Idle:100/100 ms; Shape: 1.5; Packet Size:256 Bytes; Rate: 400 kbps.	2Mbps

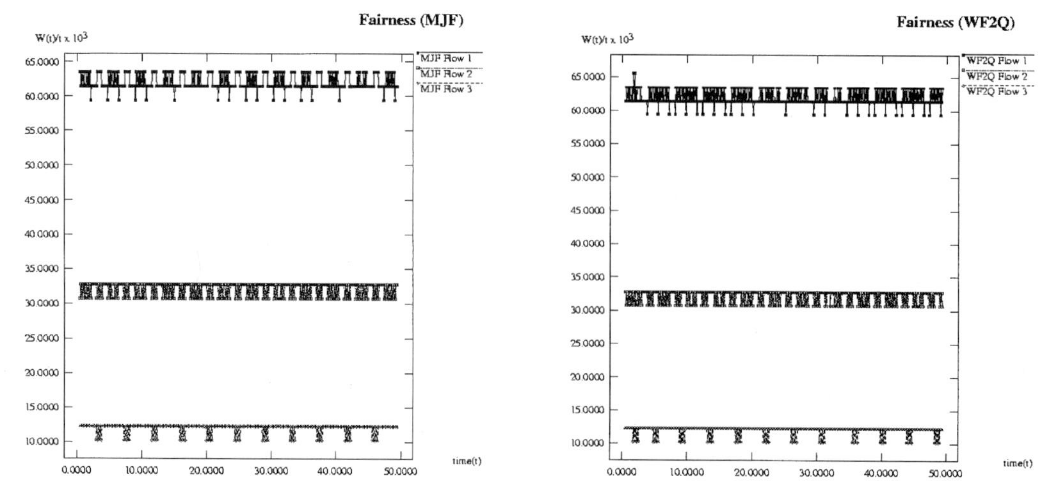

(a) JCFQ/MJF (b) WF^2Q

Fig. 8 Weighted fairness.

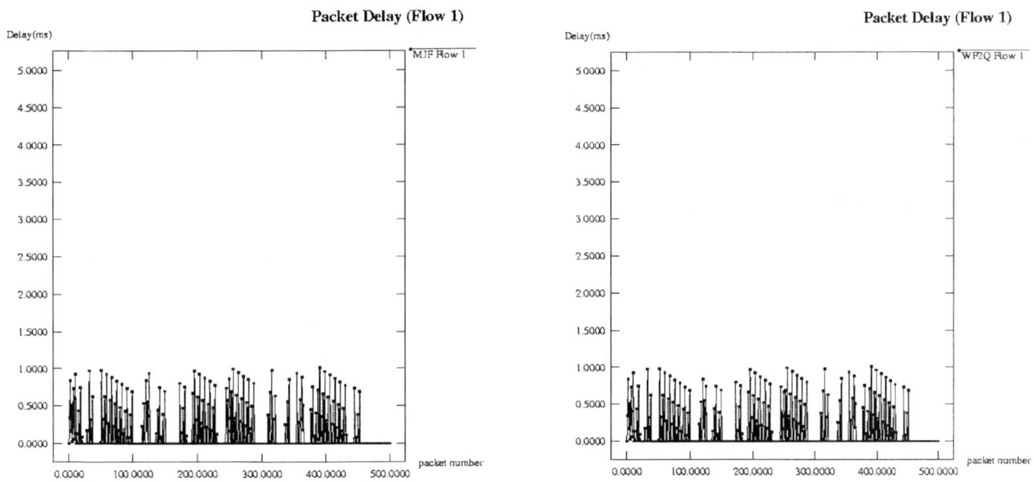

(a) JCFQ/MJF (b) WF^2Q

Fig. 9 Delay of Flow 1 ($S = 48$ Bytes).

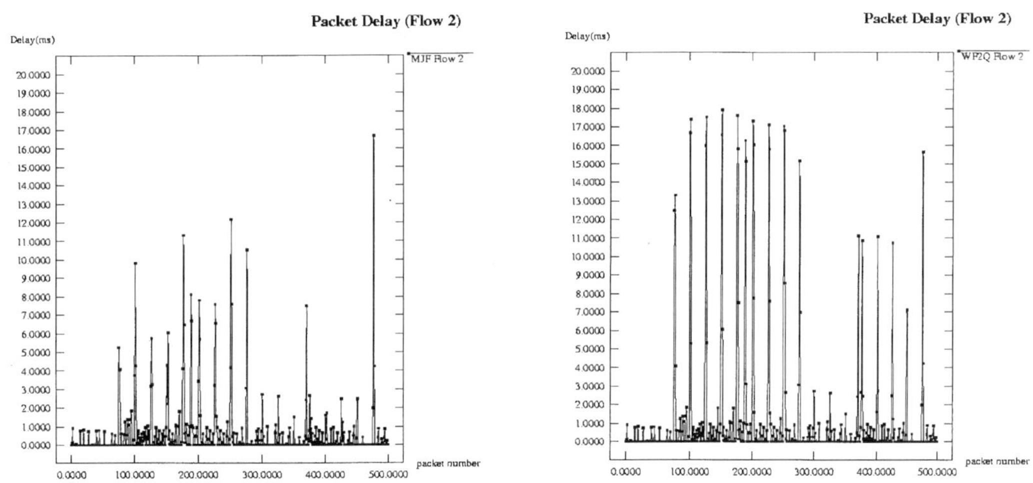

(a) JCFQ/MJF (b) WF^2Q

Fig. 10 Delay of Flow 2 ($S = 48$ Bytes).

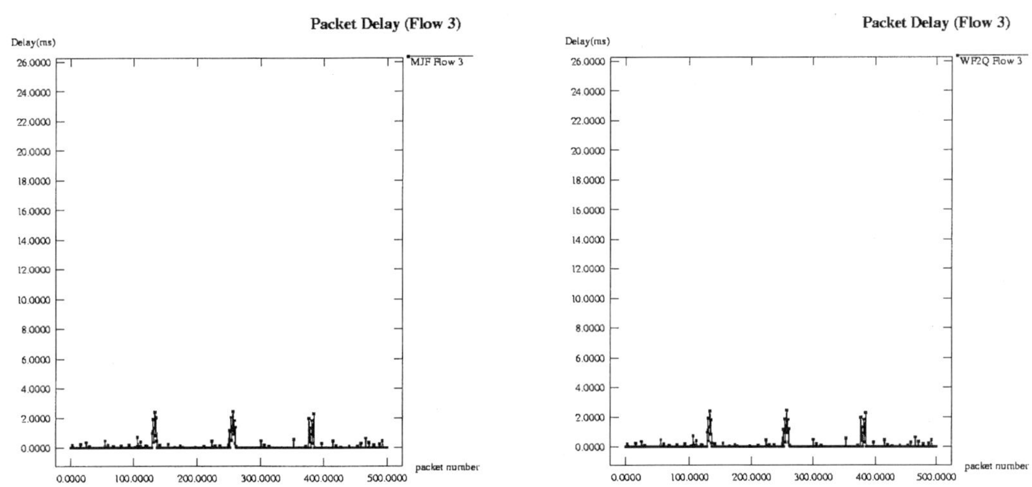

(a) JCFQ/MJF (b) WF^2Q

Fig. 11 Delay of Flow 3 ($S = 48$ Bytes).

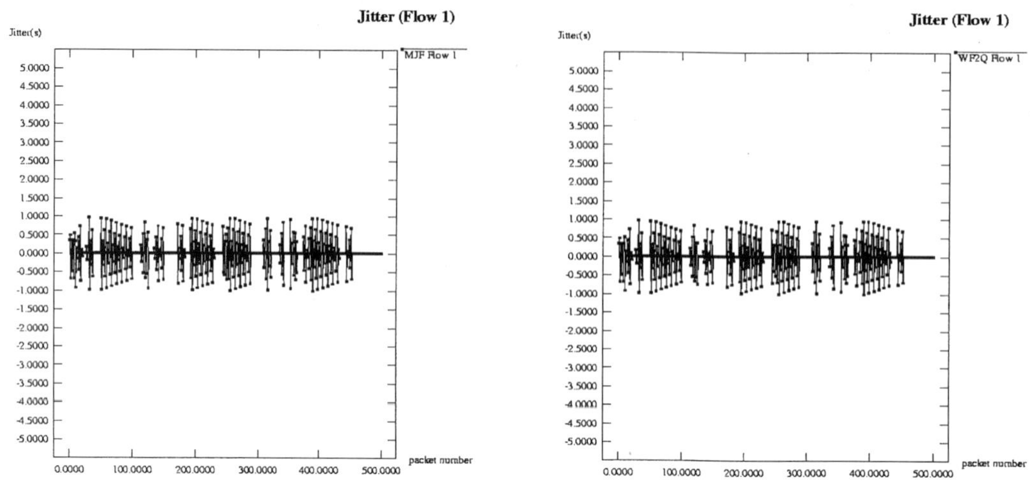

(a) JCFQ/MJF

(b) WF^2Q

Fig. 12 Jitter of Flow 1 ($S = 48$ Bytes).

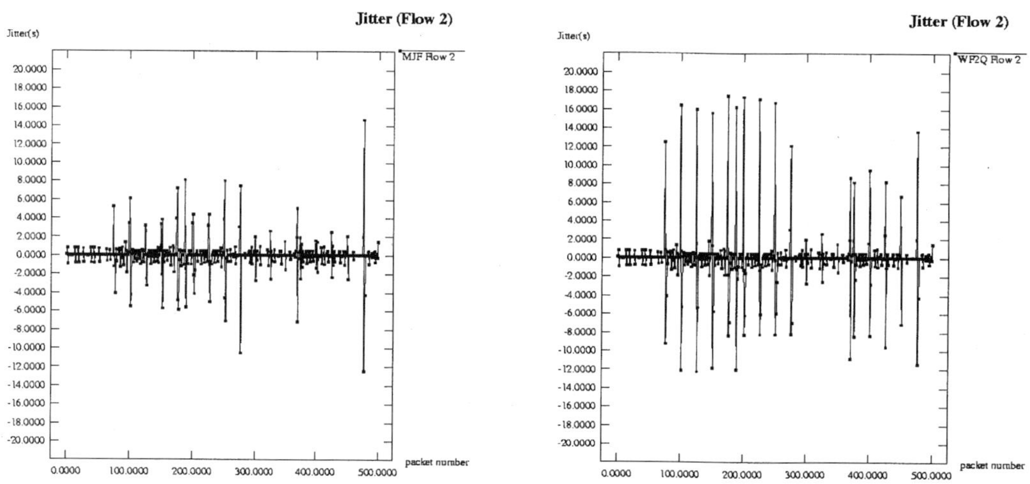

(a) JCFQ/MJF

(b) WF^2Q

Fig. 13 Jitter of Flow 2 ($S = 48$ Bytes).

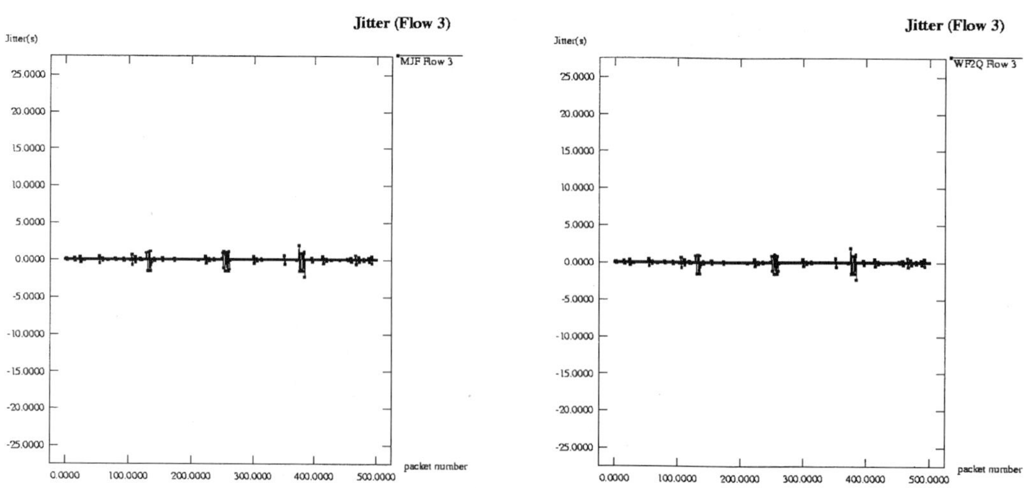

(a) JCFQ/MJF

(b) WF^2Q

Fig. 14 Jitter of Flow 3 ($S = 48$ Bytes).

Table 4 The delay and jitter statistics under JCFQ/MJF and WF^2Q.

		Average		Standard Deviation		Maximum	
		MJF	WF^2Q	MJF	WF^2Q	MJF	WF^2Q
Delay (ms)	Flow 1	0.136	0.134	0.340	0.325	4.718	4.614
	Flow 2	0.374	0.582	1.129	2.035	17.685	17.914
	Flow 3	0.402	0.405	1.951	1.984	22.716	22.716
Jitter (ms)	Flow 1	0.223	0.222	0.322	0.320	4.053	4.053
	Flow 2	0.498	0.714	1.151	1.951	17.685	17.565
	Flow 3	0.192	0.191	0.535	0.540	18.070	21.670

and jitter statistics are given in Table 4. We can see that Flows 1 and 3 exhibited similar performance under both algorithms. The delay distribution of Flow 2 under MJF was smaller than that under WF^2Q. This is an interesting observation. Note that Flow 2 was a voice connection that generated a stream of small packets and required only a comparatively small share of the bandwidth of the link. In WF^2Q, the virtual service time of packet k of flow i is defined as the ratio of the packet length to the flow's allocated weight ϕ_i, i.e., L_i^k/ϕ. The smaller the weight, the greater the virtual service time. For example, consider packets that arrive at the same time; packets belonging to flows with larger bandwidth weights may be scheduled for service before packets belonging to flows with smaller weights. As a result, the delay distribution for flows with smaller bandwidth allocation will be affected by the loads of other flows, especially those with larger allocation. In other words, when algorithms like WFQ and WF^2Q are scheduled, there is no flow isolation in terms of delay and jitter distributions. However, the flow isolation property does hold for the minimum throughput guarantee in these algorithms. In WF^2Q, the worst-case delay bound for each flow can be obtained and is independent of other flows but not the delay and jitter "distributions" [3][†]. In MJF/JCFQ because the slots are assigned at connection setup time, the maximum amount of data served between two adjacent service visits to a flow is upper bounded. Thus, the variation of packet delays and jitters is minimized. This is exactly what we observed in the simulations.

Next, we will study the delay and jitter performance of JCFQ/MJF under different slot quantum size. Three sizes will be considered here: 48 Bytes, 256 Bytes and 1024 Bytes. Tables 5 and 6 give the performance statistics of the three flows. From the tables, we can see that Flows 1 and 3 exhibited similar performance. When slot quantum was equal to the voice packet size, Flow 2 had the best performance. When the slot quantum size increased, the performance was degraded. This was mainly due to the nature of the frame-based approach. Because Flow 2 required only a small amount of bandwidth compared to the other two flows, it had a much larger increment value that determined the maximum jitter of the flow. In other words, voice packets tended to be sent in bursts in these cases.

Table 5 A comparison of delay distributions of JCFQ/MJF with different slot quantum sizes.

(a) Average delay

Average Delay	MJF S=48 Bytes	MJF S=256 Bytes	MJF S=1024 Bytes
Flow 1	0.136	0.137	0.154
Flow 2	0.374	0.803	0.959
Flow 3	0.402	0.381	0.329

(b) Standard deviation

Delay standard deviation	MJF S=48 Bytes	MJF S=256 Bytes	MJF S=1024 Bytes
Flow 1	0.340	0.337	0.431
Flow 2	1.129	3.737	4.817
Flow 3	1.951	1.894	1.652

(c)Maximum delay

Maximum Delay	MJF S=48 Bytes	MJF S=256 Bytes	MJF S=1024 Bytes
Flow 1	4.718	4.892	6.832
Flow 2	17.685	50.394	60.974
Flow 3	22.716	23.130	21.192

Table 6 A comparison of delay jitter distributions of JCFQ/MJF with different slot quantum sizes.

(a) Average delay jitter

Average Delay Jitter	MJF S=48 Bytes	MJF S=256 Bytes	MJF S=1024 Bytes
Flow 1	0.223	0.225	0.237
Flow 2	0.498	1.091	1.239
Flow 3	0.192	0.197	0.211

(b) Standard deviation

Jitter standard deviation	MJF S=48 Bytes	MJF S=256 Bytes	MJF S=1024 Bytes
Flow 1	0.322	0.322	0.372
Flow 2	1.151	3.944	4.876
Flow 3	0.535	0.577	0.727

(c)Maximum delay jitter

Maximum Delay Jitter	MJF S=48 Bytes	MJF S=256 Bytes	MJF S=1024 Bytes
Flow 1	4.053	3.028	4.053
Flow 2	17.685	50.086	60.974
Flow 3	18.070	17.878	12.960

For "heavy" flows like Flows 1 and 3, both the delay and jitter performance were less affected by the quantum size. This suggests that the slot quantum should be set to a small size to meet the jitter performance requirements of applications like IP telephony.

6. Conclusion

In this paper, we have presented a new traffic scheduling discipline called *Jitter Control Frame-based Queueing* (JCFQ) that provides not only per-flow bandwidth and delay bound guarantees as in GPS, but also an upper bound for the delay jitter. The goal of the design is to provide an upper bound for the delay jitter for rate-controlled connections, such as packet video streams and IP telephony, while guaranteeing both worst-case

[†]Theorem 2 of [3] shows the worst-case delay of flow i under WF^2Q that is bounded by the ratio of the backlog and the minimum rate. However, the delay distribution of flow i may be affected by packet arrivals of other flows.

fair weighted fairness and bounded delay and jitter as in WF^2Q. The delay jitter bound guarantee affects the smoothness of the playback quality of these applications. The algorithm achieves worst-case fair weighted fairness as in WF^2Q but with $O(1)$ complexity in selecting the next packet to serve, assuming that the number of flows is fixed. Three different slot assignment algorithms have also been proposed here. By using these algorithms, one can achieve good service order assignment in terms of minimizing the Normalized Worst-case Fair Index and Normalized Average Fair Index. The overall scheme is simple to implement and flexible in terms of slot assignment. A simple analysis of the fairness property of the proposed algorithms and their comparison with WF^2Q has been presented. It shows that if one can schedule service visits to a queue so that thay are equally spaced out in a frame table, then the zero Normalized Worst-case Fairness Index can be achieved. This is exactly the approach taken in this paper.

We have also studied the delay and jitter performance of three different types of media transmissions under JCFQ/MJF and WF^2Q via simulations. The simulation results show that flows have similar performance under both algorithms except for flows like voice connections that generate a stream of small packets and require only a comparatively small share of the bandwidth of the link. The delay distribution of the flow under JCFQ/MJF is smaller than that under WF^2Q. This is because in WF^2Q, the delay distribution for flows with smaller bandwidth allocation will be affected by the loads of other flows, especially those with larger allocation. We have also studied the delay and jitter performance of JCFQ/MJF under different slot quantum sizes. The results show that due to the nature of the frame-based approach, for "light" flows with smaller bandwidth allocation and larger increment values, when the slot quantum size increases, the performance is degraded. Both the delay and jitter performance for "heavy" flows is, however, less affected by the quantum size. This suggests that the slot quantum should be set to a small size to meet the jitter performance requirements of applications like IP telephony.

References

[1] A.K. Parekh and R.G. Gallager, "A generalized processor sharing approach to flow control—The singel-node case," Proc. INFOCOM'92, vol.2, pp.915–924, May 1992.

[2] S. Golestani, "A self-clocked fair queueing scheme for broadband applications," Proc. INFOCOM'94, pp.636–646, April 1994.

[3] J.C.R. Bennett and H. Zhang, "WF2Q: Worst-case fair weighted fair queueing," Proc. INFOCOM'96, pp.120–128, 1996.

[4] M. Katevenis, S. Sidiropoulos, and C. Courcoubetis, "Weighted round-robin cell multiplexing in a general-purpose ATM switch chip," IEEE J. Sel. Areas Commun., vol.9, pp.1265–79, Oct. 1991.

[5] M. Shreedhar and G. Varghese, "Efficient fair queueing using deficit round robin," Proc. SIGCOMM'95, pp.231–242, Sept. 1995.

[6] Y. Ohba, "QLWFQ: A queue length based weighted fair queueing algorithm in ATM networks," Proc. INFOCOM'97, pp.566–575, 1997.

[7] D. Stiliadis and A. Varma, "Design and analysis of frame-based fair queueing: A new traffic scheduling algorithm for packet-switched networks," Proc. ACM SIGMETRICS, pp.104–115, 1996.

[8] D. Ferrari and D. Verma, "A scheme for real-time channel establishment in wide-area networks," IEEE J. Sel. Areas Commun., vol.8, pp.368–379, April 1990.

[9] S. Golestani, "A framing strategy for congestion management," IEEE J. Sel. Areas Commun., vol.9, pp.1064–1077, Sept. 1991.

[10] C.R. Kalmanek, H. Kanakia, and S. Keshav, "Rate controlled servers for very high-speed networks," IEEE Globecom1990, San Diego, California, Dec. 1990.

[11] L. Zhang, "VirtualClock: A new traffic control algorithm for packet switching networks," ACM Trans. Computer System, vol.9, pp.24–110, May 1991.

[12] UCB/LBNL/VINT Network Simulator-ns (version 2), http://www-mash.cs.berkeley.edu/ns/

[13] M.W. Garrett and A. Fernandez, "MPEG-1 vide trace," Bellcore, ftp://thumper.bellcore.com/pub/vbr.video.trace /MPEG.data, 1992.

[14] R. Caputo, Cisco Packetized Voice and Data Integration, McGraw Hill, 2000.

[15] J.L. Rexford, A.G. Greenberg, and F.G. Bonomi, "Hardware-efficient fair queueing architectures for high-speed networks," Proc. INFOCOM'96, pp.638–646, 1996.

High-Performance Backbone Network Technology edited by Naoaki Yamanaka
ISBN: 0-8247-5321-6 DOI: 10.1081/(120025703)
Published 2004 by Marcel Dekker, Inc.

Adaptive Thresholds of Buffer to Solve the Beat-Down Problem of Rate Control in ATM Networks

Harry Prihanto and Kenji Nakagawa

SUMMARY ABR service is currently standardized to handle applications of data traffic in ATM network. As a flow control method, the rate-based flow control has been adopted and applied to manage the ABR service. Several control methods have been proposed, and the EPRCA is selected as one of the control methods by the ATM Forum. EPRCA is an excellent algorithm, but when the EPRCA is applied to the ATM network, several problems occur. One of the problems is the beat-down problem, which gives unfair allocation of transmission rate to connections. We propose a new control method which solves the beat-down problem. We will show that, by our proposed method, (i) the ACR is given fairly to every connection compared to the conventional method, and also (ii) the throughput is fair for both long-hop and short-hop connections, (iii) the ACR is proportional to the throughput, and finally (iv) the total throughput is larger than that of the conventional method. The fairness of the throughput in (ii) is measured by the fairness index. In (iii), being proportional means that the allocated ACR is close to the throughput and it is measured by the proportion index. The performance is evaluated by computer simulation.
key words: *flow control, rate-based, ABR traffic, EPRCA, beat-down problem, ATM network*

1. Introduction

To realize Broadband Integrated Services Digital Network (B-ISDN) which is intended to provide various services in telecommunications, Asynchronous Transfer Mode (ATM) is one of several data transfer technologies that are proposed and currently being developed [1]. In the ATM network, information is divided into a fixed length of data block, called cell. The ATM technology can provide high speed and high efficiency of the network because a simple data transfer protocol and wide bandwidth of optical fiber media are used in the network [2].

Several services are provided and defined in ATM networks. Five classes of service are defined as Constant Bit Rate (CBR), real-time Variable Bit Rate (rt-VBR), non-real-time Variable Bit Rate (nrt-VBR), Available Bit Rate (ABR), and Unspecified Bit Rate (UBR) [3]–[5]. The ABR service is designed to handle applications of data traffic in ATM network. As a flow control method, the ATM Forum has adopted and applied rate-based flow control to manage the ABR service. The rate-based flow control is a closed-loop control. Each source uses feedback information from a network to manage their transmission rate. A Resource Management (RM) cell conveys feedback information as explicit rate from switch to the source.

Several rate-based control methods have been proposed, e.g. EPRCA (Enhanced Proportional Rate Control Algorithm), CAPC (Congestion Avoidance using Proportional Control), ERICA (Explicit Rate Indication for Congestion Avoidance), etc. [4], [6]. EPRCA is a favorite control method in the ATM Forum at the present time [7]. EPRCA is an excellent algorithm, but when the EPRCA is applied in the ATM network, several problems occur [4], [6]. One of the problems is the beat-down problem, which gives unfair allocation of transmission rate to connections [10], [11]. To solve the beat-down problem, we propose a new control method based on EPRCA. We use variable thresholds of buffer to detect the congestion state of the switch.

We show through simulation results that our proposed method has better performance than conventional method. We will show that, by our proposed method, (i) the ACR is given fairly to every connection compared to the conventional method, and also (ii) the throughput is fair for both long-hop and short-hop connections, (iii) the ACR is proportional to the throughput, and finally (iv) the total throughput is larger than that of the conventional method. The fairness of the throughput in (ii) is measured by the fairness index. In (iii), being proportional means that the allocated ACR is close to the throughput and it is measured by the proportion index.

The rest of this paper is organized as follows. After describing ABR flow control in Sect. 2, we discuss an implementation of EPRCA and its problems in Sect. 3. We describe our proposed method in Sect. 4. We consider several network models and evaluate both performance of conventional and proposed method in Sect. 5. Finally, the conclusion will be described in Sect. 6.

2. ABR Flow Control

Flow control refers to the set of actions taken by the network to avoid congestion conditions. Congestion is defined as a state of network elements in which the network is not able to meet the negotiated performance objectives for already established connections [1]. The ABR flow control is operated between source end sys-

Manuscript received February 17, 1999.
Manuscript revised August 16, 1999.
† The authors are with the Department of Electrical Engineering, Nagaoka University of Technology, Nagaoka-shi, 940-2188 Japan.

IEICE TRANS. COMMUN., Vol. E83-B, No. 2, pp. 362–370, February 2000

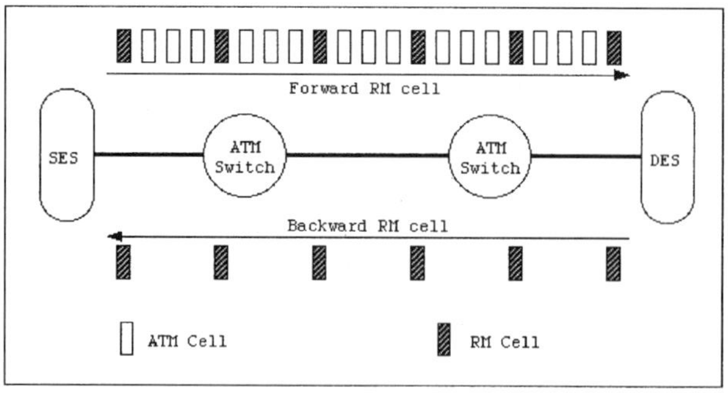

Fig. 1 Rate-based flow control.

tem (SES) and destination end system (DES). The control is a closed-loop control because there are continuous feedback of control information between the network and the SES. Generally, the ABR flow control uses control information from the network to notify the SES to set an appropriate rate. Each switch in the network computes the appropriate rate that is called explicit rate (ER) for all connections which are sharing the same link. The control information such as ER is conveyed from the network to the SES by RM cells [3]–[5]. There are two directions of the RM cell flow, forward and backward, as shown in Fig. 1.

2.1 ABR Parameters

In the ABR service, each SES negotiates several operating parameters with the network at the time of connection setup. The important parameters are peak cell rate (PCR), minimum cell rate (MCR), allowed cell rate (ACR), and so on. PCR is the maximum rate for the SES to be allowed to transmit on the link. The SES can also request the MCR, which is the guaranteed minimum rate. During transmitting data, the SES is allowed to transmit at particular rate that is defined as ACR. The ACR value is dynamically changed between MCR and PCR depending on network condition. At the connection setup or after long idle interval, the ACR is set to the initial cell rate (ICR). The list of parameters used in the ABR service are shown in the Table 1 [3].

2.2 Control Mechanism

The basic mechanism of congestion control in ABR service requires the cooperation of SES and DES to the network to avoid congestion by controlling their rates. Then the SES generates RM cells in every $Nrm - 1$ $(= 31)$ data cells to probe network condition. RM cells travel from the SES to the DES and back to the SES which carry feedback information of the network such as congestion state and rate allocation. The RM cell which travels from the SES to the DES are called forward RM (FRM) cell, and backward RM (BRM) cell from the DES to the SES. The RM cell contains a current cell rate (CCR), PCR, ER, and other parameters required by the SES. Details of the RM cell configuration can be found in [3].

Table 1 ABR parameters.

Lable	Extension	Default value
PCR	Peak cell rate	-
MCR	Minimum cell rate	0
ACR	Allowed cell rate	-
ICR	Initial cell rate	PCR
CCR	Current cell rate	-
TCR	Tagged cell rate	10 cell/s
Nrm	Number of cells between forward RM cells	32
RIF	Rate increase factor	1/16
RDF	Rate decrease factor	1/16
CDF	Cutoff decrease factor	1/16
ADTF	ACR decrease time factor	0.5 ms
TBE	Time buffer exposure	16,777,215
CRM	Missing RM-cell count	TBE/Nrm
FRTT	Fixed round-trip time	-

The switch gives feedback information to the SES in every RM cell arrival. There are three types of switch to give the feedback information[3], [4], [8]:

- "binary" switch, using explicit forward congestion indication (EFCI) bit in the header of data cells to indicate the congestion state.
- "relative rate marking" switch, using congestion indication (CI) bit and no increase (NI) bit in the payload of RM cell to notify the congestion state.
- "explicit rate (ER)" switch, using other ER field in the payload of RM cell to give the allocation rate.

The ER switch is currently used to support ABR service. When a switch receives an RM cell, the switch may reduce the ER value in FRM and/or BRM cell to the calculated fair rate.

When the SES receives a BRM cell, the SES should adapt its rate to the information carried in the RM cell. If CI bit $= 0$, the SES can linearly increase its rate by rate increase factor (RIF), where RIF is determined at call setup. The increasing rate can reach the ER value in the RM cell but should never exceed the PCR. If CI bit $= 1$, the SES must exponentially decrease its rate by using rate decrease factor (RDF), where RDF is also determined at call setup. If the rate is still greater than the returned ER, the SES must further decrease its rate to the returned ER but should not be lower than the MCR.

3. Implementation and Problems of EPRCA

Several control methods have been proposed and the EPRCA is chosen as one of the control mechanisms by the ATM Forum [3]. The EPRCA is a closed-loop rate-based flow control. The RM cells are used to transfer information from the SES to the DES and vice versa. The EPRCA is implemented at ER switch in order to achieve fairness of rate allocation [6].

3.1 EPRCA

3.1.1 Switch Behavior

The basic features of EPRCA can be described as follows [3], [4], [6], [9]. In this algorithm all the connections are assumed to share a common queue with two congestion thresholds QT and DQT. At every time an FRM cell arrives, the switch computes the mean allowed cell rate ($MACR$) using a running exponential weigth average as follows:

$$MACR = (1 - \alpha)MACR + \alpha CCR, \qquad (1)$$

where α is averaging factor and generally is chosen to be 1/16. The FRM cells sent by the SES contain a desired ER in the ER field, the CCR field set to the ACR, and the CI bit set to zero. The switch monitors its load by observing the queue length.

On arrival of BRM cell, when the queue length in the switch buffer exceeds the threshold value QT, congestion state is detected and only those connections whose CCR values are above the $fair\ share$, computed as

$$fair\ share = DPF \times MACR, \qquad (2)$$

are asked to reduce their rates. The other connections whose CCR value are below the $fair\ share$, have no adjustments. DPF is the down pressure factor of switch, and typically set to 7/8. Furthermore when the queue length in the switch buffer exceeds the second threshold value DQT, the heavy congestion is detected and all the connections are asked to reduce their rates.

3.1.2 SES Behavior

The SES adjusts its transmission rate when the SES receives BRM cells. If no congestion information is in the BRM cell, the SES can increase their rate by using the following equations to calculate their allowed cell rate (ACR):

$$ACR \leftarrow \max[MCR, \min(ACR + RIF \times PCR, ER, PCR)]. \qquad (3)$$

But if BRM cell carries some congestion information, the SES must decrease their rate with the following equation:

$$ACR \leftarrow \max(MCR, ACR - ACR \times RDF). \qquad (4)$$

Default value of the ABR parameters are shown in the Table 1.

3.2 Problems of EPRCA

Several problems can be found when we apply the EPRCA in the network, such as unfairness problem [4], oscillation of ACR [6], beat-down problem [6], and etc. In this paper we focus on the beat-down problem. According to [6], the beat-down problem occurs in the binary switch when the switch manages several type of connections which traverse through different number of switches. The connections passing more switches (long-hop connection) have a higher probability of getting their RM cells marked than those passing fewer switches (short-hop connection). Therefore the long-hop connections are given the ACR which are smaller than that of short-hop connections.

We define the beat-down problem as the difference of average ACR between short-hop and long-hop connections is too large. According to our experiment, the ER switch also have the beat-down problem when EPRCA algorithm is applied [10], [11]. Then we try to solve the problem by proposing a new method which modifies the EPRCA as described in the next section.

4. Proposed Method

The beat-down problem occurs when there are various connections which pass different number of hops, if the switch handles those connections in the same manner without taking the number of hops into consideration. We propose a new method to solve the beat-down problem. We modify the EPRCA algorithm with making longer-hop connections more advantageous by applying adaptive thresholds.

The proposed method is described as follows:

- standard threshold
  As the conventional EPRCA method, the proposed method sets 2 threshold values QT and DQT at the switch buffer to detect the congestion condition. The threshold values are set permanently in the conventional method, but the threshold values adaptively changed in the proposed method.
- adaptive threshold
  Let h and d be integers. As shown in the Fig. 2, we consider other threshold values $QT_1, \ldots, QT_h, \ldots$ and $DQT_1, \ldots, DQT_h, \ldots$, as well as the original thresholds QT and DQT. QT_h and DQT_h are defined as follows:

$$QT_h = QT + h \times d,$$

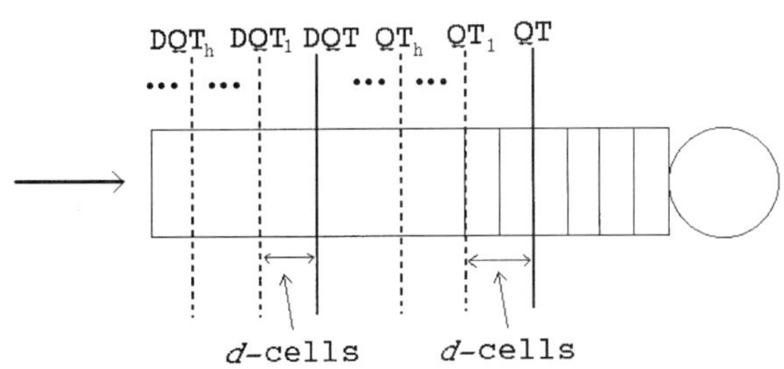

Fig. 2 Proposed method.

$$DQT_h = DQT + h \times d. \quad (5)$$

In the following algorithm, h means the number of hops that the FRM cell passed and d is the increment of threshold.

- RM cell structure
An RM cell contains the information from the SES such as PCR, MCR, CCR, ER. We add the h-field for counting the number of passed switches of the RM cell in the up stream. Then the h value can be used to decide which threshold value should be used. The initial value of the h-field is 0.
- switch behavior
When an FRM cell arrives, the switch computes the MACR and fair share using Eqs. (1) and (2). Then the switch reads the h value from the h-field of the FRM cell and computes QT_h and DQT_h by Eq. (5). The switch uses the QT_h and DQT_h as thresholds instead of QT, DQT. If the queue length Q of the switch buffer is below QT_h, the switch decides that there is no congestion and gives PCR to all connections. If $QT_h < Q \leq DQT_h$, congestion state is detected and only those connections whose CCR values are above the fairshare are asked to reduce their rates. Furthermore, if $Q > DQT_h$, the heavy congestion is detected and all the connections are asked to reduce their rate. Then, the switch increases the h value of the FRM cell by 1 and transmit it to the output link.
- source behavior
We use Eqs. (3) and (4) for increasing and decreasing rate, respectively. Other parameters are used similarly as the conventional EPRCA method.

In the conventional EPRCA, long-hop connections are given smaller ACR than short-hop connections. It is because the long-hop connections have a higher probability of decreasing rate than short-hop connections. Therefore, we use adaptive thresholds to make the rate of long-hop connections not strictly be choked at the switch. By making longer-hop connections more advantageous, we can solve the beatdown problem.

5. Simulation and Evaluation

We make simulation models to show that the proposed method provides better performance than the conventional method. We design several models to evaluate the fairness of ACR distribution and the throughput of each connection, and so on.

5.1 Simulation Model

We consider simulation models with 2, 3, and 4 switches. The 2 switch model is shown in Fig. 3(a). Each node contains 10 connections. The first switch (SW1) accommodates 10 SES's. 5 SES's are terminated to the first DES as short-connection (S1) and the other connections go to the next switch (SW2). New SES's come to the SW2 as short-hop connection (S2), they are terminated to the last DES. Other 5 SES's as a long-hop connection (L) travel from SW1 to SW2 and terminate to the final DES. The EPRCA is applied in the models and we use default EPRCA parameters as

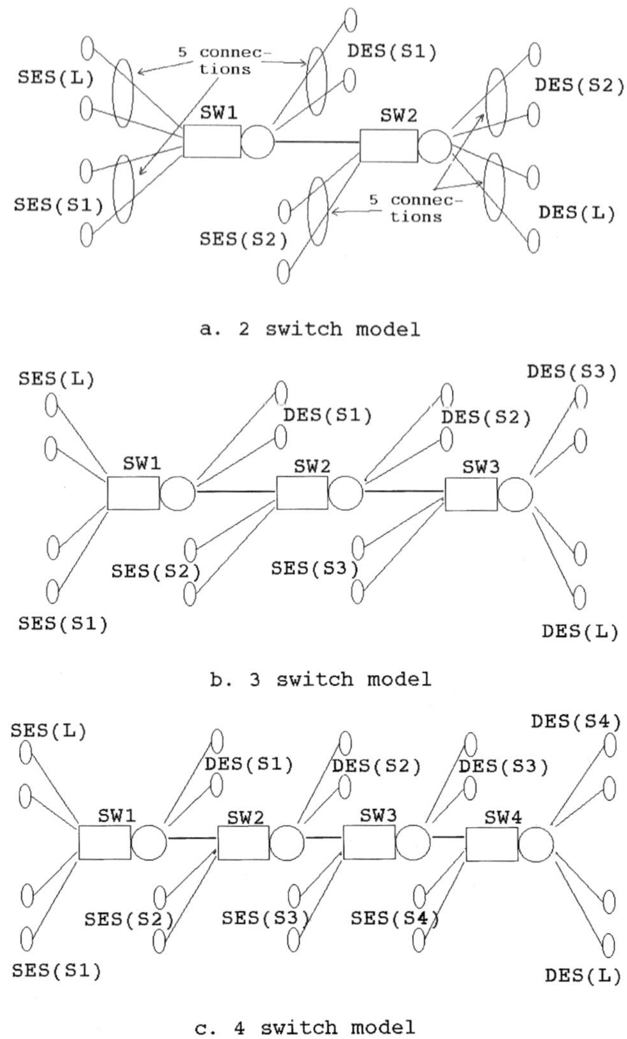

Fig. 3 Simulation models.

Table 2 Simulation condition.

Parameter	Value
PCR	150 Mbps
MCR	1.5 Mbps
Nrm	32 cells
RIF, RDF	$\frac{1}{16}, \frac{1}{16}$
α	$\frac{1}{16}$
DPF	$\frac{7}{8}$
QT, DQT	5, 10 cells
Switch buffer	35,40,45,50 cells
d for proposed method	5 cells
Load per SES	0.06-0.12
Simulation time	1,000,000 cell times

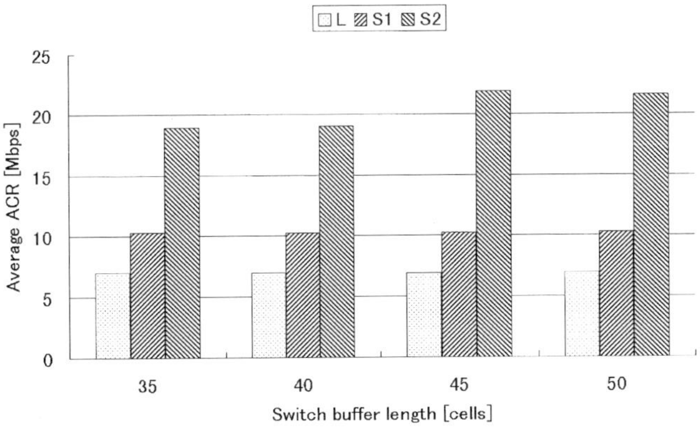

Fig. 4 Average ACR in conventional method, 2 switch model.

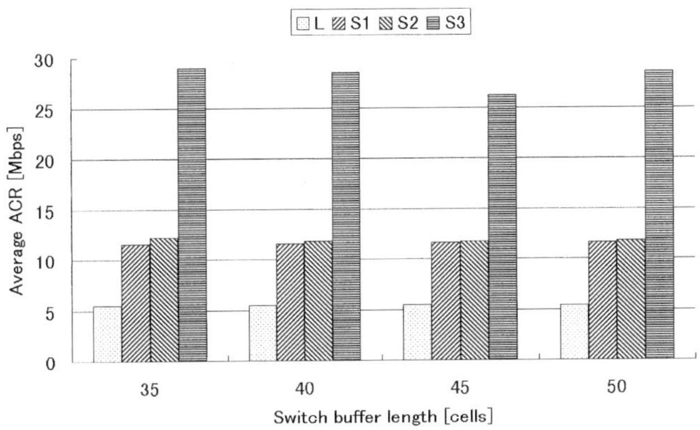

Fig. 5 Average ACR in conventional method, 3 switch model.

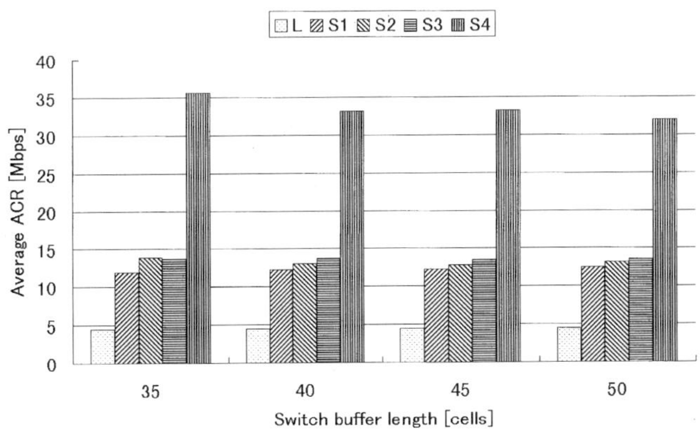

Fig. 6 Average ACR in conventional method, 4 switch model.

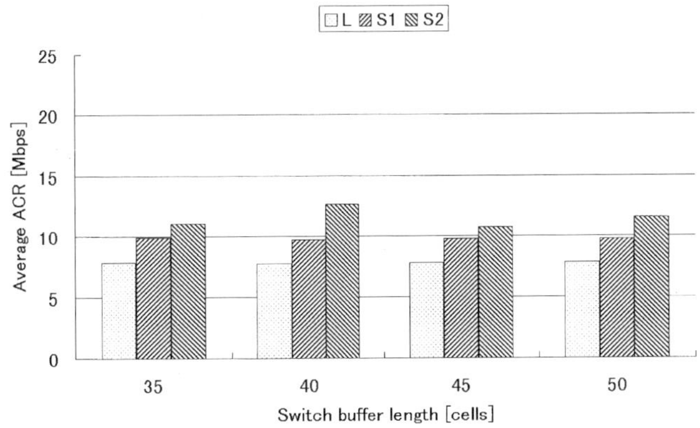

Fig. 7 Average ACR in proposed method, 2 switch model.

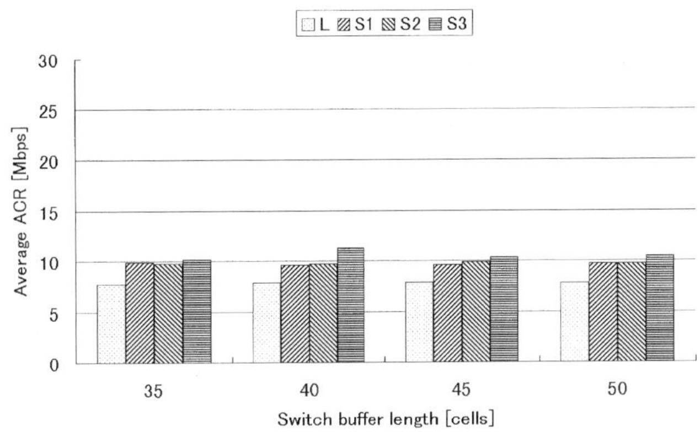

Fig. 8 Average ACR in proposed method, 3 switch model.

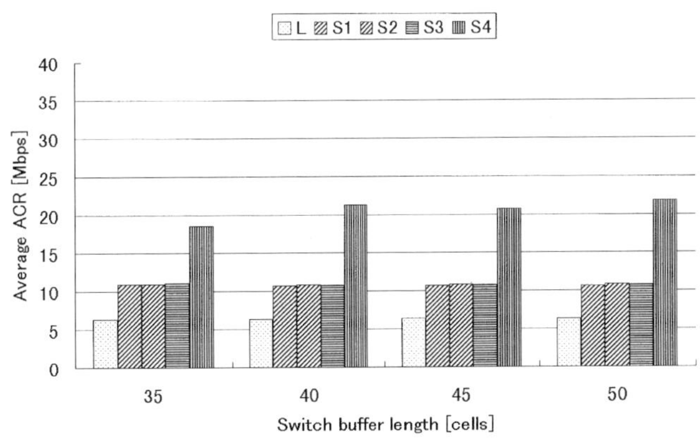

Fig. 9 Average ACR in proposed method, 4 switch model.

shown in the Table 1. We assume that the EPRCA is applied in a local area network (LAN). Because the node distances are so short, then we can assume that the transfer delay can be ignored.

It is important to decide an appropriate value of QT and DQT for preventing a beat-down problem. If we set the QT and DQT too small, the network decides that the congestion occurs very frequently, then the connections are choked and the throughput will go down. On the other hand, if the QT is too large, the network misses the congestion, then the total ACR exceeds the link capacity. Furthermore, if the DQT is too large, the beat-down problem becomes worse. This is because the connections can update their $fair\ shares$ many time before the queue length reaches DQT, so the difference of the $fair\ shares$ between a short-hop and a long-hop connections enlarges. As our experiment results, we set QT and DQT value to 5, 10 cells respectively as the best condition of EPRCA method. Other parameters are shown in the Table 2.

Similarly to 2 switch model, 3 and 4 switch models contain long-hop and short-hop connections. 5 SES's from the first node travel through all nodes to the final DES. New 5 SES's enter to each node and leave immediately for the corresponding DES. We call them short-hop connections. The simulation models can be shown in Figs. 3(b),(c).

5.2 Comparison of ACR Allocation

The average ACR is defined as the distributed ACR to each connection (SES) averaged over the simulation time. The results are shown in Figs. 4–9. We set thresholds of switch buffer $QT = 5$ cells, $DQT = 10$ cells. The buffer size of each switch is variable from 35 to 50 cells. The average load of each SES is 0.1, then the total offered load of 10 SES's is 1.0.

The beat-down problem appears as depicted in Fig. 4. The long-hop connection (L) is given smaller ACR value than the short-hop connections (S1, S2). This is because the long-hop connection passes many switches, so it has a high probability to meet the congestion at one of the switches. On the other hand, the problem are solved when we apply the proposed method as shown in the Fig. 7. Because of the advantage of L connections, they can get more ACR than the conventional method.

Similarly, for 3 switch model we have the same phenomenon. See Figs. 5 and Figs. 8. In the conventional method, the beat-down problem appears, especially at the short-hop connection S3. The S3 has the highest ACR among connections. We can solve the problem by applying the proposed method. Each connection has almost the same ACR values as seen in Fig. 8.

In the 4 switch model, we can also see that the beat-down problem still occurs in the conventional method, Fig. 6. The problem can be solved by applying the proposed method as shown in Fig. 9.

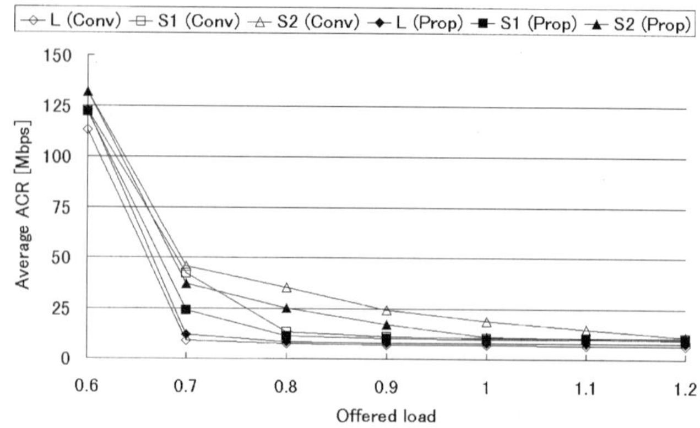

Fig. 10 Average ACR for 2 switch model.

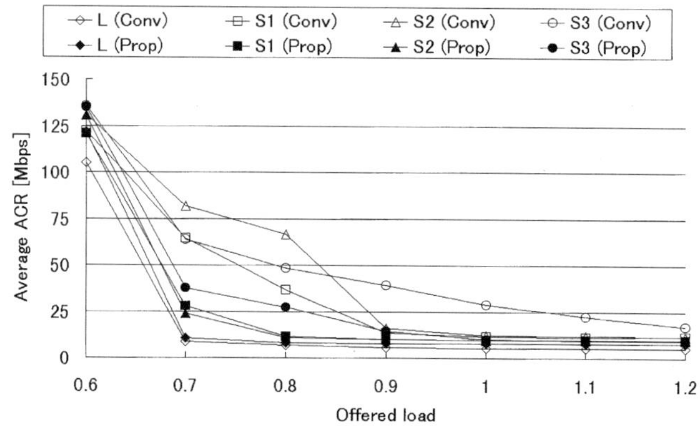

Fig. 11 Average ACR for 3 switch model.

According to the results, the conventional method distributes higher ACR than the proposed method to short-hop connections, especially to the last short-hop connections, i.e., S2 in Fig. 4, S3 in Fig. 5, and S4 in Fig. 6. On the other hand, by the proposed method, the ACR are distributed fairly to every connection.

Now we evaluate the ACR which are given to each SES versus offered load. Figure 10, Fig. 11, and Fig. 12 show the average ACR of each connection at 2, 3, and 4 switch models, respectively. We find that the large ACR are distributed when the offered load is low. The ACR are choked at the high offered load because the heavy congestion is detected. The proposed method shows better performance by appropriate ACR allocation when offered load become high. For the high offered load ≥ 0.7, the proposed method allocates the ACR to long-hop and short-hop connections fairly compared to the conventional method. Because the long-hop connection is applied higher threshold QT_h and DQT_h, the ACR of the long-hop connection is not easy to be choked.

5.3 Comparison of Throughput

Next, let us investigate the comparison of throughput

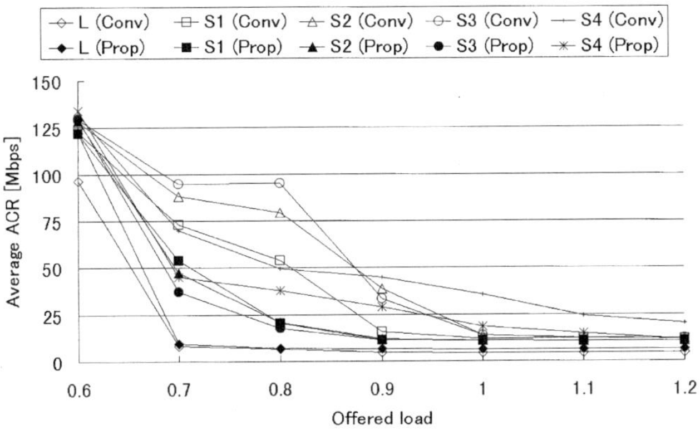

Fig. 12 Average ACR for 4 switch model.

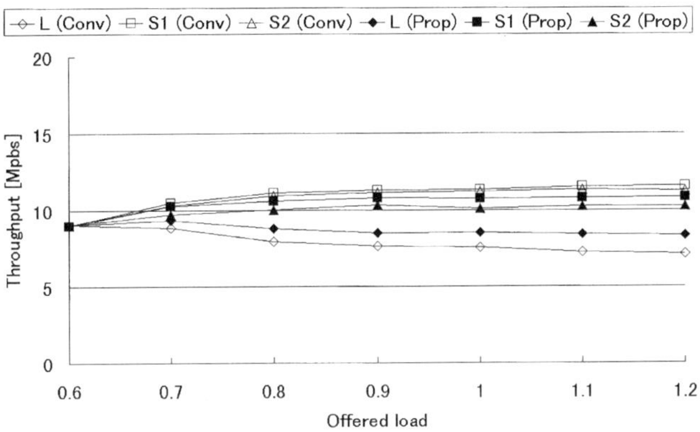

Fig. 13 Throughput for 2 switch model.

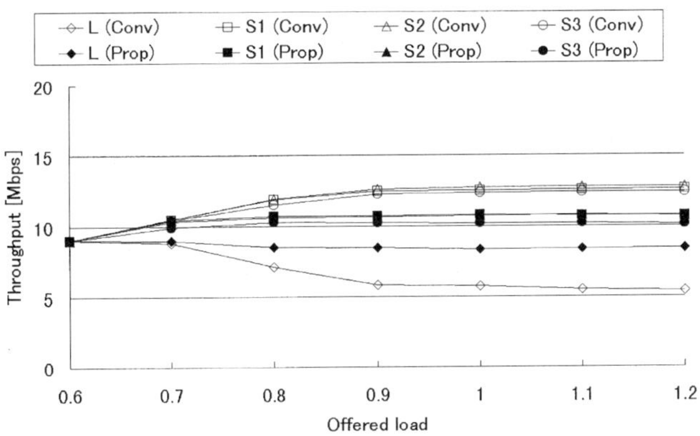

Fig. 14 Throughput for 3 switch model.

performance between the conventional EPRCA and the proposed method. We define the throughput as the average transmitted information rate during the simulation. Figure 13, Fig. 14, and Fig. 15 show the throughput of each connection at 2, 3, and 4 switch models, respectively. We see that the difference of the throughputs between long-hop and short-hop connections by the proposed method is smaller than that by the conventional method. We will examine the difference numerically with the following fairness index.

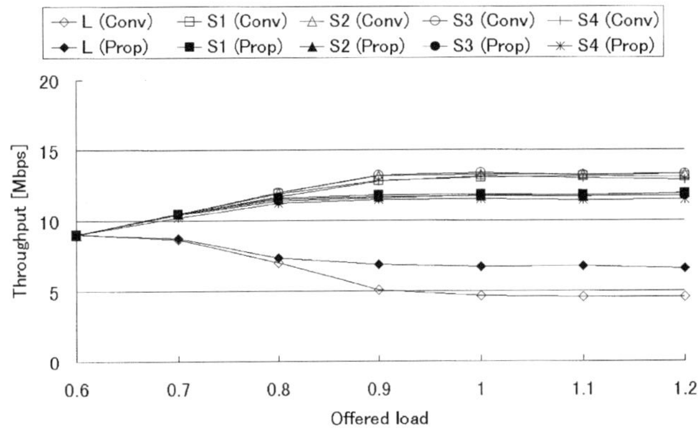

Fig. 15 Throughput for 4 switch model.

Table 3 Fairness Index for 2 switch model.

Offered load	0.6	0.7	0.8	0.9	1.0	1.1	1.2
Conventional	1.00	1.19	1.40	1.48	1.50	1.59	1.62
Proposed	1.00	1.10	1.21	1.28	1.27	1.28	1.30

Table 4 Fairness Index for 3 switch model.

Offered load	0.6	0.7	0.8	0.9	1.0	1.1	1.2
Conventional	1.00	1.20	1.68	2.16	2.22	2.33	2.37
Proposed	1.00	1.16	1.26	1.27	1.30	1.28	1.27

Table 5 Fairness Index for 4 switch model.

Offered load	0.6	0.7	0.8	0.9	1.0	1.1	1.2
Conventional	1.00	1.21	1.71	2.61	2.88	2.89	2.91
Proposed	1.00	1.20	1.58	1.71	1.75	1.74	1.82

5.3.1 Fairness Index

We here define the *fairness index* as the ratio of the maximum throughput among connections to the minimum throughput. If the fairness index is close to 1.0, we can say that the throughput of every connection is fair. We calculate the fairness indices for the conventional and proposed methods and show them in Tables 3, 4, and 5. We see from these tables that the fairness index of the proposed method is smaller than that of the conventional method. The fairness between the long-hop and short-hop connections is improved by the proposed method.

5.4 Allocation of ACR Proportional to Throughput

We define the *proportion index* as the ratio of the average ACR over all the connections to the average throughput. If the proportion index is close to 1.0, we can say that the ACR is allocated proportionally to the throughput for every connection. We show in Table 6 the proportion indices in the case of the offered load is 0.9. We see that the proportion index of the proposed method is smaller than that of the conventional method. The ACR is proportionally allocated by the

Table 6 Proportion Index.

model	2 switch	3 switch	4 switch
Conventional	1.41	1.72	2.41
Proposed	1.19	1.05	1.31

Table 7 Total throughput at each switch on average.

model	2 switch	3 switch	4 switch
Conventional [Mbps]	94.07	91.47	89.23
Proposed [Mbps]	94.69	94.66	92.12

proposed method.

5.5 Total Throughput at a Switch

Total throughput at a switch is defined as the sum of the throughputs of all the connections which enter to the switch. We calculate the total throughputs at each switch and take the average over all switches and show the average values in Table 7 in the case of the offered load is 1.0. We see that the total throughputs of the proposed method are larger than the conventional method, either in the 2, 3, or 4 switch models. Although the throughputs of short-hop connection in the proposed method are smaller than that in the conventional method, the throughputs of long-hop connection are larger. As a result, the total throughput at a switch becomes larger than that of the conventional method.

6. Conclusion

When the original EPRCA is applied in the switch, several problems occur. The beat-down problem is one of the problems that appears not only in the binary switch but also in the ER switch. We show the problem by simulation models. Then we proposed a new algorithm based on the EPRCA to solve the beat-down problem. By the simulation, the average ACR is allocated fairly either to long-hop and short-hop connections. We find that the throughputs are fair for both long-hop and short-hop connections. Moreover, the proposed method allocates the ACR proportionally to the achieved throughput. We find also that the total throughput at each switch is larger than the conventional method.

References

[1] L.G. Cuthbert and J.-C. Sapanel, "ATM the broadband telecommunications solution," IEE, London, 1994.

[2] F. Bonomi and K.W. Fendick, "The rate-based flow control framework for the available bit rate ATM service," IEEE Commun. Mag., vol.9, no.2, pp.25–39, March/April 1995.

[3] The ATM Forum, "Traffic Management Specification version 4.0," ftp://ftp.atmforum.com/pub/approved-specs/af-tm-0056.000.pdf, April 1996.

[4] R. Jain, "Congestion control and traffic management in ATM networks: Recent advances and a survey," Computer Networks and ISDN System, vol.28, pp.1723–1738, 1996.

[5] R. Jain, et al., "Source behavior for ATM ABR traffic management: An explanation," IEEE Commun. Mag., vol.34, pp.50–57, Nov. 1996.

[6] A. Arulambalam, et al., "Allocating fair rates for available bit rate service in ATM networks," IEEE Commun. Mag., vol.34, pp.92–100, Nov. 1996.

[7] A. Pitsilides, et al., "Effective control of traffic flow in ATM networks using fuzzy explicit rate marking (FERM)," IEEE J. Sel. Areas Commun., vol.15, no.2, pp.209–225, Feb. 1997.

[8] T. Kubo, "Traffic issues in ATM LAN—Principle and practice," IEICE Trans. Commun., vol.E81-B, no.5, pp.840–848, May 1998.

[9] M. Hashmani, et al., "ABR congestion control in ATM networks and proposal of EPRCAM," IEICE Trans. Commun., vol.E81-B, no.11, pp.2064–2071, Nov. 1998.

[10] H. Prihanto and K. Nakagawa, "Problems in enhanced proportional rate control algorithm (EPRCA)," IEICE Shinetsu-Shibutaikai, Oct. 1997.

[11] R. Soelistijorini, "A study of rate-based flow control in ATM networks," Bachelor Thesis, Nagaoka University of Technology, March 1998.

[12] H. Saito, et al., "Performance issues in public ABR service," IEEE Commun. Mag., vol.34, pp.40–48, Nov. 1996.

[13] H.B. Chiou and Z. Tsai, "An age priority packet discarding scheme for ATM switches on internet backbone," IEICE Trans. Commun., vol.E81-B, no.2, pp.380–391, Feb. 1998.

[14] D.H.K. Tsang and W.K.F. Wong, "A two-level flow control scheme for ABR traffic in ATM networks," IEICE Trans. Commun., vol.E79-B, no.11, pp.1633–1640, Nov. 1996.

High-Performance Backbone Network Technology edited by Naoaki Yamanaka
ISBN: 0-8247-5321-6 DOI: 10.1081/(120025704)
Published 2004 by Marcel Dekker, Inc.

Method of Implementing GFR Service in Large-Scale Networks Using ABR Control Mechanism and Its Performance Analysis

Ryoichi Kawahara, Yuki Kamado, Masaaki Omotani, and Shunsaku Nagata

SUMMARY This paper proposes implementing guaranteed frame rate (GFR) service using the available bit rate (ABR) control mechanism in large-scale networks. GFR is being standardized as a new ATM service category to provide a minimum cell rate (MCR) guarantee to each virtual channel (VC) at the frame level. Although ABR also can support MCR, a source must adjust its cell emission rate according to the network congestion indication. In contrast, GFR service is intended for users who are not equipped to comply with the source behavior rules required by ABR. It is expected that many existing users will fall into this category. As one implementation of GFR, weighted round robin (WRR) with per-VC queueing at each switch is well known. However, WRR is hard to implement in a switch supporting a large number of VCs because it needs to determine in one cell time which VC queue should be served. In addition, it may result in ineffective bandwidth utilization at the network level because its control mechanism is closed at the node level. On the other hand, progress in ABR service standardization has led to the development of some ABR control algorithms that can handle a large number of connections. Thus, we propose implementing GFR using an already developed ABR control mechanism that can cope with many connections. It consists of an explicit rate (ER) control mechanism and a virtual source/virtual destination (VS/VD) mechanism. Allocating VSs/VDs to edge switches and ER control to backbone switches enables us to apply ABR control up to the entrance of a network, which results in effective bandwidth utilization at the network level. Our method also makes it possible to share resources between GFR and ABR connections, which decreases the link cost. Through simulation analysis, we show that our method can work better than WRR under various traffic conditions.

key words: *ATM, GFR, ABR, large-scale networks, VS/VD, explicit rate control, WRR*

1. Introduction

Guaranteed frame rate (GFR) service has been attracting attention as a new asynchronous transfer mode (ATM) service category. It is intended for users who are either unable to specify traffic parameters needed to request most ATM services, or not equipped to comply with the source behavior rules required by existing ATM services. Traffic parameters for most existing applications cannot be predicted; however, to achieve good throughput, certain cell-loss requirements must be met. For these applications, a static allocation of bandwidth, such as a constant bit rate (CBR) or variable bit rate (VBR), would be inadequate or wasteful because their traffic characteristics are unpredictable. In available bit rate (ABR) service, on the other hand, bandwidth is not allocated statically. Instead, the peak cell rate (PCR) and minimum cell rate (MCR) are specified at call set-up, and the cell emission rate of each virtual channel (VC) is controlled between the PCR and MCR based on feedback from the network. This feedback control makes it possible to utilize the bandwidth effectively without cell loss when the traffic characteristics of the cell stream are unpredictable. However, a source must adjust its cell emission rate according to the network indication. Thus, GFR is being standardized as a new service to provide an MCR guarantee to VCs at the frame level, while allowing the user to send in excess of the MCR. The excess traffic will be delivered on the basis of available network resources. That is, a user can always attain MCR throughput, and may be able to achieve higher throughput if the network is underloaded; this results in effective bandwidth utilization.

The performance of various GFR implementations has been studied (e.g., [1]–[4]). These results showed that weighted round robin (WRR) with per-VC queueing at each switch had the best performance. However, WRR is hard to implement in a switch supporting a large number of VCs because it needs to determine in one cell time which VC queue should be served. In addition, it may cause ineffective bandwidth utilization at the network level because its control mechanism is closed at each node without taking account of other node conditions in the network. On the other hand, accompanied with standardization of ABR service, several kinds of ABR control algorithms have been developed, and some of these can handle a large number of VCs. Thus, in this paper, we propose implementing GFR using an already developed ABR control mechanism, which consists of virtual source/virtual destination (VS/VD) and explicit rate (ER) control

Manuscript received February 16, 1999.
Manuscript revised June 21, 1999.

†The author is with NTT Information Sharing Platform Laboratories, Musashino-shi, 180-8585 Japan.
††The authors are with NTT Network Service Systems Laboratories, Musashino-shi, 180-8585 Japan.
†††The author is with NTT Phoenix Communications Network, Inc., Tokyo, 100-0004 Japan.

IEICE TRANS. COMMUN., Vol. E82-B, No. 12, pp. 2081–2094, December 1999 mechanisms [5] that can handle many VCs. (As explained in Sect. 2, the ER control and VS/VD mechanisms are also useful and necessary to support ABR service in large-scale networks.) Allocating VSs/VDs to edge switches and ER control to backbone switches makes it possible to regulate excess traffic adequately at the entrance of the network, which results in effective bandwidth utilization at the network level. Although VSs/VDs also require per-VC queueing, like WRR, our method does not need to allocate VSs/VDs at backbone switches, where a lot of VCs are multiplexed into a high-speed link, while WRR requires per-VC queueing and scheduling at every switch. Our method can also lower the GFR implementation cost when the ABR control mechanisms have already been implemented for ABR service. Furthermore, it enables bandwidth to be shared between ABR-VCs and GFR-VCs, which also lowers the link cost.

To evaluate the effectiveness of our proposed method from the users' viewpoint, we investigated the performance of transmission control protocol (TCP) when our method is applied to ATM networks. As related works, the performance of TCP over ABR has been investigated (e.g., [12]–[17]). However, most works have assumed that ABR service is applied to each VC between the source end system and the destination end system (that is, end-to-end ABR). Though Refs. [13] and [14] have investigated the performance of TCP over ATM when ABR was applied only to backbone ATM networks, these works have neither investigated the applicability of ABR control to implementation of GFR service nor discussed the effective bandwidth utilization at the network level.

We begin by explaining ABR service and the ABR control algorithm used for our GFR implementation. Then, we describe our GFR implementation method, and compare it with WRR from the viewpoint of scalability to the number of multiplexed connections, performance, and cost. After that, by simulating various traffic conditions, we show that our method can achieve minimum throughput guarantee and fairness like WRR and more effective bandwidth utilization than WRR. In addition, we show that our method can work well even when ABR and GFR-VCs are coexisting.

2. ABR Service

In this section we describe the ABR flow control mechanism [5]. Figure 1 summarizes the behavior of ABR service. In ABR, the cell-emission rate of each VC is controlled according to the network-congestion level by using a closed-loop control indication. We use a Resource Management (RM) cell to provide this closed loop. A forward RM cell is sent every Nrm (e.g., 32) cells from a Source End System (SES) along an ATM connection to a Destination End System (DES). After reaching the DES, the forward RM cell returns to the

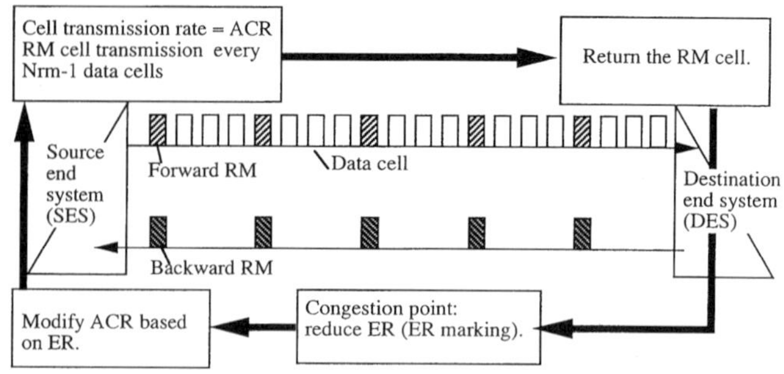

Fig. 1 ABR behavior (ER marking).

SES as a backward RM cell. During this round trip, the network congestion level is written in the RM cell and is reported to the SES through either an Explicit Forward Congestion Indication (EFCI), Relative Rate (RR), or Explicit Rate (ER) marking. Switches supporting EFCI marking only mark the EFCI bit of data cells. When the DES receives an EFCI-marked data cell, it marks the Congestion Indication (CI) bit in an RM cell returning to the SES. The ER and RR marking switches do not need to set the EFCI bit; instead they can directly process the RM cell. RR marking switches set the CI bit in the RM cells returning to the SES from the DES. Therefore, RR and EFCI marking are binary feedback that indicates whether or not the network is congested. In contrast, ER marking switches can process an Explicit Rate field in an RM cell; therefore, they can let the sources know explicitly what rate they can transmit at. In this paper, we focus on ER marking because it is the main mode of ABR service, and it can achieve better performance than others [6].

In Sects. 2.1 and 2.2, we explain source behavior and ER marking switch behavior. Section 2.3 shows VS/VD behavior, which is used later in our GFR implementation method.

2.1 Source Behavior

The cell emission rate at which the SES is allowed to transmit is called the Allowed Cell Rate (ACR). The ACR is initially set to the Initial Cell Rate (ICR).

When the SES sends a forward RM cell, the SES writes the current ACR value in the Current Cell Rate (CCR) field of the RM cell. When the SES receives a backward RM cell, the SES updates the ACR by using the following formula.

If the CI bit in the RM cell=1, then

$$ACR := ACR - ACR\ RDF, \tag{1}$$

else if the Not Increase (NI) bit in the RM cell = 0, then

$$ACR := ACR + PCR\ RIF. \tag{2}$$

$$ACR := \min(ACR, PCR). \tag{3}$$

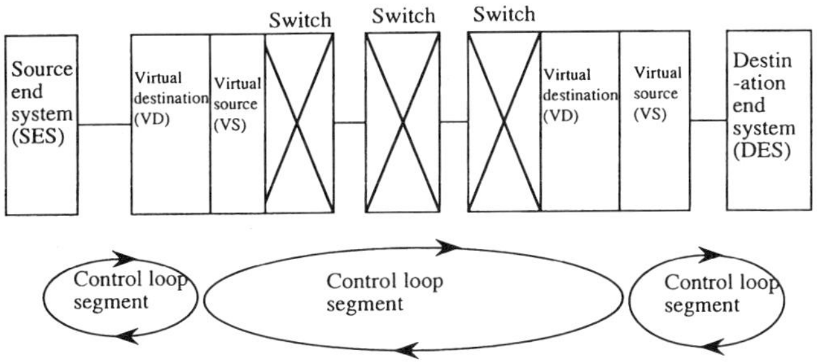

Fig. 2 Virtual source/virtual destination approach.

$$ACR := \min(ACR, ER). \quad (4)$$

$$ACR := \max(ACR, MCR). \quad (5)$$

Here, ER is a value written in the ER field of a backward RM cell, RDF is the Rate Decreasing Factor, and RIF is the Rate Increasing Factor. PCR, MCR, ICR, RDF, and RIF are negotiation parameters, specified at the call set-up. In general, ER marking switches do not set either the CI bit or the NI bit, so RDF is not used in the case of an ER marking.

2.2 ER-Switch Behavior

When an ER marking switch receives an RM cell belonging to VCi, it either calculates the ER for VCi or reads the ER value that has already been calculated. It also writes the ER to the ER field of the RM cell if it is less than the value already there.

2.3 VS/VD Behavior

We can divide the ABR control loop into segments by using the VS/VD technique, as shown in Fig. 2. Virtual sources and virtual destinations are defined along the route of the ABR connection to divide the connection into separately controlled segments. Each VS mimics the behavior of an SES; backward RM cells received by a VS are removed from the connection. Each VD mimics the behavior of a DES; forward RM cells received by a VD are returned to the virtual or actual source, and not transmitted to the next control segment of the connection. The VS/VD technique is mainly used to improve performance by reducing the delay in each control segment. In addition, it is also used, to protect users from malicious users by isolating the latter, and to improve network security by enabling the network to correct information in the RM cell fields that may have been generated by malicious users. Therefore, the VS/VD mechanism is necessary for large-scale networks, especially for public networks that assure network security [7].

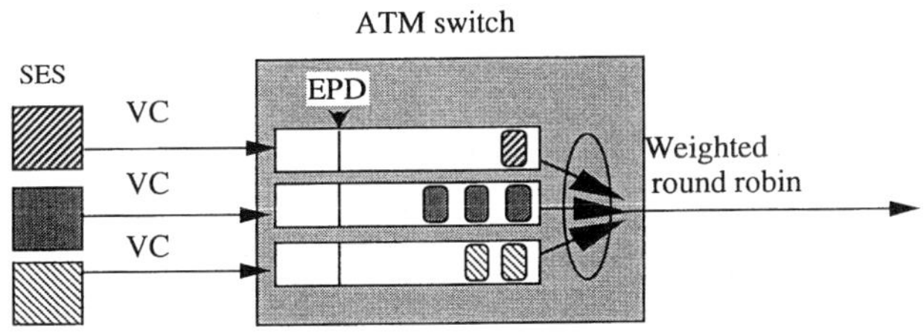

Fig. 3 Per-VC queueing with weighted round robin (WRR).

3. GFR Service and Its Conventional Implementation

In GFR, the values of PCR and MCR are specified at the call set-up. The user can always send cells at PCR, and the network must guarantee MCR throughput at the frame level. That is, the excess traffic above the MCR may be discarded at the frame level when the network is congested. On the other hand, when the network is not congested, the excess traffic will be delivered within the limits of available resources, which gives the user higher throughput than MCR. Furthermore, the service specifies that the excess traffic from each user should have access to a fair share of the available resources. Unlike ABR, cell discards may occur; instead, the GFR service need not give users explicit feedback regarding the current level of network congestion.

3.1 GFR Implementations [18]

To date, two possible implementations for GFR have been suggested [2]:

- Using per-VC queueing and per-VC scheduling (e.g., weighted round robin (WRR))
- Using single FIFO (first-in first-out) queueing and per-VC tagging.

Hereafter, we call the former "Per-VC_WRR" or simply WRR," and the latter "FIFO_tagging." References [1]–[4] reported that Per-VC_WRR has better performance than FIFO_tagging in terms of throughput, fairness, and MCR guarantee. To implement Per-VC_WRR, one must prepare a per-VC queue at each switch (see Fig. 3). Furthermore, WRR schedules every cell time which VC queue should be served, according to the MCR and the predetermined weight of each VC. Thus, Per-VC_WRR makes it possible to transmit cells from each VC at a rate of at least MCR. To discard cells at the frame level, the early packet discard (EPD) mechanism is implemented. When the first cell of a frame arrives at the VC buffer, if the queue of the VC buffer is longer than a predetermined EPD threshold, then the first cell and every subsequent cell of the frame are discarded.

On the other hand, FIFO_tagging is implemented by executing per-VC tagging at the frame level at the

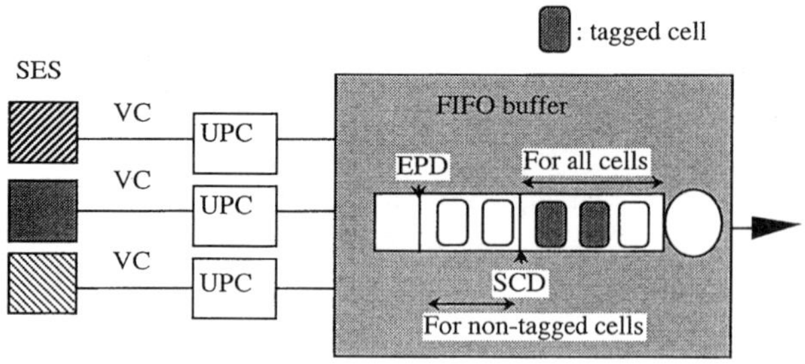

Fig. 4 FIFO queueing with per-VC tagging.

usage parameter control (UPC) point, as shown in Fig. 4. At the UPC point, a counter X is increased by one every time a cell arrives and decreased by one every time $1/MCR$ has elapsed. When the first cell of a frame arrives at the UPC point, if X is larger than a predetermined threshold L for detection of the burst in excess of MCR (which is related to MBS), the first cell and every subsequent cell of the frame are tagged. Here, MBS is the maximum burst size. Using this tagging, selective cell discard (SCD) is executed at the FIFO buffer shared among VCs (Fig. 4). When the first cell of a frame arrives at the FIFO, if the queue is longer than the EPD threshold, the first cell and every subsequent cell of the frame are discarded. If the SCD threshold < the queue length < the EPD threshold and the first cell is tagged, then the first cell and every subsequent cell of the frame are discarded. Giving priority to the acceptance of non-tagged cells enables each VC to attain MCR throughput.

3.2 Problems of WRR

Although WRR has good performance, it has the following problems, as explained in the introduction:

(1) difficulty of being implemented in a switch supporting a large number of VCs and
(2) ineffective bandwidth utilization at the network level.

The first problem is caused by per-VC queueing and scheduling. WRR must prepare a per-VC queue at each switch. In addition, WRR needs to schedule every one cell time which VC queue should be served, taking account of other VCs' conditions (e.g., whether or not the VC queue is empty). Therefore, WRR is hard to implement in a switch multiplexing a large number of VCs to a high-speed link.

We illustrate the second problem by the following example. Figure 5 consists of three WRR switches, two 20-Mbps links, and three groups of VCs; 10 VCs fro group 1 pass through both Link 1 and Link 2, 10 VCs from group 2 pass through only Link 1, and 30 VCs from group 3 pass through only Link 2. Let the offered load of each VC be 2 Mbps. Here, the weight of WRR to each VC is set to 1. In this case, switch 1 allocates 1 Mbps (= Link 1 capacity/no. of VCs multiplexed at

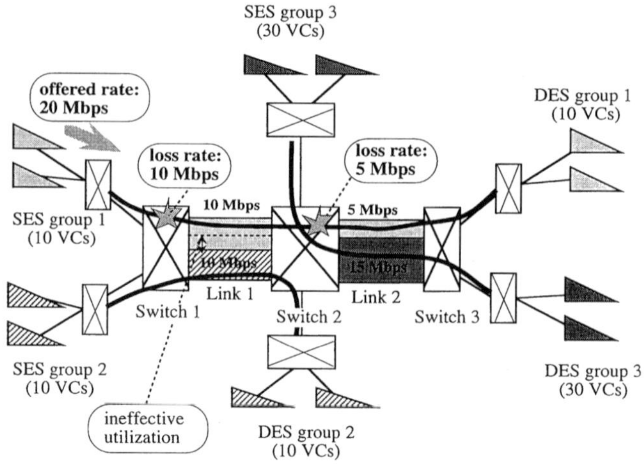

Fig. 5 Example of ineffective bandwidth utilization under WRR.

Link 1) to each VC of groups 1 and 2. Thus, the VC of group 1 transmits cells at 1 Mbps to switch 2 via Link 1. However, switch 2 allocates only 0.5 Mbps (= Link 2 capacity/no. of VCs multiplexed at Link 2) to each VC of groups 1 and 3, which causes cell loss not only at switch 1 but also at switch 2. Therefore, the VCs of group 1 utilize 5 Mbps (= (1 Mbps−0.5 Mbps)*10 VCs) ineffectively at Link 1.

To solve these problems while maintaining the same good performance as WRR, we propose the GFR implementation method described in the next section.

4. Proposed GFR Implementation Using ABR Mechanism

4.1 Proposed Method

Our method allocates a VS/VD to each VC at the entrance of the network (i.e., at the edge switches), and uses an ER marking mechanism at backbone FIFO-switches (see Fig. 6). By doing this, we can regulate excess traffic adequately at the entrance of the network, which results in effective bandwidth utilization at the network level (solving problem (2)). We illustrate this using the same example as in Sect. 3.2. Figure 7 shows the bandwidth allocation of our method. Unlike WRR (Fig. 5), our method can regulate excess traffic at the edge switch by adequately allocating bandwidth to each group. Allocating 15 Mbps from Link 1's capacity to VCs of group 2 makes it possible to utilize the bandwidth of Link 1 effectively. We will confirm this observation by simulation analysis in Sect. 6(iii).

As for problem (1), although the VS/VD method also requires per-VC queueing like in WRR, scheduling processing of each VC's cell transmission time under the VS/VD method is easier than that under WRR. This is because the VS/VD method can determine each VC's cell transmission time only based on the VC's

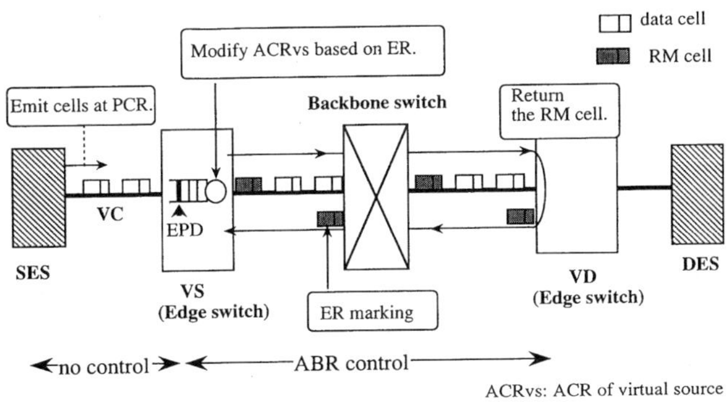

Fig. 6 Proposed GFR implementation.

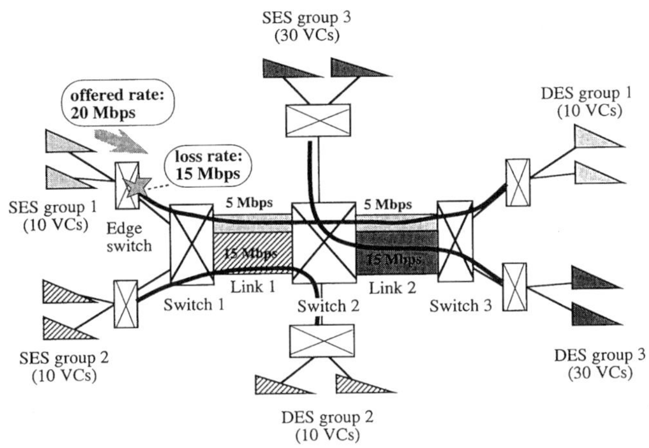

Fig. 7 Example of the effect of the proposed method.

ACR, independently of other VCs' conditions, unlike WRR (see Sect. 3.2). In addition, our method does not need to allocate VSs/VDs at backbone switches, where a lot of VCs are multiplexed into a high-speed link, while WRR requires per-VC queueing and scheduling at every switch. Another merit of our method from the implementation viewpoint is that our method can lower the amount of necessary buffer size in a network, compared with WRR (which requires per-VC queue at every switch).

In GFR, a source need not adjust its cell emission rate; therefore, cell loss may occur at the VS buffer when allowed cell rate of the VS ($ACRvs$) is decreased by the ABR control because of network congestion. Thus, we implement the EPD mechanism at the VS buffer in order to discard cells at the frame level (Fig. 6).

4.2 Choice of ER Control Algorithms

Several kinds of ER control algorithms have been proposed because ABR switch behavior is not completely specified in the international standards bodies such as ITU-T and ATM Forum. In considering GFR implementation in large-scale networks, we should choose an algorithm that can:

(1) be implemented in a switch supporting many multiplexed connections and

(2) handle various long propagation delays and numbers of connections.

In addition, to allocate bandwidth to each VC according to its relative service price, which is attained by WRR, the chosen algorithm should

(3) provide weighted bandwidth allocations.

In this paper, we chose our previously proposed ER algorithm [8] as one satisfying these three requirements, as confirmed in our simulation study [8] through comparison with other ER algorithms (e.g., CAPC2 (congestion avoidance using proportional control version 2) [9] and FMMRA (fast max-min rate allocation) [10]).

The details of our algorithm are given in Appendix A. Our algorithm calculates ER by using the total cell arrival rate ($Rate$) and mean allowed cell rate of active connections ($MACR$), which enables it to satisfy the second requirement (see Appendix B). $Rate$ is calculated by measuring the number of arriving cells N offered at the link during time T. The calculation of $MACR$ uses the ACR value in the CCR field of a forward RM cell. Every time a forward RM cell arrives at the switch, the $MACR$ is calculated using a running exponential average. Since this $MACR$ updating does not need to keep the per-VC ACR value at the switch, our algorithm is easy to implement in a large-scale network (satisfying the first requirement). The algorithms that need to store the condition of each VC, including per-VC ACR, like FMMRA [10], are hard to implement in a switch supporting a large number of VCs in large-scale networks. This is because the memory needed to store the per-VC conditions is enormous. In addition, the computational complexity is proportional to the number of VCs because the condition of each VC (e.g., active or not) must be determined and updated every time ER is calculated. Compared with these algorithms, our algorithm is easy to implement in a large-scale network, because it uses only $Rate$ and $MACR$.

We can satisfy the third requirement, by updating $MACR$ using ACR/w instead of ACR, and by multiplying ER by w, where w is a predetermined weight of a VC. (By writing w into an RM cell at the virtual source, we avoid the need to keep w of each VC at backbone switches.) The $MACR$ when ACR/w is used means the mean cell rate normalized by the weight of each VC. For example, when a forward RM cell whose ACR is 20 Mbps and weight is 2 arrives at a switch, the $MACR$ is updated after normalizing this ACR by the weight of the VC (the ACR used here for $MACR$ updating is 10 Mbps). The ER is then calculated as an explicit rate to be allocated to each VC with a weight of 1, using the $MACR$ updated above. Thus, when a backward RM cell arrives at a switch, the weight is read from the cell, and after multiplying the calculated ER by this weight, the multiplied value is written into the ER field of the backward RM cell as the weighted explicit rate for the VC. By doing this, our algorithm can provide weighted bandwidth allocation to each VC [8].

4.3 Merits of Our Method

As pointed out in Sect. 4.1, our method is
(i) easy to implement in a switch supporting a large number of VCs and
(ii) can utilize bandwidth effectively at the network level.
In addition, if we have already implemented an ER control mechanism and VS/VD mechanism for ABR services in a switch, our method
(iii) can implement GFR at a lower cost,
because we only need to add EPD mechanisms at VS buffers. Furthermore, since our method can use ABR control mechanisms not only for ABR service but also for GFR service, it enables to share link capacity between ABR-VCs and GFR-VCs, which
(iv) can lower the link cost.

5. Evaluation Criteria and Evaluation Models

In this section, we establish criteria for evaluating our method and build simulation models for evaluation. The aim of GFR service is to guarantee MCR throughput to each VC and to achieve efficient and non-discriminatory bandwidth utilization among a number of active VCs. In this paper, we evaluate these criteria by measuring each VC throughput not only at the cell level but also at the TCP level. This is because TCP-level throughput indicates effective throughput from the users' viewpoint. Here, VC throughput at the TCP level is calculated by summing the TCP throughputs of connections multiplexed at the VC; the TCP throughput is calculated by dividing the measured amount of data transmitted successfully by the measurement time (set to 30 s in this paper).

To evaluate our method in terms of the above criteria, we built three simulation models (Figs. 8, 9, and 10). In each model, some VCs are established between routers and are multiplexed to a link, and some TCP connections are multiplexed to each VC. In simulation models 1 and 2, the only VCs are GFR-VCs, while in model 3, GFR-VCs and ABR-VCs coexist. Using simulation model 1, we evaluated the fundamental characteristics of our method by comparing it with Per-VC_WRR and FIFO_tagging. Using a 2-link model (simulation model 2), we evaluated whether bandwidth was utilized at the network level, as mentioned in Sect. 4.1. Using simulation model 3, we evaluated whether our method could work well with ABR-VCs coexisting.

5.1 Source Model

In all our simulations, each SES modified Reno [11] implemented in 4.3 BSD (Berkeley Software Distribution) to simulate TCP flow control. We used an "infinite source model" to characterize application-level traffic, which emulates an application that transfers a file of infinite length. As TCP parameters, we set the maximum window size to 64 kbytes, maximum segment size (MSS) to 1518 bytes, and timer granularity to 100 ms. (Though the value of the timer granularity may have influence on the TCP performance, we set it to be 100 ms here because it is one of the generally used values in the present TCP/IP communications. The analysis of TCP parameter setting is further study.) Each SES

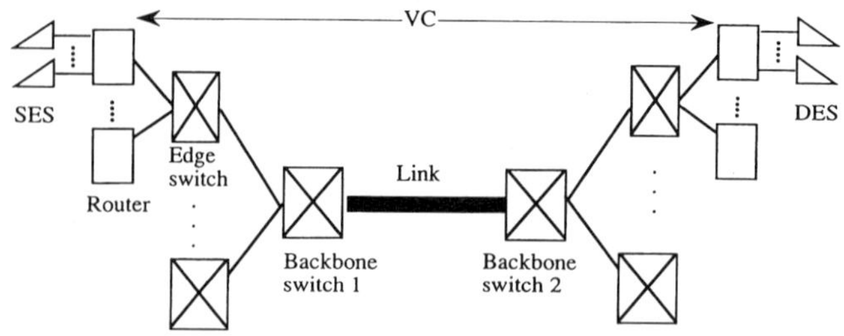

Fig. 8 The 1-link model (simulation model 1).

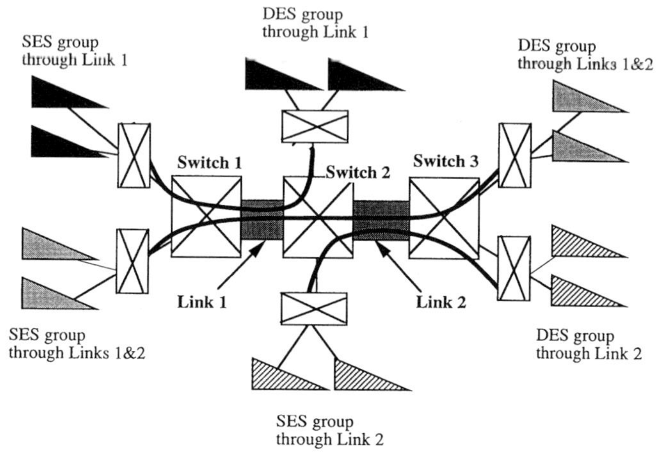

Fig. 9 The 2-link model (simulation model 2).

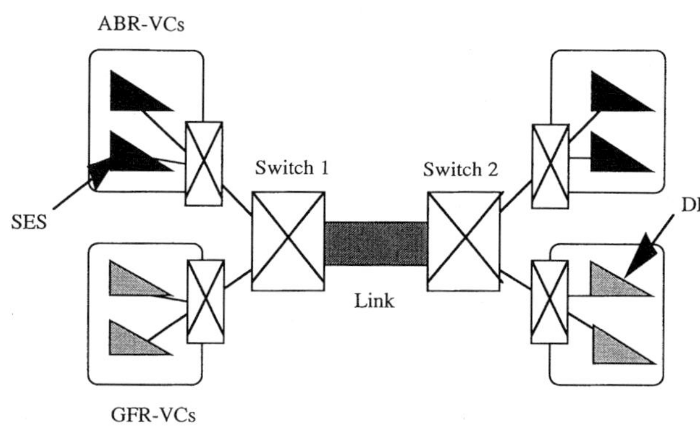

Fig. 10 The 1-link model with GFR- and ABR-VCs coexisting (simulation model 3).

generated packets according to TCP control, and transmitted them to a router, where arriving packets were segmented into cells. The router emitted cells at PCR = 6 Mbps to the edge switch.

5.2 Network Model

Simulation model 1 consisted of two backbone-switches and GFR-VCs multiplexed at one link via edge-switches. Each VC had five TCP connections. The distance between backbone-switches was 100 km; the distance between the router and an edge-switch was 10 km. To evaluate the fundamental stationary behavior under various kinds of traffic conditions, we changed the number of VCs, link capacity, and MCR value of each VC. Under each condition, we ran the simulation for 31 s and used only data in the interval from 1 to 31 s in order to remove the initial bias, such as the influence of TCP slow-start.

Simulation model 2 consisted of three backbone-switches and three groups of GFR-VCs, where each VC had five TCP connections. This three-switch model was used to evaluate network-level utilization. To evaluate it under realistic conditions, we used UDP sources for some VCs instead of TCP sources. In the present Internet, UDP traffic is increasing as the use of real-time applications increases. Unlike TCP, UDP has no flow control mechanism, which causes frequent occupation of bandwidth. Therefore, coexisting UDP traffic is considered to be a severe condition for traffic control. In model 2, 10 UDP-VCs passed through both Links 1 and 2, n TCP-VCs passed through only Link 1, and 10 VCs passed through only Link 2. By changing n from 4 to 20, we obtained several traffic conditions in model 2.

Simulation model 3 was a heterogeneous model, where both GFR-VCs and ABR-VCs coexisted. This model was used to evaluate whether our method could work well even when there was a mixture of GFR-VCs and ABR-VCs. In this model, each VC had one TCP connection. There were four groups of VCs: 10 ABR-VCs with MCR = 600 kbps, 10 GFR-VCs with MCR = 600 kbps, 15 ABR-VCs with MCR = 200 kbps, and 15 GFR-VCs with MCR = 200 kbps. The distance between backbone-switches was 100 km; the distance between the SES and an edge-switch was 10 km. In this model, we varied the link capacity from 18 to 54 Mbps.

5.3 Parameter Setting

[Per-VC_WRR] The VC buffer size at each switch was set to 512 cells, EPD threshold was set to (MBS-MSS) cells, where MBS is 512 cells and MSS is 33 cells (corresponding to 1518 bytes).

[FIFO_tagging] The threshold L for detection of the burst in excess of MCR was set to be the same as the EPD threshold at VC buffer under WRR. The FIFO buffer size was set to 15,360 cells, which corresponds to 512 cells*30 VCs. In most simulations, the number of VCs was equal to or less than 30. Therefore, this condition is considered to be fair for other methods. The threshold for EPD was 7/8*15360 cells, and that for SCD was 3/4*15360 cells.

[Our method] We set the FIFO buffer size of a backbone switch to 4096 cells, VS buffer size at an edge-switch to 512 cells, and EPD threshold at a VS buffer to (MBS-MSS) = (512–33) cells. Parameters of the ER algorithm are shown in Appendix A.

6. Simulation Results

Using the three simulation models, we evaluated whether our method could:

- guarantee minimum throughput at both the cell level and TCP level,
- attain fair bandwidth allocation to each VC,
- utilize bandwidth effectively at the network level, and
- work well with GFR-VCs and ABR-VCs coexisting.

(i) MCR throughput guarantee under homogeneous conditions

Using simulation model 1, we measured VC throughput at both the cell level and TCP level when each VC had an MCR of 600 kbps. First, to evaluate the MCR guarantee of our method, we set the link capacity to 18 Mbps and number of VCs to 30, which means that the residual bandwidth was zero. Here, residual bandwidth is defined as (link capacity–ΣMCR). The measured cell-level throughput of each VC was 581 kbps in our method, compared to 600 kbps in both WRR and FIFO_tagging. This was because our method needs to transmit RM cells every Nrm (here, 32) cells from a VS to a VD. Therefore, to guarantee MCR using our method, we need a little extra bandwidth. The frequency distribution of VC throughputs at the TCP level is shown in Fig. 11. Although the throughput was a little smaller than the ideal throughput†, each method could attain high throughput at the TCP level

†Ideal throughput is defined as the maximum TCP level throughput when MCR is allocated at the cell level to each VC. It is calculated by using MSS (= 1518 bytes), TCP/IP header (= 40 bytes), AAL 5 trailer (= 8 bytes) as follows:

$$\begin{aligned}\text{Ideal TCP throughput} &= \text{MSS}/\{(\text{MSS}\\ &\quad + \text{TCP/IP header}\\ &\quad + \text{AAL5 trailer}\\ &\quad + \text{padding})/48 * 53\} * \text{MCR}\\ &= 1518/1749 * 600\\ &= 520\,\text{kbps}\end{aligned}$$

Exactly speaking, ideal throughput is determined as min{TCP throughput determined by MCR, maximum-window-size/round-trip-time}, but the latter is larger than the former under the simulation conditions in this paper. Therefore, we used 520 kbps as ideal TCP throughput.

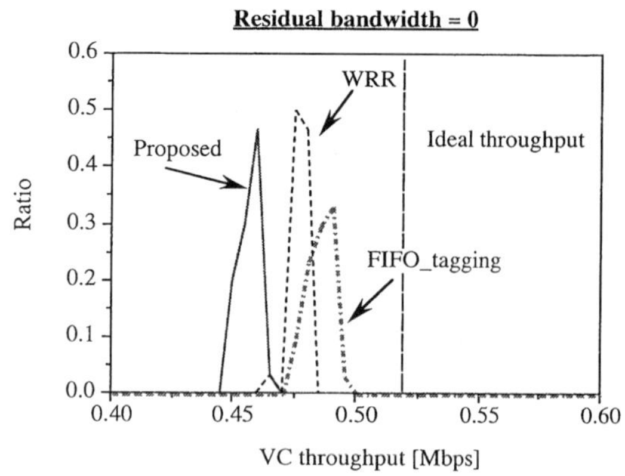

Fig. 11 Frequency distribution of throughputs for 30 VCs.

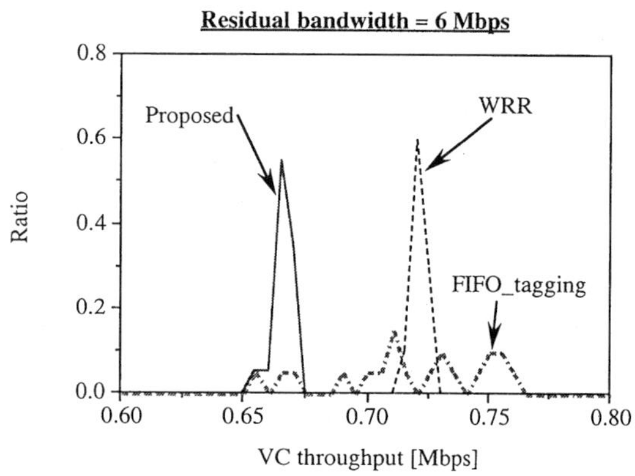

Fig. 12 Frequency distribution of throughputs for 20 VCs.

for each VC. This loss was due to TCP flow control, which caused duplicated retransmission of packets when packet losses occurred. In GFR service, packet losses (cell discards) are inevitable when Σ PCR is larger than link capacity. To guarantee minimum throughput at the TCP level, we need to take this loss into consideration. Thus, we measured TCP-level throughputs when the number of VCs was 20 (residual bandwidth = 6 Mbps), as shown in Fig. 12. In this case, both controls could guarantee minimum throughput to each VC at the TCP level. However, the variance of the throughput distribution among VCs under FIFO_tagging was a little larger, because the FIFO buffer was shared among VCs. Thus, fairness under FIFO_tagging was inferior to that under other methods.

(ii) Fairness under heterogeneous conditions **⟨Coexistence of different MCR-VCs⟩** In simulation model 1, we set the MCR of 24 VCs to 600 kbps, and that of 18 VCs to 200 kbps. Here, we set w of each VC proportional to the MCR value. By doing this, we evaluated whether bandwidth could be allocated to each VC in proportion to the MCR. Figure 13 shows average VC throughputs (at the TCP level) among VCs with the same MCR value, when

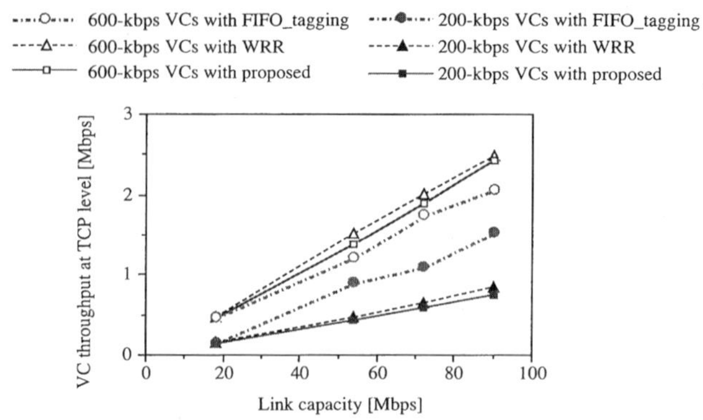

Fig. 13 VC throughput vs. link capacity.

Table 1 VC throughput ratios.

Link capacity [Mbps]	18	54	72	90
FIFO_tagging	2.909	1.361	1.591	1.350
WRR	3.178	3.149	3.070	2.916
Proposed	3.196	3.164	3.252	3.259

we varied the link capacity. The throughput ratio between VC throughput with MCR = 600 kbps and that with MCR = 200 kbps is also shown in Table 1. The throughput ratio was calculated as {average VC throughput of VCs with 600 kbps} ÷ {that of VCs with 200 kbps}. The ideal ratio is 600 kbps ÷ 200 kbps = 3 because we want to allocate bandwidth in proportion to the MCR value. As shown in Fig. 13 and Table 1, our method could maintain high throughput while achieving weighted bandwidth allocation, similar to WRR. On the other hand, the throughputs of 600-kbps VCs under FIFO_tagging were smaller than those under WRR or our method. This is because residual bandwidth was allocated equally to each VC regardless of the MCR value, because the FIFO buffer was shared among all VCs.

To evaluate fairness among the VCs with the same MCR, we also measured the frequency distribution of VC throughput (Fig. 14). When the link capacity was 18 Mbps (that is, residual bandwidth = 0), all controls could achieve fairness among the VCs with the same MCR (Fig. 14(a)). This is because each control could allocate MCR to each VC. Even when the residual bandwidth was larger than zero (the link capacity = 72 Mbps), our method could keep the variance small, like WRR (Fig. 14(b)). However, under FIFO_tagging, throughputs were more widely spread out because of FIFO queueing. Thus, FIFO_tagging had poor fairness.

Table 2 shows the sum of the throughputs. The reason throughputs in our method were slightly lower than those in WRR is due to ABR-ER rate control. In addition to the RM cell overhead mentioned in Sect. 6(i), our ER control algorithm maintains the total offered rate at the predetermined target rate R_0 (here,

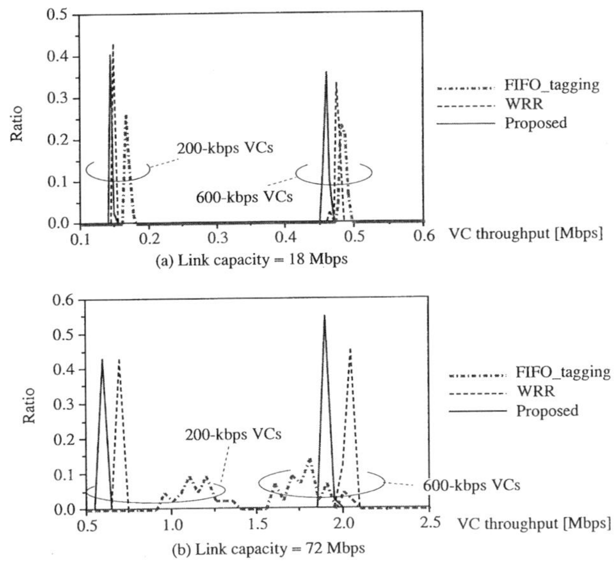

Fig. 14 Frequency distribution of throughputs.

Table 2 Sum of throughput in model 1.

Link capacity [Mbps]	18	54	72	90
FIFO_tagging	14.49	45.60	62.20	77.56
WRR	14.05	45.03	60.07	74.81
Proposed	13.55	40.84	55.93	71.58

0.95*link capacity). This 5% margin is necessary to cope with changes in traffic conditions (e.g., a rapid increase in the number of active VCs).

⟨Coexistence with UDP-VCs⟩ By changing some TCP-VCs to UDP-VCs in simulation model 1, we evaluated the fairness with coexisting UDP traffic. The number of TCP-VCs was fixed to 10, while we varied the number of UDP-VCs from 4 to 20. We used the greedy model as a UDP source, that is, each UDP-VC always emitted cells at PCR = 6 Mbps. This condition is considered to make it hard for TCP-VCs to maintain high throughput. Here, we set the MCR of each VC to 600 kbps.

Figure 15 shows average VC throughput of UDP-VCs at the UDP level and that of TCP-VCs at the TCP level. FIFO_tagging could only achieve minimum throughputs to TCP-VCs in any condition. That is, the residual bandwidth was occupied by UDP-VCs, which caused unfair bandwidth allocation between TCP-VCs and UDP-VCs. Since the FIFO buffer was shared among TCP-VCs and UDP-VCs, UDP-VCs with no control achieved higher throughput than TCP-VCs with the flow control mechanism. On the other hand, our method and WRR could achieve fair allocation between TCP-VCs and UDP-VCs, because both controls can allocate bandwidth to each VC.

(iii) Utilization at network level

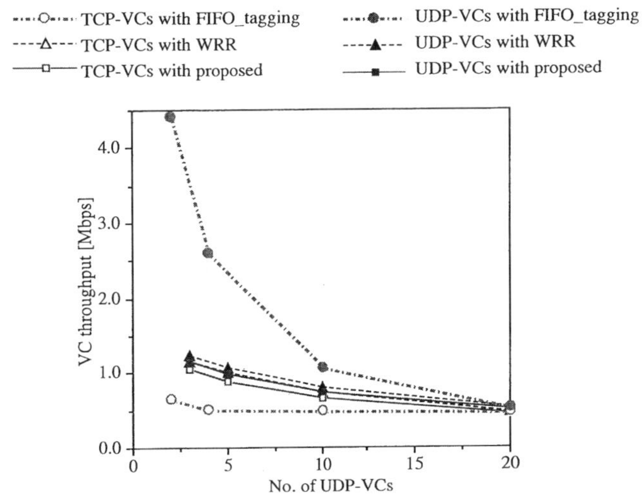

Fig. 15 VC throughputs when multiplexing UDP-VCs and TCP-VCs.

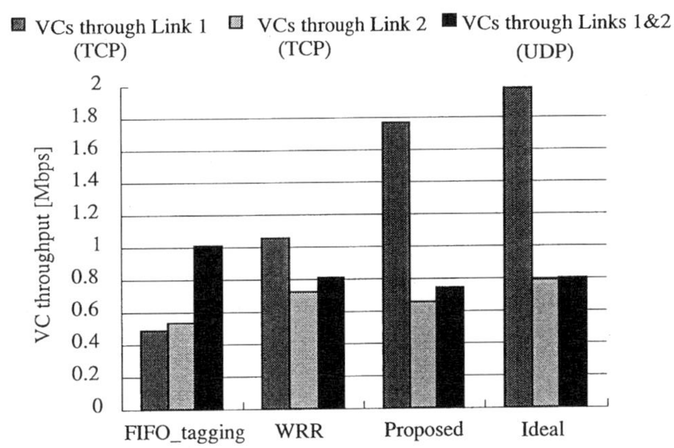

Fig. 16 VC throughputs in 2-link model.

Using simulation model 2, we measured VC throughputs in each group (see Fig. 16) when the number of UDP-VCs passing through both Links 1 and 2 was 10, that of TCP-VCs passing through only Link 2 was 10, and that of TCP-VCs passing through only Link 1 was 4. This condition means that Link 2 was more congested than Link 1 because more VCs were multiplexed at Link 2 than at Link 1. Here, we set each link capacity to 18 Mbps, and the MCR of each VC to 600 kbps. The ideal VC throughputs of each group are also shown; they were calculated as follows. First, the target cell rate was calculated as the max-min fair share, which is defined in [5] as $(C - Ccon) \div (Ntot - Ncon)$, where C is the link capacity, $Ccon$ is the capacity achievable by the constrained VCs, $Ntot$ is the number of connections, and $Ncon$ is the number of constrained VCs. In this case, the target cell rate of each VC through Link 2 was 0.9 Mbps, which was calculated by dividing 18 Mbps by the number of VCs through Link 2. That of each VC through Link 1 only was calculated to be 2.25 Mbps by dividing the unused bandwidth of Link 1 by the number of VCs through only Link 1. Next, the ideal throughput at the TCP level (or UDP level) was calculated using the target cell rate. Considering the overheads of TCP header (or UDP header) and IP header, the ideal throughput of a TCP-VC through Link 1 was calculated to be 1.95 Mbps, that of a TCP-VC through Link 2 to be 0.79 Mbps, and that of a UDP-VC through Links 1 and 2 was calculated to be 0.78 Mbps.

As shown in Fig. 16, our method could maintain high throughput close to the ideal throughput, while poor throughput was observed in the TCP-VC through Link 1 under WRR and FIFO_tagging. This was because the bandwidth of Link 1 was ineffectively utilized by UDP-VC through Links 1 and 2. The amount of Link 1 bandwidth that was ineffectively utilized by UDP-VCs was 3.85 Mbps under WRR and 4.29 Mbps under FIFO_tagging; on the other hand, none was observed under our method. This value was calculated as {(Link 1 bandwidth occupied by UDP-VCs) − (Link 2 bandwidth occupied by UDP-VCs)} because UDP-VCs need to transmit cells from the SES to the DES through both Links 1 and 2. The reason WRR or FIFO_tagging caused ineffective use of bandwidth is that cell discard occurred not only at switch 1 but also at switch 2. This was confirmed by observing the transient behavior of the queue length at UDP-VC buffers under WRR (Fig. 17(b)) and that at the FIFO buffer under FIFO_tagging (Fig. 17(a)). Even though cells from UDP-VCs were successfully transmitted to switch 2 via Link 1, these cells had a probability of being discarded at switch 2. This phenomenon caused ineffective utilization of Link 1. In contrast, under our method, cell discard occurred only at edge switches (Fig. 17(c)). (No cell discard was observed under WRR and FIFO_tagging because both methods could transmit cells from edge switches at PCR.) This is because our method could regulate excess traffic adequately at the entrance of the network, which gives the effective bandwidth utilization.

Figure 18 shows the Link 1 bandwidth that was ineffectively utilized by UDP-VCs when we varied the number of TCP-VCs through Link 1. This result shows that our method could utilize bandwidth effectively under any conditions. On the other hand, in WRR, ineffective use occurred when there were fewer VCs at Link 1 than at Link 2. The bandwidth allocated by WRR is equal to each link capacity divided by the number of active VCs at the link, if all VCs have the same MCR-value. Therefore, UDP-VCs through Links 1 and 2 wasted an amount of Link 1 bandwidth equal to {(Link 1 capacity) ÷ (no. of VCs at Link 1) − (Link 2 capacity) ÷ (no. of VCs at link 2)}.

(iv) Characteristics under coexistence of GFR- and ABR-VCs

Using simulation model 3, we evaluated the traffic characteristics of our method when ABR-VCs coexist with GFR-VCs. Although GFR does not need to feed information from the VS/VD at an edge-switch back to the SES, ABR does need to. Thus, in this paper, each VD at an edge-switch returns an RM cell sent from each SES after writing the explicit cell rate into its ER

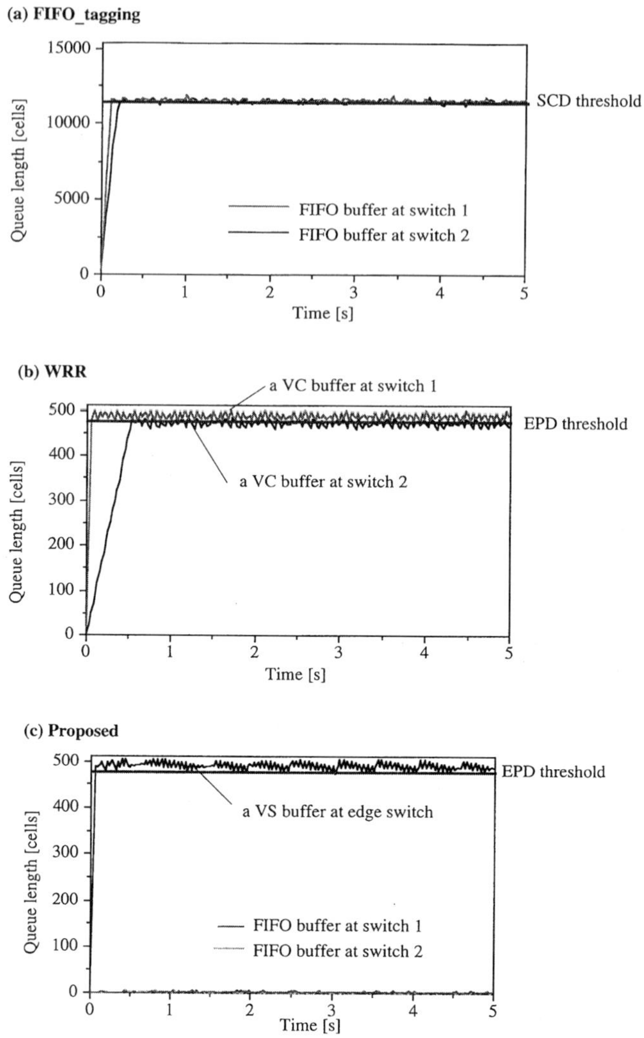

Fig. 17 Queue length behaviors.

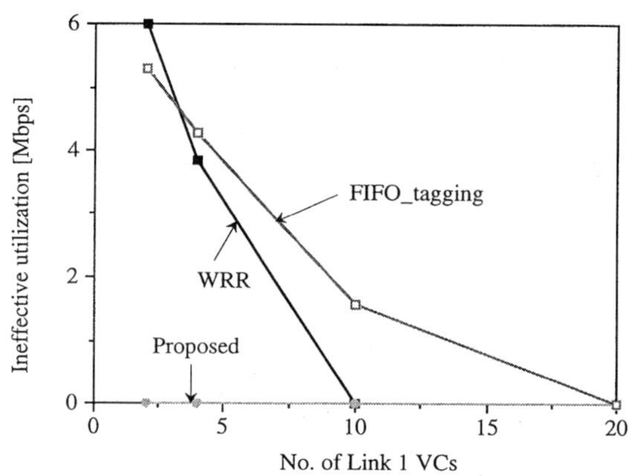

Fig. 18 Ineffective utilization of bandwidth at Link 1.

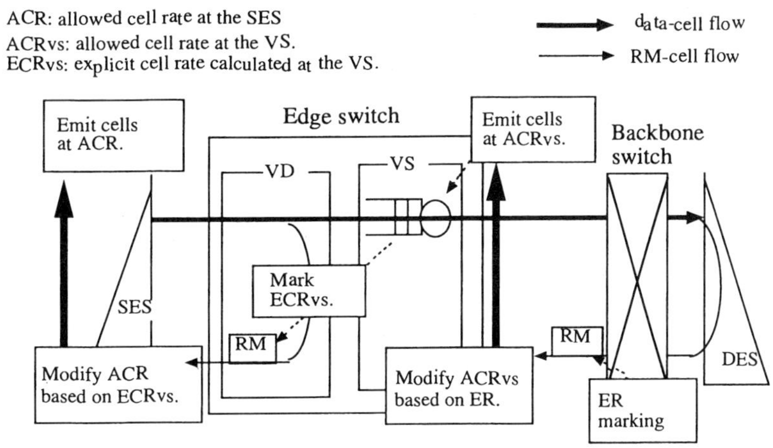

Fig. 19 VS/VD behavior under ABR service.

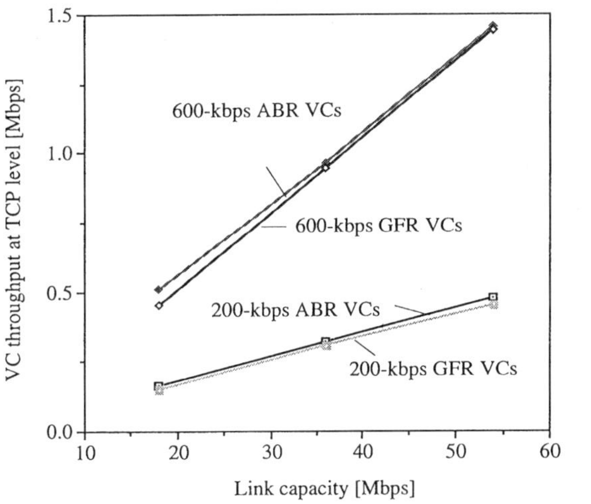

Fig. 20 VC throughputs with ABR-VCs and GFR-VCs coexisting.

Table 3 Sum of throughput in model 3.

Link capacity [Mbps]	18	36	54
TCP level	14.4	28.5	43.1
Cell level	17.5	34.0	51.3

field (see Fig. 19). The explicit cell rate ($ECRvs$) was determined as follows.

If the queue length of the virtual source < 256 cells,

$$ECRvs := \text{current } ACR - \text{value at the virtual source } (ACRvs). \quad (6)$$

Otherwise,

$$ECRvs := MCR. \quad (7)$$

Applying this algorithm to ABR-VCs, we measured the GFR-VC throughput and ABR-VC throughput when we varied the link capacity (see Fig. 20 and Table 3). Table 3 shows high utilization of bandwidth. Figure 20 shows that both ABR-VCs and GFR-VCs could attain weighted bandwidth allocation. Although the cell-level throughput was almost the same for ABR- and GFR-VCs (see Fig. 21(a)), the TCP-level throughput of ABR-VCs was higher than that of GFR-VCs (Fig. 20). This was because ABR could avoid cell loss in a VS buffer (see Fig. 21(b)) by controlling the cell emission rate of each SES adequately (see Fig. 21(c)). On the other hand, cell discard occurred at the VS buffer of a GFR-VC because it has no feedback mechanism.

Summarizing our simulation results, we clarified that our method could achieve better performance. It could:

(1) guarantee minimum throughput at both the cell level and TCP level like other methods, although it required a little extra bandwidth because ABR control is used,
(2) achieve better fairness than FIFO_tagging even under heterogeneous conditions, like WRR,
(3) utilize bandwidth more efficiently at the network level than WRR or FIFO_tagging, and
(4) work well with ABR-VCs coexisting, which enables a link to be shared between GFR-VCs and ABR-VCs, which lowers the link cost.

7. Conclusion

We have proposed a GFR implementation method using the ABR-ER control mechanism at backbone switches and VS/VD mechanism at edge switches. Unlike per-VC queueing and scheduling with weighted round robin (WRR), our method is easy to implement in large-scale networks. Simulation analysis showed that it could achieve good performance in terms of MCR guarantee, fairness, and throughput, like WRR. In addition, unlike WRR, it could also utilize bandwidth effectively at the network level and could work well with coexisting ABR connections, which may decrease the link cost because link capacity can be shared

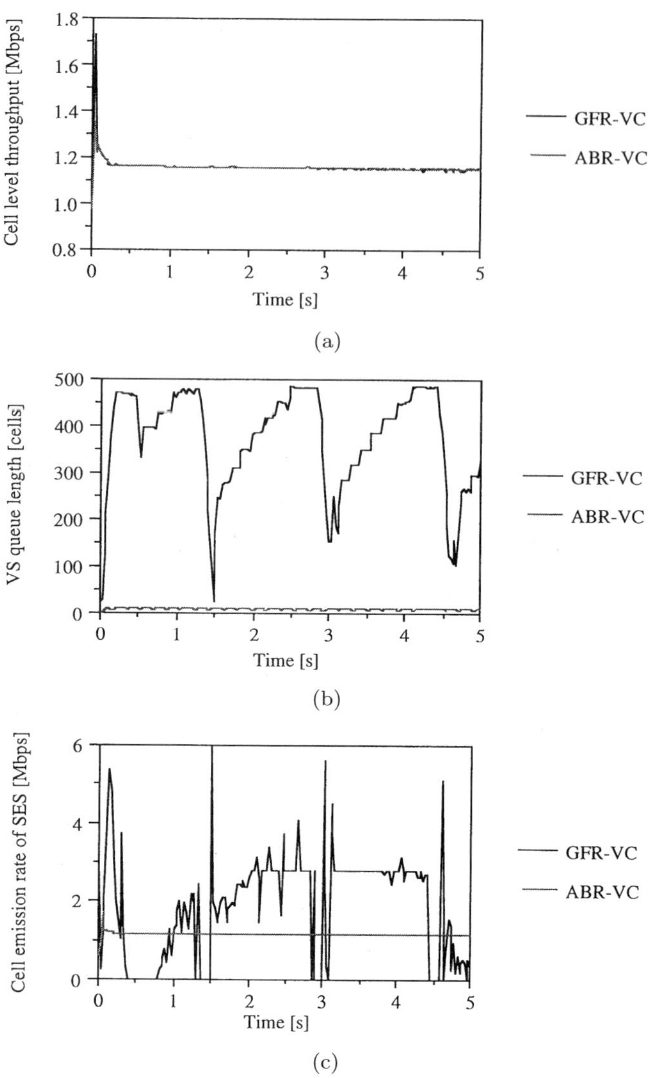

Fig. 21 Transient behaviors under GFR and ABR.

between ABR and GFR services. We will evaluate the effect of this link sharing from the cost viewpoint. Our future work will also focus on practical uses of our control method. We will consider how to apply GFR service implemented by it to various kinds of multimedia applications.

Acknowledgments

We greatly appreciate the cooperation for the simulation evaluation by Miss Mika Ishizuka of NTT Information Sharing Platform Laboratories and Miss Hiromi Noguchi of NTT Advanced Technology.

References

[1] T. Asaka, R. Kawahara, and T. Ozawa, "Performance analysis of ATM traffic control mechanisms for guaranteed throughput service," IEICE Trans. Commun., vol.J81-B-I, no.11, pp.731–740, Nov. 1998.

[2] R. Goyal, R. Jain, S. Fahmy, B. Vandalore, and S. Kalyanaraman, "Simulation experiments with guaranteed frame rate for TCP/IP traffic," ATM Forum Contribution, 97–0607, June 1997.

[3] S.K. Pappu and D. Basak, "TCP over GFR implementation with different service discipines: A simulation study," ATM Forum Contribution, 97–0310, April 1997.

[4] R. Goyal, R. Jain, S. Fahmy, B. Vandalore, and S. Kalyanaraman, "GFR—Providing rate guarantees with FIFO Buffers to TCP traffic," ATM Forum Contribution, 97–0831, Sept. 1997.

[5] The ATM Forum, Traffic Management Specification 4.0, af-tm-0056.000, April 1996.

[6] H. Saito, K. Kawashima, H. Kitazume, A. Koike, M. Ishizuka, and A. Abe, "Performance issues in public ABR services," IEEE Commun. Mag., vol.34, no.5, 1996.

[7] R. Kawahara, H. Saito, and M. Kawarasaki, "Characteristics of ABR explicit rate control algorithms in WAN environments and an ABR control algorithm suitable for public networks," Int. J. Commun. Syst., vol.11, pp.189–209, Nov. 1998.

[8] R. Kawahara and T. Ozawa, "A simple and efficient ABR control algorithm in large scale networks," IEICE Trans. Commun., vol.J81-B-I, no.11, pp.720–730, Nov. 1998.

[9] A.W. Barnhart, "Example switch algorithm for section 5.4 of TM spec.," ATM-Forum/95-0195.

[10] A. Arulambalam, X. Chen, and N. Ansari, "Allocating fair rates for available bit rate service in ATM networks," IEEE Commun. Mag., vol.34, no.11, pp.92–100, Nov. 1996.

[11] G. Wright and W.R. Stevens, TCP/IP Illustrated Volume 2: The implementation, Addison Wesley, 1995.

[12] D.P. Hong and T. Suda, "Performance of ERICA and QFC for transporting bursty TCP sources with bursty interfering traffic," INFOCOM98, pp.1341–1349, 1998.

[13] S. Kalyanaraman, R. Jain, R. Goyal, S. Fahmy, and S.C. Kim, "Performance of TCP over ABR on ATM backbone and with various VBR background traffic patterns," ICC97, pp.1035–1041, 1997.

[14] S. Jagannath and N. Yin, "End-to-end TCP performance in IP/ATM internetworks," ATM-Forum/96-1711.

[15] M. Ishizuka, A. Koike, and M. Kawarasaki, "Performance analysis of TCP over ABR in high-speed WAN environment," IEICE Trans. Commun., vol.E80-B, no.10, pp.1436–1443, Oct. 1997.

[16] B. Feng, D. Ghosal, and N. Kannappan, "Impact of ATM ABR control on the performance of TCP-tahoe and TCP-reno," GLOBCOM97, pp.1832–1837, 1997.

[17] L. An, N. Ansari, and A. Arulambalam, "TCP/IP traffic over ATM networks with ABR flow and congestion control," GLOBCOM97, pp.1845–1849, 1997.

[18] The ATM Forum, Traffic Management Specification 4.1, af-tm-00121.000, Final Ballot, March 1999.

Appendix A: An ER Control Algorithm [8]

(1) When a forward RM cell arrives at a switch, the switch reads the ACR value and weight w from the cell's fields. Then, $MACR$ is updated as follows:

if ($Rate > R1$)

$$MACR := (1-\alpha)MACR + \alpha\min\{MACR, ACR/w\} \qquad (\text{A}\cdot 1)$$

else

if ($ACR/w =< \varepsilon ER$)
  if ($Rate =< R2$)

$$MACR := (1-\alpha)MACR + \alpha\max\{MACR, ACR/w\} \quad (A\cdot 2)$$

else

$$MACR := (1-\alpha)MACR + \alpha ACR/w \quad (A\cdot 3)$$

else

$$MACR := MACR \quad (A\cdot 4)$$

(2) When $Rate$ is updated, ER is also updated as follows:
if (queue length of FIFO buffer =< threshold)

$$ER := \min\{(\gamma + (1-\gamma) * R0/Rate), ERU\} * MACR \quad (A\cdot 5)$$

else

$$ER := \min\{(\beta + (1-\beta) * R0/Rate), ERD\} * MACR \quad (A\cdot 6)$$

(3) When a backward RM cell arrives at a switch, the switch reads w from the cell. Then, it writes $w * ER$ into the cell's ER field.

Here, $R0$ is the target rate, $R1$ is the threshold for overload detection, and $R2$ is the threshold for underload detection. ε, α, β, and γ are averaging parameters. ERU and ERD are parameters for preventing a rapid increase in ER under non-congestion and congestion conditions, respectively. The values of the parameters used in the simulations were as follows: $R0 = 0.95*$ link capacity [Mbps], $R1 = 0.98*$ link capacity [Mbps], $R2 = 0.98*$ link capacity [Mbps], $\varepsilon = 1$, $\alpha = 1/16$, $\beta = 0.25$, $\gamma = 0.75$, $ERU = 2$, and $ERD = 3/4$.

Appendix B: The Effect of MACR and Rate under Various Long Propagation Delays and Number of Connections [7], [8]

Previous studies by our group [7] and [8] clarified that the ER algorithms using Rate can update ER according to the number of active connections, and that those using $MACR$ can update ER according to the propagation delay.

B.1 Effect of Rate on a Large Number of Connections

If all active VCs emit cells according to the current ER at a switch,

$$Nvc = Rate/ER, \quad (A\cdot 7)$$

where Nvc is the number of active VCs. This means that the total input cell rate to the switch implies the number of active VCs. Therefore, updating the ER relative to the $Rate$ means updating the ER according to the size of Nvc.

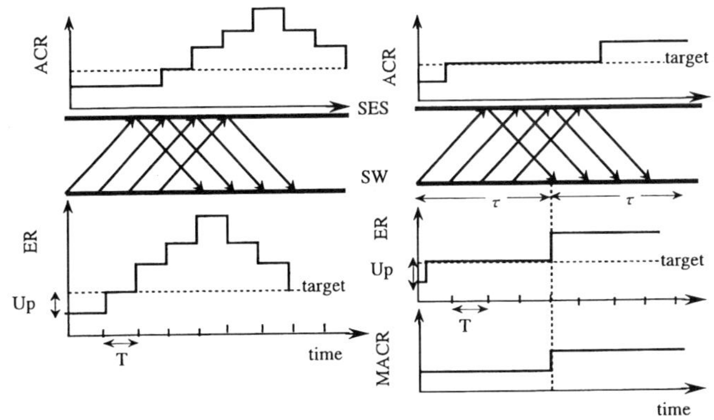

Fig. A·1 The effect of MACR on long propagation delay [8].

B.2 Effect of MACR on Long Propagation Delay

When we update ER, we need to take into account the size of propagation delay τ. An algorithm that updates ER by adding a fixed value (Up) to the current ER every time T (see the left-hand side of Fig. A·1) increases ER too much before the effect of the ER updating is reflected at the switch. In contrast, an algorithm that updates ER by adding Up to $MACR$ (right-hand side of Fig. A·1), increase ER by the size of τ because $MACR$ does not change until the effect of ER updating is reflected at the switch. Though this example is the simple case that there is a single VC, our study [7] confirmed that using $MACR$ is effective even when there are multiple VCs with different propagation delays.

High-Performance Backbone Network Technology edited by Naoaki Yamanaka
ISBN: 0-8247-5321-6 DOI: 10.1081/(120025705)
Published 2004 by Marcel Dekker, Inc.

Connection-Wise End-to-End Delay Analysis in ATM Networks

Huei-Wen Ferng and Jin-Fu Chang

SUMMARY A systematic method for connection-wise end-to-end delay analysis in asynchronous transfer mode (ATM) networks is proposed. This method consists of the followings: (i) per-stream nodal analysis; (ii) output processes characterization; and (iii) moment matching scheme. Following our previous work [1], we employ H-MMPPs/Slotted D/1 to model ATM queues. Each virtual connection (VC) in ATM networks can be regarded as a tandem configuration of such queues. In [1], the per-stream analytical results for such an H-MMPPs/Slotted D/1 queue have been provided. In this paper, not only the composite output process is exactly characterized, but also the component in an output process that corresponds to a specific traffic stream is approximated via a decomposition scheme. A moment matching scheme to emulate the per-stream output process as a two-state MMPP is further proposed. Through moment matching, we can then approximate the connection-wise end-to-end delay by recursively performing the nodal performance analysis. The connection-wise end-to-end delay is crucial to network resource decision or control problems such as call admission control (CAC) and routing.

key words: *ATM, MMPP, H-MMPPs/Slotted D/1 queue, tandem queues, end-to-end performance*

1. Introduction

Broadband integrated services digital networks (B-ISDNs) [2] provide a means of exchanging multimedia information such as voice, video, image, data, and etc. Among various transport technologies, ATM is targeted to provide such diverse services in one network. It provides different quality of service (QoS) according to service requirements. Since the QoS requirements are defined on a per-connection/stream basis, performance evaluation in connection-oriented ATM networks has the need to be done for a specific connection which is complicated by interference from other traffic streams. Past researches mostly focused on the performance analysis for an isolated node. Recently, end-to-end QoS, e.g., connection-wise end-to-end sojourn delay time, has become more and more attentioned. Knowledge of end-to-end QoS helps us to understand the interplay between traffic descriptors and QoS requirements. [3]–[5] are some of the works that have appeared in the literature to study the end-to-end performance in ATM networks. Addie and Zukerman [3] studied the performance of a tree type ATM network with discrete-time Gaussian arrivals. Kroner et al. [4] approximated the end-to-end delay jitter in ATM networks with burst silence cell streams. Ren et al. [5] also studied the end-to-end performance of ATM networks with On-Off cell streams. For a specific connection or tagged VC, we shall in this paper examine the end-to-end cell sojourn delay time. We propose an analytical method to estimate these delays. These estimates can be applied to facilitate call admission control, congestion control, routing algorithms and etc.

The output stream of a switch output port or a multiplexer becomes one of the input streams to the next stage. Research in this regard can be found in [6]–[10]. Ohba et al. [6] studied the departure process of a designated GI-stream interfered by other batch Bernoulli and IPP arrivals. Park et al. [7] studied the departure process of the lossy geometric server receiving an MMBP input. Saito [8] studied the departure process of an $N/G/1$ queue based on the embedded Markov chain at departure epochs. Takine et al. [9] studied the departure process of a discrete-time lossy deterministic server with correlated arrivals. Kleinrock [10] discussed the output process of queues with server(s) and input of exponential type. Among these past works, [7] [10] investigated the composite departure process only. Although [6] dealt with the per-stream departure process, the results obtained are not in closed-form and the model employed is perhaps too simplified. In our work, part of the accomplishment is to analyze the exact composite departure statistics of the H-MMPPs/Slotted D/1 queue. In addition, we obtain the approximate per-stream departure statistics corresponding to a specific traffic stream via a decomposition scheme. Along with a method of moment matching, we succeed in obtaining the end-to-end sojourn delay time of a tagged VC by matching the corresponding per-stream output process to a two-state MMPP at each node en route to the destination node. There are several existing moment matching techniques, e.g., [11], [12]. In [11], Heffes and Lucantoni modeled a superposition of packet arrival processes by an MMPP so that the following four statistics are matched: mean cell arrival rate, short term and long term variance-to-mean ratio of the number of arrivals,

Manuscript received April 30, 1999.
Manuscript revised September 29, 1999.

†The author is with the Department of Electrical Engineering, National Taiwan University, Taipei, Taiwan 10617 R.O.C.

††The author was with the Department of Electrical Engineering, National Taiwan University, Taipei, Taiwan 10617 R.O.C. He is now with the National Science Council, Taipei, Taiwan 10622 R.O.C.

IEICE TRANS. COMMUN., Vol. E83-B, No. 3, pp. 659–667, March 2000

and the third moment of the number of arrivals in a period. In this work, the following four departure statistics are matched: the first moment of interdeparture times, squared coefficient of variation of interdeparture times, the third moment of interdeparture times, and lag 1 covariance of interdeparture times. These four statistics are selected for the purpose of characterizing the output process. A detailed discussion about why these four statistics are selected is given in Sect. 4. We approximate the per-stream output process by a two-state MMPP if the matching scheme works; or the per-stream output process is taken to be Poisson. The failure of moment matching is due to that burstiness of the (estimated) output process has been found to be reduced to beyond the scope of MMPPs.

The rest of this paper is organized as follows. Preliminaries are given in Sect. 2. In Sect. 3, we analyze the composite and per-stream departure processes for the H-MMPPs/Slotted D/1 queue. In Sect. 4, we propose a moment matching scheme to emulate the per-stream output process as a two-state MMPP. Section 5 gives detailed description of the systematic method to evaluate connection-wise end-to-end performance measures. In Sect. 6, numerical examples are given. Computer simulations are also provided to demonstrate the accuracy of approximation. Section 7 concludes the paper.

2. Preliminaries

Following [1], we use the H-MMPPs/Slotted D/1 queue, a discrete-time deterministic server receiving heterogeneous MMPPs, to model ATM queues, e.g., a multiplexer or an output port of a switch. For the convenience of subsequent sections, we here present notation related to the MMPP [13] and results of an H-MMPPs/Slotted D/1 queue previously obtained in [1].

2.1 MMPP Traffic

The following discussions can be found in [13].

A two-state MMPP of type j is characterized by the following infinitesimal generator $\boldsymbol{Q}_j$ of the underlying Markov chain and rate matrix $\boldsymbol{\Lambda}_j$ with two Poisson arrival rates λ_{1j} and λ_{2j}, i.e.,

$$\boldsymbol{Q}_j = \begin{bmatrix} -\sigma_{1j} & \sigma_{1j} \\ \sigma_{2j} & -\sigma_{2j} \end{bmatrix}, \quad \boldsymbol{\Lambda}_j = \begin{bmatrix} \lambda_{1j} & 0 \\ 0 & \lambda_{2j} \end{bmatrix}. \tag{1}$$

Because of the property that superposition of independent MMPPs yields again an MMPP, the superposition of r different types of two-state MMPP is also MMPP with $m = 2^r$ possible states. The state space of the superposed MMPP can be described using Kronecker product and sum of matrices. In other words, the superposition of r independent two-state MMPPs parameterized by the descriptor $(\boldsymbol{Q}_j, \boldsymbol{\Lambda}_j)$, $1 \le j \le r$, can be represented by the m-state MMPP parameterized by the descriptor $(\boldsymbol{Q}, \boldsymbol{\Lambda})$ with

$$\begin{aligned} \boldsymbol{Q} &= \boldsymbol{Q}_1 \oplus \boldsymbol{Q}_2 \oplus \ldots \oplus \boldsymbol{Q}_r, \\ \boldsymbol{\Lambda} &= \boldsymbol{\Lambda}_1 \oplus \boldsymbol{\Lambda}_2 \oplus \ldots \oplus \boldsymbol{\Lambda}_r \end{aligned} \tag{2}$$

where $\oplus$ denotes the Kronecker sum operator. The rate matrix $\boldsymbol{\Lambda}$ is a diagonal matrix with $\lambda_1, \ldots, \lambda_m$ on the diagonal, i.e., $\boldsymbol{\Lambda} = Diag(\lambda_1, \lambda_2, \ldots, \lambda_m)$.

For the superposed m-state MMPP $(\boldsymbol{Q}, \boldsymbol{\Lambda})$, the steady-state probability vector $\boldsymbol{\pi} = (\pi_1, \ldots, \pi_m)$ of the underlying Markov chain is given by solving the equations $\boldsymbol{\pi Q} = \boldsymbol{0}$ and $\boldsymbol{\pi e} = 1$ where $\boldsymbol{e}$ and $\boldsymbol{0}$ are $m \times 1$ column vector of all ones and $1 \times m$ row vector of all zeros, respectively.

Define the conditional probability $P_{i,j}(n, t) \triangleq Pr\{N(t) = n, J(t) = j \mid N(0) = 0, J(0) = i\}$ where $N(t)$ and $J(t)$ denote respectively the number of arrivals during the interval $(0, t)$ and the state of the underlying Markov process at time t. From [13], the $m \times m$ matrix of probabilities $\boldsymbol{P}(n, t) \triangleq [P_{i,j}(n, t)]_{1 \le i,j \le m}$ has the probability generating function

$$\boldsymbol{P}^*(z, t) = e^{\boldsymbol{R}(z)t}, \quad |z| \le 1 \tag{3}$$

with

$$\boldsymbol{R}(z) = \boldsymbol{Q} + (z - 1)\boldsymbol{\Lambda}. \tag{4}$$

And the effective arrival rate of the superposed MMPP $(\boldsymbol{Q}, \boldsymbol{\Lambda})$ is $\lambda^* = \boldsymbol{\pi} \frac{d}{dz} \boldsymbol{R}(z) \mid_{z=1} \boldsymbol{e} = \boldsymbol{\pi \lambda}$ where $\boldsymbol{\lambda} = \boldsymbol{\Lambda e}$. The arrival rate from type j $(1 \le j \le r)$ traffic stream is independent of the other traffic streams and is given by

$$\boldsymbol{\Lambda}(j) = \boldsymbol{0} \oplus \ldots \oplus \boldsymbol{0} \oplus \boldsymbol{\Lambda}_j \oplus \boldsymbol{0} \oplus \ldots \oplus \boldsymbol{0} \tag{5}$$

where $\boldsymbol{0}$ is a 2×2 zero matrix. The effective arrival rate from type j traffic stream is $\lambda_j^* = \boldsymbol{\pi \lambda}(j)$ where $\boldsymbol{\lambda}(j) = \boldsymbol{\Lambda}(j)\boldsymbol{e}$.

2.2 System Size of the H-MMPPs/Slotted D/1 Queue at Slot Boundaries

The followings are taken from [1].

Let L_n^s denote the number of cells in the system immediately after the end of the nth slot, i.e., L_n^s includes cells arrived and accommodated into the system in the nth slot but excludes the cell departed at the end of the nth slot. Then $L_{n+1}^s = (L_n^s - 1)^+ + \tilde{A}_{n+1}$ where $\tilde{A}_n$ denotes the number of cells arrive in the nth slot and $(x)^+ = max(x, 0)$ (see the system time diagram shown in Fig. 1). Let J_n^s denote the state of the input process

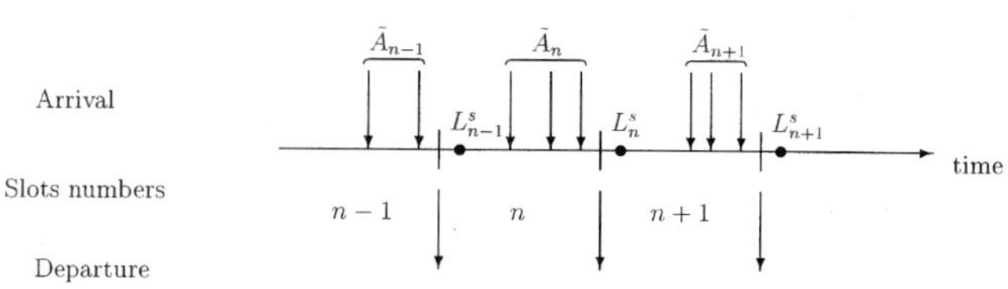

Fig. 1 System time diagram.

at the end of the nth slot. And let $\boldsymbol{A}_i \stackrel{\triangle}{=} \boldsymbol{P}(i,T)$ $(i \geq 0)$ where T denotes the duration of a slot and is used as the time unit of the time axis. $\boldsymbol{A}_i$ may be computed using the method in [13] or [14]. Then $\{(L_n^s, J_n^s) : n \geq 0\}$ forms the infinite-state Markov chain with state space $\{0,1,2,\ldots\} \times \{1,2,\ldots,m\}$ and has transition probability matrix in the following form:

$$\boldsymbol{U} = \begin{bmatrix} \boldsymbol{B}_0^s & \boldsymbol{B}_1^s & \boldsymbol{B}_2^s & \cdots \\ \boldsymbol{A}_0 & \boldsymbol{A}_1 & \boldsymbol{A}_2 & \cdots \\ \boldsymbol{0} & \boldsymbol{A}_0 & \boldsymbol{A}_1 & \cdots \\ \vdots & \vdots & \vdots & \ddots \end{bmatrix} \tag{6}$$

where $\boldsymbol{B}_i^s = \boldsymbol{A}_i$, $i \geq 0$ and $\boldsymbol{0}$ is an $m \times m$ zero matrix.

Let $\boldsymbol{x}^s = (\boldsymbol{x}_0^s, \boldsymbol{x}_1^s, \ldots, \boldsymbol{x}_n^s, \ldots)$ denote the stationary probability vector of $\boldsymbol{U}$, where $\boldsymbol{x}_i^s$ is a $1 \times m$ vector whose jth element $x_{i,j}^s = \lim_{n\to\infty} Pr\{L_n^s = i, J_n^s = j\}$. $\boldsymbol{x}^s$ satisfies $\sum_{n=0}^{\infty} \boldsymbol{x}_n^s \boldsymbol{e} = 1$ and $\boldsymbol{x}^s = \boldsymbol{x}^s \boldsymbol{U}$ or equivalently,

$$\boldsymbol{x}_i^s = \boldsymbol{x}_0^s \boldsymbol{B}_i^s + \sum_{\nu=1}^{i+1} \boldsymbol{x}_\nu^s \boldsymbol{A}_{i+1-\nu}\,, \quad i \geq 0 \tag{7}$$

where $\boldsymbol{x}_0^s$ can be calculated via resorting to [15] and [16]. Then use (7), we can obtain the system size distribution at slot boundaries.

3. Analysis of Departure Processes

3.1 Interdeparture Time Distribution of the Composite Departure Process

Following a similar philosophy done for a finite queue accepting correlated arrivals in [9], we now derive interdeparture time distribution for the composite output stream of an H-MMPPs/Slotted D/1 queue. Comparing with [9], we shall not only derive neater but more results. For example, the lag 1 covariance is not treated in [9]. Let $\boldsymbol{u}_k$ be a $1 \times m$ vector whose jth element $u_{k,j}$ is the joint stationary probability that the number of cells in the system is k and the state of the underlying Markov chain immediately after a departure is j, then $u_{k,j} = Pr\{L_n^s = k, J_n^s = j \mid L_{n-1}^s \geq 1\}$. The condition $\{L_{n-1}^s \geq 1\}$ is to ensure that the number of cells in the system immediately after the $(n-1)$st slot is at least one and there is a cell departing from the system at the end of the nth slot. Rewrite $u_{k,j}$ as follows:

$$\begin{aligned} u_{k,j} &= \frac{Pr\{L_{n-1}^s \geq 1, L_n^s = k, J_n^s = j\}}{Pr\{L_{n-1}^s \geq 1\}} \\ &= \frac{\sum_{i=1}^{k+1} Pr\{L_{n-1}^s = i, L_n^s = k, J_n^s = j\}}{1 - Pr\{L_{n-1}^s = 0\}}. \end{aligned} \tag{8}$$

Then from (7), (8), and the definition of $\boldsymbol{u}_k$,

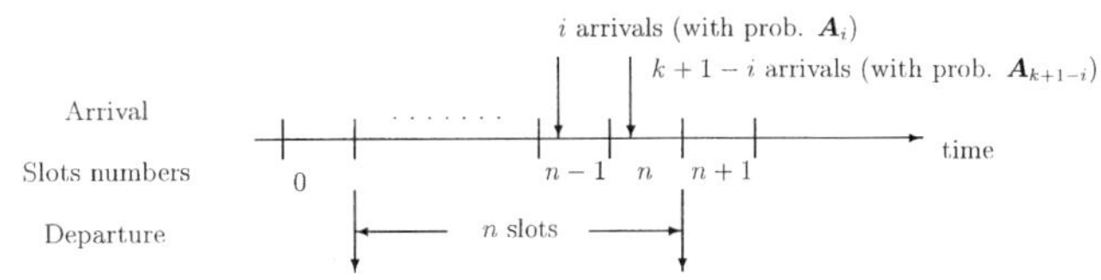

Fig. 2 Illustration of the time diagram of $\boldsymbol{d}_k^{(1)}(n)$.

$$\boldsymbol{u}_k = \frac{\sum_{i=1}^{k+1} \boldsymbol{x}_i^s \boldsymbol{A}_{k+1-i}}{1 - \boldsymbol{x}_0^s \boldsymbol{e}} = \frac{\boldsymbol{x}_k^s - \boldsymbol{x}_0^s \boldsymbol{B}_k^s}{1 - \boldsymbol{x}_0^s \boldsymbol{e}}, \quad k \geq 0. \tag{9}$$

For convenience, we number the slot that the first cell departure occurs as the 0th slot. Let D_m denote the mth interdeparture time. Let $\boldsymbol{d}_k^{(1)}(n)$ be a $1 \times m$ vector whose jth element $d_{k,j}^{(1)}(n)$ is the joint stationary probability that D_1 equals n (slots), the number of cells in the system after the second departure is k and the state of the underlying Markov chain immediately after the second departure is j, i.e., $d_{k,j}^{(1)}(n) = Pr\{D_1 = n, L_n^s = k, J_n^s = j\}$. We use the superscript $^{(1)}$ to indicate a single interdeparture time. We now derive $\boldsymbol{d}_k^{(1)}(n)$ $(n \geq 1)$ as follows. First, we consider $\boldsymbol{d}_k^{(1)}(1)$. Only when the number of cells left behind the first departure at the end of the 0th slot is at least one, say i cells, the interdeparture time becomes one. Thus, $\boldsymbol{d}_k^{(1)}(1) = \sum_{i=1}^{k+1} \boldsymbol{u}_i \boldsymbol{A}_{k+1-i}$ $(k \geq 0)$. We next consider $\boldsymbol{d}_k^{(1)}(n)$ $(n \geq 2)$. When the number of cells left behind the first departure is zero, the interdeparture time is then longer than one. If the first interdeparture time is n, there must be no arrivals between the first and the $(n-2)$nd slot since the first departure. And there must be at least one cell arriving in the $(n-1)$st slot. Thus, $\boldsymbol{d}_k^{(1)}(n) = \boldsymbol{u}_0 \boldsymbol{A}_0^{n-2} \sum_{i=1}^{k+1} \boldsymbol{A}_i \boldsymbol{A}_{k+1-i}$ $(k \geq 0,\ n \geq 2)$. The detail of $\boldsymbol{d}_k^{(1)}(n)$ $(n \geq 2)$ is illustrated in Fig. 2. Let $d^{(1)}(n)$ denote the probability that interdeparture time D_1 is of length n (slots), i.e., $d^{(1)}(n) = \sum_{k=0}^{\infty} \boldsymbol{d}_k^{(1)}(n)\boldsymbol{e}$. Then

$$\begin{aligned} d^{(1)}(1) &= 1 - \boldsymbol{u}_0 \boldsymbol{e}, \\ d^{(1)}(n) &= \boldsymbol{u}_0 \boldsymbol{A}_0^{n-2}(\boldsymbol{I} - \boldsymbol{A}_0)\boldsymbol{e}, \quad n \geq 2. \end{aligned} \tag{10}$$

Let $\tilde{D}^{(1)}(z)$ denote the probability generating function of the interdeparture time distribution. Then we have

$$\begin{aligned} \tilde{D}^{(1)}(z) &= \sum_{n=1}^{\infty} d^{(1)}(n) z^n \\ &= z(1 - \boldsymbol{u}_0 \boldsymbol{e}) \\ &\quad + z^2 \boldsymbol{u}_0 (\boldsymbol{I} - z\boldsymbol{A}_0)^{-1} (\boldsymbol{I} - \boldsymbol{A}_0)\boldsymbol{e}. \end{aligned} \tag{11}$$

Let $D^{(m)}$ denote the mth factorial moment of the interdeparture time distribution, that is, $D^{(m)} =$

$\frac{d^m}{dz^m}\tilde{D}^{(1)}(z)\mid_{z=1}$. Then

$$\begin{aligned} D^{(1)} &= 1 + \boldsymbol{u}_0(\boldsymbol{I} - \boldsymbol{A}_0)^{-1}\boldsymbol{e}, \\ D^{(m)} &= m!\boldsymbol{u}_0(\boldsymbol{I} - \boldsymbol{A}_0)^{-m}\boldsymbol{A}_0^{m-2}\boldsymbol{e}, \\ &\quad m \geq 2. \end{aligned} \tag{12}$$

From (9) and (12), $D^{(1)}$ can be further reduced to $1/(1 - \boldsymbol{x}_0^s\boldsymbol{e}) = 1/\rho = 1/\lambda^*$ where ρ denotes the utilization factor. Therefore, the mean input rate equals the mean output rate that can also be deduced from flow conservation. From the definition of factorial moment of the interdeparture time, we can obtain moments of D_1 as follows:

$$\begin{aligned} &E[D_1] = D^{(1)}, \quad E[D_1^2] = D^{(2)} + D^{(1)}, \\ &E[D_1^3] = D^{(3)} + 3D^{(2)} + D^{(1)}. \end{aligned} \tag{13}$$

3.2 Joint Distribution of Successive Interdeparture Times and Their Correlation for the Composite Departure Process

As in Sect. 3.1, let $\boldsymbol{u}_k$ denote a $1 \times m$ vector whose jth element is the joint stationary probability that the number of cells in the system is k and the state of the underlying Markov chain immediately after the departure is j. Also D_m denotes the mth interdeparture time.

Let $\boldsymbol{d}_k^{(2)}(n_1, n_2)$ denote a $1 \times m$ vector whose jth element $d_{k,j}^{(2)}(n_1, n_2)$ is the joint stationary probability of the two successive interdeparture times of length n_1 and n_2, respectively, the number of cells in the system is k and the state of the underlying Markov chain at the end of the second departure point is j, i.e., $d_{k,j}^{(2)}(n_1, n_2) = Pr\{D_1 = n_1, D_2 = n_2, L_{n_1+n_2}^s = k, J_{n_1+n_2}^s = j\}$. The derivation of $\boldsymbol{d}_k^{(2)}(n_1, n_2)$ for various n_1 and n_2 can be done for the following situations: (a) when $n_1 \geq 1$ and $n_2 = 1$: $\boldsymbol{d}_k^{(2)}(n_1, 1) = \sum_{i=1}^{k+1} \boldsymbol{d}_i^{(1)}(n_1)\boldsymbol{A}_{k+1-i}$ $(k \geq 0)$; (b) when $n_1 \geq 1$ and $n_2 \geq 2$: $\boldsymbol{d}_k^{(2)}(n_1, n_2) = \boldsymbol{d}_0^{(1)}(n_1)\boldsymbol{A}_0^{n_2-2}\sum_{i=1}^{k+1}\boldsymbol{A}_i\boldsymbol{A}_{k+1-i}$ $(k \geq 0)$. Let $d^{(2)}(n_1, n_2) \triangleq \sum_{k=0}^{\infty}\boldsymbol{d}_k^{(2)}(n_1, n_2)\boldsymbol{e}$, then we can obtain $d^{(2)}(n_1, n_2)$ for each case of n_1 and n_2 as follows:

$$\begin{aligned} d^{(2)}(1,1) &= 1 - \boldsymbol{u}_0\boldsymbol{e} - \boldsymbol{u}_1\boldsymbol{A}_0\boldsymbol{e} \\ &\quad \text{for } n_1 = n_2 = 1, \\ d^{(2)}(n_1,1) &= \boldsymbol{u}_0\boldsymbol{A}_0^{n_1-2}(\boldsymbol{I} - \boldsymbol{A}_0 - \boldsymbol{A}_1\boldsymbol{A}_0)\boldsymbol{e} \\ &\quad \text{for } n_1 \geq 2 \text{ and } n_2 = 1, \\ d^{(2)}(1,n_2) &= \boldsymbol{u}_1\boldsymbol{A}_0^{n_2-1}(\boldsymbol{I} - \boldsymbol{A}_0)\boldsymbol{e} \\ &\quad \text{for } n_1 = 1 \text{ and } n_2 \geq 2, \\ d^{(2)}(n_1,n_2) &= \boldsymbol{u}_0\boldsymbol{A}_0^{n_1-2}\boldsymbol{A}_1\boldsymbol{A}_0^{n_2-1}(\boldsymbol{I} - \boldsymbol{A}_0)\boldsymbol{e} \\ &\quad \text{for } n_1 \geq 2 \text{ and } n_2 \geq 2. \end{aligned} \tag{14}$$

Define $\tilde{D}_c(y, z) \triangleq \sum_{n_1=1}^{\infty}\sum_{n_2=1}^{\infty} d^{(2)}(n_1, n_2)y^{n_1}z^{n_2}$ and use (14), we have

$$\begin{aligned} &\tilde{D}_c(y, z) \\ &= (1 - \boldsymbol{u}_0\boldsymbol{e} - \boldsymbol{u}_1\boldsymbol{A}_0\boldsymbol{e})yz + \boldsymbol{u}_0(\boldsymbol{I} - y\boldsymbol{A}_0)^{-1} \\ &\quad \times (\boldsymbol{I} - \boldsymbol{A}_0 - \boldsymbol{A}_1\boldsymbol{A}_0)\boldsymbol{e}y^2z + \boldsymbol{u}_1\boldsymbol{A}_0 \\ &\quad \times (\boldsymbol{I} - z\boldsymbol{A}_0)^{-1}(\boldsymbol{I} - \boldsymbol{A}_0)\boldsymbol{e}yz^2 \\ &\quad + \boldsymbol{u}_0(\boldsymbol{I} - y\boldsymbol{A}_0)^{-1}\boldsymbol{A}_1\boldsymbol{A}_0 \\ &\quad \times (\boldsymbol{I} - z\boldsymbol{A}_0)^{-1}(\boldsymbol{I} - \boldsymbol{A}_0)\boldsymbol{e}y^2z^2. \end{aligned} \tag{15}$$

Then we can derive $E[D_{k-1}D_k]$ from (15) by differentiating it with respect to y and z, respectively. After tedious manipulations, we obtain

$$\begin{aligned} &E[D_{k-1}D_k] \\ &= E[D_1D_2] \\ &= \frac{\partial^2}{\partial z \partial y}\tilde{D}_c(y, z)\mid_{y=1,z=1} \\ &= 1 + (\boldsymbol{u}_0 + \boldsymbol{u}_1\boldsymbol{A}_0)(\boldsymbol{I} - \boldsymbol{A}_0)^{-1}\boldsymbol{e} \\ &\quad + \boldsymbol{u}_0(\boldsymbol{I} - \boldsymbol{A}_0)^{-2}(2\boldsymbol{I} - \boldsymbol{A}_0)\boldsymbol{A}_1\boldsymbol{A}_0 \\ &\quad \times (\boldsymbol{I} - \boldsymbol{A}_0)^{-1}\boldsymbol{e}. \end{aligned} \tag{16}$$

The covariance of D_{k-1} and D_k can be obtained via the following relation

$$\begin{aligned} Cov(D_{k-1}, D_k) &= Cov(D_1, D_2) \\ &= E[D_1D_2] - E^2[D_1]. \end{aligned} \tag{17}$$

From (12) and (17), we now have the departure statistics to characterize the composite output process.

3.3 Per-Stream Departure Process of a Tagged Traffic Stream

In order to facilitate end-to-end performance analysis in ATM networks, it is necessary to figure out the per-stream output process corresponding to a tagged cell stream. However, it is much more difficult to extract the per-stream output process corresponding to a tagged cell stream from the composite output process of the H-MMPPs/Slotted D/1 queue. Instead, we propose a heuristic decomposition scheme to obtain the approximate statistics. The usefulness and effectiveness of the decomposition scheme are checked in Sect. 6.2 through numerical examples.

We now describe the decomposition scheme as follows. Denote the tagged MMPP by MMPP_t and the other $r-1$ cross traffic streams as a whole by MMPP_c. Then the H-MMPPs arrivals are now replaced by MMPP_t+MMPP_c. The decomposition scheme is to substitute the MMPP_t+MMPP_c/Slotted D/1 queue by a decomposed MMPP_t/Slotted D_eff/1 queue† with a

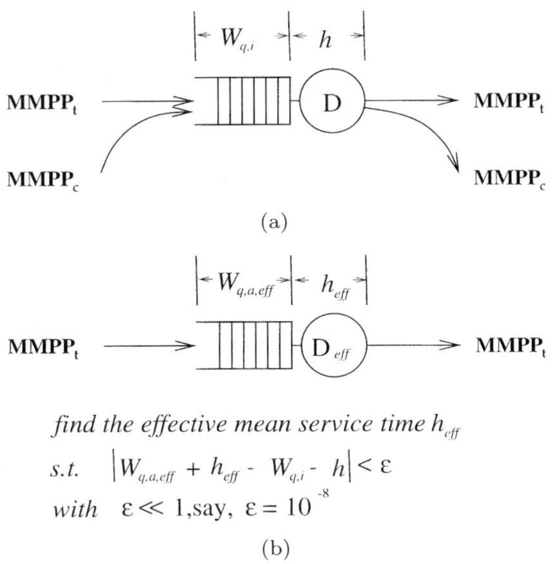

Fig. 3 The decomposition scheme: (a) the undecomposed original queueing model; (b) the replaced queueing model.

modified effective deterministic server D_{eff}. The modification of server is to reflect the interference from the cross cell streams. Since we still employ a deterministic server, it is merely necessary to determine its service time. The effective service time h_{eff} is determined via matching the per-stream sojourn delay time that $MMPP_t$ experiences in $MMPP_t$+$MMPP_c$/Slotted D/1 queue with that in the decomposed queue. That is, we first calculate the per-stream sojourn delay time $W_{q,i} + h$ ($W_{q,i}$ can be calculated using the result in [1]) for the $MMPP_t$ in the undecomposed queue. Then recursively apply binary search to find an h_{eff} over a feasible region such that the sojourn delay time $W_{q,a,eff} + h_{eff}$ ($W_{q,a,eff}$ can also be calculated using the result in [1]) of the decomposed queue satisfies $|W_{q,a,eff} + h_{eff} - W_{q,i} - h| < \epsilon$ where ϵ is a small number, say $\epsilon = 10^{-8}$, specified for stoppage of the recursion. Using the decomposition scheme, we obtain the approximate per-stream departure statistics corresponding to a tagged cell stream. This scheme is also illustrated in Fig. 3.

In this work, we again approximate the per-stream output process by a two-state MMPP through moment matching technique stated latter. Of course, we can approximate the per-stream output process by a multi-state (higher than two) MMPP to gain more accurate results, but this causes more complexities on moment matching scheme.

4. Moment Matching Scheme

For more general description, we introduce the following notation:

- $T_{A,i}$: the time between the ith and the $(i+1)$st arrivals of the matched two-state MMPP parameterized by the descriptor $(\sigma_1^{(m)}, \sigma_2^{(m)}, \lambda_1^{(m)}, \lambda_2^{(m)})$.
- $T_{D,i}$: the time between the ith and the $(i+1)$st departures of the output process.

In order to match the composite or per-stream output process as a two-state MMPP traffic source, we need to select four statistics of the output process to match. In this work, we select the following four statistics: (a) the first moment of interdeparture times, i.e., $E[T_{D,i}]$; (b) squared coefficient of variation of interdeparture times, i.e., $c^2(T_{D,i}) = Var[T_{D,i}]/E^2[T_{D,i}]$; (c) the third moment of interdeparture times, i.e., $E[T^3_{D,i}]$; (d) lag 1 covariance of interdeparture times, i.e., $Cov(T_{D,i}, T_{D,i+1})$.

We now explain why these four statistics are selected. From [17]: (i) the mean $E[T_{D,i}]$ is a measure of central tendency of the random variable $T_{D,i}$; (ii) the variance $Var[T_{D,i}]$ is a measure of the dispersion of the random variable $T_{D,i}$ with respect to (w.r.t.) the mean. Equivalently, $c^2(T_{D,i})$ measures the dispersion; (iii) the skewness $\nu = E[(T_{D,i} - E[T_{D,i}])^3]/(Var[T_{D,i}])^{\frac{3}{2}}$ is a measure of symmetry of the random variable $T_{D,i}$ w.r.t. the mean; (iv) the covariance $Cov(T_{D,i}, T_{D,i+1})$ is a measure of dependence between $T_{D,i}$ and $T_{D,i+1}$. We note that (i)–(iii) concern the shape of the distribution of the random variable $T_{D,i}$ and (iv) describes the correlation/dependence between $T_{D,i}$ and $T_{D,i+1}$. These four statistics well reflect the output process.

We obtain the four statistics after tedious algebraic manipulations as follows:

$$E[T_{A,i}] = \frac{\sigma_1^{(m)} + \sigma_2^{(m)}}{\sigma_2^{(m)}\lambda_1^{(m)} + \sigma_1^{(m)}\lambda_2^{(m)}}, \tag{18}$$

$$\begin{aligned} c^2(T_{A,i}) &= 1 + 2\sigma_1^{(m)}\sigma_2^{(m)}(\lambda_1^{(m)} - \lambda_2^{(m)})^2 \\ &\quad /\{[\sigma_2^{(m)}\lambda_1^{(m)} + \sigma_1^{(m)}\lambda_2^{(m)} + \lambda_1^{(m)}\lambda_2^{(m)}] \\ &\quad \times \sigma_1^{(m)} + \sigma_2^{(m)})^2\}, \end{aligned} \tag{19}$$

$$\begin{aligned} E[T^3_{A,i}] &= 3E[T^2_{A,i}](\sigma_1^{(m)} + \sigma_2^{(m)})/[\sigma_2^{(m)}\lambda_1^{(m)} \\ &\quad + \sigma_1^{(m)}\lambda_2^{(m)} + \lambda_1^{(m)}\lambda_2^{(m)}] + 6[(\sigma_1^{(m)} \\ &\quad + \sigma_2^{(m)})(\sigma_1^{(m)}\lambda_1^{(m)} + \sigma_2^{(m)}\lambda_2^{(m)}) \\ &\quad + \sigma_1^{(m)}\lambda_1^{(m)^2} + \sigma_2^{(m)}\lambda_2^{(m)^2}]/\{(\sigma_2^{(m)} \\ &\quad \times \lambda_1^{(m)} + \sigma_1^{(m)}\lambda_2^{(m)})(\sigma_2^{(m)}\lambda_1^{(m)} \\ &\quad + \sigma_1^{(m)}\lambda_2^{(m)} + \lambda_1^{(m)}\lambda_2^{(m)})^2\}, \end{aligned} \tag{20}$$

[†]From flow conservation, the mean output rates of $MMPP_t$ in both $MMPP_t$+$MMPP_c$/Slotted D/1 and $MMPP_t$/Slotted D_{eff}/1 queues are equal. Therefore, the mean interdeparture time is obtainable without approximation using this decomposition scheme.

$$\begin{aligned} Cov(T_{A,i},T_{A,i+1}) = \{&Var[T_{A,i}] - E^2[T_{A,i}]\} \\ &\times \lambda_1^{(m)}\lambda_2^{(m)}/[2(\sigma_2^{(m)}\lambda_1^{(m)} \\ &+ \sigma_1^{(m)}\lambda_2^{(m)} + \lambda_1^{(m)}\lambda_2^{(m)})]. \end{aligned} \tag{21}$$

Therefore, we now work with the following four matching equations:

$$\frac{\sigma_1^{(m)} + \sigma_2^{(m)}}{\sigma_2^{(m)}\lambda_1^{(m)} + \sigma_1^{(m)}\lambda_2^{(m)}} = A_d, \tag{22}$$

$$\begin{aligned} &1 + 2\sigma_1^{(m)}\sigma_2^{(m)}(\lambda_1^{(m)} - \lambda_2^{(m)})^2/\{[\sigma_2^{(m)}\lambda_1^{(m)} \\ &+\sigma_1^{(m)}\lambda_2^{(m)} + \lambda_1^{(m)}\lambda_2^{(m)}](\sigma_1^{(m)} + \sigma_2^{(m)})^2\} \\ &= B_d, \end{aligned} \tag{23}$$

$$\begin{aligned} &\frac{3C_{d,1}(\sigma_1^{(m)} + \sigma_2^{(m)})}{\sigma_2^{(m)}\lambda_1^{(m)} + \sigma_1^{(m)}\lambda_2^{(m)} + \lambda_1^{(m)}\lambda_2^{(m)}} \\ &\quad + 6[(\sigma_1^{(m)} + \sigma_2^{(m)})(\sigma_1^{(m)}\lambda_1^{(m)} + \sigma_2^{(m)}\lambda_2^{(m)}) \\ &\quad + \sigma_1^{(m)}\lambda_1^{(m)2} + \sigma_2^{(m)}\lambda_2^{(m)2}]/\{(\sigma_2^{(m)}\lambda_1^{(m)} \\ &\quad + \sigma_1^{(m)}\lambda_2^{(m)})(\sigma_2^{(m)}\lambda_1^{(m)} + \sigma_1^{(m)}\lambda_2^{(m)} \\ &\quad + \lambda_1^{(m)}\lambda_2^{(m)})^2\} = C_d, \end{aligned} \tag{24}$$

$$\frac{D_{d,1}\lambda_1^{(m)}\lambda_2^{(m)}}{\sigma_2^{(m)}\lambda_1^{(m)} + \sigma_1^{(m)}\lambda_2^{(m)} + \lambda_1^{(m)}\lambda_2^{(m)}} = D_d \tag{25}$$

where $A_d = E[T_{D,i}]$, $B_d = Var[T_{D,i}]/E^2[T_{D,i}]$, $C_d = E[T_{D,i}^3]$, $C_{d,1} = E[T_{D,i}^2]$, $D_d = Cov(T_{D,i}, T_{D,i+1})$, $D_{d,1} = (Var[T_{D,i}] - E^2[T_{D,i}])/2$.

The system of equations in (22)–(25) can be solved by the following procedure:

- Introduce another set of four variables α, β, γ, δ as follows:

$$\begin{aligned} \alpha &= \sigma_1^{(m)} + \sigma_2^{(m)}, \\ \beta &= \sigma_1^{(m)}\lambda_2^{(m)} + \sigma_2^{(m)}\lambda_1^{(m)}, \\ \gamma &= \lambda_1^{(m)}\lambda_2^{(m)}, \\ \delta &= \lambda_1^{(m)} + \lambda_2^{(m)}. \end{aligned} \tag{26}$$

- Rewrite (22)–(25) in terms of variables α, β, γ, δ. After algebraic manipulations, we have the following solution:

$$\begin{aligned} \alpha &= 6(\zeta\xi - \eta)/[C_d\zeta(\zeta + \eta)^2 - 3C_{d,1}\zeta(\zeta + \eta) \\ &\quad -6(\xi + \xi^2)], \\ \beta &= \zeta\alpha, \\ \gamma &= \eta\alpha, \\ \delta &= \zeta + \xi\alpha \end{aligned} \tag{27}$$

with $\zeta = 1/A_d$, $\eta = D_d/[A_d(D_{d,1} - D_d)]$, $\xi = [(B_d - 1)(\zeta + \eta) + 2\eta]/(2\zeta)$.

- Solving (27), we further obtain

$$\begin{aligned} \lambda_1^{(m)} &= \frac{\delta + \sqrt{\delta^2 - 4\gamma}}{2}, \\ \lambda_2^{(m)} &= \frac{\delta - \sqrt{\delta^2 - 4\gamma}}{2}, \\ \sigma_1^{(m)} &= \frac{\alpha\lambda_1^{(m)} - \beta}{\lambda_1^{(m)} - \lambda_2^{(m)}}, \\ \sigma_2^{(m)} &= \frac{\beta - \alpha\lambda_2^{(m)}}{\lambda_1^{(m)} - \lambda_2^{(m)}}. \end{aligned} \tag{28}$$

  Here we assume $\lambda_1^{(m)} \geq \lambda_2^{(m)}$. For $\lambda_1^{(m)} < \lambda_2^{(m)}$, we switch the above solutions for $\lambda_1^{(m)}$ and $\lambda_2^{(m)}$.

In order to apply the above procedure to match the output process of the H-MMPPs/Slotted D/1 queue, we note that

$$\begin{aligned} E[T_{D,i}] &= TE[D_1], \\ E[T_{D,i}^2] &= T^2E[D_1^2], \\ E[T_{D,i}^3] &= T^3E[D_1^3], \\ Cov(T_{D,i}, T_{D,i+1}) &= T^2Cov(D_{k-1}, D_k). \end{aligned} \tag{29}$$

Then using (12), (17), (27)–(29), we are able to match the output process as a two-state MMPP $(\sigma_1^{(m)}, \sigma_2^{(m)}, \lambda_1^{(m)}, \lambda_2^{(m)})$. Note that (28) may not have a feasible solution set. For such case, we then approximate the output process as a Poisson process by setting $\lambda_1^{(m)} = \lambda_2^{(m)} = \zeta$, $\sigma_1^{(m)}$ and $\sigma_2^{(m)}$ are freely specified. The reason why we select the Poisson process is that only its arrival rate is needed to specify and this parameter can be correctly obtained as mentioned earlier in Sect. 3.3, while the other output statistics are underestimated (at heavy traffic load) using the decomposition scheme (see Sect. 6.2).

5. Connection-Wise End-to-End Performance Analysis in ATM Networks

As we mentioned earlier, each VC in ATM networks can be regarded as a tandem configuration of H-MMPPs/Slotted D/1 queues. Applying the results we obtained in the previous sections, we now propose a recursive scheme outlined in Fig. 4 to evaluate the connection-wise performance for a specific VC. This scheme is done for the tandem queues shown in Fig. 5 in which all external arrival sources are assumed to be two-state MMPPs and mutually independent.

When applying our heuristic recursive calculation of the per-stream end-to-end sojourn delay time, we must modify the sojourn delay time after the first node. This is because the per-stream output processes of the first and the descendent nodes possess the discrete-time nature. Thus, if we use two-state MMPP to match the

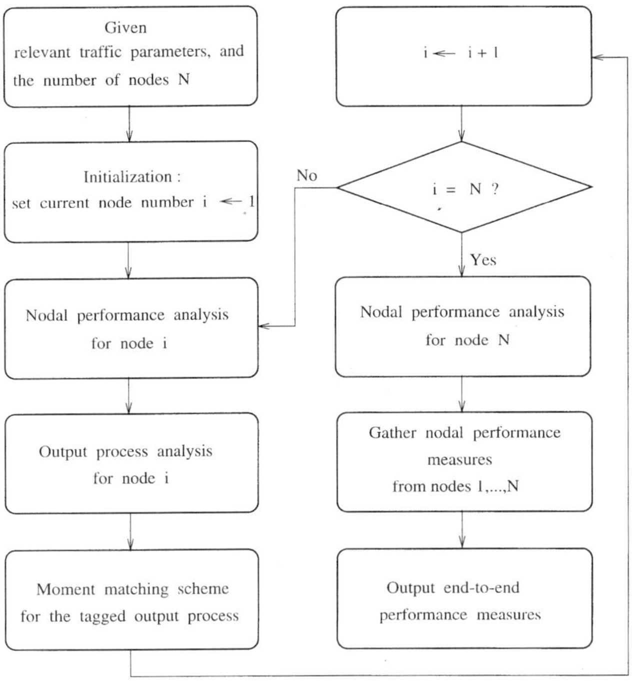

Fig. 4 A recursive scheme applied to the end-to-end performance analysis.

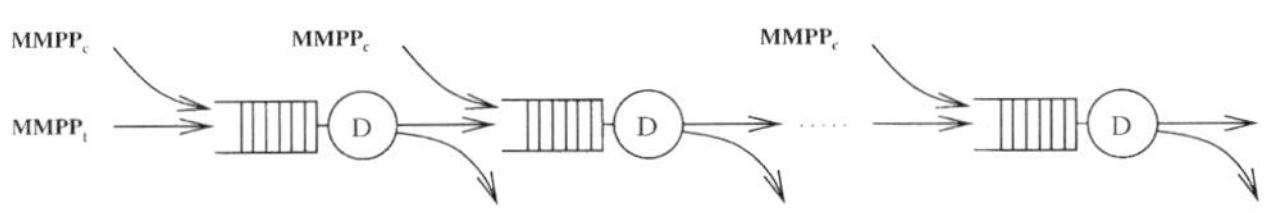

Fig. 5 The tandem queueing configuration.

per-stream output process, we must modify the results by the following interpolation scheme:

- Let the cross traffic be zero, we calculate its per-stream mean delay time drift $\triangle W_{q,i,0} = -W_{q,i,0}$. The subscript 0 denotes zero cross traffic and $W_{q,i,0}$ is the per-stream mean delay time. Note that the tagged traffic coming from the upstream node exhibits interarrival time of multiple slots. Then per-stream mean delay time should be 0. However, $W_{q,i,0}$ is not equal to 0 because we employ MMPP to match the per-stream output process. Hence we need to set $\triangle W_{q,i,0} = -W_{q,i,0}$ to compensate the per-stream mean delay time.
- Assume that the tagged traffic source has effective load ρ_i, we heuristically set $\triangle W_{q,i,1-\rho_i} = 0$, i.e., the overall traffic load equals one. Under the full load condition, the per-stream mean delay time drift becomes negligible. Hence we set $\triangle W_{q,i,1-\rho_i} = 0$.
- We use $\triangle W_{q,i,\rho_c} = -(1-\rho_i-\rho_c)W_{q,i,0}/(1-\rho_i)$ when the tagged traffic load and cross traffic load are ρ_i and ρ_c, respectively. Note that $\triangle W_{q,i,\rho_c}$ can be easily obtained via internal interpolation between $(0, \triangle W_{q,i,0})$ and $(1-\rho_i, \triangle W_{q,i,1-\rho_i})$. This is why we call this scheme an "interpolation" scheme.

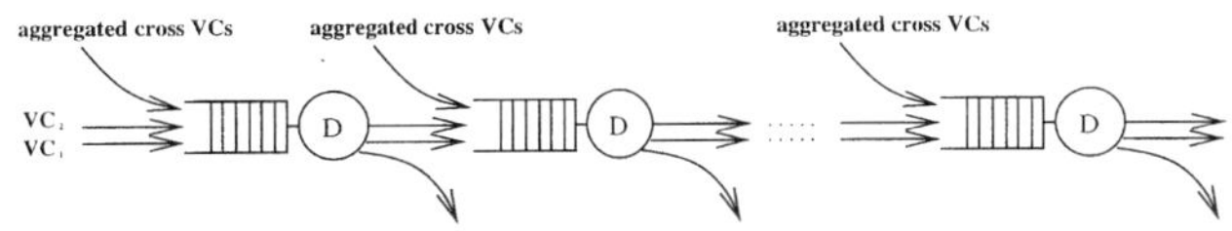

Fig. 6 An example of two VCs sharing the same route.

Using the recursive calculation together with the interpolation scheme, we can obtain a more accurate estimate of the per-stream sojourn delay time incurred at the intermediate nodes.

Note that it is possible for different VCs to use partly/entirely the same route, e.g., Fig. 6. The tandem configuration in Fig. 5 can still be employed for such case in an approximate manner if we neglect the correlation between these VCs when performing the recursion to obtain the connection-wise end-to-end performance measures. In Sect. 6.3, we also demonstrate the effectiveness and good accuracy of the approximation through numerical experiments (see Table 3).

6. Numerical Examples and Discussions

In this section, we investigate both the departure statistics and the connection-wise end-to-end performance through numerical experiments. Computer simulations are provided (with 95% confidence interval) to show the accuracy of approximation. For convenience, we use the definition of burstiness in [2], i.e., burstiness $B = Peak\ rate/\ Average\ rate$ in the following discussions. In this work, burstiness $B = \lambda_1(\sigma_1+\sigma_2)/(\sigma_2\lambda_1+\sigma_1\lambda_2)$ for a two-state MMPP with descriptor $(\sigma_1, \sigma_2, \lambda_1, \lambda_2)$ (we assume $\lambda_1 \geq \lambda_2$). It is easy to show that the mean burst duration is in proportion to $1/\sigma_1$ (or $1/\sigma_2$). In the following examples, we set service time $h = T = 1$.

6.1 Composite Departure Statistics

We consider an example of $r = 2$ under $\lambda_1^*/\lambda_2^* = 1/3$, i.e., we feed two sources which are both represented by two-state MMPPs into the slotted deterministic server. Varying total traffic load from 0.01 to 0.9, we obtain the composite departure statistics in Figs. 7(a)–(d) for three different values of burstiness (under $\sigma_{11} = \sigma_{21} = 0.01$ and $\sigma_{12} = \sigma_{22} = 0.1$). In Fig. 7(a), $E[T_{D,i}]$ for MMPPs, Poissons, and IPPs are all equal. This follows the flow conservation we mentioned earlier. In Fig. 7(b), $c^2(T_{D,i})$ decreases in a smooth manner for MMPPs and Poissons as total traffic load increases and $c^2(T_{D,i})$ for MMPPs is slightly above Poissons. For IPPs, $c^2(T_{D,i})$ increases as total traffic load goes from 0.01 to 0.58 and then decreases. We also note that if input traffic is bursty, then $c^2(T_{D,i})$ is large (as shown in Fig. 7(b): $B = 1$ for Poissons, $B = 4/3$ for MMPPs, $B = 2$ for IPPs). The difference between $c^2(T_{D,i})$ of IPPs and MMPPs (or Poissons) is considerably large. In Fig. 7(c), $E[T_{D,i}^3]$

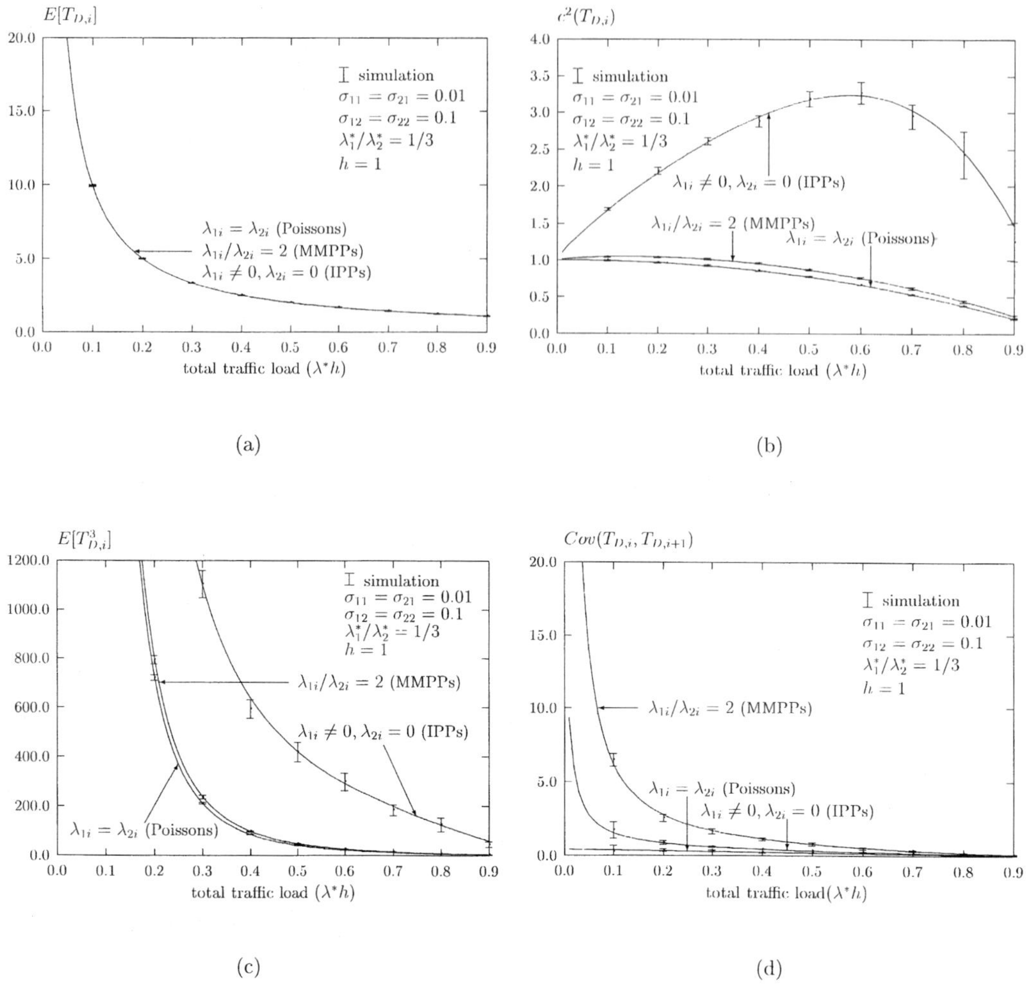

Fig. 7 Composite departure statistics under three different sets of $(\lambda_{1i}, \lambda_{2i})$ $(i = 1, 2)$: (a) $E[T_{D,i}]$; (b) $c^2(T_{D,i})$; (c) $E[T_{D,i}^3]$;(d) $Cov(T_{D,i}, T_{D,i+1})$ versus total traffic load.

decreases rather sharply for MMPPs, IPPs, and Poissons. We notice that the difference between MMPPs and Poissons is very small while the difference between IPPs and MMPPs (or Poissons) is considerably large. Both $c^2(T_{D,i})$ and $E[T_{D,i}^3]$ become larger if input traffic gets more bursty ($c^2(T_{D,i})$ and $E[T_{D,i}^3]$ for IPPs are the largest while for Poissons they are the smallest). From (21), we know that both IPP and Poisson are renewal while MMPP is not. After going through a queue, we see from Fig. 7(d) that composite output process becomes non-renewal (i.e., $Cov(T_{D,i}, T_{D,i+1}) \neq 0$) due to the discreteness of the queue. $Cov(T_{D,i}, T_{D,i+1})$ decreases as total traffic load increases (and finally behaves more like the deterministic process at heavy total traffic load). $Cov(T_{D,i}, T_{D,i+1})$ decreases almost linearly for Poissons as total traffic load increases and is below that of MMPPs. $Cov(T_{D,i}, T_{D,i+1})$ decreases rather sharply for IPPs and MMPPs.

6.2 Per-Stream Departure Statistics

In many situations, it is important to know the per-stream output process corresponding to a tagged source multiplexed with other cross traffic at the input. We fix the descriptor of the tagged traffic at $(\sigma_{11}, \sigma_{21}, \lambda_{11}, \lambda_{21}) = (0.01, 0.01, 0.12, 0.04)$ and apply our proposed heuristic decomposition scheme to find the aper-stream departure statistics corresponding to the tagged source. In Fig. 8, we compare the per-stream departure statistics corresponding to the tagged source under three different interfering MMPPs represented by $(\sigma_{12}, \sigma_{22}) = (0.002,0.008)$, $(0.02,0.08)$, and $(0.2,0.8)$. We notice the followings according to simulation. First, $E[T_{D,i}]$ maintains a constant level since we feed a fixed tagged traffic source. Second, $c^2(T_{D,i})$, $E[T_{D,i}^3]$, and $Cov(T_{D,i}, T_{D,i+1})$ almost maintain at a constant level regardless of the variation in cross traffic except they go up slightly when the cross traffic load becomes large. Third, $c^2(T_{D,i}) \geq 1$, $Cov(T_{D,i}, T_{D,i+1}) > 0$, i.e., (22)–

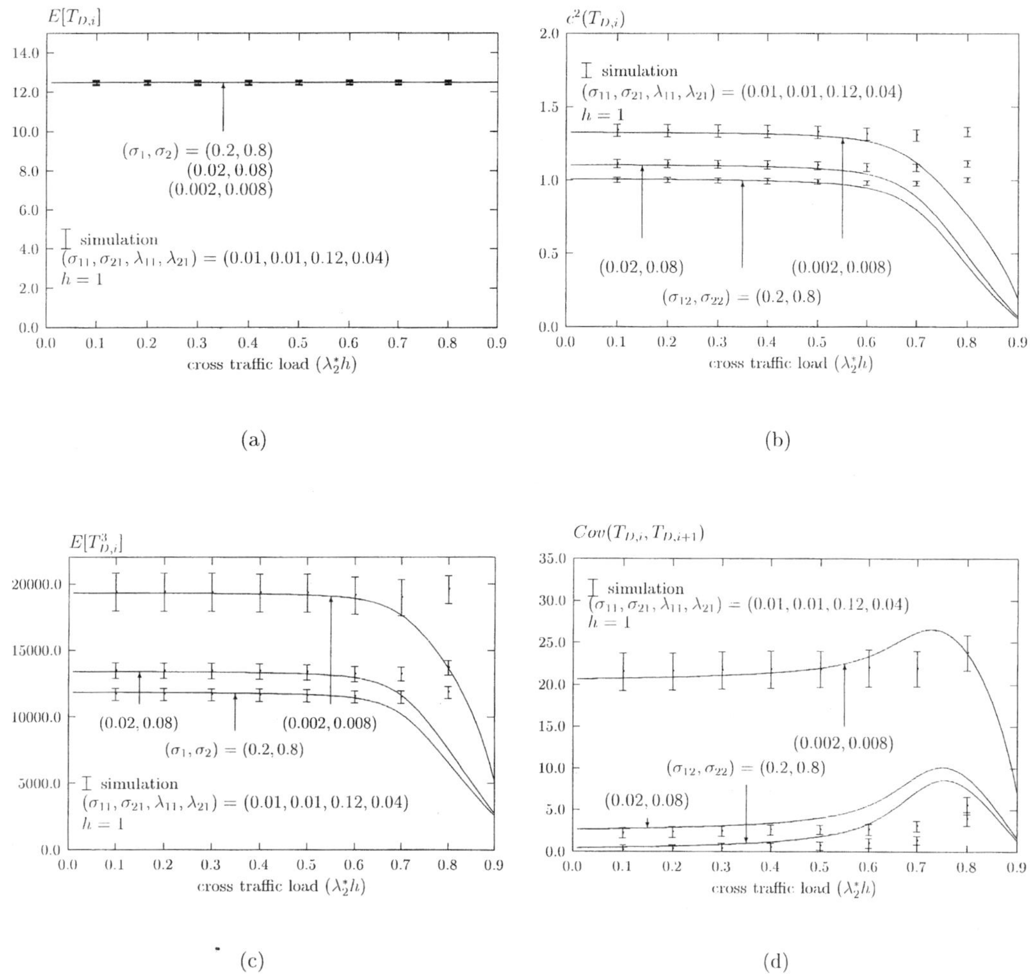

Fig. 8 Per-stream departure statistics: (a) $E[T_{D,i}]$; (b) $c^2(T_{D,i})$; (c) $E[T_{D,i}^3]$;(d) $Cov(T_{D,i}, T_{D,i+1})$ versus cross traffic load.

(25) are satisfied. Hence we can approximate the per-stream output process of a tagged traffic as a two-state MMPP. Fourth, the decomposition scheme works well except for higher utilization. It underestimates the departure statistics except the mean interdeparture time at heavy traffic load. This makes the moment matching scheme fail to work. Since only the mean interdeparture time is valid, the estimated departure process is replaced by a Poisson process.

6.3 Per-Stream Mean End-to-End Sojourn Delay Time

Due to the discrete nature of the H-MMPPs/Slotted D/1 queue, we need the interpolation scheme previously described in Sect. 5 to compensate the sojourn delay time after the first node when applying the recursive calculation of the per-stream end-to-end sojourn delay time. In the following, we show how this scheme works at the second node. Consider a two-node tandem queueing configuration in Fig. 5 in which each node is modeled by a deterministic server fed by a fixed tagged MMPP at the first node described by $(\sigma_{11}, \sigma_{21}, \lambda_{11}, \lambda_{21}) = (0.01, 0.01, 0.12, 0.04)$ and a cross MMPP at the first and second node, respectively, with traffic load varying from 0.01 to 0.9 under the fixed ratio of $\lambda_{12}/\lambda_{22} (= 2)$ (here, two different sets of cross MMPPs are arranged for comparison by setting $(\sigma_{12}, \sigma_{22})$ at (0.2,0.8) and (0.002,0.008), respectively). In Fig. 9, we show the uncompensated sojourn delay $W_{q,i} + h$ and the mean delay time drift $\triangle W_{q,i,\rho_c}$ s the cross traffic load varies from 0.01 to 0.9. Comparing with the simulation result, the uncompensated sojourn delay time indeed is overestimated except at very heavy traffic load and the overestimated value decreases as the (cross) traffic load increases. As shown in Fig. 9, $\triangle W_{q,i,\rho_c} < 0$ and $-\triangle W_{q,i,\rho_c}$ is roughly equal to the deviation of the overestimated sojourn delay from the simulated value so that the compensated sojourn delay $(W_{q,i} + h + \triangle W_{q,i,\rho_c})$ provides a more accurate result than the uncompensated one. We also note that the compensation effect contributes more at light traffic load than at heavy traffic load. This is equivalent to say the interpolation scheme contributes little when traffic is heavy. One may wonder why the sojourn delay time at heavy traffic load maintains comparatively accurate? This is due to that the decomposition scheme underestimates the per-stream departure statistics (except the mean interdeparture time) and the interpolation scheme does little help. We note that the estimated output process well matches an MMPP when the decomposition scheme works, while a Poisson process is used to replace the estimated output process when moment matching fails to work. It is the use of Poisson process which makes the estimated sojourn delay time (at an intermediate node) remains accurate, not the interpolation scheme. However, accuracy of the (compensated) sojourn delay is better at light traffic load than at heavy traffic load (see Table 1). The error of approximation reaches 5–10% at high traffic load. In the following examples, we shall show the compensated results only.

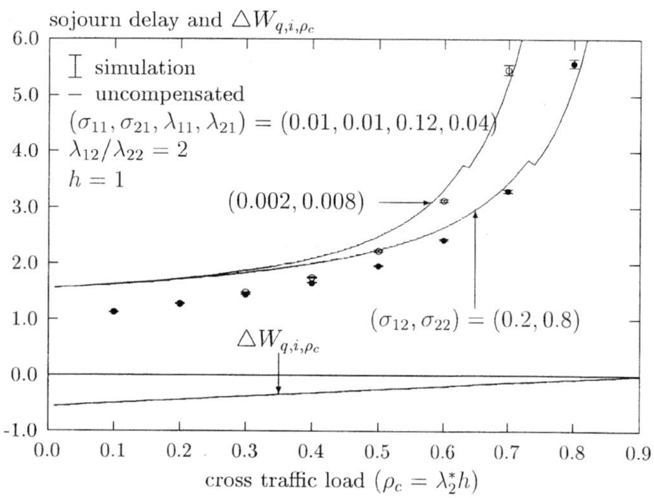

Fig. 9 Uncompensated sojourn delay $W_{q,i} + h$ and mean delay time drift $\triangle W_{q,i,\rho_c}$ at the second node.

Now we consider a tandem queueing configuration of four nodes (see Fig. 5). Each node is modeled as an H-MMPPs/Slotted D/1 queue fed by two sources (one is the tagged traffic and the other denotes the cross traffic). We assume all nodes are synchronized and all cross traffic streams are identical and mutually independent. Consider this model for the scenario of VCs in ATM networks, with tagged information flow parameterized by $(\sigma_{11}, \sigma_{21}, \lambda_{11}, \lambda_{21}) = (0.01, 0.01, 0.12, 0.04)$. Then we see how the per-stream mean end-to-end sojourn delay time $S_i^{ee} = \sum_{j=1}^{N=4} S_{i,j}$, ($N$: the number of nodes, i: connection identity, and j: intermediate node identity, $S_{i,j}$: per-stream mean sojourn delay time corresponding to connection i at intermediate node j) is affected by the cross traffic in Figs. 10(a)–(b) (when cross traffic load varies from 0.01 to 0.9). In Fig. 10(a), we change the mean burst duration of the cross MMPP by varying $\sigma_{12} = \sigma_{22}$ from 0.01 to 0.1 under $\lambda_{12}/\lambda_{22} = 2$ and $B = 4/3$. The result corresponding to Poissonian cross traffic is also included. The cross MMPP causes the per-stream mean end-to-end sojourn delay time to become larger than Poisson. Cross traffic with long mean burst duration makes the per-stream mean end-to-end sojourn delay time large. Table 1 gives the results for the case of $\sigma_{12} = \sigma_{22} = 0.01$ for detailed numerical comparison. From this table, we note that the approximation errors are about 0.8% (0.6%) for $\rho_c = 0.2$ and −5.4% (−5.0%) for $\rho_c = 0.8$ at the second node (for the total sojourn delay). In Fig. 10(b), we change the mean burst duration of the cross MMPP with $B = 10/9$ by varying $(\sigma_{12}, \sigma_{22})$ from (0.002, 0.008) to (0.2, 0.8) under $\lambda_{12}/\lambda_{22} = 2$. Results corresponding to Poissonian and IPP (with $B = 5/4$) cross traffic under $(\sigma_{12}, \sigma_{22}) = (0.02, 0.08)$ are also included. We observe that under fixed σ_{12}/σ_{22}, IPP and Poisson respectively serve as

Note that it is possible for different VCs to use

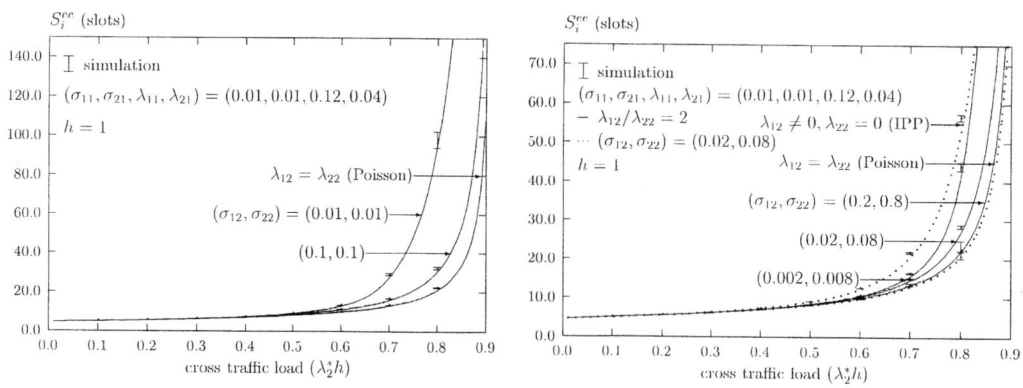

(a) (b)

Fig. 10 Per-stream mean end-to-end sojourn delay time S_i^{ee} versus cross traffic load of a four-node tandem configuration accepting two sources: (a) cross traffic with $\sigma_{12} = \sigma_{22}$; (b) cross traffic with $\sigma_{12}/\sigma_{22} = 0.25$.

Table 1 Delay at each node of the four-node tandem queues in Fig. 5.

ρ_c		Node 1	Node 2	Node 3	Node 4	Total
0.2	Simulation	1.724±0.2%	1.276±0.2%	1.273±0.5%	1.270±0.4%	5.544±0.2%
	Analysis	1.7219	1.2860	1.2853	1.2847	5.5779
0.4	Simulation	2.079±0.3%	1.743±0.8%	1.728±0.6%	1.718±0.5%	7.269±0.3%
	Analysis	2.0746	1.7559	1.7521	1.7486	7.3312
0.6	Simulation	3.552±1.5%	3.301±2.0%	3.248±1.7%	3.204±1.0%	13.305±0.9%
	Analysis	3.5062	3.2873	3.0986	3.0986	12.9908
0.8	Simulation	24.946±7.0%	24.316±5.3%	24.648±5.4%	24.095±7.1%	98.005±4.2%
	Analysis	24.1057	22.9920	22.9920	22.9920	93.0816

Table 2 Traffic parameters employed in the example of Fig. 6.

	Node 1	Node 2	Node 3	Node 4
VC_1 $(\sigma_{11}, \sigma_{21}, \lambda_{11}, \lambda_{21})$	(0.01,0.01,0.12,0.04)	—	—	—
VC_2 $(\sigma_{12}, \sigma_{22}, \lambda_{12}, \lambda_{22})$	(0.01,0.01,0.10,0.06)	—	—	—
cross VCs $(\sigma_{1c}, \sigma_{2c}, \lambda_{1c}, \lambda_{2c})$	$(0.01,0.01,\frac{4\rho_c}{3},\frac{2\rho_c}{3})$	$(0.01,0.01,\frac{4\rho_c}{3},\frac{2\rho_c}{3})$	$(0.01,0.01,\frac{4\rho_c}{3},\frac{2\rho_c}{3})$	$(0.01,0.01,\frac{4\rho_c}{3},\frac{2\rho_c}{3})$

Table 3 Delay at each node of the four-node tandem queues in Fig. 6.

ρ_c			Node 1	Node 2	Node 3	Node 4	Total
0.2	VC_1	Simulation	1.821±0.2%	1.325±0.3%	1.317±0.3%	1.314±0.4%	5.776±0.2%
		Analysis	1.8181	1.4296	1.4277	1.4266	6.1020
	VC_2	Simulation	1.800±0.1%	1.314±0.3%	1.308±0.2%	1.306±0.2%	5.729±0.1%
		Analysis	1.7997	1.4179	1.4142	1.4140	6.0459
0.4	VC_1	Simulation	2.318±0.4%	1.931±0.5%	1.907±0.5%	1.900±0.6%	8.055±0.4%
		Analysis	2.3178	2.0395	2.0321	1.9819	8.3714
	VC_2	Simulation	2.278±0.4%	1.897±0.6%	1.876±0.3%	1.874±0.5%	7.925±0.2%
		Analysis	2.2732	1.9932	1.9917	1.9819	8.2400
0.6	VC_1	Simulation	5.320±2.2%	4.990±2.0%	4.896±1.0%	4.86±2.0%5	20.071±0.7%
		Analysis	5.2587	4.6809	4.6809	4.6809	19.3015
	VC_2	Simulation	5.118±1.5%	4.832±1.8%	4.7370±1.8%	4.734±1.8%	19.422±0.8%
		Analysis	5.0606	4.6809	4.6809	4.6809	19.1034
0.8	VC_1	Simulation	101.6±15%	96.93±6.9%	88.28±16%	91.05±12%	377.8±6.2%
		Analysis	93.293	90.295	90.295	90.295	364.18
	VC_2	Simulation	99.43±13%	94.99±6.4%	87.25±13%	93.55±13%	375.2±5.3%
		Analysis	92.663	90.295	90.295	90.295	363.55

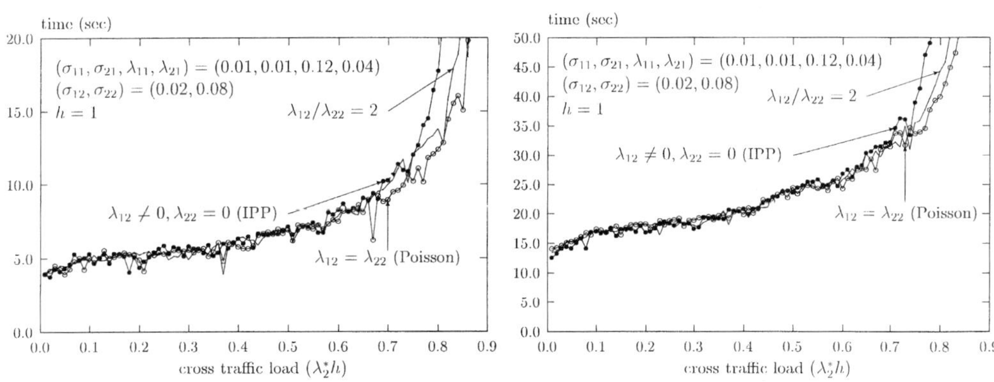

(a) (b)

Fig. 11 Computation time: (a) for the decomposition scheme; (b) for the calculation of a four-node tandem queueing configuration.

partly/entirely the same route as mentioned earlier. Hence the tandem configuration in Fig. 5 can be employed for such case in an approximate manner if we may neglect the correlation between these VCs when performing the recursion to obtain the connection-wise end-to-end performance measures. We now consider a case in Fig. 6 where two VCs share entirely the same route and all other cross VCs are assumed to be still identical and mutually independent. For this configuration with the traffic parameter sets in Table 2, we obtain delays along both routes VC_1 and VC_2 in Table 3. Table 3 says that the approximation is still good since there is merely about 8.6% error for the worst case.

The above examples have demonstrated the accuracy of our proposed method in conducting end-to-end performance analysis. Let us further examine the efficiency regarding the computation time consumed. The following examples are run on a *Sun SPARC-20E* workstation using Matlab Ver.5.1. Shown in Fig. 11(a) is the CPU time used to find the value of h_{eff} under the following parameters: the tagged traffic fixed at $(\sigma_{11}, \sigma_{21}, \lambda_{11}, \lambda_{21}) = (0.01, 0.01, 0.12, 0.04)$, at every node the cross traffic fixed only for $(\sigma_{12}, \sigma_{22}) = (0.02, 0.08)$ but the traffic load varies from 0.01 to 0.9 under three different traffic sources, i.e., MMPP $(\lambda_{21}/\lambda_{22} = 2)$, IPP $(\lambda_{12} \neq 0,\ \lambda_{22} = 0)$, and Poisson $(\lambda_{12} = \lambda_{22})$. Note that the above traffic arrangement has been employed in previous examples (i.e., in Fig. 10). The decomposition scheme needs about 4–10 seconds when the traffic load is lower than 0.7 and up to 10–50 seconds when the traffic load goes beyond 0.7. Using the above traffic arrangement, Fig. 11(b) shows the CPU time consumed for a four-node end-to-end calculation. The CPU time falls in the range 12–30 seconds when the traffic load varies from 0.01 to 0.7. At extremely heavy traffic load, the calculation even consumes more than one minute. This seems unsatisfactory for practical CAC need. However, if faster processors such as *Sun UltraSPARC* are used, the computation time may be reduced by a factor (at least five to ten) to make the proposed method usable on-line.

7. Conclusion

In this paper, we have investigated the composite output process of the H-MMPPs/Slotted D/1 queueing system and proposed a simple heuristic decomposition scheme to obtain the per-stream output process corresponding to a tagged traffic stream. This decomposition scheme can also be used in other queueing systems. A moment matching scheme used to emulate the output process as a two-state MMPP is also provided. These results can be easily extended to obtain the composite end-to-end sojourn delay time or per-stream end-to-end sojourn delay time.

Our results enable us to conduct end-to-end performance analysis for each VC in ATM networks. It can be of course extended to evaluate the overall network performance since a network consists of a collection of individual VCs. In other words, we offer a powerful analytical method to call admission control, congestion control and routing algorithms in ATM networks.

References

[1] H.W. Ferng and J.F. Chang, "Per-stream nodal delay analysis in ATM networks," Research Note, Dept. Elect. Eng., National Taiwan University, 1997.

[2] D.E. McDysan and D.L. Spohn, ATM Theory and Application, McGraw-Hill, New York, 1995.

[3] R.G. Addie and M. Zukerman, "Queueing performance of a tree type ATM network," IEEE INFOCOM'94, Toronto, Canada, pp.48–55, June 1994.

[4] H. Kroner, M. Eberspacher, T.H. Theimer, P.J. Kuhn, and U. Briem, "Approximate analysis of the end-to-end delay in ATM networks," IEEE INFOCOM'92, Florence, Italy, pp.978–986, May 1992.

[5] J.F. Ren, J.W. Mark, and J.W. Wong, "End-to-end performance in ATM networks," IEEE ICC'94, New Orleans, USA, pp.996–1002, May 1994.

[6] Y. Ohba, M. Murata, and H. Miyahara, "Analysis of interdeparture processes for bursty traffic in ATM networks," IEEE J. Sel. Areas Commun., vol.9, no.3, pp.468–476, April 1991.

[7] D. Park, H.G. Perros, and H. Yamashita, "Approximate analysis of discrete-time tandem queueing networks with bursty and correlated input traffic and customer loss," Oper. Res. Lett., vol.15, pp.95–104, 1994.

[8] H. Saito, "The departure process of an $N/G/1$ queue," Performance Evaluation, vol.11, pp.241–251, 1990.

[9] T. Takine, T. Suda, and T. Hasegawa, "Cell loss and output process analyses of a finite-buffer discrete-time ATM queueing system with correlated arrivals," IEEE Trans. Commun., vol.43, no.2/3/4, pp.1022–1037, Feb./March/April 1995.

[10] L. Kleinrock, Queueing Systems, vol.I: Theory, Wiley, New York, 1975.

[11] H. Heffes and D.M. Lucantoni, "A Markov modulated characterization of packetized voice and data traffic and related statistical multiplexer performance," IEEE J. Sel. Areas Commun., vol.SAC-4, no.6, pp.856–868, Sept. 1986.

[12] H. Heffes, "A class of data traffic processes—Covariance function characterization and related queuing results," Bell Syst. Tech. J., vol.59, no.6, pp.897–929, July-Aug. 1980.

[13] W. Fischer and K.S. Meier-Hellstern, "The Markov-modulated Poisson process (MMPP) cookbook," Performance Evaluation, vol.18, pp.149–171, 1993.

[14] D.M. Lucantoni and V. Ramaswami, "Efficient algorithms for solving the non-linear matrix equations arising in phase type queues," Commun. Statist.–Stochastic Models, vol.1, no.1, pp.29–51, 1985.

[15] M.F. Neuts, Structured Stochastic Matrices of $M/G/1$ Type and Their Applications, Dekker, New York, 1989.

[16] V. Ramaswami, "Nonlinear matrix equations in applied probability—Solution techniques and open problems," SIAM Rev., vol.30, no.2, pp.256–263, June 1988.

[17] A.M. Law and W.D. Kelton, Simulation Modeling & Analysis, 2 ed., McGraw-Hill, New York, 1991.

High-Performance Backbone Network Technology edited by Naoaki Yamanaka
ISBN: 0-8247-5321-6 DOI: 10.1081/(120025706)
Published 2004 by Marcel Dekker, Inc.

End-to-End Call Admission Control in Service Guaranteed Networks

Yung-Chung Wang and Chung-Chin Lu

SUMMARY A per-connection end-to-end call admission control (CAC) problem is solved in this paper to allocate network resources to an input session to guarantee its quality of service (Qos) requirements. In conjunction with the solution of the CAC problem, a traffic descriptor is proposed to describe the loss rate and the delay bound Qos requirements of the connection to be set up as well as the statistical characteristics of the associated input traffic which is modeled as a linear mean function plus a (zero-mean) fractional Brownian motion. The information in the traffic descriptor is sufficient to determine the allocation of channel bandwidth and buffer space to the input traffic in a network which employs leaky bucket shapers and scheduling algorithms to guarantee the Qos requirements. The CAC problem is solved by an iterative algorithm of which there are two stages in each iteration: one is responsible for the search of a candidate end-to-end routing path and the other for the verification of the legitimacy of this candidate path to meet the Qos requirements and for the allocation of resources in such a legitimate path.

key words: *service-guaranteed networks, dal admission control, traffic descriptor, lWeibul bounded burstiness process, quality of service*

1. Introduction

In many multimedia applications, the generated network traffic requires quality of service (Qos) guarantee in terms of packet loss rate and end-to-end delay. The lack of Qos guarantee in first-come-first-served service discipline has motivated the creation of leaky bucket and scheduling algorithm to shape the input traffic and to allocate the capacity of the scheduling server respectively in order to guarantee the Qos requirements. With both leaky bucket traffic shaper and scheduling algorithm employed in a service-guaranteed network, it is the aim of this paper to solve the per-connection end-to-end call admission control (CAC) problem, i.e., the problem of searching out a routing path from the caller to the callee of an input session such that the network resources, such as channel bandwidth and buffer size, can be allocated to guarantee the Qos requirements of the input session. In order to do so, we will propose a traffic descriptor to describe not only the Qos requirements of the connection to be set up but also the statistical characteristics of the associated input traffic.

There are a number of scheduling algorithms proposed in the literature for providing per-connection performance guarantee, including Virtual Clock [10], [27], [30], Packet-by Packet Generalized processor sharing (PGPS) [23], [24], Self Clocked Fair Queueing (SCFQ) [12]–[14], [17], Delay EDD [31], Jitter EDD [31], and Rate Control Static Priority Queueing (RCSP) [31]. Since a network is unlikely to be homogeneous in the type of scheduling algorithms operated in the switches or routers, we shall consider the case that an individual session in a network may be serviced by servers with distinct scheduling algorithms.

The contents of a traffic descriptor has been a key issue in service-guaranteed networks. In [1], a source traffic descriptor, including the peak cell rate, and optionally, a sustainable cell rate, maximum burst size, and minimum cell rate, is proposed for ATM networks. In [8], the flow descriptor for Internet RSVP is the combination of a flowspec and a filterspec. In [26], a filterspec, including the maximum packet size, peak rate, sustainable rate and burst tolerance is proposed for RSVP. Neither of the above two descriptors are thought as complete solutions to be used in an ATM network or in the Internet RSVP. In [15], a triple (D, T, p) is used to describe a real-time transport connection in a deterministic Qos guaranteed network, where D is the message delay bound, T is the minimum message interarrival time, and p is the maximum message processing time. In [25], sliding window and leaky bucket method are used to describe teleconferencing service over an ATM network. Both papers are lacking of providing possible statistical multiplexing gain in service-guaranteed networks.

Gaussian approximation technique has been widely used to model packet videos and to estimate congestion behavior in high speed networks, see [2], [16], [20], [29]. In [4], [18], self-similar property has been observed in VBR video traffic, ethernet traffic and ISDN traffic in high speed networks. Fractional Brownian motion (FBM) is commonly manipulated of modeling self-similar traffic since not only it is identified as promising but also its applicability has been extensively studied with surprisingly positive results, see [4], [9], [18], [21], [22]. It is then reasonable to model a generic stochastic input process in a multimedia high speed network by an FBM through Gaussian approximation. Thus

Manuscript received May 12, 1999.
Manuscript revised November 10, 1999.

[†]The authors are with the Department of Electrical Engineering, National Tsing Hua University, Hsinchu, Taiwan.

*This work was supported by the National Science Council, Taiwan, under Contract NSC88-2218-E-007-015. Parts of this manuscript was presented at ISCOM'97, Hsinchu, Taiwan.

IEICE TRANS. COMMUN., Vol. E83-B, No. 4 pp. 791–802, April 2000 in this paper, the accumulated input traffic, associated with the connection to be set up, will be modeled as a linear mean function μt plus a (zero-mean) fractional Brownian motion (FBM) $B(t; H, \sigma^2)$, where σ^2 and H are the variance constant and the Hurst parameter of the FBM.

Thus we will propose a traffic descriptor which describe the two Qos requirements of the input traffic, i.e., the tolerable data loss rate P_{loss} of the input traffic under the leaky bucket filtering and the tolerable per connection end-to-end delay D_{bound} through heterogeneous scheduling servers, as well as the three statistical parameters μ, σ^2 and H of the input traffic modeled as a linear mean function μt plus an FBM $B(t; H, \sigma^2)$.

The network architecture considered in this paper consists of access nodes and switching nodes, where each access node implements both a leaky-bucket shaper and a scheduling algorithm and each switching node implements a scheduling algorithm only. The differentiated service architecture is an example of this network architecture [5], [7]. Thus the per-connection end-to-end call admission control (CAC) problem is the problem of searching out a routing path from the access node associated with the caller of an input session to the access node associated with the callee of the input session such that the network resources, such as channel bandwidth and buffer size, can be allocated to guarantee the Qos requirements of the input session. To the best knowledge of the authors, the per-connection end-to-end call admission control problem investigated in this paper is seldom or never considered in the literature. In [11], [15], [19], the call admission control problem was investigated only on a per-node (an access or a switching node) basis. The global (per-connection end-to-end) call admission control problem was not discussed there. Furthermore, in contrast to our study, the potential statistical multiplexing gain in a service-guaranteed network was not exploited there.

The outline of this paper is as follows. In Sect. 2, we will give a brief review of main results on service guaranteed networks, together with an introduction to a newly defined Weibull bounded burstiness input processes as an extension of exponentially bounded burstiness input processes. In Sect. 3, we will present a performance analysis of a service-guaranteed network with Gaussian input processes which have stationary increments, including FBMs. We then propose a format of traffic descriptor to describe an input session in a guaranteed service network. In Sect. 4, we will show that the contents of the proposed traffic descriptor are sufficient for the calculation of the allocation of channel bandwidth and buffer space to an input session to meet its Qos requirements, when leaky bucket shapers and scheduling algorithms are employed. In Sect. 5, we will formulate the per-connection end-to-end call admission control problem as an optimization problem and solve the problem by proposing an iterative algorithm. In Sect. 6, we will give a conclusion.

2. Service Guaranteed Networks

In this section, main results on service guaranteed scheduling algorithms in the literature are reviewed as reported in [10], [14], [17], [27]. Since future networks are unlikely to be homogeneous in the type of scheduling algorithms employed by the switches or routers, we shall consider the case that an individual session in a network may be serviced by servers with distinct scheduling algorithms. We then define Weibull bounded burstiness (WBB) process as an extension of exponentially bounded burstiness process investigated in [28]. And, the tail distribution of the end-to-end packet delay for a WBB input process through a heterogeneous network of service guaranteed servers is derived for possible interest.

Consider a session s being serviced by a scheduling server, called server n. The scheduling algorithm in server n, such as VirtualClock [30], PGPS [23] or SCFQ [13], calculates the virtual finishing time $VF_{i,s}^{(n)}$ of the ith packet in server n based on the arrival time $a_{i,s}^{(n)}$ of the ith packet to server n, its packet length $l_{i,s}$ and the virtual finishing time $VF_{i-1,s}^{(n)}$ of its previous packet as follows:

$$VF_{i,s}^{(n)} = \max\{v^{(n)}(a_{i,s}^{(n)}), VF_{i-1,s}^{(n)}\} + \frac{l_{i,s}}{r_s} \tag{1}$$

where $v^{(n)}(t)$ is the virtual time function of the scheduling server n and depends on the scheduling algorithm, and r_s is the bandwidth allocated to session s. The operations of the scheduling algorithm in server n are as follows:

- use (1) to calculate the virtual finishing time of an incoming packet;
- serve packets from all sessions traversing server n in the order of virtual finishing time.

With the operations in above, the scheduling algorithm can guarantee a deadline by which a packet of a session will be transmitted. This guarantee is referred to as *service guarantee*. The service guarantee provided by the algorithm is based on the service deadline value associated with each packet. The service deadline $SD_{i,s}^{(n)}$ of the ith packet of session s in server n is given by

$$SD_{i,s}^{(n)} = \max\{a_{i,s}^{(n)}, SD_{i-1,s}^{(n)}\} + \frac{l_{i,s}}{r_s}. \tag{2}$$

The scheduling algorithm at server n is called *service-guaranteed*, if the departure time $d_{i,s}^{(n)}$ of the ith packet of session s from server n is upper bound by the service deadline $SD_{i,s}^{(n)}$ plus a constant $\Delta^{(n)}$, i.e.

$$d_{i,s}^{(n)} \le SD_{i,s}^{(n)} + \Delta^{(n)}, \tag{3}$$

where $\Delta^{(n)}$ depends on the scheduling algorithm at server n. As shown in [14], VirtualClock, PGPS and SCFQ are service-guaranteed scheduling algorithms.

The performance of service-guaranteed scheduling algorithms can be analyzed in tandem. Consider the entire route of session s whose traffic traverses N possibly heterogeneous service-guaranteed scheduling servers. Let $D_{i,s}$ be the end-to-end delay of the ith packet in session s. As proved in [10], [14], we have the following lemma:

Lemma 1:

$$D_{i,s} \le SD_{i,s}^{(1)} - a_{i,s}^{(1)} + (N-1)\frac{\max_{j\le i} l_{j,s}}{r_s} + \sum_{n=1}^{N}\Delta^{(n)} + \sum_{n=1}^{N}\Omega^{(n)},$$

where $\Omega^{(n)}$ is the propagation delay of the outgoing link at server n and $\Delta^{(n)}$ is in (3).

A more explicit bound on the end-to-end delay can be derived if a deterministic characterization of the input traffic is specified. In particular, an input traffic is said to conform to a (b, r)-leaky bucket shaper [23] if $A(s, s+t) \le b + rt$, where $A(s, s+t)$ is the accumulated amount of input traffic over the interval $[s, s+t)$. As shown in [14], we have the following theorem:

Theorem 2: If the input traffic of session s conforms to a (b_s, r_s)-leaky bucket shaper, then the end-to-end delay $D_{i,s}$ of the ith packet in session s satisfies

$$D_{i,s} \le \frac{b_s}{r_s} + (N-1)\frac{l_{\max,s}}{r_s} + \sum_{n=1}^{N}\Delta^{(n)} + \sum_{n=1}^{N}\Omega^{(n)}$$

where $l_{\max,s}$ is the maximum packet length of session s. □

On the other hand, a bound on the tail distribution of the end-to-end delay can be derived if a stochastic characterization of the input traffic can be specified. For our purposes, we define a new class of input processes as follows. Let $A(s, s+t)$ be the accumulated amount of input traffic over the time interval $[s, s+t)$. A stochastic input process is called a Weibull bounded burstiness (WBB) process with parameters (r, K, B, γ), if its accumulated amount $A(s, s+t)$ has the following property: for any $b \ge 0$,

$$Pr\{A(s, s+t) \ge rt + b\} \le K\exp(-Bb^{\gamma}),\ \forall\, s, t \ge 0. \quad (4)$$

The exponentially bounded burstiness (EBB) process, discussed in [14], [28], is a special case of WBB process with $\gamma = 1$. The WBB process can be used to describe a stochastic input process which is modeled by a fractional Brownian motion as will be seen in the next section. Fractional Brownian motion is usually used to model traffic generated in a multimedia high speed network, see [18], [21]. By using a similar argument as in [14, Theorem 4], the following theorem is proved for possible interest.

Proposition 3: If session s traverses N possibly heterogeneous service-guaranteed scheduling servers and if its input traffic is WBB with parameters $(r_s, K_s, B_s, \gamma_s)$, then the tail distribution of the end-to-end delay $D_{i,s}$ of the ith packet in session s is

$$Pr\left\{ D_{i,s} \ge x + (N-1)\frac{\max_{j\le i} l_{j,s}}{r_s} + \sum_{n=1}^{N}\Delta^{(n)} + \sum_{n=1}^{N}\Omega^{(n)} \right\} \le K_s \exp(-B_s(xr_s)^{\gamma_s}).$$

Proof: Let $k \equiv \max\{j \le i \mid SD_{j-1,s}^{(1)} < a_{j,s}^{(1)}\}$. Then by (2), we have

$$SD_{i,s}^{(1)} = a_{k,s}^{(1)} + \sum_{j=0}^{i-k}\frac{l_{k+j,s}}{r_s} = a_{k,s}^{(1)} + \frac{A(a_{k,s}^{(1)}, a_{i,s}^{(1)})}{r_s}. \quad (5)$$

Since the input traffic is WBB with parameters $(r_s, K_s, B_s, \gamma_s)$, we have

$$Pr\{A(a_{k,s}^{(1)}, a_{i,s}^{(1)}) \ge r_s(a_{i,s}^{(1)} - a_{k,s}^{(1)}) + b\} \le K_s\exp(-B_s b^{\gamma_s}),\ \forall\, b \ge 0$$

by (4). And by (5), we have

$$Pr\{SD_{i,s}^{(1)} - a_{i,s}^{(1)} \ge x\} \le K_s\exp(-B_s(xr_s)^{\gamma_s})$$

with $x = b/r_s$. Since $Pr\{Y \ge c\} \le Pr\{Z \ge c\}$ for any r.v.'s $Y \le Z$ and any constant c, the proof is completed by Lemma 1. □

3. Traffic Descriptor

In this section, we first analyze the loss behavior of a stochastic accumulated input traffic, modeled by a fractional Brownian motion, under the filtering of a leaky bucket traffic shaper in a service-guaranteed network. We then define a traffic descriptor to describe the Qos requirements and the statistical characterizations of a stochastic accumulated input traffic. The usage of this traffic descriptor will be extensively examined in the next two sections.

We shall assume that the accumulated amount $A(s, s+t)$ of an input traffic arriving at the network ingress in the interval $[s, s+t)$ is stationary in time, i.e. depending on t only and not on s. This says that the accumulated input process $A(0, t)$ has stationary increments. Let $\Lambda_t(\theta)$ be the log-moment generating function of $A(0, t)$.

Lemma 4: For any $r > 0$ and $b > 0$, we have

$$Pr\{A(s, s+t) \geq rt + b\} \leq \exp[-\Lambda^*_{t^*}(rt^* + b)]$$

where t^* is the argument of $\inf_{t>0} \Lambda^*_t(rt+b)$ and $\Lambda^*_t(rt+b) = \sup_{\theta>0}(\theta(rt+b) - \Lambda_t(\theta))$.

Proof: For any $\theta > 0$, $t > 0$, we have

$$\begin{aligned} Pr\{A(s, s+t) \geq rt + b\} &= Pr\{A(0, t) \geq rt + b\} \\ &= Pr\{e^{\theta A(0,t)} \geq e^{\theta(rt+b)}\} \\ &\leq e^{-(\theta(rt+b) - \Lambda_t(\theta))} \end{aligned}$$

by Chernoff bounding technique. And by taking the supreme of $\theta(rt+b) - \Lambda_t(\theta)$ over $\theta > 0$, we have

$$\begin{aligned} Pr\{A(s, s+t) \geq rt + b\} &\leq e^{-\sup_{\theta>0}(\theta(rt+b) - \Lambda_t(\theta))} \\ &= e^{-\Lambda^*_t(rt+b)} \\ &\leq e^{-\Lambda^*_{t^*}(rt^*+b)} \end{aligned}$$

where t^* is the argument of $\inf_{t>0} \Lambda^*_t(rt+b)$ and $\Lambda^*_t(rt+b) = \sup_{\theta>0}\{\theta(rt+b) - \Lambda_t(\theta)\}$. □

Lemma 4 can be applied to any stochastic input process whose accumulated amount has stationary increments. Suppose further that $A(0, t)$ is a Gaussian process with mean function $E[A(0, t)] = \mu t$ and variance function $Var[A(0, t)] = \sigma_t^2$. Then we have

$$\Lambda_t(\theta) = \log E[e^{\theta A(0,t)}] = \mu t\theta + \frac{\theta^2\sigma_t^2}{2},$$

$$\begin{aligned} \Lambda^*_t(rt+b) &= \sup_{\theta>0}\{\theta(rt+b) - \Lambda_t(\theta)\} \\ &= \frac{((r-\mu)t+b)^2}{2\sigma_t^2}, \end{aligned}$$

$$\begin{aligned} t^* &= arg \inf_{t>0} \Lambda^*_t(rt+b) \\ &= arg \inf_{t>0} \frac{((r-\mu)t+b)^2}{2\sigma_t^2}. \end{aligned}$$

From Lemma 4, we have the following theorem.

Theorem 5: If $A(0, t)$ is a Gaussian process with stationary increments and has mean function μt and variance function σ_t^2, then

$$Pr\{A(s, s+t) \geq rt + b\} \leq \exp\left\{-\frac{((r-\mu)t^*+b)^2}{2\sigma_{t^*}^2}\right\}$$

where t^* is the argument of $\inf_{t>0} \frac{((r-\mu)t+b)^2}{2\sigma_t^2}$. □

We will specialize our results to a class of Gaussian input processes. A Gaussian process $B(t; H, \sigma^2)$ is called a fractional Brownian motion if it has zero mean function $E[B(t; H, \sigma^2)] = 0$ and covariance function

$$\begin{aligned} C(s, t) &= E[B(s; H, \sigma^2)B(t; H, \sigma^2)] \\ &= \frac{1}{2}\sigma^2\left(t^{2H} - (t-s)^{2H} + s^{2H}\right), \\ &\forall\, 0 \leq s \leq t, \end{aligned} \tag{6}$$

where H and σ^2 are called the Hurst parameter (dimensionless) and the variance constant (with dimension $\text{bits}^2/\text{sec}^{2H}$) of the fractional Brownian motion $B(t; H, \sigma^2)$ respectively. Now consider $A(0, t)$ to be μt plus a zero mean fractional Brownian motion $B(t; H, \sigma^2)$. The variance function $Var[A(0, t)] = \sigma_t^2$ of $A(0, t)$ is $\sigma_t^2 = \sigma^2 t^{2H}$ by (6). Then we have

$$\begin{aligned} t^* &= arg \inf_{t>0} \frac{((r-\mu)t+b)^2}{2\sigma^2 t^{2H}} \\ &= \frac{Hb}{(1-H)(r-\mu)}, \end{aligned}$$

$$\begin{aligned} \Lambda^*_{t^*}(rt^*+b) &= \frac{((r-\mu)t^*+b)^2}{2\sigma^2(t^*)^{2H}} \\ &= \frac{(r-\mu)^{2H}}{2\sigma^2 H^{2H}(1-H)^{2(1-H)}} b^{2(1-H)}. \end{aligned}$$

By Theorem 5, we have the following corollary.

Corollary 6: Let $A(0, t)$ be μt plus a zero mean fractional Brownian motion with Hurst parameter H and variance constant σ^2. Then

$$\begin{aligned} &Pr\{A(s, s+t) \geq rt + b\} \\ &\leq \exp\left\{-\frac{(r-\mu)^{2H}}{2\sigma^2 H^{2H}(1-H)^{2(1-H)}} b^{2(1-H)}\right\}. \end{aligned}$$

□

The above corollary shows that a stochastic accumulated input process which is μt plus a zero mean fractional Brownian motion with Hurst parameter H and variance constant σ^2 is a WBB process with parameters r, provided $r > \mu$, $K = 1$, $B = (r-\mu)^{2H}/2\sigma^2 H^{2H}(1-H)^{2(1-H)}$, and $\gamma = 2(1-H)$.

When $A(0, t)$ is a Gaussian process with independent and stationary increments, i.e. $\sigma_t = \sigma^2 t$, the result is a special case of Corollary 6 with Hurst parameter $H = \frac{1}{2}$ and we have

$$Pr\{A(s, s+t) \geq rt + b\} \leq \exp\left\{-\frac{2(r-\mu)b}{\sigma^2}\right\},$$

which is the same result as in [28] in the discussion of exponentially bounded burstiness process, a special case of Weibull bounded burstiness process.

It has been illustrated in the literature that Gaussian approximation technique is useful to model a large class of traffics and to estimate congestion behavior in high-speed multimedia networks. For examples, Gaussian process is used to model packet video in [16], [20], [29] and a Gaussian approximation technique is used to estimate congestion behavior in an ATM network in [2]. In [4], [18], self-similar property has been observed to be possessed by VBR video traffic, ethernet traffic and ISDN traffic in high speed networks. Fractional Brownian motion is commonly manipulated of modeling self-similar traffic since it is not only identified as promising but its applicability has been extensively studied with surprisingly positive results, see [4], [18], [21], [22]. It is then reasonable to approximate a generic stochastic accumulated input process in a high-speed multimedia network by a linear mean function μt plus a (zero mean) fractional Brownian motion $B(t; H, \sigma^2)$. For the estimation of the Hurst parameter H for an accumulated input process, please see [3].

Thus we define a 5-tuple $\boldsymbol{T} = (P_{loss}, D_{bound}, \mu, \sigma^2, H)$ as the traffic descriptor for a generic stochastic input process in a high-speed multimedia network which employs leaky bucket traffic shapers and scheduling servers to guarantee Qos. The first two parameters in the traffic descriptor $\boldsymbol{T}$ are the Qos requirements of the input traffic, where P_{loss} is the tolerable data loss rate of the input traffic under the leaky bucket filtering and D_{bound} is the tolerable per connection end-to-end delay through heterogeneous scheduling servers. The last three parameters are used to characterize the statistics of the input traffic through Gaussian approximation as discussed in above, where μ is the mean rate of the input traffic, σ^2 is the variance constant and H is the Hurst parameter of the fractional Brownian motion which approximates the input process.

4. Acceptance Region

In this section, we will investigate the allocation of resources in a service-guaranteed network to an input traffic to meet its Qos requirements. More precisely, we will use the mean rate, the variance constant and the Hurst parameter described in the traffic descriptor of the input process to determine acceptable bandwidth and buffer size allocation for the leaky bucket shaper and scheduling servers to meet the packet loss rate and the delay bound requirements, also described in the traffic descriptor.

Let the traffic descriptor of the stochastic input process of session s be $(P_{s,loss}, D_{s,bound}, \mu_s, \sigma_s^2, H_s)$. If the upper bounds of the packet loss rate and the end-to-end delay in Corollary 6 and in Theorem 2 respectively are set to be no greater than $P_{s,loss}$ and $D_{s,bound}$ in the traffic descriptor respectively, the Qos requirements are guaranteed. Thus from Corollary 6 and Theorem 2, the following two inequalities

$$\exp\left(-\frac{(r_s-\mu_s)^{2H_s}}{2\sigma_s^2 H_s^{2H_s}(1-H_s)^{2(1-H_s)}} b_s^{2(1-H_s)}\right) \leq P_{s,loss}, \; 0 < H_s < 1, \mu_s \leq r_s,$$

$$\frac{b_s}{r_s} + (N-1)\frac{l_{\max,s}}{r_s} + \sum_{n=1}^{N} \Delta_n + \sum_{n=1}^{N} \Omega_n \leq D_{s,bound},$$

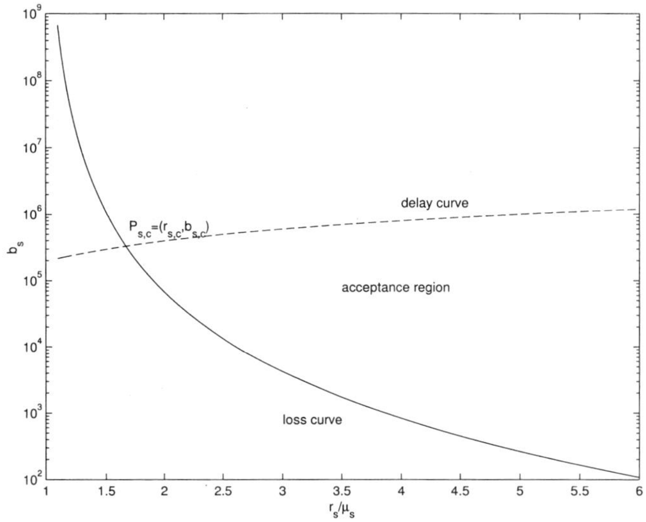

Fig. 1 The acceptance region.

can be used to determine the required bandwidth r_s and the bucket size b_s of the leaky bucket shaper for session s. After some algebraic manipulations, the above two inequalities can be rewritten as

$$b_s \geq \left(\frac{-2\sigma_s^2 H_s^{2H_s}(1-H_s)^{2(1-H_s)} \ln P_{s,loss}}{(r_s-\mu_s)^{2H_s}}\right)^{\frac{1}{2(1-H_s)}}, \tag{7}$$

$$b_s \leq \left(D_{s,bound} - \sum_{n=1}^{N}\Delta_n - \sum_{n=1}^{N}\Omega_n\right) r_s - (N-1)l_{\max,s}, \quad 0 < H_s < 1, \quad \mu_s < r_s. \tag{8}$$

Since (7) describes the loss constraint, the curve determined by the equality in (7) in the r_s-b_s plane is called the *loss curve*. And since (8) describes the delay constraint, the curve determined by the equality in (8) in the r_s-b_s plane is called the *delay curve*. It can be seen that the delay curve is a line in the r_s-b_s plane with slope $D_{s,bound} - \sum_{n=1}^{N}\Delta_n - \sum_{n=1}^{N}\Omega_n$ which is interpreted as the *delay margin* of session s and denoted as $D_{s,margin}$.

The region in between the delay and the loss curves is the region for the (r_s, b_s) pair to satisfy both the loss and the delay constraints and is called the *acceptance region*. An example of the acceptance region for a stochastic input process with traffic descriptor $(P_{s,loss}, D_{s,bound}, \mu_s, \sigma_s^2, H_s) = (10^{-8}$, 40 ms, 10 Mbps, 1 Mbps^2/sec$^{1.6}$, 0.8) is shown in Fig. 1 in a semilog scale, where the input traffic is assumed to traverses $N = 10$ possibly heterogeneous service-guaranteed scheduling servers in an ATM network. The packet length (i.e. cell size) $l_{j,s}$ is 53×8 bits. We have also assumed that both Δ_n and Ω_n are 1 ms for each of the 10 servers.

The intersection point of the delay and the loss

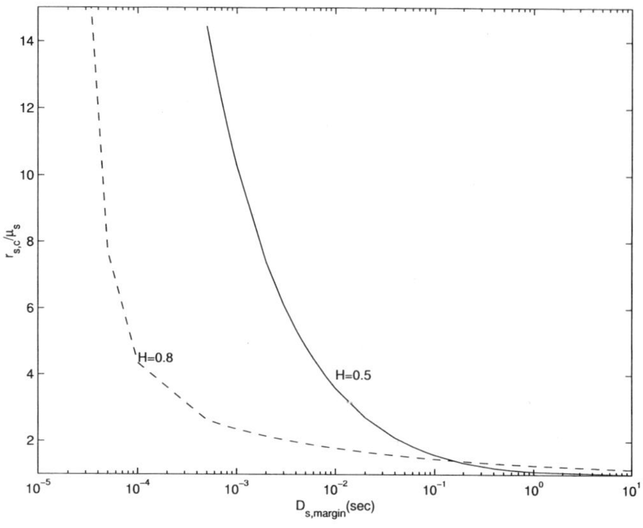

Fig. 2 The effect of delay margin on the critical bandwidth allocation with $\frac{\sigma_s}{\mu_s} = 0.1$ sec^2, $P_{s,loss} = 10^{-8}$.

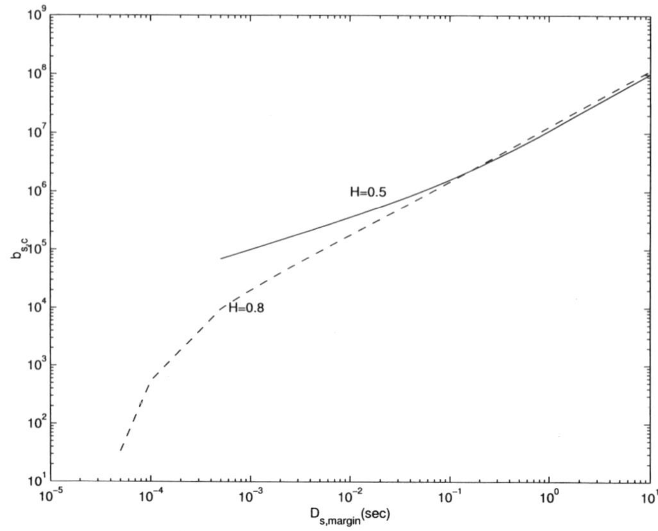

Fig. 3 The effect of delay margin on the critical buffer size allocation with $\frac{\sigma_s}{\mu_s} = 0.1$ sec^2, $P_{s,loss} = 10^{-8}$.

curves in the r_s-b_s plane is called the *critical point* of session s and is denoted by $P_{s,c} = (r_{s,c}, b_{s,c})$ as shown in Fig. 1, where $r_{s,c}$, called the *critical* bandwidth allocation, is the minimum required bandwidth allocation for session s to satisfy both the delay and the loss constraints and $b_{s,c}$ is the corresponding buffer size allocation. Any acceptable bandwidth allocation r_s for session s must be no less than $r_{s,c}$. Note that $r_{s,c} > \mu_s$ as indicated in (7).

The delay margin does not affect the loss curve but do affect the critical bandwidth allocation $r_{s,c}$. Combine (7) and (8) with equalities together, we have the following equation

$$\left(\frac{1}{D_{s,margin}}\right)(\sigma_s/\mu_s)^{1/(1-H_s)} f(H_s) \cdot (-2\ln P_{s,loss})^{1/(2(1-H_s))} = \left(\frac{r_{s,c}}{\mu_s} - (N-1)\frac{l_{\max,s}}{D_{s,margin}\mu_s}\right) \cdot \left(\frac{r_{s,c}}{\mu_s} - 1\right)^{H_s/(1-H_s)},$$

where $f(x) = \left(x^x(1-x)^{(1-x)}\right)^{1/(1-x)}$, which relates the normalized critical bandwidth allocation $r_{s,c}/\mu_s$ to the delay margin $D_{s,margin}$ as well as other parameters such as the Hurst parameter H, the normalized standard deviation constant σ_s/μ_s and the loss probability requirement P_{loss}. If the delay margin increases to infinity, (i.e no delay constraint), then the critical point $P_{s,c}$ climbs up along the loss curve so that the critical bandwidth allocation $r_{s,c}$ approaches to the mean rate μ_s as shown in Fig. 2. In the mean time, the bucket size allocation $b_{s,c}$ will approach to infinity to ensure the loss constraint as shown in Fig. 3. And if the delay margin decreases to zero, the critical point $P_{s,c}$ slides down along the loss curve so that the critical bandwidth allocation $r_{s,c}$ approaches to infinity and the critical buffer size allocation $b_{s,c}$ approaches to zero. Since bandwidth is a precious resource, we desire to have delay margin $D_{s,margin}$ as large as possible such that the critical bandwidth allocation $r_{s,c}$ is as small as possible. And since the delay margin depends on the route through which the input traffic traverses if a delay bound is give in the traffic descriptor of an input traffic, we will select a route which provides maximal possible delay margin if there are choices as will be discussed in the next section.

Observing the two curves in Fig. 2, an input process with Hurst parameter $H = 0.5$, i.e. whose accumulated process is modeled as a linear mean function μt plus a Brownian motion with stationary and independent increments, demands more bandwidth allocation than an input process with Hurst parameter $H = 0.8$, i.e. whose accumulated process is modeled as a linear mean function μt plus a Brownian motion with long-range dependence [4], [9], [18], [21], [22], when the delay margin is limited, for example less than 100 msec in Fig. 2. The difference becomes more significant if the delay margin becomes very limited. On the other hands, if the delay margin is wide, for example greater than 200 msec in Fig. 2, an input process with short-range dependence (with Hurst parameter $H = 0.5$) demands less bandwidth allocation than an input process with long-range dependence (with Hurst parameter $H = 0.8$), but the difference is not so significant. This can be observed more clearerly in Fig. 4 where limited delay margin requires limited buffer allocation to satisfy the delay constraint and then requires significantly more bandwidth allocation to accommodate the larger short-time jitter in a short-range dependent input process (with Hurst parameter close to 0.5) to

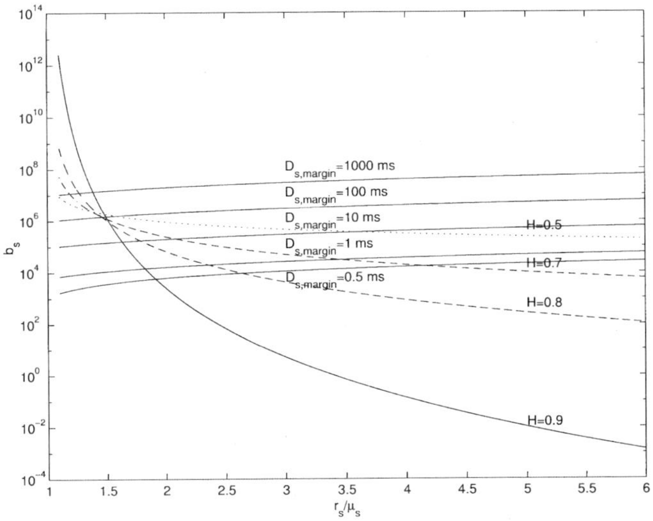

Fig. 4 The effect of Hurst parameter on the critical bandwidth and buffer size allocation with $\frac{\sigma_s}{\mu_s} = 0.1$ sec^2, $P_{s,loss} = 10^{-8}$.

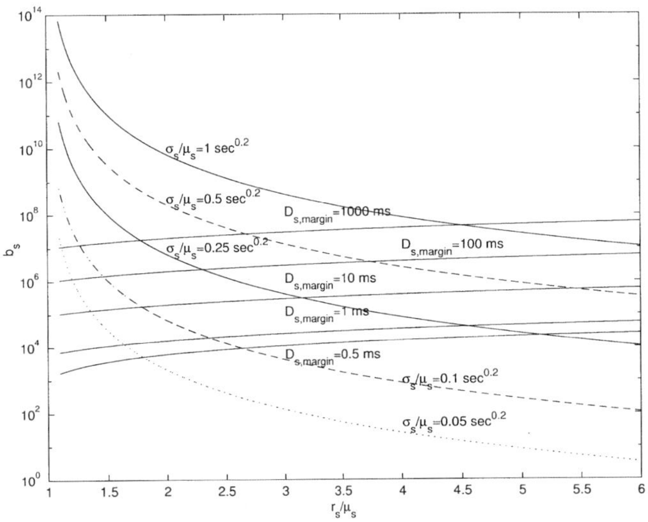

Fig. 5 The effect of the normalized standard deviation constant $\frac{\sigma_s}{\mu_s}$ of the input traffic on the critical bandwidth and buffer size allocation with $P_{s,loss} = 10^{-8}$, $H_s = 0.8$.

guarantee the loss constraint. And if delay margin is wide, larger buffer size can be allocated without violating the delay constraint such that large short-term jitter in a short-range dependent input traffic can be smoothed. Thus smaller bandwidth allocation is required for a short-range dependent input traffic. This reflects the basic difference between short-range dependent and long-range dependent input process: buffer allocation is an effective means to compensate bandwidth allocation for a short-range dependent traffic to guarantee the loss rate.

Another important parameter is the normalized standard deviation constant σ_s/μ_s of the input traffic. An input traffic with large σ_s/μ_s has large traffic jitter such that the loss curve moves up entirely. This means that both bandwidth allocation and buffer allocation increase as the normalized standard deviation constant σ_s/μ_s increases, even for a long-range dependent input traffic with Hurst parameter $H = 0.8$ as shown in Fig. 5. It can be seen that the demand of bandwidth allocation is significant if the normalized standard deviation constant σ_s/μ_s is not small, for example no less than 0.25 in Fig. 5.

Similar results can be observed in Fig. 6 where the loss curve moves up entirely when the loss rate requirement becomes tight. Thus both bandwidth allocation and buffer allocation increase as the loss probability decreases. But the change is not so significant within a wide range of delay margin and loss rate constraint.

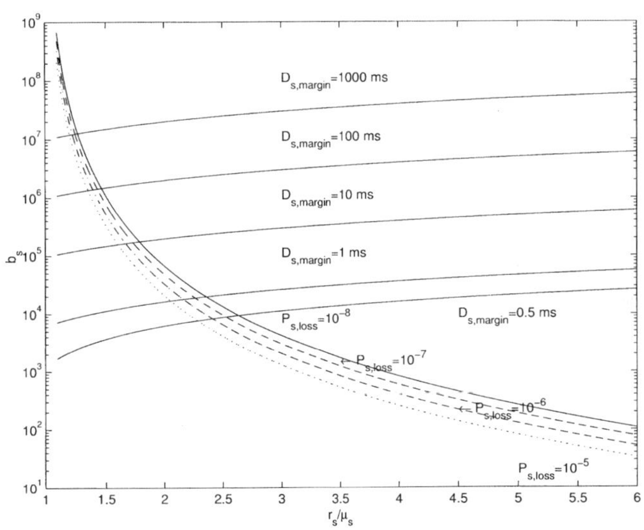

Fig. 6 The effect of loss probability requirement $P_{s,loss}$ on the critical bandwidth and buffer size allocation with $\frac{\sigma_s}{\mu_s} = 0.1$ sec^2, $H_s = 0.8$.

5. End-to-End Call Admission Control

The network architecture considered in this paper consists of access and switching nodes, where each access node implements both a leaky-bucket shaper and a scheduling algorithm and each switching node implements a scheduling algorithm only. The differentiated service architecture is an example of this network architecture [5], [7]. The scheduling algorithms implemented in access nodes and switch nodes may not be of the same type but are all assumed to be service-guaranteed as stated in Sect. 2. In this section, we will consider the per-connection end-to-end call admission control (CAC) problem of input traffic processes (called sessions) with traffic descriptor $\mathbf{T} = (P_{loss}, D_{bound}, \mu, \sigma^2, H)$ as proposed in Sect. 3. The CAC problem is said to be solved if the question of the existence of an end-to-end connection (i.e., a routing path) from the caller to the callee, which can guarantee the Qos requirements of the input session, can be determined. If the answer to the question is yes, then the call is accepted. Otherwise, the call is rejected. Furthermore, if the call can be accepted and there are more than one legitimate routing paths, then we will select the routing path which minimizes the allocation of channel bandwidth which is a precious network resource as mentioned in the last section.

To formulate the per-connection end-to-end call admission control problem, the whole network is modeled as a directed graph (abbr. as a digraph) with nodes being access nodes or switching nodes and directed links being transmission channels from a node to another node. A link (i, j) from node i to node j in the digraph is labeled by a pair $(d_{i,j}, C_{i,j}^{(r)})$ of quantities, where $d_{i,j}$, called the *delay distance* of link (i, j), is the sum $\Delta_{i,j} + \Omega_{i,j}$ of the constant scheduling delay $\Delta_{i,j}$ at node i to forward traffic to node j and the propagation delay $\Omega_{i,j}$ through the transmission channel from node i to node j, and $C_{i,j}^{(r)}$ is the unused capacity of the transmission channel from node i to node j, called the *residual capacity* of link (i, j).

Consider an input session s with traffic descriptor $\mathbf{T} = (P_{s,loss}, D_{s,bound}, \mu_s, \sigma_s^2, H_s)$ to be set up in the guaranteed service network. Let α and β be the access nodes associated with the caller and the callee of session s respectively. Let $b_{\max}$ be the maximum buffer size for a leaky-bucket shaper pre-determined by the access node α with which the caller of session s is associated. Let $\mathcal{P}_{\alpha,\beta}$ be the collection of all (acyclic) directed path from access node α to access node β in the digraph. And for a path $\Gamma = t_0 t_1 \ldots t_N$ in $\mathcal{P}_{\alpha,\beta}$, where $t_0 = \alpha$ and $t_N = \beta$, we define the delay distance $D(\Gamma)$ of path Γ as the sum of the delay distance of its component links, i.e., $D(\Gamma) = \sum_{n=1}^{N} d_{t_{n-1},t_n}$ and the residual capacity $C^{(r)}(\Gamma)$ of path Γ as the minimum of the residual capacity of its component links, i.e., $C^{(r)}(\Gamma) = \min_{1\le n\le N} C_{t_{n-1},t_n}^{(r)}$. If Γ is selected as the routing path for the input session s to guarantee its Qos requirements, the critical point $P_c(\Gamma) = (r_c(\Gamma), b_c(\Gamma))$ in the resource allocation r_s-b_s plane is determined by the delay margin $D_{margin}(\Gamma) = D_{s,bound} - D(\Gamma)$ of path Γ as well as the loss probability $P_{s,loss}$ and the three traffic statistical parameters μ_s, σ_s^2, H_s in the traffic descriptor of session s, as seen from (7) and (8). Since bandwidth is a precious resource, we desire to have a routing path Γ with delay margin $D_{margin}(\Gamma)$ as large as possible such that the critical bandwidth allocation $r_c(\Gamma)$ is as small as possible.

Now, the call admission control problem for session s amounts to the following optimization problem:

$$\text{minimize}_{\ \Gamma\in\mathcal{P}_{\alpha,\beta}} D(\Gamma)$$

$$\begin{aligned} \text{subject to}\quad & D(\Gamma) < D_{s,bound} \\ & b_c(\Gamma) \le b_{max} \end{aligned}$$

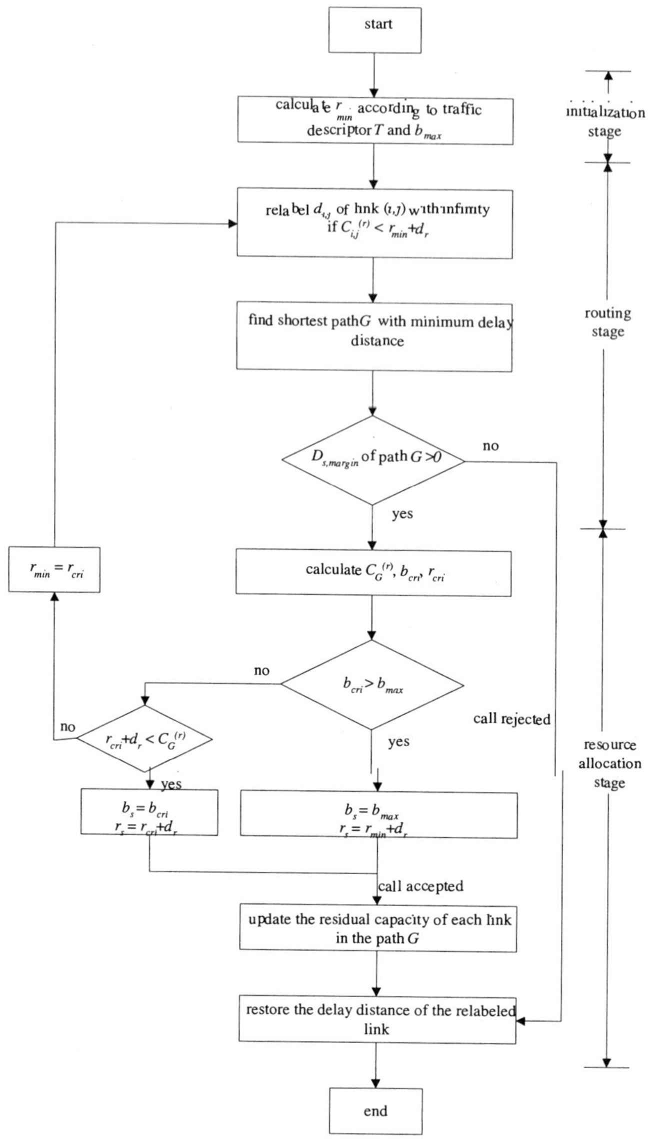

Fig. 7 The flowchart of the proposed iterative CAC algorithm.

$$r_c(\Gamma) + \delta_r \le C^{(r)}(\Gamma) \tag{9}$$

where δ_r is a bandwidth margin considered for the smooth operation of the network. If a solution of the optimization problem in (9) exists, then the call for the setup of session s is accepted, otherwise rejected. If accepted, a solution path Γ will be the routing path of session s and the resource allocation for session s is

$$b_s = b_c(\Gamma) \text{ and } r_s = r_c(\Gamma) + \delta_r.$$

It can be seen from (9) that there is a routing problem embedded in the the formulation of the CAC problem. We now propose an iterative algorithm to solve the CAC problem in (9). Except for an *initialization* stage, each iteration consists of a *routing* stage and a *resource allocation* stage as shown in Fig. 7.

In the initialization stage, the algorithm calculates

the minimum bandwidth required to set up the connection by solving the loss inequality in (7) with $b_s = b_{\max}$ and the parameters $P_{s,loss}$, μ_s, σ_s^2 and H_s in the traffic descriptor of session s, i.e.,

$$r_s \geq \mu_s + \frac{\sigma_s^{\frac{1}{H_s}} H_s(1-H_s)^{\frac{1-H_s}{H_s}}(-2\ln P_{s,loss})^{\frac{1}{2H_s}}}{b_{max}^{\frac{1-H_s}{H_s}}}. \tag{10}$$

The lower bound in the above inequality of r_s will be denoted as r_{min}. Thus the minimum required bandwidth is $r_{min} + \delta_r$, where we have added a bandwidth margin δ_r for the smoothness of network operations. Since $r_{min} + \delta_r$ is the minimum required bandwidth to set up the connection, every link in a possible route must have residual capacity no less than $r_{min} + \delta_r$. Thus, a link (i, j) from node i to node j with residual capacity $C_{i,j}^{(r)}$ less than $r_{min} + \delta_r$ will be relabeled by the infinity distance ∞. This is equivalent to delete link (i, j) from the digraph.

After the initialization stage, the algorithm goes to the first routing stage to find a path $\Gamma = t_0 t_1 \ldots t_N$ in $\mathcal{P}_{\alpha,\beta}$, where $t_0 = \alpha$ and $t_n = \beta$, which has the minimum possible delay distance $D(\Gamma)$. Then its delay margin $D_{margin}(\Gamma) = D_{s,bound} - D(\Gamma)$ would be the maximum possible. To find such a path in $\mathcal{P}_{\alpha,\beta}$, we can employ a shortest path algorithm (see Sect. 5.2 in [6]) in the relabeled digraph with delay distance as the link metric. Let G be such a shortest path from node α to node β. If the delay margin $D_{margin}(G)$ is non-positive, then the first constraint in (9) cannot be satisfied by any path in $\mathcal{P}_{\alpha,\beta}$ with residual capacity no less than the minimum required bandwidth $r_{min} + \delta_r$. Thus there exist no end-to-end connections which can guarantee the Qos requirements and the call will be rejected.

One the other hand, if the delay margin $D_{margin}(G)$ is positive, the algorithm goes to the resource allocation stage of this iteration. In this stage, the algorithm firstly calculates the residual capacity $C^{(r)}(G)$ of path G and secondly calculates the critical buffer allocation $b_c(G)$ and the critical bandwidth allocation $r_c(G)$ by finding the intersection point of the loss curve and the delay curve associated with the shortest path G.

If $b_c(G) \geq b_{max}$, then $r_c(G) \leq r_{min}$ and the point $(r_{min} + \delta_r, b_{max})$ on the r_s-b_s resource allocation plane is in the acceptance region. Since every link in path G has residual capacity no less than $r_{min} + \delta_r$, we assign

$$b_s = b_{max} \text{ and } r_s = r_{min} + \delta_r$$

and the call is admissible by using path G to set up the end-to-end connection for session s. Before exit from the algorithm, we update the residual capacity of every link in the path $G = t_0 t_1 \ldots t_N$ by

$$C_{t_{n-1},t_n}^{(r)} \longleftarrow C_{t_{n-1},t_n}^{(r)} - r_s \tag{11}$$

for all $1 \leq n \leq N$.

We now suppose that $b_c(G) < b_{max}$. If the residual capacity of path G is no less than the critical bandwidth allocation $r_c(G)$ plus a margin δ_r, i.e.

$$r_c(G) + \delta_r \leq C^{(r)}(G),$$

then we assign

$$b_s = b_c(G) \text{ and } r_s = r_c(G) + \delta_r$$

and the call is admissible by using path G to set up the end-to-end connection for session s. Before exit from the CAC algorithm, we update the residual capacity of every link in the path $G = t_0 t_1 \ldots t_N$ by (11). But if the residual capacity of path G is less than the critical bandwidth allocation $r_c(G)$ plus a margin δ_r, i.e.

$$r_c(G) + \delta_r > C^{(r)}(G),$$

then the path cannot be used to set up an end-to-end connection for session s. The iterative algorithm must find another candidate path by going to the second iteration. Since any path G' which connects the access node α and the access node b in the modified digraph has delay margin no larger than the delay margin of path G, the critical bandwidth allocation $r_c(G')$ of path G' must be no less than the critical bandwidth allocation $r_c(G)$ of path G. Thus, before going to the second iteration, we relabel a link (i, j) in the digraph with the infinity "distance" ∞ if the residual capacity $C_{i,j}^{(r)}$ of link (i, j) is less than $r_c(G) + \delta_r$ (i.e. delete link (i, j) from the digraph).

With the newly relabeled digraph, the iterative algorithm begins with the routing stage of the second iteration and so forth as shown in Fig. 7. After several iterations, the algorithm will terminate either because of the non-existence of a path with positive delay margin in the routing stage of an iteration or because of the existence of a path which guarantees Qos requirements in the resource allocation stage of an iteration.

As an illustration of the iterative algorithm described in above, consider a service-guaranteed ATM network modeled as a digraph as shown in Fig. 8(a) where each link (i, j) is labeled with the delay distance $d_{i,j} = \Delta_{i,j} + \Omega_{i,j}$ in ms and the residual capacity $C_{i,j}^{(r)}$ from node i to node j in Mbps. Suppose that there is a session s to be set up from access node 1 to access node 6 with input traffic descriptor $\mathbf{T} = (P_{s,loss}, D_{s,bound}, \mu_s, \sigma_s^2, H_s) = (10^{-8}, 40\,\text{ms}, 10\,\text{Mbps}, 1\,\text{Mbps}^2/\text{sec}^{1.6}, 0.8)$. Assume that the maximum buffer size $b_{\max}$ is 1000 cells and the bandwidth margin δ_r is set to $0.1\,\mu_s = 1\,\text{Mbps}$.

In the initialization stage, the iterative algorithm substitutes

$$P_{s,loss} = 10^{-8}, \mu_s = 10\,\text{Mbps}, \ \sigma_s = 1\,\text{Mbps/sec}^{0.8},$$

$$H_s = 0.8, \text{and } b_s = b_{\max} = 1000\,\text{cells}$$

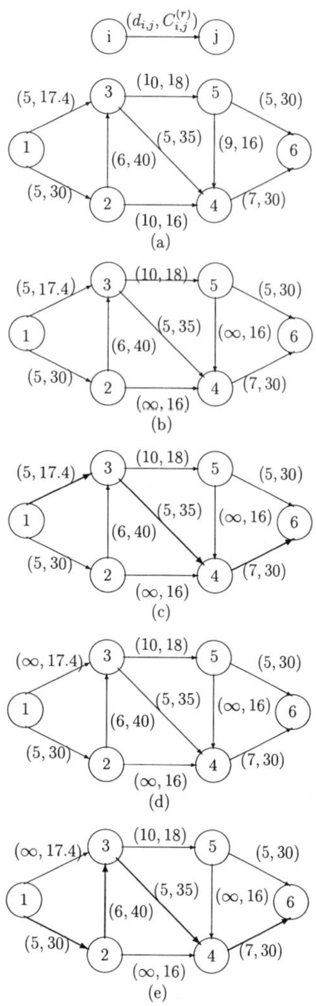

Fig. 8 An illustrative example of the CAC algorithm.

into the inequation in (10) to calculate $r_{\min} = 16.313$ Mbps. Thus, with bandwidth margin $\delta_r = 1$ Mbps, the minimum required bandwidth is $r_{min} + \delta_r = 17.313$ Mbps and every link in a possible route for session s must have residual capacity at least 17.313 Mbps. Consequently, the delay distances of links $(2,4)$ and $(5,4)$ are replaced with ∞ to delete these two links from the digraph since the residual capacity of both links is less than 17.313 as shown in Fig. 8(b).

Then in the routing stage of the first iteration, the shortest path algorithm employed by the iterative algorithm finds a path $G = 1 \rightarrow 3 \rightarrow 4 \rightarrow 6$ with a minimum path delay $D(G) = d_{1,3} + d_{3,4} + d_{4,6} = 17$ ms as shown in Fig. 8(c). Since the delay margin $D_{margin}(G) = D_{s,bound} - D(G) = 23$ ms of path G is positive, the algorithm proceeds to the resource allocation stage of the first iteration. Note that the residual capacity $C^{(r)}(G)$ of path G is 17.4 Mbps. The algorithm then determines the critical bandwidth allocation $r_c(G)$ and the critical buffer allocation $b_c(G)$ associated with the shortest path G as

$$r_c(G) = 16.5\,\text{Mbps}, \;\; b_c(G) = 892\,\text{cells}.$$

Since the residual path capacity $C^{(r)}(G)$ is less than the critical bandwidth allocation $r_c(G)$ plus the bandwidth allocation margin δ_r, path G is not legitimate. Now the link $(1,3)$ is removed from the digraph by replacing the delay distance of link $(1,3)$ with ∞ before going to the second iteration, as shown in Fig. 8(d).

In the routing stage of the second iteration, a shortest path $\Lambda = 1 \rightarrow 2 \rightarrow 3 \rightarrow 4 \rightarrow 6$ with a minimum path delay $D(\Lambda) = d_{1,2} + d_{2,3} + d_{3,4} + d_{4,6} = 23$ ms is found as shown in Fig. 8(e). Since the delay margin $D_{margin}(\Lambda) = D_{s,bound} - D(\Lambda) = 17$ ms of path Λ is positive, the algorithm proceeds to the resource allocation stage of the second iteration.

The new critical bandwidth allocation $r_c(\Lambda)$ and critical buffer allocation $b_c(\Lambda)$ associated with the shortest path Λ are

$$r_c(\Lambda) = 16.965\,\text{Mbps}, \;\; b_c(\Lambda) = 676\,\text{cells}.$$

Since the residual path capacity $C^{(r)}(\Lambda)$ is 30 Mbps which is no less than the critical bandwidth allocation $r_c(\Lambda)$ plus the bandwidth allocation margin δ_r, path Λ is legitimate. Thus the iterative algorithm admits the setup of session s with bandwidth allocation $r_s = r_c(\Lambda) + \delta_r = 17.965$ Mbps and buffer allocation $b_s = b_c(\Lambda) = 676$ cells through the route Λ. The loss rate and delay bound Qos requirements of session s is then guaranteed.

6. Conclusion

We proposed a format of traffic descriptor to describe an input session in a service-guaranteed network. The traffic descriptor $\boldsymbol{T} = (P_{loss}, D_{bound}, \mu, \sigma^2, H)$ consists of five parameters: The first two parameters in $\boldsymbol{T}$ are the Qos requirements of the input traffic, where P_{loss} is the tolerable data loss rate of the input traffic under the leaky bucket filtering and D_{bound} is the tolerable per connection end-to-end delay through heterogeneous scheduling servers. The last three parameters are used to characterize the statistics of the input traffic through Gaussian approximation, where μ is the mean rate of the input traffic, σ^2 and H are the variance constant and the Hurst parameter, respectively, of the fractional Brownian motion which characterizes the input process. The contents of such a traffic descriptor are sufficient to calculate the required allocation of bandwidth and buffer space to the input session when leaky bucket shapers and scheduling algorithms are employed to guarantee the Qos. The variation of demand of bandwidth and buffer allocation subject to the five parameters in the traffic descriptor has been extensively investigated. It is observed that the delay margin D_{margin}, the Hurst parameter H and the normalized standard deviation constant σ/μ plays the most important factors to determining the resource allocation for an input traffic. We then proposed an iterative algorithm to solve the per-connection end-to-end call admission control problem for the setup of input traffic sessions. This algorithm is developed to work with the traffic descriptor to meet the Qos requirements as well as to minimize the allocation of the precious bandwidth resource. There are two stages in each iteration of the algorithm: one responsible for the search of a candidate end-to-end routing path and the other for the verification of the legitimacy of this candidate path to meet the Qos requirements and for the allocation of resources in such a legitimate path.

Acknowledgement

The authors would like to thank the associate editor and the anonymous referees for their valuable comments and suggestions that helped improve the presentation of this paper.

References

[1] "Traffic management specification version 4.0," ATM Forum Technical Committee, April 1996.

[2] S. Abe and T. Soumiya, "A traffic control method for service quality assurance in an ATM network," IEEE J. Sel. Areas Commun., vol.12, no.2, pp.322–331, Feb. 1994.

[3] J. Beran, Statistics for Long-Memory Processes, Chapman & Hall, New York, 1994.

[4] J. Beran, R. Sherman, M.S. Taqqu, and W. Willinger, "Long-range dependence in variable-bit-rate video traffic," IEEE Trans. Commun., vol.43, no.2/3/4, pp.1566–1579, Feb./March/April 1995.

[5] Y. Bernet, J. Binder, S. Blake, M. Carlson, S. Keshav, E. Davies, B. Ohlman, D. Verma, Z. Wang, and W. Weiss, "A framework for differentiated services," Internet Draft, Oct. 1998.

[6] D. Bertsekas and R.G. Gallager, Data Networks, Prentice-Hall, Englewood Cliffs, NJ, 1987.

[7] S. Blake, D. Black, M. Carlson, E. Davies, Z. Wang, and W. Weiss, "An architecture for differentiated services," RFC 2475, Dec. 1998

[8] R. Braden, L. Zhang, S. Berson, S. Herzog, and S. Jamin, "Resource ReSerVation protocol (RSVP)—Version 1 functional specification," RFC 2205, Sept. 1997.

[9] A. Erramilli, O. Narayan, and W. Willinger, "Experimental queueing analysis with long-range dependent packet traffic," IEEE Trans. Networking, vol.4, no.2, pp.209–223, April 1996.

[10] N.R. Figueira and J. Pasquale, "An upper bound on delay for the virtual clock service discipline ," IEEE/ACM Trans. Networking, vol.3, no.4, pp.399–408, Aug. 1995.

[11] V. Firoiu, J. Kurose, and D. Towsley, "Efficient admission control for EDF schedulers," INFOCOM'97, pp.310–317.

[12] L. Georgiadis, R. Guerin, V. Peris, and K.N. Sivarjan, "Efficient network Qos provisioning based on per node traffic shaping," IEEE/ACM Trans. Networking, vol.4, no.4, pp.482–501, Aug. 1996.

[13] S.J. Golestani, "A self-clocked fair queueing scheme for high speed application," INFOCOM'94, pp.636–646.

[14] P. Goyal, S.S. Lam, and H.M. Vin, "Determining end-to-end delay bounds in heterogeneous networks," ACM/Springer-Verlag Multimedia Systems J., vol.5, no.3, pp.157–163, May 1997; also, Proc. Workshop Network and OS Support for Digital Audio Video, pp.287–298, April 1995.

[15] S. Guo and N.D. Georganas, "Resource and connection admission control in real-time transport protocols with deterministic Qos guarantees," INFOCOM'95, pp.1095–1102.

[16] D.P. Heyman, A. Tabatabai, and T.V. Lakshman, "Statistical analysis and simulation study of video teleconference traffic in ATM networks," IEEE Trans. Circuit & Syst. for Video Tech., vol.2, no.1, pp.49–59, March 1992.

[17] A. Hung and G. Kesidis, "Bandwidth scheduling for wide-area ATM networks using virtual finishing times," IEEE/ACM Trans. Networking, vol.4, no.1, pp.49–54, Feb. 1996.

[18] W.E. Leland, M.S. Taqqu, W. Willinger, and D.V. Wilson, "On the self-similar nature of ethernet traffic (extended version)," IEEE/ACM Trans. Networking, vol.2, no.1, pp.1–15, Feb. 1994.

[19] J. Liebeherr, D. Wrege, and D. Ferrari, "Exact admission control for networks with a bound delay service," IEEE/ACM Trans. Networking, vol.4, no.6, pp.885–901, Dec. 1996.

[20] B. Maglaris, D. Anastassiou, P. Sen, G. Karlsson, and J.D. Robbins, "Performance models of statistical multiplexing in packet video communications," IEEE Trans. Commun., vol.36, no.7, pp.834–844, July 1988.

[21] I. Norros, "A storage model with self-similar input," Queueing Systems, vol.16, pp.387–396, Jan. 1994.

[22] I. Norros, "On the use of fractional Brownian motion in the theory of connectionless networks," IEEE J. Sel. Areas Commun., vol.13, no.6, pp.1028–1038, Aug. 1995.

[23] A.K. Parekh and R.G. Gallager, "A generalized processor sharing approach to flow control in integrated service networks: The single node case," IEEE/ACM Trans. Networking, vol.1, no.3, pp.344–357, June 1993.

[24] A.K. Parekh and R.G. Gallager, "A generalized processor sharing approach to flow control in integrated service networks: The multiple node case," IEEE/ACM Trans. Networking, vol.2, no.2, pp.137–150, April 1994.

[25] A.R. Reibman and A.W. Berger, "Traffic descriptors for VBR video teleconferencing over ATM networks," IEEE/ACM Trans. Networking, vol.3, no.3, pp.329–339, June 1995.

[26] S. Shenker, C. Partridge, and R. Guerin, "Specification of guaranteed quality of service," RFC 2212, Sept. 1997.

[27] G. Xie and S.S. Lam, "Delay guarantee of virtual clock server," IEEE/ACM Trans. Networking, vol.3, no.6, pp.683–689, Dec. 1995.

[28] O. Yaron and M. Sidi, "Performance and stability of communication networks via robust exponential bounds," IEEE/ACM Trans. Networking, vol.1, no.3, pp.372–385, June 1993.

[29] F. Yegenoglu, B. Jabbari, and Y.Q. Zhang, "Motion-classified autoregressive modeling of variable bit rate video," IEEE Trans. Circuit & Syst. for Video Tech., vol.3, no.1, pp.42–53, Feb. 1993.

[30] L. Zhang, "VirtualClock: A new traffic control algorithm for packet switching networks," ACM Trans. Comput. Syst., vol.9, pp.101–124, May 1991.

[31] H. Zhang, "Service disciplines for guaranteed performance service in packet-switching networks," IEEE Proc., vol.83, no.10, pp.1374–1396, Oct. 1995.

High-Performance Backbone Network Technology edited by Naoaki Yamanaka
ISBN: 0-8247-5321-6 DOI: 10.1081/(120025707)
Published 2004 by Marcel Dekker, Inc.

Performance Analysis of Fast Reservation Protocols for Burst-Level Bandwidth Allocation in ATM Networks

You-Ze Cho and Alberto Leon-Garcia

SUMMARY In this paper, we investigate the performance of Fast Reservation Protocols (FRP) for burst-level bandwidth allocation in ATM networks. FRP schemes can be classified into delayed transmission (DT) and immediate transmission (IT) methods according to reservation procedure. Moreover, according to the responsibility for negative acknowledgment (NAK) cell generations when burst blocking occurs, FRP schemes can be further classified into blocking node NAK (BNAK) and destination node NAK (DNAK) schemes. We analize the FRP schemes with different reservation and NAK methods for single node and multihop network models, respectively. We then discuss the dependence of performance for each FRP scheme on propagation delay, peak rate, and the number of hops.

key words: *ATM network, burst-level bandwidth allocation, fast reservation protocol*

1. Introduction

The Asynchronous Transfer Mode (ATM) is attractive due to its bit rate flexibility and statistical multiplexing capability. In many applications of ATM networks such as distributed computing, multimedia, high-speed file transfer and image retrieval, the traffic exhibits high peak rates and long burst lengths. In particular, data traffic on ATM LANs is expected to become much more bursty as network and workstation speeds become faster [1]–[4]. However, for such highly bursty traffic, statistical multiplexing at the call level with moderate buffer size fails. To obtain a substantial bandwidth efficiency, a dynamic bandwidth management at the burst level is desirable.

The Fast Reservation Protocol (FRP) was originally proposed for efficient resource management for bursty traffic in B-ISDN [5], but the FRP is expected to be more suitable for ATM LANs due to small propagation delay [6]. Recently, a variant of FRP, called ATM Block Transfer (ABT), is being standardized for an efficient data service in B-ISDN by the ITU-T [7]. ABT capability is an ATM layer mechanism for providing a service where bandwidth is allocated dynamically on an ATM block basis. An ATM block is a group of cells of an ATM connection delineated by two resource management cells.

The FRP can be classified into two schemes according to reservation method. In the *delayed transmission (DT)* method, a burst can be transmitted only after the bandwidth on its route has been reserved explicitly through the network. On the other hand, in the *immediate transmission (IT)* method, a source can transmit a burst right after transmitting the reservation-request cell.

In traditional FRP schemes, the network allocates a bandwidth equivalent to the peak rate for each burst [4], [8]. But, in many applications of ATM LANs, the throughput for each burst could be negotiated between source and network as long as the burst transmission delay is kept within an acceptable range [6], [10], [11].

When burst blocking occurs, a negative acknowledgement (NAK) cell can be generated by either the blocking node or the destination node. In this paper, these two NAK methods will be referred to blocking node NAK (BNAK) and destination node NAK (DNAK) schemes, respectively.

During past several years, many researches have been focused on investigating the FRP performance in single hop and multihop network environments [4], [6]–[11]. In [8], the FRP/DT with the peak bandwidth allocation scheme was analyzed for multihop network based on the Erlang loss formula assuming Poisson arrivals of bursts. In [10], the FRP/DT with the negotiated bandwidth scheme was investigated in ATM LANs by simulation. In [11], the FRP/DT with the dynamic bandwidth scheme was analyzed for single hop network assuming Poisson arrivals of bursts. However, all these literatures considered only the DNAK scheme, even though the BNAK scheme can improve the FRP performance. Also, there are few results reported on comparing DT and IT reservation schemes in FRP by analysis.

In this paper, in order to investigate the impact of reservation methods on performance, we analyze the FRP/DT and FRP/IT schemes for single node and multihop network models, respectively. Moreover, we examine the impact of the NAK schemes, BNAK and DNAK, on the fast reservation performance. As performance measures, the mean burst transfer delay, the burst blocking probability, the burst throughput, and

Manuscript received January 27, 2000.
Manuscript revised August 2, 2000.

[†]The author is with the School of Electronic and Electrical Engineering of the Kyungpook National University, Taegu, 702-701, Korea.
[††]The author is with the Department of Electrical and Computer Engineering of the University of Toronto, Toronto, M5S 1A4, Canada.

[*]This work was partially supported by the Brain Korea 21 project.

IEICE TRANS. COMMUN., Vol. E84-B, No. 2, pp. 284–293, February 2001

the bandwidth waste are employed. For the single-node analysis, we use Markovian approaches as a closed queueing model assuming ON/OFF arrivals of bursts. Our multihop analysis uses similar techniques as in [8] based on the Erlang loss formula, but [8] considered only the FRP/DT scheme with the DNAK.

Following this introduction, in Sect. 2, we describe the FRP schemes considered and define notations used in this analysis. In Sect. 3, we analyze the FRP schemes for a single-node case using a closed queueing model. In Sect. 4, using the Erlang loss formula, we analyze the FRP schemes for different reservation and NAK methods in multi-hop environments. Finally, we make conclusion in Sect. 5.

2. FRP Schemes Considered and Notations

In this section, we describe two FRP schemes considered in this paper and define notations which will be used in the analysis.

2.1 FRP Schemes Considered

Fundamentally, fast reservation protocols have properties similar to random access methods such as the ALOHA scheme. A common problem of random access techniques is the waste of bandwidth due to collisions. In this paper, we consider two FRP schemes, FRP/DT and FRP/IT, in which the random backoff and the peak bandwidth allocation methods are assumed. If the peak bandwidth is not available upon request, the source tries again after an exponential backoff time.

Figure 1 illustrates the procedures of bandwidth reservations for the FRP/DT and IT for the BNAK scheme. Before transmitting a burst, the source first sends a reservation-request cell, and after transmitting the last cell of the burst it sends a release-request cell. In the FRP/DT scheme, bandwidth is reserved for each burst during a round-trip delay between source and destination plus a burst transmission time, while in the FRP/IT scheme, bandwidth is reserved just during the burst transmission time.

When burst blocking occurs, the FRP/IT scheme wastes bandwidth before receiving the NAK cell, which otherwise could be exploited by best-effort traffic. However, from the viewpoint of fast reservation traffic, both the FRP/DT and IT schemes waste bandwidth, since the bandwidth reserved until receiving the NAK cell cannot be used by other fast reservation traffic. When burst blocking occurs, the DNAK scheme therefore wastes more bandwidth than the BNAK scheme, because it takes more time for the source to receive the NAK cell.

2.2 Notations

We define following notations which will be used in the analysis:

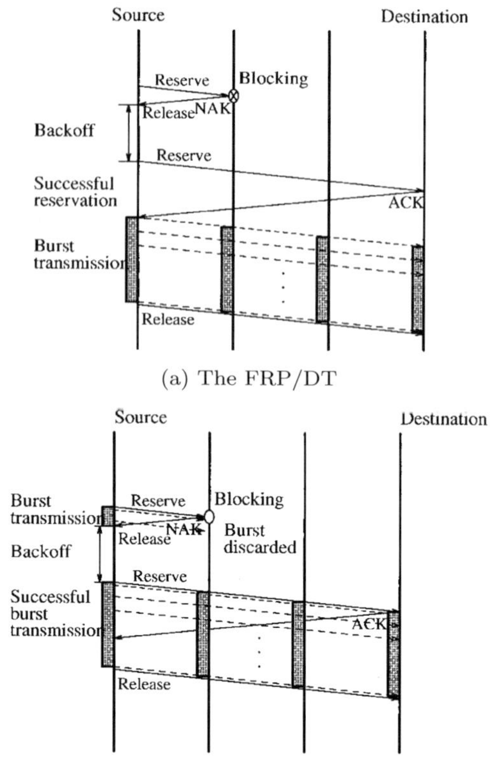

Fig. 1 Bandwidth reservation procedures of the FRP/DT and IT for the BNAK scheme.

- R_p: the peak rate for each ON/OFF source (in b/s).
- C: the output link rate (in b/s).
- T_{OFF}: the mean idle duration of each source.
- T_{ON}: the mean burst transmission time of each source.
- τ_b: the mean backoff time for each burst blocking instance.
- τ_p: the one-way propagation delay per link.
- M: the maximum number of channels which can be supported at the peak rate $= \lfloor C/R_p \rfloor$.
- W: the mean reservation delay from the instant a burst is generated until the bandwidth reservation is successful.
- U_s: the mean bandwidth holding time.
- D: the mean burst transfer delay from the first attempt until the last bit is successfully transmitted $= W + U_s$.

3. Single-Node Model Analysis

3.1 Network Model and Assumption

We first consider a single-node model connected through a single ATM LAN switch, in which distributed client computers share a single server. As shown in

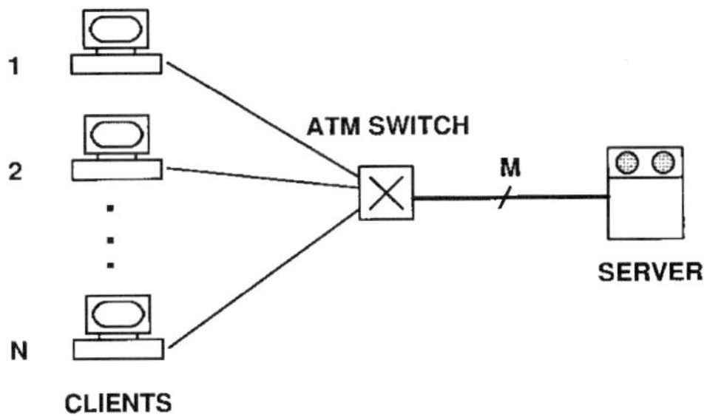

Fig. 2 A client-server model.

Fig. 2, N identical clients are connected to the ATM switch by each line and the peak bandwidth requests are assumed to be $1/M$ of the total output link capacity.

We consider an interactive data communication between N statistically identical clients and a single server as shown in Fig. 2, in which only idle clients can generate a burst. The idle duration of each ON/OFF source can be considered as the time period for waiting response from the server and thinking time in the client. In effect, this system is self-regulating, since the arrival rate of bursts becomes smaller as the system gets busier. Since we focus on the bandwidth sharing of the output link to the server by clients, we assume the BNAK scheme only. That is, if the output link is not available on request, the switch sends back the NAK cell to the corresponding client.

We assume that the burst transmission time and idle period of each ON/OFF source are exponentially distributed with means T_{ON} and T_{OFF} ($\equiv 1/\gamma$), respectively. However, in the analysis we approximate the bandwidth holding time and the additional burst delay for each blocking as being exponentially distributed with means $1/\mu$ and $1/\delta$, respectively. In the analysis, the bandwidth overhead by reservation-related cells will be neglected. Since the output channel for each burst plays the role of a server in a queueing model, the system can be considered as a multi-server queueing model. We can consider the output link to the server as consisting of M channels, each of which has the capacity of the burst peak bandwidth.

In the FRP/DT scheme, bandwidth is reserved for each burst during the round-trip propagation delay plus the burst transmission time, while in the FRP/IT scheme, bandwidth is reserved just during the burst transmission time. We can then represent the mean bandwidth holding time, U_s ($\equiv 1/\mu$), for each reservation scheme as

$$U_s = \begin{cases} 2\tau_p + T_{ON} & \text{for the FRP/DT} \\ T_{ON} & \text{for the FRP/IT.} \end{cases} \tag{1}$$

For every burst blocking instance, burst transmission is delayed by a round-trip propagation delay plus a backoff time. Hence, the mean burst delay for each blocking, $1/\delta$, is given by

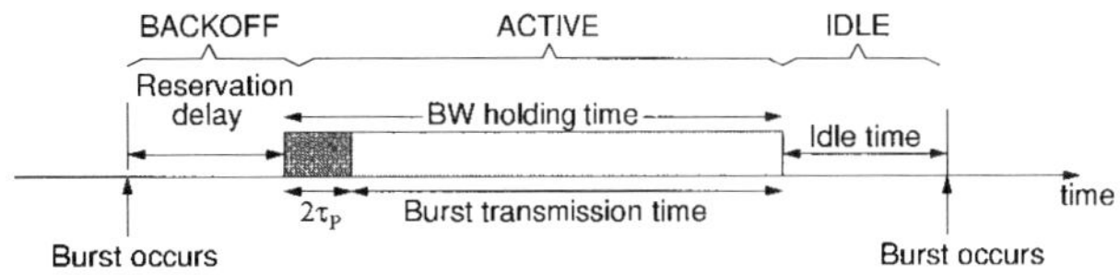

(a) The FRP/DT

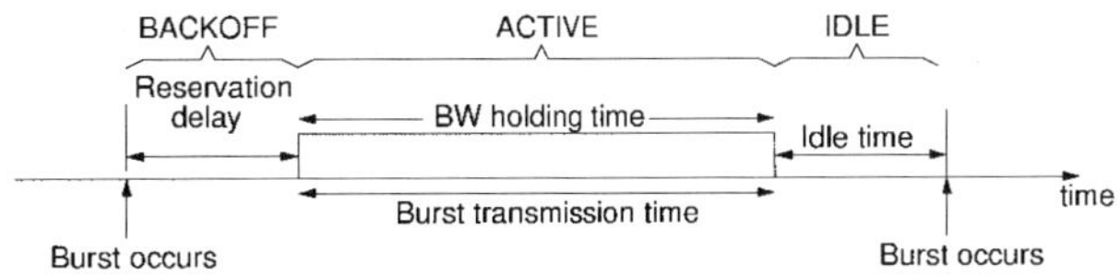

(b) The FRP/IT

Fig. 3 The system states of the FRP/DT and IT schemes.

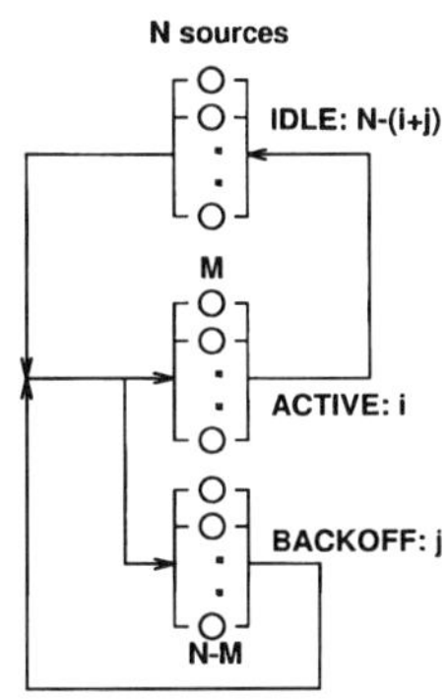

Fig. 4 Closed queueing model of the FRP/DT and IT schemes.

$$\frac{1}{\delta} = \tau_b + 2\tau_p. \tag{2}$$

3.2 Closed Queueing Network Model

Each source alternates between three states: *idle, backoff* and *active* as shown in Fig. 3. A source must wait for service in the backoff state if all M channels are busy when placing a reservation request, and then retries after an exponential backoff time. We say that a source is in the active state if it is in the period which begins with sending a reservation-request cell which is accepted successfully until completing the transmission of the release-request cell. We can then represent the FRP/DT and IT schemes as a closed queueing model as shown in Fig. 4. In this queueing model, the residence time for each source in the idle state can be considered as its idle duration, while the residence time in the active state is equivalent to the bandwidth holding time. In the backoff state, the residence time is identical to the sum of a round-trip propagation delay and a backoff time.

In the queueing model of Fig. 4, if all M channels are busy, a source from a backoff state reenters the same state after its residence time in the backoff state. But, in a continuous-time Markov chain, since the system state cannot go into the same state, we consider each

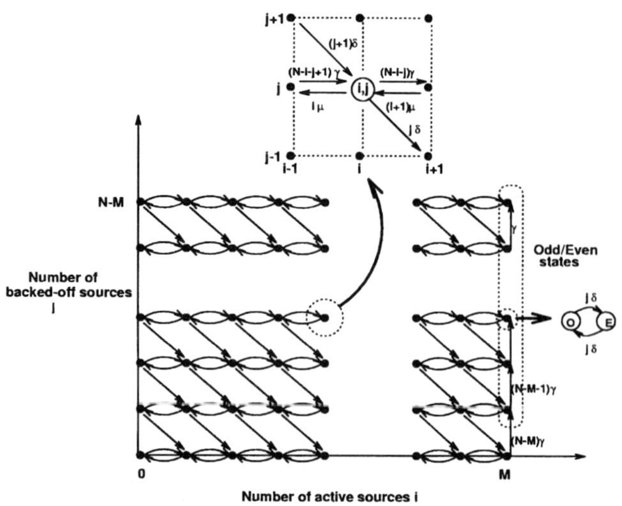

Fig. 5 The state transition diagram of the FRP/DT and IT schemes.

backoff state with all servers busy as consisting of a pair of odd/even states internally and the system as alternating between odd/even states whenever a backed-off source requests bandwidth. However, since we are concerned only with the number of backed-off sources, it doesn't matter whether the system resides in the odd or even backoff states. Therefore, we don't have to identify even and odd states and can consider them as if they were the same state. Letting the number of sources which are in the active and backoff states as i and j respectively, a pair (i,j) forms a two-dimensional Markov chain, and the total number of system states is $(N-M+1)(M+1)$ as in Fig. 5.

Denoting $\pi_{i,j}^{[N]}$ as the steady-state probability of the N-source system being in a state (i,j), we can represent the global balance equations for each state $(i,j), 0 \le i \le M, 0 \le j \le N-M$, as follows:

1) $1 \le i \le M-1, 0 \le j \le N-M-1:$

$$\begin{aligned}&[i\mu + (N-i-j)\gamma + j\delta]\pi_{i,j}^{[N]}\\ &\quad = (N-i-j+1)\gamma\pi_{i-1,j}^{[N]} + (i+1)\mu\pi_{i+1,j}^{[N]}\\ &\qquad + (j+1)\delta\pi_{i-1,j+1}^{[N]}\end{aligned}$$

2) $1 \le i \le M-1, j = N-M:$

$$\begin{aligned}&[i\mu + (M-i)\gamma + (N-M)\delta]\pi_{i,N-M}^{[N]}\\ &\quad = (M-i+1)\gamma\pi_{i-1,N-M}^{[N]} + (i+1)\mu\pi_{i+1,N-M}^{[N]}\end{aligned}$$

3) $i = 0, 1 \le j \le N-M:$

$$[(N-j)\gamma + j\delta]\pi_{0,j}^{[N]} = \mu\pi_{1,j}^{[N]}$$

4) $i = M, 0 < j \le N-M-1:$

$$\begin{aligned}&[M\mu + (N-M-j)\gamma]\pi_{M,j}^{[N]}\\ &\quad = (N-M-j+1)\gamma\pi_{M-1,j}^{[N]}\\ &\qquad + (N-M-j+1)\gamma\pi_{M,j-1}^{[N]}\\ &\qquad + (j+1)\delta\pi_{M-1,j+1}^{[N]}\end{aligned}$$

5) $i = 0, j = 0:$

$$N\gamma\pi_{0,0}^{[N]} = \mu\pi_{1,0}^{[N]}$$

6) $i = M, j = N-M:$

$$M\mu\pi_{M,N-M}^{[N]} = \gamma\pi_{M-1,N-M}^{[N]} + \gamma\pi_{M,N-M-1}^{[N]}$$

7) $i = M, j = 0:$

$$\begin{aligned}&[M\mu + (N-M)\gamma]\pi_{M,0}^{[N]}\\ &\quad = (N-M+1)\gamma\pi_{M-1,0}^{[N]} + \delta\pi_{M-1,1}^{[N]}.\end{aligned} \tag{3}$$

We note that μ has a different value for the FRP/DT and IT schemes as shown in Eq. (1).

Representing the steady-state probabilities $\{\pi_{i,j}^{[N]}\}$ as a row vector $\mathbf{\Pi}$, the global balance equations can be expressed in matrix form as

$$\mathbf{\Pi Q} = \mathbf{0} \tag{4}$$

where $\mathbf{Q}$ is a transition rate matrix of $(N-M+1)(M+1) \times (N-M+1)(M+1)$. Also, the total sum of $\pi_{i,j}^{[N]}$ must be equal to 1, hence it is given by

$$\sum_{j=0}^{N-M}\sum_{i=0}^{M}\pi_{i,j}^{[N]} = 1. \tag{5}$$

3.3 Performance Measures

The burst blocking probability P_{burst} at the first attempt for an N-source system is the same as the probability that all M channels are busy for $(N-1)$-source system at a random instant. For simplicity of notation, we let

$$\phi_i^{[N]} = \sum_{j=0}^{N-M}\pi_{i,j}^{[N]}, \tag{6}$$

where $\phi_i^{[N]}$ denotes the probability that the number of sources in the active state is equal to i for the N-source system. Hence, P_{burst} can be obtained as

$$P_{burst} = \phi_M^{[N-1]}. \tag{7}$$

If we define total burst throughput Λ as the total rate at which bursts are successfully transferred through the output link to the server, Λ (in bursts/sec) is given by

$$\Lambda = \sum_{i=1}^{M} i\mu\phi_i^{[N]}. \tag{8}$$

For a stability condition, the total burst arrival rate

Table 1 Parameters considered in Figs. 7–10 (T_{ON} and T_{OFF} are normalized to the mean burst transmission time).

Source Type	R_p	M	(T_{ON}, T_{OFF})	α
Medium peak rate	10 Mb/s	10	(1,9)	0 0.01
High peak rate	100 Mb/s	1	(1,99)	0.1 1

from sources should be equal to Λ. Hence, for an N-source system, the mean cycle time for a source is obtained from Little's formula as

$$\text{the mean cycle time} = \frac{N}{\Lambda}. \tag{9}$$

Therefore, from (9) the mean burst transfer delay D can be represented as

$$D = \frac{N}{\Lambda} - T_{OFF}. \tag{10}$$

3.4 Numerical Results and Discussion

In order to compare the FRP/DT and IT schemes for single-node case in terms of the fast reservation traffic performance, we use as performance criteria the mean burst transfer delay, the burst blocking probability and the total throughput. In all numerical examples, the output link capacity C is assumed to be 100 Mb/s for simplicity.

We first investigate the dependence of these performance parameters on the peak rate and propagation delay. We consider two different peak-rate sources with the same activity: medium peak-rate traffic $R_p = 10$ Mb/s and high peak-rate traffic $R_p = 100$ Mb/s. (T_{ON}, T_{OFF}) is selected to give the average rate—1 Mbps for each R_p. Therefore, when R_p are 10 Mbps and 100 Mbps, (T_{ON}, T_{OFF}) becomes (1,9) and (1,99), respectively. The mean backoff value τ_b is assumed to be three times of the mean burst transmission time T_{ON} if not specified otherwise [8]. The mean burst transfer delay is normalized to the mean burst transmission time, and for simplicity of notation we let $\alpha \equiv \tau_p/T_{ON}$ and $\beta \equiv \tau_b/T_{ON}$. In Table 1, we summarize the parameter values considered. The number of sources N is varied from 1 to a large number in order to get a different offered traffic or throughput.

Figure 6 shows burst delay and blocking probability characteristics for different peak rates for zero propagation delay. When the propagation delay is zero, there is no fundamental difference between the FRP/DT and IT schemes. Figure 6(b) shows that the medium peak rate traffic ($M = 10$) exhibits a much lower burst blocking probability, compared with the high peak rate traffic ($M = 1$). However, the high peak-rate traffic experiences larger burst delay and high burst blocking probability unless the network utilization is very low. From this result, we note that the performance of FRP schemes is highly dependent on

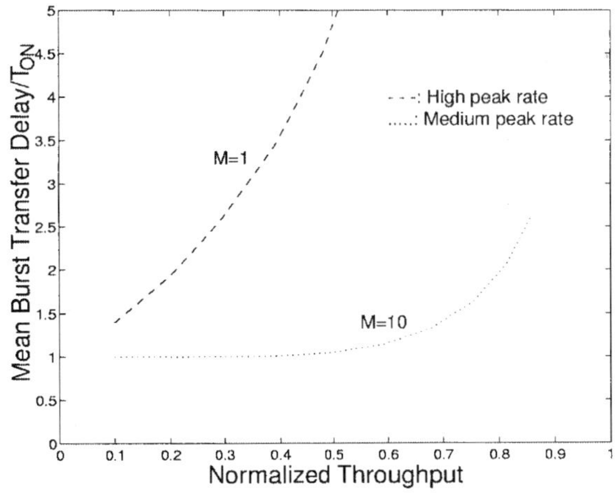

(a) Mean burst transfer delay

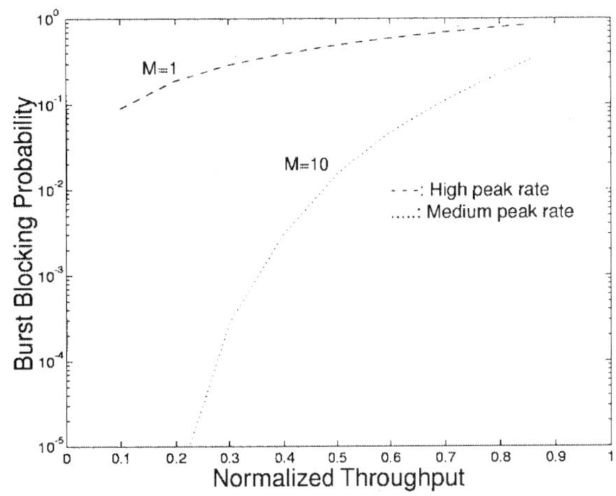

(b) Burst blocking probability

Fig. 6 Burst delay and blocking characteristics for different peak rates for zero propagation delay (α=0, β=3).

the peak rate-to-link speed ratio. Therefore, to get high utilization for high peak-rate traffic, channel or trunk grouping to provide higher output link capacity may be desirable.

Figure 7 shows burst delay characteristics with different propagation delay for the medium peak-rate traffic, respectively. In these figures, the analysis results are compared with simulation results in order to varify the assumption of an exponential round-trip propagation delay in our analysis. The circle mark indicates the simulation results by computer simulator. Also, these figures show that there is no significant difference between the FRP/DT and IT schemes for relatively low propagation delay-to-burst transmission time ratio. But, for higher α, the FRP/DT scheme exhibits limited achievable throughput since bandwidth goes unused for fast reservation traffic during the request-acknowledgment (REQ-ACK) cycle.

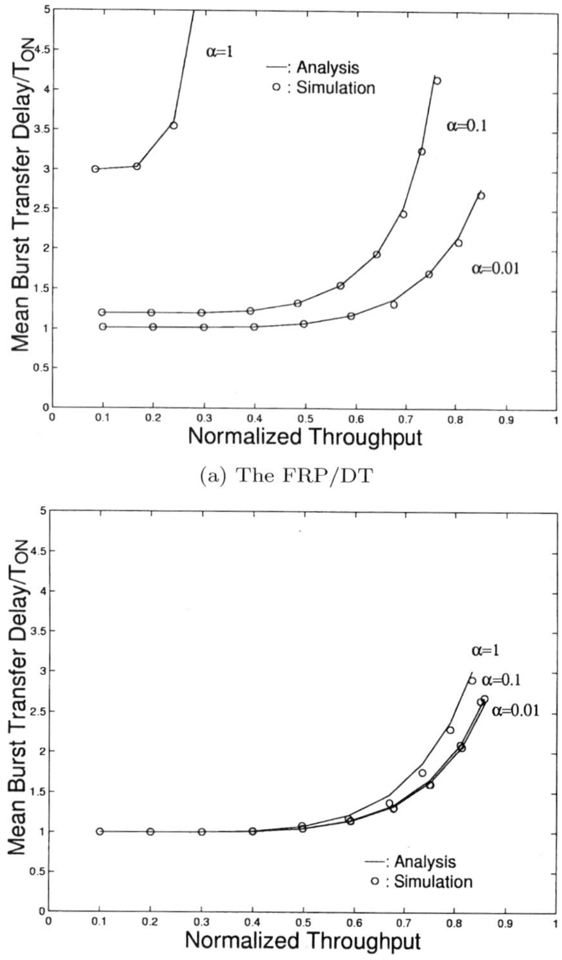

(a) The FRP/DT

(b) The FRP/IT

Fig. 7 Mean burst transfer delay versus normalized throughput for different propagation delays (M=10, β=3).

4. Multihop Model Analysis

4.1 Network Model and Assumption

Fast reservation seizes simultaneous resources along a source-destination path. The states of each link along the path, therefore, are cross correlated and the degree of correlation is dependent on network topology, routing technique, connection types and so on. In general, as the number of nodes in the network increases, the correlation between nodes weakens. For simplicity of analysis, we assume the states of nodes are independent of each other. Thus our analysis is applicable to the case where nodes are weakly coupled.

We suppose that all source-destination pairs have H hops consisting of H links and $H+1$ nodes as shown in Fig. 8. We suppose that each station is fed by a large number of bursty sources. The burst duration and idle duration of each source are assumed to be exponentially distributed with means T_{ON} and T_{OFF}, respectively. We suppose that the number of sources is sufficiently large that the aggregate new burst arrivals from ON/OFF sources can be reasonably approximated to be Poisson. We assume that if the burst peak bandwidth is not available upon request, the source tries again after a random backoff time. In reality, the blocking probability of the reattempt is correlated with the previous attempt. If we select the mean backoff time τ_b to be greater than three times of the mean burst duration, the burst blocking probabilities of new bursts and reattempts are almost independent [12]. In this paper, the mean backoff time is assumed to be three times of the mean burst duration. For simplicity of analysis, we also assume the followings:

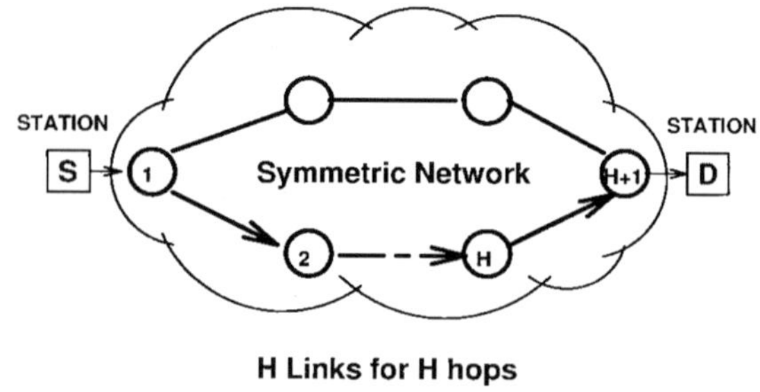

Fig. 8 A source-destination pair with H hops.

- A symmetric network: the link load is assumed to be statistically identical at all links of the subnetwork.
- The distance between successive nodes are equal.
- Poisson arrivals of the aggregate traffic at each link including new bursts and reservation reattempts.
- Propagation delay between a station and its access node is negligible.
- The link bandwidth between a station and its access node is always available.
- All sources have statistically identical characteristics and all links have the same capacity.

4.2 Performance Analysis

In this section, using an analytical model based on the Erlang loss formula, we derive the burst blocking probability, the mean burst transfer delay, and the burst throughput.

4.2.1 Burst Blocking Probability

We can consider an output link at each ATM switch as consisting of M channels, each of which has the capacity of the burst peak bandwidth. From the *Erlang B formula*, the link blocking probability P_{link} can then be obtained as

$$P_{link} \equiv B(M,a) = \frac{a^M/M!}{\sum_{j=0}^{M} a^j/j!}, \quad (11)$$

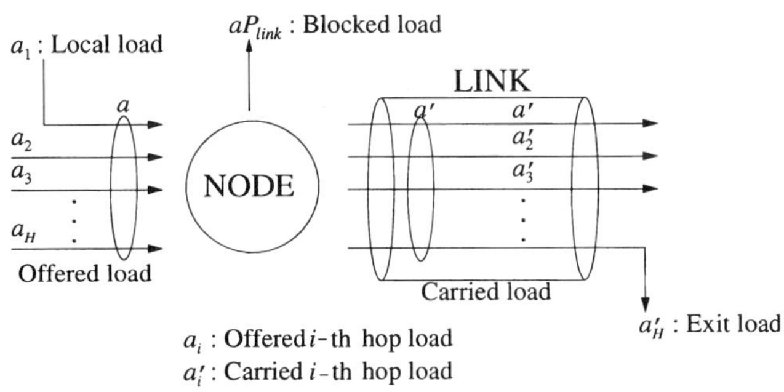

Fig. 9 Offered and carried loads of a particular link in the subnetwork.

where M is the number of servers and a is offered load (in erlangs) to a link. For H hops, a source-destination pair goes through H tandem links. Assuming that blocking probabilities at each link along a path are statistically independent, we can represent the burst blocking probability P_{burst} for H hops as

$$P_{burst} = 1 - (1 - P_{link})^H. \tag{12}$$

P_{burst} indicates the probability that a burst cannot be transmitted successfully through a network due to one or more links on the path are not available.

Let's now consider a particular link of the subnetwork. Let S be the total rate of new bursts generated from all source-destination pairs passing through the link considered, and let G be the total traffic rate from those sources including reservation reattempts. For convenience, we consider the unit of S and G as number of bursts during a mean burst duration. Since we assume that traffic load at any link in the subnetwork are statistically identical, every link has the same value of S and G. If we assume that steady state exists, S is equal to the throughput of a link and we then have the following relation:

$$S = G(1 - P_{burst}). \tag{13}$$

Let P_i be the probability that the bandwidth reservation fails for the first time at the i-th hop along a source-destination path. P_i is then expressed as

$$P_i = (1 - P_{link})^{i-1} P_{link}. \tag{14}$$

In order to calculate P_{link} for a particular link using Eq. (11), we need to find the offered load a first. As shown in Fig. 9, if we define the offered ith-hop load from all source-destination pairs passing through a particular link as a_i, the total offered load a is the sum of a_i, $i = 1, 2, ..., H$. The offered load from a given source-destination pair is reduced as it traverses each node, since the traffic carried downstream is less than the traffic offered to each node due to link blocking. It is therefore quite complicate to derive the offered load directly. We use an indirect approach in which the carried load is calculated first and then the offered load is derived from it.

The carried load a'_i on the ith-hop link for a source-destination pair contains all the bursts that succeed at all of the nodes on its path as well as those that succeed at the i-th hop but fail at the downstream nodes. Using the probability P_i that the bandwidth reservation fails for the first time at the i-th hop and the probability that a particular link belongs to the i-th hop, ($i = 1, 2, ..., H$), of a source-destination pair passing through it is $1/H$, the carried i-th hop load a'_i from all source-destination pairs through a link can be derived. The carried load a'_i is therefore dependent on reservation and NAK methods.

A. The FRP/DT Case

When a burst blocking occurs, the FRP/DT scheme releases the reserved bandwidth after receiving the NAK cell, thereby the bandwidth holding time for an unsuccessful burst depends on the location of the link considered on a path. In the BNAK scheme, the bandwidth holding time for an unsuccessful burst depends on the distance from the node considered to the blocking node while, in the DNAK scheme, it depends on the distance to the destination.

For the FRP/DT, a'_i is given as

$$a'_i = \begin{cases} \dfrac{G}{H}[(1 - P_{link})^H U_s \\ \quad + \sum_{j=i+1}^{H} 2\tau_p(j - i) \cdot P_j] \\ \qquad \text{for the BNAK,} \\ \dfrac{G}{H}[(1 - P_{link})^H U_s \\ \quad + 2\tau_p(H - i + 1) \sum_{j=i+1}^{H} P_j] \\ \qquad \text{for the DNAK.} \end{cases} \tag{15}$$

where U_s denotes the average bandwidth holding time for successful burst transmissions. U_s is given by

$$U_s = 2H\tau_p + T_{ON}. \tag{16}$$

In Eq. (15), the first term of the right hand side is the carried load that succeeds at all of the nodes on its path, and the second term is the load that succeeds at the i-th hop but fails at downstream nodes.

B. The FRP/IT Case

Since the FRP/IT scheme transmits a burst immediately after sending a bandwidth request cell, the carried load for the i-th hop link due to a burst blocking depends on the location of the blocked link and the burst duration. If the burst duration is less than the time period taken for the NAK cell to arrive at the i-th hop node, the link bandwidth is held only for the burst duration. Therefore, for the FRP/IT, a'_i can be represented as

$$a_i' = \begin{cases} \frac{G}{H}[(1-P_{link})^H U_s \\ \quad + \sum_{j=i+1}^{H} \min\{2\tau_p(j-i), T_{ON}\} \cdot P_j] \\ \qquad\qquad \text{for the BNAK,} \\ \frac{G}{H}[(1-P_{link})^H U_s \\ \quad + \min\{2\tau_p(H-i+1), T_{ON}\} \sum_{j=i+1}^{H} P_j] \\ \qquad\qquad \text{for the DNAK.} \end{cases} \tag{17}$$

For the FRP/IT, U_s is given by

$$U_s = T_{ON}. \tag{18}$$

Generally speaking, in both of the FRP/DT and FRP/IT schemes, the bandwidth overhead due to reservations and burst blocking increases as the propagation delay-to-burst duration ratio increases. However, in the FRP/IT, from Eq. (17), we note that the wasted bandwidth depends only on the burst duration irrespective of the propagation delay when the one-way propagation delay-to-burst duration ratio between the two adjacent nodes is greater than 1/2.

The carried load a', therefore, can be obtained by summation of a_i':

$$a' = \sum_{i=1}^{H} a_i'. \tag{19}$$

From the carried load a', we can get the offered load a for a link using the following relation:

$$a' = a(1 - B(M, a)). \tag{20}$$

From Eqs. (11) to (13), a is a function of G and P_{link}. For a given value G, P_{link} can then be found by iterative recalculation of Eqs. (11) through (13), until convergence is achieved. Once P_{link} is found, P_{burst} and S can be obtained using Eqs. (12) and (13).

4.2.2 Mean Burst Delay and Link Utilization

In the DNAK scheme, for each burst blocking instance, the burst transmission is delayed by $2H\tau_p + \tau_b$. But, in the BNAK scheme, the burst delay is dependent on the location of burst blocking. Because the average number of reservation reattempts per burst transmission is $(G - S)/S$, the mean reservation delay W is thus given by

$$W = \begin{cases} \frac{G-S}{S}(\sum_{j=1}^{H} 2\tau_p(j-1)P_j/P_{burst} + \tau_b) \\ \qquad\qquad \text{for the BNAK,} \\ \frac{G-S}{S}(2H\tau_p + \tau_b) \\ \qquad\qquad \text{for the DNAK.} \end{cases} \tag{21}$$

The mean burst transfer delay D therefore is given by

$$D = W + U_s = \begin{cases} W + 2H\tau_p + T_{ON} \\ \qquad\qquad \text{for the FRP/DT,} \\ W + T_{ON} \\ \qquad\qquad \text{for the FRP/IT.} \end{cases} \tag{22}$$

The actual link utilization ρ due to successful and unsuccessful bandwidth reservations is:

$$\rho = \frac{a'}{M}. \tag{23}$$

If we define the total burst throughput as the total rate at which bursts are successfully transferred through a link, the total burst throughput in a stable situation is equal to S. Letting ρ_e be the burst throughput normalized by link capacity, ρ_e then represents the effective link utilization and is given by

$$\rho_e = \frac{S}{M}. \tag{24}$$

4.3 Numerical Results and Discussion

When the propagation delay is negligible, there is no fundamental difference between the FRP/DT and the FRP/IT schemes. Moreover, when the propagation delay is zero or H=1, the performance of the DNAK scheme is the same as the BNAK scheme. In order to compare the performance of the FRP/DT and FRP/IT schemes, we use as performance criteria the mean burst transfer delay and the total throughput. We investigate the dependence of these performance parameters on peak rate, propagation delay and the number of hops. In all numerical examples, the output link capacity C is assumed to be 100 Mb/s for simplicity. And we assume the mean backoff time τ_b as three times of the mean burst duration. We let the one-way per-hop propagation delay-to-mean burst duration ratio τ_p/T_{ON} as α.

Figures 10 and 11 show the impact of propagation delay on the throughput and the mean burst transfer delay, respectively. In these figures, the lower bound indicates the minimum throughput which the FRP/IT can provide regardless of α. In these examples, we assume the medium peak-rate traffic with M=10 and the number of hops H=3. These figures show performance behaviors similar to random access methods such as the ALOHA scheme.

Figure 10 shows the offered traffic-throughput characteristics. In this figure, for a low α value below 0.01, there is no significant difference between the FRP/DT and FRP/IT schemes. But, as α value increases, the maximum achievable throughput of the FRP/DT scheme decreases much faster than the FRP/IT scheme. This is due to the fact that more bandwidth goes unused for fast reservation traffic during the REQ-ACK cycle. When α=0.1 and 0.5 with the BNAK scheme, the maximum achievable throughput are 0.49 and 0.18 for the FRP/DT, while 0.72 and 0.62 for the FRP/IT, respectively. Also, we note that the FRP/IT supports the maximum achievable throughput over 0.62 irrespective of the α value, while the maximum achievable throughput of the FRP/DT scheme decreases continuously as α increases. From Fig. 11, we

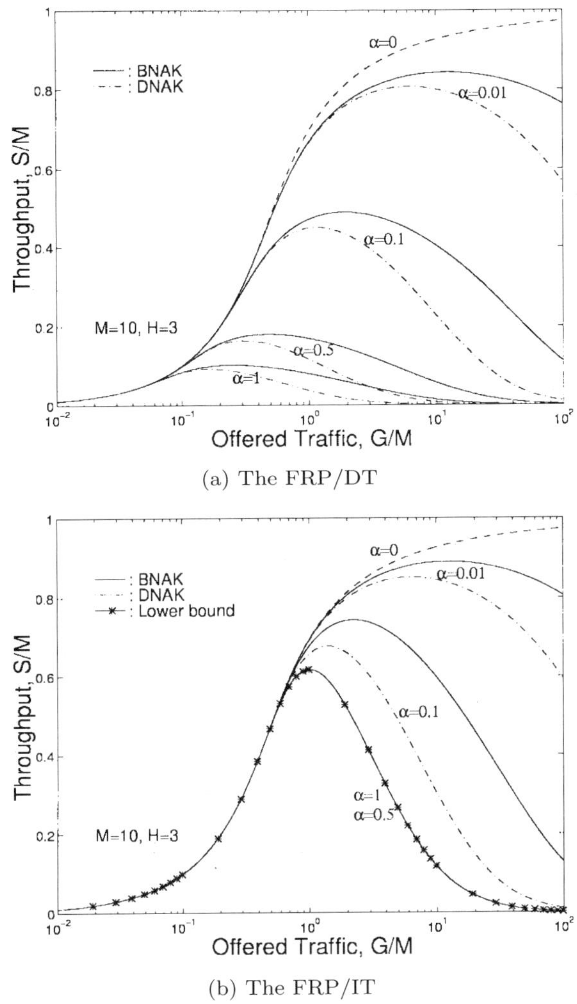

(a) The FRP/DT

(b) The FRP/IT

Fig. 10 Throughput vs. offered traffic (M=10, H=3).

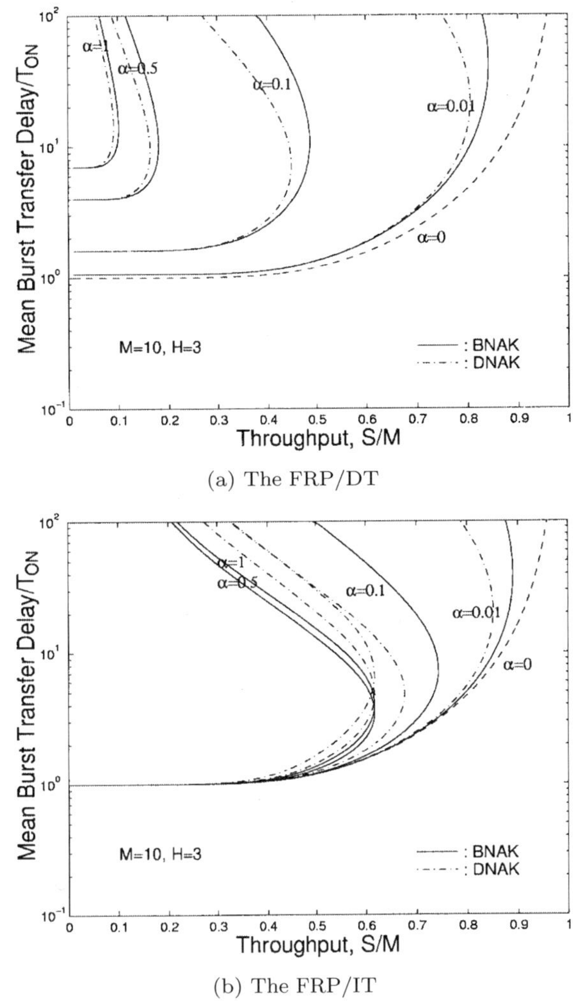

(a) The FRP/DT

(b) The FRP/IT

Fig. 11 Mean burst transfer delay vs. throughput (M=10, H=3).

can observe the same trends in performance between the FRP/DT and IT schemes.

From Figs. 10 and 11, we therefore see that the maximum achievable throughput of the FRP/IT is always greater than the FRP/DT for the same value of α, and that, for a given burst delay or blocking probability, the FRP/IT exhibits higher achievable throughput than the FRP/DT. In these figures, we also see that the BNAK scheme gives better performance than the DNAK scheme.

Figure 12 shows the impact of number of hops for the same end-to-end round-trip propagation delay. In this example, we assume the medium peak-rate traffic with M=10 and the end-to-end round-trip delay $2H\alpha$=0.1. We consider three different number of hops, H=1, 2, 5. When H=1, 2, 5 with the BNAK scheme, the maximum achievable throughput are 0.91, 0.81 and 0.71 for the FRP/DT, while 1, 0.91 and 0.81 for the FRP/IT, respectively. From this figure, we note that the maximum achievable throughput of both the FRP/DT and the FRP/IT decreases as the number of hops increases, but the FRP/IT still has the better performance. Clearly as the number of hops increases, the probability of path blocking increases. From this numerical example, we can say that minimum hop routing is highly desirable in fast reservation schemes for the same end-to-end propagation delay.

5. Conclusion

In this paper, using a closed queueing model and Erlang loss formula, we analyzed the FRP/DT and FRP/IT fast reservation schemes for single node and multihop network environments. We also investigated the effect of NAK schemes, BNAK and DNAK, on performance. To compare performance, we derived two analytical models for the mean burst transfer delay, burst blocking probability and throughput. We discussed the dependence of these performance parameters on the propagation delay, peak rate and the number of hops.

In concluding remarks, we note that the fast reservation schemes are more suitable for LAN environments

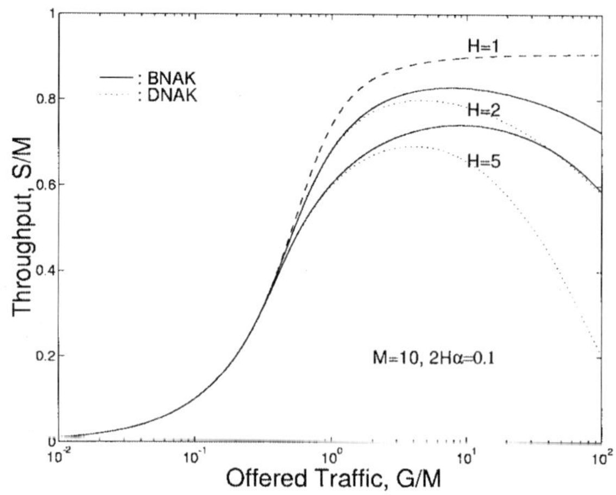

(a) The FRP/DT

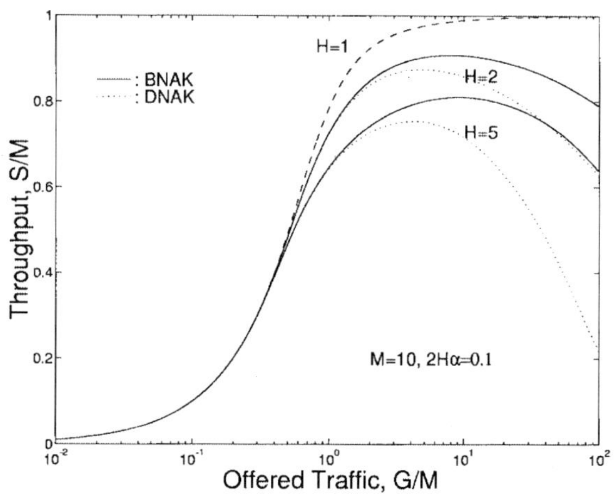

(b) The FRP/IT

Fig. 12 The impact of the number of hops for the same end-to-end round-trip delay (M=10, $2H\alpha$=0.1).

than WAN irrespective of reservation types and NAK schemes. For WAN applications, in order to achieve higher link utilization, it is desirable to concatenate small bursts into a long burst and to limit the number of routing hops. For higher propagation delay-to-burst duration ratio, the FRP/IT scheme with the BNAK is most desirable in terms of the fast reservation traffic performance. In order to stabilize the fast reservation performance, the call-level admission control is necessary in advance of the burst-level control.

References

[1] H.T. Kung, "Gigabit local area networks: A system perspective," IEEE Commun. Mag., vol.30, no.4, pp.79–89, April 1992.

[2] P. Newman, "ATM LANs: The customer premises ATM network," The 3-rd B-ISDN Workshop, April 1993.

[3] J.B. Lyles and D.C. Swinehart, "The emerging gigabit environment and the role of local ATM," IEEE Commun. Mag., vol.30, no.4, pp.52–58, April 1992.

[4] J.S. Turner, "Managing bandwidth in ATM networks with bursty traffic," IEEE Network, vol.6, no.5, pp.50–58, Sept. 1992.

[5] P.E. Boyer and D.P. Tranchier, "A reservation principle with applications to the ATM traffic control," Computer Networks and ISDN Systems 24, pp.321–334, 1992.

[6] Y.Z. Cho and A. Leon-Garcia, "Performance of burst-level bandwidth reservation in ATM LANs," IEEE INFOCOM'94, pp.812–820, 1994.

[7] ITU-T Recommunication I.371, May 1996.

[8] H. Suzuki and F.A. Tobagi, "Fast bandwidth reservation scheme with multi-link and multi-path routing in ATM networks," INFOCOM'92, pp.10A.2.1-8, 1992.

[9] D.P. Tranchier et al., "Fast bandwidth allocation in ATM networks," ISS'92, A5.2, Oct. 1992.

[10] C. Ikeda and H. Suzuki, "Adaptive congestion control schemes for ATM LANs," IEEE INFOCOM'94, pp.829–838, 1994.

[11] N. Wakamiya, K. Baba, M. Murata, and H. Miyahara, "Performance analysis of traffic control methods in multimedia ATM LAN," IEEE INFOCOM'95, pp.1027–1036, 1995.

[12] N.F. Maxemchuck, "Dispersity routing in high-speed networks," The 2nd Annual Workshop on Very High Speed Networks, March 1991.

[13] A. Pattavina, "Multichannel bandwidth allocation in a broadband packet switch," IEEE J. Sel. Areas Commun., vol.6, no.9, pp.1489–1499, Dec. 1988.

High-Performance Backbone Network Technology edited by Naoaki Yamanaka
ISBN: 0-8247-5321-6 DOI: 10.1081/(120025708)
Published 2004 by Marcel Dekker, Inc.

Diffserv-Based QoS over ATM Access Networks

Tomohiro Ishihara, Jun Tanaka, Michio Goto, and Sataro Oda

SUMMARY We have developed a new scheme to provide Diffserv-based QoS over ATM access networks. Well-known Diffserv over ATM scheme requires some extension for conventional routers with ATM interfaces. The routers must map their Diffserv classes of services into ATM QoS classes and forward IP packets into prioritized VCs based on DSCP (DiffServ Code Point). The purpose of this work is to provide Diffserv-based QoS over ATM network using conventional IP over ATM interfaces on routers. We propose DSCP snooping at ATM edge nodes, which differentiates services over a single VC between two IP domains. A prototype circuit was used to evaluate this scheme.

key words: *QoS (Quality of Services), ATM (Asynchronous Transfer Mode), IP (Internet Protocol), access network, Diffserv (Differentiated Services)*

1. Introduction

Explosive Internet growth has led to a variety of research and development on backbone networks for high-speed and efficient IP (Internet Protocol) traffic transport. Giga/Tera bit routers have been developed for routing and traffic concentration, and WDM (Wavelength Division Multiplexing) technology is being used for high-capacity transmission. These technologies provide a new direction for building carrier-scale networks using IP-based technology instead of ATM (Asynchronous Transfer Mode).

There are some questions whether the same IP-centric approach can be applied for access networks, as well. If the services are restricted to IP traffic transport, this approach may work. However, it is currently difficult for the IP-centric approach to provide leased line service or universal bearer services that require strictly guaranteed QoS (Quality of Service).

We take ATM as a base technology with regard to access networks for several reasons. First, ATM is QoS-capable. Because multi-service should be provided in an access network [1], multi-QoS for each service shall be satisfied. Although QoS-capable IP networks [2] are a very active research topic, the powerful QoS capability of ATM [3]–[5] still exceeds that of IP and has been providing commercial network services. Second, network resources can be utilized efficiently by statistical multiplexing. Bursty IP traffic is concentrated into some bandwidth by ATM cell multiplexing. In broadband access networks, both the traffic concentration and the QoS capability are essential features. The ATM technology has still advantages for access networks [6].

There may be a concern on the scalability for the connection-oriented scheme using ATM. For core networks, it may not be a practical solution because they must handle huge number of connections for each service. On the contrary, the number of required connections for access networks is not huge, less than tens of thousands. In practical, ATM-based devices (e.g. ATM edge switch) have been successfully deployed for broadband access networks. For access networks, we believe that connection-oriented approach is better than connection-less one because of its multi-service capability derived from its powerful QoS capability.

Figure 1 shows our proposed access network featuring ATM-OLTs (Optical Line Terminals). The add-drop multiplexing, grooming, statistical multiplexing, and QoS control functions are integrated into each OLT. Cost-effective access lines can be provided by adopting ATM-PON (Passive Optical Networks) technology [7], [8] between an OLT and ONTs (Optical Network Terminals). Both OLTs and ONTs can provide a

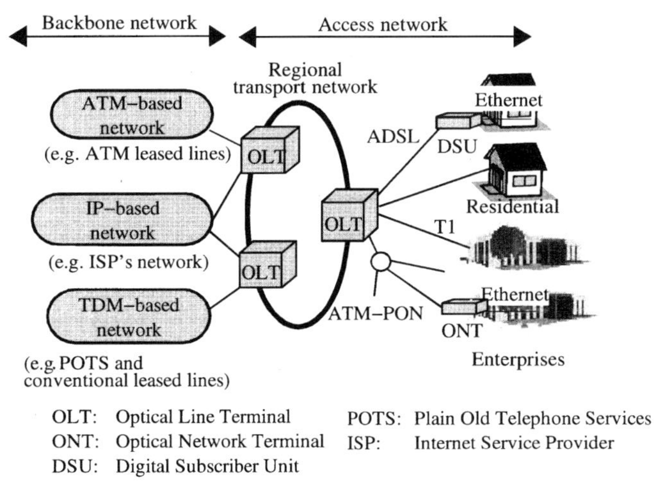

Fig. 1 Access network architecture.

Manuscript received November 6, 2000.
Manuscript revised January 16, 2001.
[†]The authors are with Fujitsu Laboratories Ltd., Kawasaki-shi, 211-8588 Japan.
[††]The authors are with Fujitsu Kyushu Digital Technology Ltd., Fukuoka-shi, 812-0011 Japan.
[*]This paper was presented at IEEE GLOBECOM 2000.

IEICE TRANS. COMMUN., Vol. E84-B, No. 6, pp. 1498–1503, June 2001

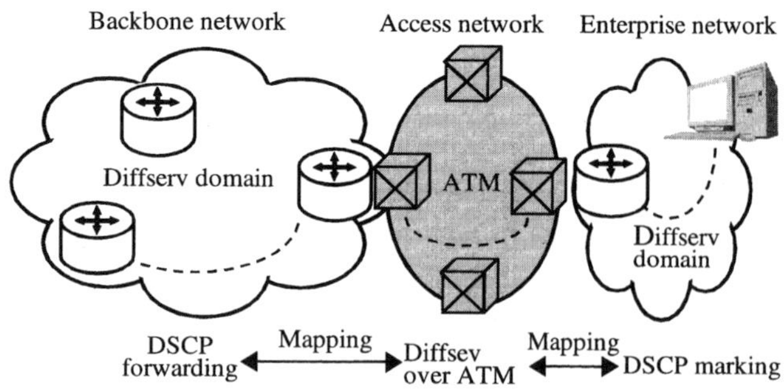

Fig. 2 Diffserv over ATM on access network.

variety of UNIs (User Network Interfaces) such as ATM, DSL (Digital Subscriber Line), T1/E1 (conventional 1.5 or 2 Mb/s line), 10/100 Base-T Ethernet, etc., depending on user's request. The Ethernet interfaces such as 10/100 Base-T are preferable as UNIs because users can connect their routers or PCs simply using their built-in Ethernet interfaces. The ATM technology should be devoted to providing paths between users and service nodes.

2. Network Architecture and Diffserv over ATM

Although we propose ATM-based access networks, the primary service is IP traffic transport. The ATM network should be IP-aware and well designed to provide IP-QoS as well. As a scalable IP-QoS architecture, Differentiated Services (Diffserv) [9] will be supported within service provider's backbone and enterprise networks. The access network that connect these two Diffserv domains should maintain the differentiated QoS between them, as shown in Fig. 2.

DiffServ over ATM was one of the main issues at the ATM-Forum. The following two approaches both provide QoS for DiffServ over ATM networks [10], [11].

- Mapping DiffServ PHBs (Per-Hop-Behaviors) such as EF (Expedited Forwarding) [12] or AF (Assured Forwarding) [13] to ATM service categories such as CBR (Constant Bit Rate) or GFR (Guaranteed Frame Rate) [5].
- Mapping DiffServ classes to relative priorities provided within ATM-UBR (Unspecified Bit Rate), e.g., AF_{1x} to AF_{4x} are assigned to four priority classes within UBR.

Both approaches require several prioritized VCs (Virtual Channels) corresponding to the Diffserv's class of service or priorities, as shown in Fig. 3(a). A drawback of this approach is that the router must have Diffserv over ATM functions. Particularly, the sending router has to forward IP packets to appropriate ATM VCs based not only on IP addresses but also on DSCP. Conventional router forwards IP packets to VCs based on only IP address. It seems small possibility that this

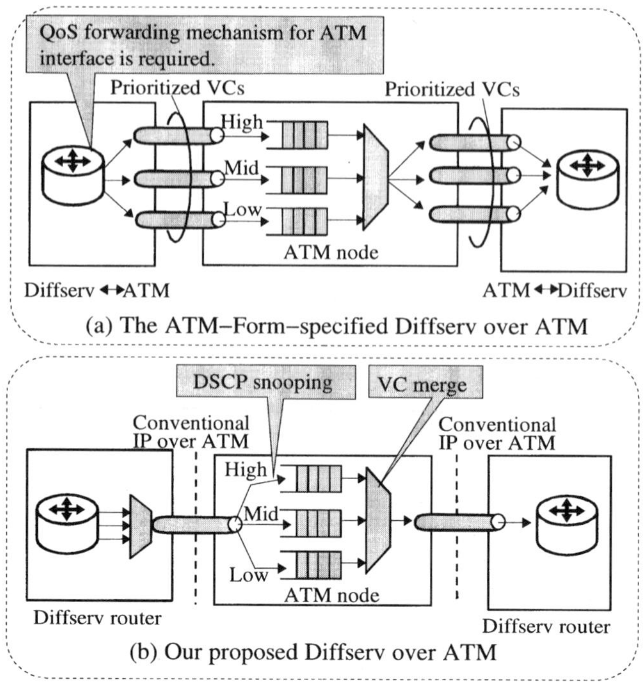

Fig. 3 Alternatives for diffserv over ATM.

extension on the QoS forwarding mechanism is implemented onto routers commonly. This is because the ATM is regarded as one of the interface alternatives from the viewpoint of IP routers, then there is little incentive to extend QoS forwarding functions just for ATM.

According to this consideration, we believe that a scheme is required which achieves Diffserv over ATM access network without any extension onto user routers.

3. Our Proposal: DSCP Snooping

For providing the Diffserv without any extension for routers, we propose a new scheme using DSCP snooping at the OLT. Figure 3(b) illustrates our approach, which requires only a single VC between routers. In the OLT, IP-level QoS information or DSCP is snooped. After QoS control is provided, packets are merged into a single VC again. The interface between routers and ATM nodes is conventional IP over ATM. Neither prioritized VCs nor extension for routers are required. The main difference between the Forum scheme and ours is whether the OLT knows cell-level QoS information a priori.

A functional diagram to achieve our scheme is shown in Fig. 4. After an AAL type-5 frame is identified, the DSCP in the first cell of each frame is snooped. Here, we assume a relative priority control, such as the class selector PHB. Referring to a pre-assigned priority table, each cell is marked with a priority same as the first cell.

Cells are stored in the frame-reassemble queues be-

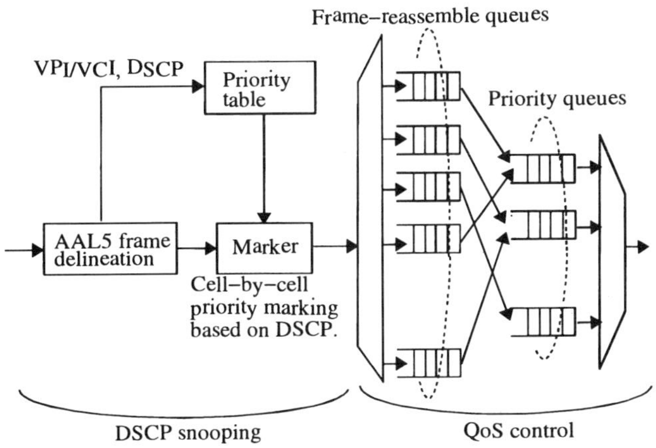

Fig. 4 Block diagram of proposed scheme.

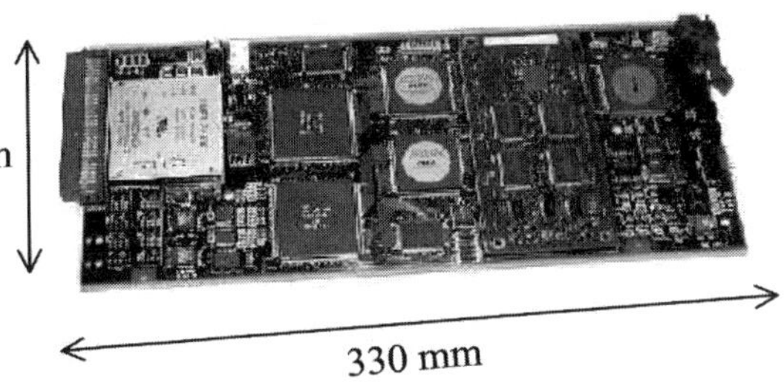

Fig. 5 Prototype card.

fore the priority control process and are read out frame-by-frame. This is because the AAL5 frame structure has to be preserved in re-multiplexing into each VC. The required length of each queue is 32-cells to store 1500-byte IP packet. It is enough because user PCs or routers are connected to the access system via Ethernet interfaces where the maximum size of IP packet is 1500 bytes.

Then cells are stored in appropriate priority queues based on their priority marks. The highest priority queue among non-empty queues pushes cells out frame-by-frame basis. This operation makes a relative priority control, or a service differentiation.

Using frame-by-frame multiplexing and priority control in each VC, the cell sequence is not maintained. This does not comply with ITU-T I.150 [14], which prescribes that ATM cell sequence integrity shall be maintained. Based on the concept of "Ordered Aggregate (OA)" [15], each VC is regarded as a unit with an ordering constraint. However, AAL5 frame-structure integrity is much more important than cell-sequence integrity for IP traffic transport service. When the UNI is an IP-based interface in particular, such as 10/100Base-T, users don't require ATM cell sequence integrity but do need IP-QoS enforcement.

Based on our proposed scheme, we developed a prototype circuit that can be installed in ATM-OLTs. The prototype card shown in Fig. 5 is connected to a switching block at a speed of 622 Mb/s, providing eight

Table 1 Specification of prototype card.

Input rate	622.04 Mb/s
Output rate	155.52 Mb/s
Priority count	8
Capable VC count	16
Total cell buffer memory capacity	7281 cells

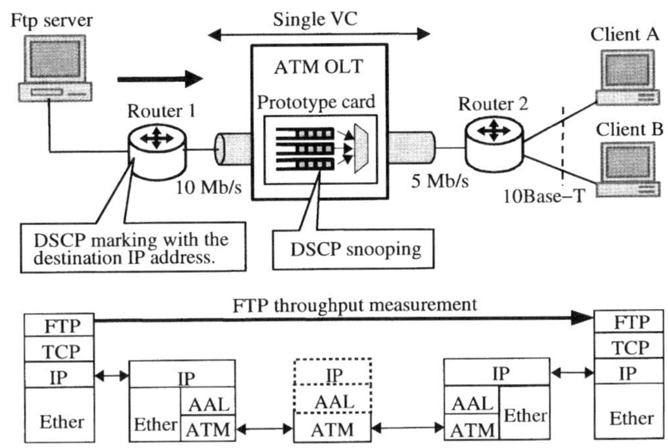

Fig. 6 Setup for performance evaluation.

levels of relative priority control. The specifications of the prototype circuit are shown in Table 1. This circuit was developed with four FPGAs. All queues, frame-reassemble queues and priority queues, were implemented into a single cell buffer memory. The queues are assigned dynamically in the memory, have no fixed boundary. This implementation scheme can achieve effective use of the cell buffer memory.

4. Performance Evaluation of Prototype Card

We evaluated the performance of the Diffserv over ATM in two ways. First, we verified the service differentiation within a single VC. Figure 6 shows the setup configuration for this performance evaluation, which uses FTP (file transfer protocol) throughput measurement. Two FTP clients download files from a server through the same VC via two routers. The router 1 marks DSCP based on the destination IP address. The prototype card snoops on this code and establishes priorities. We set 5 Mb/s as the output rate for the sum of prioritized queues. This means the two FTP flows share 5 Mb/s of bandwidth for priority control.

The results are shown in Fig. 7. First, the client A starts downloading, then the client B cuts into this communication. When A has priority over B, the traffic of B doesn't affect the traffic of A as shown in Fig. 7(a). However, when B has priority over A, it preempts the bandwidth of A as shown in Fig. 7(b). These results prove that the priority control works by snooping DSCP within a single VC. This means Diffserv can be emulated over ATM networks without any extension for the router side. In other words, Diffserv over ATM is avail-

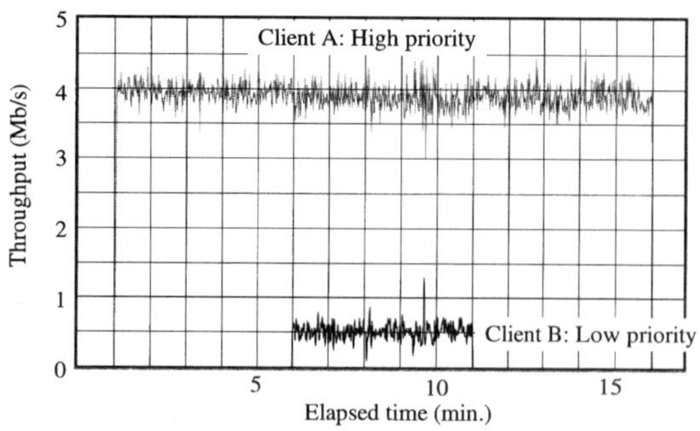

(a) Client A has priority over client B.

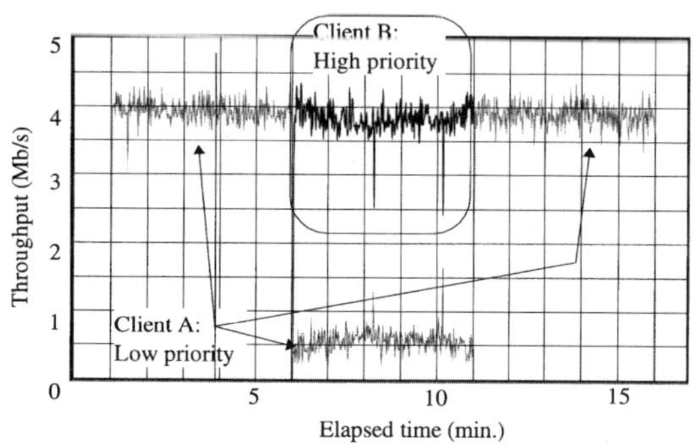

(b) Client B has priority over client A.

Fig. 7 Evaluation results.

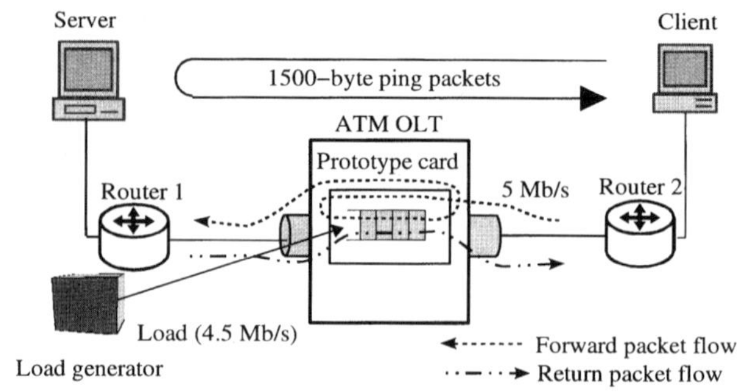

Fig. 8 Setup for performance comparison.

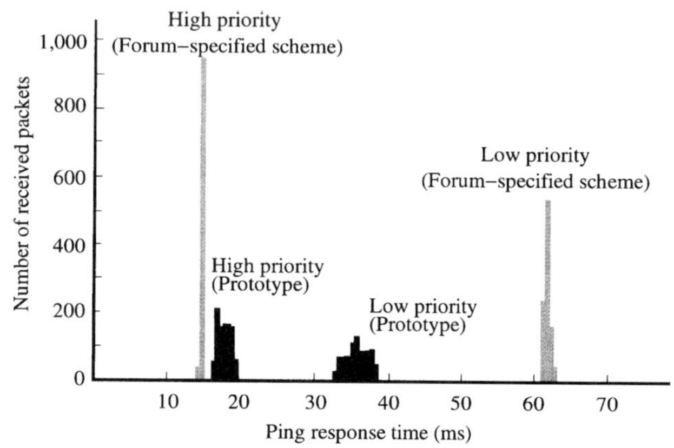

Fig. 9 Delay and jitter performance.

able with an extension inside the ATM clouds.

5. Performance Comparison

Following the initial performance evaluation, we compared the performance of the Forum-specified Diffserv over ATM and ours. The Forum scheme provides cell-by-cell QoS control using VC priority. On the other hand, our proposed scheme provides frame-by-frame control using DSCP snooping. To emulate the Forum scheme, we added a cell-multiplexing mode with VC priority to our prototype card. With this prototype card, we could evaluate both the Forum scheme and ours.

Because the multiplexing units are different from each other (cell vs. frame), differences in performance are expected in terms of delay and jitter. For services containing delay-sensitive or jitter-sensitive applications such as VoIP (Voice over IP), QoS control is a crucial issue. We compared the delay and jitter performance using the system configuration shown in Fig. 8. Test traffic and load traffic share a 5-Mb/s output link. We used ping packets to measure delay and jitter on IP layer. Also we generated 1500-byte-long ping packets to measure the performance for long packets. The load traffic has a constant bit rate and the frame size is 1500 bytes. The load-input rate is 90% of that of output link, or 4.5 Mb/s. Assigning different priorities between test and load, ping responses were measured for 1000 trials. The results are shown in Fig. 9.

Based on these results, the Forum scheme shows better delay and jitter performance than our proposed scheme when high priority is assigned to the test traffic. Because our proposed scheme adopts frame-based multiplexing, even a high priority packet must wait while a lower priority packet is read out. The maximum delay in the high priority queue, which corresponds to a 32-cell readout time, is calculated as 2.7 ms at a 5-Mb/s output rate. The maximum round-trip delay should be 5.4 ms, and the results are consistent with the calculation. When low priority is assigned to the test traffic, our scheme shows better delay performance. This is because once the low-priority load traffic started to be read, all 32 cells would be read out consecutively. However, the jitter performance of our scheme is still worse than that of the Forum scheme.

6. Discussion

Our objective is to achieve Diffserv over ATM without any extension onto routers. From this point of view, it was confirmed that our scheme worked properly as described in Sect. 4.

Looking over the results described in Sect. 5, the Forum scheme seems better than ours in terms of QoS control capability. However, we think the difference is negligible in practical conditions, because of the following reason;

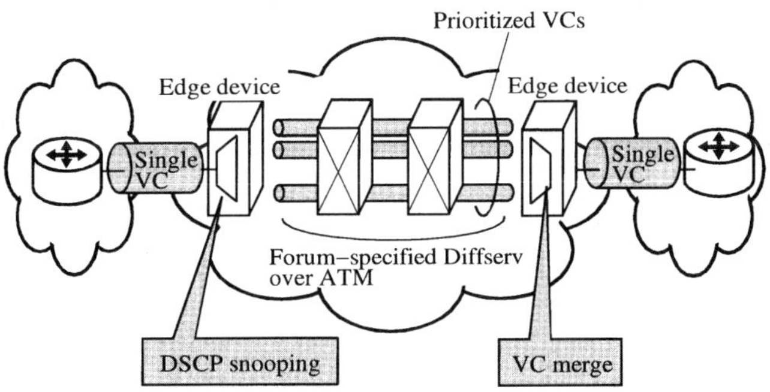

Fig. 10 Proposed diffserv over ATM architecture.

The measurement was done at the output rate of 5 Mb/s because of the measurement tool limitation. In practical system the output rate is much higher. When the output rate is 155 Mb/s, the maximum delay difference should shrink to less than 1/30, or 160 μs. Even though VoIP requires low delay and jitter, its actual one-way delay is usually distributed in a range from tens of milliseconds to hundreds of milliseconds. We regard delay and jitter of less than one millisecond as negligible for providing QoS in a Diffserv framework. Our proposed scheme will work as well as the Forum scheme.

A drawback of ours is that it requires more complexity in ATM node hardware. But it is a small drawback, because of the following reasons;

The DSCP snooping mechanism shown in Fig. 4 can be implemented onto relatively-small hardware using only one control logic shared by all incoming VCs and status memory for each VC. The frame-reassemble queues require large memory, if each queue is allocated onto the memory statically. However, dynamic allocation scheme, which has already been implemented onto the prototype, resolves this problem. The last part, priority queues are definitely required even the Forum scheme is chosen.

According to the above discussion, our proposed DSCP snooping scheme is a better solution than that of using the Forum scheme.

Our scheme is not only superior than the Forum scheme, but it co-exist with the Forum scheme when it was already implemented. As shown in Fig. 10, the Forum scheme should be adopted inside the ATM network, and the ATM edge nodes should implement our scheme. This requires the least extension for the whole network and still provides Diffserv-QoS over ATM networks.

7. Conclusion

We have developed a new scheme to provide Diffserv over ATM. Because access networks must provide multiple services, we believe ATM technology is suitable for access networks. To support Diffserv-based QoS over ATM-based access networks, we proposed a DSCP snooping scheme. This approach provides conventional IP over ATM interfaces for routers and does not require any extension for the router side. Performance evaluation with FTP showed that our scheme strictly provided relative priority control. In terms of delay and jitter performance, the Forum scheme showed better results, but the difference is effectively negligible. Our technology can provide better solution for smooth migration from current network to Diffserv-supporting network especially in terms of access networks.

References

[1] Full Service Access Network (Forum), "Full service access network requirements specification, issue 3," http://www.fsanet.net/, Aug. 1998.

[2] X. Xiao and L. Ni, "Internet QoS: A big picture," IEEE Network, pp.8–18, March/April 1999.

[3] ITU-T Recommendation I.371, "Traffic control and congestion control in B-ISDN," 1998.

[4] ITU-T Recommendation I.356, "B-ISDN ATM layer cell transfer performance," 1996.

[5] ATM Forum, "Traffic management specification 4.1," af-tm-0121.000, 1999.

[6] A. Azcorra, D. Larrabelti, E. Hernandez-Valenica, and J. Berrocal, "IP/ATM integrated services over broadband access copper technologies," IEEE Commun. Mag., vol.37, no.5, pp.90–97, May 1999.

[7] ITU-T Recommendation G.983, "High speed optical access system based on passive optical network (PON) techniques," 1999.

[8] M. Yano, K. Yamaguchi, and H. Yamashita, "Global optical access systems based on ATM-PON," FUJITSU Sci. Tech. J., vol.35, no.1, pp.56–70, July 1999.

[9] Blake, D. Black, M. Carlson, E. Davies, Z. Wang, and W. Weiss, "An architecture for differentiated services," RFC2475, 1998.

[10] S. Brim and T. Davis, "ATM support for IETF differentiated services and IEEE 802.1D," ATM Forum/98-0931, 1998.

[11] H. Suzuki, J. Heinanen, R. Vallee, D. Basak, and B. Klessig, "Solutions to support IETF diffserv and IEEE 802.1D over ATM," ATM Forum/98-0918, 1998.

[12] V. Jacobson, K. Nicols, and K. Poduri, "An expedited forwarding PHB," RFC2598, 1999.

[13] J. Heinanen, F. Baker, W. Weiss, and J. Wroclawski, "Assured forwarding PHB group," RFC2597, 1999.

[14] ITU-T Recommendation I.150, "B-ISDN asynchronous transfer mode functional characteristics," 1995.

[15] D. Grossman "New terminology for diffserv," draft-ietf-diffserv-new-terms-02.txt, 1999.

High-Performance Backbone Network Technology edited by Naoaki Yamanaka
ISBN: 0-8247-5321-6 DOI: 10.1081/(120025709)
Published 2004 by Marcel Dekker, Inc.

Integration of Scheduling Real-Time Traffic and Cell Loss Control for ATM Networks

Chuang Lin and Lijie Sheng

SUMMARY In this paper, new integrated schemes of scheduling real-time traffic and cell loss control in high speed ATM networks are proposed for multiple priorities based on variable queue length thresholds for scheduling and the Partial Buffer Sharing policy for cell loss control. In our schemes, the queues for buffering arriving cells can be constructed in two ways: one individual queue for each user connection, or one physical queue for all user connections. The proposed schemes are considered to provide guaranteed QoS for each connection and cell sequence integrity for virtual channel/path characteristics. Moreover, these new schemes are quite flexible and can realize different scheduling algorithms. This paper also provides the Stochastic Petri Net models of these integrated schemes and an approximate analysis technique, which significantly reduces the complexity of the model solution and can be applied to real ATM switch models. From the numerical results, we can see that our schemes outperform those well-known schemes such as the head-of-line (HOL) priority control and the queue length threshold (QLT) policy.

key words: *scheduling real-time traffic, del loss control, ATM networks, stochastic Petri nets, approximate performance analysis*

1. Introduction

Asynchronous Transfer Mode (ATM) is a communication architecture capable of supporting high speed multimedia applications. The key to its efficiency is the ability to manage a wide variety of traffic conditions, diverse services and certain performance requirements which include real-time traffic scheduling and cell loss control [1].

ATM networks must be able to provide Quality of Service (QoS) guarantees to connections (it's the same meaning of calls in this paper) based on users' requirements. For most real-time applications, each cell can only be delayed up to its deadline, or it is of no use to the recipient and may be considered to discard. While the scheduling algorithms can effectively control the delay of the cells by determining the sequence in which the cells from all user connections are transmitted, still, congestion due to statistical traffic fluctuations can result in cell loss in ATM networks. Some applications such as video and voice have stringent timing requirements but can tolerate some degree of cell loss. On the other hand, applications such as file transfers have low loss tolerance but large delay is not a problem. The main function of cell discarding (or loss control) mechanisms is to control the relative cell loss probabilities of different cell types in the ATM networks based on the requirements of diverse applications. Additionally, applications which contain multimedia information within a single user connection require synchronization of different types of cells during transmission. Moreover, ATM networks are connection orientated so the cells should be served in the same order as they arrive. Therefore, it is important to find schemes to satisfy all these requirements simultaneously.

For scheduling, most of the algorithms proposed and studied in the past can be categorized into three classes: rate-based, time-based, and space-based, according to the way in which they are defined and configured. Rate-based schedulers guarantee bandwidth to users connections, which implies guarantees on delay bounds. Examples of such schemes are Weight Fair Queue, Virtual Clock and other various forms of fair queuing [2], [3]. Time-based schedulers, such as Waiting-Time Priority (WTP) [4] and Earliest-Deadline-First (EDF), directly utilize a time-metric in determining the next cell to be served, which is expected to be more effective in delivering a target time-metric. EDF received a lot of attention and was well studied in the past [5], [6], based on which many extensions and variations were proposed [7], [8]. EDF is optimal in terms of minimizing the loss rate due to the violation of its deadline. However, EDF will be no longer optimal if it is considered more important to reduce the loss rate incurred by congestion. Also, the implementation of EDF is more complicated in ATM networks. Space-based schedulers make the scheduling decisions according to the conditions of buffer occupancy of logical queues. Head-of-line (HOL) priority control can be viewed as the simplest space-based scheduler. HOL always schedules delay-sensitive traffic first. Obviously, HOL degrades severely the performance of low priority traffic as the portion of high priority traffic increases. Queue Length Threshold (QLT) policy [9], which gives priority to non-delay-sensitive traffic as long as the queued non-delay-sensitive cells

Manuscript received November 27, 2000.
Manuscript revised July 6, 2001.

†The authors are with the Department of Computer Science and Technology, Tsinghua University, Beijing, 100084, P.R.China.

*This research was supported by NSFC under grant number 90104002 and 69873012, Development Plan of the State Key Fundamental Research (973) in China under grant number G1999032707, and Intel IXA University Research Plan.

IEICE TRANS. COMMUN., Vol. E85-B, No. 4, pp. 778–795, April 2002 exceed some threshold, gives a better overall tradeoff between performance and complexity. However, QLT is not adequate in differentiating the cell loss probabilities of different user connections.

There are three well-known cell discarding schemes: complete buffer partitioning (CBP), push-out (PO), and partial buffer sharing (PBS) [10]. These schemes correspond to the three buffer allocation policies: dedicated allocation, complete sharing, and partial sharing [11]. CBP uses a separate buffer queue for each class of cells. The advantage of CBP is its simplicity of implementation, as well as ensuring fairness among all classes of cells—cells of each class can only occupy the buffers dedicated to that class. However, dedicated allocation also makes the overall buffer utilization of CBP relatively low. In addition, the cells belonging to the same user connection may be assigned to different classes, so a resequencing mechanism must be used to restore the order of the cells at the destination. This destroys the advantages of the connection-oriented nature of ATM. PO uses the same buffer for all arriving cells irrespective of their priorities. An arriving high priority cell may enter a full buffer only if there exists at least one low priority cell in the buffer, and a low priority one will be discarded (pushed out) to make room for the high priority one. Thus, complex operations are needed to select and push out a low priority cell when necessary. PBS controls cell loss based on a threshold related to the overall buffer size. When buffer occupancy reaches or exceeds the given threshold, only high priority cells may enter the buffer while arriving low priority ones will be discarded. PBS is simple to implement since no selecting and pushing out operations are needed. Moreover, PBS may offer the best tradeoff between the principles of fairness and resource sharing [11].

Using only a delay priority control or a loss priority control scheme separately in an ATM network, it is difficult to effectively satisfy all the QoS requirements of diverse applications, which have different delay and loss constraints. Thus, some type of integrated priority control schemes which treats each user connection according to its particular QoS requirements is needed. In the past several years, a few results on integrated priority control schemes have been proposed in ATM networks. The real-time scheduling algorithm in [12] is optimal in terms of minimizing the weighted loss at an ATM network node. This scheme is considered an integration of real-time scheduling and cell loss control, but it may destroy the cell sequence integrity in the same user connection and it is complicated to implement. A versatile scheduling discipline, called Precedence with Partial Push-out (PPP), has been proposed for delay and loss control in ATM switches in [13]. However, in PPP only two delay and two loss priorities are considered, and nothing regarding cell sequence integrity is studied. The performance analysis of a delay and loss priority control scheme is presented in [14]. However, the analysis of the proposed scheme is not considered useful for real ATM networks since all cell arrival rates at each input queue are assumed equal. Weighted Round Robin (WRR) combined with Threshold Pushout (TP) is studied in [15]. The scheme achieves relatively low overall CLP (cell loss probability), and good fairness among connections to a certain extent, which derive from the intrinsic advantages of WRR and TP. Nevertheless, the implementation complexity of TP is still an issue that should be considered.

In this paper, we propose new integrated schemes, called Minimum Empty Buffers First (MEBF), for multiple priorities based on the variable queue length thresholds in scheduling and the PBS policy in loss control. In our schemes, one or more connections with the same delay-sensitive priority use one buffer queue for arriving cells. Each individual queue has two separate types of thresholds, one for scheduling and the other for loss control. To improve the buffer utilization of independent queues, all arriving cells can be inserted into individual logical sub-queues within one physical queue according to its real-time deadline in the one queue scheme. The proposed schemes are considered to provide guaranteed QoS for each connection and cell sequence integrity for virtual channel/path characteristics. We also provide the Stochastic Petri Net (SPN) models of these integrated schemes and an approximate analysis technique, which significantly reduces the complexity of the model solution and can be applied to real ATM switch models. The numerical results in this paper show that our schemes obtain better performance in both cell loss rate differentiation and overall weighted cell loss rate.

The focus of this paper is the performance analysis of the new integrated schemes. Performance evaluation of such schemes is basically a matter of queuing analysis. The main difficulties of such an analysis essentially arise from the fact that the corresponding Markov chain of such a model is usually multidimensional, and its state space grows exponentially with the increase of the number of queues and the size of each queue. Furthermore, it usually does not have a product form solution. Thus, it is generally very hard to solve it directly, and the computation complexities in both space and time are beyond the ability of common computers. This is the so-called "state space explosion problem." To overcome this problem, approximate methods with decomposition techniques are usually reasonable choices. The keys in applying such approximate methods are how to decompose the whole model into a number of sub-models that are loosely coupled, and how to express the relationship among these sub-models. In this paper, performance evaluation is based on Stochastic Petri Net, for it provides an intuitive yet efficient decomposition mechanism. Moreover, SPN models have been shown effective for describ-

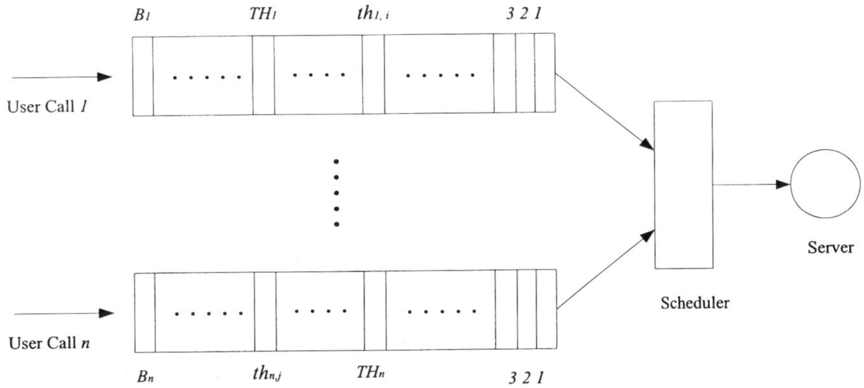

Fig. 1 The scheme of MEBF with individual input queues.

ing prioritized, stochastic, non-deterministic, concurrent and asynchronous behaviors, and they have been successfully applied to performance evaluation of some control mechanisms in ATM and IP networks [16]–[19]. The main difficulties in approximately analyzing and solving with SPN include how to refine the whole model to exploit the near-independences of the sub-models, how to transform the predicates into probabilities, how to express the import and export parameters of the sub-models with the probability distribution of steady state, and how to iterate among the sub-models. The details are presented in Sect. 3 and Sect. 5.

The rest of the paper is organized as follows. Section 2 describes the integration of scheduling and loss control based on queue length thresholds in separate queues. The models and analysis of the schemes based on SPN are presented in Sect. 3. In Sect. 4, in order to improve buffer utilization of above schemes, we present the one queue scheme which employ one physical queue. In the same way, the models and analysis of the one queue scheme are discussed in Sect. 5. Numerical examples are given in Sect. 6. Finally, Sect. 7 concludes this paper.

2. Minimum Empty Buffer First (MEBF) Scheduling Scheme with Individual Input Queues

New scheduling schemes, MEBF, are proposed based on queue length thresholds. The schemes can arrange the cells of the same user call to be served in FIFO, and are simple to implement in an ATM network.

At an ATM network node, the scheduling policy can be represented by the model with a single server and multiple buffer queues, as shown in Fig. 1. One or more user calls with the same real-time requirements share one buffer queue. In each queue there is one queue length threshold for scheduling, which is similar to QLT policy, but in our new policy the cells of the same user call are buffered in the same queue so that they can be served in FIFO. The PBS policy is accepted for cell loss control in each queue, and multiple thresholds in each buffer queue can be set for multiple loss priorities. The scheduler may shift the positions of the cells in the buffer queues in order to accommodate the arriving cells. We assume that there are n queues, and they are numbered with the priorities for real-time scheduling. The queue 1, q_1, has the highest scheduling priority. The queue n, q_n, has the lowest scheduling priority. The scheduling priority of q_i is higher than that of q_{i+1}. The buffer size of q_i is B_i and the buffer is numbered from head to tail as shown in Fig. 1. In this paper each arriving cells is assumed to have a constraint of maximum delay time (or deadline), which represents its urgent degree and scheduling priority. A cell should be dropped if its maximum delay time is violated. In an ATM environment the service time of each cell is constant. Hence, the system can be considered as being slotted and each slot is equal to the transmission time of one cell. In order to simplify the computation, we can assume the service time is 1. If the cells of user call i in q_i have the maximum delay time as d, then $B_i = \lfloor d-1 \rfloor$ ($\lfloor x \rfloor$ represents the largest positive integer which is not more than x). All arriving cells of the same user call have same maximum delay time. We have $B_1 \leq B_2 \leq \ldots \leq B_n$.

In this model the following two types of parameters are used in scheduling and loss control:

TH_i is the threshold of real-time scheduling for user call i. When the number of occupied buffers in q_i, $M(q_i)$, reaches or exceeds TH_i, and $M(q_j)(1 \leq j < i)$ is smaller than TH_j, the cells in the buffer queue i are served and sent into the network. In general, the higher the real-time scheduling priority of a user call is, the smaller the threshold value for its cell scheduling will be, and vice versa.

th_{ij} is the threshold of the cell loss priority j for user call i. Multiple thresholds in one queue can be set for Multiple loss priorities. When a cell of user call i with loss priority j is coming, if the queue length is smaller than th_{ij}, the arriving cell i will be buffered in q_i, otherwise it will be discarded. The greater the loss priority j is, the greater th_{ij}

and the smaller the loss probability of the cells with the loss priority j will be. For $\forall j, 1 \leq th_{ij} \leq B_i$. If $t < k$, then $1 \leq th_{it} < th_{ik} \leq B_i$.

Note that there are various different real-time scheduling schemes when the value of TH_i is set according to different policies. We choose the following 4 schemes for numerical examples:

(1) *Head of Line (HOL) priority scheduling*: cells with the highest real-time priority are always scheduled first, i.e.,

$$TH_1 = TH_2 = \ldots = TH_n = 1$$

When there are cells in q_i and there is no cell in $q_j (1 \leq j < i)$, the cells in q_i will be served and sent into the network first.

HOL is a static scheduling priority scheme. In HOL, the scheduling priority is not variable with the system run-time states.

(2) *Queue Length Threshold (QLT) priority scheduling*: when $M(q_i)$ reaches or exceeds TH_i and $M(q_j)(1 \leq j < i)$ is smaller than TH_j, the cells in q_i will be served and sent into the network even though $M(q_j) > 0$, i.e.,

$$1 < TH_i \leq B_i$$

In this scheme, we can regularly set: $1 < TH_1 \leq TH_2 \leq \ldots \leq TH_n \leq B_n$.

QLT is a dynamic scheduling priority scheme. The scheduling priority is variable with the queue length, but the queue length thresholds are static.

(3) *Minimum Empty Buffers First (MEBF) with priority scheduling*:

$$\text{for } 1 \leq i \leq n,\ TH_i = \begin{cases} 1, & \text{if } i \in MEBF(q); \\ B_i, & \text{otherwise.} \end{cases}$$

where $MEBF(q) = \{j \mid min(B_j - M(q_j)), \text{for } 1 \leq j \leq n,\ M(q_j) > 0\}$.

The cells in the queue with the minimum empty buffers are always scheduled first based on the scheduling priority.

The MEBF with priority scheme is a dynamic scheduling priority scheme, in which the scheduling priority and the thresholds are variable with the system run-time states. If we assumed that TH_i is not a threshold value, but is a queue scheduling probability for cells in q_i and cells in all queues are scheduled according to the queue scheduling probabilities, the scheme iii. can be modified into the scheme of Minimum Empty Buffers First (MEBF) without priority scheduling, which is given below.

(4) *Minimum Empty Buffers First (MEBF) without priority scheduling*:

for $1 \leq i \leq n$,

$$TH_i = \begin{cases} \dfrac{1}{|MEBF(q)|}, & \text{if } i \in MEBF(q); \\ 0, & \text{otherwise.} \end{cases}$$

where $MEBF(q) = \{j \mid min(B_j - M(q_j)), \text{for } 1 \leq j \leq n,\ M(q_j) > 0\}$; and $|MEBF(q)|$ is the number of elements in set $MEBF(q)$.

3. SPN Modeling and Analysis for the MEBF Schemes

Readers are assumed to have basic knowledge on Petri nets such as that given in [25]. SPNs are Petri nets augmented with the set of average (possibly marking dependent) transition rates for the exponentially distributed transition firing time. A marking in a SPN model represents a distribution of tokens in each place of the model. A transition represents a class of possible changes of markings. Such change, also called transition firing, consists of removing tokens from the input places of the transition and adding tokens to the output places of the transition according to the expressions given on the arcs. A transition is enabling whenever the predicates associated with the transition are satisfied, all input places carry enough copies of the proper tokens, and all output places have sufficient capacity to accommodate the addition of the respective copies of tokens. The state space of a model consists of the set of all markings reachable from the initial marking through occurrences of transition firing. A transition may be associated with an enabling predicate that can be expressed in terms of the token variable or place marking expressions. If a predicate of transition evaluates to be false, the transition is disabled. For more details about SPN, please refer to [22]–[24].

• Modeling Assumption

Modelings of traffic and server behavior play key roles in performance analysis of queuing systems. They have significant influences on performance metrics such as delay and buffer overflow probability. Different models may lead to quite different analysis results. We will discuss the modeling issues of these two key components first.

(a) Traffic Modeling

Traffic characterization and modeling of ATM networks is still an active area in current research. For packet formed voice, video, image, and computer-generated data traffic, both individual and multiplexed sources, various traffic models were created and analyzed [26]. Besides the traditional uniform arrival and Possion arrival models, Markov Modulated Possion Process (MMPP) and self-similar traffic models are usually used in analysis. In [19], an approximate self-similar model based on MMPP is introduced for analysis based on

SPN. However, due to the limitation of space, in this paper, only the analysis results of the most traditional Possion arrival model are presented. Comparatively, with the same average arrival rate, the performance under Possion arrivals will be better than under MMPP and self-similar arrivals.

Note that the Possion arrival (the same as MMPP arrival) is actually a point process model in which the arrival entities are abstract into points, and it will lead to Continuous-Time Markov Chain (CTMC) in the analysis. But in ATM networks, the arrival entities are cells with equal length—a "slotted" system. Thus, with exact modeling, the memoryless property of the arrival process should be translated to a geometric distribution of arrival intervals, and should lead to an analysis of a Discrete-Time Markov Chain (DTMC). However, if we choose to study the process over a period of many cell intervals, we can ignore the discreteness of the cell intervals and approximate the arrival process with Possion process. It is a commonly used technique in performance evaluation for ATM networks to approximate a discrete-time process with a continuous-time process [26]. In this paper, we also make this approximation and take the Possion arrival model. Generally, such an approximation, compared to the discrete-time process model, degrades the accuracy of analysis results by some degree [26].

(b) Modeling of Server Behavior

Since the cells in ATM networks are of equal length, the most reasonable choice for server modeling is a deterministic transition whose firing time is a constant value. A useful modeling method for such systems is Deterministic and Stochastic Petri Net (DSPN) which is a stochastic modeling formalism with graphical representation that includes both exponentially distributed and constant time transitions [27]. Some analysis methods for DSPN have been developed for both steady state and transition analysis, and the study of non-Markov SPN (of which DSPN is only a special case) has become an active research area over the last few years [28]–[31]. However, most of the currently available methods have some structure restrictions. For example, deterministic transitions cannot be enabling concurrently, nor can the predicates or the rates of deterministic transitions be marking-dependable. In this paper, we model servers with exponential transitions. Of course this does not yield complete accuracy, but it provides a low-bound for the system performance, because if the average service times of the two queuing systems—one with exponentially distributed service time, and the other with constant service time—are the same, performance of the former will be always worse than that of the latter. An analysis with constant service time, for instance based on DSPN, will be carried out in future work.

In Fig. 2, a SPN model for our schemes is given. In this model, generation and sending of cells are represented by timed transitions whose firing rates are exponentially distributed, while discarding and entry in the queue of cells are represented by immediate transitions which have zero firing time. The buffer queues are represented by places, with buffer occupancy represented by the number of markings in the place. In this model, the first subscript of the character represents the scheduling priority, while the second subscript of the character represents the loss priority. If there is only one subscript, it represents the scheduling priority.

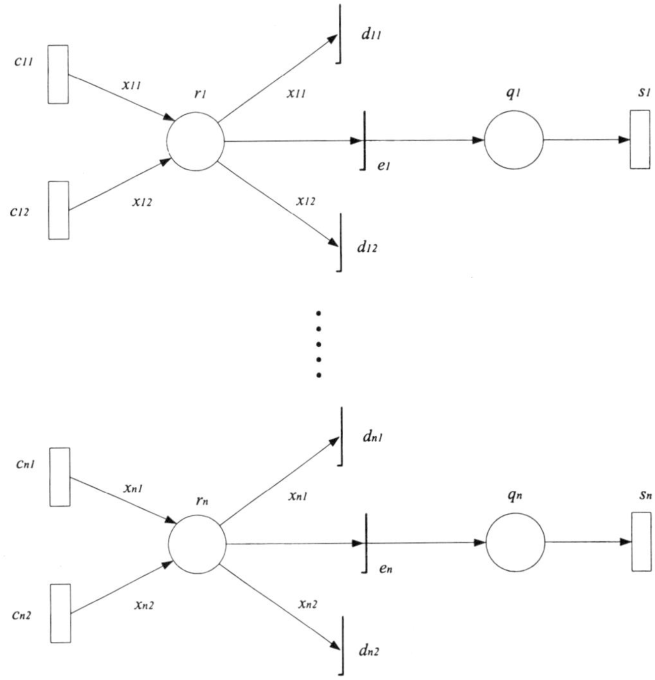

Fig. 2 The SPN model of MEBF scheduling scheme with individual input queues.

For simplicity, only the sub-models of user call 1 and user call n are given in Fig. 2. The sub-models of other user calls can be presented in the same way. For the same reason, the loss priorities have only two classes in this model. In each user call, the loss priorities may have m classes, $1 \leq m$.

The meaning of the *transitions* and *places* in Fig. 2 are described in the following:

c_{ij} represents the generation of cells with loss priority j of user call i. Its firing rate is λ_{ij}. In user call i, the generation proportion of the cells with loss priority j is $\lambda_{ij} / \sum_j \lambda_{ij}$.

d_{ij} represents discarding of cells with loss priority j of user call i.

e_i represents the entry of arriving cells of user call i.

s_i represents the server of the buffer queue i, its service rate is μ_i. In this model, $\mu_i (1 \leq i \leq n)$ has the same value since there is only one server.

r_i represents the place as temporary holding space for

arriving cells of user call i while the discarding decision is being made.

q_i represents the buffer queue of the user i, and its capacity is B_i.

In this model, the *firing priorities* of transitions are as following (the higher the value, the higher the priority):

- The immediate transition d_{ij} has priority 2.
- The immediate transition e_i has priority 1.
- The timed transitions c_{ij} and s_i have the same priority 0.

The *predicates* associated with the transitions in Fig. 2 are described in the following:

- The predicate of d_{ij} is $M(q_i) \geq th_{i,j}$, i.e., the occupancy of q_i is greater than or equal to the relevant threshold.
- The predicate of s_i for schemes (1) and (2) is

$$[(M(q_i) \geq TH_i) \wedge (\forall j, 1 \leq j < i, M(q_j) < TH_j)] \vee \left[\begin{array}{l} (\forall j, 1 \leq j < i, M(q_j) = 0) \wedge \\ (\forall j, i < j \leq n, M(q_j) < TH_j) \end{array} \right].$$

For user call 1 with the highest scheduling priority, the predicate of s_1 is simple:

$$(M(q_1) \geq TH_1) \vee (\forall j,\ 1 < j \leq n,\ M(q_j) < TH_j).$$

- The predicate of s_i for scheme (3) is

$$\left[\forall j, 1 \leq j < i, \begin{array}{l} [M(q_j) = 0] \quad \vee \\ [B_i - M(q_i) < B_j - M(q_j)] \end{array} \right] \wedge\ [\forall j, i < j \leq n, B_i - M(q_i) \leq B_j - M(q_j)].$$

For user call 1 with the highest scheduling priority, the predicate of s_1 is simple:

$$[\forall j, 1 < j \leq n, B_1 - M(q_1) \leq B_j - M(q_j)].$$

- The predicate of s_i for scheme (4) is

$$\left(\forall j, 1 \leq j \leq n, \begin{array}{l} [M(q_j) = 0] \quad \vee \\ [B_i - M(q_i) \leq B_j - M(q_j)] \end{array} \right).$$

For user call 1 with the highest scheduling priority, the predicate of s_1 for scheme (4) is the same as scheme (3):

$$[\forall j, 1 < j \leq n, B_1 - M(q_1) \leq B_j - M(q_j)].$$

For scheme (4), the rate f_i of transition s_i in Fig. 2 is dependent upon the markings:

$$\text{for } 1 \leq i \leq n,\quad f_i = \begin{cases} \dfrac{\mu_i}{|MEBF(q)|}, & \text{if } i \in MEBF(q) \\ 0, & \text{otherwise} \end{cases}$$

where $MEBF(q) = \{\, j \mid min(B_j - M(q_j)),$ for $1 \leq j \leq n,\ M(q_j) > 0\}$.

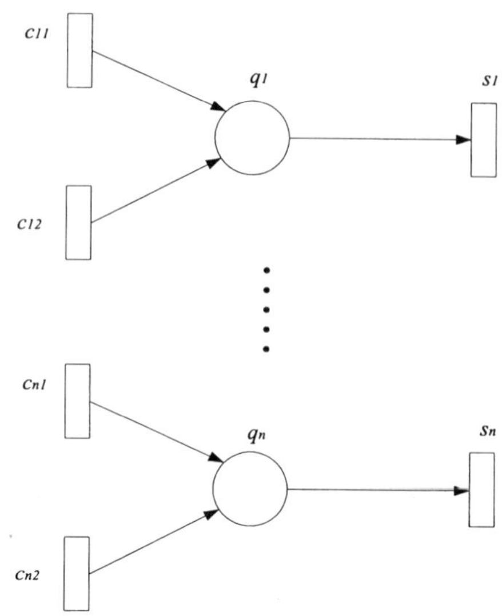

Fig. 3 The refinement of the SPN model in Fig. 2.

In these integrated schemes, the static PBS policy is accepted for cell loss control, and the value of th_{ij} is fixed in the long running time. In order to obtain higher resource utilization and smaller cell loss probability as well as a cell loss ratio guarantee, the dynamic PBS policy can be accepted and the thresholds can be dynamically varied based on the cell loss behaviors. The dynamic PBS idea has been described in our other paper [20] and is not discussed here due to the limitation of space.

To simplify the structure and to expose independence among sub-models of the SPN model in Fig. 2, we can use transition predicates to refine this model and obtain the equivalent model as shown in Fig. 3. In Fig. 3, all immediate transitions are deleted and the description of the predicate of the transition d_{ij} is moved into the predicate of the transition c_{ij}. Others have no change. The predicate of c_{ij} is $M(q_i) < th_{ij} (1 \leq th_{ij} \leq B_i)$.

There are two ways to analyze the performance of the SPN model in Fig. 3: directly solving the whole model using the corresponding Markov Chain (MC) or decomposing the model into sub-models, then approximately solving the model using iteration among sub-models.

In the first way, from the model in Fig. 3 we can construct the corresponding MC. Let us denote the marking variant b_i for the occupied capacity of the buffer queue q_i. A state (marking) is denoted by $M[b_1, \ldots, b_i, \ldots, b_n]$. Figure 4 gives the transition from any state to all reachable states and the transition conditions.

Based on the MC in Fig. 4 and the state transition rates, we can construct the state transition matrix and obtain all the steady state probabilities. We can give the performance measurements of the scheme based on the steady state probabilities. The steady state proba-

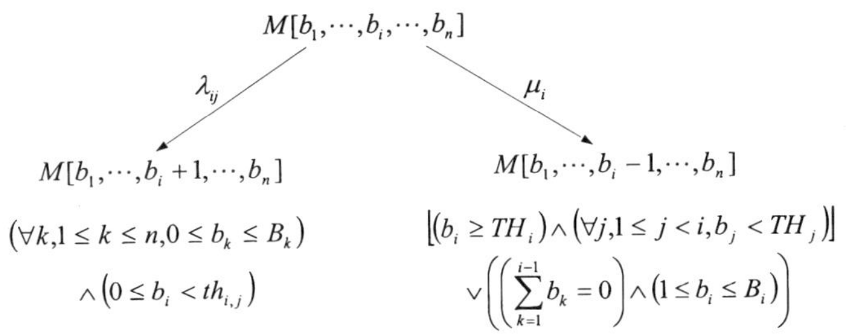

Fig. 4 The part of Markov chain of the SPN model in Fig. 3.

bility of the state M is denoted by $P[M]$.

The loss probability L_{ij} of the cells with loss priority j of user call i is given below:

$$L_{ij} = P[M(q_i) \geq th_{ij}] \tag{1}$$

In this paper, we not only consider the cell loss probabilities, but also observe the weighted loss rate for different cell loss priorities. According to Eqs. (1), the weighted loss rate WL_{ij} of the cells with loss priority j of user call i is defined by

$$WL_{ij} = w_j P[M(q_i) \geq th_{ij}] \tag{2}$$

The objective of the definition of the weighted loss rate is to provide a measurement for the overall cell loss rate in a system with multiple loss priorities. The overall loss rate WL_i of the cells of user call i is defined by

$$WL_i = \sum_j w_j P[M(q_i) \geq th_{ij}] \tag{3}$$

The different weights are selected so that the cell loss probabilities of different loss priorities for one user call can be evaluated at the same scale level. These weights correspond to the cell loss requirements of users. The overall loss rate WL of the cells of all user calls is defined by

$$WL = \sum_{i=1}^{n} WL_i \tag{4}$$

The difference in overall loss rates ($DOLR$) between user call i and user call j is defined by

$$DOLR_{ij} = |WL_i - WL_j| \tag{5}$$

In general, in (4) and (5) the weight w_j for the same loss priority is selected as the same even though they are used for the different user calls. Note that $DOLR$ and WL are not probabilities since they can be larger than 1, but they can effectively measure the fairness between user calls and the system efficiency. The scheduling scheme will be considered optimal if it can minimize $DOLR$ and WL in the system.

The throughput of the system is one of important performance measurements. The throughput $T(s_i)$ of user call i is:

$$T(s_i) = \mu_i \times \sum_{M \in E} P[M], \tag{6}$$

where E is the set of markings, under which the transition s_i is enabling, and the enabling condition is described in the above predicates of transition s_i.

Further, the throughput of the system, $T(S)$, is:

$$T(S) = \sum_{i=1}^{n} T(s_i) \tag{7}$$

When the size of the model is small, the above method can be directly accepted. But in general, the corresponding MC of the model is with n dimensions. The state space of the MC grows exponentially with the increase of B_i and n. The state space is extremely large that directly solving the state equations is impossible for practical systems.

Solving the MC with large size and multiple dimensions is a difficult problem. One possible way accepts an approximate solution based on decomposition and aggregation of the model as well as iteration among sub-models.

In general, an approximate solution follows the steps below:

(1) *Refinement*: Simplify the structure of a SPN model using transition enabling predicates and rate functions, and to exploit near-independence of the SPN model.

(2) *Decomposition*: Given an SPN model A, decompose A into sub-models A_1, A_2, ... , A_n.

(3) *Import and Export*: For each sub-model A_i, fix its import parameters from $A_j (1 \leq j \leq n)$. Also, specify its export parameters from itself after its solution.

(4) *Iteration*: Define the order for the solution of sub-models $A_j (1 \leq j \leq n)$. If A_i and A_j are in a cycle import relation, one of them must be chosen to be calculated first. In this case imports are not available and an initial guess must be provided for imports. After A_j is computed, the most recent values for the imports are used for subsequent iteration. Convergence is reached when the relative change between successive iterations in all imports is below a given tolerance.

In our scheme, the SPN model in Fig. 3 has been refined, each independent part in structure is a sub-model A_i $(1 \leq i \leq n)$. The import and export relations among sub-models have been described in the predicates of the transition s_i $(1 \leq i \leq n)$. Whether the transition s_i is enabling or not, not only depends on the marking of its queue q_i, but also depends on the markings of other queues. Note that in the SPN model, the markings are not actual import and export parameters since they are time functions. The import and export parameters among sub-models should be the probabilities of the markings in steady state. The import parameters of the sub-model A_i are the occupancy distribution probability of the queue $q_j (1 \le j < i)$ in steady state. For the sub-model $A_i (2 \le i \le n)$, the enabling probability X of the transition s_i, which depends on other sub-models, for schemes (1) and (2) is described as below:

$$X = \begin{cases} \left(\prod_{j=1}^{i-1} P[M(q_j)=0]\right) \times \left(\prod_{j=i+1}^{n} P[M(q_j) < TH_j]\right) & \text{if } M(q_i) < TH_i \\ \prod_{j=1}^{i-1} P[M(q_j) < TH_j] & \text{if } M(q_i) \ge TH_i. \end{cases} \tag{8}$$

For the sub-model A_1, the enabling probability X of the transition s_1 is

$$X = \begin{cases} \prod_{j=2}^{n} P[M(q_j) < TH_j], & \text{if } M(q_i) < TH_i \\ 1.0, & \text{if } M(q_i) \ge TH_i. \end{cases} \tag{9}$$

Let us represent $B_{ji} = B_j - B_i$. For scheme (3), in the sub-model $A_i (2 \le i \le n)$ the enabling probability X of the transition s_i is

$$X = \left(\prod_{j=1}^{i-1} P[M(q_j) < t - B_{ij}] + \prod_{j=1}^{i-1} P[M(q_j) = 0]\right) \times \prod_{j=i+1}^{n} P[M(q_j) \le B_{ji} + t], \tag{10}$$

where $M(q_i) = t$.

For the sub-model A_1, the enabling probability X of the transition s_1 is

$$X = \prod_{j=2}^{n} P[M(q_j) \le t + B_{j1}], \tag{11}$$

where $M(q_1) = t$.

For scheme (4), the scheduling probabilities can be represented in the sub-models. For $A_i (1 \le i \le n)$ the enabling probability X of the transition s_i is described as below:

$$\begin{aligned} X = {} & \prod_{j=1}^{n} P[M(q_j) < t + B_{ji}] + \\ & \vdots \\ & + \frac{1}{|S|+1} \left(\prod_{j \in S} P[M(q_j) = B_{ji} + t] \times \prod_{j \notin S} P[M(q_j) < B_{ji} + t] \right) \\ & \vdots \\ & + \frac{1}{n} \prod_{j=1}^{n} P[M(q_j) = B_{ji} + t] \end{aligned} \tag{12}$$

where $i \ne j$, $M(q_i) = t$; and for $1 \le j < i$, if $M(q_j) = 0$ then $M(q_j) < B_{ji} + t$.

Equations (8)–(12) for different schemes only describe the influence of the enabling probability X of the transition s_i from other sub-models. The competition for one server among sub-models will also result in decreasing the firing rates of the transition s_i. In the performance equivalent analysis, the firing time of transition s_i, t_i, can be obtained through the following derivation.

$$\begin{aligned} t_i = {} & \text{service time of } s_i \text{ under condition } X \\ & + \text{waiting time for other submodels service under} X; \end{aligned}$$

$$\text{service time of } s_i \text{ under condition } X = \frac{1}{X\mu_i};$$

$$\begin{aligned} T_w & = \text{waiting time for other submodels service under } X \\ & = X \times \text{ other sub-model busy probability} \times \text{ mean residual time}; \end{aligned}$$

$$\begin{aligned} & \text{other submodel busy probability} \\ & = \frac{\sum_{j \ne i} T(s_j)}{\text{server service rate}} = \frac{\sum_{j \ne i} T(s_j)}{\mu_i}; \end{aligned}$$

$$\begin{aligned} & \text{mean residual time} \\ & = \text{the complete service time according to the renewal theory} \\ & = \frac{1}{\mu_i}. \end{aligned}$$

Finally, we can get

$$t_i = \frac{1}{X\mu_i} + X\frac{\sum_{j \ne i} T(s_j)}{\mu_i} \times \frac{1}{\mu_i} = \frac{\mu_i + X^2 \sum_{j \ne i} T(s_j)}{X\mu_i^2} \tag{13}$$

During the iteration, the solution sequence of sub-models is ordered in $A_1, A_2, \ldots, A_n$. In the initial iteration, sub-model A_1 is solved first and all values of

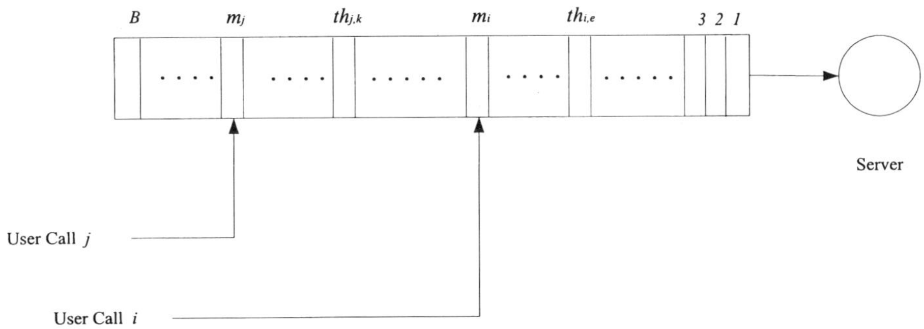

Fig. 5 The one queue scheduling model.

probabilities from other sub-models can be set as equal to or less than 1, but larger than zero.

The performance measurements of the sub-models are computed according to Eqs. (1)–(7), but the firing rate of s_i is determined by Eqs. (8)–(13) for different schemes and depends on the probabilities of the markings of sub-models in steady state.

In the new schemes, the resequencing operations are not needed in cell loss control so that it is simple for practical implementation while being sensitive to the delay constraints of process. The buffer queues can be dynamically allocated according to the requirements from user calls. The buffers are not shared among user calls with different real-time requirements, and sometimes the utilization of buffers is relative low, but this is not very important in today's hardware environment.

4. MEBF Scheduling Scheme with One Physical Input Queue

In this section, MEBF scheme with one queue is presented based on the same considerations described in above sections. In this scheme, we use a buffer queue to receive cells from all user calls so that buffers can be shared by all users even though they have different real-time requirements. At the same time, the scheme can achieve real-time scheduling and loss control as well as maintain cell sequence integrity.

In this new scheme, only one buffer queue is used to receive arriving cells from all user calls, but there are different entries in the queue for user calls with different real-time requirements as shown in Fig. 5. The user calls with the same real-time requirements have the same entry in the queue. In general, the cells that belong to the same user call have the same maximum delay time in buffer queue. In our scheme, the entrances of the buffer queue are set according to the real-time requirements, and the cells from the same user call are put into the queue as arriving sequence so that they satisfy FIFO service. PBS policy is accepted for cell loss control, and the thresholds in the queue are set for the cells with different loss priorities. The overall buffer capacity of the queue is B and the queue is numbered from head to tail as shown in Fig. 5. In fact, this queue is constructed by combination of n queues in Fig. 1 in the priority order, i.e., queue q_1 is shared by n user calls, and q_2 is shared by $n-1$ user calls and so on.

The maximum waiting time that an arriving cell may spend in the queue defines a buffer position that is the entry position for it. In order to achieve high throughput and reduce difference in cell loss probabilities, intuitively each arriving cell should be delayed as close to its deadline as possible (see the definition of m_i below). In addition to the following procedure used to admit an arbitrary cell into the queue, the following work-conserving rule is used to serve cell scheduling.

♦ *Work-conserving rule*:

> When it happens that the next-to-be-served position is empty immediately after the server finishes a service, the cell currently occupying the smallest buffer position will be served next, while the cells behind it in the queue move forward (i.e. closer to the head of the queue) by only one buffer position.

The work-conserving rule assures that the server never idles as long as there exist at least one cell waiting in the queue.

In the model of Fig. 5, the following parameters are used in scheduling and loss control:

m_i: the entry of buffer position for user call i. In order to simplify the computation, we can assume the service time is 1. If the cells of user call i have the maximum delay time in the queue as d, then $m_i = \lfloor d-1 \rfloor$, $m_i = B_i$ in the above scheme. If an arriving cell of user call i can only be put into the buffer position behind m_i, the cell is said to be useless and should be discarded since its deadline is violated.

th_{ij}: the threshold of the cell loss priority j for user call i. When a cell with the loss priority j of user call i arrives, if the buffer occupancy from the head to the cell entry position m_i is smaller than th_{ij}, the arriving cell will be put into the queue position m_i, otherwise it will be discarded. The greater the loss priority j is, the greater th_{ij} and the smaller the loss probability of the cells with the loss priority j will be. For $\forall j, 1 \leq th_{ij} \leq m_i$.

In this scheme, once the cell is admitted into the queue, it is never subject to discard. There is no selection operation needed for cell loss or resequencing for FIFO so that the implementation complexity is reduced.

In order to clearly describe our scheme, an algorithm will be given below. The following parameters will be used in the algorithm.

- $q(k)$ = content of the kth buffer position, i.e., if the kth buffer position contains cell c, then $q(k) = c$; and if the kth buffer position is empty, $q(k) = 0$.
- C_i = the buffer occupancy from the head to the entry buffer position for user call i. Whenever a cell is served, the relevant $C_i (1 \leq i \leq n)$ is decreased by 1. In the initial state, $C_i = 0$.

When an arriving cell c with loss priority j of user call i enters a network node, the algorithm 1 is applied.

Algorithm 1:

```
IF (th_ij > C_i) /* Decide whether to dropped cell c*/
THEN {
    k = m_i; x = c;
    WHILE ( q(k) ≠ 0 )
    { /* Move forward the cells in FIFO order */
        y = q(k); q(k) = x;
        x = y; k = k − 1;
    }
    q(k) = x; h = n;
    WHILE ( h > 0 AND k ≤ m_h )
    { /* update C_* /
        C_h = C_h + 1; h = h − 1;
    }
}
ELSE cell c is dropped.
END.
```

In this algorithm, only one type of operations is performed—moving forward the cells according to FIFO policy. The number of moving operations is less than or equal to m_i. Thus the number of operations in algorithm 1 is not more than B, and in the worst case the complexity of algorithm 1 is O(B).

5. SPN Modeling and Analysis for the One Queue Scheme

There are two ways to model this scheme using SPN: modeling the whole queue directly; or decomposing the whole queue into sub-queues (the entry buffer positions for user calls are the cutting points) and then modeling this constructed system. The first way leads to a complicated model and solving such a model is rather difficult. The second way will expose near-independence among sub-models and the analysis of sub-models is relative easy.

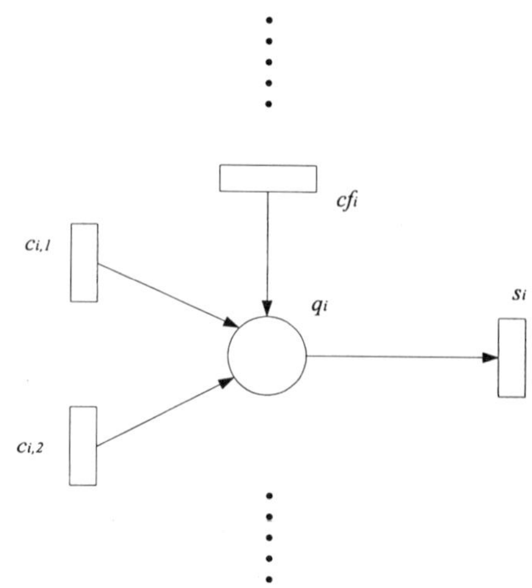

Fig. 6 The SPN model of one queue scheme.

Figure 6 shows the SPN model of our scheme, in which the second way is accepted. We can assume that there are n scheduling priorities for user calls, i.e., there are n entry buffer positions. The queue is decomposed into n sub-queues, and the size of the sub-queue i, is $B_i\,(1 \leq i \leq n)$, so that $B = \sum_i B_i$. The user call 1 has the highest scheduling priority. The user call n has the lowest scheduling priority. The scheduling priority of user call i is higher than that of user call $i+1$. In this model, the first subscript of the character represents the scheduling priority, while the second subscript of the character represents the loss priority. If the character has only one subscript, it represents the scheduling priority.

For simplicity, only the sub-model of user call i is given in Fig. 6. Sub-models for other user calls can be represented in the same way. For the same reason, the loss priorities have only two classes represented in this model. In each user call, the loss priorities may have m classes, $m \geq 1$.

The meaning of the transitions and places in Fig. 6 are described in the following:

s_i represents the server of the sub-queue i, whose service rate is μ_i. In this model, $\mu_i\,(1 \leq i \leq n)$ has the same value since there is only one server in the system.

cf_i represents the cell movement from the sub-queue $j (i < j \leq n)$ into the sub-queue i. Its firing rate corresponds to the cell arriving rate in the sub-queue $j (i < j \leq n)$.

c_{ij} represents the generation of cells with loss priority j of user call i. Its firing rate is λ_{ij}. In user call i, the generation proportion of the cells with loss priority j is $\lambda_{ij}/\sum_j \lambda_{ij}$.

q_i represents sub-queue i, and its capacity is B_i.

The predicates associated with the transitions in

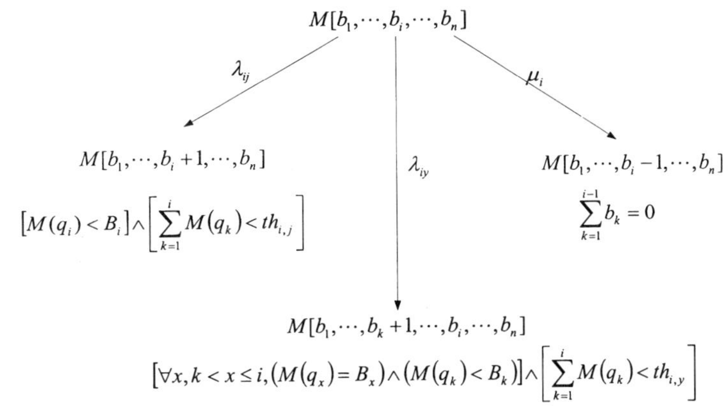

Fig. 7 The part of Markov chain of the SPN model in Fig. 6.

Fig. 6 are described in the following:

- The predicate of s_i is $\sum_{k=1}^{i-1} M(q_k) = 0$, which expresses the work-conserving rule.
- The predicate of c_{ij} is

$$[M(q_i) < B_i] \wedge \left[\sum_{k=1}^{i} M(q_k) < th_{ij}\right].$$

The first term expresses that an arriving cell is input when the sub-queue i is not full. The second term expresses that threshold is used for the cell loss control. Note that $\sum_{k=1}^{i} M(q_k)$ corresponds to the term C_i in Algorithm 1.

- The predicate of cf_i is

$$\{M(q_i) < B_i\} \wedge \left\{ \begin{array}{l} \left[\begin{array}{l}(M(q_{i+1}) = B_{i+1}) \\ \wedge \left(\sum\limits_{k=1}^{i+1} M(q_k) < th_{i+1,p}\right)\end{array}\right] \\ \vee \left[\begin{array}{l}(M(q_{i+2}) = B_{i+2}) \\ \wedge (M(q_{i+1}) = B_{i+1}) \\ \wedge \left(\sum\limits_{k=1}^{i+2} M(q_k) < th_{i+2,p}\right)\end{array}\right] \\ \vee \left[\begin{array}{l}\left(\sum\limits_{k=i+1}^{n} M(q_k) = \sum\limits_{k=i+1}^{n} B_k\right) \\ \wedge \left(\sum\limits_{k=1}^{n} M(q_k) < th_{np}\right)\end{array}\right] \end{array} \right\}$$

where p represents the arriving cell loss priority. There are two expressions in the predicate above. The first represents that there is an empty buffer in sub-queue i. Each term in the second expression represents the condition in which a cell from sub-queue $k(i < k \leq n)$ will be moved into sub-queue i. This is equivalent to that the transition $c_{kp}(i < k \leq n)$ puts a cell into sub-queue i.

In this model, the moving operation and scanning operation in the queues are not directly represented, but the work-conserving rule is represented by n server transitions. The marking function instead of buffer occupancy appears in the predicates of transitions.

From the model in Fig. 6 we can construct the corresponding MC and obtain all the steady state probabilities as we have done before. Figure 7 gives the transition from any state to all reachable states and the transition conditions.

In the steady state, the loss probability of user call i with loss priority j, L_{ij}, is

$$L_{ij} = P\left[\sum_{k=1}^{i} M(q_k) \geq th_{ij}\right] \tag{14}$$

The weighted loss rates, the overall loss rates and the difference in overall loss rates can be obtained from Eqs. (2)–(5). The throughput of transition s_i is

$$T(s_i) = \mu_i \times P\left[\left(\sum_{k=1}^{i-1} M(q_k) = 0\right) \wedge (M(q_i) > 0)\right] \tag{15}$$

The approximate solution is accepted for the model in Fig. 6 based on decomposition and aggregation of the model as well as iteration among sub-models. The SPN model in Fig. 6 has been refined; each independent part in structure is a sub-model $A_i (1 \leq i \leq n)$. The import and export relations among sub-models have been described in the predicates of the transitions c_{ij}, cf_i and s_i $(1 \leq i \leq n)$. Whether these transitions are enabling or not, not only depends on the marking of its sub-queue q_i, but also depends on the markings of other sub-queues. The import parameters of sub-model A_i are the occupancy distribution probabilities of sub-model A_j $(1 \leq j \leq i-1)$ in the steady state.

For sub-model A_i $(1 \leq i \leq n)$, the enabling probability X of the transition s_i is

$$X = \prod_{k=1}^{i-1} P[M(q_k) = 0] \tag{16}$$

Table 1 The analysis comparison for the various scheduling schemes.

schemes	parameters	exact solution	approximate solution	relative error %
HOL	*DOLR*	0.999	0.998	0.1%
	throughput	0.9674	0.9424	2.6%
QLT	*DOLR*	0.918	0.902	1.7%
	throughput	0.9780	0.9450	3.4%
MEBF with priority	*DOLR*	0.535	0.491	8.2%
	throughput	0.9791	0.9844	0.5%
MEBF without priority	*DOLR*	0.192	0.203	5.4%
	throughput	0.9778	0.9465	3.2%

The firing rate of the transition s_i can be obtained through Eq. (13) based on the same considerations. There is no predicate for s_i in sub-model A_i.

The predicate of c_{ij} is modified into $[M(q_i) < B_i]$ in sub-model A_i. The firing rate of the transition c_{ij} depends on the marking of q_i and is given by:

$$\lambda_{ij} \times P\left[\sum_{k=1}^{i-1} M(q_k) + M(q_i) < th_{ij}\right] \tag{17}$$

The predicate of cf_{ij} is modified into $[M(q_i) < B_i]$ in sub-model A_i. The firing rate of the transition cf_{ij} depends on the marking of q_i and is given by:

$$\begin{pmatrix} \lambda_{i+1,j} \times P\left[M(q_{i+1}) = B_{i+1}\right] \\ \times P\left[\sum_{k=1}^{i+1} M(q_k) < th(i+1,j)\right] \end{pmatrix} + \vdots + \begin{pmatrix} \lambda_{n,j} \quad \times \prod_{k=i+1}^{n} P\left[M(q_k) = B_k\right] \\ \times P\left[\sum_{k=1}^{n} M(q_k) < th(n,j)\right] \end{pmatrix} \tag{18}$$

Equations (13), (16), (17) and (18) describe the influence of the import parameters among sub-models. During the iteration, the solution sequence of sub-models can be ordered in $A_1, A_2, \ldots, A_n$. In the initial iteration, sub-model A_1 is solved first and all probabilities may be set as equal to or less than 1, but larger than zero.

The performance measurements of the sub-models are computed according to Eqs. (14) and (15), but the transition firing rates are determined by Eqs. (13), (16), (17) and (18) and depend on the markings of sub-models.

6. Numerical Examples

We used the software package SPNP (Stochastic Petri Net Package) developed by Trivedi etc. [21] for numeric solutions to our models. We first compare the approximation results with the exact models and discuss the complexity and relative errors. Then we present several sets of experiments where various performance measurements such as difference in overall loss rates (*DOLR*), throughputs and overall loss rate were calculated. We have also studied the impact of several system parameters including traffic load, scheduling thresholds and loss control thresholds on the system performance.

6.1 Validations

6.1.1 The Analysis Comparison for Various Scheduling Schemes with Individual Queues

The first example was based on the SPN model in Fig. 3 for various scheduling policies with two queues (two scheduling priorities) and one input transition (one loss priority) in each user call. The parameter values are given below: the cell arrival rates are $\lambda_{11} = 0.45$ and $\lambda_{21} = 0.55$; the buffer sizes are $B_1 = 20$ and $B_2 = 28$; $k_2 = B_2 - B_1 = 8$ and $\mu_1 = \mu_2 = 1.0$. The comparison between the exact models and decomposition sub-models for difference in overall loss rates (*DOLR*) and throughputs is shown in Table 1.

The state number can be regarded as the number of states in the corresponding Markov process, thus reflecting the complexity of the model. For various scheduling schemes, the state number in the exact solution is 609, but the state number in the approximate solution is 21 for sub-model A_1 and 29 for A_2 separately. The decomposition method significantly reduces the state number as well as system complexity. In QLT scheme, TH_1 and TH_2 are set to 16 and 18 separately. From Table 1, we can see that the MEBF without priority scheme have the best performance in *DOLR*. Additionally the results in Table 1 show that our approximate solution is relatively accurate.

6.1.2 Three Examples of Two Loss Priorities in Each User Call for MEBF without Priority Scheme

The three examples are based on the SPN model in Fig. 3 for the MEBF without priority scheme. In the examples, there are two scheduling priorities for two user calls and two loss priorities in each user call. The cell input rates are set as following: $\lambda_{11} = 0.2$, $\lambda_{12} = 0.25$, $\lambda_{21} = 0.25$ and $\lambda_{22} = 0.3$; the cell loss thresholds are set to $th_{11} = 4$, $th_{12} = B_1 = 5$, $th_{21} = 6$ and $th_{22} = B_2 = 7$ in the first example. In the second example the threshold values and buffer sizes are two times those of the

Table 2 Comparison for *DOLR* in the scheme of MEBF without priority.

Weights of each example	parameters	Exact solution	Approximate solution	relative error %
Example 1	State number	48	14	
$w_1 = 10^0$	*DOLR*	0.195	0.185	5.1%
$w_2 = 10^1$	throughput	0.9014	0.8808	2.3%
Example 2	state number	165	26	
$w_1 = 10^0$	*DOLR*	0.283	0.294	3.6%
$w_2 = 10^2$	Throughput	0.9494	0.9258	2.5%
Example 3	state number	609	50	
$w_1 = 10^0$	*DOLR*	0.372	0.431	13.7%
$w_2 = 10^3$	throughput	0.9740	0.9485	2.6%

Table 3 Comparison for the *DOLR* in the one-queue scheme.

Examples	parameters	exact solution	approximate solution	relative error %
Example 1	*DOLR*	0.511	0.501	2%
$th_{11} = 24$	throughput	0.9772	0.9599	1.8%
Example 2	*DOLR*	0.407	0.415	1.9%
$th_{11} = 22$	Throughput	0.9769	0.9591	1.8%
Example 3	*DOLR*	0.266	0.278	4.3%
$th_{11} = 21$	throughput	0.9767	0.9586	1.9%

first example, and in the third example the threshold values and buffer sizes are two times those of the second example; $\mu_1 = \mu_2 = 1.0$. The comparison between the exact models and decomposition sub-models is shown in Table 2.

The different weights are selected so that the cell loss probabilities of different loss priorities in one user call can be evaluated at the same scale level. In a practical system, the ratio of weights corresponds to the ratio of cell loss probabilities for different loss priorities in user requirements. For the exact and approximate solution, we use the same weights. The results from Table 2 clearly show that *DOLR* is mainly affected by the cell loss thresholds in one user call, and the throughputs may be changed with buffer sizes.

6.1.3 Three Examples for One Queue Scheme

The three examples are based on the SPN model in Fig. 6 for the one queue scheme. In the examples, there are two scheduling priorities for two user calls and two loss priorities in each user call. The cell input rates are set as following: $\lambda_{11} = 0.192$, $\lambda_{12} = 0.288$, $\lambda_{21} = 0.208$ and $\lambda_{22} = 0.312$; the cell loss thresholds are set to $th_{12} = B_1 = 30$, $th_{21} = 40$, $th_{22} = 50$ and $B_2 = 20$ in all three examples, but th_{11} is set to different values in different examples. The weights for different loss priorities are set to $w_1 = 10$ and $w_2 = 1000$ for three examples; $\mu_1 = \mu_2 = 1.0$. The comparison between the exact models and decomposition sub-models is shown in Table 3.

For the three examples, the state number in the exact solution is 651, but the state numbers in the approximate solution are 31 for sub-model A_1 and 21 for A_2 respectively. The decomposition method significantly reduces the state number as well as system complexity. The results from Table 3 show that when the threshold th_{11} changes, *DOLR* changes quickly, but the throughput does not differ much. For $th_{22} - B_2 = th_{12}$, when the difference between $th_{21} - B_2$ and th_{11} is smaller, *DOLR* is smaller.

We have validated many examples, but only some were shown here. These examples show quantitatively that our approximate analysis technique is efficient in reducing the computation complexity while keeping the accuracy of the numerical results.

6.2 Numerical Results

6.2.1 The First Example of *DOLR*s versus Input Rates

In this example, the difference in cell loss rates for different user connections was compared between MEBF policy and HOL or QLT policies. In the schemes, there are two scheduling priorities for two user calls ($n = 2$) and one loss priority in each user call. The parameter values are given below: the ratio between cell arrival rates is $\lambda_{21}/\lambda_{11} = 2/3$; the cell loss thresholds are $th_{11} = B_1$, $th_{21} = B_2$; the buffer sizes are $B_1 = 20$ and $B_2 = 40$; $k_2 = B_2 - B_1 = 20$; $\mu_1 = \mu_2 = 1.0$. The scheduling thresholds are $TH_1 = TH_2 = 1$ for HOL policy, and $TH_1 = 16$ and $TH_2 = 18$ for QLT policy. The comparison among the different policies for *DOLR* is shown in Fig. 8.

If the ratio of λ_{21} to λ_{11} is changed to $\lambda_{21}/\lambda_{11} = 31/29$ and other parameter values are not changed, we can obtain the new results in Fig. 9.

Figure 8 and Fig. 9 clearly show that in most cases the MEBF schemes have better performance in *DOLR*, specifically in high traffic load. In light traffic load, the queuing is not completely formed so that the schemes based on the queue length cannot completely exhibit their advantages. In QLT, the thresholds TH_* are

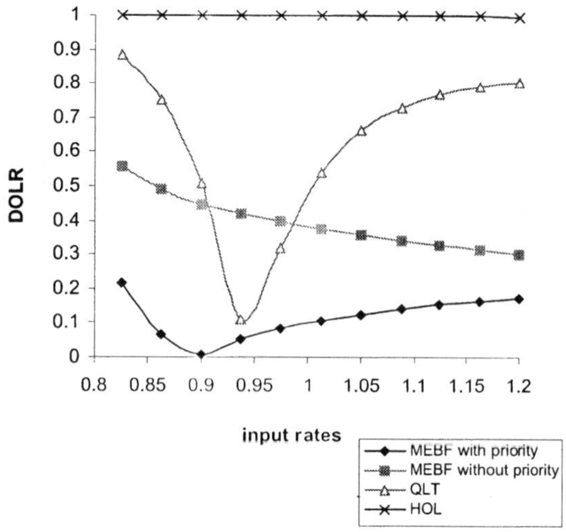

Fig. 8 Comparison among the different policies with one loss priority and $\lambda_{21}/\lambda_{11} = 2/3$.

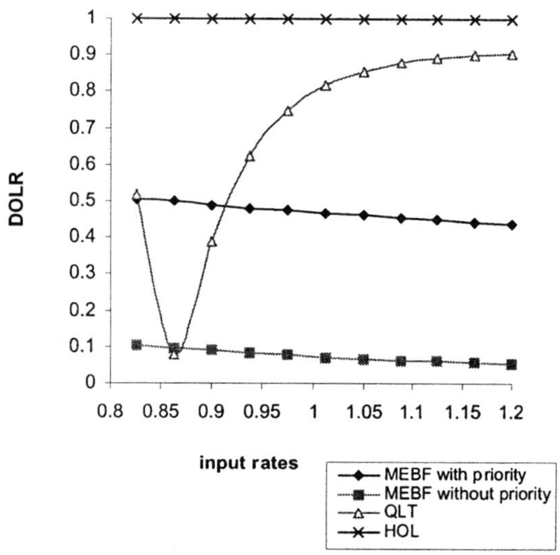

Fig. 9 Comparison among the different policies with one loss priority and $\lambda_{21}/\lambda_{11} = 31/29$.

static so that in only a few cases QLT performs better, but in MEBF the thresholds TH_* are dynamic so that in most cases MEBF can achieve better performance. Moreover, we can use the different version of MEBF schemes according to user input rates to obtain better performance. When the users with high scheduling priorities have a larger ratio of input rates, the MEBF with priority scheme can be considered to apply, since they have better performance in $DOLR$ than the schemes without priority in these cases (see Fig. 8). In other cases, the MEBF without priority scheme can be applied, since they perform better, as shown in Fig. 9. The user input rates may come from the traffic profiles that are statically set or dynamically negotiated between users and the network. The rate information may also come from the on-line measurement or prediction of users traffic.

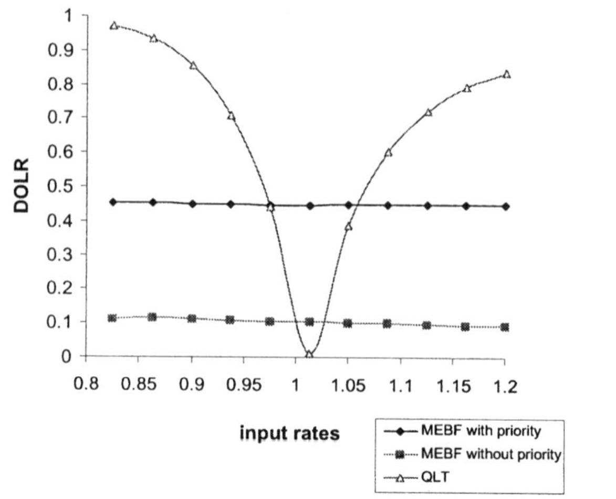

Fig. 10 Comparison among the different policies with two loss priorities.

The curves for QLT and MEBF with priority in Fig. 8 are not monotonous, nor are the QLT curves in Figs. 9 and 10. This can be explained by the definition of $DOLR$. According to Eq. (5), $DOLR$ is the absolute value of the difference in overall loss rate between two user calls. Assume the difference in loss rate between two user calls is monotonously ascending or descending with the increase of input rates (i.e. the difference is a monotonous function of input rates), and the curve crosses the zero axis. Then after we get the absolute value of the function, the curve will become non-monotone, and its lowest point should be on the zero axis, similarly to the QLT curves in the figures.

6.2.2 The Second Example of Difference in Overall Loss Rates ($DOLR$) versus Input Rates

In this example, the difference in overall loss rates for two user connections was compared between MEBF and QLT policies. In the example, there are two scheduling priorities for two user calls and two loss priorities in each user connection. The parameter values are given below: the ratio between cell arrival rates are $(\lambda_{21} + \lambda_{22})/(\lambda_{11} + \lambda_{12}) = 31/29$ and $\lambda_{21}/\lambda_{22} = \lambda_{11}/\lambda_{12} = 2/3$; the weights for the different loss priorities are set to $w_1 = 1$ and $w_2 = 100$; the cell loss thresholds are $th_{11} = 16$, $th_{12} = B_1$, $th_{21} = 36$ and $th_{22} = B_2$; the buffer sizes are $B_1 = 20$ and $B_2 = 40$; $k_2 = B_2 - B_1 = 20$; $\mu_1 = \mu_2 = 1.0$. The scheduling thresholds are $TH_1 = 16$ and $TH_2 = 18$ for QLT policy. The comparison result for $DOLR$s is shown in Fig. 10.

The results from Fig. 10 show the same conclusion from Fig. 8 and Fig. 9, and in most cases MEBF schemes have better performance in $DOLR$ than the QLT scheme even though overall loss rates instead of cell loss probabilities are compared.

If three loss priorities are set in each user call, we can get the results illustrated in Fig. 11 and Fig. 12.

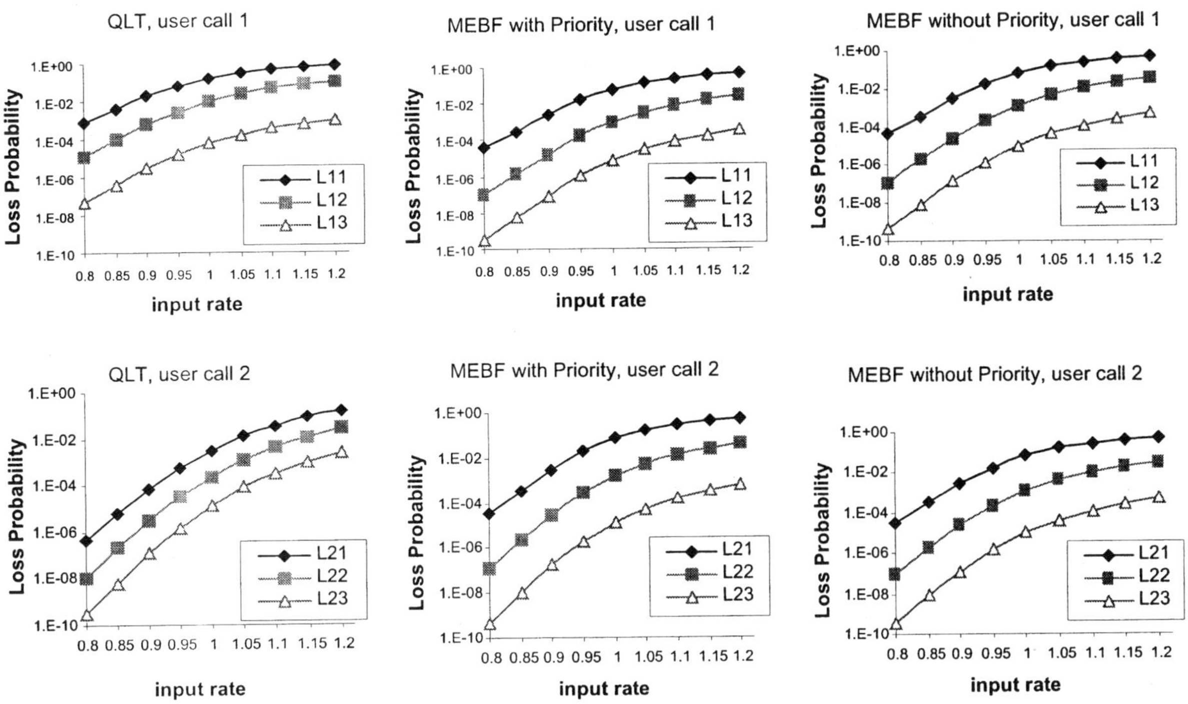

Fig. 11 Loss probabilities of the cells in two user calls with three loss priorities under the three scheduling policies.

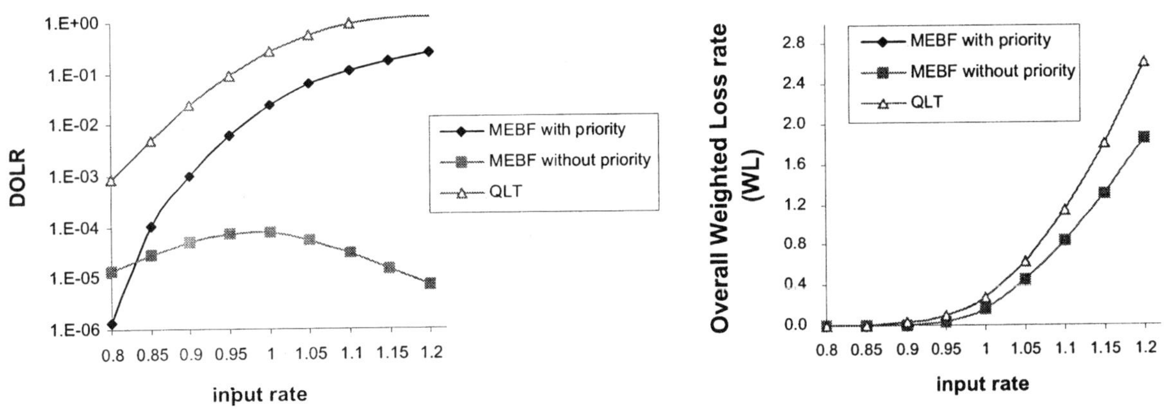

Fig. 12 Comparisons of $DOLR$ and WL among the three policies for the scenario with three loss priorities in two user calls.

The parameters are as following: the ratio between cell arrival rates are $(\lambda_{11}+\lambda_{12}+\lambda_{13})/(\lambda_{21}+\lambda_{22}+\lambda_{23}) = 1$ and $\lambda_{11} : \lambda_{12} : \lambda_{13} = \lambda_{21} : \lambda_{22} : \lambda_{23} = 3 : 3 : 4$; the weights for the different loss priorities are set to $w_1 = 1$, $w_2 = 10$ and $w_3 = 100$; the cell loss thresholds are $th_{11} = 12$, $th_{12} = 16$, $th_{13} = B_1$, $th_{21} = 32$, $th_{21} = 36$ and $th_{22} = B_2$; the buffer sizes are $B_1 = 20$ and $B_2 = 40$; $\mu_1 = \mu_2 = 1.0$; The scheduling thresholds are $TH_1 = 16$ and $TH_2 = 18$ for the QLT policy.

From Fig. 11, we can see that the loss probabilities of the three loss priorities in each user call are differentiated distinctly under all three policies, and the ratio of the loss probabilities between each pair of adjacent loss priorities in each user call is usually between 10 and 100. That means the control of loss priority in each user call is effective.

Figure 12 compares the three policies in two metrics: overall weighted loss rates (WL) and difference in overall loss rates ($DOLR$), which are defined separately in Eqs. (4) and (5). Note that the two metrics are not probabilities since they can be bigger than 1, but they can effectively measure the whole system efficiency and the fairness between user calls. From the comparison of $DOLR$s, we can see that MEBFs are better than QLT in fairness, and the policy of MEBF without priority is extremely good in this scenario. The comparison of WLs shows that MEBFs are also better than QLT in the whole system efficiency. It is valuable to especially point out that WLs of the two MEFB policies are almost the same under all the input rates, so that it is hard or impossible to distinguish their curves in the figure.

6.2.3 Comparison between the Individual-Queues Scheme and the Single-Queue Scheme

The MEBF without priority scheme and one queue scheme are compared and analyzed separately based on the SPN models in Fig. 3 and Fig. 6. In this example, there are two scheduling priorities for two user calls and two loss priorities in each user call. The parameter values are given below: the ratio between cell arrival rates are $(\lambda_{21} + \lambda_{22})/(\lambda_{11} + \lambda_{12}) = 31/29$ and $\lambda_{21}/\lambda_{22} = \lambda_{11}/\lambda_{12} = 2/3$; the weights for the different loss priorities are set to $w_1 = 1$ and $w_2 = 100$; the cell loss thresholds are $th_{11} = 16$, $th_{12} = B_1$, $th_{21} = 36$ and $th_{22} = B_2$; the buffer sizes are $B_1 = 20$ and $B_2 = 40$ for MEBF scheme, and $B_2 = 20$ for one queue scheme; $k_2 = B_2 - B_1 = 20$ for MEBF scheme; $\mu_1 = \mu_2 = 1.0$. The comparison among the different policies for $DOLR$s is shown in Fig. 13.

Figure 13 shows that one queue scheme has better performance in $DOLR$ than MEBF and its $DOLR$ is almost at zero when $th_{2,i} - B_2 = th_{1,i}$ for $i = 1, 2$. In one queue scheme, cell loss rates of user call 1 is larger than that of user call 2 since cells of user call 2 will be dropped when sub-queue q_2 is full and the buffer occupancy in sub-queue q_1 is equal to or greater than $th_{1,i}$. The possibility of cell loss rates of user call 1 being larger than that of user call 2 is very small, since the probability which sub-queue q_2 is not full and at the same time buffer occupancy in sub-queue q_1 is equal to or greater than $th_{1,i}$ is very small. One queue scheme also has better performance in buffer utilization than that of MEBF since queue q_1 for MEBF is saved, but the cost is expensive in the overall loss rates. The cell loss rates of user calls with lower scheduling priorities increases, which is not expected, since their buffers are shared by the user calls with higher scheduling priorities. Figure 14 shows the above conclusion, and the overall loss rates for all user connections are compared in the same example.

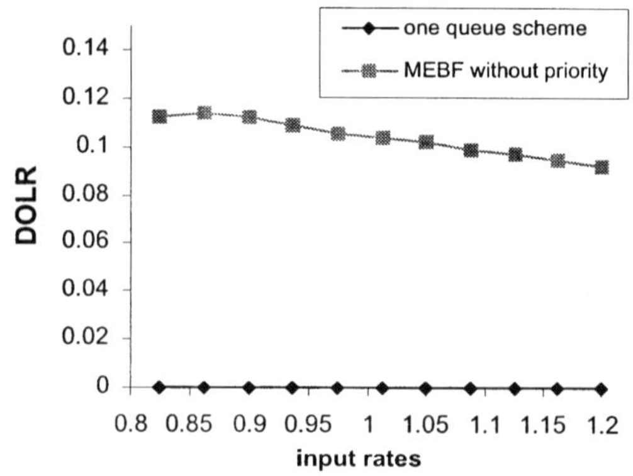

Fig. 13 Comparison of $DOLR$s for MEBF without priority and one queue scheme.

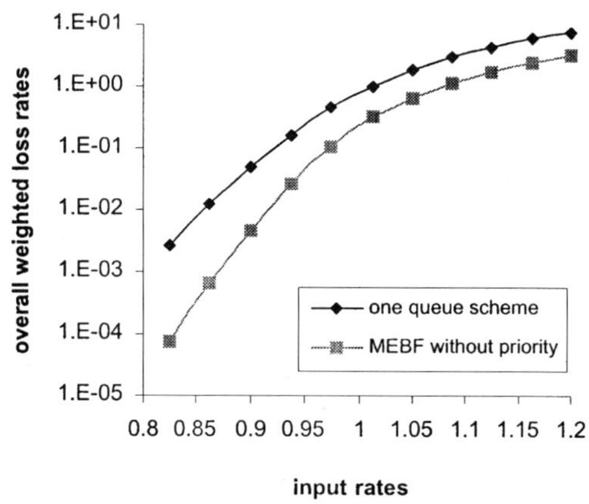

Fig. 14 Comparison of overall loss rates for MEBF without priority and one queue scheme.

7. Conclusion

In this paper, we have proposed two types of integrated schemes of real-time scheduling and cell loss control. The scheme with individual input queues is a versatile scheduling policy, which may represent various different real-time scheduling schemes. Among these schemes, MEBF policy is a dynamic real-time requirements balancing strategy, which balances the performance of all user connections according to their real-time requirements. The scheme with one physical input queue improves the buffer utilization of individual queue schemes, but increases the overall loss probability. The proposed schemes can provide guaranteed QoS for multiple priorities, maintain cell sequence integrity and are simple to implement in a software package or in a single chip. The numerical results show that our schemes significantly outperform some well-known schemes such as the head-of-line (HOL) priority control and the queue length threshold (QLT) policy. Moreover, our approximate analysis technique significantly reduces the complexity of analysis and solving to these schemes based on stochastic Petri net models and can be applied to various complex systems analyses.

Acknowledgement

We are really grateful to Yin Li, Dr. Kenneth Fields, and Lizhi Sheng for their help to improve the presentation of this paper. We would also like to thank the Associate Editor, Dr. Kenji Nakagawa, and the anonymous reviewers for their helpful suggestions.

References

[1] J.J. Bae and T. Suda, "Survey of traffic control schemes and protocols in ATM networks," Proc. IEEE, vol.7, no.2,

pp.170–189, 1991.

[2] F. Chiussi and A. Francini, "Advances in implementing fair queueing schedulers in broadband networks," 1999 IEEE International Conference on Communications (ICC99), Vancouver, British Columbia, Canada, June 1999.

[3] Y. Chen and S. Lee, "Bounded tag fair queueing for broadband packet switching networks," Computer Communications, V.23, I.1, pp.45–61, Jan. 2000.

[4] C. Dovrolis, D. Stiliadis, and P. Ramanathan, "Proportional differentiated services: Delay differentiation and packet scheduling," Proc. 1999 ACM SIGCOMM conference, Cambridge MA, Sept. 1999.

[5] M. Andrews, "Probabilistic end-to-end delay bounds for earliest deadline first scheduling," Proc. IEEE INFOCOM'2000, Tel Aviv, Israel, March 2000.

[6] V. Sivaraman and F. Chiussi, "Providing end-to-end statistical delay guarantees with earliest deadline first scheduling and per-hop traffic shaping," Proc. IEEE INFOCOM'2000, Tel Aviv, Israel, March 2000.

[7] H. Sariowan, R. Cruz, and G. Polyzos, "SCED: A generalized sceduling policy for guaranteeing quality-of-service," IEEE/ACM Trans. Networking, vol.7, no.5, pp.669–683, Oct. 1999.

[8] I. Stavrakakis and G. Chen, "Study of a scheduling policy for diverse deadline-based quality of service," Performance Evaluation, vol.4, no.1, pp.37–62, May 2000.

[9] R. Chiopalkatti, J. Kurose, and D. Towsley, "Scheduling policies for real-time and non-real-time traffic in a statistical multiplexer," Proc. IEEE INFOCOM'89, pp.774–783, Ottawa, Canada, April 1989.

[10] C. Dovrolis, D. Stiliadis, and P. Ramanathan, "Proportional ifferentiated services, Part II: Loss rate differentiation and packet dropping," Proc. 2000 International Workshop on Quality of Service (IWQoS), Pittsburgh PA, June 2000.

[11] J.W. Causey and H.S. Kim, "Comparison of buffer allocation schemes in ATM switched: Complete sharing, partial sharing and dedicated allocation," ICC, pp.1164–1168, 1994.

[12] T. Ling and N. Shroff, "Scheduling real-time traffic in ATM networks," Proc. IEEE INFOCOM'96, pp.198–205, San Francisco, California, USA, March 1996.

[13] J.M. Hah and M.C. Yuang, "A delay and loss versatile scheduling discipline in ATM switches," Proc. IEEE INFOCOM'98, pp.939–946, April 1998.

[14] J. Lee and Y. Kim, "Performance analysis of a hybrid prority control scheme for input and output queueing ATM switches," Proc. IEEE INFOCOM'98, pp.1470–1477, April 1998.

[15] N. Matsufuru, K. Nishimura, and R. Aibara, "Comparative evaluation of resource allocation strategies using weighted round robin scheduler in ATM switches," IEICE Trans. Commun., vol.E82-B, no.1, pp.60–69, Jan. 1999.

[16] C. Lin and S.T. Chanson, "ATM admission models of stochastic high level Petri nets based on hierarchical modeling," Proc. 1995 International Conf. on Network Protocols, IEEE Computer Society, pp.144–151, Tokyo, Japan, Nov. 1995.

[17] C. Lin, B. Li, and J. Wu, "Modeling ATM traffic using stochastic petri nets," Proc. ICCCN' 97, Las Vegas, USA, Sept. 1997.

[18] B. Li, C. Lin, and S.T. Chanson, "Analysis of a hybrid cutoff priority scheme for multiple classes of traffic in multimedia wireless networks," ACM J. Wireless Networks, vol.4, no.4, pp.279–290, Aug. 1998.

[19] C. Lin, L. Sheng, J. Wu, and M. Xu, "An integrative scheme of differentiated services: Modeling and performance analysis," Proc. Eighth International Symp. on Modeling, Analysis and Simulation of Computer and Telecommunication Systems (MASCOTS 2000), IEEE Computer Society, pp.441–448, San Francisco CA, USA, Aug. 2000.

[20] C. Lin, W. Luo, B. Yan, and S.T. Chanson, "A dynamic partial buffer sharing scheme for packet loss control in congested networks," Proc. International Conf. on Commun. Technology, 16th IFIP World Computer Congress (WCC2000), pp.1286–1293, IEEE Press, Publishing House of Electronics Industry, Beijing, China, Aug. 2000.

[21] G. Ciaodo, J. Muppala, and K.S. Trivedi, "SPNP: Stochastic Petri net package," Proc. Petri Nets and Performance Models, pp.142–151, Kyoto, Japan, Dec. 1989.

[22] M A. Marsan, G. Conte, and G. Balbo, "A class of generalized stochastic Petri nets for the performance evaluation of multiprocessor systems," ACM Trans. Computing Systems, vol.2, no.2, pp.93–122, May 1984.

[23] C. Lin and D.C. Marinescu, "Stochastic high level Petri nets and applications," IEEE Trans. Comput., vol.37, no.2, pp.815–825, 1988. Also in ed. K. Jensen and G. Rozenberg, High Level Petri Nets, Springer-Verlag, 1991.

[24] C. Lin and D.C. Marinescu, "On the analysis of stochastic high level Petri net models," Microelectron. Reliab., vol.31, no.4, pp.747–767, 1991.

[25] T. Murata, "Petri nets: Properties, analysis and applications," Proc. IEEE, vol.77, no.4, pp.541–580, April 1989.

[26] M. Schwartz, "Broadband integrated networks," Chapter 3, in Traffic Characterization for Broadband Services, Prentice Hall PTR, 1996.

[27] M. Ajmone Marsan, and G. Chiola, "On Petri nets with deterministic and exponentially distributed firing times," in Advances in Petri Nets 1987, Lecture Notes in Computer Science 266, ed. G. Rozenberg, pp.132–145, Springer, Berlin, 1987.

[28] C. Lindemann, "An improved numerical algorithm for calculating steady-state solutions of deterministic and stochastic petri net models," Performance Evaluation 18, pp.75–91, 1993.

[29] G. Ciardo and G. Li, "Approximate transient analysis for subclasses of deterministic and stochastic Petri nets," Performance Evaluation 35, pp.109–129, 1999.

[30] C. Lindemann and A. Thummler, "Transient analysis of deterministic and stochastic Petri nets with concurrent deterministic transitions," Performance Evaluation 36–37, pp.35–54, 1999.

[31] R. German, "Markov regenerative stochastic Petri nets with general execution policies: Supplementary variable analysis and a prototype tool," Performance Evaluation 39, pp.165–188, 2000.

Chapter 6

Traffic Control

One important function offered by ATM networks is traffic control to guarantee QoS requirements. Several studies have addressed the traffic control issues of connection admission control (CAC), usage parameter control (VPC), bandwidth allocation, traffic measurement, flow control, and traffic shaping.

Papers 6.1 and 6.2 describe the cell traffic patterns that are enforced by the VPC circuit. Paper 6.1 shows the worst cell pattern with respect to the Generic Cell Rate Algorithm (CGRA) VPC circuit. The worst pattern increases the cell loss rate of the ATM network. This realizes a form of connection admission control.

Paper 6.2 is a more general study. The cell traffic enforced by the VPC circuit is estimated in detail. The study considers the jumping window with sliding window leaky bucket VPC circuit. It is noted that negotiated cell loss rates can be guaranteed by this CAC method.

Measurement techniques are sometimes used to estimate the residual bandwidth of ATM. Paper 6.3 introduces a traffic measurement method for IP traffic over ATM. The experimental measurement results match MMPP (Markov modified Poisson process) parameter. Cell loss rate or required bandwidth of the network can be calculated by using the MMPP/G/1 model. On the other hand, Paper 6.4 examines connection admission control based on a measurement technique. Equivalent bandwidth is determined from the measurement results. The technique uses traffic characteristics such as video or voice and makes it easy to calculate the equivalent bandwidth.

Traffic shaping for VPC circuit is used to mold the user traffic to the ideal pattern. Paper 6.5 presents a GCRA traffic shaper for VBR cell streams. Mechanism details and performance studies are described.

Paper 6.6 addresses scheduling algorithms such Waited Round Robin (WRR) and static priority scheduling. Important parameters for ATM QoS are bandwidth and delay bounds. The paper points out that it is difficult to guarantee the delay bounds, even if static priority scheduling is combined with WRR. Some numerical study results are shown in detail.

H. Hasegawa et al. present an ATM flow control method and its design scheme in Paper 6.7. The ATM layer protocols are emulated at the edge node and link utilization is monitored periodically between edges. Network control is also described.

Paper 6.8 by K. Kyandoghere discusses VP control strategies for reliability. Multi-layer survivability is one of the important focus points. T. Yahara et al. propose a self-healing scheme that differentiates the bandwidth requirement for each network service on ATM networks in Paper 6.9. Simulation results suggest that the proposed scheme requires fewer spare resources while offering comparable restoration time to the conventional approach against any demand pattern.

In ATM networks, if a source violates the traffic parameters, the probability of congestion increases. The network must control the source traffic stream and detect the violating cells. M. H. Yaghmaee et al. propose an efficient fuzzy traffic policer with excellent ability to detect violating sources in Paper 6.10.

The active network provides a new approach and paradigm to networking. It departs from the traditional networking by making the network programmable and allowing computations to occur within the network. Paper 6.11 by H. K. Pung describes a design and implementation of an ATM multicast service based on programmable and active network concepts.

In IP networks, the sharing of network resources among the many flows is on a best-effort basis. T. Miyamura et al. propose a First-In-First-Out (FIFO)-based mechanism that achieves fairness in terms of throughput in Paper 6.12. They investigate its performance in terms of both fairness and link utilization through extensive simulations.

The final paper of this chapter, written by A. Koike, addresses end-to-end TCP flow control mechanisms using Available Bit Rate (ABR) information in ATM networks. Goodput comparisons and fairness issues are described.

High-Performance Backbone Network Technology edited by Naoaki Yamanaka
ISBN: 0-8247-5321-6 DOI: 10.1081/(120025710)
Published 2004 by Marcel Dekker, Inc.

The Worst Case Cell Arrival Patterns in ATM Networks

Toshiaki Tsuchiya and Hiroshi Saito

SUMMARY We introduce the concepts of conservative cell loss ratio (CLR) estimation and worst case cell arrival patterns, and apply them to cell arrival patterns that conform to the generic cell rate algorithm (GCRA). We define new sets of cell arrival patterns which contain the worst case cell arrival patterns for conforming cell arrival patterns. Based on these sets, we propose an upper bound formula using the burst tolerance as well as peak cell rate and sustainable cell rate, and develope a connection admission control method that guarantees cell loss ratio performance satisfying its objective.
key words: *ATM, CAC, GCRA, conformance*

1. Introduction

The Asynchronous Transfer Mode (ATM) is well known technology for implementing Broadband Integrated Services Digital Networks. One of major issues in ATM technology is traffic control. In ATM networks, preventive controls are important, although reactive controls such as Explicit Forward Congestion Indication and rate control are mainly used for Available Bit Rate (ABR). The concept of preventive controls is simple. A user notifies the network of traffic characteristics by the source traffic descriptor at a connection set up. The network then decides whether to admit or reject the connection based on the source traffic descriptor, cell delay variation tolerance (CDVT), quality of service (QoS) requirements, network load and the amount of network resources. This decision is called Connection Admission Control (CAC).

If the connection is established after CAC, the network monitors the connection and determines whether it satisfies the source traffic descriptor or not. When the network finds the actual traffic do not conform with the traffic characteristics reported by the source traffic descriptor, the network may impose a penalty on the connection. The conformity is detremined by using the Generic Cell Rate Algorithm (GCRA, see Fig. 1) [1]. This algorithm is essentially equivalent to the leaky bucket algorithm. The peak cell rate (PCR), the sustainable cell rate (SCR), the burst tolerance (BT) and the cell delay variation tolerance (CDVT) are included in the traffic descriptor. The connection is then tested by GCRA ($T=1/PCR$, $\tau=CDVT$) and GCRA ($T=1/SCR$, $\tau=BT+CDVT$). We assume that $CDVT=0$ for simplicity, but our results can be extended to the case in which $CDVT \neq 0$, combined with the results in [3].

A connection which satisfies the conformity requirements is called a compliant connection. The network must guarantee that the QoS objective of the compliant connection is satisfied [1]. In such a situation, an idealistic CAC goal is to keep the QoS satisfactory for all possible conforming cell streams from compliant connections. This objective is often called QoS estimation under the worst case arrival pattern. In practice, the network may provide satisfactory QoS when the network does not experience the worst case arrival pattern, because not all the admitted cell streams have a worst case arrival pattern. However, estimating QoS under the worst case cell arrival pattern is useful because it reduces the QoS monitoring and congestion control.

QoS estimation under the worst case arrival pattern presents a new field in teletraffic theory. The QoS boundaries must be evaluated for the family of arrival processes for which a part of the information is given by a source traffic descriptor or UPC parameters. Saito [2]–[4] considered this bandwidth management issue and introduced the non-parametric approach, since the complete parametric description for the arrival process can not be obtained. His study derived the upper bound of the cell loss ratio (CLR) when the stationary distribution are known for the number of cells arriving during a fixed time period, or for the maximum and average number of cells arriving during a fixed time period. The latter gives a bandwidth that

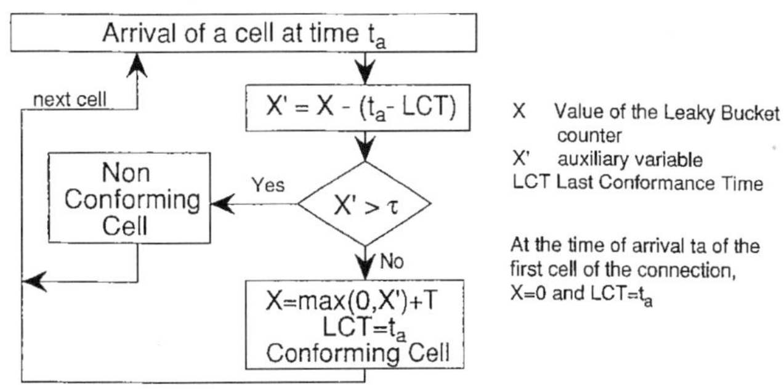

Fig. 1 Generic cell rate algorithm (T, τ) (continuous state leaky bucket algorithm).

Manuscript received September 22, 1997.
Manuscript revised December 10, 1997.
[†]The authors are with NTT Multimedia Networks Laboratories, Musashino-shi, 180–8585 Japan.
[*]This paper is an expanded version of a paper presented at GLOBECOM '95, Singapore, November 1995.

IEICE TRANS. COMMUN., Vol. E81-B, No. 5, pp. 996–1003, May 1998

accommodates the worst case arrival pattern when the sliding window and the jumping window are used as UPC. (Another formulation and a proof are shown in [2].) However, these results assumed that the burst size may be arbitrarily large. Thus, the proposed bandwidth management results in low utilization when the burst size is finite. Recently, Shioda et al. [5] extended Saito's results, and clarified the conditions for which these results are valid. Doshi [6] clearly described bandwidth management for the worst case arrival pattern and derived results based mainly on fluid flow approximation. Bonatti et al. [7] also discussed the worst case and derived some formulas. Doshi's and Bonatti's formulas are similar to Saito's, and they must be studied further as the effective bandwidth management techniques for the worst case with a finite burst size. In addition, Yamanaka et al. [8] and Worster [9] gave a non-intuitive example in which a traffic pattern alternating between off-periods and on-periods may not be the worst. In [10], the authors proposed a CLR estimation formula under the worst case arrival pattern in which burst size is finite. This paper is an extended version of [10] and includes newer numerical results, in which the impact of burst size and mixed traffic is considered.

2. Notations and Formulation

Consider a virtual path (VP) of bandwidth C and with output buffer size K. Here, the VP bandwidth is assumed to be deterministic. That is, the cell transmission of the VP is periodical and its rate is equal to the VP bandwidth when there exist cells waiting for transmission. This VP includes with virtual connections (VCs). (In the remaining of this paper, we consider a VC connection admission in a VP for simplicity. Thus, a connection implies a VC connection in the remaining.) The time axis is assumed to be divided into slots with lengths equal to the cell transmission time. In this paper, all the parameters (such as the PCR R_i and the SCR S_i in the i-th VC's source traffic descriptor) are transformed into per slot. We also assume that, for each i, the i-th VC has the same actual average cell rate as its SCR S_i. Here, the "actual" average cell rate means

$$\lim_{T\to\infty} \frac{\#\{\text{cell arrivals from the } i\text{-th VC in [0,T]}\}}{T}.$$

When the actual average cell rate does not correspond to the specified SCR, further consideration (as in [11]) is needed.

Let $\mathcal{A}_i$ be the cell arrival pattern of the i-th VC. We assume that all the cell arrival patterns are stationary, ergodic and independent of each other. Consider the cell loss ratio $CLR(\mathcal{A}_1, \mathcal{A}_2, \ldots, \mathcal{A}_N)$ of the VP at its output buffer when the cell arrival patterns $\mathcal{A}_1, \mathcal{A}_2, \ldots, \mathcal{A}_N$ are multiplexed. We define the conservative CLR estimation and the worst case cell arrival patterns.

Definition 1: A function f of cell arrival patterns $\{\mathcal{A}_i, i = 1, \ldots, N\}$ of multiplexed VCs is called a conservative CLR estimation of a VP if

$$f(\mathcal{A}_1, \ldots, \mathcal{A}_N) \geq CLR(\mathcal{A}_1, \ldots, \mathcal{A}_N). \tag{1}$$

Definition 2: A cell arrival pattern $\mathcal{A}'$ is called the worst case cell arrival pattern from a VC for a conservative CLR estimation f in the set Ω of cell arrival patterns if

$$f(\mathcal{A}', \tilde{\mathcal{A}}) \geq CLR(\mathcal{A}, \tilde{\mathcal{A}}) \qquad \text{for any } \mathcal{A} \in \Omega, \tag{2}$$

where $\tilde{\mathcal{A}}$ denotes any cell arrival pattern from other VCs.

Definition 3: Cell arrival patterns $\{\mathcal{A}'_i, i = 1, \ldots, N\}$ are called the worst case cell arrival patterns to a VP for a conservative CLR estimation f in the sets $\{\Omega_i, i = 1, \ldots, N\}$ of cell arrival patterns if

$$f(\mathcal{A}'_1, \ldots, \mathcal{A}'_N) \geq CLR(\mathcal{A}_1, \ldots, \mathcal{A}_N) \tag{3}$$

for any $\mathcal{A}_i \in \Omega_i$ $(i = 1, \ldots, N)$.

Clearly if each cell arrival pattern $\mathcal{A}'_i$ from the i-th VC is worst case, then $\{\mathcal{A}'_i, \ldots, \mathcal{A}'_N\}$ are the worst case cell arrival patterns to a VP.

By combining conservative CLR estimation and worst case cell arrival patterns, we can develop a CAC method. Suppose a new customer requires connection admission when there are N VCs multiplexed to a VP, and that the arrival process of the i-th connection is $\mathcal{A}_i \in \Omega_i$ where Ω_i is a set of cell arrival patterns conforming to the source traffic descriptor of the i-th VC. Then, the network calculates $f(\mathcal{A}'_1, \ldots, \mathcal{A}'_N, \mathcal{A}'_{N+1})$, where $\mathcal{A}'_i$ is the worst case cell arrival pattern from the i-th VC for $i = 1, \ldots, N$, and $\mathcal{A}'_{N+1}$ is the worst case cell arrival pattern from a new connection. If the calculated value remains less than the CLR required by each connection, the network accepts the new connection. If not, the new connection is denied by the newtwork and is lost. With this CAC method, the CLR required by each connection is satisfied as long as the connections maintain their conforming cell arrival patterns in $\Omega_i, i = 1, \ldots, N+1$.

3. Past Results for Conservative CLR Estimation and Worst Case Cell Arrival Patterns

Theorem 1 (CLR upper bound formula [2]): Let $X_{\mathcal{A},\gamma}$ be the number of cell arrivals from a VC in the γ time slot and $p_{\mathcal{A},\gamma}(k)$ be its stationary probability distribution of $X_{\mathcal{A},\gamma}$ when the cell arrival pattern is $\mathcal{A}$. Set $\boldsymbol{p}_{\mathcal{A},\gamma} = \{p_{\mathcal{A},\gamma}(0), p_{\mathcal{A},\gamma}(1), \ldots\}$ and define $D(\boldsymbol{p}_{\mathcal{A},\gamma}; n)$ as $\sum_{k=0}^{\infty}(k-n)_+ p_{\mathcal{A},\gamma}(k)$, where $(x)_+ = \max(0, x)$. Then, we have

$$D(\boldsymbol{p}_{\mathcal{A},\gamma}; n)/D(\boldsymbol{p}_{\mathcal{A},\gamma}; 0) \geq CLR(\mathcal{A}), \tag{4}$$

where the integer n is less than $(K+1)/2$.

Theorem 2 (Shioda and Saito [5]): In the above theorems, we can let n be less than or equal to $K+1$, when the cell arrival process $\mathcal{A}$ is ergodic.

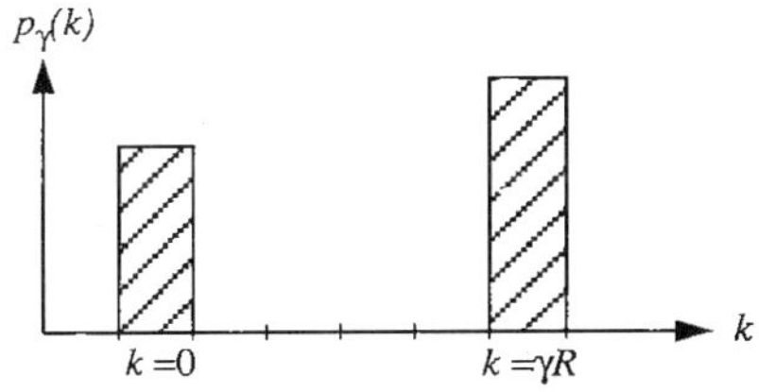

Fig. 2 Probability distribution $p_\gamma(k)$.

These two theorems conservatively estimate CLR, but they require detailed information cell arrival distribution $\boldsymbol{p}_{\mathcal{A},\gamma}$ in a determined number of slots. However, such knowledge is not realistic, especially for the call admission time period. The next theorem provides a CAC method based on the PCR and the SCR by using partial information from them. Let $\Omega_i(R_i, S_i)$ be the set of conforming cell arrival patterns from the i-th VC, of which the peak cell rate is R_i and the sustainable cell rate is S_i. Then, we have

Theorem 3 (Worst case cell arrival patterns [2]): For the i-th VC, let $\mathcal{A}_i \in \Omega_i(R_i, S_i), X_{\mathcal{A}_i,\gamma}$ and $p_{\mathcal{A}_i,\gamma}(k)$, as in theorem 1. Consider a cell arrival pattern $\mathcal{A}_i' \in \Omega_i(R_i, S_i)$ such that

$$p_{\mathcal{A}_i',\gamma}(k) = \begin{cases} 1 - S_i/R_i & k = 0 \\ S_i/R_i & k = \gamma R_i \\ 0 & \text{otherwise} \end{cases} \tag{5}$$

(see Fig. 2). Then, for any $\mathcal{A}_i \in \Omega_i(R_i, S_i)$, we have

$$\frac{D(\boldsymbol{p}_{\mathcal{A}_1',\gamma} \star \cdots \star \boldsymbol{p}_{\mathcal{A}_N',\gamma}; n)}{D(\boldsymbol{p}_{\mathcal{A}_1',\gamma} \star \cdots \star \boldsymbol{p}_{\mathcal{A}_N',\gamma}; 0)} \geq \frac{D(\boldsymbol{p}_{\mathcal{A}_1,\gamma} \star \cdots \star \boldsymbol{p}_{\mathcal{A}_N,\gamma}; n)}{D(\boldsymbol{p}_{\mathcal{A}_1,\gamma} \star \cdots \star \boldsymbol{p}_{\mathcal{A}_N,\gamma}; 0)}, \tag{6}$$

where $\star$ denotes the convolution and N is the number of VCs. In other words, $\mathcal{A}_i'$ is the worst case cell arrival pattern from the i-th VC in the conforming cell arrival pattern set $\Omega_i(R_i, S_i)$ and $\{\mathcal{A}_i', i = 1, \ldots, N\}$ are the worst case cell arrival patterns to a VP in $\{\Omega_i(R_i, S_i), i = 1, \ldots, N\}$.

In theorem 3, the worst case cell arrival pattern $\mathcal{A}_i'$ is such that the cells arrive at R_i during sufficently long intervals.

4. The Worst Case Cell Arrival Pattern under the Compliance Property

The worst case cell arrival patterns in theorem 3 use only the PCR and the SCR. Next we try to derive the worst case cell arrival pattern in the set of conforming cell arrival patterns which use the BT as well as the PCR and the SCR. Our method gives a tighter upper bound CLR formula than those in the previous section. We first focus on a particular set of cell arrival patterns. After deriving the worst case cell arrival patterns in that set, we extend the results to the conforming cell arrival patterns.

4.1 Particular Cell Arrival Patterns

Consider a set $\tilde{\Omega}(R, S, M) \subset \Omega(R, S)$ of cell arrival patterns from a VC and its subset $\tilde{\Omega}'(R, S, M)$. We assume that any cell arrival pattern $\mathcal{A} \in \tilde{\Omega}(R, S, M)$ has the following properties:

(P1) The cell arrival intervals are either $r = 1/R$ or greater than $W = Lr + l$, where r, L, l are assumed to be integers, $r > 0$, $L \geqq 0$ and $1 \leqq l \leqq r$.

(P2) The number of consecutive cell arrivals at rate R (called a burst) is at most $M (\geqq L + 1)$.

Besides (P1) and (P2), we assume that any $\mathcal{A}' \in \tilde{\Omega}'(R, S, M) \subset \tilde{\Omega}(R, S, M)$ has one more property:

(P3) The burst size is always equal to M.

Let $b_i (i = 1, 2, \ldots, M)$ be the probablity that a burst size is i when the cell arrival patterns are in $\tilde{\Omega}(R, S, M)$ and set $\boldsymbol{b} = (b_1, b_2, \ldots, b_M)$. In particular, when the cell arrival pattern is in $\tilde{\Omega}'(R, S, M)$, we have $b_i = 0$ for $i \leqq M - 1$ and $b_M = 1$. We consider the number of arriving cells $X_\mathcal{A}$ in one interval (called a window) of size W when the cell arrival pattern is $\mathcal{A} \in \tilde{\Omega}(R, S, M)$. Let the distribution of $X_\mathcal{A}$ be $p(k) = P\{X_\mathcal{A} = k\}, k = 0, 1, \ldots, L + 1$. From the burst size distribution $\boldsymbol{b}$, we have

$$p(k) = \begin{cases} C_1\Big(\{W - (k-1)r\}b_k + \sum_{m=k+1}^{M} 2rb_m\Big) \\ \qquad\qquad (1 \leqq k \leqq L - 1) \\ C_1\Big(\{W - (L-1)r\}b_L \\ \quad + \sum_{m=L+1}^{M} \{2r + (m - L - 1)(r - l)\}b_m\Big) \\ \qquad\qquad (k = L) \\ C_1 \sum_{m=L+1}^{M} (m - L)lb_m \qquad (k = L + 1) \end{cases} \tag{7}$$

(see Appendix A), where C_1 and $p(0)$ are determined so that $\sum_k kp(k) = SW$ and $\sum_k p(k) = 1$. Similarly, let $X_{\mathcal{A}'}$ be the number of arriving cells in a window when the cell arrival pattern is $\mathcal{A}' \in \tilde{\Omega}'(R, S, M)$ and $p'(k) = P\{X_{\mathcal{A}'} = k\}, n = 0, 1, \ldots, L+1$. Then we have

$$p'(k) = \begin{cases} 2rC_2 & (1 \leqq k \leqq L - 1) \\ C_2\{2r + (M - L - 1)(r - l)\} \\ \qquad\qquad (k = L) \\ C_2(M - L)l & (k = L + 1), \end{cases} \tag{8}$$

where C_2 and $p'(0)$ are determined so that $\sum_k kp'(k) =$

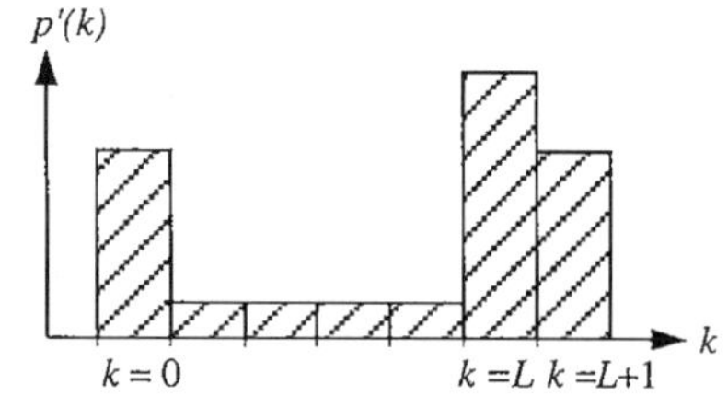

Fig. 3 Probability distribution of $X_{\mathcal{A}'}$.

SW and $\sum_k p'(k) = 1$ (Fig. 3). Note that we need to consider only a single burst, because we have assumed (P1). Between $X_{\mathcal{A}}$ and $X_{\mathcal{A}'}$, we have the following order relation

Theorem 4: It holds that $X_{\mathcal{A}} \leqq_c X_{\mathcal{A}'}$ where $\leqq_c$ means convex ordering. In other words,

$$E(X_{\mathcal{A}} - n)_+ \leqq E(X_{\mathcal{A}'} - n)_+ \quad \text{for all real } n, \tag{9}$$

where $x_+ = \max(0, x)$.

Proof: See Appendix B. □

Now, let $\mathcal{B}$ be any cell arrival pattern in some set $\hat{\Omega}$ from other VCs and $q_{\mathcal{B}}(k) = P\{X_{\mathcal{B}} = k\}$. Theorem 4 then gives

Theorem 5: For any integer n, when $\mathcal{B} \in \hat{\Omega}$, it holds that

$$\sum_k (k-n)_+ p \star q_{\mathcal{B}}(k) / \sum_k k\, p \star q_{\mathcal{B}}(k) \leqq \sum_k (k-n)_+ p' \star q_{\mathcal{B}}(k) / \sum_k k\, p' \star q_{\mathcal{B}}(k), \tag{10}$$

where $\star$ denotes the convolution.

Proof: Since convex ordering is closed for convolution, we immediately have $X_{\mathcal{A}} + X_{\mathcal{B}} \leqq_c X_{\mathcal{A}'} + X_{\mathcal{B}}$. In other words,

$$\sum_k (k-n)_+ p \star q_{\mathcal{B}}(k) \leqq \sum_k (k-n)_+ p' \star q_{\mathcal{B}}(k), \tag{11}$$

for any integer n. The two Eqs. (10) and (11) are equivalent since $\sum_k k\, p \star q_{\mathcal{B}}(k) = E(X_{\mathcal{A}} + X_{\mathcal{B}})$ and $\sum_k k\, p' \star q_{\mathcal{B}}(k) = E(X_{\mathcal{A}'} + X_{\mathcal{B}})$ are identical. □

From theorems 2 and 5, we can conclude that $\mathcal{A}'$ is the worst case cell arrival pattern in the set $\tilde{\Omega}(R, S, M)$.

4.2 Conforming Cell Arrival Patterns

Now, we can apply theorem 5 to the set $\Omega(R, S, B)$ of conforming cell arrival patterns to the source traffic descriptor that PCR$=R$, SCR$=S$ and BT$=B$. Consider one conforming cell arrival pattern $\omega \in \Omega(R, S, B)$, and the number of cell arrivals in a window X_ω. We denote the n-th cell arrival time in ω as a_n. Then, let us change the pattern ω so that the property (P1) holds for the modified pattern as follows (Fig. 4).

(Step 1)

(1.1) By moving cell arrival times, change the cell arrival pattern in $[0, a_L]$ so that the consecutive cell arrival intervals become r and the last cell arrives at time a_L, where $r = 1/R$.

(1.2) By moving cell arrival times, change the cell arrival pattern in $[a_L + 1, a_{N_1}]$ so that the consecutive cell arrival intervals become r and the first cell arrives at time $a_L + r$, where $N_1 = \min\{n : (a_{n+1} - a_L) - r(n - L) \geqq W\}$.

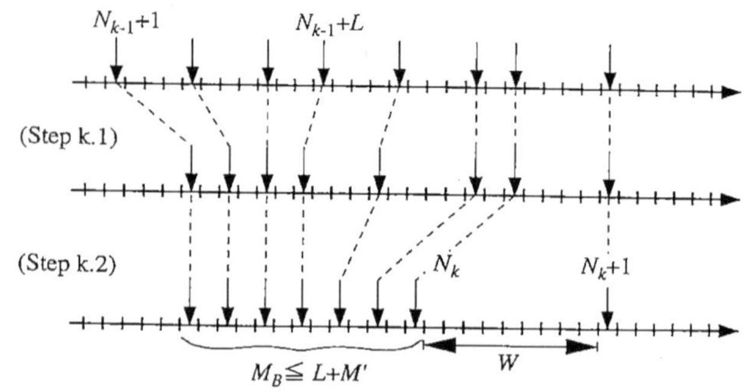

Fig. 4 Change in the pattern ω.

(Step k)

(k.1) By moving cell arrival times, change the cell arrival pattern in $[a_{N_{k-1}} + 1, a_{N_{k-1}+L}]$ so that the consecutive cell arrival intervals become r and the last cell arrives at time $a_{N_{k-1}+L}$.

(k.2) By moving cell arrival times, change the cell arrival pattern in $[a_{N_{k-1}+L} + 1, a_{N_k}]$ so that the consecutive cell arrival intervals become r and the first cell arrives at time $a_{N_{k-1}+L} + r$, where N_k is determined by

$$\min\{n : (a_{n+1} - a_{N_{k-1}+L}) - r\{n - (N_{k-1} + L)\} \geqq W\}. \tag{12}$$

Note that after the k-th step, the modified cell arrival intervals from $N_{k-1} + 1$ to N_k are equal to r and the interval length between a_{N_k} (modified) and $a_{N_k} + 1$ is equal to or greater than W. Then, consider the possible number of cells $M = N_k - (N_{k-1} + L)$ gathered in step (k.2). Let $I = a_{N_k+1} - a_{N_{k-1}+L}$ before step (k.2). From (12), we need $I - rM \geqq W$. Since ω is conforming, the maximum number of cell arrivals M' in any interval of length I is bounded by $\lfloor IS + B \rfloor$ where $\lfloor x \rfloor$ is the integer part of x. Then, $M' \leqq IS + B$, which corresponds to $I - rM' \geqq I - (IS + B)/R$. Therefore, M is bounded by $M' = \lfloor I'S + B \rfloor$ where $I' = \min\{I : I - (IS + B)/R \geqq W\}$. Consequently, we have

$$M \leqq M' = \left\lfloor \left\lceil \frac{WR + B}{R - S} \right\rceil S + B \right\rfloor, \tag{13}$$

where $\lceil x \rceil$ is the smallest integer which is larger than x. Let the modified cell arrival pattern after the i-th step be $\tilde{\omega}_i$ and the number of cell arrivals in a window for a pattern $\tilde{\omega}_i$ be $X_{\tilde{\omega}_i}$. Then, we have

$$E(X_{\tilde{\omega}_i} - m)_+ \leq E(X_{\tilde{\omega}_j} - m)_+ \tag{14}$$

for all m and $i \leq j$ (see Appendix C). Here, let $\tilde{\omega} = \lim_i \tilde{\omega}_i$ and $\tilde{X} = \lim_i X_{\tilde{\omega}_i}$. Then, we have

$$\tilde{\omega} \in \tilde{\Omega}(R, S, M_B), \tag{15}$$

where $M_B = L + M'$ and

$$X_\omega \leq_c \tilde{X}. \tag{16}$$

Equation (15) and theorem 5 imply that if we choose some $\omega' \in \tilde{\Omega}'(R, S, M_B)$, it will be the worst case arrival pattern in $\tilde{\Omega}(R, S, M_B)$. Furthermore, ω' is the worst case arrival pattern in $\Omega(R, S, B)$ since (16) holds. Finally, we have

Theorem 6: Consider the cell arrival paterns $\omega^i \in \tilde{\Omega}'(R_i, S_i, M_B^i), i = 1, \ldots, N$ in which $M_B^i = L^i + M^i$ is determined by

$$M^i = \left\lfloor \left\lceil \frac{WR_i + B_i}{R_i - S_i} \right\rceil S_i + B_i \right\rfloor. \tag{17}$$

Then, $\{\omega^i \in \tilde{\Omega}'(R_i, S_i, M_B^i), i = 1, \ldots, N\}$ are the worst case cell arrival patterns to a VP in the sets $\{\Omega(R_i, S_i, B_i), i = 1, \ldots, N\}$ for the conservative CLR estimation

$$\frac{\sum_k (k - W)_+ p^1 \star \cdots \star p^N(k)}{\sum_k k p^1 \star \cdots \star p^N(k)}, \tag{18}$$

where

$$p^i(k) = \begin{cases} 2r_i C_1^i & (1 \leq k \leq L^i - 1) \\ C_1^i\{2/R_i + (M_B^i - L^i - 1)(1/R_i - l^i)\} & (k = L^i) \\ C_1^i(M_B^i - L^i)l^i & (k = L^i + 1). \end{cases} \tag{19}$$

Here, the window size W must be equal to or less than the buffer size $+1$, l^i and L^i are non-negative integegers, $1 \leq l^i \leq 1/R_i$ and $L^i/R_i + l^i = W$. In (19), $p^i(0)$ and C_1^i are determined so that $\sum_k kp^i(k) = S_i W$ and $\sum_k p^i(k) = 1$.

Note that the worst case cell arrival patterns in theorem 6 give tighter upper bounds than those in theorem 3, because $\tilde{\Omega}_i'(R_i, S_i, M_B^i) \subset \Omega(R_i, S_i)$ for each i. The upper bound based on the theorem 6 is the minimum upper bound when the burst tolerance becomes sufficiently large. This because the upper bound based on the theorem 3 has this property [2] and because that based on the the theorem 6 becomes tighter than that based on the theorem 3.

5. CAC with Conservative CLR Estimation

Theorem 6 provides the following CAC method. Suppose that a new customer requires connection admission when N VCs are already multiplexed to the VP. For each $i \in \{1, \ldots, N\}$, R_i, S_i and B_i represent the i-th VC's PCR, SCR and BT. Also, M_B^i and $p^i(k)$ are determined by (17) and (19). Then,

1. Receive the PCR R_{N+1}, the SCR S_{N+1} and the BT B_{N+1} from the new customer.
2. Estimate the CLR upper bound by
$$\frac{\sum_{k>W} (k - W)_+ p^1 \star \cdots \star p^{N+1}(k)}{\sum_k k p^1 \star \cdots \star p^{N+1}(k)}, \tag{20}$$
where the window size W is set eqaul to the VP's output buffer size.
3. Compare the estimated value with the CLR objective.
4. If (20) is equal to or less than the CLR objective, then we accept the new cusutomer's connection as the $N+1$-st VC.
5. If (20) is greater than the objective, then we deny the new customer's connection to avoid degrating the CLR at the VP.

This CAC method guarantees the CLR required by VCs under the condition that each VC maintains its conformity.

6. Numerical Examples

We plot the CLR upper bounds for narrowband VCs (Fig. 5) and for broadband VCs (Fig. 6) as a function of VP utilization (or the number of VCs). For both figures, VP bandwidth is 150 Mbps, the buffer size for the VP is 128 and the source traffic descriptors of all VCs are the same. In Fig. 5, PCR = 1.5 Mbps and SCR = 0.5 Mbps. In Fig. 6, PCR = 15 Mbps and SCR = 5 Mbps. The upper bound is a function of the maximum burst size (MBS). Here, MBS is given by the

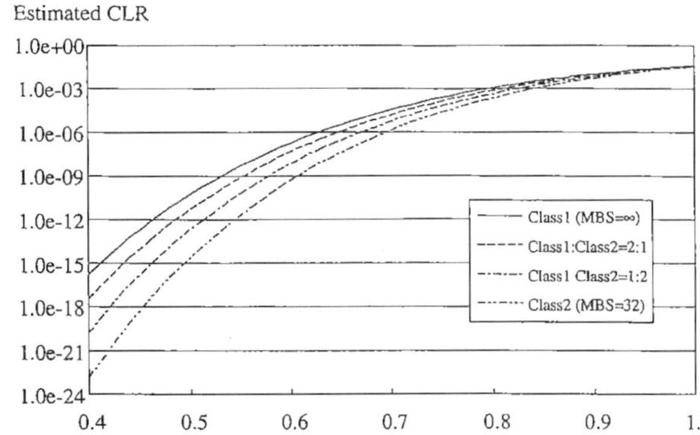

Fig. 5 Cell loss ratio (narrowband VCs).

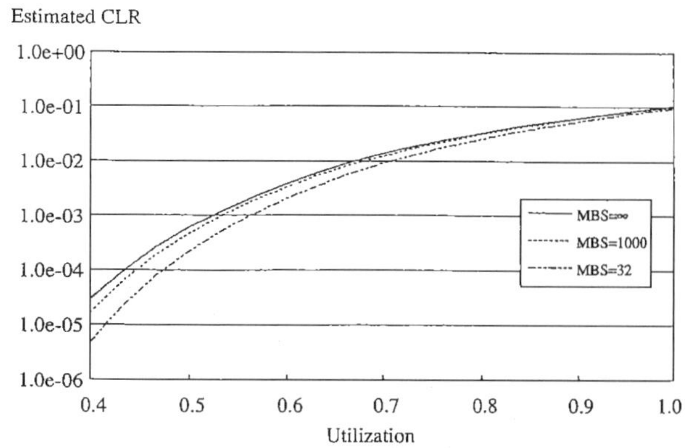

Fig. 6 Cell loss ratio (broadband VCs).

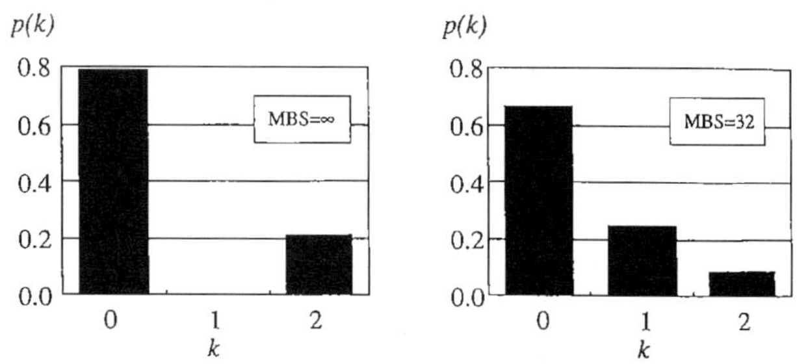

Fig. 7 Probability distribution of the number of arrivals (narrowband VCs).

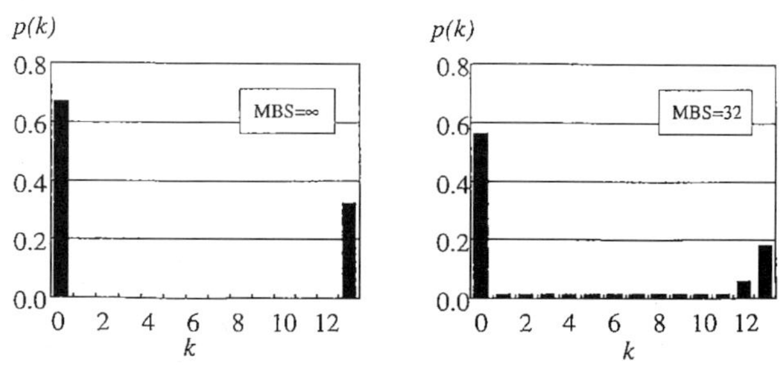

Fig. 8 Probability distribution of the number of arrivals (broadband VCs).

integer part of $(1+B/(R-S))$ where B is the burst tolerance, R is the peak cell rate and S is the sustainable cell rate.

Figures 5 and 6 show that the BT or the MBS has a larger impact for the CLR upper bound for narrowband VCs than broadband VCs. When VP utilization is 0.5, the CLR upper bound with MBS$=$32 is tighter than that with infinite MBS by about 5 orders of magnitude in Fig. 5 and by 0.5 orders of magnitude in Fig. 6.

When VP utilization is 0.7, the MBS has little impact in Fig. 6 but the upper bound with MBS$=$32 is still tighter than that with infinite MBS by about 2 orders of magnitude in Fig. 5. The reason is shown in Figs. 7 and 8. Let L be the maximum number of arriving cells of each VC. For infinite MBS, $p(k) = 0$ for $k = 1,\ldots,L-1$. For small MBS, $p(k)$ becomes positive for $k = 1,\ldots,L-1$. When PCR is large, however, $p(k)(1 \leq k \leq L-1)$ cannot be large. This is because $p(k)$ is the same for $k = 1,\ldots,L-1$, $p(1)+\cdots+p(L-1) < 1$, and L is large. Thus, the MBS has a large impact for low PCR VCs.

Let $CLR*$ be the CLR upper bound when n VCs with various MBSs are multiplexed and $CLR(x)$ be the CLR when n VCs with MBS$=x$ are multiplexed. Figure 5 shows $\sum_i n_i \log CLR(x_i)/\sum n_i$ is a good estimate of $\log CLR*$ when n_i is the number of VCs with MBS$=x_i$ and $n = \sum_i n_i$.

7. Conclusion

This paper introduced the concepts of conservative CLR estimation and worst case cell arrival patterns, and applied them to the case in which the patterns conform to GCRA. With the compliance properties, we defined a new set of cell arrival patterns. This set contained the worst case cell arrival patterns for conforming cell arrival patterns. As a result, the estimate was tighter than those with existing cell arrival patterns. Therefore, the proposed CAC with CLR estimation better utilizes the VP and can assign an adequate bandwidth as long as all the VCs are conforming. We used a numerical example to demonstrate our method's effectiveness.

References

[1] Draft Recommendation I.371, "Traffic congestion control in B-ISDN," 1994.

[2] H. Saito, "Call admission control using upper bound of cell loss probability," IEEE Trans. Commun., vol.40, no.9, pp.1512–1521, 1992.

[3] H. Saito, "Teletraffic Technologies in ATM Networks," Artech House, Boston, 1994.

[4] H. Saito, "A future traffic-dimensioning method: A non-parametric approach for cell loss-rate evaluation," Electronics and Communications in Japan (Part 1: Communications), vol.77, no.2, pp.106–119, Feb. 1994.

[5] S. Shioda and H. Saito, "Connection admission control guaranteeing cell loss ratio of the traffic through usage parameter control," ICC 95.

[6] B.T. Doshi, "Deterministic rule based traffic descriptions for broadband ISDN: Worst case behavior and connecton acceptance control," ITC14, pp.591–600, 1994.

[7] M.B. Bonatti and A.A. Gaivoronski, "Worst case analysis of ATM sources with application to access engineering of broadband multiservice networks," ITC14, pp.559–569, 1994.

[8] N. Yamanaka, Y. Sato, and K. Sato, "Performance limitation of leaky bucket algorithm for usage parameter control on bandwidth allocation methods," IEICE Trans. Commun., vol.E75-B, no.2, pp.82–86, Feb. 1992.

[9] T. Worster, "Modelling deterministic queues: The leaky buckets as an arrival process," ITC14, pp.581–590, 1994.

[10] T. Tsuchiya and H.Saito, "The worst cast cell arrival patterns that conform to the GCRA in ATM networks," GLOBECOM'95, vol.2, pp.1432–1438, Nov. 1995.

[11] M. Aida and H. Saito, "Traffic contract parameters and CAC guaranteeing cell-loss ratio in ATM networks," IEICE Trans. Commun., vol.E78-B, no.3, pp.336–343,

March 1995.

Appendix A: Number of Cell Arrivals in a Window

For convenience, we consider the case $l = r$, that is, $W = (L+1)r$. We first focus on one burst with a size of $n \leq L$. In this case, the event $\{X > 0\}$ occurs for $(n-1)r + W = r(n+L)$ times where X is the number of arriving cells in a window (Fig. A·1). For the first r windows and the last r windows, $\{X = 1\}$ (Fig. A·1). Next, from the $r+1$-st to the $2r$-th and from the $r(n+L-2)+1$-st to the $r(n+L-1)$-th windows, $\{X = 2\}$ and so on. Consequently, the total window observation results in $\{X = i\}$ for $2r$ times for each $i = 1, \ldots, n-1$ and $\{X = n\}$ for $(L-n+2)r = W-(n-1)r$ times. Similarly, when the burst size is $k \geqq L+1$, we have $\{X = i\}$ for $2r$ times for each $i = 1, \ldots, L$, and $\{X = L+1\}$ for $(n-L)r$ times. A burst length distribution $\{b_1, \ldots, b_M\}$ gives

$$P\{X = k\} \propto \begin{cases} \{W-(k-1)r\}b_k \\ \quad + \sum_{n=k+1}^{M} 2rb_n & (1 \leqq k \leqq L) \\ \sum_{n=L+1}^{M} (n-L)rb_n & (k = L+1). \end{cases} \tag{A·1}$$

Equation (A·1) corresponds to the special case of (7) when $l = r$.

When $l < r$, we need some modifications, but the argument is straightforward.

Appendix B: Proof of Theorem 5

We only check when n is an integer. First, we calculate $E(X_{\mathcal{A}} - n)_+$ in terms of $p(k)$ for $n = 1, 2, \cdots, L$.

$$E(X_{\mathcal{A}} - n)_+/C_1 = \sum_{k=n+1}^{L+1} (k-n)p(k)/C_1$$

$$= \sum_{k=n+1}^{L} (k-n)\{W-(k-1)r\}b_k$$

$$+ \sum_{k=n+1}^{L} (k-n) \sum_{m=k+1}^{M} 2rb_m$$

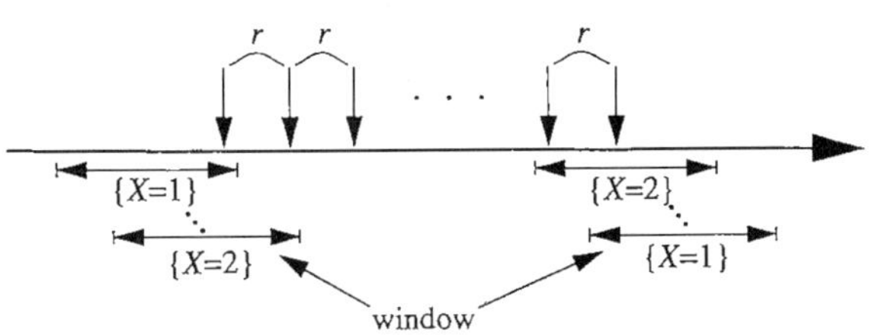

Fig. A·1 Number of cell arrivals in a window.

$$+ (L-n) \sum_{m=L+1}^{M} (m-L-1)(r-l)b_m$$

$$+ (L+1-n) \sum_{m=L+1}^{M} (m-L)lb_m$$

$$= \sum_{k=n+1}^{L} (k-n)\{W-(k-1)r\}b_k$$

$$+ \left\{ \sum_{m=n+2}^{L} \sum_{k=n+1}^{m-1} + \sum_{m=L+1}^{M} \sum_{k=n+1}^{L} \right\} 2(k-n)rb_k$$

$$+ \sum_{m=L+1}^{M} \{(L-n)(m-L-1)r + (m-n)l\}b_m$$

$$= \sum_{k=n+1}^{L} (k-n)\{W-(k-1)r\}b_k$$

$$+ \sum_{m=n+1}^{L} (m-n)(m-1-n)rb_m$$

$$+ \sum_{m=L+1}^{M} (L-n)(L+1-n)rb_m$$

$$+ \sum_{m=L+1}^{M} \{(L-n)(m-L-1)r + (m-n)l\}b_m$$

$$= (W-nr) \sum_{k=n+1}^{L} (k-n)b_k$$

$$+ \sum_{k=L+1}^{M} (k-n)\{(L-n)r + l\}b_k$$

$$= (W-nr) \sum_{k=n+1}^{M} (k-n)b_k. \tag{A·2}$$

Similarly,

$$E(X_{\mathcal{A}}) = C_1 W \sum_{k=1}^{M} kb_k = SW, \tag{A·3}$$

$$E(X_{\mathcal{A}'} - n)_+ = C_2(W-nr)(M-n), \tag{A·4}$$

and

$$E(X_{\mathcal{A}'}) = C_2 WM = SW. \tag{A·5}$$

We finally have

$$E(X_{\mathcal{A}'} - n)_+ - E(X_{\mathcal{A}} - n)_+$$
$$= C_2(W-nr)(M-n)$$
$$- C_1(W-nr) \sum_{k=n+1}^{M} (k-n)b_k$$
$$= S\frac{(W-nr)(M-n)}{M}$$

$$
\begin{aligned}
&- S\frac{(W-nr)\sum_{k=n+1}^{M}(k-n)b_k}{\sum_{k=1}^{M} kb_k} \\
&= \left\{(M-n)\sum_{k=1}^{M} kb_k - M\sum_{k=n+1}^{M}(k-n)b_k\right\} \\
&\quad \cdot \frac{(W-nr)S}{M\sum_{k=1}^{M} kb_k} \\
&= \left\{(M-n)\sum_{k=1}^{n} kb_k + n\sum_{k=n+1}^{M}(M-k)b_k\right\} \\
&\quad \cdot \frac{(W-nr)S}{M\sum_{k=1}^{M} kb_k} \\
&\geqq 0. \qquad \text{(A· 6)}
\end{aligned}
$$

Since $X_{\mathcal{A}}$ and $X_{\mathcal{A}'}$ are at most $L+1$, we have $E(X_{\mathcal{A}} - n)_+ = E(X_{\mathcal{A}'} - n)_+ = 0$ for $n \geqq L+1$. For $n \leqq 0$, $E(X_{\mathcal{A}} - n)_+ = E(X_{\mathcal{A}'} - n)_+ = SW - n$. □

Appendix C: Proof of Eq. (14)

We need to prove the steps (k.1) and (k.2) increase $X_{\tilde{\omega}_i}$ in convex order. Consider the cell arrival times $a_n, \ldots a_{n+L-1}$, where $a_n - a_{n-1} \geqq W$ and $r \leqq a_i - a_{i-1} \leqq W$ for $i = n, \ldots, n+L-1$ (Fig. A· 2). Let the arrival time of the n-th cell move to $a_{n,1}$ so that $a_{n+1} - a_{n,1} = r$. This change increases the frequency of $\{X = 0\}$ by $a_{n+1} - a_n - r$, and decreases the frequency of $\{X = 1\}$ by $a_{n+1} - a_n - r$. Moreover, several occurrence of $\{X = i\}$ should change into $\{X = i+1\}$ for some $i = 2, \ldots, L$ (note that i is not necessarily unique), since the expected value of X does not change. Those changes of the frequency increases X in the meaning of convex ordering (simple and straightforward calculation like in Appendix B gives its proof). Next, let $a_{n,1}$ and a_{n+1} move to $a_{n,2}$ and $a_{n+1,2}$ so that $a_{n+2} - a_{n+1,2} = a_{n+1,2} - a_{n,2} = r$. This also increases X in convex order. Consequently, these changes increase X in convex order, and construct step (k.1).

Next, consider the cell arrival times $a_{n+1}, \ldots, a_{n+L+1}$, where $a_{i+1} - a_i = r$ for $i = n, \ldots, n+L-1$, and $r \leqq a_{n+L+1} - a_{n+L} < W$ (Fig. A· 3). Let a_{n+L+1} move to a'_{n+L+1} so that $a'_{n+L+1} - a_{n+L} = r$. The possible decrease in frequencies of the event $\{X = L+1\}$ is at most l, although the increase in $\{X = L+1\}$ by l is

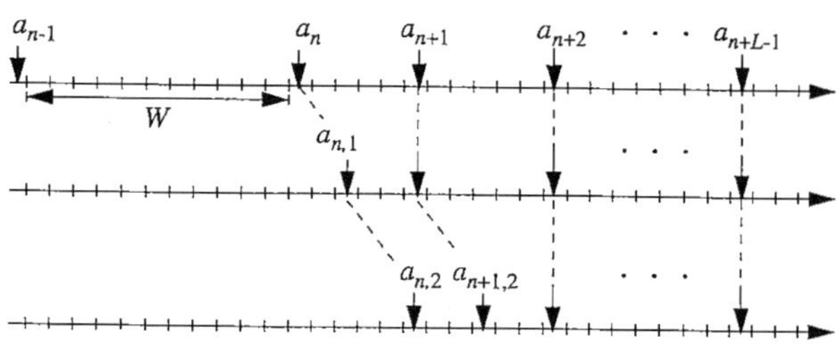

Fig. A· 2 Step (k.1).

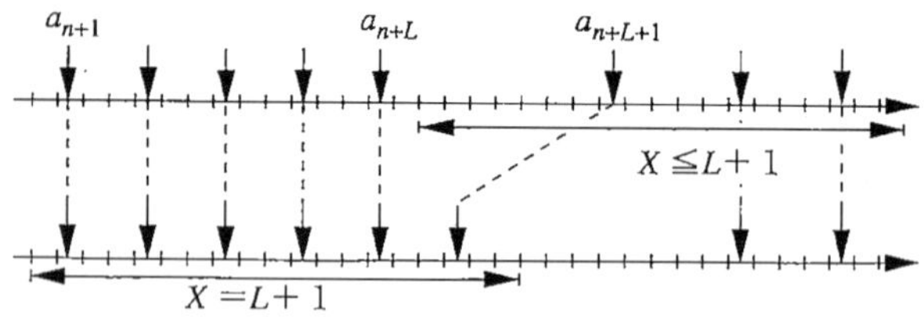

Fig. A· 3 Step (k.2).

certain, and so on. Therefore, this change also increass X in convex order and repeating this change constructs step (k.2). □

High-Performance Backbone Network Technology edited by Naoaki Yamanaka
ISBN: 0-8247-5321-6 DOI: 10.1081/(120025711)
Published 2004 by Marcel Dekker, Inc.

Connection Admission Control Guaranteeing Negotiated Cell-Loss Ratio of Cell Streams Passing Through Usage Parameter Control

Shigeo Shioda and Hiroshi Saito

SUMMARY A connection admission control (CAC) that guarantees a negotiated cell-loss ratio for all cell-streams passing through the usage parameter control (UPC) in ATM networks is proposed. In particular, the cases in which a jumping-window, sliding-window, or continuous-leaky-bucket scheme are used for peak-cell-rate policing are discussued, and the upper bound for cell-loss ratio of the cell-streams passing through each type of UPC is derived. The CACs based on the derived cell-loss-ratio upper bounds ensure the quality of service in all cases by combining the relevant UPCs. There are three possible combinations of CAC and UPC, depending on the UPC mechanism used. The impact of the choice of CAC and UPC combination on bandwidth utilization is discussed using several numerical examples.

key words: *ATM, CAC, UPC, leaky bucket, worst case performance, queueing systems*

1. Introduction

The Asynchronous Transfer Mode (ATM) is a key technology for Broadband Integrated Services Digital Networks (B-ISDNs) [2], which can provide a variety of services using a unified interface and architecture. In ATM networks, information is segmented into fixed-size blocks called cells and the necessary number of cells are transmitted for the amount of information in the network. Therefore, the traffic control methods used in existing transfer modes are not applicable to ATM networks.

The principle of ATM traffic control should be simple. At connection setup, the user specifies both quality-of-service (QOS) requirements and, using a source traffic descriptor, the anticipated traffic characteristics of the connection. Network resources for the connection are assigned on the basis of the source traffic descriptor values and the QOS requirements. If there are not enough network resources, the connection is rejected. If the connection is accepted, actual traffic characteristics are checked for conformity with the source traffic descriptor specified at connection set-up by monitoring the cell stream from the connection. If the characteristics do not conform, a penalty is imposed on the connection, e.g. some cells from the connection may be discarded. The first of these procedures, at the connection setup, is called connection admission control (CAC); the procedure after connection setup is called usage parameter control (UPC).

To simplify traffic specification by the user and monitoring by the network, the traffic descriptors are required to be observable and easily checked through some algorithmic mechanisms. Algorithmic or rule-based traffic descriptors currently being discussed in the ITU satisfy these requirements. Such traffic descriptors, however, cannot simplify the operation of connection admission control and resource management. For instance, algorithmic traffic descriptors allow a wide variety of traffic patterns to enter the network, and the network needs to guarantee the QOS requirement for all "compliant connections" that conform to the source traffic descriptor [1]. We therefore need to establish a means of evaluating the "worst" QOS of all possible cases anticipated from the information of the source traffic descriptor.

In this sense, the specification of, and the resource allocation based on, the worst traffic pattern have recently become key issues for ATM traffic control. The pioneering work in this field is Ref. [11] by Doshi. This formally addressed this problem for the case where the leaky bucket algorithm is used for conformity checks and shows that the so-called on/off pattern, which is generally believed to be the worst traffic pattern [13], is not always the worst. Based on the analysis of queue-length distribution, Yamanaka et al. [23] also gave a non-intuitive example in which the on/off source may not be the worst. Lee [14] found the worst traffic pattern of possible output processes from the leaky bucket; in the sense that it maximizes the average waiting time. However, based on the criteria of cell loss ratio (CLR), the worst case traffic pattern has not yet been clarified and a simple implementation of CAC based on the worst case traffic has not been proposed.

This paper proposes a connection admission control that guarantees CLR standards for all connections passing through the UPC, which is used for peak-cell-rate (PCR) and cell-delay-variation-tolerance (CDVT) policing. Although the conformance definitions of the PCR and CDVT are given in terms of the continuous-leaky-bucket-type algorithm, the network provider is not required to use this algorithm for the UPC. The net-

Manuscript received January 22, 1996.
Manuscript revised July 16, 1996.

[†]The authors are the with NTT Multimedia Networks Laboratories, Musashino-shi, 180 Japan, E-mail: s-sioda@hashi.tnl.ntt.co.jp

IEICE TRANS. COMMUN., Vol. E80-B, No. 3, pp. 399–411, March 1997

work may use any UPC as long as the operation of the UPC supports the QOS objectives of a compliant connection [3]. Thus network providers choose their own UPC algorithms in terms of the trade-off between implementability and resource utilization. This paper considers three different UPC mechanisms, that is, jumping window (JW), sliding window (SW), and continuous leaky bucket (LB).

After presenting some preliminaries in Sect. 2, this paper analyzes the G/D/1/K+1 system in Sect. 3, assuming that the traffic characteristics are specified by the relative frequency of the number of cell arrivals during a fixed interval. Since the relative frequency of the number of cells arriving during a fixed interval does not fully characterize the cell arrival process, cell arrival patterns specified by the same relative frequency may experience different cell-loss ratios. Thus in Sect. 3, we derive the upper bounds of CLRs of all possible patterns specified by the same relative frequency. These upper bounds are used in Sect. 4 to derive upper bounds for CLRs of cell streams passing through the UPC for the three mechanisms, JW, SW, and LB. CACs based on these upper bounds allow us to assure the QOS of all compliant connections by combining the relevant UPCs. There are three possible combinations of CAC and UPC, depending on the UPC mechanism used. We discuss the impact of the choice of CAC and UPC combination on bandwidth utilization, using several numerical examples. The proofs of several formulas used in this paper are given in the appendix.

2. Preliminaries

Of the various performance-deterioration factors, this paper concentrates on the cell-loss ratio (CLR) in an ATM node output buffer for a transmission link. The reason for choosing this focus is that the CLR depends mainly on traffic characteristics and is therefore the key factor in traffic control problems [17]. (The QOS cell-delay requirement can be satisfied by proper dimensions of the output buffer [17], so it will not be discussed in this paper.) The ATM output buffer can be regarded as a G/D/1/K+1 system with the first-in-first-out (FIFO) discipline because the cell length is fixed. The time axis is assumed to be divided into slots with a length equal to the cell transmission time L/C, where L is the cell length and C is the virtual-path bandwidth (Fig. 1). A cell arrives at the beginning of a slot and is transmitted at the end of a slot. The cell arrival process is assumed to be a stationary stochastic process. In this paper, the CLR is defined sample-path-wise as

$$\mathrm{CLR} = \lim_{t\to\infty} \frac{\text{Number of lost cells during } [0,t]}{\text{Number of cells arriving during } [0,t]}.$$

Connection Admission Control aims to satisfy the cell-loss-ratio standard assigned to each virtual path connection, as usual [18], [22]. For simplicity, the cell-loss-ratio standard is assumed to be the same for all individual connections.

The traffic parameters in a source traffic descriptor are assumed to be the peak cell rate (PCR), cell delay variation tolerance (CDVT), sustainable cell rate (SCR), and burst tolerance (BT). The PCR is defined by the inverse of the minimum of the cells' inter-arrival time; for the definitions of other traffic parameters, see Ref. [1] or [3].

The number of cells arriving in every fixed period (the definition period) is assumed to be monitored in order to determine the characteristics of the cell arrival process. This assumption is used in manipulating the UPC mechanisms, which monitor the number of cells arriving within the fixed-length period for conformity with the source traffic descriptor. Since the cell arrival process is assumed to be stationary, the relative frequency of the number of cells arriving during each definition period, $\{p(k;r)\}_{k=0}^{\infty}$, can be defined sample-path-wise as

$$p(k;r) = \lim_{N\to\infty} \frac{1}{N} \sum_{n=1}^{N} \mathbf{1}(a_n^{(r)} = k), \tag{1}$$

where $a_n^{(r)}$ is the number of cells arriving during the n-th definition period (Fig. 1), and $\mathbf{1}(x)$ is the indicator

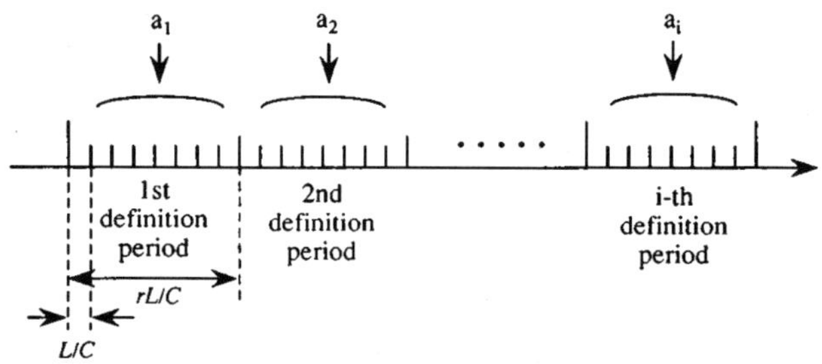

Fig. 1 Definition-period sequence.

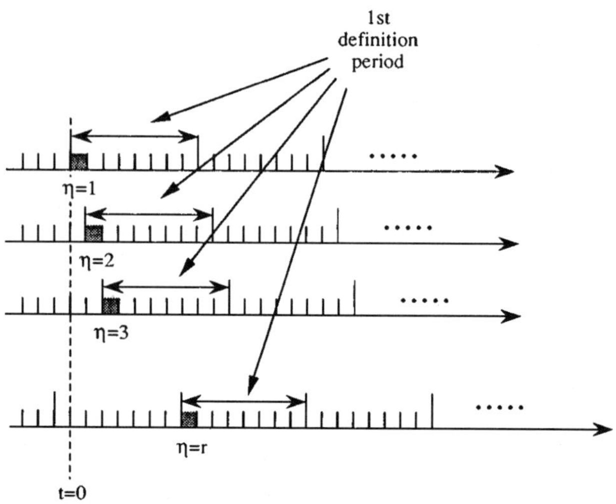

Fig. 2 Parameter η.

function [5]. The parameter r is used to denote the definition period's length in terms of L/C, i.e. the length of the definition period is equal to rL/C. Where necessary, we will use the notation $\{p(k;r,\eta)\}_{k=0}^{\infty}$ to recall that the function may depend on how the definition period sequence is defined on the time axis: the parameter η $(1 \leqq \eta \leqq r)$ denotes the sequential number of the slot at which of the first definition period starts (Fig. 2).

3. Bandwidth Allocation Based on a Source Traffic Descriptor

3.1 Upper Bounds of CLR for General Sources

This section describes several upper bounds of the CLR for the cell arrival process where traffic characteristics are specified by the relative frequencies of the number of cells arriving during the definition period. The derivation of these upper bounds is given in the Appendix.

Suppose that the ATM switch output-buffer size is K. The following inequalities concerning the CLR are satisfied for each sample path:
(1) Formula 1: for all η

$$\mathrm{CLR} \leqq B_1(p;r,\eta) \equiv \frac{\sum_k [k-\gamma_1(r)]^+ p(k;r,\eta)}{\sum_k kp(k;r,\eta)}, \quad (2)$$

where

$$\gamma_1(r) \stackrel{\mathrm{def}}{=} \min\{r,(K+2)/2\}, \qquad [x]^+ \stackrel{\mathrm{def}}{=} \begin{cases} x, & x \geqq 0, \\ 0, & \text{otherwise}. \end{cases} \quad (3)$$

(2) Formula 2: for all η

$$\mathrm{CLR} \leqq B_2(p;r,\eta) \equiv \frac{\sum_k [k-\gamma_2(r)]^+ p(k;r,\eta)}{\sum_k kp(k;r,\eta)} + q(r,\eta), \quad (4)$$

where

$$\gamma_2(r) \stackrel{\mathrm{def}}{=} \min\{r, K+1\}, \quad (5)$$

and $q(r,\eta)$ is a function of r and η (See Appendix for the definition of $q(r,\eta)$) satisfying

$$\min_\eta q(r,\eta) \leqq 0. \quad (6)$$

These upper bounds are applicable to multiplexed connections when the relative frequency of each multiplexed connection is known. If n connections are multiplexed and the relative frequency at which k cells from the i-th connection arrive during the definition period is $p_i(k;r,\eta)$, Formula 1 gives

$$\mathrm{CLR} \leqq B_1(p_1,\ldots,p_n;r,\eta) \equiv \frac{\sum_k [k-\gamma_1(r)]^+ p_1 \cdots \star p_n(k;r,\eta)}{\sum_k kp_1 \star \cdots \star p_n(k;r,\eta)}, \quad (7)$$

and Formula 2 gives

$$\mathrm{CLR} \leqq B_2(p_1,\ldots,p_n;r,\eta) \equiv \frac{\sum_k [k-\gamma_2(r)]^+ p_1 \star \cdots \star p_n(k;r,\eta)}{\sum_k kp_1 \star \cdots \star p_n(k;r,\eta)} + q(r,\eta), \quad (8)$$

where $\star$ denotes convolution.

3.2 Upper Bounds of CLR for Connections Specified by PCR and SCR

Consider the upper bounds of the CLR for the multiplexed connections specified by PCR and SCR. First we consider the full-rate case, that is, when the PCR and SCR are respectively equal to the true peak and average cell rate. (The non-full-rate case is discussed in Sect. 3.3.) In addition, for simplicity, we assume that CDVT $= 0$, but the value of BT is arbitrary. Let R^* and A^* be the PCR and SCR, respectively. The average and maximum number of cells arriving (ANA and MNA) during the definition period with length rL/C are given by

$$\hat{A}^*(r) = rLA^*/C, \qquad \hat{R}^*(r) = \lceil rLR^*/C \rceil \quad (9)$$

where $\hat{A}^*(r)$ and $\hat{R}^*(r)$ denote ANA and MNA respectively, and $\lceil x \rceil$ is the smallest integer larger than or equal to x. Let $\{\theta_i(k_i;r)\}_{k=0}^{\infty}$ be the following Bernoulli-like distribution:

$$\theta_i^*(k;r) \equiv \begin{cases} \hat{A}_i^*(r)/\hat{R}_i^*(r), & k = \hat{R}_i^*(r), \\ 1-\hat{A}_i^*(r)/\hat{R}_i^*(r), & k = 0, \\ 0, & \text{otherwise}. \end{cases} \quad (10)$$

Here $\hat{A}_i^*(r)$ and $\hat{R}_i^*(r)$ are respectively the ANA and MNA of the i-th connection. It is proved in [17] that the upper bound of the CLR can be obtained by using $\{\theta_i^*(k_i;r)\}_{k=0}^{\infty}$ as follows:

$$\mathrm{CLR} \leqq B_1(p_1,\ldots,p_n;r,\eta) \leqq B_1(\theta_1^*,\ldots,\theta_n^*;r) \equiv \frac{\sum_k [k-\gamma_1(r)]^+ \theta_1^* \star \cdots \star \theta_n^*(k;r)}{\sum_k k\theta_1^* \star \cdots \star \theta_n^*(k;r)} \quad \text{for all } \eta. \quad (11)$$

(Note that the replacement of $p(k;r,\eta)$ by $\theta^*(k;r)$ is possible for all η because the MNA and the ANA are always given by Eq. (9) regardless of how the definition-period sequence is defined on the time axis.) Similarly, from Formula 2 we have:

$$\mathrm{CLR} \leqq B_2(p_1,\ldots,p_n;r,\eta) \leqq B_2(\theta_1^*,\ldots,\theta_n^*;r,\eta) \equiv \frac{\sum_k [k-\gamma_2(r)]^+ \theta_1^* \star \cdots \star \theta_n^*(k;r)}{\sum_k k\theta_1^* \star \cdots \star \theta_n^*(k;r)} + q(r,\eta) \quad \text{for all } \eta. \quad (12)$$

By using the fact that the first term of Eq. (12) does not depend on η, Eq. (12) can be reduced to the following simpler expression:

$$\begin{aligned} \text{CLR} &\leq \min_{\eta}\left\{\frac{\sum_k [k-\gamma_2(r)]^+ \theta_1^* \star \cdots \star \theta_n^*(k;r)}{\sum_k k\theta_1^* \star \cdots \star \theta_n^*(k;r)} + q(r,\eta)\right\} \\ &= \frac{\sum_k [k-\gamma_2(r)]^+ \theta_1^* \star \cdots \star \theta_n^*(k;r)}{\sum_k k\theta_1^* \star \cdots \star \theta_n^*(k;r)} + \min_{\eta}\{q(r,\eta)\} \\ &\leq B_3(\theta_1^*,\ldots,\theta_n^*;r) \\ &\equiv \frac{\sum_k [k-\gamma_2(r)]^+ \theta_1^* \star \cdots \star \theta_n^*(k;r)}{\sum_k k\theta_1^* \star \cdots \star \theta_n^*(k;r)}. \end{aligned} \tag{13}$$

To derive the final inequality, we use the property of $q(r,\eta)$ described in Eq. (6). It is easily seen that $B_3(\theta_1^*,\ldots,\theta_n^*;r)$ is smaller than or equal to $B_1(\theta_1^*,\ldots,\theta_n^*;r)$ for all r.

The right hand side of Eq. (13) contains the parameter r, which corresponds to the length of the definition period. To obtain tight bounds, r should take a value such that rLR/C is an integer; otherwise, the peak cell rate will be overestimated (cf. Eq. (9)). When rLR/C is an integer and $r \leq K+1$, Eq. (13) is identical to the burst overflow probability proposed by Murase et al. [15].

Figure 3 shows the upper bounds of the CLRs determined using Eq. (13), together with CLRs derived from the other models, that is, the M/D/1/K and MMPP/D/1/K models [6], [9]. We consider the case where 70 connections of on/off-sources were multiplexed in a VP with a capacity of 150 Mb/s. Peak and average bit rates of each on/off-source are respectively 10 Mbps and 1 Mbps. The CLRs are evaluated by three different average burst sizes (amount of information transmitted during the on state): 1000, 5000, and 20000 Bytes. In the M/D/1/K and MMPP/D/1/K models, cell arrival processes of the multiplexed connections are modeled by the Poisson process and a two-state MMPP, respectively [7], and the CLRs are evaluated by standard queueing analysis [6], [9]. The parameter r in Eq. (13) was set at $C/(LR) = 15$, considering the case where the definition period length is equal to the inverse of the peak cell rate. Figure 3 plots the CLRs while the buffer size is changed from 1 to 100. When the buffer size in cells is larger than $r-1 = 14$, the CLRs of the MMPP/D/1/K model seems to become close to the upper bounds from Eq. (13) as the burst size approaches infinity. This is very reasonable: The derived upper bound does not assume any information about the burst size. Thus it should give the result of the case where the burst size is infinite because it is the worst case. It should be noted that, for the case where the burst size is known a priori, we can derive a tighter upper bound than with Eq. (13) [21].

When the buffer size is smaller than $r-1 = 14$, the results of Eq. (13) are rather larger than those of the M/D/1/K model and the MMPP/D/1/K model. Within this region, the cell-loss-ratio upper bound of Eq. (13) was derived in consideration of the very exceptional queueing behavior illustrated in Fig. 4, where at most $(K+1)$ cells are served during one definition period. Further discussion is needed as to whether such exceptional queueing behavior is worth considering in deriving the upper bounds of the CLR.

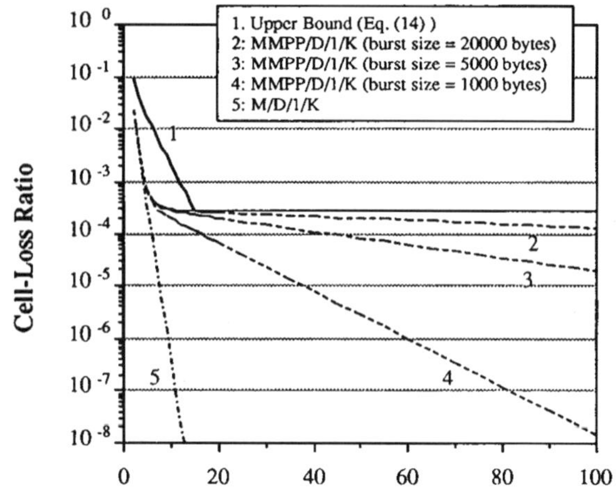

Fig. 3 Cell loss ratio versus buffer size: comparison between different analytical approaches for a reference source with peak bit rate = 10 Mb/s and average bit rate = 1 Mb/s.

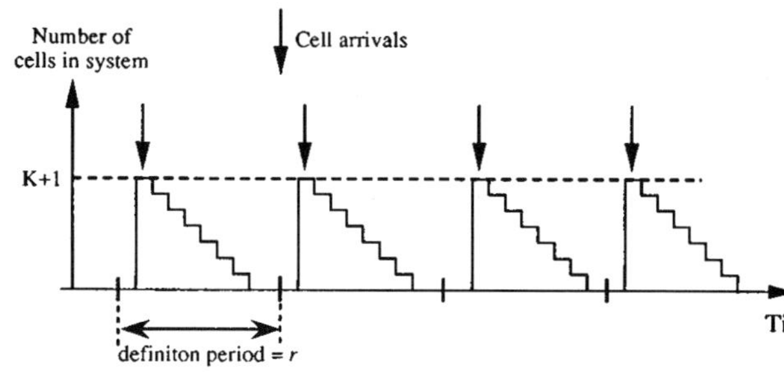

Fig. 4 Worst case queueing behavior.

3.3 Differences between Actual and Contracted Values in Traffic Parameters

In Sect. 3.2, we obtained the upper bounds of CLR for the full-rate case where each user transmits as many cells as possible while conforming to the source traffic descriptor. In practice, however, the actual peak and average cell rates of each connection can take any value up to their contracted values. We therefore need to clarify whether Eq. (13) gives the correct upper bounds when the actual rates are equal to or less than the contracted ones.

Let A_i and R_i be the actual average and peak cell

rates, respectively, of the i-th connection, and define

$$\theta_i(k;r) \equiv \begin{cases} \hat{A}_i(r)/\hat{R}_i(r), & k = \hat{R}_i(r), \\ 1 - \hat{A}_i(r)/\hat{R}_i(r), & k = 0, \\ 0, & \text{otherwise}, \end{cases} \tag{14}$$

$$\hat{A}_i(r) = rLA_i/C \qquad \hat{R}_i(r) = \lceil rLR_i/C \rceil. \tag{15}$$

Actual rates should not be greater than contracted values, that is,

$$\begin{aligned} &A_i \leqq A_i^* \quad \text{and} \quad R_i \leqq R_i^*. \\ &(\hat{A}_i(r) \leqq \hat{A}_i^*(r) \quad \text{and} \quad \hat{R}_i(r) \leqq \hat{R}_i^*(r).) \end{aligned} \tag{16}$$

We assume that A_i, the true average cell rate of the i-th connection ($i = 1, \ldots, N$), is identical to the contracted SCR A_i^*. From Eq. (13)

$$\begin{aligned} &B_3(\theta_1, \ldots, \theta_n; r), \\ &= \frac{\sum_k [k - \gamma_2(r)]^+ \theta_1 \star \theta_2 \star \cdots \star \theta_n(k;r)}{\sum_k k\theta_1 \star \theta_2 \star \cdots \star \theta_n(k;r)} \\ &= \Big(\sum_i \hat{A}_i(r)\Big)^{-1} \Big\{ \sum_k [k - \gamma_2]^+ \theta_1 \star \cdots \\ &\quad \star \theta_{n-1}(k;r)(1 - \frac{\hat{A}_n(r)}{\hat{R}_n(r)}) \\ &\quad + \sum_k [k - \gamma_2]^+ \theta_1 \star \cdots \\ &\quad \star \theta_{n-1}(k - \hat{R}_n(r); r) \frac{\hat{A}_n(r)}{\hat{R}_n(r)} \Big\} \\ &= \Big(\sum_i \hat{A}_i(r)\Big)^{-1} \Big\{ \sum_k [k - \gamma_2]^+ \theta_1 \star \cdots \star \theta_{n-1}(k;r) \\ &\quad + \sum_k ([k - \gamma_2 + \hat{R}_n(r)]^+ - [k - \gamma_2]^+) \\ &\quad \theta_1 \star \cdots \star \theta_{n-1}(k;r) \frac{\hat{A}_n(r)}{\hat{R}_n(r)} \Big\} \\ &\leqq \Big(\sum_i \hat{A}_i^*(r)\Big)^{-1} \Big\{ \sum_k [k - \gamma_2]^+ \theta_1 \star \cdots \star \theta_{n-1}(k;r) \\ &\quad + \sum_k ([k - \gamma_2 + \hat{R}_n^*(r)]^+ - [k - \gamma_2]^+) \\ &\quad \theta_1 \star \cdots \star \theta_{n-1}(k;r) \frac{\hat{A}_n^*(r)}{\hat{R}_n^*(r)} \Big\} \\ &= B_3(\theta_1, \ldots, \theta_{n-1}, \theta_n^*; r) \\ &\leqq B_3(\theta_1^*, \ldots, \theta_n^*; r). \end{aligned} \tag{17}$$

The CLR is therefore maximized when the actual peak cell rate is identical to its contracted value.

However, this is not the case for the average cell rate. The CLR is a complicated function of the average cell rate and, for arbitrary traffic conditions, cannot be a monotonic function of it. In general, the CLR is not an increasing function of the true average cell rate. We show an intuitive example in Fig. 5, where CLR increases even though the average cell rate decreases.

Aida et al. [4] showed that, if the inequality $(\hat{R}^*(r))^2/\hat{A}^*(r) \leqq r$ holds, the full-rate case is the worst;

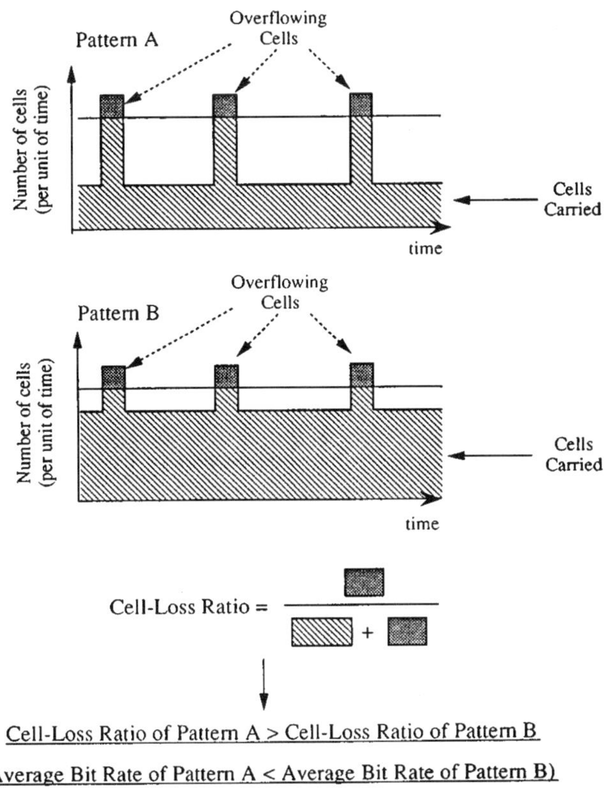

Fig. 5 Example where cell-loss ratio is not an increasing function of average cell rate.

that is, Eq. (13) calculated based on the contracted traffic parameter values (i.e. A^* and R^*) gives the correct upper bound. Even if this inequality does not hold, it can be proved that the CLR is upper bounded by Eq. (13) simply by replacing $A(r)$ by $\hat{R}^*(r))^2/r$ and by letting $\hat{R}(r) = \hat{R}^*(r)$, although this upper bound is not tight.

4. Bandwidth Allocation for Connections Passing through the UPC

4.1 UPC Parameters and Traffic Descriptors

Here we describe how to determine the parameter values of the UPC for PCR policing from the source traffic descriptor values. The UPC mechanism for average-cell-rate policing will not be explicitly considered in this paper; instead, the contracted SCR and BT are assumed to be known. Further, for simplicity, BT is assumed to be infinite. For the case in which BT is finite, see ref. [21].

(1) Jumping Window (JW) and Sliding Window (SW)
The JW and SW systems are both characterized by two parameters: window length W and the maximum number of cells accepted within that window, N [16]. The difference between JW and SW is that the former always limits the number of cells accepted in any possible interval of length W to a maximum N, but the latter cannot

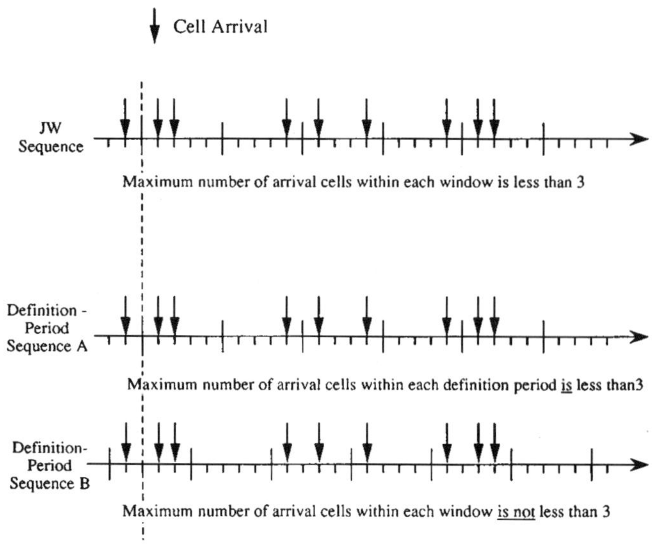

Fig. 6 JW sequence and the definition-period sequence.

do so if the window sequence is not synchronized with the cell-counting sequence. (See Fig. 6.)

Although the conformance definitions of the peak cell rate R^* and the CDVT τ are defined in terms of the Generic Cell Rate Algorithm (GCRA) or its equivalent algorithms, i.e., continuous leaky bucket, the network is not required to use these algorithms; the JW or SW algorithm can be used for PCR policing instead.

Consider the necessary condition of JW and SW for being transparent to the cells conforming two declared parameters, R^* and τ. Let $t_n(R^*,\tau)$ be the shortest period during which n consecutive cells from a compliant user can arrive. It is easily seen that t_n is given by

$$t_n(R^*,\tau) = \max\{0,(n-1)/R^* - \tau\}. \tag{18}$$

We first observe that $N(W)$, the maximum number of cells arriving during the period of length W, should satisfy

$$t_{N(W)}(R^*,\tau) \leqq W < t_{N(W)+1}(R^*,\tau). \tag{19}$$

Applying Eq. (18) to this, we have

$$\max\{0,(N(W)-1)/R^* - \tau\} \leqq W$$
$$< \max\{0, N(W)/R^* - \tau\}, \tag{20}$$

where

left-hand-side inequality
$$\Rightarrow (N(W)-1)/R^* - \tau \leqq W,$$
right-hand-side inequality
$$\Rightarrow N(W)/R^* - \tau, \tag{21}$$

which yield

$$(W+\tau)R^* < N(W) \leqq 1 + (W+\tau)R^*. \tag{22}$$

Thus the UPC parameters of JW and SW, i.e. W and N, need to satisfy

$$N = 1 + \lfloor (W+\tau)R^* \rfloor, \tag{23}$$

where $\lfloor x \rfloor$ denotes the largest integer smaller than or equal to x.

(2) Leaky Bucket (LB)

The Continuous Leaky Bucket algorithm is described in Annex 1 of ref. [1], and is specified by the parameter T and the CDV tolerance τ. The leak rate is equal to $1/T$, so T is given by

$$T = 1/R^*. \tag{24}$$

4.2 Upper Bound of CLR for the Cell Streams Passing through the UPC

(1) Jumping Window

We assume that all JWs used in the network have the same window length, $W = wL/C$. We let $\{p_i^{JW}(k;w,\eta)\}_{k=0}^{\infty}$ be the relative frequency of the number of cells arriving from the i-th connection, and assume that all connections are enforced by JW. The definition period is equal to the window length of the JW. The maximum number of cells (MNA) arriving during each definition period from each connection is equal to $N = 1 + \lfloor (W+\tau)R^* \rfloor$ if the window sequence of the JW is synchronized with the definition-period sequence. (Otherwise, MNA is not necessarily equal to N, see Fig. 6.) There exists a value of η that synchronizes the definition periods with the JW windows, which we denote as η_{ws}. Let $\{\theta_i(k;w)\}_{k=0}^{\infty}$ be the following Bernoulli-like distribution:

$$\theta_i(k;w) \equiv \begin{cases} \hat{A}_i^*(w)/N_i, & k = N_i, \\ 1 - \hat{A}_i^*(w)/N_i, & k = 0, \\ 0, & \text{otherwise}, \end{cases}$$
$$\text{and} \quad N_i = 1 + \lfloor (wL/C+\tau)R_i^* \rfloor. \tag{25}$$

By substituting Eq. (25) into Formula 1 and using the same argument as those used in sect. 3.2, we can show that the CLR of the cell stream passing through JW, CLR^{JW}, is upperbounded by

$$\begin{aligned} \mathrm{CLR}^{JW} &\leqq B_1(p_1^{JW},\ldots,p_n^{JW};w,\eta_{ws}) \\ &\leqq B_1(\theta_1,\ldots,\theta_n;w) \\ &\equiv \frac{\sum_k [k-\gamma_1(w)]^+ \theta_1 \star \cdots \star \theta_n(k;w)}{\sum_k k\theta_1 \star \cdots \star \theta_n(k;w)}. \end{aligned} \tag{26}$$

(Notice that, by the argument in Sect. 3.3, we need to replace $\hat{A}_i^*(w)$ by $(\hat{R}_i^*(w))^2/w$ in the derived upper bounds when $\hat{A}_i^*(w) < (\hat{R}_i^*(w))^2/w$.)

(2) Sliding Window

Similarly, let $\{p_i^{SW}(k;w,\eta)\}_{k=0}^{\infty}$ be the relative frequency of the i-th connection enforced by a SW whose window length, $W = wL/C$, is equal to the definition period. The maximum number of cells arriving during each definition period (MNA) from each multiplexed connection is also equal to $N_i = 1 + \lfloor (W+\tau)R_i^* \rfloor$. Unlike the JW mechanism, this is assured even if the window sequence of the SW is not synchronized with the definition-period sequence. We can therefore replace $\{p_i^{SW}(k; w, \eta)\}_{k=0}^{\infty}$ by the Bernoulli-like distribution in Eq. (25) for all η in Formula 2. By the same arguments as those used for deriving Eq. (13) in Sect. 3.2, it can be proved that the CLR of the cell streams multiplexed through the SW, CLR^{SW}, is upper-bounded by

$$\text{CLR}^{SW} \leqq B_3(\theta_1, \dots, \theta_n; w) \equiv \frac{\sum_k [k - \gamma_2(w)]^+ \theta_1 \star \cdots \star \theta_n(k; w)}{\sum_k k\theta_1 \star \cdots \star \theta_n(k; w)}, \quad (27)$$

where

$$\theta_i(k; w) \equiv \begin{cases} \hat{A}_i^*(w)/N_i, & k = N_i, \\ 1 - \hat{A}_i^*(w)/N_i, & k = 0, \\ 0, & \text{otherwise}, \end{cases}$$
$$\text{and} \quad N_i = 1 + \lfloor (wL/C + \tau) R_i^* \rfloor. \quad (28)$$

(3) Leaky Bucket

When the length of the definition period is equal to rL/C, each connection through the LB transmits at most $1 + \lfloor (rL/C + \tau)/T \rfloor$ cells during one definition period (see Eq. (23)). As with the SW, this condition is assured regardless of how the definition period sequence is defined on the time axis. We therefore arrive at the following upper bounds of the CLR, by the same argument as those used for the SW:

$$\text{CLR}^{LB} \leqq \min_r \{B_3(\theta_1, \dots, \theta_n; r)\} \equiv \min_r \left\{ \frac{\sum_k [k - \gamma_2(r)]^+ \theta_1 \star \cdots \star \theta_n(k; r)}{\sum_k k\theta_1 \star \cdots \star \theta_n(k; r)} \right\}, \quad (29)$$

where

$$\theta_i(k; r) \equiv \begin{cases} \hat{A}_i^*(r)/N_i, & k = N_i, \\ 1 - \hat{A}_i^*(r)/N_i, & k = 0, \\ 0, & \text{otherwise}, \end{cases}$$
$$\text{and} \quad N_i = 1 + \lfloor (rL/C + \tau_i)/T_i \rfloor. \quad (30)$$

The CAC based on Eq. (29) is hard to implement because it is necessary to search for the value of r minimizing the right-hand side on a call-by-call basis. Instead of directly applying Eq. (29) to the CAC, we propose to use the CLR upper bound of Eq. (29) for $r = K + 1$, that is:

$$\text{CLR}^{LB} \leqq B_3(\theta_1, \dots, \theta_n; K+1) \equiv \frac{\sum_k [k - \gamma_2(K+1)]^+ \theta_1 \star \cdots \star \theta_n(k; K+1)}{\sum_k k\theta_1 \star \cdots \star \theta_n(k; K+1)}, \quad (31)$$

where

$$\theta_i(k; K+1) \equiv \begin{cases} \hat{A}_i^*(K+1)/N_i, & k = N_i, \\ 1 - \hat{A}_i^*(K+1)/N_i, & k = 0, \\ 0, & \text{otherwise}, \end{cases}$$
$$N_i = 1 + \lfloor ((K+1)L/C + \tau_i)/T_i \rfloor. \quad (32)$$

(Several numerical experiments confirm that $B_3(\theta; K+1)$ well approximates $\min_r\{B_3(\theta; r)\}$ although no satisfactory explanation regarding it has been found.) The CLR upper bounds of Eq. (29) and Eq. (31) are numerically compared in Sect. 4.3.

It should be noted that the CACs based on the CLR upper bounds derived in Sect. 4.2 can be implemented in the same manner as that discussed in [17]. We simply outline how the CAC operates based on the derived upper bounds:

1. The current relative frequency $\theta(k; r)$ is stored in memory.
2. When a new customer requests connection admission, calculate the "worst relative frequency," $\theta_{new}(k; r)$, by Eq. (25) (for JW), (28) (for SW), or (32) (for LB).
3. Next calculate $\theta \star \theta_{new}(k; r)$ by taking the convolution and, by Eqs. (26), (27), or (31), estimate the upper bound of the CLR, assuming the case where the requested connection is further multiplexed.
4. If the resultant CLR is less than the QOS standard, a connection is established and $\theta(k; r)$ is replaced by $\theta \star \theta_{new}(k; r)$ as the new relative frequency; otherwise, the request is rejected.
5. When one of the multiplexed connections is released, deconvolute the current relative frequency and released connection, and store the resultant relative frequency in memory.

The complexity required for taking convolution and deconvolution in the above steps was disccused in [17].

4.3 Numerical Comparison of Different Combinations of CAC and UPC

4.3.1 Comparison between Actual CLRs and Upper Bounds

We numerically compare the derived upper bounds of the CLR with the actual CLR when the output process from the SW, JW, or LB is input into a single server queue whose service speed is 10 Mb/s. The output processes from each UPC used in the numerical examples are depicted in Fig. 7. The number of consecutive cell arrivals of the output process from the LB is arbitrary, so we set it so as to maximize the CLR. The results are shown in Figs. 8 (a) and (b), where the contracted PCR of each connection is assumed to be 9 Mb/s and the BT is infinite. (When BT is infinite, the value of the SCR does not influence the results of Figs. 8 (a) and (b).) The buffer length is 128 cells. First observe that the actual CLR of the JW's output process is the largest and that of the LB's output process is the smallest. This result

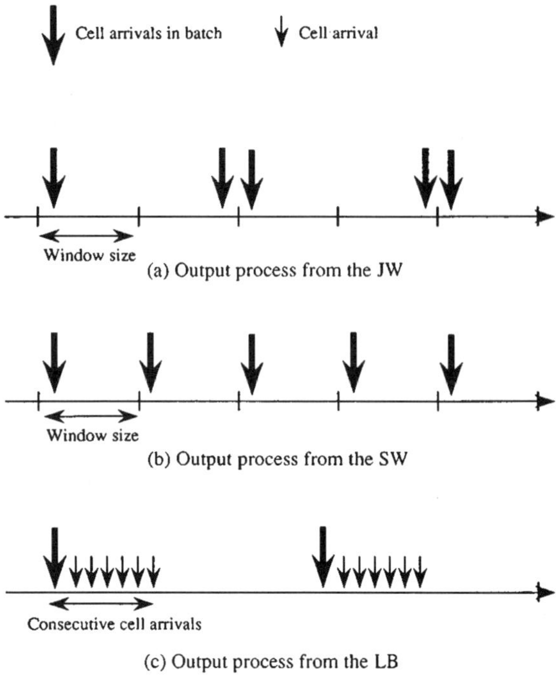

Fig. 7 Output processes from each UPC.

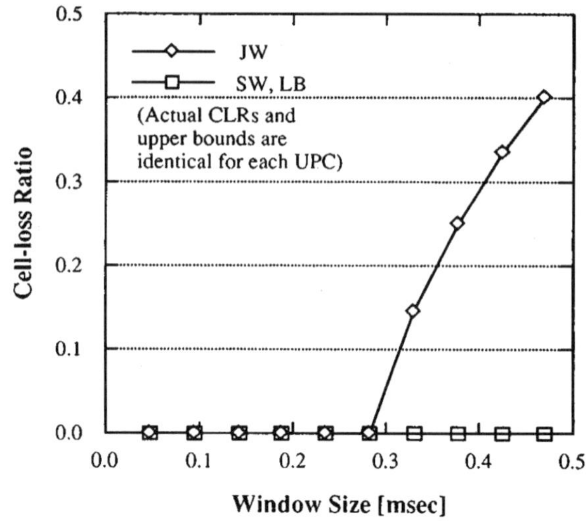

Fig. 8 (a) Cell-loss ratios of the output processes from UPCs: CDVT = 0.

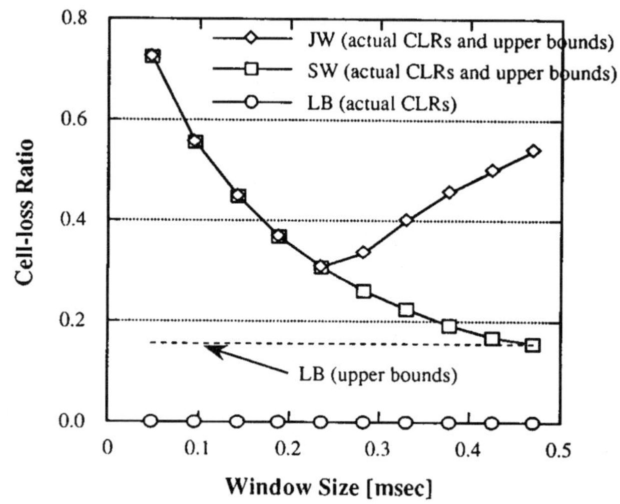

Fig. 8 (b) Cell-loss ratios of the output processes from UPCs: CDVT = 50-cell transmission time.

is reasonable and does not conflict with previous studies [12], [16]: that is, the LB is most suited to policing the LB-based traffic descriptor. The other UPC mechanism, e.g. the JW or SW, police the LB-based traffic descriptor less accurately than the LB itself, and this causes the larger cell-loss ratio. When the CDVT is equal to 0, the upper bounds of the CLRs are the same as the actual CLRs for all three UPCs (Fig. 8 (a)). When the CDVT is 50-cell transmission time (transmission speed is 150 Mb/s), the upper bounds of the CLRs for the SW and JW are still the same as the actual CLRs, but the upper bounds for the LB (Eq. (31)) overestimate the actual CLRs (Fig. 8 (b)).

To investigate the difference between the actual CLR and the upper bound of Eq. (31) when several connections are multiplexed, we carried out another numerical experiment by simulation (Fig. 9). (Here we do not compare the upper bounds and the actual CLRs for the output process of the SW and JW: these are identical for all cases of our numerical experiments.) We assume that 20 connections policed by the LB are multiplexed in a 150-Mb/s transmission path. The cells from each connection arrive as the output process of the LB depicted in Fig. 7. The LB parameters are for the case where the contracted PCR is equal to 10 Mb/s. The average cell rates of the connections are all 5 Mb/s. Figure 9 shows the actual CLRs and upper bounds by Eq. (31) for different numbers of consecutive cell arrivals. From Fig. 9 we also know that, when CDVT $= 0$, the upper bound becomes close to the actual CLR as the number of consecutive cell arrivals increases, but that the upper bounds slightly overestimate the actual CLR when CDVT > 0.

4.3.2 Influence of the Choice of CAC and UPC Combinations on Resource Assignment Efficiency

Next we compare the choice of CAC and UPC combinations in terms of resource assignment efficiency. There are three possible UPC and CAC combinations: JW and the CAC based on Eq. (26), SW and the CAC based on Eq. (27), and LB and the CAC based on Eq. (31). We show numerical examples in Fig. 10 (a) (CDVT $= 0$) and Fig. 10 (b) (CDVT $= 10L/C$) which are plots of the upper bounds of CLR obtained by Eqs. (26), (27), and (31) for different contracted peak bit rates of connections multiplexed in a 150-Mb/s transmission path. (The average bit rates of the sources are all 0.5 Mb/s.) For reference, we also show the upper bounds of Eq. (29). The length of the output buffer of

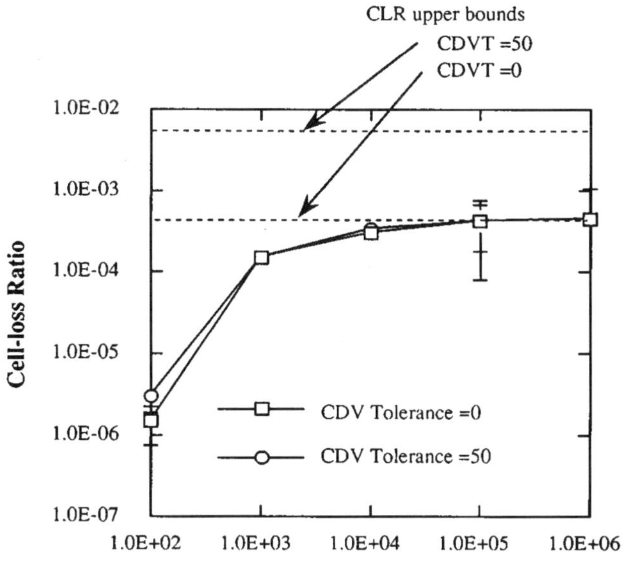

Fig. 9 Cell-loss ratios of the output processes from the LB.

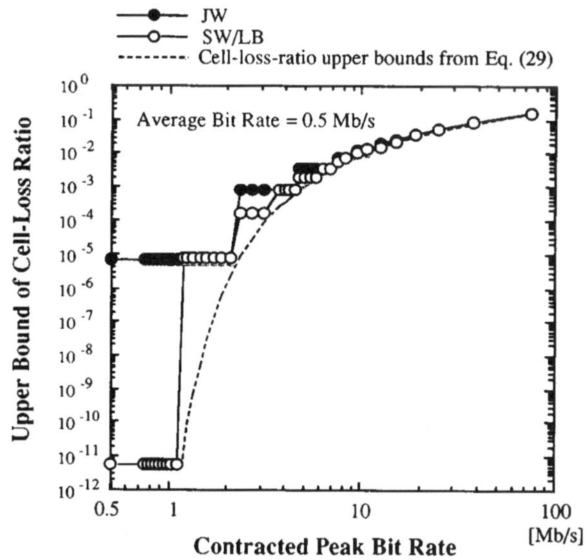

Fig. 10 (a) Upper bound of cell-loss ratio of cells passing through a UPC versus its window size: comparison between JW, SW, and LB (CDV tolerance = 0).

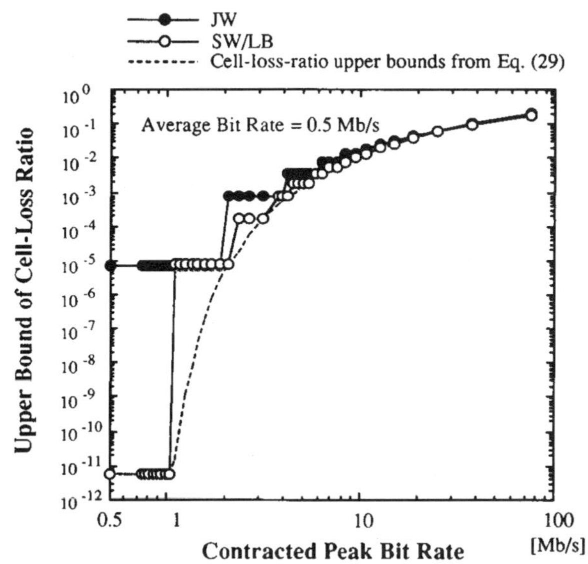

Fig. 10 (b) Upper bound of cell-loss ratio of cells passing through a UPC versus its window size: comparison between JW, SW, and LB (CDV tolerance = 10-cell transmission time).

the ATM switch is 128 cells, and the number of multiplexed connections is 200. The window lengths of the JW and SW are respectively $(K+2)/2$ and $K+1$ in units of L/C. Figures 10 (a) and (b) show the following:

1. The CLR upper bounds of the traffic through the JW given by Eq. (26) are the largest of all bounds.
2. The CLR upper bounds of the traffic through the LB given by Eq. (31) and those of the traffic through the SW given by Eq. (27) are identical for all contracted peak-bit rates.

Result 1, that is, JW is less suitable for PCR policing than the SW and the LB in terms of resource utilization, is intuitive. Result 2 is rather non-intuitive, and is due to the derived upper bound for the LB (Eq. (31)) slightly overestimating the actual CLR as seen in Sect. 4.3.1.

Next we show the influence of the choice of UPC algorithms on bandwidth resource utilization. It is assumed that a call acceptance decision is made based on the bounds derived in the previous section: The upper bound of the CLR is evaluated by Eq. (26) (for JW), Eq. (27) (for SW), or Eq. (31) (for LB) supposing that the new requesting call is further multiplexed. If and only if the upper bound of CLR is less than the CLR standard, the connection request is accepted. Figures 11 (a) and (b) show numerical examples in which bandwidth utilization was plotted by changing the peak and average bit rates of multiplexed sources while maintaining the ratio of peak bit rate to average bit rate constant. Bandwidth utilization is defined as $N_m A^* L/C$, where N_m is the maximum number of multiplexable connections for which the CLR upper bounds remain less than those specified by the QOS standard. The QOS standard of CLR was set to 1.0×10^{-6}, the virtual-path bandwidth to 150 Mb/s, and the buffer length was 128 cells. The figures show that higher bandwidth utilization is attained with the SW or LB for PCR policing than with the JW, especially for low-speed connections (less than 1.2 Mb/s).

5. Conclusion

This paper proposed a CAC that guarantees the CLR standards for all cell-stream patterns passing through a UPC. We theoretically derived upper bounds for the

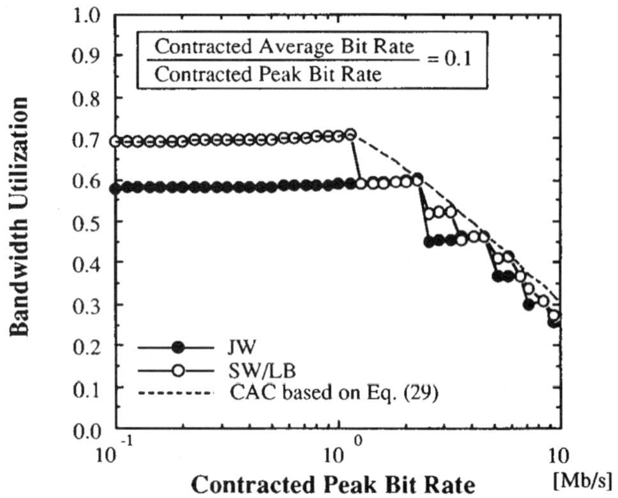

Fig. 11 (a) Bandwidth utilization ratio versus peak bit rates of multiplexed cell-streams passing through different types of UPCs: comparison between JW, SW, and LB (ratio of contracted average bit rate to peak bit rate is 0.1).

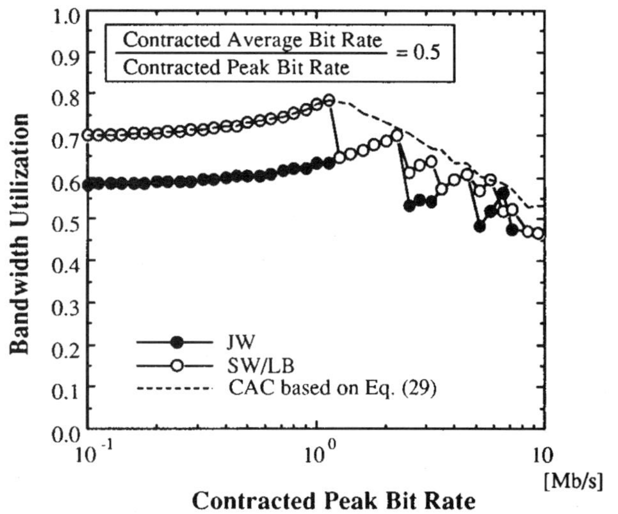

Fig. 11 (b) Bandwidth utilization ratio versus peak bit rates of multiplexed cell-streams passing through different types of UPCs: comparison between JW, SW, and LB (ratio of contracted average bit rate to peak bit rate is 0.5).

CLRs of the output process of each of three UPCs: the jumping window, sliding window, and continuous leaky bucket. The CACs based on the derived upper bounds of the CLR always ensure the QOS by combining the relevant UPCs. There are three possible combinations of CAC and UPC, depending on the UPC mechanism used. It was shown through several numerical examples that using JW makes bandwidth utilization a little lower than using SW or LB. (The derived upper bounds concerning the JW's and SW's output processes are actual bounds, so the bandwidth utilization for the JW and SW in this paper cannot be improved.) The difference in bandwidth utilization due to the choice between JW and SW is about 10% when low-speed connections are dominant, but is not significant (less than 10%) when high-speed connections are dominant. For high-speed networks, the JW might be considered to be one promising UPC algorithm because of the ease of its implementation. Otherwise, the LB and SW are more attractive.

The derived upper bounds of the output process from the LB are not tight for cases where CDVT is larger than 0. In this sense, our results imply that the CAC-UPC combination using the LB potentially attains higher bandwidth utilization than those using the SW or JW. When the output processes from the LB are multiplexed, however, whether a certain traffic pattern is the worst or not possibly depends on the traffic characteristics of other connections sharing the same multiplexer. In other words, we cannot construct the worst traffic pattern from only the source traffic descriptor of the corresponding connection. In such a case, making the upper bound tighter is questionable and the bandwidth utilization shown in this paper when the LB is used cannot be improved in practice.

Acknowledgement

The author would like to thank Mr. Toshiaki Tsuchiya, Mr. Hideaki Yoshino, and Dr. Toyofumi Takenaka of the NTT Multimedia Networks Laboratories for carefully reading the manuscript and making a number of helpful suggestions.

References

[1] Draft Recommendation I.371.
[2] Draft Recommendation I.150.
[3] The ATM Forum, Traffic management draft specification (version 4.0), 1995.
[4] M. Aida and H. Saito, "Traffic contract parameters and CAC guaranteeing cell-loss ratio in ATM networks," IEICE Trans. Commun., vol.E78-B, no.1, 1995.
[5] F. Baccelli and P. Bremaud "Elements of Queueing Theory," Springer-Verlag, 1994.
[6] A. Baiocchi, N.B. Melazzi, and A. Roveri, "Buffer dimensioning criteria for an ATM multiplexer loaded with homogeneous ON-OFF sources," ITC 13, pp.13–18, 1991.
[7] A. Baiocchi, N.B. Melazzi, M. Listanti, A. Roveri, and R. Winkler, "Loss performance analysis of an ATM multiplexer loaded with high speed on-off sources," IEEE J.S.A.C., vol.9, no.3, pp.388–393, 1991.
[8] A.A. Billingsley, "Ergodic Theory and Information," John Wiley, New York, 1965.
[9] C. Blondia, "The N/G/1 finite capacity queue," Stochastic Models, vol.5 no.2, pp.273–294, 1989.
[10] P.E. Boyer, F.M. Guillemin, M.J. Servel, and J.P. Coudreuse, "Spacing cells protects and enhances utilization of ATM networks links," IEEE Network, vol.9, pp.38–49, 1992.
[11] B.T. Doshi, "Deterministic rule based traffic descriptors for broadband ISDN: Worst case behavior and connection acceptance control," ITC 14, pp.591–600, 1994.
[12] H. Heiss and E. Wallmeier, "Performance comparison of three policing mechanisms based on their maximum throughput functions," ITC 14, pp.1405–1414, 1994.
[13] K. Kvols and S. Blaabjerg, "Bounds and approximations

for the periodic on/off queue with applications to ATM traffic control," INFOCOM '92, pp.487–494, 1992.
[14] D.C. Lee, "Effects of leaky bucket parameters on the average queue delay: worst case analysis," INFOCOM '94, pp.482–489, 1994.
[15] T. Murase, H. Suzuki, and T. Takeuchi, "A call admission control for ATM networks based on individual multiplexed traffic characteristics," ICC '91, pp.391–396, 1991.
[16] E.P. Rathgeb, "Modeling and performance comparison of policing mechanisms for ATM networks," IEEE J.S.A.C., vol.9, no.3 pp.325–334, 1991.
[17] H. Saito, "Call admission control in ATM networks using upper bound of cell loss probability," IEEE Trans. Commun., vol.40, no.9, pp.1512–1521, 1992.
[18] H. Saito, K. Kawashima, and K. Sato, "Teletraffic control in ATM netwoks," IEICE Trans., vol.E74, no.4, pp.761–771, 1991.
[19] H. Saito, "Teletraffic Technologies in ATM Networks," Artech House, 1994.
[20] S. Shioda and H. Saito, "Cell-loss-ratio analysis with insufficient knowledge of traffic characteristics," Journal of Operation Research Society of Japan, vol.40, no.1, 1997.
[21] T. Tsuchiya and H. Saito "The worst case cell arrival patterns that conform to the GCRA in ATM networks," GLOBECOM '95, 1995.
[22] F. Vakil and H. Saito, "On congestion control in ATM networks," IEEE LCS, pp.55–65, 1991.
[23] N. Yamanaka, Y. Sato, and K. Sato, "Performance limitation of the leaky bucket algorithm for ATM networks," IEEE Trans. Commun, vol.43, no.8, pp.2298–2230, 1995.

Appendix

This appendix outlines a proof of the CLR upper bounds presented in Sect. 3.1. A more detailed proof and other interesting results concerning CLR upper bounds are found in Ref. [20]. First, we introduce several notational conventions necessary for describing properties of the cell arrival process. The time axis is assumed to be divided into slots of length equal to the cell transmission time L/C, where L is the cell length and C is the virtual-path bandwidth. A cell arrives at the beginning of a slot, and is transmitted at the end of a slot. Let $a_i^{(1)}$ be the number of cells arriving during the i-th slot. Suppose that the stochastic process $\{a_i^{(1)}\}_{i=-\infty}^{\infty}$ is stationary.

We construct the definition-period sequence on the time axis (Fig. 1): denote a period of length r ($r \in \mathbb{N}$) slots as the definition period and give each definition period a sequential number. The first definition period starts on the η-th slot ($1 \leq \eta \leq r$). We introduce the following supplementary stochastic process, $\{a_n^{(r,\eta)}\}_{n=-\infty}^{\infty}$:

$$a_n^{(r,\eta)} \stackrel{\text{def}}{=} \sum_{j=0}^{r-1} a_{r(n-1)+\eta+j}. \tag{A·1}$$

Here $a_n^{(r,\eta)}$ effectively corresponds to the number of cells arriving during the n-th definition period of length rL/C. Obviously, $\{a_n^{(r,\eta)}\}_{n=-\infty}^{\infty}$ is also stationary. The ergodic theorem [8] implies that the relative frequency of $\{a_n^{(r,\eta)}\}_{n=-\infty}^{\infty}$ converges sample-path-wise, that is,

$$\frac{1}{N}\sum_{n=1}^{N} \mathbf{1}(a_n^{(r,\eta)} = k) \stackrel{P\text{-a.s.}}{\longrightarrow} p(k;r,\eta), \text{ as } N \to \infty \tag{A·2}$$

where "$P-$ a.s." indicates that the relation holds with probability 1.

Consider the busy-period structure of a discrete G/D/1/K+1 system with a FIFO service discipline. The busy period where at least one cell is lost is denoted here as the Loss Busy Period (LBP). The l-th LBP starts at the $t_s(l)$-th slot within the $n_s(l)$-th definition period and the last cell loss in l-th LBP occurs at the $t_e(l)$-th slot within the $n_e(l)$-th definition period (Fig. A·1). Let $b(l)$ be the number of lost cells within the l-th LBP. The CLR is defined in time-average meaning as follows:

$$\text{CLR} = \lim_{t\to\infty} \frac{\text{Number of lost cells during } [0,t]}{\text{Number of cells arriving during } [0,t]} \leq \lim_{L\to\infty} \frac{\sum_{l=1}^{L} b(l)}{\sum_{n=1}^{n_e(L)} a_n^{(r,\eta)}}. \tag{A·3}$$

The values of parameters, $t_s(l)$, $t_e(l)$, $n_s(l)$, $n_e(l)$, change when the definition-period sequence is shifted on the time axis. When necessary, we therefore use notation like $t_s(l;\eta)$ or $n_s(l;\eta)$ to indicate how the definition-period sequence is defined on the time axis.

Theorem 1: For all η,

$$\text{CLR} \leq \frac{\sum_k [k-\gamma_1(r)]^+ p(k;r,\eta)}{\sum_k k p(k;r,\eta)} \qquad P\text{–a.s.} \tag{A·4}$$

Proof: Consider the number of lost cells within the l-th LBP. The number of lost cells in the l-th LBP is equal to the number of cells arriving in $[(n_s(l)-1)r+t_s(l)+(\eta-1),(n_e(l)-1)r+t_e(l)+(\eta-1)]$ minus the sum of the number of cells transmitted during $[(n_s(l)-1)r+t_s(l)+(\eta-1),(n_e(l)-1)r+t_e(l)+(\eta-1)]$

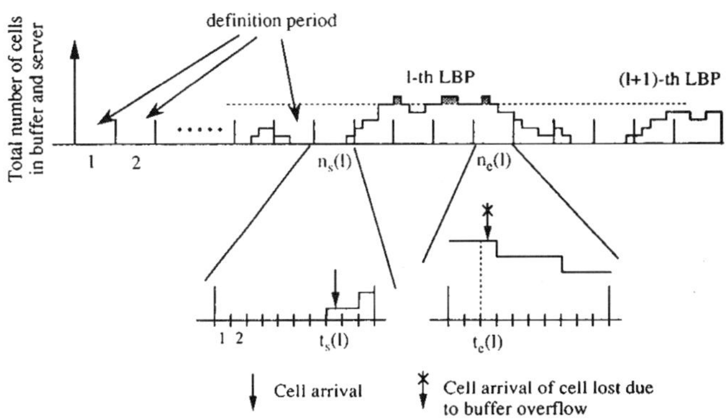

Fig. A·1 Example of LBP in the definition-period sequence.

and the number of cells waiting in the buffer at the end of the $\{(n_e(l)-1)r+t_e(l)+(\eta-1)\}$-th slot. ($[x,y]$ denotes the range from x-th to y-th slots.) Thus, we have

$$b(l) = \begin{cases} a^{(l,r,\eta)}_{n_s(l)} - \{(t_e(l)-t_s(l)+1)+K\} & : n_e(l)=n_s(l), \\ (a^{(l,r,\eta)}_{n_s(l)} + a^{(r,\eta)}_{n_s(l)+1} + \cdots + a^{(r,\eta)}_{n_e(l)-1} + a^{(l,r,\eta)}_{n_e(l)}) \\ \quad -(n_e(l)-n_s(l))r - t_e(l) + t_s(l) - K - 1 & : n_e(l) > n_s(l), \end{cases} \tag{A·5}$$

where $a^{(l,r,\eta)}_{n_s(l)}$ is the number of cells arriving during the $n_s(l)$-th definition period within the l-th LBP, and $a^{(l,r,\eta)}_{n_e(l)}$ is the number of cells arriving by the end of the $t_e(l)$-th slot of the $n_e(l)$-th definition period.

When $n_s(l)=n_e(l)$, $t_s(l) \leqq t_e(l)$, we therefore get

$$b(l) \leqq a^{(l,r,\eta)}_{n_s(l)} - (K+1). \tag{A·6}$$

For $n_s(l) < n_e(l)$, we can derive the following inequality from Eq. (A·5):

$$b(l) \leqq (a^{(l,r,\eta)}_{n_s(l)} - \gamma_1(r)) + \sum_{n=n_s(l)+1}^{n_e(l)-1} (a^{(r,\eta)}_n - \gamma_1(r)) + (a^{(l,r,\eta)}_{n_e(l)} - \gamma_1(r)). \tag{A·7}$$

Therefore, for both cases, $b(l)$ is upper-bounded by Eq. (A·7), using which we can derive the following:

$$\begin{aligned} \text{CLR} &\leqq \lim_{L\to\infty} \sum_{l=1}^{L} \{(a^{(l,r,\eta)}_{n_s(l)} - \gamma_1) + \sum_{n=n_s(l)+1}^{n_e(l)-1} (a^{(r,\eta)}_n - \gamma_1) \\ &\quad + (a^{(l,r,\eta)}_{n_e(l)} - \gamma_1)\} / \sum_{n=1}^{n_e(L)} a^{(r,\eta)}_n \\ &\leqq \lim_{L\to\infty} \frac{\sum_{l=1}^{L} \sum_{n=n_s(l)}^{n_e(l)} (a^{(r,\eta)}_n - \gamma_1(r))}{\sum_{n=1}^{n_e(L)} a^{(r,\eta)}_n} \\ &\leqq \lim_{L\to\infty} \frac{\sum_{n=1}^{n_e(L)} [a^{(r,\eta)}_n - \gamma_1(r)]^+}{\sum_{n=1}^{n_e(L)} a^{(r,\eta)}_n} \\ &= \lim_{N\to\infty} \frac{\sum_k \sum_{n=1}^{N} [k-\gamma_1(r)]^+ \mathbf{1}(a^{(r,\eta)}_n = k)}{\sum_k \sum_{n=1}^{N} k\mathbf{1}(a^{(r,\eta)}_n = k)}. \end{aligned} \tag{A·8}$$

The last line on the right-hand side of the above equation converges with probability 1 and can be rewritten with the frequency distribution $\{p(k;r,\eta)\}$ defined by Eq. (A·2). Consequently, we have the following:

$$\text{CLR} \leqq \frac{\sum_k [k-\gamma_1(r)]^+ p(k;r,\eta)}{\sum_k kp(k;r,\eta)} \quad P\text{–a.s.} \tag{A·9}$$ □

Theorem 2: For all η

$$\text{CLR} \leqq \frac{\sum_k [k-\gamma_2(r)]^+ p(k;r,\eta)}{\sum_k kp(k;r,\eta)} + q(\eta) \quad P\text{–a.s.,} \tag{A·10}$$

where

$$q(\eta) \stackrel{\text{def}}{=} \frac{\lim_{L\to\infty} \frac{1}{n_e(L)} \sum_{l=1}^{L} (t_s(l;\eta) - t_e(l;\eta))}{\sum_k kp(k;r,\eta)}. \tag{A·11}$$

Proof: From Eq. (A·5), we have

$$b(l) \leqq (a^{(l,r,\eta)}_{n_s(l)} - \gamma_2(r)) + \sum_{n=n_s(l)+1}^{n_e(l)-1} (a^{(r,\eta)}_n - \gamma_2(r)) + (a^{(l,r,\eta)}_{n_e(l)} - \gamma_2(r)) + (t_s(l) - t_e(l)). \tag{A·12}$$

Using Eq. (A·12), we thus obtain

$$\begin{aligned} \text{CLR} &\leqq \lim_{L\to\infty} \sum_{l=1}^{L} \{(a^{(l,r,\eta)}_{n_s(l)} - \gamma_2) + \sum_{n=n_s(l)+1}^{n_e(l)-1} (a^{(r,\eta)}_n - \gamma_2) \\ &\quad + (a^{(l,r,\eta)}_{n_e(l)} - \gamma_2)\} / \sum_{n=1}^{n_e(L)} a^{(r,\eta)}_n \\ &\quad + \lim_{L\to\infty} \frac{\sum_{l=1}^{L} (t_s(l) - t_e(l))}{\sum_{n=1}^{n_e(L)} a^{(r,\eta)}_n} \\ &\leqq \lim_{L\to\infty} \frac{\sum_{l=1}^{L} \sum_{n=n_s(l)}^{n_e(l)} [a^{(r,\eta)}_n - \gamma_2(r)]^+}{\sum_{n=1}^{n_e(L)} a^{(r,\eta)}_n} \\ &\quad + \lim_{L\to\infty} \frac{\sum_{l=1}^{L} (t_s(l) - t_e(l))}{\sum_{n=1}^{n_e(L)} a^{(r,\eta)}_n} \\ &\leqq \lim_{L\to\infty} \frac{\sum_{n=1}^{n_e(L)} [a^{(r,\eta)}_n - \gamma_2(r)]^+}{\sum_{n=1}^{n_e(L)} a^{(r,\eta)}_n} \end{aligned}$$

$$+\lim_{L\to\infty}\frac{n_e(L)}{\sum_{n=1}^{n_e(L)} a_n^{(r,\eta)}}\frac{\sum_{l=1}^{L}(t_s(l)-t_e(l))}{n_e(L)}$$

$$=\frac{\sum_k [k-\gamma_2(r)]^+ p(k;r,\eta)}{\sum_k kp(k;r,\eta)} + \frac{\lim_{L\to\infty}\frac{1}{n_e(L)}\sum_{l=1}^{L}(t_s(l;\eta)-t_e(l;\eta))}{\sum_k kp(k;r,\eta)}. \quad (A\cdot 13) \quad \square$$

Theorem 3:

$$\min_{\eta} q(\eta) \leqq 0. \quad (A\cdot 14)$$

Proof: We have

$$\sum_{\eta=0}^{r-1}\lim_{L\to\infty}\frac{1}{n_e(L)}\sum_{l=1}^{L}(t_s(l;\eta)-t_e(l;\eta))$$
$$=\lim_{L\to\infty}\frac{1}{n_e(L)}\sum_{l=1}^{L}\left(\sum_{\eta=1}^{r}t_s(l;\eta)-\sum_{\eta=1}^{r}t_e(l;\eta)\right)$$
$$=0. \quad (A\cdot 15)$$

To derive the final equality, we use the following relationship:

$$\sum_{\eta=1}^{r}t_s(l;\eta)=\sum_{\eta=1}^{r}t_e(l;\eta)=(r+1)r/2. \quad (A\cdot 16)$$

Thus, we have

$$\min_{\eta} q(\eta)$$
$$=\min_{\eta}\left\{\lim_{L\to\infty}\frac{1}{n_{e(L)}}\sum_{l=1}^{L}(t_s(l;\eta)-t_e(l;\eta)) \Big/ \sum_k kp(k;r,\eta)\right\}$$
$$\leq \frac{1}{r}\left\{\sum_{\eta=1}^{r}\lim_{L\to\infty}\frac{1}{n_{e(L)}}\sum_{l=1}^{L}(t_s(l;\eta)-t_e(l;\eta)) \Big/ \sum_k kp(k;r,\eta)\right\}$$
$$=\lim_{L\to\infty}\frac{1}{n_{e(L)}}\sum_{l=1}^{L}\sum_{\eta=1}^{r}\{t_s(l;\eta)-t_e(l;\eta)\} \Big/ r\sum_k kp(k;r,\eta)$$
$$=0. \quad (A\cdot 17) \quad \square$$

High-Performance Backbone Network Technology edited by Naoaki Yamanaka
ISBN: 0-8247-5321-6 DOI: 10.1081/(120025712)
Published 2004 by Marcel Dekker, Inc.

A Measured-Traffic-Based Bandwidth Dimensioning Method for Internet ATM Backbone Networks

Yuki Kamado and Kou Miyake

SUMMARY A traffic engineering method has been developed to meet the requirements for efficient bandwidth dimensioning and for a practical and consolidated network design method. It characterizes the offered-traffic burstiness on a transit link by using time-series measurement of the aggregate traffic. It estimates future traffic characteristics based on the average traffic volume at that time which is easily derived from trend analysis, i.e., an x% increase in bandwidth each year and gives the required link capacity. Simulation showed that the parameters estimated using this method fit the actual behavior of a network well. This method enables an appropriate bandwidth to be allocated to a transit link without having to estimate the specific traffic characteristics for each connection over the link. Once the burstiness parameter and its trend have been identified based on this method, it is possible to use a simple traffic measurement method to detect changes in network traffic and feed them back to the engineering procedure.

key words: *ATM, ATM bandwidth management, internet traffic, network engineering*

1. Introduction

Asynchronous transfer mode (ATM) is a promising technology for supporting computer communication networks used for legacy IP applications, such as electronic mail (E-mail), file transfer protocol (FTP), Telnet, and HTTP (web browsing), in conjunction with emerging multimedia applications requiring higher bandwidth and more stringent quality of service (QoS), such as high-speed network computing and distant learning. ATM has been widely used for high-speed computer communication network backbones, such as the Internet.

Because source traffic in such a network is quite bursty and unpredictable, the unspecified bit rate (UBR) [1] was defined as one of the ATM transfer capabilities (ATC) to support this type of traffic. ATM backbone networks should thus support UBR service. To construct and operate this kind of network cost-effectively, while maintaining such network QoS objectives as the cell loss ratio, an efficient bandwidth dimensioning method for transit links that makes use of statistical multiplexing is required. Bandwidth dimensioning means designing the required bandwidth for a given connection or connections. Furthermore, because such a network becomes larger and more complex as the number of users grows, a practical and consolidated dimensioning method is needed. We have developed a traffic engineering method to meet these requirements.

Most conventional dimensioning methods have two steps [2]: first, the detailed traffic characteristics of each connection over the transit link are estimated, then the link capacity required to meet the network QoS objective is calculated using an appropriate performance evaluation model. Expected future demand is reflected in the number of connections or in the traffic volume for each connection. However, as a network becomes larger, and as a larger number of connections having heterogeneous traffic characteristics are carried over each link, estimating the individual source traffic characteristics becomes more difficult. This limits the application of conventional methods.

In contrast, with our proposed method, such aggregate traffic characteristics on a link as the burstiness are directly estimated based on the measured data of the aggregate traffic, rather than that of the individual source traffic characteristics. The correlation between the burstiness and the mean cell rate (MCR) is then identified. This means that the burstiness parameters characterizing the offered traffic on a link can be determined from the estimated MCR representing future traffic demand. Once these correlation functions are formulated, calculating the required bandwidth for any traffic load is easy.

To determine the effectiveness of this method, we used the Markov Modulated Poisson Process (MMPP) [3] and its parameters to represent burstiness. We also used measured traffic data obtained from an actual large-scale ATM backbone network [4]. Simulation demonstrated that the parameters estimated using this method fit the actual network behavior well.

Using this method, one is able to allocate the appropriate bandwidth for a transit link without having to estimate the detailed traffic characteristics for each connection over the link. Once the burstiness parameter and its relationship with the MCR have been identified, only the MCR needs to be monitored to detect changes in network traffic. Network resources can then be reallocated accordingly.

In this paper, we first discuss the network traffic engineering methodology and then formulate the prob-

Manuscript received May 6, 1997.
Manuscript revised July 28, 1997.

† The authors are with NTT Network Service Systems Laboratories, Musashino-shi, 180-0012 Japan.

IEICE TRANS. COMMUN., Vol. E81-B, No. 2, pp. 449–458, February 1998

lem and propose an engineering method. Section 3 presents numerical results for correlation between the burstiness parameters and the MCR, and trend analysis results using actual measured traffic data. Section 4 shows simulation results confirming the effectiveness of our proposed method. Section 5 discusses a way in which our proposed method can be applied to actual network traffic engineering. Finally, we discuss the improvements needed to enhance the application of this method.

2. Formulation of Problem and New Dimensioning Method

2. 1 Network Resource Dimensioning Problem

The network configuration and traffic handling model discussed in this paper are based on the following definitions.

(1) Between two transit ATM switching nodes, a transit link is established.

(2) Several UBR connections are set up on the link. The connections are identified by their virtual channel connection (VCC) identifier. The only traffic descriptor for a UBR connection is its peak cell rate (PCR). The mean cell rate, burst size, or other traffic descriptors for that connection are not specified because the source traffic characteristics of such IP applications as E-mail, FTP, Telnet, and HTTP are quite bursty and unpredictable.

(3) The UBR connections share the common resources.

(4) To obtain a statistical multiplexing effect, the total peak cell rate of the UBR connections can be higher than the peak cell rate of the Virtual Path (VP) or physical link capacity. Buffers are provided to handle the multiplexing.

The dimensioning problem is simply defined as follows:

"Dimension the required link capacity considering the statistical multiplexing effect and required network QoS, such as cell loss ratio (CLR), cell delay variation (CDV), and cell transfer delay (CTD)."

However, to solve this problem for practical purposes, we should first forecast and identify the traffic volume and characteristics offered to the link considering the network planning interval.

In the next section, we introduce our new dimensioning method, focusing on how to take into account the burstiness of present and future multiplexed traffic. In this paper, the network QoS objective is limited to the CLR objective on the link being designed because UBR application is generally sensitive to the value of the CLR only; it is not sensitive to such real-time QoS objectives as CTD or CDV.

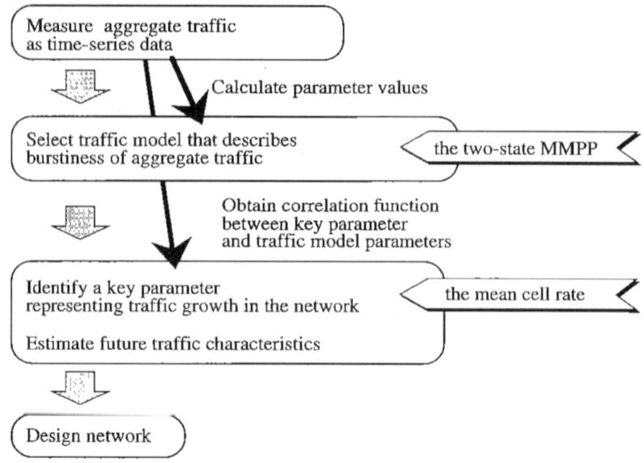

Fig. 1 Outline of proposed method.

2. 2 Outline of Proposed Dimensioning Method

As shown in Fig. 1, our proposed dimensioning method has four steps:

(1) First, the actual aggregate traffic data on the target link are measured. They are collected as time-series data.

(2) The characteristics of the time-series data are converged into traffic parameters characterizing a traffic model. We use MMPP as the traffic model. Four traffic parameters are used in this case. The parameter matching for the MMPP conforms to that in Ref. [3].

(3) The four traffic parameters are correlated with the MCR. This means that the MCR can be used to determine the four MMPP parameters and the overall characteristics of the traffic. Because the future MCR is easy to estimate, we can obtain four traffic parameters representing future traffic characteristics.

(4) The required network resources are determined according to the derived traffic.

These steps are described in more detail below.

(1) Measurement of traffic data

First, we measure the actual time-series traffic data from an ATM backbone network. To analyze the characteristics of the aggregate traffic, we count the number of cells arriving over a link during some "measurement interval." Based on the analysis of this traffic data, we estimate traffic parameters reflecting both the actual and aggregate traffic characteristics.

(2) Modeling

(2-1) Traffic model and parameter definition

To define a traffic model from which we can easily derive traffic parameters for network designing, and which can also describe the burstiness of aggregate traffic, we selected the Markov Modulated Poisson Process (MMPP) [3], which can model bursty traffic simply and requires only the chronological sequence of the arriving cells as traffic data for determining its

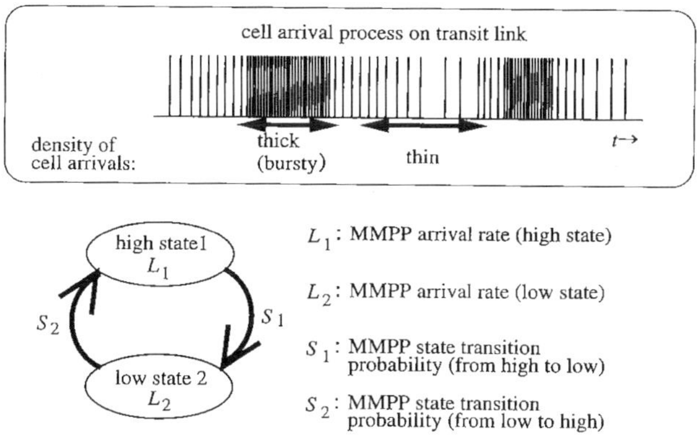

Fig. 2 Two-state MMPP model.

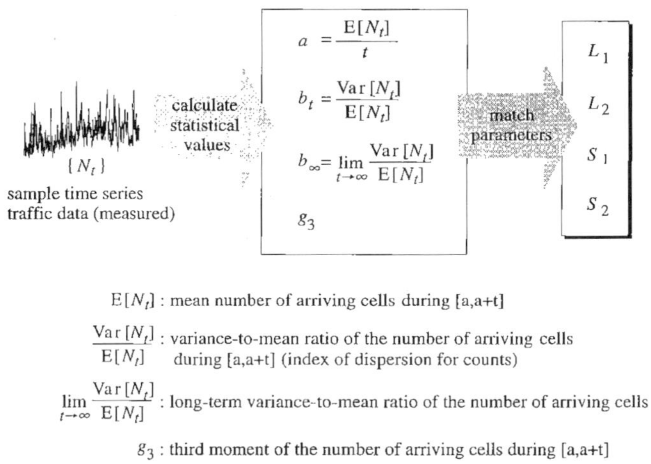

Fig. 3 MMPP parameter matching.

parameter values. Specifically, we use a two-state MMPP model (see Fig. 2). This model has two states $\{i=1, 2\}$, where the mean sojourn time in state i is S_i^{-1}, and at the end of a sojourn in state i, a transition is made to the other state. The arrival process in state i is a Poisson process of rate L_i. We used four parameters: the arrival rates at high state L_1 and at low state L_2, and the state transition probabilities from high to low S_1 and from low to high S_2.

(2-2) Parameter matching

We calculate the value of the MMPP parameters from the measured traffic. For the parameter matching (Fig. 3), one set of four MMPP parameters is derived from one sample process measured from an actual network.

Details of the parameter matching procedure will be explained in Sect. 2.3.

(3) Formulation of MMPP parameters and MCR

In MMPP-based engineering, the four parameters determine the exact traffic characteristics. They should be determined considering the growth in traffic demand. To do this, we propose analyzing the trends in the parameters based on the measured traffic, as described below.

Because the four MMPP parameters are difficult for network operators to handle directly, more convenient parameters are needed to describe future traffic characteristics. The mean cell rate (MCR) is linked to the variation in traffic demand during the network planning interval, so it is the key to estimating the characteristics of the future offered traffic. As will be shown in Sect. 3, through MMPP parameter matching of the collected chronological sequence of arriving cells, we found that these parameters and the MCR are correlated. That is, the MMPP parameters and their trend can be derived from the MCR. Having determined the correlation between the MMPP parameters and the MCR, we can easily estimate the four MMPP parameters representing the characteristics of the future offered traffic.

Because four MMPP parameters are needed, we have to provide four values as input parameters. If the three parameters in Fig. 3 other than the MCR can be described by a function correlating them to the MCR, they can be estimated from only the MCR. The candidates for these three parameters are the statistical values, b_t, b_∞ and μ_3, or the MMPP parameters themselves. There are thus two approaches to identifying the correlation between the MCR and the burstiness parameters. In Sect. 3, both approaches are evaluated.

(3-1) Approach 1 —based on primitive statistical values—

If the statistical values are selected as the parameters, we first estimate the correlations between the MCR and "the index of dispersion for counts," "the long term index of dispersion for counts," and "the third central moment of the number of arriving cells." We calculate these statistical values using the value of the given MCR and then obtain the MMPP parameters by MMPP parameter matching.

(3-2) Approach 2 —based on derived MMPP parameters—

If the MMPP parameters are selected, we obtain the MMPP parameters directly from the correlation functions between the MCR and the MMPP parameters.

(4) Dimensioning of network resources

After obtaining the set of traffic parameters, i.e., the four MMPP parameters, of the future offered traffic, we dimension the network resources analytically by using a performance evaluation model, such as MMPP/D/1/K [5].

2.3 Detailed Parameter-Matching Procedure

Assume that the number of cells arriving during some "measurement interval" t is N_t. First we calculate the statistical values based on the measured cell-arrival process $\{N_t\}$. We denote μ_3 as the third central moment of the number of arriving cells.

$$a = \frac{E[N_t]}{t} \tag{1}$$

$$b_t = \frac{\mathrm{Var}[N_t]}{\mathrm{E}[N_t]} \tag{2}$$

$$b_\infty = \lim_{t\to\infty}\left(\frac{\mathrm{Var}[N_t]}{\mathrm{E}[N_t]}\right) \tag{3}$$

$$\mu_3 = \mathrm{E}[(N_t - \mathrm{E}[N_t])^3] \tag{4}$$

Because "b_∞" cannot be directly calculated from Eq. (3), it is estimated from the values for b_{2t} ($=\mathrm{Var}[N_{2t}]/\mathrm{E}[N_{2t}]$) and b_{3t} ($=\mathrm{Var}[N_{3t}]/\mathrm{E}[N_{3t}]$) by changing their time intervals.

The third moment of the number of arriving cells is calculated according to the definition

$$g_3 = \mu_3 + 3at(at-1)b_t + at(at-1)(at-2). \tag{5}$$

From these statistical values, MMPP parameters L_1, L_2, S_1, and S_2 can be converted based on the following equations. This parameter-matching procedure conforms to that in Ref. [3].

$$L_1 = \left(\frac{ad}{S_2} - \frac{K}{S_1 - S_2}\right)\left(\frac{S_2}{S_1 + S_2}\right) \tag{6}$$

$$L_2 = \frac{K}{S_1 - S_2} + L_1 \tag{7}$$

$$S_1 = \frac{d}{2}\left(1 + \frac{1}{\sqrt{4e+1}}\right) \tag{8}$$

$$S_2 = d - S_1 \tag{9}$$

Here,

$$d = \frac{1}{t}\left(\frac{b_\infty - 1}{b_\infty - b_t}\right)(1 - \exp(-dt)) \tag{10}$$

$$e = \frac{ad^3(b_\infty - 1)}{2K^2} \tag{11}$$

$$K = \frac{A}{B} \tag{12}$$

where

$$A = g_3 - a^3t^3 - 3a^2t(b_\infty - 1)\left(t - \frac{1 - \exp(-dt)}{d}\right) \tag{13}$$

$$B = \frac{3a(b_\infty - 1)}{d^2}\left\{t(1 + \exp(-dt)) - \frac{2}{d}(1 - \exp(-dt))\right\} \tag{14}$$

To confirm the applicability and accuracy of our method, we present numerical results in the following section.

3. Numerical Results

3.1 Measured Traffic Data for Numerical Analysis

The example data for the numerical evaluation of our proposed method were collected from the largest Internet backbone in Japan, the Science Information Network Internet Backbone (SINET) [4]. This network uses ATM technology and consists of 29 large-scale ATM switches, mainly having a switching capacity of 10 Gbit/s. The link speeds between switches are 150 Mbit/s SDH, 50 Mbit/s SDH, and 6 Mbit/s PDH. Statistical multiplexing gain is obtained by using UBR service.

We measured the aggregate traffic for each virtual path between ATM switches by using a commercial ATM analyzer. About 80 sequences of cell-arrival times were obtained. A sequence comprises the time-stamps (accuracy of approximately 10^{-8}) of each cell as it arrives. We converted these data into the number of cells arriving during a fixed interval, and this number was used for the numerical evaluation. The time intervals for defining the cell-arrival process, equivalent to measurement interval t, were set to 0.5, 1, 2, 10, 20, and 50 ms.

The number of cell arrivals during one observation is always constant and only depends on the storage capacity of the ATM analyzer. In other words, the duration necessary to fill up the storage capacity varies depending on the offered load. To keep the duration in which the data were measured constant, we used a four-second duration.

Figure 4 shows a sequence of cell-arrival times

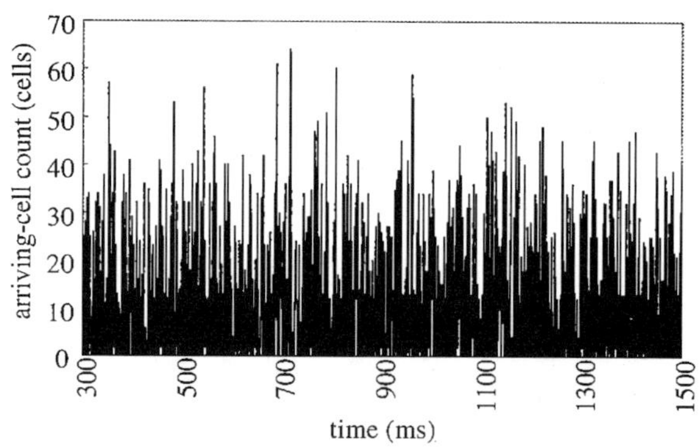

Fig. 4 Example of measured cell-arrival data (time interval = 0.5 ms).

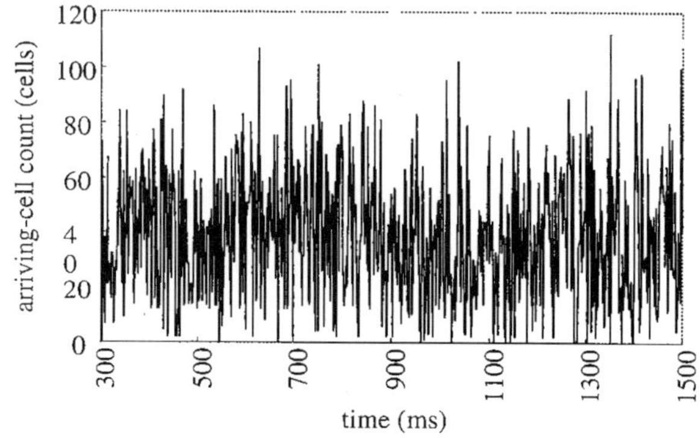

Fig. 5 Example of measured cell-arrival data (time interval = 2 ms).

converted into the number of cells arriving during a time interval 0.5 ms. Figure 5 shows the same sequence for a time interval of 2 ms.

3.2 Statistical Analysis of Measured Traffic Data

Statistical calculation of the measured data obtained as described above was performed for time intervals of 0.5, 1, 2, 10, 20, and 50 ms, respectively.

We examined the correlation between the statistical values and the MCR, m. Figures 6 to 8 show the variance in the number of arriving cells, the index of dispersion for counts, and the third central moment of the number of arriving cells against the MCR, respectively.

(1) Mean number of arriving cells
The mean number of arriving cells is the MCR itself. $E[N_t]$ is proportional to mt:

$$E[N_t] = C_0 mt, \tag{15}$$

where, C_0 is a coefficient for adjusting the dimensions; we give it the value of 1 to simplify the discussion.

(2) Variance of the number of arriving cells
Figure 6 shows the correlation between MCR and $Var[N_t]$. Both the first and second-order correlations are acceptable. $Var[N_t]$, or the second central moment of the number of arriving cells is defined as

$$Var[N_t] = E[(N_t - E[N_t])^2] = E[N_t^2] - E[N_t]^2, \tag{16}$$

where $E[N_t^2]$ is the mean number of arriving cells squared. It is clear that $E[N_t^2]$ has a second-order correlation with MCR. Also, the second term on the right side of this equation is equal to MCR squared, so this equation shows that there is a second-order correlation between MCR and $Var[N_t]$. We thus use a second-order fitting curve to represent the correlation between MCR and $Var[N_t]$.

(3) Index of dispersion for counts
Figure 7 shows the correlation between the MCR and the index of dispersion for counts (b_t). The linear correlation between b_t ($=Var[N_t]/E[N_t]$) corresponds to the second-order correlation between MCR and $Var[N_t]$. The results for b_{2t} and b_{3t} conform to that for b_t.

(4) Third central moment of the number of arriving cells
Figure 8 shows the correlation between the MCR and the third central moment of the number of arriving cells (μ_3). There is no correlation between the MCR and μ_3, although we expect a third-order correlation.

3.3 MMPP Parameter Matching

(1) MMPP arrival rate parameters
Figures 9 and 10 show that there are linear correlations between the MCR and the MMPP arrival rate parameters L_1 and L_2. Figures 9 and 10 show the relationships when the time interval is 1 and 10 ms, respectively. The longer the time interval, the smaller the difference between the MMPP high and low arrival-rates. The same relationship is also observed for other time intervals. This means that as the time interval for the arriving cell counts becomes longer, the information about burstiness in the cell arrival process becomes less clear.

(2) MMPP state transition probability parameters
Figure 11 shows there is no correlation between the

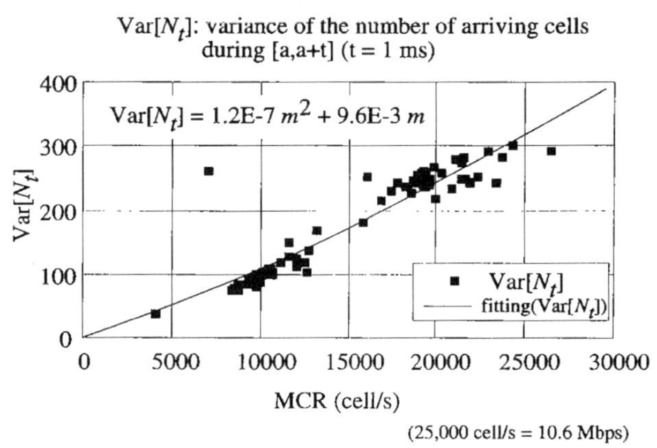

Fig. 6 Correlation between MCR and variance of the number of arriving cells (time interval = 1 ms).

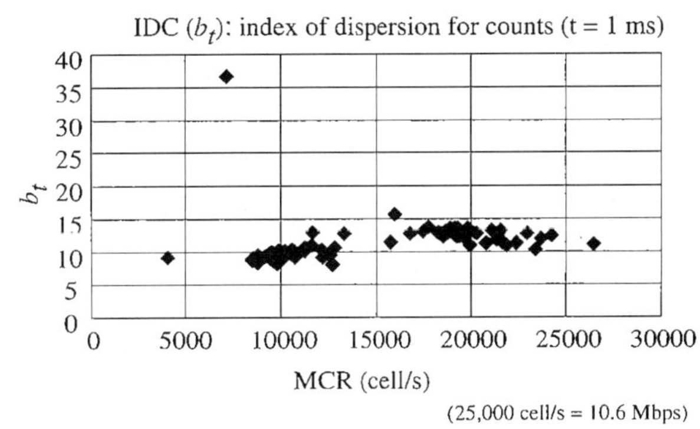

Fig. 7 Correlation between MCR and index of dispersion for counts (time interval = 1 ms).

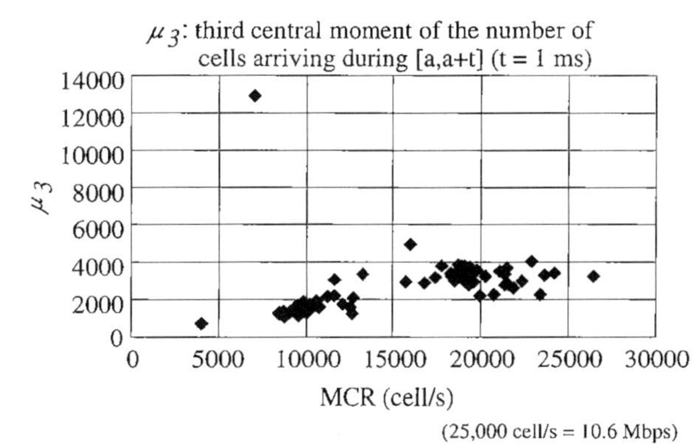

Fig. 8 Correlation between MCR and third central moment of the number of arriving cells (time interval = 1 ms).

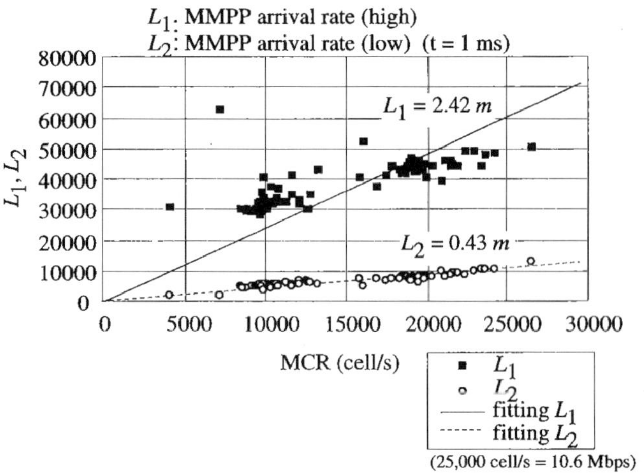

Fig. 9 Correlation between MCR and L_1 and L_2 (time interval $=1$ ms).

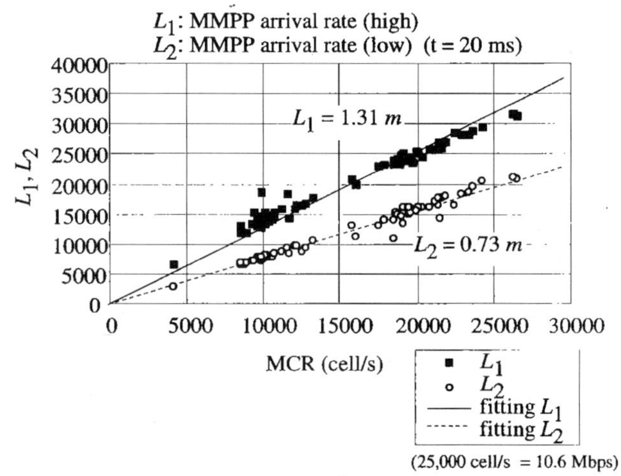

Fig. 10 Correlation between MCR and L_1 and L_2 (time interval $=20$ ms).

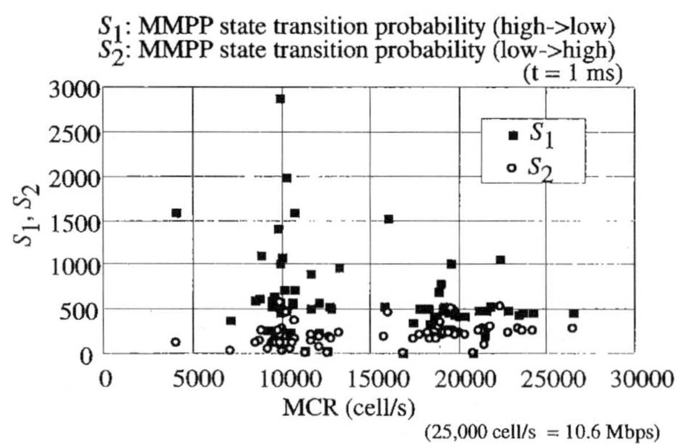

Fig. 11 Correlation between MCR and S_1 and S_2 (time interval $=1$ ms).

MCR and the MMPP state-transition probability parameters, S_1 and S_2.

3.4 Correlation between MCR and Burstiness Parameters

As described in Sect. 2.2, there are two approaches to determining the correlation between the MCR and the MMPP parameters. We evaluated approach 1, which is based on the primitive statistical values, by using the results in Sect. 3.2. This approach failed because we could not identify a correlation between the MCR and the third central moment, μ_3, as described in (3) in Sect. 3.2.

A slight modification to approach 1 would be to try to find the correlations between MCR and $E[N_t^2]$ and $E[N_t^3]$ rather than μ_3 itself, because μ_3 consists of those two primitive statistical values:

$$\mu_3=E[N_t^3]-3E[N_t]E[N_t^2]+2E[N_t]^3. \tag{17}$$

In this modified approach, there are six statistical values, a, b_t, b_{2t}, b_{3t}, $E[N_t^2]$, and $E[N_t^3]$, representing the future MCR. Although the MCR can be correlated with all six parameters, MMPP parameter-matching failed again because the MMPP arrival-rate was negative for many values of the MCR, which does not make sense for real applications.

We therefore selected approach 2, which is based on the derived MMPP parameters. Numerical evaluation revealed a clear linear correlation between the MCR and L_1 and L_2, as indicated in (1) in Sect. 3.3. We can thus estimate parameters L_1 and L_2 from the MCR:

$$L_1=\alpha m \quad \text{and} \tag{18}$$

$$L_2=\beta m, \tag{19}$$

where α and β are the fitting parameters for L_1 and L_2.

We also need S_1 and S_2 to determine the two-state MMPP model. However, no correlation is observed between the MCR and S_1 and S_2, as indicated in (2) in Sect. 3.3. As described in Ref. [3], the parameters $E[N_t]$, $\mathrm{Var}[N_t]$, L_1, L_2, S_1, and S_2 have the following relationships.

$$\frac{\mathrm{Var}[N_t]}{E[N_t]}=1+\frac{2(L_1-L_2)^2S_1S_2}{(S_1+S_2)^2(L_1S_2+L_2S_1)} -\frac{2(L_1-L_2)^2S_1S_2}{(S_1+S_2)^3(L_1S_2+L_2S_1)} \times(1-\exp(-(S_1+S_2)t)) \tag{20}$$

$$E[N_t]=\frac{L_1S_2-L_2S_1}{S_1+S_2}\,t \tag{21}$$

From Eqs. (18), (19), and (20), the sum of S_1 and S_2 can be calculated by solving the following non-linear equation.

$$\frac{\mathrm{Var}[N_t]}{E[N_t]}=1+\frac{C}{(S_1+S_2)}$$

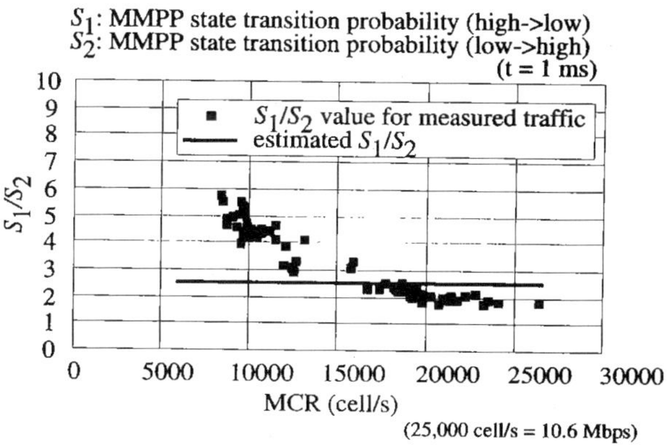

Fig. 12 Correlation between S_1/S_2 value for measured traffic and estimated S_1/S_2 (time interval =1 ms).

$$-\frac{C}{(S_1+S_2)^2 t}(1-\exp(-(S_1+S_2)t)), \tag{22}$$

where

$$C=2m(\alpha-1)(1-\beta) \tag{23}$$

From Eq. (21), as well as from (22) and (23), and the correlation between $\mathrm{Var}[N_t]$ and MCR, we can obtain S_1 and S_2.

In the course of transferring the above equations, the ratio of S_1 to S_2 can be represented by Eq. (24), which is irrelevant to the MCR. The validity of this method is demonstrated by the correlation shown in Fig. 12:

$$\frac{S_1}{S_2}=\frac{\alpha-1}{1-\beta}. \tag{24}$$

From the above correlation analysis, a clear relationship between the MCR and the two-state MMPP model parameters is recognized. We can thus establish a method for estimating the traffic parameters based on the two-state MMPP model. This is enables network resources to be dimensioned from only an input parameter representing traffic demand, i.e., the MCR.

4. Evaluation of Proposed Method

To evaluate the accuracy of our proposed method, we compared the cell loss ratio (CLR) derived from the proposed method with that derived from actual traffic data. The VP bandwidth and the buffer capacity were fixed, and the MCRs of the aggregate traffic for both methods were the same. If the CLR where the proposed method is applied is similar to the CLR where the actual traffic is offered under the condition that the MCR is the same, the proposed method should regenerate the traffic with the same characteristics as those of the actual traffic. Figure 13 shows the results.

Using the proposed method, we estimated the

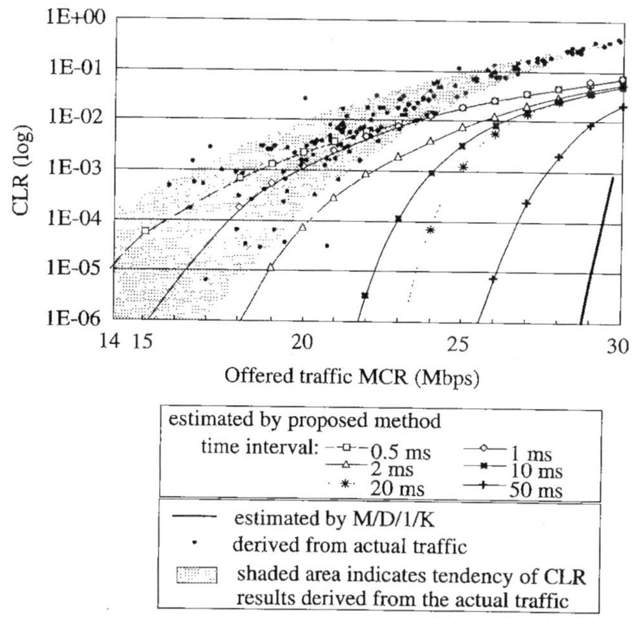

Fig. 13 Comparison of CLR as estimated using proposed method as derived from actual traffic, and from a simple M/D/1 model for a given VP bandwidth (30 Mbps).

two-state MMPP model parameters for a given MCR and calculated the CLR based on the MMPP/D/1/K performance evaluation model [5], where D was given as the VP bandwidth and K was given as the buffer size. Because the time interval for defining the basic MMPP parameters was dominant, we evaluated the proposed method by using six different time intervals. The evaluation conditions are summarized below.

For reference we also calculated the CLR by using simple computer simulation, as if the actual traffic pattern were offered to the link, by using traffic data obtained from an actual network. This was achieved by regenerating the same traffic patterns as the measured ones on the computer, then offering them to a constant-rate server model. Sequences of the cell-arrival times from the actual network were used. The VP bandwidth and buffer capacity were the same as those used for the proposed method. For the simulation, we used the data gathering three or four actual arrival processes (4-second duration) in order to make the utilization of the VP fall within the target range for the MCR, which is difficult to obtain from the actual arrival-process data. Two-hundred and forty patterns of multiplexed arrival processes were generated in this simulation. Their CLRs are shown as scattered dots in the figure because it is hard to control or select the offered load to meet the exact value of the MCR from the limited number of actual traffic patterns.

As one more comparison, the cell loss ratio calculated using the M/D/1/K performance model, in which the input traffic process is a Poisson arrival process, is also shown in Fig. 13.

The simulation conditions were as follows:

- VP bandwidth was 30 Mbps;
- buffer size k was 128 cells;
- the range of the MCR of the offered load was from 14 to 30 Mbps, i.e., about 46% to 100% of the VP bandwidth; and
- the time intervals for defining the basic MMPP parameters for the proposed method were 0.5, 1, 2, 10, 20, and 50 ms.

Note that the CLR obtained by this simulation is reliable as a reference when its value is from 1 to 10^{-3}. This is because the number of cells in the actual traffic data used in this reference is about 10^5. To compensate for the region where the CLR is less than 10^{-3}, the range of possible reference values is shown by the shadowed area.

Several observations can be made from Fig. 13.

(1) All the plots rise towards the right; in other words, the higher the MCR, the higher the CLR, because the VP bandwidth is fixed.

(2) The shorter the time interval, the closer the estimated CLR to the actual CLR. This is due to two factors. One is that the observed burstiness of the original traffic disappears and the estimated traffic becomes less bursty as the time interval becomes longer. Such traffic is estimated to have a lower CLR. The other reason comes from the experimental restriction in this particular case. As described above, the duration of the data measurement was four seconds. As the time interval becomes longer, the number of time intervals, which are the primitive data for estimating the parameters, becomes smaller. This may cause fluctuations in the statistical processes, leading to worse results. If the duration is lengthened, a longer interval can be used, improving the estimation. We will continue to study the relation between the time interval and the duration of the data measurement.

(3) Provided that the practical range of the MCR is about 60% to 80% of the VP bandwidth, the proposed method gives a good fitting when the time interval is 0.5 or 1 ms. Regarding the lower utilization area (less than 60% of the VP bandwidth or less than 10^{-3} in terms of CLR), the proposed method works well when the time interval is 0.5 or 1 ms because the shaded area covers the estimated lines for these intervals. Regarding the higher utilization area (more than 80% of the VP bandwidth or more than 10^{-2} in terms of CLR), the proposed method gives a CLR lower than the actual one, which may cause over provisioning of the traffic. This comes from the limitation of the traffic model used here, i.e., the two-state MMPP. Due to insufficient representation of burstiness in this model, the estimated traffic in the high utilization area behaves less bursty, so the derived CLR becomes lower than the actual one. The difference is at most one order of magnitude.

(4) The simple M/D/1 model gives a poor estimation because the target traffic behaves more bursty than that modeled by the Poisson process.

These results provide a guideline for how long the current VP link can be used under the same conditions, if the future MCR is given.

5. Discussion

5. 1 Actual Usage of Proposed Method

As shown in the previous section, the proposed method covers the area where the CLR is less than 10^{-3}. For a practical network, the CLR should be about 10^{-5}, even though the network supports UBR, in order to use TCP/IP efficiently. To achieve this level, the MCR of the aggregate traffic should be about 50% of the VP bandwidth according to the results of using our proposed method.

Using the proposed method, we can estimate the CLR for various MCRs if the VP bandwidth and buffer capacity are given. This is useful for determining how long the current VP link will last without increasing the bandwidth. This method can also be used to search for the appropriate VP bandwidth while maintaining the CLR at a certain value by changing the VP bandwidth. This is useful for determining the bandwidth needed to accommodate the foreseen traffic. This is because once the MCR of the target network is given, we can estimate the MMPP parameters reflecting the aggregated traffic flow in the VP link and utilize the existing MMPP/D/1/K model.

Although we chose the CLR as the key QoS parameter, the cell delay variation (CDV) could also be used because the MMPP/D/1/K model can show the delay characteristics.

5. 2 Improving the Performance of the Proposed Method

Because the parameter matching is performed based on actual traffic, the greater the amount of actual traffic data collected, the more precise the obtained parameters.

5. 3 Applicability to Other Types of Networks

We presented one example network to show the usefulness of the proposed method. Although we need more examples to verify this method further, it should be applicable to many data networks because no special-conditions are required. In this paper, we used a set of parameters to represent the future traffic . Other parameter sets can be used to adapt the proposed method to other data networks.

5. 4 Applicability to Other ATCs

In this paper, we proposed traffic estimation for UBR

ATC (ATM traffic capability). Because the target is to obtain a statistical multiplexing gain while maintaining a certain level of QoS, especially for the CLR, this method is applicable to the real-time variable bit rate (rt-VBR) and to the non-real-time variable bit rate (nrt-VBR). This method is not applicable to the constant bit rate (CBR) because the aim is to reserve the bandwidth corresponding to the declared PCR and to provide high QoS in general. The required bandwidth depends on the number of connections and their PCR values, while actual traffic behavior has nothing to do with the required bandwidth. This method is also not applicable to the available bit rate (ABR) because ABR has a mechanism which feeds back network conditions to the end system, and the bandwidth usage dynamically changes according to this, so the VP traffic conditions are always fluctuating.

6. Conclusion

When ATM is used as a backbone network for supporting a computer communication network, such as for IP applications or multimedia applications, it is difficult to design the network cost-effectively considering the future traffic trend, because it has a quite bursty and unpredictable traffic pattern and consists many connections having heterogeneous traffic characteristics. By analyzing actual traffic data collected from a large-scale ATM backbone network, we have established a traffic engineering method for such a network.

We used the two-state MMPP model to describe the burstiness of the aggregate traffic, so individual source traffic data is not needed to achieve statistical multiplexing gain. By matching the parameters of the measured and aggregate traffic data, we obtained a correlation between the MCR and the MMPP parameters. This correlation enables us to easily estimate traffic parameters in accordance with future trends in demand. We also evaluated the correlation between the QoS and traffic load on an ATM VP. Comparison between the value estimated by the proposed method and simulation results obtained from actual traffic showed good estimation in the range of 80% or less VP utilization, when the time interval for the measurement was 0.5 or 1 ms.

We also proposed a bandwidth dimensioning method for operating the network cost-effectively while maintaining the target QoS. That is, we determined how to properly allocate the bandwidth for a transit link now, and how to reallocate resources to handle future traffic flows. We do this by monitoring the MCR of the link by using a simple traffic measurement method. When a significant change in the load is detected, we measure the traffic data in detail and analyze it by using this method.

Our proposed ATM traffic engineering method could be further improved by

- finding ways other than the MMPP to model the bursty characteristics of ATM aggregate traffic,
- evaluating the usefulness of other measured data, e.g., queue occupancy, and
- developing a method for dimensioning Switched Virtual Channel (SVC) service.

Because of its ease of use and estimation accuracy, our proposed estimation method can play a key role in designing VP resources. It should thus become widely used for constructing ATM multimedia networks.

Acknowledgments

We greatly appreciate the encouragement and suggestions given by Professor Shoichiro Asano of the National Center for Science Information Systems, and by Dr. Shigehiko Suzuki, Dr. Tatsuro Takahashi, Mr. Naotaka Morita, and Mr. Toshiaki Tsuchiya of NTT.

References

[1] ATM Forum, ATM User-Network Interface Specification Version 3. 1., Sept. 1994.

[2] J. M. Pitts and J. A. Schormans, "Introduction to ATM Designing and Performance," John Wiley & Sons, 1996.

[3] H. Heffes and D. M. Lucantoni, "A Markov modulated characterization of packetized voice and data traffic and related statistical multiplexer performance," IEEE J. Sel. Areas Commun., vol. 4, no. 6, Sept. 1986.

[4] S. Asano, W. Zhao, and H. Uose, "Enhanced Academic Network in Japan," Global Information Infrastructure (GII) Evolution, record of INTERWORKING '96, Oct. 1996.

[5] A. Baiocchi and N. Blefari-Melazi, "Steady-state analysis of the MMPP/G/1/K queue," IEEE Trans. Commun., vol. 41, no. 4, April 1993.

High-Performance Backbone Network Technology edited by Naoaki Yamanaka
ISBN: 0-8247-5321-6 DOI: 10.1081/(120025713)
Published 2004 by Marcel Dekker, Inc.

Measurement-Based Real-Time Call Admission Control in ATM Networks

Cheul Shim

SUMMARY The concept of a schedulable region (SR) was introduced to characterize the capacity of a multiplexer and provide a separation between call-level and cell-level phenomena. In this paper, we present a framework and algorithm for real-time estimation of the schedulable region. A major problem associated with online estimation is that the objects of measurement are not fixed in the presence of call arrivals and departures. The *invariance* property is exploited to carry out measurements in the presence of call arrivals and departures. By virtue of it, the equivalent bandwidth could be defined on the condition of the number of each traffic class call in progress. Another important thing we consider here is that the search algorithm to estimate the effective bandwidth should be chosen depending on the arrival statistics and QOS constraints. The algorithms presented here have been implemented on an ATM switch.

key words: *online estimation, schedulable region, QOS, admission control*

1. Introduction

Efficient admission control is one of the main research challenges in multiclass broadband networks. While a conservative admission control policy is trivial to formulate, it will result in underutilization of resources. On the other hand, a lack of admission control will result in severe quality of service degradation, and possibly network breakdown. Thus admission control is a crucial functionality. As a step toward simplifying the problem, our COMET Group introduced the notion of a schedulable region (SR) for characterizing the capacity of a multiplexer [1], [2]. It is the space of possible combinations of calls that a multiplexer can handle while guaranteeing quality of service. From the point of view of the call admission controller, cell-level phenomena are completely abstracted by the schedulable region. It is an extension of the Shannon capacity concept, and as such it is a stability concept. Thus, the schedulable region is the set of all possible calls that can be handled by the multiplexer while guaranteeing per-class quality of service. The shape and size of this region are completely determined by the input traffic characteristics, buffer size(s), buffer management algorithm, scheduling algorithm, output line speed, and the QOS constraints.

Given any scheduling discipline, it is crucial to determine the minimum bandwidth required by traffic streams with QOS guarantees. The equivalent bandwidth (EB) of a traffic stream has been intended to denote it so that the problems of admission control and routing in ATM networks could be treated as the same way in circuit-switched networks. Among the first studies of EB is an analysis of bufferless system by Hui [3]. Kelly [4] presented the EB results for the buffered systems subject to either mean delay or tail constraints. The EB results expanded for Markov fluid traffic models are proposed by Eliward and Mitra [5]. These methods provide explicit results for multiplexed systems and approximated characteristics of large buffer asymptotics. An alternative approach has been to investigate large buffer asymptotics on the base of the theory of large deviation principle. The authors in [6], [7] showed the existence of EB under the large buffer assumption. Chang and Thomas introduced the EB results which could be applied to the several resources managements in the high speed digital network [8]. In particular, Chang and Zajic [9] presented the EB results of departure processes from queues with time varying capacities. Hsu and Warland [10] proposed the admission control for multi-class traffic under large buffer and many source assumption. Due to tight delay constraints, real-time traffic cannot tolerate extensive buffering in the network. They presented an accurate estimate of the overflow probability for real-time traffic based on a theorem by Bahadur and Rao. In addition, recent some results show that asymptotic EB results can be either optimistic or conservative depending on the traffic arrival statistics [11], [12]. Authors in [12] gave a warning that there may be a flaw in the EB concept because it corresponding to having no traffic smoothing from multiplexing. Finally, they concluded that the notion of EB based directly on large deviation asymptotics may not be effective for admission control. As another approach, the admission control based on the online measurement has been studied in some recent works. In [13], [14], authors proposed admission control based on the online estimation of EB. Tedijanto and Gun [15] estimated the EB of an ATM connection using a dynamically adjusted token bucket. The estimation method of delay and rate for the predictive service was suggested in [16]. Also, the authors in [17] suggested an adaptive EB estimation of the aggregated traffic via Kalman filter. A major problem associated

Manuscript received August 25, 1997.
Manuscript revised May 21, 1998.

[†]The author is with Center for Telecommunications Research, Rm. 801 Schapiro Research Bldg., Columbia University, New York, NY 10027-6699, U.S.A.

IEICE TRANS. COMMUN., Vol. E82-B, No. 9, pp. 1371–1379, September 1999

with online estimation is the change of traffic and buffer statistics as call arrive and depart. The time between such changes is insufficient to collect adequate data. This is more so due to the presence of multiple time scales and subexponentiality. We overcome this difficulty by imposing the invariance structure.

A key aspect of the schedulable region and the EB approaches that facilitate online learning is the concept of traffic classes. By limiting the load space to be finite dimensional, traffic classes give rise to the concept of operating point, much like the corresponding notion in circuit-switched networks. The main problem with online measurements is that the cell-level statistics change as calls arrive and depart. Thus, the object of estimation changes before it can be estimated. The power of the notion of traffic classes is that cell-level statistics corresponding to a given state may be estimated over successive visits of the operating point to that state. We exploit this fact to develop a search algorithm for learning the size of schedulable region.

The present work provides a framework and technique for online estimation of the schedulable region. Experimental results collected from our implementation on an Advanced Telecommunications Modules ATM switch are compared with those resulting from the EB approximation. The experiments were conducted using a real-time traffic generator which generates ATM cell-streams based on actual MPEG traces. The switch supports static priority, weighted round robin and MARS [1] scheduling policies, but the proposed techniques are not constrainted by these scheduling policies.

This paper is organized as follows: In Sect. 2, we define the concepts of traffic classes and schedulable region, and in Sect. 3, we present the search algorithm for online estimation. The fixed and floating operating concept will be explained in Sect. 4. In Sect. 5, we present a feasible implementation method based on mean-rate approximation, and the issues of reliability are treated in Sect. 6. In Sect. 7, we present results from real-time experiments.

2. Schedulable Region

The schedulable region is an abstraction of the capacity of a statistical multiplexer. This capacity depends, among other factors, on the traffic characteristics and quality of service (QOS) requirements. We devote the present section to defining these factors.

For our purposes, the traffic characteristics of calls shall be captured via a set of *traffic classes*, which represent statistical models for information streams. A traffic class is defined by a peak-rate and an additional qualitative stochastic characterization such as video, voice, or data. The latter is specified, for example, with its mean rate. The *operating point* of a multiplexer is defined as the vector of the number of calls of each traffic class in progress. It is an abstraction of the state of a multiplexer. As such, it deems all calls of a given traffic class to be statistically equivalent, a restriction that shall refer to as *invariance*.

Manuscript Figures

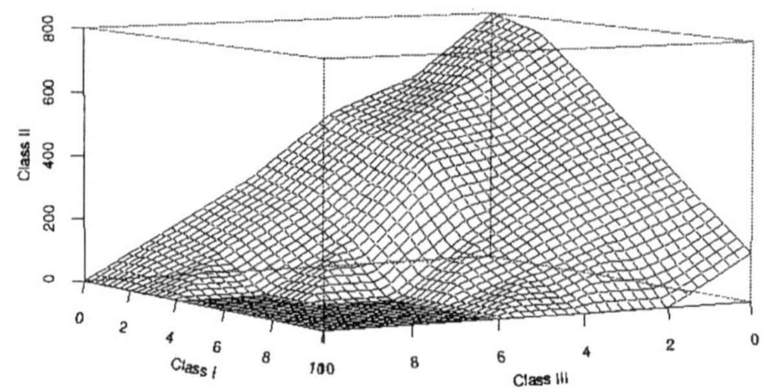

Fig. 1 Schedulable region.

Quality of service constraints refer to bounds on the probabilities of certain events related to cell delay. In this paper, we shall be content with the QOS constraint $P(D^i < d^i|X) < \varepsilon^i$ for every traffic class i and every realization of X, where the random variable D^i represents the delay distribution of class i and the random vector X represents the distribution of the operating points.

Schedulable region of a multiplexer is the space of possible combinations of operating points that the multiplexer can handle while guaranteeing QOS of real traffic. It is an extension of the concept of Shannon capacity and represents the region of stability of the multiplexer under QOS constraints. Figure 1 illustrates the concept. In order to guarantee cell-level QOS, the operating points should be within the surface of the schedulable region. Otherwise, violated.

The schedulable region is related to the equivalent bandwidth concept as follows: Denote the equivalent bandwidth of any set of calls that results in an operating point x by $\mu(x)$. Here, operating point means the vector of existing all class calls. We shall refer to it as the equivalent bandwidth of the operating point x. Such a description is justified by the invariance property. It follows directly from the definition that

$$S = \{x \in \mathcal{R}^N | \mu(x) \leq C\}$$

where S is the schedulable region, N is number of traffic classes and C is the line speed. Also, $\mathcal{R}^N$ is the positive integer set of N-dimension.

The advantage of the schedulable region is that it is the basis of a separation principle between scheduling, a cell-level task, and admission control, a call-level task. In order to guarantee cell-level QOS, the admission controller has to ensure that the operating point lies within the schedulable region.

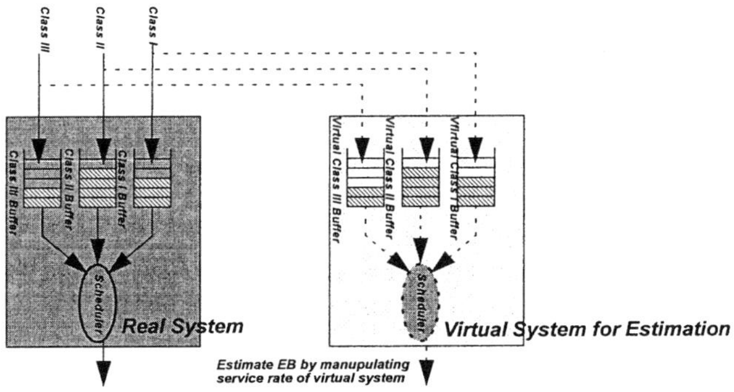

Fig. 2 Estimator's functional blocks.

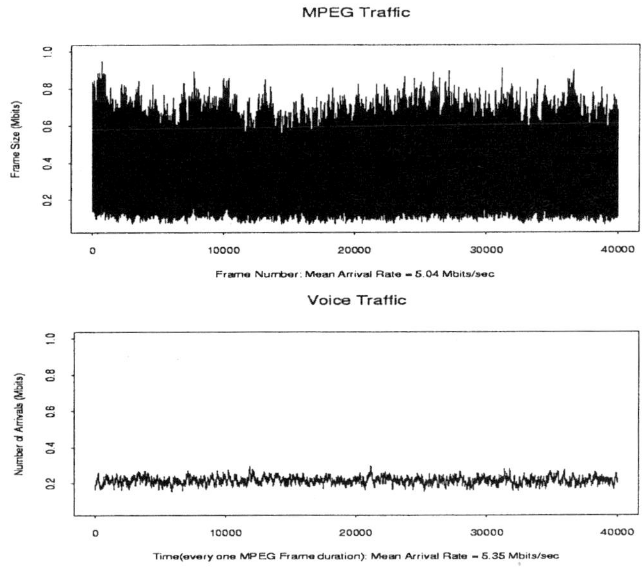

Fig. 3 Sample paths of multiplexed MPEG video and voice.

3. Algorithm for Online Estimation

The schedulable region of a multiplexer in not a constant but varies over time as traffic conditions change. Consequently, estimation of the schedulable region requires two phases. In the first, the algorithm attempts to lock onto the schedulable region, and in the second it attempts to keep track of any changes. We provide a direct approach for estimating $\mu(x)$, first under a quasi-static operating point and then under more general scenarios. Figure 2 shows the functional blocks of the implemented estimator.

3.1 The Choice of Search Algorithm

In order to estimate $\mu(x)$, an estimator has to determine the minimum line speed at which QOS is guaranteed. The technique we adopt requires the capability to alter the effective line speed and the capacity to measure QOS. The latter is already available to some extent on most ATM switches while the former is not hard to realize if we use virtual system running in parallel with real system. Given these two capabilities, the task of estimating $\mu(x)$ reduces to a search in the set $\left[0, \sum_{i=1}^{N} p^i x^i\right]$, where p^i and x^i refer to the peak arrival rate of a call of traffic class i and the ith coordinate of a point $x \in \mathcal{R}^N$ respectively. When the search algorithm is chosen, traffic statistics and QOS constraints should be considered.

Let us look at the two different sample paths shown in Fig. 3. One is for the 16 multiplexed MPEG, the other is for the 225 multiplexed voice traffic. Even though their mean arrival rate is very close to each other, their statistics are quite different. Thus, it is really difficult to find out the search algorithm which is effective for both types of traffic. It means it is necessary to choose the search algorithm depending on the arrival statistics and QOS constraints. Usually, MPEG traffic exhibits subexponentiality and multiple time scale characteristics [21]. Also, its QOS constraints are very tight. In order to estimate the EB of MPEG traffic, we should use a specific search algorithm which is quite different from the existing online measurement-based algorithms. Details are explained in the next following section.

3.2 Search Algorithms

In our framework, we focus on the two types of traffic, voice and MPEG video. Traffic statistics of voice sources could be characterized by the Markovian-type On-Off source and their QOS constraints are not tight. In order to estimate the bandwidth available to the future voice calls at a given operating point, we use linear search algorithm combined with binary search algorithm. The search algorithm for voice traffic consists of two phases. At the first phase, binary search algorithm is used to find out the convergence point in short time. After finding out convergence point, linear search algorithm will be used. Counterexample of voice traffic is MPEG video traffic with multiple time scale, subexponentiality, and very tight QOS constraints. One of the best ways to estimate EB of MPEG traffic is to keep track of the peak arrival rate and its duration with a long measurement period (multiple time scale). To provide it, we use the following search algorithms depending on the arrival statistics and QOS constraints.

3.2.1 Linear Search Algorithm

A linear search algorithm can operate in one of three modes. The *optimistic* mode starts at $\mu(x) = 0$ and increases successive estimates by some step size δ when QOS violations occur. The *pessimistic* mode start at

$\mu(x) = \sum_{i=1}^{N} p^i x^i$ and reduces successive estimates by a step size δ until a QOS violation occurs. A *mixed* mode where steps are taken in both positive and negative directions based on whether a QOS violation occurs or not is also possible. The step size δ is chosen based on the accuracy required for the estimate of $\mu(x)$, while keeping in mind that the convergence time varies inversely with δ. Convergence of the optimistic and pessimistic linear search is also guaranteed due to the monotonicity of the sequences produced by each. However, we have yet to indicate the circumstances under which any of the above limits lies in the range $\mu(x) \pm \delta$. All of the above search procedures operate with update-steps chosen a *priori*, thus assimilating information gradually with each occurrence of the relevant traffic characteristics. Hence, the sequence of estimates will converge to $\mu(x)$ only if the frequency of such occurrence is adequate for rapid learning. Example of traffic providing the requisite frequency are on-off voice and Markov-type streams.

3.2.2 Binary Search Algorithm

A binary search algorithm operates by successively bisecting the interval [0, C] after each measurement period, thus converging rapidly. To process the binary search algorithm, three parameters are required: upper-threshold, lower-threshold, and test value. First, the middle point between upper-threshold and lower-threshold would be chosen as a current test value. Once the test value could be accepted, it will be saved as a new lower-threshold. Otherwise, it will be saved as a new upper-threshold. This precess keeps going until the difference beween lower- and upper-threshold comes to be smaller than the given value. The convergence time of binary search alogrithm is much faster than the linear search algorithms, but it is not appropriate for time-continuous estimation. In our system, we use it at the initial stage just to reduce convergence time.

3.2.3 Optimistic Linear Search Algorithm with Gradient Estimates (OLG)

The prime example is compressed video which exhibits high autocorrelation and a long tail-distribution. Measurements of video traffic collected during a low bit-rate period should not be interpreted as an indication of low EB. Thus, the only search algorithm capable of handling this case is the optimistic linear search. Due to the infrequent occurrence of the high bit rates, it is important to extract the most from each such occurrence. Below we enhance the optimistic linear search by providing it with gradient estimates.

In the OLG, we shall consider even a single cell loss to be a violation of QOS, due to the infeasibility of measuring very small values of cell loss in a period of reasonable duration. Observe that

$$a_n(x) \leq \mu(x)d + \mu(x)\tau_n \tag{1}$$

where $a_n(x)$ is the number of cells that arrived during the period τ_n. d is the cell delay QOS, and τ_n is the duration of the peak arrivals. The term on the left represents the number of cells that arrive during a peak period, the first term on the right bounds the number of cells that can be buffered and the second term on the right represents the number of cells that can be transmitted during peak period. Therefore, we set the estimate of $\mu_n(x)$ made from observations preceding the nth update as

$$\mu_n(x) = max\left(\mu_{n-1}(x) + \delta, \frac{a_n(x)}{\tau_n + d}\right) \tag{2}$$

The new effective line speed $\mu_n(x)$ is an estimate of the EB of the aggregate arrival process as observed up to time n. Inequality Eq. (1) suggests that the sequence $\mu_0(x)$, $\mu_1(x)$, $\mu_2(x)$, $\cdots$ might be bounded by $\mu(x) + \delta$, provided that $\mu_0(x)$ satisfies the same property. Though this statement is not true in general, experiments show that the sequence dose not get much larger than $\mu(x) + \delta$. Essentially, the simplification presented here requires only the arrival-rate to be monitored.

4. Operating Point

A major problem associated with online estimation is that in the presence of call arrivals and departures, the objects of measurement are not fixed. Hence, there is insufficient time to take accurate measurements. We overcome this difficulty by imposing invariance structure stated in Sect. 2. Here, we introduce two types of operating points: *fixed operating point and floating operating point*.

In the fixed operating point, there are no changes of number of calls in progress. Most previous measurement-based estimations are done under this assumption. The estimation of EB at a given operating point is executed by the search algorithm chosen depending on the arrival statistics and QOS constraints (see Sect. 3). Meanwhile, in the floating operating point, we slice time corresponding to call arrival and departure epochs in order to take measurements over a period longer than a sojourn time. Incremental measurements are taken at each visit to a given point. This is also in conformance with our definition of QOS based on probabilities conditioned on the operating point. The above "time-slicing" framework is described in the Fig. 4. Every operating point has its own measurement profile (MP). Whenever operating point changes, the MP of the new operating point is loaded into the estimator while the MP of the current operating point is unloaded from the estimator. Thus, the

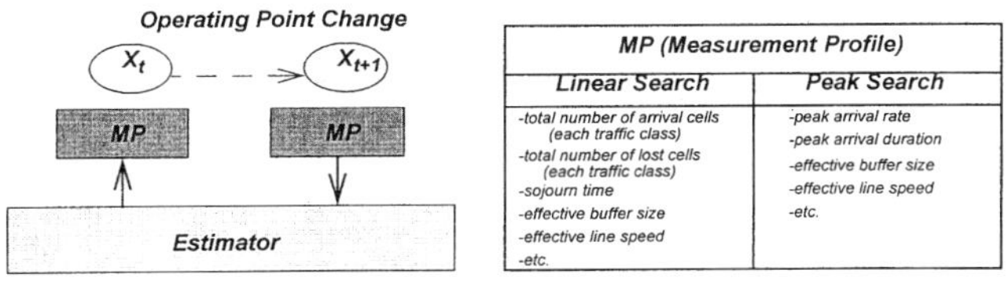

Fig. 4 Time-slicing framework for floating operating point.

operations of the estimator are the same as in the fixed operating point. Depending on the search algorithm, the contents of MP are different.

The storage space required by the above procedure is proportional to the number of operating points in the schedulable region. Moreover, the convergence time will be long compared to the fixed operating point case since the operating point spends only a fraction of the time at each point in the schedulable region. In the next section, we consider the approximation that will alleviate these concerns. This approximation can be combined to yield greater reduction in storage space requirement and convergence time.

5. Mean-Rate Approximation

It is shown in [18] that with respect to the queue-length distribution, exponential sources like voices behave asymptotically like constant bit rate sources of the same mean when they are multiplexed together with subexponential sources like MPEGs. To illustrate this approximation, consider a system with two traffic classes, class I being subexponential and class II being exponential. Then,

$$\mu((x^I, x^{II})) \approx \mu((x^I, 0)) + m^{II} x^{II} \tag{3}$$

where m^{II} is the mean rate of a class II call. Thus, we only need to keep track of the values of $\mu((0,0))$, $\mu((1,0))$, $\cdots$, $\mu((m,0))$. The convergence time also reduces since the time spent by the operating point in states of the form $(x,^*)$ is more than that spent in (x,y).

In our framework, MPEG sources are assumed to be class I and voice sources are assumed to be class II. As explained in the previous Sect. 3, the estimation of EB for voices source is based on the mixed linear search algorithm. If QOS of real traffic will be guaranteed in a given measure period, $\mu((x^I,0))$ increases by δ. Otherwise, it decreased by δ. To reduce the convergence time, binary search algorithm is used at the initial step. When operating point changes into (x^I, x^{II}), the effective line speed of the estimator will be changed by Eq. (3). Here, the effective line speed of the estimator is the same as EB of real traffic. Also, the effective buffer size of each class is given by $d^I \cdot \mu((x^I, x^{II}))$ and $d^{II} \cdot \mu((x^I, x^{II}))$ respectively. The verification of cell loss QOS for each class is done by the total number of each class arrivals and the total number of each class

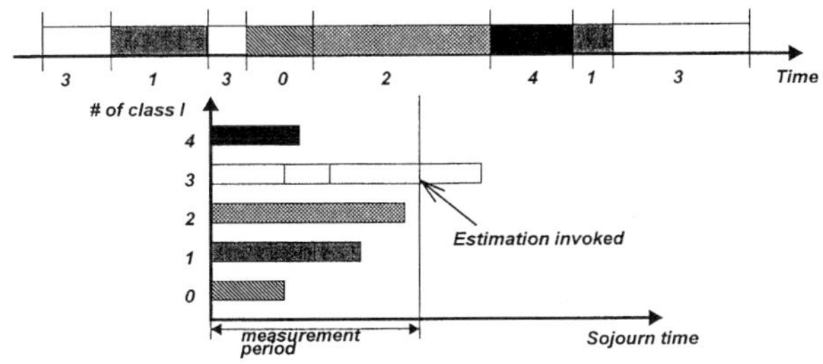

Fig. 5 Estimation timing.

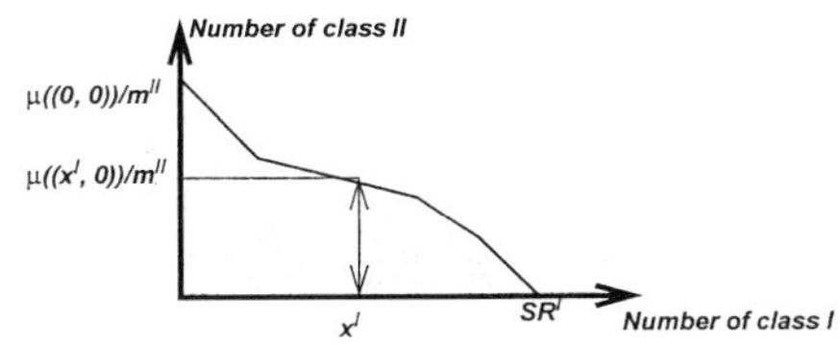

Fig. 6 Schedulable region from mean-rate approximation.

lost cells, which are kept in the MP. Note that the above update of $\mu((x^I,0))$ is done only when the total sojourn time at the operating point $(x^I,^*)$ is equal to a given measurement period (see Fig. 5). The number of MPs is the same as the number of class I calls in the schedulable region and its contents are described in Fig. 4.

The above estimation does not include the estimation of the number of class I calls in the schedulable region. That is, the above estimations are executed at the given number of class I calls. The estimation of the number of class I calls is done by OLG. Let us assume that the maximum class I call in the schedulable region is equal to SR^I and $\mu_p(x^I)$ denotes the EB estimated by OLG on the condition that x^I class I calls exist. This estimation is carried out by extracting class I traffic from the real traffic streams and inserting them into a virtual FIFO system. Initially, SR^I increase by one whenever $p^I < C - \mu((SR^I, 0))$. However, once $C < \mu_p(x^I)$, SR^I comes to be fixed by $x^I - 1$ for a long time due to multiple time scale and very tight cell loss QOS. Note that $\mu((x^I,0))$ and $\mu_p(x^I)$ are different. $\mu((x^I,0))$ is a coupled EB with multiplexing gain across traffic classes while $\mu_p(x^I)$ is a decoupled EB without multiplexing gain across traffic classes (the scheduling policy in a multiplexer has no effects on its estimation). The schedulable region is thus described as follows:

$$S = \{(x^I, x^{II}) \in \mathcal{R}^2 | \mu((x^I, x^{II})) \leq C\} \tag{4}$$

Figure 6 shows the schedulable region estimated from mean-rate approximation.

6. Reliability

When the schedulable region will be estimated from the mean-rate approximation, we should consider some reliability issues. Given a fixed measurement and a cell

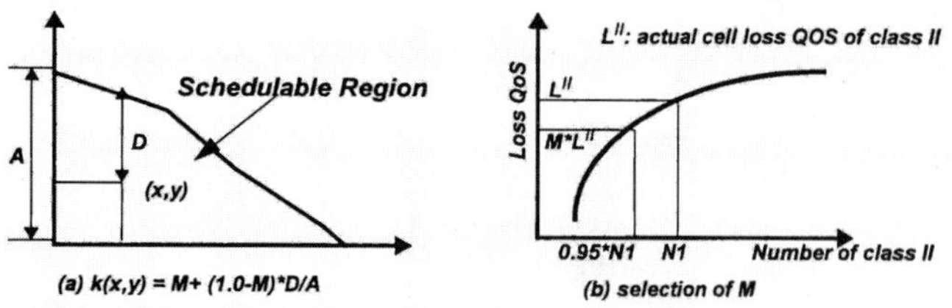

Fig. 7 Tighter QOS assignment.

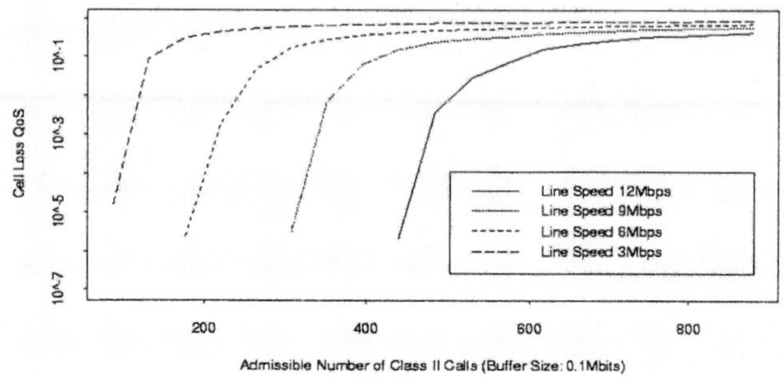

Fig. 8 Cell loss QOS and admissible number of class II calls (voice calls).

loss QOS, we should get the arrivals enough to check whether cell loss QOS is violated or not. When the current operating point is far away from the schedulable region boundary resulting from light traffic load, we do not have to estimate schedulable region during this period because many number of future calls could be acceptable and it might make EB estimation underestimated. In our framework, the estimator will be turned off if $\mu((x^I, x^{II}))$ is less than a given portion of line speed.

The mean-rate approximation causes underestimation of EB for class II when the number of class I is very small (about 0–2). To solve it, we give a tighter cell loss QOS to class II traffic than actual its cell loss QOS as operating point is close to the schedulable region boundary. Figure 7 shows how to decide tighter cell loss QOS at a given operating point (x, y). As shown in Fig. 7, $k(x, y)$ is given by $k(x, y) = M + (1.0 - M) \cdot D/A$, where A denotes the schedulable region boundary at $x = 0$ and D is the distance from schedulable region boundary in the direction of class II. Then, cell loss QOS of class II used in estimator is given by $k(x, y) \cdot L^{II}$, where L^{II} is an actual cell loss QOS of class II. In our framework, we apply the cell loss QOS which corresponds to the 95% of actual number of class II calls when the operating point is on the schedulable region boundary (see Fig. 7(b)). To get M, we use the experimental result shown in Fig. 8. For example, given 12 Mbps line speed and 10^{-3} cell loss QOS, M was decided by 0.05.

In order to include the effect of long-time past history in $\mu((x^I, 0))$, we take a moving window average whose window size is much larger than measurement

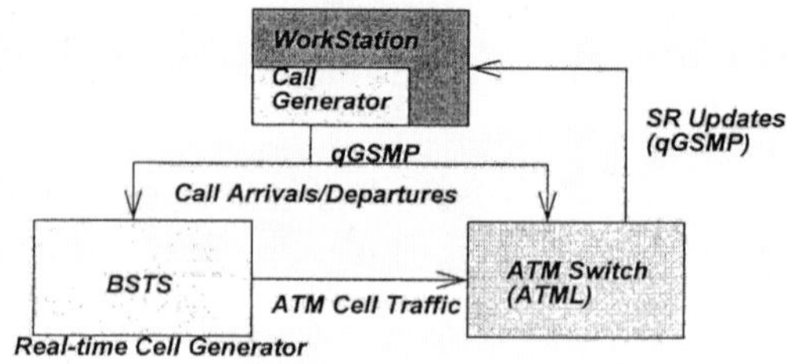

Fig. 9 Experimental setup.

period. For simplicity, it was implemented by simple exponential forecasting [19] in our next experiment. The coefficient of simple exponential forecasting is approximately given by $2/(W+1)$, where W is the window size of moving window average.

7. Experiments

7.1 Experimental Setup

The framework and algorithm presented in the previous sections have been implemented on an Advanced Telecommunications Modules Limited (ATML) switch in our laboratory. The setup of the experiments we conducted in shown in Fig. 9. It consists of an ATML ATM switch, a Hewlett-Packard Broadband Series Test System (BSTS) and an workstation. The dynamics of the operating point, i.e., call arrivals and departures, are generated by the workstation and communicated to the BSTS, which generates cell streams based on the operating point. qGSMP [22] is used to communicate call arrivals and departures to the switch and to retrieve schedulable region estimates which are displayed in real-time on the workstation. In our experiment, we used the MPEG sequences encoded at University of Wurzburg Institute of Computer Science using the UC Berkeley MPEG-I software encoder [20]. Class II calls were generated using an on-off voice model with mean talk-spurt and silence duration 3.5 ms and 6.5 ms respectively. Class I traffic consisted of MPEG video with a peak rate of 6 Mb/s and a hard delay bound of 10 ms and 0% cell loss ratio. Class II consisted of on-off voice with a peak rate of 70 kb/s (64 kb/s plus 10% for the ATM header), delay bound of 15 ms and loss ratio bound of 10^{-3}. The line speed was 12 Mb/s. Our implementation on the ATML switch supports FIFO, static priority, weighted round robin, MARS scheduling policies. The proposed estimator was also implemented inside ATML switch. The line speed seen by the scheduler is under software control.

7.2 Experimental Results

7.2.1 Estimation of Schedulable Region Boundary of Class II Call

Figure 10 shows the result of the mixed linear search al-

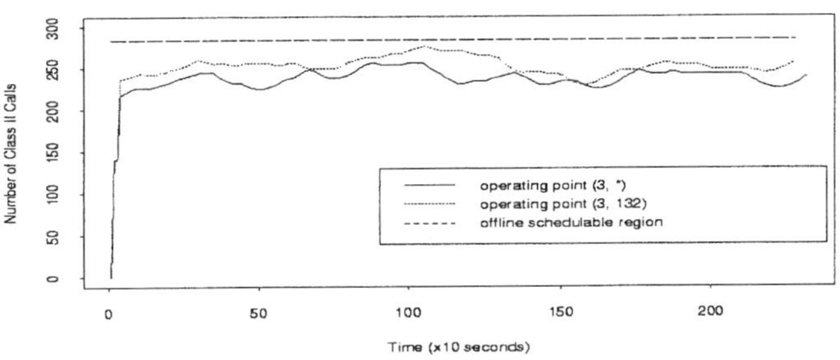

Fig. 10 The estimation of schedulable region boundary of class II calls at the fixed 3 class I calls.

gorithm combined with binary search algorithm, which is used to estimate the schedulable region boundary in the direction of voice call at the given number of MPEG traces. To get it, we used the MARS scheduler and two traffic classes (class I is MPEG and class II is voice). Measurement period is set to 10 seconds and the window size of moving average is given by 18 (about 3 minutes). In this figure, the number of class I call is fixed as 3. The dotted line represents the result at the fixed operation point, (3, 132). That is, the number of class I and class II call is 3 and 132 respectively. The solid line represents the result at the fixed 3 class I calls and moving number of class II calls within current schedulable region (melting the fixed operating point on the side of class II calls).

7.2.2 Estimation of Maximum Number of Class I Calls

To make schedulable region, we should now frezen the operating point on the side of class I call. It is important to find out the maximum number of class I calls admissible at the given line speed. To do this, we apply OLG to the class I calls as explained in Sect. 5. Figures 11 and 12 show the results of the OLG applied to MPEG video traces. We argue that for MPEG traffic with tight QOS constraints, the OLG is a good estimation method of the EB. In Fig. 11, the graph at the left side shows the sample path of a 30-minute MPEG clip from the *Worldcup Soccer final match*, and the right figure shows the online estimate of the clip's EB corresponding to a hard delay bound of 10 ms and loss bound of 10^{-10} . The dotted line shows the value computed off-line. From the top line in the right figure, each line shows the actual peak rate, offline EB, online EB, and actual mean rate.

In Fig 12, the graph at the left side shows the sample path of six 30-minute MPEG clips multiplexed. The starting instances were randomly chosen. The right figure shows the online estimate of the clip's EB. From the top line in the right figure, each line shows the sum of the individual peak rates, the sum of the individual EBs, the actual peak rate of the multiplexed streams, offline EB of the multiplexed streams, online EB of the

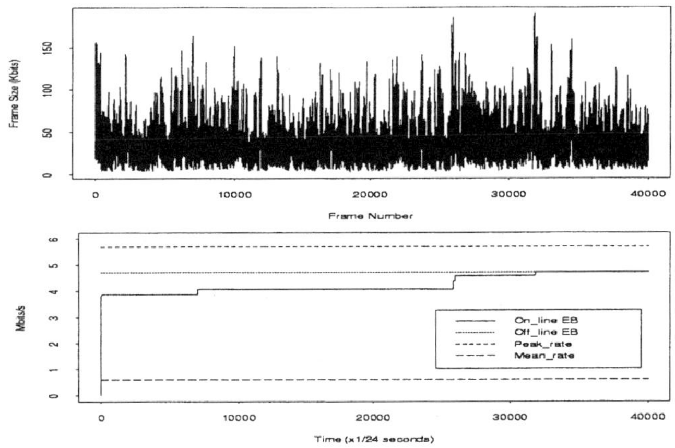

Fig. 11 OLG for single MPEG trace.

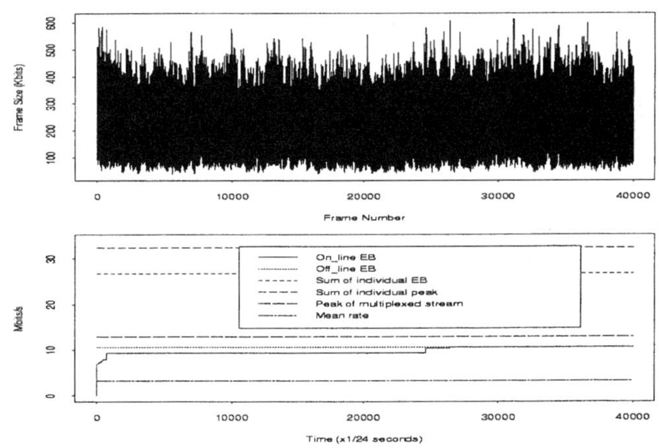

Fig. 12 OLG for 6 multiplexed MPEG traces.

multiplexed streams, and actual mean rate of the multiplexed streams.

7.2.3 The Estimates of Schedulable Region

Figure 13 shows the schedulable region estimates under static priority, weighted round robin, and MARS scheduling policies. The arrival rate and average holding time for class I call were 0.1 calls/s and 100 s, and for class II call the quantities were 3 calls/s and 30 s.

The mean-rate approximation was applied to class II traffic; the mean rate was measured at the beginning of the experiment to be about 25 kb/s (including ATM header size). The measurement period was chosen to be 10 seconds. However, increase in equivalent bandwidth estimates were made only if no QOS violations occurred for 30 seconds. The update step was chosen to be 15 kb/s for increases and 90 kb/s for decreases. The results show that the estimates rapidly expand and settle down to a fairly accurate approximation of the schedulable region in about 10 minutes.

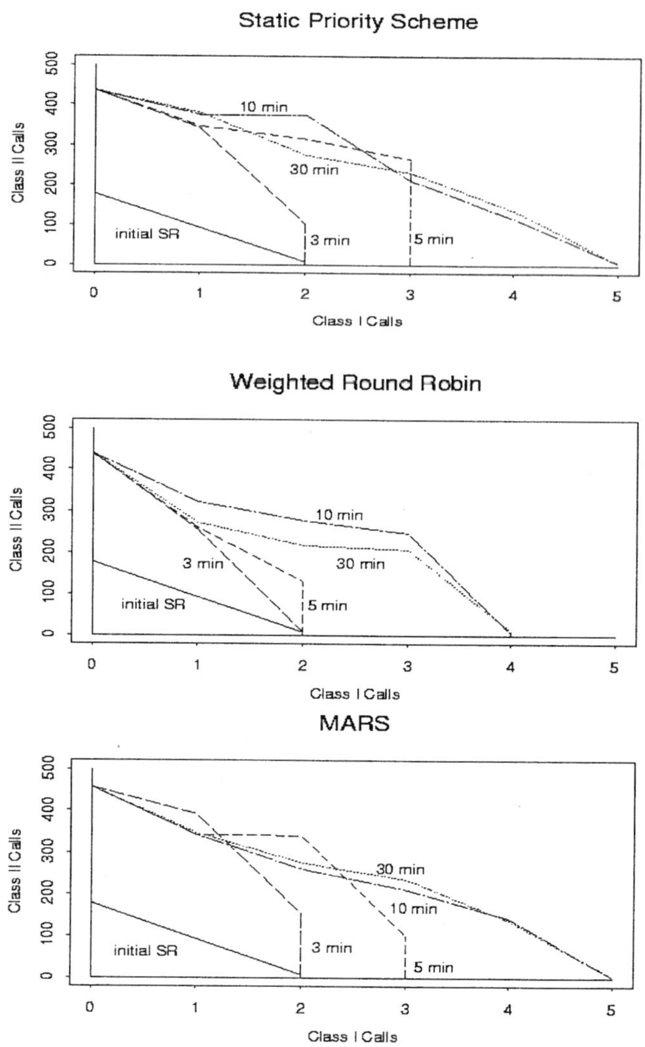

Fig. 13 The estimates of two-dimensional schedulable regions.

8. Conclusion

In this paper, we have presented direct search-methods for online estimation of equivalent bandwidths. The exact technique appropriate for estimation depends on the traffic characteristics (e.g., video, voice) and the QOS constraints. These techniques estimate the equivalent bandwidth quite accurately. In the case of compressed video, they accurately estimate the equivalent bandwidth of the arrival process as observed up to the current time, i.e., they estimate the minimum line speed required to transmit the arrival process from time zero to the present subject to QOS constraints. In the absence of a traffic model, this is the best possibility since there is no way to predict the extent of high-activity periods in the future. Though at first glance the estimation algorithms may appear complex, they are computationally very simple. As proof of concept, the estimation algorithms have been implemented on an ATM switch along static priority, weighted round robin and MARS scheduling policies. Results were provided for actual MPEG traces and on-off voice models. To overcome the difficulty arising due to call arrivals and departures, we exploited the concept of *invariance*. Under the invariance-assumption, the techniques referred to in the preceding paragraph extend readily to the case of call arrivals and departures. The main unsolved problem is the long length of time required for convergence, an issue completely overlook by many previous papers. The problem is caused by the traffic characteristics of compressed video, and significantly aggrevated by the presence of call arrivals and departures. There is an inherent trade-off between accuracy and convergence speed. Traffic models would be the ideal way out of this delemma, but no satisfactory model has come out thus far. A hybrid solution relying on both a model and measurement is probably the way out.

Acknowledgments

Author is grateful to Professor Aurel A. Lazar and Mahesan Nandikesan for their incisive comments and suggestions.

References

[1] J. Hyman, A.A. Lazar, and G. Pacifici, "Real-time scheduling with quality of service constraints," IEEE J. Select. Areas Commun., vol.9, pp.1052–1063, Sept. 1991.

[2] J. Hyman, A.A. Lazar, and G. Pacifici, "A separation principle between scheduling and admission control for broadband switching," IEEE J. Select. Areas Commun., vol.11, May 1993.

[3] J.Y. Hui, "Resource allocation for broadband networks," IEEE J. Select. Areas Commun., vol.6, pp.1598–1608, Dec. 1988.

[4] F.P. Kelly, "Effective bandwidth at multi-class queues," Queueing Systems, vol.9, pp.5–16, 1991.

[5] A.I. Elwalid and D. Mitra, "Effective bandwidth of general Markovian traffic sources," IEEE/ACM Trans. Networking, vol.1, no.3, June 1993.

[6] G. Kesidis, J. Walrand, and C.-S. Chang, "Effective bandwidths for multiclass Markov fluids and other ATM sources," IEEE/ACM Trans. Networking, vol.11, pp.424–428, Aug. 1993.

[7] C.-S. Chang, "Stability, queue length and delay of deterministic and stochastic queueing networks," IEEE Automatic Control, pp.913–931, May 1994.

[8] C.-S. Chang and J.A. Thomas, "Effective bandwidth in high-speed digital networks," IEEE J. Select. Areas Commun., vol.13, pp.1091–1100, Aug. 1995.

[9] C.-S. Chang and T. Zajic, "Effective bandwidths of departure processes from queues with time varying capacities," IEEE INFOCOM'95, pp.1001–1009, 1995.

[10] I. Hsu and J. Walrand, "Admission control for multi-class ATM traffic with overflow constraints," Computer Networks and ISDN Systems, vol.28, no.13, pp.1739–1751, 1996.

[11] D.D. Botvich and N.G. Duffield, "Large deviations, the shape of the loss curve, and economics of scale in large

multiplexers," Technical Report DIAS-APG-94-12, Dublin Institute for Advanced Studies, 1994.

[12] G. Choudhury, D. Lucantoni, and W. Whitt, "Squeezing the most out of ATM," IEEE Trans. Commun., pp.203–217, Feb. 1996.

[13] C. Courcoubetis, G. Kesidis, A. Ridder, J. Walrand, and R. Weber, "Admission Control and Routing in ATM networks using Inferences from Measured Buffer Occupancy," IEEE Trans. Commun., vol.43, pp.1778–1784, Feb./March/April 1995.

[14] C. Courcoubetis, G. Fouskas, and R. Weber, "An on-line Estimation Procedure for Cell Loss Probabilities in ATM Links," Preprint.

[15] T.E. Tedijanto and L. Gun "Effective bandwidth management mechanisms in ATM networks," IEEE INFOCOM'93, March 1993.

[16] S. Jamin, P.B. Danzig, S. Shenker, and L. Zhang, "A measurement-based admission control algorithm for integrated service packet networks," SIGCOM'95, Aug. 1995.

[17] Z. Dziong, B. Shukhman, and L. Mason, "Estimation of aggregated effective bandwidth for traffic admission in ATM networks," IEEE INFOCOM'95, pp.810–818, 1995.

[18] P.R. Jelekovic and A.A. Lazar, "Multiplexing on-off sources with subexponential on periods," CTR Technical Report 457-96-23, IEEE INFOCOM'97, Kobe Japan, April 1997.

[19] G.E.P. Box and G.M. Jenkins, Time Series Analysis, Holden-Day, 1970.

[20] O. Rose, "Statistical properties of MPEG video traffic and their impact on traffic modeling in ATM systems," Technical Report no.101, University of Wurzburg Institute of Computer Science, Feb. 1995.

[21] P.R. Jelekovic, A.A. Lazar, and N. Semret, "The effect of multiple time scales and subexponentiality of MPEG video stream on queueing behavior," IEEE J. Sel. Areas Commun., vol.15, no.6, pp.1052–1071, Aug. 1997.

[22] C.M. Adam, A.A. Lazar, and M. Nandikesan, "QOS extensions to GSMP," CTR Technical Report 471-97-05, Center for Telecommunication Research, Columbia University, New York, April 1997, Available at http://comet.ctr.columbia.edu/xbind/qGSMP.

High-Performance Backbone Network Technology edited by Naoaki Yamanaka
ISBN: 0-8247-5321-6 DOI: 10.1081/(120025714)
Published 2004 by Marcel Dekker, Inc.

Design and Analysis of the GCRA Traffic Shaper for VBR Services in ATM Networks

Mingfu Li and Zsehong Tsai

SUMMARY The proposed GCRA (Generic Cell Rate Algorithm) traffic shaper consists of a regulator and a scheduler. It can shape multiple incoming VBR (Variable Bit Rate) cell streams simultaneously to be strictly conforming according to the GCRA algorithm when the cells depart for the ATM output link. The impact of cell emission conflicts is considered and resolved by using an EDD (Earliest-Due-Date) scheduler and a feedback signal from the scheduler to the regulator. The call admission control condition and the cell delay bound are derived. Simulation results demonstrate that the output cell streams of the proposed GCRA traffic shaper do not contain any non-conforming cells and the scheduler queue size is significantly reduced. Meanwhile, the delay performance is almost not affected by the use of the feedback mechanism.

key words: *ATM networks, GCRA, traffic shaper, VBR services*

1. Introduction

Since the traffic management of ATM networks became an important research area, various traffic shapers or cell spacing devices have been proposed to reduce the burstiness of the VBR traffic [1]–[5] or to avoid violation of a traffic contract. The concept of traffic shaper or cell spacer is to store the input cells in the buffer and then output them at eligible time or theoretical reemission time. Traffic shapers can be placed in a terminal with an ATM network interface card, in the network edge ATM switch or gateway, or as an independent network device, as shown in Fig. 1. In [1], [2], traffic shapers or cell spacers are designed according to the so-called Virtual-Scheduling Algorithm (VSA) to reduce the cell delay variation (CDV) due to queueing in multiplexing nodes. In [3], Brochin employs a cell spacing device to prevent any two successive cells from being transmitted within a time shorter than a variable value. Traffic shaper is implemented by the sliding window mechanism followed by a peak bit rate reduction mechanism in [4]. In [5], two level shaping algorithm which depends on the buffer occupancy of the shaper is proposed.

In most of these papers, the assumed models include multiple traffic streams passing through separate cell spacing devices and a second-stage multiplexer before output. Since the impact of the multiplexer on the output traffic usually is not taken into account, the output cell streams may be non-conforming with respect to the traffic contract due to emission conflicts within the multiplexer, regardless the fact that the cell streams observed at the output of spacing devices are conforming. In other words, such emission conflicts are likely to lead to the unnecessary peak cell rate, sustainable cell rate, or burst tolerance violation when cells are transmitted to the ATM output link. For example, under the special case where the peak cell rate is set to be the physical link speed, no non-conforming events due to peak cell rate violation should exist. However, the cell emission conflicts can still result in large queueing delay and the CDV caused by such queueing delay can lead to the sustainable cell rate violation or burst tolerance violation. In practice, GCRA-based UPC (Usage Parameter Control) mechanism [6], [7] may be implemented at the other side of the User-Network Interface (UNI) or Network-Node Interface (NNI) (see Fig. 1). At these UPC devices, non-conforming cells may be tagged as low priority or be discarded and lead to unnecessary cell losses. It reveals that the multiplexer should be modeled as a component of the traffic shaper such that the impact of cell output conflict can be resolved. In [8], a software method is proposed for multiplexing spacer outputs on cell emissions. Unfortunately, the output

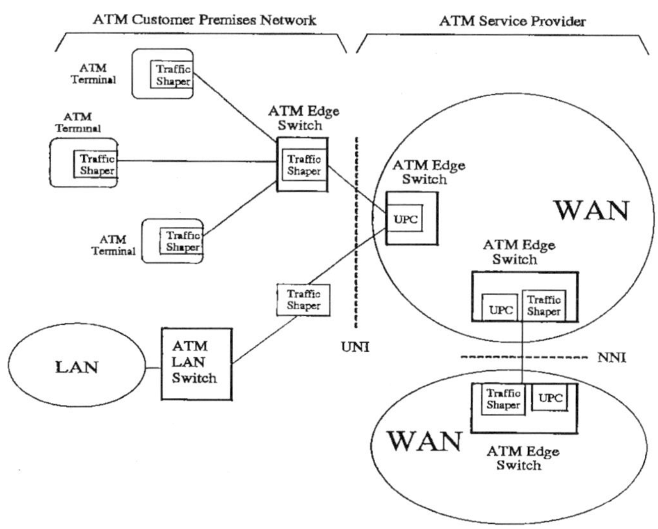

Fig. 1 The network scenario where the traffic shaper is widely applied.

Manuscript received February 13, 1997.
Manuscript revised August 16, 1997.

[†]The authors are with the Department of Electrical Engineering, National Taiwan University, Taipei, Taiwan, R.O.C.

IEICE TRANS. COMMUN., Vol. E81-B, No. 3, pp. 511–519, March 1998

cell streams of the multiplexer are still not guaranteed to be conforming. Thus, the design of traffic shapers which can regulate the peak cell rate, sustainable cell rate and burst tolerance simultaneously to be conforming is still at issue.

In this paper, we proposed the GCRA traffic shaper which can shape multiple incoming VBR traffic streams simultaneously to be strictly conforming with respect to the GCRA algorithm when the cells enter the ATM output link. Since the shaping delay is related to both the input and output traffic descriptors, we will first derive the call admission control condition to guarantee the shaper performance and then analyze the total shaping delay bound later.

The organization of this paper is as follows. In Sect. 2, the traffic model is defined. In Sect. 3, we describe the design of the GCRA traffic shaper. In Sect. 4, the delay bound of the GCRA traffic shaper is derived. In Sect. 5, the merits of the proposed GCRA traffic shaper is demonstrated by numerical examples. And finally in Sect. 6, we present our conclusions.

2. The Traffic Model

We use the concept of traffic specification in the ATM Forum UNI Specification [9] to define the traffic descriptor (X, T, τ) for each virtual connection (VC). Here, X indicates the minimum cell interarrival time, which corresponds to the peak cell rate parameter if CDV is zero, and T represents the minimum average cell interarrival time. Note that the inverse of T is equivalent to the sustainable cell rate. The burst tolerance τ is the maximal amount of time that is allowed for the actual cell arrival to be earlier than the cell arrival time theoretically predicted, under the assumption that any two consecutive cells are separated by the interval T. The latter burst tolerance parameter τ is identical to that defined in Forum's GCRA definition [9]. In other words, the traffic descriptor (X, T, τ) indicates the following traffic characteristics of a cell stream: regarding the sustainable cell rate, the cell stream conforms to GCRA (T, τ); when the peak cell rate is concerned, the cell interarrival time is strictly larger or equal to X, i.e., conforms to GCRA$(X, 0)$. On the other hand, ITU-T recommendation I.371 allows the specification of the permitted CDV value τ_p in the Customer Premises Networks in a traffic contract. Hence, it is possible to specify in a traffic contract that a cell stream conforms to GCRA(T_p, τ_p) for peak cell rate constraint and GCRA(T_s, τ_s) for sustainable cell rate constraint. To translate the GCRA-based traffic contract $\{(T_p, \tau_p), (T_s, \tau_s)\}$ to our traffic descriptor (X, T, τ), one can employ the following formula:

$$X = \max\{T_p - \tau_p, 1\}$$
$$T = T_s$$
$$\tau = \tau_s.$$

In the rest of this paper, the desired output traffic descriptor of VC j is specifically specified as (X_j, T_j, τ_j).

3. The GCRA Traffic Shaper

3.1 GCRA Traffic Shaper Model

In this section, we propose a new traffic shaper model which consists of two components: a regulator and a scheduler, as shown in Fig. 2. The regulator consists of several departure controllers. Each departure controller regulates the peak cell rate, sustainable cell rate and burst tolerance for one VC according to its individual output traffic descriptor. The operation of the traffic shaper follows a discrete time-axis, on which the normalized cell emission time is used as a time slot. If a cell arrives early, the regulator will stop it in the buffer. The latter cell then enters the scheduler at its eligible time. The scheduler is responsible for resolving the emission conflicts between cells from different VCs and transmits cells to the network. When a cell reaches the HOL (Head of Line) position in the scheduler, it can be transmitted at the beginning of the next slot. Hence, the minimum sojourn time at the scheduler for any cell is one slot. In order to prevent the cell emissions from violating the output traffic descriptor, there exists a feedback signal from the scheduler to the GCRA mechanism within the regulator. This assures that the calculation of the eligible time of the next cell emission of the same VC is always done after the receipt of the feedback signal triggered by the current emission. Therefore, there can be at most one cell in the scheduler for each VC at any time. The buffer size in the scheduler is thus determined by the maximal number of VCs that the GCRA traffic shaper is designed to support.

3.2 Algorithm in the Regulator

In order to describe the GCRA shaping algorithm in the

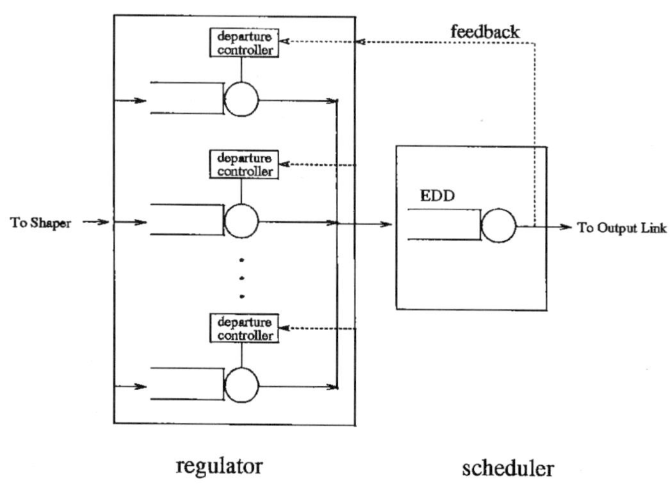

Fig. 2 The model of the GCRA traffic shaper.

regulator in details, some control variables for shaped VCs are defined as follows. The k-th cell from VC j arrives at the shaper (or regulator) at time $A_j(k)$ and is buffered until the Eligible Time $ET_j(k)$ before it enters the scheduler. It is attached an Initial Due-Date $IDD_j(k)$ when it enters the scheduler. Next, it is targeted to leave the scheduler at the Theoretical Departure Time $TDT_j(k)$. However, the time epoch it leaves the scheduler could be different. We denote the Actual Departure Time for the k-th cell of VC j as $ADT_j(k)$. During the operation of the traffic shaper, these variables are calculated by the following iterative form

$$TDT_j(k) = \max\{ADT_j(k-1),\ TDT_j(k-1)\} + T_j, \quad (1)$$

$$ET_j(k) = \max\{ADT_j(k-1) + X_j - 1,\ TDT_j(k) - \tau_j - 1,\ A_j(k)\}, \quad (2)$$

$$IDD_j(k) = \max\{TDT_j(k) - ET_j(k),\ \delta_j\}, \quad (3)$$

with initial values $TDT_j(1) = A_j(1)$, $ET_j(1) = A_j(1)$, $IDD_j(1) = \delta_j$, where δ_j is the minimal initial due-date, given by $\delta_j = \min\{T_j - X_j,\ \tau_j\} + 1$.

Consider only a specific VC, say VC j. When the k-th cell arrives at the regulator, $TDT_j(k)$ is not determined until the feedback signal triggered by the departure of the $(k-1)$-th cell is received. $TDT_j(k)$ is then set to $\max\{ADT_j(k-1),\ TDT_j(k-1)\} + T_j$ so that it meets the output sustainable cell rate constraint of VC j. In order to guarantee the output cell stream to conform to (X_j, T_j, τ_j), the k-th cell can not enter the scheduler earlier than either the time epoch $ADT_j(k-1) + X_j - 1$ or $TDT_j(k) - \tau_j - 1$; otherwise, it might violate the peak cell rate constraint or the burst tolerance constraint, respectively. By taking into account of the fact that a cell can never enter the scheduler before it arrives at the regulator, Eligible Time can be expressed as Eq. (2). Actually, Eq. (2) implies that a cell is allowed to enter the scheduler as early as possible. Thus, such a traffic shaper implementation strategy can be called the *Most Greedy Method* (*MGM*). Under ideal operation conditions, the maximal throughput of all VCs can be maintained with the use of the traffic shaper. This requires that no cell losses occur in scheduler and all cells to be emitted before their Theoretical Departure Times are reached. The first requirement can be easily met by allocating M buffers at the scheduler, where M is the maximal number of VCs that the shaper supports. To satisfy the second requirement, the difference between $TDT_j(k)$ and $ET_j(k)$ is used as the preliminary initial due-date whenever the shaper is nonempty with respect to VC j. From (1) and (2), it can be shown that $TDT_j(k) - ET_j(k)$ can not be less than δ_j unless the k-th cell of VC j observed no cells of VC j in the scheduler and regulator. The proof is left to the readers. The final initial due-date, as given by Eq. (3), is set to be the maximum of preliminary initial due-date $TDT_j(k) - ET_j(k)$ and the minimal initial due-date δ_j. The latter is used to reduce overdue condition in case of output conflicts. From (1) to (3), it can be shown that $IDD_j(k)$ satisfies $\tau_j + 1 \geq IDD_j(k) \geq \delta_j$.

3.3 Algorithm in the Scheduler

After the k-th cell of VC j enters the buffer of the scheduler, its instantaneous due-date at time t, denoted as $DD_j(k, t)$, is calculated as follows.

$$DD_j(k, ET_j(k)) = IDD_j(k) \quad (4)$$

$$DD_j(k, t+1) = \max\{DD_j(k, t) - 1,\ 0\} \quad (5)$$

for $t \geq ET_j(k)$.

In this paper, we say that a cell is *overdue* if its due-date is zero and still resides at the scheduler. One should note that overdue events may delay the operation of the regulator and increase the total queue size or cell loss ratio in the shaper. Hence, the Earliest-Due-Date (EDD) service discipline [10]–[15] is employed in the scheduler to minimize the number of overdue events. Under the EDD service discipline, the cell with the smallest due-date in the scheduler is emitted first. If multiple cells have equal due-date, they are emitted in FIFO order.

The optimality of EDD with the criterion of minimizing the number of overdue events has been established in [10]–[13]. Hence, we believe EDD is superior to other service disciplines. The disadvantage of EDD scheduling is that queued cells must be sorted according to their due-dates, requiring a search or sort operation whenever a new cell enters the scheduler. Thus, the time complexity of the proposed scheduler depends on the maximum number of cells queued in the scheduler. If software implementation based on a standard search algorithm is employed for our proposed scheduler and the maximum number of VCs is M, then the time complexity of the scheduler is $O(\log_2 M)$ [16]. As we will illustrate later, such time complexity is much smaller than that without feedback system.

Figure 3 shows one sample-path of the control variables associated with a specific VC, say VC j, with the

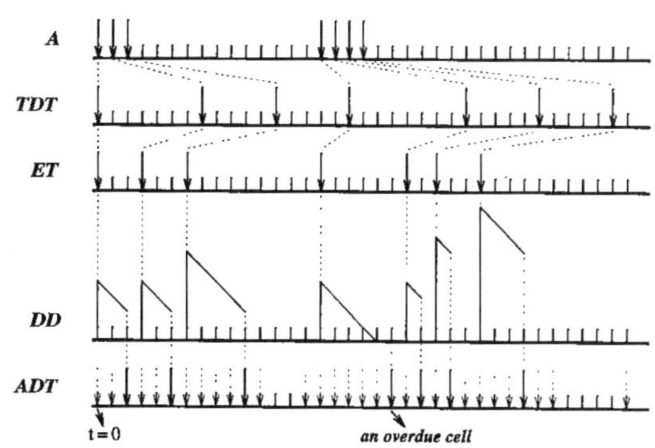

Fig. 3 Sample-path of a specific VC with traffic descriptor $(X, T, \tau) = (2, 5, 8)$, where the dotted arrows in the ADT pattern represent the cell emissions of other VCs.

output traffic descriptor $(X_j, T_j, \tau_j) = (2, 5, 8)$. The following elaborations explain the behavior of the traffic shaper, according to Fig. 3. Assume that cells of VC j arrive at the regulator at epochs $t = 0, 1, 2, 15, 16, 17, 18, \cdots$, i.e., $A_j(1) = 0$, $A_j(2) = 1$, $A_j(3) = 2$, etc. According to the algorithm in the regulator, one has $TDT_j(1) = ET_j(1) = 0$, and $IDD_j(1) = 4$. Hence, the first cell enters the scheduler at $t = 0$ immediately. Since at least 1 slot is required to schedule a cell, it can only depart at epoch $t \in [ET_j(1) + 1, \infty)$. Due to cell emission conflicts with other VCs, this cell eventually departs at $t = 2$. As a result, $ADT_j(1) = 2$. The traffic shaper then calculates $TDT_j(2) = 7$, $ET_j(2) = 3$ and $IDD_j(2) = 4$. Although $A_j(2) = 1$, the second cell enters the scheduler at $t = 3$. Next, due to cell contention within the scheduler, its departure time $ADT_j(2)$ is 5. After $t = 5$, the traffic shaper calculates $TDT_j(3) = 12$, $ET_j(3) = 6$ and $IDD_j(3) = 6$. The third cell then enters the scheduler at $t = 6$. Its departure is scheduled at $t = ADT_j(3) = 10$, again due to contentions from other VCs. Based on a similar operation, the reader should be able to trace the rest departure pattern of VC j.

4. Performance Analysis

4.1 Delay Bound in the Scheduler

Theorem 1: *Assume that the number of active VCs is g. The output traffic descriptor of VC j is (X_j, T_j, τ_j), $j = 1, \cdots, g$. Consider a target cell from VC j with initial due-date y. Then the delay bound, $d_j(y)$, of the target cell in the scheduler is given as follows.*

$$d_j(y) = \min\left\{ \max_{\Delta \geqq 0}\left\{ N_j(\Delta) + \sum_{i=1, i \neq j}^{g} N_i(y+\Delta-\delta_i) - \Delta \right\}, \; g + y - \min_{1 \leqq i \leqq g}\{\delta_i\} \right\}, \tag{6}$$

where $N_i(t) = \min\{\lfloor 1 + \frac{t}{X_i} \rfloor, \lfloor 1 + \frac{t+\tau_i}{T_i} \rfloor\}$, *for* $t \geq 0$; $N_i(t) = 0$, *otherwise.*

Proof: Consider the target cell from VC j which enters the scheduler with initial due-date y at time t. Assume the busy period that contains time t started at earlier time $(t - \Delta)$. Then the last cell from VC $i (i \neq j)$ that can delay the target cell must have arrived before time $(t + y - s)$, where $s (s \geq \delta_i)$ is the initial due-date of this cell. Hence, between time $(t - \Delta)$ and time $(t + y - s)$ the number of cells that do not belong to VC j and will be emitted before the target cell is bounded by

$$\sum_{i=1, i \neq j}^{g} N_i(y + \Delta - s) \leqq \sum_{i=1, i \neq j}^{g} N_i(y + \Delta - \delta_i).$$

Moreover, the number of cells which belong to VC j and enter the scheduler during $[t - \Delta, t]$ is bounded by $N_j(\Delta)$. Since both the cells which do not belong to VC j and enter the scheduler during $[t - \Delta, \; t + y - s]$ and the VC j cells entering the scheduler during $[t - \Delta, t]$ contribute to the delay of the target cell, we have

$$d_j(y) \leq \max_{\Delta \geqq 0}\left\{ N_j(\Delta) + \sum_{i=1, i \neq j}^{g} N_i(y+\Delta-\delta_i) - \Delta \right\}. \tag{7}$$

Since there exists a feedback signal between the scheduler and the regulator, the queue size in the scheduler can be no more than g at any time. Next, one should note that under the EDD discipline all the cells entering the scheduler after time $t + y - \min_{1 \leqq i \leqq g}\{\delta_i\}$ do not delay the target cell, because after time $t + y - \min_{1 \leqq i \leqq g}\{\delta_i\}$ the instantaneous due-date of the target cell ($< \min_{1 \leqq i \leqq g}\{\delta_i\}$) is always less than the initial due-dates of those new arrivals. Hence, under the worst case that there are g cells in the scheduler at time $t + y - \min_{1 \leqq i \leqq g}\{\delta_i\}$, the target cell should be emitted no later than time $t + y - \min_{1 \leqq i \leqq g}\{\delta_i\} + g$. It indicates that the delay of the target cell should not be larger than $g + y - \min_{1 \leqq i \leqq g}\{\delta_i\}$. Combining this result with inequality (7), we have the delay bound Eq. (6). □

4.2 Call Admission Control

The following theorem provides the sufficient condition to avoid cell overdue. One should note that if a cell is overdue, then its ADT is larger than its TDT. However, $ADT > TDT$ doesn't always imply an overdue event.

Theorem 2: *If $d_j(\delta_j) \leqq \delta_j$, then no cell of VC j will be overdue.*

Proof: Assume that the number of active VCs is g. Consider an arbitrary cell from VC j and let its initial due-date be y. If $d_j(y) \leqq y$, for all $y \geqq \delta_j$, this cell shall not experience overdue events. To show the sufficient condition, by Eq. (6) one can express $d_j(y) - y$ as

$$\begin{aligned} d_j(y) - y &= \min\left\{ \max_{\Delta \geqq 0}\left\{ N_j(\Delta) + \sum_{i=1, i \neq j}^{g} N_i(y+\Delta-\delta_i) - (y+\Delta) \right\}, \; g - \min_{1 \leqq i \leqq g}\{\delta_i\} \right\} \\ &= \min\left\{ \max_{\Delta \geqq y-\delta_j}\left\{ N_j(\Delta - y + \delta_j) + \sum_{i=1, i \neq j}^{g} N_i(\delta_j + \Delta - \delta_i) - (\delta_j + \Delta) \right\}, \; g - \min_{1 \leqq i \leqq g}\{\delta_i\} \right\}. \end{aligned}$$

For $y \geqq \delta_j$, one can write

$$d_j(y) - y \leq \min\left\{ \max_{\Delta \geq 0} \left\{ N_j(\Delta) + \sum_{i=1, i \neq j}^{g} N_i(\delta_j + \Delta - \delta_i) - (\delta_j + \Delta) \right\}, \ g - \min_{1 \leq i \leq g} \{\delta_i\} \right\}$$
$$= d_j(\delta_j) - \delta_j.$$

Thus, $d_j(y) \leq y$ for all $y \geq \delta_j$ is implied by $d_j(\delta_j) \leq \delta_j$. We then conclude $d_j(\delta_j) \leq \delta_j$ implies no cell overdue for VC j. □

The call admission control procedure operates as follows. When a new VC, say VC n, requests to be supported by the shaper, $d_j(\delta_j)$ is computed for all existing VCs and VC n. If all $d_j(\delta_j) \leq \delta_j$ are satisfied, the new request is granted. Otherwise, it is rejected.

4.3 Delay Bound of the GCRA Traffic Shaper

Lemma 1: *If no cell from VC j is overdue, then $ADT_j(k) > TDT_j(k)$ implies $IDD_j(k) = \delta_j$.*

Proof: Since no cell from VC j is overdue, the cell departure must be early enough to satisfy

$$ADT_j(k) \leq ET_j(k) + IDD_j(k). \tag{8}$$

Assume that $IDD_j(k) > \delta_j$, then from Eq. (3) one can obtain

$$TDT_j(k) = ET_j(k) + IDD_j(k) \geq ADT_j(k), \tag{9}$$

which is a conflict with $ADT_j(k) > TDT_j(k)$. Hence, $IDD_j(k) = \delta_j$ must hold. □

Theorem 3: *Assume the output and input traffic descriptors of VC j are given by (X_j, T_j, τ_j) and (X'_j, T'_j, τ'_j), where $T_j \leq T'_j$. If no overdue for VC j is guaranteed, the cell delay of VC j within the whole GCRA traffic shaper is bounded by D_j which satisfies*

$$D_j \leq \delta_j + \left\lfloor \frac{\tau'_j}{T'_j - X'_j} \right\rfloor (T_j - X'_j). \tag{10}$$

If $T_j = T'_j$, then

$$D_j \leq \delta_j + \tau'_j. \tag{11}$$

Proof: Consider an arbitrary interval $[A_j(m+1), A_j(m+L))$, where m and $L (L \geq 2)$ are integers, satisfying $ADT_j(m+1) > TDT_j(m+1)$, $ADT_j(m+n) \leq TDT_j(m+n)$ for $2 \leq n < L$ and $ADT_j(m+L) > TDT_j(m+L)$, as shown in Fig. 4. Notice that a sequence of such intervals should have represented all possible cell arrival and departure patterns. From (1), it is obvious that $TDT_j(m+2) = ADT_j(m+1) + T_j$ and $TDT_j(m+n) = TDT_j(m+n-1) + T_j$ for $3 \leq n < L$. Since $ADT_j(m+1) > TDT_j(m+1)$, from Lemma 1 $IDD_j(m+1) = \delta_j$ and one can show that $ET_j(m+1) = A_j(m+1)$. Thus, the $(m+1)$-th

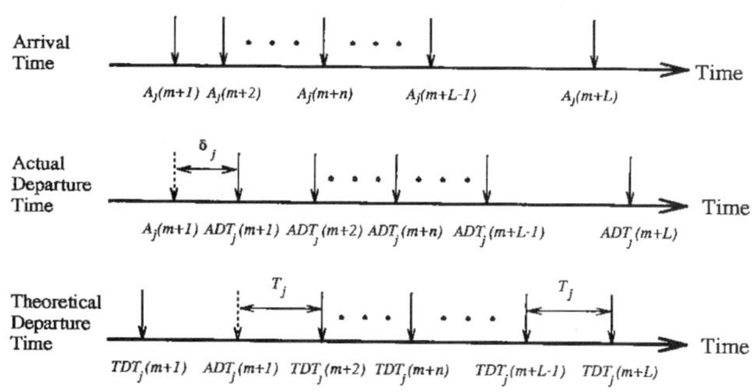

Fig. 4 A time diagram for $A_j(k)$, $ADT_j(k)$ and $TDT_j(k)$.

cell does not experience any delay in the regulator. If no cells from VC j are overdue, we conclude that the delay of the $(m+1)$-th cell can not be larger than δ_j. Hence, for $2 \leq n < L$, we have

$$TDT_j(m+n) = ADT_j(m+1) + (n-1)T_j \leq A_j(m+1) + \delta_j + (n-1)T_j. \tag{12}$$

On the other hand, with input traffic descriptor (X'_j, T'_j, τ'_j), it can be shown that

$$A_j(m+n) \geq \begin{cases} A_j(m+1) + (n-1)X'_j, & n \leq MBS'_j \\ A_j(m+1) + (n-1)T'_j - \tau'_j, & n > MBS'_j \end{cases} \tag{13}$$

where MBS'_j is the maximum burst size of the j-th input VC [9] and is equal to

$$MBS'_j = \left\lfloor 1 + \frac{\tau'_j}{T'_j - X'_j} \right\rfloor. \tag{14}$$

By definition, D_j satisfies

$$D_j = \max_{L > n \geq 1} \{ADT_j(m+n) - A_j(m+n)\}$$
$$\leq \max\{ADT_j(m+1) - A_j(m+1), \max_{L > n \geq 2} \{TDT_j(m+n) - A_j(m+n)\}\}. \tag{15}$$

By combining (12), (13) and (15), the delay bound for arbitrary L becomes

$$D_j \leq \max_{n \geq 1} \begin{cases} \delta_j + (n-1)(T_j - X'_j), & n \leq MBS'_j \\ \delta_j + (n-1)(T_j - T'_j) + \tau'_j, & n > MBS'_j \end{cases} \tag{16}$$

By substituting $n = MBS'_j$ and $n = MBS'_j + 1$ into (16), and taking maximum, we obtain (10). For the special case $T_j = T'_j$, one can obtain (11). □

5. Numerical Results

In this section, we provide examples to show the advantages of the proposed GCRA traffic shaper. We compare the proposed GCRA traffic shaper with feedback signal against three other GCRA traffic shaper designs: FIFO

Table 1 Comparison of the GCRA traffic shapers under overload traffic condition.

(a) $X_j = T_0 + j + 1$, $T_j = T_0 + 3j$, $\tau_j = 49(T_j - X_j)$, $j = 1, \cdots, 10$

GCRA Shaper	Simulation Results	T_0							
		0	1	2	4	6	9	15	27
EDD	R	0	0	0	0	0	0	0	0
With	$S(\times 10^{-2})$	97.6±0.03	83.8±0.03	74.4±0.02	61.8±0.02	53.4±0.02	44.9±0.02	34.5±0.02	24.0±0.01
Feedback	Q_s	10	10	10	10	10	10	10	10
FIFO	R	0	0	0	0	0	0	0	0
With	$S(\times 10^{-2})$	95.2±0.01	83.8±0.02	74.4±0.02	61.8±0.02	53.4±0.02	44.9±0.02	34.5±0.02	24.0±0.01
Feedback	Q_s	10	10	10	10	10	10	10	10
EDD	$R(\times 10^{-4})$	8.1±0.03	1.06±0.15	1.2±0.15	3.1±0.22	3.3±0.24	3.2±0.21	3.1±0.13	2.8±0.14
Without	$S(\times 10^{-2})$	97.6±0.02	83.8±0.02	74.4±0.02	61.8±0.01	53.4±0.01	44.9±0.01	34.5±0.01	24.0±0.01
Feedback	Q_s	275	180	113	20	10	10	10	10
FIFO	$R(\times 10^{-4})$	53.6±3.48	156.3±0.45	138.4±1.05	56.0±2.71	10.1±1.02	3.8±0.32	3.5±0.21	3.0±0.25
Without	$S(\times 10^{-2})$	97.6±0.03	83.8±0.03	74.4±0.03	61.8±0.02	53.4±0.02	44.9±0.02	34.5±0.02	24.0±0.01
Feedback	Q_s	275	180	113	20	10	10	10	10

(b) $X_j = T_0 + 2j$, $T_j = T_0 + 3j$, $\tau_j = 49(T_j - X_j)$, $j = 1, \cdots, 10$

GCRA Shaper	Simulation Results	T_0							
		0	1	2	4	6	9	15	27
EDD	R	0	0	0	0	0	0	0	0
With	$S(\times 10^{-2})$	97.6±0.02	83.8±0.02	74.4±0.02	61.8±0.02	53.4±0.01	44.9±0.01	34.5±0.01	24.0±0.01
Feedback	Q_s	10	10	10	10	10	10	10	10
FIFO	R	0	0	0	0	0	0	0	0
With	$S(\times 10^{-2})$	95.3±0.01	83.8±0.02	74.4±0.02	61.8±0.01	53.4±0.01	44.9±0.01	34.5±0.01	24.0±0.01
Feedback	Q_s	10	0	10	10	10	10	10	10
EDD	$R(\times 10^{-4})$	97.6± 2.46	11.7±0.41	3.7± 0.26	3.3±0.22	2.9±0.13	2.7±0.19	2.6±0.16	2.1±0.18
Without	$S(\times 10^{-2})$	97.6±0.02	83.8±0.02	74.4±0.02	61.8±0.01	53.4±0.01	44.9±0.01	34.5±0.01	24.0±0.01
Feedback	Q_s	150	50	11	10	10	10	10	10
FIFO	$R(\times 10^{-4})$	241.1±1.70	171.1±1.84	149.6±2.15	59.0±1.58	19.1±0.75	3.7±0.12	3.2±0.13	3.0±0.18
Without	$S(\times 10^{-2})$	97.6±0.02	83.8±0.02	74.4±0.02	61.8±0.01	53.4±0.01	44.9±0.01	34.5±0.01	24.0±0.01
Feedback	Q_s	150	50	11	10	10	10	10	10

scheduler with feedback signal, EDD scheduler without feedback signal, and FIFO scheduler without feedback signal. The regulators of these shaper designs all follow the GCRA algorithm and thus are very similar. Modification is only required for those GCRA traffic shapers without feedback signals in which no information about $ADT_j(k-1)$ is available. Therefore, in their regulators, the algorithm is modified by replacing $ADT_j(k-1)$ in Eqs. (1) and (2) with $ET_j(k-1)+1$. In the following, two simulation scenarios are considered and the number of active VCs is set to be 10.

In the first scenario, we consider the worst case that the buffer in the regulator is always nonempty for each VC. This condition occurs when the traffic shaper is overloaded. Hence, no input traffic streams need to be simulated. The performance results are summarized in Table 1, where R indicates the non-conforming cell ratio, S is the average throughput in *cell/slot*, and Q_s represents the maximum queue size observed in the scheduler. Note that the queue size in the regulator is assumed to be infinite and is thus omitted in Table 1. The output traffic descriptor of VC j $(1 \leqq j \leqq 10)$, (X_j, T_j, τ_j), satisfies $X_j = T_0 + j + 1$, $T_j = T_0 + 3j$, $\tau_j = 49(T_j - X_j)$ in Table 1 (a) and $X_j = T_0 + 2j$, $T_j = T_0 + 3j$, $\tau_j = 49(T_j - X_j)$ in Table 1 (b), where T_0 ranges from 0 to 27 to yield different output traffic loads. In the second scenario, the input traffic streams are assumed to be Poisson processes, while the output traffic descriptor of VC j $(1 \leqq j \leqq 10)$, (X_j, T_j, τ_j), satisfies $X_j = 2$, $T_j = T_0 + 10$, and $\tau_j = 49(T_0 + 8)$, where $1 \leqq T_0 \leqq 33$. The simulation results are summarized in Table 2. Two additional performance metrics are included: Q_r is the maximum queue size of each VC observed in the regulator and $\overline{D}$ indicates the average total delay (in *slots*) in the shaper. For both scenarios, the maximum queue size is observed during a simulation session larger than 90 minutes (2×10^9 *slots*).

All computer simulations, with 99% confidence intervals, confirmed that the output cell streams of the GCRA traffic shaper with feedback signal, either with EDD or FIFO scheduler, do not contain any non-conforming cells. However, the output cell streams of the traffic shapers without feedback signal do contain non-conforming cells, due to serious emission conflicts in the scheduler. The shaper performance with FIFO-based scheduler and feedback signal is found to be almost the same as that of the proposed traffic shaper, except non-negligible throughput degradation in heavy traffic load (see Table 1). This result coincides with the optimality of EDD scheduling.

Table 2 Comparison of the GCRA traffic shapers with homogeneous VCs and Possion arrivals. The output traffic descriptor (X_j, T_j, τ_j) satisfies $X_j = 2$, $T_j = T_0 + 10$, $\tau_j = 49(T_0 + 8)$, $j = 1, \cdots, 10$.

GCRA Shaper	Simulation Results	T_0							
		1	2	4	6	9	13	20	33
EDD With Feedback	R	0	0	0	0	0	0	0	0
	Q_s	9	9	9	9	9	9	9	7
	Q_r	115	152	165	173	228	264	338	493
	$\overline{D}$	6.0±0.01	6.1±0.33	9.3±0.50	16.0±0.01	32.2±0.65	69.6±3.43	181.2±5.97	591.9±30.3
FIFO With Feedback	R	0	0	0	0	0	0	0	0
	Q_s	9	9	9	9	9	9	8	7
	Q_r	115	152	165	173	228	260	338	493
	$\overline{D}$	6.0±0.01	6.0±0.01	9.3±0.50	16.0±0.01	32.2±0.65	69.1±3.12	181.2±5.97	591.9±30.3
EDD Without Feedback	$R(\times 10^{-3})$	59.7±0.25	51.6±0.08	36.8±0.15	26.9±0.08	17.7±0.16	11.1±0.09	5.5±0.05	2.0±0.03
	Q_s	56	40	25	20	15	11	10	8
	Q_r	115	152	165	173	228	260	338	493
	$\overline{D}$	6.0±0.01	6.0±0.01	9.2±0.43	15.9±0.33	32.2±0.65	68.9±3.23	181.2±5.97	591.9±30.3
FIFO Without Feedback	$R(\times 10^{-3})$	31.1±0.06	26.5±0.08	22.5±0.05	18.4±0.05	13.9±0.04	9.8±0.05	5.8±0.03	2.6±0.04
	Q_s	56	40	25	20	15	11	10	8
	Q_r	115	152	165	173	228	260	338	493
	$\overline{D}$	6.0±0.01	6.0±0.01	9.2±0.43	15.9±0.33	32.2±0.65	68.9±3.23	181.2±5.97	591.9±30.3

As one might have expected, in the scheduler where feedback signal is used, the maximum queue size is controlled to be no larger than the number of VCs. While for the other traffic shapers without feedback signal, the maximum queue size can be much larger than the number of active VCs. Hence, either the software or hardware design of the latter scheduler may be much more difficult, and the EDD scheduler should be a superior solution. If the implementation complexity of EDD algorithm is still considered not acceptable, other alternative scheduling algorithm, such as *Rotating-Priority-Queues (RPQ)* [15] which approximates EDD scheduling without sorting queued cells, can be employed in the scheduler. And we expect the resulting performance for any EDD-like scheduler should be similar. For possible hardware design alternatives, please refer to [17] for details.

Last but not least, one can observe that the average total delay is almost the same for the shaper designs with feedback and the designs without feedback, under light and normal traffic condition. It indicates that the feedback mechanism does not degrade the average delay performance while it successfully avoids non-conforming events. We thus suggest that the feedback mechanism must be included in the GCRA traffic shaper.

Since the multiplexing gain is often considered a major performance metric, we define the multiplexing gain of the GCRA shaper, G, using the formula

$$G \equiv \frac{\text{total output peak cell rate of accepted VCs}}{\text{physical link speed}} = \sum_{i=1}^{g} \frac{1}{X_i},$$

where g is the maximum number of VCs accepted under the CAC condition, and provide additional numerical results. The multiplexing gains of the proposed traffic shaper with homogeneous VCs under the CAC condition are shown in Fig. 5, where the output traffic descriptor satisfies $T_i = 100$, $X_i = X$ for all i, and τ_i ranges from 0 to 200 to yield different multiplexing gains. Note that when the burst tolerance $\tau_i \geq T_i - X_i$, the resulting multiplexing gain is a constant for any τ_i, according to the CAC condition. Only when the burst tolerance $\tau_i < T_i - X_i$, the minimum initial due-date is affected by τ_i and the multiplexing gain may be seriously degraded.

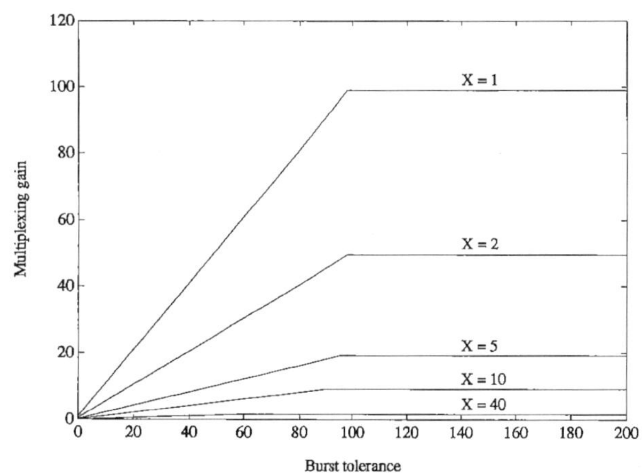

Fig. 5 The multiplexing gains of the proposed GCRA traffic shaper under CAC condition with the traffic parameters $T_i = 100$, $X_i = X$ for all i, and the burst tolerance τ_i ranges from 0 to 200.

6. Conclusions

In this paper, a new GCRA traffic shaper has been described and analyzed for supporting VBR traffic. This shaper has considered and resolved the emission conflict problem that other designs [1]–[5] did not. The two most important features of the proposed traffic shaper are the EDD scheduler and the feedback mechanism between the scheduler and the regulator. The EDD scheduler can minimize the number of overdue events and effectively maintain the shaper's operation efficiency and maximal throughput. By using a feedback signal between the regulator and the scheduler, not only it can guarantee the output cell streams to conform to the output traffic descriptor, necessary call admission control can also be easily applied. Once the call admission control policy is employed, the cell delay bound for each VC in this traffic shaper, including both the regulator and the scheduler, becomes predictable. Facing either random input or heavy traffic load, our EDD scheduler-based GCRA shaper illustrates its robustness and superior performance when it is compared with other designs. When this shaper is used as a multiplexer, the cell delay bound of a specific VC in the traffic shaper is independent of the traffic load of the other VCs, as long as CAC is applied appropriately. This feature is very different from what we observed in most existing network devices and should be another advantage of the proposed design.

When one would like to implement such a GCRA shaping function in an existing ATM network, one possible approach is to implement the proposed shaper as an independent network device and deploy it at appropriate UNI or NNI (refer to Fig. 1). However, in the long term we suggest the following approach. The proposed traffic shaper mechanism should be implemented within all ATM edge switches deployed at the UNI between ATM Customer Premises Networks and ATM WAN or at the NNI between ATM service providers. Of course, this would increase the design complexity of ATM edge switches, but we expect the improved performance, such as the reduction of non-conforming cell ratio, can justify the cost.

Acknowledgements

The authors would like to thank the National Science Council of the Republic of China for financially supporting this research under Grant NSC 87-2215-E002-024.

References

[1] F. Guillemin and W. Monin, "Management of cell delay variation in ATM networks," IEEE GLOBECOM'92, pp.128–132.
[2] P.E. Boyer, F.M. Guillemin, M.J. Servel, and J. Coudreuse, "Spacing cells protects and enhances utilization of ATM network links," IEEE Network, pp.38–49. Sept. 1992.
[3] F.M. Brochin, "A cell spacing device for congestion control in ATM networks," Performance Evaluation, pp.107–127, 1992.
[4] G. Rigolio, L. Verri, and L. Fratta, "Source control and shaping in ATM networks," IEEE GLOBECOM'91, pp.276–280.
[5] F. Bernabei, L. Gratta, M. Listanti, and M. Testa, "Analysis of two level shaping for multiplexing of ON-OFF ATM sources," IEEE ICC'93, pp.1380–1385.
[6] N. Endo, K. Miki, and N. Ogasawara, "A usage parameter control for multi-services ATM networks," IEEE GLOBECOM'94, pp.946–950.
[7] G. Ramamurthy, R.S. Doghe, and D. Raychaudhuri, "An UPC-based traffic control framework for ATM networks," IEEE GLOBECOM'94, pp.600–605.
[8] G. Mercankosk, T. Moors, and A. Gantoni, "Multiplexing spacer outputs on cell emissions," IEEE INFOCOM'95, pp.49–55.
[9] The ATM Forum, "ATM User-Network Interface Specification," Version 3.0, PTR Prentice-Hall, 1993.
[10] C.L. Liu and J.W. Layland, "Scheduling algorithms for multiprogramming in a hard-real-time environment," J. ACM, vol.3, no.1, pp.46–61, Jan. 1973.
[11] H. Saito, "Optimal queueing discipline for real-time traffic at ATM switching nodes," IEEE Trans. Commun., vol.38, no.12, pp.2131–2136, Dec. 1990.
[12] M.L. Dertouzos and A.K.-L. Mok, "Multiprocessor on-line scheduling of hard-real-time tasks," IEEE Trans. Softw. Eng., vol.15, no.12, pp.1497–1506, Dec. 1992.
[13] L. Georgiadis, R. Guerin, and A. Parekh, "Optimal multiplexing on a single link: Delay and buffer requirements," IEEE INFOCOM'94, pp.524–532.
[14] D. Ferrari and D.C. Verma, "A scheme for real-time channel establishment in wide-area networks," IEEE J. Select. Areas in Commun., vol.8, no.3, pp.368–379, April 1990.
[15] J. Liebeherr, D.E. Wrege, and D. Ferrari, "Exact admission control in networks with bounded delay services," IEEE/ACM Trans. on Networking, vol.4, no.6, pp.885–901, Dec. 1996.
[16] E. Horowitz and S. Sahni, "Fundamentals of Data Structures," Computer Science Press, Inc., 1982.
[17] H.J. Chao and N. Usun, "A VLSI sequencer chip for ATM traffic shaper and queue manager," IEEE J. Solid-State Circuits, vol.27, no.11, pp.1634–1643, Nov. 1992.

High-Performance Backbone Network Technology edited by Naoaki Yamanaka
ISBN: 0-8247-5321-6 DOI: 10.1081/(120025715)
Published 2004 by Marcel Dekker, Inc.

Performance Analysis of Weighted Round Robin Cell Scheduling and Its Improvement in ATM Networks

Hideyuki Shimonishi and Hiroshi Suzuki

SUMMARY Weighted Round Robin (WRR) scheduling is an extension of round robin scheduling. Because of its simplicity and bandwidth guarantee, WRR cell scheduling is commonly used in ATM switches. However, since cells in individual queues are sent cyclically, the delay bounds in WRR scheduling grow as the number of queues increases. Thus, static priority scheduling is often used with WRR to improve the delay bounds of real-time queues. In this paper, we show that the burstiness generated in the network is an even greater factor affecting the degradation of delay bounds. In ATM switches with per-class queueing, a number of connections are multiplexed into one class-queue. The multiplexed traffic will have a burstiness even if each connection has no burstiness, and when the multiplexed traffic is separated at the down stream switches, the separated traffic will have a burstiness even if the multiplexed traffic has been shaped in the upstream switches. In this paper, we propose a new WRR scheme, namely, WRR with Save and Borrow (WRR/SB), that helps improving the delay bound performance of WRR by taking into account the burstiness generated in the network. We analyze these cell scheduling methods to discuss their delay characteristics. Through some numerical examples, we show that delay bounds in WRR are mainly dominated by the burstiness of input traffic and, thus WRR/SP, which is a combination of WRR and static priority scheduling, is less effective in improving delay bounds. We show that WRR/SB can provide better delay bounds than WRR and that it can achieve the same target delay bound with a smaller extra bandwidth, while large extra bandwidth must be allocated for WRR.

key words: *ATM, scheduler, weighted round robin, delay bound, burstiness*

1. Introduction

ATM (Asynchronous Transfer Mode) is as a key technology in the construction of a high speed multi-media backbone network. The ATM Forum defines a number of different QoS (Quality of Service) classes to deal with the diverse requirements of different kinds of traffic [1]. Thus, ATM switches require cell scheduling methodologies that enable ATM networks to provide a suitable QoS to each service class, while retaining efficient network resource usage.

Scheduling algorithms are extensively studied and there are a lot of research on them. Of these, WFQ (Weighted Fair Queueing) [2]–[4] and WRR (Weighted Round Robin) [9] have been popular for scheduling because they guarantee bandwidth for each queue. WFQ, which is also called as PGPS (Packetized Generalized Processor Sharing), is a packetized version of GPS (Generalized Processor Sharing). Although WFQ can handle variable length packets and guarantee smaller delay bounds than WRR, it is difficult to use in high speed ATM networks because of its computational complexity and cost of implementation. SCFQ (Self-Clocked Fair Queueing) [5]–[7] is a less complex approximation of GPS and Rexford et al, proposed a hardware efficient implementation for it [8]. WRR is an extension of round robin scheduling. By allocating *Weight* to each queue, a minimum bandwidth is guaranteed according to the proportioning of weights. Though WRR has larger delay bounds than WFQ, it is more popular in ATM switches because of its computational simplicity and low implementation cost.

Since WRR scheduler takes a cell from the head of each queue cyclically, delay bounds grow as the number of queues increases [10]. To decrease delays caused by the cyclical mechanism of scheduling and to achieve smaller delay bounds for real-time traffic, static priority queueing is often used with WRR [11]. In this paper, we call this method WRR with Static Priority (WRR/SP). WRR/SP has the same mechanism to guarantee minimum bandwidth as WRR, however, cells from the head of individual queues are not sent in a cyclical manner. A priority level is assigned to each queue according to the required delay bound, and cells in higher priority queues are scheduled prior to those in lower priority queues.

In this paper, we show that the burstiness generated in the network is an another delay factor which greatly degrades the delay bounds of WRR scheduling. In ATM switches with per-class queueing, a number of connections are multiplexed into one class-queue. The multiplexed traffic will have a burstiness even if each connection has no burstiness, and when the multiplexed traffic is separated at the down stream switches, the separated traffic will have a burstiness even if the multiplexed traffic has been shaped in the upstream switches. Since WRR schedules each input traffic at a constant rate, burstiness increases maximum queue length and creates large delays, especially, large maximum delay bounds which are much larger than its mean delays. This delay bound degradation must be serious in treating real-time traffic.

To decrease delays caused by burstiness and achieve

Manuscript received September 29, 1997.
Manuscript revised December 10, 1997.

† The author is with C&C Media Laboratories, NEC Corporation, Kawasaki-shi, 216–8555 Japan.

IEICE TRANS. COMMUN., Vol. E81-B, No. 5, pp. 910–918, May 1998

smaller delay bounds for real-time traffic, we propose a new WRR scheme, which is WRR with Save and Borrow (WRR/SB). WRR/SB is mainly proposed for ATM switches having a per-class queueing discipline, in which many connections are multiplexed into one class queue and a burstiness is generated. For classes that require real-time service, WRR/SB saves some bandwidth for the future when there are input cells and borrows some bandwidth from the future when there are excess input cells for real-time classes. Thus, real-time classes are scheduled according to their burstiness by saving and/or borrowing bandwidth, not scheduled at a constant rate, so long as the burstiness of input traffic is within a pre-defined tolerance level.

Using the model in Sect. 3, we have analyzed the cell delay distribution of WRR, WRR/SP and WRR/SB. We discuss their delay characteristics to show that the burstiness of input traffic is a major factor of affecting large delays. Through some examples, we show that WRR/SB has a better maximum delay bound than WRR or WRR/SP.

2. Scheduling Methods

2.1 Weighted Round Robin (WRR) and WRR with Static Priority (WRR/SP)

WRR is an extension of round robin scheduling. Each class queue has a counter that specifies the number of cells that can be sent from it. The counter value is initially equalized to the weight value assigned to that class. Cells from various classes are sent in a cycle from the head of these class queues whose counter values are greater than zero, with one cell being sent for each unit of time. After sending a cell, the class's counter value is reduced by one. When the counter value and/or the queue length has reached zero in all classes, all counters are reset to their weight values. There is an example of WRR cell scheduling in Fig. 1, where weights of 2, 1 and 3 have respectively been allocated to queues 1, 2 and 3. In this figure, cells in queues that have a non-zero counter value are sent cyclically and the counters are reset after sending six cells. Thus, minimum bandwidths of two sixth, one sixth and three sixth are allocated for individual queues.

Figure 2 illustrates how the cyclical mechanism of scheduling induces delays. In this figure, we suppose that Class 1 requires real-time service and has a half of the total weight. Since cells of individual classes are sent cyclically, Class 1 cells are sent after every six cells as long as other classes have a non-zero counter value, though Class 1 cells should be sent after each one cell ideally. Thus, Class 1 cells have delays.

WRR/SP aims to improve the delay bound performance of WRR degraded by the cyclical mechanism of scheduling. WRR/SP has the same counters, weights and counter reset behaviors as WRR to guarantee a

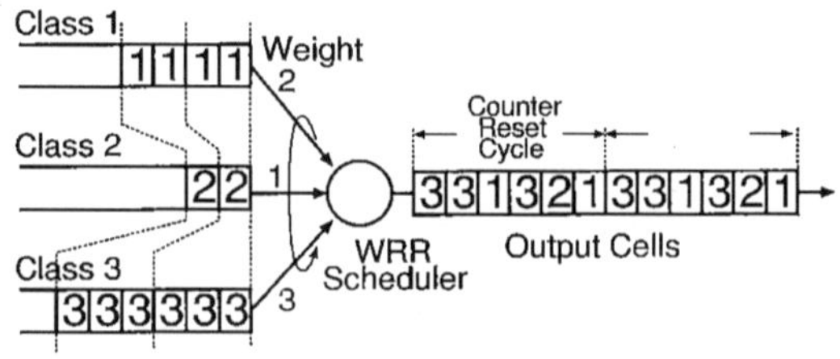

Fig. 1 Weighted round robin.

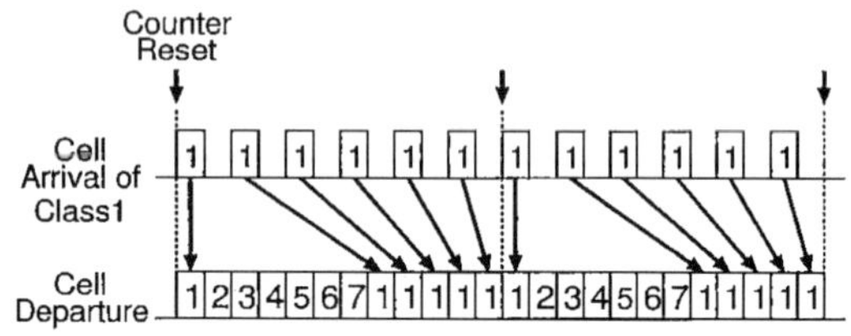

Fig. 2 Example of cell scheduling of WRR (1): Number of classes = 7: Weights of Class 1 = 6.

minimum bandwidth. The difference is that WRR/SP assigns a priority level to each class according to the required delay bound. High priority class cells are always sent prior to lower priority class cells and low priority cells are only sent when the counter value or queue length is zero in all higher priority classes. Thus, the delay bound of higher priority classes can be decreased by sacrificing of lower priority classes, while maintaining bandwidth guarantees for all queues. Figure 3 shows an example of WRR/SP scheduling under the same conditions as Fig. 2 except that Class 1 has the highest priority. In this figure, we can see that the cell delays of class 1 are reduced by sending Class 1 cells prior to other classes.

Since the way the counter is used follows WRR, throughput performance also follows WRR as long as there are no cell losses. When multiple real-time classes are required to have the same delay performance, the same priority level is assigned to these classes and round robin scheduling is done for these classes.

2.2 Weighted Round Robin with Save and Borrow (WRR/SB)

Burstiness of input traffic will induce large cell delays. Delays caused by the burstiness of input traffic are illustrated in Fig. 4. A class may at times have more arrival cells than the weight value for one counter reset cycle can support. In such a cycle, the counter value will reach zero before all the cells in the queue have been sent, and cell scheduling for this class will be suspended until the counters are reset. Cells that arrive when the counter value is less than the queue length will have to wait until the next or an even later counter reset. In Fig. 4, three cells have arrived at the Class 1 queue during one counter reset cycle, while only one weight has been allocated to this class. Thus, the second or third

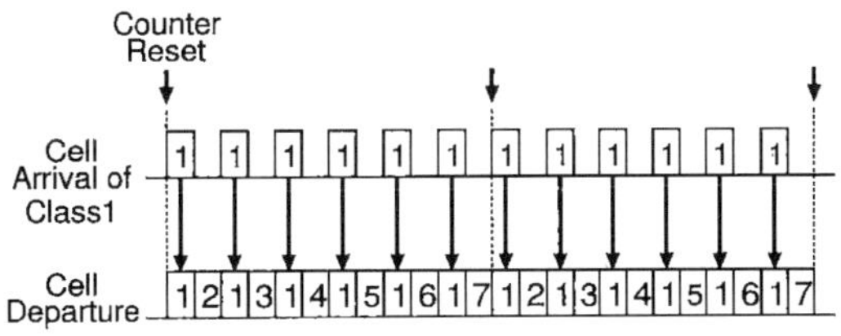

Fig. 3 Example of cell scheduling of WRR/SP: Number of classes = 7: Weights of Class 1 = 6: Class 1 has the highest priority.

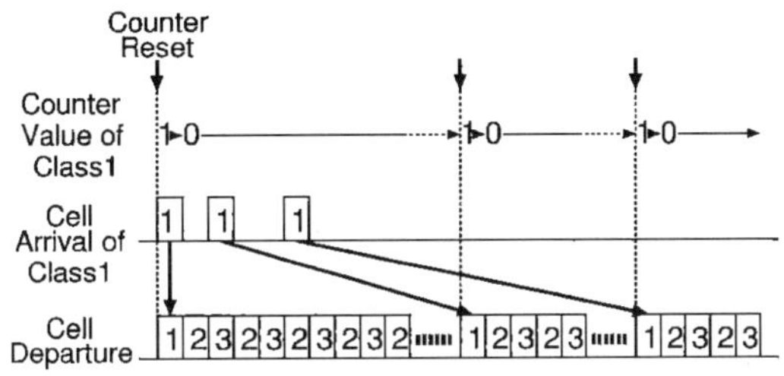

Fig. 4 Example of cell scheduling of WRR (2): Number of classes = 3: Weight of Class 1 = 1.

cells must have large delay.

Since WRR/SP only decreases delays caused by the cyclical mechanism of scheduling, we propose WRR/SB to reduce the delays induced by the burstiness of input traffic. Although WRR/SB has the same counters and weights as WRR, it has different reset behaviors. If the counter value for a class is greater than zero when counters are reset, the counter will be reset to its weight value plus the remaining counter value, up to a predetermined upper bound (see line 2, Table 1, which shows an example pseudocode for WRR/SB). That is to say, WRR/SB *saves* counter values when there are few input cells and adds them back when there are excess input cells (this approach is similar to that adopted by N. Mizukoshi, et al., for the cell scheduler of a single-chip ATM switch [11]).

Conversely, cells are permitted to be sent until the counter value reaches a predetermined lower bound of minus value (see line 3, Table 1), and when counters are reset the weight value for that class will be added to the final counter value, even if it is in fact minus. Here, then, WRR/SB *borrows* counter values from future counter resets to cope with times when there are excess cell arrivals, and it reduces the number of cells that can be sent during the next counter reset cycle by the amount borrowed. In Fig. 5, an example of WRR/SB scheduling is illustrated under the same conditions as Fig. 4 except that Class 1 is permitted to borrow the counter value. In this figure, we can see that Class 1 cells delays are reduced by borrowing counter value.

Since counter values are increased by the weight values, over the long term the total number of cells sent is dependent on the weight values and is independent of both upper bound and lower bound of the

Table 1 Pseudocode of WRR/SB scheduler.

```
If (Ci<=Wli or Qi=0) for all i
   Ci←min(Ci+Wi,Whi) for all i
While (Ci>Wli and Qi>0) {
   send a cell of class i
   Ci←Ci-1
   go to the next class
}
go to the next class

Qi= Queue length of class i
Ci= Counter value of class i
Wi= Weight of class i
Whi= Counter upper bound of class i
Wli= Counter lower bound of class i
```

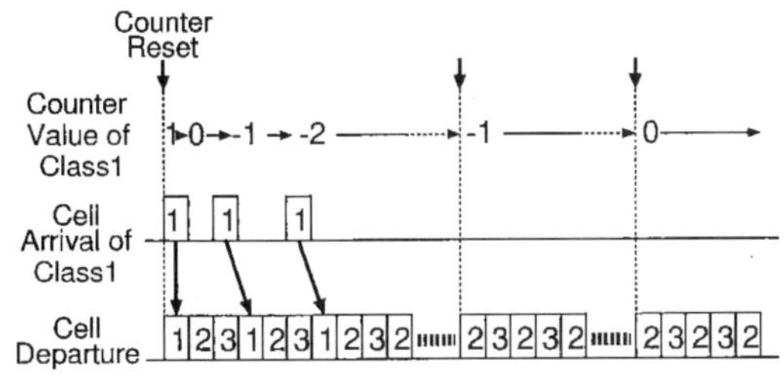

Fig. 5 Example of cell scheduling of WRR/SB: Number of classes = 3: Weight of Class 1 = 1: Counter lower bound of Class 1 = −2.

counter. This means counter saving/borrowing in one class will not degrade the throughput performance of other classes in the long run, as long as the counter upper bound and lower bound have been properly chosen and there are no cell losses. If too large counter upper bound or too small lower bound are chosen, counter saving/borrowing of one class may affect throughput performance of other classes in a small time scale. For non-real-time classes, the scheduler operates in the same way as a WRR scheduler by setting the counter upper bound to be a weight value and the counter lower bound to be zero.

3. Analysis

In this section, we discuss an analysis which describes the delay characteristics of WRR/SB. Since WRR is a special case of WRR/SB, the analysis can be applied to WRR by setting the counter upper bound to be a weight value and the lower value to be zero. The analysis can also be applied to WRR/SP with some modifications.

3.1 Model, Assumptions, and Notations

Our analytical model consists of per-class queues and a scheduler as shown in Fig. 6. We focus on the behavior of a single class, Class 1, which requires a real-time service. In WRR/SB, the counter upper bound and lower bound for other classes are chosen so that these classes

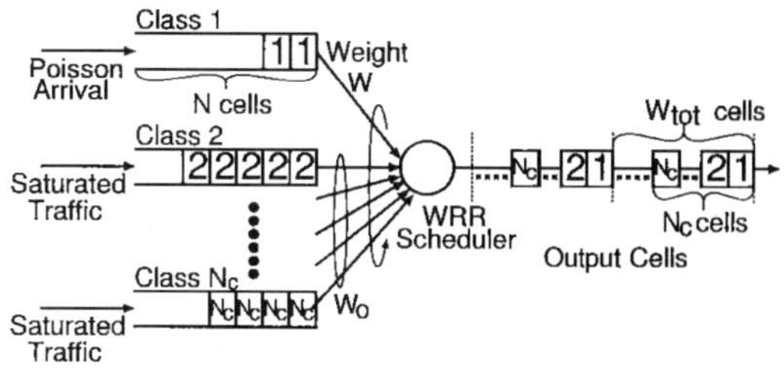

Fig. 6 Analytical model.

are scheduled in the same way as WRR, namely the counter upper bound is the same as its weights value and the counter lower bound is zero. Class 1 is given the highest priority in WRR/SP. Cell arrival to this class is assumed to be Poissonian, which means many connections having different cell rates are multiplexed into one queue. The input rate of other classes is assumed to be saturated. This means that the input rate of these classes is more than their allocated bandwidth and their queue occupancy does not become zero. When a cell arrives at a full queue, it is discarded and not considered in the evaluation of cell delay characteristics. Counter resets and the determination to which class to send a cell are done just before each cell output.

One unit time is assumed to be the time required to send one cell. Let us denote the time at which t unit times have elapsed since the counters were reset, as *slot* t. The allocated weight of Class 1 and the upper bound and the lower bound of its counter value are denoted by $W(>0)$, $W_h(\geqq W)$, and $W_l(\leqq 0)$, respectively. In WRR and WRR/SP, W_h is W and W_l is 0, and in WRR/SB, W_h is equal to or grater than W and W_l is equal to or less than 0. The sum of the weights of other classes is denoted by W_o and the sum of the weights of all classes is denoted by W_{tot} $(= W + W_o)$. The number of classes is denoted by N_c (>1). Class 1 has a queue capacity of N cells while that of other classes is infinite. The input rate of Class 1 is denoted by $\lambda(0 < \lambda < 1)$. Since cell arrival to this class is Poissonian, the probability $p(x)$, where x cells arrive in one unit time, can be denoted by

$$p(x) = e^{-\lambda}\lambda^x/x!. \tag{1}$$

3.2 Derivation of Steady State Distribution

We describe our analysis of the system by the behavior of Class 1. The states of the system are expressed in terms of the counter value w and the queue length q of Class 1 and slot t. Since we assume that the queue length of Class 1 is finite, the system can be modeled as a finite-state discrete-time Markov chain. Because of the very large number of possible states, we begin our analysis defining a simple state s of the system as

$$s = (q, w). \tag{2}$$

Thus its possible set becomes

$$S = \{(q, w); \quad 0 \leqq q \leqq N, \quad W_l \leqq w \leqq W_h\}. \tag{3}$$

Let $g(t)$ specify whether Class 1 can be scheduled or not at slot t. $g(t)$ is one if it can, otherwise, it is zero. As Class 1 can be scheduled at any time in WRR/SP since this class has the highest priority, we have

$$g(t) = 1, \qquad \text{for WRR/SP.} \tag{4}$$

For WRR and WRR/SB, we introduce the following approximation to simplify the analysis. If no Class 1 cells are in the queue when this class is selected, cell scheduling immediately shifts to Class 2 (Fig. 7 (a)), whereas in our approximated model, a cell from any one of the classes is scheduled instead of Class 1, then the cell scheduling shifts to Class 2 (Fig. 7 (b)). Thus in this approximation, the time between the cell outputs of Class 1 does not change regardless of whether or not there are cells in its queue. This approximation can be validated when the offered load to the switch is high. Thus, $g(t)$ can uniquely be determined as:

$$g(t) = \begin{cases} 1, & t \bmod N_c = 0 \\ & \quad \text{or } t - \left\lceil \dfrac{t}{N_c} \right\rceil \geqq W_o, \\ 0, & \text{otherwise.} \end{cases} \tag{5}$$
$$\text{for WRR, WRR/SB}$$

where $\lceil x \rceil$ denotes the minimum integer value which is equal to or more than x.

We define C^t as a transition probability matrix which denotes the transitions of system states s in slot t. C^t does not contain any transitions caused by counter resetting. We have

$$C^t_{(q,w)\to(q',w')} = \begin{cases} p(q'-q+1) & \\ \quad \text{if } g(t)=1, \quad q>0, \quad w>W_l & \\ \qquad q-1 \leqq q' < N, \quad w' = w-1 & \\ \sum_{k=N}^{\infty} p(k-q+1) & \\ \quad \text{if } g(t)=1, \quad q>0, \quad q'=N & \\ \qquad w>W_l, \quad w'=w-1 & \\ p(q'-q) & \\ \quad \text{if } g(t)=1, \quad (q=0 \text{ or } w=W_l), & \\ \qquad q \leqq q' < N, \quad w'=w & \\ \quad \text{or } g(t)=0, \quad q \leqq q' < N, & \\ \qquad w'=w & \\ \sum_{k=N}^{\infty} p(k-q) & \\ \quad \text{if } g(t)=1, \quad (q=0 \text{ or } w=W_l), & \\ \qquad q'=N, \quad w'=w & \\ \quad \text{or } g(t)=0, \quad q'=N, \quad w'=w & \\ 0 \quad \text{otherwise,} & \end{cases} \tag{6}$$

where w and q denotes the counter value and the queue length of Class 1 at the beginning of slot t, and w' and

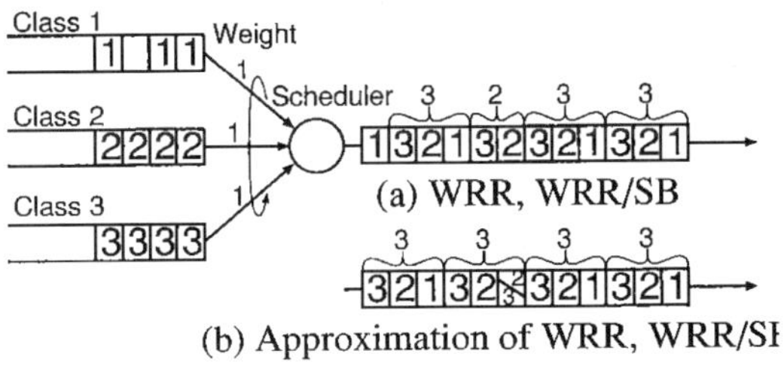

Fig. 7 Approximation of cell scheduling.

q' denote the values at the beginning of the next slot. We define A^t as the transition probability matrix that denotes transitions from slot 0, via slot $1 \ldots t$, back to slot 0. In A^t, the system starts its transition from slot 0 toward slot t without resetting counters, then counters are reset at the end of slot t and the system transits to slot 0. B^t is defined in the same way as A^t except that after slot t the system transits to slot $t+1$ without resetting counters. Thus we have A^t and B^t as

$$\begin{aligned} &A^t_{(q,w)\to(q',\min(w'+W,W_h))} \\ &= \begin{cases} (B^{t-1}C^t)_{(q,w)\to(q',w')} & \\ \qquad \text{if } (q'=0 \text{ or } w' \leq W_l), & \\ \qquad\quad t+1-(w-w')=W_o & \\ 0 \qquad \text{otherwise}, & \end{cases} \\ &B^t_{(q,w)\to(q',w')} \\ &= \begin{cases} 0 \qquad \text{if } (q'=0 \text{ or } w' \leq W_l), & \\ \qquad\quad t+1-(w-w')=W_o & \\ (B^{t-1}C^t)_{(q,w)\to(q',w')} & \\ \qquad \text{otherwise}, & \end{cases} \end{aligned} \tag{7}$$

where, now, w and q denote the counter value and the queue length at the beginning of slot 0, respectively, and w' and q' denote the values at the end of slot t. Here, we assume B^{-1} to be a unit matrix. Since Class 1 sends at most $W_h - W_l$ cells and other classes send a total of W_o cells during one counter reset cycle, the counters are reset before slot $(W_h - W_l + W_o)$.

We define $\overline{A}$ as

$$\begin{aligned} \overline{A} &= \sum_{i=0}^{\mathcal{N}-1} A^i, \\ \mathcal{N} &= W_h - W_l + W_o \end{aligned} \tag{8}$$

which is the transition probability matrix that denotes all possible transitions from slot 0 to the next slot 0. Since the state space is finite, the steady state probability distribution of the system π^t at slot t can be obtained by solving

$$\begin{aligned} (\overline{A} - I)\pi^0 = \mathbf{o}, \quad & \pi^0 \mathbf{e} = 1 \\ \pi^t = \pi^0 B^{t-1} \quad & (0 < t < \mathcal{N}), \end{aligned} \tag{9}$$

where I denotes a unit matrix and $\mathbf{e}$ denotes a unit vector. Note $\pi^t\mathbf{e}$ is equal to or less than one and $\pi^t\mathbf{e}$ denotes the probability that slot reaches t, in other words, the probability that the length of a counter reset cycle is equal to or more then t. Thus the probability $\theta(t)$ that the system is in slot t at any given time is denoted by

$$\theta(t) = \sum_{i=t}^{\mathcal{N}-2} \frac{(\pi^i\mathbf{e} - \pi^{i+1}\mathbf{e})}{i-1} + \frac{\pi^{\mathcal{N}-1}\mathbf{e}}{\mathcal{N}-2} \tag{10}$$

Next, let us describe a system using a new state $\hat{s}$, which is a combination of the simple state s and slot t. The system state $\hat{\mathbf{s}}$ and its possible set $\hat{S}$ can be denoted by

$$\begin{aligned} \hat{\mathbf{s}} &= (\mathbf{s}, w) = (q, w, t) \\ \hat{S} &= \{(q, w, t); \quad 0 \leqq q \leqq N, \\ & \qquad W_l \leqq w \leqq W_h, \ 0 \leq t < \mathcal{N}\} \end{aligned} \tag{11}$$

The steady state probability distribution $\hat{\pi}$ for $\hat{S}$ can be obtained by normalizing π^t for all slots.

$$\hat{\pi}(q, w, t) = \frac{\theta(t)}{\pi^t\mathbf{e}} \pi^t(q, w) \tag{12}$$

3.3 Derivation of Queue Length and Delay Distribution

When we denote the queue length distribution of Class 1 as $Q(i)$, we have

$$Q(i) = \sum_{j=0}^{\mathcal{N}-1} \sum_{k=W_l}^{W_h} \hat{\pi}(i, k, j). \tag{13}$$

We define $r_{\hat{\mathbf{s}}}$ as a queueing delay of a cell which arrived at the queue at state $\hat{\mathbf{s}}$. To simplify the calculation, we approximate that the length of counter reset cycle is fixed to be $\mathcal{N}$, thus, $r_{\hat{\mathbf{s}}}$ can be approximated by

$$r_{(q,w,t)} = \begin{cases} \{k | \sum_{i=t}^{k} g(i) = q+1\} - t + 1 & \\ \quad \text{if } q < w - W_l, \quad q < n(t) & \\ (W_{tot} - t) & \\ \quad +r_{(q-\min(w-W_l,n(t)),} & \\ \quad \min(w-\min(w-W_l,n(t))+W,W_h),0) & \\ \quad \text{otherwise}, & \end{cases} \tag{14}$$

where

$$n(t) = \sum_{i=t}^{\mathcal{N}-1} g(i), \tag{15}$$

and $\lfloor x \rfloor$ denotes a maximum integer values which does not exceed x. Then the delay time distribution $D(i)$ of cells which are not discarded by buffer overflow is

$$D(i) = \sum_{\substack{\hat{\mathbf{s}} \in \hat{S}' \\ r_{\hat{\mathbf{s}}}=i}} \hat{\pi}(\hat{\mathbf{s}}) / \sum_{\hat{\mathbf{s}} \in \hat{S}'} \hat{\pi}(\hat{\mathbf{s}}), \tag{16}$$

where

$$\begin{aligned} \hat{S}' = \{(q, w, t); \quad 0 \leqq q < N, \quad W_l \leqq w \leqq W_h, \\ 0 \leq t < \mathcal{N}\}. \end{aligned} \tag{17}$$

4. Numerical Examples

In this section, we discuss the delay characteristics of WRR, WRR/SP and WRR/SB using the analysis presented in the previous section. Our ATM switch model is an output-buffered switch with per-class queues as shown in Fig. 8. Each output port has five queues for five service classes. We evaluate the delay characteristics of Class 1, which requires real-time service and has a 50 cell length queue. The sum of the weights of all classes is always 100; for example, when the weight of Class 1 is 10, the sum of the weights of other classes is 90. The input rate of Class 1 is 0.1, thus this class needs a weight of at least 10. Other classes are assumed to be saturated.

4.1 Accuracy of Analysis

Let us examine the accuracy of the analysis using one typical example. The delay distributions obtained by analysis and simulation for WRR, WRR/SP and WRR/SB are shown in Fig. 9. The weight of Class 1 is 12 and the sum of the weights of other classes is 88. The counter upper bound and lower bound are 18 and −6 for WRR/SB, respectively. Class 1 has the highest priority among all classes in WRR/SP. Each simulation experiment is executed during over 10,000,000 cell times.

Figure 9 shows that the results obtained by analysis closely agree with simulation unless the distribution probability is very small, i.e. less than 10^{-5}, where the simulation results are not trustworthy. While the delay obtained by analysis and that of simulation agree in WRR/SP, there is a slight difference in WRR and WRR/SB because of the approximation of $g(t)$. However, since we are mainly concerned here with maximum delay bounds, this difference is concerned to be negligible.

4.2 Delay Distribution of WRR

In this section, the delay characteristics of WRR are discussed. Figure 10 shows the delay distributions of Class 1 for weights of 11, 12, and 14. This means 10%, 20%, and 40% of extra bandwidth is allocated to Class1. The delay distributions are shown in two parts: the distribution of *Part 1* cells and the distribution of *Part 2* cells. When a Part 1 cell arrives at the switch, the queue length is larger than the counter value. Thus this cell must wait for the next counter reset or for even more than one reset and will have large delay. Part 2 cells are other cells which will be sent before the next counter reset and will have smaller delays. Note that there are few Part 1 cells if the input traffic has no burstiness and large queueing delays of Part 1 cells are mainly caused by the burstiness of input traffic while delays of Part 2 cells are mainly result of the cyclical mechanism of scheduling.

From Fig. 10, we can see that delays of Part 2 cells occur within 50 cell times, while those of Part 1 cells occur in the hundreds. Since the worst case delay time of Part 1 cells is much longer than that of Part 2 cells, we can see that Part 1 cells dominate the maximum delay bound. In other words, the delay bound degradation in WRR is caused by the bursty nature of input traffic, rather than the cyclical mechanism of scheduling.

This figure also shows that allocating a large weight

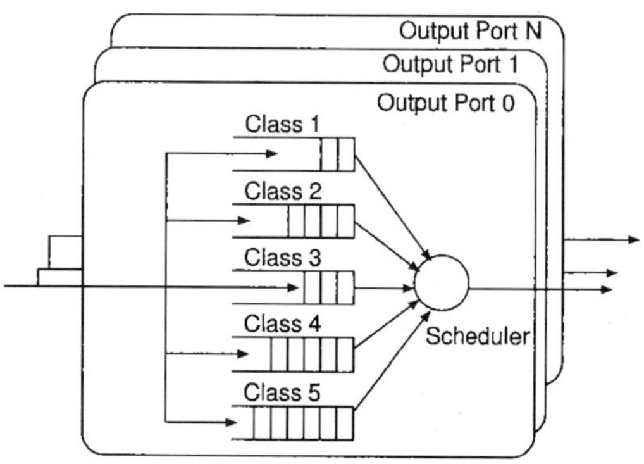

Fig. 8 Switch model.

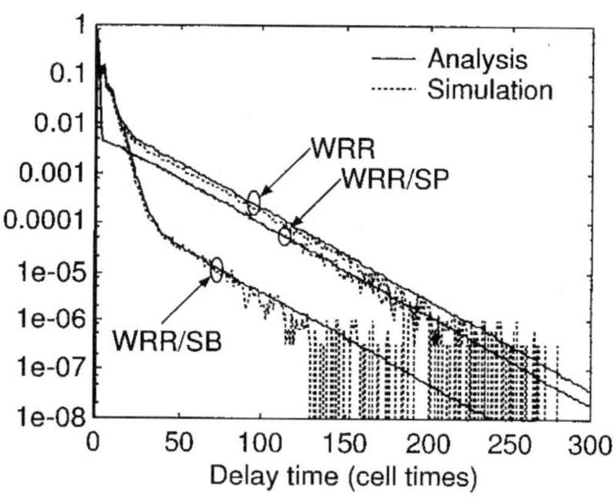

Fig. 9 Results of analysis and simulation: Input rate of Class 1 = 0.1.

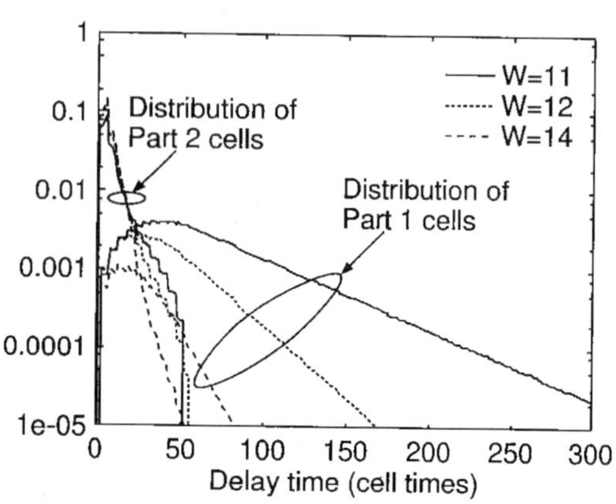

Fig. 10 Delay distribution of WRR: Input rate of Class 1 = 0.1: Weight of Class 1 = 11.

is effective in decreasing delays of Part 1 cells. While allocating a large extra bandwidth for Class 1 would reduce the number of Part 1 cells and lead to an improvement in delay bound performance, such allocation would not be desirable in terms of effective network resource usage.

4.3 Delay Distribution of WRR/SP

Figure 11 shows the delay distribution of WRR/SP with the same parameters as in Fig. 10. Class 1 has the highest priority among all classes.

From Fig. 11, we can see that the delay time of Part 2 cells is reduced while that of Part 1 cells is not significantly reduced. Since the number of cells which can be scheduled during one counter reset cycle is the same with WRR, the number of cells belonging to Part 1 cannot be reduced, instead, the delay time of Part2 cells has decreased because Class 1 cell scheduling is not obstacled by other classes.

4.4 Delay Distribution of WRR/SB

Figure 12 shows the delay distribution of WRR/SB and the effects of borrowing. The counter lower bounds we chose are −6, −11, and −22 and the counter upper bound is equal to the weight value. Counter saving and borrowing are not allowed in other classes. The weight allocated to Class 1 is 11 out of 100.

Figure 12 shows that, contrary to WRR/SP, the delay time of Part 1 cells is reduced while that of Part 2 cells is not. The delay time of Part 1 cells is improved by choosing a smaller counter lower bound, without the need to allocate a large extra bandwidth. By borrowing counter values, the scheduler attempts to schedule input cells according to their burstiness, so long as the counter value is greater than its lower bound. Thus, the number of cells that must wait for future counter resets decreases.

In WRR and WRR/SP, their delay characteristics depends on the length of a counter reset cycle, i.e. the sum of the weight values. We have confirmed that, though WRR/SB also has the same dependency, WRR/SB has still smaller delays than WRR and WRR/SP regardless of the length of a counter reset cycle.

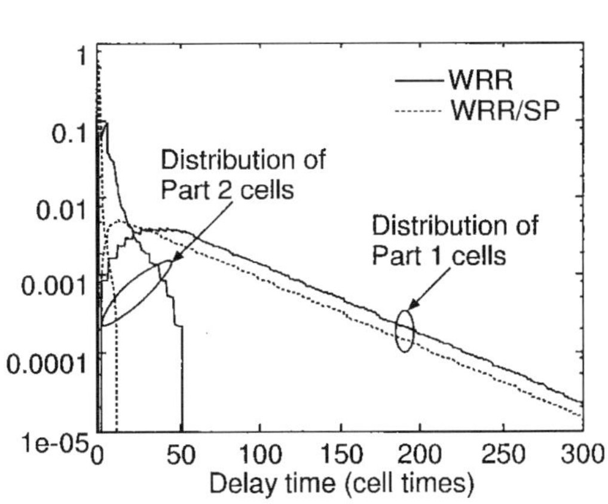

Fig. 11 Delay distribution of WRR/SP: Input rate of Class 1 = 0.1: Weight of Class 1 = 11.

4.5 99% Delay Bounds Comparison

Table 2 shows 99% delay bounds of Class 1 cells for WRR, WRR/SP and WRR/SB. A 99% delay bound means that 99% of cells are sent within this delay boundary. The input rate of Class 1 has been chosen from 0.1 and 0.2. The weight of Class 1 has been allocated so that 10% of extra bandwidth is allocated. A case with 40% of extra bandwidth is shown for WRR so that the effectiveness of large extra bandwidth can be assessed. The effects of counter saving is shown for WRR/SB; the counter upper bound is two or three times of the weight value and counter lower bound is 0.

From Table 2, we can see that the delay bounds decreased by WRR/SP is not significant when the Class 1 traffic is small. We can confirm that WRR/SB is effective in decreasing delay bounds of WRR and that WRR/SB can achieve a target delay bound with less extra bandwidth compared to WRR. For example, while a 40% extra bandwidth is needed to guarantee a 99% delay bound to be within 31 cell times in WRR, a 10% extra bandwidth is sufficient in WRR/SB by setting the counter upper bound to 33, when the input rate of Class 1 is 0.1.

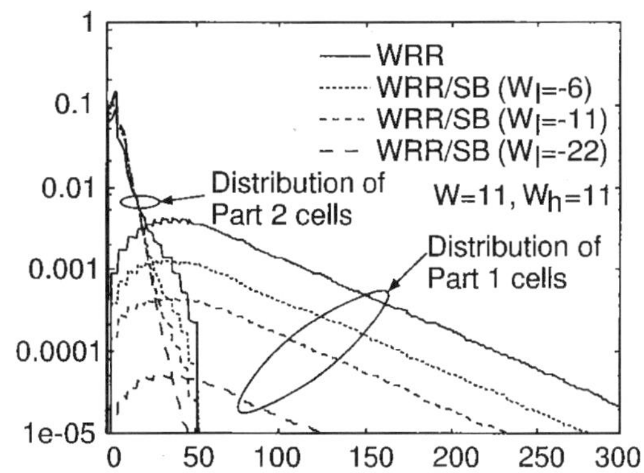

Fig. 12 Delay distribution of WRR/SB: Input rate of Class 1 = 0.1: Weight of Class 1 = 11.

Table 2 99% delay bounds dependency on input rate of Class 1 (cell times): Weight of Class 1 = Input rate × 110 (Also, input rate × 140 for WRR).

Input Rate	WRR		WRR/SP	WRR/SB ×110	
of Class 1	×110	×140	×110	$W_h = 2W$	$= 3W$
0.1	191	31	171	100	25
0.2	115	50	72	58	53

4.6 Influence of Saving/Borrowing on Other Classes

In WRR/SB, saving/borrowing of the counter value in one class may degrade the delay bound of other classes. This degradation may have little impact on the QoS of non-real-time classes, however, if there are multiple real-time classes, this degradation will be serious in other real-time classes. We ran some simulations to investigate the influence of counter saving/borrowing of one class on other classes.

Figure 13 shows 99% delay bounds of real-time class (Class 1) and non-real-time class (Class 5) are shown. The weights of each class are 20 and the input load of real-time classes is 0.19, which means about 5% of extra bandwidth is allocated for each real-time class. To examine the effects of the amount of real-time traffic, the number of real-time classes is chosen from 1 to 3 out of 5 and other classes are non-real-time. The input load of non-real-time classes is assumed to be saturated except that the input load of Class 5 is set at 0.19 to examine the delay bound of non-real-time class. In WRR/SP, all real-time classes have the same priority level, which is higher than that of non-real-time classes. The saving of counter value is only permitted for real-time classes in WRR/SB.

From this figure, we can see that delay bounds decrease as the number of real-time classes increases. This is because, about 5% extra bandwidth is allocated for each real-time class while the input traffic of the non-real-time classes are saturated, thus the total extra bandwidth increases as the number of real-time classes increases.

This figure also shows that WRR/SP is effective in reducing the delay bounds of WRR but WRR/SB is still more effective. In WRR/SB, counter saving/borrowing of one real-time class affects the delay bounds of other real-time classes and this effect becomes more significant as the total input load of real-time classes increases. However, since the delay bounds of each real-time class decreases as the number of real-time classes increases, we can see that the effect of counter saving/borrowing is less significant than the effect of increasing total extra bandwidth. This means counter upper bound and lower bound for real-time classes can be decided without considering the effects of counter saving/borrowing of other real-time classes.

To examine how counter saving/borrowing of real-time classes affects the delay characteristics of non-real-time classes, 99% delay bounds of Class 5 is also shown in Fig. 13. From this figure, we can confirm that WRR/SP and WRR/SB degrade the delay bound of Class 5. This degradation becomes more significant when the total input load of real-time classes increases and when the counter upper bound becomes high. For example, if the total input rate of real-time classes is 0.57, that is, the number of real-time classes is 3, and the counter upper bound is three times as much as the weight value in WRR/SB, the 99% delay bound of Class 5 is increased by 36% while that of real-time classes are decreased by 68%.

4.7 Network Environment

Here, we examine the effectiveness of WRR/SB in a network environment. To investigate the delay bounds of multi-hop connections, we show simulation results of 99% delay bounds for different VCs in Fig. 15. The

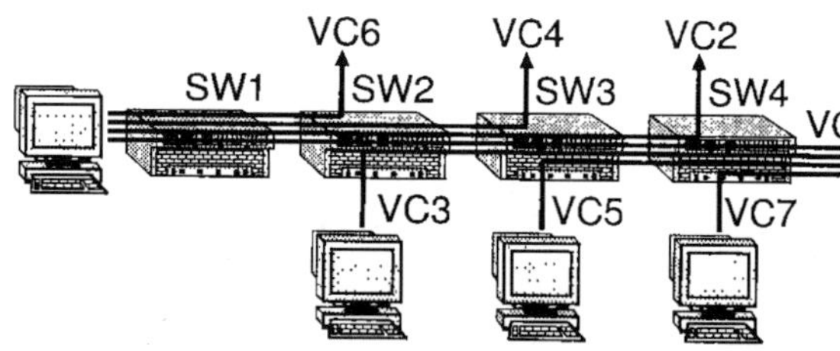

Fig. 14 Network model.

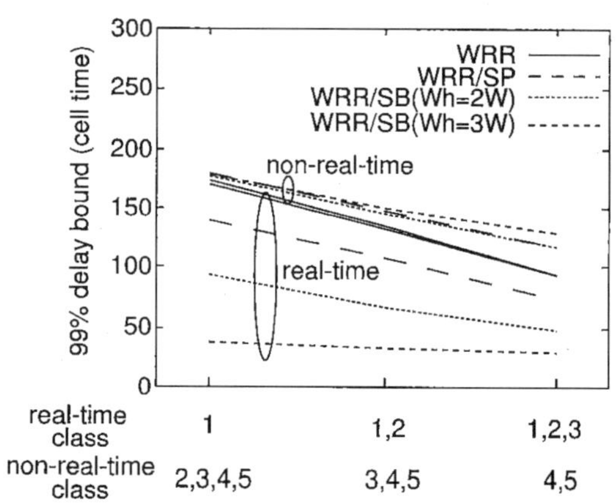

Fig. 13 99% delay bounds dependency on numbers of the real-time classes: Weight of each class = 20: Input rate of real-time classes and Class 5 = 0.19 (other classes are saturated).

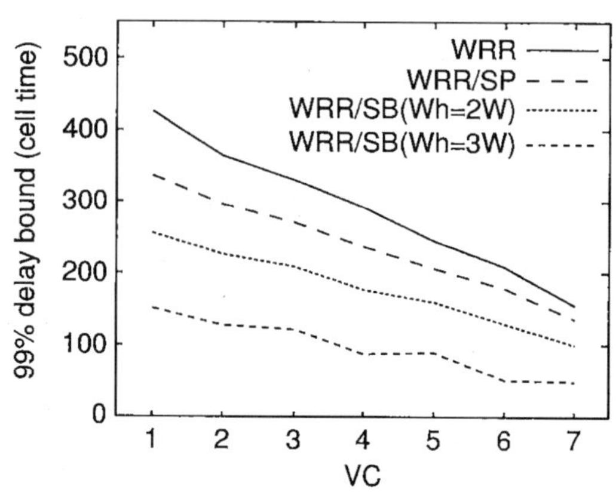

Fig. 15 99% delay bounds of different VCs: Weight of each class = 20: Input rate of one Class 1 VC = 0.0475 (other classes are saturated).

network model used is shown in Fig. 14. The weight of each class is 20 and the input load of one VC in Class 1 is 0.0475. Since all switches accommodate four Class 1 VCs, the total input rate of Class 1 becomes 0.19 at all switches. Other classes are assumed to be saturated.

In this figure, we can see that WRR/SB is effective regardless of the number of switches a VC passes through. This is because many connections are multiplexed and separated in the switches and the burstiness is generated in these switches. Since WRR/SB is robust for the burstiness of input traffic, its delay bounds are smaller than other scheduling methods. WRR/SB is generally effective in the networks in which many connections are multiplexed and separated.

5. Conclusion

In this paper, we proposed a new scheduling method, WRR/SB, to improve the delay bound performance of WRR. We studied the delay characteristics of WRR, WRR/SP and WRR/SB to show the effectiveness of WRR/SB.

Delays in WRR scheduling are induced in the cyclical mechanism of scheduling and the burstiness generated in the network. We showed that the delay bounds of WRR are mainly dominated by the burstiness rather than by the cyclical mechanism of scheduling. WRR/SB is an improvement of WRR, which takes into account the potentially bursty nature of input traffic. WRR/SB attempts to schedule cells according to the burstiness of input traffic by saving the counter value when there are few input cells and borrowing some counter values from future counter resets when there are excess input cells. We analyzed the cell-by-cell behavior of the scheduler and obtained the delay distribution of WRR, WRR/SP and WRR/SB. Our results indicate that WRR/SB is effective in improving the delay bound performance of WRR though WRR/SP is less significant in improving delay bounds.

If there are multiple real-time classes, counter saving/borrowing of one real-time class affects the delay bounds of other real-time classes. However, we showed that these effects are not significant and we can choose counter upper bound and lower bound for real-time classes without considering interference among counter saving/borrowing of real-time classes.

References

[1] ATM Forum Traffic Management Specification Version 4.0 1996.

[2] A. Demers, S. Keshav, and S. Shenker, "Analysis and simulation of a fair queueing algorithm," Internet Rs. Exper., vol.1, 1990.

[3] A.K. Parekh and R.G. Gallager, "A generalized processor sharing approach to flow control in integrated services networks: The single node case," IEEE/ACM Trans. Networking, vol.1, no.3, pp.344–357, 1994.

[4] A.K. Parekh and R.G. Gallager, "A generalized processor sharing approach to flow control in integrated services networks: The multiple node case," IEEE/ACM Trans. Networking, vol.2, no.2, pp.137–150, 1994.

[5] S.J. Golestani, "A self-clocked fair queueing scheme for broadband applications," Proc. IEEE INFOCOM'94, pp.636–646, 1994.

[6] J.W. Roberts, "Virtual spacing for flexible traffic control," Int. J. Commun. Syst., vol.7, pp.307–318, 1994.

[7] A. Hunh and G. Kesidis, "Bandwidth scheduling for wide-area ATM networks using virtual finishing times," IEEE/ACM Trans. Networking, vol.4, no.1, pp.643–641, 1996.

[8] J. Rexford, A.G. Greenberg, and F.G. Bonomi, "Hardware-efficient fair queueing architectures for high-speed networks," Proc. IEEE INFOCOM'96, pp.638–646, 1996.

[9] M. Katavenis, S. Sidiropoulos, and C. Courcoubetis, "Weighted round-robin cell multiplexing in a general-purpose ATM switch chip," IEEE J. Sel. Areas Commun., vol.9, no.8, 1991.

[10] D. Stiliadis and A. Varma, "Latency-rate servers: A general model for analysis in traffic scheduling algorithms," Proc. IEEE INFOCOM'96, pp.111–119, 1996.

[11] N. Mizukoshi, R. Fan, H. Suzuki, and M. Shinohara, "A Single-Chip Controller for 1.2 Gbps Shared Buffer ATM Switches," Proc. CICC, pp.449–452, 1997.

High-Performance Backbone Network Technology edited by Naoaki Yamanaka
ISBN: 0-8247-5321-6 DOI: 10.1081/(120025716)
Published 2004 by Marcel Dekker, Inc.

Adaptive Remote Rate Control Using Extrapolation and Correction Mechanism for Periodic Notification of Link Utilization Ratio

Haruhisa Hasegawa, Naoaki Yamanaka, and Kohei Shiomoto

SUMMARY A new adaptive rate control with congestion prediction is developed that is highly robust against long propagation delays. It minimizes the network performance degradation caused by the delay based on prediction by extrapolating past data and correction using new notification. The simulation results show that our proposed control maintains high throughput and a smaller buffer even in long propagation delay networks, like ATM-WAN.

key words: *ATM, congestion control, adaptive rate control, simulation, propagation delay*

1. Introduction

A new adaptive rate control mechanism with congestion prediction based on periodical notification of link utilization is developed that is efficient for networks where there is a long propagation delay between the observing and controlling points. Our proposed control provides a high network throughput and maintains a smaller required buffer even in long propagation delay networks. The control adopts an architecture [1] in which the remote node (Obsv node) observes the link status and notifies upstream nodes (Cntrl nodes), which control the traffic flow according to these notification messages. The remote Obsv node periodically sends the utilization ratio of its output link to the Cntrl nodes. These nodes predict the current utilization ratio by extrapolating past notification data until the next notification message arrives. It also corrects the prediction when a new notification is received. Services with conventional adaptive rate control [2] require huge buffers to prevent cell loss and/or reduce response speed to the user's request in a WAN environment [3]. A problem occurs because the rate is controlled based on the past information which is irrelevant to the current status. This paper shows that the prediction approach using extrapolation and correction improves the robustness of the rate control to diminish the propagation delay. This paper proposes this control mechanism and its basic algorithm and demonstrates its performance through simulation.

Manuscript received January 13, 1997.
Manuscript revised May 6, 1997.

† The authors are with NTT Network Service Systems Laboratories, Musashino-shi, 180 Japan.

2. Adaptive Remote Rate Control

A new adaptive rate control is proposed which is efficient for networks where there is a long propagation delay between the observing and controlling nodes (Cntrl node) such as depicted in Fig. 1.

In this figure, the remote observing nodes (Obsv nodes) notify the current link utilization ratio information (Ψ) of their output links to the Cntrl node by control cell. The control cell is sent by the Cntrl node and is looped back by the destination side edge node. The link utilization ratio is defined as the Total input traffic (T)/Output link capacity (C). Each Obsv node updates the Ψ on the control cell to the larger value of Ψ on the control cell and its own observed Ψ. Each Cntrl node adaptively varies its output rate (R) toward the Obsv node based on the notification messages. The Cntrl node updates its output rate by calculating the following two equations iteratively.

$$R = R \cdot \gamma \tag{1}$$

$$\gamma(\Psi) = P_1 \cdot \Psi + P_2 \quad P_1 < 0, P_2 >= 1.0 \tag{2}$$

If there is no delay between the Cntrl node and the Obsv node, the output rate, R, can be controlled by ratio $\gamma(\Psi)$, which is defined as a decreasing function of Ψ. For simplicity, function γ is defined as (2). When Ψ is more than 1.0, γ should be less than 1.0 to

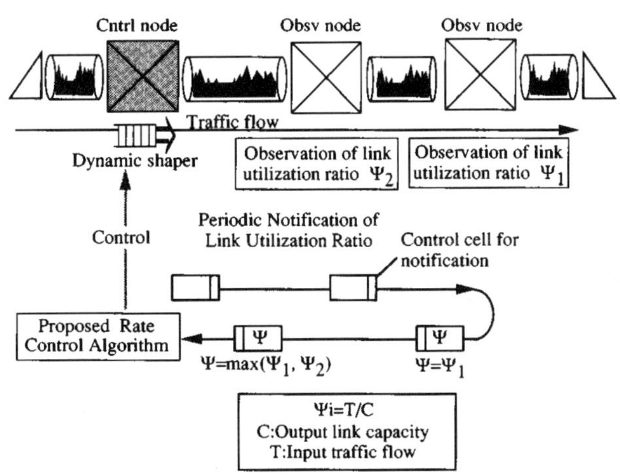

Fig. 1 Control system architecture.

IEICE TRANS. COMMUN., Vol. E80-B, No. 10, pp. 1576–1580, October 1997

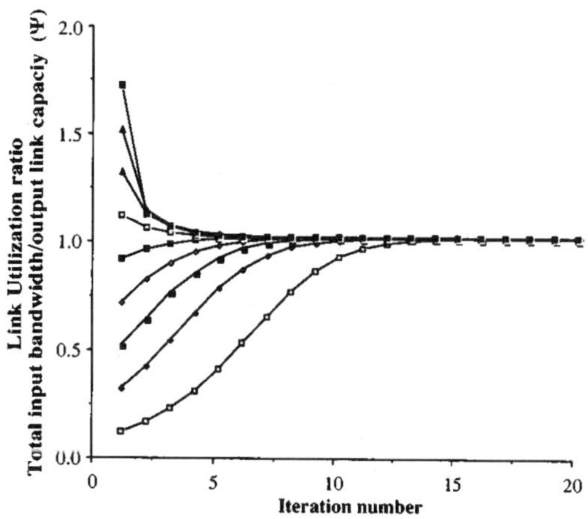

Fig. 2 Utilization rate vs. iteration number.

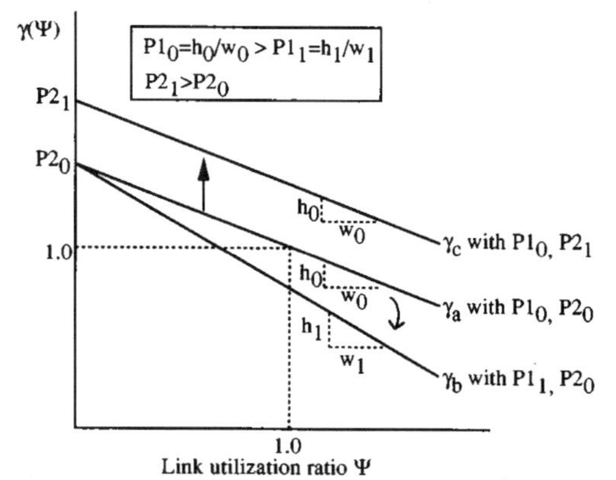

Fig. 3 Relationship between $\gamma(\Psi)$ and parameters P_1 and P_2.

reduce the traffic volume. When Ψ is less than 1.0, γ should be more than 1.0. Parameter P_2 should not be less than 1.0. In this case, the greedy traffic converges to the utilization ratio of 1.0. In this paper, "greedy" expresses a state in which a source always wants sufficient bandwidth to maintain the average traffic as high as its offered load. Figure 2 shows the link utilization ratio with iterated functions (1) and (2). In Fig. 2, the source always wants more bandwidth than it has because its total offered load is set to 2.0. Parameters P_1 and P_2 are set at -0.5 and 1.5 respectively in Fig. 2. The initial value of link utilization ratio, Ψ is changed from 0.1 to 1.7. Ψ varies from the initial value by the iteration of (1) and (2). Only when the link utilization ratio is 1.0, is $\gamma(\Psi)=1.0$. If the utilization ratio is 1.0, the link is used completely without congestion. Even if the initial utilization ratio is changed, link utilization Ψ converges it to 1.0 by the iteration of (1) and (2). Even if the control starts in a congested condition, i.e, the initial Ψ is not less than 1.0, it converges Ψ to 1.0.

Parameters P_1 and P_2 determine the characteristics of this control method. Figure 3 illustrates the relationship between $\gamma(\Psi)$ and parameters P_1 and P_2. P_1 determines the sensitivity of γ to Ψ. As P_1 approaches 0, $\gamma(\Psi)$ becomes more constant, even when Ψ changes drastically. Conversely, if P_1 is set much smaller than 0, the rate increases more quickly under lower utilization ratios and falls faster under higher loads. In Fig. 3, when P_2 is fixed, as P_1 becomes smaller, γ decreases faster for Ψ. Therefore, the rate increases relatively more at lower utilization ratios. P_2 determines the entire behavior of $\gamma(\Psi)$. If P_2 is set larger, $\gamma(\Psi)$ becomes larger as depicted in Fig. 3, and the rate tends to increase as a result. In general, the $\gamma(\Psi)$ in Fig. 3 should approach the point of (1.0, 1.0) as depicted by γ_a in Fig. 3, which is a critical point between low-load and congested conditions. If γ is 1.0 when Ψ is 1.0, the link utilization ratio is stable. Therefore, P_1 and P_2 should satisfy the next equation.

$$1.0 \approx P_1 + P_2$$

When Ψ is more than 1.0, γ should be less than 1.0 to recover from congestion. When Ψ is less than 1.0, γ should be more than 1.0 to maximize the utilization ratio.

If the current Ψ is known, it is easy to calculate the utilization ratio after the next iteration using (3) because $\gamma(\Psi)$ expresses the next maximum traffic increasing ratio.

$$\Psi_{\text{next}} = \Psi \cdot \gamma(\Psi) \tag{3}$$

If the initial utilization is given to the Cntrl nodes, they can completely predict the current utilization ratio using extrapolation. In other words, any trajectory in Fig. 2 can be calculated only by Cntrl nodes for greedy traffic sources.

If the traffic source is greedy, Ψ exactly follows the trajectories in Fig. 2 and the prediction does not need any corrections. If the source is not greedy, the link utilization may become less than Ψ_{next} in (3) since traffic sources do not always increase their output rate to the maximum value allowed by γ. In this case, the predicted Ψ should be corrected by the notification. The new notification value is the only way for Cntrl node to know the current Ψ. However, the Cntrl node should not use the raw notification information to control the buffer output rate directly, instead the effect of the delay and the notification interval must be taken into account. We propose the control algorithm below to achieve the requirements. The algorithm consists of two separate control functions.

/* Control function 1*/
If rate update timer expires {

$\gamma = P_1 \cdot \Psi_{\text{pred}} + P_2$;

$R = R \cdot \gamma$;

$\Psi_{\text{pred}} = \Psi_{\text{pred}} \cdot \gamma$;

$\pi = \pi \cdot \gamma$;

Restart timer;

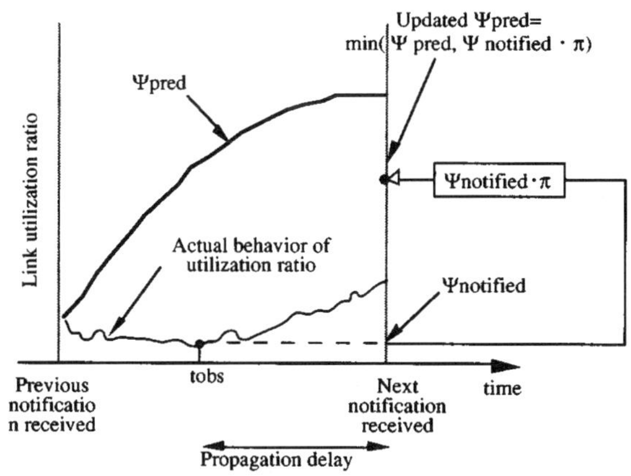

Fig. 4 Control function 2.

}
/* Control function 2*/
If a new notification of utilization ratio (Ψ_{notified}) arrives {

$\Psi_{\mathrm{pred}} = \min(\Psi_{\mathrm{pred}}, \Psi_{\mathrm{notified}} \cdot \pi)$;

$\pi = 1.0$;

}

If the rate update interval is set shorter than the notification interval, the Cntrl node must update its output rate without new information provided by the Obsv node. Control function 1 is based on the calculations (1)-(3) described above. It, therefore, predicts the current Ψ by extrapolating past Ψ assuming all the traffic sources are greedy. Ψ_{pred} in Control function 1 is the predicted value of Ψ. The prediction has error only on the safe side because greedy traffic is the heaviest of the possible loads. Control function 1 shown above updates the output rate R periodically using a timer. The variable π expresses the aggregate ratio of R since the last notification and is used in Control function 2. The initial value of π is set to 1.0.

Control function 2 represents the actions taken when the Cntrl node is notified of the latest link utilization ratio Ψ_{notified}. Figure 4 illustrates Control function 2. In control functions 1 and 2, the same variable π is commonly used. Therefore, the value of π, which is calculated by control function 1, is used first and then reset to 1.0 in control function 2.

In Fig. 4, the horizontal axis represents time and the vertical axis represents the link utilization ratio. The observed utilization ratio at time t_{obs} is sent to the Cntrl node. The Cntrl node receives it after the propagation delay. If the new Ψ_{notified}, which is the observed utilization ratio at t_{obs}, multiplied by π is less than the current Ψ_{pred}, Ψ_{pred} is updated to $\Psi_{\mathrm{notified}} \cdot \pi$. Otherwise, the current value is used. The value of $\Psi_{\mathrm{notified}} \cdot \pi$ is a safe prediction if the notification interval is longer than the propagation delay. Because both Ψ_{pred} and $\Psi_{\mathrm{notified}} \cdot \pi$ err only on the safe side, a new Ψ_{pred} can be selected from the minimum of these values.

The adaptive remote rate control proposed in this section is expected to have capabilities for network control, where the information of the link utilization ratio is received intermittently by the Obsv node with a propagation delay. The proposed adaptive rate control does not need an explicit value of propagation delay. The next section evaluates the performance of this algorithm through simulations.

This adaptive rate control is designed to function between ATM switch with dynamic rate shaper (as Cntrl node) and ones with observation function of link utilization ratio (as Obsv nodes). The source side edge switch is the Cntrl node and other below switches are Obsv nodes, as depicted in Fig. 1. If multiple Obsv nodes notify their own link utilization rates to one Cntrl node, the load of handling this information imposed on the Cntrl node increases as network size increases. In this case, the notification mechanism of link utilization can be achieved by sending one control cell from each source side edge switch per VC. The control cell is looped back by the destination side edge switch. The destination side edge switch writes its link utilization ratio Ψ. The other switches write their own Ψ to the looped back control cell only if the Ψ in the cell is smaller than their Ψ value. This mechanism allows the Cntrl node to receive aggregate information of multiple Obsv nodes, i.e., the largest Ψ on the route.

This adaptive rate control can be used between user's terminal and switches. However, UNI (User-Network Interface) should be chosen to meet the standarized services. This proposed rate control has not been standarized but can interwork with the standardized ATM service. The standardized ATM service should be used between user and edge nodes while and the proposed flow control suits ATM-WANs. In this case, the UNI should be selected to match ABR service [2], which is also an adaptive rate control. Fortunately, the Cntrl node on the source side edge is able to function as VS/VD for ABR users, because the switch calculates its output rate and is able to notify users of the rate through the ER [2], [4] field of backward RM cells. In the next section, the interaction between the proposed control method and ABR service is shown.

3. Simulation Results

The performance of the proposed adaptive remote flow control in Asynchronous Transfer Mode (ATM) networks was evaluated by simulation. The simulation model is depicted in Fig. 5. In this model, the terminal access is based on ABR service.

The traffic between switches is managed by our proposed flow control. In Fig. 5, we examined the performance concerning traffic flow from $S = \{s_{11}, s_{12}, s_{21}, s_{22}\}$ to $D = \{d_{11}, d_{12}, d_{21}, d_{22}\}$. For the traffic flow,

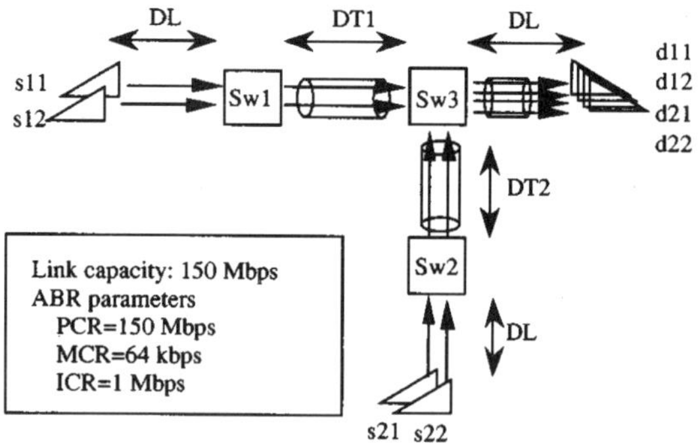

Fig. 5 Simulation model.

Sw_1 and Sw_2 are Cntrl nodes and Sw_3 is the Obsv node. Each edge switch controls its output rate and informs the terminals its own output rate using the Explicit rate (ER) field of Resource Management (RM) cells [2]. For comparison, transmission by conventional end-to-end ABR with the ER [2], [4] mode was also evaluated under the same conditions. Link capacity of all links is assumed to be 150 Mbps. The parameters P_1 and P_2 are set to -1.0 and 2.0, respectively. The initial value of variable π is set to 1.0. The interval of notification by control cell according to the mechanism described in the previous section is set to 1600 cell times. Sw_1 and Sw_2 control their output rate based on the aggregate information of Sw_3 and themselves. The simulation fixed the total offered load to 70%. The terminals generate a burst with the size following geometric distribution and the interval of exponential distribution. The propagation delay is expressed by the transmission length in this paper. The simulation fixed D_L to 0.5 km and changed $D_{T1}(=D_{T2})$ from 50 to 1000 km. In this model, the number of active terminals is 4 but the number of active VCs dynamically changes because they sends cells in a bursty manner. The mean burst length was set to 40 and 126.

Figures 6 and 7 show the network throughput and maximum queue length of Sw_3 as functions of distance, respectively. According to Fig. 6, the proposed scheme has a slightly higher throughput than ABR. Both are unaffected by distance. The maximum queue length of ABR does, however, increase with distance. The maximum queue length of the proposed scheme is not affected by distance. As a result, it can be said that the algorithm proposed in this paper successfully reduces the effect of distance.

Next, performance under various loads was evaluated using the same simulation model. The simulation fixed D_L to 0.5 km and $D_{T1}(= D_{T2})$ to 1000 km. In the simulation, the total offered load was changed from 0.2 to 1.2. Each termial can give a quarter of the total offered load on average. The mean burst length is set to 40 cells. Figures 8 and 9 show the network throughput and maximum queue length of Sw_3 as functions of total offered load, respectively.

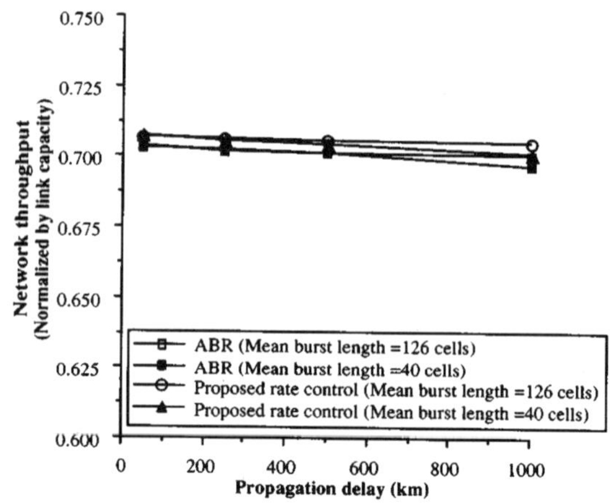

Fig. 6 Network throughput.

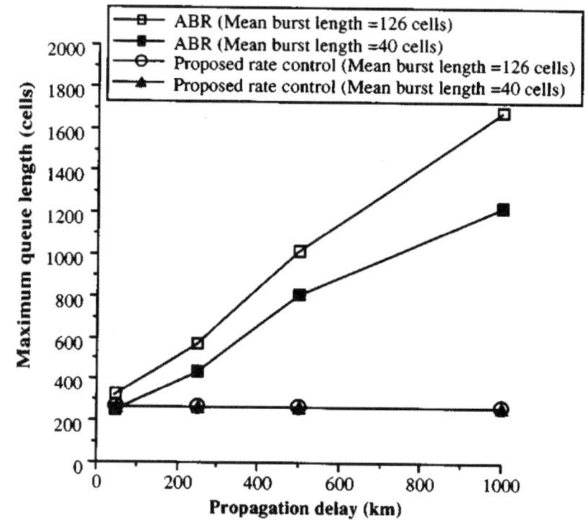

Fig. 7 Maximum queue length.

According to Fig. 8, with low loads, the network throughput of the proposed rate control method is almost the same as that of ABR. With high loads, the proposed method offers higher throughput. This difference is due to the fact that ABR reduces the rate after the input load exceeds the link capacity, i.e., congestion occurs based on the queue length information, but the proposed control method dynamically changes the rate based on the utilization information. This is also seen in the maximum queue length of Sw_3 (see Fig. 9). For total offered loads from 0.6 to 0.8, the maximum queue length of the proposed control method shows the highest value. Under higher loads, maximum queue length decreases because the absolute value of γ becomes small. On the other hand, ABR is mainly controlled by queue length information and is forced to reduce the rate more because the rate tends to be strongly decreased after the input rate exceeds the link capacity.

According to Fig. 9, the maximum queue length of the proposed control method is much smaller than ABR and it can be said again that the algorithm proposed in this paper successfully reduces the effect of

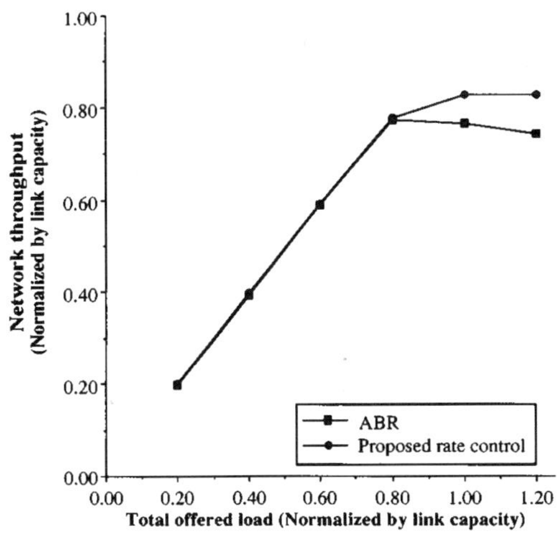

Fig. 8 Total offered load vs. network throughput.

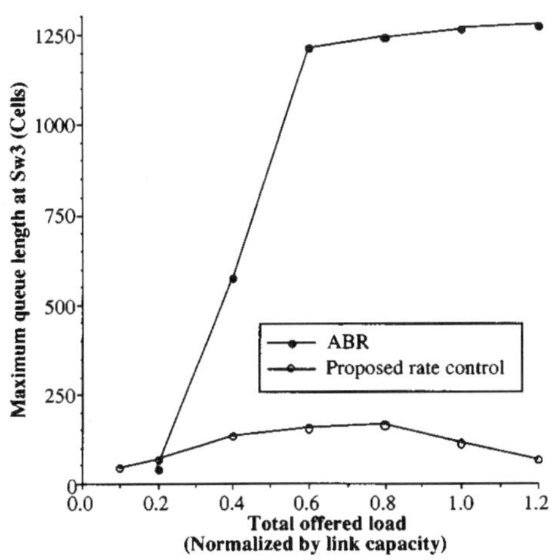

Fig. 9 Total offered load vs. maximum queue length.

distance even under various loads.

4. Conclusion

A new adaptive rate control that improves network performance and reduces hardware cost in long propagation network is proposed. Our proposed adaptive rate control does not need a large buffer even in a network with long propagation delay. The Cntrl node predicts the current utilization ratio by extrapolating the past notification until the next notification. It also corrects the prediction when a new notification is received. This paper showed that the prediction using the extrapolation and correction approach improves the robustness of the rate control affecting the propagation delay. The simulation results show our proposed control maintains a high throughput and smaller buffer even in long propagation delay networks without any information of the delay.

References

[1] N. Yamanaka, K.Shiomoto, and H. Hasegawa, "ALPEN : A simple and flexible ATM network based on multi protocol emulation at edge nodes," IEICE Trans. Commun., vol. E79-B, no. 4, pp. 611-615, April 1996.

[2] S. S. Sathaey, "Draft ABR version 4.0 of ATM forum traffic management specification version 4.0," ATM Forum/95-0013R11, March 1996.

[3] A. Koike, H. Kitazume, H. Saito, and M. Ishizuka, "On end system behavior for explicit forward congestion indication of ABR service and its performance," IEICE Trans. Commun., vol. E79-B, no. 4, pp. 605-610, April 1996.

[4] A. W. Barhnhart, "Example switch algorithm for section 5.4 of TM spec," ATM Forum/95-0195, Jan. 1995.

High-Performance Backbone Network Technology edited by Naoaki Yamanaka
ISBN: 0-8247-5321-6 DOI: 10.1081/(120025717)
Published 2004 by Marcel Dekker, Inc.

VP Control for ATM Networks with Call-Level QoS (Quality of Service) Guarantees

Kyamakya Kyandoghere

SUMMARY In this paper, a design concept that offers ATM-VP connections with different protection levels is presented. The users have the choice to select the protection level they wish, the network transport service they need, and the worst cell loss they can tolerate at call set up time, and pay accordingly. Besides, an advanced adaptive traffic control scheme that simplifies call and cell processing is also presented. Many important functions such as call admission, VC-bandwidth reservation, cell-level congestion control, etc. are efficiently performed at the boundary of the backbone network. In this way is given a suitable answer to the important question: "How can future telecommunication networks based on ATM provide services with customized availability ?" A platform that outlines the potential interaction between restoration methods and congestion avoidance schemes is also obtained.

key words: *ATM networks, QoS (Quality of Service), multiple-level availability, bandwidth control*

1. Introduction

B-ISDN based on ATM will integrate a variety of services. Besides, there are just some examples where the availability of a communication system is critical: on-line banking, military communication, signaling data, management messages, etc. In many cases however, the communication system is used in parallel with other non-critical services. Since availability requirements vary widely, several availability grades have to be realized and integrated into the framework of B-ISDN in order to achieve appropriate service costs. In the area of accounting management also, several aspects must be taken into account. When the restoration process is offered as a service to network users, possibly including the subscription to a higher availability class, rules which determine the corresponding subscription rates must be defined. A related charging adaptation should be considered to take into account the impact of network fault on the user.

A challenging question for research in the field of Telecommunication Networks Integrity is "How can future telecommunication networks based on ATM provide services with customized availability? [1]" In other words, the determination of the exact relationship between the choice of a network architecture and a set of availability objectives is still an open issue [2]. In this paper this issue is handled. In fact, an ATM network concept that allows the network carrier to provide customized survivability* keeping fairness is presented, that is, under any situation QoS (Quality of Service = protection level + cell loss level) should be equal for all users that use the same traffic class. The VP level is here the top level of traffic control. It efficiently provisions network resources to meet the bandwidth demand and satisfy the performance requirements on the VC and cell levels. Calls for which grade of service requirements cannot be met due to shortage of bandwidth at the time of arrival are provisioned to be rejected at the network boundary. Furthermore, this concept constitutes a platform that makes it possible to outline the interaction between restoration methods and congestion avoidance schemes.

It is also well known that a restoration mechanism alone is not enough to efficiently achieve a high level of survivability and to deploy a survivable network in a cost-effective manner. Thus virtual paths should be dynamically configured in response to a change in the demand patterns as well to a possible change in the underlying facility network. Given the available network resources and a traffic demand, a flow assignment determines not only the allocation of bandwidth over the network (for the different availability classes), but also the amount of demand to be rejected due to shortage of bandwidth. As such, a survivable network can be realized with less resource installation in a dynamically reconfigurable network compared to its static counterpart.

In this paper, the central question is how to design an ATM network which offers VP connections with different protection levels. The users should have the choice of selecting the protection grade they need at call set-up time. Availability classes are defined. Each class corresponds to a set of failure scenarios for which a restoration guarantee is provided. Besides, the bigger the failure scenarios set of a class is, the higher the availability provided, and the higher the priority in the network concerning VP rate assignment. Each availability

Manuscript received January 29, 1997.
Manuscript revised August 11, 1997.

[†]The author is with the Faculty of Electrical Engineering at FernUniversitaet-Hagen, FTK, ET-KS, Feithstr.142/TGZ, 58084 Hagen, Germany.

*The terms survivability, availability and restorability are used here interchangeably. Their meaning here really corresponds to what is called "Instantaneous Availability" in CCITT 8101, where it is defined as the probability that an item is in an up state at a given instant of time t.

IEICE TRANS. COMMUN., Vol. E81-B, No. 1, pp. 32–44, January 1998

class constitutes a virtual (logical) VP-subnetwork on top of a large ATM network. Given the traffic demand for each availability class, we are searching for the partition of the (physical) transmission capacities (among the different availability classes) that maximizes the total carried traffic. This partition is dynamically changed in response to changing traffic demand.

This paper is organized as follows. Section 2 explains and describes different milestones of the concept. Then the optimization problem, major properties of the optimal solution and an efficient parallel optimization algorithm are presented in Sect. 3. Numerical examples are reported in Sect. 4 along with the discussion of some simulation results. Finally, Sect. 5 presents concluding remarks.

2. The Concept's Main Features

The VP-based ATM network architecture [3], [4], in which different logical (virtual) VP-networks share the available (physical) transmission capacity is considered. Each virtual VP network corresponds to an availability class. The VP-level control provisions and performs the optimal logical assignment of VP bandwidth among different availability classes for an upcoming control period.

Under each failure scenario as well as under each non-failure scenario, the maximum rate for a VP is set (a) to satisfy end-to-end delay and cell loss requirements for the VP, (b) to maximize call and cell throughput, and (c) to satisfy the full restorability in all failure scenarios of concern for its corresponding availability class.

Therefore two important concerns are among others involved in the traffic control: (a) the network must be engineered to satisfy the maximum end-to-end delay requirements†; (b) the VP-level should collect real-time network measurements and forecasts bandwidth demand and network VC load for an upcoming control period.

The provisioning by VP-level control produces (a) the set of VP maps, (b) VP maximum rate assignment for each VP of the map sets, (c) the amount of demand to be rejected (at VC-level) at each source node for each SD (source-destination), and (d) local VP routing tables for each intermediate node under each scenario. Furthermore, the VP-level control is responsible for initiating the routing tables swap at the beginning of a control period (after completion of the optimization). If a failure occurs, the network is reconfigured in a distributed fashion (for the classes with the adequate restoration guarantee for this failure scenario) and a new control period is started. At each source node, for each incoming VC request, VC-level control reserves virtual bandwidth within the VP connection of the class with the required availability. Moreover, VPI assignments (VC-to-VP mapping), for each VC, are stored at the VP's source node.

2.1 Multiple-Availability Restoration Plan

The essentials of the multiple-availability restoration plan concept are presented in the following. An hybrid restoration strategy is considered. Given a traffic demand pattern (within a control period), VP bandwidth allocation (for all availability classes) is centraly computed not only for the non-failure scenarios but also for the failure scenarios corresponding to each availability class. Besides, routing tables for VP-crossconnection within the network are also accordingly prepared. This information is then stored in the network nodes (VP crossconnects). When a failure occurs, the failure information is broadcasted in a distributed way. Upon receiving failure information, every node carrying VPs affected by the failure activates the stored alternate paths and performs required routing tables updates. The time required for restoration consists so of the time taken to propagate the failure information to all nodes and the time taken to activate alternate paths at all nodes. This time has been estimated to be less than 60 ms even for a large size backbone network [5]. And due to optimization, even calls that are not concerned by the failure may be reallocated. In such cases a VP/VC alignment function that realizes hitless VP switching has been considered in order to solve the potential cell sequencing loss problem.

There are two general types of failures in a network: link failure and node failure. In view of many years of experience, the network operator may easily determine the most probable failure scenarios [6], [7]. As example we consider here the following failures set: (i) 1 link-failures, (ii) 2-link failures. And following availability classes are defined: (a) class $a = 0$ with no restoration guarantee; (b) class $a = 1$ with 100%-restoration guarantee for all failure scenarios (i); and (c) class $a = 2$ with 100%-restoration guarantee for all failure scenarios, i.e., (i) and (ii).

As already explained, traffic segregation is done within the VP networking on the basis of availability requirements (see Fig. 1). We suppose that each availability class has its own (logical) VP network, and these logical networks are superimposed over the facility network. If a call requires more than one availability level, for example in multimedia communications, routing may be performed on multiple VP networks.

Fundamentally, restoration is (here) guaranteed by providing redundant capacity and precomputing alternate paths for all VPs for the failure scenarios of concern according to their respective availability classes. The logical spare capacity assignment is done within

†This is done by properly sizing the outgoing link buffers (see Fig. 3) and limiting the maximum number of hops of the VP sets.

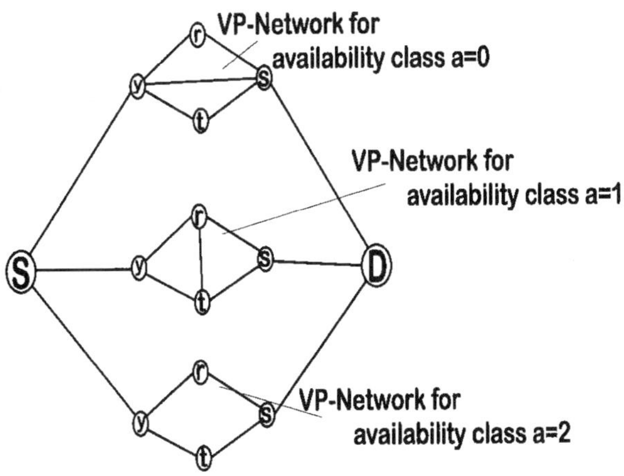

Fig. 1 VP-networks for multiple availability levels.

the frame of the joint optimization problem presented in Sect. 3. Considerations made above show that the optimal spare capacity reservation follows directly from network admission and bandwidth allocation decisions. Alternate paths may be computed according to source-destination, local-destination, and local rerouting [5]. But we'll concentrate on source-destination rerouting since this has been proven to be the best one concerning (limited) resources (spare capacity and VPI number) utilization [8]. Its drawbacks (a more complex rerouting decision and longer restoration times) are considerably reduced (throttled) by the here proposed precomputing of restoration paths.

2.2 Simplified Traffic Control Architecture

A layered architecture for adaptive traffic control which utilizes the VP concept in ATM backbone networks is proposed. The suggested control guarantees quality of service and simplifies call and cell processing through advanced allocation of network resources and distribution of traffic control functions to network edges. This simplification is done at the expense of some bandwidth utilization which is but optimized in the constraints of the suggested architecture. In this way, the proposed control complies with standard recommendations ITU-T I.371. The control is organized in three levels: a VP-level, a VC-level, and a cell-level. We use the terms VC and call interchangeably.

For each availability class there is at least one VP between each SD (Source-Destination) pair. This is an achievement of the so called diversity concept, that we briefly explain. Traffic demand between two nodes is divided and carried out on different disjoint routes. In doing so, a failure on one of the routes will affect only a portion of the total traffic under consideration. Hence diversity ensures some connectivity between the two nodes, but it does not protect all the traffic.

Three types of network transport services (TS) on VC-level are considered: (a) express service (ES), appropriated for real-time-traffic; (b) first-class-service (AS), appropriated for non-real-time traffic; and (c) second-class-service (BS), appropriated for calls that generate very long bursts of non-real-time traffic separated by silent periods (see [9]). The BS call uses bandwidth reservation only during the burst. Within an availability class there is bandwidth sharing between all TS. This is called multiservice bandwidth sharing (MBS) strategy. It is located at VC-level control. The parameters of the MBS strategy may be periodically adjusted at the beginning of each control period.

The VC-level control performs call admission and processing according to the provisioning specified by the VP-level control for an upcoming period. The control makes all call admission and routing decisions and bandwidth reservations at the access node independently of all other nodes. Call admission and routing decisions within an availability class are done taking into account the following elements: (a) traffic descriptors incorporated in the connection request ID; (b) the amount of bandwidth currently reserved on the VP set for the call's SD pair; and (c) maximum permitted VP rates assigned to real-time and non-real-time traffic for the VP set. Note that the MBS is performed according to either threshold or profitability strategies.

According to the user's choice, the VC-level control assigns a TS for the admitted call. To provide the required quality of service for the TS, the VC-level control reserves the peak rate for ES calls and BS bursts and some rate between the peak and guaranteed rates for AS calls. The rate reserved for AS calls depends on the selected guaranteed rates and is determined by the network provider to satisfy the average delay requirement. Due to the bursty nature of AS traffic, the scheme can use statistical multiplexing to utilize network resources more efficiently. For BS calls, within a session, if the user requests permission to send a burst, the VC-level control, in cooperation with the cell-level control, checks to see if there is sufficient VP bandwidth to handle the burst with the requested quality of service. When the bandwidth is available, the control reserves, at the source node, the burst peak rate on the VP and sends a positive acknowledgment. The user then transmits the burst. The last cell of the burst is a token that acknowledges the source node with the end of the burst transmission on the outgoing link. Then the reservation is released. The bandwidth reservation for each burst is done at the access node. No other backbone nodes are involved in the process. This leads to a small reservation overhead. It also makes the reservation process independent of propagation delay, which may be quite substantial in wide area networks (see a similar mechanism called ATM Block Transfer - Immediate Transmission (ABT-IT) in ITU-T (I.371)). Traffic control schemes in which the control is not done only at the source node result in a higher overhead and slower burst admission in the network [9].

The cell-level control handles cell processing in the inter-office network and is the fastest. One focuses on cell-level Virtual Path Connection Control (VPCC) and do not consider other cell-level control functions such as policing, cell retransmission, etc. The cell-level VPCC is executed for each VP in the Front Server of the access backbone node. This may be realized with the help of intelligent controllers (neural networks + fuzzy logic, learning automata, etc. [10], [17], [22]. The main goals of this regulation are: (a) to maintain, at the VP local access, the total peak rate of AS cells below the maximum permitted rate assigned to this TS; (b) to save bandwidth by reserving for AS calls the rate between their peak and guaranteed rates; and (c) perform cell loss adaptation: note that cell loss requirements are satisfied since they are uniformly limited for all classes by the VP logical bandwidth assignment (see Eq. (6)). But some calls may tolerate a higher cell loss rate. The cell level control may adapt their cell rates in order to reach their respective tolerable cell loss rates. We save this way additional bandwidth.

Since for ES calls and BS-bursts the peak rate is reserved, the total MPR on the VPs is not exceeded. This precludes congestion inside each VP logical pipe and sustains the quality of service. To avoid the VP local access overflow, the cell-level control chokes AS users if the VP access for AS cells is getting congested. In the absence of congestion AS traffic will be transmitted at peak rate. On the other hand, when congestion at the access node is detected, the transmission rate of AS traffic has to be reduced to the guaranteed rate. The detection is performed by observing the queue of the AS cells locally accessing the VP source node. No additional cell-level congestion control is applied at the intermediate nodes, and no information exchange between the source and other nodes is required.

A simplified traffic control architecture is presented in Fig. 2. Thus the connection ID request contains at least the following information: (a) desired availability level; (b) network transport service (TS) specification; (c) traffic descriptors; and (d) acceptable value of the cell loss ratio. On the basis of the ID request content it will be decided, whether the connection may be accepted or not.

In the suggested scheme VP assignment allocates bandwidth for many calls for an upcoming control period and is not done on a per-call basis. To accommodate the traffic going through the failed links, or to adjust to a large change in demand, VP bandwidth is optimally reassigned to all VPs (that survive) and not only for the VPs affected by the failure. In such cases, even calls that are not using the failed links may be reallocated, but no calls will be dropped, according to the protection level. A cell sequencing loss becomes possible, since the new route(s) may be shorter than the past one(s). To solve this potential problem, a VP/VC alignment function is introduced in the backbone node. This function is shared by all VPs switched at the node and allows a hitless path switching on demand. A realization example of such a function is presented in [10].

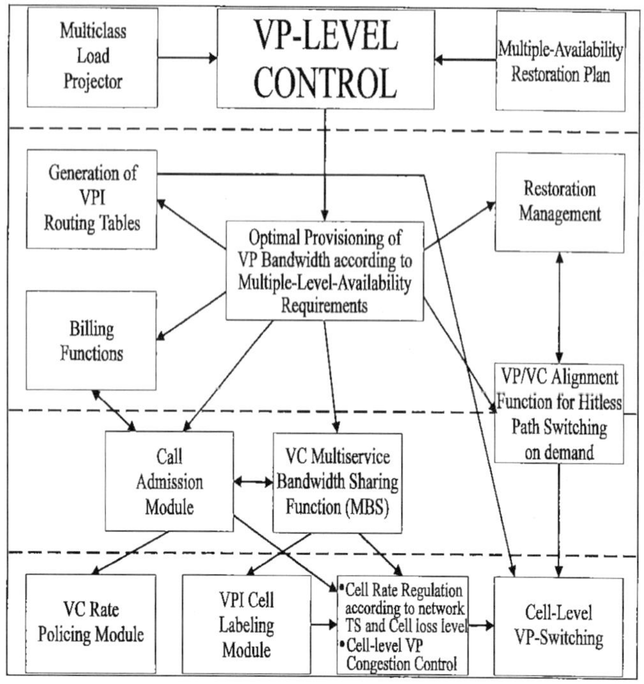

Fig. 2 Simplified traffic control architecture.

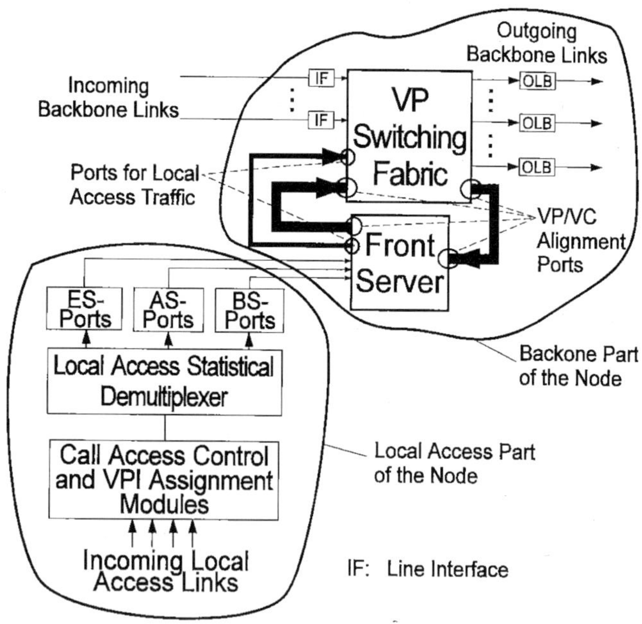

Fig. 3 Proposed architecture of an ATM backbone node.

The proposed backbone node architecture, for its part, consists of following building blocks (see Fig. 3): (a) Call admission, VC-rate policing, VPI cell labeling, TS calls segregation and MBS-VC-bandwidth reservation functions are located in the local access part of the node architecture; (b) Cell-rate-regulation according to MBS, bandwidth adaptation according to call cell loss level tolerance, and the VP/VC alignment functions are put together in a "Front-Server" that may be included in the backbone part of the architecture without major modification of existing ATM node concepts; and (c) the cell-level VP-switching is realized by the VP switching fabric in the backbone part of the node.

Furthermore, the VP set (number of hops (NH), Output Link Buffer (OLB, see Fig. 3) sizes) is engineered to meet the maximum and average VP delay requirements for ES traffic [11], [12]. All OLBs have the same size. The selected buffer size is also used to determine the VP Maximum Permitted Rate (MPR) assignment satisfying the cell loss requirements (see Sect. 3).

Concerning demand projection, the VP-level control collects real-time network measurements and forecasts bandwidth demand and network VC load for an upcoming control period. The length of the control period is to be chosen empirically to permit the adaptation of VP maximum permitted rates (MPR) to the dynamics of bandwidth demand and network VC load. It also represents a tradeoff between processing and measurement costs incurred in obtaining the required information and the resulting savings in bandwidth or improvements in throughput. If bandwidth is inexpensive and demand is relatively stable, the control period can be made long, for instance it can be set hourly.

The optimal VP network dimensioning presented in Sect. 3 requires knowledge of the multiclass (availability classes) Source-Destination (SD) bandwidth demand for a given control period. The projection of this bandwidth demand requires real-time measurements and may be realized by several prediction techniques such as: (a) queuing models [13], [14]; (b) regressor and Normalized Least Mean Square (NLMS) parameter estimation [15]; and even (c) neuronal networks [16], [17].

3. Optimal Virtual Path Bandwidth Allocation and Restoration

Since multiple point-to-point connection demand represents only a fraction of the total traffic—what is a realistic assumption of the situation in backbone networks—the bandwidth efficiency of multipoint connections becomes less important [18]. Therefore one focuses on point-to-point traffic only, assuming that multipoint demand within the interoffice network is allocated using multiple point-to-point connection strategy.

3.1 Joint Optimization Problem

For simplicity of presentation, it is assumed that at the beginning of the control period, the network is intact. Let us now introduce some notations:

$K = \{k \in K\}$: is the set of all possible failure scenarios considered.
$k = 0$: denotes a normal network operation without failure.
K_i, $(i = 0, 1, 2)$: subsets of K. The index $i = 0$ designates the subset containing the sole non-failure scenario; the index $i = 1$ designates the subset containing all one-link failure scenarios; and the index $i = 2$ designates the subset containing all two-link failure scenarios.
Availability classes[†] examples:

$$a = 0 \Rightarrow A_0 = K_0;$$
$$a = 1 \Rightarrow A_1 = K_0 \cup K_1;$$
$$a = 2 \Rightarrow A_2 = K_0 \cup K_1 \cup K_2.$$

The set $A_i (i = 0, 1, 2)$ is the restorability space of an availability class i, i.e., a VP of availability class i requires 100% restoration for all failure scenarios of the set A_i.

$$\Gamma_a^k = \begin{cases} 1, & \text{if failure scenario } k \in A_a \\ 0, & \text{otherwise} \end{cases}$$
$$D^k = \{a \mid \Gamma_a^k = 1\}.$$

L^k: a set of non failed links under failure scenario k.
C_l: capacity of link l.
p: cell loss requirement for a VP. To simplify the protocol on VP-level, we impose—from the VP-level point of view—the same cell loss requirement for all classes of traffic and set it to the strictest loss requirement among them[††]. In ATM Networks, the cell loss requirement is required to be small. This sets a limit to the sum of the maximum permitted rates (MPR) of all VPs passing through a link that is tighter than the limit imposed by the link capacity (see also Eq. (5)).
$S^k = \{(i, j)\}$: a set of SD (source-destination) pairs of non-failed nodes (i and j) that can be connected, under scenario k.
R_{ij}^{ak}: the VP set for the backbone SD pair (i, j) in S^k with availability level a; R_{ij}^{ak} does not contain loops.
R_{ij}^{ask}: route for the VP of index s, with $s \in R_{ij}^{ak}$.
Z_{ij}^{as}: the total bandwidth that has been logically reserved for all VCs allocated on the sth VP for $s \in R_{ij}^{a0}$ $(s > 0)$ and has not been released by the beginning of the upcoming control period. We designate the path with index $s = 0$ as a slack route for rejected demand of each SD pair within an availability class (see Fig. 4).

The slack path (route) is just imaginary. It has unlimited capacity to be used for allocation of the rejected traffic and it is so far useful, within the frame of the joint optimization problem, since we don't know a priori how much demand is going to be rejected. The

[†] Note: the higer a, the higher the priority in the network concerning VP rate assignment.

[††] In the numerical examples presented in Sect. 4, we take (arbitrary) $p = 10^{-17}$ (which is really small enough).

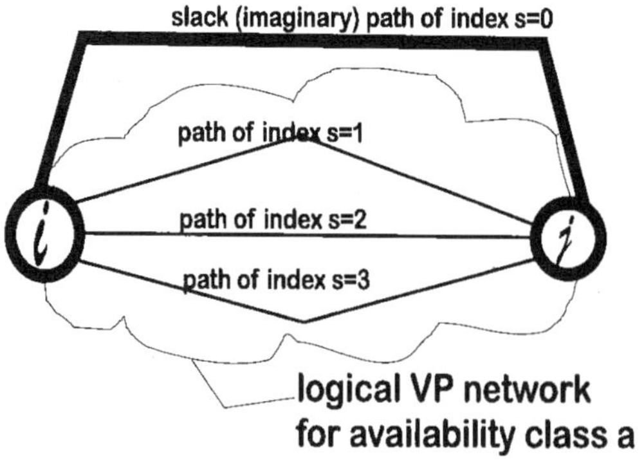

Fig. 4 Symbolic representation of the slack path.

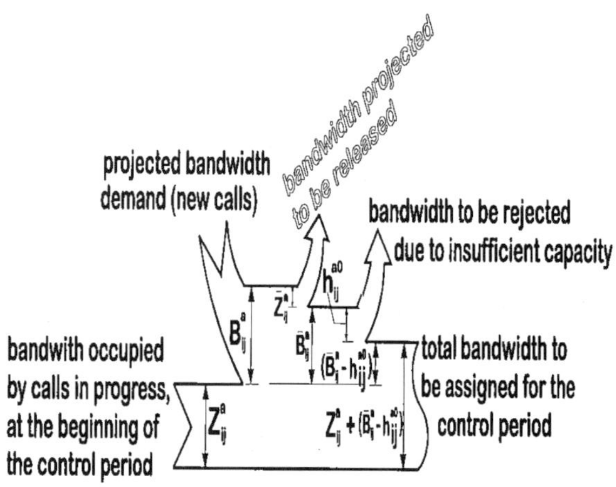

Fig. 5 Bandwidth balance within a control period.

total cell loss along this slack path is the maximum cell loss p.

$Z^a_{ij} = \sum_{s \in R^{a0}_{ij}} Z^{as}_{ij}$: the total bandwidth, for availability class a, that has been logically reserved and has not been released by the beginning of the upcoming control period for SD pair $(i, j) \in S^0$.

$\overline{Z}^a_{ij}$: portion of the total bandwidth Z^a_{ij} projected to be released during the control period and thus can be reused for bandwidth reservation.

B^a_{ij}: total projected bandwidth demand (for the upcoming period) for logical bandwidth reservation for SD pair (i, j) within availability class a.

$\overline{B}^a_{ij}$: a portion of the total projected demand B^a_{ij} that exceeds the total released (for the period) bandwidth $\overline{Z}^a_{ij}$, $\overline{B}^a_{ij} = \max(0, B^a_{ij} - \overline{Z}^a_{ij})$.

h^{a0}_{ij}: portion of demand $\overline{B}^a_{ij}$ that must be rejected due to insufficient capacity on VP set R^{ak}_{ij} for at least one $k \in A_a$.

With these notations, the total bandwidth to be assigned for an upcoming control period for SD pair (i, j) within an availability class a is (see Fig. 5) equals to $Z^a_{ij} + (\overline{B}^a_{ij} - h^{a0}_{ij})$. The first term describes the bandwidth that was logically reserved by the beginning of the control period and thus has to be reserved for the upcoming control period (to protect the calls in progress)[†]. The second term describes the portion of the bandwidth $\overline{B}^a_{ij}$ to be allocated during the upcoming control period (new calls).

h^{ask}_{ij}: MPR provisioned for VP of availability a of index s (between SD-pair (i, j)) under failure scenario k, for allocation of portion of bandwidth $Z^a_{ij} + (\overline{B}^a_{ij} - h^{a0}_{ij})$ for the upcoming control period.

Thus following flow conservation and restorability constraints hold:

$$\sum_{s>0} h^{ask}_{ij} = Z^a_{ij} + (\overline{B}^a_{ij} - h^{a0}_{ij}),\ \forall k \in A_a;$$

$$\overline{B}^a_{ij} \geqq h^{a0}_{ij} \geqq 0;$$

$$h^{ask}_{ij} \geqq 0,\ s \in R^{ak}_{ij}. \tag{1}$$

X^k_l: the sum of all MPRs, to be assigned for the upcoming control period under scenario k, to VPs that pass through link l,

$$X^k_l = \sum_{i,j} \sum_a \sum_{s \in R^{ak}_{ij}} h^{ask}_{ij} \delta^{ask}_{ij}(l), \quad \text{where}$$

$$\delta^{ask}_{ij}(l) = \begin{cases} 1, & \text{if } R^{ask}_{ij} \text{ passes through link } l \in L^k \\ 0, & \text{otherwise.} \end{cases} \tag{2}$$

$q_l(X)$: steady state probability of the cell loss on link l as a function of the VP rate X_l assigned to link l. Contrary to the circuit-switching-based network architecture, in which time slots are assigned to individual circuits, the link equipment transmits all cells on a FIFO basis regardless of their VPIs. Since the physical link is shared by all VPs passing through the link, the sum of the MPRs for all VPs passing through the link cannot exceed its capacity.

As such the VPs are statistical multiplexed on the link. Statistics of cell arrivals for new services depend on information transfer content and are often unknown. This feature and the high speed of ATM networks make it very important to obtain a reliable estimate of cell loss probability q_l in order to guarantee quality of service. A robust and conservative method based on Refs. [11], [12] derives the following upper bound on cell loss probability q_l:

$$q_l(r) \leq [\beta_l(L)(L-2)]^r \frac{1-\rho_l}{1-\frac{\rho_l}{L}}, \tag{3}$$

where r is the OLB size, L is the maximum cell burst size and is selected sufficiently large ($L > 2$; any network design and sizing imposes restrictions on L[††]; there is

[†]For simplicity we consider in the numerical examples $Z^a_{ij} = 0 = \overline{Z}^a_{ij}$.

[††]In the numerical examples we consider (arbitrary) $L = 4$.

no more than one burst per slot), $\rho_l = \frac{X_l}{C_l(1-q_l)} \approx \frac{X_l}{C_l}$, $\beta_l(L) = \frac{\rho_l/L}{1-\rho_l/L}$, $\beta_l(L)(L-2) < 1$, and $u_l = \frac{\rho_l}{L}$ is the probability of arrival of burst on link l in a time slot (which is equal to cell transmission time on physical link).

$F_A = \sum_{i,j}\sum_{a} h_{ij}^{a0}$: a portion of the total projected demand $\sum_{i,j}\sum_{a} \overline{B}_{ij}^{a}$ which cannot be allocated for the period.

Q_{ij}^{ask}: the sum of cell loss probabilities on VP $s \in R_{ij}^{ak}$. Q_{ij}^{ask} represents an upper bound of the cell loss probability on VP s:

$$Q_{ij}^{ask} = \sum_{l} q_l(X_l^k)\delta_{ij}^{ask}(l). \tag{4}$$

With these notations, the capacity constraints and cell loss requirements can be written as follows:

$$X_l^k \le C_l,\ k \in K \tag{5}$$

$$\text{if } h_{ij}^{ask} > 0, \text{ then } Q_{ij}^{ask} \le p,\ k \in K \tag{6}$$

Since: (a) the statistics for some services (e.g. live video) depend on the instantaneous information contents, (b) the high speed network environment requires robust and reliable cell loss probability estimation (used to control the quality of service), (c) statistics for many future services are unknown, we propose to use Q_{ij}^{ask} which is an upper bound on the cell loss probability in constraint (6). Thus Z_{ij}^{as} is feasible.

Let,

$$F_L(H) = \sum_{l \in L^k} \int_0^{X_l^k} q_l(y_l)dy_l, \tag{7}$$

where $H = (\{h_{ij}^{ask}\}, \{h_{ij}^{a0}\})$. H represents the total traffic demand and consists naturally of two parts: the part to be admitted (represented by $\{h_{ij}^{ask}\}$) and a part to be rejected (represented by $\{h_{ij}^{a0}\}$).

Since the goal is, among others, to minimize both the cell loss upon the admitted demand and the rejected demand, an objective function that adequately expresses this feature is needed. We introduce the following objective function for the optimal VP bandwidth assignment and restoration problem:

$$F_0(H) = \sum_{k \in K} F_L(H) + pF_A. \tag{8}$$

The function $F_0(H)$ is a weighted sum of the total rejected demand F_A and monotonously increasing convex function of the total load assigned to each link (considering all failure scenarios). The first and the second term of Eq. (8) represent respectively the cell loss on the admitted traffic and on the traffic to be rejected. Minimization of function $F_0(H)$ results in, and corresponds to the minimization of its two components $(\sum_{k \in K} F_L(H))$ and (pF_A). A bandwidth allocation that minimizes $(\sum_{k \in K} F_L(H))$ minimizes the cell loss on the admitted traffic too. And this optimal bandwidth allocation guarantees the QoS. Further, a minimization of (pF_A) corresponds to a minimization of F_A (the rejected traffic demand), since p is a constant. And a minimization of the rejected traffic demand corresponds by all means to an optimization (maximization) of the call throughput. So it can be sayed that the minimization of function $F_0(H)$ results in optimization of bandwidth allocation and call throughput under all failure scenarios.

Calls in progress allocated on the failed links may be optimally reassigned for all VPs that must survive (according to the availability classes) and not only for the VPs affected by the failure. Thus the optimal VP rate assignment optimization problem can be formulated as follows:

> "Given network topology S_0, demand $\overline{B}_{ij}^{a}$, VP sets R_{ij}^{ak}, link capacities C_l, minimize $F_0(H)$ subject to flow conservation and multiple-restorability constraints (1), capacity constraints (5), and cell loss requirements (6)."

Since q_l is a monotonously increasing function of X_l and $p \ll 1$, the capacity constraint (5) follows from the stronger requirement (6). Thus constraint (5) may be omitted in the formulation.

The following length L_{ij}^{sk} is assigned to each VP $s \in R_{ij}^{ak}$:

$$L_{ij}^{ask} = \frac{\partial F_0}{\partial h_{ij}^{ask}} = \begin{cases} Q_{ij}^{ask}, & \text{if } s > 0 \\ p, & \text{if } s = 0 \end{cases} \tag{9}$$

$$\overline{Q}_{ij}^{ak} = \min_{s>0} Q_{ij}^{ask} \tag{10}$$

With these notations, we can formulate some properties of the optimal solution.

3.2 Equilibrium Problem and Properties of the Optimal Solution

In fact, we have work to do with a case of equilibrium programming. The system consists of many components, the different source-destination (SD) pairs (within all availability classes). Each component has its own sub-problem and is continually resolving it given the most current information on the action of the others. In other words, each component pursues its own objectives given the opportunities and limitations either created by other components or relative to the constrains of the problem (restorability, capacity constraints, cell loss, and flow conservation). Equilibrium programming is really (fortunately) a well documented problem in the relevant literature (see [19] for instance).

Theorem 1: The optimal solution of the VP rate assignment problem has the following properties (provided that connections with higher priority are not disturbed, see Steps 4 and 5 of the algorithm below):

- h_{ij}^{ask} can be allocated on VP if the VP length L_{ij}^{ask} is minimal and satisfies the cell loss requirements

$$L_{ij}^{ask} = \overline{Q}_{ij}^{ak} \leqq p. \tag{11}$$

- No bandwidth demand is rejected under following conditions:

$$h_{ij}^{a0} = 0 \quad \text{if} \quad \max_{k \in A_a} \overline{Q}_{ij}^{ak} < p. \tag{12}$$

- Bandwidth can be rejected ($h_{ij}^{a0} \neq 0$) if there is a failure scenario $k \in A_a$ for which

$$\min_{k \in A_a} \overline{Q}_{ij}^{ak} \geqq p. \tag{13}$$

- Bandwidth is totally rejected ($h_{ij}^{a0} = \overline{B}_{ij}^{a}$) if there is a failure scenario $k \in A_a$ for which the bandwidth allocation will lead to an excessive loss, i.e.,

$$\min_{k \in A_a} \overline{Q}_{ij}^{ak} > p. \tag{14}$$

- h_{ij}^{ask} can be allocated on a VP, although $k \notin A_a$, as long as the VPs with higher priorities are not disturbed, see parameter $V_{ij}^{ak}(m)$ in Eq. (16) below.

Proof
The objective function (8) is a convex, separable function of flow X_l^k. The flow conservation and restorability constraints are linear. The result follows directly from Kuhn-Tucker optimality conditions which, in this case, are necessary and sufficient. See theorem 9.6.2 and 3.C.2 in [19].
The theorem shows (see Eq. (11)) that the optimal VP rate assignment allocates the demand satisfying the cell loss requirements (6). Thus, constraint (5) may be omitted in the optimization algorithm. From the theorem it also follows that the optimal solution equalizes the cell loss on the VP set of each SD pair within an availability class. This feature of the optimal solution assures fairness with respect to cell loss. □

3.3 Optimization Algorithm for VP Rate Assignment

In this subsection, a parallel algorithm for solving the VP rate assignment optimization problem is presented. Parallel processing is used to overcome combinatorial complexity introduced by the necessity to simultaneously consider a large variety of failure scenarios. The algorithm assigns to the VP set between two SD nodes (i, j) within an availability class a, the virtual bandwidth sufficient to accommodate as much of bandwidth demand $\overline{B}_{ij}^{a}$ as feasible. It also calculates the portion of demand for each availability class, to be rejected at the source node.

3.3.1 Algorithm Description

The quintessence of the algorithm presented below results from: (a) an application of the properties of the optimal solution as formulated in the theorem above; (b) a formulation of the prioritization of traffic belonging to high availability classes (at the disadvantage of low availability ones); this is done with the help of some indicator functions $P_{ij}^{a}(m)$ and $V_{ij}^{ak}(m)$ defined in the Step 5 of the algorithm).

The iterative parallel algorithm is composed from subprocesses defined for each scenario $k \in K$. At each iteration each subprocess shifts a small portion of each SD demand, within an availability class, on the VP of the minimum length from all other VPs. Communication between the subprocesses is invoked when one of the subprocesses finds that some of the demand cannot be accommodated due to excessive cell loss or to priority constraints (see Step 4).

The algorithm consists of three phases and is similar to the "Accelerated Flow Deviation Algorithm" described in [20]:

1. The first phase optimizes the objective function and is based only upon the calculation of the total link flows. This phase does not require the calculation of multicommodity flows (i.e., the h_{ij}^{ask}) at each iteration, which is a major computational effort if the number of commodities is large. The execution of the second phase starts only when a δ-neighborhood of the optimal solution is reached. The task of the first phase consists in redistribution of the flows to the minimum length VPs of a class a (Step 2–Step 5);

2. The second phase finds the optimal multicommodity flows h_{ij}^{ask} by means of averages upon the indicator functions $T_{ij}^{a0}(t)$, $T_{ij}^{ask}(t)$, $V_{ij}^{ask}(t)$, and $P_{ij}^{a}(t)$. The task of this phase consists in the calculation of the VP maximum permitted rates (MPR) for paths of the map sets;

3. The third phase consists in computation of the amount (if any) of rejected demand for each SD pair within an availability class (Step 7).

From the convexity of the optimization problem (8), it is easy to see that if one is in the neighborhood of the optimum, one remains there after the averaging. This consideration guarantees the convergence of the algorithm. The algorithm goes to the second phase (δ-neighborhood of the optimal solution is reached) when the change in the minimum length VPs is less than a specified value for a given number of iterations τ. We call it stability requirement (Step 6).

Let $\Delta_{ij}^{a} = \max_{t-\tau<n\leq t} \overline{Q}_{ij}^{a}(n) - \min_{t-\tau<n\leq t} \overline{Q}_{ij}^{a}(n)$. The stability requirement is expressed by the following function:

$$stabf(t) = \max_{ij,\, a} \frac{\triangle^a_{ij}}{\overline{Q}^a_{ij}(t)} \leq \epsilon \tag{15}$$

The stability requirement utilizes the following property of the optimal solution: the cell loss is equalized, and thus no VP of a smaller length can be found.

3.3.2 The Algorithm, Subprocess k (corresponding to failure scenario k):

- **Step 0:** Set iteration counter $m = 0$.
- **Step 1:** Set all $X^k_l(m) = \overline{X}_l$, where $\overline{X}_l$ is the value of X^k_l at the beginning of the control period and accommodate the bandwidth $\overline{B}^a_{ij}$ on the slack VPs.
- **Step 2:** Compute $Q^{ask}_{ij}(m)$ as a function of $X^k_l(m)$ using Eq. (4).
- **Step 3:** Find $\overline{Q}^{ak}_{ij}(m)$ and the minimum length VP for each SD pair using Eqs. (3) and (10) by executing a shortest path algorithm.
Let
$$T^{ask}_{ij}(m) = \begin{cases} 1 & \text{if } L^{ask}_{ij}(m) = \overline{Q}^{ak}_{ij}(m),\ s > 0 \\ 0 & \text{otherwise} \end{cases},$$
with $\sum_{s>0} T^{ask}_{ij}(m) = 1,\ \forall k \in K$.
- **Step 4:** If $\overline{Q}^{ak}_{ij} > p$ do:
    1. if $\Gamma^k_a = 1$, broadcast $\overline{Q}^{ak}_{ij} > p$ to all subprocesses $k \in A_a$
    2. $\forall i_+ j_+ \in S^k$ and $\forall a_+ \in A_+ = \{a_+ \mid \exists (a_+ < a,\ l_+ \in L^k,\ s_+ > 0)$ so that $q_{l_+}(X^k_{l_+}(m)) \geqq \theta$, and $T^{ask}_{ij}(m)\delta^{ask}_{ij}(l_+)\delta^{a_+ s_+ k}_{i_+ j_+}(l_+)T^{a_+ s_+ k}_{i_+ j_+}(m) = 1\}$, set $E^{aa_+ k}_{i_+ j_+}(m) = 1$. Note that $\theta < p$ and is set in relation with the accuracy of Eq. (3) and priority needs.
    3. if $a_+ \in D^k$ and $\exists (E^{aa_+ k}_{i_+ j_+}(m) = 1,\ i_+ j_+ \in S^k)$ broadcast $E^{aa_+ k}_{i_+ j_+}(m) = 1$ to all subprocesses $k \in A_{a_+}$.
- **Step 5:** Update $X^k_l(m)$ by the following formula:

$$\begin{aligned} X^k_l(m+1) &= (1-\gamma)X^k_l(m) \\ &\quad + \gamma \sum_a \sum_{ij} \sum_{s \in R^{ak}_{ij}} \{Z^a_{ij} \\ &\quad + [1 - T^{a0}_{ij}(m)]\overline{B}^a_{ij} P^a_{ij}(m)\} \\ &\quad \cdot T^{ask}_{ij}(m) V^{ak}_{ij}(m) \delta^{ask}_{ij}(l), \end{aligned} \tag{16}$$

where $0 < \gamma < 1$ and:

$$P^a_{ij}(m) = \begin{cases} 0, & \text{if } \Gamma^k_a = 1 \text{ and } \exists a_+ > a \\ & \mid E^{a_+ ak_+}_{i_+ j_+}(m) = 1, \\ & \text{with } k_+ \in A_a, i_+ j_+ \in S^{k_+} \\ 1, & \text{otherwise} \end{cases},$$

$$T^{a0}_{ij}(m) = \begin{cases} 1, & \text{if } \max_{k \in A_a} \overline{Q}^{ak}_{ij}(m) > p, \\ & \text{for } \Gamma^k_a = 1 \\ 1, & \text{if } \min_{s>0} Q^{ask}_{ij} > p, \\ & \text{for } \Gamma^k_a = 0 \\ 0, & \text{otherwise} \end{cases}$$

$$V^{ak}_{ij}(m) = \begin{cases} 0, & \text{if } \Gamma^k_a = 0, \text{ and } \exists \{a_+ \in D^k) \\ & \text{or } (a_+ \notin D^k, \\ & a_+ > a)\} \mid E^{a_+ ak}_{i_+ j_+} = 1, \\ & \text{with } i_+ j_+ \in S^k, \\ & \text{and } k \text{ is the number of} \\ & \text{this subprocess} \\ 1, & \text{otherwise} \end{cases}$$

The values of $T^{a0}_{ij}(m)$ and P^a_{ij} are known from Step 3 of the subprocess or from the broadcast (Step 4) of subprocesses for other k.

- **Step 6:** Perform the stability test. If passed GOTO Step 7, otherwise increment m and GOTO Step 2.
- **Step 7:** Compute h^{ask}_{ij} for $s > 0$ by averaging over n iterations:

$$\begin{aligned} h^{ask}_{ij} &= \frac{1}{n} \sum_{u=1}^{n} T^{ask}_{ij}(m+u) V^{ak}_{ij}(m+u)[Z^a_{ij} \\ &\quad + (1 - T^{a0}_{ij}(m+u))\overline{B}^a_{ij} P^a_{ij}(m)],\ n \geqq 1 \end{aligned} \tag{17}$$

$$h^{a0}_{ij} = (Z^a_{ij} + \overline{B}^a_{ij}) - \sum_{s>0} h^{ask}_{ij},\ k \in A_a \tag{18}$$

The number of iterations used in Step 7 for calculating h^{ask}_{ij} should be sufficiently large so that the values of h^{ask}_{ij} and X^k_l are in agreement, i.e. Eq. (2) holds with the desired accuracy. Similarly to [20] we use a variable step size γ in (16), starting with a large γ and decreasing it when approaching the optimal solution. The VPs with higher availability level have a higher priority in the network. This is expressed through the parameters $V^{ak}_{ij}(m)$[†] and $P^a_{ij}(m)$. See also a similar way of thinking in [21]. Note that all subprocesses are executed independently and in parallel, with the exception of the broadcast in Step 4. The broadcast ties the subprocesses together.

The outcome of the algorithm consists of the optimal VP bandwidth allocation (one for the non failure and one for each failure scenario) by the VP-level control. These assignments are stored at each VP's source node. Since reconfiguration is done on the VP level by changing local VP routing tables, only local VP routing tables (one for the non-failure and one for each failure scenario) are needed to be stored at each node. In addition, each source node needs to store only one VC-to-VP

† In the numerical examples (see Sect. 4) $V^{ak}_{ij}(m)$ was simplified and taken equals to Γ^k_a.

mapping. Thus, the amount of storage required for implementation of the described reconfiguration scheme is not excessive.

4. Numerical Examples

The proposed algorithm was coded in Java and the programs were executed on a Sun Supersparc workstation. Throughout the experiments a sample network was used: a 5 node network with 8 arcs, see Fig. 6.

This sparse and small network was chosen in regards both of the high computation power requirements of the algorithm and of the big data quantity to be managed. But it satisfies the desired goals, since our first objective is to outline the main features of the concept presented. Moreover, a sparse network topology is typically found in long-haul (or long distance) networks. In the heart of the calculations, a repeated application of the shortest path algorithm (Dijikstra's) is involved, and the overall complexity is polynomial. Though the computational effort required is relative heavy. In order to reduce the computational time, a pre-calculated table of the restoration routes is used. We have the following availability classes constellation: (a) class $a = 0$: no restoration guarantee; (b) class $a = 1$: restoration guarantee for all 8 possible one-link failure scenarios; and (c) class $a = 2$: restoration guarantee for all 8 possible one link-failure and all 28 two-link-failure scenarios. For restoration, we use the source-destination rerouting strategy. So the parallel optimization algorithm consists in a total of 37 subprocesses (each corresponding to a failure scenario + the no-failure case) that communicates in a way already described. When not expressly indicated, B_{ij}^{a} was taken uniformly equals 100 BU (Bandwidth Units) between each SD-pair in each availability class. Throughout the experiments we examined the following aspects: (a) convergence of the algorithm; (b) effect of some parameters on the performance (throughputs for different classes); (c) price of availability enhancement, in comparison to the case where nothing is done to assure a certain survivability (class $a = 0$).

4.1 Convergence of the Algorithm

The stability requirement has been defined in Eq. (15) by function stabf (t). A typical progress of this function is presented in Fig. 7. For a value of $\epsilon \approx 1$, a δ-neighborhood of the optimal solution is reached within less than 20 iterations. Recall that a variable step size γ is used in the flow deviation algorithm, starting with a large γ and decreasing it when approaching the optimal solution. For the case presented in Fig. 7, the value of γ should be decreased for the iterations above number 15, and the second phase of the algorithm should be initiated.

4.2 Effect of Some Simulation Parameters

The effect of following parameters is examined: iterations number, traffic demand, and availability classes combination.

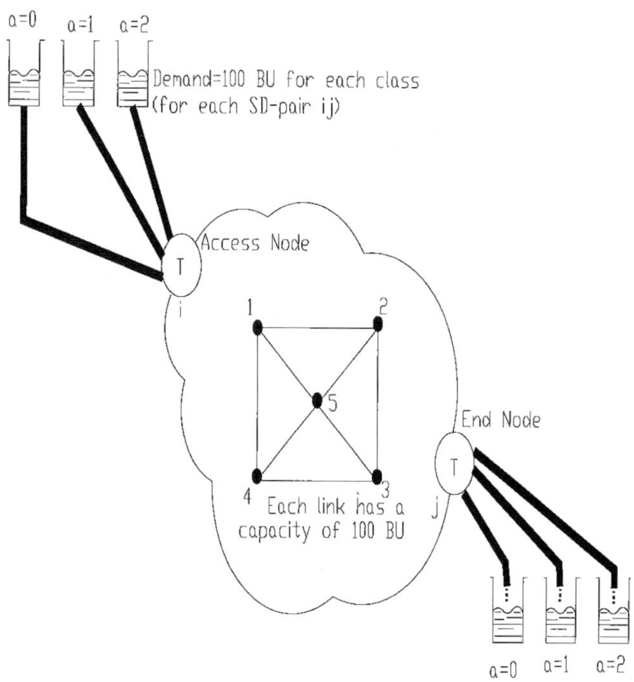

Fig. 6 Network topology considered for the numerical examples presented in this paper. BU: Bandwidth Units; SD: Source-Destination.

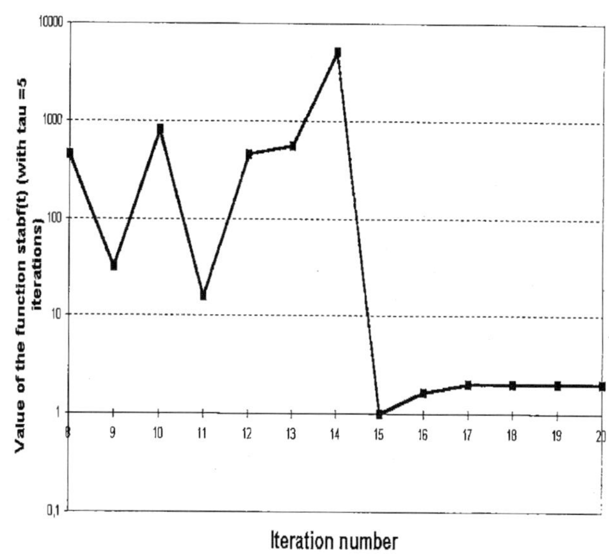

Fig. 7 Convergence illustration of the parallel algorithm by a plot sample of the function stabf (t) defined in Eq. (15), with $\tau = 5$. Classes scenario considered: 0 & 1; For the step size in the accelerated flow deviation algorithm we took $\gamma = 0.04$ in the first phase, and 0.01 in the second phase; and θ wa taken equals 10^{-18}.

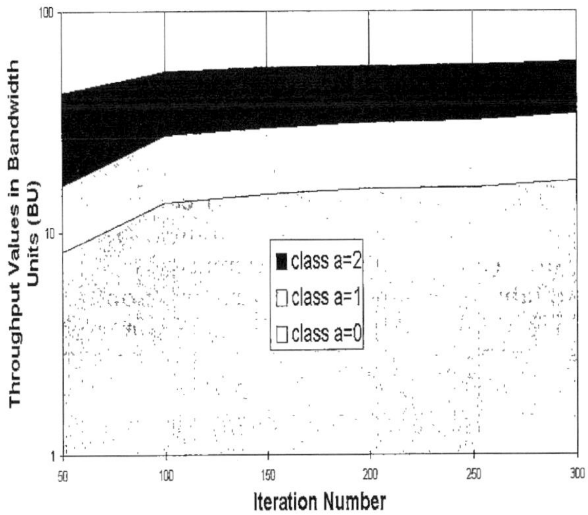

Fig. 8 Effect of Iterations Number on the Throughputs obtained for the different Classes; Availability Classes Scenario: 0, 1 & 2. For the step size in the accelerated flow deviation algorithm we took $\gamma = 0.04$ in the first phase, and 0.01 in the second phase; and θ was taken equals 10^{-18}.

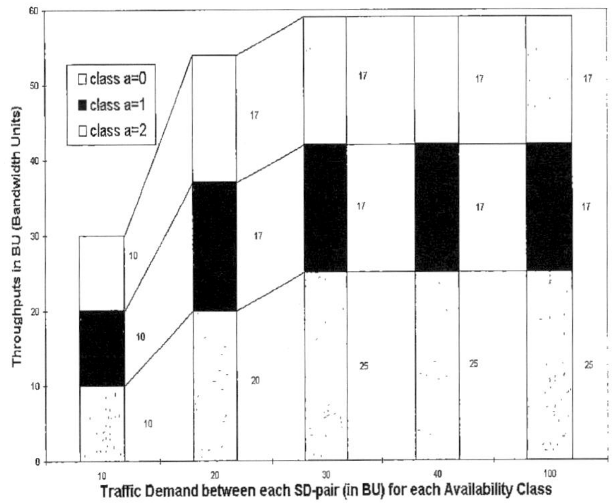

Fig. 9 Effect of Traffic Demand for different Availability Classes on their respective Throughputs.

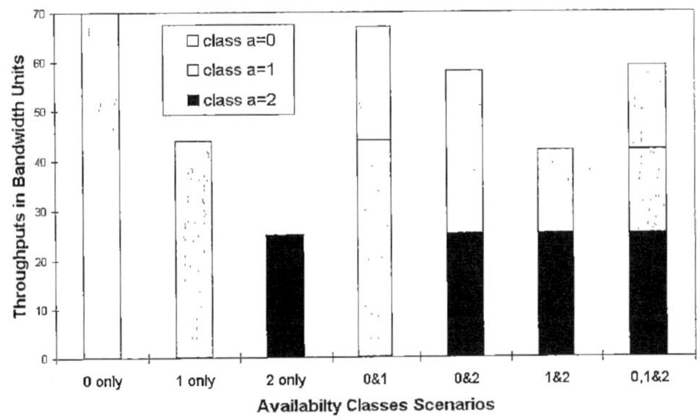

Fig. 10 Demand Throughputs for different Availability Classes Scenarios.

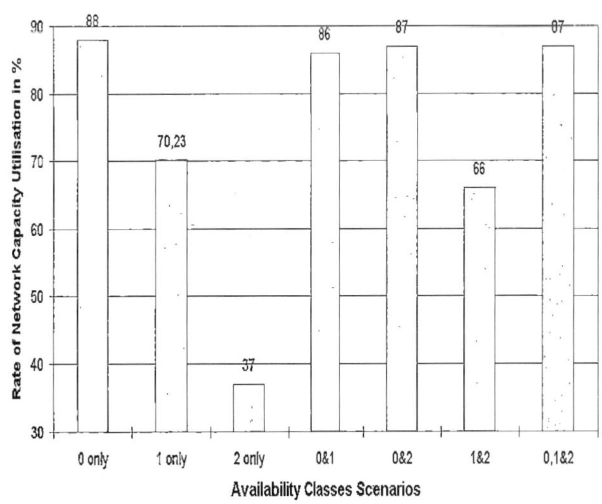

Fig. 11 Rate of network capacity utilization for different availability classes scenarios.

4.2.1 Iterations Number

As shown in Fig. 8, the number of iterations used in Step 7 of the algorithm for calculating h_{ij}^{ask} should be sufficiently large so that the values of h_{ij}^{ask} and X_l^k are in agreement, i.e., Eq. (2) holds with the desired accuracy. We see that a saturation is reached by 100 iterations. So we may say that about 100 iterations are enough to reach a relative good accuracy of the throughputs. The plots show the respective throughputs of the three availability classes that compete for the limited network capacity. The class $a = 2$, which has the highest priority chokes at the beginning the other ones. These last ones are only served after that class $a = 2$ has satisfied its bandwidth requirements.

4.2.2 Effect of Traffic Demand

For simplicity, a uniform traffic demand is considered for all availability classes and it is varied from 10 BU to 100 BU. The evolution of the throughputs is presented in Fig. 9. We see that for low demand, fairness between the different classes is provided. But as soon as network capacity begins to be meager (compared to the demand), the high priority classes begin to choke the low priority ones.

4.2.3 Effect of Availability Classes Scenarios

The availability classes configuration choice has a great influence on obtainable throughputs, as may be seen on Fig. 10. The following general rule prevails: "the higher the availability required, the lower the throughput." In constellations where many availability classes are defined, the lower availability classes just use the capacity left unused from the higher priority classes. And Fig. 11 shows how the different classes share the available network capacity.

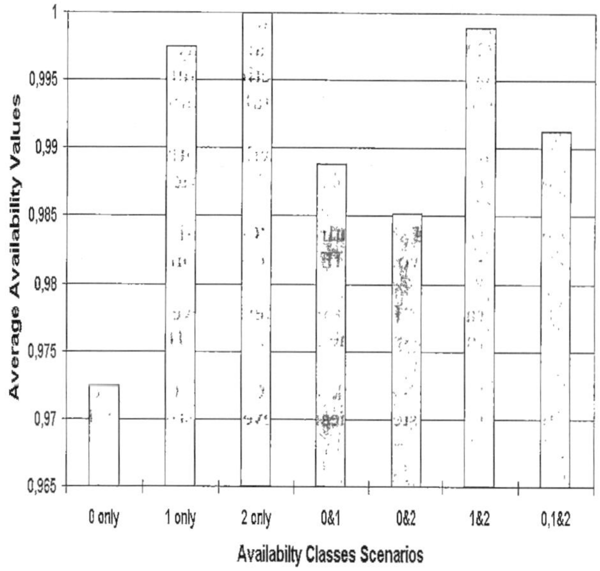

Fig. 12 Average availability for different classes scenarios.

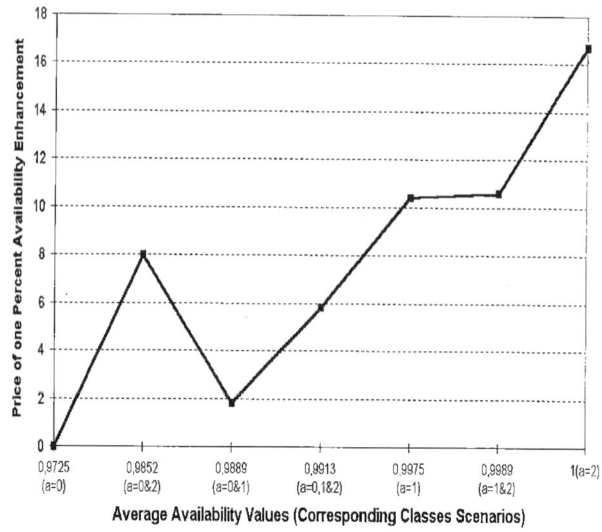

Fig. 13 Price in Bandwidth Units (BU) of one percent (average) availability enhancement.

4.3 Price of Availability Enhancement

As may be seen from Fig. 10, availability enhancement beyond the one provided by no restoration guarantee costs something, since bandwidth must logically be reserved so that a 100% survivability is assured for the failure scenarios of concern for a given availability class. For an illustrative evaluation of this price, we make some assumptions. We suppose, for simplicity, that only two independent failure events may happen: (a) one failure case with a probability pf1 = 0.025; (b) two failure case, with a probability of pf2 = (pf1)/10. Under these conditions, the probability of no-failure case is 1-pf1-pf2. We then obtain the following availability values for the different classes: (a) class $a = 0$: 0.9725; (b) class $a = 1$: 0.9975; (c) class $a = 2$: 1. For a scenario where many availability classes are considered, an average availability may be calculated taking the respective throughputs in account. The results are summarized in Fig. 12.

Now we want to evaluate the price of one percent average availability enhancement, evaluated in BU (Bandwidth Units). This price is given by the ratio between the throughput difference (compared to the case $a = 0$ only) and the average availability difference. If we process the obtained results for different scenarios, we obtain the curve plotted in Fig. 13. Figure 13 teaches us two things. First, we see that the average availability enhancement price is not constant at all. The higher the required availability level is, the higher the bandwidth price for it. Second, we establish that one of the availability classes combinations is not good. The classes combination 0 & 2 ($a = 0 + a = 2$) conducts to an unacceptable (price) - to - (average availability enhancement) ratio as depicted in Fig. 13. Thus this classes combination should be avoided.

5. Conclusion

An optimal virtual path control strategy for ATM-VP survivable networks that maximizes network throughput and provides customized (instantaneous) availability guaranteeing call-level QoS was proposed. Besides, a layered architecture for adaptive traffic control in ATM networks was proposed. The traffic control simplifies call and cell processing. Many important functions, such as call admission and routing decisions, VC bandwidth reservation, and cell-level congestion control, are efficiently performed at the boundary of the backbone network. These functions do not require the source node to communicate with other nodes. The concept presented integrates multiple-availability-requirements, the dynamic of call admission, VP bandwidth allocation and logical spare capacity assignment for maximizing network throughput while satisfying quality-of-service requirements. The joint optimization problem of demand admission and multiple-level-restorability bandwidth allocation was formulated and an efficient parallel algorithm suitable for solving large-scale network problems was presented. This constitutes also a powerful tool to evaluate restoration schemes, through the network throughputs they make possible. This concepts is also a platform that outlines the interaction between restoration methods and congestion avoidance schemes Several numerical experiments have shown not only the effectiveness of the proposed algorithm, but also that offering different levels of network availability saves network resources and allows the network operator, by an appropriate configuration/choice of the availability classes, to provide its customers with services at the appropriate cost. Future work consists in evaluating the effects of network scale, topology variations and different rerouting strategies on the performance of the concept presented.

Aknowledgment

The research work presented was carried out at FTK-Research Institute for Telecommunications, Hagen-Germany, under supervision of Prof. Kaderali. And we thank the DAAD (Deutscher Akademischer AusstauschDienst) for its financial support.

References

[1] R. Kawamura and I. Tokizawa, "Self-healing virtual path architecture in ATM networks," IEEE Communications Magazine, vol.33 no.9, pp.72–79, Sept. 1995.

[2] M. To and P. Neusy, "Unavailability analysis of long-haul networks," IEEE J. Sel. Areas Commun., vol.12, no.1, Jan. 1994.

[3] K. Sato and I. Tokizawa, "Flexible asynchronous transfer mode networks utilizing virtual paths," Proc. ICC'90, April 1990.

[4] T. Kanada, K. Sato, and T. Tsuboi, "An ATM based transport network architecture," Int. Workshop Burst/Packetized Multimedia Commununications, Osaka, Japan, Nov. 1987.

[5] J. Anderson, B.T. Doshi, S. Dravida, and P. Harshavardana, "Fast restoration of ATM networks," IEEE J. Sel. Areas Commun., vol.12, no.1, Jan. 1994.

[6] W.C. Marra and J. Schesser, "Africa ONE: The africa optical network," IEEE Communication Magazine, Feb. 1996.

[7] D.S. Kobayashi and M. Tesfaye, "Availability of bidirectional line switched rings," Rep.T1X1.5/91-070, April 1991.

[8] K. Kyandoghere, "Survivability functions as performance measure tool for rerouting strategies in an ATM/VP mesh transport network," Proc. of the Joint IEEE ICCS'96/ISPACS'96, Singapore, pp.25–29, Nov. 1996.

[9] A. Gersht, A. Shulman, J. Vucetic, and J. Keilson, "Dynamic bandwidth allocation, routing, and access control in ATM networks," in Network Management and Control, vol.2, eds. I.T. Frisch et al., Plenum Press, New York, 1994.

[10] B. Edmaier, J. Eberspaecher, W. Fischer, and A. Klug, "Alignment server for hitless path-switching in ATM networks," XV International Switching Symposium ISS'95, Proceedings, vol.2, pp.403–407, 1995.

[11] X. Gu, K. Sohraby, and D.R. Vaman, "Control and Performance in Packet, Circuit and ATM Networks," Kluwer Academic Publishers, 1995.

[12] H. Saito, "Teletraffic Technologies in ATM Networks," Artech House, 1994.

[13] S. Kheradpir, "PARS: A predictive access control and routing strategy for real-time control of telecommunications networks," Network Management and Control, ed. A. Kershenbaum, 1990.

[14] A. Gersht and S. Kheradpir, "Integrated traffic management in SONET-based multi-service networks," ITC-13, ed. A. Jensen and V.B. Iversen, 1991.

[15] C.R. Johnson, "Lectures on Adaptive Parameter Estimation," Prentice Hall, 1988.

[16] Z. Fan and P. Mars, "ATM Traffic Prediction Using FIR Neural Networks," in the 2nd IFIP Volume on Performance Modelling and Evaluation of ATM Networks, ed. D. Kouvatsos, University of Bradford, 1996.

[17] S.V. Fuzzy, "Logic Neural Network in Communications: Concepts Applications," Tutorial and Workshops, SUPERCOM ICC'94, New Orleans, May 1994.

[18] K.J. Lee, "Multipoint connection routing," International Journal of Analog and Digital Communication Systems, vol.3., pp.177–186, 1990.

[19] C.B. Garcia and W.I. Zangwill, "Pathways to solutions, Fixed Points and Equilibria," Prentice-Hall Inc., Englewood Cliffs, 1981.

[20] A. Gersht and A. Shulman, "Optimal Routing in Circuit Switched Networks," IEEE Trans. Commun., vol.37, Nov. 1989.

[21] M. Peyravian, "Providing different levels of network availability in high-speed networks," Proc. IEEE GLOBECOM, 1994.

[22] S. Hu, "ATM Network Flow Control Using Learning Automaton," Masters Thesis, Stevens Institute of Technology, 1995.

High-Performance Backbone Network Technology edited by Naoaki Yamanaka
ISBN: 0-8247-5321-6 DOI: 10.1081/(120025718)
Published 2004 by Marcel Dekker, Inc.

New Self-Healing Scheme that Realizes Differentiated Bandwidth Requirement on ATM Networks

Taishi Yahara, Ryutaro Kawamura, and Satoru Ohta

SUMMARY This paper proposes a new self-healing scheme that differentiates the bandwidth requirement for each network service on ATM networks. First, we show the necessity of our proposed scheme. In the future network, we must satisfy two demands, rapid restoration from failure and differentiated bandwidth requirements. The conventional restoration scheme, called the self-healing scheme, realizes rapid restoration, but does not support bandwidth differentiation; the new self-healing scheme proposed herein does. We also show that the proposed scheme reduces the spare resources required for backup. The scheme can be realized as a simple extension of the conventional self-healing scheme. Finally, simulations show that the proposed scheme requires fewer spare resources while offering comparable restoration time to the conventional approach against any demand pattern.

key words: *ATM, restoration, self-healing, differentiated bandwidth requirement*

1. Introduction

Future networks must offer extremely high capacity. To this end, backbone networks based on Asynchronous Transfer Mode (ATM) networks are being established to provide sufficient network resources. In these networks, transmission paths tend to be large, so one failure causes enormous data loss. Furthermore, society will become highly dependent on data communication. Mission critical network services include electric commerce and tele-medicine. Even one failure will significantly damage the information society politically and economically, so rapid restoration from failure is required. Self-healing schemes were developed to achieve this goal. They realize rapid restoration by distributing the control mechanism among network elements, and many variants have been proposed [1]-[18].

Other characteristic of future networks is that each network service will have a different bandwidth requirement according to the network service concept. Some high reliability services should be assured of a Peak Cell Rate (PCR) to fully satisfy customer demands even if failure occurs. For these services, sufficient spare resources must be reserved to ensure restoration. On the other hand, the bandwidth of low reliability services can be decreased to a predefined Minimum Cell Rate (MCR) for a short time to decrease cost. In view of user cost and application, it is possible to define the following service classes according to failure state.

(1) PCR guaranteed regardless of failure state
(2) Rate is allowed to fall to MCR only when failure happens
(3) Rate is allowed to fall to MCR regardless of failure state

Manuscript received November 6, 1998.
Manuscript revised July 25, 1999.
† The authors are with NTT Network Innovation Laboratories, Yokosuka-shi, 239-0847 Japan.

In ATM networks, the same network can carry these services. Given reasonable failure patterns, spare resources can be reduced by forcing service classes (2) and (3) to MCR as needed. This realizes cost effective network management with fewer spare resources needed for backup. However, no conventional self-healing scheme provides an adequate framework to achieve these states.

This paper proposes a new self-healing scheme that realizes differentiated bandwidth management. It is based on the flexible bandwidth control permitted by the Virtual Path (VP) layer in ATM networks. This paper also shows that the scheme realizes more cost effective network management than the conventional restoration scheme.

The remainder of the paper is organized as follows. In the next section, we describe the conventional self-healing scheme and its problems in realizing bandwidth differentiation. Section 3 describes how to realize the new self-healing scheme. Section 4 quantitatively evaluates the proposed scheme through simulations. Concluding remarks appear in Sect. 5.

2. Conventional Self-Healing Scheme

2.1 Backup VP Scheme

Existing self-healing schemes are classified as the flooding type [2], [8]-[12] and the backup VP type [3]-[7]. Flooding type restoration looks for the backup route by broadcasting messages after a failure is detected, so this scheme can detect a backup route flexibly regardless of the failure pattern. But, it takes much restoration time. On the other hand, backup restoration preassigns backup routes. This approach simplifies the restoration process. Backup restoration also suits ATM networks (the advantages are described later). We have developed a scheme of this type called the Backup VP scheme for ATM networks [3]-[5]. It is realized in the VP layer in ATM networks. The control and flexibility possible with ATM realizes effective restoration.

In this scheme, Backup VPs are preassigned with zero bandwidth when established. This is feasible because ATM networks allow route establishment to be independent of bandwidth setting [19]. The backup VPs accommodated in that link and are managed by the node connected to the link shares

IEICE TRANS. COMMUN., Vol. E83-B, No. 3, pp. 672–679, March 2000

the spare resources of a link. If a network failure occurs, the appropriate bandwidth is captured by distributed message communication between nodes on the backup VP after the failure is detected by a VP failure detection message, such as VP-AIS[20]. After appropriate bandwidth is captured in all links on the backup VP, the failed working VP is switched to the backup VP.

The Backup VP scheme realizes rapid restoration by pre-assigning backup routes [3]. It also decreases cost by sharing spare resources and switching at termination nodes unlike restoration schemes that utilize dedicated spare resources and switching at switching nodes [4], [21]-[24]. Furthermore, utilizing the Backup VP scheme allows us to estimate required spare resources against foreseeable failure patterns (i.e. 1 link/node failure). This simplifies network resource (bandwidth) management. The maximum bandwidth required to counter all foreseeable failure patterns is established in all links [4].

2.2 Problems of Conventional Self-Healing Scheme and its Solution

In the conventional Backup VP scheme and other self-healing schemes, all network services are assumed to have the same bandwidth requirement, so the spare resources are established to cope with the most stringent requirement of all services carried. Thus conventional schemes require much spare resources.

This paper extends the Backup VP scheme to solve this problem. We define the 3 VP service classes introduced in chapter 1.

- High reliability VP (H-VP): Guaranteed PCR regardless of failure
- Middle reliability VP (M-VP): Decrease the bandwidth to MCR in case of own failure
- Low reliability VP (L-VP): Decrease the bandwidth to MCR in case of own or other failure as necessary.

We extend the conventional self-healing scheme to support and utilize these VP services. The proposed scheme realizes cost-effective networks that require fewer spare resources by following below concepts.

- M-VPs and L-VPs captures MCR under own failure.
- Bandwidth released from L-VPs is utilized for restoration.

3. Proposed Scheme

This section details the proposed scheme. To provide a background to the proposed scheme, we first explain how to establish spare resources and node databases. We then explain how to extend the Backup VP scheme.

3.1 Establishment of Spare Resources

3.1.1 Estimation of Spare Resources

The proposed scheme is based on the Backup VP scheme, so it is able to estimate the spare resources that fulfill all bandwidth requirements of each VP against all foreseeable failure patterns [4]. In the same manner, we can estimate the spare resources for the 3 VP service classes. This means that sufficient spare resources are established in all links when H-VPs capture PCR, and M-VPs and L-VPs capture MCR in all foreseeable failure patterns. Furthermore, the proposed scheme utilizes the bandwidth of L-VPs to decrease the bandwidth needed for restoration. These two technique decreases the spare resources required.

Spare resources required in link l (SR_l) are given by equation (1). If SR_l is below zero, spare resources are not required in this link. That is, all failed VPs are restored by the bandwidth released from L-VPs in foreseeable failure patterns.

$$SR_l = Max\,(WR_{l,m}) - \sum_{vp}^{l} (PCR_{l,vp} - MCR_{l,vp}) \qquad (1)$$

$WR_{l,m}$: *Total bandwidth required in link l to restore all failed VPs when optional link m (≠l) fails*
$PCR_{l,vp}$: *PCR of low reliability VP in link l*
$MCR_{l,vp}$: *MCR of low reliability VP in link l*

3.1.2 Minimization of Spare Resources

Spare resources required in each link depend on the route of each working and Backup VP. To roughly minimize required spare resources, we utilize the procedure in Fig. 1 to decide the route of a working VP (VP_a) and a backup VP (BVP_a). It presumes that VP_a and BVP_a are disjoint except termination nodes.

In ***Seq3***, condition (1) and (2) means that the link has sufficient bandwidth (captured from L-VPs) such that all backup VPs in the link can be restored against all foreseeable failure patterns. If condition (1) and (2) are satisfied, it is useless to increase spare resources. So, such a link will be selected as a candidate of backup VP first. If there are plural candidates that satisfies two conditions or there are no candidates that fulfill the two conditions in ***Seq3***, ***Subseq*** (in Fig. 1(b)) is executed. In ***Subseq***, the link that increases spare resources least is selected, that means the link satisfies condition (3). When estimating spare resources after including VP_a and link i as elements of BVP_a, some links may need more spare resources to guarantee the bandwidth requirements than before including VP_a, while other links may not due to the sharing of spare resources. To reduce total spare resources in whole network, we prefer the link that does not need more spare resources.

One advantage of the proposed scheme is that it does not affect the routes of existing working VPs and backup VPs. To achieve complete optimization, we must change all routes drastically, but this is not realistic because it would increase node complexity and stop services. The proposed scheme does not require route reestablishment, and so yields an approximately optimal solution.

3.2 Link and VP Information

To realize the proposed scheme in a distributed manner, all or some nodes must manage information about the bandwidth utilization in the links, and VP characteristics accommodated in the links. Thus, at least one node connected to each link must obtain the information described above. In this paper, we define such nodes as "Management nodes." A Management node obtains the Link information described in Table 1, and the VP information containing all data of working VPs and backup VPs in the link. This information (see Table 2) is utilized by the proposed scheme.

3.3 Capture MCR Bandwidth in Failure State

When failure occurs, the trigger message for capturing bandwidth is sent. Management nodes on the backup VP capture bandwidth from spare resources upon receiving the trigger message by referring to Bandwidth in VP information. This information is established based on the service requirement of the VP. That is, H-VP captures PCR and L-VP and M-VP capture MCR in failure state (see Table 2).

3.4 Utilization of the Bandwidth Released from L-VPs

A bumping algorithm that releases all the bandwidth of some VPs and utilizes the released bandwidth for restoration has been proposed [25]. As an extension of the Bumping algorithm, we realize a restoration scheme that releases the bandwidth of L-VP to MCR and utilizes released bandwidth for backup. Releasing the bandwidth is executed any time the spare resources are insufficient when failure occurs. Two mechanisms that release the bandwidth are described below.

(1) Release the needed bandwidth exactly.
(2) Force the bandwidth to MCR.

Requests to release bandwidth may come in groups, since mechanism (1) would force each request to initiate bandwidth

Definition:
LE = {Set of links not utilized as a route of a working VP}
CBVP = {Set of links that can be utilized to carry backup VP}

Seq1. Select the shortest path between two termination nodes for the route of VP_a. If BVP_a is not selected (e.g. returned back from ***Seq5***), the VP_a is exchanged to the next shortest one.

Seq2. Initialize two sets of links ***LE*** and ***CBVP***

LE = {all links that are not utilized to carry VP_a }
CBVP = *null* (it can be utilized to carry BVP_a)

Seq3. Select link *m* while satisfying the following two conditions

(1) SR_m is less than zero
(2) SR_m is the smallest among all elements in ***LE***

The selected link *m* is added to ***CBVP*** and deleted from ***LE***. If there are plural candidates (e.g. their SR_m are same and satisfy conditions (1) and (2)) or there are no candidates, (e.g. there is no link that satisfies condition (1) and (2)) execute ***Subseq*** in fig. 1 (b) and select the link *m* that increases the required bandwidth the least among candidates against all foreseeable failure patterns.

Seq4. If it is able to establish the route of BVP_a using only elements in *CBVP*, select this route and finish this procedure.

Seq5. Repeat processes 3 and 4 until the route of BVP_a is selected. If BVP_a is not selected when *LE* = *null*, go back to ***Seq1*** and continue with a new VP_a.

(a)

Definition:
CLS= {Set of links that are candidates of link *m*}

Subseq1. Initialize a set of links, *CLS* . If there are plural candidates that satisfy condition (1) and (2), *CLS* is initialized as the set of links that fulfill two conditions and have same SR_m in ***Seq3***. If not, *CLS* is initialized as the set of all links in *LE*.

Subseq2. Select the optional link *i* from *CLS* that fulfills condition (3)

(3) The result of ($After\_BW_i$ - $Before\_BW_i$) is the smallest among all element in *CLS*

$Before\_BW_i$ represents SR_i assuming all failure patterns without VP_a, and $After\_BW_i$ represents the SR_i assuming all failure patterns with VP_a and link *i* to be an element of BVP_a.

(b)

Fig. 1 Procedure to establish virtual path network.

Table 1 Link information.

Resources utilized for working VPs
Spare resources

Table 2 VP information.

Data	Contents	
VP classification	Working VP / Backup VP	
VP Identifier(VPI)	Specific value for the VP in a link[3]	
Bandwidth	Working VP	PCR and MCR of the VP (MCR is equal to PCR in H-VP and M-VP)
	Backup VP	Bandwidth required when failure happens (H-VP : PCR, M-VP,L-VP : MCR)
VP reliability value	A working VP in L-VP is set to 1, Other one is set to an integral number over 1	
Reserve Condition	Working VP	Null(Initial state) / Reserved*(Reserved for release the bandwidth) / Executing* (Now executed bandwidth release) / Finish*(Finish the bandwidth release)
	Backup VP	Null(Initial state)/Release(Reserve the bandwidth of L-VPs for restoration)
Reserve Infomation	Working VP	Null(Initial state)/VPI of a Backup VP which reserves this VP and reserved bandwidth **(Reserve condition = "Reserved")
	Backup VP	Null(Initial state)/VPI of reserved Low reliability VPs(Reserve condition ="Release")

*, ** : these are for L-VPs

release which increase restoration time [25] and complicate the restoration algorithm. Restoration time is also important factor, so we adopt (2) because it minimizes the increase in restoration time and simplify the restoration algorithm.

Figure 2 describes the algorithm for releasing the bandwidth of L-VPs and restoring failed VPs by utilizing the bandwidth so released. In Fig. 2, restoration starts using BVP_m, but the link "2" connecting node "B" and node "C" lacks sufficient spare resources, so L-VPs (VP_n) are released to provide enough free bandwidth. The termination node located downstream that detects the failure by VP alarm detection (ex. VP-AIS[20]) is defined as the Start node, while that located upstream is the Destination node. Node "A" manages Link "1," Node "B" manages Link "2" and Node "C" manages Link "3" about the VP and Link information in each link.

Phase 1: Bandwidth Capture Phase

The Start node sends a Capture_message to the next node on the backup VP until that message reaches the Destination node. If there are sufficient spare resources for backup, the Management node captures spare resources and sends a Capture_message to the next node. If not, the Management node captures all spare resources first and selects some L-VPs until adequate resources are captured for backup and reserves their bandwidth. The bandwidth is not released in this phase to eliminate useless bandwidth release (e.g. restoration can not completed after some L-VPs were bumped).

In the Capture_message, The information *Rel*, which represents whether the bandwidth of L-VP is reserved or not in all upstream nodes, is included. *Rel* is initialized to 0 at Start node and updated when the bandwidth of L-VPs is reserved. When updating, *Rel* is set to 1.

When Management nodes (node "A," "B" and "C" in this figure) receive a Capture_message, they execute the procedure in Fig. 3. Management nodes select one or more VPs that have VP reliability value 1(e.g. L-VP(s)) that ReserveCondition is "Null" or ReserveCondition is "Reserved" but all bandwidth are not reserved at this time. ReserveInfomation includes information about reserved bandwidth in a L-VP, so it enables us to check whether all upstream links have already been reserved or not. To reserve appropriate bandwidth, one or more L-VPs are reserved. Finally, if the appropriate bandwidth for restoration is successfully reserved on all links on the backup VP, the Capture_message reaches the Destination node.

If a Management node does not capture or reserve the appropriate bandwidth, it sends a Cancel_message back to

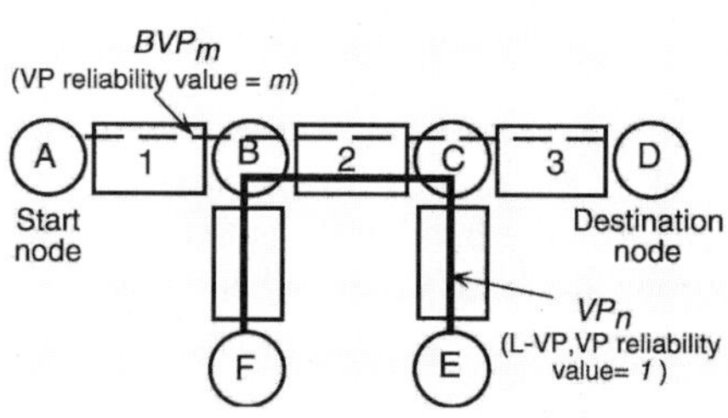

(a)

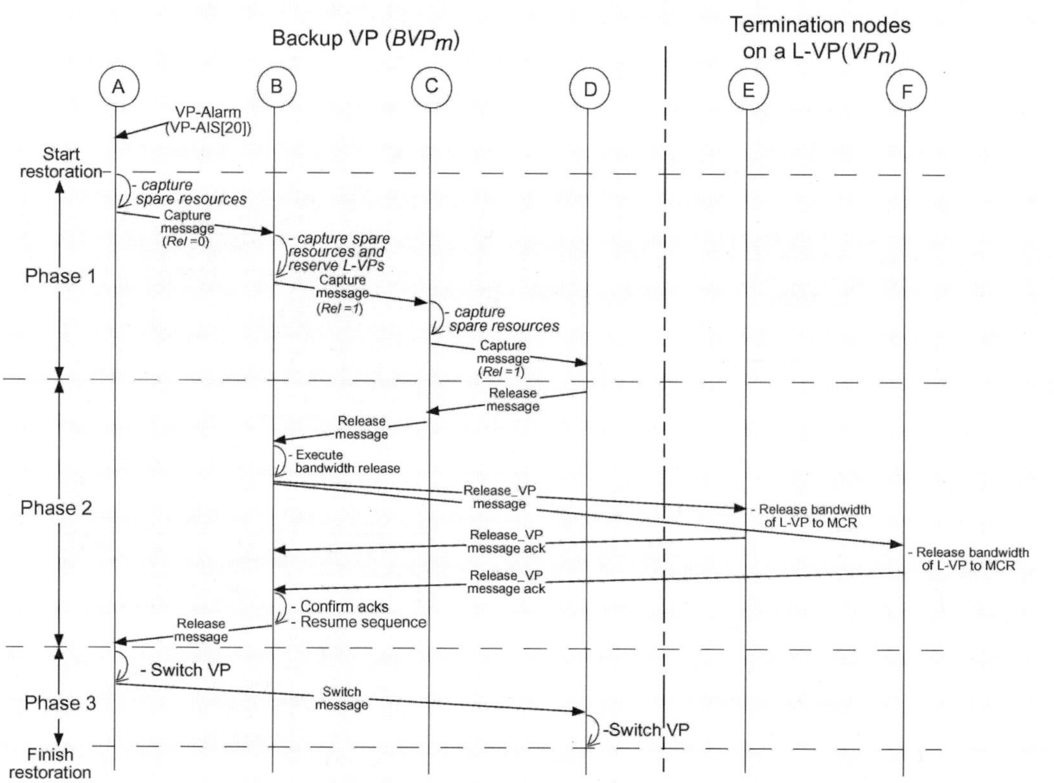

(b)

Fig. 2 Restoration sequence utilizing bandwidth releasing mechanism.

the Start node. Management nodes that receive the Cancel_message initialize the updated Link information and VP information. This procedure assumes unforeseeable failure patterns, such as multiple link/node failure or more.

Phase 2: Bandwidth Release Phase

If it is not necessary to release any L-VPs (e.g. $Rel = 0$), there is no need to release the bandwidth, so the Destination node switches the failed working VP to the backup VP and send a Switch_message to the Start node as in the conventional Backup VP scheme [4]. Upon receiving the Switch_message, the Start node switches the failed working VP to the backup VP that finishes the restoration sequence.

If it is necessary to release bandwidth (e.g. $Rel = 1$, in Fig. 2), it sends a Release_message to the Start node via the backup VP. Upon receiving the Release_message, the Management nodes execute the procedure in Fig. 4.

This procedure is needed to prevent cell overflow from the released L-VPs. If we simply release a VP and reassign its bandwidth in one link, cells will continue to arrive at the upstream node that may degrade the cell loss quality of other working VPs. To realize a safe sequence, we use ack for confirmation.

```
if (there are sufficient spare resources) then
  - Update the Link information and capture appropriate
    bandwidth;
  - Send the Capture_message to the next node;
else if (it is able to restore by utilizing released
         bandwidth from L-VPs) then
     - Update the Link information and capture all spare
       resources;
     - Select L-VPs for bandwidth release and update
       VP information both Backup VP and L-VPs;
         VP information of the Backup VP:
            ReserveCondition  = "Release";
            ReserveInfomation = VPI of L-VPs which this
                                backup VP reserves;
         VP information of reserved L-VPs:
            ReserveCondition  = "Reserved";
            ReserveInfomation = VPI of reserved Backup
                                VP and reserved
                                bandwidth;
     - Update Rel to "1" and send Capture_message to the
       next node;
  else - Send back Cancel_message to the Start node;
```

Fig. 3 Procedure for bandwidth capture phase.

```
if (ReserveCondition of the Backup VP = "Release") then
   - Send Release_VP_message to both termination nodes
     of reserved L-VPs whose condition are
        ReserveCondition = "Reserved"
        ReserveInfo = "VPI of the Backup VP";
   - Update VP information of above L-VPs
        ReserveCondition = "Executing";
   - When confirming acks from a L-VP, update VP
     information of this L-VP:
        ReserveCondition = "Finish";
   - Send Release_message to the next node after
     ReserveCondition of all reserved L-VPs become
     "Finish";
else
   - Send Release_message to the next node
```

Fig. 4 Procedure for bandwidth release phase.

In this procedure, when one L-VP(VP_b) is reserved by multiple backup VPs, the Management node receiving Release_message first switches the ReserveCondition of VP_b to "Executing." When the Management node receives another Release_message requesting bandwidth release of VP_b, it does not execute the sequence for bandwidth release. This is because the first disposal releases the bandwidth until MCR, so all reserved bandwidth are released at one time. This method decreases the total number of message communications to release bandwidth and simplifies the restoration procedure.

After updating the ReserveCondition of all reserved L-VPs to "Finish," Management node sends Release_message to the next node.

Phase 3: Switch Phase

Upon receiving Release_message, the Start node switches the failed working VP to the backup VP using the captured bandwidth and sends a Switch_message to the Destination node. The Destination node then switches the failed VP to its backup VP and restoration finishes.

4. Evaluation of Proposed Scheme

This section quantitatively evaluates the effectiveness of our proposed scheme in terms of spare resources and restoration time by simulation. Three schemes are compared, Backup VP scheme, proposed scheme that establish the route of working and backup VP by the concept of shortest path (proposed1), and proposed scheme that establish routes by the sequence proposed in Sect. 3.1.2 (proposed2). The simulation conditions are described below.

-Network topology:
1. Metropolitan LATA network (15 nodes joined by 28 links (Fig. 5(a)))
2. Japan long haul network (56 nodes joined by 85 links (Fig. 5(b)))

-Foreseeable failure: 1 link failure
-PCR: all VPs have same PCR.
-VP routing:
Backup VP scheme, Proposed1 (using proposed scheme): Working VPs occupy the shortest path. Backup VPs occupy the shortest path that is not common (except termination nodes) with working VP.
Proposed2: Working VPs and Backup VPs are established based on the scheme described in Sect. 3.1.2 with the aim of minimizing spare resources.

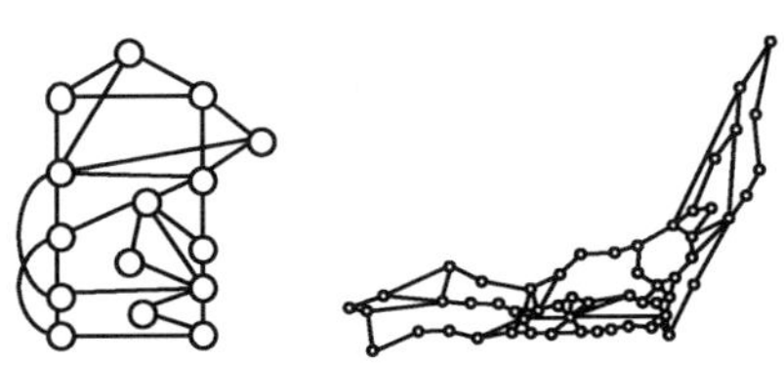

(a) LATA (b) Japan long haul

Fig. 5 Network topology.

-Message processing time: 10ms
-Failure detection delay: transmission delay (hop number from the point where failure occurs to termination node)
-Alarm detection delay: 10ms (failure generation time is set to 0)

4.1 Evaluation of Spare Resources

Figures 6 and 7 show a comparison of total spare resources required versus MCR/PCR ratio. In Figs. 6 (a) and 7 (a), one H-VP, M-VP and L-VP are established between all nodes with the same bandwidth in proposed1 and proposed2. In (b), demand pattern in each node pair differs. The same numbers of 3 class VP (H-VP, M-VP and L-VP) are established in proportion to the total number of links in connecting each node pair. This demand pattern nearly reflects the real demand pattern. In Backup VP scheme, the same numbers of VPs are established compared with proposed1 and proposed2. All VPs must offer PCR. Spare resources are established to satisfy all bandwidth requirements against any foreseeable failure pattern, and the spare resources that satisfy this requirement are set to 1. The value in the graph is relative to the total spare resources of the Backup VP scheme. From both graphs, spare resources are proportionally decreased depending on MCR. Furthermore, In Fig. 6 proposed2 requires 20% less spare resources than proposed1. However, proposed2 does not work well in the Japan long haul network. This is because there is no alternative backup VPs in that network topology, so the results almost the same

Figures 8 and 9 shows evaluations of spare resources required when the demand pattern in each class varies. The number of M-VP and L-VP are proportionally variable in all node pairs one VP in all node pairs carries H-VP. MCR is set to half of PCR. In both networks, the graph shows that the same number of H-VP, regardless of M-VP, are required to achieve a 50% decrease in spare resources. Furthermore, quite a lot less spare resources are needed when there are a lot of L-VPs. Proposed 2 takes into account the releasing of L-VP, so it realizes network with lower spare resource requirements.

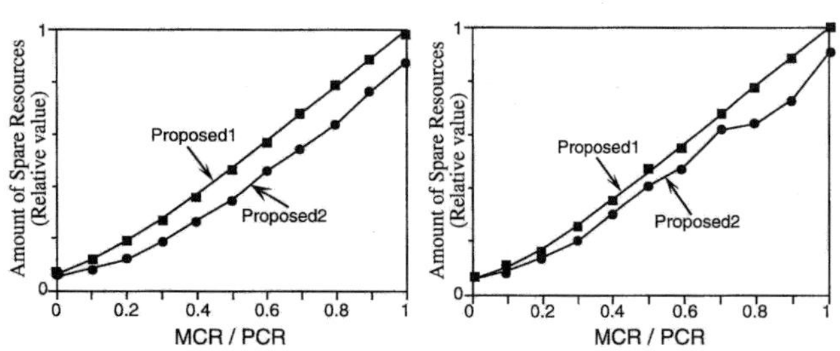

(a) Uniform demand pattern (b) Variable demand pattern

Fig. 6 Evaluation of the spare resources with variable MCR. (Metropolitan LATA network, H-VP : M-VP : L-VP = 1 : 1 : 1)

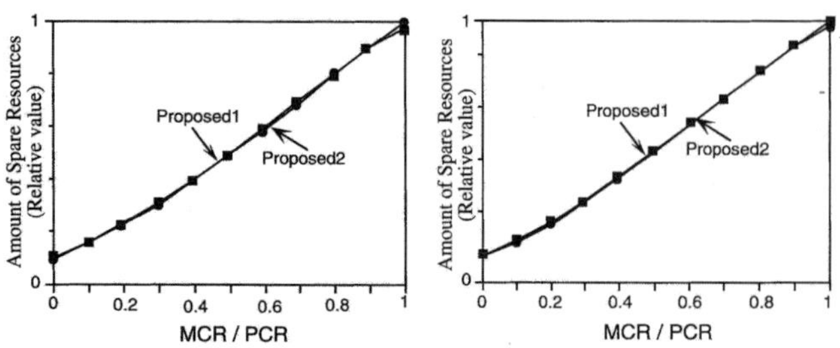

(a) Uniform demand pattern (b) Variable demand pattern

Fig. 7 Evaluation of the spare resources with variable MCR. (Japan long haul network, H-VP : M-VP : L-VP = 1 : 1 : 1)

4.2 Evaluation of Restoration Time

When restoration involves the release of bandwidth, it tends to increase the restoration time [25]. Restoration time is also an important factor, so the increase must be as small as possible. Figure 10 shows evaluations of restoration time for each network topology with variable MCR. Settings are the same

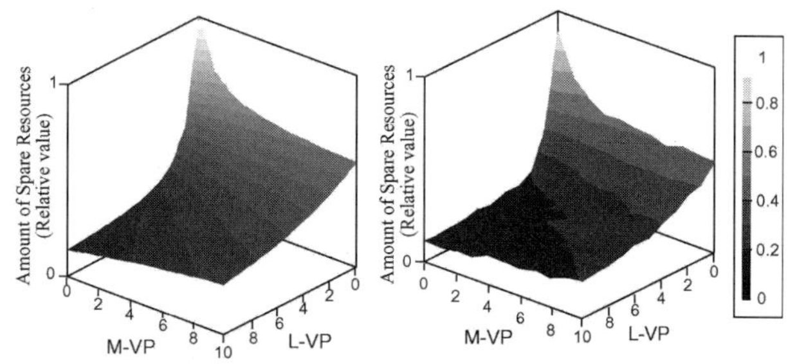

(a) Proposed1 (b)Proposed2

Fig. 8 Evaluation of the spare resources with variable demand pattern in each service class. (Metropolitan LATA network, MCR / PCR = 0.5, the total VP number of H-VP is set to 1)

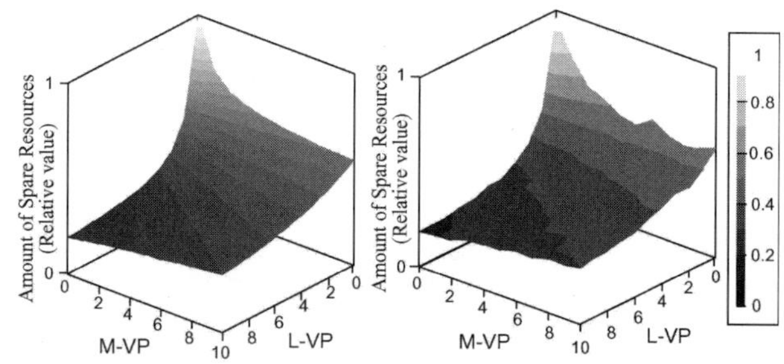

(a) Proposed1 (b)Proposed2

Fig. 9 Evaluation of the spare resources with variable demand pattern in each service class. (Japan long haul network, MCR / PCR = 0.5, the total VP number of H-VP is set to 1)

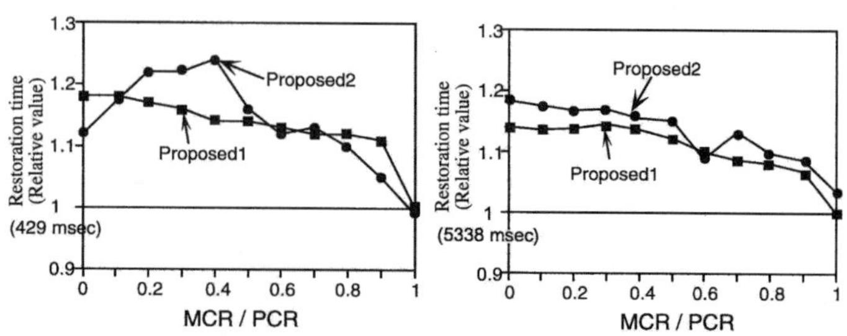

(a) Metropolitan LATA network (b) Japan long haul network

Fig. 10 Evaluation of the restoration time. (Uniform demand pattern, H-VP : M-VP : L-VP = 1 : 1 : 1) as in Figs 6 (a) and 7 (a) The values are relative to the average restoration time when utilizing the Backup VP scheme against all foreseeable failure patterns. Average restoration time using the Backup VP scheme is 429 msec in Fig. 10 (a), and 5,338 msec in Fig. 10(b).

The figures show that proposed schemes require more time compared to the Backup VP scheme when using bandwidth release for restoration, but the increase is relatively small. This is because the sequence for bandwidth release is executed in a distributed manner, so the time spent is dispersed among the nodes involved. This result shows that the proposed scheme decreases the spare resource requirements without increasing the restoration time significantly. The restoration time of proposed2 is sometimes less than that of proposed1 even though proposed2 requires fewer spare resources. This is because message processing does not concentrate at specific nodes. It is a by-product of the routing scheme adopted in proposed2.

5. Conclusion

We have proposed a new self-healing scheme that realizes differentiated bandwidth requirement in each network service on ATM networks. The proposed scheme decreases the bandwidth assigned to MCR for backing up M-VP and L-VPs and utilizes bandwidth released from L-VPs to restore failed VPs. We have also proposed ways to establish spare resources and node databases.

Simulation results show that the proposed scheme requires fewer resources than conventional restoration schemes and that it increases the restoration time by a relatively small degree for foreseeable failure in any demand pattern. These results show that it is now feasible to construct effective networks that require fewer spare resources without degrading restoration time significantly. This means that the proposed scheme realizes networks that fulfill the rapidity and differentiated bandwidth requirement for the restoration required in future networks.

References

[1] T.-H.Wu, Fiber Network Service Survivability, Artech House, 1992.

[2] W.D.Grover, "The selfhealing network," Proc. IEEE Globecom, pp.1090-1095, 1987.

[3] R.Kawamura and I.Tokizawa, "Self-healing virtual path architecture in ATM networks," IEEE Commun. Mag., vol.33, no.9, pp.72-79, Sept. 1995.

[4] R.Kawamura, K.Sato, and I.Tokizawa, "Self-healing ATM networks based on virtual path concept," IEEE J-SAC, vol.12, no.1, pp.120-127, Jan. 1994.

[5] R.Kawamura and S.Ohta, "SELF-HEALING ATM NETWORK ARCHITECTURE; Implementation Results from NTT's Nationwide Testbed ATM Network," Proc. Networks, pp.717-722, 1996.

[6] P.Veitch and I.Hawker, "Administration of restorable virtual path mesh networks," IEEE Commun. Mag., vol.34, no.12, pp.96-101, Dec. 1996.

[7] J.Carlier and D.Nace, "A new distributed restoration algorithm for telecommunication networks," Proc. Networks, pp.377-380, 1994.

[8] H.Komine, T.Chujo, T.Ogura, K.Miyazaki, and T.Soejima, "A distributed restoration algorithm for multiple-link and node failures of transport networks," Proc. IEEE Globecom, pp.459-463, 1990.

[9] K.Wipusitwarakun, H.Tode, and H.Ikeda, "Fast failure restoration algorithm with reduced messages based on flooding mechanism," IEICE Trans. Commun., vol.E80-B, no.4, pp.564-572, April 1997.

[10] H.G.Song, H.Lee, B.Moon, and S.J.Chung, "Dynamic rerouting for ATM virtual path restoration," IEEE Globecom, 1997.

[11] H.Saito, M.M.Slominski, and M.Yoshida, "An improved guided restoration algorithm for ATM crossconnect networks," Proc. NOMS, pp.225-234, 1996.

[12] E.Ayanoglu, "A fast topology update algorithm for restoration under multiple failures in broadband networks," Proc. IEEE ICC, pp.1295-1299, 1993.

[13] J.Anderson, B.T.Doshi, S.Dravida, and P.Harshavardhana, "Fast restoration of ATM networks," IEEE J-SAC, vol.12, no.1, pp.128-138, Jan. 1994.

[14] M.T.Busche, C.M.Lockhart, and C.Olszewski, "Dynamic K-shortest path (DKSP) facility restoration algorithm," Proc. IEEE Globecom, pp.536-542, 1994.

[15] K.Struyve and P.Demeester, "Escalation between recovery schemes: Beyond ATM backup recovery," Proc. IEEE Globecom, pp.755-761, 1996.

[16] K.W.Ko and K.W.Cheung, "A local map based (LMB) self-healing scheme for arbitrary topology networks," IEEE Globecom, 1997.

[17] D.A.Dunn, W.D.Grover, and M.H.MacGregor, "Comparison of k-shortest paths and maximum flow routing for network facility restoration," IEEE J-SAC, vol.12, no.1, pp.88-99, Jan. 1994.

[18] P.Veitch and D.Johnson, "ATM network resilience, " IEEE Network, pp.26-33, Sept./Oct. 1997.

[19] K.Sato and I.Tokizawa, "Flexible asynchronous transfer mode networks utilizing virtual paths," Proc. IEEE ICC, pp.831-838, 1990.

[20] ITU-T Recommendation I.610, B-ISDN operation and maintenance principles and functions, 1992.

[21] S.Ohta and K.Nunokawa, "Comparison of alternative service protection mechanism for wide area ATM networks," Proc. Networks, 1996.

[22] D.Johnson, "Survivability strategies for broadband networks," IEEE Globecom, pp.452-456, 1996.

[23] M.Herzberg and A.Utano, "Optimal assignment of spare capacity to back-up VPs in survivable ATM networks," Proc. Networks, pp.665-670, 1996.

[24] K.Murakami and H.S.Kim, "Comparative study on restoration schemes of survivable ATM networks," Proc. IEEE INFOCOM, 1997.

[25] T.Yahara and R.Kawamura, "Virtual path self-healing scheme based on multi-reliability ATM network concept," Proc. IEEE Globecom, 1997.

High-Performance Backbone Network Technology edited by Naoaki Yamanaka
ISBN: 0-8247-5321-6 DOI: 10.1081/(120025719)
Published 2004 by Marcel Dekker, Inc.

An Efficient Fuzzy Based Traffic Policer for ATM Networks

Mohammad Hossien Yaghmaee, Mostafa Safavi, and Mohammad Bager Menhaj

SUMMARY In Asynchronous Transfer Mode (ATM) networks, congestion can be caused by unpredictable statistical fluctuations of traffic flows and fault conditions within the network. If congestion happens, then the network performance for the already established connection will decrease. ATM networks use the preventive congestion control mechanisms such as Usage Parameter Control (UPC) and Connection Admission Control (CAC) to avoid congested conditions. Knowing that many sources in ATM networks have variable traffic stream with different QoS characteristics, traffic management functions become necessary to control the traffic flows within the network. By using the signaling procedures at the call setup phase, the network and source reach an agreement for some traffic characteristic parameters. If the source violates the traffic parameters, then the probability of congestion increases. So the network must control the source traffic streams and detect well the violating cells. Therefore, fast detection of any violating source is one of the most important characteristics of a good traffic policer. In this paper we propose a fuzzy traffic policer with high ability in detection of violating sources. Our proposed fuzzy controller has two inputs, estimated passed mean cell rate and the current state of the counter. If the output of fuzzy controller is 1, then the input cell is detected as violating cell, otherwise it is a non-violating cell. Simulation results obtained from two traffic sources, show that the proposed traffic policer has better selectivity than the conventional leaky bucket. It is observed that our proposed traffic policer has better ability for mean cell rate control, peak cell rate control and burst duration control. Furthermore, it is observed that the proposed traffic policer outperforms the conventional leaky bucket specially when the dynamic behavior is considered.
key words: *ATM networks, traffic control, fuzzy logic control*

1. Introduction

Asynchronous Transfer Mode (ATM) has been accepted by ITU-T as the new technology for implementation of future B-ISDN services. In ATM networks the user information is conveyed by small size (53 bytes) packets named cells. In ATM networks, many traffic sources with different traffic parameters and Quality of Service (QoS), share the network resources through statistical multiplexing. Therefore a set of traffic management functions will become necessary to offer the customers a desired QoS. The most important ATM traffic management functions include [1]: Resource Management (RM), Connection Admission Control (CAC), Usage/Network Parameter Control (UPC/NPC) and priority control.

The CAC function is invoked at the call setup, and its purpose is to decide whether a requested connection can be accepted or not. The new connection is only accepted if the network has enough resources to provide the QoS requirements of the new connection without affecting the QoS of the existing connections. Two important issues for designing the proper CAC mechanism are: 1) determining the bandwidth required for the new connection, and 2) assuring that QoS of the existing connections are not affected by the new connection. The most important CAC algorithms which have been proposed in literature are: equivalent capacity [2], [3], heavy traffic approximation [4], upper bounds of the cell loss probability [5], [6], fast buffer/bandwidth allocation [7], [8] and time windows [9]–[11].

The ATM network is connection oriented. By using the signaling procedures at the call setup phase, user sends its traffic parameters and also its requested QoS to the network. The traffic parameters and the QoS value agreed by both user and the network form a traffic contract. After the call setup phase, it is necessary for the network to monitor and control the traffic in terms of conformity with the agreed traffic contract. The UPC does this function by detecting the violating sources and taking appropriate actions, such as discarding or tagging any violating cells. So the main purpose of UPC is to protect network resources from the user unintentional misbehavior.

The specification of complex and non-linear relationships can be simplified by using the fuzzy logic methods [15]. A fuzzy logic controller (FLC) can be conceived as a nonlinear controller of which the input output relationship can be expressed by using a set of linguistic rules or relational expressions. During the past years, the fuzzy logic has found many applications in telecommunications networks. Ghosh et al. [16] present a survey of recent advances in fuzzy logic in telecommunication applications. Fuzzy logic has been very successful in the traffic control and congestion control of high speed networks [17]–[21].

In [17] a fuzzy traffic controller has been designed that incorporates call admission control and congestion control simultaneously. The simulation results show that their proposed fuzzy admission controller and also

Manuscript received December 24, 1998.
Manuscript revised June 1, 1999.
[†]The authors are with the Department of Electrical Engineering Amir-kabir University of Technology, Hafez Ave., Tehran, Iran.

IEICE TRANS. COMMUN., Vol. E83-B, No. 1, pp. 1–9, January 2000

the fuzzy congestion controller have better performance than the conventional non-fuzzy mechanisms. In [18], by using the fuzzy expert systems, a fuzzy traffic controller has been designed which has better selectivity and dynamic behavior than the traditional UPC mechanisms. In [19], Ndousse propose a fuzzy leaky bucket which is based on conventional virtual leaky bucket. By using the QoS feedback obtained from the network, the proposed fuzzy controller can reduce the rate of dropping cells but it does not show any performance improvement in the inherence efficiency of the leaky bucket mechanism. In [20], Douligeris and Develekos propose a fuzzy rule-base system which is transparent to the traffic source. In [21], a novel Fuzzy Backward Congestion Notification (FBCN) has been proposed which has high ability for congestion control in ATM networks.

In this paper, we have extended our previous research activities on fuzzy ATM traffic control [22], [23]. we have employed fuzzy logic concepts to design an efficient Fuzzy Traffic Policer (FTP) for ATM networks. As we mentioned above, combining the fuzzy logic into the traffic policing has already proposed in [18]. They proposed a fuzzy model of jumping window mechanism for only mean cell rate control of traffic sources. Note that traffic policing must be done in real time, but the fuzzy rule base in [18] consists of many rules which consumes a lot of CPU time. Unlike [18], our proposed FTP is based on leaky bucket mechanism, which is one of the most popular UPC algorithms. Our proposed FTP has a high ability in detecting any kind of violation from traffic parameters including mean cell rate, peak cell rate and burst duration. It also has better dynamic behavior than conventional UPC mechanisms. To design the FTP, we attempted to avoid any complexity. The rule base of our proposed FTP contains only 6 rules so it needs a small memory and also it is very suitable for real time traffic policing. The inputs of FTP are 1) the state of bucket and 2) the normalized estimated passed mean cell rate. Note that these two inputs represent the instantaneous behavior and the past statistical information of traffic source, respectively. By using the linguistic rules stored in the fuzzy rule base and according to the values of inputs, the FTP decides to accept or reject the input cells.

The rest of this paper is organized as follows: in Sect. 2 we describe the principles of UPC and then describe two important UPC mechanisms. In Sect. 3, we present a brief introduction to fuzzy logic controllers and then describe the details of proposed FTP mechanism. In Sect. 4 by using the simulation results, we compare the selectivity and dynamic behavior of the proposed FTP with those of the traditional UPC mechanisms. Section 5 concludes the paper.

2. Usage Parameter Control (UPC)

As mentioned already in the introduction, the main function of UPC mechanisms is to ensure, both for user and the network, that the traffic contract is not violated. Any UPC algorithm should have some desirable features including:

- Slow time-response in detection of the violating traffic stream (good dynamic behavior),
- Ability of detecting the violating traffic stream (high selectivity),
- Simplicity of implementation (low cost).

The violation probability of an ideal UPC mechanism is determined as:

$$P_v = \frac{\sigma - 1}{\sigma}, \tag{1}$$

where $\sigma \geq 1$ is the normalized actual mean cell rate of the traffic source. The ideal UPC policer has zero violation probability up to the negotiated parameter and a violation probability corresponding to the percentage of excess cells beyond the negotiation parameter point.

Several UPC mechanisms have been proposed including: Leaky Bucket (LB) [12]–[14], Jumping Window (JW) [14], Triggered Jumping Window (TJW) [14], Exponentially Weighted Moving Average (EWMA) and Moving Window (MW) [14]. In [14] the performance characteristics of these mechanisms is calculated and is compared with each other. The LB and EWMA mechanisms are two popular UPC mechanisms in the ATM networks.

2.1 LB Mechanism

The LB algorithm is based on the concept of pseudo-queue. It consists of a counter which is increased by the arrival of each cell and is decreased periodically by a constant leaky rate. When the counter reaches a given threshold N, then the arriving cell is detected as a violating cell. If m_p and m_0 show the controlled mean cell rate and the nominal mean cell rate of the traffic source, respectively then to have more flexibility in dimensioning the LB, an over-dimensioning factor $C \geq 1$ is defined as: $C = \frac{m_p}{m_0}$ [14]. In the case of mean cell rate control, the leaky rate is equal to $lr = m_p = C \cdot m_0$. It is well known that for actual mean cell rate control of traffic source, the over-dimensioning factor C must be set to 1. By increasing C at a constant N, the ability of LB to detect long-term parameter violations will be decreased. For the well-behavior ON/OFF sources, to achieve a cell loss probability less than P_0 the counter size N of the LB mechanism is calculated as bellow:

$$N = \frac{\ln\left(\frac{\left(\frac{b_0 - lr}{b_0}\right)\left(1 - \frac{\lambda_2}{\lambda_1}\right)}{P_0} + \frac{\lambda_2}{\lambda_1}\right)}{(\lambda_1 - \lambda_2) \cdot cell\_size}, \qquad (2)$$

$$\lambda_1 = \frac{1}{h_0 \cdot (b_0 - lr)}, \lambda_2 = \frac{1}{k_0 \cdot lr},$$

where b_0, h_0, k_0 and $cell\_size$ are the nominal peak cell rate, the burst duration, the silence length and the cell size, respectively. This equation has been obtained by using the results in [13] and doing some mathematical manipulations. It has been assumed that both the burst and silence length of traffic source have exponential distribution.

2.2 EWMA Mechanism

In the EWMA mechanism, the number of arriving cells in the non-overlapping consecutive time windows, is counted. The maximum number of accepted cells in the ith window (N_i), depends on a fixed threshold N and an exponentially weighted sum of the number of accepted cells in the previous windows (X_i). To calculate N_i, the following formula is used [14]:

$$N_i = \frac{N - \gamma \cdot S_{i-1}}{1 - \gamma},$$
$$S_{i-1} = (1 - \gamma) \cdot X_{i-1} + \gamma \cdot S_{i-2}$$
$$0 \leq \gamma < 1 \qquad (3)$$

where γ is a flexibility factor.

3. The Proposed Fuzzy Traffic Policer (FTP)

In this section, we describe our proposed FTP, which is designed by combining the conventional LB with the fuzzy logic. At the beginning of this section, we introduce a brief introduction to fuzzy logic controllers and then describe our proposed FTP.

3.1 Fuzzy Logic Controllers (FLC)

Figure 1 shows a basic configuration of a FLC, which consists of four important parts includung: fuzzification unit, knowledge base, decision making unit and defuzzification unit.

The fuzzification unit, converts the input crisp parameters into suitable linguistic values which are needed in the decision making unit. The knowledge base consists of data base and linguistic control rule base. In the rule base a set of fuzzy control rules are defined in a linguistic way. These linguistic rules characterize the dynamic behavior of the fuzzy system. For example, in the case of two-input-single-output fuzzy system (which we use to design FTP), the fuzzy rules have the following general form:

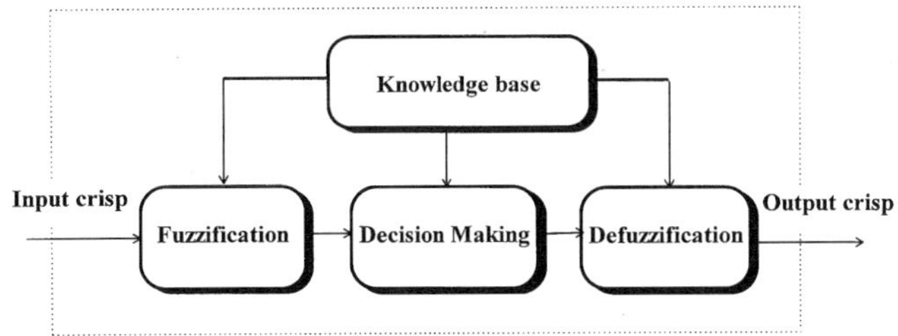

Fig. 1 Block diagram of a fuzzy logic controller.

Rule 1: if x is A_1 and y is B_1 then z is C_1,
Rule 2: if x is A_2 and y is B_2 then z is C_2,
...,
Rule n: if x is A_n and y is B_n then z is C_n.

where x, y are inputs linguistic variables and z is the output linguistic variable. A_i, B_i and C_i are linguistic values of the linguistic variables x, y and z, respectively. In the above rule base, the terms "x is A_i and y is B_i" is the premise of the rule and "z is C_i" is the conclusion. The ith fuzzy control rule is implemented as a fuzzy relation R_i in $U \times V \times W$ and is defined as: $R_i = (A_i \ and \ B_i) \to C_i$. U, V and W are the universes discourse of the linguistic variables x, y and z, respectively and $\to$ denotes a fuzzy implication function [24]. Let $\mu_{A_i}: U \to [0,1]$ and $\mu_{B_i}: V \to [0,1]$ are the membership functions which characterize the fuzzy sets A_i and B_i, respectively. Suppose that an "and" of input variables in a fuzzy relation is implemented by the product $\mu_{A_i \times B_i}(x, y) = \min\{\mu_{A_i}(x), \mu_{B_i}(y)\}$ of the membership functions. For each rule the firing strength is calculated as: $\alpha_i = \min(\mu_{A_i}(x), \mu_{B_i}(y))$. By using the Mamdanis method [25] for fuzzy reasoning, the ith rule leads to the control decision: $\mu_{C_i'}(w) = \min(\alpha_i, \mu_{C_i}(w))$. The membership function $\mu_C(w)$ of the inferred consequence C is given by: $\mu_C(w) = \max(\mu_{C_i'}(w))$, $i = 1, 2, \cdots, n$. After decision making, the final crisp output is generated by using a defuzzification method. Defuzzification is a mapping from a space of fuzzy control actions which are deffined by output universe of discourse into a space of crisp control actions. In our proposed FTP, we use the Center of Gravity (COG) method [25] as a defuzzification technique. In this method the final output crisp z_0 is calculated as bellow:

$$z_0 = \frac{\sum_{j=1}^{m} \mu_z(w_j) \cdot w_j}{\sum_{j=1}^{m} \mu_z(w_j)} \qquad (4)$$

where m is the number of quantization levels of the output.

3.2 FTP Model

Figure 2 shows a simple block diagram of the proposed FTP. As shown in this figure, the FTP consists of three parts which are: the up/down counter, the normalized passed mean cell rate estimator and a fuzzy logic controller.

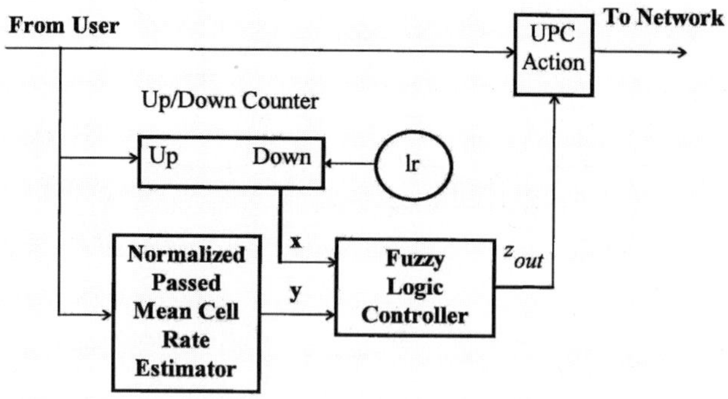

Fig. 2 A simple block diagram of the proposed FTP.

Table 1 Rule base of the proposed FTP.

Rule Number	Description
1	If estimator's output is Low then the source has Very Low Violation.
2	If estimator's output is Medium and the state of counter is Very Low then the source has Very Low Violation.
3	If estimator's output is Medium and the state of counter is Low then the source has Low Violation.
4	If estimator's output is Medium and the state of counter is Medium then the source has Medium Violation.
5	If estimator's output is Medium and the state of counter is High then the source has High Violation.
6	If estimator's output is High then the source has Very High Violation.

The up/down counter is increased by arriving each input cell and is decreased (if positive) periodically by a constant rate lr. In the fuzzy logic controller we have used two linguistic inputs: the current state of counter (x) and the normalized estimated passed mean cell rate (y). The x and y inputs show the short and long term behavior of the traffic source, respectively. The fuzzy logic controller has a single binary output, which represents the violation or non-violation of the input cell. The inputs are first fuzzified, manipulated with rules in the rule base, and then defuzzified to yield the desired crisp output. The term sets of x and y are defined as bellow:

T(x) = {Very Low (VL), Low (Lo), Medium (Me), High (Hi)},

T(y) = {Low (L), Medium (M), High (H)}.

The term set of the output control action z is also defined as:

T(z) = {Very Low Violation (VLV), Low Violation (LV), Medium Violation (MV), High Violation (HV), Very High Violation (VHV)}.

Table 1 shows the rule base of proposed FTP. These linguistic rules were obtained from human experts and also from the previous knowledge about leaky bucket mechanism and voice traffic source.

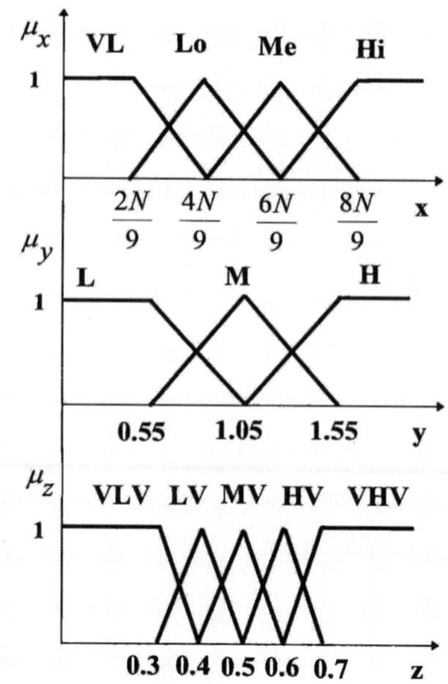

Fig. 3 Membership functions of the x, y, z variables.

To obtain the optimum linguistic rules and also the best parameter values for membership functions, many independent simulation trials were performed and for each simulation trial the performance of FTP was measured. After doing many simulation trials, the optimum fuzzy rules and also the best membership functions were obtained. Figure 3 shows the selected membership functions of the inputs x, y and the output z. The number of terms in a term set is selected as a compromise between the complexity of the fuzzy logic controller and the controlled performance.

Simulation results, show that the proposed FTP with the rule base demonstrated in Table 1 and membership functions shown in Fig. 3, has better performance than the traditional UPC mechanisms. We use the Mamdanis inference method in the inference engine.

We also use the center of gravity method to implement defuzzification strategy. After the defuzzification operation, the final binary output is calculated as bellow:

$$z_{out} = \begin{cases} 0 & \text{if} \quad z_0 \leq 0.5 \\ 1 & \text{if} \quad z_0 > 0.5 \end{cases}, \tag{5}$$

$$z_0 = \frac{\sum_{j=1}^{m} \mu_z(w_j) \cdot w_j}{\sum_{j=1}^{m} \mu_z(w_j)},$$

where m is the number of output quantization levels. The input cell is detected as violating cell if the binary output z_{out} is 1 otherwise it is considered as a non-violating cell.

3.3 Mean Estimator

To estimate the source mean cell rate, the number of

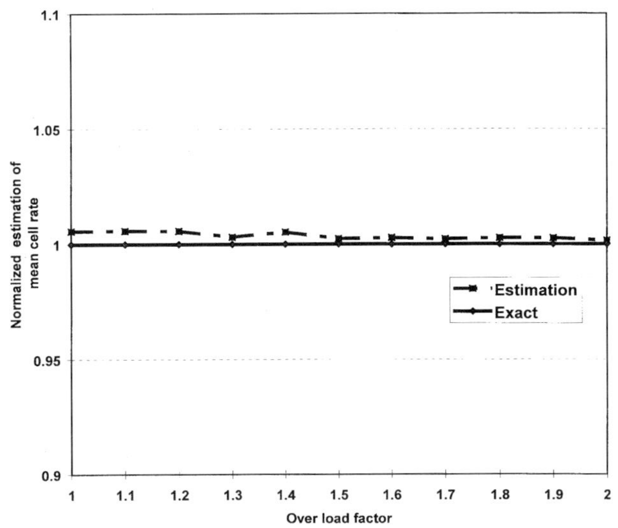

Fig. 4 Accuracy curve of the mean cell rate estimator.

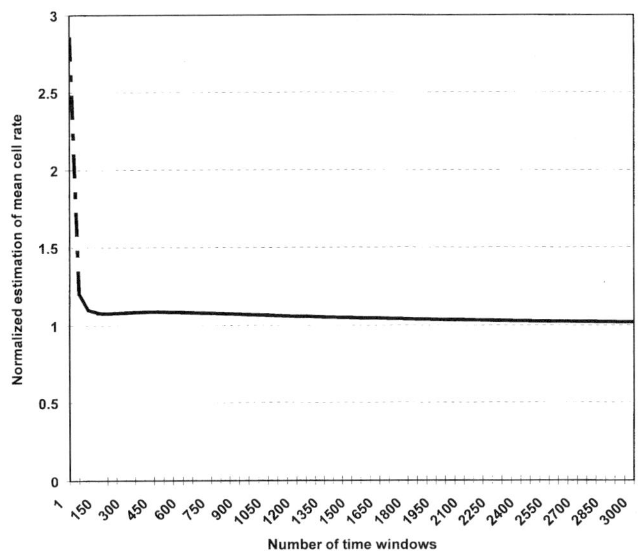

Fig. 5 Time behavior of the mean cell rate estimator.

input cells in a constant time window T, is counted. If N_i represents the number of input cells in ith time window then the estimated mean cell rate $\hat{m}_k$ is calculated by the following recursive formula:

$$\hat{m}_k = \frac{\sum_{i=1}^{k} N_i}{kT} = \frac{(k-1)}{k}\hat{m}_{k-1} + \frac{N_k}{kT}, \tag{6}$$

As the number of time windows goes to infinity, then the second term in the above formula will be zero and we have: $\lim_{k\to\infty} \hat{m}_k = \lim_{k\to\infty} \hat{m}_{k-1} = m_\infty = m_0$, where m_0 is the source long term mean cell rate. The passed mean cell rate (The average cell rate, which is passed through the FTP) is calculated like above formula except that in this case N_i represents the number of non-violating cells in the ith time window. At the end of each time window, the new value of the normalized output cell rate is calculated and it will be constant until the end of next time window. Figure 4 shows the accuracy of the mean cell rate estimator for a packetized voice source [26].

The window size, on average corresponds to two burst/silence patterns.

It is clear that the accuracy of the proposed mean cell rate estimator is very high. The time behavior of mean cell estimator is shown in Fig. 5. At the beginning of estimation due to a few sampling windows, the accuracy of mean estimator is low, but after some time windows the estimator can properly estimate the mean cell rate. So, in our simulation software, we set the first time window enough high to avoid the initial error.

4. Simulation Results

In this section by using simulation results, we compare the performance of the proposed FTP with that of the conventional LB mechanism. We simulated two types of packetized voice sources with traffic parameters shown in Table 2.

Table 2 Traffic parameters of two voice traffic sources.

Traffic Source	Peak Bit Rate	Mean Bit Rate	E(x)	E(s)
Source 1	32000 b/s	11200 b/s	5 cells	0.14772 s
Source 2	32000 b/s	11200 b/s	22 cells	0.65000 s

The number of cells per burst has a geometric distribution with a mean of E[x] and the silence phase has an exponential distribution with a mean of E[s]. The cell inter-arrival time in the burst phase is $\Delta = 16$ ms.

The traffic parameters are based on 32 kb/s ADPCM with a 64 bytes information field. We developed a simulation software in C language at the SunOS environment. For mean cell rate estimator, we assume a window size on average corresponds to two burst/silence pattern. We use the Mamdani's inference method and center of gravity defuzzification technique. The simulation results were obtained by several repetition of experiment and with a 95% confidence interval of the true value but as it is so slight, only the average values are given in the figures. The counter size N, is dimensioned so that the violation probability of a well behavior source is less than 10^{-9}. The selectivity curve of ideal UPC mechanism is obtained by using Eq. (1).

4.1 Peak Cell Rate Control

By decreasing the cell inter-arrival time Δ during the burst phase, the peak cell rate of traffic source will be varied. The mean burst and mean silence times are assumed to be constant. In this case, the leaky rate must be near the nominal peak cell rate and also the counter size N must be equal to 1 [13]. To allow cell delay variation between the customer premises and the

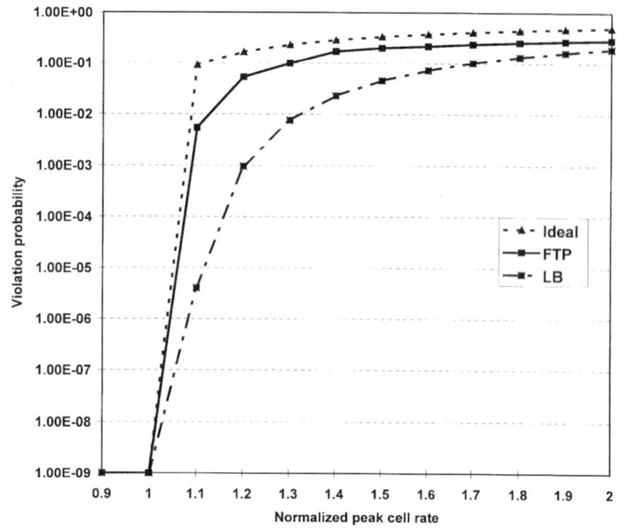

Fig. 6 Performance of the FTP and LB for the peak cell rate control of traffic source 1.

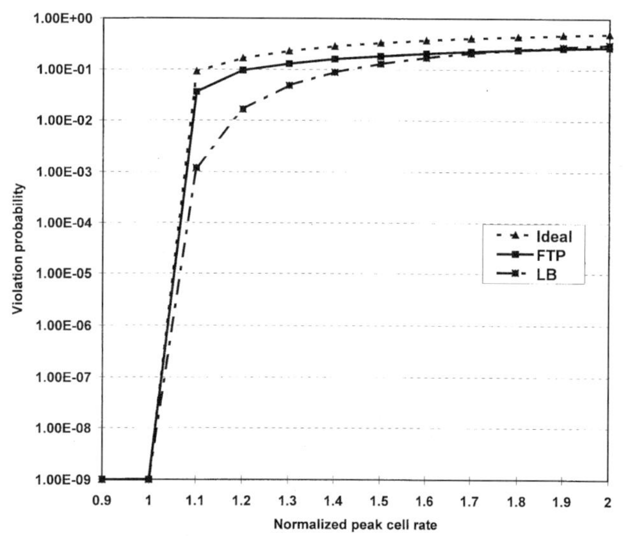

Fig. 7 Performance of the FTP and LB for the peak cell rate control of traffic source 2.

network access point, the values $N > 1$ are selected for counter size. Figures 6 and 7 show the selectivity curves of the proposed FTP and the conventional LB for two traffic source 1 and 2, respectively. For both traffic sources 1 and 2, lr was set to 500 cell/s which is equal to peak cell rate of traffic sources.

For two traffic sources 1 and 2, the counter size N was set to 9 and 18, respectively. Note that as traffic source 1 has smoother traffic than traffic source 2, the value of counter size N for traffic source 1 is less than that of traffic source 2. As shown in these figures, the proposed FTP has zero false-alarm probability for $\sigma \leq 1$ and it has a probability of violation very close to the ideal policer for $\sigma > 1$.

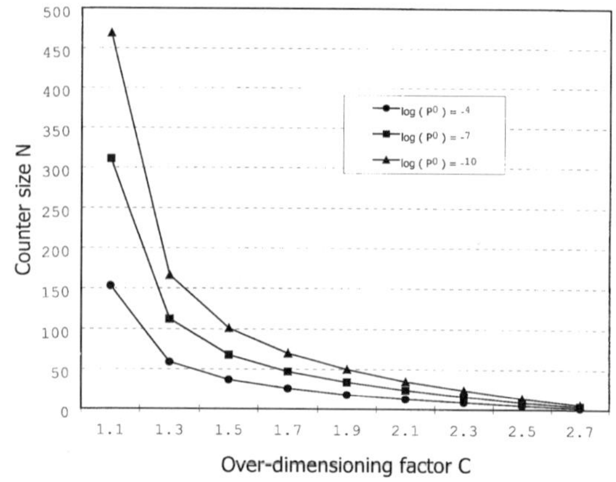

Fig. 8 The required counter size N versus different values of C (for the traffic source 1).

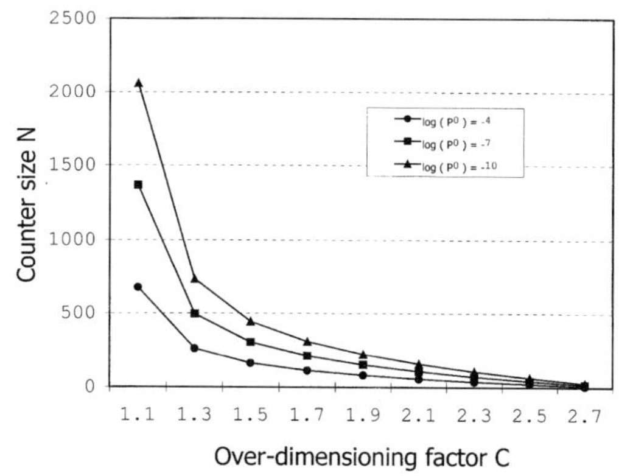

Fig. 9 The required counter size N versus different values of C (for the traffic source 2).

4.2 Mean Cell Rate Control

The mean cell rate of a traffic source with mean burst N_0 and mean silence time k_0 and peak cell rate b_0 is calculated as $m_0 = \frac{h_0}{h_0+k_0}b_0$. According to this formula, by changing the values of h_0 or k_0 at a constant peak cell rate, the source mean cell rate will be varied. To obtain a mean cell rate which is σ times the nominal mean cell rate ($m = \sigma \cdot m_0$), the new mean burst time h' at a constant mean silence time k_0 is given by: $h' = \frac{k_0 \cdot \sigma \cdot h_0}{(1-\sigma)\cdot h_0 + k_0}$. If the mean cell rate varies by changing the mean silence time and keeping the mean burst time constant, then the new mean silence time k' becomes: $k' = \frac{(1-\sigma)\cdot h_0 + k_0}{\sigma}$. For mean cell rate control, the leaky rate must be near the nominal mean cell rate and counter size must be enough high [13]. For both traffic sources, the over-dimensioning factor C was set to $C = 1.15$ so the lr is equal to $lr = 1.15m_0 = 201$ cell/s. In Figs. 8 and 9 by using Eq. (2), at different values of P_0 and for two traffic sources 1 and 2, the requested counter size N has been plotted versus over-dimensioning factor C, respectively.

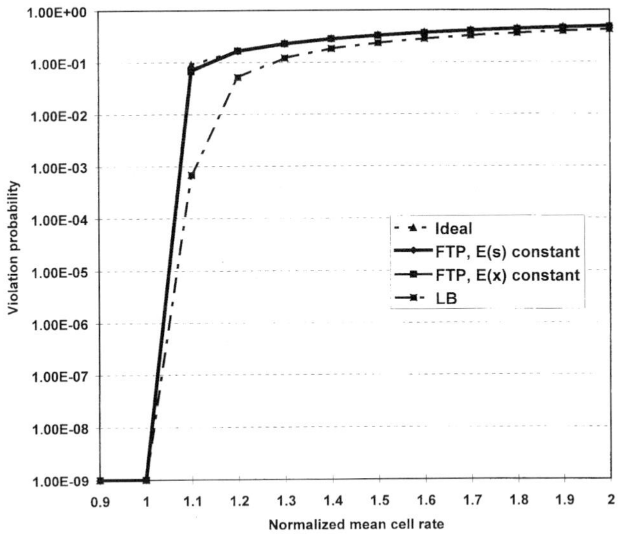

Fig. 10 Performance of the FTP and LB for the mean cell rate control of traffic source 1.

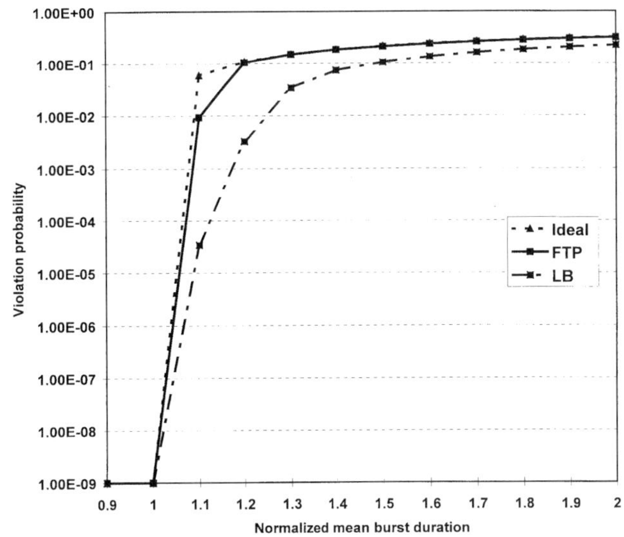

Fig. 12 Performance of the FTP and LB for the burst duration rate control of traffic source 1.

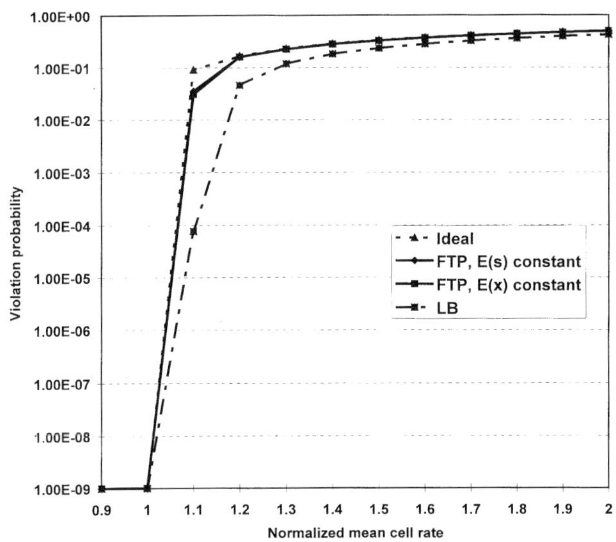

Fig. 11 Performance of the FTP and LB for the mean cell rate control of traffic source 2.

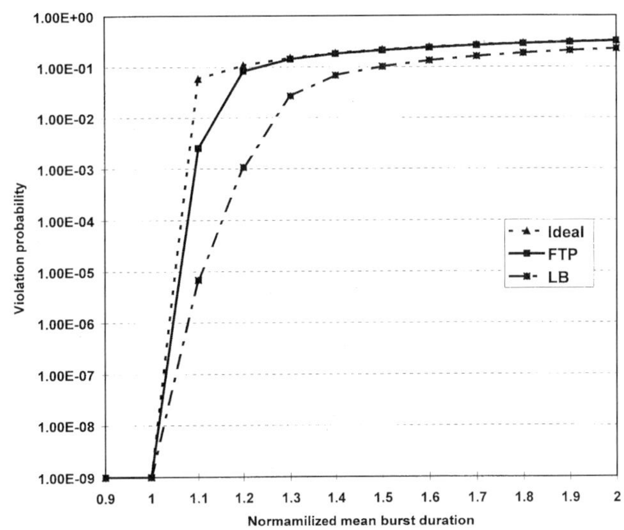

Fig. 13 Performance of the FTP and LB for the burst duration rate control of traffic source 2.

According to these figures, at an over-dimensioning factor $C = 1.15$ to achieve a cell loss probability less than 10^{-9}, for traffic source 1 and 2 the counter size N must be equal to 400 and 2000, respectively.

Figures 10 and 11, for two traffic source 1 and 2, show the selectivity curves of the proposed FTP and the conventional LB, respectively. As shown in these figures the proposed FTP is very efficient regardless of the kind of violation.

4.3 Mean Burst Time Control

For a traffic source with mean burst time h' which is j times the nominal value ($h' = j \cdot h_0$), the mean cell rate m at a constant values of k_0, b_0, is calculated as: $m = \frac{j \cdot (h_0 + k_0)}{j \cdot h_0 + k_0} m_0$. To control the mean burst time, the leaky rate must be near the nominal mean cell rate and the counter size must be high [13].

The selectivity curves for mean burst control of two traffic sources 1 and 2 are shown in Figs. 12 and 13, respectively.

The values of C, and N are the same values used for mean cell rate control. These figures show that the proposed FTP is more efficient than the conventional LB and it is very close to the ideal policer.

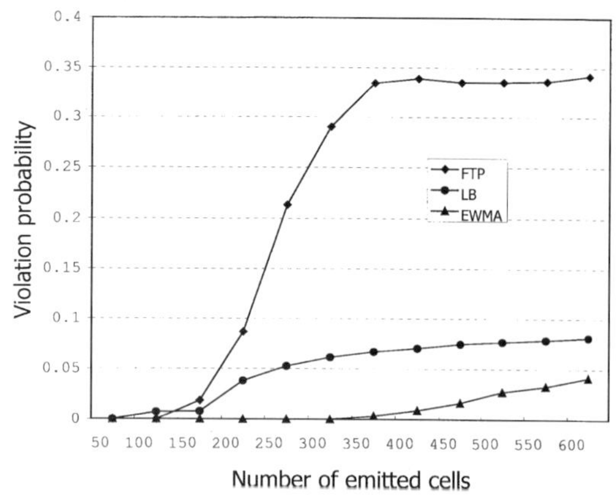

Fig. 14 The dynamic behavior of FTP and LB (for the traffic source 1).

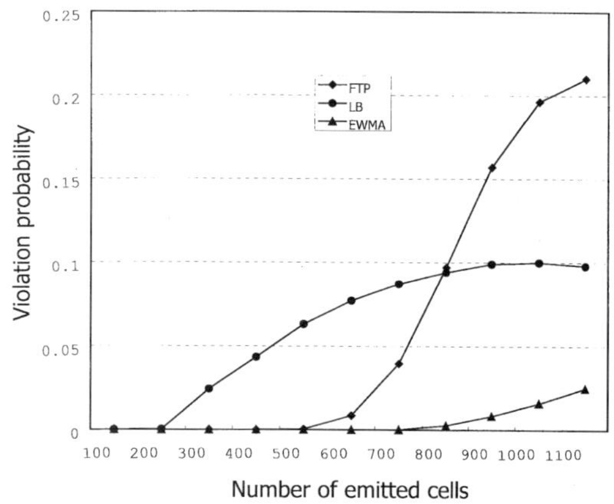

Fig. 15 The dynamic behavior of FTP and LB (for the traffic source 2).

4.4 Dynamic Behavior

An ideal traffic policer should detect any violation as quickly as possible and take appropriate actions to protect the network from congestion. The dynamic behavior of the proposed FTP, for two traffic source 1 and 2 are shown in Figs. 14 and 15, respectively and compared with those of the conventional LB and EWMA mechanisms. All mechanisms have been dimensioned for the same violation probability. The traffic source generates 50% overload by increasing the mean burst time. For all UPC mechanisms, the over-dimensioning factor C was set to $C = 1.42$. The flexibility factor γ of EWMA is equal to $\gamma = 0.9$. The initial value of the EWMA S_0 was set to zero. For two traffic sources 1 and 2, the counter size N was set to 40 and 100, respectively.

As shown in these figures, although the LB start detecting violation sooner than the proposed FTP, but the violation probability of the proposed FTP grows very fast and the percentage of cells detected as excessive is high in comparison with the conventional LB and EWMA mechanisms.

These figures show that for both traffic sources 1 and 2, the EWMA mechanism has the worst dynamic behavior. This is because for the traffic sources 1 and 2, in the first interval the actual counter limit of EWMA mechanism N_1 starts with a value of 200 and 500, respectively which is greater than fixed threshold N, so it takes some window periods until the overload can be detected.

5. Conclusion

In ATM networks, usage parameter control is necessary to protect network resources from misbehavior of traffic sources which can affect the quality of service of the existing connections. An ideal UPC mechanism must detect any violation from nominal traffic parameters and it must take an appropriate action to violating cells. In this paper we proposed an efficient fuzzy based traffic policer for ATM networks. The performance of our proposed FTP was compared with those of the conventional leaky bucket and EWMA mechanisms. Simulation results obtained from two packetized voice traffic sources show that for well-behavior sources the proposed FTP has zero false-alarm violation probability. The proposed FTP can detect any violation from nominal traffic parameters better than the conventional leaky bucket, especially for mean cell rate control. The fuzzy rule base of FTP contains only a few rules so it is suitable for real time traffic policing in ATM networks. We also examined the dynamic behavior of the proposed FTP and we observed that our fuzzy traffic policer can detect violating sources sooner than the conventional LB and EWMA mechanisms.

References

[1] ITU-T recommendation I.371, "Traffic control and congestion control in B-ISDN."

[2] R. Guerin, H. Ahmadi, and M. Naghshineh, "Quivalent capacity and its application to bandwidth allocation in high speed networks," IEEE J. Sel. Areas. Commun., vol.9, pp.968–981, 1991.

[3] D. Anick, D. Mitra, and M.M. Sondhi, "Stochastic theory of a data handling system with multiple sources," Bell System Technical Journal, vol.61, pp.1871–1894, 1982.

[4] K. Sohraby, "On asymptotic behavior of heterogeneous statistical multiplexer with applications," Proc. INFOCOM'92, pp.839–847, 1992.

[5] H. Saito, "Call admission control in an ATM network using upper bound of cell loss probability," IEEE Trans. Commun., vol.40, pp.1512–1521, 1992.

[6] C. Rasmussen, J.H. Sresen, K.S. Kvols, and S.B. Jacobsen, "Source independent call acceptance procedures in ATM networks," IEEE J. Sel. Areas. Commun., vol.9, pp.351–358, 1991.

[7] J.S. Turner, "A proposed bandwidth management and congestion control scheme for multicast ATM networks," Tech. Rep., Comp. and Commun. Res. Ctr., Washington Univ., 1991.

[8] J.S. Turner, "Bandwidth management in ATM networks using fast buffer reservation," Proc. Australian Broadband

Switching and Services Symp., Melbourne, Australia, July 1992.

[9] S.J. Golestani, "Congestion-free communication in broadband packet networks," IEEE Trans. Commun., vol.39, pp.1802–1812, 1991.

[10] T. Faber and L. Landweber, "Dynamic time windows: Packet admission control with feedback," Proc. SIGCOMM'92, pp.124–135, 1992.

[11] F. Vakil and R.P. Singh, "Shutter: A flow control scheme for ATM networks," Proc. 7th ITC Spec. Sem., Morristown, NJ, Oct. 1990.

[12] J.S. Turner, "New direction in communication (or which way in the information age?)," Proc. Int. Zurich Sem. Digit. Communication, pp.25–32, Zurich, Switzerland, March 1986.

[13] M. Butto and E. Cavallero, "Effectiveness of the leaky bucket policing mechanism in ATM networks," IEEE J. Sel. Areas. Commun., vol.9, no.3, pp.335–342, 1991.

[14] E.P. Rathgeb, "Modeling and performance comparison of policing mechanisms for ATM networks," IEEE J. Sel. Areas. Commun., vol.9, no.3, pp.325–334, 1991.

[15] B. Kosko, ed., Neural Networks and Fuzzy Systems, Prentice-Hall, 1992.

[16] S. Ghosh, Q. Razouqi, H.J. Schumacher, and A. Celmins, "A survey of recent advances in fuzzy logic in telecommunications networks and new challenges," IEEE Trans. Fuzzy Systems, vol.6, no.3, pp.443–447, Aug. 1998.

[17] R.-G. Cheng and C.-J. Chang, "Design of a fuzzy traffic controller for ATM networks," IEEE Trans. Networking, vol.4, pp.460–469, June 1996.

[18] V. Catania, G. Ficili, S. Palazzo, and D. Panno, "A comparative analysis of "fuzzy" versus "conventional" policing mechanism for ATM networks," IEEE/ACM Trans. Networking, vol.4, pp.449–459, 1996.

[19] T.D. Ndousse, "Fuzzy neural control of voice cells in ATM networks," IEEE J. Sel. Areas. Commun., vol.12, no.9, pp.1488–1494, Dec. 1994.

[20] C. Douligeris and G. Develekos, "A fuzzy logic approach to congestion control in ATM networks," Proc. IEEE Int. Conf. Commun., Seattle, WA, vol.3, pp.1969–1973, June 1995.

[21] A. Pitsillides and Y.A. Sekercioglu, "Fuzzy logic control of cell flow in asynchronous transfer mode(ATM)," Australian Telecommunication Network Application Conference, pp.249–254, Clayton, Australia, Dec. 1994.

[22] M.H. Yaghmaee, M.B. Menhaj, and M. Safavi, "A simple model of fuzzy leaky bucket for ATM traffic control," Proc. IASTED International Conference on Intelligent Systems and Control (ISC'98), pp.173–176, Halifax, Nova Scotia, Canada, June 1998.

[23] M.H. Yaghmaee, M. Safavi, and M.B. Menhaj, "A novel fuzzy traffic policer for ATM traffic control," Proc. IEEE International Workshop on Intelligent Signal Processing and Communication Systems (ISPACS98), Melbourne, Australia, Nov. 1998.

[24] C.C. Lee, "Fuzzy logic in control systems: Fuzzy logic controller—Part I," IEEE Trans. Syst., Man. & Cybern., vol.20, no.2, pp.404–418, March/April 1990.

[25] C.C. Lee, "Fuzzy logic in control systems: Fuzzy logic controller, part II," IEEE Trans. Syst., Man. & Cybern., vol.20, no.2, pp.419–432, March/April 1990.

[26] H. Heffes and D. Lucantoni, "A markov modulated characteristization of packetized voice and data traffic and related statistical multiplexer performance," IEEE J. Sel. Areas. Commun., vol.SAC-4, no.6, pp.856–868, Sept. 1986.

High-Performance Backbone Network Technology edited by Naoaki Yamanaka
ISBN: 0-8247-5321-6 DOI: 10.1081/(120025720)
Published 2004 by Marcel Dekker, Inc.

A Programmable ATM Multicast Service with Congestion Control

Hung Keng Pung and Naftali Bajrach

SUMMARY This paper presents a design and implementation of a ATM multicast service based on programmable and active network concepts. It aims to address the design and implementation issues of creating new network services—multicast in this case—through a set of corba-based network interfaces, and with a java based user codes injection mechanism for supporting customization of network services. We demonstrate the feasibility of our prototype through the implementation of a wavelet video multicast application with active filters implanted at intermediate nodes for supporting heterogeneous receivers and the implementation of a congestion control scheme. The performance of the prototype over an ATM test-bed is measured and evaluated.

key words: *programmable networks, active networks, ATM multicast, congestion control, protocol implementation*

1. Introduction

Active network provides a new approach and paradigm to networking [1]. It departs from the traditional networking by making the network programmable and allowing computations to occur within the network. There are at least two possible approaches to programmable networking—quasi-static and dynamic [2]. The quasi-static approach allows applications and middleware to manipulate low-level network protocols and resources to construct and manage new services through a set of standardized network programming interfaces. This is the approach pursued by the newly formed IEEE Project 1520 [3]. The dynamic approach is to have the executable contents delivered to the network nodes as packets or 'capsules' (integrated approach), or via out-of-band signaling (discrete approach). The dynamic approach is potentially more flexible but requires a secured and robust codes delivery and injection environment.

We began to investigate the application of Active Network (AN) concept in the multicasting problem of ATM [4]–[6], as it is a network service which AN technique could be effectively applied. Moreover, ATM multicast itself is an interesting and challenging research problem, especially in the leaf-initiated join/leave multipoint-to-point and multipoint-to-multipoint connections with QoS support [7]. Our work was therefore not intended to provide a complete framework of active network solutions in ATM. Instead, it focuses on mechanisms for setting up and management of the multicast-tree with QoS support, and mechanisms for dynamic injections of user codes into network nodes for customization of network services. Our approach is a hybrid of the P1520 and the active networking. Our multicast service architecture is similar in principle with that of the P1520, while our mechanism for customization of application specific network services is using the out-of-band discrete approach of active networking. This makes our work unique in comparison with other related works (See Sect. 6).

The remaining paper is organized as follow. To facilitate our discussion, Sect. 2 presents the Extended Multicast Service Architecture (EAMSA) for ATM. Section 3 describes the design and implementation of EAMSA for supporting dynamic code injection. The performance measurements of our new system are presented in Sect. 4. Section 5 describes a customized video multicast application over our ATM network. The application demonstrates how a network can adapt a video stream to meet the presentation requirements of different receivers. A simple but effective congestion control mechanism for video multicast is also presented. Section 6 reviews some related work in ATM multicast, congestion control and media scaling and filtering. Finally, Sect. 7 concludes the paper.

2. Extended Active Multicast Service Architecture (EAMSA) and Operations

EAMSA is functionally divided into two levels. Level one provides the basic services of multicast connection setup and management, which is partly based on our previous work of Active Multicast Service Architecture (AMSA). It uses an opening signaling consisting of a set of primitives. These primitives are invoked through a set of multicast related APIs for requests of joining or leaving a multicast, sending data, receiving data and etc. Level 2 provides means to program the network. These include injection of customized programs into specific nodes of the network, as well as supporting management of on-going sessions such as QoS renegotiation.

EAMSA is modeled as a distribution of producers (senders), consumers (receivers), brokers (for con-

Manuscript received May 28, 1999.
Manuscript revised August 24, 1999.

[†]The authors are with School of Computing, National University of Singapore, 3, Science Drive 2, Singapore 117543.

IEICE TRANS. COMMUN., Vol. E83-B, No. 2, pp. 253–263, February 2000

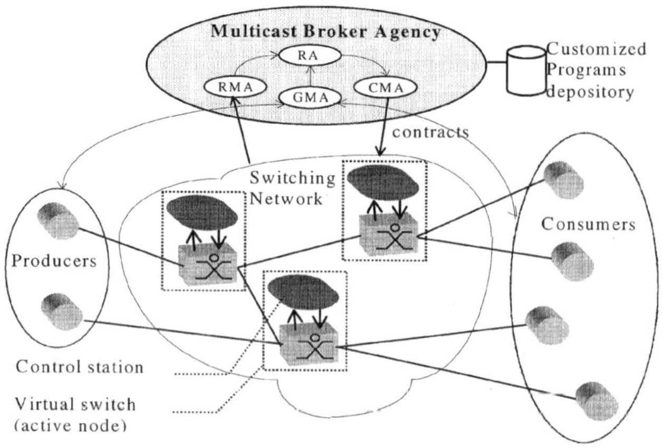

Fig. 1 Active multicast service architecture.

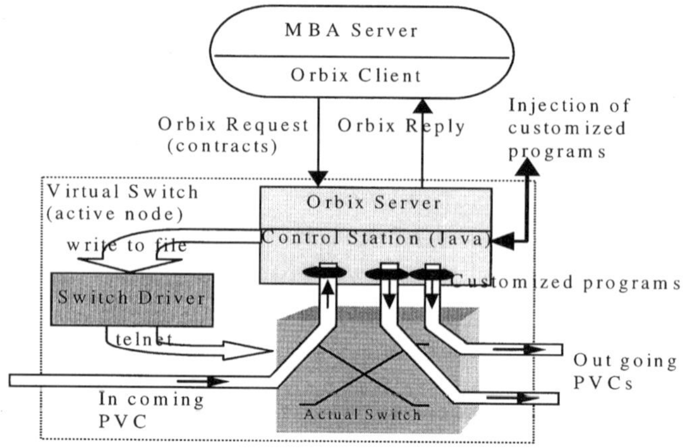

Fig. 2 Virtual switch and its interaction with MBA.

trol, monitoring and other management services), virtual switches and programs depositor as illustrated in Fig. 1. It uses a Multicast Broker Agency (MBA) to perform most of its operations. The MBA consists of four agents: (i) the Resource Monitoring Agent (RMA) which collects link state and topology information of the network, (ii) the Group Management Agent (GMA) which maintains the group membership status; manages the joining or leaving of members and negotiates agreements (including QoS parameters) between the source and the members, (iii) the Routing Agent (RA) computes a multicast tree that connects sources and receivers as well as deciding where to inject a customization program. It does these by using the link state and topology information collected from RMA, as well as the group membership information and the negotiated agreements from GMA. RA may also support different routing protocol options, (iv) the Connection Management Agent (CMA) which setups the actual multicast tree by sending contracts (as part of our multicast signaling) to the designated virtual switches via the links between MBA and the switches. In summary, MBA manages connections only and hence plays no role in the actual transfer of user's data. The functions of the virtual switch will be elaborated on later.

2.1 Multicast Operations

EAMSA allows dynamic leaf initiated joining and leaving of members. The MBA supports point-to-point, point-to-multiple-point and multipoint-to-multipoint connections, as well as QoS negotiations and re-negotiations [7]. A new member joins an existing multicast session through a *join* operation. MBA receives this request and processes it; GMA updates its group membership table. After which, processing involves computing a bi-directional path from the source to the joining host by RA. The path is merged with the existing multicast tree. The merging process starts from the joining node up to the source, allocating sufficient resources along the way. RA actually works out contracts instructing network nodes to set up VCs. A contract specifies the following key information: the 'switched' PVC connection to be set up between two adjacent switches, some link QoS parameters and the user programs to be used, including their locations.

Up till now all computations and changes have been performed logically within MBA; and no connection has been setup physically. When the logical multicast tree is ready, CMA takes over. The necessary new links are then created, and existing links to logically merge with some of the new links (as they are the common links) may have to re-adjust their QoS. All these operations are performed concurrently. The actions taken for a leave operation is the reverse of a join operation. Switches and links no longer needed due to a leaving user will have their resources released and pruned from the tree. For convenience of route computation, the current implementation employs the shortest path route from the source to the joining host.

2.2 Virtual Switch (Active Node)

Most of the ATM switches in the market are not programmable and lacking the functionalities and structure to support customization of network services as required by EAMSA. To overcome this limitation, we introduce a control station (a software residing in a workstation) to every switch. The control-station together with the physical switch form the virtual switch (Fig. 2). Data flow into the physical switch is directed up to the control station then back down to the switch. Hence, changes required would be done in the control station as if the switch were programmable even though its firmware could not be modified (i.e., we use virtual switches to emulate the capability and activities of a programmable ATM switch.).

The virtual switch is implemented as an Orbix server object (Fig. 2) in EAMSA. MBA, which is the Orbix client, would send a request to the control station (implemented in Java) via Orbix links. The virtual switch controls the actual switch (mainly to manipulate the routing table and to initiate set-up/tear down of PVCs) via the ATM network management port (this is one way to access the ATM black-box). The virtual switch can support functions like point-to-multipoint multicasting, multipoint-to-multipoint multicasting, input/output filtering and cut through forwarding [7]. It can also have access to the link state information and set/reset routing table in the switch. The advantages of this scheme are in controlling the switch via standard protocols without tempering with the ATM switch firmware, and the flexibility of implementing any new control functions. This is of course achieved with additional communication and signaling overheads but adequate for our prototyping purposes. However, these extra steps may be eliminated if internal resources and functionalities of switches are accessible through standard interfaces, as proposed in the P1520.

2.3 Programs Injection

The injection of programs into the active nodes is done out-of-band. When an active node is requested (via the contract) to use a customized program, it first checks if the program is available in its local environment, either in the local cache or in the local file system. Otherwise the active node finds and downloads the customized program at a remote location as specified in the contract (we have not implemented the checking of the version number of the program in our prototype). Since program fragments are not encapsulated within the data packets, the amount of data traveling along the network links is greatly reduced. In general, the programs to be injected can reside in a repository, which can be part of the MBA or in a designated computer. This approach also makes it easier for applying security policies to the programming of the network.

2.4 Some Design Issues

The virtual switch model in Fig. 2 is limited by our implementation constraints. Since each ATM switch always maintains an up-to-date link-state and topology information as part of the PNNI requirements, it is in principle possible for MBA to have access to this information through its local ATM switch. This will eliminate the need for collecting states information of links and switches by RMA, and hence effectively remove the potential bottleneck that could be created by these traffic. The potential scalability and reliability problems of MBA being a centralized service can be addressed in two ways. First is to restructure the multicast membership management of MBA in a hierarchical manner. As in the case of PNNI, a large scale network can be divided into peer groups, each managed by a physical MBA. Logical MBAs are formed to facilitate the exchange of multicast membership information as well as multicast routing between peer groups. The basic principle of hierarchical multicast routing is similar to that of PNNI routing—the logical MBAs allow multicast links to be set up between peer groups whose physical MBAs in turn take care of detailed multicast routing within each peer group. Second, one can also take advantage of the strength of Corba to support a distributed implementation of physical MBA of each peer group. This is achieved at the cost of extra communication overheads for better reliability.

```
typedef struct {
int command;                to create/delete , join/leave a session or modify QOS
int unicast;                Specifies unicast or multicast
char group_name[64];
char host_name[64];
// send channel
trafficQoS qos_s;           QoS over ATM connection
int traffic_type_s;         traffic type
int bandwidth_s;            required bandwidth
//receive channel
trafficQoS qos_r;           QoS over ATM connection
int traffic_type_r;         traffic type
int bandwidth_r;            required bandwidth
// active network
CharURL_address[64];        address of the HTTP server from where the customized
                            programs can be downloaded
char functionname[64];      name of the customized program to be used the
                            receiver/user
char option[64];            String specifying the option values in the customized
                            programs resulting in certain QOS over the active service
} Saa_cmd;
```

Fig. 3 The request form.

3. Implementation

This section focuses on the implementation of the active node of EAMSA and the mechanism for injection of user codes. However, it is worth elaborating first the interactions between the end-hosts and the MBA.

The MBA server waits for requests (e.g. join or leave a multicast) of end-users via Orbix (which relies on TCP/IP). A data structure (Fig. 3) is passed as part of the request.

The user can customize the multicast service by specifying in the request its own choices of programs/service and QoS for useover its connection, or rely on the default setups of the initiator of the session. Contracts are then worked out by RA (data structure not shown here due to the limitation of space) and sent to the designated network nodes by CMA. Finally the MBA sends a response (containing a data structure with information of VPI/VCI, QoS to use, etc.) to the user indicating the result of its last request, also via Orbix.

3.1 The Active Node

The virtual switch was implemented in Java as a software switching element with central queuing. The mechanism is based on three threads that are running simultaneously and continuously (Fig. 4).

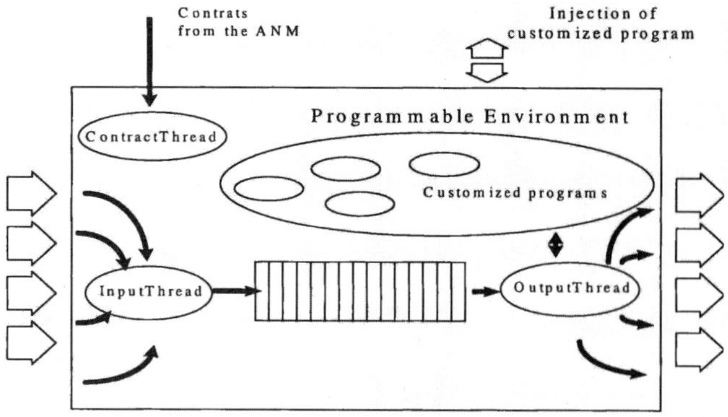

Fig. 4 Threads in active node implementation.

The InputThread reads data from Inlinks. It then writes the data to the central queue. For every selected Outlink of a particular connection, the OutputThread passes the data to the designated customized program for processing (i.e. each leaf may has its own customized program). Finally, it sends the processed data onto the outgoing link(s) (via the native ATM's built-in 'copy' function in the case of duplicating data of the same QoS requirement). The InputThread and the OutputThread are scheduled through an event driven process; they operate asynchronously.

The ContractThread is activated every few seconds to check if new contracts have arrived for processing. It can then create/delete connections or pass information to the customized programs. The ContractThread is scheduled following a time base, and has been made as lightweight as possible to always give priority to the two other threads.

The injected programs can be called by the active node only through two methods: compute() and setOption(). The data can be modified by the designated customized program; it can also be passed on to a shared environment for sharing with processed by other customized programs. The modified data may then be sent on the ATM link by the OutputThread. SetOptions() takes a string as argument that contains specific information for the program like QoS parameters or special options. The information in the string is first assembled at the MBA. The active node passes this string to the program without modification. We also added two other methods: congestion() and restoreOptions()—for a congestion control scheme. They are used to activate/deactivate a mechanism in the customized programs that modify the data in certain ways to reduce the load on the network.

The active node was completely developed in JAVA which ensures interoperability of the mobile code through the Java VM. The Java APIs includes multi-level support for network communications. Low-level: sockets connections for IP; high-level: distributed objects, remote connections. However, no Java APIs exist for support of ATM connections during our development. The only way to integrate native ATM APIs into the active node was to write them in another language (C in this case) and use the Java native-code interface (Java JNI) to link the implementation in Java to the native-code. Java ATM APIs were written for the active node prototype, as "native methods" to support the setup of ATM links, as well as methods to send and receive from those links. It should be mentioned that native-code interface in Java is complicated, and can quickly lead to fragile or difficult-to-maintain codes.

```
method loadClass() {
Look up the requested class in the local cache
IF not found
  Use the method findSystemClass() to try to locate the class using
  the standard java CLASSPATH mechanism that lists directories on
  the local file system
 IF not found
   Try to load the bytes for the class from the specified URL by
   calling the method loadClassData()
 IF successful
     Call the method defineClass() to create an instance of Class
     out of the loaded bytes , then
     install the new class in the local cache
 ELSE
     Throw a ClassNotFoundException exception to notify the active
node
     that the requested class could not be found, it allows the system
     to take some action
IF the class needs to be resolved (i.e. if it mentions other classes
  that are not yet known)
  Call the method resolveClass() to resolve it.
Return the class
}
```

Fig. 5 The network class loader.

3.2 The NetworkClassLoader

The mechanism for the injection of customized programs at runtime is based on the support of Java for dynamic class loading. It enables a Java system to load classes mentioned in a Java program at runtime. A class known as NetworkClassLoader has been implemented in our prototype. The NetworkClassLoader loads classes that are located at URLs on HTTP servers. It defines and implements a stream-based class loader. This is accomplished partly by changing the semantics of the string argument to the loadClass() method from the ClassLoader class. The ClassLoader class defines this argument as the name of the class being requested; the NetworkClassLoader class changes the definition to include class "locators" in general. A class locator may be a URL, a host/port/filename combination, or some other means for addressing a class located on the network, or be anywhere else accessible via an input stream. In our case, a class locator is expected to be in the form of a valid URL.

When loading a class, an instance (an object) of the NetworkClassLoader class goes through the steps as shown in Fig. 5.

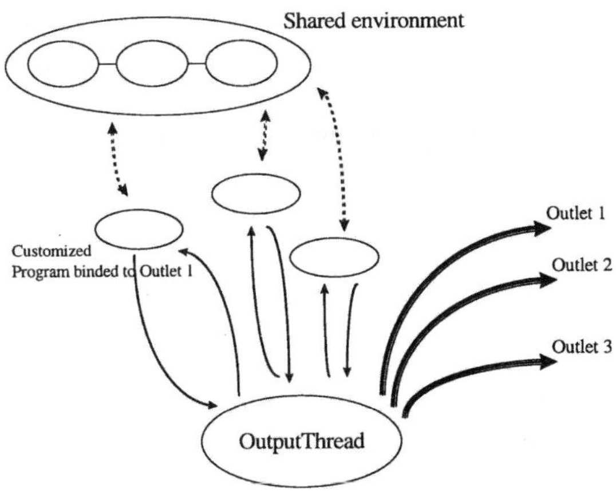

Fig. 6 Active processing.

3.3 Customized Programs

All the customized programs are in Java and are defined by the "Active_Component's" abstract interface. This interface defines different methods that the Output-Thread can call for active processing. The customized programs must implement these methods.

Interface Active_Component{
void setOptions (String option, Shared_Env shared_env_linkedlist);
void compute(Element packet, Shared_Env shared_env_linkedlist);
void Congestion();
void restoreOption();}

The roles of these methods were described earlier in this paper. The shared environment is passed as a whole to each customized program. An active component can refer to other classes on remote locations. The customized programs, once are downloaded, might need to be resolved. This can result in a whole chain of other programs, if necessary, be downloaded to the Active node.

The shared environment (Fig. 6) is more than just a shared memory, to be explained as follow:

The NodeTable class defines a root to the linked list that represents the shared environment for the customized programs. This shared environment is implemented as a linked list of instances of the Shared_Env class. An instance of Shared_Env is added to the linked list for every new Outlink, but the linked list can be accessed as a whole by any customized program.

The "envID" is used to find the right element in the linked list. Every element of the linked list holds an instance "space" of the class Object. The Object class is defined in Java as a superclass of all classes. Java allows the creation of "space" at runtime, an instance of any subclass of Object. It allows "space" to be an instance of any class; most probably provided as part of the service and downloaded with the customized program when it is resolved. This class can define any kind of data structure and methods to operate on them. Hence, the shared environment can be made to be much more complex than just a data repository.

Class NodeTable{
ConnectionEntry conn_linkedlist;
Shared_env shared_env_linkedlist;}

Class Shared_Env{
Int envID;
Object space;
Shared_Env next;}

4. Performance of the AN Prototype

This section presents some measurements of the performance of our prototype. The machines used in measurement setup are:

Workstations

"SPARC1": Sun Ultra-1 with one 167 MHz Ultra SPARC Processor, Solaris Software 2.5.1 Desktop Media

"SPARC2": Sun Ultra-2 with two 300 MHz Ultra SPARC-2 Processor, Solaris Software 2.5.1 Media for Server

ATM switch

Two ATM Switches: FORE System ASX 200 BX, 155 Mbps per Port, full duplex.

Each workstation connecting to an ATM switch creates and runs an active virtual switch environment(Fig. 2) independently. The SPARC2 in addition, ran the MBA and acted as an http server for depositing of customized programs.

The performance measurements have been performed in two parts. First concerns the signaling in the AN, e.g. the delay to setup/tear down connections and the delay to modify option values in the injected programs. Second concerns the performance of the transmission of data, e.g. the latency and throughput of an active node.

4.1 Signaling

4.1.1 Delay for Processing Queries at the MBA

The measurements (Table 1) were obtained with the MBA running on SPARC1, for a network configuration of two ATM switches.

The delays include working out contracts, computing routes, modifying status information in the MBA database, and generating the contacts to their designated network nodes. The delays become larger when the communication delays of the CORBA system(used to link the MBA with the active nodes) and the TCP/IP system (used between the MBA and end-users) are included.

Table 1 Query processing delay.

query	delay
set up a link	50 ms
leave a session	25 ms
set option	25 ms

Table 2 Delay for download codes to active node.

injected program	size	time to download SPARC1	time to download SPARC2
class_A	544 bytes	120 ms	100 ms
class_B	2800 bytes	120 ms	100 ms
class_C	776 bytes	400 ms	300 ms
sub_A	534 bytes		
sub_B	534 bytes		
sub_C	534 bytes		

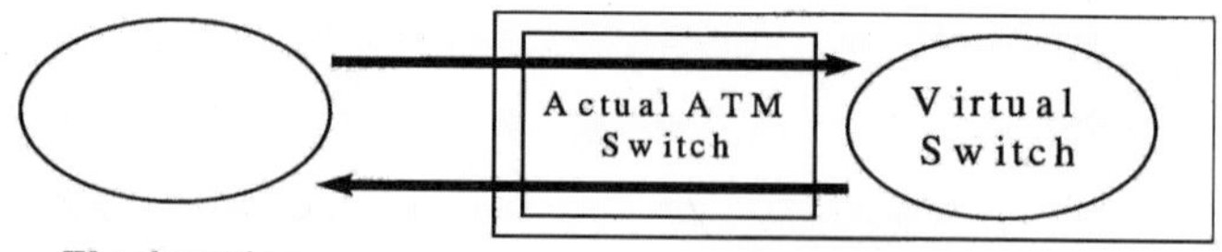

Fig. 7 Experiment setup to measure additional latency.

Table 3 Additional per-hop delay introduced by the virtual switch.

size of AAL packets	SPARC1	SPARC2
8k	536.5 ms	370.2 ms
7k	480.0 ms	366.2 ms
6k	407.6 ms	288.6 ms
5k	432.6 ms	249.0 ms
4k	314.4 ms	194.0 ms
3k	223.8 ms	157.4 ms
2k	162.9 ms	106.3 ms
1k	94.4 ms	57.7 ms

4.1.2 Time Needed to Download a Program from the HTTP Server to the Active Node

To download a program from the HTTP server, a connection is setup to the server specified by the URL. This takes approximately 430 ms on SPARC1 and 280 ms on SPARC2. Different programs of different sizes and complexities were being downloaded to evaluate the injection mechanism. Table 2 shows the delays for downloading of codes to the active node (exclude the connection setup time). Class_C needs to be resolved by the class loader, since it refers to other classes i.e. subA, subB and subC. The class loader fetches those sub's from the HTTP server too.

The above measurements were obtained when other Unix users were not allowed to access the server during the experiments. Once a class is downloaded, creating a new instance takes less than 1 ms. The class needs to be downloaded only once. Then it is seen as being local to the system, and new instances can be created easily as for native system classes.

4.1.3 Processing Delay of Contracts by the Active Node

Different contracts loaded from MBA will be processed at the active node, resulting in: setup/tear down of connections, adding/removing branches to existing connections, and passing information to the customized programs to modify options/parameters. The time needed for processing contracts (including connections creation/deleting or adding/removing) is typically 4 ms on SPARC1 and 2 ms on SPARC2.

The time needed to process contracts that modify option values in customized programs is 1 ms on SPARC1 and less than 1 ms on SPARC2. The processing operations include passing the information to the customized program and waiting for it to be processed.

4.1.4 Summary on the Performance of Signaling

Our measurements show that the processing delays at the MBA and the active nodes are reasonably low. Most of the delays are contributed by the downloading of programs from the HTTP server. Fortunately, these delays are incurred only during the connection set up phase (i.e. joining phase) and hence virtually have little impact on subsequent data transfers.

4.2 Performance of Data Transmission

4.2.1 Per-hop Delay Overhead due to Active Node

The following setup (Fig. 7) was used to measure the additional per-hop delay. The time a packet taken to travel from a host and back is measured for the case of a conventional ATM switch and for an active node respectively. These two values are subtracted to reflect the additional per-hop delay introduced at a node by the virtual switch (average over 1000 packets) (Table 3).

These additional delays were contributed by the following processing: redirecting the cells through the ATM switch to the virtual switch, reading from the ATM link (reassembling the ATM cells at the ATM Adaptation Layer into AAL5 packets), switching the packet through the virtual switch, sending the packet on the ATM link (segmentation of the AAL5 packet at the ATM Adaptation Layer into ATM cells), and redirecting the cells through the ATM switch to the host.

The delay introduced by the virtual switch's Java application was also evaluated (Table 4). In the experiment setup(Fig. 8), the following may occur when a packet travels the virtual switch: writing the data to the central queue, reading from the queue, transfer control to a 'dummy' customized program that returns

Table 4 Delay introduced by Java.

Size of the AAL packets	*SPARC1*	*SPARC2*
8k	33 ms	18 ms
7k	28 ms	16 ms
6k	25 ms	14 ms
5k	21 ms	12 ms
4k	17 ms	10 ms
3k	13 ms	8 ms
2k	9 ms	5 ms
1k	5 ms	3 ms

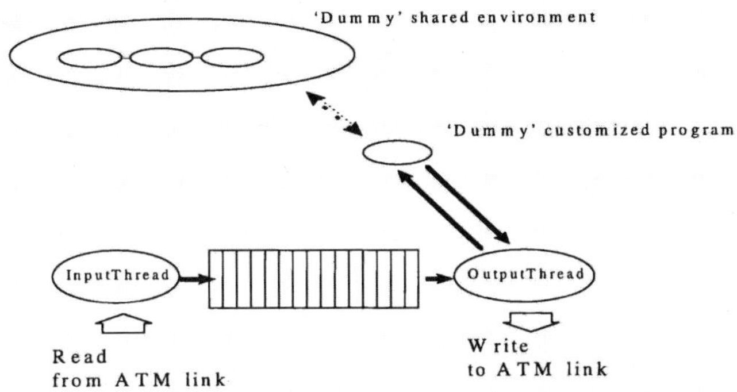

Fig. 8 Packet traversing the Java virtual switch.

the data without modification, and finally preparing the data for sending.

In summary most of the data transfer delays, as expected, are attributed to the active node due to the virtual switch software layer. We expect such delays to be reduced greatly if the virtual switch could be implemented within the actual switch. We also observed that, from the entire additional per-hop delay, only 4 to 6% of the delay is introduced in the Java application itself.

4.2.2 The Throughput of an Active Node

Figure 9 is the experiment setup used to measure the throughput of an active node. The source send's data at a certain rate through the active node to a receiver. Under a certain rate, the receiver receives exactly the same rate as the sender's rate. When the source rate is increased beyond a 'saturation rate,' the receiver's rate remains unchanged at this saturation rate. This constant value is said to be the maximum throughput of the active node. Notice that the active node will not lose data until the central queue buffer is overflowed.

A unicast connection was used for the following measurements (Table 5). The active node reads from one Inlink and sends to one Outlink. A 'dummy' customized program emulates the active processing.

The same experiment was repeated for a multicast connection, with one Inlink and four Outlinks (Table 6). The OutputThread makes three copies of the data before it can be processed and sent on the different Outlinks. In this case, the active node has more work to do

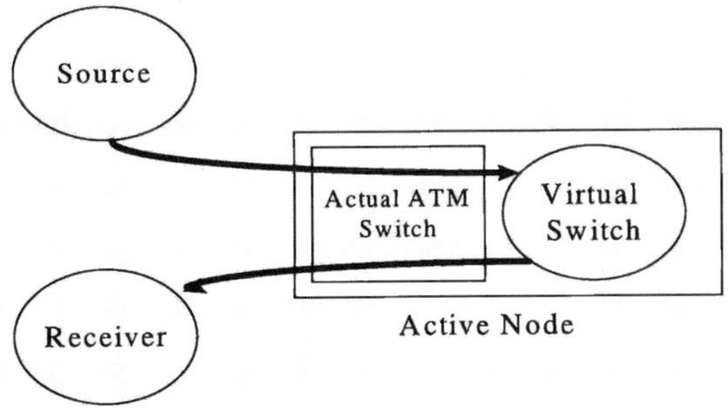

Fig. 9 Experiment setup to measure throughput.

Table 5 Throughput of an active node with a unicast connection.

Size of the AAL packets	SPARC1 throughput	SPARC2 throughput
8k	1.77 Mbps	3.24 Mbps
7k	1.75 Mbps	3.22 Mbps
6k	1.73 Mbps	3.20 Mbps
5k	1.71 Mbps	3.20 Mbps
4k	1.66 Mbps	3.12 Mbps
3k	1.62 Mbps	3.04 Mbps
2k	1.55 Mbps	2.93 Mbps
1k	1.36 Mbps	2.60 Mbps

Table 6 Throughput of an active node with a multicast connection.

Size of AAL packets	*SPARC1 throughput*	*SPARC2 throughput*
8k	1.46 Mbps	2.67 Mbps
5k	1.44 Mbps	2.69 Mbps
1k	1.20 Mbps	2.27 Mbps

for each incoming packet. Hence, the rate it can handle is slightly lower.

In summary the major concern regarding the prototype performance is obviously the throughput bottleneck of the virtual switch. However it could be seen that the current throughput rate of the active node is still adequate for real-time transmissions, as in the showcase video stream multicast application to be described in next section.

5. Video Multicast and Congestion Control

We have developed a real-time video multicast application to demonstrate the feasibility of our prototype in meeting different receiver presentation requirements. In the application, a video-stream encoded using Wavelet transform is multicast from one source to multiple receivers. The stream sent by the source has a fixed image quality and a frame-rate. During the session, receivers may request MBA to modify the quality of the service. The necessary steps are then taken to identify how and where to modify the filtering functions. The MBA then sends contracts to the specific

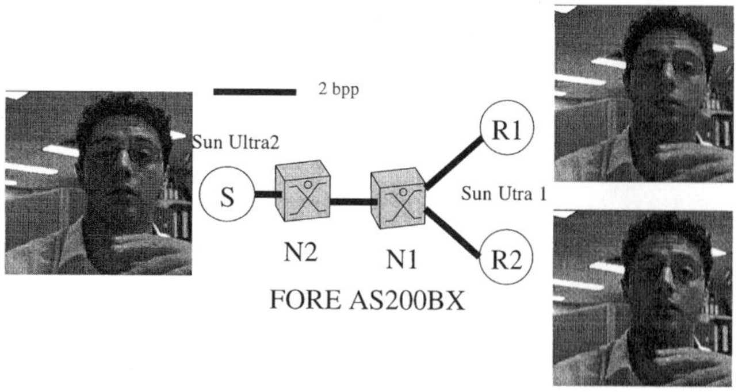

Fig. 10 An instance of a session.

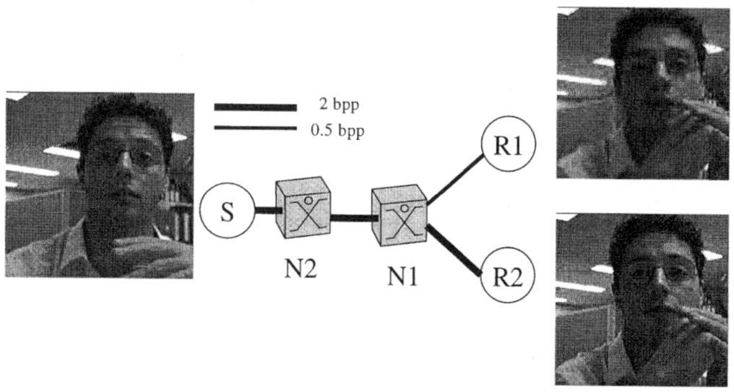

Fig. 11 R1 reduces presentation QoS to 0.5 bpp.

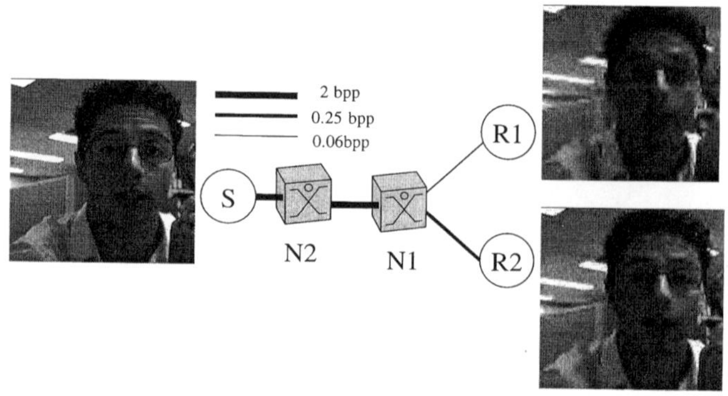

Fig. 12 R1 reduces presentation QoS to 0.06 bpp and R2 to 0.25 bpp.

nodes to realize those modifications in the video compression scheme. The multi-layered property of the data allows filtering of unwanted parts of the video, resulting in a graceful degradation of the image quality.

For instance, the session might look like Fig. 10 at a certain moment. The effects of changing QoS are clearly shown in Fig. 11 and Fig. 12. When receiver R1 requests to modify the quality of the image from 2.0 bpp to 0.5 bpp, the filter function in node N1 will be modified to filter the excess of data flowing to receiver R1 (Fig. 11). When receiver R2 requests to modify the quality of the image from 2.0 bpp to 0.25 bpp and R1 to 0.06 bpp, the filter function in node N2 will be modified to filter the excess of data (to 0.25 bpp) flowing to node N1. The filter in N1 will then filter the excessdata flowing to receiver R1, but not the data flowing to receiver R2, since it already has the right quality (Fig. 12).

We have also developed a simple congestion control program which is activated at a switch when a traffic level exceeds a predefined threshold [8]. The control program can also be de-activated when the congestion threshold returns to the normal level. The scheme is somewhat application data specific as a prior knowledge of the source data structure is required to trim data of less significant when detecting a local congestion. Our scheme is particularly useful for congestion control of video and audio multicast streams that tolerate small loss of data.

It should be pointed that our emphasis here is the programming platform to allow different congestion control schemes be invoked from the repository to reside within switches, and to ensure they are correctly bound during the connection setup or QoS renegotiation phase. Clearly it is the designer responsibility to take care of the QoS interface to the users, the mapping of QoS requirements into congestion control parameters and the definition and implementation of the congestion control protocols.

We could address user QoS requirements in our simple congestion scheme as follow (not implemented in the prototype). Users could specify the acceptable video presentation quality from best to poor during connection set-up. This would then be translated into different thresholds of congestion tolerance. Data from connections with higher congestion tolerance are more likely to be discarded in times of congestion being developed in a switch.

As the congestion control is done within the network automatically by the active nodes, the latency for responding to congestion situations is as short as possible. It also provides a more satisfactory mechanism than rate-based schemes that require at least a round trip delay for the congestion control to take effect. In the case of point-to-multi-point connections, only the parts of the network that experience congestion will perceive a reduction in the amount of data, but the other parts of the network will still get the full rate as sent by the source.

6. Related Work

Multicast in ATM has been an active topic of research and standardization in recent years. The focus of these works is on connection management using conventional networking techniques. Most of the research work reported are proposals with simulation results without physical implementation.

The ATM Forum's UNI 3.1 [9] together with PNNI1.0 [10] support unidirectional source initialized (i.e. source is responsible for the connection setups) point-to-multipoint (PTM) connections. This solution does not support receivers with heterogeneous quality of service requirements. The multicast capability of ATM network services could be enhanced with the introduction of UNI 4.0 [11] and PNNI 2.0 [12]. They allow leaf initiated joining (LIJ) PTM connections but support only homogeneous receiver QoS as in UNI3.1/PNNI-1.

Few protocols have been specified to support multipoint-to-multipoint (MTM) multicast service utilizing existing ATM services and legacy protocols. IP multicast over ATM networks [13] proposed by IETF supports MTM multicast utilizing UNI 3.1 PTM connections with MARS. MARS model supports both source-rooted tree (full mesh) and shared tree (multicast server) architecture. It is not scalable because it utilizes sender-controlled joining. MARS model can only be used in each Logical IP Subnet (LIS); IP multicast routers are required to support IP multicast between LIS's. Since ATM is used as the data link layer of IP, IP multicast over ATM provides only best-effort service and hence does not support QoS.

Work on native ATM MTM connection begins to emerge recently. Two better known examples are SEAM [14] and SMART [15]. The Shared Many-to-Many ATM Reservation (SMART) is a protocol for controlling a shared ATM multicast tree supporting many-to-many communication. The protocol supports one or several ATM VCCs of the MTM type. The main contributions of SMART are in dealing with cell interleaving and resource management (of which SEAM lacks), instead of MTM connection setup (it supports merging and splitting of tree).

SEAM supports MTP connections with the use of a shared tree, and is conceptually very similar to the Core-based-tree (CBT) for internet. Inter-leaving of AAL5's cells at switches at 'merging points' are avoided through the use of a cut-through VC merging method, which is similar to that in [7]. The signaling protocol for the setting up of the share-tree is also similar to that of UNI 4.0 with some extensions for handling multiple-senders. The performance of SEAM is dependent on the location of the core. The criteria and algorithm for selecting a core to yield an optimal share-tree have not been addressed. QoS resource management for multiple senders over a MTM shared tree remains an open issue; simulation of SEAM was carried out but no actual prototyping was reported.

Work on network protocols over internet platform using active networking techniques includes the Smart-Packets of University of Kansas who demonstrated the implementation of a RIP-like router protocol [16] the reliable multicast project at MIT [17].

Various sender- and receiver-based media scaling and media filtering techniques have been proposed for meeting heterogeneous receiver requirements and better network traffic and resource management [18]–[21]. These techniques are still limited in their ability to detect changes in congestion in a timely manner and receive the best possible quality. A couple of recent works report the performance gains for active network-adaptation of multicast audio and video distributions [22], [23].

The use of active networking in network congestion control was also reported in [24] which considered application-specific processing of user data within the network at congested nodes. They used a programmable switch approach by applying transformations and other techniques to reduce the quantity of data at a congestion point.

7. Conclusion

Programmable and active network technologies are still in its infancy with many basic issues unresolved. The most controversial issue of active networking is the organization and placement of 'active' functions within the network that do not seem to blend well with the established 'end-to-end' design principle of networks [25]. The prospectus of dynamic AN in the upcoming ultra high speed optical networks remains an open question. Similarly, the overheads associated with binding and dynamic invocation in programmable networking are also a matter of great concern. Though the proliferation of these technologies are still remote, we believe they can provide effective solutions in selective areas of networking, such as in network management and multicast.

This paper proposes an alternative design of an ATM multicast scheme based on a hybrid programmable-active networking approach. This design includes a programmable network architecture consisting a set of Corba-based network interfaces, and a Java based user codes injection mechanism. The feasibility of the proposed solution is demonstrated through a realization of a customizable video multicast application and an application specific congestion control. The video multicast shows how receivers can customize the video stream to match their own presentation requirement. The congestion control allows active nodes to automatically exercise local congestion control to prevent congestion by reducing traffic through filtering of the video stream.

The performance of the prototype was measured and evaluated through a series of experiments. The data transfer delay between the actual switch and the virtual switch, and the loading of user programs from the http server to a virtual switch were found to be most time consuming. The former can range between 58 ms to 537 ms and is incurred for each data packet switching through the virtual switch. This delay can be greatly reduced if the virtual switch functions were actually executed within the switch rather than in an external host as it is done now. The program down loading time is between 100 ms to 400 ms which occurs once per injection of customized program. Once the program is residing in the virtual switch, the subsequent processing delay per switch hop is negligible.

The prototype was implemented through the use of general-purpose Java and Iona's Orbix for Unix. We found these platforms and the development tools though less efficient than the native tools and OS environment, are useful and adequate for our prototyping purposes.

Acknowledgements

This work was supported by the research grant RP960349 from the Ministry of Education Academic Research Fund and the National Science and Technology Board of Singapore. The authors gratefully acknowledge their support.

References

[1] D.L. Tennenhouse and D.J. Wetherall, A Survey of Acive Network Research, Telemedia Networks and Systems Group, Laboratory for Computer Science, MIT, 1997.

[2] T.M. Chen and A.W. Jackson, "Active and programmable networks, guest editorial note, " IEEE Network, Special Issue in Active and Programmable Networks, vol.12, no.3, pp.10–11, May/June 1998.

[3] J. Biswas, A.A. Lazar, J.F. Huard, K. Lim, S. Mahjoub, L.F. Pau, M. Suzuki, S. Torstensson, W. Wang, and S. Weinstein, "The IEEE P1520 standards initiative for programmable network interfaces," IEEE Commun. Mag., vol.36, no.10, p.64, Oct. 1998.

[4] H. Li, L.H. Ngoh, and H.K. Pung, "Simulating QoS guaranteed ATM multicast in CORBA environment," Proc. Eleventh International Conference on Systems Engineering, pp.503–508, University of Nevada, Las Vegas, USA, July 1996.

[5] H. Li, L.H. Ngoh, H.K. Pung, and Y.Q. Liang, "A QoS guaranteed ATM multicast service in supporting selective multimedia data transmission," Fifth International Conference on Computer Communications and Networks, pp.293–296, Washington D.C., USA, Oct. 1996.

[6] H.Y. Li, H.K. Pung, and L.H. Ngoh, "Active multicast service architecture for user customised multimedia transmission over ATM network," SPIE Proceedings on Multimedia Computing and Networking 1997 (MMCN97), pp.17–28, USA, 10–12 Feb. 1997.

[7] H.K. Pung and A.S. Leow, "A multipoint-to-multipoint multicast over ATM supporting heterogeneous QoS," International Conference on Computer Communications and Networks '98, Lafayette Hilton and Towers, Lafayete, Louisiana USA. Oct. 12–15 1998.

[8] H.K. Pung and N. Bajrach, "An active congestion control in ATM multicast," IEEE ATM Workshop'99, Kochi, Japan, 24–27 May 1999.

[9] ATM Forum, ATM User Network Interface (UNI) Specification Version 3.1. Englewood Cliffs, Prentice Hall, NJ, Sept. 1995.

[10] ATM Forum Technical Committee, "Private Network-Network Interface Specification Version 1.0 (PNNI1.0)," ATM Forum, March 1996.

[11] ATM Forum, ATM User Network Interface (UNI) Signaling Specification Version 4.0, ⟨af-sig-0061.000⟩, July 1996.

[12] ATM Forum, Private Network-Network Interface (PNNI) Version 2.0 Specification, Draft, ⟨BTD-PNNI-02.00⟩, Sept. 1997.

[13] J. Grenville, "IP multicasting over ATM networks," IEEE J. Sel. Areas Commun., vol.15, no.3, April 1997.

[14] M. Grossglauser and K.K. Ramakrishnan, "SEAM: An architecture for scalable and efficient ATM multipoint-to-multipoint communication," 15th International Teletraffic Congress (ITC), Washington DC, USA, June 1997.

[15] E. Gauthier, J.-Y. Le Boudec, Ph Oechslin, SMART: A many-to-many multicast protocol for ATM, http://lrceee.epfl.ch/PS_files/SMARTpaper.ps.gz.

[16] A.B. Kulkarni, G.J. Minden, R. Hill, Y. Wijata, A. Gopinath, S. Sheth, F. Wahhab, H. Pindi, and A. Nagarajan, "Implementation of a prototype active network," OpenArch98.

[17] L.H. Lehman, S.J. Garland, and D.L. Tennenhouse, "Active reliable multicast," Proceedings INFOCOM'98, vol.2, pp.581–589, 1988.

[18] T. Kappner and L. Wolf, "Media scaling in distributed multimedia object services," in Proc. Multimedia Advanced Teleservices and High Speed Communication Architecture, ed. R. Steinmetz, Springer LNCS 868, pp.34–43, 1995.

[19] S.R. McCanne, "Scalable compression and transmission of internet multicast vide," Ph.D. Thesis, report no.UCB/CSD-96-928, Computer Science Division, UC Berkeley, 1996.

[20] N. Yeodon, A. Mauthe, D. Hutchison, and F. Garcia, "QoS filters: Addressing the heterogeneity gap," in Proc Interactive Distributed Multimedia Systems, eds. B. Butcher, E. Moller, and H. Pusch, Springer LNCS 1045, Berlin, pp.227–244, 1996.

[21] B.J. Vickers, M. Lee, and T. Suda, "Feedback control mechanisms for real-time multipoint video services," IEEE J. Sel. Areas Commun., vol.15, no.3, pp.512–530, April 1997.

[22] A. Banchs, W. Effelsberg, C. Tschudin, and V. Turau, "Multicasting multimedia streams with active networks," International Computer Science Institute, Berkeley, TR-97-050.

[23] S. Bhattacharjee, K.L. Calvert, and E.W. Zegura, "Network support for multicast video," Technical Report, GIT-CC-98-16, College of Computing, Georgia Institute of Technology, 1998.

[24] S. Bhattacharjee, K.L. Calvert, and E.W. Zegura, "On active networking and congestion," Technical Report, GIT-CC-96-02, College of Computing, Georgia Institute of Technology, 1996.

[25] T.M. Chen and A.W. Jackson, "Commentaries on active networking and end-to-end arguments" IEEE Networks, vol.12, no.3, p.66, May/June, 1998.

High-Performance Backbone Network Technology edited by Naoaki Yamanaka
ISBN: 0-8247-5321-6 DOI: 10.1081/(120025721)
Published 2004 by Marcel Dekker, Inc.

A Buffer Management Mechanism for Achieving Approximately Fair Bandwidth Allocation in High-Speed Networks

Takashi Miyamura, Takashi Kurimoto, Kenji Nakagawa, Prasad Dhananjaya, Michihiro Aoki, and Naoaki Yamanaka

SUMMARY We propose a buffer management mechanism, called V-WFQ (Virtual Weighted Fair Queueing), for achieving an approximately fair allocation of bandwidth with a small amount of hardware in a high-speed network. The basic process for the allocation of bandwidth uses selective packet dropping that compares the measured input rate of the flow with an estimated fair share of bandwidth. Although V-WFQ is a hardware-efficient FIFO-based algorithm, it achieves almost ideal fairness in bandwidth allocation. V-WFQ can be implemented in the high-speed core routers of today's IP backbone networks to provide various high-quality services. We have investigated V-WFQ's performance in terms of fairness and link utilization through extensive simulation. The results of simulation show that V-WFQ achieves a good balance between fairness and link utilization under various simulation conditions.

key words: *fairness, QoS, Diff-serv, high-speed network, bandwidth management*

1. Introduction

In today's Internet, the sharing of network resources among the many flows is on a best-effort basis. A malicious user of a network may thus be able to get more bandwidth than a well-behaved user. Protecting bandwidth for well-behaved flows from ill-behaved flows is thus important in the Internet environment. One way to stop ill-behaved flows from adversely affecting other flows is to allocate a fair bandwidth to each flow.

Two types of algorithms can be used to achieve fair bandwidth allocation: scheduling-based algorithms and preferential-dropping-based algorithms.

Scheduling-based algorithms (e.g., Weighted Fair Queueing (WFQ) [4] and its variants [5]–[7]) are known to be ideal as mechanisms allocating bandwidth in a fair way and providing guaranteed QoS. In these approaches, however, must be maintained a separate queue for each flow and state is maintained on a per-flow basis, so that hardware efficiency is out of the question. More precisely, WFQ has a computational complexity of $O(\log(n))$, where n is the number of flows currently queued at a router. WFQ is hard to implement in high-speed backbone routers with trunks that carry large numbers of flows.

Core-Stateless Fair Queueing (CSFQ), proposed by Stoica et al. [9], is a well-known preferential-dropping-based algorithm. The CSFQ mechanism achieves almost ideal fairness without using the per-flow states of the core routers. To avoid maintaining the per-flow state for each router, they use a distributed algorithm in which per-flow state is only maintained for edge routers and is not maintained for core (non-edge) routers. An edge router estimates the arrival rate for each flow and this information is carried by a label in each packet's header to the core router, which utilizes this information in deciding whether to discard or queue each arriving packet. Thus, in terms of the core routers, CSFQ is simpler and easier to implement than WFQ. However, the architecture of the edge routers is still complicated. Although it is easy to extend CSFQ to support flows that have different weights, the algorithm is incapable of accommodating situations where the relative weights of flows differ from router to router.

Recently, Cao et al. [10] have proposed Rainbow Fair Queueing (RFQ), which, like CSFQ, may be used to achieve approximately fair sharing. The approach here is to divide each flow into a set of layers, based on rate. At the edge routers, the packets in a flow are marked with layer labels rather than with explicit rates of flows. While the calculation of fair share in CSFQ requires exponential averaging, the core routers in RFQ only need to perform threshold-based dropping; the routers are thus simple and amenable to hardware implementation. However, the approach applied in the edge routers in their remains complicated. It is important to note that the application of either CSFQ or RFQ requires to the Internet of today extension of the IP packet's header.

We have proposed the basic architecture of an active queue-control scheme, called V-WFQ (Virtual Weighted Fair Queueing), which achieves almost fair

Manuscript received February 1, 2002.
Manuscript revised April 10, 2002.

†The authors are with NTT Network Service System Laboratories, NTT Corporation, Musashino-shi, 180-8585 Japan.
††The authors are with Electrical Department, Nagaoka University of Technology, Nagaoka-shi, 940-2188 Japan.
†††The author is with NTT Network Innovation Laboratories, NTT Corporation, Musashino-shi, 180-8585 Japan.

IEICE TRANS. COMMUN., Vol. E85-B, No. 8, pp. 1434–1441, August 2002

allocation of bandwidth and minimum bandwidth guarantees in high-speed networks [1], [2]. The word *"Virtual"* here indicates that V-WFQ emulates the function and performance of WFQ without using sophisticated per-flow queueing. In this paper, we present the details of the V-WFQ algorithm and an investigation of its performance through extensive experiments by simulation.

This algorithm's basic process for the fair allocation of bandwidth uses preferential packet dropping to compare the measured input rate of the flow with an estimated fair share of bandwidth. Although V-WFQ is a hardware-efficient FIFO-based algorithm, it is capable of achieving almost ideal fair bandwidth allocation. Unlike CSFQ and RFQ, V-WFQ does not require any extensions to the headers of IP packets and does not need to distinguish between the core and edge routers of the network, so it is easy to apply to today's Internet backbone networks.

2. Proposed Algorithm

2.1 An Overview of V-WFQ

The buffer architecture of a V-WFQ router is shown in Fig. 1. The proposed mechanism: (1) estimates the input rates of the flows at each router in the network, (2) calculates the bandwidth allocated to each flow at each router, and (3) uses FIFO buffering with preferential dropping.

We avoid the need for per-flow buffer management and scheduling by using FIFO buffers with preferential packet discarding on arrival. Thus, as well as providing improved fairness, our proposed mechanism is suited to efficient implementation in high-speed core routers.

A V-WFQ router measures the input rates of the flows it is processing. When each packet arrives, the V-WFQ router calculates a fair bandwidth for the corresponding flow on the basis of the degree of output-link congestion. The router decides whether or not to discard a given packet by comparing the estimated input rate with the calculated result for allocated bandwidth. If the packet is not discarded, it is enqueued in the

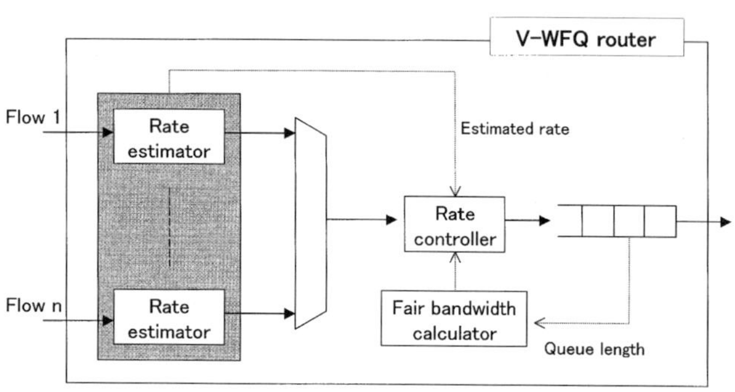

Fig. 1 Buffer architecture of a V-WFQ router.

buffer of the output link.

2.2 The Algorithm of V-WFQ

2.2.1 Estimating Flow Rate

To estimate the packet-arrival rate for a flow, we use the simple 'jumping window' method. This requires only two states for each flow and involves no complex computation to estimate the arrival rate. Let R_i be the estimated rate of flow i and T_w be the window size for using with the method. Let Ct_i be the cumulative packet length of the flow i. At the beginning of the window, Ct_i is set to 0. When a packet of flow i arrives, Ct_i is renewed in the following way:

$$Ct_i = Ct_i + l_i,$$

where l_i is the packet length. At the end of the window, the estimated rate R_i is calculated as

$$R_i = \frac{Ct_i}{T_w}.$$

Ct_i is then reset to 0.

2.2.2 Deciding to Drop or Enqueue a Packet

The V-WFQ router estimates the degree of output-link congestion from the variation in queue length. The following terms are used in calculating the allocated rate:

- N: number of sources,
- $weight_i$: weight of flow i,
- R_i: estimated arrival rate of packets in flow i,
- AR: the fair share for a flow,
- AR_i: the bandwidth allocated to flow i,
- AR_{max}: maximum allocatable bandwidth,
- MGR: per-flow bandwidth achieved in an ideal situation,
- t_{update}: time interval for the updating of AR,
- C: capacity of the output link.

The flowchart and pseudocode of the V-WFQ router algorithm are given in Figs. 2 and 3, respectively. The router updates the fair share AR according to the current congestion status of the output link at the end of every time interval t_{update}. When the average queue length $q_{average}$ reaches the threshold q_{max}, the output link is assumed to be congested and AR should be decreased. If $q_{average}$ is below the threshold q_{min}, the output link is assumed to be underutilized and AR should be increased to raise the degree of link utilization.

When a packet arrives, the router calculates the bandwidth AR_i which is allocated to the flow.

$$AR_i = weight_i \cdot AR$$

Then, the V-WFQ router decides whether or not to discard the arriving packet. We use the random-dropping

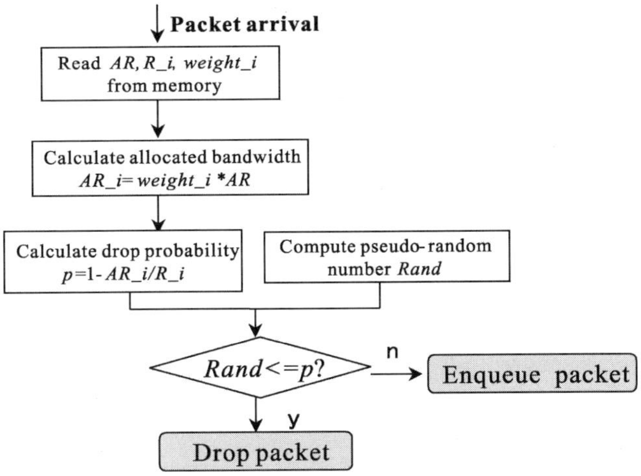

Fig. 2 The V-WFQ algorithm.

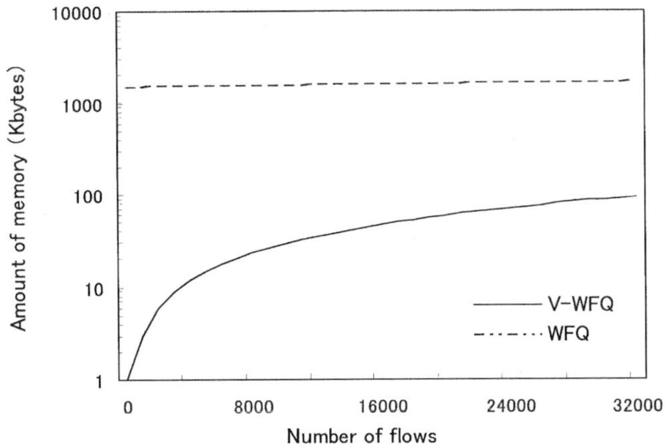

Fig. 4 Comparison of V-WFQ with WFQ in terms of required memory.

```
Every time interval t_update
If(q_average > q_max){
        AR = MGR;
}
Else if(q_average< q_min){
        AR = AR+0.1AR;
}
Else{
        AR = AR/2;
}
 AR=min{AR, AR_max};
/* Updating q_average */
 q_average=α*q_now+(1− α)*q_average;


On packet arrival (flowID =i):
/* Calculating the bandwidth allocated AR_i */
 AR_i= weight_i *AR;
/* Calculating dropping probability p_drop */
If(R_i ≦ AR_i){
        p_drop= 0;
}
Else if{
        p_drop= (R_i−AR_i)/R_i;
}

If(p_drop≧ Rand){
        drop packet;
}
Enqueue packet;
```

Fig. 3 Algorithm of packet dropping decision.

approach. The probability of the arriving packet being dropped p_{drop} is given by

$$p_{drop} = \begin{cases} \frac{R_i - AR_i}{R_i} & \text{if } R_i \geq AR_i \\ 0 & \text{if } R_i < AR_i \end{cases}$$

If one ill-behaved flow keeps sending packets at a higher rate than its fair share, the V-WFQ router discards packets of this flow on a probabilistic basis and provides only that bandwidth which corresponds to the calculated fair share for the ill-behaved flow. This mechanism can ensure a fair allocation of bandwidth for all flows that share a given link's resources. Furthermore, by dropping packets long before the buffer becomes full, this approach avoids the congestion collapse and provides low levels of latency.

2.2.3 Hardware Implementation

Here we compare V-WFQ with WFQ in terms of the cost of hardware implementation. In VLSI, this is directly dependent on the chip area that is required, which in turn is dominated by the requirement for memory [12].

Figure 4 is a comparison of the amount of memory required by V-WFQ and WFQ across a range of numbers of flows. Much less memory is required by V-WFQ than is required by WFQ. For example, V-WFQ only requires about 11 kbytes for 4000 flows, while WFQ requires about 1560 kbytes. V-WFQ is thus a more hardware-efficient algorithm than is WFQ.

However, V-WFQ needs to measure the input rate of each flow and compute a fair bandwidth. The calculation of the fair share consists of simple operations and its computational complexity does not depend on the number of flows. Per-flow traffic measurement is now becoming common in high-speed routers that support line speeds of OC-48 and higher [17]. With today's technology, it is feasible to implement V-WFQ to support several thousands of flows at line speeds of OC-48 and higher.

2.2.4 A New Network Service Based on V-WFQ

We have proposed several new network services that are based on the use of V-WFQ routers [3]. Here we introduce one proposed service. A network which consists of V-WFQ routers is capable of providing a new VPN service as well as a high-quality IP service. Figure 5 shows an isolated-VPN service where the bandwidth is determined on the basis of the tariff concept. The word "isolated" here means that the bandwidth for each VPN is not affected by the number of VPN connections. In

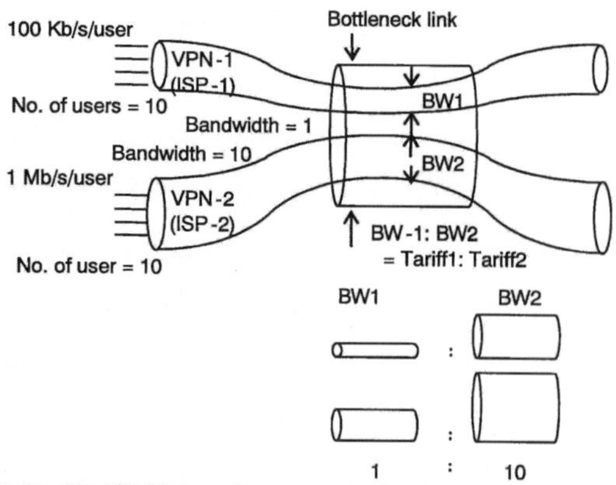

Fig. 5 Example of a new VPN service (BW indicates bandwidth).

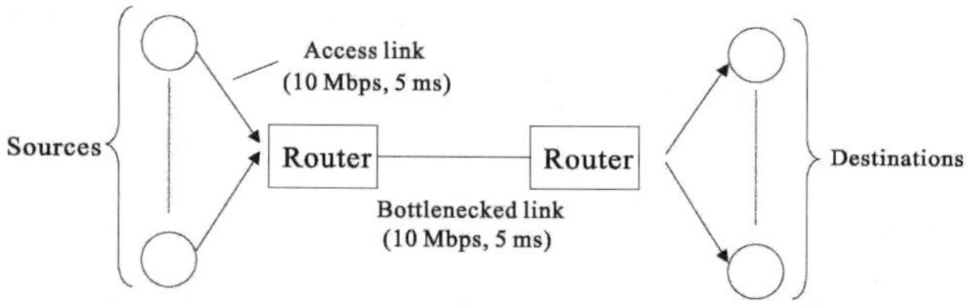

Fig. 6 One-link network model.

other words, the bandwidth for each VPN is provided independently of the other VPNs with which it shares link resources and is determined by the VPN's connection weight alone. Conventional Diff-serv routers can guarantee only a minimum bandwidth. Excess bandwidth is then equally apportioned among the VPNs. In the proposed service, on the other hand, bandwidth in excess of the minimum guaranteed bandwidth is allocated to each of the VPNs according to its weight.

The example shows two ISPs connected to a network and taking advantage of the isolated-VPN service. VPN Service providers can charge each VPN (or ISP) according to the bandwidth with which it is actually provided. For example, the tariff for VPN-1 is \$100/month while that for VPN-2 is \$1000/month; accordingly, VPN-2 has better quality and gets always more bandwidth than VPN-1. Each user can be assured of the required bandwidth even when all go through the same bottleneck link. This service will become the basis of next-generation ISP services and of premium IP services.

3. Performance Evaluation

3.1 Configuration of the Simulation

To investigate the performance of V-WFQ when there is a single congested link in the network, we considered the network configuration shown in Fig. 6. The congested link is between two routers. The link, which has a capacity of 10 Mbps, is shared by m TCP flows and n UDP flows. Each end host is connected to the router via a 10-Mbps link and one end host persistently transmits packets to the other end host. We compared the performance of V-WFQ with Random Early Discard (RED) [13], Flow Random Early Discard (FRED) [7], Deficit Round Robin (DRR) [14], CSFQ, and DropTail.

RED monitors the average queue length and starts to drop arriving packets on a probabilistic basis when the average queue length exceeds a certain threshold. To make a router start to drop packets before its buffer becomes full, RED gives an early congestion indication to each flow and avoids congestion collapse. However, RED does not provide fairness among competing flows. FRED brings fairness of a bandwidth allocation to RED by maintaining a per-flow state. FRED drops packets from flows that have had many packets dropped in the past and from flows that have longer queues than the average queue length.

DRR is a well-known implementation of WFQ that has a computational complexity $O(1)$. The buffer management scheme of DRR requires a sophisticated per-flow queueing mechanism, while the other algorithms applied in our simulation only require simple FIFO-buffer mechanisms. DRR achieves a higher degree of fairness than the other algorithms, so we used DRR as the benchmark for the fair allocation of bandwidth.

All simulations were performed in ns-2 [15]. The simulation code for the DRR, RED, FRED, and CSFQ algorithms is supplied with the ns-2 package. In all simulations, unless otherwise stated, we used the following assumptions and parameters.

- All TCP sources were persistent TCPs and started at the same time. For a TCP flow, the maximum segment size (MSS) was 1000 bytes and the version of TCP was Reno. The TCP hosts transmitted files of infinite size by FTP to the corresponding end hosts.
- The UDP hosts sent packets at a constant bit rate (CBR) of k kbps, where k is a variable, and all started sending at the same time. The packet size of the UDP flows was set to 500 bytes.
- The output buffer size was 128 KB.
- All flows had the same weight values.
- The following parameters were used with V-WFQ. The jumping window size was 50 ms. q_{max} and q_{min} were 128 KB and 8 KB, respectively. Fair sharing MGR was 625 kbps and AR_{max} was 10 Mbps. α was 0.05.

3.2 Indices of Performance

We evaluated the performance of our algorithms in terms of fairness and efficiency. Let N be the number of TCP sources, C be the capacity of the bottleneck link, x_i be the measured throughput of the i-th TCP source, and z_i be the ideal throughput of the i-th TCP source. We define efficiency as

$$efficiency = \frac{\sum_{i=1}^{N} x_i}{C}.$$

We measured the fairness of algorithms in terms of the *fairness index* (FI) [16], which is defined as

$$fairness\ index = \frac{\left(\sum_{i=1}^{N} \frac{z_i}{x_i}\right)^2}{N \times \sum_{i=1} \left(\frac{z_i}{x_i}\right)^2}.$$

3.3 Results of Simulation

3.3.1 Fair Allocation for UDP Flows

In the first experiment, we evaluated the performance of V-WFQ when all hosts were transmitting an infinite amount of data according to the UDP. Each of the 16 UDP sources sent packets at $i \times \frac{10}{16}$ kbps, where $i\,(1,\ldots,16)$ is the flow ID. Each UDP source constantly had data to transmit so the backbone link was always severely congested. Under max-min fairness [16], each flow should have achieved an average throughput of 626 kbps. Figure 7 shows the average throughput achieved by the 16 UDP sources sharing the bottleneck link which was configured with V-WFQ, DRR, CSFQ, FRED, RED or DropTail. Table 1 shows the index of fairness and the level of efficiency for each algorithm used in this simulation. The throughput achieved by RED or DropTail was almost exactly in proportion to the input rate of each flow, while the throughput achieved by V-WFQ, DRR, or CSFQ was nearly constant at around 626 kbps. These results show that RED and DropTail completely failed to allocate fair amounts of bandwidth to the flows during periods of congestion while DRR and CSFQ achieved almost ideal fairness in sharing bandwidth among competing flows. V-WFQ

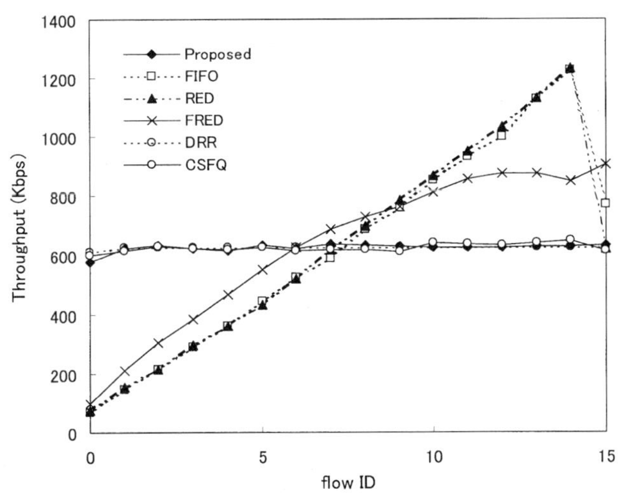

Fig. 7 Throughput for 16 UDP flows sharing a 10-Mbps bottleneck link. Each UDP flow i sends packets at i times its fair allocation (626 kbps).

had the same performance as DRR and much better performance than FRED.

3.3.2 A Single Misbehaving Flow

In the second experiment, we examined the levels of protection provided by the various algorithms for well-behaved flows in the presence of a single misbehaving flow. In this simulation, we assumed that 15 TCP flows and 1 UDP flow shared a 10-Mbps bottleneck link. The UDP source sent packets at the constant bit rate of 10 Mbps during the 30 seconds of simulation. Figure 8 shows the normalized throughput, which is defined as the throughput/fair-share ratio. Note that the normalized throughput for each flow should equal 1.0 in the optimal situation. Table 2 shows the fairness index and level of efficiency achieved by each algorithm used in this simulation.

Table 1 Comparison of fairness indices and efficiency.

	Efficiency	Fairness
V-WFQ	0.999	0.999
DRR	0.999	0.999
CSFQ	0.999	0.999
FRED	0.999	0.859
RED	0.999	0.766
DropTail	0.999	0.768

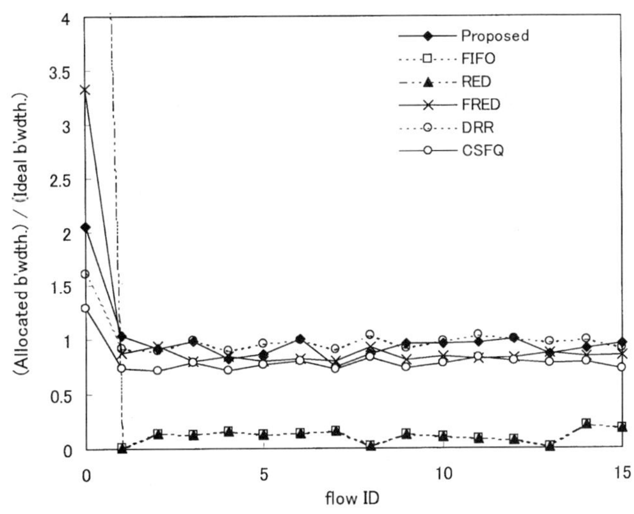

Fig. 8 Throughput for 1 UDP flow and 15 TCP flows sharing a 10-Mbps bottleneck link. The UDP flow i sends packets at 10 Mbps.

Table 2 Indices of efficiency and fairness for a configuration with a single misbehaving UDP flow and 15 TCP flows.

	Efficiency	Fairness
V-WFQ	0.995	0.926
DRR	1.00	0.973
CSFQ	0.996	0.952
FRED	1.00	0.735
RED	1.00	0.0778
DropTail	1.00	0.0778

Under FIFO and RED, the misbehaving UDP flow got more than 8 Mbps, while each of the other TCP flows received less than 30 percent of fair share. Under V-WFQ, the UDP flow got about twice the fair bandwidth, but most of the TCP flows got more than 85 percent of the fair bandwidth. CSFQ shows similar tendencies. Under DRR, most of the flows got almost the ideal bandwidth. These results show that RED and DropTail failed to provide fair share to the well-behaved TCP flows while DRR, CSFQ, and V-WFQ protected the TCP flows against the over-transmitting UDP flow and achieved almost ideal fairness. On fairness index, CSFQ slightly outperformed V-WFQ. Both CSFQ and V-WFQ are FIFO-based algorithms and require per-flow traffic measurement. In V-WFQ, the simple jumping-window method is used to measure per-flow traffic, while CSFQ requires complex exponential averaging at a high-speed core router. The difference in performance is traded-off against the cost of implementation.

3.3.3 Flows with Different Weights

Next, we investigated the performance of V-WFQ for flows with different weights. We compared the performance of V-WFQ with DRR in terms of throughput achieved for each flow and of queue dynamics. Here, the number of source-destination pairs was 5. The weight $weight_i$ is given by

$$weight_1 : weight_2 : \cdots : weight_5 = 1 : 2 : \cdots : 5.$$

In this simulation, the sources were all TCP sources. During the 30-s simulation, each TCP source had an infinite amount of data to transmit so the backbone link was always severely congested. The ideal result is the allocation of bandwidth to each flow in proportion to its weight.

Figure 9 shows the average throughput achieved for the 5 TCP sources over a 30-s interval. Under RED and FIFO, the allocation of shared bandwidth to each flow was independent of its weight. The throughput achieved by V-WFQ or DRR was proportional to its weight. RED and FIFO failed to ensure fairness in terms of throughput, while V-WFQ and DRR achieved almost ideal fairness.

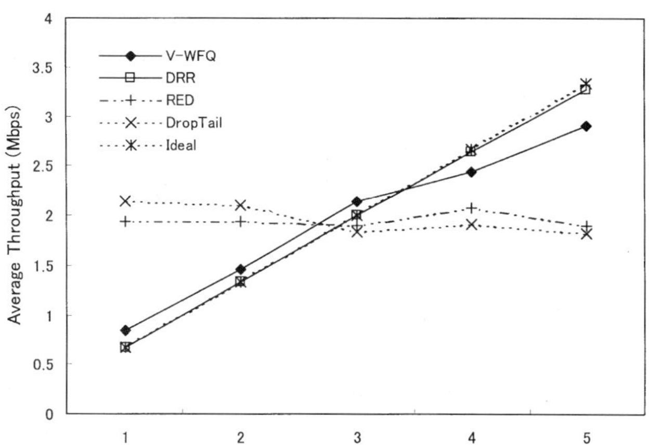

Fig. 9 Average throughput achieved by each of the 5 TCP flows with different weights.

Figure 10 compares the queue dynamics for V-WFQ and DRR. The queue length of DRR is the sum of the length of the separate per-flow queues. The queue length for V-WFQ fluctuated greatly between 0 kbytes and 60 kbytes, while the length for DRR tended to stabilize at a higher level of buffer occupancy, i.e., around 90 kbytes. V-WFQ estimated its fair allocations for each flow from the queue length, hence the rate control of a V-WFQ router ceases to operate the queue length is not becoming longer. The average queue length was much shorter for V-WFQ than for DRR. V-WFQ, like RED, starts to drop packets before the buffer becomes full, so it gives an early congestion indication to each flow and hence can provide much lower latency than DRR.

4. Concluding Remarks

We have proposed a FIFO-based mechanism called V-WFQ that achieves fairness in throughput and have investigated its performance in terms of both fairness and link utilization through extensive simulations. In this investigation, we compared the performance of V-WFQ with DRR and other algorithms. DRR requires a sophisticated per-flow queueing mechanism, while V-WFQ only requires simple FIFO-buffer mechanisms. DRR achieved a higher degree of fairness than the other algorithms, so we used DRR as the benchmark for the fair allocation of bandwidth. The results of simulation show that V-WFQ achieves almost ideal fairness under various simulated conditions. In particular, we should note that it showed almost the same levels of performance as DRR and achieved a much higher degree of fairness than FRED, which, like V-WFQ, is a preferential dropping algorithm.

It is also possible to use the V-WFQ architecture to provide minimum bandwidth guarantees for flows. One example of a new application which V-WFQ allows is

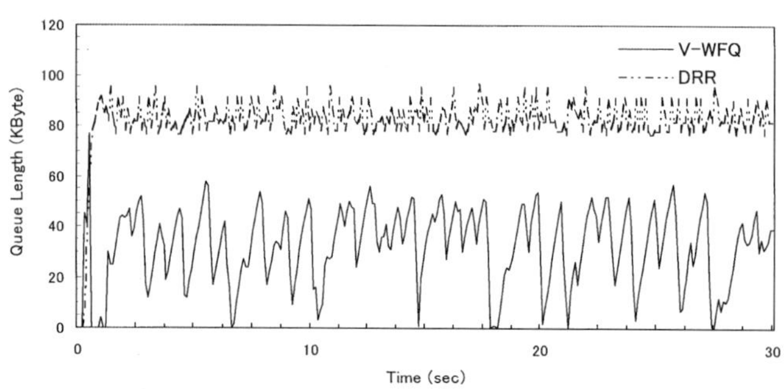

Fig. 10 Queue dynamics for V-WFQ and DRR.

an IP VPN backbone network that provides minimum bandwidth guarantees to each of multiple VPNs.

References

[1] T. Miyamura, T. Kurimoto, K. Nakagawa, P. Dhananjaya, M. Aoki, and N. Yamanaka, "A new queueing mechanism for achieving fair bandwidth allocation in high-speed networks," Proc. APCC 2001, vol.E84-B, pp.473–476, 2001.

[2] T. Miyamura, T. Kurimoto, K. Nakagawa, P. Dhananjaya, M. Aoki, and N. Yamanaka, "Active queue control scheme for achieving approximately fair bandwidth allocation," Proc. IEEE ICC, vol.2, pp.1269–1273, 2002.

[3] N. Yamanaka, T. Kurimoto, T. Miyamura, and M. Aoki, "MSN Type-X: Next generation Internet backbone switch/router architecture," Proc. IEEE ICC, vol.4, pp.2085–2099, 2002.

[4] A. Parekh and G. Gallager, "A generalized processor sharing approach to flow control in integrated services networks: The single-node case," IEEE/ACM Trans. Networking, vol.1, no.3, pp.344–357, 1993.

[5] S.J. Golestani, "A self-clocked fair queueing scheme for broadband applications," Proc. IEEE INFOCOM, vol.2, pp.636–645, 1994.

[6] S. Suri and G. Varghese, "Leap forward virtual clock: A new fair queueing scheme with guaranteed delays and throughput fairness," Proc. IEEE INFOCOM, vol.3, pp.557–565, 1997.

[7] M. Schreedhar and G. Varghese, "Efficient fair queueing using deficit round robin," Proc. SIGCOMM'95, Sept. 1995.

[8] M. Katevenis, S. Sidiropoulos, and C. Courcoubetis, "Weighted round-robin cell multiplexing in a general-purpose ATM switch chip," IEEE J. Sel. Areas Commun., vol.9, no.8, pp.1265–1279, 1991.

[9] I. Stoica, S. Shenker, and H. Zhang, "*Core*-stateless fair queueing: Achieving approximately fair bandwidth allocations in high-speed networks," Proc. ACM SIGCOMM'98, pp.118–130, Sept. 1998.

[10] Z. Cao, Z. Wang, and E. Zegura, "Rainbow fair queueing: Fair bandwidth sharing without per-flow state," Proc. IEEE INFOCOM 2000, vol.2, pp.922–931, March 2000.

[11] B. Braden, D. Clark, J. Crowcroft, B. Davie, S. Deering, D. Estrin, S. Floyd, V. Jacobson, G. Minshall, C. Partridge, L. Peterson, K. Ramakrishnan, S. Shenker, J. Wroclawski, and L. Zhang, "Recommendations on queue and congestion avoidance in the Internet," IETF, RFC2309, April 1998.

[12] M. Grossglauser and S. Keshav, "On CBR service," Proc. IEEE INFOCOM, pp.129–137, 1996.

[13] S. Floyd and V. Jacobson, "Random early detection gateways for congestion avoidance," IEEE/ACM Trans. Networking, vol.1, no.4, pp.397–413, 1993.

[14] D. Lin and R. Morris, "Dynamics of random early detection," Proc. SIGCOMM'97, pp.127–137, Oct. 1997.

[15] UCB/LBNL/VINT, "Network Simulator - NS (Version 2)," http://www-mash.cs.berkley.edu/ns/

[16] R. Jain, A. Durresi, and G. Babic, "Throughput fairness index: An explanation," ATM Forum, 99-0045, 1999.

[17] Cisco Systems Inc., "Cisco Documentation," http://www.cisco.com /univercd/home/home.htm

High-Performance Backbone Network Technology edited by Naoaki Yamanaka
ISBN: 0-8247-5321-6 DOI: 10.1081/(120025722)
Published 2004 by Marcel Dekker, Inc.

Controlling End-to-End TCP Flow Using ABR Rate Information

Arata Koike

SUMMARY Using Available Bit Rate (ABR) service of Asynchronous Transfer Mode (ATM) at a backbone link will provide a lossless network. It, however, causes congestion at an edge-router and is reported to show poor end-to-end TCP performance. We consider how to improve the performance of TCP over a backbone ABR network. By using ABR feedback information (ACR) at edge-routers, we can adjust intervals of TCP Acknowledgement packets. This adjustment couples the ABR rate-based control loop and the TCP window-based control loop. Based on this scheme, we can achieve a good end-to-end TCP performance as well as good ATM layer performance over backbone ABR. This paper describes comprehensive study on the proposed mechanism.
key words: *interworking, ABR, TCP, acknowledgement, edge-router*

1. Introduction

Assurance of end-to-end communication is a basic issue for computer communication and one of the most important ones. So far TCP/IP [1] does provide this assurance even if we use Asynchronous Transfer Mode (ATM) as an underlying layer. If we consider the performance of TCP/IP over ATM, cell loss is an important factor. The loss of one cell means the loss of a whole packet. To reduce the cell loss ratio, the ATM Forum has specified the Available Bit Rate (ABR) service category [2]. So far, many simulation results have been shown to reveal various aspects of ABR. Some studies aimed to show the ATM layer performance while others aimed to discuss TCP over ABR.

Recently, the IP/ATM interworking issue is highlighted regarding to the ABR flow control characteristics. In this case, legacy networks, such as ethernet, are connected to a router and the router is the ATM edge device. We show this reference configuration in Fig. 1. There are two independent flow control loops. The ABR control loop is limited in the ATM segment while the TCP loop spans end-to-end. There is no interworking function at the router.

As shown by previous studies [3]–[7], ABR mitigates switch congestion and successfully reduces cell loss at the ATM layer. However, it induces ATM edge-device level congestion. When we use ABR as the ATM layer service category, congestion in ATM switches is simply moved to the ATM edge-device. And since this has a limited buffer capacity, this congestion may cause buffer overflow at the edge level. This results in poor TCP performance even if we achieve good ATM layer performance [3]–[7].

In order to improve this situation, some interworking function for flow control between ATM layer and upper layer protocols is required [8]. We have proposed flow control primitives for the native ATM API that can be applicable to this situation [9], [10].

In this paper, we propose a TCP-ATM interworking function in a router. Our approach does not require any modification of TCP end-hosts. The key mechanism of our approach is to adjust the Acknowledgment (Ack) interval based on the Allowed Cell Rate (ACR) value. Using this mechanism, we achieve zero packet loss at the edge-router, and zero cell loss at the ATM layer; thus achieving a good end-to-end TCP performance as well as good ATM layer performance.

In the next section, we briefly summarize TCP and ABR. We also explain the background of our proposal and related works. In Sect. 3, we describe the algorithm of the proposal. In Sect. 4, we explain the simulation model and various parameters. We present simulation results in Sects. 5 and 6, and demonstrate how our proposed approach improves end-to-end TCP performance. Section 7 gives a conclusion.

2. Motivation of TCP and ABR Interworking and Related Works

ABR is a rate-based ATM layer protocol. It features a closed feedback control loop by using a special cell,

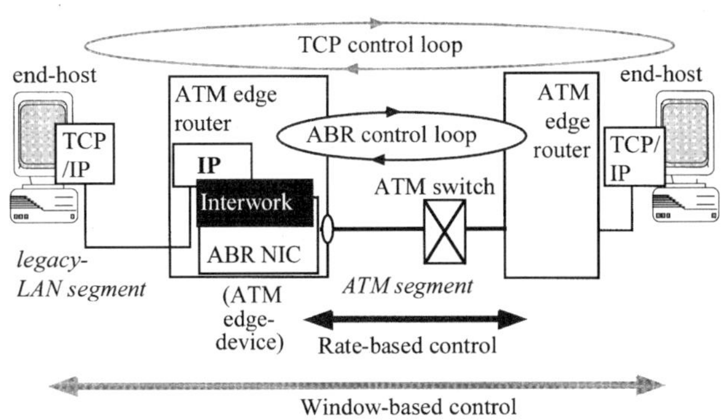

Fig. 1 TCP-ABR reference configuration.

Manuscript received January 25, 2001.
Manuscript revised March 16, 2001.
†The author is with Traffic Research Center, NTT Advanced Technology Corp., Musashino-shi, 180-0006 Japan.

IEICE TRANS. COMMUN., Vol. E84-B, No. 8, pp. 2105–2114, August 2001

called Resource Management (RM) cell. The allowed source transmission rate, ACR, is determined by feedback information which reflects the status of intermediate network nodes or the destination. On the other hand, TCP is a window-based transport layer protocol. A source host sends data packets and the destination host sends back acknowledgment (Ack) packets to the source host. Detection of network congestion at a source host is based on the time out due to packet discard or duplicate Ack arrival. If a source host does not receive an Ack packet within a specific time or receives duplicate Ack packets, the source will think that the packet has been lost and it will retransmit the packet. Since TCP is based on end-to-end control, it does not assume any congestion control mechanism for the underlying layers.

We are focusing on a legacy-LAN and ABR backbone network configuration in this paper. As described in the previous section, the lack of an interworking function yields poor TCP performance due to packet discard at the router even if we use a lossless ATM-ABR network. Our basic motivation is to introduce an interworking function at a router and to relate these two control loops. The main problems of this interworking are how to influence end-to-end basis TCP control and how to utilize ABR rate information.

There are several different ideas to seek TCP and ABR interworking. Jagannath and Yin [11] proposed a sophisticated queueing strategy using ACR information to handle this situation. They fully utilize a TCP congestion detection mechanism by dropping packets based on ACR information. They show that their approach improves overall TCP performance over ABR backbone network. However, it discards packets at the edge-router. For high-speed networks, or long-delay networks, packet discard for TCP may cause a big degradation of performance and fairness. This is because the TCP source has to wait a long time to perceive packet loss. The more we use high speed network, the more this becomes significant. We will analyze this later in this paper.

Since the ABR source implements a very complex scheduling method, we would like to seek a way to keep the performance achieved at the ATM layer, extending it end-to-end, rather than discarding packets at the edge of ABR.

A preliminary version of this paper was presented at the ATM Forum meeting [12],[13]. This paper presents comprehensive study on this mechanism. Our approach couples ABR rate information with Ack packets interval. This is achieved by creating a simple FIFO extra buffer for Ack packets at an ATM edge router and controlling the output interval of Ack packet based on ACR information.

Similar ideas of delaying TCP Ack packets coupling ABR rate information were independently proposed by [14],[15]. In [14], they emulate TCP source behaviour to know the dynamics of congestion window size at the edge of ABR. They show that their algorithm improves the TCP performance as good as infinite forwarding buffer at the ABR edge. However, their method requires emulation of TCP source behaviour at the ABR edge and thus it requires complex implementation to realization their mechanism. They made little concern about fairness among TCP sources.

In [15], it also suggested delaying TCP Ack packets based on ABR rate information but no algorithm and performance results were shown.

3. Algorithm

As described in the previous section, ABR is a rate-based control while TCP is an end-to-end window-based control. The end-to-end basis of the TCP mechanism means that there is limited room for intermediate network elements, such as routers, to explicitly contribute to the TCP control. One such explicit contribution could be achieved by sending an ICMP (Internet Control Message Protocol) source quench message when a connection's ACR falls below a specific threshold value (such as 0) [16]. According to the TCP/IP implementation, this will cause a TCP slow start. However, the use of source quench message causes additional side effects. These disadvantages are summarized in Ref. [17].

The other way to influence TCP control by a router is implicit contribution [16]. The key idea of this approach is based on the fact that TCP is a window-based control. This could be done by scheduling forward and/or backward packets according to the ACR value. This will cause a response delay and will limit new packet transmission from the TCP source since the number of packets traveling without being acknowledged is determined by the TCP window size. This approach enables us to relate rate-based control at the ATM layer and window-based control at the transmission layer. ATM layer rate information contained in ACR is interpreted as an Ack packet interval, which controls the transmission speed of TCP packets (Fig. 2). We want to emphasis that our approach does not require any modification of TCP end-hosts.

Here, we only consider the case delaying Ack packets. Comparing the effects of adjusting the forward packet interval and backward Ack packet interval, we find that Ack packet interval adjustment has two benefits. The first one is the length of control loops. If we adjust the forward packet interval, we have to wait one whole round-trip-time (RTT) period until we observe the effects at the router. On the other hand, if we adjust Ack packet interval, we only need to wait the RTT of the legacy segment which may be much smaller than the whole RTT. The second point is the buffer requirement. Adjusting the Ack packet interval requires additional Ack packet buffers in a router. However since

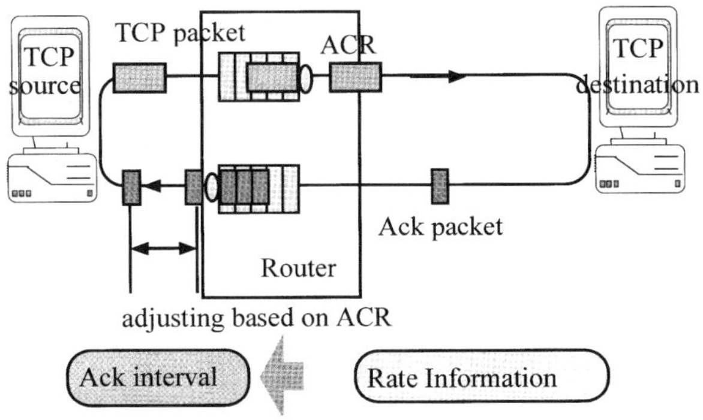

Fig. 2 Adjustment ACK interval.

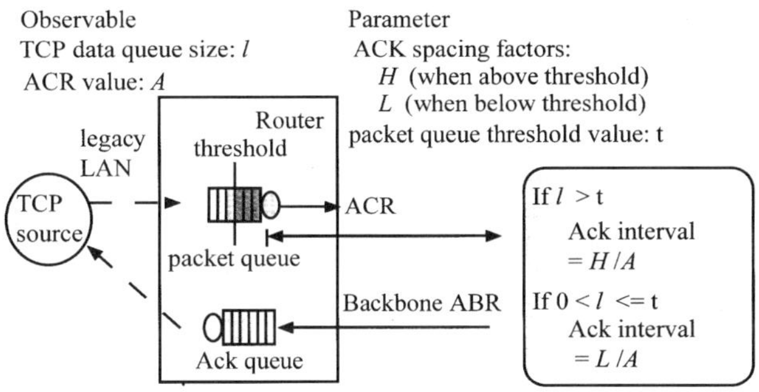

Fig. 3 Proposed algorithm.

Ack packets only contain a sequence number in their payload field, the net buffer requirement in a router is extremely lower.

In order to extend ACR information to legacy segments, we introduce a simple algorithm (Fig. 3). If the queue length l of a buffer for the forward direction in a router exceeds a certain threshold, say t, we schedule the Ack packet of the backward direction based on the rate ACR/H, where H is a parameter. If the queue length is less than or equal to the threshold t but greater than 0, we schedule the Ack packet based on the rate ACR/L, where L is also a parameter. This is our entire algorithm. This interworking function only needs ACR information from the ATM layer.

4. Simulation Setup

4.1 Simulation Model

The simulation model is described in Fig. 4. There are five routers and upto five TCP sources at the source side. Each TCP source is connected to a router. The same configuration appears at the destination side. We call these segments legacy segments (source side: LAN1, destination side: LAN4). Each router is assumed to be an ABR edge-device. Routers are connected to a switch and are aggregated to a backbone link. Then they are distributed to destination side routers at the opposite side switch. Router-switch (switch-router) segments are ATM segments (source side: LAN2, destination side: LAN3).

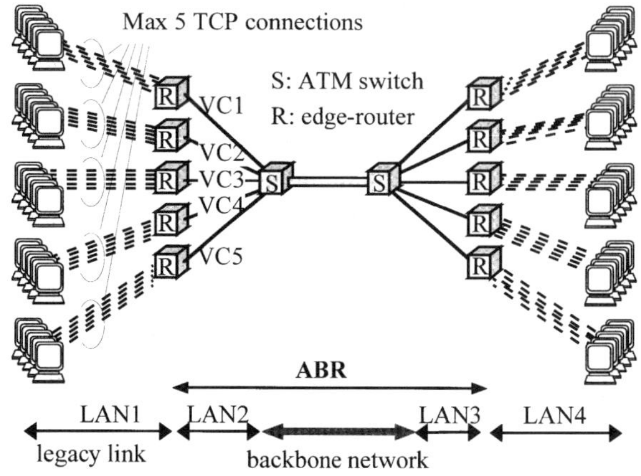

Fig. 4 Simulation configuration.

4.2 Simulation Parameters

We use the following simulation parameters.

• Network parameters:

We assume Minimum Cell Rate (MCR) is 0 Mbit/s. We investigate three Peak Cell Rates (PCR) of each Virtual Connection (VC). We also use three link rates for legacy segments. Distances in legacy segments (LAN1 and LAN4) are 1 km, and ATM-LAN segments (LAN2 and LAN3) are also 1 km. Three backbone link (BL) lengths are used (10 km, 100 km, 1000 km). The light velocity is assumed to be 200,000 km/s. The ABR switch algorithm we used is CAPC2 [18] and we assume 95% utilization.

• TCP parameters:

We use TCP-Reno, which includes fast retransmission and fast recovery. The TCP timer granularity is 200 ms. We examined two different maximum TCP window sizes: 64 and 128 kbytes (we explicitly note when we use 64 kbyte window size). Each packet size is 9188 bytes. We also assume the TCP destination packet processing delay obeys an exponential distribution with a 1-ms mean. Each TCP source transmits data 10 seconds and all sources start to transmit simultaneously (except Sect. 6.4). Note that in order to eliminate the effect of initial connection setup, we use the time interval (0.5, 10.5) for throughput accumulation.

• Router parameters:

Data buffer size is 10 packets (Sect. 5) or 15 packets (Sect. 6). The threshold of data queue is 3 packets and H (when $l > t$) = 2.08 and L (when $0 < l = t$) = 0.5. The Ack buffer size is unlimited. This final assumption is validated because Ack packets only contain a sequence number in their payload so the storage requirement is much smaller than for TCP datagram packets.

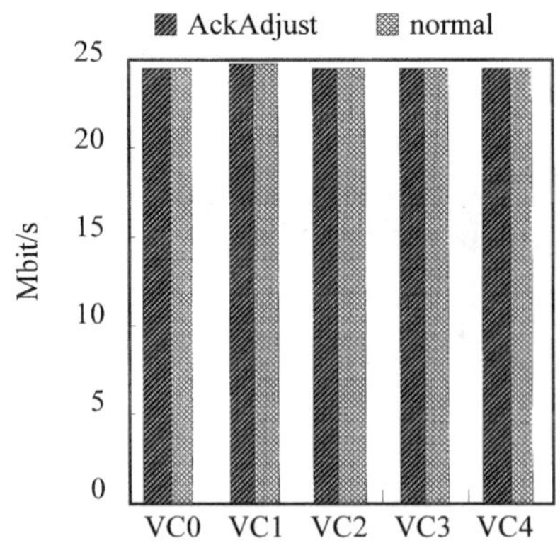

BL=10Km, Window size=64Kbyte, legacy link rate=150Mbit/s

Fig. 5 TCP goodput.

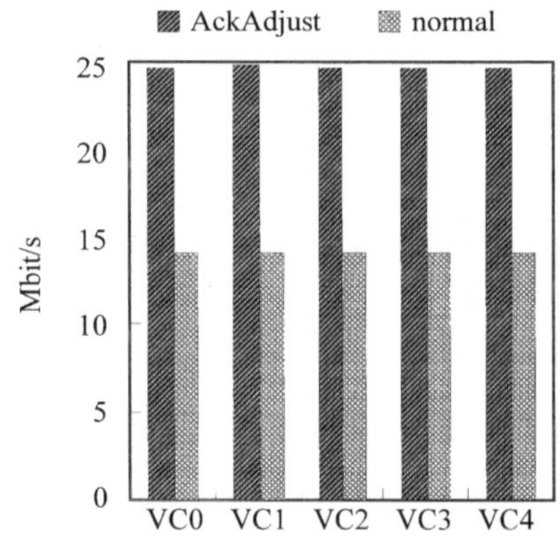

BL=10Km, Window size=128Kbyte, legacy link rate=150Mbit/s

Fig. 6 TCP goodput.

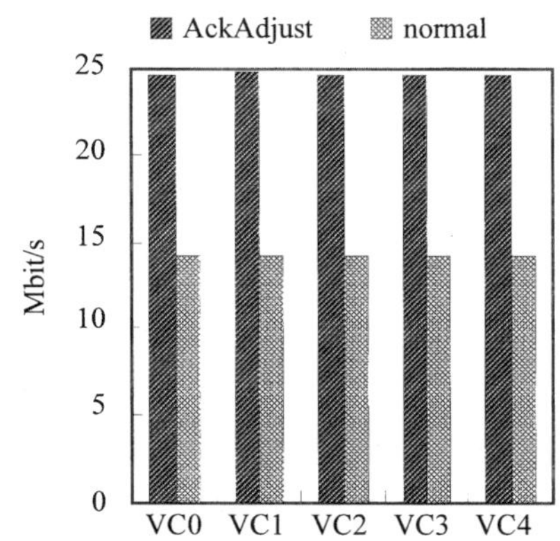

BL=100Km, Window size=128Kbyte, legacy link rate=150Mbit/s

Fig. 7 TCP goodput.

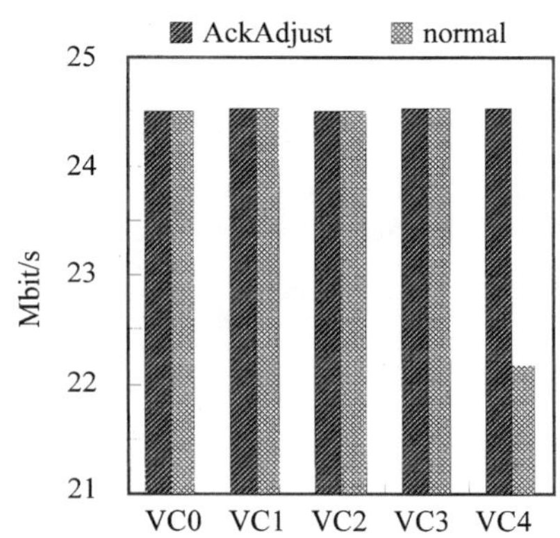

BL=1000Km, Window size=128Kbyte, legacy link rate=150Mbit/s

Fig. 8 TCP goodput.

5. Single TCP Connection

In this section, we consider VCs with a single TCP connection, i.e., a single TCP connection is connected to a router.

5.1 Algorithm Overhead

The first point we want to look at is the overheads of our algorithm. In our simulation, we assume the buffer size at a router is 10 packets. If we assume a maximum TCP window size of 64 kbytes, which corresponds to seven TCP packets, then no packet dropping occurs at the router. In this case, we observe little difference between the usual router configuration (denoted "normal" in the figures) and our algorithm (denoted "Ac kAdjust") (Fig. 5). We can see that our algorithm shows slightly lower throughput for all cases if we look at the figure through high precision magnifier. This is because it regulates the Ack interval based on a threshold, so there is some period of over-regulation but this overhead is obviously negligible.

5.2 Effect of Ackadjust Algorithm

We show the case for a maximum TCP window size of 128 kbytes. This corresponds to 14 TCP packets, so buffer overflow at a router is inevitable. Without our algorithm, the queue length of the buffer would grow and repeatedly exceed the buffer size. This causes packet retransmission and greatly degrades the end-to-end throughput performance. On the other hand, with our algorithm there is no packet discard at the router. The data queue length grows but never exceeds the buffer size. This results in higher throughput. We can see this result in many different parameter configurations (Figs. 6–11). Note that we did not observe any cell drop events in any of our simulations. In Figs. 12–15, we show snapshots of ACR and packet buffer transition of a VC. We specifically look at the 10 km backbone link length case with a 150-Mbit/s legacy link rate. Figures 12 and 13 show the buffer occupancy level for our algorithm and for the normal case, respectively. Our algorithm limits the occupancy level to seven packets

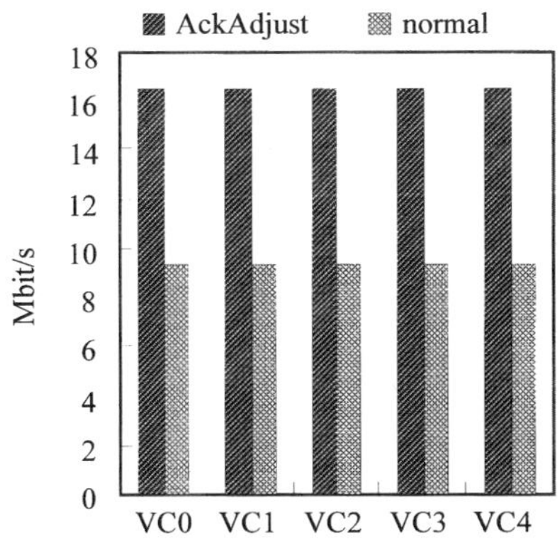

BL=10Km, Window size=128Kbyte, legacy link rate=100Mbit/s

Fig. 9 TCP goodput.

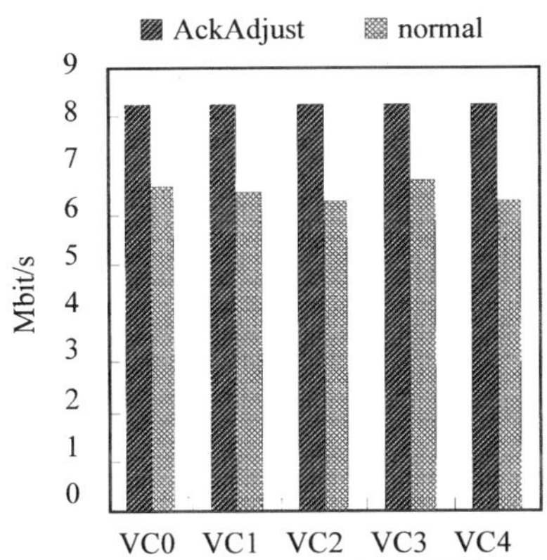

BL=10Km, Window size=128Kbyte, legacy link rate=50Mbit/s

Fig. 10 TCP goodput.

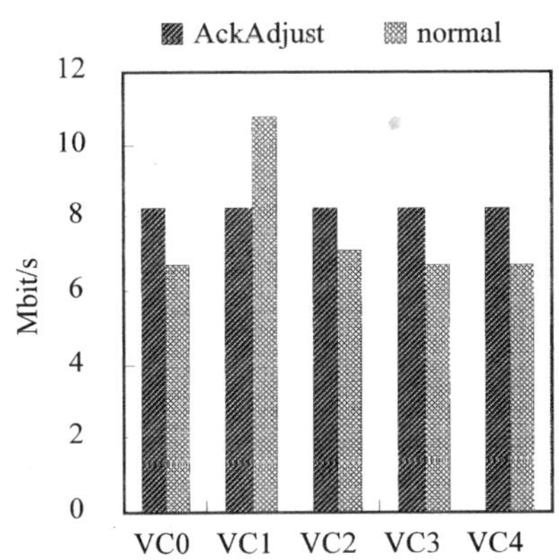

BL=1000Km, Window size=128Kbyte, legacy link rate=50Mbit/s

Fig. 11 TCP goodput.

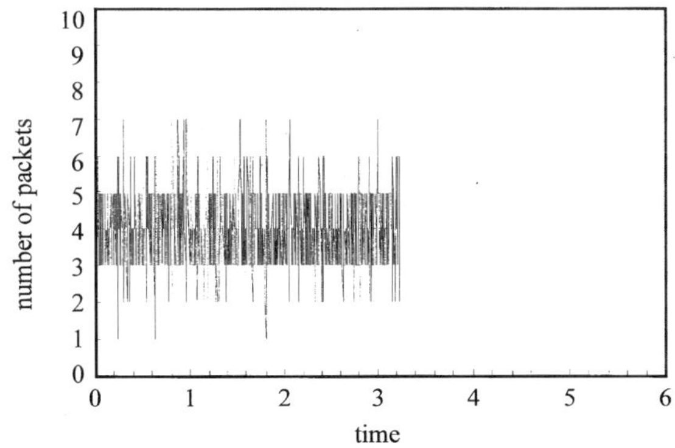

BL=10Km, Window size=128Kbyte, legacy link rate=150Mbit/s

Fig. 12 Packet queue transition for Ackadjust algorithm.

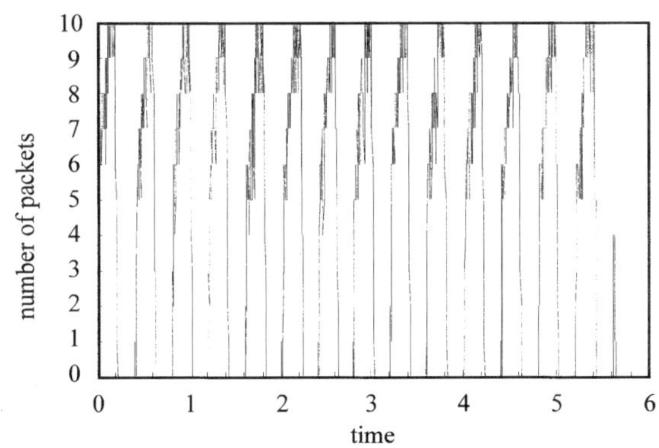

BL=10Km, Window size=128Kbyte, legacy link rate=150Mbit/s

Fig. 13 Packet queue transition for normal algorithm.

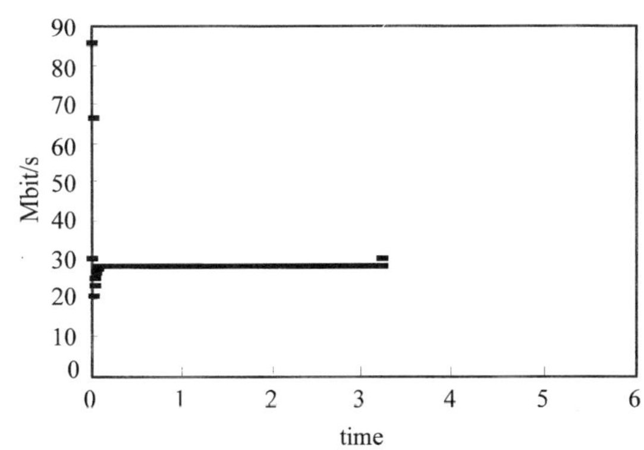

BL=10Km, Window size=128Kbyte, legacy link rate=150Mbit/s

Fig. 14 ACR for AckAdjust algorithm.

but the normal case repeatedly causes packet overflow. After a packet drop event, there is a silence periods while waiting for TCP time-out. Therefore, the packet buffer becomes empty, which results in low throughput. Packet discard also causes oscillation of ACR values (Figs. 14,15). This will be harmful to other connections sharing the same backbone link.

We also note that since we set the target utilization for the ABR switch at 95%, we can see that our approach fully utilizes the ABR link. For example, in the case of the 150 Mbit/s-link sharing by five connections, the ideal fare share of the throughput is 25.68 Mbit/s. This is calculated as follows:

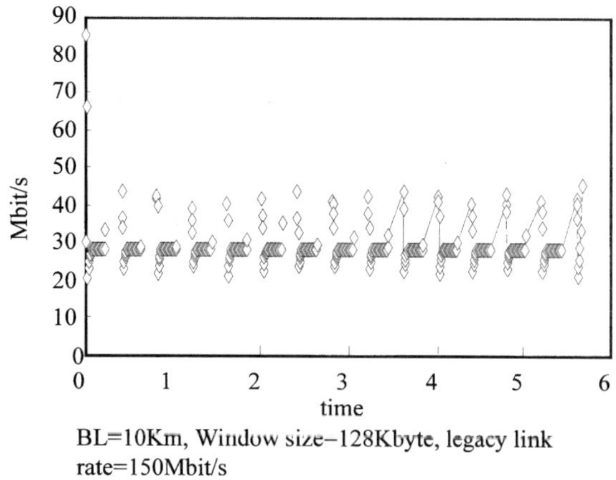

Fig. 15 ACR for normal algorithm.

$$150 * (48/53) * (9140/9188) * 0.95 * 1/5 = 25.68.$$

5.3 Consideration on RTT

We also want to point out the effects of backbone link distance. RTT delay is an important factor for control schemes using a feedback mechanism such as TCP or ABR. In this paper, we show the throughput performance for three categories of backbone link (BL) distances (10 km: Fig. 6, 100 km: Fig. 7, and 1000 km: Fig. 8). TCP uses a window-based approach. The number of packets sent without an Ack packet is limited by the current window size. Our approach keeps Ack packets in an edge-router to adjust TCP source transmission events. Without our algorithm, TCP data packets would be stored in the router. Notice that the more RTT increases, the more TCP packets or TCP Ack packets we can store in the whole network. This means that if we increase the backbone link distance, the number of TCP packets or Ack packets contained in the link increases. Since the number of TCP packets traveling in the network is limited, an increase in the number of packets in the link mean a decrease in the number of packets in the router. Therefore, as you can see in the Figs. 6–8, we can observe our algorithm is more effective for small backbone link distances. For 1000 km link length with a legacy link rate of 150 Mbit/s, only one of the VCs causes packet loss and degrades its performance for normal case. On the other hand, our algorithm keeps fairness among connections (Fig. 8).

5.4 Consideration on Link Speed

For a low legacy link rate, the packet interval is larger than the higher link rate. Therefore even if a packet is discarded, the silence period is only a small proportion of the whole transmission period. This improves the ratio of throughput for a normal router for ideal fair-share throughput (Figs. 6, 9, and 10). On the other hand, it increases unfairness between connections (Figs. 10,11).

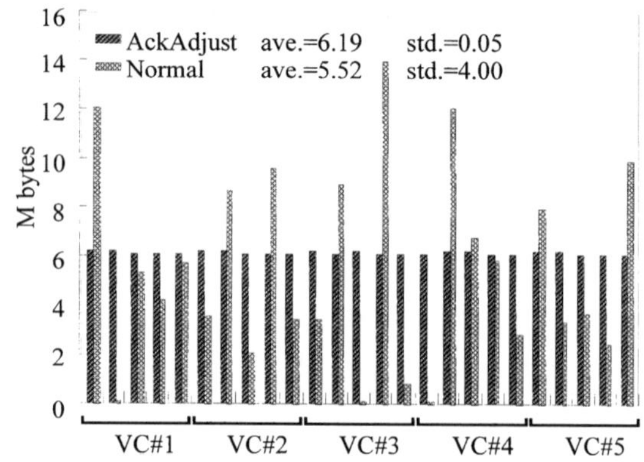

Fig. 16 Goodput for BL=10 km and PCR=150 Mbit/s.

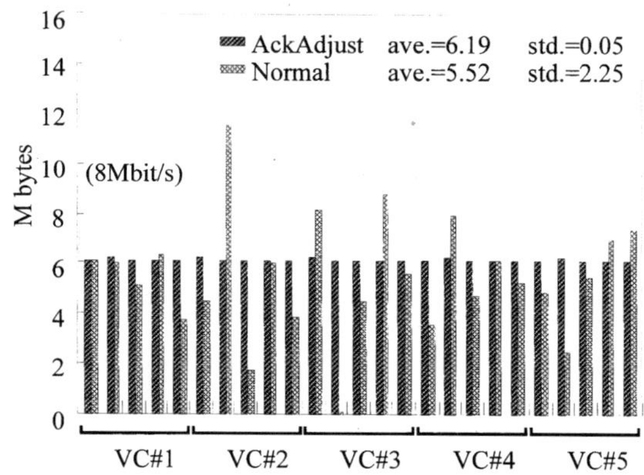

Fig. 17 Goodput for BL=1000 km and PCR=150 Mbit/s.

Our algorithm keeps fairness among connections for all cases in our simulation.

6. Multiple TCP Connections

6.1 Multiple TCP Configuration

In the previous section, we presented results for a single TCP source connected to a router (VCs with a single TCP connection). Usually a router handles multiple TCP connections simultaneously (VCs with multiple TCP connections). We will show that even in this case, our proposed algorithm will work well since it stores mainly Ack packets in the routers, so no data in the forward direction will be discarded.

Now, we change the network configuration to the following multiple TCP connections: five TCP connections are aggregated at a router, and totally we have 25 TCP connections. In this section, we assume the legacy LAN speed is 100 Mbit/s and PCRs for the ATM segments are 150 Mbit/s.

6.2 Performance for High-Speed Network

In Figs. 16, 17, we show the results for 10 km and 1000 km backbone link length with 150 Mbit/s PCR.

The horizontal axis represents different TCP connections per-VC, and vertical axis represents TCP goodput which is the amounts of data a TCP connection could send during 10 seconds. Thus indication of "10 Mbyte" at the vertical axis means that the TCP source could send data on the average rate at 8 Mbit/s. "ave." in the figures shows goodput average of all TCP connections and "std." shows standard deviation of goodput for all TCP connections.

As we can see from the both of figures, without our algorithm, packet dropping at a router causes serious fairness degradation as well as goodput degradation. ABR is intended to keep fairness among VCs but TCP disturbs this characteristics and we observes fairness violation for not only among TCP connections but also among VCs. On the other hand, when we use our "Ackadjust" algorithm, we observes superior fairness as well as high goodput. In general, the ABR control only works for ATM layer. It does not care what kind of upper layer protocols are used, nor how many TCP connections are aggregated to a VC. However, if we look at the figures, by using "Ackadjust" algorithm, we find that we can achieve not only fairness among VCs but also fairness among TCP connections. As described in the previous section, "Ackadjust" uses a simple FIFO buffer without distinguishing aggregated TCP connections. We must emphasis that our approach accomplish excellent fairness among TCP connections even for such simple queueing method.

6.3 Heterogeneous Configuration

We further study this issue by changing number of aggregated TCP connections to a VC: for VC $\#n$, we aggregate n TCP connections. We show the results in Fig. 18. We can see that "Ackadjust" algorithm accomplish fairness among TCP connections within a VC. If we compare the aggregated goodput for each VC, we also find that it keeps fairness among VCs. On the other hand, without using our algorithm, we observe fairness violation not only among TCP connections in the same VC but also among the VCs which were expected to be maintained fairness resulted from the ABR control. Therefore "Ackadjust" is a very simple algorithm but we confirm that it can keep fairness even for TCP connections.

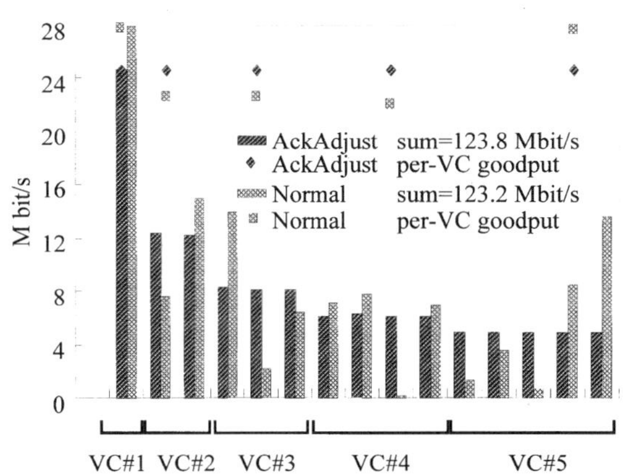

Fig. 18 Goodput for different number of TCP connections.

6.4 Fairness for Newly Joined Connection

We so far considered greedy TCP sources that start sending data simultaneously. In this subsection, however, we first consider TCP sources that start at different times. We studied how fairly a TCP source activated later will be treated among other already existing TCP sources. As in the previous section, we assume that there are five routers and each router serves five TCP connections. Five TCP sources are activated with initial delay of one second each, i.e., TCP connection #1 in a router starts sending data at time zero-second, TCP connection #2 initiated at time one-second and so on.

In Figs. 19 and 20, we show cumulative transferred data of each TCP connections of the router connected to VC#1 for both with and without the proposed algorithm. By using the Ac kAdjust algorithm, the gra-

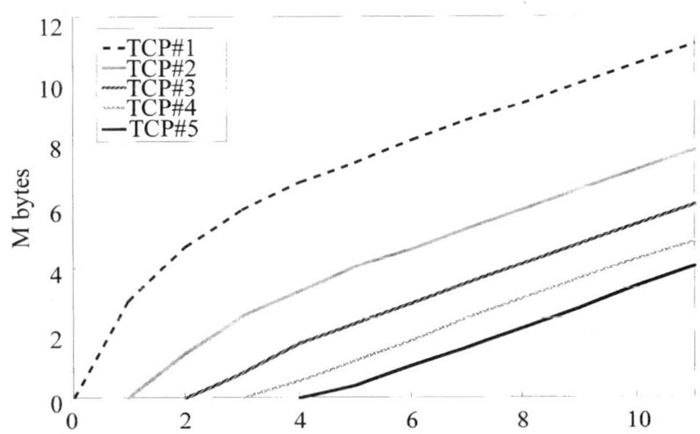

Fig. 19 Time sequence of cumulative goodput for TCPs in router #1 with AckAdjust.

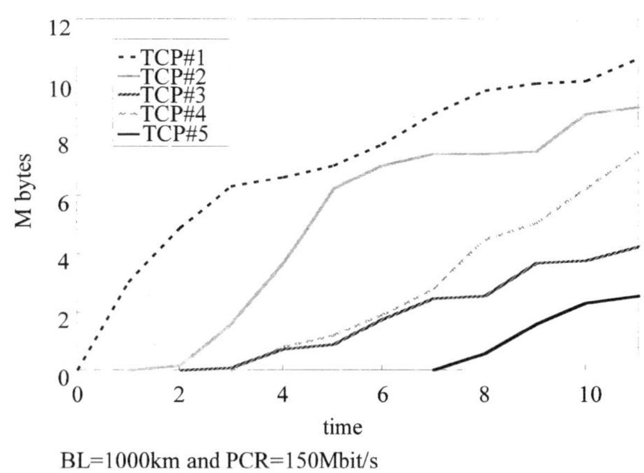

Fig. 20 Time sequence of cumulative goodput for TCPs in router #1 without AckAdjust.

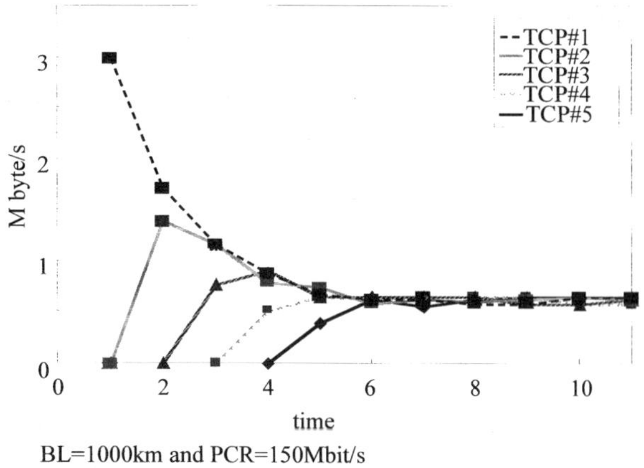

Fig. 21 Time sequence of data transmission for TCPs in router #1 with AckAdjust.

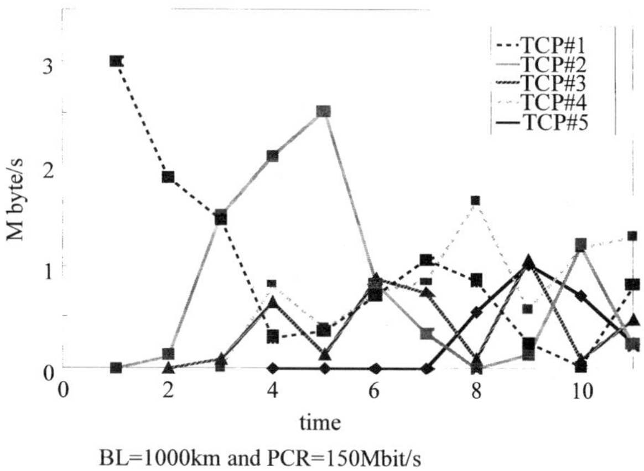

Fig. 22 Time sequence of data transmission for TCPs in router #1 without AckAdjust.

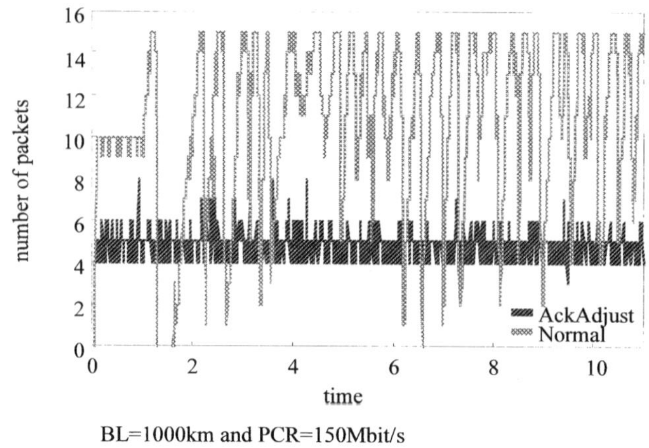

Fig. 23 Forward queue transition at VC #1's router.

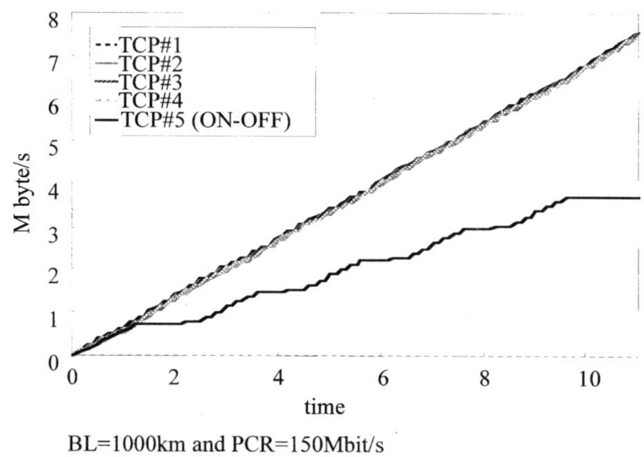

Fig. 24 Time sequence of cumulative goodput for TCPs in router #1 with AckAdjust.

dient of the graph changes when a new connection is activated. We can see that finally all TCP connections show the same slope. To clarify this, we show the time sequence of transferred data during past one second for these TCP connections in Figs. 21 and 22. Each mark in the figure indicates the amount of data transferred during past one second. Figure 21 shows that the newly activated TCP connection obtains smaller share than that of the others during first one second. This is because newly joined TCP connections uses "slow-start" mode at the initial period and their window size is smaller than that of the others. But at the next one-second duration, the amount of transferred data for newly joined TCP connection is the same as the other ones. And thus we could achieve fairness among TCP connections at this stage.

On the other hand, without using the algorithm, the buffer at the router overflows repeatedly (Fig. 23) and thus the goodputs and the transferred rates are unstable. Therefore we could not achieve fairness as in the previous section.

The second simulation scenario focuses on the ON-OFF TCP sources. We assume that one of five TCP sources in a router is ON-OFF source and others are greedy. We discuss the performance of ON-OFF TCP sources comparing to greedy ones. In this scenario we assume ON-OFF sources start every two-second and transfers 750 kbyte data. This quantity is carried approximately one-second if 150 Mbit/s link is fairly shared by simultaneous 25 TCP users. We differentiated the wake-up time of ON-OFF source depending on the router.

In Fig. 24, we show cumulative transferred data of each TCP connections of the router connected to VC#1 with the proposed algorithm. As described above, due to TCP slow-start mechanism, the ON-OFF TCP source receives less bandwidth share at the very initial period. But we can see from the figure that it soon reach the comparative gradient to the others. Thus the proposed algorithm is effective for both greedy and ON-OFF TCP sources except for the cases with very short durations.

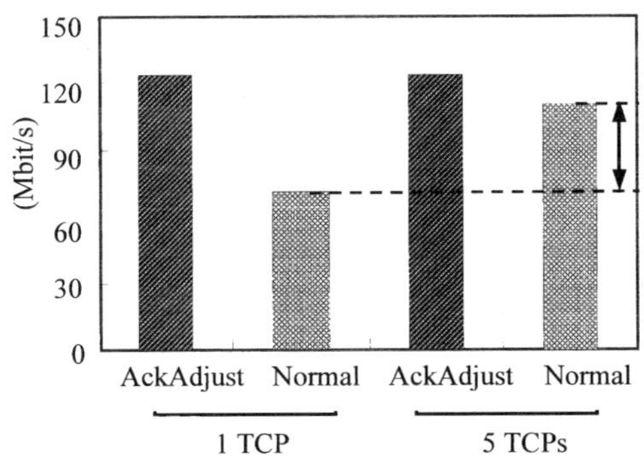

Fig. 25 Effect of TCP aggregation.

6.5 Aggregated Goodput Comparison

In Fig. 25, we compare aggregated goodput for VCs with multiple TCP connections and for VCs with single TCP connection which is described in the previous section. Here, the backbone length is 10 km and the PCR is 150 Mbit/s. We can see no difference between VCs with single TCP connection and VCs with multiple TCP connections for "Ackadjust" algorithm. This is because even the case VCs with single TCP connection, it can fully utilize the backbone bandwidth and there are no room for performance improvement. We, however, observe a big goodput improvement for "normal" by aggregating multiple TCP connections. Without using "Ackadjust" algorithm, many packets will be dropped in a router. A TCP connection which suffered losses will stop transmission until it resumes retransmission. ABR mechanism allocates the network resources in terms of ACR for a VC and if the input rate to the VC from TCP sources is less than the ACR, the backbone utilization rate will decrease. When we increase the number of aggregated TCP connections to a VC, several TCP connections, however, will survive and exclusively utilize the link and thus keep their goodputs. Therefore overall utilization of backbone links is much improved compared to the case for single TCP connection. On the other hand, connections suffering packet discard at a router will cause packet retransmission and their goodputs will be low. This will increase the fairness problem among connections.

7. Conclusions

In order to achieve TCP-ABR interworking at a router, we propose using Ack adjustment. This approach achieves zero packet loss in a router as well as zero cell loss at the ATM layer and thus achieves good end-to-end TCP performance. We consider this performance for both VCs with single TCP connection and VCs with multiple TCP connections configuration.

The first issues are goodput comparison. For low speed backbone network, we see little difference in TCP goodput. For high speed backbone network, however, normal router causes serious goodput degradation since TCP retransmission mechanism itself cannot adapt the network speed. On the other hand, our proposed algorithm, "Ackadjust," does not cause packet discard and thus can extend the ATM layer performance to the end-to-end.

The second issues are fairness. Unfairness among TCP connections are one of the well known problem. By using Ackadjust algorithm, we can achieve fairness at TCP layer. We must note that this fairness is realized by using only a simple FIFO buffer for Ack packets and relate the Ack interval with ACR information. No special per-TCP-connection handlings are required.

This clearly shows using network information enables us to achieve controls which could not be realized only by end-to-end control. We will further extend the study of utilization of network capability and information.

Acknowledgments

We thank Ms. K. Takeishi and Ms. M. Ishizuka for their supporting of simulation works and fruitful discussion.

References

[1] W.R. Stevens, TCP/IP Illustrated, vol.1, Addison Wesley, 1994.
[2] The ATM Forum, Traffic Management Specification 4.0, af-tm-0056.000, April 1996.
[3] N. Yin and S. Jagannath, "End-to-end traffic management in IP/ATM internetworks," ATM Forum/96-1406, Oct. 1996.
[4] S. Jagannath and N. Yin, "End-to-end TCP performance in IP/ATM internetworks," ATM Forum/96-1711, Dec. 1996.
[5] S. Jagannath and N. Yin, "End-to-end traffic management in IP/ATM internetworks," IEEE ATM'97 Workshop Proceedings, pp.83–89, May 1997.
[6] W. Ren, K.-Y. Siu, H. Suzuki, and G. Ramamurth, "Performance of TCP over ATM with legacy LAN," ATM Forum/96-1716, Dec. 1996.
[7] S. Jagannath and N. Yin, "More results on end-to-end traffic management in IP/ATM internetworks," ATM Forum/97-0653, July 1997.
[8] A. Koike and H. Saito, "Utilization of ACR across ATM API," ATM Forum/97-0342, April 1997.
[9] A. Koike and S. Sasyan, "Revised proposal of flow control management primitives," ATM Forum/97-0754, Sept. 1997.
[10] The ATM Forum, API Semantics for Native ATM Services using UNI 4.0, af-saa-0108.000, Feb. 1999.
[11] S. Jagannath and N. Yin, "End-to-end traffic management issues in IP/ATM interworks," Internet Draft, draft-jagan-e2e-traf-mgmt-00, Aug. 1997.
[12] A. Koike, "TCP flow control with ACR information," ATM Forum/97-0758, Sept. 1997.
[13] A. Koike, "TCP/ABR interworking," ATM Forum/97-0998, Dec. 1997.
[14] P. Narváez and K.-Y. Siu, "Acknowledgment bucket scheme for regulating TCP flow over ATM," Computer Networks and ISDN Systems, vol.30, pp.1775–1791, 1998.
[15] J. Ma, "Interworking between TCP and ATM flow controls," ATM Forum/97-0960, Dec. 1997.

[16] A. Koike, "AF-SAA-TM: Re: AF-SAA-TM: 97-0342 utilization of ACR across ATM API," ATM Forum e-mail exploder, April 17 1997.
[17] S. Floyd, "TCP and explicit congestion notification," ACM Computer Communication Review, vol.24, no.5, pp.10–23, Oct. 1994.
[18] W. Barnhart, "Example switch algorithm for section 5.4 of TM spec.," ATM Forum/95-0195, Feb. 1995.

Chapter 7

Photonic Network

The following three chapters describe advanced technologies with especial emphasis on photonic technologies. Three chapters are basically dedicated to networks, architectures, and devices. Very advanced devices are needed to realize photonic systems, so we must optimize the combination of architectures and devices. Chapter 7 describes photonic networks.

Photonic technologies are critical to realizing the future backbone network. Recent progress in wavelength division multiplexing (WDM) has led to a breakthrough in network performance. Paper 7.1 shows an IP over WDM network architecture. Electrical IP routers are connected by fat and robust optical WDM paths. WDM networks provide very high-speed (Tb/s) transmission capability. IP packets are transferred by using layer-2 and optical layer routing. Cost estimation results confirm the effectiveness of the proposed architecture. The optical time-division technique is used for not only LANs and MANs, but also WANs. An ultra-fast bus architecture using TDM technology is proposed in Paper 7.2. The remote bridge technique is used in a hierarchical WAN architecture. Some optical TDM transmission experimental results are also presented.

The optical technology requirements to ensure application to the backbone and those related to the access network are quite different. Paper 7.3 describes a backbone crossconnect and Paper 7.4 describes one for a local network. Paper 7.3 proposes a free-space micromachine optical crossconnect system. This micromachine switch has sub-millisecond switching time, and offers transparency to all WDM signals.

Paper 7.4 examines an architectural approach to the local area network. The architecture is based on a broadcast and select mechanism. 128 channels consisting of 8 wavelengths with 16 time slots are demonstrated. The basic architecture for this paper is also presented in Paper 1.7. An experimental prototype system is shown to achieve channel speeds from 20 ~ 30 Mb/s to 155 Mb/s with both point-to-point and multicasting functions.

T. Miyazaki et al. report in Paper 7.5 the optical bridge and switch function for protection networks. This function is used in the ADM (Add Drop Multiplexer) system. An experimental system is realized by using a AWG (Arrayed Waveguide Filter) and fiber amplifiers. The architecture is applicable to the WDM restoration ADM network.

I. Saeki et al. address the all-optical code division switch in Paper 7.6. While other papers consider either TDM or WDM, this paper uses the CDM architecture. In addition, self-routing is realized by CDM. Performance simulations are also reported.

Networks carrying huge transmission loads must ensure survivability, since even a very short outage will cause a large amount of information to be lost. Papers 7.7, and 7.8 address network survivability and reliability. Y. Miyao describes a λ-ring system, based on a survivable wavelength-based self-healing-ring, in Paper 7.7. C. Bowornttummarat et al. present in Paper 7.8 a comprehensive investigation of three optical WDM mesh network protection approaches, namely minimal cost, single link basis and disjoint path approaches. The operation of each approach is described and their performances are extensively evaluated and compared.

M. Kashima et al. study an optimum network configuration of optical cross connect (OXC) systems and optical add-drop-multiplexer (OADM) systems in Paper 7.9. They describe a test-bed system used to evaluate various problems associated with network configurations.

T. Yasui presents in Paper 7.10 a photonic switching system called WAPS (wavelength assignment photonic switching system). The principle of the system is to assign or allot a wavelength to a call when it is originated. He proposes an IP network configured over WAPS. The proposed architecture provides semi-permanent and flow-by-flow connections, and the edge routes can use either of these connections according to the required services.

The final paper of this chapter, written by M. Ohta et al., addresses the issue of IP over glass (IOG). IOG is a link layer protocol specially designed for the high speed and bandwidth efficient transmission of IPv4 and IPv6 datagrams over optical fibers.

High-Performance Backbone Network Technology edited by Naoaki Yamanaka
ISBN: 0-8247-5321-6 DOI: 10.1081/(120025723)
Published 2004 by Marcel Dekker, Inc.

Robust IP Backbone Network Utilizing WDM Optical Paths

Atsushi Watanabe, Satoru Okamoto, and Ken-ichi Sato

SUMMARY A wavelength division multiplexing (WDM) optical path-based Internet protocol (IP) backbone network is proposed as a cost-effective way of realizing robust IP-over-photonic systems. The WDM optical path is based on WDM transmission and wavelength routing. Between end-to-end IP backbone routers, the WDM optical path, a fat and robust optical pipe, is defined across photonic transport systems (PTS's). Tera-bit class PTS's will be required for the future IP backbone network and this level of performance is achievable. Optical layer routing is done at intermediate nodes, so the electrical packet-by-packet routing required by existing systems is eliminated. An optical signal format that permits cost-effective IP packet transmission is presented. WDM optical paths directly accommodate the IP packets via layer-2 frames. The cost-effectiveness of the proposed system, especially for heavy traffic, is demonstrated from the viewpoint of the overall network traffic transport capability and network node cost. The proposed system is as robust as existing systems; e.g. fault/degradation localization mechanism and optical layer network protection one are implemented. Thus the proposed IP-over-photonic system will create cost-effective and robust IP backbone networks.

key words: *optical communication, optical path, Internet protocol*

1. Introduction

Due to the recent explosion in transmission control protocol over Internet protocol (TCP/IP) applications such as the world-wide-web (WWW), both Internet host number and the traffic in the IP backbone network have been increasing rapidly each year. A recent Internet domain survey indicated that the number of Internet hosts as of July 1988 and July 1998 was 33,000 and 36,739,000, respectively [1]. This means that the number of Internet hosts has increased over one thousand times in the last ten years. There is a problem in that IP backbone systems, which are being created around synchronous optical network (SONET), synchronous digital hierarchy (SDH), and asynchronous transfer mode (ATM) technologies, will not yield the bandwidth abundant and effective IP backbone networks needed in the near future. The main problem stems from the use of packet-by-packet electrical routing which causes high equipment cost and volume, poor system throughput scalability, expensive and complicated network management, and increased end-to-end delay time. This paper demonstrates that the introduction of the wavelength division multiplexing (WDM) optical path network [2], which is based on WDM transmission and wavelength routing, effectively avoids these problems. This paper also describes how the WDM optical path transport network will yield the cost-effective and robust IP backbone networks needed in the future.

The performance of the IP backbone network depends on three components: the photonic transport system (PTS), the IP router, and the way in which IP packets are carried by the WDM optical path transport network. PTS performance is described in a later section. It was recently announced that IP routers with a few hundred giga-bit per second (Gbit/s) IP packet switching performance will be commercially available in the near future. IP routers with tera-bit per second (Tbit/s) IP packet switching performance have also been demonstrated in exhibitions (e.g., Networld+Interop 98 Atlanta). Therefore, IP router performance appears satisfactory. The last component is the focus of this paper. This paper presents a new physical layer (layer 1) technology for constructing IP-over-photonic systems. This paper first describes the schematic of the WDM optical path-based IP backbone network, PTS performance for the long-haul network, and an optical signal format for transporting IP packets. Recent advances in electrical circuit technologies are also addressed. Network layer structure and cost-effectiveness are then discussed in a comparison with other physical layer technologies.

2. WDM Optical Path-Based IP Backbone Network

2.1 Network Structure

Figure 1 shows a schematic of the proposed IP backbone network. The WDM optical path transport network is based on WDM transmission and wavelength routing [2]. It enhances various network functions including transmission capacity and node throughput cost-effectively. Each optical path is established between two PTS's, which offer optical path termination, and routing is based on wavelength. Each PTS also has WDM and wavelength routing functions, and provides flexible and cost-effective access capability to the large bandwidth transmission system. By establishing optical paths between all PTS's, a full-mesh connection can be realized in the IP backbone network. Therefore, a fat and robust optical pipe is defined between end-to-end IP routers; electrical level routing (e.g. packet-by-packet IP routing) is unnecessary at each intermediate node. Thus end-to-end delay time will be decreased. The IP packets are accommodated in the optical path payload via layer-2 frames (e.g., point to point protocol

Manuscript received December 22, 1998.

† The authors are with NTT Network Innovation Laboratories, Yokosuka-shi, 239-0847 Japan.

* This paper is also published in the IEICE Trans. Electron., Vol.E82-C, No.8, pp.1381-1386, August 1999.

IEICE TRANS. COMMUN., Vol. E82-B, No. 8, pp. 1115–1120, August 1999

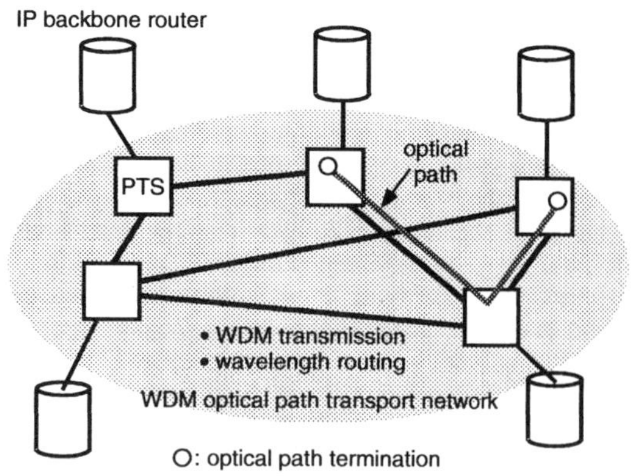

Fig. 1 WDM optical path-based IP backbone network. (PTS: photonic transport system)

(PPP) [3], high-level data link control (HDLC) [4], classical IP over ATM [5], multiprotocol over ATM (MPOA) [6], frame relay (FR) [7], or simplified data link (SDL) [8]). The detailed optical signal format and layer structure are described in Sects. 2.3 and 3.1, respectively. The layer-2 frame plays a key role in forming stream type data from the burst IP packet data.

2.2 Tera-Bit Class PTS

PTS throughput must be of Tbit/s class to support the full-mesh connection between all IP backbone routers. In the IP router world, IP routers with more than 640 Gbit/s IP packet switching performance are generally called tera-bit class IP routers. Following this classification, this paper defines tera-bit class PTS's as those with more than 640 Gbit/s throughput. Japan's long-distance network, which consists of 30 nodes, is shown in Fig. 2(a). Each node is composed of an IP router and a PTS. The following was assumed in this assessment. Optical paths are realized using the virtual wavelength path scheme [2] in which the wavelength of each optical path is converted node-by-node as required. 1:1 (one for one) optical path protection with link-distinct routes between each working and protection path pair is prepared against link failure, and the upstream-downstream path pair are assigned the same route. The traffic pattern was determined to be the bi-directional full-mesh configuration, and the number of wavelengths multiplexed into each fiber was 16. Figure 2(b) shows a PTS fiber pair histogram as determined by an optical path accommodation design tool [9]. The maximum number of PTS fiber pairs was 40. As shown in Fig. 2(b), 26 nodes require a tera-bit class PTS if each optical path has 2.5 Gbit/s capacity. In this example, the required maximum PTS throughput is 1.6 Tbit/s (40-fiber pairs × 16 wavelengths × 2.5 Gbit/s optical path capacity). The realization of the tera-bit class PTS is discussed next. A 320-Gbit/s throughput PTS with 16 fiber pairs, 8 wavelengths and 2.5 Gbit/s per optical path has been already developed [10]. A 900-Gbit/s throughput architecture with 45 fiber pairs, 8 wavelengths and 2.5 Gbit/s per optical path was proposed in [11] and is a simple extension of an earlier development. Therefore, the tera-bit

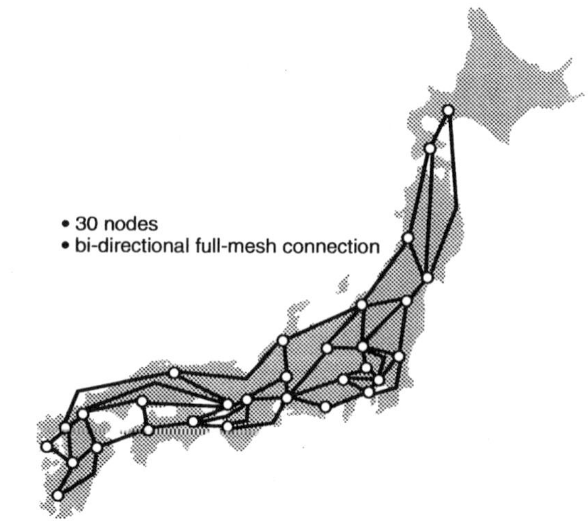

(a) network model

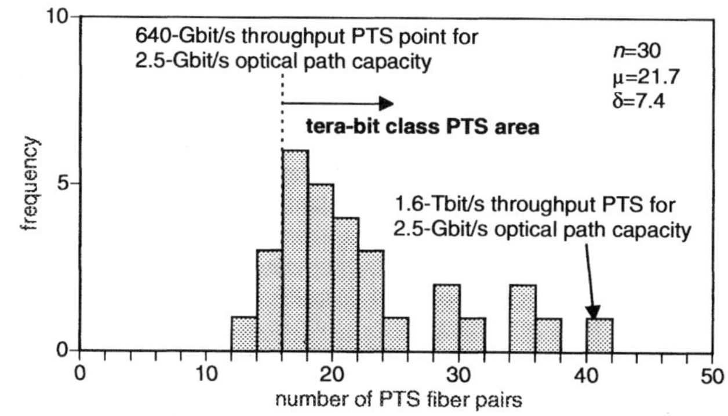

(b) histogram of number of required PTS fiber pairs

Fig. 2 Number of required PTS fiber pairs. (n: the number of samples, μ: average, σ: standard deviation)

class PTS is now within our grasp. In the second system architecture with 45 fiber pairs, a 1.8-Tbit/s throughput PTS can be constructed by upgrading the wavelength number to 16. The 1.8-Tbit/s PTS's can realize the WDM optical path-based IP backbone network for Japan's long-distance network shown in Fig. 2(a). Moreover, by upgrading optical path capacity to 10-Gbit/s, a 7-Tbit/s throughput PTS can be realized.

2.3 Optical Signal Format

The main details of our previously proposed optical transport unit (OTU) format [12] for the WDM optical path transport network, i.e. optical signal format, are shown in Fig. 3. The format is intended to map the IP packets via layer-2 frames in a cost-effective manner. Each OTU consists of optical path overhead and optical path payload. Here, two kinds of optical path overhead are utilized: pilot tone modulation of each path wavelength and in-frame overhead. IP packets, via layer-2 frames, are accommodated in this optical path payload area. The time division multiplexing (TDM) frame structure shown in Fig. 3 retains maximum commonality with the synchronous transport module (STM)-N (N=16, 64, ...)

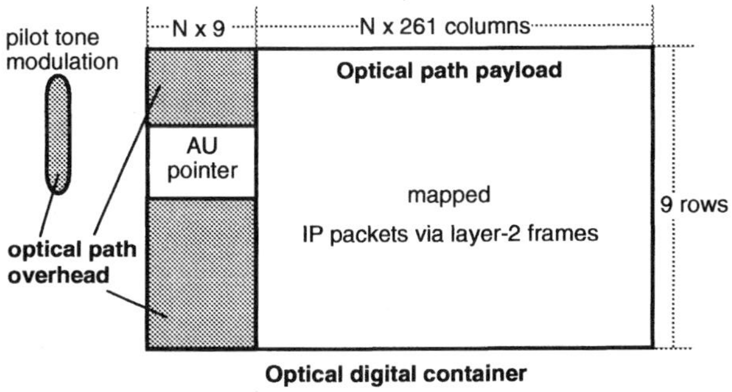

Fig. 3 Optical transport unit (OTU). (AU: administrative unit, N=16, 64, ...)

frame structure of the SDH network. The optical path overhead region and optical path payload region in the frame correspond to the SDH section overhead region and STM payload region in the STM-N frame, respectively. Therefore, by allocating in-frame optical path overhead bytes appropriately [12], already developed SDH section overhead processing large scale integration (LSI) devices can be used for optical path overhead processing. This feature enables the most cost-effective system introduction. The cost and size of SONET/SDH section terminating circuits are continually decreasing. We have recently developed a compact 2.5-Gbit/s SONET/SDH section terminating circuit [13] that supports regenerator section and multiplex section termination functions. This circuit uses low-power bipolar LSI technologies and multi-chip-module (MCM) technology. Module size and power consumption are 33 mm × 33 mm and 2.2 watts, respectively. These are just 1/10th and 1/6th of the corresponding values of currently available modules with the same functions. This technology will support optical path overhead processing, and will surely realize compact optical path terminating circuits.

By utilizing optical path overhead information, robustness, e.g. fault/degradation location identification and optical path automatic protection switching, similar to the benefits offered by SONET/SDH and ATM systems, can be realized in the optical layer [9].

3. Benefits of the Proposed System

3.1 Simplified Signal Mapping Structure

Figure 4 shows the layer structures of existing and newly proposed physical layer technologies for transporting IP packets. The structure shown in Fig. 4(a) is based on SONET/SDH technologies and is called the IP-over-SONET/SDH-over-WDM system (sometimes, IP-over-WDM system). The IP packets, via layer-2 frames, are mapped into SONET/SDH synchronous payload envelopes (SPE's) [14]. WDM technology is utilized for point-to-point WDM transmission in order to provide large transmission capacity. Electrical level routing (e.g. IP routing) is needed at each node. The optical layer has only the multiplexing function, so networking functions such as automatic protection switching are provided by the SONET/SDH system. Quality of service (QoS) is assured

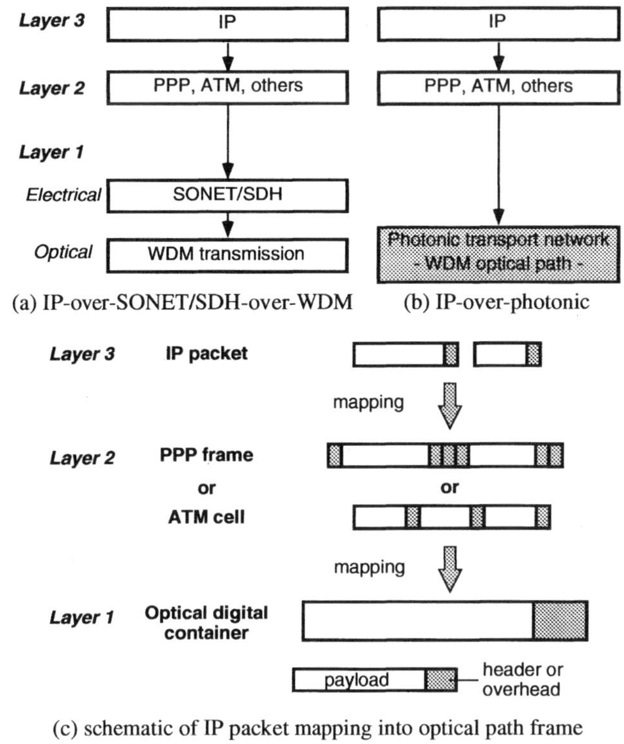

Fig. 4 Network layered model.

Table 1 IP-over-SONET/SDH-over-WDM vs. IP-over-photonic systems.

Items	IP-over-SONET/SDH-over-WDM	IP-over-photonic
optical pipe type	physically adjacent IP routers (point-to-point)	logically adjacent IP routers (end-to-end)
QoS	point-to-point	end-to-end
end-to-end delay time	large	small
protection	SONET/SDH	optical

only at point-to-point nodes.

The layer structure of the proposed network system, shown in Fig. 4(b), is simpler. As shown in Fig. 4(c), the IP packets are directly mapped into the optical path payload via layer-2 frames. If other frames including those mentioned in Sect. 2.1 are utilized in layer-2, they can also be accommodated in the optical path payload. In this layer structure, networking functions such as routing and protection are realized in the optical layer, as described above. QoS is assured at end-to-end nodes by the optical path which acts as a fat and robust optical pipe. Wavelength routing at each intermediate node is done in the optical layer, so end-to-end delay time is minimized; packet-by-packet IP level routing at intermediate nodes is eliminated. The simplified structure brings about not only simple signal mapping relations, but also cost-effective network systems including the minimization of management and networking function duplication. According to these discussions, the characteristics of these two structures are compared in Table 1. Overall, the proposed IP backbone network will provide large bandwidth in the most cost-effective manner.

3.2 Cost-Effectiveness

This section assesses the cost-effectiveness of the proposed IP backbone network.

In discussing the cost-effectiveness of the proposed IP-over-photonic network, the relation between the overall network traffic transport capability and necessary IP backbone router capacity is important. The relation between overall network node cost, IP router capacity and traffic in the network is discussed with reference to Fig. 5. Figure 5(a) shows the IP-over-SONET/SDH-over-WDM system. It is assumed that the terminating traffic in each IP backbone router is *C2* and the necessary IP router capacity is *C1* (*C1* = *C2*, generally speaking *C1* > *C2*). The difference between *C1* and *C2* (*C1*-*C2*), represents the through traffic. This through traffic wastes the routing capacity of the intermediate IP backbone routers in IP-over-SONET/SDH-over-WDM systems. Given the same traffic, IP-over-photonic system can use smaller IP backbone routers with just *C2* capacity, as shown in Fig. 5(b).

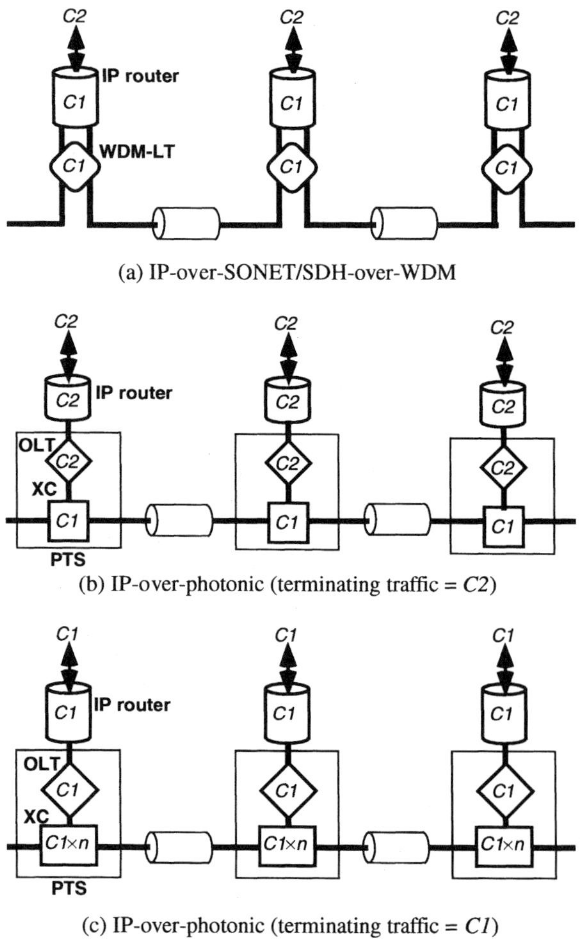

(a) IP-over-SONET/SDH-over-WDM

(b) IP-over-photonic (terminating traffic = *C2*)

(c) IP-over-photonic (terminating traffic = *C1*)

Fig. 5 Relation of IP router capacity and traffic amount. (*C1* and *C2* indicate capacity or traffic amount. WDM-LT: WDM line terminal, OLT: optical line terminal, XC: cross-connect)

This is because only the terminating-traffic is handled by IP backbone routers in the IP-over-photonic system. The through-traffic is routed in the optical layer. It is noted that PTS offers the optical path cross-connect function (PTS-XC) and the optical path termination function (PTS-OLT). If IP backbone routers with *C1* capacity are utilized in IP-over-photonic system, the amount of terminating traffic in each node can be *C1*. Therefore, if IP backbone routers with the same capacity are utilized in IP-over-SONET/SDH-over-WDM and IP-over-photonic systems (shown in Fig. 5(c)), the IP-over-photonic system can support more traffic (*n* times) than the IP-over-SONET/SDH-over-WDM system.

End-to-end cost is evaluated first. Figure 6 shows two IP backbone network models; (a) IP-over-SONET/SDH-over-WDM, and (b) IP-over-photonic (optical path). The same traffic is assumed for the architectures shown in Figs. 5(a) and (b). For evaluation simplicity, a 2.5-Gbit/s IP router interface speed (optical carrier (OC)-48c or STM-16) and 20-Gbit/s transmission speed (8 wavelengths) were assumed for both architectures. Therefore, only end-to-end node cost is highlighted; transmission line cost for both architectures is the same. The costs for PTS-XC, PTS-OLT, IP router (with capacities of *C1* and *C2*), and WDM line terminal (WDM-LT) per 2.5-Gbit/s capacity are shown in Table 2, six patterns were considered. When the traffic is large, PTS-XC cost is smaller than IP router cost since PTS-XC performs optical path routing without electrical packet-by-packet processing (a 320-Gbit/s throughput PTS has been realized that occupies only

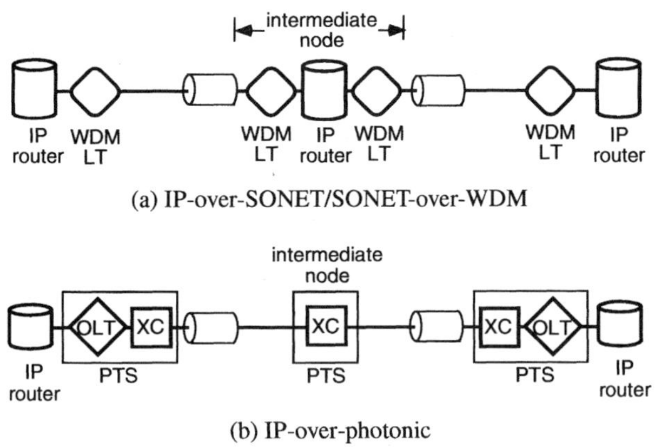

(a) IP-over-SONET/SONET-over-WDM

(b) IP-over-photonic

Fig. 6 End-to-end cost evaluation model. (2.5-Gbit/s (OC-48c or STM-16) IP router interface, and 20 Gbit/s (8 wavelengths) WDM transmission)

Table 2 Node cost ratio.

pattern no.	Cost ratio per 2.5-Gbit/s capacity				
	PTS-XC	PTS-OLT	IP router (capacity=*C1*)	IP router (capacity=*C2*)	WDM-LT
1	1.0	0.7	3.0	2.0	0.5
2	1.0	1.5	3.0	2.0	1.0
3	1.0	2.5	3.0	2.0	2.0
4	1.0	0.7	6.0	4.0	0.5
5	1.0	1.5	6.0	4.0	1.0
6	1.0	2.5	6.0	4.0	2.0

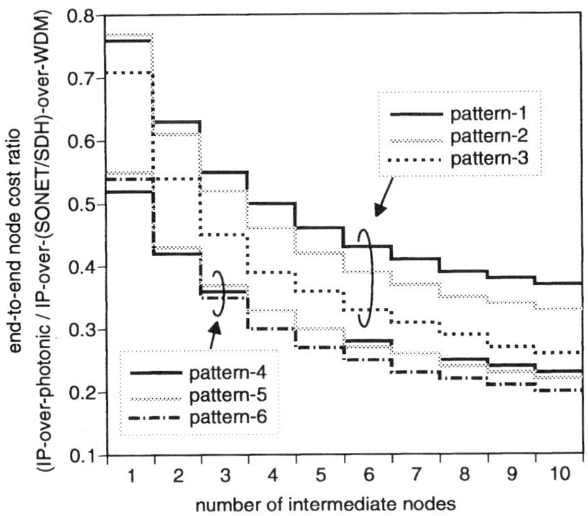

Fig. 7 End-to-end total node cost.

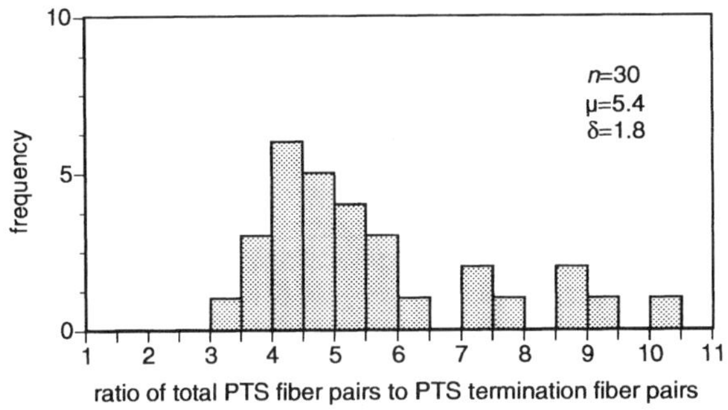

Fig. 8 Ratio of PTS total fiber pairs to terminating fiber pairs. (n: the number of samples, μ: average, σ: standard deviation)

one cabinet [10]). PTS-OLT cost is larger than WDM-LT cost since PTS-OLT must offer operation, administration and maintenance (OA&M) functions such as network protection while WDM-LT does not. Figure 7 shows the end-to-end total node cost ratio for 2.5-Gbit/s capacity. It is clear that the IP-over-photonic system is always cheaper than the IP-over-SONET/SDH-over-WDM system. As the number of intermediate nodes increases, this cost advantage increases. This is due to the increase in intermediate node cost with IP-over-SONET/SDH-over-WDM. For each intermediate node, the IP-over-photonic system needs only one PTS-XC while the IP-over-SONET/SDH-over-WDM system needs one IP router and two WDM-LT's.

The cost-effectiveness of the proposed system will be demonstrated for the physical network topology shown in Fig. 2(a). IP router cost utilizes the $C1$ capacity values listed in Table 2. Each node in this network example has the configuration shown in Fig. 5(a) or (c). That is, identical IP router capacity (= $C1$) is assumed for comparing IP-over-SONET/SDH-over-WDM and IP-over-photonic systems. The node cost ratio is the same as in the above cost evaluation. Assuming full-mesh connection in the network, see Fig. 2(a), a histogram of the ratio of the amount of incoming traffic (terminating and through traffic) to terminating traffic is shown in

Table 3 Overall network node cost ratio per bit. (overall network traffic of IP-over-photonic is five times that of IP-over-SONET/SDH-over-WDM)

pattern no.	network node cost ratio per bit of IP-over-photonic to IP-over-SONET/SDH-over-WDM
1	0.50
2	0.48
3	0.42
4	0.36
5	0.36
6	0.34

Fig. 8. The average is about 5:1 ($C1=C2\times5$). For evaluation simplicity, this ratio was assumed in all network nodes. Under this condition, PTS-XC capacity in Fig. 5(c) must be $C1\times5$ (n=5). Table 3 shows, for each cost pattern shown in Table 2, the overall network node cost per bit of the IP-over-photonic over that of the IP-over-SONET/SDH-over-WDM. It is clear that, in this network and node cost model, the proposed IP-over-photonic system lets the same network support five times more traffic than IP-over-SONET/SDH-over-WDM, with less than half the cost per bit. The proposed IP-over-photonic system is thus shown to realize large bandwidth IP backbone networks cost-effectively.

4. Conclusion

This paper proposed the WDM optical path-based IP backbone network. This proposal will realize IP-over-photonic systems in which the IP packets are directly accommodated in the optical path payload via layer-2 frames. A fat and robust optical pipe (optical path) is defined between logically adjacent IP backbone routers and wavelength routing in the optical layer is done at each intermediate node which eliminates electrical packet-by-packet routing at intermediate nodes. It was demonstrated that tera-bit class PTS's are required for constructing a long-haul network, and that they are now within our grasp. The proposed IP backbone network was shown to have the following advantages; (i) simple layer structure, (ii) end-to-end QOS capability, and (iii) cost-effective realization. From the view of IP router capacity, the proposed IP backbone network can be realized by lower capacity IP backbone routers than the IP-over-SONET/SDH-over-WDM network for the same network traffic demands. This is because the proposed system handles through traffic at the optical level, and only the terminating-traffic is handled by IP backbone routers. The cost-effectiveness against large traffic was demonstrated using Japan's long distance network model. Robustness similar to that offered by existing systems can be realized in the optical layer. The proposed physical layer technology will play a key role in constructing large bandwidth and robust IP backbone networks.

Acknowledgment

The authors are grateful to Mr. Naohide Nagatsu who carried out the estimation of PTS fiber pair requirements described in this paper.

References

[1] Network Wizards, “Internet domain survey,” Online available http://www.nw.com/, July 1998.
[2] K.Sato, S.Okamoto, and H.Hadama, “Network performance and integrity enhancement with optical path layer technologies,” IEEE J. Select. Areas Commun., vol.12, no.1, pp.159-170, Jan. 1994.
[3] W.Simpson, “PPP in HDLC-like framing,” IETF RFC 1662, July 1994.
[4] “Information technology — Telecommunications and information exchange between systems — High-level data link control (HDLC) procedures — Frame structure,” ISO/IEC 3309:1993, 1993.
[5] M.Laubach and J.Halpern, “Classical IP and ARP over ATM,” IETF RFC 2225, April 1998.
[6] J.Heinanen, “Multiprotocol encapsulation over ATM adaptation layer 5,” IETF RFC 1483, July 1993.
[7] C.Brown and A.Malis, “Multiprotocol interconnect over frame relay,” IETF RFC 2427, Sept. 1998.
[8] J.Carlson, P.Langner, and J.Manchester, “PPP over Simple Data Link (SDL) using SONET/SDH with ATM-like framing,” Internet Draft, <draft-ietf-pppext-sdl-00.txt>, Nov. 1998.
[9] N.Nagatsu, S.Okamoto, and K.Sato, “Optical path cross-connect system scale evaluation using path accommodation design for restricted wavelength multiplexing,” IEEE J. Select. Areas Commun., vol.14, no.5, pp.893-902, June 1996.
[10] M.Koga, A.Watanabe, S.Okamoto, K.Sato, H.Takahashi, and M.Okuno, “Optical path cross-connect demonstrator designed to achieve 320-Gbit/s,” Proc. ECOC ’96 , ThC.3.1, Sept. 1996.
[11] A.Watanabe, S.Okamoto, and K.Sato, “Optical path cross-connect system architecture suitable for large scale expansion,” IEEE J. Lightwave Technol., vol.14, no.10, pp.2162-2172, Oct. 1996.
[12] S.Okamoto and K.Sato, “Inter-network interface for photonic transport networks and SDH transport networks,” Proc. GLOBECOM ’97, pp.850-855, Nov. 1997.
[13] K.Kawai, K.Koike, M.Koga, Y.Takei, and H.Ichino, “2.5-Gbit/s SDH/SONET terminating circuit that uses low-power bipolar LSI technologies and multi-chip module technology,” IEEE Bipolar/BiCMOS circuits and technology meeting, paper 8.3, Sept. 1998.
[14] W. Simpson, “PPP over SONET/SDH,” *IETF RFC 1619*, May 1994.

High-Performance Backbone Network Technology edited by Naoaki Yamanaka
ISBN: 0-8247-5321-6 DOI: 10.1081/(120025724)
Published 2004 by Marcel Dekker, Inc.

Ultrafast Optical TDM Networking: Extension to the Wide Area

John D. Moores, Jeff Korn, Katherine L. Hall, Steven G. Finn, and Kristin A. Rauschenbach

SUMMARY Recent work in the area of ultrafast optical time-division multiplexed (OTDM) networking at MIT Lincoln Laboratory is presented. A scalable helical local area network or HLAN architecture, presented elsewhere as an architecture well-suited to ultrafast OTDM LANs and MANs, is considered in the context of wide area networking. Two issues arise in scaling HLAN to the wide area. The first is protocol extension, and the second is supporting the required bandwidth on the long-haul links. In this paper we discuss these challenges and describe progress made in both architecture and technologies required for scaling HLAN to the wide area.

key words: *optical network, optical time division multiplexing (OTDM), wide area network, transmission, regeneration, quality of service, bandwidth, protocol, packet, scalable, long-haul, experiment, numerical simulation*

1. Introduction

Growing demands for capacity have stimulated the development of high-speed optical shared-media networks. At present, wavelength division multiplexed (WDM) networks with WDM channel rates of up to approximately 10 Gb/s per channel and, limited by available electronics, are being aggressively developed and deployed. Ultrafast optical time-division multiplexed (OTDM) networks [1]–[5] with single channel rates in excess of 100 Gb/s may offer advantages over WDM networks, depending upon the nature of the users and traffic. Many advantages stem from using a single fast channel and digital processing at the 100 Gb/s rate. Specifically, OTDM advantages include: simpler management and control (single stream vs. many wavelength channels); the provision of truly flexible bandwidth on demand (BOD) service; packet switching or self-routing of packets; scalability in number of users (packet address space); the use of statistical multiplexing; the ability to support variable quality of service (QoS); and the use of digital regeneration, digital buffering, coding and encryption.

The acceptance and deployment of OTDM systems has been limited because the technologies needed to implement these networks are less mature than those needed for WDM networks, and also because significant challenges exist in scaling these systems to the wide area. Much of our previous work [1], [2] has centered on maturing the technology base, and significant progress has been made. Here we present some of the key challenges surrounding scaling OTDM to wide areas. One major issue is determining architectures that function well given the OTDM component limitations—limited buffering, switching and logic at 100 Gb/s rates. Another issue is the challenge of the underlying high capacity link technology. These issues are related in that the choice of the underlying link technology impacts the choices available in the architecture.

In this paper we begin by reviewing the salient features of our ultrafast OTDM architecture suitable for local and metropolitan areas (LANs and MANs). Then we describe some of the challenges in extending this ultrafast system to the wide area, and present some architecture options. Finally, we discuss recent progress in, and future possibilities for, high-capacity transmission over long distances that enable such extensions.

2. Ultrafast OTDM LAN/MAN

Within the realm of OTDM networks, multiplexing may be at the bit level, in which data from several sources, each at a rate less than the full channel rate, are multiplexed up to the channel rate. This is often referred to as bit-interleaved TDM [3], [4]. From the perspective of the network, this multiplexing approach is analogous to WDM, where each low-rate OTDM channel is comparable to each WDM wavelength. An alternative method that has significant advantages over bit-interleaved TDM or WDM is slotted TDM [1], [2], [5]. In a slotted TDM network, time is partitioned into slots of many bits, e.g. 10–100 kbits per slot. Users insert packets into these slots as dictated by the network protocols. The data rate within the packets can be as high as the channel rate, or as high as the aggregate rate in a WDM or bit-interleaved TDM system. Using this packet-based approach, the inherent statistical multiplexing improves network performance with regard to throughput, access time, and delay.

In previous work [1], [2], we have described the Helical Local Area Network (HLAN) architecture. This architecture was originally intended for slotted TDM

Manuscript received November 24, 1998.

[†]The authors are with Lincoln Laboratory, Massachusetts Institute of Technology, 244 Wood Street, Lexington, MA, USA.

*This paper is also published in IEICE Trans. Electron., Vol.E82-C, No.2, pp.157–169, February 1999.

IEICE TRANS. COMMUN., Vol. E82-B, No. 2, pp. 209–221, February 1999

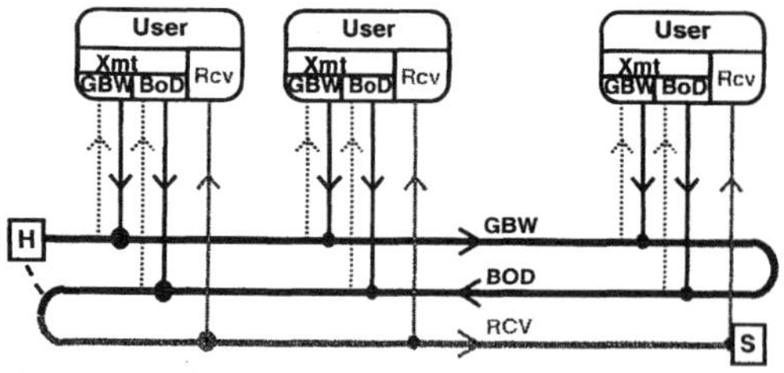

Fig. 1 HLAN topology. Unidirectional double-folded bus. Head-end (H) generates "empty" slots. Users with guaranteed bandwidth service (GBW) monitor the availability of slots and may transmit packets into empty pre-allocated GBW slots on the GBW segment. Users with bandwidth on demand (BOD) service monitor the availability of slots and, if a user has sufficient credits (see text), that user may transmit packets into empty slots on the BOD segment. Users read the headers of packets on the receive (RCV) segment of the bus and download those packets that are addressed to them. A slot sink (S) terminates the bus.

LANs or MANs, with channel rates of 100+ Gb/s, supporting a heterogeneous user population including certain very high-speed users operating at the majority of the full channel capacity in a bursty manner, as well as aggregates of lower rate users. The immaturity of all-optical processing technology dictates that the protocols should require a minimum of ultrafast processing, and should preferably avoid read-write-modify operations. Such operations are not required in HLAN. Both flexible bandwidth-on-demand (BOD) service and guaranteed bandwidth (GBW) service are required. HLAN is the only reported multi-access architecture that can provide both GBW and BOD in a high latency environment.

In one implementation of HLAN, the topology is a unidirectional double folded bus, as depicted in Fig. 1. The bus consists of three segments. The first is a guaranteed bandwidth segment on which users with guaranteed bandwidth service may inject packets at regular intervals with a guaranteed rate. The second is a bandwidth-on-demand segment on which users may inject packets when bandwidth is available, subject to a flow control strategy. On the third, or receive (RCV) segment, users receive packets but no packets are inserted. Users monitor the channel for available slots, with sufficient delay in the channel that the user may transmit a packet into the empty slot immediately after identifying that the slot is empty. The bus has a head-end with the global slot clock and with control and management electronics. The head-end monitors traffic on the network at the beginning of the receive segment, where the cable of the bus is in physical proximity to the head-end. Although there is a global slot clock, there need not be a global data clock at the channel data rate. Packets in neighboring slots may be asynchronous, at least in terms of bit phase, if not bit rate. A slot sink terminates the bus.

The architecture supports both GBW and BOD services and provides fair sharing of bandwidth under heavy loading, but is simple enough that minimal all-optical processing at the full channel rate need be performed. The architecture is frame-based. A frame consists of a number of slots, with the number being selected based upon the user population. Each slot is provided with a slot marker with three components. There is a self-synchronization code for time-alignment, there is a frame-start marker for those slots that are first in their frames, and there are credit markers for flow control, in certain slots selected by the head-end. A certain number of slots within each frame are reserved for the GBW users. Additional slots within a frame are provided for BOD use. Furthermore, any unused GBW slots are available for BOD use.

To provide fairness on the BOD segment, the head-end implements a credit allocation based medium access control scheme. The head-end generates credit markers that are sent out with the empty slots. Users have credit counters that are incremented, up to a finite maximum, when passing credit markers are detected. When a user transmits, filling a slot, that user's credit counter is decremented. If a user has no credits, that user cannot transmit. A flow control algorithm based upon the monitored traffic determines the rate at which the head-end distributes credits.

Under low traffic loads, HLAN mimics the greedy algorithm. The greedy algorithm is highly efficient and simple, but unfair under heavy loading. The greedy algorithm permits any user to transmit data whenever an empty slot passes by. Thus, the user nearest the head-end has an unfair advantage. Under light loads, this is optimal. With HLAN, under heavy loads, the rate of credit distribution is reduced, limiting the rate at which any user may transmit. Under heavy loading, the HLAN protocol approaches min-max fairness, where each user is given the same maximum fraction of the available bandwidth. Under overloaded conditions, low rate BOD users should have access to their full bandwidths, while the high rate users may be flow controlled.

In HLAN, each packet contains a header, in addition to the payload. The header contains addressing information and may contain additional information. The packet duration is less than a slot time. Guard intervals may be allocated to the leading and trailing edges of packets, to allow for switching of packets utilizing switching technologies that are not capable of switching in less than a bit interval. For example, electro-optic switches with, e.g., 20 GHz bandwidth might be utilized for switching entire packets, in which case it would be necessary to allot a minimum of a 50 ps interval between packets. Furthermore, packets in different slots need not be synchronous at the bit level.

Certain hardware is required in order to implement HLAN. For example, a GBW user may be assigned a particular slot within the framing structure in which to inject packets. That GBW user must count slots and also keep track of frame boundaries in order to inject packets into the proper slots. BOD users perform more complicated monitoring of the slots passing by. They read credit markers from the slot markers, and increment their credit counters. Furthermore, BOD users must determine whether slots are empty or occupied. This determination is performed upstream of the transmitter, so that a packet may be immediately injected into a slot that has been identified as empty. After transmitting on the BOD segment, the credit counter is decremented. On the receive segment, users tap off a fraction of the power in the bus. They must perform header recognition on every packet in order to determine if the packet is intended for that receiver. In the general case in which packets in different slots are asynchronous at the bit level, receivers must perform clock recovery on each packet. If the packet is identified as being destined for that receiver, the receiver feeds the packet into an ultrafast buffer or demultiplexes the packet and feeds it into parallel lower-speed buffers. Receivers have queues in order to accommodate packet influx exceeding packet readout rate.

The HLAN architecture provides both GBW and BOD service with QoS guarantees and with fairness. HLAN is suitable for high latency environments, where the packet duration is much less than the network propagation time, as might be the case in networks with ultrafast bit rates. The simplicity of the topology simplifies the protocols and hardware. The protocols are effective despite their simplicity, and the protocol simplicity keeps the all-optical ultrafast gate count to a manageably small number.

3. Extension of HLAN to Wide Area Networks

As with any multiple access system, there are a number of ways to extend HLAN to the wide area. The choice of the approach is not unrelated to the underlying link technologies. Today, the only method for achieving more than 100 Gb/s capacity over long distance is through the use of WDM. In this case, the transmission is provided in channels of up to about 10 Gb/s. Therefore, the HLAN stream would have to be inverse multiplexed at the physical layer in order to propagate over the long haul. As such, the only architectural option is to terminate HLAN connections and interface to another networking technology, e.g. WDM SONET, for transmission through a non-HLAN WAN. We refer to this as the WDM-based gateway architecture. This option is complicated by the need to preserve bit order from end to end, which is difficult given fiber-dispersion-induced delay among the channels as well as other electronic sources of delay jitter among different channels.

Still, the emergence of WDM transmission systems makes a gateway approach the most likely early method to wide-area extension of HLAN. In previous work [2] we have described the pulse width to wavelength conversion technology that would be required to interface the physical layer at the local and wide-area boundary. The buffering and variable delay required for bit-level reassembly would be handled with high-speed electronics in each channel at the receive-end local-to-wide-area boundary. In a later section, we describe in more detail the status of WDM transmission, and some recent work toward improving the performance high capacity WDM systems.

There are several limitations in the WDM-based gateway-architecture HLAN extension. First is the complexity of the gateway and inverse multiplexing systems. Second is that HLAN QoS and fairness are not preserved because of the protocol termination. Other methods for HLAN extension do not necessarily terminate the HLAN protocol. However, these options require that the long-haul transmission be supported in a single fast channel. In these cases, the single fast channel implies that no inverse multiplexing and bit-level reassembly are required.

Conceptually, the simplest architecture to use in a single fast channel case is a single HLAN that spans the wide area. Note that delay in the feedback used to allocate credits in HLAN could be long. The HLAN algorithm ensures stability despite the slow feedback, but the control would be sluggish. One major ramification is that fairness could only be guaranteed over long (i.e. seconds) time periods. This approach also requires that accurate slot timing be guaranteed over the wide area, which is difficult.

A second, somewhat more complex approach is to bridge together a hierarchy of HLANs, as shown in Fig. 2(a). Bridges would interface the different HLANs, aligning timing but still utilizing the HLAN protocol, including the packet and framing format. While this approach alleviates some of the timing issues, it still suffers from the sluggishness of the control channel. We note it is possible to modify HLAN to address the control system latency. One method would insert more feedback points within the network. These points could feed back into a centralized HLAN controller as in the original design of the HLAN, or, for even faster response, they could be used in conjunction with a distributed HLAN controller. We are currently studying such approaches. The challenge is to work within the limitations imposed by the technology—limited buffering and fast switching and logic capability.

A third architecture option for extending HLAN using single channel transmission is to use remote bridging, as depicted in Fig. 2(b). This is similar to the bridged hierarchy, but the HLAN protocol terminates on the remote bridges, and ultrafast links interconnect the remote bridges. By terminating the HLAN protocol, the problem of long feedback delays can be reduced. However, if bridges interconnect more than two links, then they must also perform routing functions,

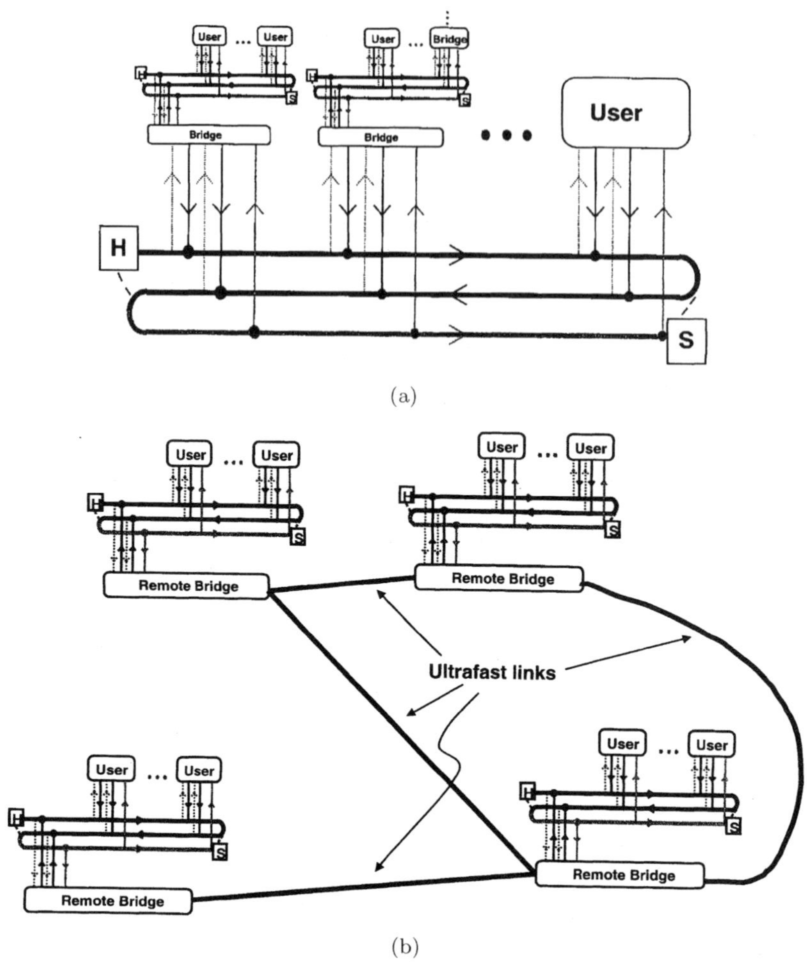

Fig. 2 (a) A WAN in the form of a bridged HLAN hierarchy. (b) A WAN in the form of an HLAN hierarchy with remote bridging.

and will therefore place considerable requirements on the ultrafast buffering technology. As a result, we feel the bridged hierarchy of HLANs is a better choice for extending HLAN to wide areas when a single fast channel is available. In a later section, we discuss the potential for realizing single fast channel transmission over wide areas and describe enabling technology.

In summary, we have discussed some of the architectures for extending HLAN to wide areas. When only WDM transmission is available, a gateway approach is required. Because WDM is currently the only demonstrated method to achieve the required transmission capacity, we feel this approach will be the most likely near term method of wide area interconnection. However, difficult inverse multiplexing, and the fact that the protocol must terminate at the local-wide area boundary are drawbacks. In the future, as fast single channel transmission systems emerge, we feel a bridged hierarchy of HLANs is a good architectural choice. We are currently exploring methods to improve HLAN for the bridged hierarchy architecture without severely impacting the simplicity and data rate scalability of the current protocol.

4. Transmission Technologies

As discussed above, one major issue in extending HLAN to the wide area is the requirement of a high capacity link technology. The required capacities have already been demonstrated using WDM technologies. However, there are still challenges in WDM transmission to provide systems with lower cost and higher capacity. One major advantage of the WDM systems is that the lower channel rates allow the use of electronic regeneration, on a per channel basis. Regeneration allows very long-reach links. However, electronic repeaters are costly, and a separate repeater is needed for each channel in the system. As a result, one current challenge for WDM links is to reduce the number of electronic repeaters necessary to reduce the cost. We describe work toward the goal in Sects. 4.1 and 4.2.

Although WDM is near-term, in a slotted OTDM network, high-speed single channel transmission is more desirable. Single channel systems are conceptually simpler, and, as discussed above, enable HLAN physical

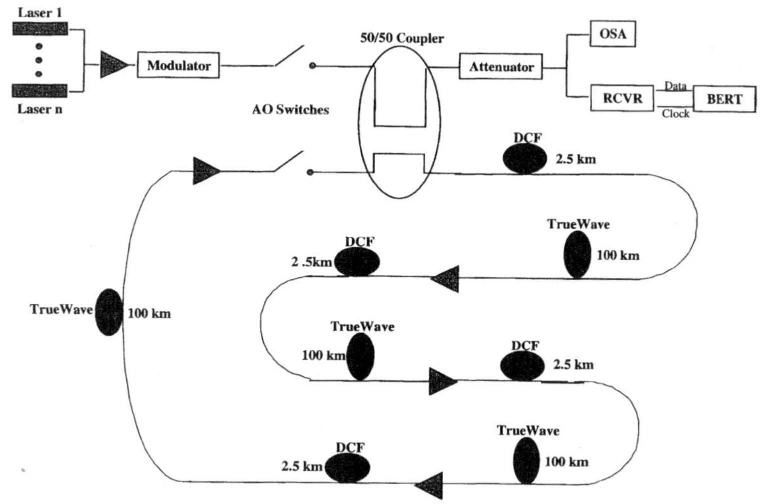

Fig. 3 Schematic of 400 km recirculating loop test bed as configured for NRZ experiments.

and protocol extensions over the wide area. However, high-speed transmission is very difficult due to the effects of noise accumulation, group velocity dispersion (GVD and dispersion slope, nonlinearities, polarization mode dispersion (PMD), polarization-dependent loss (PDL), and fiber amplifier gain non-flatness. All-optical regeneration is a potential solution to mitigate these effects, as we discuss in Sect. 4.3.

4.1 High-Speed Non-return-to-Zero WDM Systems

Terrestrial WDM transmission systems are rapidly being developed and deployed. For these systems, non-return-to-zero, NRZ, is the data format of choice. The 70 to 100 km spacing of the regeneration huts in terrestrial systems, however, limits the total propagation length of these systems [6]. WDM "hero" experiments [7]–[11] tend to utilize shorter amplifier spacings, improving noise performance. Current WDM commercial offerings regenerate every 400 km. We recently demonstrated error-free transmission of 12 WDM NRZ channels at 10 Gb/s each over 1600 km in a circulating loop using TrueWave^{TM}(TW) and dispersion compensating fiber (DCF) with a 100 km erbium-doped fiber amplifier (EDFA) spacing. This represents a four-fold improvement in the electronic regenerator spacing, and therefore, a substantial reduction in total system cost. In our experiment, the dispersion compensation is performed in each span. Moreover, we find [21] that placing the DCF immediately after each single-stage EDFA significantly reduces the nonlinear penalty [12].

The experimental setup is shown in Fig. 3. The transmitter consists of one external-cavity diode, eight DFB, and three DBR lasers star-coupled into a fiber. The wavelengths range from 1546.23 to 1555.02 nm, with 100 GHz spacing. The data stream is a 10 Gb/s NRZ pseudo-random bit stream (PRBS) of length $2^{23}-1$ bits, encoded via a $LiNbO_3$ Mach-Zehnder modulator. The light enters the transmission loop [13] via an acousto-optic (AO) switch and a 50/50 coupler. The loop itself consists of four 100-km spans of TW fiber. Each span has the requisite amount of DCF to bring the path averaged dispersion zero to 1550 nm. The net dispersion slope is 0.07 ps/(km-nm^2). The span losses (TW + DCF) range from 23.5 to 25.6 dB. The receiver uses an erbium-doped fiber preamplifier, followed by a tunable 18 GHz 3-dB bandwidth Fabry-Perot filter, an 18 GHz *pin* detector and a low-noise 12 GHz preamplifier. The signal is split for clock-recovery and then ac-coupled to a decision circuit.

Figures 4 and 5 show the effect of ordering the DCF relative to the TW in each span. Eleven channels were propagated through EDFA#1 and one span of TW+DCF, then through EDFA#2 and sent into the optical spectrum analyzer (OSA). When the TW is placed after the EDFA and the DCF follows the TW (post-compensation), four-wave mixing (FWM) is evident as indicated by the arrows in Fig. 4(a). Figure 4(b) shows the output spectrum with the DCF first (pre-compensation). FWM is greatly reduced in this configuration. Figure 5 was obtained by launching twelve 3-dBm channels into the loop and observing the spectrum after the 4 th round-trip (1600-km). For post-compensation, Fig. 5(a), spectral broadening is evident. The pre-compensation scheme mitigates broadening, Fig. 5(b). The difference between pre- and post-compensation can be understood from two facts. First, the DCF has a high loss ($\sim$ 4 dB per segment). Second, the local dispersion of DCF is high (-100 ps/nm-km). Thus, when the DCF is placed first, it not only attenuates the light prior to entering the TW fiber, but it also induces a large phase-mismatch, both between the different wavelength channels and within each channel, rapidly dispersing the pulses. It is anticipated that sys-

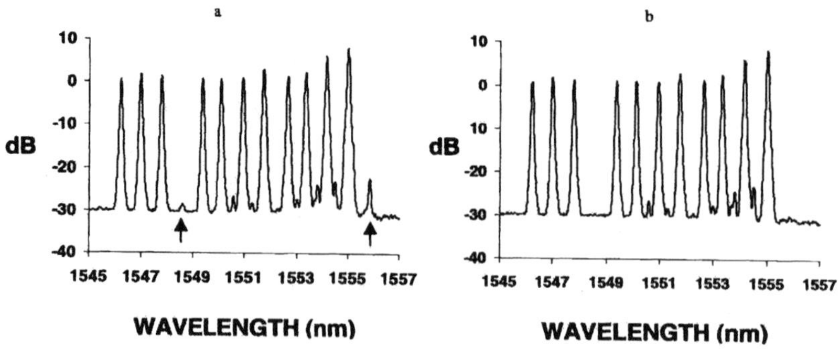

Fig. 4 Spectrum after a single span consisting of EDFA #1 TW and DCF: (a) TrueWave followed by DCF, (b) DCF followed by TrueWave. The arrows indicate FWM products.

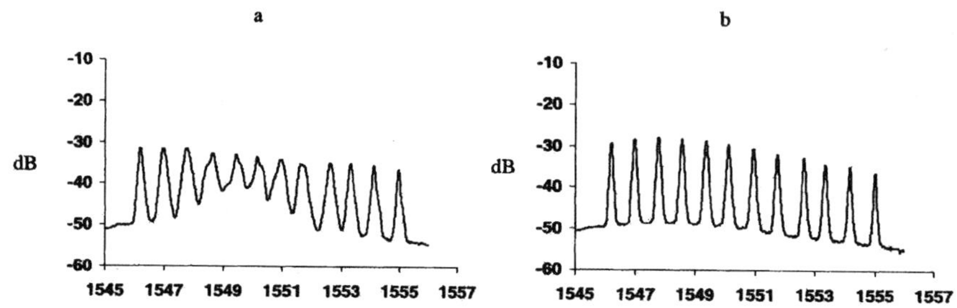

Fig. 5 Spectrum (resolution bandwidth = 0.1 nm) after 4 round-trips (1600 km) with: (a) TrueWave followed by DCF, (b) DCF followed by TrueWave. All channels have the same launched power of 3 dBm.

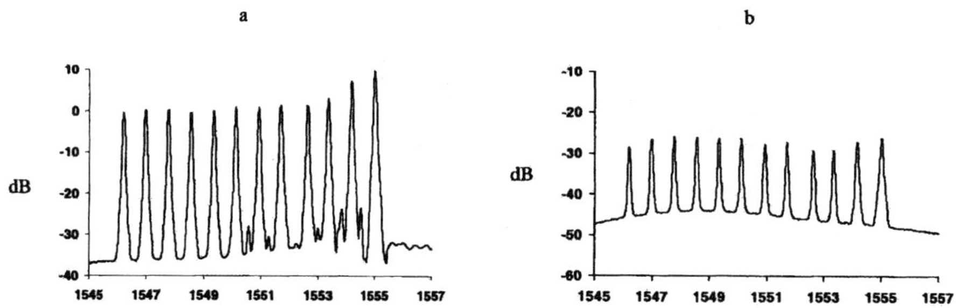

Fig. 6 (a) Spectrum launched into the first span of the loop, (b) spectrum after 4 round-trips (1600 km). Resolution bandwidth = 0.1 nm.

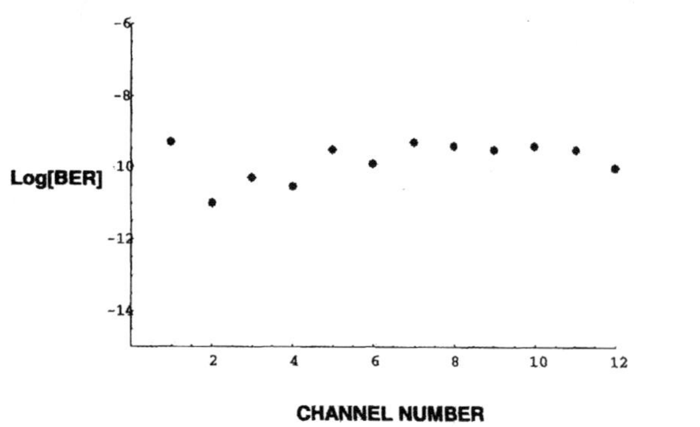

Fig. 7 Log[BER] versus Channel #. Channel 1 = 1546.23 nm. Channel 12 = 1555.02 nm.

tem performance could be further improved by using two-stage amplifiers with the DCF between stages.

Figure 6(a) shows the launched spectrum for a different experiment. Here, the two longest wavelength channels were pre-emphasized to 7.6 and 10 dBm, while the remaining ten channels had powers from 0 to 2.5 dBm. Pre-emphasis makes up for the lower EDFA gain for the longer channels. Figure 6(b) shows the spectrum after 4 round-trips. The BER for each channel was less than 10^{-9} as indicated by Fig. 7.

Because of the clear evidence of nonlinearity in these links, the use of return-to-zero (RZ), or pulsed, data formats may be beneficial in achieving long propagation distances. Traditionally, pulsed or RZ data streams have utilized soliton effects in links with anomalous path-averaged dispersion. For example, long-haul RZ transmission has been demonstrated for undersea systems, as discussed further in the next section. However, preliminary numerical studies suggest

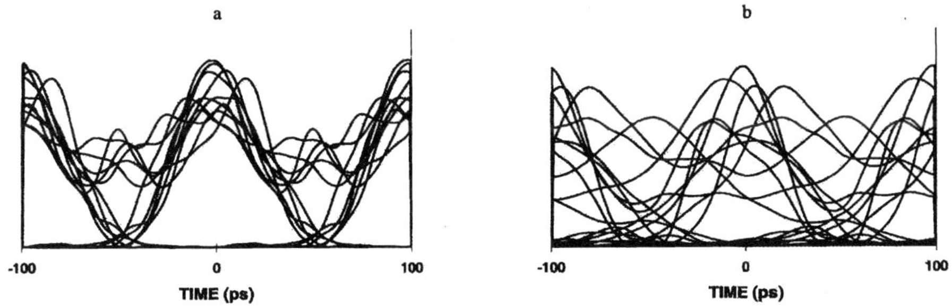

Fig. 8 Simulated optical eye diagram for: (a) 2000-km propagation of RZ transmission format, and (b) 1600-km propagation for NRZ.

that pulsed or RZ data streams may also offer transmission advantages in certain links where the path-averaged dispersion is positive, or normal. Figure 8 shows a comparison of simulated RZ and NRZ propagation for a 1547.8 nm channel in our transmission test-bed. The simulation used the same parameters as the loop experiment with 10 channels modulated at 10 Gb/s. The RZ pulse full width at half-maximum intensity (FWHM) was 30 ps and the NRZ pattern had 30 ps rise and fall times. In both cases the data stream peak power was 7.8 dBm. As can be seen qualitatively in Fig. 8, the RZ data survives 2000 km with a clean optical eye, whereas the NRZ eye is closed after 1600 km.

In summary, we have shown the importance of appropriate ordering of DCF and TW fiber in long-haul transmission links. Correct ordering may mitigate nonlinear effects in the scenario where compensation is performed at each amplifier. We have also demonstrated error-free performance of 120 Gb/s over 1600 km with an amplifier spacing of 100 km. Simulations suggest even greater link lengths may be achieved using RZ formatted data, in both dispersion regimes.

4.2 High-Speed Return-to-Zero TDM Experiments

Over the past decade, several major telecommunications companies around the world have supported ongoing efforts to push the envelope of optical fiber telecommunications [14]–[17]. These "hero" experiments have for the most part been aimed at undersea systems. Undersea systems have a number of attributes that distinguish them from terrestrial systems, such as inaccessibility of nodes after installation, simple topologies (lines or rings, with almost no branching, and without demultiplexing at the majority of nodes), and more difficult (remote) monitoring. Another distinction is the willingness of the major international carriers to employ relatively short amplifier spacings in undersea systems, with a corresponding increase in cost per unit length of cable, greatly improving the noise performance of the networks and improving survivability in the event of a repeater failure.

In terrestrial networks, however, short amplifier spacings may not easily be accommodated. As a result,

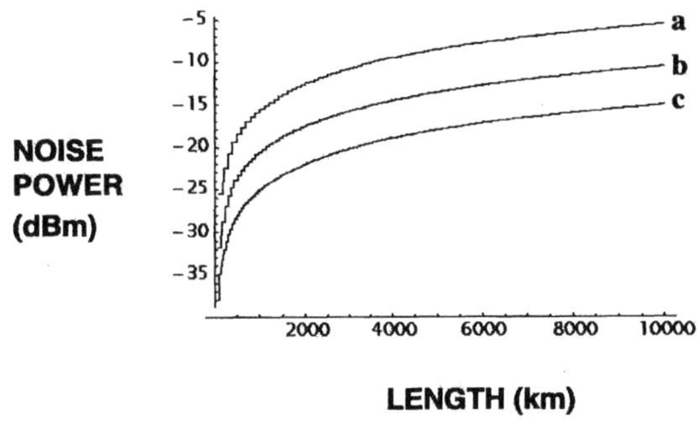

Fig. 9 Noise power accumulation (NEB = 0.2 nm) after + each EDFA in a chain of EDFAs for 3 different amplifier spacing: (a) 100 km (−25 dB span loss), (b) 75 km (−18.75 dB), and (c) 50 km (−12.5 dB).

our experimental transmission testbed, which models a terrestrial network, uses relatively long amplifier spacings. In addition, the loop was designed to have the dispersion-compensating period equal to the loss-compensating period, which is potentially advantageous in huts with add/drop capability. The 100 km spacing of the optical amplifiers, however, limits the total transparent propagation length due to noise accumulation [6]. By comparison, undersea links [13] typically use an amplifier spacing of 33 to 45 km, as do many of the dispersion-managed and path-averaged soliton experiments [18], [19]. Figure 9 compares the noise power accumulation for the situations where the amplifier spacing is 100 km, 75 km, and 50 km (assuming fiber loss is 0.25 dB/km). Longer amplifier spacing necessitates higher signal power for the same propagation length, which may lead to nonlinear impairments. Recently Marcerou et al. [20] reported 10 Gb/s transmission over 8200 km with 81 km amplifier spacing and sliding-guiding filtering. As discussed above, we have achieved error-free WDM transmission of 12 channels each at 10 Gb/s over 1600 km with 100-km amplifier spacing, using the NRZ format [21]. In that experiment, the transmission length was limited by nonlinearity, while the total number of channels was limited by gain non-flatness. Using a similar loop configuration we have used dispersion-managed [22], [23] RZ propagation to extend the error-free distance (EFD—which

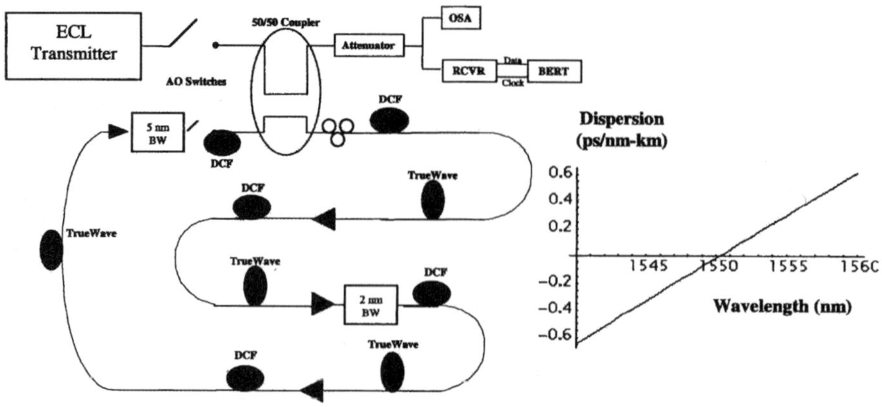

Fig. 10 Experimental setup and measured path averaged (per span, TW+DCF) dispersion for RZ transmission experiments.

we define to be the longest distance for which the BER $< 10^{-9}$). We obtain what we believe to be a record 8800-km error-free distance with 100-km EDFA spacing and without the use of controls (e.g., sliding-guiding filtering or synchronous in-line modulation).

The experimental setup and measured path average dispersion are shown in Fig. 10. The transmitter consists of a mode-locked tunable external-cavity semiconductor laser with 10 ps FWHM pulse width. The data consist of a 10 Gb/s PRBS of length $2^{15} - 1$ bits, encoded via a $LiNbO_3$ Mach-Zehnder modulator. The light enters the loop via an acousto-optic (AO) switch and a 50/50 coupler. The loop itself consists of four periods of the dispersion map. Each period contains 1.25 km of dispersion-compensating fiber (DCF) followed by 100-km of anomalous TrueWave (TW) fiber (2.7 ps/nm-km), 1 EDFA with a 4.5 dB noise figure, and another section of DCF to bring the path averaged dispersion zero to 1550 nm. The insertion loss of the DCF is approximately 3 dB, including splice losses. The net dispersion slope is 0.063 ps/(km-nm^2). Thus, a study of propagation in different dispersion regimes can be performed by changing the wavelength of the light source. Two- and five-nm bandpass filters were placed after EDFAs #2 and 4, respectively, to reduce ASE noise. The receiver includes an EDF preamplifier, followed by either a tunable 64 GHz 3-dB bandwidth fiber Fabry-Perot filter or a 0.8-nm BW filter (used only for the 1550 nm wavelength), and an HP lightwave converter with an electrical bandwidth of 11 GHz.

Experiments were performed separately at seven different wavelengths: 1558, 1554, 1552, 1551, 1550, 1548, and 1544 nm, in order to investigate propagation in the net anomalous-, zero-, and normal group velocity dispersion (GVD) regimes. The path averaged dispersions were 0.5, 0.25, 0.12, 0.06, 0, −0.25, and −0.5 ps/(nm-km) respectively. Figure 11 summarizes the results. As expected, the pulses with center wavelengths in the net normal dispersion regime do not propagate as far as those with center wavelengths in the net anomalous dispersion regime as a result of self-phase modulation and accumulated dispersion.

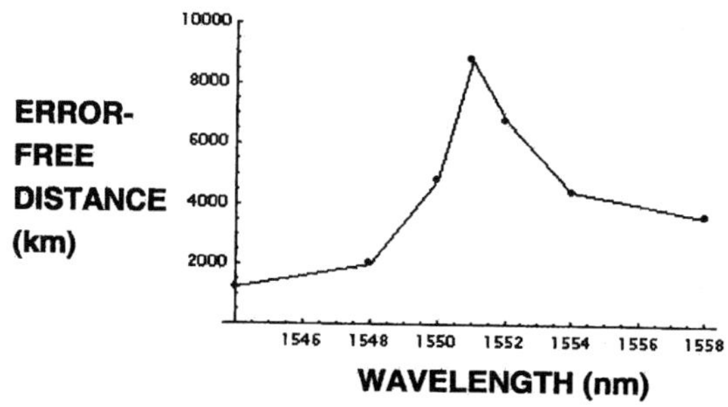

Fig. 11 Error-free distance for each wavelength.

The longest error-free distance was obtained at the 1551 nm wavelength, which survived error-free for 8800 km when the output power from the EDFAs was set to 8 dBm. The evolution of the eye and spectrum for this wavelength are shown in Fig. 12. After an initial period of spectral broadening, temporal broadening occurs followed by a damped oscillation to an apparent steady state.

In summary, a record propagation distance of 8800 km for 10 Gb/s optical data transmission has been set with 100-km EDFA spacing. Narrow (10 ps) RZ pulses are transmitted, suggesting similar results might be attainable at higher data rates. Also, we have studied propagation in the net normal-, zero-, and anomalous dispersion regimes. Best performance was achieved in the weakly anomalously dispersive regime.

4.3 Optical Regeneration

As we look to the goal of 100 Gb/s single stream links, physical layer considerations become severe. These considerations include polarization mode dispersion (PMD), additive noise, nonlinearities, dispersion slope, and noise-induced timing jitter, all of which tend to become more problematic as data rates increase. Crosstalk, including coherent crosstalk, and nonlinear

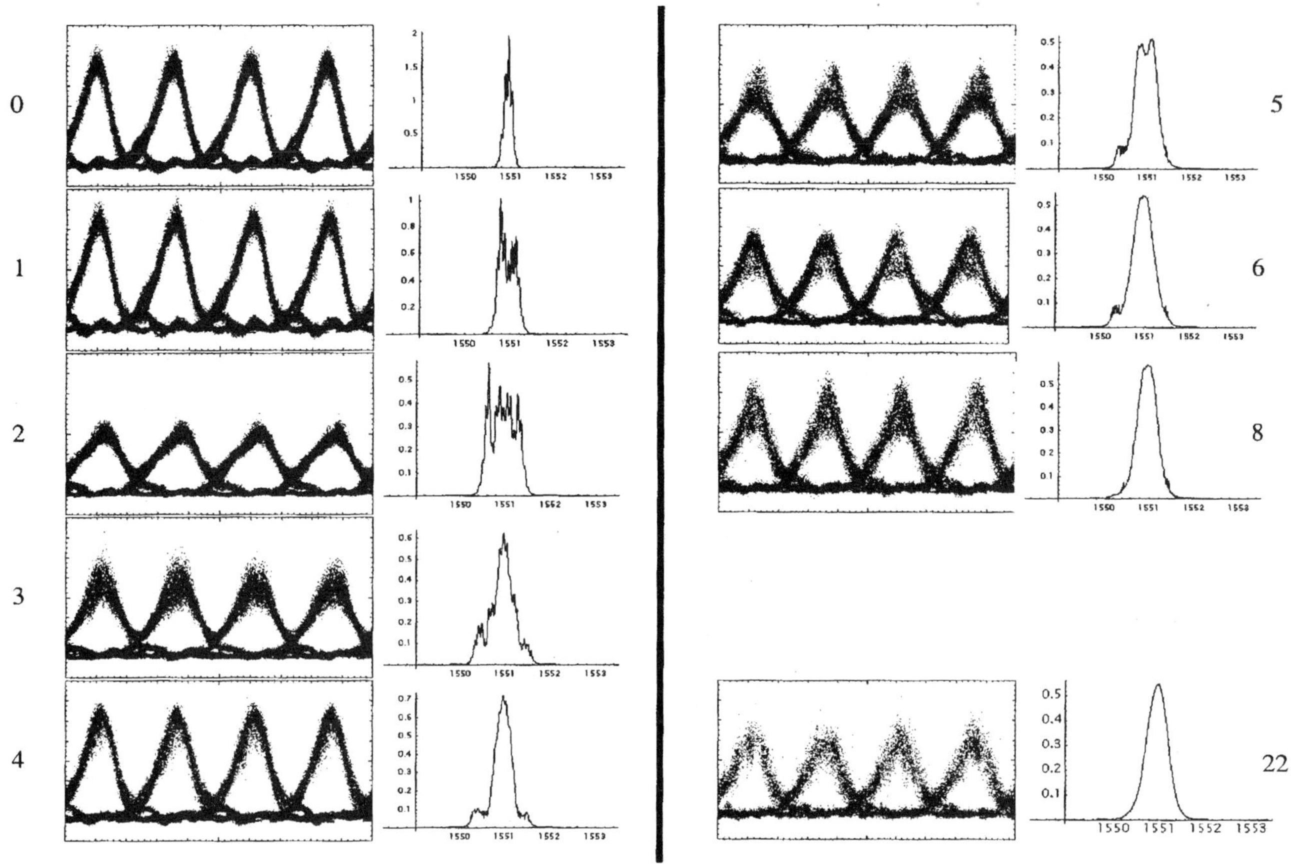

Fig. 12 Evolution of the 'eyes' and spectra for RZ propagation at the 1551 nm wavelength. The spectra are plotted on a linear scale. The number of round-trips is indicated beside each sub-figure. 1 round-trip = 400 km.

crosstalk in the transmission fiber, impose serious physical limitations. In these cases, regeneration becomes an important technique. In systems where the single channel rate is too fast for electronic regeneration, the regeneration may be performed all-optically.

For a proof-of-concept 100 Gb/s transmission experiment, short amplifier spacings (e.g., 25–45 km) could be utilized in order to minimize the impact of noise accumulation. However, such short amplifier spacings would not accurately reflect a typical terrestrial link, as described above. Rather, a realistic transmission experiment should have amplifier spacings on the order of 100 km. Such large spacings are problematic because the higher amplified spontaneous emission (ASE) noise levels necessitate high launched powers, exacerbating nonlinear effects. From a system perspective, at some point it is more cost effective to use optical regenerators (one or more in the network—not necessarily a regenerator at every amplifier) and longer amplifier spacings, rather than to use shorter amplifier spacings and no regenerators.

Numerical simulations demonstrate the tremendous potential benefits of regeneration. Figure 13 illustrates the distortion of a 32 bit 100 Gb/s optical data pattern without the use of regeneration or other types of compensation. In these results, propagation is at the zero dispersion wavelength of the fiber with third-order dispersion k‴ = 0.1 ps^3/km, the PMD is assumed compensated, and the launched pulses are 1 mW peak 3.64 ps FWHM raised cosines. The loss is 0.2 dB/km, the amplifier spontaneous emission factor is 3, and the amplifier spacing is 50 km. Third-order dispersion is the primary effect that washes out the pattern. By including numerical models for all-optical regenerators

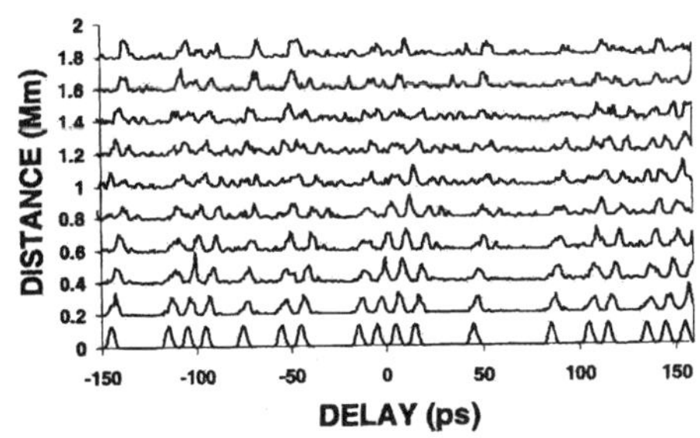

Fig. 13 Unregenerated simulation. 100 Gb/s, 1 mW peak 3.64 ps raised cosines, 50 km amplifier spacing, k‴=0.1 ps^3/km. Third-order dispersion is uncompensated and is the primary effect that washes out the data pattern.

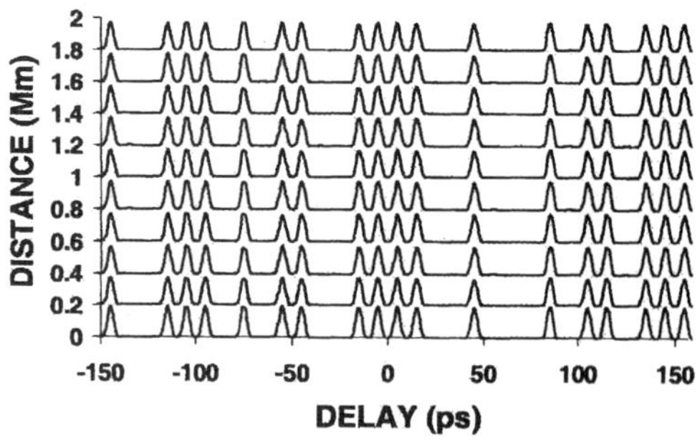

Fig. 14 Same as previous figure, but with optical regenerators every 200 km.

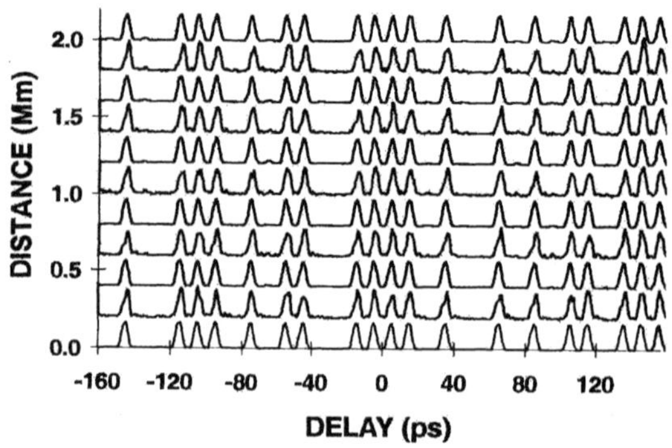

Fig. 15 2 Mm, 100 Gb/s, 1 mW 3.64 ps raised cosine transmission with 100 km amplifier spacings, 400 km regenerator spacings, D=0, k'''=0.1 ps^3/km, using lumped third-order dispersion compensation after each amplifier.

with the squared sinusoidal response of an interferometric switch every 200 km, and looking at the pattern right after each regenerator, we obtain Fig. 14. In this case, the data are much better preserved. Even better performance can be achieved using ideal third-order dispersion compensation, which might be achieved using specialty fiber [24] or a chirped grating [25]. This is illustrated in Fig. 15, where the result is of a simulation in which the amplifier spacing has been doubled to 100 km and the regenerator spacing has also been doubled to 400 km. The data pattern is well preserved at 2 Mm. These three simulations can be thought of not only in the context of long-haul transmission within a network, but also from the standpoint of data traveling between networks, with regenerators at the bridges or gateways between the networks.

To realize a regenerator at 100 Gb/s stream rates requires optical processing. One necessary element which has recently been demonstrated is optical clock recovery (possibly using a hybrid optoelectronic device) [26], [27]. Note that it is desirable for the local optical source to run at 100 Gb/s rates [28]. Clock recovery is used to synchronize the incoming data and local optical source. Subsequently, the incoming data can be used to modulate (on and off) the locally generated clock stream via an all-optical switch.

There are several choices available for optical switching, with interferometric optical switches showing the greatest promise for ultra-high speed operation [29]–[32]. A drawback of such switches for regenerator applications is the squared sinusoidal response [29], [33], which is much softer than the response a digital electronic switch, and therefore provides less accurate logic level restoration. Some softness is tolerable in that the action of successive regenerators will tend to restore logic levels asymptotically. For the same reason, better restoration can be achieved in each regenerator by cascading more than one switch [33] within the regenerator. Furthermore, cascading permits the use of optical frequency-shifting switches internal to the regenerator, with the original frequency restored at the output. In practice, these optical regenerators should be polarization-insensitive, and such switches have been fabricated [34]. Furthermore, these switches demand input pulses within a certain energy range, and the output pulse energies may have to be tailored to the output links, so the devices would in general have to be tuned to the particular input and output links. Optical regenerators of the type described here are suited to return-to-zero (RZ) data, and are compatible with non-soliton or soliton data. Optical semiconductor switches have also been utilized for regeneration [35]. NRZ optical regeneration has been demonstrated in an interferometric wavelength converter [36]. Regenerative memories have also been developed [37].

With an optical regenerator of the type considered here, using a local optical clock source, ultrafast optical clock recovery, and an optical interferometric switch, the 3 Rs of regeneration can be approximated. By using a timing-jitter insensitive optical switch, the regenerator can perform retiming [38], [39]. Reshaping is automatic, because clean locally-generated clock pulses are replacing the incoming data pulses, but inaccuracies in switching can lead to minor shape modifications. Reamplification is provided because of the replacement of pulses with the local source, but the soft response of the switching can give rise to energy fluctuations.

Recent literature contains work that applies the term "regeneration" to certain other optical data preservation schemes. These schemes typically involve clock recovery, followed by modulation (phase or amplitude). Optical bandpass filtering may be introduced as well. The clock recovery permits synchronization of the modulation with the incoming data. However, there is no local optical clock source in these schemes. Rather, the modulation (often combined with passive optical bandpass filtering) primarily provides a restoring force for the timing of the data, and may also provide some reshaping. But all three R's are typically comparatively weak in this approach. Lack of nonlinearity in the "regenerator" limits the capability for logic level restoration. Nonetheless, data preservation over very long distances has been achieved using these techniques

[40]–[42].

Choice of network topology is impacted by the amount of optical processing required. Ultra-high speed optical processing is costly and relatively immature. All-optical processors are practically limited to a few gates. Single buffers have been demonstrated at the research level but queues have not been demonstrated. There is no convenient low-loss variable optical delay, and there is no optical equivalent of a digital latch (similar to an optical regenerator, and there is no practical true 3R all-optical regenerator). Lack of such functionality makes efficient mesh topologies difficult to implement without optical-to-electronic conversion. For example, alignment of packets at each node is difficult to achieve. While it would be most straightforward to utilize optical regenerators in simpler topologies such as HLAN, there are proponents of regeneration for arbitrary mesh networks [43], requiring asynchronous regenerators with sufficient dynamic range to account for path length differences, as well as contention resolution. Slotted TDM systems in simple topologies such as buses and rings have the advantage that much of the processing can be done in electronics at the slot rate, and that variable delays are unnecessary. Regeneration can be comparatively easily introduced within such a network to increase the span of the network. Ultrafast clock recovery can be performed on the payload of each packet if necessary.

Whatever form of regeneration is utilized, it will certainly enable the increase of bit rate—distance products in optical networks. Regeneration can play a key role in providing robust interfaces between networks. All-optical regeneration in particular will be a powerful enabler for ultrafast TDM networks.

5. Conclusion

We have described architectures and technologies applicable to future ultrafast optical networks. We have focused on slotted TDM networks. The HLAN architecture, described in previous work, is well suited to LANs and MANs. Here we have discussed possibilities for extending HLAN into the wide area. When the underlying link technology must be channelized at less that the HLAN operating rate, as is the case with WDM systems, a gateway termination for both the physical layer and the protocol is appropriate. More options exist when high-speed single channels are available. One attractive option is a bridged hierarchy of HLANs.

We also discussed transmission limitations, with emphasis on terrestrial networks. The trend with terrestrial networks is to use long amplifier spacings on the order of 100 km, which imposes a severe noise penalty. We presented recent experimental results obtained with a 400 km recirculating loop with 100 km amplifier spacing. Error-free performance of 12×10 Gb/s over 1600 km, using the NRZ format, was obtained. Ordering DCF at the beginning of the link is important to mitigate nonlinear effects. Return-to-zero experiments have also been performed with this recirculating loop. RZ data at 10 Gb/s, with the narrow pulse width of 10 ps, have been propagated over 8.8 Mm in a 400 km recirculating loop of Lucent TrueWave^{TM} (anomalous dispersion) and dispersion compensating fiber (DCF).

Optical regeneration is an essential element in wide area single channel systems because of the transmission impairments suffered by the optical data. The advantage of utilizing a high-speed single channel is greater architectural flexibility in extending HLAN over wide areas. Ultrafast all-optical regenerators are feasible, but have yet to be demonstrated as full scale systems operating at 100 Gb/s rates. They are expected to be important enablers for wide area hierarchical HLAN-based ultrafast TDM networks.

References

[1] S.G. Finn, "HLAN—An architecture for optical multi-access networks," IEEE LEOS Summer Topical Meeting Digest, vol.45, 1995.

[2] R.A. Barry, V.W.S. Chan, K.L. Hall, E.S. Kintzer, J.D. Moores, K.A. Rauschenbach, E.A. Swanson, L.E. Adams, C.R. Doerr, S.G. Finn, H.A. Haus, E.P. Ippen, W.S. Wong, and M. Haner, "All-optical network consortium—Ultrafast TDM networks," IEEE J. Sel. Areas Commun., vol.14, no.5, 999, June 1996.

[3] B.Y. Yu, P. Tollver, R.J. Runser, D. Kung-Li, D. Zhou, I. Glesk, and P.R. Prucnal, "Packet-switched optical networks," IEEE Micro, vol.18, no.1, pp.28–38, Jan.–Feb. 1998.

[4] D. Cotter, J.K. Lucek, and D.D. Marcenac, "Ultra-high-bit-rate networking: From the transcontinental backbone to the desktop," IEEE Communications Magazine, vol.35, no.4, pp.90–95, April 1997.

[5] A. Hasegawa and H. Toda, "A feasible all optical soliton based inter-LAN network using time division multiplexing," IEICE Trans. Commun., vol.E81-B, no.8, pp.1681–1686, Aug. 1998.

[6] E. Desurvire, "Erbium-Doped Fiber Amplifiers," Wiley, New York, 1994.

[7] A.K. Srivastava, Y. Sun, J.W. Sulhoff, C. Wolf, M. Zirngibl, R. Monnard, A.R. Chraplyvy, A.A. Abramov, R.P. Espindola, T.A. Strasser, J.R. Pedrazzani, A.M. Vengsarkar, J.L. Zyskind, J. Zhou, D.A. Ferrand, P.F. Wysocki, J.B. Judkins, and Y.P. Li, "1 Tb/s transmission of 100 WDM 10 Gb/s channels over 400 km of TrueWave^{TM} fiber," Optical Fiber Communications Conference, vol.2, Paper PD10, 1998.

[8] S. Aisawa, T. Sakamoto, M. Fukui, J. Kani, M. Jinno, and K. Oguchi, "Ultra-wide band, long distance WDM transmission demonstration: 1 Tb/s (50×20 Gb/s), 600 km transmission using 1550 and 1580 nm wavelength bands," Optical Fiber Communications Conference, vol.2, Paper PD11, 1998.

[9] N.S. Bergano, C.R. Davidson, M. Ma, A. Pilipetskii, S.G. Evangelides, H.D. Kidorf, J.M. Darcie, E. Golovchenko, K. Rottwitt, P.C. Corbett, R. Menges, M.A. Mills, B. Pedersen, D. Peckham, A.A. Abramov, and A.M. Vengsarkar, "320 Gb/s WDM transmission (64×5 Gb/s) over 7200 km using large mode fiber spans and chirped return-to-zero signals," Optical Fiber Communications

Conference, vol.2, Paper PD12, 1998.

[10] H. Taga, N. Edagawa, M. Suzuki, N. Takeda, K. Imai, S. Yamamoto, and S. Akiba, "213 Gbit/s (20×10.66 Gbit/s), over 9000 km transmission experiment using dispersion slope compensator," Optical Fiber Communications Conference, vol.2, Paper PD13, 1998.

[11] N.S. Bergano and C.R. Davidson, "Circulating loop transmission experiments for the study of long-haul transmission systems using erbium-doped fiber amplifiers," J. Lightwave Tech., vol.13, p.879, 1995.

[12] G.P. Agrawal, "Fiber-Optic Communication Systems," 2nd ed., John Wiley & Sons Inc., 1997.

[13] N.S. Bergano, "Undersea Amplified Lightwave Systems Design," in "Optical Fiber Communications IIIA," eds. I.P. Kaminow and T.L. Koch, pp.302-335, Academic Press, Boston, 1997.

[14] M. Nakazawa, E. Yoshida, T. Yamamoto, E. Yamada, and A. Sahara, "TDM single channel 640 Gbit/s transmission experiment over 60 km using a 400 fs pulse train and a walk-off free, dispersion-flattened nonlinear optical loop mirror," Optical Fiber Communications Conference, vol.2, Paper PD14, 1998.

[15] S. Kawanishi, "Ultrafast time-division multiplexed optical communications," Optical Fiber Communications Conference, vol.2, Paper ThN1, 1998.

[16] S. Bigo, P Brindel, O. Leclerc, E. Brun-Maunand, and E. Desurvire, "Error-free 20-Gbit/s soliton transmission over 7150 km through all-optical synchronous phase modulation," Optical Fiber Communications Conference, vol.6, Paper WH1, 1997.

[17] L. Dong, M.J. Cole, A.D. Ellis, M. Durkin, M. Ibsen, V. Gusmeroli, and R.I. Laming, "40 Gbit/s 1.55 μm transmission over 109 km of non-dispersion shifted fibre with long continuously chirped fibre gratings," Optical Fiber Communications Conference, vol.6, Paper PD, 1997.

[18] J.M. Jacob and G.M. Carter, "Error-free transmission of dispersion-managed solitons at 10 Gbit/s over 24500 km without sliding-frequency filters," Electron. Lett., vol.33, no.13, p.1128, 1997.

[19] L.F. Mollenauer, J.P. Gordon, and P.V. Mamyshev, "Solitons in High Bit-Rate, Long-Distance Transmission," in "Optical Fiber Communications IIIA," eds. I.P Kaminow and T.L. Koch, pp.373–460, Academic Press, Boston, 1997.

[20] J.F. Marcerou, G. Bassier, J.P. Collet, S. Cohen, and N. Robin, "10-Gbit/s transmission over 8200 km with 81-km spacing using nonlinear non-soliton modulation format," Optical Fiber Communications Conference, vol.7, Paper TuI3, 1998.

[21] J. Korn and J.D. Moores, "12×10 Gb/s WDM transmission over 1600 km in a circulating loop using TrueWave^{TM} and DCF," Optical Amplifiers and their Applications Conference 1998, Paper TuD13, 1998.

[22] N.J. Smith, N.J. Doran, F.M. Knox, and W. Forysiak, "Energy-scaling characteristics of solitons in strongly dispersion-managed fibers," Opt. Lett., vol.21, p.1981, 1996.

[23] M. Nakazawa, H. Kubota, A. Sahara, and K. Tamura, "Marked increase in the power margin through the use of a dispersion-allocated soliton," IEEE Photon. Technol. Lett., vol.8, p.1088, 1996.

[24] K. Yonenaga, A. Matsuura, S. Kuwahara, M. Yoneyama, Y. Miyamoto, K. Hagimoto, and N. Noguchi, "Dispersion-compensation-free 40-Gbit/s×4-channel WDM transmission experiment using zero-dispersion-flattened transmission line," Optical Fiber Communications Conference, vol.2, Paper PD20, 1998.

[25] M. Durkin, M. Ibsen, M.J. Cole, and R.I. Laming, "1 m long continuously-written fibre Bragg gratings for combined second- and third-order dispersion compensation," Electronics Letters, vol.33, no.22, pp.1891–1893, Oct. 23, 1997.

[26] K.L. Hall, K.A. Rauschenbach, E.A. Swanson, S.R. Chinn, and G. Raybon, "Picosecond-accuracy all-optical bit phase sensing using a nonlinear optical loop mirror," IEEE Photonics Technology Letters, vol.7, no.8, pp.935–937, Aug. 1995.

[27] O. Kamatani and S. Kawanishi, "Ultrahigh-speed clock recovery with phase lock loop based on four-wave mixing in a traveling-wave laser diode amplifier," Journal of Lightwave Technology, vol.14, no.8, pp. 1757–1767, Aug. 1996.

[28] E.A. Swanson, S.R. Chinn, K. Hall, K.A. Rauschenbach, R.S. Bondurant, and J.W. Miller, "100-GHz soliton pulse train generation using soliton compression of two phase side bands from a single DFB laser," IEEE Photonics Technology Letters, vol.6, no.10, pp.1194–1196, Oct. 1994.

[29] K.J. Blow, N.J. Doran, and B.K. Nayar, "Experimental demonstration of optical soliton switching in an all-fiber nonlinear Sagnac interferometer," Optics Letters, vol.14, no.14, pp.754–756, July 15, 1989.

[30] N.S. Patel, K.L. Hall, and K.A. Rauschenbach, "Interferometric all-optical switches for ultrafast signal processing," Applied Optics, vol.37, no.14, pp.2831–2842, May 10, 1998.

[31] K.L. Hall and K.A. Rauschenbach, "100-Gbit/s bitwise logic," Optics Letters, vol.23, pp.1271–1273, Aug. 15, 1998.

[32] I.D. Phillips, A. Gloag, P.N. Kean, N.J. Doran, I. Bennion, and A.D. Ellis, "Simultaneous demultiplexing, data regeneration, and clock recovery with a single semiconductor optical amplifier-based nonlinear-optical loop mirror," Optics Letters, vol.22, no.17, pp.1326–1328, Sept. 1, 1997.

[33] B.K. Nayar, N. Finlayson, and N.J. Doran, "Concatenated all-optical loop mirror switches," Journal Modern Optics, vol.40, no.12, pp.2327–2332, Dec. 1993.

[34] K. Uchiyama, H. Takara, S. Kawanishi, T. Morioka, and M. Saruwatari, "Ultrafast polarisation-independent all-optical switching using a polarisation diversity scheme in the nonlinear optical loop mirror," Electronics Letters, vol.28, no.20, pp.1864-1866, Sept. 24, 1992.

[35] D. Wolfson, S.L. Danielsen, H.N. Poulsen, P.B. Hansen, and K.E. Stubkjaer, "Experimental and theoretical investigation of the regenerative capabilities of electrooptic and all-optical interferometric wavelength converters," IEEE Photonics Technology Letters, vol.10, no.10, pp.1413–1415, Oct. 1998.

[36] D. Chiaroni, B. Lavigne, L. Hamon, C. Janz, and A. Jourdan, "10 Gbit/s optically regenerated NRZ transmission experiment over 20,000 kms with 140 km repeater spacing," Optical Fiber Communications Conference, vol.2, Paper PD15, 1998.

[37] A.J. Poustie, K.J. Blow, and R.J. Manning, "Storage threshold and amplitude restoration in an all-optical regenerative memory," Optics Communications, vol.146, no.1-6, pp.262–267, Jan. 15, 1998.

[38] J.K. Lucek and K. Smith, "All-optical signal regenerator," Optics Letters, vol.18, no.15, pp.1226–1228, Aug. 1, 1993.

[39] J.D. Moores, K. Bergman, H.A. Haus, and E.P. Ippen, "Optical switching using fiber ring reflectors," Journal of the Optical Society of America B (Optical Physics), vol.8, no.3, pp.594–601, March 1991.

[40] M. Nakazawa, K. Suzuki, H. Kubota, A. Sahara, and E. Yamada, "160 Gbit/s WDM (20 Gbit/s∗8 channels) soliton transmission over 10000 km using in-line synchronous modulation and optical filtering," Electronics Letters, vol.34, no.1, pp.103–104, Jan. 8, 1998.

[41] S. Bigo, O. Leclerc, P. Brindel, G. Vendrome, E. Desurvire, P. Doussiere, and T. Ducellier, "All-optical regenerator for

20 Gbit/s transoceanic transmission," Electronics Letters, vol.33, no.11, pp.975-976, May 22, 1997.

[42] S. Bigo, O. Leclerc, and E. Desurvire, "All-optical fiber signal processing and regeneration for soliton communications," IEEE J. Selected Topics in Quantum Electronics, vol.3, no.5, pp.1208–1223, Oct. 1997.

[43] D. Cotter, J.K. Lucek, P. Gunning, A.J. Poustie, K.J. Blow, and R.J. Manning, "Ultrafast photonic self-routing," 11th International Conference on Integrated Optics and Optical Fibre Communications 23rd European Conference on Optical Communications IOOC-ECOC 97 (Conf. Publ. No.448). IEE. Part vol.2, pp.67–68, 1997.

High-Performance Backbone Network Technology edited by Naoaki Yamanaka
ISBN: 0-8247-5321-6 DOI: 10.1081/(120025725)
Published 2004 by Marcel Dekker, Inc.

Multiwavelength Opaque Optical-Crossconnect Networks

Evan L. Goldstein, Lih Y. Lin, and Robert W. Tkach

SUMMARY Over roughly the past decade, the lightwave-research community has converged upon a broad architectural vision of the emerging national-scale core network. The vision has been that of a transparent, reconfigurable, wavelength-routed network, in which signals propagate from source to destination through a sequence of intervening nodes without optoelectronic conversion. Broad benefits have been envisioned. Despite the spare elegance of this vision, it is steadily becoming clear that due to the performance, cost, management, and multivendor-interoperability obstacles attending transparency, the needs of civilian communications will not drive the core network to transparency on anything like a national scale. Instead, they will drive it to 'opaque' form, with critical reliance on optoelectronic conversion via transponders. Transponder-based network architectures in fact not only offer broad transmission and manageability benefits. They also make networking at the optical layer possible by offering to the nodes managed and performance-engineered standard-interface signals that can then be reconfigured for provisioning and restoration purposes by optical-layer elements. Because of this, the more pressing challenges in lightwave networking are steadily shifting towards the mechanisms that will be used for provisioning and restoration. Among these are mechanisms based on free-space micromachined optical crossconnects. We describe recent progress on these new devices and the architectures into which they fit, and summarize the reasons why they appear to be particularly well-matched to the task of provisioning and restoring opaque multiwavelength core long-haul networks.

key words: *optical networks, optical crossconnects, optical switching, WDM network architecture*

1. Introduction

WDM point-to-point transmission is fast emerging as the structural cornerstone upon which national-scale core communications will be built. Less clear, though, is whether the nodes of these WDM networks will be *transparent*—whether signals will propagate from source to destination through intervening nodes without optoelectronic conversion. Notwithstanding the past decade's momentum in favor of transparent WDM networks, we argue that civilian communications needs will in fact drive the core network not to transparent, but to *opaque* form [1], with critical reliance on the emerging class of small, inexpensive optoelectronic converter known as the transponder.

Manuscript received December 25, 1998.
Manuscript revised February 15, 1999.

†The authors are with AT&T Labs-Research, 100 Schulz Drive, Red Bank, N.J. 07701, U.S.A.

*This paper is also published in the IEICE Trans. Electron., Vol. E82-C, No.8, pp.1361–1370, August 1999.

There are two broad reasons for this: (1) transparency is on balance a liability in national-scale networks; (2) the alternative to transparency (call it 'opacity') is on balance a virtue in such networks. These facts, once appreciated, lead one naturally to a broad class of opaque optical-crossconnect networks whose optical-transmission, network-management, and interoperability features make them, unlike transparent systems, eminently engineerable on a national scale.

2. Opaque Networks

There are two long-standing arguments for building transparent WDM networks. The first concerns *format-independence*: transparent WDM networks can transport analog or digital or unspecified modulation formats on separate wavelengths of a single network. This is clearly true in principle, and it constitutes an elegant feature of WDM optics. It is our working assumption, however, that the case for network-transparency will not rest on this form of format-independence. Global momentum towards digital intensity-modulated systems in general, and towards SONET and SDH protocols and bit rates in particular, has fast become overwhelming. What is critical instead is that a single network be capable of transporting a diversity of services—SONET, ATM, and IP-based—at a restricted set of specified standard bit rates.

The second argument concerns *node-bypass*: transparent WDM networks readily allow express signals to bypass extensive electronic processing in intermediate nodes. This is clearly true for nodes consisting of a transparent fixed wavelength-add/drop-multiplexer, as shown in Fig. 1(a). It is a relatively small step to add reconfigurability, as in Fig. 1(b). From a transmission point of view, it is a substantial yet natural step to consider networks of transparent, reconfigurable wavelength-selective crossconnects, as in Fig. 11(c). All of the depicted network elements provide node-bypass, with its attendant prospects for cost-reduction.

Transparency is thus an appealing feature: it allows signals to propagate at high power levels through long chains of losses. But transparency also lets small degradations do likewise. This opens the system up to a familiar, but ever-expanding, menagerie of cumulative performance-degradations due to chromatic and polarization-mode dispersion, optical-fiber

IEICE TRANS. COMMUN., Vol. E82-B, No. 8, pp. 1095–1104, August 1999

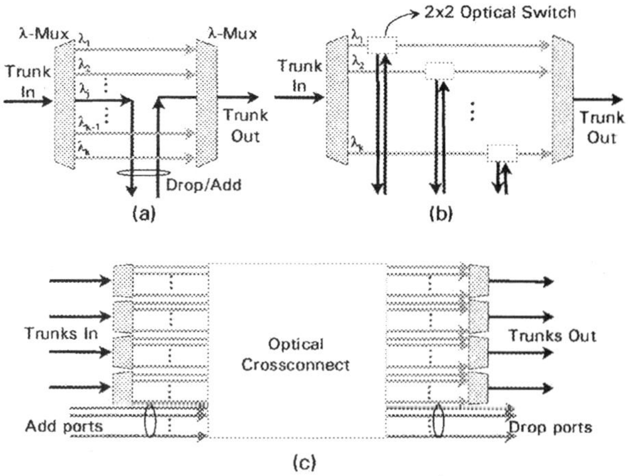

Fig. 1 Transparent implementations of a fixed wavelength-add/drop-multiplexer (a); a reconfigurable wavelength-add/drop-multiplexer (b); and a reconfigurable wavelength-selective cross-connect (c). All offer the benefits of node-bypass.

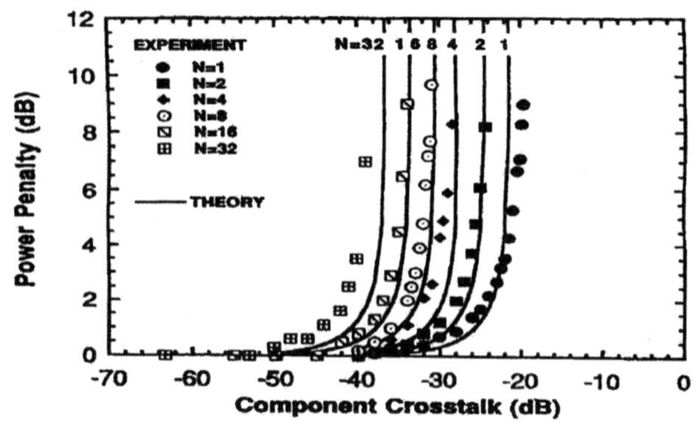

Fig. 2 Scaling of in-band component-crosstalk requirements with number of leakage paths N. Experimental points, from Ref. [7], were obtained for signal and crosstalk having matched polarization and optical center frequency, at a per-wavelength bit rate of 2.5 Gb/s. Curves result from simplified Gaussian theory. Sources of degradation other than crosstalk are excluded.

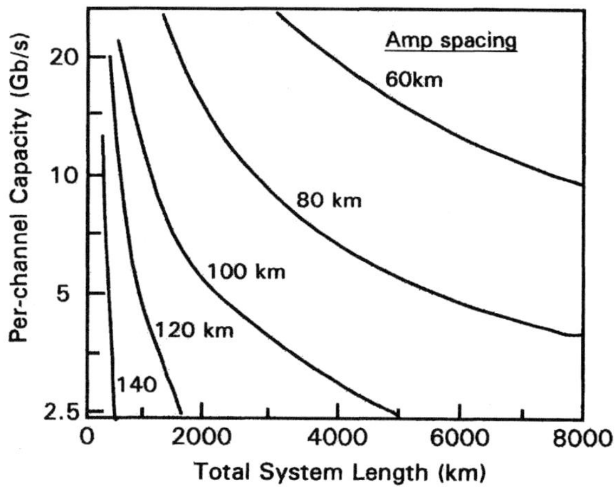

Fig. 3 Decline of supportable per-channel capacity with total system length, due to amplifier noise, in a transparent network with ideal optical amplifiers (after Ref. [6]). The amplifiers are assumed to have ideal 3-dB noise figures, flat gain precisely compensating inter-amplifier loss, perfectly equalized signal-input-power levels, and output power-levels of +7 dBm/channel.

nonlinearities, polarization-dependent loss, multipath interference, wavelength-misalignment of lasers and WDM filters, WDM-filter passband-narrowing, component crosstalk, noise-accumulation in ideal, flat-gain optical amplifiers, and a variety of gain-shape, cross-saturation, and compression effects in fiber amplifiers.

These accumulating impairments have lately been amply chronicled by a voluminous literature (e.g., [2]–[6]), to which we add three observations. First, even when considered in isolation from one another, the above effects impose on the components of transparent networks stringent requirements that scale with reach. In-band crosstalk alone, for example, in a network corrupted by a single leakage path and operating at 2.5 Gb/s per wavelength, requires component crosstalk values smaller than −20 dB [4]. This is itself not a trivial demand, but it steadily tightens with system reach. Thus, a network corrupted by 32 leakage paths—not a large number in a national-scale system—requires component crosstalk values smaller than −40 dB (Fig. 2) [7]. For each of the other impairments listed above, a transparent network will suffer similar accumulation, and thus a similarly relentless tightening of component requirements.

Transparency in WDM networks in general carries either performance costs or dollar costs or both, and these are offset only by savings in optoelectronic converters. This is evident from the accumulation of spontaneous-emission noise alone. Consider a transparent network whose amplifiers are ideal, with quantum-limited noise figures of 3 dB, flat gain spectra that precisely compensate fiber loss, perfectly equalized input-signal power-levels, and fairly aggressive amplifier output-power levels of +7 dBm/channel. Assume an operating margin of 6 dB. In such an idealized system, due to spontaneous noise-accumulation alone, the supportable per-channel capacity declines with system length as shown in Fig. 3. Thus, a network with ideal amplifiers spaced at 80 km, operating at 10 Gb/s per channel, will suffer intolerable noise-build-up after about 3,200 km if the nodes are loss-free; finite node loss will necessitate proportionate reach-reduction. Further cumulative impairments suffered by real amplifiers, including transient cross-saturation, intolerance to loss-variations, and non-flat gain spectra, will tighten these constraints [6].

If one wishes to scale a transparent network beyond these limits, there are only two options. The first option is to accept degraded noise performance. The second option is to decrease the repeater spacing below 80 km, as indicated in Fig. 3. Network operators have thus far resisted this on cost grounds.

Finally, the above comments naïvely consider only

a single impairment acting alone. In a deployed national-scale transparent network, however, a small zoo of cumulative impairments will act in concert. Recent research shows that this raises two serious concerns.

First, it has been found that two types of cumulative degradation that are individually benign can, when acting in concert, impose unexpectedly severe impairments. For example, the nonlinear-index-induced broadening in a long-haul network-segment, when transparently passed to a conventional-fiber local-exchange network, has been observed to produce severe dispersive error floors. These arise even in a system operating at less than 15% of its theoretical dispersion limit [8]. Moreover, such floors have been observed even for signals whose nonlinear broadening was wholly indiscernible using laboratory-grade optical spectrum-analyzers. This latter observation suggests that it will be difficult to erect signal-quality interface standards that guarantee performance in transparent long-haul networks, while at the same time being verifiable using field-deployable optical diagnostic equipment. If this is so, it is unclear how national-scale transparent networking would be made to work in practice.

Transparency in fact constrains the upgradability of the network in two ways. First, as noted above, a transparent network has by definition lower capacity than a regenerated network. Second, a transparent network must standardize technology choices at the outset, and cannot exploit, for example, developments in broader amplifier bandwidths or narrower channel spacings as they become available. Opaque networks by contrast can employ the best technology available in each system that is added, with no barrier to interconnecting systems with different technologies. The various transmission systems entering a node need not use the same channel wavelengths, number of channels, or even modulation formats. This graceful adoption of new technology while continuing to use the old—a sort of 'technology transparency'—represents one of the most powerful features of opaque optical networking.

We expect the above considerations to have the following impact. Both fixed and reconfigurable varieties of transparent wavelength-add/drop-multiplexing (Figs. 1(a) and (b) are currently on the brink of deployment over limited (not national-scale) domains. Reconfigurable wavelength-selective crossconnect networks (Fig. 1(c)), however, even without wavelength-translation, are expected to prove engineerable on a national scale only by sacrificing either network cost or network performance or both. Moreover, if transparent wavelength-translation is needed in such networks, this will require adding a feature that has not yet been demonstrated in plausible form in the laboratory.

It is an obvious but too-often-overlooked fact that the full variety of functional network elements in Fig. 1 is readily implemented in opaque form, by incorporating transponders at the interfaces of each transmission span. Opacity in this form exhibits exactly one bad feature: it requires numerous transponders. Notwithstanding their virtues, transponders are admittedly less than ideal in two respects. First, they are not free, although their precipitously dropping price trajectories, due partly to miniaturization, show no signs yet of leveling off. Second, transponders retime and reshape and are thus bit-rate-specific: increasing the bit rate requires transponder upgrade.

In all other respects, opacity is no vice in optical-crossconnect networks. It is instead a positive virtue. Opaque crossconnects arrest cumulative transmission impairments through effectively complete signal-waveform clean-up. They facilitate performance-monitoring and fault-detection by monitoring, as desired, any of various SONET signal-quality indicators (e.g., loss-of-signal and loss-of-frame) and parity-check bytes, and communicating the results to the optical-crossconnect system. Moreover, opaque optical crossconnects ease the task of fault-location since, by contrast with trans-parent architectures, their transponders offer a means of generating the alarm-insertion (AIS) signals upon which downstream-alarm-suppression ordinarily relies [9]–[11]. Beyond this, transponders not only ease the task of topology self-identification, by providing a way to insert identifiers; they also provide the function of wavelength-translation at no incremental cost [1].

Finally, and perhaps more important than all of the preceding considerations, opaque networks provide standard, nonproprietary interfaces. Since such open interfaces arise at the boundary between each WDM transport system and the optical crossconnects on which its endpoints terminate, the critical issue of multi-vendor inter-operability effectively vanishes. Thus, for a broad variety of reasons, of which physical-level transmission impairments are only the most obvious, it grows increasingly clear that practical core-long-haul networks will be deployed in opaque form.

3. Basic Restoration Architecture

Having resolved to build long-haul networks in opaque form, one still faces the far more complex challenge of choosing a restoration architecture. In multiwavelength networks, there are two fundamentally divergent ways of restoring failures, distinguished by the locations at which one places the restoration-switching elements. One approach is to respond to failures by rerouting individual wavelengths, or *wavelength paths*. This is achieved by placing switching elements on the "node side" of the network's wavelength-multiplexers, where they operate only on single-wavelength paths [11], [12]. The various affected wavelengths in a failed fiber may then be rerouted over a variety of restoration routes. Alternatively, one can respond to failures by

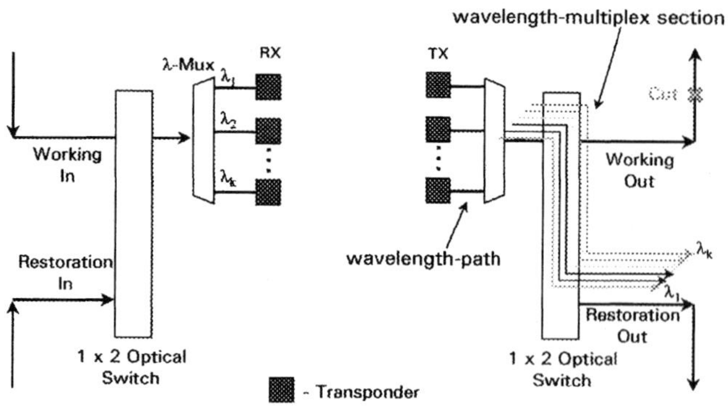

Fig. 4 Wavelength-multiplex-section-level restoration vs. wavelength-path-level restoration.

rerouting full sections of wavelength-multiplexed signals, or *wavelength-multiplex sections*, as depicted in Fig. 4. In this section we examine the potential benefits of restoration at the wavelength-multiplex-section (WMS) level, and point out the operational difficulties that face this approach. We focus in particular on the cost of such an approach, compared with wavelength-path-level (WP) restoration, when applied to national-scale long-haul WDM net-works.

WMS-level restoration in general offers two strengths, independent of the network's geographical scale. First, traffic is restored in large bundles (full wavelength-multiplex sections), thus reducing both computational complexity and restoration switch-fabric size. Second, restoration is carried out "on the network side" of the wavelength-multiplexer, thus avoiding the expense of duplicated transmitter and receiver line cards.

However, WMS-level restoration also faces three obstacles: (1) due to its coarse-grained nature, the approach necessarily uses restoration capacity less efficiently than do WP-level alternatives; (2) WMS-level approaches create optically transparent (unregenerated) domains within which it will be difficult to deploy multi-vendor transmission equipment; (3) WMS-level restoration paths, being both unregenerated and long in reach, will require the insertion of additional regeneration in order to satisfy transmission-engineering constraints. Thus, in general, restoring at the WP-level economizes on transmission facilities, including fiber amplifiers, while restoring at the WMS-level economizes on the aggregate costs of transmitters and receivers as well as cross-connects.

3.1 Network Assumptions

We examine a WDM mesh network whose topology is representative of the national-scale long-haul network, with approximately 450 switching offices (nodes) interconnected by 550 links. The links employ optical transport systems (OTS) supporting eight wavelengths, each

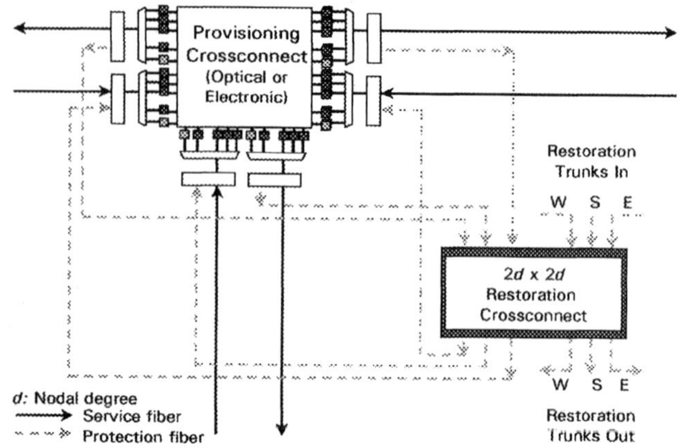

Fig. 5 Sharing of restoration facilities at the wavelength-multiplex-section level.

modulated at 2.5 Gb/s. Both the transmit and receive sides of the OTS are assumed to terminate on transponders. Transmission-engineering rules are assumed to permit 360-km transmission reach for systems built on 120-km amplifier spacings, and 560-km reach for systems built on 80-km repeater spacings. The network is assumed to support traffic projections for the year 2002.

In the WMS-level approach, OTS failures are detected by transponders, and result in reconfiguration of WMS-level restoration cross-connects, which then provide an alternate path between end-points of the failed OTS. The restoration crossconnect at each office, as shown in Fig. 5, has a size of $2d \times 2d$, where d is the number of optical transport systems terminating at this office. The restoration crossconnect can connect any of the failed service fibers to restoration facilities. These crossconnects are configured to allow the restoration facilities to be shared among multiple optical transport systems.

By contrast, we assume that WP-level restoration reroutes each affected connection between its end-points employing what has come to be called *end-to-end* or *source-based* rerouting [13]. We consider only single OTS failures, and assume that any such failure must be restorable.

3.2 Numerical Results

Under the above assumptions, WMS-level restoration is, as expected, found to be wasteful of transmission facilities. It on average requires 1.27 km of restoration fiber for each km of service fiber for the national-scale mesh network. WP-level restoration, by contrast, consumes 0.54 restoration km per service km. This finding is consistent with previous studies of WP-level restoration using end-to-end rerouting [11], [12]. The reason for this difference is granularity. WP-level restoration is able to utilize capacity more efficiently both because it can extract restoration capacity from underused service fibers, and because it can employ bifurcated routing techniques [14].

In addition, WMS-level restoration consumes additional regenerators, used to satisfy transmission-engineering rules on the long WMS restoration paths. We assume that these regenerators can only be placed at the offices. For the national-scale network outlined above, an average of 1.1 regeneration points are required in systems built on 80-km repeater spacings, and 2.1 such points per restoration path in systems built on 120-km repeater spacings.

These numbers can be reduced, however, by engineering the network so as to allow restoration paths corresponding to link-disjoint service OTSs to share regenerators. We employed a greedy algorithm to increase the sharing of regenerator locations, subject to transmission-engineering constraints [14]. The algorithm starts by assigning a regenerator to the location that can be shared by the largest number of paths. For those paths that have a regenerator assignment containing the selected location, all other assignments that do not include the selected node are discarded. The algorithm continues to select regenerator locations by assigning them to the location with largest sharing until, eventually, all paths satisfy transmission-engineering rules. Computationally, this turns on solving the Maximum Independent Set Problem [15], as outlined in Ref. [14]

When the greedy algorithm is applied to WMS-level restoration, one obtains a 31% reduction in regeneration points for systems employing a 120-km repeater spacing. By contrast, this declines to a 19% reduction for systems with 80-km spacings. The improvement is larger for 120-km spacings because one has more alternatives for assigning regeneration points to a given path, and thus more options for sharing.

On the other hand, WP-level restoration requires transponders at the receive and transmit sides of each office traversed by the restoration path, and it is this approach that consumes very large numbers of transponders. By comparison, WMS-level restoration reduces the number of transponders on restoration paths by 89% at 80-km repeater spacings, and by 83% at 120-km repeater spacings.

Given the above results, together with equipment costs, the relative economic merits of WMS-level and WP-level end-to-end restoration are readily calculated. Equipment costs are normalized to the cost of an optical amplifier (=1). Link costs include only the costs of lighting, with optical amplifiers, a fiber that is presumed to be already available. We assume that the same basic technology is used for both the WP-level and WMS-level cross-connect, so that costs per port for the two approaches are identical. With these assumptions, the total network restoration equipment cost is plotted in Fig. 6 as a function of the unit transponder

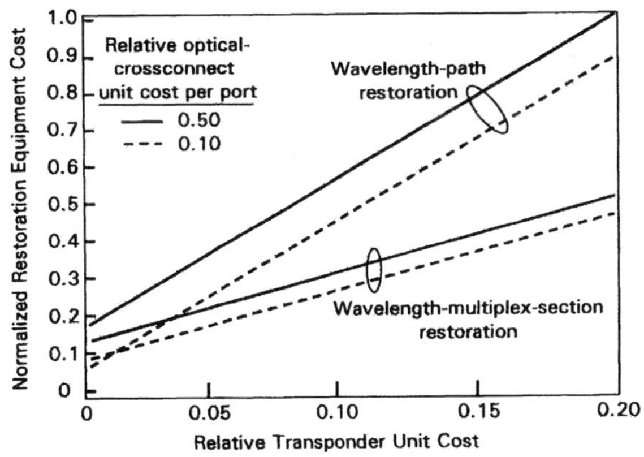

Fig. 6 Normalized restoration equipment cost as a function of unit transponder cost.

cost per port. Total cost is normalized to that of the most costly system plotted, namely WP-level restoration with unit transponder and cross-connect port costs of 0.2 and 0.5, respectively.

Because transponder costs are so rapidly dropping, only parametrized trend statements are possible. At transponder costs on the order of 0.2 to 0.4, roughly their value as of early 1999, aggregate system cost is seen to be dominated by the transponders. Thus, WMS-level restoration offers the short-term promise of substantial cost advantages. However, should transponder unit costs drop by an order of magnitude, as miniaturization trends would appear to suggest, this advantage largely disappears. Thus, neglecting the operational liabilities of WMS-level restoration summarized above, this approach offers the prospect of significant short-term savings. However, particularly given the swiftly dropping trajectories of transponder size and cost, the manageability, interoperability, and upgradeability associated with wavelength-path-based approaches make their adoption appear inevitable.

4. Free-Space Micromachined Optical Crossconnects

The immediate implication of the above is that the swiftly changing landscape of WDM networking, and the precipitous rise in aggregate per-fiber bit rates, are rapidly imposing the need for a new network element: a high-port count optical or optical-layer crossconnect. These network elements will likely be chiefly used for network restoration, to begin with. Substantial provisioning value will likely emerge thereafter. Optical switching technologies offer the potential advantages of bit-rate transparency, low power consumption, small volume, and low cost. Nevertheless, the requirements in port count (on the order of 1,000 ports in 3 to 5 years) and loss budget represent deep challenges that have not yet been met by any current photonic switching technology. Although conventional mechanical switches can achieve high optical quality, they are large in size and mass, and are thus relatively slow in switching speed. On the other hand, guided-wave solid-state switches, though compact, generally have high loss and high crosstalk. The inherent disadvantages of these technologies thus appear to limit their expandability to the port counts mentioned above.

By contrast, micromachined free-space optical-switching technology holds particular appeal in this application because it combines the advantages of free-space interconnection—low loss and high optical quality—with those of monolithic integrated optics, namely, compactness. Various small-scale (2×2) micromachined switches [16], [17] utilizing sliding micromirrors have been demonstrated. In addition, collimating optics and rotating micro-mirrors have also been proposed as a means of achieving high-density optical switches [18]–[20]. For the application of restoration and provisioning in core-transport lightwave networks, free-space micromachined optical switches (FS-MOS) with free-rotating hinged micro-mirrors are particularly attractive. This is because such applications do not require frequent switching, but do require very high reliability even for switch mirrors that remain in one switching state for extended periods on the order of years. Furthermore, the sub-millisecond switching times exhibited by FS-MOS devices are well-matched to the needs of restoration and provisioning in core-transport lightwave-communications networks. In the remainder of this paper, we review the design and performance of FS-MOS devices, and their particularly strong match to the application of optical-network restoration and provisioning.

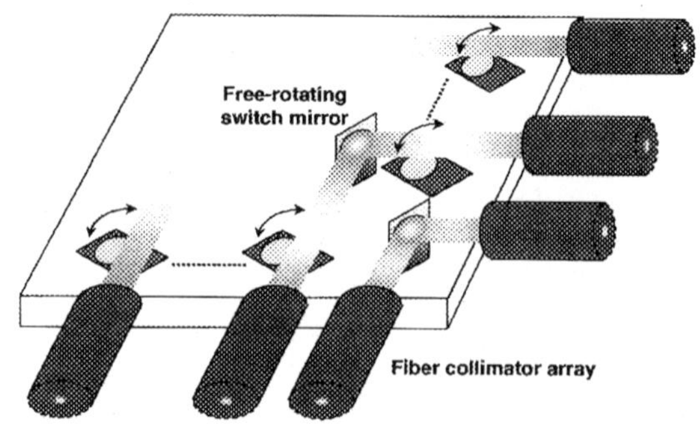

Fig. 7 Schematic drawing of the matrix free-space micromachined optical switch (FS-MOS).

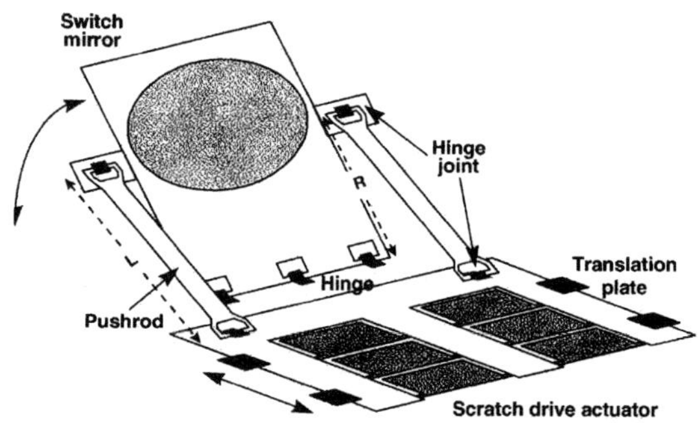

Fig. 8 Schematic drawing of the micro-actuated free-rotating switch mirror.

4.1 Device Design and Performance

The working principle of a matrix FS-MOS is shown in Fig. 7. The micro-actuated free-rotating mirrors are monolithically integrated on a silicon chip by means of surface-micromachining techniques. The collimated light is switched to the desired output port by rotating a selected mirror. Figure 8 contains a schematic drawing of the actuated mirror. Micro-fabricated hinges [21] anchor the mirror on the Si substrate. The modified interleaved hinges constitute the hinge joints, through which the mirror is connected to the translation stage by push rods. The translation stage is actuated by arrays of scratch-drive actuators (SDA's) [22].

FS-MOS chips utilizing mirrors of the above-described structure have been fabricated using the MCNC MUMPs fabrication process [23]. The epitaxial layers consist of one Si_3N_4 layer for insulation; three polysilicon layers for ground-plane purposes (poly-0) and for mechanical structures (poly-1 and poly-2); two phosphosilicate glass layers for use as sacrificial material; and one gold layer for use in fabricating electrical contacts and mirror coatings. The mirror and the translation plate are built on the second polysilicon (poly-1) layer, and anchored to the substrate through hinge and guiding rail structures built on the third polysilicon layer (poly-2). The SDA's are L-shaped polysilicon plates formed on poly-2. The hinge joints consist of interleaved poly-1/poly-2 structures, and are connected to the pushrods built on poly-2. After epitaxial growth, the sacrificial material is selectively etched by immersing the device in hydrofluoric acid, thus releasing the mechanical structures from the Si substrate. Figure 9 shows a top-view photograph of an 8×8 switch. The switch fabric occupies a 1×1 cm^2 chip.

The free-rotating micro-mirror structures described above offer precisely the kinds of switching speeds that core-transport-network crossconnects demand. Device measurements have shown sub-millisecond switching times when the SDA's are actuated with a ±100 V square wave at a frequency of 500 kHz. The maximum mirror-rotation angle is defined by the dimension of the ground electrode under the translation stage. Figure 10 shows the measured results for switching time. The switching time is 500 μs for rotating the mirror from the OFF position to the ON position. To this, one must add a 200-μs delay between the application of the switching voltage and the onset of measurable optical-switching action. For

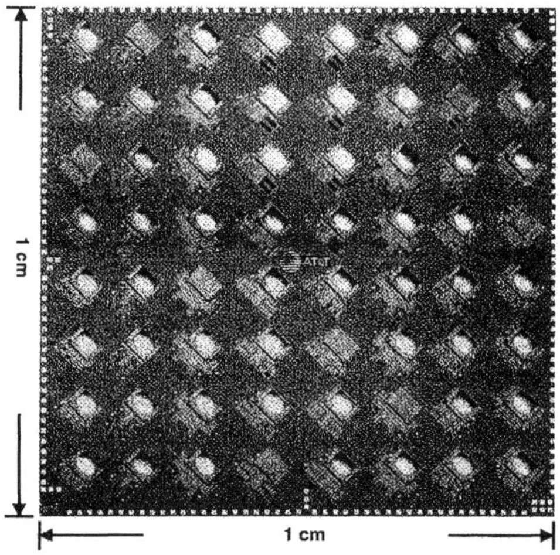

Fig. 9 Top view photograph of an 8×8 FS-MOS.

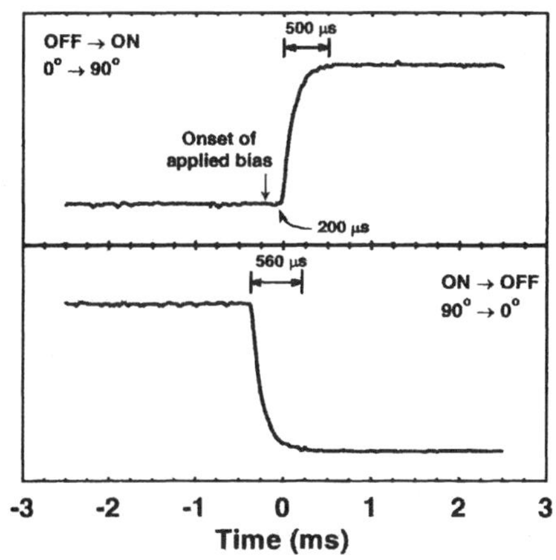

Fig. 10 Switching response of the FS-MOS.

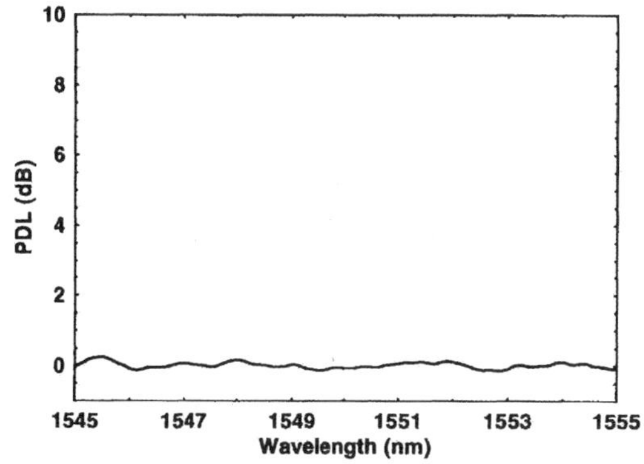

Fig. 11 Polarization-dependent loss (PDL) of FS-MOS versus wavelength. The device's PDL is below the noise floor of the measurement system.

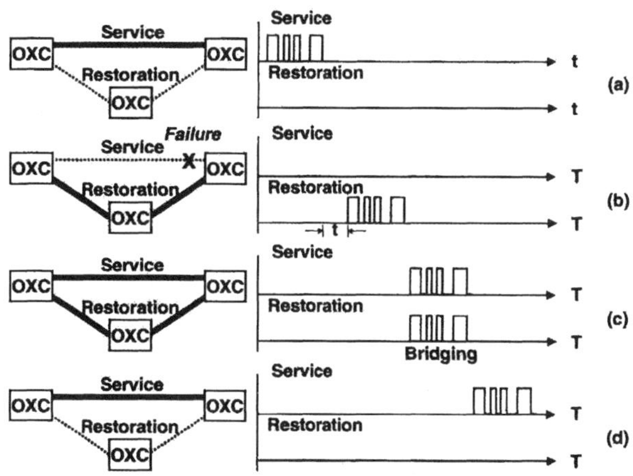

Fig. 12 Schematic illustration of the bridging operation in core-transport networks.

the first demonstration, the mirror is rotated down by pulling the translation plate back with polysilicon springs, resulting in an ON→OFF switching time of 560 μs. In practical systems, the springs would likely be replaced by bi-directional SDA's. The device's switching curve shows an extinction ratio of more than 60 dB. By employing fiber collimators for input and output coupling, losses as low as 3.1 to 3.5 dB are achieved for the shortest and longest optical paths in an 8×8 switch, respectively. Unlike guided-wave approaches, these devices tend to exhibit very low crosstalk values (<-60 dB) between adjacent channels without resorting to dilated architectures. Moreover, one substantial advantage of the FS-MOS is its low polarization-dependent loss (PDL). Figure 11 shows PDL as a function of the wavelength used to illuminate the switch ports. The device's PDL is seen to be smaller than the measurement system's noise floor of about 0.3 dB. In general, the high optical quality of these switching structures highlight the inherent virtues of free-space interconnection approaches.

4.2 Incorporating Enhanced Functionality: Bridging

One particularly appealing feature of free-space micromachined crossconnects is the ease with which rather complicated functionality can be incorporated into them. This is well-illustrated by the function of bridging. Optical-switching elements used for core-transport restoration must provide the critical function of bridging. This is illustrated in Fig. 12. Under normal operation (Fig. 12(a)), traffic traverses the service link. Upon link failure (Fig. 12(b)), it is switched to the restoration path, sustaining an interruption of duration T. Upon repair, traffic must be *bridged* to *both* service and restoration links (Fig. 12(c)), in order to avoid a second interruption due to imperfectly synchronized head- and tail-end switching. What is critical is that the above bridging capability be incorporated into the optical switch with sufficient simplicity to avoid exacerbating the already-challenging scaling demands incumbent on it.

This is readily accomplished for devices of the FS-MOS type. By integrating into the FS-MOS a few sim-

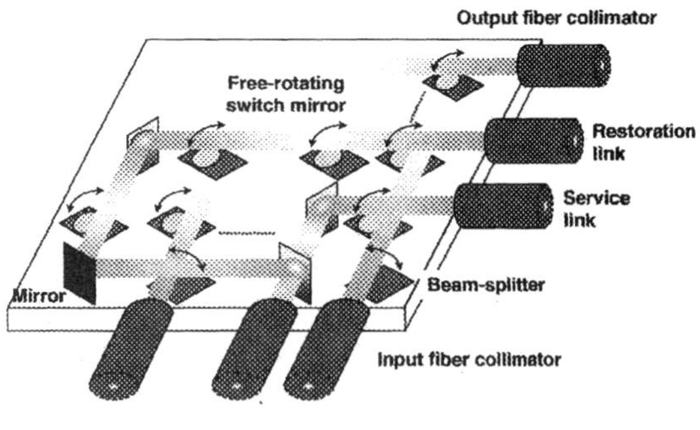

Fig. 13 Schematic of the FS-MOS incorporating the bridging function.

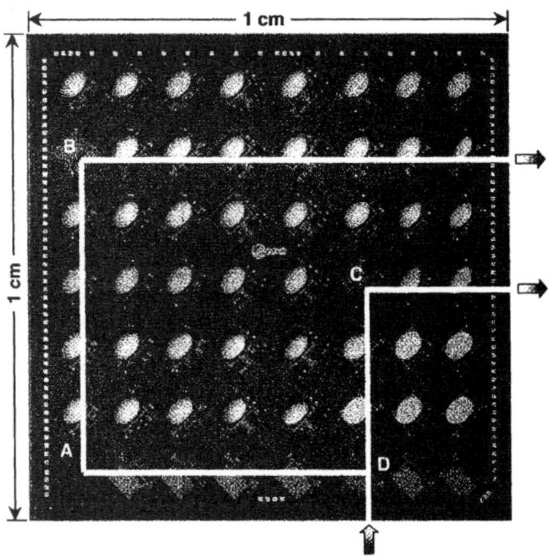

Fig. 14 Top-view photograph of the FS-MOS configured to execute bridging.

ple additional micro-optical elements, advanced functions like bridging can be incorporated at little cost in insertion loss or, ultimately, in scalability. Figure 13 contains a schematic drawing of the FS-MOS with bridging. An additional row of beam-splitters, an additional column of free-rotating mirrors, and a fixed mirror are incorporated into the switch fabric to achieve the any-one-to-any-two bridging function. To accommodate multiple simultaneous bridging operations, multiple copies of these elements could be integrated on a substrate. The beam-splitters are fabricated using the same process as the switch mirrors, except that they consist of a 1.5-μm polysilicon plate without any reflective gold coating.

The device's operation is most readily apparent from Fig. 14. This is a top-view photograph of the FS-MOS, with array elements configured to execute bridging. The corner mirror (A) has been assembled and is held perpendicular to the substrate by spring latches [21] and precision side-latches [24]. Switch mirrors B and C and beam-splitter D have also been rotated up, perpendicular to the substrate. During normal operation, only switch mirror C is rotated up, providing a

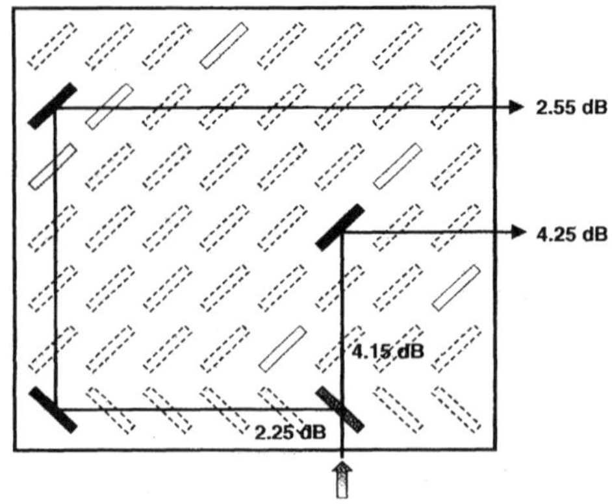

Fig. 15 Measured loss at 1.55 μm through the beam-splitter and mirrors of the FS-MOS.

light path from the input to the output service link. Upon service-link failure, A, B, and D rotate up, thus directing traffic to the restoration link. With C upright as well, traffic is bridged onto both service and restoration links. Finally, to carry out recovery, A, B, and D are lowered, with C left upright, so that traffic again flows on the service link only.

Among the chief determinants of scalability in these devices is loss. Here, the added losses due to the bridging structure are characterized. Insertion losses through beamsplitters and mirrors, for both service and restoration paths, are measured by coupling light at 1550 nm into the switch fabric using a fiber collimator. Losses along the optical paths were then measured with a large-area detector. Figure 15 shows the results. Insertion losses along the beamsplitter's transmission and reflection paths were 4.15 dB and 2.25 dB, respectively. The service path's switch mirror adds an additional 0.1-dB loss. Along the restoration path, the loss from beam-splitter to output is 0.3 dB. The total excess loss for the two paths is 0.31 dB, of which 0.09 dB results from scattering and absorption in the beam-splitter itself. This level of performance, though already reasonable, is likely to be substantially improved as these devices progress beyond the preliminary research stages. The results are suggestive of the kinds of sophisticated function that can be incorporated into such free-space devices at incremental cost in device complexity and optical performance.

5. Conclusion

The considerations outlined above, taken together, comprise an argument that core long-haul WDM networks will be built in opaque form, that they will likely restore at the wavelength-path level, that this raises great challenges for optical crossconnection, and that the resulting challenges are already beginning to spawn substantial new switching-device innovation.

Even when considered individually, the transmission-performance-, cost-, and network-management obstacles associated with network-transparency make it unlikely that a transparent, reconfigurable wavelength-selective crossconnect network could be engineered on anything approaching a national scale. When these considerations are taken collectively, the case seems quite overwhelming. By contrast, the corresponding virtues of transponder-based systems in general, and of networks with open interfaces in particular, suggest that opaque WDM optical-crossconnect-based networks will likely provide the functional value of their transparent counterparts while being engineerable on a national or global scale.

Still, having decided to build opaque networks, one can nonetheless carry out restoration in two fundamentally divergent ways: by restoring individual wavelength-paths, or by restoring full wavelength-multiplex sections. The two approaches offer largely orthogonal sets of virtues. Wavelength-multiplex-section level restoration appears to offer the prospect of significant economic benefits in national-scale WDM mesh networks in the short-term. These benefits appear poised to vanish, however, as transponder unit costs steadily decline. Since wavelength-multiplex-section restoration in addition carries with it the broad array of operational impediments associated with transparency, it appears that wavelength-path restoration will in the long-haul network be inescapable.

However, this imposes a need for large crossconnects whose port counts far outstrip current technologies. In response to this challenge, both optical and electronic switching technologies that have long been in quiet repose are beginning to stir. Indeed, on the optical side, broad new classes of switching technology have begun to arise, among which are the free-space micromachined optical switches described here. This technology is seen to achieve low loss, low crosstalk, low polarization-dependence, and fast switching speed. Moreover, such complex functions as bridging can be readily incorporated at marginal cost in optical performance and device complexity. In view of these virtues, together with the compactness conferred by integrated micro-fabrication and the high optical quality derived from free-space interconnection, devices of the type described here appear to hold substantial promise as restoration and provisioning vehicles in high-capacity core-transport lightwave networks to come.

Acknowledgement

The approach described here owes much to the thinking of Tom Afferton, Krishna Bala, Sid Chaudhuri, Bruce Cortez, Bob Doverspike, Ezhan Karasan, Leda Lunardi, Jon Nagel, Jane Simmons, and John Strand.

References

[1] K. Bala, R.R. Cordell, and E.L. Goldstein, "The case for opaque multiwavelength lightwave networks," Proc. IEEE/LEOS Summer Topical Meeting on Global Information Infrastructure, Keystone, CO, Aug. 1995.

[2] D. Marcuse, A.R. Chraplyvy, and R.W. Tkach, "Effect of fiber nonlinearity on long-distance transmission," IEEE J. Lightwave Technol., vol.9, pp.121–128, 1991.

[3] R.W. Tkach, A.R. Chraplyvy, F. Forghieri, A.H. Gnauck, and R.M. Derosier, "Four-photon mixing and high-speed WDM systems," IEEE J. Lightwave Technol., vol.13, pp.841–849, 1995.

[4] P.J. Legg, D.K. Hunter, I. Andonovic, and P.E. Barnsley, "Interchannel crosstalk phenomena in optical time-division-multiplexed switching networks," IEEE Photon. Technol. Lett., vol.6, pp.661–663, 1994.

[5] N.N. Khrais, A.F. Elrefaie, R.E. Wagner, and S. Ahmed, "Performance of cascaded misaligned optical demultiplexers in multiwavelength optical networks," IEEE Photon. Technol. Lett., vol.8, pp.1073–1075, 1996.

[6] J.L. Zyskind, J.A. Nagel, and H.D. Kidorf, "Erbium-doped fiber amplifiers for optical communications," in Optical Fiber Telecommunications IIIB, ed. I. Kaminow and T. Koch, pp.13–68, Academic, New York, 1997.

[7] E.L. Goldstein, L. Eskildsen, C. Lin, and Y. Silberberg, "Polarization statistics of crosstalk-induced noise in transparent lightwave networks," IEEE Photon. Technol. Lett., vol.7, pp.1345–1347, 1995.

[8] L.D. Garrett, R.S. Vodhanel, S.H. Patel, R.W. Tkach, and A.R. Chraplyvy, "Dispersion management in optical networks," Proc. Eur. Conf. on Optical Commun., Edinburgh, Scotland, 1997.

[9] A. McGuire, "Architectural models for supervisory and maintenance aspects of optical transport networks," Proc. Third IEEE Workshop on Network Management and Control, International Conf. on Commun., Montreal, Canada, 1997.

[10] K. Sato, "Photonic transport network OAM technologies," IEEE Commun. Mag., vol.34, pp.86–94, 1996.

[11] S. Okamoto, A. Watanabe, and K.-I. Sato, "Optical path crossconnect architecture for photonic transport networks," IEEE J. Lightwave Technol., vol.14, pp.1410–1422, 1996.

[12] Y. Hamazumi, N. Nagatsu, S. Okamoto, and K. Sato, "Number of wavelengths required for constructing optical path networks considering restoration," Electronics and Communications in Japan, Part 1, vol.78, pp.30–41, 1995.

[13] J. Anderson, B.T. Doshi, S. Dravida, and P. Harshavardhana, "Fast restoration of ATM networks," IEEE J. Select. Areas Commun., vol.12, pp.128–138, 1994.

[14] E. Karasan and E.L. Goldstein, "Optical restoration at the wavelength-multiplex-section level in WDM mesh networks," IEEE Photon. Technol. Lett., Oct. 1998.

[15] M.S. Garey and D.S. Johnson, Computers and intractability: A guide to the theory of NP-Completeness, W.H. Freeman, New York, 1979.

[16] S.S. Lee, E. Motamedi, and M.C. Wu, "Surface-micromachined free-space fiber optic switches with integrated microactuators for optical fiber communication systems," Transducers'97, Chicago, IL, June 16–19, 1997.

[17] C. Marxer, M.-A. Crétillat, N.F.de Rooij, R. Bättig, O. Anthamattern, B. Valk, and P. Vogel, "Vertical mirror fabricated by reactive ion etching for fiber optical switching applications," IEEE Workshop on Micro Electro Mechanical Systems, Nagoya, Japan, Jan. 26–30, 1997.

[18] H. Toshiyoshi and H. Fujita, "Electrostatic micro torsion

mirrors for an optical switch matrix," J. Microelectromechanical Systems, vol.5, no.4, pp.231–237, 1996.
[19] L.Y. Lin, E.L. Goldstein, and R.W. Tkach, "Free-space micromachined optical switches with submillisecond switching time for large-scale optical crossconnects," IEEE Photon. Technol. Lett., vol.10, no.4, pp.525–527, 1998.
[20] L.Y. Lin, E.L. Goldstein, J.M. Simmons, and R.W. Tkach, "High-density connection-symmetric free-space micromachined polygon optical crossconnects with low loss for WDM networks," Conference on Optical Fiber Communications, Postdeadline paper, San Jose, CA, Feb. 22–27, 1998.
[21] K.S.J. Pister, M.W. Judy, S.R. Burgett, and R.S. Fearing, "Microfabricated hinges," Sensors and Actuators A, vol.33, pp.249 256, 1992.
[22] T. Akiyama and H. Fujita, "A quantitative analysis of scratch drive actuator using buckling motion," IEEE Workshop on Micro Electro Mechanical Systems, Amsterdam, the Netherlands, Jan. 29–Feb. 2, 1995.
[23] http://www.mcnc.org.
[24] L.Y. Lin, S.S. Lee, K.S.J. Pister, and M.C. Wu, "Micromachined three-dimensional micro-optics for integrated free-space optical system," IEEE Photon. Technol. Lett., vol.6, no.12, pp.1445–1447, 1994.

High-Performance Backbone Network Technology edited by Naoaki Yamanaka
ISBN: 0-8247-5321-6 DOI: 10.1081/(120025726)
Published 2004 by Marcel Dekker, Inc.

Feasibility Demonstrations of Hyper-Media Photonic Information Networks Using Prototype WDM Broadcast-and-Select Local Network Systems

Masanori Ogawara, Atsushi Hiramatsu, Jun Nishikido, Masayuki Yanagiya, Masato Tsukada, and Ken-ichi Yukimatsu

SUMMARY This paper describes the implementation and demonstration of local networks for the hyper-media photonic information network (HM-PIN), a candidate for the information service platform offering broadcast and telecommunication services. In addition, the feasibility of the HM-PIN is also demonstrated using prototype local network systems. This local network adopts architecture based on wavelength-division-multiplexing (WDM) and broadcast-and-select (B&S) switching, and supports all HM-PIN services except inter-local-network communication. The major issues of this proposed network are the technologies that support many broadcast channels and reduce channel selection cost. This paper also considers the combination of WDM technology and three alternatives: electrical TDM, subcarrier multiplexing (SCM or electrical FDM), and optical TDM (O-TDM). Three 128 ch (8 wavelengths × 16 channels) WDM B&S prototype systems are built to demonstrate the feasibility of the proposed HM-PIN. In WDM/SCM, 30 and 20 Mb/s channels are realized as 16-QAM and 64-QAM, and 155 Mb/s channels are realized by WDM/TDM. Moreover, these three prototypes were connected to form a small HM-PIN and applications such as video distribution and IP datagram cut-through are demonstrated. Furthermore, the delay and throughput of the HM-PIN are evaluated by connecting a local network to a 200-km WDM-ring backbone network. Our discussions and demonstrations confirm the impact and feasibility of the proposed hyper-media photonic information network.

key words: *hyper-media, photonic switching, broadcast-and-select, WDM/TDM, WDM/SCM*

1. Introduction

The concept of a hyper-media photonic information network (HM-PIN) was proposed recently [1]–[3]. "Hyper-media" means a novel communication platform supporting various functions of the current communication media like TV broadcast, newspapers, magazines, and voice and data communications. This network concept has two big targets:

(1) establishing a universal infrastructure for integrating future information services with current telecommunication and broadcast services, and
(2) building a network technology by which the user channel capacities can be increased in concert with optical transmission technology evolution.

The HM-PIN consists of access links, local networks, and a backbone network, as shown in Fig. 1. Access links connect user terminals to a local network by using either a wireless or an optical fiber link depending on user's requirements. Their basic function is to establish point-to-point connections with a certain capacity and security. At this moment, the asynchronous transfer mode-passive optical network (ATM-PON) seems

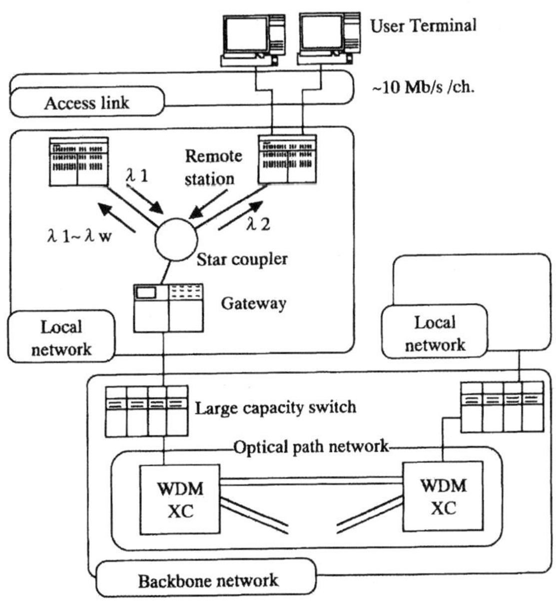

Fig. 1 Hyper-media photonic information network architecture.

Manuscript received June 21, 1998.
Manuscript revised September 17, 1998.

[†]The authors are with NTT Optical Network Systems Laboratories, Yokosuka-shi, 239-0847 Japan.
[††]The author is with NTT Technology Department, Tokyo, 163-8019 Japan.
[†††]The author is with NTT Network Service Systems Laboratories, Musashino-shi, 180-0012 Japan.
[††††]The author is with NTT PC Communications, Inc., Tokyo, 105-0004 Japan.
[†††††]The author is with NTT Telecommunication Network Laboratory Group, Tokyo, 160-0022 Japan.

*This paper is also published in IEICE Trans. Electron., Vol.E82-C, No.2, pp.265–273, February 1999.

IEICE TRANS. COMMUN., Vol. E82-B, No. 2, pp. 317–325, February 1999

suitable for supporting broadband bi-directional links [4].

Local networks are self-contained and support all network services in the HM-PIN except inter-local-network communication. They are based on the broadcast-and-select (B&S) switching architecture to achieve connection variety (e.g. point-to-point, multicast, broadcast). Each user connected to a local network can use one of the broadband broadcast channels and send his information to any other user in the network. Each user can also receive information through the selector located in a local network after the user is authenticated. In the HM-PIN, the local network technology is a key for supporting a variety of attractive network services.

The backbone network is a broadband information transfer network that connects all the local networks. The backbone network architecture is simplified and dedicated to simple point-to-point connections because of cost effectiveness and reliability. The capacity needed to carry all the traffic among local networks must be very large. The wavelength-division-multiplexing (WDM) optical path technology will be indispensable for realizing such high capacity easily and flexibly [5]. The ATM switching technology will be also used for flexible division of the optical path bandwidth to support many links between local networks [6], [7].

This paper discusses issues of implementing the photonic B&S local networks and presents the prototype systems. In addition, we show the results of feasibility demonstrations of the HM-PIN using these prototype systems. After briefly showing the architecture of local network in Sect. 2, alternative technologies for local networks are discussed in Sect. 3. Sections 4 and 5 show prototype local network systems and feasibility demonstrations of the HM-PIN using prototype systems, respectively. Conclusions are given in Sect. 6.

2. Architecture of Local Network

2.1 Changes in the Role of Switching Nodes

Due to the transmission cost reduction induced by optical technology, the role of the switching systems in a local network has started to change. Originally the switching systems were developed to reduce the transmission link cost by enabling the expensive link capacity to be shared by many users.

We note that while the cost of transmission links is rapidly falling, switching system cost reductions have not kept pace. As a result, the main role of a switching system is shifting from how efficiently it uses the transmission links to how effectively it shares the transmission capacity among users by using a variety of connection service functions. It is very important that switching network designs must recognize this change. Otherwise, the switching system may become a bottleneck in the future communication network. This means that the end-users see no improvements in service functions or quality even when the network capacity increases.

2.2 All-Optical Network Trials

Many recent projects and testbeds are pursuing the potential of WDM technology in long distance transmission [8]–[10]. But the target of these trials is not just to increase the transmission capacity but to establish technologies making a transparent switching system whole throughput automatically improves as the transmission technology improves. However, since the number of available wavelength is not large enough with the current technology, they try to make only the ultra high-speed point-to-point links for global backbone networks or very special users like super computer interconnections.

2.3 Photonic B&S Switching

In the HM-PIN, local switching is based on photonic switching in order to build a transparent switching system for ordinary users, including small business and home users. The key technology is WDM B&S switching, which has been applied to computer LAN/WAN [11] and TV studio distribution [12]. In a WDM B&S switching system, all information in the network is multiplexed on an optical fiber by a combination of electrical multiplexing technologies and WDM technologies, and distributed to all users as a WDM signal. Each user receives the intended information from the distributed signal by choosing the optical signal wavelength and the electrical channel.

The core of photonic B&S switching system consists of passive optical couplers and amplifiers, and is transparent. The optical signal is not directly provided to users and the capacity of an optical signal is divided into a number of channels that are assigned to different users. The channel capacity can be easily increased by replacing the multiplexers and selectors on the same core, when the transmission technology improves.

2.4 Structure of Local Networks in the HM-PIN

A local network consists of a star coupler, remote stations (RSs) and a gateway (GW). Because the B&S switching architecture is adopted, the local network can freely provide (logically) a full-mesh topology and a variety of connections. The RS is composed of a transmitter (including a pre-assigned-wavelength laser and a multiplexer), a distributor (an optical amplifier and a splitter), selectors (including tunable optical filters (TOFs), optical receivers, and electrical channel selectors (SELs)) for users, and terminal interfaces (Fig. 2).

The transmitter multiplexes information sent from users by means of an electrical multiplexing technique,

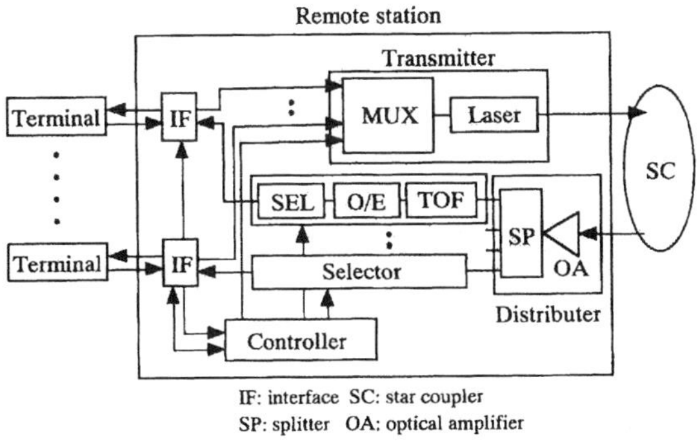

Fig. 2 Structure of a remote station.

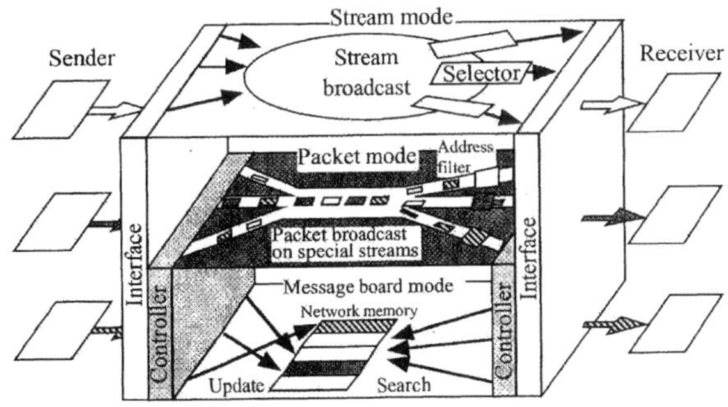

Fig. 3 Information transfer modes in the HM-PIN.

allowing users in each RS to share a single laser. The star coupler combines optical signals into WDM signals and distributes the WDM signals to remote stations. At the distributor in the RS, the WDM signals are amplified and distributed to all selectors so that each user can independently select a channel. Each user can receive his desired channel by demultiplexing the signals using a TOF and a SEL at the selector. Before selecting the channel, the controller authenticates the user's right-to-receive for the channel. The physical layer of the selector control mechanism can never be accessed by users; this gives the system inherent resistance to unauthorized access.

A GW is a special RS that has an interface with the backbone network. The idea of inter-local-network communication is to simply copy the broadcast channel in a local network on the broadcast channel in another local network. In the destination local network, the copied channel is handled as a local channel.

2.5 Information Transfer Mode

A local network based on B&S switching can provide a novel information service platform that supports three information transfer modes: stream, packet, and message board (Fig. 3). By combining these transfer modes, users can create a variety of information services.

The stream mode means communication on a broadcast channel. The channel bandwidth and quality of service (QoS) are guaranteed. Most broadcast channels are used in the stream mode. The expected applications are real-time voice/video transmission and connections between computers or routers. The capacity of each stream should be more than 10 Mb/s, which is sufficient for MPEG-2 video applications.

The packet mode is for receiving a small amount of information from many senders in parallel. This mode is achieved by multiplexing the data from the various senders on common broadcast channels. When a user receives the common channel, he can receive packets from all other users. The transfer delay and the bandwidth are not guaranteed because so many users share a common resource. This mode is useful for exchanging small amounts of data between computers or for sending signals for stream control.

The message board mode means the message broadcast via network common memory. A certain space in the memory is allocated to each user, and the user can write a message in this space; other users can search this memory space and find messages of interest. This mode is useful for applications based on broadcast messages, such as informing anonymous users of a sender's channel contents or user's conditions.

3. Alternative Technologies for Local Networks

3.1 Technology Requirements

There are three alternatives for WDM B&S architecture: electrical TDM, subcarrier multiplexing (SCM or electrical FDM), and optical TDM (O-TDM). We compare these alternatives from the viewpoints of scalability, upgradability, and cost. Scalability means the maximum system capacity achieved by the technology. Upgradability means how easily system throughput can be upgraded with each transmission technology advance.

Here it is important to note the difference between point-to-point transmission application and the B&S switching application. In point-to-point transmission, there is only one multiplexer and demultiplexer pair, but in the B&S system, there is one multiplexer with many demultiplexers. Thus the requirements for the cost reduction in the demultiplexer or selector circuits is much more severe in the B&S switching system than in the transmission system. Also problems occur when upgrading the system. Upgrading a transmission system means just replacing individual transmitter and receiver pairs. But in the B&S switching system, we have to replace many receivers at one time. Thus the comparison results here will differ from those for point-to-point transmission.

In order to reduce system cost, reducing the cost of

optical amplifiers and optical filters is very important, but these are common in all the alternatives examined. Accordingly, the following discussion concentrates only on the electrical circuits.

3.2 WDM/TDM Technology

The combination of WDM and electrical TDM is well established. 2.6-Tb/s transmission experiments have been achieved using this combination [13]. Thus there is a possibility of building a B&S switching system with more than 1-Tb/s throughput. System design is rather easy because the power budget, the optical filter crosstalk, and optical amplifier wavelength dependence are the main issues to be considered. The required optical power for a receiver is much smaller than that of a WDM/SCM system. Also it is easy to support high-speed channels.

The problems of this alternative are the cost and power consumption of the high-speed electrical circuits. For example, giving a 10-Gb/s receiver to each user and packing many receiver circuits on a small board seem impractical with the current technology. Also system upgrades are not easy because the TDM multiplexer/demultiplexers are incompatible if the clock speeds differ. That is, upgrading a 2.4 Gb/s/wave system to 10 Gb/s/wave forces us to upgrade all the receivers at the same time.

Thus this WDM/TDM combination is suitable for high-speed channel applications where many channels are required from initial installation.

3.3 WDM/SCM Technology

The combination of WDM and SCM is under development. SCM has been mainly applied to analog CATV distribution [14] and has been thoroughly investigated [15]. SCM has also been applied to distribute digital TV programs based on 16-, 64-, or 256-quadrature-amplitude-modulation (QAM), and a 10 Gb/s/wave capacity was achieved in a 16-QAM 20 Mb/s × 500 ch system [16]. As for the combination of WDM and SCM, experiments into 50 ch × 4 waves have been reported [17].

The advantage of WDM/SCM technology is the possibility of a dramatic receiver cost reduction by using the inexpensive electrical circuits developed for digital TV. VLSI technology for QAM decoder has been developed with the aim of digital TV set-top box [18]. The power required for this circuit would be much smaller than that of the TDM circuit. In addition, it is possible to upgrade the multiplexer (digital-analog modulator, up converter and electrical combiner) and the demultiplexers. Thus initial cost of this system would be small.

It is more complicated to design the WDM/SCM system than the WDM/TDM system because the former is more sensitive to transmission noise and distortion. System parameters must be controlled include carrier-to-noise ratio (CNR) and desired-to-undesired signal ratio (DUR).

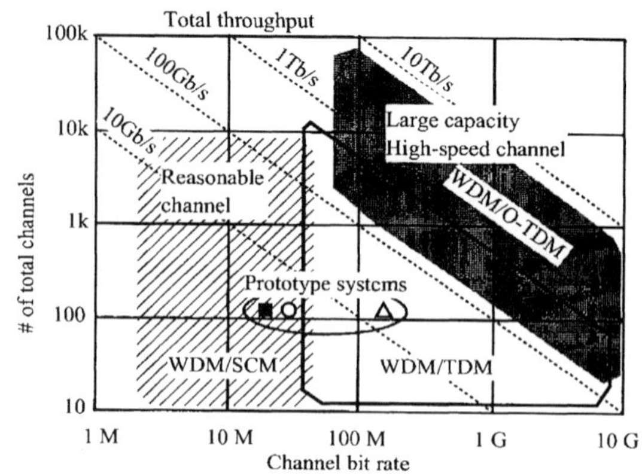

Fig. 4 Target areas of alternative technologies.

3.4 WDM/O-TDM System Design

The combination of WDM and O-TDM has great potential in building a very large B&S switching system with more than 1 Tb/s. It has been found recently that this technology is very practical even for low-speed channel demultiplexing [19]. This is because the received optical power decreases with the channel bit rate and very small non-linear optical effect can be utilized. Although this technology is still immature, a 100 Mb/s × 10,000 ch system does seem feasible with this technology. Short pulse generation, timing control, and synchronization are the main issues remaining.

3.5 Targets of the Technology Alternatives

By summarizing the above discussions, we reached the following conclusions:

(1) Although the alternatives offer different maximum channel bit-rates, all can support more than ten thousand channels.
(2) For channels of 20–30 Mb/s, WDM/SCM has high potential achieving very low system cost since the low-cost electrical circuits for digital TV set-top box can be applied and the initial cost is also small; WDM/TDM is necessary if high-speed channels are to be supported.
(3) WDM/O-TDM is the only technology that can achieve more than Tb/s throughput at reasonable cost. Figure 4 summarizes the target areas of these alternatives for a WDM B&S system.

4. Prototype B&S Local Network

4.1 General Description of the Systems

As the first step of the feasibility study of local network, three prototype systems, 128 ch (8 wavelengths × 16 channels) were fabricated as shown in Fig. 5 [20]. One was based on WDM/TDM, and the others on WDM/SCM. Various implementation alternatives were tested in these systems. The important parameters of the three systems are summarized in Table 1. For each system, two RSs were actually built and connected to an 8 × 8 star coupler by single mode fibers.

In all systems, the 1.5-μm LDs (DFB-LDs in the two SCM systems and DBR-LD in the TDM system) wavelength were controlled only by temperature, while the power level was maintained by an APC circuit. The typical deviations of the wavelength and power lay in the range of ± 0.01 nm and ± 0.3 dB, respectively. The wavelength spacing depends on the optical filters used in each system, and was 200 GHz for dielectric filters and grating filters and 400 GHz for waveguide acousto-optic (AO) filters [21]. For reducing the effect of multiple reflections, SC connectors of the advanced-PC type were used in all systems to keep the return loss more than 40 dB.

Fig. 5 Prototype systems.

The basic characteristics like bit error rate (BER), noise level, and channel control functions were assessed in each system, and the expected operations were successfully confirmed. In the following sections, only the important features of each system are described in detail.

4.2 WDM/TDM System

A WDM/TDM system based on Synchronous Transfer Mode level 16 (STM-16) technology was fabricated in order to confirm high-speed channel (STM-1) support. The system design was easy because the required optical power for a receiver is rather small and noise is not important. However, each user interface requires an STM-16 receiver module and an STM-16 to STM-1 demultiplexer.

The user interface is STM-1-based ATM, and all cells from the user interface are broadcast on an assigned STM-1 channel in the network. The channel selection commands from a user is sent as ATM adaptation layer (AAL0) cells with a special virtual path identifier/virtual channel identifier (VPI/VCI).

An RS multiplexes 16 STM-1 channels into STM-16 format and converts it into a DBR-LD light signal

Table 1 Specifications of prototype systems.

Multiplexing method	WDM/TDM (STM-1)	WDM/SCM (16QAM)	WDM/SCM (64QAM)
# of wavelengths (spacing)	8 (400GHz)	8 (200GHz)	8 (200GHz)
# of channels on a wavelength	16	16	16
Total throughtput	19.17 Gb/s	4 Gb/s	2.6 Gb/s
Wavelength assigned	1542.42 nm, 1545.60 nm	1552.00 nm, 1553.61 nm	1560.08 nm, 1561.70 nm
LD output power	-2 dBm	+4.5 dBm	+4.5 dBm
Optical modulation index	-	7 %	9 %
Splitting ratio in RS	4	16	8
Type of optical filter	Waveguide acousto-optic filter	Dielectric film optical filter	Rotatable grating filter
Optical filter crosstalk	> 20dB	> 25 dB	> 25 dB
Optical filter FWHM	1 nm	0.4 nm	0.8 nm
Optical filter insertion loss	10 dB	2.5 dB	2 dB
Optical filter tuning speed	~ 20 μsec	~ 500 msec	~ 5 msec
O-AMP gain	> 11 dB	> 13.5 dB	> 15.8 dB
O-AMP gain deviation	3 dB	2.5 dB	5 dB
Subcarrier frequency	-	1.0 ~ 1.5 GHz	450 ~ 660 MHz
Subcarrier bandwidth	-	12.38 MHz	5.26 MHz
Symbol rate	-	8.2075 Mbaud	4.384 Mbaud
Subcarrier spacing	-	24 MHz	12 MHz
Channel bit rate	149.76 Mb/s	31.56 Mb/s	20.48 Mb/s
Bit error rate	$< 10^{-12}$	$< 10^{-10}$	$< 10^{-11}$
CNR	-	> 34 dB	> 37 dB
CTB	-	< -50 dB/15carriers	< -60 dB/2carriers

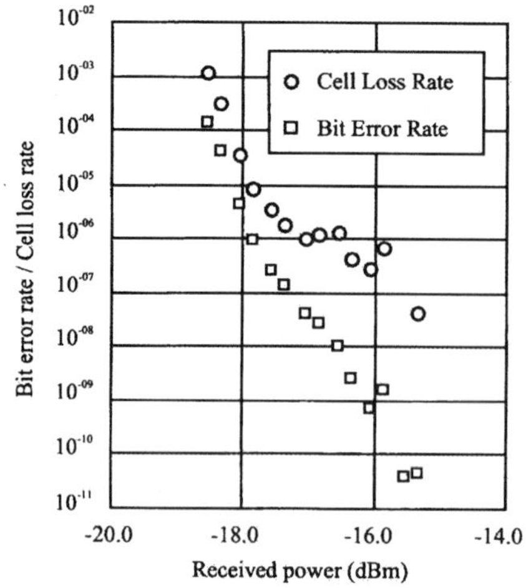

Fig. 6 Bit error rate and cell loss rate.

with external electro-absorption modulator. The optical receiver module was a standard STM-16 device. The wavelength spacing was 400 GHz because of the crosstalk characteristics of the AO filter. Channel selection could be completed in about 100 msec. This period is much longer than the AO filter switching speed of about 20 μsec because the time required to establish the synchronization with the newly selected STM-16 signal is much longer than the AO filter response.

4.3 WDM/SCM (16QAM) System

One of the WDM/SCM systems was based on 16-QAM. 16-QAM can tolerate higher noise levels than 64-QAM. In this system, forward error correction is not employed and the target BER was set to 10^{-8}. The modem can transmit 31.56 Mb/s in the 12.38 MHz subcarrier bandwidth. The user interface is STM-1 ATM, and cells from users are once buffered in the interface board and then transmitted one-by-one at the rate of 31.56 Mb/s on a subcarrier.

Theoretical CNR is 22 dB in 16-QAM. Considering the 4 dB loss of the actual 16-QAM modem circuit and 2.5 dB of margin, the target total CNR was 28.5 dB. In this system, it was allocated equally to CNR and DUR, and the optical modulation index and the minimum received power were set to 7% and -15 dBm, respectively. Here the subcarrier frequencies were allocated within the octave bandwidth in order to escape the second order inter-modulation noise (CSO), and only CTB was considered. Double cavity dielectric filters with a linear-motor actuator and optical amplifiers with auto gain control [12] were used.

Typically, the received power was -9.5 dBm, and CNR and CTB were 34 dB and 50 dB, respectively, on the worst optical path. Since this noise level was far better than the requirements, the BER was better than 10^{-10}. The ATM cell level performance measured at lower received powers is shown in Fig. 6. Cell loss occurs when bit error occurs in the header part, and the bit means the BER in the body of successfully received ATM cells.

4.4 WDM/SCM (64QAM) System

To more efficiently use the carrier bandwidth, 64-QAM modulation was tried in the other WDM/SCM system. Since 64-QAM requires much better CNR than 16-QAM, the target BER was set to 10^{-6} and Reed-Solomon encoding was employed to achieve 10^{-11} after forward error correction. The modem could transmit 20.48 Mb/s in the 5.26 MHz subcarrier bandwidth.

Theoretical CNR is 26.5 dB in 64-QAM. Considering modem performance and margins, the target CNR and DUR were set to 30.5 dB and 38 dB, respectively. The optical modulation index and the minimum received power were 9% and -12.5 dBm, respectively. The subcarrier frequencies were also allocated within the octave bandwidth. Rotating grating filters with a step-motor controller [22] were used.

Actually, the received power on the worst optical path was -12.5 dBm, and CNR and CTB were 37 dB and 60 dB, respectively. Since this noise level was far better than the requirements, the BER after forward error correction was better than 10^{-11}.

5. Feasibility Demonstration of the HM-PIN Using Prototype Systems

5.1 Demonstration of Inter-Local-Network Communication

An experimental HM-PIN that has minimum functions was built by interconnecting the user interfaces of the three prototype systems (local networks) in Fig. 7. This means that through the interconnection links the output from a selector in one system was transferred to a broadcast channel in another system.

In the experimental HM-PIN, the selectors in WDM/SCM (64-QAM) system were preset to establish a fixed connection between workstation (WS) #1 and the WDM/SCM (16-QAM) system. Therefore, WS #1 worked as a terminal directly connected to the WDM/SCM (16-QAM) system. Thus the experimental HM-PIN consists of two local networks (Domain #1 and #2) each with three workstations. In the experimental network, no datagram/signal channel was supported, so signals for connection set-up were exchanged via Ethernet.

A domain master in each domain managed the selector for the interconnection link in the domain, and the two domain masters communicated to establish connections. For example, when WS #2 tried to receive information from WS #5, WS #2 sent a request to domain master #1 through the Ethernet. Domain master #1 passed the request to domain master #2,

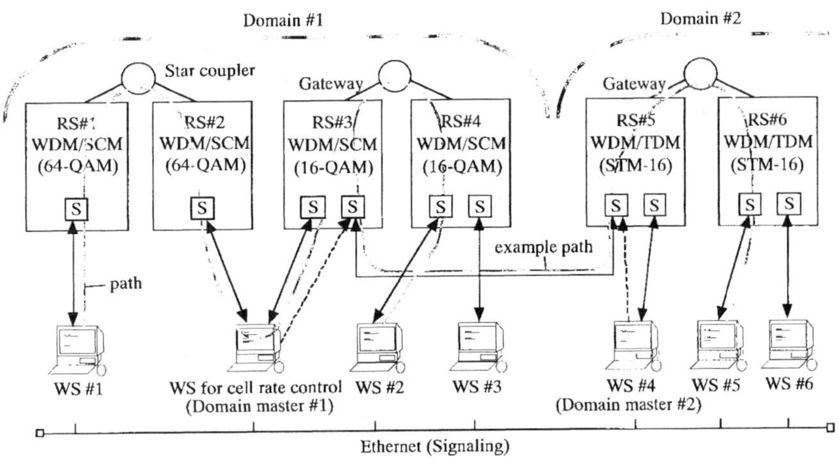

Fig. 7 Configuration of experimental HM-PIN inter-connecting three prototype systems.

which then controlled the selector in RS #5 via the logical control line (broken line). The information from WS #5 was selected and transferred to RS #3. Finally, domain master #1 passed the channel ID for the interconnection link to WS #2, which sent a command to select the channel to the private selector for WS #2.

By using this primitive function, the distribution of real-time video at about 20 Mb/s (simple compression was used) was demonstrated, and also some connection control functions were tested. For example, there is a function that allows the video transmitter terminal to determine which receivers are allowed to receive his channel.

As a datagram service example, an IP cut-through service was demonstrated. Usually IP datagrams are exchanged over Ethernet or Internet, but when the terminal tries to send a large number of IP datagrams to one destination, a direct connection to the destination is established through this HM-PIN to enhance transfer speed. This function was implemented in the network interface driver. The driver always checks the proper IP destination for the direct connection, and automatically sets up the channel. Thus users do not have to worry about which datagrams should be cut through.

5.2 Connection with WDM-Ring Backbone Network

To demonstrate wide area communication with the HM-PIN, we connected a prototype local network system to a WDM-ring backbone network. Figure 8 shows the experimental network configuration using a WDM/SCM (16-QAM) system. Two RSs were connected to an ADM in a WDM-ring backbone network (a 200-km long field test system in Kanagawa, Japan [23]) through electrical ATM switches and a 2.4-Gb/s transmission line. The electrical ATM switches were used just for physical layer conversion and speed conversion, and the Ethernet was used for exchanging the

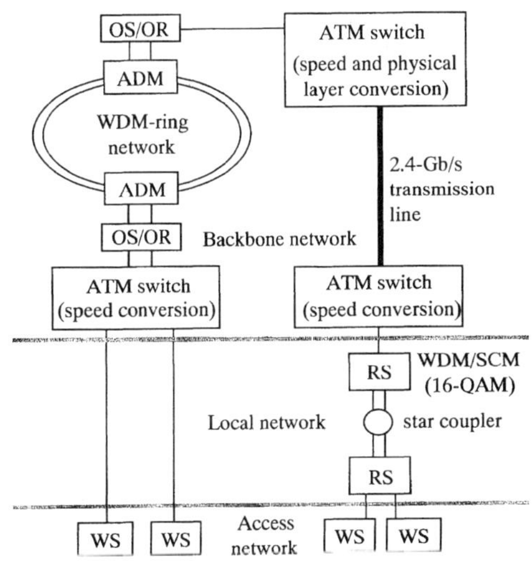

Fig. 8 Experimental network configuration with a WDM-ring backbone network.

connection set-up signals. This experimental network enabled us to estimate end-to-end throughput and delay characteristics of the HM-PIN.

Before investigating the experimental system, we evaluated the influence of the additional ATM switches in the network by measuring delays when the WDM-ring network and 2.4-Gb/s transmission lines were omitted from the experimental network. The total delay as measured by the ping function of the WS for three electrical ATM switches was only 0.4 msec. On the other hand, the throughput of the electrical ATM switch was so high compared with system throughput that was neglected in the experiment.

The measured delay between the terminals through the whole experimental network was 2.1 msec when the ATM switch delay (0.4 msec) was omitted. The throughput of the experimental network was evaluated by measuring file-transfer speeds between end terminals using file-transfer-protocol (ftp) commands of the WS. The transferred file was a video image of 10.5 Mbytes (binary mode). The measured throughput was 2–3 Mbyte/s (corresponding to 16–24 Mb/s); the interface speed was 155 Mb/s and the RS channel speed was 32 Mb/s. The difference between the system throughput and the speed of the RS was caused by the overhead of ftp protocol processing.

6. Conclusions

We implemented three prototype systems to confirm local network feasibility for the HM-PIN. Moreover, the feasibility of the HM-PIN was demonstrated by connecting the prototype local networks.

Combining electrical TDM/SCM with WDM is the key to support more than ten thousand broadband channels. Each of the alternative technologies has its own application area: For channels of 20–30 Mb/s, WDM/SCM has great possibility of achieving very low system cost since low cost electrical circuits can be applied, while WDM/TDM is suitable for supporting high-speed channels. WDM/O-TDM is the best way of realizing more than Tb/s throughput with reasonable cost.

Three prototypes of 128 ch (8 wavelengths × 16 channels) WDM B&S switching systems were fabricated. The WDM/SCM prototype used 16-QAM and 64-QAM to realize 30 and 20 Mb/s channels; 155 Mb/s channels were created by WDM/TDM. Moreover, we connected these three prototypes to build a small HM-PIN and applications such as video distribution and IP datagram cut-through were successfully demonstrated. Furthermore, the delay and throughput of the HM-PIN were evaluated by connecting a local network to a 200-km WDM ring backbone network.

To develop practical systems, we need further studies such as detailed examinations on the available number of the channels, cost comparison of SCM/WDM and TDM/WDM systems, network services in HM-PIN, and applications of B&S architecture.

Acknowledgement

The authors wish to thank Kimio Oguchi for his support of the ADMs and WDM-ring network, and Tohru Matsunaga and Masayasu Yamaguchi for their helpful discussions.

References

[1] A. Hiramatsu, M. Yanagiya, M. Ogawara, K. Hirabayashi, and K. Yukimatsu, "Hyper-media photonic information networks based on WDM/SCM broadcast-and-select switching," LEOS '96, ThG1, Nov. 1996.

[2] K. Yukimatsu and A. Hiramatsu, "Hypermedia Photonic Information Networks," ed. G. Prati, Photonic Networks, Springer, pp.67–78, 1997.

[3] A. Hiramatsu, M. Tsukada, T. Matsunaga, and K. Yukimatsu, "Hyper-media photonic information networks: WDM-based solutions and prototype systems," ISS '97, IS2-20, Toronto, Sept. 1997.

[4] Y. Takigawa, S. Aoyagi, and E. Maekawa, "ATM based passive double star system offering B-ISDN, N-ISDN and POTS," GLOBECOM '93, pp.14–18, Houston, Nov. 1993.

[5] N. Nagatsu, S. Okamoto, and K. Sato, "Optical path cross-connect system scale evaluation using path accommodation design for restricted wavelength multiplexing," IEEE J. Select. Areas Commun., vol.14, pp.893–902, 1996.

[6] M. Renaud and F. Masetti, "Network and system concepts for optical packet switching," IEEE Commun. Mag., vol.35, no.4, pp.96–102, 1997.

[7] T. Matsunaga, K. Habara, A. Misawa, Y. Yamada, T. Okugawa, M. Tsukada, S. Hino, K. Sasayama, and K. Yukimatsu, "Design of photonic ATM switch and a rack-mounted prototype," Conference Record of ICC '97, pp.1292–1297, Montreal, 1997.

[8] G. Chang, G. Ellinas, J.K. Gamelin, M.Z. Iqbal, and C.A. Brackett, "Multiwavelength reconfigurable WDM/ATM/SONET network testbed," IEEE J. Lightwave Technol., vol.14, pp.1320–1340, 1996.

[9] R.E. Wagner, R.C. Alferness, A.A. Saleh, and M.S. Goodman, "MONET: Multiwavelength optical networking," IEEE J. Lightwave Technol., vol.14, pp.1349–1355, 1996.

[10] K. Oda, H. Toda, and T. Matsumoto, "Application of WDM technology to regional optical networks," OECC '96, 17B1-2, Makuhari, July 1996.

[11] E. Hall, J. Kravitz, R. Ramaswami, M. Halvorson, S. Tenbrink, and R. Thomse, "The Rainbou-II Gigabit optical network," IEEE J. Select. Areas Commun., vol.14, pp.814–823, 1996.

[12] M. Misono, N. Shimosaka, T. Shiozawa, M. Murakami, and M. Fujiwara, "Practical demonstration of WD/TD optical network for broadcasting station," Proc. ECOC '95, Tu.B.1.6, Brussels, Sept. 1995.

[13] Y. Yano, T. Ono, K. Fukuchi, T. Ito, H. Yamazaki, M. Yamaguchi, and K. Emura, "2.6 Terabit/s WDM transmission experiment using optical duobinary coding," ECOC '96, ThB.3.1, Oslo, Sept. 1996.

[14] E. Yoneda, K. Suto, K. Kikushima, and H. Yoshinaga, "All-fiber video distribution (AFVD) systems using SCM and EDFA techniques," IEEE J. Lightwave Technol., vol.11, pp.128–137, 1993.

[15] W.I. Way, "Subcarrier multiplexed lightwave system design considerations for subscriber loop applications," IEEE J. Lightwave Technol., vol.7, pp.1806–1818, 1989.

[16] K. Maeda, M. Fuse, and K. Fujito, "Ultrahigh channel capacity optical CATV systems," OFC '96, WN3, San Jose, Feb. 1996.

[17] T. Uno, M. Mitsuda, and J. Ohya, "Low-distortion characteristics in amplified 4 × 50-channel WDM AM SCM transmission," OFC '94, WM 2, San Jose, Feb. 1994.

[18] Y. Ishizawa, Y. Yanai, and I. Nakayama, "64/256 QAM chip-set for digital CATV application," Proc. Int'l Conf. Consumer Elect., WAM 3.2, Rosemont, June 1995.

[19] A. Keating, M. Tsukada, and A. Hiramatsu, "All-optical selection of 160-Mb/s channels from a 51.2-Gb/s datastream," PS '97, PthA4-1, Stochholm, April 1997.

[20] K. Yukimatsu, T. Matsunaga, and A. Hiramatsu, "Hypermedia photonic information network," OSA, Photonic in Switching, 1997 Tech. Digest series, vol.10, pp.62–65, 1997.

[21] T. Fukida, R. Watanabe, M. Kashima, H. Okayama, and M. Kawahara, "New thermal stabilizing control of acousto-optic tunable filter using digital feedback technique," OFC '97, WL 61, Dallas, Feb. 1997.

[22] M. Iida and H. Asakura, "A computer controlled narrow bandpass optical tunable filter using a Fourier diffraction grating in the range of 1.3–1.55 μm," IEEE J. Lightwave Technol., vol.13, pp.2343–2348, 1995.

[23] M. Fukui, K. Oda, K. Shimano, A. Umeda, T. Sakamoto, and M. Fukutoku, "Field experiment of all-optical WDM ring network up to 70-Gb/s capacity employing 198-km installed dispersion shifted fiber," ECOC '96, ThC.3.2, Oslo, Sept. 1996.

High-Performance Backbone Network Technology edited by Naoaki Yamanaka
ISBN: 0-8247-5321-6 DOI: 10.1081/(120025727)
Published 2004 by Marcel Dekker, Inc.

A Demonstration of an Optical Switch Circuit with "Bridge and Switch" Function in WDM Four-Fiber Ring Networks

Tetsuya Miyazaki, Toshio Kato, and Shu Yamamoto

SUMMARY We propose and demonstrate for the first time in our knowledge, an optical switch circuit architecture furnishing with the "Bridge and Switch" function, conforming to ITUT-T Recommendation G.841 Annex A for optical Add-Drop Multiplexers (ADMs) in WDM four-fiber ring networks. This function enables optical ADMs to revert automatically from the switching state to their idle state just after the recovery of failure, that is indispensable for the extra traffic accommodation to enhance efficiency of the network operation. We have developed the optical ADM nodes employing the proposed optical switch circuit for each wavelength, arrayed-waveguide gratings (AWGs) and Er-doped fiber amplifiers. In the demonstration, transmission characteristics of the cascaded optical ADM nodes without regenerative repeaters have been verified at first. We have confirmed the ring protection and the automatic protection switching (APS) sequence which includes the automatic reversion in the optical ADM nodes with proposed optical switch circuits.

key words: *optical switch, optical ADM, WDM, four-fiber-ring networks*

1. Introduction

Optical ring add/drop multiplexers (ADMs) [1]-[4] employing optical switches are promising network protection equipment in growing demand of SDH/SONET-based ring optical networks at multigigabit capacity more than STM-16, in stead of lower rate electrical ADMs. In wavelength division multiplexing (WDM) networks, the implementation of the optical switch circuit for each wavelength between wavelength multiplexers/demultiplexers (MUXs/DMUXs) offers protection path against network failure to each wavelength path.

Four-fiber ring networks which comprises bi-directional working and protection fiber pairs, can effectively accommodate all paths for a full mesh interconnection with the fewer number of wavelength [5], [6] in WDM ring networks. The protection fiber pair is shared by the several wavelength paths which are assigned with an identical wavelength without color clash in the working fiber pair. Furthermore, lower priority traffics (extra traffics) is frequently assigned with the identical wavelength in the protection fiber pairs for the efficient network operation. Naturally, the restoration strategy in the WDM Four-fiber ring networks should take the 1 : N protection scheme [7] in the wavelength path layer if the application of the wavelength conversion technique [8] in the restoration is not considered.

The ITUT-T Recommendation G.841 Annex A [9] defines automatic protection switching (APS) sequence with a "Bridge and Switch" function for electrical ADMs in four-fiber ring networks with a single wavelength. This function offers the extra traffic support and the self-healing by the shortest path protection of the each SDH tributary path between the end to end nodes.

Our main concern is to expand the application of the "Bridge and Switch" function to the optical ADMs [10] in WDM four-fiber ring networks with the extra traffic support and wavelength path protections. Therefore the optical switch architecture in optical ADMs with "Bridge and Switch" function becomes a key issue.

Figure 1 shows the schematics of traffic flow in an idle state (Fig. 1(a)), and a failure state (Fig. 1(b)), with the APS sequence in the recommendation. Dual feed protection scheme (1+1 trail protection [9]) is also shown in Fig. 1(c) for a comparison. The Bridge function is to permanently supply the working signal to the working path by dividing the working signal. This gives a trigger to revert from the failure state to the idle state. The other portion of the divided working signal is introduced into a switch to select the output signal to the protection path. The selection of the working signal into the protection path only in the case of the failure

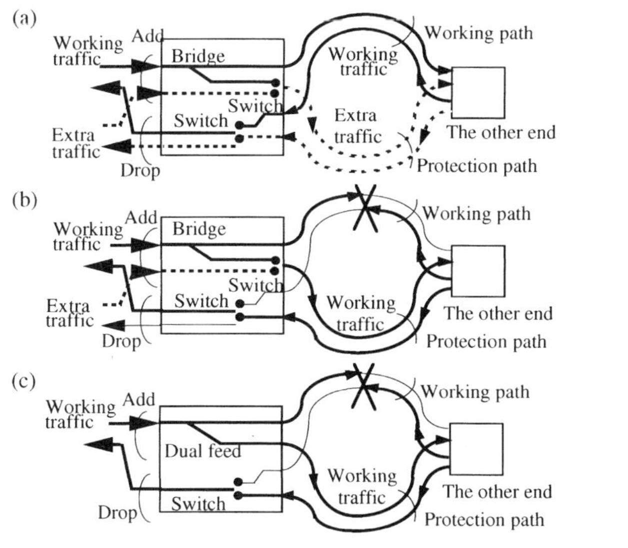

Fig. 1 Traffic flows in the ITUT-Recommendation G.841 Annex A (a) idle state, (b) failure state, (c) dual feed protection scheme.

Manuscript received June 14, 1998.
Manuscript revised September 8, 1998.

† The authors are with KDD R&D Laboratories Inc., Kamifukuoka-shi, 356-8502 Japan.

* This paper is also published in IEICE Trans. Electron., Vol. E82-C, No. 2, pp. 274-282, February 1999.

IEICE TRANS. COMMUN., Vol. E82-B, No. 2, pp. 326–334, February 1999

allows the accommodation of the extra traffic in the protection path during the idle state (Fig. 1(a)). The extra traffics will be preempted by the restoration of working traffics until the recovery of filler (Fig. 1(b)). The switch function means to switch the drop port in each state. In the dual feed configuration (Fig. 1(c)), the APS sequence is not necessary because the working traffic is always supplied to both ways and selected the survived path at the drop port. However the extra traffic support is no longer possible and much more wavelength number is required to establish full mesh wavelength paths in WDM networks.

In this paper, we propose and demonstrate an optical switch circuit architecture with "Bridge and Switch" function for the optical ADMs to realize capacity-enhanced and reliable WDM four-fiber ring networks. In Sect. 2 we introduce the "Bridge and Switch" function in WDM Four-fiber ring networks and propose the architecture of the optical switch circuit with the function. Fabrication results of the optical ADMs employing the proposed optical switch circuits and arrayed waveguide gratings (AWGs) [11] as MUXs/DMUXs, are shown in Sect. 3. In Sect. 4, the transmission characteristics of the cascaded optical ADMs without regeneration are clarified at first. We demonstrate the ring protection operation of the optical switch circuits in the four-fiber ring network. We also demonstrate the APS sequence which includes automatic reversion of the optical switch circuits after the recovery of failure. Section 5 contains a conclusion.

2. Optical Switch Architecture

2.1 "Bridge and Switch" Function

Figure 2 shows a schematic diagram of an optical ADM node. The optical ADM consists of eight EDFAs, four sets of wavelength multiplexers (MUXs) and demultiplexers (DMUXs), optical switch circuit for each wavelength. Regenerative repeaters should be employed between MUXs or DMUXs and optical switch circuit in the case of long haul transmission networks.

Figure 3 shows self-healing WDM four-fiber ring network in the idle state (Fig. 3(a)), in the cable failure state between node A and D (Fig. 3(b)). The four-fiber ring network comprises bi-directional working (W) and protection (P) fiber pairs, which are indicated by thin solid line between

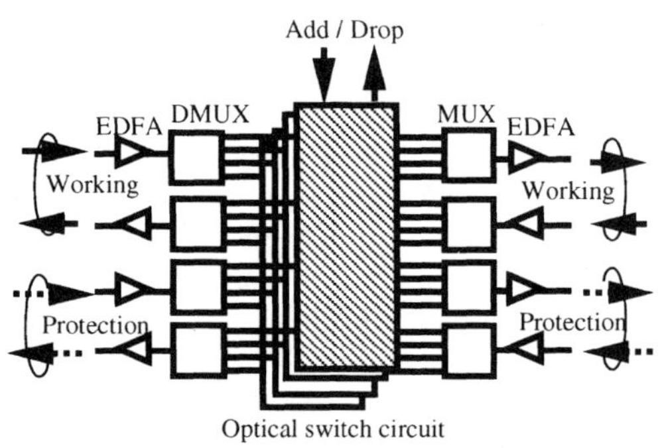

Fig. 2 Schematic diagram of optical ADM node.

node A - D. A thick solid lines in the working fiber pair, show the working traffics with the assigned identical wavelength between node A and D, node B and C. Thick dotted lines in the protection fiber pair between node A and B, node C and D indicate the extra traffics using the identical wavelength with the working traffics. When a cable failure occurs in the span between node A and D, added optical signal in these two nodes are divided ("Bridge") into the short path in the working fiber pair, and the long detour path in the protection fiber pair as shown in Fig. 3(b). At the same time, dropped paths are switched from the working fiber pair to the protection fiber pair in these nodes ("Switch"). Consequently the working traffic is survived by the detour path and the extra traffics are preempted. The divided working optical signals into the failure short path allows optical ADMs in these nodes to detect the recovery of the failure for reversion of the optical switches from the cable failure state with activating the "Bridge and Switch" function (Fig. 3(b)) to the idle state (Fig. 3(a)). The extra traffic resume after the failure is recovered. This sequence can be automatically accomplished, by detecting alarm indication signal (AIS) or loss of signal (LOS) at optical layer [12] for example.

2.2 Optical Switch Circuit Configuration

Figure 4 shows the proposed optical switch circuit architecture which is applied to node A for communication with D in various state, (a): idle state, and (b): the cable failure state. There are four input and output ports for inter-node traffics (To / From node B and D), and four add and drop ports for

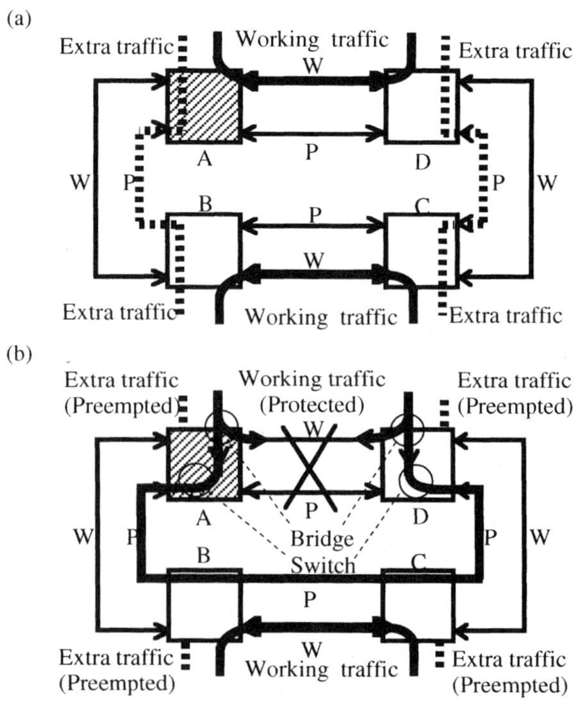

Fig. 3 Self-healing four-fiber ring network.
(a) Self-healing WDM four-fiber ring network in an idle state.
(b) Cable failure state between node A and D.

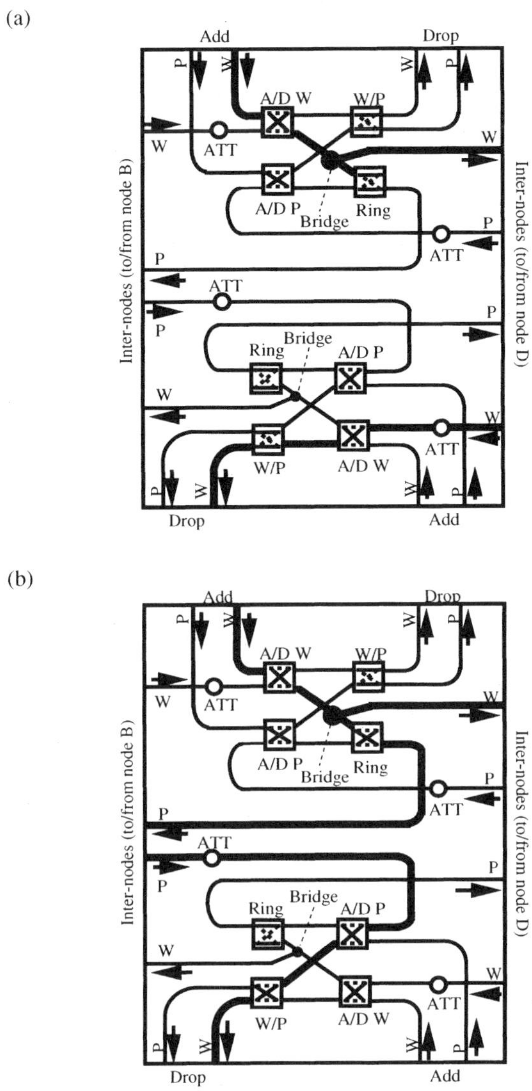

Fig. 4 The proposed optical switch circuit architecture which is applied to node A for communication with D.
(a) idle state, (b) the cable failure state.

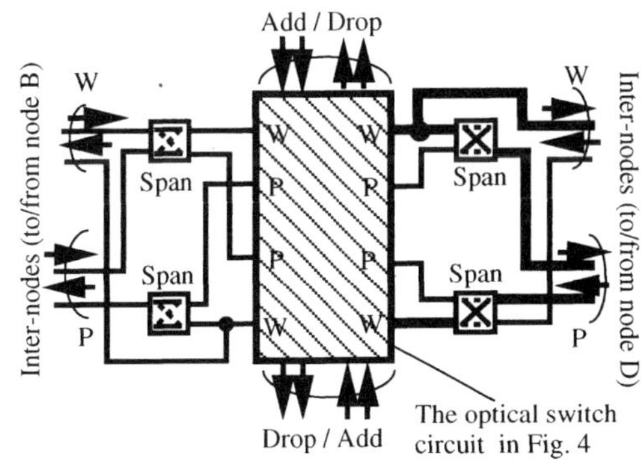

Fig. 5 The proposed optical switch circuit employing span switches. The hatched part shows the optical switch circuit in Fig. 4.

local traffics in node A. Thick lines in Fig. 4 indicate the paths for the working traffics between node A and D in each state. The paths for the extra traffics are omitted here. Four 2×2 optical switches are employed for each direction (total eight optical switches) to cope with various occasions such as add/drop for working and protection (A/D W, A/D P), working/protection switch for dropping signal (W/P), ring switch for detour protection fiber (Ring). Optical power splitter (Bridge) which is employed between A/D W and Ring switches permanently provide added working traffic towards short working path to node D. Mechanically adjustable optical attenuators (ATTs) are also fabricated in the four inter-node input ports to compensate optical power variances among incoming wavelength paths. In idle state (Fig. 4(a)), only A/D W and A/D P switches are in cross state ("×"). In the cable failure state (Fig. 4(b)), working signal to node D goes to both short working path and detour protection path by setting the Ring switch in cross state ("×"), then the signal to node D is "Bridged." Similarly, a working signal from node D to node A is "Bridged" in node D. Simultaneously, W/P switch becomes in cross state (Fig. 4(b)) to receive the protected service signal from node D, then "Bridge and Switch" function in node A is established (vice versa in node D).

Figure 5 shows the proposed optical switch circuit architecture for node A which additionally employs span switches for unidirectional signal failure (span failure) [9]. The hatched part shows the optical switch circuit in Fig. 4. When the span failure occurs, the added optical signal to D is "Bridged" by setting the span switch in cross state in node A. According to the ITU-T Recommendation G.841 Annex A, the sequence of the protection and restoration for the span failure is almost the same as in the cable failure case, except the existing of the intermediate state "Bridged state " before "Bridged and Switched." The employment of span switch for every wavelength also allows the optical ADMs to cope with the node equipment failure such as regenerative repeater failure for each working wavelength path.

3. Fabrication

3.1 Optical Switch Package

The proposed optical switch circuit as shown in Fig. 4 is fabricated in a package with two 3-dB fiber couplers for "Bridge" function and a switch driver circuit, a switch control interface, an optical interface, a power feeding interface. Here, we adopted 2×2 mechanical optical switches. The typical switching time is 5 ms, and cross talk is less than –50 dB in each optical switch.

Table 1 shows the typical optical insertion loss for various paths in the package. The insertion loss variances among the packages in each path are within ±1.0 dB. These values in the table also contain insertion loss of optical attenuator with loss minimum setting and 3-dB fiber coupler. Very low optical insertion losses at around 3-dB have been realized except two paths which contains the loss by 3-dB coupler and attenuator or both. Although the loop-back path from working to protection gives the highest loss of 6.7 dB, which contains the attenuator, two optical switches and the 3-dB

Table 1 The typical optical insertion loss for various path in the package.

In \ Out		Inter-node [dB]		Drop [dB]	
		W	P	W	P
Inter-node [dB]	W	5.3	6.7	3.4	3.2
	P	---	2.8	3.1	3.3
Add [dB]	W	4.4	5.4	2.0	2.1
	P	---	2.1	2.7	2.9

Table 2 The performance of the AWG package.

Channel spacing	0.8 nm
Wavelength	1530.0 ~ 1560.8
Loss	4.0dB (@1548nm)
Loss imbalance	1.5 dB
3-dB BW	0.35 nm
20-dB BW	0.93 nm
Cross talk	-28 dB (@adjacent channel)
TE-TM shift	0.04 nm

fiber coupler, the path will not be used in the transoceanic application where the distances between the nodes on the ring exceed 1500 km [9]. The status of each optical switch, bar or cross, can be monitored from both a PC controller and LEDs in the front panel of the package.

3.2 Arrayed Waveguide Grating (AWG) Filter Package

A 1×32 AWG is fabricated in a module which contains a thermal insulator for heater temperature control. An AWG filter package consists of the AWG module, a temperature control circuit, an optical interface, a power feeding interface. There is an LED temperature alarm indicator in the front panel of the package.

Figure 6 shows the 32-channels measured spectral responses of the AWG packages. The performance of the AWG package is shown in Table 2. The insertion loss variances among the packages at each wavelength port are within ±1.5 dB. Low optical insertion loss at around 4 dB is obtained by the peak top type filtering characteristics.

3.3 Optical ADM Cabinet

Optical ADM nodes employing the optical switch packages, the AWG packages, and also EDFA packages are mounted in cabinets which have four Hi-PAS standard shelves as shown in Fig. 7. One shelf can accommodate ten AWG packages, or the switch packages for twelve bi-directional wavelength paths, or eighteen EDFA packages. We can examine various network topologies including ring networks by arranging the

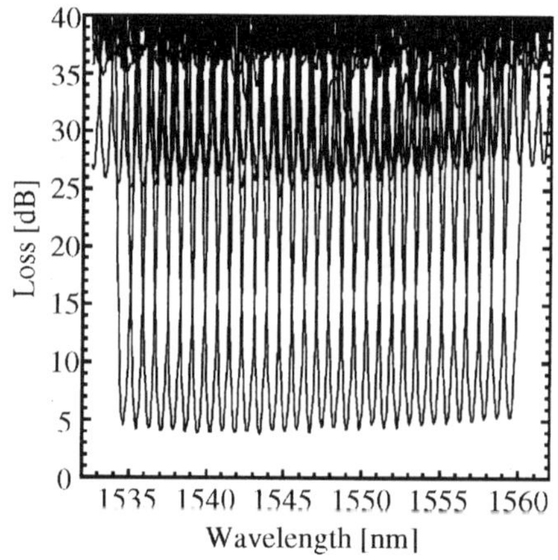

Fig. 6 The 32-channels measured spectral responses of the AWG packages.

Fig. 7 The optical ADM cabinets.

optical connection between these packages.

4. Demonstration

4.1 Cascadability

Figure 8 shows the experimental setup for the cascadability measurement. The aim of this measurement is to clarify the transmission characteristics through the optical ADM nodes without regenerative repeaters. Three optical ADM nodes (A - C) were connected in the four-fiber ring network. Here, we exclude EDFAs for output signals from a node configuration in Fig. 2. In the each span between the nodes, a 10-dB optical attenuator was inserted to simulate a transmission. Wavelengths of the eight WDM signals ($\lambda 1$ - $\lambda 8$) at 0.8 nm wavelength spacing were assigned to correspond to the spectral responses of the AWGs. The multi-wavelength fiber ring-laser (MFRL) [13] was used as a light sources for the shorter four wavelength signals (λ1 - λ4). Additional four DFB-LD modules (4TRs) were used as light sources for the longer four wavelength signals (λ5 - λ8). The eight WDM signals can experience six optical ADM nodes transit without regeneration by switching their route from the working fiber in clockwise direction to the protection fiber in anti-clockwise direction in node The all optical node cascadability was evaluated by the Bit error rate (BER) characteristics and the Q-factor [14] measurement by dropping λ6 at six points in optical switch circuits as shown in Fig. 8.

Figure 9 shows the level diagram of λ6 which is measured at output ports of nodes (C, B, A, C, B, A), attenuators (ATT), EDFAs, optical switch circuits (- Drop). Average optical loss by node transit was 11.5 dB, maximum and minimum values were 12.5 dB and 10.9 dB, respectively. Closed circles shows the power level at the drop port output of the optical switch circuit where the BER characteristics and Q-factors were measured.

Figure 10 shows the BER characteristics of l6 signal at various bit rate (Fig. 10(a): 5 Gbit/s, Fig. 10(b): 10 Gbit/s) with PRBS of length $2^{15}–1$, when the transit node number is one (open circles) and six (closed circles). The power penalty at 10^{-9} is 1 dB at 5 Gbit/s, 3 dB at 10 Gbit/s, respectively.

Figure 11 shows the Q-factors versus the number of transit nodes. The Qs at the six nodes transit at 5 Gbit/s (open circles) and 10 Gbit/s (closed circles) are 26 dB and 18 dB, respectively. The signal to noise ratio (SNR) degradation by pass through optical loss, the additional 10-dB span loss, and the noise figure of EDFAs (8 dB nominal) are dominant degradation factor of Qs in the case of 5 Gbit/s. The all optical cascadability of the optical ADM nodes up to six nodes at 5 Gbit/s has been confirmed even if there is a 10 dB span optical loss between the nodes.

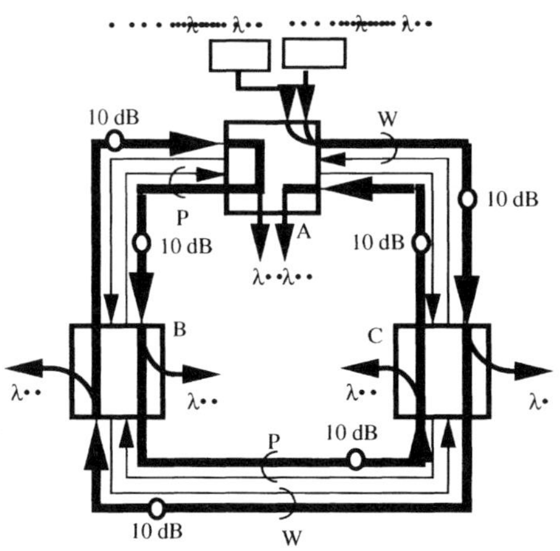

Fig. 8 The experimental setup for the cascadability measurement. MFRL: multi-wavelength fiber ring-laser, 4TRs: Four transmitters, Open circles; 10 dB optical attenuator.

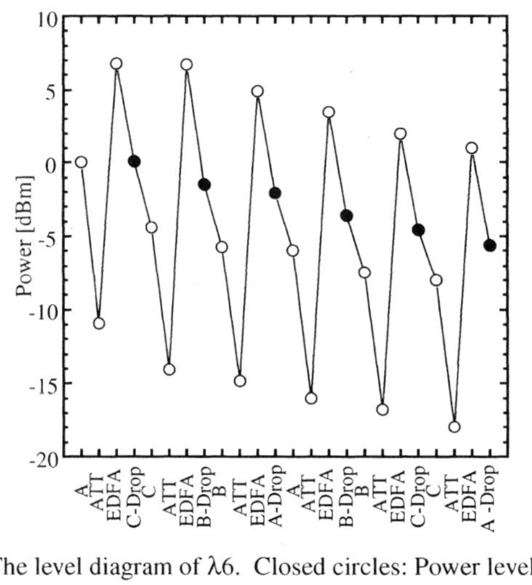

Fig. 9 The level diagram of λ6. Closed circles: Power level at the BER characteristics and Q-factors were measured.

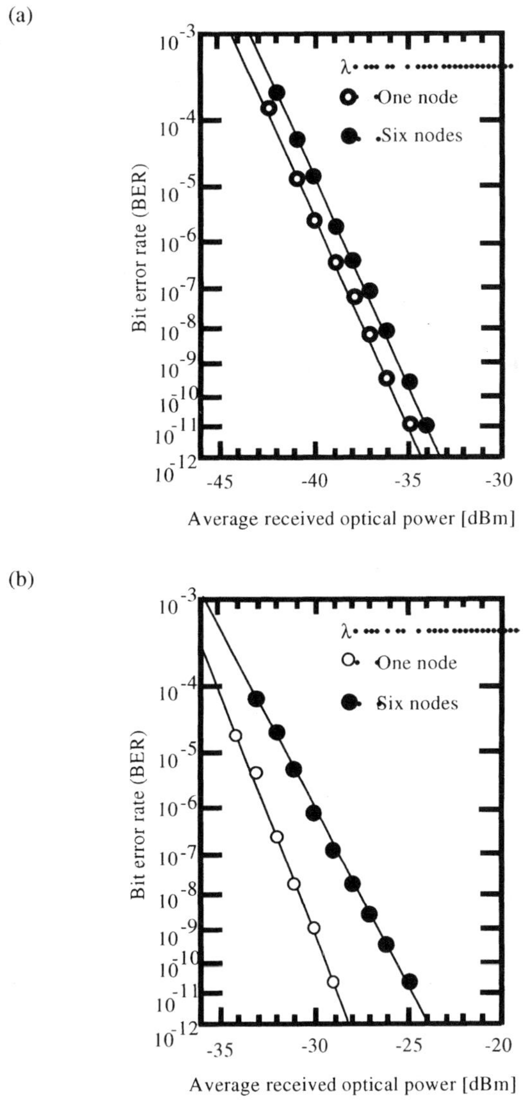

Fig. 10 The BER characteristics of λ6 at various bit rate when the transit node number is one (open circles) and six (closed circles). (a) 5 Gbit/s, (b) 10 Gbit/s.

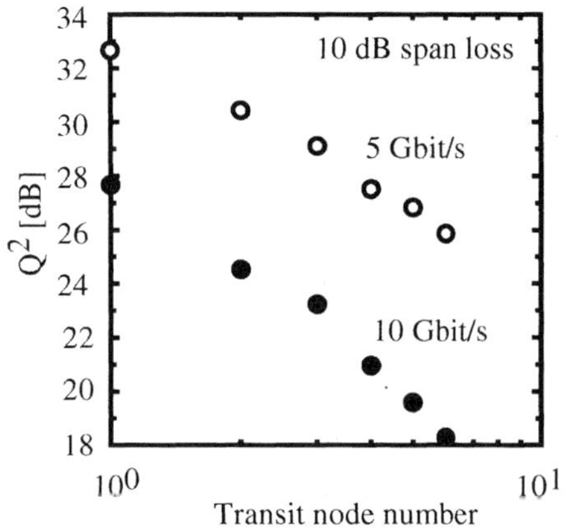

Fig. 11 The Q-factors versus the number of transit node. Open circles: 5 Gbit/s, closed circles: 10 Gbit/s.

In the case of 10 Gbit/s, bandwidth narrowing effect becomes serious, in addition to the SNR degradation. Further more polarization fluctuation and miss aligning by temperature control to AWGs become critical factors in the signal degradation.

4.2 The Ring Protection Demonstration

The optical switch operation for the ring protection has been confirmed in optical spectrum observation at various points in the ring network. Here, the ring switch in the node A is controlled by sending commands from the controller.

Figure 12 shows the signal flow (Fig. 12(a)) and optical spectrum at the dropped port for λ6 in node B (Fig. 12(b)), and dropped port for λ5 in node C (Fig. 12(c)) in the idle state. The longer four wavelength signals λ5 - λ8 were introduced into the working fiber in clockwise direction from node A. λ5 was assigned for wavelength path between node A and C, and l6 was assigned for wavelength path between node A and B in the clockwise direction. The shorter four wavelength signals λ1 - λ4 were added only into the protection fiber of anti-clockwise direction. This allows us to know whether the dropped signals experience the working path, or the protection path by the optical spectrum measurement. The output optical power of each wavelength signal from node A was adjusted at around 0 dBm, in the both working fiber and the protection fiber. The variation of dropped optical power of λ7, λ8 in Fig. 12(b) and λ6, λ7, λ8 in Fig. 12(c) is caused by the wavelength dependent cross talk suppression in transit AWGs. The complete disappearance of λ5 in node B (Fig. 12(b)) offers the proof of the ideal cross talk performance in the optical switch circuit.

Figure 13 shows the signal flow (Fig. 13(a)) and optical spectrum at the dropped port for λ6 in node B (Fig. 13(b)), and dropped port for λ5 in node C (Fig. 13(c)) in the cable failure state between node A and C. This time, the shorter four wavelength signals are included in the optical spectrum at the dropped ports in node B and C as shown in Fig. 13(b) and Fig. 13(c).

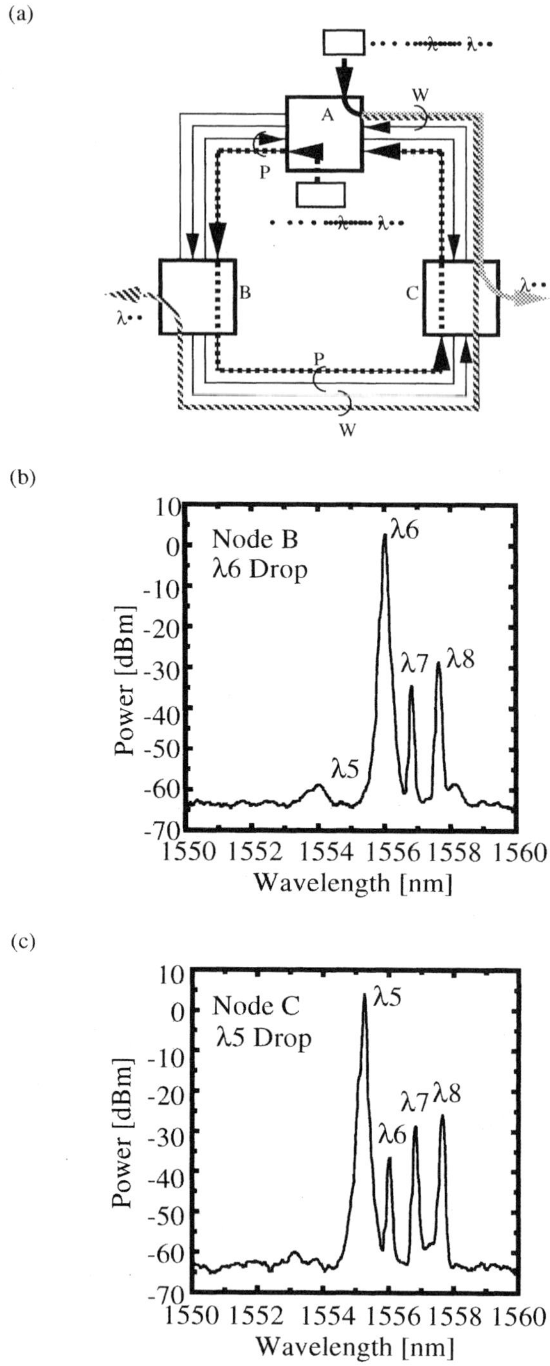

Fig. 12 The ring protection demonstration.
(a) The signal flow in the idle state.
(b) Optical spectrum at dropped port for λ6 in node B.
(c) Optical spectrum at dropped port for λ5 in node C.

4.3 The Automatic Protection and Reversion

The automatic protection and reversion for a working signal from node A to node C are demonstrated in the case of cable failure and recovery between these nodes. Since the switch control of node B is not considered here, the equivalent net-

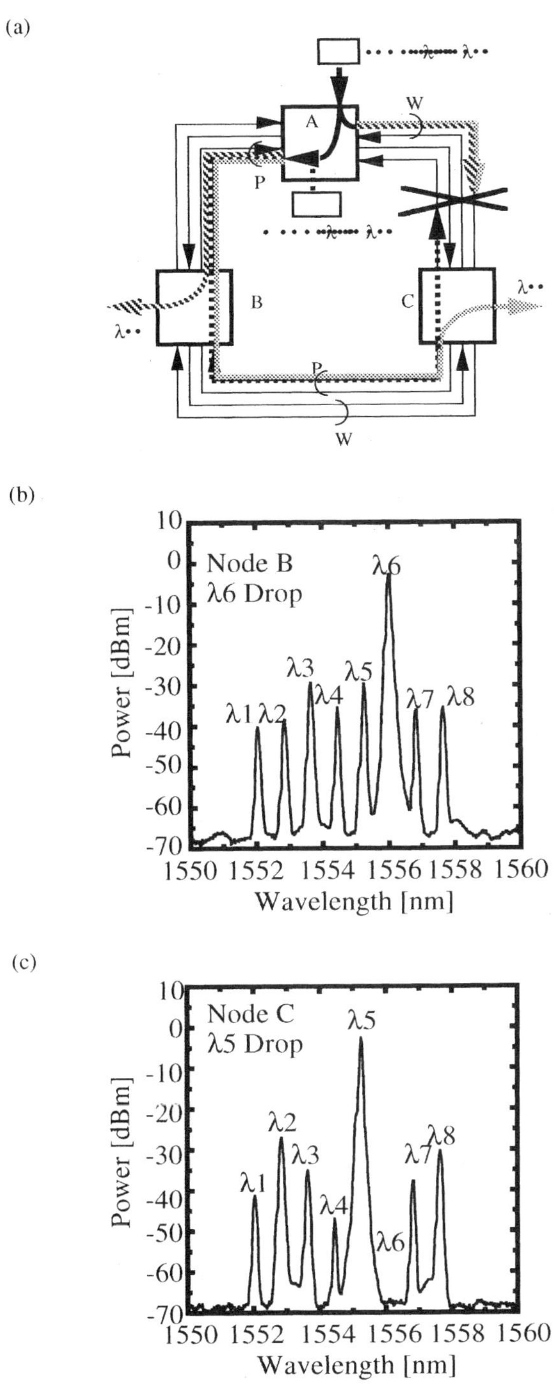

Fig. 13 The ring protection demonstration.
(a) The signal flow in the cable failure state.
(b) Optical spectrum at the dropped port for λ6 in node B.
(c) Optical spectrum at the dropped port for λ5 in node C.

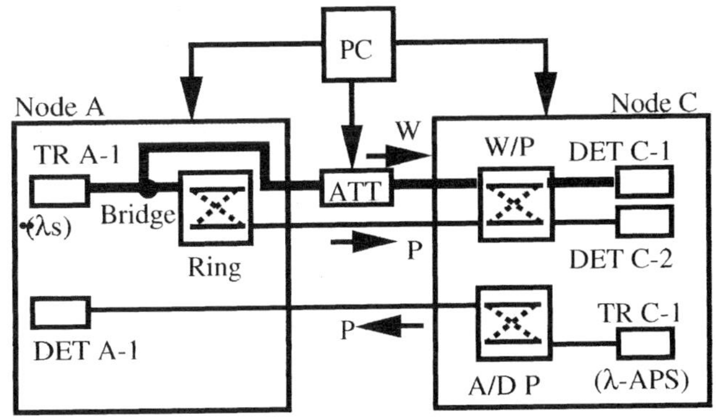

Fig. 14 The automatic protection and restoration demonstration. TR: Transmitter, DET: Detector.

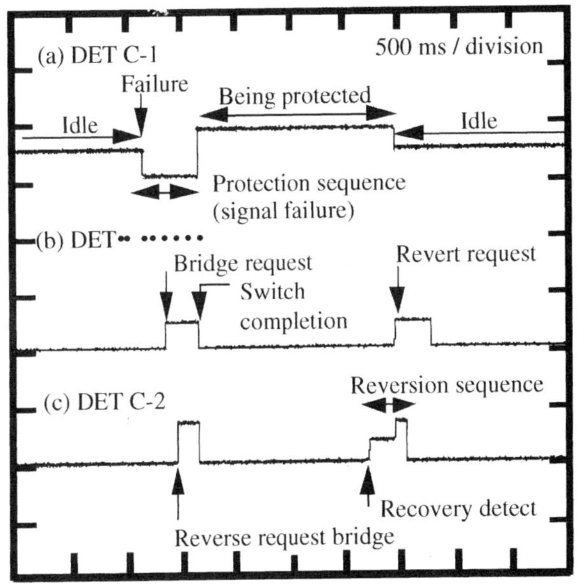

Fig. 15 The detected waveform in detectors in the automatic protection and reversion.
(a) DET C-1(λs), (b) DET A-1 (λ-APS), (c) DET C-2 (λs).

work configuration for the demonstration, consequently becomes linear network as shown in Fig. 14.

A continuos wave (CW) light sources was employed as an optical APS signal source in node C (λ-APS) to proceed the automatic protection and reversion sequence which corresponds to the ITU-T Recommendation G.841 Annex A. An optical attenuator was employed between node A and C to simulate a cable cut failure and a recovery by gating the working signal λs from node A. The simple threshold decision of the optical power detectors in both nodes (DET-A1, DET-C1 - C2) for the λs and λ-APS offers the trigger to control optical switches. Therefore the gating of λ-APS by the optical switch (A/D P) in node C can inform node A of the "Bridge request" or "Switch completion." The APS sequence was accomplished by connecting these optical power detectors to the PC controller with IEEE 488 busses as well as the optical switch circuits in both nodes. Indeed this "λ-APS" technique requires further consideration to implement into real systems, for example, about the threshold decision in low SNR condition and the effect on the transmission characteristics of other WDM signals. Here we adopt this technique to demonstrate the APS performance of the optical switch circuits without using any SDH equipment to extract APS bytes in the experiment.

Figures 15(a) - (c) show the detected waveform in the

automatic protection and reversion operation, for DET C-1 (λs), DET A-1 (λ-APS), DET C-2 (λs), respectively. The excess loss in the working path through the optical attenuator helps to show clearly the path switching between the working and the protection path. One division of the time scale is 500 ms. Due to the bridge function of the proposed optical circuit, the signal failure period in the reversion sequence (from "Being protected" to "Idle") is only determined by the switching time of the W/P optical switch (~ 5 ms) which can not be seen in Fig. 15(a). The obtained protection and reversion sequence time (< 500 ms) here are not essential, since the PC control through the buses is the dominant delaying factor in our demonstration setup. The implementation of the control sequence into the firmware in each node will improve the speed of the sequence by the order of few times of the switching time of the optical switches.

5. Conclusion

We have proposed and demonstrated an optical switch circuit architecture with "Bridge and Switch" function, which is compatible with ITU-T Recommendation G.841 Annex A, in WDM four-fiber ring optical ADM nodes. This function is essential for WDM four-fiber ring networks with 1 : N protection scheme in wavelength path layer, especially for supporting the extra traffics. The all optical cascadability, the ring protection operation and the APS operation including the automatic reversion, have been successfully demonstrated. The proposed optical switch circuit architecture will provide reliable and efficient operation in growing demand WDM four-fiber ring networks.

Acknowledgment

The Authors wish to acknowledge Dr. Y. Matsushima, Dr. S. Akiba, Dr. K. Suzuki, and Dr. T. Muratani, for their encouragement and Mr. Y. Horiuchi for his help.

References

[1] S.Johansson, "Transport network involving a reconfigurable WDM network layer—A European demonstration," IEEE J. Lightwave Technol., vol.14, no.6, pp.1341-1348, June 1996.

[2] G.Chang, G.Ellinas, J.K.Gamelin, M.Z.Iqbal, and C.A.Brackett, "Multiwavelength reconfigurable WDM/ATM/SONET network testbed," IEEE J. Lightwave Technol., vol.14, no.6, pp.1320-1340, June 1996.

[3] M.Koga, Y.Hamazumi, A.Watanabe, S.Okamoto, H.Obara, K.Sato, M.Okuno, and S.Suzuki, "Design and performance of an optical path cross-connect system based on wavelength path concept," IEEE J. Lightwave Technol., vol.14, no.6, pp.1106-1119, June 1996.

[4] T.Miyazaki, S.Yamamoto, and S.Akiba, "Blocking-free optical cross-connect employing a pair of arrayed-waveguide-grating filter," Proc. Opto-electronics conference (OECC'97), 10B4-2, pp.428-429, 1997.

[5] G.Ellinas, K.Bala, and G.C.Chang, "Scalability of a novel wavelength assignment algorithm for WDM shared protection rings," Proc. OFC '98, ThU4, pp.363-364, 1998.

[6] D.K.Hunter and D.Marcenac, "Optimal mesh routing in four-fiber WDM rings," Electron. Lett., vol.34, no.8, pp.796-797, Jan. 1990.

[7] D.Uehara, K.Asahi, Y.Nakabayashi, C.Yamashita, C.Konoshi, and S.Fujita, "Highly reliable and economical WDM ring with optical self-healing and 1 : N wavelength protection," Proc. ECOC '97, pp.65-68, Sept. 1997.

[8] S.L.Danielsen, B.Mikkelsen, C.Joergrnsen, T.Durhuus, and K.E.Stubkjaer, "WDM packet switch architectures and analysis of the influence of tunable wavelength converters on the performance," IEEE J. Lightwave Technol., vol.15, no.2, pp.216-226, Feb. 1997.

[9] ITU-T Recommendation G.841, "Types and characteristics of SDH network protection architectures," COM15-62-E, pp.87-92, 1997.

[10] T.Miyazaki and S.Yamamoto, "A proposal of the optical switch circuit architecture with "Bridge and Switch" function for WDM four-fiber ring network," Proc. OECC '98, 16A4-4, 1998.

[11] H.Uetsuka, K.Akiba, K.Morosawa, H.Okano, S.Takasugi, and K.Inaba, "Wavelength division multi / demultiplexer with arrayed-waveguide grating," IEICE Trans. Electron. vol.E80-C, no.5, pp.1205-1208, May 1997.

[12] H.Takeshita, S.Takahashi, T.Kato, S.Kitamura, H.Harano, and N.Henmi, "A demonstration of an optical cross-connect system for a self-healing optical network," Proc. ECOC '97, pp.335-338, 1997.

[13] T.Miyazaki, N.Edagawa, S.Yamamoto, and S.Akiba, "A multi-wavelength fiber ring-laser employing a pair of silica-based arrayed-waveguide-gratings," IEEE Photonics Technol. Lett., vol.9, no.7, pp.910-912, 1997.

[14] N.S.Bergano, F.W.Kerfoot, and C.R.Davidson, "Margin measurements in optical amplifier systems," IEEE Photonics Technol. Lett., vol.5, no.3, pp.304-306, 1993.

High-Performance Backbone Network Technology edited by Naoaki Yamanaka
ISBN: 0-8247-5321-6 DOI: 10.1081/(120025728)
Published 2004 by Marcel Dekker, Inc.

All-Optical Code Division Multiplexing Switching Network Based on Self-Routing Principle

Isamu Saeki, Shouhei Nishi, and Koso Murakami

SUMMARY The tera-bit order capacity of ultrahigh-speed and wide-band networks will become necessary to provide highly advanced multimedia services. In conventional networks, electronic circuits limit the speed capability of the networks. Consequently, all-optical networks are essential to realize ultrahigh-speed and wide-band communications. In this paper, we propose the configuration of an all-optical code division multiplexing (CDM) switching network based on self-routing principles and the structure of a nonlinear all-optical switching device as one of the key components for the network. We show that the required performances of the optical devices used in the CDM switching fabric are lower than those used in the TDM and illustrate the basic transmission characteristics of the switching device utilizing FD-BPM. To evaluate the multiplexing performance, we demonstrate the maximum number of channels under an error-free condition and the BER characteristics when the Gold sequence is applied as one of the CDM code sets, and show that the network of the sub-tera-bit order capacity is realizable by adopting TDM, WDM and CDM technologies. We also illustrate the packet assembly method suitable for self-routing transmissions and one of network architectures where the proposed switching fabric can be exploited.

key words: *labptic al switch, code division multiplexing, self-routing network, FD-BPM*

1. Introduction

In the highly advanced information processing society of the 21st century, the tera-bit order capacity of ultrahigh-speed and wide-band networks will become necessary to realize advanced multimedia communications, for example, video on demand, telemedicine and virtual reality. In conventional networks, optical fibers are used as transmission lines, but switching nodes are composed of complex electronic circuits. Therefore, O/E and E/O conversions are inevitable at each node and electronic circuits limit the speed capability of the networks. Consequently, all-optical networks, where optical signals are switched by optical control signals, are essential for realizing near-future networks of the tera-bit order capacity [1]. Ultrahigh-speed all-optical networks require autonomous routing principles, because call setup functions and signal synchronization in each node make switching node configurations complicated and expensive.

As for optical multiplexing methods, the time division multiplexing (TDM) and the wavelength division multiplexing (WDM) methods have mainly been studied. The former utilizes high-speed and large-capacity optical memory devices, and also needs strict time synchronization between communicating nodes [2]–[4]. The latter does not require internode time synchronization, but necessitates complicated optical devices such as wavelength converters and filters. Moreover, even if both TDM and WDM technologies are adopted, the realizable multiplexing capacity is insufficient.

On the other hand, the optical code division multiplexing access (CDMA) method has drawn researchers' attention [5]–[8]. Although the CDMA is characterized by the on-demand access method, this study will deal with the connection oriented method, called CDM hereafter. The outstanding features of the optical CDM method are listed below. Firstly, optical CDM encoders and decoders can be realized with much simpler optical devices compared to the optical TDM and WDM methods. Secondly, it does not require time synchronization control systems, unlike the TDM method. Lastly, it will be possible to connect wireless and wired networks seamlessly [2], [3], for the CDMA method has been practically deployed especially in mobile wireless network systems.

Thus far, optical CDM encoding and decoding systems have mainly been studied, such as the experiment on the coherent CDM code converter [9]. On the other hand, the photonic switch based on CDM was reported [10], but there are very few reports that consider its application to large-sized public networks [17].

In this paper, we propose the configuration of an all-optical CDM switching network based on self-routing principles and the structure of a nonlinear all-optical switching device as one of the key components for the network. In Sect. 2, we illustrate the packet assembly method suitable for self-routing transmissions, the configuration of the all-optical CDM switching fabric and the network architecture where the proposed switching fabric can be exploited. We also show that the required performances of the optical devices used

Manuscript received June 18, 1998.
Manuscript revised September 21, 1998.

†The author is with the Computation Center, Osaka University, Ibaraki-shi, 567-0047 Japan.
††The authors are with the Department of Information Systems Engineering, Graduate School of Engineering, Osaka University, Suita-shi, 565-0871 Japan.

*This paper is also published in IEICE Trans. Electron., Vol.E82-C, No.2, pp.187–193, February 1999.

IEICE TRANS. COMMUN., Vol. E82-B, No. 2, pp. 239–245, February 1999

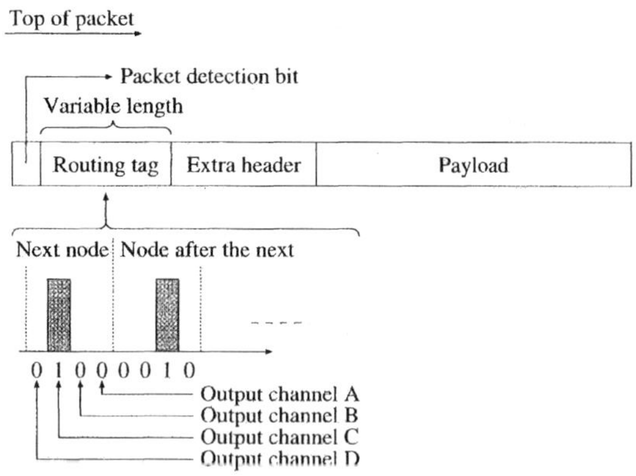

Fig. 1 Optical packet format.

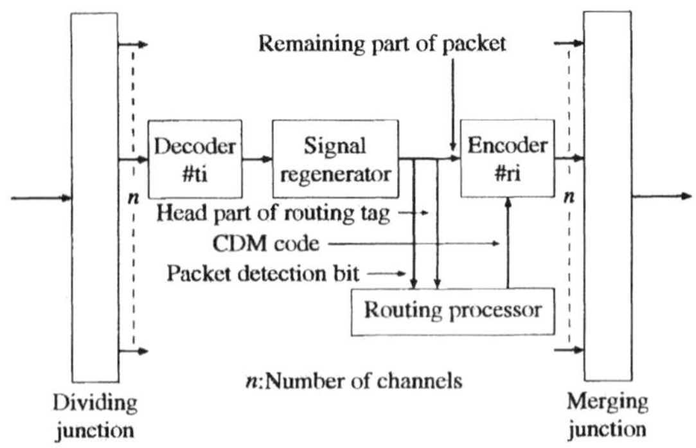

Fig. 2 Configuration of all-optical CDM switching fabric.

in the CDM switching fabric are lower than those used in the TDM. In proposing new systems, the feasibility of constructing devices should be taken into consideration. Therefore, in Sect. 3, we propose the model of the all-optical switching device and show the basic transmission characteristics utilizing FD-BPM. In Sect. 4, to evaluate the multiplexing performance, we demonstrate the maximum number of channels under an error-free condition and the bit error rate (BER) characteristics when the Gold sequence is applied as one of the CDM code sets. We also show that the network of the sub-tera-bit order capacity is realizable by adopting TDM, WDM and CDM technologies.

2. Switch Configuration and Network Architecture

2.1 Packet Assembly Method

Figure 1 shows the proposed optical packet format suitable for all-optical self-routing transmissions. The variable-length routing tag contains complete routing information, that is, a set of output port numbers of all passing nodes to the destination. To simplify the routing processing, each bit referred by switches corresponds to each output channel of switches, namely, only one bit representing the desired output channel is set to '1,' and the others are set to '0.' When packets are sent out from switches, referred bits at the head of routing tags are discarded, which increases the occupation rate of payloads in packets and ensures that the routing information for the next node is always located in the forefront of routing tags.

In the upper part of Fig. 1, we show the packet as a sequence of labeled fields. Note that the packet is transmitted beginning with the leftmost field. The packet detection bit indicates that the packet is transmitted beginning with this bit. In this format, the packet detection bit is assumed to be one bit '1,' but several bits can also be adopted considering the accuracy. The extra header contains information tags such as source addresses or higher layer tags.

2.2 All-Optical CDM Switching Fabric

Figure 2 shows the proposed configuration of the all-optical CDM switching fabric, where packets described in the previous subsection are switched based on self-routing principles. The switch consists of the following modules; the junctions, the decoders, the signal regenerators, the routing processors and the encoders [9]. Basically, CDM switching is realized by converting the CDM codes of input signals to those of output signals according to routing tags.

In Fig. 2, input signals multiplexed by the CDM codes are divided by the junction to n lines, where n is the number of channels. The CDM decoders demodulate the divided signals by tapped-delay lines. The decoded signals are sent to the signal regenerators to remove interfering noises by optical hard-limiters. Signals of the packet detection bit are extracted from regenerated signals and sent to the routing processors. The regenerated signals, except those of the packet detection bit, are divided into the head parts of routing tags and other parts, which are sent to the routing processors and the encoders, respectively. To distribute the bit sequences, we can utilize either optical switches or dividers with controls for receiving messages. At the inputs of the encoders, signals of the packet detection bit are added to the packet signals. Referring to the head parts of routing tags, the routing processors generate the CDM codes and drive the encoders. Encoded signals are merged by the junction and sent to the output of switches.

Figure 3 shows the configuration of the all-optical CDM encoder. The encoder has a configuration similar to that of a bipolar optical encoder [9], except that switches or phase shifters are controlled by optical control signals from the routing processors. Note that the decoder can be realized by inverting the input and the output of the encoder and setting the switches or phase shifters statically according to the CDM code.

Figure 4 shows the configuration of the all-optical CDM routing processor. Signals of the head part of

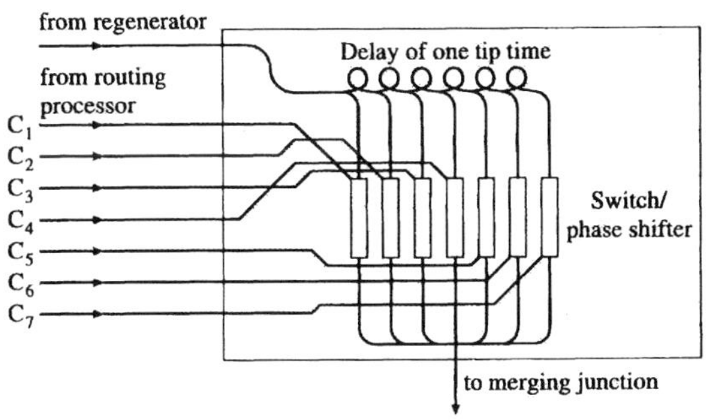

Fig. 3 Configuration of all-optical CDM encoder.

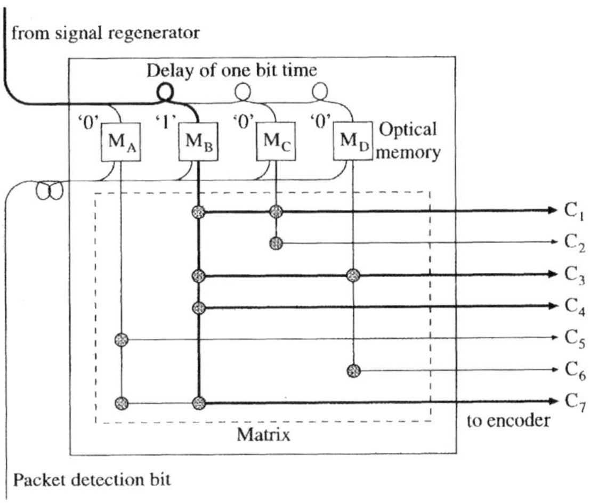

Fig. 4 Configuration of all-optical CDM routing processor.

routing tags are separated into individual bit signals by optical delaylines and memories. To memorize these signals for the time length of incoming packets, optical memories are controlled by signals of the packet detection bit, which are delayed by optical delaylines for an appropriate time length. Output signals from optical memories are sent to the optical matrix and converted to those of the CDM code bit for the desired output channels. Note that vertical and horizontal lines are statically connected, and correspond to output channels and the CDM code bits, respectively. The CDM code bit signals are discarded after driving the CDM encoders.

The proposed all-optical self-routing CDM switching fabric is composed of optical junctions, hard-limiters, delaylines, memories and switches. The important feature of the CDM switching fabric is that the required performances of the optical devices are lower than those used in the TDM; for example, the necessary capacity of the memories is much smaller and phase synchronization is simpler. This feature is especially fascinating for realizing ultrahigh-speed and wide-band networks in the nearfuture.

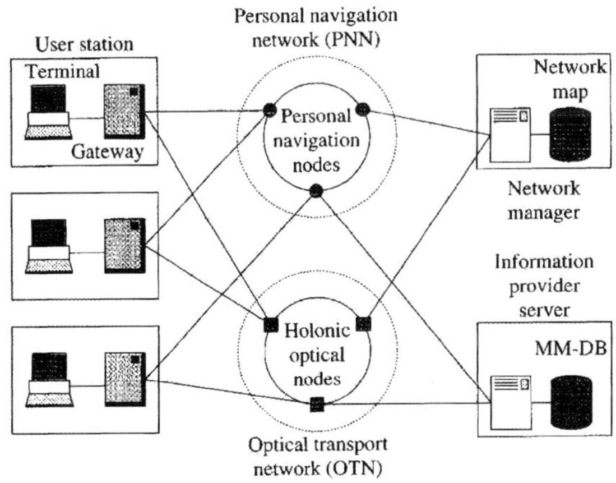

Fig. 5 Network architecture.

2.3 Network Architecture

In this subsection, we show one of the all-optical self-routing network configurations, where the proposed switching fabric can be exploited. The holonic optical network [11] was proposed as one of the highly advanced multimedia network systems. The network is characterized by distributed cooperative networking based on autonomous management and all-optical tera-bit order networks.

Figure 5 shows the configuration of the network. The network consists of an optical transport network (OTN) and a personal navigation network (PNN). The OTN is an all-optical self-routing network with simple structures, that facilitate complex controls with optical logical gates. In the OTN, information is transferred and switched between the multiplexers of sending nodes and the demultiplexers of receiving nodes, integrating TDM, WDM and CDM technologies. The PNN is an intelligent electronic network that compensates for simple functions of the OTN. Sending stations obtain complete routing information from the PNN, which refers to the network map and sets connections while avoiding congestion and contention at the beginning of communications. Thus, user terminals can assemble packets including entire routing information. Each optical switching node autonomously performs the routing function while referring only to the routing tags in packets, that is, without exchanging routing information between switching nodes. See [11] for the detailed concept of the Holonic network.

3. All-Optical Switching Device

Among the optical component devices listed in Sect. 2.2, optical junctions, hard-limiters and delaylines are relatively easier to realize than optical memories and switches. Because optical memories in the proposed switch can be constructed by optical switches

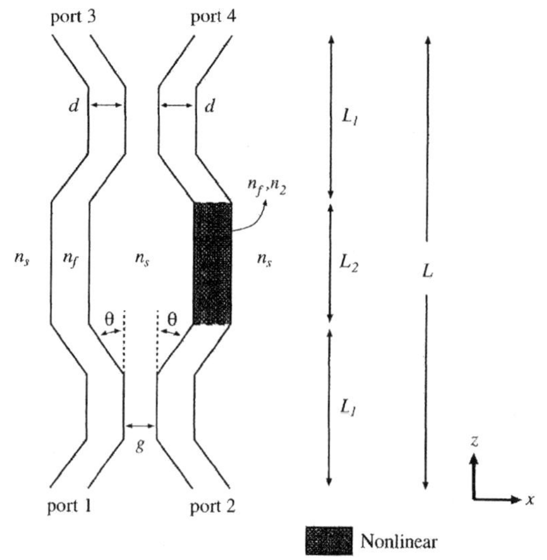

Fig. 6 Structure of nonlinear Mach-Zehnder all-optical switching device.

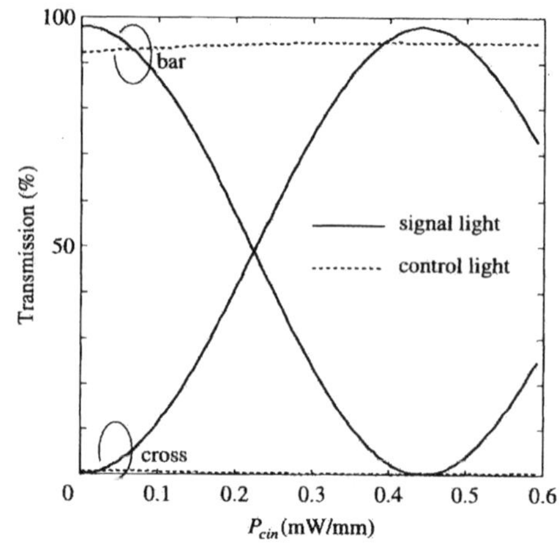

Fig. 7 Transmission characteristics of signal and control light for P_{cin}.

utilizing bistable optical materials, we focus on optical switches as the key components.

Optical switching devices for the switching fabric require the following properties. First, the switching devices must be controlled by optical signals. To control optical signals optically, we must make use of nonlinear optical effects, which are caused by the nonlinear response of the polarization. Second, the switching devices must show ultrafast response characteristics. Basically, since nonlinear optical effects show a faster response speed and optical pulse widths can be shortened much more than electronic ones, we can expect that all-optical switching devices exhibit much faster switching speed than those controlled by electronic signals. In a recent study of an all-optical switching device utilizing phase modulation, the operation speed was reported to be as high as 160 Gbps [12].

In this section, we propose the model of the all-optical switching device using phase modulation in the nonlinear Mach-Zehnder interferometer. To demonstrate the primary switching characteristics of the device, we numerically analyze the proposed structure utilizing the iterative finite difference beam propagation method (iterative FD-BPM) [13] with the transparent boundary condition [14], [15].

3.1 Nonlinear Mach-Zehnder All-Optical Switching Device

Figure 6 shows the structure of the nonlinear Mach-Zehnder all-optical switching device. We simulate the propagation characteristics of the device with continuous lightwave inputs.

The wavelengths of signal and control lights are $\lambda_s = 0.9\,\mu$m and $\lambda_c = 0.515\,\mu$m, respectively. Although the wavelengths are different, the refractive indices and the nonlinear coefficient for the control light are assumed to be equal to those for the signal light for simplicity. The structure is uniform in the y-direction and the nonlinear dielectric region (shaded) has only Kerr-like nonlinearity. Two 3-dB couplers (length $L_1 = 3000\lambda_s$) are jointed by parallel straight cores (length $L_2 = 2200\lambda_s$). The width of cores, the smallest distance between two cores and the bent angle of cores are $d = 4\lambda_s$, $g = 4\lambda_s$ and $\theta = 0.25°$, respectively. The refractive indices of cores and claddings are $n_f = 1.55$ and $n_s = 1.545$, respectively. The refractive index and the nonlinear coefficient of the nonlinear dielectric region are n_f and $n_2 = 10^{-9}\,\mathrm{m}^2/\mathrm{W}$, respectively. Here, we assume the nonlinear material to be the liquid crystal MBBA, which has a relatively large nonlinear dielectric coefficient.

The signal and the control lights are launched into port 1 and port 2 with input powers $P_{sin} = 2.1\,\mathrm{mW/mm}$ and P_{cin}, respectively. The signal light will be equally divided by the input 3-dB coupler (L_1) and then they will propagate along the straight cores (L_2) without coupling. Two guided waves will be coupled again by the output 3-dB coupler (L_1). On the other hand, the control light does not couple in the input 3-dB coupler because of its short wavelength, and propagates along the right core. Since the L_2 region of the right core is nonlinear, two guided signal lights will be coupled differently and the transmission coefficient in each output port will vary with the input power of the control light.

3.2 Numerical Results

Figure 7 shows the transmission characteristics of the signal and the control lights in each output port for P_{cin}. In Fig. 7, the transmission coefficients of the signal light vary periodically with P_{cin}, whereas those of the control light keep almost constant values. The control power needed to change the paths of the signal light

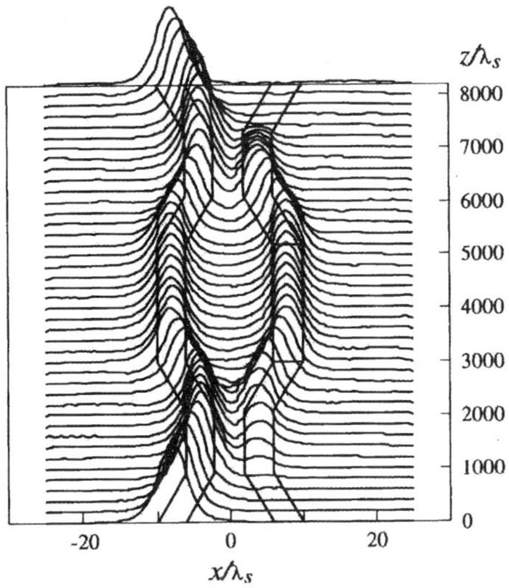

(a) $P_{cin} = 0$

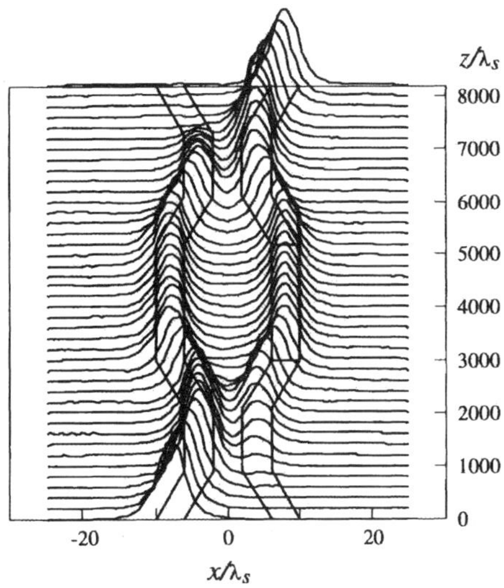

(b) $P_{cin} = 0.44$ (mW/mm)

Fig. 8 Field amplitude evolution of signal light.

is estimated to be 0.44 mW/mm, which is considerably small compared to the input power of the signal light of 2.1 mW/mm.

Figures 8(a) and 8(b) show the field amplitude evolutions of the signal light for $P_{cin} = 0$ and $P_{cin} = 0.44$ mW/mm, respectively. These figures illustrate well that the paths of the signal light are all-optically controlled.

Figure 9 shows the field amplitude evolutions of the control light for $P_{cin} = 0.44$ mW/mm. Since the wavelengths of the signal and the control lights are different, the control light launched into port 2 does not couple in the 3-dB coupler and propagates along the right nonlinear core. Therefore, the paths of the signal light are switched by the relatively lowpower of the control light.

In this analysis, we selected the wavelengths of 0.9 μm and 0.515 μm, but even if the wavelengths are

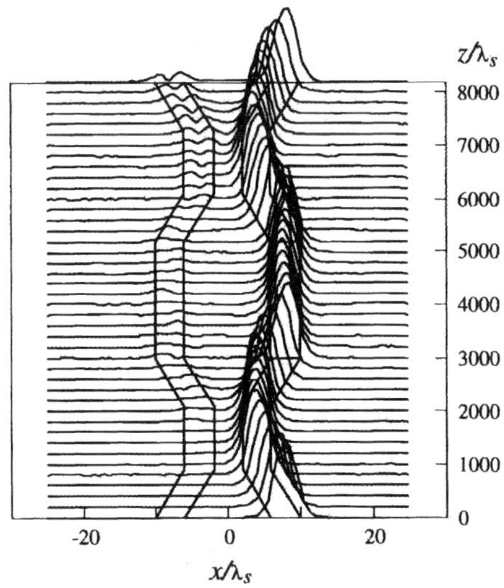

Fig. 9 Field amplitude evolution of control light.

shifted by the same ratio, we can expect basic transfer characteristics similar to those shown in Figs. 7–9. To analyze the accurate transfer characteristics for the shifted wavelengths, we must measure the optical constants.

4. Performance Analysis

In general, the selection of a set of CDM codes influences the BER performance of network systems. To increase the number of channels in CDM network systems, we must choose a class of CDM codes that have less cross- and auto-correlation properties as well as a large family size. The Gold sequence [16] and the M sequence are two major CDM codes. The Gold sequence is easily generated and has a much larger family size than the M sequence. In this section, to evaluate the capacity of the proposed network system, we calculate the maximum number of channels under an error-free condition and the BER characteristics using the Gold sequence.

4.1 Maximum Number of Channels under Error-Free Condition

First of all, we assume that there is no quantum and thermal noise. According to the following algorithm, we calculate the maximum numbers of channels for the Gold sequence generated by the n-bit shift register.

1. Take two of the generated M sequences. Generate the $(2^n - 1)$ Gold sequences by calculating XOR for the pair with all phase differences.
2. Take two of the Gold sequences generated in 1. Calculate the maximum and minimum cross-correlations of the pair.
3. Perform 2, for all of the Gold sequences generated in 1, and select the maximum and mini-

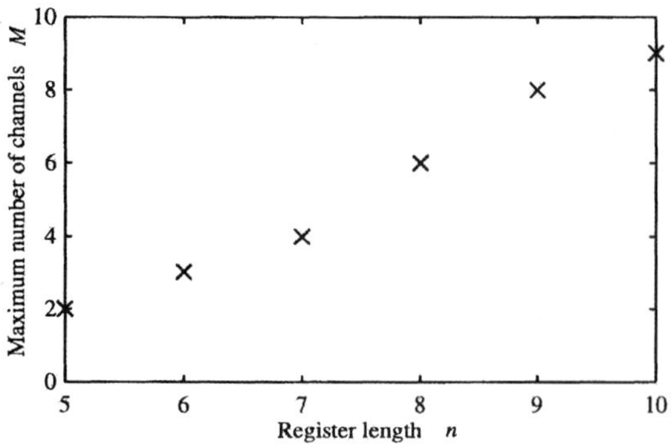

Fig. 10 Maximum number of channels with Gold sequence.

mum cross-correlations. Calculate the difference between them.

4. Perform 3, for all pairs of the M sequences and define C as the smallest value of the difference.
5. To achieve an error-free condition, the received signal level should be greater than the maximum summation of all correlation noises from other channels, which can be expressed as

$$C(M-1) < 1, \tag{1}$$

where M is the number of channels. Thus, the maximum number of channels under the error-free condition M_{max} is

$$M_{max} = \left\lfloor \frac{1}{C} + 1 \right\rfloor, \tag{2}$$

where $\lfloor x \rfloor$ is the integer part of x.

Figure 10 shows the calculated maximum number of channels versus register length for $5 \leq n \leq 10$. The figure illustrates that the maximum number of channels increases with the register length. If the register length is 10, the achievable number of channels under the error-free condition is 9, where the period of sequences is 1023 chips. Assuming that one physical link has a 100GHz band width, the link can carry 9 CDM channels of $100\,\text{GHz}/1023\,\text{bits}^{-1} \simeq 100\,\text{Mbps}$ signals, which can also be multiplexed by the TDM method. Since the band width of optical fibers is about 10THz, the entire data transfer rate with one optical fiber is estimated at $100\,\text{Mbps} \times 9 \times (10\,\text{THz}/100\,\text{GHz}) = 90\,\text{Gbps}$, adopting the TDM, WDM and CDM methods in the proposed network system.

4.2 BER Characteristics

To evaluate BER characteristics, we first calculate the probability density function $P_C(u)$ of cross-correlation for the Gold sequences generated in the previous subsection. Assuming that m transmitter and receiver pairs communicate continuously with the on-off keying, the probability density function of the interference from $(m-1)$ transmitters $P_{C_m}(u)$ is given by

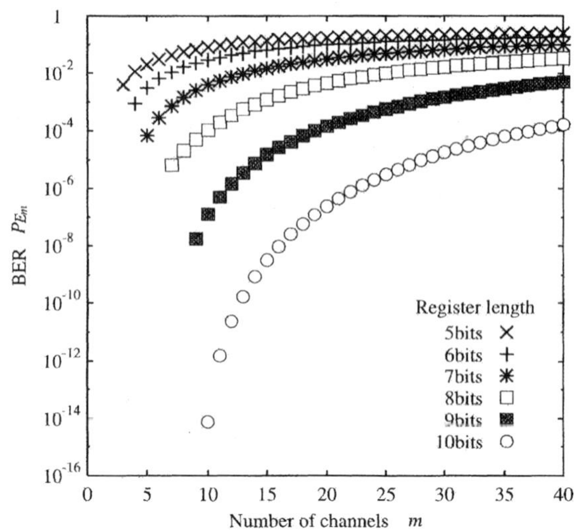

Fig. 11 BER performances with Gold sequence.

$$P_{C_m}(u) = \int_{-\infty}^{\infty} P_{C_{m-1}}(u-x)\{P_0 \cdot \delta(u) + P_1 \cdot P_C(x)\}dx \tag{3}$$

and

$$P_{C_1}(u) = \delta(u), \tag{4}$$

where $\delta(x)$ is Dirac's delta function and P_0 and P_1 are the probabilities with which transmitters send signals of '0' and '1,' respectively. Supposing that $P_0 = P_1 = 1/2$, the BER P_{E_m} is given by

$$P_{E_m} = \min_{-\infty \leq Th \leq \infty} \left\{ P_0 \int_{Th}^{\infty} P_{C_m}(u)du + P_1 \int_{-\infty}^{Th-1} P_{C_m}(u)du \right\}, \tag{5}$$

where Th is the threshold of decoders.

Figure 11 shows the calculated BER characteristics for $5 \leq n \leq 10$. Taking the same consideration as that described in the previous subsection, the entire data transfer rate is estimated to be $100\,\text{Mbps} \times 12 \times (10\,\text{THz}/100\,\text{GHz}) = 120\,\text{Gbps}$, under the condition that $n = 10$ and $P_{E_m} < 10^{-10}$.

As a result of the analysis in this section, we showed that the network of sub-tera-bit order capacity is realizable by adopting TDM, WDM and CDM technologies. Although the data transfer rate estimated in the analysis is sufficiently large to meet the demand of networks for the time being, developments in CDM sequences will further enlarge network capacities, the detailed examination of which will be the subject for future studies.

5. Conclusion

We proposed the configuration of the all-optical CDM

switching network based on self-routing principles and the structure of the nonlinear all-optical switching device as one of the key components for the network. The main feature of the proposed network is the fact that the required performances of the optical devices are lower than those used in the TDM and that it does not require time synchronization control systems.

We then demonstrated the basic switching characteristics of the device by simulating the propagation characteristics utilizing FD-BPM. We also evaluated the multiplexing performance of the CDM switching system by demonstrating the maximum number of channels under the error-free condition and the BER characteristics when the Gold sequence is applied as one of the CDM code sets. We showed that the network of the sub-tera-bit order capacity is realizable by adopting TDM, WDM and CDM technologies. Moreover, we illustrated the packet assembly method suitable for self-routing transmissions and one of the network configurations where the proposed switching fabric can be exploited. In the future, we will confirm the behavior of the proposed switching fabric by experiments.

References

[1] S.B. Alexander, R.S. Bondurant, D. Byrne, V.W.S. Chan, S.G. Finn, R. Gallger, B.S. Glace, H.A. Haus, P. Humblet, R. Jain, I.P. Kaminow, M. Karol, R.S. Kennedy, A. Kirby, H.Q. Le, A.A.M. Saleh, B.A. Schofield, J.H. Shapiro, N.K. Shankaranarayanan, R.E. Thomas, R.C. Williamson, and R.W. Wilson, "A precompetitive consortium on wide-band all-optical networks," IEEE J. Lightwave Technol., vol.11, no.5/6, pp.714–735, 1993.

[2] S. Park, K. Tsukamoto, and S. Komaki, "Proposal of direct optical switching CDMA for cable-to-the-air system and its performance analysis," IEICE Trans. Commun., vol.E81-B, no.6, pp.1188–1196, June 1998.

[3] S. Park, K. Tsukamoto, and S. Komaki, "Radio highway networks using optical CDMA with optical polarity reversed correlator," Proc. MWP'97, vol.1, no.1, pp.227–230, Sept. 1997.

[4] H. Harada, S. Kajiya, K. Tukamoto, S. Komaki, and N. Morinaga, "TDM intercell connection fiber-optic bus link for personal radio communication systems," IEICE Trans. Commun., vol.E78-B, no.9, pp.1287–1294, Sept. 1995.

[5] J.A. Salehi, "Code division multiple-access techniques in optical fiber networks—Part I: Fundamental principles," IEEE Trans. Commun., vol.37, pp.824–833, Aug. 1989.

[6] J.A. Salehi and C.A. Brackett, "Code division multiple-access techniques in optical fiber networks—Part II: Systems performance analysis," IEEE Trans. Commun., vol.37, pp.834–842, Aug. 1989.

[7] T. Ohtsuki, "Performance analysis of direct-detection optical asynchronous CDMA system with double optical hard-limiters," IEICE Technical Report, CS96-46, July 1996.

[8] H.M.H. Shalaby, "Performance analysis of optical synchronous CDMA communication systems with PPM signaling," IEEE Trans. Commun., vol.43, no.2/3/4, pp.624–634, Feb./March/April 1995.

[9] K. Kitayama, "Error-free 1.24Gb/s 4-chip coherent CDM code converter," ECOC Conference Publication, no.448, pp.117–120, Sept. 1997.

[10] Y. Takushima and K. Kikuchi, "Photonic switching using spread spectrum technique," Electron. Lett., vol.30, no.5, pp.436–438, March 1994.

[11] K. Murakami, S. Uno, K. Kinoshita, M. Higashida, and T. Takine, "Distributed cooperative networking architecture for multimedia communications based on autonomous routing," Proc. ITC-OSCC'96, vol.1, pp.83–86.

[12] K. Suzuki, K. Iwasaki, S. Nishi, and M. Saruwatari, "Error-free demultiplexing of 160 Gbit/s pulse signal using optical loop mirror including semiconductor laser amplifier," Electron. Lett., vol.30, no.18, pp.1501–1503, Sept. 1994.

[13] H. Yokota, K. Kimura, and S. Kurazono, "Numerical analysis of an optical X coupler with a nonlinear dielectric region," IEICE Trans. Electron., vol.E78-C, no.1, pp.61–66, Jan. 1995.

[14] G.R. Hardley, "Transparent boundary condition for beam propagation," Opt. Lett., vol.16, no.9, pp.624–626, May 1991.

[15] G.R. Hardley, "Transparent boundary condition for the beam propagation method," IEEE J. Quantum Electron., vol.28, no.1, pp.363–370, Jan. 1992.

[16] W. Huang and K. Kitayama, "Optical pulse code division multiple access utilizing coherent correlation detection," Proc. ICT'97, pp.1221–1226, April 1997.

[17] S. Nishi, I. Saeki, and K. Murakami, "All-optical CDMA self-routing switch for holonic network," Proc. ITC-CSCC'98, vol.1, pp.347–350, July 1998.

High-Performance Backbone Network Technology edited by Naoaki Yamanaka
ISBN: 0-8247-5321-6 DOI: 10.1081/(120025729)
Published 2004 by Marcel Dekker, Inc.

λ-Ring System: An Application in Survivable WDM Networks of Interconnected Self-Healing Ring Systems

Yasuhiro Miyao

SUMMARY This paper proposes a λ-ring system that is a wavelength-based self-healing-ring application unlike ordinary fiber-based ones. To design survivable networks of interconnected such self-healing ring systems, a virtual mesh network scheme is used, in which wavelength assignment in virtual links can be considered according to the λ-ring-system or fiber-ring-system applications of the bidirectional wavelength-path switched architecture. Integer-programming-based design problems are then formulated that minimize the total fiber length in these self-healing-ring applications. Numerical examples show that the λ-ring-system application is always superior to 4-fiber and 2-fiber-ring-system applications and 1+1 end-to-end path protection.

key words: *network protection, self-healing ring, OADM*

1. Introduction

Wavelength division multiplexing (WDM) based transport systems are being widely used in backbone networks to satisfy increased demand for higher bandwidth, caused by the growth of the Internet [1]. The transmission capacity of a single fiber that multiplexes 32 WDM-channels of OC-192 is 320 Gbps. Networks with such huge transmission systems must have a high degree of survivability since even a very short outage can cause a substantial amount of information to be lost. There are two approaches to designing highly survivable WDM-based transport networks, namely mesh and ring approaches [2]. In a WDM-based network, the minimum unit of resource allocation is an optical channel (OCh), which is specified by being assigned a route and a single wavelength within each fiber on the route. In the mesh approach, each OCh affected by a failure can be restored by using optical cross connects (OXCs), if sufficient wavelengths are provided as spare bandwidths. To optimize routing and resource allocation in the mesh approach, we have proposed a design scheme based on mathematical programming [3].

Meanwhile in the ring approach, when OChs are affected by a failure, WDM self-healing ring systems that consist of fibers and optical add-drop multiplexers (OADMs) perform protection switching to recover the failed OChs. It should be noted here that in some cases this paper distinguish *ring* system from *ring* which is used as a word meaning the topology of a circle on which ring systems are placed. The architectures for this ring approach can be categorized into the bi-directional self-healing ring (BSHR) and uni-directional self-healing ring (USHR) like those in SONET self-healing architectures [2]. They can be divided further according into the protection switching scheme they use (i.e., either loop-back based or path based). The BSHR architecture based on loop-back protection switching is referred to as bidirectional fiber-line switched ring (BFLSR), and that based on path protection switching, as bidirectional wavelength-path switched ring (BW-PSR) [4]. Both BFLSR and BWPSR have two applications: 4-fiber ring and 2-fiber ring systems. The USHR architecture based on path protection switching is referred to UWPSR [4]. This is an application of 1+1 path protection to a ring, while the above two BSHR architectures can share protection bandwidth among different failure scenarios. In T1X1 standardization process, UWPSR is referred to as OCh dedicated protection ring (DPRing), BFLSR, as OMS shared protection ring (SPRing), and BWPSR, as OCh SPRing [5].

Ordinary applications for the above self-healing ring architectures are fiber-ring systems as in SONET ring systems, and we have reported comparison between the mesh approach and the fiber-ring approach in terms of overall facility cost in [6]. In this paper, we propose a new application of the self-healing architecture, referred to as a λ-ring system. This system is not per-fiber based like SONET 4(2)-fiber-ring systems, but wavelength (or λ) based. Table 1 summarizes the architectures and applications of WDM self-healing ring systems. UWPSR and BWPSR can be directly applied

Table 1 Classification of WDM self-healing ring systems.

Application \ Architecture		USHR Loop-back based	USHR path based	BSHR Loop-back based	BSHR path based
Fiber-ring system	4-fiber	–	–	O	O
	2-fiber	O	O	O	O
λ-ring system	4-λ	–	–	O (4-fiber SONET BLSR)	O
	2-λ	O	O	O (2-fiber SONET BLSR)	–

UWPSR (OCh DPRing) — BFLSR (OMS SPRing) — BWPSR (OCh SPRing)

Manuscript received October 24, 2000.
Manuscript revised December 28, 2000.
[†]The author is with Development Laboratories, NEC Networks, NEC Corporation, Abiko-shi, 270-1198 Japan.

IEICE TRANS. COMMUN., Vol. E84-B, No. 6, pp. 1596–1604, June 2001

to λ-ring systems, because their protection-switching schemes are path based.

In addition, this paper proposes a mathematical design scheme for networks of interconnected λ-ring systems. When the ring approach is applied to large backbone networks such as those of U.S. carriers [7], interconnection of multiple ring systems is indispensable to these networks. This is because the maximum number of nodes within a single ring system is limited when the BSHR scheme is used (16 for SONET BSHR protocol [2]), and the circumference of a single ring system should also be limited to save the time required for completing protection switching based on the BSHR scheme. To achieve mathematical design of networks of interconnected λ-ring systems, we use a virtual mesh network scheme proposed in [6], in which wavelength assignment in virtual links can be specified according to λ-ring-system application. We then formulate an integer-programming based optimal design problem. By solving this, we can determine where and how many λ-ring systems are to be established so as to minimize total fiber length. Total facility cost that is adopted as the design objective in [3] and [6] is not considered here, since using this measure will require sufficiently precise modeling of node cost in each protection scheme.

Finally, we compare the ring-system applications to a mesh approach in which each working OChs has a dedicated protection channel with the reserved bandwidth (i.e. a 1+1 end-to-end path protection scheme). We consider this protection scheme because it provides the fastest recovery [8], [9]. It corresponds to *trail protection* in [10], but is hereafter simply referred to as 1+1 protection. In order to compare these protection schemes in terms of resource usage, we take advantage of the proposed optimal design scheme for networks of interconnected λ-ring systems, along with previously reported optimal design schemes for mesh networks [3] and networks of interconnected fiber-ring systems [6]. Another mesh approach of shared path restoration that utilizes OXCs to share spare bandwidths [3] is not considered here, since it has smaller total fiber length while it requires longer restoration time than any other protection scheme.

The reminder of this paper is organized as follows: Sect. 2 discusses the λ-ring application and its characteristics by comparing it to ordinary fiber-ring systems. Section 3 discusses design models including the network model, facility configuration model, hub-based path routing model, and wavelength assignment model. Section 4 describes a mathematical programming-based design for networks based on either the fiber-ring systems or λ-ring systems. Section 5 compares the ring-system applications and 1+1 protection, numerically. Section 6 offers conclusions.

2. λ-Ring System

In the following discussion, we will restrict discussions to the BWPSR architecture because it is more advantageous than other ring architectures [4]. For simplicity, in some cases, we refer to the 4(2)-fiber BWPSR system as a 4(2)-fiber ring system, and the 4-λ BWPSR system, as a λ-ring system.

Within the fiber-ring system, the number of fibers between each pair of OADMs is the same. These fibers are divided, half are working ones, and half, protection. Figure 1 shows a 4-fiber ring system, consisting of fibers and OADMs. Between each pair of OADMs there is a group of four fibers. Within the fiber-ring system, working rings exist in the clockwise and counterclockwise directions. Similarly, protection rings also exists in both directions. Figure 2 shows the nodal structure of an OADM. It consists of element switches that perform protection switching per for the OCh. Unlike TDM-based ADMs for SONET fiber-ring systems, segregating an OADM per-wavelength based is possible in WDM-based fiber-ring systems. This fact has led to work aimed at developing a ring system that is per-wavelength based. This is shown in Fig. 3. We define a λ ring as a ring system which is located on a ring and consists of *wavelength segments* that are assigned wavelengths from fibers and established between pairs of element switches. A 4-λ BWPSR system or simply *λ-ring system* is then defined as a group of four λ

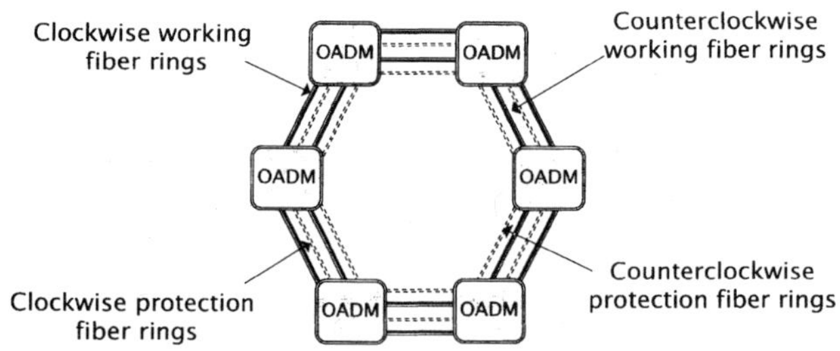

Fig. 1 4-fiber BWPSR system.

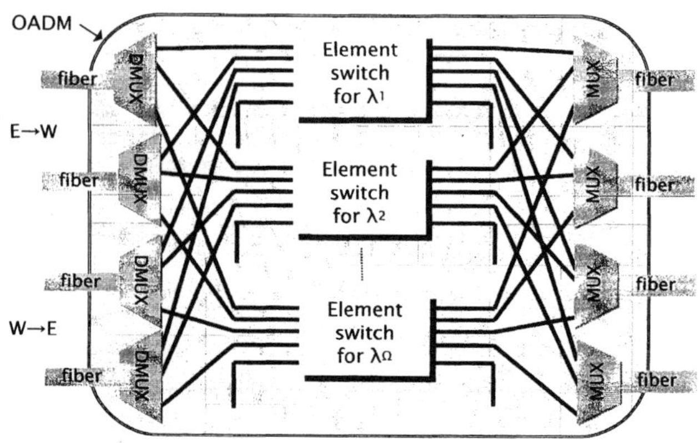

Fig. 2 Nodal structure of an OADM.

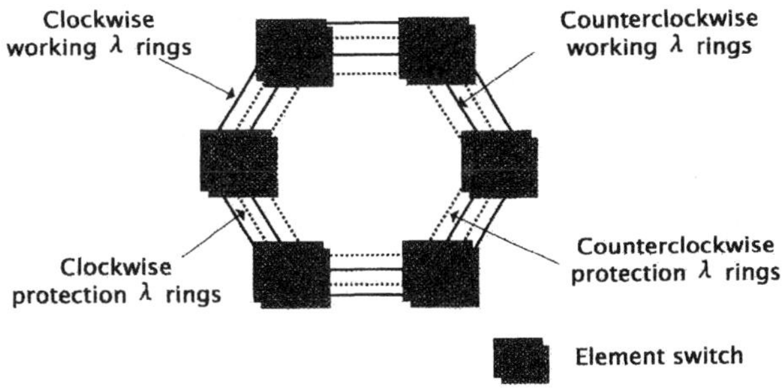

Fig. 3 4-λ BWPSR system.

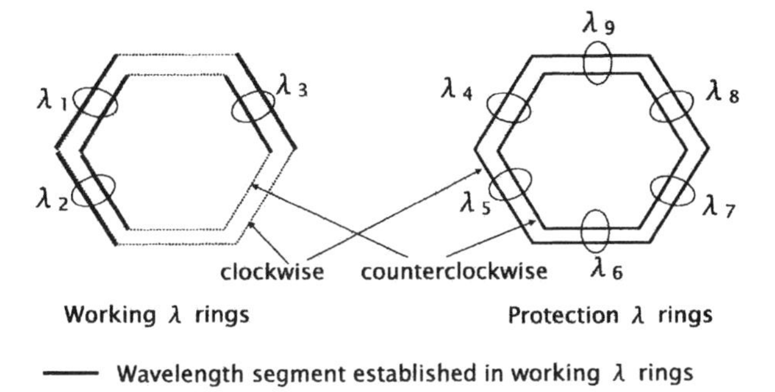

Fig. 4 Establishment of wavelength segments in a 4-λ BWPSR system.

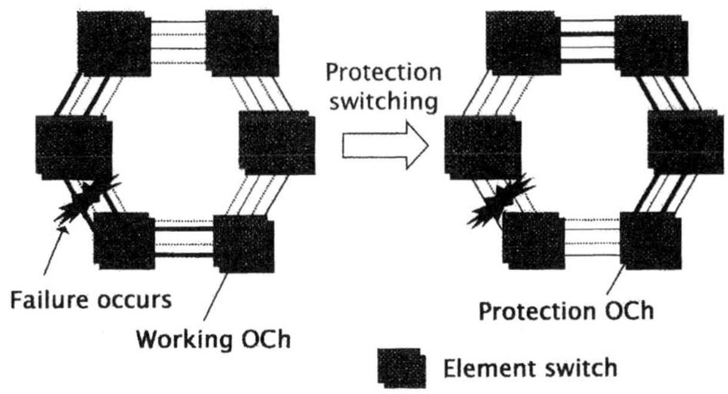

Fig. 5 Protection switching in a 4-λ BWPSR system.

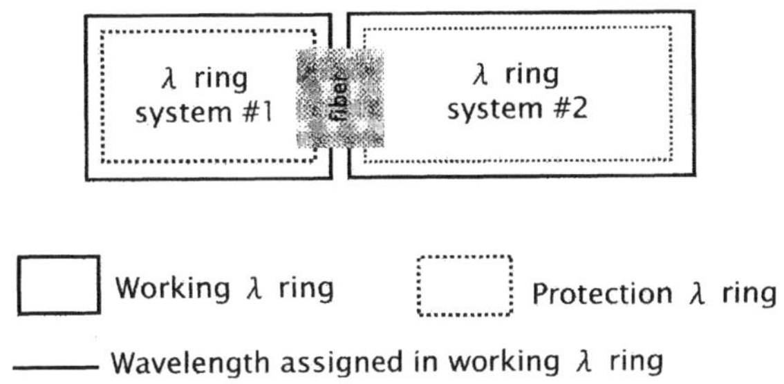

Fig. 6 Accommodation of 4-λ BWPSR systems to fiber.

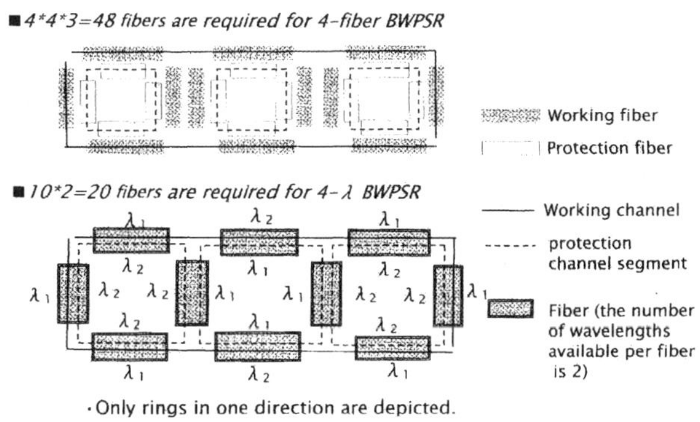

Fig. 7 4-fiber vs. 4-λ BWPSR systems.

rings of working and protection variety, in the clockwise and counterclockwise directions, connected by element switches that perform add/drop and protection switching for a bi-directional OCh. Let Ω denote the maximum number of wavelengths available per fiber. It can then be said that a 4-fiber ring system consists of $\Omega \times \lambda$-ring systems that are located on an identical ring and have a common OADM in each node on the ring.

Figure 4 shows a way in which wavelength segments are established within a λ-ring system. Within a λ ring, wavelength segments with different wavelengths may be established between individual pairs of element switches. In the clockwise and counterclockwise λ rings of working or protection λ rings, wavelength segments established between the same pair of element switches have the same wavelength. Within the two working λ rings in the both directions, wavelengths are only established between pairs of element switches that a working OCh passes through or terminate the OCh. Figure 5 shows path-based protection switching in a λ-ring system. If an OCh on working λ rings is affected by a network failure, an alarm indication signal is transmitted to the originating/terminating node of the OCh, and then the OCh is switched to the far-side path on the associated protection λ rings. This protection switching is performed by element switches along the route of the protection OCh.

Figure 6 shows a situation in which λ-ring systems are accommodated by fibers. This situation is different from that for 4-fiber-ring-system applications in the following ways.

(1) Working and protection λ rings belonging to the same λ-ring system can be accommodated by the same fiber.

(2) λ rings belonging to different λ-ring systems can be accommodated by the same fiber.

These two features, combined with the assumption that wavelength segments which are just used by an OCh are only established, can greatly reduces the number of required fibers. Figure 7 demonstrates this by comparing a λ-ring system with a 4-fiber ring system. If we assume the number of wavelengths available per fiber is 2, 20 fibers are required for λ-ring systems. By contrast, 48 fibers are required for 4-fiber ring systems.

The λ-ring-system application has another advantage. It can coexist with SONET 4(2)-fiber BLSR sys-

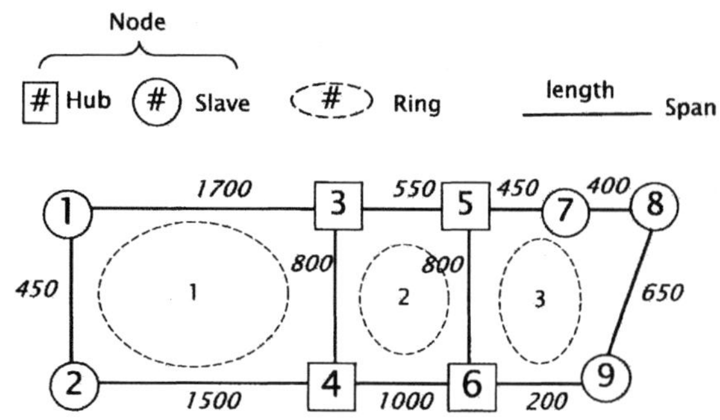

Fig. 8 Network model (9, 11).

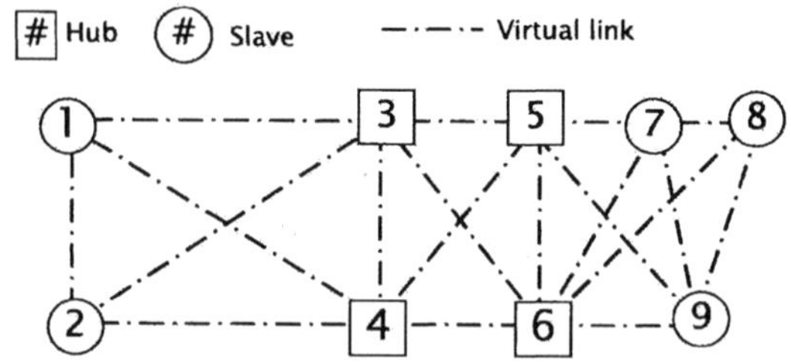

Fig. 9 Virtual network for network model (9, 11).

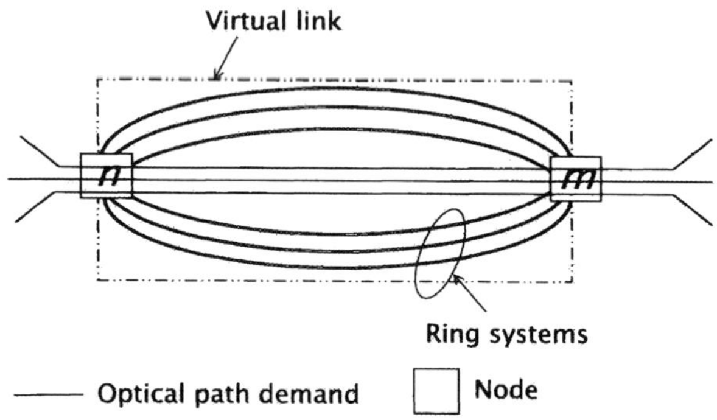

Fig. 10 Demand and capacity assignment to a virtual link.

tems in a WDM network, because they can simply be regarded as a 4(2)-λ-ring systems with a loop-back protection switching architecture, as shown in Table 1.

3. Design Model for Interconnected Self-Healing Ring Systems

3.1 Network and Facility Configuration Model

As illustrated in Fig. 8, a network consists of spans and nodes. Fibers are placed on spans, while OADMs and OXCs are placed on nodes. We divide nodes into hubs and slaves. In hubs, OChs traversing different ring systems are dropped from one ring system, cross-connected by an OXC, and then added to another ring system. In slaves, OChs either pass through or are added and dropped by OADMs or element switches. As mentioned in the introduction, a ring system differs from a ring, which is defined as a set of spans and nodes that form a circle. Multiple ring systems can exist on the same ring, and they may be called stacked rings.

We assume that none of the OADMs or element switches perform wavelength conversion on OChs in each node on a ring system. We also assume that all OChs that traverse different ring systems must pass through a hub, in which wavelength conversion may be performed. We disregard dual-ring-interworking [11] for the sake of model simplicity.

3.2 Path Routing Model

Hereafter a bundle of OChs sharing a common route is referred to as an optical path (OP) [3]. The capacity of an OP is determined by the number of OChs it includes. We assume that each source-destination (s-d) pair may use multiple OPs with different routes to distribute loads over the network. In networks comprised of self-healing ring systems, an OP is used as a unit of routing over multiple adjacent rings. Since we have assumed that an OCh must pass through different ring systems via a hub, we specify the route of each OP by using a list of these hubs. To simplify such hub-based routing, we transform the original network into a *virtual mesh network* [6]. In this transformation, we set a *virtual link* between each pair of nodes that are included in at least one ring. Figure 9 shows the virtual mesh network transformed from the original network in Fig. 8. In this virtual network, only OP routes that do not include a slave as a transit node are allowed. This is, again, due to our hub constraint. The result is two possible OP-route-patterns. One connects two nodes directly with a virtual link. In the other, the OP route has one or more hubs as transit nodes. If insufficient virtual links are created within a virtual network, no OP can be provided for a given s-d pair. In this case, more rings must be generated in the original network.

Figure 10 shows the demand and capacity assignment in a virtual link. The demand assigned to each virtual link is given by the sum of the capacities of the individual OPs that pass through. Here, we define a λ *segment* as that which is established between a pair of nodes within a ring system in order to set up a working OCh. This λ *segment* corresponds to a sub-network connection (SNC) in T1X1 standardization [10]. An OCh is given by a concatenation of a single λ segment in each virtual link on the route of the OCh. λ segments in each virtual link can thus be regarded as a pool of wavelength resources to set up OCs. We define the capacity of a virtual link as the total number of λ segments established on the virtual link.

3.3 Wavelength Assignment Model

Since in Sect. 3.1 we assume that none of OADMs or element switches perform wavelength conversion on an OCh passing through the associated node, individual

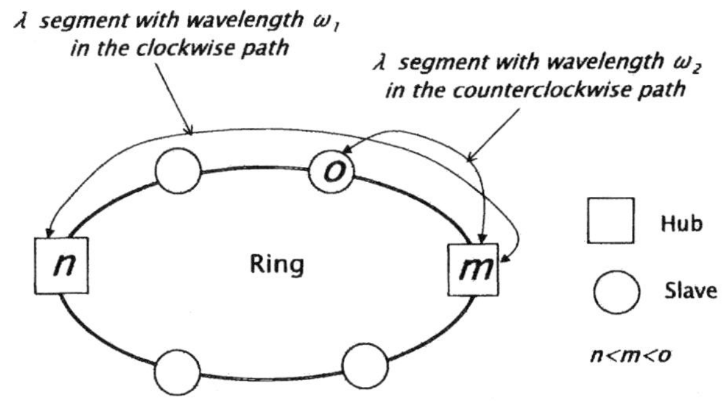

Fig. 11 Path and wavelength assignment to a λ segment.

wavelength segments within a λ ring must have the same wavelength. Each λ segment is then assigned an identical wavelength from wavelength segments in all spans along the path of the λ segment. The λ segment is hence specified by both a wavelength assigned in the normal state and a path in either the clockwise or counterclockwise direction in the associated ring, as shown in Fig. 11. We refer to a path in the clockwise (counterclockwise) direction, from node n to node m, on a ring, where $n < m$, as a clockwise (counterclockwise) path, as shown in Fig. 11.

Within a fiber-ring system, we assume all wavelength segments within each fiber are established, which means that a fiber is required in such a span that even no OCh passes through. A significant constraint on the system is that one wavelength can be assigned from a wavelength segment to only one λ segment at a time in each span, which is not modeled in [2], [12], and [13]. When a failure occurs, the λ segment is re-assigned an identical wavelength within another fiber in the opposite direction in a 4-fiber ring system, and a different wavelength within another fiber in the opposite direction in a 2-fiber one.

As for a λ-ring system, we assume that working and protection λ rings belonging to the same λ-ring system may have different wavelengths from each other. As we assumed in Sect. 2, within a working λ-ring, wavelength segments are only established on the spans that a working OCh passes through. Meanwhile within a protection λ ring, we assume that wavelength segments are established on all the spans in the associated ring.

4. Design Problem Formulation

In this section, we describe a mathematical programming scheme for designing a network of interconnected multiple self-healing ring systems. In it, we minimize total fiber length.

4.1 Formulation for 4(2)-Fiber BWPSR Systems

Since in [6] we have already formulated mathematical programming problems whose objective function is overall facility cost that includes the term of the total fiber length, design problems to minimize the total fiber length can be derived by omitting unnecessary terms from the formulation of the above problems. We, however, present here the formulation to minimize the total fiber length so that it can be compared with that for 4-λ BWPSR systems.

The following items are used in our discussion.

$n, m = 1 \dots N$: Number assigned to each node. To prevent unnecessary duplication, we restrict the range of subscripts for a node pair (n, m) to satisfy $n < m$, without loss of generality.

$s = 1 \dots S$: Number assigned to each span.

$r = 1 \dots R$: Number assigned to each ring.

$p = 1 \dots P$: Number assigned to each s-d pair.

$l_p = 1 \dots L_p$: Number assigned to each OP for s-d pair p.

$k = 1 \dots K$: Number assigned to each individual 4(2)-fiber-ring system that is established on each ring. For the λ-ring-system application, K corresponds to the upper limit on the number of the ring systems established on each ring.

$h(l_p) = 1 \dots H(l_p)$: Number assigned to each individual hop of a virtual link along the route of OP l_p.

The design parameters are as follows.

a_s: Length of Span s.

v_p: Demand for s-d pair p, in terms of the number of OChs.

Ω: Maximum number of wavelengths available per fiber.

D: Fiber count (i.e., the maximum allowable number of unidirectional fibers in each span).

The following are binary indicators determined by the rings and routes of the OPs.

$g_{r,n,m}^{cw(ct),s}$: Indicator that is 1 if the clockwise (counterclockwise) path between Nodes n and m included in Ring r goes through Span s; otherwise, it is 0.

$\delta_{h(l_p)}^{n,m}$: Indicator that is 1 if the hth hop in OP l_p, $h(l_p)$, includes Nodes n and m; otherwise, it is 0.

c_{l_p}: Capacity to be allocated to OP l_p.

$d_{r,k}$: Indicator that is 1 if the kth fiber-ring system is established on Ring r; otherwise, it is 0.

$u_{r,k,n,m}^{\omega,cw(ct)}$: Indicator that is 1 if in the kth fiber-ring system on Ring r, a λ segment is established between Nodes n and m, and if the clockwise (counterclockwise) path and Wavelength ω are assigned; otherwise, it is 0. Using this variable, we can specify the wavelength assignment from wavelength segments established within each individual fiber-based BWPSR system.

$b_{r,k,n,m}$: Total number of λ segments between Nodes n and m within the kth fiber-ring system on Ring r.

Finally, we define the following sets.

Q_r: The set of nodes included in Ring r.

Using the above notation, we have the following integer-programming problem.

Minimize

$$4(2)\sum_{r=1}^{R}\sum_{k=1}^{K}\sum_{s=1}^{S} a_s g_r^s d_{r,k} \quad (1)$$

Subject to

$$\sum_{l_p=1}^{L_p} c_{l_p} = v_p \quad (p = 1 \dots P) \quad (2)$$

$$\sum_{p=1}^{P}\sum_{l_p=1}^{L_p}\sum_{h(l_p)=1}^{H(l_p)} \delta_{h(l_p)}^{n,m} c_{l_p} = \sum_{r=1}^{R}\sum_{k=1}^{K}\sum_{\omega=1}^{\Omega}\sum_{i=(cw,ct)} u_{r,k,n,m}^{\omega,i} \left(n = 1 \dots N, m = n+1 \dots N \quad \text{s.t.} \sum_{p=1}^{P}\sum_{l_p=1}^{L_p} \delta_{h(l_p)}^{n,m} \geq 1 \right) \quad (3)$$

$$\sum_{i=(cw,ct)} \sum_{(n,m)\in Q_r} g_{r,n,m}^{i,s} u_{r,k,n,m}^{\omega,i} \leq d_{r,k} \quad (r = 1 \dots R, s = 1 \dots S, \text{ s.t. } g_s^r = 1, k = 1 \dots K, \omega = 1 \dots \Omega) \quad (4)$$

$$d_{r,k} \geq d_{r,k+1} \ (r = 1 \dots R, k = 1 \dots K-1) \quad (5)$$

$$4(2)\sum_{r=1}^{R}\sum_{k=1}^{K} g_r^s d_{r,k} \leq D \ (s = 1 \dots S) \quad (6)$$

$$0 \leq c_{l_p} \leq v_p, \ d_{r,k} = 0 \text{ or } 1, u_{r,k,n,m}^{\omega,i} = 0 \text{ or } 1$$

Equation (1) is the total fiber length to be minimized. Multiplier 4 is used for a network only 4-fiber ring systems are used, and 2, for 2-fiber ring systems. Equation (2) ensures that the total sum of the capacities allocated to the individual OPs for an s-d pair is equal to the demand for that pair. Equation (3) ensures that the demand offered to a virtual link, through which at least one OP candidate passes, is equal to its capacity which is given by the total number of λ segments in that link (see Sect. 3.2). Equation (4) ensures that if a fiber-ring system is established on a ring ($d_{r,k} = 1$), one wavelength is assigned from a wavelength segment to only one λ segment at a time in each span (see Sect. 3.3).

Equation (5) ensures that if the kth fiber-ring system does not exist in a ring ($d_{r,k} = 0$), then the $(k+1)$th fiber-ring system is, likewise, not established ($d_{r,k+1} = 0$). This constraint does not affect the optimal value, but is very effective at reducing the search space when this integer-programming problem is solved using the branch-and-bound method [14]. Equation (6) ensures that the total number of fibers required in each span does not exceed the fiber count.

4.2 Formulation for 4-λ BWPSR Systems

The following variables are used to formulate integer programming problems, which are, in turn, used to design a network of interconnected 4-λ BWPSR systems.

h_s: Number of fibers required in Span s.

$x_{r,n,m}^{v,cw(ct)}$: Total number of λ segments that are established between Node n and m on ring r and are assigned a clockwise (counterclockwise) path and Wavelength ω. Setting this variable is based on Sect. 3.3.

$y_{r,s,\omega}$: Number of λ segments that are established on Ring r, passes through Span s, and are assigned Wavelength ω.

$q_{r,\omega,z}$: Number of protection λ rings that have Wavelength z, and whose corresponding working λ rings have Wavelength ω.

Using the above, we arrive at the following integer programming problem.

Minimize

$$2\sum_{s=1}^{S} a_s h_s \quad (7)$$

Subject to

$$\sum_{l_p=1}^{L_p} c_{l_p} = v_p \quad (p = 1 \dots P) \quad (8)$$

$$\sum_{p=1}^{P}\sum_{l_p=1}^{L_p}\sum_{h(l_p)=1}^{H(l_p)} \delta_{h(l_p)}^{(n,m)} c_{l_p} = \sum_{r=1}^{R}\sum_{\omega=1}^{\Omega}\sum_{i=(cw,ct)} x_{r,n,m}^{\omega,i} \left(n = 1 \dots N, m = n+1 \dots N \quad \text{s.t.} \sum_{p=1}^{P}\sum_{l_p=1}^{L_p} \delta_{h(l_p)}^{n,m} \geq 1 \right) \quad (9)$$

$$y_{r,s,\omega} = \sum_{(n,m)\in Q} \sum_{i=(cw,ct)} g_{r,s}^{n,m,a} x_{r,n,m}^{\omega,i} \quad (r = 1 \dots R, s = 1 \dots S, \text{ s.t. } g_s^r = 1, \omega = 1 \dots \Omega) \quad (10)$$

$$y_{r,s,\omega} \leq \sum_{z=1}^{\Omega} q_{r,\omega,z}$$
$$(r = 1 \ldots R, s = 1 \ldots S, \text{ s.t. } g_s^r = 1,$$
$$\omega = 1 \ldots \Omega) \qquad (11)$$

$$\sum_{r=1}^{R} g_r^s y_{r,s,\omega} + \sum_{r=1}^{R}\sum_{z=1}^{\Omega} g_r^s q_{r,z,\omega} \leq h_s$$
$$(s = 1 \ldots S, \omega = 1 \ldots \Omega)$$
$$0 \leq h_s \leq D,\ 0 \leq c_{l_p} \leq v_p,\ x_{r,n,m}^{\omega,i} \geq 0,$$
$$0 \leq y_{r,s,\omega} \leq K,\ 0 \leq q_{r,\omega,z} \leq K \qquad (12)$$

Equation (7) is the total fiber length to be minimized. Equation (8) ensures that the sum of the capacities allocated to the individual OPs of an s-d pair is equal to the demand for that pair. Equation (9) ensures that the demand offered to such a virtual link that at least one OP candidate passes through is equal to its capacity which is given by the total number of λ segments in that link (see Sect. 3.2). Equation (10) ensures that the total number of λ segments with a wavelength in each span on a ring is given by the sum of the number of λ segments that are established between a node pairs along a path on the ring, have that wavelength, and pass through that span. Equation (11) ensures that the total working bandwidth requirement does not exceed the total protection bandwidths for each wavelength in each span. Here an inequality constraint is adopted because within working λ rings, wavelength segments are only established in necessary spans (see Sect. 3.3). Equation (12) ensures that the sum of the working and protection bandwidth requirements with each wavelength does not exceed the number of fibers in each span.

5. Numerical Example

Figures 8 and 12 show the network models considered. The former is a smaller one that consists of 9 nodes and 11 spans. The latter is a larger network that consists of 18 nodes and 24 spans. They are denoted by (9, 11) and (18, 24), respectively. The (9, 11) network model is a sub-set of the (18, 24) network model, which is derived from [7]. We assume that $\Omega = 32$ in both network models. In the smaller one, we assumed a fully meshed demand pattern with the same demand, and in the larger, the basic pattern shown in Fig. 13, based on [15]. Since these networks are not complex, the hubs and rings are arbitrarily chosen. We set the OP routes according the kth shortest path over a virtual mesh network in which the length of each virtual link is given by the shortest circumference among all the rings included in it.

For each of the two network models, we solved the design problems based on the interconnected fiber-ring systems, interconnected λ-ring systems, and 1+1 protection. To design the networks based on 1+1 protection, we applied the optimal design scheme for a mesh-dedicated model proposed in [3] with a minor modification made necessary by the unidirectional-fiber assumption. In this 1+1 protection-based network design, we also assume that: (1) for any single span failure, each working OP is fully protected by a single backup OP that is span-disjointed from the working OP. (2) wavelength may be converted in each node on the route of each OP. To solve the integer-programming problems, we used CPLEX 6.5† with the branch-and-bound method [13]. Computations were done on a DEC Model53au2.

In the following, we describe the design and evaluation results for the (9, 11) network model. Figure 14 shows the optimized total fiber length as a function of the demand. In the 4-fiber-ring-system application, total fiber lengths for low demand such as at $v_p = 1$ or 2 are the same. In these cases, each ring has just a single ring system, which is sufficient to accommodate a demand of up to 2. For the other architectures, the graph increases monotonically. For all levels of demand, the order of length, from largest to smallest, is (i) the 4-fiber-ring-system application, (ii) the 2-fiber-

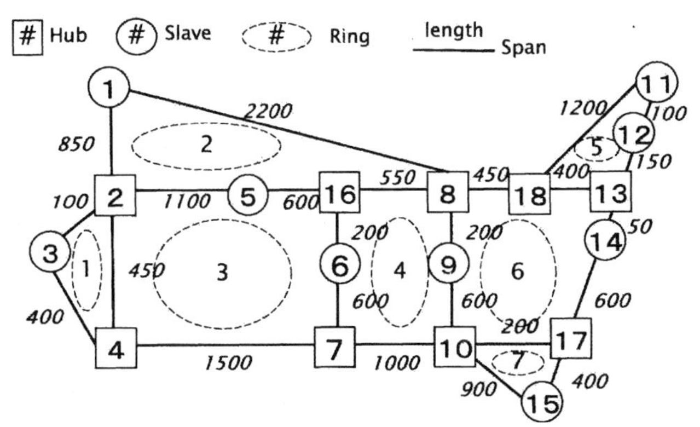

Fig. 12 Network model (18, 24).

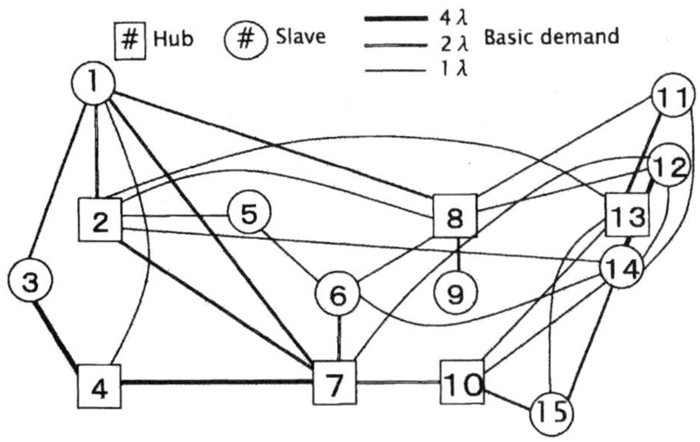

Fig. 13 Basic demand pattern in network model (18, 24).

†http://www.cplex.com

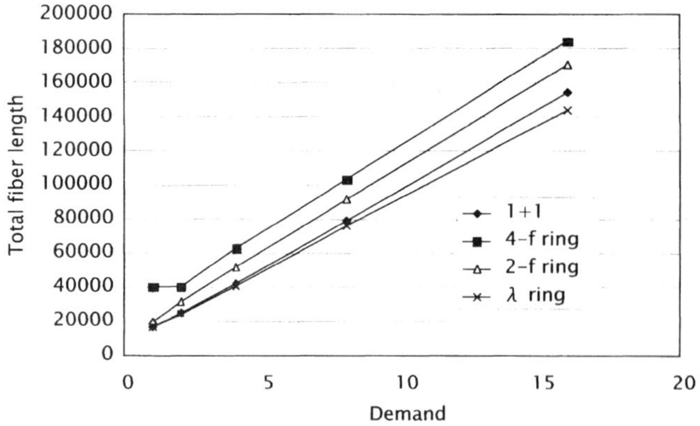

Fig. 14 Total fiber length vs. demand for (9, 11).

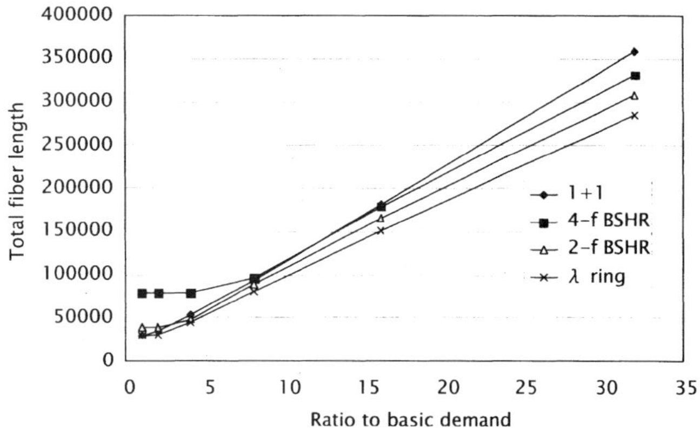

Fig. 15 Total fiber length vs. ratio to basic demand for (18, 24).

ring-system application, (iii) 1+1 protection, and (iv) the λ-ring-system application. This suggests that the λ-ring-system application is the most cost-effective for small starts.

Next we describe the design results for the (18, 24) network model. Figure 15 shows the optimized total fiber length as a function of the ratio to basic demand. For the 4-fiber-system application, the total fiber lengths for ratios of 1, 2, and 4 are the same. For the 2-fiber-ring-system application, the total fiber lengths for ratios of 1 and 2 are the same. For other levels of ratio and applications, the graphs increase monotonically. Moreover, for low demand, the order for fiber lengths, from largest to smallest, is: (i) the 4-fiber-ring-system application, (ii) the 2-fiber-ring-system application, (iii) 1+1 protection, and (iv) the λ-ring-system application. For sufficiently large demand, however, the order changes to: (i) 1+1 protection, (ii) the 4-fiber-ring-system application, (iii) the 2-fiber-ring-system application, and (iv) the λ-ring-system application. This also suggests that the λ-ring-system application is the most cost-effective for small starts.

Note that, for sufficiently large demand, 4-fiber and 2-fiber-ring-system applications are inferior to 1+1 protection in model (9, 11), but that these two ring system applications are superior to 1+1 protection in model (18, 24).

The order of total fiber length for individual ring-system applications depends on the granularity of the resource allocation. This granularity is highest in the λ-ring-system application, followed by the 2-fiber-ring-system application, and finally the 4-fiber-ring-system application. The total fiber length obtained by calculations follow the reverse order of the granularity.

Based on the solutions obtained, we evaluated overall utilization, defined as

$$(\text{Overall utililzation}) = \frac{(\text{total OCh length})}{(\text{the number of working wavelengths per fiber}) \times (\text{total fiber miles})}, \tag{13}$$

where total OCh length is the sum of the length of all wavelength segments used by all OChs and that of protection wavelength segments that will be used if the OChs fails. Overall utilization for 4(2)-fiber-ring systems becomes

$$\left\{ \sum_{r=1}^{R}\sum_{k=1}^{K}\sum_{\omega=1}^{\Omega(\Omega/2)}\sum_{i=(cw,ct)}\sum_{(n,m)\in Q_r}\sum_{s=1}^{S} g_{r,n,m}^{i,s} a_s u_{r,k,n,m}^{\omega,i} + \sum_{r=1}^{R}\sum_{k=1}^{K}\sum_{\omega=1}^{\Omega(\Omega/2)}\sum_{s=1}^{S} g_r^s a_s \left(\text{if} \sum_{(n,m)\in Q_r}\sum_{i=(cw,ct)} u_{r,k,n,m}^{\omega,i} \geq 1 \text{ then } 1 \text{ else } 0 \right) \right\} \Bigg/ \left\{ 2 \times \Omega(\text{or } \Omega/2) \times \sum_{r=1}^{R}\sum_{k=1}^{K}\sum_{s=1}^{S} g_r^s a_s d_{r,k} \right\}, \tag{14}$$

in which we assume that the protection λ ring exists if at least one working wavelength segment is established within a λ-ring system. Ω is applied to 4-fiber ring systems, and $\Omega/2$, to 2-fiber ring systems.

Overall utilization for the λ-ring-system application is calculated by using Eq. (12) as

$$\frac{\sum_{r=1}^{R}\sum_{s=1}^{S}\sum_{\omega=1}^{\Omega} g_r^s a_s y_{r,s,\omega} + \sum_{r=1}^{R}\sum_{z=1}^{\Omega}\sum_{\omega=1}^{\Omega}\sum_{s=1}^{S} g_r^s a_s q_{r,z,\omega}}{\Omega \sum_{r=1}^{R}\sum_{s=1}^{S} a_s h_s}. \tag{15}$$

Figure 16 shows overall utilization as a function of demand for network model (9, 11), and Fig. 17 shows overall utilization as a function of the ratio to basic demand for model (18, 24). Utilization increases rapidly as demand increases for all ring system applications and for 1+1 protection.

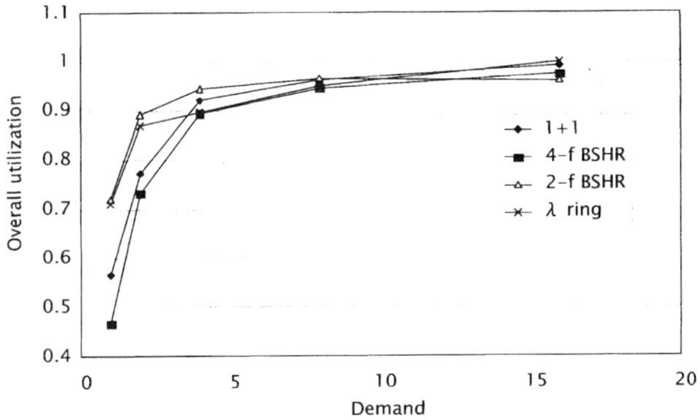

Fig. 16 Overall utilization vs. demand for (9, 11).

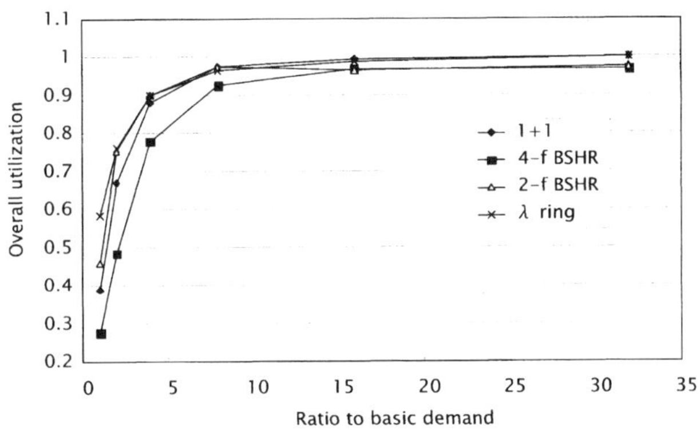

Fig. 17 Overall utilization vs. ratio to basic demand for (18, 24).

For both network models, we obtained the following results:
(1) For sufficiently low demand, the 2-fiber-ring-system and the λ-ring-system applications are superior to both the 4-fiber-ring-system application and 1+1 protection with respect to overall utilization.
(2) For sufficiently high demand, the λ-ring-system application and 1+1 protection have utilizations close to unity, and they are superior to the 4-fiber and 2-fiber ring-system applications.

6. Conclusion

We proposed a λ-ring systems as a new application for WDM self-healing ring architecture whose ordinary applications are 4-fiber and 2-fiber ring systems. The optimal design schemes for networks of interconnected 4-λ BWPSR systems and 4-fiber and 2-fiber BWPSR systems are also described. We compared these ring-system applications and 1+1 end-to-end path protection in terms of total fiber length. Numerical examples show that the λ-ring-system application has the shortest total fiber length, and compared with the 4-fiber and 2-fiber-ring-system applications, the λ-ring-system application is more suitable for introducing WDM self-healing ring systems into networks with low demand.

Acknowledgment

The author would like to thank S. Hasegawa and T. Shiragaki for their helpful discussions and comments.

References

[1] J.P. Ryan, "WDM: North American deployment trends," IEEE Commun. Mag., vol.36, no.2, pp.40–44, Feb. 1998.
[2] T.H. Wu, Fiber Network Service Survivability, Artech House, 1992.
[3] Y. Miyao and H. Saito, "Design and evaluation of survivable optical transport network," IEEE J. Sel. Areas Commun., vol.16, no.7, pp.1190–1198, Sept. 1998.
[4] T. Shiragaki, H. Takeshita, H. Shimomura, S. Nakamura, and N. Henmi, "Network resource advantages of bidirectional wavelength-path switched ring," IEEE Photon. Technol. Lett., vol.11, no.10, pp.1325–1327, Oct. 1999.
[5] M. Cvijetic, T. Shiragaki, and A.J. Weissberger, "OCh shared protection ring," T1X1.5/99-178, July 1999.
[6] Y. Miyao and H. Saito, "Optimal design of survivable photonic transport networks of interconnected WDM self-healing ring systems," IEICE Trans. Commun., vol.E83-B, no.10, pp.2261–2269, Oct. 2000.
[7] R.K. Butler and D.R. Polson, "Wave-division multiplexing in the Sprint long distance network," IEEE Commun. Mag., vol.36, no.2, pp.52–55, Feb. 1998.
[8] S. Nathan, "The impact of IP traffic patterns of today's backbone network," LIGHTWAVE, Dec. 1999.
[9] S. Hardy, "Qwest unveils all-optical-network plans," LIGHTWAVE, Jan. 2000.
[10] T. Shiragaki and M. Cvijetic, "Protection architecture for OCh SPRing and required functions in the OChT overhead," T1X1.5/2000-45, Jan. 2000.
[11] B. Doshi and P. Harshavardhana, "Broadband network infrastructure of the future: Roles of network design tools in technology deployment strategies," IEEE Commun. Mag., vol.36, no.5, pp.60–71, May 1998.
[12] L. Wuttisittikulkij and M.J. O'Mahony, "Design of a WDM network using a multiple ring approach," GLOBECOM'97, pp.551–555, 1997.
[13] S. Bortolon, H.M.F. Tavares, R.V. Ribeiro, E. Quaglia, and M.A. Bergamaschi, "A methodology to SDH networks design using optimization tools," GLOBECOM'96, pp.1867–1871, 1996.
[14] H.P. Williams, Model building in mathematical programming third revised, John Wiley & Sons, 1993.
[15] http://boardwatch.internet.com/isp/summer99/bb/sprint.jpg

High-Performance Backbone Network Technology edited by Naoaki Yamanaka
ISBN: 0-8247-5321-6 DOI: 10.1081/(120025730)
Published 2004 by Marcel Dekker, Inc.

A Comparison on Capacity Requirement of Optical WDM Mesh Network Protection Strategies

Charoenchai Boworntummarat, and Lunchakorn Wuttisittikulkij

SUMMARY This paper presents a comprehensive investigation of three optical wavelength-division multiplexed (WDM) mesh network protection approaches, namely minimal cost, single link basis and disjoint path approaches. The operation of each approach is described and their performances are extensively evaluated and compared. Key aspects that are taken into the consideration and comparison of the designs include a spare capacity requirement, ease of operation and practical feasibility. A mathematical model based on integer linear programming is introduced to obtain a lower bound on the spare capacity requirement for full protection against all single link failures. Two heuristic algorithms have also been developed to perform wavelength resource allocation under both normal and failure conditions for both systems with and without wavelength conversion capability. It is shown that the minimal cost approach can accomplish the lowest extra cost requirement for protection, but this approach is considered not appropriate for practical applications due to complicated restoration and management. The single link basis scheme is on the other hand more practical and very cost efficient. For the disjoint path technique, the cost for spare capacity is generally slightly greater than that of the single link basis scheme. Its main advantages lie in the simple re-configuration and inherent protection against node failure for in-transit traffic. Finally, a new framework for obtaining a good spare capacity cost estimate of a mesh restorable network is presented.

key words: *wavelength division multiplexing, mesh design, lower bound on network cost, heuristic approach, network connectivity*

1. Introduction

The wavelength division multiplexed (WDM) technique has been seen as a promising technology for realizing future broadband networks to support the increasing bandwidth demands of various emerging applications, such as multimedia and Internet web-browsing [1]–[3]. In such networks, a number of multiple data streams can be multiplexed into a single fiber, each operating at a few Gbit/s. Therefore, the aggregate throughput of this type of network is expected to be in the order of Tbit/s. A basic element of a multi-wavelength network is a set of optical-cross connects (OXCs) which can be interconnected by fiber links to form an arbitrary network topology. A traffic connection demand is supported by an optical path which is established from a source node to a destination passing through intermediate OXC nodes in a circuit-switched fashion.

Manuscript received December 11, 2000.
Manuscript revised November 22, 2001.

[†]The authors are with the Department of Electrical Engineering, Faculty of Engineering, Chulalongkorn University, Bangkok, 10330 Thailand.

Since the WDM networks can support very high bit-rate services, a single element (link or node) failure in the network could lead to a large amount of data loss, causing serious service disruption. Consequently, it is considered essential that certain network protection measures must be provided at the network design and dimensioning stage, so that service continuity can be maintained despite failures. In this paper, we examine and explore several network protection techniques suitable for multi-wavelength network systems. The main objective is to identify an appropriate protection scheme for full restoration against all single link failures. Major design aspects that are taken into the design consideration include a spare capacity requirement, ease of operation and practical feasibility.

In general, network protections for link failures can be classified into two categories, namely link restoration and path restoration. For link restoration, the corrupted traffic is rerouted around the failed link whereas for path restoration the interrupted traffic can be rerouted on a new entire physical route from the source to destination nodes, i.e., end-to-end restoration. The key advantage of link restoration is that restoration can be performed locally and is transparent to the source and destination nodes, hence its restoration process is relatively faster than the path restoration counterpart. This restoration technique is however not suitable for a multi-wavelength network architecture, because of some difficulties in allocating spare wavelength resources which arise due to a wavelength continuity constraint. Nevertheless, the link restoration can become a viable protection alternative, if either wavelength conversion is provided or spare capacity is allocated with the granularity of fibers. For the latter, the protection problem will become identical to the problem encountered in conventional electronic networks and it has been extensively addressed in the literature [4]–[7]. Therefore, the protection techniques investigated in this paper are all based on path restoration.

Although there have been a number of interesting studies on multi-wavelength network design and dimensioning over the past few years, the design requirements of each study varied. For example, references [8]–[11] focused on minimizing the number of wavelength requirements by assuming one optical fiber per link. This objective may not be appropriate for realistic network systems where the number of fibers allocated to each link is not necessarily the same and usually more cost-effective designs can be achieved when allowing fiber allocation as well as wavelength allocation. Therefore, in this paper we study how routes, wavelengths, and fibers are allocated to support traffic demands while providing full protection against all single link failures.

In this study, two different systems are considered namely wavelength path (WP) and virtual wavelength path (VWP) systems [12], [13]. The VWP system differs from the WP system in that its node configurations include an additional wavelength conversion capability. Therefore, when setting up an optical path for a connection, the VWP can assign wavelengths on a link-by-link basis, whereas the WP must choose a single wavelength for all links along the entire physical route. At present, the number of wavelengths that can be multiplexed into a single fiber is still limited to a few wavelengths and it is dictated by optical technology. Accordingly, it is important to understand how this factor will influent the network design outcomes and thus this issue is addressed here.

In the WDM network design with WP system, as described in [13] there are two methods that can be considered for assigning the wavelength to the restoration path. In the first method, the restoration path and its corresponding active path are assigned with the same wavelength. On the other hand, in the second method, the wavelength assigned to the restoration path is not necessarily the same wavelength as its corresponding active path. In this paper, we adopt the second method, because it is much more cost effective and practically feasible.

The remainder of this paper is organized as follows. In Sect. 2, the protection strategies against a single link failure are elaborated. In Sect. 3, a mathematical model for protection is proposed. The heuristic algorithms of wavelength routing and protection methods are described in Sect. 4. In Sect. 5, the simulation results of a few network problems are given and discussed. A comparison between each of the protection approaches is also displayed in this section. A new framework for determining the spare capacity requirement is also presented and discussed in this section. Finally, the major findings in this paper are summarized in Sect. 6.

2. Protection Strategies

The network protection strategies in this study are classified into three categories according to the way in which corrupted traffic is restored: minimal cost approach, single link basis approach and disjoint path approach.

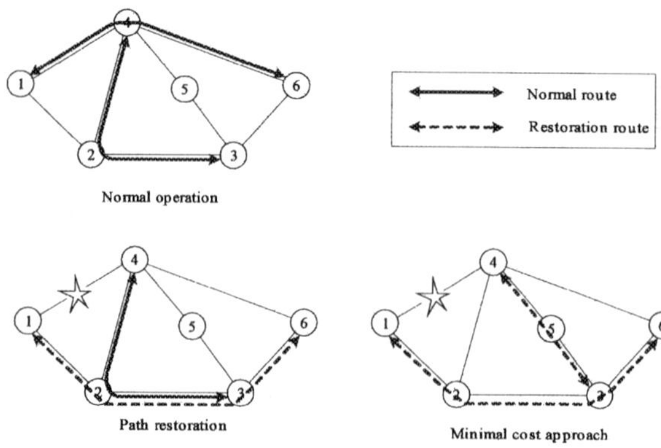

Fig. 1 An example illustrating the difference between the minimal cost protection and the commonly known path restoration scheme.

2.1 Minimal Cost (MC)

For the minimal cost protection approach, in the events of a link failure all optical connections are released and subject to be rearranged, eventhough they may not be directly interrupted by the failure. Accordingly, in principle this approach allows the design to be very efficient and results in minimal network cost. It should be noted that in the commonly known path restoration technique, only interrupted traffic connections are rerouted and other connections remain unchanged.

In order to illustrate the difference between the minimal cost approach and the commonly known path restoration scheme, an example of a network scenario is given in Fig. 1. In this example, under normal operation two connections (1,6) and (3,4) are set up along the physical routes of 1-4-6 and 3-2-4 respectively. Now consider a situation where a failure occurs on the link between nodes 1 and 4. In the conventional path restoration scheme, only the connection (1,6) is subject to rerouting as it is directly affected by the failure and in this example the new route chosen is through nodes 1-2-3-6 resulting in an extra wavelength channel being needed on link 2-3. In contrast, for the minimal cost approach, both connections are re-arranged and it appears that no additional wavelength is required.

Although this technique is very cost-efficient, it may not be suitable for practical use due to two reasons. Firstly, it requires all optical connections to be re-arranged, causing unnecessary disturbance to all ongoing traffic in the entire network, which is extremely undesirable. Secondly, if this protection technique is to be employed, the network management system will have to maintain a huge database keeping all the restoration paths for each active path in all failure scenarios. However, the design outcomes of the MC protection technique are useful for comparisons in terms of network costs with other more practical network protection strategies.

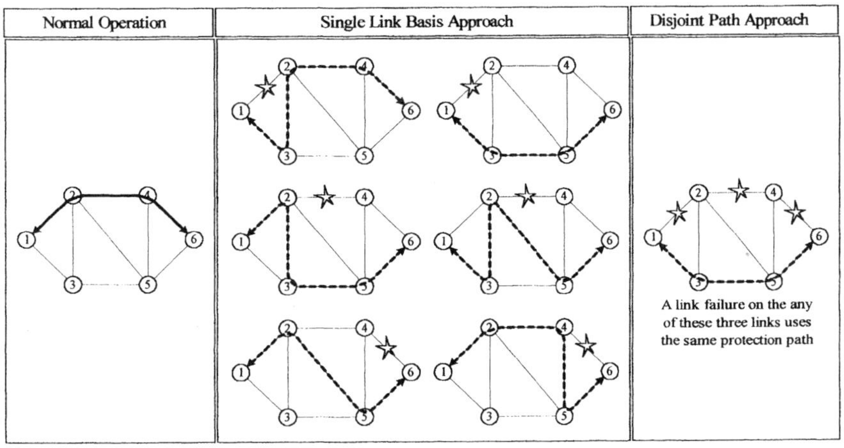

Fig. 2 An example illustrating the difference between the SLB and DJP techniques.

2.2 Disjoint Path (DJP)

In this approach, when a connection is to be established between a source-destination node pair, two physically disjoint paths are carefully chosen. One of the paths is selected as an active path for supporting the traffic demands in normal operation whereas another path is considered as a backup for the active path, i.e., the restoration path. Since these two paths are physically disjoint, failure on any link along the active path will not affect the protection path. Notice that this protection technique in effect provides a conventional 1:1 protection for all single link failure events. However, the amount of spare capacity required is not necessarily 100% as the capacity reserved as restoration paths can be shared among active paths as long as these active paths are physically disjoint.

In fact, disjoint paths can be divided into two classes, i.e., node disjoint paths and link disjoint paths. The difference between them is that the node disjoint path must not have both links and nodes in common with the working path, whereas the link disjoint path only requires disjoint links. Therefore, the choice of a node disjoint path is more limited, but it has benefits over link disjoint paths; the node disjoint scheme can also protect in-transit traffic from node failures. In this paper, we concentrate on the node disjoint technique, as in fact the differences between these two techniques are rather marginal. In contrast to the MC approach, when a link failure occurs in the network, only the interrupted paths are involved in the restoration process. Consequently, the network management of the DJP protection method is much simpler and more practical than that of the MC method. Moreover, the optical nodes can immediately switch from the interrupted paths to the protection paths after the network management system detects an interruption of certain active paths without knowing the exact location of the failure.

Table 1 The advantages and disadvantages of three protection strategies.

Protection Strategy	Routing Flexibility	Restoration Complexity	Network Cost	Size of Routing database
MC	High	High	Low	High
SLB	Moderate	Moderate	Relatively low	Moderate
DJP	Low	Low	High	Low

2.3 Single Link Basis (SLB)

Like the DJP approach, the single link basis (SLB) approach [6], [7], [13] demands the re-arrangements of only the active paths that are disrupted while other paths are left undisturbed. However, in the SLB scheme, each active path can have different restoration paths depending on the place of the failed link, whereas in the DJP scheme each active path can have only one physical backup path. For instance, as shown in Fig. 2, a connection is to be established between the nodes 1 and 6. The working path is established along the path 1-2-4-6. For the DJP approach, there is only one possible restoration path, that is the path 1-3-5-6. This restoration path can protect the working path when a failure happens on any link along the working path. Conversely, the SLB approach allows at least two paths for surviving the working path for each failure event. For example, if a failed link occurs in link 1-2, the restoration path is 1-3-2-4-6 or 1-3-5-6. Intuitively, it is expected that the network capacity requirements for the SLB approach are less than for the DJP approach since the SLB approach can utilize the network resources more flexibly and efficiently.

For three protection strategies as discussed above, we can summarize the advantages and disadvantages of each of protection strategy with respect to other strategies as shown in Table 1.

3. Lower Bound on Protection Cost

In this section, we describe a mathematical model based on the integer linear programming (ILP) to obtain a lower bound on the protection cost.

Given a set of traffic demands and a physical network topology, the network is modeled as an undirected graph $G(\nu, \varepsilon)$, where ν is a set of nodes with size N and ε is a set of edges with size L. Let M be the number of wavelengths multiplexed into a fiber. Let X_i be the number of fibers (or their capacity) on the ith edge and each fiber is assumed to be bi-directional and can only contain one wavelength ($M = 1$). The objective function of this model is

$$\text{Min:} \sum X_i \quad \text{where} \quad i \in \varepsilon \tag{1}$$

The constraint to formulate the equation is derived from the minimum flow on a cut set which is defined as follows. The network graph is partitioned into two parts, G_1 and G_2, with cut set ε_c, which corresponds to a set of edges that has one endpoint in G_1 and the other in G_2. The possible minimum traffic of a cut set, $f_{min}(c)$, is the summation of each traffic demand which has a source and a destination lying on different parts. As a result, the capacity of any cut set must be larger than or equal to the possible minimum traffic:

$$\sum X_i \geq f_{min}(c) \quad \text{where} \quad i \in \varepsilon_c \tag{2}$$

Now, we can determine the network cost from the linear formulations related to every possible cut set of the network graph. This cost only provides for the traffic demands in normal operation. In order to extend the model to cover the protection requirements against all single link failures, additional constraints are included as follows. Edges are sequentially removed from the network graph, simulating each link failure event. Given a network graph with an edge removed, the same formulation of the minimum flow on a cut set is applied, producing a new set of required constraints. By applying this procedure to all edges, a total of L additional new sets of constraints would result, covering all link failure scenarios. All these constraints are sufficient to form a model for finding the lower bound on the network cost with full protection and the solution can be obtained by solving this mathematical formulation using a known integer linear programming algorithm.

4. Heuristic Algorithms

As will be shown in the result section, the execution time of the ILP formulation is computationally expensive for large or even medium network sizes. Therefore, the ILP technique is only suitable for relatively small networks. Remember also that this model can only be for the lower bound; no actual route or wavelength assignments are performed. As a result, there exists a need to develop an efficient heuristic algorithm to perform and optimize the network resource allocation. This paper introduces two heuristic algorithms, one for the MC and another for the DJP scheme. For the SLB protection scheme, a known efficient algorithm exists and can be found in [13]. Hence this algorithm is adopted for this study.

We shall describe the path allocation and wavelength assignment design algorithms for the optical paths below (no protection is considered).

4.1 Path Accommodation Design and the Wavelength Assignment Algorithms

Due to the complexity of the network capacity allocation problem, the problem is divided into two procedures: path accommodation and wavelength assignment. Note that the wavelength assignment in the VWP system is considered on a link-by-link basis. Therefore, only the path accommodation procedure is required. We will first explain the path accommodation procedure, then followed by the wavelength assignment. An outline of the path accommodation is described as follows:

Step 1: All possible routes between each node pair are identified.

Step 2: Randomly select an appropriate physical route for each traffic connection.

Step 3: After all connections have been established, the next step is an attempt to rearrange the existing allocation, so that the overall network cost is reduced. To accomplish this, each connection is reconsidered in turn to see whether they can be redirected on another route that leads to more efficient wavelength resource utilization.

Step 4: As the VWP system assigns wavelengths on a link-by-link basis, the number of wavelength channels required on each link is equal to the number of optical connections passing through the corresponding link. The number of fibers required on each link can be obtained from $\lceil P/M \rceil$ where P is the number of optical paths going through the link and M is the number of wavelengths multiplexed in one fiber. Note that $\lceil X \rceil$ is the smallest integer greater than or equal to X. Consequently, the network cost can be obtained from the summation of the number of required fibers of every link times the number of wavelengths.

As mentioned earlier that the wavelength resource allocation in the WP system is more complex than that of the VWP, care must be taken in choosing a wavelength for the optical path. Here is the outline of the algorithm for wavelength assignment.

Step 1: The same procedure for physical route selection as for VWP is applied for each traffic connection.

Step 2: Divide those connections into groups according

to the length of the route; connections of the same route length are grouped together.

Step 3: A matrix $\boldsymbol{W}$ is defined. All elements are set to zero. Apply the following actions to all connections in each group. The groups that have longer route lengths are considered before those that have shorter routes. Among the routes in the same group, the routes are randomly considered.

3.1) If a connection in the group being considered goes through link j the value at w_{1j} is set to 1, i.e., marking element w_{1j}. However, if the connection goes through a link that has been marked, this connection is skipped and will be considered in the next row of the matrix.

3.2) Repeat step 3.1. Both steps are applied until all connections in the group are assigned.

3.3) Apply the same rules of 3.1 and 3.2 to all the remaining groups.

Step 4: Assign λ_1 for the connections marked in the first row of matrix $\boldsymbol{W}$, λ_2 for the connections in the second row and so on. The number of fibers required in link i can now be obtained.

4.2 MC Heuristic Algorithm

This algorithm guarantees 100% protection against any single link failure in the network and it can be applied to both VWP and WP systems.

Step 1: A fiber link i is removed from the original structure for simulating a link failure event i. The resulting network is referred to here as an incomplete network i. Note that there exists a total of L different incomplete networks.

Step 2: Apply the algorithms of VWP or WP to the incomplete network i and determine the number of fibers required for each fiber link, producing a total of $L-1$ results. These numbers are then stored in row i of an $L \times L$ matrix $\boldsymbol{R}$. Each row i in the matrix $\boldsymbol{R}$ contains results from each failure event i. Each column j contains the number of fibers required on link j for each failure event.

Step 3: Repeat step 1 until all link failure events have been considered.

Step 4: The maximum value of column j in the matrix $\boldsymbol{R}$ is taken as the number of fibers required to be placed on link j. Therefore, the total number of fibers required for full protection is the summation of the largest value in each column.

Step 5: The matrix $\boldsymbol{R}$ obtained from the above steps is used as an initial result for further improvement. Since the network cost depends on the largest value in each column, it is useful to reduce the largest value of a certain column while keeping the largest values in all other columns unchanged. This technique can be repeated iteratively, until no further cost reduction is observed.

4.3 DJP Heuristic Algorithm

This algorithm can be applied to both the VWP and WP systems. The details are described as follows:

Step 1: In this step, the working paths are established for satisfying traffic demands by applying the path accommodation design as described above. Simultaneously, we also assign the restoration paths for surviving the active paths in the failure event. Note that, for the WP system, there is a single wavelength for all links along the entire route. Thus, for the active and restoration paths, the wavelength assignment algorithm is used for allocating the wavelengths. After that, the network cost of each link is calculated as the initial cost, which will be employed in next step.

Step 2: A link in the network is defined as the reroute link. The restoration routes that contain the reroute link are re-established. As in the MC heuristic algorithm, the restoration route will be changed to a new restoration route, if the new route can reduce the initial network cost of the reroute link without increasing the cost of other links.

Step 3: It is possible that the network cost can be further decreased. Therefore, step 2 is repeated iteratively, until no further cost reducing is observed.

5. Simulation Results and Discussion

5.1 Execution Time Requirement of the ILP Technique

In this first section, we shall evaluate and discuss the execution time requirement of the ILP technique and identify the extent to which this model can be used in practice. A chart describing the computational time requirement of the ILP technique is summarized in Fig. 3. The execution time results here are based on numerous test samples with an assumption of uniform traffic patterns. Experimental network problems are presented in terms of the number of nodes and the network connectivity, which are normalized by the maximum value, i.e., $N-1$, where N is the number of nodes. Referring to Fig. 3, the dark areas represent those network problems that cannot be solved within 30 minutes (using a commonly available personal computer) whereas the white areas represents the problems that can be solved almost

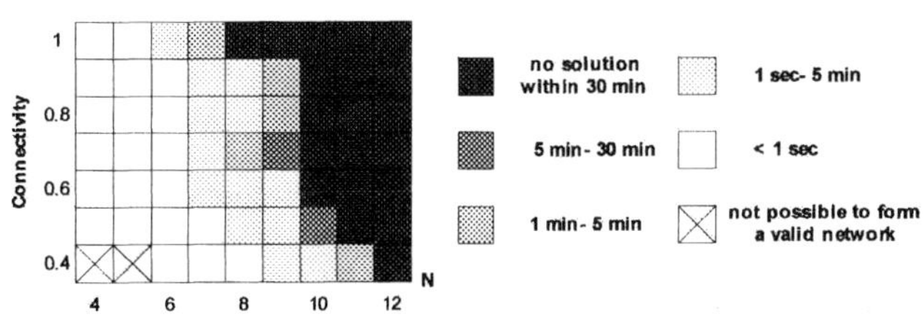

Fig. 3 The computational time for the ILP technique.

Table 2 Results of ILP and MC heuristic solutions at $M = 1$.

N	L	C	No Protection		Protection	
			ILP	MC	ILP	MC
7	7	2.0	42	42	84	84
7	8	2.29	38	38	76	76
7	9	2.57	37	37	70	70
7	10	2.86	35	35	68	68
7	11	3.14	31	31	53	54
7	12	3.43	30	30	42	45
7	13	3.71	29	29	46	48
7	14	4.0	28	28	43	45
7	15	4.29	27	27	34	36
7	16	4.57	26	26	34	35
7	17	4.86	25	25	35	35
7	18	5.14	24	24	28	32
7	19	5.43	23	23	28	31
7	20	5.71	22	22	28	30
7	21	6.0	21	21	28	30

instantly. As we can see, the ILP technique is not applicable to any network that is larger than 11 nodes and connectivity of 0.4. For the networks of smaller sizes in the range of 8–11 nodes, the ILP can still require prohibitive processing time if the connectivity is high. The ILP can complete and find the solution within a reasonable time for only those networks that are smaller than 8 nodes. These results show that the ILP technique can be used to determine a lower bound of the network costs only for small network problems.

5.2 ILP Technique vs. the Heuristic MC

In this section, we shall analyze and compare the performance of the ILP technique with the heuristic MC technique. Due to the execution time restriction of the ILP model, it is not possible to conduct a comparison for large network problems. Accordingly, networks of 7 nodes are considered appropriate and used here. In order to understand their performance under various network connectivities, an experiment is carried out as follows: a number of 7-node networks with the number of links L varied from 7 to 21 corresponding to a connectivity C of 2 to 6 are generated, simulating all possible levels of connectivity. For $C = 2$ the sample network becomes a ring structure, whereas for $C = 6$ the network is fully interconnected. In each of these generated networks, the network costs for systems with and without protection, assuming uniform traffic requests, are determined using the ILP technique and the MC heuristic with M set to 1. Note that when $M = 1$ there is no difference between the WP and VWP systems. The results are summarized in Table 2. The first two columns show the number of nodes (N) and the number of links (L), respectively. The number of nodes is fixed at $N = 7$. The third column shows the network connectivity (C) which is defined as the average number of links adjacent to a node. The last two columns display the resulting network costs with and without protection, respectively, and in each column the results are obtained from the ILP and MC heuristic algorithms.

Comparing the results where no protection is provided, it appears that both results are identical for all network problems. Each of these costs are in fact truly optimal, as in the case of the MC technique where the network costs are obtained from the sum of the lowest hop counts of all node pairs. Now considering the protection cases, results from these two techniques are not always equal. In some cases, the MC design demands more capacity as compared to the ILP technique, which serves as the lower bound. However, the differences are not substantial in most cases. Therefore, based on these limited results it is conceivable that the MC algorithm may be able to offer near-optimal solutions for other network problems.

5.3 Comparison of Protection Strategies

In this section, the performances of all three protection schemes are evaluated, analyzed and compared with respect to the network cost requirements. The two proposed heuristic algorithms for the MC and DJP schemes and the SLB algorithm described in [13] are extensively used to numerically determine the protection costs of each scheme for various network scenarios. Other important design aspects including the influences of the number of wavelengths, the level of connectivity, and traffic growth on the system performance are also discussed.

A realistic network structure, referred to as a European optical network (EON), is used here as an experimental network for discussion purposes. The network consists of 18 nodes and 35 bi-directional links, see its physical topology in Fig. 4. The traffic demands are assumed to be uniform with five levels of traffic volume ranging from 1 to 5. The traffic volume of level i is defined as i times the total number of active paths among all node pairs, i.e., $i \cdot N(N-1)/2$ active paths.

In this paragraph, we shall examine the effect of the number of wavelengths multiplexed (M) on the resulting network costs and the characteristics of fiber and wavelength resource allocation. An experiment is carried out as follows: the traffic demand is kept constant at one connection per node pair and the number of wavelengths used is varied from 1 to 72. Then we ap-

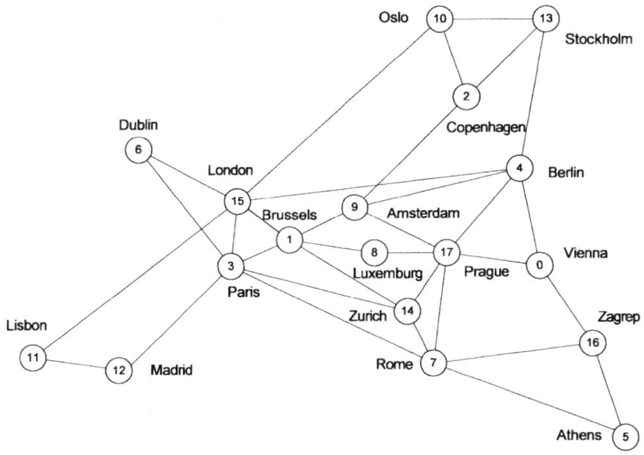

Fig. 4 A European optical network (EON), $N = 18$, $L = 35$.

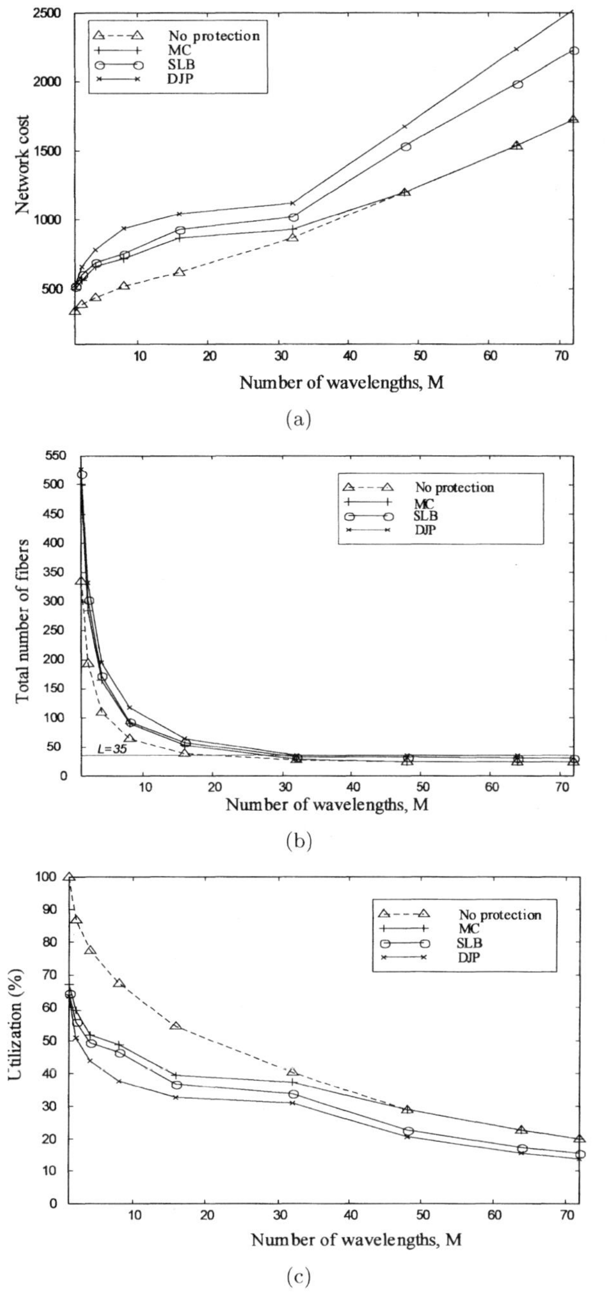

Fig. 5 Numerical results of the EON network for various values of M.

ply all the algorithms to determine the network costs, the total number of fibers and the link utilization as a function of M. The results are summarized in Fig. 5. Let first consider the case where no protection is provided: it is apparent that the network costs increase with M. Note that at a certain point ($M = 32$) the increment of the network costs with M becomes a linear relation. Referring to Fig. 5(b), the total number of fibers required at $M = 1$ is relatively high, i.e., 336, which is equivalent to installing over 9.6 (336/35) fibers per link on average. At greater values of M, the total number of fibers decreases rapidly until $M = 8$ and after that the decrement is saturated. Larger values of M mean a greater capacity in each fiber, hence more traffic connections can be accommodated. When the number of wavelengths multiplexed in each fiber is increased to 32, the total number of fibers becomes equal to the number of links in the network, i.e., 35. At this point, only one fiber is needed on each link. An increase in the value of M beyond this point produces further small reduction in the number of fibers, implying that some links in the network topology are not at all utilized. In summary, the simulation results indicate clearly that networks with a small number of wavelengths are more capacity efficient and can be better utilized; see Fig. 5(c) for the details of link utilization.

Now considering the results when protections are included, it is seen from Fig. 5(a) that the MC scheme is still the most capacity efficient for all values of M, while the capacity requirements of the SLB and the DJP schemes are about 3–7% and 5–20% over the MC scheme, respectively, for the values of M smaller than 16. It should be pointed out at this stage that systems using a value of M greater than 32 are considered to be overproviding network capacity, as this assumed amount of traffic will never fill up each fiber even when only one fiber is placed on each link. This can be observed from curves illustrating the network costs and the link utilization in Figs. 5(a) and 5(c), respectively; all results for M greater than 32 change linearly with M. Therefore, our discussion will focus on the results when M is smaller than 32. It is clear that although the number of fibers required becomes fewer as M gets large, the total network costs still tend to increase with M, signifying that using a greater number of multiplexed wavelengths generally causes the wavelength resource allocation to be less efficient and leads to low link utilization.

Here we shall investigate how wavelength conversion will affect the resource allocation and hence the

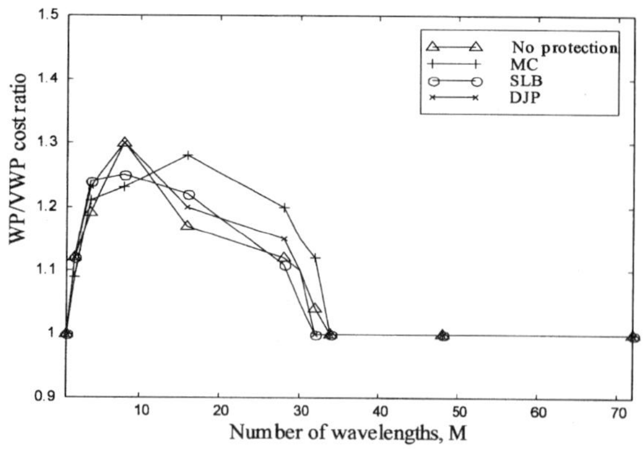

Fig. 6 Difference between the WP and VWP systems.

system performance. Figure 6 shows a performance comparison between WP and VWP systems. The results are demonstrated in the form of the WP/VWP network cost ratio. At $M = 1$, the network cost of the WP and VWP systems are equal (in other word, the WP/VWP cost ratio is one), confirming the equivalence of these two systems. Therefore, conversion is not an issue in this case. However, at other values of M, wavelength conversion is found to be useful, as there exist cost savings ranging from 10% to 30% depending on the number of wavelengths multiplexed (see detail in Fig. 6). However, when M is greater than 32, no differences between them are observed. This is because when M greater than 32 the network capacity is already over provided. Based on many other experiments, it is concluded that wavelength conversion can sometimes be useful to facilitate network resource allocation and hence effective network resource utilization can result.

5.4 The Network Connectivity and the Average Distance of Shortest Paths

In the previous section, a comparison of the network costs based on the three protection techniques and the influence of the maximum number of wavelengths per fiber on the spare network costs have been described. In this section, we will look into other important and influential factors that can be used to estimate and determine the spare capacity requirement given a network topology. In most previous research studies on mesh restorable networks, the relation between the spare capacity requirement and the network connectivity has been a subject of intense study. It is generally perceived that the lower the connectivity the higher the spare capacity requirement. However, in some experimental cases, the reverse is true and no explicit explanation has been given. For example, one study [10] experimented on a large number of different network structures of equal connectivity and performed network resource allocation to determine the spare capacity requirements. It was found that the spare capacity requirements of these networks varied considerably; no definite conclusion can be drawn. This investigation implies that knowing only the connectivity of the network will not be enough to obtain a good estimate of the spare capacity requirement.

Here we will show that the spare capacity requirement can be better estimated by considering both the network connectivity and the average distance of the shortest hop counts of all node pairs simultaneously. In order to illustrate this new concept, an experiment on networks of 14 nodes is given and detailed. A large number of sample networks of 14 nodes are generated with the following two properties: i) the connectivity C of 2.5, 3, 3.5, 4, 4.5 and 5 and ii) the average distances d_{avg} of 1.5, 1.75, 2.0, 2.25, 2.5, 2.75 and 3.0. For each sample network generated, the total network costs for all three protection techniques are determined using the heuristic algorithms with $M = 1$ and the results measured in terms of spare capacity requirements (%) are presented in Table 3. Note that an important constraint imposed on the way in which a network sample is selected for this experiment is that the number of links adjacent to every node in the network should be made equal or almost the same. This is to avoid any discrepancy that may arise due to unbalanced connectivity of each node.

According to the results in Table 3, the spare capacity requirements of all sample networks using the three protection strategies exhibit similar trend as all the experimental results shown before. The MC scheme is always the most cost effective, whereas the SLB and DJP schemes are more expensive on average by 4–5% and 7–8%, respectively.

From Table 3, consider the results with a fixed level of connectivity, say at $C = 4$. It is seen that the spare capacity requirements tend to increase with the average distance. For example, in the MC scheme the protection cost is only 20.98% for the average distance of 1.75 but this cost is increased up to 81.41% for the average distance of 2.5. These results indicate and confirm clearly that knowing only the network connectivity will not be sufficient to make any good cost estimate of the spare capacity requirement.

Now we will explain why the average distance is another influential factor in determining the spare capacity. For the same number of nodes and number of links (the same level of connectivity), the link locations interconnecting nodes dictate the average distance of the network; see three network structure examples in Fig. 7. These three networks have the same number of nodes and the same level of connectivity, but they do not have the same values of the average distance. If we look closely at all structures, we will notice that links in the network that have a short average distance are placed in such a way that all links can be used as backup for most interrupted traffic in the event of failure equally and evenly (see Fig. 7(c)). This is not the

Table 3 The influence of the network connectivity (C) and the average distance of the shortest path between all node pairs (d_{avg}) on the spare network cost for full protection. The results are percentages of the spare capacity over the normal condition.

d_{avg}	1.5			1.75			2.0			2.25			2.5			2.75			3.0		
C	MC	SLB	DJP	MC	SLB	DJP	MC	SLB	DJP	MC	SLB	DJP	MC	SLB	DJP	MC	SLB	DJP	MC	SLB	DJP
2.5													48.02	48.45	51.10	91.30	92.05	92.88	97.09	97.09	97.81
3										36.18	38.91	38.91	50.65	52.83	53.28	93.17	93.98	94.38	95.41	96.18	96.56
3.5							41.84	42.93	45.63	51.94	58.74	59.22	76.75	79.82	81.14	85.36	86.58	86.99			
4				20.98	24.69	24.54	35.52	35.52	36.61	66.34	66.83	67.32	81.41	84.51	84.51						
4.5	21.71	22.36	23.68	21.02	24.20	24.84	51.67	55	57.22	65.84	68.31	70.79									
5	21.37	25.52	25.52	27.92	31.17	33.17	56.83	61.20	61.20												

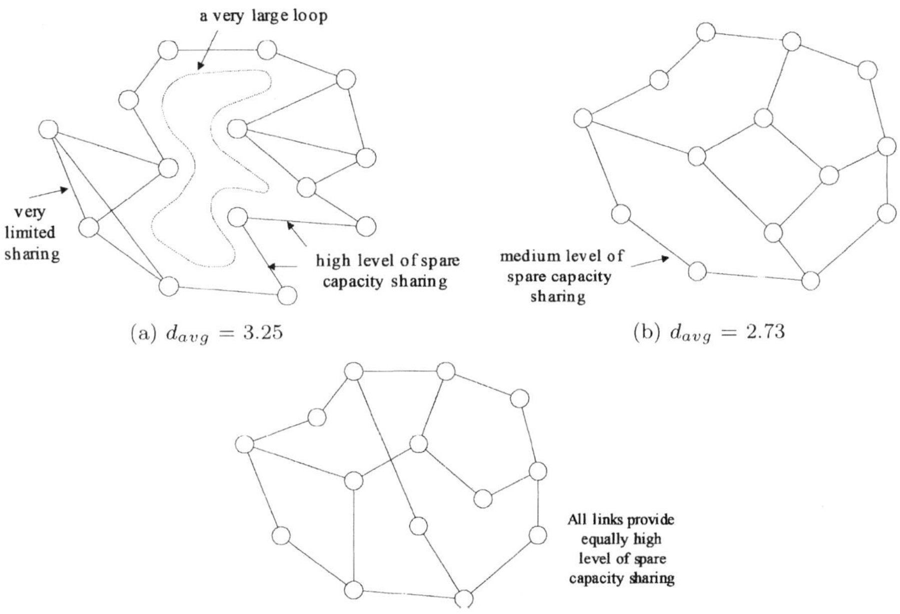

(a) $d_{avg} = 3.25$ (b) $d_{avg} = 2.73$

(c) $d_{avg} = 2.46$

Fig. 7 Three distinct network structures of the same number of nodes and links.

case in the network that has a long average distance, because only some links are useful to restore most disrupted traffic scenarios in the event of failure, while the others offer limited assistance (see Fig. 7(a)), i.e., only nearby interrupted traffic can be efficiently supported. An implication of this observation is that the sharing of space capacity on all links from failures is maximized with short average distance networks whereas the level of sharing can only be partially achieved with long average distance networks.

Another interesting observation is that the network structure that generates large loops such as the network in Fig. 7(a) causes expensive spare capacity requirement. For example if any link on the loop fails, the traffic between the two nodes adjacent to the failed link will have to traverse over a very long alternate route, i.e., 11 times longer than the original route in this case, resulting a very expensive backup path, 1100% extra cost. However, the total spare capacity over the entire network is not over 100%, because these spare capacities can be shared among other failure events. As the structures that have low average distance tend not to form very long loops as in Figs. 7(b) and 7(c), these excessive long backup paths are not usually seen. It is now completely clear why the network structure in Fig. 7(c) requires very low spare capacity, i.e., 54.46% compared to 73.09% and 97.29% for the structures in Figs. 7(b) and 7(a) respectively. This is because such a structure does not contain large loops and it also provides equally high level of spare capacity sharing among mutual failure events.

6. Conclusion

This paper has presented and compared three protection strategies for mesh restorable multi-wavelength networks against a single link failure, namely minimal cost (MC), single link basis (SLB) and disjoint path (DJP) techniques. An integer linear programming formulation is proposed to identify the lower bound of total network costs considering network protections. It is shown that this formulation offers a useful lower bound for small network problems, but when the problem gets large the computational time becomes excessive. As a result, heuristic algorithms are considered more practical for large network system designs. Two new heuristic algorithms are introduced to optimize working and spare capacity allocations for the MC and DJP protection strategies. Simulation results show that among the three protection schemes the MC scheme is the most cost efficient, as it always requires the least spare capacity. However, its operation is not applicable to practical environments, due to the unnecessary disturbance of all traffic connections in the entire network in the event of a failure. The SLB scheme is clearly a more realistic approach, as only interrupted connections are subject to re-arrangement, leaving other connections undisturbed. It is in fact a commonly known technique and has been extensively discussed in the literature. An interesting new aspect that this paper has added to the study of this protection technique is to show that its spare capacity requirement is not at all far from the minimal requirement. Therefore, the SLB technique can indeed be considered very cost effective and suitable for real applications. The DJP scheme is found the most expensive to put into practice. However, it offers other advantages, such as the simple re-configuration and inherent protection against node failure for in-transit traffic.

The number of wavelengths multiplexed into a fiber (M) is an influential parameter to the network system design performance in all protection schemes. In general, large values of M mean less flexibility in assigning wavelength resources, causing more unused capacity and hence resulting in more expensive system design. However, good and effective capacity allocation algorithm can minimize the total network cost for various values of M. Then, this paper studies the effect of the wavelength conversion to the network cost. Based on the simulation results, the network cost reduction can be achieved when the wavelength conversions are available in the OXCs.

Finally, we considers the influence of the network connectivity and the average distance of the shortest path between all node pairs to the spare network cost requirement for protection. From the results, it is found that the network connectivity alone is not sufficient to provide a good estimate of the spare capacity requirements. When the average distance is included, it is shown that the estimation can become more precise. In general, for the same level of connectivity the spare capacity requirement increases with the average distance.

Acknowledgments

The authors would like to thank Thailand Graduate Institute of Science and Technology (TGIST), National Science and Technology Development Agency for the scholarship. The authors also wish to gratefully acknowledge Chulalongkorn University for supporting this research work.

References

[1] L.D. Garrett, et al., "The MONET New Jersey network demonstration," IEEE J. Sel. Areas Commun., vol.16, pp.1199–1219, Sept. 1998.

[2] M.W. Chbat, et al., "Toward wide-scale all optical transparent networking the ACTS optical pan-european network (OPEN) project," IEEE J. Sel. Areas Commun., vol.16, pp.1226–1244, Sept. 1998.

[3] A. Jourdan, et al., "Fully reconfigurable WDM optical crossconnect: Feasibility validation and preparation of prototype crossconnect for ACTS "OPEN" field trials," Proc. ECOC'97, pp.55–58, 1997.

[4] H. Sakauchi, Y. Nishimura, and S. Hasegawa, "A self-healing network with an economical spare-channel assignment," Proc. GLOBECOM'90, pp.47.3.1–47.3.5, 1990.

[5] R.R. Iraschko, M.H. MacGregor, and W.D. Grover, "Optimal capacity placement for path restoration in STM or ATM mesh-survivable networks," IEEE/ACM Trans. Networking, vol.14, pp.893–902, June 1996.

[6] L. Wuttisittikulkij, C. Boworntummarat, and T. Iamvasant, "A comparative study of mesh and multi-ring designs for survivable WDM networks," IEICE Trans. Commun., vol.E83-B, no.10, pp.2270–2277, Oct. 2000.

[7] C. Boworntummarat and L. Wuttisittikulkij, "Design and analysis of 100% resilient all-optical WDM networks based on mesh and multi-ring approaches using joint optimization technique," Proc. IEEE HPSR2002, May 2002.

[8] D. Banerjee and B. Mukherjee, "A practical approach for routing and wavelength assignment in large wavelength-routed optical networks," IEEE J. Sel. Areas Commun., vol.14, pp.903–908, June 1996.

[9] L. Wuttisittikulkij, et al., "Routing and wavelength allocation in multi-wavelength all-optical ring networks," Proc. ICC'99, Vancouver, Canada, June 1999.

[10] S. Baroni, P. Bayvel, and R.J. Gibbens, "On the number of wavelengths in arbitrarily-connected wavelength-routed optical networks," OSA Trends in Optics and Photonics Series (TOPS): Optical Network and Their Applications, vol.20, pp.195–204, July 1998.

[11] L. Wuttisittikulkij and M.J. O'Mahony, "Use of spare wavelength for traffic restoration in multi-wavelength transport network," Proc. ICC'96, pp.1778–1782, Dallas, Texas, 1996.

[12] N. Nagatsu, K. Sato, and H. Hadama, "Network performance and integrity enhancement with optical path layer technologies," IEEE J. Sel. Areas Commun., vol.12, no.1, pp.159–169, Jan. 1994.

[13] N. Nagatsu, S. Okamoto, and K. Sato, "Optical path crossconnect system scale evaluation using accommodation design for restricted wavelength multiplexing," IEEE J. Sel. Areas Commun., vol.14, no.5, pp.893–902, June 1996.

High-Performance Backbone Network Technology edited by Naoaki Yamanaka
ISBN: 0-8247-5321-6 DOI: 10.1081/(120025731)
Published 2004 by Marcel Dekker, Inc.

A Realistically Architecture of WDM Ring Network Using OXC and OADM

Masayuki Kashima, Naoki Minato, Satoko Kutsuzawa, and Saeko Oshiba

SUMMARY A configuration capable of wavelength routing is indispensable in constructing an optical network that has the IP-over-WDM capability. A ring network based on WDM is one of the configurations that can make wavelength routing possible. As the nodes used to construct a WDM ring network, we have the optical ADM system (OADM) and optical cross connect system (OXC). In this paper, in order to make ring network realistic, we examined a wavelength routing way using the number of possible wavelengths and the number of Node-Connections. A wavelength routing way placement on a lattice letter logically, and the all paths forward by 1 hop or 2 hops. As the parameters for determining the number of nodes and the distance of transmission, we evaluated the deterioration resulting from coherent crosstalk and OSNR. As a result of evaluation, the number of node-passes for 1 hop transmission amounts to less than 20. In addition, when we made a test bed and made evaluations, the results almost coincided with theoretical values.
key words: *WDM, OADM, OXC, coherent crosstalk, OSNR*

1. Introduction

With data traffic rapidly increasing in recent years, data communication (IP network) systems are required to increase their transmission capacity and processing speeds. Conventional IP networks are based on hop-by-hop communication and wavelength division multiplexing (WDM) technology is used for transmission lines between routers to increase transmission capacity and meet increasing traffic demands. Because in this network configuration, each router processes destinations by taking in all IP packets passing through it, the processing speeds of routers must be increased as traffic increases. Considering that routers are limited in their processing speeds, they have difficulty meeting the increasing traffic demand. In addition, because routers perform their tasks in layer 3, a considerable processing delay occurs; although no problem occurs during ordinary data communication, handling of real-time services will be beyond the capability of the presently used system. The type of routers that can perform their processing tasks in layer 2 are being studied, but there still exists a considerable delay since intermediate routers (not destination routers) must be installed.

Wavelength routing (changing a destination based on wavelengths), which one of the features of WDM technology, is under consideration as a solution to this problem and forming a network capable of transferring data to a destination router without routing it via unnecessary routers (cut-through) by using this wavelength routing is being studied. Specifically, this wavelength routing is the process of adding and dropping necessary optical wavelengths and letting unnecessary wavelengths pass through (OADM: Optical Add/Drop MUX). Studies in the development of this OADM system and networks based on this OADM are now actively being pursued [1]–[3].

There are three typical OADM configurations:
(1) FBG (fiber bragg gratings) filters
(2) Tunable-wavelength AO (acousto optic) filters
(3) The wavelength MUX/DEMUX and optical switches

The characteristics (pass-through loss, crosstalk and others) of each configuration are varied and many. By referring to these parameters, the level of signal deterioration must be calculated and the most appropriate network must be designed.

An optical cross connection system (OXC) is being studied as a backbone node of an optical network. OXC configurations include the WP type which performs switching by the wavelength plane alone and the VWP type which switches all paths irrespective of wavelength. Because a large-scale OXC system of the VWP type is impracticable using the present technology, a study of how this system can be configured is being made [4], [5].

In this paper, we studied an optimum configuration with the objective of realizing an optical network using the OADM and OXC systems. By evaluating the level of signal deterioration, we designed the system by using as parameters a distance of transmission, a distance between nodes and the number of nodes. Last of all, we report the results of making a test bed and evaluation.

2. Network Configuration

2.1 Physical Configuration

The physical network connection is ring-shaped. If the network is to be configured by using one ring, the

Manuscript received February 28, 2000.
Manuscript revised June 23, 2000.
[†]The authors are with Oki Electric Industry Co., Ltd., Network Systems Development Division, Hachioji-shi, 193-8550 Japan.

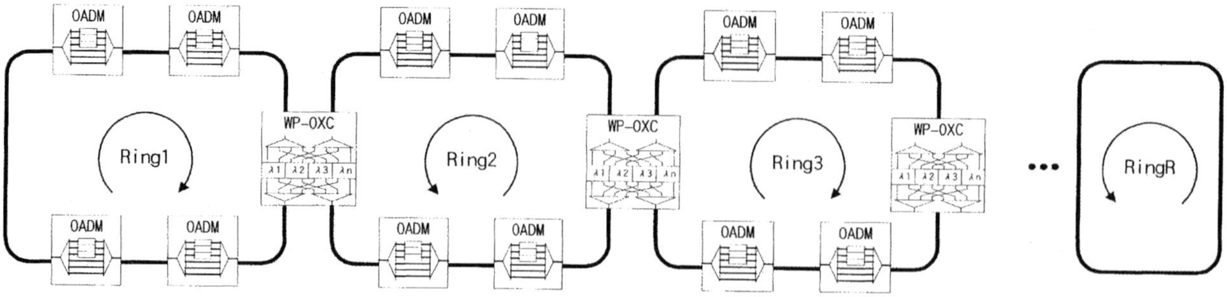

Fig. 1 Network configuration.

Table 1 Comparison of network configurations.

	Number of wavelengths		Expandability	Cost	Number of nodes	
OADM alone	many	$\frac{N(N-1)}{2}$	difficult	low	small	$N_{TH}+2$
OADM+ WP-OXC	few	$\frac{N(N-1)}{2}-\frac{N(N-2)}{8}$	easy	low	medium	$\frac{2R(N_{TH}+2)}{R+2}$
OADM+ VWP-OXC	few	Same as above	easy	high	large	$(N_{TH}+2)R$

OADM system alone can be used to realize the configuration. In this case, however, the number of pass-through nodes is limited due to the number of wavelengths and signal deterioration. It is estimated, as shown in Fig. 1, that the number of wavelengths to be used decreases and the number of nodes to be connected increases if multiple rings are used and rings are connected to the OXC. The number of wavelengths and the number of nodes used in these configurations are compared as follows.

In the connection shown in Fig. 1, the communication route of odd-number rings runs in the reverse direction to that of even-number rings.

Table 1 shows a comparison between a network configuration using OADM alone (1 ring) and a network configuration using OADM and OXC (2 or more rings) in relation to the number of wavelengths. These configurations assume that the same number of nodes are used and the path is set up in a full mesh. When multiple rings are used, the rings to be connected inside rings are distinguished from those to be connected outside. Because data goes through the same route outside rings, different wavelengths must be allocated. Inside rings, data does not pass through other rings and, therefore, the same wavelengths can be allocated to rings. Table 1 shows the number of wavelengths calculated based on this principle. In the formula, N is the number of nodes and $N \geq 4$. Accordingly, the number of wavelengths decreases when multiple rings are used. Although the WP and VWP types may use different types of wavelengths, the number of wavelengths that they use is the same.

Then the number of nodes is compared based on the principle that the maximum number of nodes passing through each configuration must be the same. Multiple rings are used and the number of nodes when nodes are connected with WP-OXC is calculated. Communication over a maximum distance is the transfer from one end to the other and data passes through the intermediate rings only halfway round. Providing that the number of nodes in a network is N_{NW}, the number of rings is R and the number of nodes through which data can pass is N_{TH}, we obtain $N_{NW} = 2R(N_{TH}+2)/(R+2)$ and $R \geq 2$. When $R = 2$, the case is the same as when one ring (OADM alone) is used and the number of nodes remain unchanged. When $R \geq 3$, the number of nodes in a network increases though the number of them through which data can pass is the same.

In the case of the VWP/OXC connection using multiple rings, VWP/OXC performs wavelength conversion. If VWP/OXC is designed to perform not only this wavelength conversion but also 3R, wavelengths can be shaped each time they pass through OXC and, therefore, the number of nodes will become several times larger than the number of rings. Table 1 shows the results of a comparison made of the number of nodes.

As a result of the comparative study made by including other items, as shown in Table 1, the network configurations can be classified as follows:

(1) Large-scale high cost: OADM + VWP/OXC
(2) Medium-scale low cost: OADM + WP/OXC
(3) Small-scale low cost: OADM alone

The system is designed to form a configuration that matches the target network structure (backbone, local network and access network).

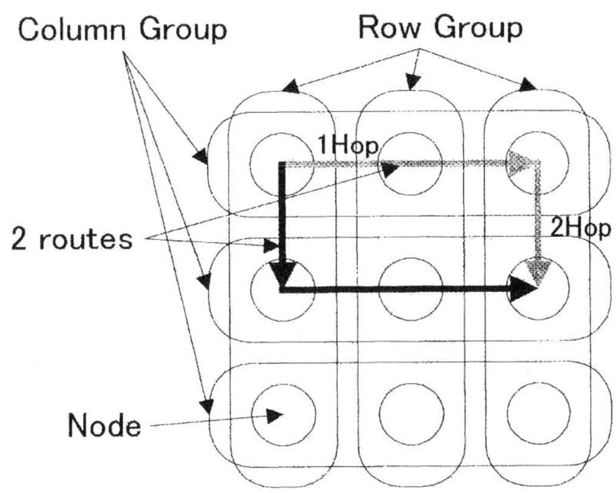

Fig. 2 RookNet configuration.

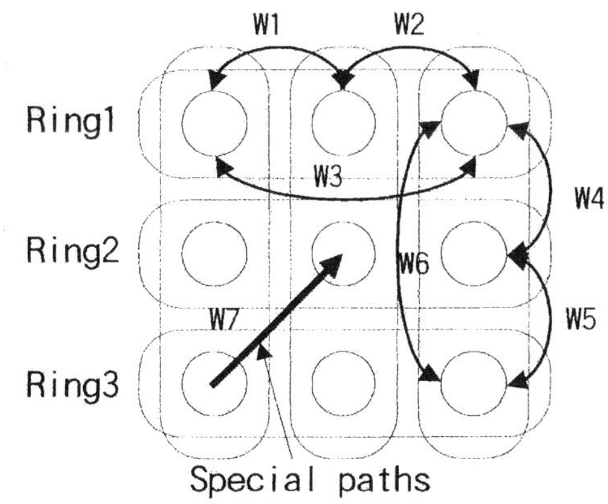

Fig. 3 Allocation of wavelengths.

Table 2 Comparison of the number of wavelengths used.

	Uni-directional	Bi-directional
Full mesh	$N(N-1)/2$	$(N^2-1)/8$
2 hops	$N(\sqrt{N}-1)$	$(N-1)\sqrt{N}/4$

2.2 Logical Configuration

When multiple rings are used, the path control inside rings can be done for each individual ring if VWP/OXC is used for connection. The path control between rings, however, must be done for the entire system. Controlling all paths by 1-hop transfer (without routing via other routers) is considered impracticable using the number of wavelengths that can be used at the present time. If 1-hop transfer is mixed with multiple-hop transfer, the path control will be complicated.

As a solution to this technical difficulty, a study is being made to form a network (RookNet [6], [7]) in which nodes are placed in a lattice and each node is grouped in column and row directions. This RookNet allows 1-hop transfer of data inside a group and 2-hop transfer of data outside a group via the node where a group of the originating nodes intersects a group of terminating nodes (see Fig. 2). Therefore, all paths can be transferred by 2 or less hops. Because in the case of the 2-hop transfer, two routes are available, routes can be changed logically and quickly, depending on the condition of congestion, allowing the network throughput to increase.

This RookNet is applied to the WDM ring network. With regard to wavelengths, each node need not have a full-mesh path for all nodes, i.e. each node is allowed to have a full-mesh path for each node in a group and consequently a smaller number of wavelengths (see Table 2).

Inside a group, wavelengths are allocated in a full mesh, i.e. wavelengths are allocated to each path. Because each node is physically connected, the same wavelength cannot be used by different groups (see Fig. 3). If the network is configured using OXC and multiple rings, the same wavelength can be used by different column groups by allocating column groups to each ring. Accordingly, the number of wavelengths shown in Table 2 further decreases.

Regarding the case that a lattice cannot be formed due to a difference in the number of nodes of each ring, the third-row group has only Ring 1 if Ring 2 and Ring 3 have only 2 nodes in a 3×3 lattice, as shown in Fig. 3. The RingX in Fig. 3 means the RingX in Fig. 1. The one ring of physical configuration is applied the one group of logical configuration. This node can transfer data in only one direction. This node will decrease the network throughput when the traffic to be transferred to groups other than the column group increases. In this case, the 1-hop path can be connected as a special path directly to a node in the adjacent row group. Because a wavelength that this node uses in the Row group is allocated, this particular wavelength must be used. Namely, a new wavelength need not be added as a special path.

3. System Design

To compensate for the transmission loss of optical signals, an optical amplifier must be established somewhere in the transmission line. A loss also occurs with OADM and the intensity of each wavelength being transmitted varies if the transmission intensity of each node is made uniform, causing signal deterioration due to crosstalk to worsen. In order to prevent crosstalk from causing signal deterioration, the intensity of each wavelength being transmitted must be made uniform. Although this can be theoretically achieved by monitoring and controlling the intensity of signals passing through each node, it is difficult in practice. Therefore, a method of placing an optical amplifier inside the OADM node and adjusting the intensity of optical signals that pass through and that of optical signals being transmitted to a uniform level is generally used. The optical amplifier used here amplifies multiple wavelengths simultaneously and generates optical noise which is added each time it passes through the optical amplifier. The number of passing OADM nodes is limited by signal degradation due to crosstalk and optical noise. The following sections describe the results of evaluating crosstalk and optical noise.

3.1 Evaluation of Crosstalk

Errors occurring inside a node are caused by the crosstalk resulting from light leaks. Crosstalk includes signal crosstalk (the effect of other wavelengths on the wavelength of a main signal) and coherent crosstalk (the effect of wavelengths that have the same wavelength as a main signal). With the FBG and AO types, coherent crosstalk occurs during Add/Drop. With the SW type, it is caused by extinction ratios of optical switches during Add/Drop and they are also caused by the repetition of MUX and DEMUX of AWG. With the latter case, addition takes place by the number of AWGs that pass through during the period of Add to Drop. With the SW type, therefore, the number of nodes that pass through is limited if the amount of coherent crosstalk caused by AWG is large. Also if the port of unused wavelengths are set to a through state, an infinite amount of coherent crosstalk is added. To avoid this, some nodes at this port must be set to the Add/Drop states. The crosstalk caused by AWG is calculated on the condition that all ports are dropped after signals make one circle.

The amounts of crosstalk of a filter which has certain characteristics is defined as follows: On the long wavelength side $= P_1, P_2, P_3, \cdots$, on the short wavelength side $= P_1', P_2', P_3', \cdots$. The coherent-crosstalk (XT_c) can be calculated by wavelength division and synthesis with the following formula:

$$XT_c = 10\log\left(K \times \sum_{j=1}^{n}\left(10^{P_j/5} + 10^{P_j'/5}\right)\right) \quad (1)$$

providing that the number of nodes to pass through is K.

Supposing that adjacent crosstalk which produces the strongest effect is 15 dB and the number of nodes to pass through only in the area affected by this crosstalk is 5, the coherent crosstalk becomes 23 dB. When the number of nodes to pass through is 10, it becomes 20 dB.

Because the extinction ratio (P_s) is further added during Add/Drop, the formula (1) can be rewritten as follows:

$$XT_c = 10\log\left(K \times \sum_{j=1}^{n}\left(10^{P_j/5} + 10^{P_j'/5}\right) + 2 \times 10^{P_s/10}\right) \quad (2)$$

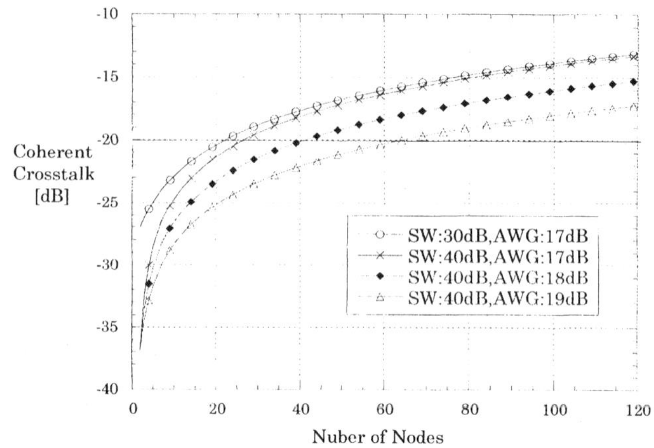

Fig. 4 Coherent crosstalk in relation to the number of nodes.

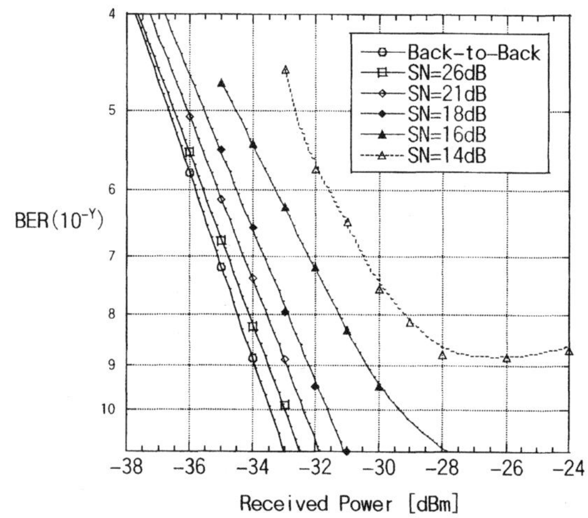

Fig. 5 BER in relation to coherent crosstalk.

Using the formula (2), the amount of coherent crosstalk of the system can be calculated. Figure 4 shows the relationship between the number of nodes and coherent crosstalk which is calculated taking a total of the extinction ratios of optical switches and the amount of AWG crosstalk as parameters.

Then the results of measurement made to verify what effects this amount of coherent crosstalk produces are shown. Figure 5 shows the BER characteristics in relation to the amount of coherent crosstalk. Coherent crosstalk is calculated relative to optical signals at 26 dB, 21 dB, 18 dB, 16 dB and 14 dB. The results of this measurement show that the power penalty is less than 1 dB when the crosstalk is more than 21 dB and it is about 0.5 dB when the crosstalk is more than 26 dB. If the power penalty resulting from coherent crosstalk is permitted to be 1 dB or less, the optimum system design should be based on the principle that the calculation result obtained by using the formula (2) is more than 21 dB. In this case, the number of nodes is approximately 30, as shown in Fig. 4 (SW:40 dB, AWG:18 dB).

3.2 Evaluation of OSNR

OSNR is defined as a ratio of optical signal to optical noise occurring in the range of 0.1 nm from the center of optical signal. This optical noise is caused by the loss that occurs when signals pass through optical modules (optical switches and filters) and optical fibers. To compensate this loss, an optical amplifier is used. The optical amplifier (EDFA) amplifies multiple wavelengths simultaneously, i.e. it amplifies (ASE) wavelengths other than signals, resulting in a situation that these undesired wavelengths and noise (NF) resulting from gains together become optical noise. As signals pass through the optical amplifier, the ASE and NF noise is added and the transmission distance and the number of pass-through steps are limited. This limitation determines the physical configuration of a network. OSNR can be calculated as follows:

The OADM node is equipped with a preamplifier for compensating for transmission loss at the first stage and a postamplifier for compensating the loss inside a node. Using this transmission loss and the number of stages of the amplifier to pass through as parameters, OSNR is calculated with the following formula:

$$1/SNR = 1/SNR_0 + h\nu\Delta\nu \left\{ \sum_{k=1}^{N-1} NF_1/P_1^{(k)} + \sum_{k=2}^{N} NF_2/P_2^{(k)} \right\} \quad (3)$$

Here, h is the blank constant, ν is the optical frequency, SNR0 is the SNR of a transmitter, NF_1 is the noise of a post-amplifier, NF_2 is the noise of a pre-amplifier and P_1 and P_2 are the input power of each amplifier. If the gain deviation, which is the variation in output power, must be considered, the following formula is used:

$$P_{total} = 10^{G/10} \left\{ P_{I1} + P_{I2} \times 10^{\Delta G/10(N-1)} + \cdots + P_{IN} \times 10^{(N-1)\Delta G/10(N-1)} \right\} \quad (4)$$

Here, ΔG is the gain deviation, N is the number of wavelengths and P_{I1} to P_{IN} are input power of each wavelength.

Figure 6 shows the relationship between the number of nodes and OSNR. In this figure, the gain is defined as 20 dB and the distance between nodes is changed in the order of 5 km, 10 km, 20 km, 40 km, 60 km and 80 km. OSNR for the scale of the network can be calculated from Fig. 6.

Figure 7 shows the relationship between OSNR and error rates. This figure shows the results of OSNR measurements based on 20.7 dB, 26.8 dB and Back-to-Back. The power penalties are 26.8 dB: 0.4 dB and 20.7 dB: 1 dB. If OSNR was allowed to 20 dB, the number of a Node-Connections amounts to less than 20.

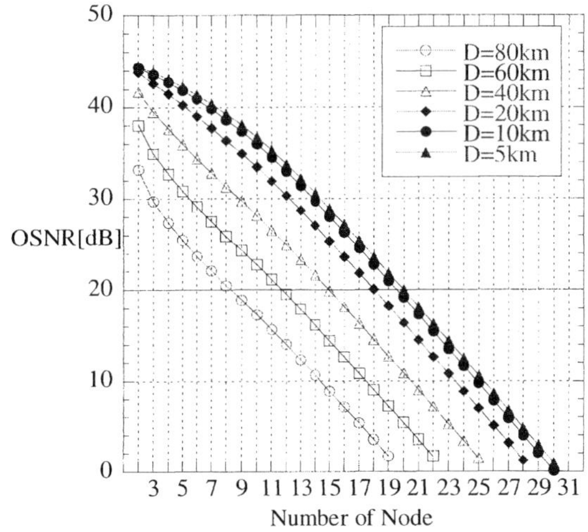

Fig. 6 OSNR in relation to the number of nodes.

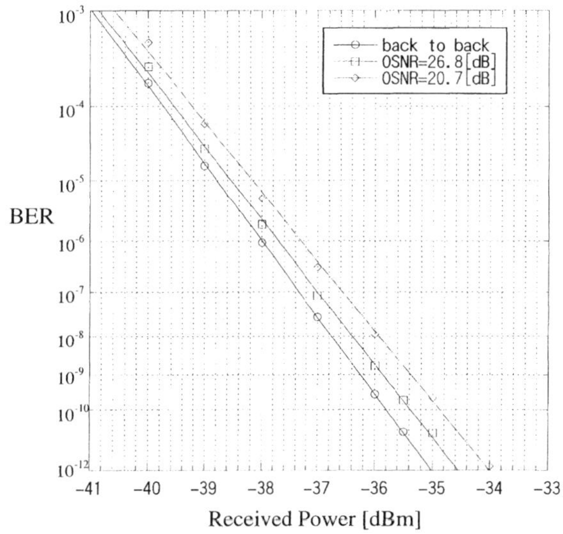

Fig. 7 BER in relation to OSNR.

The result of evaluation, OADM nodes is applied to the metro-ring that a distance of between nodes is no longer.

4. Test Bed

We made a test bed to evaluate various problems of the optical network. Figure 8 shows the configuration of this test bed and Fig. 9 shows the appearance. Add/Drop is configured by the wavelength MUX/DEMUX (AWG) and 2×2 TO-SW. To adjust the intensity of optical signals, optical amplifiers (EDFA) are set at the input and output. After the wavelength DEMUX, optical signals are branched by an asymmetrical coupler to monitor them. The NE-Control controls the switching of TO-SW, monitors optical signals and collects system data. The PC boards include the EDFA board, AWG board, SW board, monitor board

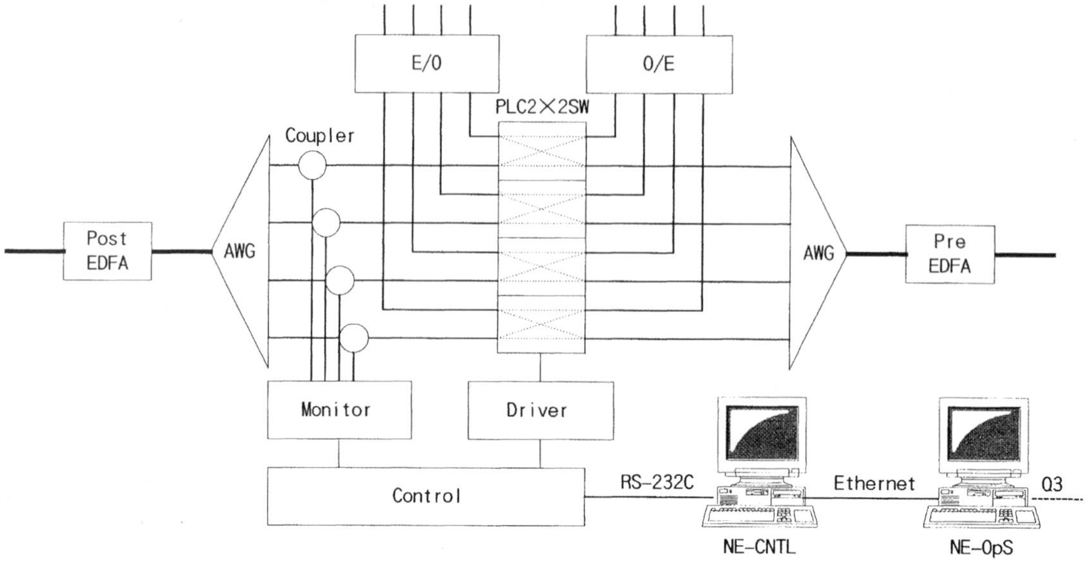

Fig. 8 Configuration of the OADM system.

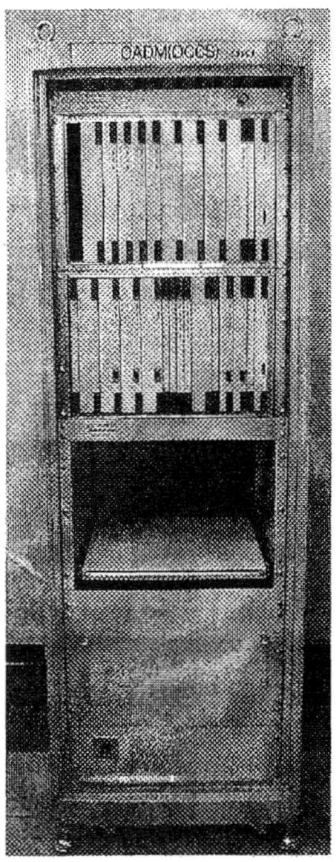

Fig. 9 Appearance of the OADM/OXC system.

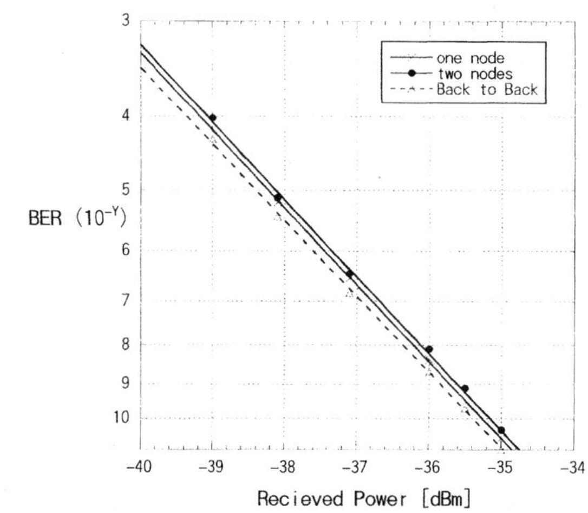

Fig. 10 BER in coherent crosstalk by relation to passed the number of nodes.

and OEEO board for Add/Drop. The monitor board has a simple CPU and a function to interface with the NE-Control. RS-232C is used as an interface with the NE-Control. The NE-Control is connected to NE-OpS via Ethernet. We used a personal computer as NE-Control for the present test bed to make the configuration simple. In the future, a CPU board will be used. This configuration enables us to control the Add/Drop wavelengths of OADM and monitor the conditions of optical signals from NE-OpS via NE-Control.

By replacing the optical switch section (SW-PKG) with the switch for cross connection, this system can be used as the OXC system. Therefore, this test bed using 4 different systems can be used as a platform for testing and evaluating various network configurations, namely, single-ring and double-ring systems, ring connections using OXC and others.

Then we evaluated system performance when a test bed was ADM systems in ring networks. First of all, the result of measurement is shown effect this amount of coherent crosstalk which occur by a repetition of wavelength MUX/DEMUX. Figure 10 shows the BER characteristics in relation to the number of nodes passed. This measurement is for the case of one passed node and two passed nodes. The results of this measurement show that the power penalty is about 0.2 dB when the two nodes are passed. Here,

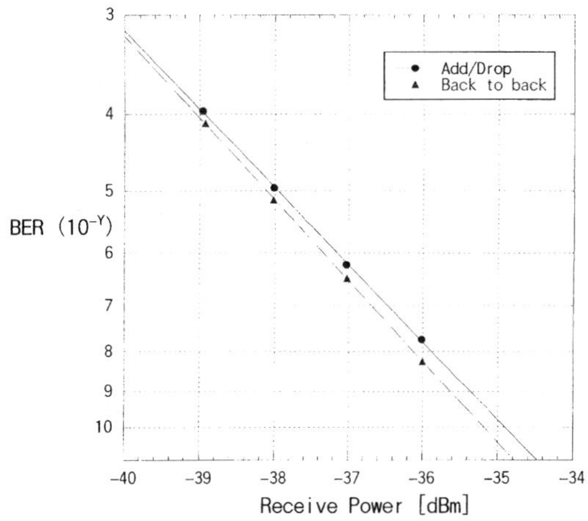

Fig. 11 BER in coherent crosstalk of Add/Drop.

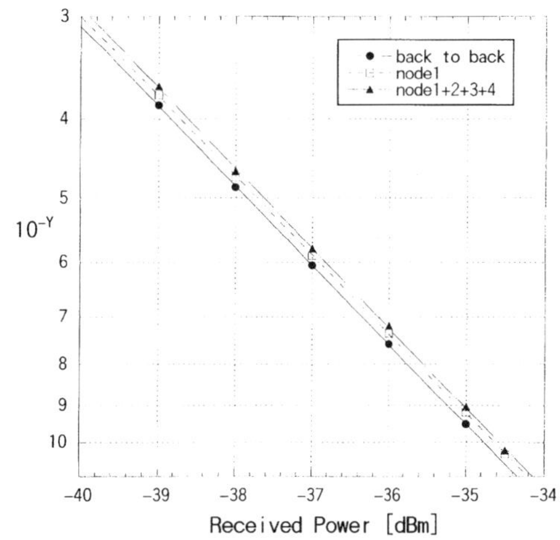

Fig. 12 BER in OSNR by relation to passed the number of nodes.

near side crosstalk of an AWG is 19 dB, other crosstalk is 36 dB. Then the result of measurement is shown effects this amount of coherent crosstalk which occur by Add/Drop. Figure 11 shows the BER characteristics in Add/Drop. The results of this measurement show that power penalty is about 0.4 dB. Here, extinction ratio of an optical switch is 38 dB to 40 dB. These results almost coincide with the theoretical values shown in Sect. 3.1. Then the result of measurement is shown about influence of OSNR. Figure 12 shows the BER characteristics in relation to the number of nodes passed. This measurement is the case of passed an one node and the fore nodes. The results of this measurement show that the power penalty is about 0.8 dB when the fore nodes passed. Here, NF of an EDFA is 5.0. This value of OSNR is about 27 dB. This result almost also coincides with the theoretical value shown in Sect. 3.2.

5. Conclusion

We studied the optimum configuration of the WDM ring network in which OADM/OXC is used. Because the maximum number of nodes to pass through in the OADM system is approximately 20, using multiple rings to form a large-scale network is recommended. Also, to make the control of these multiple rings easy, it was proposed that nodes be logically arranged in a lattice to control communication paths and wavelengths. In addition, when we made a test bed and made evaluations, the results almost coincide with the theoretical values. As the next step of our studies, we intend to evaluate an actual system constructed based on this test bed and to develop a software program for controlling this lattice network.

References

[1] H. Toba, K. Oda, K. Inoue, K. Nosu, and T. Kitoh, "An optical FDM-based self-healing ring network employing arrayed waveguide grating filters and EDFA's with level equalizers," IEEE J. Sel. Areas Commun., vol.14, no.5, pp.800–813, June 1996.

[2] M. Sharma, H. Ibe, and T. Ozeki, "WDM ring network using a centralized multiwavelength light source and add-drop multiplexing filters," IEEE J. Lightwave Technol., vol.15, no.6, pp.917–929, June 1997.

[3] S. Hamada, K. Asahi, T. Hosoi, K. Nakaya, D. Uehara, C. Konishi, and S. Fujita, "WDM four-fiber ring with add/drop acousto-optic tunable filter and 4×4 optical switch," OFC'97, vol.6, ThO2, 1997.

[4] S. Okamoto, A. Watanabe, and K. Sato, "Optical path cross-connect node architectures for photonic transport network," IEEE J. Lightwave Technol., vol.14, no.6, pp.1410–1422, June 1996.

[5] I. Nakajima, T. Maeda, T. Nishi, I. Tsuyama, and S. Kuroyanagi, "Prototype WP-based optical path cross-connect node using piloss optical switches," ECOC'98, vol.1, pp.251–254, Sept. 1998.

[6] Y. Oie, Y. Sasaki, and H. Miyahara, "RookNet: A switching network for high speed communication," ICC'93, vol.2/3, pp.1225–1230, May 1993.

[7] M. Kashima, R. Watanabe, R. Konuma, H. Inoue, and Y. Shiraishi, "RookNet: A rearrangeable optical network with high speed relay processing," IEICE Trans. Commun., vol.E78-B, no.5, pp.679–685, May 1995.

High-Performance Backbone Network Technology edited by Naoaki Yamanaka
ISBN: 0-8247-5321-6 DOI: 10.1081/(120025732)
Published 2004 by Marcel Dekker, Inc.

Internetworking Based on Wavelength Assignment Photonic Switching System (WAPS)

Tadahiko Yasui, Kumio Kasahara, and Yoshiaki Nakano

SUMMARY Wavelength Assignment Photonic Switching System (WAPS) provides a wavelength for an end-to-end communication. In this way the features of wavelength can be fully utilized by users. We will give an architectural proposal in which two types of connections over WAPS network are provided and are adaptively used according to service demand by customers. One is a connection established semi-permanently between edge routers and relay-routers and the other is a connection established on flow-by-flow basis between edge routers. We will compare with conventional router networks in terms of data-transfer time. Pre-processing time is a crucial issue in connection-oriented networks, and this is very much reduced thanks to the WAPS network structure.

key words: *WDM, photonic network, routers, IP, cut through, photonic switches*

1. Introduction

It is almost nonsense to design a new telecommunication network without considering an explosion of the Internet communications. It is said that in near future the Internet traffic will surely exceed that of voice.

The present Internet, however, is said to be 'Send and Pray' model or 'Best Effort Service,' and various problems are pointed out.

(1) The Internet is essentially a store and forward system, which cannot guarantee the time interval between packets. The maximum delay time or the maximum loss of packets cannot be specified. There is no guarantee for QoS (Quality of Services) or reliability of a whole network. Specifically, this network is not suited for real time traffic such as voice, music or live TV broadcasting.

(2) A secure wide area enterprise closed network cannot be constructed using the Internet because it cannot exclude malicious traffic to disturb the enterprise network.

(3) Information transferred through the Internet is basically open to the public and their security is not easy to be achieved.

Many solutions are presented to above problems. General trends of the technologies for improving IP network quality are functional partitioning, distribution and network-wide traffic management. This naturally results in the more connection-oriented-like network architecture.

Cut-through mechanism is an example of functional partitioning [1]. RSVP (Reservation Protocol) [2] secures an end-to-end path communication bandwidth with cooperation of the relevant routers. Studies on protocols for multicast service are being made. Ipv6 enlarges address space. Differentiated Services implementation [3] will improve router architecture.

MPOA (Multi-Protocol over ATM (Asynchronous TransferMode)) performs a network cut-through with the network kept as it is. GMN-CL (GigaMedia Network-Connectionless) proposed by NTT [4] is a kind of a huge router comprised of optical access networks, edge nodes and a core network based on ATM backbone network. The GMN-CL aims at a network-wide solution of Internet problems from a network carrier side. Traffic engineering attempt by MPLS (Multiprotocol Label Switching) [5] can be said that it tries to realize the network like connection-oriented as much as possible.

We proposed already in the previous papers [6], [7] a photonic switching system called WAPS (Wavelength Assignment Photonic Switching System). The basic principle of the system is to assign or allot a wavelength to a call originated. In this paper we will propose an IP (Internet Protocol) network configured over WAPS in order to solve IP network problems mentioned above. The proposed IP network will guarantee QoS by securing high reliability, allowable delay time and allowable packet losses. IP over WAPS realizes an IP network by providing both semi-permanent connections like PVC (Permanent Virtual Circuit) in ATM network and flow-by-flow connections like SVC (Switched Virtual Circuit) in ATM network. The former connection mode provides a network similar to the conventional router network. The latter can provide a QoS guaranteed network.

Compared with the conventional router networks, the most crucial problem in the flow-by-flow connection is the pre-processing time that is essentially needed in order to establish a path before sending data. In this paper this is discussed and analyzed in detail and it will be clarified that WAPS network structure is beneficial

Manuscript received February 26, 2000.
Manuscript revised May 18, 2000.

[†]The authors are with Mitsubishi Electric Corporation, Kamakura-shi, 247-8501 Japan.
[††]The author is with the Department of Electronic Engineering, The University of Tokyo, Tokyo, 113-0052 Japan.

IEICE TRANS. COMMUN., Vol. E83-B, No. 10, pp. 2312–2320, October 2000

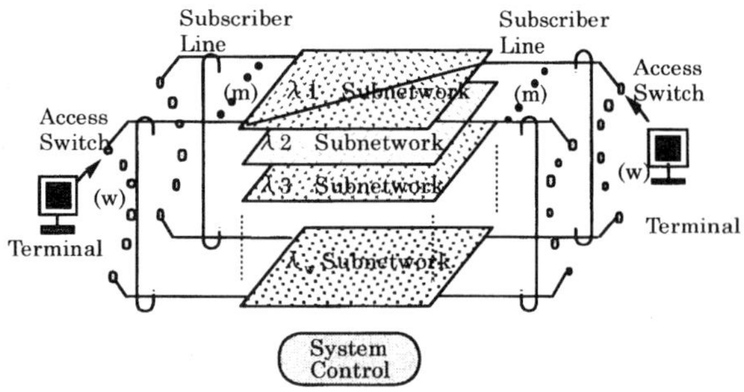

Fig. 1 WAPS basic concept.

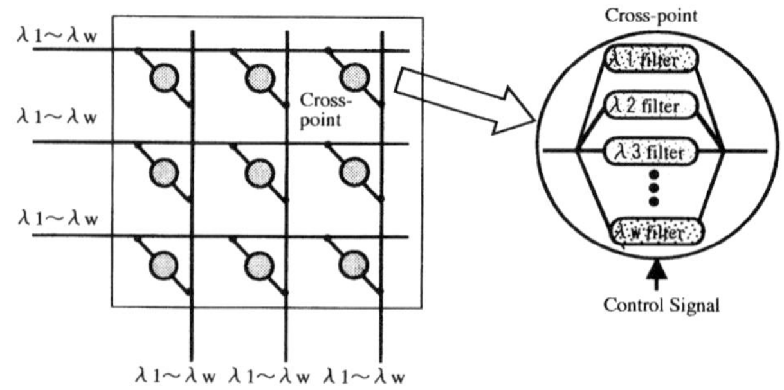

Fig. 2 Wavelength switch fabric.

to shorten the preprocessing time.

2. WAPS Architecture

The basic concept of WAPS is to assign a wavelength for a connection on call-by-call basis with the resource management done by the system control. With this architecture subscribers can fully utilize features of an optical wavelength, and will be able to enjoy various services such as bitrate-free (variable bitrate) and protocol-free services that can not be provided by electronic systems.

Logical configuration of a WAPS network is given in Fig. 1. Each wavelength has its own dedicated subnetwork. Terminals select a subnetwork, and space switches dedicated to the subnetwork perform connections within the subnetwork. There is no connection among different wavelengths; i.e. subnetworks can be controlled independently of others. Node capacity is very large, but substantially we only need to design a switching node of small capacity, because the whole node consists of w (number of wavelengths) identical small nodes. This is cost effective in view of switch cost, because the cost is roughly proportional to m^2 (m = number of switch ports), and this will give better transmission characteristics. Each subnetwork is separated logically but is combined into one physical network; the transmission lines, switches and subscriber lines are all wavelength multiplexed. An access function is required to realize wavelength selection and traffic concentration. A network control administrates and manages such network resources as transmission lines and switch fabrics.

Figure 2 shows the basic concept of a WAPS switch. Logically the WAPS switch consists of w identical small space division switches. A WAPS switch is incomplete in the sense that there are some input/output terminal pairs with no connection path. A switch crosspoint must filtrates any wavelength according to switch control signal. There are many candidates to be used for the WAPS switch such as AOTF (Acousto-Optic Tunable Filter) [8], a transversal filter [9], but in view of present state of technologies a space division (SD) switch [10], [11] solution seems most suitable for the

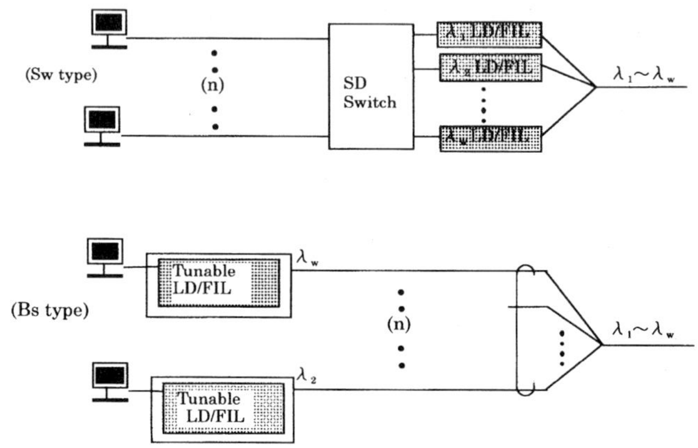

Fig. 3 Access switch configuration.

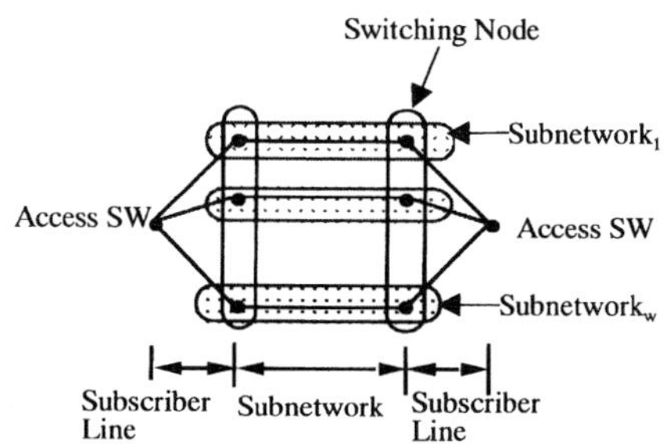

Fig. 4 WAPS channel graph.

time being.

As shown in Fig. 3, there are roughly two methods for implementing access function; Sw (Switch) and Bs (Bus) types. In Sw type, space switches are used for access to wavelength and traffic concentration. Sw type subscriber network consisting of switches, subscriber lines and terminals can be constructed independently of the backbone network architecture. In Bs type the access function is distributed to terminals, and thus space switches are eliminated. Terminals in this type of subscriber network must be able to originate any wavelength designated by the network. LAN or loop topology can be used for this type of subscriber network.

Channel graph of WAPS network is shown in Fig. 4. When a subnetwork is represented by a link,

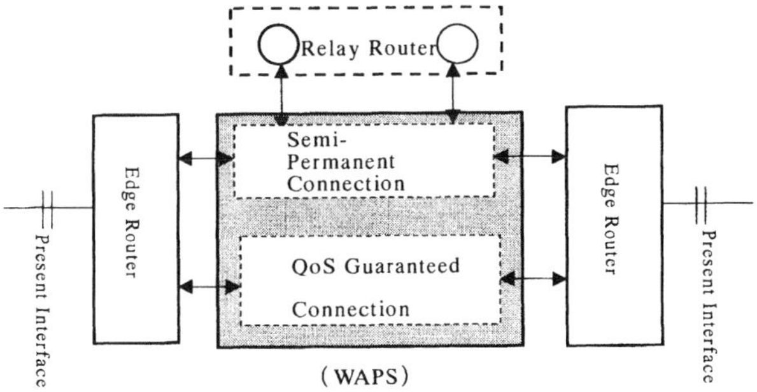

Fig. 5 IP network over WAPS.

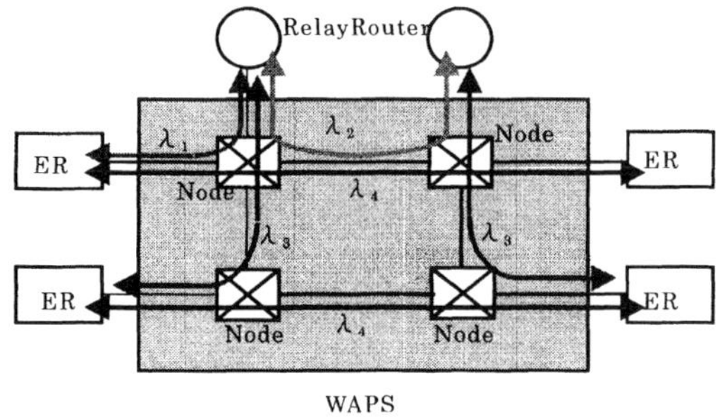

Fig. 6 IP over WAPS by semi-permanent connections.

the channel graph is of a serial-parallel type with no connection among subnetworks. Thus, a hierarchical structure consisting of system and subnetwork controls is possible, in which the subnetwork resource management is carried out by the subnetwork control and the wavelength administration of the whole network is done by the system control.

3. Concept of IP Network over WAPS

Because of its open nature the Internet has resulted in a disorderly and non-systematic communication network, which raises various problems and increases complexity against both network and traffic growth. Among these, the biggest problem is a degradation of transferring capacity owing to mixture of routing and data-transfer control. A cut-through is one of the solutions to this problem, which performs data-transfer by a cut-through switch without disturbing forwarding control. The Internet is going towards more connection-oriented networks.

In its nature WAPS is a connection-oriented network establishing a connection for every call/flow. Data-transfer control and connection control is essentially separated, and therefore, there is no disturbance of data-transfer on connection control process. A WAPS network can easily provide either network or node cut-through function, and thereby can give solutions to problems of the Internet. The architecture of an IP network over WAPS is shown in Fig. 5. Following two kinds of connections provide IP over WAPS.
(a) semi-permanent connections.
(b) flow-by-flow connections.

These two connections are to be adaptively used depending on traffic and services characteristics demanded (TCP/IP, UTP, FTP or QoS guaranteed service). Short data packets frequently changing IP addresses are to be carried by connection (a), and long, latency sensitive and time critical packets are to be carried by connection (b). Traffic suitable for (a) and (b) will change time to time, but in future most of the Internet traffic will surely be carried by the flow-by-flow connections. This architecture of adaptive usage of these connections will provide optimal IP network not only for the present situation but also for the future.

4. IP over WAPS with Semi-Permanent Connections

As shown in Fig. 6, WAPS provides semi-permanent optical paths between ERs (Edge Routers), RRs (Relay Routers), and between RRs and ERs. Not through conventional digital circuits but through optical wavelength paths can a router network be provided. A WAPS node of this configuration is quite similar to a photonic crossconnect [12], [13] in that switch crosspoints are semi-permanently closed and are not operated on flow by flow basis. Usually they are operated manually, or through a management system. Photonic paths of this network are terminated by ERs or RRs, where wavelengths are converted to electronics and vice versa. Electronic switches in relay routers cannot be replaced with photonic switches, because it is impossible at present to distinguish IP headers at optical level.

The difference against the conventional router network is that it enables flexible network reconfiguration and guaranteed provisioning of assured service grade using a network-wide management and administration system. In this network it is possible to adopt a proprietary but orderly and systematic addressing scheme which enables an easy and quick address resolution. This configuration also makes it possible to construct ERs and RRs particularly specified for WAPS network. It will be possible to make hierarchical grouping of RRs. Routing systems with default forwarders [14] as in GMN-CL can be configured, which leads to quicker address resolution and easier routing.

The optical path provided by the WAPS network is a wavelength, and it is used to carry aggregate of flows or single flow depending on the characteristics of IP packets. It is easy to implement CUG (Closed User Group) service using this connection mode which never suffers from the attack by malicious traffic.

By using LSRs (Label Switch Routers) for relay routers, and by implementing MPLS (Multi-Protocol Label Switching) edge functions for ERs, we can easily construct an MPLS network, in which the WAPS net-

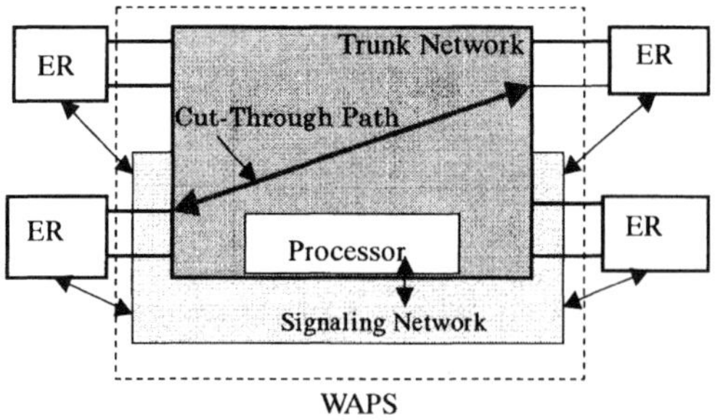

Fig. 7 Network cut-through over WAPS.

work only provide wavelength paths. But, RR network configured over WAPS with a proprietary and efficient address system has more possibility to provide a higher speed and more cost effective IP network than MPLS network. In MPLS network as well as IP over WAPS with semi-permanent connection, there is still a possibility to have some delay time variations and packet losses owing to processor over-loads and/or packet multiplexing. But through WAPS cut-through configuration, we can provide a completely QoS guaranteed IP network with no delay variation and packet loss.

5. IP over WAPS with Flow-by-Flow Connections

In this mode of connection a connection path is established for every flow. As shown in Fig. 7, the data-transfer and routing/forwarding functions are separated and header information is distributed through a signaling network in order to control switch fabrics in the nodes. Thus, IPs are carried from originating ER to destination ER in a cut-through mode. This enables to provide,

① fixed delay independent from network traffic variation,
② routing without packet loss,
③ secure provisioning of required transmission capacity, and
④ security against interference.

Through these provisioning, QoS of IP over WAPS is to be fully guaranteed.

5.1 Data-Transfer Time

This method establishes a connection path with necessary capacity for every new flow origination. The biggest problem is that IP packets cannot be sent out to destination immediately after the end of address resolution, because the packets should wait for the end-to-end path to be established. In the present router network, packets can be sent right after the address resolution with no path established beforehand. The time difference between these two methods is to be roughly studied.

Figure 8 shows a data-transfer model of a router

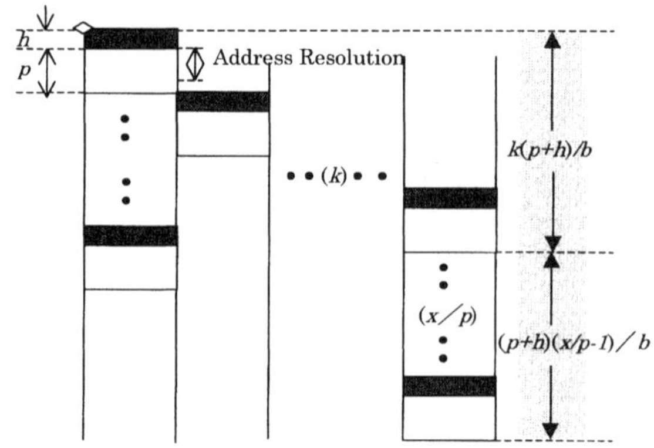

Fig. 8 IP packets transfer model.

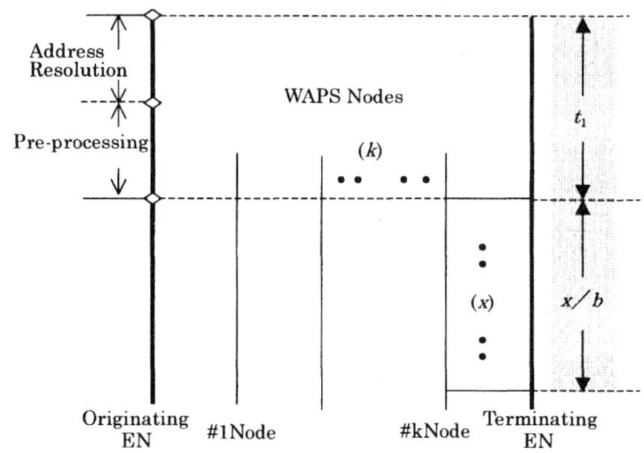

Fig. 9 IP packets transfer model over WAPS.

network. When we assume that the address resolution time is less than the IP payload receiving time, total transmission time t_{IP} is given below.

$$t_{IP} = \frac{1}{b}\left(kp + kh - p - h + x + \frac{hx}{p_{IP}}\right) \tag{1}$$

where,
x: total information length (bit),
p_{IP}: information part length of an internet packet (bit),
h: header length of a internet packet (bit),
b: transmission bitrate (bit/sec), and
k: number of routers traversed.

The data-transfer model of IP over WAPS is shown in Fig. 9. The total transfer time t_{WAPS} in this model is given,

$$t_{WAPS} = t_1 + \frac{1}{b}\left(x + \frac{hx}{p_{WAPS}}\right) \tag{2}$$

where t_1 is preprocessing time consumed before start of data sending and p_{WAPS} denotes packet length used in a WAPS network. t_1 is a parameter proper to the IP over WAPS. If p_{IP} is optimized for IP routers and $p_{IP} = p_{WAPS}$, then, from formulas (1) and (2) the time difference Δ is given below,

$$\Delta = t_1 - \frac{1}{b}\left(\sqrt{hx(k-1)} + h(k-1)\right). \tag{3}$$

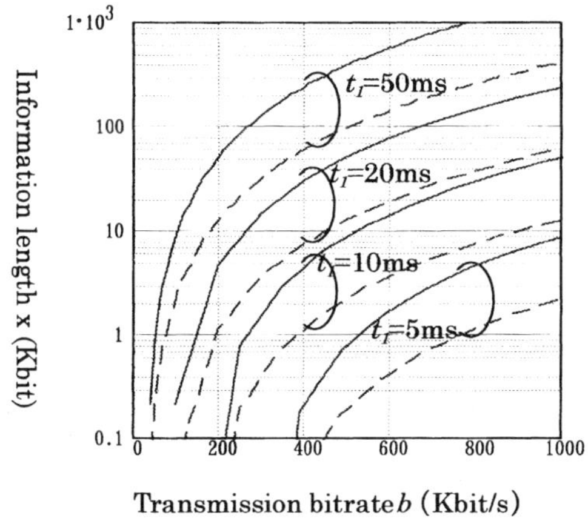

Fig. 10 Application area of WAPS cut-through (Rigid: Packet length not optimized for WAPS, Broken: Packet length optimized for WAPS).

If p_{IP} is optimized for IP routers and p_{WAPS} is for WAPS, the difference is given below.

$$\Delta = t_1 - \frac{1}{b}\left(2\sqrt{hx(k-1)} + h(k-1)\right). \tag{4}$$

As is clear from formulas (3) and (4), the optimization of the WAPS packet length contributes to reduce the trade-off information length (x) by 1/4 time. When $\Delta = 0$, relations between x (kbits) and b (kbps) for

$$t_1 = 5\,\text{ms},\ 10\,\text{ms},\ 20\,\text{ms},\ 50\,\text{ms},$$
$$k = 10\,\text{hops}$$
$$h = 20\,\text{bytes}$$

are given in Fig. 10 for both cases. It is important to note that in order to realize IP over WAPS the preprocessing time t_1 should never exceed roughly 100 ms.

5.2 Path Connection Time t_1

The control scheme of the network cut-through in IP over WAPS is shown in Fig. 11, and the basic process flow from the end of receiving all input data to the beginning of data-transfer is shown in Fig. 12. Table 1 explains each process and consumed time notations in Fig. 12.

In this configuration SC administrates the wavelength busy/idle state at the output links of ERs, and SNC maintains busy/idle state of subnetwork transmission links and switching node links. Sub-Node can be structured to administrate the switch fabric link busy/idle states. But here, only the partitioning between SNC and SC is considered.

The most important feature of WAPS from a viewpoint of establishing a lightwave path is that network path searches can be done in parallel because subnetworks have no interconnection with other subnetworks. Inquiry data is sent to all SNCs from SC in parallel whose incoming and outgoing link pairs are idle. This

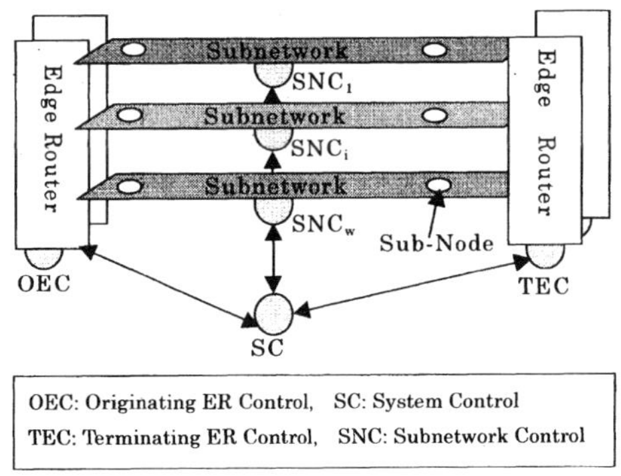

Fig. 11 Network cut-through control scheme.

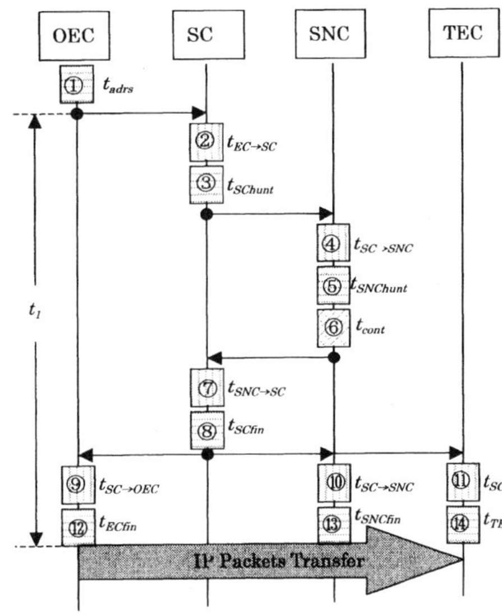

Fig. 12 Process flow of WAPS network cut-through.

means the total set-up time can be reduced to a great extent, because it is almost equal to the path hunting time of one subnetwork. Therefore, if the number of usable wavelength increases with the number of terminals fixed, the processing time become shorter accordingly, because the subnetwork size becomes smaller.

Assuming,

$$t_{SNCfin} + t_{SC\to SNC2} \text{ and}$$
$$t_{TECfin} + t_{SC\to TEC} < t_{OECfin} + t_{SC\to OEC}, \tag{5}$$

preprocessing time t_1 is expressed below.

$$t_1 = t_{OEC\to SC} + t_{SChunt} + t_{SC\to SNC} + t_{SNChunt} + t_{cont} + t_{SNC\to NC} + t_{SCfin} + t_{NC\to OEC} + t_{OECfin} \tag{6}$$

We assume $t_{SNChunt}$ is roughly proportional to number of WAPS nodes (n) traversed by IP packets, and

$$t_{SChunt} = t_{SNChunt}/n = t_{SCfin} = t_{OECfin} = t_2,$$
$$t_{EC\to SC} = t_{SC\to SNC} = t_{SNC\to NC} = t_{NC\to EC} = t_3 \tag{7}$$

Table 1 Process and consumed time.

No.	Process	Time consumed
①	IP address resolution	t_{adrs}
②	Receiving of header information from OEC	$t_{OEC \to SC}$
③	Determining of wavelengths available between caller and called and making of inquiry data to SNC for path hunting within the subnetworks	t_{SChunt}
④	Receiving of inquiry data from SC	$t_{SC \to SNC}$
⑤	Determining/reserving of an idle path between terminals designated by SC	$t_{SNChunt}$
⑥	Switching equipment control	t_{cont}
⑦	Receiving of data from SNC	$t_{SNC \to SC}$
⑧	Dicision of a path used, making data for OEC,TEC and SNC ot inform the decision	t_{SCfin}
⑨	Receiving of data from SC	$t_{SC \to OEC}$
⑩	Receiving of data from SC	$t_{SC \to TEC}$
⑪	Receiving of data from SC	$t_{SC \to SNC2}$
⑫	Preparation for data-transfer	t_{OECfin}
⑬	Receiving reparation of IP packets from OEC	t_{TECfin}
⑭	Shutting down of wavelength path reserved but not used.	t_{SNCfin}

Firstly, we assume the WAPS network configuration is the same with that studied in the previous paper [7]. The network is two-layered and the nodes at the top layer are connected in mesh and the nodes at the lower layer are connected to the specified upper layer nodes in star topology. Therefore maximum n is 4.

Secondly, we assume $t_2 = 1.5$ ms. This value is based on the study of Habara et al. [15], who calculated the software dynamic steps of a dial-tone connection process in an electronic switching system with 3400 Erlangs. It is about 1500 steps in an assembler language. We assumed executing time of a dynamic step is 1 μs. With a WAPS node capacity of about 1500 Erlangs, the estimation of $t_2 = 1.5$ ms is appropriate.

The selection of a signaling network is important. Even without considering transmission delay, a conventional 64 kb/s common channel signaling network takes 6 ms ($= t_3$) to send 48 bytes (15 bytes for control information and 33 bytes for data). This means total data-transfer time is 24 ms, which is too large for the signaling network for WAPS cut-through mode. One of the alternatives for signaling networks is ATM backbone network. Another is a WAPS crossconnect network configured on the semi-permanent connections for the WAPS router network. Both alternatives have capability of realizing 10 or 100 Mb/s signaling network, but we postulate 1 Mb/s signaling network here. Then, $t_3 = 0.75$ ms.

We assume thermo-optical PLC switches [10] or MEMS optical switches [11] are used for WAPS SD nodes because they are technologically mature. In this case, roughly t_{cont}=10 ms.

From above, we have preprocessing time $t_1 =$ 23.5 ms. From Fig. 10, if $b = 600$ kb/s, IP over WAPS will be better than conventional IP networks for the information length x of more than 10 kbytes. If t_1 becomes 2 ms, WAPS will be better for around 100-byte information length. In order to realize this, switching time (t_{cont}) and processor instruction executing time should be less. Semiconductor optical switches can be used in the future [16]. As for the executing time of a processor it is very easy to improve by more than one order.

Bearing in mind things mentioned above, and considering that in the conventional router model no waiting time in traversing routers is assumed, we could conclude that from the viewpoint of data-transfer time the IP network over WAPS cut-through mode can be feasible in the very near future.

5.3 Waiting Time and Necessary Buffer Capacity

There is a chance for a flow to be unable to reach to destination, because of the busy conditions of terminating links and/or the internal blocking of a subnetwork. Following two measures are considered.

(1) Discarding of the flow.
Discarding rate can be less than 10^{-4} [7], but whether or not this is allowable cannot be judged because there is no criteria for IP packet discarding.

(2) Waiting for the path to become idle.
Let BL_{E-E} be blocking probability in the network. Waiting time is t_1, if the flow is not blocked. Waiting time is t_w, if a connection request is blocked. Then, average total waiting time T_w is given below.

$$T_w = t_1(1 - BL_{E-E}) + t_w BL_{E-E} \tag{8}$$

As is mentioned, it is possible to make BL_{E-E} sufficiently low, i.e. less than 10^{-4}. If so, T_w is almost equal to t_1. An ER with $T_w = 50$ ms, 100 wavelengths, and 200 Mb/s transmission bitrate per wavelength needs about 125 Mbytes of buffer capacity. No problems exist for waiting when packets are blocked. t_1 (therefore, necessary buffer capacity) will decrease according to progress of processing or signaling network

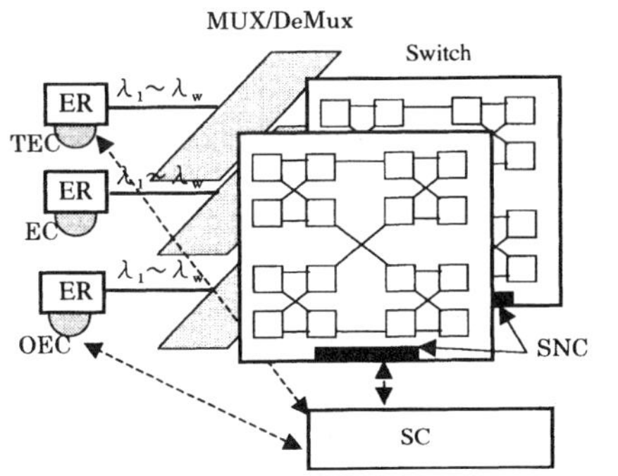

Fig. 13 Node cut-through processing flow.

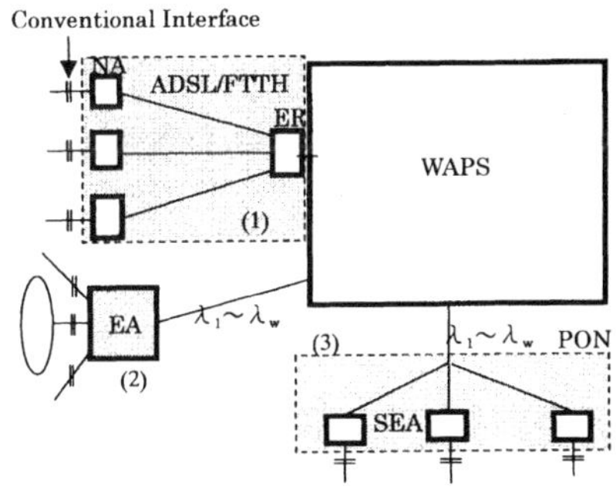

Fig. 14 Subscriber network structure.

technologies.

6. Node Cut-Through

When we consider the introduction scenario of IP over WAPS, the first step is to introduce WAPS nodes in limited and independent areas, and these nodes will be gradually integrated to form IP network over WAPS. A WAPS node can be used as an IP router as shown in Fig. 13. This IP node can be made to have considerably large capacity. Throughput of a WAPS node with 2000 ports, 100 wavelengths and 200 Mb/s per wavelength would be 40 Tb/s, which is still one of the biggest routers at present.

The control configuration is almost the same with the network cut-through configuration. The difference is that the subnetwork is not a network but a switching fabric ($n = 1$), and there is no need to have signaling network so long as the ERs are in the vicinity of the WAPS node. In this case, therefore, $t_{SC \to SNC}$, $t_{SNC \to SC}$, $t_{EC \to NC}$, and $t_{NC \to EC}$ can be roughly eliminated. Pre-processing time t_1 is given below.

$$t_1 = t_{SChunt} + t_{SChunt} + t_{cont} + t_{SCfin} + t_{OECfin} \tag{9}$$

The same condition with the former calculation gives preprocessing time $t_1 = 16$ ms. Figure 10 can be also applied in this case for the relation of transmission bit rate (b kb/s) vs. information length (x kbits). The situation does not improve very much compared with the network cut-through mode, because the switching time (10 ms) is dominant. A processor with processing time of 10 times faster than postulated, and switches with 10 times faster switching time make $t_1 = 1.6$ ms, which will make WAPS node better than a router network even in the case of information length of 50 bytes.

The WAPS node has a large but blocking switch fabric, while OXCs• Optical Cross Connects), which is considered to be LSRs in Multi-Protocol Lambda Switching (MPλS), usually has a small and non-blocking switch fabric. A WAPS node in a node cut-through mode, although it is different from OXC in many points, can be an optical LSR which satisfies requirements of MPλS. When we construct a WAPS network of cut-through mode with more than one WAPS nodes, the whole WAPS network is considered one large LSR with the system control maintaining the WAPS network resources. Therefore, in IP over WAPS network with flow-by-flow connections, very limited set of MPλS functions is only necessary. There is no need for label distribution for example.

7. Subscriber Network

As mentioned earlier, configurations of subscriber network are categorized into two, i.e. Sw and Bs types. Above discussions are made on the basis of Sw type subscriber network. In Fig. 14(1), distributed NAs (Network Adapters) provide standardized interfaces, and ERs terminates the terminal protocols and adapt these standardized interfaces to the WAPS network protocol or give core addresses of the WAPS network to these adapted packets. Subscriber networks can be constructed using copper wires (xSDL: Subscriber Digital Loop) or optical fibers (FTTH: Fiber To The Home). In Fig. 14(2), an EA (Edge Adapter) integrates all functions of both ER and NA still having conventional interfaces. In Fig. 14(3) an EA is divided into SEAs (Sub-Edge Adapter), which are similar to terminal adapters but still have functions of ER + NA. The last configuration of Fig. 14(3) (this is PON (Passive Optical Network) configuration) is eventually Bs type subscriber network of WAPS concept. When ER is situated remotely from a WAPS node as Figs. 14(2) and (3), signaling system and delay consideration will be needed.

8. Conclusion

We proposed an IP network configured over WAPS, in which WAPS provides two types of connection, semi-permanent and flow-by-flow connections, and the edge routers can use either of these connections according to the required services. The pre-processing time in the flow-by-flow connection is crucial and is discussed in details. The parallelism of the WAPS network architecture contributes to shortening the time very much, and we can conclude that IP network configured over WAPS network with WAPS subscriber network is sufficiently feasible. Requirements are clarified for a signaling network, switching time of WAPS node and a processing time of processors used in subnetworks and a system controller.

A network provider can give a reliable and QoS guaranteed IP network using WAPS, but at some extra cost. At present the demand for more inexpensive networks is so strong, and present router networks with some further improvement are preferable for the time being, but complaints against the real time services or broadband picture services will lead to gradual introduction of IP network over WAPS.

References

[1] P. Newman, G. Minshall, and T. Lyon, "IP switching: ATM under IP," IEEE/ACM Trans. Networking, vol.6, no.2, April 1998.

[2] P.P. White, "RSVP and integrated services in the internet: A tutorial," IEEE Commun. Mag., vol.35, no.5, pp.100–106, May 1997.

[3] V.P. Kumar, T.V. Lakshman, and D. Stiliadis, "Beyond best effort: Router architectures for the differentiated services of tomorrow's internet," IEEE Commun. Mag., vol.35, no.5, pp.152–164, May 1998.

[4] H. Sugioka, K. Koyanagi, T. Adachi, and T. Chiba, "A technological initiative towards next-generation computer networks," NTT R&D, vol.47, no.4, pp.411–414, 1998.

[5] T. Li, "MPLS and the evolving internet architecture," IEEE Commun. Mag., vol.37, no.12, pp.38–41, Dec. 1999.

[6] T. Yasui and A. Uemura, "A proposal of a new photonic FDM switching system FASP—Frequency assign photonic switching system," IEICE Trans. Commun., vol.E77-B, no.2, pp.174–183, Feb. 1994.

[7] T. Yasui and Y. Nakano, "Switching node consideration from the aspect of transmission characteristics in wavelength assignment photonic network (WAPN)," IEICE Trans. Commun., vol.E82-B, no.2, pp.306–316, Feb. 1999.

[8] D.A. Smith, R.H. Chakravarthy, Z. Bao, J.E. Baran, J.L. Jackel, A. d'Alessandro, D.J. Fritz, S.H. Huang, X.Y. Zou, S.-M. Hwang, A.E. Willner, and K.D. Li, "Evolution of the acousto-optic wavelength routing switch," J. Lightwave Technol., vol.14, no.6, pp.1005–1019, June 1996.

[9] K. Sasayama, M. Okuno, and K. Habara, "FDM multichannel selector using a coherent optical transversal filter," Proc. OFC, pp.10–11, 1992.

[10] A. Himeno, T. Goh, M. Okuno, H. Takahashi, and K. Hattori, "Silica-based low loss and high extinction ratio 8 × 8 thermo-optic matrix switch with path-independent loss arrangement using double Mach-Zehnder interferometer switching units," Proc. ECOC'96, vol.4, pp.149–152, Oslo, 1996.

[11] L.Y. Lin, E.L. Goldstein, and R.W. Tkach, "Free-space micromachined optical switches with sub-millisecond switching time for large-scale optical crossconnects," IEEE Photon. Technol. Lett., vol.10, no.4, pp.525–527, 1998.

[12] S. Okamoto, A. Watanabe, and K. Sato, "Optical path cross-connect node architectures for photonic transport network," J. Lightwave Technol., vol.14, no.6, pp.1410–1422, June 1996.

[13] A. Jourdan, F. Masetti, M. Garnot, G. Soulage, and M. Sotom, "Design and implementation of a fully reconfigurable all-optical crossconnect for high capacity multiwavelength transport networks," J. Lightwave Technol., vol.14, no.6, pp.1198–1206, June 1996.

[14] N. Watanabe, H. Ichikawa, J. Aramomi, M. Tanikawa, H. Hara, S. Kato, H. Kitazumi, and J. Murayama, "Backbone network architecture of the next-generation computer network," NTT R&D, vol.47, no.4, pp.435–444, 1998.

[15] K. Habara, M. Kajiwara, M. Eguchi, and H. Ogiwara, "Information theoretical analysis of switching processing," IECEJ Trans., vol.E60, no.9, pp.459–466, Sept. 1977.

[16] S. Takahashi, S. Araki, Y. Suemura, A. Tajima, Y. Maeno, H. Takahashi, K. Saito, K. Matsuda, T. Tamanuki, H. Hatakeyama, O. Nakakubo, S. Dohmae, and N. Henmi, "Terabit/s throughput opto-electronic ATM switch prototype with PVC cell switching operation," Proc. ECOC1999, Post Deadline Paper, pp.50–51, 1999.

High-Performance Backbone Network Technology edited by Naoaki Yamanaka
ISBN: 0-8247-5321-6 DOI: 10.1081/(120025733)
Published 2004 by Marcel Dekker, Inc.

IOG: A Protocol for IP over Glass

Masataka Ohta, Hideaki Oonaka, Kazuyuki Sato, Shinichi Aoki, Shigeyuki Takayama, and Akio Iijima

SUMMARY IOG is a link layer protocol specifically designed for the high speed and bandwidth efficient transmission of IPv4 and IPv6 datagrams over optical fibers. That is, IOG is a simple point-to-point packetization protocol over a bit stream with a low bit error rate. MTU of IOG is 1535, which is long enough for the Internet with IPv6 multicast packets and Ethernet frames. IOG has a framing structure of fixed length (2048 bytes) for synchronization, CRCC (Cyclic Redundancy Check Code) and scrambling. A frame consists of 4 bytes of a frame header, 2040 bytes of a frame payload and 4 bytes of a frame trailer for CRCC. CRCC is also used for scrambling. A frame header consists of a 21 bit flag sequence ("011111111111111111110") and a 11 bit packet boundary pointer. A packet has an 11 bit length field and a 21 bit label field. The label field contains an Ethertype or a link layer label. Packets are packed continuously in frame payload. A packet is at least 20 and at most 1535 bytes long. If there are no packets to send, 20 byte packets of Ethertype 0 are sent, which is ignored by the receiver. A packet may be included in two adjacent frames. The packet boundary pointer in a frame header of a frame points to the first packet boundary of the frame, which means that once a frame synchronization is established, packet synchronization is also established. IOG is designed to allow high-speed implementations to enable 32 or 64 bit parallel processing. CRC polynomial of $X^{32} + X^2 + X + 1$ is newly designed for high-speed 32 or 64 bit parallel calculation and frame wise SECDED (Single Error Correction, Double Error Detection). IOG is applicable to long haul transmission of IP datagrams in Internet backbone and to inter-chip or inter-module transmission of IP datagrams in parallel IP routers.

key words: *internet, optical fiber, WDM*

1. Introduction

IOG is a link layer protocol specifically designed for the high speed and efficient transmission of IPv4 and IPv6 datagrams over optical fibers.

With the increasing demand on the speed of the Internet backbone, the backbone needs more bandwidth than that available with electric circuit. It is necessary to use multiple electric circuit connected by parallel optical transport multiplexed at the physical, photonic, layer with multiple fibers, multiple wave length etc. As IP traffic can use up all the bandwidth available with electric circuit, it is meaningless to use electric multiplexing protocol such as SONET/SDH or ATM to share the bandwidth of electric circuit between IP and non-IP protocols. Instead, it is necessary to have a data link protocol tuned specifically to transmit IP packets over some optical transport. The protocol should be simple to be able to extract as much speed of electric circuit as possible without complex electric operations. The protocol should be efficient to pass as much bandwidth of electric circuit as possible to the IP layer without a lot of data link headers and trailers. The protocol, on the other hand, does not need to have complex features to support non-IP communication. For example, no multiplexing other than packet multiplexing is necessary, which makes global clock synchronization for TDM relaying unnecessary.

The simplicity and efficiency of IOG makes it applicable for not only long haul transmission of packets at the Internet backbone but also short optical wiring of high speed equipment such as massively parallel routers.

Manuscript received March 6, 2000.
Manuscript revised May 31, 2000.

†The author is with Tokyo Institute of Technology, Tokyo, 152-8550 Japan.
††The author is with Japan Music Agency Company Limited, Tokyo, 150-0012 Japan.
†††The author is with World Axle Corporation, Tokyo, 100-8166 Japan.
††††The authors are with NEC Shizuoka, Ltd., Kakegawa-shi, 436-8501 Japan.
†††††The author is with NEC Corporation, Abiko-shi, 270-1198 Japan.

2. Physical Layer Considerations

Physical characteristics of communication over optical fiber are that it has low bit error rate.

Moreover, optical fiber should be used only for point to point communication.

Optical fiber is a physical layer entity to connect a pair of end points. However, it is not impossible to use optical couplers to physically couple multiple fibers to form a multi-access optical physical layer (Fig. 1(a)). With the multi-access physical layer, the number of interfaces to convert electric and optical signals can be reduced. For example, to connect 3 terminal stations to a central station, 6 interfaces are necessary with point to point configuration (Fig. 1(b)). On the other hand, with optical coupling, the number of interfaces can be reduced to 4 (Fig. 1(a)). But, with the optical coupling, each interface must support bandwidth for all the communication in the physical layer. If the requirement were to connect 10 terminals to the central station at 64 kbps, the required bandwidth of 640 kbps had not

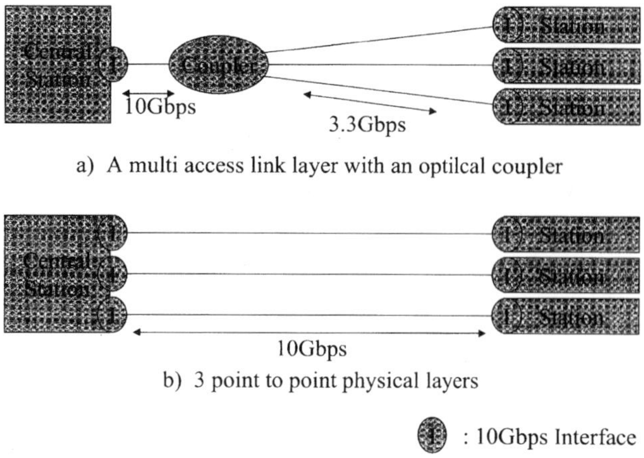

Fig. 1 Multi access and point to point physical layers.

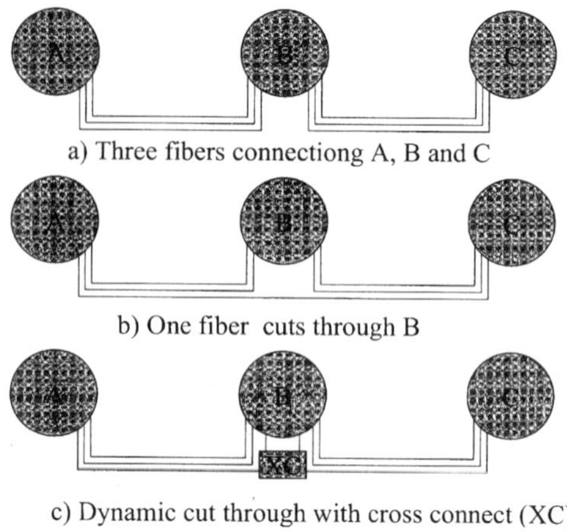

Fig. 2 Physical layer cut through and cross connect.

been a problem. However, as is discussed in Sect. 1, the requirement is to have more bandwidth than that available with electric circuit. With the optical coupling of Fig. 1(a)), the average bandwidth available to each terminal station is reduced to one third of that of the point to point case of Fig. 1(b)). That is, for really high speed optical transmission, interfaces to convert electric and optical signals must optically be connected point to point. It is, of course, possible to use optical multiplexing (such as WDM) to have multiple optical point to point connections over a single optical fiber.

It is difficult to estimate the maximum bandwidth available with electric circuit. However, as it is not economically feasible to transmit DC to 300 MHz signal between a pair of LSIs or to have more than 32 parallel transmission lines, 10 Gbps seems to be a good estimate on the maximum bandwidth and it is really unlikely that 100 Gbps can be available with electric circuit.

Thus, IOG is designed assuming point to point physical layer with low bit error rate and very high speed of 10 or 100 Gbps.

With the current optical technology, it is not difficult to modulate light with 100 different wave length each at 10 Gbps and multiplex them into a single optical fiber to transmit date over the fiber at 1 Tbps.

If there are 1,000 parallel fibers laid between two points, it is easy to make the 1,000 fibers a Peta bit backbone of the Internet.

3. Data Link Layer Considerations

In this section, it is argued that data link layers over optical layers for Internet backbone should be point to point.

Even though physical layer of optical network is point to point, it is possible to construct multi-access data link by connecting point to point physical layers by bridges. Typical reasons are flexibility, speed, traffic engineering or peering. However, it is not beneficial to do so.

Compared to a point to point data link, a multi-access data link layer is complex and need extra overhead for complex header information such as MAC addresses. Such information were useful, if there had been various protocols used at the Internetworking layer. However, in the Internet backbone, all the information is carried as IP packets, headers of which contain all the information to relay the packets to the destination. So, data link layer headers are redundant. Worse, to support IP multicast or QoS (Quality of Service) assurance, multi-access data link layer must have its own mechanism of multicast or QoS assurance and, in addition, mapping between the data link and the Internetworking layer multicast or QoS assurance must be defined. On the other hand, if IP routers are connected by point to point data link, data link layer multicast is no different from data link layer unicast and QoS is assured by proper queuing at the output of the IP routers.

It is often misunderstood that data link layer bridging is faster than Internetworking layer routing. However, data link layer bridges do not operate much faster than the internetworking layer routers, because the performance of bridges and routers are both bounded by the speed of memory look up to determine the next hop bridges or routers [1].

It may be argued that data link layer labels are necessary for traffic engineering with MPLS (Multi-Protocol Label Switching). However, as is argued in [2], physical layer labels (spatial, time slot, wave length, code labels are all physical layer labels) can also be used for label switching. For example, as shown in Fig. 2(a)), if locations A, B and C are connected by 3 parallel fibers but there are a lot of traffic between A and C, as shown in Fig. 2(b)), it is possible to physically connect a fiber at B to physically cut through Internetworking and data link layer processing. Unlike data link layer bridging, physical layer cut through automatically operate at the wire speed. If the amount of traffic between A and C varies dynamically, there may be spatial cross connect at B (Fig. 2(c)). MPLS protocols can control

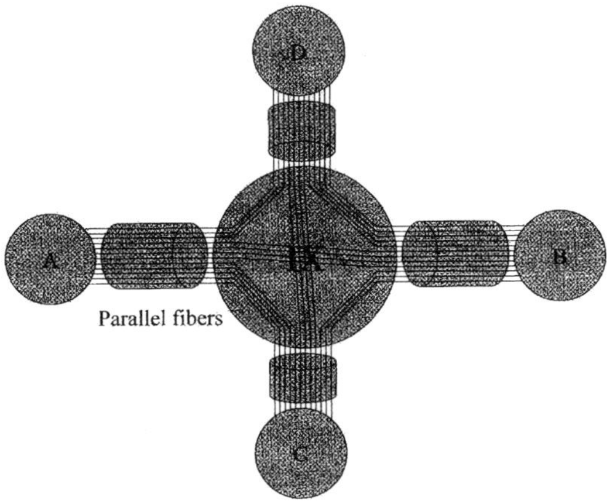

Fig. 3 Peering at IX with multiple interfaces.

such physical layer cross connects.

Assuming 10 Gbps electric interfaces, a 1 Tbps backbone has 100 physical layer labels and a 1 Pbps backbone has 100,0000 labels that they are more than enough for complex traffic engineering.

Finally, it may be argued that multi-access data link is useful for peering at IXes (Internet Exchanges), because only one interface is necessary for a router of an ISP (Internet Service Providers) to peer with N other routers of other ISPs. However, as there are a lot of parallel electric interfaces multiplexed to form a Tera or Peta bit backbone, they may be used to peer with other ISPs (Fig. 3).

That is, the data link layer of the optical Internet backbone should be point to point.

The other functionality of data link layer is multiplexing. Internet needs packet multiplexing only. Though IPv4 [3] and IPv6 [4] packets can be distinguished by the first 4 bits of IP packets, IPv4 and IPv6 packets are also distinguished by data link layer labels [5], the reason of which is that some buggy IPv4 hosts do not check the first 4 bits and may crash or behave poorly if IPv6 packets are passed to the Internetworking layer of the hosts. As a result, information in the first 4 bits of IP packets is not used and, in the future, the field may be used for other purposes. So, it is safe to distinguish IP versions with data link layer headers.

Labels in data link layer are also useful for QoS assurance on massively parallel routers [6].

Finally, MTU (Maximum Transfer Unit) is an important property of a data link. The larger the MTU, the smaller the overhead to process packets. However, for real time communication, the larger MTU, the longer to accumulate data stuffed in a packet. With IPv4, minimum MTU is 68 bytes and larger packets can be fragmented into smaller packets at the source or on intermediate routers. However, it is recognized that fragmentation is a complex operation and badly degrade router performance. So, with IPv6, fragmentation is allowed only at the source. Instead, minimum

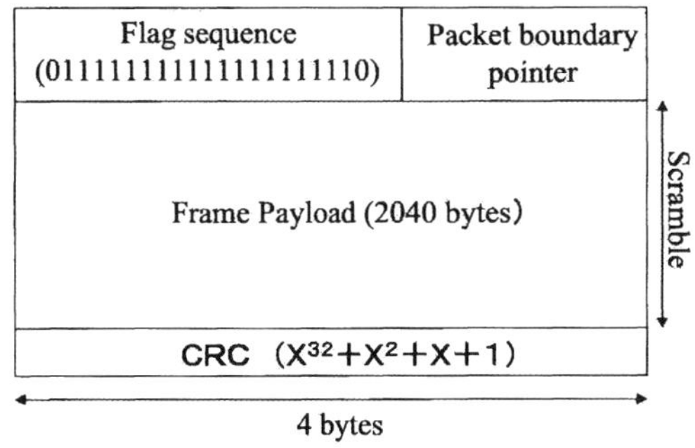

Fig. 4 Frame format of IOG.

MTU is increased to 1280 bytes, which is considered to be small enough even if a packets is transmitted over Ethernet (MTU of Ethernet is 1500 byte) with doubly nested tunneling. IPv6 hosts cannot send packets larger than 1280 bytes, unless it knows that all the data link between it and the destination has larger MTU. While there is a mechanism to discover path MTU [7], the minimum MTU of links between a source and a destination, it does not scale and can not be used for multicast. That is, considering IPv6 minimum MTU, it is desirable to support MTU larger than or equal to 1500 bytes but, considering IPv6 multicast, it is not meaningful to support MTU much larger than 1280 bytes.

4. Design of IOG

IOG is designed for efficient transmission of IP packets over high speed point to point data link layer with low bit error rate.

IOG has a fixed length frame structure for synchronization, scrambling and CRC (Cyclic Redundancy Check).

Variable length packets are stuffed continuously in payload fields of frames with MTU of 1535. Packets must at least be 20 byte long, which is not a problem for IP, because shortest IPv4 packet is 20 byte long and shortest IPv6 packet is 40 byte long.

Bit ordering of IOG is bigendean.

The header and CRC overhead of a frame is 8 bytes for 2040 bytes of data and is less than 0.5%. Each packet, further, has a 4 byte header.

4.1 Frame Structure of IOG

A frame of IOG is 2048 byte long. As shown in Fig. 4, a frame begins with 32 bits of a frame header (21 bits of a flag sequence ("011111111111111111110") followed by 11 bits of a packet boundary pointer), 2040 bytes of frame payload and 4 bytes of CRCC.

Frame payload is first scrambled and CRCC is calculated over the scrambled payload.

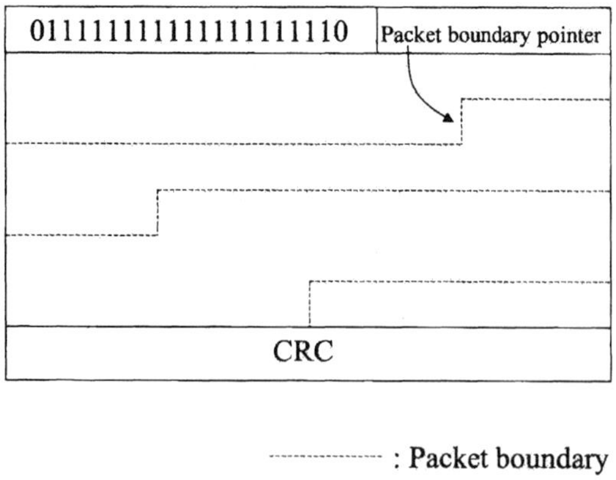

Fig. 5 Packet boundaries and packet boundary pointer.

4.1.1 Frame and Packet Synchronization of IOG

To establish frame synchronization, the 21 bit pattern of the flag sequence is sought. If CRCC check of 2048 bytes of data from the flag sequence is successful (single bit errors are not tolerated), the frame synchronization is established immediately and packets are extracted from the frame payload.

Packets are tightly stuffed within a frame payload. A packet may be contained in two consecutive frames.

The first packet boundary in a frame payload is designated by the packet boundary pointer in the frame header. Remaining packet boundaries are calculated using packet length fields in packet headers (Fig. 5).

The probability of false synchronization is 2^{-53}.

4.1.2 Scrambling of IOG

To prevent bit patterns (such as a long sequence of 0's) harmful to optical devices (laser diode can not be modulated by DC), payload field of IOG is scrambled.

Frame payload is XORed with a reset scrambler with generating polynomial of $X^{15} + X + 1$ and CRCC of previous frame replicated every 32 bits.

Even if a packet contains bit pattern identical to the reset scrambler, the probability that the result of scrambling becomes a long sequence of 0's is 2^{-43}.

4.1.3 CRC of IOG

CRCC of a frame is computed over CRCC of the previous frame and the current frame.

CRC polynomial is $X^{32} + X^2 + X + 1$, which is factorized into the following four irreducible polynomials:

$$
\begin{aligned}
&X + 1 \\
&X^3 + X^2 + 1 \\
&X^7 + X^6 + X^5 + X^4 + 1 \\
&X^{21} + X^{20} + X^{19} + X^{16} + X^{15} \\
&\quad + X^{13} + X^{10} + X^6 + 1
\end{aligned}
$$

The first factor, $X + 1$, and the last factor, which is a primitive polynomial of 21st degree, makes the hamming distance of the code generated by the CRC polynomial 4 for frames less than 2^{21} bit long. So, it is possible to detect all the triple bit errors. Or, it is possible to perform SECDED (Single Error Correction, Double Error Detection).

The relationship between raw bit error rates and the probabilities to drop a frame is summarized in Tables 1 and 2.

Note that the argument in [8] that CRC polynomial must be chosen carefully is important to choose a general purpose CRC but is not applicable if, like IOG, frame size is large or raw bit error rate is low.

4.1.4 Parallel Processing of Frames

IOG is designed for efficient parallel processing in mind.

For 32 (64) bit parallel processing of frames, 64 (128) bit input shift register should be provided.

To establish frame synchronization, flag sequence should be matched in parallel starting at 32 (64) different bit positions of the input shift register. Then, 32 (64) bit barrel shifter should be used to align bit positions to 32 (64) bit boundary.

Pattern of reset scrambler is fixed and CRC of the previous frame is already 32 bit wide (which may be replicated twice) that 32 (64) bit parallel scrambling is easy.

So far, operations are serialized with no loop that time consuming operations can be serially pipelined for better performance.

However, the remaining operation of CRC computation involves a loop. So, the latency of the loop limits the performance.

CRC polynomial is chosen to minimize the fan in of XOR for 32 (64) bit parallel computation of CRC. XOR with large fan in needs cascaded XORs with smaller fan in, which means a large amount of delay. For example, at least 5 stages of 2 input XORs are necessary to construct 17 input XOR.

Let $C(i)$ i-th bit of CRC and $I(i)$ i-th bit of 32 (64) bit parallel input.

For 32 bit parallel input, CRC with polynomial of $X^{32} + X^2 + X + 1$ can be computed in parallel as

$$
\begin{aligned}
C(0) &\leftarrow C(0) \wedge C(30) \wedge C(31) \wedge I(0) \\
C(1) &\leftarrow C(0) \wedge C(1) \wedge C(30) \wedge I(1) \\
C(2) &\leftarrow C(0) \wedge C(1) \wedge C(2) \wedge C(30) \wedge I(2) \\
C(3) &\leftarrow C(1) \wedge C(2) \wedge C(3) \wedge C(31) \wedge I(3) \\
C(4) &\leftarrow C(2) \wedge C(3) \wedge C(4) \wedge I(4) \\
&\ \ \vdots \\
C(31) &\leftarrow C(29) \wedge C(30) \wedge C(31) \wedge I(31)
\end{aligned}
$$

Table 1 Probability to drop a frame with triple bit error detection.

Raw Bit Error Rate	10^{-8}	10^{-9}	10^{-10}	10^{-11}
Probability of Drop	$1.64 * 10^{-4}$	$1.64 * 10^{-5}$	$1.64 * 10^{-6}$	$1.64 * 10^{-7}$

Table 2 Probability to drop a frame with SECDED.

Raw Bit Error Rate	10^{-8}	10^{-9}	10^{-10}	10^{-11}
Probability of drop	$1.34 * 10^{-8}$	$1.34 * 10^{-10}$	$1.34 * 10^{-12}$	$1.34 * 10^{-14}$

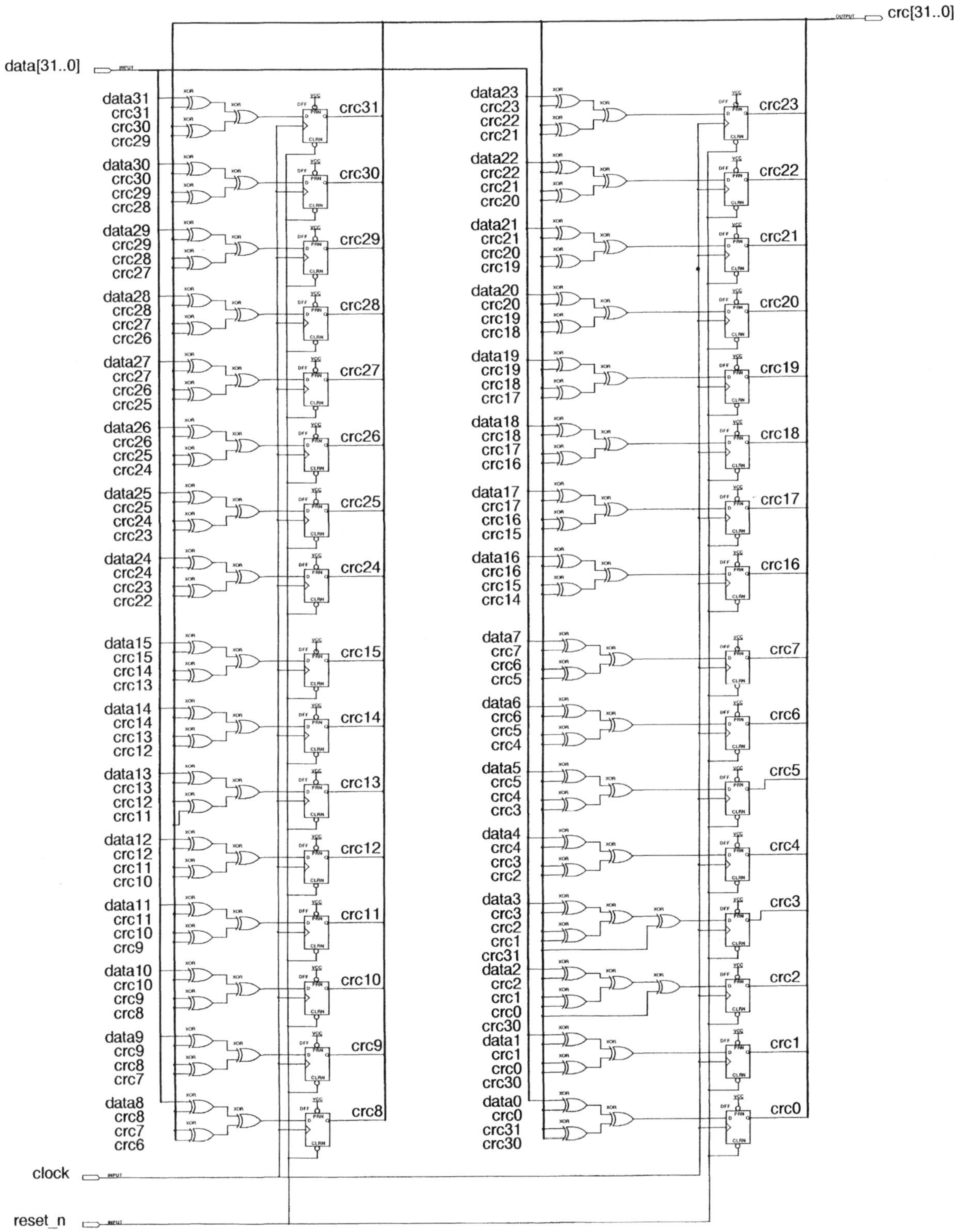

Fig. 6 32 bit parallel CRC computation circuit of IOG.

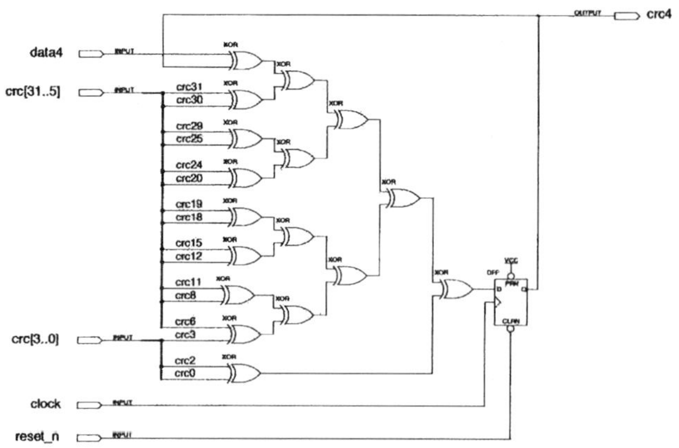

Fig. 7 Forth bit of 32 bit parallel CRC32 computation circuit.

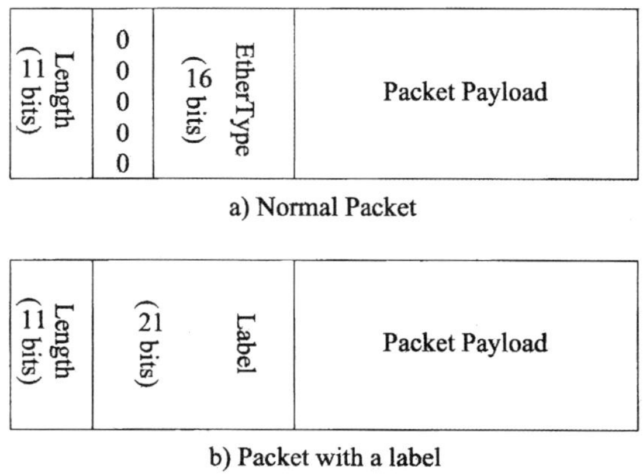

Fig. 8 Packet format of IOG.

which requires fan in of XORs of at most 5 (Fig. 6).

Similarly, for 64 bit parallel input, the CRC computation requires fan in of XORs of at most 9.

Other CRC polynomial needs more fan in. For example, 32 bit parallel computation of CRC32 requires fan in of XOR of 18 at a bit of 4th degree (Fig. 7).

With a timing model of gate arrays of 0.35 μm process, 32 bit parallel CRC computation is the bottleneck of the IOG protocol. CRC32 can be computed at 179 MHz (5.7 Gbps) while the CRC of IOG can be computed at 248 MHz (7.9 Gbps). That is, the performance improvement is critical for leading edge implementations at 10 or 100 Gbps.

4.2 Packet Structure of IOG

A packet of IOG is of variable length and from 24 bytes to 1539 bytes long. As shown in Fig. 8, a packet begins with 32 bits of a frame header (11 bits of a packet length field followed by a 21 bit packet label) and packet payload. The packet length field gives a byte count of the packet payload. If the label is 0, the packet is a 20 byte long space filler stuffed by the sender, if there is no packet to send. The receiver ignores it. If the upper 5 bits of the label is 0, the remaining 16 bits of the label designates the Ethertype of the packet payload (Fig. 8(a)). Otherwise, the label may be used for data link dependent special meaning (Fig. 8(b)). The packet payload is a packet at the Internetworking layer.

The length of frames and packets are chosen to simplify processing. With the restriction on the length of the packet, a packet may be contained in one or two frames. With 32 or 64 bit parallel processing, there is at most one packet boundary in a sequence of 32 or 64 bits.

5. OAM and IOG

IOG has no OAM by itself. As routers are connected point to point, ICMP or routing protocols are the OAM at the Internetworking layer and no additional OAM is necessary at the data link layer. Physical layer OAM may be necessary if a point to point physical layer is complex consisting with photonic elements such as EDFA, spatial XC or WDM XC. However, with photonic elements, it is practically impossible to add OAM information to data stream at intermediate devices that physically separate channel for physical layer OAM should be provided with its own protocol, which is independent to data link layer protocol such as IOG.

6. Prototype Implementation of IOG

To confirm the validity of the design of IOG, prototype implementation of IOG was manufactured with FPGA and 32 bit parallel processing at OC-12 speed to bridge eight pairs of 100 Mbps Ethernet interfaces.

The implementation experience was fed back to refine the design.

The prototype implementation worked successfully and was demonstrated at Interop+Networld Tokyo 1999 as part of high speed backbone network of the show.

7. Comparison to Other Protocols

Performance of routers or bridges is limited by the speed of routing table look up. To forward ATM cells, a routing table must be looked up for every 48 bytes, which is about 10 times slower than packet forwarding of IP routers [1]. So, anything related to ATM is not a choice for high speed backbone.

Other than ATM, IP over PPP over HDLC over SONET/SDH [9], [10] is a major protocol currently used to transmit IP over optical fiber. However, SONET/SDH layer has about 3% of overhead. HDLC is a byte based protocol with byte wise comparison, insertion and deletion that there is no room for more parallel processing, waste 1% (average) to 50% (worst case) of bandwidth (because some byte values can not be used without an additional escape byte). Worse, as SONET/SDH use a scrambler sequence of a period of

127 bits, a malicious user can send a packet containing the same bit pattern as the scrambler to create an optical output of long sequences of 0's once every 127 packets, which is often fatal for proper network management.

There are several proposals to overcome the problem of PPP over SONET/SDH.

IETF added a scrambling to the specification of PPP over SONET/SDH to solve the last problem, but other problems remain unchanged [11].

SDL [12] is a completely new protocol. With SDL, a variable length packet is preceded by a 32 bit header containing packet length and 16 bit CRC for single bit error correction and followed by a 32 bit CRC. Packet boundaries are detected by searching a 32 bit sequence with valid header CRC. Compared to IOG, the detection of the packet boundaries is time consuming involving XOR operations. Worse, with SDL, bit errors in packet content are not corrected. There may be some traffic for which bit errors are tolerated. However, for the Internet use, bit errors are serious errors. IPv4 packets have their own checksums and packets with bit errors are discarded on routers. IPv6 packets do not have checksums at all that bit errors in IP headers may cause the packets delivered to wrong transport layer entities. SDL uses CRC32, which is slow to compute.

8. Conclusions

IOG is a protocol for IP over WDM tuned for high speed transmission of IP datagram over optical fiber. It assumes point to point data link with low bit error rate. CRC is specially designed for high speed parallel computation. Its effectiveness is proven both in theory and in practice.

Some of the authors are now considering to implement IOG at much higher speed, say 10 Gbps.

Acknowledgements

The Authors are grateful to NEDO for funding the project to develop IOG protocol with 600 Mbps prototype implementation. IOG was developed using an open mailing list to which a lot of people subscribed and gave responses. Especially, comments from Kenji Fujikawa of Kyoto University and Ryota Motobayashi of NEC Corporation was quite helpful.

References

[1] M. Ohta, "Toward the real internet—The farewell to cells," IPSJ Magazine, vol.41, no.1, pp.66–67, Jan. 2000.

[2] D. Awduche, J. Malcolm, J. Agogbua, M. O'Dell, and J. McManus, "Requirements for traffic engineering over MPLS," RFC2702, Sept. 1999.

[3] J. Postel, "Internet protocol," RFC791, Sept. 1981.

[4] S. Deering and R. Hinden, "Internet protocol, version 6 (IPv6) specification," RFC2460, Dec. 1998.

[5] M. Crawford, "Transmission of IPv6 packets over ethernet networks," RFC2464, Dec. 1998.

[6] M. Ohta, M. Sola, K. Fujikawa, A. Kojima, H. Fukumori, and Y. Muraoka, "Hash parallel and label parallel routing for high performance multicast router with fine grain Qos control," Proc. Internet Workshop '99, Feb. 1999.

[7] J. McCann, S. Deering, and J. Mogul, "Path MTU discovery for IP version 6," RFC1981, Aug. 1996.

[8] "A comparison of some error detecting CRC code standards," IEEE Trans., vol.COM-33, no.9, Sept. 1985.

[9] W. Simpson, "The Point-to-Point Protocol (PPP)," STD51, RFC 1661, July 1994.

[10] W. Simpson, "PPP over SONET/SDH," RFC1619, May 1994.

[11] A. Malis and W. Simpson, "PPP over SONET/SDH," RFC2615, June 1999.

[12] B. Doshi, S. Dravida, E. Hernandez-Valencia, W. Matragi, M. Qureshi, J. Anderson, and J. Manchester, "A simple data link protocol for high speed packet networks," Bell Lab. Tech. J., vol.4 no.1, pp.85–104, Jan.-March 1999.

Chapter 8

Photonic Switch

Chapter 8 describes a photonic switch architecture system and its development. Photonic switches are the key to realizing high-performance systems. Photonic switch architectures can be classified into three types: space division, time division packet, and Add Drop Multiplexer (ADM). Paper 8.1 reports an architectural evaluation for photonic ATM switches, and focuses on the number of optical devices. Wavelength division routing is more suitable for cell switching than space division switching. In addition, the ATM cell buffer configuration is discussed. The architecture, system configuration, and implementation method are examined.

Paper 8.2 is another architectural paper. The paper provides design and performance study results for free space photonic switches. The free space photonic switches overcome the interconnection limitation. A switch system is constructed by 2 × 2 space switch elements and a multi-stage configuration. Its architecture is optimized in terms of interconnection simplicity. In addition, a connection setup algorithm is discussed for point-to-point and point-to-multipoint connections. Finally, the paper details implementation examples based on birefringence plates, polarization controllers, and Exciton Absorption Reflection Switch (EARS) devices.

T. Yasui et al. provide an architectural performance study for a wavelength division multiplexed network in Paper 8.3. In the system, each cell is assigned a wavelength. The network is very efficient in terms of its S/N and Q factor. Numerical evaluation results show that a 2000-km wide area network with 50-Mb/s users can be realized.

S. Araki et al. present a developed prototype of a photonic core node based on 2.56-Tbit/s opto-electronic switching fabric in Paper 8.4. Their architecture drastically reduces the amount of optical switching elements, by 7/8, compared to a conventional optical switching fabric. Generally speaking, optical devices are very expensive. So architectural approaches to reduce cost as well as to achieve scalability is very important. The technology will be a key to the installation of future broadband networks.

K. Habara et al. describe a large-capacity photonic packet switch prototype based on wavelength routing techniques in Paper 8.5. The switch has a maximum throughput of 320 Gbit/s. It features a WDM star-based switch architecture, an electrical control circuit layer, and a high-speed optical WDM layer.

Paper 8.6 by H. Sotobayashi et al. describes all-optical label swapping for the photonic label switching router (LSR). The optical code routing photonic LSR, in which label is mapped onto and optical code, is one of the most promising photonic network technologies.

S. Nishi et al. present an optical code division multiplexing (CDM) self-routing network system in Paper 8.7. In the proposed system, a set of multiple-encoding CDM codes is utilized as routing information. They evaluate the bit error rate (BER) characteristics of the multiple-encoding CDM system by simulations. It is suitable for a photonic IP switching system.

K. Yamakoshi et al. present an experimental 5 Tbit/s packet-by-packet wavelength switching system in Paper 8.8. The developed switching system is a non-blocking three-stage switch using optical WDM links and dynamic bandwidth sharing. Details of the hardware architecture are described.

The final paper of this chapter studies optical label switching. K. Tanaka et al. present a novel concept for an optical label switching network and the configuration and function of the core nodes in paper 8.9. The concept combines wavelength with multiplexed pilot tone frequencies. Experimental results show the feasibility of the proposed optical label discrimination process and optical label conversion process.

High-Performance Backbone Network Technology edited by Naoaki Yamanaka
ISBN: 0-8247-5321-6 DOI: 10.1081/(120025734)
Published 2004 by Marcel Dekker, Inc.

Comparative Evaluation of Photonic ATM Switch Architectures

Yoshihiro Nakahira, Hideki Sunahara, and Yuji Oie

SUMMARY In this paper, we discuss configurations of photonic ATM (Asynchronous Transfer Mode) switches and their advantages in terms of the number of optical switching devices to be implemented on the system, the number of wavelengths, throughput, broadcast function etc. In particular, we focus on photonic ATM switch architectures which can be built in the near future; that is, with presently available optical and electrical devices. For example, we assume the optical devices such as optical gate switches with 40 dB on/off ratio. In this context, we evaluate 17 types of photonic ATM switches; they are 6 types of input buffer type switches, 6 types of output buffer type switches, 4 types of shared buffer switches, and 1 proposed type. From our evaluation, for cell switching, wavelength division switching technologies are desirable compared with space division switching technologies in the sense that the former enables us to build a photonic ATM switch with the less number of optical gate switches. Furthermore, we propose a switch architecture equipped with optical delay line buffers on outputs and electric buffers on inputs. We show that our switch architecture is superior in the number of required optical gate switch elements under the given conditions.
key words: *photonic ATM, throughput, WDM, optical switch, optical memory*

1. Introduction

Recently, a lot of photonic ATM switch architectures have been proposed and prototypes of photonic ATM switches have been developed [1]–[11].

About for 10 years, a lot of studies have been done to clarify performance of electric ATM switches depending upon the buffering styles, i.e. input buffer types or output buffer types [12]. The studies to realize useful functions like a multicast [13] also have been done. The electric ATM switch technologies have been promoted by these studies.

However, it seems that little study has been done on comparative evaluation of photonic ATM switch architectures with regard to performance and the number of required optical switching devices to be implemented. Therefore, we will evaluate the performance and the number of devices required through simulations to realize a practical photonic ATM switch with presently available devices.

Manuscript received May 6, 1997.
Manuscript revised July 28, 1997.

† The authors are with NAIST: Nara Institute of Science and Technology, Takayama, Ikoma-shi, 630–0101 Japan.

†† The author is with Faculty of Computer Science and System Engineering, Kyushu Institute of Technology, Iizuka-shi, 820–8502 Japan.

In Sect. 2, we describe hypotheses of optical switching devices for system evaluation, and we also address conditions for throughput simulations.

In Sect. 3, we discuss advantages and disadvantages of 16 photonic ATM switch architectures. Then we propose a photonic ATM switch architecture with input electric buffers and output optical delay line buffers.

2. Device Technologies and Conditions for Throughput Simulation

In this section, we describe relevant device technologies used in photonic ATM switches, which will be evaluated in the following section.

2.1 Target Photonic ATM Switch

We evaluate photonic ATM switches expected to be realized in the near future, which are defined as the switches realized with devices available now.

In our opinion, it is not yet easy to build an all optical ATM switching system with all optical header converter, all optical call admission control unit, all optical cell switching, all optical buffering, etc. with the state-of-the-art devices.

In some studies, relevant ideas and experimental results of optical intelligent processing have been reported: for example, optical header recognitions [14], [15] or an optical self routing [5]. However, in these experiments, realized functions are limited among a lot of functions to be implemented on the ATM switch system.

Therefore, in this paper, we evaluate photonic ATM switches with both optical cell switching functions and electric processing functions as shown in Fig. 1 [3].

It is very likely that high speed (for example 10 Gbit/sec) cell transmission in cell switching (routing) stage is carried out in an optical way, and control and header conversion requiring intelligent processing are carried out in an electric way with relatively low speed (for example 42.4 nsec/cell) parallel processing. With regard to buffering, we deal with both optical buffers and electric buffers.

We assume the photonic ATM switch with 8–32 input ports of 10 Gbit/s. 10 Gbit/s is treated as just an example of the highest available transmission speed in practical use.

IEICE TRANS. COMMUN., Vol. E81-B, No. 2, pp. 473–481, February 1998

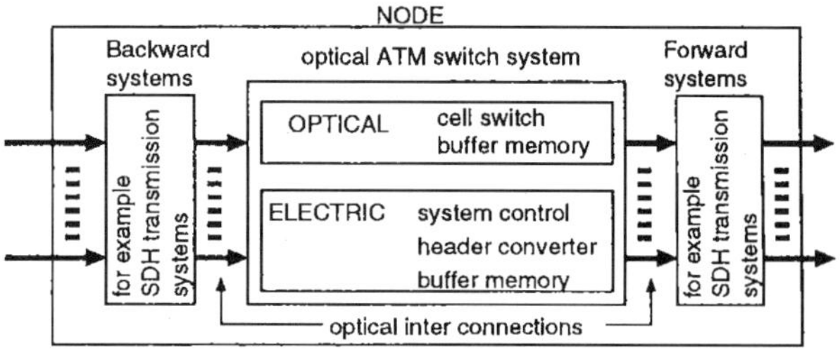

Fig. 1 A photonic ATM switch with optical and electric layers.

Sometimes, it is said that a study on photonic ATM switches is a challenging theme [16] or it is difficult to realize photonic ATM switches which are superior to electric ATM switches. We do not explicitly address this issue by comparing these switches. Whereas we believe our results will also help to clarify how to build practical photonic ATM switches.

2.2 Space Division Switching Devices

In this section, we describe available optical SD (space division) switching devices for photonic ATM switches meeting some requirements on switching speed and crosstalk ratio.

Firstly, we discuss the required switching speed for devices from a viewpoint of system requirements. For the photonic ATM switches described in the above section, optical space division switches should be switched from a state to the other state within an interval between the end of a cell and the start of the following cell as shown in Fig. 2 (a) [2]. We call this interval time "cell switching guard time" or just "guard time". In the guard time, cells are not sent, and thus a large guard time deteriorates the performance of ATM switches. The speedup of cell transmission within ATM switches can avoid the deterioration.

For example, when the line speed is 10 Gbit/s, one time slot of cell transmission is 42.4 nsec. If cell transmission speed is increased to 10.5 Gbit/s, cell transmission time is about 40.4 nsec, and guard time is 2 nsec. We believe a few nano seconds will be available as a guard time.

On the other hand, in case of present electric ATM switches, switching devices have to be switched within a time between a bit and the following bit in cell data stream. Therefore, very high speed switching or parallel processing with tight skew is required. Therefore, some problems arise from difficulties of achieving required device speed, heat, and interconnections in enlarging switch size.

Secondly, we discuss required on/off ratio of optical switches for switch networks. Now we assume two types of 8×8–32×32 switching networks. The one is a rearrangeable nonblocking network with 2×2 switch elements as shown in Fig. 2 (b); the other is a crossbar network with gate switch elements as shown in Fig. 2 (c).

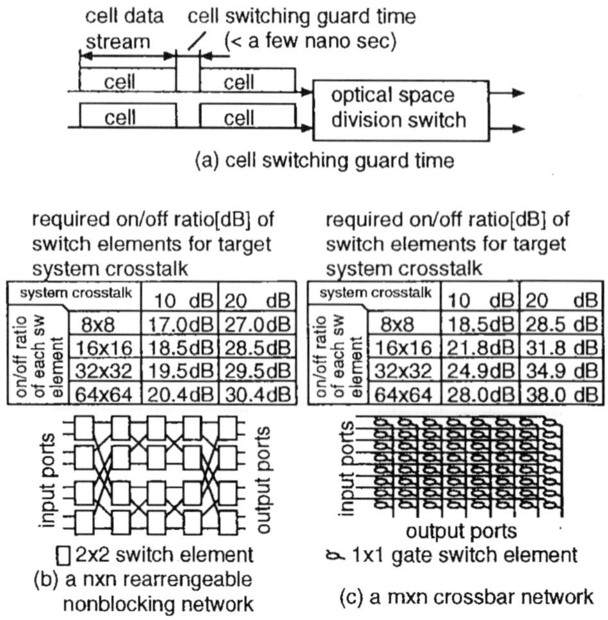

Fig. 2 Optical space division switches.

For the 10 dB of signal/crosstalk (S/X) ratio at output ports of the 8×8–32×32 switch networks, to build the former networks, 17.0 dB–19.5 dB or more of S/X ratio is required for each 2×2 switch element. To make the latter networks, 18.5 dB–24.9 dB or more of on/off ratio is required for each gate switch element. Experimental results show that S/X of 10 dB results in less than 1 dB power penalty; when crosstalk is more than 10 dB, power penalty is increased drastically [17]. S/X ratio of rearrangeable nonblocking network R_{RNET} is calculated as follows. (Crosstalk denotes a total of crosstalk of switch elements at each stage; Appendix I)

$R_{rnet} \cong R_{sw}/(2\log_2 n - 1)$

R_{RNET} [dB] $= 10\log_{10} R_{rnet}$

R_{SW} [dB] $= 10\log_{10} R_{sw}$

R_{rnet}: (Signal/Crosstalk) of switch network

R_{sw}: S/X ratio of switch elements

n: the number of I/O ports of switch network

S/X ratio of crossbar network R_{CNET} is calculated as follows.

$R_{cnet} = R_{gsw}/(n-1)$

R_{CNET} [dB] $= 10\log_{10} R_{cnet}$

R_{gsw}: on/off ratio of switch elements

It seems that there is no 2×2 switch element with S/X ratio of 19.5 dB and a few nano seconds of switching speed in present products.

There are two types of gate switch elements with these required performance. They are SOA (Semiconductor Optical Amplifier) gate switches and EA (Electro Absorption) gate switches. For the former switch elements, 40–60 dB of on/off ratio was achieved [18], [19], and for the latter switch elements, 35 dB of on/off ratio was obtained [20]. In this paper, we assume the optical crossbar switches built with gate switches of 40 dB on/off ratio regardless of wavelength, power and polarization. Actually, 60 dB of gate switch on/off ratio can be expected, we can use additional 20 dB is used for another system margin like the margin for optical buffer memories. We assume optical insertion loss can be recovered by optical amplifier without any power penalty.

2.3 Wavelength Division Switching Devices

Relevant wavelength division (WD) switching techniques have been developed. One is a technique enabling a wavelength switching laser with a few nano seconds switching speed [8], [21]. That is done by two wavelength tunable laser diodes (TLD) connected to high speed optical gate switches as shown in Fig. 3 (a). For example, the TLD source sequentially sends out optical cells with wavelengths λ1, λ3, and λ2 in consecutive 3 timeslots. During the first timeslot, TLD1 is emitting a beam with a wavelength λ1, and TLD2 is tuning to a wavelength λ3; the optical gate switch connected to TLD1 allows the optical beam to pass through (through state), and the other does not allow the beam to pass through (stop state). Then, in the following timeslot, TLD2 is emitting with a wavelength λ3, TLD1 is tuning its wavelength from λ1 to λ2. 42.4 n seconds of tuning time is allowed. With a AWGF (arrayed waveguide grating filter) router as shown in Fig. 3 (b), it can perform a self routing switch [7], [8].

A technique of a high speed tunable filter (TFL) [18] has been developed as shown in Fig. 3 (c). When wavelength division multiplex (WDM) ed signals are sent to this TFL, a wavelength which is sent to the gate switch with through state is forwarded to output. The AWGF can be also used as a WDMux (wavelength division multiplexer) or a WMux (wavelength multiplexer). The number of optical gate switches corresponds to the number of wavelengths required to make a TFL shown in Fig. 3 (c). A configuration of TFL shown in Fig. 3 (d) requires the less number of gate switches than the configuration of Fig. 3 (c). This TFL consists of 2 stages of wavelengths selection.

The number of gate switches required to make this TFL is calculated as follows (It is about square root of the number of wavelengths).

number of gate switches $= n_{s1} + n_{s2}$

where

number of wavelengths $= n_{s1} \times n_{s2}$

n_{s1}: num. of wavelengths groups in first stage

n_{s2}: num. of wavelengths groups in second stage

A few nano second switching speed is available in both techniques. In this paper, we will also consider photonic ATM switches with TLD and TFL techniques. Whereas it seems that TLD technique is more difficult to implement than TFL technique because, for the former, stable thermal control is required.

With regard to the switch size, the 32 × 32 (64 × 64) AWGF with 2.1 dB (3.1 dB) of insertion loss and −28 dB (−27 dB) of channel crosstalk is developed [18]. With the gate switches with 40 dB on/off ratio, system signal/crosstalk ratio is more than 10 dB for the 32 × 32 WD switch with TFL and FLD (fixed wavelength laser diode).

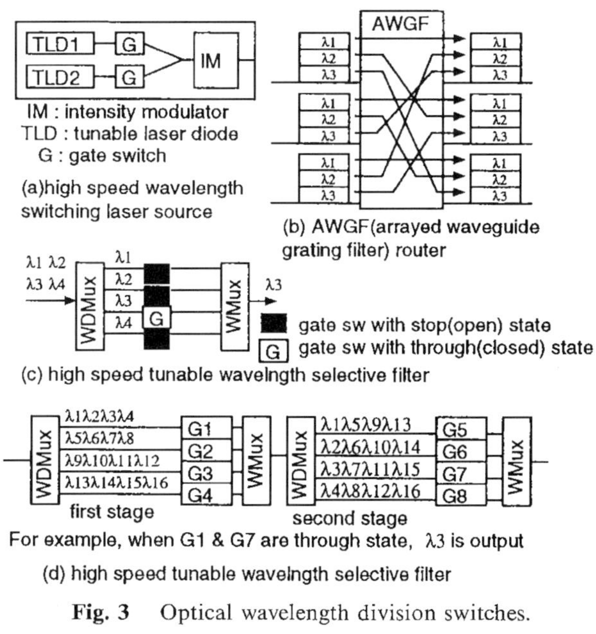

Fig. 3 Optical wavelength division switches.

2.4 Buffer Memory Devices

We assume two types of buffer memory devices; the first is an electric buffer, and the other is a tap type optical delay line buffer [9].

The latter is shown in Fig. 4 (b). Several prototypes of photonic ATM switches with this optical buffers have been reported [6], [10]. As another possible technology, there is a loop type optical buffer memory [22]. It has logically infinite memorizing time, whereas the technology to overcome optical loss and noise in the loop and switches is required.

We evaluate photonic ATM switches with 8 stages and 64 stages tap buffers in the following section. With conventional technologies, the possible buffer size may be restricted to 8–16 because of coherent crosstalk [23] (Appendix II) when on/off ratio of SOAG is 40 dB. The system crosstalk of the tap buffer should be less

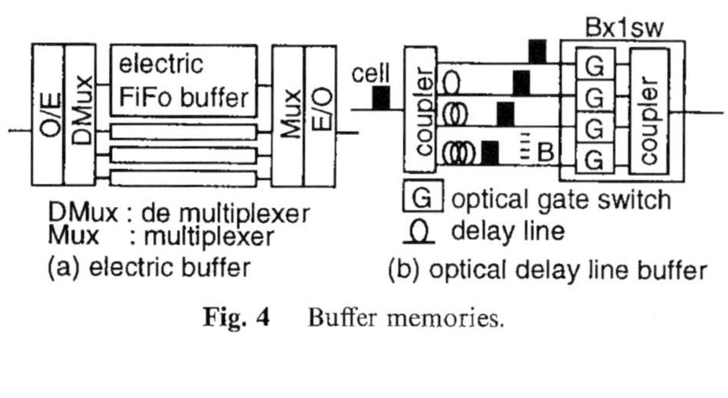

Fig. 4 Buffer memories.

than $-25\,\mathrm{dB}$ [17]. On the other hand, with a differential receiver, a tap type buffer can be implemented with less power penalty [23], we can expect to realize a tap buffer with 64 stages. We evaluate both of these cases.

2.5 Conditions of Throughput Simulations

Simulations are carried out under the following conditions.

- Destination of each cell is decided at random.
- Arrival pattern is in accordance with Poisson process.
- Capacity of buffers is 8 cells/line or 64 cells/line.
- The number of input ports (n1) and output ports (n2) is 8, 16 and 32.

3. Modeling and Evaluation of Photonic ATM Switches

In this section, we evaluate 17 kinds of photonic ATM switches with regard to the number of optical switching devices, throughput, and useful functions.

3.1 Input Buffer Type Photonic ATM Switches

First, we discuss the advantages and disadvantages of input buffer type photonic ATM switches. Figure 5 shows models of their configurations, with the simulation results of cell loss ratio to the random traffic load.

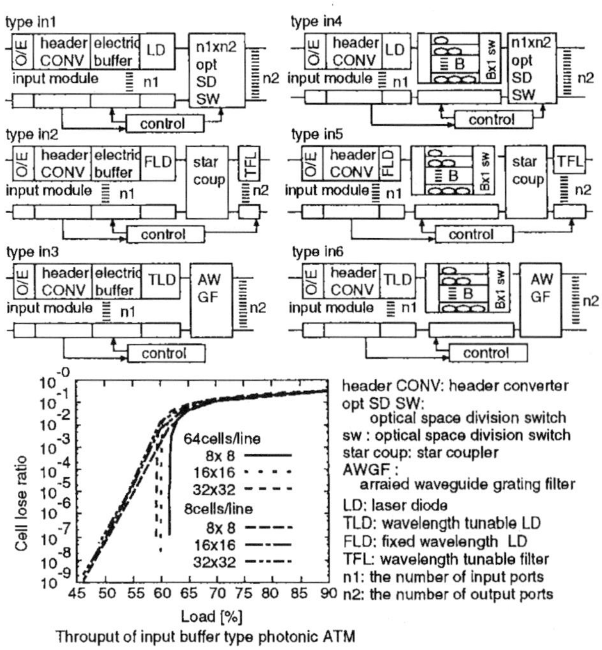

Fig. 5 Photonic ATM switches with input buffers.

A prototype photonic ATM switch with input electric buffers as shown in Fig. 5 (type in1) has been proposed [2]. In this switch, arriving cells from the input ports are processed electrically to analyze their headers in the header converter, and then, cells are stored in electric buffer memories. During this period, E/O or O/E conversion is not required. In header converters and electric buffers, cells are processed in parallel at the same speed. Cells read out from buffers are converted to optical signals by laser diodes (LD) and sent to destination outputs by an optical space division switch (opt SD SW). When there are cells to be sent to the same destination, cells with lower priority is not read out from buffers in this time slot. In our simulations, the higher priority is given to the cell in the buffer in which a larger number of cells are stored.

The number of required gate switch elements, wavelengths, and O/E converters are shown in Table 1. The number of required switch elements equals (n1) $\times$ (n2), where the number of input and output lines are n1 and n2. The number of O/E converters is only n1. As discussed above, it seems that there are not serious technological problems associated with devices in building a in1 type of switch, whereas its performance should be further improved.

From the load cell-loss-ratio curve shown in Fig. 5, the maximum load is about 58% at 10^{-8} of cell loss ratio when electric buffers for 64 cell/line are adopted. This is caused by HOL (head of line) blocking problem. To improve the cell loss ratio, some techniques have been proposed for electric input buffer type ATM switches [24]. For example, there are scheduling control methods like Nemawashi [25], and a way to avoid blocking due to output contention by speeding up cell transmission within the ATM switches.

In what follows, we compare various types of switches in terms of the number of gate switch elements required to achieve a total throughput of 320 Gbit/s with 10^{-8} of cell loss ratio. For example, to attain the above performance, the ATM switch shown in Fig. 5 (type in1) requires 56 input/output ports, i.e., $10\,\mathrm{Gbit/s} \times 56$ ports $\times\, 58\% = 320\,\mathrm{G}$. The related information, e.g., the required switch size and the number of gate switch elements etc., is listed in Table 1 (see the 11th to 15th columns marked with $*1 \sim *5$).

In the model shown in Fig. 5 (type in2), the header of cell is analyzed in the header converter; information of the header is sent to the control unit, and arbitration for output is done there. Then, cells are sent to electric buffers, and after being read out, they are again converted to optical signals by FLD. Then, they are mixed and distributed by a star coupler. The TFL connected to the destination output selects a cell destined for it among them. In this model, the number of gate switches is less than that in the type in1 switch, whereas laser sources must be adjusted to emit at their exact intended wavelengths. Note that the gate switches are

Table 1 The number of required devices and the performance of each ATM switch.

(a) input buffer type

		input buffer type					
		in1	in2	in3	in4	in5	In6
	Num.of gate sw 32x32	n1xn2 1024	2SQR(n1) x n2 384	2xn1 64	(B+n2)xn1 1280 (B=8) 3072(B=64)	2SQR(n1) xn2+ n1xB 640(B=8) 2432(B=64)	Bxn1+ 2xn1 320(B=8) 2112(B=64)
	Num.of wavelengths	1	n1	n2	1	n1	n2
	Num.of TLDs	0	0	2xn1	0	0	2xn1
	Num.of O/Es	n1	n1	n1	n1	n1	n1
	Num. of electric FIFo buffers	n1	n1	n1	0	0	0
	Load at 10E-8 cell loss ratio	59%	59%	59% (when capacity of buffer is 64cell/line)	47% (B=8) 59% (B=64)	47% (B=8) 59% (B=64)	47% (B=8) 59% (B=64)
	multicast	△	△	△	△	△	△
	distributed control	×	×	×	×	×	×
	normalized evaluation						
*1	required system scale	54x54	54x54	54x54	68x68 (B=8)	68x68 (B=8)	68x68 (B=8)
*2	Num.of gate sw	2916	810	108	5168	1700	680
*3	Num.of TLDs	0	0	108	0	0	136
*4	Num. of electric FIFo buffers	54	54	54 (64cell/line, total capacity is 64x54 cells)	0	0	0
*5	Num.of wavelengths	1	54	54	1	68	68

(b) output buffer type

		output buffer type					
		out1	out2	out3	out4	out5	out6
	Num.of gate sw 32x32	n1xn2xn3 32768 (N3=32)	2SQR(n1) x n2 x n3 12288 (N3=32)	2xn2 64	n1x(Bxn2) 8192(B=8) 32768(B=32)	Bxn2+ 2SQR(n1) xn2 640(B=8) 2432(B=64)	2xn1+Bxn2+ 2SQR(n1) xn2 704(B=8) 2496(B=64)
	Num.of wavelengths	1	n1	n2	1	n1	n2
	Num.of TLD	0	0	2xn1	0	0	2xn1
	Num.of O/E	n1+ n2xn3	n1+ n2xn3	n1+ n2xn3	n1	n1	n1
	Num.of electric FIFo buffers	n2xn3	n2xn3	n2xn1	0	0	0
	Load at 10E-8 cell loss ratio	87%	87% (64cell/line)	>87% (64xn1 cell/line)	25%(B=8) 87%(B=64)	25%(B=8) 87%(B=64)	25%(B=8) 87%(B=64)
	multicast	○	○	△	○	○	△
	distributed control	×	×	○	×	×	○
	normalized evaluation						
*1	required system scale	37x37	37x37	37x37	128x128 (B=8) 37x37(B=64)	128x128 (B=8) 37x37(B=64)	128x128 (B=8) 37x37(B=64)
*2	Num.of gate sw	50653	17797	74	131072(B=8) 87616(B=64)	3968 (B=8) 2849 (B=64)	4224 (B=8) 2923 (B=64)
*3	Num.of TLDs	0	0	74	0	0	256
*4	Num.of electric FIFo buffers	1369	1369 (64cell/line,total 64x37cell)	1369 (total 64x1369)	0	0	0
*5	Num.of wavelengths	1	37	37	1	128 (B=8) 37 (B=64)	128 (B=8) 37 (B=64)

(c) shared buffer type, proposed type and electric developed switch

		shared buffer type				proposed	electric developed
		shared1	shared2	shared3	shared4	inout1	
	Num.of gate sw 32x32	n1xn3 +n3xn2 2048 (n3=32)	2SQR(n1) x (n3 + n2) 768(n3=32)	2xn1+ 2xn3 128(n3=32)	Bxn2+ 2SQR(n1) xn2 640(B=8) 2432(B=64)	Bxn2+ 2SQR(n1) xn2 640(B=8)	cross point sw for 10Gb/sx2 line speed
	Num.of wavelengths	1	n1 or n3	n1 or n3	n1	n1	7W/pcs
	Num.of TLDs	0	0	2xn1+2xn3	0	0	
	Num.of O/Es	n3	n3	n3	n1	n1	
	Num.of electric FIFo buffers	n3 RAM	n3 RAM	n3 RAM	0	n1	
	Load at 10E-8 cell loss ratio	88%	88%	88% (when total capacity of RAM is 8xn1 cells)	25%(B=8) 87%(B=64)	82% (opt: B=8, ele: 64cell)	
	multicast	○	○	△	○	△	△
	distributed control	×	×	×	×	×	×
	normalized evaluation						
*1	required system scale	37x37	37x37	37x37	128x128 (B=8) 37x37(B=64)	40x40 (B=8)	for 32x32 ATM sw, 1024 pcs. of this sw are needed
*2	Num.of gate sw	2738	962	148	3968 (B=8) 2849(B=64)	840	
*3	Num.of TLDs	0	0	148	0	0	
*4	Num. of electric FIFo buffers	37 RAM	37 RAM	37 RAM (total capacity : 8x37cells)	0	40 (total capa : 40x64cells)	
*5	Num.of wavelengths	1	37	37	128 (B=8) 37 (B=64)	40	

*1∼*5 for the same performance of the 32x32 switch with 10E-8 of cell lose ratio at 100% load

supposed to be included in TFLs in Fig. 5 (type in2).

In the model shown in Fig. 5 (type in3), cells from TLDs are sent to their destination output ports through an AWGF router according to their wavelengths. In this model, TLDs are required, while only a small number of optical gate switches are enough. The switches are included in TLD sources in Fig. 5 (type in3).

Next, we consider a model using optical delay line buffers as input buffers as shown in Fig. 5 (type in4). A cell from header converter is converted from electric to optical signals and distributed to delay lines. If there is no competing cell, the cell is read out from the delay line with 0 cell time delay which is the uppermost line in Fig. 5 (type in4). If there is a competing cell, the cell with low priority is not read in this time slot; at the next time slot, the cell is read out from one cell time delay line.

WD cell switching technologies are used in the configurations as shown in Fig. 5 (type in5, in6). In these cases, the number of gate switches are reduced compared with that of type in4.

Then, we will discuss whether the broadcast function and the distributed processing can be allowed. These are of practical interest because the former is crucial in providing multimedia services, and the latter enables us to design a switch in a rather easy way. In the input buffer type of switches, broadcast and multicast communication can be allowed in a way to copy the cell to be broadcasted and multicasted for each destination in the input module, while this results in the increase of the number of cells transmitted. Implementation of distributed processing is not easy because each input module has to communicate with the control unit for cell arbitration.

If the capacity of optical delay line buffer is 8, the throughput is not good (about 47% for 10^{-8}). Even if tap buffers with 64 stages are possible, a lot of gate switches are required for optical delay line buffer and only throughput of 58% will be obtained (see Table 1). Compared with the electric buffer type (type in1, in2 or in3), for the same capacity of 320 Gbit/s, a larger number of optical gate switch elements are required as shown in the 12th line of Table 1 (marked with *2). Because in input buffer type switches, optical gate switches have to be implemented in optical delay line buffer; on the other hand, additional optical gate switches or additional O/E converters are not required when electric buffers are implemented.

For the above reasons, in case of the input buffer type ATM switches, photonic ATM switches with electric buffers are better than photonic ATM switches with optical buffers.

3.2 Output Buffer Type Photonic ATM Switches

Figure 6 shows configuration models of output buffer type photonic ATM switches and performance of these switches.

In case of output buffer type with electric buffers

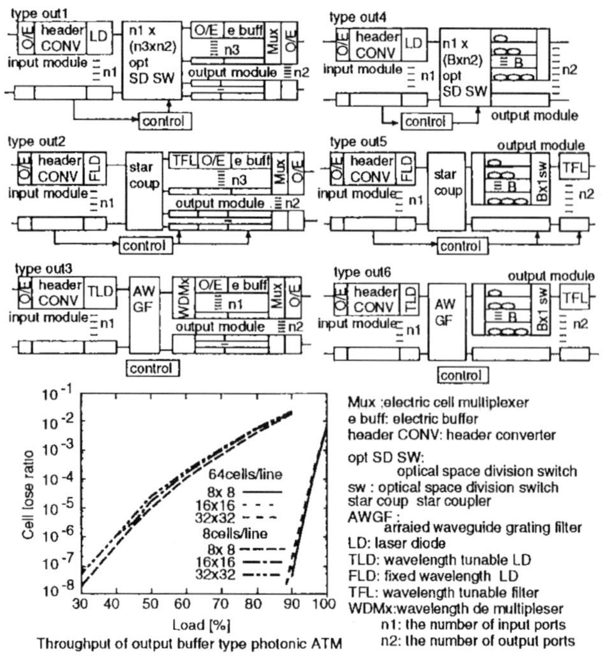

Fig. 6 Photonic ATM switches with output buffers.

as shown in Fig. 6 (type out1), after the electric header recognition and conversion in header converter, an input electric cell is converted to an optical cell and sent to an optical space division switch (opt SD SW). The cell is then sent to an O/E converter and further forwarded to an electric buffer connected to the destination output port. There is no problem due to output contention when less than n3 cells in input ports are sent to the same output port in a time. In many electric ATM switch implementations, to prevent cell loss in switching (routing) stage, n3 is equal to n1. In our throughput simulations, we assume $n1 = n2 = n3$. When there are cells in buffers connected to the same output port, the cell with a high priority is selected and forwarded to the output.

In case of type out2, an input cell is sent to all outputs of a star coupler, and then selected by the tunable wavelength filters (TFL) connected to the destination output port. In case of type out3, a cell from TLD is sent to a destination by AWGF router. In case of type out3, it is not possible to share the buffers in an output module. Therefore, type out3 needs larger capacity of each buffer to achieve the same performance.

Compared with the input buffer type, in achieving a throughput of 320 Gbit/s, the number of switch elements required for type out1 or type out2 is larger than that for type in1 or type in2, while the number of switch elements for type out3 is less than that of type in3.

The increase of the number of optical gate switches is not desirable because it in turn increases both power consumption and cost as well.

Another critical problem is related to the multiplexers (Mux) located on the outputs of buffers. Since the multiplexer is actually an $n3 \times 1$ high speed electric switch, it seems that it is not easy to implement. Type out1, out2 and out3 switches can be used at a load of 87% or less with the cell loss ratio of 10^{-8}.

Configurations of output buffer type photonic ATM switches with optical buffers are shown in Fig. 6 (type out4, 5, 6). In case of type out4 switch, if there is no competing cell, the cell is sent to the delay line with 0 cell time delay connected to destination output port. If there are cells to be sent to the same output port, the cell with high priority is sent to the delay line with 0 cell time delay, and the cell with low priority is sent to the delay lines with 1 cell time delay.

Type out5, 6 are called FRONTIER-NET [8], [10]; prototype photonic ATM switches, FRONTIER-NET, have been developed by Misawa et al. [10]. In case of type out5, all cells are distributed to all delay lines. The cell to be read out is selected from delay lines by a $B \times 1$ opt SD SW with WD multiplexed other cells in the same time slot; then, the cell to be read out is selected by the TFL among the multiplexed cells in the same time slot. In this configuration, multicast or broadcast function is easy to be implemented because cells are distributed to all outputs by the star coupler.

In case of type out6, distributed processing is possible. In each input module, header conversion and wavelength switching can be carried out independently. In each output module, buffering and arbitration can be carried out independently as well. There is no need to communicate with other input modules or other output modules.

Load cell-loss-ratio curves obtained by throughput simulations are also shown in Fig. 6. The photonic ATM switches with output buffers are able to have high throughput when a lot of buffers are equipped; however, throughput of ATM switches with a small capacity of buffers is very poor.

FRONTIER-NET has an excellent advantage like a multicast function or a distributed control; the problem is how to realize buffers with large capacity.

3.3 Shared Buffer Type Photonic ATM Switches

Figure 7 (type shared1, 2, 3) shows configuration models of shared buffer type photonic ATM switches with electric buffers. They are designed to resolve memory bus speed problems. Optical space division switching or wavelength division switching is used as a technique of switching bus from input modules to buffers or from buffers to destination outputs. The problems are how to realize high speed RAM buffers and how to realize a buffer management system. In our throughput simulation, we assumed $n1 = n2 = n3$, and buffers are not FiFos but RAMs. Obtained load cell-loss-ratio curves are shown in Fig. 7 as well. Cell loss ratio is excellent even with a small capacity of RAMs; about 88%

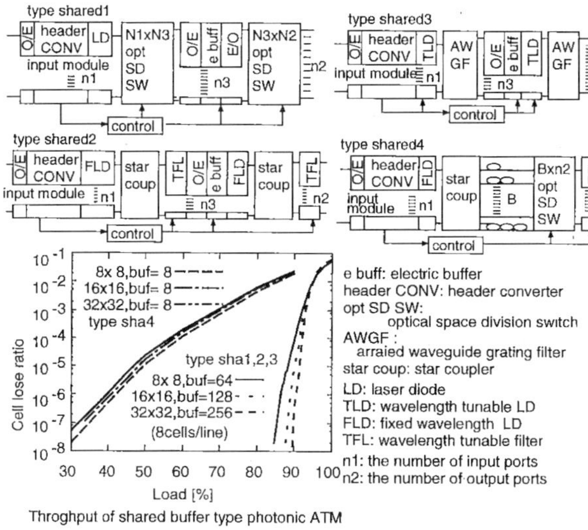

Fig. 7 Photonic ATM switches with shared buffers.

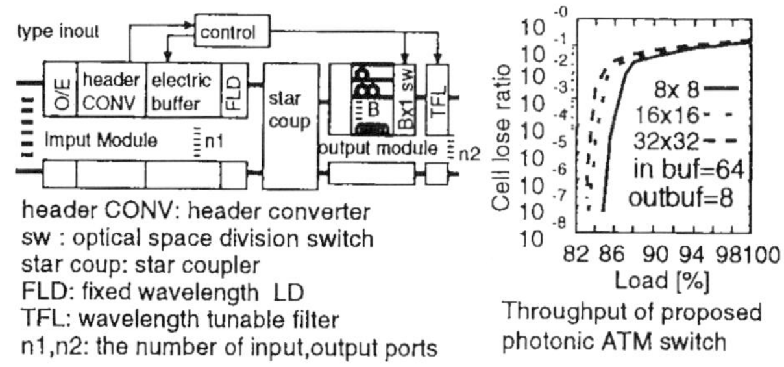

Fig. 8 Photonic ATM switches with input electric buffers and output optical buffers.

of maximum load can be achieved at cell loss ratio of 10^{-8} with 8 cells/line buffer.

The photonic ATM switch with optical delay line buffers shown in Fig. 7 (type shared4) has been proposed and developed by Chiaroni et al. [6]. Cells from input modules are distributed to all delay lines by the star coupler. If there is no contention cell, wavelength multiplexed cells in a 0 cell time delay line are sent to all TFLs being connected to destination outputs by an opt SD SW. Please note that cells can be sent to desirable plural output ports by the optical space division switch considered here (crossbar switch as shown in Fig. 3 (c)). When there are cells to be sent to the same output, the cell with higher priority is selected by the TFL in this time slot. In the next time slot, the cell with lower priority is sent to the output from one cell delay line and selected by TFL. In this ATM switch, optical delay line buffers are shared by all output lines, but, the capacity is not $B \times n1$ but B for each output port; the throughput is the same as that of type out4 switch with output buffers of B stages.

3.4 16 Types of Photonic ATM Switches, and a Proposed ATM Switch

In the above discussions, each photonic ATM switch has both advantages and disadvantages.

In this subsection, we discuss only the number of optical gate switches to be implemented. Optical ATM switches treated here are ones whose consumption electric power and cost are reduced with the number of optical gate switches.

Assumed WD switching technology needs a variety of device technologies, whereas, the number of devices must be reduced drastically as shown in Table 1. If tunable laser diode (TLD) technology is available, TLD and AWGF router technologies are very useful to reduce the number of optical gate switches compared with FLD and TFL technologies. As mentioned above, TLD technology needs more stable thermal control for LD than TFL technology.

If TLD technology is not available but FLDs and TFLs (with optical gates and WDMuxs) are available, the FLD and TFL technology can reduce the number of optical gate switches compared with the space division cell switching technology.

Next, we consider advantages and disadvantages of photonic ATM switches depending on the buffering styles. Photonic ATM switches with input buffer have a problem of HOL blocking. Whereas in case of input buffer type switches, a number of components like O/E converters or FiFos are not increased when electric buffer is implemented; it is an advantageous point. An advantageous point of output buffer type photonic ATM switches is to achieve high throughput. Output buffer type switches have a problem to realize optical buffers with large capacity, or a problem that a lot of O/E converters and a lot of FiFos are required for electric output buffers. Shared buffer type switches have a problem to realize both high speed RAM buffers and a high speed memory management system.

Therefore, we propose a configuration in which large capacity electric buffers are placed on the input side and small capacity optical delay line buffers are on the output side as shown in Fig. 8. The input electric buffers are equipped to provide large capacity buffers without increase of the number of O/Es and optical gate switches. The output optical delay line buffers are equipped for prevention of HOL blocking,

In the proposed configuration, cells from FLD are distributed to all outputs of the star coupler. When there is no competing cell, the cell for each destination is read out from the 0 cell time delay line by $B \times 1$ opt SD SW. A delay line buffer contains a number of wavelength multiplexed cells. Then, the cell to be output should be selected among them by TFL. If there are competing cells, the cell with lower priority is read out from the one cell time delay line on the next time slot. If there is no space in the output buffers, cells are buffered in the input electric buffer. As shown in the figure, the number of electric buffers and O/E converters are very small. Nevertheless the throughput performance is not bad; it works on 82% or less of traffic load with 10^{-8} of cell loss ratio.

With regard to the number of optical gate switches, if we can not use TLD technology but TFL technology, type inout shown in Fig. 8 requires the least number of optical gate switches with relatively a small number of required wavelengths. (see Table 1).

4. A Comparison with an Electric ATM Switching Technology

A very high performance electric ATM switch LSI has been developed [27]. It is a cross point switch for a ATM switch with 10 Gbit/s input line speed. To prevent HOL blocking problem, the switch operates at a speed of 20 Gbit/s. Consumption electric power of this LSI is 7 W/piece [26]. For 32×32 ATM switch, $32 \times 32 \times 7 + \alpha = 7168 + \alpha$ [W] is required. "$+\alpha$" is consumption electric power in a control part, input/output buffers, header converters, etc. In case of type inout in Fig. 8, for the same performance, switch size of 40×40, $40 \times (8 + 13) = 840$ pieces of optical gate switches and 40 pieces of FLD are required. In this system, the number of tap stages is 8, 13 pieces of optical gate switches are implemented on each TFL for 40 wavelengths (first stage 8 pieces, second stage 5 pieces). When an optical gate switch needs 245 mW (3.5 V, 70 mA), a switch driver needs 1.5 W, and a thermal control with a cooler needs 1.5 W, the system with the same performance totally needs $(0.245 + 1.5 + 1.5) \times 840 + \beta = 2727 + \beta$ [W]. "$+\beta$" is electric consumption on EDFAs ($10\,W \times 40$ lines $\times$ (2 or 3)), input buffers, header converters, O/Es, etc. Of course, electric and optical technologies are being made progress still now so that the comparison of both technologies at the moment may not be true in the near future. Nevertheless, our comparison shows that a photonic ATM technology can be competitive from some point of view as mentioned above.

5. Conclusion

In this paper, we discuss configurations of photonic ATM switches and their advantages with regard to the number of optical switching devices to be implemented on the system, the number of wavelengths, throughput, broadcast function etc.

With regard to device technologies to be implemented on photonic ATM switches, if possible, for cell switching, wavelength division switching technologies are desirable compared with space division switching technologies to make a photonic ATM switch with a less number of optical gate switches.

Our evaluation of switch configurations shows that each switch has both inherent advantages and disadvantages. Among switches examined here, with regard to the number of optical gate switch elements, our proposed configuration with optical delay line buffers on outputs and the electric buffers on inputs is superior to others under the given conditions.

References

[1] K.Y. Eng, "A photonic knockout switch for high speed packet networks," Proc. of GCOM'87, 47.2.1, pp.1861–1865, 1987.

[2] E. Amada, Y. Takahashi, and H. Inoue, "A photonic approach to ATM switching," Proc. of GCOM'89, pp.1910–1814, 1989.

[3] T. Matsunaga, "Photonic switching for ATM networks," Proc. of ISS'90, Stockholm, Sweden, no.poster18, pp.105–110, 1990.

[4] B. Bostica, P. Cinato, A.de Bosio, E. Garetti, and E. Vezzoni, "Electro optical ATM digital cross connect system based on cell aggregation and compression," Proc. of ISS'92, Yokohama, Japan, no.B10.3 pp.422–426, Oct. 1992.

[5] M. Nishio, S. Suzuki, K. Takagi, I. Ogawa, T. Numai, K. Kasahara, and K. Kaede, "Photonic ATM switch using vertical to surface transmission electro-photonic devices (VSTEPs)," Proc.of ISS'92, Yokohama, Japan, no.B10.4, pp.427–431, Oct. 1992.

[6] D. Chiaroni, P.G. Morin, P.A. Perrier, S. Ruggeri, S. Gauchard, D. Bouard, J.C. Jacquinot, P. Doussiere, M. Monnot, E. Grard, D. Laclerc, M. Sotom, J.M. Gabriagues, and J. Benoit, "Rack-mounted 2.5 Gbit/s ATM photonic switch demonstrator," Proc. of ECOC'93, no.ThP12.7, pp.77–80, 1993.

[7] T. Ozeki, "Ultra opt electronic devices for photonic ATM switching systems with Tera bits/sec throughput," IEICE Trans. Commun., vol.E77-B, no.2, pp.100–109, Feb. 1994.

[8] K. Sasayama, Y. Yamada, F. Fruh, K. Habara, K. Yukimatsu, S. Suzuki, and H. Ishii, "Demonstration of a photonic frequency routing type time division interconnection network FRONTIERNET and performance analysis of FDM output buffers," Proc. of ISS'95, Berlin, Germaney, no.C8.3, pp.452–456, June 1995.

[9] A. Misawa, Y. Yamada, M. Tsukada, K. Habara, and T. Matsunaga, "20 Gbit/s photonic ATM switching with differential packet receiver," Electron. Lett., vol.32, no.19, pp.1798–1799, Sept. 1996.

[10] T. Matsunaga, K. Habara, A. Misawa, Y. Yamada, T. Ogawa, M. Tsukada, S. Hino, K. Sasayama, and K. Yukimatsu, "Design of photonic ATM switch and rack mounted prototype," Proc. of ICC'97, Montreal, Canada, session 31.6, pp.1292–1297, June 1997.

[11] S. Kuroyanagi, K. Hironishi, T. Maeda, and A. Hakata, "Photonic ATM switches with optical buffers using wavelength division multiplexing," Proc. of PS'96, Sendai, Japan, no.PWC5, pp.94–95, April 1996.

[12] M.G. Hluchy and M.J. Karol, "Queueing in high performance packet switching," J. Selected Areas in Commun., vol.6, no.9, pp.1587–1597, Sept. 1988.

[13] R. Bubenik and J. Turner, "Performance of a broadcast packet switch," IEEE Trans. Commun., vol.37, no.1, Jan. 1989.

[14] J. Nishikido, M. Okuno, and A. Himeno, "Experiments with 4 bit optical pattern matching circuit," Proc. IEICE Spring Conf.'90, B-963, March 1990.

[15] D. Cotter, M.C. Tatham, J.K. Shabeer, K. Smith, D.C. Rogers, D. Nesset, and P. Gunning, "Photonic address header recognition and self routing in ultrafast packet networks," Proc. of PS'96, Sendai, Japan, no.PMA4, pp.8–9,

April 1996.
[16] J. Mizusawa, "Foreword of special issue on photonic switching technologies," IEICE Trans. Commun., vol.E77-B, no.2, p.95, Feb. 1994.
[17] M. Koga, Y. Hamazumi, A. Watanabe, S. Okamoto, and S. Suzuki, "Design and performance of an optical path cross connect system based on wavelength path concept," J. Lightwave Technol., vol.14, no.6, pp.1106–1119, June 1996.
[18] Y. Tachikawa, Y. Inoue, M. Ishii, and T. Nozawa, "Arrayed waveguide grating multiplexer with loop back optical paths and its applications," IEEE J. Lightwave Technol., vol.14, no.6, pp.977–983, June 1996.
[19] S. Takahashi, T. Kato, H. Takeshita, S. Kitamura, and N. Henmi, "10 Gbps/ch SD optical cell switching with 8 × 8 gate type switch matrix employing gate turn on delay compensator," Proc. of PS '96, Sendai, Japan, no.PTHC1 (post deadline), pp.12–15, April 1996.
[20] T. Ido, M. Koizumi, S. Tanaka, M. Suzuki, and H. Inoue, "Polarization and wavelength insensitive MQW electro absorption optical cate switches for WDM applications," Proc. of PS '96, Sendai, Japan, no.PWB4, pp.80–81, April 1996.
[21] Y. Yoshikuni, "Tunable laser dynamics and their realization," Proc. of PS '96, Sendai, Japan, no.PWC1, pp.26–27, April 1996.
[22] P. Gambini, G. Bendelli, B. Bostica, and M. Puleo, "Photonic packet switching based on fiber memories," Proc. of PS '96, Sendai, Japan, no.PMA1, pp.2–3, April 1996.
[23] T. Okugawa, Y. Yamada, K. Habara, and T. Matsunaga, "Reduction of optical coherent crosstalk induced degradation by manchester cording," Proc. IEICE Spring Conf. '97, Osaka, Japan, B-6-28, p.28, March 1997.
[24] Y. Oie, T. Suda, M. Murata, and H. Miyahara, "Survey of switching techniques in high-speed networks and their performance," International Journal of Satellite Communications, vol.9., no.5, pp.285–303, Sept.-Oct. 1991.
[25] N. Arakawa, A. Noiri, and H. Inoue, "ATM switch for multimedia switching system," Proc. of ISS '90, Stockholm, Sweden, no.A7, vol.5, pp.9–14, May 1990.
[26] S. Kawamura, H. Ichino, M. Suzuki, K. Genda, and Y. Doi, "An over 20 Gb/s throughput Si bipolar ATM crosspoint switch LSI," Proc. IEICE Spring Conf. '94, B-548, March 1994.
[27] N. Yamanaka, "Interconnection and packaging technologies in ultra high speed ATM switching system," Proc. of PS '96, Sendai, Japan, no.PTuA2, pp.44–45, April 1996.

Appendix I

Exactly, R_{rnet} is calculated as follows.

$$R_{rnet} = \frac{1}{1 - \left(1 - \dfrac{1}{R_{sw}}\right)^m}$$

where $m = 2\log_2 n - 1$.

Appendix II

From paper [23] written by Okugawa, optical power detected in receiver is calculated as follows.

$$I = S + 2S\sqrt{\xi}\cos(\theta_s - \theta_c) + \xi S \qquad \text{(A· 1)}$$

I: optical power detected in receiver
where the electric field of signal is $\sqrt{S}\cos(\omega t + \theta_s)$, the electric field of crosstalk is $\sqrt{\xi S}\cos(\omega t + \theta_c)$.

The second term of (A· 1) is related to a coherent crosstalk, which can increase bit error ratio. For example, when $\xi = 0.01$, the first term of (A· 1) is S and the second term of (A· 1) is $0.2S\cos(\theta_s - \theta_c)$.

High-Performance Backbone Network Technology edited by Naoaki Yamanaka
ISBN: 0-8247-5321-6 DOI: 10.1081/(120025735)
Published 2004 by Marcel Dekker, Inc.

A High-Performance Switch Architecture for Free-Space Photonic Switching Systems

Shigeo Urushidani, Masayasu Yamaguchi, and Tsuyoshi Yamamoto

SUMMARY Design and evaluation of a high-performance switch architecture for free-space photonic switching systems is described. The switch is constructed of 2×2 switching elements and employs special multistage interconnection patterns. The connection setup algorithm and the control procedure at the switching elements are based on a "rerouting algorithm." Performance analysis shows that the blocking probability of the switch is easily controlled by increasing the number of switching stages. Example implementations of this switch are shown in which birefringent plates, polarization controllers, etc. are used.
key words: *free-space photonic switching, multistage interconnection, self-routing, rerouting*

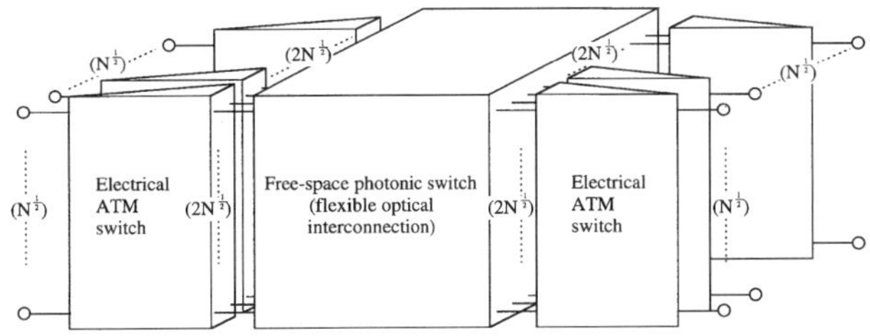

Fig. 1 Example architecture of very-large-scale ATM switch.

1. Introduction

Free-space photonic switching technology is considered very promising for compact packaging of large-scale photonic switching systems and cross-connect systems [1]-[6] and for flexible two-dimensional optical interconnection in large-scale ATM switching systems [7]. Recently we have become interested in its application to ATM switching systems of one Tera-bit/s to several tens of Tera-bit/s. Figure 1 shows an example of ATM switch architecture for this purpose. The switch has multiple electrical ATM switches used as sub-switches and a free-space photonic switch used as a flexile optical interconnection fabric between the electrical ATM switches. The free-space photonic switch has a greater number of links (four times; see Fig. 1) than the total number of input links to the sub-switches, to enhance switching performance without link speedup. For example, a 1024×1024 free-space photonic switch is required for a 256×256 ATM switch. When the link speed is 10 Gbit/s, the switch has a capacity of 2.5 Tbit/s. For more capacity, a larger-scale photonic switch is required. The free-space photonic switch also changes the connection patterns between the sub-switches according to the traffic flow and volume. It should change the connection patterns between the sub-switches without influencing the existing connections and should have a low blocking probability for new connection setup requests.

Several switch architectures have been proposed for free-space switching. The crossbar-type switch and the three-stage Clos switch perform adequately, but they require a very large number of switching stages [2]. The banyan network requires a very small number of stages and can employ a simple connection setup algorithm but has a high probability of blocking especially for large switch sizes. Extra-stage architectures [3], [4] and parallel architectures [5] have been proposed to alleviate the problem, but these architectures require undesirable connection rearrangement or complicate the switching control and packaging.

This paper proposes a high-performance switch architecture that attains a low blocking probability with a small number of stages for arbitrary switch sizes. The performance is enhanced by a novel connection setup algorithm applied to a special multistage interconnection network. The switch is constructed of 2×2 switching elements for an arbitrary switch size, enables a "rerouting" within the switch [9]-[12], and embeds bypass links toward the outputs. The switch can perform point-to-multipoint connection as well as point-to-point connection. Performance analysis shows that the blocking probability for an arbitrary switch size can be controlled easily by increasing the number of switching stages. We also show that this switch can be implemented with a combination of birefringent plates, polarization controllers, EARS [8], and so on.

2. Switch Design

2.1 Switch Architecture

Figure 2 shows our proposed switch architecture. In this figure, the switch size N $(= 2^n)$ is 16 and the total number of switching stages m is 13, where m can be arbitrary. This switch is composed of routing links, bypass links, switching elements, and dropping elements.

To connect a specified input line to a specified output line, this switch holds routing links until the connection

Manuscript received June 21, 1998.
Manuscript revised September 21, 1998.
† The author is with NTT Network Service Switching Systems Laboratories, Musashino-shi, 180-8585 Japan.
†† The authors are with NTT Optical Network Systems Laboratories, Yokosuka-shi, 239-0847 Japan.
* This paper is also published in IEICE Trans. Electron., Vol. E82-C, No. 2, pp. 246-253, February 1999.

IEICE TRANS. COMMUN., Vol. E82-B, No. 2, pp. 298–305, February 1999

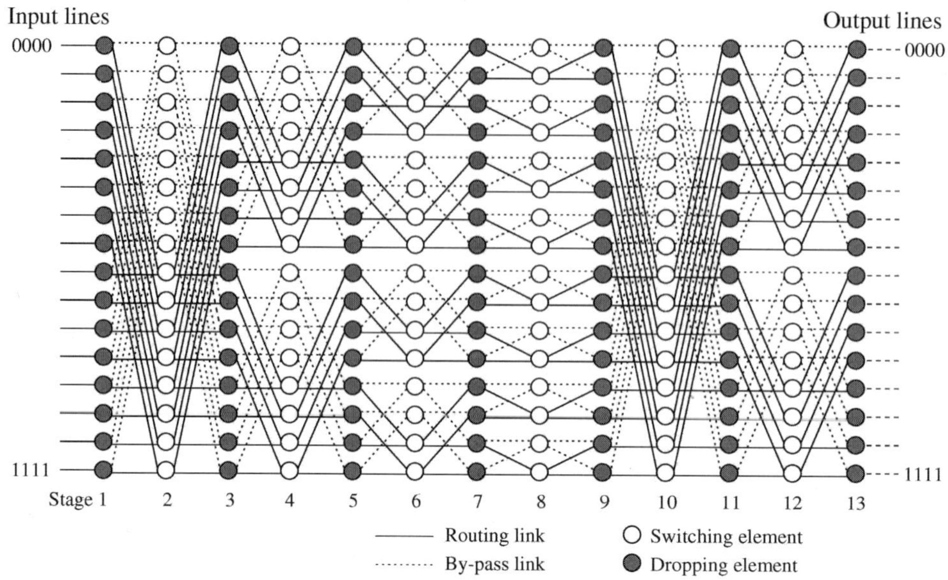

Fig. 2 Proposed switch architecture.

reaches the specified position and holds bypass links from the stage that the connection reaches the specified position. If a desired output routing link has been occupied by another connection, i.e., an internal conflict occurs, the connection holds the opposite routing link and restarts the connection setup from the next stage. The connection setup can be performed in the same way no matter how many times the connection experiences internal conflict.

Each switching element for routing links exchanges the routing links. Each switching element for bypass links is usually set to the bar state. Each dropping element connects the input routing link to either the output bypass link or the output routing link.

When an inlet or an outlet position is numbered in binary $(p_n \cdots p_1 p_0)$ regardless of routing link or bypass link, the interconnection rule between the adjacent stages $(2k-1)$ and $(2k)$ $(k \geq 1)$ or stages $(2k)$ and $(2k+1)$ is represented by the following mapping rules X_k.

$$X_k[(p_n \cdots p_1 p_0)] = (p_n \cdots p_{j+1} p_0 p_{j-1} \cdots p_1 p_j)$$

where $j = n - (k-1)\text{mod}(n)$

This interconnection rule is suitable for two-dimensional arrangement of the switching elements, because the link length between the adjacent stages can be drastically shorter. Figure 3 is an example of a two-dimensional arrangement of the switching elements for switch size $N = 16$. This shows that the maximum link length can be reduced to a distance of two switching elements, while the one-dimensional arrangement requires a link length distance of eight switching elements. As the switch size becomes larger, the effect of the two-dimensional arrangement of the switching elements becomes larger. For example, a 1024×1024 switch requires only the maximum link length distance of 16 switching elements.

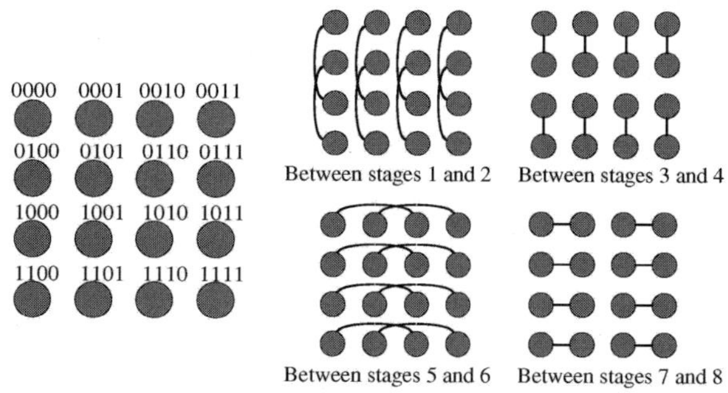

Fig. 3 Interconnection patterns in the proposed switch.

2.2 Connection Setup Algorithm

In this switch, the internal links can be held stage by stage like a self-routing switch. The control information for connection setup consists of the destination address of $(d_n \cdots d_1)$ in binary and the remaining number of stages (abbreviated to RNS). The RNS indicates the number of stages through which the connection should pass to reach the specified position. The initial value of RNS at an input line is given as follows. When a source address is represented by $(s_n \cdots s_1)$ in binary, if $(s_h \cdots s_1) = (d_h \cdots d_1)$ $(1 \leq h \leq n)$ the initial RNS is set to $(n-h)$; otherwise the value is set to n. The switching algorithms for the switching elements and dropping elements are as follows.

(1) Dropping element

This element at stage $(2k-1)$ $(k \geq 1)$ connects the routing inlet to the routing outlet if $RNS \geq 1$ or to the bypass outlet if $RNS = 0$. The bypass inlet, which includes the active signal, is connected to the bypass outlet. The *RNS* retains the same value at this switching element.

(2) Switching element for routing links

This element at stage $2k$ $(k \geq 1)$ connects the inlet to the upper outlet if $d_j = 0$ or to the lower outlet if $d_j = 1$, where $j = n - (k-1)\mathrm{mod}(n)$. Then *RNS* is decreased by one. If an internal conflict occurs, i.e., another connection has occupied the desired outlet, this element connects the inlet to the opposite outlet and resets the *RNS* to n.

(3) Switching element for bypass links

This element at stage $2k$ ($k \geq 1$) always connects an inlet to the same position outlet. Thus the switching elements are kept in the bar state.

2.3 Connection Setup Examples

Figures 4(a) and (b) show connection setup examples and clarify how the switch works. Figure 4(a) is an example of connection from input line (0000) to output line (1010). The thick solid line represents the route where no internal conflict occurs and the thick broken line represents the route where a conflict for the outlet occurs at stage 2. The connection information is initially composed of a destination address $(d_4d_3d_2d_1) = (1010)$ and an *RNS* of 3. The *RNS* has been decreased by one from 4 because the least significant bits are the same between the source address (0000) and the destination address (1010).

The route toward the destination is determined as follows for the case of the thick solid line.

(Stage 1) To the routing link because $RNS = 3$ (≥ 1).
(Stage 2) To the lower routing link because $d_4 = 1$ ($j = 4 - 0\mathrm{mod}4 = 4$) and decrease *RNS* by one.
(Stage 3) To the routing link because $RNS = 2$ (≥ 1).
(Stage 4) To the upper routing link because $d_3 = 0$ ($j = 4 - 1\mathrm{mod}4 = 3$) and decrease *RNS* by one.
(Stage 5) To the routing link because $RNS = 1$ (≥ 1).
(Stage 6) To the lower routing link because $d_2 = 1$ ($j = 4 - 2\mathrm{mod}4 = 2$) and decrease *RNS* by one.
(Stage 7) To the bypass link because $RNS = 0$.

From stage 7 on, the route is kept on the bypass links.

Next consider the route for the case of the broken lines.

(Stage 2) To the upper routing link due to the conflict for the lower outlet and reset *RNS* to 4.
(Stage 3) To the routing link because $RNS = 4$ (≥ 1).
(Stage 4) To the upper routing link because $d_3 = 0$ ($j = 4 - 1\mathrm{mod}4 = 3$) and decrease *RNS* by one.
(Stage 5) To the routing link because $RNS = 3$ (≥ 1).
(Stage 6) To the lower routing link because $d_2 = 1$ ($j = 4 - 2\mathrm{mod}4 = 2$) and decrease *RNS* by one.
(Stage 7) To the routing link because $RNS = 2$ (≥ 1).
(Stage 8) To the upper routing link because $d_1 = 0$ ($j = 4 - 3\mathrm{mod}4 = 1$) and decrease *RNS* by one.
(Stage 9) To the routing link because $RNS = 1$ (≥ 1).
(Stage 10) To the lower routing link because $d_4 = 1$ ($j = 4 - 0\mathrm{mod}4 = 4$) and decrease *RNS* by one.
(Stage 11) To the bypass link because $RNS = 0$.

From stage 11 on, the route is kept on the bypass links.

This switch can perform point-to-multipoint connection using a similar connection setup algorithm. The connection information is composed of the destination addresses, the RNSs, and the multipoint group number (abbreviated to MGN). Figure 4(b) shows a point-to-multipoint connection example from input line (0000) to output lines (0000), (0100), (1010), and (1111). First the connection from (0000) to (0000) is established. The route is kept on the bypass links and the dropping elements remember the MGN. Next the connection from (0000) to (0100) is established. The dropping element at stage 1 copies the input signal to the routing outlet by receiving the connection setup request with the same MGN as the existing connection. From stage 2 the route is different from the existing connection, so the switching elements and dropping elements on the route remember the MGN. For the connection from (0000) to (1010), the existing route is split at the switching element at stage 2. For the connection from (0000) to (1111), the existing route is split at the switching element at stage 4. Note that in the case of the connection to the contiguous addresses, the address interval and its splitting algorithm [12] can be employed.

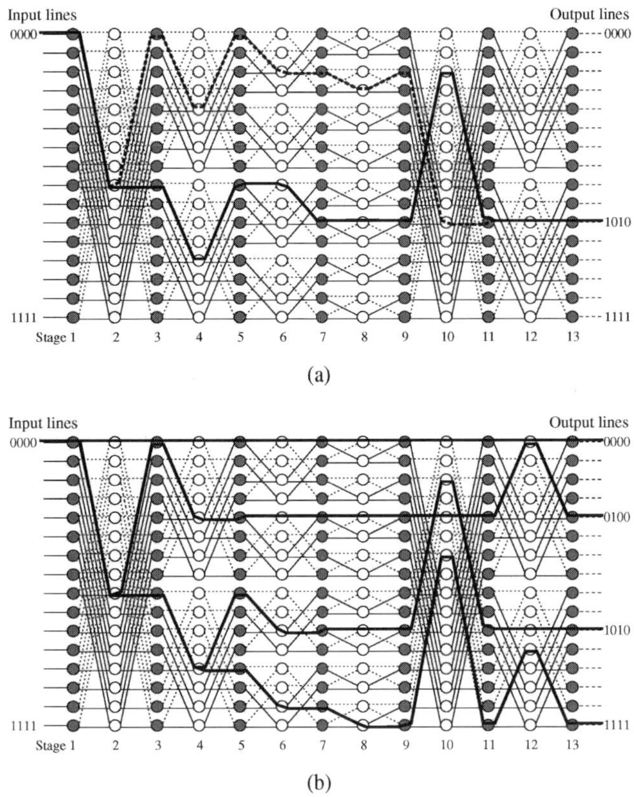

Fig. 4 (a) Example of point-to-point connection setup. (b) Example of point-to-multipoint connection setup.

3. Switch Performance

The performance of the proposed switch is enhanced as the number of stages is increased, so the relationship between performance, especially the blocking probability, and the number of switching stages is studied.

3.1 Performance Analysis

Let us assume that the existing connections have been uniformly set up, i.e., internal loads have been uniformly distributed within the switch. We then calculate the existing loads by setting all the connections simultaneously. When a conflict occurs between the connections, we give preference to the connection that has the smallest RNS.

For the offered load on an input line we define two symbols, as follows.

ρ_0: Total offered load on an input line.

$\rho_{0p}(d, r)$: Offered load of input line p $(0 \le p \le N-1)$ toward output line d $(0 \le d \le N-1)$ with an RNS of r $(0 \le r \le n)$.

For the internal loads at stage t, some symbols for routing links are defined as follows, where position p is numbered from the top.

$\rho_{tip}(d, s, r)$: Load of inlet p toward output line d from input line s with an RNS of r

$\rho_{top}(d, s, r)$: Load of outlet p toward output line of d from input line s with an RNS of r

$\sigma_{tip}(0, r)$: Sum of loads of inlet p with an RNS of r, aimed toward the upper outlet.

$\sigma_{tip}(1, r)$: Sum of loads of inlet p with an RNS of r, aimed toward the lower outlet.

$dd(z,q,y)$: Destination address, determined by z $(0 \le z \le 2^{n-j}-1)$, q $(= 0, 1)$, and y $(0 \le y \le 2^{j-1}-1)$ and expressed by $z_n \cdots z_{j+1} q y_{j-1} \cdots y_1$ in binary, where $j = n - (k-1)\mathrm{mod}(n)$, $z = z_n \cdots z_{j+1}$, and $y = y_{j-1} \cdots y_1$.

At a switching element of stage $(2k-1)$ $(k \ge 1)$, the load for the routing outlet is obtained as follows.

$$\rho_{(2k-1)op}(d, s, 0) = 0$$

$$\rho_{(2k-1)op}(d, s, r) = \rho_{(2k-1)ip}(d, s, r) \quad (r \ge 1) \tag{1}$$

Next consider a switching element at stage $2k$ $(k \ge 1)$ that has two routing inlets and outlets labeled $2c$ and $(2c+1)$ $(0 \le c \le 2^{n-1}-1)$.

The load that becomes $RNS = 0$ at stage $2k$ and reaches the specified position dd_q $(q = 0, 1)$ at stage $2k+1$, $\rho_{(2k)o(2c+q)}(dd_q, s, 0)$ is obtained from the input loads with $RNS = 1$ as follows.

$$\rho_{(2k)o(2c+q)}(dd_q, s, 0) = \sum_{x=0}^{1} \rho_{(2k)i(2c+x)}(dd_q, s, 1) \tag{2}$$

The loads toward an address $dd(z, q, y)$ $(0 \le z \le (2^{n-j}-1)$, $q = 0, 1$, $0 \le y \le (2^{j-1}-1))$ with an RNS of $(r-1)$ $(2 \le r \le n)$ and leave stage $2k$ are calculated as follows.

$$\rho_{(2k)o(2c+q)}(dd(z,q,y), s, r-1) = \sum_{x=0}^{1}\Big\{\rho_{(2k)i(2c+x)}(dd(z,q,y), s, r) \cdot \Big(1 - \sum_{h=1}^{r-1} \sigma_{(2k)i(2c+1-x)}(q,h) - \frac{1}{2}\sigma_{(2k)i(2c+1-x)}(q,r)\Big)\Big\} \tag{3}$$

where

$$\sigma_{(2k)i(2c+q)}(q,1) = \sum_{s=0}^{N-1} \rho_{(2k)i(2c+q)}(dd_q, s, 1)$$

$$\sigma_{(2k)i(2c+q)}(q,r) = \sum_{s=0}^{N-1} \sum_{z=0}^{2^{n-j}-1} \sum_{y=0}^{2^{j-1}-1} \rho_{(2k)i(2c+q)}(dd(z,q,y), s, r)$$

$$(r \ge 2)$$

The loads that have their RNS reset to n at stage $2k$ are calculated as follows.

$$\rho_{(2k)o(2c+q)}(dd(z,1-q,y), s, n) = \sum_{x=0}^{1}\Big\{\sum_{r=2}^{n} \rho_{(2k)i(2c+x)}(dd(z,1-q,y), s, r) \cdot \Big(\sum_{h=1}^{r-1} \sigma_{(2k)i(2c+1-x)}(1-q,h) + \frac{1}{2}\sigma_{(2k)i(2c+1-x)}(1-q,r)\Big)\Big\} \tag{4}$$

Then we obtain the total outlet loads at stage $2k$, $\sigma\sigma_{(2k)o(2c+q)}$ from Eqs. (1) to (4).

$$\sigma\sigma_{(2k)o(2c+q)} = \sum_{s=0}^{N-1} \rho_{(2k)o(2c+q)}(2c+q, s, 0) + \sum_{s=0}^{N-1} \sum_{r=2}^{n} \sum_{z=0}^{2^{n-j}-1} \sum_{y=0}^{2^{j-1}-1} \rho_{(2k)o(2c+q)}(dd(z,q,y), s, r-1) + \sum_{s=0}^{N-1} \sum_{z=0}^{2^{n-j}-1} \sum_{y=0}^{2^{j-1}-1} \rho_{(2k)o(2c+q)}(dd(z,1-q,y), s, n) \tag{5}$$

Consider the setup of a new connection on the load condition of Eq. (5). When a connection between input line a and output line b is set up, the existing internal loads are calculated by eliminating the loads from input line a and the loads toward output line b. The loads, $\sigma\sigma_{(2k)o(2c+q)}(-b, -a)$, are given as follows.

$$\sigma\sigma_{(2k)o(2c+q)}(-b,-a) = \sigma\sigma_{(2k)o(2c+q)} - \sum_{d=0}^{N-1} \sum_{r=0}^{n} \rho_{(2k)o(2c+q)}(d, a, r) - \sum_{s \ne a} \sum_{r=0}^{n} \rho_{(2k)o(2c+q)}(b, s, r) \tag{6}$$

As a result, the probabilities that the new connection from a to b gets the RNS of $r-1$ $(1 \le r \le n)$ at stage $2k$, $P_{(2k)o(2c+b_j)}(b,a,r-1)$, and the probabilities that it gets the RNS of n at stage $2k$, $P_{(2k)o(2c+1-b_j)}(b,a,n)$, are obtained as follows, where, $b = b_n \cdots b_j \cdots b_1$.

$$P_{(2k)o(2c+b_j)}(b,a,r-1) = \sum_{x=0}^{1} P_{(2k)i(2c+x)}(b,a,r) \cdot \left(1 - \sigma\sigma_{(2k)o(2c+b_j)}(-b,-a)\right)$$

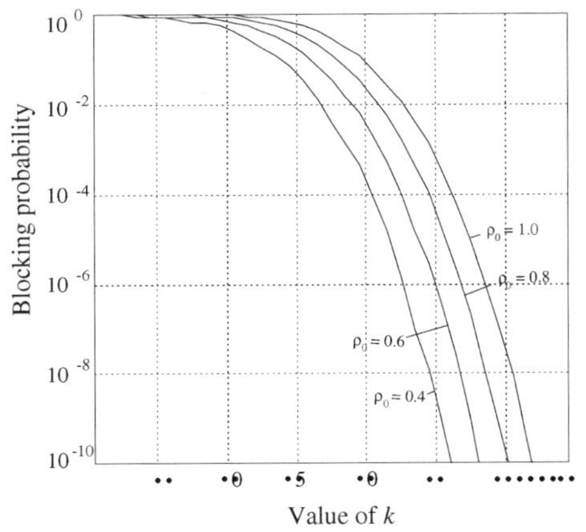

Fig. 5 Blocking probability for switch size $N = 64$.

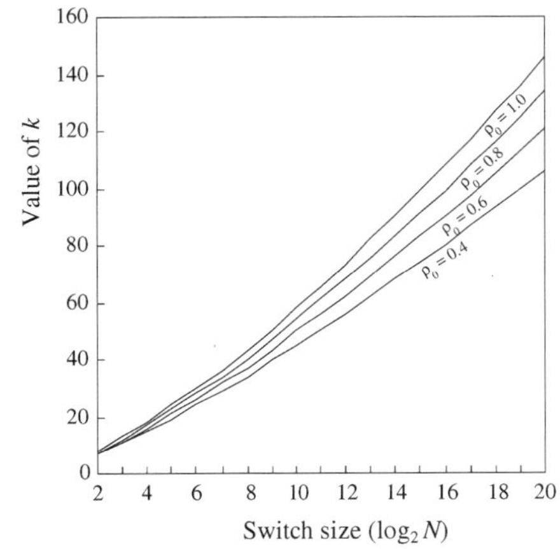

Fig. 6 Required number of stages.

$$P_{(2k)o(2c+1-b_j)}(b,a,n) = \sum_{x=0}^{1}\sum_{r=1}^{n} P_{(2k)i(2c+x)}(b,a,r)\cdot\sigma\sigma_{(2k)o(2c+b_j)}(-b,-a) \tag{7}$$

Accordingly, the blocking probabilities, P_{loss}, are given by the following equation:

$$P_{loss} = \frac{\sum_{b=0}^{N-1}\sum_{a=0}^{N-1}\sum_{r=1}^{n} P_{mob}(b,a,r)}{N^2} \tag{8}$$

where m is the last stage number.

Using Eqs. (1) through (8), the performance is evaluated for $N = 64$. Figure 5 shows the blocking probabilities against the value k of $(2k–1)$ for different loadings. The offered load varied from 0.6 to full loading. The blocking probabilities decreases slowly at small values of k but sharply at large values of k.

3.2 Number of Stages Required

The number of stages required for a given blocking probability for any switch size is investigated. The previously mentioned analysis gives accurate values but the number of variables becomes very large as the switch size increases. A simplified analysis is therefore used to calculate the number of stages. The simplified method gives good approximate values, as shown in the appendix.

The simplified analysis ignores the difference between the parameters p, d, and s. When $\rho_{ti}(r)$ and $\rho_{to}(r)$ are defined as the loads with $RNS = r$ of inlet and outlet at stage t, the loads that leave stage $2k$ and have $RNS = 0$, $r-1$ $(2 \le r \le n)$, and n are derived as follows.

$$\rho_{(2k)o}(0) = \rho_{(2k)i}(1)$$

$$\rho_{(2k)o}(r-1) = \rho_{(2k)i}(r)\cdot\left\{1-\frac{1}{2}\sum_{h=1}^{r-1}\rho_{(2k)i}(h)-\frac{1}{4}\rho_{(2k)i}(r)\right\}$$

$$\rho_{(2k)o}(n) = \sum_{r=2}^{n}\rho_{(2k)i}(r)\cdot\left\{\frac{1}{2}\sum_{h=1}^{r-1}\rho_{(2k)i}(h)+\frac{1}{4}\rho_{(2k)i}(r)\right\} \tag{9}$$

From this the probabilities that a new connection gets $RNS = r-1$ $(2 \le r \le n)$ or n are derived as follows.

$$P_{(2k)o}(r-1) = P_{(2k)i}(r)\cdot\left\{1-\frac{1}{2}\sum_{h=0}^{n}\rho_{(2k)o}(h)\right\}$$

$$P_{(2k)o}(n) = \sum_{r=1}^{n}P_{(2k)i}(r)\cdot\left\{\frac{1}{2}\sum_{h=0}^{n}\rho_{(2k)o}(h)\right\} \tag{10}$$

Accordingly, the blocking probability, P_{loss}, is given in the following equations.

$$P_{loss} = \sum_{r=1}^{n} P_{mo}(r) \tag{11}$$

where m is the last stage number.

Figure 6 shows the value of k required for a 10^{-8} blocking probability with the offered load from 0.4 to full loading. It can be seen that the relationship between the required number of stages and switch size n $(= \log_2 N)$ is nearly linear. This means that the required number of stages is of $O(\log_2 N)$ for a switch size N. This switch can therefore attain high performance with acceptable hardware overhead regardless of its size.

4. Implementation Examples

As mentioned in the previous section, the proposed switch is effective for two-dimensional arrangement. For example, switching stages 1 to 5 in Fig. 2 can be transformed into a four-plane structure as shown in Fig. 7(a). Each plane has a five-stage, four-input, four-output structure. Switching stages 5 to 9 can also be transformed into a four-plane structure as shown in Fig. 7(b). Thus the switch can be constructed by cascading these planes alternately until the required number of stages is obtained.

Figure 8 shows an example implementation of the structure of Fig. 7(a). Each switching element for bypass links is separated into two elements in this implementation, as the element always stays in the bar state. The switching stages are composed of birefringent plates and polarization controllers [2]. Optical beams with orthogonal polarization (an ordinary beam and an extraordinary beam) and birefringent plates are used to obtain the interconnection patterns. The ordinary beam proceeds straight into the birefringent plates and the extraordinary beam proceeds at a slanting into the birefringent plates. The width of the birefringent plate can change the position that the angled beam reaches. A 2×2 switching operation is obtained by the rotation (0 or 90 degrees) of the polarization state of the input beam at the polarization controller. The 0-degree polarization rotation maintains the polarization states, while the 90-degree polarization rotation changes the polarization states, i.e., from ordinary to extraordinary or vice versa. This changes the direction of beam propagation in the rear birefringent plate. Figure 9 shows a view of the packaging of the switch with 13 switching stages, 16 inputs, and 16 outputs. Using these optical devices, we have previously obtained fine experimental results for an 8-stage, 1024×256 concentrator [2], so we believe that a relatively large photonic switch can be created in the near future.

The proposed switch can also use signal regeneration devices such as the exciton absorption reflection switch (EARS) to produce a digital free-space switch [1], [8]. An example is shown in Fig. 10. Birefringent plates and patterned half-wave plates are employed to obtain the interconnection pattern. In the birefringent plate, a circularly polarized input beam is divided into two beams: an s-polarized beam that proceeds straight and a p-polarized beam that is angled. The polarization of beams incident on the gray-colored part of the patterned half-wave plate is converted from p to s or vice versa. An EARS array operates as an optically controllable, one-input, one-output gate switch and has a level regeneration function for digital-signal operation [8]. An s-polarized control light beam enters a polarizing beam splitter (PBS) from the bottom, is reflected by the PBS, and is then modulated and reflected by the EARS. Using these devices, we have previously obtained experimental results for a two-stage 16×16 switch (four sets of 4×4 switches) [1]. Although the devices require future enhancement, especially in operating speed, we believe that these signal regeneration technologies

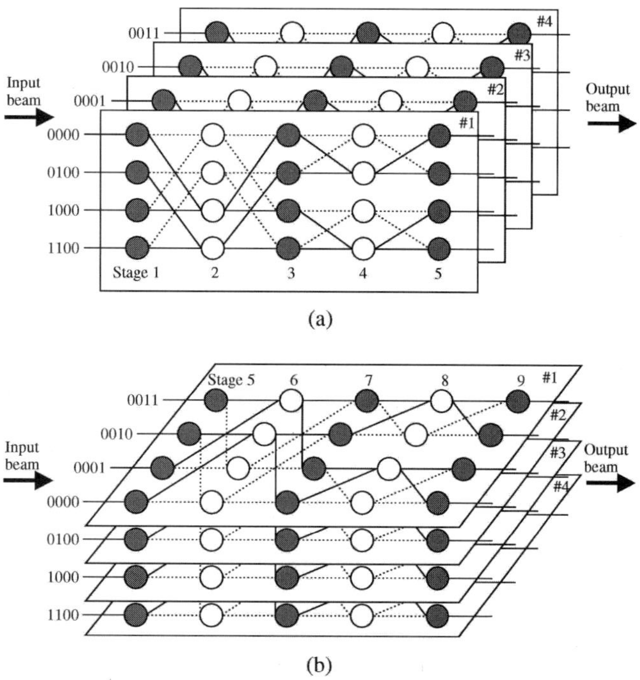

Fig. 7 (a) Four-plane structure of stages 1 to 5. (b) Four-plane structure of stages 5 to 9.

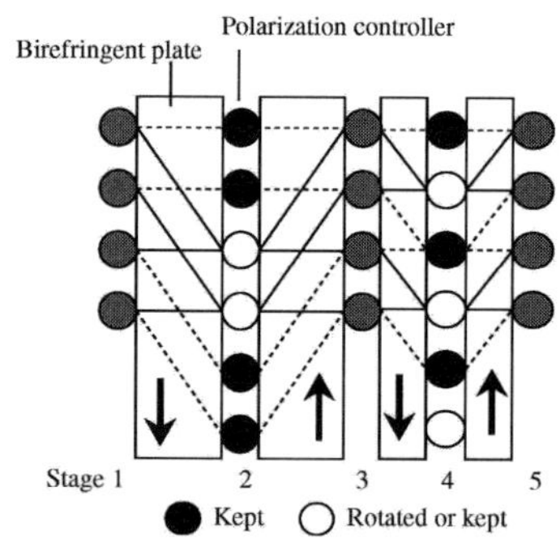

Fig. 8 Optical implementation using birefringent plates.

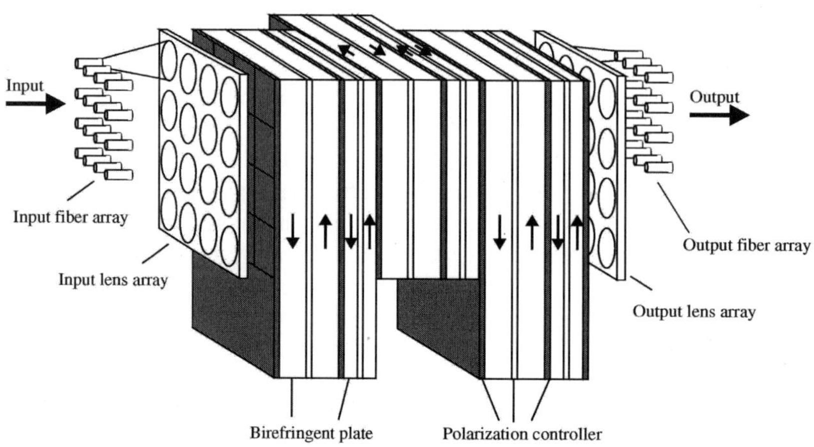

Fig. 9 Packaging of the proposed switch.

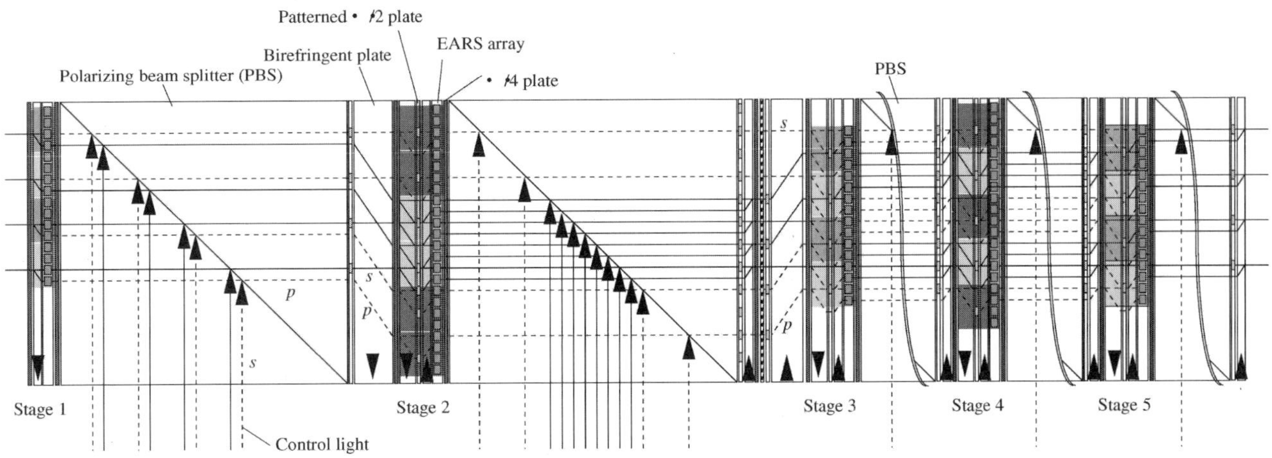

Fig. 10 Optical implementation using EARS.

open the door to the development of a full-scale over-Terabit/s ATM switch.

5. Conclusion

We have described a high-performance switch architecture for free-space photonic switching systems. The switch is constructed of 2×2 switching elements and employs a rerouting algorithm for a connection setup within special multistage interconnection patterns. Performance evaluation of the switch through analysis showed that its blocking probability is easily controlled by increasing the number of switching stages and that the switch can attain high performance with acceptable hardware overhead regardless of its size. Implementation examples with various combination of birefringent plates, polarization controllers, and EARS, were shown and indicated the possibility of producing a relatively large photonic switch.

Acknowledgments

We thank Dr. Tatsuro Takahashi and Dr. Kou Miyake of the NTT Network Service Systems Laboratories and Dr. Kenichi Yukimatsu of NTT Network Innovation Project and Mr. Tohru Matsunaga of the NTT Optical Network Systems Laboratories.

References

[1] M.Yamaguchi, T.Yamamoto, K.Hirabayashi, and K.Koyabu, "High-density digital free-space photonic-switching fabrics using exciton absorption reflection-switch (EARS) arrays and microbeam optical interconnections," IEEE J. Select. Topics Quantum Elec., vol.2, no.1, pp.47-53, April 1996.

[2] M.Yamaguchi, T.Matsunaga, S.Shirai, and K.Yukimatsu, "Analog free-space optical switch structure based on cascaded beam shifters," IEICE Trans. Commun., vol.E77-B, no.2, pp.163-173, Feb. 1994.

[3] K.Noguchi, K.Hogari, T.Sakano, and T.Matsumoto, "A rearrangeable multichannel free-space optical switch using polarization-multiplexing technique," PS'90, 13C-5, Kobe, 1990.

[4] H.Itoh, T.Yamamoto, and T.Nakagami, "A polarization-switched optically reconfigurable interconnection network," PS'90,13C-6, Kobe, 1990.

[5] T.J.Cloonan, F.B.McCormick, M.J.Herron, F.A.P.Tooley, G.W.Richards, E.Kerbis, J.L.Brubaker, and A.L.Lentine, "A 3D crossover switching network based on S-Seed arrays," PS'90, 14D-2, Kobe, 1990.

[6] K.Kimura, K.Noguchi, and T.Sakano, "A study on multichannel liquid-crystal free-space optical switch for optical digital cross-connect systems," IEICE Technical Report, SSE93-151, March 1994.

[7] K.Hirabayashi, T.Yamamoto, S.Hino, Y.Kohama, and K.Tateno, "Optical beam direction compensating system for board-to-board free space optical interconnection in high-capacity ATM switch," IEEE J. Lightwave Technol., vol.15, no.5, pp.874-882, May 1997.

[8] S.Matsuo, C.Amano, and T.Kurokawa, "Operation characteristics of three-terminal hybrid structure with multiple-quantum-well reflection modular and heterojunction phototransister," IEEE Photon. Tech. Lett., vol.3, pp.330-332, 1991.

[9] S.Urushidani, H.Imagawa, and K.Yukimatsu, "The rerouting banyan network," ISS'90 Proc., vol.V, pp.27-31, May 1990.

[10] S.Hino, S.Urushidani, J.Nishikido, and K.Yamasaki, "Implementation issues for very-large-capacity ATM switching fabrics optically intraconnected," GLOBECOM '92 Proc., vol.1, pp.208-212, 1992.

[11] S.Hino, S.Yamaguchi, Y.Ohtomo, S.Yasuda, S.Urushidani, and K.Yamasaki, "An implementation of a high-performance ATM switching fabric," ISS '95 Proc., vol.1, pp.399-403, 1995.

[12] S.Urushidani, S.Hino, Y.Ohtomo, and S.Yasuda, "A high-performance multicast switch and its feasibility study," IEICE Trans. Commun., vol.E81-B, no.2, pp.284-296, Feb. 1998.

[13] Y.Ohtomo, S.Yasuda, M.Nogawa, J.Inoue, K.Yamakoshi, H.Sawada, M.Ino, S.Hino, Y.Sato, Y.Takei, T.Watanabe, and K.Takeya, "A 40-Gb/s 8×8 ATM switch LSI using 0.25-mm CMOS/SIMOX," ISSCC '97 Proc., pp.154-155, 1997.

[14] J.H.Patel, "Performance of processor-memory interconnection for multi-processors," IEEE Trans. Comput., vol.C-30, no.10, pp.771-780, Oct. 1981.

Appendix

This appendix compares values calculated using the analysis of Eqs. (1) to (8) and the simplified analysis of Eqs. (9) to (11). Figure A·1 compares the blocking probability between

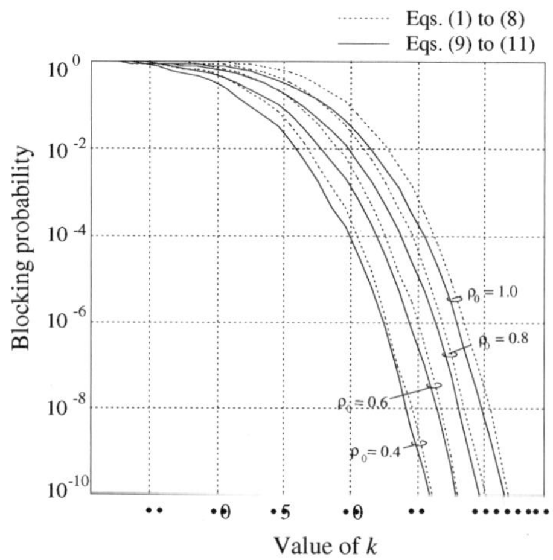

Fig. A·1 Blocking probability comparison.

two analyses. Compared with that of detailed analysis, the value yielded by the simplified analysis is somewhat optimistic, but is nonetheless a good approximation.

High-Performance Backbone Network Technology edited by Naoaki Yamanaka
ISBN: 0-8247-5321-6 DOI: 10.1081/(120025737)
Published 2004 by Marcel Dekker, Inc.

Switching Node Consideration from the Aspect of Transmission Characteristics in Wavelength Assignment Photonic Network (WAPN)

Tadahiko Yasui and Yoshiaki Nakano

SUMMARY By adopting a network architecture in which not only a calling but also a called terminal can select a wavelength, a novel WDM network becomes possible. This we call Wavelength Assignment Photonic Network (WAPN). In this network wavelengths are a kind of network resources and according to requests from terminals, wavelengths are allocated or assigned to calls. In the system a wavelength used for a call is to be used for another call after the call is terminated. By supplying wavelengths to the home, a bitrate-free, protocol free or even transmission method free network can be realized. In this paper, from a viewpoint of S/N or Q factor, WAPN is evaluated with special focus on the node architecture—i.e., from the viewpoint of node size, number of switching stages, crosstalk level,and losses, because the allowable node size is the crucial issue to decide the whole network capacity. After brief explanations of this proposed system, the model for system evaluations will be established and a node system is to be evaluated for some practical parameter values considering especially traffic characteristics of a node. As a result of this study a node system with capacity more than 100 thousands erl (about 20 Tbps throughput) can be constructed using present available technologies, which will enable us to construct large WAPN network with radius of 2,000 km and subscribers of about 50 millions.
key words: *photonic network, switching matrix, transmission characteristics, traffic performance*

1. Introduction

Wavelength Division Multiplexing (WDM) transmission technologies are accelerating R&D activities of devices and systems such as fiber transmission technologies, wideband fiber amplifiers, and multiwavelength LDs etc. Thanks to these activities, the increase of the transmission capacity is remarkable. Not only wavelengths are used for the increase of a network capacity, but also WDM technology is being used for an enhancement of network functionalities by enabling wavelength re-routing in the case of transmission line failures and by enabling network reconfigurations for traffic flow changes. There are many reports on experimental optical crossconnect systems [1]–[3]. Photonic crossconnects are being studied in various countries not only on terrestrial transmission but also on submarine transmission systems stimulated by WDM technologies.

In this paper we will propose a network in which the traffic calls are carried by wavelength on call by call basis. The network based on this concept can be tailored according to the available number of wavelengths without changing the number of subscribers.

Photonic cross-connect systems aim at network-wide utilization of wavelengths, but it is aimed to enhance flexibility of transmission systems and limited to within trunk lines. The proposed network WAPN (Wavelength Assignment Photonic Network) does go beyond to the subscriber network, and moreover wavelengths are switched call by call basis. Therefore, even if the apparent architecture is similar to crossconect systems, essential concept is quite different. Explicit differences are as follows.

1. Switching networks of the WAPN can accommodate more than 100,000 ports.
2. Switching networks may have connection loss, while crossconnects are not allowed to have connection loss because they should be always ready to reroute transmission lines in case transmission failures happen. Non blocking network with 100,000 ports is practically too large to be realized.
3. One wavelength is assigned to one call. The maximum bitrate from one terminal will not exceed 100 Mbps even in far future. This will ease the requirement for the network transmission specification very much. On the contrary, the bit-rate of cross-connects should exceed 10 Gbps or more.

By lowering the bitrate of the information carried on one wavelength (end to end bitrate is 100 Mbps at most), resultant transmission margins are diverted to the crosstalk and loss margins of the switching system. Moreover by dividing a switching network of large capacity into identical small capacity switching networks through the subnetwork concept and by adopting end to end channel matching scheme, a construction of a very large switching network is made possible. These factors have led to a possibility of realizing a nationwide photonic telecommunication network with a node capacity of 100,000 erl or 20 Tbps throughput.

Manuscript received August 5, 1998.
Manuscript revised October 13, 1998.
†The author is with Mitsubishi Electric Corporation Kamakura-shi, 247-8501 Japan.
††The author is with the Faculty of Electronic Engineering, The University of Tokyo, Tokyo, 113-8654 Japan.
*This paper is also published in IEICE Trans. Electron., Vol.E82-C, No.2, pp.254–264, February 1999.

IEICE TRANS. COMMUN., Vol. E82-B, No. 2, pp. 306–316, February 1999

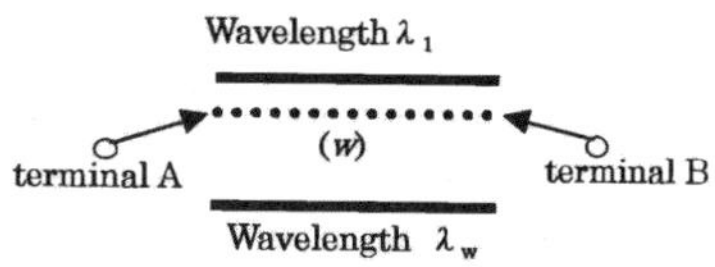

Fig. 1 WAPN basic idea.

A lot of study will be necessary to investigate the feasibility of WAPN system, but in this paper a special focus is given to the node transmission characteristics, because estimation of how large network is to be constructed in WAPN is the most important. At present it is not clear whether the nationwide WDM telecommunication network including large capacity switching nodes is feasible or not from the technological aspects, especially from the viewpoint of transmission characteristics.

The composition of this paper is as follows. Section 2 is devoted to rough system description. In Sect. 3 the formulas to evaluate the concept are introduced and in Sect. 4, giving the typical values for the introduced parameters, some of the calculated data will be shown and at the same time discussions are given concerning the calculated data. Finally in Sect. 5 the problems which this system has and must be studied further will be clarified.

2. System Concept

The principle of WAPN is quite simple [4]. Basic idea of WAPN is illustrated in Fig. 1. Network has wavelengths as one of network resources like time slots in conventional digital switching systems and assign an idle wavelength to a call originated. The switching node in WAPN administrates idle/busy condition of wavelengths. The proposed system is to be bitrate independent, protocol-free, and even transmission method free. WAPN needs no clocking devices for the transmission purpose and the system needs essentially no digital equipment.

A terminal communicates with a system control and tells the control the address of the subscriber he wants to talk with. According to the request from the terminal the system control searches for wavelengths not used (idle wavelengths), and inform them of the idle wavelength with which they should communicate. The controller sets up a path between them. The calling terminal generates the designated wavelength and the called terminal adjusts his filter to the wavelength designated. After termination of the talk, the communication path is cleared and the wavelength becomes idle again and is ready for another call.

Number of calls $Msim$ simultaneously connected is given in Eq. (1) with N as number of subscribers and b as calling rate of a subscriber.

$$Msim = N \cdot b \tag{1}$$

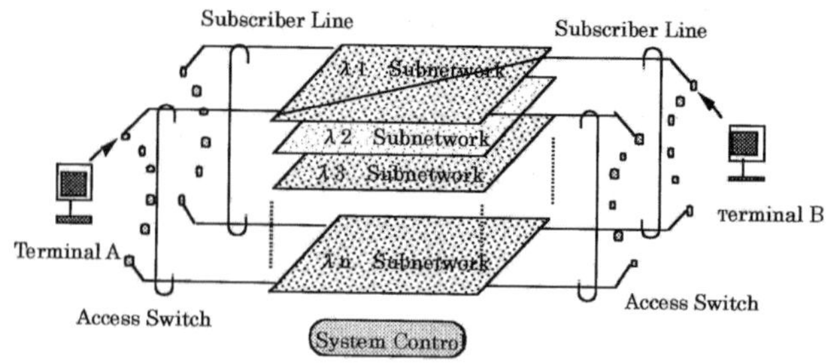

Fig. 2 WAPN concept.

If we were to realize the system in which every connection should use different wavelengths, required number of wavelength is 0.5 million (N = 50,000,000 and $b = 0.01$). A nation-wide network is not realistic even for far future because it is impossible to prepare 0.5 million wavelengths. But by using space division switches, required number of wavelength is reduced to practical number. The principle of the system is illustrated in Fig. 2.

Suppose w wavelengths are available, and construct a small subnetwork of m subscribers with space division switches and a wavelength, and then make $(w - 1)$ subnetworks identical with it. The result is a network with identical w subnetworks, each of which is dedicated to one specific wavelength. Each subnetwork has m ports, and total available ports $Msim$ can be written in Eq. (2).

$$Msim = w \cdot m \tag{2}$$

Therefore from Eq. (1),

$$w = Msim/m = N \cdot b/m. \tag{3}$$

Equation (3) indicates that w becomes smaller with N and b fixed, as the number of ports m of each subnetwork increases. In other words, if w is fixed, m can be adjusted for a given $Msim$, i.e. can be adjusted to match the number of available wavelength w. Number of wavelength is not a limiting factor of this proposed network. If $m = Nb$, then $w = 1$, which is the present network.

It is very important to note that there is no connection between subnetworks. This means that we only need to design a small capacity switching network to realize a large capacity node, which is key to make feasible a large WAPN. The penalty of this idea is that network-wide path hunting is usually needed in order to decrease the call blocking rate of the network and to decrease the path hunting time.

Each terminal must be able to access to every subnetwork, i.e. can access to every wavelength of the network. Further, as is clear in Eq. (2), $Msim$ is smaller than N, so it is necessary to have a mechanism to match these two numbers, which is called traffic concentration. These access and concentration functions are illustrated in Fig. 2 as "access switch." The access switch performs

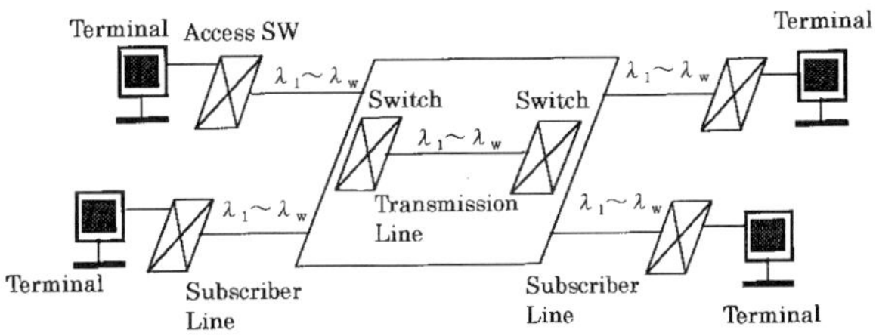

Fig. 3 Physical structure of WAPN.

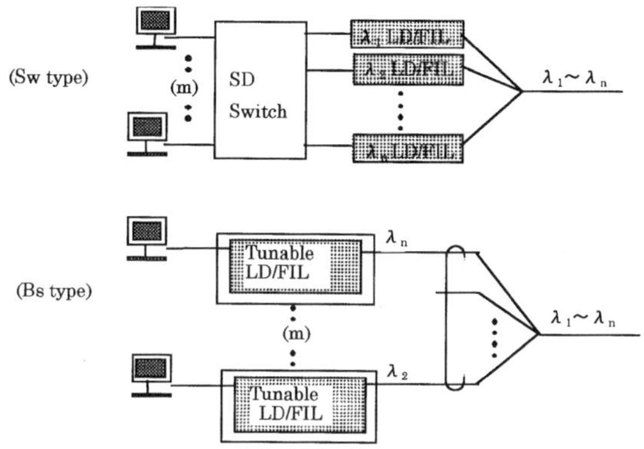

Fig. 4 WAPN subscriber network structure.

any connection between N/m subscribers and $Msim/m$ subnetwork ports.

The physical configuration of WAPN is illustrated in Fig. 3. Each subnetwork which is completely separated logically (interconnections between subnetworks are not allowed) is reduced to a single physical network, in which switches, transmission links and even subscriber lines are wavelength multiplexed. The access switch concentrates traffic from terminals and selects the necessary wavelengths for communications. System control administrates these whole call processing.

The feasibility of WAPN should be investigated first by estimating possible maximum physical network size, maximum number of subscribers accommodated and number of wavelengths multiplexed. In order to do it, formulas should be deduced to evaluate system transmission performance, taking into accounts optical amplifier gain, ASE (amplified spontaneous emission) noise, fiber transmission link loss, loss and crosstalk of switching nodes. Next section is given to formulate WAPN from the transmission point of view.

3. System Evaluation Model

3.1 Network Model

A WDM system is essentially an analog system, and at present it is not possible to implement 3R functions into the system and the system should be evaluated in view of total overall end-end performance. In order to do this, network model should be identified. There are two network models depending on structure of access switches. The subscriber network structure is shown in Fig. 4

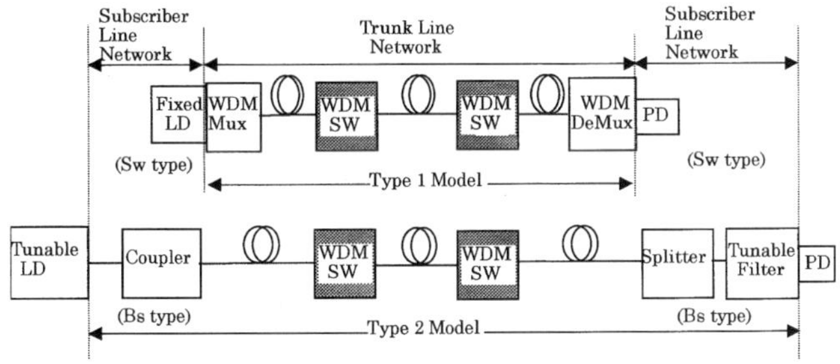

Fig. 5 WAPN network model.

Sw type is one that has physical switches for traffic concentration and wavelength selection. In Sw type, there is no special requirement for the switches used for access switch and even conventional electronic switches can be used. This is very important, because network can be designed without considering complicated conditions of the subscriber network. In this configuration each wavelength of LDs and filters is fixed.

Bs type is the final goal of the subscriber network of the proposed WAPN network. All the wavelengths are on a fiber, and terminals can access to all of them. In a sense this is a kind of a fiber bus. There is no switches. In Bs type configuration, every terminal must have a tunable LD and a tunable filter and should generate and select any wavelength designated by the system control. The switching functions that are performed by switches in Sw type reside in every terminal. In other words distributed switching is performed in this type. Distances between subscribers and a node are different and input levels for a node or input power level for terminals differ very much. These conditions of the subscriber network should be taken into consideration when designing WAPN based on the Bs type subscriber network.

According to the configuration of subscriber networks, two types of network models are considered as shown in Fig. 5. One is Type 1 model whose subscriber network is Sw type. The other is Type 2 model whose subscriber network consists of Bs type.

In this WAPN as is stated before, a path for a call is found out by the system control from a group of idle wavelengths, so even when a terminal communicates with the same terminal the wavelength used for a connection is different from the former one. Depending on the destination, number of nodes and transmission distance and links for a call to pass through varies. Therefore in order to make evaluation easy, we assume every WAPN path for a call is composed of arbitrary cascaded connections of the section modeled in Fig. 6. A section is configured so that loss generated in the section are to be compensated by an optical amplifier at the output, but noises added in the section are to be output without compensation. If noise accumulation exceeds

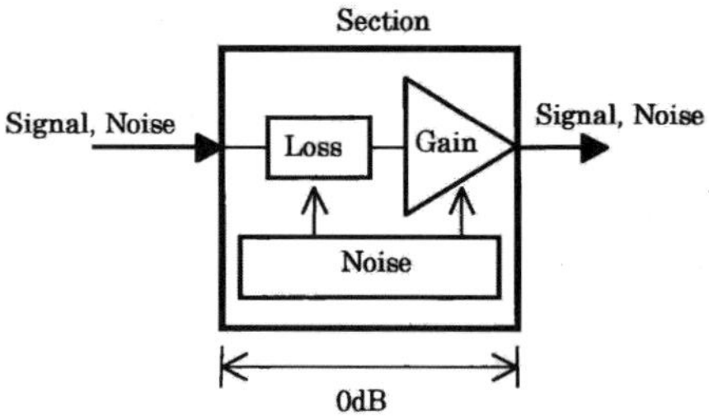

Fig. 6 Section model.

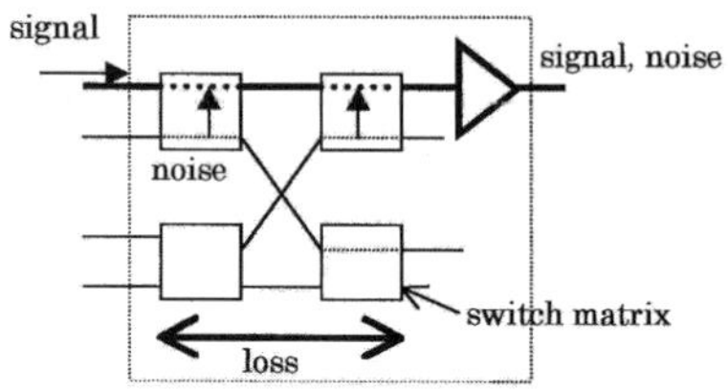

Fig. 7 Node section illustration.

allowed value, this is the system limit. For example, a switching node is expressed like Fig. 7 from a transmission characteristics point of view. A switching node consists of meshed switch matrices, which have losses and suffer from crosstalks from other channels. The optical amplifier compensates switch losses but crosstalk is accumulated.

3.2 Deduction of Evaluation Formulas

In this section we will calculate end-to-end bit error rate on the assumption every connection is a cascade of the block shown in Fig. 6.

(1) Cascade Connection

We will deduce transmission characteristics of K sections cascade connection shown in Fig. 8. Parameters to characterize the ith section are shown in Fig. 9. In this figure, $n_{out(i)}$ denotes output photon number per unit time, $n_{sp(i)}$ is spontaneous emission coefficient of optical amplifier, G_i denotes gain of amplifier, L_i is loss of components and $n_{xt(i)}$ denotes crosstalk. Average values are expressed by $\langle\ \rangle$. Let

$$\begin{array}{l} U_i = G_i L_i \\ Msig = U_K U_{K-1} \cdots U_2 U_1 \langle n_{out(0)} \rangle \\ Msp_i = U_K U_{K-1} \cdots U_{i+1} (G_i - 1) n_{sp(i)} \\ Mxt_i = U_K U_{K-1} \cdots U_{i+1} U_i n_{xt(i)} = Msig Xt_{xt(i)}, \end{array} \quad (4)$$

where $Xt_{(i)}$ is crosstalk attenuation ratio of ith section. Further, we consider p polarizations and bandwidth of optical filter Δf, and also assume that there is no correlation between crosstalks and signal. Then, from the master equation we obtain the total output photon number per unit time after K sections cascade as follows [5]–[8].

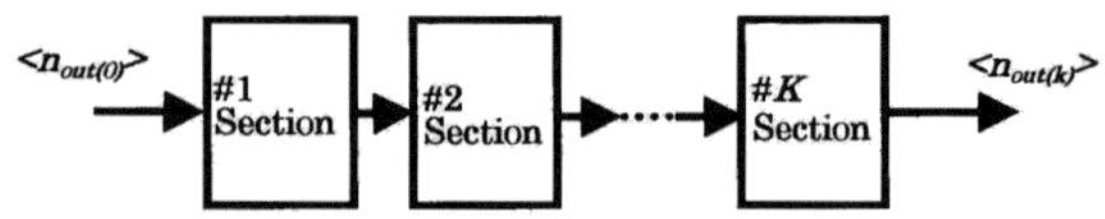

Fig. 8 Cascade connection.

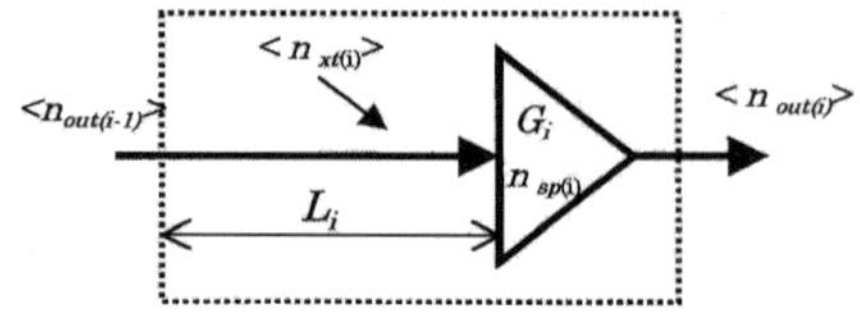

Fig. 9 Section i parameter notation.

$$\langle n_{out(K)} \rangle = Msig + \sum_{i=1}^{K} Mxt_i + p\Delta f \sum_{i=1}^{K} Msp_i \quad (5)$$

Assuming Poisson distribution for both $n_{out(0)}$, and $n_{xt(i)}$ for every i, spectral density of output photon flux fluctuation after K sections cascade $\sigma^2_{out(K)}$ is given below by again solving master equation.

$$\begin{aligned} &\sigma^2_{out(K)} \\ &= Msig + \sum_{i=1}^{K} Mxt_i + p\Delta \sum_{i=1}^{K} Msp_i \\ &+ 2Msig \sum_{i=1}^{K} Msp_i + p \left(\sum_{i=1}^{K} Msp_i \right)^2 \Delta f \\ &+ 2 \sum_{i=1}^{K} \left(Mxt_i \sum_{j=1}^{K} Msp_i \right) \end{aligned} \quad (6)$$

(2) Bit Error Rate

As is well known, noise power spectrum density *Nnoise* (power per a unit frequency or power consumed by a unit resister: A^2/Hz) is given below [6].

$$N_{noise} = 2e^2\sigma^2 \quad (7)$$

where σ^2 is spectral density of photon flux fluctuation at the input of the optical receiver. Here, we must consider quantum efficiency η, and this can be treated as loss η which is connected further to the last section (K+1 section).

Assuming signals are modulated in ASK and signal format is NRZ, we can calculate mark/space current and total noise on mark signal and space signal. Let *Pin* be system input power, r be on/off (or mark/space) ratio of input power, and occurrence of mark/space be equal. Then, the mark signal current *Imark*, space signal current *Ispace*, crosstalk current on mark signal Ixt_{mark} and crosstalk current on space signal Ixt_{space} are given in the following equations with one photon energy $h\nu$ (h: Planck constant, ν: frequency of the signal light).

$$\begin{aligned} Imark &= e\eta Msig_{mark} \\ &= U_K U_{K-1} \cdots U_2 U_1 \frac{\eta e}{h\nu} \frac{2r}{1+r} Pin \end{aligned} \tag{8}$$

$$\begin{aligned} Ispace &= e\eta Msig_{space} \\ &= U_K U_{K-1} \cdots U_2 U_1 \frac{\eta e}{h\nu} \frac{2}{1+r} Pin \end{aligned} \tag{9}$$

$$Ixt_{mark} = Imark \sum_{i=1}^{K} Xt_{(i)} \tag{10}$$

$$Ixt_{space} = Ispace \sum_{i=1}^{K} Xt_{(i)} \tag{11}$$

where, $Msig_{mark}$ is $Msig$ for mark signal, and $Msig_{space}$ is $Msig$ for space signal. From Eq. (6) and Eq. (7), the noise power on mark signal $Nmark$ and the noise power on space signal $Nspace$ are given in the following equations, with optical receiver bandwidth B, spectral density of photon flux fluctuation for mark signal in the optical receiver σ^2_{mark} and spectral density of photon flux fluctuation for space signal in the optical receiver σ^2_{space}.

$$\begin{aligned} Nmark &= 2e^2\sigma^2_{mark}B \\ &= 2e^2B\Bigg[\eta Msig_{mark} + \eta\sum_{i=1}^{K} Mxt_{(i)} \\ &+ p\Bigg\{\eta^2\left(\sum_{i=1}^{K} Msp_i\right)^2 + \eta\sum_{i=1}^{K} Msp_i\Bigg\}\Delta f \\ &+ \eta^2\Bigg\{2Msig_{mark}\sum_{i=1}^{K} Msp_i \\ &+ 2\sum_{i=1}^{K}\left(Msp_i\sum_{j=1}^{i} Mxt_j\right)\Bigg\}\Bigg] \end{aligned} \tag{12}$$

$$\begin{aligned} Nspace &= 2e^2\sigma^2_{space}B \\ &= 2e^2B\Bigg[\eta Msig_{space} + \eta\sum_{i=1}^{K} Mxt_{(i)} \\ &+ p\Bigg\{\eta^2\left(\sum_{i=1}^{K} Msp_i\right)^2 + \eta\sum_{i=1}^{K} Msp_i\Bigg\}\Delta f \\ &+ \eta^2\Bigg\{2Msig_{space}\sum_{i=1}^{K} Msp_i \\ &+ 2\sum_{i=1}^{K}\left(Msp_i\sum_{j=1}^{i} Mxt_j\right)\Bigg\}\Bigg] \end{aligned} \tag{13}$$

The Q factor is given below, assuming the Gaussian distribution for $Nmark$ and $Nspace$ [9] and the crosstalk currents degrade directly eye-opening of the signals. It is assumed that crosstalk adds to signal lights in the worst manner.

$$Q = \frac{Imark - Ispace - (Ixt_{mark} - Ixt_{space})}{\sqrt{Nmark} + \sqrt{Nspace}} \tag{14}$$

Finally bit error rate BER is given by the following formula.

$$BER = \frac{1}{2}\left(1 - \int_{-\frac{Q}{\sqrt{2}}}^{\frac{Q}{\sqrt{2}}} \frac{1}{\sqrt{\pi}} \exp\left(-x^2\right) dx\right) \tag{15}$$

Equations (14) and (15) have one to one correspondences, and so we treat Q factor as an evaluation tool of the system.

4. Switching Node Consideration

Using the formulas given in the previous section, switching node considerations from the aspect of transmission characteristics are to be made.

4.1 Type 1 Model

In Type 1 model, only trunk lines and switching nodes are taken into account. All optical amplifiers used for nodes are equal (n_{spsw}: spontaneous emission coefficient, G_{sw}: gain), transmission characteristics of node switches are equal (L_{sw}: loss and n_{xtsw}: crosstalk), and number of nodes is K_{sw} and

$$G_{sw}L_{sw} = 1 \tag{16}$$

is assumed. All optical amplifiers used for trunk lines are assumed equal (n_{sptr}: spontaneous emission coefficient, G_{tr}: gain), transmission characteristics of transmission lines are equal (L_{tr}: loss and n_{xttr}: crosstalk), and number of transmission line sections is K_{tr}, and

$$G_{tr}L_{tr} = 1 \tag{17}$$

is assumed. Further, we assume crosstalks occur only in switching nodes.

Applying these relations to Eq. (14), Q factor is given for Type 1 model. Figure 10 shows the switching network loss L_{sw} versus Q factor for $Pin = -8$, -10, -12 dBm and $n_{xtsw} = -15$, -20, -25 dB, when $G_{tr} = 20$ dB and $K_{tr} = 20$. This means that if the fiber loss is 0.2 dB/km, the optical amplifier spacing on transmission lines is 100 km and total transmission line length is 2000 km. Other parameter values for the calculation of Fig. 10 is shown in Table 1. Figure 10 indicates various important characteristics of the proposed network. 25 dB to 35 dB loss/node will only be allowed with crosstalk around -20 dB. Given this condition it needs to estimate roughly the size or capacity of the node to be realized.

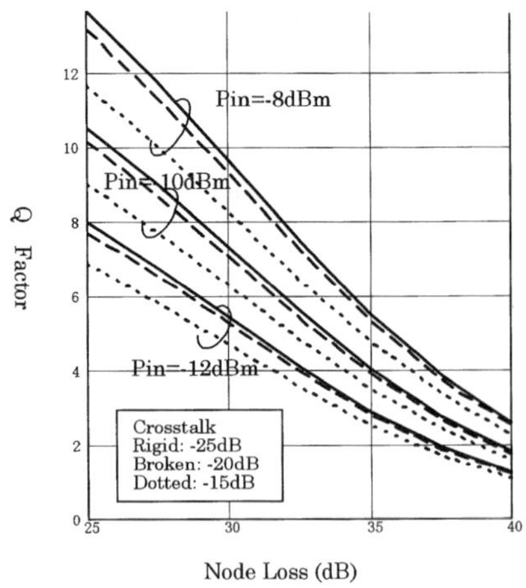

Fig. 10 Calculated Q factor vs. node loss.

Table 1 Paramerer values used.

NRZ bitrate of signal carried B (Mbps)	250
On-off Ratio of input signal r (dB)	13
Spontaneous Emission Coefficient n_{spsw}, n_{sptr}	2.0
No.of Switching stage K_{sw}	4
Optical Filter Bandwidth Δf (nm)	0.1
Optical Frequency ν (nm)	1558

In order to get a rough sketch of possible node size, we here assume simple traffic condition and network structure. All the traffic is homogeneous, and the network structure is as in Fig. 11, in which the network is 2 layered (A: upper layer, B: lower layer) and B layer adopts star topology and for A layer, mesh topology is assumed. Mesh topology means that each node extends its lines to every node in A layer networks. In this hierarchical network maximum number of nodes for a call to pass through is 4. We assume traffic flow in each A, B node area as in Fig. 12. $(100-c)\%$ of a B node traffic goes back to subscribers of the same node, and other c% traffic goes equally to A node. $1/M$ of one A node traffic incoming from B nodes belonging to it goes to every node in the A layer.

Under this condition, total subscriber traffic N_B of B node, and total transit traffic M_B of an A node are given below.

$$N_B = \frac{N \cdot b}{S \cdot w}\, erl \qquad M_B = \frac{N \cdot b}{M \cdot w} \cdot \frac{c}{100}\, erl \tag{18}$$

where, N: number of subscribers, M: number of A node, S: number of B node and c: outgoing traffic rate of a B node, and b: calling rate of a subscriber. As for the number of the nodes in each layer, taking into account the Japanese present telecommunication network, following numbers are assumed.

$$\begin{aligned} &N = 50\,\text{million},\ M = 10\,\text{nodes},\\ &S = 50\,\text{nodes},\ c = 30,\ b = 0.1 \end{aligned} \tag{19}$$

From these, N_B and M_B are given below.

$$N_B = \frac{100{,}000}{w}\, erl \qquad M_B = \frac{150{,}000}{w}\, erl \tag{20}$$

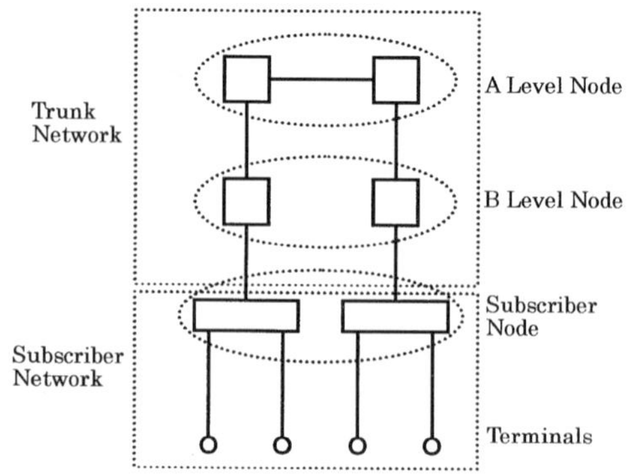

Fig. 11 Network structure.

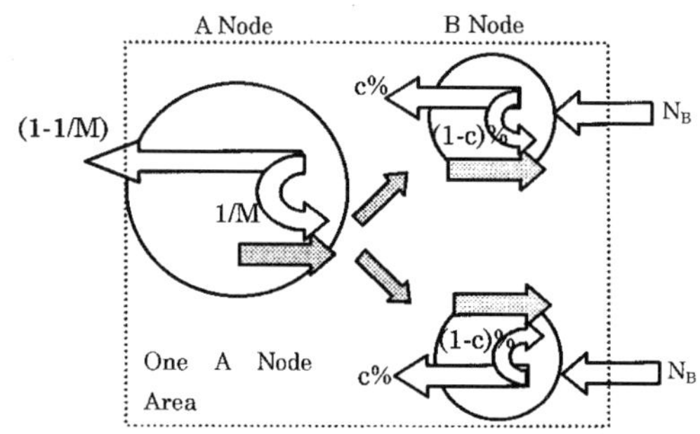

Fig. 12 Postulated traffic flow pattern.

If w=100, then N_B =1,000, and M_B=1,500.

Usually blocking rate of a switching network is less than 0.001, and this is a severe requirement especially for a large capacity-switching network. In WAPN, each subnetwork is completely independent, which means there is no need for connections between different subnetworks. Therefore design target capacity of a switching network reduces to $1/w$, and as a result, end to end channel matching (or route finding) of one subnetwork becomes easy to be realized, which further eases design requirement for switching network blocking rate. End-end channel matching means to find out end-end idle path by knowing the busy/idle state of the transmission lines and switch jumpers of a subnetwork, while link-link channel matching finds out end-end idle path not knowing the busy/idle states of the next stages. This is illustrated in Fig. 13. In link by link method the controller of #1 node do not know the jumper busy/idle of #2 node. In WAPN in order to relax blocking rate requirement for a switching network as much as possible and to decrease number of switching network stages, the end-end channel matching algorithm is considered better although the control becomes complicated a little.

For a case of 4 nodes, channel graph of end-end channel matching is shown in Fig. 14, and that of link by link is illustrated in Fig. 15. Blocking rate $p_{E-E}(n)$ of end to end and p_{L-L} (n) of link by link for a subnetwork are given below,

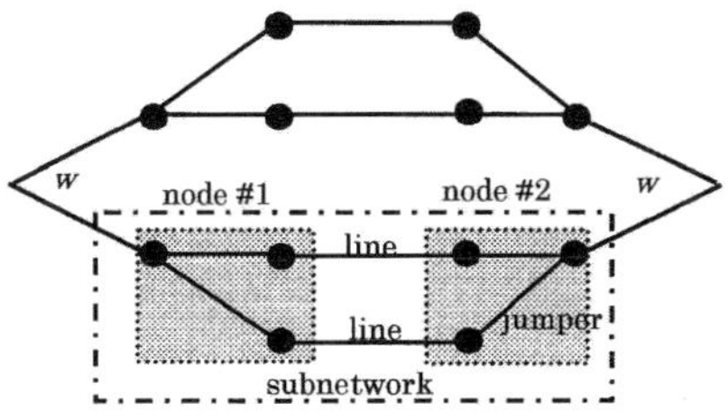

Fig. 13 Network channel graph.

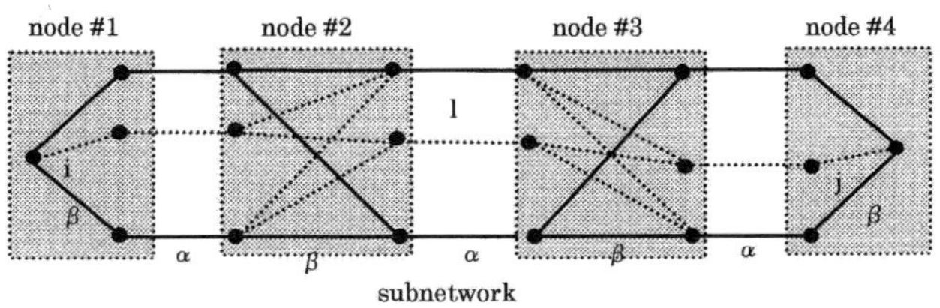

Fig. 14 4 Stage node subnetwork channel graph (end to end).

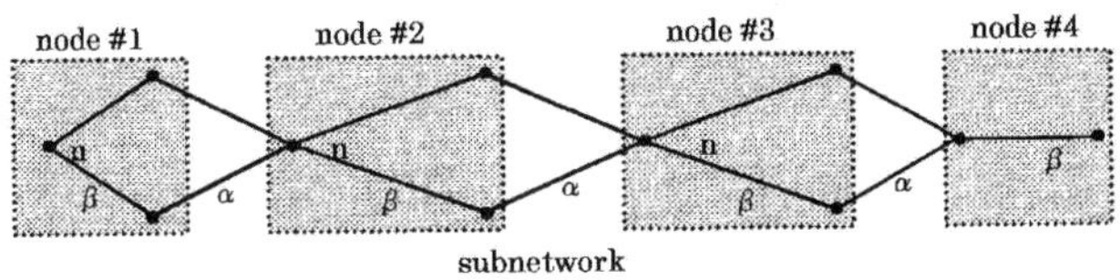

Fig. 15 4 Stage node subnetwork channel graph (link by link).

$$
\begin{aligned}
p_{E-E}(n) &= \sum_{l=0}^{n}\sum_{i=0}^{n}\sum_{j=0}^{n} C_l^n C_i^n C_j^n \\
&\alpha^{n-l}(1-\alpha)^l \gamma^{n-i}(1-\gamma)^i \gamma^{n-j}(1-\gamma)^j \\
&\left(\beta^{i\cdot l} + \beta^{j\cdot l} - \beta^{(i+j)l}\right)
\end{aligned}
\tag{21}
$$

$$
p_{L-L}(n) = 1 - (1-\gamma^n)^3 (1-\beta) \tag{22}
$$

where, n: the number of output transmission lines available in each subnetwork, and β: usage rate of one link path of a switching node (equivalent switching network jumper usage), α: usage rate of one transmission line and

$$
\gamma = 1 - (1-\alpha)(1-\beta) \tag{23}
$$

Total blocking rate of the whole network is given in the following formula.

$$
BL_{E-E} = \left\{1 - (1-\alpha)^2(1-p_{E-E}(n))\right\}^w \tag{24}
$$

$$
BL_{L-L} = \left\{1 - (1-\alpha)^2(1-p_{L-L}(n))\right\}^w \tag{25}
$$

where, BL_{E-E} and BL_{L-L} indicate total end-end blocking rate and total link by link blocking rate respectively. BL_{E-E} and BL_{L-L} versus equivalent switch jumper usage β are shown in Fig. 16, for $n = 5, 10, 20, 40, 100$, α=0.7 and $w = 100$. Figure 16 shows that BL_{E-E} is always better than BL_{L-L}.

Let us consider the switching network of which channel graph is shown in Fig. 17 with jumper usage: a, number of links: s and multiplicity: t. The equivalent jumper usage β is given below.

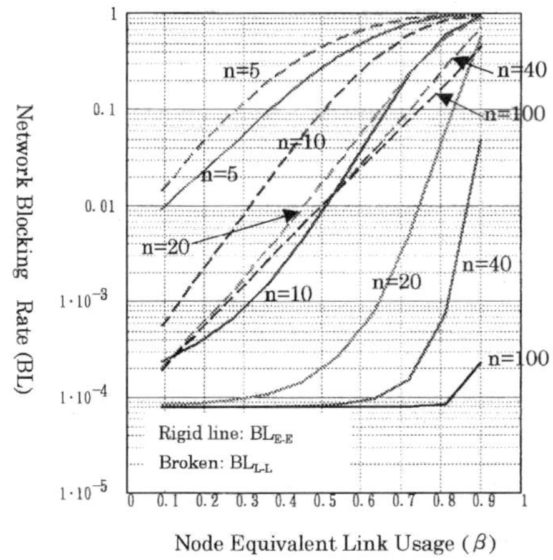

Fig. 16 Calculated network blocking rate.

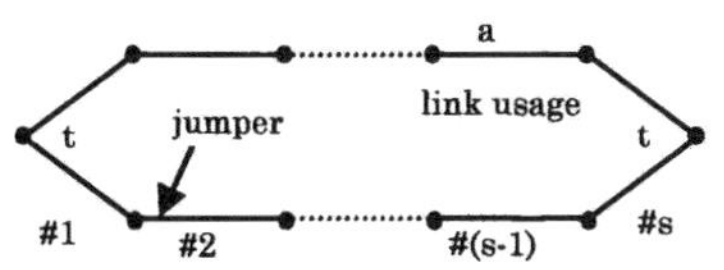

Fig. 17 Postulated channel graph.

$$
\beta = \{1 - (1-a)^s\}^t \tag{26}
$$

From Eq. (20) if w is 100, about 1500 erl switching network capacity is required, which means 2,000 ($a = 0.75$) to 3,000 ($a = 0.5$) ports switching network. If 8×8 switch matrixes are to be used for every stage and $s \leq 3$, it is obvious that we cannot construct the network whose ports are more than 2000, because $t \leq 2$. Figure 18 shows the relations between a and β for $s = 2, 3, 4, 5$ and $t = 8$. From the view point of transmission characteristics, the number of stages are better to be as small as possible. When we assume s=4 (5 stage switching network), we can configure switching network shown in Fig. 19 with 4096 ports in which $t = 8$. From Fig. 18, we can take $a = 0.5$ (2048 erl) and $\beta = 0.7$. In order to construct an economical switching network, jumper usage is better to be high. But from the view point of flexibility against the traffic fluctuations and variations, a should be around 0.5 even if BL is less than 0.0001 for $n \geq 100$.

If we build this type of switching network with space division switch matrices, the architecture shown in Fig. 20 must be taken. The node should have wavelength demultiplexers and multiplexers at the input and at the output of the switching network, and these have also losses (arrayed waveguide filters have about 5 to 7 dB). Figure 10 shows only 25 to 35 dB loss is allowed for 1 node stage, so that losses less than 3 dB to 5 dB must be attained for each 8×8 switch matrix.

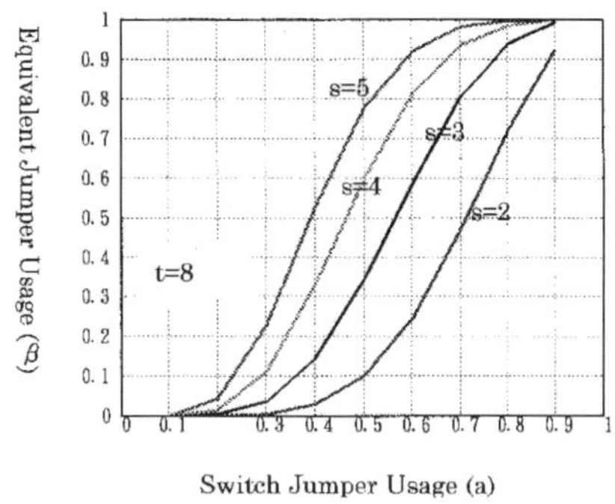

Fig. 18 Switch jumper usage vs. multiplicity.

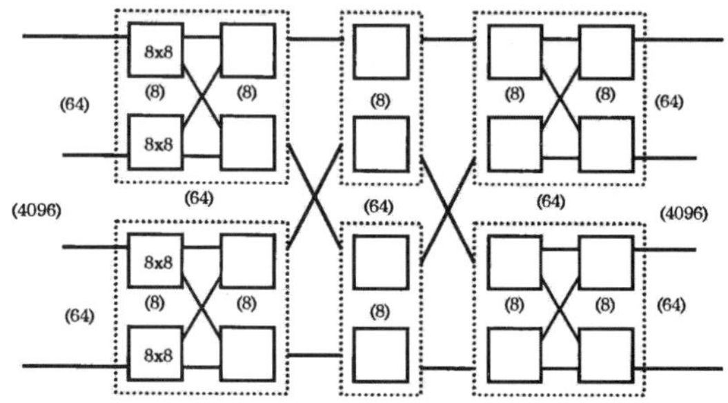

Fig. 19 Switching network of a node.

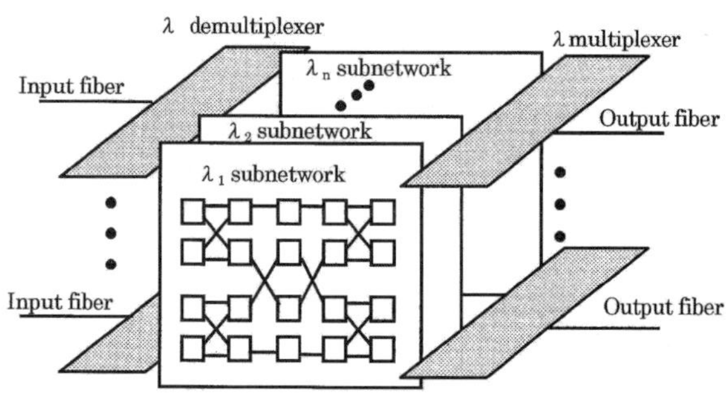

Fig. 20 Node architecture.

As for the crosstalk, −20 dB is only allowed under these conditions. When an 8×8 switch matrix has a crosstalk attenuation ratio *Xt* dB, an $s+1$ stage switching network will have a total crosstalk *Xteq* given by following equation with m_{sw} as number of crosstalk sources. We assume that all the crosstalks *Xt*s of each switch matrix and mux/demux constituting the switching node shown in Fig. 20 are equal.

$$Xt(dB) = Xt_{eq}(dB) - 10\log(m_{sw}\cdot(s+3)) \quad (27)$$

From this equation *Xt* is −37 dB, when *Xteq*=−20 dB, m_{sw} =7 and $s = 4$.

As far as we know, the only 8×8 switch matrix that might achieve 3–5 dB loss and −37 dB crosstalk is PLC (Planar Lightwave Circuit) switch matrix [10].

Table 2 Parameter values used.

Parameter	Value
NRZ bitrate of signal carried B (Mbps)	100
On-off Ratio of input signal r (dB)	13
Spontaneous Emission Coefficient n_{spsw}, n_{sptr}	2.0
No.of Switching stage K_{sw}	4
Optical Filter Bandwidth Δf (nm)	0.1
Optical Frequency ν (nm)	1558

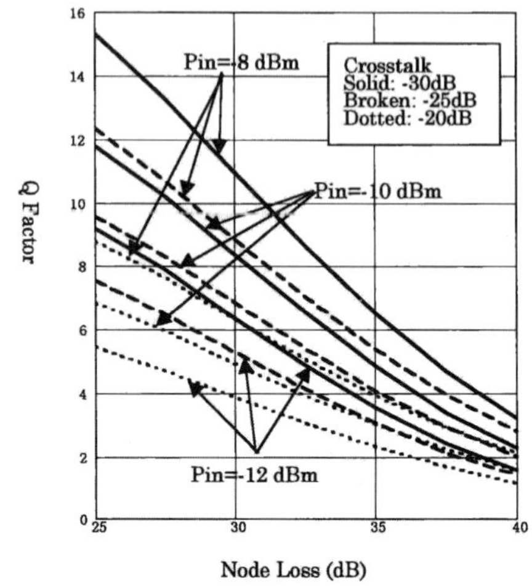

Fig. 21 Q-node loss.

4.2 Type 2 Model

Type 2 model is one in which subscriber networks are added to both sides of Type 1 model. All parameters for the subscriber network of a call originator have suffix $_{sub1}$, and parameters for the subscriber network of a called terminal are suffixed $_{sub2}$. The length of a subscriber line depends on location of a subscriber, so losses of each subscriber line differ and it can not be assumed that loss is completely compensated by an optical amplifier that is used commonly by several subscribers. But it will be compensated to some extent by a common amplifier in the subscriber network of an originating call. In the subscriber network of a called terminal, loss cannot be compensated by a common amplifier. Parameters U_{sub1} $(= G_{sub1}L_{sub1})$, U_{sub2} $(= G_{sub2}L_{sub2})$ can not be 1. Thus, assuming

$$U_{sub1} = 0.8,\ n_{spsub1} = 2.0,$$
$$G_{sub1} = 20\,\text{dB},\ n_{xtsub1} = -\infty$$
$$U_{sub2} = 0.005(= -23\,\text{dB}),$$
$$G_{sub2} = 1\ \text{(No optical amplifier)},$$

and the parameter values shown in Table 2, switching node loss versus Q factor is obtained in Fig. 21.

Compared with Type 1 model, the information bitrate should be lowered from 250 Mbps to 100 Mbps and the allowable switching node crosstalk should be higher than Type 1. Because there is no pre-amplifier in the called subscriber network, S/N improvement cannot be expected. As is shown, more than 5 dB crosstalk improvement is required. The loss increase owing to both

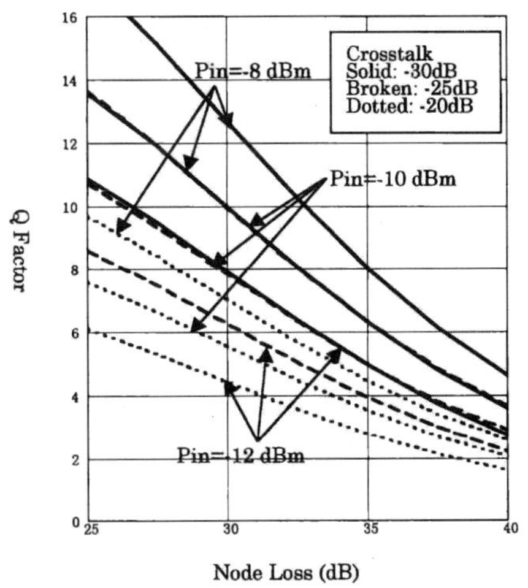

Fig. 22 Q-node loss for Δf=0.01 nm, r=20 dB.

subscriber networks does not affect the loss requirement for the switching network. Allowable loss becomes rather large in the high input power region. This is because ASE increase of both subscriber networks is not so great to affect the requirement.

Crosstalk requirement is severe in Type 2 model, and there may be many ways to improve this situation.

1. Loss suppression of switching matrix (e.g. 2 dB to 3 dB/8×8 switch matrix)
   Loss suppression results in fairly large improvement in crosstalk requirement but this seems not possible from the present state of arts.
2. Lower signal bitrate (e.g. 100 Mbps → 50 Mbps)
   This is easy to implement and very effective but this will cause the subscriber's discontent.
3. Narrower optical filter (e.g. 0.1 nm → 0.01 nm)
   The bandwidth of the above mentioned example is rather wide. Because carried bitrate is low, so this may be a practical solution for this problem. But this imposes severe requirement for the wavelength mux/demux.
4. Increase of on-off ratio of the input signal (e.g. 13 dB → 20 dB)
   This also contributes to enhance system performance. Because the required bitrate is 100 Mbps, requirement for the LDs are not so severe. This is one of the alternatives.

Those alternatives strongly depends on the device technologies, so optimum solutions comes from practical device performances. Figure 22 shows Q-Loss relations for $\Delta f = 0.01$ nm (1 Gbps interval) and $r = 20$ dB (on/off ratio of the input signal). About 2–3 dB relaxation in a node loss requirement is expected.

5. Considerations and Discussions

In order to make more detailed and precise analysis and evaluation, following should be studied further.

(1) Loss Variations (Ui)
In the analysis of Type 1 and Type 2 models, amplifiers compensate all the losses caused by switches and transmission lines. That is to say, Ui's of node and transmission line sections are assumed unity. In fact this is not always true. Following are the reasons.

- Transmission line loss drift and loss variations depending on the wavelength
- Gain drift and wavelength dependent loss variations
- Connection pattern dependent loss variations
- Wavelength dependent loss variation of switches and filters

These factors can be taken into considerations by giving Ui values other than unity. We have to calculate loss and gain variations and estimate the impact which have on the system requirement from various data of the devices and systems such as switches, filters, amplifiers, fibers etc.

(2) Crosstalk (n_{xt})
We considered two kinds of noises, one from amplifiers and the other (we called crosstalk) from noisy devices such as switch matrix and filters. The noises from these materials are considered not correlated or independent of the signal carried, but in fact the correlated noises such as coherent or bursty noises may exist.

We have also to consider the crosstalk from the signals of other wavelength in the optical amplifiers caused by the input power variations of wavelengths [11]. Wavelengths are dynamically assigned to calls, and the number of active wavelengths varies time to time. These inter-carrier modulations should be avoided as much as possible.

(3) Transmission Lines Non-linearity
In this paper we focused our discussion on the node architecture from the traffic characteristics. But in an analog network like this proposed system, transmission line characteristics is very important and more discussion should be done concerning mainly not random noise but the degradation of pulse shapes by fiber nonlinear distortions and chromatic dispersions.

(4) Input Power Level
General discussion says total power of an input to an optical amplifier must not exceed amplifiers saturation level. When number of wavelength is w, amplifier saturation output is $pout_{max}$, minimum input power is pin_{min}, allowable range of an input power is given below.

$$\frac{pin_{min}}{L} \leq p_{out} \leq \frac{pout_{max}}{w} \tag{28}$$

where L is transmission loss and w is the number of wavelength. In this case, L=−20 dB (1/100), and w=100 (20 dB), so following simple relation is given for the input power to the system.

$$pin_{min} + 20\,\text{dB} \leq p_{out} \leq pout_{max} - 20\,\text{dB} \tag{29}$$

If we want to use the transmitter output of −10 dBm,

and,

$$pout_{max} = 10\,\text{dBm}$$

power sensitivity of an amplifier should be less than $-30\,\text{dBm}$, and in order to have margins the sensitivity of amplifiers should be less than $-40\,\text{dBm}$.

(5) Switch Matrix

In the discussions, we only considered an architecture that is composed of wavelength demultiplexers and switch matrix and wavelength multiplexers. There are many switching devices considered for crossconnects [12]–[14] and they have their own transmission characteristics. It may be especially interesting to study switches which has SOAs [13] in the path. In these switches the ASE noise will be generated from switch devices and it imposes another problems on the evaluation.

(6) Structure of Subscriber Network

The feasibility of WAPN wholly depends on the subscriber network feasibility and one of the problems of realizing the network is stabilization of the frequencies generated by terminals. A terminal generates a wavelength designated by the network and called terminal should be able to select any wavelength designated by the network. This means tunable LDs and tunable filters are essential in a WAPN subscriber network. Not relative but absolute frequency preciseness is important and if this is not realized the result is too much fluctuation of Ui's, which will make WAPN unrealistic. The way to secure stabilization and accuracy of wavelengths will be the most urgent problem to be solved.

(7) Slow Wavelength Switching

Required switching time of a node is not so high and several msecs are enough. But this wavelength switching at that speed may pose some trouble on the optical fiber amplifiers. This also must be studied further.

6. Conclusion

So far the deduction of formulas and evaluation of WAPN with focus on a node switching network are intensively made, and feasibility of an example which is aiming at introduction into the area with 2000 km radius and 50million terminals is ascertained to some extent. We have shown a feasibility of a photonic switching node with 100,000 erl or 20 Tbps throughput. But still there are many problems to be studied and solved.

Acknowledgement

Authors are grateful for many useful suggestions and advises given by Prof. Kikuchi of Univ. of Tokyo and Prof. Ozeki of Sophia Univ. Especially Prof. Kikuchi has given us many useful comments on deducing the formulas for the system evaluation, which are central parts of this paper. Also we are thankful to Mr. Kitayama, manager of the lightwave communication department and members of the department for giving us many informations on the practical lightwave transmission technologies.

References

[1] S. Johansson, "Transport network involving a reconfigurable WDM network layer—A European demonstration," IEEE J. Lightwave Technol., vol.14, no.6, pp.1341–1348, June 1996.

[2] G.-K. Chang, et al., "Multiwavelength reconfigurable WDM/ATM/SONET network testbed," IEEE J. Lightwave Technol., vol.14, no.6, pp.1320–1340, June 1996.

[3] M. Teshima, et al., "Demonstration of virtual wavelength path cross-connect," Proc. ECOC'97, vol.3, pp.59–62, Sept. 1997.

[4] T. Yasui and A. Uemura, "A proposal of a new photonic FDM switching system FAPS—Frequency assign photonic switching system," IEICE Trans. Commun., vol.E77-B, no.2, pp.174–183, Feb. 1994.

[5] K. Kikuchi, "Theory of noise in optical amplifier," Fiber and Integrated Optics, vol.12, pp.369–380, 1993.

[6] H. Ishio, et al., "Optical Amplifier and Its Applications," Ohmusha, 1992.

[7] K. Kikuchi, "Foundations of Optical Fiber Communication," Shokodo, 1997.

[8] S. Yamashita and T. Okoshi, "Analysis of optical fiber communications systems with optical amplifiers," IEICE Technical Report, OCS90-20, 1990.

[9] S.B. Alexander, "Optical communication receiver design," SPIE Tutorial Texts in Optical Engineering, vol.TT22, 1997.

[10] A. Himeno, T. Goh, M. Okuno, H. Takahashi, and K. Hattori, "Silica-based low loss and high extinction ratio 8×8 thermo-optic matrix switch with path-independent loss arrangement using double mach-zehnder interferometer switching units," Proc. ECOC'96, vol.4, Oslo, pp.149–152, 1996.

[11] K. Motoshima, K. Shimizu, and K. Takano, "Automatic gain control of erbium-doped fiber amplifiers for WDM transmission systems," IEICE Trans. Commun., vol.E80-B, no.9, Sept. 1997.

[12] D.A. Smith, et al., "Evolution of the acousto-optic wavelength routing switch," IEEE J. Lightwave Technol., vol.14, no.6, pp.1005–1019, June 1996.

[13] G. Soulage, A. Jourdan, P. Doussière, M. Backman, J.Y. Emery, J. Da Loura, and M. Sotom, "4×4 space-switch based on clamped-gain semiconductor optical amplifiers in a 16×10 Gbit/s WDM experiment," Proc. ECOC'96, paper ThD.2.1, Oslo, 1996.

[14] M. Koga, Y. Hamazumi, A. Watanabe, S. Okamoto, H. Obara, K. Sato, M. Okuno, and S. Suzuki, "Design and performance of an optical path cross-connect system based on wavelength path concept," IEEE J. Lightwave Technol., vol.14, no.6, pp.1106–1119, June 1996.

High-Performance Backbone Network Technology edited by Naoaki Yamanaka
ISBN: 0-8247-5321-6 DOI: 10.1081/(120025738)
Published 2004 by Marcel Dekker, Inc.

Photonic Core Node Based on a 2.56-Terabit/s Opto-Electronic Switching Fabric

Soichiro Araki, Naoya Henmi, Yoshiharu Maeno, Kazuhiko Mtsuda, Osamu Nakakubo, Masayuki Shinohara, Yoshihiko Suemura, Akio Tajima, Hiroaki Takahashi, Seigo Takahashi, Hiromi Koganemaru, and Ken-ichi Saisho

SUMMARY This paper proposes Photonic Core Node based on a 2.56-Terabit/s opto-electronic switching fabric, which can economically handle the rapidly increasing multimedia traffics, such as Internet traffic. We have successfully developed the first prototype of Photonic Core Node. The prototype consists of a single-stage full-crossbar opto-electronic switching fabric, super-packet buffers for input queuing, and a desynchronized-round-robin scheduler. The switching fabric is upgradable up to 2.56 Tb/s, and employs wavelength-division-multiplexing techniques, which dramatically reduce the total number of optical switching elements down to one-eighth the number of those used in a conventional switching fabric. The super-packet buffer assembles 16 ATM cells routed to the same output port into a single fixed-length packet. The super-packet-switching scheme drastically reduces the overhead of optical switching from 32 to 2.9%, although it tends to decrease effective throughput. The desynchronized-round-robin scheduler maintains nearly 100% effective throughput for random traffic, recursively resolving the contention of connection requests in one scheduling routine while keeping fairness in a round robin manner. The proposed Photonic Core Node can accommodate not only ATM switching but also WDM optical path grooming/multiplexing, and IP routing by using IP input buffer interfaces, because optical switches are bit-rate/format-independent.

key words: *optical packet switch, WDM, ATM, IP, scheduler*

1. Introduction

The tremendous growth in Internet traffic and the continually increasing demand for broadband-data services have generated a need for increased transmission capacity and switching system throughput of more than Terabit/s in core networks [1], [2]. The next-generation communication networks should economically support various types of services at a throughput of more than Tb/s. In other words, a Tb/s-throughput Asynchronous Transfer Mode (ATM) core switch and a Tb/s-throughput Internet Protocol (IP) router are required for the next generation of broadband multimedia services. The maximum throughput of conventional electrical switching fabrics still remains a few hundreds of Gb/s even with the state-of-the-art CMOS technology [2], [3]. Optical switching technology is a key to build up a single-stage full crossbar switch with a throughput of more than 1 Tb/s and accommodate various types of services, taking advantages of the bit-rate/format-independent and broad-bandwidth features. An optical switching fabric is also advantageous to achieve large scalability, small power consumption, small volume and negligible EMI effects. Among previously proposed optical switching devices [4], a semiconductor optical amplifier gate (SOAG) has the advantages of switching time around 1 ns, loss compensation, high on/off ratio, and possible integration [5].

The use of optical switching devices, however, does pose some problems. One problem is their relatively high cost compared with electrical switching devices. Another problem is their lack of a bit-synchronization function and a frame-synchronization function within themselves. The other problem is the difficulties in implementing buffers and a contention resolution logic circuit employing only optical technology. We therefore aim to resolve these problems and construct an opto-electronic switching fabric providing throughput greater than 1 Tb/s. We have developed wavelength-division/space-division (WD/SD) optical switch architecture that needs far fewer SOAGs than that of conventional space-division (SD) architecture and can provide throughput of 2.56 Tb/s [6], [7]. We have also developed an optical receiver with a fast bit-synchronization circuit [8] and optical interfaces with an instantaneous-frame-reception and automatic-frame-alignment technologies [9], which enables optical packet reception at a bit rate of 10 Gb/s with short guard time.

Our proposed Photonic Core Node consists of an optical crossbar switch, electrical input queues and a scheduler [10]. An input-queued switch does not severely require high-speed buffer memory for enlargement of a switch size and speedup of a line rate, so

Manuscript received November 21, 2000.

†The authors are with Computer & Communication Media Research, NEC Corporation, Kawasaki-shi, 211-8666 Japan.

††The authors are with Development Laboratories, NEC Networks, NEC Corporation, Abiko-shi, 270-1198 Japan.

†††The authors are with Network Node Division, NEC Networks, NEC Corporation, Fukuoka-shi, 814-8560 Japan.

††††The author is with NEC Communication Systems Kyushu, Ltd., Fukuoka-shi, 814-8560 Japan.

†††††The author is with IP Network Division, NEC Networks, NEC Corporation, Abiko-shi, 270-1198 Japan.

*This paper is also published in the IEICE Trans. Electron., Vol.E84-C, No.5, pp.485–492, May 2001.

that it can be electronically implemented. However, it requires severe performance of a scheduler.

In this paper, we propose Photonic Core Node, which accommodates WDM optical path grooming/multiplexing, ATM switching and IP routing functions. We have successfully built the first prototype of Photonic Core Node equipped with ATM switching function, employing a super-packet-switching scheme and a desynchronized-round-robin (DRR) scheduler. We have chosen ATM for the first prototype implementation, because ATM has the most severe switching performance requirements among all the types of networks. The developed 2 × 2 Photonic Core Node prototype consists of a partially installed 256 × 256 upgradable opto-electronic switching fabric, but it is fully functional with super-packet buffers, and a DRR scheduler. We have demonstrated error-free and no-cell-loss switching and stable operation by downloading MPEG video streams over ATM for as long as nine hours.

2. Photonic Core Node

2.1 Concept and Architecture

Figure 1 shows the Photonic Core Node concept model. Photonic Core Node exhibits optical cross-connecting function and/or IP core routing/ATM core switching function simultaneously, sharing the ultra-large-scale opto-electronic switching fabric. The optical cross-connecting function is realized by simply connecting the WDM transponders to the opto-electronic switching fabric. An IP routing/ATM switching function requires data buffering and contention-resolution logic, for which the optics has not yet matured. To implement an IP routing/ATM switching function, an opto-electronic switching fabric should be used as the juncture switch for expanding the capacity of the ATM switches and IP routers. Therefore, in Photonic Core Node, the data buffering is performed using electronic buffers in each switch and router. The contention-resolution logic is performed by an electronic switch controller, which resolves traffic contention among switches and routers.

The first prototype of Photonic Core Node (Fig. 2) consists of a crossbar-equivalent opto-electronic switching fabric, super-packet buffers, and a DRR scheduler, a contention resolution logic circuit.

2.2 2.56-Terabit/s Opto-Electronic Switching Fabric

The opto-electronic switching fabric is based on our wavelength-division/space-division (WD/SD) optical switch architecture which is scalable up to 2.56 Tb/s capacity as shown in Fig. 3. It utilizes WDM technology in order to reduce the switching fabric's physical size and power consumption. Among previously proposed optical switching devices [4], we employ semiconductor optical amplifier gates (SOAGs) for a fast switching time around 1ns, loss compensation, high on/off ratio, and possible integration [5]. The WD/SD optical switch architecture has a full-crossbar-equivalent function and multicasting capability, and requires far fewer SOAGs and higher modularity than conventional space-division (SD) architecture [11], which is shown in Fig. 4.

A conventional SD optical switch works as follows: 256 input signals are broadcast through two hundred fifty-six EDFAs and two hundred fifty-six 1 × 256 splitters to every 256 × 1 selector, which consists of 256 SOAGs, and the selector chooses one signal by turning on one SOAG. Therefore, this architecture needs 65536 SOAGs and 256 EDFAs for a 256 × 256 switch.

Compared to a conventional SD optical switch, our proposed WD/SD optical switch architecture can drastically reduce the number of devices. A schematic diagram of our proposed 256 × 256 WD/SD optical switch architecture using 16 wavelength-channels is shown in Fig. 3. On the input side, each of 256 input ports is pre-assigned to one of 16 wavelength-channels of 16 spatial-groups. Each input signal is broadcast to every 256 × 1 WD/SD selector via a 16 × 1 wavelength multiplexer (λ-MUX), an Erbium-doped fiber amplifier (EDFA) to boost the optical signal strength and a 1 × 256 splitter, implemented as 1 × 16 splitters in a two-stage tree-structure. Boosting the signal strength ensures that enough optical power is distributed to each

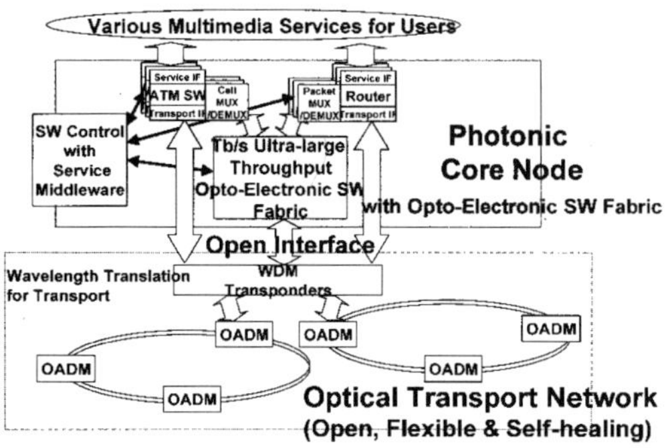

Fig. 1 Photonic core node concept model.

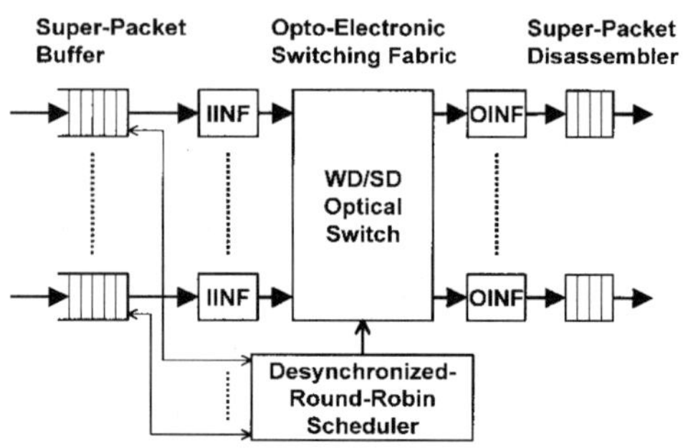

Fig. 2 Photonic core node adapted to ATM.

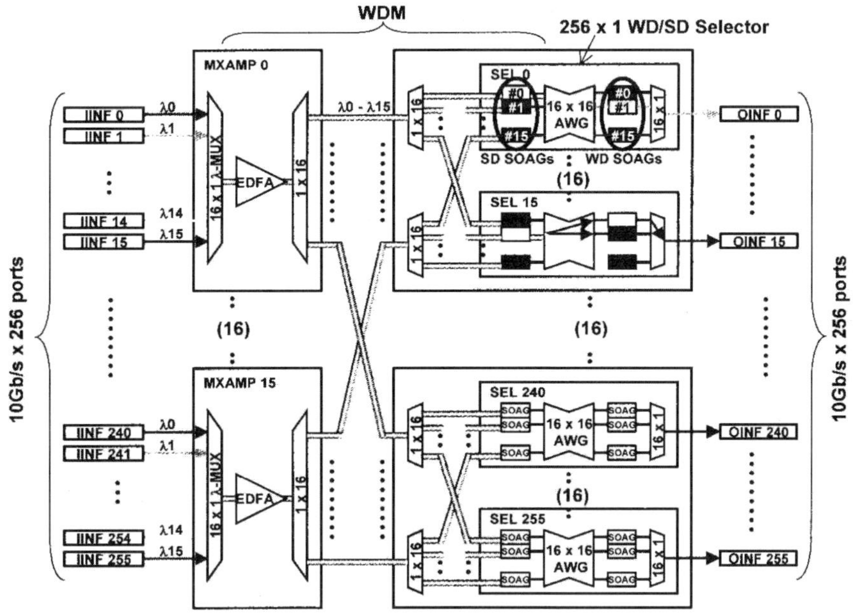

Fig. 3 Opto-electronic switching fabric (WD/SD optical switch architecture).

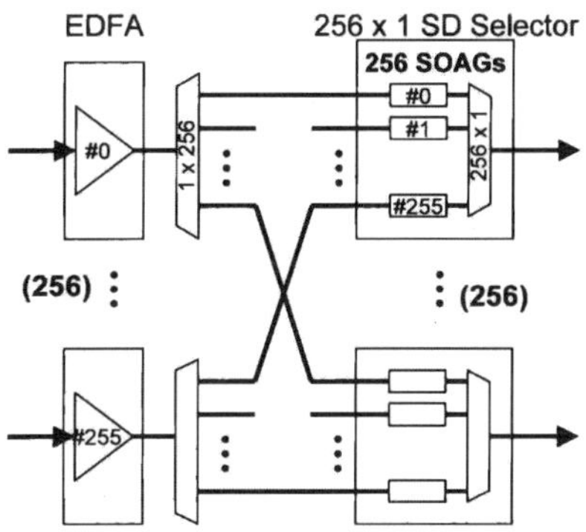

Fig. 4 Conventional space division (SD) optical switch.

WD/SD selector. Since the EDFA simultaneously amplifies 16 signals, the number of required EDFAs is reduced to 16 (256 for a SD optical switch). On the output side, each output port equips a WD/SD selector, which consists of 32 SOAGs (two arrays of 16 SOAGs), a 16 × 16 AWG router and a 16 × 1 passive combiner. The first-stage SOAG array of each 256 × 1 WD/SD selector chooses one spatial-group signal containing 16 wavelength-channels. Then the WDM signal of the chosen group is demultiplexed as it passes through an arrayed waveguide grating (AWG) router [12]. The second-stage SOAG array selects one wavelength-channel out of 16 wavelength-channels. By using WDM technologies and the wavelength routing function of the AWG, the number of SOAGs required is reduced to 32 (256 for a SD optical switch), and the loss due to combining between the first-stage and the second-stage SOAG arrays is reduced by more than 12 dB. Therefore, WD/SD optical switch architecture requires only 8192 SOAGs to implement a 256 × 256 switch, and this number is 1/8 the amount of those used for a SD switch.

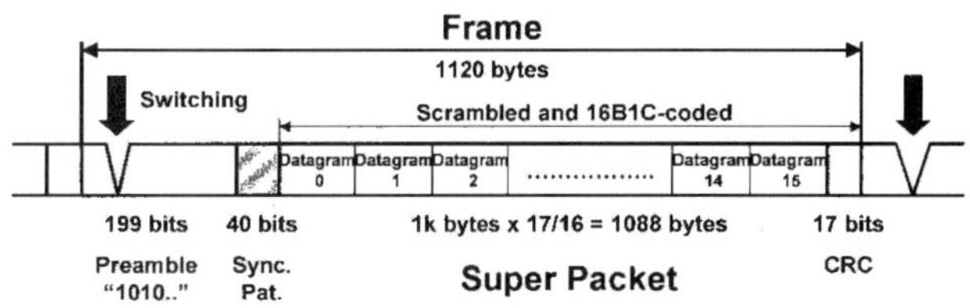

Fig. 5 Frame format in opto-electronic switching fabric.

In our architecture, each selector independently selects one channel from among the 256 input channels, so the architecture is logically equivalent to a strictly internally non-blocking full-crossbar switch. It also supports multicasting capability. The number of input/output ports can be easily increased in terms of modularity, by adding more optical interfaces, EDFA boosters, and optical-selector packages.

2.3 Super-Packet Switching with a Desynchronized-Round-Robin Scheduler

The opto-electronic switching-fabric works along with super-packet buffers as input queues and with a desynchronized-round-robin (DRR) scheduler. A super-packet buffer is equipped with Virtual Output Queues (VOQs) [13] in order to avoid head-of-line (HOL) blocking [14], which limits the throughput of an input-queued packet switch to 58.6%. A super-packet buffer assembles 16 ATM cells routed to the same output port into a 1120-byte frame. Figure 5 shows the frame format in the opto-electronic switching fabric. The 199-bit preamble pattern is needed to supply enough guard time of optical switching and bit-synchronization at an optical receiver. The 40-bit synchronization pattern is used for frame synchronization. The 17-bit 16B1C-coded Cyclic Redundancy Check (CRC) bytes are used for transmission error detection. The total overhead in a frame is 256 bits (32 bytes). The payload, we call a super packet, accommodates sixteen 68-byte 16B1C-coded datagrams. The ratio of the overhead to the whole frame is 32B/(32B+68B*16), 2.9%. If the payload contains only one datagram, the ratio of the overhead to the whole frame inflates up to 32B/(32B+68B), 32%. The super-packet-switching scheme drastically can, therefore, reduce the overhead of optical switching from 32 to 2.9%.

Figure 6 shows a schematic diagram of the DRR scheduler. The scheduler has request queues and two stages of round-robin arbiters, maintaining fairness in a round-robin manner. And it recursively resolves the contention of requests from input ports to output ports in one scheduling routine as described in Fig. 7. Request queues are arranged so as to represent the state of occupancy for each VOQ. The request queues receive information of assigned input ports and output ports after each sub-scheduling routine. The request queues do not send a request to the assigned output ports. This protocol avoids the HOL blocking, and gains nearly 100% effective throughput.

A round robin arbiter operation cycle is estimated at 68 ns for a 256-port arbiter, employing 0.35-μm CMOS technology. That means the scheduler requires at least 136 ns for a single sub-scheduling routine.

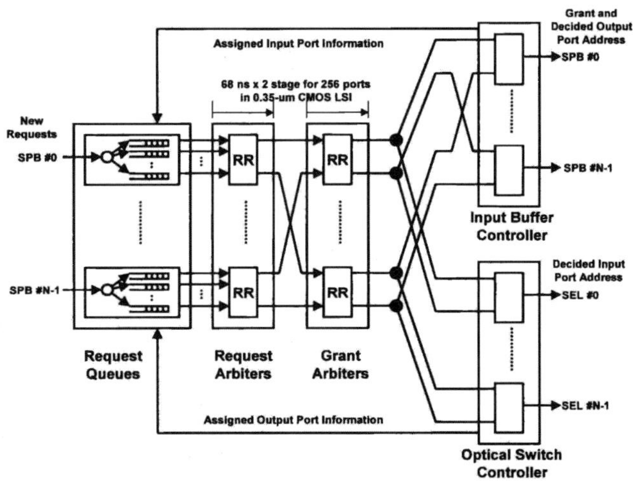

Fig. 6 Desynchronized-round-robin (DRR) scheduler.

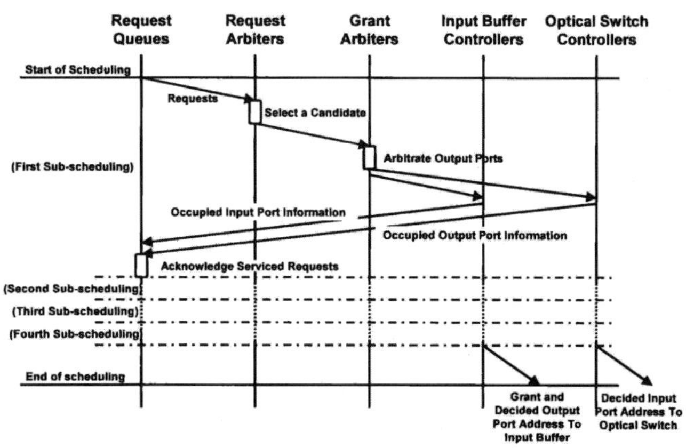

Fig. 7 DRR scheduling procedure.

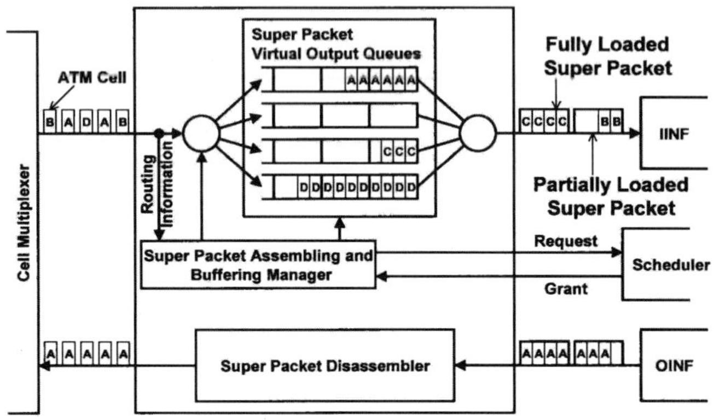

Fig. 8 Super packet assembling in a super packet buffer.

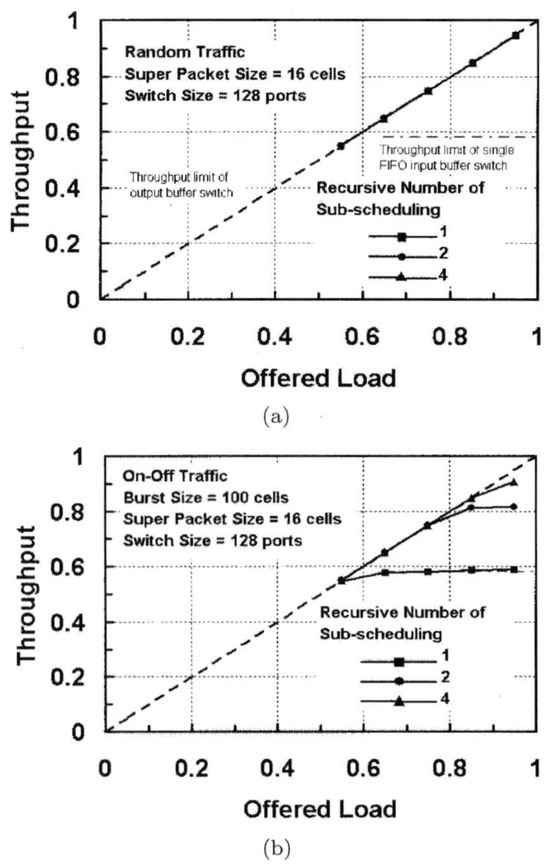

Fig. 9 (a) Throughput performance of DRR scheduler (random traffic). (b) Throughput performance of DRR scheduler (burst traffic).

Meanwhile, 1 time slot of an ATM cell is 44.2 ns at a line rate of 10 Gb/s (OC-192). Therefore, a 256-port scheduler cannot complete even one sub-scheduling routine in an ATM time slot at a line rate of 10 Gb/s. However, a super-packet time slot is 707.2 ns, which is enough for the scheduler to repeat a sub-scheduling routine four times.

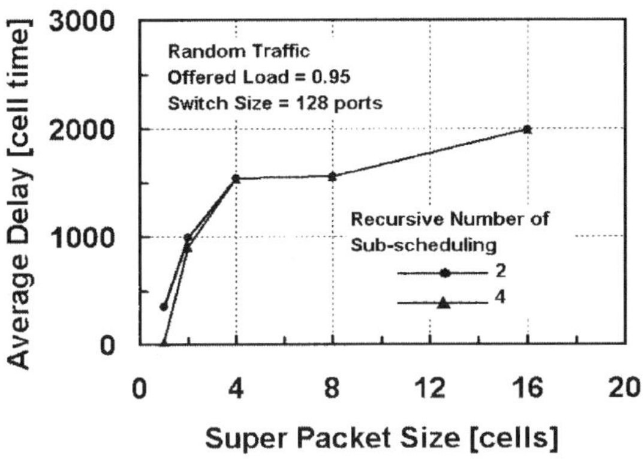

Fig. 10 Delay performance of DRR scheduler.

2.4 Performance Analysis of Super-Packet Switching

As we mentioned in the previous section, the super-packet-switching scheme drastically reduces the overhead of optical switching, and gains enough time to schedule 256 ports. However, the super-packet switching scheme tends to decrease effective throughput. In order to avoid unacceptable long latency, a scheduler occasionally permits a super-packet buffer to send a partially loaded super packet as described in Fig. 8. Even though this operation can affect the switch performance, the DRR scheduler maintains high effective throughput.

We evaluated the throughput performance for a DRR scheduler in computer simulation. We assume that the size of the super-packet buffer is 4000 cells

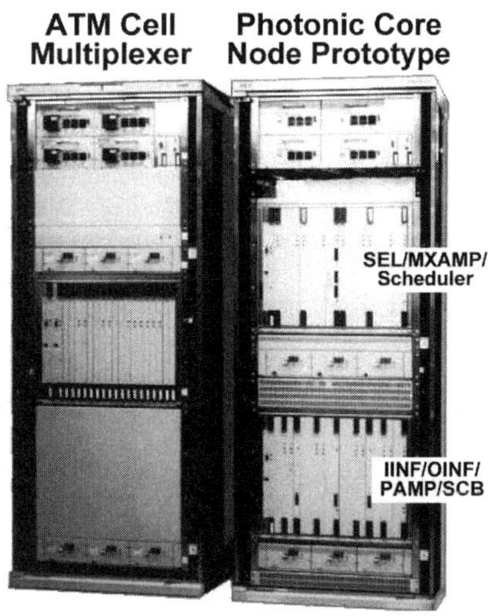

(a)

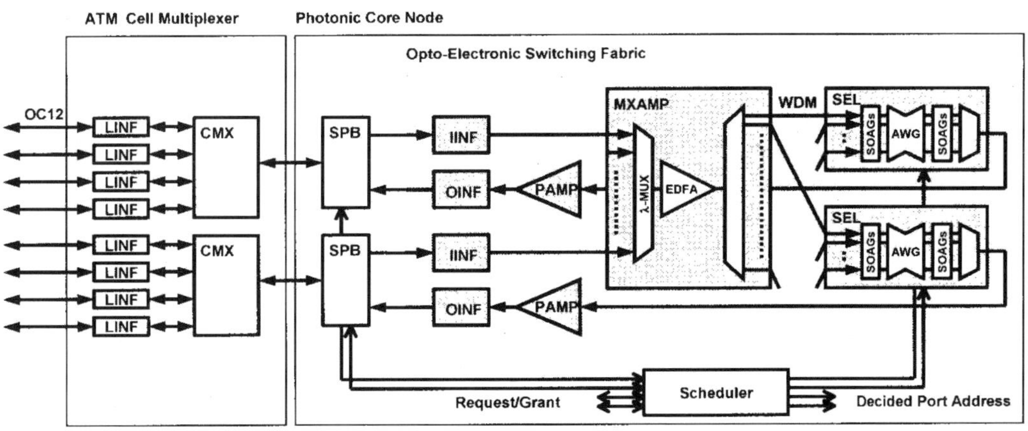

(b)

Fig. 11 (a) 2×2 photonic core node prototype (cabinet: H 1800 × W 750 × D 600 mm^3). (b) Configuration of the developed prototype system.

per port and the size of the switch matrix is 128 ports. The simulation was performed for 10^6 cell time. Figures 9(a) and 9(b) show the simulation results for random traffic and burst traffic respectively. The recursive number of sub-scheduling in a scheduling routine is varied from 1 to 4 in this simulation. The scheduler achieves effective throughput of nearly 100% with random traffic for every recursive number. With burst traffic, the performance drastically improves in proportion to the recursive number. When the recursive number of sub-scheduling is one, the throughput performance remains less than 60%. When the recursive number is increased to four, it achieves 85% throughput. More times of sub-scheduling can avoid the head-of-line blocking, when the destinations of super packets are concentrated to a part of the output ports.

We also estimated the delay performance of the DRR scheduler in the simulation. Figure 10 shows the average DRR-scheduler delay as a function of super packet size offering a 95% load of random traffic. In our super packet format, the super packet size is equal to 16 cells. The delay performance of 16-cell super packet is 2000 cells. A 2000-cell delay corresponds to less than 90 μs at a line rate of 10 Gb/s. This delay performance is acceptable for a practical use.

3. Prototype System

Figure 11(a) shows a photograph of the developed prototype system. A configuration of the prototype system is shown in Fig. 11(b). There are two cabinets. One is for a Photonic Core Node prototype, which has the 2×2 WD/SD opto-electronic switching fabric, a 4 × 4 DRR scheduler and two super-packet buffers. The other is for an ATM cell multiplexer as an interface for ATM networks. The interface speed between Photonic Core Node and the cell multiplexer was set to 5 Gb/s obeying the cell muliplexer's clock speed.

One of the most important parts in the WD/SD opto-electronic switching fabric is our newly developed 256 × 1 WD/SD selector. Figures 12(a) and 12(b) show a photograph and a configuration of the 256×1 WD/SD selector. The selector includes two arrays of sixteen SOAG modules, an arrayed waveguide grating (AWG), flexible optical fiber sheets and electrical control circuits. All the optical components are densely and well orderly mounted as never been done before.

We have successfully demonstrated the Photonic Core Node prototype system, and confirmed the stable operation of the system by downloading MPEG video streams through the system for as long as nine hours, and also confirmed error-free and no-cell-loss operation with an ATM analyzer.

4. Conclusions

The first prototype of Photonic Core Node based on

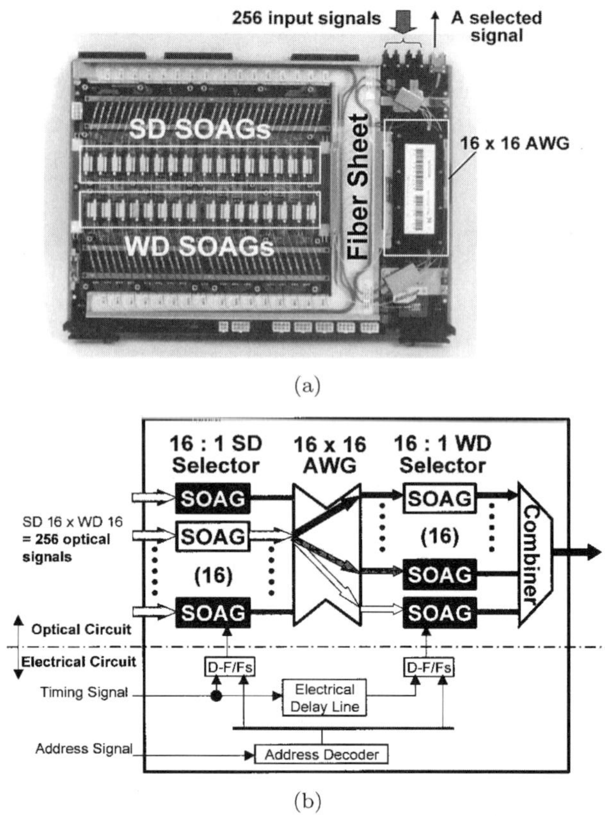

Fig. 12 (a) 256 × 1 WD/SD selector package (W 414 × H 222 mm^2). (b) Configuration of the WD/SD selector.

a 2.56-Tb/s opto-electronic switching fabric has been successfully developed, employing super-packet switching scheme and a desynchronized-round-robin scheduler. The prototype system has operated stably. Our WD/SD architecture drastically reduced the amount of optical switching elements to 1/8 that of a conventional optical switching fabric. From computer simulation results, we have verified that the DRR scheduler can achieve nearly 100% effective throughput avoiding the HOL blocking problem.

The proposed Photonic Core Node will accommodate not only ATM switching but also WDM optical path grooming/multiplexing, and IP routing by supplying it with IP input buffer interfaces, because optical switches are bit-rate/format-independent. We believe this technology will be a key to the installation of future broadband networks.

References

[1] N. Henmi, "Towards Tb/s throughput optical switching network," Tech. Digest of OECC'98, 16A3-1, 1998.
[2] N. McKeown, M. Izzard, A. Mekkittikul, W. Ellersick, and M. Horowitz, "The tiny tera: A packet switch core," IEEE Micro, pp.26–33, Jan./Feb. 1997.
[3] N. Yamanaka, S. Yasukawa, E. Oki, T. Kurimoto, T. Kawamura, and T. Matsumura, "OPTIMA: Tb/s ATM switching system architecture based on highly statistical op-

tical WDM interconnection," International Switching Symposium (ISS'97), IS-02.8, pp.255–260, 1997.

[4] "Special issue on photonic packet switching technologies, and systems," IEEE J. Lightwave Technol., vol.16, no.12, Dec. 1998, see also "Joint special issue on photonics in switching: Systems and devices," IEICE Trans. Commun., vol.E82-B, no.2, Feb. 1999.

[5] S. Kitamura, H. Hatakeyama, T. Kato, N. Kimura, M. Yamaguchi, and K. Komatsu, "Very-low-operating-current SOA-gate modules for optical matrix switches," Conf. Optical Amplifiers and Their Applications, TuC3-1, Victoria, BC, Canada, 1997.

[6] Y. Maeno, Y. Suemura, A. Tajima, and N. Henmi, "A 2.56-Tb/s multiwavelength and scalable switch-fabric for fast packet-switching networks," IEEE Photon. Tech. Lett., vol.10, pp.1180–1182, 1998.

[7] S. Araki, S. Takahashi, Y. Maeno, Y. Suemura, A. Tajima, H. Takahashi, K. Matsuda, T. Tanamuki, S. Dohmae, and N. Henmi, "A 2.56 Tb/s throughput packet/cell-based optical switch-fabric demonstrator," Proc. ECOC, Post Deadline Paper, p.127, Madrid, 1998.

[8] A. Tajima, H. Takahashi, S. Araki, and N. Henmi, "A 10-Gb/s optical asynchronous cell/packet receiver with a fast bit-synchronization circuit," Tech. Digest of OFC, San Diego, TuI6, 1999.

[9] Y. Maeno, Y. Suemura, S. Araki, S. Takahashi, A. Tajima, H. Takahashi, and N. Henmi, "A bandwidth-effective Terabit/s opto-electronic packet-switch with skew-insensitive synchronisation scheme," Proc. ECOC, TuA1.2, Nice, 1999.

[10] S. Araki, N. Henmi, Y. Maeno, K. Matsuda, K. Saisho, M. Shinohara, Y. Suemura, A. Tajima, H. Takahashi, and S. Takahashi, "Photonic core node based on a 2.56-Terabit/s opto-electronic switching fabric," WTC/ISS2000, 2000.

[11] S. Takahashi, T. Kato, H. Takeshita, S. Kitamura, and N. Henmi, "10 Gbps/ch space-division optical cell switching with 8×8 gate type switch matrix employing gate turn-on-delay compensator," Tech. Digest of the Photonics in Switching, PThC1, 1996.

[12] H. Takahashi, K. Oda, H. Toba, and Y. Inoue, "Transmission characteristics of arrayed waveguide $N \times N$ wavelength multiplexer," IEEE J. Lightwave Technol., vol.13, no.3, pp.447–455, 1995.

[13] Y. Tamir and G. Frazier, "High-performance multi-queue buffers for VLSI communication switches," Proc. 15th Annu. Int. Symp. Comput. Architecture, pp.343–354, 1988.

[14] M. Karol, M. Hluchyj, and S. Morgan, "Input versus output queueing on a space-division packet switch," IEEE Trans. Commun., vol.COM-35, no.12, pp.1347–1356, 1987.

High-Performance Backbone Network Technology edited by Naoaki Yamanaka
ISBN: 0-8247-5321-6 DOI: 10.1081/(120025739)
Published 2004 by Marcel Dekker, Inc.

Large-Capacity Photonic Packet Switch Prototype Using Wavelength Routing Techniques

Keishi Habara, Hiroaki Sanjo, Hideki Nishizawa, Yoshiaki Yamada, Shigeki Hino, Ikuo Ogawa, and Yasumasa Suzaki

SUMMARY A rack-mounted prototype packet switch that makes use of wavelength-division-multiplexing (WDM) interconnect techniques has been developed. The switch has a maximum throughput of 320 Gbit/s. It features a WDM star-based switch architecture, an electrical control circuit layer and a broad-bandwidth optical WDM layer. The basic characteristics of the broad bandwidth WDM layer, such as level diagram, transmission characteristics, 32-wavelength-channel switching, and high-speed optical gating within a 1.6-ns guard time, are described. Experimental results demonstrated that the switch can perform practical self-routing switch operations, such as address-extraction, optical buffering, and filtering for packet speeds of up to 10 Gbit/s. The switch is promising for such applications as a terabit-per-second switching node in future WDM transport networks.

key words: *photonic packet switch, WDM broadcast-and-select switch, optical buffer, wavelength selector, optical burst mode receiver*

1. Introduction

The current rapid growth of data traffic, such as Internet traffic, has highlighted the need for Tbit/s wavelength-division-multiplexing (WDM) transmission systems that have intelligent IP routers. In the near future, large-scale packet switching nodes will have to have a total throughput of over 1 Tbit/s. High signal speed due to transparency to the data rate and a photonic bandwidth over several tens of terahertz are the main features of photonic switching. Thus, application of photonic technologies to the switching systems has been intensively studied, and several switch architectures have been reported [1]–[5]. In particular, WDM-based systems have drawn much attention.

It is essential that the switching operation be done packet-by-packet, instead of bit-by-bit switching, in photonic switching systems because this packet-rate photonic switching technology can handle very high-capacity packet transmission. On the other hand, photonics are not yet practical for logic functions such as packet header handling, routing control and precise synchronization. Terabit-per-second packet switches therefore require an efficient combination of electronics and photonics. Thus, high-speed electrical circuits capable of the complex processing required in large-scale photonic packet switches are needed as well as equipment for the photonic switching layer itself.

We have already demonstrated WDM photonic switching systems that have throughputs of several tens of Gbit/s [6], [7]. Our aim with those systems was to confirm the basic operating principles of photonic packet switching systems and not to test the maximum throughput capability of photonic switching. In addition, a control circuit for the routing process was not implemented. The basic design for a WDM photonic switching system that has a larger throughput and a routing process function has been discussed in Ref. [8]. However, the actual implementation of such a system has not yet been realized.

This paper describes a large-capacity packet switch prototype that makes use of WDM interconnect techniques. The switch has 32-channel 10-Gbit/s inputs and outputs and thus, has a maximum throughput of 320 Gbit/s. First, the WDM star-based broadcast-and-select output-buffer-type switch architecture and prototype switch configurations are described. An electrical control circuit layer, as well as a broadband optical WDM layer, was implemented in the prototype. The ability to process the address information contained in the high-speed packet required development of LSIs having 10-Gbit/s serial I/Os. Some optical component modules, such as the 32-channel wavelength channel selector, were developed to reduce the size of the WDM layer. Next, the basic characteristics of the broad-bandwidth WDM layer, such as its transmission characteristics, 32-wavelength-channel switching operation, and high-speed optical gating within a 1.6-ns guard time, are described. Finally, we review experiments that demonstrated that the switch can perform practical self-routing system operations, such as address extraction, optical buffering, and filtering for packet speeds of up to 10 Gbit/s.

2. Switch Architecture

Figure 1 shows the block diagram of our star architecture for a large-scale packet switch. The switch architecture is a broadcast-and-select WDM output-buffer type, and it has two separate layers: an optical WDM

Manuscript received March 1, 2000.
Manuscript revised May 17, 2000.

[†]The authors are with NTT Network Innovation Laboratories, Yokosuka-shi, 239-0847 Japan.
[††]The authors are with NTT Photonics Laboratories, Ibaraki-ken, 319-1193 Japan.

IEICE TRANS. COMMUN., Vol. E83-B, No. 10, pp. 2304–2311, October 2000

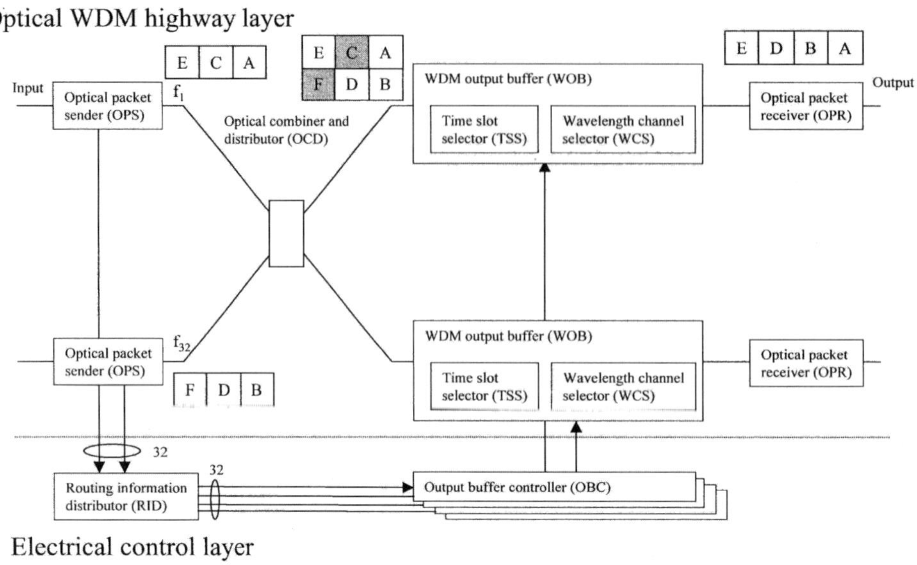

Fig. 1 $N \times N$ broadcast-and-select output-buffer switch architecture ($N = 32$).

highway layer and an electrical control layer. The advantages of the architecture are that its WDM output-buffering scheme overcomes the bottlenecks that can occur when receiving and storing concurrent high-speed cells and that there is no need to deal with contention resolution between input highways.

The WDM highway layer consists of four parts: N fixed wavelength optical packet senders (OPSs), an optical combiner and distributor (OCD), N WDM output buffers (WOBs) and N optical packet receivers (OPRs). The optical wavelengths of the input packets are converted by the OPSs, wavelength-division-multiplexed (WDMed) and distributed by the OCD, buffered by the WOBs, then received and output to the output lines by the OPRs. The WOB converts the WDMed packets from parallel input into sequential output, so only one packet is received by the OPR in one time slot. The electrical control layer consists of a routing information distributor (RID), and N output buffer controllers (OBCs).

Each OPS is assigned a fixed wavelength from f_1 to f_N. Packets arriving in the same time slot are WDMed by the OCD and broadcast to the WOBs as stacks of WDM packets. Since the switch uses the output buffered architecture and thus, the WOBs work independently of each other, the control circuits, that is OBC, can be placed in each WOB.

Each WOB consists of a time-slot selector (TSS) and a wavelength-channel selector (WCS). The TSS selects one of the time slots, in which many packets are WDMed in a stack, by holding the in-coming stacks in optical fiber delay-lines and reading out one stack. The WCS selects one wavelength in each packet stack output from the TSS.

The packets are aligned at each WOB input. All switching operations, that is, WOB resetting and OPR synchronization establishment, are completed within a guard time, which is assumed to be a few nano-seconds. The destination address information of each incoming packet is extracted packet-by-packet in the OPS by using electric LSIs having high-speed O/E- and E/O-serial interfaces.

Table 1 320-Gbit/s system specifications.

Architecture	Broadcast-and-select with WDM output buffers
Line speed	9.95328 Gbit/s (STM-64)
Number of inputs/outputs	32/32
Wavelength-channel	f1: 192.0 THz~f32:195.1 THz, ITU-T grid (1536-1561 nm)
Wavelength-channel span	100 GHz
Packet format	64 bytes fixed length (including 2-byte guard time)
Buffer size per output	15 packets
Internal transmission code	IM/Manchester code
Packet frame clock	19.44 MHz
Frame synchronization	Optical frame pulse (f0:191.9 THz) detection method

This switch architecture features non-block switching, and the switch works as a large-capacity unit-switch element. Therefore, terabit-per-second switching node can be realized using a multi-stage switching network configuration.

3. Prototype Switch Configuration

The prototype is a 10-Gbit/s and 32-wavelength-channel rack-mounted switch. The system specifications and configuration are shown in Table 1 and Fig. 2. The signals were modulated at a rate of 9.95328 Gbit/s (rate of STM-64). The wavelength channels were spaced 100 GHz apart and set to $f_1 = 1561.42$ nm (192000 GHz) and $f_{32} = 1536.61$ nm (195100 GHz). In contrast with the architecture in Fig. 1, the routed packet selector (RPS) is placed after the OCD. The RPS has the same structure as the WCS, and it selects only the packets destined to a specific output. The RPS

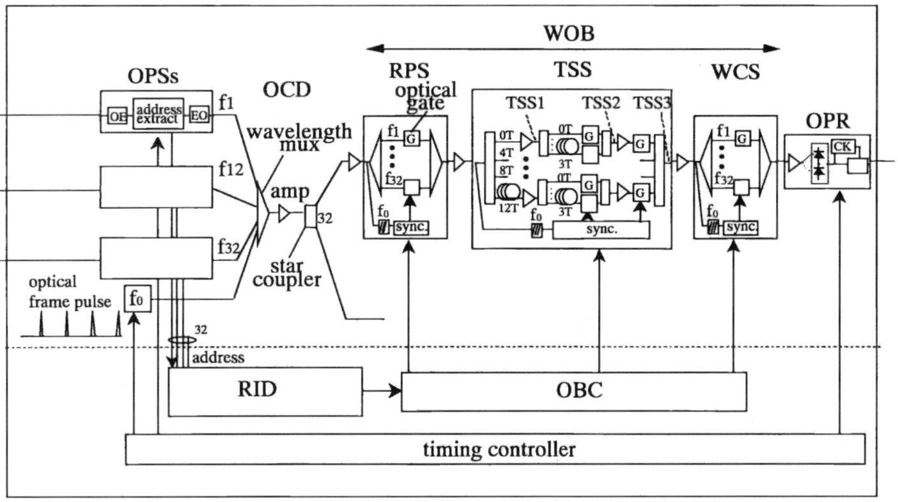

Fig. 2 Configuration of prototype system.

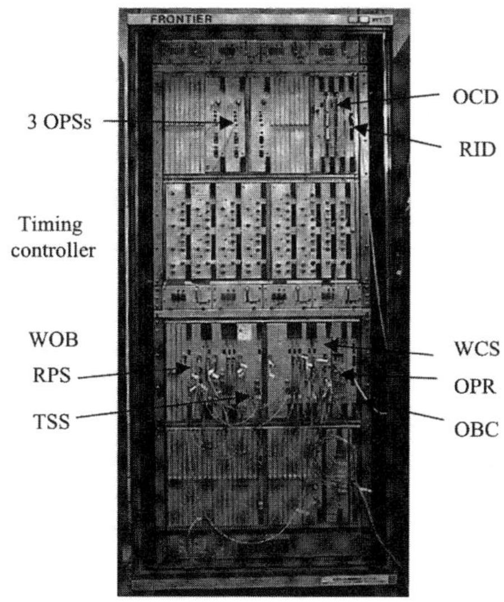

Fig. 3 Photograph of prototype system.

reduces the average number of multiplexed wavelengths from 32 to about 1 and helps increase the signal to noise ratio (SNR) [9].

Figure 3 shows a photograph of the prototype. Three OPSs (f_1, f_{12} and f_{32}) and one WOB were mounted in the rack. For the electrical control circuits, one RID board and one OBC board were implemented. One RID board was capable of controlling up to 32 OPSs and 4 WOBs, although only one WOB was actually mounted.

The packet format is shown in Fig. 4. The packet length is fixed at 64-bytes. The format includes a 1.6-ns guard time for optical gating, 7-byte synchronization patterns for instantaneous clock and frame synchronization, 2-byte internal routing information, and 53-byte payload. The 2-byte routing information contains a 5-bit internal routing address (32 ports), 1-bit ACTive for indicating packet activity (ACT), 1-bit BroadCast mode (BC), 1-bit MultiCast mode (MC), 1-bit Test mode (T), and 7 bits for further applications.

Figure 5 shows the configuration of the OPS. The OPS consists of a PIN-PD, a DFB-LD, an LN intensity modulator and electrical circuits for routing information extraction. The input packets are converted to electrical signals and routing information is electrically extracted packet by packet. The packets are then transmitted at a fixed wavelength. We developed LSIs using Si-bipolar and GaAs-MESFET technologies to achieve the 10-Gbit/s OPS function [10]. The LSIs also have a packet frame synchronization function.

The OCD was originally intended to function as a 32×32 combiner and distributor, and thus, in principle, it can be made from a simple star coupler. However, the configuration had to be modified from its original form to meet the required signal to noise ratio, number of EDFAs, and number of parallel optical fibers and to simplify production. Consequently, we selected the OCD, which consisted of a 32-channel wavelength multiplexer, an EDFA, and a 32-port coupler, shown in Fig. 2. The calculated Q-factor of this configuration was increased by about 7 dB over that of the original 32 × 32 star coupler configuration [9].

Each RPS or WCS consists of a 32-channel wavelength demultiplexer and multiplexer and 32 optical gates [11]. Semiconductor laser diode optical gates (SLGs) were used for the optical gates, and 32 SLGs were implemented using hybrid-integrated PLC technologies. These technologies make it possible to fabricate small selector modules and to equalize the optical path lengths of different wavelength channels.

The TSS consists of splitters, fiber delay-lines, optical gates, and combiners. The delay time was set to 0T–15T [12], where T is the unit delay time corresponding to the fixed packet length. Multiple wavelength channels can be delayed in a single fiber delay line, and thus, the number of required optical components can be

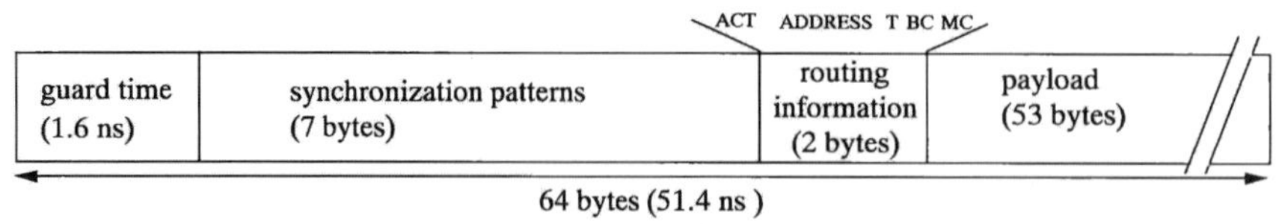

Fig. 4 Packet format.

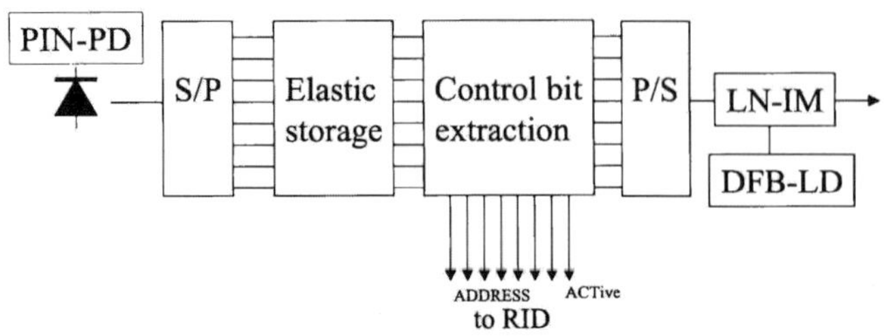

Fig. 5 Optical packet sender (OPS).

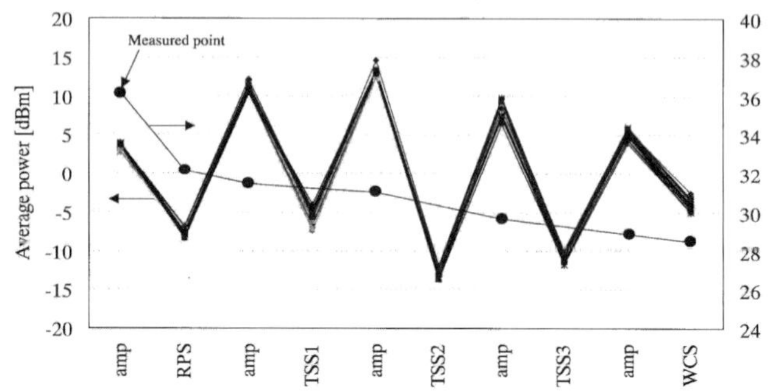

Fig. 6 Level diagram and calculated SNR.

reduced. Electro-absorption gates were used for the optical gates because of their high-power gain-saturation characteristics. If the SLGs are used, the signals are degraded because of the non-linear saturation characteristics of the SLGs. The maximum delay time of 15T is not sufficient from the viewpoint of traffic characteristics, but a cascaded configuration of the TSSs or a composite electrical/optical buffer configuration could solve this problem [13].

Manchester-coded signals were used for transmission coding, and a differential receiver was used in the OPR [14]. The use of Manchester-code and differential receiver in combination is tolerant to packet-by-packet level fluctuations and coherent crosstalk [15]. In addition, the waveform distortion produced by the SLGs in the gain-saturated condition can be suppressed by using Manchester coding [16]. A hybrid-integrated PLC technology was also used to reduce the size of the photoreceiver module [17]. The OPR can receive optical burst packets having more than 9-dB packet-by-packet level fluctuations. The OPR also includes an instantaneous clock recovery circuit and can receive optical burst packets that have packet-by-packet phase fluctuations [18].

In the electrical control layer, the RID multiplexes and converts the routing information from 32 OPSs into bit-map data, where each bit corresponds to the input port number, and delivers them to the OBCs. The OBC resolves the contention problem and creates on-off signals to drive the optical gates in the WOB. The RID and OBC are composed of field programmable gate arrays (FPGAs), and each of them was mounted on a single printed circuit board. One RID board can control 32 OPSs and 4 OBCs. The signal rate of the bit-map data between RID and OBC was 622 Mbit/s (= 32 bits in a 19.44-MHz frame), and optical interconnection was used for the transmission.

Another important issue is the frame synchronization between the optical WDM layer and the electrical control layer. We used the optical frame-pulse detection method in the prototype system. The optical frame pulses (pulse width = 1 ns) of f_0 (191900 GHz) were multiplexed with a precise time position and were detected at the inputs of the RPS, TSS and WCS. The detected pulses trigger the gate on-off signals from the OBCs as shown in Fig. 2. By using this method, once the optical frame pulse position is initially set, the electrical control signals can always control the optical gates at the proper timing even when the length of the optical path changes due to temperature fluctuations, or other causes.

4. Experiments

In this section, several experimental results are described. First, the basic characteristics of the broad bandwidth WDM layer, such as level diagram, transmission characteristics, 32-wavelength-channel switching, and high-speed optical gating within a 1.6-ns guard time, are evaluated. Then, experimental results demonstrate that the switch can perform practical self-routing switch operations, such as address-extraction, optical buffering, and filtering.

The level diagram and transmission characteristics were measured to evaluate the broad bandwidth of the optical WDM layer. Figure 6 shows the measured level diagram of the 32 wavelength channels in the WOB and the calculated SNR. During the measurements, the optical path in the TSS was fixed to T0. The measuring points in the TSS were set to TSS1, TSS2 and TSS3, as indicated in Fig. 2. The lowest power level was −13.8 dBm at TSS2, and it determined the total transmission characteristics of the whole system. The SNR at the output of the amplifier just before the RPS was measured. The SNRs at the following modules were cal-

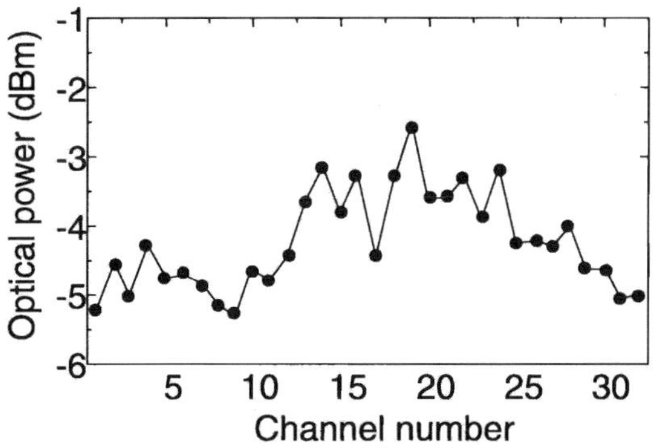

Fig. 7 Wavelength-channel power dependence of WCS output power.

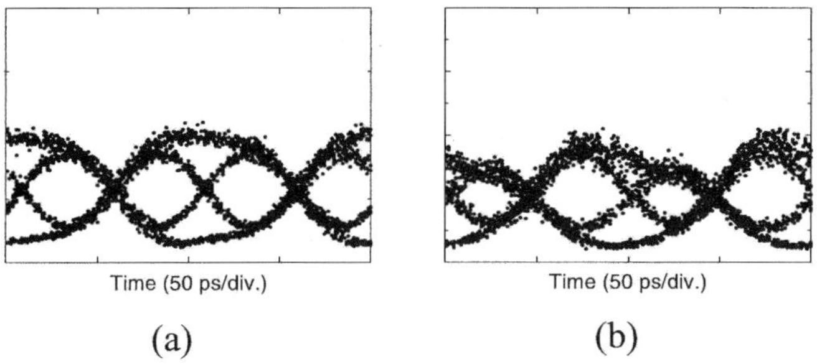

Fig. 8 Observed waveforms of Manchester code signals: (a) OPS output, (b) WCS output.

culated using device parameters. The calculated SNR at the WCS output was greater than 28 dB, which is high enough to obtain a bit-error-rate (BER) of less than 10^{-12}. In the experiment, when the power level was intentionally attenuated at TSS2, the power margin was measured to be about 10 dB.

Figure 7 shows the wavelength-channel power dependence at the WCS output. The figure is for the case when the optical path in the TSS was fixed to T0. In spite of the polarization and wavelength dependence of the optical devices, level differences between wavelength channels were less than 3 dB. This result is mainly because the SLGs in RPS and WCS function as power equalizers [16]. The polarization dependence was measured to be less than 1 dB at the WCS output, and the path dependence in the TSS (0T–15T) was less than 1 dB. Therefore, the total power level fluctuation was less than 5 dB including wavelength, polarization and optical path dependencies. It was much less than the tolerance of the OPR.

Figure 8 shows the optical waveforms at the OPS output and WCS output. The payload was a PRBS of 2^7-1. By using Manchester coding, a good eye-opening was observed and measured power penalty caused by the waveform distortion was less than 1 dB even after the signals passed through the 2-stage SLGs in the RPS and WCS. Figure 9 shows the measured BER characteristics for f_1, f_{12} and f_{32}. A PRBS of 2^7-1 was used for the payload data, and a variable attenuator was in-

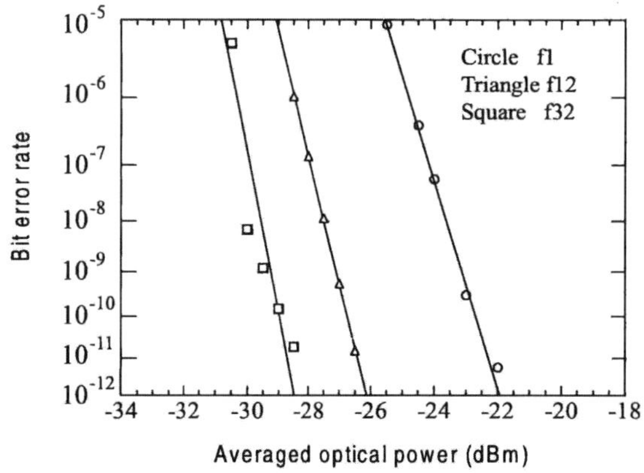

Fig. 9 Bit error rate.

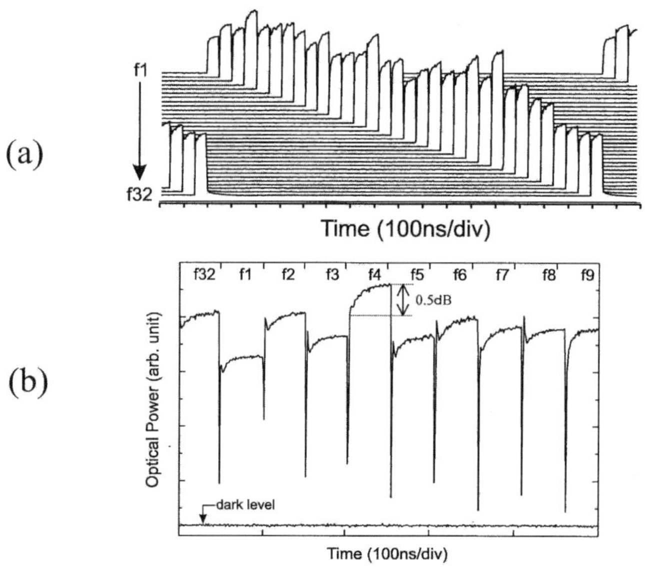

Fig. 10 32-wavelength-channel switching waveforms: (a) observed after WDM demultiplexing, (b) observed without WDM demultiplexing.

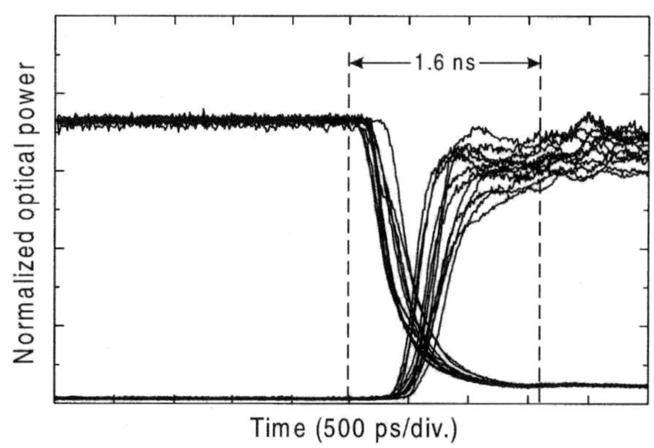

Fig. 11 Switching waveforms of RPS and WCS.

serted just before the OPR. A BER of less than 10^{-12} was achieved. The receiver sensitivity was -21.8 dBm at the BER of 10^{-12} for f_1. It has large enough power

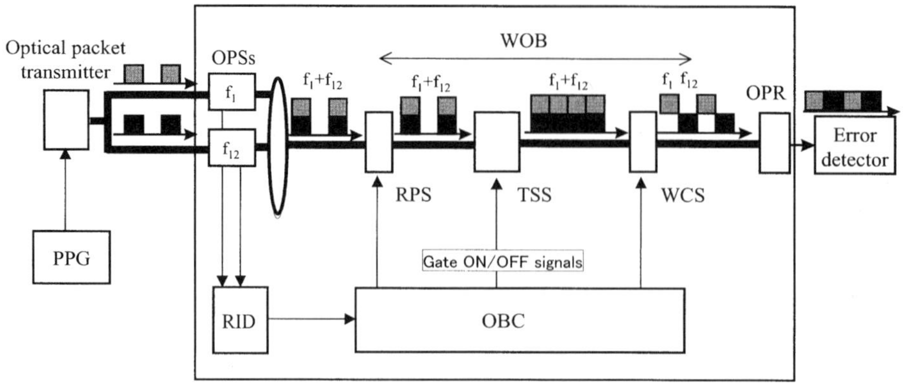

Fig. 12 Experimental setup for self-routing switch operation.

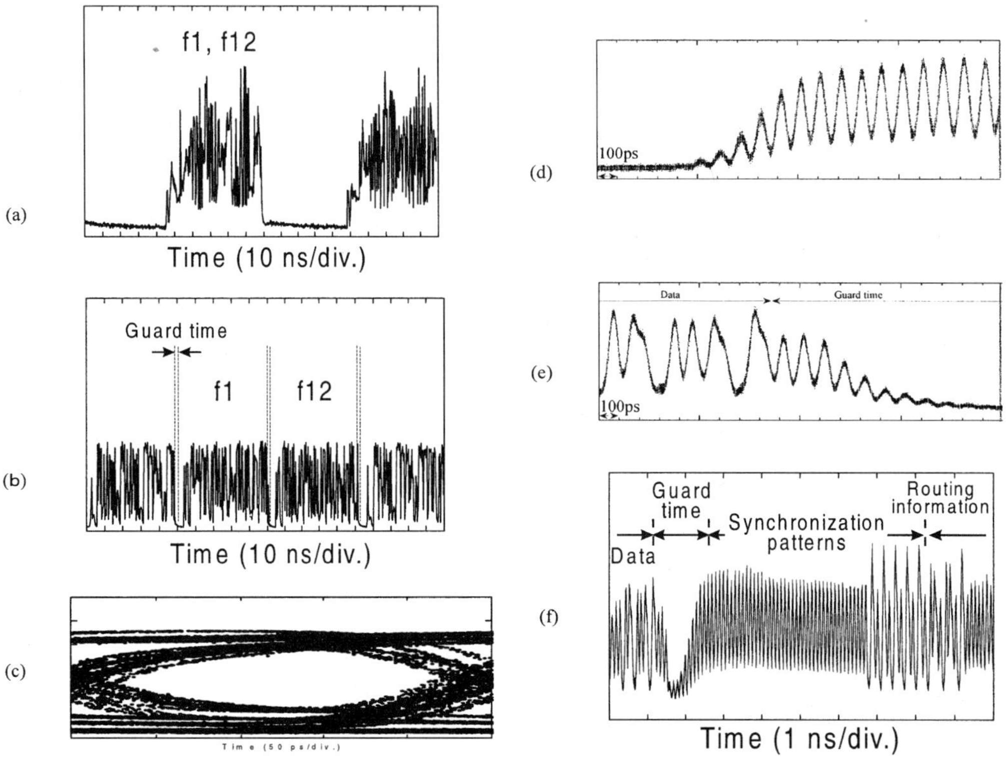

Fig. 13 Self-routing switch experiments: (a) RPS output (Manchester code), (b) OPR output (NRZ), (c) Eye diagram of OPR output (f_1 and f_{12} superimposed), (d) Guard time (rise), (e) Guard time (fall), (f) Header.

margin compared with the output power shown in the level diagram of Fig. 6. The power penalty difference between the wavelengths is mainly caused by the characteristics of the pre-amplifier EDFA in the OPR at the optical power level of $-22 \sim -28$ dBm. However, the wavelength dependence of the pre-amplifier EDFA characteristics is much smaller at the pre-amplifier's operating power level of $-2 \sim -5$ dBm as previously shown in Fig. 6.

A switching experiment of 32-wavelength channels was performed to confirm the switching operation over a wide range of wavelengths. Figure 10 shows the switched waveforms of the 32 channels at the WCS output. Thirty-two channel DC lights were injected at the OCD using a 32-channel DFB laser light source. The signals were observed after the 32-channel wavelength demultiplexing (Fig. 10(a)) or without demultiplexing (Fig. 10(b)). Packets were switched every 64 bytes (51.4 ns), and the waveform distortion within a packet was less than 0.5 dB, which was small enough for the OPR to receive the signal correctly.

The switching times of the RPS and WCS were evaluated. Figure 11 shows the switching waveforms, where the waveforms of different wavelength channels are superimposed. Both of the SLGs in the RPS and WCS were switched at the same time. As shown in the figure, switching of all the channels can be executed within a 1.6-ns guard time.

A self-routing switching operation was performed as demonstration of a practical switching operation, Figure 12 shows the experimental setup. Packets of f_1 and f_{12} were launched into the switch. The routing information of the two packets was extracted in the OPSs and analyzed in the RID and OBC. The contention problem was resolved by the WOB, and the packets were output sequentially. In the experiments, packets of the two wavelength channels were launched simultaneously every two time slots as shown in the figure. Figure 13(a) shows the WDMed f_1 and f_{12} (Manchester code) observed at the output of the RPS. Figures 13(b) and 13(c) show the contention-resolved output (NRZ) of the OPR and its eye-diagram. Figures 13(d) to 13(f) show the switching waveforms in the guard time and header part. This switching experiment also achieved a BER of less than 10^{-12}.

5. Conclusion

A 10-Gbit/s and 32-wavelength-channel photonic packet switch was developed. Electrical control circuits for self-routing operation, as well as equipment for an optical WDM layer, were implemented. The basic characteristics of the optical layer were measured. The total power level fluctuation was less than 5 dB including wavelength, polarization and optical path dependencies. We also developed a differential receiver for Manchester coding. The receiver sensitivity was -21.8 dBm, and a BER of less than 10^{-12} was achieved. Thirty-two-channel wavelength selector was also developed using hybrid-integrated PLC technologies. A switching experiment of 32-wavelength channels was successfully performed within a 1.6-ns guard time. The self-routing switching experiment demonstrated practical system operations, such as address extraction, optical buffering, and filtering for packet speeds of up to 10 Gbit/s. A BER of less than 10^{-12} was achieved in the self-routing demonstration.

References

[1] D.J. Blumenthal, P.R. Prucnal, and J.R. Sauer, "Photonic packet switches: Architectures and experimental implementations," Proc. IEEE, vol.82, no.11, pp.1650–1657, Nov. 1994.

[2] M. Renaud, F. Masetti, C. Guillemot, and B. Bostica, "Network and system concepts for optical packet switching," IEEE Commun. Mag., vol.35, no.4, pp.96–102, April 1997.

[3] M. Tsukada, W.D. Zhong, T. Matsunaga, M. Asobe, and T. Oohara, "An ultrafast photonic ATM switch based on bit-interleave multiplexing," IEEE J. Lightwave Technol., vol.14, no.9, pp.1979–1985, Sept. 1996.

[4] K. Sasayama, Y. Yamada, K. Habara, and K. Yukimatsu, "FRONTIERNET: Frequency-routing-type time-division interconnection network," IEEE J. Lightwave Technol., vol.15, no.3, pp.417–429, March 1997.

[5] A. Misawa and M. Tsukada, "Broadcast-and-select photonic ATM switch with frequency division multiplexed output buffers," IEEE J. Lightwave Technol., vol.15, no.10, pp.1769–1777, Oct. 1997.

[6] Y. Yamada, K. Sasayama, K. Habara, A. Misawa, M. Tsukada, T. Matsunaga, and K. Yukimatsu, "Optical output buffered ATM switch prototype based on FRONTIERNET architecture," IEEE J. Sel. Areas Commun., vol.16, no.7, pp.1298–1308, Sept. 1998.

[7] A. Misawa, Y. Yamada, M. Tsukada, K. Sasayama, K. Habara, T. Matsunaga, and K. Yukimatsu, "A prototype broadcast-and-select photonic ATM switch prototype with a WDM output buffer," IEEE J. Lightwave Technol., vol.16, no.12, pp.2202–2211, Dec. 1998.

[8] K. Habara, T. Matsunaga, and K. Yukimatsu, "Large-scale WDM star-based photonic ATM switches," IEEE J. Lightwave Technol., vol.16, no.12, pp.2191–2201, Dec. 1998.

[9] S. Mino, T. Matsunaga, Y. Shibata, A. Misawa, Y. Yamada, and K. Habara, "Q-factor-based level design for photonic ATM switches," IEICE Trans. Commun., vol.E82-B, no.2, pp.288–297, Feb. 1999.

[10] S. Hino, M. Togashi, and Y. Shibata, "High-speed electronic control circuits with 10 Gb/s serial I/Os for photonic ATM switching fabric," Proc. XVI World Telecom Congress, ISS'97, IS-03-14, Toronto, Sept. 1997.

[11] F. Ebisawa, I. Ogawa, Y. Akahori, K. Takiguchi, Y. Tamura, T. Hashimoto, A. Sugita, Y. Yamada, Y. Suzaki, N. Yoshimoto, Y. Tohmori, S. Mino, T. Ito, K. Magari, Y. Kawaguchi, A. Himeno, and K. Kato, "High-speed 32-channel optical wavelength selector using PLC hybrid integration," OFC'99, ThB1, 1999.

[12] S. Kitajima, H. Takano, and M. Kobayashi, "Traveling type optical cell buffer with small variation of output cell level," IEICE Trans. Commun., vol.E82-B, no.2, pp.281–287, Feb. 1999.

[13] T. Okugawa, K. Habara, and T. Matsunaga, "Composite optical/electrical buffer configuration for photonic ATM switching systems," Electronic. Lett., vol.33, no.16, pp.1398–1400, July 1997.

[14] Y. Yamada, Y. Shibata, K. Habara, and N. Yoshimoto, "High level-fluctuation tolerant optical receiver for optical packet switch and WDM cross-connect," IEEE J. Lightwave Technol., vol.16, no.12, pp.2220–2227, Dec. 1998.

[15] T. Okugawa, Y. Yamada, T. Matsunaga, and K. Habara, "Reduction of optical coherent-crosstalk-induced degradation by Manchester coding," Asia Pacific Commun. Conf., APCC'97, MoB, pp.23–27, 1997.

[16] Y. Shibata, Y. Yamada, K. Habara, and N. Yoshimoto, "Semiconductor laser diode optical amplifiers/gates in photonic packet switching," IEEE J. Lightwave Technol., vol.16, no.12, pp.2228–2235, 1998.

[17] Y. Akahori, T. Ohyama, M. Oguma, K. Kato, and Y. Yamada, "Hybrid integrated differential photoreceiver module for photonic packet switching systems using planar lightwave circuit platform," IEEE Photon. Technol. Lett., vol.10, no.6, pp.869–871, June 1998.

[18] Y. Yamada, S. Mino, and K. Habara, "Ultra-fast clock recovery for burst-mode optical packet communication," OFC '99, TuI7, 1999.

High-Performance Backbone Network Technology edited by Naoaki Yamanaka
ISBN: 0-8247-5321-6 DOI: 10.1081/(120025740)
Published 2004 by Marcel Dekker, Inc.

Optical Code Based Label Swapping for Photonic Routing

Hideyuki Sotobayashi and Ken-ichi Kitayama

SUMMARY This paper describes an all-optical label swapping for the photonic label switching router (LSR). The optical code routing photonic LSR in which label is mapped onto an optical code is one of the most promising photonic network technologies. It utilizes such unique features of optical code division multiplexing (OCDM) as asynchronous transmission, tell-and-go access protocol, and high degree of scalability. In practical photonic LSRs, all optical code conversion will play an important role. All-optical code conversion of 10 Gbit/s binary phase-shift keying (BPSK) codes by use of cross-phase modulation (XPM) in an optical fiber without wavelength-shift is proposed for the photonic LSR and experimentally demonstrated.

key words: *label swapping, photonic router, optical code division multiplexing, photonic processing, optical soliton*

1. Introduction

Networks with fast packet switching, supported by native optical transmission are necessary to cope with the continued rapid increase in traffic levels. Routing and switching technologies currently remain electronic based. With the advent of advanced optical devices, a new class of optically routed network is becoming feasible. The efficient utilization of the bandwidth of optical fiber is a major issue in the design of high-speed photonic networks. The two primary techniques for multiplexing data signals onto the channel of a single fiber are currently time division multiplexing (TDM) and wavelength division multiplexing (WDM). OCDM is an alternative method. A proper choice of optical codes allows signals from all connected network nodes to be carried without interference between signals. Simultaneous multiple access can thus be achieved without a complex network protocol to coordinate data transfer among the communicating nodes [1], [2]. Therefore, OCDM can provide certain real advantages when applied to photonic networks, due to its unique combination of qualities: asynchronous transmission, a high level of information security [1], tell-and-go access protocol [3], and high degree of scalability. The applications range from point-to-point transmission [4], multiple access [5], and an optical path network [6]. In transport networks, the optical code path, which is defined as the logical path and is determined by the optical code, has been proposed as a core concept for OCDM networks [6]. Thus, photonic IP routing based on OCDM, where the IP address is mapped onto an optical code, has been outlined recently [7]. Multihop, optical code routed, all-optical packet networks have the potential to realize high-throughput in a scalable form and data-rate based photonic routing in future photonic networks. Photonic LSRs that process headers and forward packets to either local hosts or further routers are a critical network element. An important feature of the future photonic LSR will be the ability to process labels, switch optical packets, and replace packet headers for label swapping, all in real time. Label swapping is an important technique that promises to allow simple, high-performance packet forwarding while decoupling packet routing and forwarding, and thus to support multiple routing services.

In this paper, we propose and experimentally demonstrate all-optical label swapping for a photonic LSR. We demonstrate 10 Gbit/s BPSK optical code conversion without a wavelength-shift by using XPM in an optical fiber.

2. Optical Code Routed Photonic Label Switching Router (LSR)

2.1 Principle of OCDM Operation

As shown in Fig. 1, a time spread/despread technique is adopted for our system [2]–[8], [10]. Time spread BPSK pulse codes are used as optical codes and optical transversal filters are used as optical encoders and decoders. A short optical pulse, with a frequency spectrum much higher than the data bandwidth, is spread over a one-one-bit period by the encoding process. The encoded signal is time-despread by the decoder on the receiver side, which is used as the counterpart of the transversal filter on the transmitter side. The output shows the correlation waveform, that is, the matched filtering response in the time domain. When the receiver's optical code matches the transmitter's optical code, an autocorrelation waveform, with a sharp peak at its center, is produced. On the other hand, when the codes don't match, the cross-correlation waveform is produced. When the threshold is set appropriately,

Manuscript received February 25, 2000.
Manuscript revised May 22, 2000.

[†]The author is with Communications Research Laboratory, Ministry of Posts and Telecommunications, Koganei-shi, 184-8795 Japan.
[††]The author is with the Osaka University, Suita-shi, 565-0871 Japan.

IEICE TRANS. COMMUN., Vol. E83-B, No. 10, pp. 2341–2347, October 2000

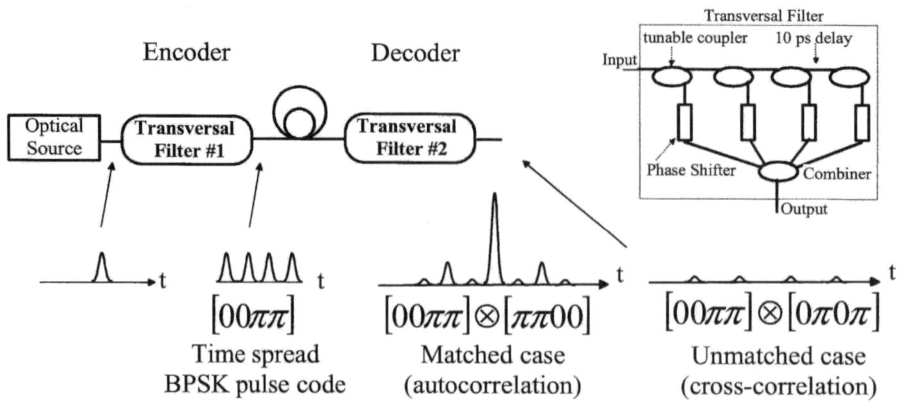

Fig. 1 Principle of OCDM based on time spread/despread basis.

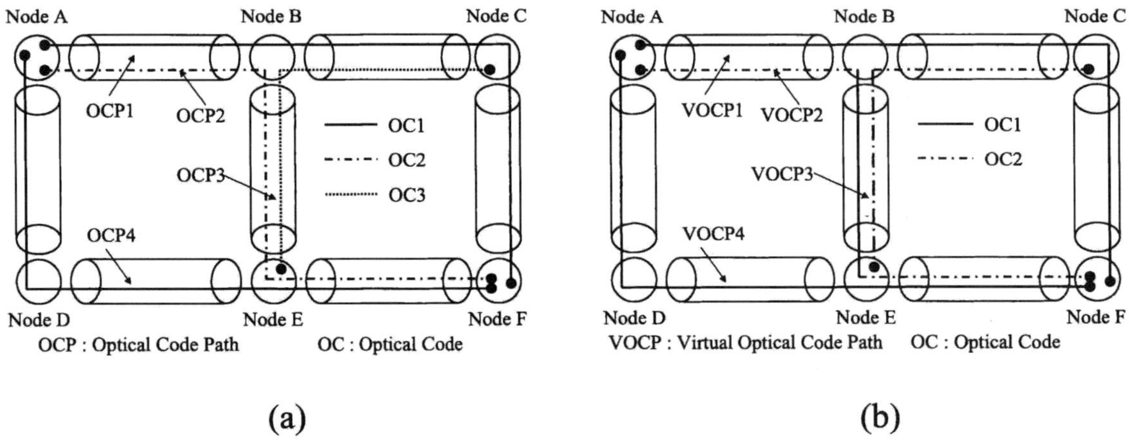

Fig. 2 Schematics of (a) optical code path (OCP) and (b) virtual optical code path (VOCP).

only matched receivers can detect the signal.

The transversal filter used in our system consists of tunable taps, delay lines, optical phase shifters, and a combiner, all of which are monolithically integrated in a planar lightwave circuit (PLC). The taps are tuned so that they produce an equi-amplitude four-pulse sequence. Each delay line has a 10 ps delay. The phase of each tapped pulse is binary changed by 0, or π by the optical phase shifter. Consequently, at the output, a train of four pulses with 10 ps separation that carries a BPSK code is generated. The BPSK optical code is denoted in the text and figures as, for example, $[00\pi\pi]$ and so on.

2.2 Optical Code Routing

In OCDM transport networks, an optical code path (OCP) is determined by the optical code and allocated link by link [6]. In a similar way of the WDM transport layers, there are two approaches to assign optical codes: one based on optical code routing and another on optical code conversion. In optical code routing, as shown in Fig. 2(a), an optical code is assigned and persists along the entire optical path, i.e., the optical path is wholly identified by OCP. Provision of optical paths on the basis of optical code conversion is referred to as a virtual optical code path (VOCP). In such a case, as shown in Fig. 2(b), an optical code is allocated link by link, and the path is decomposed into several OCPs under the control the optical code conversion. For full network functionality, optical code conversion is the key technology for OCDM transport networks and is analogous to the wavelength conversion for virtual wavelength path (VWP) of WDM transport networks [9]. On the example route shown in Fig. 2, to establish four OCPs (Fig. 2(a)) and four VOCPs (Fig. 2(b)) with the same routes, an OCP scheme requires three optical codes. On the other hand, a VOCP scheme requires only two. The introduction of VOCP potentially solves the optical code path assignment problems, which may limit the network and optical path expansion. Thus, optical code conversion is needed for VOCP cross-connect.

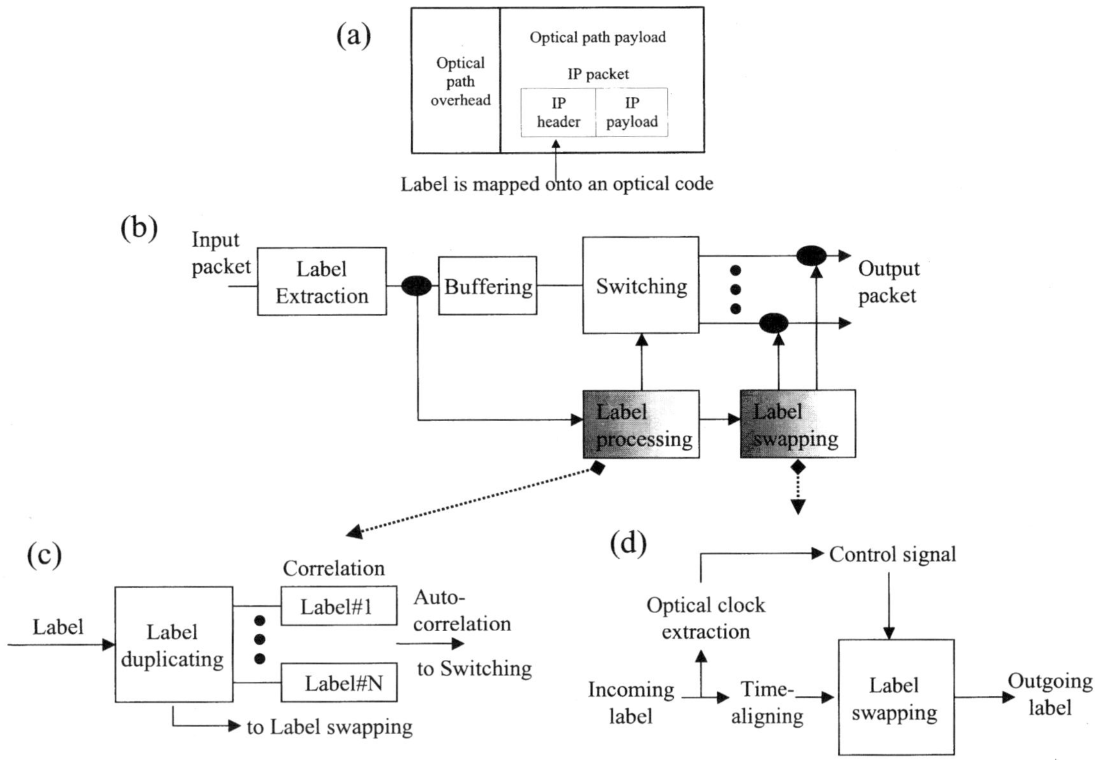

Fig. 3 (a) Optical frame in which a label is mapped on an optical code, (b) architecture of photonic label switching router (LSR), (c) architecture of photonic label processing, and (d) architecture of photonic label swapping.

2.3 Photonic LSR

Photonic IP routing is a crucial technology for IP over photonic networks. A type of photonic LSR based on parallel processing has been proposed recently [7]. As shown in Fig. 3(a), each label is mapped onto a single optical code, and the IP packet is encapsulated in an optical frame. The architecture of the photonic LSR is shown in Fig. 3(b). The label processing, $1 \times N$ optical gate switching, and label swapping are the key functions. As shown in Fig. 3(c), label processing is based on optical correlation in the time domain between the input optical code and the codes in the address bank. A unique feature is that this label processing of the incoming label with all the labels in the address bank can be performed in parallel [7]. An autocorrelation peak during the bit period only appears at a given output port when the incoming code matches the corresponding code in the address bank. The optical frame is gated for routing to the desired port in the switch fabric. As shown in Fig. 3(d), the label has to be swapped at the output of the LSR according to the address bank. In order to swap labels, the control signal has to be time-aligned within a chip rate precision. Therefore, the control signal is generated by the optical clock, which is extracted by the label itself using injection locked mode-locked semiconductor laser diode (MLLD) [2]. After time-aligning the front of the label to the control signal, the label is swapped by the control signal. Label swapping is achieved by optical conversion, while the resultant wavelength maintains to be the same. The target performance of the photonic LSR is discussed in [7], [10].

3. All-Optical Label Swapping by the Optical Code Conversion

3.1 Principle

Ultrafast photonic processing is expected to play a major role in future photonic networks. At data rates near or above 40 Gbit/s, electronics imposes severe technology and economic constraints, which ultrafast photonic processing could advantageously remove. The quasi-instantaneous response of the Kerr nonlinearity in fibers makes it a most attractive effect for the overcoming of bandwidth limitations. The refractive index corresponding to the signal light is modified through XPM in proportion to the instantaneous power envelope of the control light, resulting in modulation of the signal phase [11]. Figure 4 shows the XPM induced phase shift and frequency chirping. Figure 4(a) shows the situation when there is no walkoff between the control pulse and the signal pulse. The amount of phase shift is maximized. However, the shape of the phase

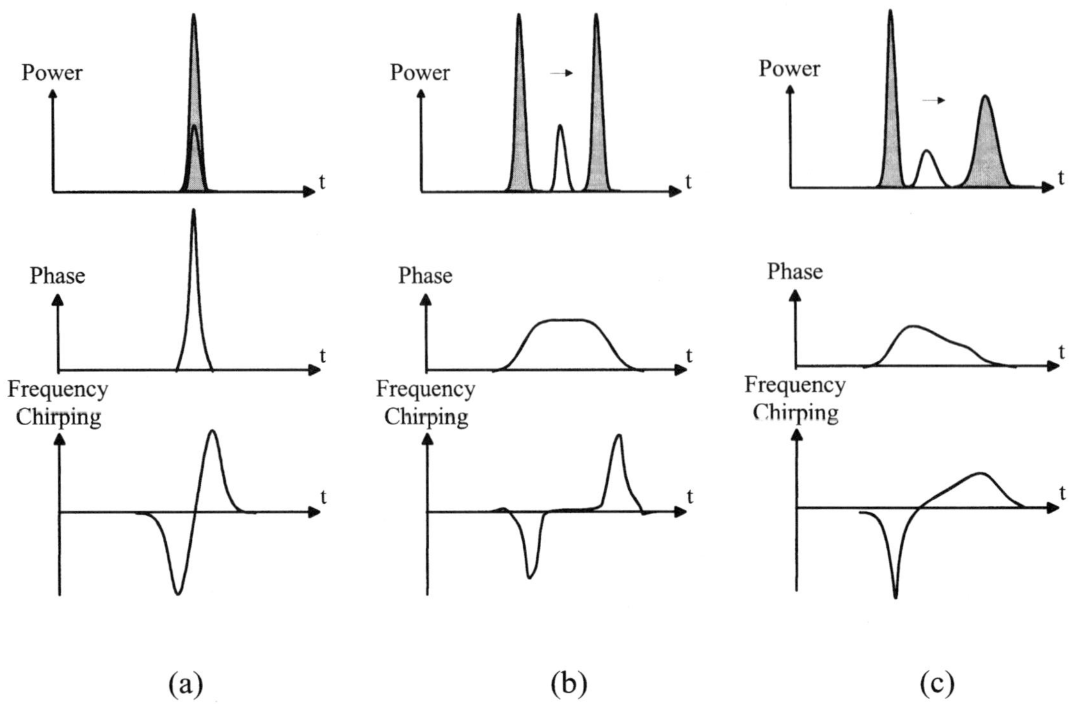

Fig. 4 XPM induced phase change and frequency chirping (a) without walkoff, (b) with total walkoff in an ideal case, and (c) with total walkoff in a real case.

shift is proportional to the shape of signal pulse. Therefore, the induced phase shift is not constant over the whole signal pulse, and consequently, the wavelength is shifted. Figure 4(b) shows the case for total walkoff without pulse broadening. In this case, the induced phase shift is constant over the whole signal pulse, and there is no resulting wavelength-shift. These features are ideal for optical code conversion. Figure 4(c) shows a real case with the total walkoff. Both the control pulse and signal pulse are broadened because of the group velocity dispersion (GVD). In this occasion, there is less phase shift on the trailing edge than on the leading edge, and then the wavelength is shifted.

As shown in Fig. 5, the total phase shift during the interaction length must be π to convert the optical BPSK code by using XPM. And to avoid shifting the wavelength, the control pulse has to pass through the entire signal pulse utilizing the GVD. When the fundamental optical soliton is used as a control pulse, the difference in phase shift between the leading and tailing edges is minimized, because the shape of the control pulse keeps unchanged.

The nonlinear phase shift through XPM is the local index change over the entire fiber length L as

$$\phi(t) = \frac{4\pi n_2}{\lambda_s A_{eff}} \int_0^L P(t - \tau z)dz, \tag{1}$$

where n_2 is the nonlinear index coefficient, A_{eff} is the effective core area, λ_s is the wavelength of the signal, $P(t)$ is the control pulse envelope, and τ is the relative delay between the signal and control pulse in a unit length given by

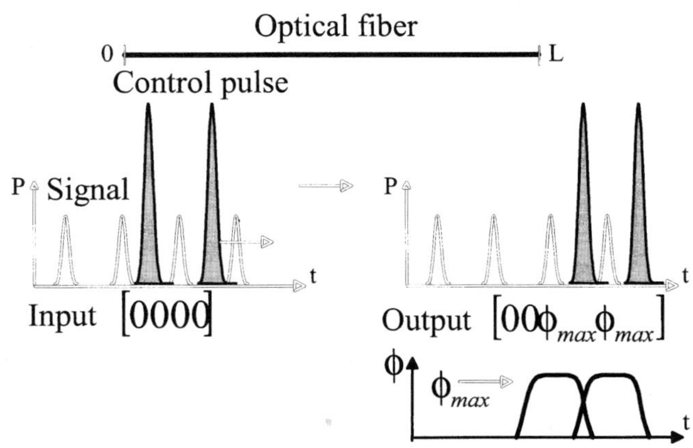

Fig. 5 XPM induced optical code conversion.

$$\tau = \frac{1}{\nu_c} - \frac{1}{\nu_s}, \tag{2}$$

where ν_c is the group velocity of the control pulse with a wavelength of λ_c, and ν_s is the group velocity of the signal pulse. When a soliton pulse is used as the control pulse Eq. (1) becomes

$$\phi(t) = \frac{4\pi n_2}{\lambda_s A_{eff}} \int_0^L P_0 \text{sech}^2\left(\frac{t}{t_0} - \frac{\tau_z}{t_0}\right) dz, \tag{3}$$

where t_0 is the half-width at 1/e-intensity point, and P_0 is the peak power of the soliton pulse. The phase

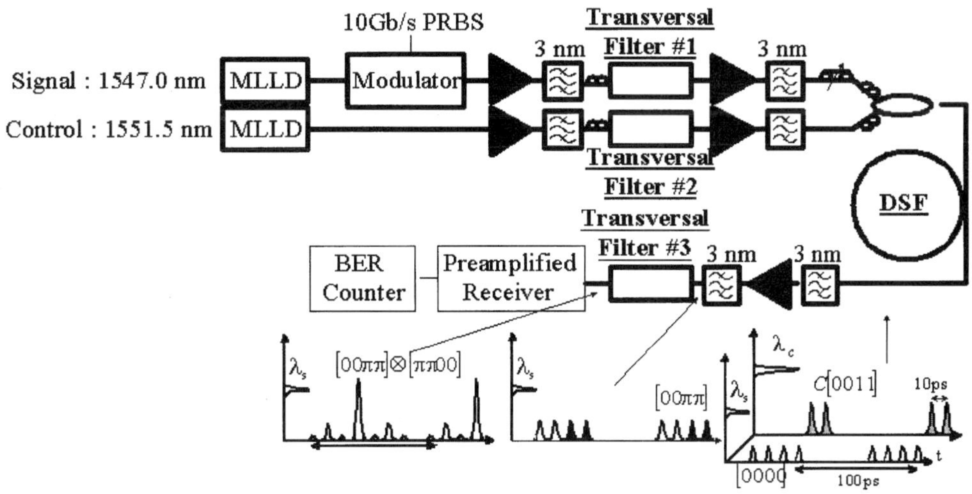

Fig. 6 Experimental setup for the optical code conversion.

shift profile becomes broader and the peak phase shift $\phi_{\max}$ decreases with an increase in the relative delay τ, because of walkoff. When the control pulse does not initially overlap the signal pulse at the input, and then walks off the entire signal pulse at the output, as shown in Fig. 4, the control pulse induced maximum phase shift becomes

$$\phi_{\max} = \frac{4\pi n_2}{\lambda_s A_{eff}} \frac{2P_0 t_0}{\tau} = \frac{4\pi n_2}{\lambda_s A_{eff}} \frac{1.14 P_0 t_{FWHM}}{\tau}. \tag{4}$$

This means that the maximum phase shift in the co-propagating signal is proportional to the soliton pulse's energy and inversely proportional to the relative delay τ, and it is independent of the fiber length in the total walkoff case. To convert the BPSK code, the XPM induced phase shift of $\phi_{\max}$ should be π.

3.2 Experiments and Discussions

Figure 6 shows the experimental setup. Transversal filter #1 is used as the optical encoder that generates the time-spread BPSK optical code, which is used as the label. Here, the payload data has been omitted for simplicity. The signal BPSK pulse code sequences and the control pulse sequences are combined and launched into a dispersion-shifted fiber (DSF). The optical code is converted by co-propagating the control pulse in the DSF fiber through XPM. The output signal pulse sequence is matched-filtered in the time domain by Transversal Filter #3. The output becomes the correlation waveform, which results in a train of seven pulses. By observing the shape of the correlation waveform with various phase shift patterns in Transversal Filter #3, the converted BPSK code can be identified, so that it enables the verification of the code conversion, that is, the label swapping.

The signal and pump sources are MLLDs that operate at a repetition rate of 10 GHz. The full width at half maximum of the both laser pulses are 1.5 ps, and the center wavelength of the signal is ($\lambda_s =$) 1547.0 nm and of the control is ($\lambda_c =$) 1551.5 nm, respectively. As shown in Fig. 7(a), the dispersion values of DSF at λ_s and λ_c are -0.006 ps/nm/km and 0.292 ps/nm/km, respectively. In order to avoid signal pulse waveform distortion, the signal wavelength λ_s is set to the zero-dispersion wavelength. The length is 4.0 km and the loss is 1.8 dB. As shown in Fig. 7(b), the group delay difference is 3.20 ps. The DSF is longer than is necessary to allow the control pulse to pass through the signal pulse. By considering the pulse width and the group delay difference, time-aligning range should be within 300 fs.

The signal pulses are externally modulated using a 10 Gbit/s pseudo-random bit stream ($2^{23} - 1$ PRBS). The control pulse sequence is generated by using the same type of the transversal filter. However, in this case, taps are tuned to generate the control pulses at the specific positions in order to convert the signal's optical code. The control pulse sequence is denoted in the text and figures as, for example, C(0011), and so on. After the polarization and timing adjustment, both pulse sequences are combined together and launched into the DSF. By using Eq. (1), the peak power for the XPM induced π phase shift is calculated to be 292 mW. At the control wavelength of λ_c, the peak power required to form the fundamental soliton is 196 mW and consequently, the soliton order is calculated to be 1.22. The control pulse for the π phase shift is almost nearly the fundamental soliton. The signal BPSK code is converted by co-propagating the control pulse. After rejecting the control pulse by using two 3 nm optical filters, the output stream is matched-filtered in the time domain by Transversal Filter #3.

Figure 8(a) shows the output of Transversal Filter #3 when the control pulse is not injected. The signal

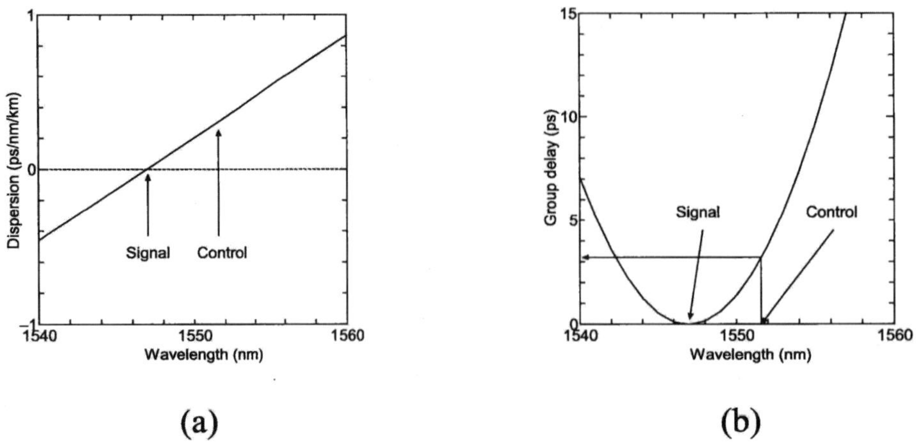

Fig. 7 (a) Measured wavelength dispersion value and (b) measured group delay difference versus the wavelength of the DSF.

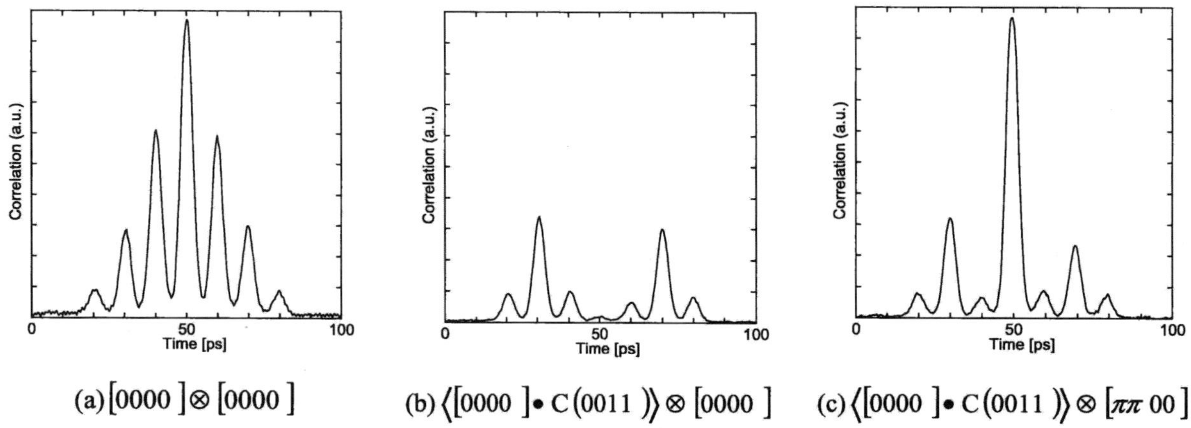

Fig. 8 Output correlation waveforms of Transversal Filter #3 for the signal code of [0000]. (a) without control pulse, (b)(c) with control pulse.

code is [0000] and the phase shifts in Transversal Filter #3 are [0000], resulting in an autocorrelation trace. Figure 8(b) shows the output when the control pulse of C(0011) is co-propagated with the signal code in the same setup as Fig. 8(a). The cross-correlation with the power distribution of (1,4,1,0,1,4,1) is clearly visible because the BPSK code is converted to $[00\pi\pi]$ by XPM. Figure 8(c) shows the output when the phase shift in Transversal Filter #3 is $[\pi\pi 00]$ with the same signal code and the control pulse as in Fig. 8(b). An autocorrelation with the power distribution of (1,4,1,16,1,4,1) is visible. The shape of the autocorrelation is different from that in Fig. 8(a) because the BPSK codes are converted by XPM. The shapes of these cross- and autocorrelations are in good agreement with the theoretical predictions.

Figure 9 shows the measured BER for a matched case with an optical code of $[00\pi\pi]$, and with and without code conversion. The latter case corresponds to Fig. 8(c). Figure 9 shows that the optical code is successfully converted at a BER of $< 10^{-10}$ with almost no power penalty. In a practical system, the clock, which is

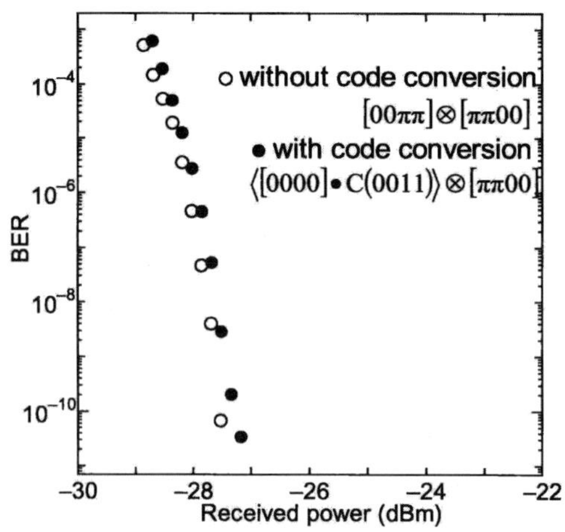

Fig. 9 Measured BER in the matched case with and without optical code conversion.

used as the control pulse as shown in Fig. 3(d), should be extracted by the waveform distorted and S/N degraded signals. These results show that label swapping can be achieved without lowering the signal quality using this method when the clock extraction works ideally.

4. Conclusion

We have proposed and experimentally demonstrated the all-optical label swapping by optical code conversion. 10 Gbit/s BPSK code conversion without wavelength-shift by using XPM was successfully demonstrated at a BER of $< 10^{-10}$. It needs further consideration about the total system performance including link design, S/N degradation, number of hops, and so on. These results will enhance the flexibility of the future photonic LSR based on OCDM.

Acknowledgement

The authors wish to thank W. Chujo and N. Wada of the Communications Laboratory for their useful discussions and invaluable comments. The authors would also like to thank T. Iida and T. Itabe of the Communications Research Laboratory for their encouragement.

References

[1] D.D. Sampson, G.J. Pendock, and R.A. Griffin, "Photonic code-division multiple-access communications," Fiber and Int. Opt., vol.16, pp.129–157, 1997.

[2] K. Kitayama, H. Sotobayashi, and N. Wada, "Optical code division multiplexing (OCDM) and its application to photonic networks," IEICE Trans. Fundamentals, vol.E82-A, no.12, pp.2616–2626, Dec. 1999.

[3] P.E. Green, Fiber optic networks, chapter 13, Prentice-Hall Inc., New Jersey, 1993.

[4] N. Wada, H. Sotobayashi, and K. Kitayama, "Error-free 100 km transmission of 10 Gbit/s optical code division multiplexing using BPSK picoseconds-pulse code sequence with novel time-gating detection," IEE Electron. Lett., vol.35, no.10, pp.833–834, 1999.

[5] H. Sotobayashi and K. Kitayama, "10 Gb/s OCDM/WDM multiple access using spectrum-sliced supercontinuum BPSK pulse code sequences," Optical Amplifiers and their Applications (OAA '99), PD7, Nara, June 1999.

[6] K. Kitayama, "Code division multiplexing lightwave networks based upon optical code conversion," IEEE J. Sel. Areas Commun., vol.16, no.7, pp.1309–1319, 1998.

[7] K. Kitayama and N. Wada, "Photonic IP routing," IEEE Photon. Technol. Lett., vol.11, no.12, pp.1689–1691, 1999.

[8] N. Wada and K. Kitayama, "10 Gbit/s optical code division multiplexing using 8-chip optical bipolar code and coherent detection," IEEE/OSA J. Lightwave Technol., vol.17, no.10, pp.1758–1765, 1999.

[9] K. Sato, Advances in transport network technologies, Artech house Publisher, Boston, 1996.

[10] K. Kitayama, N. Wada, and H. Sotobayashi, "Architectural considerations for photonic IP router based upon optical code correlation," to appear in Lightwave Technol., vol.18, no.12, 2000.

[11] M. Jinno and T. Matsumoto, "Nonlinear Sagnac interferometer switch and its application," IEEE J. Quantum Electron., vol.28, no.4, pp.875–882, 1992.

High-Performance Backbone Network Technology edited by Naoaki Yamanaka
ISBN: 0-8247-5321-6 DOI: 10.1081/(120025741)
Published 2004 by Marcel Dekker, Inc.

A Photonic IP Switching Technique Using Code Division Multiplexing

Shouhei Nishi, Isamu Saeki, Hideki Tode, and Koso Murakami

SUMMARY Increasing the capacity and intelligence of the next-generation Internet requires the application of optical technologies to switching nodes as well as transmission lines, and the development of advanced network architectures with end-to-end connection setup processing at the source node and autonomous routing at intermediate nodes. In the present paper, we design a new CDM-based switching scheme and node configurations that are suitable for a photonic IP switching system, in which a set of multiple-encoding CDM codes is utilized as routing information. In addition, we calculate the BER characteristics of the multiple-encoding CDM system by simulation. Under the condition that the chip duration of a certain code is a multiple of that of another code, the BER characteristics of the multiple-encoding system are shown to coincide with that of the single-encoding system by the longer code.

key words: *multiple-encoding, optical network, self-routing, code division multiplexing, IP switching network*

1. Introduction

In order to promote quantitative and qualitative changes in the communication environment and improve multimedia communication services such as audio, visual and data communications, the Internet must achieve higher speed and become more intelligent.

Therefore, optical links and fast transport technologies, e.g., SONET and SDH, are applied to backbone networks. The maximum transport rate achieved is currently 1 Tbits/s. This was achieved experimentally using optical fibers and WDM technology. The theoretical maximum transport rate using optical fibers may be 50–75 Tbits/s. However, high-speed switching nodes capable of utilizing the wide-band and high-speed characteristics of optical fibers have not been developed yet. Data transferred on the Internet is divided into packets. In order to decide the desired output port, switches search a routing table for a destination address in the packet header. Consequently, fast switching operation is required for any future high-speed network. Multiprotocol Label Switching routing (MPLS) technologies such as Multiprotocol Over ATM (MPOA), tag switch [1], IP switch [2] and Cell Switch Router (CSR) [3] have been proposed as fast IP switching methods. These methods shorten and unify the packet transport time, but bottlenecks based on software operation still exist, making realization of an extremely high-speed next-generation Internet backbone have been difficult. Although optical switches that are capable of processing optical packet signals that are controlled by electronical signals have been realized, optical switching technologies without E/O and O/E conversions are required for a next-generation network, because the speed capability of electronical circuits may be restricted.

Secondly, reduction of routing tables is one the major obstacles that must be overcome in order to relieve the bottleneck of routing and exchanging routing information on the Internet. However, maintaining the routing table of host groups will be required for multicast communication in addition to unicast communication. Moreover, in order to guarantee QoS such as delay and packet drop rate, modification of routes complying with QoS requests will be essential. Consequently, in order to accommodate the increasing number of connections, it is necessary to maintain a huge routing table. The main bottleneck that is hindering improvement of network performance is the analysis of the packet header and referencing the routing table for every router. In addition, it is impractical to construct the same complicated logic using optical devices as electronical devices for these routing operations. Thus network systems based on a new paradigm for optical node operations must be studied. In addition, distributed routing operations are performed by exchanging the routing tables between nodes on the current Internet. However, reliability and throughput over the entire network cannot be guaranteed, because no end-to-end systematic controls are performed.

Consequently, in order to construct future high-speed multimedia networks, a network architecture in which end-to-end connections are processed integratedly and autonomously must be investigated.

As typical fast IP networking technologies, IP-over-ATM, IP-over-SONET and IP-over-WDM methods have mainly been studied. IP-over-ATM and IP-

Manuscript received March 1, 2000.
Manuscript revised April 17, 2000.

† The authors are with the Department of Information Systems Engineering, Graduate School of Engineering, Osaka University, Suita-shi, 565-0871 Japan.

†† The author is with the Computation Center, Osaka University, Ibaraki-shi, 567-0047 Japan.

* Presently, with the Department of Letters, Konan Women's University.

** In this paper, the term "switching" is defined as exchanging signals or packets in the switch and is distinguished from network level "routing."

over-SONET are considered to be a kind of Dynamic Assigned TDM, in which a time-slot is assigned to every IP packet. The TDM is a basic method for digital communication. However, the TDM utilizes optical memory devices, which are still in the experimental stage, and requires strict time synchronization between communicating nodes [4]–[6]. Also, the TDM method is not very suitable for an IP network, because variable length IP packets must be treated in the IP network.

In addition, the IP-over-WDM is considered to be a kind of Dynamic Assigned WDM, in which wavelength is allocated to every IP packet. The WDM does not require inter-node time synchronization, but does require complicated optical devices such as wavelength converters and filters. Moreover, if each route is expressed by wavelength, a great deal of wavelength is required.

On the other hand, the optical code division multiplexing (CDM) method has been examined extensively by researchers because of the following four features [7]–[10]. First, optical CDM encoders and decoders can be realized using much simpler optical devices compared to the optical TDM and WDM methods. In particular, CDM encoders and decoders do not require optical memories, which is required in the TDM method, wavelength filters and converters for the WDM. Secondly, in the CDM method, the frame synchronization for receiving is required, however, unlike the TDM method, control systems for time synchronization between communicating nodes are not forcibly required. Therefore, the synchronization for the CDM method is easier than that for the TDM method. Thirdly, seamless connection of wireless and wired networks will be possible. The CDM method has been practically deployed, particularly in mobile wireless network systems. Finally, the available wavelength and the time slots in the WDM and TDM, respectively, which can be used for routing information, should be less than the maximum number of multiplexing channels for implementation. While, the CDM codes can be used much more than such number of channels independent of implementation constraints. Consequently, the CDM codes are suitable for expressing routing information. Thus for, optical CDM encoding and decoding systems have been the primary focus of research, e.g., experiment is on the coherent CDM code converter [11]. However, a efficient switching network based on optical CDM has not yet been reported.

In this paper, in order to deal with huge network demands in near future, we propose the optical CDM self-routing network system as a photonic IP switching method [12], [13]. In the source nodes for the proposed system, the packet signals are multiple-encoded by sets of CDM codes corresponding to routes obtained by the conventional source routing scheme before packet signal transmission. Intermediate switches decide the destination output ports by decoding the incoming packet signals. In Sect. 2, we illustrate the principle of the proposed method. In Sect. 3, we show the structure of the source, intermediate and destination nodes in the proposed network. In Sect. 4, in order to show the feasibility of the multiple-encoding method, which is the core technology of the proposed network, we calculate the BER characteristics of the multiple-encoding CDM system by simulation.

2. Photonic IP Switching Scheme Using Optical Multiple-Encoding

2.1 Network Configuration

In Sect. 1, we have shown that the network architecture in which end-to-end connections are processed integratedly and autonomously is required for future high-speed multimedia networks. Consequently, we propose that the network is separated to an electronical routing processing layer and an all-optical transport layer. Such a network has already been proposed, as the holonic optical network [14].

The holonic network is characterized by distributed cooperative networking technology based on autonomous management and all-optical tera-bit order transport networks. Figure 1 shows the configuration of the network. The network consists of the optical transport network (OTN) and the personal navigation network (PNN). The OTN is an all-optical self-routing network having simple structures, which alleviates complex controls using optical logical gates. The PNN is an intelligent electronic network, which compensates for the simplicity of the OTN. The OTN performs self-routing without direct controls from PNN, and communicates without the limitations associated with the speed of the electronical circuit.

In Fig. 1, sending stations obtain complete routing information from the PNN, which references the network map and sets connections at the beginning of communications, and store obtained routing informa-

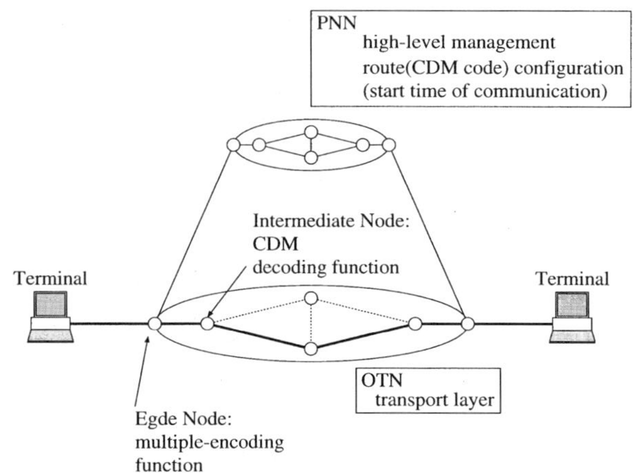

Fig. 1 Configuration of the holonic optical network.

tion into the packets. Consequently, intermediate nodes in the OTN can switch packets using only the routing information in the packets.

2.2 Basic Concepts

The multiple-encoding scheme assumes that an adequate route of a generated connection is determined by a routing algorithm beforehand in the PNN. In order to put generated routing information into packets, transmitting packets are encoded repeatedly in the source nodes. The sequence of CDM codes for encoding corresponds to the routing information which is generated in the PNN, that is, sequence of output ports on the traversed intermediate nodes. The order of encoding is the reverse of the order of packet traversing. These packet signals are then transferred to the network. In intermediate nodes of the OTN, all incoming photonic signals that indicate packet streams are merged and then broadcast to all output modules. Each output module decodes the merged signals by using the preassigned inherent code. If certain packets are destined for the target outgoing link, these packets will be safely decoded, and then transferred directly to the outgoing link without having the buffering operation. As for the assignment of CDM code sets for multiple-encoding, if code length in each encoding operation at the source node is not same, it is required to use different CDM code sets for each encoding. However, if code length in each encoding operation is same, multiple-encoding signals, in which same combination of CDM codes is used, might not be distinguished each other. Therefore, in this paper, different CDM code sets are used for each encoding operation at the source node.

Figure 2 illustrates the proposed scheme and its example, and shows only OTN part of holonic optical network. We assume that packets are sent to the destination node from the source node, and that different CDM codes are allocated if, at least, one of the information is not the same; node ID, output link ID, and residual encoding times until destination nodes (residual number of hops). The PNN calculates the route for transferring packets. In this examples, we assume that calculated transferring order of packets is the intermediate nodes 1, the intermediate node 3 and the destination node. First, an incoming IP packet signal is first-encoded by code E for multiplexing on the link between the intermediate node 3 and the destination node. In addition, the first-encoded signal is second-encoded by code D, which is used for the second-encoding of transferring packets from the intermediate node 3 to the destination node. Finally, the second-encoded signal is third-encoded by the code B similarly. The PNN gives CDM codes for multiple-encoding as routing information to the source node.

At each intermediate switching node, incoming packet signals are output to the port corresponding to the code that was used for the last decoding. In this example, at the intermediate node 1, the received signals of IP packets are output to the intermediate node 3, because this signal is the last encoded by the code B. After decoding of code B, the signal of the output packet is regarded as being in the same condition as what is the last encoded by code D. In intermediate node 3, the signal of the IP packet is transported to the destination node corresponding to code D. Consequently, in the proposed method, complicated routing operation, such as conventional table lookup, is removed, and the intermediate nodes can be realized by the configuration in which decoders and output ports are statically connected.

2.3 Optical Multiple-Encoding

Figure 3 shows the principle of optical multiple-encoding using optical delay-lines, which are widely used in the optical CDM currently. In the optical multiple-encoding, transferred signals are encoded repeatedly. N is the maximum number of encoding repetitions. The degree of encoding, n, is an integer and $1 \leq n \leq N$. M_n is the code period of the nth-encoding. B and $T = T_0 = 1/B$ are the bit rate of the transmitting signal and the bit duration, respectively, and $T_n = T_{n-1}/M_n$ is the chip duration in the

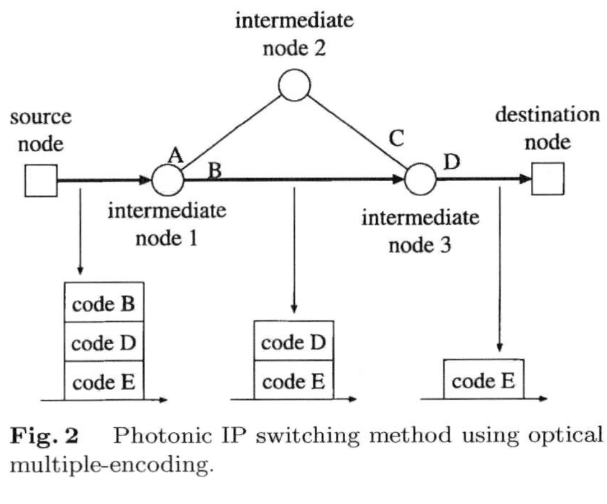

Fig. 2 Photonic IP switching method using optical multiple-encoding.

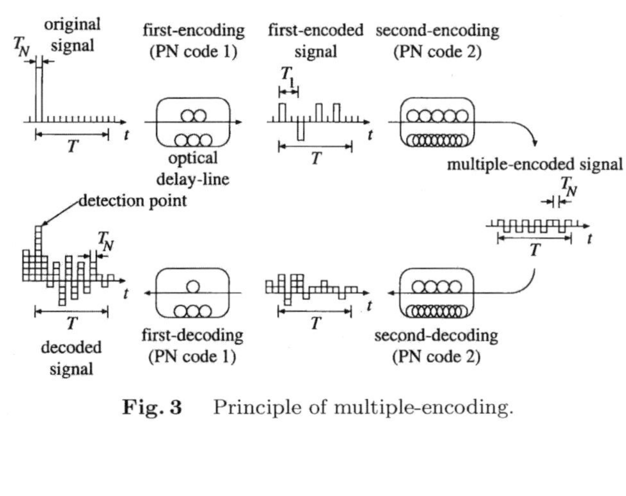

Fig. 3 Principle of multiple-encoding.

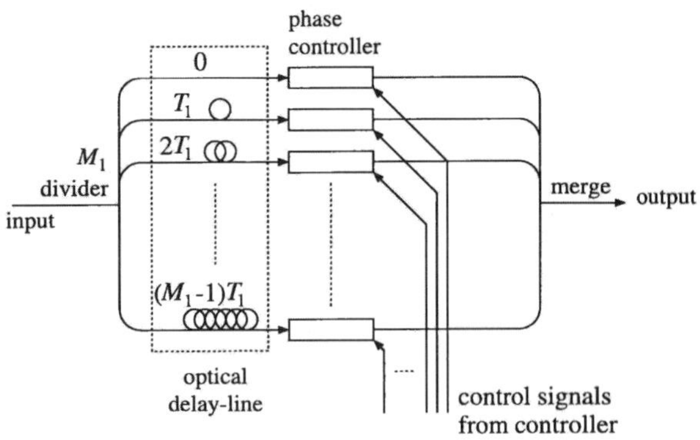

Fig. 4 Configuration of the encoder.

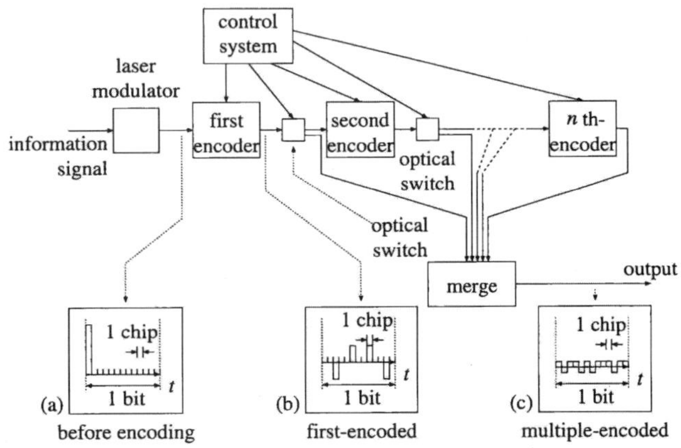

Fig. 5 Configuration of the source node.

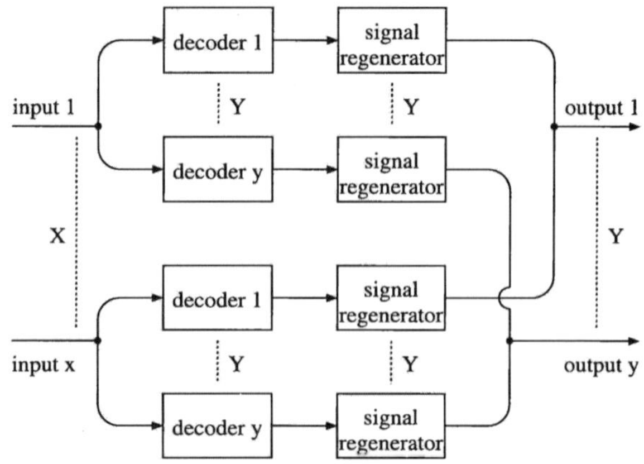

Fig. 6 Configuration of the intermediate node.

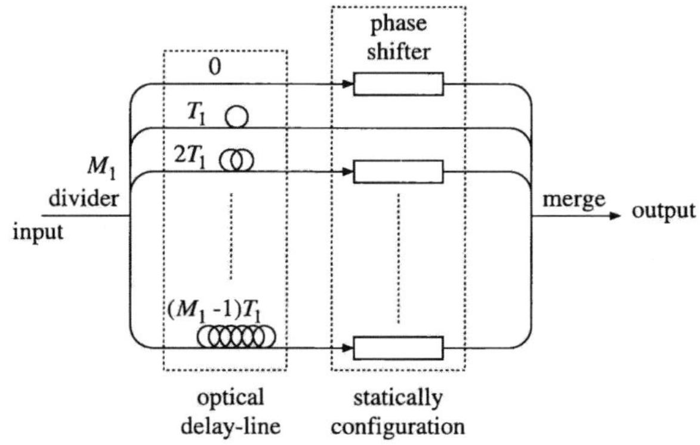

Fig. 7 Configuration of the decoder.

nth-encoding. First, in the source node, a transmitting signal is modulated using rectangular pulses, the duration and time interval of which are T_N and T, respectively. Next, the modulated signal is first-encoded by the encoder shown in Fig. 4. The encoder consists of optical delay-lines and phase-controllers. An input signal is divided into M_1 lines, and each signal is delayed for mT_1, where m is an integer satisfying $0 \leq m < M_1$. In addition, the phase of some of the delayed signals is shifted by π in the phase controller. Signals from the phase controller are merged into the output of encoder. The CDM code corresponds to whether the phase of each signal is shifted or not. After this encoding, similar processes are performed in the nth-encoder.

3. Node Configurations

3.1 Source Nodes

The source node needs to multiple-encode the optical packet signals according to routing information before transmission. Figure 5 shows the configuration of the proposed source node.

In the laser modulator, information packets are converted to optical packet signals, which consist of optical pulses, the duration of witch is the maximum modulated chip duration (Fig. 5(a)). The packet signals are converted to spread signals by the first encoder, as shown in Fig. 5(b). The source node repeats the encoding operation. Total number of encoding operations in the source node is equal the number of intermediate nodes plus one. The optical switches control the encoding times according to the determined number of codes.

3.2 Intermediate Nodes

Figure 6 illustrates the configuration of the intermediate nodes. The intermediate nodes decode incoming IP packets, and decoded packets are output to the desired output port corresponding to the CDM codes for decoding. First, input signals are divided into a number of lines equal to the number of output ports. The signals are then decoded by the corresponding output codes and sent to the signal regenerators in order to remove interference noise. We consider that signal regenerators are realized by optical hard-limiters [15], [16]. Consequently, only a signal that is encoded by the CDM code corresponding to an desired output port can go through the switching node. The peak level detection of the decoded signal is adopted for signal filtering.

Figure 7 shows the configuration of the decoders.

The decoder has a configuration similar to that of

the encoder, except that the impulse response is reversed in the time dimension. The input signal is divided, and each divided signal is delayed for multiples of chip duration, respectively. After which, phase shifters shift the phase of some of the delayed signal, in compliance with the CDM code. Unlike the encoders, the operation of the phase shifters in the decoders is fixed beforehand. Thus, these phase shifters can be replaced by delay-lines. Consequently, the intermediate node is composed of optical delay-lines, junctions and signal regenerators.

In the destination nodes, a decoder shown in Fig. 7 decodes the received signals. Each bit is discriminated via peak detection.

3.3 Required Devices

Required optical devices for realizing the CDM network with the proposed method are lasers, phase controllers, phase shifters, optical switches and optical delay-lines. We consider that the required speed of lasers using the CDM method is less than that using the TDM method and more than that using the WDM method. However, multi-wavelength lasers are not required unlike the WDM method. Phase controllers and phase shifters are only required for the CDM method. The former is realizable using optical switches and phase shifters, and, the later is realizable with optical delay-lines. It is considered that the required speed of optical switches is similar to that of lasers.

4. Characteristics of Optical Multiple-Encoding

In this section, in order to examine the effect of multiplexing signals in the proposed multiple-encoding scheme, we evaluate the BER characteristics.

4.1 Evaluation Model

The evaluation model is shown in Fig. 8. In this model, first, we assume the ideal condition without quantum and thermal noise, because the treatment of the nonlinear characteristics of optical devices is difficult.

N is the number of source nodes. In these source nodes, optical packet signals are encoded $M+1$ times, that is, packets are forwarded through M intermediate nodes. We assume that times of encoding in all source nodes are same. n is an integer and $1 \leq n \leq N$. d_l^n is the binary data sequence of the nth transmitter ($d_l^n \in \{-1, 1\}$ for each l). T is the bit duration. $d_n(t)$ is the nth baseband data signal given by

$$d_n(t) = \sum_{l=-\infty}^{\infty} d_l^n \cdot P_T(t - Tl), \tag{1}$$

where $P_\tau(x)$ is an unit rectangular pulse defined by

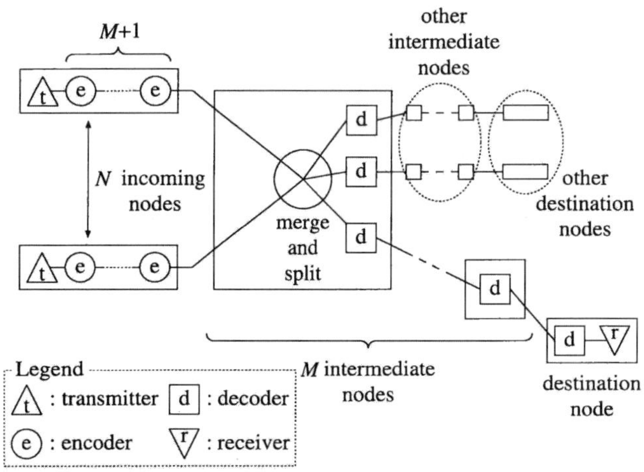

Fig. 8 Model of the optical multiple-encoding.

$$P_\tau(x) = \begin{cases} 1 & (0 \leq x < \tau) \\ 0 & (\text{otherwise}) \end{cases} . \tag{2}$$

$d_n(t)$ is multiplied by $\sqrt{2P}\cos(\omega_c t + \theta_n)$ as the carrier for modulating, where P and ω_c are commonly defined as the power and carrier frequency, respectively, of each transmitter's signal. θ_n denotes the carrier phase shift of the nth transmitter. Moreover, the nth modulated signal is multiplied by $M+1$ code signals $c_n^i(t)(1 \leq i \leq M+1)$ for multiple-encoding. Transmitted signal $S_n(t)$ of nth transmitter is

$$S_n(t) = \sqrt{2P} \cdot d_n(t) \cdot \cos(\omega_c t + \theta_n) \cdot \prod_{i=1}^{M+1} c_n^i(t). \tag{3}$$

$c_n^i(t)$ is the ith-encoding CDM code signal of nth transmitter given by

$$c_n^i(t) = \sum_{k=-\infty}^{\infty} \sum_{l=0}^{M_i-1} C_n^i[l] \cdot P_{T_i}(t - Tk - T_i l), \tag{4}$$

where $C_n^i[l]$ is a periodic binary sequence, the elements of which have the value of $\{+1, -1\}$, M_i and $T_i = T/M_i$ are code period and chip duration, respectively, and l is an integer satisfying $0 \leq l \leq M_i - 1$. N transmitted signals are sent to focused intermediate node. In this intermediate node, these signals are delayed for τ_n and merged into each other. Finally, merged signal $M(t)$ is given by

$$\begin{aligned} M(t) &= \sum_{n=1}^{N} S_n(t - \tau_n) \\ &= \sqrt{2P} \sum_{n=1}^{N} d_n(t - \tau_n) \\ &\quad \cdot \cos(\omega_c(t - \tau_n) + \theta_n) \\ &\quad \cdot \prod_{i=1}^{M+1} c_n^i(t - \tau_n). \end{aligned} \tag{5}$$

We assume that only nth transmitted signal is gone through a certain port of this node. For decoding, the merged signal $M(t)$ is multiplied by CDM code signal $c_n^{M+1}(t)$ that matches the phase to the nth transmitter's signal. The decoded signal $M(t) \cdot c_n^{M+1}(t-\tau_n)$ is sent to next intermediate node. After nth signal is repeatedly decoded in M intermediate nodes, the decoded signal is sent to the destination node. Accordingly, the decoded signal $r_n(t)$ is given by

$$r_n(t) = M(t) \cdot \prod_{i=2}^{M+1} c_n^i(t-\tau_n). \tag{6}$$

In the destination node, received signal $r_n(t)$ is multiplied by the corresponding CDM code signal $c_n^1(t)$ and carrier $\sqrt{2P}\cos(\omega_c - t + \theta_n)$. In addition, the multiplied signal is integrated over bit duration T. The integrated value R_l^n for bit d_l^n is given by

$$\begin{aligned} R_l^n &= \int_{Tl+\tau_n}^{T(l+1)+\tau_n} r(t) \cdot c_n^1(t-\tau_n) \\ &\quad \cdot \cos(\omega_c(t-\tau_n)+\theta_n)dt \\ &= \sqrt{2P} \int_{Tl+\tau_n}^{T(l+1)+\tau_n} \left\{ \sum_{m=1}^{N} d_m(t-\tau_m) \right. \\ &\quad \cdot \cos(\omega_c(t-\tau_m)+\theta_m) \\ &\quad \left. \cdot \prod_{i=1}^{M+1} c_m^i(t-\tau_m) \right\} \cdot \prod_{i=1}^{M+1} c_n^i(t-\tau_n) \\ &\quad \cdot \cos(\omega_c(t-\tau_n)+\theta_n)dt. \end{aligned} \tag{7}$$

In order to identify the bit d_l^n, R_l^n is compared with the threshold value 0. P_{el}^n is the lth bit error rate of the nth transmitter given by

$$P_{el}^n = P(R_l^n < 0 | d_l^n = 1) + P(R_l^n > 0 | d_l^n = -1). \tag{8}$$

Since we are concerned with relative phase shift modulo 2π and relative time delay modulo T, this case is equivalent to the case assuming $\phi_n = \theta_n - \omega_c\tau_n$ and considering only $0 \le \phi_n < 2\pi$ and $0 \le \tau_n < T$ in generality. In addition, the generality of this analysis is preserved even if we assume $n = 1$, $l = 0$, $\phi_1 = 0$ and $\tau_1 = 0$. Thus, we obtain

$$\begin{aligned} R_0^1 &= \sqrt{2P} \int_0^T \left\{ \sum_{n=1}^{N} d_n(t-\tau_n) \cdot \cos(\omega_c t + \phi_n) \right. \\ &\quad \left. \cdot \prod_{i=1}^{M+1} c_n^i(t-\tau_n) \right\} \cdot \prod_{i=1}^{M+1} c_1^i(t) \\ &\quad \cdot \cos(\omega_c t)\, dt, \end{aligned} \tag{9}$$

where m is replaced by n. Because $\cos x \cdot \cos y$ is equal to $(\cos(x+y) + \cos(x-y))/2$,

$$\begin{aligned} R_0^1 &= \sqrt{\frac{P}{2}} \sum_{n=1}^{N} \int_0^T d_n(t-\tau_n) \\ &\quad \cdot \{\cos(2\omega_c t + \phi_n) + \cos\phi_n\} \\ &\quad \cdot \prod_{i=1}^{M+1} \{c_n^i(t-\tau_n) \cdot c_1^i(t)\}\, dt. \end{aligned} \tag{10}$$

We can assume $T \gg 1/\omega_c$, taking into account the frequency response of realistic hardware implementation of the receiver. Therefore, the double frequency component $\cos(2\omega_c t + \phi_n)$ can be ignored. Accordingly, we obtain

$$\begin{aligned} R_0^1 &= \sqrt{\frac{P}{2}} \sum_{n=1}^{N} \cos\phi_n \left[\int_0^T d_n(t-\tau_n) \right. \\ &\quad \left. \cdot \prod_{i=1}^{M+1} \{c_n^i(t-\tau_n) \cdot c_1^i(t)\}\, dt \right] \\ &= \sqrt{\frac{P}{2}} \left[\int_0^T d_1(t) \cdot \prod_{i=1}^{M+1} \{c_1^i(t)\}^2 dt \right. \\ &\quad + \sum_{n=2}^{N} \cos\phi_n \left[\int_0^T d_n(t-\tau_n) \right. \\ &\quad \left. \left. \cdot \prod_{i=1}^{M+1} \{c_n^i(t-\tau_n) \cdot c_1^i(t)\}\, dt \right] \right] \end{aligned} \tag{11}$$

Because $c_n^i(t)$ is always $+1$ or -1 and $d_1(t) = d_0^1 (0 \le t < T)$, the integrated value R_0^1 can be expressed as

$$\begin{aligned} R_0^1 &= \sqrt{\frac{P}{2}} \left[d_0^1 T + \sum_{n=2}^{N} \cos\phi_n \left[\int_0^T d_n(t-\tau_n) \right. \right. \\ &\quad \left. \left. \cdot \prod_{i=1}^{M+1} \{c_n^i(t-\tau_n) \cdot c_1^i(t)\}\, dt \right] \right] \end{aligned} \tag{12}$$

We evaluate the BER characteristics by applying Eqs. (8) and (12) to the computer simulation. Where, the CDM code for evaluation is the Gold sequence which consists of preferred paired m-sequences, because the Gold sequence is generally used as CDM codes and easily generated even if the family size of the code set is large. Generally, to achieve more channels in CDM network, we must choose a class of the CDM codes which has less cross and auto correlation properties as well as a large family size.

4.2 Numerical Results

Figure 9 shows the BER characteristics in the case of multiple-encoding when the chip duration of one encoding is not a multiple of the other. We evaluated the BER characteristics under single-encoding where $M_1 = 31$ and $M_1 = 63$, under double-encoding with two CDM codes having the same periods and under double-encoding in which the periods of the codes for first-encoding and second-encoding are 31 and 63, respectively. As shown in Fig. 9, the BER characteristics

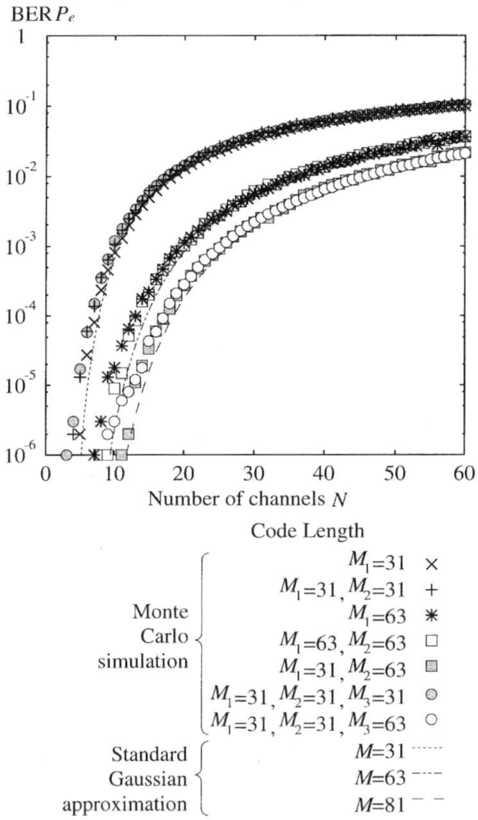

Fig. 9 BER characteristics under multiple-encoding when chip durations are not multiples of each other.

under double-encoding with same-period codes are similar to those under single-encoding.

On the other hand, under the appropriate Gaussian assumption the BER can be calculated as P_e, given by

$$P_e = Q\left(\sqrt{\frac{3M}{N-1}}\right), \tag{13}$$

where

$$Q(x) = \frac{1}{\sqrt{2\pi}}\int_x^{\infty} e^{-t^2/2}dt. \tag{14}$$

Here, N and M are the number of channels and the code period, respectively [17]–[20]. In order to verify simulation results under the single-encoding, we compare them with the BER characteristics obtained by Eq. (13), where $M = 31$ and $M = 63$. From Fig. 9, for the cases of the single-encoding and multiple-encoding with same-period codes, the approximate curve fits the BER results by simulation very well. From Eq. (13), the achievable number of channels is roughly proportional to the code length. On the other hand, the bandwidth of encoded signal is proportional to the code length too. Therefore, the number of channels per unit bandwidths becomes nearly constant.

On the other hand, the BER under double-encoding in which the periods of codes are 31 and 63

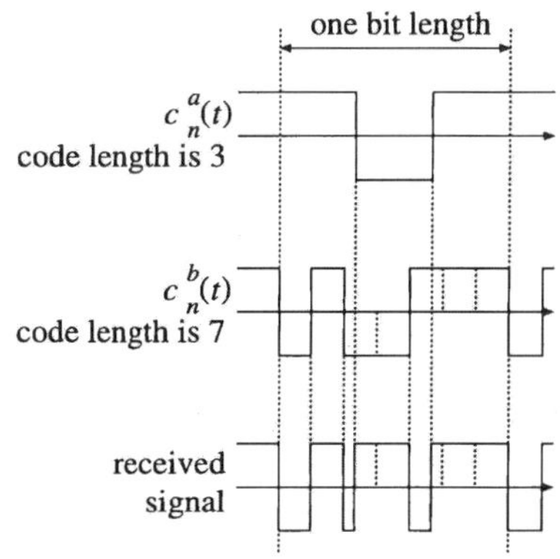

Fig. 10 Signal waveform when chip durations are not multiples of each other.

is less than that under $M_1 = 63$. This result indicates that under the same target BER, the former encoding can accommodate more channels than the latter. In addition, the result for $M_1 = 31, M_2 = 63$ coincides with that of the Standard Gaussian approximation for $M = 81$ which is calculated by heuristic fitting. From the point of view of bandwidth and "spread ratio," which is defined as bandwidth of encoded signals per signal symbol rate, BER is considered to depend on the relationship between the chip duration of first-encoding and that of the second-encoding.

Under the evaluation model shown in Sect. 4.1, spread ratio, bandwidth and BER for single-encoding are similar to these for double-encoding with the same the code period. However, this is not the case when M_2/M_1 is not an integer, that is, when the chip duration of one encoding is not a multiple of the other. In this case, as shown in Fig. 10, the boundaries of the chips do not match between the two codes. Because the number of fluctuation points in the final transferred signal becomes grater than those of the CDM codes, the bandwidth of transferred signal is enlarged. As a result, the BER is decreased. For example, in Fig. 10, the code periods for first-encoding and second-encoding are 3 and 7, respectively. The minimum width of pulse of the transferred signal is the same as the CDM codes having periods of $3 \times 7 = 21$. Similarly, when the code periods for first-encoding and second-encoding are 31 and 63, respectively, the BER is similar to that of the single-encoding system for which the code period is 81. Using single-encoding where $M_1 = 31$, the achievable number of channels is 13 under the condition that the BER is less than $< 10^{-4}$. On the other hand, the achievable number of channels with double-encoding, where $M_1 = 31$ and $M_2 = 63$, is 17 under the same BER condition. However, bandwidth of single-encoded signals is 31 times as wide as baseband signals, but, bandwidth of double-encoded signals is $1953 (= 31 \times 63)$ times. Namely, the number of channels per unit bandwidth of single-encoded and double-encoded signals are $13/31 \approx 0.42$ and $17/1953 \approx 8.7 \times 10^{-3}$, respectively.

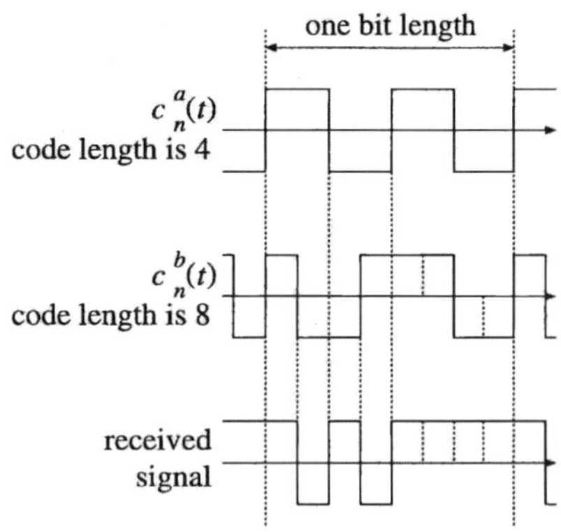

Fig. 11 Signal waveform when chip durations are multiples of each other.

Consequently, the number of channels per unit bandwidths decreases as the multiple-encoding times are increasing.

The signals when the chip duration of one encoding is a multiple of the other encoding is shown in Fig. 11, the code periods of which are 4 and 8, respectively. In this case, because the minimum duration between fraction points of double-encoded signals is as same as that of single-encoded signals with narrower chip duration, double-encoded signals can be treated as single-encoded signals approximately. Consequently the BER characteristics of double-encoded signals are as same as that of single-encoded signals. Therefore, the number of channels per unit bandwidths might be similar to that under single-encoding. In order to verify this result, one chip was appended to two m-sequences for the Gold sequence in Fig. 9, and, we calculated the BER characteristics under the double-encoding when chip durations are multiples of each other. Figure 12 shows that the BER under double-encoding with code periods of 32 and 64 is similar to that of single-encoding. In particular, the BER characteristics of double-encoding signals with same period codes are as same as single-encoded signals.

In this simulation, when the chip duration of one encoding is not a multiple of the other, the BER appears to be less than that under single-encoding, while, the bandwidth of the double-encoded signal is wider than that of the single-encoded signal under the same code period. Consequently, the number of channels per unit bandwidths decreases. Therefore, the condition such that the chip duration of one code is not a multiple of another should be avoided. However, when the chip durations are multiples of each other, the BER under the double-encoding is similar to that under single-encoding using the code for which the chip duration is as long as the shortest duration code of the double-encoding. In particular, when the chip durations in the double-encoding are the same, the BER of the double-encoding coincides with that of the single-encoding.

Similarly, the BER characteristics under the triple-

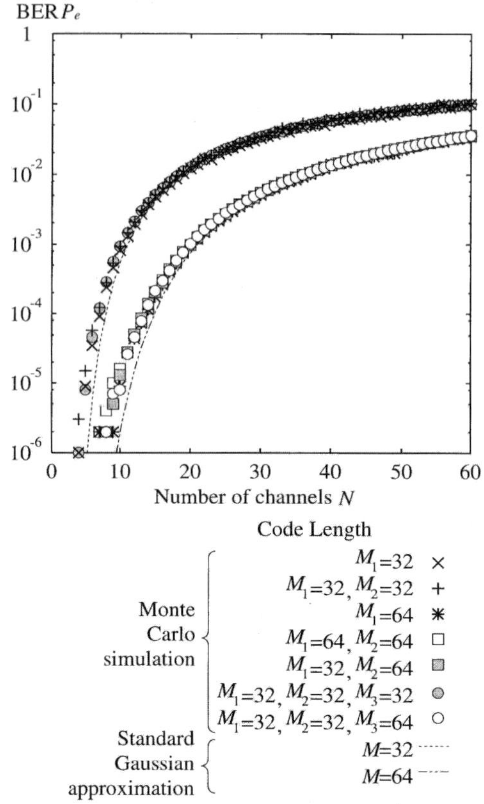

Fig. 12 BER characteristics under multiple-encoding when chip durations are multiples of each other.

encoding is similar to that under the single-encoding using the CDM code whose chip duration is shortest, when the chip durations are multiples of each other, like, for example, the case of $M_1 = 32$, $M_2 = 32$ and $M_3 = 64$ in Fig. 12. If the chip durations are not multiples, like, for example, the case of $M_1 = 31$, $M_2 = 31$ and $M_3 = 63$ in Fig. 9, the BER characteristics is less than that under the single-encoding. In particular, when all chip durations for the triple-encoding are same, BER characteristics are close to that the under single-encoding. Consequently, the BER characteristics under the generic multiple-encoding is considered to be similar to that under the double-encoding or the single-encoding.

In the ideal condition, there may be no limitation for the number of the encoding. However, the practical limitation is less than the theoretical one by the loss and distortion in devices and the limitation for generating short optical pulses. We can expect that developments in CDM sequences and the increase of the code period will enlarge the limitation of the number of encodings. The evaluation in this practical condition will be further research issue.

The achievable number of channels with the double-encoding, where $M_1 = M_2 = 64$, is 8 under the condition that BER is less than $< 10^{-6}$. Assuming that the data transfer rate is 1 Gbps, the bandwidth of the encoded signal is 64 GHz. As the band width of optical fibers is about 10 THz, the whole data transfer rate with one optical fiber is estimated at $10\,\text{THz} \times 8 \times 1\,\text{Gbps}/64\,\text{GHz} = 1.25\,\text{Tbps}$. We can expect that developments in CDM sequences and error correction method will enlarge network capacities. Furthermore, the bandwidth is enlarged, the increase of the code period improves the BER characteristics.

5. Conclusion

In this paper, in order to deal with huge network demands in near future, we proposed an optical CDM self-routing network system, in which a set of multiple-encoding CDM codes is utilized as routing information. In addition, we illustrated source, intermediate and destination nodes in the proposed network. In particular, we have shown that the intermediate nodes can be realized by the configuration in which decoders and output ports are statically connected.

Moreover, in order to clarify the feasibility of the multiple-encoding method, which is the fundamental technology of the proposed network, we calculated the BER characteristics of the multiple-encoding CDM system by Monte Carlo simulation. As a results, we have found that under the condition for which the chip duration of a code is a multiple of another, the BER characteristics of the multiple-encoding system coincides with that of the single-encoding system using the code for which the chip duration is as long as the shortest duration code of the multiple-encoding.

In the future, we will evaluate BER characteristics under multiple-encoding in which the number of encodings on each channel is not the same. In addition, we intend to design the detail routing method and investigate an adequate network topology.

References

[1] S.J. Vaughan-Nichols, "Switching to a faster Internet," Computer, vol.30, no.1, pp.31–32, Jan. 1997.

[2] G. Copperwheat, "Layer-3 Internet switching—The next step for LAN infrastructure," Electron. Eng., vol.70, no.861, pp.17–19, Oct. 1998.

[3] Y. Katsube, K. Nagami, S. Matsuzawa, and H. Esaki, "Internetworking based on cell switch router—Architecture and protocol overview," Proc. IEEE, vol.85, no.12, pp.1998–2006, Dec. 1997.

[4] S. Park, S. Kajiya, K. Tsukamoto, and S. Komaki, "Multimedia mobile radio highway networks using optical CDMA multiplexing method," Proc. Multi-Dimensional Mobile Communication (MDMC), vol.1, no.1, pp.838–892, July 1996.

[5] S. Park, K. Tsukamoto, and S. Komaki, "Radio highway networks using optical CDMA with optical polarity reversed correlator," Microwave photonics (MWP)'97, vol.1, no.1, pp.227–230, Sept. 1997.

[6] H. Harada, S. Kajiya, K. Tukamoto, S. Komaki, and N. Morinaga, "TDM intercell connection fiber-optic bus link for personal radio communication systems," IEICE Trans. Commun., vol.E78-B, no.9, pp.1287–1294, Sept. 1995.

[7] J.A. Salehi, "Code division multiple-access techniques in optical fiber networks—Part I: Fundamental principles," IEEE Trans. Commun., vol.37, no.8, pp.824–833, Aug. 1989.

[8] J.A. Salehi and C.A. Brackett, "Code division multiple-access techniques in optical fiber networks—Part II: Systems performance analysis," IEEE Trans. Commun., vol.37, no.8, pp.834–842, Aug. 1989.

[9] T. Ohtsuki, "Performance analysis of direct-detection optical asynchronous CDMA system with double optical hard-limiters," IEICE Technical Report, CS96-46, July 1996.

[10] H.M.H. Shalaby, "Performance analysis of optical synchronous CDMA communication systems with PPM signaling," IEEE Trans. Commun., vol.43, no.2/3/4, pp.624–634, Feb./March/April 1995.

[11] K. Kitayama, "Error-free 1.24 Gb/s 4-chip coherent CDM code converter," ECOC Conference Publication, no.448, pp.117–120, Sept. 1997.

[12] S. Nishi, I. Saeki, H. Tode, and K. Murakami, "A configuration of optical network with self-routing switch," IEICE Techincal Report, OCS99-39, PS99-13, OFT99-16, July 1999.

[13] S. Nishi, Z. Tianwei, I. Saeki, H. Tode, and K. Murakami, "Photonic IP switching scheme using optical multiple encoding," IEICE Techincal Report, SSE99-112, PS99-36, Dec. 1999.

[14] K. Murakami, S. Uno, K. Kinoshita, M. Higashida, and T. Takine, "Distributed cooperative networking architecture for multimedia communications based on autonomous routing," Proc. ITC-CSCC'96, vol.1, pp.83–86, 1996.

[15] T. Ohtsuki, "Direct-detecition optical asynchronous CDMA systems with double optical hard-limiters: APD noise and thermal noise," IEICE Trans. Commun., vol.E81-B, no.7, pp.1491–1499, July 1998.

[16] A.A Sawchuk and T.C. Strand, "Degital optical computing," Proc. IEEE, Special Issue on Optical Computing, pp.758–779, July 1984.

[17] R.K. Morrow, Jr. and J.S. Lehnert, "Bit-to-bit error dependence in slotted DS/SSMA packet system with random signature sequence," IEEE Trans. Commun., vol.37, no.10, pp.1052–1061, Oct. 1989.

[18] M.B. Pursley, "Performance evaluation for phase-coded spread-spectrum multiple-access communication—Part I: System analysis," IEEE Trans. Commun., vol.COM-25, no.8, pp.795–799, Aug. 1977.

[19] K. Yao, "Error probability of asynchronous spread spectrum multiple access communication systems," IEEE Trans. Commun., vol.COM-25, no.8, pp.803–809, Aug. 1977.

[20] J.M. Holtman, "A simple, accurate method to calculate spread-spectrum multiple-access error probabilities," IEEE Trans. Commun., vol.40, no.3, pp.461–464, March 1992.

High-Performance Backbone Network Technology edited by Naoaki Yamanaka
ISBN: 0-8247-5321-6 DOI: 10.1081/(120025742)
Published 2004 by Marcel Dekker, Inc.

Experimental 5-Tb/s Packet-by-Packet Wavelength Switching System Using 2.5-Gb/s × 8-λ WDM Links

Kimihiro Yamakoshi, Nobuaki Matsuura, Kohei Nakai, Eiji Oki, Naoaki Yamanaka, Takaharu Ohyama, and Yuji Akahori

SUMMARY We have developed an experimental 5-Tb/s packet-by-packet wavelength switching system, OPTIMA-2. This paper describes its hardware architecture. OPTIMA-2 is a non-blocking 3-stage switch using optical wavelength division multiplexing (WDM) links and dynamic bandwidth-sharing. A new scheduling algorithm for variable-length packets is used for the receiver ports of WDM links and simulation results show that it can suppress short-packet delay while keeping high throughput. An implementation of the WDM link using field programable gate arrays and a compact planar lightwave circuit platform is described. Experimental results for the basic operation of optical wavelength switching are also presented.

key words: *packet, switch, WDM, AWG*

1. Introduction

Recently, the broadband access services such as ADSL, FTTH, wireless LAN, have been spread and the volume of Internet Protocol (IP) data traffic across the access networks are growing year-by-year. The growth of the access networks will also drive the increase of data traffic of backbone network. Moreover, new services using contents delivery network (CDN) and pear-to-pear have started and these services will also increase the data traffic of the backbone network. As a result, the volume of Internet Protocol (IP) data traffic is growing 2–3 times by year, as shown in Fig. 1. On the other hand, the size of commercial-base switch system for the backbone network is now hundreds of Gb/s. This means that a Tb/s-class switching system [1], [2] for the backbone network will be required in the near future if the data traffic are going to increase at the same pace. The throughput of conventional switches has increased as the process technology for VLSI has progressed according to Moore's law, but IP traffic has increased even faster. Therefore, some breakthrough technologies are required.

To increase the switch size, there are two approaches. One is to enlarge the size of a basic switch-element by using VLSIs fabricated by deep sub-micron process technology and the high density packaging technologies such as ball grid array (BGA), chip-scale package (CSP), and multi-chip module (MCM) to assemble switch VLSIs. Another approach is to scale the switch size by using the basic switch-elements as many as required using a scalable multi-stage architecture such as a clos-network.

Both approaches have limitations as shown in Fig. 2. In the first approach, a cooling limitation will arise. High-density packaging technologies has resulted in high power consumptions over 1 kW, so cooling systems using a cooling fan or a liquid coolant with a radiator have been required. To deal with much higher power consumption, we propose a distributed switch architecture in order to disperse the heat of the basic switch-element.

Another limitation is the interconnection between basic switch-elements in a scalable architecture. High-speed interconnections are required if the multi-stage switch architecture is used. Electrical signals are likely to suffer from noise or reflections when the signal speed exceeds 1 Gb/s over tens of cm. Moreover, it becomes difficult to deal with the large number of electrical cables used for interconnection between switch-elements. An optical WDM interconnection is superior to an electrical interconnection in that it can propagate high-speed signals over long distances. There-

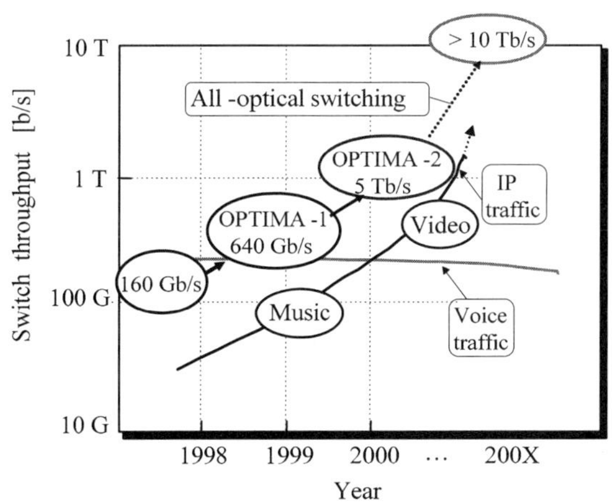

Fig. 1 Trends of data traffic and backbone network.

Manuscript received October 17, 2001.
Manuscript revised December 13, 2001.

†The authors are with NTT Network Service Systems Laboratories, NTT Corporation, Musashino-shi, 180-8585 Japan.
††The authors are with NTT Network Innovation Laboratories, NTT Corporation, Musashino-shi, 180-8585 Japan.
†††The authors are with NTT Photonics Laboratories, NTT Corporation, Ibaraki-ken, 319-1193 Japan.

IEICE TRANS. COMMUN., Vol. E85-B, No. 7, pp. 1293–1301, July 2002

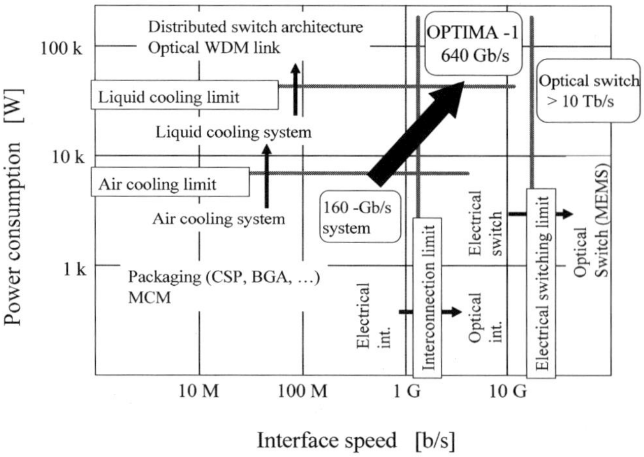

Fig. 2 Breakthrough technologies for large-throughput switching system.

fore we have proposed a distributed switch architecture combined with optical WDM interconnections for Tb/s-class switch system in order to break the above limitations.

As a first step towards Tb/s switch system, we have already developed a 640-Gb/s switching system called OPTIMA-1, which stands for optically interconnected distributed multi-stage Tb/s switching network architecture. OPTIMA-1 has a non-blocking 3-stage switch architecture composed of eight 80-Gb/s multi-chip module (MCM)-based basic switch-elements at each stage, interconnected by optical wavelength division multiplexing (WDM) links [3].

To expand the OPTIMA-1 architecture, we have proposed a 5-Tb/s switching system, OPTIMA-2, that can better handle IP packets [4]. OPTIMA-2 also has a 3-stage switch architecture and employs optical WDM links and dynamic bandwidth sharing to expand the total throughput to 5 Tb/s. The WDM links enable the switching system to expand throughput decreasing the number of cables needed to interconnect switch-elements. OPTIMA-2 has only 1/8 the number of cables that a system using conventional links would need. Moreover, the 2.5-Gb/s maximum bandwidth of each wavelength enables dynamic bandwidth sharing over a total bandwidth of 10 Gb/s at the input or output ports of a switch-element. This makes it possible to expand the total switch throughput to 5 Tb/s without decreasing the statistical multiplexing gain. The bandwidth of each wavelength is controlled dynamically according to the data traffic so that packet-by-packet optical wavelength switching is performed.

Most high-speed packet switching systems including IP routers adopt a fixed-sized cell in the switch fabric. Variable-length packets are segmented into several fixed-sized cells when they arrive, are switched through the switch fabric, and are reassembled into packets before they depart [5]. On the other hand, the cost of reassembling cells into packets is considerable high, because the cells switched based on cell-based scheme are likely to be interleaved. To reduce the reassembly cost, OPTIMA-2 has a packet-based switching scheme as well as cell-based switching. In the packet-based switching scheme, the switched cells belonging to a packet is not interleaved. We call such cells as concatenated-cells. A conventional scheduler for variable-length packets, such as deficit round-robin (DRR) [6], can provide the max-min fair share independent of packet length and are widely used for its simplicity. However, it cannot support both short-packet delay requirement and high throughput simultaneously because the granularity for the deficit counter is fixed. The degradation of short packet dalay is likely to affect the quality of real time services such as VoIP. In stead of a conventional DRR scheduler, a new scheduling algorithm, called dynamic deficit round-robin (DDRR), is introduced in order to resolve the above drawbacks. DDRR is based on DRR but modified by employing a technique in which the granularity for the deficit counter in the scheduler is varied adaptively according to the packet lengths at the heads of queues. It is shown that DDRR can satisfy both high throughput and small delay requirements for short packets while providing the max-min fair share independent of packet length.

For a switching system to be cost effective, it should be implemented compactly. Along with the basic switch elements, as many as 128 WDM links are used to fully implement a 5-Tb/s system. Therefore the E/O and O/E modules of the WDM link are made using compact planar lightwave circuit (PLC) platforms combined with by hybrid integration.

This paper briefly describes the 5-Tb/s switch architecture and switching operation in Sect. 2. Section 3 describes the architecture of the WDM link and the scheduling algorithm for variable-length packets, along with simulation results showing the efficiency of the scheduling algorithm. Section 4 describes an implementation of the sender and receiver ports for the WDM links composed of electrical boards and optical modules using field programable gate arrays (FPGAs) and PLC platforms. Section 5 presents experimental results for the basic operation of the WDM link system of optical wavelength switching. Finally, Sect. 6 concludes the paper.

2. 5-Tb/s Switch Architecture

This section briefly describes the architecture and operation of OPTIMA-2 [9], [10]. Figure 3 shows the architecture of OPTIMA-2, which has a 3-stage switch architecture eight times larger than that of OPTIMA-1 [3]. Each stage has 64 electrical switch-elements in an 8×8 matrix. A switch-element has 8 input $\times$ 8 output ports of 10 Gb/s each, giving it a total through-

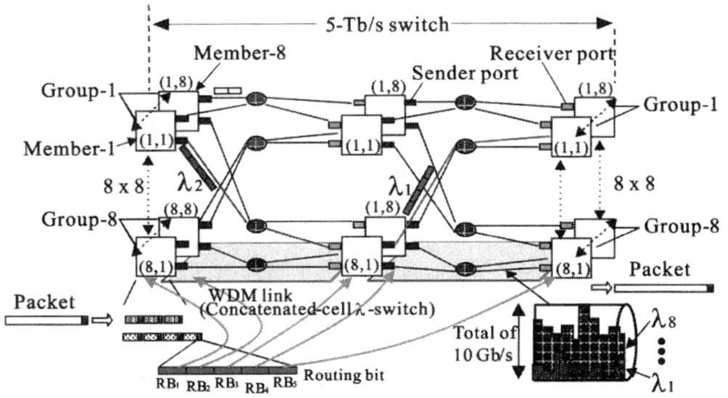

Fig. 3 Architecture of OPTIMA-2: 5-Tb/s switching system.

put of 80 Gb/s [7], [8]. The switch-elements in the 1st stage are interconnected to those in the 2nd stage by optical WDM links using eight wavelengths. The switch-elements in the 2nd and 3rd stages are similarly interconnected. An optical WDM link consists of sender ports, receiver ports, and a wavelength arrayed waveguide grating (AWG) router. To expand the total throughput of the switching system, we divide the 10-Gb/s bandwidth of the output port of a switch-element into one of the eight WDM wavelengths at the sender port.

The link speed for each wavelength, S_{λ_i} (λ_i : wavelength), should be set between 1.25 Gb/s (1/8 of 10 Gb/s) to 10 Gb/s so as to keep the statistical multiplexing gain. On the other hand, the bandwidth for each wavelength, B_{λ_i}, is limited by S_{λ_i}. Therefore the B_{λ_i} satisfies the following formula.

$$B_{\lambda_i} \leqq S_{\lambda_i}, \quad i = 1, \cdots, 8,$$

$$\sum_{i=1}^{8} B_{\lambda_i} \leqq 10 \text{ Gb/s.}$$

B_{λ_i} is dynamically changed from 0 to S_{λ_i} as the data traffic varies. In this way, the total 10-Gb/s bandwidth of a switch port is shared among the eight wavelengths. The bandwidths of these wavelengths are fed to the receiver port via the wavelength AWG router. The scheduler in the receiver port at the input port of a switch-element at the next stage arbitrates cells with or without concatenation for all eight wavelengths. S_{λ_i} should be chosen to be as slow as possible under the worst traffic condition [9]. In OPTIMA-2, S_{λ_i} is cost-effectively set to 2.5 Gb/s.

To explain the switching operation, we define a switch-element as follows. A switch-element belongs to a particular group numbered from 1 to 8, as shown in Fig. 3. Within a group, each switch-element is a member numbered from 1 to 8. Therefore, we denote the m-th member (switch-element) of the g-th group at the n-th stage as $S_n(g, m)$. Routing bits of a cell header, RB_i ($1 \leqq i \leqq 5$) are used for switching the cell inside the system. Here, we explain the case when a cell is switched from switch-element $S_1(g, m)$ at the 1st stage

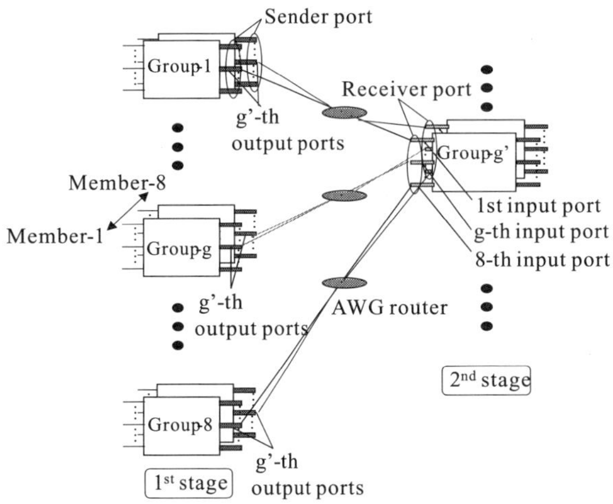

Fig. 4 Interconnection among neighboring-stage groups.

Table 1 Wavelength specifications for optical routing.

		Rx							
		m'_1	m'_2	m'_3	m'_4	m'_5	m'_6	m'_7	m'_8
Tx	m_1	λ_2	λ_3	λ_4	λ_5	λ_6	λ_7	λ_8	λ_1
	m_2	λ_3	λ_4	λ_5	λ_6	λ_7	λ_8	λ_1	λ_2
	m_3	λ_4	λ_5	λ_6	λ_7	λ_8	λ_1	λ_2	λ_3
	m_4	λ_5	λ_6	λ_7	λ_8	λ_1	λ_2	λ_3	λ_4
	m_5	λ_6	λ_7	λ_9	λ_1	λ_2	λ_3	λ_4	λ_5
	m_6	λ_7	λ_8	λ_1	λ_2	λ_3	λ_4	λ_5	λ_6
	m_7	λ_8	λ_1	λ_2	λ_3	λ_4	λ_5	λ_6	λ_7
	m_8	λ_1	λ_2	λ_3	λ_4	λ_5	λ_6	λ_7	λ_8

to $S_2(g', m')$ at the 2nd stage. The g'-th output ports of all members of the g-th group at the 1st stage are connected to the g-th input ports of all members of the g'-th group at the 2nd stage via the wavelength AWG router, as shown in Fig. 4. Thus, a cell is switched to the g'-th output port at the 1st-stage switch-element in order to switch it to a switch-element belonging to the g'-th group at the 2nd stage. First, the cell is electrically switched at $S_1(g, m)$ to the g'-th output port according to the routing bit RB_1. Then, an appropriate wavelength $\lambda_{m,m'}$ is selected to route the cell from the m-th member at the 1st stage to the m'-th member at the 2nd stage according to the routing bit RB_2. Table 1 shows the wavelength rules for routing from member m_i at the 1st stage (i.e. sender side) to member m'_j at the 2nd stage (i.e. receiver side). The wavelengths vary from 1536 to 1564 nm at a spacing of 500 GHz. In this way, cell-based switching from the g-th group to the g'-th group is performed electrically by the 1st-stage switch-element, and wavelength routing from m_i to m'_j is performed optically by the optical WDM link using wavelength λ_{m_i,m'_j}. Switching from a member at the 2nd stage to one at the 3rd stage is performed using the routing bits, RB_3, RB_4.

When a connection request is made, such as from $S_1(g, m)$ to $S_3(g'', m'')$, the system selects an appropriate 2nd-stage switch-element so as to equalize the average load in each wavelength and establishes the con-

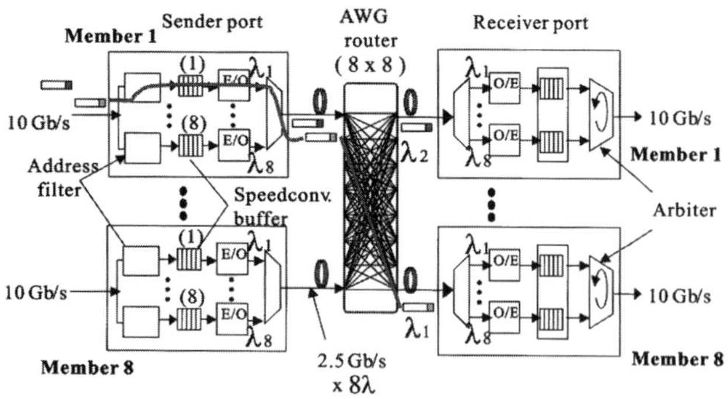

Fig. 5 Architecture of WDM link.

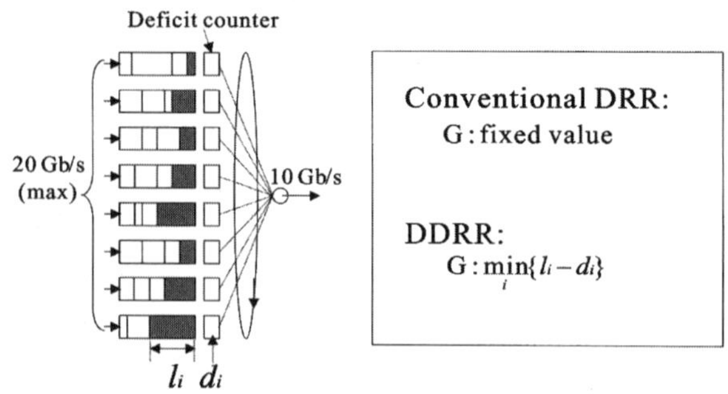

Fig. 6 Arbiter for variable-length packets.

nection from the 1st stage to the 3rd stage. To monitor the data traffic in each wavelength, the system uses operations, administration, and maintenance (OAM) cells [10]. Figure 3 shows a cell being switched from $S_1(1,1)$ to $S_3(1,8)$ via $S_2(8,1)$. This represents the system selecting $S_2(8,1)$ at the 2nd stage. The cell is first electrically switched to the 8th output-port in $S_1(1,1)$ at the 1st stage and then optically switched to $S_2(8,1)$ by using wavelength $\lambda_{m_1,m'_1} = \lambda_2$. At $S_2(8,1)$, the cell is electrically switched to the 1st output-port and then optically switched to $S_3(1,8)$ using wavelength $\lambda_{m_1,m'_8} = \lambda_1$.

3. Design of WDM Link

Cells switched by a switch element at the third stage are reassembled into an IP packet at a line-interface card. In a cell-based switching system, cells belonging to a packet are switched in an interleaved manner. This results in a high reassembly cost at the line interface card. In OPTIMA-2, a variable-length packet is segmented to variable number of fixed-sized cells. Here, we say the variable number of fixed-sized cells as variable-length concatenated-cells. The WDM link is designed to switch the cells belonging to a variable-length packet as variable-length concatenated-cells to reduce the reassembly cost at the line-interface card.

The optical WDM link consists of sender ports, receiver ports, and an AWG router as shown in Fig. 5. The cells from an output port of a switch element are fed to the sender ports. The sender port assigns a definite wavelength to a cell according to the routing bit in the cell header in order to send the cell to the destination receiver port. Cells are distributed through an address filter into a speed conversion buffer corresponding to the specific wavelength.

Optical cells of eight wavelengths are multiplexed at the output of the sender port and each cell is optically routed at the AWG router. At the receiver port, optical cells of eight wavelengths are converted to electrical signals and arbitrated to be fed to the 10 Gb/s input port of the switch element at the next stage. Arbitration is done at each receiver port independently in order to output variable-length concatenated cells to an input port of the next-stage switch-element. In this way, packet-by-packet wavelength switching is performed in the WDM link.

3.1 Sender Port

The sender port has an 8-channel structure of address filters, speed conversion buffers, and an electro-optical (E/O) module with eight wavelengths. Each channel in the sender port corresponds to one wavelength, and the eight wavelength signals are multiplexed onto a single optical fiber by the AWG coupler in the E/O module. After the routing bits have been checked by the address filters, the cells are accumulated in the speed conversion buffers in order to absorb the link-speed difference between 10 and 2.5 Gb/s, and the cells are sent from the sender port with the specific wavelength corresponding to the routing bit. For optical transmission, scrambling is also employed to avoid long sequences of the same bit in the bit stream of the optical signal. The discontinuity of the same bit sequences ensures operation of the clock data recovery (CDR) circuit in the receiver port. The 2.5-Gb/s × 8-ch electrical signals are converted by the E/O module into optical signals that have a wavelength range from 1536 to 1564 nm with a spacing of 500 GHz.

3.2 Receiver Port

The receiver port consists of an opto-electrical (O/E) module with eight wavelengths, buffers for received cells, and a scheduler for concatenated-cells in eight queues. The eight-wavelength WDM signals are first optically demultiplexed into individual wavelengths by a decoupling-AWG and then converted into electrical signals by the O/E module. Data electrically converted from optical signals of each wavelengths are descrambled, and cells accumulated in the buffers are arbitrated in order to be sent out to a 10-Gb/s bandwidth input port of the next-stage switch-element.

Figure 6 shows a DRR-based arbiter for variable-length packets. As described in Sect. 1, a conventional DRR cannot suppress short-packet delay while keeping

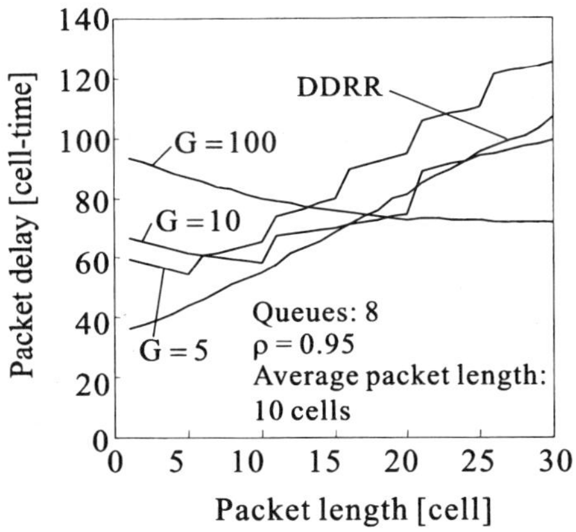

Fig. 7 Packet delay dependency on packet length.

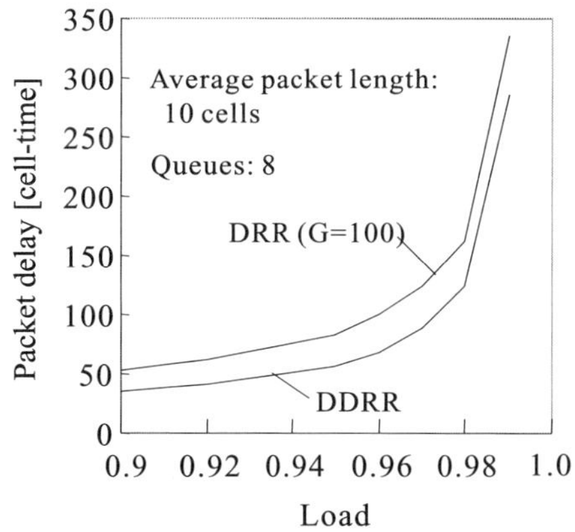

Fig. 9 Average packet delay.

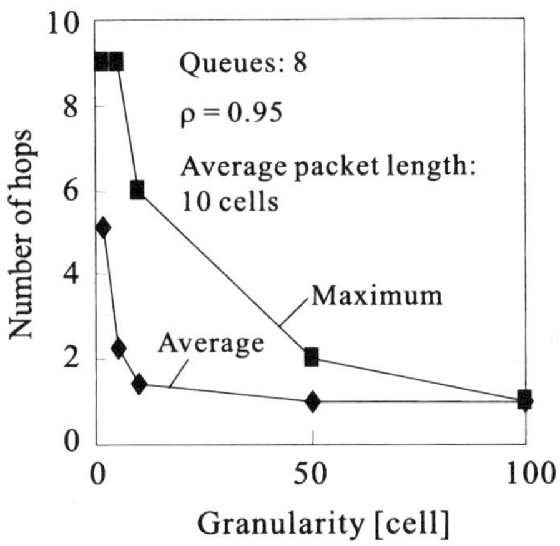

Fig. 8 Number of hops in DRR.

high-throughput. Figure 7 shows the average packet delay dependency on packet length in the case of DRR with granularity of 5, 10, and 100 cells. Here we assumed that the packet length was normalized to an integer number of cell times. The length of arriving packets was assumed to have an exponential distribution with an average length of ten cells. The probability of packet arrival was assumed to have a Poisson distribution. The short packets are delayed as the granularity becomes large ($G = 100$). It is necessary to use a small granularity ($G \leqq 10$) in order to suppress the delay for short packets. On the other hand, the number of hops traversing queues needed to decide the output queue becomes large as the granularity becomes small. This causes the throughput to become low, because there is an upper limit of the hop number at a given line speed. In Fig. 8, though the average number of hops is just 1.43 in the case of $G = 10$, the maximum number of hops is 6 for $G = 10$. If G is set to more than 50 cells, the maxmum number as well as the average number of hops is merely one. But the short packets have a large delay than long packets, as in Fig. 7. Therefore, it is difficult to suppress short packet delay while keeping high throughput when we use a conventional DRR.

To resolve the above bottleneck, we used a scheduler with a variable granularity counter value for arbitration. Its algorithm is as follows.

Step 1. Calculate the difference $m_i = l_i - d_i$, where l_i and d_i are the packet length at the head of channel i ($1 \leqq i \leqq N$) and the counter value for the channel, respectively.
Step 2. Determine the minimum value m_0 from m_i.
Step 3. Add the minimum value m_0 to the counter for each channel other than the selected channel.

All d_i are initialized to 0 before these steps. In step 1, m_i is calculated only when there is a packet at the head of the queue in channel i. Three steps must be done within one packet-time in order to meet the line speed. When the number of channels is 2^N, only N operations determining the smaller of two values m_i, m_j are required to decide m_0 in step 2. For example, when the number of channels is eight, three operations are enough to determine the minimum value among any m_i ($1 \leqq i \leqq 8$). This resolves the timing margin bottleneck of DRR with small granularity ($G \leqq 10$).

We estimated the packet delay by simulation for the case of an 8-channel scheduler. The conditions for the simulation are the same as in the case of Fig. 7. Figure 9 shows the simulation results for average packet delay for our new scheduler and the conventional DRR scheduler. Our scheduler has a smaller packet delay than the conventional DRR scheduler especially under a heavy load larger than $\rho = 0.9$. Here, the granularity of DRR was assumed to be 100 cells. As Fig. 7 shows, packet delay decreases linearly with decreasing packet length in our scheduling method. Moreover, the delays of packets shorter than 15 cells are smaller than that in DRR for any value of granularity. Therefore, our scheduler can satisfy both high throughput and small delay for short packets especially under a heavy load.

4. Implementation of WDM Link

We experimentally implemented the functions of the

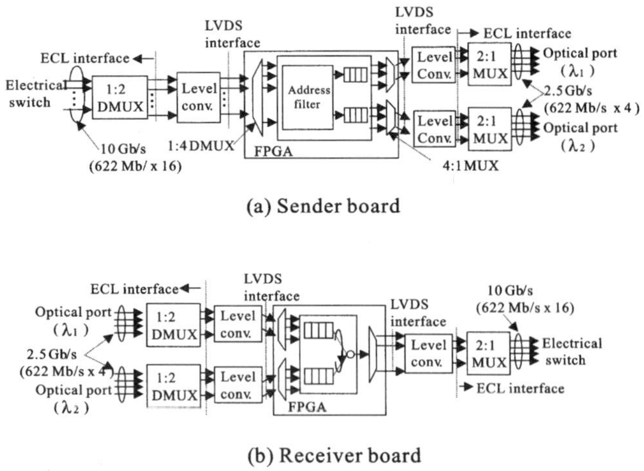

Fig. 10 Block diagrams of the electrical boards for the sender/receiver ports.

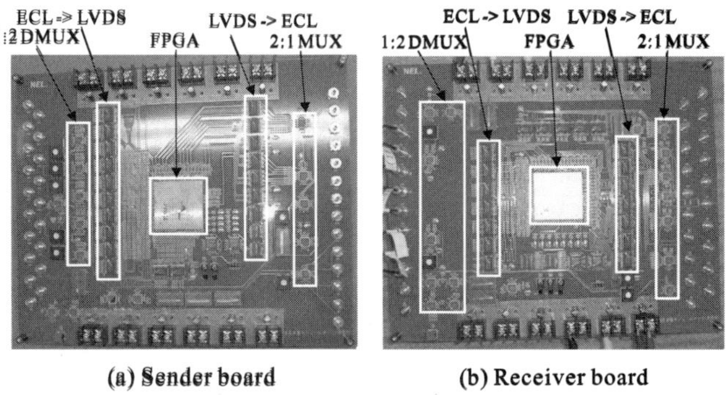

Fig. 11 Overview of electrical boards for the sender/receiver ports.

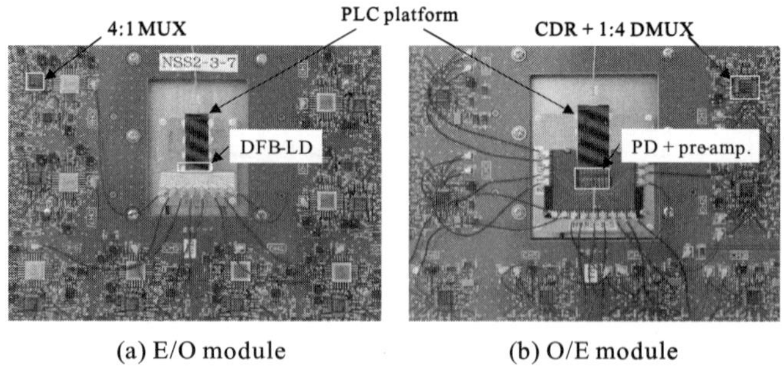

Fig. 12 Overview of E/O and O/E modules.

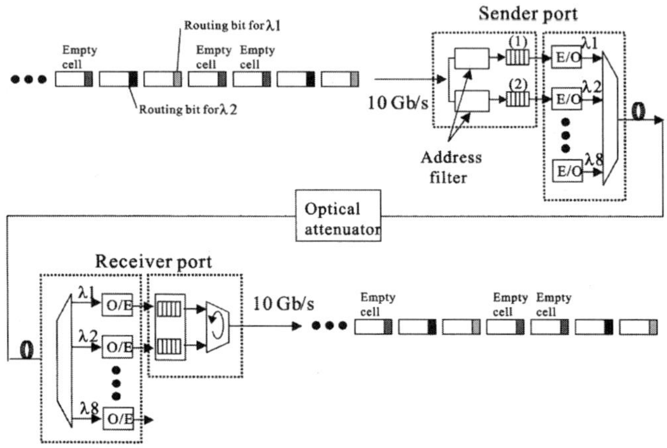

Fig. 13 Experimental system for WDM link.

sender and receiver ports for the WDM link. Each port consists of an optical (E/O or O/E) module and an electrical board [11].

4.1 Sender Port

The interface of the electrical board can handle a total of 10 Gb/s from an output port of the switch-element in the form of 622 Mb/s × 16. The 622-Mb/s signals are first down-converted to 311 Mb/s and then reduced again by a factor of 4 in order to enable a FPGA to operate at 78 MHz, as shown in Fig. 10(a). Signals processed by the FPGA are up-converted to 622-Mb/s. Figure 11 shows an overview of the electrical boards of the sender and receiver ports. The lengths of parallel interconnects on the boards are designed to have variation of less than 2 cm to reduce skews between parallel signals and to ensure a timing margin.

The four parallel 622-Mb/s signals are converted to a 2.5-Gb/s optical signal of the specific wavelength and optically multiplexed onto a single fiber at the E/O module. Figure 12 shows an overview of the E/O and O/E modules of the sender and receiver ports. The distributed feed-back laser diode (DFB-LD) and optical multiplexer provided by the AWG for eight wavelengths were fabricated on a compact PLC platform by hybrid integration [12]. The PLC platform was maintained at a constant temperature of 25°C using a Peltier device in order to prevent wavelength shifts of the DFB-LD.

4.2 Receiver Port

As with the E/O module of the sender port, the AWG-based optical module of demultiplexer, photodetectors (PDs), and pre-amplifiers for O/E module of the receiver port are fabricated on a PLC platform [12], and the temperature of the PLC platform is also maintained at a constant 25°C. Each 2.5-Gb/s electrical signal is converted to the form of 622 Mb/s × 4 parallel by the CDR circuit and then down-converted to 78 Mb/s, just as for the sender port to ensure the operation of the FPGA. The processed data in the FPGA are again up-converted to 622 Mb/s × 16 and input to a switch-element at the next stage.

5. Measurements

To demonstrate the feasibility of wavelength switching of the WDM link, we performed an experiment in the environment shown in Fig. 13. The output of the sender port was connected to the input port of the receiver port via an optical attenuator, which had an optical

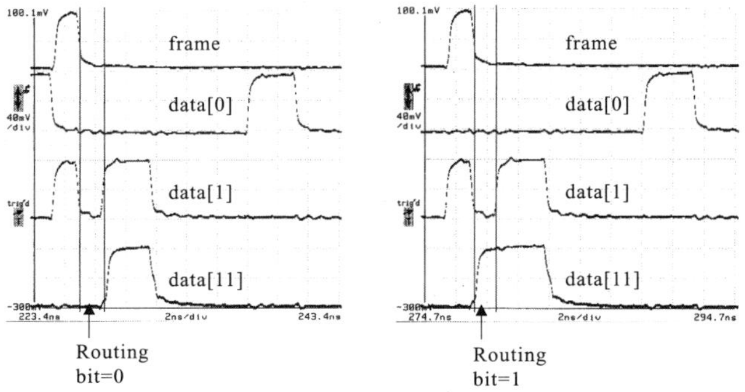

(a) Waveform from port 0 (b) Waveform from port 1

Fig. 14 Measured waveforms for receiver port.

loss corresponding to that of the AWG router. In our system, the AWG router has an average loss of about 7 dB. Therefore we set the loss of the optical attenuator to 10 dB to represent the worst-case scenario for the optical loss of the AWG router.

To estimate the performance of the WDM link, we measured the electrical waveform of the output of a receiver port. We examined the wavelength switching of cells for two wavelengths. Cells corresponding to λ_1, λ_2 and two dummy cells were periodically sent to the input port of the sender port. Figure 14 shows the measured waveforms for the output of the receiver port. Figure 14(a) corresponds to the received cells of wavelength λ_1, which had a routing bit of 0, while Fig. 14(b) corresponds to wavelength λ_2 with a routing bit of 1. In this way, optical wavelength switching was performed according to the routing bit in the cell header.

6. Conclusion

An experimental system of 5-Tb/s packet-by-packet wavelength switch, OPTIMA-2, was developed. Its hardware architecture and efficiency for handling variable-length packets was presented and the architecture of the optical WDM link was described. The dynamic bandwidth sharing technique of the WDM link enables a total throughput up to 5-Tb/s while keeping the statistical multiplexing gain. A packet-based wavelength switching scheme was used for the receiver port of the WDM link to reduce the cell-assembly cost at the line-interface card. A new scheduling algorithm for variable-length packets was used to resolve the bottleneck of conventional DRR. Simulation results showed that our algorithm can reduce the delay for short packets while keeping high throughput. We described as implementation of an experimental optical WDM link system, composed of electrical boards and optical (E/O and O/E) modules for the sender and receiver ports fabricated using FPGAs and compact PLC platforms. The basic operation of the WDM link for optical wavelength switching was demonstrated in experiments.

References

[1] J. Turner and N. Yamanaka, "Architectural choices in large scale ATM switches," IEICE Trans. Commun., vol.E81-B, no.2, pp.120–137, Feb. 1998.

[2] T. Chaney, J.A. Fingerhut, M. Flucke, and J.S. Turner, "Design of a gigabit ATM switch," Proc. IEEE INFOCOM'97, pp.2–11.

[3] N. Yamanaka, E. Oki, S. Yasukawa, R. Kawano, and K. Okazaki, "OPTIMA: Scalable, multi-stage, 640-Gbit/s ATM switching system based on advanced electronic and optical WDM technologies," IEICE Trans. Commun., vol.E83-B, no.7, pp.1488–1496, July 2000.

[4] K. Nakai, E. Oki, and N. Yamanaka, "Optical WDM grouped links and dynamic bandwidth sharing for scalable 3-stage ATM systems," Proc. IEEE ICC'99, vol.3, pp.1473–1477.

[5] H.J. Chao, C.H. Lam, and E. Oki, Broadband Packet Switching Technologies: A Practical Guide to ATM Switches and IP Routers, John Wiley & Sons, New York, 2001.

[6] M. Shreedhar and G. Varghese, "Efficient fair queuing using deficit round-robin," IEEE/ACM Trans. Networking, vol.4, no.3, pp.375–385, June 1996.

[7] E. Oki, N. Yamanaka, and Y. Ohtomo, "A 10 Gb/s (1.25 Gb/s × 8) 4 × 2 CMOS/SIMOX ATM switch," Proc. IEEE ISSCC'99, pp.172–173.

[8] E. Oki, N. Yamanaka, Y. Ohtomo, K. Okazaki, and R. Kawano, "A 10 Gb/s (1.25 Gb/s × 8) 4 × 2 0.25 μ CMOS/SIMOX ATM switch based on scalable distributed arbitration," Proc. IEEE Trans. Solid-State Circuit, vol.34, pp.1921–1934, Dec. 1999.

[9] K. Nakai, E. Oki, and N. Yamanaka, "Scalable 3-stage ATM switch architecture using optical WDM grouped links based on dynamic bandwidth sharing," IEICE Trans. Commun., vol.E82-B, no.2, pp.265–270, Feb. 1999.

[10] K. Nakai, E. Oki, and N. Yamanaka, "A distributed traffic control scheme for large-scale multi-state ATM switching systems," IEICE Trans. Commun., vol.E83-B, no.2, pp.231–237, Feb. 2000.

[11] K. Yamakoshi, K. Nakai, N. Matsuura, E. Oki, R. Kawano, and N. Yamanaka, "5-Tbit/s frame-based ATM switching system using 2.5-Gbit/s × 8 optical WDM links," Proc. IEEE ICC'01, pp.3117–3121.

[12] Y. Akahori, T. Ohyama, T. Yamada, S. Kamei, M. Ishii, R. Kasahara, M. Nakamura, M. Okayasu, H. Oohashi, and K. Yamakoshi, "WDM interconnection using PLC hybrid technology for 5-Tb/s electrical switching system," Proc. IEEE ECOC 2001 We.A.2.5.

High-Performance Backbone Network Technology edited by Naoaki Yamanaka
ISBN: 0-8247-5321-6 DOI: 10.1081/(120025743)
Published 2004 by Marcel Dekker, Inc.

Optical Label Switching Using Optical Label Based on Wavelength and Pilot Tone Frequency

Kiyoshi Tanaka, Katsuhiro Shimano, Kyo Inoue, Shigeru Kuwano, Takeshi Kitagawa, and Kimio Oguchi

SUMMARY This paper describes a new optical label switching technique; wavelength and pilot tone frequency are combined to form labels that are used to control transport network routing. This technique is very attractive for achieving simple nodes that offer extremely rapid forwarding. Experimental results on the discrimination of optical labels and all-optical label conversion are also presented.

key words: *optical label, pilot tone frequency, discrimination, all-optical label conversion*

1. Introduction

The variety of attractive contents on the Internet and the falling costs of high performance computers are leading to a dramatic increase not only in the number of Internet users, but also in the amount of Internet protocol (IP) data traffic. One critical research goal is to achieve IP network nodes with processing speeds that match the speeds now available from the optical links connecting these nodes.

The processing speed of IP network nodes has been increased by the use of hardware switches based on application specific integrated circuits (ASIC) [1], and link control methods such as multi-protocol label switching (MPLS) [2], [3]. Unfortunately, these speeds fail to match the tera-bit link capacities achieved with wavelength division multiplexing (WDM) [4], [5]. The key weakness of current IP network nodes is their continued reliance on the header analysis needed since the forwarding process lies in the electric domain. Several optical label techniques based on subcarrier multiplexing (SCM) [6], [7] and optical code division multiplexing (OCDM) [8], have been proposed. The former places the address information in an overhead signal that is transferred by subcarriers but header information still must be processed in the electric domain. While the latter offers the possibility of high-speed IP forwarding, its signal decision circuits are complicated.

This paper introduces a novel optical label (OL) switching method that combines WDM with pilot tone frequency (PTF) multiplexing. This method yields a simple node configuration because it performs forwarding by discriminating the packet's optical label; header information is not analyzed. In addition to explaining this OL switching method, we investigate the PTF discrimination offered by logic ICs with the goal of rapidly generating switching trigger signals; to this end we examine gate switching and all-optical OL conversion simultaneously in a semiconductor optical amplifier (SOA).

2. Optical Label Switching Network

Advanced IP-based WDM networks will be realized only if we can ease the crushing loads placed on current IP routers by implementing a transport network that offers extremely fast packet routing and forwarding. Each node in the network must switch huge packet volumes with low latency, and provide some important network functions such as quality of service and security at lower layers. OL switching has the possibility to realize the above requirements, since it has the potential to eliminate electrical buffering and routing processing. Here, our definition of the word "optical label" is physical layer information used to route optical packets and control network functions.

Figure 1 shows the concept of the OL switching network. The network consists of edge nodes (ENs), core nodes (CNs), and WDM links connecting these nodes. In the network, prior to optical packet transmission, the transmission route and the relationship between input and output OLs at each CN are established by negotiation between ENs. The optical packets are transmitted along the decided route, and OLs are converted node by node as agreed. In the EN, the network management function decides an appropriate optical label by referring to a label table. The label table is based on the header information of the received packet, such as destination network address, security level, quality class of service, etc. The OL is then rewritten with appropriate value and the optical packet is transmitted to

Manuscript received October 24, 2000.
Manuscript revised December 20, 2000.

[†]The authors are with NTT Network Innovation Laboratories, Yokosuka-shi, 239-0847 Japan.
[††]The author is with NTT Basic Research Laboratories, Atsugi-shi, 243-0198 Japan.
[†††]The author is with NTT Photonics Laboratories, Ibaraki-ken, 319-1193 Japan.
[††††]The author is with NTT Science and Core Technology Laboratory Group, Atsugi-shi, 243-0198 Japan.

*This paper is also published in the IEICE Trans. Electron., Vol.E84-C, No.5, pp.501–508, May 2001.

IEICE TRANS. COMMUN., Vol. E84-B, No. 5, pp. 1127–1134, May 2001

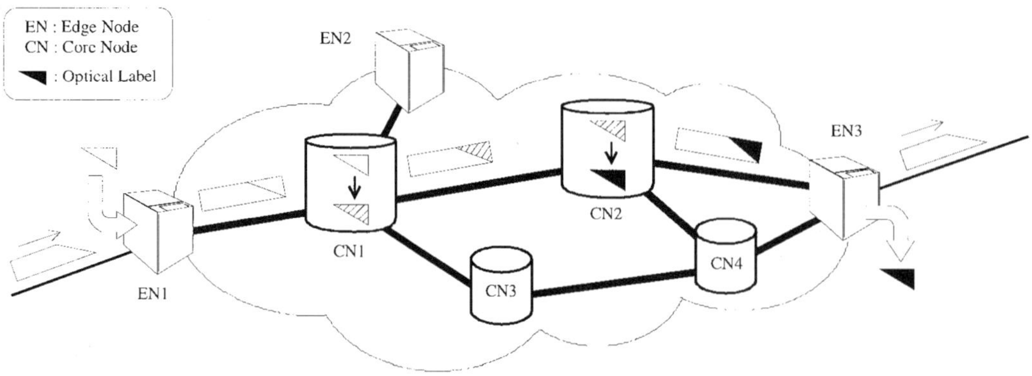

Fig. 1 The model of optical label switching network.

Table 1 Example of label table for edge node.

Current Node	Destination Node	Quality	Next Node	Optical label: WL	Optical label: Pilot tone
EN1	EN2	High	CN1	λ1	f1
		Low		λ4	f1
	EN3	High		λ2	f2
		Low		λ4	f2
	EN2&3	High		λ3	f1
		Low		λ4	f1 and f2

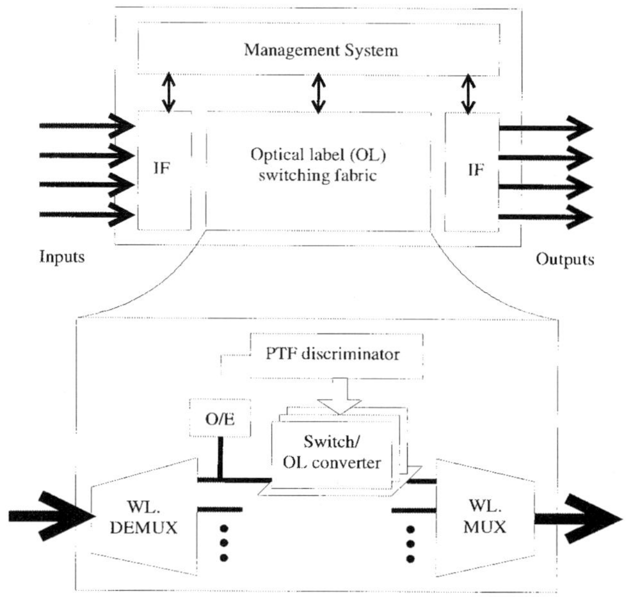

Fig. 2 Configuration of optical label switching core node.

the next node. Table 1 shows an example of the label table for EN1 in which the parameters are destination node and service quality. The CNs forward each optical packet to the appropriate node depending on its OL, and alter the OL passively.

One prior instance of OL usage is address identification, which uses wavelength as OL; it was proposed for optical cross-connect (OXC) systems and optical add/drop multiplexing (OADM) systems. Our idea is that the OL can also be used to indicate quality class or application identifier in addition to the network address. The applied area of our network is taken to be a regional area with 20 ENs for example. For a network with 1,350 addresses (unifying unicast, multicast assuming the combination up to 3 ENs, and broadcast), 2 quality classes, and 2 security levels, and 2 application (port number), 10,000 labels are required. As the current achievable number of wavelengths in a fiber is around 1,000 [9], [10], it seems that there are too few wavelengths. Obviously some multiplex technology is needed to overcome this shortage. Among the many options available for multiplexing [6]–[8], [11], [12], we choose wavelength and PTF multiplexing because it offers the possibility of fast discrimination, sufficient numbers of OLs, and simple node configuration. If there are M wavelengths and N PTFs, the number of OLs is given by $M \times N$ because they are mutually independent. Moreover, if the multiplexed PTF can be treated as a binary bit string, the available number of PTF labels becomes $2^N - 1$, where the condition of no PTF exclude from the PTF label. For instance, by thinking of three PTFs as a binary bit string such as $\{101\}$, the state number is $2^3 - 1$. It follows that the number of OLs created by combining WDM and multiplexed PTF is given by $M \times (2^N - 1)$. For example, 10,000 OLs can be obtained with 700 wavelengths and 4 PTFs, or 1400 wavelengths and 3 PTFs.

Figure 2 shows the CN configuration. The CN consists of input/output interfaces, OL switching fabric, and management system. The OL switching fabric consists of a wavelength multiplexing/demultiplexing section, a PTF discriminating section, and switching and OL converting section. When a WDM optical packet enters this CN, it first undergoes wavelength discrimination at wavelength demultiplexer. And then, the packet is divided into two parts: switching and OL con-

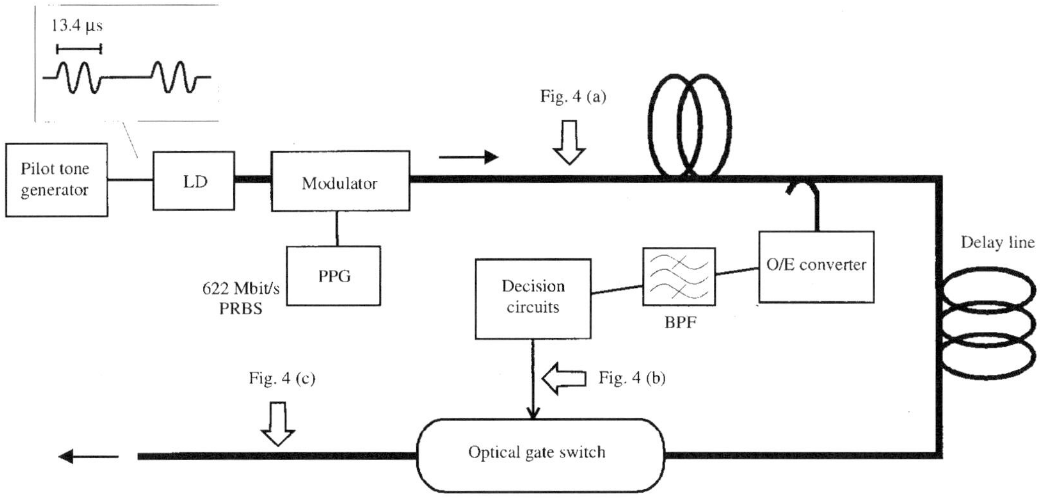

Fig. 3 Experimental setup for pilot tone frequency discrimination and optical gate switching.

verting section, and a PTF discriminating section that involve an optical-electrical (O/E) converter. Next, label switching is performed using the OL as the trigger in the switching section, and OL conversion is performed. At the end of this process, the optical packet is transmitted to the next node after wavelength multiplexing. In this CN, PTF discrimination, optical gate switching, and OL conversion are the key operations in the forwarding process. Accordingly, we conducted experiments on OL switching and confirmed the techniques needed for OL discrimination and conversion.

3. Experiments on Optical Label Switching

3.1 PTF Discrimination and Switching

Wavelength can be discriminated easily and speedily by a wavelength demultiplexer such as an arrayed-waveguide grating (AWG). Though several papers discussed using pilot tones [13]–[15] to transfer supervisory information, there is, to the best of our knowledge, no method that can generate an electrical switching trigger directly from PTF discrimination. In the proposed SCM method, since header detection and switching are performed by digital signal processing using regenerated bit sequence, a high-speed processor or procedure are required. In our method, on the other hand, the electric waveform obtained after PTF discrimination is immediately used to drive the switching section; neither signal analysis nor buffering is used so significant enhancement of forwarding speed can be expected. We confirmed the ability of PTF discriminator to generate the switching trigger directly from the detected PTF. The discriminating section consists of an O/E converter, band-pass filters (BPFs), and decision circuits. The decision circuits, which are the key to PTF discrimination, use commercial logic ICs, XNOR and D-FF, because these have high-speed operation (up to 20 GHz). The discrimination proceeds as follows. At first, a part of the optical signal is converted into an electric signal by an O/E converter, and then the PTF's waveforms are regenerated within the BPFs. Next, the power of the PTF waveforms is detected by XNOR of the BPF outputs, and the switching trigger waveform is generated by using a D-FF as a comparator. By using wide band logic ICs, a high-speed rise-time of a few nanoseconds is achieved, and the guard time, which includes the time needed for PTF discrimination, is reduced.

Figure 3 shows the experimental setup used to examine the proposed PTF discriminating section. The optical signal was modulated by an intermittent PTF signal (a PTF of 590 MHz with a period of about 13.4 μs), and then intensity modulated by a random pulse signal at 622 Mbit/s. The period of 13.4 μs corresponds to the packet length of 1,040 bytes at 622 Mbit/s. The optical signal was tapped by an optical splitter; one part passed to the optical gate switch via an optical delay line, while the other was passed to the discriminating section. As the optical gate switch, we adopted a semiconductor optical amplifier (SOA) because its gating time is within several hundred picoseconds, and the insertion loss of this element is compensated by its self-amplification. Figure 4 shows a modulated optical signal, the trigger signal generated by the PTF discriminating section, and the optical signal output as gated by the SOA. These waveforms were observed at measurement points (a), (b), and (c), respectively (see Fig. 3). These results confirm the capability of passive OL switching.

Increasing the PTF modulation index makes discrimination easier, but it also closes the eye opening of the main signal because it is regarded as the noise at the mark level. Here, we define the modulation index

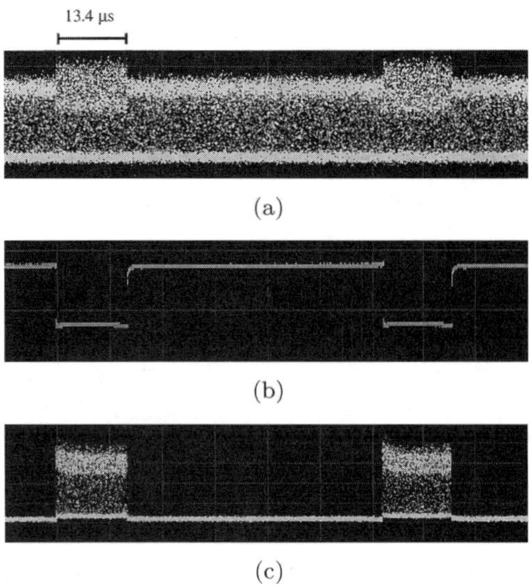

Fig. 4 Experimental results of PTF discrimination and gate switching. (a) Optical waveform before gating, (b) Electrical waveform generated by PTF discriminating section, (c) Optical waveform after gating.

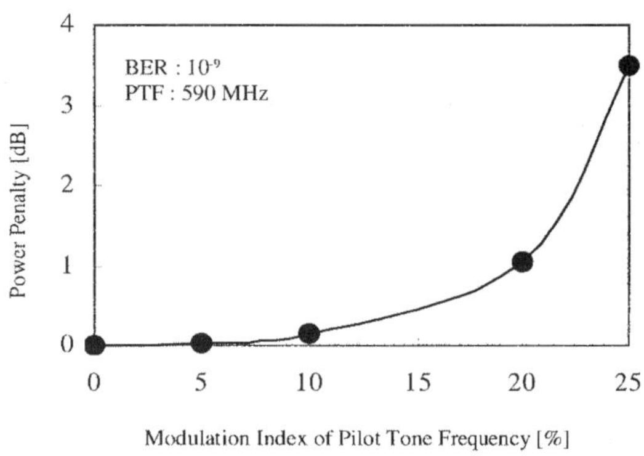

Fig. 5 Power penalty vs. modulation index of pirot tone frequency.

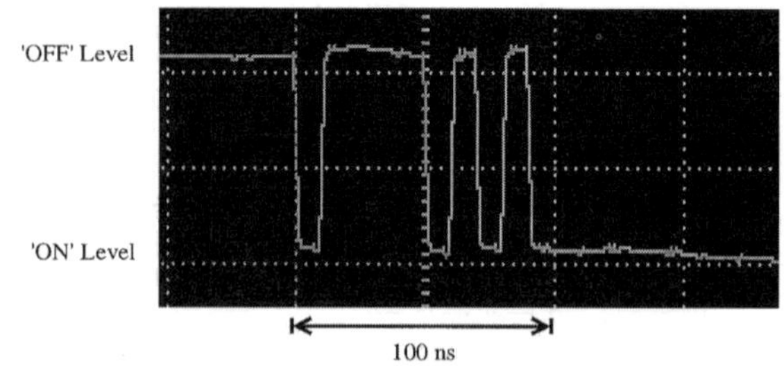

Fig. 6 Transitional phenomenon of switching trigger waveform. The pirot tone frequency was set to 590 MHz and 20% modulation index. It is shown that the chattering time was less than 100 ns in this experimental system.

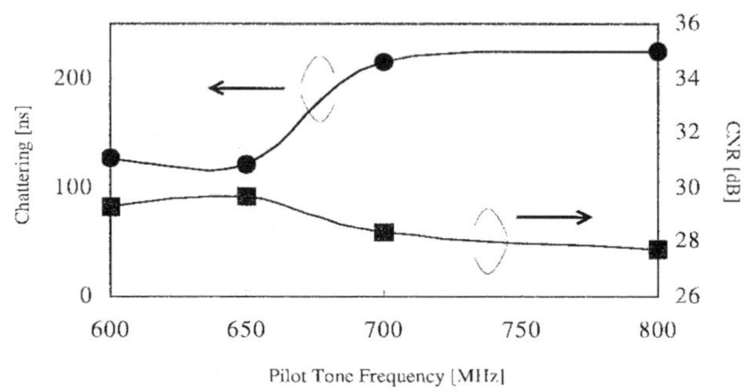

Fig. 7 The relation of pilot tone frequency, CNR and chattering.

as the percentage of the amplitude of power fluctuation caused by the pilot tone against the maximum power of the signal. We investigated the effect of the PTF modulation index on bit error rate (BER) characteristics of the main signal because the upper limit of the modulation index is determined by the BER requirement for the main signal. The relationship between the power penalty at the BER of 10^{-9} and the PTF modulation index of 590 MHz is shown in Fig. 5. The results show that in order to hold the power penalty to under 1 dB, the modulation index should be less than 20% in this experimental system.

Next, we examined a switching speed of the proposed method. In this experimental system, since a chattering, a transitional phenomenon as shown in Fig. 6, occur in the logic ICs, PTF discrimination is the slowest of all processes. Here, the PTF was set to 590 MHz and 20% modulation index. Though the rise-time of the decision circuits is fast, less than 1 ns, unstable behavior extending over hundreds of ns, was observed in the switching trigger waveform. To determine the cause, we examined the carrier-to-noise ratio (CNR) of the PTF experimentally; the results are shown in Fig. 7. In this experiment, chattering time and CNR, for several PTFs (all with the same modulation index), were measured using a one-shot oscilloscope and a spectrum analyzer, respectively. The results indicate that increasing the CNR reduces the chattering time, and that to increase the CNR, the used PTFs must suit the spectrum shape of the main signal. For example, if the main signal bit rate is 622 Mbit/s PRBS, the PTF should be set around at 622 MHz or several giga-hertz, at which values of the spectrum density is very small, for the purpose of achieving large CNR. Hence, to perform high-speed switching (equal to a short chattering time), we must not only optimize the PTF discriminator, but also investigate signal system that maximize CNR of PTF over a wide band without exceeding the modulation index.

3.2 Conversion of Optical Label

For constructing an optical label switching core node that offers high-speed and small latency time, it is important to implement all-optical OL conversion. We examined a conversion method that performs wavelength and PTF conversion in one go, by using the saturation effect based on cross-gain modulation (XGM) [16] in an

SOA. The principle of this all-optical OL conversion is shown in Fig. 8; the relation between input light power and converted light power of the SOA is sketched using a linear scale. The input signal light (λ_1, PTF_1) is input to the SOA in one direction while the new signal light (λ_2, PTF_2) is input in the other direction. Within the SOA, the input signal light modulates the pre conversion light by the effect of XGM, thus wavelength conversion from the input light λ_1 to the new light λ_2 is achieved; the new signal light is then passed to the circulator. The XGM characteristic ensures that the converted signal light power rapidly decreases as the input signal light power increases when input signal light power inserted to the SOA is low, but it changes little if the input signal power is high. This suggests that the PTF on '1'-level in the input signal light is suppressed by XGM. An advantage of our configuration, which inputs the original and converted lights in the opposite direction and uses a low-loss circulator, is that the input signal light is fully utilized in order to suppress the impact of PTF_1 on '1'-level. Moreover, the converted light is easily distinguished from the input light even when their wavelengths are close. According to this principle, the modulation pattern of the input signal light is reproduced in the converted light, but not the PTF; the PTF of the pre conversion light is retained because the PTF of the input signal light is suppressed by the XGM of the SOA. Hence, all-optical OL conversion from (λ_1, PTF_1) to (λ_2, PTF_2) is achieved in one device.

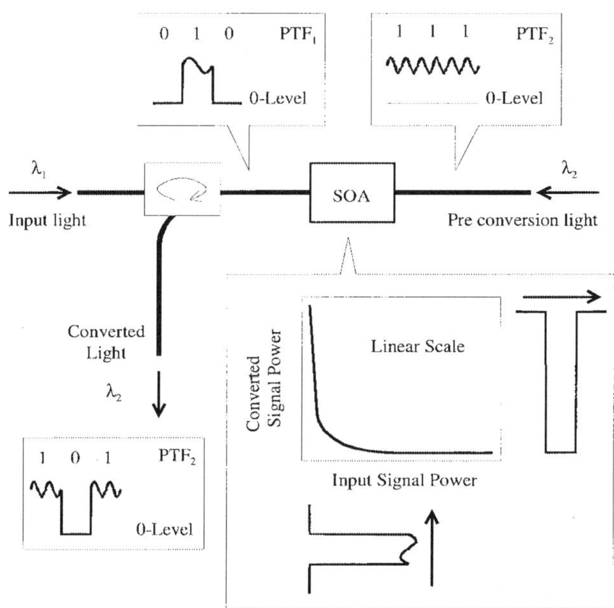

Fig. 8 Principle of all-optical optical label conversion.

3.2.1 Experiments on All-Optical Optical Label Conversion

The experimental setup used to confirm the principle of all-optical OL conversion is shown in Fig. 9. The input signal light was modulated by PRBS at 622 Mbit/s; its wavelength and PTF frequency were 1552 nm and 900 MHz, respectively. Those of the converted light were set at 1511 nm and 1.1 GHz, respectively. An optical BPF was used for the purpose of reducing the crosstalk. The results are shown in Fig. 10. Figure 10(a) shows the spectra of the input signal, and (b) shows the spectra of the converted signal, where the left and right figures are the optical spectra and electrical RF spectra, respectively. These results show that all-optical wavelength conversion and PTF conversion can be performed simultaneously in the same SOA. Here, the input PTF was suppressed more than 15 dB by the cross-gain saturation effect of the SOA, and optical crosstalk in the wavelength conversion process was about 50 dB due to the use of the optical circulator and optical BPF.

It is noted here that the SOA also realizes optical gate switching, as described above. Thus, gating and OL conversion are achieved simultaneously in the same SOA.

3.2.2 Dependency of Pilot Tone Frequency Suppression on Input Light Power

Since insufficient PTF suppression may cause an error in OL switching at the next node, it is important to investigate how well PTF is suppressed in the SOA. The

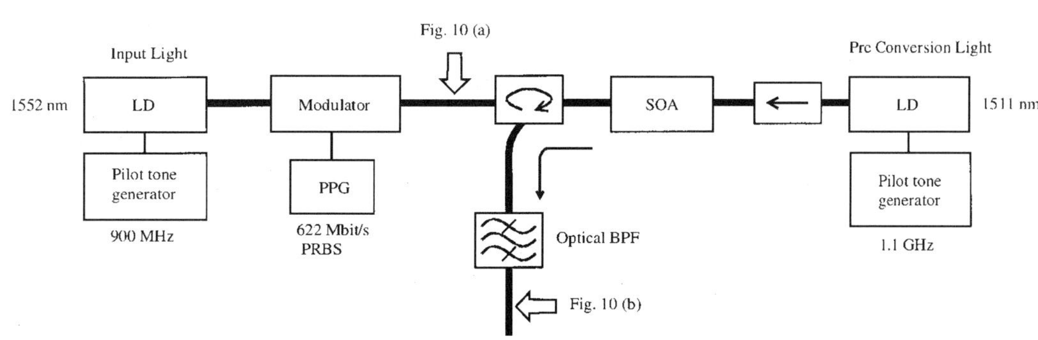

Fig. 9 Experimental setup to confirm the principle of all-optical optical label conversion.

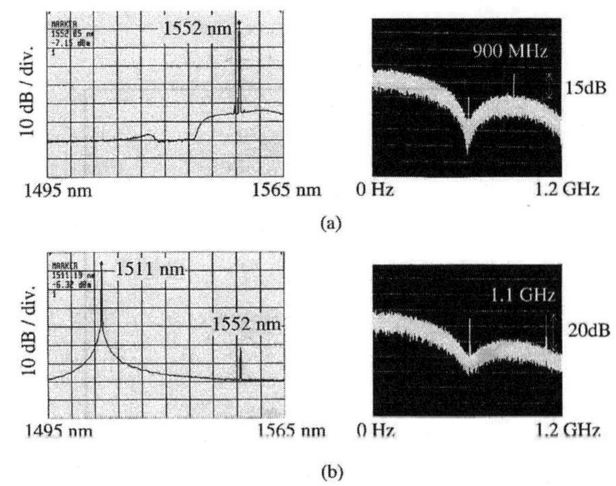

Fig. 10 Experimental results of optical label conversion. Left and right figures are the optical spectra and RF spectra, respectively. (a) The spectra of before conversion, (b) The spectra of after conversion.

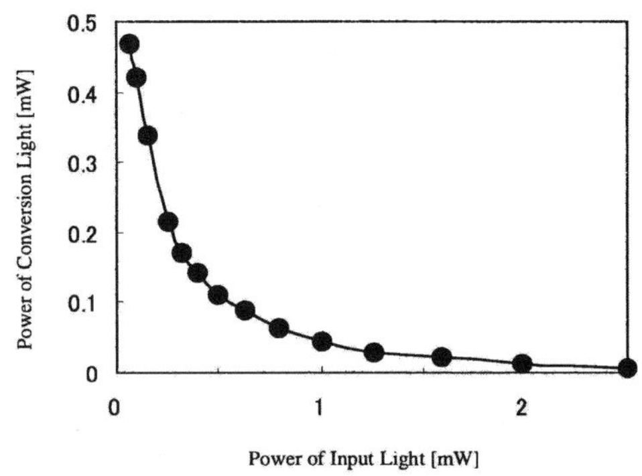

Fig. 11 Input-output power characteristic of SOA used in this paper.

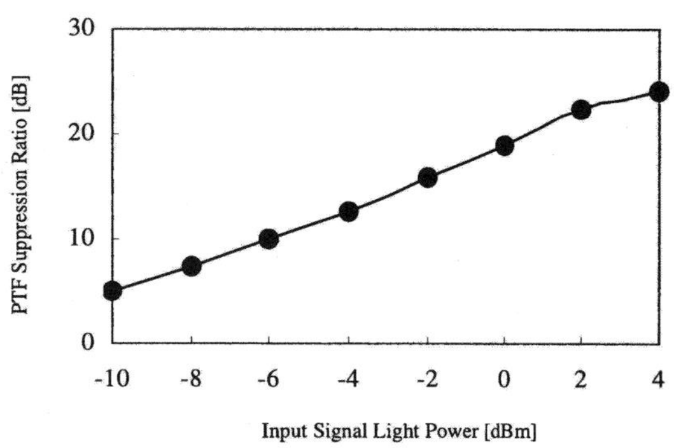

Fig. 12 Suppression ratio of pilot tone frequency vs. input signal light power. The conversion light power into SoA was 0 dBm.

cross-gain saturation effect we described above is not linear; the fluctuation in the '1'-level of the input signal light due to the PTF decreases as the signal light power input to the SOA is increased. The characteristic of the SOA used this paper is shown in Fig. 11 where the pre conversion light power was set at −2 dBm. We examined the relation between PTF suppression and input signal light power. The pre conversion light contained no pilot tone, and its input power was kept at 0 dBm. The results, shown in Fig. 12, indicate that to realize the suppression index of 20 dB, the input signal power must exceed 0 dBm for this experimental system.

4. Discussion

We discuss technical limitation of the proposed optical label switching technique in terms of PTF allocation, bit rate, and packet switching guard time.

First, we consider the frequency allocation of PTF. There are two limitation factors for this topic; the CNR at PTF and the frequency characteristics of used devices. The former, as mentioned above, is affected by the spectrum of main signal, this is to be a lower frequency limit. Since the spectrum of the main signal limits the CNR at PTFs, it is thought inserting electrical low-pass filter (LPF) at an electrical input port of optical modulator to improve the CNR. As a result, PTFs have only to be set at the out-band of main signal from the viewpoint of maximizing CNR. On the other hand, the upper limit is decided by the frequency characteristics of used devices. Since the upper limit of logic ICs marketed now is 20 GHz, it is possible to set PTFs up to 20 GHz.

Second, as for bit rate of main signal, it is necessary to set to 20 Gbit/s or less by considering PTF allocation according to above discussion. Because the carrier lifetime in a SOA performed XGM is slower than risetime of logic ICs (several hundreds pico-seconds), the upper limit of bit rate is decided by a carrier lifetime in the SOA. Although we used a bit rate of 622 Mbit/s, the wavelength conversion at 2.5 Gbit/s is possible using XGM in SOA.

On the other hand, as a Third limitation factor, packet guard time is determined by considering a transmission delay and jitter. Since a transparent bandwidth of the electrical BPF is chosen as several percents of the PTF, the delay time at the PTF discriminator depends on the bandwidth. Although a small delay is obtained at a high PTF, the discriminator needs to operate at a higher speed. In our experiment, we used relative lower PTF and narrower filter bandwidth of 20 MHz in order to relax the requirement to the device. This bandwidth corresponds to delay time of 50 ns. In addition, a chattering shown in Fig. 6 and Fig. 7 also extends the guard time to 100 ns as a jitter. These results indicated that a guard time of about 150 ns is necessary for 622 Mbit/s. However, employing higher bit rate and higher PTF, reduction in the guard time is possible.

Finally, on the basis of above discussion, all optical label switching at a bit rate of several Gbit/s with a packet guard time of less than 150 ns is achievable. Assuming a long packet of about 1,500 bytes, the guard time corresponds to less than 2.5 percents of the packet duration. Therefore, reduction of packet switching efficiency due to the guard time is negligible.

5. Conclusion

We presented a novel concept for an optical label switching network and the configuration and function of the core nodes. The concept combines wavelength with multiplexed pilot tone frequencies. Experiments confirmed the feasibility of the proposed optical label discrimination process and optical label conversion process. For the former, we experimentally examined the optical packet forwarding processes such as pilot tone frequency discrimination, switching trigger generation, and optical gate switching. For the latter, we explained the principle of optical label conversion, and introduced results that showed PTF suppression of more than 15 dB. Therefore, the feasibility of using OLs and simple node configuration to achieve low latency switching has been confirmed.

References

[1] L.L. Zhang, B. Beacham, M.R. Hashemi, P. Chow, and A. Leon-Garcia, "A scheduler ASIC for a programmable packet switch," IEEE Micro, pp.42–48, Jan.–Feb. 2000.

[2] A. Viswanathan, N. Feldman, Z. Wang, and R. Callon, "Evolution of multiprotocol label switching," IEEE Commun. Mag., vol.36, no.5, pp.165–173, 1998.

[3] C. Metz, "Ingredients for better routing? Read the label," IEEE Internet Computing, pp.10–15, Sept.–Oct. 1998.

[4] K. Oguchi, M. Jinno, J. Kani, K. Tanaka, and T. Kitagawa, "Multi-channel ultra-wide band WDM network and its supporting technologies," Proc. ICAPT2000, NA-II.2, 2000.

[5] T. Ito, K. Fukuchi, K. Sekiya, D. Ogasahara, R. Ohhira, and T. Ono, "6.4 Tb/s (140 × 40 Gb/s) WDM transmission experiment with 0.8 bit/s/Hz spectral efficiency," ECOC 2000, PD. 1.1, 2000.

[6] D.J. Blumenthal, A. Carena, L. Rau, V. Curri, and S. Humphries, "All-optical label swapping with wavelength conversion for WDM-IP networks with subcarrier multiplexed addressing," IEEE Photon. Technol. Lett., vol.11, no.11, pp.1497–1499, 1999.

[7] B.H. Wang, K.Y. Yen, and W.I. Way, "Demonstration of gigabit WDMA systems using parallelly processed subcarrier pilot-tone signaling technique," OFC'96, TuE, pp.21–22, 1996.

[8] K. Kitayama and N. Wada, "Photonic IP routing," IEEE Photon. Technol. Lett., vol.11, no.12, pp.1689–1691, 1999.

[9] H. Takara, T. Ohara, K. Mori, K. Sato, E. Yamada, K. Jinguji, Y. Inoue, T. Shibata, T. Morioka, and K.-I. Sato, "Over 1000 channel optical frequency chain generation from a single supercontinuum source with 12.5 GHz channel spacing for DWDM and frequency standards," ECOE 2000, PD. 3.1, 2000.

[10] K. Takada, M. Abe, T. Shibata, M. Ishii, Y. Inoue, H. Yamada, Y. Hibino, and K. Okamoto, "10 GHz-spaced 1010-channel AWG filter achieved by tandem connection of primary and secondary AWGs," ECOE 2000, PD. 3.8, 2000.

[11] D.J. Blumenthal, P.R. Prucnal, and J.R. Sauer, "Photonic packet switches: Architectures and experimental implementations," Proc. IEEE, vol.82, no.11, pp.1650–1667, 1994.

[12] N. Wada, J. Harai, W. Chujo, and F. Kubota, "Photonic packet routing based on multi-wavelength label switch using fiber bragg gratings," Proc. ECOE2000, vol.4, pp.71–72, 2000.

[13] M. Murakami, T. Imai, and M. Aoyama, "A remote supervisory system based on subcarrier overmodulation for submarine optical amplifier systems," J. Lightwave Technol., vol.14, no.5, pp.671–667, 1996.

[14] Y. Hamazumi and M. Koga, "Transmission capacity of optical path overhead transfer scheme using pilot tone for optical path network," J. Lightwave Technol., vol.15, no.12, pp.2197–2205, 1997.

[15] C.L. Lu, T.K. Fong, R.T. Hofmeister, P. Poggiolini, and L.G. Kazovsky, "CORD—A WDM optical network: Design and experiment of fast data synchronization by pilot-tone transport," IEEE Photon. Technol. Lett., vol.8, no.8, pp.1070–1072, 1996.

[16] K. Inoue, "Noise transfer characteristics in wavelength conversion based on cross-gain saturation in a semiconductor optical amplifier," IEEE Photon. Technol. Lett., vol.8, no.7, pp.888–890, 1996.

Chapter 9

Photonic and Advanced Switching Devices

The chapter covers hardware technologies to realize backbone networks. There are several contributions, all of which use state-of-the-art technologies. This chapter describes advanced switching devices and focuses on photonic devices.

For realizing a very high-speed backbone system, photonic devices are the key components and optical wavelength conversion is the next breakthrough needed. Paper 9.1 presents a review of wavelength conversion techniques. For speeds of up to 2.5 Gb/s, optical to electrical conversion is effective. However, the demand is for speeds of 10 Gb/s and which necessitates all-optical wavelength conversion.

Most of the details available on wavelength converters are given in Papers 9.2 and 9.3. R. Sato et al. present a developed wavelength converter based on a planar lightwave circuit (PLC) technologies in Paper 9.2. K. Tsuji et al. describe a simple polarization-independent wavelength conversion method using four-wave mixing in single-mode fibers pumped with cross-polarized high frequency pulses in Paper 9.3. The wavelength conversion technology is now rapidly changing, so one challenge is to identify the next breakthrough technology.

T. Kato et al. report a 4 × 4 broadcast and select type optical switch element in Paper 9.4. The architecture is well-known, but the paper also includes very advanced experimental results in addition to the hardware technologies. The 4 × 4 switch integrates an optical gate, realized by an optical amplifier and a 1:4 splitter, and a 4:1 multiplexer, realized as a silica waveguide. The flip chip technique is used to realize a compact hybrid optical switch module. A 10 Gb/s photonic cell switch prototype is demonstrated.

More on the system level, B. Bostica et al. present a Tb/s ATM switch architecture that uses both electrical and photonic technologies in Paper 9.5. The paper shows that the space division approach is important. Some experimental 2.5 Gb/s link techniques are also reported.

Paper 9.6, reports a 10 Gb/s very high-speed packet receiver module. 10 Gb/s bit-synchronization and synchronization is realized within the module whose dimensions are 170 mm × 130 mm × 10 mm. In addition, only 9 bits are needed for synchronization. Fabrication results for a high-speed Si-bipolar device and experimental results are also reported. It provides effective information on both device technology and transmission technology such as synchronization technology.

Temperature insensitive add/drop filters and wavelength tunable lasers are indispensable devices for future optical networks. T. Amano et al. present the design of micromachined temperature-insensitive filters in Paper 9.7. The experimental results show a temperature dependence of better than +0.01 nm/K and wavelength trimming of over 20nm.

This decade has seen a continuous increase in research of VCSELs (vertical cavity surface emitting laser diodes). Papers 9.8 and 9.9 address VCSEL technologies. K. Katsura et al. describe a very advanced optical parallel interconnection technology in Paper 9.8. The interconnection module employs a 40 channel VCSEL array for optional source. To increasing interconnection speed, optical WDM and parallel interconnection are being implemented using VCSELs. As is well known, WDM technologies are suitable for long-haul transmission. However, for short-reach transmission, they run into a cost limitation because multiple laser sources are expensive. Therefore, optical parallel interconnection is the key to cost-effective short-reach transmission. In addition, innovative bare fiber connectors are described. An interconnection module with 850-nm optical source is detailed that offers 28 Gb/s throughput (100 Mb/s × 40 channels). R. Michalzik et al. discuss several promising application areas for optical data links based on high-performance VCSELs in Paper 9.9.

Photonic crystals (PCs) have seen major advances in the past few years. C. Weisbuch et al. present experiments on a variety of structures and devices, as well as modeling tools in Paper 9.10. They show that 2D PCs etched-through waveguides supported by substrates are a viable approach to realizing high-performance PC-based photonic integrated circuits.

The final paper of this chapter, written by Y. Hibino, reviews the recent progress made by those working on optical filters and switches for photonic networks based on dense WDM. Three kinds of optical filter; thin-film interference filters, fiber gratings and AWGs are described in detail. The AWG technique on planar-type lightwave circuits is an important breakthrough technology since optical functions can be integrated into an integrated circuit. While this paper offers a review of the field, it also presents state-of-the-art development results.

High-Performance Backbone Network Technology edited by Naoaki Yamanaka
ISBN: 0-8247-5321-6 DOI: 10.1081/(120025744)
Published 2004 by Marcel Dekker, Inc.

Wavelength Converter Technology

Kristian E. Stubkjaer, Allan Kloch, Peter Bukhave Hansen, Henrik N. Poulsen, David Wolfson, Kim Stokholm Jepsen, Anders Thomas Clausen, Emmanuel Limal, and Alvaro Buxens

SUMMARY Wavelength conversion is important since it ensures full flexibility of the WDM network layer. Progress in optical wavelength converter technology is reviewed with emphasis on all-optical wavelength converter types based on semiconductor optical amplifiers.

key words: *wavelength converter, cross connect, WDM, optical network, optical processing*

1. Introduction

Wavelength division multiplexing (WDM) is considered one of the most feasible ways to upgrade point-to-point transmission links and to meet the huge demand for transmission capacity. Next step in the rapid evolution is the introduction of WDM networks for efficient transport of information. As the network technology evolves we want the possibility of wavelength conversion or translation within the network or at its interfaces. The motivation is to achieve an optical layer providing dynamic transport reconfiguration, high level restoration and efficient use of the available fiber bandwidth the same way as the synchronous digital hierarchy (SDH) layer provides efficiency to the transport layer by time division multiplexing, time slot grooming and high speed protection.

The aim of this paper is to briefly review the current state of optical wavelength conversion technology with emphasis on all-optical wavelength converters. Although WDM networks have the attractive feature that different signal formats can be transported in different wavelength layers, our focus will be on binary signal formats that are by far dominant for telecommunication traffic.

The paper will start by giving a short motivation for wavelength converters in WDM networks in Sect. 2. The various types of wavelength converters are examined in Sect. 3, while Sect. 4 gives examples of new development areas for wavelength converters. Finally, a conclusion is given in Sect. 5.

Manuscript received October 6, 1998.

[†]The authors are with COM Center, Bldg. 349, Techn. Univ. of Denmark, DK-2800 Lyngby, Denmark.

[*]This paper is also published in IEICE Trans. Electron., Vol.E82-C, No.2, pp.338–348, February 1999.

2. Wavelength Converters in WDM Networks

Several field trials with WDM networks that feature wavelength switching and routing have been reported, e.g. [1]–[3]. Although the work on WDM network architectures is in the initial phase, it seems that for full flexibility it is very attractive to be able to translate the channel wavelengths in an easy way [4], [5]. This is motivated by the resulting capacity increase, easier failure recovery, and reduced network complexity, e.g. [4]–[13]. The converters make it possible to assign wavelengths on a link by link or on a sub-network basis thereby relaxing the requirements to the wavelength precision throughout the whole network [4]. Moreover, wavelength conversion eases the recovery from link or node failures by allowing for local rather than global reconfigurations in the network. Networks will also become more scaleable by use of converters and a slight capacity increase will be achieved.

Wavelength converters will be placed in the network nodes where they can be operated with either a fixed input wavelength and a variable output wavelength or a variable input and a fixed output wavelength. Only special cases will require variable in- and output wavelengths at the same time. A converter with a fixed output wavelength is simpler than one with a variable, but the complexity of the wavelength converter is traded for that of other parts of the optical cross connects [14]. The converters will under all circumstances add complexity to the hardware. Therefore it is still an open issue if converters should be used in, e.g., all cross connects or at network segment interfaces.

The first trials with wavelength converters in switch blocks have just begun. A fine example is the optical cross connects made by the ACTS project OPEN. It has 4 ports with 4 wavelengths per port and has recently been tested in a field trial between Norway and Denmark [1]. Another impressive optical cross connect has been demonstrated by NTT. It is a 16 × 16 switch with 8 wavelengths employing tuneable wavelength converters at the input [15].

Even more advanced switch blocks with wavelength converters are investigated for optical packet switching. In this type of switch block the converters can be used to address space in optical delay-line buffers [16]–[18] or to address output ports in the switch [18], [19]. Examples of packet switch block demonstrators with wavelength converters are reported in [20], [21].

So, network solutions with wavelength conversion are emerging. Besides, it should be noted that wavelength conversion is already used in commercial WDM point-to-point systems in transponders that ensure correct channel wavelengths irrespective of the transmission equipment used in the tributary links.

The requirements to the converters will clearly be application dependent, but desired features include:

- Transparency to the signal format (typically IM-NRZ)
- Bit rate capabilities of at least 10 Gbit/s
- High dynamic range (because signal levels in the switch blocks will be path dependent)
- Signal reshaping (because other optical network elements can corrupt the signals)
- Cascadability of a few stages of converters (because multihop connections will be needed in most networks)
- Low electrical power consumption
- Operation at moderate optical power levels
- Easy implementation (to enable low price and reliability)

In the following we review the progress in wavelength converters towards these goals.

3. Wavelength Converter Types

The converters fall into four groups: 1) Opto-electronic converters; 2) Laser converters; 3) Coherent converters (four wave mixing and difference frequency generation); and 4) Converters based on optically controlled optical gates.

Opto-electronic conversion is a straightforward solution that is already used in WDM point to point links as mentioned above. Of the last three converter types, which are all-optical, those based on optically controlled gates are currently the most promising. We will therefore put most emphasis on the description of these.

3.1 Opto-Electronic Converters

The opto-electronic (O/E/O) converter is conceptually simple since it consists of a detector followed by amplification or regeneration and transmitter stages as shown in Fig. 1. The approach has the big advantage that it can be implemented with "off the shelf" technology. So implementation is relatively simple. The opto-electronic converters have low optical power requirements and potentially a large input power dynamic range and they are polarization independent. Depending on the use of simple electrical amplifiers or electronic regenerators they will even reshape the signal.

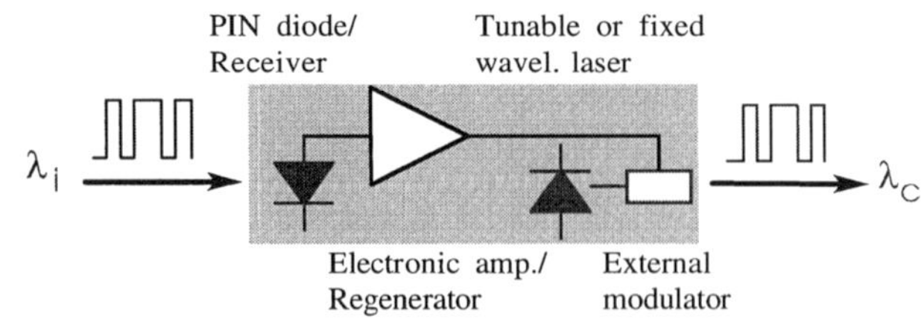

Fig. 1 Schematic of opto-electronic converter.

Demonstrators with such converters have given impressive results. As an example a 4-channel 2.5 Gbit/s network experiment with opto-electronic converters in three nodes has been reported [22]. Reference [15] reports an 8-channel 16×16 wavelength translating switch block with line rates of 2.5 Gbit/s. The O/E/O converters in this experiment are built into compact boards.

Recently, very compact units made directly for the purpose of wavelength conversion have been reported [23]. The compact, reconfigurable 2.5 Gbit/s O/E/O converter uses an integrated pin/HBT receiver for driving the modulator section of an opto-electronic integrated array laser containing up to 12 selective channels spaced 100 GHz apart.

Most opto-electronic converters reported till now operate at 2.5 Gbit/s. For wavelength conversion in high speed networks operating at 10 Gbit/s and in the future possibly at 100 Gbit/s, the power consumption of the opto-electronic converter will be high and bandwidth limitations of electronic circuitry may be encountered. Therefore, all-optical converters are highly interesting.

3.2 Laser Converters

All-optical wavelength conversion can be performed in a very simple way by optical control of single frequency lasers as shown in Fig. 2. The input signal (λ_i) to be converted is launched into the laser where it causes gain saturation that controls the oscillation of the laser. The result can be either IM or CPFSK output formats depending on the operation of the laser [24]–[26]. The lasing wavelength (λ_c) is either fixed or electronically tuneable depending on the system requirements.

With steep input-output transfer functions (see Fig. 2) the IM output format mode can achieve fine signal waveforms that even allow for cascading of a few converter stages [27]–[29]. The IM output mode is, however, associated with chirp [24], [30] that will limit transmission on non dispersion shifted optical fibers. The problems related to chirp could be overcome by operating the laser well above threshold in the CPFSK

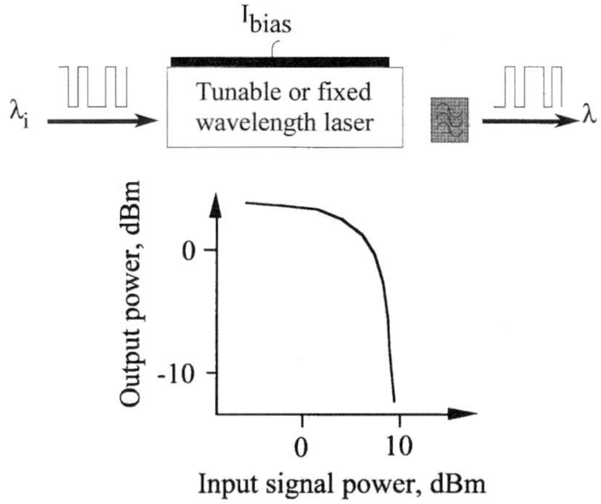

Fig. 2 Principle of wavelength converters based on semiconductor lasers together with schematic of output versus input power characteristic for laser converter.

output mode. In this case a frequency discriminator/filter is, however, needed at the output to obtain an IM signal format [24], [31].

Note, that the laser converter basically consists of a single component, so it is conceptually simple. Its optical input power requirements are 0–10 dBm, but unless a special waveguide design is implemented for the gain section of the laser, it is polarization dependent. Moreover, the maximum bit rate, determined by the laser's resonance frequency, is limited to around 10 Gbit/s [26]. Because of the speed limitation, laser converters are less interesting.

Work on laser converters exhibiting optical bistability for regeneration has also been pursued, e.g. [32]–[34]. The bistable element does, however, limit the conversion speed to the order of 1 Gbit/s, making the bistable converters unattractive.

3.3 Coherent Converters

Converters that rely on four wave mixing (FWM) have been extensively investigated using both optical fibers, e.g., [35] and semiconductor optical amplifiers (SOAs), e.g., [36]–[39], as nonlinear elements. The scheme is inherently fast for both fiber and semiconductor nonlinear elements. The converters can handle all signal modulation formats in contrast to other types of converters that are more suited for IM input signals. The transparency to signal format also implies that many WDM channels can be converted simultaneously [40].

The conversion efficiency is normally low (typically around −20 dB), so optical power levels of ~10 dBm have to be used for the pump of SOA converters while 10–20 dBm is needed for fiber based converters. Because of the low conversion efficiencies the signal-to-noise ratio for the converted signals needs attention, especially if converters have to be cascaded. Experiments using SOA converters with very long cavities

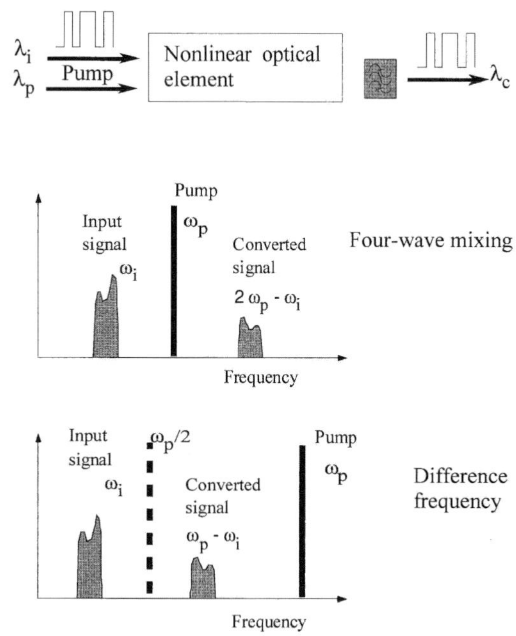

Fig. 3 Schematic of conversion in nonlinear element using four-wave mixing and difference frequency generation.

have, however, resulted in conversion efficiencies approaching 0 dB [41] thereby making FWM more attractive.

A serious drawback for the FWM converter is the dependency of the output wavelength on both the pump (λ_p) and the input signal (λ_i) wavelengths, so the pump must be tuneable even for converters with fixed output wavelength. Moreover, two pumps will be needed to ensure polarization insensitive operation [40], [42].

Because of the relatively complex pumping scheme, FWM converters will probably only be used for conversion of ultra fast signals at bit rates above 100 Gbit/s. Another potentially important application for FWM converters is mid-span spectral inversion for dispersion compensation. Here the optical phase conjugation of the converted signal is used to advantage. For mid-span spectral inversion a fixed pump wavelength can be used, so the set-up is simpler than that of a converter. Impressive results have been reported using 2 mm long SOAs [43]. For these S/N ratios of 45 dB in 0.1 nm bandwidth were achieved and a 40 Gbit/s signal successfully transmitted over 406 km of NDS fiber. Similarly, an impressive mid-span spectral inversion experiment relying on four wave mixing in highly nonlinear dispersion shifted fiber has been reported in [44]. Here simultaneous conversion of five 40 Gbit/s channels together with transmission over 105 km of standard single mode fiber is achieved.

Wavelength conversion based on difference frequency generation (see Fig. 3) in periodically loaded waveguide structures of $LiNbO_3$ [45] or AlGaAs [46] have also been reported. One advantage is polarization insensitivity of the scheme, otherwise it features

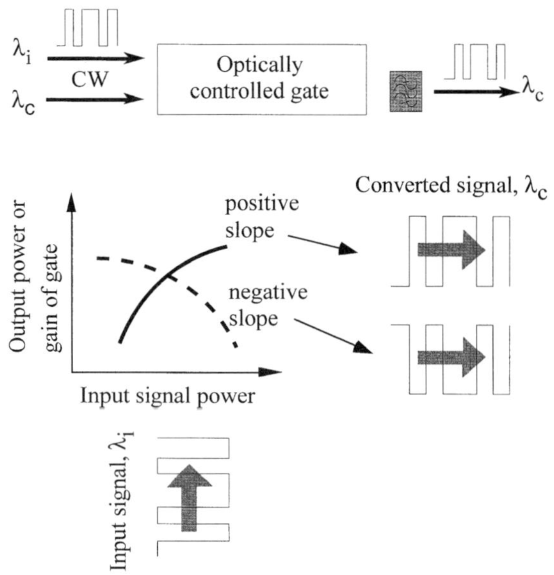

Fig. 4 Schematic of wavelength converter based on optically controlled gate and the dependency of the converted signal on the transfer characteristic of the gate.

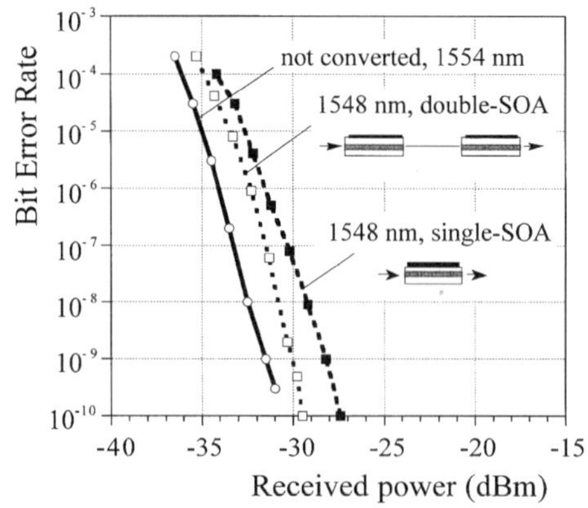

Fig. 5 Bit error rate for 40 Gbit/s signals: 1554 nm (input) and converted signals at 1548 nm using a single and double SOA converter [56].

almost the same advantages and disadvantages as four wave mixing. Conversion efficiencies are in the −17 dB range therefore requiring relatively high optical power levels.

3.4 Converters Based on Optically Controlled Gates

Converters made from optically controlled gates appear at the moment to be the most promising all-optical converter types. As illustrated in Fig. 4, the principle is to let the input power at λ_i control the gating of CW light at λ_c. Thereby the data are converted from λ_i to λ_c. The CW light originates from a light source with either a fixed or a tunable output wavelength depending on the application of the converter. Clearly, the transfer function of the gate should be as steep as possible and depending on its positive or negative slope the converted signal will be in-phase or inverted relative to the input.

3.4.1 XGM Gate

A simple optical gate is realized by a semiconductor optical amplifier (SOA). Its gain saturation due to an optical input signal is controlling the gain and thereby the state of the gate [47], [48]. The resulting converter, also called a cross gain modulated (XGM) converter, is extremely simple to assemble. It is polarization insensitive because of polarization independent SOA gain and is very power efficient. It has, however, a number of shortcomings: The signal is inverted relative to the input signal (negative slope) and the extinction ratio for the converted signal may degrade going from shorter to longer wavelength [49]. Moreover, the converted signal has a relatively large frequency chirp. Despite the nonideal properties, the converter has been used with fine results in a number of switch block experiments, e.g. [19], [50], and it remains attractive for many applications because of its simplicity. The shortcomings may also be eliminated if an interferometric converter (see next section) follows the XGM converter.

The conversion speed is determined by the carrier dynamics that are governed by the relative slow interband carrier recombinations [51]. The detailed analyses in [52] and [53] show that gain saturation plays a significant role for obtaining the high bit rate. Simple guidelines for achieving high bit rate conversion is to operate the SOAs with:

1) Large current injection
2) High optical power levels

Moreover, the SOA waveguides should have:

3) Large optical confinement factors
4) Large differential gain

Since the allowed injection current is limited the conversion speed can also be increased using longer cavity lengths. Direct coupling of two SOAs is an approach to a longer cavity [54].

For 1.2 mm long specially designed SOAs [55] 40 Gbit/s conversion has been achieved as shown in Fig. 5 [56]. The penalties are only ~4 dB for a single SOA converter and ~2 dB for two directly coupled SOAs used as a converter unit. Recently impressive 100 Gbit/s wavelength conversion has been achieved by XGM conversion in a 2 mm long SOA followed by a grating for FM to AM conversion of the chirped output to extend the conversion bandwidth [57].

Obviously, the SOA converters add spontaneous emission noise to the converted signals. Still, with noise figures of 8–10 dB (including 2–4 dB input coupling loss), it is possible to cascade a few SOA-XGM converters as shown in, e.g., [58].

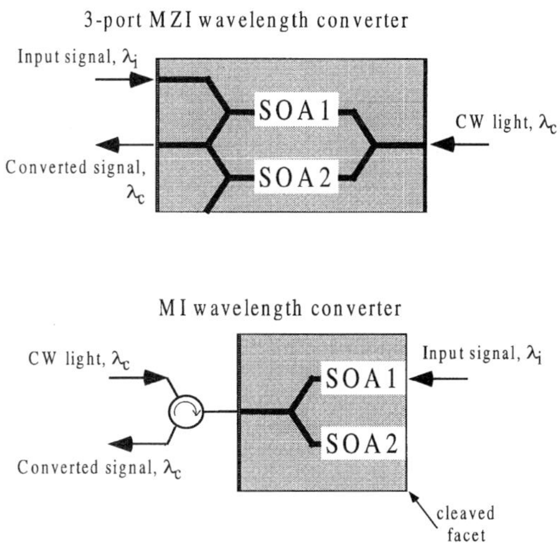

Fig. 6 Wavelength conversion principle using MZI and MI structures with SOAs.

3.4.2 XPM Gates

Better wavelength converters are obtained by integration of two or more SOAs into interferometric waveguide configurations (Fig. 6). In these converters, the optical input signal controls the phase difference between the interferometer arms through the relation between the carrier density and the refractive index in the SOAs (cross phase modulation, XPM); thereby a CW light is modulated [59].

For stable operation the XPM converters must be integrated. An impressive activity on opto-electronic integration of interferometric converters has taken place resulting in rapid progress towards compact and efficient monolithically integrated converters [60]–[67]. Several approaches are taken:

MZI-converters:
The Mach-Zehnder structures have been developed into structures with a separate waveguide for coupling the input signal into only one of the SOAs as shown in Fig. 6 (a). The idea is that the input signal depletes the carrier concentration in only one of the SOAs thereby creating the needed phase difference between the two interferometer arms in a very efficient way [63]–[66].

It should be noted that the work on Mach-Zehnder interferometric converters started with two-port devices [60], [61].

MI-converters:
Michelson interferometric converters have a simpler structure since they offer direct access for the input signal to the SOAs [62], cf. Fig. 6 (b). The high optical power levels due to the direct coupling will enable a high modulation bandwidth. The MI converter has a reflective facet making it a folded version of the MZI converter. Therefore it works very similarly. It should be noted, however, that the MZI allows the CW-light and the input signal to be launched counter-directionally, whereas, for the MI, a part of the original signal will always pass through the converter and be transmitted along with the converted signal. This prohibits the input and output wavelengths to be identical due to interference crosstalk.

NOLM-converters:
Nonlinear loop mirror configurations with SOAs are also reported [67], [68]. The converter is inherently balanced, but only RZ signals can be handled and the signal bit-rate is determined by the placement of the SOA in the loop. NOLM-converters with fiber as the nonlinear element are also reported, e.g. [69].

Different approaches are also taken regarding integration of the SOAs into the inter-ferometer:

All-active: Both splitters and gain sections have the same material composition. The gains in the different parts of the structure are determined by the bias current to the electrodes. This approach has the advantage of being simple and very compact [63]. Initially concern was raised that the pumped in- and output couplers would contribute significantly to the noise level. Experiments show that this is not so.

Active-passive: The active SOA sections are coupled to passive waveguides leading to a simpler electrode structure and to more well defined gain sections [64]–[66].

Hybrid active-passive: Hybrid integration of SOAs on PLCs has also been reported [70]. The approach is interesting since waveguide technology in Si is relatively mature.

More work is still needed to determine which approach should be taken. It should be noted that almost polarization independent operation can be achieved for correct design of the interferometers.

The interferometric converters have the advantage of very steep transfer functions (see Fig. 7) enabling extinction ratio regeneration of the converted signal. As also seen, the conversion can take place on either the positive or the negative slope of the interferometer curve, where clearly the positive slope is the most attractive since the converted signal is non-inverted. Only small input signals are needed to introduce a π phase difference between the interferometer arms, so very efficient conversion is obtained almost independent of wavelength. Because of the small modulation associated with the π phase shift, the frequency chirp of the output signal will also be small compared to, e.g., the XGM converter [62], [72], [73]. Moreover, for conversion on the positive slope the resulting chirp compensates

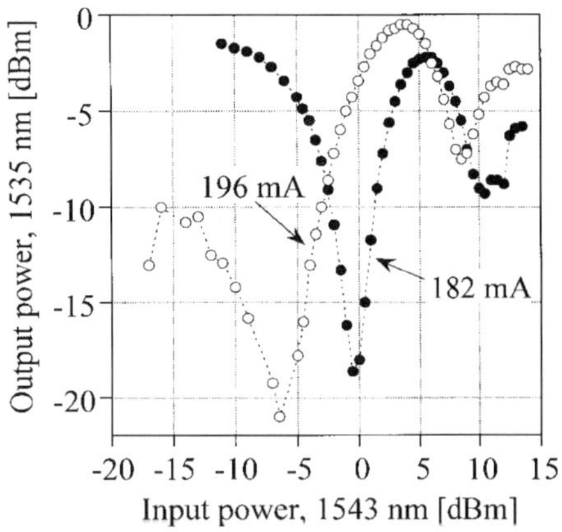

Fig. 7 Examples of static characteristics of active-passive MZI converter for two different bias currents the SOA in one of the interferometer arms [71].

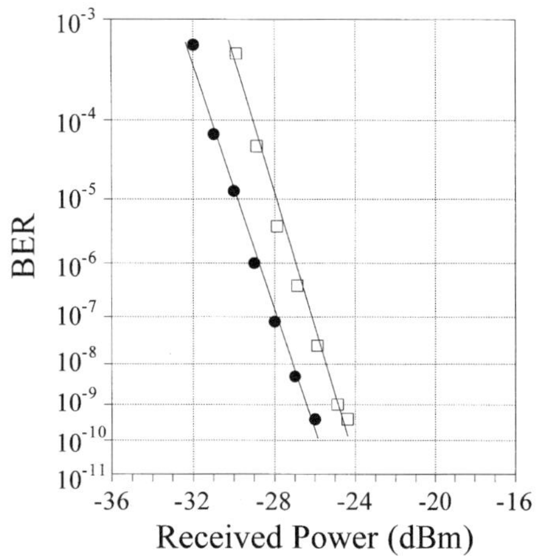

Fig. 8 BER performance of 40 Gbit/s wavelength conversion with all-active MI-converter. Results are after conversion (" ") and back-to-back (! |) [80].

against the influence of fiber dispersion.

Besides signal waveform and spectral reshaping, the interferometric converters have excellent noise properties with optical signal-to-noise ratios for the converted signals as high as 30 dB (measured in 1 nm spectral bandwidth). Moreover, the noise is redistributed due to the transfer function [74], [75]. As a result, the noise is accumulating less rapidly than for a chain of optical amplifiers. This allows for cascading of several converters [76]. The regenerative properties of the interferometric converters are important for construction of all-optical cross connects in which other components (e.g., amplifiers and gates) may degrade the signal quality.

The interferometric converters will typically have < 1 dB penalty for bit rates of 10 Gbit/s as has been reported by many groups, e.g., [62], [77]–[79]. Like the XGM converters, the interferometric XPM converters can also achieve 40 Gbit/s as seen in Fig. 8 [80]. The

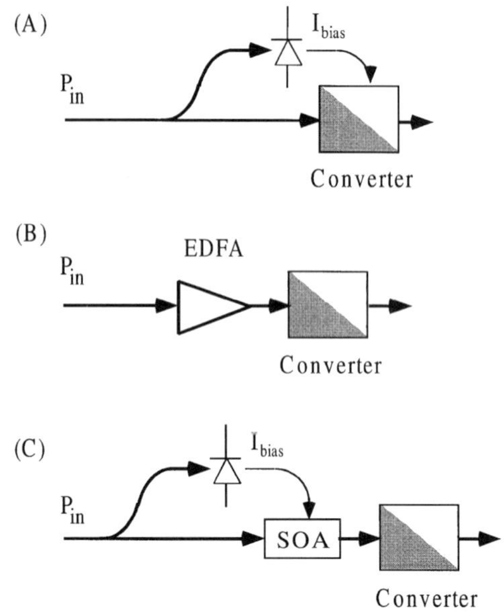

Fig. 9 Schemes for enhancement of the input power dynamic range of interferometric wavelength converters [84].

same guidelines as for XGM converters apply for achieving high bit rate. Especially coupling losses should be kept small to enable high optical power levels. Therefore the MI structure with direct coupling of the input signal has achieved the highest conversion bandwidth so far. The MI converter used in the 40 Gbit/s experiment is realized using multiquantum-well based layer structures with a 10 well tensile strained InGaAs/InGaAsP waveguide core. It consists of an active 3 dB coupler followed by two 600 μm long amplifier sections, resulting in a total device length of only 1.3 mm [77]. It should be noted, however, that differential switching schemes (e.g. [81], [82]) can be used for penalty free conversion of RZ-signals at 40 Gbit/s with potential for much higher bit rates [83].

Progress toward development of practical interferometric converters has been achieved with the realization of packaged all-active MZI structures [1]. These packaged converters have been tested in field trials. A practical obstacle for the interferometric converters is their small input power dynamic range of typically 2–3 dB. This can, however, be significantly improved by simple compensation schemes as shown in Fig. 9 [84]. In 10 Gbit/s experiments it was shown that Scheme A with simple power monitoring and electronic feed forward gives an 8 dB dynamic range, while the use of an EDFA preamplifier operated in saturation, as shown in Scheme B, can give a 40 dB dynamic range. It may be a disadvantage that the EDFA has a response time in the millisecond range. Use of a SOA preamplifier with electronic gain control as shown in Scheme C is much faster and can give a 28 dB dynamic range.

Another solution for a high input power dynamic range is the combination of a XGM and a MI-XPM converter as discussed in [85]. The scheme has the additional advantage of a relatively fast conversion bandwidth combined with possibility for conversion to the same wavelength.

Other very interesting advances in the field of interferometric converters include the integration of optical preamplifiers on the same wafer [86]. Recently MZI converters with a bi-modal waveguide structure have also been reported [87]. They allow the input signal and the probe signal to co-propagate in the structure using different spatial modes. Therefore, conversion to the same wavelength is possible without use of counter-propagation, which normally results in a smaller conversion bandwidth.

3.4.3 XAM Gate

Briefly it should also be noted that very promising results have been achieved for cross absorption modulation in electro-absorption modulators [88]. It has been shown that the technique works up to at least 40 Gbit/s, that extinction ratio enhancement may be achieved and that 30 nm operation bandwidths are possible.

4. All-Optical 3-R Regenerators

The interferometric (XPM) wavelength converters have 2-R regenerating capabilities. It is, however, very exciting that XGM and XPM wavelength converters combined with optical clocks lead to 3-R regenerative capabilities [89]–[91]. A fine example of the construction of such a regenerator is given in [90], where an XGM converter is used in the first stage to sample the input signal with extracted clock pulses and to equalize input power fluctuations. In the second stage an XPM converter follows. It regenerates the extinction ratio resulting in complete regeneration. The scheme has been demonstrated for cascading a number of 140 km links in a 10 Gbit/s loop experiment thereby allowing for transmission over more than 200,000 km of fiber [92].

5. Conclusions

Progress in the field of optical wavelength conversion has been assessed. The development and test of different converter technologies is still in progress, so final conclusions can not be made. It appears, however, that at bit rates of 2.5 Gbit/s the O/E/O converters will remain unchallenged because of their fine properties and well known technology.

At 10 Gbit/s and above all-optical conversion schemes may become competitive as they feature high bit-rate operation. This is important especially for cascading of many converters. Coherent converter types will probably be restricted in use to extremely high bit rates since they require tunable pumps and more complicated set-ups for polarization insensitive operation. A very interesting application is, however, mid-span spectral inversion for dispersion compensation.

Interferometric converters have shown excellent properties with respect to bit rate capability and signal quality for the converted signal. Importantly work on packaging of these relatively complicated devices has been reported and research on "second generation" devices is under way. Attention should also be given to XGM and XAM based converters that may be very interesting for a number of applications.

Moreover, the work on wavelength converters seems to lead to exciting new system blocks such as all-optical 3-R regenerators.

Acknowledgement

Many of the reported results are generated within the ACTS projects KEOPS, OPEN and HIGHWAY. We are grateful to our colleagues within the projects for their help. In particular we should mention P. Doussiere, M. Schilling, and C. Janz of Alcatel, S. Bouchoule of CNET and R. Hess and H. Melchior of ETH. We also thank N. Vodjdani and F. Ratovelomanana of Thomson CSF, our colleague R.J. Pedersen as well as our old colleagues B. Mikkelsen (now with Lucent Technologies), M. Vaa (now with Tyco), and C. Joergensen and S.L. Danielsen (now with L.M. Ericsson).

References

[1] A. Jourdan, L. Berthelon, P. Bonno, F. Bruyér, M. Chbat, C. Coeurjolly, J.Y. Emery, P. Gavignet, E. Grard, C. Janz, A. Noury, G. Soulage, and T. Zami, "Fully reconfigurable WDM optical crossconnect: Feasibility validation and preparation of prototype crossconnect for ACTS "OPEN" field trials," Proc. ECOC'97, vol.3, paper We1D, Edinburgh, UK, Sept. 1997.

[2] R.E. Wagner, R.C. Alferness, A.A.M. Saleh, and M.S. Goodman, "MONET: Multiwavelength optical networking (invited paper)," IEEE J. Lightwave Technol., vol.14, no.6, pp.1349–1355, 1996.

[3] M. Berger, M. Chbat, P. Demester, P. Godsvang, B. Hein, M. Huber, A. Jourdan, A. Leclert, R. Marz, T. Olsen, M. Sotom, G. Tobolka, B. Van Caenegem, and T. Van den Broeck, "Pan-European optical networking using wavelength division multiplexing," IEEE Commun. Mag., vol.35, no.4, pp.82–89, 1997.

[4] K. Sato, S. Okamoto, and H. Hadama, "Network Performance and integrity enhancement with optical path layer technologies," IEEE J. Selected Areas Commun., vol.12, no.1, pp.159–170, Jan. 1994.

[5] A.A.M. Saleh, "Transparent optical networks for the next generation information infrastructure," Techn. Dig. of OFC'95, p.241, San Diego, California, Feb. 1995.

[6] N. Wauters and P. Demester, "Wavelength routing algorithms for transparent optical networks," Proc. ECOC '95, vol.2, pp.855–858, Brussels, Belgium, Sept. 1995.

[7] S. Subramaniam, M. Azizoglu, and K. Somani, "Effect of wavelength converter density on the blocking perfomance of all-optical networks," Proc. LEOS Annual Meeting 95,

vol.1, pp.210–211, San Francisco, California, Oct.–Nov. 1995.

[8] R.A. Barry and P.A. Humblet, "Models of blocking probability in all-optical networks with and without wavelength changers," Proc. IEEE INFOCOM'95, vol.1, pp.402–412, Boston, Massachusetts, April 1995.

[9] M. Kovacevic and A. Acampora, "On wavelength translation in all-optical networks," Proc. IEEE INFOCOM'95, vol.1, pp.413–422, Boston, Massachusetts, April 1995.

[10] F. Derr, M.N. Huber, and G. Kettler, "Key issues of an optical FDM transport network," Techn. Dig. of OFC'95, pp.271–272, San Diego, California, Feb. 1995.

[11] W.D. Zhong, J.P.R. Lacey, R.S. Tucker, "New wavelength interchangeable cross-connect architectures with reduced complexity," J. Lightwave Techn., vol.15, no.11, pp.2029–2037, 1997

[12] E. Karasan and E. Ayanoglu, "Effects of wavelength routing and selection algorithms on wavelength conversion gain in WDM optical networks," Digest IEEE/LEOS 1996 Summer Topical Meetings, pp.43–44, Keystone, Colorado, Aug. 1996.

[13] C. Qiạo and Y. Mei, "Wavelength reservation under distributed control," Digest of IEEE/ LEOS 1996 Summer Topical Meetings, pp.45–46, Keystone, Colorado, Aug. 1996.

[14] B. Mikkelsen, H.N. Poulsen, S. Danielsen, C. Joergensen, and K.E. Stubkjaer, "Interferometric wavelength converters for internal routing and wavelength slot interchange," Proc. Photonics in Switching, pp.30–31, Sendai, Japan, April 1996.

[15] M. Teshima, T. Kawai, N. Sakaida, H. Ishii, and M. Koga, "Demonstration of virtual wavelength path cross-connect," Proc. IOOC/ECOC '97, vol.3, pp.59–62, Edinburgh, UK, Sept. 1997.

[16] K. Wünstel, M. Schilling, W. Idler, G. Eilenberger, S. Bunse, and W. Rehm, "Multidimensional optical switching with advanced key components," Proc. ECOC '93, vol.2, pp.89–92, Montreux, Switzerland, Sept. 1993.

[17] S.L. Danielsen, B. Mikkelsen, C. Joergensen, T. Durhuus, and K.E. Stubkjaer, "Improved performance of a WDM photonic packet switch with tuneable wavelength converters under bursty traffic conditions," Proc. IOOC'95, vol.2, pp.80–81, Hong Kong, June 1995.

[18] C. Guillemot and F. Clérot, "Optical packet switching for WDM IP gigabit routers," Proc. ECOC '98, pp.433–434, Madrid, Spain, Sept. 1998.

[19] D. Chiaroni, P. Gavignet-Morin, P.A. Perrier, S. Ruggeri, S. Gauchard, D. Boudard, J.J. Jacquinot, C. Chauzat, J. Jacquet, P. Doussiere, M. Monot, E. Grard, D. Leclert, M. Sotom, J.M. Gabriagues, and J. Benoit, "Rack-mounted 2.5 Gbit/s ATM photonic switch demonstrator," Proc. ECOC '93, vol.3, pp.77–80, Montreux, Switzerland, Sept. 1993.

[20] D. Chiaroni, et al. "New regeneration functionalities of all-optical wavelength converters for packet-switching applications," Proc. Photonics in Switching, paper PMA5, pp.10–11, Sendai, Japan, April 1996.

[21] A. Misawa, "Broadcast-and-select photonic ATM switch with differential receiver," Proc. Photonics in Switching, paper PThC2, p.16, Sendai, Japan, April 1996.

[22] L. Bertheleon, C. Caeurjolly, P.A. Perier, A. Noury, S. Ruggeri, P. Gavignet, V. Havard, S. Gauchard, and H. Fevrier, "Over 40.000 km path across layered network by recirculation through experimental WDM ring network," Proc. ECOC '96, vol.1, pp.23–26, Oslo, Norway, Sept. 1996.

[23] G. Raybon, U. Koren, B.I. Miller, M. Chien, K. Dreyer, S. Chandrasekhar, R.E. Behringer, D. Tennant, and K. Feder, "Reconfigurable optoelectronic wavelength translation based on an integrated electroabsorption modulated laser array," IEEE Photon. Techn. Lett., vol.10, pp.215–217, 1998.

[24] B. Mikkelsen, R.J.S. Pedersen, T. Durhuus, C. Braagaard, C. Joergensen, and K.E. Stubkjaer, "Wavelength conversion of high speed data signals," Electron. Lett., vol.19, no.19, pp.1716–1718, Sept. 1993.

[25] E. Lach, D. Baums, K. Daub, T. Feeser, W. Idler, G. Laube, G. Luz, M. Schilling, and K. Wünstel, "5 Gbit/s wavelength conversion with simultaneous regeneration of extinction ratio using Y-lasers," Proc. ECOC'93, Montreux, Switzerland, vol.2, pp.137–140, Sept. 1993.

[26] R.J.S. Pedersen, B. Mikkelsen, T. Durhuus, C. Braagaard, C. Joergensen, K.E. Stubkjaer, M. Öberg, and S. Nilsson, "Simple wavelength conversion for bitrate independent operation up to 10 Gbit/s," Proc. OFC'94, San Jose, paper ThQ3, Feb. 1994.

[27] P. Ottolenghi, A. Jourdan, and J. Jacquet, "All-optical wavelength conversion with extinction ratio enhancement using tuneable DBR laser," Proc. ECOC '93, vol.2, pp.141–144, Montreux, Switzerland, Sept. 1993.

[28] G. Gurib and A. Jourdan, "High extinction-ratio conservation up to 7.5 Gbit/s in cascaded all-optical wavelength converters," Techn. Dig. of OFC '95, paper TuO3, San Diego, California, Feb. 1995.

[29] H. Yasaka, H. Sanjoh, H. Ishii, Y. Yoshikuni, and K. Oe, "Repeated wavelength conversion of 10 Gb/s signals and converted signal gating using wavelength-tunable semiconductor lasers," IEEE J. Lightwave Technol., vol.14, pp.1042–1047, 1996.

[30] H. Sanjoh, H. Mawatari, H. Ishii, H. Yasaka, Y. Yoshikuni, and K. Oe, "Wavelength chirping in wavelength conversion of 10 Gbit/s signal with semiconductor laser converter," IEEE Photon. Technol. Lett., vol.8, pp.296–298, Feb. 1996.

[31] K. Inoue, H. Takahashi, S. Suzuki, and K. Oda, "Wavelength conversion using a light injected DFB-LD and a Mach-Zehnder filter with a ring resonator," IEEE Photon. Technol. Lett., vol.7, pp.998–1000, Sept. 1995.

[32] J. Jacquet, F. Serre, E. Gaumot-Goarin, C. Labourie, O. le Gouzigou, F. Gaborit, C. Fortin, and P. Pagnod, "2.5 Gbit/s optical triggering with signal reshaping using bistable laser with proton bombarded saturable absorber," Proc. ECOC '93, vol.2, pp.293–296, Montreux, Switzerland, Sept. 1993.

[33] K. Nonaka, "Digital signal regeneration with side-injection-light-controlled bistable laser diode as a wavelength converter," IEEE Photon. Technol. Lett., vol. 7, pp.29–31, Jan. 1995.

[34] K. Inoue and M. Yoshino, "Bistability and waveform reshaping in a DFB-LD with side-mode light injection," IEEE Photon. Technol. Lett., vol.7, pp.164–166, Feb. 1995.

[35] K. Inoue, T. Hasegawa, and H. Toba, "Influence of stimulated Brillouin scattering on optimum length in fiber four-wave mixing wavelength conversion," IEEE Photon. Technol. Lett., vol.7, pp.327–329, March 1995.

[36] G. Grosskopf, R. Ludwig, R. Schnabel, and H.G. Weber, "Frequency conversion with semiconductor laser amplifiers for coherent optical frequency division switching," Proc. of IOOC '89, paper 19C4-4, Kobe, July 1989.

[37] M.C. Tatham, "20 nm optical wavelength conversion using nondegenerate four-wave mixing," IEEE Photon. Technol. Lett., vol.5, pp.1303–1306, Nov. 1993.

[38] R. Ludwig and G. Raybon, "BER measurements of frequency converted signals using four-wave mixing in a semiconductor laser amplifier at 1, 2.5, 5 and 10 Gbit/s," Electron. Lett., vol.30, pp.338–339, Jan. 1994.

[39] H.G. Weber, et al. "Four-wave mixing for photonic switching," Proc. Photonics in Switching, pp.160–161, Sendai, Japan, April 1996.

[40] R. Schnabel, U. Hilbk, T. Hermes, P. Meissner, C. Helmolt, K. Magari, F. Raub, F. W. Pieper, F.J. Westphal, R. Ludwig, L. Kuller, and H.G. Weber, "Polarisation insensitive frequency conversion of a 10-channel OFDM signal using four-wave mixing in a semiconductor laser amplifier," IEEE Photon. Technol. Lett., vol.6, pp.56–58, Jan. 1994.

[41] F. Martelli, "Semiconductor optical amplifiers for very efficient frequency conversion by four-wave mixing," Proc. IQEC '96, paper ThN2, Sydney, Australia, July 1996.

[42] R.M. Jopson and R.E. Tench, "Polarisation-independent phase conjugation of lightwave signals," Electron. Lett., vol.29, pp.2216–2217, Dec. 1993.

[43] D.D. Marcenac, D. Nesset, A.E. Kelly, M. Brierley, A.D. Ellis, D. Moodie, and C.W. Ford, "40 Gbit/s transmission over 406 km of NDSF using mid-span spectral inversion by four-wave-mixing in a 2 mm long semiconductor optical amplifier," Electron. Lett., vol.33, pp.879–880, 1997.

[44] S. Watanabe, S. Takeda, G. Ishikawa, H. Ooi, J.G. Nielsen, and C. Sonne, "Simultaneous Wavelength conversion and optical phase conjugation of 200 Gb/s (5×40 Gb/s) WDM signal using a highly nonlinear fiber four-wave mixer," Proc. IOOC/ECOC '97, vol.5, pp.1–4, Edinburgh, UK, Sept. 1997.

[45] C.Q. Xu, H. Okayama, and M. Kawahara, "Efficient broadband wavelength converter for WDM optical communication systems," Techn. Dig. of OFC '94, San Jose, paper ThQ4, Feb. 1994.

[46] S.J.B. Yoo, C. Caneau, R. Bhat, and M.A. Koza, "Transparent wavelength conversion by difference frequency generation in AlGaAs waveguides," Techn. Digest of OFC'96, San Jose, CA, paper WG7, Feb. 1996.

[47] T. Durhuus, B. Fernier, P. Garabedian, F. Leblond, J.L. Lafragette, B. Mikkelsen, C.G. Joergensen, and K.E. Stubkjaer, "High speed all-optical gating using two-section semiconductor optical amplifier structure," Proc. CLEO'92, Anaheim, paper CThS4, 1992.

[48] B. Glance, J.M. Wiesenfeld, U. Koren, A.H. Gnauck, H.M. Presby, and A. Jourdan, "High performance optical wavelength shifter," Electron. Lett., vol.28, no.18, pp.1714–1715, Aug. 1992.

[49] M. Brierley, A.D. Ellis, D. Moodie, and C.W. Ford, "40 Gbit/s transmission over 406 km of NDSF using mid-span spectral inversion by four-wave-mixing in a 2 mm long semiconductor optical amplifier," Electron. Lett., "Optical wavelength converters," Proc. ECOC '94, Firenze, Italy, vol.2, pp.635–642, Sept. 1994.

[50] A. Jourdan, G. Soulage, G. Da Loura, B. Clesca, P. Doussiere, C. Duchet, D. Leclert, J.F. Vinchart, and M. Sotom, "Experimental assessment of a 4×4 four-wavelength all-optical cross-connect at a 10-Gbit/s line rate," Techn. Dig. of OFC'95, San Diego, California, pp.277–278, Feb. 1995.

[51] N. Storkfelt, B. Mikkelsen, D.S. Olesen, M. Yamaguchi, and K.E. Stubkjaer, "Measurement of carrier lifetime and linewidth enhancement factor for 1.5-μm ridge-waveguide laser amplifier," IEEE Photon. Technol. Lett., vol.3, pp.632–634, July 1991.

[52] T. Durhuus, B. Mikkelsen, C.G. Joergensen, and K.E. Stubkjaer, "All optical wavelength conversion by semiconductor optical amplifiers," IEEE J. Lightwave Techn., vol.14, pp.942–954, June 1996.

[53] C. Joergensen, S.L. Danielsen, K.E. Stubkjaer, M. Schilling, K. Daub, P. Doussiere, F. Pommerau, P.B. Hansen, H.N. Poulsen, A. Kloch, M. Vaa, B. Mikkelsen, E. Lach, G. Laube, W. Idler, and K. Wunstel, "All-optical wavelength conversion at bit-rates above 10 Gbit/s using semiconductor optical amplifiers," IEEE J. Selected Topics in Quantum Electron., vol.3, pp.1168–1180, Oct. 1997.

[54] D.D. Marcenac, A.E. Kelly, D. Nesset, and D.A.O. Davies, "Bandwidth enhancement of wavelength conversion via cross-gain modulation by optical amplifier cascade," Electron. Lett., vol.31, pp.1442–1443, Aug. 1995.

[55] P. Doussiere, P. Garabedian, C.G. Graver, B. Bonnevie, T. Fillion, E. Derouin, M. Monnot, J.G. Provost, D. Leclerc, and M. Klenk, "1.55 μm polarisation independent semiconductor optical amplifier 25 dB fiber to fiber gain," IEEE Photon. Technol. Lett., vol.6, no.2, pp.170–172, Feb. 1994.

[56] S.L. Danielsen, C.G. Joergensen, M. Vaa, B. Mikkelsen, K.E. Stubkjaer, P. Doussiere, F. Pommerau, L. Goldstein, R. Ngo, and M Goix, "Bit error assessment of 40 Git/s all-optical polarisation independent wavelength converter," Electron. Lett., vol.32, pp.1688–1689, Aug. 1996.

[57] A.D. Ellis, A.E. Kelly, D. Nesset, D. Pitcher, D.G. Moodie, and R. Kashyap, "Error free 100 Gbit/s wavelength conversion using grating assisted cross gain modulation in a 2 mm long semiconductor amplifier," Electron Lett., vol.34, pp.1958–1959, 1998.

[58] J.M Wiesenfeld and B. Glance, "Cascadability and fanout of semiconductor optical amplifier wavelength shifter," IEEE Photon. Technol. Lett., vol.4, pp.1168–1171, 1992.

[59] T. Durhuus, C.G. Joergensen, B. Mikkelsen, R.J.S. Pedersen, and K.E. Stubkjaer, "All optical wavelength conversion by SOA's in a Mach-Zehnder configuration," IEEE Photon. Technol. Lett., vol.6, pp.53–55, Jan. 1994.

[60] M. Schilling, T. Durhuus, C.G. Joergensen, E. Lach, D. Baums, K. Daub, W. Idler, D. Laube, K. Stubkjaer, and K. Wünstel, "Monolithic Mach-Zehnder interferometer based optical wavelength converter operated at 2.5 Gbit/s with extinction ratio improvement and low penalty," Proc. ECOC '94, Firenze, Italy, vol.2, pp.647–650, Sept. 1994.

[61] N. Vodjdani, F. Ratovelomanana, A. Enard, G.Glastre, D. Rondi, R. Blondeau, T.Durhuus, C.G. Joergensen, B. Mikkelsen, and K.E. Stubkjaer, "All optical wavelength conversion with SOA's monolithically integrated in a passive Mach-Zehnder interferometer," Proc. ECOC'94, Firenze, Italy, vol.4, pp.95–98, Sept. 1994.

[62] B. Mikkelsen, T. Durhuus, C. Joergensen, R.J.S Pedersen, S.L. Danielsen, K.E. Stubkjaer, M. Gustavsson, W. van Berlo, and M. Janson, "10 Gbit/s wavelength converter realised by monolithic integration of semiconductor optical amplifiers and Michelson interferometer," Proc. ECOC '94, Firenze, Italy, vol.4, pp.67–70, Sept. 1994.

[63] M. Schilling, K. Daub, W. Idler, D. Baums, U. Koerner, E. Lach, G. Laube, and K. Wunstel, "Wavelength conversion based on integrated all-active three-port Mach-Zehnder interferometer," Electron. Lett., vol.30, pp.2128–2130, Dec. 1994.

[64] N. Vodjdani, F. Ratovelomanana, A. Enard, G.Glastre, D. Rondi, R. Blondeau, C. Joergensen, S.L Danielsen, T.Durhuus, B. Mikkelsen, and K.E. Stubkjaer, "Integrated optics all optical wavelength converter," Proc. ECIO'95, Delft, The Netherlands, April 1995.

[65] F. Ratovelomanana, N. Vodjdani, A. Enard, G.GA.D. Ellis, A.E. Kelly, D. Nesset, D. Pitcher, D.G. Moodie, and R. Kashyap, "Error free 100 Gbit/s wavelength conve: An all-optical wavelength-converter with semiconductor optical amplifiers monolithically integrated in a asymmetric passive Mach-Zehnder interferometer," IEEE Photon. Technol. Lett., vol.7, pp.992–994, Sept. 1995.

[66] X. Pan, J.M. Wiesenfeld, J.S. Perino, T.L. Koch, G. Raybon, U. Koren, M. Chien, M. Young, B.I. Miller, and

C.A. Burrus, "Dynamic operation of a three-port, integrated Mach-Zehnder wavelength converter," IEEE Photon. Technol. Lett., vol.7, pp.995–997, Sept. 1995.

[67] E. Jahn, N. Agrawal, W. Pieper, H.-J. Ehrke, D. Franke, W. Fürst, and C.M. Weinert, "Monolithically integrated nonlinear Sagnac interferometer and its application as a 20 Gbit/s all-optical demultiplexer," Electron. Lett., vol.32, pp.782–783, 1996.

[68] D.A.O. Davies, A.D. Ellis, and G. Sherlock, "Regenerative 20 Gbit/s wavelength conversion and demultiplexing using a semiconductor laser amplifier nonlinear loop mirror," Electron. Lett., vol.31, pp.1000–1001, June 1995.

[69] K.A. Rauschenbach, K.L. Hall, J.C. Livas, and G. Raybon, "All-optical pulse width and wavelength conversion at 10 Gb/s using a nonlinear optical loop mirror," IEEE Photon. Technol. Lett., vol.6, pp.1130–1132, Sept. 1994.

[70] R. Sato, Y. Sakai, S. Sekine, Y. Tohmori, Y. Inoue, K. Shuto, M. Yamada, and T. Kanamori, "Hybrid integrated wavelength converter module using spot-size converter integrated semiconductor optical amplifier array on a PLC platform," Proc. OECC'97, paper 9C3-3, Seoul, Korea, July 1997.

[71] C.G. Joergensen, S.L. Danielsen, T. Durhuus, B. Mikkelsen, K.E. Stubkjaer, N. Vodjdani, F. Ratovelomanana, A. Enard, G. Glastre, D. Rondi, and R. Blondeau, "Wavelength conversion by optimised monolithic integrated Mach-Zehnder interferometer," IEEE Photon. Technol. Lett., vol.8, pp.521–524, April 1996.

[72] T. Durhuus, C. Joergensen, B. Mikkelsen, K.E. Stubkjaer, F. Ratovelomanana, A. Enard, G. Glastre, N. Vodjani, and P. Pagnod, "Monolithic integrated Mach-Zehnder wavelength converter: Conversion and transmission experiments at 5 Gbit/s," Techn. Dig. of OFC '95, paper TuO6, San Diego, Feb. 1995.

[73] W. Idler, M. Schilling, K. Daub, D. Baums, U. Korner, E. Lach, G. Laube, and K. Wunstel, "Signal quality and BER performance improvement by wavelength conversion with an integrated three-port Mach-Zehnder interferometer," Electron. Lett., vol.31, pp.454–455, March 1995.

[74] X. Pan and T.L. Koch, "Intensity noise characteristics of a Mach-Zehnder wavelength converter," IEEE Photon. Techn. Lett., vol.7, pp.1276–1278, Nov. 1995.

[75] B. Mikkelsen, S.L. Danielsen, C.G. Joergensen, R.J.S. Pedersen, H.N. Poulsen, and K.E. Stubkjaer, "All-optical noise reduction capability of interferometric wavelength converters," Electron. Lett., vol.32, pp.566–567, March 1996.

[76] B. Mikkelsen, R.J.S. Pedersen, M. Nissov, H.N. Poulsen, C.G. Joergensen, S.L. Danielsen, K.E. Stubkjaer, M. Gustavsson, W. van Berlo, and M. Janson, "Transmission through 10 all-optical interferometric wavelength converter spans," Techn. Dig. of OFC'96, Post Deadline Paper PD13, San Jose, California, Feb. 1996.

[77] M. Schilling, W. Idler, G. Laube, K. Daub, K. Dütting, E. Lach, D. Baums, and K. Wünstel, "10 Gbit/s monolithic MQW-based wavelength converter in Michelson interferometer configuration," Techn. Digest of OFC'96, San Jose, CA, paper WG2, Feb. 1996.

[78] F. Ratovelomanana, N. Vodjdani, E. Enard, G. Glastre, D. Rondi, R. Blondeau, and A. Jourdan, "Monolithic integration of a Michelson all-optical wavelength converter," Techn. Dig. of OFC'96, pp. 124-125, San Jose, California, Feb. 1996.

[79] C.G. Joergensen, S.L. Danielsen, P.B. Hansen, K.E. Stubkjaer, M. Schilling, K. Daub, E. Lach, G. Laube, W. Idler, and K. Wunstel, "All-optical 40 Gbit/s compact integrated interferometric wavelength converters," Techn. Digest of OFC'97, Dallas, Texas, paper TuO1, Feb. 1997.

[80] B. Mikkelsen, M. Vaa, H.N. Poulsen, S.L. Danielsen, C. Joergensen, A. Kloch, P.B. Hansen, K.E. Stubkjaer, K. Wunstel, K. Daub, E. Lach, G. Laube, W. Idler, M. Schilling, and S. Bouchoule, "40 Gbit/s all-optical wavelength converter and RZ-to-NRZ format adapter realised by monolithic integrated active Michelson interferometer," Electron. Lett., vol.33, pp.133–134, Jan. 1997.

[81] S. Nakamura, K. Tajima, and S. Nakamura, "Experimental investigation on high speed switching characteristics of a novel symmetric Mach-Zehnder all-optical switch," Appl. Phys Lett., vol.65, pp.283–285, 1994.

[82] C.G. Joergensen, S.L. Danielsen, T. Durhuus, B. Mikkelsen, K.E. Stubkjaer, N. Vodjdani, F. Ratovelomanana, A. Enard, G. Glastre, D. Rondi, R. Blondeau, and T. Du, "Monolithic integrated Michelson interferometer with SOA's for high speed all-optical signal processing," Proc. of OFC'95, San Diego, CA, paper TuH4 Feb. 1995.

[83] B. Mikkelsen, K.S. Jepsen, M. Vaa, H.N. Poulsen, K.E. Stubkjaer, R. Hess, M. Duelk, W. Vogt, E. Gamper, E. Gini, P.A. Besse, H. Melchior, S. Bouchoule, and F. Devaux, "All-optical wavelength converter scheme for high speed RZ signal formats," Electron. Lett., vol.33, pp.2137–2139, 1997.

[84] P.B. Hansen, S.L. Danielsen, C.G. Joergensen, and K.E. Stubkjaer, "All optical wavelength conversion schemes for increased input power dynamic range," Techn. Digest of Photonics in Switching '97, pp.132–135, Stockholm, Sweden, April 1997.

[85] A. Kloch, P.B. Hansen, D. Wolfson, S.L. Danielsen, K.E. Stubkjaer, J.Y. Emery, F. Pommereau, M. Renaud, and M. Schilling, "Assessment of dual-stage wavelength converter in OXC at 20 Gbit/s," Proc. ECOC'98, pp.659–660, Madrid, Spain, Sept. 1998.

[86] L.H. Spiekman, J.M. Wiesenfeld, U. Koren, B.I. Miller, and M.D. Chien, "All-Optical Mach-Zehnder Wavelength Converter with Monolithically Integrated Preamplifiers," IEEE Photonics Technology Letters, vol.10, no.8, pp.1115–1117, 1998.

[87] C. Janz, F. Poingt, F. Pommereau, F. Gaborit, D. Ottenwlder, I. Guillemot, B. Dagens, and M. Renaud, "All-active dual-order mode Mach-Zehnder wavelength converter for power-efficient, co-propagative operation," Proc. ECOC'98, pp.661–662, Madrid, Spain, Sept. 1998.

[88] N. Edagawa, M. Suzuki, and S. Yamamoto, "Novel wavelength converter using an electroabsorption modulator," IEICE Trans. Electron., vol.E81-C, pp.1251–1257, 1998.

[89] L. Billes, J.C. Somon, B. Kowalski, M. Henry, G. Michaud, P. Lamuler, and F. Alard, "20 Gbit/s optical 3R regeneration using SOA based Mach-Zehnder interferometer gate," Proc. IOOC/ECOC '97, Edinburgh, UK, vol.2, pp.269–272, Sept. 1997.

[90] D. Chiaroni, B. Lavigne, A. Jourdan, L. Hamon, C. Janz, and M. Renauld, "New 10 Gbit/s 3R NRZ optical regenerative interface based on semiconductor optical amplifiers for all-optical networks," Proc. IOOC/ECOC '97, Edinburgh, UK, vol.5, pp.41–44, Sept. 1997.

[91] K.S. Jepsen, A.T. Clausen, B. Mikkelsen, H.N. Poulsen, and K.E. Stubkjaer, "All-optical network interface for bit synchronisation and regeneration," Proc. IOOC/ECOC '97, Edinburgh, UK, vol.5, pp.89–92, Sept. 1997.

[92] B. Lavigne, D. Chiaroni, L. Hamon, C. Janz, and A. Jourdan, "Performance and system margins at 10 Gbit/s of an optical repeater for long haul NRZ transmission," Proc. ECOC '98, Madrid, Spain, pp.559–560, Sept. 1998.

High-Performance Backbone Network Technology edited by Naoaki Yamanaka
ISBN: 0-8247-5321-6 DOI: 10.1081/(120025745)
Published 2004 by Marcel Dekker, Inc.

Polarization Insensitive SOA-PLC Hybrid Integrated Michelson Interferometric Wavelength Converter and Its Application to DWDM Networks

Rieko Sato, Toshio Ito, Katsuaki Magari, Akira Okada, Manabu Oguma, Yasumasa Suzaki, Yoshihiro Kawaguchi, Yasuhiro Suzuki, Akira Himeno, and Noboru Ishihara

SUMMARY We fabricated a 1.55-μm polarization insensitive Michelson interferometric wavelength converter (MI-WC). The MI-WC consists of a two-channel spot-size converter integrated semiconductor optical amplifier (SS-SOA) on a planar lightwave circuit (PLC) platform. Clear eye opening and no power penalty in the back-to-back condition were obtained at 10 Gb/s modulation. We also confirmed the polarization insensitive operation on the input signal. Moreover, for an application of the MI-WC to DWDM networks, we demonstrated the selective wavelength conversion of 2.5 G/s optical packets from Fabry-Perot laser diode (FP-LD) light to four ITU-T grid wavelengths. We confirmed the good feasibility of this technique for use in DWDM networks. The wavelength conversion we describe here is indispensable for future all-optical networks, in which optical signal sources without wavelength control will be used at user-end terminals.

key words: *wavelength conversion, cross-phase modulation, ITU-T grid, hybrid integration, polarization insensitive*

1. Introduction

Dense wavelength-division-multiplexing (DWDM) photonic technology is currently of great interest as a way to increase the capacity, flexibility, and scalability of future photonic networks. Wavelength converters are indispensable for constructing scalable photonic networks [1]. For example, they are applicable to cross-connect systems in DWDM based photonic networks using virtual wavelength paths (VWP). They are used as wavelength translators to avoid wavelength blocking between nodes and also as all-optical reshaping circuits [2]. The simultaneous wavelength conversion of multi-DWDM channels is also expected to be a key technology for connecting networks with different bands (C-band to S-band or L-band) or channel spacing [3]. In those systems, all of the wavelengths used are ITU-T grid wavelengths that are controlled precisely.

On the other hand, such DWDM based photonic networks can be expanded to user-end terminals in future all-optical networks. Figure 1 shows an example of an optical network in which each user transmits and receives optical packets via a DWDM network. Each optical packet transmitted from the user-end terminals is routed to the destination node according to i ts wavelength. In terms of the network cost reduction, a signal source that does not require wavelength control, such as a Fabry-Perot laser-diode (FP-LD) or an uncooled distributed-feedback laser-diode (DFB-LD), is attractive as an optical transmitter at user-end terminals. However, if such devices are to be used in DWDM systems, they must be able to selectively convert imprecise wavelengths to exact wavelengths (ITU-T grid wavelengths) at network nodes.

So far, various types of all-optical wavelength converters have already been reported. Among them, the cross-phase modulation (XPM) wavelength converters based on interferometric configuration with semiconductor optical amplifiers (SOAs) have been vigorously investigated [4]–[9]. This is because XPM wavelength converters have certain total advantages compared with converters based on other methods (cross-gain modulation (XGM) or four-wave mixing (FWM)). These advantages include regeneration capability [2], high-speed operation [5], noise suppression [10], and less carrier modulation, which leads to reduced chirp for transmission along single-mode fiber (SMF) [11].

To realize XPM wavelength converters, hybrid integration technologies employing planar-lightwave cir-

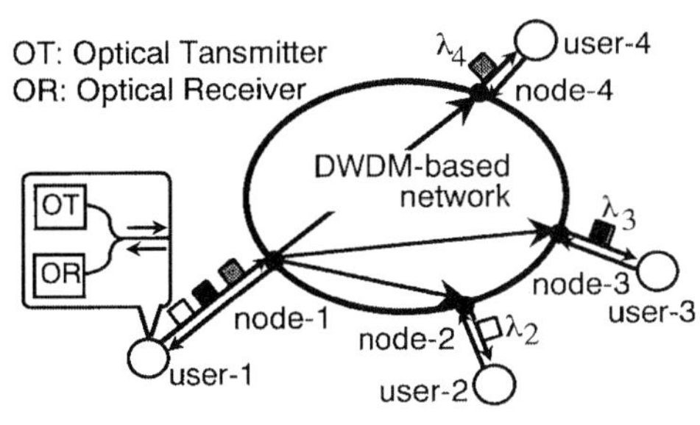

Fig. 1 An example DWDM-based all-optical network.

Manuscript received September 29, 2000.
Manuscript revised December 13, 2000.

[†]The authors are with NTT Photonics Laboratories, Atsugi-shi, 243-0198 Japan.
[††]The authors are with NTT Photonics Laboratories, Ibaraki-ken, 319-1193 Japan.

[*]This paper is also published in the IEICE Trans. Electron., Vol.E84-C, No.5, pp.571–578, May 2001.

IEICE TRANS. COMMUN., Vol. E84-B, No. 5, pp. 1197–1204, May 2001

cuit (PLC) platforms are attractive for low-cost optical modules that use fiber pigtails, i.e., for optical access applications [12]. Hybrid integrated optical modules are also expected in which an active device and a passive waveguide circuit can be optimized independently and their fabrication reproducibility is expected to be improved. In addition, PLC-PLC direct attachment makes it possible to realize a further integrated optical module, which leads to multi channel wavelength conversion [3].

Previously, we have reported on a 1.55-μm SOA-PLC hybrid integrated Michelson interferometric wavelength converter (MI-WC) that uses a two-channel spot-size converter integrated semiconductor optical amplifier (SS-SOA) on a PLC platform [13], [14]. We obtained good results at 2.5 Gb/s modulation with high conversion efficiency (-0.2 dB) and low insertion loss (0.1 dB) for wide wavelength differences of 46 nm. However, we utilized polarization control for the measurement in order to compensate the polarization dependence of the gain due to the use of the multi-quantum-well (MQW) based active region in the SS-SOA. Polarization control is unacceptable for practical purposes.

In this paper, we report on an SOA-PLC hybrid integrated MI-WC employing polarization insensitive SS-SOAs on a PLC platform. We obtained good characteristics at 10 Gb/s modulation without polarization control of input signal. We also describe the selective wavelength conversion from imprecise wavelengths to four ITU-T grid wavelengths, by using the MI-WC and an arrayed-waveguide grating (AWG) router. Using an FP-LD as an example of a signal source that does not require wavelength control, we obtained a clear eye opening at 2.5 Gb/s and confirmed no signal degradation by the wavelength conversion. We successfully transmitted FP-LD packets 500 m with a power penalty of less than 1 dB. We also successfully transmitted converted packets 80 km without using an optical fiber amplifier.

2. SOA-PLC Hybrid Integrated MI-WC

Figure 2 shows the schematic structure of a MI-WC that consists of a polarization insensitive two-channel SS-SOA on a PLC platform. The MI-WC operates as follows. A continuous-wave (CW) light with a wavelength of λ_c is injected through port 3. The input CW-light is divided into two interferometric arms by the 3-dB coupler and is injected into both sides of the SS-SOA, and reflected back to port 2. When the relative phase difference between the two interferometric arms, $\Delta\phi$, is $2n\pi$ ($n = 0, 1 \cdots$), the output power at wavelength λ_c, $P_{c\text{-}out}$, is set to the on-state. When $\Delta\phi$, is $(2n+1)\pi$ ($n = 0, 1 \cdots$), $P_{c\text{-}out}$ is set to the off-state. $\Delta\phi$ is adjusted by the bias currents to the SS-SOAs. Here we describe an example where the on-state is the initial condition. A signal light with a wavelength of

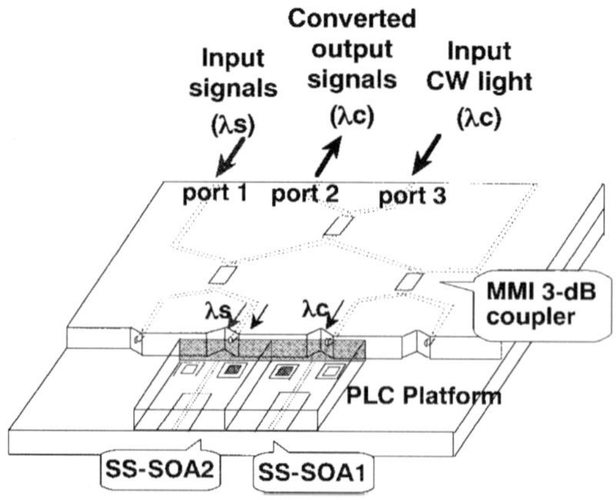

Fig. 2 Schematic structure of MI-WC.

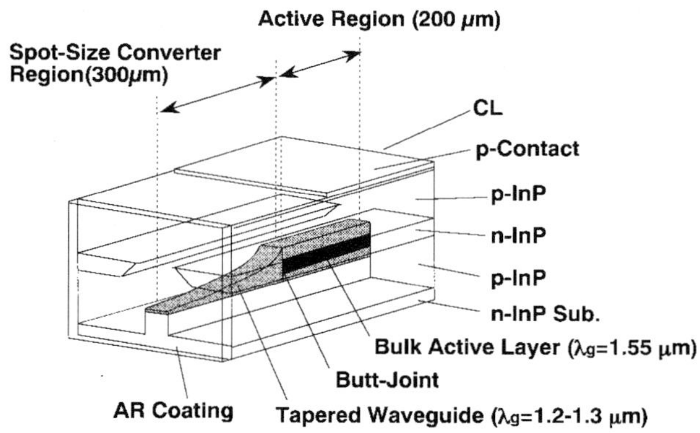

Fig. 3 Structure of SS-SOA.

λ_s is injected into one side of the SOA through port 1. The input signal light causes a depletion of the carrier density in the SOA, and thereby an increase in the refractive index of the SOA. When the relative phase difference of the interferometer, $\Delta\phi$, changes from $2n\pi$ to $(2n+1)\pi$ as a result of the increase in the refractive index of the SOA, $P_{c\text{-}out}$ decreases to the minimum Off-state value. Consequently, the amplitude of the injected signal light determines the interferometric output power, $P_{c\text{-}out}$, from the on-state to the off-state. In this way, the wavelength of a signal can be converted from λ_s to λ_c. Depending on the initial condition (on-state or off-state), the wavelength converter operates in either the inverting or non-inverting mode for input signal patterns. Although the signal light we used Sect. 4 oscillates in multiple modes, the optical power level, which is input to the SOA, controls the refractive index in the SOA.

Figure 3 shows the structure of the polarization insensitive SS-SOA we used [15]. The spot-size converter was butt-joined to the 1.55-μm bulk active region of the SOA. The SS-SOA consisted of a 200-μm bulk active region and a 300-μm SS tapered bulk passive region. The waveguide was 0.4 μm thick at the butt joint, and 0.2 μm thick at the front facet, which was treated with

Fig. 4 Photograph of MI-WC.

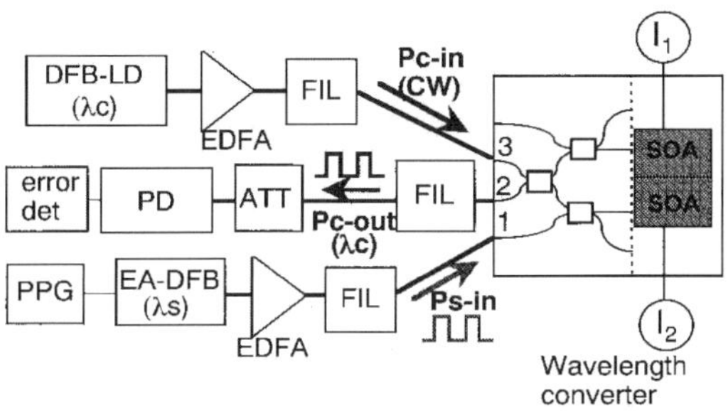

Fig. 5 Measurement setup for 10 Gb/s operation.

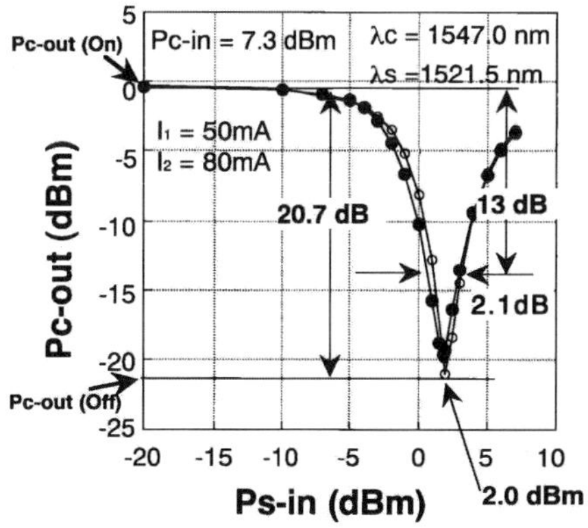

Fig. 6 An example of static switching curves.

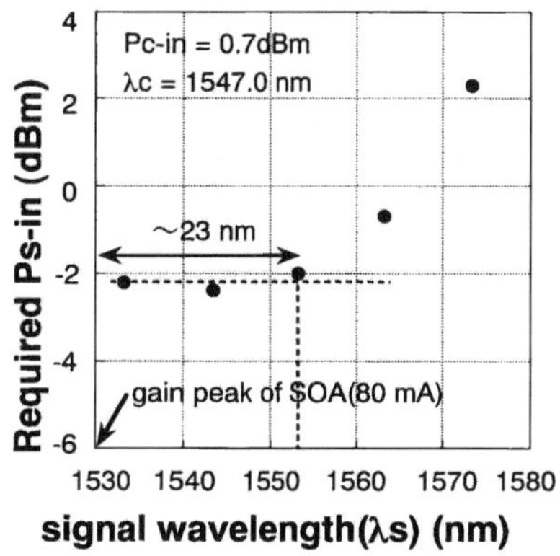

Fig. 7 Dependence of the required optical power for switching on input signal wavelength.

an anti-reflection (AR) coating. The PLC platform had a silica-on-terraced-silicon structure [16]. We flip-chip bonded the SS-SOA array to the PLC platform using a passive alignment technique [17]. The average coupling loss between the SS-SOA and the PLC waveguide, which facet was angled, was estimated as about 4 dB. The total loss from input fiber port 2 to the SS-SOA, including the coupling loss between the SS-SOA and the PLC waveguide, was estimated as about 11 dB.

We utilized multimode interference (MMI) couplers to couple the input signal. The waveguide had a refractive index difference of 0.75%. The size of the PLC chip was $13 \times 3\,\text{mm}^2$.

Figure 4 shows a photograph of our MI-WC. The PLC chip was directly attached to a single-mode fiber array by using UV-curable adhesive.

3. 10 Gb/s Operation of MI-WC

Figure 5 shows the setup we used for measuring the MI-WC when it was operating at 10 Gb/s. The converted light, $P_{c\text{-}out}$ was obtained from port 2 through the 1-nm band-pass optical filter. To measure dynamic characteristics, we used an electro-absorption modulator integrated with a DFB-LD (EA-DFB-LD) as the signal source, which we modulated at 10 Gb/s.

Figure 6 shows an example of static wavelength switching curves. The wavelength of the CW light, λ_c, was 1547.0 nm. $P_{s\text{-}in}$ shows the input power of the signal light, $P_{c\text{-}in}$ shows the input power of the CW light at port 3, and $P_{c\text{-}out}$ shows the converted output power of the CW light after an optical filter. The bias currents of the SS-SOA were $I_1 = 50\,\text{mA}$ and $I_2 = 80\,\text{mA}$, respectively. The wavelength converter was set in the inverted operation mode. Black and white circles indicate the switching curves for the TE/TM polarization dependence of $P_{s\text{-}in}$ at a wavelength of 1521.5 nm. The switching curves show the polarization insensitive characteristics of this module. In fact, we didn't need any polarization control for an input signal in the actual measurement. As the figure shows, the converted output power, $P_{c\text{-}out}$, decreased steeply and it reached a minimum value at a signal input power of 2.0 dBm (required $P_{s\text{-}in}$). Under this condition, the extinction ratio ($10\log(P_{c\text{-}out}\ (\text{On})/P_{c\text{-}out}\ (\text{Off}))$) of the $P_{c\text{-}out}$ was 20.7 dB and the conversion efficiency ($10\log(P_{c\text{-}out}$ (On)/required $P_{s\text{-}in}$)) was −2.3 dB. We define the allowable $P_{s\text{-}in}$ range as the range of the input light signal power in which the $P_{c\text{-}out}$ extinction ratio remains at 13 dB. In Fig. 6, the allowable $P_{s\text{-}in}$ range was 2.1 dB, which corresponds to the tolerance for the input power of optical signals.

Figure 7 shows the dependence of the required $P_{s\text{-}in}$ (CW light) on input signal wavelength, λ_s. The required $P_{s\text{-}in}$ is determined by the bias current to the SS-SOA into which signal light is input and $P_{c\text{-}in}$. In Fig. 7, the required $P_{s\text{-}in}$ changed by less than 0.5 dB

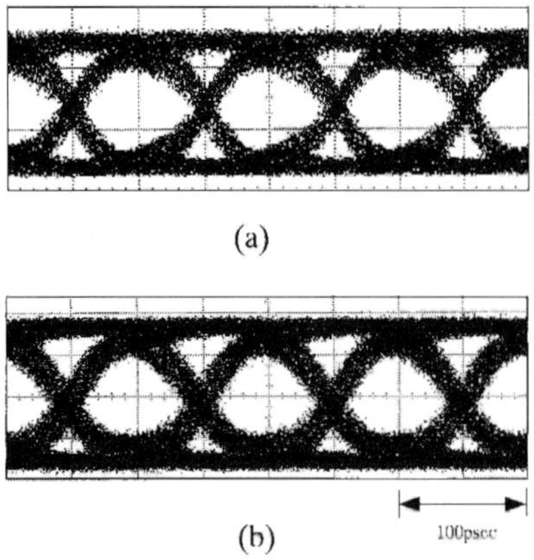

Fig. 8 Input (a) and output (b) eye patterns at 10 Gb/s modulation.

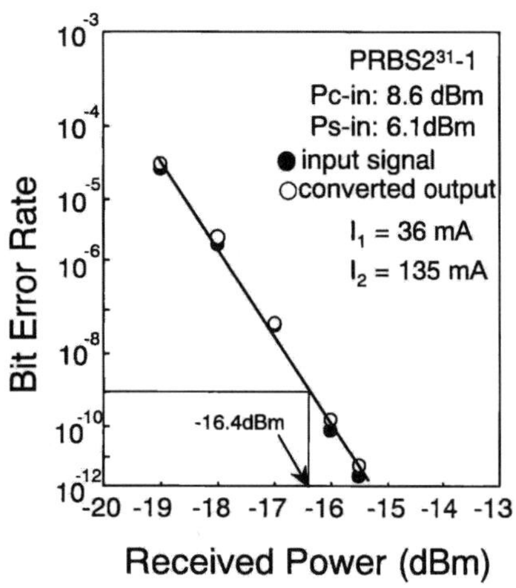

Fig. 9 BER curves of input signals and converted output signals.

for λ_s up to 1553.3 nm and showed a larger increase at more than 1563.3 nm. The center wavelength of the SOA was around 1530 nm. If we assume that the gain spectrum of the SOA is symmetric at the center wavelength, these results show that we can expect the wavelength tolerance of input signals to be as wide as 46 nm.

Figures 8(a) and (b) show the input and output eye patterns at 10 Gb/s modulation with NRZ formats respectively. The wavelength of signal, λ_s, was 1554.4 nm. The operating conditions were $I_1 = 36$ mA, $I_2 = 135$ mA, and $P_{c\text{-}in} = 8.6$ dBm, which were different from those in Fig. 7. At these conditions, the gain peak of the SOA2 must be shifted to a longer wavelength than in Fig. 7. However, as it is supposed that the wavelength tolerance for the input signal is same as in Fig. 7, the wavelength of the signal, 1554.4 nm, is included within the range. We observed a clear eye opening for the converted output signal in an inverted mode.

Figure 9 shows the BER curves of input signals and converted output signals. There was no power penalty

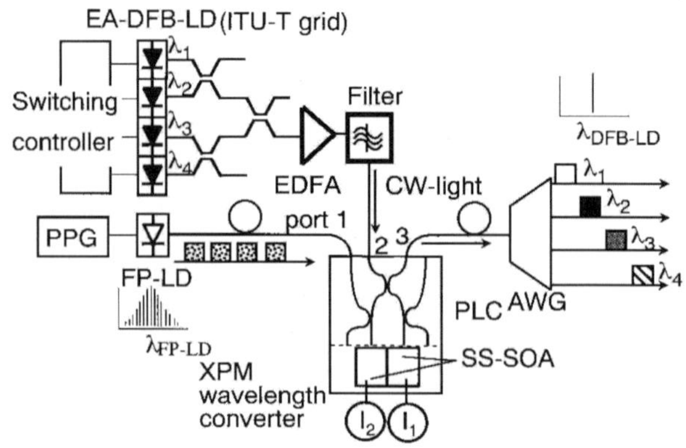

Fig. 10 Experimental setup for selective wavelength conversion.

for the converted output signals. The receiver sensitivity at a bit-error rate of 10^{-9} for the both input and output signals was −16.4 dBm.

4. Selective Wavelength Conversion Using a Hybrid Wavelength Converter

4.1 Experimental Setup

In this section we describe selective wavelength conversion using the MI-WC described in Sect. 3. Figure 10 shows the experimental setup we used for selective wavelength conversion. Optical packets at 2.5 Gb/s were input into port 1, and continuous-wave (CW) light at an ITU-T grid wavelength was input into port 2. The wavelength of the FP-LD was converted to the exact wavelength of the CW light, and the converted packets were output via port 3. The converted output packets were then routed to the AWG output ports according to their wavelengths. The AWG also acted as a filter and eliminated the input signal light of the FP-LD and the amplified spontaneous emission (ASE) of the semiconductor optical amplifiers (SOAs) in the wavelength converter.

The FP-LD was directly modulated at 2.5 Gb/s by a pulse-pattern generator (PPG). We used an optical packet length of 60 bytes with a guard time of 12.8 nsec (4 bytes), and the payload patterns were PRBS $2^{31}-1$. The average input power at port 1 was 0.7 dBm and polarization control was not used. The bias currents of the SS-SOAs were $I_1 = 55.0$ mA and $I_2 = 80.7$ mA. Under these conditions, the wavelength converter operates in a non-inverting mode. The bias currents were not adjusted for the four different CW lights that were input. CW light sources at ITU-T grid wavelengths of 1552.5, 1553.3, 1554.1, and 1554.9 nm (100 GHz spacing) were selected by switching four modules, each containing an electro-absorption modulator integrated with a DFB-LD (EA-DFB-LD). We used a fiber amplifier to increase the CW power.

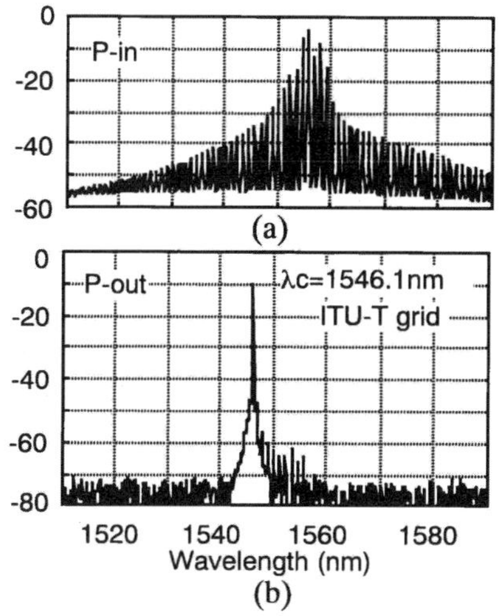

Fig. 11 Input and output light spectra, (a) Input light (FP-LD), (b) Example of converted output light (DFB-LD).

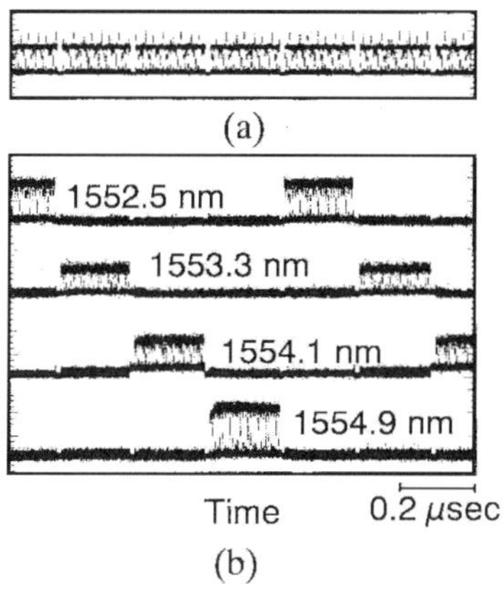

Fig. 12 Input and converted output packets, (a) Input optical packets generated by FP-LD, (b) Converted output packets at each port of AWG.

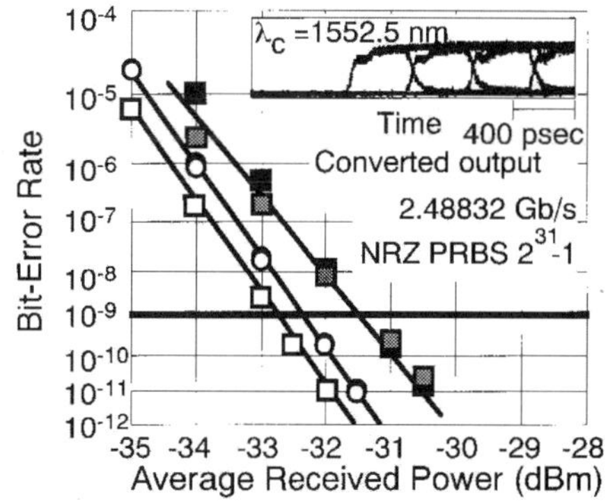

Fig. 13 BER curves and eye patterns. Black circles: Input FP-LD packets before wavelength conversion. White circles: Converted output packets (back-to-back, $\lambda_c = 1552.5$ nm), Black squares: Input FP-LD packets for 500 m transmission through SMF, Gray squares: Converted output packets for 500 m transmission through SMF before wavelength conversion ($\lambda_c = 1552.5$ nm), White squares: Converted output packets after 80 km transmission through SMF after wavelength conversion ($\lambda_c = 1552.5$ nm).

4.2 Experimental Results

Figures 11(a) and (b) show the spectra of the input light from the FP-LD and an example of the converted output light after filtering at a wavelength of 1546.1 nm (ITU-T grid wavelength). Oscillations in multiple modes were converted to oscillation in a single mode. Figure 12(a) shows the 2.5 Gb/s optical packets generated by the FP-LD. Figure 12(b) shows the converted output packets at each output port of the AWG. We successfully achieved the selective wavelength conversion from the wavelengths of the FP-LD to four exact ITU-T grid ones. The average output power after conversion at each output port of the AWG was −8 dBm. The insertion loss of the AWG was 4.5 dB.

Figure 13 shows bit-error rate (BER) curves for the back-to-back condition and for the transmission experiments. The black circles indicate the BER curves for the input FP-LD packets before wavelength conversion. The white circles indicate the BER curves for the converted output packets at a wavelength of 1552.5 nm (back-to-back). The inset eye-patterns were produced by the converted output packets. We obtained a clear eye opening and no power penalty in the back-to-back condition. Furthermore, the BER curves of converted output packets transmitted 80 km along single-mode fiber (SMF) after back-to-back wavelength conversion are also shown (white squares in Fig. 13). No optical fiber amplifier was used in the transmission span and a slight improvement of sensitivity was observed. We assume this to be the dispersion compensated effect that occurs when a signal in a non-inverting mode is transmitted through SMF [3].

Experimental results for transmission of input FP-LD packets along SMF are also shown in Fig. 13. The black squares show the BER curves of FP-LD packets after a 500 m transmission. It is clear that the signal was degraded because of the wide-band spectrum of the FP-LD. A 500 m transmission distance is possible at a power penalty of less than 1 dB with a bit-error rate of 10^{-9}. When we take the transmission distance of the FP-LD packets into consideration, it will be possible to apply the FP-LD signal source to local area networks (LANs) in buildings in which the transmission distance between each user and a wavelength converter at a node is less than 500 m. We confirmed that there was no power penalty for the BER curves of output packets at a wavelength of 1552.5 nm converted from FP-LD packets after 500 m transmission (gray squares in Fig. 13).

These experimental results—no signal degradation caused by the wavelength conversion and successful 80-km transmission after wavelength conversion along SMF without an optical amplifier—show the feasibility of this technique for use in DWDM networks.

These experiments using a 1.55-μm FP-LD as an

optical transmitter were obtained under the worst experimental condition. There are other applicable signal sources that require no wavelength control in addition to the 1.55-μm FP-LD. One is a 1.55-μm uncooled DFB-LD. We expect that the use of this DFB-LD will allow the transmission distance along SMF to be extended beyond that is achievable by a 1.55-μm FP-LD. In addition to 1.55-μm optical signal sources, a 1.3-μm FP-LD and a DFB-LD are applicable to optical LANs operating at more than 1 Gb/s. It is expected that a 2.5 Gb/s optical signal modulated from the 1.3-μm FP-LD can transmits 8–9 km along SMF, although the wavelength conversion of light between 1.3 and 1.55-μm [18] will be necessary.

These results indicate that signal sources with a wavelength that is not precisely controlled (not only FP-LDs but also uncooled DFB-LDs) are applicable to DWDM based optical networks when selective wavelength conversion is used to create ITU-T grid wavelengths.

5. Conclusion

We fabricated a hybrid integrated MI-WC that consists of a two-channel SS-SOAs on a PLC platform. We obtained good characteristics at 10 Gb/s modulation. The MI-WC had polarization insensitive characteristics. Using our MI-WC, we demonstrated the selective wavelength conversion from FP-LD light (light of signal source without wavelength control) to four ITU-T grid wavelengths. We confirmed the successful wavelength conversion among 2.5-Gb/s optical packets without any signal degradation. The selective wavelength conversion described here is one of the applications of the MI-WC for DWDM networks and is a candidate technology for designing future optical networks.

Acknowledgements

The authors thank Dr. S. Tohno for helpful discussions and encouragement.

References

[1] K. Sato, S. Okamoto, and A. Watanabe, "Optical paths and realization technologies," Proc. GLOBECOM'94, pp.1513–1520, San Francisco, Nov. 1994.

[2] F. Ratovelomanana, N. Vodjdani, A. Enard, G. Glastre, D. Rondi, R. Blondeau, A. Dupas, L. Billès, and J.C. Simon, "Regeneration improvement in all-optical wavelength converter, based on a Mach-Zehnder interferometer, by means of phase shifter section," Electron. Lett., vol.33, pp.1629–1630, 1997.

[3] T. Ito, I. Ogawa, Y. Suzaki, K. Magari, and Y. Kawaguchi, "Simultaneous 4-channel wavelength conversion using four-wave mixing in a PLC-SOA hybrid wavelength selector," Proc. Optical Amplifiers and Their Applications (OAA), FB2, Nara, 1999.

[4] T. Durhuus, B. Mikkelsen, C. Joergensen, S.L. Danielsen, and K.E. Stubkjaer, "All-optical wavelength conversion by semiconductor optical amplifiers," J. Lightwave Technol., vol.14, pp.942–954, 1996.

[5] C. Joergensen, S.L. Danielsen, P.B. Hansen, K.E. Stubkjaer, M. Schilling, K. Daub, E. Lach, G. Laube, W. Idler, and K. Wünstel, "All-optical 40-Gb/s compact integrated interferometric wavelength converter," Proc. Optical Fiber Communication'97, Dallas, TuO1, pp.72–73, 1997.

[6] L.H. Spiekman, U. Koren, M.D. Chien, B.I. Miller, J.M. Wiesenfeld, and J.S. Perino, "All optical Mach-Zehnder wavelength converter with monolithically integrated DFB probe source," IEEE Photon. Technol. Lett., vol.9, pp.1349–1351, 1997.

[7] R. Hess, M. Caraccia-Gross, W. Vogt, E. Gamper, P.A. Besse, M. Duelk, E. Gini, H. Melchior, B. Mikkelsen, M. Vaa, K.S. Jepsen, K.E. Stubkjaer, and S. Bouchoule, "All-optical demultiplexing of 80 to 10 Gb/s signals with monolithic integrated high-performance Mach-Zehnder interferometer," IEEE Photon. Technol. Lett., vol.10, pp.165–167, 1998.

[8] C. Janz, B. Dagens, A. Bisson, F. Poingt, F. Pommereau, F. Gaborit, I. Guillemot, and M. Renaud, "Integrated all-active Mach-Zehnder wavelength converter with record signal sensitivity and large dynamic range at 10 Gb/s," Proc. Optical Fiber Communication'99, FB3, pp.30–32, San Diego, 1999.

[9] K. Tajima, S. Nakamura, Y. Ueno, J. Sasaki, T. Sugimoto, T. Kato, T. Shimoda, M. Itoh, H. Hatakeyama, T. Tamanuki, and T. Sasaki, "Hybrid integrated symmetric Mach-Zehnder all-optical switch with ultrafast, high extinction switching," Electron. Lett., vol.35, pp.2030–2031, 1999.

[10] B. Mikkelsen, S.L. Danielsen, C. Joergensen, R.J.S. Pedersen, H.L. Poulsen, and K.E. Stubkjaer, "All-optical noise reduction capability of interferometric wavelength converter," Electron. Lett., vol.32, pp.566–567, 1996.

[11] W. Idler, M. Schilling, K. Daub, D. Baums, U. Körner, E. Lach, G. Laube, and K. Wünstel, "Signal quality and BER performance improvement by wavelength conversion with an integrated three-port Mach-Zehnder interferometer," Electron. Lett, vol.31, pp.454–455, 1995.

[12] N. Uchida, Y. Hibino, Y. Suzuki, T. Kurosaki, N. Ishihara, M. Nakamura, T. Hashimoto, Y. Akahori, Y. Inoue, K. Moriwaki, Y. Yamada, K. Kato, Y. Tohmori, M. Wada, and T. Sugie, "Passively aligned hybrid WDM module integrated with a spot-size converted laser diode and waveguide photodiode on a PLC platform for fiber-to-the-home," Proc. Optical Fiber Communication'96, PD-15, San Jose, 1996.

[13] R. Sato, Y. Suzuki, N. Yoshimoto, I. Ogawa, T. Hashimoto, T. Ito, A. Sugita, Y. Tohmori, and H. Toba, "Wide temperature operation of hybrid integrated wavelength converter module," Proc. Optical Amplifiers and Their Applications (OAA), WB4, Vail, 1998.

[14] R. Sato, Y. Suzuki, N. Yoshimoto, I. Ogawa, T. Hashimoto, T. Ito, A. Sugita, Y. Tohmori, and H. Toba, "1.55-μm hybrid integrated wavelength converter module using spot-size converter integrated semiconductor optical amplifiers on a planar-lightwave-circuit platform," IEICE Trans. Electron., vol.E82-C, pp.1487–1493, 1999.

[15] N. Yoshimoto, K. Magari, Y. Kawaguchi, K. Kishi, Y. Kondo, Y. Kadota, T. Ito, O. Mitomi, Y. Tohmori, Y. Yoshikuni, O. Nakajima, and Y. Hasumi, "Four-channel polarization-insensitive SOA gate array integrated with butt-joined spot-size converters," Electron. Lett., vol.33, pp.2045–2046, 1997.

[16] Y. Yamada, I. Takagi, I. Ogawa, M. Kawachi, and M. Kobayashi, "Silica-based optical waveguide on terraced sil-

icon substrate as hybrid integration platform," Electron. Lett., vol.29, pp.444–445, 1993.

[17] Y. Nakasuga, T. Hashimoto, Y. Yamada, M. Yanagisawa, K. Moriwaki, Y. Akahori, Y. Tohmori, K. Kato, S. Sekine, and M. Horiguchi, "Multi-chip hybrid integration on PLC platform using passive alignment technique," Proc. Electronic Components and Technology Conference'96, pp.20–25, Orlando, 1996.

[18] S. Diez, R. Ludwig, and H.G. Weber, "High-speed all-optical signal processing by switching devices based on semiconductor optical amplifiers," Proc. IEEE Laser and Electro-Optics Society (LEOS), FP2, 1998.

High-Performance Backbone Network Technology edited by Naoaki Yamanaka
ISBN: 0-8247-5321-6 DOI: 10.1081/(120025746)
Published 2004 by Marcel Dekker, Inc.

Polarization-Independent Wavelength Conversion Using Four-Wave Mixing in Single-Mode Fibers Pumped with Cross-Polarized High Frequency Pulses

Kenichiro Tsuji, Hideaki Yokota, and Masatoshi Saruwatari

SUMMARY This paper describes a simple polarization-independent wavelength conversion method using degenerated four-wave mixing (FWM) in single-mode fibers pumped with cross-polarized high frequency, saw-tooth pulses from a single pump source. Successful polarization-independent wavelength conversion is experimentally confirmed with less than ±12% and ±5.6% variation using a gain-switched LD pumping and a mode-locked fiber laser pumping, respectively. We clarify that the interference effect between two orthogonal pump pulses must be taken into account to achieve a good polarization-insensitive operation, since even the small pulse edges bring about the large polarization fluctuations when they are interfered. Furthermore, it is reveal that the shorter pump pulse broadens its own spectrum due to the self-phase modulation in fibers, resulting in poor FWM efficiency. Finally, possibility of high-speed operation is discussed taking into account the pump pulse conditions.

key words: *four-wave mixing, wavelength conversion, polarization-independent, cross-polarized pulse*

1. Introduction

All-optical wavelength conversion is the key to realizing future WDM based optical communications networks and various techniques utilizing optical nonlinearities both in fibers [1] and laser-diodes [2], [3] have been reported. Wavelength conversion based on the fiber four-wave mixing (FWM) is one of the most attractive techniques because of its superior characteristics including high conversion efficiency, fast response and simple configuration [4], [5]. FWM efficiency in fibers, however, strongly depends on the polarization states of pump and signal lights [6]. Therefore, polarization-independent operation is essential to apply the fiber FWM based wavelength conversion to practical use. This is because the optical signals transmitted through long fibers usually have an uncontrollable state of polarization.

To date, there are several reports that demonstrate the polarization-independent FWM operation in fibers; they include a polarization diversity method using a polarization-rotating fiber loop together with a polarization beam splitter [7], [8], or nondegenerated FWM operation using a conventional fibers pumped by two orthogonal CW lights of different wavelengths [9]. These methods, however, have drawbacks that the former requires the special dispersion-shifted PANDA fibers to achieve stable operation and that the latter uses two optical sources leading to the conversion with different wavelengths according to the signal polarization state.

In this paper, we propose a simple polarization-independent wavelength conversion technique that utilizes degenerated FWM in conventional dispersion-shifted single-mode fibers pumped with cross-polarized high frequency, saw-tooth pulses from a single pump source. Although the wavelength-converted signals have the saw-tooth profiles within a signal time slot, the converted power of one-bit signal is constant regardless of the input signal polarization state. First, we briefly introduce the principle of the polarization-independent wavelength conversion using FWM, and describe an experimental setup of the proposed method. Next, experimental results on polarization dependence of FWM efficiency are compared with each other both for the proposed cross-polarized pumping and the conventional single-polarization pumping. Finally, polarization dependence using various optical pulse sources and how the pump pulse width affects the polarization stability are also discussed.

2. Principle of Polarization-Independent FWM Generation

The principle of the proposed FWM-based polarization-independent wavelength conversion is shown in Fig. 1. A top Fig. 1(a) shows the cross-polarized high frequency pump pulses consisting of mutually orthogonal pulse trains with the same power. It is noted that P- and S-components are temporally separated each other. Figures 1(b) and 1(d) show signal pulse trains with P- and S-polarization, respectively. The pump repetition frequency is set to be much higher than that of a signal bit-rate as shown in the figure. When the signal light is P- or S-polarization, P- or S-polarized FWM pulse is generated as shown in Figs. 1(c) and 1(e). Since the

Manuscript received September 28, 2000.
Manuscript revised January 9, 2001.

†The authors are with the Department of Communications Engineering, National Defense Academy, Yokosuka-shi, 239-8686 Japan.

*This paper is also published in the IEICE Trans. Electron., Vol.E84-C, No.5, pp.564–570, May 2001.

IEICE TRANS. COMMUN., Vol. E84-B, No. 5, pp. 1190–1196, May 2001

FWM efficiency is the same for each polarization state, the overall FWM power is constant regardless of signal polarization state. Thus the polarization independent wavelength conversion can be achieved when the generated FWM signal is averaged over the signal time slot.

3. Experimental Setup

The experimental setup using cross-polarized pumping is shown in Fig. 2. The signal light from a DFB-LD (λ_s =1545.0 nm) is modulated by an electro-absorption (EA) modulator [10] at f_0=50 MHz. The pump light from a wavelength-tunable external-cavity LD is modulated by a $LiNbO_3$ (LN) intensity-modulator at several GHz, and then input into a 133-m long polarization-maintaining PANDA fiber with a 188 ps polarization walk-off via a polarization-controller (PC2), to equally generate the orthogonal pump pulses. The relative delay between orthogonal pump pulses is properly set to suppress the interference as much as possible. With this setup, the intensity of the pump light component along any polarization directions is the same throughout the long FWM fiber. Figure 3 shows the measured pulse waveforms of the cross-polarized and single-polarized pumps. The signal and pump lights are then combined with a coupler, and input into a 6.3-km dispersion-shifted (DS) wavelength-conversion fiber (zero-dispersion wavelength of 1553.8 nm) after being amplified by an EDFA. The launched power (time-averaged) of the signal and pump lights into the DS fiber is set to 8.9 mW and 10.1 mW, respectively, in the FWM experiments described in Sect. 4. The state of polarization (SOP) of the signal is controlled by the polarization controller (PC1), consisting of a half-wave plate and a quarter-wave plate, to evaluate the polarization dependence of the FWM efficiency. The pump light wavelength (λ_p=1553.8 nm) is selected to be the zero-dispersion wavelength of the DS fiber used so as to achieve high conversion efficiency [11].

Output power spectra from the FWM fiber were measured with an optical spectrum analyzer with a 0.1-nm resolution after being divided by a 3-dB coupler. The temporal waveform of the wavelength-converted FWM light was observed with a PIN photo-detector followed by a sampling oscilloscope after the FWM light being extracted by a 3-nm bandwidth optical filter.

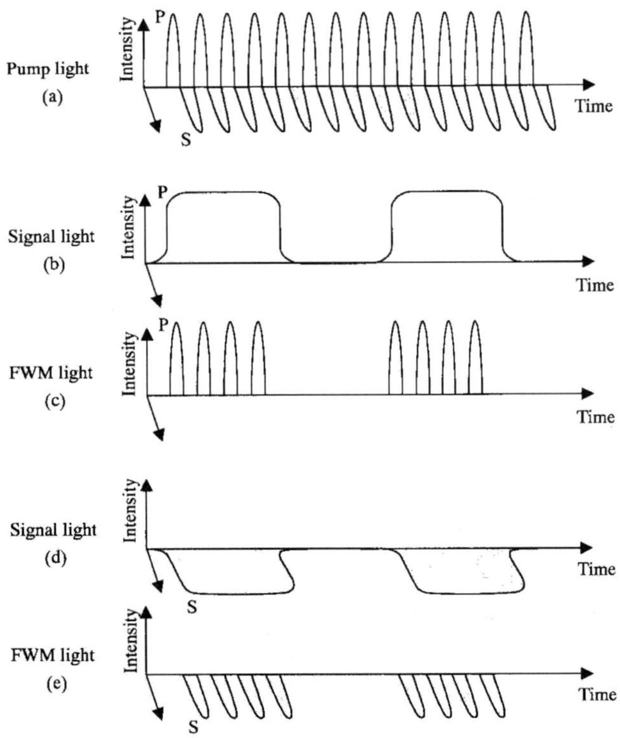

Fig. 1 Principle of polarization-independent operation.

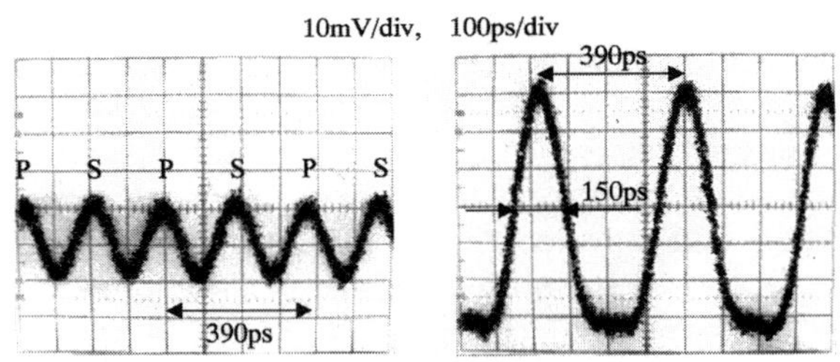

Fig. 3 Measured pulse waveforms of the cross-polarized (left) and single-polarized pumps (right).

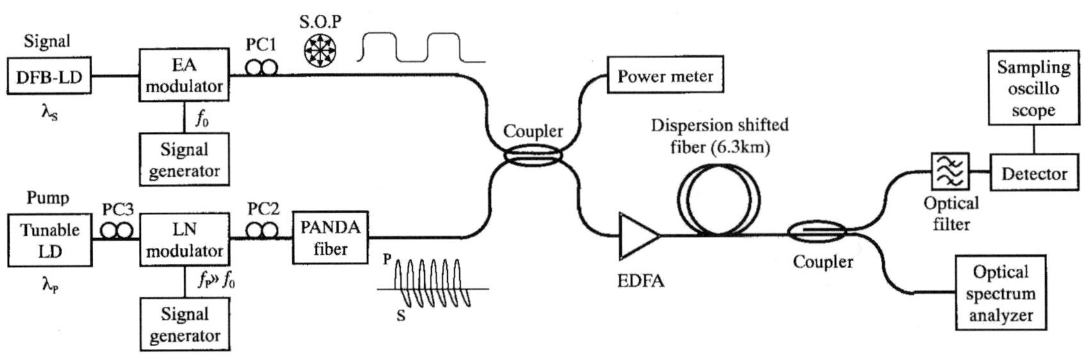

Fig. 2 Experimental setup.

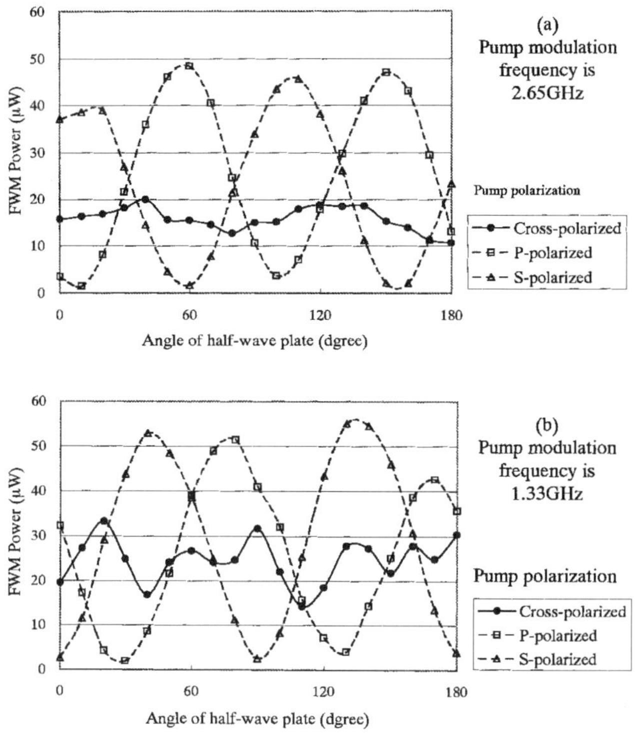

Fig. 4 FWM power as a function of the signal polarization state.

4. Experimental Results

FWM output power versus the signal polarization direction was measured by rotating the half-wave plate of the PC1. Figure 4(a) shows the FWM power measured with an optical spectrum analyzer as a function of the rotation angle of the half-wave plate when using the pump repetition frequency of 2.65 GHz. This frequency corresponds to the case where the mutually orthogonal pulses are positioned alternately with the same temporal spacing. Solid circles represent the polarization dependence of FWM output when the pump is the cross-polarized light with equal polarization components. For comparison, the cases for the single-polarization pump, P-polarized and S-polarized pumps, are also shown as open squares and open triangles, respectively. As you can see, large polarization dependence for the single-polarization pump was considerably suppressed by using the cross-polarization pumping, although there still remains polarization dependence of less than ±30%. FWM efficiency (time-averaged) of the cross-polarization pumping is as small as 0.25–0.45%, corresponding to one-fourth of the maximum efficiency for the single-polarization pumping as is expected.

Figure 4(b) shows the polarization dependence of FWM output when the pump repetition frequency is set to 1.33 GHz in which orthogonal polarization pulses are partially overlapped each other. Each mark represents the same pump condition as that for Fig. 4(a).

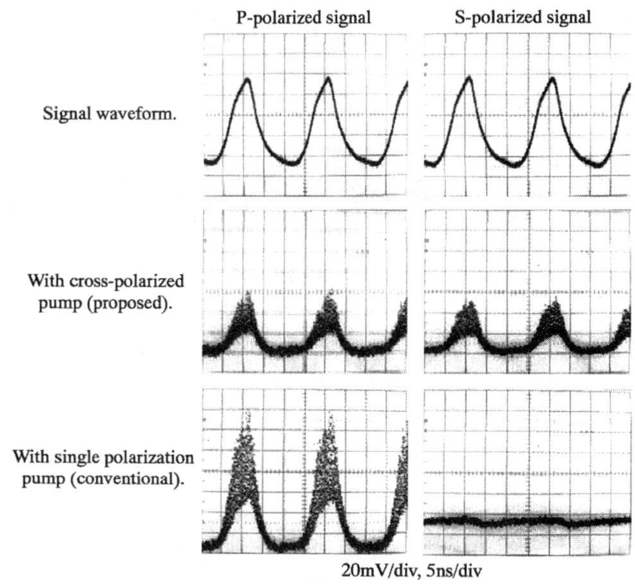

Fig. 5 Original signal waveforms (top) and wavelength-converted signal waveforms generated by a FWM process with cross- (middle) and single-polarized pumping when the signal polarization state is P- (left row) and S- (right row) polarized.

Relatively large polarization dependence of more than ±40% has been observed for this case.

Figure 5 shows the original signal waveforms (top) and wavelength-converted signal waveforms generated by a FWM process when cross-polarized pump (middle) and single-polarization pump (bottom) are adopted. Left and right figures compare the wavelength-converted signals when the input signal polarization direction is rotated by 90 degrees with PC1. As shown here, polarization-independent wavelength conversion has been confirmed for the pulse waveform when using cross-polarized pump. The generated FWM signal has a high frequency component of the pump pulses that can be seen like a noise in the waveform pictures. This component can be removed by using optical or electrical filters since its frequency is much higher than the signal frequency. Thus the deterioration of the signal-to-noise ratio of the wavelength-converted signal is not so much if an appropriate filter is adopted.

In the above measurement, the original pulses having relatively long tails were used for generating orthogonal pump pulses because the pulses were generated by an LN intensity modulator driven by a sinusoidal signal. This results in the relatively large interference between orthogonal pulses, leading to the pump polarization fluctuations.

We have measured the temporal fluctuations of the FWM power to estimate the influence effect of the of the pump pulses. The signal polarization state is set to 45 degrees to the one of orthogonal pump pulses. This is the most sensitive condition with respect to the pump polarization fluctuations. Figure 6 shows the high frequency components of the FWM light corresponding

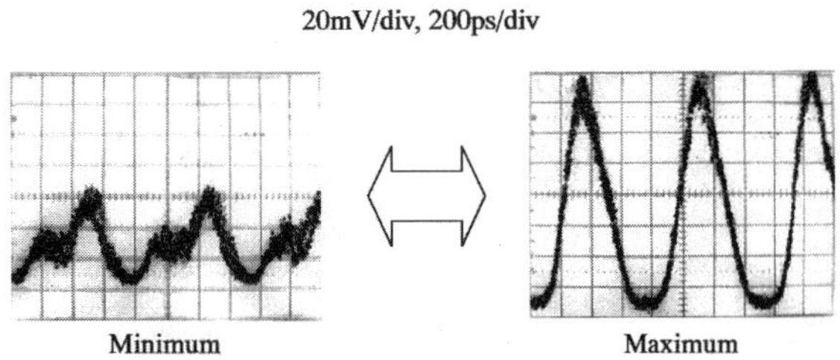

Fig. 6 Time variation of FWM waveform when pump frequency is 1.33 GHz.

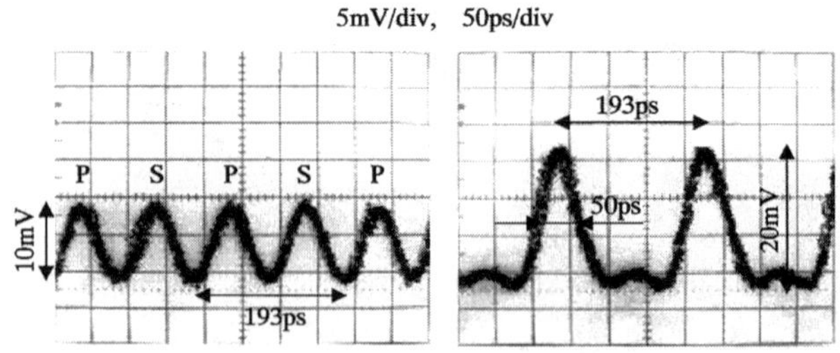

Fig. 7 Measured pump pulse waveforms generated by a gain-switched LD. (Left: Cross-polarized pump, Right: Single-polarized pump.)

to the maximum and the minimum FWM outputs, respectively, when the pump frequency is 1.33 GHz. In a few minutes, the FWM output has periodically changed from the maximum to the minimum states. This can be attributed to the interference effect of the overlapped orthogonal pulses; the polarization state of the overlapped portion is rotated by the temperature-induced birefringence change of the PANDA fiber. This has been confirmed by the intentional temperature change of the PANDA fiber, leading to fast transitions between both extreme states in Fig. 6. To further clarify the interference effect, the temporal variations for the optimum 2.65 GHz pump were also measured for the same signal polarization condition. As is expected, relatively small power fluctuations have been found as compared to the 1.33 GHz pump condition.

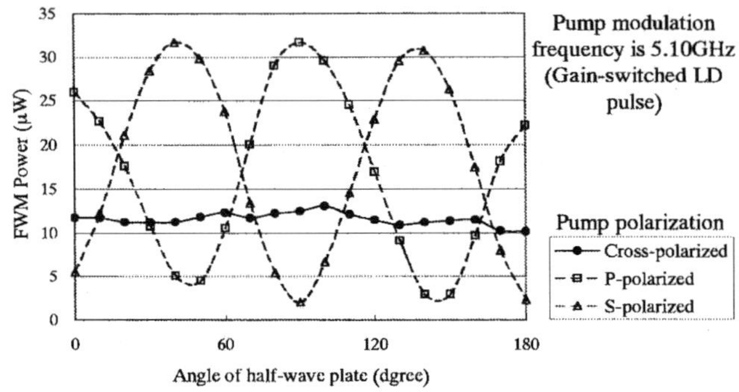

Fig. 8 Polarization dependence of the generated FWM power when using the gain-switched LD pulse.

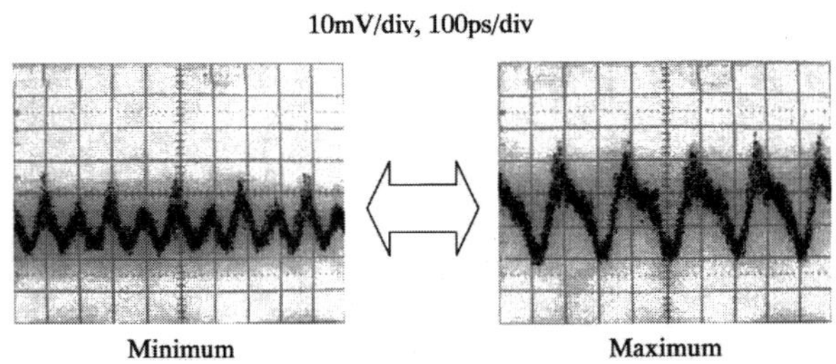

Fig. 9 Measured temporal variation of the FWM output waveform for the gain-switched LD pump.

5. Discussion

5.1 Polarization Dependence for Gain-Switched LD Pulse Pumping

To further suppress the polarization dependence, we have used gain-switched LD [12] as a pump source. The repetition frequency was selected to be 5.1 GHz, corresponding to the cavity-mode spacing, to obtain the shortest pulse duration of 50 ps. A 65-m PANDA fiber with a 98 ps polarization walk-off was used to generate orthogonal pump pulses with equal temporal spacing. The signal and pump powers coupled into the FWM fiber are 9.4 and 10.8 mW, respectively. Figure 7 shows the measured temporal waveforms of the cross- and single-polarization pumps. We can see little pulse overlap between cross-polarized adjacent pulses because the gain-switched pulse has a small duty factor of 26%.

Figure 8 show the polarization dependence of the generated FWM power when using the gain-switched LD pulse. Each mark and horizontal and vertical axes represent the same parameters as that in Fig. 4. As is expected, polarization dependence is further suppressed to be less than ±12% variation, that is, the FWM efficiency of 0.21–0.27%. These results prove that the FWM output fluctuations are mainly caused by the overlap of orthogonal pump pulses. The remaining small variations are due to the residual overlap of the orthogonal pulses, pump polarization-fluctuation before the PANDA fibers, and the polarization-dependent loss/sensitivity of optical components or measurement apparatuses including fiber couplers, optical filters, optical isolators, an optical spectrum analyzer, and so on.

Temporal variation of the FWM output waveform has been measured for the gain-switched LD pump, as shown in Fig. 9. Although the waveform was changed between the maximum and the minimum states, the time-averaged power was almost kept constant, leading to no polarization dependence of the FWM output within a signal time-slot. Moreover, the transition time between two extreme states is fairly large as compared with the case of Fig. 6. This is because the transition is not caused by the fast polarization change in the interfered portion, but is caused by the slow change in polarization difference between signal and pump lights.

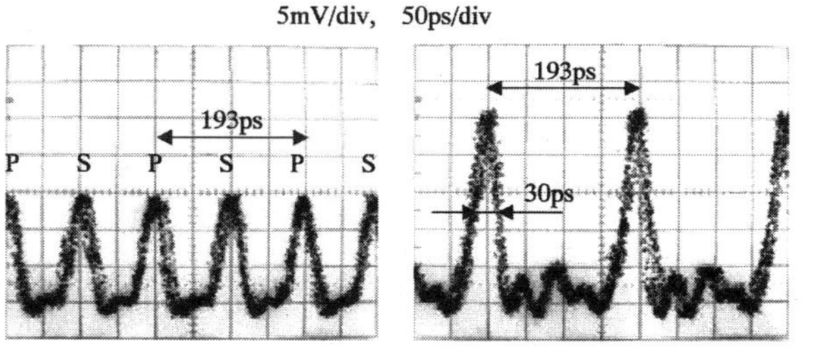

Fig. 10 Measured pump pulse waveforms generated by the mode-locked fiber laser pulse. (Left: Cross-polarized pump, Right: Single-polarized pump.)

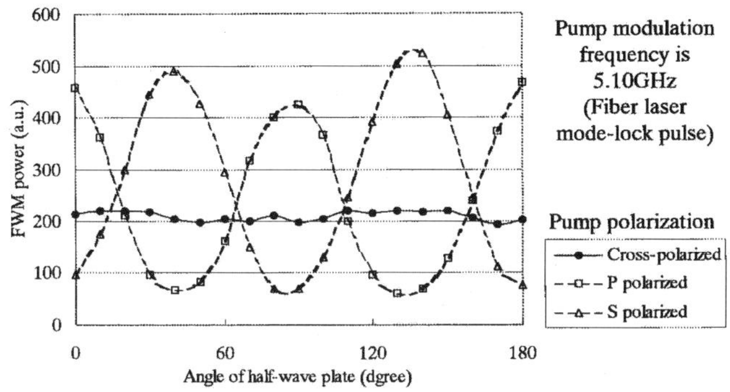

Fig. 11 Polarization dependence of the generated FWM power when using the mode-locked fiber laser pulse.

5.2 Polarization Dependence for Mode-Locked Fiber Laser Pulse Pumping

To further study pump pulse dependence, we have conducted the same FWM experiment using a mode-locked Er-doped fiber laser [13] as a pump source. The repetition was set to the same frequency of 5.1 GHz as the previous measurement. Since the mode-locked fiber laser pulse has a wide spectrum width corresponding to the short pulse width of less than 10 ps, the FWM output has also the spread spectrum. This decreases the peak FWM spectrum in an optical domain even if the same FWM power is generated. To correctly measure the FWM power, the electrical signal at modulation frequency was measured with an electrical spectrum analyzer after the O-E conversion by a PIN photodiode. The temporal waveform of the mode-locked fiber laser is shown in Fig. 10. The pulse width measured by an autocorrelator is estimated to be less than 5 ps.

Figure 11 shows polarization dependent FWM output when a mode-locked fiber laser is used. A vertical axis indicates the electrical signal components in the generated FWM signal. The signal frequency was set to 250 MHz to avoid an electrical noise of the detection system. The output fluctuation for cross-polarized pump was within $\pm 5.6\%$. The temporal waveform of the FWM light is shown in Fig. 12. The polarization-independent operation can be clearly seen from the waveform measurement.

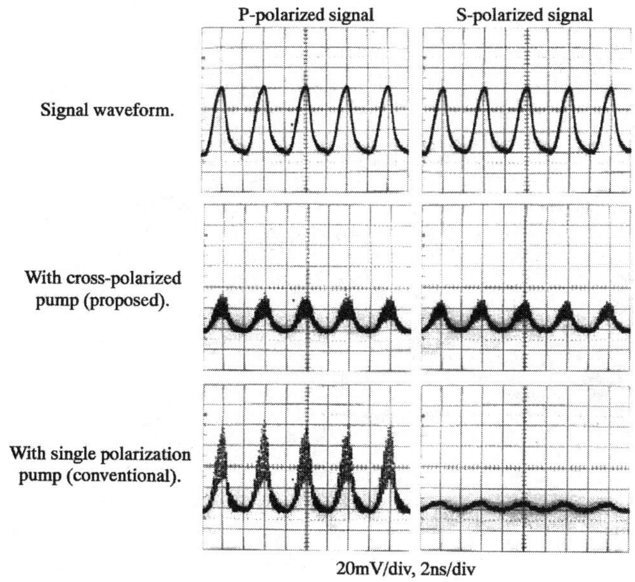

Fig. 12 Original signal waveforms (top) and wavelength-converted signal waveforms generated by a FWM process with cross- (middle) and single-polarized pumping when using the mode-locked fiber laser pulse. The left and right rows show waveforms for P- and S-polarized signal, respectively.

It is expected that the wavelength-conversion efficiency can be improved by increasing the pump power. Contrary to our expectation, the increase in pump power from the mode-locked fiber laser has drastically decreased the FWM output. By measuring the output spectra from the DS fiber, it has been clarified that the large spectrum broadening is easily induced by the self-phase modulation of the ultrashort pulse from a mode-locked fiber laser [14]. The broadened spectrum like a supercontinuum decreases the peak intensity of the pump spectra and increases the wavelength crosstalk, resulting in poor conversion efficiency and low signal-to-noise ratio. In the experiments, we had to adjust the pump power level under 5 mW not to occur the spectrum broadening, which limited the FWM efficiency.

5.3 Possibility of High-Speed Operation

In order to confirm the polarization independent operation of the proposed method, we have conducted the FWM experiment using 50 and 250 MHz modulation signals. To increase the operation speed, the following factors must be clarified; what the optimum ratio between the pump and signal modulation speeds is, what the optimum pulse width for the pump is, and how fast the pump pulse can be modulated. The minimum ratio of the pump and signal frequencies can be roughly determined taking into account the required polarization dependency. By assuming a hyperbolic secant signal waveform with a 30% duty factor, the polarization dependencies for the ratios of 3 and 5 can be calculated to be 27 and 3.2%, respectively, when the relative position between the pump and signal, and the signal polarization, are changed. Therefore the ratio 5 is sufficient to suppress the polarization dependence within 5%.

Optimum pulse width is determined by evaluating polarization fluctuations in the overlapped cross-polarized pulses. Judging from the experiment in Sect. 5.1, the 50 ps pulse width is almost the maximally tolerable duration for the present 5 GHz pump pulse train. Thus the optimum pulse width corresponding to the pulse train with a 25% duty factor. Taking into account the minimum ratio of 5 between pump and signal frequency, the polarization independent wavelength conversion can be expected at up to 1 Gbit/s signal using the present experimental setup.

The operation speed of the proposed method depends on the repetition frequency of the pump pulse train. Recently, very high-speed optical pulse generation techniques up to 640 GHz have been reported using an optical time-division multiplexing (OTDM) technique employing PLC multipliers together with supercontinuum generation or soliton pulse compression techniques. Therefore polarization-independent wavelength conversion for high-speed signals over 40 Gbit/s could be possible by applying the OTDM pump pulse source to our FWM method.

6. Conclusion

Polarization-independent wavelength conversion was proposed using degenerated four-wave mixing in single-mode fibers pumped with cross-polarized high frequency short pulses. Successful polarization-independent wavelength conversion was experimentally confirmed with less than ±12% and ±5.6% variation using a gain switched LD and a mode-locked fiber laser as a pump, respectively. It was also clarified that the interference effect between two orthogonal pump pulses must be taken into account to achieve a good polarization-insensitive operation. This is because even the small pulse edges bring about the large polarization fluctuations when they are interfered. Furthermore, it was revealed that the shorter pump pulse broadens its own spectrum due to the self-phase modulation in fibers, resulting in poor FWM efficiency. Finally, possibility of high-speed operation is discussed taking into account the pump pulse conditions.

References

[1] K. Inoue and H. Toba, "Wavelength conversion experiment using fiber four-wave mixing," IEEE Photon. Technol. Lett., vol.4, no.1, pp.69–72, Jan. 1992.

[2] R. Schnabel, W. Pieper, R. Ludwig, and H.G. Weber, "Multiterahertz frequency conversion of a picosecond pulse train using nonlinear gain dynamics in a 1.5 μm MQW semiconductor laser amplifier," Electron. Lett., vol.29, no.9, pp.821–822, April 1993.

[3] M.C. Tatham, G. Sherlock, and L.D. Westbrook, "20-nm optical wavelength conversion using nondegenerate four-wave mixing," IEEE Photon. Technol. Lett., vol.5, no.11, pp.1303–1306, Nov. 1993.

[4] G.P. Agrawal, Nonlinear fiber optics, Academic Press, 1989.

[5] P.A. Andrekson, N.A. Olsson, J.R. Simpson, T. Tanbun-Ek, R.A. Logan, and M. Haner, "16 Gbit/s all-optical demultiplexing using four-wave mixing," Electron. Lett., vol.27, no.11, p.922, May 1991.

[6] K. Inoue, "Polarization effect on four-wave mixing efficiency in a single-mode fiber," IEEE J. Quantum Electron., vol.28, no.4, pp.883–894, April 1992.

[7] T. Hasegawa, K. Inoue, and K. Oda, "Polarization independent frequency conversion by fiber four-wave mixing with a polarization diversity technique," IEEE Photon. Technol. Lett., vol.5, no.8, pp.947–949, Aug. 1993.

[8] T. Morioka, S. Kawanishi, K. Uchiyama, H. Takara, and M. Saruwatari, "Polarization-independent 100 Gbit/s all-optical demultiplexer using four-wave mixing in a polarization-maintaining fiber loop," Electron. Lett., vol.30, no.7, pp.591–592, March 1994.

[9] K. Inoue, "Polarization independent wavelength conversion using fiber four-wave mixing with two orthogonal pump lights of different frequencies," J. Lightwave Technol., vol.12, no.11, pp.1916–1920, Nov. 1994.

[10] M. Suzuki, H. Tanaka, K. Utaka, N. Edagawa, and Y. Matsusima, "Transform-limited 14 ps optical pulse generation with 15 GHz repetition rate by InGaAsP electroabsorption modulator," Electron. Lett., vol.28, no.11, pp.1007–1008, May 1992.

[11] K. Inoue, "Four-wave mixing in an optical fiber in the zero-dispersion wavelength region," J. Lightwave Technol., vol.10, no.11, pp.1553–1561, Nov. 1992.

[12] K.Y. Lau, "Gain switching of semiconductor injection laser," Appl. Phys. Lett., vol.52, no.4, pp.257–259, Jan. 1988.

[13] H. Takara, S. Kawanishi, and M. Saruwatari, "20 GHz transform-limited optical pulse generation and bit-error-free operation using a tunable, actively mode-locked Er-doped fiber ring laser," Electron. Lett., vol.29, no.13, pp.1149–1150, June 1993.

[14] T. Morioka, S. Kawanishi, K. Mori, and M. Saruwatari, "Nearly penalty-free, < 4 ps supercontinuum Gbit/s pulse generation over 1535–1560 nm," Electron. Lett., vol.30, no.10, pp.790–791, May 1994.

High-Performance Backbone Network Technology edited by Naoaki Yamanaka
ISBN: 0-8247-5321-6 DOI: 10.1081/(120025747)
Published 2004 by Marcel Dekker, Inc.

Hybrid Integrated 4 × 4 Optical Matrix Switch Module on Silica Based Planar Waveguide Platform

Tomoaki Kato, Jun-ichi Sasaki, Tsuyoshi Shimoda, Hiroshi Hatakeyama, Takemasa Tamanuki, Shotaro Kitamura, Masayuki Yamaguchi, Tatsuya Sasaki, Keiro Komatsu, Mitsuhiro Kitamura, and Masataka Itoh

SUMMARY The hybrid electrical/optical multi-chip integration technique for optical modules for optical network system has been developed. Employing the technique, a 4×4 broadcast-and-select type optical matrix switch module has been realized. The module consists of four sets of silica waveguide 1 : 4 splitters/4 : 1 combiners, four 4-channel arrays of polarization insensitive semiconductor optical amplifiers with spot-size converters as optical gates, printed wiring chips for electrical wiring and single mode fibers for optical signal interface on planar waveguide platform fabricated by atmospheric pressure chemical vapor deposition. All the gates and the wiring chips were mounted precisely onto the platform at once in flip-chip manner by self-align technique using AuSn solder bumps. Coupling loss between the waveguide and the SOA gate was estimated to be 4.5 dB. Averaged fiber-to-fiber signal gain, on-off ratio and polarization dependent loss for each of the signal paths was 7 dB ± 2 dB, more than 40 dB and 0.5 dB, respectively. High speed 10 Gb/s photonic cell switching as short as 2 nsec has been successfully achieved.

key words: *hybrid photonic integration, semiconductor optical amplifier, planar lightwave circuit, optical gate switch, flip-chip bonding, spot-size conversion, optical switching, photonic ATM switch, optical fiber communication, optical network*

1. Introduction

Large capacity switching systems based on photonic technologies are attractive for achieving high throughput telecommunication network for a large traffic increase [1]–[4]. These systems require a number of optical components of multiple-channel photonic functional devices, such as optical matrix switches. The switching systems should satisfy scalability, which strongly depends on the performances of the optical components such as insertion loss, module size, cost, etc. Low crosstalk switching performances are essential for the optical matrix switches [5]. Non-blocking feature and control scheme of the switching operation are important issues for realizing the flexible and robust optical network. In high throughput photonic cell switching systems, even signal propagation delay in the switch fabric could not be negligible. Polarization management is also indispensable for light sources and optical filters. These requirements, however, could hardly be satisfied with conventional bulky, lens-coupled modules.

From the viewpoint of these requirements, a gate type optical switch fabric based on the broadcast-and-select architecture becomes to be attractive [1],[2],[6]. Semiconductor optical amplifier (SOA) is expected to be a promising candidate as the optical gate because of its potential such as high extinction ratio (low crosstalk), optical signal gain and high speed switching. Moreover, these remarkable features can be controlled by only a single injection current [7]. Hybrid integration of semiconductor photonic devices onto a silicon (Si) platform with silica based planar lightwave circuit (PLC) seems to be the most practical way to assemble such gate type switch fabric into a compact module [8]–[11]. To realize such hybrid photonic integration, a waveguide platform for mounting the optical/electrical devices, a high precision multiple chip assembly technique and a high efficient coupling structure between different waveguide components are essential.

This paper describes design concepts, module fabrication and performances of the hybrid 4×4 gate type optical matrix switch module by our hybrid photonic integration technologies [12]. Its applicability was proved successfully through 10 Gb/s photonic cell switching demonstration.

2. Design Concept

Figure 1 is schematic of the developed 4×4 optical matrix switch module [12]. The module consists of silica based PLC platform on which four 4-channel arrays of SOA gates, seven electrical wiring devices and optical fibers are mounted. To realize the optical module in the hybrid manner, we designed the device structure and then developed our original hybrid photonic integration technologies based on the following concepts.

Manuscript received September 24, 1998.

†The authors are with Opto-Electronics and High Frequency Device Research Laboratories, NEC Corporation, Kawasaki-shi, 216-8555 Japan.

††The author is with Compound Semiconductor Device Division, NEC Corporation, Kawasaki-shi, 211-8666 Japan.

†††The author is with ULSI Device Development Laboratories, NEC Corporation, Ohtsu-shi, 520-8555 Janpan.

††††The author is with Transmission Device Division, NEC Corporation, Kawasaki-shi, 211-8666 Japan.

*This paper is also published in IEICE Trans. Electron., Vol.E82-C, No.2, pp.305–312, February 1999.

IEICE TRANS. COMMUN., Vol. E82-B, No. 2, pp. 357–364, February 1999

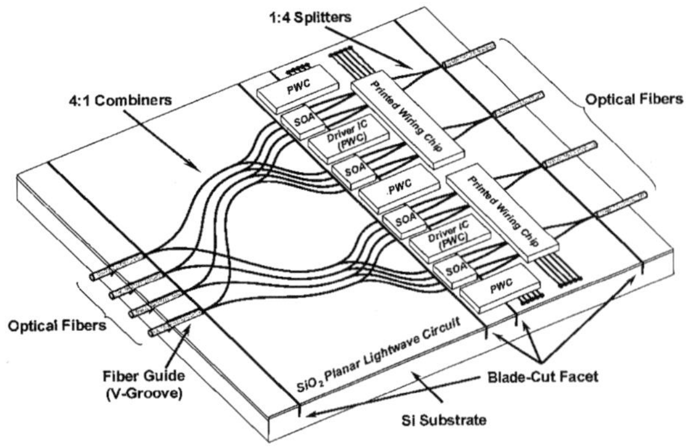

Fig. 1 A schematic of the hybrid 4 × 4 optical matrix module structure.

2.1 Hybrid Optical/Electrical Platform

At first, the hybrid photonic integration technology was expected to be a key enabler of low cost optical modules for realizing optical access network. In case of the switch module applications, however, the waveguide platform itself as a 'rigid' fabric is considered to be more important advantage of such hybridization that enables both the path synchronization and the polarization management. The waveguide platform can also simplify the complicated fiber-based fabrics with a number of surplus pigtail fibers, splicing and connectors.

In case of the 4×4 optical matrix module, over 20 electrical wiring patterns for the signal lines and the ground lines for driving the SOA gates should be located in limited space on the platform. To enlarge the chip integration scale of the hybrid photonic module, a multi-layered structure of the electrical signal lines is necessary for the platform to improve their layout flexibility. Flame hydrolysis deposition (FHD) method has been a popular way for fabricating the silica waveguide devices. The technique, however, could not provide the waveguide platform with the inner electrode layer due to its high deposition temperature exceeding 1000°C. Beside, it is not so suitable to fabricate low birefringence waveguide devices.

We therefore adopted atmospheric pressure chemical vapor deposition (AP-CVD) using tetraethoxysilane/ozone (TEOS/O_3) for fabricating the silica waveguide layers. The method has promising features that it can provide a low-loss and low-birefringence SiO_2 layer with deposition temperature as low as 400°C [13], which therefore enables us to integrate some conductive materials such as WSi for the inner signal lines in the silica waveguide layers. Besides, WSi can be charged as a masking material in fabricating V-grooves on the Si substrate for fiber alignment guides.

2.2 Simultaneous and High Precision Multiple Chip Self-Alignment

In case of the hybrid 4×4 optical switch module, the four 4-channel arrays of the SOA gates and several electrical devices are mounted onto where the silica layers of the waveguide platform were removed partially. Visual alignment has been a popular way for mounting an optical device onto a waveguide platform by flip-chip manner [8]–[11]. Employing the technique, however, each of the devices should be handled piece-by-piece, hence it restricts the assembling throughput and then the alignment accuracy. To realize the photonic multi-chip module by the hybrid manner, a precise and simultaneous multi-chip alignment technology is indispensable.

We therefore adopted the self-align technique using AuSn solder bumps which is expected to be a promising alternative because of its potentially simple and simultaneous features [14]–[16]. This technique was also applicable to the electrical wiring for the platform instead of the conventional wire bonding [12],[17]. Printed wiring chip (PWC) can connect a number of broken patterns all at once and be mounted with the other photonic devices together by the self-align manner. Such three-dimensional electrical wiring scheme is promising to potentially improve the chip layout flexibility and responses of the electrical signals. This role can be charged by some electronics such as driver IC chips for the photonic devices in near future.

2.3 High Efficient Coupling between Different Waveguide Elements with Spot-Size Converter

In case of the hybrid optical switch module, there are two waveguide discontinuities via gaps at both input and output ends of the SOA gate for each of the signal paths. To reduce such losses effectively, a spot size conversion technique is indispensable to match the mode fields as possible. In general, the core size of the silica waveguide is much larger than that of the SOA gate, then the SOA should have a structure so as to expand its spot size.

We therefore adopted selective MOVPE, now generally known as a key technology of in-plane bandgap energy control to realize monolithic photonic integration, for fabricating the SSC structures [18],[19]. The technique enables the integration of thickness-tapered waveguide structure, which acts as the spot-size converter, smoothly connected to active stripe of the SOA gate by a single-step epitaxial growth [20].

3. Module Fabrication

Figure 2 is a schematic of the PLC platform. The platform is 12 mm in width and 40 mm in length. The plat-

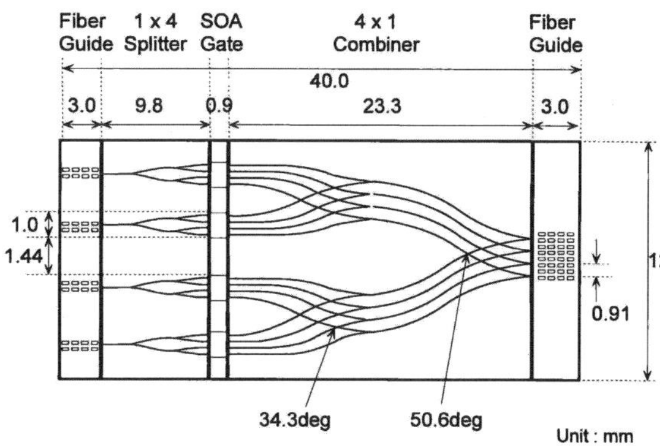

Fig. 2 A schematic of the waveguide platform of the hybrid 4×4 optical matrix module.

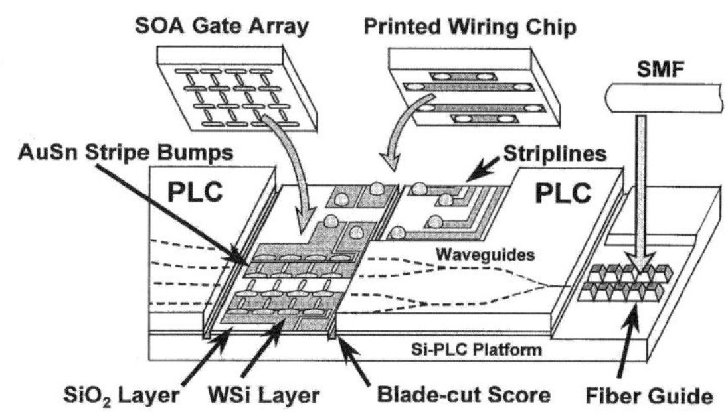

Fig. 3 A schematic of the module assembly of the hybrid optical matrix switch module.

form consists of four sets of silica optical waveguide 1 : 4 splitters/4 : 1 combiners, electrical signal lines and fiber guides on a Si substrate.

First, a WSi layer of 400 nm thick was sputtered onto thermal oxide layer of 1 μm thick on the Si wafer, and then patterned into both the electrical signal lines and the mask of the fiber guides by wet chemical etching. The silica waveguide layer was formed over the WSi layer by AP-CVD. The 1 : 4 splitters/4 : 1 combiners were made up from 2-stage waveguide Y-branches so as to adjust the signal propagation skews of all the 16 signal paths in the 4×4 switch fabric. Length and channel spacing at branched ends are 9.8 mm and 250 μm for the splitters, 23.3 mm and 2400 μm for the combiners, respectively. The waveguides have a Ge-doped square core (5 μm × 5 μm, Δn=0.5%) fabricated by reactive ion etching. Thickness of both under and upper cladding is 15 μm and 10 μm, respectively. Minimum bending radius of the waveguide is 10 mm. Minimum angle of the waveguide intersection is designed at 34° to suppress inter-port crosstalk less then −40 dB [21]. The silica waveguide layers were annealed at 800°C for 30 minute in O_2 atmosphere so as to improve their optical quality. Facets of the waveguides, where will be coupling ends to the SOA gates, were formed by dicing blade of 50 μm thick due to its practical throughput and low cost capability. The 4-channel fiber guides of 3 mm long were located at both input and output ends of the platform. The guides consist of our original segmented Si V-grooves formed by anisotropic wet chemical etching [13]. This structure can make the fiber guides less insensitive against mis-aligned patterning to crystal orientation of the Si wafer. Pitches of the fiber guides were 2.4 mm for the splitter end and 0.91 mm for the combiner end, respectively. Averaged insertion loss of the 1 : 4 splitters/4 : 1 combiners was estimated to be about 7.0 dB.

On this platform, the SOA gates and the PWCs were mounted by flip-chip manner as shown in Fig. 3. Figure 4 shows the process flow of the self-alignment technique schematically. A proper volume of the solder tip was punched out mechanically from a solder ribbon, and then pressed onto a solder-wettable metal pad on the PLC platform heated at 150°C [16]. In case of the module, we used Au-80%/Sn-20% eutectic alloy as the solder, and diameter and thickness of the solder tip were 80 μm and 20 μm, respectively. The pad metal is Au(500 nm)/Pt(100 nm)/Ti(50 nm) formed onto the WSi layer by sputtering, and precisely patterned into stripe shape of 25 μm wide and 140 μm long by stepper and electron-cyclotron-resonance (ECR) plasma etching. Such stripe shape pads can reduce the influence of the solder volume deviation effectively, then enable high precision alignment less than ±1 μm [22]. The solder tips on the pads were coated with flux and then melt into the bumps by first-step re-flow at 300°C in N_2 atmosphere. By second-step fluxless re-flow at 300°C, all of the chips temporarily placed within 10 μm displacement over the solder bumps on the platform were drawn toward center of the pads, and then aligned on them precisely all at once due to surface tension of the solder melt. Height of the AuSn bumps after the self-align process was about 17.5 μm.

In case of the module, two different sizes of the PWCs were used for the electrical wiring. The PWCs were grass chips (Corning 7059) of 300 μm thick with several patterned signal lines. The smaller one was 1 mm long and 1 mm wide, which connects the signal lines over the blade-cut score due to the facet forming. The bigger one was 5 mm long and 2 mm wide, which connects the signal lines over the PLC waveguides. Figure 5 shows a photograph of the mounted chips on to the PLC platform by the self-alignment technique.

Flat-ended single mode fibers for both optical signal input and output were aligned passively along the Si V-groove fiber guides, and then fixed with UV-curable adhesive. Typical coupling loss between the silica waveguide and the fiber was estimated less than 0.5 dB.

Figure 6 shows schematic of the polarization-insensitive SOA gates with spot-size converter. (PI-SSC-SOA gates) [23]. The gate consists of a bulk InGaAsP active stripe (λ =1550 nm) and thickness-tapered SSC regions of 250 μm long for both coupling ends, which were grown by the selective MOVPE at

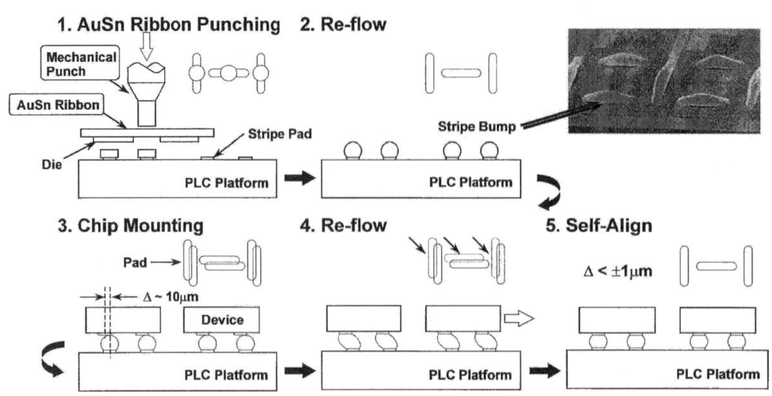

Fig. 4 A schematic of the process flow of the self-alignment technique using AuSn solder bump.

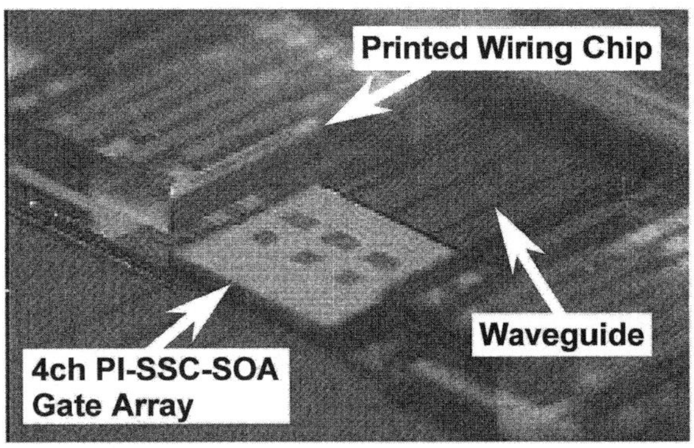

Fig. 5 A photograph of mounted chips on the PLC platform by the self-alignment technique.

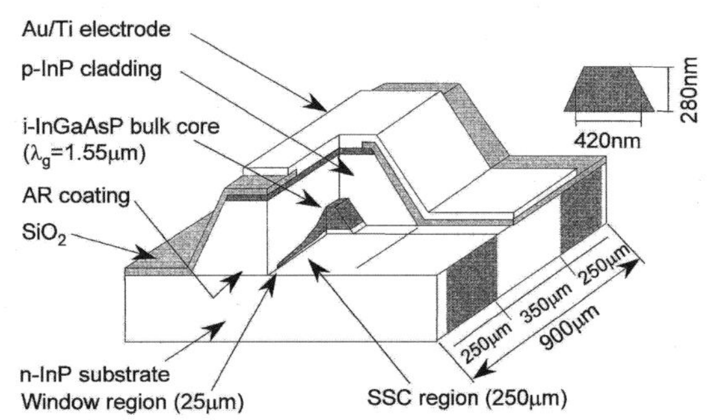

Fig. 6 A schematic of the polarization insensitive semi conductor optical amplifier gate with spot-size converter.

once and buried in InP layers. Window regions of 25 μm long and low-reflection coatings were formed at each of the facets so as to suppress residual facet reflectivity for improving the optical signal gain. Radiation divergence was 13–14° (FWHM) in both azimuth and elevation planes. Figure 7 shows measured coupling coefficient versus gap between the gate and the PLC waveguide. The loss increases for about 0.9 dB/10 μm as the gap increases. Typical coupling loss at the gap of 10 μm is estimated to be 4.5 dB. Figure 8 shows the measured coupling coefficient versus lateral displacement at the gap of 10 μm. The chip alignment tolerance with 1 dB excess loss were about $\pm 2.5\,\mu$m for both horizontal and vertical axes. These values are smaller than typical alignment error of the self-alignment, which therefore support the applicability of both the spot size converter and the self-align technique for mounting the SOA gates on the platform. Averaged saturation power and the polarization dependent loss of the SOA gates is about +4 dBm and less than 0.5 dB, respectively. Typical on/off ratio of the SOA gates was more than 40 dB for 40 mA injection current, which is almost enough to suppress the influences of the coherent crosstalk.

Figure 9 shows the photograph of the hybrid 4×4 optical switch module. The module is 54 mm long and 22 mm wide.

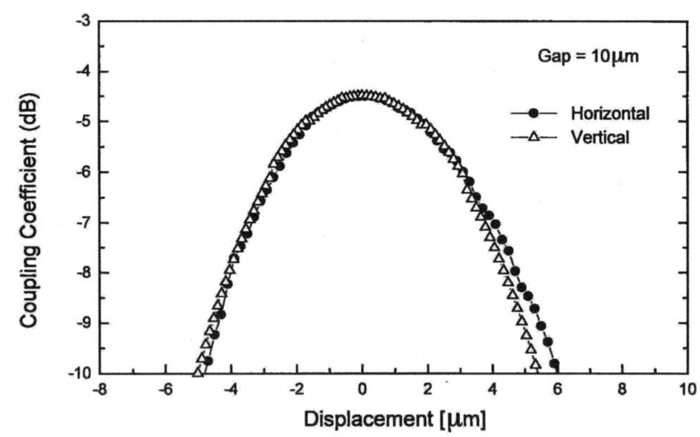

Fig. 7 Coupling coefficient versus gap between the SOA gate and the PLC waveguide.

4. Module Performances

4.1 Static Characteristics

Static optical gating performances of the module were evaluated [12]. Typical fiber-to-fiber gain as a function of injection current into the SOA gate is depicted in Fig. 10. The input signal wavelength was 1.54 μm and

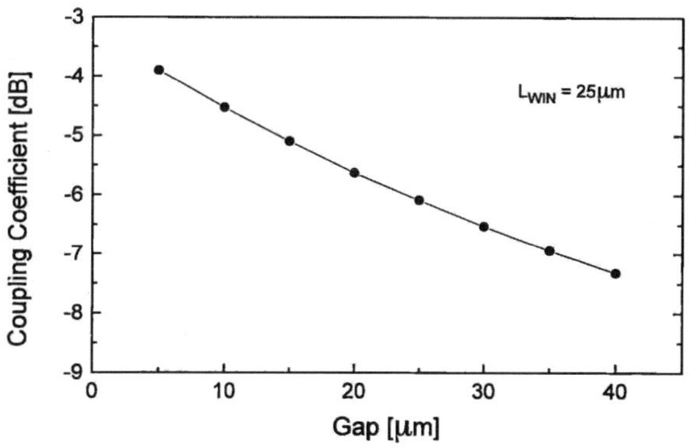

Fig. 8 Coupling coefficient versus lateral displacement.

Fig. 9 A photograph of the hybrid 4×4 optical switch module.

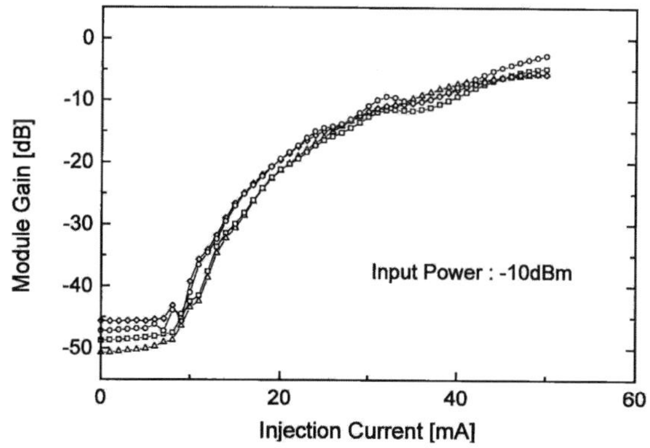

Fig. 10 Typical fiber-to-fiber gain as a function of injection current.

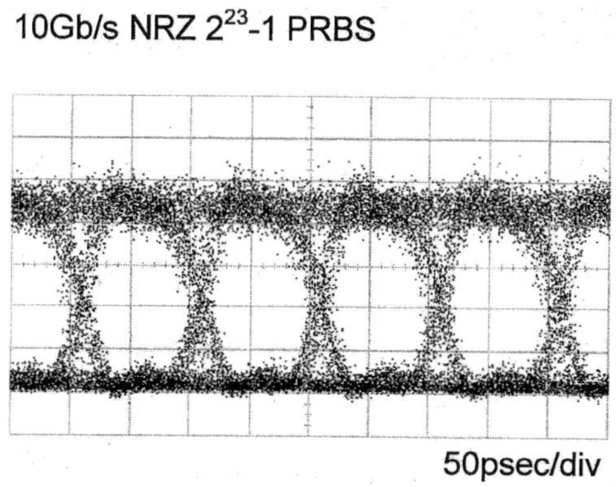

Fig. 11 10 Gb/s optical signal waveform on module output data cell under cell switching.

the signal power is −10 dBm, respectively. The insertion losses of each signal path at 50 mA of injection current were 7±2 dB. Such losses are mainly due to splitting loss of the waveguide splitters/combiners, hence they can be reduced efficiently by employing single-to-multi mode combiners instead of the single mode combiners [24]. Total polarization dependent loss of the module was less than 0.5 dB for each path. Averaged extinction ratio as high as 40 dB was obtained for 40 mA injection current, which is sufficient as a optical matrix switch for suppressing the coherent crosstalk [6]. Small ripples on the gain characteristics were due to reflection at the facets of the PLC waveguides. Such influences can be suppressed effectively by filling the gap with a polymer of proper refractive index.

4.2 High Speed Cell Switching Characteristics

To confirm its potential for the photonic cell switching applications, high speed switching performances were evaluated [12].

The cell switching is realized by alternating the signal paths. One of two input signal cells which propagate along two paths (input port 1 → output port 1′, input port 4 → output port 1′) of the module was selected alternately by the gate driving current. The input signals were split form a common light source (λ=1.54 μm), then modulated at 10 Gb/s independently. Peak power of each input signal cell was settled at −5 dBm, and then the cells were launched into the input ports in TE polarized states. The cell length and the period were 51.2 nsec (512 bits/cell) and 102.4 nsec, respectively. The gate driving current was 0–35 mA.

Figure 11 shows the optical signal waveform on the output data cell under the cell switching. Clear eye opening of the gated data cell was observed. Figure 12 shows the 10 Gb/s photonic cell waveforms under the cell switching. Transient optical waveform at the cell switching was so smooth without degradation, which indicates the ideal high on/off switching performances of the SOA gates enough to suppress the coherent crosstalk. Rise and fall time of the switched signal cell were 0.9 nsec and 1.1 nsec, respectively. Minimum switching time as short as 2 nsec were achieved under the cell switching.

These results support the promising features of the proposed hybrid photonic integration technique, which will provide a better solution to fabricate the optical modules in the future high throughput photonic switching applications.

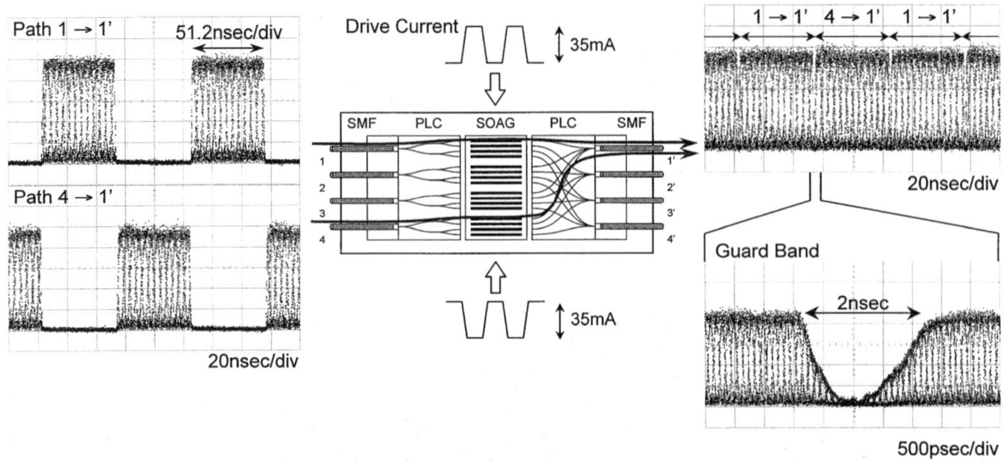

Fig. 12 10 Gb/s photonic cell waveforms under cell switching.

5. Conclusion

We developed the hybrid 4×4 gate type optical matrix switch module. The matrix switch module integrates four 4-channel arrays of PI-SSC-SOA gates, printed wiring chips and optical fibers on a Si platform with silica waveguide elements. Such hybrid integration has been enabled successfully by our original multiple chip self-align assembling technology, AP-CVD based Si-PLC platform technology and the selective MOVPE technology. Its applicability was confirmed through the 10 Gb/s photonic cell switching demonstration. This technique was proved to be quite useful to fabricate a great variety of optical modules for next generation photonic network systems.

Acknowledgement

The authors would like to thank M. Ogawa, I. Mito, K. Kobayashi, and M. Fujiwara for their continuous encouragement throughout this work. We are also grateful to N. Henmi for fruitful discussions, N. Kida, K. Yasuda, S. Sone, and Y. Akimoto for their support of the device processing, and S. Takahashi for arranging the measurement equipment.

References

[1] S. Takahashi, T. Kato, H. Takeshita, S. Kitamura, and N. Henmi, "10 Gbps/ch space-division optical cell switching with 8×8 gate type switch matrix employing gate turn-on-delay compensator," Technical Digest of Photonics in Switching '96, vol.2, Post Deadline Paper, PThC1, pp.12–15, 1996.

[2] H. Takeshita, S. Takahashi, T. Kato, S. Kitamura, H. Harano, and N. Henmi, "A demonstration of an optical cross-connect system for a self-healing optical network," Proceedings of ECOC/IOOC '97, vol.3, WE4C-2, pp.335–338, 1997.

[3] K. Sasayama, Y. Yamada, F. Frueh, K. Habara, K. Yukimatsu, S. Suzuki, and H. Ishii, "Demonstration of a photonic ATM switch using a frequency-routing-type time-division interconnection network (FRONTIERNET)," Proceedings of ECOC '94, vol.2, pp.533–536, 1994.

[4] D. Chiaroni, D. De Bouard, P. Doussiere, C. Chauzat, and M. Sotom, "High performance semiconductor optical amplifier gate for fast WDM packet switching," Proceedings of ECOC '95, vol.1, no.L.3.5, pp.115–119, 1995.

[5] D.J. Blumenthal, P. Granestrand, and L. Thylén, "BER floors due to heterodyne coherent crosstalk in space photonic switches for WDM networks," IEEE Photonics Technol. Lett., vol.8, no.2, pp.284–286, 1996.

[6] T. Kato, S. Takahashi, H. Takeshita, S. Kitamura, and N. Henmi, "High speed and low crosstalk switching characteristics of 2×2 gate type optical switch," Technical Digest of IEICE '96 Spring Conference, B-1106, 1996.

[7] S. Kitamura, K. Komatsu, and M. Kitamura, "Very low power consumption semiconductor optical amplifier array," IEEE Photonics Technol. Lett., vol.7, no.2, pp.147–148, 1995.

[8] Y. Yamada, H. Terui, Y. Ohmori, M. Yamada, A. Himeno, and M. Kobayashi, "Hybrid-integrated 4×4 optical gate matrix switch using silica-based optical waveguides and LD array chips," IEEE Journal of Lightwave Technol., vol.10, no.3, pp.383–390, 1992.

[9] G.H. Henein, D.J. Muehlner, J. Shmulovich, L. Gomez, M.A. Capuzzo, E.J. Laskowski, R. Yang, and J.V. Gates, "Hybrid integration for low-cost OE packaging PLC transceiver," Proceedings of LEOS '97, vol.1, ThB1, pp.297–298, 1997.

[10] I. Ogawa, T. Hashimoto, T. Ohyama, K. Shuto, M. Yanagisawa, Y. Yamada, Y. Akahori, A. Himeno, F. Ebisawa, K. Kato, Y. Tohmori, and O. Nakajima, "Hybrid integration of spot-size converted semiconductor amplifier on planar lightwave circuit platform using passive alignment technique," Proceedings of CLEO/Pacific-Rim '97, Post Deadline Paper PD2.3, 1997.

[11] T. Ito, I. Ogawa, N. Yoshimoto, F. Ebisawa, K. Magari, K. Shuto, Y. Kawaguchi, M. Yanagisawa, O. Mitomi, F. Hanawa, Y. Tohmori, Y. Yamada, Y. Yoshikuni, and Y. Hasumi, "Ultra-wide-band high-speed wavelength selector using a hybrid integrated gate module: A 4-channel SS-SOA gate array on PLC platform," Proceedings of ECOC/IOOC '97, vol.5, Post Deadline Paper, TH3B-6, pp.53–56, 1997.

[12] T. Kato, J. Sasaki, T. Shimoda, T. Tamanuki, H. Hatakeyama, M. Yamaguchi, M. Kitamura, and M. Itoh,

"10 Gb/s photonic cell switching with hybrid 4×4 optical matrix switch module on planar waveguide platform," Technical Digest of OFC '98, Post Deadline Paper, PD3, 1998.

[13] N. Kitamura, S. Mizuta, T. Shimoda, and M. Kitamura, "Silica based optical waveguide devices, with novel fiber guide structure for alignment-free fiber coupling," Technical Digest of Integrated Photonics Research (IPR) '96, IThB2-1, 1996.

[14] M.J. Wale and C. Edge, "Self-aligned flip-chip assembly of photonic devices with electrical and optical connections," IEEE Trans. on Components, Hybrids, and Manufacturing Technol., vol.13, no.4, pp.780–786, 1990.

[15] K. Katsura, T. Hayashi, and F. Ohira, "A micro interconnection technique using solder bumps for high-speed optical devices," Proceedings of Integrated Microelectronics Conference (IMC) '90, pp.105–108, 1990.

[16] M. Itoh, J. Sasaki, A. Uda, I. Yoneda, H. Honmou, and K. Fukushima, "Use of solder bumps in three-dimensional passive aligned packaging of LD/PD arrays on Si optical benches," Proceedings of Electronic Components & Technology Conference (ECTC) '96, pp.1–7, 1996.

[17] J. Sasaki, H. Hatakeyama, T. Tamanuki, M. Yamaguchi, N. Kitamura, T. Shimoda, M. Kitamura, T. Kato, and M. Itoh, "Hybrid 4×4 optical matrix switch module on silica based planar light wave circuit by self-align multiple chip bonding technique," Technical Digest of Integrated Photonics Research (IPR) '98, ITuC3-1, pp.210–211, 1998.

[18] T. Kato, S. Sasaki, N. Kida, K. Komatsu, and I. Mito, "Novel MQW DFB laser diode/modulator integrated light source using bandgap energy control epitaxial growth technique," Proceedings of ECOC/IOOC '91, vol.2, WeB7-1, pp.429–432, 1991.

[19] M. Aoki, H. Sano, M. Suzuki, M. Takahashi, K. Uomi, and A. Takai, "Novel structure MQW electroabsorption modulator/DFB laser integrated device fabrication by selective area MOCVD growth," Electron. Lett., vol.27, pp.2138–2140, 1992.

[20] H. Kobayashi, M. Ekawa, N. Okazaki, O. Aoki, S. Ogita, and H. Soda, "Tapered thickness MQW waveguide BH MQW lasers," IEEE Photon. Technol. Lett., vol.6, pp.1080–1081, Sept. 1994.

[21] T. Shimoda, J. Sasaki, T. Kato, H. Hatakeyama, T. Tamanuki, N. Kitamura, M. Yamaguchi, M. Itoh, M. Kitamura, and S. Kitamura, "Hybrid integrated 4×4 matrix optical switch module using silica based planar lightwave circuit," Proceedings of CLEO/Europe-EQEC '98, CThO2, p.350, 1998.

[22] J. Sasaki, H. Honmou, M. Itoh, A. Uda, and T. Torikai, "Self-aligned assembly technology for laser diode modules using stripe-type AuSn solder bump flip-chip bonding," Proceedings of LEOS '95, vol.1, OPMR 2.3, pp.234–235, 1995.

[23] H. Hatakeyama, S. Kitamura, K. Hamamoto, M. Yamaguchi, and K. Komatsu, "Spot size converter integrated semiconductor optical amplifier gates with low power consumption," Technical Digest of Optoelectronics and Communications Conference (OECC) '97, 9C3-2, 1997.

[24] A. Tajima, N. Kitamura, S. Takahashi, S. Kitamura, Y. Maeno, Y. Suemura, and N. Henmi, "10-Gb/s/port gated divider passive combiner optical switch with single-mode-to-multimode combiner," IEEE Photonics Technol. Lett., vol.10, no.1, pp.162–164, 1998.

High-Performance Backbone Network Technology edited by Naoaki Yamanaka
ISBN: 0-8247-5321-6 DOI: 10.1081/(120025748)
Published 2004 by Marcel Dekker, Inc.

Merging Electronics and Photonics Towards the Terabit/s ATM Switching

Bruno Bostica and Luigi Licciardi

SUMMARY The paper is focused on the architectural and technological solutions that will allow the transition from small to huge capacity ATM Switching Systems. This path starts from the industrial nodes available today and will arrive at the photonic switching architecture. The progressive introduction of photonics has already started with the use of optical interconnections in ATM nodes of hundreds of Gbit/s. A balanced use of microelectronics and photonics is the correct answer to the Terabit/s switching system challenge. After presenting a modular ATM Switching System, some technological solutions like Multichip Modules and Optical Interconnections are presented in order to explain how node capacity can be expanded. Some results of the research activity on photonic Switching are finally shown in order to exploit the great attitude of this technique to obtain very high throughput nodes.

key words: *ATM switching, terabit/s capacity, electronic switching, ATM switching fabric, photonic switching, multichip module, optical interconnections*

1. Introduction

Today we are witnessing the realisation of ATM networks with the aim of verifying the feasibility of the ATM technique and of offering basic services to the users. Most Telecom system manufacturers have developed their own equipments and are working on improvements both in terms of functionalities and capabilities. A step-by-step approach is the following: initially Cross Connects with low capacity (10–20 Gbit/s) are realised, then the transition to full Switch follows, granted by the introduction of signalling [1]. In terms of capacity, flexibility and modularity this pattern seems to be a practical key of success. Maintaining the same architecture from low to high capacity is a very important item to save investments and guarantee proper answers to the different market requests. Up to now, the request in terms of capacity is not too demanding but it is easy to foresee that when the services, the infrastructure and the backbone are available switches node of high capacity will be necessary. Most of the state of the art nodes are now based on electronic switches, CMOS or BiCMOS integrated circuits (IC) are used for the basic components of the switching systems and bipolar ICs for drivers and receivers. What about interconnections? Inside the node high capacity electronic switches require optical interconnections, therefore we are witnessing the necessity to merge electronic devices with optical ones. This attitude is increasing and nodes of several hundred Gigabit/s use this technique together with new high density packaging techniques like Multichip Module. And what about the challenge of the Terabit/s Switching Systems? The technological enhancement will play a strategic role: even if microelectronics is in continuous evolution, this capacity might be the threshold between the electronic and the optical switches, that are now object of advanced research. We will try to cover the technological path that starting with the state of the art (an industrial Cross Connect architecture) passes through the introduction of photonics and new enabling technologies in interconnections and packaging to finally arrive at the optical switching. This path requires a very good mastery in merging electronics with photonics defining the best architecture and partitioning the system properly.

Manuscript received February 27, 1997.
Manuscript revised July 16, 1997.
† The authors are with CSELT-Centro Studi e Laboratori Telecomunicazioni, V. Reiss Romoli 274, 10148 Torino, Italy.

2. The Electronic Switch Reference Architecture

The industrial reference architecture here presented is the Italtel UT-FastNet ATM node (the Cross-Connect UT-XC and the Switching node UT-BB): it is a modular architecture that can actually achieve throughput from 5 Gbit/s to 160 Gbit/s just adding modules. It is an electronic VLSI based switching solution that can use both electrical and optical interconnections depending on the capacity (electrical are used till 20 Gbit/s).

2.1 Features and Architecture

The UT-XC [2] basic functionalities are: semipermanent Virtual Path (VP) and Virtual Channel (VC) management, VPI and VCI translation, Cell multiplexing and de multiplexing preserving cell sequence integrity, Usage and Network Parameter Control (UPC, NPC), OAM F4 and F5 flow monitoring, VPs and VCs release under the Network Management control, Cell Header error control generation and extraction. It manages both the UNI and NNI interfaces (SDH at 155.52 Mbit/s and PDH at 34 Mbit/s).

The overall architecture (Fig. 1) is based on ATM

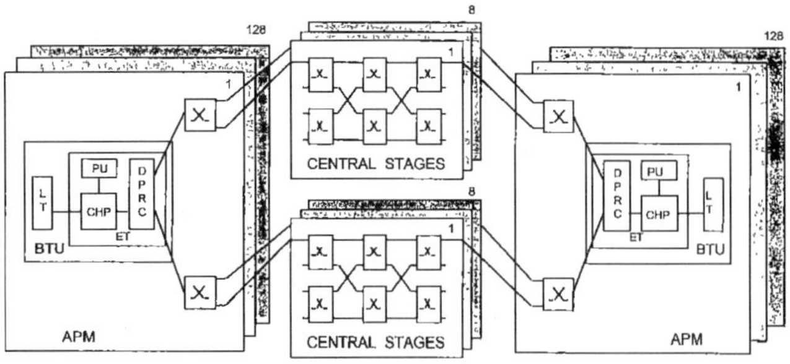

Fig. 1 The 160 Gbit/s switching architecture.

Peripheral Modules (APM), a Central ATM Switching Fabric (ASF), a Supervision Module and some optional blocks, such as Connectionless and/or Multipoint Servers and local concentrators. Modules are linked through a duplicated Ethernet for Interprocessor communication. Internally the APM a Broadband Termination Unit (BTU) acts as an interface for a single access line to the first stage of the ASF. The BTU contains both the Physical Layer interface and the ATM switch interface. Up to 8 BTU can be present in an APM module, where an 8×8 switching element distributes the traffic of 8 incoming flows towards the central ASF.

The ATM switch interface, called Exchange Termination (ET), is realised by a set of VLSI components: the Cell Header Processor (CHP[†]) that performs the ATM layer functions and transforms the standard ATM cell format (53 bytes) into an internal one (64 bytes) containing the proprietary information and vice versa removing them, the Policing Unit (PU[†]) that implements both the Usage and Network Parameter Control applying a generalised Leaky Bucket Algorithm, the Duplicated Path Recombining Controller (DPRC[†]) that realises a redundancy principle based on a duplicated path and implements a resequencing algorithm based on maximum delay equalisation.

The ASF is duplicated in order to guarantee a good robustness against faults, the ASF internal speed is 311 Mbit/s, providing a speed advantage of about 2 between the access and the internal links and hence improving the global traffic performance. The ASF is a multistage *distribution and routing* network (one connection cells are distributed in the network and follows different paths). Depending on the configuration and the node capacity the number of stages can be 3 or 5 for each duplicated plane.

The small configuration (up to 20 Gbit/s) is based on a single APM with a maximum number of 64 ports and a 3 stage network, the complete one (up to 160 Gbit/s) is based on several APMs with a maximum number of 1024 ports and a 5 stage network. The ASF is based on a 8×8 Broadband ATM Switching Element (BASE[†]) in the small configuration, on a 16 ×16 one in the complete one, i.e. using the MCM-

Table 1 The ET component features.

Circuit	CHP	PU	DPRC
Technology	0.7 μm CMOS	1.2 μm CMOS	0.7 μm CMOS
Package	257 CPGA	120 CPGA	208 CPGA
I/O signals	177	89	131
Max. freq.	40 MHz	20 MHz	40 MHz
Mos number	480000	32000	640000
Area	150 mm^2	48 mm^2	150 mm^2
Power	2.5 W	1 W	3 W

BASE16 described in Sect. 3. The BASE element performs the switching function allowing different routing (virtual, random distributed and distributed with link exclusion).

2.2 The VLSI Components

Four basic VLSI components were developed in the UT-FastNet project: three for the ET part (Table 1) and one for the Switching Fabric. CMOS technology was used in order to have high integration density, low power dissipation and low cost.

The Broad band ATM switching element is the key element of the ASF, increasing its capacity is the major effort to improve the system throughput. The BASE component can switch N (8 or 16) high speed, broadband data inputs with a variable rate of up to 311 Mbit/s onto N (8 or 16) high speed data outputs, according to the detailed routing information laid down in the header of the data stream.

The component internally provides the alignment of the incoming cells at the maximum internal frequency. The BASE can switch cells according to a deterministic routing or a random distribution with or without exclusion of one output link. A cell loss priority principle has been implemented as well as a back pressure function in order to avoid traffic congestion. The BASE device is based on a shared buffer architecture, controlled by a set of output queues. Queues are implemented as a complete sharing of a large memory properly addressed in order to realise the space switching function. The internal shared buffer is tailored in size (128) such as to maintain a cell loss probability of 10^{-10}, supposing a uniformly distributed and non-correlated input traffic, with a load of approximately 85%. The component also adapts the outgoing data from the internal frequency to the proper output link speed, that is programmable by microprocessor. In the 8×8 version (BASE8) the component has a global capacity of 2.5 Gbit/s. A specific logic was implemented in the IC in order to allow an extended use as a sub stage of a more complex 16×16 or 32×32

[†] Patents Pending.

single stage Switching Element.

3. High Performance Interconnections Techniques

Interconnections and packaging are the first technological aspects to face in order to allow the ATM nodes to increase their overall throughput from several tens of Gbit/s, which represent the actual limit using conventional technologies, up to several hundreds of Gbit/s. In fact, moving towards these capacities, interconnections become a real problem as the physical dimension of the switching fabric increases as well as the connectivity requirements, in terms of number of links and speed among the stages. Up to now, electrical interconnections have been widely adopted within the switching systems, mostly to support the exchange of data in parallel format, allowing the use of low speed, low cost technologies (e.g. CMOS) and simplified transmission schemes.

Higher node capacities push the request for a higher level of packaging of the switching blocks, which in turn imply a larger number of I/Os at both IC and board level. On the other hand the growth of the overall system physical dimensions implies longer interstage link lengths on which high speed multiplexed data flows should be transmitted to reduce the number of interconnection links. In this perspective the introduction of the Multichip Module (MCM) and Optical Interconnections (OIC) technologies gives since now the possibility to achieve higher system performance.

Optical Interconnections technology is becoming more available, thanks to the recent progress in the development of low power, small dimensions and high reliable electro-optical devices, e.g. very low threshold (hundreds of μA) Vertical Cavity Surface Emitting Lasers (VCSEL), which have made practical and relatively cheap the developments of parallel optical interconnections [3]. Therefore the first step to increase the ATM switching node capacity up to several hundreds of Gbit/s is the combined use of microelectronics and optical fibres.

3. 1 Multichip Modules

Multichip Module is a promising technique that allows to reduce area, improve integration density and increase the system global performance. It consists in mounting and interconnecting bare dice on a single substrate. In the UT-XC its use was driven by the need of increasing the Switching capacity from 8×8 links at 311 Mbit/s to 16×16 at the same speed, obtaining a global throughput of 4.98 Gbit/s per single board.

Three MCMs were developed: two low cost MCM-Ls (8 layers) performing respectively a 8 links output driver and a 8 links input receiver and one MCM-C realising the 16×16 Switching Element. The

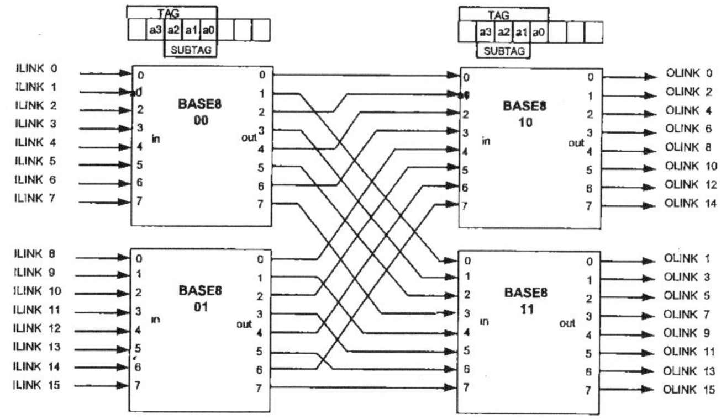

Fig. 2 16×16 switching element architecture.

16×16 final board, that can perfectly replace the 8×8 one in UT-XC, contains one 16×16 Switching Element, two Drivers and Receivers besides the other components.

The drivers and receivers MCM-Ls allow to adapt the reduced CMOS swing (0–1.5 V) of the Switching Element output/inputs to balanced PECL levels and vice versa. The output driver (MCM-DR48) can drive 48 balanced lines (ECL or PECL) up to 320 MHz, with an area of about 800 mm^2 and a power dissipation of 2.4 W. The input receiver (MCM-RC48) adapts 48 balanced PECL inputs to 48 single TTL outputs up to 100 MHz, with an area of about 800 mm^2 and a power dissipation of 2 W. Both circuits take up less than 1/6 of the space occupied on the board by using standard packaged components.

The 16×16 Switching Element (MCM-BASE16[†]) [4] is a Multichip Module that uses 4 Switching Elements (BASE8) properly connected in order to avoid blocking problems and in order to allow one routing tag to drive the proper routing inside the MCM-BASE16, that is managed as a single Switching Element (Fig. 2).

The MCM-BASE16 has 4 shared memories each of 128 cells, therefore its global capacity is 512 cells, each stage allowing 256 cells contemporaneously stored, this allows to drastically reduce cell loss probability with respect to other single chip Switching Elements based on a Shared Buffer architecture (Fig. 3).

This 16×16 Switching Element, designed by CSELT and realised as a 11 layers MCM-C (co-fired ceramic) by IBM, has a complexity of about 2,800,000 transistors, works at 78 MHz and dissipates about 10 W.

The use of CCGA (Ceramic Column Grid Array) package with 1073 pins distributed below the whole area allows to have a component smaller than the BASE8 Element, packaged in a standard 256 CPGA. Moreover, with this kind of package the problems related to the mechanical strengths are relaxed, providing an high level of reliability.

† Patents Pending.

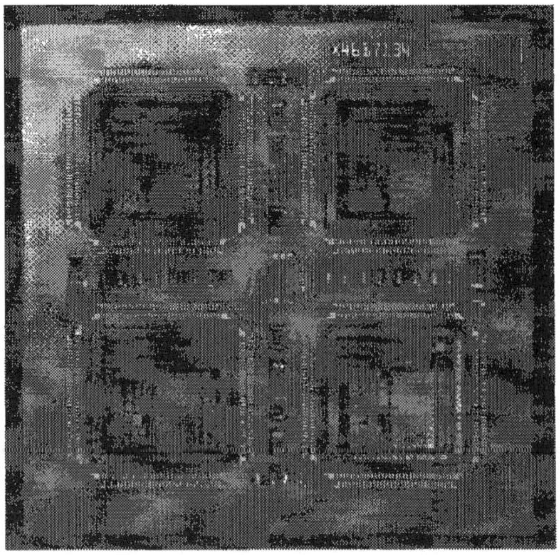

Fig. 3 MCM-BASE16 die photo.

The evolution of packaging techniques deserves great attention from Telecom system suppliers in order to improve performance and reduce costs. Having so many I/Os available allows to increase the number of links managed and, thanks to the use of parallel I/Os, to scale down the frequency inside the IC. Furthermore using thermal vias to transfer the heat on the board is very useful to improve power dissipation

The architecture presented above is very effective and allows to increase the switching capacity also to 32 ×32 or even more if using a 16×16 single Switching Element (SE). It is a general solution that can be used to increase the single SE switching capacity of a factor of 2 or 4 respectively using 4 or 12 basic IC properly connected.

3. 2 Optical Interconnections

The use of optical interconnections in ATM Switching systems is quite mandatory for high capacity nodes, where connection length is relevant (i.e. the connections between the peripheral modules and the central stages).

The activity on point-to-point optical interconnections was focused on the development of two components (Fig. 4), both realized by using on 0.7 μm CMOS technology: the first one converts parallel data into serial and vice versa (ASTRID: ATM Serial Transmitter Receiver Integrated circuit for Data links) and the other includes the laser driver and the pin diode receiver circuits (MINUET: Multipurpose Interconnection Unit for Electro optic Tranceivers).

ASTRID† has a parallel interface according to the UT-XC format (4/8 data, data clock, cell clock) at 77. 8 MHz/38.9 MHz and a serial interface working on scrambled data at 311 Mbit/s (PECL levels). It works on cells in the proprietary format (64 bytes) and the serial to parallel conversion includes cell alignment

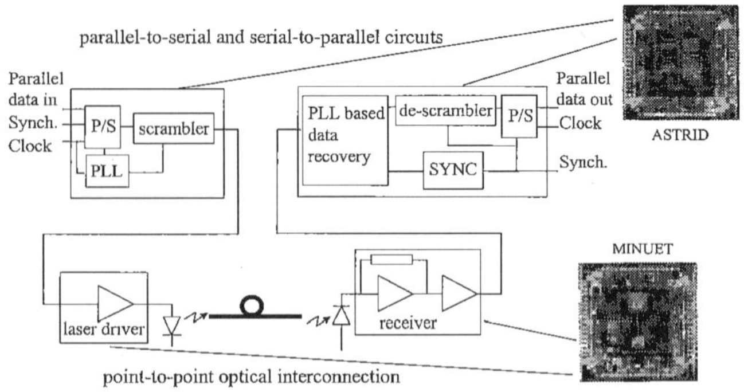

Fig. 4 Point-to-point serial optical interconnection.

(PLL based) and cell start detection. The power dissipation is 1.1 W at the working target frequency.

The optical transceiver† module uses two MINUET circuits (transmitter and receiver) and a laser diode made by CSELT. Transmitter and receiver are housed in a single low profile 20 pin metallic flat pack, the laser diode, pin diode and the fibre pigtails are directly glued on the IC. The operating bitrate is 311 Mbit/s, using PECL electrical interfaces, the receiver sensitivity is −20 dBm with a dynamic range of 12 dB, the interconnection optical power margin is 7 dB and the power dissipation is 100 mW for the transmitter and 150 mW for the receiver.

A contemporary activity on parallel optical interconnections is in progress in order to minimise the space occupancy on the board and on the backplane connector. Two 8 channels parallel transmitter and receiver modules are under development, based on laser driver and photo diode receiver arrays having the same characteristics of the previously described single point-to-point devices. The modules include a multimode fibre ribbon pigtail terminated with a MPO connector which is fully compatible within standard electrical high density metrical pitch connectorization systems.

Interstage interconnections cabling becomes in this context an important issue. According to the classical interconnection pattern among the stages of a multistage switching fabric the links from one switching module need to be re-distributed towards all the modules of the next stage causing heavy and poorly manageable cabling structures (Fig. 5(a)).

Parallel optical interconnections can be used with great advantage when an optical interconnection rearranging blocks is adopted to realise in a concentrated small size unit the interstage connection pattern. This block can be based either on fibre flex-foil technology or on optical wave guide technology. All the connectorized ribbon can be terminated on it, allowing a very light, easily re-arrangeable interconnection structure (Fig. 5(b)). In fact this solution allows the

† Patents Pending.

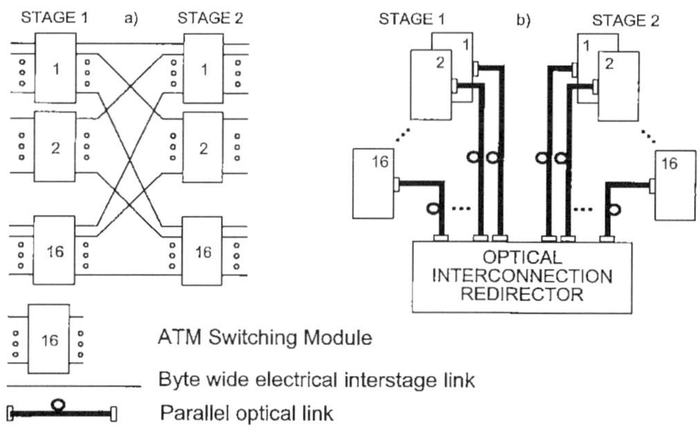

Fig. 5 Interstage interconnections in a high throughput ATM switching fabric.

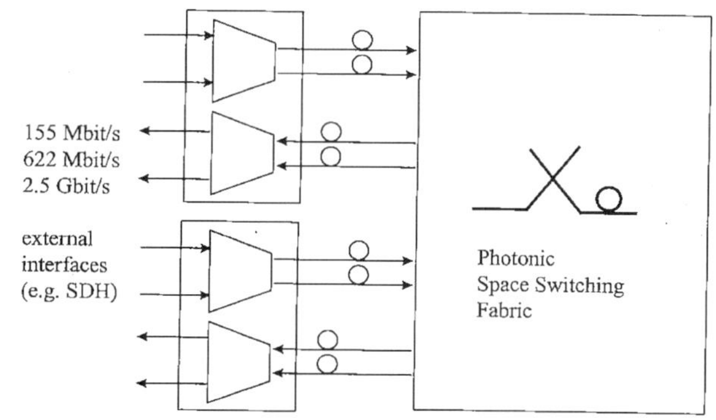

Fig. 6 Tbit/s ATM switch based on ATM electrical access switches and photonic switch core.

system growth or changes in the interconnection pattern just changing the optical interconnection rearranging block, without modifying the existing cabling of the fibre ribbons.

4. Architectural and Technological Trends towards the Terabit/s Switches

After the evaluation phase both in equipments and services, it is easy to foresee a more demanding request in terms of nodes capacity. Therefore the great capacity nodes will be asked for within 2–3 years. Technological improvements will be done in microelectronics: new CMOS and BiCMOS processes at 0.15 μm and 1 V will be available, new high I/O packaging and interconnections techniques will be industrially usable and diffused. Up to now nodes at 160 Gbit/s are available merging electronics and photonics using a digital VLSI Switching Element (SE) and optics for interconnections. What is the limit of architectures and solutions using SE based on electronic switching? It is a difficult question, the feeling is that speed of 622 MHz per link can be achieved and 32×32 SE can be realised on a single package using a single IC or multichip modules. Therefore SE of 20 Gbit/s capacity should be available within 1–2 years. It means a factor of 4 with respect of the 160 Gbit/s system previously presented (based on 311 Mbit/s and 16×16 SE) and so a capacity of 640 Gbit/s seems to be realistic. The same technological improvements should be applied in the ATM Peripheral Modules which represent a huge part of the whole ATM node hardware. The trend is to increase the level of integration by developing integrated circuits having higher levels of functionality: for instance interfaces at 622 Mbit/s or 2.5 Gbit/s, collapsing 4 or 16 ATM flows, can use a single integrated circuit to perform the ATM layer functionality.

What about the Terabit/s? It is not so far away, using both Multichip Modules and high density ICs, taking great care of power dissipation that is a very important constraint and using new techniques to refrigerate the system. This target seems to be, in a medium term perspective, challenging but feasible. It seems anyway that 1.28 Terabit/s is the upper limit for switching systems based on electronic Switching Elements. To go further (i.e. 10 Terabit/s) switching systems using photonic technologies should be the most promising solution: many studies and research activities are in progress all over the world.

5. Photonic Switching in ATM Nodes

Two different basic architectural approaches are emerging in this research field. One, *short term* oriented, basically extends the concept discussed above, where optical interconnections were introduced among electronic ATM stages, and suggests the use of a photonic cross connect to provide connectivity among high capacity (up to tens Gbit/s capacity) electronic access ATM switches [5]. The advantage of this solution is the reuse of electronic ATM switches (as well as the access interfaces), which are connected together in a very flexible and efficient way to get very high throughputs. In this solution (Fig. 6) the photonic switch provides only the routing of high speed (2.5 or 10 Gbit/s) ATM flows and does not operates at cell level. The routing across the space division photonic switch is managed by an electronic control which can reconfigure the connection pattern among the peripheral ATM access switches according to the traffic needs. An extensive use of parallel high speed OICs is suitable and generally foreseen in these architectures allowing, very compact physical structure.

The second approach is more oriented towards an extensive use of photonic technology. In this case it is assumed that photonics will evolve providing mass production, reliability and low power consumption, which are key factors for the telecom system requirements.

This concept foresees full optical high capacity ATM switches based on the exploitation of the wavelength domain for the high speed cell routing and/or for the ATM cell buffering, electronics being only used

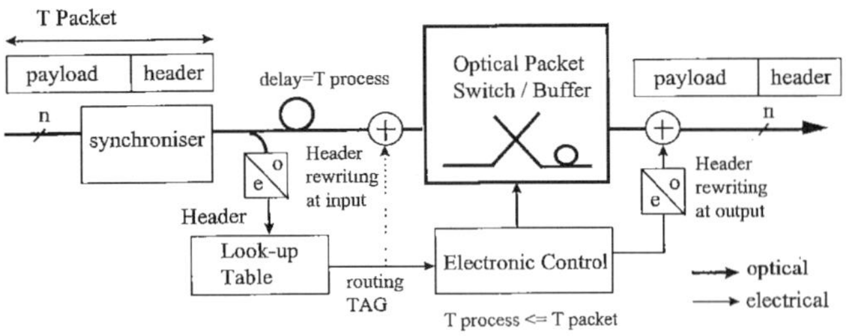

Fig. 7 Block diagram of a Tbit/s photonic ATM switch.

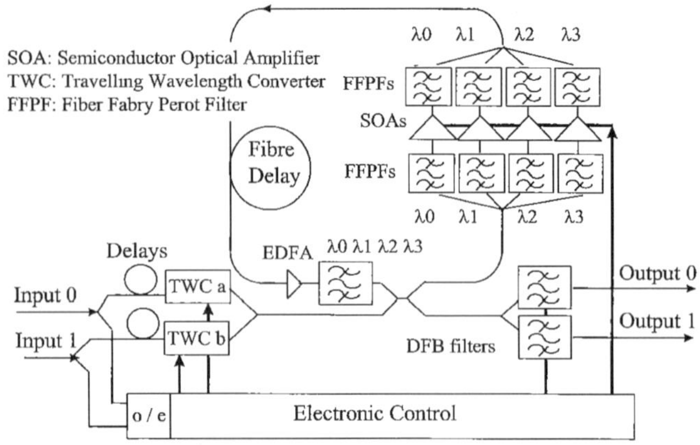

Fig. 8 Schematic of the Fibre loop memory ATM photonic switch demonstrator.

for the functional blocks requiring high level of low speed processing, like the switch control. In this field several studies have been carried out and different architectures have been proposed and demonstrated particularly in Japan and in Europe, mainly in the RACE II ATMOS (ATM Optical Switching) [6], [7] and ACTS KEOPS (Keys to Photonic Switching) european projects.

The general scheme for a photonic ATM switch is presented in Fig. 7. Optical buffers needed to solve the cell contention have been based, up to now, on fibre delays loops, as optical technology has not yet been able to develop true large scale integration optical memories. Different architectures which minimise the complexity of these optical buffers (input, output, shared buffers, use of wavelengths as cell memory positions, according to the traffic figures and to the system performance required) have been studied and experimented. Packets are stored in fibre optical memories as a whole entity: therefore processing at bit or byte level is avoided, allowing the optical nodes to operate in a transparent way even at bit and coding scheme level. The optical node electronic control manages the cell buffering/routing function on the basis of information extracted from the packet headers.

The header rewriting, either at the node input or output, is needed for the VP/VC (Virtual Path/Channel) change according to the look-up tables (electrical memories) which are updated at the connection set-up and which also provide the switch routing information. The synchronous mode of operation of the photonic ATM switches requires the incoming cells on the n switch inputs to be aligned with the local time slot. Therefore synchronisers are needed at each input to perform this function that implies packet delineation, as well.

These significant experiments (i.e. Fig. 8) have verified the feasibility of optical ATM switching nodes and the great chance that the photonic technology offers to operate beyond the Terabit/s range throughput.

A lot of work is still required especially in the technology field, where photonic devices and integration techniques are still in their earlier phase. These experiments have also led to the conclusion that the ATM cell length is not well suited to operate at speeds beyond 2.5 Gbit/s per link, where optical technologies could be conveniently introduced. New packet formats are under study, to better exploit photonic technologies, together with new solutions for the optical packet networks. In this direction new concepts like optical *transparent* packet nodes and networks are now emerging [8] based on optical transparent packets of fixed duration (transparency) applies to the payload, while the packet header is maintained at a fixed bit rate, low enough to be processed by the electronic node control.

6. Conclusions

A modular and flexible architecture for high capacity ATM Switching has been presented. Starting from a reference electronic system, some technical solutions and architectures, mainly based on Multichip Modules and optical interconnections, are proposed in order to improve the global throughput of the system. We are witnessing the progressive introduction of photonics to face the challenge of the Terabit/s Switching Systems. Results of advanced projects on photonic switching have been reported and we can finally conclude that photonics and electronics need to operate in a complementary way, each exploiting its best performance. The ATM node architectures can take advantage from the extremely wide bandwidth and the availability of the wavelength domain of photonics, while electronics can still exploit its high level of functionality and integrability.

Acknowledgements

We wish to sincerely thank all the researchers both of Italtel and CSELT involved in the UT-FastNet ATM node project and all the colleagues involved in the ATM Switching System evolution and in the ATMOS-RACE and KEOPS-ACTS projects.

References

[1] G. P. Balboni, M. Collivignarelli, L. Licciardi, A. Paglialunga, G. Rigoglio, and F. Zizza, "From transport backbone to service platform: Facing the broadband switch evolution," Proc. ISS '95, Berlin, Germany, pp. 216–220, April 1995.

[2] M. Collivignarelli, A. Daniele, P. De Nicola, L. Licciardi, M. Turolla, and A. Zappalorto, "A complete set of VLSI circuits for ATM switching," Globecom 1994, San Francisco, USA, pp. 134–138, Nov. 1994.

[3] K. J. Ebeling, et al. "Recent advances in semiconductor vertical cavity lasers for optical communications and optical interconnects," ECOC'96, OSLO, pp. 281–287.

[4] L. Licciardi, M. Peretti, L. Pilati, J. J. Ichai, F. Martins, and P .Y. Urena, "A multichip module solution for high performance ATM switching," MCMC 1996, Santa Cruz, Feb. 1996.

[5] M. Burgard, "A flexible Tbit/s ATM switch with photonic switching," ISS '95, Berlin, pp. 442–446, April 1995.

[6] B. Bostica, P. Cinato, M. Calzavara, A. Cavaciuti, P. Gambini, M. Puleo, and E. Vezzoni, "Photonics application in switching: An ATM optical node," Fiber&Integrated Optics, vol. 13.2, pp. 185–198, 1994.

[7] F. Masetti, et al., "ATMOS (ATM Optical Switching): Results and conclusions of the RACE 2039 project," ECOC'95, Bruxelles, pp. 645–652, Sept. 1995.

[8] C. Guillemot, et al., "A two stage transparent packet switch architecture based on wavelength conversion," ECOC'95, pp. 765–768, Bruxelles, Sept. 1995.

High-Performance Backbone Network Technology edited by Naoaki Yamanaka
ISBN: 0-8247-5321-6 DOI: 10.1081/(120025749)
Published 2004 by Marcel Dekker, Inc.

A 10-Gb/s Optical Asynchronous Packet Receiver with a Fast Bit-Synchronization Circuit

Akio Tajima, Hiroaki Takahashi, Yoshiharu Maeno, Soichiro Araki, and Naoya Henmi

SUMMARY A novel 10-Gb/s fast acquisition bit-synchronization circuit for use in a Tb/s throughput optical packet switch has been developed. The circuit is a best-sampled-data-select type based on multiple phase-clocks, and it processes the asynchronous input packets into a synchronous data stream in a serial manner, which is advantageous in terms of circuit scale and consumption power compared with the parallel processing type. The circuit was developed using Si-bipolar ultrahigh-speed gate arrays and it was used to develop a 10-Gb/s optical asynchronous packet receiver module. The core logic of this circuit module required about 100 gates, consume 6 W, and the size of the module was reduced to only 170 mm (W) × 130 mm (D) × 10 mm (H). Using the receiver module, a fast acquisition time of 9 bits and receiver sensitivity penalty of less than 1.5 dB due to re-synchronization were measured.
key words: *optical packet switch, bit-synchronization, asynchronous packet receiver*

1. Introduction

A Tb/s-throughput asynchronous transfer mode (ATM) core switch and an internet protocol (IP) router are required for the next generation of broadband multimedia services [1]. The maximum throughput of conventional electrical switching fabrics still remains a few hundreds of Gb/s even with the state-of-the-art CMOS technology [2], [3]. At present, one of the most attractive candidate technologies for achieving a throughput of more than 1 Tb/s is to use an optical switching fabric having functionally equivalent to a full cross-bar switch. We have proposed and demonstrated more than a Tb/s-throughput wavelength/space division (WD/SD) cross-bar-equivalent optical packet-switching architecture that has 10-Gb/s throughput per port and can be scaled up to 256 × 256 [4], [5]. A conventional packet switching core is a synchronous system, which means input and output ports of a packet switching core operate under a global clock. To achieve a synchronous system, it is necessary to control clock and signal skews below one-bit time. However, as the signal and clock lines become more than 100 m in the 256 × 256 scale switch and wiring length equivalent to one-bit-time of 10-Gb/s throughput is a few cm, it is impossible to draw such a long line equally within such an accuracy [6]. Therefore, in such a large scale, high-speed switching fabric, asynchronous operation is needed. As a consequence, the bit phase of incoming optical packet becomes random packet-by-packet at the output port. Since downstream functional circuits of the output ports including buffering, data decoding and framing ones, must operate synchronously, in order to collaborate downstream functional circuits, it is necessary to resynchronize optical packets from the switch input port into a synchronized output data stream with a bit-synchronization circuit. A short acquisition time within 10 bits, low sensitivity penalty less than 4 dB, an operation at 10 Gb/s, and a high jitter rejection are required for the bit-synchronization circuit, however, to date there has not been any adequate technology suitable for applying to those systems [7]–[10].

We proposed a novel fast-acquisition bit-synchronization circuit configuration which can re-time incoming random-phase optical packets into a synchronized data stream. We developed the circuit using Si-bipolar ultrahigh-speed gate arrays and with these circuits we developed a 10-Gb/s optical asynchronous packet receiver module.

Section 2 describes the bit-synchronization circuit in a Tb/s optical packet switch. Based on a comparison of serial and parallel processing types, we select the serial type. Section 3 describes the configuration, the operation, and an actual implementation of our proposed best-sampled-data-select type bit-synchronization circuit. Section 4 evaluates the bit-synchronization circuit and the optical asynchronous packet receiver using this circuit.

2. Bit-Synchronization Circuit in a Tb/s Optical Packet Switch

Figure 1(a) shows a configuration of a Tb/s throughput optical packet switch [4], [5], which has 10-Gb/s throughput per port and a switch scale of 256 × 256. This switch consists of 256 pair input/output ports and optical switching fabric. Figure 1(b) shows a schematic diagram of a Tb/s throughput optical packet switch with an optical asynchronous packet receiver. In this architecture, the optical switching fabric has function-

Manuscript received January 7, 1999.
Manuscript revised March 10, 1999.
[†]The authors are with C&C Media Research Laboratories, NEC Corporation, Kawasaki-shi, 216-8555 Japan.
*This paper is also published in the IEICE Trans. Electron., Vol. E82-C, No.8, pp.1387–1392, August 1999.

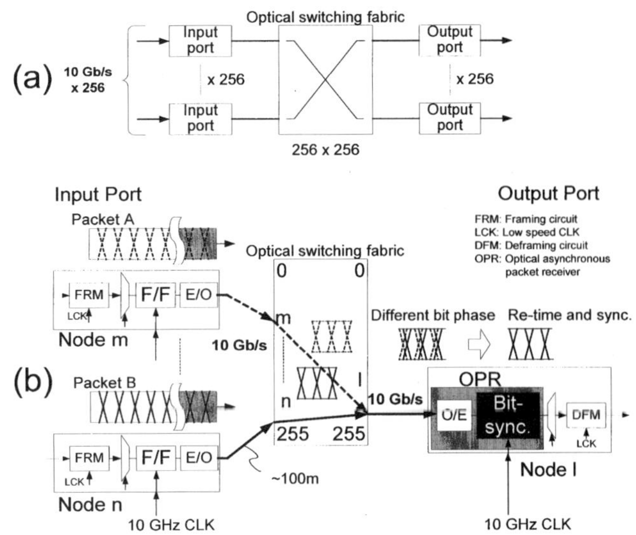

Fig. 1 (a) A Tb/s optical packet switch, (b) A Tb/s optical packet switch with an optical asynchronous packet receiver.

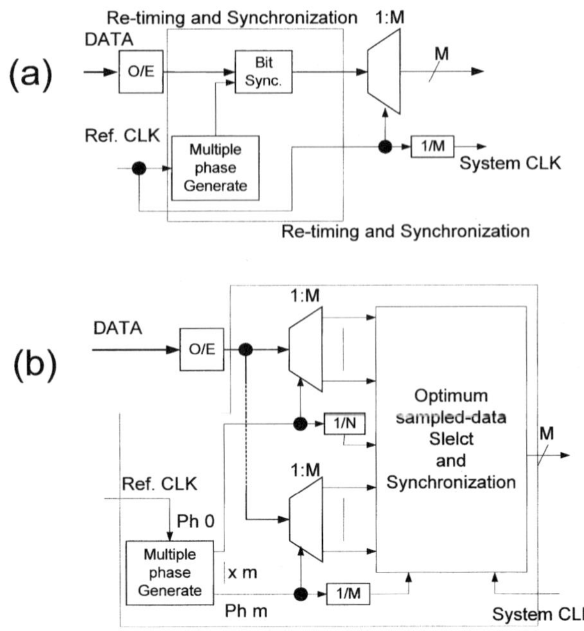

Fig. 2 (a) Serial and (b) parallel processing type bit-synchronization circuit.

ality equivalent to a full cross-bar circuit switch without input/output buffer, so it requires additional retiming function of the incoming packets, that arrive with random phases, into a synchronized output data stream by using the fast acquisition capability of a bit-synchronization circuit (Fig. 1(b)). There are two main approaches to the synchronization function. One is to process the asynchronous input data stream into a synchronous stream in a serial manner (at 10 Gb/s), and the other is to process the synchronization after demultiplexing (serial-to-parallel conversion) the incoming data to a lower speed (a few hundred Mb/s) [7] (Fig. 2). In the parallel processing, an ultra-low power CMOS technology is only the way to practically make it, but does not meet the high-speed operation with the state-of-the-art, the circuit scale and power consumption become large. We roughly estimate the circuit scale and power consumption of the serially processing to be around 1/5 of those for parallel processing. Therefore, in this paper, we concentrate on the design and evaluation of the serial approach.

3. Best-Sampled-Data-Select Type Bit-Synchronization Circuit

Retiming circuits based on a multiple phase clocks [8] are particularly well-suited for Tb/s throughput optical packet switches because of their short acquisition time, high jitter rejection, and operation at high frequencies. Especially, best-sampled-data-select type bit-synchronization circuit is suited for operation at high frequencies, because of its simple logic circuit configuration, less element and periodical layout. So, we designed and developed best-sampled-data-select type bit-synchronization circuit. There are three important issues to develop a 10-Gb/s class bit synchronization IC/module even if we utilize currently available state-of-the-art Si bipolar process and packaging technologies; (1) low sensitivity penalty (2) less power consumption and smaller circuit size, and (3) fast acquisition. To cope with (1) and (2) we optimized the number of clock phases, and to cope with (2) and (3), we devised a new circuit that can simplify the phase selection process.

3.1 Proposed Simplified Phase Selection Process

Figure 3 shows the circuit configuration in the case of m-phase clocks by using our proposed simplified phase selection process. The process is as follows: a) By the m-phase clocks with $360/m$-degree spacing, m-times-over-sampling is conducted. (over-sampling block) b) Phase comparators are used to monitor the difference between the phase of each clock and the leading edge of the data stream on a lag(0) or lead(1) basis (phase comparator block). c) The optimum data phase is determined at the boundary between lag and lead by the selection logic block. Figure 4 shows the operation flow of the best phase selection process in the case of $m = 4$. In this case, CLK 0 & 1 are in the lag-phase and CLK 2 & 3 are in the lead-phase, so CLK1 is selected as the optimized phase, and the output of Sampling 1 in the over-sampling block, which is driven by CLK1, is determined as the output data. (See decision point marked with arrows in Fig. 4).

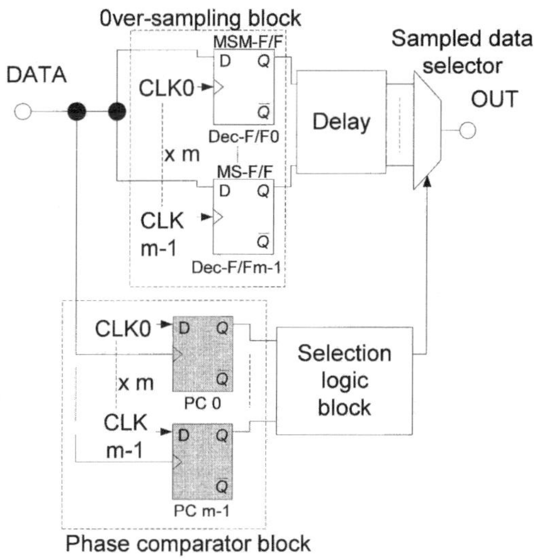

Fig. 3 Configuration of a m-times-over-sampling bit-synchronization circuit.

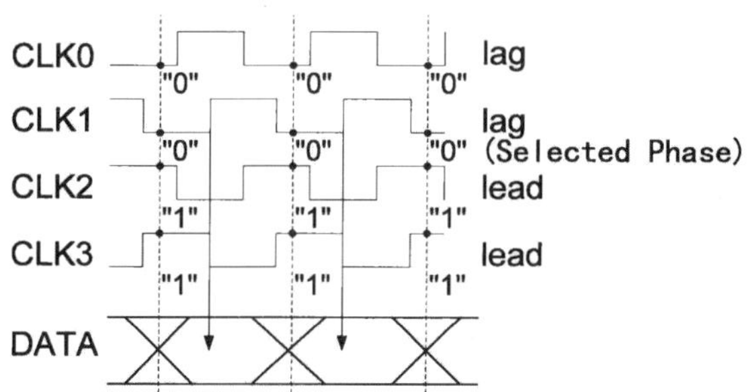

Fig. 4 Operation of selection logic block.

3.2 Number of the Clock Phases

A multi-phase clock type bit-synchronization circuit involves the following trade-off: a larger number of clock phases reduces the sensitivity penalty of the receiver, while a larger circuit scale increases the difficulty of operating at high frequencies. In order to determine the optimum number of clock phases, we measured receiver sensitivity vs. input clock timing for the decision circuit. Figure 5 shows the experimental setup. Figure 6 shows input clock timing vs. sensitivity penalty. Even optical data transmitted through a semiconductor optical amplifier gate (SOAG) [11], which is the switching element of our Tb/s switching fabric, there is a phase allowance of 90-degrees including a 20% margin which retains the sensitivity penalty within 1.5 dB (i.e. a minimum of four clock phases are necessary to keep the penalty within 1.5 dB).

3.3 Reduction of Module Size

As our estimation in Sect. 3.2 indicates that four phase

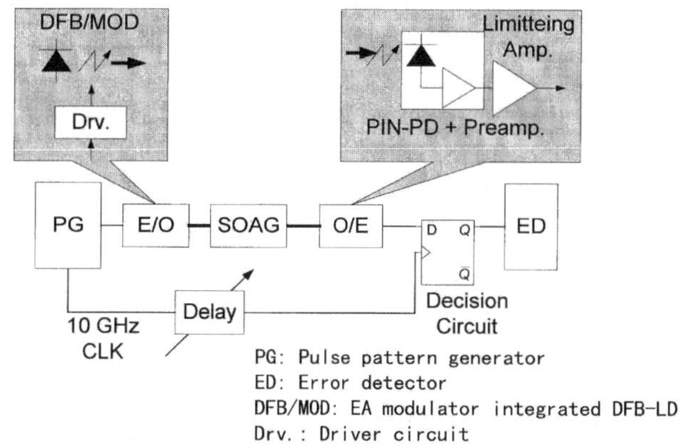

Fig. 5 Experimental setup for determining the optimum number of clock phases.

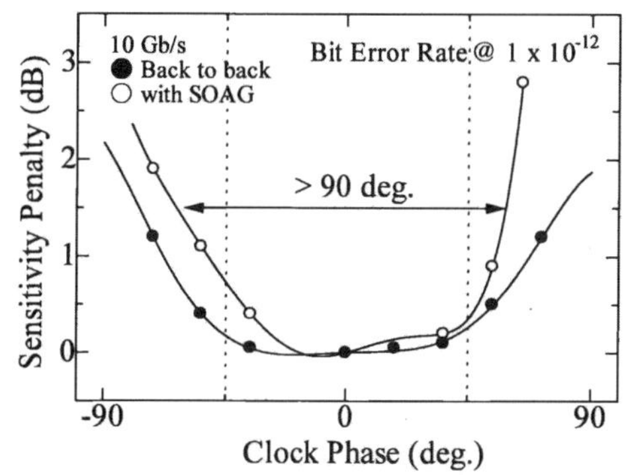

Fig. 6 Clock phase vs. sensitivity penalty.

clocks are sufficient to minify the sensitivity penalty, we designed and developed the circuit of four phase clocks. In the real implementation, we used the following techniques to reduce the module size. We used 1) the leading and trailing edges of CLK0 and CLK1 to produce four different phases in order to reduce the circuit size; 2) a simple D-F/F type phase comparator which decides the optimized data to get fast phase acquisition; 3) a combination of F/F output delay (180-degree) and delay lines (90-degree) to eliminate the buffering circuit element in order to synchronize the output data.

We have developed a bit-synchronization IC for 180-degree interval sampling using a Si-bipolar ultrahigh-speed gate array which cut-off frequency is 40 GHz. In this process, maximum operating frequency of a F/F is 13 GHz and a F/F has wide phase margin more than 300 degrees. About 50 gates were used in the implementation. By using two of these IC chips, we made a 4x-over-sampling module, whose size was reduced to only 170 mm (W) × 130 mm (D) × 10 mm (H) (See Fig. 7).

4. Experiments

To evaluate the functions of the circuit, the circuit mod-

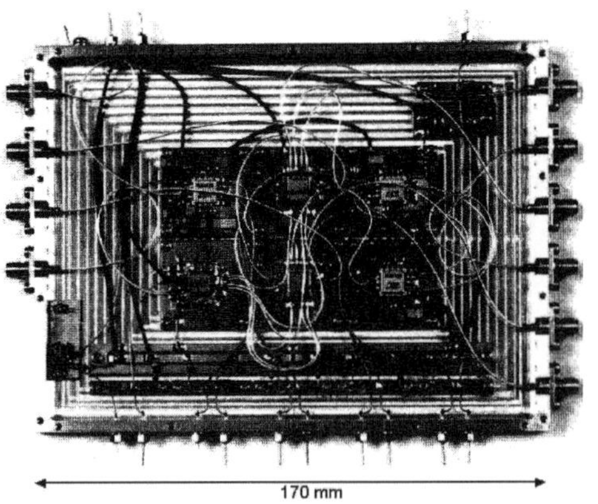

Fig. 7 Photograph of a 4x-over-sampling bit-synchronization circuit module.

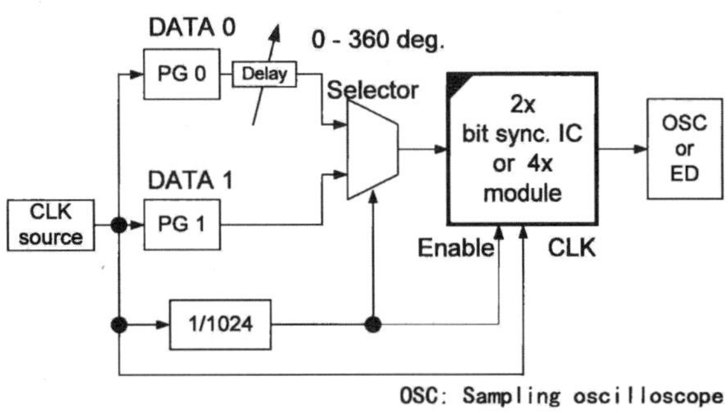

Fig. 8 Setup for acquisition experiments.

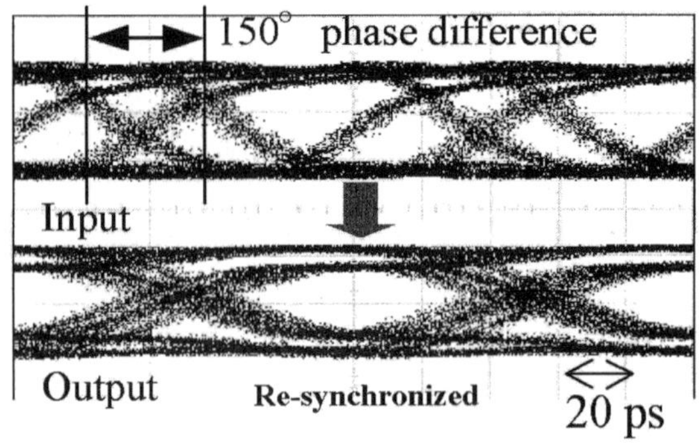

Fig. 9 Eye diagram.

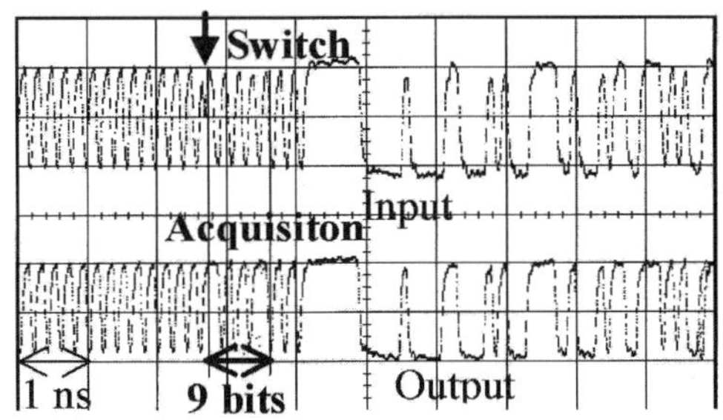

Fig. 10 Pattern synchronization waveform.

ule, and the characteristics of an optical receiver for use in a Tb/s optical packet switch, we carried out the following experiments.

4.1 IC Function Evaluation

The maximum operating speed of the developed IC was measured to be 11 Gb/s and rise/fall time was 40/35 ps. These are sufficient performance for 10-Gb/s operation. The consumption power of the IC was as small as 3 W. Figure 8 shows the experimental setup for demonstrating the bit-synchronization capability of the developed IC. Two sets of 10-Gb/s data, DATA0 and DATA1, were switched by an ultrahigh-speed selector IC, (maximum operating frequency of 18 GHz), and a 2x-over-sampling bit-synchronization IC received the switched packet. Figure 9 shows an eye-diagram of the 150-degree-phase-difference incoming signals and a synchronized eye-diagram of the out-going signal. The IC has an acquisition time of 1 bit and re-synchronized data was observed, even when the phase difference was as high as 150 degrees.

4.2 Evaluation of 10-Gb/s Optical Asynchronous Packet Receiver Characteristics

4.2.1 Evaluation of Re-Synchronization Function

Using the setup shown in Fig. 8, which is the same as that for IC evaluation, we carried out re-synchronization function experiments. Varying the phase difference between DATA0 and DATA1 from 0 to 360 degrees continuously, we measured the error through the 4x-over-sampling bit-synchronization circuit using an ANRITSU MP1762A error detector. We observed error-free operation at all phase differences and confirmed dead-band free operation.

4.2.2 Evaluation of Acquisition Time

Figure 10 shows the waveform of the input and output data of the 4x-over-sampling bit-synchronization circuit module using the setup shown in Fig. 8. The module has an acquisition time of 9 bits (Fig. 10).

4.2.3 Evaluation of Optical Burst Receiver Characteristics

In order to confirm the viability of the optical burst receiver module for a Tb/s optical packet switch, we measured the receiver sensitivity vs. input packet timing for

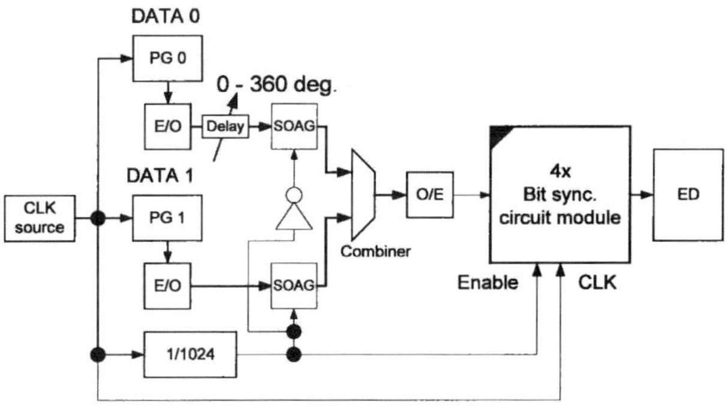

Fig. 11 Setup for optical packet switching experiments.

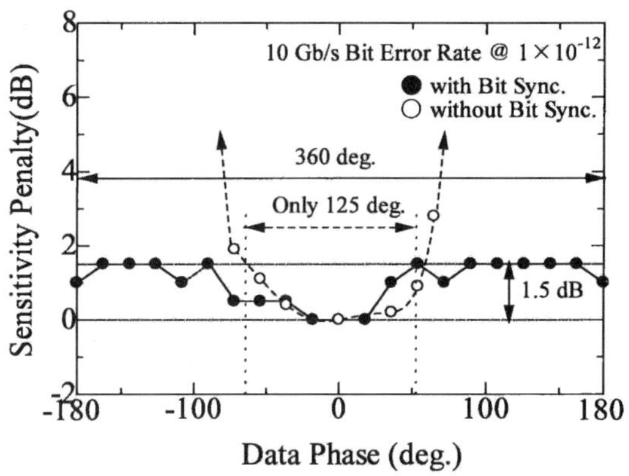

Fig. 12 Sensitivity penalty vs. packet phase.

burst data reception. Figure 11 shows the experimental setup. Optical packets were generated using two kinds of 1550-nm 10-Gb/s optical data, which were periodically switched by the semiconductor optical amplifier gates [11] used in our proposed optical cross-bar switching fabric. The optical receiver consists of a PIN-PD 2R module and a 4x-over-sampling bit-synchronization circuit module. Varying the phase difference between DATA0 and DATA1 from 0 to 360 degrees with variable optical delay line, we measured the sensitivity penalty to be less than 1.5 dB (Fig. 12).

5. Discussion

The core logic of this circuit module required about 100 gates and consume 6 W. Small scale and consumption power circuit was realized. By integrating these ICs in a one-chip IC using full-custom process, smaller size and less power consumption will be realized. The acquisition time of the bit-synchronization IC was 1 bit and that of the 4x-over-sampling bit-synchronization circuit module was 9 bits due to the delay in the cable interconnected ICs. Since an acquisition time of 9 bits corresponds to a few percent of an ATM cell size, however, these experiments show effective utilization of channel capacity.

We obtained sensitivity penalty of 1.5 dB. Without deviation, sensitivity penalty of 4x-over-sampling circuit is less than 1 dB, excess penalty of 0.5 dB is due to clock phase deviation in the circuit. From Fig. 6, penalty of 0.5 dB equivalent to 20% clock phase deviation in the circuit. This clock phase deviation is due to duty ratio variation and limited cell layout in the gate array. Since each F/F has wide phase margin more than 300 degrees, dead-band did not appear in spite of clock phase deviation. However, the sensitivity penalty of less than 1.5 dB is within the loss margin of around 4 dB for a 256×256 switch with port capacity of 10 Gb/s per port [4].

6. Conclusion

We have developed an optical asynchronous packet receiver with a 10-Gb/s fast acquisition bit-synchronization circuit. Through 10-Gb/s/port optical packet switching experiments, we found the acquisition time to be 9 bits and evaluated the sensitivity penalty to be less than 1.5 dB at any phase of the optical packets. These indicate the feasibility of scaling the size of the switch up to 256 × 256 with 10 Gb/s per port. The results show that our proposed technology will enable a Tb/s optical packet switch to be made in the near future.

References

[1] N. Henmi, "Towards Tb/s throughput optical switching network," Optoelectronics and Communication Conf. (OECC '98), 16A3-1, pp.424–425, 1998.

[2] N. McKeown, M. Izzard, A. Mekkittikul, W. Ellersick, and M. Horowitz, "The tiny tera: A packet switch core," IEEE Micro. pp.26–33, Jan./Feb. 1997.

[3] N. Yamanaka, S. Yasukawa, E. Oki, T. Kurimoto, T. Kawamura, and T. Matsumura, "OPTIMA: Tb/s ATM switching system architecture based on highly statistical optical WDM interconnection," International Switching Symposium (ISS '97), IS-02.8, pp.255–260, 1997.

[4] Y. Maeno, Y. Suemura, S. Kitamura, and N. Henmi, "10 Gbit/s/port and 256 ports optical crossbar switch with reduced hardware and enhanced modularity," European Conference on Optical Communication (IOOC-ECOC97), WE1D, pp.67–70, 1997.

[5] S. Araki, S. Takahashi, Y. Maeno, Y. Suemura, A. Tajima, H. Takahashi, K. Matsuda, T. Tamanuki, S. Dohmae, and N. Henmi, "A 2.56 Tb/s throughput packet/cell-based optical switch-fabric demonstrator," European Conference on Optical Communication (ECOC98), Post-Deadline Paper, 1998.

[6] Y. Maeno, A. Tajima, Y. Suemura, and N. Henmi, "8.5 Gbit/s/port synchronous optical packet-switch," Proc. Massively Parallel Processing Using Optical Interconnections (MPPOI'97), pp.114–119, June 1997.

[7] C.-K. Yang, R. Farjad-Rad, and M. Horowitz, "0.5-μm CMOS 4.0-Gbit/s serial link transceiver with data recovery using oversampling," IEEE J. Solid-State Circuits, vol.33, no.5, pp.713–722, 1998.

[8] R. Kakinuma, "A study on clock phase synchronization method for broadband user-network interface," IEICE Spring Conference, B-746, 1990.

[9] M. Banu and A. Dunlop, "A 660 Mb/s CMOS clock recovery circuit with instantaneous locking for NRZ data and

burst-mode transmission," IEEE International Solid-Sate Circuits Conference (ISSCC '93), TA6.4, 1993.

[10] Y. Otsuka, K. Hiraide, Y. Oikawa, and K. Murakami, "Bit synchronization circuit for high-speed switching systems," IEICE Trans., vol.J74-B-I, no.4, pp.304–312, April 1991.

[11] S. Kitamura, N. Kimura, K. Kurata, and M. Yamaguchi, "Low-switching-current SOA gate module with wide optical input dynamic range," Optical Amplifiers and Their Applications (OAA'97), TuC3-1, 1997.

High-Performance Backbone Network Technology edited by Naoaki Yamanaka
ISBN: 0-8247-5321-6 DOI: 10.1081/(120025750)
Published 2004 by Marcel Dekker, Inc.

Temperature Insensitive Micromachined GaAlAs/GaAs Vertical Cavity Wavelength Filter

Takeru Amano, Fumio Koyama, Nobuhiko Nishiyama, Akihiro Matsutani, and Kenichi Iga

SUMMARY A novel temperature insensitive wavelength filter consisting of GaAlAs/GaAs distributed Bragg reflectors (DBRs) has been demonstrated. This micromachined DBR is mechanically tuned by differential thermal expansion. The strain-induced displacement of one mirror can generate wavelength tuning and trimming functions with an adjustable temperature dependence. We succeeded in the control of temperature dependence in this micromachined semiconductor filter by properly designing a vertical cavity structure. The achieved temperature dependence was as small as +0.01 nm/K, which is one-tenth of that of conventional semiconductor based optical filters. Also, a wavelength trimming of over 20 nm was demonstrated after completing the device fabrication. In addition, we demonstrated a 4×4 multiple wavelength micromachined vertical cavity filter array. The multi-wavelength filter array with a wavelength span of 45 nm was fabricated by partially etching off a GaAs wavelength control layer loaded on the top surface of device.

key words: *wavelength tuning, micromachining, semiconductor DBR, wavelength trimming, vertical cavity filter, multiple wavelength*

1. Introduction

Temperature insensitive add/drop filters and wavelength tunable lasers are key devices for future optical networks [1]–[3]. The wavelength shift induced by temperature change is a problem of causing channel crosstalk for dense wavelength division multiplexing (WDM) networks, resulting in the need of precise temperature control of WDM devices. However, for use in low cost applications such as local area networks (LANs) and wide area networks (WANs), it is needed to avoid temperature controllers. Although there have been some reports on temperature insensitive waveguide filters based on athermal waveguides [4], there have been no reports on temperature insensitive semiconductor-based devices.

An add/drop filter and surface emitting laser (VCSEL) with micromachined semiconductor distributed Bragg reflectors (DBRs) can provide us various advantages including wide wavelength tuning range, polarization insensitive operation, two-dimensional array integration, and so on [5], [6]. GaAs based filters and VCSELs with a micromachined DBR have been demonstrated, which showed a large tuning wavelength range exceeding 40 nm. However, the temperature dependence of a resonant wavelength has remained as a problem. We previously proposed a temperature insensitive add/drop filter and VCSEL using a micromachined DBR mirror tuned by differential thermal expansion [7], [8]. We also demonstrated some preliminary results of temperature insensitive operation as well as wavelength trimming [9], [10].

In this paper, we present the design and some experiments of micromachined temperature insensitive filters. We realized a temperature dependence of less than +0.01 nm/K and wavelength trimming of over 20 nm. Also, we demonstrated a possibility of a large scale 2-dimensional multiple wavelength filter array based on the same principle. We realized a 2×2 monolithically integrated filter array with a wavelength span of 12 nm.

2. Design of Temperature Insensitive Filter

Figure 1 shows the schematic structure of the proposed filter. The upper GaAlAs/GaAs distributed Bragg reflector (DBR) is suspended above the substrate. An air gap is formed between the upper and bottom GaAlAs/GaAs DBRs. In the micromachined filter, when voltage is applied at the air gap, the electrostatic attraction causes a bending of the cantilever toward the substrate side, which enables wide wavelength tuning. Our structure is similar to that devel-

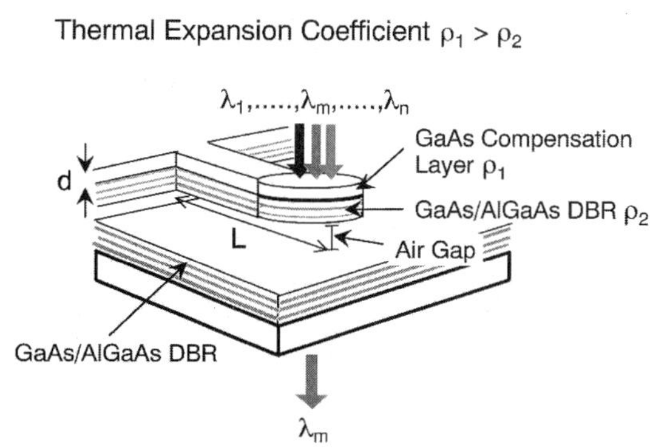

Fig. 1 Schematic structure of micromachined vertical cavity filter.

Manuscript received October 5, 2000.
Manuscript revised January 9, 2001.

[†]The authors are with Microsystem Research Center Precision & Intelligence Laboratory, Tokyo Institute of Technology, Yokohama-shi, 226-8503 Japan.

*This paper is also published in the IEICE Trans. Electron., Vol.E84-C, No.5, pp.678–684, May 2001.

IEICE TRANS. COMMUN., Vol. E84-B, No. 5, pp. 1304–1310, May 2001

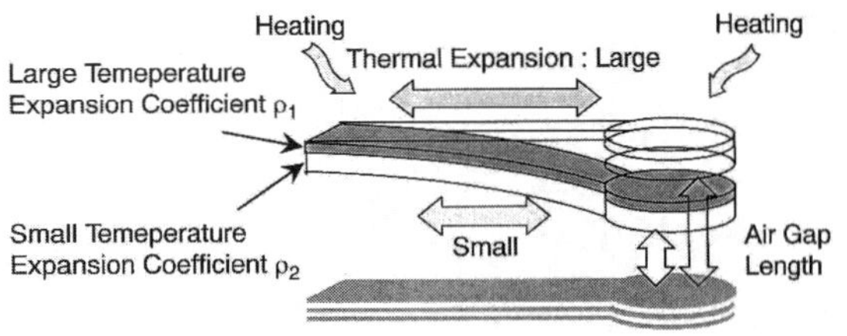

Fig. 2 Principle of temperature insensitive operation.

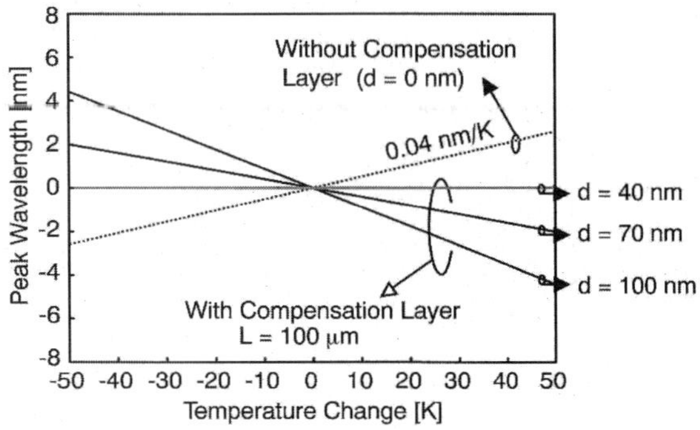

Fig. 3 Temperature dependence of filter wavelength for the thickness of GaAs temperature compensation layer.

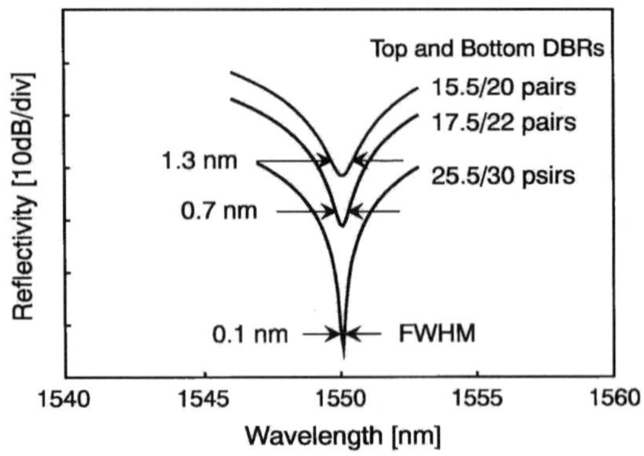

Fig. 4 Calculated filter performance of micromachined GaAlAs/GaAs vertical cavity filter.

oped for tunable wavelength VCSELs with a micromachined cantilever [5], [6]. A difference in our device is one additional GaAs layer on the upper DBR with a larger thermal expansion coefficient ρ_1 than the average one ρ_2 of the GaAlAs/GaAs DBR. Figure 2 shows the principle of temperature insensitive operation. Temperature increase causes the displacement of the DBR toward the substrate side, resulting in the blue shift of the resonant wavelength. This means the compensation of the increase of transmission wavelength caused by the temperature increase. The amount of compensation can be controlled by some structural parameters, which are the arm length L, the difference in thermal expansion coefficient, and the thickness of the GaAs compensation layer d. We used a simplified two layer model where the bottom layer is the GaAlAs/GaAs DBR with an average thermal expansion coefficient ρ_2 smaller than that of the GaAs compensation layer [8]. We calculate the mechanical displacement of the top DBR, which is induced by thermal strain. We obtain the resonant wavelength of a micromachined filter by using a standard matrix formula, including the mechanical displacement. A temperature insensitive tunable filter can be realized by controlling the thickness of the GaAs compensation layer d and the arm length L.

Figure 3 shows the calculated temperature dependence of filter wavelength as a function of the thickness of GaAs temperature compensation layer. The cantilever length is $100\,\mu$m and the air gap length is $1.55\,\mu$m. The top DBR is 15.5-pairs. When the thickness of the GaAs compensation layer is 40 nm, we can obtain temperature insensitive operations. Also, when the thickness of the GaAs compensation layer is 100 nm, a negative temperature dependence of over -0.08 nm/K can be obtained.

The calculated reflection spectra of a Fabry-Perot filter is discussed. Figure 4 shows the filter performance of a micromachined GaAlAs/GaAs vertical cavity filter. The $1.55\,\mu$m air gap is inserted between the upper and lower DBRs. Figure 4 shows the reflection spectra for three different cavity designs with DBR pairs of 15.5/20 pairs, 17.5/22 pairs and 25.5/30 pairs, respectively. When DBRs pairs are these three cases, the full width at a half of the maximum (FWHM) is 1.3 nm, 0.7 nm and 0.1 nm, respectively. We can expect a narrow band optical filter by increasing the number of pairs of DBRs.

3. Fabrication of Temperature Insensitive Micromachined Filter

We fabricated a cantilever structure by using selective oxidation of GaAlAs followed by selective wet etching and a drying process. The structure was grown on a GaAs (100) substrate by low-pressure metal-organic chemical vapor deposition. A $1.56\,\mu$m thick $Ga_{0.02}Al_{0.98}As$ spacer layer was inserted between the upper and bottom DBRs. The upper and bottom DBRs consist of 15.5-pair and 20-pair $Ga_{0.2}Al_{0.8}As$/GaAs, respectively. A top GaAs compensation layer is loaded on the upper DBR. We fabricated a mesa structure by using inductively coupled plasma (ICP) dry etching [11]. Then we carried out a wet oxidation process at 480°C to oxidize the $Ga_{0.02}Al_{0.98}As$ layer for 90 min. The oxidized $Ga_{0.02}$ $Al_{0.98}As$ (Al_xO_y) spacer layer was removed by buffered HF to form a cantilever structure [12]. After removing the oxidized layer, we dried up this filter with boiled acetone. When this filter is dried with water, the top DBR is stuck with the bottom DBR due to surface tension. Thus, we can avoid a sticking problem by drying process with boiled acetone.

The SEM photograph of the fabricated filter is shown in Fig. 5. Using selective wet oxidation followed

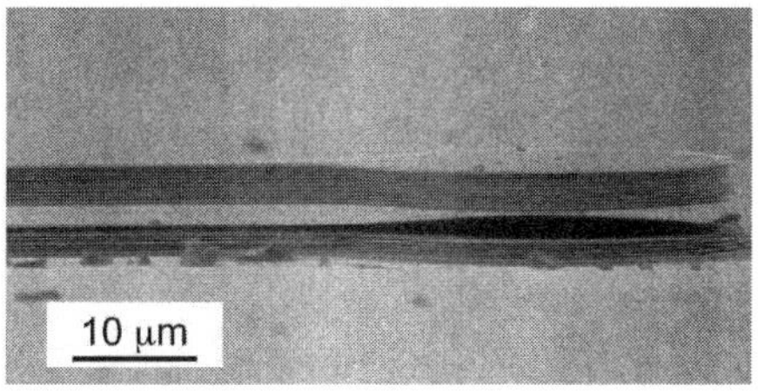

Fig. 5 SEM image of fabricated GaAlAs/GaAs filter.

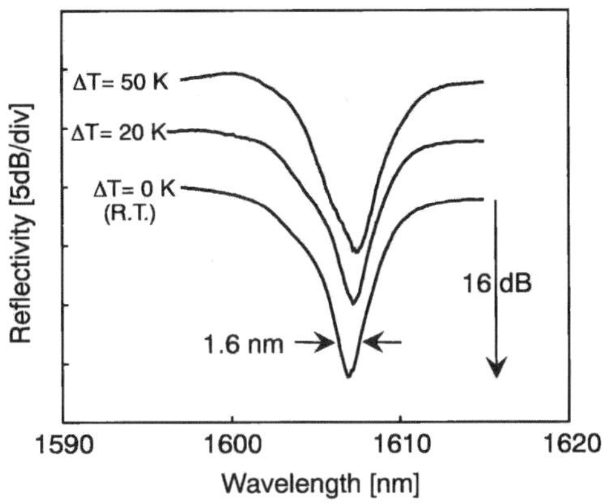

Fig. 6 Reflection spectra for various temperatures.

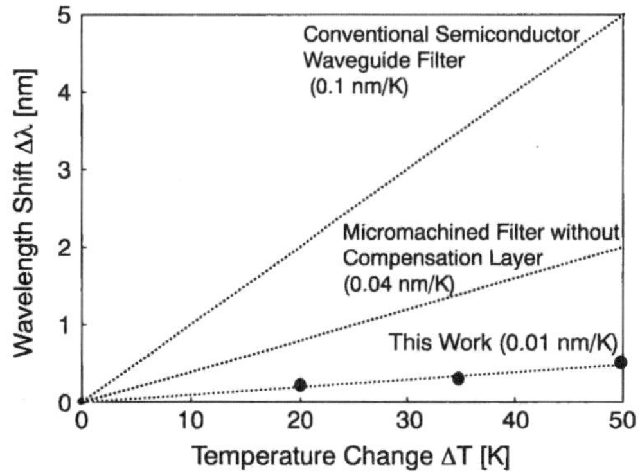

Fig. 7 Temperature dependence of resonant wavelength. The closed circles show the experimental data.

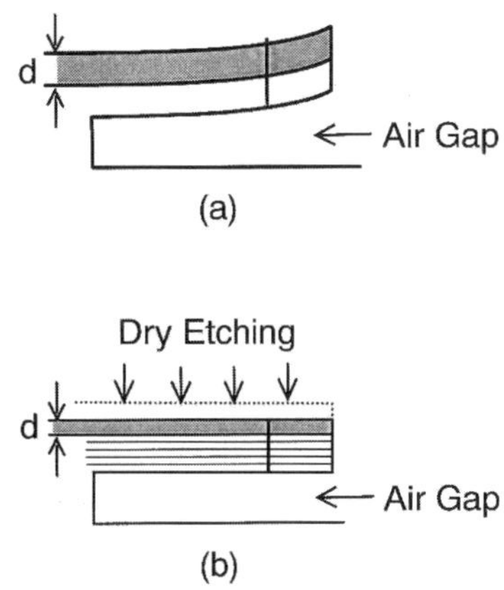

Fig. 8 Principle of wavelength trimming technique. (a) Before etching. (b) After etching.

by wet selective etching process, we can get a flat surface of an exposed GaAs surface because of a large difference in etch rate by buffered HF between Al_xO_y and GaAs. The arm length of a cantilever is 96 μm and the thickness of the top GaAs compensation layer is 40 nm. These numbers are chosen from the modeling of a structure to give a temperature insensitive operation.

4. Characteristic of Temperature Insensitive Filter

We measured the temperature dependence of the resonant wavelength of a fabricated filter. The spectra of reflectivity were measured by using broadband amplifier spontaneous emission from an Er-doped fiber amplifier coupled to the device through a lensed fiber under various temperatures. Figure 6 shows the reflection spectra for various temperatures. The reflection dip for temperature change $\Delta T = 0$ K, 20 K, and 50 K is 1607 nm, 1607.2 nm, and 1607.5 nm, respectively. Thus the temperature dependence was as small as +0.01 nm/K. The depth of the dip and the full width at a half of maximum (FWHM) are 16 dB and ~1.6 nm, respectively. These are in good agreement with the expected value from the theoretical calculation. The filter bandwidth can be reduced to 0.1 nm by increasing the number of the pair of DBRs as shown previously [12]. Figure 7 shows the wavelength shift versus temperature change. We observed a small red shift of +0.01 nm/K with increasing the temperature, while conventional semiconductor waveguide filters typically show a red shift of ~+0.1 nm/K [13]. The calculated temperature dependence of the micromachined filter without the compensation layer is +0.04 nm/K as shown in Fig. 7. Also, we have demonstrated a negative temperature coefficient (−0.44 nm/K) in the filter with compensation layer thickness of 500 nm in our previous report [9]. When the compensation layer thickness d is 40 nm, the filter should be temperature insensitive. However, the experimental result shows a small remaining temperature dependence which would be due to the deviation of GaAs layer thickness, cantilever length and DBR thickness from the designed structure.

5. Wavelength Trimming

In our filter we can adjust the filter wavelength even after completing the device, which is so-called wavelength trimming technology. The precise wavelength trimming is desirable for use in dense WDM applications. Figure 8 shows the concept of the proposed scheme using a strain for wavelength trimming. The arm length of the cantilever is 96 μm and the thickness of the top compensation GaAs layer is 500 nm. By etching off the GaAs compensation layer, the fil-

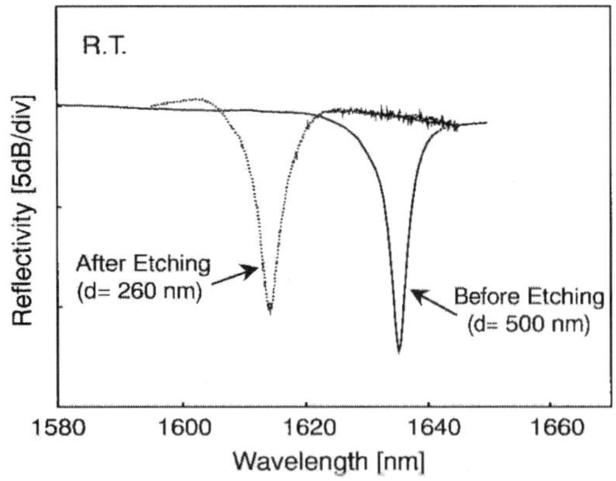

Fig. 9 Reflection spectra before and after wavelength trimming.

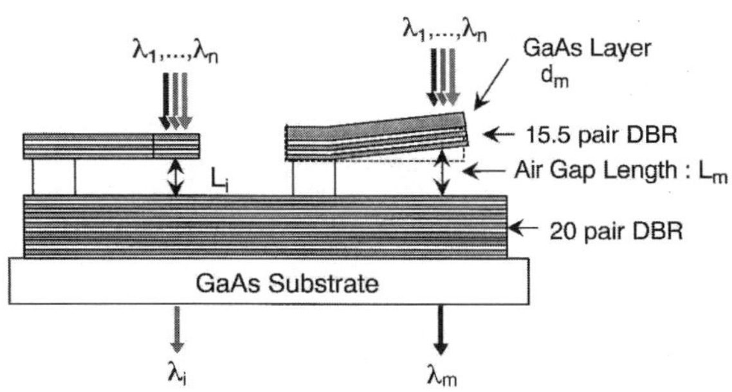

Fig. 10 Principle of multi-wavelength micromachined vertical cavity filter array.

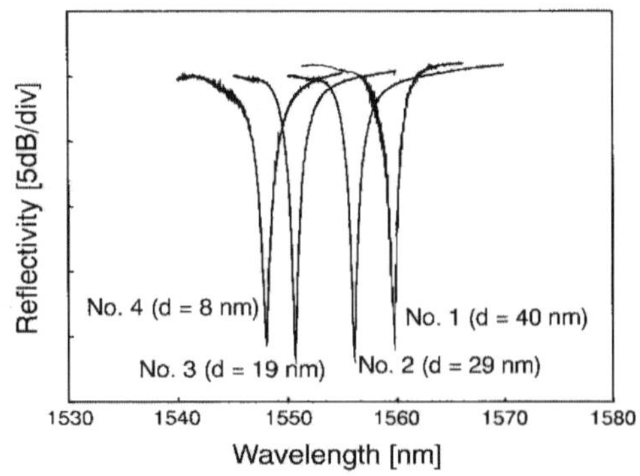

Fig. 11 Reflection spectra of the 2×2 multi-wavelength micromachined vertical cavity filter array.

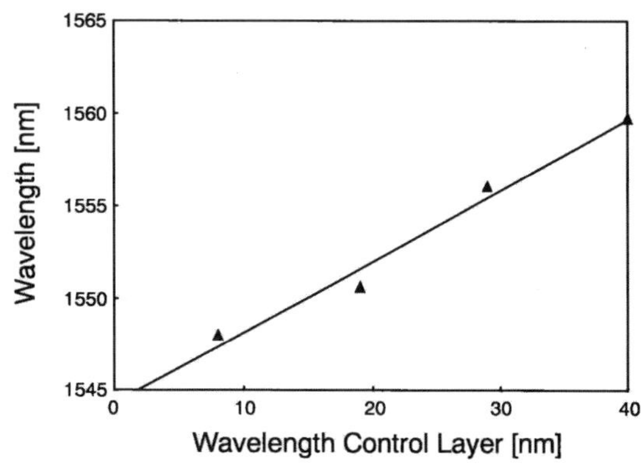

Fig. 12 Resonant wavelength as a function of the thickness of the GaAs wavelength control layer. Closed circles show experimental data.

ter wavelength can be decreased. We carried out ICP etching for an entire surface of the fabricated filter. The thickness of an etched layer was 240 nm, thus, the thickness of the remaining compensation layer was 260 nm. Figure 9 shows the reflection spectra before and after etching. We observed a large blue shift of 21 nm which shows a potential of wavelength trimming with a large wavelength span. However, we cannot keep insensitive operation with wavelength trimming because of changing the thickness of the GaAs layer. However we can realize wavelength trimming technique in temperature insensitive filters if we control both lattice constant and thermal expansion independently.

6. Multi-Wavelength Filter

We can integrate multi-wavelength filters using the same idea of wavelength trimming. Adjusting the thickness of the top GaAs layer can precisely control the wavelength of the filter. Figure 10 shows the conceptual diagram for the multi-wavelength filter array. The air-gap width L_m should be longer when the wavelength control layer d_m is loaded because of using lattice strain. The filter wavelengths λ_i and λ_m are determined by the thickness of the wavelength control layer d_m. By adjusting the thickness of the wavelength control layer by etching, a multiple wavelength vertical cavity filter array integrated on one wafer can be realized. Before we fabricate a cantilever structure, we prepared four different thicknesses of the top GaAs layer by applying two photolithography and etching steps. The thickness of the top GaAs layer is either 8 nm, 19 nm, 29 nm or 40 nm.

We measured the reflection spectra of a fabricated filter array. Figure 11 shows the reflection spectra of the 2×2 multi-wavelength micromachined vertical cavity filter array. In this structure, the top and bottom DBR is 15.5/20 pair. The resonant wavelength of the 2×2 multi-wavelength micromachined vertical cavity filter array is 1559.8 nm, 1556.1 nm, 1550.7 nm or 1548.0 nm. The depth of the dip and the full-width at half maximum (FWHM) are ~20 dB and ~0.4 nm, respectively. Figure 12 shows the resonant wavelength as a function of the thickness of the top GaAs layer. By etching off the top GaAs layer, the filter wavelength can be decreased. The $\Delta\lambda/\Delta d$ coefficient is +0.35. This is in good agreement with the expected value from the theoretical calculation represented by the solid line. When the thickness of the top GaAs layer is increased, we can increase the wavelength spacing, resulting in a larger-scale 2D filter array. We can easily extend the dimension of the array. We fabricated 4×4 multiple wave-

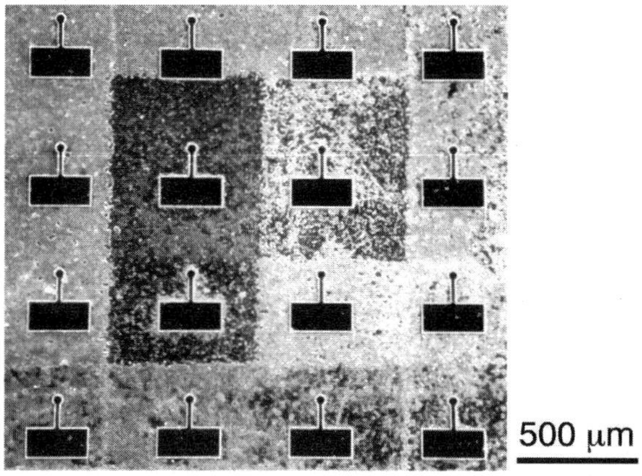

Fig. 13 SEM image of the fabricated 4 × 4 multi-wavelength micromachined vertical cavity filter array.

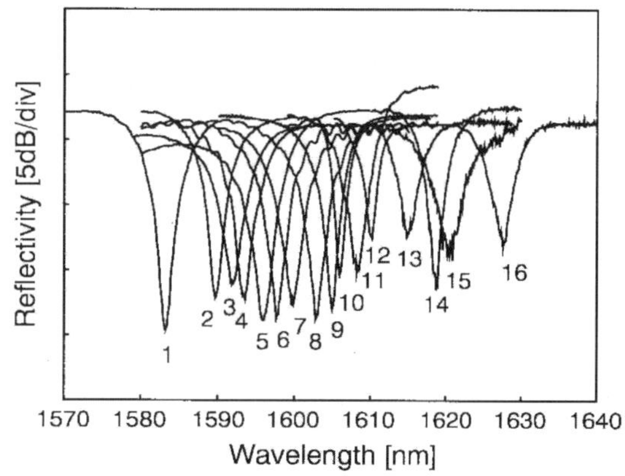

Fig. 14 Reflection spectra of the 4 × 4 multi-wavelength filter array.

length filter array as shown in Fig. 13. We used 4 step lithography and etching before the fabrication of an air-gap structure. The thickness of the top GaAs layer is from 8 nm to 40 nm. The length of micromachined cantilever is 130 μm. Figure 14 shows the filter response of a 4 × 4 filter array. In this structure, top and bottom DBR is 15.5/20 pair. Figure 15 shows the resonant wavelength as a function of the etching depth of the top GaAs layer. We could realize the multi-wavelength filter array with a wavelength span of 45 nm. The experimental result is in good agreement with the expected value from the theoretical calculation shown by the solid line. By improving the accuracy of etching process in the wavelength control layer, we may expect a precisely controlled multi-wavelength filter array with narrow channel spacing.

7. Conclusion

We presented the design and experiment of a GaAs based-micromachined filter. We have demonstrated a temperature insensitive micromachined semiconductor filter by using differential thermal expansion. We

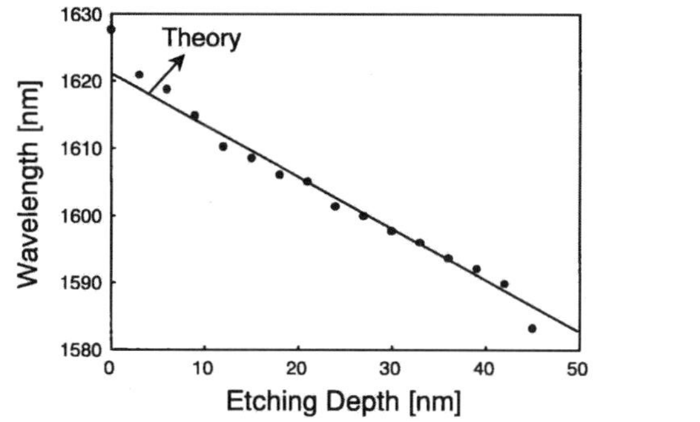

Fig. 15 Resonant wavelength as a function of the etching depth of the GaAs wavelength control layer in 4 × 4 multiple wavelength filter array. Closed circles show experimental data.

have achieved a temperature dependence as small as +0.01 nm/K in a fabricated filter. Also, wavelength trimming of over 20 nm was demonstrated after completing the device fabrication. The obtained result shows a possibility of a compact and temperature insensitive tunable filter. In addition, we realized a monolithically integrated multi-wavelength filter array with a wavelength span of over 40 nm. The technology presented in this paper may enable us to realize temperature insensitive multi-wavelength VCSELs.

Acknowledgment

The authors would like to thank S. Shinada and M. Arai for their help in experiments. This study was support by Grant-in-Aid for COE Research from the Ministry of Education, Science, Sport and Culture (#07CE2003, "Ultra-parallel Optoelectronics").

References

[1] H. Takahashi, O. Ishida, K. Oda, and H. Toba, "Anticrosstalk arrayed-waveguide add-drop multiplexer with foldback paths for penalty free transmission," Electron. Lett., vol.30, no.24, pp.2053–2055, 1994.

[2] T. Morioka, H. Takara, S. Kawanishi, O. Kamatani, K. Takiguchi, K. Uchiyama, M. Saruwatari, H. Takahashi, M. Yamada, T. Kanamori, and H. Ono, "1 Tbit/s (100 Gbit/s × 10 channel) OTDM/WDM transmission using a single supercontinuum WDM source," Electron. Lett., vol.32, no.10, pp.906–907, 1996.

[3] K. Kato, A. Okada, Y. Sakai, K. Noguchi, T. Sakamoto, S. Suzuki, A. Takahara, S. Kamei, A. Kaneko, and M. Matsuoka, "32 × 32 full-mesh (1024 path) wavelength-routing WDM network based on uniform-loss cyclic-frequency arrayed-waveguide grating," Electron. Lett., vol.36, no.15, pp.1294–1296, 2000.

[4] Y. Kokubun, S. Yoneda, and H. Tanaka, "Temperature-independent narrowband optical filter at 1.3 μm wavelength by an athermal waveguide," Electron. Lett., vol.32, no.21, pp.1998–2000, 1996.

[5] M.C. Larson, B. Pezeshki, and J.S. Harris, Jr., "Vertical coupled-cavity micro interferometer on GaAs with deformable membarance top mirror," IEEE Photon. Technol. Lett., vol.7, no.4, pp.382–384, 1995.

[6] M.S. Wu, E.C. Vail, G.S. Li, W. Yuen, and C.J. Chang-Hasnain, "GaAs micromachined widely tunable Fabry-Perot filter," Electron. Lett., vol.31, no.19, pp.1671–1672, 1995.

[7] F. Koyama and K. Iga, "Wavelength stabilization and trimming technologies for vertical-cavity surface emitting laser arrays," OSA Technical Digest Series of Quantum Optoelectronics, vol.9, QTh-14, pp.90–92, 1997.

[8] F. Koyama, T. Amano, N. Furukawa, N. Nishiyama, M. Arai, and K. Iga, "Micromachined semiconductor vertical cavity for temperature insensitive surface emitting lasers and optical filters," Jpn. J. Appl. Phys., vol.39, no.3B, pp.1542–1545, 2000.

[9] T. Amano, F. Koyama, N. Furukawa, N. Nishiyama, A. Mastutani, and K. Iga, "Micromachined AlGaAs/GaAs vertical cavity filter with adjustable temperature dependence and wavelength trimming ability," Electron. Lett., vol.36, no.1, pp.74–76, 2000.

[10] T. Amano, F. Koyama, N. Nishiyama, and K. Iga, "Temperature-insensitive micromachined AlGaAs-GaAs vertical cavity filter," IEEE Photon. Technol. Lett., vol.12, no.5, pp.510–512, 2000.

[11] A. Matsutani, H. Ohtuki, F. Koyama, and K. Iga, "Vertical and smooth etching of InP by Cl2/Xe inductively coupled plasma," Jpn. J. Appl. Phys., vol.38, no.7A, pp.4260–4261, 1999.

[12] N. Furukawa, F. Koyama, N. Nishiyama, M. Arai, T. Miyamoto, and K. Iga, "Micromachined tunable GaAlAs/GaAs DBR for temperature insensitive vertical cavity add/drop filter," Microoptics Conference, MD2, pp.70–73, 1999.

[13] E. Gini, W. Hunziker, and H. Melchior, "Polarization independent InP WDM multiplexer/demultiplexer module," IEEE J. Lightwave Technol., vol.16, no.4, pp.625–630, 1998.

High-Performance Backbone Network Technology edited by Naoaki Yamanaka
ISBN: 0-8247-5321-6 DOI: 10.1081/(120025751)
Published 2004 by Marcel Dekker, Inc.

Para BIT: Parallel Optical Interconnection for Large-Capacity ATM Switching Systems

Kosuke Katsura, Yasuhiro Ando, Mitsuo Usui, Akira Ohki, Nobuo Sato, Nobuaki Matsuura, Nobuyuki Tanaka, Toshiaki Kagawa, and Makoto Hikita

SUMMARY We have been working on a project called ParaBIT (for parallel inter-board optical interconnection technology) to achieve large-capacity switching systems. The ParaBIT module being developed as the first step in this project is a front-end module with 40 channels providing throughput of 28 Gb/s, cost-effectiveness and compactness. To realize the module, this project has developed five novel technologies: (1) 850-nm 10-ch Vertical-cavity Surface-emitting laser (VCSEL) arrays as very cost-effective light sources, (2) new high-density multiport bare fiber connectors that do not need a ferrule and spring, (3) passive optical alignment using polymeric optical waveguide film with a 45-degree mirror for coupling to the optical array chips and the waveguide, (4) transferred multichip bonding to mount optical array chips on a substrate with a positioning error of only a few micrometers, and (5) simple electronic circuits with a fixed-decision-level receiver and an APC-less transmitter, and low power consumption. Experimental results show that the design targets of throughput of 700 Mb/s per channel and a compact and cost-effectiveness structure were met. Thus, ParaBIT is a promising technology for large-capacity switching systems.

key words: *optical interconnection, ATM switch, VCSEL, optical connector, optical waveguide, multichip*

1. Introduction

The inevitable expansion in the use of information services and the Internet in homes and offices will cause an explosive increase in data transmitted over networks after the year 2000. This will prompt investigation and development of sub-Tb/s- and 1-Tb/s-class large-capacity asynchronous transfer mode (ATM) switching systems as an infrastructure to sustain such enormous data traffic [1]. In these systems, the throughput per board is expected to be 80–160 Gb/s. It is clear that electrical inter-board and inter-cabinet interconnections will surely be a system performance bottleneck, preventing an increase in the throughput of LSIs and packaging boards. To overcome this bottleneck, optical interconnections have been introduced into the input/output transmission lines of boards, replacing metal cables. It certainly offers a way to extract the maximum performance from systems.

Practically, on the other hand, optical modules are much more expensive than electrical ones because of difficulty of optical alignments. This problem causes a barrier to adopt optical interconnection in the systems, even understanding the above interconnection crisis. Putting it to practical use for switching systems will require cost-effectiveness as well as high throughput.

NTT is currently conducting a project called ParaBIT (for parallel inter-board optical interconnection technology) to achieve large-capacity switching systems after the year 2000. It aims to widely apply optical interconnection for board input/output in sub-Tb/s- and 1-Tb/s-class ATM switching systems, in massively parallel computers, and in information processing systems such as high-resolution image processing apparatus. A prototype ParaBIT module has been developed; it is a front-end module with 28-Gb/s throughput (700 Mb/s×40 ch) [2]–[6].

In this paper, we describe the development concept for ParaBIT, the major technologies used for the prototype ParaBIT module, and its performance.

2. Concept and Specifications

At first, cost effectiveness was the most important target of ParaBIT. A low cost is essential in order to apply optical interconnection even in the case of large-capacity switching systems. We aimed to use cost-effective devices and develop packaging technologies that would allow the economies of mass production, because packaging accounts for most of the total module cost in optical module fabrication.

The technical target of ParaBIT is to achieve high throughput, and high-density optical interconnection, in a word, to raise the throughput per unit volume. We chose a technical direction of increasing the channel number to be compatible with low consumption power. To set the target specifications for parallel transmission performance, we considered:

(1) Transmission speed per channel

To decide the transmission speed per channel, we assumed a synchronous digital hierarchy (SDH) interface

Manuscript received October 7, 1998.

[†]The authors are with NTT Opto-electronics Laboratories, Musashino-shi, 180-8585 Japan.

[††]The author is with NTT Opto-electronics Laboratories, Atsugi-shi, 243-0122 Japan.

[†††]The author is with NTT Opto-electronics Laboratories, Ibaraki-ken, 319-1193 Japan.

*This paper is also published in IEICE Trans. Electron., Vol.E82-C, No.2, pp.360–369, February 1999.

IEICE TRANS. COMMUN., Vol. E82-B, No. 2, pp. 412–421, February 1999

Table 1 Design targets for ParaBIT module.

Number of channels	40 (or 20 for Tx and 20 for Rx)
Data rate per channel	DC - 700 Mb/s
Power dissipation	200 mW/ch
Electrical interface	Differential ECL
Power supply voltage	-4.5 V
Module dimensions	48 (L)×47.2 (W)×8.5 (H) mm
Transmission length	> 100 m
Bit-error rate	10^{-19}

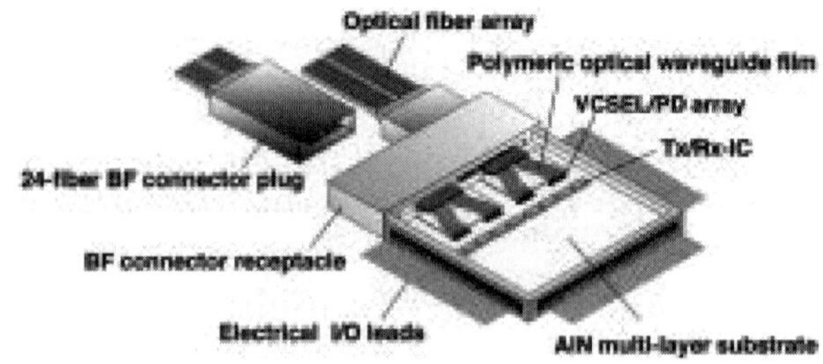

Fig. 1 Structure of prototype ParaBIT module.

and various conditions for using the ParaBIT module. For the prototype ParaBIT module, we chose a transmission speed of 700 Mb/s per channel, which can transmit signals at 622 Mb/s (STM-4) for the SDH transmission interface with some additional header information for routing in the switching LSI, and can be handled by the printed circuit boards on which we intended to mount the ParaBIT module. Furthermore, in anticipation of widespread use of optical interconnection, we decided that it should be able to operate from DC so as to handle signals consisting of an infinite number of consecutive 0's or 1's.

(2) Number of parallel channels
To make the best use of parallel interconnection, we considered the number of channels for directly transmitting parallel signals with various data widths of the switching LSI, such as 8, 16, or 32 bits. In the ParaBIT module, we chose a 40-channel configuration to be able to transmit parallel 32-wide data plus some frame synchronizing signals.

Thus, total throughput of 28 Gb/s (700 Mb/s×40 ch) was selected as a target of development. The design targets of ParaBIT are summarized in Table 1. The target transmission distance was more than 100 m. The bit error rate was set to less than 10^{-19} to establish error-free transmission for 10 years, which is calculated from total throughput of 28 Gb/s.

In order to achieve the above target, five key technologies were developed:

(1) 850-nm 10-ch vertical-cavity surface-emitting laser (VCSEL) arrays on p-type GaAs substrate, (2) a novel high-density multiport bare fiber (BF) connector, (3) passive optical alignment using a polymeric optical waveguide film, (4) transferred multichip bonding, and (5) simple electronic circuits with a fixed-decision-level receiver and an APC-less transmitter that have low power consumption.

In the next section, we describe the module configuration and the details of five key technologies developed for the prototype ParaBIT module.

3. Technologies

3.1 Module Configuration

Figure 1 illustrates the structure of the prototype ParaBIT module, and Fig. 2 is a block diagram of the prototype ParaBIT transmitter (Tx) and receiver (Rx) modules. To handle 40 channels within one module, the Tx module consists of eight 5-ch Tx-ICs, four 10-ch VCSEL arrays, two 24-ch polymeric optical waveguide films, and a BF connector receptacle on an AlN multilayer substrate which was selected for its low thermal resistance. Similarly, the Rx module consists of eight 5-ch Rx-ICs, four 10-ch PD arrays, two 24-ch polymeric optical waveguide films, and a BF connector receptacle on an AlN multilayer substrate. The module is composed of two blocks for 20-ch operation, so a 20-ch transceiver module (20-ch Tx and 20-ch Rx) is also available.

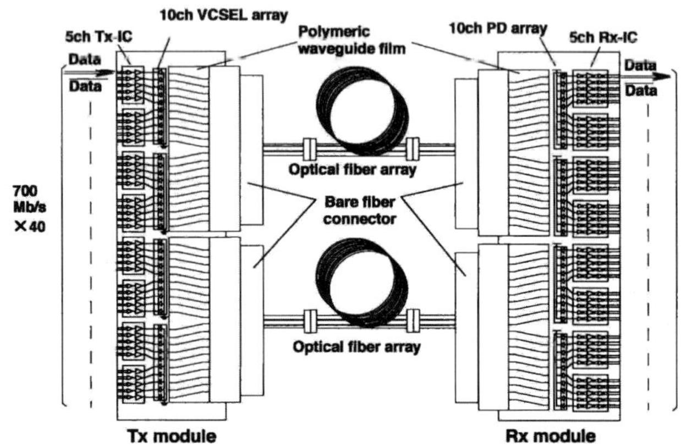

Fig. 2 Block diagram of the prototype ParaBIT Tx and Rx modules.

Two 24-fiber multimode optical fiber (graded index 50/125 μm) arrays terminated by a pair of two 24-ch BF connector plugs are used to connect Tx and Rx modules. In ParaBIT, a multimode optical transmission system was chosen, because it allows easy optical coupling enabling them to reduce packaging cost.

3.2 10-ch VCSEL Arrays

In the development of ParaBIT, one of the most significant and remarkable points is to select VCSELs as light sources. VCSELs have various advantages in terms of cost effectiveness: (1) high coupling efficiency with optical waveguides and fibers enabling them to emit circular, low-NA beams; (2) low power consumption enabling them to operate with a low threshold current; (3) high production yield in order to be able to eval-

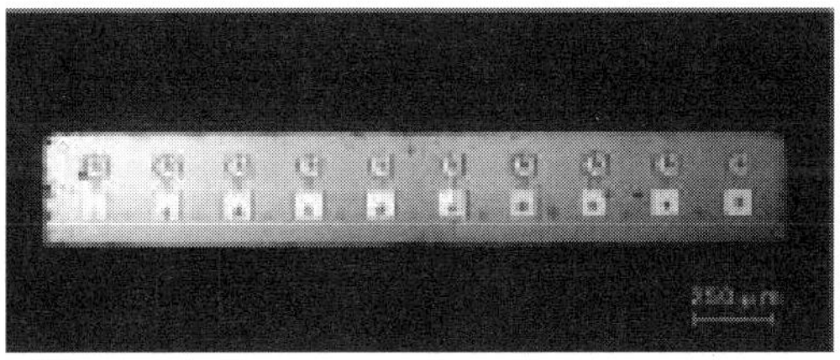

Fig. 3 Top view photograph of 250-μm-pitch 10-ch VCSEL array on p-type GaAs substrate used in Tx module.

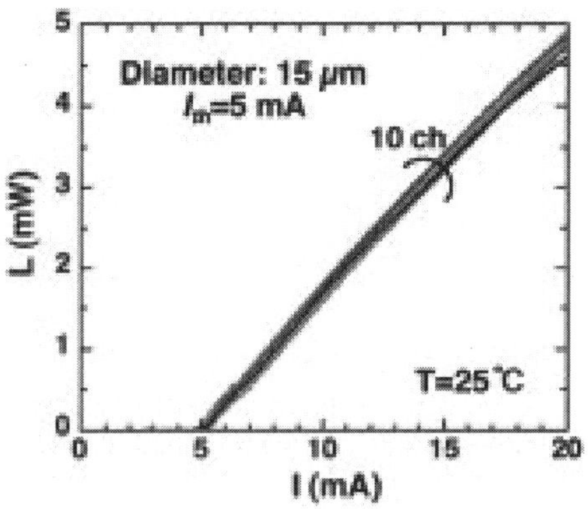

Fig. 4 I-L characteristics of 10-ch VCSEL array.

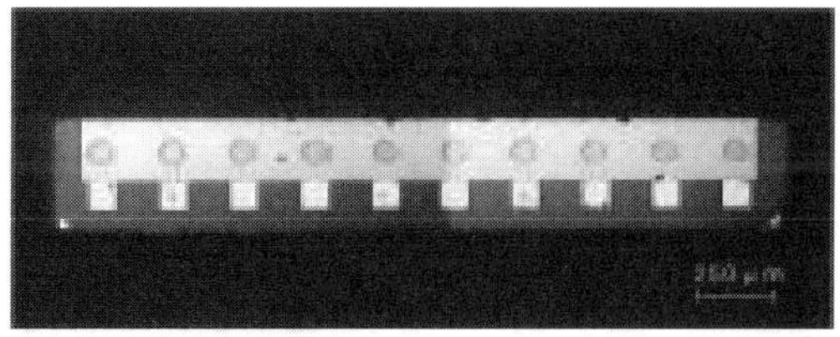

Fig. 5 Top view photograph of 250-μm-pitch 10-ch pin-PD array on p-type GaAs substrate used in Rx module.

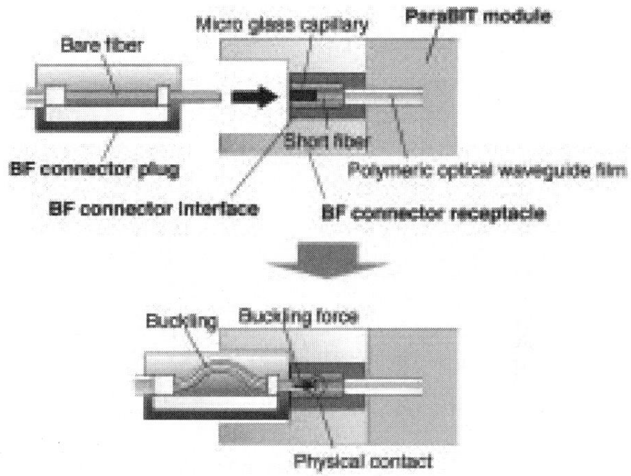

Fig. 6 Structure and mechanism of the 24-fiber BF connector exploiting the flexibility of bare optical fibers.

uate chips in the wafer state; (4) no need to cleave a wafer to fabricate chips; (5) high productivity enabling them to be made on a large GaAs wafer; and (6) the same packaging structure of Tx and Rx modules. Using VCSEL, therefore, is a very advantageous way to reduce production cost of optical interconnection modules.

Figure 3 shows a top view photograph of the 10-ch VCSEL array we developed for the ParaBIT Tx module. The VCSELs with a 250-μm pitch are grown on a p-type GaAs substrate [7], which type is available to match the electrical interface to output of Tx-ICs. The operating wavelength is around 850 nm. Figure 4 shows the I-L characteristics of the 10-ch VCSEL array. A typical threshold current of the 15-μm-diameter VCSEL is 5 mA, and the optical output power is greater than 3 mW at 15 mA. On the other hand, Fig. 5 shows a top view photograph of the 10-ch PD array used in the Rx module. The PD array is a 250-μm-pitch GaAs pin-PD array, and the chip size of 415 μm (L) $\times$ 2600 μm (W) $\times$ 300 μm (T) is the same as in the VCSEL array. Therefore, the packaging of the Tx and Rx modules can have the same packaging structure.

3.3 High-Density Multiport Bare Fiber (BF) Connector

An optical connector is a key device to accomplish super-multichannel performance, such as 40 channels, in the ParaBIT module. To make it easier to assembly the boards, it is necessary to be able to mount the modules together with other electrical components on a printed circuit board by reflow soldering. So, we chose to use a connector rather than a fiber pigtail as the optical interface of ParaBIT. Using a conventional multiport fiber connector, however, raises a definite problem, because pitch size errors between channels accumulate as the number of channels increases.

To make a super-multichannel connector, we newly developed a high-density multiport bare fiber (BF) connector. Figure 6 shows the structure and mechanism of the 24-fiber BF connectors, which exploit the flexibility of bare optical fibers [8], [9]. The BF connector can be joined by direct physical contact between short bare multimode optical fibers (graded index 50/125 μm) embedded in micro glass capillaries with a 250-μm pitch. The buckling forces of the optical fibers themselves secure the physical contact. The short fibers protect the polymeric optical waveguides from the direct pressure of buckling forces. This connector does not need a costly ferrule and spring unlike conventional physical-contact connectors. This structure results in stable high-density optical contact as long as the fibers can be inserted into the capillaries, so it should easily be possible to make even 100-fiber connectors.

Figure 7 shows the assembly of a 24-fiber BF connector designed for the 40-ch ParaBIT module. The ParaBIT module has two 24-fiber BF connector plugs and a BF connector receptacle as the optical interface. The 24-fiber BF connector contains a pair of 12-fiber

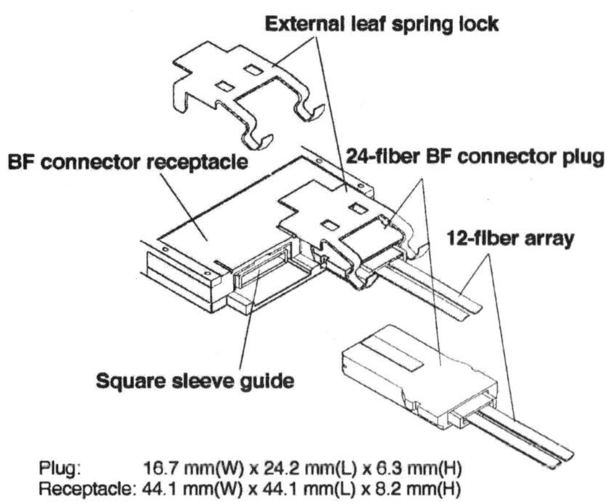

Fig. 7 Assembly of 24-fiber BF connector designed for the 40-ch ParaBIT module.

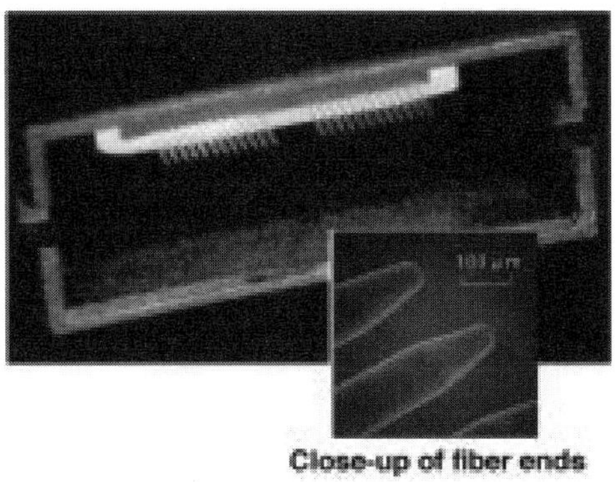

Fig. 8 Photograph of the BF plug end with 24-fiber tips.

arrays with a 250-μm pitch, in which 10-fibers are available for data and the outer two fibers are used for testing. For miniaturization, it uses an external leaf spring lock. Figure 8 shows a photograph of the BF plug end with 24-fiber tips. The fiber-ends in the plug were fabricated to a conical shape with a cleaved end. In the receptacle, the entrances of the micro-capillaries were also cut into cone shapes to guide the bare fiber tips. As a result, it possible to smoothly insert the 24 fibers of the BF plug into the micro-capillaries of the BF receptacle with a large tolerance in the pitch. Thus, the BF connector has great potential for cost-effectiveness, because it does not need expensive ferrule and does not need high accuracy to align fiber-ends of the plug.

Figure 9 shows the insertion and return loss for each port of a 24-fiber BF connector. They result in an average insertion loss of 0.04 dB and an average return loss of over 36 dB at 850 nm (the return loss value was limited by the performance of the measuring equipment). After a 10-time mating test, the deviations among the ports were small. Therefore, the 24-fiber BF connector has high optical performance.

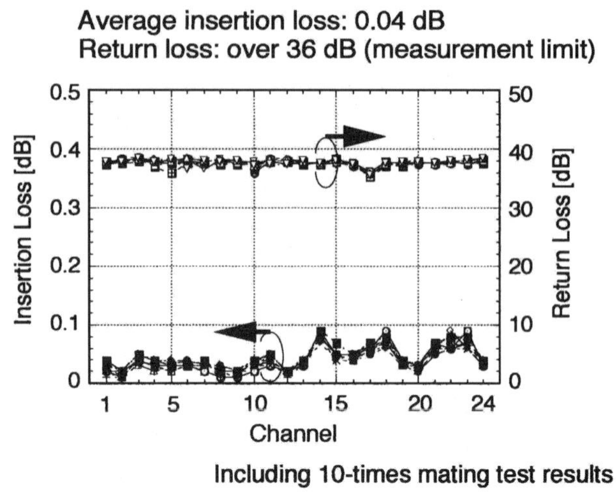

Fig. 9 Insertion loss and return loss for each port of 24-fiber BF connector.

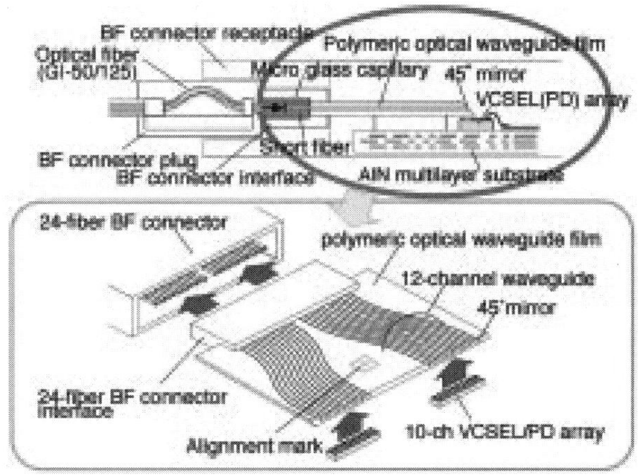

Fig. 10 Structure of the optical coupling system using polymeric optical waveguide film with 45 degree mirror.

3.4 Passive Optical Alignment Using Polymeric Optical Waveguide Film

Figure 10 shows the structure of the optical coupling system using a polymeric optical waveguide film with the 45-degree mirror and the 24-fiber BF connector interface [10]–[13]. The VCSEL/PD arrays are coupled to the optical waveguides through the 45-degree mirror. A pair of 12-ch core lines with 250-μm pitch is symmetrically positioned in order to connect two 12-fiber arrays to two 10-ch VCSEL/PD arrays. Each waveguide have an S-shaped curve, because it is also a pitch transformer between the BF connector and VCSEL/PD arrays. Alignment marks formed together with the core lines in the films enable two VCSEL/PD arrays to be easily coupled simultaneously by passive alignment.

Figure 11 shows a close-up photograph of the polymeric optical waveguide film. The film is 13 mm by 16 mm, excluding the BF connector interface. The 45-

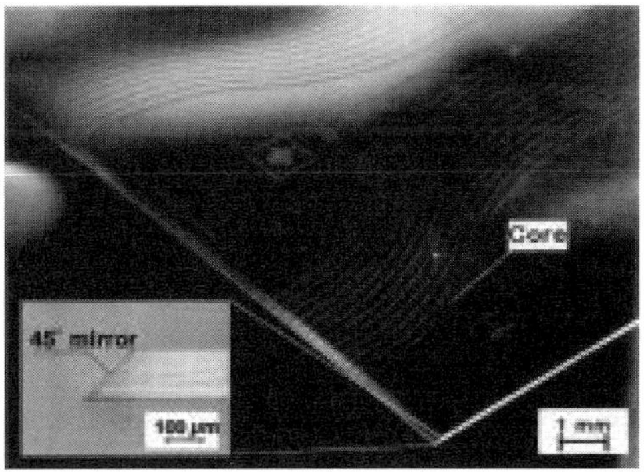

Fig. 11 Close-up photograph of polymeric optical waveguide film.

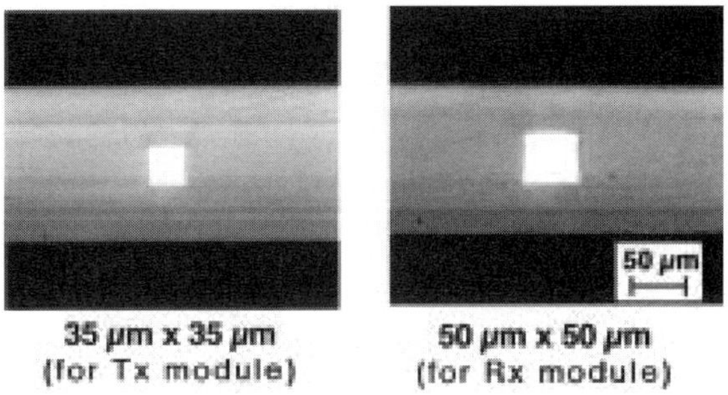

Fig. 12 Cross section of two kinds of the waveguide film.

degree mirrors were formed at the ends of the waveguides by using a smooth diamond saw with a 90-degree V-shaped tip. The polymers used were deuterated polymethylmethacrylate (d-PMMA) for the core and UV-cured resin for the cladding. The waveguide film has two kinds of core sizes as shown in Fig. 12. They were fabricated with core sizes of 35 μm by 35 μm for the Tx modules and 50 μm by 50 μm for the Rx modules, in order to reduce the coupling losses between the waveguide and 50-μm diameter core multimode fibers. This is because the light propagation path in the ParaBIT modules is always fixed in one direction from VCSEL to fiber for a Tx module or from fiber to PD for a Rx module.

Figure 13 shows a close-up top view of the optical coupling system in a ParaBIT Tx module. It includes the polymeric optical waveguide film with 45-degree mirrors formed at the ends of the waveguides, a 10-channel VCSEL array wired with electrical interconnection lines on an AlN substrate. The 45-degree mirrors are precisely aligned with the VCSELs, using the inner 10 cores of the 12-channel waveguide. The alignment could be easily done only in the plane direction by direct observation using a microscope. Thus, the above passive alignment to realize simultaneous coupling with 24 channels also has an advantage in terms of cost-effectiveness.

Figure 14 shows the insertion losses of the polymeric optical waveguide films. They include the losses

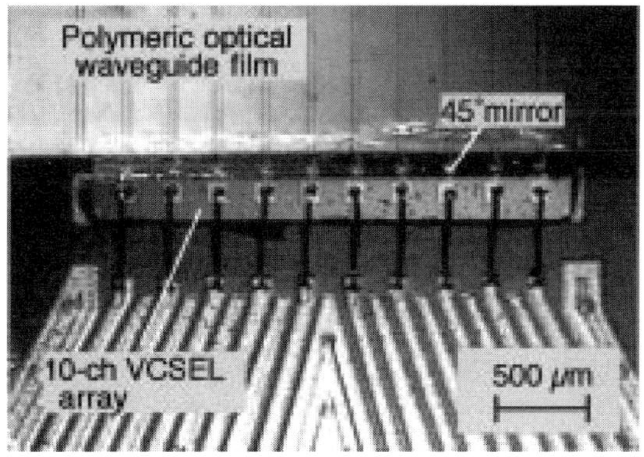

Fig. 13 Close-up top view of optical coupling system in a ParaBIT Tx module.

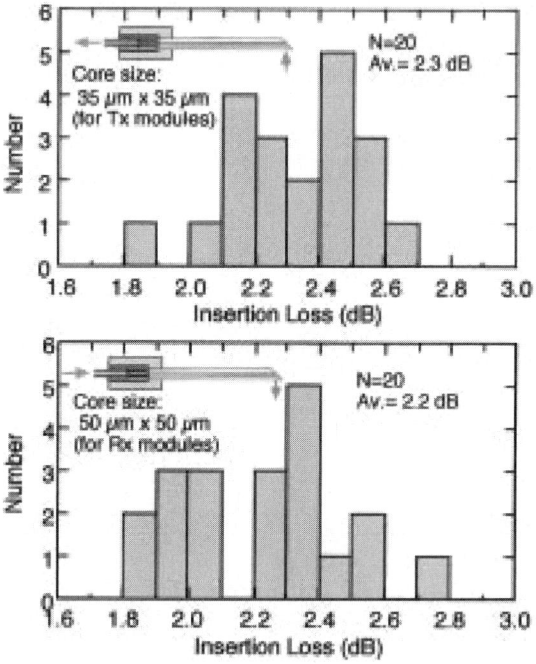

Fig. 14 Insertion losses of polymeric optical waveguide film including losses at 45 degree mirrors and BF connectors in Tx and Rx.

at the 45-degree mirror and BF connector. The average insertion losses for Tx and Rx were 2.3 and 2.2 dB, respectively. The deviations among the channels were within $\pm$0.5 dB for both. Therefore, these alignment using polymeric optical waveguide film have high potential for multi-channel optical interconnects in parallel optical interconnection modules.

3.5 Transferred Multichip Bonding (TMB)

In the ParaBIT module fabrication, diebonding of the VCSEL/PD array chips onto a substrate is a laborious procedure that requires high positional accuracy. To obtain the maximum optical coupling efficiency with the optical waveguide film, the diebonding requires accurate positioning of the chips at the designed posi-

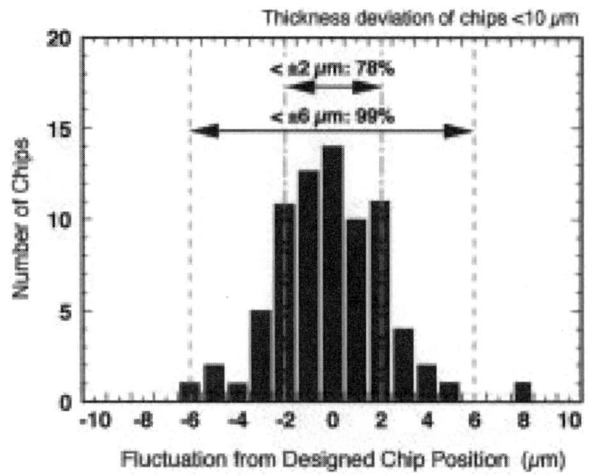

Fig. 15 Histogram of deviations from designed chip positions for VCSEL/PD array chips mounted by TMB.

tion. In particular, it is necessary to align not the external shape of the chip itself, but each element of the four 10-channel optical array chips in a row on a substrate. Moreover, to reduce thermal damage to the devices and to avoid getting out of accurate position of the chips during diebonding, all array device chips must be diebonded simultaneously to shorten the time that the devices are kept at a temperature higher than the melting point of solder. Transferred Multichip Bonding (TMB) was developed to satisfy these requirements [14], [15].

The TMB process is as follows: each VCSEL/PD array chip is adhered using paste onto a glass plate with alignment marks precisely representing the relative positions of elements on the four VCSEL/PD array chips, after alignment. And the chips attached to the plate are kept in contact with the substrate heated to the soldering temperature through a solder sheet for a short time within one minute. As a result, the four VCSEL/PD array chips are diebonded and transferred onto the substrate, and simultaneously diebonded exactly onto the desired positions.

In the prototype ParaBIT module fabrication, the target was to suppress an increase of the coupling loss due to chip positioning errors to below 0.5 dB, meaning the chip positioning errors had to be less than 6 μm. After bonding the chips by TMB, the positioning errors of the mounted chips were estimated using the microscope. Figure 15 shows a histogram of deviations from the designed chip positions for VCSEL/PD array chips mounted by TMB. Chip positioning errors less than 6 μm were achieved in most cases (more than 99%), and even a positioning error of less than 2 μm was achieved in specimens of more than 78%, when the thickness deviations of chips were less than 10 μm. The values are sufficiently low for diebonding of the VCSEL/PD array chips.

Hence, it is clear that TMB is excellent diebonding technique to realize accurate positioning of optical array chips, not to damage to devices, and to offer cost-effective module assembly for a multichannel optoelectronic module.

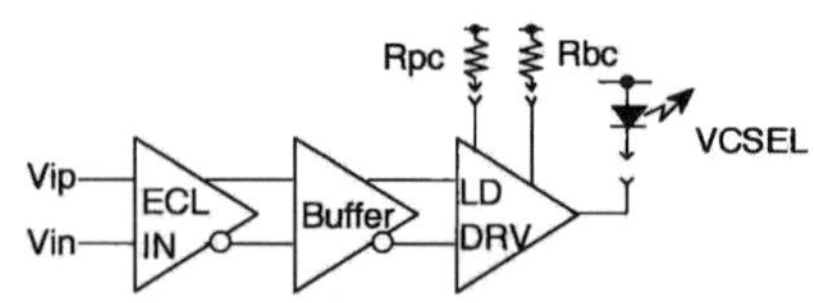

Output bias current (Ib) range	10 - 20 mA
Output pulse current (Ip) range	0 - 25 mA
Bandwidth	DC - 1 Gb/s
Supply voltage	- 4.5 V
Power dissipation	416 mW @ Ib=0, Ip=10 mA
Chip size	1.60 mm x 3.36 mm

Fig. 16 Diagram of Tx-IC.

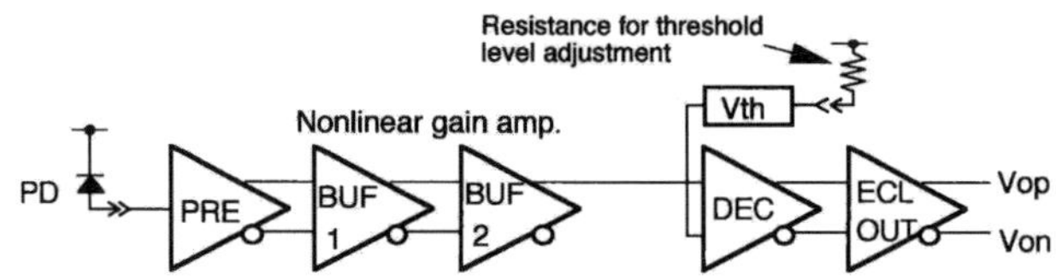

Received signal range	9.6 - 208 μA
Bandwidth	DC - 700 Mb/s
Power supply	- 4.5 V
Power dissipation	400 mW
Chip size	1.60 mm x 3.36 mm

Fig. 17 Diagram of Rx-IC.

3.6 5-Ch Tx/Rx IC

The transmitter (Tx) and receiver (Rx) ICs used in ParaBIT must exhibit high-speed operation from DC to 700 Mb/s, compact size, and low power consumption [16]. To satisfy these demands, the Tx-IC and Rx-IC with simple circuit configuration have been developed. Fabricated by a conventional 0.8-μm Si bipolar process, they have differential ECL interfaces in order to raise their noise immunity under a single −4.5 V voltage supply. For compactness and easy handling, moreover, five circuits were integrated within the chip size of 1.60 mm×3.36 mm.

Figure 16 shows a diagram of the Tx-IC circuits. To simplify the circuit configuration and to handle signals having an infinite series of consecutive 0's or 1's, the APC circuits were not used, and fixed bias current circuits and fixed pulse current circuits with external level adjustment resistors were chosen. Figure 17 shows a diagram of the Rx-IC circuits. DC coupled circuits were also chosen, and fixed-decision-level-circuits with external level adjustment resistors to ensure the low power dissipation were applied. To insure a wide dynamic range and to keep a noise margin for low-level

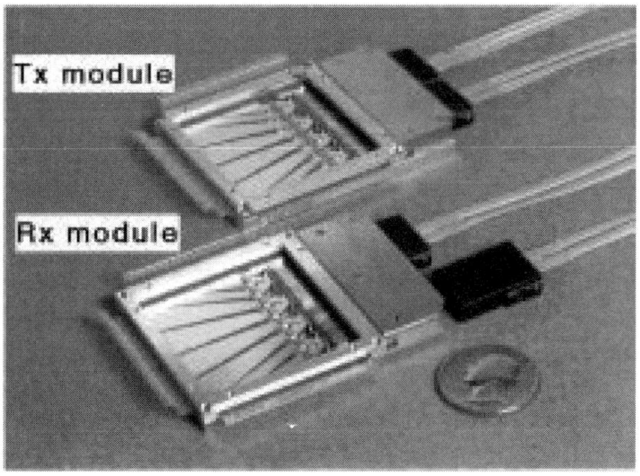

Fig. 18 Photograph of the prototype ParaBIT Tx and Rx modules.

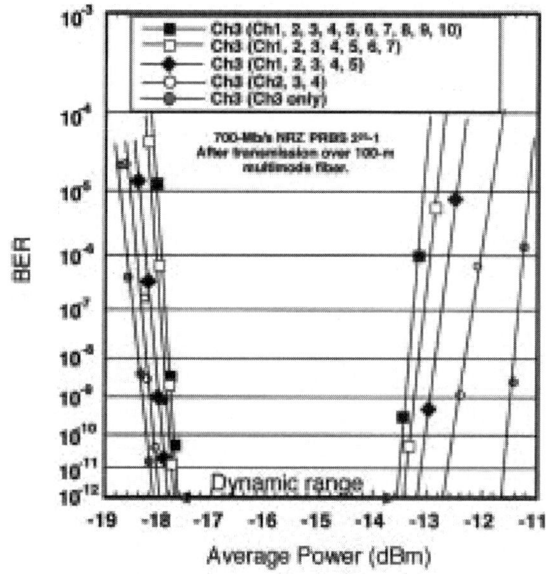

Fig. 19 Results of the bit error rate measurements for every number of operating channels.

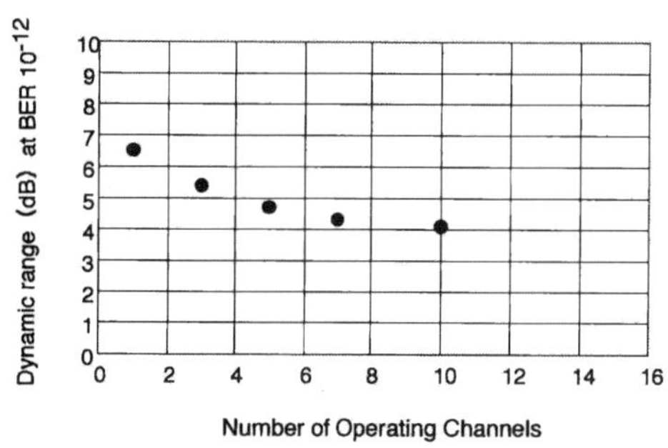

Fig. 20 Dynamic ranges at a bit error rate of 10^{-12} for every number of operating channels.

input signals, furthermore, the Rx-IC contains nonlinear gain amplifiers, which make the amplifier gain small for high-level input signals and large for low-level input signals. The 5-ch Tx and Rx ICs dissipate only 416 mW (83.2 mW/ch) when $I_b = 0$ and $I_p = 10$ mA, and 400 mW (80.0 mW/ch), respectively. The values were adequately small. The Tx-IC and Rx-IC have good advantages of low power consumption and low device cost.

Figure 18 shows a photograph of the prototype ParaBIT Tx and Rx modules concentrating the abovementioned components and technologies. Each module is a compact surface-mounting package with flat leads, 48.0 mm long, 47.2 mm wide, and 8.5 mm high, excluding the BF connectors, for easy board assembly. As VCSEL arrays are used as light sources, it clearly shows that the packaging of the Tx and Rx modules have the same packaging structure.

4. Performance of Prototype ParaBIT Module

Preliminary measurements were done by operating the Tx and Rx modules with 10 channels of 700-Mb/s NRZ $2^{23} - 1$ pseudo random bit sequence (PRBS) data. Figure 19 shows the results of the bit error rate measurements under single-channel operation (channel 3), 3-channel (channels 2–4) operation, 5-channel (channels 1–5) operation, 7-channel (channels 1–7), and 10-channel (channels 1–10) operation. It shows the results for electrical output from channel 3 of the Rx module after 105 m of multimode optical fiber (GI-50/125 μm) arrays. The 10 channels in the Tx/Rx module were ones from the same VCSEL/PD array chip. In this figure, there are two groups of lines. This results from the use of fixed-decision-level-circuits: the left lines mainly relate to the duty fluctuation near the decision limit of "1" level and the right lines mainly relate to the duty fluctuation near the decision limit of "0." In the measurement, to evaluate all the characteristics of both sides, we attenuated the power of optical signals from the Tx module with optical attenuators before they were input to the Rx module. In practical use, both optical output power and decision levels can be adjusted using external adjustment resistors.

As a result of using nonlinear gain amplifiers, dynamic range of 4.1 dB was obtained even for 10-channel operation at the bit error rate of 10^{-12}. Figure 20 shows the obtained dynamic ranges at a bit error rate of 10^{-12} for every number of operating channels. Although the dynamic ranges narrowed as the number of channels increased, this narrowing tended to saturate when the number of channels exceeded five. For over 10-channels operation, therefore, it is presumed that the dynamic ranges of around 4 dB will be obtained.

This confirmed that the design target of 700 Mb/s per channel over 100 meters was achieved in the prototype ParaBIT module.

5. Conclusion

In this paper, we have introduced and described the parallel optical interconnection: ParaBIT, which we have been developing, and its five major technologies to realize a prototype front-end module with 28-Gb/s throughput (700 Mb/s×40 ch). We have succeeded in developing the prototype ParaBIT module as a first step toward 40-channel parallel optical interconnection. The design targets of 700 Mb/s per channel with cost-effectiveness and compactness as well as high throughput, were all met. ParaBIT is a very promising solution that should make possible inter-board and inter-cabinet interconnection in large-capacity switching systems, such as sub-Tb/s- and 1-Tbit/s-class ATM switching systems, in the near future.

Optical interconnections are starting to be used for longer interconnections, such as inter-cabinet, inter-rack, and inter-board, in a packaging hierarchy. In the future, switching systems with throughput of a few terabits per second will appear. To extract the maximum performance from the systems, it will certainly become necessary to use shorter distance optical interconnetions such as inter-MCM and inter-chip interconnections as well as inter-board optical interconnections. At that time, therefore, there will be demands for higher throughput, and more compact higher-density optical interconnections using excellent optoelectronics technologies such as miniaturized O/E and E/O convertors, smart pixels, and so on.

Acknowledgements

We thank Takaaki Ohsaki and Saburo Imamura for their encouragement during this project. We also thank Masato Nakamura, Fuminori Ishitsuka, Masakaze Hosoya, and all of our colleagues involved in the ParaBIT project at NTT Opto-electronics Laboratories.

References

[1] N. Yamanaka, S. Yasukawa, E. Oki, and T. Kawamura, "OPTIMA: Tb/s ATM switching system architecture: Based on highly statistical optical WDM interconnection," Proc. ISS'97, pp.285–289, Sept. 1997.

[2] Y. Ando, "Lecture 5: Current progress in optical interconnections and their impact on packaging technologies," Proc. International Symposium of the JPCA, pp.43–50, Feb. 1998.

[3] Y. Ando and K. Katsura, "ParaBIT: An over-25-Gb/s-throughput 40-ch optical interconnection (1)," Proc. The 1997 Communications Society Conference of IEICE, vol.2, B-10-130, p.429, Tokyo, Sept. 1997.

[4] Y. Ando, "An over-25-Gb/s-throughput optical interconnection for high-speed ATM switching systems," Proc. IEEE Lasers and Electro-Optics Society 1997 (LEOS'97) 10th Annual Meeting, vol.1, pp.126–127, Nov. 1997.

[5] M. Usui, N. Matsuura, N. Sato, M. Nakamura, N. Tanaka, A. Ohki, M. Hikita, R. Yoshimura, K. Tateno, K. Katsura, and Y. Ando, "700-Mb/s × 40-channel parallel optical interconnection module using VCSEL arrays and bare fiber connectors," Proc. of IEEE Lasers and Electro-Optics Society 1997 (LEOS'97) 10th Annual Meeting, vol.1, pp.51–52, Nov. 1997.

[6] K. Katsura, M. Usui, N. Sato, A. Ohki, N. Tanaka, N. Matsuura, T. Kagawa, K. Tateno, M. Hikita, R. Yoshimura, and Y. Ando, "Packaging for a 40-channel parallel optical interconnection module with over 25-Gb/s throughput," Proc. 48th ECTC, pp.740–746, May 1998.

[7] Y. Kohama, Y. Ohiso, K. Tateno, and T. Kurokawa, "0.85-μm vertical-cavity surface-emitting laser diode arrays grown on p-type GaAs substrate," IEEE Photonics Tech. Lett., vol.9, no.3, pp.280–281, 1997.

[8] Y. Ando and I. Nishi, "Spring-retentive V-groove splice for joining groups of optical fibers," Appl. Opt., vol.21, no.15, pp.2689–2692, 1982.

[9] M. Kobayashi, S. Iwano, R. Nagase, and S. Mitachi, "A new physical contact connection method using the buckling force of optical fiber," IEICE Trans. Electron., vol.E80-C, no.2, 1997.

[10] M. Usui, M. Hikita, R. Yoshimura, N. Matsuura, N. Sato, A. Ohki, T. Kagawa, K. Tateno, K. Katsura, and Y. Ando, "An optical coupling technique for parallel optical interconnection modules using polymeric optical waveguide films," Proc. of 2nd IEMT/IMC Symposium, pp.127–132, April 1998.

[11] M. Usui, M. Hikita, R. Yoshimura, S. Tomaru, S. Imamura, K. Katsura, and Y. Ando, "Basic evaluation of polymeric optical waveguide films applied to optical interconnections," IEICE Trans. Electron., vol.E81-C, no.7, pp.1027–1033, July 1998.

[12] M. Hikita, R. Yoshimura, A. Kaneko, M. Usui, S. Tomaru, and S. Imamura, "Free-standing polymeric optical waveguide films for optical interconnections," Proc 11th Integrated Optics and Optical Fibre Communications/23rd European Conference of Optical Communications, vol.2, pp.285–288, Sept. 1997.

[13] R. Yoshimura, M. Hikita, M. Usui, S. Tomaru, and S. Imamura, "Polymeric optical waveguide films with 45° mirrors formed with a 90° V-shaped diamond blade," Electron. Lett., vol.33, no.15, pp.1311–1312, 1997.

[14] A. Ohki, M. Usui, N. Sato, N. Matsuura, K. Katsura, and Y. Ando, "Fabrication of a parallel inter-board optical interconnection module using transferred multichip bonding," Proc. of 2nd IEMT/IMC Symposium, pp.123–126, April 1998.

[15] A. Ohki, M. Usui, N. Sato, N. Matsuura, K. Katsura, and Y. Ando, "Multi-channel optical coupling between VCSEL arrays and multimode optical fibers for a 40-channel parallel optical interconnection module," Digest of 1998 IEEE/LEOS Summer Topical Meetings, Smart Pixels, FA3, pp.47–18, July 1998.

[16] N. Tanaka, M. Nakamura, F. Ishitsuka, K. Katsura, and Y. Ando, "ParaBIT: An over-25-Gb/s-throughput 40-ch optical interconnection (3)," Proc. The 1997 Communications Society Conference of IEICE, vol.2, B-10-132, p.431, Tokyo, Sept. 1997.

High-Performance Backbone Network Technology edited by Naoaki Yamanaka
ISBN: 0-8247-5321-6 DOI: 10.1081/(120025752)
Published 2004 by Marcel Dekker, Inc.

High-Performance VCSELs for Optical Data Links

Rainer Michalzik, Karl Joachim Ebeling, Max Kicherer, Felix Mederer, Roger King, Heiko Unold, and Roland Jäger

SUMMARY The present paper discusses several promising application areas for optical data links based on high-performance vertical-cavity surface-emitting laser diodes (VCSELs). Both 850 and 980 nm emission wavelength devices realized in the GaAs-AlGaAs or InGaAs-AlGaAs material systems are considered. We show data transmission results of 10 Gb/s signals at 830 nm wavelength over a new high-bandwidth multimode silica fiber of up to 1.6 km length. The same fiber type is employed to demonstrate the first 40 Gb/s transport over 300 m distance by means of a 4-channel coarse wavelength-division multiplexing approach. A first 1×10 linear VCSEL array capable of 10 Gb/s per channel operation is presented for use in next generation parallel optical modules. To improve the singlemode emission characteristics for output power in the 5 mW range we introduce a new device concept incorporating a long monolithic cavity. For low-cost short-distance data links we investigate graded-index polymer optical fibers and report on up to 9 Gb/s transmission over a length of 100 m. Polymer waveguides are also used in an optical layer of a hybrid electrical-optical printed circuit board. Transmitted 10 Gb/s optical data over a prototype board show the potential of this new technology. Finally we present two-dimensional VCSEL arrays for highly parallel data transport on a CMOS chip level. Both 980 and 850 nm bottom emitting devices with modulation capabilities up to 12.5 Gb/s are discussed.
key words: *surface-emitting lasers, optical interconnections, 10-Gigabit Ethernet, wavelength-division multiplexing, polymer waveguides*

1. Introduction

For about ten years already a continuous increase of research into VCSELs is witnessed [1]. On the other hand, in recent years quite a number of start-up companies have been established to economically exploit the benefits of these new microlasers. Generally speaking, the most attractive features of VCSELs include on-wafer testing capability, mounting technology familiar from the low-cost light emitting diode market, circularly symmetric beam profiles for ease of light focusing, high-speed modulation with low driving currents, temperature insensitive operation characteristics, and obvious forming of one- or two-dimensional arrays. Besides numerous attempts to realize 1.3 or 1.55 μm telecom wavelength VCSELs, commercial efforts mainly focus on the data communications field, where 850 nm short-wavelength transceivers are today's commodity. Data rates of 1 Gb/s as in Gigabit Ethernet (GbE) or Fiber Channel links are most widely spread while next generation devices operating at 2.5 Gb/s are appearing on the market. Recently, an option for 850 nm serial transmission over graded-index multimode fiber (MMF) was adopted into the 10-Gigabit Ethernet standards draft**, so that a lot of future VCSEL work will concentrate on the realization of reliable 10 Gb/s sources.

In this paper we report on applications of high-performance short-wavelength VCSELs on different hierarchical levels of optical interconnects. We consider optical links based on high-bandwidth graded-index silica and plastic optical multimode fibers, space-parallel data transmission over fiber bundles or polymer waveguides integrated into a conventional electrical printed circuit board, and two-dimensional guided-wave interconnections between CMOS chips. In each case, data rates in the 10 Gb/s range are demonstrated.

The layout and general dynamic and noise characteristics of the well-established conventional high-speed VCSELs are described in some detail in [2]. Since transverse singlemode rather than multimode oscillation has some inherent advantages but is often difficult to achieve, in this paper we restrict ourselves to a more detailed description of a new extended cavity device concept enabling spectrally pure emission while permitting larger active areas.

2. Long-Distance and High Throughput Multimode Fiber Links

In the upcoming 10-GbE standard, the completion of which is expected for about the beginning of the year 2002, one of the physical layer options for in-building cabling will be short-wavelength serial data transmission over up to 300 m of a new generation silica MMF. In early 2000, Lucent Technologies was the first company to introduce a 50 μm core diameter MMF with an increased bandwidth-distance product of 2.2 GHz·km for 850 nm wavelength. Fiber optimization consists of a reproducible and improved realization of a close to parabolic refractive index grading over the target wave-

Manuscript received November 23, 2000.
[†]The authors are with the Optoelectronics Department, University of Ulm, D-89069 Ulm, Germany.
*This paper is also published in the IEICE Trans. Electron., Vol.E84-C, No.5, pp.629–638, May 2001.

**See URL http://grouper.ieee.org/groups/802/3/ae

IEICE TRANS. COMMUN., Vol. E84-B, No. 5, pp. 1225–1264, May 2001

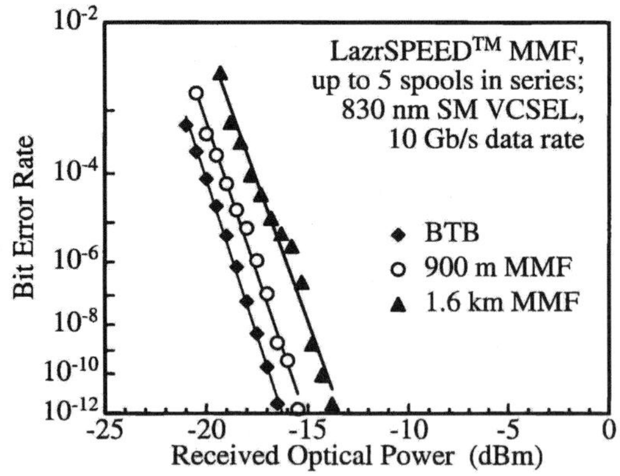

Fig. 1 Bit error rate characteristics for 10 Gb/s data transmission with an 830 nm wavelength transverse singlemode (SM) VCSEL over up to 1.6 km of a high-performance multimode fiber as well as for back-to-back (BTB) operation [3].

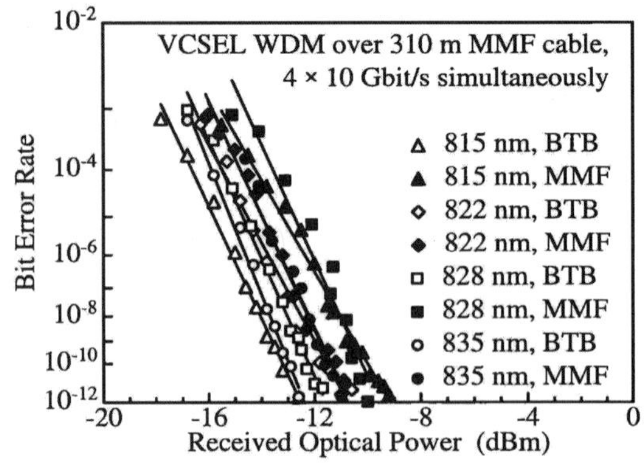

Fig. 2 Bit error rates in a 4-channel CWDM experiment showing 4 × 10 Gb/s data transmission over 310 m of a 50 μm core diameter MMF cable [7].

length regime, while further on satisfying the bandwidth requirement of 1.3 μm wavelength GbE signals[†].

Figure 1 displays bit error rate measurements for 10 Gb/s data transmission with an 830 nm VCSEL source over this fiber medium [3]. Error-free operation over up to 1.6 km distance is achieved with only 3 dB power penalty. Five connectorized fiber spools of 500, 400, 300, 300, and 100 m length are used in the experiment. Further including a 2 km long fiber piece, transmission over even 2.8 km was shown to be possible [4]. Although in these two demonstrations singlemode emission turned out to be an essential feature due to chromatic dispersion limitations, highly *multimoded* VCSELs can as well be employed for transmission over only a 300 m distance, as required for 10-GbE.

It is generally agreed upon that a further increase of data rates beyond 10 Gb/s in MMF systems will most realistically be accomplished by resorting to the coarse wavelength-division multiplexing (CWDM) technique. With a channel spacing of several nm, the requirements for wavelength stabilization of optical sources as well as on the demultiplexing process can be considerably relaxed compared to dense WDM in telecommunications. CWDM modules based on 840 nm VCSELs [5] or 1310 nm DFB lasers [6] for 4 × 2.5 Gb/s transmission over 100 m MMF or 300 m MMF, respectively, both with 62.5 μm core diameter have already been demonstrated.

For a first laboratory realization of a 40 Gb/s MMF system [7] we have selected oxide-confined high-speed transverse singlemode VCSELs in the 3 to 4 μm active diameter range with emission wavelengths of 815, 822, 828, and 835 nm. VCSEL beams are combined through polarizing beam splitters and a MMF 3-dB coupler and a grating is used for free-space demultiplexing.

The transmission experiment is performed by driving all VCSELs simultaneously with decorrelated 10 Gb/s pseudo-random bit sequences from a pattern generator, whereas only one of the channels is detected. In Fig. 2 we show the measured BERs before and after the insertion of a 310 m long high-bandwidth MMF cable, where error-free operation is achieved for all channels. Power penalties between 1 and 3 dB are mainly caused by polarization noise generated during beam multiplexing and a below 15 dB adjacent wavelength suppression in the present demultiplexing setup.

3. Space-Division Multiplexed Data Transmission

An alternative approach to the realization of high-throughput optical interconnects is the space-division multiplexing technique, where optical signals are transmitted in parallel through a MMF ribbon cable. Due to relatively high cost of this waveguide, especially if interchannel skew is to be minimized, parallel links are currently competitive only for several tens of meter transmission length. State-of-the-art modules operate at 2.5 Gb/s channel data rate and thus achieve 30 Gb/s throughput for a 12-channel system [8]. Obviously 10 Gb/s individual data rates are targeted for next generation devices.

Figure 3 displays the threshold current distribution of a linear 1 × 10 singlemode 850 nm VCSEL array to be incorporated into a parallel optical module. Devices are arranged on a 250 μm pitch and average threshold current and standard deviation are 340 μA and 14 μA, respectively.

For data transmission, identical bias currents of 1.65 mA and modulation voltages of 0.65 V_{pp} were applied for every channel. Figure 4 shows that the bit error rate curves thus obtained for back-to-back operation almost overlap. Received optical power required for BER $= 10^{-9}$ is below -15 dBm for an on-off ratio of approximately 6 dB. These results confirm that 10 Gb/s-compatible prototype VCSEL arrays are available today, leaving the main challenges for the realization of complete modules to the dense integration of

[†]See IEEE 802.3z Gigabit Ethernet standard, 1000BASE-SX physical layer.

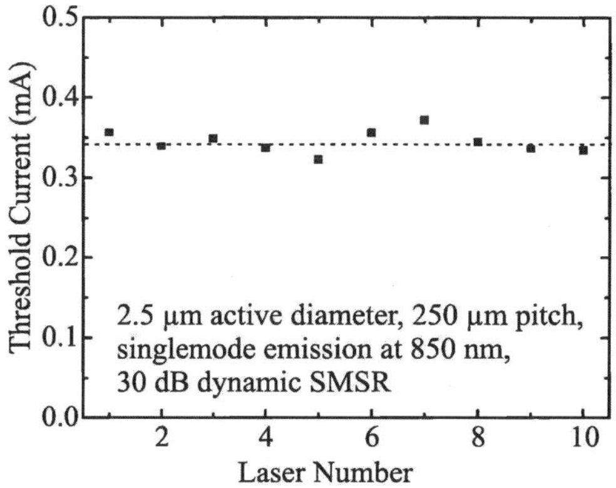

Fig. 3 Threshold current distribution of a 1×10 VCSEL array for fiber-parallel 10 Gb/s data transmission. Devices emit in a single transverse mode with above 30 dB sidemode suppression ratio (SMSR).

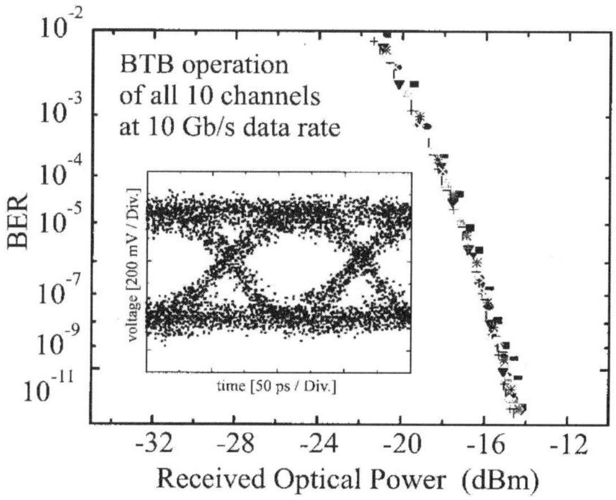

Fig. 4 10 Gb/s back-to-back data transmission of all 10 channels of the linear VCSEL array from Fig. 3 with a representative eye diagram (on a 200 ps time scale) at error-free operation in the inset.

high-speed electronics and sensitive detector arrays.

4. Large-Area Singlemode VCSELs

Standard singlemode selectively oxidized VCSELs of, e.g., 3 μm diameter suffer from comparatively large differential resistances of typically about 250 Ω, which is due to the current bottleneck resulting from the oxide aperture. Smaller VCSELs moreover tend to have higher threshold current densities which, together with an increased thermal resistance, lead to pronounced device heating. These effects may contribute to a reduced laser lifetime.

In order to overcome the problems, various modified VCSEL structures have been investigated. In VCSELs with surface relief for transverse mode filtering we observe singlemode operation in 7 μm diameter devices, however, positioning and precise layout of the surface relief is crucial [9]. More promising are VCSELs with a long monolithic cavity as shown in Fig. 5, where an ex-

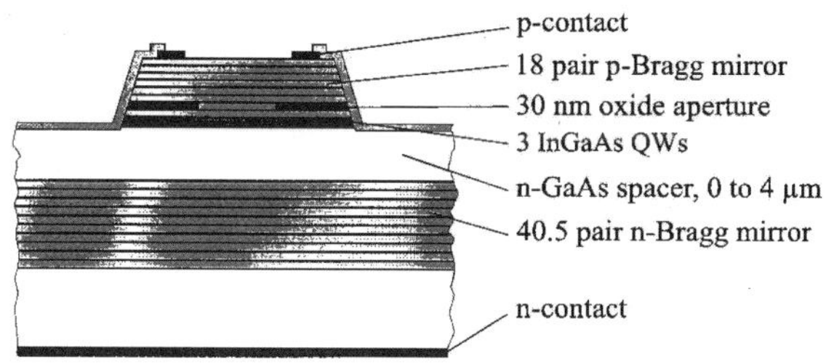

Fig. 5 Extended cavity VCSEL structure for high-power singlemode emission.

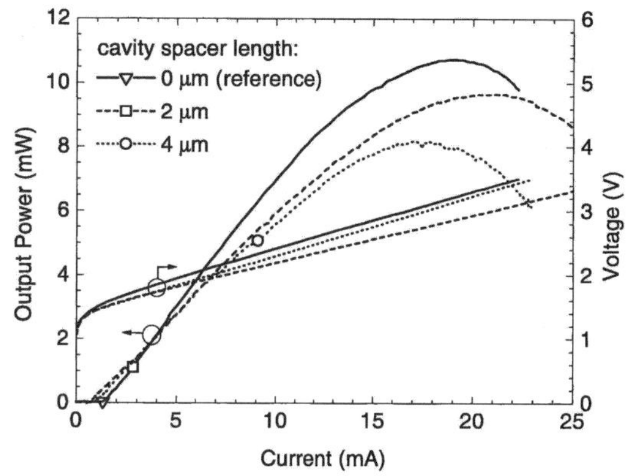

Fig. 6 Operation characteristics of regular as well as 2 and 4 μm extended cavity VCSELs. Markers indicate limits of singlemode emission.

tra n-GaAs spacer layer is inserted between the bottom Bragg reflector and the inner cavity [10]. Advantages concerning the modal structure of such a device layout are expected from increased diffraction losses for higher order transverse modes. Since the active area then can correspondingly be made larger, reductions of both electrical and thermal resistances should favorably add to improved manufacturability as well as reliability.

Figure 6 compares output characteristics of 7 μm aperture devices for various spacers of 0, 2, and 4 μm thickness. From the emission spectra in Fig. 7 for the 4 μm spacer device and the corresponding output characteristics in Fig. 6 it is seen that 5 mW singlemode output power is obtained at 9 mA driving current, which is among the highest values obtained to date [11], [12]. The differential resistance is as low as 98 Ω. The 2 μm spacer device remains singlemode up to 3 mA current or 1 mW output power, whereas the emission of the device without spacer is always on several transverse modes. Obviously, 7 μm size singlemode VCSELs are favorable for butt-coupling to standard singlemode fiber where coupling efficiencies of 80% are easily achieved.

In even longer VCSELs with 8 μm spacer, switching between neighboring longitudinal modes is observed [10], which is an unknown effect in regular VCSEL designs. Near- and far-field measurements as well as transfer matrix calculations have been employed in [13] to reveal that mode switching depends on the cavity de-

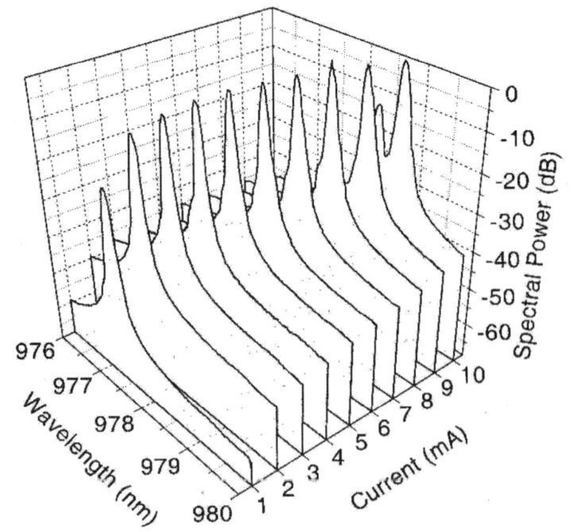

Fig. 7 Current dependent emission spectra of a 4 μm spacer long monolithic cavity VCSEL from Fig. 6.

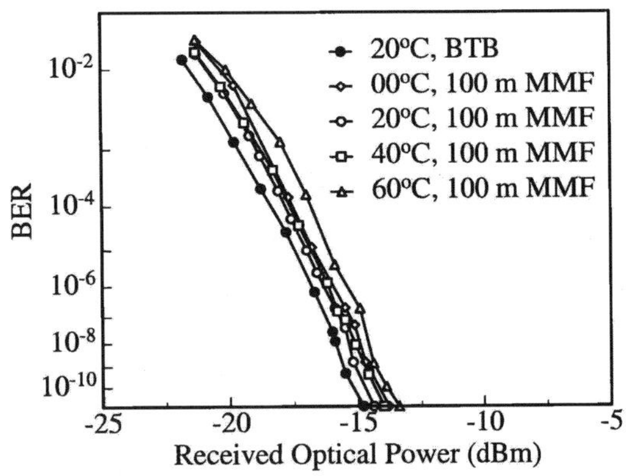

Fig. 8 10 Gb/s data transmission results at 990 nm wavelength obtained with a 2 μm spacer extended cavity singlemode VCSEL for a temperature range from 0 to 60°C.

tuning and originates form the interplay between oxide-induced index-guiding and thermal lensing.

Although the increase of spacer length results in longer photon lifetimes in the resonator, the dynamical properties of the VCSELs do not suffer dramatically. With the current design, a 3-dB small-signal frequency of 5 GHz was measured for the 2 μm spacer device. Figure 8 shows data transmission results at 10 Gb/s over 100 m of silica multimode fiber for a temperature range from 0 to 60°C. Certainly, wide temperature range operation at high bit rates is increasingly harder to achieve [14], however even results at 85°C have already been reported [15].

5. Data Transmission over Plastic Optical Fibers

Plastic optical fibers (POFs) today are widely used in low-cost applications like optical interconnection of consumer electronics equipment, image transmission, or lighting [16]. In recent years a new low-loss graded-index (GI) POF has been introduced, so that even high-speed data communications has become feasible [17].

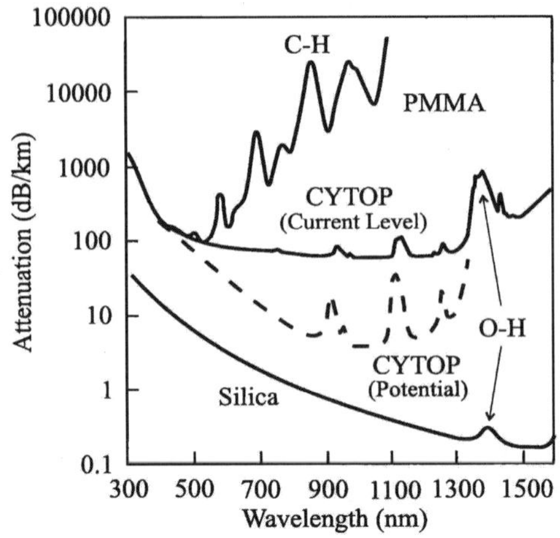

Fig. 9 Currently achieved as well as potential attenuation characteristics of a perfluorinated polymer (CYTOP) fiber compared with standard silica glass and polymethyl methacrylate (PMMA) material [18].

One of the most important benchmarks for an optical fiber is its spectral attenuation characteristics. Figure 9 shows such spectra for the well-known silica glass fiber and the very low-cost PMMA step-index POF. For data communication wavelengths of around 900 nm, these fiber types show a huge difference of attenuation coefficients of about 1 dB/km and 10^4 dB/km, respectively. The various attenuation peaks in the PMMA spectrum are mainly caused by resonances of carbon-hydrogen bonds. Data transmission experiments at 2.5 Gb/s or 1 Gb/s over 2.5 m or 15 m of such a POF with 120 μm or 1 mm core diameter are reported in [19] and [20], respectively. In the perfluorinated POF, basically hydrogen is replaced by heavier fluorine atoms, thus effectively suppressing most of the resonances and allowing to achieve intermediate (on a logarithmic dB-scale) attenuation coefficients of about 100 dB/km. The so-called Lucina^{TM} fiber, fabricated of CYTOP material by Asahi Glass Co., Japan, has a parabolic-like graded refractive index profile and offers a large over-filled launch bandwidth in the range of 1 GHz·km and losses below 80 dB/km throughout the entire data communications window from 700 to 1300 nm [18], as seen in Fig. 10. Minimum losses down to 10 dB/km seem to be in close reach.

In what follows we report on experiments employing two different types of GI POF with 130 or 155 μm core diameter, respectively. This large mode-guiding area is favorable for alignment tolerant coupling and especially allows for the fabrication of injection molded connectors, contributing to low connectorization costs. Applications for this type of fiber might be found in premises networks, short-reach (< 100 m) datacom, and computer interconnections. Data transmission over 100 m perfluorinated POF at up to 11 Gb/s with a 1300 nm wavelength Fabry-Perot laser has already been published [21].

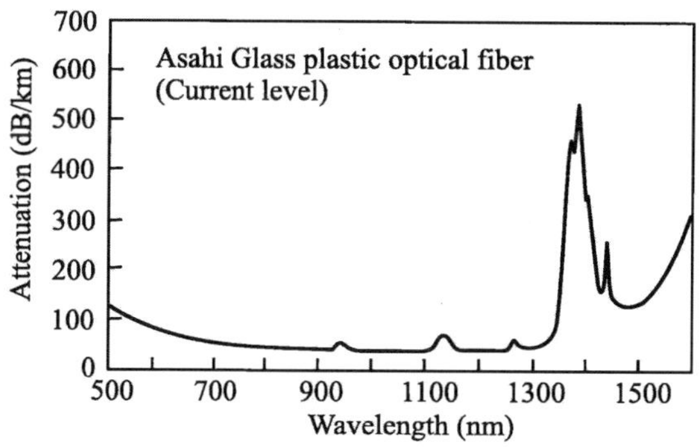

Fig. 10 Attenuation characteristics of a today's graded-index POF fabricated by Asahi Glass Co.

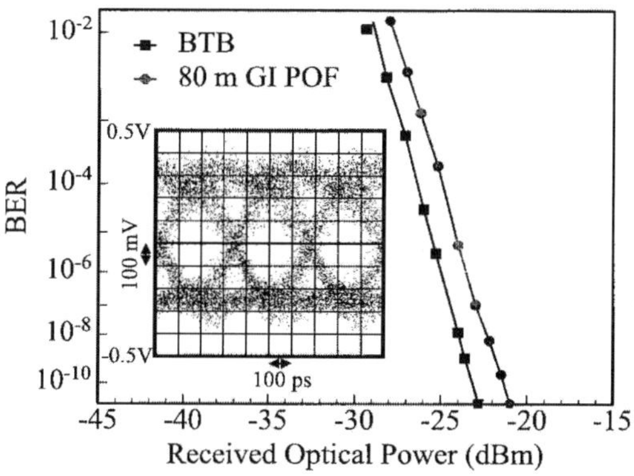

Fig. 11 Bit error rates and eye diagram for 3 Gb/s data transmission at 935 nm wavelength over 80 m of a 155 μm core diameter GI POF.

The first experiment uses an 80 m-long piece of GI POF with 155 μm core and 250 μm cladding diameter [22]. The InGaAs quantum well based VCSEL emits in a single transverse mode at about 935 nm wavelength. The threshold current is 1.3 mA and for modulation the laser is biased at 5 mA, corresponding to 3 mW output power. Light is launched into the fiber through butt-coupling and at the output end is fed into a 50 μm core diameter silica fiber pigtailed germanium avalanche photodiode using demagnifying optics. Figure 11 shows the bit error rates at 3 Gb/s for back-to-back as well as fiber transmission together with an eye diagram. The received optical power required for BER $= 10^{-9}$ is about -23.5 dBm with additional 2 dB power penalty when inserting the fiber.

In the second experiment we employ a singlemode 830 nm wavelength GaAs quantum well based VCSEL for 100 m distance transmission at 9 Gb/s data rate through an improved 130 μm core diameter GI POF [23]. The laser shows a threshold current of 0.8 mA and delivers an optical power of 1 mW at 3.2 mA bias current. Free-space optics is used both to couple the laser beam into the POF and to focus the light onto a 70 μm diameter pin-type detector. As before, the size mismatch between fiber and detector was not fully compensated by the demagnification optics. Using an 850 nm

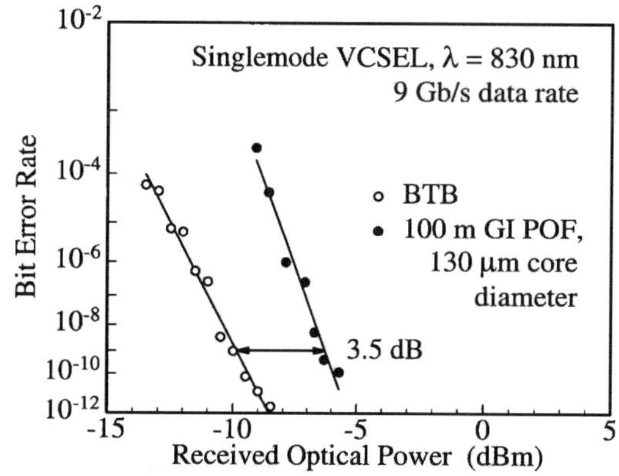

Fig. 12 Bit error rate curves for 9 Gb/s data transmission over a 100 m length graded-index plastic optical fiber using a singlemode oxide-confined VCSEL emitting at 830 nm wavelength.

LED source and a broad-area detector, we measured a total attenuation of 4.4 dB for the fiber, corresponding to a specific attenuation coefficient of only about 44 dB/km at the given wavelength. For comparison, the same fiber has a loss of 33 dB/km at 1.3 μm wavelength [21], confirming the very favorable flat attenuation characteristics of the perfluorinated material as seen in Fig. 10. Figure 12 shows bit error rate curves for back-to-back and 100 m length GI POF data transmission at 9 Gb/s. Quasi error-free operation is achieved in both cases. The relatively high power penalty of 3.5 dB at BER $= 10^{-9}$ is caused by modal bandwidth restrictions of the fiber, modal noise contributions, as well as the relatively low source extinction ratio, not exceeding 5 dB in the present case.

The receiving end remains to be the most demanding part of a high-speed 10 Gb/s POF link. First of all, the diameter of the photodiode has to be chosen large enough in order to minimize modal noise penalties, which incurs bandwidth limitations through the increased diode capacitance. On the other hand, the higher fiber attenuation compared to silica decreases the available link power margin, so that a high sensitivity receiver has to be employed, especially if the link is intended to be operated within the eye safety margin of the short-wavelength regime.

6. Intra-Circuit Board Optical Interconnects

The first step of the penetration of optical data links into a single-box computer system most likely will be the introduction of optical backplanes to interconnect the multitude of printed circuit boards (PCBs) [24]. For this task, VCSELs obviously are highly attractive light sources. Nevertheless, even if the efficient board-to-board signal transfer was guaranteed, the distribution of highly parallel electrical signals within a single board would continue to impose a significant technological problem already for clock rates as low as 200 MHz. A particularly elegant way for optics to contribute to the

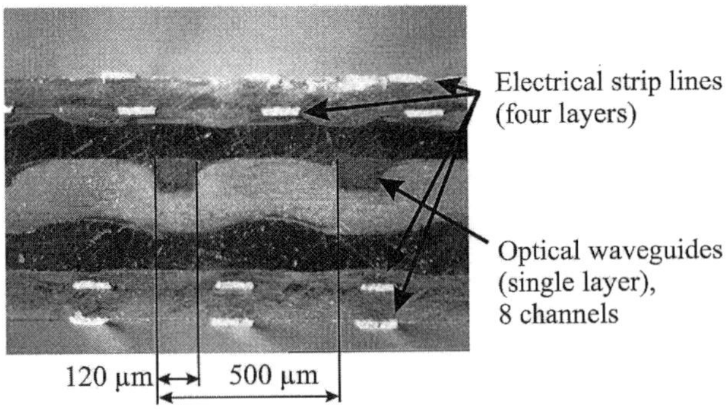

Fig. 13 Cross-sectional view of a hybrid printed circuit board containing a single polymer optical waveguide layer and four layers of electrical strip lines.

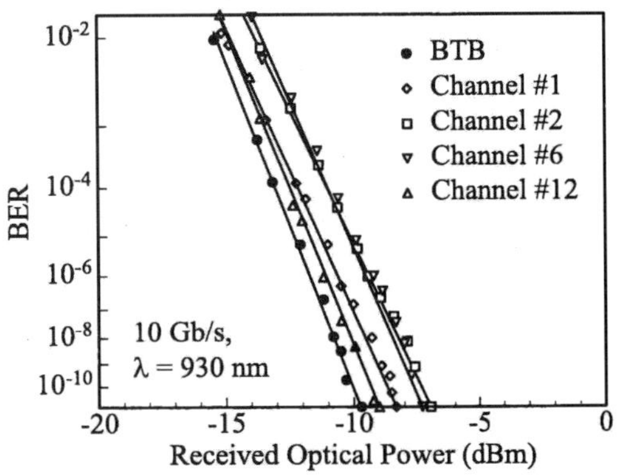

Fig. 14 Bit error rates for 10 Gb/s back-to-back operation and transmission over four representative channels of a 4.3 cm long hybrid electro-optical printed circuit board.

removal of the intraboard interconnect bottleneck could be the realization of hybrid electrical-optical PCBs into which polymer waveguides are integrated as an additional optical layer [25]–[27].

In the example in Fig. 13, hot-embossing of a polymer foil is used to arrange a total of 8 waveguides with an individual cross-sectional area of about $120 \times 120\,\mu\text{m}^2$ on a $500\,\mu$m pitch. Minimum waveguide attenuation of the first prototypes is about 0.5 dB/cm. In a recent experiment we have used a long monolithic cavity VCSEL with $2\,\mu$m spacer thickness and single-mode emission at 930 nm for 10 Gb/s data transmission over a 4.3 cm long PCB. Light is launched into the edge of the board and coupled into a $50\,\mu$m core diameter MMF pigtail of a 10 Gb/s-compatible InGaAs pin-photodiode based receiver. Figure 14 illustrates the error rate characteristics for four representative channels of the waveguide array. Power penalties between about 1 and 3 dB at bit error rates of 10^{-11} are observed. These power penalties are mainly attributed to modal noise induced by the size mismatch between waveguide and silica pick-up fiber. Slight variations in waveguide geometry and associated scattering as well as varying launch and fiber coupling conditions contribute to the penalty differences. Equivalently to the transmission over large diameter POFs, the receiving end appears to be more challenging to implement.

In an advanced configuration of the given experiment, VCSELs and photodiodes in the future will be mounted on the board together with electronic devices, and waveguide coupling will be achieved through via holes and 45° beam deflection.

7. Two-Dimensional VCSEL Arrays for Interchip Interconnections

The use of two-dimensional (2-D) VCSEL arrays for data transport is especially attractive in cases where high-speed electrical signals are *generated* on a small pitch. In such a case there is no need to map spatially distributed data into a dense 2-D arrangement which on the one hand is space-consuming and on the other hand imposes considerable crosstalk problems. The above scenario is found today in CMOS chip-to-chip interconnections where the data bottleneck is predicted to become severe in several years due to general limitations of electrical data lines and an increase of both clock speed and integration density[†]. As a partner in a joint European research project named *Optically Interconnected Integrated Circuits* (OIIC)[††] [28] we are fabricating two-dimensional VCSEL chips containing a total of 32 emitters in a rectangular 4×8 grid [29]. The goal is to demonstrate direct high-density optical interconnection between 2-D optoelectronic transmitter and receiver arrays which are hybridized onto silicon CMOS circuits by flip-chip soldering. The bulk of VCSEL arrays delivered to the project for incorporation into various technology demonstrators is designed for bottom emission in the 980 nm wavelength regime and features coplanar, bondable p- and n-contacts contained in a $250 \times 250\,\mu\text{m}^2$ size unit cell.

In the top part of Fig. 15, optical output power, driving voltage, and power conversion efficiency for all 32 elements of such a VCSEL array are summarized. A $6\,\mu$m individual active device diameter is defined through selective oxidation of a single AlAs current aperture layer. Excellent homogeneity is achieved by molecular beam epitaxial growth and devices show average threshold currents and voltages of 0.8 mA and 1.45 V, respectively. Optical output power of 1 mW is reached at only 2 mA driving current and a peak power conversion efficiency of 37% results for output power of about 3 mW. Direct microwave probing has already been used for 12.5 Gb/s data transmission over 100 m multimode fiber or 1.6 km standard singlemode fiber [29]. In the present application, guided-wave op-

[†]See the Semiconductor Industry Association's (SIA's) *National Technology Roadmap for Semiconductors, 1999 edition*, URL http://public.itrs.net/

[††]See URL http://intecsrv.rug.ac.be/OIIC/

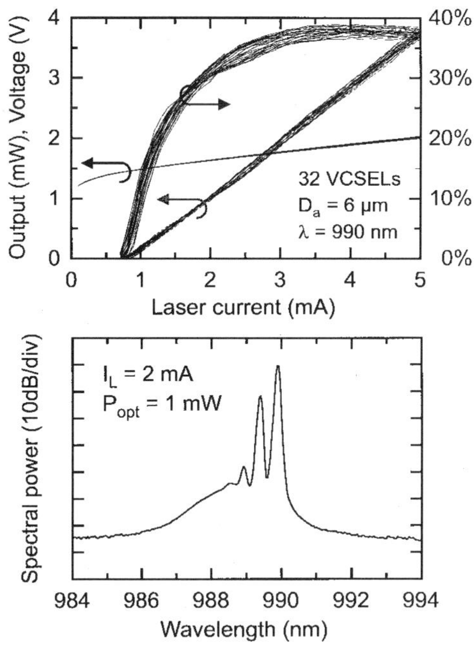

Fig. 15 Laser characteristics of a 990 nm wavelength, 4 × 8 elements bottom-emitting VCSEL array (top) and optical spectrum of a single device at 2 mA driving current, corresponding to 1 mW optical output power (bottom).

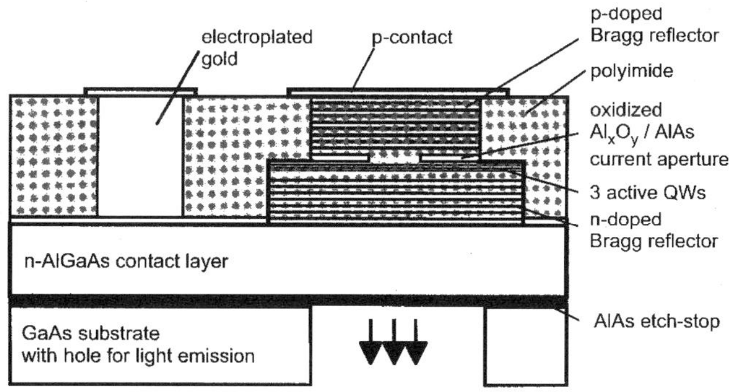

Fig. 16 Schematic layout of an 850 nm VCSEL with recessed GaAs substrate to enable bottom emission.

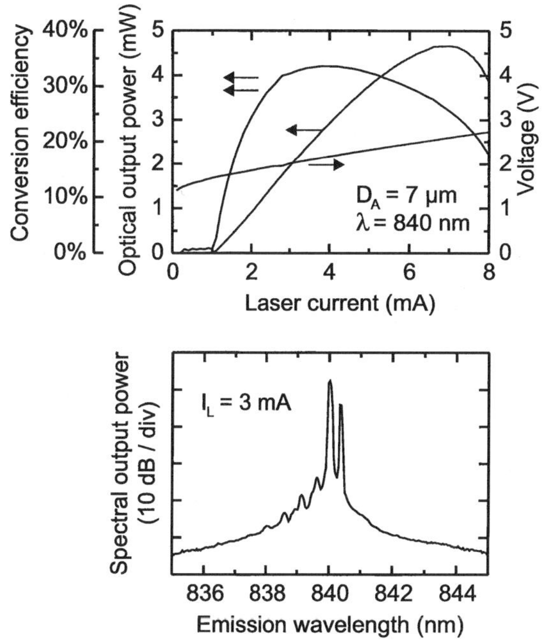

Fig. 17 Operation characteristics of a 7 μm active diameter bottom-emitting VCSEL (top) and optical spectrum at 3 mA driving current (bottom).

tical interconnection between laser and detector arrays is accomplished through 120 or 62 μm core diameter step-index POF bundles placed on top of the chips without additional optics, so that transverse multimode VCSEL operation, as seen in the optical spectrum in the bottom part of Fig. 15, is perfectly acceptable. If required, transverse singlemode operation can be achieved with smaller active diameter arrays. Devices with about 3 μm diameter delivering up to 3 mW fundamental mode output power at 30 dB sidemode suppression ratio have been demonstrated before [29].

Since most of the short-wavelength VCSEL work nowadays concentrates on the 850 nm spectral region, direct compatibility of on-chip interconnects with, e.g., the Gigabit Ethernet world would certainly be highly attractive. Unfortunately, since the GaAs substrate is highly absorptive at 850 nm, the 980 nm device concept cannot be simply transferred to the shorter wavelength regime. Bottom-emitting VCSEL arrays have been realized before by means of wafer bonding to a transparent substrate [30], [31]. Alternatively one can remove the opaque substrate after flip-chip bonding and leave only a thin film of epitaxially grown material or even individual islands. These concepts have been demonstrated with LEDs [32] or optical modulators [33]. As sketched in Fig. 16, our approach to realize bottom-emitting VCSEL arrays is to process about 50 μm diameter holes into the GaAs substrate by means of dry and subsequent wet chemical etching.

The device layout is equivalent to that of the 980 nm arrays [29], except for an additional transparent n-type AlGaAs contact layer of about 8 μm thickness and an AlAs etch-stop layer.

As seen in Fig. 17, VCSELs with 7 μm active diameter show threshold currents and voltages of 1.1 mA and 1.7 V, respectively. Maximum output power of 4.7 mW is obtained at about 7 mA current. In the linear light versus current regime the differential quantum efficiency equals 70%. The peak conversion efficiency from electrical to optical power of 34% is reached at about 4 mA current and 2.9 mW output power. The optical spectrum is centered around 840 nm and shows emission in several transverse modes. Although the devices have not been characterized dynamically yet, 10 Gb/s high-speed modulation capability is also to be expected, as has been achieved before with similar layout 980 nm VCSEL arrays [29].

8. Conclusions

We have reported on some future applications of 10 Gb/s high-speed optical interconnects based on 850 or 980 nm VCSELs. Through the development of a new generation multimode fiber optimized for the 850 nm wavelength region, VCSELs will further be the key element for the fabrication of low-cost transceivers for upcoming local area network (LAN) standards like 10-GbE. We have demonstrated 10 Gb/s transmission over 1.6 km and 4 × 10 Gb/s over 310 m of such a fiber. CWDM with higher channel count might be able to satisfy the bandwidth demand of in-building backbones at least for a while.

For fiber-parallel links up to several 10 m distance we have fabricated first 1 × 10 VCSEL arrays enabling 100 Gb/s aggregate data throughput. Data transfer within computer clusters will continue to benefit from next generation modules employing this technology.

In order to overcome the problems caused by high electrical resistances in small-area singlemode VCSELs, we have devised a new design route relying on the incorporation of a long monolithic cavity. First devices deliver 5 mW of singlemode output power for 7 μm active diameter.

Taking advantage of the rapid performance progress of plastic optical fibers, we have shown 830 nm data transmission at a rate of 9 Gb/s over 100 m length of a low-loss perfluorinated type of GI POF. With some further reductions of the fiber loss and successfully passed environmental and stability tests, these fibers can position themselves as a viable alternative to silica MMFs for a broad range of applications.

An attractive chance to improve the intraboard parallel high-speed data transport could be the introduction of hybrid PCBs, serving as a substrate for both electronic and optoelectronic components. As a first demonstration of the bandwidth potential, we have sent 10 Gb/s data along a short prototype board. Waveguide loss reductions and the integration of beam deflectors are among the challenges to be met in the near future.

Two-dimensional VCSEL arrays are especially useful for inter-MCM or interchip interconnections. We have fabricated 32-elements 980 nm arrays with excellent operation characteristics and up to 12.5 Gb/s per channel modulation capability. Successful transfer of the technology to the standardized 850 nm spectral region was shown and opens up new perspectives for direct chip-to-LAN communications.

Acknowledgements

This work was supported by the Deutsche Forschungsgemeinschaft, the Bundesministerium für Bildung und Forschung, as well as the European Commission (contract No. 22641, OIIC). We wish to thank Picometrix Inc., USA and Asahi Glass Co., Japan for providing a prototype MMF receiver module and high-performance POF samples, respectively. The LazrSPEED^{TM} fiber experiments were carried out at Bell Labs, Lucent Technologies, Holmdel, NJ and would not have been possible without the help of G. Giaretta, A.J. Ritger, K.W. Goossen, J.A. Walker, A.L. Lentine, and M.C. Nuss. S. Lehmacher and A. Neyer from the Arbeitsgebiet Mikrostrukturtechnik, Universität Dortmund, Germany have fabricated the polymer optical layers which were integrated into PCBs by ILFA in Hannover, Germany.

References

[1] C. Wilmsen, H. Temkin, and L.A. Coldren, eds., Vertical-Cavity Surface-Emitting Lasers, Cambridge University Press, Cambridge, 1999.

[2] D. Wiedenmann, R. King, C. Jung, R. Jäger, R. Michalzik, P. Schnitzer, M. Kicherer, and K.J. Ebeling, "Design and analysis of single-mode oxidized VCSEL's for high-speed optical interconnects," IEEE J. Sel. Top. Quantum Electron., vol.5, pp.503–511, 1999.

[3] R. Michalzik, R. King, G. Giaretta, R. Jäger, and K.J. Ebeling, "VCSEL arrays for 10 Gb/s multimode fiber optical interconnects," in Optoelectronic Interconnects VII, eds., S. Tang and J.P. Bristow, Proc. SPIE 3952A, pp.124–133, 2000.

[4] G. Giaretta, R. Michalzik, and A.J. Ritger, "Long distance (2.8 km), short wavelength (0.85 μm) data transmission at 10 Gb/sec over new generation high bandwidth multimode fiber," Conf. Lasers and Electro-Opt., CLEO 2000, postdeadline paper CPD13, San Francisco, CA, USA, May 2000.

[5] L.B. Aronson, B.E. Lemoff, L.A. Buckman, and D.W. Dolfi, "Low-cost multimode WDM for local area networks up to 10 Gb/s," IEEE Photon. Technol. Lett., vol.10, pp.1489–1491, 1998.

[6] B.E. Lemoff, L.A. Buckman, A.J. Schmit, and D.W. Dolfi, "A compact, low-cost WDM transceiver for the LAN," Proc. 50th Electron. Comp. & Technol. Conf., ECTC, pp.711–716, Las Vegas, NV, USA, May 2000.

[7] R. Michalzik, G. Giaretta, K.W. Goossen, J.A. Walker, and M.C. Nuss, "40 Gb/s coarse WDM data transmission with 825 nm wavelength VCSELs over 310 m of high-performance multimode fiber," Proc. 26th Europ. Conf. on Opt. Commun., ECOC 2000, vol.4, pp.33–34, Munich, Germany, Sept. 2000.

[8] K. Drögemüller, D. Kuhl, J. Blank, M. Ehlert, T. Kraeker, J. Höhn, D. Klix, V. Plickert, L. Melchior, I. Schmale, P. Hildebrandt, M. Heinemann, F.P. Schiefelbein, L. Leininger, H.-D. Wolf, T. Wipiejewski, and A. Ebberg, "Current progress of advanced high speed parallel optical links for computer clusters and switching systems," Proc. 50th Electron. Comp. & Technol. Conf., ECTC, pp.1227–1235, Las Vegas, NV, USA, May 2000.

[9] H. Unold, M. Grabherr, F. Eberhard, F. Mederer, R. Jäger, M. Riedl, and K.J. Ebeling, "Increased-area oxidised single-fundamental mode VCSEL with self-aligned shallow etched surface relief," Electron. Lett., vol.35, pp.1340–1341, 1999.

[10] H.J. Unold, S.W.Z. Mahmoud, R. Jäger, M. Kicherer, M.C. Riedl, and K.J., Ebeling, "Improving single-mode VCSEL performance by introducing a long monolithic cavity," IEEE Photon. Technol. Lett., vol.10, pp.939–941, 2000.

[11] C. Jung, R. Jäger, M. Grabherr, P. Schnitzer, R. Michalzik, B. Weigl, S. Müller, and K.J. Ebeling, "4.8 mW singlemode oxide confined top-surface emitting vertical-cavity laser diodes," Electron. Lett., vol.33, pp.1790–1791, 1997.

[12] K.D. Choquette, A.J. Fischer, K.M. Geib, G.R. Hadley, A.A. Allerman, and J.J. Hindi, "High single mode operation from hybrid ion implanted/selectively oxidized VCSELs," Proc. IEEE Int.'l Semicond. Laser Conf., pp.59–60, Monterey, CA, USA, Sept. 2000.

[13] S.W.Z. Mahmoud, H.J. Unold, W. Schmid, R. Jäger, R. Michalzik, and K.J. Ebeling, "Analysis of longitudinal mode waveguiding in vertical-cavity surface-emitting lasers with long monolithic cavity," Appl. Phys. Lett., vol.78, pp.586–588, 2001.

[14] F. Mederer, M. Kicherer, R. Jäger, H. Unold, and K.J. Ebeling, "2.5 Gbit/s data transmission with singlemode GaAs VCSELs over -20 to $100°$C temperature range," Electron. Lett., vol.36, pp.236–238, 2000.

[15] M. Webster, R.V. Penty, I.H. White, M.R.T. Tan, and S.W. Corzine, "Performance of 10 Gbit/s VCSELs operating at 85°C," Proc. Conf. on Lasers and Electro-Opt., CLEO 2000, pp.201–202, San Francisco, CA, USA, May 2000.

[16] J. Marcou, ed., Plastic Optical Fibres: Practical Applications, J. Wiley & Sons, Chichester, 1997.

[17] See, e.g., Proc. International Plastic Optical Fibres Conferences, Berlin, Germany, Oct. 1998 or Makuhari Messe, Chiba, Japan, July 1999.

[18] T. Onishi, H. Murofushi, Y. Watanabe, Y. Takano, R. Yoshida, and M. Naritomi, "Recent progress of perfluorinated GI POF," Proc. 7th International Plastic Optical Fibres Conference'98, pp.39–42, Berlin, Germany, Oct. 1998.

[19] P. Schnitzer, M. Grabherr, R. Jäger, F. Mederer, R. Michalzik, D. Wiedenmann, and K.J. Ebeling, "GaAs VCSEL's at $\lambda = 780$ and 835 nm for short-distance 2.5-Gb/s plastic optical fiber data links," IEEE Photon. Technol. Lett., vol.11, pp.767–769, 1999.

[20] M. Kicherer, F. Mederer, H. Unold, K.J. Ebeling, S. Lehmacher, E. Griese, and M. Naritomi, "VCSEL based high-speed data transmission over polymer optical fibers and circuit board integrated waveguides," Proc. International Optoelectronics Symposium, pp.7–8, Kyoto, Japan, June 2000.

[21] G. Giaretta, W. White, M. Wegmuller, and T. Onishi, "High-speed (11 Gb/sec) data transmission using perfluorinated graded-index polymer optical fibers for short interconnects (< 100 m)," IEEE Photon. Technol. Lett., vol.12, pp.347–349, 2000.

[22] F. Mederer, R. Jäger, P. Schnitzer, H. Unold, M. Kicherer, K.J. Ebeling, M. Naritomi, and R. Yoshida, "Multi-Gb/s graded-index POF data link with butt-coupled single-mode InGaAs VCSEL," IEEE Photon. Technol. Lett., vol.12, pp.199–201, 2000.

[23] G. Giaretta, F. Mederer, R. Michalzik, W. White, R. Jäger, G. Shevchuk, T. Onishi, M. Naritomi, R. Yoshida, P. Schnitzer, H. Unold, M. Kicherer, K. Al-Hemyari, J.A. Valdmanis, M. Nuss, X. Quan, and K.J. Ebeling, "Demonstration of 500 nm-wide transmission window at multi-Gb/s data rates in low-loss plastic optical fiber," Proc. 25th Europ. Conf. on Opt. Commun., ECOC'99, vol.II, pp.240–241, Nice, France, Sept. 1999.

[24] J. Moisel, J. Guttmann, H.-P. Huber, O. Krumpholz, M. Rode, R. Bogenberger, and K.-P. Kuhn, "Optical backplanes with integrated polymer waveguides," Opt. Eng., vol.39, pp.673–679, 2000.

[25] E. Griese, "Reducing EMC problems through an electrical/optical interconnection technology," IEEE Trans. Electromagnetic Compatibility, vol.41, pp.502–509, 1999.

[26] D. Krabe, F. Ebling, N. Arndt-Staufenbiel, G. Lang, and W. Scheel, "New technology for electrical/optical systems on module and board level: The EOCB approach," Proc. 50th Electron. Comp. & Technol. Conf., ECTC, pp.970–974, Las Vegas, NV, USA, May 2000.

[27] M. Kicherer, F. Mederer, R. Jäger, H. Unold, K.J. Ebeling, S. Lehmacher, A. Neyer, and E. Griese, "Data transmission at 3 Gbit/s over intraboard polymer waveguides with GaAs VCSELs," Proc. 26th Europ. Conf. on Opt. Commun., ECOC 2000, vol.3, pp.289–290, Munich, Germany, Sept. 2000.

[28] L. Vanwassenhove, R. Baets, M. Brunfaut, J. Van Campenhout, J. Hall, K. Ebeling, H. Melchior, A. Neyer, H. Thienpont, R. Vounckx, J. Van Koetsem, P. Heremans, F.-T. Lentes, and D. Litaize, "Two-dimensional optical interconnect between CMOS IC's," Proc. 50th Electron. Comp. & Technol. Conf., ECTC, pp.231–237, Las Vegas, NV, USA, May 2000.

[29] R. King, R. Michalzik, D. Wiedenmann, R. Jäger, P. Schnitzer, T. Knödl, and K.J. Ebeling, "2D VCSEL arrays for chip-level optical interconnects," in Optoelectronic Interconnects VI, eds., J.P. Bristow and S. Tang, Proc. SPIE 3632, pp.363–372, 1999.

[30] K.D. Choquette, K.M. Geib, B. Roberds, H.Q. Hou, R.D. Twesten, and B.E. Hammons, "Short wavelength bottom-emitting vertical cavity lasers fabricated using wafer bonding," Electron. Lett., vol.34, pp.1404–1405, 1998.

[31] C.-K. Lin, S.-W. Ryu, and P.D. Dapkus, "High-performance wafer-bonded bottom-emitting 850-nm VCSEL's on undoped GaP and sapphire substrates," IEEE Photon. Technol. Lett., vol.11, pp.1542–1544, 1999.

[32] P. Heremans, M. Kuijk, S. Peeters, R. Windisch, D. Filkins, and S. Borghs, "Thin-film 850 nm light emitters on fiber optic face plate," Proc. 25th Europ. Conf. on Opt. Commun., ECOC'99, pp.II-304–II-305. Nice, France, Sept. 1999.

[33] A.V. Krishnamoorthy and K.W. Goossen, "Optoelectronic-VLSI: Photonics integrated with VLSI circuits," IEEE J. Sel. Top. Quantum Electron., vol.4, pp.899–912, 1998.

High-Performance Backbone Network Technology edited by Naoaki Yamanaka
ISBN: 0-8247-5321-6 DOI: 10.1081/(120025753)
Published 2004 by Marcel Dekker, Inc.

3D Control of Light in Waveguide-Based Two-Dimensional Photonic Crystals

Claude Weisbuch, Henri Benisty, Ségolène Olivier, Maxime Rattier, Christopher J.M. Smith, and Thomas F. Krauss

SUMMARY Photonic crystals have seen major advances in the past few years in the optical range. The association of in-plane waveguiding and two-dimensional photonic crystals (PCs) in thin-slab or waveguide structures leads to good 3D confinement with easy fabrication. Such structures, much easier to fabricate than 3D PCs open many exciting opportunities in optoelectronic devices and integrated optics. We present experiments on a variety of structures and devices, as well as modelling tools, which show that 2D PCs etched through waveguides supported by substrates are a viable route to high-performance PC-based photonic integrated circuits (PICs). In particular, they exhibit low out-of-plane diffraction losses. Low-loss waveguides, high finesse microcavities, and their mutual coupling are demonstrated. PACS: 42.70 QS, 42.55 Sa, 42.82 m, 42.50-p.

key words: *photonic crystals, integrated optics, photonic integrated circuits, optical waveguides, microcavities*

1. Introduction: The Case for Waveguide-Based 2D Photonic Crystals

In three-dimensional periodic dielectric structures, the existence of photonic transmission and forbidden bands has led to contemplate new avenues to tailor the light-matter interaction, particularly spontaneous emission. Equally important, these structures should also allow to design new photonic systems, such as integrated optics components, with superior properties [1].

At optical wavelengths, three-dimensional structures have been recently fabricated such as "woodpiles," stacked dielectric spheres, "inverted opals," self-organized structures, etc., [1] but they have a number of drawbacks: reproducibility and quality problems, difficulty of precise quantitative measurements (one uses approximately-collimated optical beams to measure finite structures) and, more importantly, they lack the design versatility required to implement functional devices: it will be difficult to insert the needed photonic defects, active species, electrodes in self-organized systems in a deterministic way.

One should still be impressed by the magnificent results obtained over the years in 3D systems [1]: first on the theoretical side, after some initial wanderings around cubic structures, due to the consideration of a scalar propagation equation, the diamond structure (to which the "Yablonovite" structure belongs) emerged as a natural candidate, dictated by the properties of the vector electromagnetic propagation equation. While such a structure appeared difficult to fabricate, several other structures having the full 3D bandgap property were proposed, easier to fabricate, such as the woodpile structure or the "MIT" structure. Even more recently, the inverse opal structure appeared, with considerable success.

While basic dispersion properties of full 3D bandgaps have been observed, the impact on optical properties is so far rather limited (with the exception of the "giant" dispersion effect [2], however actually a 2D effect). Whereas one expects in well-designed systems to be able to control the spontaneous emission pattern, or even more the spontaneous emission lifetime, only limited modifications of such quantities are observed [3]: this is due to the fact that the emitting species for which one expects changes (dye molecules adsorbed on the PC surfaces for instance) are distributed randomly in position and orientation, which averages the expected effects compared to their maximum values, and, more importantly, the observed emission mainly originates from those molecules situated near the PC-air interface, for which the emission is less "forbidden" by the crystal, but which then exhibit less of a control of their emission.

One has then to consider resorting to structures more amenable to fabrication along design. 2D structures allow this, owing to existing mature nanofabrication technology. The question which then arises is whether there is an unacceptable loss of photon control as compared to 3D PCs. Taking as a similar issue the architecture of integrated circuits, one immediately notices that the electron motion is in a plane (not exactly 2D in the quantum sense, as electron motion is not quantized in the vertical direction, but remember that if that motion is really 2D the transistors perform even better, as shown by the success of HEMTs). Such a

Manuscript received December 1, 2000.

[†]The authors are with Laboratoire de Physique de la Matière Condensée, Ecole Polytechnique, 91120-Palaiseau, France.

[††]The authors are with the Department of Electronics and Electrical Engineering, Glasgow University, G12 8LT, Glasgow, UK.

[†††]The author is with the School of Physics and Astronomy, University of St Andrews, St Andrews, KY16 9SS, UK.

[*]This paper is also published in the IEICE Trans. Electron., Vol.E84-C, No.5, pp.660–668, May 2001.

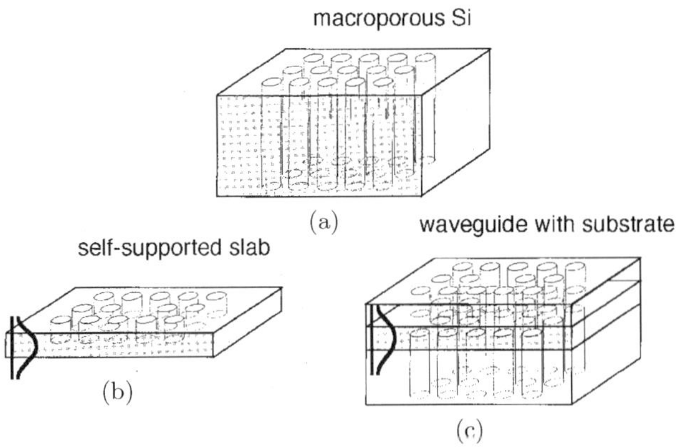

Fig. 1 Schematic of 2D photonic crystals: (a) the homogeneous (here macroporous Silicon), (b) slab and (c) waveguide-supported types.

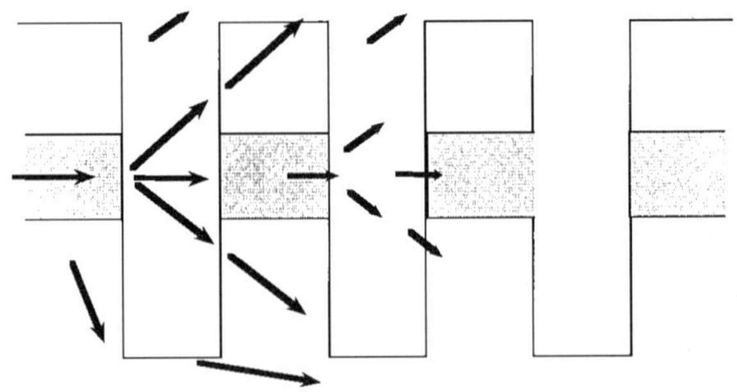

Fig. 2 Schematic of diffraction at air holes.

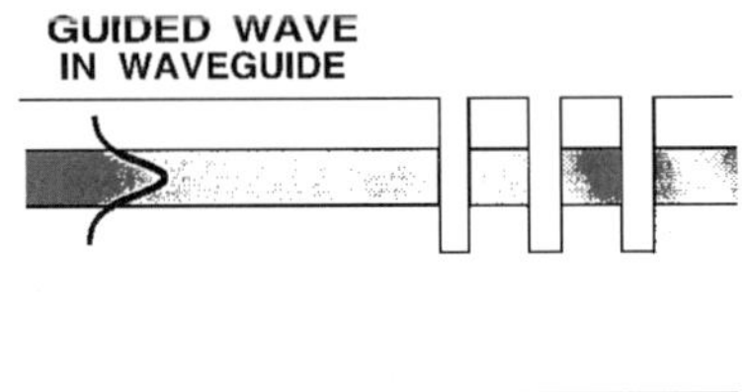

Fig. 3 Schematics of the optical quasi-2D system where vertical confinement is obtained by heterostructure waveguiding, and horizontal confinement by light control by a 2D PC of air holes etched into the semiconductor heterostructure.

2D architecture provides plenty of functionalities, as is obvious. The many past attempts to achieve 3D IC architectures have failed because the gain in performance (integration density) was not so great, sometimes at the expense of single-device performance loss as the 3D material cannot be high-quality, all this with a dramatic increase in fabrication difficulty. The same line of reasoning could well apply to photonic integrated circuits (PICs): As a consequence, 2D-PCs are being widely studied [1].

2D PCs come in three main forms: one is the 2D array of air holes (Fig. 1(a)) in a dielectric or dielectric pillars in air. A quasi 2D system is then obtained for extremely high aspect ratio if the optical beams to be controlled are reasonably collimated within the 2D plane, perpendicular to the structure's axis. A second one is the slab waveguide-based 2D PC, consisting of a 2D array of holes perforated through a thin membrane [1], [4] (Fig. 1(b)); the third one consists in holes drilled through a classic heterostructure waveguide similar to those of heterostructure lasers (Fig. 1(c)) [5]. The first type of 2D PC is not easily used as the device size has to be limited in all 3D because of beam divergence and collimation problems. Due to the natural indefinite collimation occuring in a waveguide plane, the two latter ones are quite useful: while the slab approach has some advantages, in particular the ease of the etching procedure as only a thin slab has to be pierced, the moderate size of self-supported structures jeopardizes the fabrication of the large scale photonic integrated circuits (PICs) required for the next generation of high-throughput optical communications systems. The latter solution, conversely, allows for large-scale PICs. It however makes the aspect ratio of these structures a crucial issue, as the holes have to overlap completely with the guided wave in order to display 2D PC features. It also raises the issue of light losses to the substrate which was long regarded as precluding this approach. On the other hand, it is well-known that such structures are compatible with electrodes positioned on the waveguide top, therefore allowing electrically active devices along this scheme.

It is the purpose of this paper to review the properties achieved by our group on such waveguide-based structures, which amply demonstrate the viability of this approach. One main concern in the past has been that of the radiation loss due to the air holes perforating the waveguide: the vertical waveguide index contrast is small compared to that of the air holes with the semiconductors. It has therefore been long thought that the air holes would destroy the waveguiding effect by diffraction loss occuring at each hole (Fig. 2). We have shown that the planar waveguiding provided by the heterostructure, although appearing as "weak" in the first approximation, is on the contrary very robust against the perturbation introduced by the PC air holes. We have also shown that the presence of the substrate does not alter the PC performance, provided that the holes are indeed etched deep enough.

2. Design and Experimental Procedures

One creates a quasi-2D photonic structure, analogous to quantum wells for electrons, by using waveguided light propagation [5]. In such a case, the optical beam is perfectly defined and collimated and its interaction with 2D PCs etched across the waveguide can be seen as providing the frame of a well-defined 2D-physics experiment. In addition the waveguide naturally freezes the third degree of freedom, which can lead to excellent 3D optical confinement whenever in-plane 2D confinement is provided by a 2D PC structure (Fig. 3).

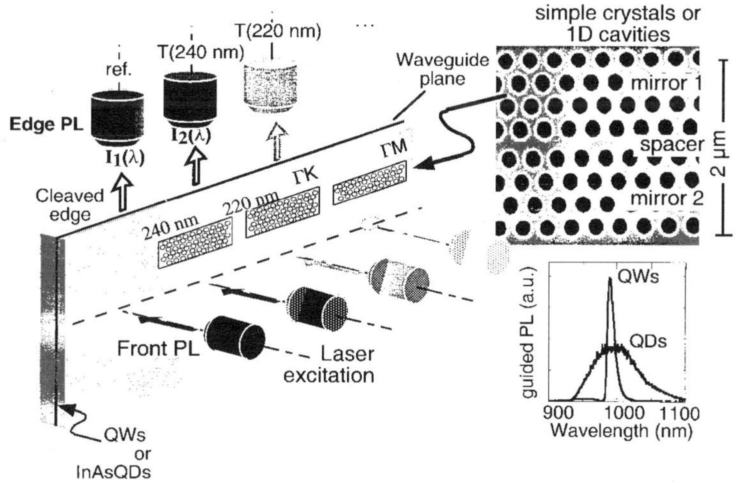

Fig. 4 Schematic of the experimental set-up for quantitative measurements of 2D photonic crystals, using internal light sources in the waveguide. Insets: PL spectra of imbedded InGaAs quantum wells or InAs quantum dots; schematics of 1D microcavity defined by two PC mirrors.

Semiconductor-based PC structures offer the opportunity to include active materials, therefore allowing to integrate both active and passive components. Here, for our measurement purposes, the spontaneous photoluminescence (PL) of quantum wells or quantum dots embedded in the waveguide produces an internal quasi-2D light beam which then probes the 2D-PCs (Fig. 4) [5]. As PL produces only a finite range of wavelengths λ, patterns with various lattice constants a allow to explore the various gaps and bands of PCs as a function of the reduced parameter a/λ. Figure 4 shows the schematics of the sample layout and experimental setup. Using a variety of layouts, one explores both spectral and directional properties. The ratio of light intensity for beams having propagated through patterned or unpatterned waveguides provides the absolute transmission coefficient of the pattern. For patterns located near the cleaved sample edge, the observed spectral fringes, due the cavity effect occuring between the mirrors provided by the cleaved edge and the pattern, allows to determine at once both the transmission and reflection coefficients of the pattern, as the reflectivity of the cleaved edge is well known. In some other geometry where the exciting light is placed between the cleaved edge and the PC pattern, in-plane diffraction by the PC can be observed.

Such measurements are quantitative with a relative precision on the order of 5% over up to 3 decades. Uncalibrated intensity measurements can also be observed in "front" photoluminescence measurements (Fig. 4). In that case, the signal is provided by light which "leaks out" of the waveguide plane into air. The efficiency of this emission mechanism is hard to quantify as it depends on many parameters (angle, aperture, etching fluctuations, ...). Frontal measurements sometimes provide however absolute spectral informations: they allow the determination of cavity energy spectra, and more importantly their quality factor Q, an absolute measurement for integrated losses (see below Sect. 3.2).

3. PC Structure Properties

3.1 Photonic Crystals Parameters

A first series of studies established the basic kinematic properties of 2D PCs such as transmission, reflection and diffraction coefficients (Fig. 5). It is now well established that, for a variety of structures and materials, excellent, quasi-intrinsic properties can be obtained. The measured features are in line with infinite 2D system calculations, which underlines the correctness of the quasi-2D description of such systems [5].

Beyond reflection and transmission, other kinematic properties are of high interest. High diffraction efficiencies can be observed for the incidence along ΓK direction, as high as 90%! This is an example of the importance of dynamical diffraction effects which lead to the dominance of the build-up of a backwards diffracted wave at the expense of the specular-reflected wave. Giant dispersion effects could also be of paramount importance in such structures.

The measurements of the fine oscillatory structure of the transmission spectra, due to the finiteness of the PC slab, are made possible by using the broad emission line of quantum dots to probe PCs (Fig. 6). Besides reproducing the fine structure predicted by theory (see Fig. 5), the sharpness of the "dielectric" band edge (i.e. corresponding to lightwave modes mainly localized in the PC dielectric) emphasizes the excellent and intrinsic behaviour of the bandgap; in other terms that there is no effect of fabrication fluctuations. In contrast, a smooth band is edge observed for the "air band." We

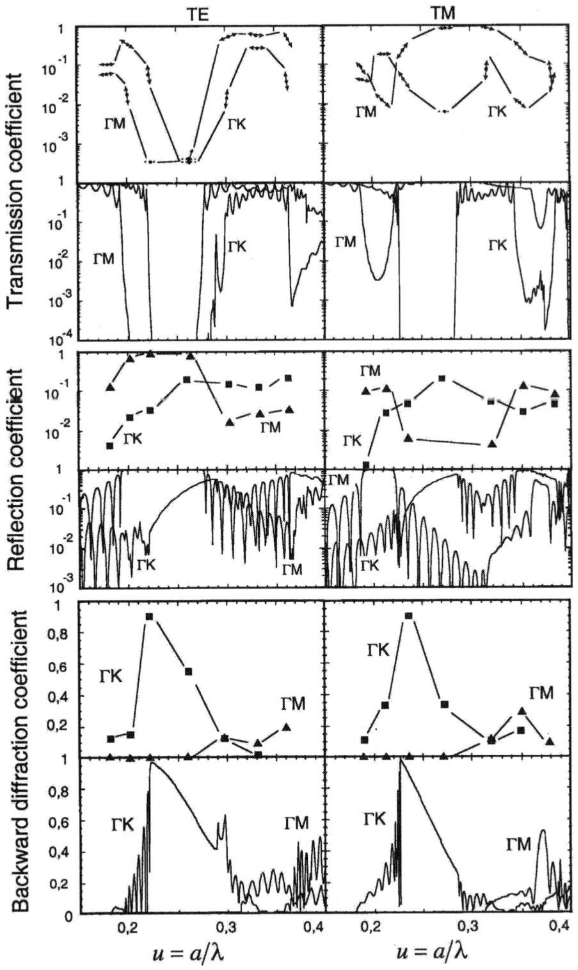

Fig. 5 Transmission, reflection, diffraction measurements for TE and TM polarizations and propagation along ΓM or ΓK directions (using QW light source for 15 rows PC patterns with a period between 180 and 300 nm).

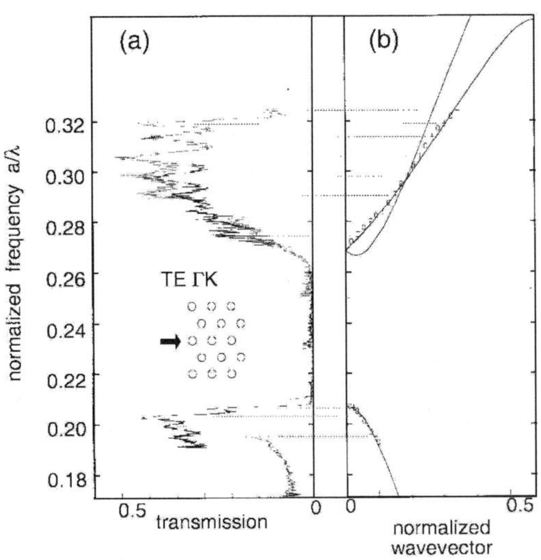

Fig. 6 Transmission experiments using QD light source for 8 rows PC patterns with a period between 180 and 300 nm (a). Photonic crystal dispersion curve deduced from the analysis of the transmission fine structure (b).

shall show that its origin is also intrinsic. We shall describe it in part 4 as due to radiation losses.

3.2 PC-Based Microcavities

High-performance microcavities are at the heart of many of the foreseen applications of PCs, such as lasers, filters, etc. They also serve to evaluate PC parameters: not only is low transmission desired in the gap, but even more desirable is high reflectivity (specular or not). While it is always difficult to calibrate reflectivities close to unity, it is obvious that measuring reflection and transmission peaks of Fabry-Perot cavities yield the most precise information regarding the mirror reflectivities and losses. An horizontal cavity analogous to the planar Fabry-Perot cavities can be created by the introduction of a spacer between two PCs, or in other words, by introducing a line defect in the form of a larger separation between two of the rows of a PC (inset of Fig. 4). Q-factors up to 200 have been measured for such structures with spacers of the order of the wavelength.

Another type of horizontal cavity, with *circular boundaries*, allowed us to confine quasi-radial modes (QRMs) instead of the whispering-gallery modes (WGMs) of microdisks (Fig. 7, left) [5]. Quasi radial modes are much more amenable to controlled radiation outside the cavity than WGMs.

A still more desirable configuration, however, is to confine all the modes of a given area by means of a 2D PC boundary that achieves omnidirectional reflection. We designed such hexagonal in-plane cavities carved into an otherwise regular triangular PC. Remarkably high Q-factors are observed, around 1000 and most certainly limited by spectrometer resolution. Such performance points to the low losses of the PC-in waveguide approach (Fig. 7, right).

4. The Issue of Losses in Waveguide-Based 2D PCs

Beyond the basic "kinematic" properties of PCs it is essential to evaluate their other properties, as these will determine the potential applications of PCs. For instance, radiation losses in the substrate or superstrate around the slab represent an unwanted loss mechanism for resonator or integrated optics purposes, or a welcome extraction mechanism for light in LEDs.

There is quite some debate about which 2D structures would be of sufficiently low loss to allow significant integrated optoelectronic systems to be made. Membrane-supported crystals have in principle some modes which lie below the light cone, therefore being uncoupled to outside radiative modes. This is only true for infinite systems, and there is a need to quantify the properties of such systems when dealing with real-world, finite-size structures. Also, the coupling ef-

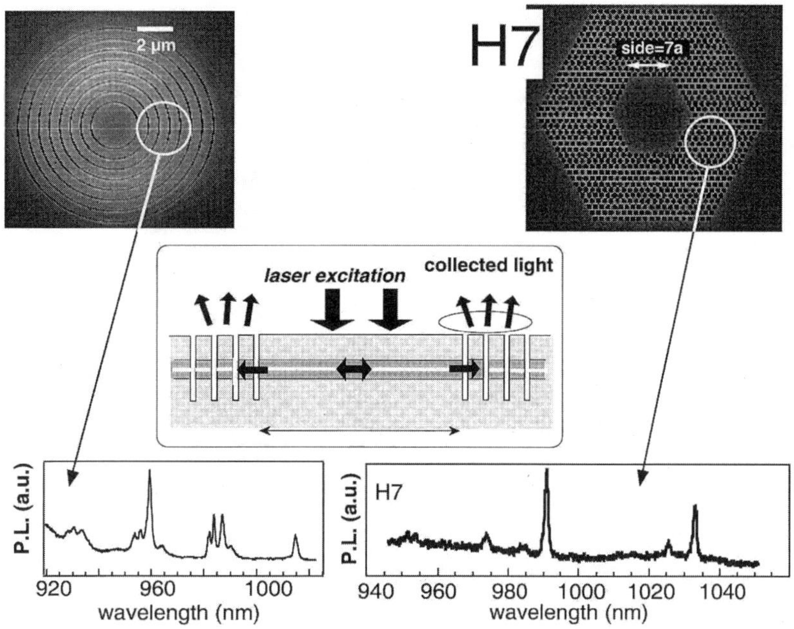

Fig. 7 Top: micrographs of PC-bounded microcavities, with circular bragg mirror (left) or 2D triangular PC (right); Middle: schematics of the measurement; Bottom: observed luminescence spectra, displaying sharp modes.

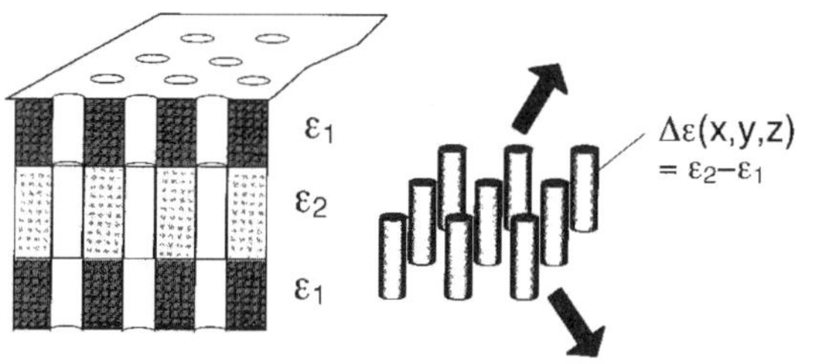

Fig. 8 Schematics of a separable-variable system for in-plane propagation (left), which has no propagation loss, and the dielectric plugs representing the difference with the real-world system obtained when etching air holes through a waveguide.

ficiency of outside modes into such modes needs to be evaluated. Nevertheless, such low-loss modes have recently been observed [6].

In our substrate-supported 2D PCs we also observe low losses. We have developed a perturbation method that gives a realistic estimation of such radiation losses [7]. It is based on the comparison of the *actual system (supposedly with infinitely deep etched holes)* with an *idealized lossless system* based on the separation of in-plane and vertical index, variations leading to a separate-variable equation for light propagation (Fig. 8). The difference between the real and the lossless system is represented by dielectric "plugs" of dielectric constant $\varepsilon_2 - \varepsilon_1$ in the air holes ($\varepsilon_2 - \varepsilon_1$ is the dielectric core-cladding contrast of the waveguide). This result is remarkable as it does not depend on the hole index, which would lead to a large index difference with the waveguide materials. It is easy to calculate the radiative losses in a perturbative spirit as the radiation of these plugs irradiated by the unperturbed, separable field [7]. We find a reasonable agreement with the value obtained from fitting experimental results by introducing in a purely 2D in-plane calculation an appropriate imaginary dielectric constant ε'' in the air holes.

Recently, we have performed more exact 3D calculations of our waveguide-based 2D PCs, taking into account the *finite depth of the holes* [8]. We find that the value for ε'' in the experimental fit of Fig. 9 depends on the PC period a, itself related with the hole depth (as we design the structures at constant air filling factor, the hole diameter changes with the period, and also the hole depth). The agreement for the observed loss is then excellent when considering that the holes have a period-dependent depth $h(a)$ of the form $h = constant \times a + h_o$. This is good news as it means that in most of our structures, losses are mainly due to the lack of hole depth, and are not yet intrinsic to the structure of 2D PCs made on substrate-supported waveguides; the perturbative approach of [7] is likely overestimating the losses.

Such determinations of loss parameters should provide a universal tool to make predictions and improve 2D PCs structures as it is a simple way to incorporate the effect of the third dimension (as radiation losses) into reasonably simple 2D models.

5. Integrated Optics Components

5.1 Waveguides

Photonic crystals allow waveguiding between "photonic

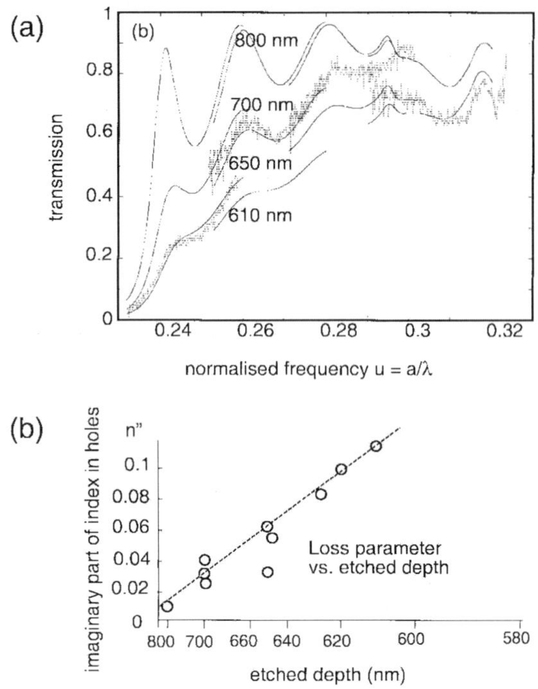

Fig. 9 (a) Measurements and 3D modelling of the "air" band-edge. The hole etching depth is taken as a parameter. As is clear, holes for larger periods a have lower loss, due to their larger depth. (b) loss, as represented in a 2D model by the imaginary part of dielectric constant for air in the holes, as a function of hole depth, assuming a period-dependent depth of the form $h = constant \times a + h_o$.

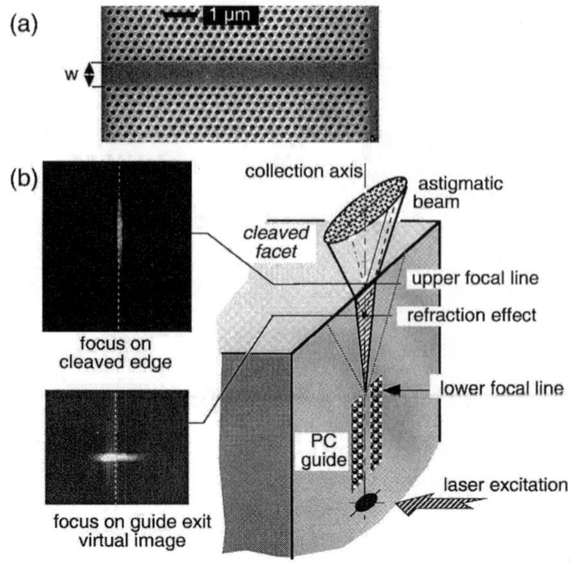

Fig. 10 Principle of measurement of PC waveguides. (a) micrograph of the waveguide; (b) schematics of the measurement procedure and images of the transmitted beam for different focus.

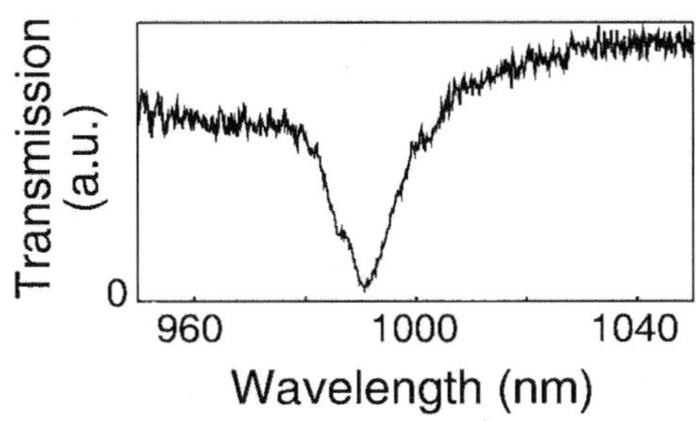

Fig. 11 Measured transmission curve for a PC waveguide displaying a mini stop-band due to Bragg mode coupling by the corrugated waveguide boundary.

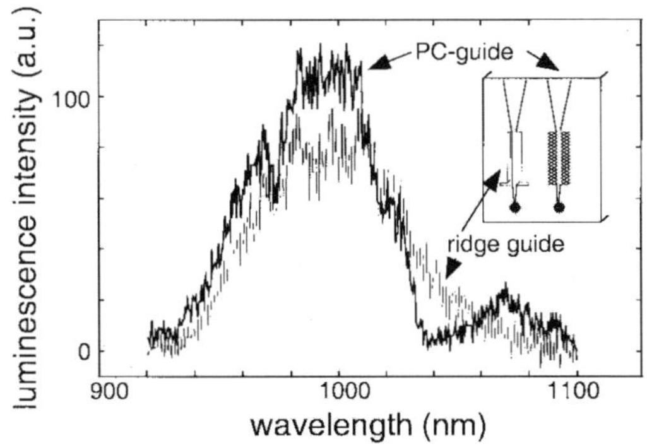

Fig. 12 Comparisons of the raw transmission spectra of a ridge waveguide and a PC bounded waveguide, made in the same run. One observes a higher transmission of the PC-bounded waveguide, except in the spectral region (around 1040 nm) where the mini stopband occurs.

walls" made of crystals used in the bandgap frequency range. Such PC waveguides are very advantageous as any coupling between the waveguide modes and outside modes occurs only through evanescent modes within the PC bandgap. Therefore, such couplings induce no loss, just a mode expansion. This is very useful in order to design very compact waveguides, usually prone to losses by defect or non-adiabatic coupling to radiative modes in narrow or bent waveguides. We recently measured the transmission properties of PC waveguides [9] (Fig. 10). They display a high transmission efficiency, except within a "mini stopband" due to the corrugated nature of the waveguide boundary which induces a Bragg reflexion, equivalently described as an anti-crossing of folded waveguide modes (Fig. 11). Simulations indeed show the anti-crossing behaviour of such mini stopbands. Loss measurements on long waveguides (up to 240 periods) show a lower loss (below 40 cm^{-1}, our measurement limit) than for deeply-etched ridge-waveguides made in the same etching run (Fig. 12).

5.2 Coupled Cavity and Waveguide

As a first demonstration of coupled PC elements we designed coupled PC hexagonal microcavities and PC waveguides separated by some number of PC rows [7]

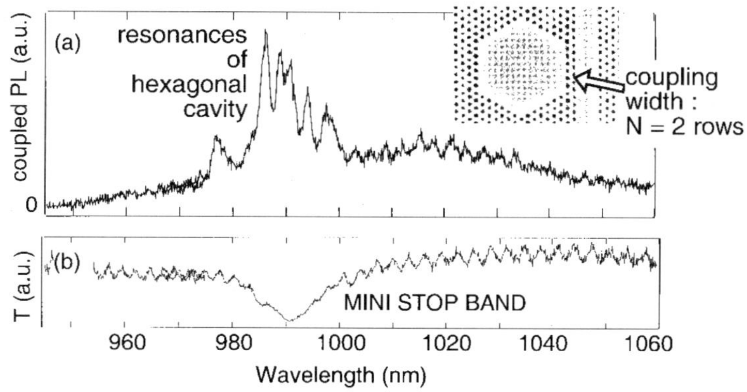

Fig. 13 Coupled PC-based cavity and waveguide (wg) (see micrograph as inset): the measured transmission of the wg (bottom curve) is roughly unity, except in the mini stopband originating in the periodic nature of the wg (see Fig. 11). Sharp resonances observed at the wg end when exciting the cavity (top curve) are to be seen in the mini stopband spectral region, indicating a coupling between cavity and wg modes through "radial" cavity modes [5]. The sharpness of the observed resonances indicates a rather lossless coupling mechanism.

(Fig. 13). We observe rather lossless coupling (as revealed by the high Q-factor of the observed modes). As the coupling occurs in the spectral region of the waveguide "mini stopband," it indicates that the coupling occurs through highly-transverse modes in the waveguide, not surprisingly as one indeed expects such waves to easily tunnel through the PC barrier between the cavity and the waveguide. Such an observation is highly demonstrative of the possibility to cascade PC elements in complex integrated optics designs.

5.3 Coupled Resonators

Another possibility provided by the omnidirectional confinement of 2D-PC is to tailor waveguide by means of chained coupled resonators, the so-called "CROWs" (Coupled Resonator Optical Waveguides) [10]. The modest coupling between the resonators makes it possible to determine, accurately enough, important properties of the waveguide from those of a single resonator. As a step towards such a system, we designed the series of coupled hexagonal cavities of Fig. 14(a) [11]. We indeed observed that the light transmitted by such a system (in the same way as [9]) exhibit marked features, attributed to minibands and minigaps, Fig. 14(b). The 2D simulation, based on the supercell method, of dispersion relation of the infinite system, shown in Fig. 14(c), is in good agreement with the observed behavior. This is an interesting case as it paves the way to an easy implementation of integrated optics waveguide with deterministic spectral shape of the transmission.

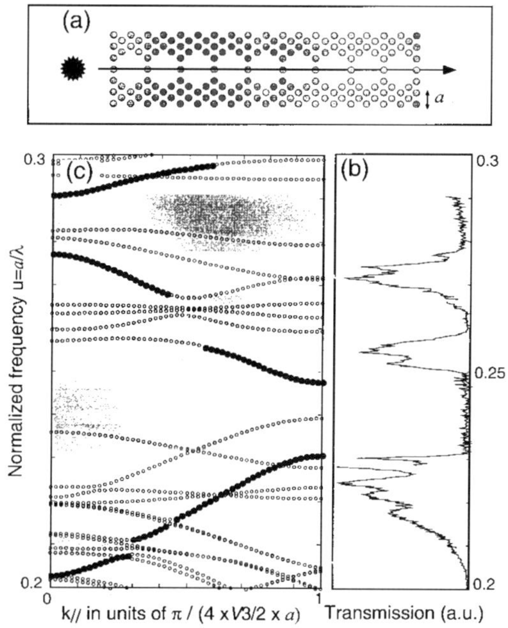

Fig. 14 (a) Scheme of coupled hexagonal cavities etched in a waveguide heterostructure; (b) Transmission spectrum (a.u.) of a system consisting of 9 cavities; (c) dispersion relation of the periodic cavity lattice (the "CROW"), showing the minibands and minigaps.

6. Conclusion: Towards PC-Based Photonic Integrated Circuits (PICs)

The focus of this paper has been the excellent properties of III-V semiconductor-based PCs in a configuration where vertical confinement is provided by a standard heterostructure waveguide and in-plane propagation control by 2D PC elements. They should provide a useful approach to the various functions needed by monolithic active and passive PICs. The excellent features of PC-based PICs can be obtained with three advantages over usual PICs: (i) a very high device density, (ii) the fabrication of multiple functionalities through a single lithographic step, (iii) the use of new functionalities allowed by PCs (such as diffraction or the giant prism effect). There is clearly a bright future for 2D PCs in integrated optics.

Acknowledgments

This work is partially supported by the European Union projects SMILED and PCIC.

References

[1] For a recent overview, see the various papers in the special issue of the IEEE Journal of Lightwave Technology 17, p.1928, 1999, see also the special issue of Optical and Quantum Electronics, 2001.
[2] see, e.g. H. Kosaka, et al., ibid. vol.1, p.2032, 1999.
[3] M. Megens, J.E.G.J. Wijnhoven, A. Lagendjik, and W.M. Vos, Phys. Rev. A59, p.4727, 1999; see also M. Megens, H.P. Schriemer, A. Lagendjik, and W.M. Vos, Phys. Rev. Lett. 83, p.5401, 1999.
[4] S.G. Johnson, S. Fan, P.R. Villeneuve, J.D. Joannopoulos, and L.A. Kolodziejski, Phys. Rev. B 60, p.5751, 1999.
[5] For a review see H. Benisty, et al., J. Lightwave Tech. 17, p.2063, 1999.
[6] E. Chow, et al., Nature 407, p.983, 2000.
[7] H. Benisty, D. Labilloy, C. Weisbuch, C.J.M. Smith, T.F. Krauss, A. Béraud, D. Cassagne, and C. Jouanin, Appl. Phys. Lett., vol.76, p.532, 2000.
[8] H. Benisty, et al., special issue of Optical and Quantum Electronics, 2001.
[9] C.J.M. Smith, et al., Appl. Phys. Lett., vol.77, p.2813, 2000.
[10] A. Yariv, Y. Xu, R.K. Lee, and A. Scherer, Optics lett., vol.24, p.711, 1999.
[11] S. Olivier, et al., to be published.

High-Performance Backbone Network Technology edited by Naoaki Yamanaka
ISBN: 0-8247-5321-6 DOI: 10.1081/(120025754)
Published 2004 by Marcel Dekker, Inc.

Passive Optical Devices for Photonic Networks

Yoshinori Hibino

SUMMARY This paper reviews the recent progress made by those working on optical filters and switches for photonic networks based on dense wavelength division multiplexing (WDM). While various kinds of optical devices have been developed for flexible and large-capacity networks, the key components for the WDM networks are narrow-band filters and switches. Three kinds of optical filter are described in this paper: thin-film interference filters, fiber grating filters and arrayed-waveguide-grating (AWG) filters. The optical switches reviewed here are mechanical fiber-type switches, thermo-optic switches made using planar-lightwave-circuit technologies, total-internal-reflection switches and micro-electromechanical-system switches. Each device has its own advantages, and has been or will be used in point-to-point WDM, optical add/drop mupliplexing systems and optical crossconnect systems. Further advances in optical components and technologies are expected to contribute greatly to the construction of future photonic networks.

key words: *photonic network, optical filter, optical switch, PLC*

1. Introduction

The rapid and global spread of the Internet and multimedia communications is accelerating the growth of optical communications networks. Photonic networks based on wavelength division multiplexing (WDM) systems [1]–[3] have played a key role in increasing the capacity and flexibility of these communications networks. Typical photonic networks are schematically shown in Fig. 1. It started from point-to-point WDM transmission systems, and is now evolving to ring networks. Various optical devices have been developed for WDM-based photonic networks, and some of them have already been installed in commercial communication systems. Among those devices, filter and optical switches are two of the most important passive devices used in the networks. The filters are needed wherever signals with different wavelengths propagating in a fiber must be multiplexed or demultiplexed in WDM networks. Various kinds of optical switches in optical add/drop multiplexing (OADM) systems [4], [5] and optical cross connect (OXC) systems [6] are needed for making the network more flexible. Moreover, the rapid progress made on networks has led to the demands for more channels at a reduced cost. This means we must develop larger scale optical devices those are less expensive.

This paper reviews the recent progress made on filter and switch devices for WDM-based photonic networks. Several bulk-type and fiber-type optical components have already been installed in practical systems, and waveguide-type devices have also been playing an important role recently, because of their suitability for large-scale integration and mass production. This paper therefore focuses on bulk-type, fiber-type and planar-waveguide-type devices.

Manuscript received April 27, 2000.

[†]The author is with NTT Photonics Laboratories, Ibaraki-ken, 319-1193 Japan.

2. Filters

Narrow-band filters are essential as multi/demultiplexers and OADM filters in WDM-based photonic networks. Figure 2 shows a typical spectrum of a WDM filter used in a system with a 100 GHz channel spacing. Since the wavelengths of the signals are densely set at frequencies standardized by the International Telecommunication Union (ITU), the wavelength of the bandpass filters must precisely match the ITU grid frequencies. In addition to the wavelength accuracy, we must achieve high levels of performance including low loss, low crosstalk and a wide passband. Furthermore, the demand is growing for filters with even narrower passbands because network operators want to increase the capacity within the fiber-amplifier-active region, which is typically 1530–1570 nm. While various kind of filter have been developed in response to this demand, there are three major types of filers used in practical systems: bulk-type thin film interference filters, fiber gratings and planar-lightwave-circuit (PLC) types. This section reviews the characteristics of these three types of filter.

2.1 Thin-Film Interference Filters

Narrow-band filters for WDM applications have been developed using thin-film multilayer interference filters, whose principle has been long understood [7]. The structure of the WDM thin-film filter is based on that of the Fabry-Perot (F-P) etalon, which is composed of a cavity and mirrors, and which acts as a bandpass filter. The center wavelength of the passband, which must be precisely controlled to the ITU grid frequency, is determined by the cavity length. A signal at the passband wavelength passes through the filter, and signals at other wavelengths are reflected with a very high reflectivity. A narrow-band thin-film interference filter consists of more than two F-P cavities separated by

IEICE TRANS. COMMUN., Vol. E83-B, No. 10, pp. 2178–2190, October 2001

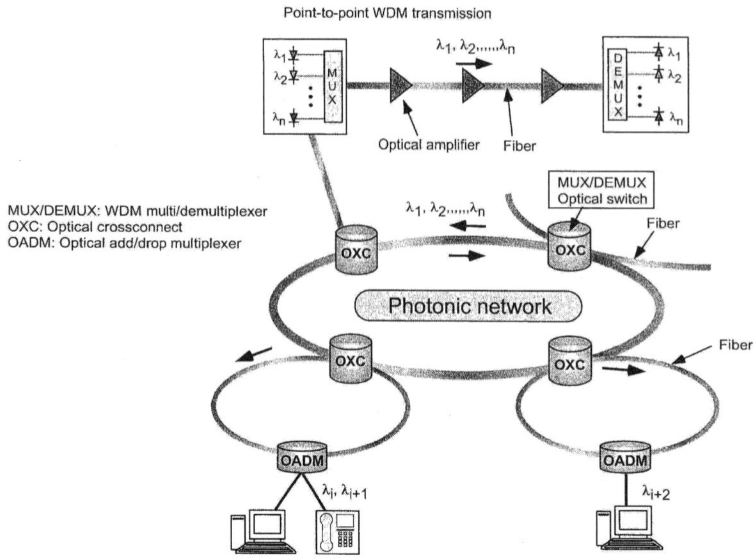

Fig. 1 Typical photonic network configuration.

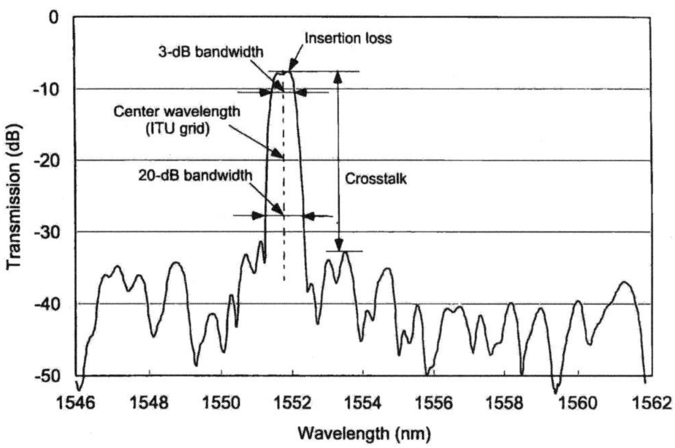

Fig. 2 Typical spectrum of narrow-band optical filter for DWDM systems.

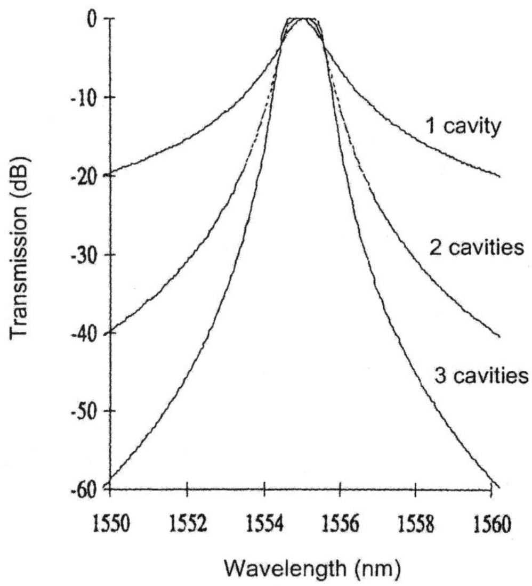

Fig. 3 Spectra of thin-film interference filters with multi cavities.

dielectric reflection layers, and each cavity contains a multilayer structure with more than 50 layers [8].

A thin-film interference filter is fabricated by using micro-plasma techniques to accumulate two kind of dense dielectric layer on a glass substrate. The two kinds of layers are deposited alternately with a unit thickness of half the target wavelength, and the materials used for the layers are usually SiO_2 and TiO_2 because of the large difference between their refractive indices. The spectra of multicavity thin-film filters for WDM application are shown in Fig. 3. The passband shape becomes better for a WDM filter as the number of cavities increases; That is, the top of the passband become flatter and the sides become steeper as shown in Fig. 3. It has thus become possible to obtain filters with a loss of less than 0.5 dB, a wide band and a low crosstalk.

A typical structure of a multi/demultiplexer formed by using thin-film filters is shown in Fig. 4.

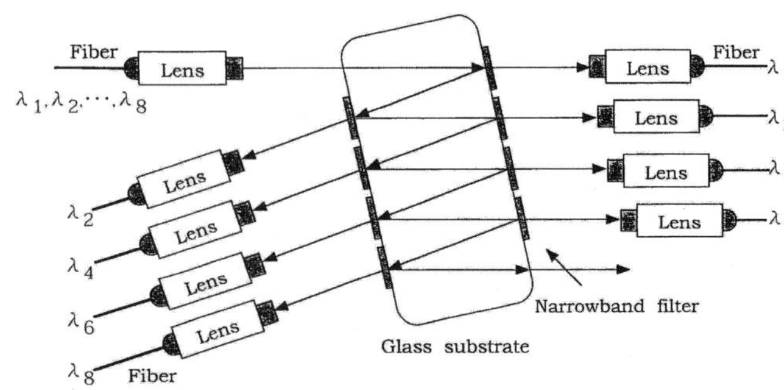

Fig. 4 Configuration of multi/demultiplexer formed with thin-film interference filters.

Thin-film filters with different wavelengths are cascaded in series. The number of filters corresponds to the number of WDM channels. A WDM signal passing through the filter is received by an output fiber with a focusing lens. The characteristics of a multi/demultiplexer made of thin-film filters are a low loss, a low crosstalk, a wide band and low polarization dependence. However, it is rather difficult to reduce the wavelength spacing and increase the port count of this type of multi/demultiplexer because making a thin-film filter with a narrow bandpass becomes more difficult, the assembly cost increases, and the insertion loss accumulates at each stage.

2.2 Fiber Grating Filters

Fiber gratings are attractive devices that can be used for a variety of applications, including filtering, add/drop functions and for dispersion compensation in the system. Their main advantages are a low loss, ease of coupling (with other fibers), polarization insensitivity and simple packaging. It has been reported that the refractive index of GeO_2-doped silica glass is increased by ultraviolet (UV) laser irradiation [9]. This phenomenon, which is known as photosensitivity, has been used to fabricate two kinds of fiber grating: Bragg (short-period) [10] and long-period gratings [11], [12].

2.2.1 Fiber Bragg Gratings

Bragg gratings with short-period changes in refractive index are formed in standard single-mode optical fibers with GeO_2-doped silica cores. The period of the refractive index change in a fiber Bragg grating (FBG) is of the order of half a wavelength, typically around 0.5 μm, and the FBG couples a forward mode propagating in the waveguide to a backward mode. The Bragg wavelength λ_B is given by

$$\lambda_B = 2n_e\Lambda \tag{1}$$

where n_e is the effective refractive index of the FBG and Λ is the pitch of the refractive index change in the FBG. The passband width can be controlled by adjusting the length and/or amount of the FBG index change. Consequently, devices containing FBGs can be fabricated with an extremely low loss (0.1 dB), low polarization dependent loss (PDL) and with low-cost packaging.

FBGs are made by focusing a UV laser with a periodic intensity distribution on the side of an optical fiber to induce a periodic refractive index change. The periodic intensity distribution of the laser can be formed with a phase mask and a KrF excimer laser as shown in Fig. 5 [13]. This method is relatively easy and widely used. Since the phase mask can be set close to the side of the fiber with this approach, an FBG with good characteristics can be formed simply with an excimer laser,

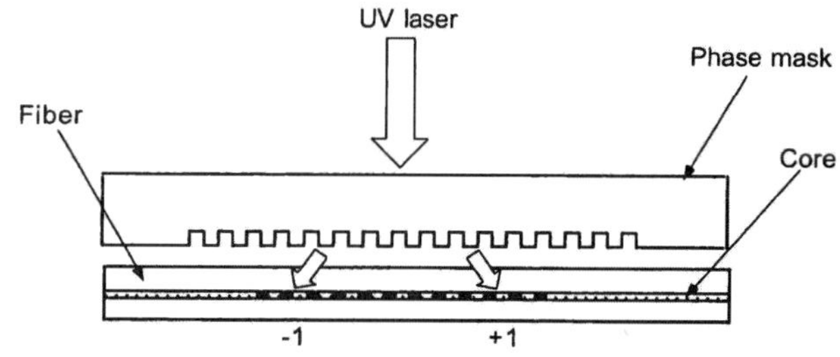

Fig. 5 Fabrication of fiber Bragg grating using a UV laser and a phase mask.

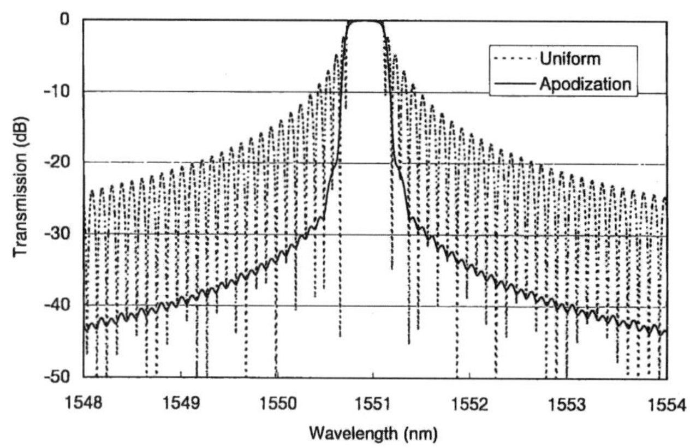

Fig. 6 Typical fiber Bragg grating spectra. The solid and dashed curves represent spectra with and without apodization, respectively.

which usually does not have good spatial coherence. Moreover, an effective way to increase the photosensitivity and thus the refractive index change is to load the optical fibers with hydrogen under high pressure at room temperature. After an optical fiber has been kept in a chamber with a hydrogen pressure of 10 MPa for two weeks, UV laser irradiation can induce a refractive index change greater than 1×10^{-3} [14].

It has been reported that this photo-induced refractive index change can be stabilized by annealing the irradiated fiber at a high temperature [15]. This is because the photo-induced change is the result of point defects in the glass network. The index change decays thermally over time, but this decay can be removed by annealing the FBG at a temperature over 100°C [15]. Moreover, the Bragg wavelength of the FBG shifts about 0.01 nm/°C because of the temperature dependence of the refractive index of silica glass. This dependence can be compensated for by using a special package that applies stress to the fiber [16].

If the amplitude of the refractive index change in the FBG is uniform along the propagation direction, many side peaks are observed in the grating spectra (Fig. 6). This type of the FBG is not suitable for use in a dense WDM network. The side peaks can be eliminated by using an apodization method to provide the amplitude of the refractive index change in the fiber with a nonuniform distribution [17]. A typi-

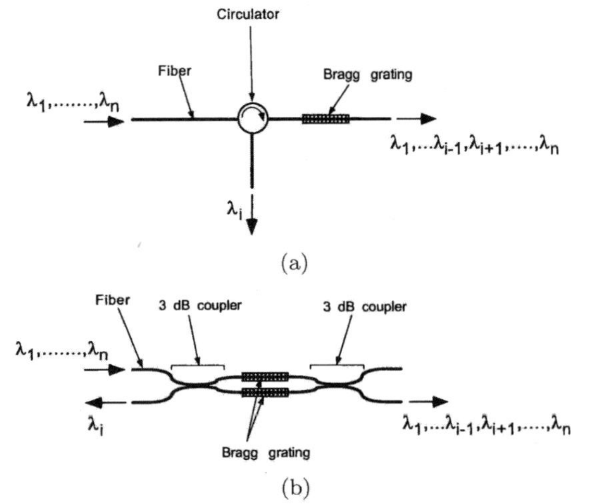

Fig. 7 Schematic configuration of OADM filters constructed using FBGs. (a) Circulator type, (b) Mach-Zehnder interferometer type.

cal amplitude distribution for reducing the side peaks is Gaussian-like, that is, the center region of the amplitude distribution is higher than the outer region. The side peaks of the FBG can be suppressed by using this apodization method as shown in Fig. 6. The apodization method enables us to use the FBG as a dense WDM filter with a low loss, a wide passband and a low neighboring crosstalk.

Two configurations of FBG filters for WDM systems are shown in Fig. 7. In one, a single FBG is combined with a circulator, and in the other, two FBGs are installed in the two arms of a Mach-Zehnder interferometer consisting of two fiber couplers [18]. Signals at Bragg wavelengths are reflected by filters with these configurations while signals at other wavelengths pass through them with a low loss. The FBG filter can therefore be used as an OADM filter with a low loss and a wide flat-topped passband [19]. Furthermore, a multi/demultiplexer can be constructed by cascading the unit structures shown in Fig. 7.

2.2.2 Long-Period Fiber Gratings

Long-period fiber gratings (LPGs) are fabricated in the same way as FBGs. Namely, a UV laser is used to irradiate optical fibers through spatial filters. These gratings are used primarily for adjusting the non-flat gain profile of erbium-doped fiber amplifiers.

In LPGs, the forward propagating mode in the core is coupled to other forward propagating modes in the cladding. These cladding modes are extremely lossy, and decay rapidly as they propagate along the fiber due to losses at the cladding-to-air interface. The coupling occurs between a core mode and a cladding mode through an LPG, whose phase-matching condition is expressed as follows;

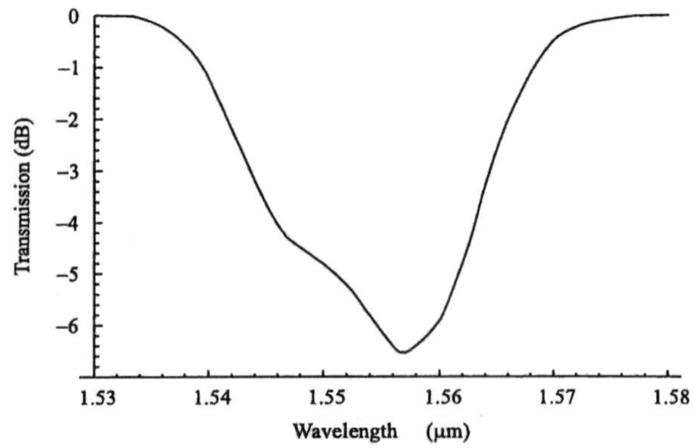

Fig. 8 Typical spectrum of long-period fiber grating.

$$\beta - \beta_{cl}^{p} = 2\pi/\Lambda \tag{2}$$

where β is the propagation constant of the single mode in the core, β_{cl}^{p} is the propagation constant of the pth-order cladding mode, and Λ is the LPG period. Since the difference between the propagation constants of two forward propagating modes is small, Λ is fairly large, usually a few hundred μm. Equation (2) can be rewritten using the relation $\beta = 2\pi n_{eff}/\lambda$ as;

$$\lambda = \Lambda(n_{eff} - n_{eff}^{p}) \tag{3}$$

If we know the effective refractive indices of the core and cladding modes, we can design the LPG spectrum. The typical transmission spectrum of the LPG is shown in Fig. 8. The amount of wavelength-dependent loss can be controlled by adjusting the UV exposure time during fabrication. LPGs are expected to be used widely in WDM systems, and various kinds of LPGs have recently been developed [12].

2.3 Planar Waveguide Filters

Planar lightwave circuits (PLCs), in which fiber-matched silica-based waveguides are integrated, can provide various key devices for optical networks. This is because they are suitable for large-scale integration, offer long-term stability, and can be mass produced. These PLC devices include wavelength $N \times N$ multi/demultiplexers, optical add/drop or cross-connect switches, multi-wavelength light sources for WDM transport networks, programmable filters [20], [21] for high-speed transmission systems, $1/N$ optical power splitters, optical couplers, and 1.3/1.55 μm WDM optical modules for access networks. Of these devices, the arrayed waveguide grating (AWG) multi/demultiplexer is one of the most successful optical filters, and is a key component of the photonic network. This subsection therefore focuses on the AWG as a planar-waveguide-type filter.

Silica-based optical waveguides are fabricated on silicon substrates by a combination of flame hydrolysis deposition (FHD) and reactive ion etching (RIE) [22].

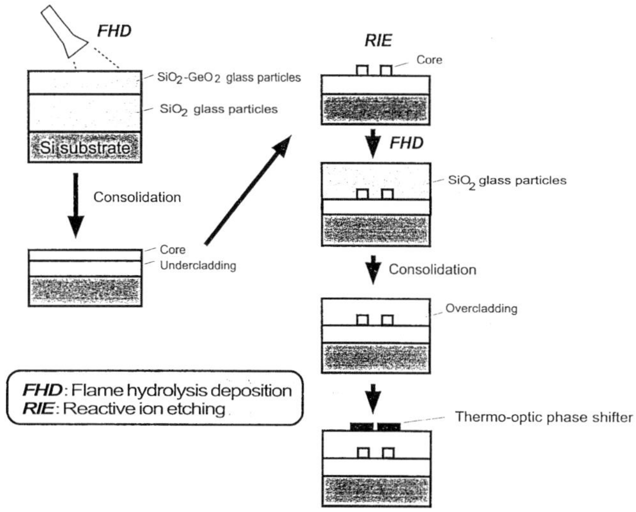

Fig. 9 Fabrication process of silica-based planar lightwave circuits.

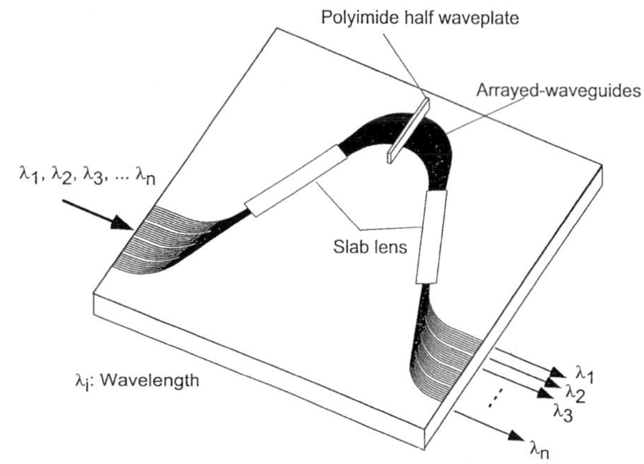

Fig. 10 Configuration of $N \times N$ AWG multiplexer.

The waveguide fabrication process is shown schematically in Fig. 9. The first step is to use FHD to deposit two successive glass particle layers that serve as the undercladding and the core. After deposition, the substrate with these two porous glass layers is heated to about 1300°C for consolidation. The waveguide core ridges are then formed by photolithography and RIE. Finally FHD is used again to cover the core ridges with an overcladding. The advantages of this type of waveguide are its low loss [23] and its low coupling loss to optical fibers. Specifically, a propagation loss of less than 0.02 dB/cm has been attained in a 10-m long waveguide [24]. These waveguides with a well-defined structure and a low loss enable us to design complicated circuits by using various numerical simulation techniques. Moreover, the reliability of the PLC devices has been confirmed based on the Bellcore reliability requirements for passive optical devices [25], [26].

Silica-based waveguides have been employed to develop AWGs for use as dense WDM filters [27]–[29]. The configuration of an $N \times N$ AWG multiplexer is shown in Fig. 10. The multiplexer consists of N input/output waveguides, two focusing slab waveguides and arrayed-waveguides with a constant path length difference ΔL between neighboring waveguides. The input light is launched into the first slab waveguide and then excites the arrayed waveguides. After traveling through the arrayed waveguides, the light beam interferes constructively at one focal point in the second slab. The location of the focal point depends on the signal wavelength λ [30] because the relative phase delay in each arrayed waveguide is given by $2\pi\Delta L/\lambda$.

In silica-based AWGs, the equivalent refractive index n_c is different for the orthogonal polarization modes due to stress-induced birefringence. The focusing position is therefore different for the TE and TM polarizations. The TE/TM mode conversion method, in which a thin half-waveplate is inserted at the center of the arrayed-waveguides, is the simplest and most practical way to correct for the different focal positions [31]. In the AWG whose configuration shown in Fig. 10, a 14.5 μm polyimide half-waveplate is inserted and fixed in a 18 μm wide groove with a UV curable adhesive. The insertion loss increment caused by the half-waveplate is less than 0.4 dB.

Various kinds of AWG multiplexers have been fabricated ranging from a 15-nm spacing 8-channel AWG [32] to a 25-GHz spacing 256-channel AWG [33]. The grating parameters and achieved performance of the fabricated AWGs are listed in Table 1. The on-chip loss at the center port of the AWGs ranges from 2.1 to 3.5 dB. The 3-dB bandwidths obtained experimentally agree very well with the theoretical values calculated by using the beam propagation method (BPM). The configuration of a 25-GHz spacing 256-channel AWG with a chip size of 55×75 mm is shown in Fig. 11. The demultiplexing characteristics for all 256 output ports are shown in Fig. 12. The background crosstalk of less than −37 dB is obtained even for this large AWG. Its on-chip loss ranges from 3.5 dB for the central outputs port to 5.9 dB for the peripheral outputs port.

The spectral shape of normal AWGs such as those listed in Table 1 are Gaussian because the spectra reflect the beam intensity distribution in the waveguide. However, flat-top spectra are desired for WDM applications. A flat passband has been obtained by using parabolic waveguide horns in the input waveguides [34]. The loss spectra of a 16 channel flat-passband AWG with a 100 GHz spacing are shown in Fig. 13. A 3-dB bandwidth of about 80 GHz was achieved. Moreover, the insertion loss ranged from 8 to 8.5 dB and the worst crosstalk was about −37 dB.

The center wavelength of a conventional AWG shifts to a longer wavelength by about 0.1 nm/°C. This is because the refractive index of a silica-based waveguide depends on temperature. The AWG temperature must therefore be controlled with a heater or a Peltier

Table 1 Experimental performance of fabricated multiplexers.

Parameters	Experimental results					
Channel spacing	15 nm	2 nm	0.8 nm (100 GHz)	0.4 nm (50 GHz)	0.2 nm (25 GHz)	0.2 nm (25 GHz)
Number of channels	8	8	32	64	128	256
Center wavelength	1.55 μm					
Path difference	12.8 μm	50.3 μm	63.0μm	63.0 μm	63.0 μm	27.7 μm
Focal length	2.38 mm	5.68 mm	11.35 mm	24.2 mm	36.3 mm	41.1 mm
Diffraction order	12	47	59	59	59	26
On-chip loss	2.4 dB	6.1 dB	2.1 dB	2.8 dB	3.5 dB	2.7 dB
3-dB Bandwidth (BPM simulation)	6.3 nm (6.3 nm)	0.74 nm (0.75 nm)	40 GHz (37 GHz)	19 GHz (21 GHz)	11 GHz (9.5 GHz)	14.4 GHz (12.5 GHz)
Channel crosstalk	< -28 dB	< -29 dB	< -28 dB	< -27 dB	< -16 dB	< -33 dB

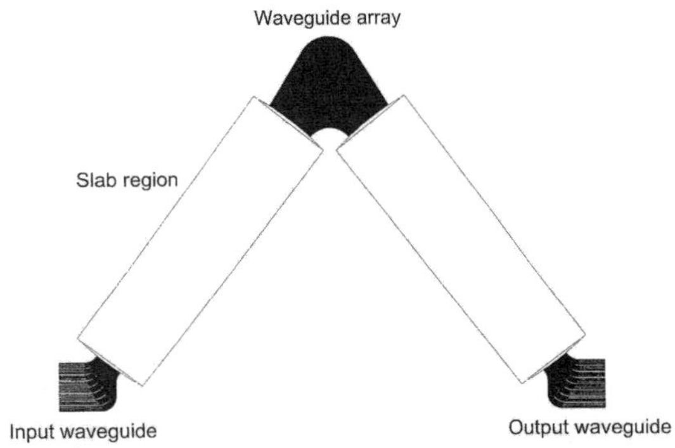

Fig. 11 Waveguide layout of 256 × 256 AWG multiplexer with 25 GHz channel spacing.

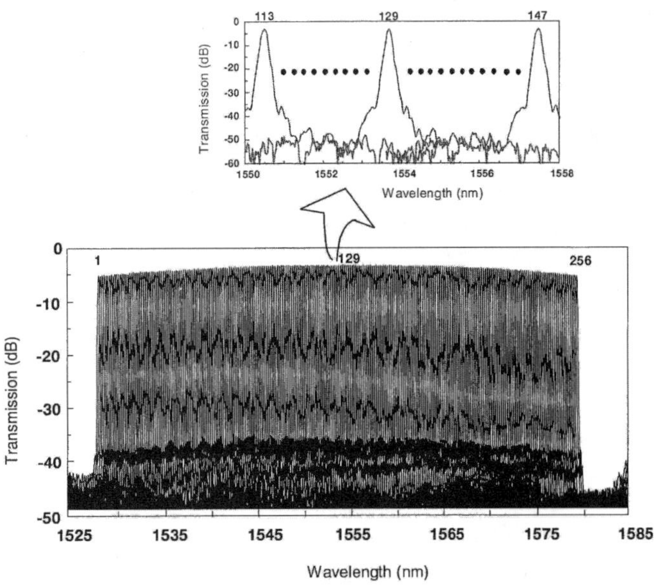

Fig. 12 Demultiplexing characteristics of 256 channel 25 GHz spacing AWG.

device to stabilize the center wavelength. By contrast, an athermal AWG has also been shown to expand the application area and reduce the cost [35]. Its configu-

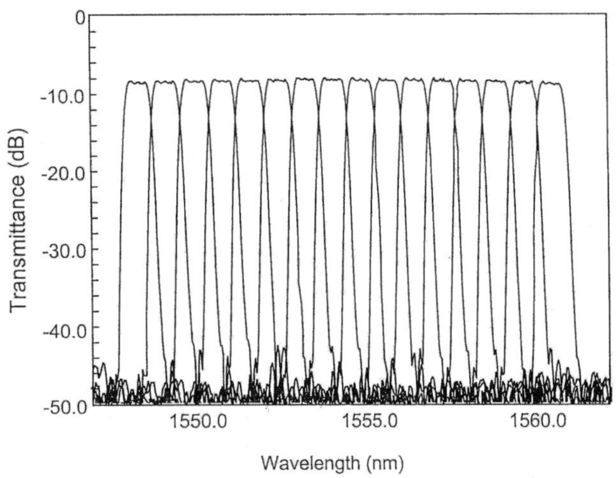

Fig. 13 Transmission spectra of flat-passband low-crosstalk AWG.

ration is shown schematically in Fig. 14. The temperature dependent optical path difference in silica-based waveguides is compensated for by employing a triangular groove filled with silicone adhesive which has a negative thermal coefficient. Since the negative optical path change in the silicone is several tens of times larger than in silica-based waveguides, the negative optical path change in the silicone can compensate for the positive optical path changes in the silica thereby achieving temperature insensitive operation with small silicone gaps which cause only a slight loss increment.

The temperature-dependent transmission spectra of conventional and athermal AWGs are shown in Figs. 15(a) and (b), respectively. The thermal behavior of the athermal AWG at various temperatures is so similar that no difference can be distinguished in Fig. 15(b). The wavelength shift observed over a 0 to 85°C temperature range was 0.95 nm for the conventional AWG , but only 0.05 nm for the athermal AWG. The fiber-to-fiber insertion losses for eight output ports ranged from 3.9 to 4.2 dB, and the excess loss caused by

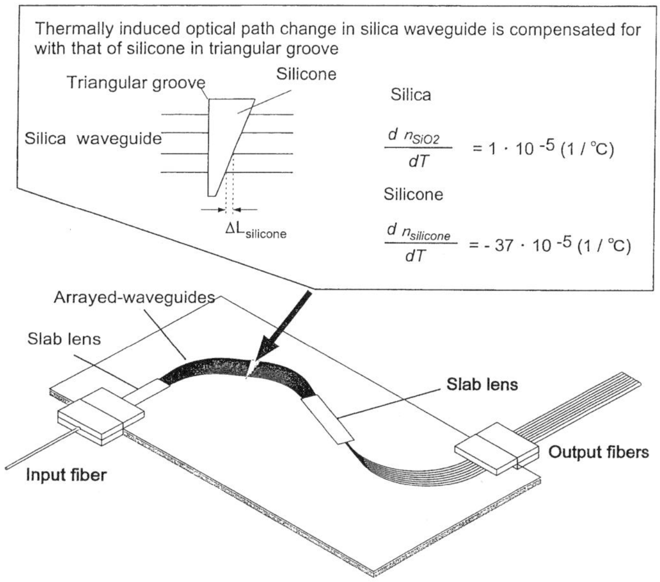

Fig. 14 Configuration of athermal AWG multiplexer.

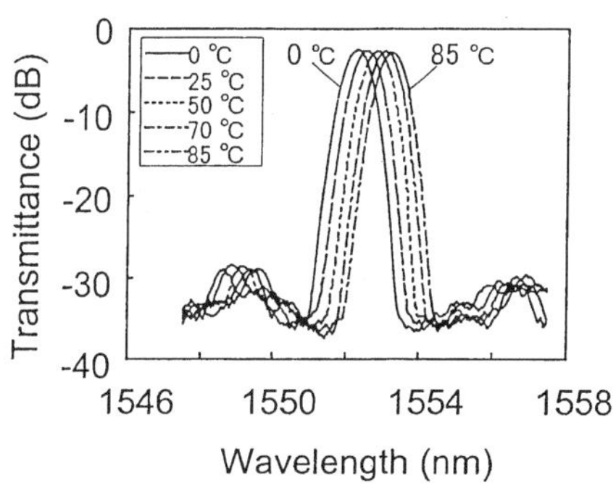

(a)

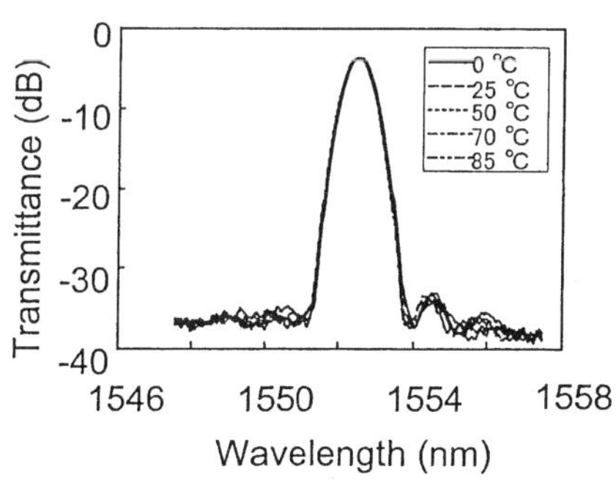

(b)

Fig. 15 Temperature dependent transmission spectra. (a) Conventional silica AWG multiplexer. (b) Athermal AWG multiplexer.

the silicone groove was estimated to be 2 dB. The wavelength change over 5000 hours at 75°C and at 90% relative humidity was only 0.02 nm. This result confirms that the athermal AWG is stable enough for practical use.

A wide variety of PLC-type filters have been developed as described above, and actually applied to WDM systems. Since they are suitable for mass production and can be integrated with active devices, the application of these PLC-type devices will undoubtedly expand.

3. Optical Switches

As already mentioned, optical switches are used in photonic networks for a variety of applications including OADM and OXC. The main feature distinguishing these applications is the required switching time, which varies from a few milliseconds to subnanosecond levels. Passive optical switches are mainly used in for provisioning of lightpaths and protection of systems because the switching speed of the passive switches is usually only of millisecond [36]. For the provisioning application, the switches are used inside optical crossconnects to reconfigure them to support new lightpaths. The challenge here is to realize large switches. In the protection application, the switches are used to change the traffic stream from a primary fiber to another fiber should the primary fiber fail. Small 2 × 2 switches are usually sufficient for this purpose.

In addition to the switching time, other important parameters used to characterize the suitability of a switch for optical networking applications are insertion loss, the extinction ratio between on and off states, crosstalk, PDL and wavelength dependence. These characteristics are reviewed here for mechanical fiber-type switches, PLC-type switches and microelectromechanical-system (MEMS)-type switches. The state of integration of optical switches is considerably less than that of electric switches. A 16 × 16 optical switch is currently considered a large switch.

3.1 Mechanical Fiber-Type Switch

Mechanical switches are constructed by moving an optical fiber directly or moving a micro-optics component in front of fibers with focusing lenses. They have the advantages of a low insertion loss, a low PDL and a low crosstalk, and are relatively inexpensive devices. Mechanical switches are usually latching type switches, and therefore need switching power only when actually operating. Their switching speeds are usually of the order of a few ms, and sufficient for the reconfiguration of optical paths.

A typical configuration of a 1 × 2 latching-type fiber switch is shown in Fig. 16 [37], [38]. The switch is composed of a movable fiber to which a micro magnetic alloy pipe is fixed, two stationary fibers fixed in half cylindrical V-grooved ferrules, a solenoid coil and a pair of rectangular magnets. Ceramic ferules with V-grooves and a slit sleeve are used for the precise alignment of the fibers, offering excellent levels of optical and mechanical switching performance, and a simple assembly process. This configuration has no bulk op-

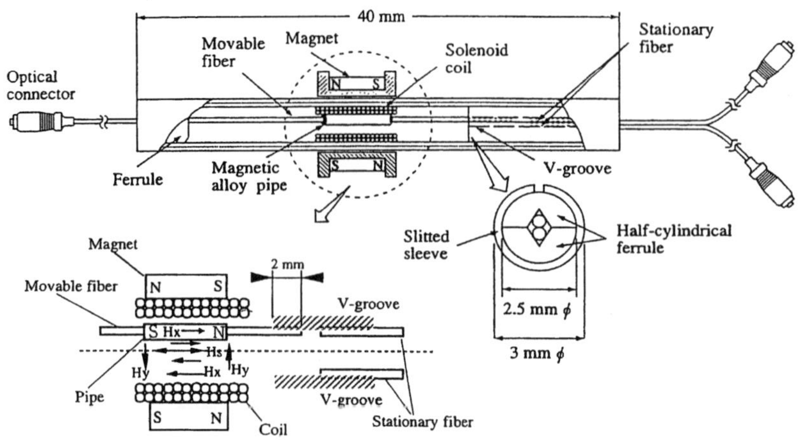

Fig. 16 Schematic view of 1 × 2 latching-type fiber optical switch.

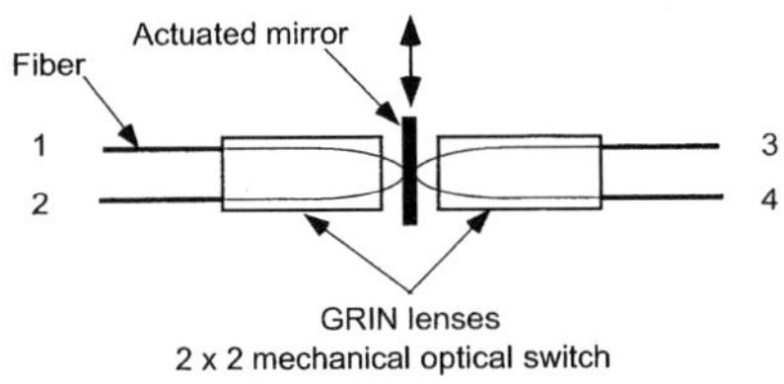

Fig. 17 Configuration of 2 × 2 mechanical optical switch.

tics such as lens as or prisms, so the switch size and power can be reduced. The insertion loss of this switch is lower than 1.0 dB, and the crosstalk and switching time are less than −70 dB and 6 ms, respectively. This high level of performance is sufficient for OADM applications and lightpath protection, and various versions of this switch type are commercially available.

A mechanical switch with a 2 × 2 configuration is also available. As shown in Fig. 17, the switching is realized by moving a mirror into and out of the optical path. This mechanical switch also performs well. However, it is rather difficult to construct larger switches by cascading these types of mechanical-fiber switch, and it is not cost effective.

3.2 Planar Waveguide Type

3.2.1 Thermo-Optic Switches

Thermo-optic (TO) switches can be made by utilizing the thermally induced refractive change of silica-based waveguides [39]. The basic configuration of a 2 × 2 switching unit is shown in Fig. 18. The unit is a Mach-Zehnder interferometer (MZI) consisting of two 3-dB directional couplers and two waveguide arms equipped with a thin film heater which operates as a TO phase shifter. The unit is in the cross state when no heating power is supplied. It can be switched to the bar state by adjusting the heating power supplied to the TO phase shifter. The switching time is less than 4 ms. This performance is acceptable for the majority of WDM applications such as OADM or crossconnect switches.

Strictly non-blocking matrix switches with scales up to 16 × 16 [40]–[42] have been fabricated using silica-based PLCs. The configurations of the switching unit and the $N \times N$ matrix switch are shown in Fig. 19. A double-gate switching unit is used to improve the extinction ratio as shown in Fig. 19(a). It consists of two asymmetrical MZIs with TO phase shifters and an intersection. Theoretically this unit can double the extinction ratio of a single MZI unit. The logical arrangement of an $N \times N$ matrix switch with a path-independent insertion loss arrangement is shown in Fig. 19(b) [43]. This arrangement is useful for reducing the insertion loss because the total waveguide length is shorter than that in a conventional crossbar matrix switch.

The measured losses and extinction ratios of the 16 × 16 matrix switch for all 256 possible input and output connection patterns are shown in Fig. 20. The insertion loss ranges from 6.0 to 8.0 dB with an average value of 6.6 dB. The extinction ratio ranges from 45 to 67 dB with an average value of 55 dB. The extinction ratio was considerably improved by adopting a double gate switching unit. The electric power needed for operating this switching unit is about 1.1 W. The total power for the 16 × 16 matrix switch is 17 W.

Recently an integrated OADM filter consisting of AWGs and TO switches has been developed [44]. This device provides simultaneous access to all wavelength channels in a WDM communication system. The configuration of a 16 channel optical OADM is shown in Fig. 21. Four AWGs and sixteen double-gate switches are integrated in a wafer. All these AWGs have the same grating parameters; a channel spacing of 100 GHz and a free spectral range of 3300 GHz in the 1.55 μm wavelength region. When switches SW_2, SW_4, SW_6, SW_7, SW_9, SW_{12}, SW_{13}, and SW_{15}, for example, are turned to the on state, the selected signals λ_2, λ_4, λ_6, λ_7, λ_9, λ_{12}, λ_{13} and λ_{15} are extracted from the main output port and led to the drop port as shown in

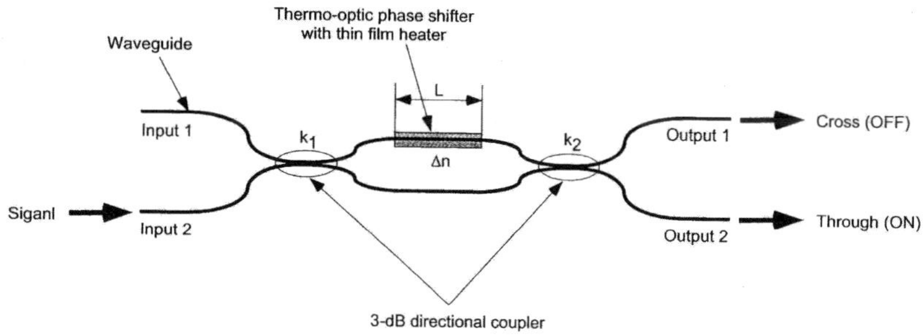

Fig. 18 Basic configuration of 2 × 2 thermo-optic switching unit.

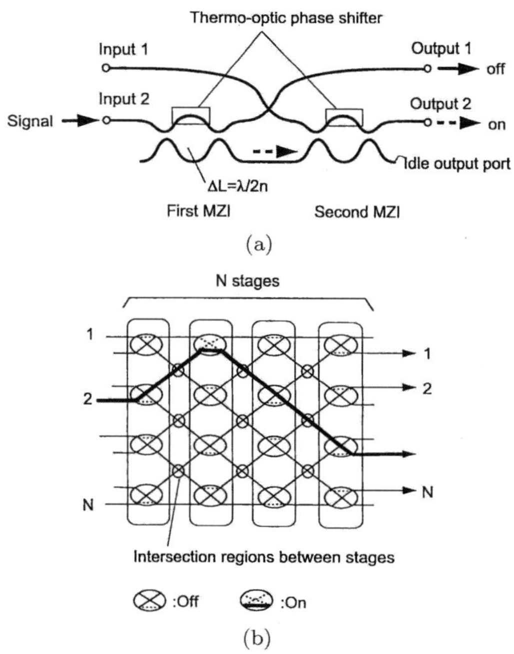

Fig. 19 Configuration of $N \times N$ matrix switch. (a) Configuration of double-gate switching unit. (b) Logical arrangement of matrix switch.

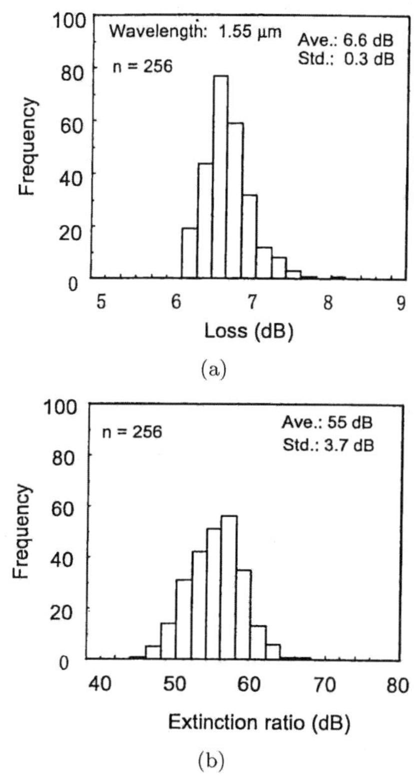

Fig. 20 Optical characteristics of 16 × 16 matrix switch. (a) Insertion loss. (b) Extinction ratio.

Fig. 22. The on-off crosstalk is less than −28 dB with on-chip losses of 8–10 dB. This OADM filter is attractive for all optical WDM routing systems and allows the network to be transparent in terms of signal format and bit rate.

3.2.2 Total-Internal-Reflection-Type Switches

PLC-type switches utilizing total-internal reflection at a groove formed in crossed waveguides have recently been developed for constructing large-scale optical switches [45]–[48]. The on/off function is achieved by moving index matching oil thermally or creating a bubble in the groove by heating. A typical structure of the PLC-type oil switch is shown in Fig. 23. A narrow rectangular groove is formed perpendicularly at the cross point of the waveguides. A small amount of index matching oil is sealed in the groove, and a thin-film metal heater is deposited around the ends of the groove. The oil moves along the groove when heated. The groove is covered with a top glass plate.

The position of the oil in the groove is controlled by the thermal capillary phenomenon. When the oil is at the cross point of the waveguides, signals pass straight through the groove. The oil can be moved by supplying an electric power to the heater near the oil. Once the oil has moved to the other side, its position can be maintained without the power. When the oil is not at the cross point, the signal is reflected totally at the groove wall, and switches to the crossed waveguide. The PLC-type switch, which uses the total-internal reflection, has good optical characteristics such as a low loss, a low PDL, a wide operating wavelength region and a high extinction ratio. Moreover, excellent isolation is attained by the large angle of the crossed

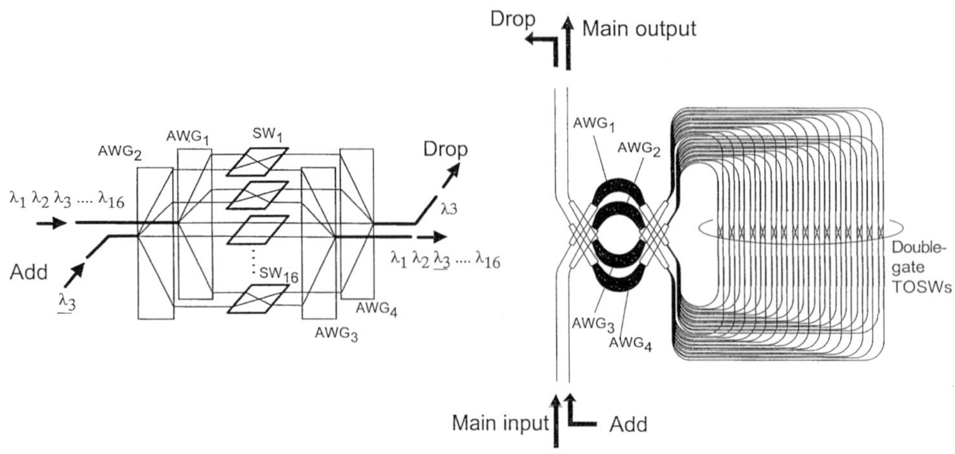

(a) (b)

Fig. 21 Schematic configuration of 16 channel optical ADM with double-gate TO switches. (a) Logical configuration. (b) Waveguide layout.

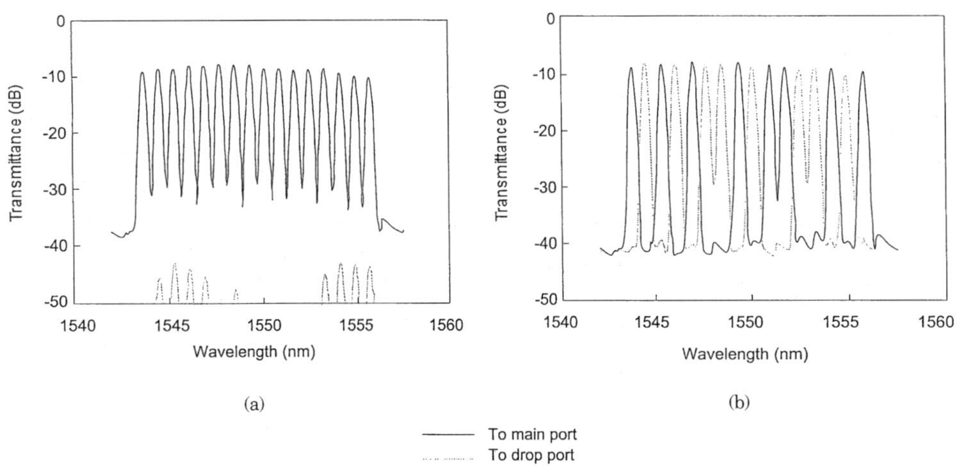

Fig. 22 Transmission spectra from main input port to main output port. (a) All switches are off. (b) TO switches SW_2, SW_4, SW_6, SW_7, SW_9, SW_{12}, SW_{13}, SW_{15} are on.

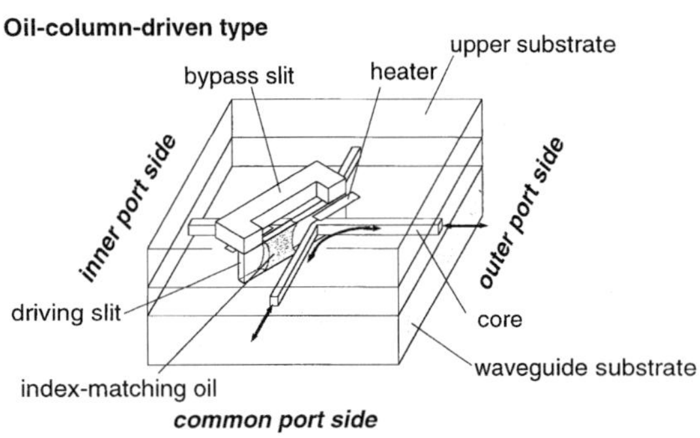

Fig. 23 Configuration of PLC total-internal-reflection-type switch. Index-matching oil is moved by heating along a groove formed at the waveguide intersection.

waveguides. A reflection loss of less than 1.0 dB can be obtained at the groove through the use of very precise etching technologies.

NTT have applied the above configuration using total-internal reflection to developing 1 × 8 [47] and 16 × 16 switches [48] called "Olive." Insertion losses of less than 1.0 and 3.0 dB were obtained for transmitted and reflected lights, respectively, with the 1 × 8 switches. The on/off ratio and the isolation were both less than −50 dB. The switching power was about 0.1 W/channel, and the switching time was of the order of 10 ms. Although the insertion loss increased up to 10 dB in the 16 × 16 switches, the other characteristics were the same as those of the 1 × 8 switches.

Another type of the total internal reflection switch called "Champagne" has also been proposed by Agilent Technologies [49]. The waveguide structure is almost the same as that of Olive. The switching is realized by displacing the liquid in the groove with a small bubble which is generated at the crosspoint by an inkjet technology-based thermal actuator. The actuator was

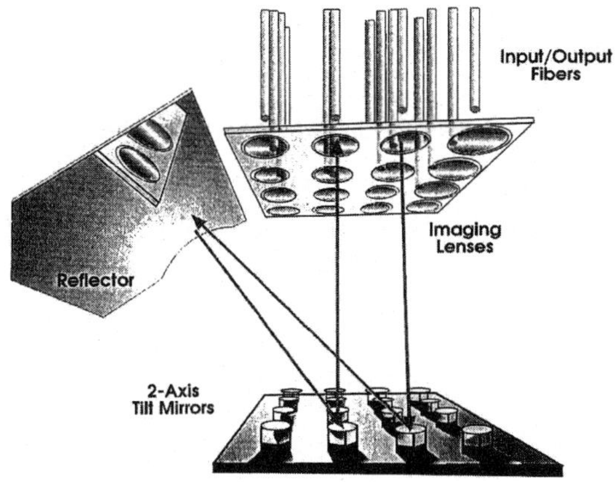

Fig. 24 Basic configuration of free-space 2-dimension MEMS switch. In the switch, each micromirror tilts independently and arbitrarily.

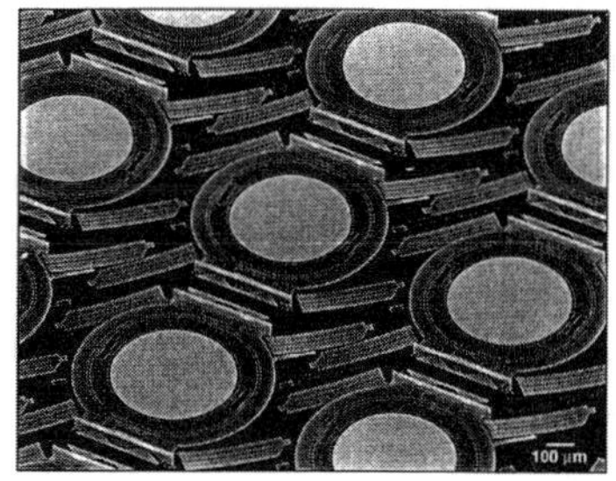

Fig. 25 Photo of MEMS mirrors.

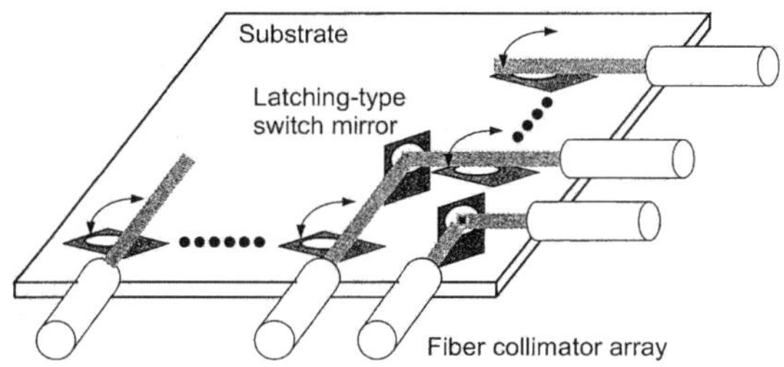

Fig. 26 Schematic drawing of matrix micromachined optical switch.

fabricated on a separate chip bonded to the waveguide substrate. In the presence of a bubble, light undergoes total internal reflection, and is switched to the crossed waveguide. The optical characteristics are almost the same as those of the Olive switch, and large switches of up to 32×32 have been demonstrated [49].

3.3 MEMS Switches

Transparent optical crossconnects (OXCs) have certain advantages when compared wtih electronic crossconnects since optical networks are characterized by rapidly increasing data rates and evolving data formats. The dramatic expansion of network size and capacity requires these optical crossconnects to be large port counts, $N > 1000$. Optical switches with a structure of 2-dimension free space optics have been proposed as a way of achieving these large port count OXCs. In this configuration, the active component scale is reduced to the order of N, while the switch elements needed to construct a planar-type matrix switch are of the order of N^2 [50].

The basic concept of the free-space 2-dimension switch is shown in Fig. 24. The cross connect function in this switch is achieved using two-axis beam steering micro-mechanical mirrors made with microelectromechanical system (MEMS) technology. The switch system consists of a 2-dimension array of MEMS mirrors, a fiber array with a collimating microlens array and a folding mirror to allow one MEMS array and fiber array to be used for both input and output. The distance between the MEMS mirror and fiber arrays depends on the area size and moving angle of the MEMS mirror, and the port counts. A 112×112 OXC switch has been demonstrated recently with this configuration [50]. The MEMS mirrors used are shown in Fig. 25. The distance between the MEMS mirror and fiber arrays was about 10 cm. The typical connection switching time and insertion loss were reported to be less than 10 ms and less than 10 dB at 1550 nm, respectively. The crosstalk was below -50 dB. Although the detailed performance and long-term stability of the MEMS OXC switch have not been confirmed yet, large OXC switch with a low cost per channel is expected to be realized using MEMS technologies.

Matrix free-space MEMS switches using vertical mirrors formed on a planar substrate have been also developed [51]–[55]. The schematic configuration of this switch is show in Fig. 26. The mircoactuated free-rotating mirrors are monolithically integrated on a silicon chip with MEMS technologies, and are controlled with electrostatic force. The collimated light from the input fiber is switched to the desired output port by rotating a selected mirror. The fiber collimators consist of GRIN lenses attached to single-mode fibers. The MEMS chip size of the 8×8 matrix switch is $1 \times 1\ \text{cm}^2$ and the on/off switching time is about 0.5 ms. The insertion loss, extinction ratio, crosstalk and PDL of the 8×8 switch are about 3.5 dB, less than -60 dB, less than -60 dB and 0.3 dB, respectively. The switch has advantages of compactness and high switching speed, and so is expected to be used in high-capacity core-transport lightwave networks.

4. Conclusion

This paper has reviewed recent progress on optical filters and switches for WDM-based photonic networks, which are expanding both rapidly and globally. While various kinds of optical devices have been developed to construct flexible and large-capacity networks, narrow-band filters and switches are key components for the DWDM networks. Three kinds of optical filter; thin-film interference filters, fiber gratings and AWGs are described in this paper. As regards optical switches, this paper reviews mechanical fiber switches, PLC-type TOSWs, PLC-type total-internal-reflection switches and MEMS switches. Each device has been or will be installed in a suitable application in WDM systems. The next step is to meet demands for lower cost, lager scale and expanded functions for both types of device. Further advances in optical components and related technologies will contribute greatly to the construction of photonic networks.

Acknowledgements

The author thanks Dr. Katsunari Okamoto, Dr. Tetsuo Miya, Dr. Tohru Maruno, Dr. Yasuji Ohmori and Dr. Akira Himeno for their useful discussions and encouragement.

References

[1] H. Toba, K. Oda, K. Nakanishi, N. Shibata, K. Nosu, N. Takato, and M. Fukuda, "A 100-ch optical WDM transmission/distribution at 622 Mbits/s over 50 km," IEEE J. Lightwave Technol., vol.8, pp.1396–1401, 1990.

[2] I.P. Kaminow, "A wideband all-optical WDM network," IEEE J. Sel. Areas Commun., vol.14, pp.780–799, 1996.

[3] T. Miki, "Photonic transport networks," Proc. IEEE, vol.81, pp.1594–1601, 1993.

[4] N. Nagatsu, S. Okamoto, K. Koga, and K. Sato, "Flexible OADM architecture and its impact on WDM ring evolution for robust and large-scale optical transport networks," IEICE Trans. Commun., vol.E82-B, pp.1105–1114, 1999.

[5] H. Toba, K. Oda, K. Inoue, K. Nosu, and T. Kitoh, "An optical FDM-based self-healing ring network employing arrayed-waveguide grating filters and EDFA's with level equalizers," IEEE J. Sel. Areas Commun., vol.14, pp.800–813, 1996.

[6] M. Koga, Y. Hamazumi, A. Watanabe, S. Okamoto, H. Obara, K. Sato, M. Okuno, and S. Suzuki, "Design and performance of an optical path cross-connect system based on wavelength path connect," IEEE J. Lightwave Technol., vol.14, pp.1106–1119, 1996.

[7] H. Toba, K. Oda, K. Inoue, K. Nosu, and T. Kitoh, "Demonstration of optical FDM based self-healing ring network employing arrayed waveguide-grating ADM filters and EDFA," Proc. ECOC'94, pp.263–265, 1994.

[8] M.A. Scobey and D.E. Spock, "Passive DWDM components using MicroPlasma optical interference filters," OFC'96 Technical Digest, pp.242–243, 1996.

[9] K.O. Hill, Y. Fujii, D.C. Johnson, and B.S. Kawasaki, "Photosensitivity in optical fiber waveguides: Application to reflector filter fabrication," Appl. Phys. Lett., vol.32, pp.647–649, 1978.

[10] G. Meltz, W. Morey, and W.H. Glenn, "Formation of Bragg gratings in optical fibers by a transverse holographic method," Opt. Lett., vol.29, pp.823–825, 1989.

[11] A.M. Vengsarkar, J.R. Pedrazzani, J.B. Judkins, P. Lemaire, N.S. Bergano, and C.R. Davidson, "Long-period fiber-grating-based gain equalizers," Opt. Lett., vol.21, pp.336–338, 1996.

[12] A.M. Vengsarkar, P. Lemaire, J.B. Judkins, V. Bhatia, T. Erdogan, and J.E. Sipe, "Long-period fiber gratings as band-rejection filters," J. Lightwave Technol., vol.14, pp.58–65, 1996.

[13] K.O. Hill, B. Maro, F. Bilodeau, D.C. Johnson, and J. Albert, "Bragg gratings fabricated in monomode photosensitive optical fiber by UV exposure through a phase mask," Appl. Phys. Lett., vol.62, pp.1035–1037, 1993.

[14] P.J. Lemaire, R.M. Atkins, V. Mizrahi, and W.A. Reed, "High pressure H_2 loading as a technique for achieving ultrahigh UV photosensitivity and thermal sensitivity in GeO_2 doped optical fibers," Electron. Lett., vol.29, pp.1191–1193, 1993.

[15] T. Erdogan, V. Mizrahi, P.J. Lemaire, and D. Monroe, "Decay of unltraviolet-induced fiber Bragg garings," J. Appl. Phys., vol.76, pp.73–80, 1994.

[16] G.W. Yoffe, P.A. Krug, F. Ouellette, and D. Thorncraft, "Temperature-compensated optical-fiber Bragg gratings," OFC'95 Technical Digest, pp.134–135, 1995.

[17] B. Maro, S. Theriault, D.C. Johnson, F. Bilodeau, J. Albert, and K.O. Hill, "Apodized in-fiber Bragg grating reflector photoimprinted using a phase mask," Electron. Lett., vol.31, pp.223–225, 1995.

[18] D.C. Johnson, K.O. Hill, F. Bilodeau, and S. Faucher, "New design concept for a narrowband wavelength-selective optical tap and combiner," Electron. Lett., vol.23, pp.668–669, 1987.

[19] K. Hattori, M. Abe, J. Albert, F. Bilodeau, K.O. Hill, Y. Hibino, T. Kitagawa, and K. Oguchi, "Coherent crosstalk of an optical add/drop filter with Bragg gratings in a PLC Mach-Zehnder interferometer for optical LAN," IEEE Photo. Technol. Lett., vol.11, pp.272–274, 1999.

[20] K. Jinguji and M. Kawachi, "Synthesis of coherent two-port lattice-form optical delay-line circuit," IEEE J. Lightwave Technol., vol.13, pp.73–82, 1995.

[21] K. Takiguchi, K. Jinguji, and Y. Ohmori, "Variable group-delay dispersion equaliser based on a lattice-form programmable optical filter," IEE Electron. Lett., vol.31, pp.1240–1241, 1995.

[22] M. Kawachi, "Silica waveguides on silicon and their application to integrated components," Opt. & Quantum Electron., vol.22, pp.391–416, 1990.

[23] Y. Hibino, H. Okazaki, Y. Hida, and Y. Ohmori, "Propagation loss characteristics of long silica-based optical waveguides on 5 inch Si wafers," IEE Electron. Lett., vol.29, pp.1847–1848, 1993.

[24] Y. Hida, Y. Hibino, H. Okazaki, and Y. Ohmori, "10 m long silica-based waveguide with a loss of 1.7 dB/m," Proc. IPR, Dana Point, paper IthC6, 1995.

[25] Y. Hibino, F. Hanawa, H. Nakagome, M. Ishii, and N. Takato, "High reliability optical splitters composed of silica-based planar lightwave circuits," IEEE J. Lightwave Technol., vol.13, pp.1728–1735, 1995.

[26] H. Hanafusa, F. Hanawa, Y. Hibino, and T. Nozawa, "Reliability estimation for PLC-type optical splitters," IEE Electron. Lett., vol.33, pp.238–239, 1997.

[27] H. Takahashi, S. Suzuki, K. Kato, and I. Nishi, "Arrayed waveguide grating for wavelength division multi/demultiplexer with nanometer resolution," IEE Electron. Lett., vol.26, pp.87–88, 1990.

[28] M.K. Smit, "New focusing and dispersive planar component based on an optical phased array," IEE Electron. Lett., vol.24, pp.385–386, 1988.

[29] C. Dragone, C.A. Edwards, and R.C. Kistler, "Integrated optics $N \times N$ multiplexer on silicon," IEEE Photon. Tehnol. Lett., vol.3, pp.896–899, 1991.

[30] H. Takahashi, K. Oda, H. Toba, and Y. Inoue, "Transmission characteristics of arrayed-waveguide $N \times N$ wavelength multiplexer," IEEE J. Lightwave Technol., vol.13, pp.447–455, 1995.

[31] Y. Inoue, H. Takahashi, S. Ando, T. Sawada, A. Himeno, and M. Kawachi, "Elimination of polarization sensitivity in silica-based wavelength division multiplexer using a polyimide half waveplate," IEEE J. Lightwave Technol., vol.15, pp.1947–1957, 1997.
[32] K. Okamoto, "Planar lightwave circuits PLCs," in Photonic Network, ed. G. Prati, pp.118–132, Springer-Verlag, London, 1987.
[33] Y. Hibino, Y. Hida, A. Kaneko, M. Ishii, M. Itoh, T. Goh, A. Sugita, T. Saida, A. Himeno, and Y. Ohmori, "Fabrication of silica-on-Si waveguide with higher index difference and its application to 256 channel arrayed-waveguide multi/demultiplexer," Technical Digest of OFC2000, Paper WH2, 2000.
[34] K. Okamoto and A. Sugita, "Flat spectral response arrayed-waveguide grating multiplexer with parabolic waveguide horns," IEE Electron. Lett., vol.32, pp.1661–1662, 1996.
[35] Y. Inoue, A. Kaneko, F. Hanawa, H. Takahashi, K. Hattori, and S. Sumida, "Athermal silica-based arrayed-waveguide grating multiplexer," IEE Electron. Lett., vol.33, pp.1945–1946, 1997.
[36] K. Sato, "Photonic transport network OAM technologies," IEEE Commun. Mag., Special issue on "Operation and management of broadband networks," vol.34, pp.86–94, 1996.
[37] S. Nagaoka, "Latching type single-mode fiber switch," Electron. Lett., vol.26, pp.744–745, 1990.
[38] S. Nagaoka, "Compact latching-type single-mode fiber switches fabricated by a fiber-micromachining technique and their practical applications," J. Select. Topics Quant. Electron., vol.5, pp.36–45 1999.
[39] M. Okuno, N. Takato, T. Kitoh, and A. Sugita, "Silica-based thermo-optic switches," NTT Review, vol.7, pp.57–63, 1995.
[40] T. Kominato, A. Himeno, M. Ishii, T. Goh, K. Hattori, F. Hanawa, and K. Kato, "Silica-based optical switch with a single-facet array fiber connection," Proc. CLEO Passific Rim, paper FH2, Makuhari, Japan, 1997.
[41] T. Goh, A. Himeno, M. Okuno, H. Takahashi, and K. Hattori, "High extinction ratio and low loss silica-based 8×8 thermo-optic matrix switch," IEEE Photon. Technol. Lett., vol.10, pp.358–360, 1998.
[42] T. Goh, M. Yasu, K. Hattori, A. Himeno, and Y. Ohmori, "Low loss and high extinction ratio silica-based strictly non-blocking 16×16 thermo-optic matrix switch," IEEE Photon. Technol. Lett., vol.10, pp.810–812, 1998.
[43] T. Nishi, T. Yamamoto, and S. Kuroyanagi, "A polarization-controlled free-space photonic switch based on a PI-LOSS switch," IEEE Photon. Technol. Lett., vol.5, pp.1104–1106, 1993.
[44] K. Okamoto, M. Okuno, A. Himeno, and Y. Ohmori, "16-ch optical add/drop multiplexer consisting of arrayed waveguide gratings and double-gate switches," IEE Electron. Lett., vol.32, pp.1471–1472, 1996.
[45] M. Sato, F. Shimokawa, M. Makihara, and Y. Nishida, "Two types of thermo-capillary optical switches," Technical Digest of MOEMS'97, p.238, 1997.
[46] M. Sato, F. Shimokawa, M. Makihara, and Y. Nishida, "Self-latching waveguide optical switch based on thermo-capillary," Technical Digest of 23rd ECOC, pp.73–76, 1997.
[47] M. Sato, F. Shimokawa, S. Inagaki, and Y. Nishida, "Waveguide optical switch for 8:1 standby system of optical line terminals," Technical Digest of OFC1998, pp.194–195, 1998.
[48] M. Makihara, F. Shimokawa, and K. Kaneko, "Strictly non-blocking $N \times N$ thermo-capillary optical matrix switch using silica-based waveguide," Technical Digest of OFC2000, TuM2, 2000.
[49] J.E. Fouquet, "Compact optical cross-connect switch based on total internal reflection in a fluid-containing planar lightwave circuit," Technical Digest of OFC2000, TuM1, 2000.
[50] D.T. Neilson, V.A. Aksyuk, S. Arney, N.R. Basavanhally, K.S. Bhalla, D.J. Bishop, B.A. Boie, C.A. Bolle, J.V. Gates, A.M. Gottieb, J.P. Hickey, N.A. Jackman, P.R. Kolodner, S.K. Horotky, B. Mikkelsen, F. Oardo, G. Raybon, R. Ruel, R.E. Scotti, T.W. Van Blarcum, L. Zhang, and C.R. Giles, "Fully provisioned 112×112 micro-mechanical optical cross-connect with 35.8 Tb/s demonstrated capacity," Technical Digest of OFC2000, PD12, 2000.
[51] L.Y. Lin, E.L. Goldstein, J.M. Simmons, and R.W. Tkach, "High-density micromachined polygon optical crossconnects exploiting network connection-symmetry," IEEE Photon. Technol. Lett., vol.10, pp.1425–1427, 1998.
[52] L.Y. Lin, E.L. Goldstein, and R.W. Tkach, "Angular-precision enhancement in free-space micromachined optical switches," IEEE Photon. Technol. Lett., vol.11, pp.1253–1255, 1999.
[53] R.T. Chen, H. Nguyen, and M.C. Wu, "A high-speed low-voltage stress-induced micromachined 2×2 optical switch," IEEE Photon. Technol. Lett., vol.11, pp.1396–1398, 1999.
[54] L.Y. Lin, E.L. Goldstein, and R.W. Tkach, "Free-space micromachined optical switches for optical networking," IEEE J. Select. Topics Quant. Electron., vol.5, pp.4–9, 1999.
[55] H. Toshiyoshi, D. Miyauchi, and H. Fujita, "Electromagnetic torsion mirrors for self-aligned fiber optic crossconnectors by silicon micromachining," IEEE J. Select. Topics Quant. Electron., vol.5, pp.10–17, 1999.

Author Index

Subject Index